NEW!

Mastering A&P
for Marieb's HAP Lab Manual

D0146975

MasteringA&P™ is an online learning and assessment system proven to help students learn and designed to help instructors teach more efficiently. It helps instructors maximize lab time with customizable, easy-to-assign, automatically graded assessments that motivate students to learn outside of class and arrive prepared for lab. The powerful gradebook provides unique insight into student and class performance, even before the first lab exam. As a result, instructors can spend valuable lab time where students need it most. The Mastering system empowers students to take charge of their learning through activities aimed at different learning styles and engages them in learning A&P through practice and step-by-step guidance—at their convenience, 24/7.

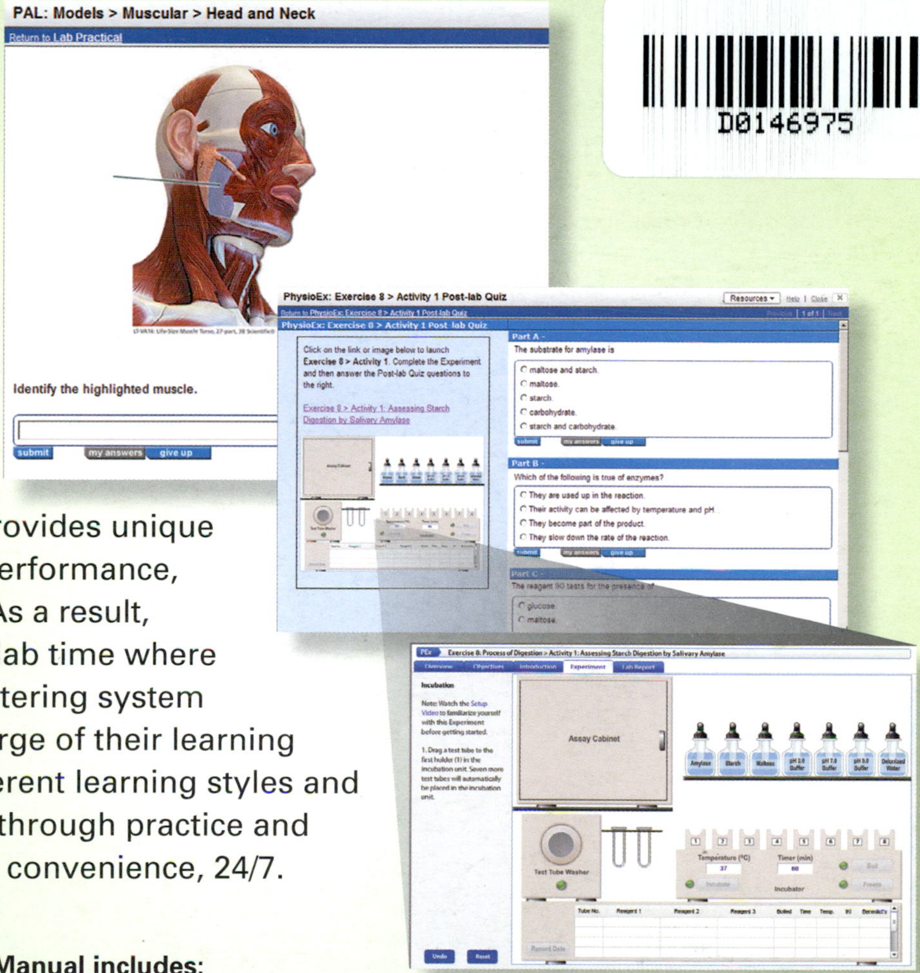
PAL: Models > Muscular > Head and Neck
Return to Lab Practical
Identify the highlighted muscle.
submit
my answers
give up
PhysioEx: Exercise 8 > Activity 1 Post-lab Quiz
Resources
Help
Close
PhysioEx: Exercise 8 > Activity 1 Post-lab Quiz
Click on the link or image below to launch Exercise 8 > Activity 1. Complete the Experiment and then answer the Post-lab Quiz questions to the right.
Exercise 8 > Activity 1: Assessing Starch Digestion by Salivary Amylase
Part A -
The substrate for amylase is
maltose and starch.
maltose.
starch.
carbohydrate.
starch and carbohydrate.
Part B -
Which of the following is true of enzymes?
They are used up in the reaction.
Their activity can be affected by temperature and pH.
They become part of the product.
They slow down the rate of the reaction.
Part C -
The reagent IKI tests for the presence of
glucose.
maltose.
Exercise 8: Process of Digestion > Activity 1: Assessing Starch Digestion by Salivary Amylase
Overview
Objectives
Introduction
Experiment
Lab Report
Incubation
Assay Cabinet
Amylase
Starch
Maltose
Test Tube Washer
Temperature (°C)
37
Timer (min)
60
Incubator
Boil
Freeze
Incubate
Tube No.
Reagent 1
Reagent 2
Reagent 3
Record Data
Undo
Reset

MasteringA&P for Marieb's HAP Lab Manual includes:

- **Access to pre-lab and post-lab quizzes for each of the 46 lab exercises, art labeling activities, PAL™ 3.0, and PhysioEx™ 9.0** in the MasteringA&P Study Area. Note: These lab manual quizzes are different than those found in the printed lab manual or in the assignments for MasteringA&P.
- **Assignable pre-and post-lab quizzes** for each of the 46 lab exercises in the lab manual.
- **Assignable quizzes and lab practicals from PAL 3.0 Testbank.**
- **Assignable pre-lab quizzes and post-lab quizzes from PhysioEx 9.0.**

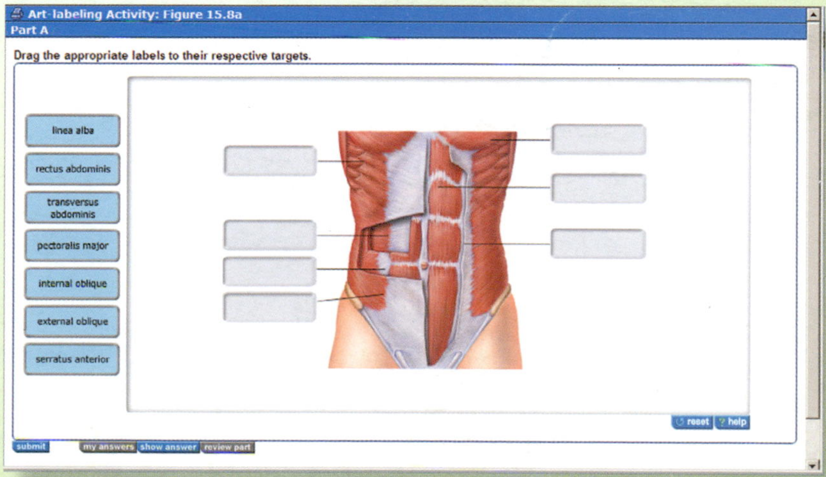
Art-labeling Activity: Figure 15.8a
Part A
Drag the appropriate labels to their respective targets.
linea alba
rectus abdominis
transversus abdominis
pectoralis major
internal oblique
external oblique
serratus anterior
reset
help
submit
my answers
show answer
review part

NEW!

Practice Anatomy Lab™ (PAL™) 3.0

PAL 3.0 is an indispensable virtual anatomy study and practice tool that gives students 24/7 access to the most widely used lab specimens, including human cadaver, anatomical models, histology, cat, and fetal pig.

NEW! Interactive Cadaver

Carefully prepared dissections show **nerves, veins,** and **arteries across body systems.**

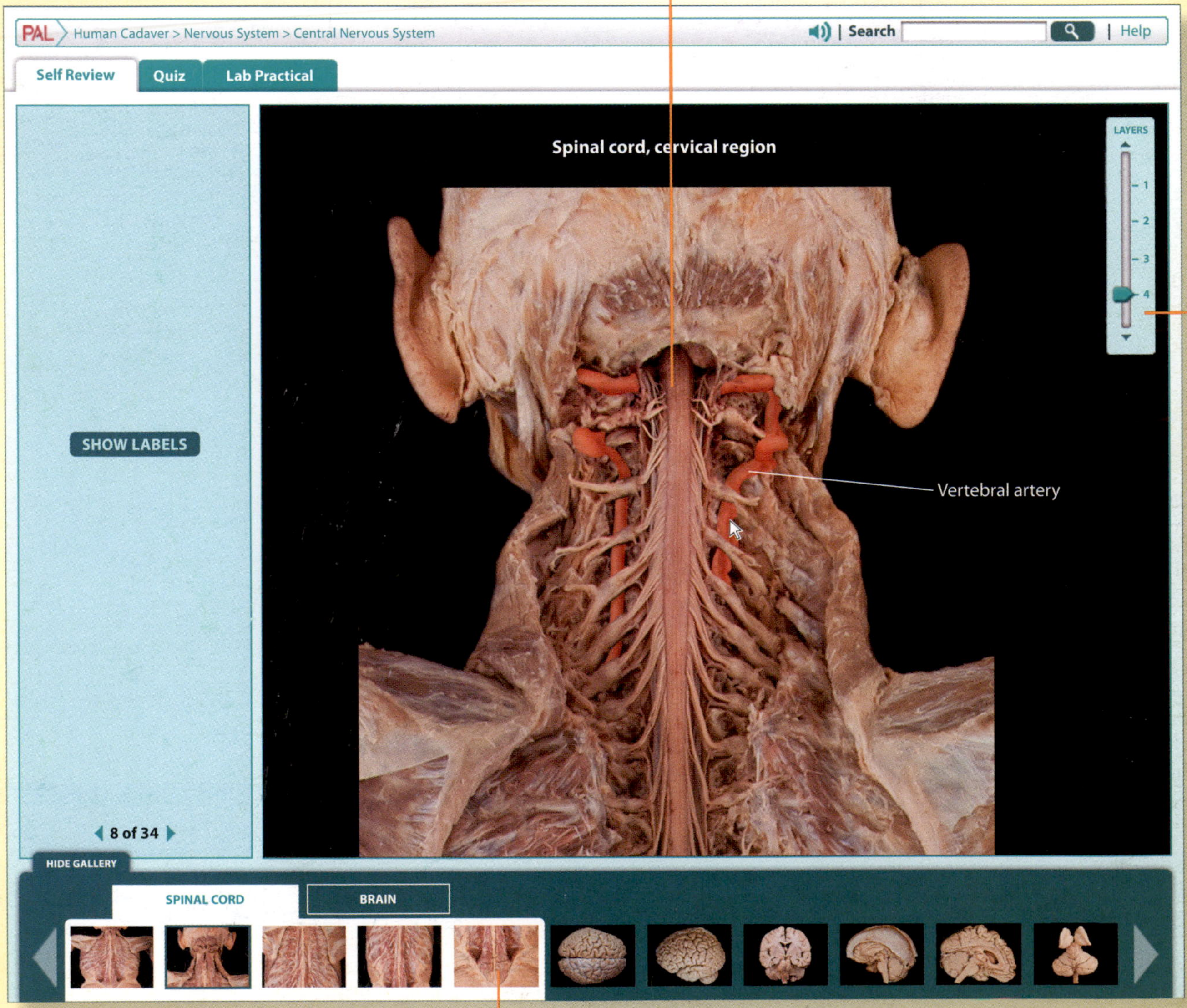

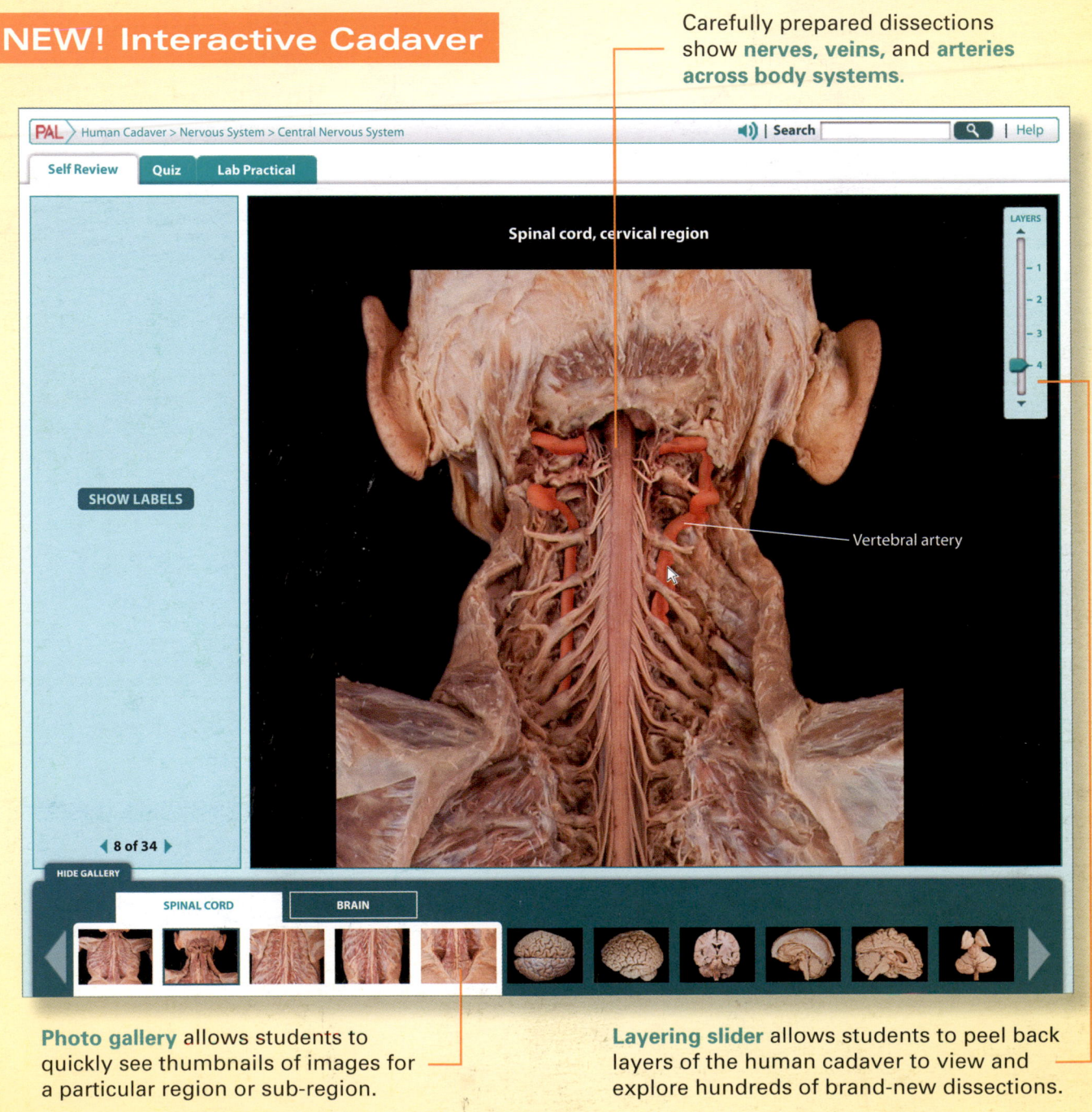

PAL
Human Cadaver > Nervous System > Central Nervous System
Search
Help
Self Review
Quiz
Lab Practical
Spinal cord, cervical region
LAYERS
1
2
3
4
SHOW LABELS
Vertebral artery
8 of 34
HIDE GALLERY
SPINAL CORD
BRAIN

Photo gallery allows students to quickly see thumbnails of images for a particular region or sub-region.

Layering slider allows students to peel back layers of the human cadaver to view and explore hundreds of brand-new dissections.

PAL 3.0 is available in MasteringA&P (www.masteringaandp.com).
The PAL 3.0 DVD is available with every new copy of this book for no additional charge.

NEW! Interactive Histology

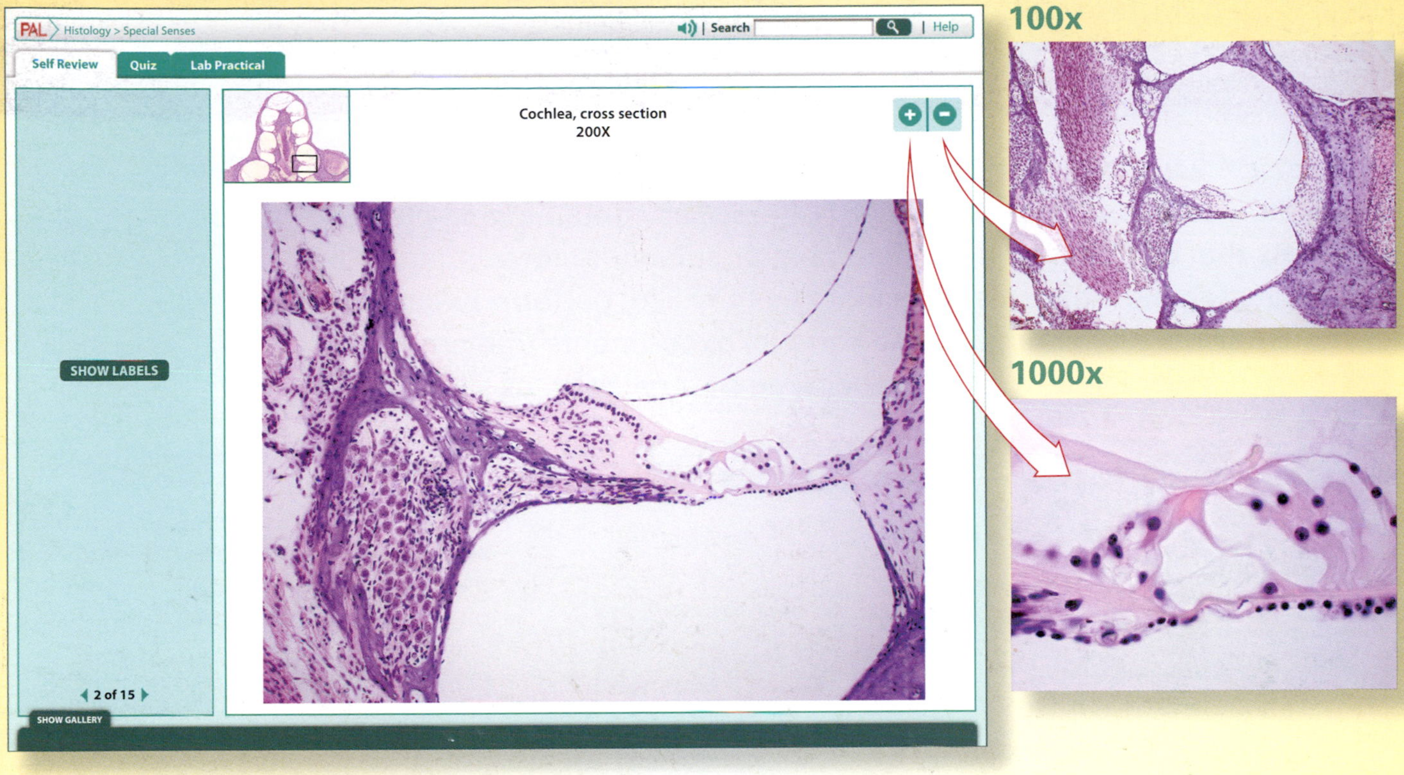

PAL
Histology > Special Senses
Search
Help
Self Review
Quiz
Lab Practical
Cochlea, cross section
200X
SHOW LABELS
2 of 15
SHOW GALLERY
100x
1000x

3-D Anatomy Animations

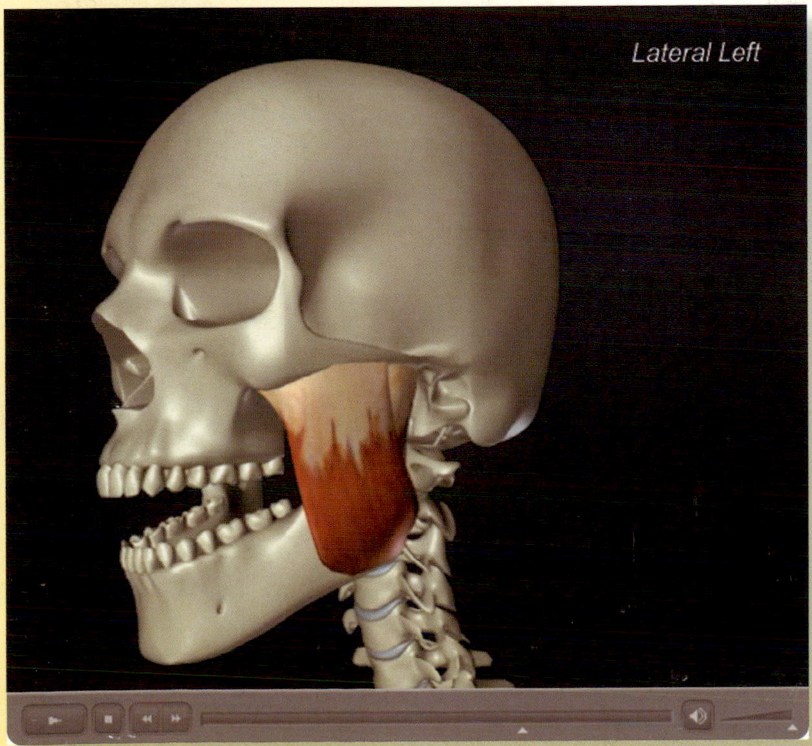

Lateral Left

PAL 3.0 also includes:

- NEW! Question randomization
- NEW! Hundreds of images and views
- NEW! IRDVD with Test Bank for PAL 3.0

SEE FOR YOURSELF!
Check out the new **PAL 3.0** at
www.masteringaandp.com

NEW!

PhysioEx™ 9.0

PhysioEx™ 9.0: Laboratory Simulations in Physiology is easy-to-use laboratory simulation software that consists of 12 exercises containing 63 physiology lab activities that can be used to supplement or substitute for wet labs. PhysioEx allows students to repeat labs as often as they like, perform experiments without harming live animals, and conduct experiments that are difficult to perform in a wet lab environment because of time, cost, or safety concerns.

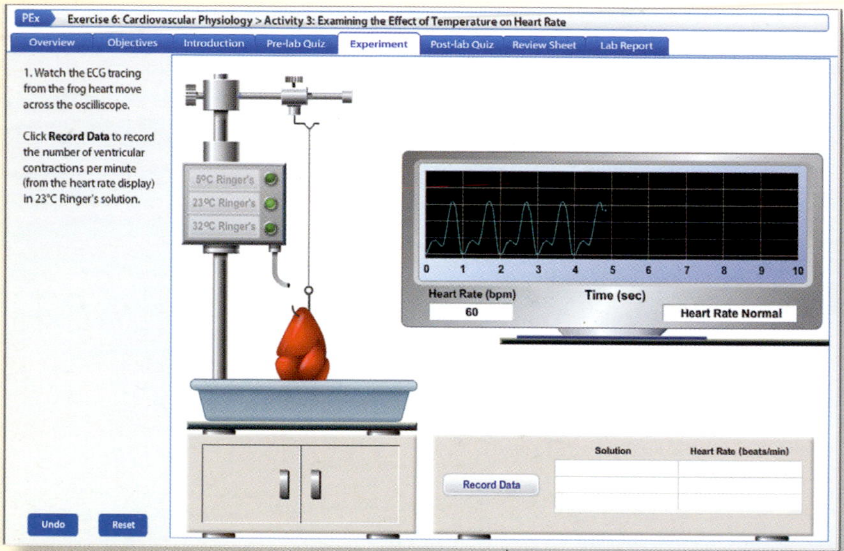

PEx
Exercise 6: Cardiovascular Physiology > Activity 3: Examining the Effect of Temperature on Heart Rate
Overview
Objectives
Introduction
Pre-lab Quiz
Experiment
Post-lab Quiz
Review Sheet
Lab Report
1. Watch the ECG tracing from the frog heart move across the oscilliscope.
Click Record Data to record the number of ventricular contractions per minute (from the heart rate display) in 23°C Ringer's solution.
5°C Ringer's
23°C Ringer's
32°C Ringer's
0 1 2 3 4 5 6 7 8 9 10
Heart Rate (bpm)
60
Time (sec)
Heart Rate Normal
Solution
Heart Rate (beats/min)
Record Data
Undo
Reset

The PhysioEx 9.0 software features:

- **Completely new online format** with **easy step-by-step instructions** so everything the student needs to do is located in one convenient place.
- **Pre-lab, Post-lab Quizzes**, and short-answer **Review sheets** for each activity give the student lots of opportunities to assess their understanding.
- **Greater data variability in the results** reflects more realistic results students would encounter in real wet-lab experiments.

- **Stop & Think** and **Predict Questions** within the steps of each experiment help students make the connection between the activities and the physiological concepts they demonstrate.
- **A Lab Report** that includes the student's answers to all of the questions and their results from the experiments. Students **can save their Lab Report** as a PDF, which they can print and/or email to their instructor.
- **Test Bank of pre-lab and post-lab quizzes** is easily assignable in **MasteringA&P**. The Test Bank is also available in TestGen® and Blackboard format.

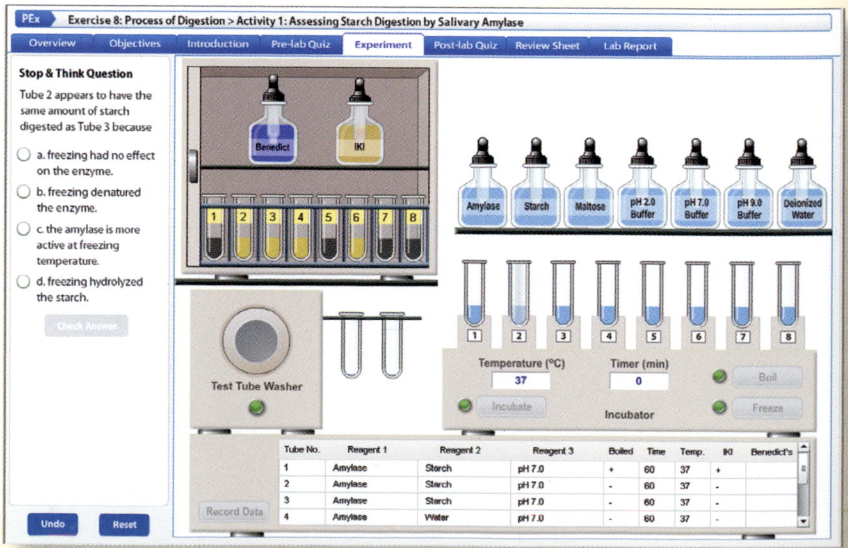

PEx
Exercise 8: Process of Digestion > Activity 1: Assessing Starch Digestion by Salivary Amylase
Overview
Objectives
Introduction
Pre-lab Quiz
Experiment
Post-lab Quiz
Review Sheet
Lab Report
Stop & Think Question
Tube 2 appears to have the same amount of starch digested as Tube 3 because
a. freezing had no effect on the enzyme.
b. freezing denatured the enzyme.
c. the amylase is more active at freezing temperature.
d. freezing hydrolyzed the starch.
Check Answer
Benedict
IKI
Amylase
Starch
Maltose
pH 2.0 Buffer
pH 7.0 Buffer
pH 9.0 Buffer
Deionized Water
Test Tube Washer
Temperature (°C)
37
Timer (min)
0
Incubate
Incubator
Boil
Freeze
Record Data
Tube No. | Reagent 1 | Reagent 2 | Reagent 3 | Boiled | Time | Temp. | IKI | Benedict's
1 | Amylase | Starch | pH 7.0 | + | 60 | 37 | +
2 | Amylase | Starch | pH 7.0 | - | 60 | 37 | -
3 | Amylase | Starch | pH 7.0 | - | 60 | 37 | -
4 | Amylase | Water | pH 7.0 | - | 60 | 37 | -
Undo
Reset

The **PhysioEx 9.0 CD-ROM** comes packaged with every new copy of the lab manual. **PhysioEx 9.0** is also available in the Study Area of MasteringA&P at www.masteringaandp.com.

Human Anatomy & Physiology Laboratory Manual

FETAL PIG VERSION

Tenth Edition Update

Elaine N. Marieb, R.N., Ph.D.
Holyoke Community College

Susan J. Mitchell, Ph.D.
Onondaga Community College

PhysioEx™ Version 9.0 authored by

Peter Z. Zao
North Idaho College

Timothy Stabler, Ph.D.
Indiana University Northwest

Lori Smith, Ph.D.
American River College

Andrew Lokuta, Ph.D.
University of Wisconsin–Madison

Edwin Griff, Ph.D.
University of Cincinnati

Benjamin Cummings

Boston Columbus Indianapolis New York San Francisco Upper Saddle River
Amsterdam Cape Town Dubai London Madrid Milan Munich Paris Montréal Toronto
Delhi Mexico City São Paulo Sydney Hong Kong Seoul Singapore Taipei Tokyo

Editor-in-Chief: Serina Beauparlant
Acquisitions Editor: Gretchen Puttkamer
Associate Project Editor: Shannon Cutt
Development Manager: Barbara Yien
Art Development Manager: Laura Southworth
Associate Project Editor for PAL 3.0: Nicole Graziano
Editorial Assistant: John Maas
Managing Editor: Deborah Cogan
Production Manager: Michele Mangelli
Production Supervisor: Janet Vail
Art Editor: Karen Gulliver
Art and Photo Coordinator: Jean Lake
Photo Researcher: Kristin Piljay
Interior and Cover Designer: Riezebos Holzbaur Design Group
Copyeditor: Antonio Padial
PhysioEx 9 Copyeditor: Anita Wagner
Compositor: S4Carlisle Publishing Services
Media Producers: Erik Fortier and Sarah Young Dualan
PhysioEx Developer: BinaryLabs, Inc.
Senior Manufacturing Buyer: Stacey Weinberger
Marketing Manager: Derek Perrigo
Cover photograph credit: Masterfile

Credits and acknowledgments for illustrations and photographs borrowed from other sources and reproduced, with permission, in this textbook appear on the appropriate page within the text or on p. BM-2.

Copyright © 2012, 2011, 2009 Pearson Education, Inc., publishing as Benjamin Cummings, 1301 Sansome St., San Francisco, CA 94111. All rights reserved. Manufactured in the United States of America. This publication is protected by Copyright and permission should be obtained from the publisher prior to any prohibited reproduction, storage in a retrieval system, or transmission in any form or by any means, electronic, mechanical, photocopying, recording, or likewise. To obtain permission(s) to use material from this work, please submit a written request to Pearson Education, Inc., Permissions Department, 1900 E. Lake Ave., Glenview, IL 60025. For information regarding permissions, call (847) 486-2635.

Many of the designations used by manufacturers and sellers to distinguish their products are claimed as trademarks. Where those designations appear in this book, and the publisher was aware of a trademark claim, the designations have been printed in initial caps or all caps.

The Authors and Publisher believe that the lab experiments described in this publication, when conducted in conformity with the safety precautions described herein and according to the school's laboratory safety procedures, are reasonably safe for the student to whom this manual is directed. Nonetheless, many of the described experiments are accompanied by some degree of risk, including human error, the failure or misuses of laboratory or electrical equipment, mismeasurement, chemical spills, and exposure to sharp objects, heat, bodily fluids, blood, or other biologics. The Authors and Publisher disclaim any liability arising from such risks in connection with any of the experiments contained in this manual. If students have any questions or problems with materials, procedures, or instructions on any experiment, they should always ask their instructor for help before proceeding.

The Benjamin Cummings Series in Human Anatomy & Physiology

By R.A. Chase
The Bassett Atlas of Human Anatomy (1989)

By Kapit/Elson
The Anatomy Coloring Book, Third Edition (2002)

By Kapit/Macey/Meisami
The Physiology Coloring Book, Second Edition (2000)

By E.N. Marieb and S.J. Mitchell
Human Anatomy & Physiology Laboratory Manual, Main Version, Ninth Edition Update (2012)

Human Anatomy & Physiology Laboratory Manual, Cat Version, Tenth Edition Update (2012)

Human Anatomy & Physiology Laboratory Manual, Fetal Pig Version, Tenth Edition Update (2012)

Human Anatomy & Physiology Laboratory Manual, Rat Version (2011)

Human Anatomy Laboratory Manual with Cat Dissections, Sixth Edition (2011)

By E.N. Marieb
The A&P Coloring Workbook: A Complete Study Guide, Tenth Edition (2012)

Laboratory Manual for Anatomy & Physiology, Fourth Edition (2011)

By Lafferty/Panella
A.D.A.M.® Interactive Anatomy Student Package, Third Edition, with Windows DVD or with Windows CD-ROM (2005)

A.D.A.M.® Interactive Anatomy Student Lab Guide, Third Edition (2005)

Benjamin Cummings
is an imprint of

PEARSON

www.pearsonhighered.com

ISBN 10: 0-321-76559-1 (student edition)
ISBN 13: 978-0-321-76559-8 (student edition)

ISBN 10: 0-321-73324-X (professional copy)
ISBN 13: 978-0-321-73324-5 (professional copy)

1 2 3 4 5 6 7 8 9 10—QGV—15 14 13 12 11

Contents

THE ENDOCRINE SYSTEM

THE CIRCULATORY SYSTEM

THE RESPIRATORY SYSTEM

THE DIGESTIVE SYSTEM

THE URINARY SYSTEM

THE REPRODUCTIVE SYSTEM, DEVELOPMENT, AND HEREDITY

SURFACE ANATOMY

DISSECTION EXERCISES

PHYSIOEX™ 9.0 COMPUTER SIMULATIONS

HISTOLOGY FIGURES

THE MICROSCOPE AND ITS USES

THE CELL

HISTOLOGY: BASIC TISSUES OF THE BODY

THE INTEGUMENTARY SYSTEM AND BODY MEMBRANES

THE SKELETAL SYSTEM

THE MUSCULAR SYSTEM

THE NERVOUS SYSTEM

THE ENDOCRINE SYSTEM

THE CIRCULATORY SYSTEM

THE RESPIRATORY SYSTEM

THE DIGESTIVE SYSTEM

THE URINARY SYSTEM

THE REPRODUCTIVE SYSTEM, DEVELOPMENT, AND HEREDITY

Preface to the Instructor

The philosophy behind the tenth edition update of this manual mirrors that of all earlier editions. It reflects a still-developing sensibility for the way teachers teach and students learn, engendered by years of teaching the subject and by listening to the suggestions of other instructors as well as those of students enrolled in multifaceted health-care programs. *Human Anatomy & Physiology Laboratory Manual, Fetal Pig Version* was originally developed to facilitate and enrich the laboratory experience for both teachers and students. This, its tenth edition update, retains those same goals.

This manual, intended for students in introductory human anatomy and physiology courses, presents a wide range of laboratory experiences for students concentrating in nursing, physical therapy, dental hygiene, pharmacology, respiratory therapy, and health and physical education, as well as biology and premedical programs. It differs from *Human Anatomy & Physiology Laboratory Manual, Main Version Update* (2012) in that it contains detailed guidelines for dissecting a laboratory animal. The manual's coverage is intentionally broad, allowing it to serve both one- and two-semester courses.

BASIC APPROACH AND FEATURES

The generous variety of experiments in this manual provides flexibility that enables instructors to gear their laboratory approach to specific academic programs, or to their own teaching preferences. The manual is still independent of any textbook, so it contains the background discussions and terminology necessary to perform all experiments. Such a self-contained learning aid eliminates the need for students to bring a textbook into the laboratory.

Each of the 46 exercises leads students toward a coherent understanding of the structure and function of the human body. The manual begins with anatomical terminology and an orientation to the body, which together provide the necessary tools for studying the various body systems. The exercises that follow reflect the dual focus of the manual—both anatomical and physiological aspects receive considerable attention. As the various organ systems of the body are introduced, the initial exercises focus on organization, from the cellular to the organ system level. As indicated by the table of contents, the anatomical exercises are usually followed by physiological experiments that familiarize students with various aspects of body functioning and promote the critical understanding that function follows structure. Homeostasis is continually emphasized as a requirement for optimal health. Pathological conditions are viewed as a loss of homeostasis; these discussions can be recognized by the homeostatic imbalance logo within the descriptive material of each exercise. This holistic approach encourages an integrated understanding of the human body.

Features

- The numerous physiological experiments for each organ system range from simple experiments that can be performed without specialized tools to more complex experiments using laboratory equipment, computers, and instrumentation techniques.
- The laboratory Review Sheets following each exercise are designed to accompany that lab exercise. The Review Sheets provide space for recording and interpreting experimental results and require students to label diagrams and answer multiple-choice and short-answer questions.
- In addition to the figures, isolated animal organs such as the sheep heart and pig kidney are employed because of their exceptional similarity to human organs.
- All exercises involving body fluids (blood, urine, saliva) incorporate current Centers for Disease Control and Prevention (CDC) guidelines for handling human body fluids. Because it is important that nursing students, in particular, learn how to safely handle bloodstained articles, the human focus has been retained. However, the decision to allow testing of human (student) blood or to use animal blood in the laboratory is left to the discretion of the instructor in accordance with institutional guidelines. The CDC guidelines for handling body fluids are reinforced by the laboratory safety procedures described on the inside front cover of this text, in Exercise 29: Blood, and in the *Instructor Guide*. You can photocopy the inside front cover and post it in the lab to help students become well versed in laboratory safety.
- Five logos alert students to special features or instructions. These include:

The dissection scissors icon appears at the beginning of activities that entail the dissection of isolated animal organs.

The homeostatic imbalance icon directs the student's attention to conditions representing a loss of homeostasis.

A safety icon notifies students that specific safety precautions must be observed when using certain equipment or conducting particular lab procedures. (For example, when working with ether, a hood is to be used, or when handling body fluids such as blood, urine, or saliva, gloves are to be worn.)

BIOPAC® The BIOPAC icon in the materials list for an exercise clearly identifies use of the BIOPAC Student Lab System and alerts you to the equipment needed. BIOPAC is used in Exercises 16, 20, 21, 22, 31, 33, 34, and 37. The instructions in the lab manual are for use with the BIOPAC MP36 (or MP35/30) data acquisition unit. New versions of the BSL software (3.7.5 and higher for Windows, 3.7.4 and higher for Mac OS X) require different channel settings and collection strategies. Instructions for their use can be found in the Instructor Resources at MasteringA&P. There you can also find instructions for the use of the new 2-channel data acquisition unit, the MP45.

PEx The PhysioEx icon at the end of the materials list for an exercise directs students to the corresponding PhysioEx computer simulation exercise found in the back of the lab manual.

- Other data aquisition instructions are available in MasteringA&P, including:

PowerLab Instructions

For Exercises 16, 22, 31, 33, 34, and 37, instructors using PowerLab equipment may print these exercises for student handouts.

iWorx Instructions

For Exercises 16, 20, 22, 31, 33, 34, and 37, instructors using iWorx equipment in their laboratory may print these exercises for student handouts.

Intelitool Instructions

Four physiological experiments (Exercises 16i, 22i, 31i, and 37i) using Intelitool® equipment are available. Instructors using Intelitool equipment in their laboratory may print these exercises for student handouts.

WHAT'S NEW

In this revision, we have continued to try to respond to reviewers' and users' feedback concerning trends that are having an impact on the anatomy and physiology laboratory experience, most importantly:

- The growing demand for student-based experimentation
- The increased use of computers in the laboratory and in students' homes, and hence the continuing desire for more computer simulation and practice exercises
- The replacement of older recording equipment with computerized data acquisition and analysis systems
- The continued importance of visual learning for today's student
- The need to reinforce writing, computation, and critical thinking skills across the curriculum

The specific changes implemented to address these trends are described next.

Pre-lab Quizzes

Brand new pre-lab quizzes at the beginning of each exercise motivate your students to prepare for lab by asking them basic information they should know before doing the lab. These quizzes are different than those found on MasteringA&P and those in the Instuctor Test Bank.

All-New Art Program

This brand-new art program uses three-dimensional, realistic styles with dramatic views and perspectives, and rich, vibrant colors. The art includes key anatomy figures that are rendered with detail, depth, and a clear focus on key anatomical structures. Ten new and improved histology images have been added. Images from the Histology Atlas have been integrated into the lab exercises so that students can review relevant histology images all in one place. See pages vi–vii for a complete listing of histology figures.

Also included are five new cadaver photos and all-new surface anatomy photos showing superb muscle definition and clear surface landmarks for skeletal, muscular, and vascular structures.

Customization Options

With Integrate, our custom laboratory publishing program for anatomy and physiology, you have the freedom to choose exercises from our library of highly regarded Pearson Benjamin Cummings lab manuals and other collections to build the right manual for your course. You'll find excellent, class-tested exercises—many printed in full color—for one- and two-semester anatomy and physiology laboratories, and one-semester human anatomy or physiology courses. Use our online Book-Build system to select just the exercises you need, in the sequence you want—your students pay only for the exercises you choose. You can also add your own original exercises. With Integrate, you're in control. For more information, visit our Integrate website at www.pearsoncustom/integrate, or contact your Pearson sales representative for details.

SUPPLEMENTS FOR THE STUDENT

New! Practice Anatomy Lab 3.0

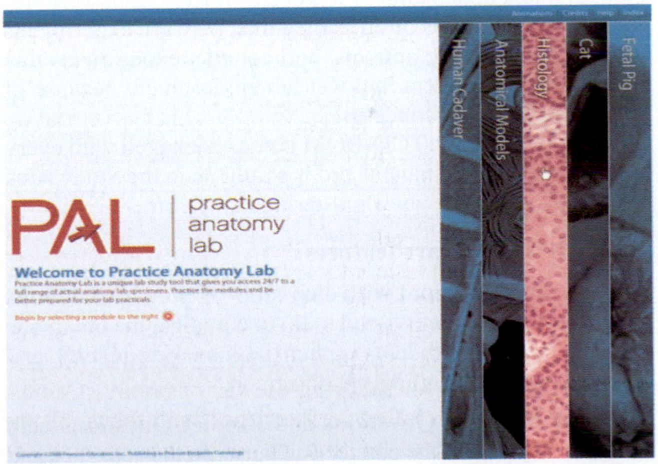

Human Cadaver
Anatomical Models
Histology
Cat
Fetal Pig
PAL
practice anatomy lab
Welcome to Practice Anatomy Lab

Practice Anatomy Lab (PAL) 3.0 is an indispensable virtual anatomy study and practice tool that gives students 24/7 access to the most widely used lab specimens including **human cadaver, anatomical models, histology, cat,** and **fetal pig.**

PAL 3.0 features:

- **A whole new interactive cadaver** that allows students to peel back layers of the human cadaver and view hundreds of brand-new dissection photographs specially commissioned for 3.0
- **New interactive histology** allows students to view the same tissue slide at varying magnifications, thereby helping the student identify structures and their characteristics.

- **Expanded, randomizable quizzing** gives students more opportunity to practice to assess learning—each time the student takes a quiz or lab practical exam, a new set of questions is generated.
- **New integration of nerves, arteries, and veins** across body systems.
- **Integrated muscle animations** of the origin, insertion, action, and innervations of key muscles.
- **Rotatable bones** help students appreciate the three-dimensionality of bone structures.

The Instructor's Resource DVD for PAL 3.0 includes images in PowerPoint with editable labels and leader lines, labeled and unlabeled images in JPEG and PowerPoint format, JPEGs with single structure highlight and leader for quizzing and testing, and PowerPoint image slides featuring embedded links to relevant 3D anatomy animations and bone rotations. Test Bank includes more than 4000 multiple-choice quiz and fill-in-the-blank lab practical exam questions available in PDF and TestGen format.

PAL 3.0 is available in the Study Area of MasteringA&P at www.masteringaandp.com. The PAL 3.0 DVD can also be packaged with Marieb's lab manual at no additional charge.

NEW! PhysioEx™ 9.0 CD-ROM

The PhysioEx CD-ROM is easy-to-use laboratory simulation software and lab exercises that consist of 12 exercises containing a total of 63 physiology lab activities. It can be used to supplement or substitute for wet labs. PhysioEx allows students to repeat labs as often as they like, perform experiments without harming live animals, and conduct experiments that are difficult to perform in a wet lab environment because of time, cost, or safety concerns.

The PhysioEx 9.0 CD-ROM comes packaged with every new copy of the lab manual and is available in the Study Area of MasteringA&P at www.masteringaandp.com.

PhysioEx 9.0 software features:

- **New online format with easy step-by-step instructions** puts everything students need to do to complete the lab in one convenient place. Students gather data, analyze results, and check their understanding, all on screen.
- **Stop & Think Questions** and **Predict Questions** help students think about the connection between the activities and the physiological concepts they demonstrate.
- **Greater data variability in the results** reflects more realistically the results that students would encounter in a wet lab experiment.
- **New Pre-lab and Post-lab Quizzes** and short-answer **Review Sheets** are offered for every activity.
- **Students can save their Lab Report as a PDF,** which they can print and/or email to their instructor.
- **A Test Bank of assignable pre-lab and post-lab quizzes** for use with TestGen® or its course management system is provided for instructors.
- **Seven videos of lab experiments** demonstrate the actual experiments simulated on screen, making it easy for students to understand and visualize the context of the simulations. Videos demonstrate the following experiments: Skeletal Muscle, Blood Typing, Cardiovascular Physiology, Use of a Water-Filled Spirometer, Nerve Impulses, BMR Measurement, and Cell Transport.
- **Convenient online access** is available at the companion website for this lab manual at www.masteringaandp.com. (To log onto the site, use the access kit that may have come with your lab manual or purchase access at the website.)

PhysioEx 9.0 topics include:

- Exercise 1: *Cell Transport Mechanisms and Permeability.* Explores how substances cross the cell's membrane. Topics covered include: simple and facilitated diffusion, osmosis, filtration, and active transport.
- Exercise 2: *Skeletal Muscle Physiology.* Provides insights into the complex physiology of skeletal muscle. Topics include: electrical stimulation, isometric contractions, and isotonic contractions.
- Exercise 3: *Neurophysiology of Nerve Impulses.* Investigates stimuli that elicit action potentials, stimuli that inhibit action potentials, and factors affecting the conduction velocity of an action potential.
- Exercise 4: *Endocrine System Physiology.* Investigates the relationship between hormones and metabolism; the effect of estrogen replacement therapy; the diagnosis of diabetes; and the relationship between the levels of cortisol and adrenocorticotropic hormone and a variety of endocrine disorders.
- Exercise 5: *Cardiovascular Dynamics.* Allows students to perform experiments that would be difficult if not impossible to do in a traditional laboratory. Topics include vessel resistance and pump (heart) mechanics.
- Exercise 6: *Cardiovascular Physiology.* Examines variables influencing heart activity. Topics include: setting up and recording baseline heart activity, the refractory period of cardiac muscle, and an investigation of factors that affect heart rate and contractility.
- Exercise 7: *Respiratory System Mechanics.* Investigates physical and chemical aspects of pulmonary function. Students collect data simulating normal lung volumes. Other activities examine factors such as airway resistance and the effect of surfactant on lung function.
- Exercise 8: *Chemical and Physical Processes of Digestion.* Examines factors that affect enzyme activity by manipulating (in compressed time) enzymes, reagents, and incubation conditions.
- Exercise 9: *Renal System Physiology.* Simulates the function of a single nephron. Topics include: factors influencing glomerular filtration, the effect of hormones on urine function, and glucose transport maximum.
- Exercise 10: *Acid-Base Balance.* Topics include: respiratory and metabolic acidosis/alkalosis, and renal and respiratory compensation.
- Exercise 11: *Blood Analysis.* Topics include: hematocrit determination, erythrocyte sedimentation rate determination, hemoglobin determination, blood typing, and total cholesterol determination.
- Exercise 12: *Serological Testing.* Investigates antigen-antibody reactions and their role in clinical tests used to diagnose a disease or an infection.

New! MasteringA&P™

See description in next section.

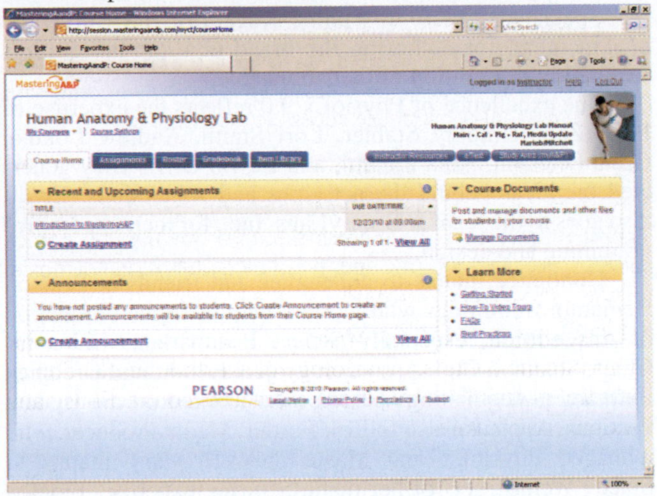
Human Anatomy & Physiology Lab
Recent and Upcoming Assignments
Course Documents
Announcements
Learn More
PEARSON

Also Available

A Brief Atlas of the Human Body, Second Edition (0-321-66261-X)

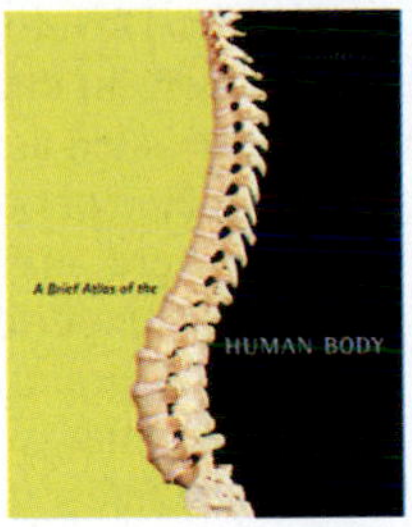
A Brief Atlas of the
HUMAN BODY

This full-color atlas includes 107 bone and 47 soft-tissue photographs with easy-to-read labels. This edition of the atlas contains a brand-new, comprehensive histology photomicrograph section with more than 50 slides of basic tissue and organ systems. Featuring photos taken by renowned biomedical photographer Ralph Hutchings, this high-quality photographic atlas makes an excellent resource for the classroom and laboratory. Available for purchase from Pearson Education.

SUPPLEMENTS FOR THE INSTRUCTOR

New! MasteringA&P™

MasteringA&P is an online learning and assessment system proven to help students learn. It helps instructors maximize lab time with customizable, easy-to-assign, automatically graded assessments that motivate students to learn outside of class and arrive prepared for lab. The powerful gradebook provides unique insight into student and class performance, even before the first lab exam. As a result, instructors can spend valuable time where students need it most. The Mastering system empowers students to take charge of their learning through activities aimed at different learning styles and engages them in learning A&P through practice and step-by-step guidance—at their convenience, 24/7.

MasteringA&P for Marieb's HAP Lab Manual includes:

- **Access to PAL 3.0, PhysioEx 9.0, pre-lab and post-lab quizzes for each of the 46 lab exercises, and art labeling activities** in the MasteringA&P Study Area. Note: These quizzes found in the Study Area differ from those found in the printed lab manual, as well as the electronic test bank, to allow for a variety of practice.
- **Assignable pre- and post-lab quizzes** that differ from the printed lab manual for each of the 46 lab exercises in the lab manual.
- **Assignable quizzes and lab practical exams from PAL 3.0 Test Bank.**
- **Assignable pre-lab quizzes and post-lab quizzes from PhysioEx 9.0.**
- **Assignable bone and dissection videos with assessments.**
- Instructor access to the Instructor Guide and procedures for using **Powerlab®, iWorx®, and Intelitool®.**

A 24-month subscription to this powerful website is included with each new copy of the lab manual. Access directions, along with access codes, are included in the access kit.

New! Electronic Test Bank of Pre-lab and Post-lab Quizzes

This brand-new electronic Test Bank of pre-lab and post-lab quizzes, different than those found in the lab manual and in the Study Area of MasteringA&P, allows you to assign these quizzes for homework or extra credit. These quizzes can be assigned in MasteringA&P or downloaded from the Instructor Resources section of MasteringA&P and imported into **Blackboard, CourseCompass,** and other course-management systems.

New! Instructor Resource DVD with Test Bank for PAL 3.0

This DVD includes everything an instructor needs for lecture and lab presentations and assessments using PAL 3.0.

The DVD includes:

- Images in PowerPoint with editable labels and leader lines.
- Labeled and unlabeled images in JPEG and PowerPoint format.
- JPEGs with single structure highlight and leader for quizzing and testing.
- PowerPoint image slides featuring embedded links to relevant 3D anatomy animations and bone rotations.
- Test Bank that includes more than 4000 multiple-choice quiz and fill-in-the-blank lab practical exam questions available in PDF and TestGen format. Includes files for importing the Test Bank into Blackboard and other course management systems.
- An index of anatomical structures covered in PAL 3.0.

MasteringA&P™ Instructor Resources

Instructor Resources include the Instructor Guide, procedures for using **PowerLab®, iWorx®,** and **Intelitool®** equipment with certain lab exercises, and instructions for BIOPAC exercises using the newest software (3.7.5 for Windows and 3.7.4 for Mac) and data acquisition unit (MP45).

Instructor Guide (0-321-75225-2)

This guide accompanies all versions of the *Human Anatomy & Physiology Laboratory Manual* and contains a wealth of information for instructors, including answers to the new pre-lab quizzes and the new PhysioEx 9.0 pre-lab quizzes, activity questions, and review sheets. Instructors can find help in planning the experiments, ordering equipment and supplies, anticipating pitfalls and problem areas, and locating audiovisual material. The probable in-class time required for each lab is indicated by an hourglass icon. Another useful resource is the Trends in Instrumentation section, which describes the latest laboratory equipment and technological teaching tools available.

ACKNOWLEDGMENTS

We wish to thank the following reviewers for their contributions to this edition: C. Thomas G. Appleton, University of Western Ontario; Richard Connett, Monroe Community College; Deanna Denault, New Hampshire Community Technical College-Claremont; Smruti Desai, Lone Star College CyFair; Jose Fierro, Florida State College at Jacksonville; Katelijne Flies, Central New Mexico Community College; Lynn Gargan, Tarrant County College Northeast; Lori Garrett, Parkland College; Fran Hardin, Ivy Tech Community College—Kokomo; DJ Hennager, Kirkwood Community College; William Hoover, Bunker Hill Community College; Sandra Hsu, Skyline College; William Huber, St. Louis Community College; Jesse Lang, Holyoke Community College; Stephen Lebsack, Linn Benton Community College; Linda Mackie, St. Johns River Community College; Cherie McKeever, Montana State University—Great Falls, College of Technology; Judy Megaw, Indian River Community College; Ellen Ott-Reeves, Blinn Community College; Steve Perry, Liberty University; Jean Revie, South Mountain Community College; Laura Ritt, Burlington Community College; Josephine Rogers, University of Cincinnati; Dale Smoak, Piedmont Technical College; Pam Strong, Quincy College; Diane Teter, South Texas College; Harriett Tresham, Kennesaw Community College; Maureen Tubbiola, St. Cloud State; Wendy Waters, Wharton Junior College.

We would like to extend a special thank you to the following authors and contributors of Practice Anatomy Lab 3.0: Ruth E. Heisler, University of Colorado at Boulder; Nora Hebert, Red Rocks Community College; Jett Chinn, Cañada College and College of Marin; Karen M. Krabbenhoft, University of Wisconsin-Madison; Olga Malakhova, University of Florida at Gainesville, College of Medicine; Lisa M.J. Lee, The Ohio State University College of Medicine; Larry DeLay, Waubonsee Community College; Patricia B. Wilhelm, Community College of Rhode Island-Warwick; Leslie C. Hendon, University of Alabama-Birmingham; Samuel Chen, Moraine Valley Community College; Leif Saul, University of Colorado at Boulder; Eric Howell, Red Rocks Community College; Steve Downing, University of Minnesota Medical School-Duluth; Yvonne Baptiste-Syzmanski, Niagara County Community College; Charles Venglarik, Jefferson State Community College; Nina Zanetti, Siena College.

Special thanks to Josephine Rogers of the University of Cincinnati for authoring the brand new pre-lab quizzes featured in this edition of the lab manual.

The excellence of PhysioEx 9.0 reflects the expertise of Peter Zao, Timothy Stabler, Lori Smith, Andrew Lokuta, Greta Peterson, Nina Zanetti, and Edwin Griff. They generated the ideas behind the activities and simulations. Credit also goes to the team at BinaryLabs, Inc., for their expert programming and design.

Continued thanks to colleagues and friends at Pearson Benjamin Cummings who worked with us in the production of this edition, especially Serina Beauparlant, Editor-in-Chief; Shannon Cutt, Associate Project Editor; and Gretchen Puttkamer, Acquisitions Editor, who worked on the Update versions. Applause also to Erik Fortier, Media Producer, who managed PhysioEx and MasteringA&P. Many thanks to Stacey Weinberger for her manufacturing expertise. Finally, our Marketing Manager, Derek Perrigo, has efficiently kept us in touch with the pulse of the market place.

Kudos also to Michele Mangelli and her production team, who did their usual great job. Janet Vail, production editor for this project, got the job done in jig time. Laura Southworth, Art Development Manager, and Karen Gulliver, Art Editor, were in charge of overseeing the entire art program. Jean Lake acted as art and photo coordinator, and Kristin Piljay conducted photo research. Our fabulous interior and cover designs were created by Yvo Riezebos. Antonio Padial brought his experience to copyediting the text.

We are grateful to the team at BIOPAC, especially to Jocelyn Kremer, who was extremely helpful in making sure we had the latest updates and answering all of our questions.

Last but not least, thank you to Robert Sullivan of Marist College for authoring the rat dissection exercises found in the brand-new Rat Version of this lab manual.

Elaine N. Marieb

Susan J. Mitchell

Elaine N. Marieb and
Susan J. Mitchell
Anatomy and Physiology
Pearson Benjamin Cummings
1301 Sansome Street
San Francisco, CA 94111

Preface to the Student

We hope you will enjoy your laboratory experiences. As with any unfamiliar experience, it really helps if you know in advance what to expect and what will be expected of you.

LABORATORY ACTIVITIES

The A&P laboratory exercises in this manual are designed to help you gain a broad understanding of both anatomy and physiology. You can anticipate examining models, dissecting a specimen, and using a microscope to look at tissue slides (anatomical approaches). You will also investigate chemical interactions in both living and nonliving systems, manipulate variables in computer simulations, and conduct experiments that examine responses of living organisms to various stimuli (physiological approaches).

Preserved organ specimens used in the anatomy and physiology labs are *not* harvested from animals raised specifically for dissection purposes. Organs that are of no use to the meat packing industry (such as the brain, heart, or lungs) are sent from slaughterhouses to biological supply houses for preparation.

Included with this edition is a companion website designed to help you practice outside of class for lab practical exams. Also included with the manual is the PhysioEx CD-ROM, with 12 simulation exercises that allow you to convert a computer into a virtual laboratory. You will be able to manipulate variables to investigate physiological phenomena.

There is little doubt that computer simulations offer many advantages; however, an animated frog muscle or heart on a computer screen is not really a substitute for observing the responses of actual muscle tissue. Consequently, living animal experiments remain an important part of the approach of this manual to the study of human anatomy and physiology. However, we use the minimum number of animals needed.

If you use living animals for experiments, you are required to handle them humanely. Inconsiderate treatment of laboratory animals will not be tolerated in your anatomy and physiology laboratory.

ICONS/VISUAL MNEMONICS

We have tried to make this manual easy for you to use by employing different icons (visual mnemonics) throughout:

The *Dissection* head is orange and is accompanied by the **dissection scissors icon** at the beginning of activities that require you to dissect isolated animal organs.

The **homeostatic imbalance icon** directs your attention to conditions representing a disruption of homeostasis.

The **safety icon** alerts you to special precautions that you should take when handling lab equipment or conducting certain procedures. For example, it alerts you to use a ventilating hood when using volatile chemicals and signifies that you should take special measures to protect yourself when handling blood or other body fluids (e.g., saliva, urine).

BIOPAC® The **BIOPAC icon** in the materials list for an exercise clearly identifies use of the BIOPAC Student Lab System. BIOPAC is used in Exercises 16, 20, 21, 22, 31, 33, 34, and 37.

PEx The **PhysioEx icon** at the end of the materials list for an exercise directs you to the corresponding PhysioEx computer simulation exercise found in the back of the lab manual.

HINTS FOR SUCCESS IN THE LABORATORY

Most students can use helpful hints and guidelines to ensure that they have successful lab experiences.

1. Perhaps the best bit of advice is to attend all your scheduled labs and to participate in all the assigned exercises. Learning is an *active* process.

2. Scan the scheduled lab exercise and the questions in the Review Sheet following it *before* going to lab, then complete the pre-lab quiz.

3. Be on time. Most instructors explain what the lab is about, pitfalls to avoid, and the sequence or format to be followed at the beginning of the lab session. If you are late, not only will you miss this information, you also will not endear yourself to the instructor.

4. Follow the instructions in the order in which they are given. If you do not understand a direction, ask for help.

5. Review your lab notes after completing the lab session to help you focus on and remember the important concepts.

6. Keep your work area clean and neat. Move books and coats out of the way. This reduces confusion and accidents.

7. Assume that all lab chemicals and equipment are sources of potential danger to you. Follow directions for equipment use and observe the laboratory safety guidelines provided inside the front cover of this manual.

8. Keep in mind the real value of the laboratory experience—a place for you to observe, manipulate, and experience hands-on activities that will dramatically enhance your understanding of the lecture presentations.

We hope that this lab manual makes learning about intricate structures and functions of the human body a fun and rewarding process. We are always open to constructive criticism and suggestions for improvement in future editions. If you have any, please write to us.

Elaine N. Marieb and
Susan J. Mitchell
Anatomy and Physiology
Pearson Benjamin Cummings
1301 Sansome Street
San Francisco, CA 94111

Getting Started—What to Expect, The Scientific Method, and Metrics

Two hundred years ago science was largely a plaything of wealthy patrons, but today's world is dominated by science and its technology. Whether or not we believe that such domination is desirable, we all have a responsibility to try to understand the goals and methods of science that have seeded this knowledge and technological explosion.

The biosciences are very special and exciting because they open the doors to an understanding of all the wondrous workings of living things. A course in human anatomy and physiology (a minute subdivision of bioscience) provides such insights in relation to your own body. Although some experience in scientific studies is helpful when beginning a study of anatomy and physiology, perhaps the single most important prerequisite is curiosity.

Gaining an understanding of science is a little like becoming acquainted with another person. Even though a written description can provide a good deal of information about the person, you can never really know another unless there is personal contact. And so it is with science—if you are to know it well, you must deal with it intimately.

The laboratory is the setting for "intimate contact" with science. It is where scientists test their ideas (do research), the essential purpose of which is to provide a basis from which predictions about scientific phenomena can be made. Likewise, it will be the site of your "intimate contact" with the subject of human anatomy and physiology as you are introduced to the methods and instruments used in biological research.

For many students, human anatomy and physiology is taken as an introductory-level course; and their scientific background exists, at best, as a dim memory. If this is your predicament, this prologue may be just what you need to fill in a few gaps and to get you started on the right track before your actual laboratory experiences begin. So—let's get to it!

THE SCIENTIFIC METHOD

Science would quickly stagnate if new knowledge were not continually derived from and added to it. The approach commonly used by scientists when they investigate various aspects of their respective disciplines is called the **scientific method.** This method is *not* a single rigorous technique that must be followed in a lockstep manner. It is nothing more or less than a logical, practical, and reliable way of approaching and solving problems of every kind—scientific or otherwise—to gain knowledge. It comprises five major steps.

Step 1: Observation of Phenomena

The crucial first step involves observation of some phenomenon of interest. In other words, before a scientist can investigate anything, he or she must decide on a *problem* or focus for the investigation. In most college laboratory experiments, the problem or focus has been decided for you. However, to illustrate this important step, we will assume that you want to investigate the true nature of apples, particularly green apples. In such a case you would begin your studies by making a number of different observations concerning apples.

Step 2: Statement of the Hypothesis

Once you have decided on a focus of concern, the next step is to design a significant question to be answered. Such a question is usually posed in the form of a **hypothesis,** an unproven conclusion that attempts to explain some phenomenon. (At its crudest level, a hypothesis can be considered to be a "guess" or an intuitive hunch that tentatively explains some observation.) Generally, scientists do not restrict themselves to a single hypothesis; instead, they usually pose several and then test each one systematically.

We will assume that, to accomplish step 1, you go to the supermarket and randomly select apples from several bins. When you later eat the apples, you find that the green apples are sour, but the red and yellow apples are sweet. From this observation, you might conclude *(hypothesize)* that "green apples are sour." This statement would represent your current understanding of green apples. You might also reasonably predict that if you were to buy more apples, any green ones you buy will be sour. Thus, you would have gone beyond your initial observation that "these" green apples are sour to the prediction that "all" green apples are sour.

Any good hypothesis must meet several criteria. First, *it must be testable.* This characteristic is far more important than its being correct. The test data may or may not support the hypothesis, or new information may require that the hypothesis be modified. Clearly the accuracy of a prediction in any scientific study depends on the accuracy of the initial information on which it is based.

In our example, no great harm will come from an inaccurate prediction—that is, were we to find that some green apples are sweet. However, in some cases human life may depend on the accuracy of the prediction; thus: (1) Repeated testing of scientific ideas is important, particularly because scientists working on the same problem do not always agree in their conclusions. (2) Careful observation is essential, even at the very outset of a study, because conclusions drawn from scientific tests are only as accurate as the information on which they are based.

A second criterion is that, even though hypotheses are guesses of a sort, *they must be based on measurable, describable facts. No mysticism can be theorized.* We cannot conjure up, to support our hypothesis, forces that have not been shown to exist. For example, as scientists, we cannot say that the tooth fairy took Johnny's tooth unless we can prove that the tooth fairy exists!

Third, a hypothesis *must not be anthropomorphic.* Human beings tend to anthropomorphize—that is, to relate all experiences to human experience. Whereas we could state that bears instinctively protect their young, it would be anthropomorphic to say that bears love their young, because love is a human emotional response. Thus, the initial hypothesis must be stated without interpretation.

Step 3: Data Collection

Once the initial hypothesis has been stated, scientists plan experiments that will provide data (or evidence) to support or disprove their hypotheses—that is, they *test* their hypotheses. Data are accumulated by making qualitative or quantitative observations of some sort. The observations are often aided by the use of various types of equipment such as cameras, microscopes, stimulators, or various electronic devices that allow chemical and physiological measurements to be taken.

Observations referred to as **qualitative** are those we can make with our senses—that is, by using our vision, hearing, or sense of taste, smell, or touch. For some quick practice in qualitative observation, compare and contrast* an orange and an apple.

Whereas the differences between an apple and an orange are obvious, this is not always the case in biological observations. Quite often a scientist tries to detect very subtle differences that cannot be determined by qualitative observations; data must be derived from measurements. Such observations based on precise measurements of one type or another are **quantitative observations**. Examples of quantitative observations include careful measurements of body or organ dimensions such as mass, size, and volume; measurement of volumes of oxygen consumed during metabolic studies; determination of the concentration of glucose in urine; and determination of the differences in blood pressure and pulse under conditions of rest and exercise. An apple and an orange could be compared quantitatively by analyzing the relative amounts of sugar and water in a given volume of fruit flesh, the pigments and vitamins present in the apple skin and orange peel, and so on.

A valuable part of data gathering is the use of experiments to support or disprove a hypothesis. An **experiment** is a procedure designed to describe the factors in a given situation that affect one another (that is, to discover cause and effect) under certain conditions.

Two general rules govern experimentation. The first of these rules is that the experiment(s) should be conducted in such a manner that every **variable** (any factor that might affect the outcome of the experiment) is under the control of the experimenter. The **independent variables** are manipulated by the experimenter. For example, if the goal is to determine the effect of body temperature on breathing rate, the independent variable is body temperature. The effect observed or value measured (in this case breathing rate) is called the **dependent** or **response variable**. Its value "depends" on the value chosen for the independent variable. The ideal way to perform such an experiment is to set up and run a series of tests that are all identical, except for one specific factor that is varied.

* *Compare* means to emphasize the similarities between two things, whereas *contrast* means that the differences are to be emphasized.

One specimen (or group of specimens) is used as the **control** against which all other experimental samples are compared. The importance of the control sample cannot be overemphasized. The control group provides the "normal standard" against which all other samples are compared relative to the dependent variable. Taking our example one step further, if we wanted to investigate the effects of body temperature (the independent variable) on breathing rate (the dependent variable), we could collect data on the breathing rate of individuals with "normal" body temperature (the implicit control group), and compare these data to breathing-rate measurements obtained from groups of individuals with higher and lower body temperatures.

The second rule governing experimentation is that valid results require that testing be done on large numbers of subjects. It is essential to understand that it is nearly impossible to control all possible variables in biological tests. Indeed, there is a bit of scientific wisdom that mirrors this truth—that is, that laboratory animals, even in the most rigidly controlled and carefully designed experiments, "will do as they damn well please." Thus, stating that the testing of a drug for its painkilling effects was successful after having tested it on only one postoperative patient would be scientific suicide. Large numbers of patients would have to receive the drug and be monitored for a decrease in postoperative pain before such a statement could have any scientific validity. Then, other researchers would have to be able to uphold those conclusions by running similar experiments. *Repeatability* is an important part of the scientific method and is the primary basis for support or rejection of many hypotheses.

During experimentation and observation, data must be carefully recorded. Usually, such initial, or raw, data are recorded in table form. The table should be labeled to show the variables investigated and the results for each sample. At this point, *accurate recording* of observations is the primary concern. Later, these raw data will be reorganized and manipulated to show more explicitly the outcome of the experimentation.

Some of the observations that you will be asked to make in the anatomy and physiology laboratory will require that a drawing be made. Don't panic! The purpose of making drawings (in addition to providing a record) is to force you to observe things very closely. You need not be an artist (most biological drawings are simple outline drawings), but you do need to be neat and as accurate as possible. It is advisable to use a 4H pencil to do your drawings because it is easily erased and doesn't smudge. Before beginning to draw, you should examine your specimen closely, studying it as though you were going to have to draw it from memory. For example, when looking at cells you should ask yourself questions such as "What is their shape—the relationship of length and width? How are they joined together?" Then decide precisely what you are going to show and how large the drawing must be to show the necessary detail. After making the drawing, add labels in the margins and connect them by straight lines (leader lines) to the structures being named.

Step 4: Manipulation and Analysis of Data

The form of the final data varies, depending on the nature of the data collected. Usually, the final data represent information converted from the original measured values (raw data) to some other form. This may mean that averaging or some other statistical treatment must be applied, or it may require conversions from one kind of units to another. In other cases, graphs may be needed to display the data.

Elementary Treatment of Data

Only very elementary statistical treatment of data is required in this manual. For example, you will be expected to understand and/or compute an average (mean), percentages, and a range.

Two of these statistics, the mean and the range, are useful in describing the *typical* case among a large number of samples evaluated. Let us use a simple example. We will assume that the following heart rates (in beats/min) were recorded during an experiment: 64, 70, 82, 94, 85, 75, 72, 78. If you put these numbers in numerical order, the **range** is easily computed, because the range is the difference between the highest and lowest numbers obtained (highest number minus lowest number). The **mean** is obtained by summing the items and dividing the sum by the number of items. What is the range and the mean for the set of numbers just provided?

1. ______________________________________ *

The word *percent* comes from the Latin meaning "for 100"; thus *percent,* indicated by the percent sign, %, means parts per 100 parts. Thus, if we say that 45% of Americans have type O blood, what we are really saying is that among each group of 100 Americans, 45 (45/100) can be expected to have type O blood. Any ratio can be converted to a percent by multiplying by 100 and adding the percent sign.

$.25 \times 100 = 25\% \qquad 5 \times 100 = 500\%$

It is very easy to convert any number (including decimals) to a percent. The rule is to move the decimal point two places to the right and add the percent sign. If no decimal point appears, it is *assumed* to be at the end of the number; and zeros are added to fill any empty spaces. Two examples follow:

0.25 = 0.25 = 25%

5 = 5 = 500%

Change the following to percents:

2. 38 = __________ **4.** 1.6 = __________

3. .75 = __________

Note that although you are being asked here to convert numbers to percents, percents by themselves are meaningless. We always speak in terms of a percentage *of* something.

To change a percent to decimal form, remove the percent sign, and divide by 100. Change the following percents to whole numbers or decimals:

5. 800% = __________ **6.** 0.05% = __________

Making and Reading Line Graphs

For some laboratory experiments you will be required to show your data (or part of them) graphically. Simple line graphs allow relationships within the data to be shown interestingly and allow trends (or patterns) in the data to be demonstrated. An advantage of properly drawn graphs is that they save the reader's time because the essential meaning of a large amount of statistical data can be seen at a glance.

To aid in making accurate graphs, graph paper (or a printed grid in the manual) is used. Line graphs have both horizontal (X) and vertical (Y) axes with scales. Each scale should have uniform intervals—that is, each unit measured on the scale should require the same distance along the scale as any other. Variations from this rule may be misleading and result in false interpretations of the data. By convention, the condition that is manipulated (the independent variable) in the experimental series is plotted on the X-axis (the horizontal axis); and the value that we then measure (the dependent variable) is plotted on the Y-axis (the vertical axis). To plot the data, a dot or a small **x** is placed at the precise point where the two variables (measured for each sample) meet; and then a line (this is called the **curve**) is drawn to connect the plotted points.

Sometimes, you will see the curve on a line graph extended beyond the last plotted point. This is (supposedly) done to predict "what comes next." When you see this done, be skeptical. The information provided by such a technique is only slightly more accurate than that provided by a crystal ball! When constructing a graph, be sure to label the X-axis and Y-axis and give the graph a legend (see Figure G.1).

To read a line graph, pick any point on the line, and match it with the information directly below on the X-axis and with that directly to the left of it on the Y-axis. Figure G.1 is a graph that illustrates the relationship between breaths per minute (respiratory rate) and body temperature. Answer the following questions about this graph:

7. What was the respiratory rate at a body temperature of 96°F? ______________

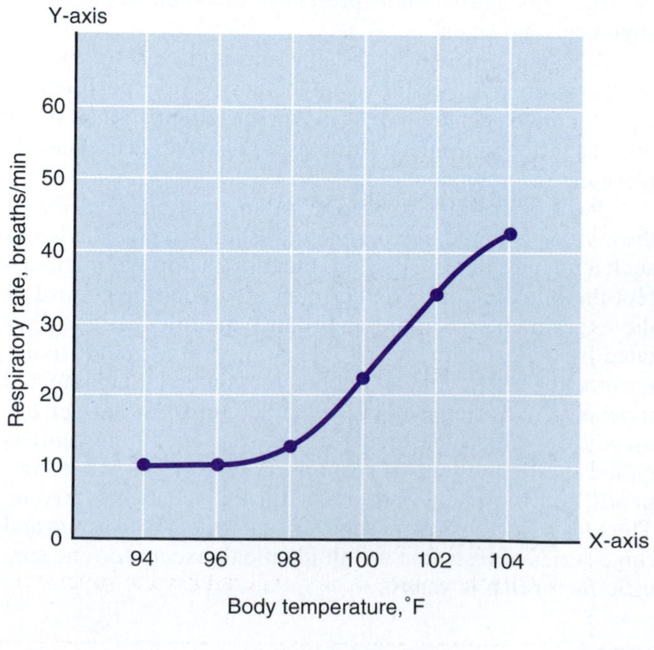
Y-axis
60
50
40
30
20
10
0
Respiratory rate, breaths/min
94
96
98
100
102
104
X-axis
Body temperature, °F

FIGURE G.1 Example of graphically presented data. Respiratory rate as a function of body temperature.

* Answers are given on page xx.

8. Between which two body temperature readings was the increase in breaths per minute greatest? __________

Step 5: Reporting Conclusions of the Study

Drawings, tables, and graphs alone do not suffice as the final presentation of scientific results. The final step requires that you provide a straightforward description of the conclusions drawn from your results. If possible, your findings should be compared to those of other investigators working on the same problem. (For laboratory investigations conducted by students, these comparative figures are provided by classmates.)

It is important to realize that scientific investigations do not always yield the anticipated results. If there are discrepancies between your results and those of others, or what you expected to find based on your class notes or textbook readings, this is the place to try to explain those discrepancies.

Results are often only as good as the observation techniques used. Depending on the type of experiment conducted,

LAB REPORT

Cover Page

- Title of Experiment
- Author's Name
- Course
- Instructor
- Date

Introduction

- Provide background information.
- Describe any relevant observations.
- State hypotheses clearly.

Materials and Methods

- List equipment or supplies needed.
- Provide step-by-step directions for conducting the experiment.

Results

- Present data using a drawing (figure), table, or graph.
- Analyze data.
- Summarize findings briefly.

Discussion and Conclusions

- Conclude whether data gathered support or do not support hypotheses.
- Include relevant information from other sources.
- Explain any uncontrolled variables or unexpected difficulties.
- Make suggestions for further experimentation.

Reference List

- Cite the source of any material used to support this report.

you may need to answer several questions. Did you weigh the specimen carefully enough? Did you balance the scale first? Was the subject's blood pressure actually as high as you recorded it, or did you record it inaccurately? If you did record it accurately, is it possible that the subject was emotionally upset about something, which might have given falsely high data for the variable being investigated? Attempting to explain an unexpected result will often teach you more than you would have learned from anticipated results.

When the experiment produces results that are consistent with the hypothesis, then the hypothesis can be said to have reached a higher level of certainty. The probability that the hypothesis is correct is greater.

A hypothesis that has been validated by many different investigators is called a **theory.** Theories are useful in two important ways. First, they link sets of data; and second, they make predictions that may lead to additional avenues of investigation. (OK, we know this with a high degree of certainty; what's next?)

When a theory has been repeatedly verified and appears to have wide applicability in biology, it may assume the status of a **biological principle.** A principle is a statement that applies with a high degree of probability to a range of events. For example, "Living matter is made of cells or cell products" is a principle stated in many biology texts. It is a sound and useful principle, and will continue to be used as such—unless new findings prove it wrong.

We have been through quite a bit of background concerning the scientific method and what its use entails. Because it is important that you remember the phases of the scientific method, they are summarized here:

1. Observation of some phenomenon
2. Statement of a hypothesis (based on the observations)
3. Collection of data (testing the hypothesis with controlled experiments)
4. Manipulation and analysis of the data
5. Reporting of the conclusions of the study (routinely done by preparing a lab report—see page xvii)

Writing a Lab Report Based on the Scientific Method

A laboratory report is not the same as a scientific paper, but it has some of the same elements and is a formal way to report the results of a scientific experiment. The report should have a cover page that includes the title of the experiment, the author's name, the name of the course, the instructor, and the date. The report should include five separate, clearly marked sections: Introduction, Materials and Methods, Results, Discussion and Conclusions, and References. Use the previous template to guide you through writing a lab report.

METRICS

No matter how highly developed our ability to observe, observations have scientific value only if we can communicate them to others. Without measurement, we would be limited to qualitative description. For precise and repeatable communication of information, the agreed-upon system of measurement used by scientists is the **metric system.**

A major advantage of the metric system is that it is based on units of 10. This allows rapid conversion to workable numbers so that neither very large nor very small figures need be used in calculations. Fractions or multiples of the standard units of length, volume, mass, time, and temperature have been assigned specific names. Table G.1 shows the commonly used units of the metric system, along with the prefixes used to designate fractions and multiples thereof.

To change from smaller units to larger units, you must *divide* by the appropriate factor of 10 (because there are fewer of the larger units). For example, a milliunit (*milli* = one-thousandth), such as a millimeter, is one step smaller than a centiunit (*centi* = one-hundredth), such as a centimeter. Thus to change milliunits to centiunits, you must divide by 10. On the other hand, when converting from larger units to smaller ones, you must *multiply* by the appropriate factor of 10. A partial scheme for conversions between the metric units is shown on the next page.

TABLE G.1 Metric System

A. Commonly used units		B. Fractions and their multiples		
Measurement	**Unit**	**Fraction or multiple**	**Prefix**	**Symbol**
Length	Meter (m)	10^6 one million	mega	M
Volume	Liter (L; l with prefix)	10^3 one thousand	kilo	k
Mass	Gram (g)	10^{-1} one-tenth	deci	d
Time*	Second (s)	10^{-2} one-hundredth	centi	c
Temperature	Degree Celsius (°C)	10^{-3} one-thousandth	milli	m
		10^{-6} one-millionth	micro	μ
		10^{-9} one-billionth	nano	n

* The accepted standard for time is the second; and thus hours and minutes are used in scientific, as well as everyday, measurement of time. The only prefixes generally used are those indicating *fractional portions* of seconds—for example, millisecond and microsecond.

$$\text{microunit} \underset{\times 1000}{\overset{\div 1000}{\rightleftharpoons}} \text{milliunit} \underset{\times 10}{\overset{\div 10}{\rightleftharpoons}} \text{centiunit} \underset{\times 100}{\overset{\div 100}{\rightleftharpoons}} \text{unit} \underset{\times 1000}{\overset{\div 1000}{\rightleftharpoons}} \text{kilounit}$$

$$\text{smallest} \longleftrightarrow \text{largest}$$

The objectives of the sections that follow are to provide a brief overview of the most-used measurements in science or health professions and to help you gain some measure of confidence in dealing with them. (A listing of the most frequently used conversion factors, for conversions between British and metric system units, is provided in the appendix.)

Length Measurements

The metric unit of length is the **meter (m).** Smaller objects are measured in centimeters or millimeters. Subcellular structures are measured in micrometers.

To help you picture these units of length, some equivalents follow:

One meter (m) is slightly longer than one yard (1 m = 39.37 in.).

One centimeter (cm) is approximately the width of a piece of chalk. (Note: There are 2.54 cm in 1 in.)

One millimeter (mm) is approximately the thickness of the wire of a paper clip or of a mark made by a No. 2 pencil lead.

One micrometer (μm) is extremely tiny and can be measured only microscopically.

Make the following conversions between metric units of length:

9. 12 cm = __________ mm

10. 2000 μm = __________ mm

Now, circle the answer that would make the most sense in each of the following statements:

11. A match (in a matchbook) is (0.3, 3, 30) cm long.

12. A standard-size American car is about 4 (mm, cm, m, km) long.

Volume Measurements

The metric unit of volume is the liter. A **liter** (l, or sometimes L, especially without a prefix) is slightly more than a quart (1 L = 1.057 quarts). Liquid volumes measured out for lab experiments are usually measured in milliliters (ml). (The terms *ml* and *cc,* cubic centimeter, are used interchangeably in laboratory and medical settings.)

To help you visualize metric volumes, the equivalents of some common substances follow:

A 12-oz can of soda is a little less than 360 ml.

A fluid ounce is about 30 (it's 29.57) ml (cc).

A teaspoon of vanilla is about 5 ml (cc).

Compute the following:

13. How many 5-ml injections can be prepared from 1 liter of a medicine? __________

14. A 450-ml volume of alcohol is __________ L.

Mass Measurements

Although many people use the terms *mass* and *weight* interchangeably, this usage is inaccurate. **Mass** is the amount of matter in an object; and an object has a constant mass, regardless of where it is—that is, on earth, or in outer space. However, weight varies with gravitational pull; the greater the gravitational pull, the greater the weight. Thus, our astronauts are said to be weightless* when in outer space, but they still have the same mass as they do on earth.

The metric unit of mass is the **gram (g).** Medical dosages are usually prescribed in milligrams (mg) or micrograms (μg); and in the clinical agency, body weight (particularly of infants) is typically specified in kilograms (kg) (1 kg = 2.2 lb).

The following examples are provided to help you become familiar with the masses of some common objects:

Two aspirin tablets have a mass of approximately 1 g.

A nickel has a mass of 5 g.

The mass of an average woman (132 lb) is 60 kg.

Make the following conversions:

15. 300 g = __________ mg = __________ μg

16. 4000 μg = __________ mg = __________ g

17. A nurse must administer to her patient, Mrs. Smith, 5 mg of a drug per kg of body mass. Mrs. Smith weighs 140 lb. How many grams of the drug should the nurse administer to her patient?

__________ g

Temperature Measurements

In the laboratory and in the clinical agency, temperature is measured both in metric units (degrees Celsius, °C) and in British units (degrees Fahrenheit, °F). Thus it helps to be familiar with both temperature scales.

The temperatures of boiling and freezing water can be used to compare the two scales:

The freezing point of water is 0°C and 32°F.

The boiling point of water is 100°C and 212°F.

* Astronauts are not *really* weightless. It is just that they and their surroundings are being pulled toward the earth at the same speed; and so, in reference to their environment, they appear to float.

As you can see, the range from the freezing point to the boiling point of water on the Celsius scale is 100 degrees, whereas the comparable range on the Fahrenheit scale is 180 degrees. Hence, one degree on the Celsius scale represents a greater change in temperature. Normal body temperature is approximately 98.6°F or 37°C.

To convert from the Celsius scale to the Fahrenheit scale, the following equation is used:

$$°C = \frac{5(°F - 32)}{9}$$

To convert from the Fahrenheit scale to the Celsius scale, the following equation is used:

$$°F = (9/5\ °C) + 32$$

Perform the following temperature conversions:

18. Convert 38°C to °F: ____________

19. Convert 158°F to °C: ____________

Answers

1. range of 94–64 or 30 beats/min; mean 77.5

2. 3800%

3. 75%

4. 160%

5. 8

6. 0.0005

7. 10 breaths/min

8. interval between 100–102° (went from 22 to 36 breaths/min)

9. 12 cm = 120 mm

10. 2000 μm = 2 mm

11. 3 cm long

12. 4 m long

13. 200

14. 0.45 L

15. 300 g = $\underline{3 \times 10^5}$ mg= $\underline{3 \times 10^8}$ μg

16. 4000 μg = $\underline{4}$ mg= $\underline{4 \times 10^{-3}}$ g (0.004 g)

17. 0.32 g

18. 100.4°F

19. 70°C

EXERCISE 1

The Language of Anatomy

MATERIALS

- ☐ Human torso model (dissectible)
- ☐ Human skeleton
- ☐ Demonstration: sectioned and labeled kidneys [three separate kidneys uncut or cut so that (a) entire, (b) transverse sectional, and (c) longitudinal sectional views are visible]
- ☐ Gelatin-spaghetti molds
- ☐ Scalpel

OBJECTIVES

1. To describe the anatomical position verbally or by demonstration, and to explain its importance.
2. To use proper anatomical terminology to describe body directions, planes, and surfaces.
3. To name the body cavities and indicate the important organs in each.

PRE-LAB QUIZ

1. Circle True or False. In anatomical position, the body is recumbent (lying down).
2. Circle the correct term. With regard to surface anatomy, abdominal / axial refers to the division relating to the head, neck, and trunk of the body.
3. The term *superficial* refers to a structure that is:
   a. attached near the trunk of the body
   b. toward or at the body surface
   c. toward the head
   d. toward the midline
4. The ___________ plane runs longitudinally and divides the body into right and left parts.
   a. frontal
   b. sagittal
   c. transverse
   d. ventral
5. Circle the correct terms. The dorsal body cavity can be divided into the cranial / thoracic cavity, which contains the brain, and the sural / vertebral cavity, which contains the spinal cord.
6. What organ would you expect to find in the thoracic cavity surrounded by the pericardium? ___________________

Most of us are naturally curious about our bodies. This fact is amply demonstrated by infants, who are fascinated with their own waving hands or their mother's nose. Unlike the infant, however, the student of anatomy must learn to observe and identify the dissectible body structures formally.

When beginning the study of any science, the student is often initially overcome by jargon unique to the subject. The study of anatomy is no exception. But without this specialized terminology, confusion is inevitable. For example, what do *over, on top of, superficial to, above,* and *behind* mean in reference to the human body? Anatomists have an accepted set of reference terms that are universally understood. These allow body structures to be located and identified with a minimum of words and a high degree of clarity.

This exercise presents some of the most important anatomical terminology used to describe the body and introduces you to basic concepts of **gross anatomy,** the study of body structures visible to the naked eye.

MasteringA&P™
Access practice quizzes and more in the Study Area at www.masteringaandp.com.

Anatomical Position

When anatomists or doctors refer to specific areas of the human body, they do so in accordance with a universally accepted standard position called the **anatomical position.** It is essential to understand this position because much of the body terminology employed in this book refers to this body positioning, regardless of the position the body happens to be in. In the anatomical position the human body is erect, with the feet only slightly apart, head and toes pointed forward, and arms hanging at the sides with palms facing forward (Figure 1.1a).

- Assume the anatomical position, and notice that it is not particularly comfortable. The hands are held unnaturally forward rather than hanging partially cupped toward the thighs.

Surface Anatomy

Body surfaces provide a wealth of visible landmarks for study. There are two major divisions of the body:

Axial: Relating to head, neck, and trunk, the axis of the body

Appendicular: Relating to limbs and their attachments to the axis

Anterior Body Landmarks

Note the following regions in Figure 1.1a:

Abdominal: Pertaining to the anterior body trunk region inferior to the ribs

Acromial: Pertaining to the point of the shoulder

Antebrachial: Pertaining to the forearm

Antecubital: Pertaining to the anterior surface of the elbow

Axillary: Pertaining to the armpit

Brachial: Pertaining to the arm

Buccal: Pertaining to the cheek

Carpal: Pertaining to the wrist

Cephalic: Pertaining to the head

Cervical: Pertaining to the neck region

Coxal: Pertaining to the hip

Crural: Pertaining to the leg

Digital: Pertaining to the fingers or toes

Femoral: Pertaining to the thigh

Fibular (peroneal): Pertaining to the side of the leg

Frontal: Pertaining to the forehead

Hallux: Pertaining to the great toe

Inguinal: Pertaining to the groin

Mammary: Pertaining to the breast

Manus: Pertaining to the hand

Mental: Pertaining to the chin

Nasal: Pertaining to the nose

Oral: Pertaining to the mouth

Orbital: Pertaining to the bony eye socket (orbit)

Palmar: Pertaining to the palm of the hand

Patellar: Pertaining to the anterior knee (kneecap) region

Pedal: Pertaining to the foot

Pelvic: Pertaining to the pelvis region

Pollex: Pertaining to the thumb

Pubic: Pertaining to the genital region

Sternal: Pertaining to the region of the breastbone

Tarsal: Pertaining to the ankle

Thoracic: Pertaining to the chest

Umbilical: Pertaining to the navel

Posterior Body Landmarks

Note the following body surface regions in Figure 1.1b:

Acromial: Pertaining to the point of the shoulder

Brachial: Pertaining to the arm

Calcaneal: Pertaining to the heel of the foot

Cephalic: Pertaining to the head

Dorsum: Pertaining to the back

Femoral: Pertaining to the thigh

Gluteal: Pertaining to the buttocks or rump

Lumbar: Pertaining to the area of the back between the ribs and hips; the loin

Manus: Pertaining to the hand

Occipital: Pertaining to the posterior aspect of the head or base of the skull

Olecranal: Pertaining to the posterior aspect of the elbow

Otic: Pertaining to the ear

Pedal: Pertaining to the foot

Perineal: Pertaining to the region between the anus and external genitalia

Plantar: Pertaining to the sole of the foot

Popliteal: Pertaining to the back of the knee

Sacral: Pertaining to the region between the hips (overlying the sacrum)

Scapular: Pertaining to the scapula or shoulder blade area

Sural: Pertaining to the calf or posterior surface of the leg

Vertebral: Pertaining to the area of the spinal column

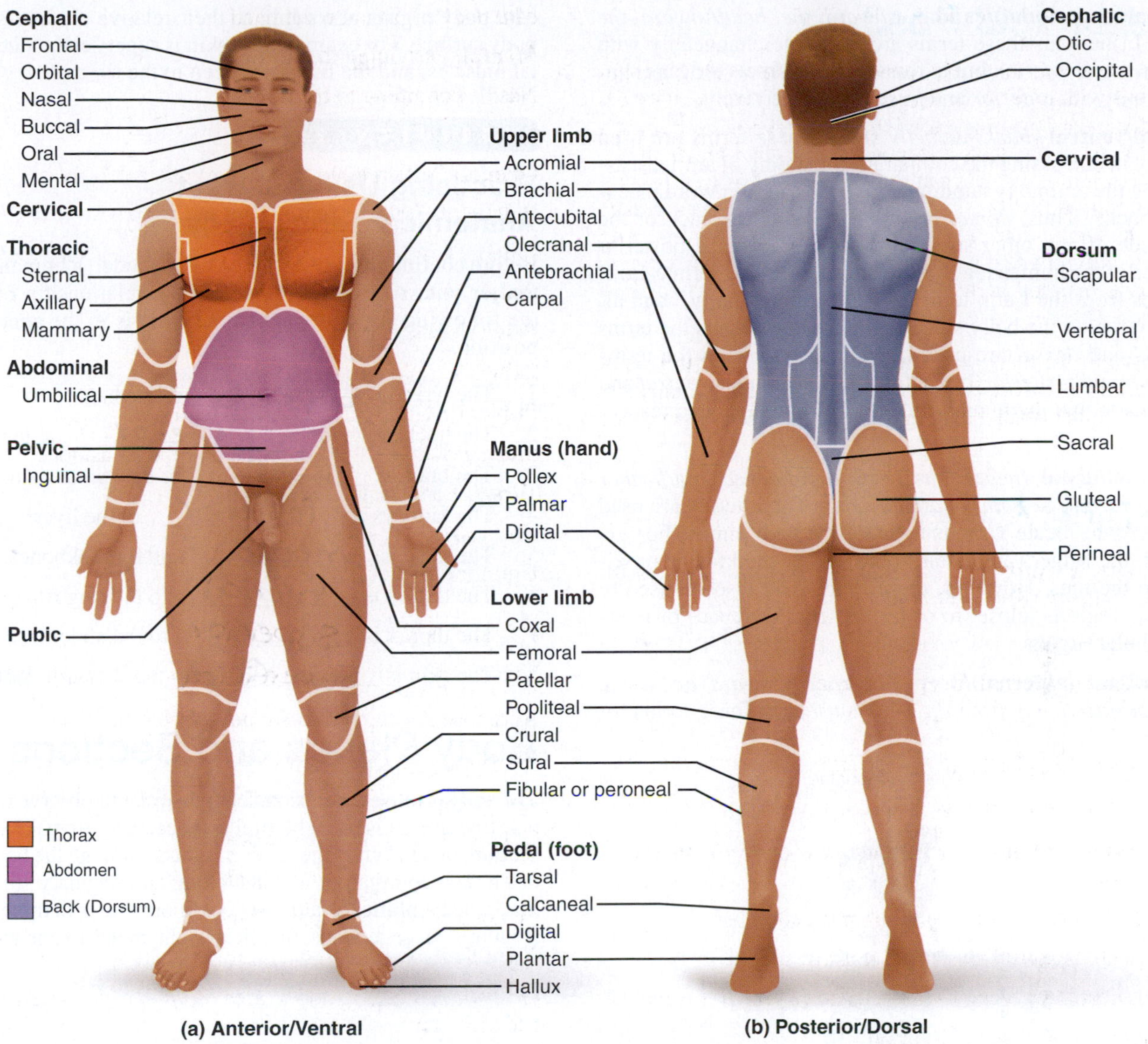
Cephalic
Frontal
Orbital
Nasal
Buccal
Oral
Mental
Cervical
Thoracic
Sternal
Axillary
Mammary
Abdominal
Umbilical
Pelvic
Inguinal
Pubic
Upper limb
Acromial
Brachial
Antecubital
Olecranal
Antebrachial
Carpal
Manus (hand)
Pollex
Palmar
Digital
Lower limb
Coxal
Femoral
Patellar
Popliteal
Crural
Sural
Fibular or peroneal
Pedal (foot)
Tarsal
Calcaneal
Digital
Plantar
Hallux
Cephalic
Otic
Occipital
Cervical
Dorsum
Scapular
Vertebral
Lumbar
Sacral
Gluteal
Perineal
Thorax
Abdomen
Back (Dorsum)
(a) Anterior/Ventral
(b) Posterior/Dorsal

FIGURE 1.1 Surface anatomy. (a) Anatomical position. **(b)** Heels are raised to illustrate the plantar surface of the foot.

ACTIVITY 1

Locating Body Regions

Locate the anterior and posterior body landmarks on yourself, your lab partner, and a human torso model before continuing.

Body Orientation and Direction

Study the terms below, referring to Figure 1.2. Notice that certain terms have a different meaning for a four-legged animal (quadruped) than they do for a human (biped).

Superior/inferior *(above/below):* These terms refer to placement of a structure along the long axis of the body. Superior structures always appear above other structures, and inferior structures are always below other structures. For example, the nose is superior to the mouth, and the abdomen is inferior to the chest.

Anterior/posterior *(front/back):* In humans the most anterior structures are those that are most forward—the face, chest, and abdomen. Posterior structures are those toward the backside of the body. For instance, the spine is posterior to the heart.

Medial/lateral *(toward the midline/away from the midline or median plane):* The sternum (breastbone) is medial to the ribs; the ear is lateral to the nose.

The terms of position just described assume the person is in the anatomical position. The next four term pairs are more absolute. Their applicability is not relative to a particular body position, and they consistently have the same meaning in all vertebrate animals.

Cephalad (cranial)/caudal *(toward the head/toward the tail):* In humans these terms are used interchangeably with *superior* and *inferior,* but in four-legged animals they are synonymous with *anterior* and *posterior,* respectively.

Dorsal/ventral *(backside/belly side):* These terms are used chiefly in discussing the comparative anatomy of animals, assuming the animal is standing. *Dorsum* is a Latin word meaning "back." Thus, *dorsal* refers to the animal's back or the *back*side of any other structures; for example, the posterior surface of the human leg is its dorsal surface. The term *ventral* derives from the Latin term *venter,* meaning "belly," and always refers to the belly side of animals. In humans the terms *ventral* and *dorsal* are used interchangeably with the terms *anterior* and *posterior,* but in four-legged animals *ventral* and *dorsal* are synonymous with *inferior* and *superior,* respectively.

Proximal/distal *(nearer the trunk or attached end/farther from the trunk or point of attachment):* These terms are used primarily to locate various areas of the body limbs. For example, the fingers are distal to the elbow; the knee is proximal to the toes. However, these terms may also be used to indicate regions (closer to or farther from the head) of internal tubular organs.

Superficial (external)/deep (internal) *(toward or at the body surface/away from the body surface):* These terms locate body organs according to their relative closeness to the body surface. For example, the skin is superficial to the skeletal muscles, and the lungs are deep to the rib cage.

ACTIVITY 2

Practicing Using Correct Anatomical Terminology

Before continuing, use a human torso model, a human skeleton, or your own body to specify the relationship between the following structures when the body is in the anatomical position.

1. The wrist is Superior to the hand.
2. The trachea (windpipe) is Anterior to the spine.
3. The brain is Superior to the spinal cord.
4. The kidneys are Posterior to the liver.
5. The nose is medial to the cheekbones.
6. The thumb is laterial to the ring finger.
7. The thorax is Superior to the abdomen.
8. The skin is Superficial to the skeleton.

Body Planes and Sections

The body is three-dimensional, and in order to observe its internal structures, it is often helpful and necessary to make use of a **section,** or cut. When the section is made through the body wall or through an organ, it is made along an imaginary surface or line called a **plane.** Anatomists commonly refer to three planes (Figure 1.3), or sections, that lie at right angles to one another.

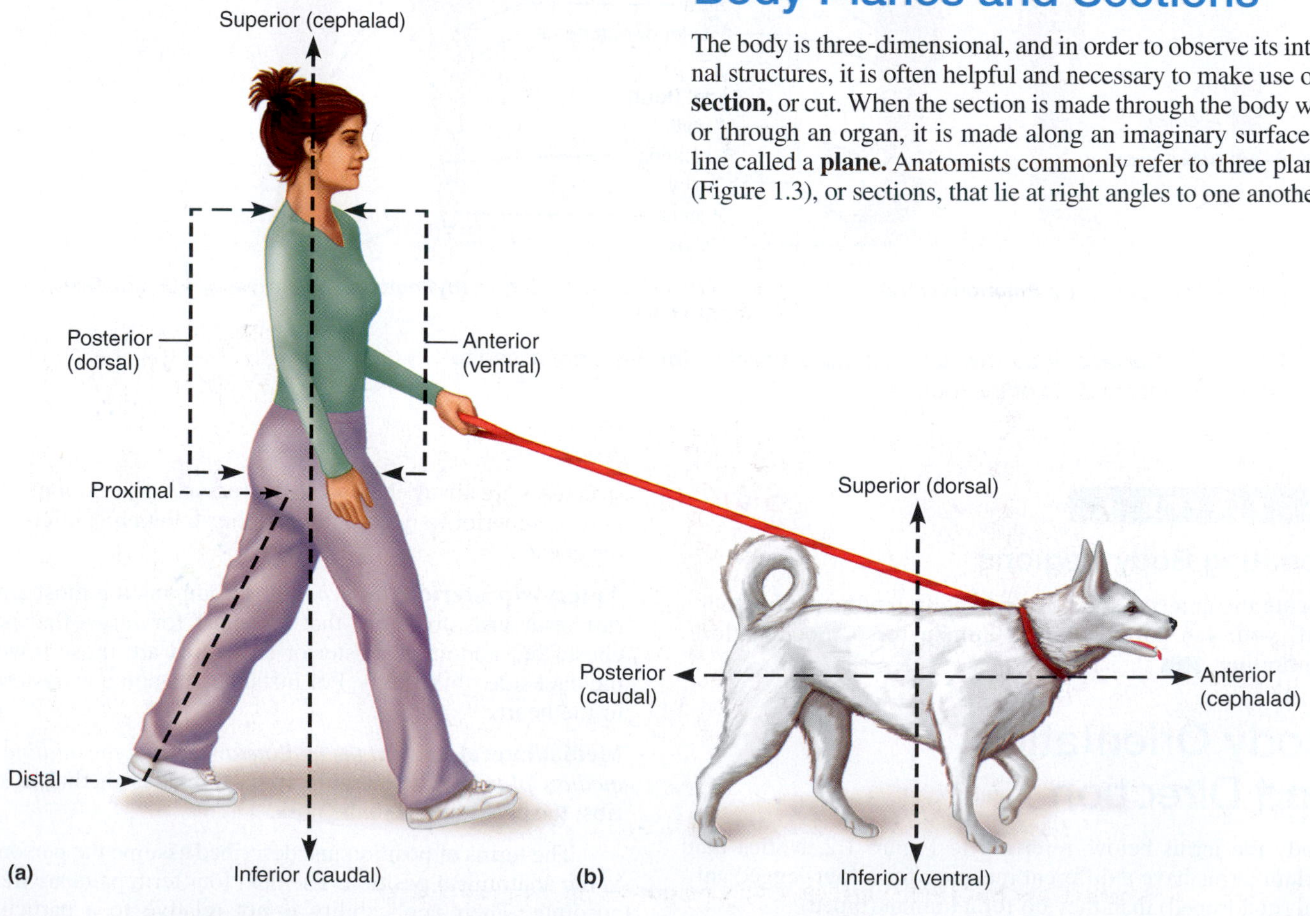
Superior (cephalad)
Posterior (dorsal)
Anterior (ventral)
Proximal
Distal
(a)
Inferior (caudal)
(b)
Superior (dorsal)
Posterior (caudal)
Anterior (cephalad)
Inferior (ventral)

FIGURE 1.2 Anatomical terminology describing body orientation and direction. **(a)** With reference to a human. **(b)** With reference to a four-legged animal.

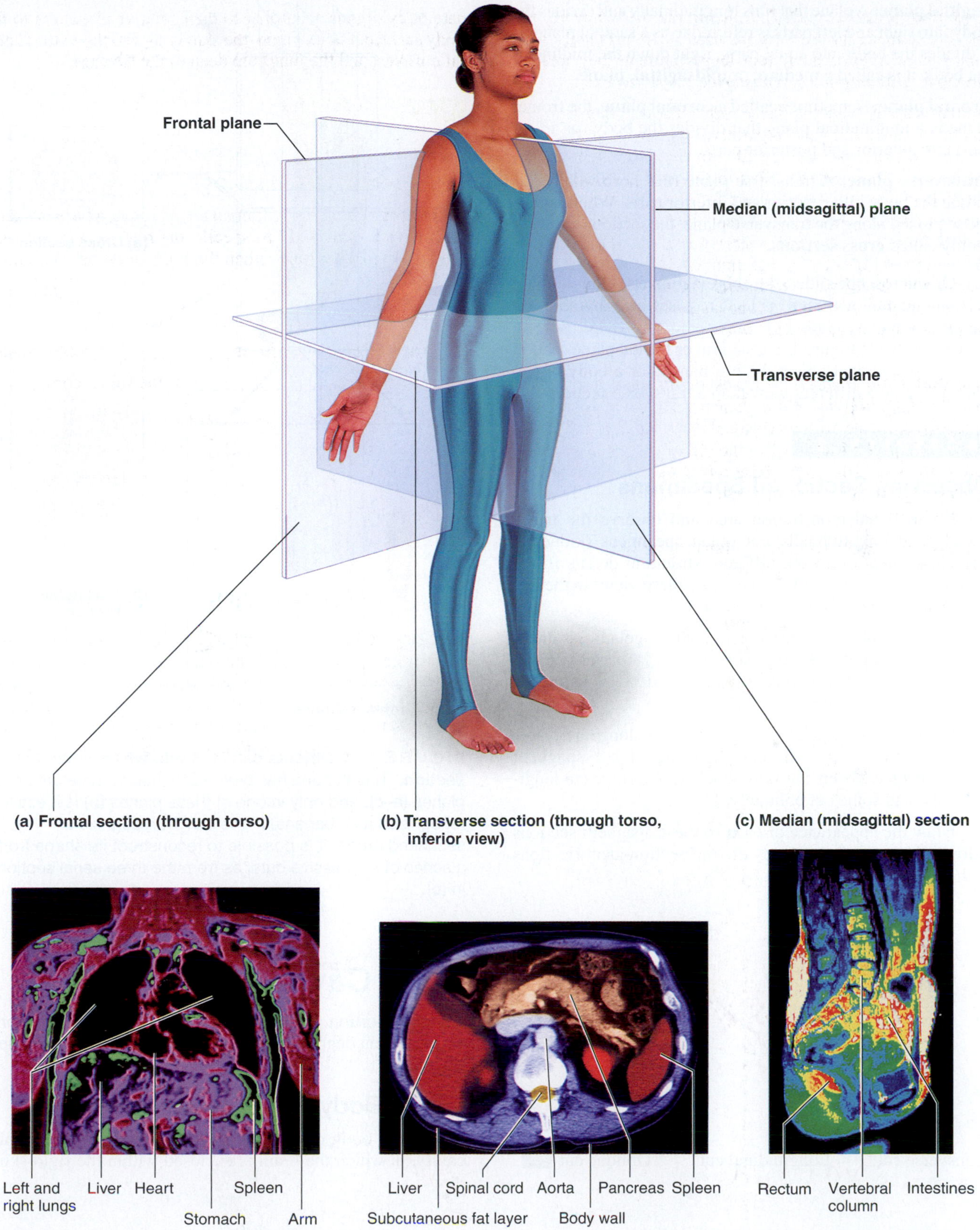
Frontal plane
Median (midsagittal) plane
Transverse plane
(a) Frontal section (through torso)
(b) Transverse section (through torso, inferior view)
(c) Median (midsagittal) section
Left and right lungs
Liver
Heart
Spleen
Stomach
Arm
Liver
Spinal cord
Aorta
Pancreas
Spleen
Subcutaneous fat layer
Body wall
Rectum
Vertebral column
Intestines

FIGURE 1.3 Planes of the body with corresponding magnetic resonance imaging (MRI) scans.

Sagittal plane: A plane that runs longitudinally and divides the body into right and left parts is referred to as a sagittal plane. If it divides the body into equal parts, right down the midline of the body, it is called a **median,** or **midsagittal, plane.**

Frontal plane: Sometimes called a **coronal plane,** the frontal plane is a longitudinal plane that divides the body (or an organ) into anterior and posterior parts.

Transverse plane: A transverse plane runs horizontally, dividing the body into superior and inferior parts. When organs are sectioned along the transverse plane, the sections are commonly called **cross sections.**

On microscope slides, the abbreviation for a longitudinal section (sagittal or frontal) is l.s. Cross sections are abbreviated x.s. or c.s.

As shown in Figure 1.4, a sagittal or frontal plane section of any nonspherical object, be it a banana or a body organ, provides quite a different view than a transverse section.

ACTIVITY 3

Observing Sectioned Specimens

1. Go to the demonstration area and observe the transversely and longitudinally cut organ specimens (kidneys). Pay close attention to the different structural details in the samples because you will need to draw these views in the Review Sheet at the end of this exercise.

2. After completing instruction 1, obtain a gelatin-spaghetti mold and a scalpel and bring them to your laboratory bench. (Essentially, this is just cooked spaghetti added to warm gelatin, which is then allowed to gel.)

3. Cut through the gelatin-spaghetti mold along any plane, and examine the cut surfaces. You should see spaghetti strands that have been cut transversely (x.s.), some cut longitudinally, and some cut obliquely.

4. Draw the appearance of each of these spaghetti sections below, and verify the accuracy of your section identifications with your instructor.

Transverse cut   Longitudinal cut   Oblique cut

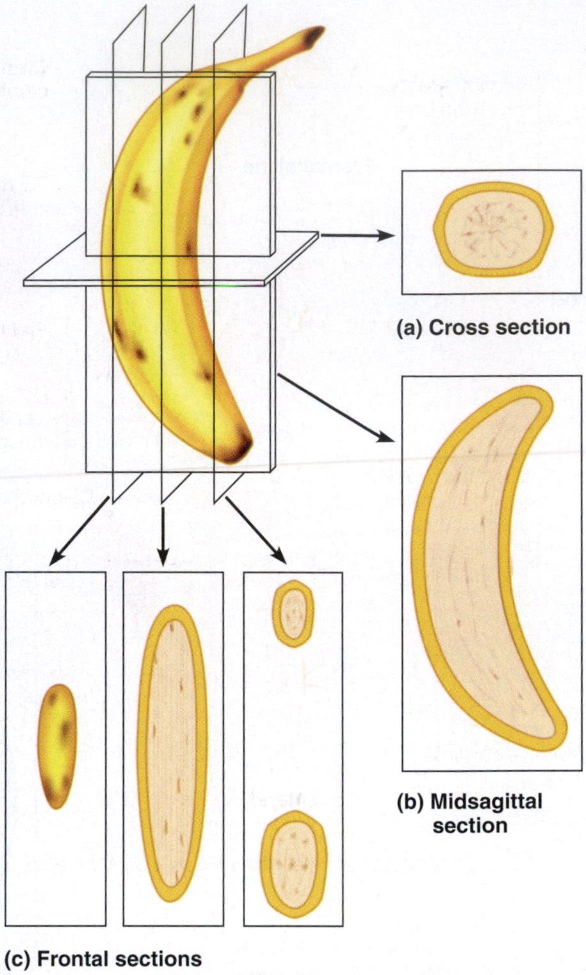
(a) Cross section
(b) Midsagittal section
(c) Frontal sections

FIGURE 1.4 Objects can look odd when viewed in section. This banana has been sectioned in three different planes **(a–c),** and only in one of these planes **(b)** is it easily recognized as a banana. If one cannot recognize a sectioned organ, it is possible to reconstruct its shape from a series of successive cuts, as from the three serial sections in **(c).**

Body Cavities

The axial portion of the body has two large cavities that provide different degrees of protection to the organs within them (Figure 1.5).

Dorsal Body Cavity

The dorsal body cavity can be subdivided into the **cranial cavity,** in which the brain is enclosed within the rigid skull,

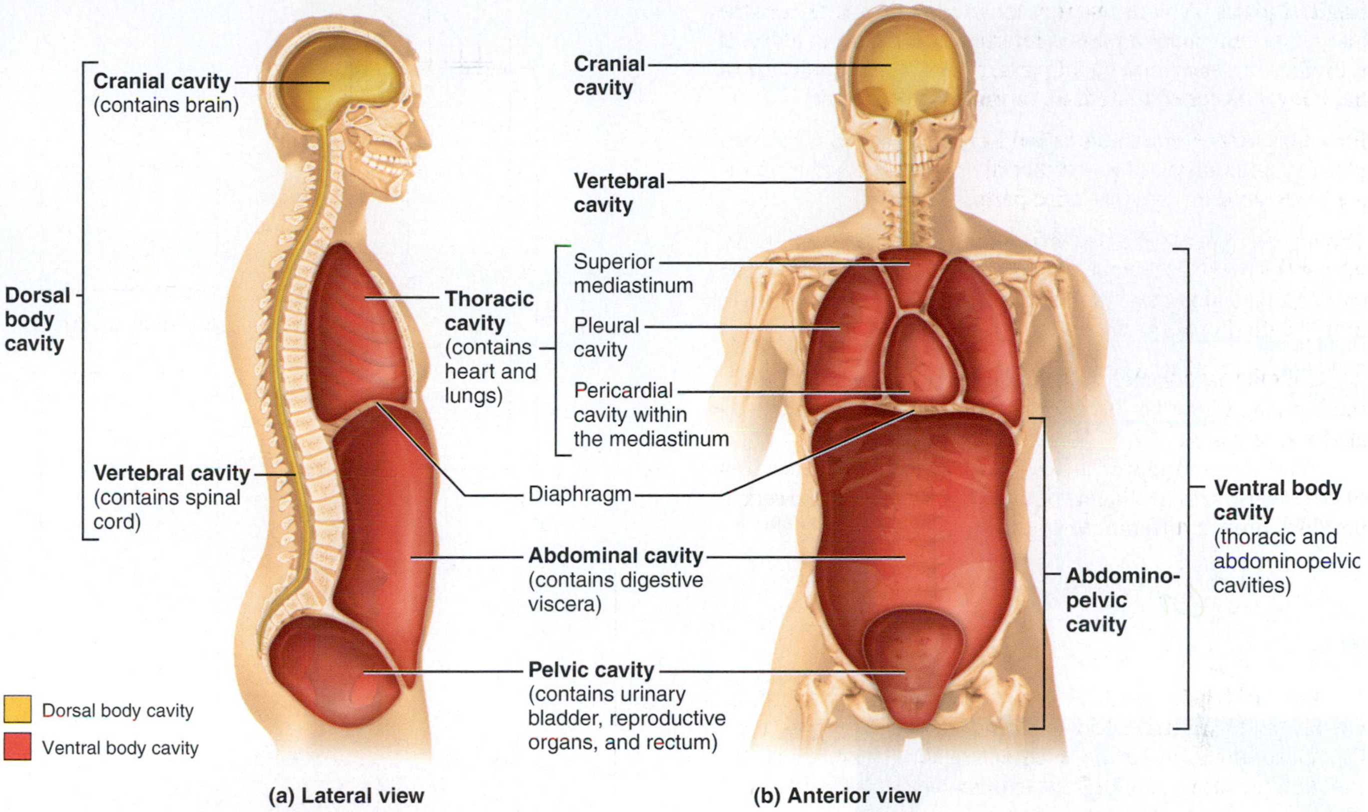

Cranial cavity (contains brain)
Dorsal body cavity
Vertebral cavity (contains spinal cord)
Thoracic cavity (contains heart and lungs)
Superior mediastinum
Pleural cavity
Pericardial cavity within the mediastinum
Diaphragm
Abdominal cavity (contains digestive viscera)
Pelvic cavity (contains urinary bladder, reproductive organs, and rectum)
Dorsal body cavity
Ventral body cavity
(a) Lateral view
Cranial cavity
Vertebral cavity
Ventral body cavity (thoracic and abdominopelvic cavities)
Abdomino-pelvic cavity
(b) Anterior view

FIGURE 1.5 Dorsal and ventral body cavities and their subdivisions.

and the **vertebral** (or **spinal**) **cavity,** within which the delicate spinal cord is protected by the bony vertebral column. Because the spinal cord is a continuation of the brain, these cavities are continuous with each other.

Ventral Body Cavity

Like the dorsal cavity, the ventral body cavity is subdivided. The superior **thoracic cavity** is separated from the rest of the ventral cavity by the dome-shaped diaphragm. The heart and lungs, located in the thoracic cavity, are afforded some measure of protection by the bony rib cage. The cavity inferior to the diaphragm is often referred to as the **abdominopelvic cavity.** Although there is no further physical separation of the ventral cavity, some prefer to describe the abdominopelvic cavity in terms of a superior **abdominal cavity,** the area that houses the stomach, intestines, liver, and other organs, and an inferior **pelvic cavity,** the region that is partially enclosed by the bony pelvis and contains the reproductive organs, bladder, and rectum. Notice in Figure 1.5 that the abdominal and pelvic cavities are not continuous with each other in a straight plane but that the pelvic cavity is tipped away from the perpendicular.

Serous Membranes of the Ventral Body Cavity

The walls of the ventral body cavity and the outer surfaces of the organs it contains are covered with an exceedingly thin, double-layered membrane called the **serosa,** or **serous membrane.** The part of the membrane lining the cavity walls is referred to as the **parietal serosa,** and it is continuous with a similar membrane, the **visceral serosa,** covering the external surface of the organs within the cavity. These membranes produce a thin lubricating fluid that allows the visceral organs to slide over one another or to rub against the body wall with minimal friction. Serous membranes also compartmentalize the various organs so that infection of one organ is prevented from spreading to others.

The specific names of the serous membranes depend on the structures they envelop. Thus the serosa lining the abdominal cavity and covering its organs is the **peritoneum,** that enclosing the lungs is the **pleura,** and that around the heart is the **pericardium** (see Figure 8.1).

Abdominopelvic Quadrants and Regions

Because the abdominopelvic cavity is quite large and contains many organs, it is helpful to divide it up into smaller areas for discussion or study.

A scheme used by most physicians and nurses divides the abdominal surface (and the abdominopelvic cavity deep to it) into four approximately equal regions called **quadrants.** These quadrants are named according to their relative position—that is, *right upper quadrant, right lower quadrant, left upper quadrant,* and *left lower quadrant* (see Figure 1.6a). Note that the terms left and right refer to the left and right of the figure, not your own. The left and right of the figure are referred to as **anatomical left and right.**

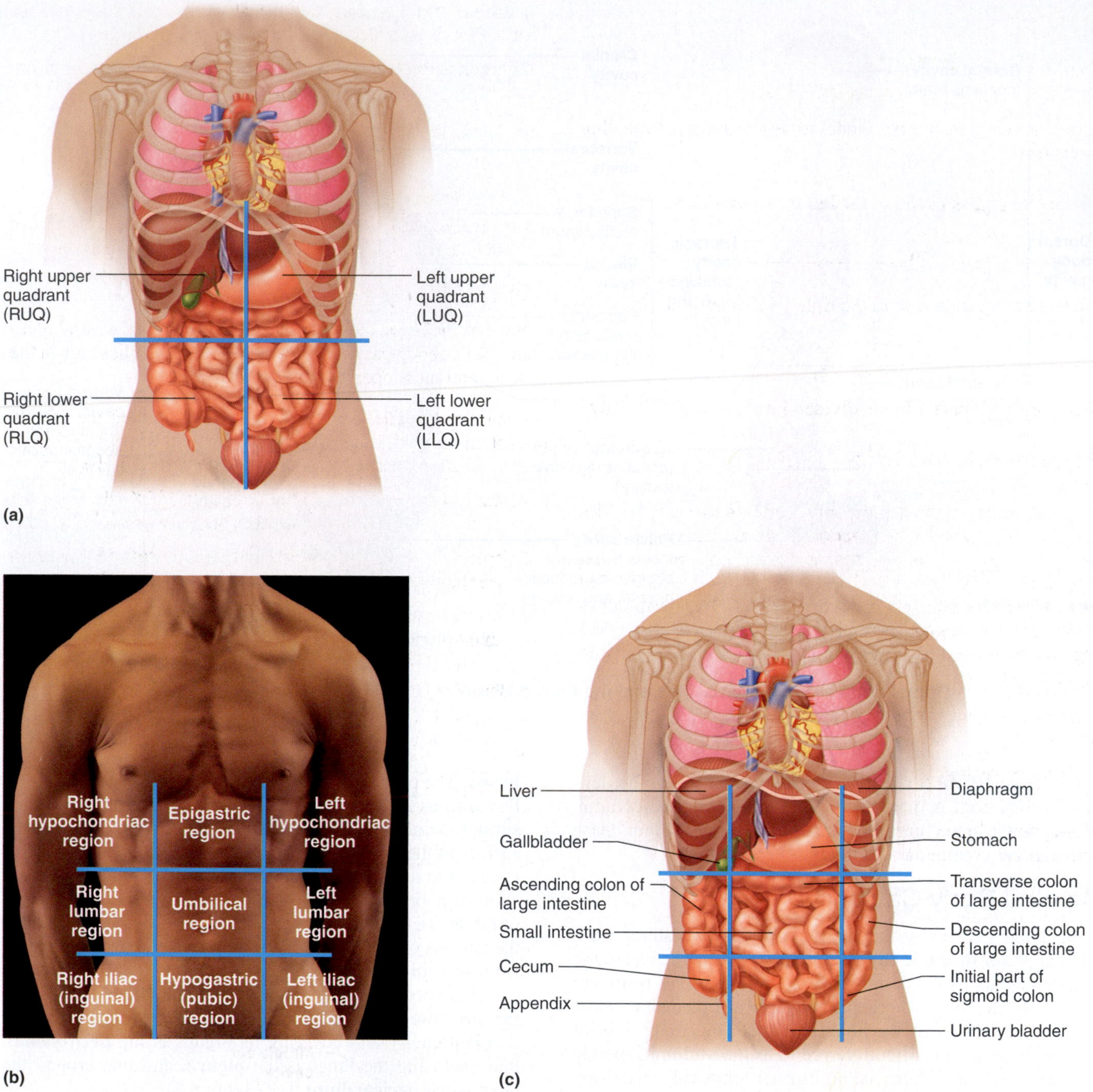
Right upper quadrant (RUQ)
Left upper quadrant (LUQ)
Right lower quadrant (RLQ)
Left lower quadrant (LLQ)
(a)
Right hypochondriac region
Epigastric region
Left hypochondriac region
Right lumbar region
Umbilical region
Left lumbar region
Right iliac (inguinal) region
Hypogastric (pubic) region
Left iliac (inguinal) region
(b)
Liver
Gallbladder
Ascending colon of large intestine
Small intestine
Cecum
Appendix
Diaphragm
Stomach
Transverse colon of large intestine
Descending colon of large intestine
Initial part of sigmoid colon
Urinary bladder
(c)

FIGURE 1.6 Abdominopelvic surface and cavity. (a) The four quadrants, showing superficial organs in each quadrant. **(b)** Nine regions delineated by four planes. The superior horizontal plane is just inferior to the ribs; the inferior horizontal plane is at the superior aspect of the hip bones. The vertical planes are just medial to the nipples. **(c)** Anterior view of the abdominopelvic cavity showing superficial organs.

ACTIVITY 4

Identifying Organs in the Abdominopelvic Cavity

Examine the human torso model to respond to the following questions.

Name two organs found in the left upper quadrant.

Stomach and ______________

Name two organs found in the right lower quadrant.

Cecum and Appendix

What organ (Figure 1.6a) is divided into identical halves by the median plane line? Urinary bladder ■

A different scheme commonly used by anatomists divides the abdominal surface and abdominopelvic cavity into nine separate regions by four planes, as shown in Figure 1.6b. Although the names of these nine regions are unfamiliar to you now, with a little patience and study they will become easier to remember. As you read through the descriptions of these nine regions and locate them in Figure 1.6b, also look at Figure 1.6c to note the organs the regions contain.

Umbilical region: The centermost region, which includes the umbilicus

Epigastric region: Immediately superior to the umbilical region; overlies most of the stomach

Hypogastric (pubic) region: Immediately inferior to the umbilical region; encompasses the pubic area

Iliac (inguinal) regions: Lateral to the hypogastric region and overlying the superior parts of the hip bones

Lumbar regions: Between the ribs and the flaring portions of the hip bones; lateral to the umbilical region

Hypochondriac regions: Flanking the epigastric region laterally and overlying the lower ribs

ACTIVITY 5

Locating Abdominal Surface Regions

Locate the regions of the abdominal surface on a human torso model and on yourself before continuing. ■

Other Body Cavities

Besides the large, closed body cavities, there are several types of smaller body cavities (Figure 1.7). Many of these are in the head, and most open to the body exterior.

Oral cavity: The oral cavity, commonly called the mouth, contains the tongue and teeth. It is continuous with the rest of the digestive tube, which opens to the exterior at the anus.

Nasal cavity: Located within and posterior to the nose, the nasal cavity is part of the passages of the respiratory system.

Orbital cavities: The orbital cavities (orbits) in the skull house the eyes and present them in an anterior position.

Middle ear cavities: Each middle ear cavity lies just medial to an eardrum and is carved into the bony skull. These cavities contain tiny bones that transmit sound vibrations to the organ of hearing in the inner ears.

Synovial (sĭ-no′ve-al) **cavities:** Synovial cavities are joint cavities—they are enclosed within fibrous capsules that surround the freely movable joints of the body, such as those between the vertebrae and the knee and hip joints. Like the serous membranes of the ventral body cavity, membranes lining the synovial cavities secrete a lubricating fluid that reduces friction as the enclosed structures move across one another.

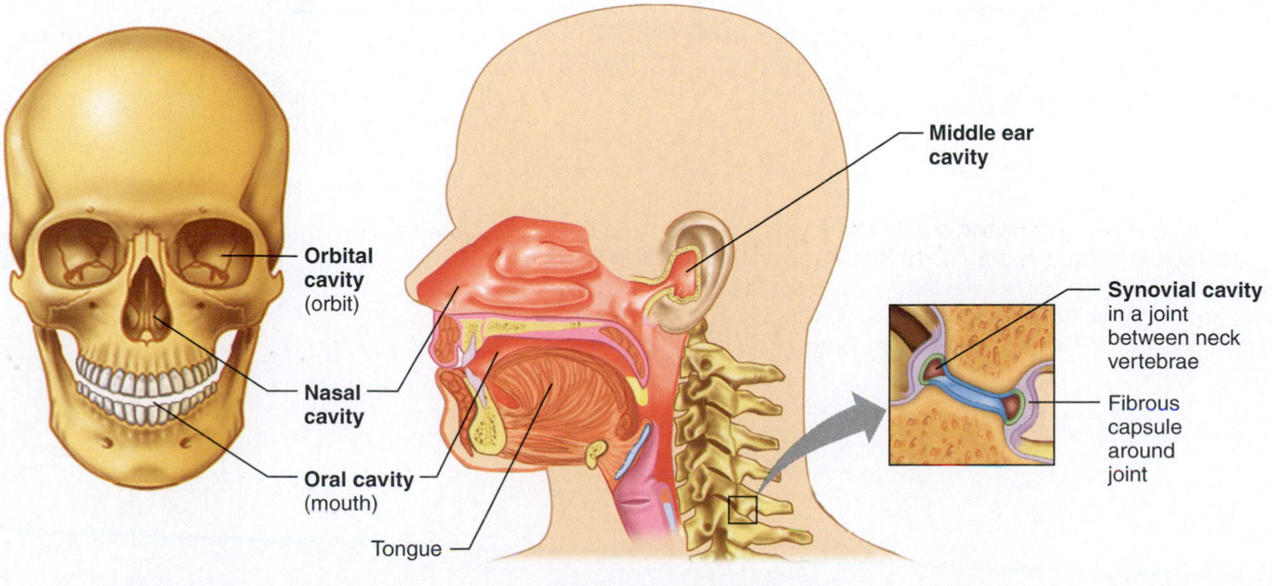
Orbital cavity (orbit)
Nasal cavity
Oral cavity (mouth)
Tongue
Middle ear cavity
Synovial cavity in a joint between neck vertebrae
Fibrous capsule around joint

FIGURE 1.7 Other body cavities. The oral, nasal, orbital, and middle ear cavities are located in the head and open to the body exterior. Synovial cavities are found in joints between many bones such as the vertebrae of the spine, and at the knee, shoulder, and hip.

E X E R C I S E 2

Organ Systems Overview

MATERIALS

- ☐ Freshly killed or preserved rat [predissected by instructor as a demonstration or for student dissection (one rat for every two to four students)] or predissected human cadaver
- ☐ Dissection trays
- ☐ Twine or large dissecting pins
- ☐ Scissors
- ☐ Probes
- ☐ Forceps
- ☐ Disposable gloves
- ☐ Human torso model (dissectible)

OBJECTIVES

1. To name the human organ systems and indicate the major functions of each.
2. To list two or three organs of each system and categorize the various organs by organ system.
3. To identify these organs in a dissected rat or human cadaver or on a dissectible human torso model.
4. To identify the correct organ system for each organ when presented with a list of organs (as studied in the laboratory).

PRE-LAB QUIZ

1. Name the basic unit or building block of all living things. ____________
2. The small intestine is an example of a(n) ___________, because it is composed of two or more tissue types performing a specific function for the body.
   a. epithelial tissue
   b. muscular tissue
   c. organ
   d. organ system
3. The ________________ system is responsible for maintaining homeostasis of the body via rapid transmission of electrical signals.
4. The _______________________ system keeps the blood continuously supplied with oxygen while removing carbon dioxide.
5. The kidneys are part of the ________________ system.
6. The thin muscle that separates the thoracic and abdominal cavities is the ________________.

The basic unit or building block of all living things is the **cell.** Cells fall into four different categories according to their structures and functions. Each of these corresponds to one of the four tissue types: epithelial, muscular, nervous, and connective. A **tissue** is a group of cells that are similar in structure and function. An **organ** is a structure composed of two or more tissue types that performs a specific function for the body. For example, the small intestine, which digests and absorbs nutrients, is composed of all four tissue types.

An **organ system** is a group of organs that act together to perform a particular body function. For example, the organs of the digestive system work together to break down foods moving through the digestive system and absorb the end products into the bloodstream to provide nutrients and fuel for all the body's cells. In all, there are 11 organ systems, which are described in Table 2.1. The lymphatic system also encompasses a *functional system* called the immune system, which is composed of an army of mobile *cells* that act to protect the body from foreign substances.

Read through this summary of the body's organ systems before beginning your rat dissection or examination of the predissected human cadaver. If a human cadaver is not available, Figures 2.3 through 2.6 will serve as a partial replacement.

MasteringA&P™
Access practice quizzes and more in the Study Area at www.masteringaandp.com.

TABLE 2.1 Overview of Organ Systems of the Body

Organ system	Major component organs	Function
Integumentary (Skin)	Epidermal and dermal regions; cutaneous sense organs and glands	• Protects deeper organs from mechanical, chemical, and bacterial injury, and desiccation (drying out) • Excretes salts and urea • Aids in regulation of body temperature • Produces vitamin D
Skeletal	Bones, cartilages, tendons, ligaments, and joints	• Body support and protection of internal organs • Provides levers for muscular action • Cavities provide a site for blood cell formation
Muscular	Muscles attached to the skeleton	• Primary function is to contract or shorten; in doing so, skeletal muscles allow locomotion (running, walking, etc.), grasping and manipulation of the environment, and facial expression • Generates heat
Nervous	Brain, spinal cord, nerves, and sensory receptors	• Allows body to detect changes in its internal and external environment and to respond to such information by activating appropriate muscles or glands • Helps maintain homeostasis of the body via rapid transmission of electrical signals
Endocrine	Pituitary, thymus, thyroid, parathyroid, adrenal, and pineal glands; ovaries, testes, and pancreas	• Helps maintain body homeostasis, promotes growth and development; produces chemical "messengers" (hormones) that travel in the blood to exert their effect(s) on various "target organs" of the body
Cardiovascular	Heart, blood vessels, and blood	• Primarily a transport system that carries blood containing oxygen, carbon dioxide, nutrients, wastes, ions, hormones, and other substances to and from the tissue cells where exchanges are made; blood is propelled through the blood vessels by the pumping action of the heart • Antibodies and other protein molecules in the blood act to protect the body
Lymphatic/ Immunity	Lymphatic vessels, lymph nodes, spleen, thymus, tonsils, and scattered collections of lymphoid tissue	• Picks up fluid leaked from the blood vessels and returns it to the blood • Cleanses blood of pathogens and other debris • Houses lymphocytes that act via the immune response to protect the body from foreign substances (antigens)
Respiratory	Nasal passages, pharynx, larynx, trachea, bronchi, and lungs	• Keeps the blood continuously supplied with oxygen while removing carbon dioxide • Contributes to the acid-base balance of the blood via its carbonic acid–bicarbonate buffer system.
Digestive	Oral cavity, esophagus, stomach, small and large intestines, and accessory structures (teeth, salivary glands, liver, and pancreas)	• Breaks down ingested foods to minute particles, which can be absorbed into the blood for delivery to the body cells • Undigested residue removed from the body as feces
Urinary	Kidneys, ureters, bladder, and urethra	• Rids the body of nitrogen-containing wastes (urea, uric acid, and ammonia), which result from the breakdown of proteins and nucleic acids by body cells • Maintains water, electrolyte, and acid-base balance of blood
Reproductive	Male: testes, prostate gland, scrotum, penis, and duct system, which carries sperm to the body exterior	• Provides germ cells (sperm) for perpetuation of the species
	Female: ovaries, uterine tubes, uterus, mammary glands, and vagina	• Provides germ cells (eggs); the female uterus houses the developing fetus until birth; mammary glands provide nutrition for the infant

DISSECTION AND IDENTIFICATION:

The Organ Systems of the Rat

Many of the external and internal structures of the rat are quite similar in structure and function to those of the human, so a study of the gross anatomy of the rat should help you understand our own physical structure. The following instructions complement and direct your dissection and observation of a rat, but the descriptions for organ observations in Activity 4 ("Examining the Ventral Body Cavity," which begins on page 18) also apply to superficial observations of a previously dissected human cadaver. In addition, the general instructions for observing external structures can easily be extrapolated to serve human cadaver observations. The photographs in Figures 2.3 to 2.6 will provide visual aids.

Note that four of the organ systems listed in Table 2.1 (integumentary, skeletal, muscular, and nervous) will not be studied at this time, as they require microscopic study or more detailed dissection.

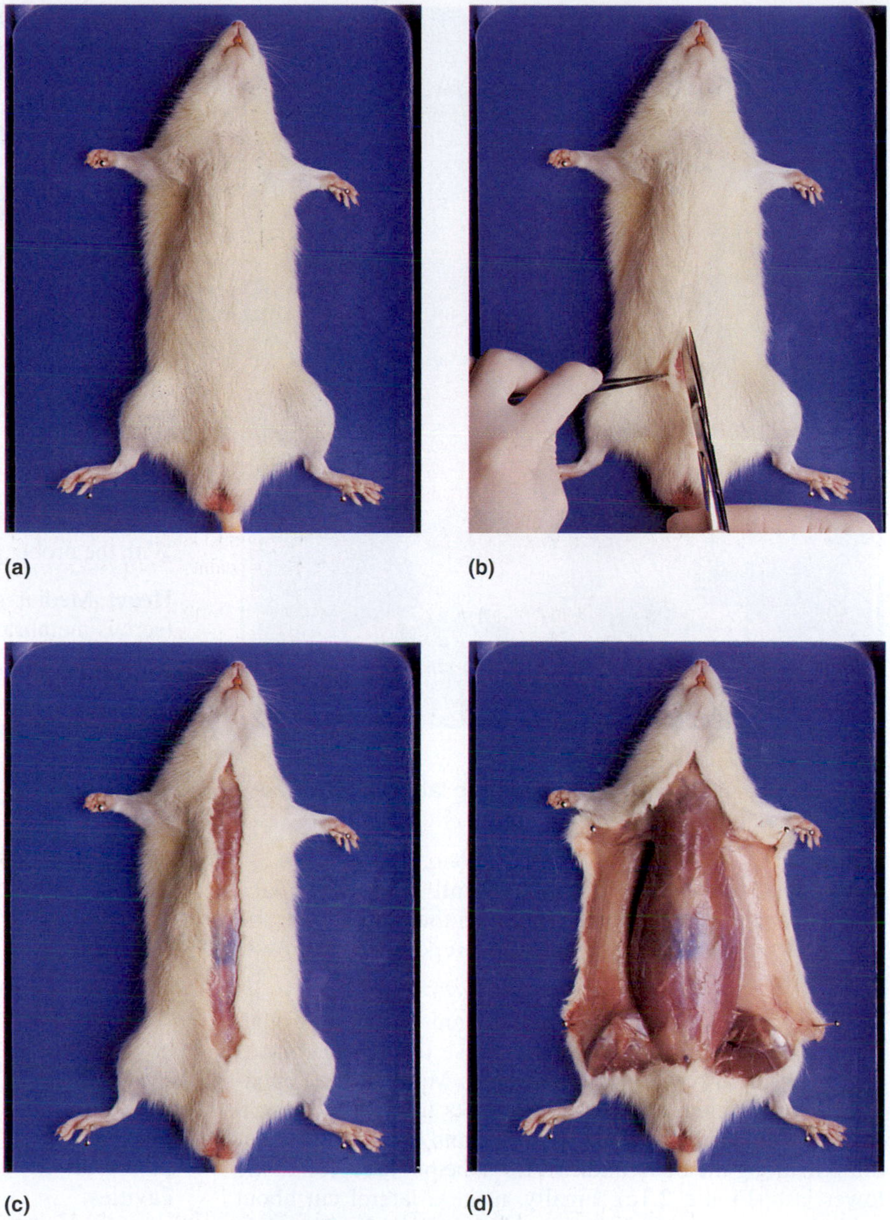
(a)
(b)
(c)
(d)

FIGURE 2.1 Rat dissection: Securing for dissection and the initial incision. (a) Securing the rat to the dissection tray with dissecting pins. **(b)** Using scissors to make the incision on the median line of the abdominal region. **(c)** Completed incision from the pelvic region to the lower jaw. **(d)** Reflection (folding back) of the skin to expose the underlying muscles.

ACTIVITY 1

Observing External Structures

1. If your instructor has provided a predissected rat, go to the demonstration area to make your observations. Alternatively, if you and/or members of your group will be dissecting the specimen, obtain a preserved or freshly killed rat (one for every two to four students), a dissecting tray, dissecting pins or twine, scissors, probe, forceps, and disposable gloves, and bring them to your laboratory bench.

If a predissected human cadaver is available, obtain a probe, forceps, and disposable gloves before going to the demonstration area.

! 2. Don the gloves before beginning your observations. This precaution is particularly important when handling freshly killed animals, which may harbor internal parasites.

3. Observe the major divisions of the body—head, trunk, and extremities. If you are examining a rat, compare these divisions to those of humans. ■

ACTIVITY 2

Examining the Oral Cavity

Examine the structures of the oral cavity. Identify the teeth and tongue. Observe the extent of the hard palate (the portion underlain by bone) and the soft palate (immediately posterior to the hard palate, with no bony support). Notice that the posterior end of the oral cavity leads into the throat, or pharynx, a passageway used by both the digestive and respiratory systems. ■

ACTIVITY 3

Opening the Ventral Body Cavity

1. Pin the animal to the wax of the dissecting tray by placing its dorsal side down and securing its extremities to the wax with large dissecting pins as shown in Figure 2.1a.

If the dissecting tray is not waxed, you will need to secure the animal with twine as follows. (Some may prefer

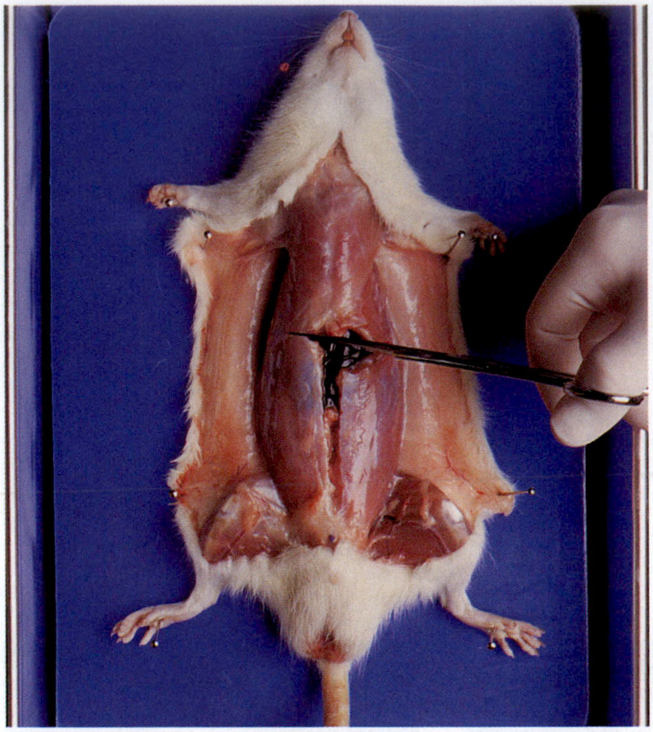

FIGURE 2.2 Rat dissection: Making lateral cuts at the base of the rib cage.

this method in any case.) Obtain the roll of twine. Make a loop knot around one upper limb, pass the twine under the tray, and secure the opposing limb. Repeat for the lower extremities.

2. Lift the abdominal skin with a forceps, and cut through it with the scissors (Figure 2.1b). Close the scissor blades and insert them flat under the cut skin. Moving in a cephalad direction, open and close the blades to loosen the skin from the underlying connective tissue and muscle. Now, cut the skin along the body midline, from the pubic region to the lower jaw (Figure 2.1c). Finally, make a lateral cut about halfway down the ventral surface of each limb. Complete the job of freeing the skin with the scissor tips, and pin the flaps to the tray (Figure 2.1d). The underlying tissue that is now exposed is the skeletal musculature of the body wall and limbs. It allows voluntary body movement. Notice that the muscles are packaged in sheets of pearly white connective tissue (fascia), which protect the muscles and bind them together.

3. Carefully cut through the muscles of the abdominal wall in the pubic region, avoiding the underlying organs. Remember, to *dissect* means "to separate"—not mutilate! Now, hold and lift the muscle layer with a forceps and cut through the muscle layer from the pubic region to the bottom of the rib cage. Make two lateral cuts at the base of the rib cage (Figure 2.2). A thin membrane attached to the inferior boundary of the rib cage should be obvious; this is the **diaphragm,** which separates the thoracic and abdominal cavities. Cut the diaphragm where it attaches to the ventral ribs to loosen the rib cage. Cut through the rib cage on either side. You can now lift the ribs to view the contents of the thoracic cavity. Cut across the flap, at the level of the neck, and remove it. ■

ACTIVITY 4

Examining the Ventral Body Cavity

1. Starting with the most superficial structures and working deeper, examine the structures of the thoracic cavity. Refer to Figure 2.3, which shows the superficial organs, as you work. Choose the appropriate view depending on whether you are examining a rat (a) or a human cadaver (b).

Thymus: An irregular mass of glandular tissue overlying the heart (not illustrated in the human cadaver photograph).

With the probe, push the thymus to the side to view the heart.

Heart: Medial oval structure enclosed within the pericardium (serous membrane sac).

Lungs: Flanking the heart on either side.

Now observe the throat region to identify the trachea.

Trachea: Tubelike "windpipe" running medially down the throat; part of the respiratory system.

Follow the trachea into the thoracic cavity; notice where it divides into two branches. These are the bronchi.

Bronchi: Two passageways that plunge laterally into the tissue of the two lungs.

To expose the esophagus, push the trachea to one side.

Esophagus: A food chute; the part of the digestive system that transports food from the pharynx (throat) to the stomach.

Diaphragm: A thin muscle attached to the inferior boundary of the rib cage; separates the thoracic and abdominal cavities.

Follow the esophagus through the diaphragm to its junction with the stomach.

Stomach: A curved organ important in food digestion and temporary food storage.

2. Examine the superficial structures of the abdominopelvic cavity. Lift the **greater omentum,** an extension of the peritoneum that covers the abdominal viscera. Continuing from the stomach, trace the rest of the digestive tract (Figure 2.4).

Small intestine: Connected to the stomach and ending just before the saclike cecum.

Large intestine: A large muscular tube connected to the small intestine and ending at the anus.

Cecum: The initial portion of the large intestine.

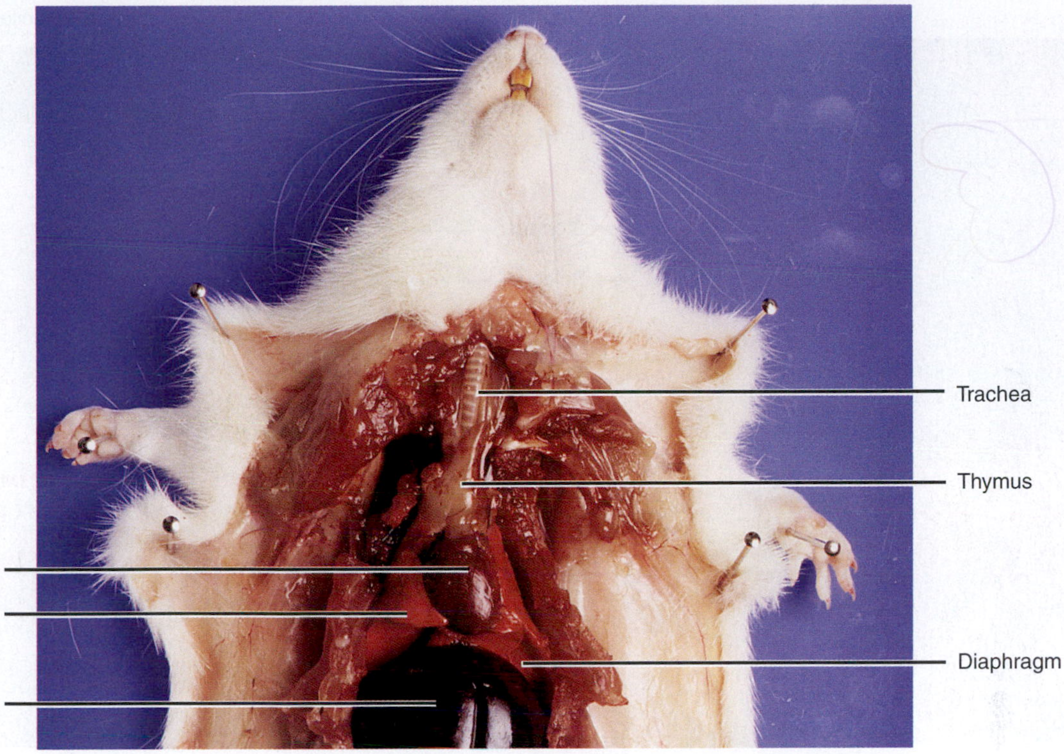
Trachea
Thymus
Heart
Lung
Diaphragm
Liver

(a)

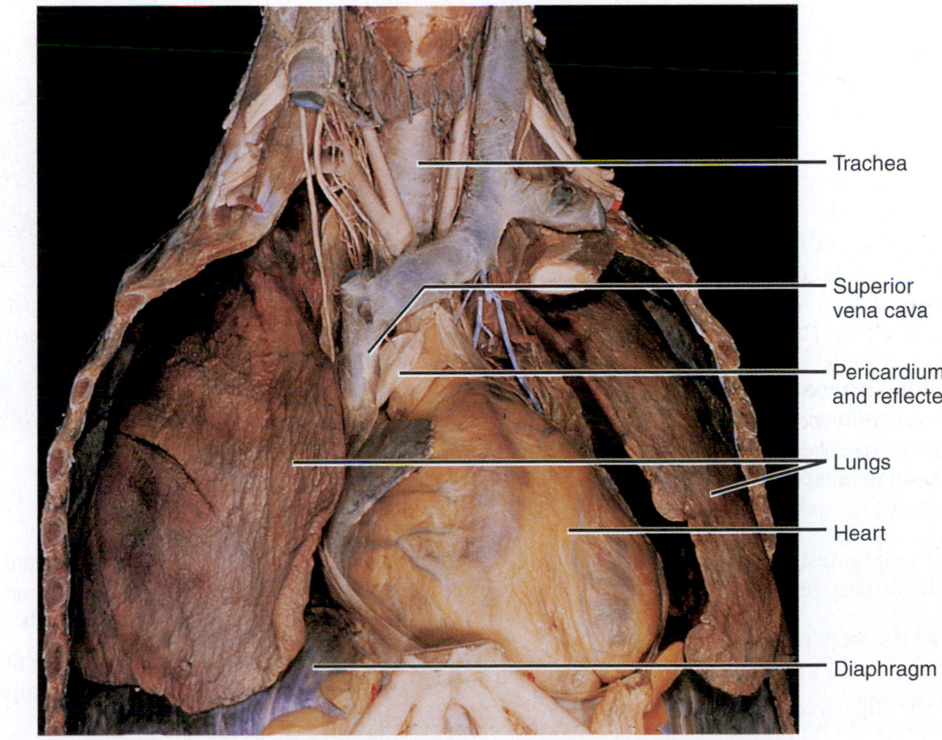
Trachea
Superior vena cava
Pericardium (cut and reflected)
Lungs
Heart
Diaphragm

(b)

FIGURE 2.3 Superficial organs of the thoracic cavity. (a) Dissected rat. **(b)** Human cadaver.

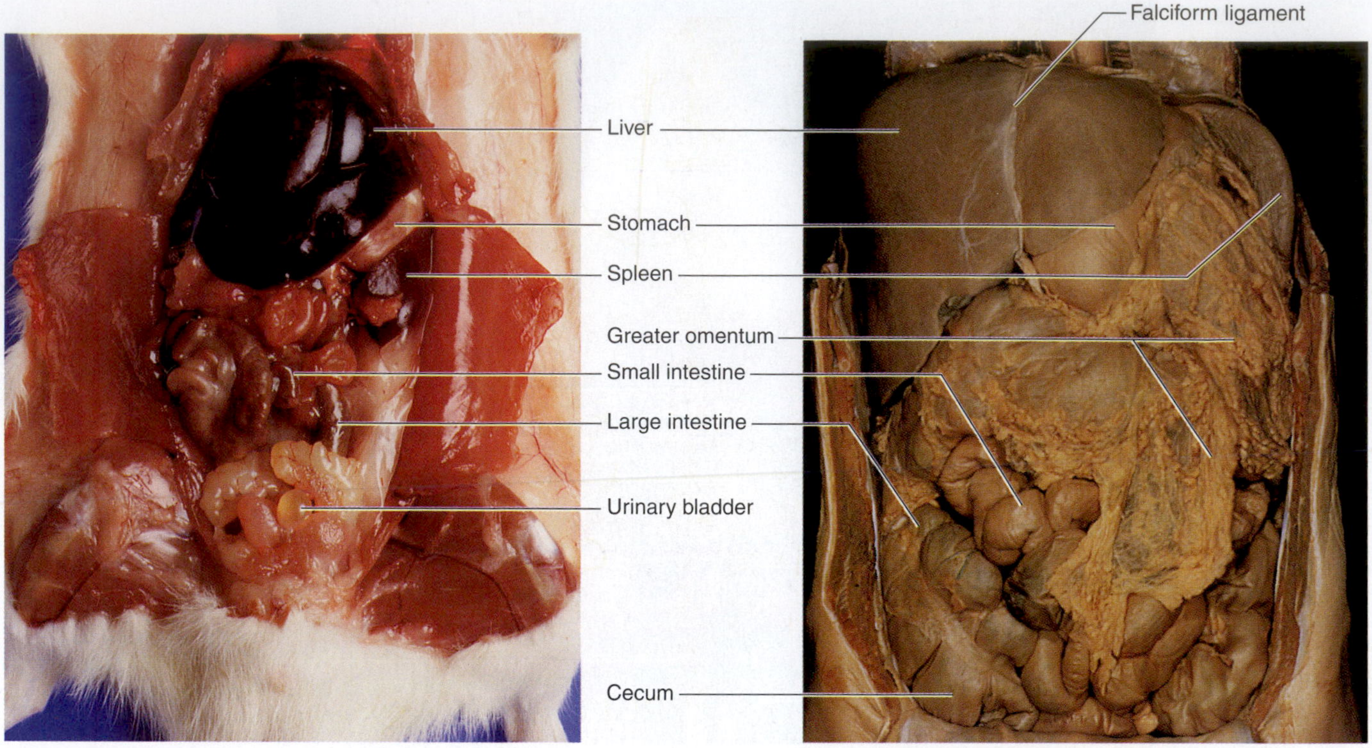
Falciform ligament
Liver
Stomach
Spleen
Greater omentum
Small intestine
Large intestine
Urinary bladder
Cecum

(a)

(b)

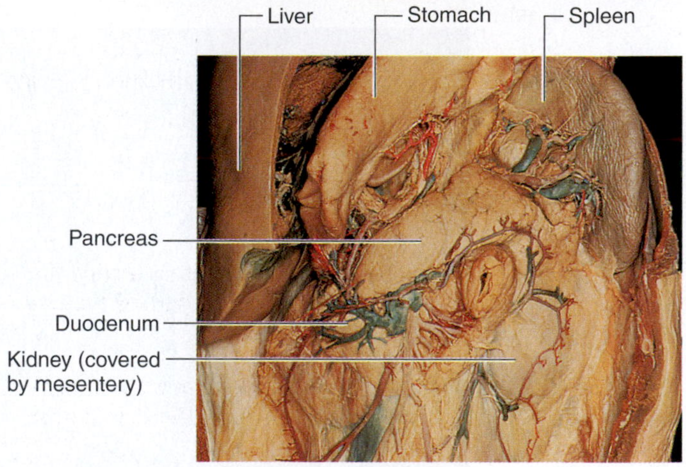
Liver
Stomach
Spleen
Pancreas
Duodenum
Kidney (covered by mesentery)

(c)

FIGURE 2.4 Abdominal organs. (a) Dissected rat, superficial view. **(b)** Human cadaver, superficial view. **(c)** Human cadaver, intermediate view. Stomach reflected superiorly to reveal the pancreas, spleen, and duodenum.

Follow the course of the large intestine to the rectum, which is partially covered by the urinary bladder (Figure 2.5).

Rectum: Terminal part of the large intestine; continuous with the anal canal.

Anus: The opening of the digestive tract (through the anal canal) to the exterior.

Now lift the small intestine with the forceps to view the mesentery.

Mesentery: An apronlike serous membrane; suspends many of the digestive organs in the abdominal cavity. Notice that it is heavily invested with blood vessels and, more likely than not, riddled with large fat deposits.

Locate the remaining abdominal structures.

Pancreas: A diffuse gland; rests dorsal to and in the mesentery between the first portion of the small intestine and the stomach. You will need to lift the stomach to view the pancreas.

Spleen: A dark red organ curving around the left lateral side of the stomach; considered part of the lymphatic system and often called the red blood cell graveyard.

Liver: Large and brownish red; the most superior organ in the abdominal cavity, directly beneath the diaphragm.

3. To locate the deeper structures of the abdominopelvic cavity, move the stomach and the intestines to one side with the probe.

Examine the posterior wall of the abdominal cavity to locate the two kidneys (Figure 2.5).

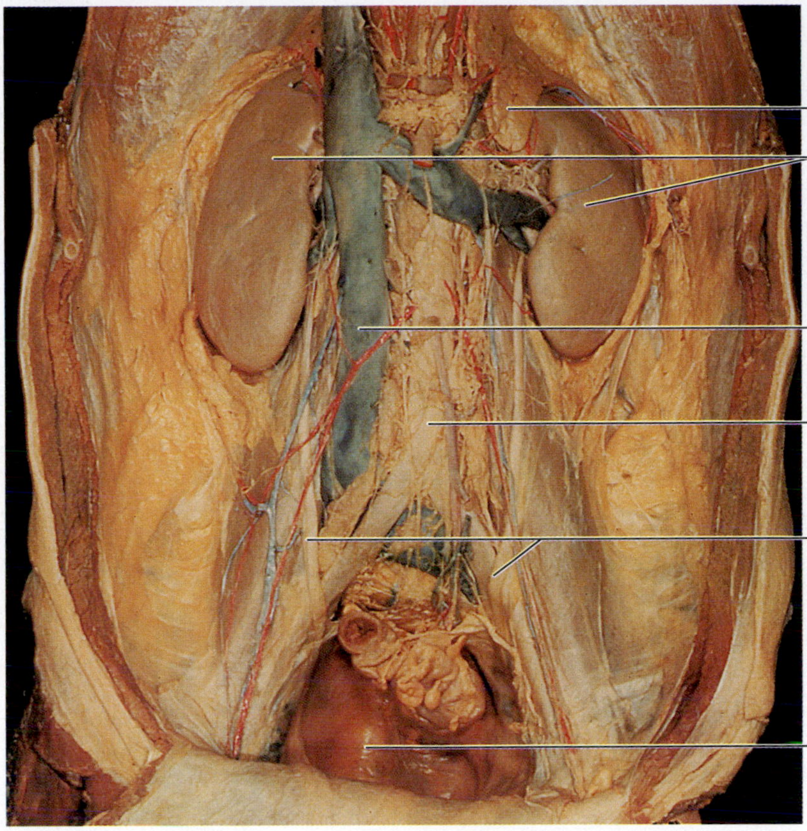
Adrenal gland
Kidneys
Inferior vena cava
Descending aorta
Ureters
Urinary bladder

(a)

FIGURE 2.5 Deep structures of the abdominopelvic cavity. (a) Human cadaver.

(Figure continues on page 22.)

Kidneys: Bean-shaped organs; retroperitoneal (behind the peritoneum).

Adrenal glands: Large endocrine glands that sit astride the superior margin of each kidney; considered part of the endocrine system.

Carefully strip away part of the peritoneum with forceps and attempt to follow the course of one of the ureters to the bladder.

Ureter: Tube running from the indented region of a kidney to the urinary bladder.

Urinary bladder: The sac that serves as a reservoir for urine.

4. In the midline of the body cavity lying between the kidneys are the two principal abdominal blood vessels. Identify each.

Inferior vena cava: The large vein that returns blood to the heart from the lower regions of the body.

Descending aorta: Deep to the inferior vena cava; the largest artery of the body; carries blood away from the heart down the midline of the body.

5. Only a cursory examination of reproductive organs will be done. If you are working with a rat, first determine if the animal is a male or female. Observe the ventral body surface beneath the tail. If a saclike scrotum and an opening for the anus are visible, the animal is a male. If three body openings—urethral, vaginal, and anal—are present, it is a female.

Male Animal

Make a shallow incision into the **scrotum.** Loosen and lift out one oval **testis.** Exert a gentle pull on the testis to identify the slender **ductus deferens,** or **vas deferens,** which carries sperm from the testis superiorly into the abdominal cavity and joins with the urethra. The urethra runs through the penis and carries both urine and sperm out of the body. Identify the **penis,** extending from the bladder to the ventral body wall. Figure 2.5b indicates other glands of the male rat's reproductive system, but they need not be identified at this time.

Female Animal

Inspect the pelvic cavity to identify the Y-shaped **uterus** lying against the dorsal body wall and superior to the bladder (Figure 2.5c). Follow one of the uterine horns superiorly to identify an **ovary,** a small oval structure at the end of the uterine horn. (The rat uterus is quite different from the uterus of a human female, which is a single-chambered organ about the size and shape of a pear.) The inferior undivided part of the rat uterus is continuous with the **vagina,** which leads to the body exterior. Identify the **vaginal orifice** (external vaginal opening).

If you are working with a human cadaver, proceed as indicated next.

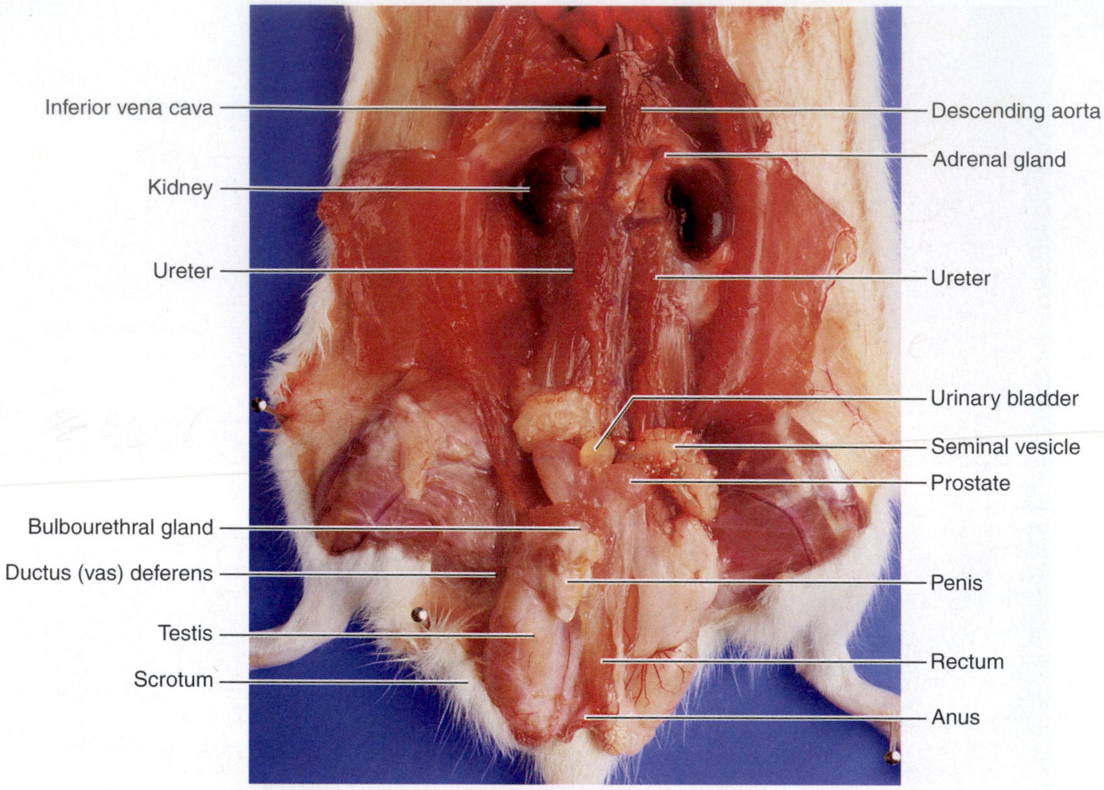
Inferior vena cava
Kidney
Ureter
Bulbourethral gland
Ductus (vas) deferens
Testis
Scrotum
Descending aorta
Adrenal gland
Ureter
Urinary bladder
Seminal vesicle
Prostate
Penis
Rectum
Anus

(b)

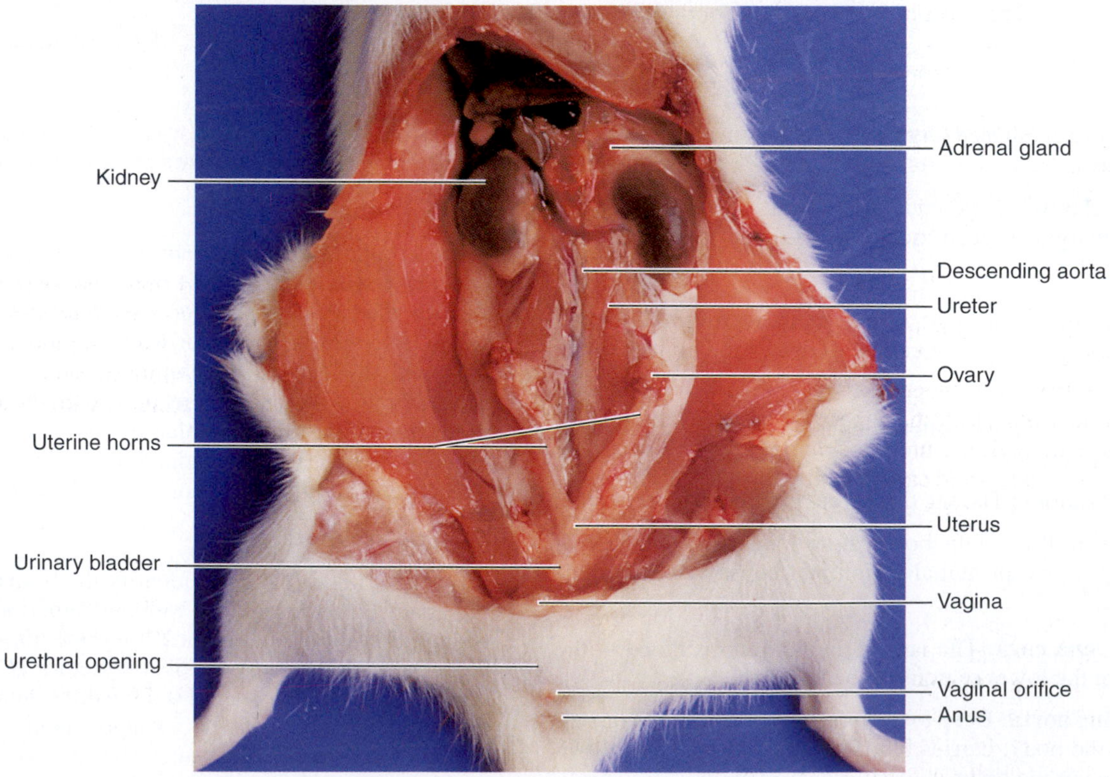
Kidney
Uterine horns
Urinary bladder
Urethral opening
Adrenal gland
Descending aorta
Ureter
Ovary
Uterus
Vagina
Vaginal orifice
Anus

(c)

FIGURE 2.5 (*continued*) Deep structures of the abdominopelvic cavity. **(b)** Dissected male rat. (Some reproductive structures also shown.) **(c)** Dissected female rat. (Some reproductive structures also shown.)

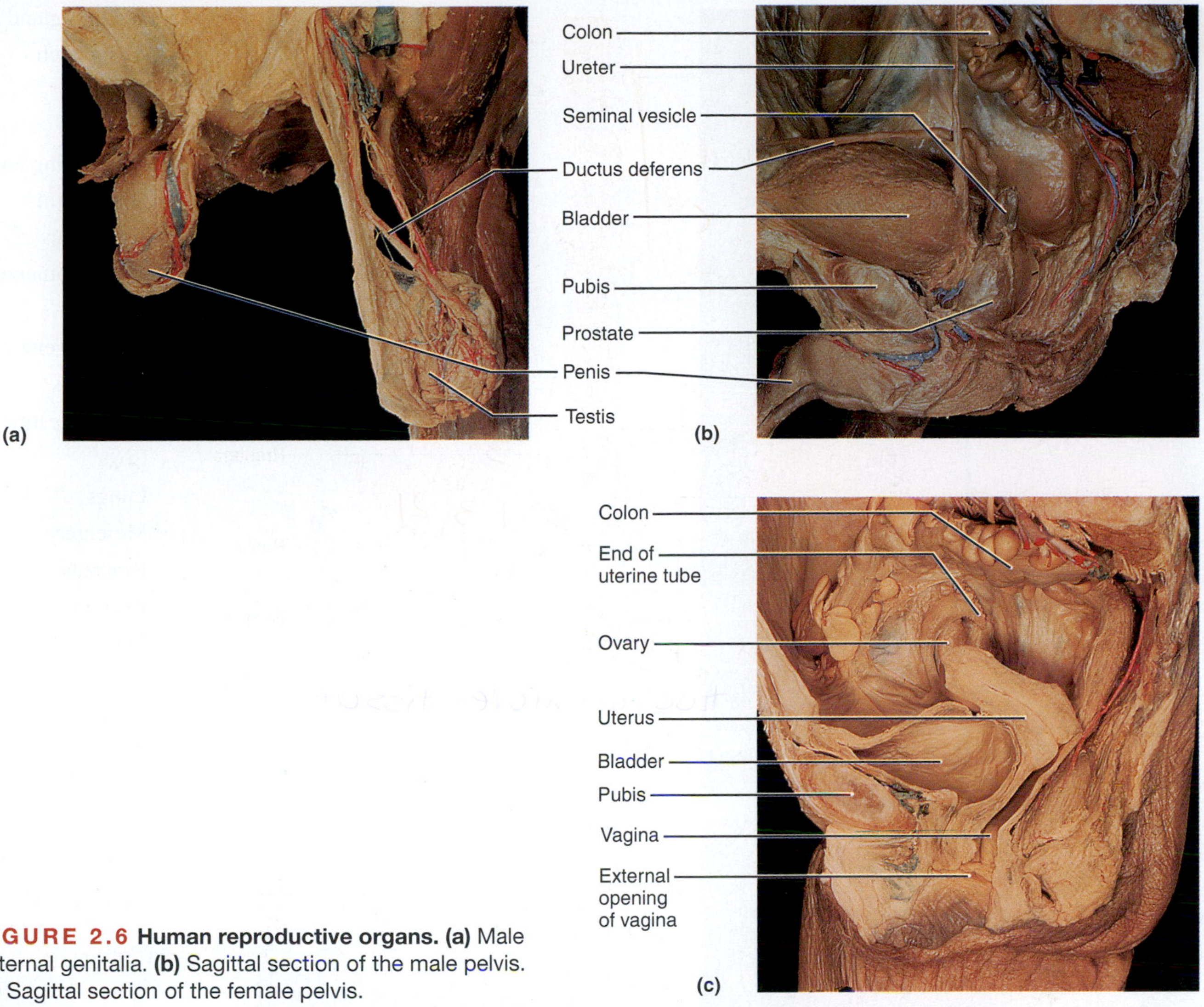

(a)
(b)
Colon
Ureter
Seminal vesicle
Ductus deferens
Bladder
Pubis
Prostate
Penis
Testis
(c)
Colon
End of uterine tube
Ovary
Uterus
Bladder
Pubis
Vagina
External opening of vagina

FIGURE 2.6 Human reproductive organs. (a) Male external genitalia. **(b)** Sagittal section of the male pelvis. **(c)** Sagittal section of the female pelvis.

Male Cadaver

Make a shallow incision into the **scrotum** (Figure 2.6a). Loosen and lift out the oval **testis.** Exert a gentle pull on the testis to identify the slender **ductus (vas) deferens,** which carries sperm from the testis superiorly into the abdominal cavity and joins with the urethra (Figure 2.6b). The urethra runs through the penis and carries both urine and sperm out of the body. Identify the **penis,** extending from the bladder to the ventral body wall.

Female Cadaver

Inspect the pelvic cavity to identify the pear-shaped **uterus** lying against the dorsal body wall and superior to the bladder. Follow one of the **uterine tubes** superiorly to identify an **ovary,** a small oval structure at the end of the uterine tube (Figure 2.6c). The inferior part of the uterus is continuous with the **vagina,** which leads to the body exterior. Identify the **vaginal orifice** (external vaginal opening).

6. When you have finished your observations, rewrap or store the dissection animal or cadaver according to your instructor's directions. Wash the dissecting tools and equipment with laboratory detergent. Dispose of the gloves. Then wash and dry your hands before continuing with the examination of the human torso model. ■

ACTIVITY 5

Examining the Human Torso Model

1. Examine a human torso model to identify the organs listed adjacent to Figure 2.7. Some model organs will have to be removed to see the deeper organs. If a torso model is not available, Figure 2.7 may be used for this part of the exercise.

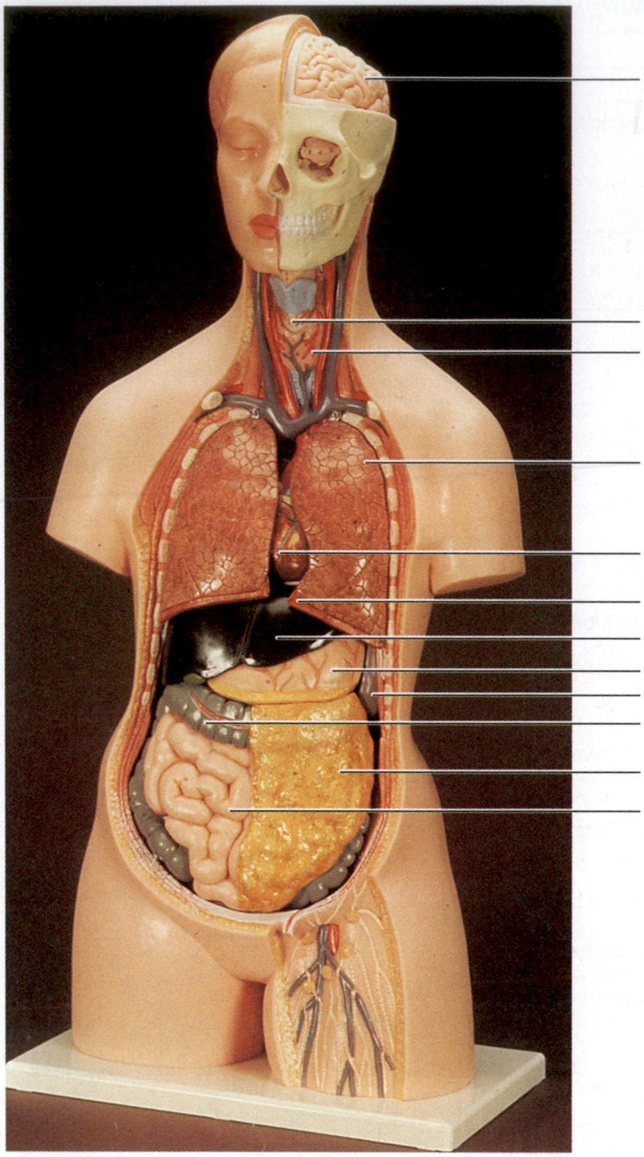

Adrenal gland
Aortic arch
Brain
Bronchi
Descending aorta
Diaphragm
Esophagus
Greater omentum
Heart
Inferior vena cava
Kidneys
Large intestine
Liver
Lungs
Mesentery
Pancreas
Rectum
Small intestine
Spinal cord
Spleen
Stomach
Thyroid gland
Trachea
Ureters
Urinary bladder

FIGURE 2.7 Human torso model.

2. Using the terms at the right of Figure 2.7, label each organ supplied with a leader line in Figure 2.7.

3. Place each of the listed organs in the correct body cavity or cavities. For organs found in the abdominopelvic cavity, also indicate which quadrant they occupy.

Dorsal body cavity ______________________

Thoracic cavity ______________________

____

Abdominopelvic cavity ______________________

____

____

4. Now, assign each of the organs to one of the organ system categories listed below.

Digestive: ______________________

____

Urinary: ______________________

Cardiovascular: ______________________

____

Endocrine: ______________________

Reproductive: ______________________

Respiratory: ______________________

Lymphatic/Immunity: ______________________

Nervous: ______________________

EXERCISE 3

The Microscope

MATERIALS

- ☐ Compound microscope
- ☐ Millimeter ruler
- ☐ Prepared slides of the letter *e* or newsprint
- ☐ Immersion oil
- ☐ Lens paper
- ☐ Prepared slide of grid ruled in millimeters (grid slide)
- ☐ Prepared slide of three crossed colored threads
- ☐ Clean microscope slide and coverslip
- ☐ Toothpicks (flat-tipped)
- ☐ Physiologic saline in a dropper bottle
- ☐ Iodine or methylene blue stain (dilute) in a dropper bottle
- ☐ Filter paper or paper towels
- ☐ Beaker containing fresh 10% household bleach solution for wet mount disposal
- ☐ Disposable autoclave bag
- ☐ Prepared slide of cheek epithelial cells

Note to the Instructor: The slides and coverslips used for viewing cheek cells are to be soaked for 2 hours (or longer) in 10% bleach solution and then drained. The slides and disposable autoclave bag (containing coverslips, lens paper, and used toothpicks) are to be autoclaved for 15 min at 121°C and 15 pounds pressure to ensure sterility. After autoclaving, the disposable autoclave bag may be discarded in any disposal facility, and the slides and glassware washed with laboratory detergent and reprepared for use. These instructions apply as well to any bloodstained glassware or disposable items used in other experimental procedures.

MasteringA&P™

Access practice quizzes and more in the Study Area at www.masteringaandp.com.

OBJECTIVES

1. To identify the parts of the microscope and list the function of each.
2. To describe and demonstrate the proper techniques for care of the microscope.
3. To define *total magnification* and *resolution*.
4. To demonstrate proper focusing technique.
5. To define *parfocal, field,* and *depth of field*.
6. To estimate the size of objects in a field.

PRE-LAB QUIZ

1. The microscope slide rests on the ___________ while being viewed.
   a. base
   b. condenser
   c. iris
   d. stage
2. Your lab microscope is *parfocal*. This means that:
   a. The specimen is clearly in focus at this depth.
   b. The slide should be in focus at higher magnifications once it is properly focused at lower magnifications.
   c. You can easily discriminate two close objects as separate.
3. If the ocular lens magnifies a specimen 10×, and the objective lens used magnifies the specimen 35×, what is the total magnification being used to observe the specimen? ___________
4. How do you clean the lenses of your microscope?
   a. with a paper towel
   b. with soap and water
   c. with special lens paper and cleaner
5. Circle True or False. You should always start your observation of specimens with the oil-immersion lens.
6. Circle True or False. It is always necessary to use a coverslip with a wet mount to prevent soiling or damaging the objective lens.

With the invention of the microscope, biologists gained a valuable tool to observe and study structures (like cells) that are too small to be seen by the unaided eye. The information gained helped in establishing many of the theories basic to the understanding of biological sciences. This exercise will familiarize you with the workhorse of microscopes—the compound microscope—and provide you with the necessary instructions for its proper use.

Care and Structure of the Compound Microscope

The **compound microscope** is a precision instrument and should always be handled with care. *At all times you must observe the following rules for its transport, cleaning, use, and storage:*

- When transporting the microscope, hold it in an upright position with one hand on its arm and the other supporting its base. Avoid swinging the instrument during its transport and jarring the instrument when setting it down.
- Use only special grit-free lens paper to clean the lenses. Use a circular motion to wipe the lenses, and clean all lenses before and after use.
- Always begin the focusing process with the lowest-power objective lens in position, changing to the higher-power lenses as necessary.
- Use the coarse adjustment knob only with the lowest-power lens.
- Always use a coverslip with temporary (wet mount) preparations.
- Before putting the microscope in the storage cabinet, remove the slide from the stage, rotate the lowest-power objective lens into position, wrap the cord neatly around the base, and replace the dust cover or return the microscope to the appropriate storage area.
- Never remove any parts from the microscope; inform your instructor of any mechanical problems that arise.

ACTIVITY 1

Identifying the Parts of a Microscope

1. Obtain a microscope and bring it to the laboratory bench. (Use the proper transport technique!)

- Record the number of your microscope in the summary chart on page 30.

  Compare your microscope with the illustration in Figure 3.1 and identify the following microscope parts:

  Base: Supports the microscope. (**Note:** Some microscopes are provided with an inclination joint, which allows the instrument to be tilted backward for viewing dry preparations.)

  Substage light or **mirror:** Located in the base. In microscopes with a substage light source, the light passes directly upward through the microscope: light controls are located on the microscope base. If a mirror is used, light must be reflected from a separate free-standing lamp.

  Stage: The platform the slide rests on while being viewed. The stage has a hole in it to permit light to pass through both it and the specimen. Some microscopes have a stage equipped with *spring clips;* others have a clamp-type *mechanical stage* as shown in Figure 3.1. Both hold the slide in position for viewing; in addition, the mechanical stage has two adjustable knobs that control precise movement of the specimen.

  Condenser: Small substage lens that concentrates the light on the specimen. The condenser may have a *rack and pinion knob* that raises and lowers the condenser to vary light delivery. Generally, the best position for the condenser is close to the inferior surface of the stage.

  Iris diaphragm lever: Arm attached to the base of the condenser that regulates the amount of light passing through the condenser. The iris diaphragm permits the best possible contrast when viewing the specimen.

  Coarse adjustment knob: Used to focus on the specimen.

  Fine adjustment knob: Used for precise focusing once coarse focusing has been completed.

  Head or **body tube:** Supports the objective lens system (which is mounted on a movable nosepiece) and the ocular lens or lenses.

  Arm: Vertical portion of the microscope connecting the base and head.

  Ocular (or *eyepiece*): Depending on the microscope, there are one or two lenses at the superior end of the head or body tube. Observations are made through the ocular(s). An ocular lens has a magnification of 10×. (It increases the apparent size of the object by ten times or ten diameters.) If your microscope has a **pointer** (used to indicate a specific area of the viewed specimen), it is attached to one ocular and can be positioned by rotating the ocular lens.

  Nosepiece: Rotating mechanism at the base of the head. Generally carries three or four objective lenses and permits sequential positioning of these lenses over the light beam passing through the hole in the stage. Use the nosepiece to change the objective lenses. Do not directly grab the lenses.

  Objective lenses: Adjustable lens system that permits the use of a **scanning lens,** a **low-power lens,** a **high-power lens,** or an **oil immersion lens.** The objective lenses have different magnifying and resolving powers.

2. Examine the objective lenses carefully; note their relative lengths and the numbers inscribed on their sides. On many microscopes, the scanning lens, with a magnification between 4× and 5×, is the shortest lens. If there is no scanning lens, the low-power objective lens is the shortest and typically has a magnification of 10×. The high-power objective lens is of intermediate length and has a magnification range from 40× to 50×, depending on the microscope. The oil immersion objective lens is usually the longest of the objective lenses and has a magnifying power of 95× to 100×. Some microscopes lack the oil immersion lens.

- Record the magnification of each objective lens of your microscope in the first row of the summary chart on page 30. Also, cross out the column relating to a lens that your microscope does not have. Plan on using the same microscope for all microscopic studies.

3. Rotate the lowest-power objective lens until it clicks into position, and turn the coarse adjustment knob about 180 degrees. Notice how far the stage (or objective lens) travels during this adjustment. Move the fine adjustment knob 180 degrees, noting again the distance that the stage (or the objective lens) moves. ■

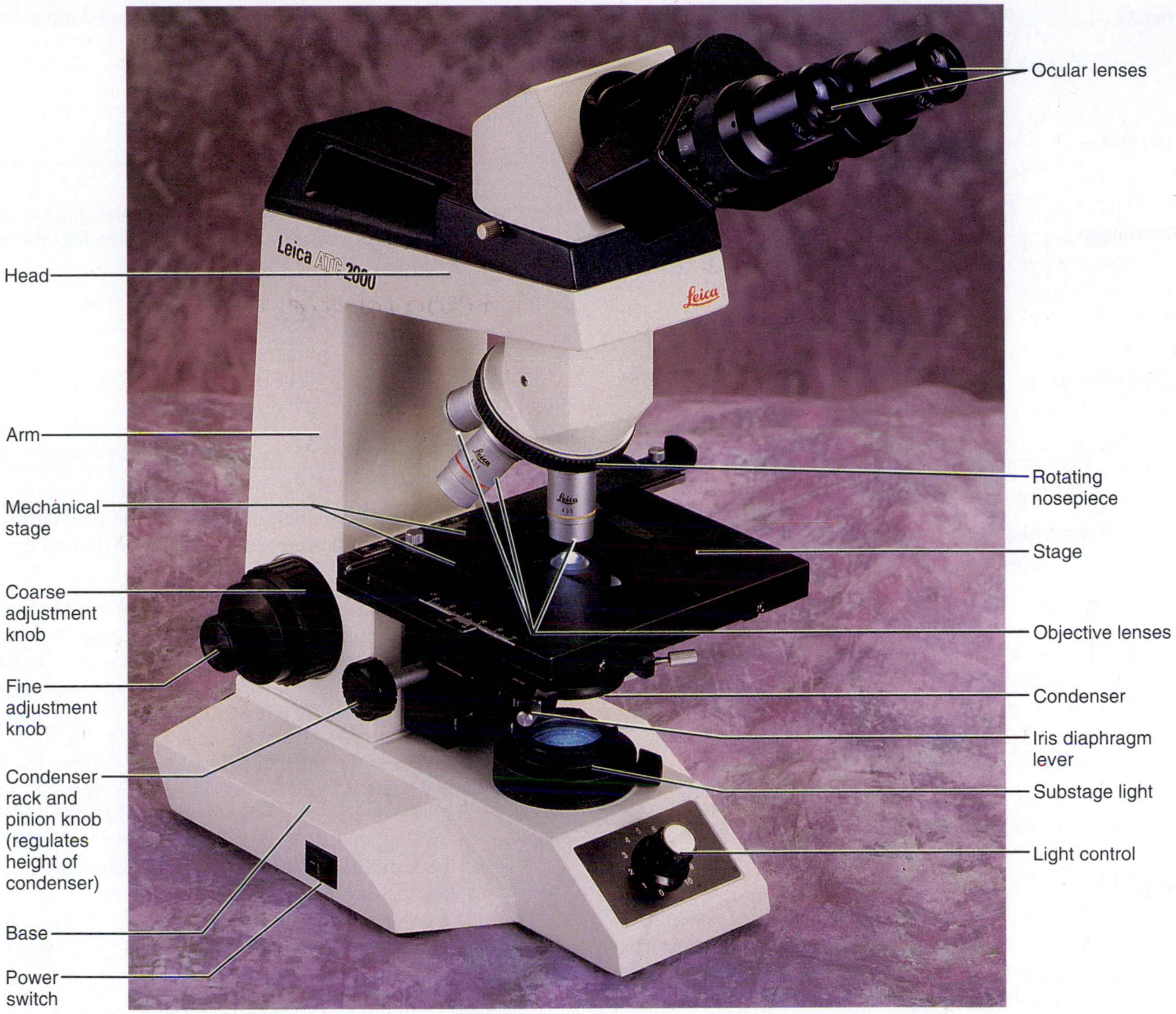
Ocular lenses
Head
Arm
Mechanical stage
Coarse adjustment knob
Fine adjustment knob
Condenser rack and pinion knob (regulates height of condenser)
Base
Power switch
Rotating nosepiece
Stage
Objective lenses
Condenser
Iris diaphragm lever
Substage light
Light control

FIGURE 3.1 Compound microscope and its parts.

Magnification and Resolution

The microscope is an instrument of magnification. In the compound microscope, magnification is achieved through the interplay of two lenses—the ocular lens and the objective lens. The objective lens magnifies the specimen to produce a **real image** that is projected to the ocular. This real image is magnified by the ocular lens to produce the **virtual image** seen by your eye (Figure 3.2).

The **total magnification** (TM) of any specimen being viewed is equal to the power of the ocular lens multiplied by the power of the objective lens used. For example, if the ocular lens magnifies 10× and the objective lens being used magnifies 45×, the total magnification is 450× (or 10 × 45).

- Determine the total magnification you may achieve with each of the objectives on your microscope, and record the figures on the third row of the chart.

The compound light microscope has certain limitations. Although the level of magnification is almost limitless, the **resolution** (or resolving power), that is, the ability to discriminate two close objects as separate, is not. The human eye can resolve objects about 100 μm apart, but the compound microscope has a resolution of 0.2 μm under ideal conditions. Objects closer than 0.2 μm are seen as a single fused image.

Resolving power is determined by the amount and physical properties of the visible light that enters the microscope. In general, the more light delivered to the objective lens, the greater the resolution. The size of the objective lens aperture (opening) decreases with increasing magnification, allowing less light to enter the objective. Thus, you will probably find it necessary to increase the light intensity at the higher magnifications.

ACTIVITY 2

Viewing Objects Through the Microscope

1. Obtain a millimeter ruler, a prepared slide of the letter *e* or newsprint, a dropper bottle of immersion oil, and some lens paper. Adjust the condenser to its highest position and switch

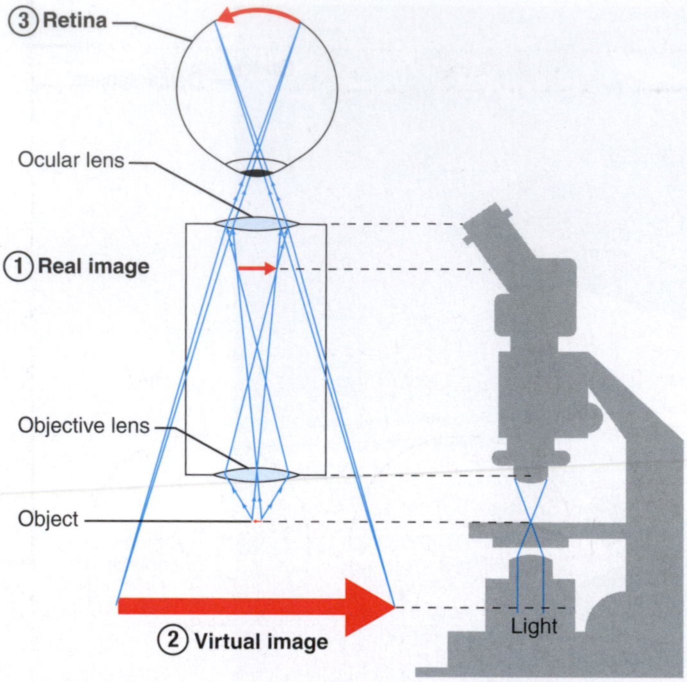
③ Retina
Ocular lens
① Real image
Objective lens
Object
② Virtual image
Light

FIGURE 3.2 Image formation in light microscopy. **Step** ① The objective lens magnifies the object, forming the real image. **Step** ② The ocular lens magnifies the real image, forming the virtual image. **Step** ③ The virtual image passes through the lens of the eye and is focused on the retina.

on the light source of your microscope. (If the light source is not built into the base, use the curved surface of the mirror to reflect the light up into the microscope.)

2. Secure the slide on the stage so that you can read the slide label and the letter *e* is centered over the light beam passing through the stage. If you are using a microscope with spring clips, make sure the slide is secured at both ends. If your microscope has a mechanical stage, open the jaws of its slide retainer (holder) by using the control lever, typically located at the rear left corner of the mechanical stage. Insert the slide squarely within the confines of the slide retainer. Check that the slide is resting on the stage (and not on the mechanical stage frame) before releasing the control lever.

3. With your lowest-power (scanning or low-power) objective lens in position over the stage, use the coarse adjustment knob to bring the objective lens and stage as close together as possible.

4. Look through the ocular lens and adjust the light for comfort using the iris diaphragm. Now use the coarse adjustment knob to focus slowly away from the *e* until it is as clearly focused as possible. Complete the focusing with the fine adjustment knob.

5. Sketch the letter *e* in the circle on the summary chart on page 31 just as it appears in the **field** (the area you see through the microscope).

How far is the bottom of the objective lens from the specimen? In other words, what is the **working distance**? Use a millimeter ruler to make this measurement.

Record the working distance in the summary chart.

How has the apparent orientation of the *e* changed top to bottom, right to left, and so on?

Upside down

6. Move the slide slowly away from you on the stage as you view it through the ocular lens. In what direction does the image move?

toward me

Move the slide to the left. In what direction does the image move?

right

At first this change in orientation may confuse you, but with practice you will learn to move the slide in the desired direction with no problem.

7. Today most good laboratory microscopes are **parfocal;** that is, the slide should be in focus (or nearly so) at the higher magnifications once you have properly focused. *Without touching the focusing knobs,* increase the magnification by rotating the next higher magnification lens (low power or high power) into position over the stage. Make sure it clicks into position. Using the fine adjustment only, sharpen the focus.* Note the decrease in working distance. As you can see, focusing with the coarse adjustment knob could drive the objective lens through the slide, breaking the slide and possibly damaging the lens. Sketch the letter *e* in the summary chart. What new details become clear?

the ink spots

As best you can, measure the distance between the objective and the slide.

Record the working distance in the summary chart.

Is the image larger or smaller? larger

Approximately how much of the letter *e* is visible now?

some of it

Is the field larger or smaller? larger

Why is it necessary to center your object (or the portion of the slide you wish to view) before changing to a higher power?

because the focus will be more centered

*If you are unable to focus with a new lens, your microscope is not parfocal. Do not try to force the lens into position. Consult your instructor.

Summary Chart for Microscope # 4				
	Scanning	Low power	High power	Oil immersion
Magnification of objective lens	4 ×	10 ×	40 ×	100 ×
Magnification of ocular lens	10 ×	10 ×	10 ×	10 ×
Total magnification	40 ×	100 ×	400 ×	1000 ×
Working distance	2.5 mm	.8 mm	.1 mm	0 mm
Detail observed Letter *e*	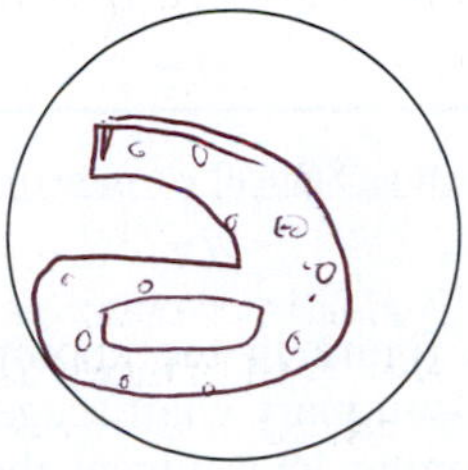	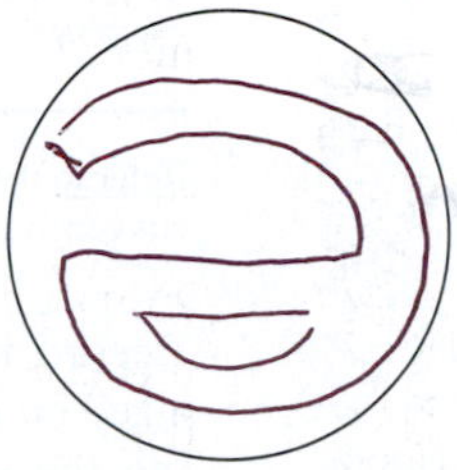	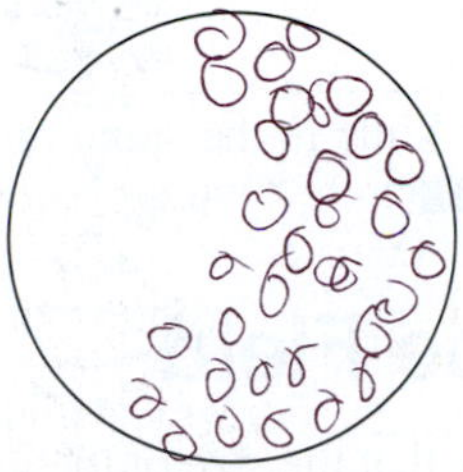	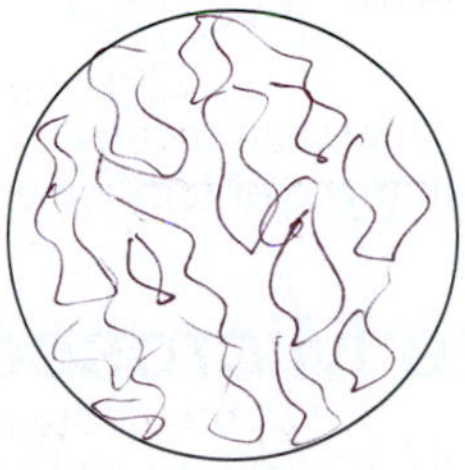
Field size (diameter)	.10 mm 100 µm	.22 mm 220 µm	.65 mm 650 µm	1.25 mm 1250 µm

Move the iris diaphragm lever while observing the field. What happens?

light decreases + increases

Is it more desirable to increase *or* decrease the light when changing to a higher magnification?

decrease Why? the light gets more intence when you get closer

8. If you have just been using the low-power objective, repeat the steps given in direction 7 using the high-power objective lens. What new details become clear?

You can see the seperation and pigment of the ink

Record the working distance in the summary chart.

9. Without touching the focusing knob, rotate the high-power lens out of position so that the area of the slide over the opening in the stage is unobstructed. Place a drop of immersion oil over the *e* on the slide and rotate the oil immersion lens into position. Set the condenser at its highest point (closest to the stage), and open the diaphragm fully. Adjust the fine focus and fine-tune the light for the best possible resolution.

Note: If for some reason the specimen does not come into view after adjusting the fine focus, do not go back to the 40× lens to recenter. You do not want oil from the oil immersion lens to cloud the 40× lens. Turn the revolving nosepiece in the other direction to the low-power lens and recenter and refocus the object. Then move the immersion lens back into position, again avoiding the 40× lens. Sketch the letter *e* in the summary chart, What new details become clear?

pigment + ink

Is the field again decreased in size? Yes

As best you can, estimate the working distance, and record it in the summary chart. Is the working distance less *or* greater than it was when the high-power lens was focused?

less

Compare your observations on the relative working distances of the objective lenses with the illustration in Figure 3.3. Explain why it is desirable to begin the focusing process at the lowest power.

You might be able to see the Object better at a lower power

10. Rotate the oil immersion lens slightly to the side and remove the slide. Clean the oil immersion lens carefully with

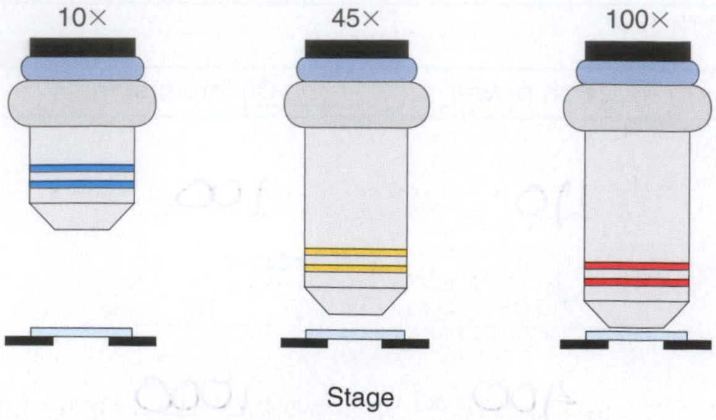
10×
45×
100×
Stage

FIGURE 3.3 Relative working distances of the 10×, 45×, and 100× objectives.

lens paper, and then clean the slide in the same manner with a fresh piece of lens paper.

The Microscope Field

By this time you should know that the size of the microscope field decreases with increasing magnification. For future microscope work, it will be useful to determine the diameter of each of the microscope fields. This information will allow you to make a fairly accurate estimate of the size of the objects you view in any field. For example, if you have calculated the field diameter to be 4 mm and the object being observed extends across half this diameter, you can estimate that the length of the object is approximately 2 mm.

Microscopic specimens are usually measured in micrometers and millimeters, both units of the metric system. You can get an idea of the relationship and meaning of these units from Table 3.1. A more detailed treatment appears in Appendix A.

ACTIVITY 3

Estimating the Diameter of the Microscope Field

1. Obtain a grid slide (a slide prepared with graph paper ruled in millimeters). Each of the squares in the grid is 1 mm on each side. Use your lowest-power objective to bring the grid lines into focus.

2. Move the slide so that one grid line touches the edge of the field on one side, and then count the number of squares you can see across the diameter of the field. If you can see only part of a square, as in the accompanying diagram, estimate the part of a millimeter that the partial square represents.

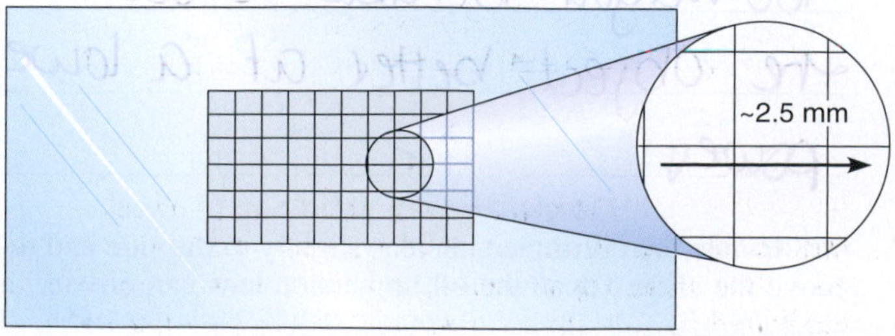
~2.5 mm

TABLE 3.1 Comparison of Metric Units of Length

Metric unit	Abbreviation	Equivalent
Meter	m	(about 39.3 in.)
Centimeter	cm	10^{-2} m
Millimeter	mm	10^{-3} m
Micrometer (or micron)	μm (μ)	10^{-6} m
Nanometer (or millimicrometeror millimicron)	nm (mμ)	10^{-9} m
Ångstrom	Å	10^{-10} m

Refer to the "Getting Started" exercise (page xv) for tips on metric conversions.

Record this figure in the appropriate space marked "field size" on the summary chart (page 31). (If you have been using the scanning lens, repeat the procedure with the low-power objective lens.)

Complete the chart by computing the approximate diameter of the high-power and oil immersion fields. The general formula for calculating the unknown field diameter is:

$$\text{Diameter of field } A \times \text{total magnification of field } A = \text{diameter of field } B \times \text{total magnification of field } B$$

where *A* represents the known or measured field and *B* represents the unknown field. This can be simplified to

$$\text{Diameter of field } B = \frac{\text{diameter of field } A \times \text{total magnification of field } A}{\text{total magnification of field } B}$$

For example, if the diameter of the low-power field (field *A*) is 2 mm and the total magnification is 50×, you would compute the diameter of the high-power field (field *B*) with a total magnification of 100 × as follows:

$$\text{Field diameter } B = (2\text{mm} \times 50)/100$$
$$\text{Field diameter } B = 1\text{mm}$$

3. Estimate the length (longest dimension) of the following microscopic objects. *Base your calculations on the field sizes you have determined for your microscope.*

Object seen in low-power field:

approximate length:

________ mm

Object seen in high-power field:

approximate length:

________ mm

or ________ μm

Object seen in oil immersion field:

approximate length:

__________ µm

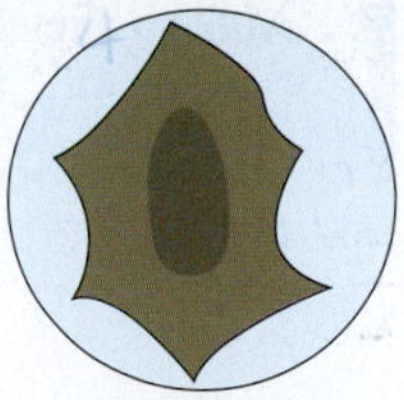

4. If an object viewed with the oil immersion lens looked as it does in the field depicted just below, could you determine its approximate size from this view?

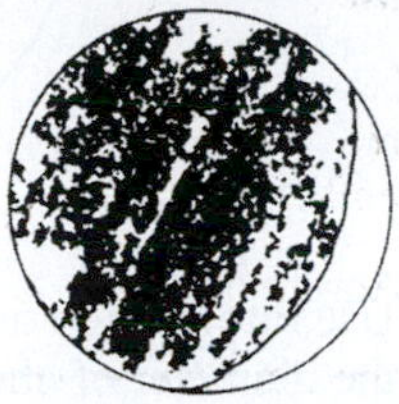

If not, then how could you determine it? ______________

____

____

____

Perceiving Depth

Any microscopic specimen has depth as well as length and width; it is rare indeed to view a tissue slide with just one layer of cells. Normally you can see two or three cell thicknesses. Therefore, it is important to learn how to determine relative depth with your microscope. In microscope work the **depth of field** (the depth of the specimen clearly in focus) is greater at lower magnifications.

ACTIVITY 4

Perceiving Depth

1. Obtain a slide with colored crossed threads. Focusing at low magnification, locate the point where the three threads cross each other.

2. Use the iris diaphragm lever to greatly reduce the light, thus increasing the contrast. Focus down with the coarse adjustment until the threads are out of focus, then slowly focus upward again, noting which thread comes into clear focus first. (You will see two or even all three threads, so you must be very careful in determining which one first comes into clear focus.) Observe: As you rotate the adjustment knob forward (away from you), does the stage rise or fall? If the stage rises, then the first clearly focused thread is the top one; the last clearly focused thread is the bottom one.

If the stage descends, how is the order affected? __________

____

Record your observations, relative to which color of thread is uppermost, middle, or lowest:

Top thread ______________________________________

Middle thread ____________________________________

Bottom thread ____________________________________

Viewing Cells Under the Microscope

There are various ways to prepare cells for viewing under a microscope. Cells and tissues can look very different with different stains and preparation techniques. One method of preparation is to mix the cells in physiologic saline (called a wet mount) and stain them with methylene blue stain.

If you are not instructed to prepare your own wet mount, obtain a prepared slide of epithelial cells to make the observations in step 10 of Activity 5.

ACTIVITY 5

Preparing and Observing a Wet Mount

1. Obtain the following: a clean microscope slide and coverslip, two flat-tipped toothpicks, a dropper bottle of physiologic saline, a dropper bottle of iodine or methylene blue stain, and filter paper (or paper towels). Handle only your own slides throughout the procedure.

2. Place a drop of physiologic saline in the center of the slide. Using the flat end of the toothpick, *gently* scrape the inner lining of your cheek. Transfer your cheek scrapings to the slide by agitating the end of the toothpick in the drop of saline (Figure 3.4a).

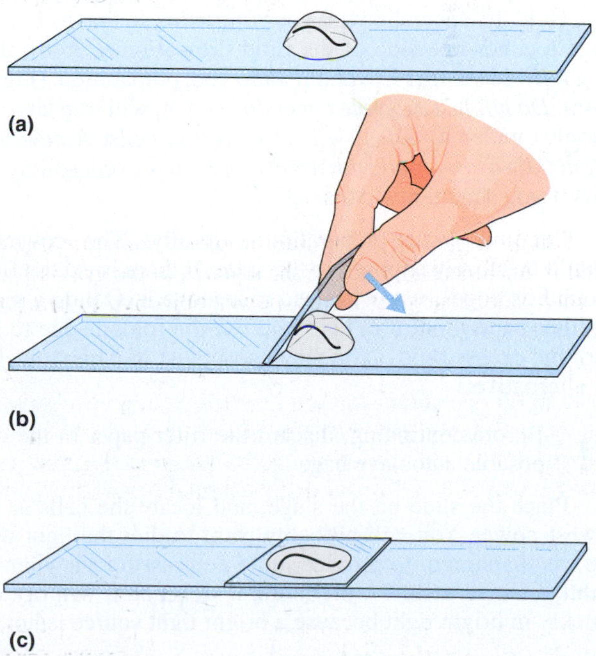
(a)
(b)
(c)

FIGURE 3.4 Procedure for preparation of a wet mount. (a) The object is placed in a drop of water (or saline) on a clean slide, **(b)** a coverslip is held at a 45° angle with the fingertips, and **(c)** it is lowered carefully over the water and the object.

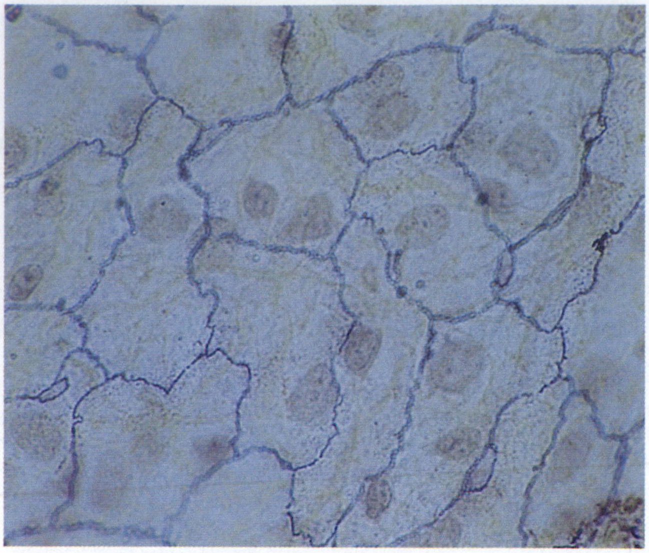

FIGURE 3.5 **Epithelial cells of the cheek cavity (surface view, 750×).**

⚠ *Immediately* discard the used toothpick in the disposable autoclave bag provided at the supplies area.

3. Add a tiny drop of the iodine or methylene blue stain to the preparation. (These epithelial cells are nearly transparent and thus difficult to see without the stain, which colors the nuclei of the cells and makes them look much darker than the cytoplasm.) Stir again.

⚠ *Immediately* discard the used toothpick in the disposable autoclave bag provided at the supplies area.

4. Hold the coverslip with your fingertips so that its bottom edge touches one side of the fluid drop (Figure 3.4b), then *carefully* lower the coverslip onto the preparation (Figure 3.4c). *Do not just drop the coverslip,* or you will trap large air bubbles under it, which will obscure the cells. *A coverslip should always be used with a wet mount* to prevent soiling the lens if you should misfocus.

5. Examine your preparation carefully. The coverslip should be closely apposed to the slide. If there is excess fluid around its edges, you will need to remove it. Obtain a piece of filter paper, fold it in half, and use the folded edge to absorb the excess fluid. (You may use a twist of paper towel as an alternative.)

⚠ Before continuing, discard the filter paper in the disposable autoclave bag.

6. Place the slide on the stage, and locate the cells at the lowest power. You will probably want to dim the light with the iris diaphragm to provide more contrast for viewing the lightly stained cells. Furthermore, a wet mount will dry out quickly in bright light because a bright light source is hot.

7. Cheek epithelial cells are very thin, six-sided cells. In the cheek, they provide a smooth, tilelike lining, as shown in Figure 3.5. Move to high power to examine the cells more closely.

8. Make a sketch of the epithelial cells that you observe.

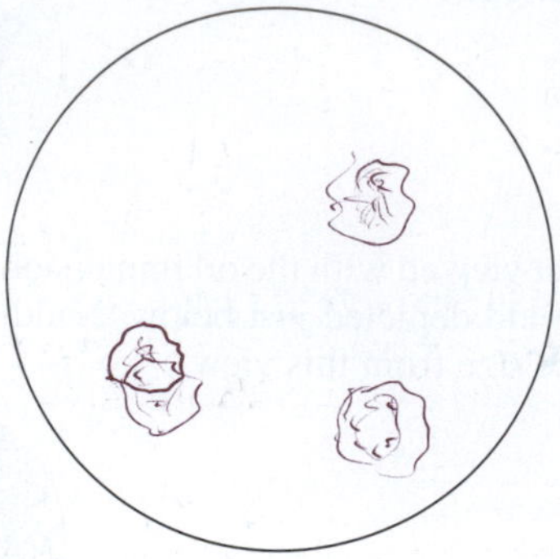

Use information on your summary chart (page 31) to estimate the diameter of cheek epithelial cells.

.1 mm

Why do *your* cheek cells look different than those illustrated in Figure 3.5? (Hint: What did you have to *do* to your cheek to obtain them?)

they arent stained and there arent as many

⚠ 9. When you complete your observations of the wet mount, dispose of your wet mount preparation in the beaker of bleach solution, and put the coverslips in an autoclave bag.

10. Obtain a prepared slide of cheek epithelial cells, and view them under the microscope.

Estimate the diameter of one of these cheek epithelial cells using information from the summary chart (page 31).

________ mm

Why are these cells more similar to those seen in Figure 3.5 and easier to measure than those of the wet mount?

____

____

____

11. Before leaving the laboratory, make sure all other materials are properly discarded or returned to the appropriate laboratory station. Clean the microscope lenses and put the dust cover on the microscope before you return it to the storage cabinet. ■

EXERCISE 4

The Cell: Anatomy and Division

MATERIALS

- ☐ Three-dimensional model of the "composite" animal cell or laboratory chart of cell anatomy
- ☐ Compound microscope
- ☐ Prepared slides of simple squamous epithelium, teased smooth muscle (l.s.), human blood cell smear, and sperm
- ☐ Animation/video of mitosis
- ☐ Three-dimensional models of mitotic stages
- ☐ Prepared slides of whitefish blastulae

Note to the Instructor: See directions for handling wet mount preparations and disposable supplies on page 33, Exercise 3. For suggestions on the animation/video of mitosis, see the Instructor Guide.

OBJECTIVES

1. To define *cell, organelle,* and *inclusion.*
2. To identify on a cell model or diagram the following cellular regions and to list the major function of each: nucleus, cytoplasm, and plasma membrane.
3. To identify and list the major functions of the various organelles studied.
4. To compare and contrast specialized cells with the concept of the "generalized cell."
5. To define *interphase, mitosis,* and *cytokinesis.*
6. To list the stages of mitosis and describe the events of each stage.
7. To identify the mitotic phases on slides or appropriate diagrams.
8. To explain the importance of mitotic cell division and its product.

PRE-LAB QUIZ

1. Define *cell.* ____

   ____
2. When a cell is not dividing, the genetic material is loosely dispersed throughout the nucleus in a threadlike form called:
   a. chromatin　　c. cytosol
   b. chromosomes　　d. ribosomes
3. The plasma membrane not only provides a protective barrier for the cell but also determines which substances enter or exit the cell. We call this characteristic:
   a. diffusion　　c. osmosis
   b. membrane potential　　d. selective permeability
4. Name the organelles that are the sites of protein synthesis.

   ____
5. Because these organelles are responsible for providing the bulk of ATP needed by the cell, they are often referred to as the "powerhouses" of the cell. They are the:
   a. centrioles　　c. mitochondria
   b. lysosomes　　d. ribosomes
6. Circle the correct term. During cytokinesis / interphase the cell grows and carries out its usual activities.
7. Circle True or False. The end product of mitosis is four genetically identical daughter cells.
8. How many stages of mitosis are there? __________
9. DNA replication occurs during:
   a. cytokinesis　　c. metaphase
   b. interphase　　d. prophase
10. Circle True or False. All animal cells have a cell wall.

MasteringA&P™

Access practice quizzes and more in the Study Area at www.masteringaandp.com.

The **cell,** the structural and functional unit of all living things, is a complex entity. The cells of the human body are highly diverse, and their differences in size, shape, and internal composition reflect their specific roles in the body. Nonetheless, cells do have many common anatomical features, and all cells must carry out certain functions to sustain life. For example, all cells can maintain their boundaries, metabolize, digest nutrients and dispose of wastes, grow and reproduce, move, and respond to a stimulus. Most of these functions are considered in detail in later exercises. This exercise focuses on structural similarities that typify the "composite," or "generalized," cell and considers only the function of cell reproduction (cell division). Transport mechanisms (the means by which substances cross a cell's external membrane) are dealt with separately in Exercise 5.

Anatomy of the Composite Cell

In general, all cells have three major regions, or parts, that can readily be identified with a light microscope: the **nucleus,** the **plasma membrane,** and the **cytoplasm.** The nucleus is typically a round or oval structure near the center of the cell. It is surrounded by cytoplasm, which in turn is enclosed by the plasma membrane. Since the advent of the electron microscope, even smaller cell structures—organelles—have been identified. Figure 4.1a is a diagrammatic representation of the fine structure of the composite cell; Figure 4.1b depicts cellular structure (particularly that of the nucleus) as revealed by the electron microscope.

Nucleus

The nucleus contains the genetic material, DNA, sections of which are called "genes." Often described as the control center of the cell, the nucleus is necessary for cell reproduction. A cell that has lost or ejected its nucleus (for whatever reason) is programmed to die.

When the cell is not dividing, the genetic material is loosely dispersed throughout the nucleus in a threadlike form called **chromatin.** When the cell is in the process of dividing to form daughter cells, the chromatin coils and condenses, forming dense, darkly staining rodlike bodies called **chromosomes**—much in the way a stretched spring becomes shorter and thicker when it is released. (Cell division is discussed later in this exercise.) Carefully note the appearance of the nucleus—it is somewhat nondescript when a cell is healthy. A dark nucleus and clumped chromatin are indications that the cell is dying and undergoing degeneration.

The nucleus also contains one or more small round bodies, called **nucleoli,** composed primarily of proteins and ribonucleic acid (RNA). The nucleoli are assembly sites for ribosomal particles (particularly abundant in the cytoplasm), which are the actual protein-synthesizing "factories."

The nucleus is bound by a double-layered porous membrane, the **nuclear envelope.** The nuclear envelope is similar in composition to other cellular membranes, but it is distinguished by its large **nuclear pores.** Although they are spanned by diaphragms, these pores permit easy passage of protein and RNA molecules.

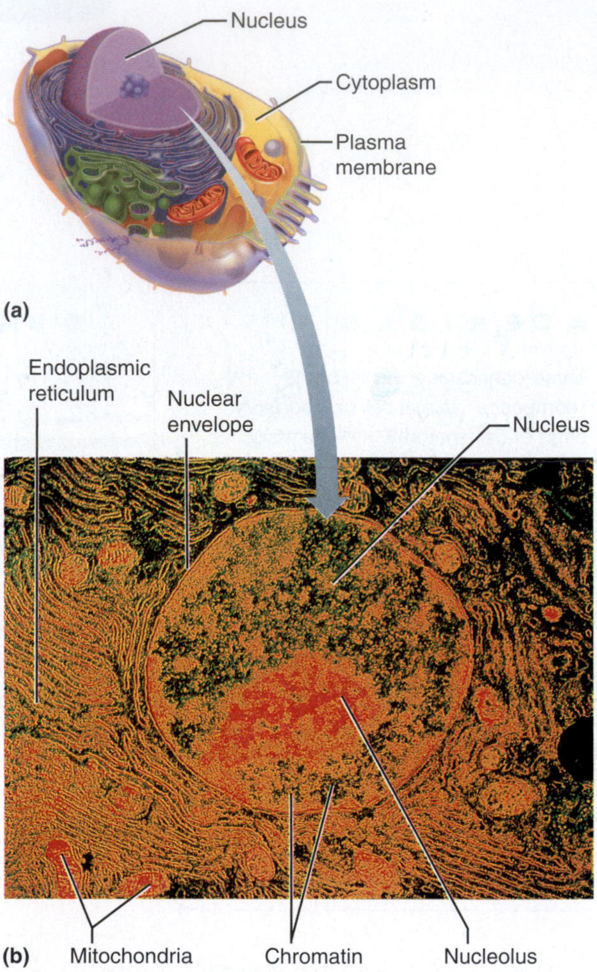
Nucleus
Cytoplasm
Plasma membrane
(a)
Endoplasmic reticulum
Nuclear envelope
Nucleus
(b)
Mitochondria
Chromatin
Nucleolus

FIGURE 4.1 Anatomy of the composite animal cell. (a) Diagrammatic view. **(b)** Transmission electron micrograph (10,000×).

ACTIVITY 1

Identifying Parts of a Cell

As able, identify the nuclear envelope, chromatin, nucleolus, and the nuclear pores in Figure 4.1a and b and Figure 4.3.

Plasma Membrane

The **plasma membrane** separates cell contents from the surrounding environment. Its main structural building blocks are phospholipids (fats) and globular protein molecules, but some of the externally facing proteins and lipids have sugar (carbohydrate) side chains attached to them that are important in cellular interactions (Figure 4.2). Described by the fluid-mosaic model, the membrane is a bilayer of phospholipid molecules in which the protein molecules float. Occasional cholesterol molecules dispersed in the fluid phospholipid bilayer help stabilize it.

Besides providing a protective barrier for the cell, the plasma membrane plays an active role in determining which substances may enter or leave the cell and in what quantity. Because of its molecular composition, the plasma membrane is selective about what passes through it. It allows nutrients to

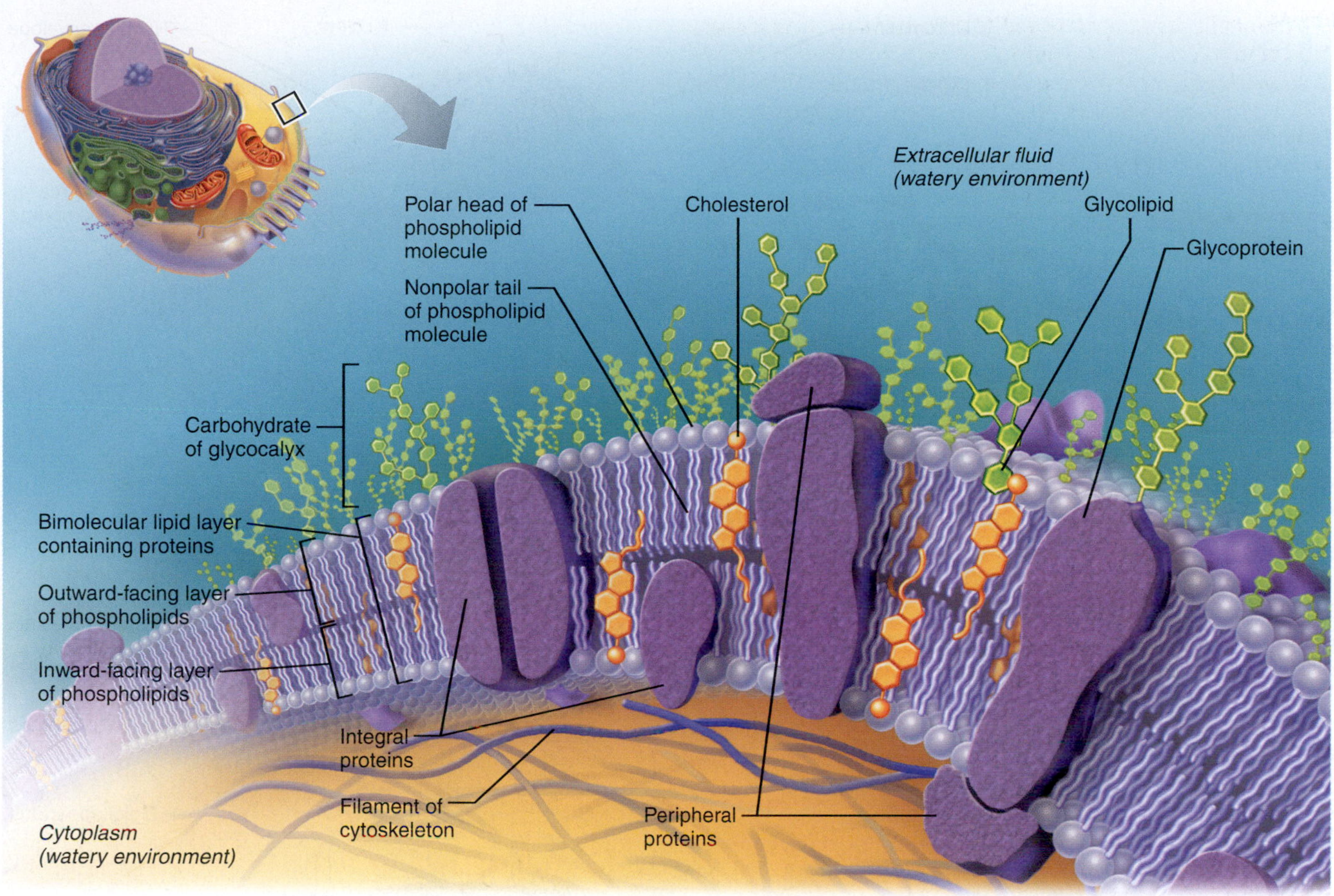
Extracellular fluid
(watery environment)
Polar head of phospholipid molecule
Nonpolar tail of phospholipid molecule
Cholesterol
Glycolipid
Glycoprotein
Carbohydrate of glycocalyx
Bimolecular lipid layer containing proteins
Outward-facing layer of phospholipids
Inward-facing layer of phospholipids
Integral proteins
Filament of cytoskeleton
Peripheral proteins
Cytoplasm
(watery environment)

FIGURE 4.2 Structural details of the plasma membrane.

enter the cell but keeps out undesirable substances. By the same token, valuable cell proteins and other substances are kept within the cell, and excreta or wastes pass to the exterior. This property is known as **selective permeability.** Transport through the plasma membrane occurs in two basic ways. In *active transport,* the cell must provide energy (adenosine triphosphate, ATP) to power the transport process. In *passive transport,* the transport process is driven by concentration or pressure differences.

Additionally, the plasma membrane maintains a resting potential that is essential to normal functioning of excitable cells, such as neurons and muscle cells, and plays a vital role in cell signaling and cell-to-cell interactions. In some cells the membrane is thrown into minute fingerlike projections or folds called **microvilli** (see Figure 4.3 on page 42). Microvilli greatly increase the surface area of the cell available for absorption or passage of materials and for the binding of signaling molecules.

ACTIVITY 2

Identifying Components of a Plasma Membrane

Identify the phospholipid and protein portions of the plasma membrane in Figure 4.2. Also locate the sugar (*glyco* = carbohydrate) side chains and cholesterol molecules. Identify the microvilli in the orientation diagram at the top of Figure 4.2 and in Figure 4.3. ■

Cytoplasm and Organelles

The cytoplasm consists of the cell contents between the nucleus and plasma membrane. It is the major site of most activities carried out by the cell. Suspended in the **cytosol,** the fluid cytoplasmic material, are many small structures called **organelles** (literally, "small organs"). The organelles are the metabolic machinery of the cell, and they are highly organized to carry out specific functions for the cell as a whole. The organelles include the ribosomes, endoplasmic reticulum, Golgi apparatus, lysosomes, peroxisomes, mitochondria, cytoskeletal elements, and centrioles.

ACTIVITY 3

Locating Organelles

Each organelle type is summarized in Table 4.1 and described briefly below. Read through this material and then, as best you can, locate the organelles in both Figures 4.1b and 4.3. ■

- **Ribosomes** are densely staining, roughly spherical bodies composed of RNA and protein. They are the actual sites of

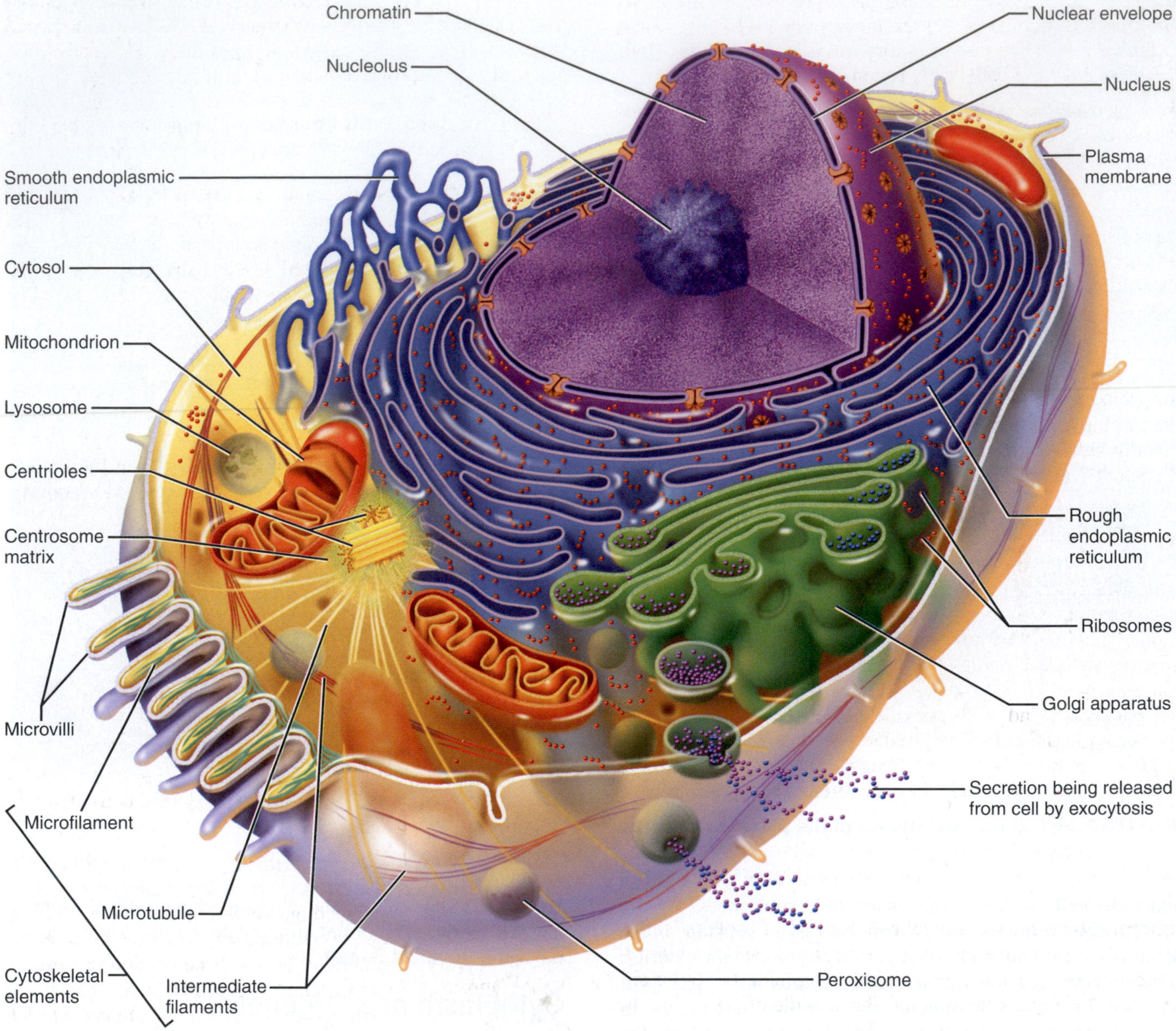
Chromatin
Nucleolus
Nuclear envelope
Nucleus
Plasma membrane
Smooth endoplasmic reticulum
Cytosol
Mitochondrion
Lysosome
Centrioles
Centrosome matrix
Rough endoplasmic reticulum
Ribosomes
Golgi apparatus
Microvilli
Secretion being released from cell by exocytosis
Microfilament
Microtubule
Peroxisome
Cytoskeletal elements
Intermediate filaments

FIGURE 4.3 Structure of the generalized cell. No cell is exactly like this one, but this composite illustrates features common to many human cells. Not all organelles are drawn to the same scale in this illustration.

protein synthesis. They are seen floating free in the cytoplasm or attached to a membranous structure. When they are attached, the whole ribosome-membrane complex is called the *rough endoplasmic reticulum.*

- The **endoplasmic reticulum (ER)** is a highly folded system of membranous tubules and cisternae (sacs) that extends throughout the cytoplasm. The ER is continuous with the nuclear envelope. Thus, it is assumed that the ER provides a system of channels for the transport of cellular substances (primarily proteins) from one part of the cell to another. The ER exists in two forms; a particular cell may have both or only one, depending on its specific functions. The **rough ER,** as noted earlier, is studded with ribosomes. Its cisternae modify and store the newly formed proteins and dispatch them to other areas of the cell. The external face of the rough ER is involved in phospholipid and cholesterol synthesis. The amount of rough ER is closely correlated with the amount of protein a cell manufactures and is especially abundant in cells that make protein products for export—for example, the pancreas cells that produce digestive enzymes destined for the small intestine.

  The **smooth ER** does not participate in protein synthesis but is present in conspicuous amounts in cells that produce steroid-based hormones—for example, the interstitial cells of the testes, which produce testosterone. Smooth ER is also abundant in cells that are active in lipid metabolism and drug detoxification activities—liver cells, for instance.

- The **Golgi apparatus** is a stack of flattened sacs with bulbous ends that is generally found close to the nucleus. Within its cisternae, the proteins delivered to it by transport vesicles from the rough ER are modified (by attachment of sugar groups), segregated, and packaged into membranous vesicles that ultimately (1) are incorporated into the plasma membrane, (2) become secretory vesicles that release their contents from the cell, or (3) become lysosomes.

- **Lysosomes,** which appear in various sizes, are membrane-bound sacs containing an array of powerful digestive enzymes. A product of the packaging activities of the Golgi apparatus, the lysosomes contain *acid hydrolases,* enzymes capable of digesting worn-out cell structures and foreign substances that enter the cell via vesicle formation through phagocytosis or endocytosis (see Exercise 5). Because they have the capacity of total cell destruction, the lysosomes are often referred to as the "suicide sacs" of the cell.

- **Peroxisomes,** like lysosomes, are enzyme-containing sacs. However, their *oxidases* have a different task. Using oxygen, they detoxify a number of harmful substances, most importantly free radicals. Peroxisomes are particularly abundant in kidney and liver cells, cells that are actively involved in detoxification.

- **Mitochondria** are generally rod-shaped bodies with a double-membrane wall; the inner membrane is thrown into folds, or *cristae.* Oxidative enzymes on or within the mitochondria catalyze the reactions of the Krebs cycle and the electron transport chain (collectively called oxidative respiration), in which end products of food digestion are broken down to produce energy. The released energy is captured in the bonds of ATP molecules, which are then transported out of the mitochondria to provide a ready energy supply to power the cell. Every living cell requires a constant supply of ATP for its many activities. Because the mitochondria provide the bulk of this ATP, they are referred to as the "powerhouses" of the cell.

- **Cytoskeletal elements** ramify throughout the cytoplasm, forming an internal scaffolding called the *cytoskeleton* that supports and moves substances within the cell. The **microtubules** are slender tubules formed of proteins called *tubulins,* which have the ability to aggregate and then disaggregate spontaneously. Microtubules organize the cytoskeleton and direct formation of the spindle formed by the centrioles during cell division. They also act in the transport of substances down the length of elongated cells (such as neurons), suspend organelles, and help maintain cell shape by providing rigidity to the soft cellular substance. **Intermediate filaments** are *stable* proteinaceous cytoskeletal elements that act as internal guy wires to resist mechanical (pulling) forces acting on cells. **Microfilaments,** ribbon or cordlike elements, are formed of contractile proteins, primarily *actin.* Because of their ability to shorten and then relax to assume a more elongated form, these are important in cell mobility and are very conspicuous in cells that are specialized to contract (such as muscle cells). A cross-linked network of *lamin* microfilaments braces and strengthens the internal face of the plasma membrane.

The cytoskeletal structures are changeable and minute. With the exception of the microtubules of the spindle, which are very obvious during cell division (see pages 46–47), and the microfilaments of skeletal muscle cells, they are rarely seen, even in electron micrographs, and are not depicted in Figure 4.1b. However, special stains can reveal the plentiful supply of these important organelles.

- The paired **centrioles** lie close to the nucleus in all animal cells capable of reproducing themselves. They are rod-shaped bodies that lie at right angles to each other. Internally each centriole is composed of nine triplets of microtubules. During cell division, the centrioles direct the formation of the mitotic spindle. Centrioles also form the cell projections called cilia and flagella, and in that role are called basal bodies.

The cell cytoplasm contains various other substances and structures, including stored foods (glycogen granules and lipid droplets), pigment granules, crystals of various types, water vacuoles, and ingested foreign materials. However, these are not part of the active metabolic machinery of the cell and are therefore called **inclusions.**

ACTIVITY 4

Examining the Cell Model

Once you have located all of these structures in Figure 4.3, examine the cell model (or cell chart) to repeat and reinforce your identifications.

Differences and Similarities in Cell Structure

ACTIVITY 5

Observing Various Cell Structures

1. Obtain a compound microscope and prepared slides of simple squamous epithelium, smooth muscle cells (teased), human blood, and sperm.

2. Observe each slide under the microscope, carefully noting similarities and differences in the cells. See photomicrographs for simple squamous epithelium (Figure 3.5 in Exercise 3) and teased smooth muscle (Figure 6.7c in Exercise 6). (The oil immersion lens will be needed to observe blood and sperm.) Distinguish the limits of the individual cells, and notice the shape and position of the nucleus in each case. When you look at the human blood smear, direct your attention to the red blood cells, the pink-stained cells that are most numerous. The color photomicrographs illustrating a blood smear (Figure 29.3 in Exercise 29) and sperm (Figure 43.3 in Exercise 43) may be helpful in this cell structure study. Sketch your observations in the circles provided on page 45.

3. Measure the length or diameter of each cell, and record below the appropriate sketch.

4. How do these four cell types differ in shape and size?

Squamous epithelium - many diff cells, very close to eachother

smooth muscle cells - squigly lines, and Black and purple dots

Blood (sickle cell) - small pink squares

Sperm

TABLE 4.1 Summary of Structure and Function of Cytoplasmic Organelles

Organelle		Location and function
Ribosomes		Tiny spherical bodies composed of RNA and protein; actual sites of protein synthesis; floating free or attached to a membranous structure (the rough ER) in the cytoplasm
Endoplasmic reticulum (ER)		Membranous system of tubules that extends throughout the cytoplasm; two varieties—rough ER is studded with ribosomes (tubules of the rough ER provide an area for storage and transport of the proteins made on the ribosomes to other cell areas; external face synthesizes phospholipids and cholesterol) and smooth ER, which has no function in protein synthesis (rather it is a site of steroid and lipid synthesis, lipid metabolism, and drug detoxification)
Golgi apparatus		Stack of flattened sacs with bulbous ends and associated small vesicles; found close to the nucleus; plays a role in packaging proteins or other substances for export from the cell or incorporation into the plasma membrane and in packaging lysosomal enzymes
Lysosomes	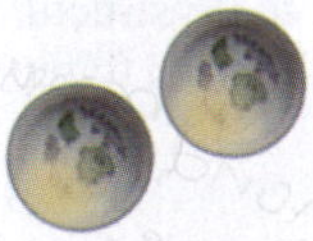	Various-sized membranous sacs containing digestive enzymes (acid hydrolases); function to digest worn-out cell organelles and foreign substances that enter the cell; have the capacity of total cell destruction if ruptured
Peroxisomes		Small lysosome-like membranous sacs containing oxidase enzymes that detoxify alcohol, hydrogen peroxide, and other harmful chemicals
Mitochondria		Generally rod-shaped bodies with a double-membrane wall; inner membrane is thrown into folds, or cristae; contain enzymes that oxidize foodstuffs to produce cellular energy (ATP); often referred to as "powerhouses of the cell"
Centrioles		Paired, cylindrical bodies lie at right angles to each other, close to the nucleus; direct the formation of the mitotic spindle during cell division; form the bases of cilia and flagella
Cytoskeletal elements: microfilaments, intermediate filaments, and microtubules		Provide cellular support; function in intracellular transport; microfilaments are formed largely of actin, a contractile protein, and thus are important in cell mobility (particularly in muscle cells); intermediate filaments are stable elements composed of a variety of proteins and resist mechanical forces acting on cells; microtubules form the internal structure of the centrioles and help determine cell shape

How might cell shape affect cell function?

____

____

____

____

____

Which cells have visible projections? ______________

____

How do these projections relate to the function of these cells?

____

____

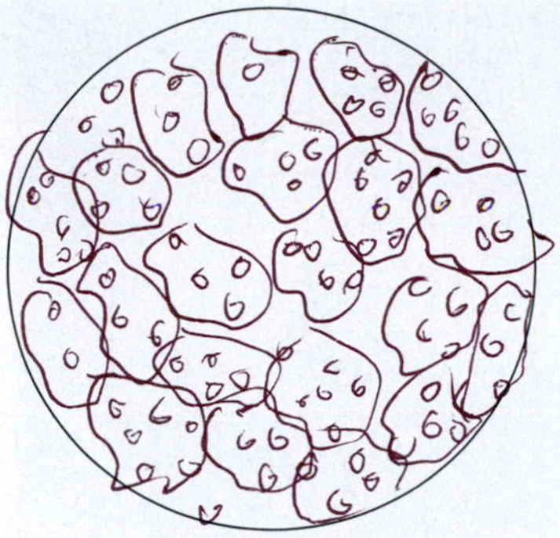

Simple squamous epithelium
Diameter __________

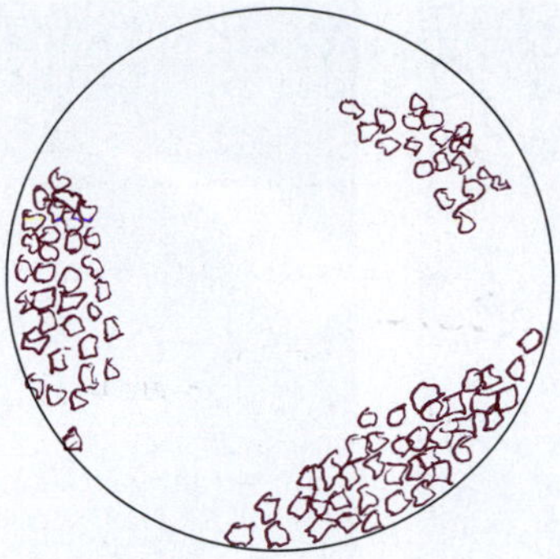

Human red blood cells
Diameter __________

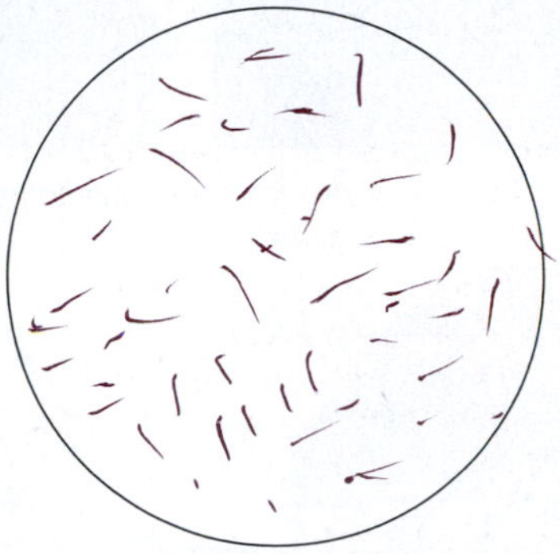

Sperm cells
Length __________

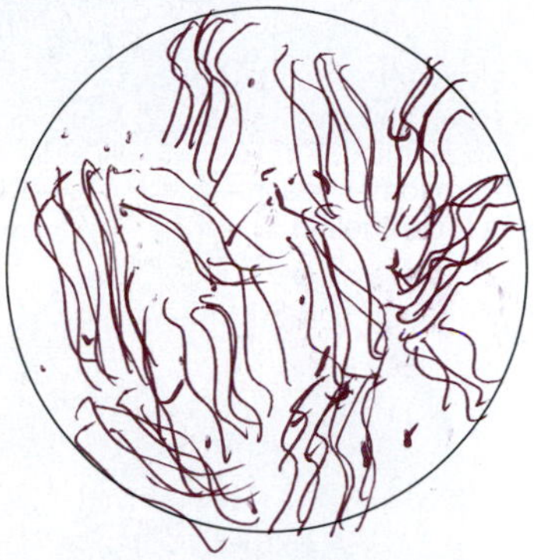

Teased smooth muscle cells
Length __________

Do any of these cells lack a plasma membrane? Sperm

A nucleus? NO

In the cells with a nucleus, can you discern nucleoli?

____

Were you able to observe any of the organelles in these cells?

________ Why or why not? __________________________

____

____

____

____

Cell Division: Mitosis and Cytokinesis

A cell's *life cycle* is the series of changes it goes through from the time it is formed until it reproduces itself. It encompasses two stages—**interphase,** the longer period during which the cell grows and carries out its usual activities (Figure 4.4a), and **cell division,** when the cell reproduces itself by dividing. In an interphase cell about to divide, the genetic material (DNA) is replicated (duplicated exactly). Once this important event has occurred, cell division ensues.

Cell division in all cells other than bacteria consists of a series of events collectively called mitosis and cytokinesis. **Mitosis** is nuclear division; **cytokinesis** is the division of the cytoplasm, which begins after mitosis is nearly complete. Although mitosis is usually accompanied by cytokinesis, in some instances cytoplasmic division does not occur, leading to the formation of binucleate (or multinucleate) cells. This is relatively common in the human liver.

The product of **mitosis** is two daughter nuclei that are genetically identical to the mother nucleus. This distinguishes mitosis from **meiosis,** a specialized type of nuclear division (covered in Exercise 43) that occurs only in the reproductive organs (testes or ovaries). Meiosis, which yields four daughter nuclei that differ genetically in composition from the mother nucleus, is used only for the production of gametes (eggs and sperm) for sexual reproduction. The function of cell division, including mitosis and cytokinesis in the body, is to increase the number of cells for growth and repair while maintaining their genetic heritage.

The stages of mitosis illustrated in Figure 4.4 include the following events:

Prophase (Figure 4.4b and c): At the onset of cell division, the chromatin threads coil and shorten to form densely staining, short, barlike **chromosomes.** By the middle of prophase the chromosomes appear as double-stranded structures (each strand is a **chromatid**) connected by a small median body called a **centromere** and an adhesive protein called *cohesin.* The centrioles separate from one another and act as focal points for the assembly of two systems of microtubules: the

Text continues on page 48.

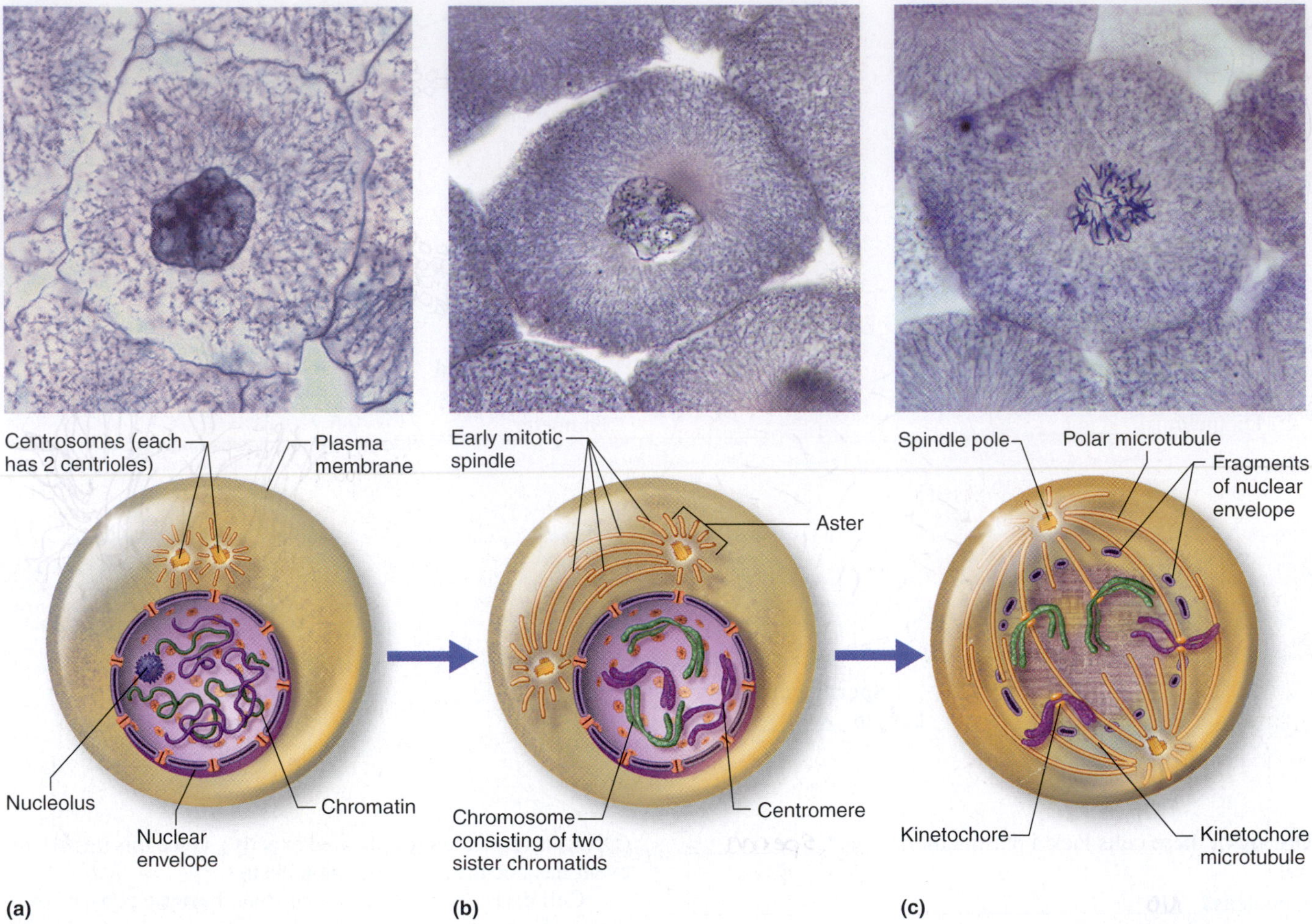
Centrosomes (each has 2 centrioles)
Plasma membrane
Nucleolus
Nuclear envelope
Chromatin
Early mitotic spindle
Aster
Chromosome consisting of two sister chromatids
Centromere
Spindle pole
Polar microtubule
Fragments of nuclear envelope
Kinetochore
Kinetochore microtubule
(a)
(b)
(c)

Interphase

Interphase is the period of a cell's life when it carries out its normal metabolic activities and grows.

- During interphase, the DNA-containing material is in the form of chromatin. The nuclear envelope and one or more nucleoli are intact and visible.
- There are three distinct periods of this phase:
  G_1: The centrioles begin replicating.
  S: DNA is replicated.
  G_2: Final preparations for mitosis are completed and centrioles finish replicating.

Early Prophase

- The chromatin condenses, forming barlike *chromosomes* that are visible with a light microscope.
- Each duplicated chromosome appears as two identical threads, now called *sister chromatids*, held together at a small, constricted region called a *centromere*. (After the chromatids separate, each is considered a new chromosome.)
- As the chromosomes appear, the nucleoli disappear, and the two centrosomes separate from one another.
- The centrosomes act as focal points for growth of a microtubule assembly called the **mitotic spindle**. As these microtubules lengthen, they propel the centrosomes toward opposite ends (poles) of the cell.
- Microtubule arrays called *asters* ("stars") are seen extending from the matrix around the centrosomes.

Late Prophase

- While the centrosomes are still moving apart, the nuclear envelope fragments, allowing the spindle to interact with the chromosomes.
- Some of the growing spindle microtubules attach to *kinetochores* (ki-ne′-to-korz), special protein structures at each chromosome's centromere. Such microtubules are called *kinetochore microtubules*.
- The remaining spindle microtubules (not attached to any chromosomes) are called *polar microtubules*. The microtubules slide past each other, forcing the poles apart.
- The kinetochore microtubules pull on each chromosome from both poles in a kind of tug-of-war that ultimately draws the chromosomes to the exact center or equator of the cell.

FIGURE 4.4 The interphase cell and the stages of mitosis. The cells shown are from an early embryo of a whitefish. Photomicrographs are above; corresponding diagrams are below. (Micrographs approximately 1210×).

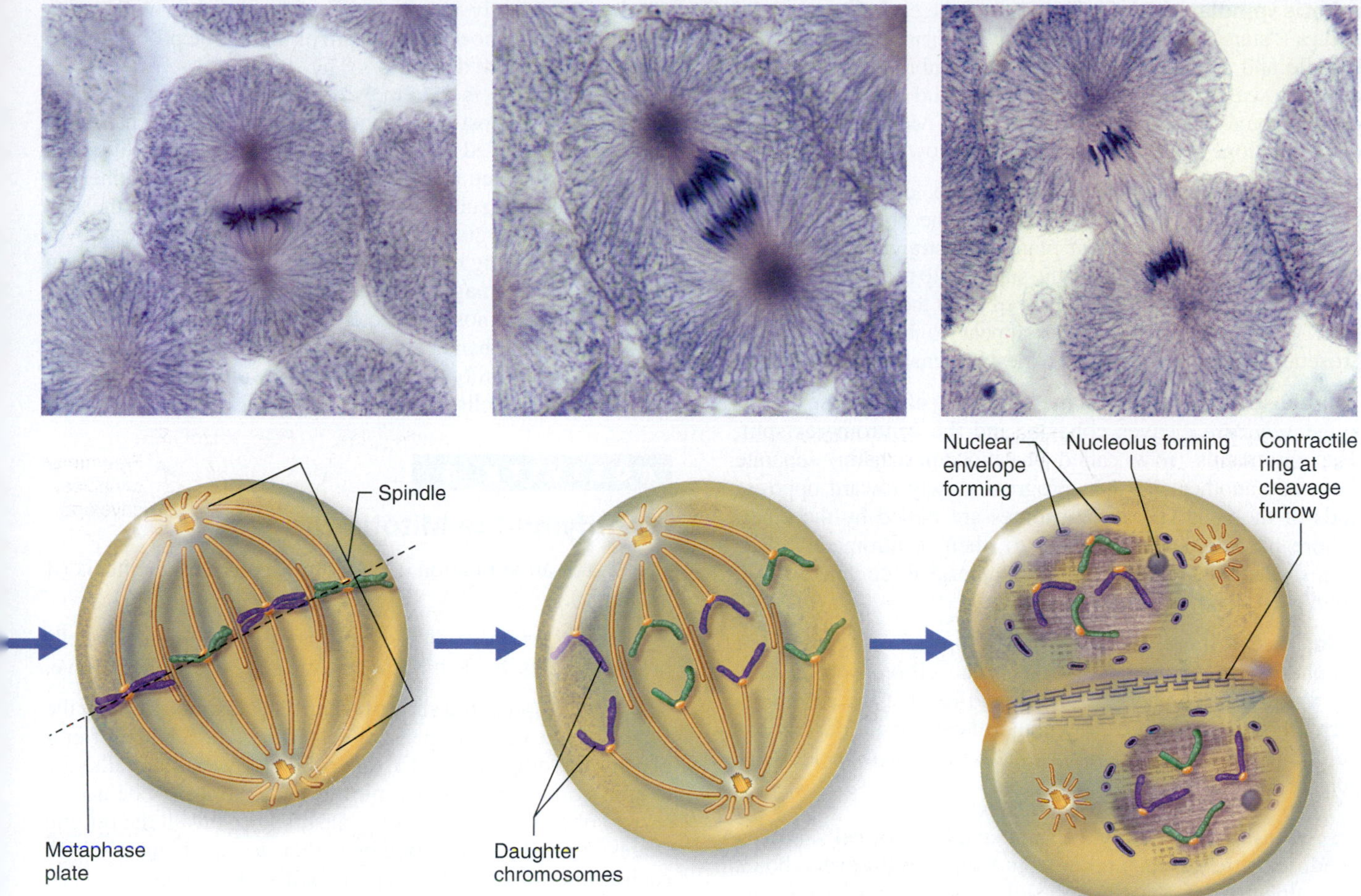
Spindle
Metaphase plate
Daughter chromosomes
Nuclear envelope forming
Nucleolus forming
Contractile ring at cleavage furrow

(d)

Metaphase
Metaphase is the second phase of mitosis.

- The two centrosomes are at opposite poles of the cell.
- The chromosomes cluster at the middle of the cell, with their centromeres precisely aligned at the *equator* of the spindle. This imaginary plane midway between the poles is called the *metaphase plate*.

(e)

Anaphase
Anaphase is the third and shortest phase of mitosis. Anaphase begins abruptly as the centromeres of the chromosomes split simultaneously. Each chromatid now becomes a chromosome in its own right.

- The kinetochore microtubules, moved along by motor proteins in the kinetochores, gradually pull each chromosome toward the pole it faces.
- At the same time, the polar microtubules slide past each other, lengthen, and push the two poles of the cell apart.
- Anaphase is easy to recognize because the moving chromosomes look V-shaped. The centromeres lead the way, and the chromosomal "arms" dangle behind them.
- This process of moving and separating the chromosomes is helped by the fact that the chromosomes are short, compact bodies. Diffuse threads of chromatin would tangle, trail, and break, resulting in imprecise "parceling out" to the daughter cells.

(f)

Telophase
Telophase begins as soon as chromosomal movement stops. This final phase is like prophase in reverse.

- The identical sets of chromosomes at the opposite poles of the cell uncoil and resume their threadlike chromatin form.
- A new nuclear envelope forms around each chromatin mass, nucleoli reappear within the nuclei, and the spindle breaks down and disappears.
- Mitosis is now ended. The cell, for just a brief period, is binucleate (has two nuclei) and each new nucleus is identical to the original mother nucleus.

Cytokinesis
- As a rule, as mitosis draws to a close, cytokinesis completes the division of the cell into two identical daughter cells. Cytokinesis occurs as a contractile ring of actin microfilaments forms the *cleavage furrow* and pinches the cell apart. It begins during late anaphase and continues through and beyond telophase.

FIGURE 4.4 ***(continued)***

mitotic spindle, which forms between the centrioles, and the **asters** ("stars"), which radiate outward from the ends of the spindle and anchor it to the plasma membrane. The spindle acts as a scaffolding for the attachment and movement of the chromosomes during later mitotic stages. Meanwhile, the nuclear envelope and the nucleolus break down and disappear.

Metaphase (Figure 4.4d): A brief stage, during which the chromosomes migrate to the central plane or equator of the spindle and align along that plane in a straight line (the so-called *metaphase plate*) from the superior to the inferior region of the spindle (lateral view). Viewed from the poles of the cell (end view), the chromosomes appear to be arranged in a "rosette," or circle, around the widest dimension of the spindle.

Anaphase (Figure 4.4e): At the beginning of anaphase the enzyme *separase* cleaves cohesin, and the centromeres split. The chromatids (now called chromosomes again) separate from one another and then progress slowly toward opposite ends of the cell. The chromosomes are pulled by the kinetochore microtubules attached to their centromeres, their "arms" dangling behind them. Anaphase is complete when poleward movement ceases.

Telophase (Figure 4.4f): During telophase, the events of prophase are essentially reversed. The chromosomes clustered at the poles begin to uncoil and resume the chromatin form, the spindle breaks down and disappears, a nuclear envelope forms around each chromatin mass, and nucleoli appear in each of the daughter nuclei.

Mitosis is essentially the same in all animal cells, but depending on the tissue, it takes from 5 minutes to several hours to complete. In most cells, centriole replication occurs during interphase of the next cell cycle.

Cytokinesis, or the division of the cytoplasmic mass, begins during late anaphase and continues through telophase (Figure 4.4f). In animal cells, a *cleavage furrow* begins to form approximately over the spindle equator and eventually splits or pinches the original cytoplasmic mass into two portions. Thus at the end of cell division, two daughter cells exist—each with a smaller cytoplasmic mass than the mother cell but genetically identical to it. The daughter cells grow and carry out the normal spectrum of metabolic processes until it is their turn to divide.

Cell division is extremely important during the body's growth period. Most cells divide until puberty, when normal body size is achieved and overall body growth ceases. After this time in life, only certain cells carry out cell division routinely—for example, cells subjected to abrasion (epithelium of the skin and lining of the gut). Other cell populations—such as liver cells—stop dividing but retain this ability should some of them be removed or damaged. Skeletal muscle, cardiac muscle, and mature neurons almost completely lose this ability to divide and thus are severely handicapped by injury. Throughout life, the body retains its ability to repair cuts and wounds and to replace some of its aged cells.

ACTIVITY 6

Identifying the Mitotic Stages

1. Watch an animation or video presentation of mitosis (if available).

2. Using the three-dimensional models of dividing cells provided, identify each of the mitotic states described above.

3. Obtain a prepared slide of whitefish blastulae to study the stages of mitosis. The cells of each *blastula* (a stage of embryonic development consisting of a hollow ball of cells) are at approximately the same mitotic stage, so it may be necessary to observe more than one blastula to view all the mitotic stages. A good analogy for a blastula is a soccer ball in which each leather piece making up the ball's surface represents an embryonic cell. The exceptionally high rate of mitosis observed in this tissue is typical of embryos, but if it occurs in specialized tissues it can indicate cancerous cells, which also have an extraordinarily high mitotic rate. Examine the slide carefully, identifying the four mitotic stages and the process of cytokinesis. Compare your observations with Figure 4.4, and verify your identifications with your instructor. ■

EXERCISE 5

The Cell: Transport Mechanisms and Cell Permeability

MATERIALS

Passive Processes

Diffusion of Dye Through Agar Gel

- ☐ Petri dish containing 12 ml of 1.5% agar-agar
- ☐ Millimeter-ruled graph paper
- ☐ Wax marking pencil
- ☐ 3.5% methylene blue solution (approximately 0.1 *M*) in dropper bottles
- ☐ 1.6% potassium permanganate solution (approximately 0.1 *M*) in dropper bottles
- ☐ Medicine dropper

Diffusion and Osmosis Through Nonliving Membranes

- ☐ Four dialysis sacs or small Hefty® "alligator" sandwich bags
- ☐ Small funnel
- ☐ 25-ml graduated cylinder
- ☐ Wax marking pencil
- ☐ Fine twine or dialysis tubing clamps
- ☐ 250-ml beakers
- ☐ Distilled water
- ☐ 40% glucose solution
- ☐ 10% sodium chloride (NaCl) solution
- ☐ 40% sucrose solution colored with Congo red dye
- ☐ Laboratory balance
- ☐ Paper towels
- ☐ Hot plate and large beaker for hot water bath
- ☐ Benedict's solution in dropper bottle
- ☐ Silver nitrate ($AgNO_3$) in dropper bottle
- ☐ Test tubes in rack, test tube holder

Text continues on next page.

MasteringA&P™

Access practice quizzes and more in the Study Area at www.masteringaandp.com.

PEx

This lab corresponds to PhysioEx Exercise 1. see p. PEx-3.

OBJECTIVES

1. To define *differential permeability* and explain the difference between *active* and *passive processes* of cellular transport.
2. To define *diffusion (simple diffusion)* and *osmosis; isotonic, hypotonic,* and *hypertonic solutions; active transport; vesicular transport;* and *exocytosis, phagocytosis,* and *pinocytosis.*
3. To describe the processes that account for the movement of substances across the plasma membrane and to indicate the driving force for each.
4. To determine which way substances will move passively through a differentially permeable membrane (given appropriate information on concentration differences).

PRE-LAB QUIZ

1. Circle the correct term. A passive process, diffusion / osmosis is the movement of solute molecules from an area of greater concentration to an area of lesser concentration.
2. A solution surrounding a cell is *hypertonic* if:
   a. it contains fewer nonpenetrating solute particles than the interior of the cell.
   b. it contains more nonpenetrating solute particles than the interior of the cell.
   c. it contains the same amount of nonpenetrating solute particles as the interior of the cell.
3. Which of the following would require an input of energy?
   a. diffusion
   b. filtration
   c. osmosis
   d. vesicular transport
4. Circle the correct term. In pinocytosis / phagocytosis, parts of the plasma membrane and cytoplasm expand and flow around a relatively large or solid material and engulf it.
5. Circle the correct term. In active / passive processes, the cell provides energy in the form of ATP to power the transport process.

Because of its molecular composition, the plasma membrane is selective about what passes through it. It allows nutrients to enter the cell but keeps out undesirable substances. By the same token, valuable cell proteins and other substances are kept within the cell, and excreta or wastes pass to the exterior. This property is known as **differential,** or **selective, permeability.** Transport through the plasma membrane occurs in two basic ways. In **passive processes,** concentration or pressure differences drive the movement. In **active processes,** the cell provides energy (ATP) to power the transport process.

(materials list continued)

Experiment 1

- ☐ Deshelled eggs
- ☐ 400-ml beakers
- ☐ Wax marking pencil
- ☐ Distilled water
- ☐ 30% sucrose solution
- ☐ Laboratory balance
- ☐ Paper towels
- ☐ Graph paper
- ☐ Weigh boat

Experiment 2

- ☐ Clean microscope slides and coverslips
- ☐ Medicine dropper
- ☐ Compound microscope
- ☐ Vials of animal (mammalian) blood obtained from a biological supply house or veterinarian—at option of instructor
- ☐ Freshly prepared physiologic (mammalian) saline solution in dropper bottle
- ☐ 5% sodium chloride solution in dropper bottle
- ☐ Distilled water
- ☐ Filter paper
- ☐ Disposable gloves
- ☐ Basin and wash bottles containing 10% household bleach solution
- ☐ Disposable autoclave bag
- ☐ Paper towels

Diffusion Demonstrations

1: Diffusion of a dye through water

Prepared the morning of the laboratory session with setup time noted. Potassium permanganate crystals are placed in a 1000-ml graduated cylinder, and distilled water is added slowly and with as little turbulence as possible to fill to the 1000-ml mark.

2: Osmometer

Just before the laboratory begins, the broad end of a thistle tube is closed with a differentially permeable dialysis membrane, and the tube is secured to a ring stand. Molasses is added to approximately 5 cm above the thistle tube bulb, and the bulb is immersed in a beaker of distilled water. At the beginning of the lab session, the level of the molasses in the tube is marked with a wax pencil.

Filtration

- ☐ Ring stand, ring, clamp
- ☐ Filter paper, funnel
- ☐ Solution containing a mixture of uncooked starch, powdered charcoal, and copper sulfate ($CuSO_4$)
- ☐ 10-ml graduated cylinder
- ☐ 100-ml beaker
- ☐ Lugol's iodine in a dropper bottle

Active Processes

- ☐ Videotape showing phagocytosis (if available)
- ☐ Videotape viewing box

Note to the Instructor: See directions for handling wet mount preparations and disposable supplies on page 33, Exercise 3.

Passive Processes

The two important passive processes of membrane transport are *diffusion* and *filtration*. Diffusion is an important transport process for every cell in the body. By contrast, filtration usually occurs only across capillary walls.

Recall that all molecules possess *kinetic energy* and are in constant motion. At a specific temperature, given molecules have about the same average kinetic energy. Since kinetic energy is directly related to both mass and velocity (KE = $\frac{1}{2}mv^2$), smaller molecules tend to move faster. As molecules move about randomly at high speeds, they collide and ricochet off one another, changing direction with each collision (Figure 5.1).

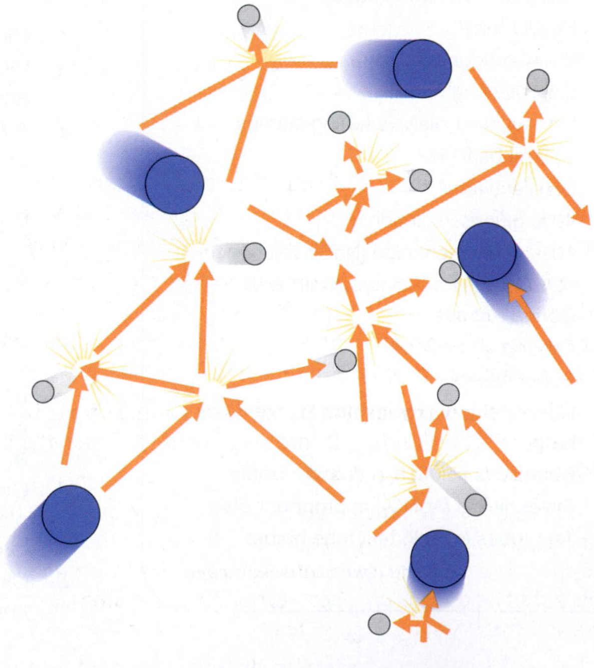

FIGURE 5.1 Random movement and numerous collisions cause molecules to become evenly distributed. The small spheres represent water molecules; the large spheres represent glucose molecules.

Diffusion

When a **concentration gradient** (difference in concentration) exists, the net effect of this random molecular movement is that the molecules eventually become evenly distributed throughout the environment; that is, the process called diffusion occurs. Hence, **diffusion** is the movement of molecules from a region of their higher concentration to a region of their lower concentration. Its driving force is the kinetic energy of the molecules themselves.

There are many examples of diffusion in nonliving systems. For example, if a bottle of ether was uncorked at the front of the laboratory, very shortly thereafter you would be nodding as the ether molecules became distributed throughout the room. The ability to smell a friend's cologne shortly after he or she has entered the room is another example.

The diffusion of particles into and out of cells is modified by the plasma membrane, which constitutes a physical barrier. In general, molecules diffuse passively through the plasma membrane if they can dissolve in the lipid portion of the membrane (for example CO_2 and O_2). The unassisted diffusion of solutes (dissolved substances) through a differentially permeable membrane is called **simple diffusion.**

Certain molecules (for example, glucose) are transported across the plasma membrane with the assistance of a protein carrier molecule. The glucose binds to the carrier and is ferried across the membrane. Small ions cross the membrane by moving through water-filled protein channels. In both cases, the substances move by a passive transport process called **facilitated diffusion.** As with simple diffusion, the substances move from an area of higher concentration to one of lower concentration, that is, down their concentration gradients.

Osmosis

The flow of water across a differentially permeable membrane is called **osmosis.** During osmosis, water moves down its concentration gradient. The concentration of water is inversely related to the concentration of solutes. If the solutes can diffuse across the membrane, both water and solutes will move down their concentration gradients through the membrane. If the particles in solution are nonpenetrating solutes (prevented from crossing the membrane), water alone will move by osmosis and in doing so will cause changes in the volume of the compartments on either side of the membrane.

Diffusion of Dye Through Agar Gel and Water

The relationship between molecular weight and the rate of diffusion can be examined easily by observing the diffusion of two different types of dye molecules through an agar gel. The dyes used in this experiment are methylene blue, which has a molecular weight of 320 and is deep blue in color, and potassium permanganate, a purple dye with a molecular weight of 158. Although the agar gel appears quite solid, it is primarily (98.5%) water and allows free movement of the dye molecules through it.

ACTIVITY 1

Observing Diffusion of Dye Through Agar Gel

1. Work with members of your group to formulate a hypothesis about the rates of diffusion of methylene blue and potassium permanganate through the agar gel. Justify your hypothesis.
2. Obtain a petri dish containing agar gel, a piece of millimeter-ruled graph paper, a wax marking pencil, dropper bottles of methylene blue and potassium permanganate, and a medicine dropper. (See Figure 5.2.)
3. Using the wax marking pencil, draw a line on the bottom of the petri dish dividing it into two sections. Place the petri dish on the ruled graph paper.
4. Create a well in the center of each section using the medicine dropper. To do this, squeeze the bulb of the medicine dropper, and push it down into the agar. Release the bulb as you slowly pull the dropper vertically out of the agar. This should remove an agar plug, leaving a well in the agar.
5. Carefully fill one well with the methylene blue solution and the other well with the potassium permanganate solution.

Record the time. _________

6. At 15-minute intervals, measure the distance the dye has diffused from each well. Continue these observations for 1 hour, and record the results in the adjacent chart.

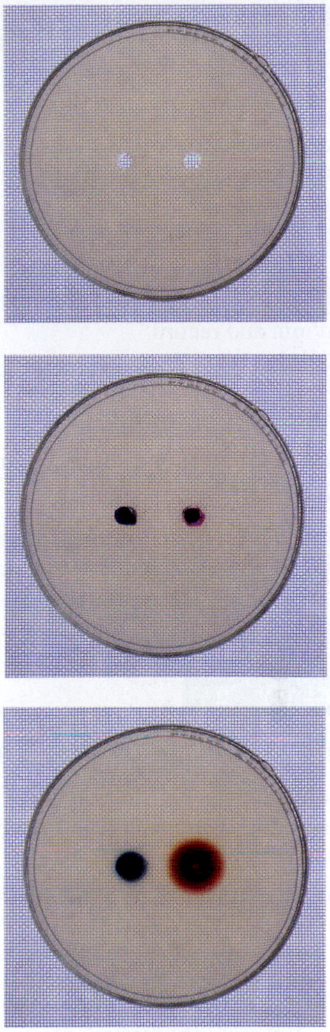

(a)

(b)

(c)

FIGURE 5.2 Comparing diffusion rates. Agar-plated petri dish as it appears after the diffusion of 0.1 *M* methylene blue placed in one well and 0.1 *M* potassium permanganate placed in another.

Time (min)	Diffusion of methylene blue (mm)	Diffusion of potassium permanganate (mm)
15		
30		
45		
60		

Which dye diffused more rapidly? _________________

What is the relationship between molecular weight and rate of molecular movement (diffusion)?

____

Why did the dye molecules move? ______________________

____

Compute the rate of diffusion of the potassium permanganate molecules in millimeters per minute (mm/min) and record.

________ mm/min

Compute the rate of diffusion of the methylene blue molecules in mm/min and record.

________ mm/min

7. Prepare a lab report for these experiments. (See Getting Started, page xv.)

Make a mental note to yourself to go to demonstration area 1 at the end of the laboratory session to observe the extent of diffusion of the potassium permanganate dye through water. At that time, follow the directions given next.

ACTIVITY 2

Observing Diffusion of Dye Through Water

1. Go to diffusion demonstration area 1, and observe the cylinder containing dye crystals and water set up at the beginning of the lab.
2. Measure the number of millimeters the dye has diffused from the bottom of the graduated cylinder and record.

________ mm

3. Record the time the demonstration was set up and the time of your observation. Then compute the rate of the dye's diffusion through water and record below.

Time of setup ________

Time of observation ________

Rate of diffusion ________ mm/min

4. Does the potassium permanganate dye move (diffuse) more rapidly through water or the agar gel? (Explain your answer.)

____

____

ACTIVITY 3

Observing Diffusion and Osmosis Through Nonliving Membranes

The following experiment provides information on the movement of water and solutes through differentially permeable membranes called dialysis sacs. Dialysis sacs have pores of a particular size. The selectivity of living membranes depends on more than just pore size, but using the dialysis sacs will allow you to examine selectivity due to this factor.

1. Read through the experiments in this activity, and develop a hypothesis for each part.
2. Obtain four dialysis sacs, a small funnel, a 25-ml graduated cylinder, a wax marking pencil, fine twine or dialysis tubing clamps, and four beakers (250 ml). Number the beakers 1 to 4 with the wax marking pencil, and half fill all of them with distilled water except beaker 2, to which you should add 40% glucose solution.
3. Prepare the dialysis sacs one at a time. Using the funnel, half fill each with 20 ml of the specified liquid (see below). Press out the air, fold over the open end of the sac, and tie it securely with fine twine or clamp it. Before proceeding to the next sac, rinse it under the tap, and quickly and carefully blot the sac dry by rolling it on a paper towel. Weigh it with a

Data from Experiments on Diffusion and Osmosis Through Nonliving Membranes

Beaker	Contents of sac	Initial weight	Final weight	Weight change	Tests—beaker fluid	Tests—sac fluid
Beaker 1 ½ filled with distilled water	Sac 1, 20 ml of 40% glucose solution				Benedict's test:	Benedict's test:
Beaker 2 ½ filled with 40% glucose solution	Sac 2, 20 ml of 40% glucose solution					
Beaker 3 ½ filled with distilled water	Sac 3, 20 ml of 10% NaCl solution				$AgNO_3$ test:	
Beaker 4 ½ filled with distilled water	Sac 4, 20 ml of 40% sucrose solution containing Congo red dye				Benedict's test:	

laboratory balance. Record the weight in the data chart on page 56, and then drop the sac into the corresponding beaker. Be sure the sac is completely covered by the beaker solution, adding more solution if necessary.

- Sac 1: 40% glucose solution
- Sac 2: 40% glucose solution
- Sac 3: 10% NaCl solution
- Sac 4: Congo red dye in 40% sucrose solution

Allow sacs to remain undisturbed in the beakers for 1 hour. (Use this time to continue with other experiments.)

4. After an hour, boil a beaker of water on the hot plate. Obtain the supplies you will need to determine your experimental results: dropper bottles of Benedict's solution and silver nitrate solution, a test tube rack, four test tubes, and a test tube holder.

5. Quickly and gently blot sac 1 dry and weigh it. (**Note:** Do not squeeze the sac during the blotting process.) Record the weight in the data chart.

Has there been any change in weight? ____________

Conclusions: ____________

____

____

Place 5 ml of Benedict's solution in each of two test tubes. Put 4 ml of the beaker fluid into one test tube and 4 ml of the sac fluid into the other. Mark the tubes for identification and then place them in a beaker containing boiling water. Boil 2 minutes. Cool slowly. If a green, yellow, or rusty red precipitate forms, the test is positive, meaning that glucose is present. If the solution remains the original blue color, the test is negative. Record results in the data chart.

Was glucose still present in the sac? ____________

Was glucose present in the beaker? ____________

Conclusions: ____________

____

____

6. Blot gently and weigh sac 2. Record the weight in the data chart.

Was there an *increase* or *decrease* in weight? ____________

With 40% glucose in the sac and 40% glucose in the beaker, would you expect to see any net movement of water (osmosis) or of glucose molecules (simple diffusion)?

____________ Why or why not? ____________

____

7. Blot gently and weigh sac 3. Record the weight in the data chart.

Was there any change in weight? ____________

Conclusions: ____________

____

Take a 5-ml sample of beaker 3 solution and put it in a clean test tube. Add a drop of silver nitrate. The appearance of a white precipitate or cloudiness indicates the presence of silver chloride (AgCl), which is formed by the reaction of $AgNO_3$ with NaCl (sodium chloride). Record results in the data chart.

Results: ____________

Conclusions: ____________

8. Blot gently and weigh sac 4. Record the weight in the data chart.

Was there any change in weight? ____________

Did the beaker water turn pink? ____________

Conclusions: ____________

____

Take a 1-ml sample of beaker 4 solution and put the test tube in boiling water in a hot water bath. Add 5 drops of Benedict's solution to the tube and boil for 5 minutes. The presence of glucose (one of the hydrolysis products of sucrose) in the bath water is indicated by the presence of a green, yellow, or rusty colored precipitate.

Did sucrose diffuse from the sac into the bath water? ____________

Conclusions: ____________

____

9. In which of the test situations did net osmosis occur?

____

In which of the test situations did net simple diffusion occur?

____

What conclusions can you make about the relative size of glucose, sucrose, Congo red dye, NaCl, and water molecules?

____

____

____

With what cell structure can the dialysis sac be compared?

____

10. Prepare a lab report for the experiment. (See Getting Started, page xv.) Be sure to include in your discussion the answers to the questions proposed in this activity.

ACTIVITY 4

Observing Osmometer Results

Before leaving the laboratory, observe demonstration 2, the *osmometer demonstration* set up before the laboratory session to follow the movement of water through a membrane (osmosis). Measure the distance the water column has moved during the laboratory period and record below. (The position of the meniscus [the surface of the water column] in the thistle tube at the beginning of the laboratory period is marked with wax pencil.)

Distance the meniscus has moved: __________ mm

Did net osmosis occur? Why or why not?

____

____

____

ACTIVITY 5

Investigating Diffusion and Osmosis Through Living Membranes

To examine permeability properties of plasma membranes, conduct the following experiments. As you read through the experiments in this activity, develop a hypothesis for each part.

Experiment 1

1. Obtain two deshelled eggs and two 400-ml beakers. Note that the relative concentration of solutes in deshelled eggs is about 14%. Number the beakers 1 and 2 with the wax marking pencil. Half fill beaker 1 with distilled water and beaker 2 with 30% sucrose.

2. Carefully blot each egg by rolling it gently on a paper towel. Place a weight boat on a laboratory balance and tare the balance (that is, make sure the scale reads 0.0 with the weigh boat on the scale). Weigh egg 1 in the weigh boat, record the initial weight in the data chart below, and gently place it into beaker 1. Repeat for egg 2, placing it in beaker 2.

3. After 20 minutes, remove egg 1 and gently blot it and weigh it. Record the weight, and replace it into beaker 1. Repeat for egg 2, placing it into beaker 2. Repeat this procedure at 40 minutes and 60 minutes.

4. Calculate the change in weight of each egg at each time period, and enter that number in the data chart below. Also calculate the percent change in weight for each time period and enter that number in the data table.

How has the weight of each egg changed?

Egg 1 ______________________________

Egg 2 ______________________________

Make a graph of your data by plotting the percent change in weight for each egg versus time.

How has the appearance of each egg changed?

Egg 1 ______________________________

Egg 2 ______________________________

A solution surrounding a cell is **hypertonic** if it contains more nonpenetrating solute particles than the interior of the cell. Water moves from the interior of the cell into a surrounding hypertonic solution by osmosis. A solution surrounding a cell is **hypotonic** if it contains fewer nonpenetrating solute particles than the interior of the cell. Water moves from a hypotonic solution into the cell by osmosis. In both cases, water moved down its concentration gradient. Indicate in your conclusions whether distilled water was a hypotonic or hypertonic solution and whether 30% sucrose was hypotonic or hypertonic.

Data from Experiment 1 on Diffusion and Osmosis Through Living Membranes

Time	Egg 1 (in distilled H_2O)	Weight change	% Change	Egg 2 (in 30% sucrose)	Weight change	% Change
Initial weight (g)		—	—		—	—
20 min.						
40 min.						
60 min.						

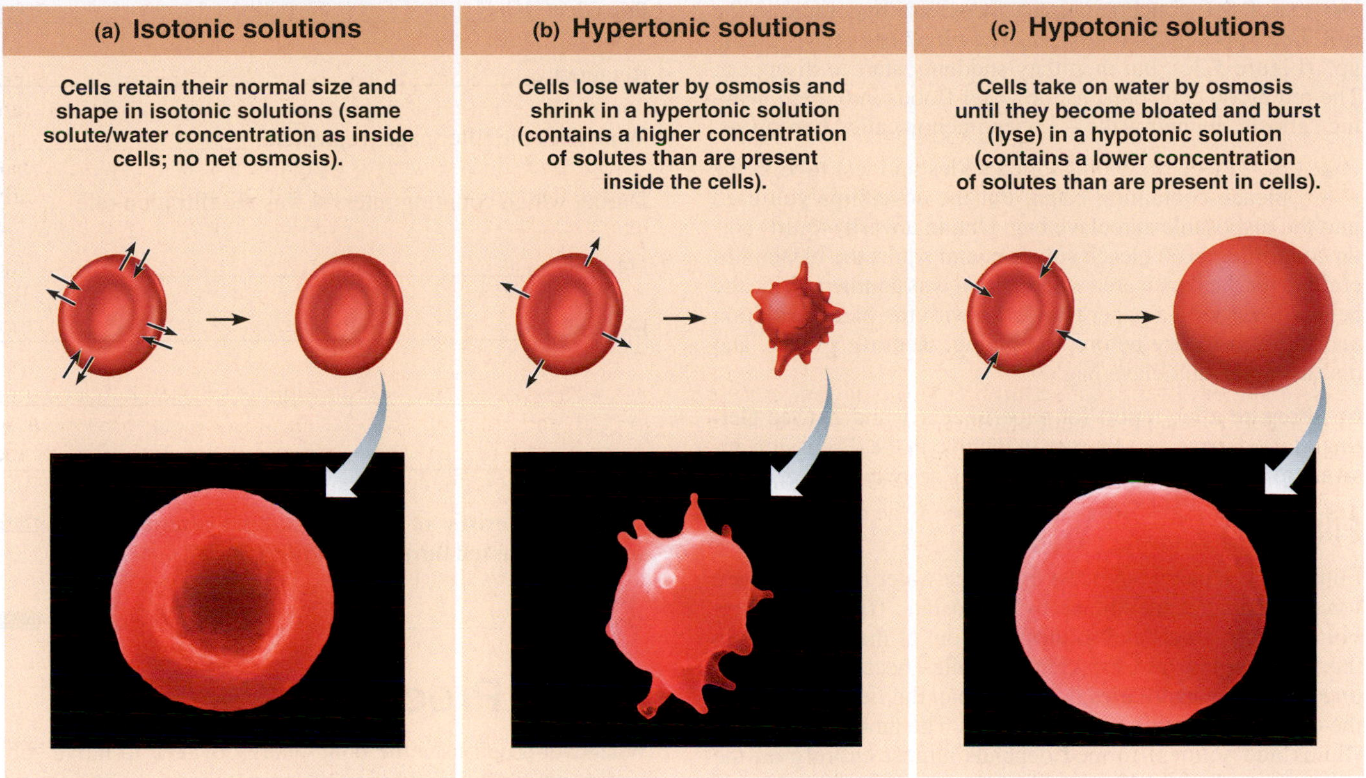
(a) Isotonic solutions
Cells retain their normal size and shape in isotonic solutions (same solute/water concentration as inside cells; no net osmosis).
(b) Hypertonic solutions
Cells lose water by osmosis and shrink in a hypertonic solution (contains a higher concentration of solutes than are present inside the cells).
(c) Hypotonic solutions
Cells take on water by osmosis until they become bloated and burst (lyse) in a hypotonic solution (contains a lower concentration of solutes than are present in cells).

FIGURE 5.3 Influence of isotonic, hypertonic, and hypotonic solutions on red blood cells.

Conclusions: ______________________________

____

____

Experiment 2

Now you will conduct a microscopic study of red blood cells suspended in solutions of varying tonicities. The objective is to determine if these solutions have any effect on cell shape by promoting net osmosis.

1. The following supplies should be available at your laboratory bench to conduct this experimental series: two clean slides and coverslips, a vial of animal blood, a medicine dropper, physiologic saline, 5% sodium chloride solution, distilled water, filter paper, and disposable gloves.

⚠ Wear disposable gloves at all times when handling blood (steps 2–5).

2. Place a very small drop of physiologic saline on a slide. Using the medicine dropper, add a small drop of animal blood to the saline on the slide. Tilt the slide to mix, cover with a coverslip, and immediately examine the preparation under the high-power lens. Notice that the red blood cells retain their normal smooth disclike shape (see Figure 5.3a). This is because the physiologic saline is **isotonic** to the cells. That is, it contains a concentration of nonpenetrating solutes (e.g., proteins and some ions) equal to that in the cells (same solute/water concentration). Consequently, the cells neither gain nor lose water by osmosis. Set this slide aside.

3. Prepare another wet mount of animal blood, but this time use 5% sodium chloride (saline) solution as the suspending medium. Carefully observe the red blood cells under high power. What is happening to the normally smooth disc shape of the red blood cells?

____

____

____

This crinkling-up process, called **crenation,** is due to the fact that the 5% sodium chloride solution is hypertonic to the cytosol of the red blood cell. Under these circumstances, water tends to leave the cells by osmosis. Compare your observations to Figure 5.3b.

4. Add a drop of distilled water to the edge of the coverslip. Fold a piece of filter paper in half and place its folded edge at the opposite edge of the coverslip; it will absorb the saline solution and draw the distilled water across the cells. Watch the red blood cells as they float across the field. Describe the change in their appearance.

____

____

Distilled water contains *no* solutes (it is 100% water). Distilled water and *very* dilute solutions (that is, those containing less than 0.9% nonpenetrating solutes) are hypotonic to the cell. In a hypotonic solution, the red blood cells first "plump up" (Figure 5.3c), but then they suddenly start to disappear. The red blood cells burst as the water floods into them, leaving "ghosts" in their wake—a phenomenon called **hemolysis.**

5. Place the blood-soiled slides and test tube in the bleach-containing basin. Put the coverslips you used into the disposable autoclave bag. Obtain a wash (squirt) bottle containing 10% bleach solution, and squirt the bleach liberally over the bench area where blood was handled. Wipe the bench down with a paper towel wet with the bleach solution and allow it to dry before continuing. Remove gloves, and discard in the autoclave bag.

6. Prepare a lab report for experiments 1 and 2. (See Getting Started, page xv.) Be sure to include in the discussion answers to the questions proposed in this activity.

Filtration

Filtration is a passive process by which water and solutes are forced through a membrane by hydrostatic (fluid) pressure. For example, fluids and solutes filter out of the capillaries in the kidneys and into the kidney tubules because the blood pressure in the capillaries is greater than the fluid pressure in the tubules. Filtration is not selective. The amount of filtrate (fluids and solutes) formed depends almost entirely on the pressure gradient (difference in pressure on the two sides of the membrane) and on the size of the membrane pores.

ACTIVITY 6

Observing the Process of Filtration

1. Obtain the following equipment: a ring stand, ring, and ring clamp; a funnel; a piece of filter paper; a beaker; a 10-ml graduated cylinder; a solution containing uncooked starch, powdered charcoal, and copper sulfate; and a dropper bottle of Lugol's iodine. Attach the ring to the ring stand with the clamp.

2. Fold the filter paper in half twice, open it into a cone, and place it in a funnel. Place the funnel in the ring of the ring stand and place a beaker under the funnel. Shake the starch solution, and fill the funnel with it to just below the top of the filter paper. When the steady stream of filtrate changes to countable filtrate drops, count the number of drops formed in 10 seconds and record.

___________ drops

When the funnel is half empty, again count the number of drops formed in 10 seconds and record the count.

___________ drops

3. After all the fluid has passed through the filter, check the filtrate and paper to see which materials were retained by the paper. (Note: If the filtrate is blue, the copper sulfate passed. Check both the paper and filtrate for black particles to see whether the charcoal passed. Finally, using a 10-ml graduated cylinder, put a 2-ml filtrate sample into a test tube. Add several drops of Lugol's iodine. If the sample turns blue/black when iodine is added, starch is present in the filtrate.)

Passed: ___________

Retained: ___________

What does the filter paper represent? ___________

During which counting interval was the filtration rate

greatest? ___________

Explain: ___________

____

____

What characteristic of the three solutes determined whether or not they passed through the filter paper?

____

Active Processes

Whenever a cell uses the bond energy of ATP to move substances across its boundaries, the process is an *active process.* Substances moved by active means are generally unable to pass by diffusion. They may not be lipid soluble; they may be too large to pass through the membrane channels; or they may have to move against rather than with a concentration gradient. There are two types of active processes: *active transport* and *vesicular transport.*

Active Transport

Like one form of facilitated diffusion, **active transport** requires carrier proteins that combine specifically with the transported substance. Active transport may be primary, driven directly by hydrolysis of ATP, or secondary, driven indirectly by energy stored in ionic gradients. In most cases the substances move against concentration or electrochemical gradients or both. Some of the substances that are moved into the cells by such carriers are amino acids and some sugars. Both solutes are insoluble in lipid and too large to pass through membrane channels but are necessary for cell life. However, sodium ions (Na^+) are ejected from cells by active transport. Carrier-mediated active transport is difficult to study in an A&P laboratory and will not be considered further here.

Vesicular Transport

Large particles and molecules are transported across the membrane by **vesicular transport.** Movement may be into the cell **(endocytosis)** or out of the cell **(exocytosis)**.

Most types of endocytosis utilize clathrin protein–coated pits to engulf the substance to be carried into the cell. Once engulfed, the substance is transported in the cell within a clathrin-coated vesicle.

In **phagocytosis** (cell eating), parts of the plasma membrane and cytoplasm expand and flow around a relatively large or solid material (for example, bacteria or cell debris) and engulf it (Figure 5.4a). The membranous sac thus

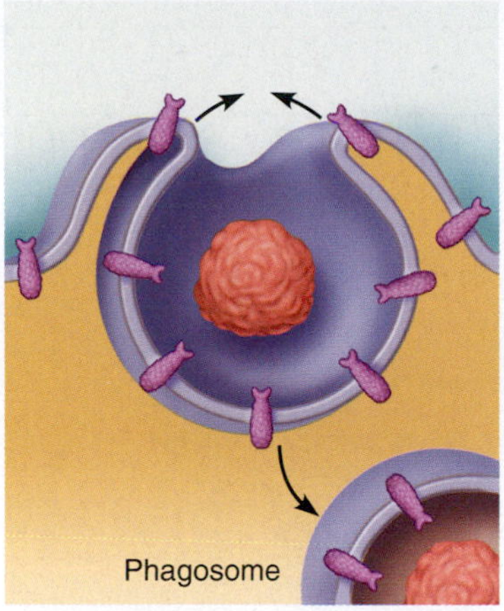
Phagosome

(a) Phagocytosis

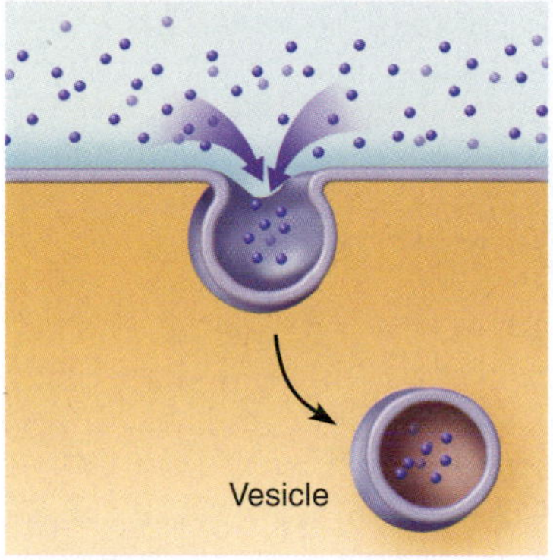
Vesicle

(b) Pinocytosis

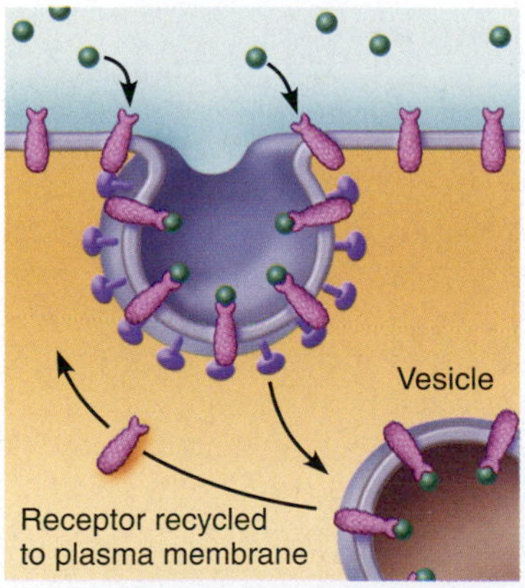
Vesicle
Receptor recycled to plasma membrane

(c) Receptor-mediated endocytosis

FIGURE 5.4 Three types of endocytosis. (a) In phagocytosis, cellular extensions (pseudopodia) flow around the external particle and enclose it within a vacuole. **(b)** In pinocytosis, dissolved proteins gather on the external surface of the plasma membrane, causing the membrane to invaginate and to incorporate a droplet of the fluid in a tiny vesicle. Most vesicles are protein coated. **(c)** In receptor-mediated endocytosis, plasma membrane proteins bind only with certain substances in regions of coated pits. The pits form protein-coated vesicles.

formed, called a *phagosome,* is then fused with a lysosome and its contents are digested. In the human body, phagocytic cells are mainly found among the white blood cells and macrophages that act as scavengers and help protect the body from disease-causing microorganisms and cancer cells.

In **pinocytosis,** also called **fluid-phase endocytosis,** the cell membrane sinks beneath the material to form a small vesicle, which then pinches off into the cell interior (see Figure 5.4b). Pinocytosis is most common for taking in liquids containing protein or fat.

A more selective type of endocytosis uses plasma membrane receptors and is called **receptor-mediated endocytosis** (Figure 5.4c). As opposed to the phagocytosis used by the body's scavenger cells, this type of endocytosis is exquisitely selective and is used primarily for cellular uptake of specific molecules, such as cholesterol, iron, and some hormones, and for transfer of substances from one side of the cell to the other.

ACTIVITY 7

Observing Phagocytosis

Go to the videotape viewing area and watch the videotape demonstration of phagocytosis (if available).

Note: If you have not already done so, complete Activity 2 ("Observing Diffusion of Dye Through Water," page 56), and Activity 4 ("Observing Osmometer Results," page 58).

NAME ______________________________

LAB TIME/DATE ______________________

REVIEW SHEET

EXERCISE 5

The Cell: Transport Mechanisms and Permeability

Choose all answers that apply to questions 1 and 2, and place their letters on the response blanks to the right.

1. Molecular motion ______________________________.

a. reflects the kinetic energy of molecules

b. reflects the potential energy of molecules

c. is ordered and predictable

d. is random and erratic

2. Velocity of molecular movement ______________________________.

a. is higher in larger molecules

b. is lower in larger molecules

c. increases with increasing temperature

d. decreases with increasing temperature

e. reflects kinetic energy

3. Summarize the results of Activity 3, diffusion and osmosis through nonliving membranes, below. List and explain your observations relative to tests used to identify diffusing substances, and changes in sac weight observed.

Sac 1 containing 40% glucose suspended in distilled water

____

____

Sac 2 containing 40% glucose suspended in 40% glucose

____

____

Sac 3 containing 10% NaCl suspended in distilled water

____

____

Sac 4 containing 40% sucrose and Congo red dye suspended in distilled water

____

____

4. What single characteristic of the differentially permeable membranes *used in the laboratory* determines the substances that can pass through them? ______

In addition to this characteristic, what other factors influence the passage of substances through living membranes?

______

______

5. A semipermeable sac containing 4% NaCl, 9% glucose and 10% albumin is suspended in a solution with the following composition: 10% NaCl, 10% glucose, and 40% albumin. Assume that the sac is permeable to all substances except albumin. State whether each of the following will (a) move into the sac, (b) move out of the sac, or (c) not move.

glucose: ______ albumin: ______

water: ______ NaCl: ______

6. Summarize the results of Activity 5, Experiment 1 (diffusion and osmosis through living membranes—the egg), below. List and explain your observations.

Egg 1 in distilled water: ______

______

______

Egg 2 in 30% sucrose: ______

______

______

7. The diagrams below represent three microscope fields containing red blood cells. Arrows show the direction of net osmosis.

Which field contains a hypertonic solution? ______ The cells in this field are said to be ______. Which field contains an isotonic bathing solution? ______ Which field contains a hypotonic solution? ______ What is happening to the cells in this field? ______

______

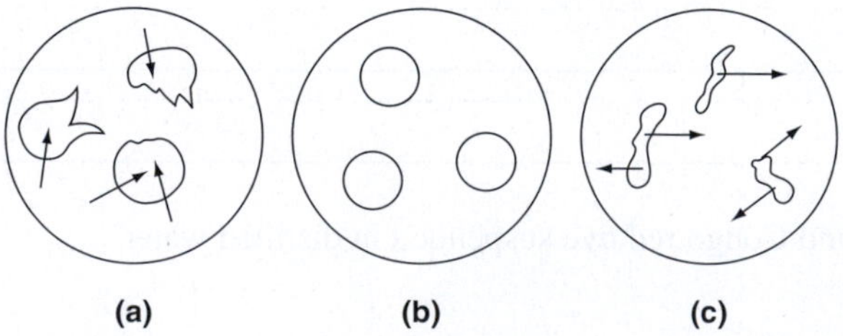
(a)
(b)
(c)

8. Assume you are conducting the experiment illustrated in the next figure. Both hydrochloric acid (HCl) with a molecular weight of about 36.5 and ammonium hydroxide (NH_4OH) with a molecular weight of 35 are volatile and easily enter the gaseous state. When they meet, the following reaction will occur:

$$HCl + NH_4OH \rightarrow H_2O + NH_4Cl$$

Ammonium chloride (NH_4Cl) will be deposited on the glass tubing as a smoky precipitate where the two gases meet. Predict which gas will diffuse more quickly and indicate to which end of the tube the smoky precipitate will be closer.

a. The faster-diffusing gas is ________________.

b. The precipitate forms closer to the ________________ end.

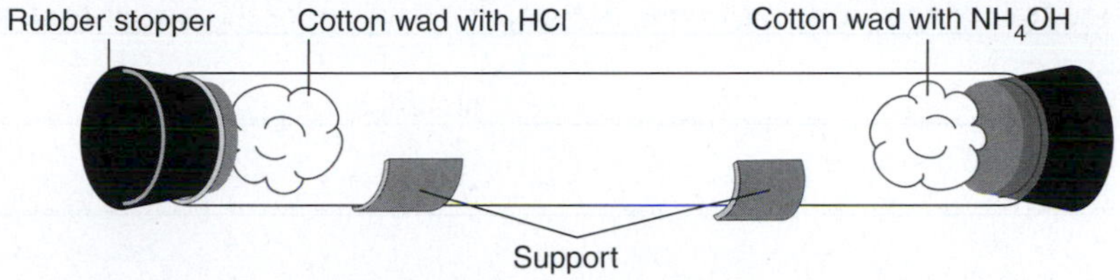
Rubber stopper
Cotton wad with HCl
Cotton wad with NH_4OH
Support

9. What determines whether a transport process is active or passive? ______________________________

____

10. Characterize membrane transport as fully as possible by choosing all the phrases that apply and inserting their letters on the answer blanks.

Passive processes: ________________ Active processes: ________________

a. account for the movement of fats and respiratory gases through the plasma membrane

b. explain solute pumping, phagocytosis, and pinocytosis

c. include osmosis, simple diffusion, and filtration

d. may occur against concentration and/or electrical gradients

e. use hydrostatic pressure or molecular energy as the driving force

f. move ions, amino acids, and some sugars across the plasma membrane

11. For the osmometer demonstration (Activity 4), explain why the level of the water column rose during the laboratory session.

____

____

12. Define the following terms.

diffusion: ______________________________

____

osmosis: ______________________________

____

simple diffusion: ______________________________

____

filtration: ______________________________

____

active transport: ______________________________

____

phagocytosis: ______________________________

____

fluid-phase endocytosis: ______________________________

____

EXERCISE 6

Classification of Tissues

MATERIALS

- ☐ Compound microscope
- ☐ Immersion oil
- ☐ Prepared slides of simple squamous, simple cuboidal, simple columnar, stratified squamous (nonkeratinized), stratified cuboidal, stratified columnar, pseudostratified ciliated columnar, and transitional epithelium
- ☐ Prepared slides of mesenchyme; of adipose, areolar, reticular, and dense (both regular and irregular connective tissues); of hyaline and elastic cartilage; of fibrocartilage; of bone (x.s.); and of blood
- ☐ Prepared slides of skeletal, cardiac, and smooth muscle (l.s.)
- ☐ Prepared slide of nervous tissue (spinal cord smear)

MasteringA&P™

Access practice quizzes and more in the Study Area at www.masteringaandp.com.

PAL

For access to anatomical models and more, check out Practice Anatomy Lab.

OBJECTIVES

1. To name the four primary tissue types in the human body and the major subcategories of each.
2. To identify the tissue subcategories through microscopic inspection or inspection of an appropriate diagram or projected slide.
3. To state the location of the various tissue types in the body.
4. To list the general functions and structural characteristics of each of the four major tissue types.

PRE-LAB QUIZ

1. Groups of cells that are similar in structure and function are called:
   a. organ systems c. organs
   b. organisms d. tissues
2. How many primary tissue types are found in the human body? ________
3. Circle True or False. Endocrine and exocrine glands are classified as epithelium because they usually develop from epithelial membranes.
4. Epithelial tissues can be classified according to cell shape. ________ epithelial cells are scalelike and flattened.
   a. Columnar c. Squamous
   b. Cuboidal d. Transitional
5. This type of epithelium lines the digestive tract from stomach to anus.
   a. simple cuboidal c. stratified squamous
   b. simple columnar d. transitional
6. All connective tissue is derived from an embryonic tissue known as:
   a. cartilage c. mesenchyme
   b. ground substance d. reticular
7. All the following are examples of connective tissue except:
   a. bones c. neurons
   b. ligaments d. tendons
8. Circle True or False. Blood is a type of connective tissue.
9. Circle the correct term. Of the two major cell populations in nervous tissue, neurons / neuroglial cells are highly specialized to receive stimuli and conduct waves of excitation to all parts of the body.
10. How many basic types of muscle tissue are there? ________
11. This type of muscle tissue is found in the walls of hollow organs. It has no striations, and its cells are spindle shaped. It is:
    a. cardiac muscle
    b. skeletal muscle
    c. smooth muscle

Exercise 4 describes cells as the building blocks of life and the all-inclusive functional units of unicellular organisms. However, in higher organisms, cells do not usually operate as isolated, independent entities. In humans and other multicellular organisms, cells depend on one another and cooperate to maintain homeostasis in the body.

With a few exceptions, even the most complex animal starts out as a single cell, the fertilized egg, which divides almost endlessly. The trillions of cells that result become specialized for a particular function; some become supportive bone, others the transparent lens of the eye, still others skin cells, and so on. Thus a division of labor exists, with certain groups of cells highly specialized to perform functions that benefit the organism as a whole. Cell specialization carries with it certain hazards, because when a small specific group of cells is indispensable, any inability to function on its part can paralyze or destroy the entire body.

Groups of cells that are similar in structure and function are called **tissues.** The four primary tissue types—epithelium, connective tissue, nervous tissue, and muscle—have distinctive structures, patterns, and functions. The four primary tissues are further divided into subcategories, as described shortly.

To perform specific body functions, the tissues are organized into **organs** such as the heart, kidneys, and lungs. Most organs contain several representatives of the primary tissues, and the arrangement of these tissues determines the organ's structure and function. Thus **histology,** the study of tissues, complements a study of gross anatomy and provides the structural basis for a study of organ physiology.

The main objective of this exercise is to familiarize you with the major similarities and dissimilarities of the primary tissues, so that when the tissue composition of an organ is described, you will be able to more easily understand (and perhaps even predict) the organ's major function. Because epithelium and some types of connective tissue will not be considered again, they are emphasized more than muscle, nervous tissue, and bone (a connective tissue), which are covered in more depth in later exercises.

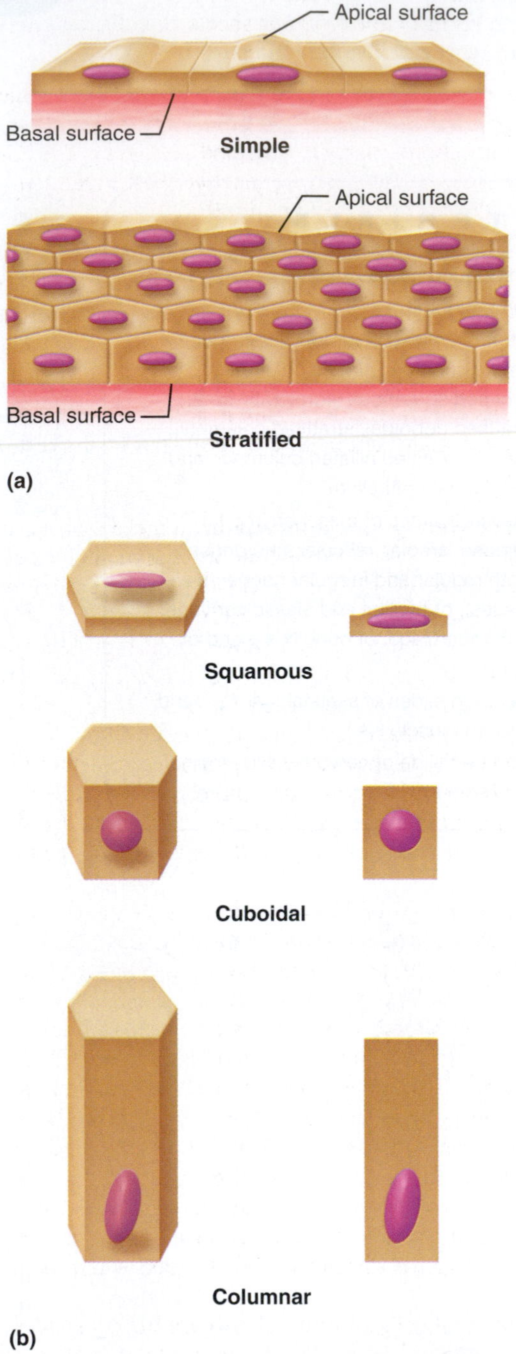
Apical surface
Basal surface
Simple
Apical surface
Basal surface
Stratified
(a)
Squamous
Cuboidal
Columnar
(b)

FIGURE 6.1 Classification of epithelia. **(a)** Classification on the basis of arrangement (relative number of layers). **(b)** Classification on the basis of cell shape. For each category, a whole cell is shown on the left and a longitudinal section is shown on the right.

Epithelial Tissue

Epithelial tissue, or **epithelium,** covers surfaces. For example, epithelium covers the external body surface (as the epidermis), lines its cavities and tubules, and generally marks off our "insides" from our outsides. Since the various endocrine (hormone-producing) and exocrine glands almost invariably develop from epithelial membranes, glands, too, are logically classed as epithelium.

Epithelial functions include protection, absorption, filtration, excretion, secretion, and sensory reception. For example, the epithelium covering the body surface protects against bacterial invasion and chemical damage; that lining the respiratory tract is ciliated to sweep dust and other foreign particles away from the lungs. Epithelium specialized to absorb substances lines the stomach and small intestine. In the kidney tubules, the epithelium absorbs, secretes, and filters. Secretion is a specialty of the glands.

The following characteristics distinguish epithelial tissues from other types:

- Polarity. The membranes always have one free surface, called the *apical surface,* and typically that surface is significantly different from the *basal surface.*

- Cellularity and specialized contacts. Cells fit closely together to form membranes, or sheets of cells, and are bound together by specialized junctions.
- Supported by connective tissue. The cells are attached to and supported by an adhesive **basement membrane,** which is an amorphous material secreted partly by the epithelial cells (*basal lamina*) and connective tissue cells (*reticular lamina*) that lie adjacent to each other.
- Avascularity. Epithelial tissues have no blood supply of their own (are avascular) but instead depend on diffusion of nutrients from the underlying connective tissue. (Glandular epithelia, however, are very vascular.)
- Regeneration. If well nourished, epithelial cells can easily regenerate themselves. This is an important characteristic because many epithelia are subjected to a good deal of friction.

The covering and lining epithelia are classified according to two criteria—arrangement or relative number of layers and cell shape (Figure 6.1). On the basis of arrangement, there are **simple** epithelia, consisting of one layer of cells attached to the basement membrane, and **stratified** epithelia, consisting of two or more layers of cells. The general types based on shape are **squamous** (scalelike), **cuboidal** (cubelike), and **columnar** (column-shaped) epithelial cells. The terms denoting shape and arrangement of the epithelial cells are combined to describe the epithelium fully. *Stratified epithelia are named according to the cells at the apical surface of the epithelial membrane,* not those resting on the basement membrane.

There are, in addition, two less easily categorized types of epithelia. **Pseudostratified epithelium** is actually a simple columnar epithelium (one layer of cells), but because its cells vary in height and the nuclei lie at different levels above the basement membrane, it gives the false appearance of being stratified. This epithelium is often ciliated. **Transitional epithelium** is a rather peculiar stratified squamous epithelium formed of rounded, or "plump," cells with the ability to slide over one another to allow the organ to be stretched. Transitional epithelium is found only in urinary system organs subjected to periodic distension, such as the bladder. The superficial cells are flattened (like true squamous cells) when the organ is distended and rounded when the organ is empty.

Epithelial cells forming glands are highly specialized to remove materials from the blood and to manufacture them into new materials, which they then secrete. There are two types of glands, as shown in Figure 6.2. **Endocrine glands** lose their surface connection (duct) as they develop; thus they are referred to as ductless glands. Their secretions (all hormones) are released into the extracellular fluid, from which they enter the blood or the lymphatic vessels that weave through the glands. **Exocrine glands** retain their ducts, and their secretions empty through these ducts to an epithelial surface. The exocrine glands—including the sweat and oil glands, liver, and pancreas—are both external and internal; they will be discussed in conjunction with the organ systems to which their products are functionally related.

The most common types of epithelia, their characteristic locations in the body, and their functions are described in Figure 6.3.

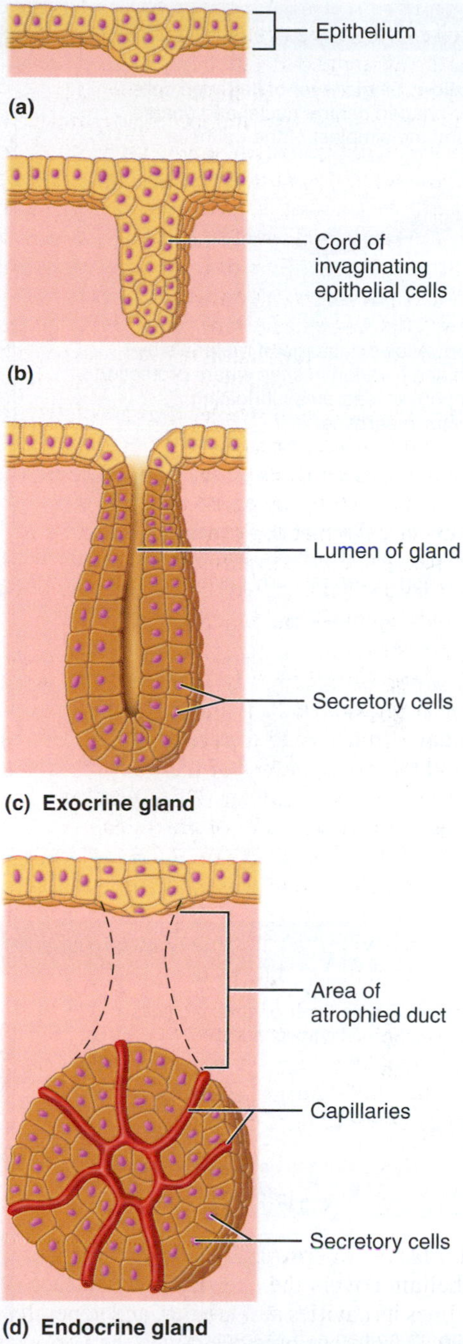
Epithelium
(a)
Cord of invaginating epithelial cells
(b)
Lumen of gland
Secretory cells
(c) Exocrine gland
Area of atrophied duct
Capillaries
Secretory cells
(d) Endocrine gland

FIGURE 6.2 Formation of endocrine and exocrine glands from epithelial sheets. (a) Epithelial cells grow and push into the underlying tissue. **(b)** A cord of epithelial cells forms. **(c)** In an exocrine gland, a lumen (cavity) forms. The inner cells form the duct, the outer cells produce the secretion. **(d)** In a forming endocrine gland, the connecting duct cells atrophy, leaving the secretory cells with no connection to the epithelial surface. However, they do become heavily invested with blood and lymphatic vessels that receive the secretions.

Text continues on page 74.

(a) Simple squamous epithelium

Description: Single layer of flattened cells with disc-shaped central nuclei and sparse cytoplasm; the simplest of the epithelia.

Function: Allows passage of materials by diffusion and filtration in sites where protection is not important; secretes lubricating substances in serosae.

Location: Kidney glomeruli; air sacs of lungs; lining of heart, blood vessels, and lymphatic vessels; lining of ventral body cavity (serosae).

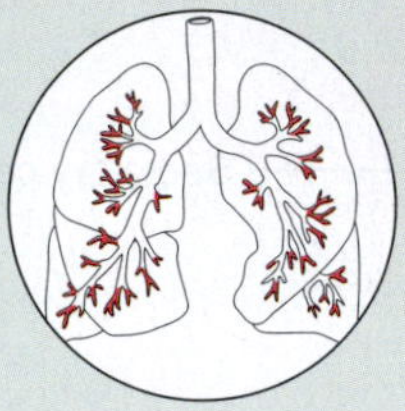

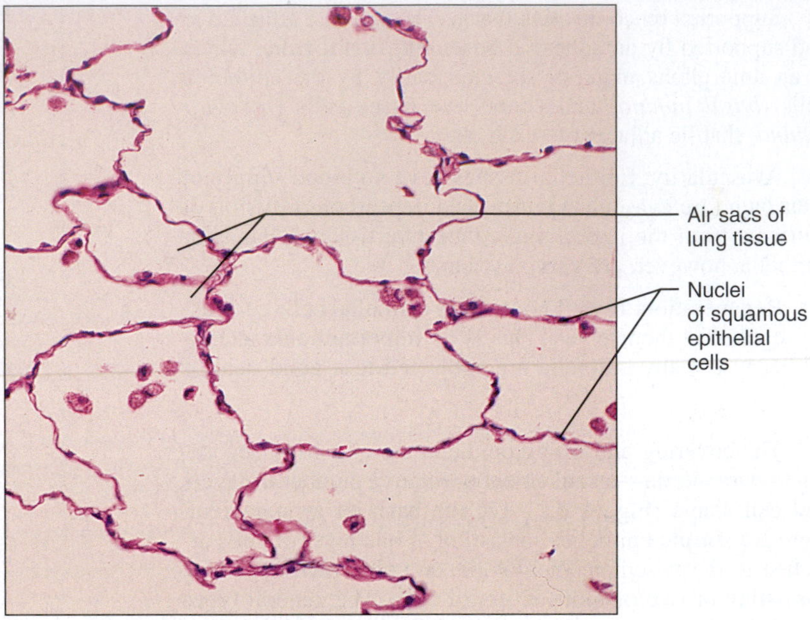
Air sacs of lung tissue
Nuclei of squamous epithelial cells

Photomicrograph: Simple squamous epithelium forming part of the alveolar (air sac) walls (125×).

(b) Simple cuboidal epithelium

Description: Single layer of cubelike cells with large, spherical central nuclei.

Function: Secretion and absorption.

Location: Kidney tubules; ducts and secretory portions of small glands; ovary surface.

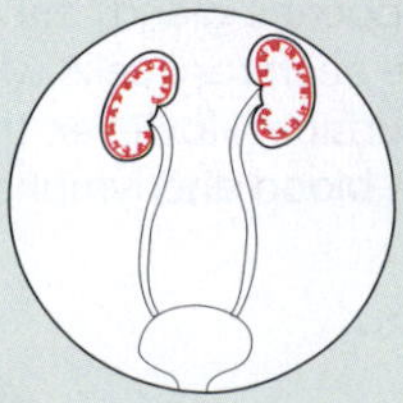

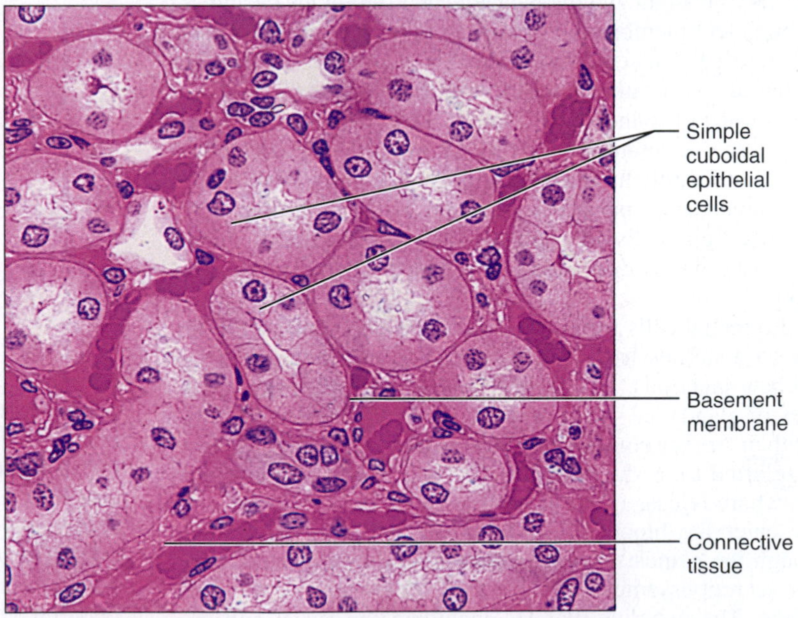
Simple cuboidal epithelial cells
Basement membrane
Connective tissue

Photomicrograph: Simple cuboidal epithelium in kidney tubules (430×).

FIGURE 6.3 Epithelial tissues. Simple epithelia **(a** and **b)**.

(c) Simple columnar epithelium

Description: Single layer of tall cells with *round* to *oval* nuclei; some cells bear cilia; layer may contain mucus-secreting unicellular glands (goblet cells).

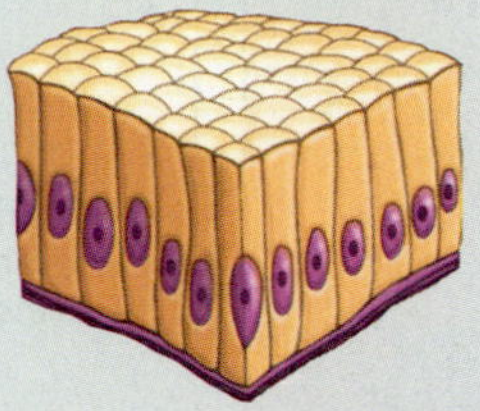

Function: Absorption; secretion of mucus, enzymes, and other substances; ciliated type propels mucus (or reproductive cells) by ciliary action.

Location: Nonciliated type lines most of the digestive tract (stomach to anal canal), gallbladder, and excretory ducts of some glands; ciliated variety lines small bronchi, uterine tubes, and some regions of the uterus.

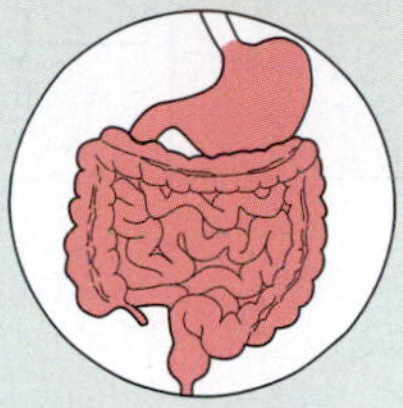

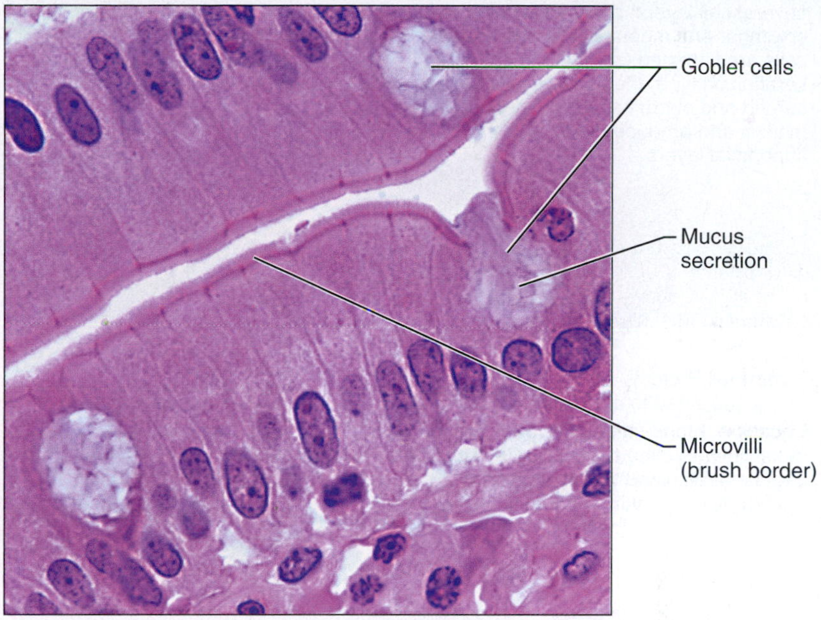
Goblet cells
Mucus secretion
Microvilli (brush border)

Photomicrograph: Simple columnar epithelium containing goblet cells from the small intestine (1150×).

(d) Pseudostratified columnar epithelium

Description: Single layer of cells of differing heights, some not reaching the free surface; nuclei seen at different levels; may contain mucus-secreting cells and bear cilia.

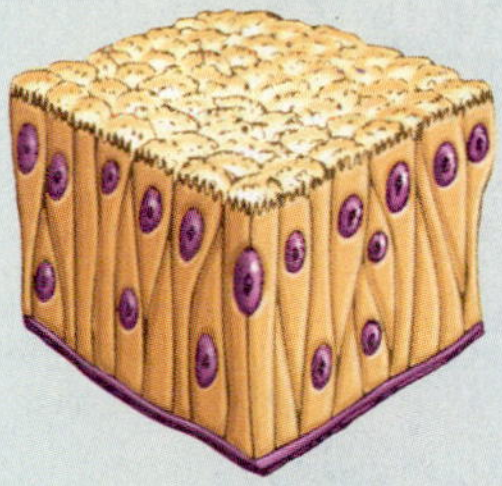

Function: Secretion, particularly of mucus; propulsion of mucus by ciliary action.

Location: Nonciliated type in male's sperm-carrying ducts and ducts of large glands; ciliated variety lines the trachea, most of the upper respiratory tract.

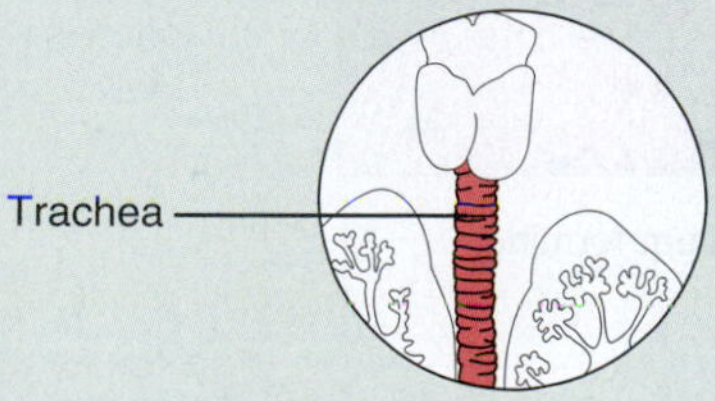
Trachea

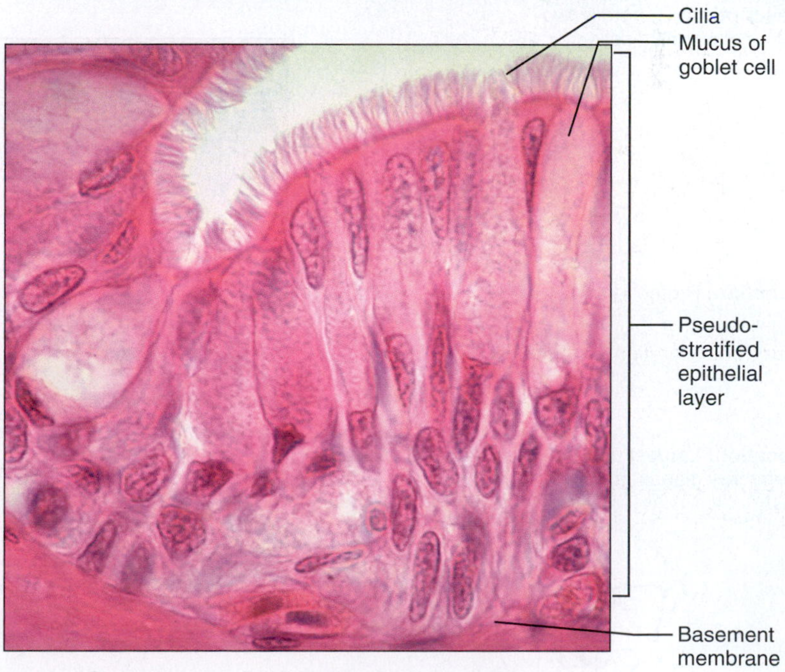
Cilia
Mucus of goblet cell
Pseudo-stratified epithelial layer
Basement membrane

Photomicrograph: Pseudostratified ciliated columnar epithelium lining the human trachea (570×).

FIGURE 6.3 ***(continued)*** Simple epithelia (**c** and **d**).

(e) Stratified squamous epithelium

Description: Thick membrane composed of several cell layers; basal cells are cuboidal or columnar and metabolically active; surface cells are flattened (squamous); in the keratinized type, the surface cells are full of keratin and dead; basal cells are active in mitosis and produce the cells of the more superficial layers.

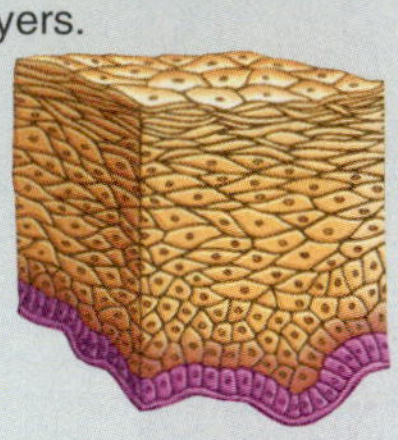

Function: Protects underlying tissues in areas subjected to abrasion.

Location: Nonkeratinized type forms the moist linings of the esophagus, mouth, and vagina; keratinized variety forms the epidermis of the skin, a dry membrane.

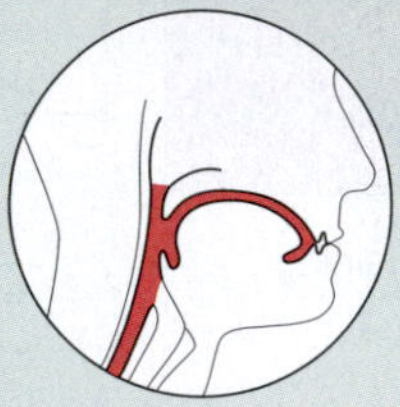

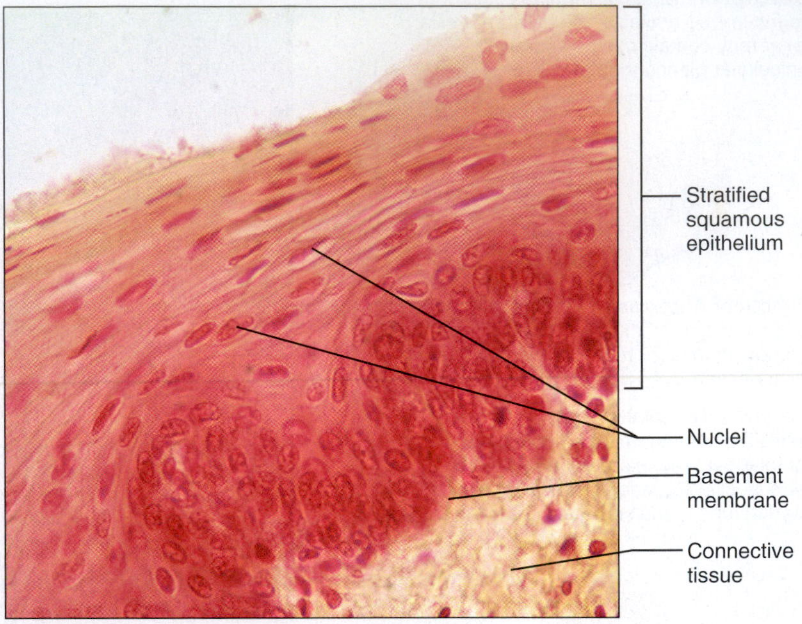
Stratified squamous epithelium
Nuclei
Basement membrane
Connective tissue

Photomicrograph: Stratified squamous epithelium lining the esophagus (285×).

(f) Stratified cuboidal epithelium

Description: Generally two layers of cubelike cells.

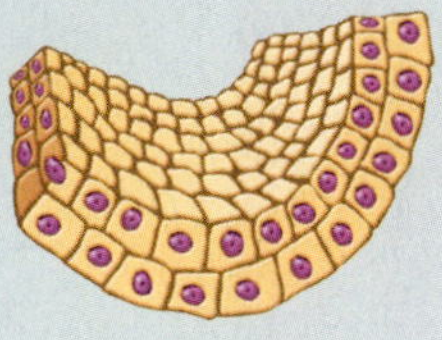

Function: Protection

Location: Largest ducts of sweat glands, mammary glands, and salivary glands.

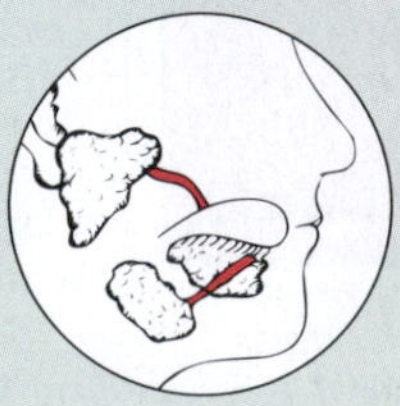

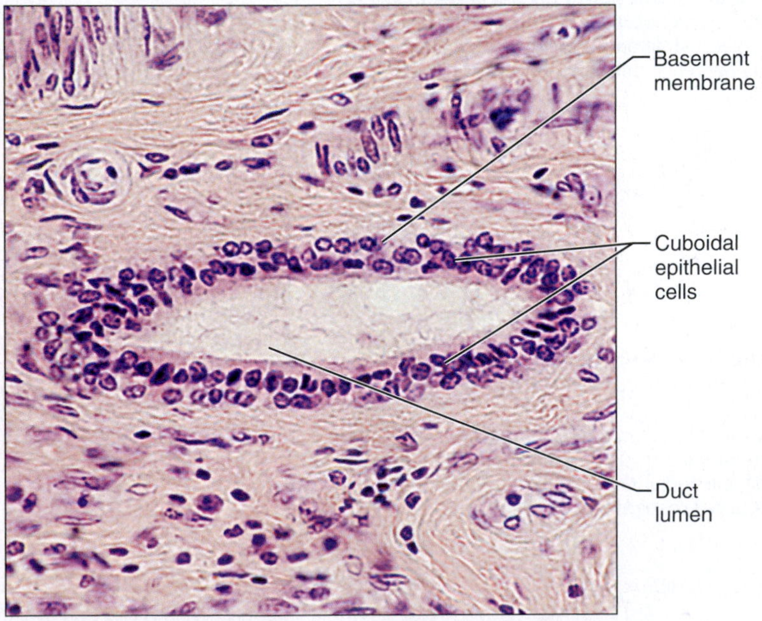
Basement membrane
Cuboidal epithelial cells
Duct lumen

Photomicrograph: Stratified cuboidal epithelium forming a salivary gland duct (285×).

FIGURE 6.3 *(continued)* Epithelial tissues. Stratified epithelia (**e** and **f**).

(g) Stratified columnar epithelium

Description: Several cell layers; basal cells usually cuboidal; superficial cells elongated and columnar.

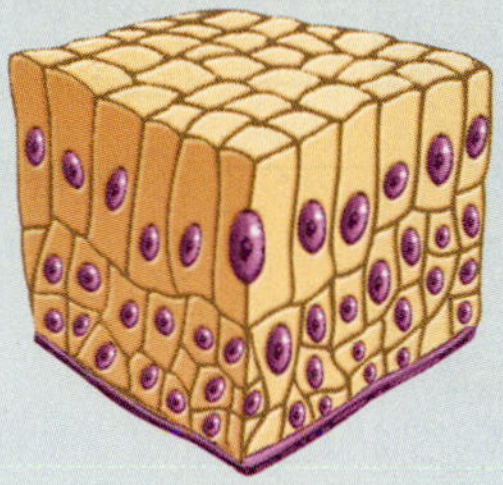

Function: Protection; secretion.

Location: Rare in the body; small amounts in male urethra and in large ducts of some glands.

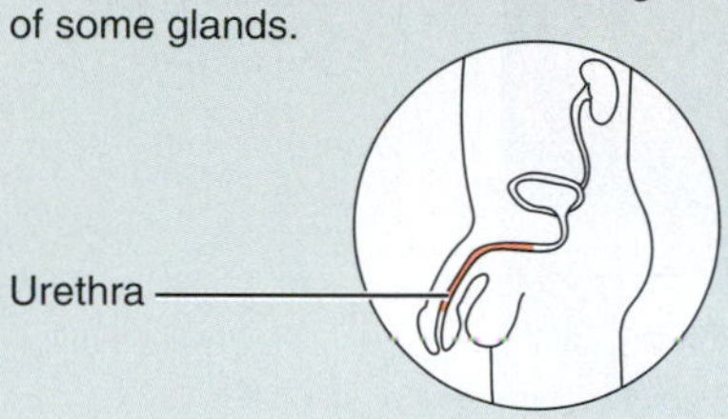
Urethra

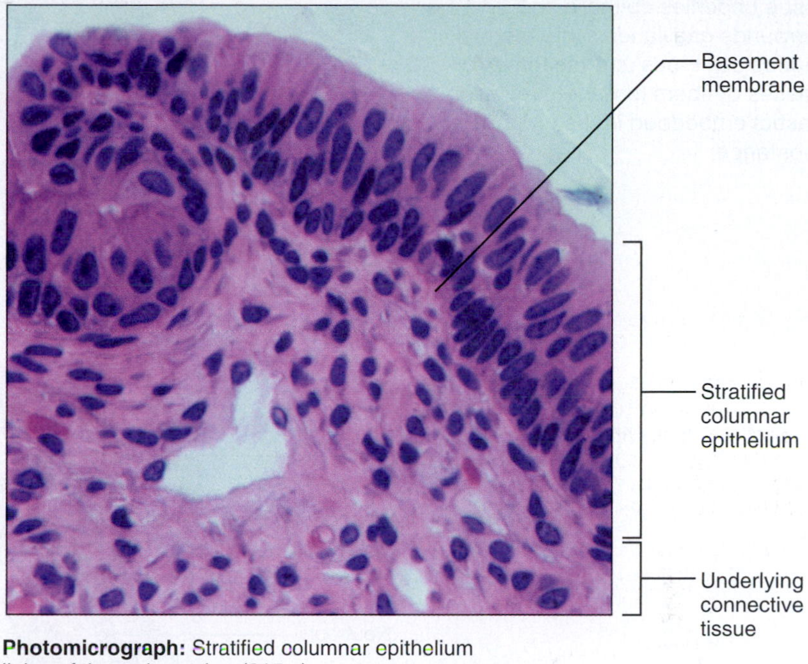
Basement membrane
Stratified columnar epithelium
Underlying connective tissue

Photomicrograph: Stratified columnar epithelium lining of the male urethra (315×).

(h) Transitional epithelium

Description: Resembles both stratified squamous and stratified cuboidal; basal cells cuboidal or columnar; surface cells dome shaped or squamouslike, depending on degree of organ stretch.

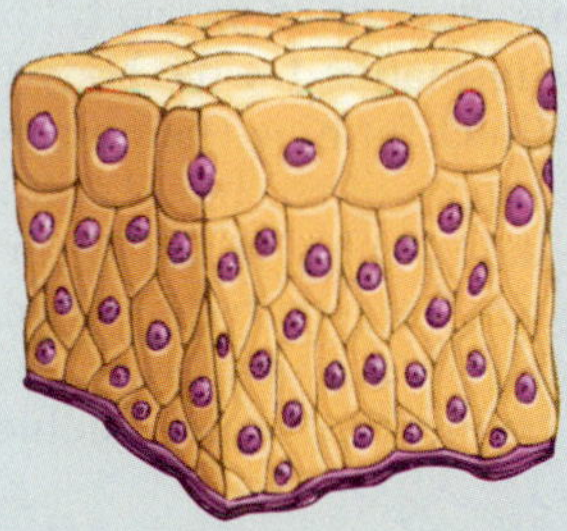

Function: Stretches readily and permits distension of urinary organ by contained urine.

Location: Lines the ureters, urinary bladder, and part of the urethra.

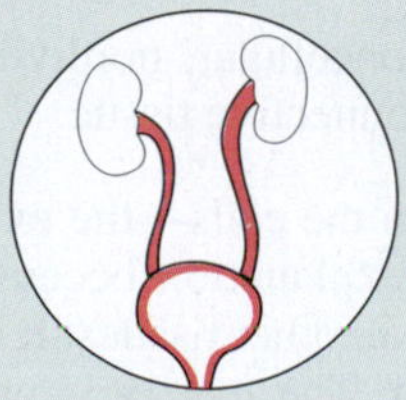

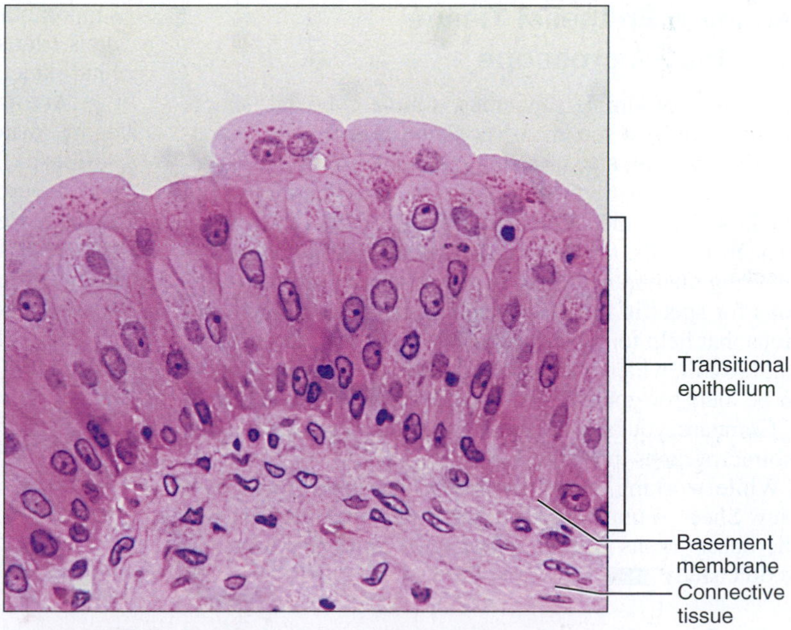
Transitional epithelium
Basement membrane
Connective tissue

Photomicrograph: Transitional epithelium lining the urinary bladder, relaxed state (360×); note the bulbous, or rounded, appearance of the cells at the surface; these cells flatten and become elongated when the bladder is filled with urine.

FIGURE 6.3 *(continued)* Stratified epithelia **(g and h)**.

Source of (g): Kessel and Kardon/Visuals Unlimited.

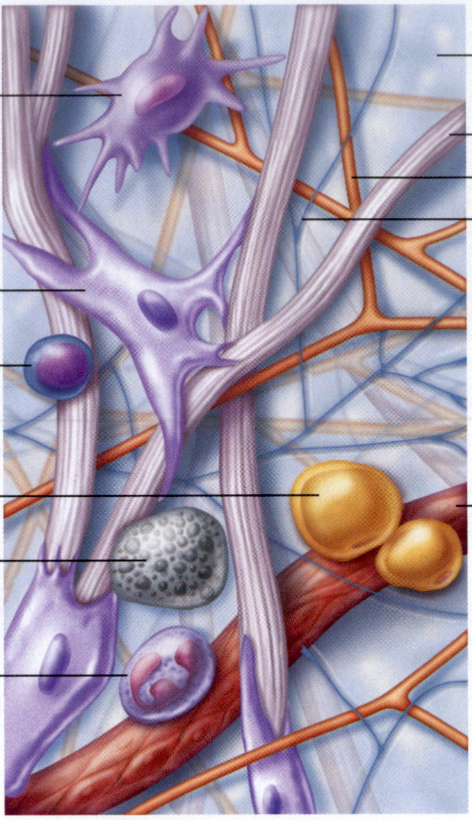

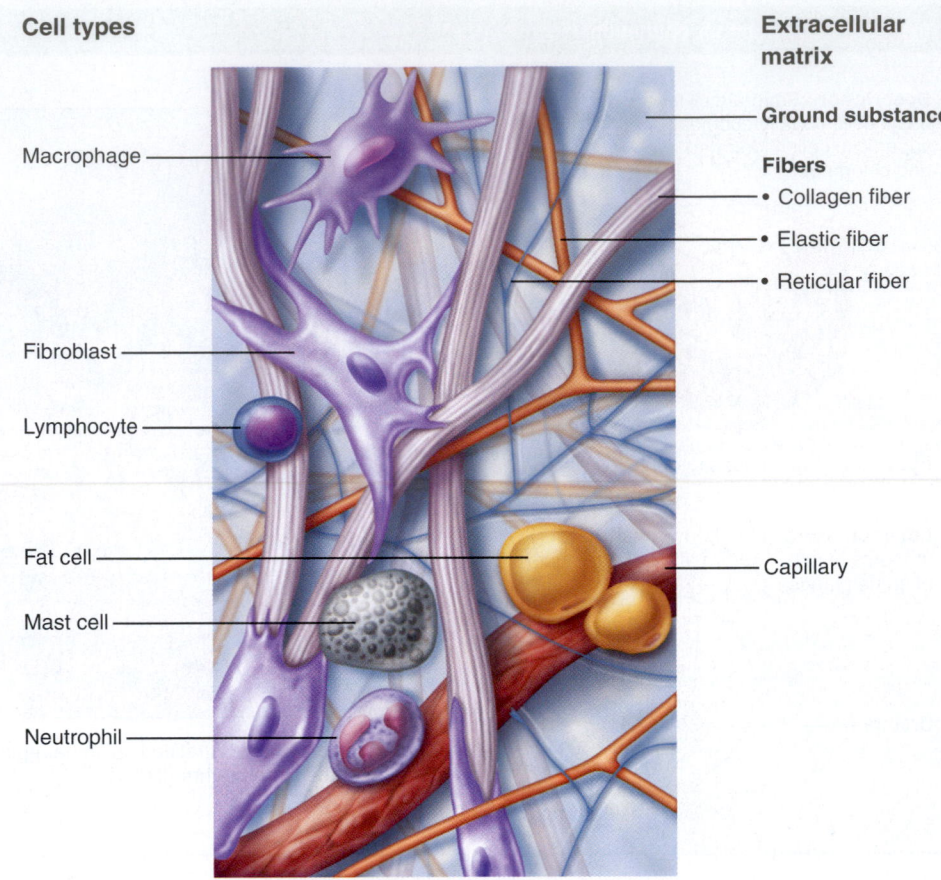

Cell types
Extracellular matrix
Ground substance
Fibers
• Collagen fiber
• Elastic fiber
• Reticular fiber
Macrophage
Fibroblast
Lymphocyte
Fat cell
Capillary
Mast cell
Neutrophil

FIGURE 6.4 Areolar connective tissue: A prototype (model) connective tissue. This tissue underlies epithelia and surrounds capillaries. Note the various cell types and the three classes of fibers (collagen, reticular, elastic) embedded in the ground substance.

ACTIVITY 1

Examining Epithelial Tissue Under the Microscope

Obtain slides of simple squamous, simple cuboidal, simple columnar, stratified squamous (nonkeratinized), pseudostratified ciliated columnar, stratified cuboidal, stratified columnar, and transitional epithelia. Examine each carefully, and notice how the epithelial cells fit closely together to form intact sheets of cells, a necessity for a tissue that forms linings or covering membranes. Scan each epithelial type for modifications for specific functions, such as cilia (motile cell projections that help to move substances along the cell surface), and microvilli, which increase the surface area for absorption. Also be alert for goblet cells, which secrete lubricating mucus. Compare your observations with the descriptions and photomicrographs in Figure 6.3.

While working, check the questions in the Exercise 6 Review Sheet. A number of the questions there refer to some of the observations you are asked to make during your microscopic study. ■

Connective Tissue

Connective tissue is found in all parts of the body as discrete structures or as part of various body organs. It is the most abundant and widely distributed of the tissue types.

Connective tissues perform a variety of functions, but they primarily protect, support, and bind together other tissues of the body. For example, bones are composed of connective tissue (**bone,** or **osseous tissue**), and they protect and support other body tissues and organs. The ligaments and tendons (**dense connective tissue**) bind the bones together or bind skeletal muscles to bones.

Areolar connective tissue (Figure 6.4) is a soft packaging material that cushions and protects body organs. **Adipose** (fat) tissue provides insulation for the body tissues and a source of stored food. Blood-forming (**hematopoietic**) tissue replenishes the body's supply of red blood cells. Connective tissue also serves a vital function in the repair of all body tissues, since many wounds are repaired by connective tissue in the form of scar tissue.

The characteristics of connective tissue include the following:

- With a few exceptions (cartilages, tendons, and ligaments, which are poorly vascularized), connective tissues have a rich supply of blood vessels.
- Connective tissues are composed of many types of cells.
- There is a great deal of noncellular, nonliving material (matrix) between the cells of connective tissue.

The nonliving material between the cells—the **extracellular matrix**—deserves a bit more explanation because it distinguishes connective tissue from all other tissues. It is produced by the cells and then extruded. The matrix is primarily responsible for the strength associated with connective tissue, but there is variation. At one extreme, adipose tissue is composed mostly of cells. At the opposite extreme, bone and cartilage have few cells and large amounts of matrix.

(a) Embryonic connective tissue: Mesenchyme

Description: Embryonic connective tissue; gel-like ground substance containing fibers; star-shaped mesenchymal cells.

Function: Gives rise to all other connective tissue types.

Location: Primarily in embryo.

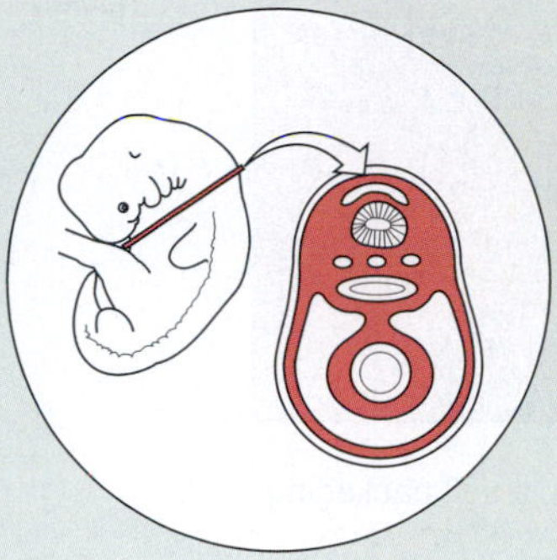

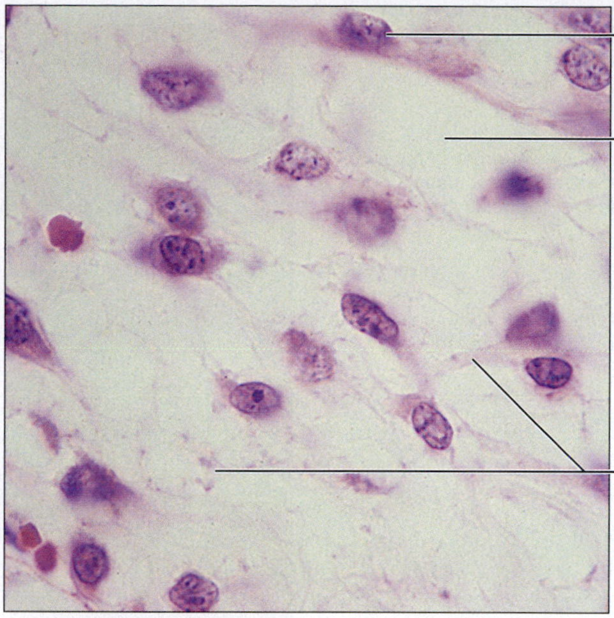
Mesenchymal cell
Ground substance
Fibers

Photomicrograph: Mesenchymal tissue, an embryonic connective tissue (600×); the clear-appearing background is the fluid ground substance of the matrix; notice the fine, sparse fibers.

FIGURE 6.5 Connective tissues. Embryonic connective tissue **(a)**.

The matrix has two components—ground substance and fibers. The **ground substance** is composed chiefly of interstitial fluid, cell adhesion proteins, and proteoglycans. Depending on its specific composition, the ground substance may be liquid, semisolid, gel-like, or very hard. When the matrix is firm, as in cartilage and bone, the connective tissue cells reside in cavities in the matrix called *lacunae.* The fibers, which provide support, include **collagen** (white) **fibers, elastic** (yellow) **fibers,** and **reticular** (fine collagen) **fibers.** Of these, the collagen fibers are most abundant.

Generally speaking, the ground substance functions as a molecular sieve, or medium, through which nutrients and other dissolved substances can diffuse between the blood capillaries and the cells. The fibers in the matrix hinder diffusion somewhat and make the ground substance less pliable. The properties of the connective tissue cells and the makeup and arrangement of their matrix elements vary tremendously, accounting for the amazing diversity of this tissue type. Nonetheless, the connective tissues have a common structural plan seen best in *areolar connective tissue* (Figure 6.4), a soft packing tissue that occurs throughout the body. Since all other connective tissues are variations of areolar, it is considered the model or prototype of the connective tissues. Notice in Figure 6.4 that areolar tissue has all three varieties of fibers, but they are sparsely arranged in its transparent gel-like ground substance. The cell type that secretes its matrix is the *fibroblast,* but a wide variety of other cells (including phagocytic cells like macrophages and certain white blood cells and mast cells that act in the inflammatory response) are present as well. The more durable connective tissues, such as bone, cartilage, and the dense fibrous varieties, characteristically have a firm ground substance and many more fibers.

There are four main types of adult connective tissue, all of which typically have large amounts of matrix. These are **connective tissue proper** (which includes areolar, adipose, reticular, and dense [fibrous] connective tissues), **cartilage, bone,** and **blood.** All of these derive from an embryonic tissue called *mesenchyme.* Figure 6.5 lists the general characteristics, location, and function of some of the connective tissues found in the body.

ACTIVITY 2

Examining Connective Tissue Under the Microscope

Obtain prepared slides of mesenchyme; of adipose, areolar, reticular, dense regular, elastic, and irregular connective tissue; of hyaline and elastic cartilage and fibrocartilage; of osseous connective tissue (bone); and of blood. Compare your observations with the views illustrated in Figure 6.5.

Distinguish between the living cells and the matrix and pay particular attention to the denseness and arrangement of the matrix. For example, notice how the matrix of the dense regular and irregular connective tissues, respectively making up tendons and the dermis of the skin, is packed with collagen fibers. Note also that in the *regular* variety (tendon), the

Text continues on page 81.

(b) Connective tissue proper: loose connective tissue, areolar

Description: Gel-like matrix with all three fiber types; cells: fibroblasts, macrophages, mast cells, and some white blood cells.

Function: Wraps and cushions organs; its macrophages phagocytize bacteria; plays important role in inflammation; holds and conveys tissue fluid.

Location: Widely distributed under epithelia of body, e.g., forms lamina propria of mucous membranes; packages organs; surrounds capillaries.

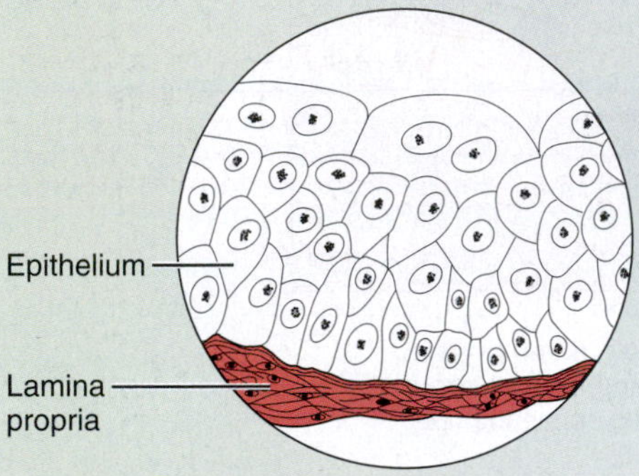
Epithelium
Lamina propria

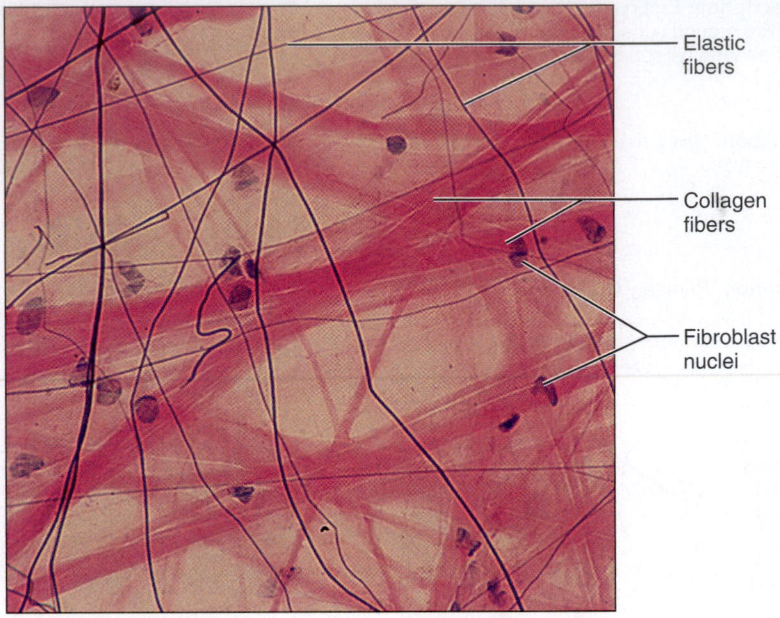
Elastic fibers
Collagen fibers
Fibroblast nuclei

Photomicrograph: Areolar connective tissue, a soft packaging tissue of the body (300×).

(c) Connective tissue proper: loose connective tissue, adipose

Description: Matrix as in areolar, but very sparse; closely packed adipocytes, or fat cells, have nucleus pushed to the side by large fat droplet.

Function: Provides reserve fuel; insulates against heat loss; supports and protects organs.

Location: Under skin; around kidneys and eyeballs; within abdomen; in breasts.

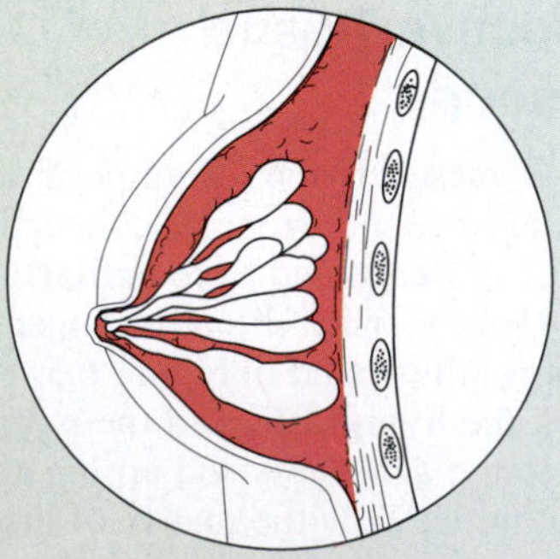

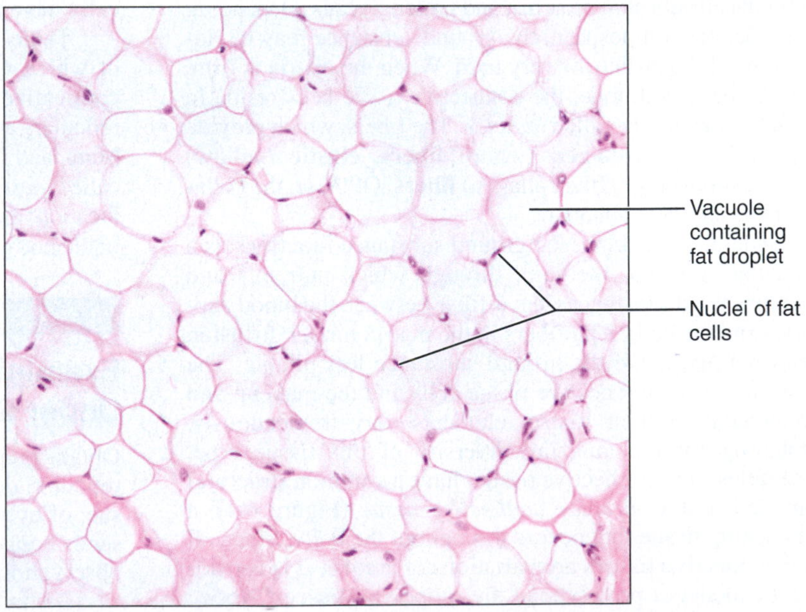
Vacuole containing fat droplet
Nuclei of fat cells

Photomicrograph: Adipose tissue from the subcutaneous layer under the skin (350×).

FIGURE 6.5 *(continued)* Connective tissues. Connective tissue proper **(b** and **c)**.

(d) Connective tissue proper: loose connective tissue, reticular

Description: Network of reticular fibers in a typical loose ground substance; reticular cells lie on the network.

Function: Fibers form a soft internal skeleton (stroma) that supports other cell types, including white blood cells, mast cells, and macrophages.

Location: Lymphoid organs (lymph nodes, bone marrow, and spleen).

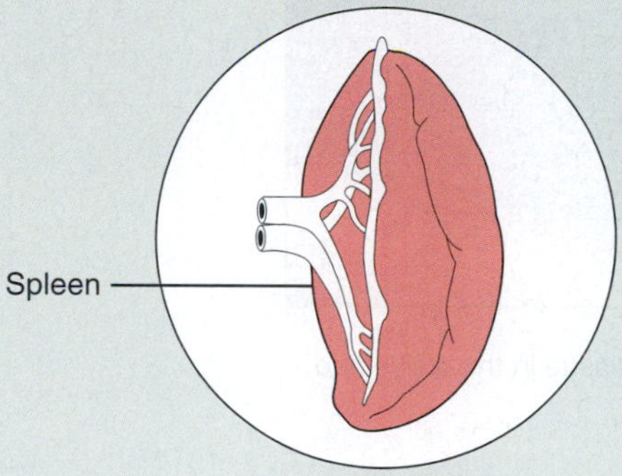
Spleen

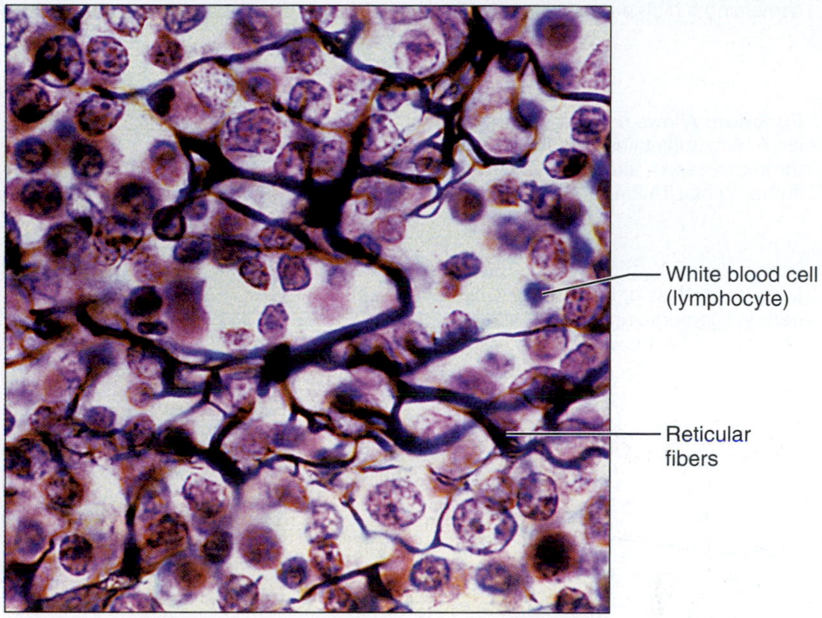
White blood cell (lymphocyte)
Reticular fibers

Photomicrograph: Dark-staining network of reticular connective tissue fibers forming the internal skeleton of the spleen (350×).

(e) Connective tissue proper: dense connective tissue, dense regular

Description: Primarily parallel collagen fibers; a few elastic fibers; major cell type is the fibroblast.

Function: Attaches muscles to bones or to muscles; attaches bones to bones; withstands great tensile stress when pulling force is applied in one direction.

Location: Tendons, most ligaments, aponeuroses.

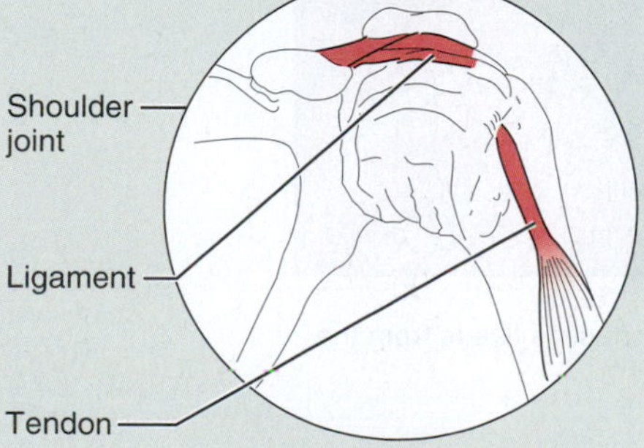
Shoulder joint
Ligament
Tendon

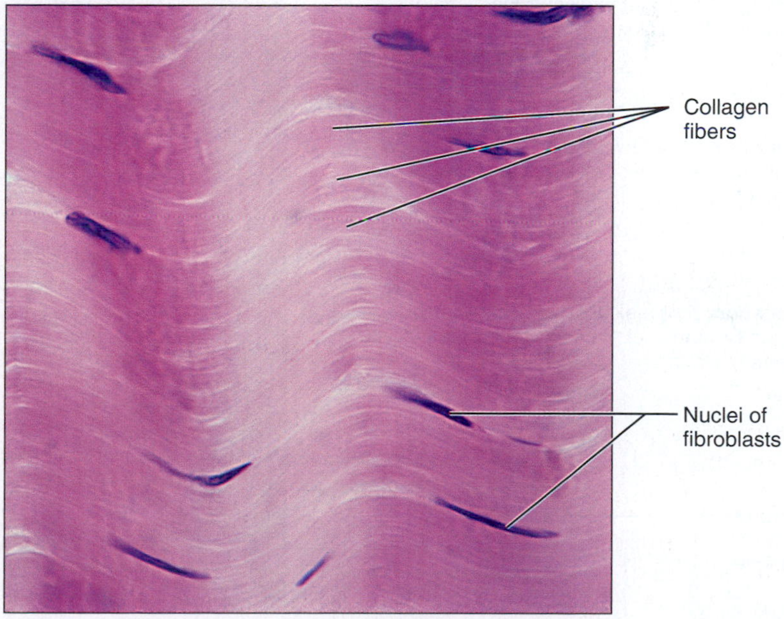
Collagen fibers
Nuclei of fibroblasts

Photomicrograph: Dense regular connective tissue from a tendon (500×).

FIGURE 6.5 *(continued)* Connective tissue proper **(d** and **e)**.

(f) Connective tissue proper: dense connective tissue, elastic

Description: Dense regular connective tissue containing a high proportion of elastic fibers.

Function: Allows recoil of tissue following stretching; maintains pulsatile flow of blood through arteries; aids passive recoil of lungs following inspiration.

Location: Walls of large arteries; within certain ligaments associated with the vertebral column; within the walls of the bronchial tubes.

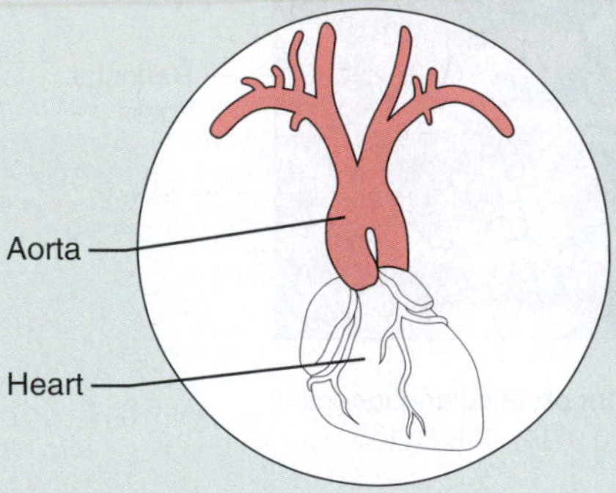
Aorta
Heart

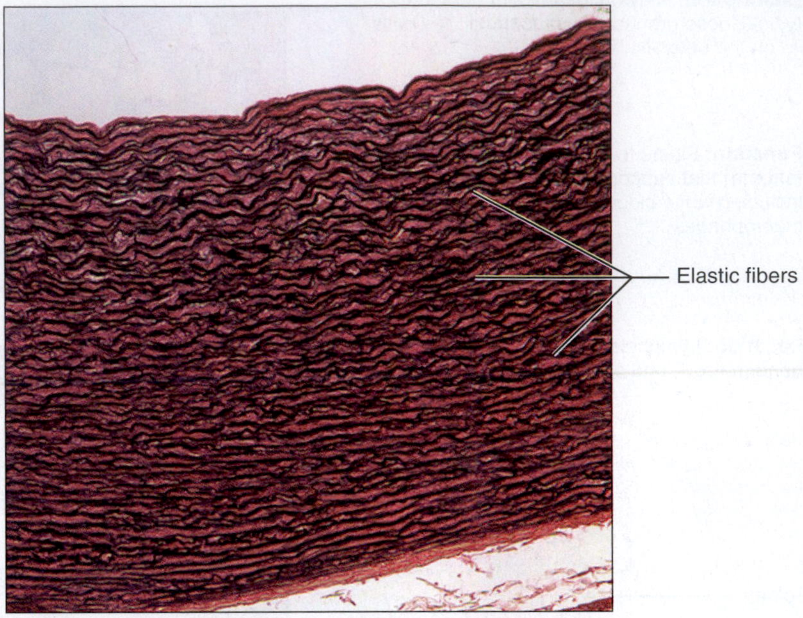
Elastic fibers

Photomicrograph: Elastic connective tissue in the wall of the aorta (250×).

(g) Connective tissue proper: dense connective tissue, dense irregular

Description: Primarily irregularly arranged collagen fibers; some elastic fibers; major cell type is the fibroblast.

Function: Able to withstand tension exerted in many directions; provides structural strength.

Location: Fibrous capsules of organs and of joints; dermis of the skin; submucosa of digestive tract.

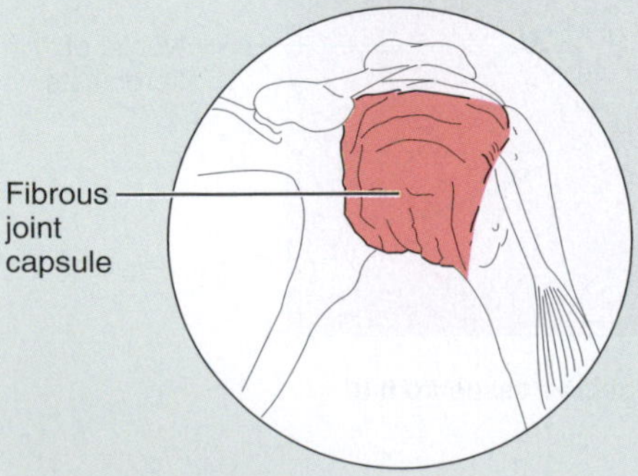
Fibrous joint capsule

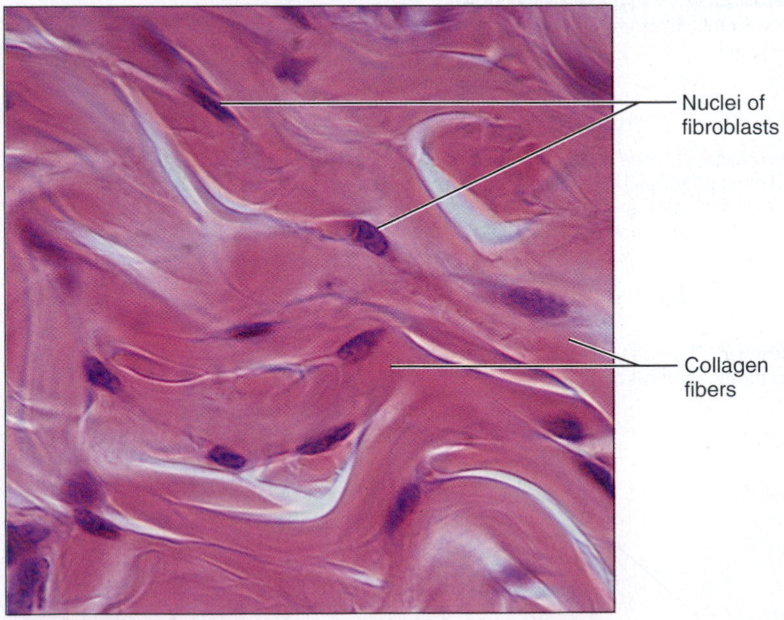
Nuclei of fibroblasts
Collagen fibers

Photomicrograph: Dense irregular connective tissue from the dermis of the skin (400×).

FIGURE 6.5 *(continued)* **Connective tissues.** Connective tissue proper **(f)** and **(g)**.

(h) Cartilage: hyaline

Description: Amorphous but firm matrix; collagen fibers form an imperceptible network; chondroblasts produce the matrix and when mature (chondrocytes) lie in lacunae.

Function: Supports and reinforces; has resilient cushioning properties; resists compressive stress.

Location: Forms most of the embryonic skeleton; covers the ends of long bones in joint cavities; forms costal cartilages of the ribs; cartilages of the nose, trachea, and larynx.

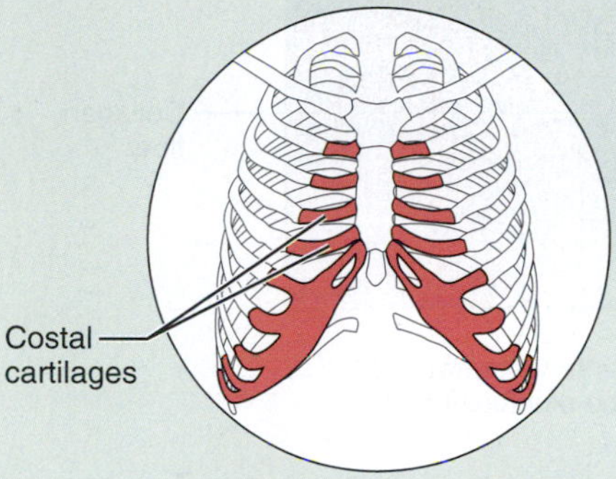
Costal cartilages

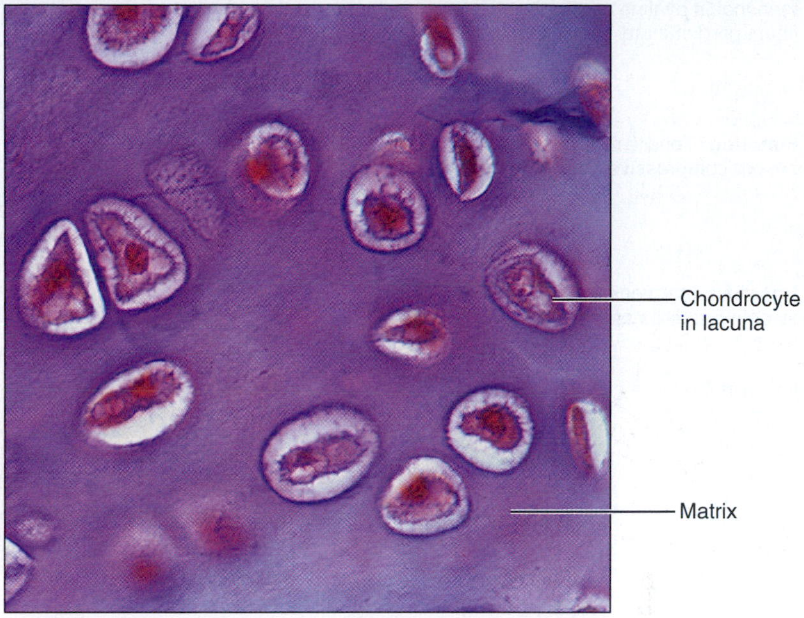
Chondrocyte in lacuna
Matrix

Photomicrograph: Hyaline cartilage from the trachea (750×).

(i) Cartilage: elastic

Description: Similar to hyaline cartilage, but more elastic fibers in matrix.

Function: Maintains the shape of a structure while allowing great flexibility.

Location: Supports the external ear (pinna); epiglottis.

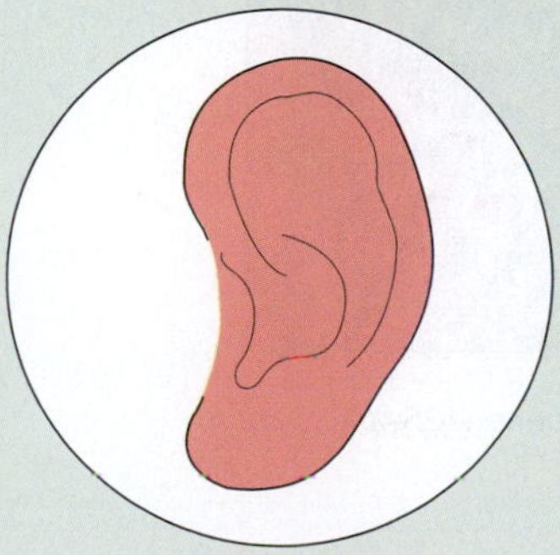

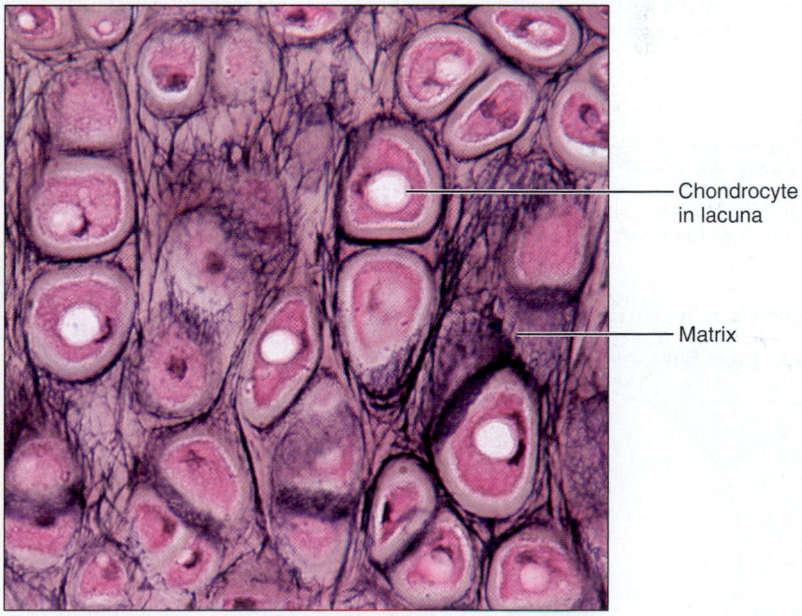
Chondrocyte in lacuna
Matrix

Photomicrograph: Elastic cartilage from the human ear pinna; forms the flexible skeleton of the ear (800×).

FIGURE 6.5 ***(continued)*** Cartilage (**h** and **i**).

(j) Cartilage: fibrocartilage

Description: Matrix similar to but less firm than that in hyaline cartilage; thick collagen fibers predominate.

Function: Tensile strength with the ability to absorb compressive shock.

Location: Intervertebral discs; pubic symphysis; discs of knee joint.

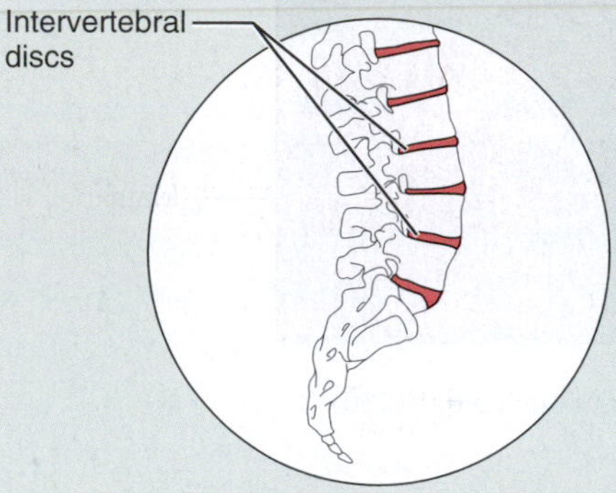
Intervertebral discs

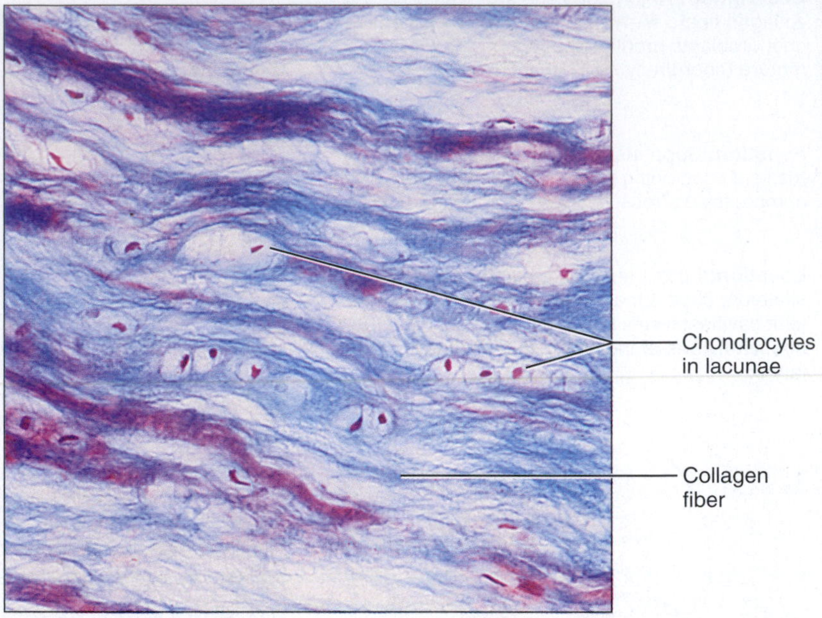
Chondrocytes in lacunae
Collagen fiber

Photomicrograph: Fibrocartilage of an intervertebral disc (125×). Special staining produced the blue color seen.

(k) Bones (osseous tissue)

Description: Hard, calcified matrix containing many collagen fibers; osteocytes lie in lacunae. Very well vascularized.

Function: Bone supports and protects (by enclosing); provides levers for the muscles to act on; stores calcium and other minerals and fat; marrow inside bones is the site for blood cell formation (hematopoiesis).

Location: Bones

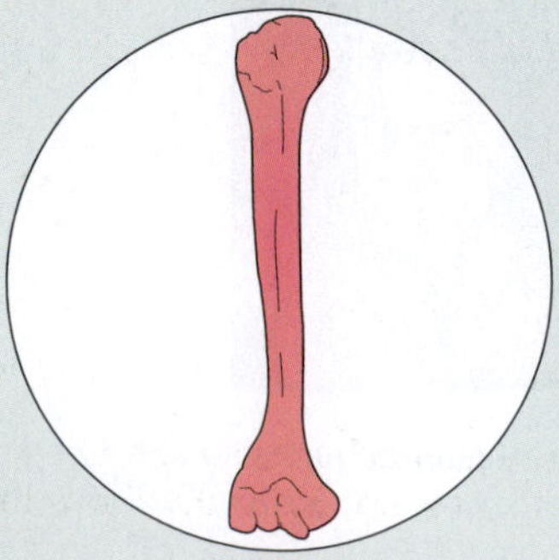

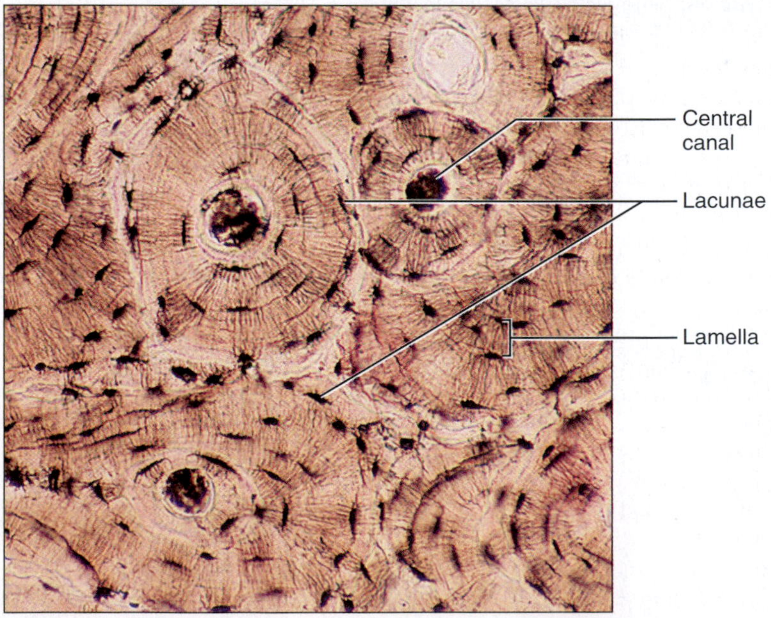
Central canal
Lacunae
Lamella

Photomicrograph: Cross-sectional view of bone (125×).

FIGURE 6.5 *(continued)* **Connective tissues.** Cartilage **(j)** and bone **(k)**.

(l) Blood

Description: Red and white blood cells in a fluid matrix (plasma).

Function: Transport of respiratory gases, nutrients, wastes, and other substances.

Location: Contained within blood vessels.

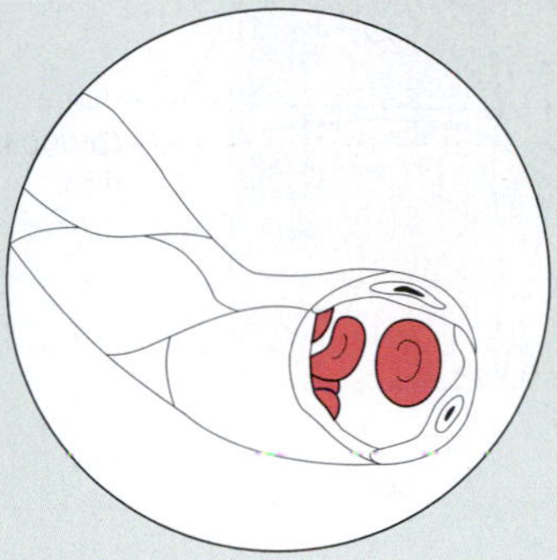

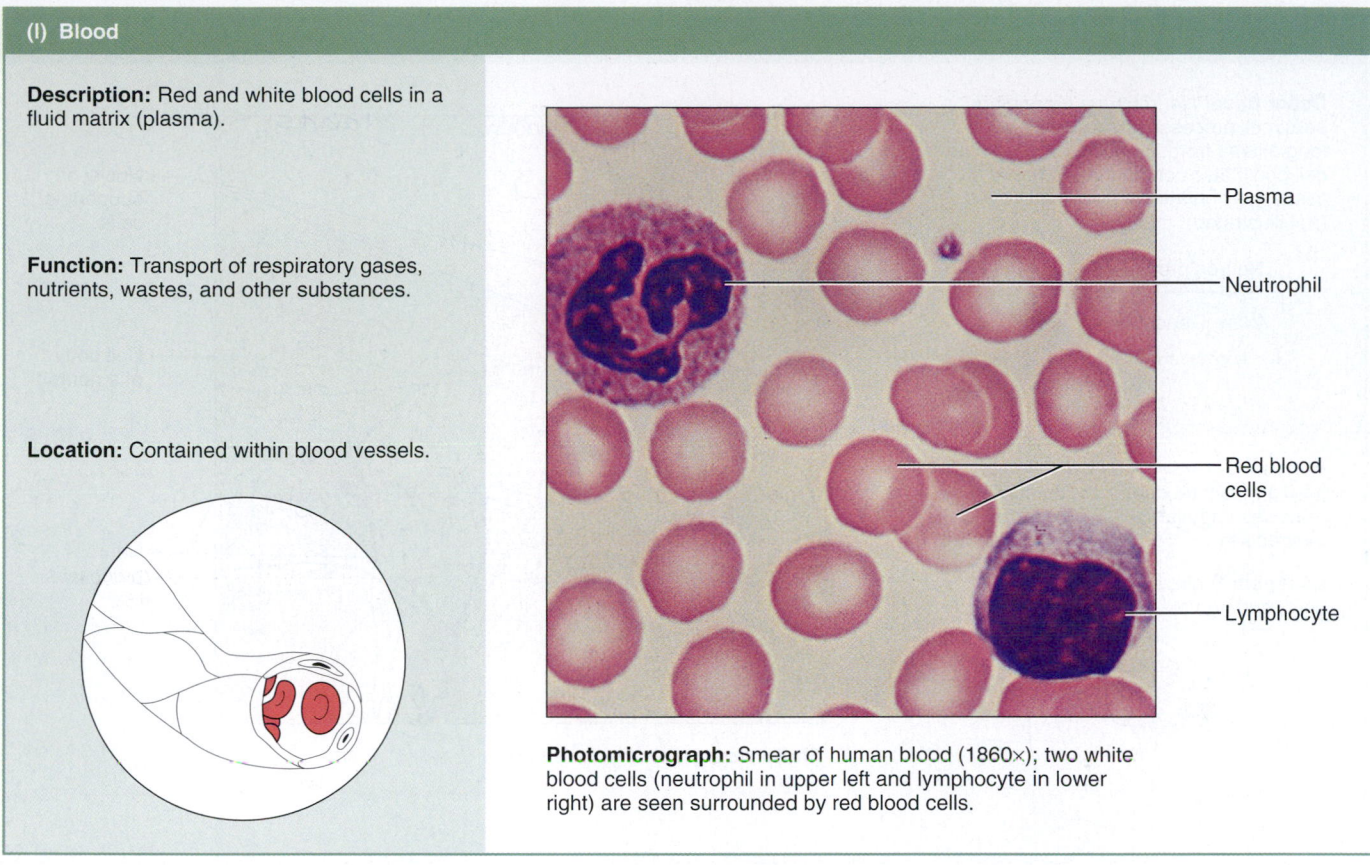

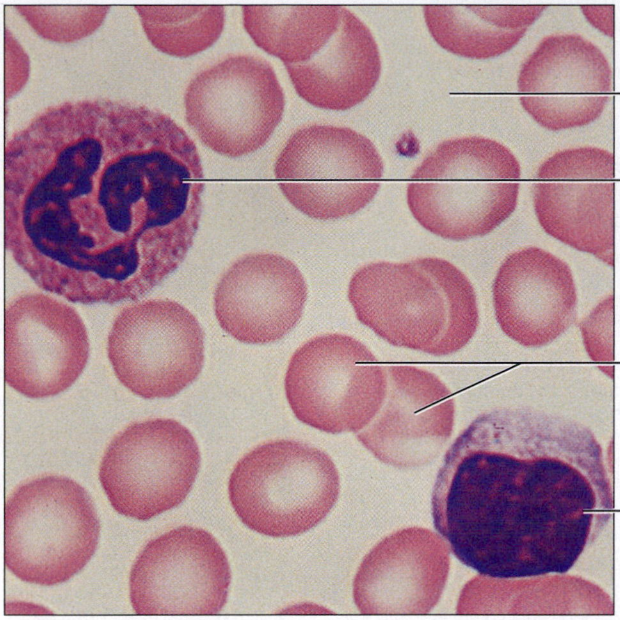
Plasma
Neutrophil
Red blood cells
Lymphocyte

Photomicrograph: Smear of human blood (1860×); two white blood cells (neutrophil in upper left and lymphocyte in lower right) are seen surrounded by red blood cells.

FIGURE 6.5 *(continued)* Blood **(l)**.

fibers are all running in the same direction, whereas in the dermis they appear to be running in many directions.

While examining the areolar connective tissue, notice how much empty space there appears to be (*areol* = small empty space), and distinguish between the collagen fibers and the coiled elastic fibers. Identify the starlike fibroblasts. Also, try to locate a **mast cell,** which has large, darkly staining granules in its cytoplasm (*mast* = stuffed full of granules). This cell type releases histamine, which makes capillaries more permeable during inflammatory reactions and allergies and thus is partially responsible for that "runny nose" of some allergies.

In adipose tissue, locate a "signet ring" cell, a fat cell in which the nucleus can be seen pushed to one side by the large, fat-filled vacuole that appears to be a large empty space. Also notice how little matrix there is in adipose (fat) tissue. Distinguish between the living cells and the matrix in the dense fibrous, bone, and hyaline cartilage preparations.

Scan the blood slide at low and then high power to examine the general shape of the red blood cells. Then, switch to the oil immersion lens for a closer look at the various types of white blood cells. How does blood differ from all other connective tissues?

____

____

Nervous Tissue

Nervous tissue is composed of two major cell populations. The **neuroglia** are special supporting cells that protect, support, and insulate the more delicate neurons. The **neurons** are highly specialized to receive stimuli (irritability) and to conduct waves of excitation, or impulses, to all parts of the body (conductivity). They are the cells that are most often associated with nervous system functioning.

The structure of neurons is markedly different from that of all other body cells. They all have a nucleus-containing cell body, and their cytoplasm is drawn out into long extensions (cell processes)—sometimes as long as 1 m (about 3 feet), which allows a single neuron to conduct an impulse over relatively long distances. More detail about the anatomy of the different classes of neurons and neuroglia appears in Exercise 17.

ACTIVITY 3

Examining Nervous Tissue Under the Microscope

Obtain a prepared slide of a spinal cord smear. Locate a neuron and compare it to Figure 6.6. Keep the light dim—this

Nervous tissue

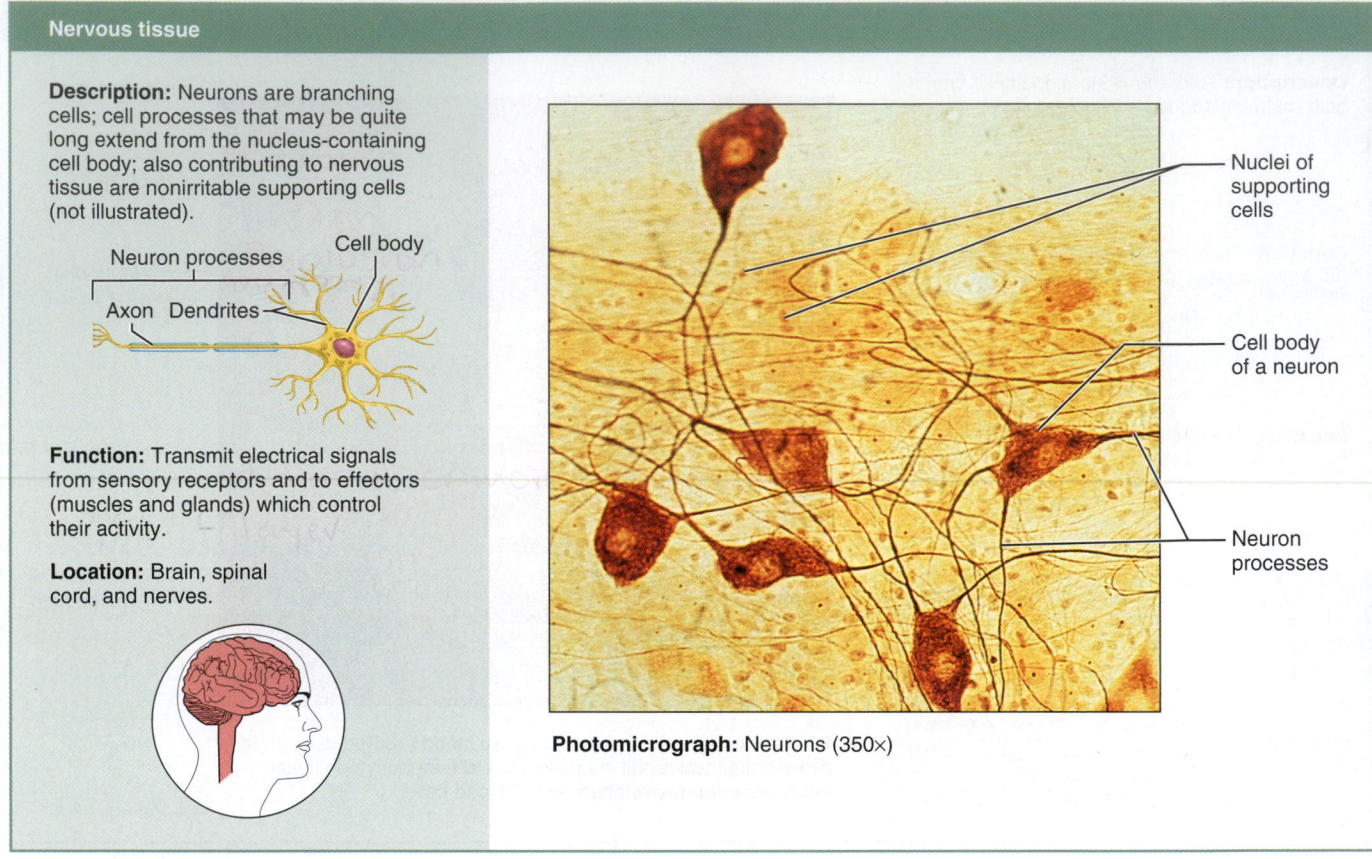

Description: Neurons are branching cells; cell processes that may be quite long extend from the nucleus-containing cell body; also contributing to nervous tissue are nonirritable supporting cells (not illustrated).

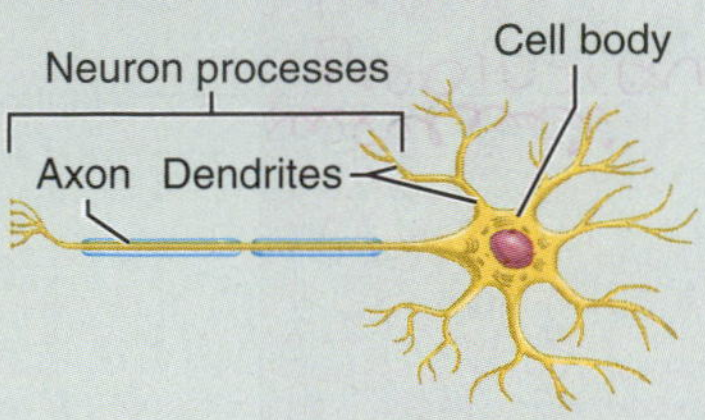
Neuron processes
Cell body
Axon
Dendrites

Function: Transmit electrical signals from sensory receptors and to effectors (muscles and glands) which control their activity.

Location: Brain, spinal cord, and nerves.

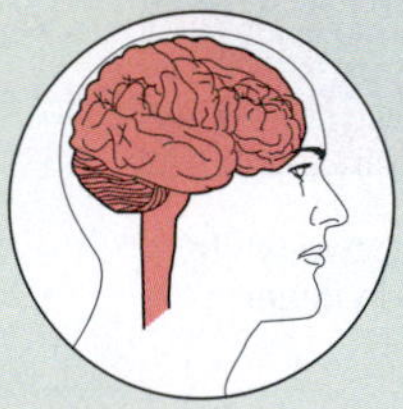

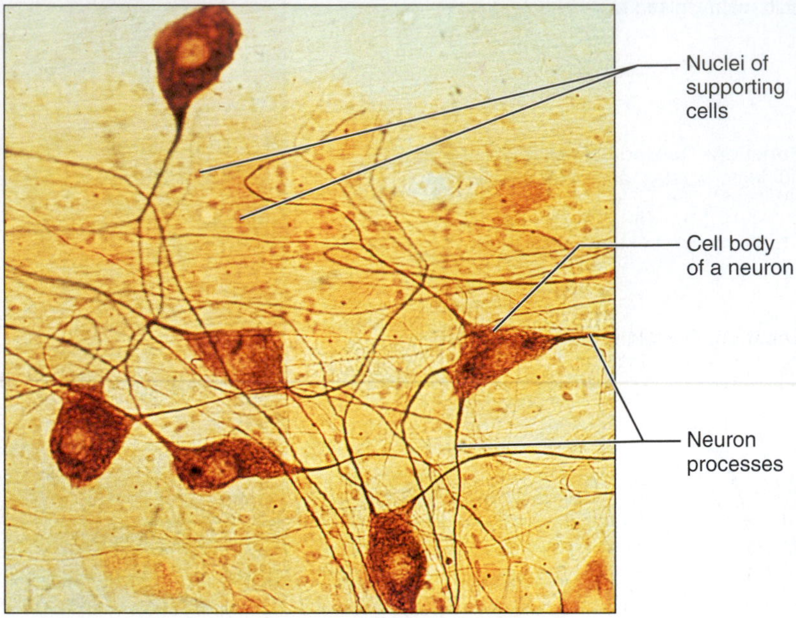
Nuclei of supporting cells
Cell body of a neuron
Neuron processes

Photomicrograph: Neurons (350×)

FIGURE 6.6 **Nervous tissue.**

will help you see the cellular extensions of the neurons. See also Figure 17.2 in Exercise 17.

Muscle Tissue

Muscle tissue (Figure 6.7) is highly specialized to contract and produces most types of body movement. As you might expect, muscle cells tend to be elongated, providing a long axis for contraction. The three basic types of muscle tissue are described briefly here. Cardiac and skeletal muscles are treated more completely in later exercises.

Skeletal muscle, the "meat," or flesh, of the body, is attached to the skeleton. It is under voluntary control (consciously controlled), and its contraction moves the limbs and other external body parts. The cells of skeletal muscles are long, cylindrical, and multinucleate (several nuclei per cell), with the nuclei pushed to the periphery of the cells; they have obvious *striations* (stripes).

Cardiac muscle is found only in the heart. As it contracts, the heart acts as a pump, propelling the blood into the blood vessels. Cardiac muscle, like skeletal muscle, has striations, but cardiac cells are branching uninucleate cells that interdigitate (fit together) at junctions called **intercalated discs.** These structural modifications allow the cardiac muscle to act as a unit. Cardiac muscle is under involuntary control, which means that we cannot voluntarily or consciously control the operation of the heart.

Smooth muscle, or *visceral muscle,* is found mainly in the walls of hollow organs (digestive and urinary tract organs, uterus, blood vessels). Typically it has two layers that run at right angles to each other; consequently its contraction can constrict or dilate the lumen (cavity) of an organ and propel substances along predetermined pathways. Smooth muscle cells are quite different in appearance from those of skeletal or cardiac muscle. No striations are visible, and the uninucleate smooth muscle cells are spindle-shaped.

ACTIVITY 4

Examining Muscle Tissue Under the Microscope

Obtain and examine prepared slides of skeletal, cardiac, and smooth muscle. Notice their similarities and dissimilarities in your observations and in the illustrations in Figure 6.7.

(a) Skeletal muscle

Description: Long, cylindrical, multinucleate cells; obvious striations.

Function: Voluntary movement; locomotion; manipulation of the environment; facial expression; voluntary control.

Location: In skeletal muscles attached to bones or occasionally to skin.

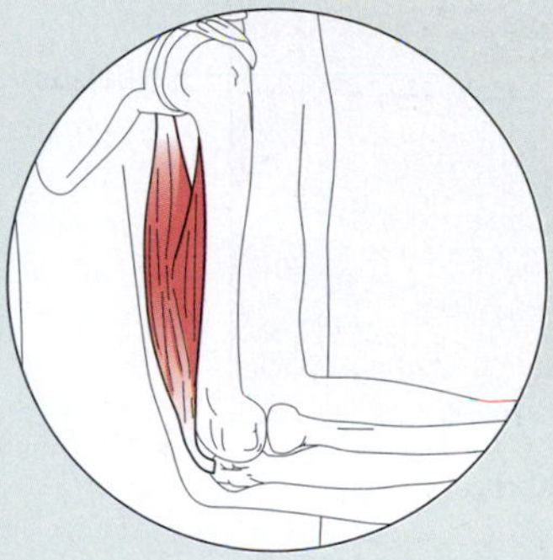

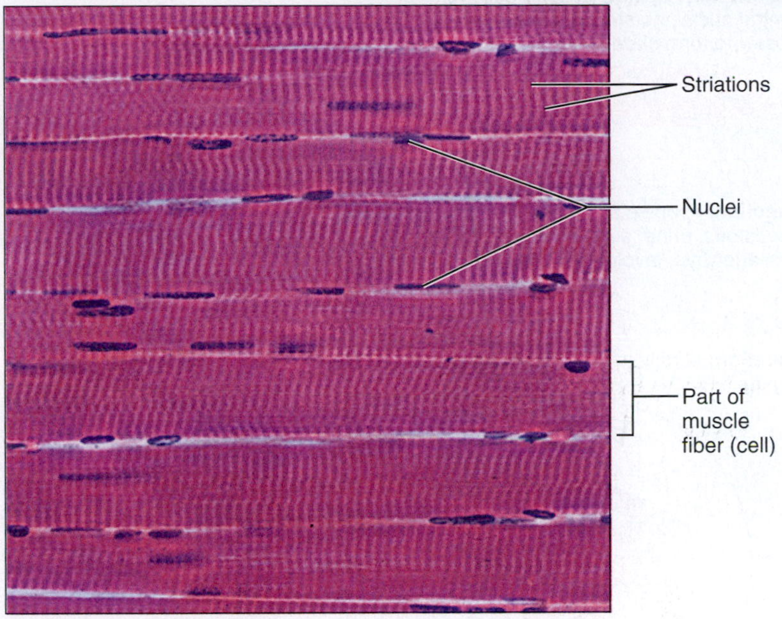
Striations
Nuclei
Part of muscle fiber (cell)

Photomicrograph: Skeletal muscle (approx. 460×). Notice the obvious banding pattern and the fact that these large cells are multinucleate.

(b) Cardiac muscle

Description: Branching, striated, generally uninucleate cells that interdigitate at specialized junctions (intercalated discs).

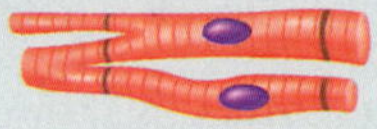

Function: As it contracts, it propels blood into the circulation; involuntary control.

Location: The walls of the heart.

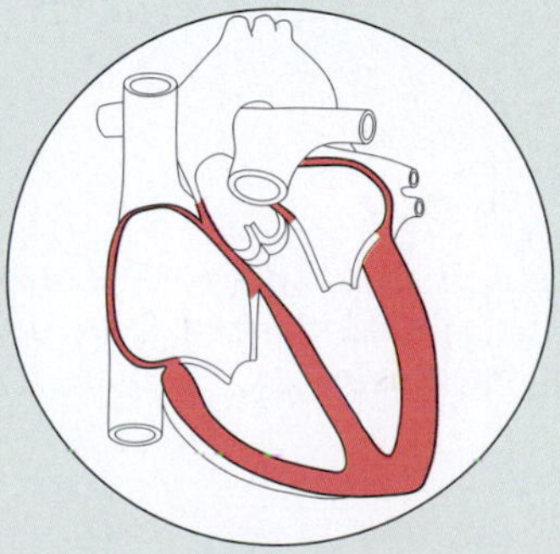

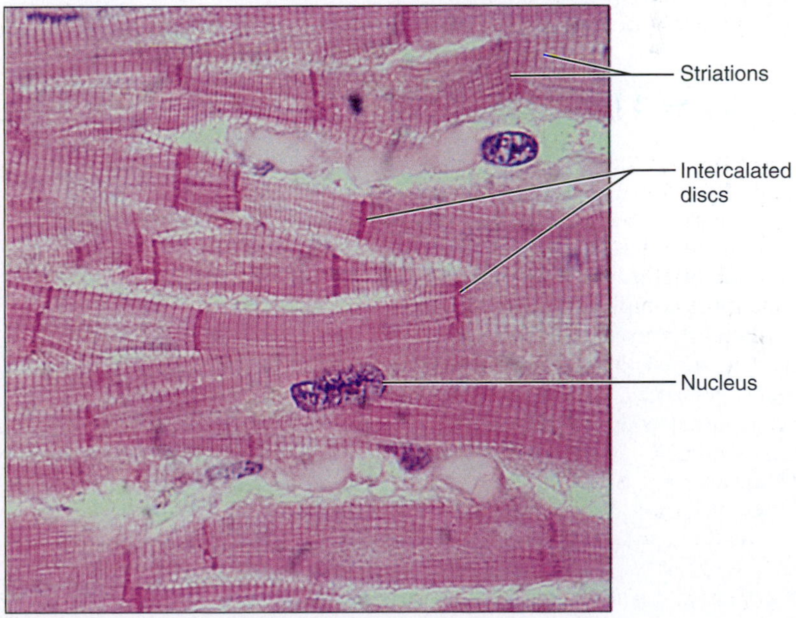
Striations
Intercalated discs
Nucleus

Photomicrograph: Cardiac muscle (500×); notice the striations, branching of cells, and the intercalated discs.

FIGURE 6.7 Muscle tissues. Skeletal muscle **(a)** and cardiac muscle **(b)**.

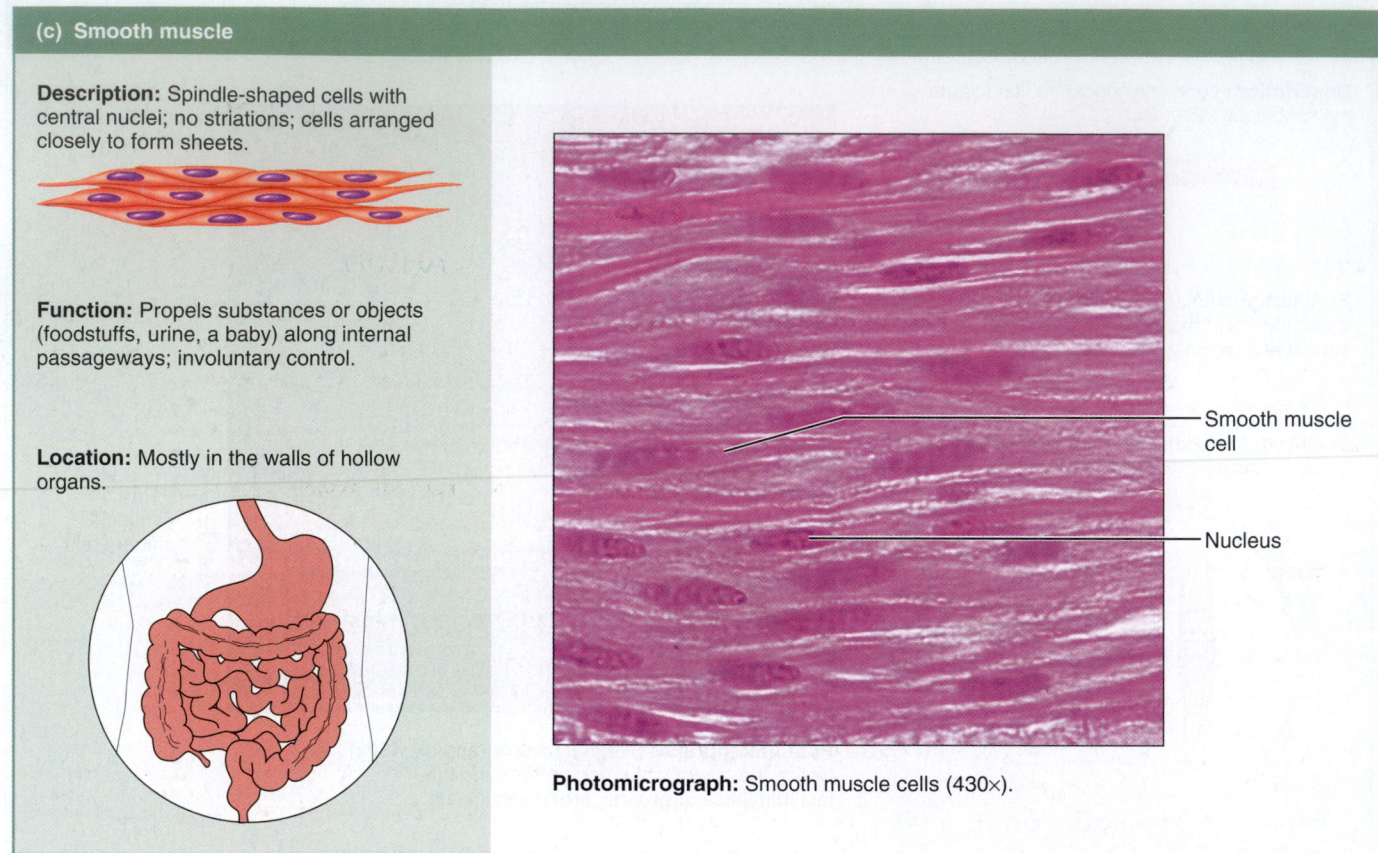

(c) Smooth muscle

Description: Spindle-shaped cells with central nuclei; no striations; cells arranged closely to form sheets.

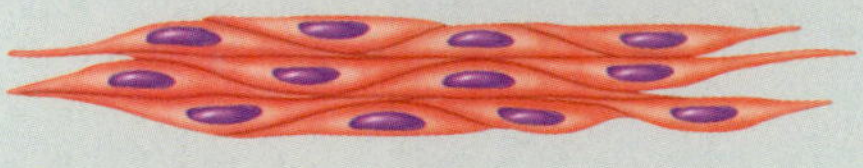

Function: Propels substances or objects (foodstuffs, urine, a baby) along internal passageways; involuntary control.

Location: Mostly in the walls of hollow organs.

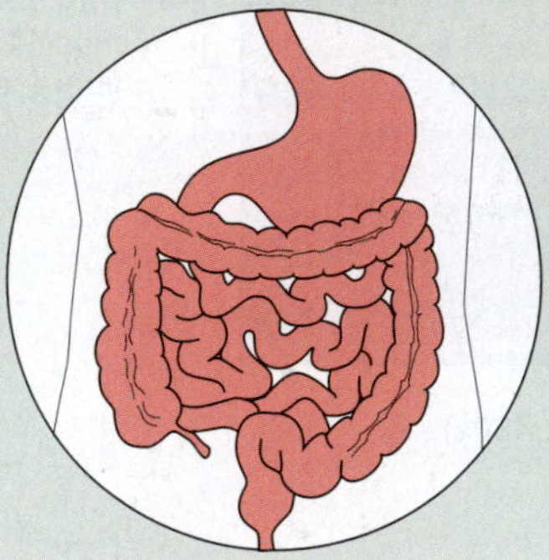

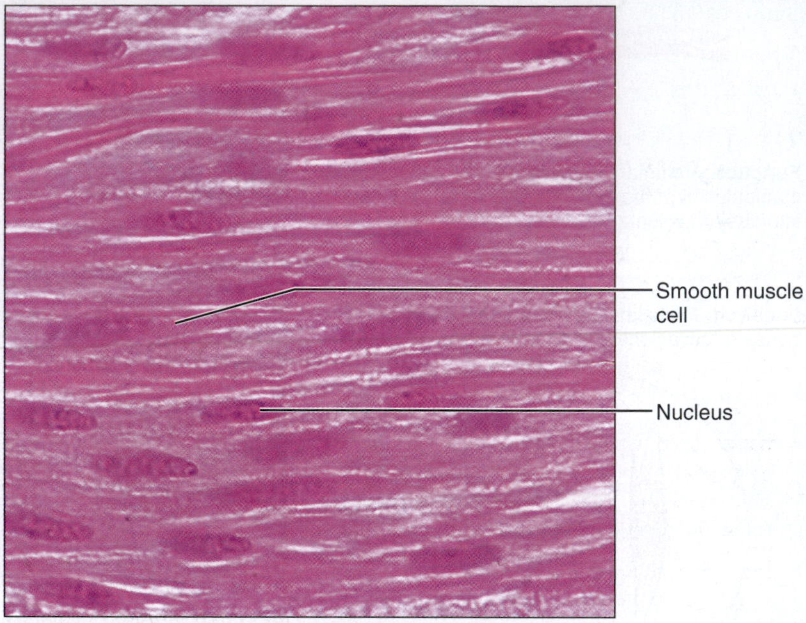

Smooth muscle cell
Nucleus

Photomicrograph: Smooth muscle cells (430×).

FIGURE 6.7 *(continued)* **Muscle tissues.** Smooth muscle **(c)**.

EXERCISE 7

The Integumentary System

MATERIALS

- ☐ Skin model (three-dimensional, if available)
- ☐ Compound microscope
- ☐ Prepared slide of human scalp
- ☐ Prepared slide of skin of palm or sole
- ☐ Sheet of #20 bond paper ruled to mark off cm^2 areas
- ☐ Scissors
- ☐ Betadine® swabs, or Lugol's iodine and cotton swabs
- ☐ Adhesive tape
- ☐ Disposable gloves
- ☐ Data collection sheet for plotting distribution of sweat glands
- ☐ Porelon fingerprint pad or portable inking foils
- ☐ Ink cleaner towelettes
- ☐ Index cards (4 in. × 6 in.)
- ☐ Magnifying glasses

OBJECTIVES

1. To list several important functions of the skin, or integumentary system.
2. To recognize and name during observation of an appropriate model, diagram, projected slide, or microscopic specimen the following skin structures: epidermis, dermis (papillary and reticular layers), hair follicles and hair, sebaceous glands, and sweat glands.
3. To name the layers of the epidermis and describe the characteristics of each.
4. To compare the properties of the epidermis to those of the dermis.
5. To describe the distribution and function of the skin derivatives—sebaceous glands, sweat glands, and hairs.
6. To differentiate between eccrine and apocrine sweat glands.
7. To enumerate the factors determining skin color.
8. To describe the function of melanin.
9. To identify the major regions of nails.

PRE-LAB QUIZ

1. All the following are functions of the skin except:
   a. excretion of body wastes
   b. insulation
   c. protection from mechanical damage
   d. site of vitamin A synthesis
2. The skin has two distinct regions. The superficial layer is the ______________________ and the underlying connective tissue, the ______________________.
3. These cells produce a brown-to-black pigment that colors the skin and protects DNA from the damaging effects of ultraviolet radiation. The cells are:
   a. epidermal dendritic cells
   b. keratinocytes
   c. melanocytes
   d. tactile cells
4. Circle True or False. Nails are hornlike derivatives of the epidermis.
5. The portion of a hair that you see that projects from the scalp surface is known as the __________.
   a. bulb
   b. matrix
   c. root
   d. shaft
6. Circle the correct term. The ducts of sebaceous / sweat glands usually empty into a hair follicle but may also open directly on the skin surface.
7. Circle the correct term. Eccrine / Apocrine glands are found primarily in the genital and axillary areas.

MasteringA&P™

Access practice quizzes and more in the Study Area at www.masteringaandp.com.

PAL

For access to anatomical models and more, check out Practice Anatomy Lab.

The **skin,** or **integument,** is considered an organ system because of its extent and complexity. It is much more than an external body covering; architecturally the skin is a marvel. It is tough yet pliable, a characteristic that enables it to withstand constant insult from outside agents.

The skin has many functions, most (but not all) concerned with protection. It insulates and cushions the underlying body tissues and protects the entire body from mechanical damage (bumps and cuts), chemical damage (acids, alkalis, and the like), thermal damage (heat), and bacterial invasion (by virtue of its acid mantle and continuous surface). The hardened uppermost layer of the skin (the cornified layer) prevents water loss from the body surface. The skin's abundant capillary network (under the control of the nervous system) plays an important role in regulating heat loss from the body surface.

The skin has other functions as well. For example, it acts as a mini-excretory system; urea, salts, and water are lost through the skin pores in sweat. The skin also has important metabolic duties. For example, like liver cells, it carries out some chemical conversions that activate or inactivate certain drugs and hormones, and it is the site of vitamin D synthesis for the body. Finally, the cutaneous sense organs are located in the dermis.

Basic Structure of the Skin

The skin has two distinct regions—the superficial *epidermis* composed of epithelium and an underlying connective tissue *dermis* (Figure 7.1). These layers are firmly "cemented" together along an undulating border. But friction, such as the rubbing of a poorly fitting shoe, may cause them to separate, resulting in a blister. Immediately deep to the dermis is the **hypodermis,** or **superficial fascia** (primarily adipose tissue), which is not considered part of the skin. The main skin areas and structures are described below.

ACTIVITY 1

Locating Structures on a Skin Model

As you read, locate the following structures in Figure 7.1 and on a skin model.

Epidermis

Structurally, the avascular epidermis is a keratinized stratified squamous epithelium consisting of four distinct cell types and four or five distinct layers.

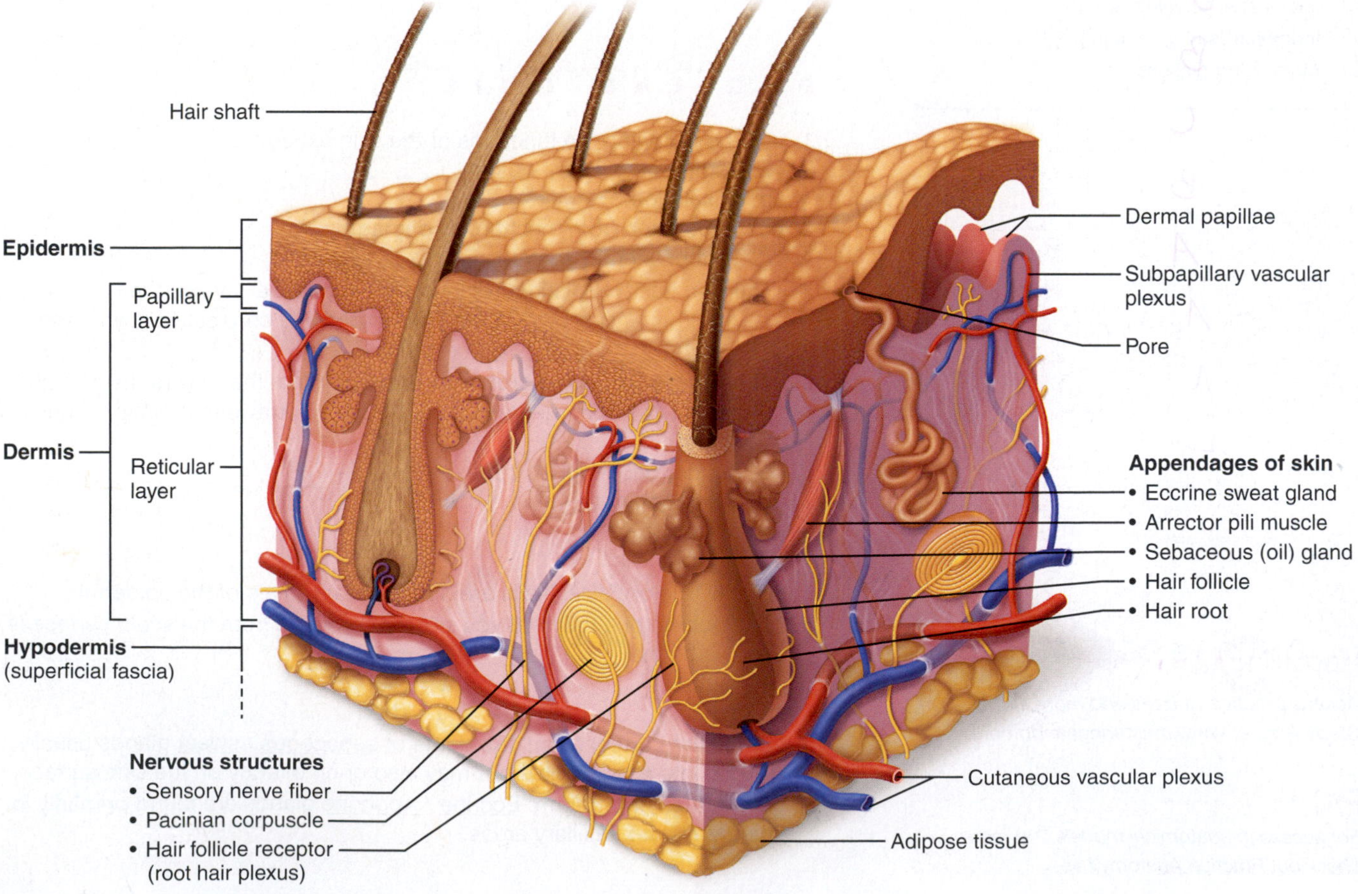
Hair shaft
Epidermis
Papillary layer
Dermis
Reticular layer
Hypodermis (superficial fascia)
Nervous structures
• Sensory nerve fiber
• Pacinian corpuscle
• Hair follicle receptor (root hair plexus)
Dermal papillae
Subpapillary vascular plexus
Pore
Appendages of skin
• Eccrine sweat gland
• Arrector pili muscle
• Sebaceous (oil) gland
• Hair follicle
• Hair root
Cutaneous vascular plexus
Adipose tissue

FIGURE 7.1 Skin structure. Three-dimensional view of the skin and the underlying hypodermis. The epidermis and dermis have been pulled apart at the right corner to reveal the dermal papillae.

Cells of the Epidermis

- **Keratinocytes** (literally, keratin cells): The most abundant epidermal cells, they function mainly to produce keratin fibrils. **Keratin** is a fibrous protein that gives the epidermis its durability and protective capabilities. Keratinocytes are tightly connected to each other by desmosomes.

Far less numerous are the following types of epidermal cells (Figure 7.2):

- **Melanocytes:** Spidery black cells that produce the brown-to-black pigment called **melanin.** The skin tans because melanin production increases when the skin is exposed to sunlight. The melanin provides a protective pigment umbrella over the nuclei of the cells in the deeper epidermal layers, thus shielding their genetic material (deoxyribonucleic acid, DNA) from the damaging effects of ultraviolet radiation. A concentration of melanin in one spot is called a *freckle.*

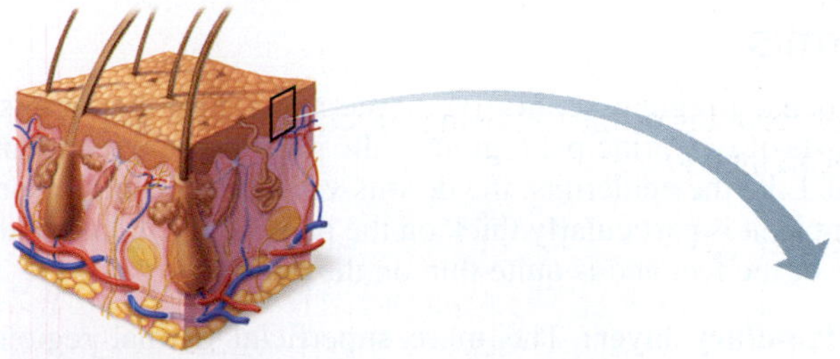

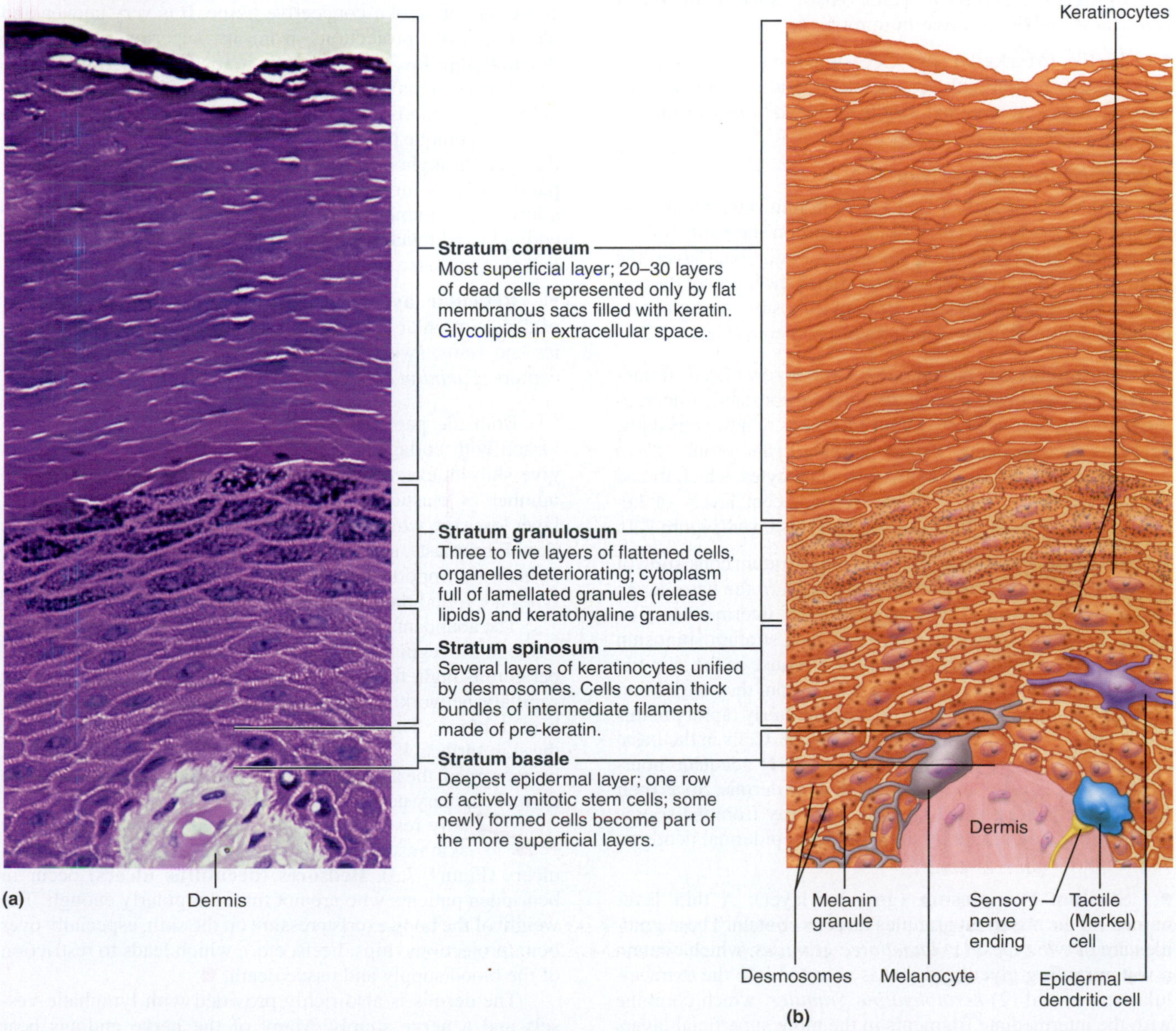
Keratinocytes
Stratum corneum
Most superficial layer; 20–30 layers of dead cells represented only by flat membranous sacs filled with keratin. Glycolipids in extracellular space.
Stratum granulosum
Three to five layers of flattened cells, organelles deteriorating; cytoplasm full of lamellated granules (release lipids) and keratohyaline granules.
Stratum spinosum
Several layers of keratinocytes unified by desmosomes. Cells contain thick bundles of intermediate filaments made of pre-keratin.
Stratum basale
Deepest epidermal layer; one row of actively mitotic stem cells; some newly formed cells become part of the more superficial layers.
(a)
Dermis
Dermis
Melanin granule
Sensory nerve ending
Tactile (Merkel) cell
Desmosomes
Melanocyte
Epidermal dendritic cell
(b)

FIGURE 7.2 The main structural features in epidermis of thin skin. (a) Photomicrograph depicting the four major epidermal layers (200 ×). **(b)** Diagram showing the layers and relative distribution of the different cell types. Keratinocytes (orange), melanocytes (gray), epidermal dendrite cells (purple), and tactile cells (blue). A sensory nerve ending (yellow) extending from the dermis is associated with a tactile cell forming a tactile disc (touch receptor). Notice that the keratinocytes are joined by numerous desmosomes. The stratum lucidum, present in thick skin, is not illustrated here.

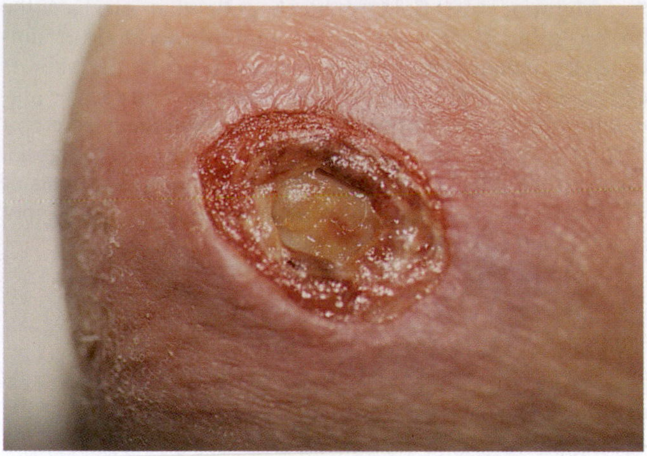

FIGURE 7.3 Photograph of a deep (stage III) decubitus ulcer.

- **Epidermal dendritic cells:** Also called *Langerhans cells,* these cells play a role in immunity.
- **Tactile (Merkel) cells:** Occasional spiky hemispheres that, in conjunction with sensory nerve endings, form sensitive touch receptors called *tactile* or *Merkel discs* located at the epidermal-dermal junction.

Layers of the Epidermis

The epidermis consists of four layers in thin skin, which covers most of the body. Thick skin, found on the palms of the hands and soles of the feet, contains an additional layer, the stratum lucidum. From deep to superficial, the layers of the epidermis are the stratum basale, stratum spinosum, stratum granulosum, stratum lucidum, and stratum corneum (Figure 7.2).

- **Stratum basale** (basal layer): A single row of cells immediately adjacent to the dermis. Its cells are constantly undergoing mitotic cell division to produce millions of new cells daily, hence its alternate name *stratum germinativum.* From 10% to 25% of the cells in this stratum are melanocytes, which thread their processes through this and the adjacent layers of keratinocytes. Note also the tactile cells of this layer (Figure 7.2).
- **Stratum spinosum** (spiny layer): A stratum consisting of several cell layers immediately superficial to the basal layer. Its cells contain thick weblike bundles of intermediate filaments made of a pre-keratin protein. The stratum spinosum cells appear spiky (hence their name) because as the skin tissue is prepared for histological examination, they shrink but their desmosomes hold tight. Cells divide fairly rapidly in this layer, but less so than in the stratum basale. Cells in the basal and spiny layers are the only ones to receive adequate nourishment via diffusion of nutrients from the dermis. So as their daughter cells are pushed upward and away from the source of nutrition, they gradually die. Note the epidermal dendritic cell (Figure 7.2).
- **Stratum granulosum** (granular layer): A thin layer named for the abundant granules its cells contain. These granules are of two types: (1) *lamellated granules,* which contain a waterproofing glycolipid that is secreted into the extracellular space; and (2) *keratohyaline granules,* which combine with the intermediate filaments in the more superficial layers to form the keratin fibrils. At the upper border of this layer, the cells are beginning to die.
- **Stratum lucidum** (clear layer): A very thin translucent band of flattened dead keratinocytes with indistinct boundaries. It is not present in regions of thin skin.
- **Stratum corneum** (horny layer): This outermost epidermal layer consists of some 20 to 30 cell layers, and accounts for the bulk of the epidermal thickness. Cells in this layer, like those in the stratum lucidum (where it exists), are dead, and their flattened scalelike remnants are fully keratinized. They are constantly rubbing off and being replaced by division of the deeper cells.

Dermis

The dense irregular connective tissue making up the dermis consists of two principal regions—the papillary and reticular areas. Like the epidermis, the dermis varies in thickness. For example, it is particularly thick on the palms of the hands and soles of the feet and is quite thin on the eyelids.

- **Papillary layer:** The more superficial dermal region composed of areolar connective tissue. It is very uneven and has fingerlike projections from its superior surface, the **dermal papillae,** which attach it to the epidermis above. These projections lie on top of the larger dermal ridges. In the palms of the hands and soles of the feet, they produce the *fingerprints,* unique patterns of *epidermal ridges* that remain unchanged throughout life. Abundant capillary networks in the papillary layer furnish nutrients for the epidermal layers and allow heat to radiate to the skin surface. The pain (free nerve endings) and touch receptors (*Meissner's corpuscles* in hairless skin) are also found here.
- **Reticular layer:** The deepest skin layer. It is composed of dense irregular connective tissue and contains many arteries and veins, sweat and sebaceous glands, and pressure receptors *(Pacinian corpuscles).*

Both the papillary and reticular layers are heavily invested with collagenic and elastic fibers. The elastic fibers give skin its exceptional elasticity in youth. In old age, the number of elastic fibers decreases, and the subcutaneous layer loses fat, which leads to wrinkling and inelasticity of the skin. Fibroblasts, adipose cells, various types of macrophages (which are important in the body's defense), and other cell types are found throughout the dermis.

The abundant dermal blood supply allows the skin to play a role in the regulation of body temperature. When body temperature is high, the arterioles serving the skin dilate, and the capillary network of the dermis becomes engorged with the heated blood. Thus body heat is allowed to radiate from the skin surface. If the environment is cool and body heat must be conserved, the arterioles constrict so that blood bypasses the dermal capillary networks temporarily.

Any restriction of the normal blood supply to the skin results in cell death and, if severe enough, skin ulcers (Figure 7.3). **Bedsores (decubitus ulcers)** occur in bedridden patients who are not turned regularly enough. The weight of the body exerts pressure on the skin, especially over bony projections (hips, heels, etc.), which leads to restriction of the blood supply and tissue death. ■

The dermis is also richly provided with lymphatic vessels and a nerve supply. Many of the nerve endings bear highly specialized receptor organs that, when stimulated by environmental changes, transmit messages to the central nervous system for interpretation. Some of these receptors—free nerve endings (pain receptors), a Pacinian corpuscle, and a hair follicle receptor (also called a *root hair plexus*)—are shown in Figure 7.1. (These receptors are discussed in depth in Exercise 23.)

Skin Color

Skin color is a result of the relative amount of melanin in skin, the relative amount of carotene in skin, and the degree of oxygenation of the blood. People who produce large amounts of melanin have brown-toned skin. In light-skinned people, who have less melanin pigment, the dermal blood supply flushes through the rather transparent cell layers above, giving the skin a rosy glow. *Carotene* is a yellow-orange pigment present primarily in the stratum corneum and in the adipose tissue of the hypodermis. Its presence is most noticeable when large amounts of carotene-rich foods (carrots, for instance) are eaten.

Skin color may be an important diagnostic tool. For example, flushed skin may indicate hypertension, fever, or embarrassment, whereas pale skin is typically seen in anemic individuals. When the blood is inadequately oxygenated, as during asphyxiation and serious lung disease, both the blood and the skin take on a bluish or cyanotic cast. **Jaundice,** in which the tissues become yellowed, is almost always diagnostic for liver disease, whereas a bronzing of the skin hints that a person's adrenal cortex is hypoactive **(Addison's disease).** ■

Accessory Organs of the Skin

The accessory organs of the skin—cutaneous glands, hair, and nails—are all derivatives of the epidermis, but they reside in the dermis. They originate from the stratum basale and grow downward into the deeper skin regions.

Nails

Nails are hornlike derivatives of the epidermis (Figure 7.4). Their named parts are:

- **Body:** The visible attached portion.
- **Free edge:** The portion of the nail that grows out away from the body.
- **Root:** The part that is embedded in the skin and adheres to an epithelial nail bed.
- **Nail folds:** Skin folds that overlap the borders of the nail.
- **Eponychium:** The thick proximal nail fold commonly called the cuticle.
- **Nail bed:** Extension of the stratum basale beneath the nail.
- **Nail matrix:** The thickened proximal part of the nail bed containing germinal cells responsible for nail growth. As the matrix produces the nail cells, they become heavily keratinized and die. Thus nails, like hairs, are mostly nonliving material.
- **Lunule:** The proximal region of the thickened nail matrix, which appears as a white crescent. Everywhere else,

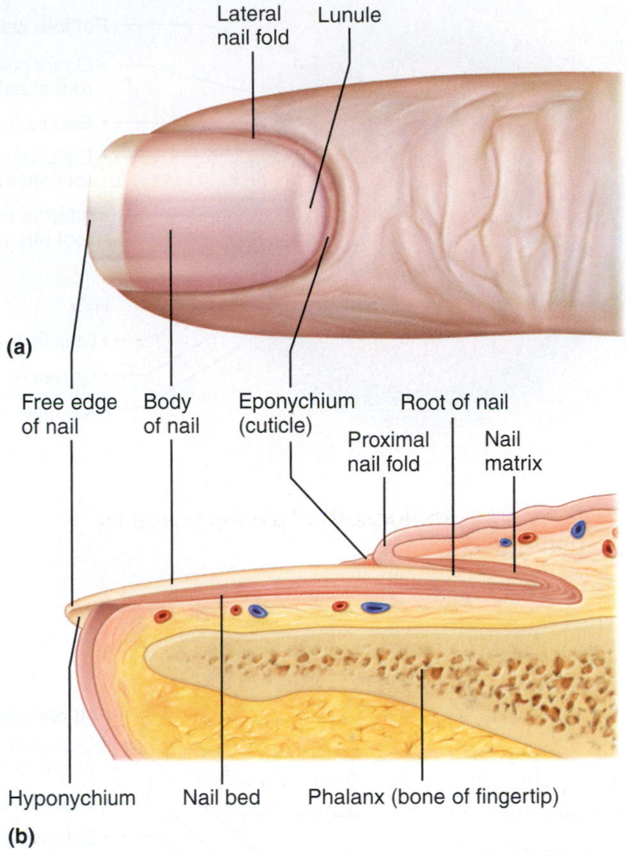
Lateral nail fold
Lunule
(a)
Free edge of nail
Body of nail
Eponychium (cuticle)
Root of nail
Proximal nail fold
Nail matrix
Hyponychium
Nail bed
Phalanx (bone of fingertip)
(b)

FIGURE 7.4 Structure of a nail. (a) Surface view of the distal part of a finger showing nail parts. The nail matrix that forms the nail lies beneath the lunule; the epidermis of the nail bed underlies the nail. **(b)** Sagittal section of the fingertip.

nails are transparent and nearly colorless, but they appear pink because of the blood supply in the underlying dermis. When someone is cyanotic because of a lack of oxygen in the blood, the nail beds take on a blue cast.

ACTIVITY 2

Identifying Nail Structures

Identify the nail structures shown in Figure 7.4 on yourself or your lab partner. ■

Hairs and Associated Structures

Hairs, enclosed in hair follicles, are found all over the entire body surface, except for thick-skinned areas (the palms of the hands and the soles of the feet), parts of the external genitalia, the nipples, and the lips.

- **Hair:** Structure consisting of a medulla, a central region surrounded first by the *cortex* and then by a protective *cuticle* (Figure 7.5). Abrasion of the cuticle results in split ends. Hair color is a manifestation of the amount and kind of melanin pigment within the hair cortex. The portion of the hair enclosed within the follicle is called the **root;** that portion projecting from the scalp surface is called the **shaft.** The **hair bulb** is a collection of well-nourished germinal epithelial

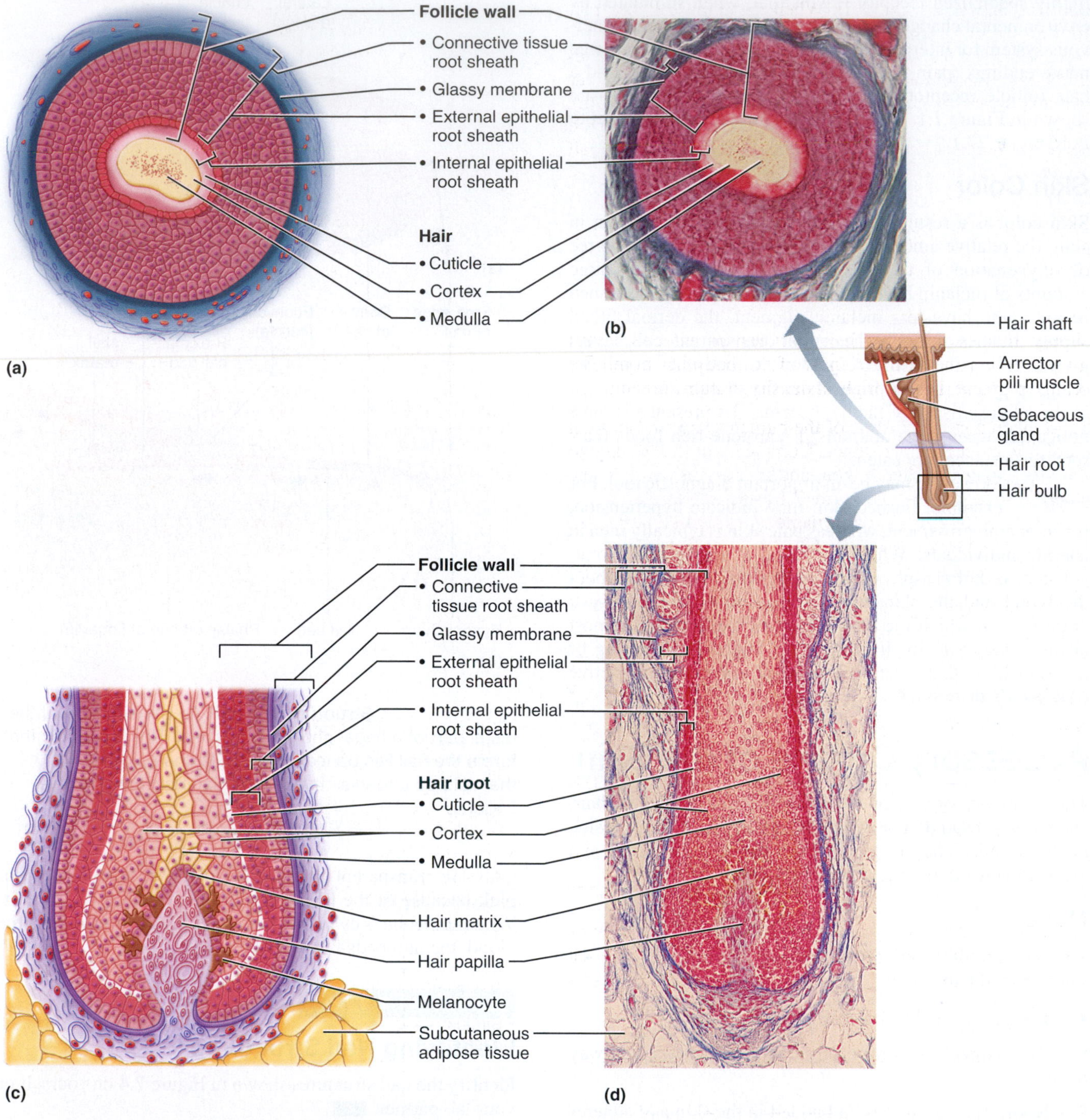
Follicle wall
• Connective tissue root sheath
• Glassy membrane
• External epithelial root sheath
• Internal epithelial root sheath
Hair
• Cuticle
• Cortex
• Medulla
(a)
(b)
Hair shaft
Arrector pili muscle
Sebaceous gland
Hair root
Hair bulb
Follicle wall
• Connective tissue root sheath
• Glassy membrane
• External epithelial root sheath
• Internal epithelial root sheath
Hair root
• Cuticle
• Cortex
• Medulla
Hair matrix
Hair papilla
Melanocyte
Subcutaneous adipose tissue
(c)
(d)

FIGURE 7.5 Structure of a hair and hair follicle. (a) Diagram of a cross section of a hair within its follicle. **(b)** Photomicrograph of a cross section of a hair and hair follicle (145×). **(c)** Diagram of a longitudinal view of the expanded hair bulb of the follicle, which encloses the matrix, the actively dividing epithelial cells that produce the hair. **(d)** Photomicrograph of longitudinal view of the hair bulb in the follicle (105×).

cells at the basal end of the follicle. As the daughter cells are pushed farther away from the growing region, they die and become keratinized; thus the bulk of the hair shaft, like the bulk of the epidermis, is dead material.

- **Follicle:** A structure formed from both epidermal and dermal cells (see Figure 7.5). Its inner epithelial root sheath, with two parts (internal and external), is enclosed by a thickened basement membrane, the glassy membrane, and a connective tissue root sheath, which is essentially dermal tissue. A small nipple of dermal tissue that protrudes into the hair bulb from the connective tissue sheath and provides nutrition to the growing hair is called the **papilla.**

- **Arrector pili muscle:** Small bands of smooth muscle cells connect each hair follicle to the papillary layer of the dermis (Figures 7.1 and 7.5). When these muscles contract (during cold or fright), the slanted hair follicle is pulled upright, dimpling the skin surface with goose bumps. This phenomenon is especially dramatic in a scared cat, whose fur actually stands on end to increase its apparent size. The activity of the arrector pili muscles also exerts pressure on the sebaceous glands surrounding the follicle, causing a small amount of sebum to be released.

ACTIVITY 3

Comparison of Hairy and Relatively Hair-Free Skin Microscopically

While thick skin has no hair follicles or sebaceous (oil) glands, thin skin typical of most of the body has both. The scalp, of course, has the highest density of hair follicles.

1. Obtain a prepared slide of the human scalp, and study it carefully under the microscope. Compare your tissue slide to the view shown in Figure 7.6a, and identify as many of the structures diagrammed in Figure 7.1 as possible.

How is this stratified squamous epithelium different from that observed in Exercise 6?

____

____

How do these differences relate to the functions of these two similar epithelia?

____

____

____

2. Obtain a prepared slide of hairless skin of the palm or sole (Figure 7.6b). Compare the slide to Figure 7.6a. In what ways does the thick skin of the palm or sole differ from the thin skin of the scalp?

____

____

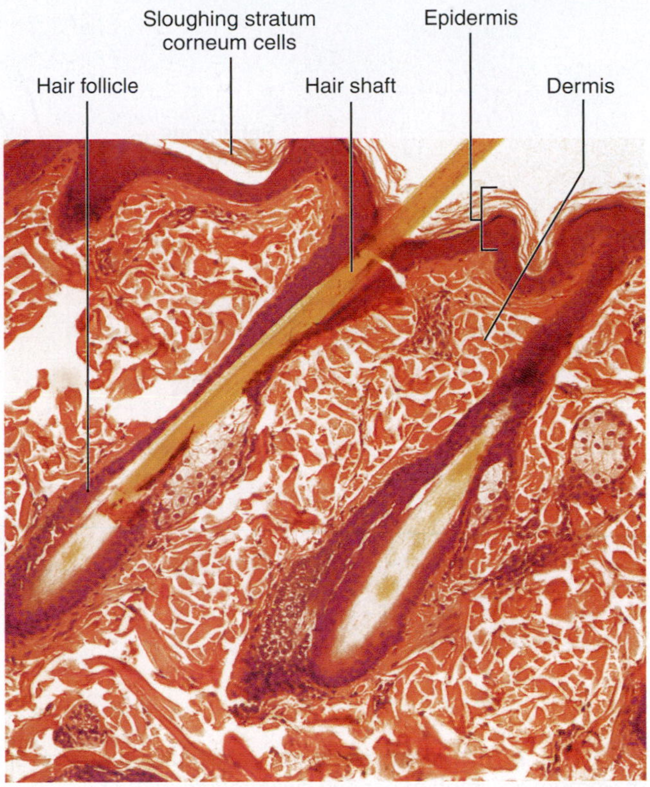

Sloughing stratum corneum cells
Epidermis
Hair follicle
Hair shaft
Dermis

(a)

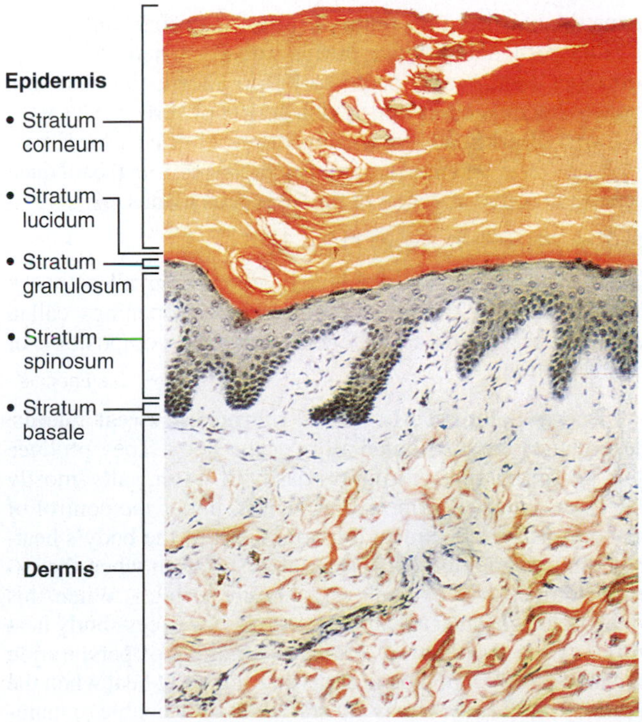

Epidermis
• Stratum corneum
• Stratum lucidum
• Stratum granulosum
• Stratum spinosum
• Stratum basale
Dermis

(b)

FIGURE 7.6 Photomicrographs of skin. (a) Thin skin with hairs (115×). **(b)** Thick hairless skin (120×).

Cutaneous Glands

The cutaneous glands fall primarily into two categories: the sebaceous glands and the sweat glands (Figure 7.1 and Figure 7.7).

Sebaceous (Oil) Glands

The sebaceous glands are found nearly all over the skin, except for the palms of the hands and the soles of the feet. Their ducts usually empty into a hair follicle, but some open directly on the skin surface.

Sebum is the product of sebaceous glands. It is a mixture of oily substances and fragmented cells that acts as a lubricant to keep the skin soft and moist (a natural skin cream) and

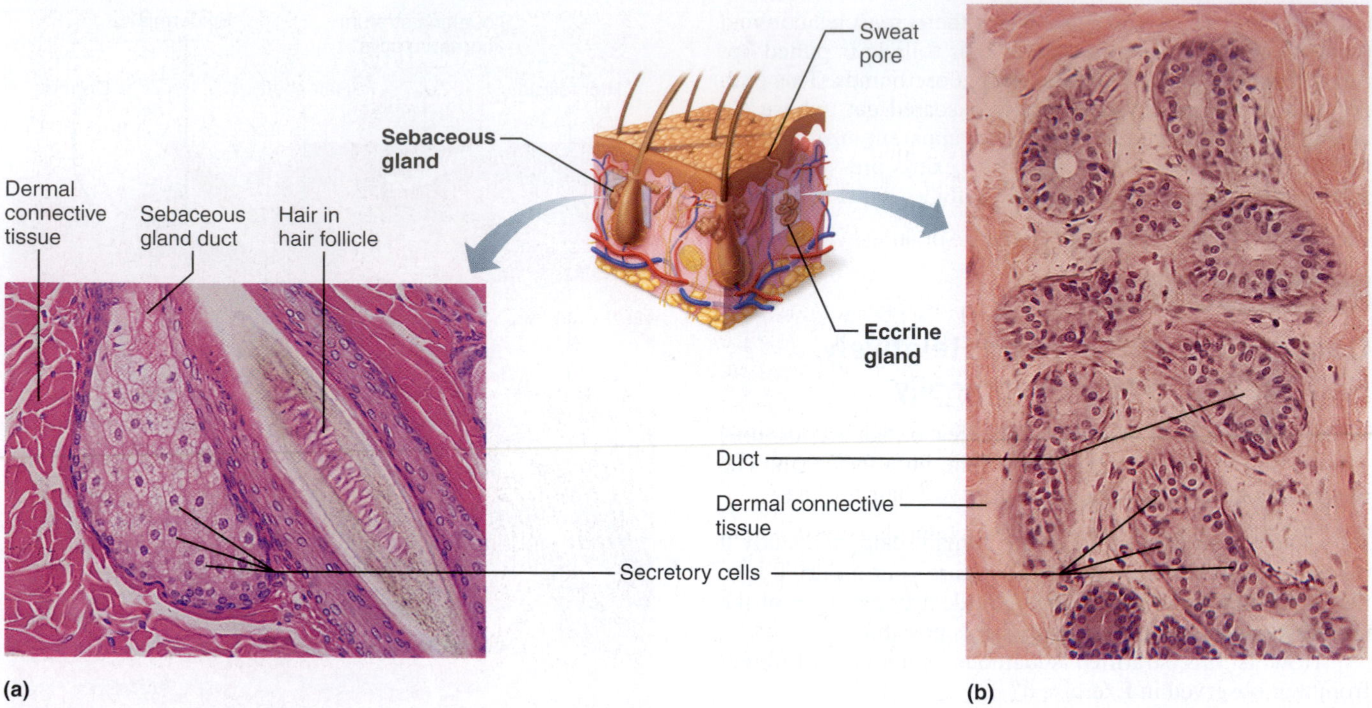
Dermal connective tissue
Sebaceous gland duct
Hair in hair follicle
Sebaceous gland
Sweat pore
Eccrine gland
Duct
Dermal connective tissue
Secretory cells
(a)
(b)

FIGURE 7.7 Cutaneous glands. (a) Photomicrograph of a sebaceous gland (220×). **(b)** Photomicrograph of eccrine sweat gland (220×).

keeps the hair from becoming brittle. The sebaceous glands become particularly active during puberty when more male hormones (androgens) begin to be produced; thus the skin tends to become oilier during this period of life.

Blackheads are accumulations of dried sebum, bacteria, and melanin from epithelial cells in the oil duct. **Acne** is an active infection of the sebaceous glands. ■

Sweat (Sudoriferous) Glands

These exocrine glands are widely distributed all over the skin. Outlets for the glands are epithelial openings called *pores*. Sweat glands are categorized by the composition of their secretions.

- **Eccrine glands:** Also called **merocrine sweat glands,** these glands are distributed all over the body. They produce clear perspiration consisting primarily of water, salts (mostly NaCl), and urea. Eccrine sweat glands, under the control of the nervous system, are an important part of the body's heat-regulating apparatus. They secrete perspiration when the external temperature or body temperature is high. When this water-based substance evaporates, it carries excess body heat with it. Thus evaporation of greater amounts of perspiration provides an efficient means of dissipating body heat when the capillary cooling system is not sufficient or is unable to maintain body temperature homeostasis.

- **Apocrine glands:** Found predominantly in the axillary and genital areas, these glands secrete a milky protein- and fat-rich substance (also containing water, salts, and urea) that is an excellent nutrient medium for the microorganisms typically found on the skin. Because these glands enlarge and recede with the phases of a woman's menstrual cycle, the apocrine glands may be analogous to the pheromone-producing scent glands of other animals.

ACTIVITY 4

Differentiating Sebaceous and Sweat Glands Microscopically

Using the slide *thin skin with hairs*, and Figure 7.7 as a guide, identify sebaceous and eccrine sweat glands. What characteristics relating to location or gland structure allow you to differentiate these glands?

____

____

____ ■

ACTIVITY 5

Plotting the Distribution of Sweat Glands

1. Form a hypothesis about the relative distribution of sweat glands on the palm and forearm. Justify your hypothesis.

2. For this simple experiment you will need two squares of bond paper (each 1 cm × 1 cm), adhesive tape, and a Betadine (iodine) swab *or* Lugol's iodine and a cotton-tipped swab. (The bond paper has been preruled in cm^2—put on disposable gloves and cut along the lines to obtain the required squares.)

3. Paint an area of the medial aspect of your left palm (avoid the crease lines) and a region of your left forearm with the iodine solution, and allow it to dry thoroughly. The painted area in each case should be slightly larger than the paper squares to be used.

4. Have your lab partner *securely* tape a square of bond paper over each iodine-painted area, and leave them in place for 20 minutes. (If it is very warm in the laboratory while this test is being conducted, good results may be obtained within 10 to 15 minutes.)

5. After 20 minutes, remove the paper squares, and count the number of blue-black dots on each square. The presence of a blue-black dot on the paper indicates an active sweat gland. (The iodine in the pore is dissolved in the sweat and reacts chemically with the starch in the bond paper to produce the blue-black color.) Thus "sweat maps" have been produced for the two skin areas.

6. Which skin area tested has the greater density of sweat glands?

___

___

___

7. Tape your results (bond paper squares) to a data collection sheet labeled "palm" and "forearm" at the front of the lab. Be sure to put your paper squares in the correct columns on the data sheet.

8. Once all the data has been collected, review the class results.

9. Prepare a lab report for the experiment. (See Getting Started, page xv.)

Dermography: Fingerprinting

As noted previously, each of us has a unique genetically determined set of fingerprints. Because of the usefulness of fingerprinting for identifying and apprehending criminals, most people associate this craft solely with criminal investigations. However, civil fingerprints are invaluable in quickly identifying amnesia victims, missing persons, and unknown deceased such as those killed in major disasters.

The friction ridges responsible for fingerprints appear in several patterns, which are clearest when the fingertips are inked and then pressed against white paper. Impressions are also made when perspiration or any foreign material such as blood, dirt, or grease adheres to the ridges and the fingers are then pressed against a smooth, nonabsorbent surface. The three most common patterns are *arches, loops,* and *whorls* (Figure 7.8). The *pattern area* in loops and whorls is the only area of the print used in identification, and it is delineated by the *type lines*—specifically the two innermost ridges that start parallel, diverge, and/or surround or tend to surround the pattern area.

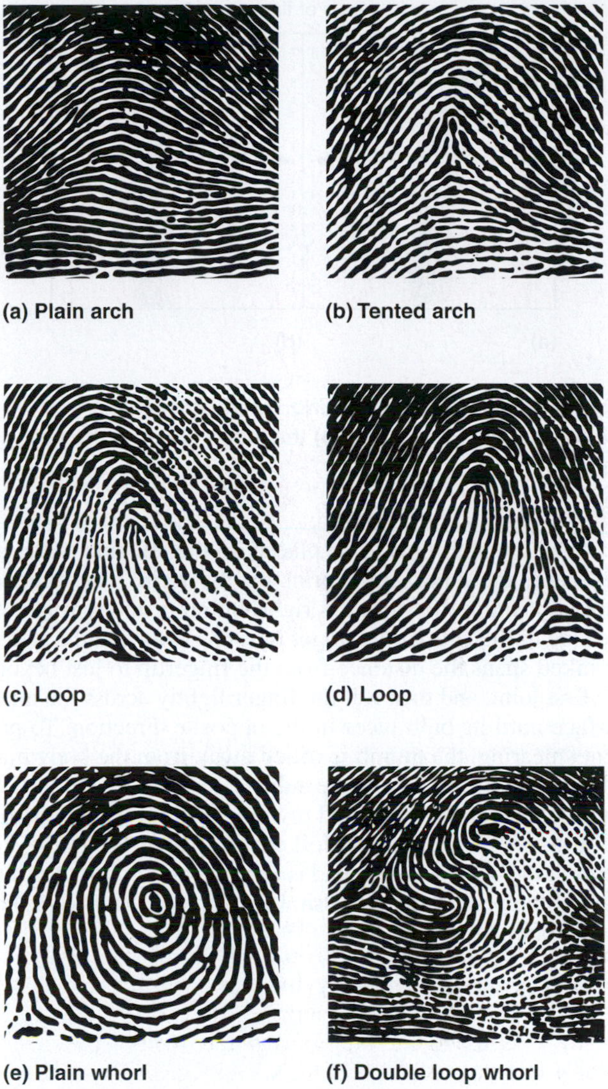
(a) Plain arch
(b) Tented arch
(c) Loop
(d) Loop
(e) Plain whorl
(f) Double loop whorl

FIGURE 7.8 Main types of fingerprint patterns. **(a–b)** Arches. **(c–d)** Loops. **(e–f)** Whorls.

ACTIVITY 6

Taking and Identifying Inked Fingerprints

For this activity, you will be working as a group with your lab partners. Though the equipment for professional fingerprinting is fairly basic, consisting of a glass or metal inking plate, printer's ink (a heavy black paste), ink roller, and standard 8 in. × 8 in. cards, you will be using supplies that are even easier to handle. Each student will prepare two index cards, each bearing his or her thumbprint and index fingerprint of the right hand.

1. Obtain the following supplies and bring them to your bench: two 4 in. × 6 in. index cards per student, Porelon fingerprint pad or portable inking foils, ink cleaner towelettes, and a magnifying glass.

2. The subject should wash and dry the hands. Open the ink pad or peel back the covering over the ink foil, and position it close to the edge of the laboratory bench. The subject should position himself or herself at arm's length from the bench edge and inking object.

3. A second student, called the *operator,* stands to the left of the subject and with two hands holds and directs movement of the subject's fingertip. During this process, the subject should look away, try to relax, and refrain from trying to help the operator.

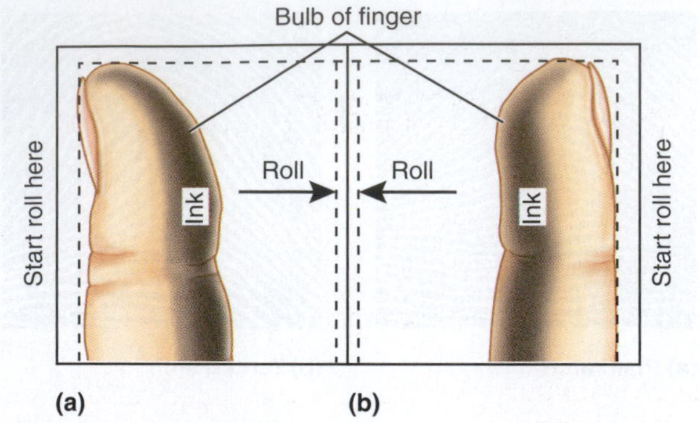
Bulb of finger
Start roll here
Roll
Ink
Roll
Ink
Start roll here
(a)
(b)

FIGURE 7.9 Fingerprinting. Method of inking and printing **(a)** the thumb and **(b)** the index finger of the right hand.

4. The thumbprint is to be placed on the left side of the index card, the index fingerprint on the right. The operator should position the subject's right thumb or index finger on the side of the bulb of the finger in such a way that the area to be inked spans the distance from the fingertip to just beyond the first joint, and then roll the finger lightly across the inked surface until its bulb faces in the opposite direction. To prevent smearing, the thumb is rolled away from the body midline (from left to right as the subject sees it; see Figure 7.9) and the index finger is rolled toward the body midline (from right to left). The same ink foil can be reused for all the students at the bench; the ink pad is good for thousands of prints. Repeat the procedure (still using the subject's right hand) on the second index card.

5. If the prints are too light, too dark, or smeary, repeat the procedure.

6. While subsequent members are making clear prints of their thumb and index finger, those who have completed that activity should clean their inked fingers with a towelette and attempt to classify their own prints as arches, loops, or whorls. Use the magnifying glass as necessary to see ridge details.

7. When all members at a bench have completed the above steps, they are to write their names on the backs of their index cards, then combine their cards and shuffle them before transferring them to the bench opposite for classification of pattern and identification of prints made by the same individuals.

How difficult was it to classify the prints into one of the three categories given?

____

Why do you think this is so?

____

Was it easy or difficult to identify the prints made by the same individual?

____

Why do you think this was so?

____

NAME ______________________________

LAB TIME/DATE ______________________

REVIEW SHEET

EXERCISE 7

The Integumentary System

Basic Structure of the Skin

1. Complete the following statements by writing the appropriate word or phrase on the correspondingly numbered blank:

The two basic tissues of which the skin is composed are dense irregular connective tissue, which makes up the dermis, and __1__, which forms the epidermis. The tough water-repellent protein found in the epidermal cells is called __2__. The pigments melanin and __3__ contribute to skin color. A localized concentration of melanin is referred to as a __4__.

1. ______________________
2. ______________________
3. ______________________
4. ______________________

2. Four protective functions of the skin are

a. ______________________ c. ______________________

b. ______________________ d. ______________________

3. Using the key choices, choose all responses that apply to the following descriptions.

Key:

a. stratum basale	d. stratum lucidum	g. reticular layer
b. stratum corneum	e. stratum spinosum	h. epidermis as a whole
c. stratum granulosum	f. papillary layer	i. dermis as a whole

______________ 1. translucent cells in thick skin containing keratin fibrils

______________ 2. dead cells

______________ 3. dermal layer responsible for fingerprints

______________ 4. vascular region

______________ 5. major skin area that produces derivatives (nails and hair)

______________ 6. epidermal region exhibiting the most rapid cell division

______________ 7. scalelike dead cells, full of keratin, that constantly slough off

______________ 8. mitotic cells filled with intermediate filaments

______________ 9. has abundant elastic and collagenic fibers

______________ 10. location of melanocytes and tactile (Merkel) cells

______________ 11. area where weblike pre-keratin filaments first appear

______________ 12. region of areolar connective tissue

4. Label the skin structures and areas indicated in the accompanying diagram of thin skin. Then, complete the statements that follow.

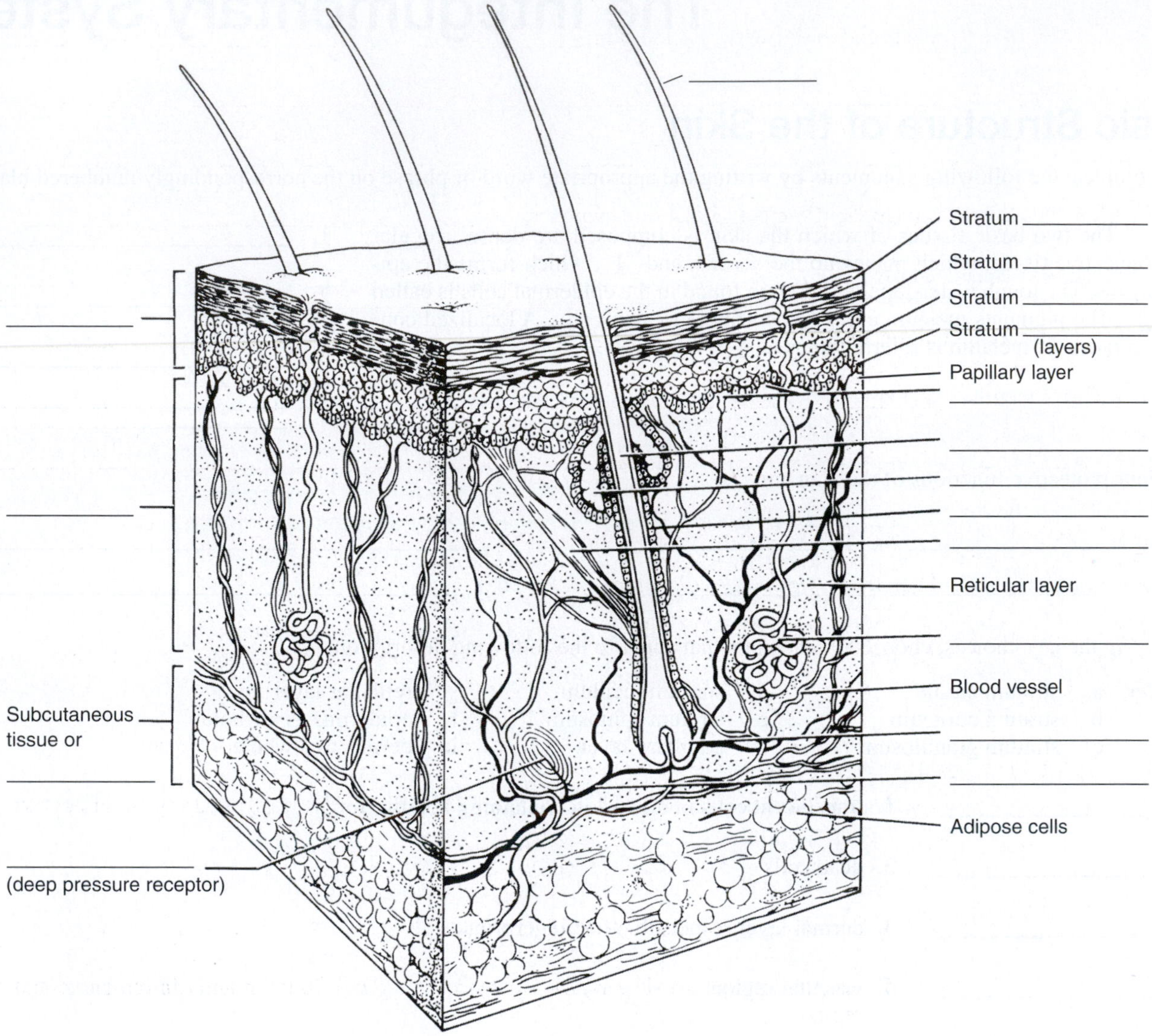
Stratum
Stratum
Stratum
Stratum
(layers)
Papillary layer
Reticular layer
Blood vessel
Adipose cells
Subcutaneous tissue or
(deep pressure receptor)

a. ______________ granules extruded from the keratinocytes prevent water loss by diffusion through the epidermis.

b. Fibers in the dermis are produced by ______________.

c. Glands that respond to rising androgen levels are the ______________ glands.

d. Phagocytic cells that occupy the epidermis are called ______________.

e. A unique touch receptor formed from a stratum basale cell and a nerve fiber is a ______________.

f. What layer is present in thick skin but not in thin skin? ______________

g. What cell-to-cell structures hold the cells of the stratum spinosum tightly together? ______________

5. What substance is manufactured in the skin that plays a role in calcium absorption elsewhere in the body?

____

6. List the sensory receptors found in the dermis of the skin. ______________________________

____

____

7. A nurse tells a doctor that a patient is cyanotic. Define *cyanosis*. ______________________________

What does its presence imply? ______________________________

8. What is a bedsore (decubitus ulcer)? ______________________________

Why does it occur? ______________________________

____

Accessory Organs of the Skin

9. Match the key choices with the appropriate descriptions.

Key:

a. arrector pili
b. cutaneous receptors
c. hair
d. hair follicle
e. nail
f. sebaceous glands
g. sweat gland—apocrine
h. sweat gland—eccrine

__________ 1. produces an accumulation of oily material that is known as a blackhead

__________ 2. tiny muscles, attached to hair follicles, that pull the hair upright during fright or cold

__________ 3. perspiration glands with a role in temperature control

__________ 4. sheath formed of both epithelial and connective tissues

__________ 5. less numerous type of perspiration-producing gland; found mainly in the pubic and axillary regions

__________ 6. found everywhere on the body except the palms of hands and soles of feet

__________ 7. primarily dead/keratinized cells

__________ 8. specialized nerve endings that respond to temperature, touch, etc.

__________ 9. secretes a lubricant for hair and skin

__________ 10. "sports" a lunule and a cuticle

10. Describe two integumentary system mechanisms that help in regulating body temperature. ______________________

____

____

11. Several structures or skin regions are listed below. Identify each by matching its letter with the appropriate area on the figure.

a. adipose cells

b. dermis

c. epidermis

d. hair follicle

e. hair shaft

f. sloughing stratum corneum cells

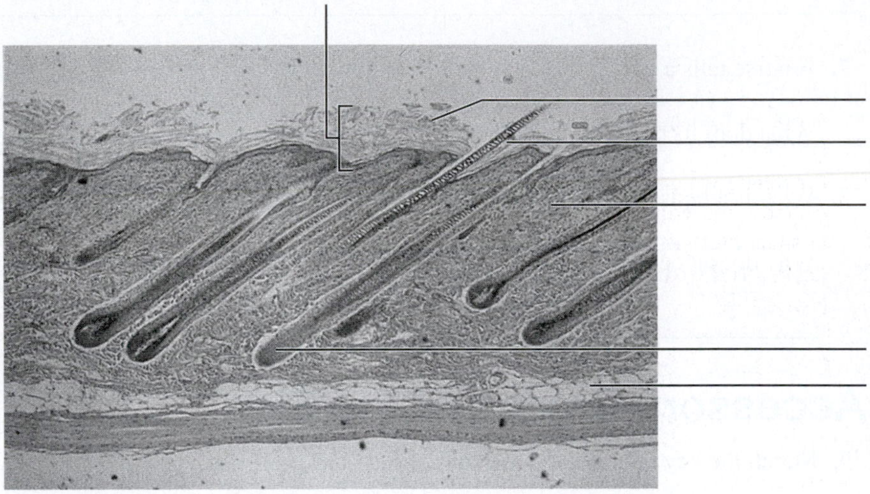

Plotting the Distribution of Sweat Glands

12. With what substance in the bond paper does the iodine painted on the skin react? ______________________

13. Based on class data, which skin area—the forearm or palm of hand—has more sweat glands? ______________________

Was this an expected result? __________ Explain. ____

____

Which other body areas would, if tested, prove to have a high density of sweat glands? ______________________

____

14. What organ system controls the activity of the eccrine sweat glands? ______________________

Dermography: Fingerprinting

15. Why can fingerprints be used to identify individuals?

____

16. Name the three common fingerprint patterns.

______________________, ______________________, and ______________________

EXERCISE 8

Classification of Covering and Lining Membranes

MATERIALS

- ☐ Compound microscope
- ☐ Prepared slides of trachea (x.s.), esophagus (x.s.), and small intestine (x.s.)
- ☐ Prepared slide of serous membrane (for example, mesentery artery and vein [x.s.] or small intestine [x.s.])
- ☐ Longitudinally cut fresh beef joint (if available)

OBJECTIVES

1. To compare the structure and function of the major membrane types.
2. To list the general functions of each membrane type and indicate its location in the body.
3. To recognize by microscopic examination cutaneous, mucous, and serous membranes.

PRE-LAB QUIZ

1. Body membranes fall into two major categories: _____________ and _____________.
   a. epithelial and cutaneous
   b. epithelial and synovial
   c. connective and cutaneous
   d. serous and synovial
2. The _____________ membrane is the skin, a dry membrane with a keratinizing epithelium.
   a. cutaneous
   b. mucous
   c. serous
3. The mucous membranes are made up of a layer of epithelial cells resting on a layer of loose connective tissue called the _____________ _____________.
   a. goblet cells
   b. lamina propria
   c. visceral pericardium
4. Circle True or False. Mucous membranes line body cavities that open to the surface.
5. Circle the correct terms. The serous membranes generally occur in twos and are actually continuous. The parietal / visceral layer lines a body cavity, and the parietal / visceral layer covers the outside of the organs in that cavity.
6. Unlike mucous and serous membranes, _____________ membranes contain no epithelial cells and are composed entirely of connective tissue.
   a. cutaneous c. synovial
   b. parietal d. visceral

The body membranes, which cover surfaces, line body cavities, and form protective (and often lubricating) sheets around organs, fall into two major categories. These may be designated as *epithelial membranes* and *synovial membranes.*

Epithelial Membranes

Most of the covering and lining epithelia take part in forming one of the three common varieties of epithelial membranes: cutaneous, mucous, or serous. The term "epithelial membrane" is used in various ways. Here we will define an **epithelial membrane** as a simple organ consisting of an epithelial sheet bound to an underlying layer of connective tissue proper.

MasteringA&P™

Access practice quizzes and more in the Study Area at www.masteringaandp.com.

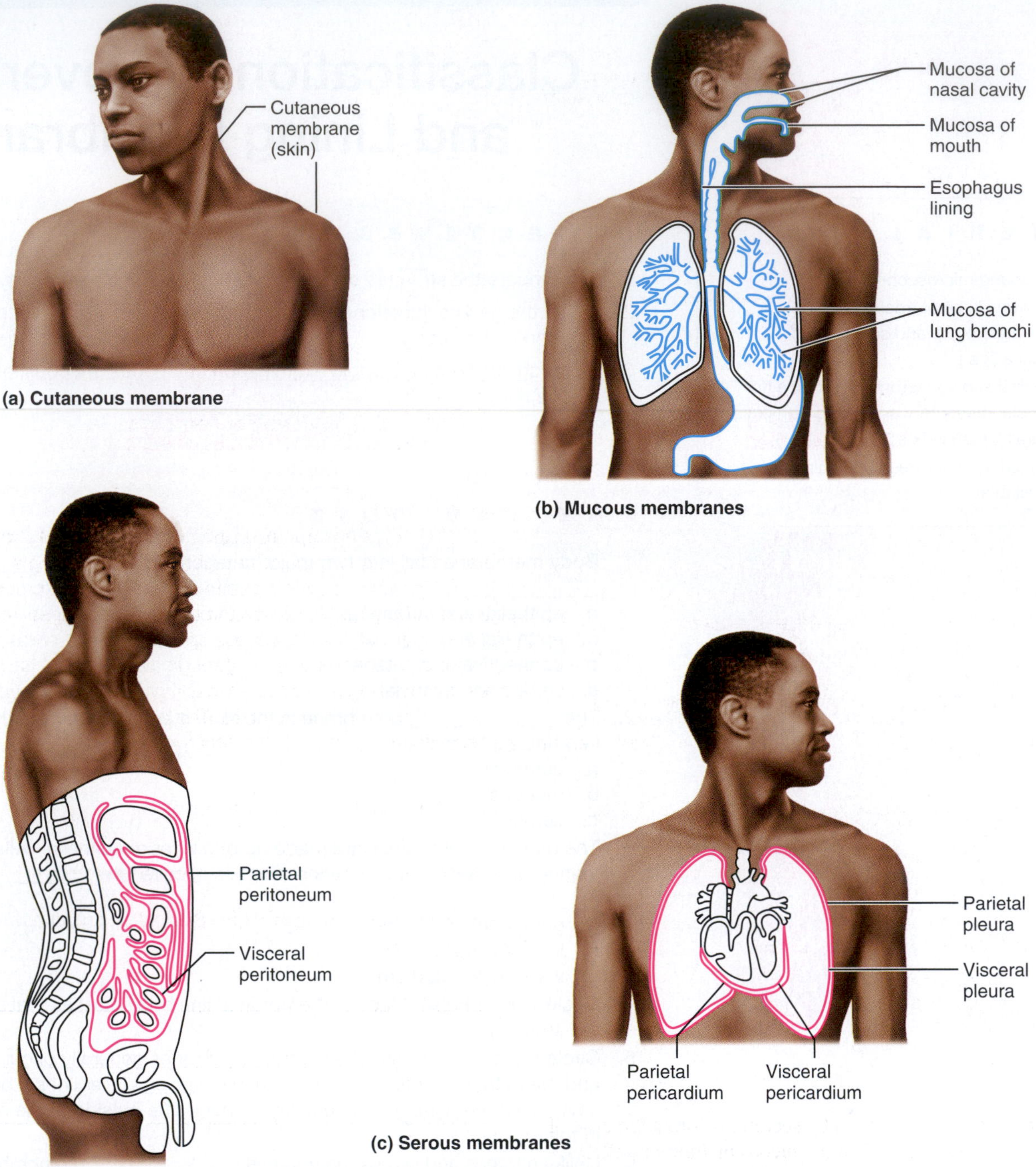
Cutaneous
membrane
(skin)
(a) Cutaneous membrane
Mucosa of
nasal cavity
Mucosa of
mouth
Esophagus
lining
Mucosa of
lung bronchi
(b) Mucous membranes
Parietal
peritoneum
Visceral
peritoneum
Parietal
pleura
Visceral
pleura
Parietal
pericardium
Visceral
pericardium
(c) Serous membranes

FIGURE 8.1 Classes of epithelial membranes. Epithelial membranes are composite membranes with epithelial and connective tissue elements. **(a)** The cutaneous membrane, or skin, covers and protects the body surface. **(b)** Mucous membranes line body cavities (hollow organs) that open to the exterior. **(c)** Serous membranes line the closed ventral cavity of the body. Three examples, the peritoneums, pericardia, and pleurae, are illustrated here.

The **cutaneous membrane** (Figure 8.1a) is the skin, a dry membrane with a keratinizing epithelium (the epidermis). Since the skin is discussed in some detail in Exercise 7, we focus on the mucous and serous membranes here.

Mucous Membranes

The **mucous membranes (mucosae)*** are composed of epithelial cells resting on a layer of loose connective tissue called the **lamina propria.** They line all body cavities that open to the body exterior—the respiratory, digestive (Figure 8.1b), and urogenital tracts. In most cases mucosae are "wet" membranes,

*Notice the spelling difference between *mucous*, an adjective describing the membrane type, and *mucus*, a noun indicating the product of glands.

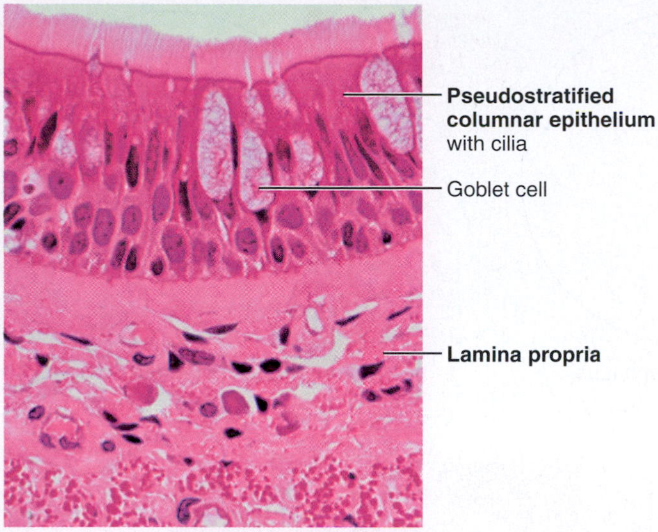
Pseudostratified columnar epithelium with cilia
Goblet cell
Lamina propria

(a)

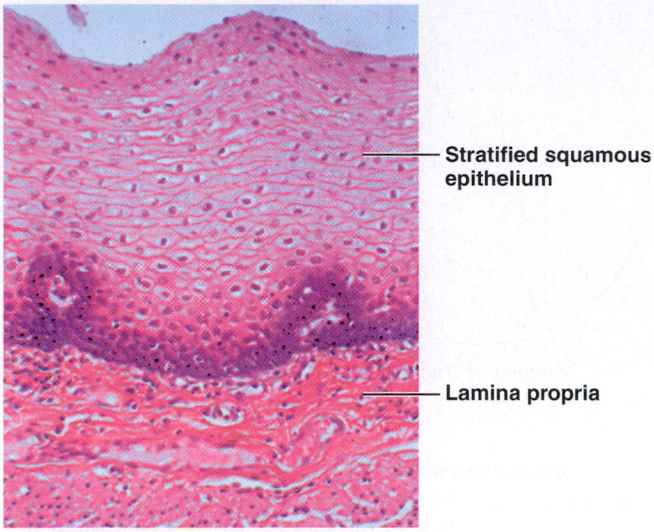
Stratified squamous epithelium
Lamina propria

(b)

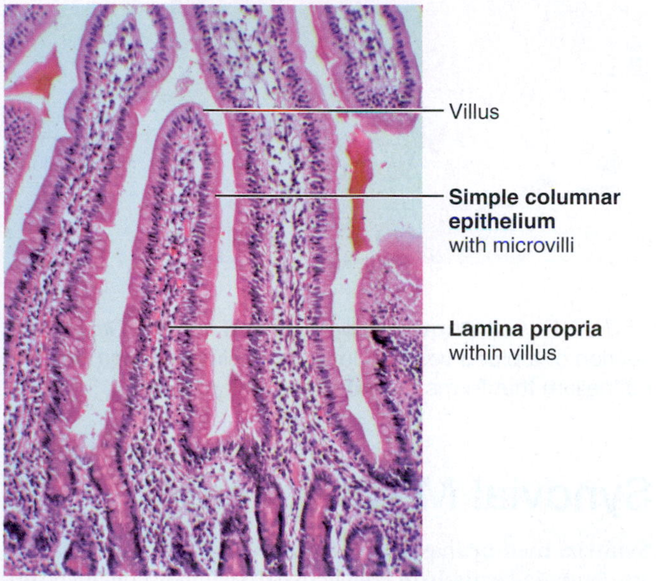
Villus
Simple columnar epithelium with microvilli
Lamina propria within villus

(c)

FIGURE 8.2 Photomicrographs of mucosae. (a) Cross section through the mucosa of the trachea showing a pseudostratified columnar epithelium with lamina propria beneath (315×). **(b)** Section through the esophageal mucosa showing a stratified squamous epithelium with lamina propria beneath (140×). **(c)** Section through several villi of the small intestine. Intestinal mucosa has a simple columnar epithelium with microvilli and goblet cells. The lamina propria lies between the epithelia in the center of the villus (30X).

which are continuously bathed by secretions (or, in the case of urinary mucosa, urine). Although mucous membranes often secrete mucus, this is not a requirement. The mucous membranes of both the digestive and respiratory tracts secrete mucus; that of the urinary tract does not.

ACTIVITY 1

Examining the Microscopic Structure of Mucous Membranes

Using Figure 8.2 as a guide, examine slides made from cross sections of the trachea, esophagus, and small intestine. Draw the mucosa of each in the appropriate circle, and fully identify each epithelial type. Remember to look for the epithelial cells at the free surface. Also search the epithelial sheets for **goblet cells**—columnar epithelial cells with a large mucus-containing vacuole (goblet) in their apical cytoplasm.

Which mucosae contain goblet cells?

___

Compare and contrast the roles of these three mucous membranes.

___

___

___

___

Serous Membranes

The **serous membranes (serosae)** are also epithelial membranes (Figure 8.1c). They are composed of a layer of simple squamous epithelium on a scant amount of areolar connective tissue. The serous membranes generally occur in twos and are actually continuous. The *parietal layer* lines a body cavity, and the *visceral layer* covers the outside of the organs in that cavity. In contrast to the mucous membranes, which line open body cavities, the serous membranes line body cavities that are closed to the exterior (with the exception of the female peritoneal cavity and the dorsal body cavity). These double-layered serosae secrete a thin fluid (serous fluid) that lubricates the organs and body walls and thus reduces friction as the organs slide across one another and against the body cavity walls. Inflammation of these membranes typically results in less serous fluid being produced and excruciating pain.

Mucosa of trachea

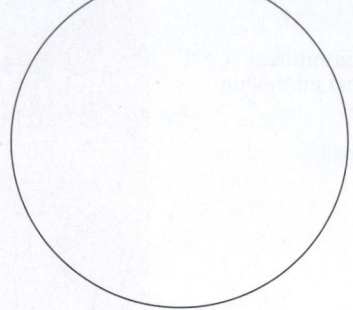

Mucosa of esophagus

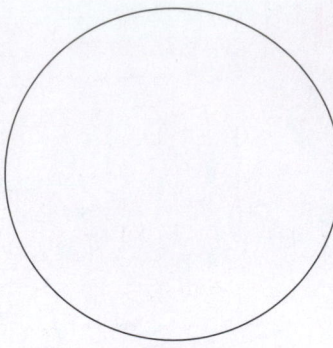

Mucosa of small intestine

A serous membrane also lines the interior of blood vessels (endothelium) and the heart (endocardium). In capillaries, the entire wall is composed of serosa that serves as a selectively permeable membrane between the blood and the tissue fluid of the body.

ACTIVITY 2

Examining the Microscopic Structure of a Serous Membrane

Using Figure 8.3 as a guide, examine a prepared slide of a serous membrane and diagram it in the circle provided here.

Serosa of blood vessel

What are the specific names of the serous membranes covering the heart and lining the cavity in which it resides (respectively)?

______________________ and ______________________

Of the abdominal viscera and visceral cavity (respectively)?

______________________ and ______________________

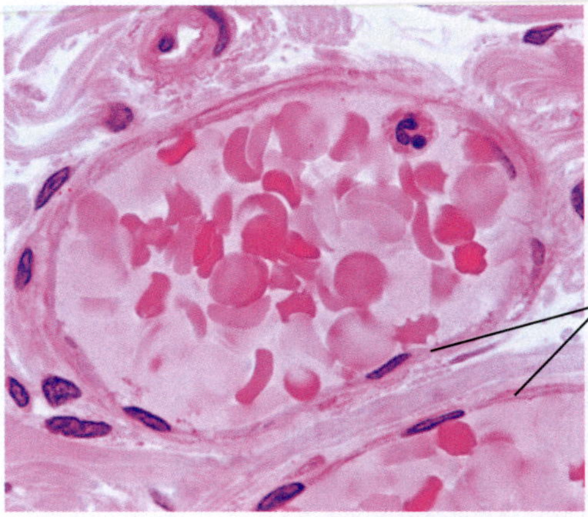
Simple squamous epithelium (endothelium)

FIGURE 8.3 Photomicrograph of a serosa. Cross section of a blood vessel showing the simple squamous epithelium that forms its endothelial lining (200×).

Synovial Membranes

Synovial membranes, unlike mucous and serous membranes, are composed entirely of connective tissue; they contain no epithelial cells. These membranes line the cavities surrounding the joints, providing a smooth surface and secreting a lubricating fluid. They also line smaller sacs of connective tissue (bursae and tendon sheaths), which cushion structures moving against each other, as during muscle activity. See Figure 13.2C, which illustrates a synovial membrane in the joint cavity.

ACTIVITY 3

Examining the Gross Structure of a Synovial Membrane

If a freshly sawed beef joint is available, visually examine the interior surface of the joint capsule to observe the smooth texture of the synovial membrane. A more detailed study of synovial membranes is done in Exercise 13.

NAME ______________________________

LAB TIME/DATE ______________________

REVIEW SHEET

EXERCISE **8**

Classification of Covering and Lining Membranes

1. Complete the following chart.

Membrane	Tissue types: membrane composition (epithelial/connective)	Common locations	General functions
cutaneous			
mucous			
serous			
synovial			

2. Respond to the following statements by choosing an answer from the key.

Key: a. cutaneous b. mucous c. serous d. synovial

_______________ 1. membrane type in joints, bursae, and tendon sheaths

_______________ 2. epithelium of this membrane is always simple squamous epithelium

_______, _______ 3. membrane types *not* found in the ventral body cavity

_______________ 4. the only membrane type in which goblet cells are found

_______________ 5. the dry membrane with keratinizing epithelium

_____, _____, _____ 6. “wet” membranes

_______________ 7. adapted for absorption and secretion

_______________ 8. has parietal and visceral layers

3. Using terms from the key above the figure, identify the different types of body membranes (cutaneous, mucous, and serous) by writing the terms at the end of the leader lines.

Key: a.	cutaneous membrane (skin)	g.	parietal pericardium
b.	esophageal mucosa	h.	parietal pleura
c.	gastric mucosa	i.	tracheal mucosa
d.	mucosa of lung bronchi	j.	visceral pericardium
e.	nasal mucosa	k.	visceral pleura
f.	oral mucosa		

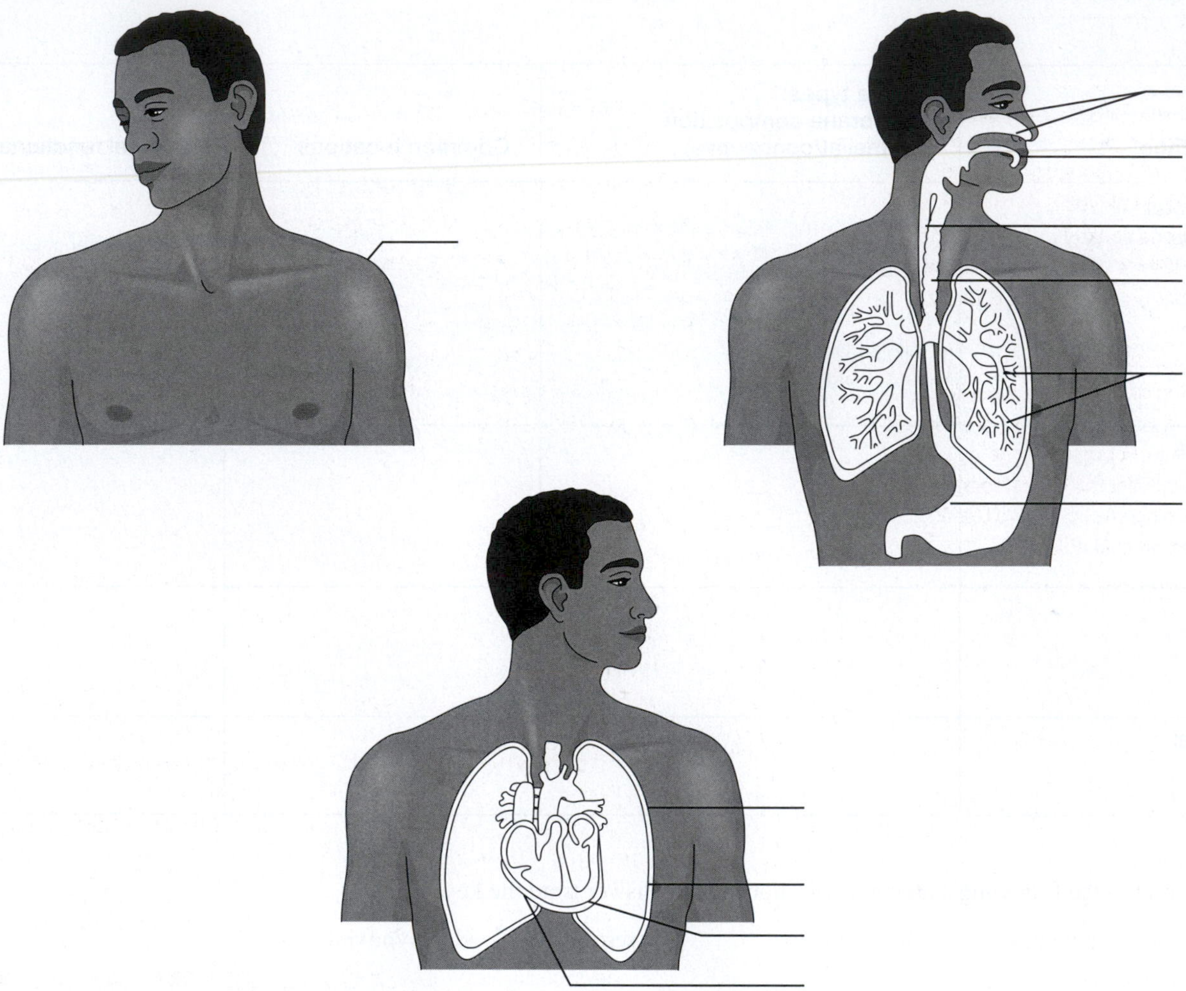

4. The suffix *-itis* means "inflammation of." What do peritonitis, pleurisy, and pericarditis (pathological conditions) have in common?

____

5. Why are these conditions accompanied by a great deal of pain? ______________________________

____

EXERCISE 9

Overview of the Skeleton: Classification and Structure of Bones and Cartilages

MATERIALS

- ☐ Long bone sawed longitudinally (beef bone from a slaughterhouse, if possible, or prepared laboratory specimen)
- ☐ Disposable gloves
- ☐ Long bone soaked in 10% hydrochloric acid (HCl) (or vinegar) until flexible
- ☐ Long bone baked at 250°F for more than 2 hours
- ☐ Compound microscope
- ☐ Prepared slide of ground bone (x.s.)
- ☐ Three-dimensional model of microscopic structure of compact bone
- ☐ Prepared slide of a developing long bone undergoing endochondral ossification
- ☐ Articulated skeleton

MasteringA&P™

Access practice quizzes and more in the Study Area at www.masteringaandp.com.

PAL

For access to anatomical models and more, check out Practice Anatomy Lab.

OBJECTIVES

1. To list five functions of the skeletal system.
2. To locate and identify the three major types of skeletal cartilages.
3. To identify the four main groups of bones.
4. To identify surface bone markings and functions.
5. To identify the major anatomical areas on a longitudinally cut long bone (or diagram of one).
6. To identify the major regions and structures of an osteon in a histologic specimen of compact bone (or diagram of one).
7. To explain the role of the inorganic salts and organic matrix in providing flexibility and hardness to bone.

PRE-LAB QUIZ

1. All the following are functions of the skeleton except:
   a. attachment for muscles
   b. production of melanin
   c. site of red blood cell formation
   d. storage of lipids
2. Circle the correct term. The axial / appendicular skeleton consists of bones that surround the body's center of gravity.
3. This type of cartilage has great tensile strength and is found in the knee joint and intervertebral discs.
   a. elastic
   b. fibrocartilage
   c. hyaline
4. Circle the correct term. Compact / Spongy bone looks smooth and homogeneous.
5. _________ bones are generally thin, with two waferlike layers of compact bone sandwiching a layer of spongy bone.
   a. Flat
   b. Irregular
   c. Long
   d. Short
6. The femur is an example of a(n) _________ type of bone.
   a. flat
   b. irregular
   c. long
   d. short
7. Circle the correct term. The shaft of a long bone is known as the epiphysis / diaphysis.
8. A central canal and all the concentric lamellae surrounding it are referred to as:
   a. an osteon
   b. canaliculi
   c. lacunae
9. Circle True or False. Embryonic skeletons consist primarily of elastic cartilage, which is gradually replaced by bone during development and growth.

The **skeleton,** the body's framework, is constructed of two of the most supportive tissues found in the human body—cartilage and bone. In embryos, the skeleton is predominantly composed of hyaline cartilage, but in the adult, most of the cartilage is replaced by more rigid bone. Cartilage persists only in such isolated areas as the external ear, bridge of the nose, larynx, trachea, joints, and parts of the rib cage (see Figure 9.2).

Besides supporting and protecting the body as an internal framework, the skeleton provides a system of levers with which the skeletal muscles work to move the body. In addition, the bones store lipids and many minerals (most importantly calcium). Finally, the red marrow cavities of bones provide a site for hematopoiesis (blood cell formation).

The skeleton is made up of bones that are connected at *joints,* or *articulations.* The skeleton is subdivided into two

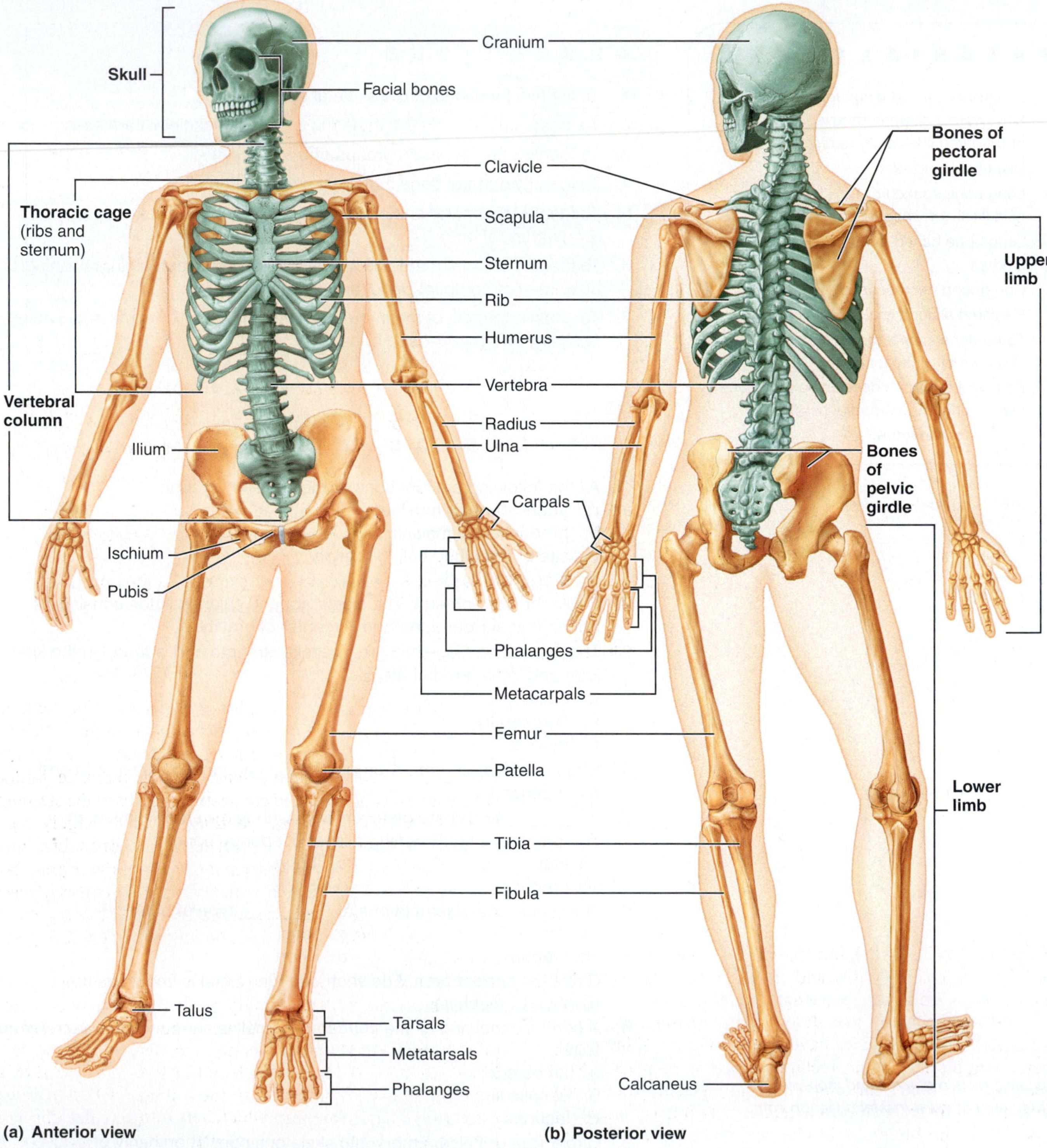
Skull
Cranium
Facial bones
Thoracic cage (ribs and sternum)
Clavicle
Scapula
Sternum
Rib
Humerus
Vertebra
Vertebral column
Radius
Ulna
Ilium
Carpals
Ischium
Pubis
Phalanges
Metacarpals
Femur
Patella
Tibia
Fibula
Talus
Tarsals
Metatarsals
Phalanges
Calcaneus
Bones of pectoral girdle
Upper limb
Bones of pelvic girdle
Lower limb
(a) Anterior view
(b) Posterior view

FIGURE 9.1 The human skeleton. The bones of the axial skeleton are colored green to distinguish them from the bones of the appendicular skeleton.

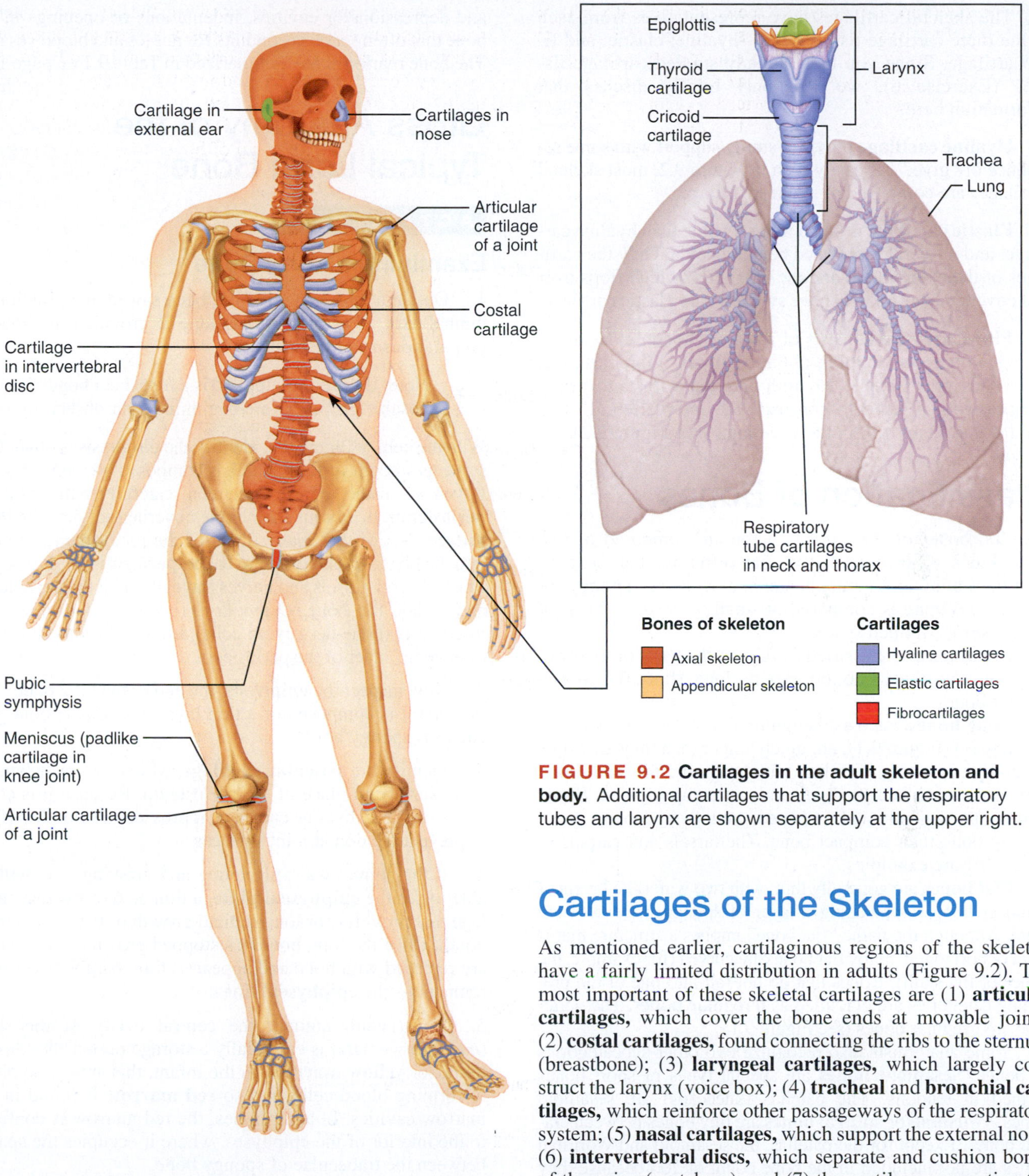

Cartilage in external ear
Cartilages in nose
Articular cartilage of a joint
Costal cartilage
Cartilage in intervertebral disc
Pubic symphysis
Meniscus (padlike cartilage in knee joint)
Articular cartilage of a joint
Epiglottis
Thyroid cartilage
Cricoid cartilage
Larynx
Trachea
Lung
Respiratory tube cartilages in neck and thorax
Bones of skeleton
Axial skeleton
Appendicular skeleton
Cartilages
Hyaline cartilages
Elastic cartilages
Fibrocartilages

FIGURE 9.2 Cartilages in the adult skeleton and body. Additional cartilages that support the respiratory tubes and larynx are shown separately at the upper right.

divisions: the **axial skeleton** (those bones that lie around the body's center of gravity) and the **appendicular skeleton** (bones of the limbs, or appendages) (Figure 9.1).

Before beginning your study of the skeleton, imagine for a moment that your bones have turned to putty. What if you were running when this metamorphosis took place? Now imagine your bones forming a continuous metal framework within your body, somewhat like a network of plumbing pipes. What problems could you envision with this arrangement? These images should help you understand how well the skeletal system provides support and protection, as well as facilitating movement.

Cartilages of the Skeleton

As mentioned earlier, cartilaginous regions of the skeleton have a fairly limited distribution in adults (Figure 9.2). The most important of these skeletal cartilages are (1) **articular cartilages,** which cover the bone ends at movable joints; (2) **costal cartilages,** found connecting the ribs to the sternum (breastbone); (3) **laryngeal cartilages,** which largely construct the larynx (voice box); (4) **tracheal** and **bronchial cartilages,** which reinforce other passageways of the respiratory system; (5) **nasal cartilages,** which support the external nose; (6) **intervertebral discs,** which separate and cushion bones of the spine (vertebrae); and (7) the cartilage supporting the external ear.

The skeletal cartilages consist of some variety of *cartilage tissue,* which typically consists primarily of water and is fairly resilient. Cartilage tissues are also distinguished by the fact that they contain no nerves and very few blood vessels. Like bones, each cartilage is surrounded by a covering of dense connective tissue, called a *perichondrium* (rather than a periosteum), which acts to resist distortion of the cartilage when it is subjected to pressure, and plays a role in cartilage growth and repair.

The skeletal cartilages have representatives from each of the three cartilage tissue types—hyaline, elastic, and fibrocartilage. Since you have already studied cartilage tissues (Exercise 6), we will only briefly discuss that information here.

- **Hyaline cartilage** provides sturdy support with some resilience or "give." As easily seen in Figure 9.2, most skeletal cartilages are composed of hyaline cartilage.
- **Elastic cartilage** is much more flexible than hyaline cartilage, and it tolerates repeated bending better. Only the cartilages of the external ear and the epiglottis (which flops over and covers the larynx when we swallow) are elastic cartilage.
- **Fibrocartilage** consists of rows of chondrocytes alternating with rows of thick collagen fibers. Fibrocartilage, which has great tensile strength and can withstand heavy compression, is used to construct the intervertebral discs and the cartilages within the knee joint (see Figure 9.2).

Classification of Bones

The 206 bones of the adult skeleton are composed of two basic kinds of osseous tissue that differ in their texture. **Compact bone** looks smooth and homogeneous; **spongy** (or *cancellous*) **bone** is composed of small *trabeculae* (bars) of bone and lots of open space.

Bones may be classified further on the basis of their relative gross anatomy into four groups: long, short, flat, and irregular bones.

Long bones, such as the femur and phalanges (bones of the fingers) (Figure 9.1), are much longer than they are wide, generally consisting of a shaft with heads at either end. Long bones are composed predominantly of compact bone. **Short bones** are typically cube shaped, and they contain more spongy bone than compact bone. The tarsals and carpals in Figure 9.1 are examples.

Flat bones are generally thin, with two waferlike layers of compact bone sandwiching a layer of spongy bone between them. Although the name "flat bone" implies a structure that is level or horizontal, many flat bones are curved (for example, the bones of the skull). Bones that do not fall into one of the preceding categories are classified as **irregular bones.** The vertebrae are irregular bones (see Figure 9.1).

Some anatomists also recognize two other subcategories of bones. **Sesamoid bones** are special types of short bones formed in tendons. The patellas (kneecaps) are sesamoid bones. **Wormian** or **sutural bones** are tiny bones between cranial bones. Except for the patellas, the sesamoid and Wormian bones are not included in the bone count of 206 because they vary in number and location in different individuals.

Bone Markings

Even a casual observation of the bones will reveal that bone surfaces are not featureless smooth areas but are scarred with an array of bumps, holes, and ridges. These **bone markings** reveal where bones form joints with other bones, where muscles, tendons, and ligaments were attached, and where blood vessels and nerves passed. Bone markings fall into two categories: projections, or processes that grow out from the bone and serve as sites of muscle attachment or help form joints; and depressions or cavities, indentations or openings in the bone that often serve as conduits for nerves and blood vessels. The bone markings are summarized in Table 9.1 on page 116.

Gross Anatomy of the Typical Long Bone

ACTIVITY 1

Examining a Long Bone

1. Obtain a long bone that has been sawed along its longitudinal axis. If a cleaned dry bone is provided, no special preparations need be made.

! Note: If the bone supplied is a fresh beef bone, don disposable gloves before beginning your observations.

With the help of Figure 9.3, identify the **diaphysis** or shaft. Observe its smooth surface, which is composed of compact bone. If you are using a fresh specimen, carefully pull away the **periosteum,** or fibrous membrane covering, to view the bone surface. Notice that many fibers of the periosteum penetrate into the bone. These fibers are called **perforating (Sharpey's) fibers.** Blood vessels and nerves travel through the periosteum and invade the bone. *Osteoblasts* (bone-forming cells) and *osteoclasts* (bone-destroying cells) are found on the inner, or osteogenic, layer of the periosteum.

2. Now inspect the **epiphysis,** the end of the long bone. Notice that it is composed of a thin layer of compact bone that encloses spongy bone.

3. Identify the **articular cartilage,** which covers the epiphyseal surface in place of the periosteum. Because it is composed of glassy hyaline cartilage, it provides a smooth surface to prevent friction at joint surfaces.

4. If the animal was still young and growing, you will be able to see the **epiphyseal plate,** a thin area of hyaline cartilage that provides for longitudinal growth of the bone during youth. Once the long bone has stopped growing, these areas are replaced with bone and appear as thin, barely discernible remnants—the **epiphyseal lines.**

5. In an adult animal, the central cavity of the shaft *(medullary cavity)* is essentially a storage region for adipose tissue, or **yellow marrow.** In the infant, this area is involved in forming blood cells, and so **red marrow** is found in the marrow cavities. In adult bones, the red marrow is confined to the interior of the epiphyses, where it occupies the spaces between the trabeculae of spongy bone.

6. If you are examining a fresh bone, look carefully to see if you can distinguish the delicate **endosteum** lining the shaft. The endosteum also covers the trabeculae of spongy bone and lines the canals of compact bone. Like the periosteum, the endosteum contains both osteoblasts and osteoclasts. As the bone grows in diameter on its external surface, it is constantly being broken down on its inner surface. Thus the thickness of the compact bone layer composing the shaft remains relatively constant.

! 7. If you have been working with a fresh bone specimen, return it to the appropriate area and properly dispose of your gloves, as designated by your instructor. Wash your hands before continuing to the microscope study.

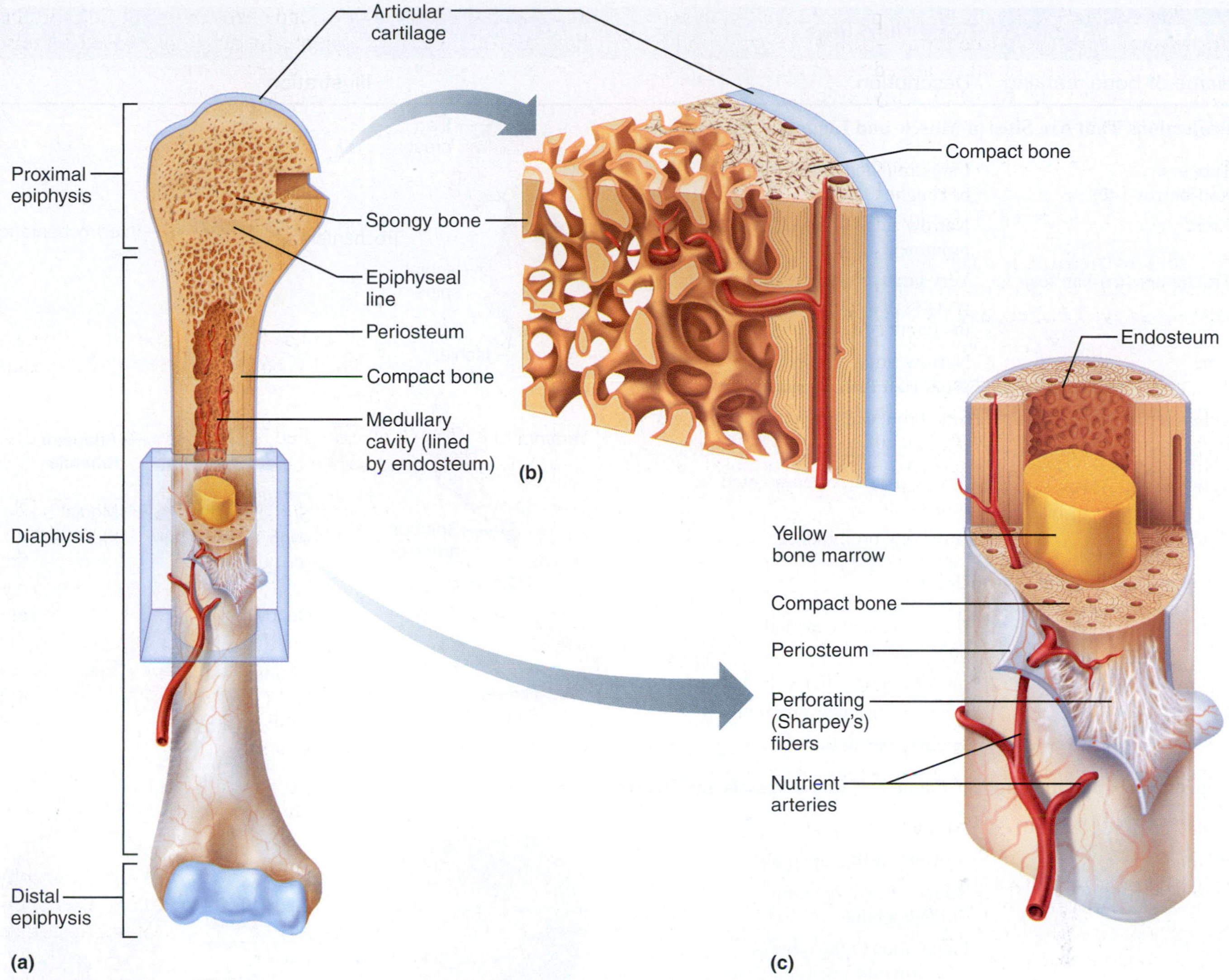
Articular cartilage
Proximal epiphysis
Spongy bone
Epiphyseal line
Periosteum
Compact bone
Medullary cavity (lined by endosteum)
Diaphysis
Distal epiphysis
(a)
Compact bone
(b)
Endosteum
Yellow bone marrow
Compact bone
Periosteum
Perforating (Sharpey's) fibers
Nutrient arteries
(c)

FIGURE 9.3 The structure of a long bone (humerus of the arm). (a) Anterior view with longitudinal section cut away at the proximal end. **(b)** Pie-shaped, three-dimensional view of spongy bone and compact bone of the epiphysis. **(c)** Cross section of diaphysis (shaft). Note that the external surface of the diaphysis is covered by a periosteum, but the articular surface of the epiphysis is covered with hyaline cartilage.

Longitudinal bone growth at epiphyseal plates (growth plates) follows a predictable sequence and provides a reliable indicator of the age of children exhibiting normal growth. In cases in which problems of long-bone growth are suspected (for example, pituitary dwarfism), X rays are taken to view the width of the growth plates. An abnormally thin epiphyseal plate indicates growth retardation. ■

Chemical Composition of Bone

Bone is one of the hardest materials in the body. Although relatively light, bone has a remarkable ability to resist tension and shear forces that continually act on it. An engineer would tell you that a cylinder (like a long bone) is one of the strongest structures for its mass. Thus nature has given us an extremely strong, exceptionally simple (almost crude), and flexible supporting system without sacrificing mobility.

The hardness of bone is due to the inorganic calcium salts deposited in its ground substance. Its flexibility comes from the organic elements of the matrix, particularly the collagen fibers.

ACTIVITY 2

Examining the Effects of Heat and Hydrochloric Acid on Bones

Obtain a bone sample that has been soaked in hydrochloric acid (HCl) (or in vinegar) and one that has been baked. Heating removes the organic part of bone, while acid dissolves out the minerals. Do the treated bones retain the structure of untreated specimens?

____

TABLE 9.1 Bone Markings

Name of bone marking	Description
Projections That Are Sites of Muscle and Ligament Attachment	
Tuberosity (too″be-ros′ĭ-tē)	Large rounded projection; may be roughened
Crest	Narrow ridge of bone; usually prominent
Trochanter (tro-kan′ter)	Very large, blunt, irregularly shaped process (the only examples are on the femur)
Line	Narrow ridge of bone; less prominent than a crest
Tubercle (too′ber-kl)	Small rounded projection or process
Epicondyle (ep″ĭ-kon′dīl)	Raised area on or above a condyle
Spine	Sharp, slender, often pointed projection
Process	Any bony prominence
Projections That Help Form Joints	
Head	Bony expansion carried on a narrow neck
Facet	Smooth, nearly flat articular surface
Condyle (kon′dīl)	Rounded articular projection
Ramus (ra′mus)	Armlike bar of bone
Depressions and Openings for Passage of Blood Vessels and Nerves	
Groove	Furrow
Fissure	Narrow, slitlike opening
Foramen (fo-ra′men)	Round or oval opening through a bone
Notch	Indentation at the edge of a structure
Others	
Meatus (mē-a′tus)	Canal-like passageway
Sinus	Bone cavity, filled with air and lined with mucous membrane
Fossa (fos′ah)	Shallow basinlike depression in a bone, often serving as an articular surface

Illustration

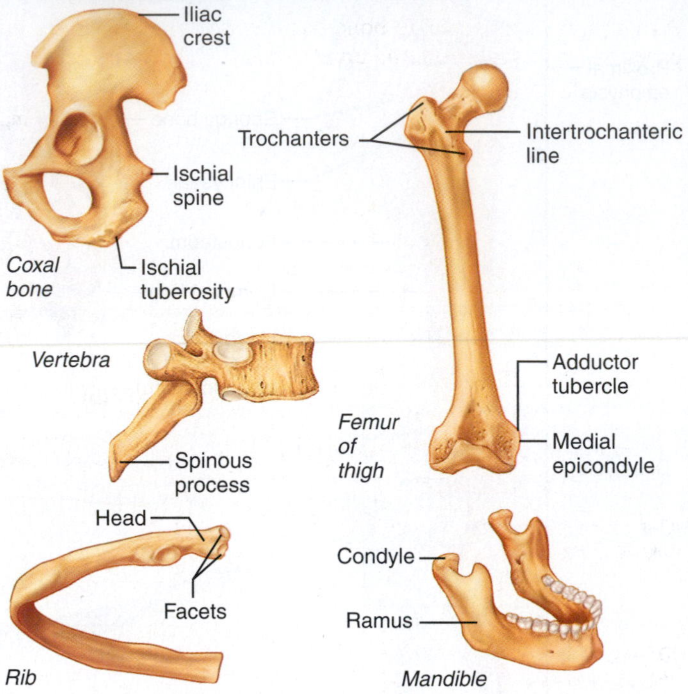
Iliac crest
Ischial spine
Coxal bone
Ischial tuberosity
Trochanters
Intertrochanteric line
Vertebra
Spinous process
Adductor tubercle
Femur of thigh
Medial epicondyle
Head
Facets
Rib
Condyle
Ramus
Mandible

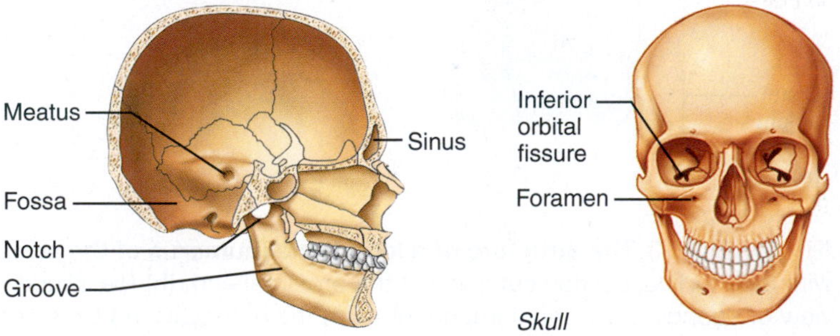
Meatus
Sinus
Fossa
Notch
Groove
Inferior orbital fissure
Foramen
Skull

Gently apply pressure to each bone sample. What happens to the heated bone?

____

____

What happens to the bone treated with acid?

____

What does the acid appear to remove from the bone?

____

What does baking appear to do to the bone?

____

____

In rickets, the bones are not properly calcified. Which of the demonstration specimens would more closely resemble the bones of a child with rickets?

____

Microscopic Structure of Compact Bone

As you have seen, spongy bone has a spiky, open-work appearance, resulting from the arrangement of the **trabeculae** that compose it, whereas compact bone appears to be dense and homogeneous. However, microscopic examination of compact bone reveals that it is riddled with passageways carrying blood vessels, nerves, and lymphatic vessels that provide the living bone cells with needed substances and a way to eliminate wastes. Indeed, bone histology is much easier to

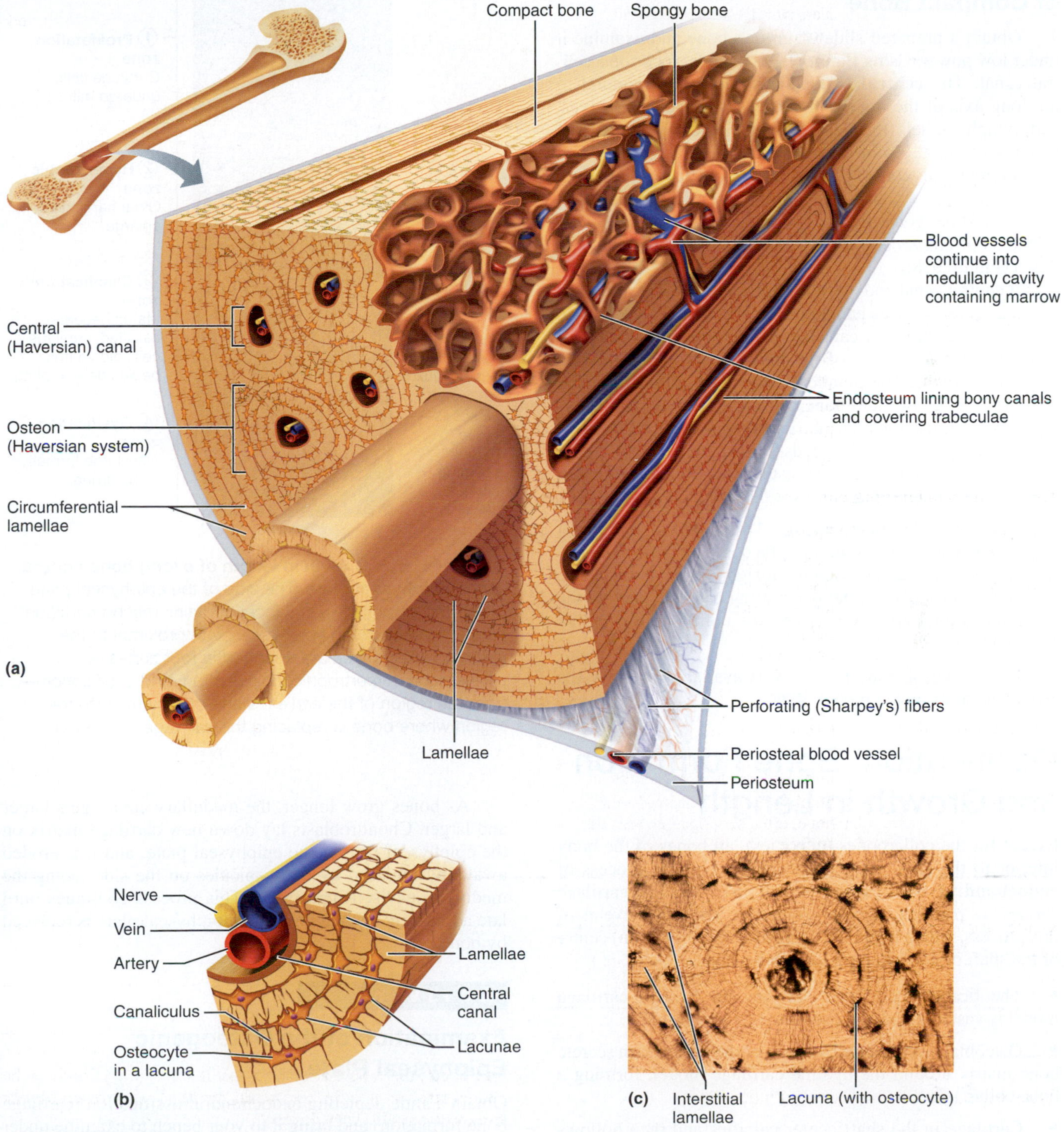

Compact bone
Spongy bone
Blood vessels continue into medullary cavity containing marrow
Central (Haversian) canal
Osteon (Haversian system)
Endosteum lining bony canals and covering trabeculae
Circumferential lamellae
(a)
Lamellae
Perforating (Sharpey's) fibers
Periosteal blood vessel
Periosteum
Nerve
Vein
Artery
Canaliculus
Osteocyte in a lacuna
Lamellae
Central canal
Lacunae
(b)
(c)
Interstitial lamellae
Lacuna (with osteocyte)

FIGURE 9.4 Microscopic structure of compact bone. (a) Diagrammatic view of a pie-shaped segment of compact bone, illustrating its structural units (osteons). **(b)** Higher-magnification view of a portion of one osteon. Note the position of osteocytes in lacunae. **(c)** Photomicrograph of a cross-sectional view of an osteon (160×). Veins, lymphatic vessels, and nerves that also occupy bone canals are not illustrated.

understand when you recognize that bone tissue is organized around its blood supply.

ACTIVITY 3

Examining the Microscopic Structure of Compact Bone

1. Obtain a prepared slide of ground bone and examine it under low power. Using Figure 9.4 as a guide, focus on a central canal. The **central** (Haversian) **canal** runs parallel to the long axis of the bone and carries blood vessels, nerves, and lymph vessels through the bony matrix. Identify the **osteocytes** (mature bone cells) in **lacunae** (chambers), which are arranged in concentric circles **(concentric lamellae)** around the central canal. Because bone remodeling is going on all the time, you will also see some *interstitial lamellae,* remnants of *circumferential lamellae* that have been broken down (Figure 9.4c).

A central canal and all the concentric lamellae surrounding it are referred to as an **osteon,** or Haversian system. Also identify **canaliculi,** tiny canals radiating outward from a central canal to the lacunae of the first lamella and then from lamella to lamella. The canaliculi form a dense transportation network through the hard bone matrix, connecting all the living cells of the osteon to the nutrient supply. The canaliculi allow each cell to take what it needs for nourishment and to pass along the excess to the next osteocyte. You may need a higher-power magnification to see the fine canaliculi.

2. Also note the **perforating (Volkmann's) canals** in Figure 9.4. These canals run into the compact bone and marrow cavity from the periosteum, at right angles to the shaft. With the central canals, the perforating canals complete the communication pathway between the bone interior and its external surface.

3. If a model of bone histology is available, identify the same structures on the model. ■

Ossification: Bone Formation and Growth in Length

Except for the collarbones (clavicles), all bones of the body inferior to the skull form in the embryo by the process of **endochondral ossification,** which uses hyaline cartilage "bones" as patterns for bone formation. The major events of this process, which begins in the (primary ossification) center of the shaft of a developing long bone, are:

- The fibrous membrane covering the hyaline cartilage model is vascularized and converted to a periosteum.
- Osteoblasts at the inner surface of the periosteum secrete bone matrix around the hyaline cartilage model, forming a bone collar.
- Cartilage in the shaft center calcifies and then hollows out, forming an internal cavity.

A *periosteal bud* (blood vessels, nerves, red marrow elements, osteoblasts, and osteoclasts) invades the cavity, which becomes the medullary cavity. This process proceeds in both directions from the *primary ossification center.*

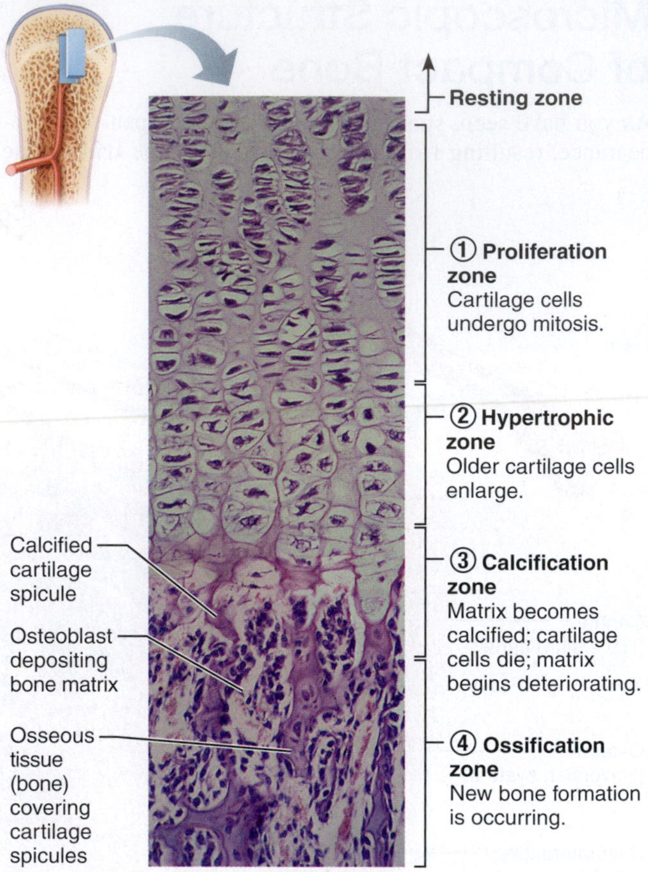

Resting zone
① Proliferation zone
Cartilage cells undergo mitosis.
② Hypertrophic zone
Older cartilage cells enlarge.
③ Calcification zone
Matrix becomes calcified; cartilage cells die; matrix begins deteriorating.
④ Ossification zone
New bone formation is occurring.
Calcified cartilage spicule
Osteoblast depositing bone matrix
Osseous tissue (bone) covering cartilage spicules

FIGURE 9.5 Growth in length of a long bone occurs at the epiphyseal plate. The side of the epiphyseal plate facing the epiphysis (distal face) contains resting cartilage cells. The cells of the epiphyseal plate proximal to the resting cartilage area are arranged in four zones—proliferation, hypertrophic, calcification, and ossification—from the region of the earliest stage of growth ① to the region where bone is replacing the cartilage ④ (150×).

As bones grow longer, the medullary cavity gets larger and larger. Chondroblasts lay down new cartilage matrix on the epiphyseal face of the epiphyseal plate, and it is eroded away and replaced by bony spicules on the side facing the medullary cavity (Figure 9.5). This process continues until late adolescence when the entire epiphyseal plate is replaced by bone.

ACTIVITY 4

Examination of the Osteogenic Epiphyseal Plate

Obtain a slide depicting endochondral ossification (cartilage bone formation) and bring it to your bench to examine under the microscope. Using Figure 9.5 as a reference, identify the growth, hypertrophic, calcification, and ossification zones of the epiphyseal plate. Then, also identify the area of resting cartilage cells distal to the growth zone, some hypertrophied chondrocytes, bony spicules, the periosteal bone collar, and the medullary cavity. ■

NAME ____________________________

LAB TIME/DATE ______________________

REVIEW SHEET

EXERCISE 9

Overview of the Skeleton: Classification and Structure of Bones and Cartilages

Bone Markings

1. Match the terms in column B with the appropriate description in column A.

Column A		**Column B**
______________	1. sharp, slender process*	a. condyle
______________	2. small rounded projection*	b. crest
______________	3. narrow ridge of bone*	c. epicondyle
______________	4. large rounded projection*	d. facet
______________	5. structure supported on neck†	e. fissure
______________	6. armlike projection†	f. foramen
______________	7. rounded, articular projection†	g. fossa
______________	8. narrow opening‡	h. head
______________	9. canal-like structure	i. meatus
______________	10. round or oval opening through a bone‡	j. process
______________	11. shallow depression	k. ramus
______________	12. air-filled cavity	l. sinus
______________	13. large, irregularly shaped projection*	m. spine
______________	14. raised area on or above a condyle*	n. trochanter
______________	15. projection or prominence	o. tubercle
______________	16. smooth, nearly flat articular surface†	p. tuberosity

*a site of muscle and ligament attachment
†takes part in joint formation
‡a passageway for nerves or blood vessels

Classification of Bones

2. The four major anatomical classifications of bones are long, short, flat, and irregular. Which category has the least amount of spongy bone relative to its total volume? ____________

3. Place the name of each labeled bone in Figure 9.1, page 112, into the appropriate column of the chart here.

Long	Short	Flat	Irregular

Gross Anatomy of the Typical Long Bone

4. Use the terms below to identify the structures marked by leader lines and braces in the diagrams (some terms are used more than once).

Key:
a. articular cartilage
b. compact bone
c. diaphysis
d. endosteum
e. epiphyseal line
f. epiphysis
g. medullary cavity
h. nutrient artery
i. periosteum
j. red marrow cavity
k. trabeculae of spongy bone
l. yellow marrow

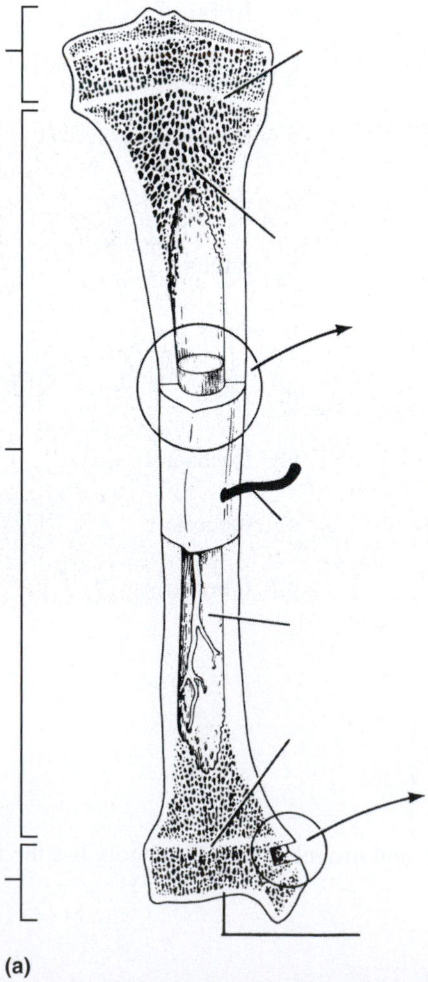

(a)

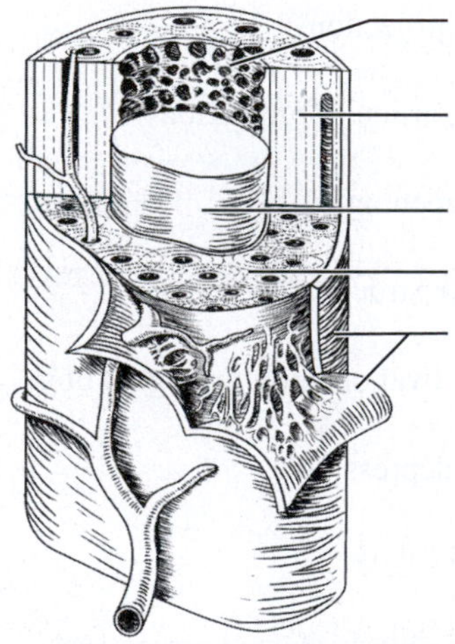

(b)

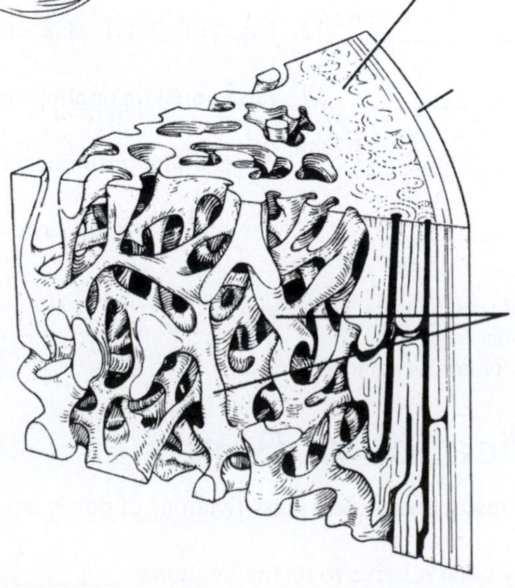

(c)

5. Match the terms in question 4 with the information below.

_________ 1. contains spongy bone in adults

_________ 2. made of compact bone

_________ 3. site of blood cell formation

_________ 4. major submembranous site of osteoclasts

_________ 5. scientific term for bone shaft

_________ 6. contains fat in adult bones

_________ 7. growth plate remnant

_________ 8. major submembranous site of osteoblasts

6. What differences between compact and spongy bone can be seen with the naked eye? ____________________

____

7. What is the function of the periosteum? ____________________

____

Microscopic Structure of Compact Bone

8. Trace the route taken by nutrients through a bone, starting with the periosteum and ending with an osteocyte in a lacuna.

Periosteum → ____________________ → ____________________

____________________ → ____________________ → ________ osteocyte

9. Several descriptions of bone structure are given below. Identify the structure involved by choosing the appropriate term from the key and placing its letter in the blank. Then, on the photomicrograph of bone on the right (365×), identify all structures named in the key and bracket an osteon.

Key: a. canaliculi b. central canal c. concentric lamellae d. lacunae e. matrix

_________ 1. layers of bony matrix around a central canal

_________ 2. site of osteocytes

_________ 3. longitudinal canal carrying blood vessels, lymphatics, and nerves

_________ 4. minute canals connecting osteocytes of an osteon

_________ 5. inorganic salts deposited in organic ground substance

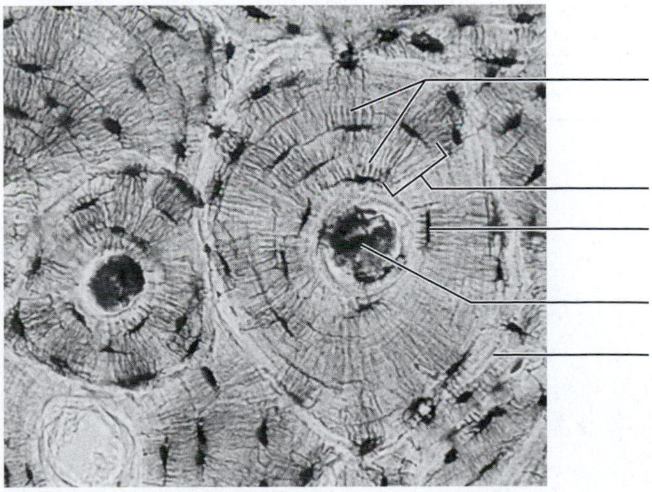

Chemical Composition of Bone

10. What is the function of the organic matrix in bone? ____________________

11. Name the important organic bone components. ____________________

12. Calcium salts form the bulk of the inorganic material in bone. What is the function of the calcium salts?

____

13. Baking removes ____________ from bone. Soaking bone in acid removes ____________.

Ossification: Bone Formation and Growth in Length

14. Compare and contrast events occurring on the epiphyseal and diaphyseal faces of the epiphyseal plate.

Epiphyseal face: ____________________

Diaphyseal face: ____________________

Cartilages of the Skeleton

15. Using the key choices, identify each type of cartilage described (in terms of its body location or function) below.

Key: a. elastic b. fibrocartilage c. hyaline

____________ 1. supports the external ear

____________ 2. between the vertebrae

____________ 3. forms the walls of the voice box (larynx)

____________ 4. the epiglottis

____________ 5. articular cartilages

____________ 6. meniscus in a knee joint

____________ 7. connects the ribs to the sternum

____________ 8. most effective at resisting compression

____________ 9. most springy and flexible

____________ 10. most abundant

EXERCISE 10

The Axial Skeleton

MATERIALS

- ☐ Intact skull and Beauchene skull
- ☐ X rays of individuals with scoliosis, lordosis, and kyphosis (if available)
- ☐ Articulated skeleton, articulated vertebral column, removable intervertebral discs
- ☐ Isolated cervical, thoracic, and lumbar vertebrae, sacrum, and coccyx

OBJECTIVES

1. To identify the three bone groups composing the axial skeleton.
2. To identify the bones composing the axial skeleton, either by examining isolated bones or by pointing them out on an articulated skeleton or a skull, and to name the important bone markings on each.
3. To distinguish the different types of vertebrae.
4. To discuss the importance of intervertebral discs and spinal curvatures.
5. To distinguish three abnormal spinal curvatures.

PRE-LAB QUIZ

1. The axial skeleton can be divided into the skull, the vertebral column, and the _________.
   a. bony thorax
   b. femur
   c. hip bones
   d. humerus
2. Eight bones make up the _________, which encloses and protects the brain.
   a. cranium
   b. face
   c. skull
3. How many bones compose the face? _________
4. Circle the correct term. The lower jawbone, or maxilla / mandible, articulates with the temporal bones in the only freely movable joints in the skull.
5. Circle the correct term. The body / spinous process of a typical vertebra forms the rounded, central portion that faces anteriorly in the human vertebral column.
6. The seven bones of the neck are called _________ vertebrae.
   a. cervical
   b. lumbar
   c. spinal
   d. thoracic
7. The _________ vertebrae articulate with the corresponding ribs.
   a. cervical
   b. lumbar
   c. spinal
   d. thoracic
8. The _________, commonly referred to as the breastbone, is a flat bone formed by the fusion of three bones: the manubrium, the body, and the xiphoid process.
   a. coccyx
   b. sacrum
   c. sternum
9. Circle True or False. The first seven pairs of ribs are called floating ribs because they have only indirect cartilage attachments to the sternum.

MasteringA&P™
Access practice quizzes and more in the Study Area at www.masteringaandp.com.

PAL
For access to anatomical models and more, check out Practice Anatomy Lab.

The **axial skeleton** (the green portion of Figure 9.1 on page 112) can be divided into three parts: the skull, the vertebral column, and the bony thorax.

The Skull

The **skull** is composed of two sets of bones. Those of the **cranium** enclose and protect the fragile brain tissue. The **facial bones** present the eyes in an anterior position and form the base for the facial muscles, which make it possible for us to present our feelings to the world. All but one of the bones of the skull are joined by interlocking joints called *sutures.* The mandible, or lower jawbone, is attached to the rest of the skull by a freely movable joint.

ACTIVITY 1

Identifying the Bones of the Skull

The bones of the skull, shown in Figures 10.1 through 10.7, are described below. As you read through this material, identify each bone on an intact and/or Beauchene skull (see Figure 10.6c).

Note: Important bone markings are listed beneath the bones on which they appear, and a color-coded dot before each bone name corresponds to the bone color in the figures.

The Cranium

The cranium may be divided into two major areas for study—the **cranial vault** or **calvaria,** forming the superior, lateral, and posterior walls of the skull, and the **cranial floor** or **base,** forming the skull bottom. Internally, the cranial floor has three distinct concavities, the **anterior, middle,** and **posterior cranial fossae** (see Figure 10.3). The brain sits in these fossae, completely enclosed by the cranial vault.

Eight bones construct the cranium. *With the exception of two paired bones (the parietals and the temporals), all are single bones.* Sometimes the six ossicles of the middle ear are also considered part of the cranium. Because the ossicles are functionally part of the hearing apparatus, their consideration is deferred to Exercise 25, Special Senses: Hearing and Equilibrium.

● ***Frontal Bone*** See Figures 10.1, 10.3, 10.6, and 10.8. Anterior portion of cranium; forms the forehead, superior part of the orbit, and floor of anterior cranial fossa.

Supraorbital foramen (notch): Opening above each orbit allowing blood vessels and nerves to pass.

Glabella: Smooth area between the eyes.

● ***Parietal Bone*** See Figures 10.1 and 10.6. Posterolateral to the frontal bone, forming sides of cranium.

Sagittal suture: Midline articulation point of the two parietal bones.

Coronal suture: Point of articulation of parietals with frontal bone.

● ***Temporal Bone*** See Figures 10.1 through 10.3 and 10.6. Inferior to parietal bone on lateral skull. The temporals can be divided into four major parts: the **squamous region** abuts the parietals; the **tympanic region** surrounds the external ear opening; the **mastoid region** is the area posterior to the ear; and the **petrous region** forms the lateral portion of the skull base.

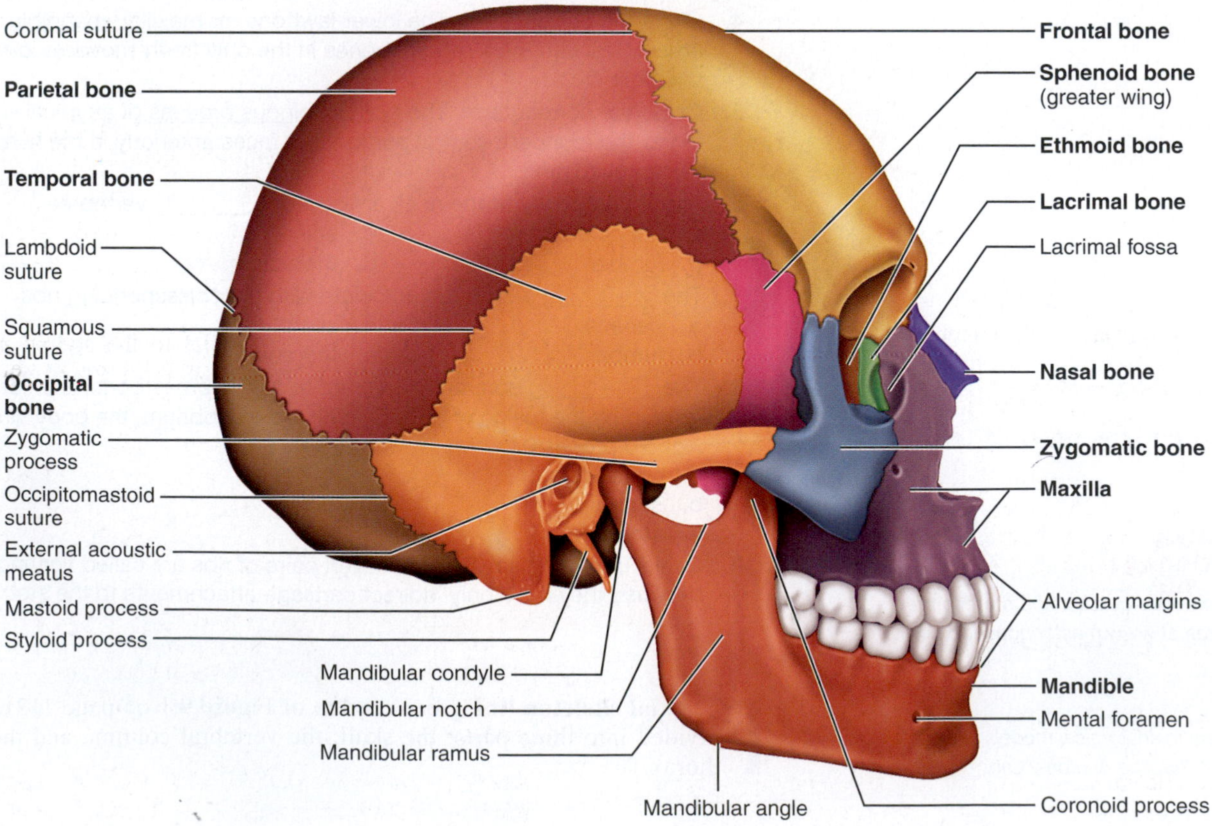

Coronal suture
Parietal bone
Temporal bone
Lambdoid suture
Squamous suture
Occipital bone
Zygomatic process
Occipitomastoid suture
External acoustic meatus
Mastoid process
Styloid process
Mandibular condyle
Mandibular notch
Mandibular ramus
Mandibular angle
Frontal bone
Sphenoid bone (greater wing)
Ethmoid bone
Lacrimal bone
Lacrimal fossa
Nasal bone
Zygomatic bone
Maxilla
Alveolar margins
Mandible
Mental foramen
Coronoid process

FIGURE 10.1 **External anatomy of the right lateral aspect of the skull.**

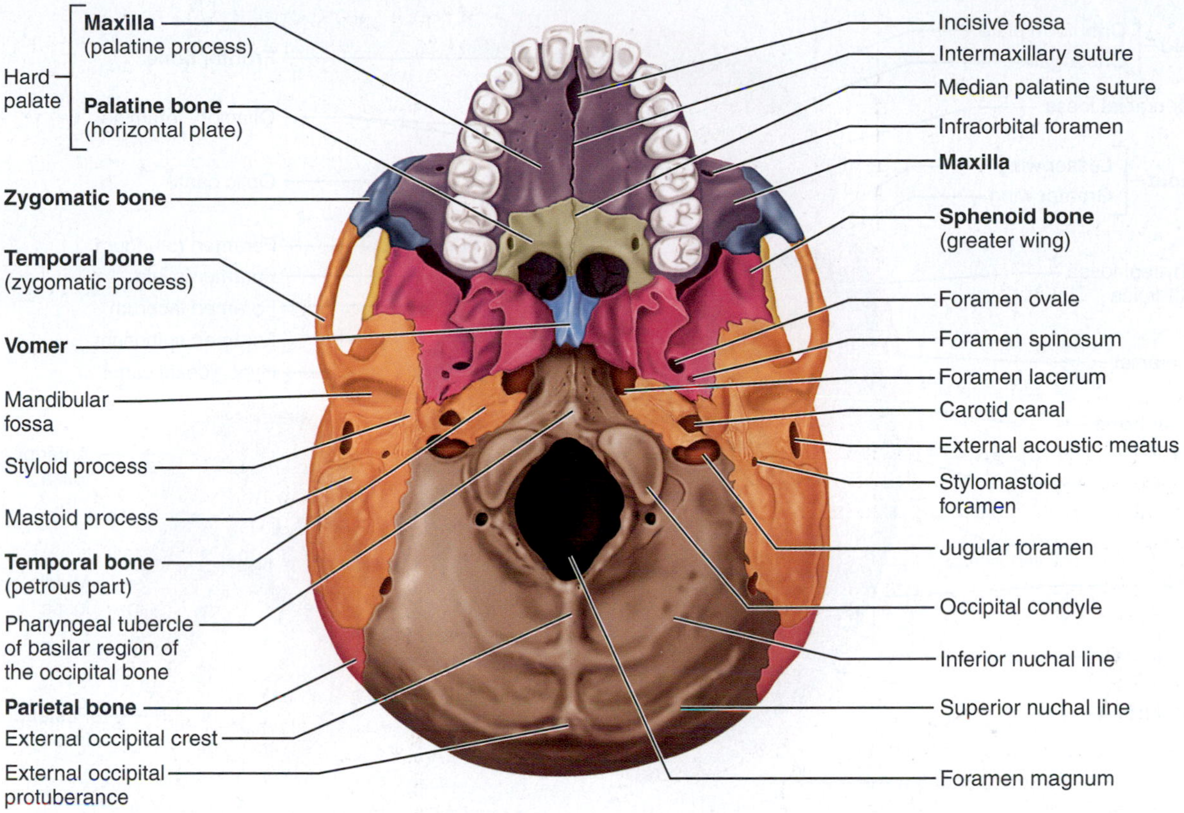

Hard palate
Maxilla (palatine process)
Palatine bone (horizontal plate)
Zygomatic bone
Temporal bone (zygomatic process)
Vomer
Mandibular fossa
Styloid process
Mastoid process
Temporal bone (petrous part)
Pharyngeal tubercle of basilar region of the occipital bone
Parietal bone
External occipital crest
External occipital protuberance
Incisive fossa
Intermaxillary suture
Median palatine suture
Infraorbital foramen
Maxilla
Sphenoid bone (greater wing)
Foramen ovale
Foramen spinosum
Foramen lacerum
Carotid canal
External acoustic meatus
Stylomastoid foramen
Jugular foramen
Occipital condyle
Inferior nuchal line
Superior nuchal line
Foramen magnum

FIGURE 10.2 Inferior view of the skull, mandible removed.

Important markings associated with the flaring squamous region (Figures 10.1 and 10.2) include:

Squamous suture: Point of articulation of the temporal bone with the parietal bone.

Zygomatic process: A bridgelike projection joining the zygomatic bone (cheekbone) anteriorly. Together these two bones form the *zygomatic arch.*

Mandibular fossa: Rounded depression on the inferior surface of the zygomatic process (anterior to the ear); forms the socket for the mandibular condyle, the point where the mandible (lower jaw) joins the cranium.

Tympanic region markings (Figures 10.1 and 10.2) include:

External acoustic meatus: Canal leading to eardrum and middle ear.

Styloid (*stylo*=stake, pointed object) **process:** Needlelike projection inferior to external acoustic meatus; attachment point for muscles and ligaments of the neck. This process is often broken off demonstration skulls.

Prominent structures in the mastoid region (Figures 10.1 and 10.2) are:

Mastoid process: Rough projection inferior and posterior to external acoustic meatus; attachment site for muscles.

The mastoid process, full of air cavities and so close to the middle ear—a trouble spot for infections—often becomes infected too, a condition referred to as **mastoiditis.** Because the mastoid area is separated from the brain by only a thin layer of bone, an ear infection that has spread to the mastoid process can inflame the brain coverings, or the meninges. The latter condition is known as **meningitis.** ■

Stylomastoid foramen: Tiny opening between the mastoid and styloid processes through which cranial nerve VII leaves the cranium.

The petrous part (Figures 10.2 and 10.3), which helps form the middle and posterior cranial fossae, exhibits several obvious foramina with important functions:

Jugular foramen: Opening medial to the styloid process through which the internal jugular vein and cranial nerves IX, X, and XI pass.

Carotid canal: Opening medial to the styloid process through which the internal carotid artery passes into the cranial cavity.

Internal acoustic meatus: Opening on posterior aspect (petrous region) of temporal bone allowing passage of cranial nerves VII and VIII (Figure 10.3).

Foramen lacerum: A jagged opening between the petrous temporal bone and the sphenoid providing passage for a number of small nerves and for the internal carotid artery to enter the middle cranial fossa (after it passes through part of the temporal bone).

● ***Occipital Bone*** See Figures 10.1, 10.2, 10.3, and 10.6. Most posterior bone of cranium—forms floor and back wall. Joins sphenoid bone anteriorly via its narrow basioccipital region.

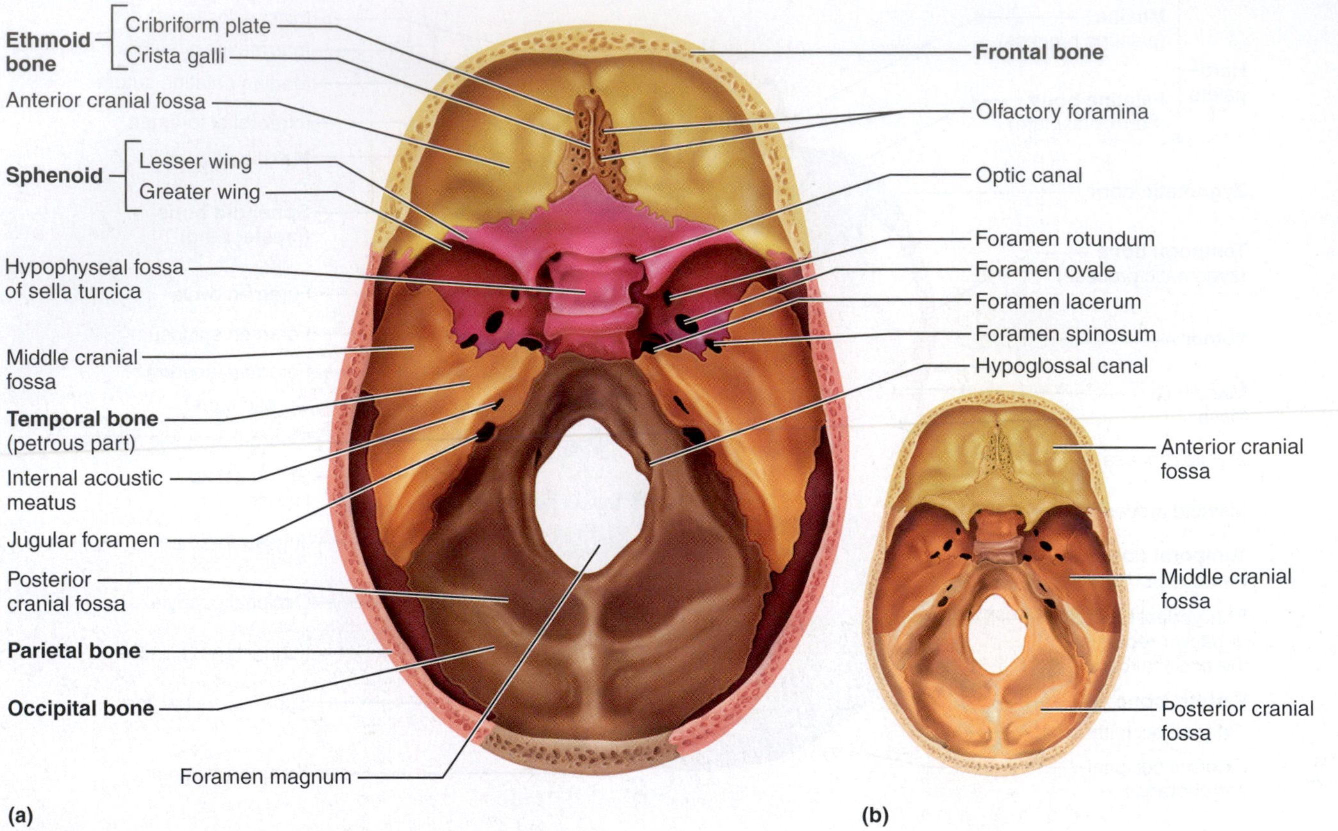

Ethmoid bone
Cribriform plate
Crista galli
Anterior cranial fossa
Sphenoid
Lesser wing
Greater wing
Hypophyseal fossa of sella turcica
Middle cranial fossa
Temporal bone (petrous part)
Internal acoustic meatus
Jugular foramen
Posterior cranial fossa
Parietal bone
Occipital bone
Foramen magnum
Frontal bone
Olfactory foramina
Optic canal
Foramen rotundum
Foramen ovale
Foramen lacerum
Foramen spinosum
Hypoglossal canal
(a)
Anterior cranial fossa
Middle cranial fossa
Posterior cranial fossa
(b)

FIGURE 10.3 Internal anatomy of the inferior portion of the skull. (a) Superior view of the floor of the cranial cavity, calvaria removed. **(b)** Schematic view of the cranial cavity floor showing the extent of its major fossae.

Lambdoid suture: Site of articulation of occipital bone and parietal bones.

Foramen magnum: Large opening in base of occipital, which allows the spinal cord to join with the brain.

Occipital condyles: Rounded projections lateral to the foramen magnum that articulate with the first cervical vertebra (atlas).

Hypoglossal canal: Opening medial and superior to the occipital condyle through which the hypoglossal nerve (cranial nerve XII) passes.

External occipital crest and protuberance: Midline prominences posterior to the foramen magnum.

- ***Sphenoid Bone*** See Figures 10.1 through 10.4, 10.6, and 10.8. Bat-shaped bone forming the anterior plateau of the middle cranial fossa across the width of the skull. The sphenoid bone is the keystone of the cranium because it articulates with all other cranial bones.

Greater wings: Portions of the sphenoid seen exteriorly anterior to the temporal and forming a part of the eye orbits.

Pterygoid processes: Inferiorly directed trough-shaped projections from the junction of the body and the greater wings.

Superior orbital fissures: Jagged openings in orbits providing passage for cranial nerves III, IV, V, and VI to enter the orbit where they serve the eye.

The sphenoid bone can be seen in its entire width if the top of the cranium (calvaria) is removed (Figure 10.3).

Sella turcica (Turk's saddle): A saddle-shaped region in the sphenoid midline. The seat of this saddle, called the **hypophyseal fossa,** surrounds the pituitary gland (hypophysis).

Lesser wings: Bat-shaped portions of the sphenoid anterior to the sella turcica.

Optic canals: Openings in the bases of the lesser wings through which the optic nerves enter the orbits to serve the eyes.

Foramen rotundum: Opening lateral to the sella turcica providing passage for a branch of the fifth cranial nerve. (This foramen is not visible on an inferior view of the skull.)

Foramen ovale: Opening posterior to the sella turcica that allows passage of a branch of the fifth cranial nerve.

- ***Ethmoid Bone*** See Figures 10.1, 10.3, 10.5, 10.6, and 10.8. Irregularly shaped bone anterior to the sphenoid. Forms the roof of the nasal cavity, upper nasal septum, and part of the medial orbit walls.

Crista galli (cock's comb): Vertical projection providing a point of attachment for the dura mater, helping to secure the brain within the skull.

Cribriform plates: Bony plates lateral to the crista galli through which olfactory fibers pass to the brain from the nasal

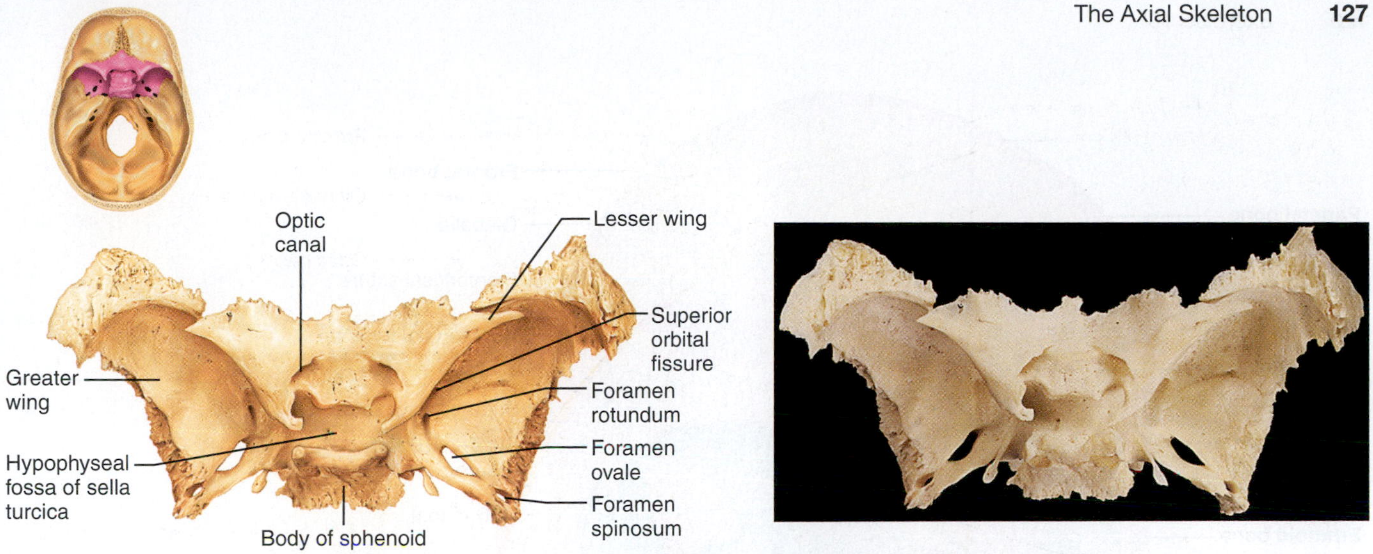
Optic
canal
Lesser wing
Superior
orbital
fissure
Greater
wing
Foramen
rotundum
Hypophyseal
fossa of sella
turcica
Foramen
ovale
Foramen
spinosum
Body of sphenoid

(a) Superior view

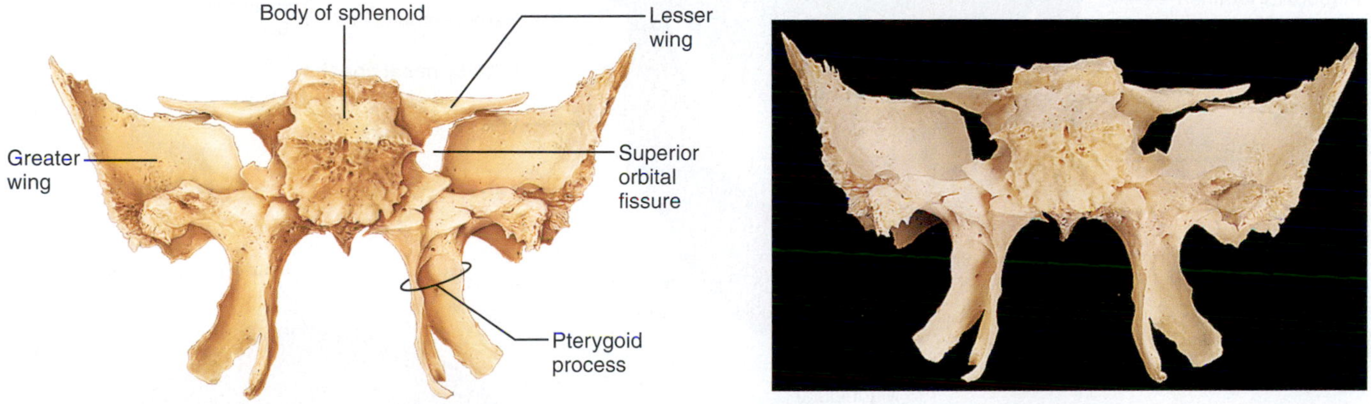
Body of sphenoid
Lesser
wing
Greater
wing
Superior
orbital
fissure
Pterygoid
process

(b) Posterior view

FIGURE 10.4 The sphenoid bone.

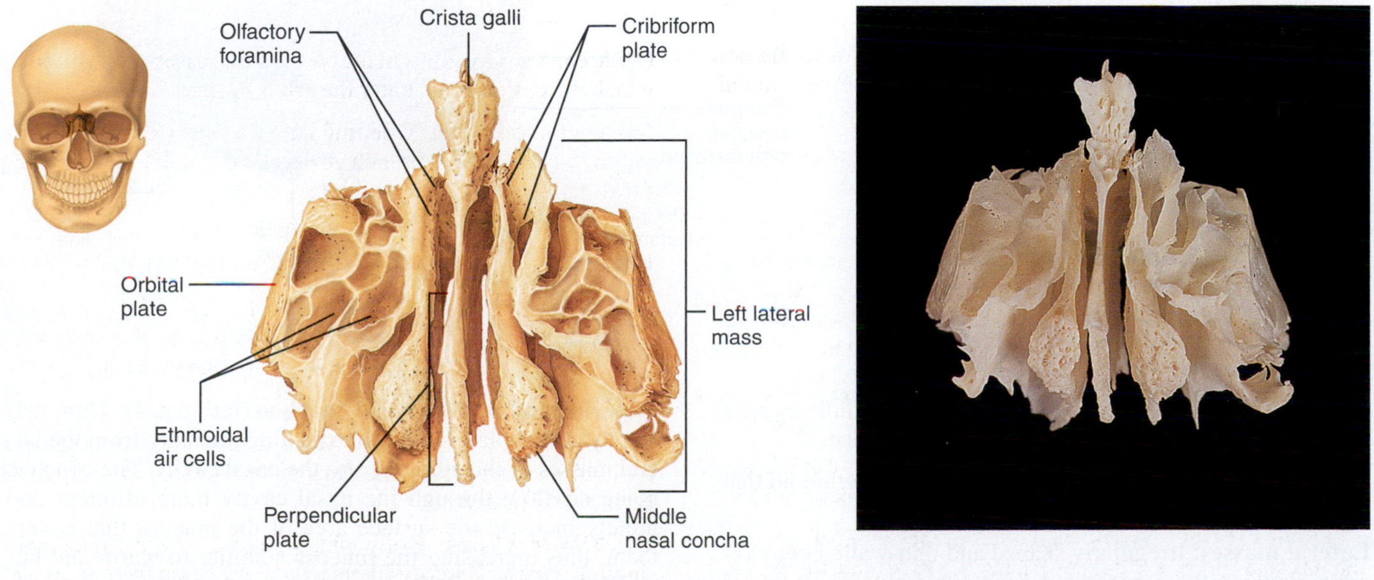
Olfactory
foramina
Crista galli
Cribriform
plate
Orbital
plate
Left lateral
mass
Ethmoidal
air cells
Perpendicular
plate
Middle
nasal concha

FIGURE 10.5 The ethmoid bone. Anterior view.

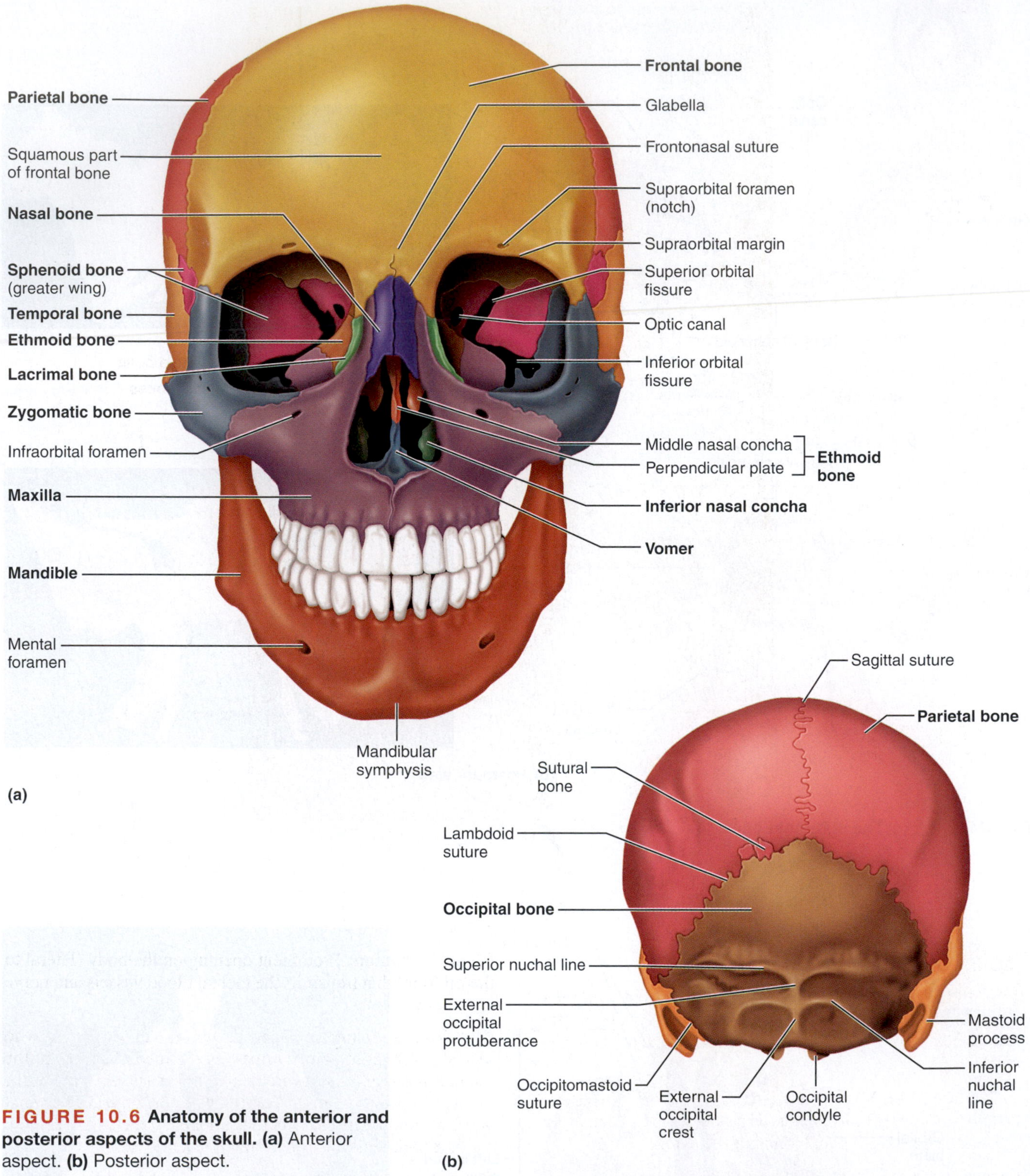
Parietal bone
Squamous part of frontal bone
Nasal bone
Sphenoid bone (greater wing)
Temporal bone
Ethmoid bone
Lacrimal bone
Zygomatic bone
Infraorbital foramen
Maxilla
Mandible
Mental foramen
Mandibular symphysis
Frontal bone
Glabella
Frontonasal suture
Supraorbital foramen (notch)
Supraorbital margin
Superior orbital fissure
Optic canal
Inferior orbital fissure
Middle nasal concha
Perpendicular plate
Ethmoid bone
Inferior nasal concha
Vomer
(a)
Sagittal suture
Parietal bone
Sutural bone
Lambdoid suture
Occipital bone
Superior nuchal line
External occipital protuberance
Mastoid process
Inferior nuchal line
Occipitomastoid suture
External occipital crest
Occipital condyle
(b)

FIGURE 10.6 Anatomy of the anterior and posterior aspects of the skull. (a) Anterior aspect. **(b)** Posterior aspect.

mucosa. Together the cribriform plates and the midline crista galli form the *horizontal plate* of the ethmoid bone.

Perpendicular plate: Inferior projection of the ethmoid that forms the superior part of the nasal septum.

Lateral masses: Irregularly shaped and thin-walled bony regions flanking the perpendicular plate laterally. Their lateral surfaces (*orbital plates*) shape part of the medial orbit wall.

Superior and middle nasal conchae (turbinates): Thin, delicately coiled plates of bone extending medially from the lateral masses of the ethmoid into the nasal cavity. The conchae make air flow through the nasal cavity more efficient and greatly increase the surface area of the mucosa that covers them, thus increasing the mucosa's ability to warm and humidify incoming air.

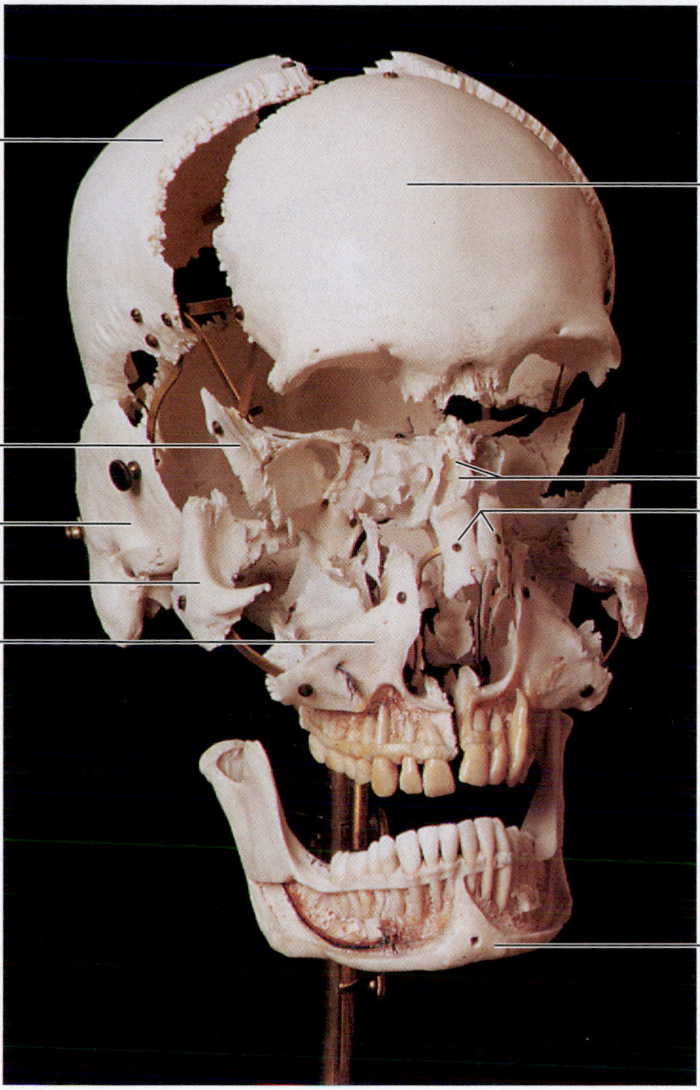
Parietal bone
Frontal bone
Sphenoid bone
Ethmoid bone
Nasal bones
Temporal bone
Zygomatic bone
Maxilla
Mandible

(c)

FIGURE 10.6 *(continued)* **(c)** Frontal view of the Beauchene skull.

Facial Bones

Of the 14 bones composing the face, 12 are paired. *Only the mandible and vomer are single bones.* An additional bone, the hyoid bone, although not a facial bone, is considered here because of its location.

- ***Mandible*** See Figures 10.1, 10.6, and 10.7. The lower jawbone, which articulates with the temporal bones in the only freely movable joints of the skull.

Mandibular body: Horizontal portion; forms the chin.

Mandibular ramus: Vertical extension of the body on either side.

Mandibular condyle: Articulation point of the mandible with the mandibular fossa of the temporal bone.

Coronoid process: Jutting anterior portion of the ramus; site of muscle attachment.

Mandibular angle: Posterior point at which ramus meets the body.

Mental foramen: Prominent opening on the body (lateral to the midline) that transmits the mental blood vessels and nerve to the lower jaw.

Mandibular foramen: Open the lower jaw of the skull to identify this prominent foramen on the medial aspect of the mandibular ramus. This foramen permits passage of the nerve involved with tooth sensation (mandibular branch of cranial nerve V) and is the site where the dentist injects Novocain to prevent pain while working on the lower teeth.

Alveolar margin: Superior margin of mandible; contains sockets in which the teeth lie.

Mandibular symphysis: Anterior median depression indicating point of mandibular fusion.

- ***Maxillae*** See Figures 10.1, 10.2, 10.6, and 10.7. Two bones fused in a median suture; form the upper jawbone and part of the orbits. All facial bones, except the mandible, join the maxillae. Thus they are the main, or keystone, bones of the face.

FIGURE 10.7 Detailed anatomy of the mandible and maxilla.

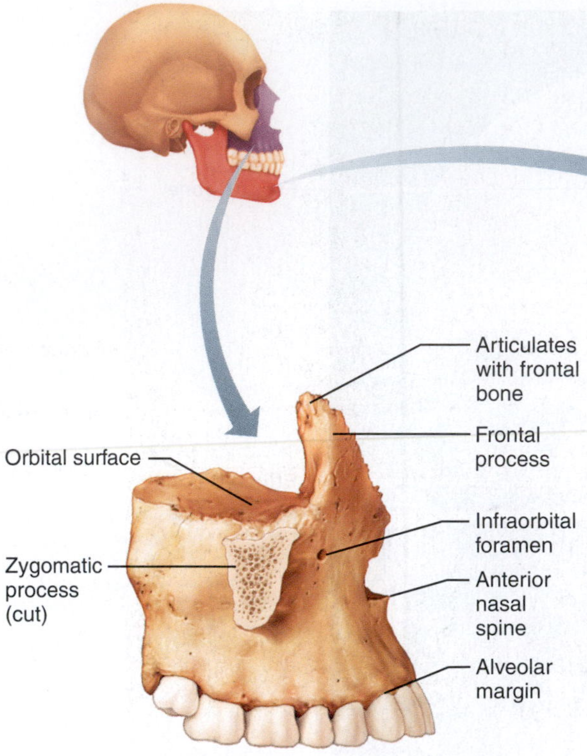
Articulates
with frontal
bone
Frontal
process
Orbital surface
Infraorbital
foramen
Zygomatic
process
(cut)
Anterior
nasal
spine
Alveolar
margin

(a) Maxilla, right lateral view

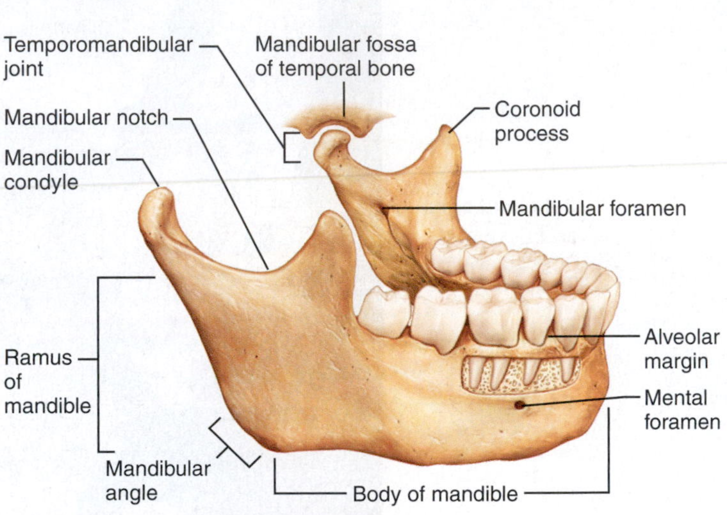
Temporomandibular
joint
Mandibular fossa
of temporal bone
Mandibular notch
Coronoid
process
Mandibular
condyle
Mandibular foramen
Ramus
of
mandible
Alveolar
margin
Mental
foramen
Mandibular
angle
Body of mandible

(b) Mandible, right lateral view

Alveolar margin: Inferior margin containing sockets (alveoli) in which teeth lie.

Palatine processes: Form the anterior hard palate; meet medially in the intermaxillary suture.

Infraorbital foramen: Opening under the orbit carrying the infraorbital nerves and blood vessels to the nasal region.

Incisive fossa: Large bilateral opening located posterior to the central incisor tooth of the maxilla and piercing the hard palate; transmits the nasopalatine arteries and blood vessels.

● ***Lacrimal Bone*** See Figures 10.1 and 10.6a. Fingernail-sized bones forming a part of the medial orbit walls between the maxilla and the ethmoid. Each lacrimal bone is

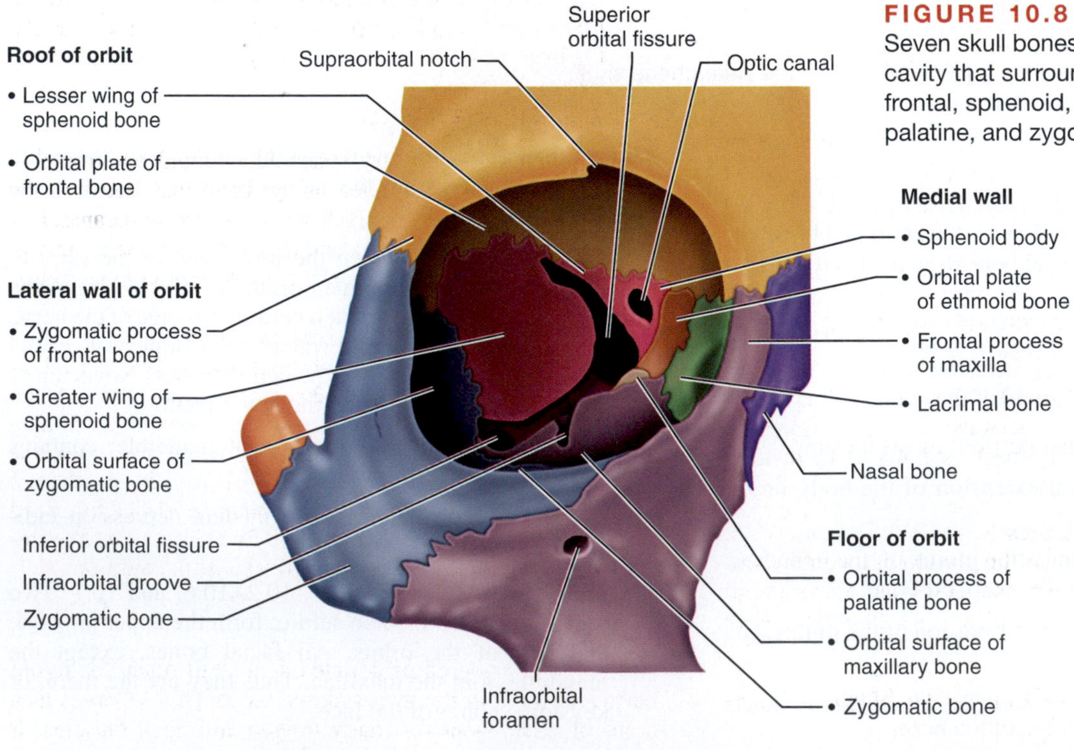
Roof of orbit
• Lesser wing of
sphenoid bone
• Orbital plate of
frontal bone
Supraorbital notch
Superior
orbital fissure
Optic canal
Lateral wall of orbit
• Zygomatic process
of frontal bone
• Greater wing of
sphenoid bone
• Orbital surface of
zygomatic bone
Inferior orbital fissure
Infraorbital groove
Zygomatic bone
Infraorbital
foramen
Medial wall
• Sphenoid body
• Orbital plate
of ethmoid bone
• Frontal process
of maxilla
• Lacrimal bone
Nasal bone
Floor of orbit
• Orbital process of
palatine bone
• Orbital surface of
maxillary bone
• Zygomatic bone

FIGURE 10.8 Bones that form the orbit. Seven skull bones form the orbit, the bony cavity that surrounds the eye. They are frontal, sphenoid, ethmoid, lacrimal, maxilla, palatine, and zygomatic.

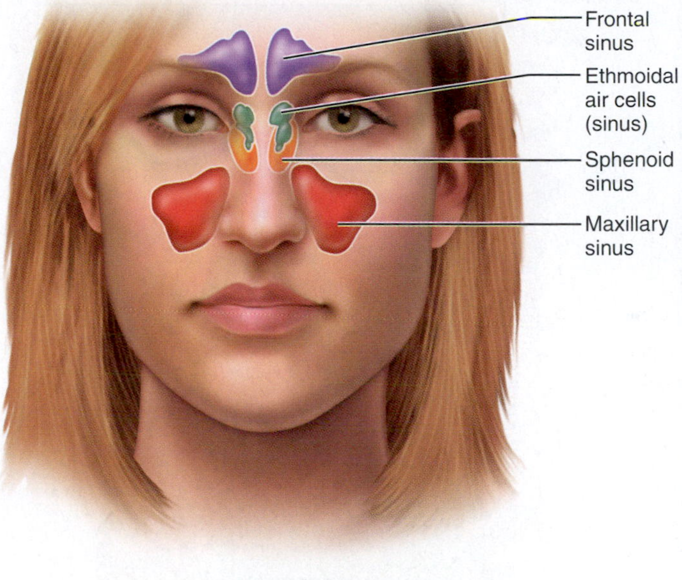

Frontal sinus
Ethmoidal air cells (sinus)
Sphenoid sinus
Maxillary sinus

(a)

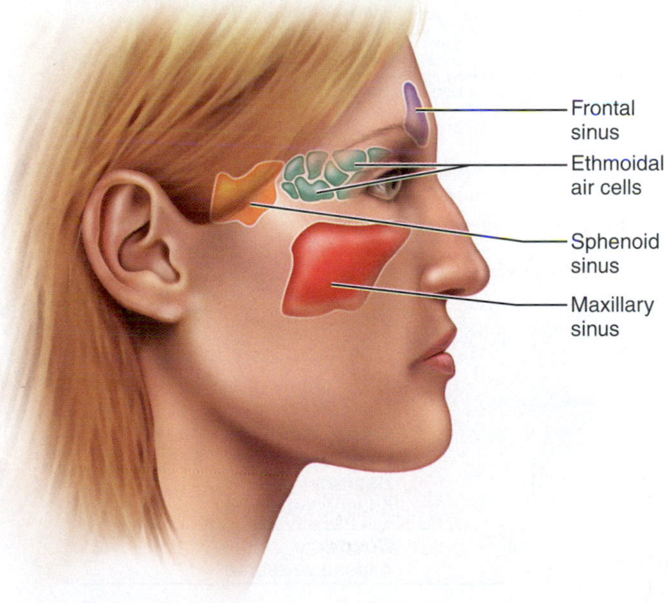

Frontal sinus
Ethmoidal air cells
Sphenoid sinus
Maxillary sinus

(b)

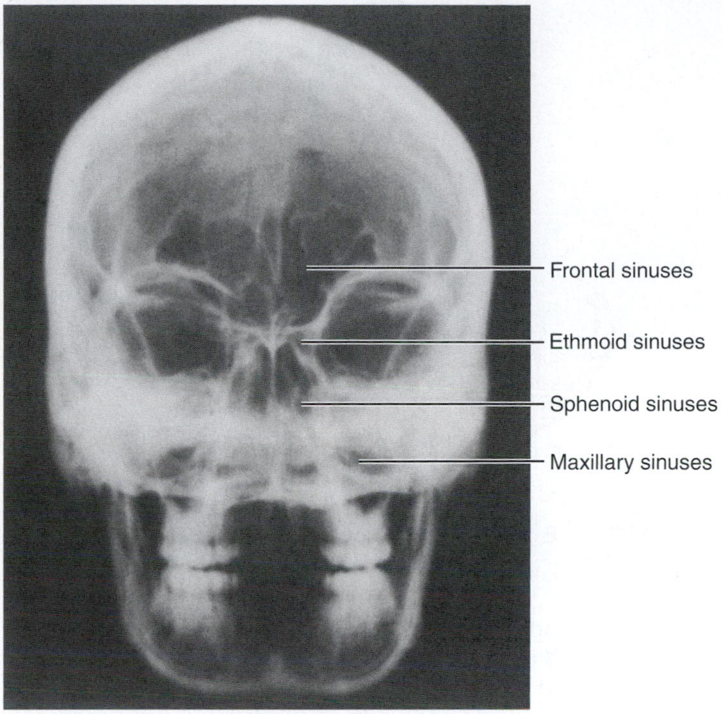

Frontal sinuses
Ethmoid sinuses
Sphenoid sinuses
Maxillary sinuses

(c)

FIGURE 10.9 Paranasal sinuses. (a) Anterior aspect. **(b)** Medial aspect. **(c)** Skull X ray showing the paranasal sinuses, anterior view.

pierced by an opening, the **lacrimal fossa,** which serves as a passageway for tears (*lacrima* = tear).

- ***Palatine Bone*** See Figure 10.2 and 10.8. Paired bones posterior to the palatine processes; form posterior hard palate and part of the orbit; meet medially at the median palatine suture.

- ***Zygomatic Bone*** See Figures 10.1, 10.2, 10.6, and 10.8. Lateral to the maxilla; forms the portion of the face commonly called the cheekbone, and forms part of the lateral orbit. Its three processes are named for the bones with which they articulate.

- ***Nasal Bone*** See Figures 10.1 and 10.6. Small rectangular bones forming the bridge of the nose.

- ***Vomer*** See Figures 10.2 and 10.6. Blade-shaped bone (*vomer* = plow) in median plane of nasal cavity that forms the posterior and inferior nasal septum.

- ***Inferior Nasal Conchae (Turbinates)*** See Figure 10.6a. Thin curved bones protruding medially from the lateral walls of the nasal cavity; serve the same purpose as the turbinate portions of the ethmoid bone (described earlier).

Paranasal Sinuses

Four skull bones—maxillary, sphenoid, ethmoid, and frontal—contain sinuses (mucosa-lined air cavities) that lead into the nasal passages (see Figures 10.5 and 10.9). These paranasal sinuses lighten the facial bones and may act as resonance chambers for speech. The maxillary sinus is the largest of the sinuses found in the skull.

Sinusitis, or inflammation of the sinuses, sometimes occurs as a result of an allergy or bacterial invasion of the sinus cavities. In such cases, some of the connecting passageways between the sinuses and nasal passages may become blocked with thick mucus or infectious material. Then, as the air in the sinus cavities is absorbed, a partial vacuum forms. The result is a sinus headache localized over the inflamed sinus area. Severe sinus infections may require surgical drainage to relieve this painful condition. ■

Hyoid Bone

Not really considered or counted as a skull bone, the hyoid bone is located in the throat above the larynx. It serves as a point of attachment for many tongue and neck muscles. It

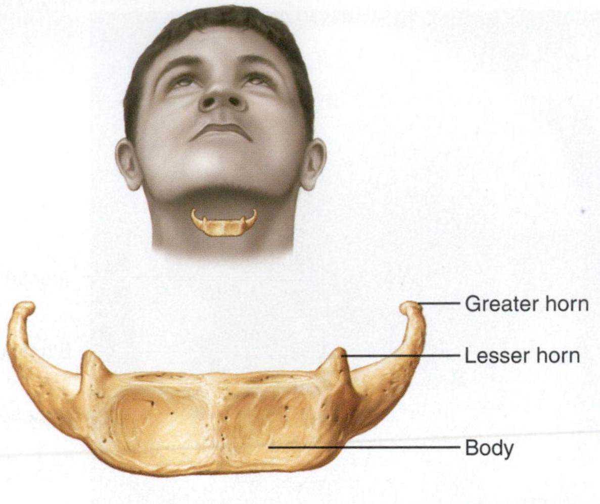
Greater horn
Lesser horn
Body

FIGURE 10.10 Hyoid bone.

does not articulate with any other bone and is thus unique. It is horseshoe-shaped with a body and two pairs of **horns,** or **cornua.** (Figure 10.10)

ACTIVITY 2

Palpating Skull Markings

Palpate the following areas on yourself:

- Zygomatic bone and arch. (The most prominent part of your cheek is your zygomatic bone. Follow the posterior course of the zygomatic arch to its junction with your temporal bone.)
- Mastoid process (the rough area behind your ear).
- Temporomandibular joints. (Open and close your jaws to locate these.)
- Greater wing of sphenoid. (Find the indentation posterior to the orbit and superior to the zygomatic arch on your lateral skull.)
- Supraorbital foramen. (Apply firm pressure along the superior orbital margin to find the indentation resulting from this foramen.)
- Infraorbital foramen. (Apply firm pressure just inferior to the inferomedial border of the orbit to locate this large foramen.)
- Mandibular angle (most inferior and posterior aspect of the mandible).
- Mandibular symphysis (midline of chin).
- Nasal bones. (Run your index finger and thumb along opposite sides of the bridge of your nose until they "slip" medially at the inferior end of the nasal bones.)
- External occipital protuberance. (This midline projection is easily felt by running your fingers up the furrow at the back of your neck to the skull.)
- Hyoid bone. (Place a thumb and index finger beneath the chin just anterior to the mandibular angles, and squeeze gently. Exert pressure with the thumb, and feel the horn of the hyoid with the index finger.)

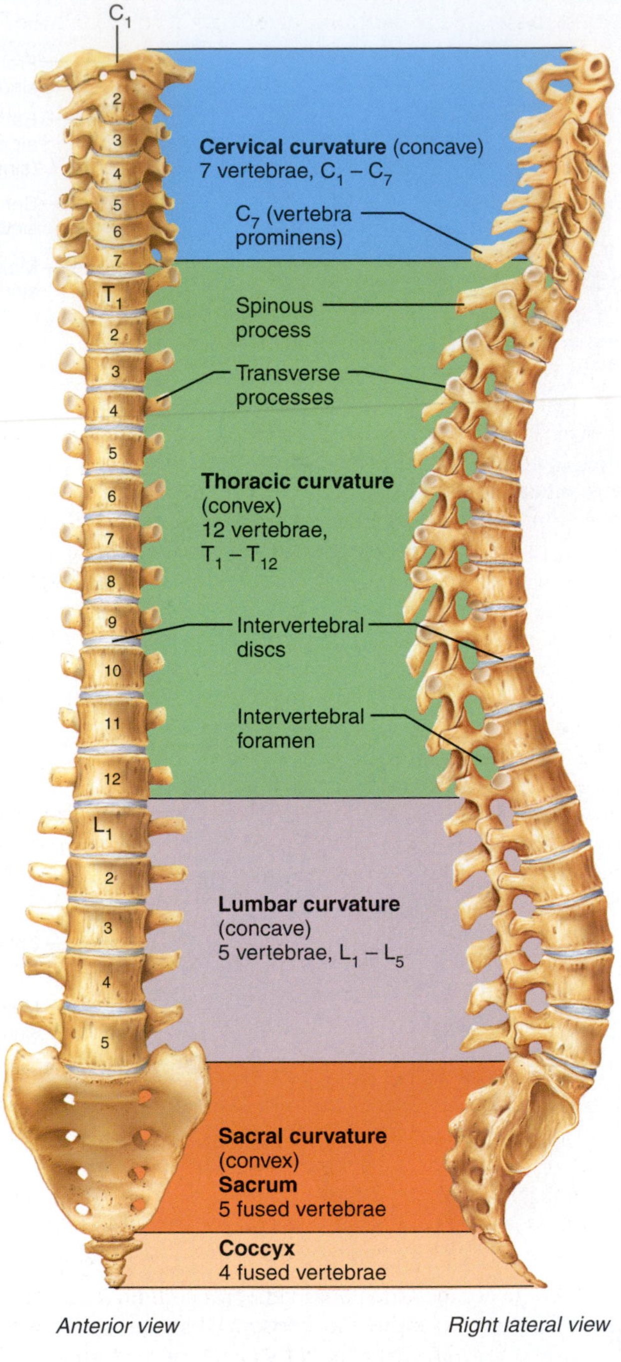
C_1
Cervical curvature (concave)
7 vertebrae, $C_1 - C_7$
C_7 (vertebra prominens)
Spinous process
Transverse processes
Thoracic curvature (convex)
12 vertebrae, $T_1 - T_{12}$
Intervertebral discs
Intervertebral foramen
Lumbar curvature (concave)
5 vertebrae, $L_1 - L_5$
Sacral curvature (convex)
Sacrum
5 fused vertebrae
Coccyx
4 fused vertebrae
Anterior view
Right lateral view

FIGURE 10.11 The vertebral column. Notice the curvatures in the lateral view. (The terms *convex* and *concave* refer to the curvature of the posterior aspect of the vertebral column.)

The Vertebral Column

The **vertebral column,** extending from the skull to the pelvis, forms the body's major axial support. Additionally, it surrounds and protects the delicate spinal cord while allowing the spinal nerves to issue from the cord via openings between adjacent vertebrae. The term *vertebral column* might suggest a rather rigid supporting rod, but this is far from the truth. The vertebral column consists of 24 single bones called **vertebrae** and two composite, or fused, bones (the sacrum and coccyx) that are connected in such a way as to provide a flexible curved structure (Figure 10.11). Of the 24 single vertebrae, the seven bones of the neck are called *cervical vertebrae;* the next 12 are *thoracic vertebrae;* and the 5 supporting the lower back are *lumbar vertebrae.* Remembering common mealtimes for breakfast, lunch, and dinner (7 A.M., 12 noon, and 5 P.M.) may help you to remember the number of bones in each region.

The vertebrae are separated by pads of fibrocartilage, **intervertebral discs,** that cushion the vertebrae and absorb shocks. Each disc is composed of two major regions, a central gelatinous *nucleus pulposus* that behaves like a fluid, and an outer ring of encircling collagen fibers called the *annulus fibrosus* that stabilizes the disc and contains the pulposus.

As a person ages, the water content of the discs decreases (as it does in other tissues throughout the body), and the discs become thinner and less compressible. This situation, along with other degenerative changes such as weakening of the ligaments and tendons of the vertebral column, predisposes older people to **ruptured discs.** In a ruptured disc, the nucleus pulposus herniates through the annulus portion and typically compresses adjacent nerves. ■

The presence of the discs and the S-shaped or spring-like construction of the vertebral column prevent shock to the head in walking and running and provide flexibility to the body trunk. The thoracic and sacral curvatures of the spine are referred to as *primary curvatures* because they are present and well developed at birth. Later the *secondary curvatures* are formed. The cervical curvature becomes prominent when the baby begins to hold its head up independently, and the lumbar curvature develops when the baby begins to walk.

ACTIVITY 3

Examining Spinal Curvatures

1. Observe the normal curvature of the vertebral column in the articulated vertebral column or laboratory skeleton, and compare it to Figure 10.11. Then examine Figure 10.12, which depicts three abnormal spinal curvatures—*scoliosis, kyphosis,* and *lordosis.* These abnormalities may result from disease or poor posture. Also examine X-rays, if they are available, showing these same conditions in a living patient.

2. Then, using the articulated vertebral column (or an articulated skeleton), examine the freedom of movement between two lumbar vertebrae separated by an intervertebral disc.

When the fibrous disc is properly positioned, are the spinal cord or peripheral nerves impaired in any way?

____

Remove the disc and put the two vertebrae back together. What happens to the nerve?

____

What would happen to the spinal nerves in areas of malpositioned or "slipped" discs?

______________________________ ■

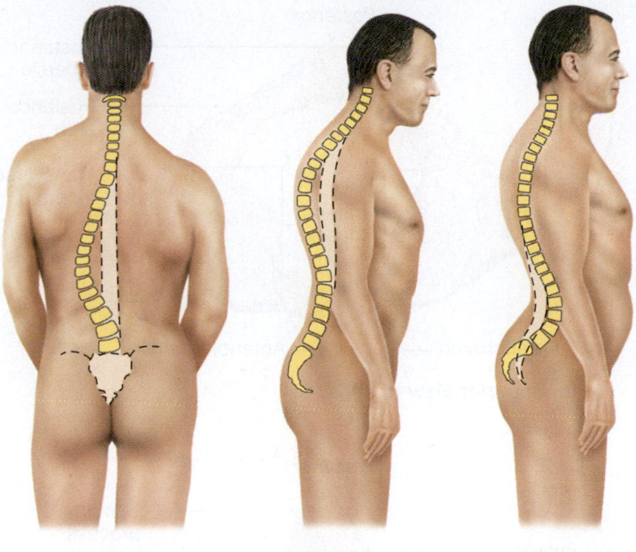
Scoliosis
Kyphosis
Lordosis

FIGURE 10.12 Abnormal spinal curvatures

Structure of a Typical Vertebra

Although they differ in size and specific features, all vertebrae have some features in common (Figure 10.13).

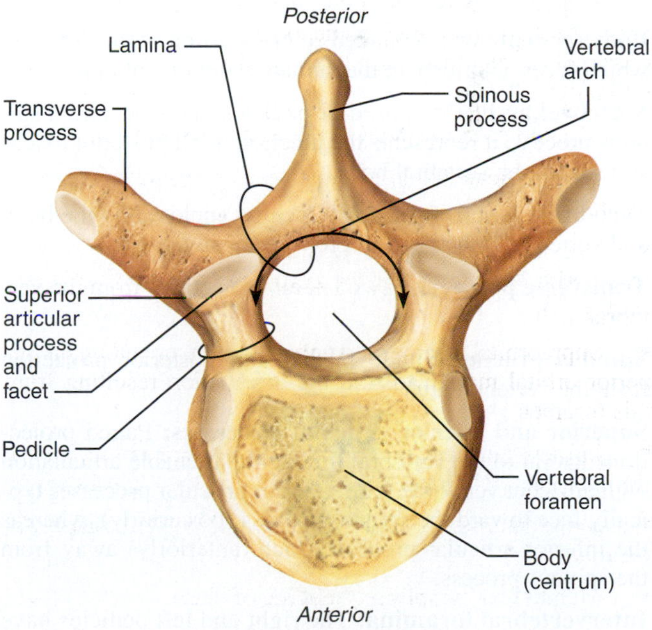
Posterior
Lamina
Vertebral arch
Transverse process
Spinous process
Superior articular process and facet
Pedicle
Vertebral foramen
Body (centrum)
Anterior

FIGURE 10.13 A typical vertebra, superior view. Inferior articulating surfaces not shown.

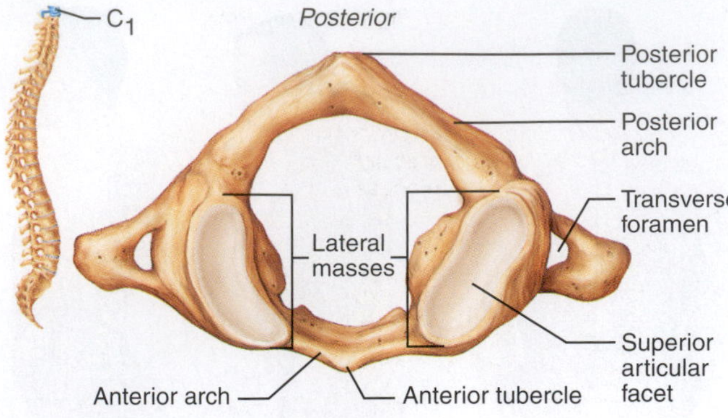

C_1
Posterior
Posterior tubercle
Posterior arch
Transverse foramen
Lateral masses
Superior articular facet
Anterior arch
Anterior tubercle

(a) Superior view of atlas (C_1)

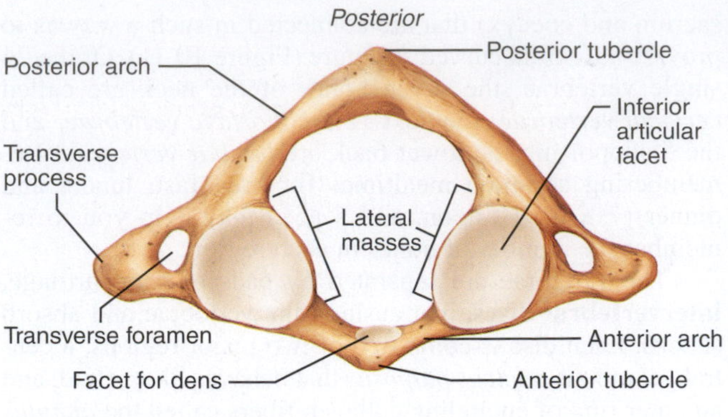

Posterior
Posterior arch
Posterior tubercle
Inferior articular facet
Transverse process
Lateral masses
Transverse foramen
Anterior arch
Facet for dens
Anterior tubercle

(b) Inferior view of atlas (C_1)

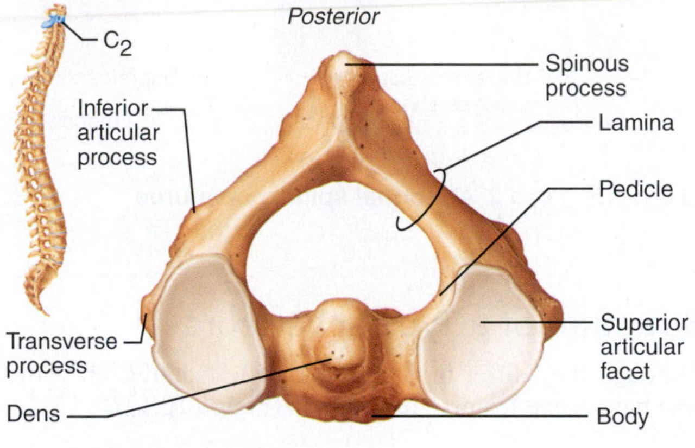

C_2
Posterior
Spinous process
Lamina
Inferior articular process
Pedicle
Transverse process
Superior articular facet
Dens
Body

(c) Superior view of axis (C_2)

FIGURE 10.14 The first and second cervical vertebrae.

Body (or **centrum**): Rounded central portion of the vertebra, which faces anteriorly in the human vertebral column.

Vertebral arch: Composed of pedicles, laminae, and a spinous process, it represents the junction of all posterior extensions from the vertebral body.

Vertebral (spinal) foramen: Opening enclosed by the body and vertebral arch; a conduit for the spinal cord.

Transverse processes: Two lateral projections from the vertebral arch.

Spinous process: Single medial and posterior projection from the vertebral arch.

Superior and inferior articular processes: Paired projections lateral to the vertebral foramen that enable articulation with adjacent vertebrae. The superior articular processes typically face toward the spinous process (posteriorly), whereas the inferior articular processes face (anteriorly) away from the spinous process.

Intervertebral foramina: The right and left pedicles have notches (see Figure 10.15) on their inferior and superior surfaces that create openings, the intervertebral foramina, for spinal nerves to leave the spinal cord between adjacent vertebrae.

Figures 10.14 through 10.16 and Table 10.1 show how specific vertebrae differ; refer to them as you read the following sections.

Cervical Vertebrae

The seven cervical vertebrae (referred to as C_1 through C_7) form the neck portion of the vertebral column. The first two cervical vertebrae (atlas and axis) are highly modified to perform special functions (see Figure 10.14). The **atlas** (C_1) lacks a body, and its lateral processes contain large concave depressions on their superior surfaces that receive the occipital condyles of the skull. This joint enables you to nod "yes." The **axis** (C_2) acts as a pivot for the rotation of the atlas (and skull) above. It bears a large vertical process, the **dens,** or **odontoid process,** that serves as the pivot point. The articulation between C_1 and C_2 allows you to rotate your head from side to side to indicate "no."

The more typical cervical vertebrae (C_3 through C_7) are distinguished from the thoracic and lumbar vertebrae by several features (see Table 10.1 and Figure 10.15). They are the smallest, lightest vertebrae, and the vertebral foramen is triangular. The spinous process is short and often bifurcated (divided into two branches). The spinous process of C_7 is not branched, however, and is substantially longer than that of the other cervical vertebrae. Because the spinous process of C_7 is visible through the skin, it is called the *vertebra prominens* (Figure 10.11) and is used as a landmark for counting the vertebrae. Transverse processes of the cervical vertebrae are wide, and they contain foramina through which the vertebral arteries pass superiorly on their way to the brain. Any time you see these foramina in a vertebra, you can be sure that it is a cervical vertebra.

- Palpate your vertebra prominens.

Thoracic Vertebrae

The 12 thoracic vertebrae (referred to as T_1 through T_{12}) may be recognized by the following structural characteristics. As shown in Figure 10.15, they have a larger body than the cervical vertebrae. The body is somewhat heart-shaped, with two small articulating surfaces, or **costal facets,** on each side (one superior, the other inferior) close to the origin of the vertebral arch. Sometimes referred to as *costal demifacets* because of their small size, these facets articulate with the heads of the corresponding ribs. The vertebral foramen is oval or round, and the spinous process is long, with a sharp downward hook. The closer the thoracic vertebra is to the lumbar region, the less sharp and shorter the spinous process. Articular facets on the transverse processes articulate with the tubercles of the

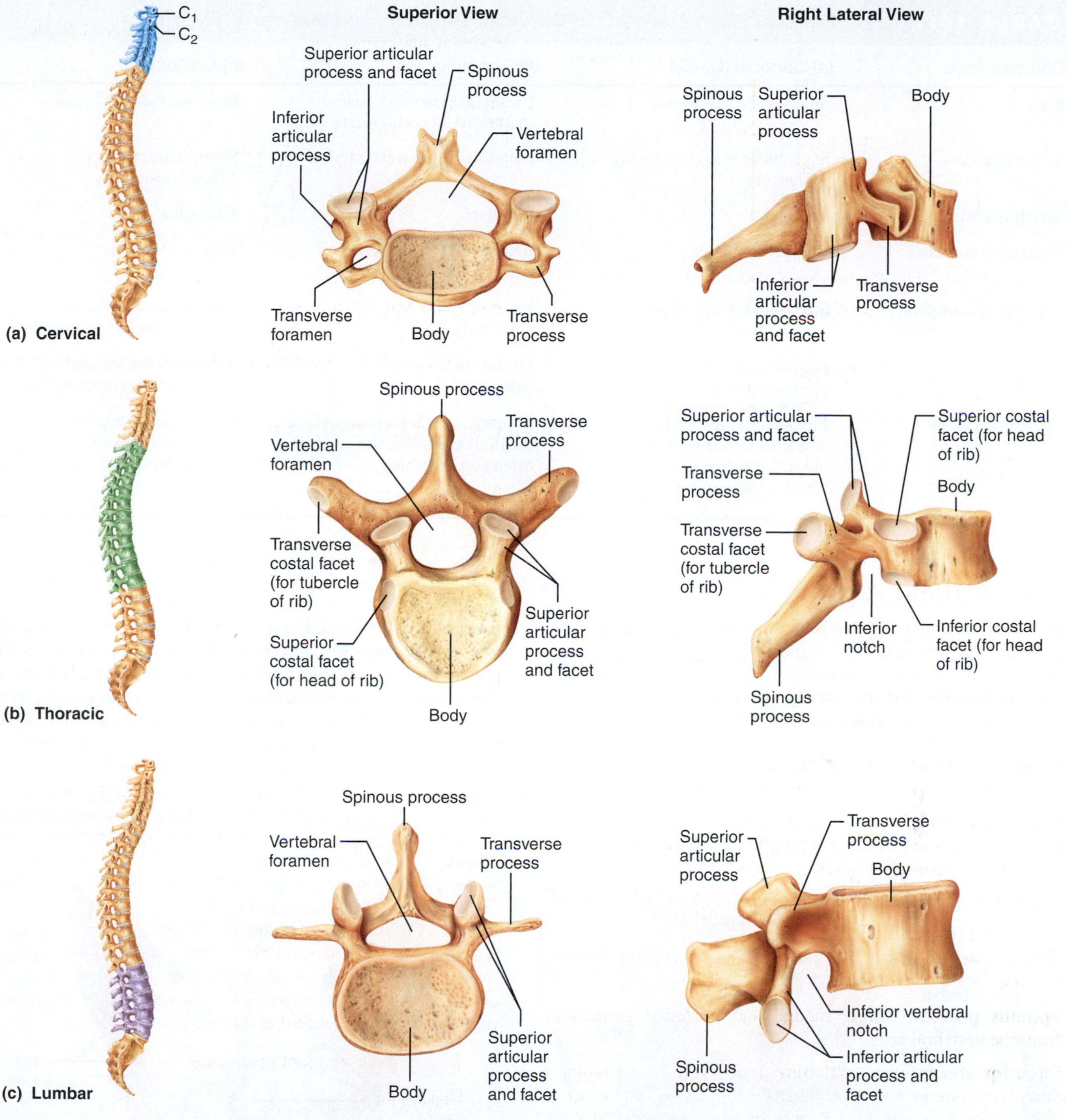
C1
C2
Superior View
Right Lateral View
Superior articular process and facet
Spinous process
Inferior articular process
Vertebral foramen
Transverse foramen
Body
Transverse process
Spinous process
Superior articular process
Body
Inferior articular process and facet
Transverse process
(a) Cervical
Spinous process
Transverse process
Vertebral foramen
Transverse costal facet (for tubercle of rib)
Superior costal facet (for head of rib)
Superior articular process and facet
Body
Superior articular process and facet
Superior costal facet (for head of rib)
Transverse process
Body
Transverse costal facet (for tubercle of rib)
Inferior notch
Inferior costal facet (for head of rib)
Spinous process
(b) Thoracic
Spinous process
Vertebral foramen
Transverse process
Superior articular process and facet
Body
Superior articular process
Transverse process
Body
Inferior vertebral notch
Inferior articular process and facet
Spinous process
(c) Lumbar

FIGURE 10.15 Superior and right lateral views of typical vertebrae. (a) Cervical. **(b)** Thoracic. **(c)** Lumbar.

ribs. Besides forming the thoracic part of the spine, these vertebrae form the posterior aspect of the bony thorax. Indeed, they are the only vertebrae that articulate with the ribs.

Lumbar Vertebrae

The five lumbar vertebrae (L_1 through L_5) have massive blocklike bodies and short, thick, hatchet-shaped spinous processes extending directly backward (see Table 10.1 and Figure 10.15). The superior articular facets face posteromedially; the inferior ones are directed anterolaterally. These structural features reduce the mobility of the lumbar region of the spine. Since most stress on the vertebral column occurs in the lumbar region, these are also the sturdiest of the vertebrae.

The spinal cord ends at the superior edge of L_2, but the outer covering of the cord, filled with cerebrospinal fluid, extends an appreciable distance beyond. Thus a *lumbar puncture* (for examination of the cerebrospinal fluid) or the administration of "saddle block" anesthesia for childbirth is normally done between L_3 and L_4 or L_4 and L_5, where there is little or no chance of injuring the delicate spinal cord.

TABLE 10.1	Regional Characteristics of Cervical, Thoracic, and Lumbar Vertebrae		
Characteristic	**(a) Cervical (C_3–C_7)**	**(b) Thoracic**	**(c) Lumbar**
Body	Small, wide side to side	Larger than cervical; heart-shaped; bears costal facets	Massive; kidney-shaped
Spinous process	Short; bifid; projects directly posteriorly	Long; sharp; projects inferiorly	Short; blunt; projects directly posteriorly
Vertebral foramen	Triangular	Circular	Triangular
Transverse processes	Contain foramina	Bear facets for ribs (except T_{11} and T_{12})	Thin and tapered
Superior and inferior articulating processes	Superior facets directed superoposteriorly	Superior facets directed posteriorly	Superior facets directed posteromedially (or medially)
	Inferior facets directed inferoanteriorly	Inferior facets directed anteriorly	Inferior facets directed anterolaterally (or laterally)
Movements allowed	Flexion and extension; lateral flexion; rotation; the spine region with the greatest range of movement	Rotation; lateral flexion possible but limited by ribs; flexion and extension prevented	Flexion and extension; some lateral flexion; rotation prevented

The Sacrum

The **sacrum** (Figure 10.16) is a composite bone formed from the fusion of five vertebrae. Superiorly it articulates with L_5, and inferiorly it connects with the coccyx. The **median sacral crest** is a remnant of the spinous processes of the fused vertebrae. The winglike **alae,** formed by fusion of the transverse processes, articulate laterally with the hip bones. The sacrum is concave anteriorly and forms the posterior border of the pelvis. Four ridges (lines of fusion) cross the anterior part of the sacrum, and **sacral foramina** are located at either end of these ridges. These foramina allow blood vessels and nerves

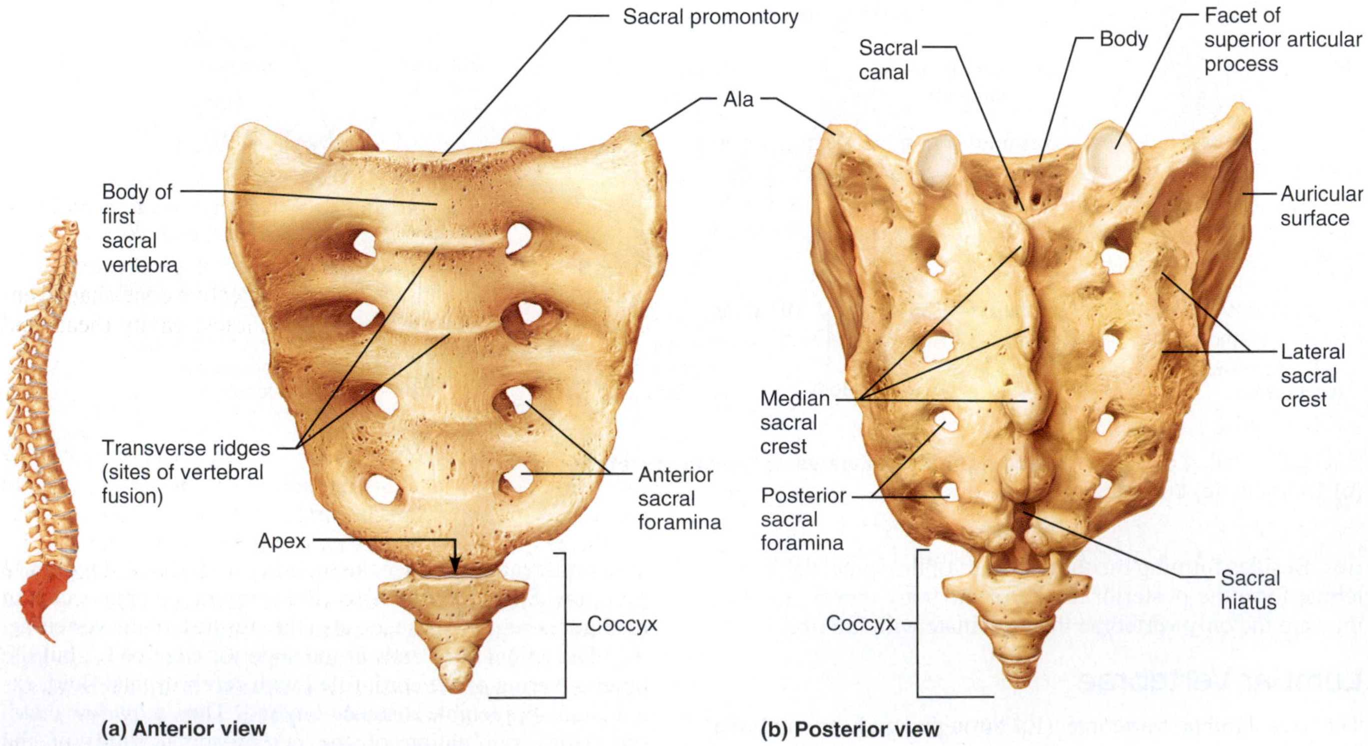

Sacral promontory
Ala
Body of first sacral vertebra
Transverse ridges (sites of vertebral fusion)
Anterior sacral foramina
Apex
Coccyx
(a) Anterior view
Sacral canal
Body
Facet of superior articular process
Auricular surface
Lateral sacral crest
Median sacral crest
Posterior sacral foramina
Sacral hiatus
Coccyx
(b) Posterior view

FIGURE 10.16 Sacrum and coccyx.

FIGURE 10.17 The thoracic cage. (a) Skeleton of the bony thorax, anterior view (costal cartilages are shown in blue). **(b)** Midsagittal section of the thorax, illustrating the relationship of the surface anatomical landmarks of the thorax to the vertebral column (thoracic portion).

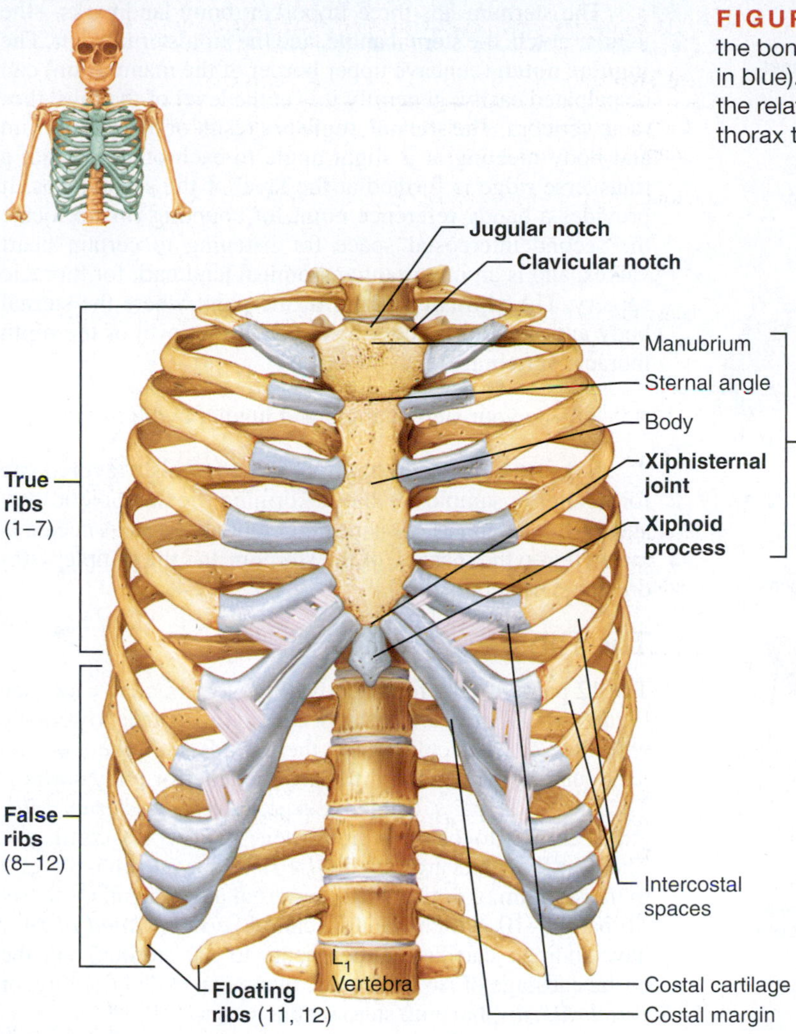

Jugular notch
Clavicular notch
Manubrium
Sternal angle
Body
Xiphisternal joint
Xiphoid process
Sternum
True ribs (1–7)
False ribs (8–12)
Intercostal spaces
L1 Vertebra
Floating ribs (11, 12)
Costal cartilage
Costal margin

(a)

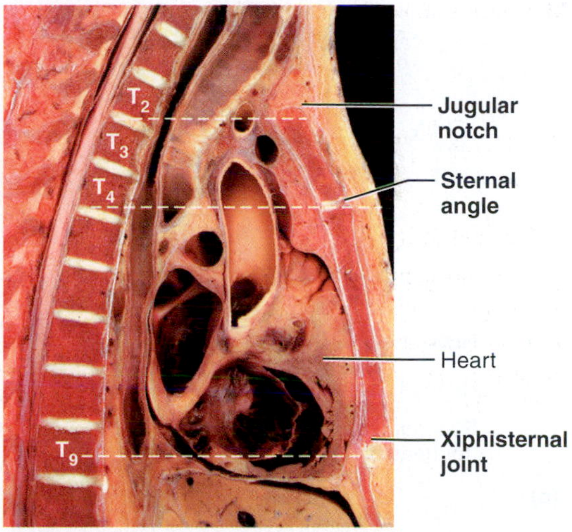

T2
T3
T4
T9
Jugular notch
Sternal angle
Heart
Xiphisternal joint

(b)

to pass. The vertebral canal continues inside the sacrum as the **sacral canal** and terminates near the coccyx via an enlarged opening called the **sacral hiatus.** The **sacral promontory** (anterior border of the body of S_1) is an important anatomical landmark for obstetricians.

- Attempt to palpate the median sacral crest of your sacrum. (This is more easily done by thin people and obviously in privacy.)

The Coccyx

The **coccyx** (see Figure 10.16) is formed from the fusion of three to five small irregularly shaped vertebrae. It is literally the human tailbone, a vestige of the tail that other vertebrates have. The coccyx is attached to the sacrum by ligaments.

ACTIVITY 4

Examining Vertebral Structure

Obtain examples of each type of vertebra and examine them carefully, comparing them to Figures 10.14, 10.15, 10.16 and Table 10.1, and to each other. ■

The Thoracic Cage

The bony thorax is composed of the sternum, ribs, and thoracic vertebrae (Figure 10.17). Combined with the costal cartilages, it is referred to as the **thoracic cage** because of its appearance and because it forms a protective cone-shaped enclosure around the organs of the thoracic cavity (heart and lungs, for example).

The Sternum

The **sternum** (breastbone), a typical flat bone, is a result of the fusion of three bones—the manubrium, body, and xiphoid process. It is attached to the first seven pairs of ribs. The superiormost **manubrium** looks like the knot of a tie; it articulates with the clavicle (collarbone) laterally. The **body (gladiolus)** forms the bulk of the sternum. The **xiphoid process** constructs the inferior end of the sternum and lies at the level of the fifth intercostal space. Although it is made of hyaline cartilage in children, it is usually ossified in adults.

In some people, the xiphoid process projects dorsally. This may present a problem because physical trauma to the chest can push such a xiphoid into the heart or liver (both immediately deep to the process), causing massive hemorrhage. ■

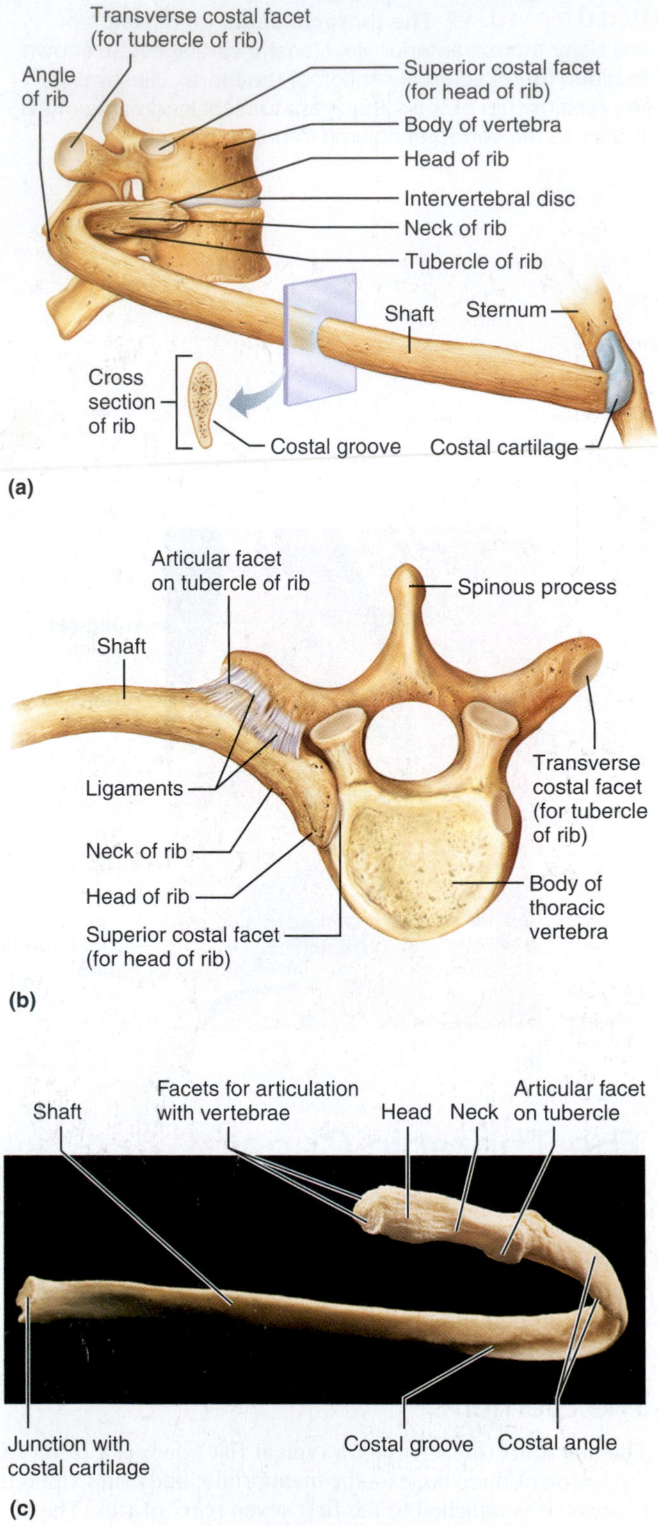
Transverse costal facet
(for tubercle of rib)
Angle
of rib
Superior costal facet
(for head of rib)
Body of vertebra
Head of rib
Intervertebral disc
Neck of rib
Tubercle of rib
Shaft
Sternum
Cross
section
of rib
Costal groove
Costal cartilage
(a)
Articular facet
on tubercle of rib
Spinous process
Shaft
Ligaments
Neck of rib
Head of rib
Superior costal facet
(for head of rib)
Transverse
costal facet
(for tubercle
of rib)
Body of
thoracic
vertebra
(b)
Shaft
Facets for articulation
with vertebrae
Head
Neck
Articular facet
on tubercle
Junction with
costal cartilage
Costal groove
Costal angle
(c)

FIGURE 10.18 Structure of a typical true rib and its articulations. (a) Vertebral and sternal articulations of a typical true rib. **(b)** Superior view of the articulation between a rib and a thoracic vertebra, with costovertebral ligaments shown on left side only. **(c)** Right rib 6, posterior view.

The sternum has three important bony landmarks—the jugular notch, the sternal angle, and the xiphisternal joint. The **jugular notch** (concave upper border of the manubrium) can be palpated easily; generally it is at the level of the third thoracic vertebra. The **sternal angle** is a result of the manubrium and body meeting at a slight angle to each other, so that a transverse ridge is formed at the level of the second ribs. It provides a handy reference point for counting ribs to locate the second intercostal space for listening to certain heart valves, and is an important anatomical landmark for thoracic surgery. The **xiphisternal joint,** the point where the sternal body and xiphoid process fuse, lies at the level of the ninth thoracic vertebra.

- Palpate your sternal angle and jugular notch.

Because of its accessibility, the sternum is a favored site for obtaining samples of blood-forming (hematopoietic) tissue for the diagnosis of suspected blood diseases. A needle is inserted into the marrow of the sternum and the sample withdrawn (sternal puncture).

The Ribs

The 12 pairs of **ribs** form the walls of the thoracic cage (see Figures 10.17 and 10.18). All of the ribs articulate posteriorly with the vertebral column via their heads and tubercles and then curve downward and toward the anterior body surface. The first seven pairs, called the *true,* or *vertebrosternal, ribs,* attach directly to the sternum by their "own" costal cartilages. The next five pairs are called *false ribs;* they attach indirectly to the sternum or entirely lack a sternal attachment. Of these, rib pairs 8–10, which are also called *vertebrochondral ribs,* have indirect cartilage attachments to the sternum via the costal cartilage of rib 7. The last two pairs, called *floating,* or *vertebral, ribs,* have no sternal attachment.

ACTIVITY 5

Examining the Relationship Between Ribs and Vertebrae

First take a deep breath to expand your chest. Notice how your ribs seem to move outward and how your sternum rises. Then examine an articulated skeleton to observe the relationship between the ribs and the vertebrae.

Refer to Activity 3 (Palpating Landmarks of the Trunk) and Activity 4 (Palpating Landmarks of the Abdomen) in Exercise 46, Surface Anatomy Roundup.

EXERCISE 11

The Appendicular Skeleton

MATERIALS

- ☐ Articulated skeletons
- ☐ Disarticulated skeletons (complete)
- ☐ Articulated pelves (male and female for comparative study)
- ☐ X rays of bones of the appendicular skeleton

OBJECTIVES

1. To identify on an articulated skeleton the bones of the pectoral and pelvic girdles and their attached limbs.
2. To arrange unmarked, disarticulated bones in proper relative position to form the entire skeleton.
3. To differentiate between a male and a female pelvis.
4. To discuss the common features of the human appendicular girdles (pectoral and pelvic), and to note how their structure relates to their specialized functions.
5. To identify specific bone markings in the appendicular skeleton.

PRE-LAB QUIZ

1. The ______________________ skeleton is composed of 126 bones of the appendages and pectoral and pelvic girdles.
2. Circle the correct term. The pectoral / pelvic girdle attaches the upper limb to the axial skeleton.
3. The ______________________ or shoulder blades are generally triangular in shape. They have no direct attachment to the axial skeleton but are held in place by trunk muscles.
4. The arm consists of one long bone, the ________.
   a. femur
   b. humerus
   c. tibia
   d. ulna
5. The hand consists of three groups of bones. The carpals make up the wrist. The ______________________ make up the palm, and the phalanges make up the fingers.
6. You are studying a pelvis that is wide and shallow. The acetabula are small and far apart. The pubic arch/angle is rounded and greater than 90°. It appears to be tilted forward, with a wide, short sacrum. Is this a male or a female pelvis? ______________
7. The strongest, heaviest bone of the body is in the thigh. It is the
   a. femur
   b. fibula
   c. tibia
8. The ________________, or "knee cap," is a sesamoid bone that is enclosed in the quadriceps tendon. It guards the knee joint anteriorly and improves leverage of the thigh muscles acting across the knee joint.
9. Circle True or False. The fingers of the hand and the toes of the foot—with the exception of the great toe and the thumb—each have three phalanges.

MasteringA&P™

Access practice quizzes and more in the Study Area at www.masteringaandp.com.

PAL

For access to anatomical models and more, check out Practice Anatomy Lab.

The **appendicular skeleton** (the gold-colored portion of Figure 9.1) is composed of the 126 bones of the appendages and the pectoral and pelvic girdles, which attach the limbs to the axial skeleton. Although the upper and lower limbs differ in their functions and mobility, they have the same fundamental plan, with each limb composed of three major segments connected together by freely movable joints.

ACTIVITY 1

Examining and Identifying Bones of the Appendicular Skeleton

Carefully examine each of the bones described throughout this exercise and identify the characteristic bone markings of each. The markings aid in determining whether a bone is the right or left member of its pair; for example, the glenoid cavity is on the lateral aspect of the scapula and the spine is on its posterior aspect. *This is a very important instruction because you will be constructing your own skeleton to finish this laboratory exercise.* Additionally, when corresponding X rays are available, compare the actual bone specimen to its X-ray image.

Bones of the Pectoral Girdle and Upper Limb

The Pectoral (Shoulder) Girdle

The paired **pectoral,** or **shoulder, girdles** (Figure 11.1) each consist of two bones—the anterior clavicle and the posterior scapula. The shoulder girdles function to attach the upper limbs to the axial skeleton and provide attachment points for many trunk and neck muscles.

The **clavicle,** or collarbone, is a slender doubly curved bone—convex forward on its medial two-thirds and concave laterally. Its *sternal* (medial) *end,* which attaches to the sternal manubrium, is rounded or triangular in cross section. The sternal end projects above the manubrium and can be felt and (usually) seen forming the lateral walls of the *jugular notch* (see Figure 10.17, page 137). The *acromial* (lateral) *end* of the clavicle is flattened where it articulates with the scapula to form part of the shoulder joint. On its posteroinferior surface is the prominent **conoid tubercle** (Figure 11.2b). This projection anchors a ligament and provides a handy landmark for determining whether a given clavicle is from the right or left side of the body. The clavicle serves as an anterior brace, or strut, to hold the arm away from the top of the thorax.

The **scapulae** (Figure 11.2c–e), or shoulder blades, are generally triangular and are commonly called the "wings" of humans. Each scapula has a flattened body and two important processes—the **acromion** (the enlarged, roughened end of the spine of the scapula) and the beaklike **coracoid process** (*corac* = crow, raven). The acromion connects with the clavicle; the coracoid process points anteriorly over the tip of the shoulder joint and serves as an attachment point for some of the upper limb muscles. The **suprascapular notch** at the base of the coracoid process allows nerves to pass. The scapula has no direct attachment to the axial skeleton but is loosely held in place by trunk muscles.

The scapula has three angles: superior, inferior, and lateral. The inferior angle provides a landmark for auscultating (listening to) lung sounds. The **glenoid cavity,** a shallow socket that receives the head of the arm bone (humerus), is located in the blunted lateral angle. The scapula also has three named borders: superior, medial (vertebral), and lateral (axillary). Several shallow depressions (fossae) appear on both sides of the scapula and are named according to location; there are the anterior *subscapular fossa* and the posterior *infraspinous* and *supraspinous fossae.*

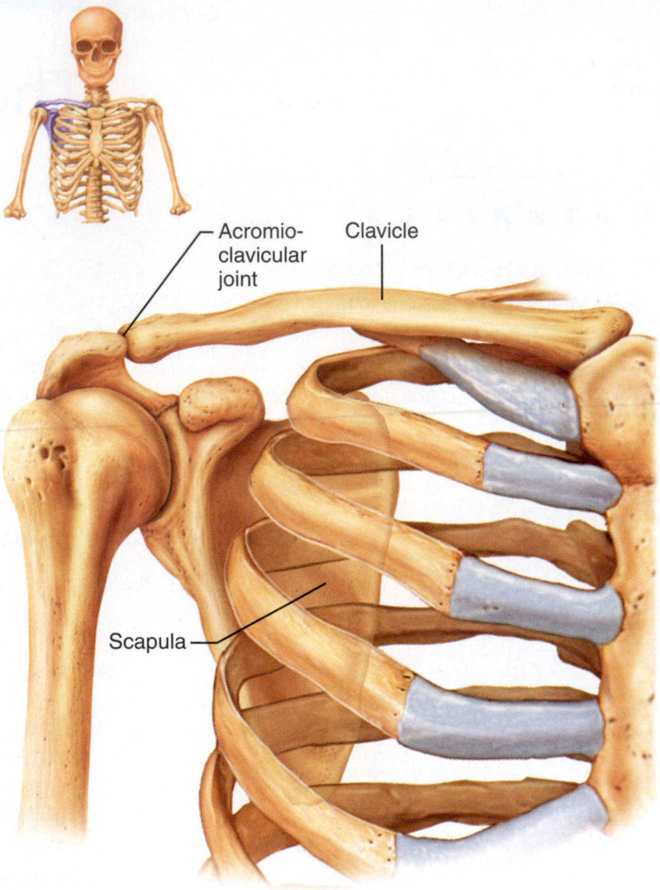
Acromio-
clavicular
joint
Clavicle
Scapula

FIGURE 11.1 Articulated bones of the pectoral (shoulder) girdle. The right pectoral girdle is articulated to show the relationship of the girdle to the bones of the thorax and arm.

The shoulder girdle is exceptionally light and allows the upper limb a degree of mobility not seen anywhere else in the body. This is due to the following factors:

- The sternoclavicular joints are the *only* site of attachment of the shoulder girdles to the axial skeleton.
- The relative looseness of the scapular attachment allows it to slide back and forth against the thorax with muscular activity.
- The glenoid cavity is shallow and does little to stabilize the shoulder joint.

However, this exceptional flexibility exacts a price: the arm bone (humerus) is very susceptible to dislocation, and fracture of the clavicle disables the entire upper limb.

The Arm

The arm (Figure 11.3) consists of a single bone—the **humerus,** a typical long bone. Proximally its rounded *head* fits into the shallow glenoid cavity of the scapula. The head is separated from the shaft by the *anatomical neck* and the more constricted *surgical neck,* which is a common site of fracture. Opposite the head are two prominences, the **greater** and **lesser tubercles** (from lateral to medial aspect), separated by

Text continues on page 148.

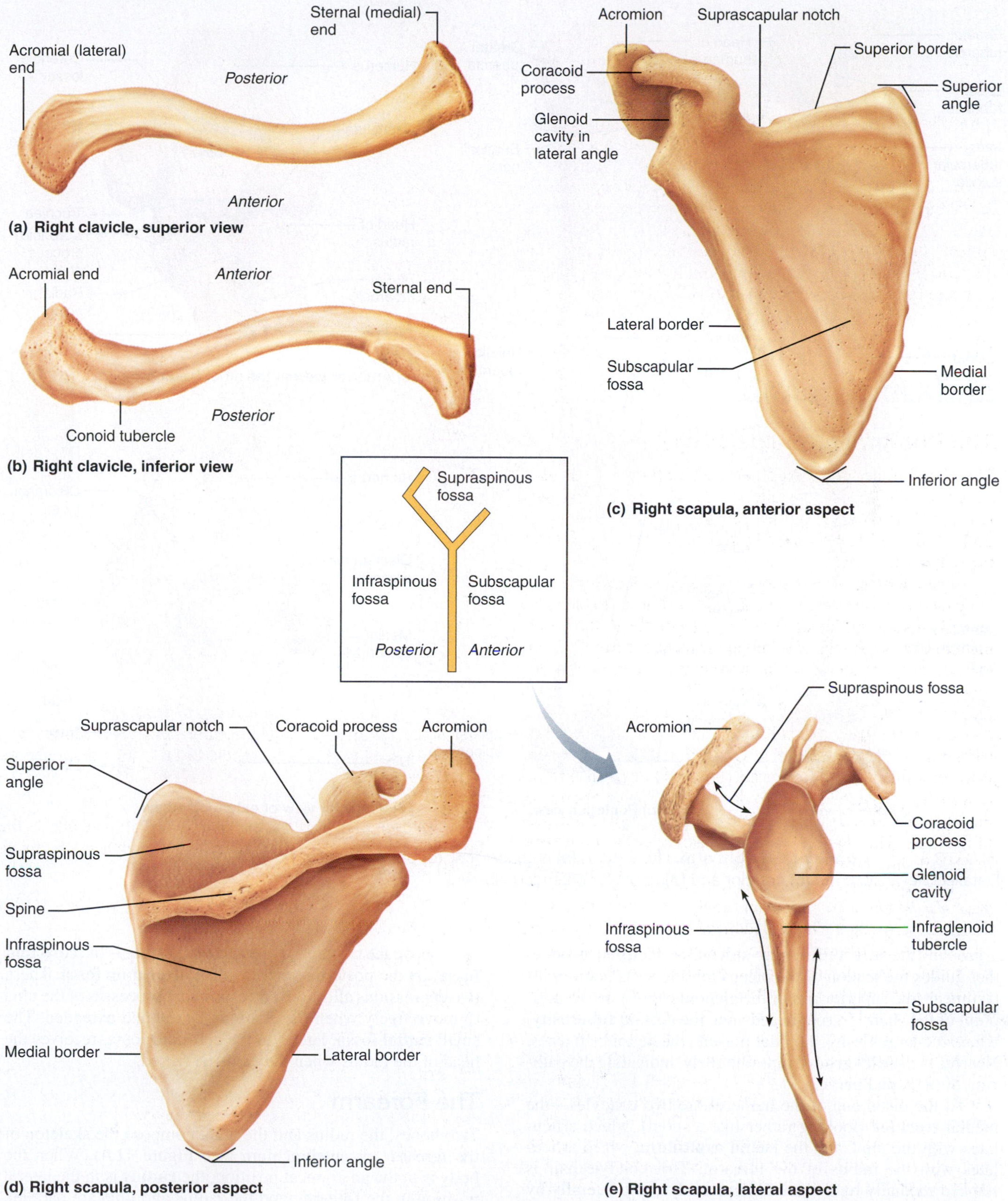
Acromial (lateral) end
Posterior
Sternal (medial) end
Anterior
(a) Right clavicle, superior view
Acromial end
Anterior
Sternal end
Posterior
Conoid tubercle
(b) Right clavicle, inferior view
Acromion
Suprascapular notch
Superior border
Coracoid process
Superior angle
Glenoid cavity in lateral angle
Lateral border
Subscapular fossa
Medial border
Inferior angle
(c) Right scapula, anterior aspect
Supraspinous fossa
Infraspinous fossa
Subscapular fossa
Posterior
Anterior
Suprascapular notch
Coracoid process
Acromion
Superior angle
Supraspinous fossa
Spine
Infraspinous fossa
Medial border
Lateral border
Inferior angle
(d) Right scapula, posterior aspect
Supraspinous fossa
Acromion
Coracoid process
Glenoid cavity
Infraspinous fossa
Infraglenoid tubercle
Subscapular fossa
(e) Right scapula, lateral aspect

FIGURE 11.2 Individual bones of the pectoral (shoulder) girdle. View **(e)** is accompanied by a schematic representation of its orientation.

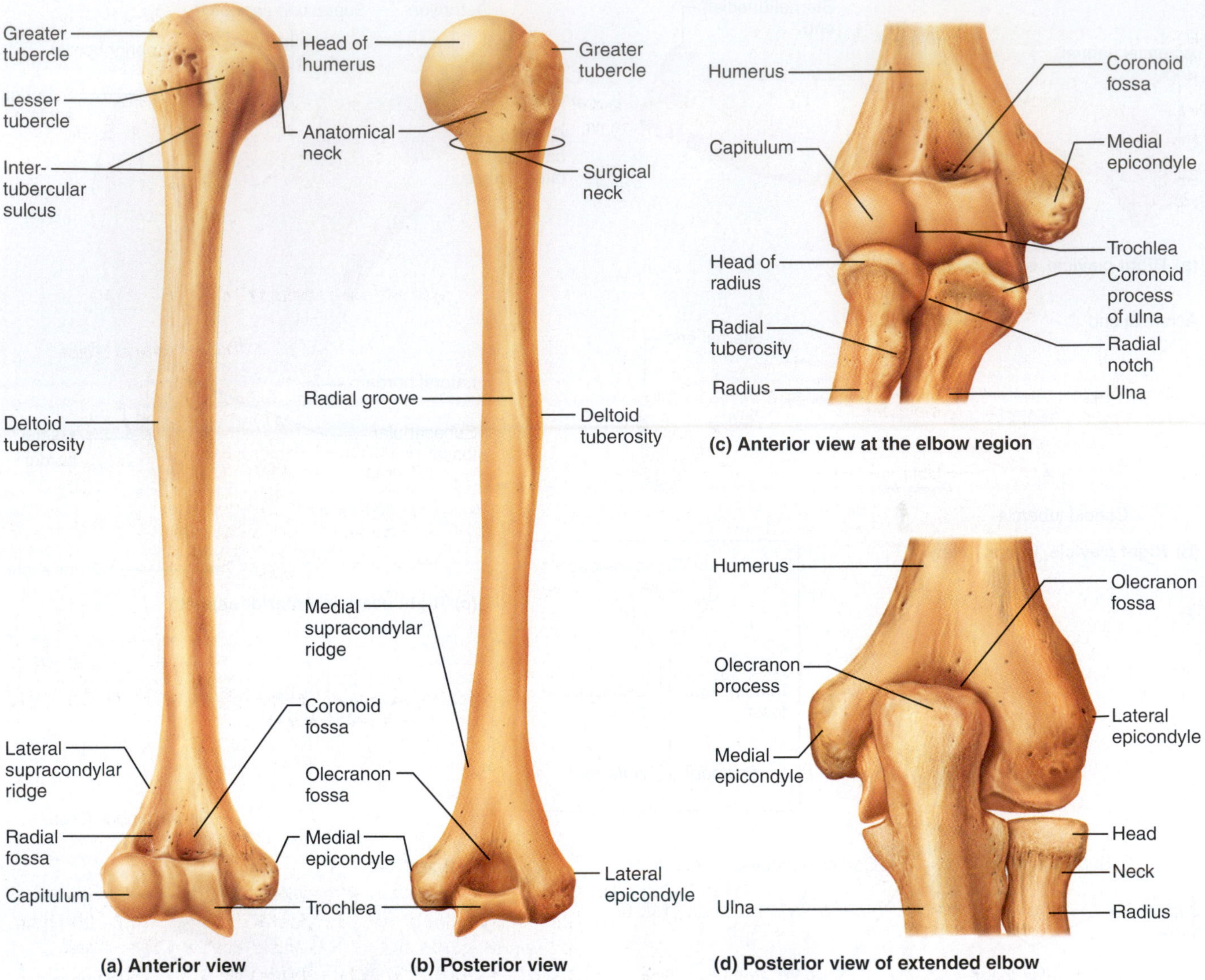
Greater tubercle
Lesser tubercle
Inter-tubercular sulcus
Head of humerus
Anatomical neck
Greater tubercle
Surgical neck
Radial groove
Deltoid tuberosity
Deltoid tuberosity
Medial supracondylar ridge
Coronoid fossa
Olecranon fossa
Medial epicondyle
Lateral supracondylar ridge
Radial fossa
Capitulum
Trochlea
Lateral epicondyle
(a) Anterior view
(b) Posterior view
Humerus
Coronoid fossa
Capitulum
Medial epicondyle
Trochlea
Head of radius
Coronoid process of ulna
Radial tuberosity
Radial notch
Radius
Ulna
(c) Anterior view at the elbow region
Humerus
Olecranon fossa
Olecranon process
Lateral epicondyle
Medial epicondyle
Head
Neck
Ulna
Radius
(d) Posterior view of extended elbow

FIGURE 11.3 Bone of the right arm. Humerus in **(a)** anterior view, **(b)** posterior view. Detailed views illustrate **(c)** anterior and **(d)** posterior extended elbow.

a groove (the **intertubercular sulcus,** or **bicipital groove**) that guides the tendon of the biceps muscle to its point of attachment (the superior rim of the glenoid cavity). In the midpoint of the shaft is a roughened area, the **deltoid tuberosity**, where the large fleshy shoulder muscle, the deltoid, attaches. Nearby, the **radial groove** runs obliquely, indicating the pathway of the radial nerve.

At the distal end of the humerus are two condyles—the medial **trochlea** (looking rather like a spool), which articulates with the ulna, and the lateral **capitulum,** which articulates with the radius of the forearm. This condyle pair is flanked medially by the **medial epicondyle** and laterally by the **lateral epicondyle.**

The medial epicondyle is commonly called the "funny bone." The ulnar nerve runs in a groove beneath the medial epicondyle, and when this region is sharply bumped, we are likely to experience a temporary, but excruciatingly painful, tingling sensation. This event is called "hitting the funny bone," a strange expression, because it is certainly *not* funny!

Above the trochlea on the anterior surface is the **coronoid fossa;** on the posterior surface is the **olecranon fossa.** These two depressions allow the corresponding processes of the ulna to move freely when the elbow is flexed and extended. The small **radial fossa,** lateral to the coronoid fossa, receives the head of the radius when the elbow is flexed.

The Forearm

Two bones, the radius and the ulna, compose the skeleton of the forearm, or antebrachium (see Figure 11.4). When the body is in the anatomical position, the **radius** is in the lateral position in the forearm, and the radius and ulna are parallel. Proximally, the disc-shaped head of the radius articulates with the capitulum of the humerus. Just below the head, on the medial aspect of the shaft, is a prominence called the **radial tuberosity,** the point of attachment for the tendon of the biceps muscle of the arm. Distally, the small **ulnar notch** reveals where it articulates with the end of the ulna.

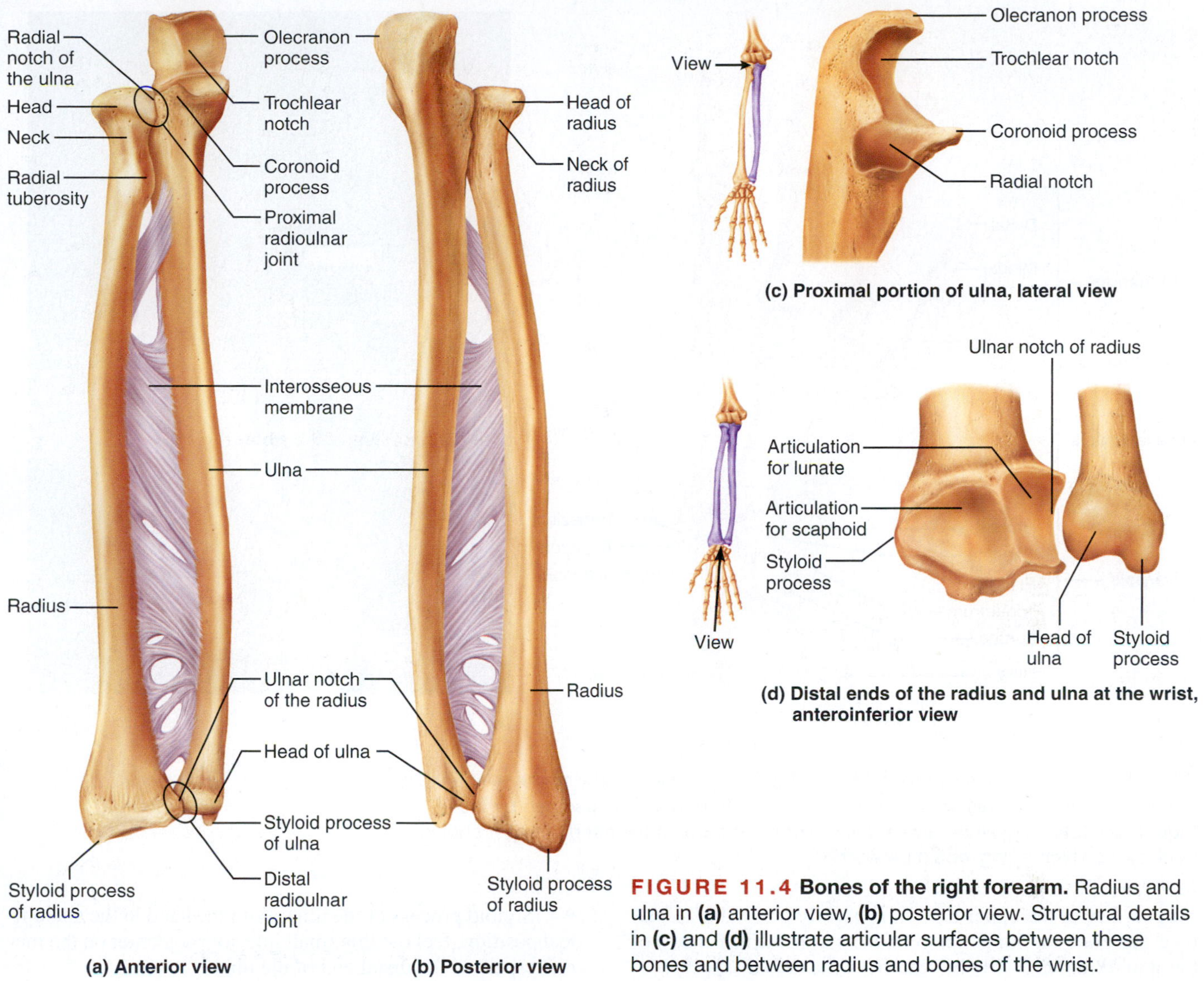

Radial notch of the ulna
Head
Neck
Radial tuberosity
Olecranon process
Trochlear notch
Coronoid process
Proximal radioulnar joint
Head of radius
Neck of radius
Interosseous membrane
Ulna
Radius
Ulnar notch of the radius
Head of ulna
Styloid process of ulna
Distal radioulnar joint
Styloid process of radius
Radius
Styloid process of radius
(a) Anterior view
(b) Posterior view
View
Olecranon process
Trochlear notch
Coronoid process
Radial notch
(c) Proximal portion of ulna, lateral view
Ulnar notch of radius
Articulation for lunate
Articulation for scaphoid
Styloid process
View
Head of ulna
Styloid process
(d) Distal ends of the radius and ulna at the wrist, anteroinferior view

FIGURE 11.4 Bones of the right forearm. Radius and ulna in **(a)** anterior view, **(b)** posterior view. Structural details in **(c)** and **(d)** illustrate articular surfaces between these bones and between radius and bones of the wrist.

The **ulna** is the medial bone of the forearm. Its proximal end bears the anterior **coronoid process** and the posterior **olecranon process,** which are separated by the **trochlear notch.** Together these processes grip the trochlea of the humerus in a plierslike joint. The small **radial notch** on the lateral side of the coronoid process articulates with the head of the radius. The slimmer distal end, the ulnar **head,** bears a small medial **styloid process,** which serves as a point of attachment for the ligaments of the wrist.

The Hand

The skeleton of the hand, or manus (Figure 11.5), includes three groups of bones, those of the carpus (wrist), the metacarpals (bones of the palm), and the phalanges (bones of the fingers).

The wrist is the proximal portion of the hand. It is referred to anatomically as the **carpus;** the eight bones composing it are the **carpals.** (So you actually wear your wristwatch over the distal part of your forearm.) The carpals are arranged in two irregular rows of four bones each, which are illustrated in Figure 11.5. In the proximal row (lateral to medial) are the *scaphoid, lunate, triquetral,* and *pisiform bones;* the scaphoid and lunate articulate with the distal end of the radius. In the distal row are the *trapezium, trapezoid, capitate,* and *hamate*. The carpals are bound closely together by ligaments, which restrict movements between them.

The **metacarpals,** numbered 1 to 5 from the thumb side of the hand toward the little finger, radiate out from the wrist like spokes to form the palm of the hand. The *bases* of the metacarpals articulate with the carpals of the wrist; their more bulbous *heads* articulate with the phalanges of the fingers distally. When the fist is clenched, the heads of the metacarpals become prominent as the knuckles.

Like the bones of the palm, the fingers are numbered from 1 to 5, beginning from the thumb *(pollex)* side of the hand. The 14 bones of the fingers, or digits, are miniature long bones, called **phalanges** (singular: *phalanx*) as noted above. Each finger contains three phalanges (proximal, middle, and distal) except the thumb, which has only two (proximal and distal).

ACTIVITY 2

Palpating the Surface Anatomy of the Pectoral Girdle and the Upper Limb

Before continuing on to study the bones of the pelvic girdle, take the time to identify the following bone markings on the

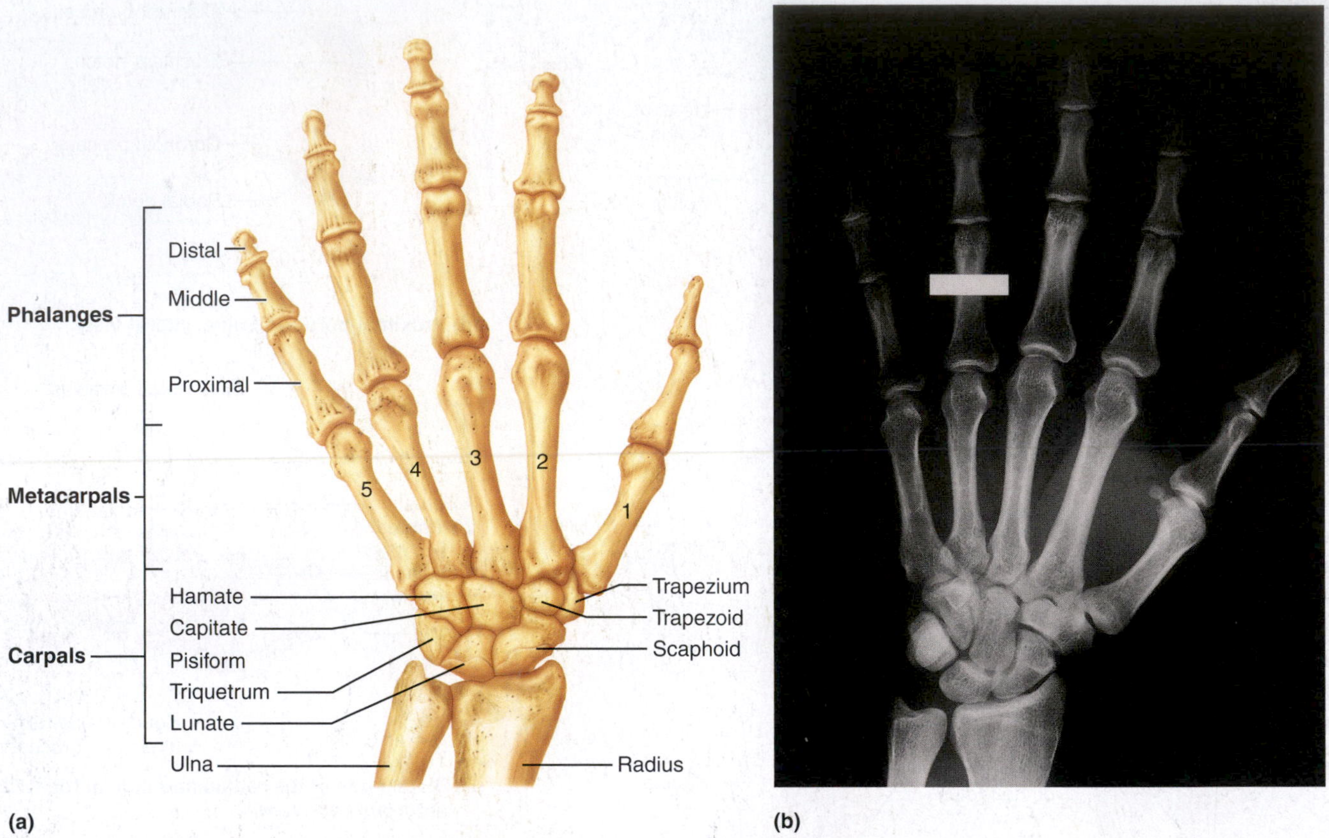
Distal
Middle
Phalanges
Proximal
Metacarpals
5
4
3
2
1
Hamate
Capitate
Carpals
Pisiform
Triquetrum
Lunate
Ulna
Trapezium
Trapezoid
Scaphoid
Radius
(a)
(b)

FIGURE 11.5 Bones of the left hand. (a) Posterior view showing the relationships of the carpals, metacarpals, and phalanges. The pisiform bone is obscured by the triquetrum. **(b)** X ray. White bar on the proximal phalanx of the left ring finger shows the position at which a ring would be worn.

skin surface of the upper limb. It is usually preferable to palpate the bone markings on your lab partner since many of these markings can only be seen from the dorsal aspect.

- Clavicle: Palpate the clavicle along its entire length from sternum to shoulder.
- Acromioclavicular joint: The high point of the shoulder, which represents the junction point between the clavicle and the acromion of the scapular spine.
- Spine of the scapula: Extend your arm at the shoulder so that your scapula moves posteriorly. As you do this, your scapular spine will be seen as a winglike protrusion on your dorsal thorax and can be easily palpated by your lab partner.
- Lateral epicondyle of the humerus: The inferiormost projection at the lateral aspect of the distal humerus. After you have located the epicondyle, run your finger posteriorly into the hollow immediately dorsal to the epicondyle. This is the site where the extensor muscles of the hand are attached and is a common site of the excruciating pain of tennis elbow, a condition in which those muscles and their tendons are abused physically.
- Medial epicondyle of the humerus: Feel this medial projection at the distal end of the humerus.
- Olecranon process of the ulna: Work your elbow—flexing and extending—as you palpate its dorsal aspect to feel the olecranon process of the ulna moving into and out of the olecranon fossa on the dorsal aspect of the humerus.
- Styloid process of the ulna: With the hand in the anatomical position, feel out this small inferior projection on the medial aspect of the distal end of the ulna.
- Styloid process of the radius: Find this projection at the distal end of the radius (lateral aspect). It is most easily located by moving the hand medially at the wrist. Once you have palpated the styloid process, move your fingers just medially onto the anterior wrist. Press firmly and then let up slightly on the pressure. You should be able to feel your pulse at this pressure point, which lies over the radial artery (radial pulse).
- Pisiform: Just distal to the styloid process of the ulna, feel the rounded pealike pisiform bone.
- Metacarpophalangeal joints (knuckles): Clench your fist and find the first set of flexed-joint protrusions beyond the wrist—these are your metacarpophalangeal joints.

Bones of the Pelvic Girdle and Lower Limb

The Pelvic (Hip) Girdle

As with the bones of the pectoral girdle and upper limb, pay particular attention to bone markings needed to identify right and left bones.

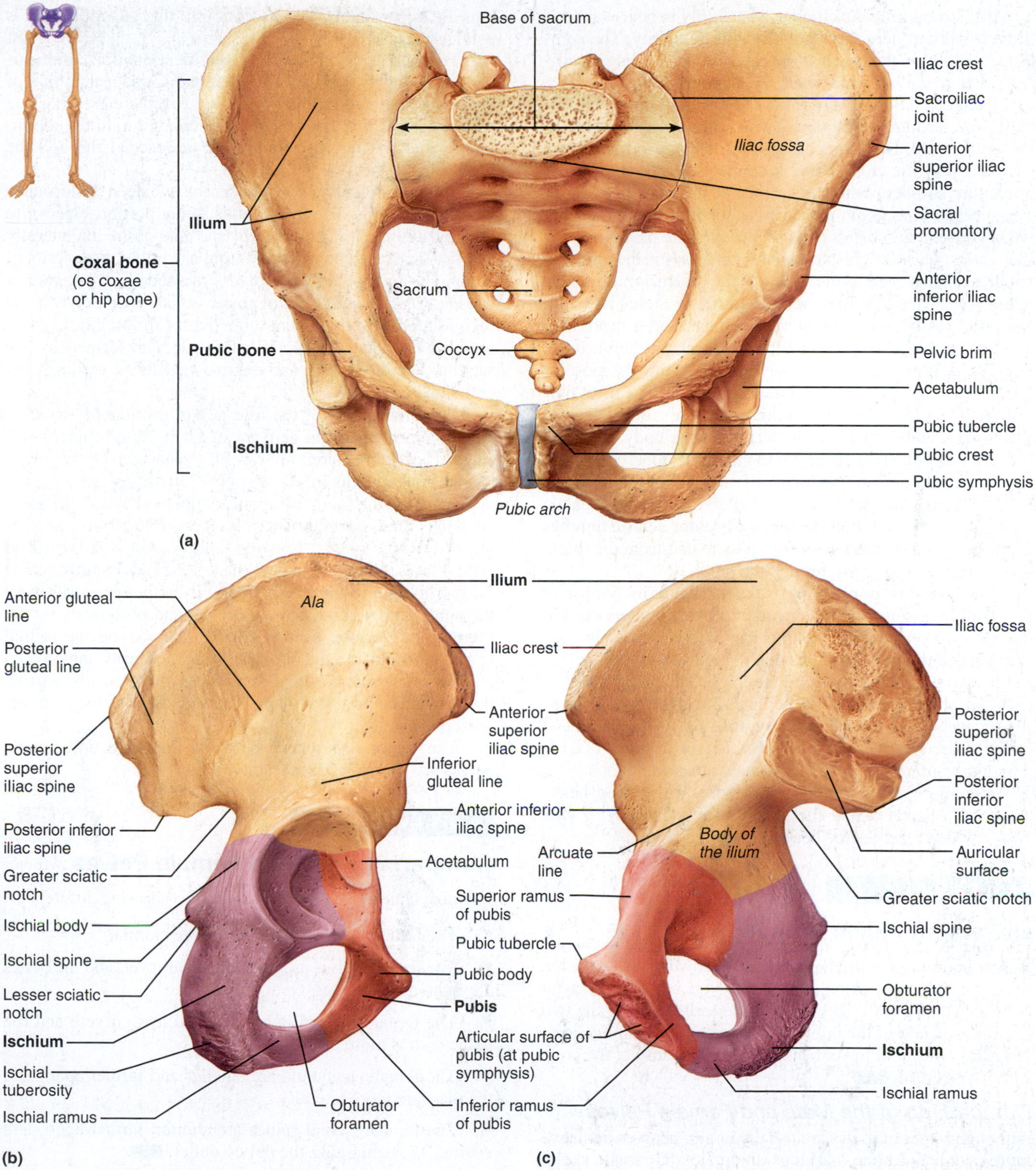
Base of sacrum
Iliac crest
Sacroiliac joint
Iliac fossa
Anterior superior iliac spine
Sacral promontory
Ilium
Coxal bone (os coxae or hip bone)
Sacrum
Anterior inferior iliac spine
Pubic bone
Coccyx
Pelvic brim
Acetabulum
Pubic tubercle
Ischium
Pubic crest
Pubic symphysis
Pubic arch
(a)
Ilium
Ala
Anterior gluteal line
Posterior gluteal line
Iliac crest
Iliac fossa
Anterior superior iliac spine
Posterior superior iliac spine
Inferior gluteal line
Posterior superior iliac spine
Posterior inferior iliac spine
Anterior inferior iliac spine
Posterior inferior iliac spine
Arcuate line
Body of the ilium
Auricular surface
Acetabulum
Greater sciatic notch
Greater sciatic notch
Superior ramus of pubis
Ischial body
Ischial spine
Pubic tubercle
Ischial spine
Pubic body
Lesser sciatic notch
Obturator foramen
Pubis
Ischium
Articular surface of pubis (at pubic symphysis)
Ischium
Ischial tuberosity
Ischial ramus
Ischial ramus
Obturator foramen
Inferior ramus of pubis
(b)
(c)

FIGURE 11.6 Bones of the pelvic girdle. (a) Articulated bony pelvis, showing the two hip bones (coxal bones), which together comprise the pelvic girdle, the sacrum, and the coccyx. **(b)** Right hip bone, lateral view, showing the point of fusion of the ilium, ischium, and pubic bones. **(c)** Right hip bone, medial view.

The **pelvic girdle,** or **hip girdle** (Figure 11.6), is formed by the two **coxal** (*coxa* = hip) **bones** (also called the **ossa coxae,** or hip bones). The two coxal bones together with the sacrum and coccyx form the **bony pelvis.** In contrast to the bones of the shoulder girdle, those of the pelvic girdle are heavy and massive, and they attach securely to the axial skeleton. The sockets for the heads of the femurs (thigh bones) are deep and heavily reinforced by ligaments to ensure a stable, strong limb attachment. The ability to bear weight is more important here than mobility and flexibility. The combined weight of the upper body rests on the pelvis (specifically, where the hip bones meet the sacrum).

Each coxal bone is a result of the fusion of three bones—the ilium, ischium, and pubis—which are distinguishable in the young child. The **ilium,** a large flaring bone, forms the major portion of the coxal bone. It connects posteriorly, via its **auricular surface,** with the sacrum at the **sacroiliac joint.** The superior margin of the iliac bone, the **iliac crest,** is rough; when you rest your hands on your hips, you are palpating your iliac crests. The iliac crest terminates anteriorly in the **anterior superior spine** and posteriorly in the **posterior superior spine.** Two inferior spines are located below these. The shallow **iliac fossa** marks its internal surface, and a prominent ridge, the **arcuate line,** outlines the pelvic inlet, or pelvic brim.

The **ischium** is the "sit-down" bone, forming the most inferior and posterior portion of the coxal bone. The most outstanding marking on the ischium is the rough **ischial tuberosity,** which receives the weight of the body when sitting. The **ischial spine,** superior to the ischial tuberosity, is an important anatomical landmark of the pelvic cavity. (See Comparison of the Male and Female Pelves, Table 11.1 on the facing page.) The obvious **lesser** and **greater sciatic notches** allow nerves and blood vessels to pass to and from the thigh. The sciatic nerve passes through the latter.

The **pubis,** or **pubic bone,** is the most anterior portion of the coxal bone. Fusion of the **rami** of the pubis anteriorly and the ischium posteriorly forms a bar of bone enclosing the **obturator foramen,** through which blood vessels and nerves run from the pelvic cavity into the thigh. The pubic bones of each hip bone meet anteriorly at the **pubic crest** to form a cartilaginous joint called the **pubic symphysis.** At the lateral end of the pubic crest is the *pubic tubercle* (see Figure 11.6c) to which the important *inguinal ligament* attaches.

The ilium, ischium, and pubis fuse at the deep hemispherical socket called the **acetabulum** (literally, "wine cup"), which receives the head of the thigh bone.

ACTIVITY 3

Observing Pelvic Articulations

Before continuing with the bones of the lower limbs, take the time to examine an articulated pelvis. Notice how each coxal bone articulates with the sacrum posteriorly and how the two coxal bones join at the pubic symphysis. The sacroiliac joint is a common site of lower back problems because of the pressure it must bear. ■

Comparison of the Male and Female Pelves

Although bones of males are usually larger, heavier, and have more prominent bone markings, the male and female skeletons are very similar. The exception to this generalization is pelvic structure.

The female pelvis reflects modifications for childbearing—it is wider, shallower, lighter, and rounder than that of the male. Not only must her pelvis support the increasing size of a fetus, but it must also be large enough to allow the infant's head (its largest dimension) to descend through the birth canal at birth.

To describe pelvic sex differences, we need to introduce a few more terms. The **false pelvis** is that portion superior to the arcuate line; it is bounded by the alae of the ilia laterally and the sacral promontory and lumbar vertebrae posteriorly. Although the false pelvis supports the abdominal viscera, it does not restrict childbirth in any way. The **true pelvis** is the region inferior to the arcuate line that is almost entirely surrounded by bone. Its posterior boundary is formed by the sacrum. The ilia, ischia, and pubic bones define its limits laterally and anteriorly.

The dimensions of the true pelvis, particularly its inlet and outlet, are critical if delivery of a baby is to be uncomplicated. These dimensions are carefully measured by the obstetrician. The **pelvic inlet,** or **pelvic brim,** is the opening delineated by the sacral promontory posteriorly and the arcuate lines of the ilia anterolaterally. It is the superiormost margin of the true pelvis. Its widest dimension is from left to right, that is, along the frontal plane. The **pelvic outlet** is the inferior margin of the true pelvis. It is bounded anteriorly by the pubic arch, laterally by the ischia, and posteriorly by the sacrum and coccyx. Since both the coccyx and the ischial spines protrude into the outlet opening, a sharply angled coccyx or large, sharp ischial spines can dramatically narrow the outlet. The largest dimension of the outlet is the anterior-posterior diameter.

The major differences between the male and female pelves are summarized in Table 11.1.

ACTIVITY 4

Comparing Male and Female Pelves

Examine male and female pelves for the following differences:

- The female inlet is larger and more circular.
- The female pelvis as a whole is shallower, and the bones are lighter and thinner.
- The female sacrum is broader and less curved, and the pubic arch is more rounded.
- The female acetabula are smaller and farther apart, and the ilia flare more laterally.
- The female ischial spines are shorter, farther apart, and everted, thus enlarging the pelvic outlet. ■

Text continues on page 154.

TABLE 11.1 Comparison of the Male and Female Pelves

Characteristic	Female	Male
General structure and functional modifications	Tilted forward; adapted for childbearing; true pelvis defines the birth canal; cavity of the true pelvis is broad, shallow, and has a greater capacity	Tilted less far forward; adapted for support of a male's heavier build and stronger muscles; cavity of the true pelvis is narrow and deep
Bone thickness	Less; bones lighter, thinner, and smoother	Greater; bones heavier and thicker, and markings are more prominent
Acetabula	Smaller; farther apart	Larger; closer
Pubic angle/arch	Broader (80°–90°); more rounded	More acute (50°–60°)
Anterior view		

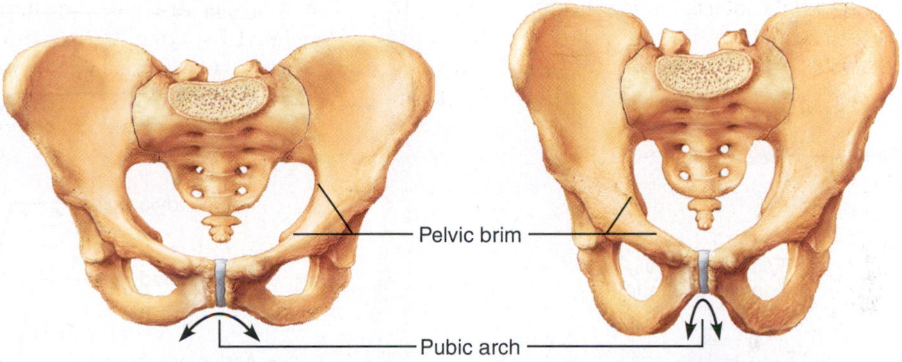
Pelvic brim
Pubic arch

Characteristic	Female	Male
Sacrum	Wider; shorter; sacrum is less curved	Narrow; longer; sacral promontory more ventral
Coccyx	More movable; straighter	Less movable; curves ventrally
Left lateral view		

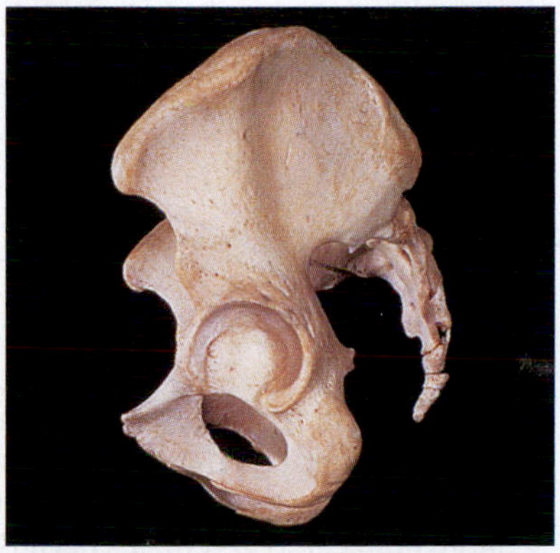
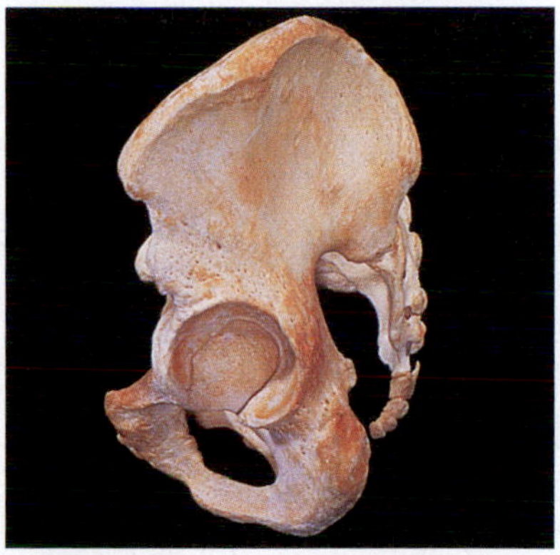

Characteristic	Female	Male
Pelvic inlet (brim)	Wider; oval from side to side	Narrow; basically heart shaped
Pelvic outlet	Wider; ischial spines shorter, farther apart, and everted	Narrower; ischial spines longer, sharper, and point more medially
Posteroinferior view		

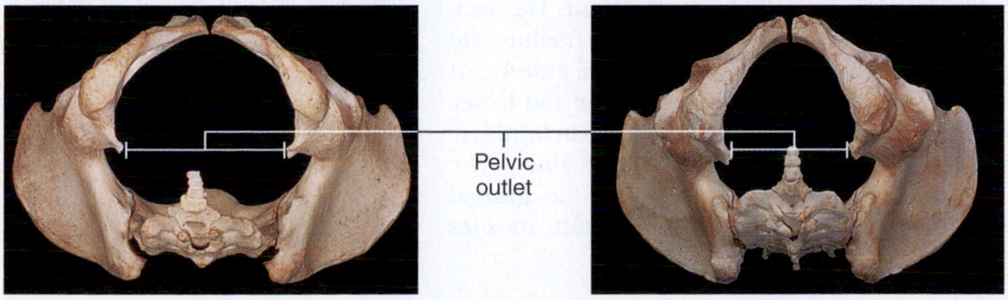
Pelvic outlet

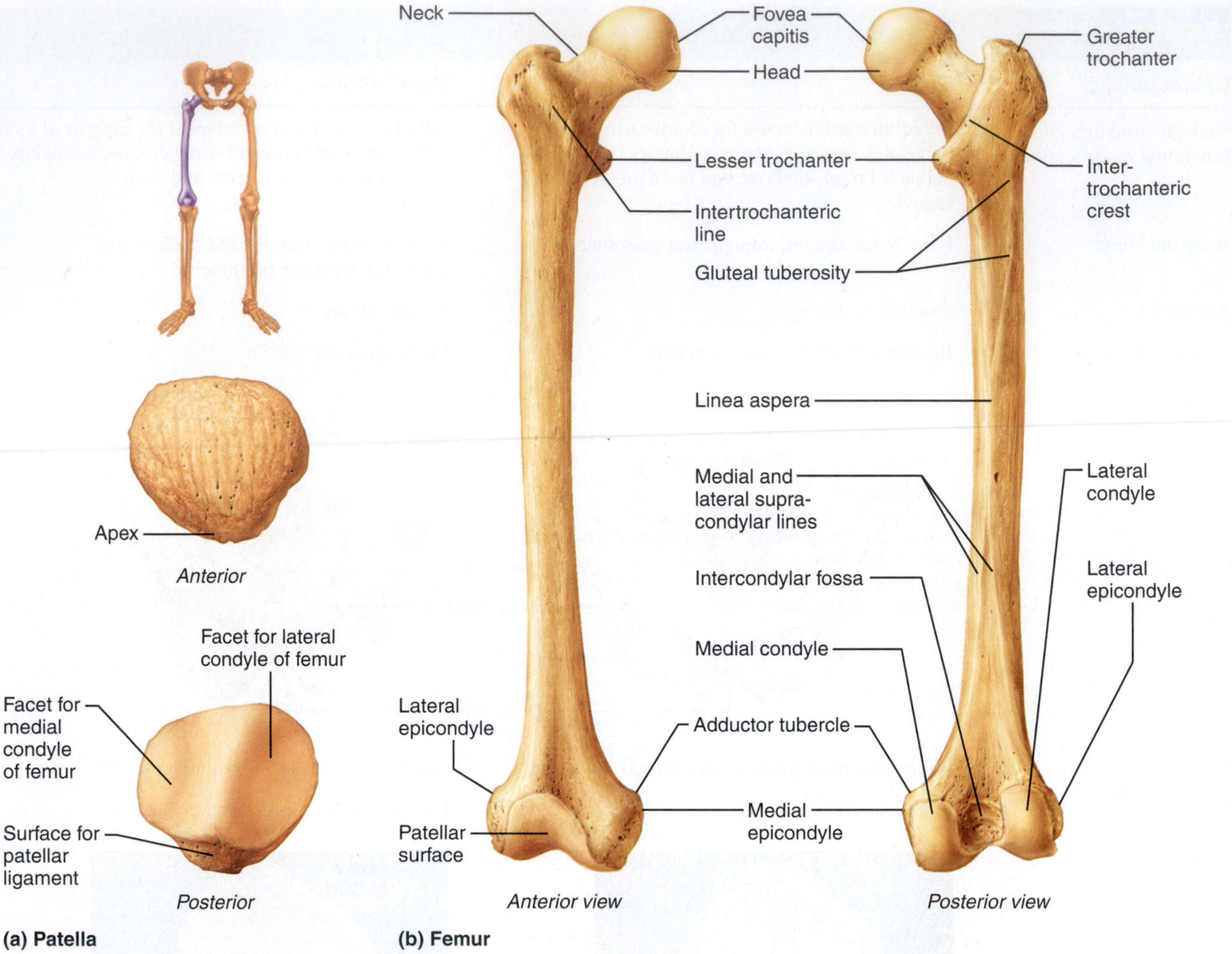
Apex
Anterior
Facet for lateral condyle of femur
Facet for medial condyle of femur
Surface for patellar ligament
Posterior
(a) Patella
Neck
Fovea capitis
Head
Greater trochanter
Lesser trochanter
Inter-trochanteric crest
Intertrochanteric line
Gluteal tuberosity
Linea aspera
Medial and lateral supra-condylar lines
Lateral condyle
Intercondylar fossa
Lateral epicondyle
Medial condyle
Lateral epicondyle
Adductor tubercle
Medial epicondyle
Patellar surface
Anterior view
Posterior view
(b) Femur

FIGURE 11.7 Bones of the right knee and thigh. (a) The patella (kneecap). **(b)** The femur (thigh bone).

The Thigh

The **femur,** or thigh bone (Figure 11.7b), is the sole bone of the thigh. It is the heaviest, strongest bone in the body. The ball-like head of the femur articulates with the hip bone via the deep, secure socket of the acetabulum. Obvious in the femur's head is a small central pit called the **fovea capitis** ("pit of the head"), from which a small ligament runs to the acetabulum. The head of the femur is carried on a short, constricted *neck,* which angles laterally to join the shaft. The neck is the weakest part of the femur and is a common fracture site (an injury called a broken hip), particularly in the elderly. At the junction of the shaft and neck are the **greater** and **lesser trochanters** (separated posteriorly by the **intertrochanteric crest** and anteriorly by the **intertrochanteric line**). The trochanters and trochanteric crest, as well as the **gluteal tuberosity** and the **linea aspera** located on the shaft, are sites of muscle attachment.

The femur inclines medially as it runs downward to the leg bones; this brings the knees in line with the body's center of gravity, or maximum weight. The medial course of the femur is more noticeable in females because of the wider female pelvis.

Distally, the femur terminates in the **lateral** and **medial condyles,** which articulate with the tibia below, and the **patellar surface,** which forms a joint with the patella (kneecap) anteriorly. The **lateral** and **medial epicondyles,** just superior to the condyles, are separated by the **intercondylar fossa.** On the superior part of the medial epicondyle is a bump, the **adductor tubercle,** to which the large adductor magnus muscle attaches.

The **patella** (Figure 11.7a) is a triangular sesamoid bone enclosed in the (quadriceps) tendon that secures the anterior thigh muscles to the tibia. It guards the knee joint anteriorly and improves the leverage of the thigh muscles acting across the knee joint.

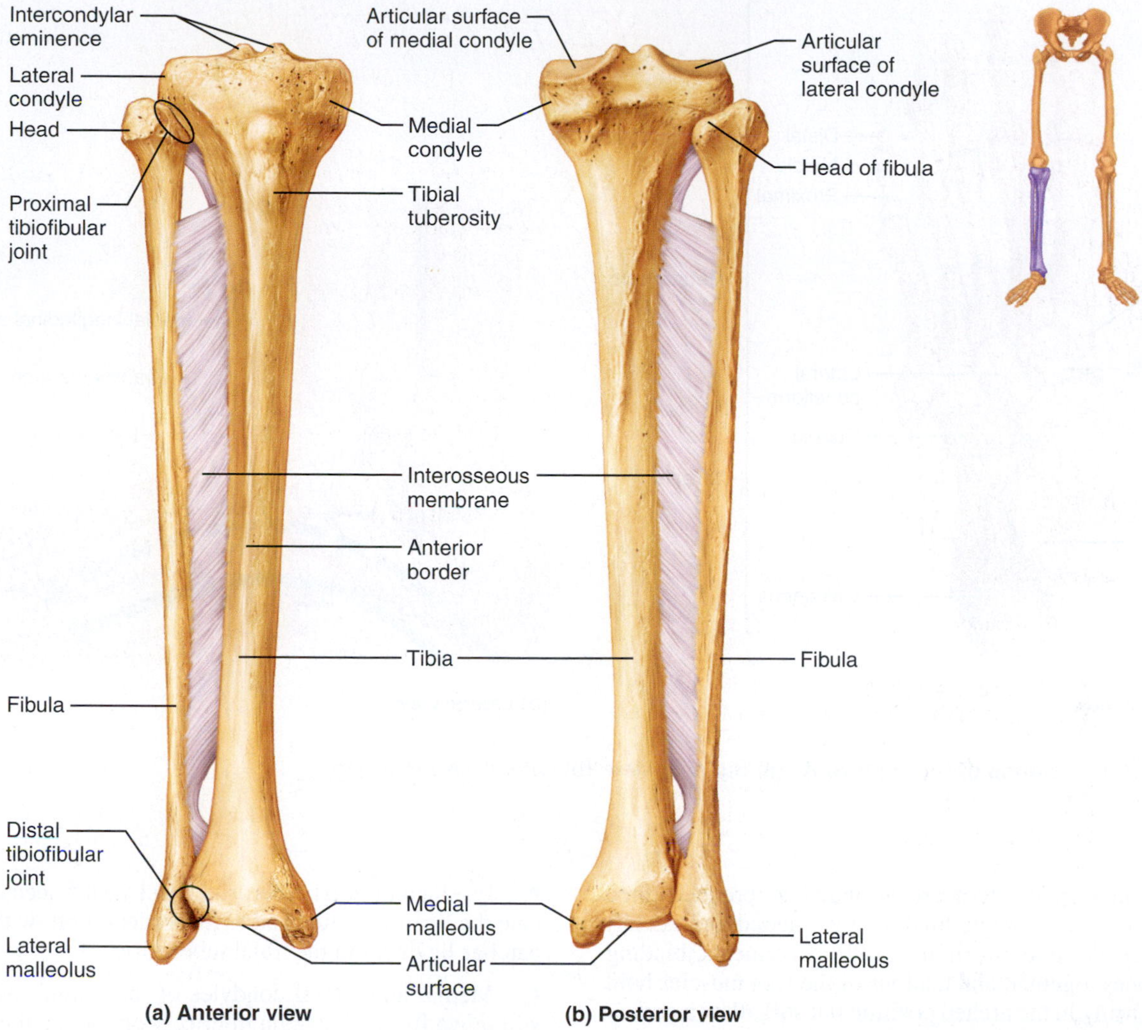

Intercondylar eminence
Lateral condyle
Head
Proximal tibiofibular joint
Articular surface of medial condyle
Medial condyle
Tibial tuberosity
Articular surface of lateral condyle
Head of fibula
Interosseous membrane
Anterior border
Tibia
Fibula
Fibula
Distal tibiofibular joint
Lateral malleolus
Medial malleolus
Articular surface
Lateral malleolus
(a) Anterior view
(b) Posterior view

FIGURE 11.8 Bones of the right leg. Tibia and fibula, anterior view on left; posterior view on right.

The Leg

Two bones, the tibia and the fibula, form the skeleton of the leg (see Figure 11.8). The **tibia,** or *shinbone,* is the larger and more medial of the two leg bones. At the proximal end, the **medial** and **lateral condyles** (separated by the **intercondylar eminence**) receive the distal end of the femur to form the knee joint. The **tibial tuberosity,** a roughened protrusion on the anterior tibial surface (just below the condyles), is the site of attachment of the patellar (kneecap) ligament. Small facets on the superior and inferior surface of the lateral condyle of the tibia articulate with the fibula. Distally, a process called the **medial malleolus** forms the inner (medial) bulge of the ankle. The more medial articular surface of the tibia articulates with the talus bone of the foot. The anterior surface of the tibia bears a sharpened ridge that is relatively unprotected by muscles. This so-called **anterior border** is easily felt beneath the skin.

The **fibula,** which lies parallel to the tibia, takes no part in forming the knee joint. Its proximal head articulates with the lateral condyle of the tibia. The fibula is thin and stick-like with a sharp anterior crest. It terminates distally in the **lateral malleolus,** which forms the outer part, or lateral bulge, of the ankle.

The Foot

The bones of the foot include the 7 **tarsal** bones, 5 **metatarsals,** which form the instep, and 14 **phalanges,** which form the toes (see Figure 11.9). Body weight is concentrated on the two largest tarsals, which form the posterior aspect of the foot. These are the *calcaneus* (heel bone) and the *talus,* which lies between the tibia and the calcaneus. The other tarsals are named and identified in Figure 11.9. The metatarsals are numbered 1 through 5, medial to lateral. Like the fingers of the hand, each toe has three phalanges except the great toe, which has two.

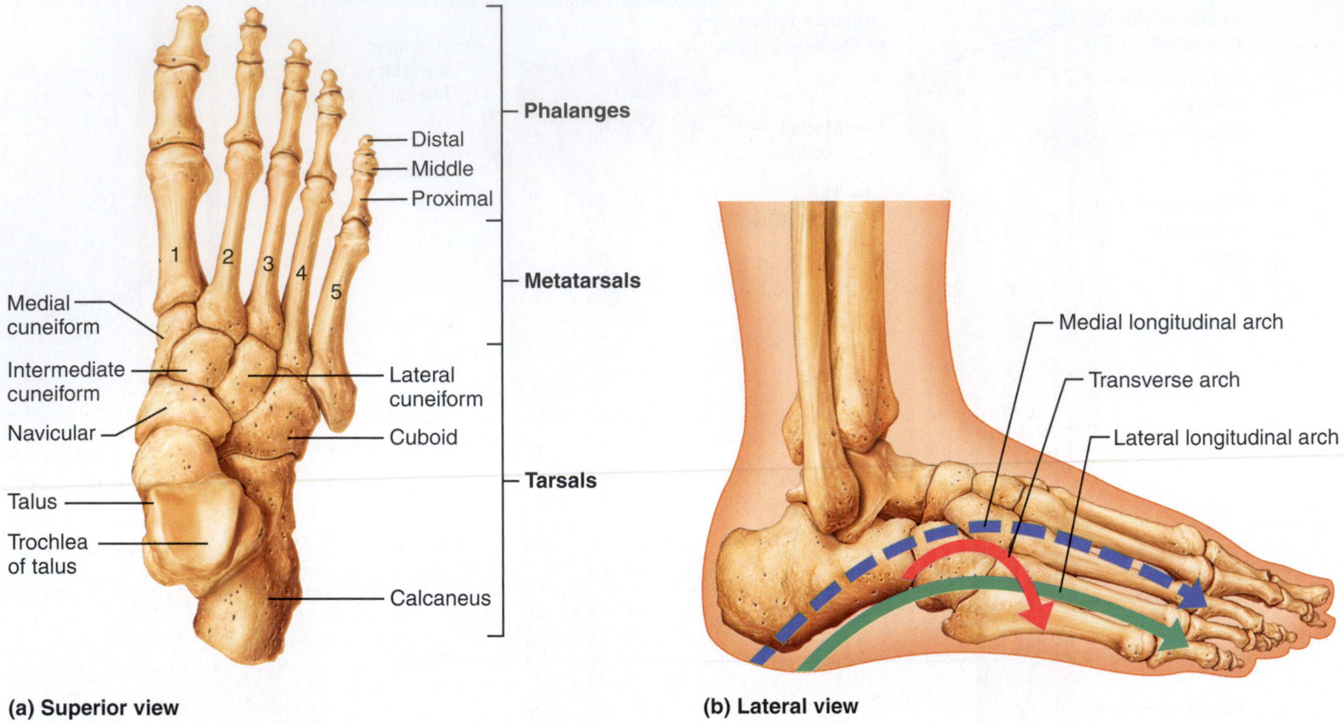
Phalanges
Distal
Middle
Proximal
Metatarsals
1
2
3
4
5
Medial cuneiform
Intermediate cuneiform
Navicular
Lateral cuneiform
Cuboid
Tarsals
Talus
Trochlea of talus
Calcaneus
(a) Superior view
Medial longitudinal arch
Transverse arch
Lateral longitudinal arch
(b) Lateral view

FIGURE 11.9 Bones of the right foot. **(a)** Superior view. **(b)** Lateral view showing arches of the foot.

The bones in the foot are arranged to produce three strong arches—two longitudinal arches (medial and lateral) and one transverse arch (Figure 11.9b). Ligaments, binding the foot bones together, and tendons of the foot muscles hold the bones firmly in the arched position but still allow a certain degree of give. Weakened arches are referred to as fallen arches or flat feet.

ACTIVITY 5

Palpating the Surface Anatomy of the Pelvic Girdle and Lower Limb

Locate and palpate the following bone markings on yourself and/or your lab partner.

- Iliac crest and anterior superior iliac spine: Rest your hands on your hips—they will be overlying the iliac crests. Trace the crest as far posteriorly as you can and then follow it anteriorly to the anterior superior iliac spine. This latter bone marking is easily felt in almost everyone and is clearly visible through the skin (and perhaps the clothing) of very slim people. (The posterior superior iliac spine is much less obvious and is usually indicated only by a dimple in the overlying skin. Check it out in the mirror tonight.)
- Greater trochanter of the femur: This is easier to locate in females than in males because of the wider female pelvis; also it is more likely to be clothed by bulky muscles in males. Try to locate it on yourself as the most lateral point of the proximal femur. It typically lies about 6 to 8 inches below the iliac crest.
- Patella and tibial tuberosity: Feel your kneecap and palpate the ligaments attached to its borders. Follow the inferior patellar ligament to the tibial tuberosity.
- Medial and lateral condyles of the femur and tibia: As you move from the patella inferiorly on the medial (and then the lateral) knee surface, you will feel first the femoral and then the tibial condyle.
- Medial malleolus: Feel the medial protrusion of your ankle, the medial malleolus of the distal tibia.
- Lateral malleolus: Feel the bulge of the lateral aspect of your ankle, the lateral malleolus of the fibula.
- Calcaneus: Attempt to follow the extent of your calcaneus or heel bone. ■

ACTIVITY 6

Constructing a Skeleton

1. When you finish examining yourself and the disarticulated bones of the appendicular skeleton, work with your lab partner to arrange the disarticulated bones on the laboratory bench in their proper relative positions to form an entire skeleton. Careful observation of bone markings should help you distinguish between right and left members of bone pairs.

2. When you believe that you have accomplished this task correctly, ask the instructor to check your arrangement to ensure that it is correct. If it is not, go to the articulated skeleton and check your bone arrangements. Also review the descriptions of the bone markings as necessary to correct your bone arrangement. ■

EXERCISE 12

The Fetal Skeleton

MATERIALS

- ☐ Isolated fetal skull
- ☐ Fetal skeleton
- ☐ Adult skeleton

OBJECTIVES

1. To define *fontanel* and discuss the function and fate of fontanels in the fetus.
2. To demonstrate important differences between the fetal and adult skeletons.

PRE-LAB QUIZ

1. Circle the correct term. The fetal skeleton has more / fewer bones than the adult skeleton.
2. A fontanel
   a. is an indentation between bones of the fetal skull
   b. is a fibrous membrane
   c. allows for compression of the skull during birth
   d. all of the above
3. The fetal skull has _________ fontanel(s).
   a. 1　　c. 3
   b. 2　　d. 4 or more
4. Ossification is
   a. the process of bone formation
   b. the development of fontanels in the skull
   c. the process of red blood cell formation
   d. the process of white blood cell formation

MasteringA&P™
Access practice quizzes and more in the Study Area at www.masteringaandp.com.

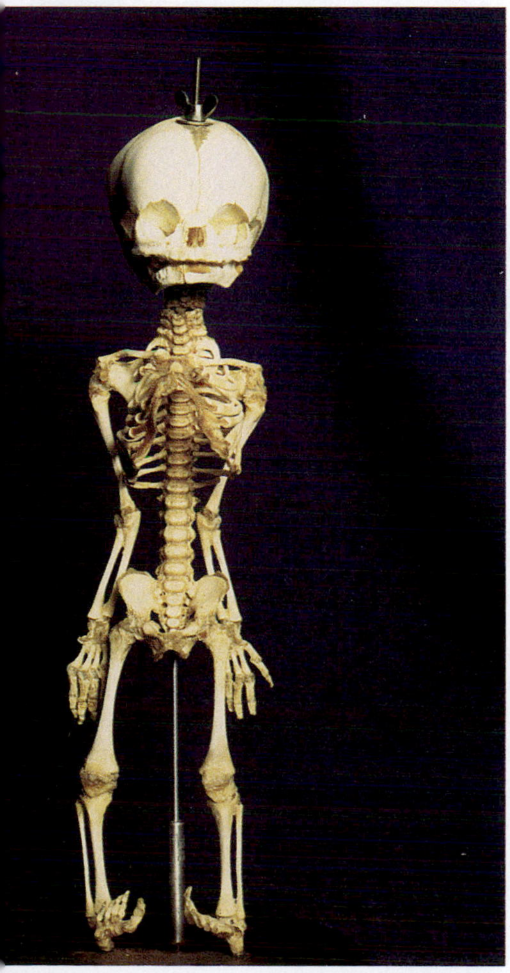

FIGURE 12.1 The fetal skeleton. Anterior view.

A human fetus about to be born has 275 bones, many more than the 206 bones found in the adult skeleton. This is because many of the bones described as single bones in the adult skeleton (for example, the coxal bone, sternum, and sacrum) have not yet fully ossified and fused in the fetus.

ACTIVITY

Examining a Fetal Skeleton and Skull

1. Obtain a fetal skeleton or use Figure 12.1, and examine it carefully, noting differences between it and an adult skeleton. Pay particular attention to the vertebrae, sternum, frontal bone of the cranium, patellae (kneecaps), coxal bones, carpals and tarsals, and thoracic cage.

2. Obtain a fetal skull and study it carefully. Make observations as needed to answer the questions on page 166.

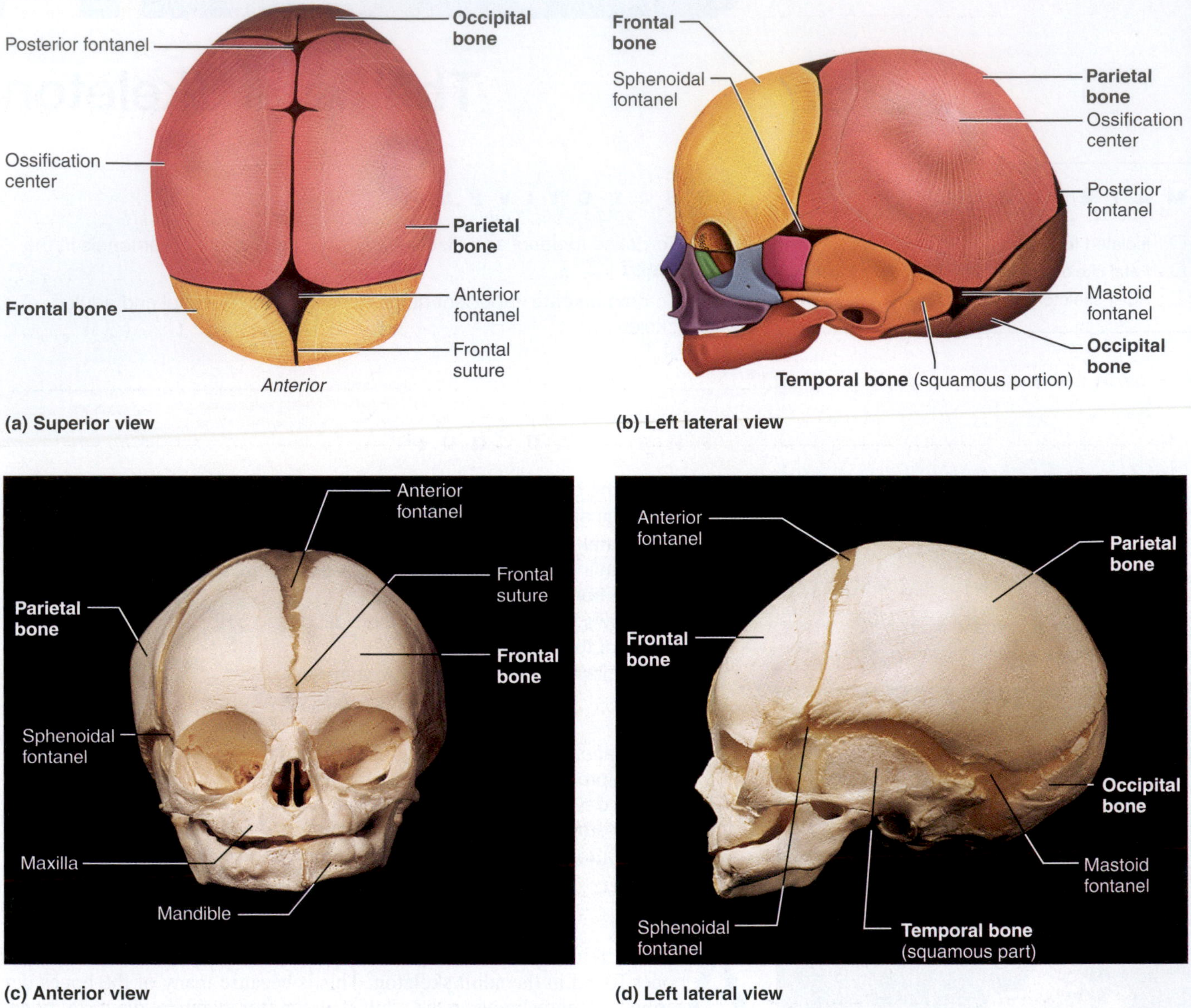
Posterior fontanel
Occipital bone
Ossification center
Parietal bone
Frontal bone
Anterior fontanel
Frontal suture
Anterior
(a) Superior view
Frontal bone
Sphenoidal fontanel
Parietal bone
Ossification center
Posterior fontanel
Mastoid fontanel
Occipital bone
Temporal bone (squamous portion)
(b) Left lateral view
Anterior fontanel
Frontal suture
Parietal bone
Frontal bone
Sphenoidal fontanel
Maxilla
Mandible
(c) Anterior view
Anterior fontanel
Parietal bone
Frontal bone
Occipital bone
Mastoid fontanel
Sphenoidal fontanel
Temporal bone (squamous part)
(d) Left lateral view

FIGURE 12.2 Skull of a newborn.

- Does it have the same bones as the adult skull?
- How does the size of the fetal face relate to the cranium?
- How does this compare to what is seen in the adult?

3. Indentations between the bones of the fetal skull, called **fontanels,** are fibrous membranes. These areas will become bony (ossify) as the infant ages, completing the process by the age of 20 to 22 months. The fontanels allow the fetal skull to be compressed slightly during birth and also allow for brain growth during late fetal life. Locate the following fontanels on the fetal skull with the aid of Figure 12.2: *anterior* (or *frontal*) *fontanel, mastoid fontanel, sphenoidal fontanel,* and *posterior* (or *occipital*) *fontanel.*

4. Notice that some of the cranial bones have conical protrusions. These are **ossification (growth) centers.** Notice also that the frontal bone is still bipartite, and the temporal bone is incompletely ossified, little more than a ring of bone.

5. Before completing this study, check the questions on the review sheet at the end of this exercise to ensure that you have made all of the necessary observations.

NAME ______________________________

LAB TIME/DATE ______________________

REVIEW SHEET

EXERCISE 12

The Fetal Skeleton

1. Are the same skull bones seen in the adult also found in the fetal skull? ______________________

2. How does the size of the fetal face compare to its cranium? ______________________

____

How does this compare to the adult skull? ______________________

____

____

3. What are the outward conical projections on some of the fetal cranial bones? ______________________

4. What is a fontanel? ______________________

What is its fate? ______________________

What is the function of the fontanels in the fetal skull? ______________________

____

5. Describe how the fetal skeleton compares with the adult skeleton in the following areas:

vertebrae: ______________________

____

coxal bones: ______________________

____

carpals and tarsals: ______________________

sternum: ______________________

frontal bone: ______________________

patella: ______________________

thoracic cage: ______________________

____

6. How does the size of the fetus's head compare to the size of its body?________________________

____

7. Using the terms listed, identify each of the fontanels shown on the fetal skull below.

Key:

a. anterior fontanel

b. mastoid fontanel

c. posterior fontanel

d. sphenoidal fontanel

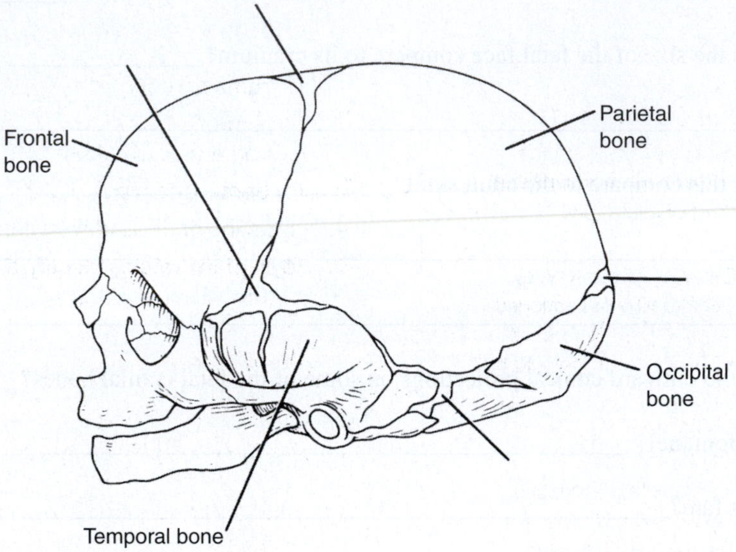
Frontal bone
Parietal bone
Occipital bone
Temporal bone

EXERCISE 13

Articulations and Body Movements

MATERIALS

- ☐ Skull
- ☐ Articulated skeleton
- ☐ X ray of a child's bone showing the cartilaginous growth plate (if available)
- ☐ Anatomical chart of joint types (if available)
- ☐ Diarthrotic joint (fresh or preserved), preferably a beef knee joint sectioned sagittally (Alternatively, pig's feet with phalanges sectioned frontally could be used)
- ☐ Disposable gloves
- ☐ Water balloons and clamps
- ☐ Functional models of hip, knee, and shoulder joints (if available)
- ☐ X-rays of normal and arthritic joints (if available)

OBJECTIVES

1. To name and describe the three functional categories of joints.
2. To name and describe the three structural categories of joints, and to compare their structure and mobility.
3. To identify the types of synovial joints.
4. To define origin and insertion of muscles.
5. To demonstrate or identify the various body movements.

PRE-LAB QUIZ

1. Name one of the two functions of an articulation, or joint. ____________

   ____
2. The functional classification of joints is based on
   a. a joint cavity
   b. amount of connective tissue
   c. amount of movement allowed by the joint
3. Structural classification of joints includes fibrous, cartilaginous, and __________________, which have a fluid-filled cavity between articulating bones.
4. Circle the correct term. Sutures, which have their irregular edges of bone interlocked by short fibers of connective tissue, are an example of fibrous / cartilaginous joints.
5. Circle True or False. All synovial joints are diarthroses, or freely movable joints.
6. Circle the correct term. Every muscle of the body is attached to a bone or other connective tissue structure at two points. The origin / insertion is the more movable attachment.
7. The hip joint is an example of a _________ synovial joint.
   a. ball-and-socket
   b. hinge
   c. pivot
   d. plane
8. Movement of a limb *away* from the midline or median plane of the body in the frontal plane is known as
   a. abduction
   b. eversion
   c. extension
   d. rotation
9. Circle the correct term. This type of movement is common in ball-and-socket joints and can be described as the movement of a bone around its longitudinal axis. It is rotation / flexion.
10. Circle True or False. The knee joint is the most freely movable joint in the body.

MasteringA&P™

Access practice quizzes and more in the Study Area at www.masteringaandp.com.

PAL

For access to anatomical models and more, check out Practice Anatomy Lab.

With rare exceptions, every bone in the body is connected to, or forms a joint with, at least one other bone. **Articulations,** or joints, perform two functions for the body. They (1) hold the bones together and (2) allow the rigid skeletal system some flexibility so that gross body movements can occur.

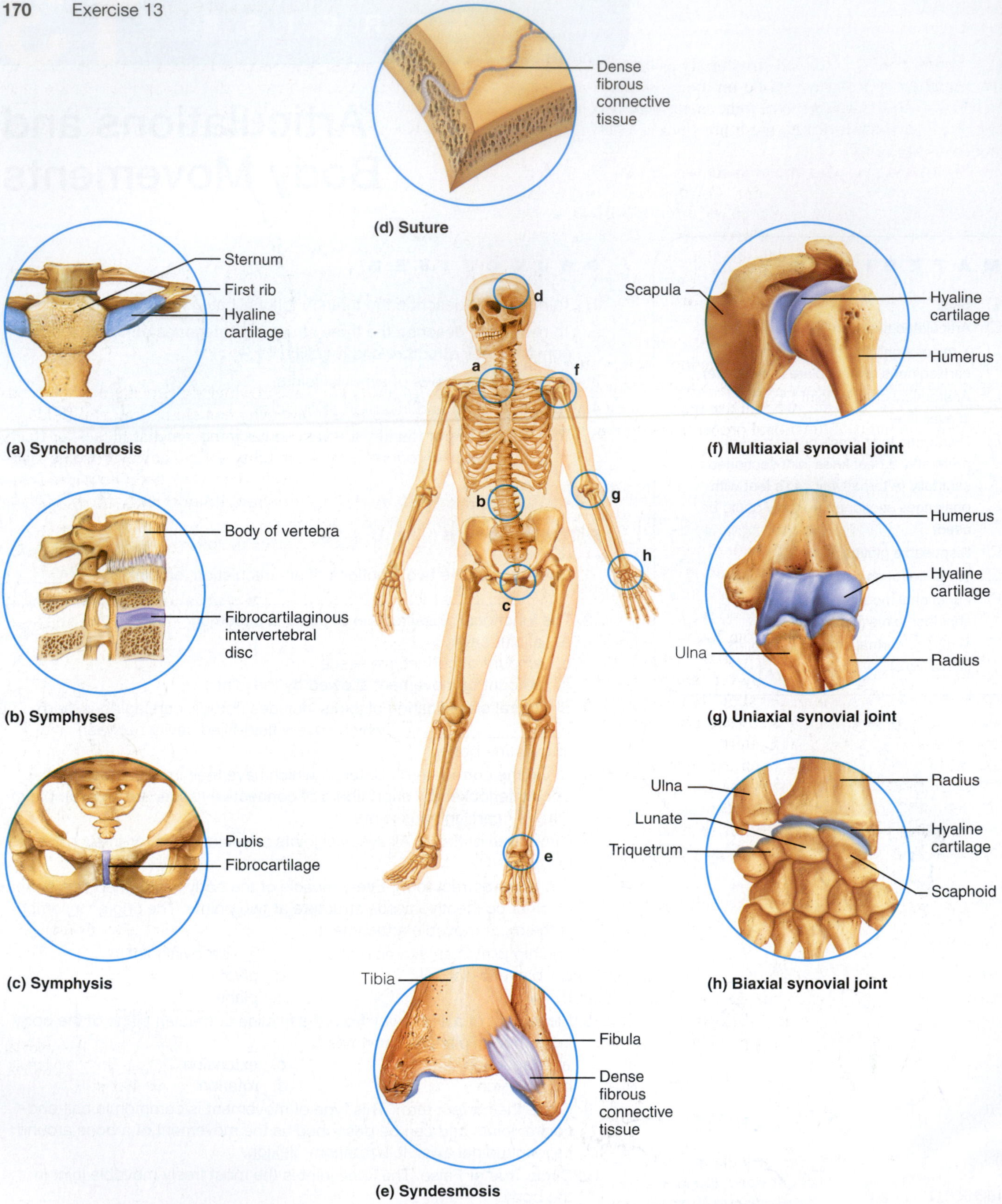

Dense fibrous connective tissue
(d) Suture
Sternum
First rib
Hyaline cartilage
(a) Synchondrosis
Body of vertebra
Fibrocartilaginous intervertebral disc
(b) Symphyses
Pubis
Fibrocartilage
(c) Symphysis
a
b
c
d
e
f
g
h
Scapula
Hyaline cartilage
Humerus
(f) Multiaxial synovial joint
Humerus
Hyaline cartilage
Ulna
Radius
(g) Uniaxial synovial joint
Ulna
Radius
Lunate
Triquetrum
Hyaline cartilage
Scaphoid
(h) Biaxial synovial joint
Tibia
Fibula
Dense fibrous connective tissue
(e) Syndesmosis

FIGURE 13.1 Types of joints. Joints to the left of the skeleton are cartilaginous joints; joints above and below the skeleton are fibrous joints; joints to the right of the skeleton are synovial joints. **(a)** Synchondrosis (joint between costal cartilage of rib 1 and the sternum). **(b)** Symphyses (intervertebral discs of fibrocartilage connecting adjacent vertebrae). **(c)** Symphysis (fibrocartilaginous pubic symphysis connecting the pubic bones anteriorly). **(d)** Suture (dense fibrous connective tissue connecting interlocking skull bones). **(e)** Syndesmosis (ligament of dense fibrous connective tissue connecting the distal ends of the tibia and fibula). **(f)** Synovial joint (multiaxial shoulder joint). **(g)** Synovial joint (uniaxial elbow joint). **(h)** Synovial joint (biaxial radiocarpal joint of the hand).

Joints may be classified structurally or functionally. The structural classification is based on the presence of connective tissue fiber, cartilage, or a joint cavity between the articulating bones. Structurally, there are *fibrous, cartilaginous,* and *synovial joints.*

The functional classification focuses on the amount of movement allowed at the joint. On this basis, there are **synarthroses,** or immovable joints; **amphiarthroses,** or slightly movable joints; and **diarthroses,** or freely movable joints. Freely movable joints predominate in the limbs, whereas immovable and slightly movable joints are largely restricted to the axial skeleton, where firm bony attachments and protection of enclosed organs are priorities.

As a general rule, fibrous joints are immovable, and synovial joints are freely movable. Cartilaginous joints offer both rigid and slightly movable examples. Since the structural categories are more clear-cut, we will use the structural classification here and indicate functional properties as appropriate.

Fibrous Joints

In **fibrous joints,** the bones are joined by fibrous tissue. No joint cavity is present. The amount of movement allowed depends on the length of the fibers uniting the bones. Although some fibrous joints are slightly movable, most are synarthrotic and permit virtually no movement.

The two major types of fibrous joints are sutures and syndesmoses. In **sutures** (Figure 13.1d) the irregular edges of the bones interlock and are united by very short connective tissue fibers, as in most joints of the skull. In **syndesmoses** the articulating bones are connected by short ligaments of dense fibrous tissue; the bones do not interlock. The joint at the distal end of the tibia and fibula is an example of a syndesmosis (Figure 13.1e). Although this syndesmosis allows some give, it is classed functionally as a synarthrosis. Not illustrated here is a **gomphosis,** in which a tooth is secured in a bony socket by the periodontal ligament (see Figure 38.12).

ACTIVITY 1

Identifying Fibrous Joints

Examine a human skull again. Notice that adjacent bone surfaces do not actually touch but are separated by fibrous connective tissue. Also examine a skeleton and anatomical chart of joint types and Table 13.1 (pp. 176–177) for examples of fibrous joints. ■

Cartilaginous Joints

In **cartilaginous joints,** the articulating bone ends are connected by a plate or pad of cartilage. No joint cavity is present. The two major types of cartilaginous joints are synchondroses and symphyses. Although there is variation, most cartilaginous joints are *slightly movable* (amphiarthrotic) functionally. In **symphyses** (*symphysis* = a growing together), the bones are connected by a broad, flat disc of fibrocartilage. The intervertebral joints and the pubic symphysis of the pelvis are symphyses (see Figure 13.1b and c). In **synchondroses** the bony portions are united by hyaline cartilage. The articulation of the costal cartilage of the first rib with the sternum (Figure 13.1a) is a synchondrosis, but perhaps the best examples of synchondroses are the epiphyseal plates seen in the long bones of growing children. View an X ray of the cartilaginous growth plate (epiphyseal disc) of a child's bone if one is available. The epiphyseal plates are flexible during childhood, but eventually they are totally ossified.

ACTIVITY 2

Identifying Cartilaginous Joints

Identify the cartilaginous joints on a human skeleton, Table 13.1, and on an anatomical chart of joint types. ■

Synovial Joints

Synovial joints are those in which the articulating bone ends are separated by a joint cavity containing synovial fluid (see Figure 13.1f–h). All synovial joints are diarthroses, or freely movable joints. Their mobility varies, however; some synovial joints can move in only one plane, and others can move in several directions (multiaxial movement). Most joints in the body are synovial joints.

All synovial joints have the following structural characteristics (Figure 13.2):

- The joint surfaces are enclosed by a two-layered *articular capsule* (a sleeve of connective tissue), creating a joint cavity.

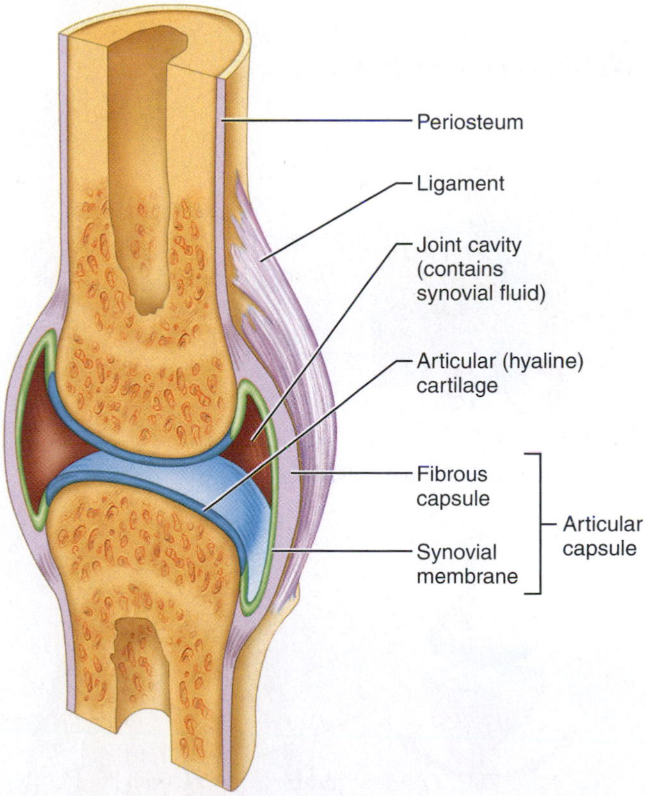
Periosteum
Ligament
Joint cavity (contains synovial fluid)
Articular (hyaline) cartilage
Fibrous capsule
Synovial membrane
Articular capsule

FIGURE 13.2 General structure of a synovial joint. The articulating bone ends are covered with articular cartilage, and enclosed within an articular capsule that is typically reinforced by ligaments externally. Internally the fibrous capsule is lined with a smooth synovial membrane that secretes synovial fluid.

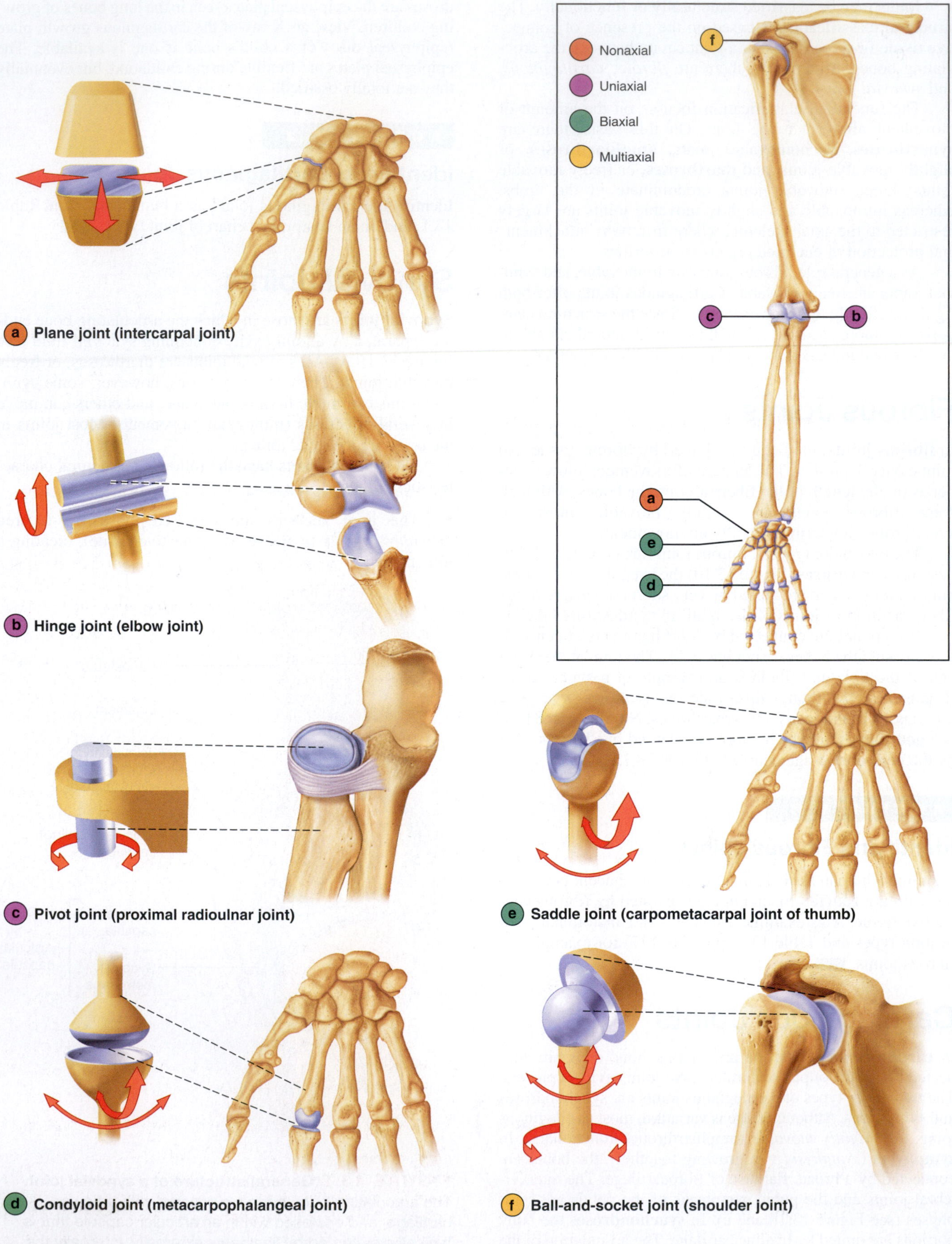
Nonaxial
Uniaxial
Biaxial
Multiaxial
a Plane joint (intercarpal joint)
b Hinge joint (elbow joint)
c Pivot joint (proximal radioulnar joint)
d Condyloid joint (metacarpophalangeal joint)
e Saddle joint (carpometacarpal joint of thumb)
f Ball-and-socket joint (shoulder joint)

FIGURE 13.3 Types of synovial joints. Dashed lines indicate the articulating bones.

- The inner layer is a smooth connective tissue membrane, called the *synovial membrane,* which produces a lubricating fluid (synovial fluid) that reduces friction. The outer layer, or *fibrous capsule,* is dense irregular connective tissue.
- *Articular* (hyaline) *cartilage* covers the surfaces of the bones forming the joint.
- The articular capsule is typically reinforced with ligaments and may contain *bursae* (fluid-filled sacs that reduce friction where tendons cross bone).
- Fibrocartilage pads *(articular discs)* may be present within the capsule.

ACTIVITY 3

Examining Synovial Joint Structure

Examine a beef or pig joint to identify the general structural features of diarthrotic joints as listed above.

! If the joint is freshly obtained from the slaughterhouse and you will be handling it, don disposable gloves before beginning your observations.

ACTIVITY 4

Demonstrating the Importance of Friction-Reducing Structures

1. Obtain a small water balloon and clamp. Partially fill the balloon with water (it should still be flaccid), and clamp it closed.

2. Position the balloon atop one of your fists and press down on its top surface with the other fist. Push on the balloon until your two fists touch and move your fists back and forth over one another. Assess the amount of friction generated.

3. Unclamp the balloon and add more water. The goal is to get just enough water in the balloon so that your fists cannot come into contact with one another, but instead remain separated by a thin water layer when pressure is applied to the balloon.

4. Repeat the movements in step 2 to assess the amount of friction generated.

How does the presence of a sac containing fluid influence the amount of friction generated?

___

What anatomical structure(s) does the water-containing balloon mimic?

___

What anatomical structures might be represented by your fists?

___

Types of Synovial Joints

Because there are so many types of synovial joints, they have been divided into the following subcategories on the basis of movements allowed (Figure 13.3):

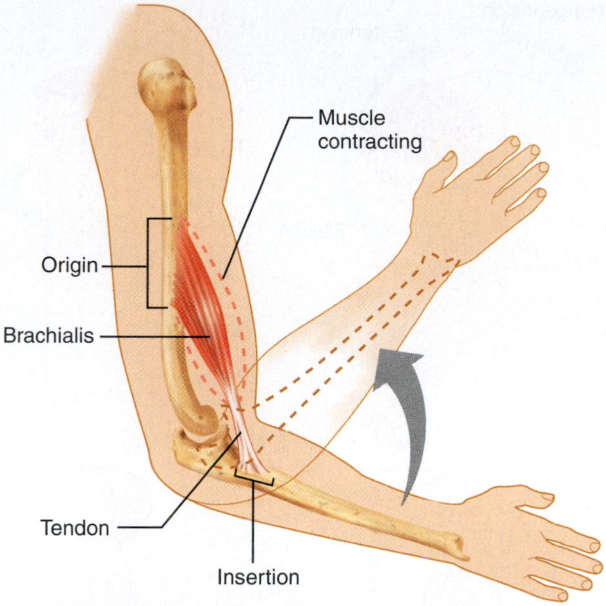
Muscle contracting
Origin
Brachialis
Tendon
Insertion

FIGURE 13.4 Muscle attachments (origin and insertion). When a skeletal muscle contracts, its insertion moves toward its origin.

- Plane (Gliding): Articulating surfaces are flat or slightly curved, allowing sliding movements in one or two planes. Examples are the intercarpal and intertarsal joints and the vertebrocostal joints of ribs 2–7.
- Hinge: The rounded process of one bone fits into the concave surface of another to allow movement in one plane (uniaxial), usually flexion and extension. Examples are the elbow and interphalangeal joints.
- Pivot: The rounded or conical surface of one bone articulates with a shallow depression or foramen in another bone. Pivot joints allow uniaxial rotation, as in the proximal radioulnar joint and the joint between the atlas and axis (C_1 and C_2).
- Condyloid (Ellipsoidal): The oval condyle of one bone fits into an ellipsoidal depression in another bone, allowing biaxial (two-way) movement. The radiocarpal (wrist) joint and the metacarpophalangeal joints (knuckles) are examples.
- Saddle: Articulating surfaces are saddle-shaped; the articulating surface of one bone is convex, and the reciprocal surface is concave. Saddle joints, which are biaxial, include the joint between the thumb metacarpal and the trapezium of the wrist.
- Ball and socket: The ball-shaped head of one bone fits into a cuplike depression of another. These are multiaxial joints, allowing movement in all directions and pivotal rotation. Examples are the shoulder and hip joints.

Movements Allowed by Synovial Joints

Every muscle of the body is attached to bone (or other connective tissue structures) at two points—the **origin** (the stationary, immovable, or less movable attachment) and the **insertion** (the movable attachment). Body movement occurs when muscles contract across diarthrotic synovial joints (Figure 13.4). When the muscle contracts and its fibers shorten, the insertion moves toward the origin. The type of movement depends on the construction of the joint (uniaxial,

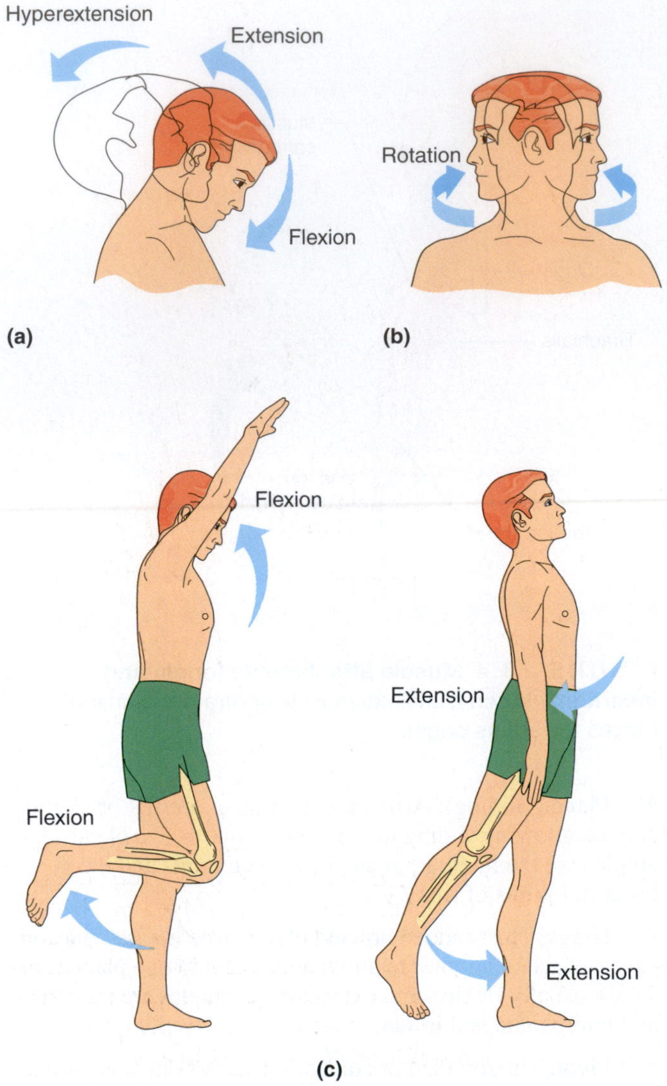
Hyperextension
Extension
Flexion
(a)
Rotation
(b)
Flexion
Flexion
Extension
Extension
(c)

FIGURE 13.5 Movements occurring at synovial joints of the body. **(a)** Flexion and extension of the head. **(b)** Rotation of the head. **(c)** Flexion and extension of the knee and shoulder.

biaxial, or multiaxial) and on the placement of the muscle relative to the joint. The most common types of body movements are described below and illustrated in Figure 13.5.

ACTIVITY 5

Demonstrating Movements of Synovial Joints

Attempt to demonstrate each movement as you read through the following material:

Flexion (Figure 13.5a and c): A movement, generally in the sagittal plane, that decreases the angle of the joint and reduces the distance between the two bones. Flexion is typical of hinge joints (bending the knee or elbow) but is also common at ball-and-socket joints (bending forward at the hip).

Extension (Figure 13.5a and c): A movement that increases the angle of a joint and the distance between two bones or parts of the body (straightening the knee or elbow); the opposite of flexion. If extension is greater than 180 degrees (bending the trunk backward), it is termed *hyperextension.*

Abduction (Figure 13.5d): Movement of a limb away from the midline or median plane of the body, generally on the frontal plane, or the fanning movement of fingers or toes when they are spread apart.

Adduction (Figure 13.5d): Movement of a limb toward the midline of the body or drawing the fingers or toes together; the opposite of abduction.

Rotation (Figure 13.5b and e): Movement of a bone around its longitudinal axis without lateral or medial displacement. Rotation, a common movement of ball-and-socket joints, also describes the movement of the atlas around the odontoid process of the axis.

Circumduction (Figure 13.5e): A combination of flexion, extension, abduction, and adduction commonly observed in ball-and-socket joints like the shoulder. The proximal end of the limb remains stationary, and the distal end moves in a circle. The limb as a whole outlines a cone. Condyloid and saddle joints also allow circumduction.

Pronation (Figure 13.5f): Movement of the palm of the hand from an anterior or upward-facing position to a posterior or downward-facing position. The distal end of the radius moves across the ulna.

Supination (Figure 13.5f): Movement of the palm from a posterior position to an anterior position (the anatomical position); the opposite of pronation. During supination, the radius and ulna are parallel.

The last four terms refer to movements of the foot:

Inversion (Figure 13.5g): A movement that results in the medial turning of the sole of the foot.

Eversion (Figure 13.5g): A movement that results in the lateral turning of the sole of the foot; the opposite of inversion.

Dorsiflexion (Figure 13.5h): A movement of the ankle joint in a dorsal direction (standing on one's heels).

Plantar flexion (Figure 13.5h): A movement of the ankle joint in which the foot is flexed downward (standing on one's toes or pointing the toes). ■

ACTIVITY 6

Demonstrating Uniaxial, Biaxial, and Multiaxial Movements

Using the information gained in the previous activity, perform the following demonstrations and complete the three charts.

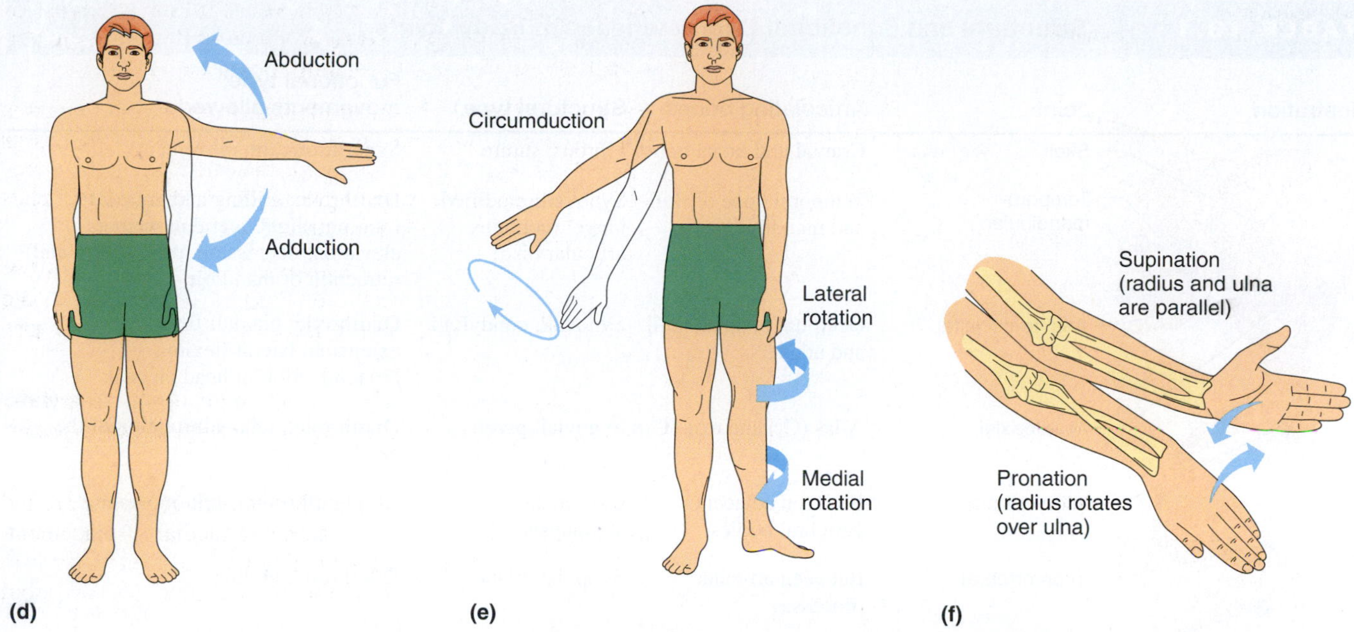
Abduction
Adduction
Circumduction
Lateral rotation
Medial rotation
Supination (radius and ulna are parallel)
Pronation (radius rotates over ulna)
(d)
(e)
(f)

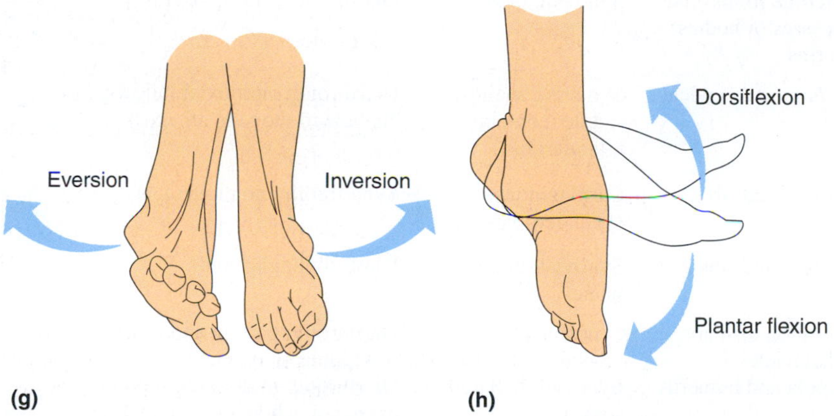
Eversion
Inversion
Dorsiflexion
Plantar flexion
(g)
(h)

FIGURE 13.5 *(continued)* **(d)** Abduction and adduction of the upper limb. **(e)** Circumduction of the upper limb and lateral and medial rotation of the lower limb around its long axis. **(f)** Supination and pronation of the forearm. **(g)** Eversion and inversion of the foot. **(h)** Dorsiflexion and plantar flexion of the foot.

1. Demonstrate movement at two joints that are uniaxial.

Name of joint	Movement allowed

2. Demonstrate movement at two joints that are biaxial.

Name of joint	Movement allowed	Movement allowed

3. Demonstrate movement at two joints that are multiaxial.

Name of joint	Movement allowed	Movement allowed	Movement allowed

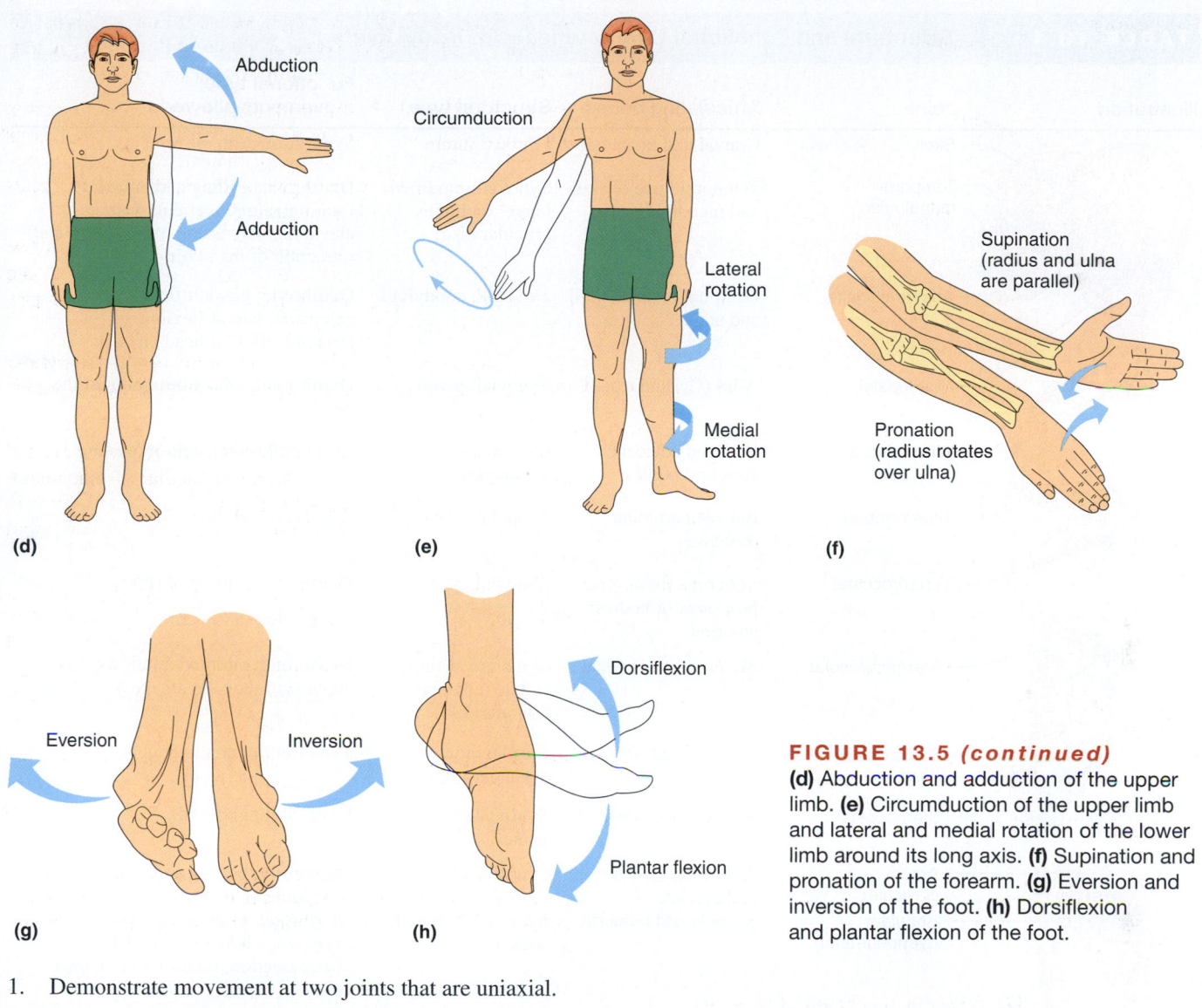

TABLE 13.1 Structural and Functional Characteristics of Body Joints

Illustration	Joint	Articulating bones	Structural type*	Functional type; movements allowed
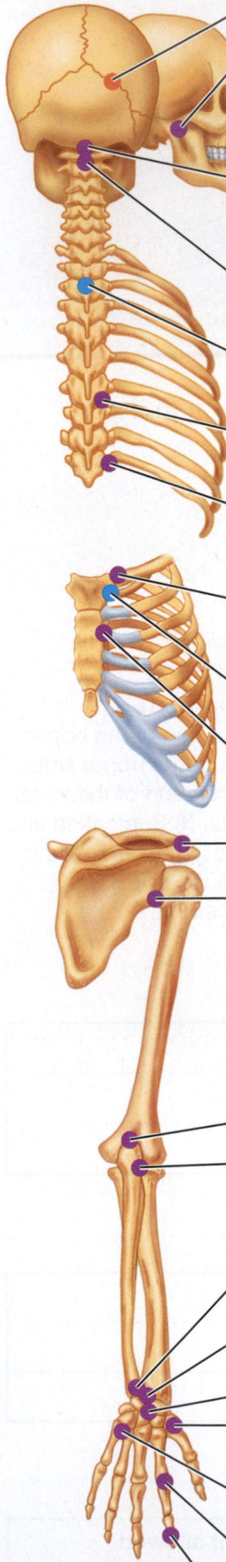	Skull	Cranial and facial bones	Fibrous; suture	Synarthrotic; no movement
	Temporo-mandibular	Temporal bone of skull and mandible	Synovial; modified hinge† (contains articular disc)	Diarthrotic; gliding and uniaxial rotation; slight lateral movement, elevation, depression, protraction, and retraction of mandible
	Atlanto-occipital	Occipital bone of skull and atlas	Synovial; condyloid	Diarthrotic; biaxial; flexion, extension, lateral flexion, circumduction of head on neck
	Atlantoaxial	Atlas (C_1) and axis (C_2)	Synovial; pivot	Diarthrotic; uniaxial; rotation of the head
	Intervertebral	Between adjacent vertebral bodies	Cartilaginous; symphysis	Amphiarthrotic; slight movement
	Intervertebral	Between articular processes	Synovial; plane	Diarthrotic; gliding
	Vertebrocostal	Vertebrae (transverse processes or bodies) and ribs	Synovial; plane	Diarthrotic; gliding of ribs
	Sternoclavicular	Sternum and clavicle	Synovial; shallow saddle (contains articular disc)	Diarthrotic; multiaxial (allows clavicle to move in all axes)
	Sternocostal (first)	Sternum and rib 1	Cartilaginous; synchondrosis	Synarthrotic; no movement
	Sternocostal	Sternum and ribs 2–7	Synovial; double plane	Diarthrotic; gliding
	Acromio-clavicular	Acromion of scapula and clavicle	Synovial; plane (contains articular disc)	Diarthrotic; gliding and rotation of scapula on clavicle
	Shoulder (glenohumeral)	Scapula and humerus	Synovial; ball and socket	Diarthrotic; multiaxial; flexion, extension, abduction, adduction, circumduction, rotation of humerus
	Elbow	Ulna (and radius) with humerus	Synovial; hinge	Diarthrotic; uniaxial; flexion, extension of forearm
	Radioulnar (proximal)	Radius and ulna	Synovial; pivot	Diarthrotic; uniaxial; rotation of radius around long axis of forearm to allow pronation and supination
	Radioulnar (distal)	Radius and ulna	Synovial; pivot (contains articular disc)	Diarthrotic; uniaxial; rotation (radius moves around ulna in ulnar notch of radius)
	Wrist (radiocarpal)	Radius and proximal carpals	Synovial; condyloid	Diarthrotic; biaxial; flexion, extension, abduction, adduction, circumduction of hand
	Intercarpal	Adjacent carpals	Synovial; plane	Diarthrotic; gliding
	Carpometacarpal of digit 1 (thumb)	Carpal (trapezium) and metacarpal 1	Synovial; saddle	Diarthrotic; biaxial; flexion, extension, abduction, adduction, circumduction, opposition of metacarpal 1
	Carpometacarpal of digits 2–5	Carpal(s) and metacarpal(s)	Synovial; plane	Diarthrotic; gliding of metacarpals
	Knuckle (metacarpo-phalangeal)	Metacarpal and proximal phalanx	Synovial; condyloid	Diarthrotic; biaxial; flexion, extension, abduction, adduction, circumduction of fingers
	Finger (interphalangeal)	Adjacent phalanges	Synovial; hinge	Diarthrotic; uniaxial; flexion, extension of fingers

TABLE 13.1 *(continued)*

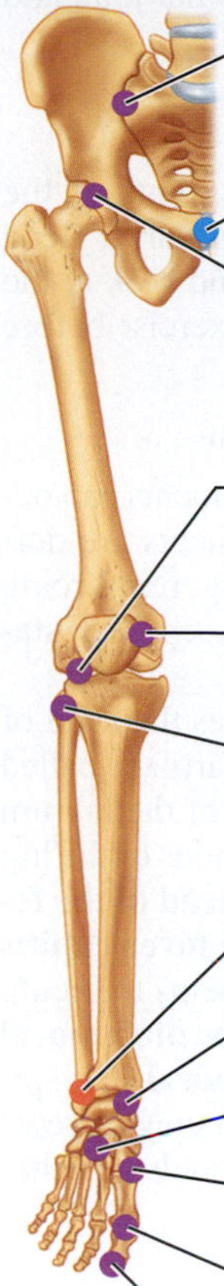

Illustration	Joint	Articulating bones	Structural type*	Functional type; movements allowed
	Sacroiliac	Sacrum and coxal bone	Synovial; plane	Diarthrotic; little movement, slight gliding possible (more during pregnancy)
	Pubic symphysis	Pubic bones	Cartilaginous; symphysis	Amphiarthrotic; slight movement (enhanced during pregnancy)
	Hip (coxal)	Hip bone and femur	Synovial; ball and socket	Diarthrotic; multiaxial; flexion, extension, abduction, adduction, rotation, circumduction of thigh
	Knee (tibiofemoral)	Femur and tibia	Synovial; modified hinge† (contains articular discs)	Diarthrotic; biaxial; flexion, extension of leg, some rotation allowed
	Knee (femoropatellar)	Femur and patella	Synovial; plane	Diarthrotic; gliding of patella
	Tibiofibular (proximal)	Tibia and fibula (proximally)	Synovial; plane	Diarthrotic; gliding of fibula
	Tibiofibular (distal)	Tibia and fibula (distally)	Fibrous; syndesmosis	Synarthrotic; slight "give" during dorsiflexion
	Ankle	Tibia and fibula with talus	Synovial; hinge	Diarthrotic; uniaxial; dorsiflexion, and plantar flexion of foot
	Intertarsal	Adjacent tarsals	Synovial; plane	Diarthrotic; gliding; inversion and eversion of foot
	Tarsometatarsal	Tarsal(s) and metatarsal(s)	Synovial; plane	Diarthrotic; gliding of metatarsals
	Metatarso-phalangeal	Metatarsal and proximal phalanx	Synovial; condyloid	Diarthrotic; biaxial; flexion extension, abduction, adduction, circumduction of great toe
	Toe (interpha-langeal)	Adjacent phalanges	Synovial; hinge	Diarthrotic; uniaxial; flexion, extension of toes

*__Fibrous joint__ indicated by orange circles; **cartilaginous joints**, by blue circles; **synovial joints**, by purple circles.

†These modified hinge joints are structurally bicondylar.

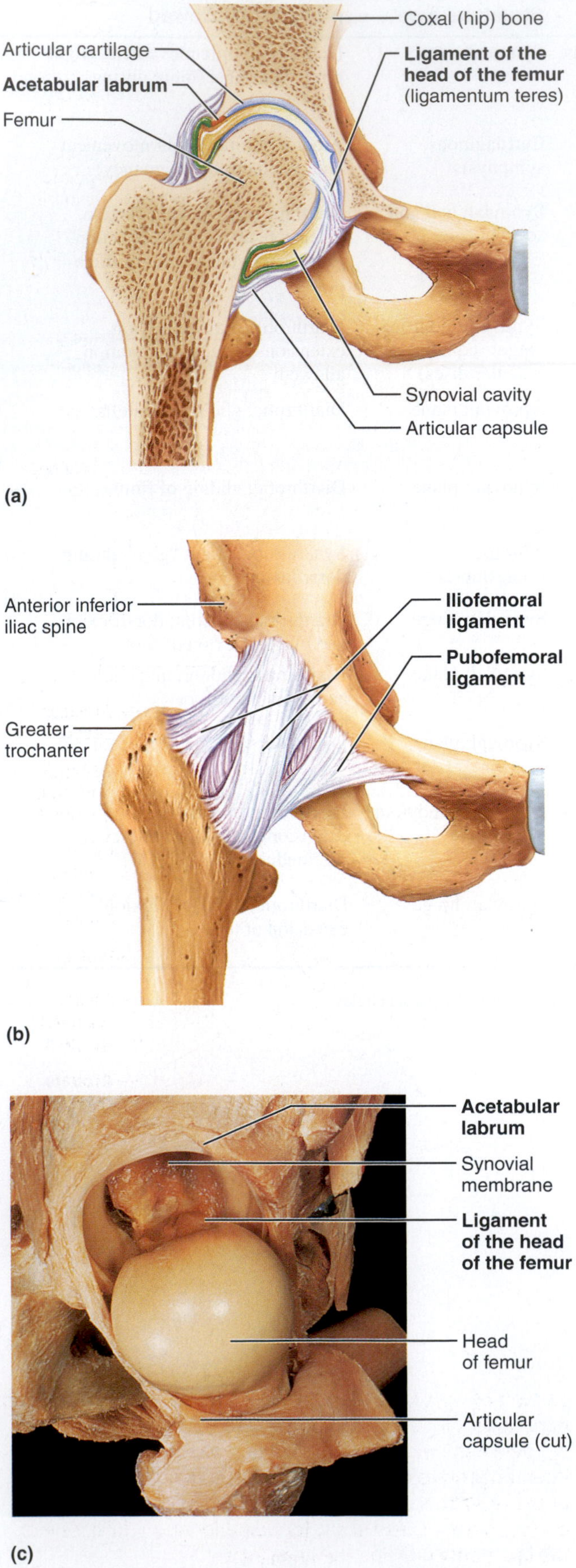
Coxal (hip) bone
Articular cartilage
Ligament of the head of the femur (ligamentum teres)
Acetabular labrum
Femur
Synovial cavity
Articular capsule
(a)
Anterior inferior iliac spine
Iliofemoral ligament
Pubofemoral ligament
Greater trochanter
(b)
Acetabular labrum
Synovial membrane
Ligament of the head of the femur
Head of femur
Articular capsule (cut)
(c)

Selected Synovial Joints

Now you will have the opportunity to compare and contrast the structure of the hip and knee joints and to investigate the structure and movements of the temporomandibular joint and shoulder joint.

The Hip and Knee Joints

Both of these joints are large weight-bearing joints of the lower limb, but they differ substantially in their security. Read through the brief descriptive material below, and look at the questions in the review sheet at the end of this exercise before beginning your comparison.

The Hip Joint

The hip joint is a ball-and-socket joint, so movements can occur in all possible planes. However, its movements are definitely limited by its deep socket and strong reinforcing ligaments, the two factors that account for its exceptional stability (Figure 13.6).

The deeply cupped acetabulum that receives the head of the femur is enhanced by a circular rim of fibrocartilage called the **acetabular labrum.** Because the diameter of the labrum is smaller than that of the femur's head, dislocations of the hip are rare. A short ligament, the **ligament of the head of the femur** or **ligamentum teres,** runs from the pitlike **fovea capitis** on the femur head to the acetabulum where it helps to secure the femur. Several strong ligaments, including the **iliofemoral** and **pubofemoral** anteriorly and the **ischiofemoral** that spirals posteriorly (not shown), are arranged so that they "screw" the femur head into the socket when a person stands upright.

ACTIVITY 7

Demonstrating Actions at the Hip Joint

If a functional hip joint model is available, identify the joint parts and manipulate it to demonstrate the following movements: flexion, extension, abduction, and inner and outer rotation that can occur at this joint.

Reread the information on what movements the associated ligaments restrict, and verify that information during your joint manipulations.

The Knee Joint

The knee is the largest and most complex joint in the body. Three joints in one (Figure 13.7), it allows extension, flexion, and a little rotation. The **tibiofemoral joint,** actually a duplex joint between the femoral condyles above and the **menisci** (semilunar cartilages) of the tibia below, is functionally a hinge joint, a very unstable one made slightly more secure by the menisci (Figure 13.7b and d). Some rotation occurs when the knee is partly flexed, but during extension, rotation and side-to-side movements are counteracted by the menisci and ligaments. The other joint is the **femoropatellar joint,** the intermediate joint anteriorly (Figure 13.7a and c).

FIGURE 13.6 Hip joint relationships. (a) Frontal section through the right hip joint. **(b)** Anterior superficial view of the right hip joint. **(c)** Photograph of the interior of the hip joint, lateral view.

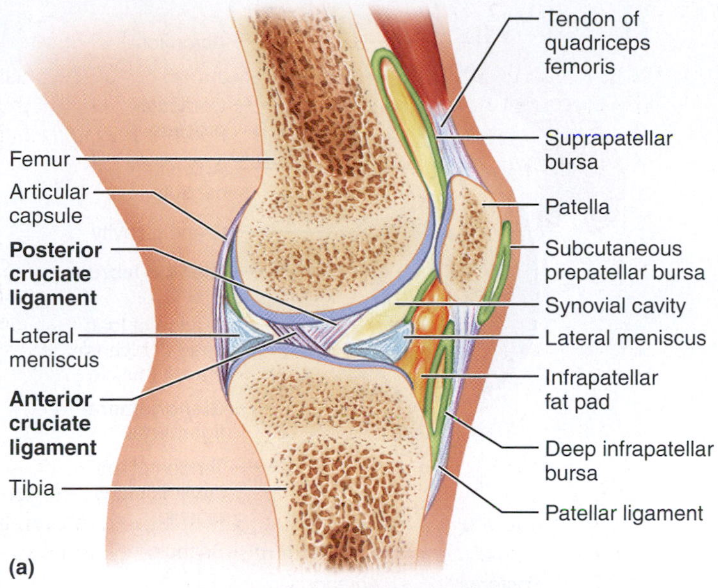
Tendon of quadriceps femoris
Femur
Suprapatellar bursa
Articular capsule
Patella
Posterior cruciate ligament
Subcutaneous prepatellar bursa
Synovial cavity
Lateral meniscus
Lateral meniscus
Anterior cruciate ligament
Infrapatellar fat pad
Deep infrapatellar bursa
Tibia
Patellar ligament
(a)

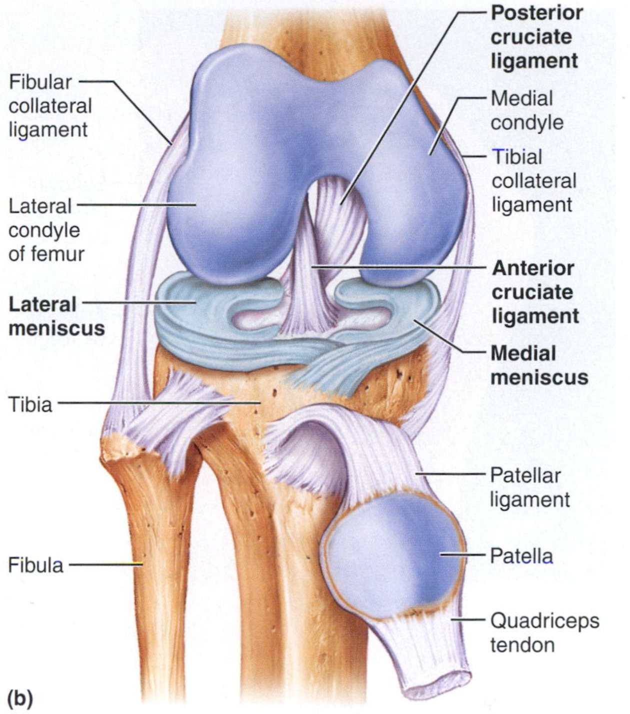
Posterior cruciate ligament
Fibular collateral ligament
Medial condyle
Tibial collateral ligament
Lateral condyle of femur
Anterior cruciate ligament
Lateral meniscus
Medial meniscus
Tibia
Patellar ligament
Fibula
Patella
Quadriceps tendon
(b)

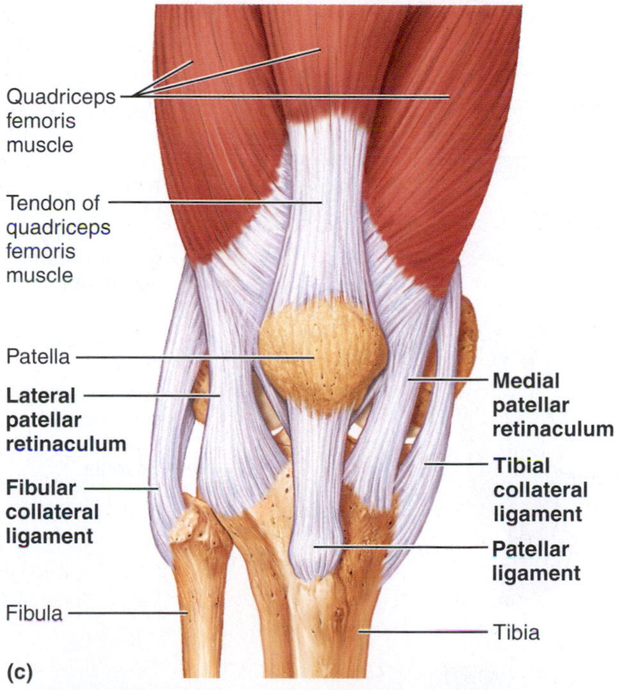
Quadriceps femoris muscle
Tendon of quadriceps femoris muscle
Patella
Medial patellar retinaculum
Lateral patellar retinaculum
Tibial collateral ligament
Fibular collateral ligament
Patellar ligament
Fibula
Tibia
(c)

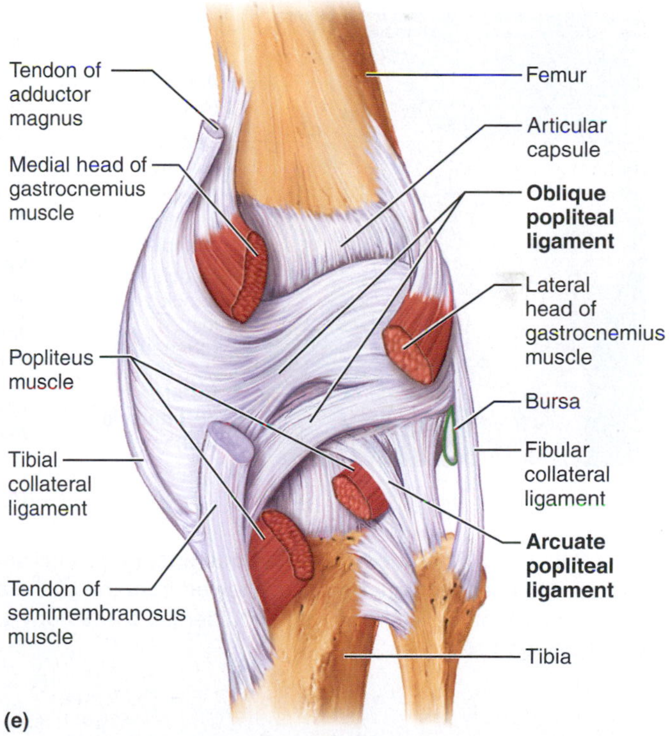
Tendon of adductor magnus
Femur
Articular capsule
Medial head of gastrocnemius muscle
Oblique popliteal ligament
Lateral head of gastrocnemius muscle
Popliteus muscle
Bursa
Tibial collateral ligament
Fibular collateral ligament
Arcuate popliteal ligament
Tendon of semimembranosus muscle
Tibia
(e)

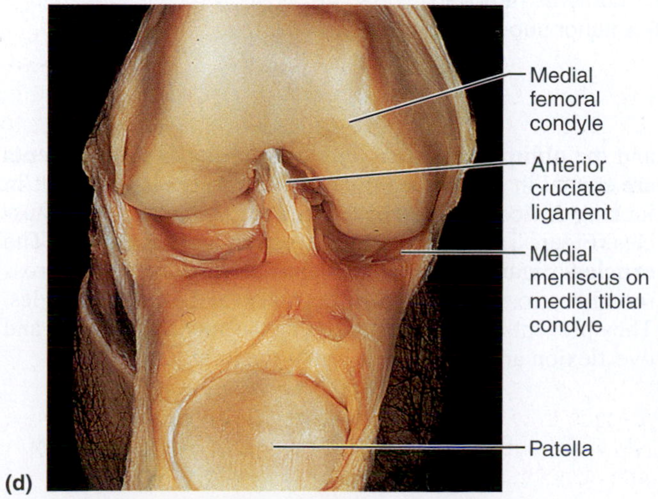
Medial femoral condyle
Anterior cruciate ligament
Medial meniscus on medial tibial condyle
Patella
(d)

FIGURE 13.7 Knee joint relationships. (a) Midsagittal section of right knee joint. **(b)** Anterior view of slightly flexed right knee joint showing the cruciate ligaments. Articular capsule has been removed; the quadriceps tendon has been cut and reflected distally. **(c)** Anterior superficial view of the right knee. **(d)** Photograph of an opened knee joint corresponds to view in (b). **(e)** Posterior superficial view of the ligaments clothing the knee joint.

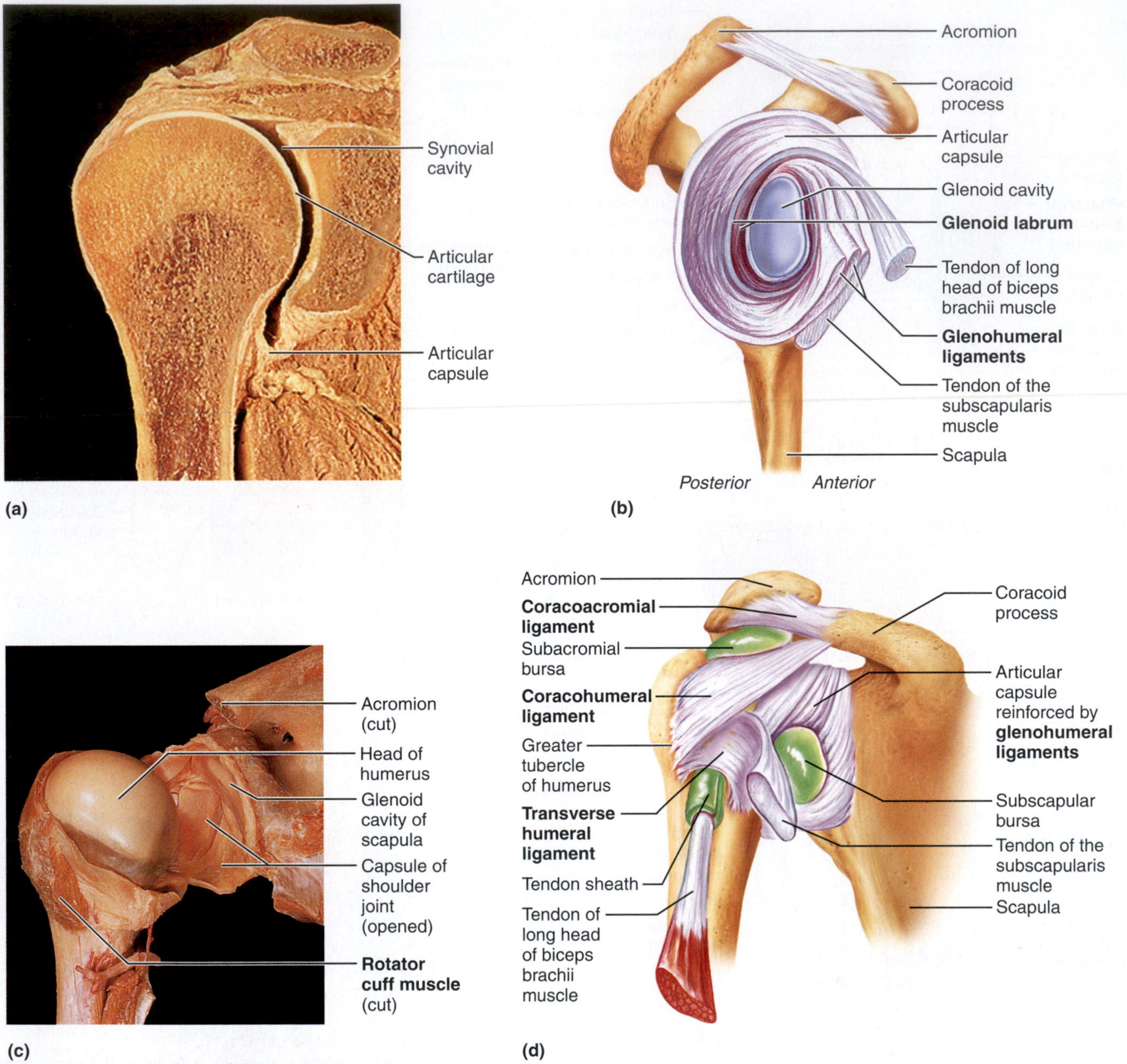
Synovial cavity
Articular cartilage
Articular capsule
(a)
Acromion
Coracoid process
Articular capsule
Glenoid cavity
Glenoid labrum
Tendon of long head of biceps brachii muscle
Glenohumeral ligaments
Tendon of the subscapularis muscle
Scapula
Posterior
Anterior
(b)
Acromion (cut)
Head of humerus
Glenoid cavity of scapula
Capsule of shoulder joint (opened)
Rotator cuff muscle (cut)
(c)
Acromion
Coracoacromial ligament
Subacromial bursa
Coracohumeral ligament
Greater tubercle of humerus
Transverse humeral ligament
Tendon sheath
Tendon of long head of biceps brachii muscle
Coracoid process
Articular capsule reinforced by glenohumeral ligaments
Subscapular bursa
Tendon of the subscapularis muscle
Scapula
(d)

FIGURE 13.8 Shoulder joint relationships. (a) Frontal section through the shoulder. **(b)** Right shoulder joint, cut open and viewed from the lateral aspect; humerus removed. **(c)** Photograph of the interior of the shoulder joint, anterior view. **(d)** Anterior superficial view of the right shoulder.

The knee is unique in that it is only partly enclosed by an articular capsule. Anteriorly, where the capsule is absent, are three broad ligaments, the **patellar ligament** and the **medial** and **lateral patellar retinacula** (retainers), which run from the patella to the tibia below and merge with the capsule on either side.

Capsular ligaments including the **fibular** and **tibial collateral ligaments** (which prevent rotation during extension) and the **oblique popliteal** and **arcuate popliteal ligaments** are crucial in reinforcing the knee. The knees have a built-in locking device that must be "unlocked" by the popliteus muscles (Figure 13.7e) before the knees can be flexed again. The **cruciate ligaments** are intracapsular ligaments that cross (cruci = cross) in the notch between the femoral condyles. They prevent anterior-posterior displacement of the joint and overflexion and hyperextension of the joint.

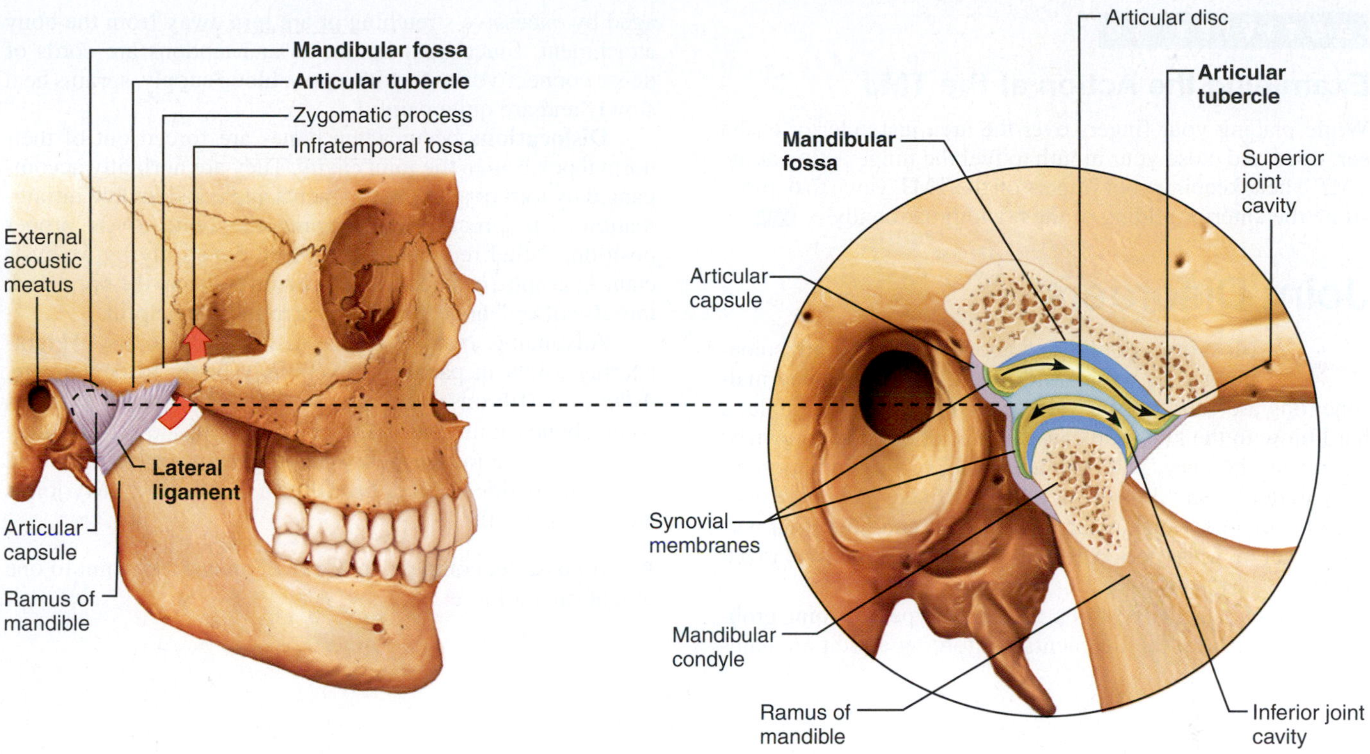
Mandibular fossa
Articular tubercle
Zygomatic process
Infratemporal fossa
External acoustic meatus
Lateral ligament
Articular capsule
Ramus of mandible
Articular disc
Articular tubercle
Mandibular fossa
Superior joint cavity
Articular capsule
Synovial membranes
Mandibular condyle
Ramus of mandible
Inferior joint cavity

(a) Location of the joint in the skull

(b) Enlargement of a sagittal section through the joint

FIGURE 13.9 The temporomandibular (jaw) joint relationships. Note that the superior and inferior compartments of the joint cavity allow different movements indicated by arrows.

ACTIVITY 8

Demonstrating Actions at the Knee Joint

If a functional model of a knee joint is available, identify the joint parts and manipulate it to illustrate the following movements: flexion, extension, and inner and outer rotation.

Reread the information on what movements the various associated ligaments restrict, and verify that information during your joint manipulations.

The Shoulder Joint

The shoulder joint or **glenohumeral joint** is the most freely moving joint of the body. The rounded head of the humerus fits the shallow glenoid cavity of the scapula (Figure 13.8). A rim of fibrocartilage, the **glenoid labrum,** deepens the cavity slightly.

The articular capsule enclosing the joint is thin and loose, contributing to ease of movement. Few ligaments reinforce the shoulder, most of them located anteriorly. The **coracohumeral ligament** helps support the weight of the upper limb, and three weak **glenohumeral ligaments** strengthen the front of the capsule. In some people they are absent. Muscle tendons from the biceps brachii and **rotator cuff** muscles (subscapularis, supraspinatus, infraspinatus, and teres minor) contribute most to shoulder stability.

ACTIVITY 9

Demonstrating Actions at the Shoulder Joint

If a functional shoulder joint model is available, identify the joint parts and manipulate the model to demonstrate the following movements: flexion, extension, abduction, adduction, circumduction, and inner and outer rotation.

Note where the joint is weakest and verify the most common direction of a dislocated humerus.

The Temporomandibular Joint

The **temporomandibular joint (TMJ)** lies just anterior to the ear (Figure 13.9), where the egg-shaped condyle of the mandible articulates with the inferior surface of the squamous region of the temporal bone. The temporal bone joint surface has a complicated shape: posteriorly is the **mandibular fossa** and anteriorly is a bony knob called the **articular tubercle.** The joint's articular capsule, though strengthened by the **lateral ligament,** is slack; an articular disc divides the joint cavity into superior and inferior compartments. Typically, the mandibular condyle–mandibular fossa connection allows the familiar hingelike movements of elevating and depressing the mandible to open and close the mouth. However, when the mouth is opened wide, the mandibular head glides anteriorly and is braced against the dense bone of the articular tubercle so that the mandible is not forced superiorly when we bite hard foods.

ACTIVITY 10

Examining the Action at the TMJ

While placing your fingers over the area just anterior to the ear, open and close your mouth to feel the hinge action at the TMJ. Then, keeping your fingers on the TMJ, yawn to demonstrate the anterior gliding of the mandibular condyle.

Joint Disorders

Most of us don't think about our joints until something goes wrong with them. Joint pains and malfunctions are caused by a variety of things. For example, a hard blow to the knee can cause a painful bursitis, known as "water on the knee," due to damage to, or inflammation of, the patellar bursa. Slippage of a fibrocartilage pad or the tearing of a ligament may result in a painful condition that persists over a long period, since these poorly vascularized structures heal so slowly.

Sprains and dislocations are other types of joint problems. In a **sprain,** the ligaments reinforcing a joint are damaged by excessive stretching or are torn away from the bony attachment. Since both ligaments and tendons are cords of dense connective tissue with a poor blood supply, sprains heal slowly and are quite painful.

Dislocations occur when bones are forced out of their normal position in the joint cavity. They are normally accompanied by torn or stressed ligaments and considerable inflammation. The process of returning the bone to its proper position, called reduction, should be done only by a physician. Attempts by the untrained person to "snap the bone back into its socket" are often more harmful than helpful.

Advancing years also take their toll on joints. Weight-bearing joints in particular eventually begin to degenerate. *Adhesions* (fibrous bands) may form between the surfaces where bones join, and extraneous bone tissue (*spurs*) may grow along the joint edges. Such degenerative changes lead to the complaint so often heard from the elderly: "My joints are getting so stiff. . . ."

- If possible, compare an X-ray of an arthritic joint to one of a normal joint.

EXERCISE 14

Microscopic Anatomy and Organization of Skeletal Muscle

MATERIALS

- ☐ Three-dimensional model of skeletal muscle cells (if available)
- ☐ Forceps
- ☐ Dissecting needles
- ☐ Clean microscope slides and coverslips
- ☐ 0.9% saline solution in dropper bottles
- ☐ Chicken breast or thigh muscle (freshly obtained from the meat market)
- ☐ Compound microscope
- ☐ Prepared slides of skeletal muscle (l.s. and x.s. views) and skeletal muscle showing neuromuscular junctions
- ☐ Three-dimensional model of skeletal muscle showing neuromuscular junction (if available)

OBJECTIVES

1. To describe the structure of skeletal muscle from gross to microscopic levels.
2. To define and explain the role of the following:

fiber	aponeurosis
myofibril	tendon
myofilament	epimysium
actin	perimysium
myosin	endomysium

3. To describe the structure of a neuromuscular junction and to explain its role in muscle function.

PRE-LAB QUIZ

1. Which is *not* true of skeletal muscle?
   a. It enables you to manipulate your environment.
   b. It influences the body's contours and shape.
   c. It is one of the major components of hollow organs.
   d. It provides a means of locomotion.
2. Circle the correct term. Because the cells of skeletal muscle are relatively large and cylindrical in shape, they are also known as fibers / tubules.
3. Circle True or False. Skeletal muscle cells have more than one nucleus.
4. The two contractile proteins that make up the myofilaments of skeletal muscle are ________________ and ________________.
5. Each muscle cell is surrounded by thin connective tissue called the
   a. aponeuroses
   b. endomysium
   c. epimysium
   d. perimysium
6. A strong cordlike structure that connects a muscle to another muscle or bone is
   a. a fascicle
   b. a tendon
   c. deep fascia
7. The junction between a nerve fiber and a muscle cell is called a ________________.
8. Circle True or False. The neuron and muscle fiber membranes do not actually touch but are separated by a fluid-filled gap.

MasteringA&P™

Access practice quizzes and more in the Study Area at www.masteringaandp.com.

PAL

For access to anatomical models and more, check out Practice Anatomy Lab.

The bulk of the body's muscle is called **skeletal muscle** because it is attached to the skeleton (or associated connective tissue structures). Skeletal muscle influences body contours and shape, allows you to grin and frown, provides a means of locomotion, and enables you to manipulate the environment. The balance of the body's muscle—smooth and cardiac muscle, the major components of the walls of hollow organs and the heart, respectively—is involved with the transport of materials within the body.

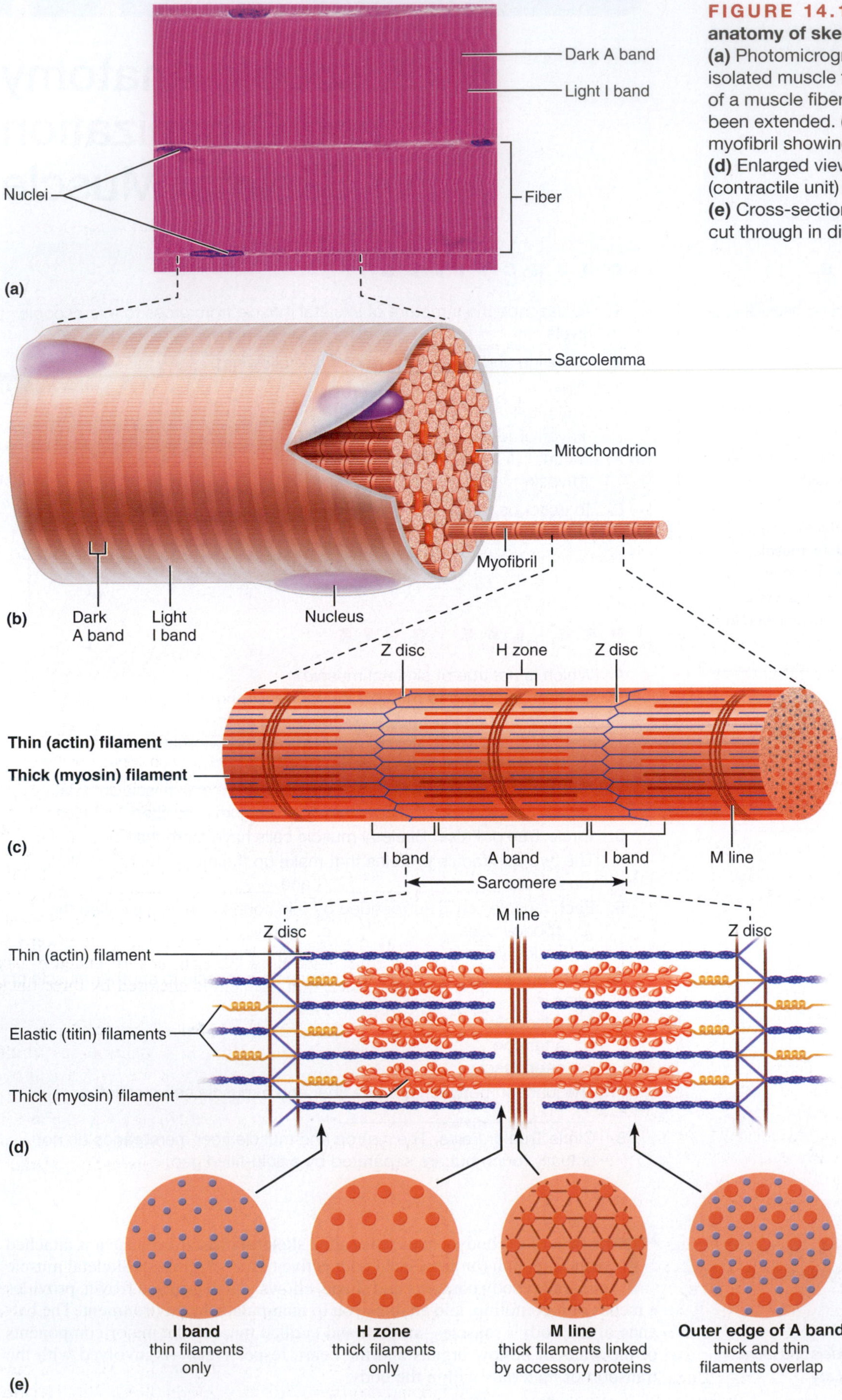
Dark A band
Light I band
Nuclei
Fiber
(a)
Sarcolemma
Mitochondrion
Myofibril
(b)
Dark A band
Light I band
Nucleus
Z disc
H zone
Z disc
Thin (actin) filament
Thick (myosin) filament
(c)
I band
A band
I band
M line
Sarcomere
M line
Z disc
Z disc
Thin (actin) filament
Elastic (titin) filaments
Thick (myosin) filament
(d)
I band
thin filaments only
H zone
thick filaments only
M line
thick filaments linked by accessory proteins
Outer edge of A band
thick and thin filaments overlap
(e)

FIGURE 14.1 Microscopic anatomy of skeletal muscle. **(a)** Photomicrograph of portions of two isolated muscle fibers (700×). **(b)** Part of a muscle fiber. One myofibril has been extended. **(c)** Enlarged view of a myofibril showing its banding pattern. **(d)** Enlarged view of one sarcomere (contractile unit) of a myofibril. **(e)** Cross-sectional view of a sarcomere cut through in different areas.

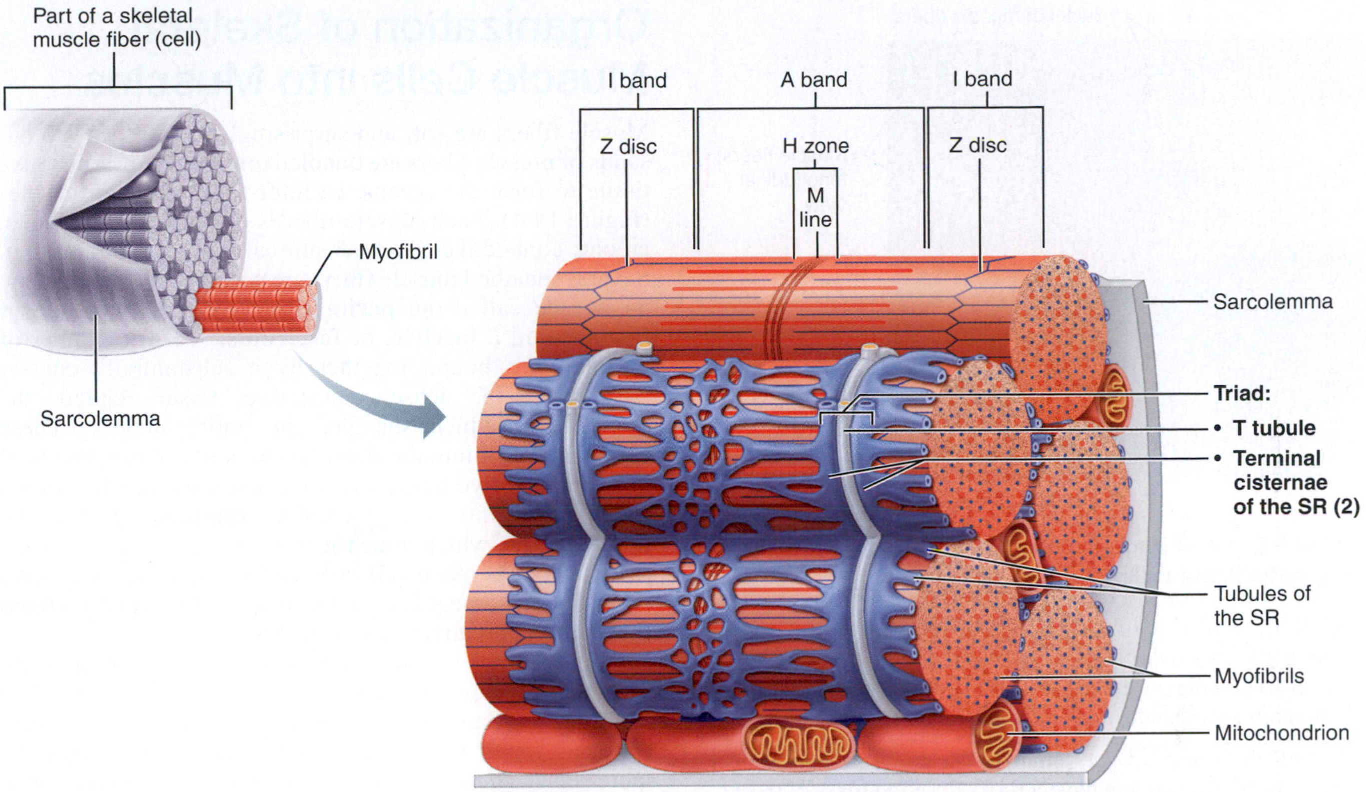

Part of a skeletal muscle fiber (cell)
Myofibril
Sarcolemma
I band
A band
I band
Z disc
H zone
Z disc
M line
Sarcolemma
Triad:
• T tubule
• Terminal cisternae of the SR (2)
Tubules of the SR
Myofibrils
Mitochondrion

FIGURE 14.2 Relationship of the sarcoplasmic reticulum and T tubules to the myofibrils of skeletal muscle.

Each of the three muscle types has a structure and function uniquely suited to its task in the body. However, because the term *muscular system* applies specifically to skeletal muscle, the primary objective of this unit is to investigate the structure and function of skeletal muscle.

Skeletal muscle is also known as *voluntary muscle* (because it can be consciously controlled) and as *striated muscle* (because it appears to be striped). As you might guess from both of these alternative names, skeletal muscle has some special characteristics. Thus an investigation of skeletal muscle should begin at the cellular level.

The Cells of Skeletal Muscle

Skeletal muscle is composed of relatively large, long cylindrical cells, sometimes called **fibers,** ranging from 10 to 100 μm in diameter and up to 6 cm in length. However, the cells of large, hard-working muscles like the antigravity muscles of the hip are extremely coarse, ranging up to 25 cm in length, and can be seen with the naked eye.

Skeletal muscle cells (Figure 14.1a and b) are multinucleate; multiple oval nuclei can be seen just beneath the plasma membrane (called the *sarcolemma* in these cells). The nuclei are pushed peripherally by the longitudinally arranged **myofibrils,** which nearly fill the sarcoplasm. Alternating light (I) and dark (A) bands along the length of the perfectly aligned myofibrils give the muscle fiber as a whole its striped appearance.

Electron microscope studies have revealed that the myofibrils are made up of even smaller threadlike structures called **myofilaments** (Figure 14.1d). The myofilaments are composed largely of two varieties of contractile proteins—**actin** and **myosin**—which slide past each other during muscle activity to bring about shortening or contraction of the muscle cells. It is the highly specific arrangement of the myofilaments within the myofibrils that is responsible for the banding pattern in skeletal muscle. The actual contractile units of muscle, called **sarcomeres,** extend from the middle of one I band (its Z disc) to the middle of the next along the length of the myofibrils (Figure 14.1c and d.) Cross sections of the sarcomere in areas where **thick** and **thin filaments** overlap show that each thick filament is surrounded by six thin filaments; each thin filament is enclosed by three thick filaments (Figure 14.1e).

At each junction of the A and I bands, the sarcolemma indents into the muscle cell, forming a **transverse tubule (T tubule).** These tubules run deep into the muscle cell between cross channels, or **terminal cisternae,** of the elaborate smooth endoplasmic reticulum called the **sarcoplasmic reticulum (SR)** (Figure 14.2). Regions where the SR terminal cisternae abut a T tubule on each side are called **triads.**

ACTIVITY 1

Examining Skeletal Muscle Cell Anatomy

1. Look at the three-dimensional model of skeletal muscle cells, noting the relative shape and size of the cells. Identify the nuclei, myofibrils, and light and dark bands.

2. Obtain forceps, two dissecting needles, slide and coverslip, and a dropper bottle of saline solution. With forceps,

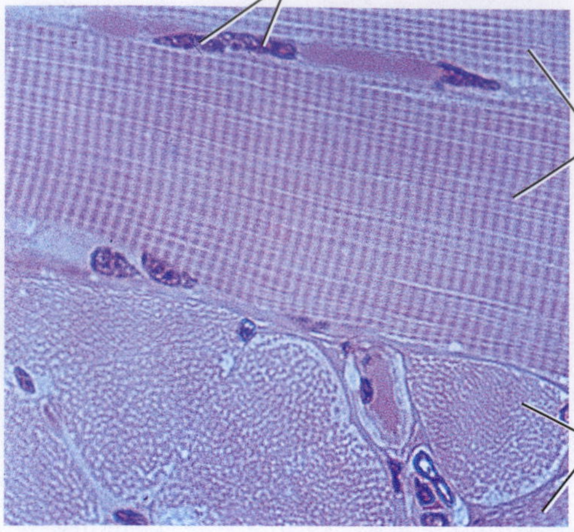
Nuclei of muscle fibers
Muscle fibers, longitudinal view
Muscle fibers, cross-sectional view

FIGURE 14.3 Photomicrograph of muscle fibers, longitudinal and cross sections (400×).

remove a very small piece of muscle (about 1 mm diameter) from a fresh chicken breast (or thigh). Place the tissue on a clean microscope slide, and add a drop of the saline solution.

3. Pull the muscle fibers apart (tease them) with the dissecting needles until you have a fluffy-looking mass of tissue. Cover the teased tissue with a coverslip, and observe under the high-power lens of a compound microscope. Look for the banding pattern by examining muscle fibers isolated at the edge of the tissue mass. Regulate the light carefully to obtain the highest possible contrast.

4. Now compare your observations with Figure 14.3 and with what can be seen with professionally prepared muscle tissue. Obtain a slide of skeletal muscle (longitudinal section), and view it under high power. From your observations, draw a small section of a muscle fiber in the space provided below. Label the nuclei, sarcolemma, and A and I bands.

What structural details become apparent with the prepared slide?

____

____

Organization of Skeletal Muscle Cells into Muscles

Muscle fibers are soft and surprisingly fragile. Thus thousands of muscle fibers are bundled together with connective tissue to form the organs we refer to as skeletal muscles (Figure 14.4). Each muscle fiber is enclosed in a delicate, areolar connective tissue sheath called the **endomysium.** Several sheathed muscle fibers are wrapped by a collagenic membrane called the **perimysium,** forming a bundle of fibers called a **fascicle,** or **fasciculus.** A large number of fascicles are bound together by a substantially coarser "overcoat" of dense connective tissue called the **epimysium,** which sheathes the entire muscle. These epimysia blend into the **deep fascia,** still coarser sheets of dense connective tissue that bind muscles into functional groups, and into strong cordlike **tendons** or sheetlike **aponeuroses,** which attach muscles to each other or indirectly to bones. As noted in Exercise 13, a muscle's more movable attachment is called its *insertion* whereas its fixed (or immovable) attachment is the *origin.*

Tendons perform several functions, two of the most important being to provide durability and to conserve space. Because tendons are tough collagenic connective tissue, they can span rough bony prominences that would destroy the more delicate muscle tissues. Because of their relatively small size, more tendons than fleshy muscles can pass over a joint.

In addition to supporting and binding the muscle fibers, and providing strength to the muscle as a whole, the connective tissue wrappings provide a route for the entry and exit of nerves and blood vessels that serve the muscle fibers. The larger, more powerful muscles have relatively more connective tissue than muscles involved in fine or delicate movements.

As we age, the mass of the muscle fibers decreases, and the amount of connective tissue increases; thus the skeletal muscles gradually become more sinewy, or "stringier."

ACTIVITY 2

Observing the Histological Structure of a Skeletal Muscle

Obtain a slide showing a cross section of skeletal muscle tissue. Using Figure 14.4 as a reference, identify the muscle fibers, their peripherally located nuclei, and their connective tissue wrappings, (the endomysium, perimysium, and epimysium, if visible).

The Neuromuscular Junction

The voluntary skeletal muscle cells are always stimulated by motor neurons via nerve impulses. The junction between a nerve fiber (axon) and a muscle cell is called a **neuromuscular,** or **myoneural, junction** (Figure 14.5).

FIGURE 14.4 Connective tissue coverings of skeletal muscle. (a) Diagrammatic view. **(b)** Photomicrograph of a cross section of skeletal muscle (150×).

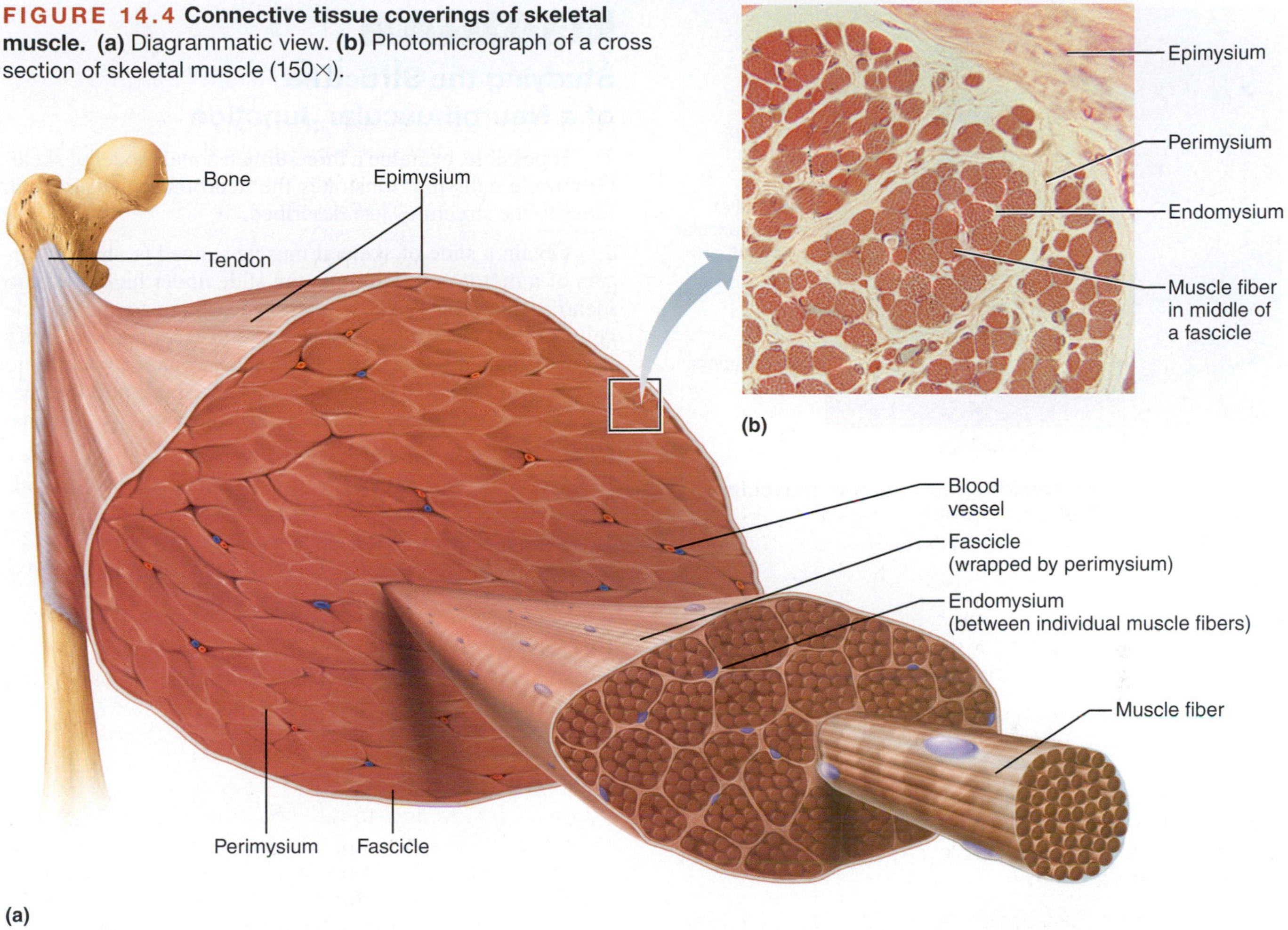
Epimysium
Perimysium
Endomysium
Muscle fiber in middle of a fascicle
(b)
Bone
Epimysium
Tendon
Blood vessel
Fascicle (wrapped by perimysium)
Endomysium (between individual muscle fibers)
Muscle fiber
Perimysium
Fascicle
(a)

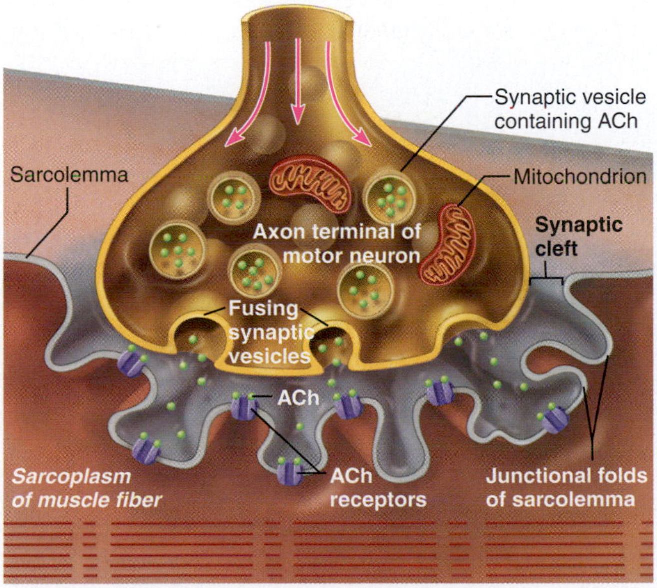
Synaptic vesicle containing ACh
Sarcolemma
Mitochondrion
Axon terminal of motor neuron
Synaptic cleft
Fusing synaptic vesicles
ACh
Sarcoplasm of muscle fiber
ACh receptors
Junctional folds of sarcolemma

FIGURE 14.5 The neuromuscular junction. Red arrows indicate arrival of the nerve impulse (action potential), which ultimately causes vesicles to release ACh. The ACh receptor is part of the ion channel that opens briefly, causing depolarization of the sarcolemma.

Each axon of the motor neuron breaks up into many branches called *axon terminals* as it approaches the muscle, and each of these branches participates in forming a neuromuscular junction with a single muscle cell. Thus a single neuron may stimulate many muscle fibers. Together, a neuron and all the muscle fibers it stimulates make up the functional structure called the **motor unit.** Part of a motor unit, showing two neuromuscular junctions, is shown in Fig. 14.6. The neuron and muscle fiber membranes, close as they are, do not actually touch. They are separated by a small fluid-filled gap called the **synaptic cleft** (see Figure 14.5).

Within the axon terminals are many mitochondria and vesicles containing a neurotransmitter chemical called acetylcholine (ACh). When a nerve impulse reaches the axon terminal, some of these vesicles release their contents into the synaptic cleft. The ACh rapidly diffuses across the junction and combines with the receptors on the sarcolemma. When receptors bind ACh, a change in the permeability of the sarcolemma occurs. Channels that allow both sodium (Na^+) and potassium (K^+) ions to pass open briefly. Because more Na^+ diffuses into the muscle fiber than K^+ diffuses out, depolarization of the sarcolemma and subsequent contraction of the muscle fiber occurs.

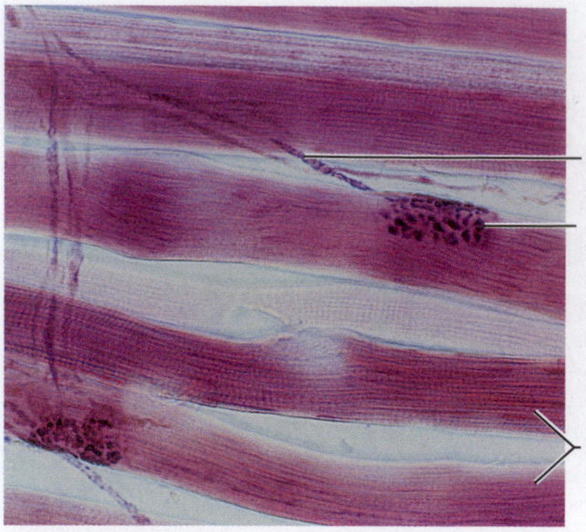
Branch of an axon
Axon terminal at neuromuscular junctions
Muscle fibers

FIGURE 14.6 Photomicrograph of neuromuscular junctions (160×).

ACTIVITY 3

Studying the Structure of a Neuromuscular Junction

1. If possible, examine a three-dimensional model of skeletal muscle cells that illustrates the neuromuscular junction. Identify the structures just described.

2. Obtain a slide of skeletal muscle stained to show a portion of a motor unit. Examine the slide under high power to identify the axon fibers extending leashlike to the muscle cells. Follow one of the axon fibers to its terminus to identify the oval-shaped axon terminal. Compare your observations to Figure 14.6. Sketch a small section in the space provided below. Label the axon of the motor neuron, its terminal branches, and muscle fibers.

NAME ____________________

LAB TIME/DATE ____________________

REVIEW SHEET

EXERCISE 14

Microscopic Anatomy and Organization of Skeletal Muscle

Skeletal Muscle Cells and Their Packaging into Muscles

1. Use the items in the key to correctly identify the structures described below.

Key:

a. endomysium
b. epimysium
c. fascicle
d. fiber
e. myofibril
f. myofilament
g. perimysium
h. sarcolemma
i. sarcomere
j. sarcoplasm
k. tendon

__________ 1. connective tissue ensheathing a bundle of muscle cells

__________ 2. bundle of muscle cells

__________ 3. contractile unit of muscle

__________ 4. a muscle cell

__________ 5. thin reticular connective tissue surrounding each muscle cell

__________ 6. plasma membrane of the muscle fiber

__________ 7. a long filamentous organelle with a banded appearance found within muscle cells

__________ 8. actin- or myosin-containing structure

__________ 9. cord of collagen fibers that attaches a muscle to a bone

2. List three reasons why the connective tissue wrappings of skeletal muscle are important.

____

____

____

3. Why are there more indirect—that is, tendinous—muscle attachments to bone than there are direct attachments?

____

____

____

4. How does an aponeurosis differ from a tendon structurally? ____________________

____

How is an aponeurosis functionally similar to a tendon? ____________________

____

5. The diagram illustrates a small portion of several myofibrils. Using letters from the key, correctly identify each structure indicated by a leader line or a bracket.

Key: a. A band
b. actin filament
c. I band
d. myosin filament
e. T tubule
f. terminal cisterna
g. triad
h. sarcomere
i. Z disc

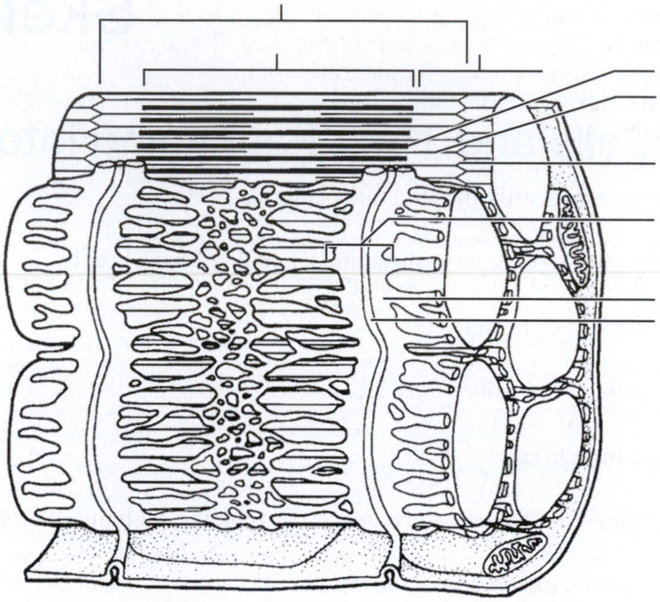

6. On the following figure, label a blood vessel, endomysium, epimysium, a fascicle, a muscle cell, perimysium, and the tendon.

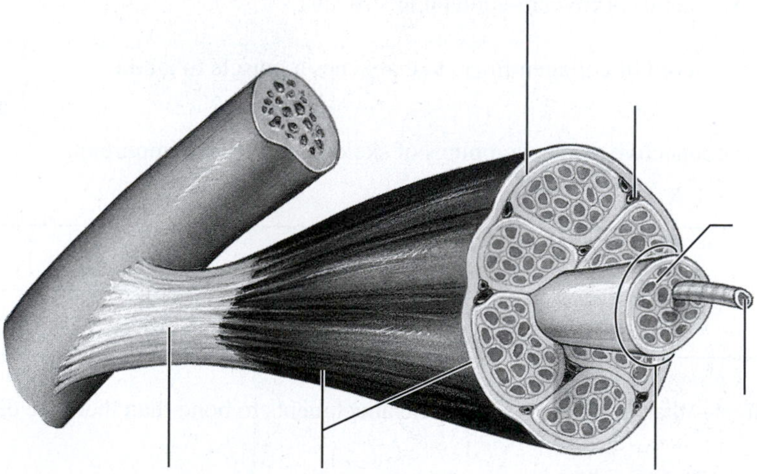

The Neuromuscular Junction

7. Complete the following statements:

The junction between a motor neuron's axon and the muscle cell membrane is called __1__ junction. A motor neuron and all of the skeletal muscle cells it stimulates is called a __2__. The actual gap between the axon terminal and the muscle cell is called a __3__. Within the axon terminal are many small vesicles containing a neurotransmitter substance called __4__. When the __5__ reaches the ends of the axon, the neurotransmitter is released and diffuses to the muscle cell membrane to combine with receptors there. The combining of the neurotransmitter with the muscle membrane receptors causes the membrane to become permeable to both sodium and potassium. The greater influx of sodium ions results in __6__ of the membrane. Then contraction of the muscle cell occurs.

1. ____________________
2. ____________________
3. ____________________
4. ____________________
5. ____________________
6. ____________________

8. The events that occur at a neuromuscular junction are depicted below. Identify by labeling every structure provided with a leader line.

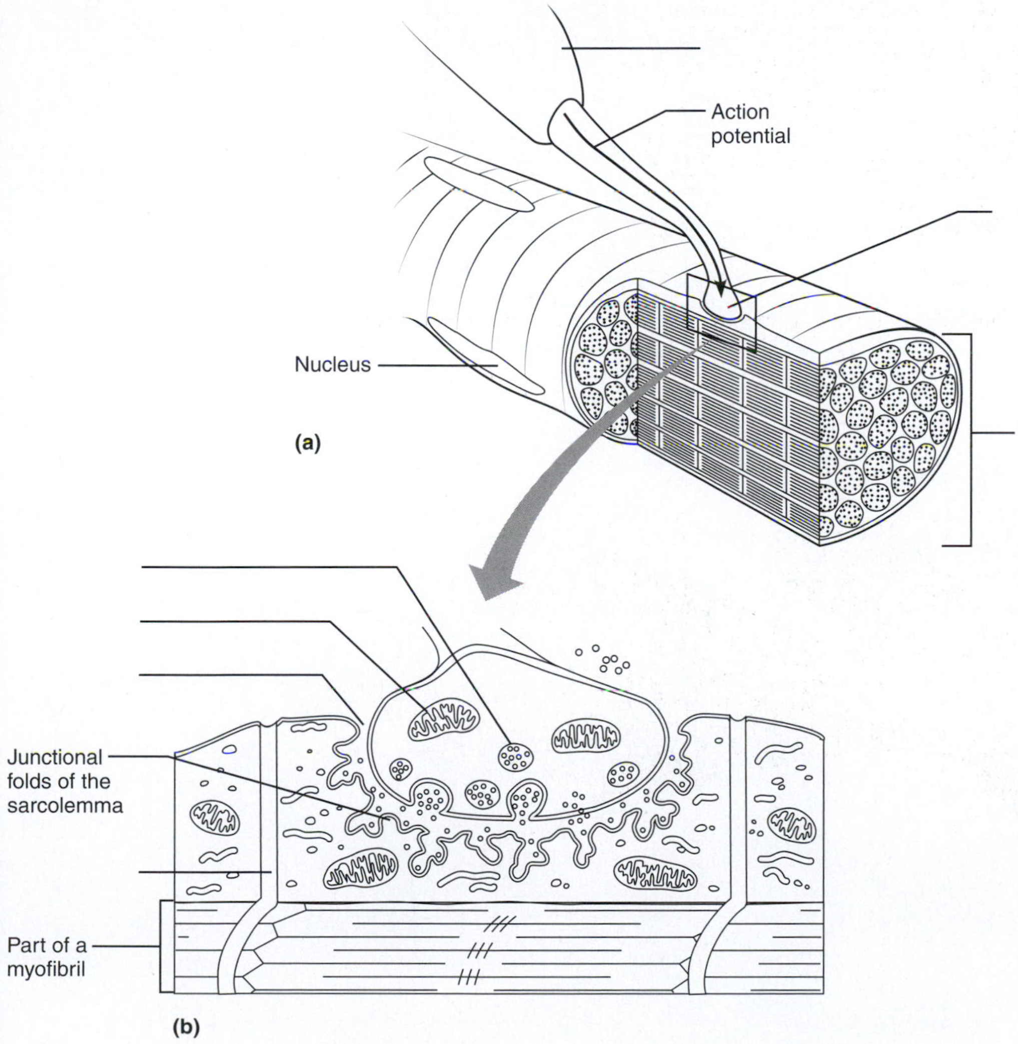
Action potential
Nucleus
(a)
Junctional folds of the sarcolemma
Part of a myofibril
(b)

Key:

a. axon terminal
b. mitochondrion
c. muscle fiber
d. myelinated axon
e. sarcolemma
f. synaptic cleft
g. T tubule
h. vesicle containing ACh

EXERCISE 15

Gross Anatomy of the Muscular System

MATERIALS

- ☐ Human torso model or large anatomical chart showing human musculature
- ☐ Human cadaver for demonstration (if available)
- ☐ Disposable gloves
- ☐ *Human Musculature* videotape
- ☐ Tubes of body (or face) paint
- ☐ 1″ wide artist's brushes

For instructions on animal dissections, see the dissection exercises starting on page 697 in the cat, fetal pig, and rat editions of this manual.

OBJECTIVES

1. To define *agonist* (prime mover), *antagonist, synergist, fixator, origin,* and *insertion.*
2. To cite criteria used in naming skeletal muscles.
3. To name and locate the major muscles of the human body (on a torso model, a human cadaver, lab chart, or diagram) and state the action of each.
4. To explain how muscle actions are related to their location.
5. To name muscle origins and insertions as required by the instructor.
6. To identify antagonists of the major prime movers.

PRE-LAB QUIZ

1. A prime mover or _________ is responsible for producing a particular type of movement.
   a. agonist
   b. antagonist
   c. fixator
   d. synergist
2. Skeletal muscles are named on the basis of many criteria. Name one.

   ____
3. Circle True or False. Muscles of facial expression differ from most skeletal muscles because they insert into the skin or other muscles rather than into bone.
4. The _________ musculature includes muscles that move the vertebral column and muscles that move the ribs.
   a. head and neck b. lower limb c. trunk
5. Muscles that act on the _________ cause movement at the hip, knee, and foot joints.
   a. lower limb b. trunk c. upper limb
6. This two-headed muscle bulges when the forearm is flexed. It is the most familiar muscle of the anterior humerus. It is the
   a. biceps brachii
   b. extensor digitorum
   c. flexor carpii radialis
   d. triceps brachii
7. These abdominal muscles are responsible for giving me my "six-pack." They also stabilize my pelvis when walking. They are the _________ muscles.
   a. internal intercostal
   b. quadriceps
   c. rectus abdominis
   d. triceps femoris
8. Circle the correct term. This lower limb muscle, which attaches to the calcaneus via the calcaneal tendon and plantar flexes the foot when the knee is extended, is the sartorius / gastrocnemius.
9. The _________ is the largest and most superficial of the gluteal muscles.
   a. gluteus internus
   b. gluteus maximus
   c. gluteus medius
   d. gluteus minimus

MasteringA&P™

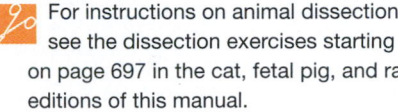

Access practice quizzes and more in the Study Area at www.masteringaandp.com.

PAL

For access to anatomical models and more, check out Practice Anatomy Lab.

Skeletal muscles cause movement. Among the movements are smiling, frowning, speaking, singing, breathing, dancing, running, and playing a musical instrument. Most often, purposeful movements require the coordinated action of several skeletal muscles.

Classification of Skeletal Muscles

Types of Muscles

Muscles that are primarily responsible for producing a particular movement are called **prime movers,** or **agonists.** Muscles that oppose or reverse a movement are called **antagonists.** When a prime mover is active, the fibers of the antagonist are stretched and in the relaxed state. The antagonist can also regulate the prime mover by providing some resistance, to prevent overshoot or to stop its action. Antagonists can be prime movers in their own right. For example, the biceps muscle of the arm (a prime mover of elbow flexion) is antagonized by the triceps (a prime mover of elbow extension).

Synergists aid the action of agonists either by assisting with the same movement or by reducing undesirable or unnecessary movement. Contraction of a muscle crossing two or more joints would cause movement at all joints spanned if the synergists were not there to stabilize them. For example, you can make a fist without bending your wrist only because synergist muscles stabilize the wrist joint and allow the prime mover to exert its force at the finger joints.

Fixators, or fixation muscles, are specialized synergists. They immobilize the origin of a prime mover so that all the tension is exerted at the insertion. Muscles that help maintain posture are fixators; so too are muscles of the back that stabilize or "fix" the scapula during arm movements.

Naming Skeletal Muscles

Remembering the names of the skeletal muscles is a monumental task, but certain clues help. Muscles are named on the basis of the following criteria:

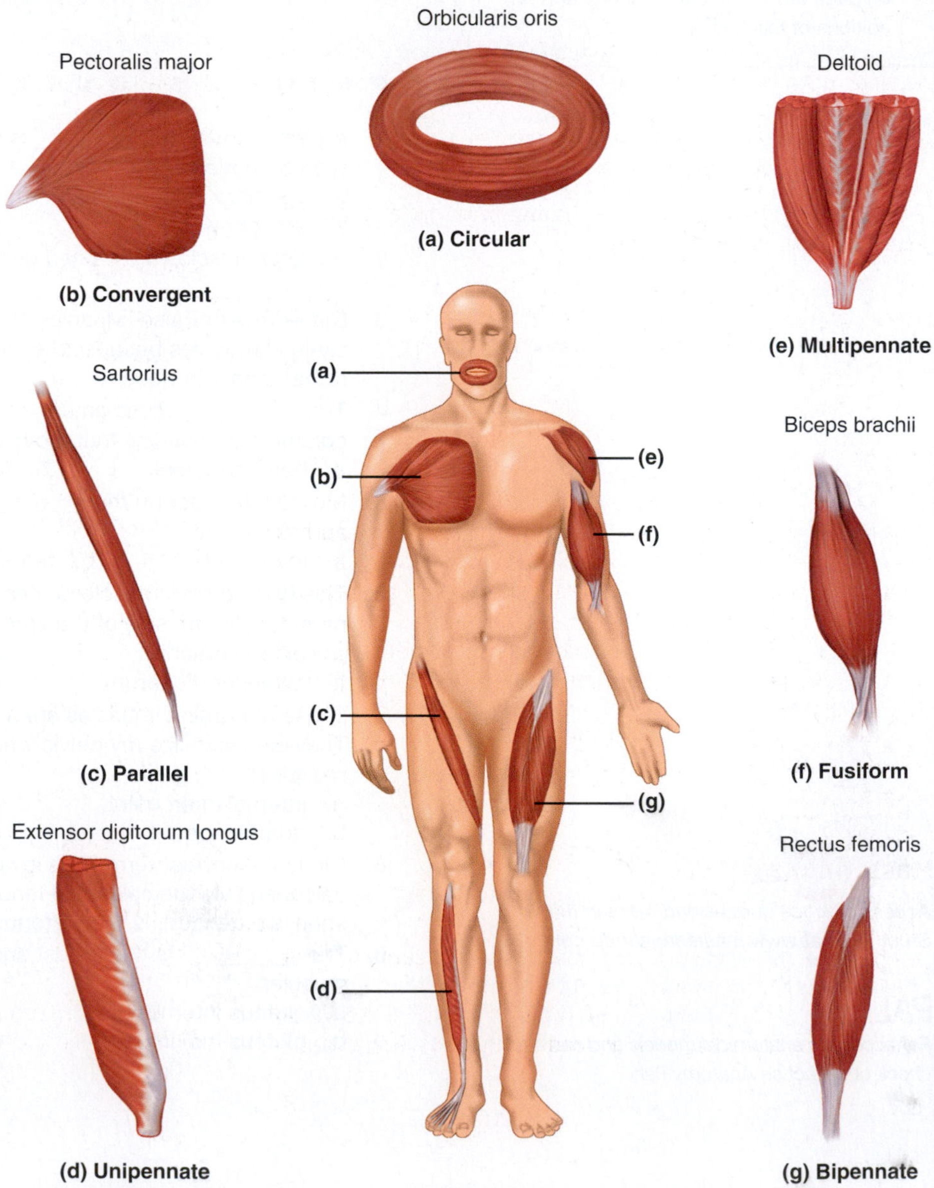

Orbicularis oris
(a) Circular
Pectoralis major
(b) Convergent
Deltoid
(e) Multipennate
Sartorius
(c) Parallel
Biceps brachii
(f) Fusiform
Extensor digitorum longus
(d) Unipennate
Rectus femoris
(g) Bipennate
(a)
(b)
(c)
(d)
(e)
(f)
(g)

FIGURE 15.1 Patterns of fascicle arrangement in muscles.

- **Direction of muscle fibers:** Some muscles are named in reference to some imaginary line, usually the midline of the body or the longitudinal axis of a limb bone. A muscle with fibers (and fascicles) running parallel to that imaginary line will have the term *rectus* (straight) in its name. For example, the rectus abdominis is the straight muscle of the abdomen. Likewise, the terms *transverse* and *oblique* indicate that the muscle fibers run at right angles and obliquely (respectively) to the imaginary line. Figure 15.1 shows how muscle structure is determined by fascicle arrangement.
- **Relative size of the muscle:** Terms such as *maximus* (largest), *minimus* (smallest), *longus* (long), and *brevis* (short) are often used in naming muscles—as in gluteus maximus and gluteus minimus.
- **Location of the muscle:** Some muscles are named for the bone with which they are associated. For example, the temporalis muscle overlies the temporal bone.
- **Number of origins:** When the term *biceps, triceps,* or *quadriceps* forms part of a muscle name, you can generally assume that the muscle has two, three, or four origins (respectively). For example, the biceps muscle of the arm has two heads, or origins.
- **Location of the muscle's origin and insertion:** For example, the sternocleidomastoid muscle has its origin on the sternum *(sterno)* and clavicle *(cleido)*, and inserts on the mastoid process of the temporal bone.
- **Shape of the muscle:** For example, the deltoid muscle is roughly triangular (*deltoid* = triangle), and the trapezius muscle resembles a trapezoid.
- **Action of the muscle:** For example, all the adductor muscles of the anterior thigh bring about its adduction, and all the extensor muscles of the wrist extend the wrist.

Identification of Human Muscles

While reading the tables and identifying the various human muscles in the figures, try to visualize what happens when the muscle contracts. Since muscles often have many actions, we have indicated the primary action of each muscle in blue type in the tables. Then, use a torso model or an anatomical chart to again identify as many of these muscles as possible. (If a human cadaver is available for observation, your instructor will provide specific instructions for muscle examination.) Then carry out the instructions for demonstrating and palpating muscles.

Muscles of the Head and Neck

The muscles of the head serve many specific functions. For instance, the muscles of facial expression differ from most skeletal muscles because they insert into the skin (or other muscles) rather than into bone. As a result, they move the facial skin, allowing a wide range of emotions to be shown on the face. Other muscles of the head are the muscles of mastication, which manipulate the mandible during chewing, and the six extrinsic eye muscles located within the orbit, which aim the eye. (Orbital muscles are studied in Exercise 24.)

ACTIVITY 1

Identifying Head and Neck Muscles

Neck muscles are primarily concerned with the movement of the head and shoulder girdle. Figures 15.2 and 15.3 are summary figures illustrating the superficial musculature of the body as a whole. Head and neck muscles are discussed in Tables 15.1 and 15.2 and shown in Figures 15.4 and 15.5.

Demonstrating Operations of Head Muscles

1. Raise your eyebrow to wrinkle your forehead. You are using the *frontal belly* of the *epicranius* muscle.
2. Blink your eyes; wink. You are contracting *orbicularis oculi.*
3. Close your lips and pucker up. This requires contraction of *orbicularis oris.*
4. Smile. You are using *zygomaticus.*
5. To demonstrate the temporalis, place your hands on your temples and clench your teeth. The masseter can also be palpated now at the angle of the jaw. ■

Muscles of the Trunk

The trunk musculature includes muscles that move the vertebral column; anterior thorax muscles that act to move ribs, head, and arms; and muscles of the abdominal wall that play a role in the movement of the vertebral column but more importantly form the "natural girdle," or the major portion of the abdominal body wall.

ACTIVITY 2

Identifying Muscles of the Trunk

The trunk muscles are described in Tables 15.3 and 15.4 and shown in Figures 15.6 through 15.9. As before, identify the muscles in the figure as you read the tabular descriptions and then identify them on the torso or laboratory chart.

Demonstrating Operation of Trunk Muscles

Now, work with a partner to demonstrate the operation of the following muscles. One of you can demonstrate the movement (the following steps are addressed to this partner). The other can supply resistance and palpate the muscle being tested.

1. Fully abduct the arm and extend the elbow. Now adduct the arm against resistance. You are using the *latissimus dorsi.*
2. To observe the *deltoid,* try to abduct your arm against resistance. Now attempt to elevate your shoulder against resistance; you are contracting the upper portion of the *trapezius.*
3. The *pectoralis major* is used when you press your hands together at chest level with your elbows widely abducted. ■

Text continues on page 213.

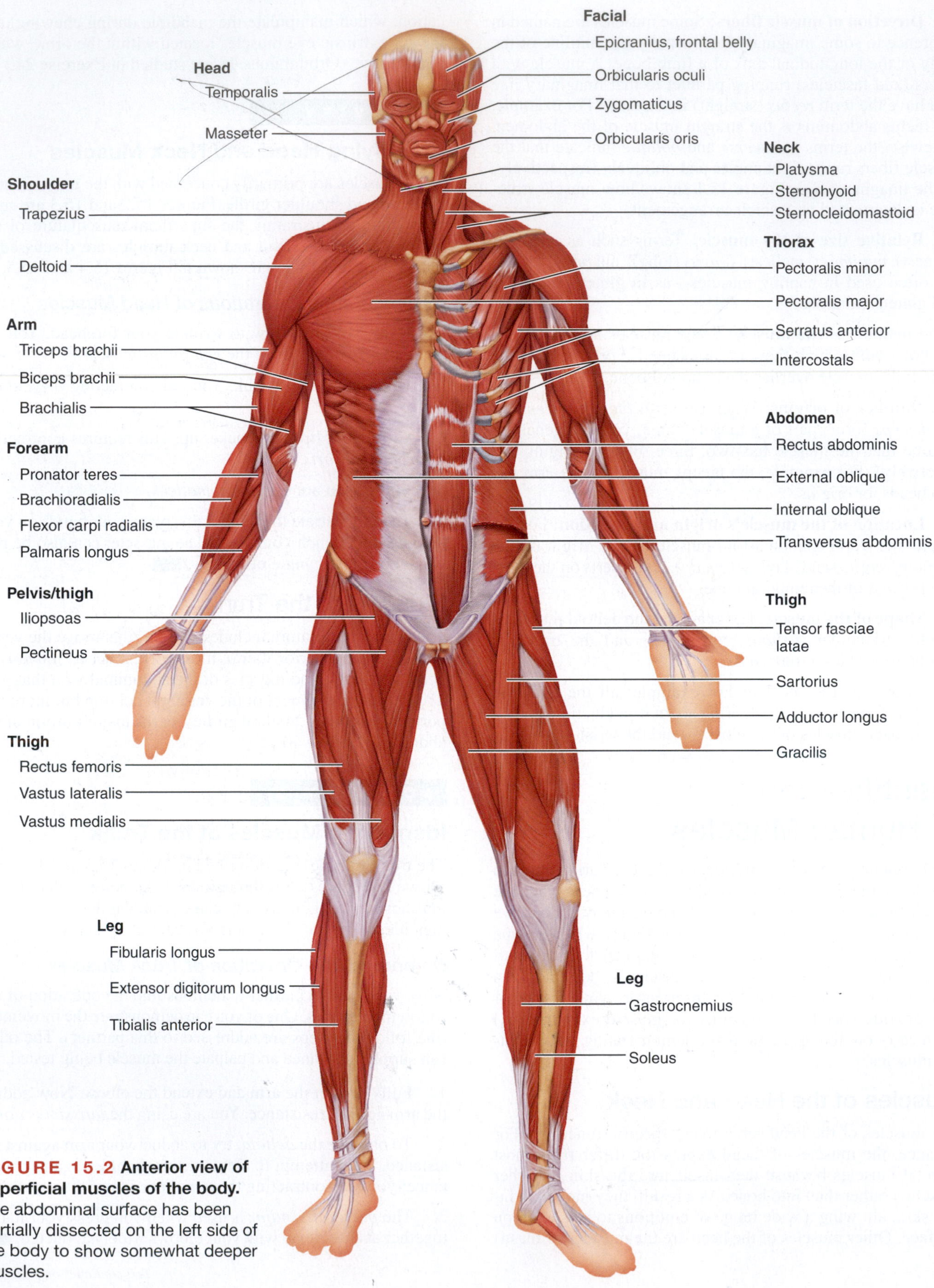
Facial
Epicranius, frontal belly
Orbicularis oculi
Zygomaticus
Orbicularis oris
Head
Temporalis
Masseter
Neck
Platysma
Sternohyoid
Sternocleidomastoid
Shoulder
Trapezius
Deltoid
Thorax
Pectoralis minor
Pectoralis major
Serratus anterior
Intercostals
Arm
Triceps brachii
Biceps brachii
Brachialis
Abdomen
Rectus abdominis
External oblique
Internal oblique
Transversus abdominis
Forearm
Pronator teres
Brachioradialis
Flexor carpi radialis
Palmaris longus
Pelvis/thigh
Iliopsoas
Pectineus
Thigh
Tensor fasciae latae
Sartorius
Adductor longus
Gracilis
Thigh
Rectus femoris
Vastus lateralis
Vastus medialis
Leg
Fibularis longus
Extensor digitorum longus
Tibialis anterior
Leg
Gastrocnemius
Soleus

FIGURE 15.2 Anterior view of superficial muscles of the body. The abdominal surface has been partially dissected on the left side of the body to show somewhat deeper muscles.

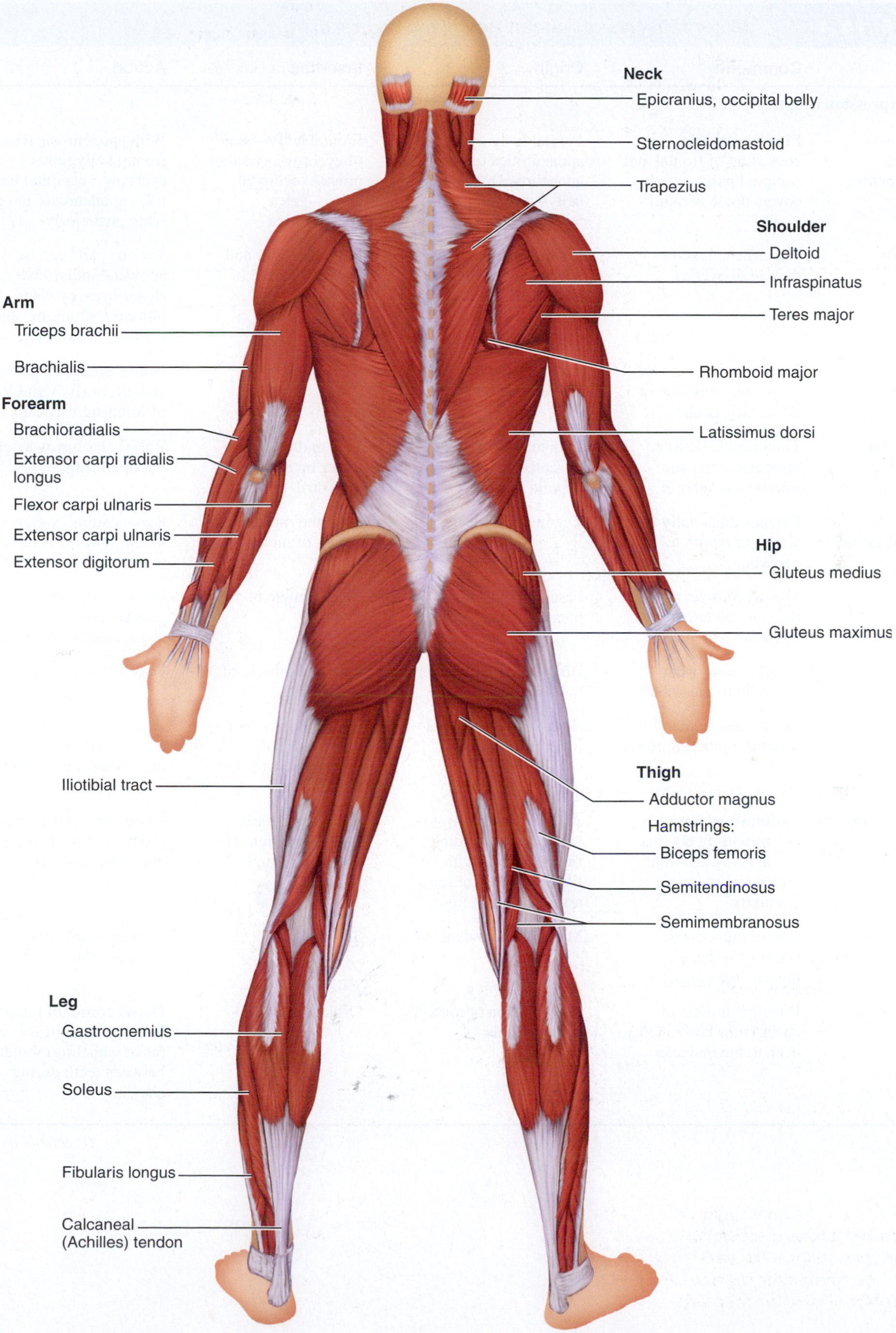
Neck
Epicranius, occipital belly
Sternocleidomastoid
Trapezius
Shoulder
Deltoid
Infraspinatus
Teres major
Rhomboid major
Latissimus dorsi
Arm
Triceps brachii
Brachialis
Forearm
Brachioradialis
Extensor carpi radialis longus
Flexor carpi ulnaris
Extensor carpi ulnaris
Extensor digitorum
Hip
Gluteus medius
Gluteus maximus
Iliotibial tract
Thigh
Adductor magnus
Hamstrings:
Biceps femoris
Semitendinosus
Semimembranosus
Leg
Gastrocnemius
Soleus
Fibularis longus
Calcaneal (Achilles) tendon

FIGURE 15.3 Posterior view of superficial muscles of the body.

TABLE 15.1 Major Muscles of Human Head (see Figure 15.4)

Muscle	Comments	Origin	Insertion	Action
Facial Expression (Figure 15.4a)				
Epicranius—frontal and occipital bellies	Bipartite muscle consisting of frontal and occipital parts, which covers dome of skull	Frontal belly—galea aponeurotica (cranial aponeurosis); occipital belly—occipital and temporal bones	Frontal belly—skin of eyebrows and root of nose; occipital belly—galea aponeurotica	With aponeurosis fixed, frontal belly raises eyebrows; occipital belly fixes aponeurosis and pulls scalp posteriorly
Orbicularis oculi	Tripartite sphincter muscle of eyelids	Frontal and maxillary bones and ligaments around orbit	Encircles orbit and inserts in tissue of eyelid	Various parts can be activated individually; closes eyes, produces blinking, squinting, and draws eyebrows inferiorly
Corrugator supercilii	Small muscle; activity associated with that of orbicularis oculi	Arch of frontal bone above nasal bone	Skin of eyebrow	Draws eyebrows medially and inferiorly; wrinkles skin of forehead vertically
Levator labii superioris	Thin muscle between orbicularis oris and inferior eye margin	Zygomatic bone and infraorbital margin of maxilla	Skin and muscle of upper lip and border of nostril	Raises and furrows upper lip; opens lips
Zygomaticus—major and minor	Extends diagonally from corner of mouth to cheekbone	Zygomatic bone	Skin and muscle at corner of mouth	Raises lateral corners of mouth upward (smiling muscle)
Risorius	Slender muscle; runs inferior and lateral to zygomaticus	Fascia of masseter muscle	Skin at angle of mouth	Draws corner of lip laterally; tenses lip; zygomaticus synergist
Depressor labii inferioris	Small muscle from lower lip to mandible	Body of mandible lateral to its midline	Skin and muscle of lower lip	Draws lower lip inferiorly
Depressor anguli oris	Small muscle lateral to depressor labii inferioris	Body of mandible below incisors	Skin and muscle at angle of mouth below insertion of zygomaticus	Zygomaticus antagonist; draws corners of mouth downward and laterally
Orbicularis oris	Multilayered muscle of lips with fibers that run in many different directions; most run circularly	Arises indirectly from maxilla and mandible; fibers blended with fibers of other muscles associated with lips	Encircles mouth; inserts into muscle and skin at angles of mouth	Closes lips; purses and protrudes lips (kissing and whistling muscle)
Mentalis	One of muscle pair forming V-shaped muscle mass on chin	Mandible below incisors	Skin of chin	Protrudes lower lip; wrinkles chin
Buccinator	Principal muscle of cheek; runs horizontally, deep to the masseter	Molar region of maxilla and mandible	Orbicularis oris	Draws corner of mouth laterally; compresses cheek (as in whistling); holds food between teeth during chewing

(Continues on page 204.)

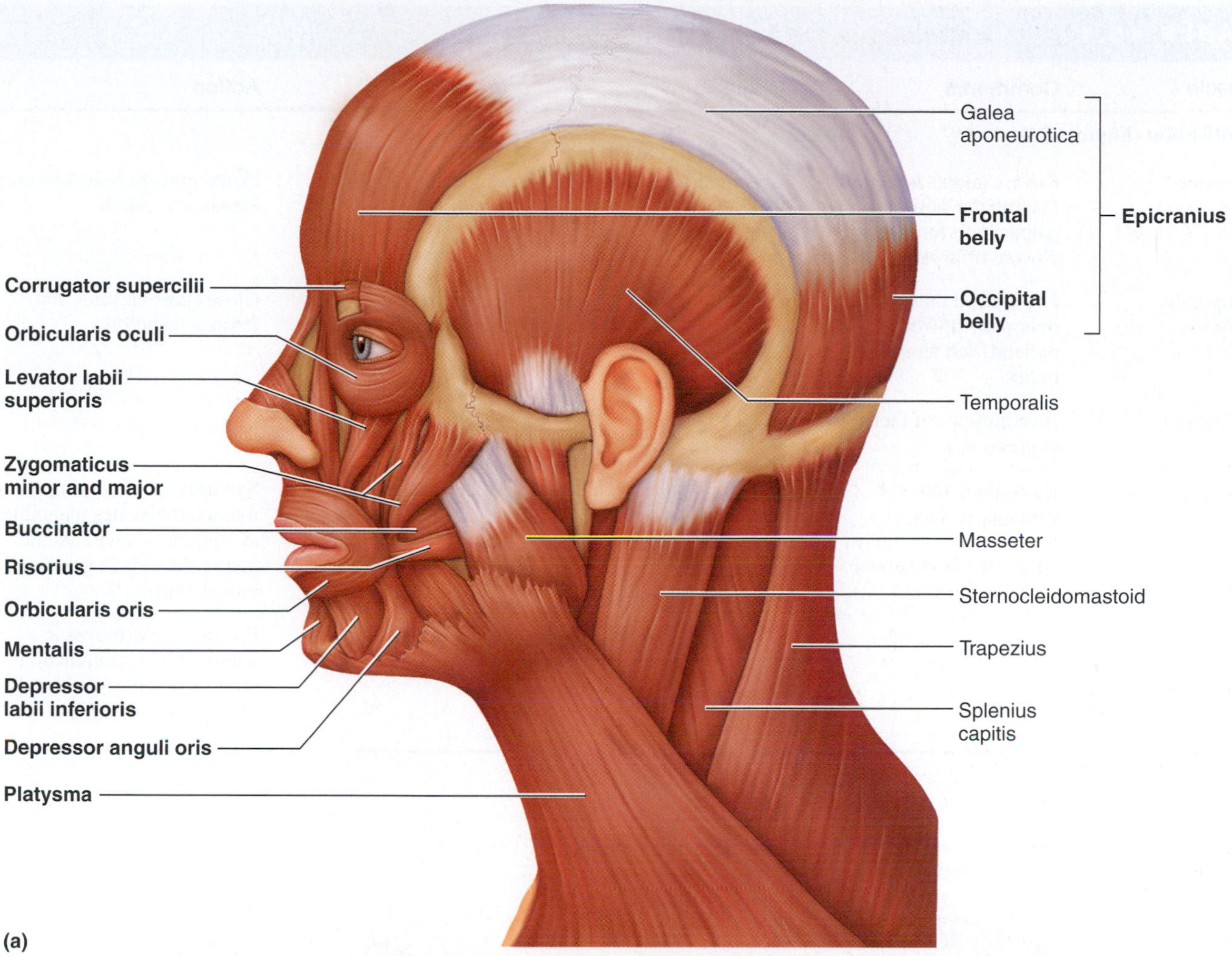
Galea aponeurotica
Frontal belly
Epicranius
Occipital belly
Temporalis
Masseter
Sternocleidomastoid
Trapezius
Splenius capitis
Corrugator supercilii
Orbicularis oculi
Levator labii superioris
Zygomaticus minor and major
Buccinator
Risorius
Orbicularis oris
Mentalis
Depressor labii inferioris
Depressor anguli oris
Platysma
(a)

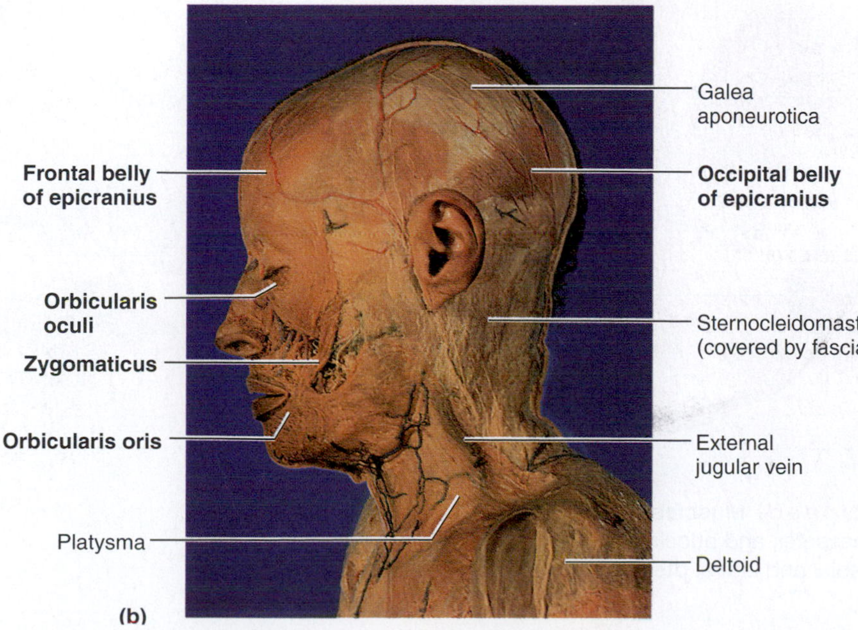
Galea aponeurotica
Frontal belly of epicranius
Occipital belly of epicranius
Orbicularis oculi
Zygomaticus
Sternocleidomastoid (covered by fascia)
Orbicularis oris
External jugular vein
Platysma
Deltoid
(b)

FIGURE 15.4 Muscles of the head (left lateral view). (a) Superficial muscles. **(b)** Photo of superficial structures of head and neck.

TABLE 15.1 Major Muscles of Human Head *(continued)*

Muscle	Comments	Origin	Insertion	Action
Mastication (Figure 15.4c,d)				
Masseter	Covers lateral aspect of mandibular ramus; can be palpated on forcible closure of jaws	Zygomatic arch and maxilla	Angle and ramus of mandible	Prime mover of jaw closure; elevates mandible
Temporalis	Fan-shaped muscle lying over parts of frontal, parietal, and temporal bones	Temporal fossa	Coronoid process of mandible	Closes jaw; elevates and retracts mandible
Buccinator	(See muscles of facial expression.)			
Medial pterygoid	Runs along internal (medial) surface of mandible (thus largely concealed by that bone)	Sphenoid, palatine, and maxillary bones	Medial surface of mandible, near its angle	Synergist of temporalis and masseter; elevates mandible; in conjunction with lateral pterygoid, aids in grinding movements
Lateral pterygoid	Superior to medial pterygoid	Greater wing of sphenoid bone	Mandibular condyle	Protracts jaw (moves it anteriorly); in conjunction with medial pterygoid, aids in grinding movements of teeth

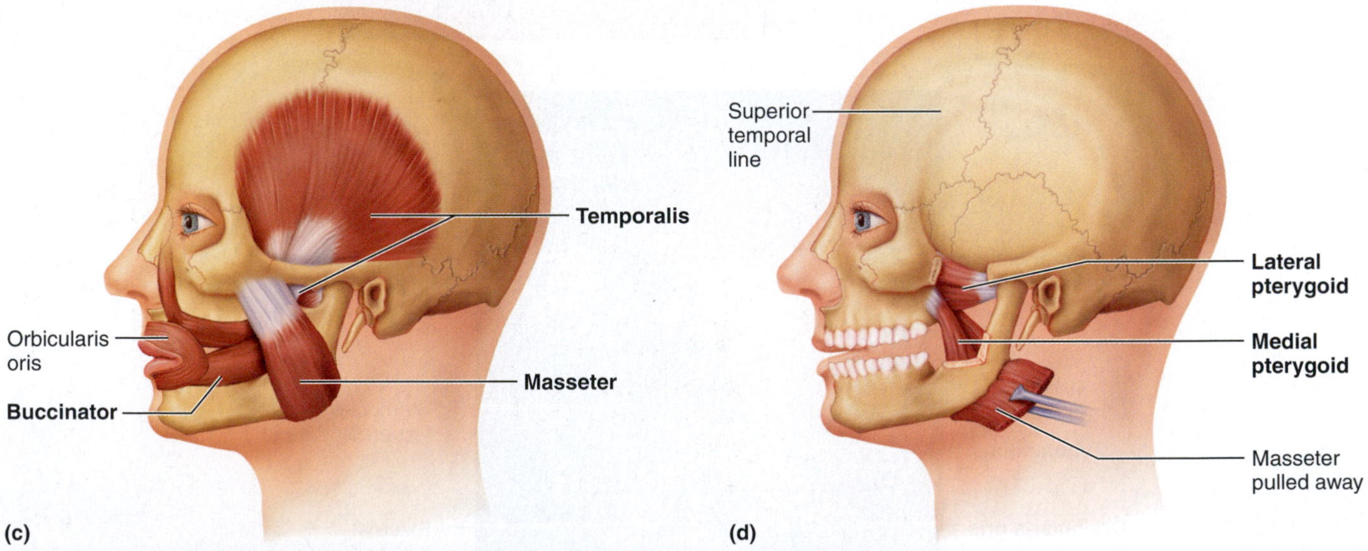
Temporalis
Orbicularis oris
Buccinator
Masseter
(c)
Superior temporal line
Lateral pterygoid
Medial pterygoid
Masseter pulled away
(d)

FIGURE 15.4 ***(continued)*** **Muscles of the head: mastication. (c)** Lateral view of the temporalis, masseter, and buccinator muscles. **(d)** Lateral view of the deep chewing muscles, the medial and lateral pterygoid muscles.

TABLE 15.2 **Anterolateral Muscles of Human Neck (see Figure 15.5)**

Muscle	Comments	Origin	Insertion	Action
Superficial				
Platysma	Unpaired muscle: thin, sheetlike superficial neck muscle, not strictly a head muscle but plays role in facial expression (see also Fig. 15.4a)	Fascia of chest (over pectoral muscles) and deltoid	Lower margin of mandible, skin, and muscle at corner of mouth	Tenses skin of neck; depresses mandible; pulls lower lip back and down (i.e., produces downward sag of the mouth)
Sternocleidomastoid	Two-headed muscle located deep to platysma on anterolateral surface of neck; fleshy parts on either side indicate limits of anterior and posterior triangles of neck	Manubrium of sternum and medial portion of clavicle	Mastoid process of temporal bone and superior nuchal line of occipital bone	Simultaneous contraction of both muscles of pair causes flexion of neck forward, generally against resistance (as when lying on the back); acting independently, rotate head toward shoulder on opposite side
Scalenes—anterior, middle, and posterior	Located more on lateral than anterior neck; deep to platysma and sternocleidomastoid (see Fig. 15.5c)	Transverse processes of cervical vertebrae	Anterolaterally on ribs 1–2	Flex and slightly rotate neck; elevate ribs 1–2 (aid in inspiration)

(continues)

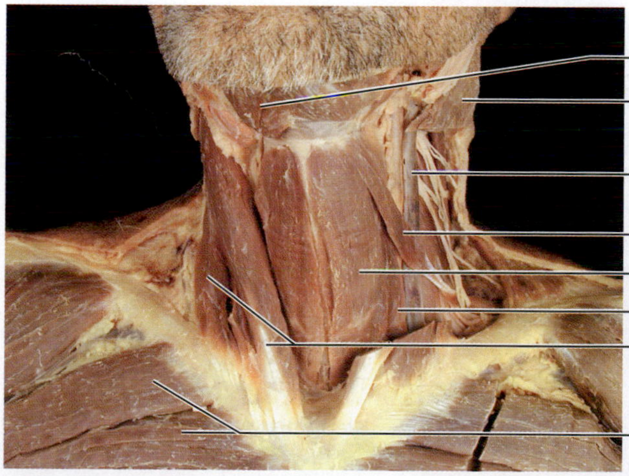
Platysma (cut)
Sternocleidomastoid (cut)
Internal jugular vein
Omohyoid
Sternohyoid
Sternothyroid
Sternocleidomastoid
Pectoralis major

(a)

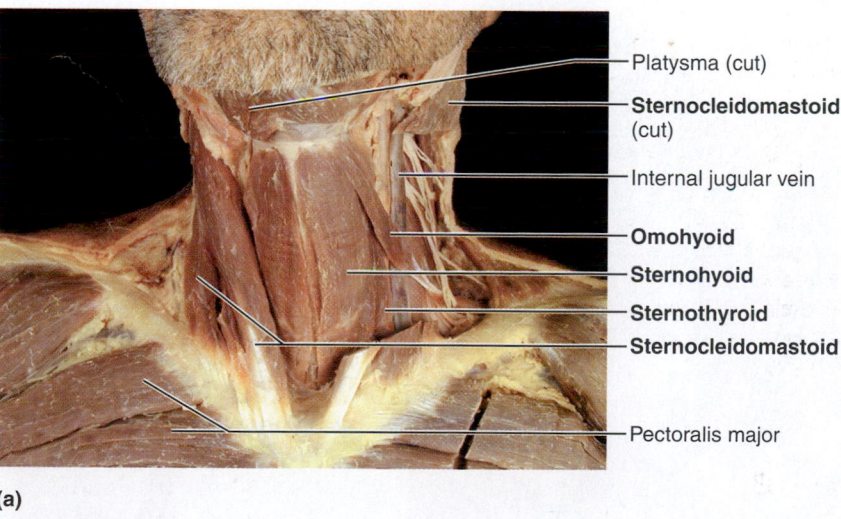

FIGURE 15.5 Muscles of the anterolateral neck and throat. (a) Cadaver photo of the anterior and lateral regions of the neck.

TABLE 15.2 Anterolateral Muscles of Human Neck *(continued)*

Muscle	Comments	Origin	Insertion	Action
Deep (Figure 15.5a,b)				
Digastric	Consists of two bellies united by an intermediate tendon; assumes a V-shaped configuration under chin	Lower margin of mandible (anterior belly) and mastoid process (posterior belly)	By a connective tissue loop to hyoid bone	Acting in concert, elevate hyoid bone; open mouth and depress mandible
Stylohyoid	Slender muscle parallels posterior border of digastric; below angle of jaw	Styloid process of temporal	Hyoid bone	Elevates and retracts hyoid bone
Mylohyoid	Just deep to digastric; forms floor of mouth	Medial surface of mandible	Hyoid bone and median raphe	Elevates hyoid bone and base of tongue during swallowing
Sternohyoid	Runs most medially along neck; straplike	Manubrium and medial end of clavicle	Lower margin of body of hyoid bone	Acting with sternothyroid and omohyoid depresses larynx and hyoid bone if mandible is fixed; may also flex skull
Sternothyroid	Lateral and deep to sternohyoid	Posterior surface of manubrium	Thyroid cartilage of larynx	(See Sternohyoid above)
Omohyoid	Straplike with two bellies; lateral to sternohyoid	Superior surface of scapula	Hyoid bone; inferior border	(See Sternohyoid above)
Thyrohyoid	Appears as a superior continuation of sternothyroid muscle	Thyroid cartilage	Hyoid bone	Depresses hyoid bone; elevates larynx if hyoid is fixed

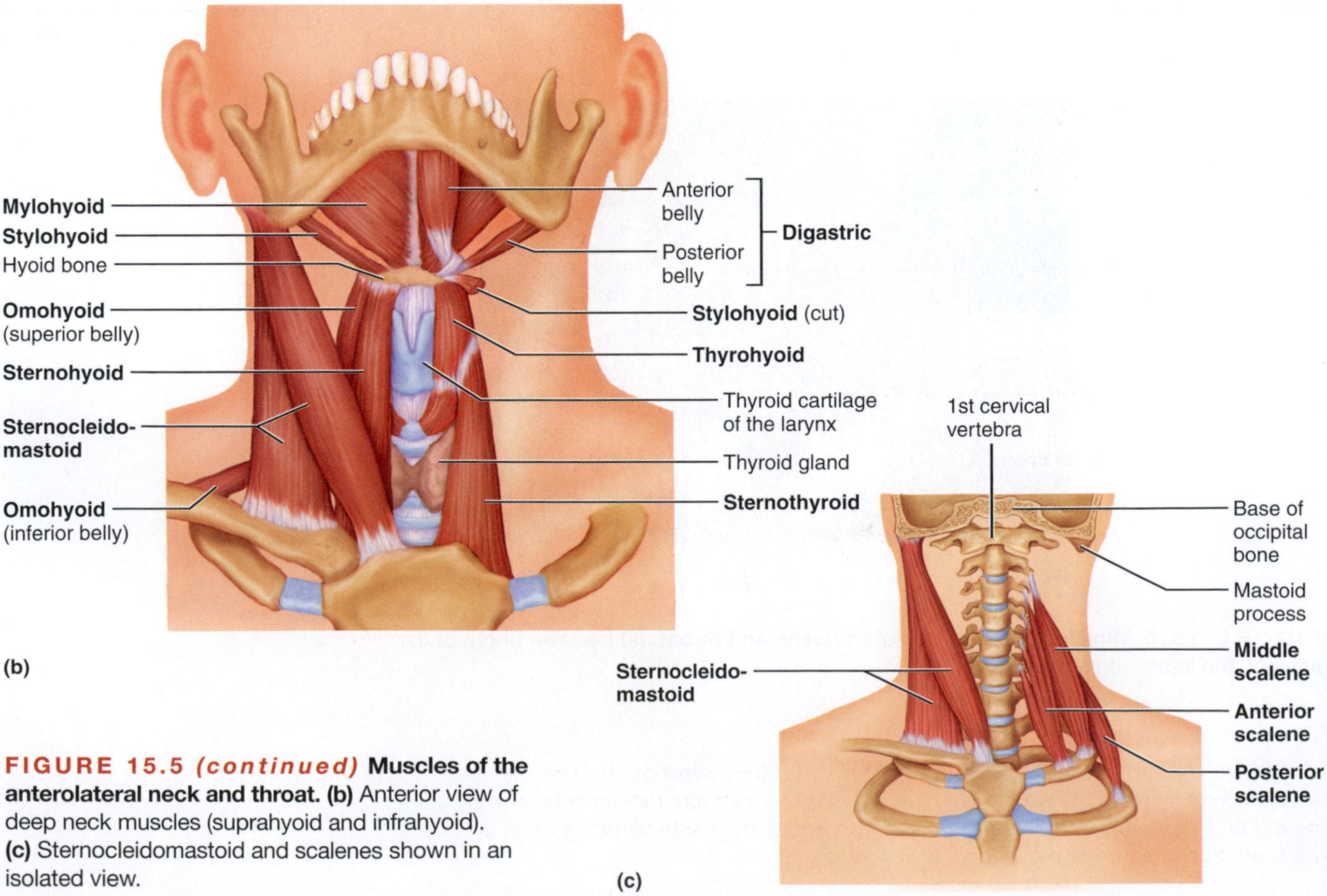
Mylohyoid
Stylohyoid
Hyoid bone
Omohyoid (superior belly)
Sternohyoid
Sternocleido-mastoid
Omohyoid (inferior belly)
Anterior belly
Posterior belly
Digastric
Stylohyoid (cut)
Thyrohyoid
Thyroid cartilage of the larynx
Thyroid gland
Sternothyroid
(b)
1st cervical vertebra
Base of occipital bone
Mastoid process
Middle scalene
Anterior scalene
Posterior scalene
Sternocleido-mastoid
(c)

FIGURE 15.5 *(continued)* Muscles of the anterolateral neck and throat. (b) Anterior view of deep neck muscles (suprahyoid and infrahyoid). **(c)** Sternocleidomastoid and scalenes shown in an isolated view.

TABLE 15.3 Anterior Muscles of Human Thorax, Shoulder, and Abdominal Wall (see Figures 15.6, 15.7, and 15.8)

Muscle	Comments	Origin	Insertion	Action
Thorax and Shoulder, Superficial (Figure 15.6)				
Pectoralis major	Large fan-shaped muscle covering upper portion of chest	Clavicle, sternum, cartilage of ribs 1–6 (or 7), and aponeurosis of external oblique muscle	Fibers converge to insert by short tendon into intertubercular sulcus of humerus	Prime mover of arm flexion; adducts, medially rotates arm; with arm fixed, pulls chest upward (thus also acts in forced inspiration)
Serratus anterior	Fan-shaped muscle deep to scapula; beneath and inferior to pectoral muscles on lateral rib cage	Lateral aspect of ribs 1–8 (or 9)	Vertebral border of anterior surface of scapula	Prime mover to protract and hold scapula against chest wall; rotates scapula, causing inferior angle to move laterally and upward; essential to raising arm; fixes scapula for arm abduction
Deltoid	Fleshy triangular muscle forming shoulder muscle mass; intramuscular injection site	Lateral ⅓ of clavicle; acromion and spine of scapula	Deltoid tuberosity of humerus	Acting as a whole, prime mover of arm abduction; when only specific fibers are active, can aid in flexion, extension, and rotation of humerus

(continues)

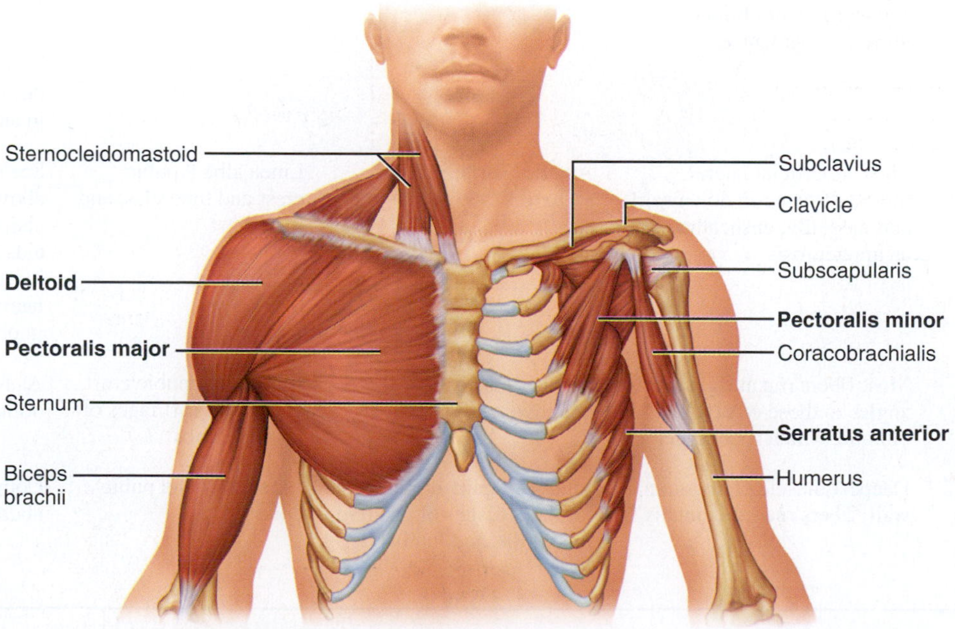

Sternocleidomastoid
Deltoid
Pectoralis major
Sternum
Biceps brachii
Subclavius
Clavicle
Subscapularis
Pectoralis minor
Coracobrachialis
Serratus anterior
Humerus

FIGURE 15.6 Muscles of the thorax and shoulder acting on the scapula and arm (anterior view). The superficial muscles, which effect arm movements, are shown on the left. These muscles have been removed on the right side of the figure to show the muscles that stabilize or move the pectoral girdle.

TABLE 15.3 Anterior Muscles of Human Thorax, Shoulder, and Abdominal Wall *(continued)*

Muscle	Comments	Origin	Insertion	Action
Thorax and Shoulder, Superficial *(continued)*				
Pectoralis minor	Flat, thin muscle directly beneath and obscured by pectoralis major	Anterior surface of ribs 3–5, near their costal cartilages	Coracoid process of scapula	With ribs fixed, draws scapula forward and inferiorly; with scapula fixed, draws rib cage superiorly
Thorax, Deep: Muscles of Respiration (Figure 15.7)				
External intercostals	11 pairs lie between ribs; fibers run obliquely downward and forward toward sternum	Inferior border of rib above (not shown in figure)	Superior border of rib below	Pull ribs toward one another to elevate rib cage; aid in inspiration
Internal intercostals	11 pairs lie between ribs; fibers run deep and at right angles to those of external intercostals	Superior border of rib below	Inferior border of rib above (not shown in figure)	Draw ribs together to depress rib cage; aid in forced expiration; antagonistic to external intercostals
Diaphragm	Broad muscle; forms floor of thoracic cavity; dome-shaped in relaxed state; fibers converge from margins of thoracic cage toward a central tendon	Inferior border of rib and sternum, costal cartilages of last six ribs and lumbar vertebrae	Central tendon	Prime mover of inspiration flattens on contraction, increasing vertical dimensions of thorax; increases intra-abdominal pressure
Abdominal Wall (Figure 15.8a and b)				
Rectus abdominis	Medial superficial muscle, extends from pubis to rib cage; ensheathed by aponeuroses of oblique muscles; segmented	Pubic crest and symphysis	Xiphoid process and costal cartilages of ribs 5–7	Flexes and rotates vertebral column; increases abdominal pressure; fixes and depresses ribs; stabilizes pelvis during walking; used in sit-ups and curls
External oblique	Most superficial lateral muscle; fibers run downward and medially; ensheathed by an aponeurosis	Anterior surface of last eight ribs	Linea alba,* pubic crest and tubercles, and iliac crest	See rectus abdominis, above; compresses abdominal wall; also aids muscles of back in trunk rotation and lateral flexion; used in oblique curls
Internal oblique	Most fibers run at right angles to those of external oblique, which it underlies	Lumbar fascia, iliac crest, and inguinal ligament	Linea alba, pubic crest, and costal cartilages of last three ribs	As for external oblique
Transversus abdominis	Deepest muscle of abdominal wall; fibers run horizontally	Inguinal ligament, iliac crest, cartilages of last five or six ribs, and lumbar fascia	Linea alba and pubic crest	Compresses abdominal contents

*The linea alba (white line) is a narrow, tendinous sheath that runs along the middle of the abdomen from the sternum to the pubic symphysis. It is formed by the fusion of the aponeurosis of the external oblique and transversus muscles.

FIGURE 15.7 Deep muscles of the thorax: muscles of respiration. (a) The external intercostals (inspiratory muscles) are shown on the left and the internal intercostals (expiratory muscles) are shown on the right. These two muscle layers run obliquely and at right angles to each other. **(b)** Inferior view of the diaphragm, the prime mover of inspiration. Notice that its muscle fibers converge toward a central tendon, an arrangement that causes the diaphragm to flatten and move inferiorly as it contracts. The diaphragm and its tendon are pierced by the great vessels (aorta and inferior vena cava) and the esophagus.

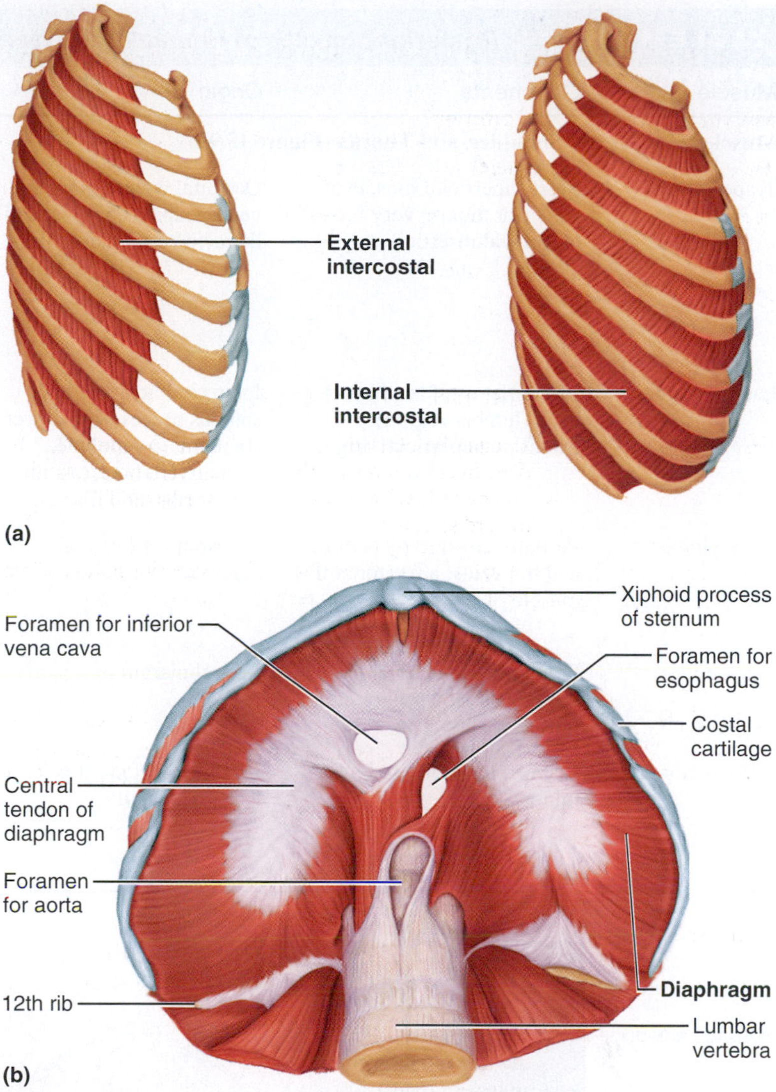
External intercostal
Internal intercostal
(a)
Foramen for inferior vena cava
Xiphoid process of sternum
Foramen for esophagus
Costal cartilage
Central tendon of diaphragm
Foramen for aorta
12th rib
Diaphragm
Lumbar vertebra
(b)

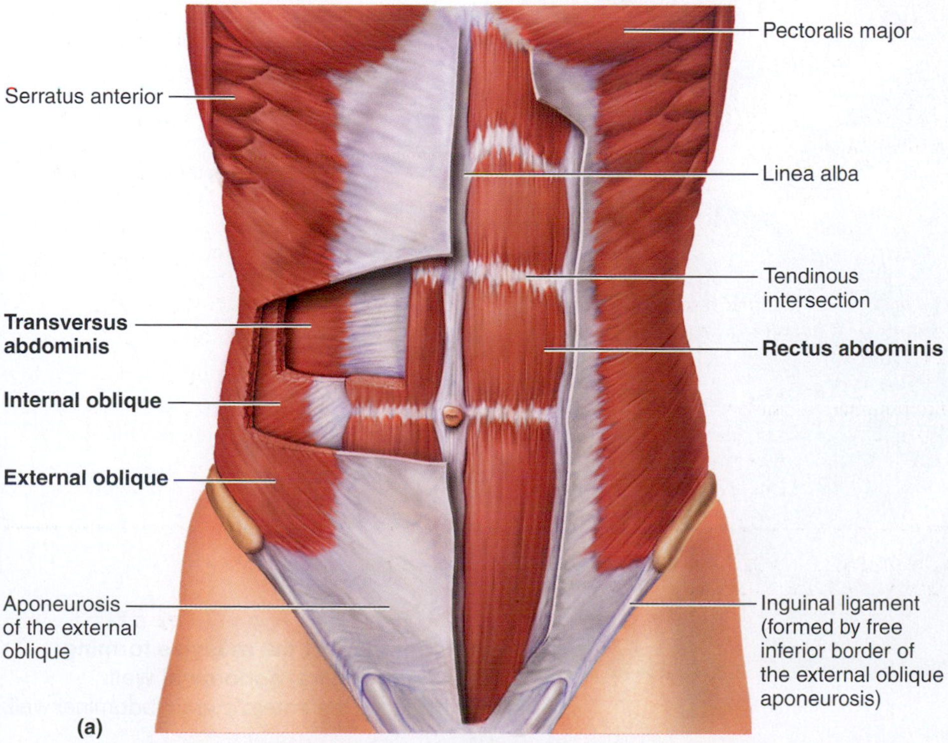
Pectoralis major
Serratus anterior
Linea alba
Tendinous intersection
Transversus abdominis
Rectus abdominis
Internal oblique
External oblique
Aponeurosis of the external oblique
Inguinal ligament (formed by free inferior border of the external oblique aponeurosis)
(a)

FIGURE 15.8 Anterior view of the muscles forming the anterolateral abdominal wall. (a) The superficial muscles have been partially cut away on the left side of the diagram to reveal the deeper internal oblique and transversus abdominis muscles.

TABLE 15.4 Posterior Muscles of Human Trunk (see Figure 15.9)

Muscle	Comments	Origin	Insertion	Action
Muscles of the Neck, Shoulder, and Thorax (Figure 15.9a)				
Trapezius	Most superficial muscle of posterior thorax; very broad origin and insertion	Occipital bone; ligamentum nuchae; spines of C_7 and all thoracic vertebrae	Acromion and spinous process of scapula; lateral third of clavicle	Extends head; raises, rotates, and retracts (adducts) scapula and stabilizes it; superior fibers elevate scapula (as in shrugging the shoulders); inferior fibers depress it
Latissimus dorsi	Broad flat muscle of lower back (lumbar region); extensive superficial origins	Indirect attachment to spinous processes of lower six thoracic vertebrae, lumbar vertebrae, last three to four ribs, and iliac crest	Floor of intertubercular sulcus of humerus	Prime mover of arm extension; adducts and medially rotates arm; brings arm down in power stroke, as in striking a blow
Infraspinatus	Partially covered by deltoid and trapezius; a rotator cuff muscle	Infraspinous fossa of scapula	Greater tubercle of humerus	Lateral rotation of humerus; helps hold head of humerus in glenoid cavity; stabilizes shoulder
Teres minor	Small muscle inferior to infraspinatus; a rotator cuff muscle	Lateral margin of scapula	Greater tubercle of humerus	As for infraspinatus
Teres major	Located inferiorly to teres minor	Posterior surface at inferior angle of scapula	Intertubercular sulcus of humerus	Extends, medially rotates, and adducts humerus; synergist of latissimus dorsi

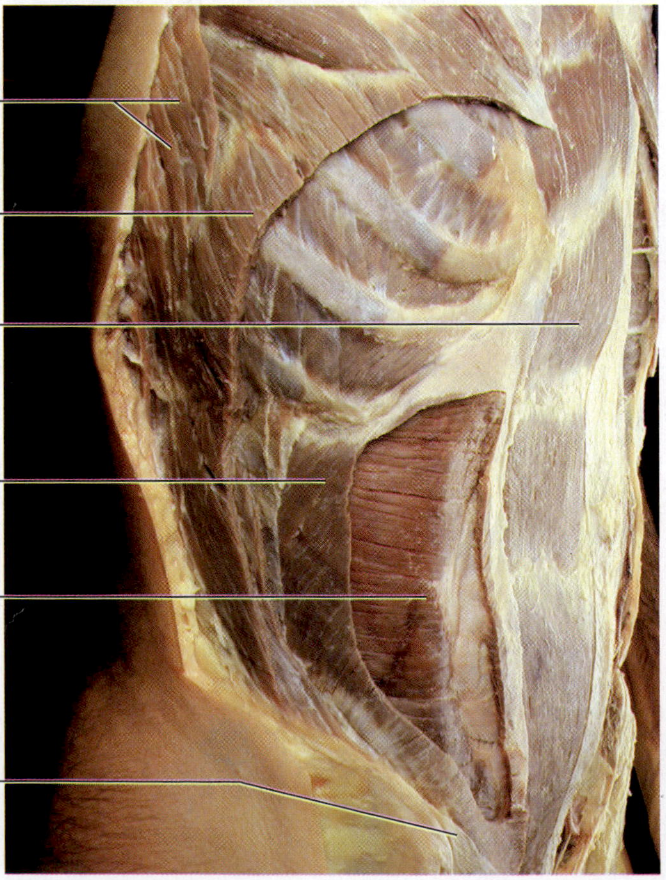
Serratus anterior
External oblique
Rectus abdominis
Internal oblique
Transversus abdominis
Inguinal ligament
(b)

FIGURE 15.8 ***(continued)*** **Anterior view of the muscles forming the anterolateral abdominal wall.**
(b) Photo of the anterolateral abdominal wall.

TABLE 15.4 *(continued)*

Muscle	Comments	Origin	Insertion	Action
Supraspinatus	Obscured by trapezius; a rotator cuff muscle	Supraspinous fossa of scapula	Greater tubercle of humerus	Initiates abduction of humerus; stabilizes shoulder joint
Levator scapulae	Located at back and side of neck, deep to trapezius	Transverse processes of C_1–C_4	Medial border of scapula superior to spine	Elevates and adducts scapula; with fixed scapula, laterally flexes neck to the same side
Rhomboids—major and minor	Beneath trapezius and inferior to levator scapulae; rhomboid minor is the more superior muscle	Spinous processes of C_7 and T_1–T_5	Medial border of scapula	Pull scapula medially (retraction); stabilize scapula; rotate glenoid cavity downward
Muscles Associated with the Vertebral Column (Figure 15.9b)				
Semispinalis	Deep composite muscle of the back—thoracis, cervicis, and capitis portions	Transverse processes of C_7–T_{12}	Occipital bone and spinous processes of cervical vertebrae and T_1–T_4	Acting together, extend head and vertebral column; acting independently (right vs. left) causes rotation toward the opposite side

(continues)

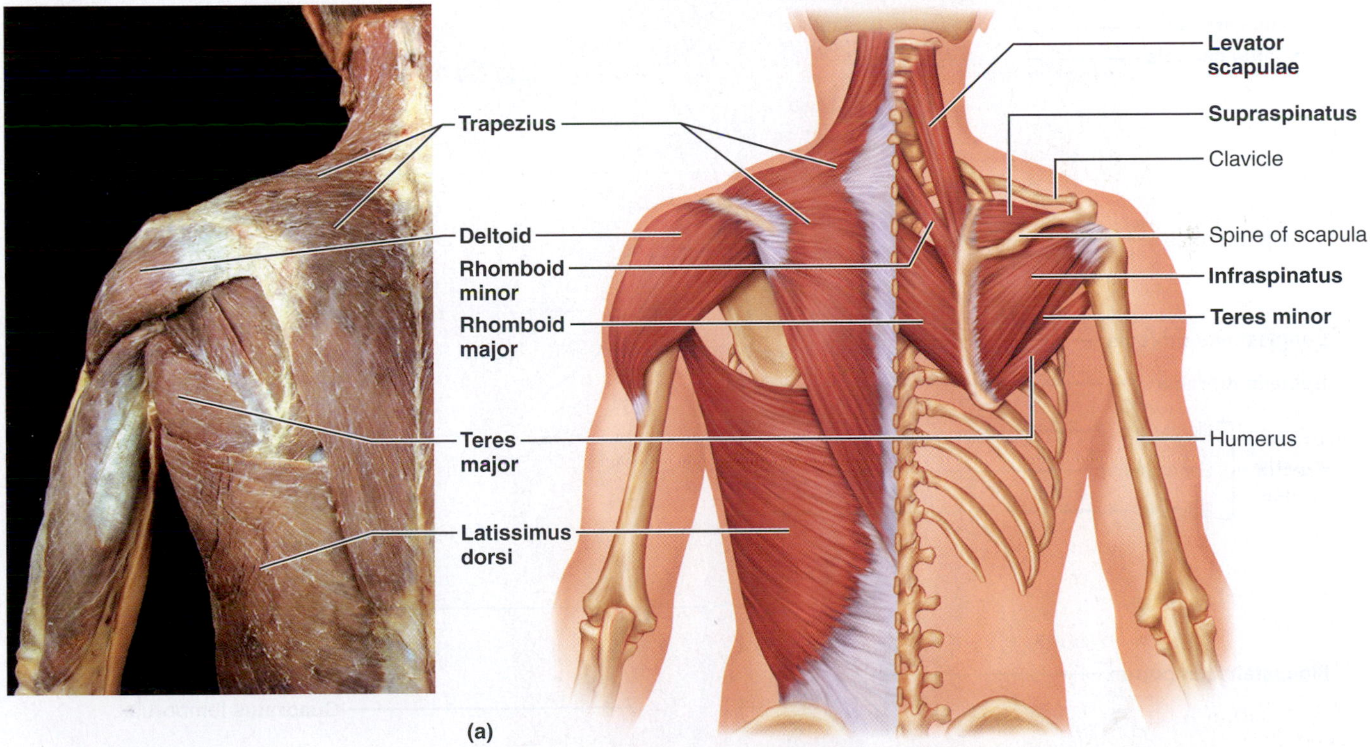
Trapezius
Deltoid
Rhomboid minor
Rhomboid major
Teres major
Latissimus dorsi
Levator scapulae
Supraspinatus
Clavicle
Spine of scapula
Infraspinatus
Teres minor
Humerus
(a)

FIGURE 15.9 Muscles of the neck, shoulder, and thorax (posterior view). **(a)** The superficial muscles of the back are shown for the left side of the body, with a corresponding photograph. The superficial muscles are removed on the right side of the illustration to reveal the deeper muscles acting on the scapula and the rotator cuff muscles that help to stabilize the shoulder joint.

TABLE 15.4 Posterior Muscles of Human Trunk *(continued)*

Muscle	Comments	Origin	Insertion	Action
Erector spinae	A long tripartite muscle composed of iliocostalis (lateral), longissimus, and spinalis (medial) muscle columns; superficial to semispinalis muscles; extends from pelvis to head	Iliac crest, transverse processes of lumbar, thoracic, and cervical vertebrae, and/or ribs 3–6 depending on specific part	Ribs and transverse processes of vertebrae about six segments above origin; longissimus also inserts into mastoid process	Extend and bend the vertebral column laterally; fibers of the longissimus also extend head

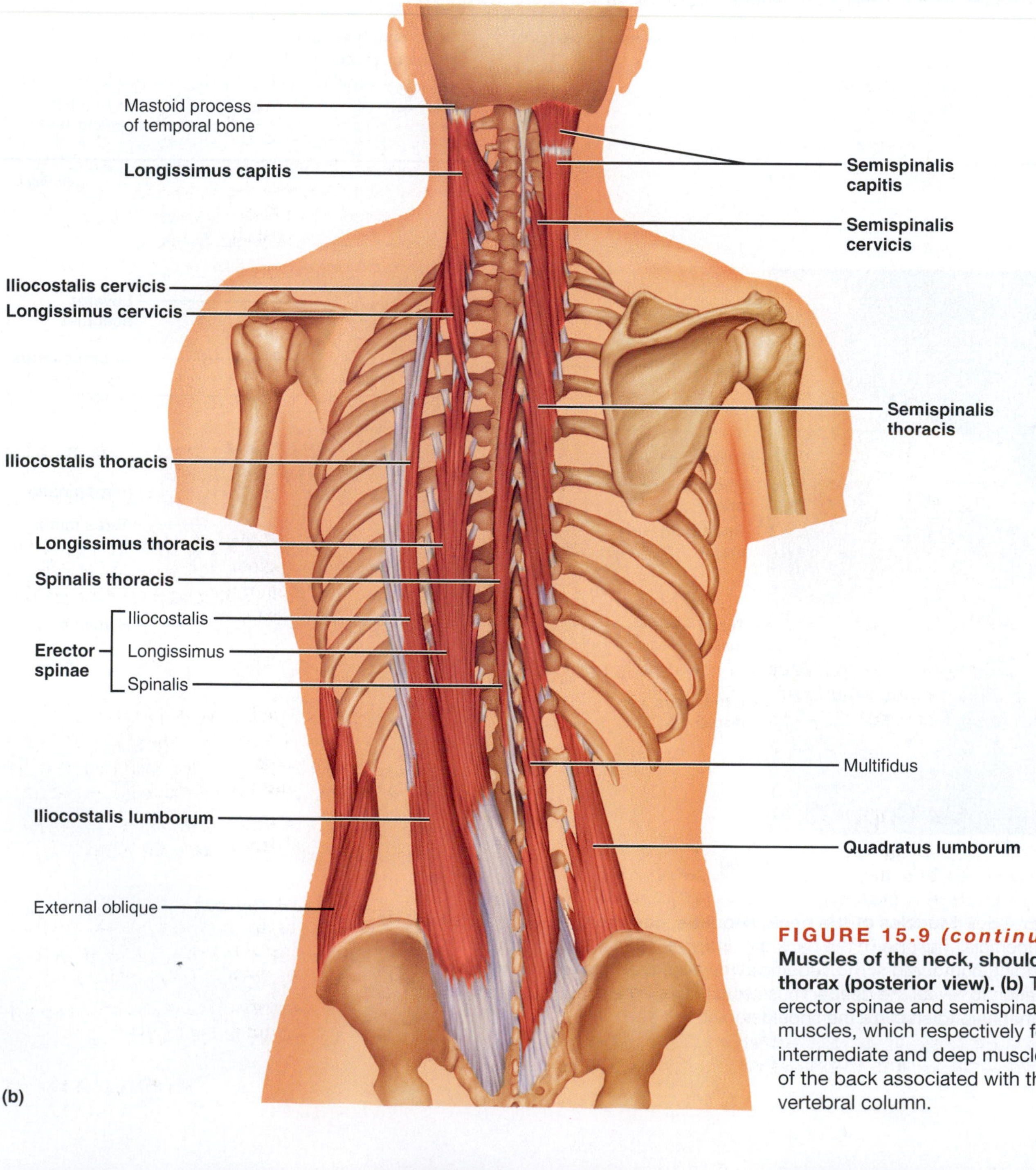

Mastoid process of temporal bone
Longissimus capitis
Semispinalis capitis
Semispinalis cervicis
Iliocostalis cervicis
Longissimus cervicis
Semispinalis thoracis
Iliocostalis thoracis
Longissimus thoracis
Spinalis thoracis
Erector spinae
Iliocostalis
Longissimus
Spinalis
Multifidus
Iliocostalis lumborum
Quadratus lumborum
External oblique
(b)

FIGURE 15.9 ***(continued)*** **Muscles of the neck, shoulder, and thorax (posterior view). (b)** The erector spinae and semispinalis muscles, which respectively form the intermediate and deep muscle layers of the back associated with the vertebral column.

TABLE 15.4 *(continued)*

Muscle	Comments	Origin	Insertion	Action
Splenius (see Figure 15.9c)	Superficial muscle (capitis and cervicis parts) extending from upper thoracic region to skull	Ligamentum nuchae and spinous processes of C_7–T_6	Mastoid process, occipital bone, and transverse processes of C_2–C_4	As a group, extend or hyperextend head; when only one side is active, head is rotated and bent toward the same side
Quadratus lumborum	Forms greater portion of posterior abdominal wall	Iliac crest and lumbar fascia	Inferior border of rib 12; transverse processes of lumbar vertebrae	Each flexes vertebral column laterally; together extend the lumbar spine and fix rib 12; maintains upright posture

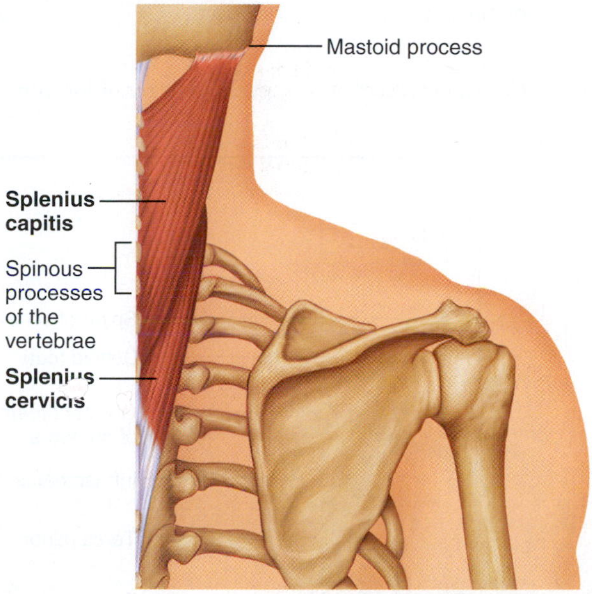
Mastoid process
Splenius capitis
Spinous processes of the vertebrae
Splenius cervicis

(c)

FIGURE 15.9 *(continued)* Muscles of the neck, shoulder, and thorax (posterior view). (c) Deep (splenius) muscles of the posterior neck. Superficial muscles have been removed.

Muscles of the Upper Limb

The muscles that act on the upper limb fall into three groups: those that move the arm, those causing movement at the elbow, and those effecting movements of the wrist and hand.

The muscles that cross the shoulder joint to insert on the humerus and move the arm (subscapularis, supraspinatus and infraspinatus, deltoid, and so on) are primarily trunk muscles that originate on the axial skeleton or shoulder girdle. These muscles are included with the trunk muscles.

The second group of muscles, which cross the elbow joint and move the forearm, consists of muscles forming the musculature of the humerus. These muscles arise primarily from the humerus and insert in forearm bones. They are responsible for flexion, extension, pronation, and supination.

The third group composes the musculature of the forearm. For the most part, these muscles insert on the digits and produce movements at the wrist and fingers.

ACTIVITY 3

Identifying Muscles of the Upper Limb

The origins, insertions, and actions of muscles that move the forearm are summarized in Table 15.5 and the muscles are shown in Figure 15.10.

In general, muscles acting on the wrist and hand are more easily identified if their insertion tendons are located first. These muscles are described in Table 15.6 and illustrated in Figure 15.11.

First study the tables and figures, then see if you can identify these muscles on a torso model, anatomical chart, or cadaver. Complete this portion of the exercise with palpation demonstrations as outlined next.

Demonstrating Operations of Upper Limb Muscles

1. To observe the *biceps brachii,* attempt to flex your forearm (hand supinated) against resistance. The insertion tendon of this biceps muscle can also be felt in the lateral aspect of the antecubital fossa (where it runs toward the radius to attach).

2. If you acutely flex your elbow and then try to extend it against resistance, you can demonstrate the action of your *triceps brachii.*

3. Strongly flex your wrist and make a fist. Palpate your contracting wrist flexor muscles (which originate from the medial epicondyle of the humerus) and their insertion tendons, which can be easily felt at the anterior aspect of the wrist.

4. Flare your fingers to identify the tendons of the *extensor digitorum* muscle on the dorsum of your hand.

Text continues on page 218.

TABLE 15.5 Muscles of Human Humerus That Act on the Forearm (see Figure 15.10)

Muscle	Comments	Origin	Insertion	Action
Triceps brachii	Sole, large fleshy muscle of posterior humerus; three-headed origin	Long head—inferior margin of glenoid cavity; lateral head— posterior humerus; medial head—distal radial groove on posterior humerus	Olecranon process of ulna	Powerful forearm extensor; antagonist of forearm flexors (brachialis and biceps brachii)
Anconeus	Short triangular muscle blended with triceps	Lateral epicondyle of humerus	Lateral aspect of olecranon process of ulna	Abducts ulna during forearm pronation; extends elbow
Biceps brachii	Most familiar muscle of anterior humerus because this two-headed muscle bulges when forearm is flexed	Short head: coracoid process; tendon of long head runs in intertubercular sulcus and within capsule of shoulder joint	Radial tuberosity	Flexion (powerful) of elbow and supination of forearm; "it turns the corkscrew and pulls the cork"; weak arm flexor
Brachioradialis	Superficial muscle of lateral forearm; forms lateral boundary of antecubital fossa	Lateral ridge at distal end of humerus	Base of styloid process of radius	Synergist in forearm flexion
Brachialis	Immediately deep to biceps brachii	Distal portion of anterior humerus	Coronoid process of ulna	A major flexor of forearm

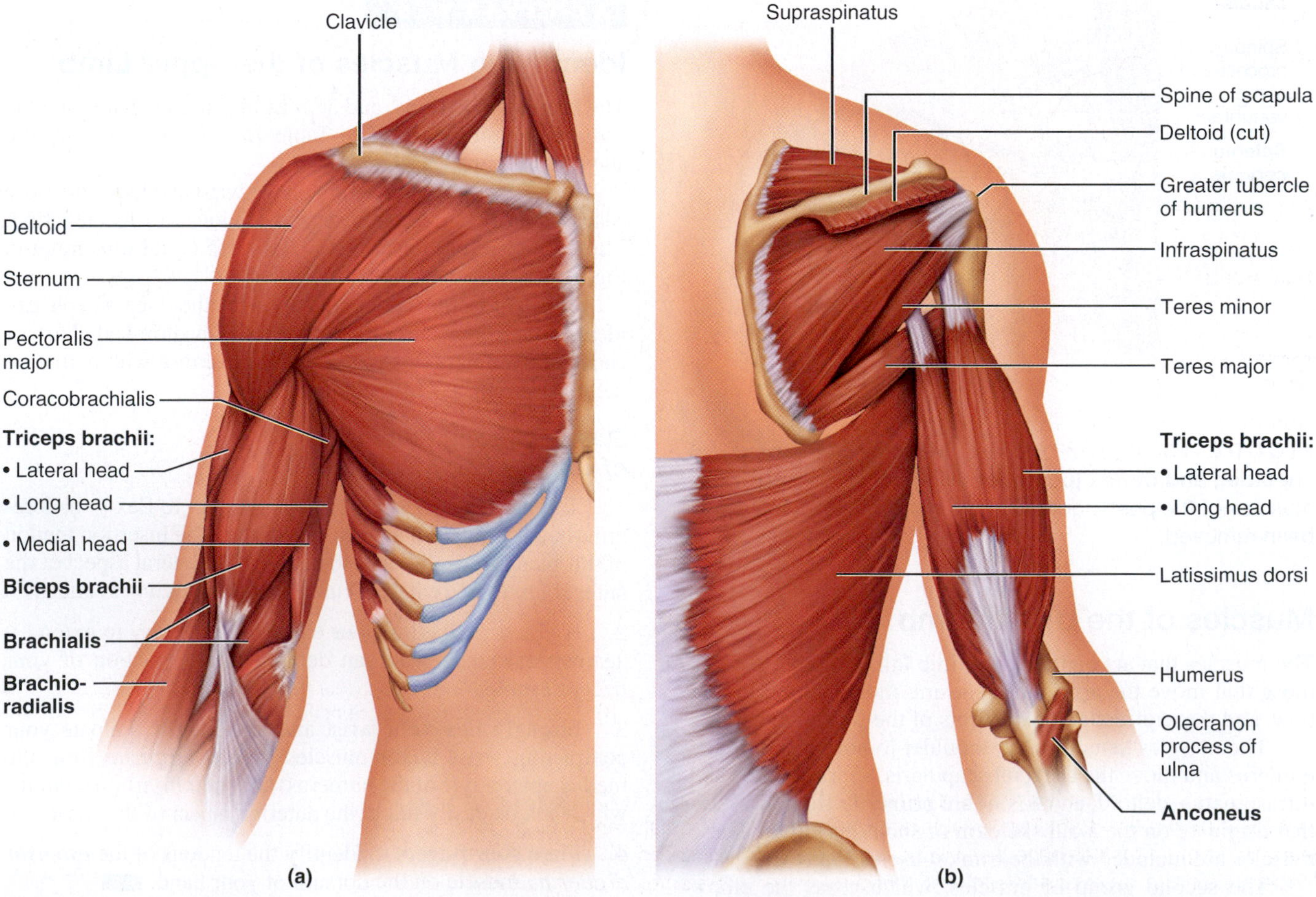
Clavicle
Deltoid
Sternum
Pectoralis major
Coracobrachialis
Triceps brachii:
• Lateral head
• Long head
• Medial head
Biceps brachii
Brachialis
Brachio-radialis
(a)
Supraspinatus
Spine of scapula
Deltoid (cut)
Greater tubercle of humerus
Infraspinatus
Teres minor
Teres major
Triceps brachii:
• Lateral head
• Long head
Latissimus dorsi
Humerus
Olecranon process of ulna
Anconeus
(b)

FIGURE 15.10 Muscles causing movements of the arm and forearm. (a) Superficial muscles of the anterior thorax, shoulder, and arm, anterior view. **(b)** Posterior aspect of the arm showing the lateral and long heads of the triceps brachii muscle.

TABLE 15.6 Muscles of Human Forearm That Act on Hand and Fingers (see Figure 15.11)

Muscle	Comments	Origin	Insertion	Action
Anterior Compartment (Figure 15.11a, b, c)				
Superficial				
Pronator teres	Seen in a superficial view between proximal margins of brachioradialis and flexor carpi radialis	Medial epicondyle of humerus and coronoid process of ulna	Midshaft of radius	Acts synergistically with pronator quadratus to pronate forearm; weak elbow flexor
Flexor carpi radialis	Superficial; runs diagonally across forearm	Medial epicondyle of humerus	Base of metacarpals 2 and 3	Powerful flexor of wrist; abducts hand
Palmaris longus	Small fleshy muscle with a long tendon; medial to flexor carpi radialis	Medial epicondyle of humerus	Palmar aponeurosis; skin and fascia of palm	Flexes wrist (weak); tenses skin and fascia of palm

(continues)

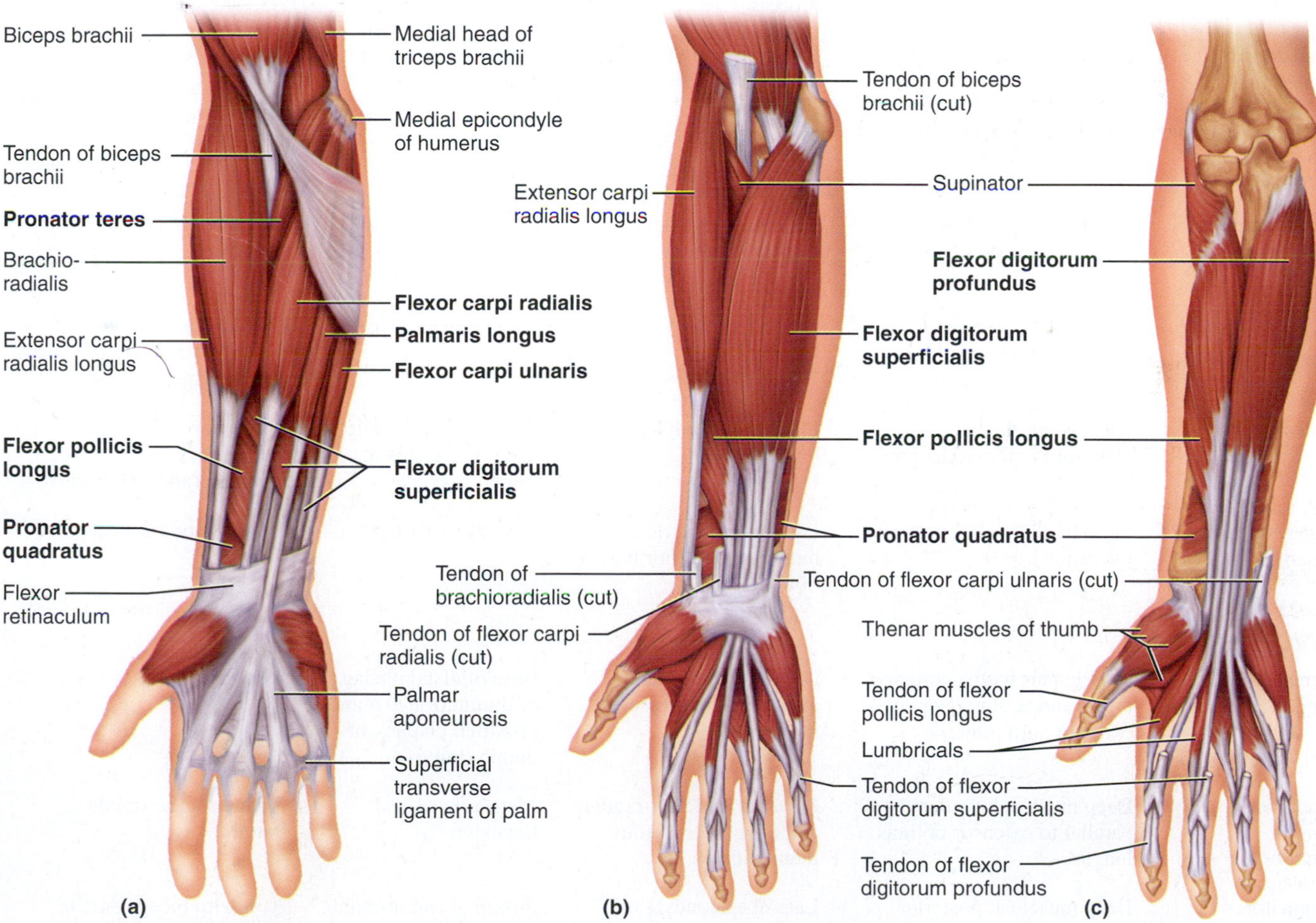

Biceps brachii
Medial head of triceps brachii
Medial epicondyle of humerus
Tendon of biceps brachii
Pronator teres
Brachio-radialis
Extensor carpi radialis longus
Flexor carpi radialis
Palmaris longus
Flexor carpi ulnaris
Flexor pollicis longus
Flexor digitorum superficialis
Pronator quadratus
Flexor retinaculum
Palmar aponeurosis
Superficial transverse ligament of palm
(a)
Tendon of biceps brachii (cut)
Extensor carpi radialis longus
Supinator
Flexor digitorum profundus
Flexor digitorum superficialis
Flexor pollicis longus
Pronator quadratus
Tendon of brachioradialis (cut)
Tendon of flexor carpi ulnaris (cut)
Tendon of flexor carpi radialis (cut)
Thenar muscles of thumb
Tendon of flexor pollicis longus
Lumbricals
Tendon of flexor digitorum superficialis
Tendon of flexor digitorum profundus
(b)
(c)

FIGURE 15.11 Muscles of the forearm and wrist. (a) Superficial anterior view of right forearm and hand. **(b)** The brachioradialis, flexors carpi radialis and ulnaris, and palmaris longus muscles have been removed to reveal the position of the somewhat deeper flexor digitorum superficialis. **(c)** Deep muscles of the anterior compartment. Superficial muscles have been removed. *Note:* The thenar muscles of the thumb and the lumbricals that help move the fingers are illustrated here but are not described in Table 15.6.

TABLE 15.6 Muscles of Human Forearm That Act on Hand and Fingers *(continued)*

Muscle	Comments	Origin	Insertion	Action
Flexor carpi ulnaris	Superficial; medial to palmaris longus	Medial epicondyle of humerus and olecranon process and posterior surface of ulna	Base of metacarpal 5; pisiform and hamate bones	Powerful flexor of wrist; adducts hand
Flexor digitorum superficialis	Deeper muscle (deep to muscles named above); visible at distal end of forearm	Medial epicondyle of humerus, coronoid process of ulna, and shaft of radius	Middle phalanges of fingers 2–5	Flexes wrist and middle phalanges of fingers 2–5
Deep				
Flexor pollicis longus	Deep muscle of anterior forearm; distal to and paralleling lower margin of flexor digitorum superficialis	Anterior surface of radius, and interosseous membrane	Distal phalanx of thumb	Flexes thumb (*pollex* is Latin for "thumb")
Flexor digitorum profundus	Deep muscle; overlain entirely by flexor digitorum superficialis	Anteromedial surface of ulna, interosseous membrane, and coronoid process	Distal phalanges of fingers 2–5	Sole muscle that flexes distal phalanges; assists in wrist flexion
Pronator quadratus	Deepest muscle of distal forearm	Distal portion of anterior ulnar surface	Anterior surface of radius, distal end	Pronates forearm
Posterior Compartment (Figure 15.11d, e, f)				
Superficial				
Extensor carpi radialis longus	Superficial; parallels brachioradialis on lateral forearm	Lateral supracondylar ridge of humerus	Base of metacarpal 2	Extends and abducts wrist
Extensor carpi radialis brevis	Deep to extensor carpi radialis longus	Lateral epicondyle of humerus	Base of metacarpal 3	Extends and abducts wrist; steadies wrist during finger flexion
Extensor digitorum	Superficial; medial to extensor carpi radialis brevis	Lateral epicondyle of humerus	By four tendons into distal phalanges of fingers 2–5	Prime mover of finger extension; extends wrist; can flare (abduct) fingers
Extensor carpi ulnaris	Superficial; medial posterior forearm	Lateral epicondyle of humerus; posterior border of ulna	Base of metacarpal 5	Extends and adducts wrist
Deep				
Extensor pollicis longus and brevis	Muscle pair with a common origin and action; deep to extensor carpi ulnaris	Dorsal shaft of ulna and radius, interosseous membrane	Base of distal phalanx of thumb (longus) and proximal phalanx of thumb (brevis)	Extend thumb
Abductor pollicis longus	Deep muscle; lateral and parallel to extensor pollicis longus	Posterior surface of radius and ulna; interosseous membrane	Metacarpal 1 and trapezium	Abducts and extends thumb
Supinator	Deep muscle at posterior aspect of elbow	Lateral epicondyle of humerus; proximal ulna	Proximal end of radius	Acts with biceps brachii to supinate forearm; antagonist of pronator muscles

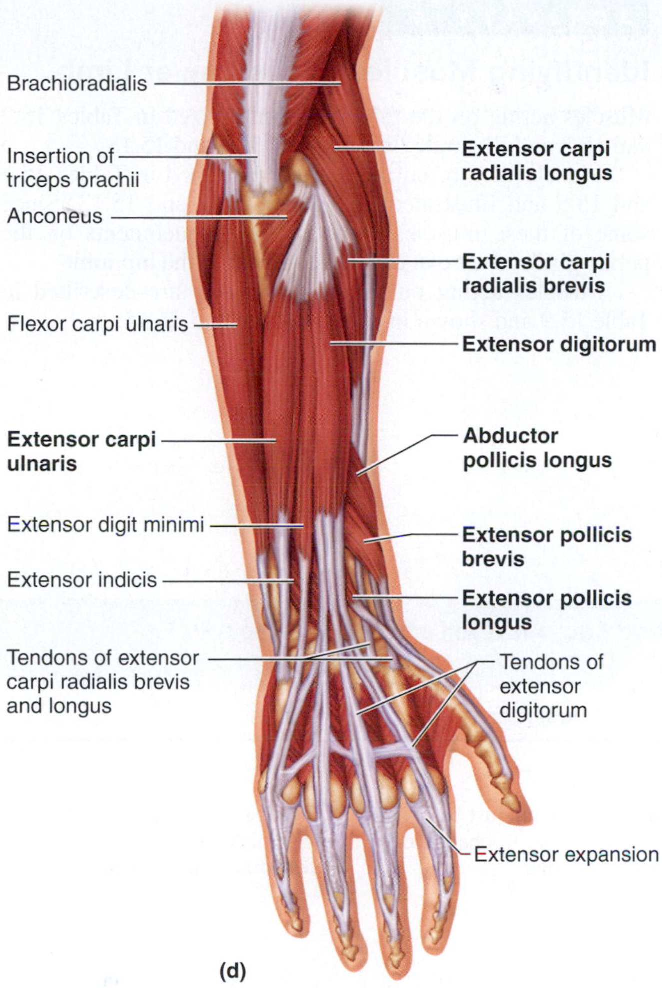

Brachioradialis
Insertion of triceps brachii
Anconeus
Flexor carpi ulnaris
Extensor carpi ulnaris
Extensor digit minimi
Extensor indicis
Tendons of extensor carpi radialis brevis and longus
Extensor carpi radialis longus
Extensor carpi radialis brevis
Extensor digitorum
Abductor pollicis longus
Extensor pollicis brevis
Extensor pollicis longus
Tendons of extensor digitorum
Extensor expansion
(d)

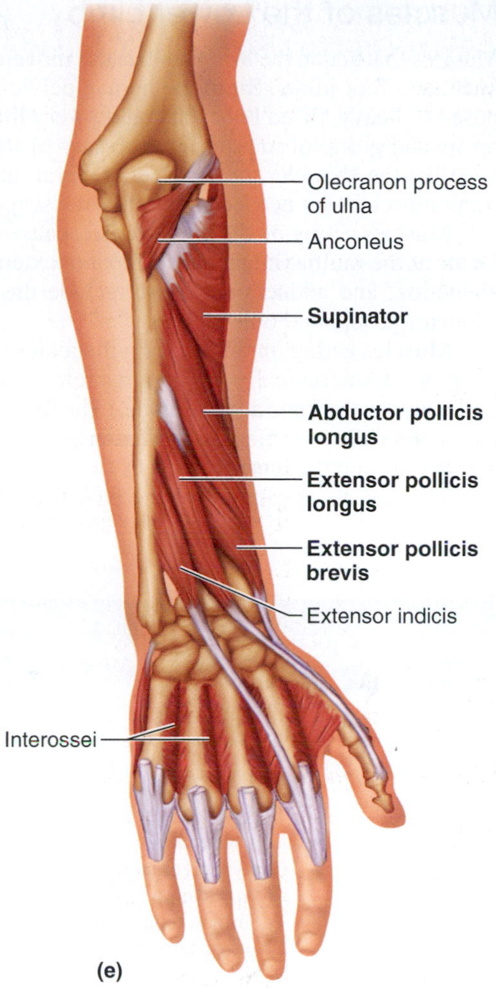

Olecranon process of ulna
Anconeus
Supinator
Abductor pollicis longus
Extensor pollicis longus
Extensor pollicis brevis
Extensor indicis
Interossei
(e)

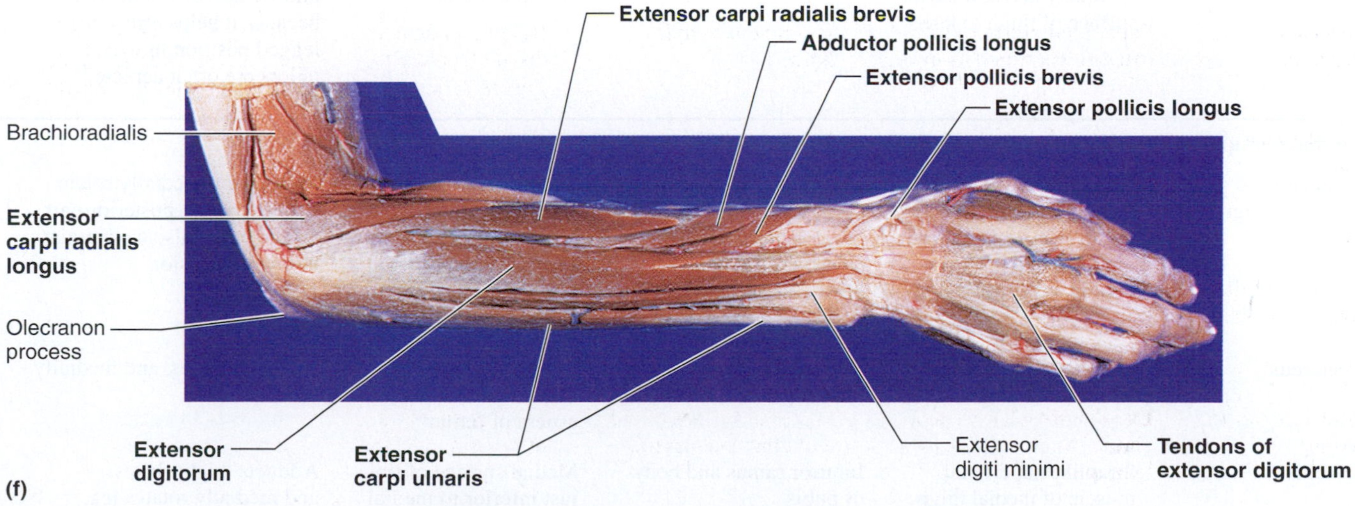

Extensor carpi radialis brevis
Abductor pollicis longus
Extensor pollicis brevis
Extensor pollicis longus
Brachioradialis
Extensor carpi radialis longus
Olecranon process
Extensor digitorum
Extensor carpi ulnaris
Extensor digiti minimi
Tendons of extensor digitorum
(f)

FIGURE 15.11 ***(continued)*** **Muscles of the forearm and wrist. (d)** Superficial muscles, posterior view. **(e)** Deep posterior muscles; superficial muscles have been removed. The interossei, the deepest layer of instrinsic hand muscles, are also illustrated. **(f)** Photo of deep posterior muscles of the right forearm. The superficial muscles have been removed.

Muscles of the Lower Limb

Muscles that act on the lower limb cause movement at the hip, knee, and foot joints. Since the human pelvic girdle is composed of heavy, fused bones that allow very little movement, no special group of muscles is necessary to stabilize it. This is unlike the shoulder girdle, where several muscles (mainly trunk muscles) are needed to stabilize the scapulae.

Muscles acting on the thigh (femur) cause various movements at the multiaxial hip joint (flexion, extension, rotation, abduction, and adduction). These include the iliopsoas, the adductor group, and others.

Muscles acting on the leg form the major musculature of the thigh. (Anatomically the term *leg* refers only to that portion between the knee and the ankle.) The thigh muscles cross the knee to allow its flexion and extension. They include the hamstrings and the quadriceps.

The muscles originating on the leg act on the foot and toes.

ACTIVITY 4

Identifying Muscles of the Lower Limb

Muscles acting on the thigh are summarized in Tables 15.7 and 15.8 and illustrated in Figures 15.12 and 15.13.

Muscles acting on the leg are described in Tables 15.7 and 15.8 and illustrated in Figures 15.12 and 15.13. Since some of these muscles can also have attachments on the pelvic girdle, they can cause movement at the hip joint.

Muscles acting on the foot and toes are described in Table 15.9 and shown in Figures 15.14 and 15.15.

TABLE 15.7 Muscles Acting on Human Thigh and Leg, Anterior and Medial Aspects (see Figure 15.12)

Muscle	Comments	Origin	Insertion	Action
Origin on the Pelvis				
Iliopsoas—iliacus and psoas major	Two closely related muscles; fibers pass under inguinal ligament to insert into femur via a common tendon; iliacus is more lateral	Iliacus—iliac fossa and crest, lateral sacrum; psoas major—transverse processes, bodies, and discs of T_{12} and lumbar vertebrae	On and just below lesser trochanter of femur	Flex trunk on thigh; flex thigh; lateral flexion of vertebral column (psoas)
Sartorius	Straplike superficial muscle running obliquely across anterior surface of thigh to knee	Anterior superior iliac spine	By an aponeurosis into medial aspect of proximal tibia	Flexes, abducts, and laterally rotates thigh; flexes knee; known as "tailor's muscle" because it helps effect cross-legged position in which tailors are often depicted
Medial Compartment				
Adductors—magnus, longus, and brevis	Large muscle mass forming medial aspect of thigh; arise from front of pelvis and insert at various levels on femur	Magnus—ischial and pubic rami and ischial tuberosity; longus—pubis near pubic symphysis; brevis—body and inferior ramus of pubis	Magnus—linea aspera and adductor tubercle of femur; longus and brevis—linea aspera	Adduct and medially rotate and flex thigh; posterior part of magnus is also a synergist in thigh extension
Pectineus	Overlies adductor brevis on proximal thigh	Pectineal line of pubis (and superior ramus)	Inferior from lesser trochanter to linea aspera of femur	Adducts, flexes, and medially rotates thigh
Gracilis	Straplike superficial muscle of medial thigh	Inferior ramus and body of pubis	Medial surface of tibia just inferior to medial condyle	Adducts thigh; flexes and medially rotates leg, especially during walking

(Continues on page 220.)

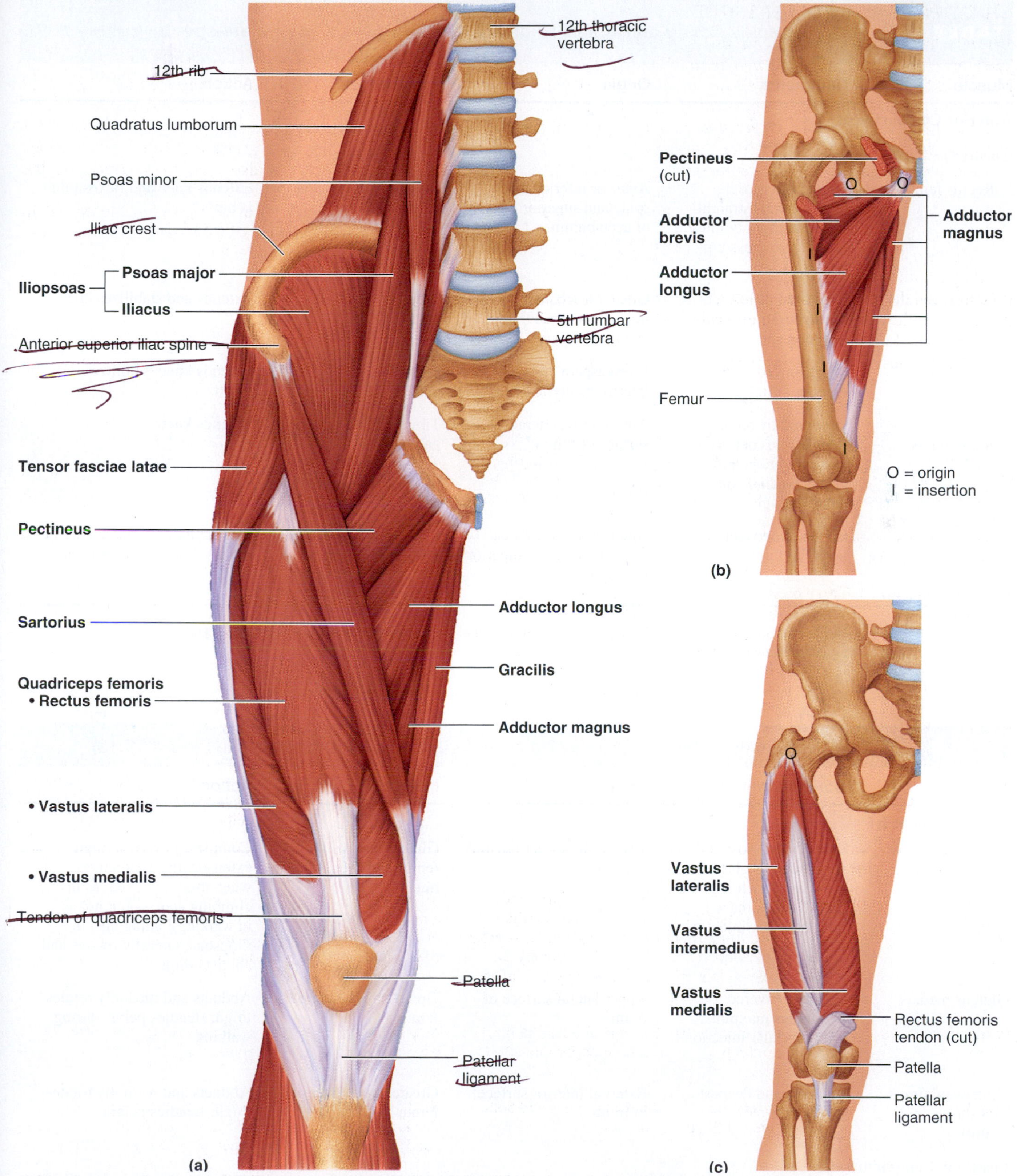
12th thoracic vertebra
12th rib
Quadratus lumborum
Psoas minor
Iliac crest
Iliopsoas
Psoas major
Iliacus
5th lumbar vertebra
Anterior superior iliac spine
Tensor fasciae latae
Pectineus
Sartorius
Adductor longus
Gracilis
Quadriceps femoris
• Rectus femoris
Adductor magnus
• Vastus lateralis
• Vastus medialis
Tendon of quadriceps femoris
Patella
Patellar ligament
(a)
Pectineus
(cut)
Adductor brevis
Adductor magnus
Adductor longus
Femur
O = origin
I = insertion
(b)
Vastus lateralis
Vastus intermedius
Vastus medialis
Rectus femoris tendon (cut)
Patella
Patellar ligament
(c)

FIGURE 15.12 Anterior and medial muscles promoting movements of the thigh and leg. (a) Anterior view of the deep muscles of the pelvis and superficial muscles of the right thigh. **(b)** Adductor muscles of the medial compartment of the thigh. **(c)** The vastus muscles (isolated) of the quadriceps group.

TABLE 15.7 Muscles Acting on Human Thigh and Leg, Anterior and Medial Aspects *(continued)*

Muscle	Comments	Origin	Insertion	Action
Anterior Compartment				
Quadriceps femoris*				
Rectus femoris	Superficial muscle of thigh; runs straight down thigh; only muscle of group to cross hip joint	Anterior inferior iliac spine and superior margin of acetabulum	Tibial tuberosity and patella	Extends knee and flexes thigh at hip
Vastus lateralis	Forms lateral aspect of thigh; intramuscular injection site	Greater trochanter, intertrochanteric line, and linea aspera	Tibial tuberosity and patella	Extends and stabilizes knee
Vastus medialis	Forms inferomedial aspect of thigh	Linea aspera and intertrochanteric line	Tibial tuberosity and patella	Extends knee; stabilizes patella
Vastus intermedius	Obscured by rectus femoris; lies between vastus lateralis and vastus medialis on anterior thigh	Anterior and lateral surface of femur	Tibial tuberosity and patella	Extends knee
Tensor fasciae latae	Enclosed between fascia layers of thigh	Anterior aspect of iliac crest and anterior superior iliac spine	Iliotibial tract (lateral portion of fascia lata)	Flexes, abducts, and medially rotates thigh; steadies trunk

*The quadriceps form the flesh of the anterior thigh and have a common insertion in the tibial tuberosity via the patellar tendon. They are powerful leg extensors, enabling humans to kick a football, for example.

TABLE 15.8 Muscles Acting on Human Thigh and Leg, Posterior Aspect (see Figure 15.13)

Muscle	Comments	Origin	Insertion	Action
Origin on the Pelvis				
Gluteus maximus	Largest and most superficial of gluteal muscles (which form buttock mass); intramuscular injection site	Dorsal ilium, sacrum, and coccyx	Gluteal tuberosity of femur and iliotibial tract*	Complex, powerful thigh extensor (most effective when thigh is flexed, as in climbing stairs—but not as in walking); antagonist of iliopsoas; laterally rotates and abducts thigh
Gluteus medius	Partially covered by gluteus maximus; intramuscular injection site	Upper lateral surface of ilium	Greater trochanter of femur	Abducts and medially rotates thigh; steadies pelvis during walking
Gluteus minimus (not shown in figure)	Smallest and deepest gluteal muscle	External inferior surface of ilium	Greater trochanter of femur	Abducts and medially rotates thigh; steadies pelvis
Posterior Compartment				
Hamstrings†				
Biceps femoris	Most lateral muscle of group; arises from two heads	Ischial tuberosity (long head); linea aspera and distal femur (short head)	Tendon passes laterally to insert into head of fibula and lateral condyle of tibia	Extends thigh; laterally rotates leg; flexes knee

(continues)

TABLE 15.8	(continued)			
Muscle	**Comments**	**Origin**	**Insertion**	**Action**
Semitendinosus	Medial to biceps femoris	Ischial tuberosity	Medial aspect of upper tibial shaft	Extends thigh; flexes knee; medially rotates leg
Semimembranosus	Deep to semitendinosus	Ischial tuberosity	Medial condyle of tibia; lateral condyle of femur	Extends thigh; flexes knee; medially rotates leg

* The iliotibial tract, a thickened lateral portion of the fascia lata, ensheathes all the muscles of the thigh. It extends as a tendinous band from the iliac crest to the knee.

† The hamstrings are the fleshy muscles of the posterior thigh. The name comes from the butchers' practice of using the tendons of these muscles to hang hams for smoking. As a group, they are strong extensors of the hip; they counteract the powerful quadriceps by stabilizing the knee joint when standing.

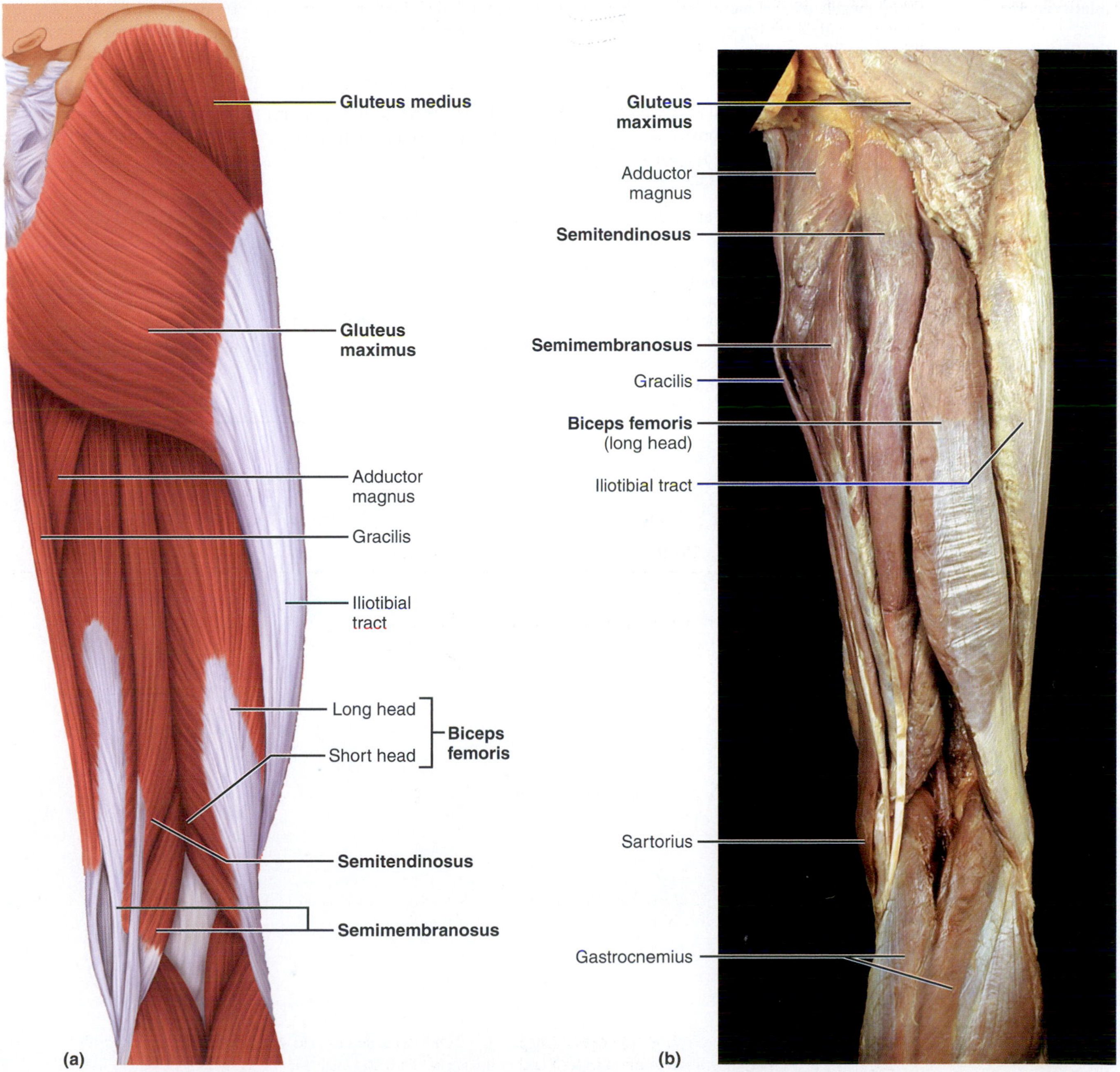
Gluteus medius
Gluteus maximus
Adductor magnus
Gracilis
Iliotibial tract
Long head
Short head
Biceps femoris
Semitendinosus
Semimembranosus
(a)
Gluteus maximus
Adductor magnus
Semitendinosus
Semimembranosus
Gracilis
Biceps femoris (long head)
Iliotibial tract
Sartorius
Gastrocnemius
(b)

FIGURE 15.13 Muscles of the posterior aspect of the right hip and thigh. **(a)** Superficial view showing the gluteus muscles of the buttock and hamstring muscles of the thigh. **(b)** Photo of muscles of the posterior thigh.

TABLE 15.9 Muscles Acting on Human Foot and Ankle (see Figures 15.14 and 15.15)

Muscle	Comments	Origin	Insertion	Action
Lateral Compartment (Figure 15.14a, b and Figure 15.15b)				
Fibularis (peroneus) longus	Superficial lateral muscle; overlies fibula	Head and upper portion of fibula	By long tendon under foot to metatarsal 1 and medial cuneiform	Plantar flexes and everts foot; helps keep foot flat on ground
Fibularis (peroneus) brevis	Smaller muscle; deep to fibularis longus	Distal portion of fibula shaft	By tendon running behind lateral malleolus to insert on proximal end of metatarsal 5	Plantar flexes and everts foot, as part of fibularis group
Anterior Compartment (Figure 15.14a, b)				
Tibialis anterior	Superficial muscle of anterior leg; parallels sharp anterior margin of tibia	Lateral condyle and upper ⅔ of tibia; interosseous membrane	By tendon into inferior surface of first cuneiform and metatarsal 1	Prime mover of dorsiflexion; inverts foot; supports longitudinal arch of foot
Extensor digitorum longus	Anterolateral surface of leg; lateral to tibialis anterior	Lateral condyle of tibia; proximal ¾ of fibula; interosseous membrane	Tendon divides into four parts; inserts into middle and distal phalanges of toes 2–5	Prime mover of toe extension; dorsiflexes foot
Fibularis (peroneus) tertius	Small muscle; often fused to distal part of extensor digitorum longus	Distal anterior surface of fibula and interosseous membrane	Tendon inserts on dorsum of metatarsal 5	Dorsiflexes and everts foot
Extensor hallucis longus	Deep to extensor digitorum longus and tibialis anterior	Anteromedial shaft of fibula and interosseous membrane	Tendon inserts on distal phalanx of great toe	Extends great toe; dorsiflexes foot

(Continues on page 224.)

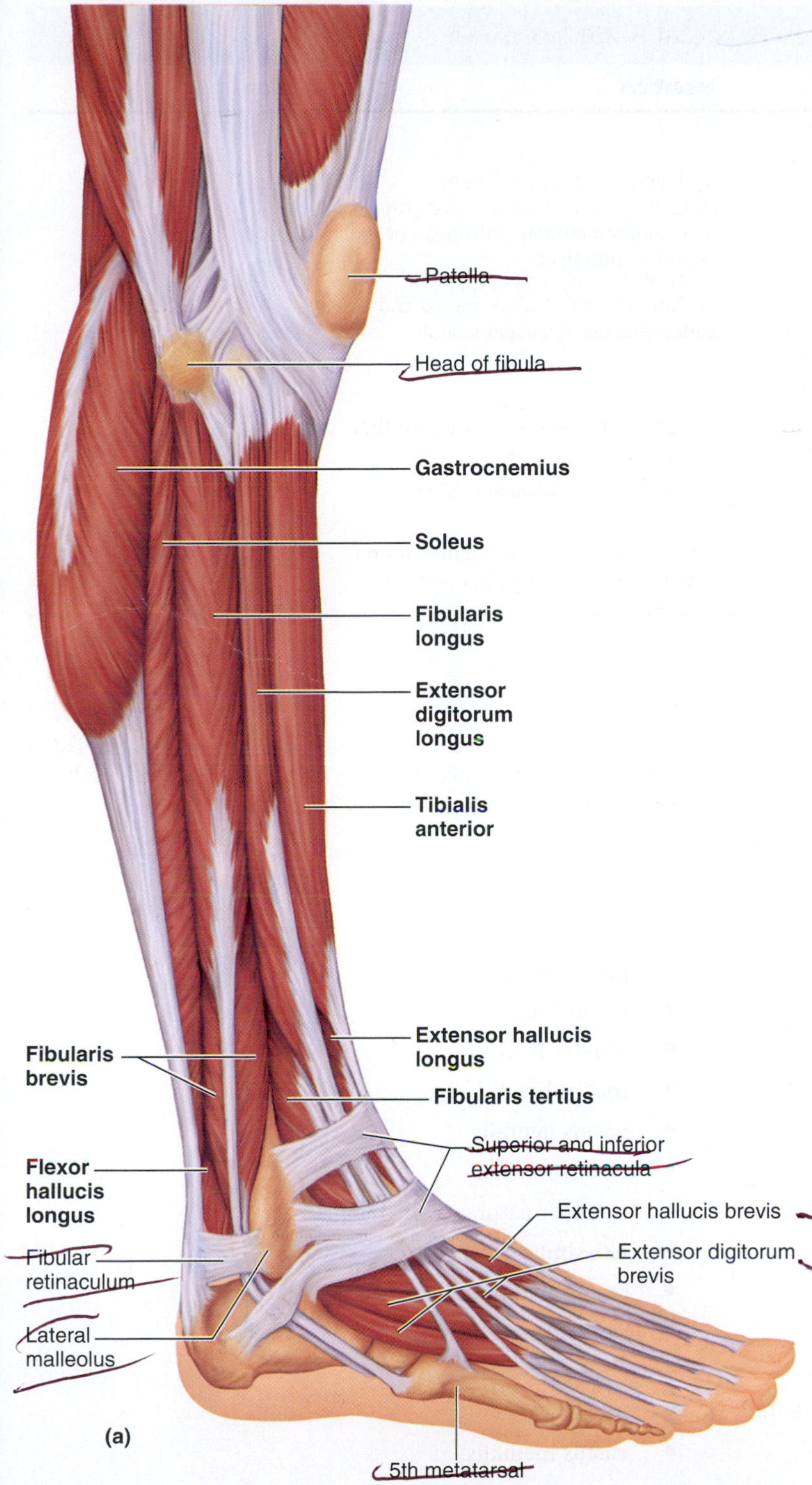
Patella
Head of fibula
Gastrocnemius
Soleus
Fibularis longus
Extensor digitorum longus
Tibialis anterior
Fibularis brevis
Extensor hallucis longus
Fibularis tertius
Superior and inferior extensor retinacula
Flexor hallucis longus
Extensor hallucis brevis
Fibular retinaculum
Extensor digitorum brevis
Lateral malleolus
5th metatarsal
(a)

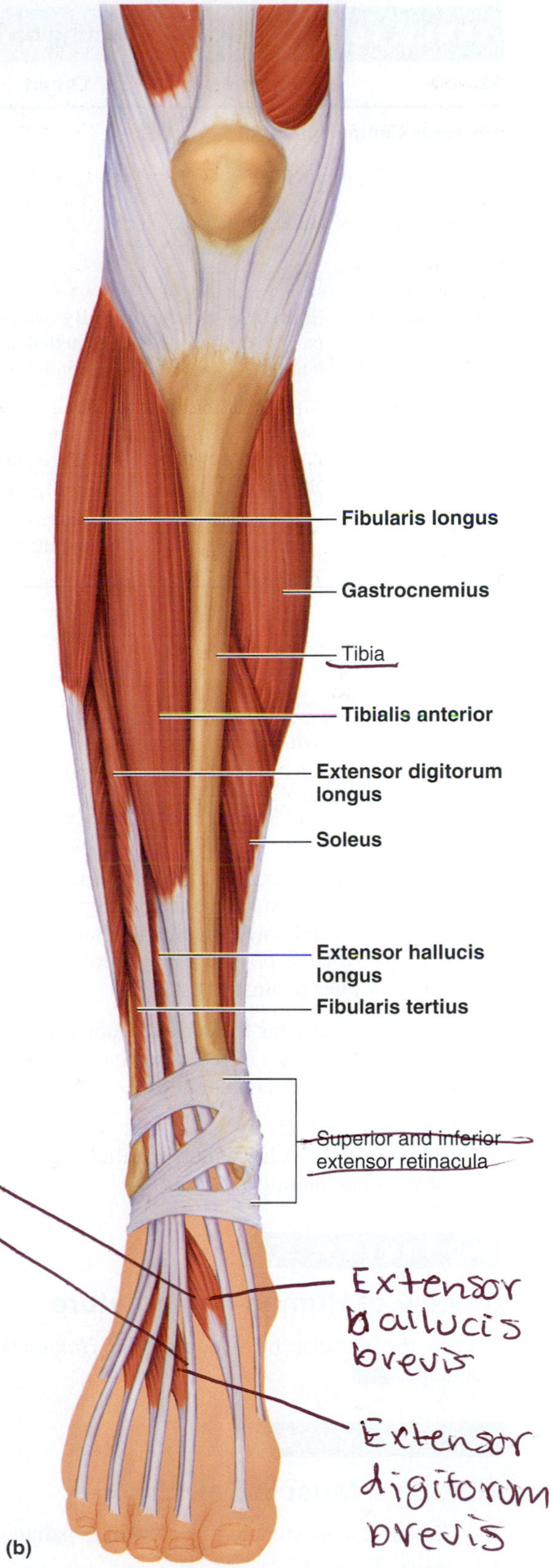
Fibularis longus
Gastrocnemius
Tibia
Tibialis anterior
Extensor digitorum longus
Soleus
Extensor hallucis longus
Fibularis tertius
Superior and inferior extensor retinacula
Extensor hallucis brevis
Extensor digitorum brevis
(b)

FIGURE 15.14 Muscles of the anterolateral aspect of the right leg. **(a)** Superficial view of lateral aspect of the leg, illustrating the positioning of the lateral compartment muscles (fibularis longus and brevis) relative to anterior and posterior leg muscles. **(b)** Superficial view of anterior leg muscles.

TABLE 15.9 Muscles Acting on Human Foot and Ankle *(continued)*

Muscle	Comments	Origin	Insertion	Action
Posterior Compartment				
Superficial (Figure 15.15a)				
Triceps surae	Refers to muscle pair below that shapes posterior calf		Via common tendon (calcaneal, or Achilles) into heel	Plantar flex foot
Gastrocnemius	Superficial muscle of pair; two prominent bellies	By two heads from medial and lateral condyles of femur	Calcaneus via calcaneal tendon	Plantar flexes foot when knee is extended; crosses knee joint; thus can flex knee (when foot is dorsiflexed)
Soleus	Deep to gastrocnemius	Proximal portion of tiba and fibula; interosseous membrane	Calcaneus via calcaneal tendon	Plantar flexion; is an important muscle for locomotion

(Continues on page 226.)

Demonstrating Operations of Lower Limb Muscles

1. Go into a deep knee bend and palpate your own *gluteus maximus* muscle as you extend your hip to resume the upright posture.

2. Demonstrate the contraction of the anterior *quadriceps femoris* by trying to extend your knee against resistance. Do this while seated and note how the patellar tendon reacts. The *biceps femoris* of the posterior thigh comes into play when you flex your knee against resistance.

3. Now stand on your toes. Have your partner palpate the lateral and medial heads of the *gastrocnemius* and follow it to its insertion in the calcaneal tendon.

4. Dorsiflex and invert your foot while palpating your *tibialis anterior* muscle (which parallels the sharp anterior crest of the tibia laterally). ■

ACTIVITY 5

Review of Human Musculature

Review the muscles by watching the *Human Musculature* videotape. ■

ACTIVITY 6

Making a Muscle Painting

1. Choose a male student to be "muscle painted."

2. Obtain brushes and water-based paints from the supply area while the "volunteer" removes his shirt and rolls up his pant legs (if necessary).

3. Using different colored paints, identify the muscles listed below by painting his skin. If a muscle covers a large body area, you may opt to paint only its borders.

- biceps brachii
- deltoid
- erector spinae
- pectoralis major
- rectus femoris
- tibialis anterior
- triceps brachii
- vastus lateralis
- biceps femoris
- extensor carpi radialis longus
- latissimus dorsi
- rectus abdominis
- sternocleidomastoid
- trapezius
- triceps surae
- vastus medialis

4. Check your "human painting" with your instructor before cleaning your bench and leaving the laboratory. ■

For instructions on animal dissections, see the dissection exercises starting on page 697 in the cat, fetal pig, and rat editions of this manual.

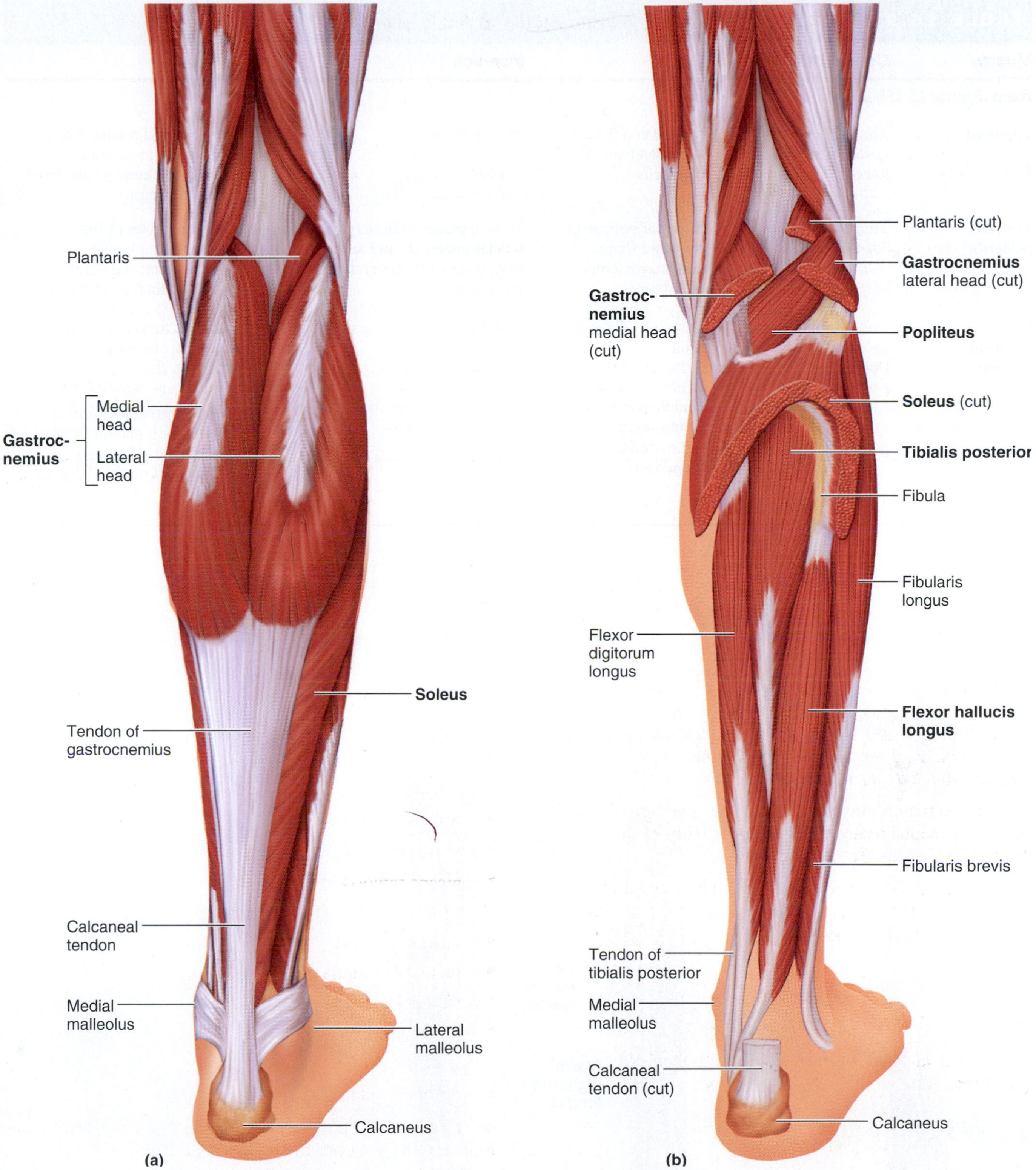
Plantaris
Medial head
Gastrocnemius
Lateral head
Soleus
Tendon of gastrocnemius
Calcaneal tendon
Medial malleolus
Lateral malleolus
Calcaneus
(a)
Plantaris (cut)
Gastrocnemius lateral head (cut)
Gastrocnemius medial head (cut)
Popliteus
Soleus (cut)
Tibialis posterior
Fibula
Fibularis longus
Flexor digitorum longus
Flexor hallucis longus
Fibularis brevis
Tendon of tibialis posterior
Medial malleolus
Calcaneal tendon (cut)
Calcaneus
(b)

FIGURE 15.15 Muscles of the posterior aspect of the right leg. (a) Superficial view of the posterior leg. **(b)** The triceps surae has been removed to show the deep muscles of the posterior compartment.

TABLE 15.9 Muscles Acting on Human Foot and Ankle *(continued)*

Muscle	Comments	Origin	Insertion	Action
Deep (Figure 15.15b–e)				
Popliteus	Thin muscle at posterior aspect of knee	Lateral condyle of femur and lateral meniscus	Proximal tibia	Flexes and rotates leg medially to "unlock" extended knee when knee flexion begins
Tibialis posterior	Thick muscle deep to soleus	Superior portion of tibia and fibula and interosseous membrane	Tendon passes obliquely behind medial malleolus and under arch of foot; inserts into several tarsals and metatarsals 2–4	Prime mover of foot inversion; plantar flexes foot; stabilizes longitudinal arch of foot
Flexor digitorum longus	Runs medial to and partially overlies tibialis posterior	Posterior surface of tibia	Distal phalanges of toes 2–5	Flexes toes; plantar flexes and inverts foot
Flexor hallucis longus (see also Figure 15.14a)	Lies lateral to inferior aspect of tibialis posterior	Middle portion of fibula shaft; interosseous membrane	Tendon runs under foot to distal phalanx of great toe	Flexes great toe (*hallux* = great toe); plantar flexes and inverts foot; the "push-off muscle" during walking

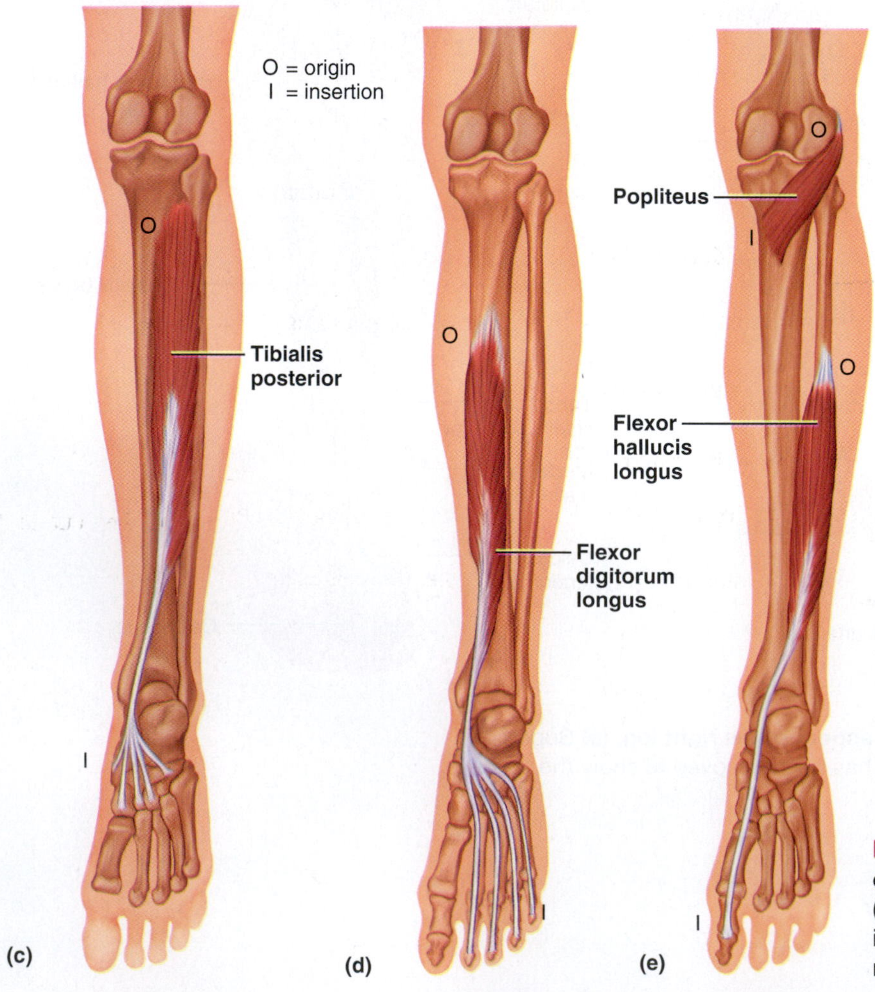
O = origin
I = insertion
O
Tibialis posterior
I
(c)
O
Flexor digitorum longus
I
(d)
O
Popliteus
I
O
Flexor hallucis longus
I
(e)

FIGURE 15.15 *(continued)* Muscles of the posterior aspect of the right leg. (c–e) Individual muscles are shown in isolation so that their origins and insertions may be visualized.

EXERCISE 16

Skeletal Muscle Physiology: Frogs and Human Subjects

MATERIALS

- ☐ ATP muscle kits (glycerinated rabbit psoas muscle;* ATP and salt solutions obtainable from Carolina Biological Supply)
- ☐ Petri dishes
- ☐ Microscope slides
- ☐ Coverslips
- ☐ Millimeter ruler
- ☐ Compound microscope
- ☐ Stereomicroscope
- ☐ Pointed glass probes (teasing needles)
- ☐ Small beakers (50 ml)
- ☐ Distilled water
- ☐ Glass-marking pencil
- ☐ Textbooks or other heavy books
- ☐ Watch or timer
- ☐ Ringer's solution (frog)
- ☐ Scissors
- ☐ Metal needle probes
- ☐ Medicine dropper
- ☐ Cotton thread
- ☐ Forceps
- ☐ Disposable gloves
- ☐ Glass or porcelain plate
- ☐ Pithed bullfrog†

Notes to the Instructor:

*At the beginning of the lab, the muscle bundle should be removed from the test tube and cut into approximately 2-cm lengths. Both the cut muscle segments and the entubed glycerol should be put into a petri dish. One muscle *segment* is sufficient for every two to four students making observations.

†Bullfrogs to be pithed by lab instructor as needed for student experimentation. (If instructor prefers that students pith their own specimens, an instructional sheet on that procedure suitable for copying for student handouts is provided in the Instructor's Guide.)

Text continues on next page.

MasteringA&P™

Access practice quizzes and more in the Study Area at www.masteringaandp.com.

PEx

This lab corresponds to PhysioEx Exercise 2. See p. PEx-17.

OBJECTIVES

1. To observe muscle contraction on the microscopic level and describe the role of ATP and various ions in muscle contraction.
2. To define and explain the physiological basis of the following:

   depolarization
   repolarization
   action potential
   absolute and relative refractory periods
   subthreshold stimulus
   threshold stimulus
   maximal stimulus
   wave summation
   tetanus
   muscle fatigue

3. To trace the events that result from the electrical stimulation of a muscle.
4. To recognize that a graded response of skeletal muscle is a function of the number of muscle fibers stimulated and the frequency of the stimulus.
5. To name and describe the phases of a muscle twitch.
6. To distinguish between a muscle twitch and a sustained (tetanic) contraction and to describe their importance in normal muscle activity.
7. To demonstrate how a computer or physiograph can be used to obtain pertinent and representative recordings of various physiological events of skeletal muscle activity.
8. To explain the significance of muscle tracings obtained during experimentation.

PRE-LAB QUIZ

1. Circle the correct term. The potential difference, or voltage, across the plasma membrane is the result of the difference in membrane permeability to anions / cations, most importantly Na^+ and K^+.
2. Depolarization of skeletal muscle is caused by a(n)
   a. influx of Ca^+
   b. influx of K^+
   c. influx of Na^+
   d. release of Na^+
3. The _________ wave follows the depolarization wave across the sarcolemma.
   a. hyperpolarization
   b. refraction
   c. repolarization
4. Circle the correct term. Contraction of a skeletal muscle that results from a single stimulus is called twitch / tetanus.
5. Circle True or False. The voltage at which the first noticeable contractile response is obtained is called the threshold stimulus.
6. Circle True or False. A single muscle is made up of many motor units, and the gradual activation of these motor units results in a graded contraction of the whole muscle.
7. A sustained, smooth, muscle contraction that is a result of high-frequency stimulation is
   a. tetanus b. tonus c. twitch

Text continues on next page.

- ☐ Apparatus A or B:‡
  A: physiograph, physiograph paper and ink, force transducer, pin electrodes, stimulator, stimulator output extension cable, transducer stand and cable, straight pins, frog board, laboratory stand, clamp
  BIOPAC® B: BIOPAC® MP36 (or MP35/30) data acquisition unit, PC or Mac computer, BIOPAC Student Lab electrode lead set, hand dynamometer, headphones, metric tape measure, disposable vinyl electrodes, and conduction gel

New versions of the BSL software 3.7.5 and higher for Windows, and 3.7.4 and higher for Mac) require slightly different channel settings and collection strategies. Instructions for using the newer software with the MP36/35 data acquisition units can be found on MasteringA&P. In addition, instructions for use of the new 2-channel data acquisition unit, the MP45, may also be found on MasteringA&P.

‡Additionally, instructions for Activity 3 using a kymograph can be found in the Instructor Guide. Instructions for using PowerLab® equipment can be found on MasteringA&P.

8. Muscle fatigue, the loss of the ability to contract, may be a result of
   a. oxygen debt in the tissue after prolonged activity
   b. potassium buildup in the tissue after prolonged activity
   c. sodium debt in the tissue after prolonged activity
   d. the accumulation of ATP in the tissue after prolonged activity

The contraction of skeletal and cardiac muscle fibers can be considered in terms of three events—electrical excitation of the muscle cell, excitation-contraction coupling, and shortening of the muscle cell due to sliding of the myofilaments within it.

At rest, all cells maintain a potential difference, or voltage, across their plasma membrane; the inner face of the membrane is approximately –60 to –90 millivolts (mV) compared with the cell exterior. This potential difference is a result of differences in membrane permeability to cations, most importantly sodium (Na^+) and potassium (K^+) ions. Intracellular potassium concentration is much greater than its extracellular concentration, and intracellular sodium concentration is considerably less than its extracellular concentration. Hence, steep concentration gradients across the membrane exist for both cations. Because the plasma membrane is more permeable to K^+ than to Na^+, the cell's **resting membrane potential** is more negative inside than outside. The resting membrane potential is of particular interest in excitable cells, like muscle cells and neurons, because changes in that voltage underlie their ability to do work (to contract and/or issue electrical signals).

Action Potential

When a muscle cell is stimulated, the sarcolemma becomes temporarily more permeable to Na^+, which enters the cell. This sudden influx of Na^+ alters the membrane potential. That is, the cell interior becomes less negatively charged at that point, an event called **depolarization.** When depolarization reaches a certain level and the sarcolemma momentarily changes its polarity, a depolarization wave travels along the sarcolemma. Even as the influx of Na^+ occurs, the sarcolemma becomes less permeable to Na^+ and more permeable to K^+. Consequently, K^+ ions move out of the cell, restoring the resting membrane potential (but not the original ionic conditions), an event called **repolarization.** The repolarization wave follows the depolarization wave across the sarcolemma. This rapid depolarization and repolarization of the membrane that is propagated along the entire membrane from the point of stimulation is called the **action potential.**

The **absolute refractory period** is the period of time when Na^+ permeability of the sarcolemma is rapidly changing and maximal, and the following period when Na^+ permeability becomes restricted. During this period there is no possibility of generating another action potential. As Na^+ permeability is gradually restored to resting levels during repolarization, an especially strong stimulus to the muscle cell may provoke another action potential. This period of time is the **relative refractory period.** Repolarization restores the muscle cell's normal excitability. If the muscle cell is stimulated to contract rapidly again and again, the changes in Na^+ and K^+ concentrations near the membrane begin to reduce its ability to respond. Therefore, the sodium-potassium pump, which actively transports K^+ into the cell and Na^+ out of the cell, must be "revved up" (become more active) to reestablish the ionic concentrations of the resting state.

Contraction

Propagation of the action potential along the sarcolemma causes the release of calcium ions (Ca^{2+}) from storage in the sarcoplasmic reticulum within the muscle cell. When the calcium ions bind to regulatory proteins on the actin myofilaments, they act as an ionic trigger that initiates contraction, and the actin and myosin filaments slide past each other. Once the action potential ends, the calcium ions are almost immediately transported back into the sarcoplasmic reticulum. Instantly the muscle cell relaxes.

The events of the contraction process can most simply be summarized as follows: Muscle cell contraction is initiated by generation and transmission of an action potential along the sarcolemma. This electrical event is coupled to the sliding of the myofilaments—contraction—by the release of calcium ions (Ca^{2+}). Keep in mind this sequence of events as you conduct the experiments.

ACTIVITY 1

Observing Muscle Fiber Contraction

In this simple observational experiment, you will have the opportunity to review your understanding of muscle cell anatomy and to watch fibers respond to the presence of ATP and/or a solution of K^+ and magnesium ions (Mg^{2+}).

This experiment uses preparations of glycerinated muscle. The glycerination process denatures troponin and tropomyosin. Consequently, calcium, so critical for contraction in vivo, is not necessary here. The role of magnesium and potassium salts as cofactors in the contraction process is not well understood, but magnesium and potassium salts seem to be required for ATPase activity in this system.

1. Talk with other members of your lab group to develop a hypothesis about requirements for muscle fiber contraction for this experiment. The hypothesis should have three parts: (1) salts only, (2) ATP only, and (3) salts and ATP.

2. Obtain the following materials from the supply area: two glass teasing needles; six glass microscope slides and six coverslips; millimeter ruler; dropper bottles containing the following solutions: (a) 0.25% ATP in triply distilled water, (b) 0.25% ATP plus 0.05 *M* KCl plus 0.001 *M* $MgCl_2$ in distilled water, and (c) 0.05 *M* KCl plus 0.001 *M* $MgCl_2$ in distilled water; a petri dish; a beaker of distilled water; a glass-marking pencil; and a small portion of a previously cut muscle bundle segment. While you are at the supply area, place the muscle fibers in the petri dish and pour a small amount of glycerol (the fluid in the supply petri dish) over your muscle cells. Also obtain both a compound and a stereo microscope and bring them to your laboratory bench.

3. Using clean fine glass needles, tease the muscle segment to separate its fibers. The objective is to isolate *single* muscle cells or fibers for observation. Be patient and work carefully so that the fibers do not get torn during this isolation procedure.

4. Transfer one or more of the fibers (or the thinnest strands you have obtained) onto a clean microscope slide with a glass needle, and cover it with a coverslip. Examine the fiber under the compound microscope at low- and then high-power magnifications to observe the striations and the smoothness of the fibers when they are in the relaxed state.

5. Clean three microscope slides well and rinse in distilled water. Label the slides A, B, and C.

6. Transfer three or four fibers to microscope slide A with a glass needle. Using the needle as a prod, carefully position the fibers so that they are parallel to one another and as straight as possible. Place this slide under a *stereomicroscope* and measure the length of each fiber by holding a millimeter ruler adjacent to it. Alternatively, you can rest the microscope slide *on* the millimeter ruler to make your length determinations. Record the data on the chart below.

7. Flood the fibers (situated under the stereomicroscope) with several drops of the solution containing ATP, K^+, and Mg^{2+}. Watch the reaction of the fibers after adding the solution. After 30 seconds (or slightly longer), remeasure each fiber and record the observed lengths on the chart. Also, observe the fibers to see if any width changes have occurred. Calculate the degree (or percentage) of contraction by using the simple formula on the next page, and record this data on the chart below.

Salts and ATP, slide A	Muscle fiber 1	Muscle fiber 2	Muscle fiber 3	Average
Initial length (mm)				
Contracted length (mm)				
% Contraction				
ATP only, slide B				
Initial length (mm)				
Contracted length (mm)				
% Contraction				
Salts only, slide C				
Initial length (mm)				
Contracted length (mm)				
% Contraction				

$$\frac{\text{Initial}}{\text{length (mm)}} - \frac{\text{contracted}}{\text{length (mm)}} = \frac{\text{degree of}}{\text{contraction (mm)}}$$

then:

$$\frac{\text{Degree of contraction(mm)}}{\text{initial length(mm)}} \times 100 = \_\_\_\_\_\%\ \text{contraction}$$

8. Carefully transfer one of the contracted fibers to a clean, unmarked microscope slide, cover with a coverslip, and observe with the compound microscope. Mentally compare your initial observations with the view you are observing now. What differences do you see? (Be specific.)

___

___

___

What zones (or bands) have disappeared?

___

9. Repeat steps 6 through 8 twice more, using clean slides and fresh muscle cells. On slide B use the solution of ATP in distilled water (no salts). Then, on slide C use the solution containing only salts (no ATP) for the third series. Record data on the chart on page 237.

10. Collect the data from all the groups in your laboratory and use these data to prepare a lab report. (See Getting Started, page xv.) Include in your discussion the following questions:

What degree of contraction was observed when ATP was applied in the absence of K^+ and Mg^{2+}?

___

What degree of contraction was observed when the muscle fibers were flooded with a solution containing K^+ Mg^{2+}, and lacking ATP?

___

What conclusions can you draw about the importance of ATP, K^+, and Mg^{2+} to the contractile process?

___

___

Can you draw exactly the same conclusions from the data provided by each group? List some variables that might have been introduced into the procedure and that might account for any differences.

___

___

ACTIVITY 2

Demonstrating Muscle Fatigue in Humans

1. Work in small groups. In each group select a subject, a timer, and a recorder.

2. Obtain a copy of the laboratory manual and a copy of the textbook. Weigh each book separately, and then record the weight of each in the chart in step 6.

3. The subject is to extend an upper limb straight out in front of him or her, holding the position until the arm shakes or the muscles begin to ache. Record the time to fatigue on the chart.

4. Allow the subject to rest for several minutes. Now ask the subject to hold the laboratory manual while keeping the arm and forearm in the same position as in step 3 above. Record the time to fatigue on the chart.

5. Allow the subject to rest again for several minutes. Now ask the subject to hold the textbook while keeping the upper limb in the same position as in steps 3 and 4 above. Record the time to fatigue on the chart.

6. Each person in the group should take a turn as the subject, and all data should be recorded in the chart below.

Load	Weight of object	Time elapsed Until fatigue		
		Subject 1	Subject 2	Subject 3
Appendage	N/A			
Lab Manual				
Textbook				

7. What can you conclude about the effect of load on muscle fatigue? Explain.

___

___

___

ACTIVITY 3

Inducing Contraction in the Frog Gastrocnemius Muscle

Physiologists have learned a great deal about the way muscles function by isolating muscles from laboratory animals and then stimulating these muscles to observe their responses. Various stimuli—electrical shock, temperature changes, extremes of pH, certain chemicals—elicit muscle activity, but laboratory experiments of this type typically use electrical shock. This is because it is easier to control the onset and cessation of electrical shock, as well as the strength of the stimulus.

Various types of apparatus are used to record muscle contraction. All include a way to mark time intervals, a way to indicate exactly when the stimulus was applied, and a way to

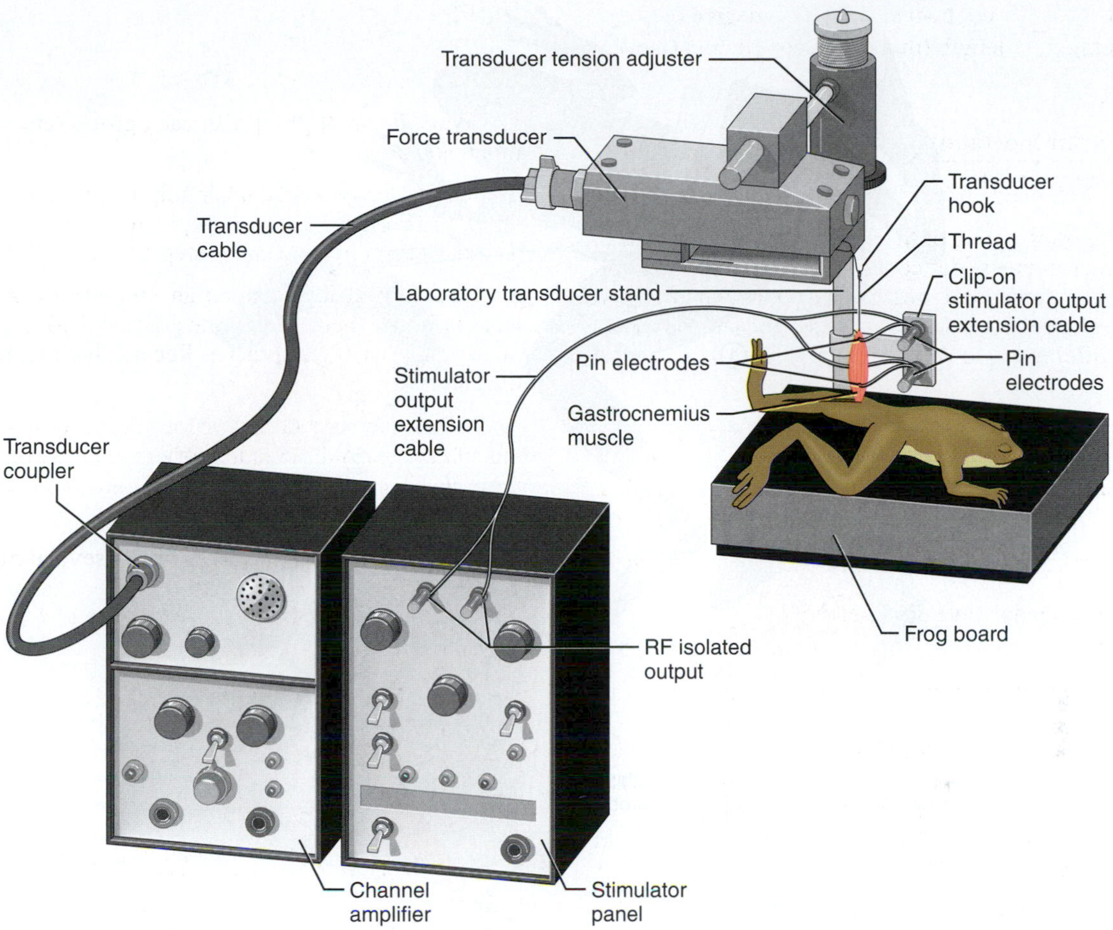
Transducer tension adjuster
Force transducer
Transducer cable
Transducer hook
Thread
Laboratory transducer stand
Clip-on stimulator output extension cable
Pin electrodes
Pin electrodes
Stimulator output extension cable
Gastrocnemius muscle
Transducer coupler
RF isolated output
Frog board
Channel amplifier
Stimulator panel

FIGURE 16.1 Physiograph setup for frog gastrocnemius experiments.

measure the magnitude of the contractile response. Instructions are provided here for setting up a physiograph apparatus (Figure 16.1). Specific instructions for use of recording apparatus during recording will be provided by your instructor.

Preparing a Muscle for Experimentation

The preparatory work that precedes the recording of muscle activity tends to be quite time-consuming. If you work in teams of two or three, the work can be divided. While one of you is setting up the recording apparatus, one or two students can dissect the frog leg (Figure 16.2). Experimentation should begin as soon as the dissection is completed.

Materials

Channel amplifier and transducer cable

Stimulator panel and stimulator output extension cable

Force transducer

Transducer tension adjuster

Transducer stand

Two pin electrodes

Frog board and straight pins

Prepared frog (gastrocnemius muscle freed and calcaneal tendon ligated)

Frog Ringer's solution

Procedure

1. Connect transducer to tranducer stand and attach frog board to stand.
2. Attach tranducer cable to transducer and to input connection on channel amplifier.
3. Attach stimulator output extension cable to output on stimulator panel (red to red, black to black).
4. Using clip at opposite end of extension cable, attach cable to bottom of transducer stand adjacent to frog board.
5. Attach two pin electrodes securely to electrodes on clip.
6. Place knee of prepared frog in clip-on frog board and secure by inserting a straight pin through tissues of frog. Keep frog muscle moistened with Ringer's solution.
7. Attach thread from the calcaneal tendon of frog to transducer spring hook.
8. Adjust position of tranducer on stand to produce a constant tension on thread attached to muscle (taut but not tight). Gastrocnemius muscle should hang vertically directly below hook.
9. Insert free ends of pin electrodes into the muscle, one at proximal end and the other at distal end.

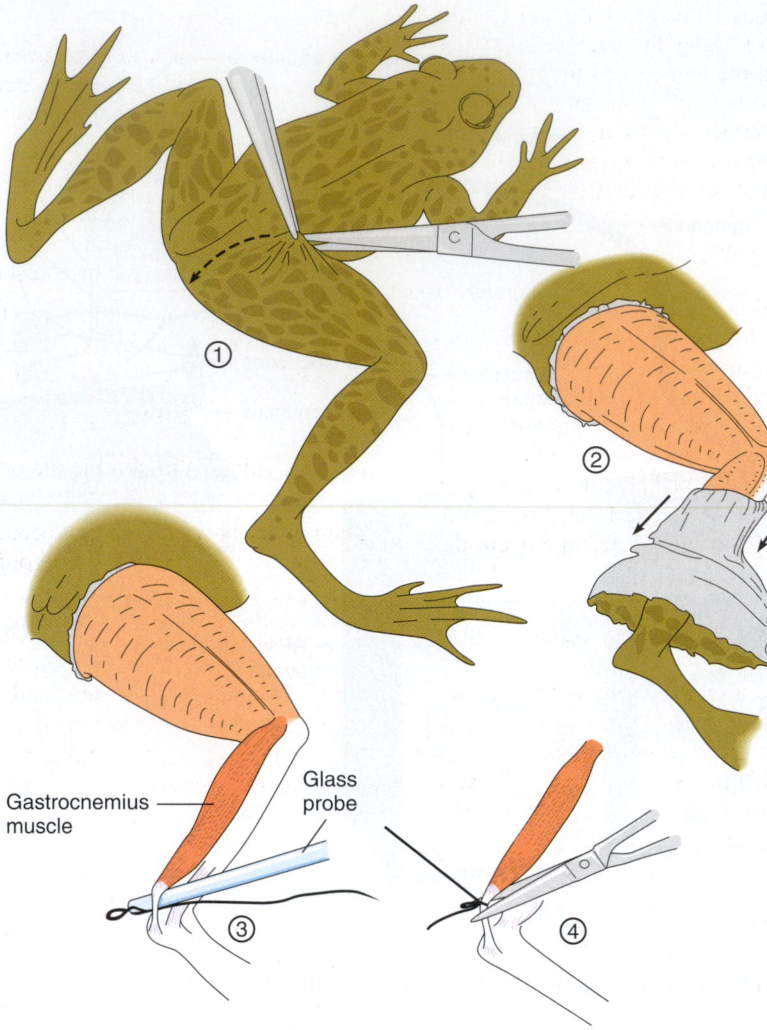

①
②
Gastrocnemius muscle
Glass probe
③
④

FIGURE 16.2 Preparation of the frog gastrocnemius muscle. Numbers indicate the sequence of manipulation.

DISSECTION: Frog Hind Limb

1. Before beginning the frog dissection, have the following supplies ready at your laboratory bench: a small beaker containing 20 to 30 ml of frog Ringer's solution, scissors, a metal needle probe, a glass probe with a pointed tip, a medicine dropper, cotton thread, forceps, a glass or porcelain plate, and disposable gloves. While these supplies are being accumulated, one member of your team should notify the instructor that you are ready to begin experimentation, so that a frog can be prepared (pithed). Preparation of a frog in this manner renders it unable to feel pain and prevents reflex movements (like hopping) that would interfere with the experiments.

2. All students who will be handling the frog should don disposable gloves. Obtain a pithed frog and place it ventral surface down on the glass plate. Make an incision into the skin approximately midthigh (Figure 16.2), and then continue the cut completely around the thigh. Grasp the skin with the forceps and strip it from the leg and hindfoot. The skin adheres more at the joints, but a careful, persistent pulling motion—somewhat like pulling off a nylon stocking—will enable you to remove it in one piece. *From this point on, the exposed muscle tissue should be kept moistened with the Ringer's solution* to prevent spontaneous twitches.

3. Identify the gastrocnemius muscle (the fleshy muscle of the posterior calf) and the calcaneal (Achilles) tendon that secures it to the heel.

4. Slip a glass probe under the gastrocnemius muscle and run it along the entire length and under the calcaneal tendon to free them from the underlying tissues.

5. Cut a piece of thread about 10 inches long and use the glass probe to slide the thread under the calcaneal tendon. Knot the thread firmly around the tendon and then sever the tendon distal to the thread. Alternatively, you can bend a common pin into a Z-shape and insert the pin securely into the tendon. The thread is then attached to the opposite end of the pin. Once the tendon has been tied or pinned, the frog is ready for experimentation (see Figure 16.2).

Recording Muscle Activity

Skeletal muscles consist of thousands of muscle cells and react to stimuli with graded responses. Thus muscle contractions can be weak or vigorous, depending on the requirements of the task. Graded responses (different degrees of shortening) of a skeletal muscle depend on the number of muscle cells being stimulated. In the intact organism, the number of motor units firing at any one time determines how many muscle cells will be stimulated. In this laboratory, the frequency and strength of an electrical current determines the response.

A single contraction of skeletal muscle is called a **muscle twitch.** A tracing of a muscle twitch (Figure 16.3) shows three distinct phases: latent, contraction, and relaxation. The **latent phase** is the interval from stimulus application until the muscle begins to shorten. Although no activity is indicated on the tracing during this phase, important electrical and chemical changes are occurring within the muscle. During the **contraction phase,** the muscle fibers shorten; the tracing shows an increasingly higher needle deflection and the tracing peaks. During the **relaxation phase,** represented by a downward curve of the tracing, the muscle fibers relax and lengthen. On a slowly moving recording surface, the single muscle twitch appears as a spike (rather than a bell-shaped curve, as in Figure 16.3), but on a rapidly moving recording surface, the three distinct phases just described become recognizable.

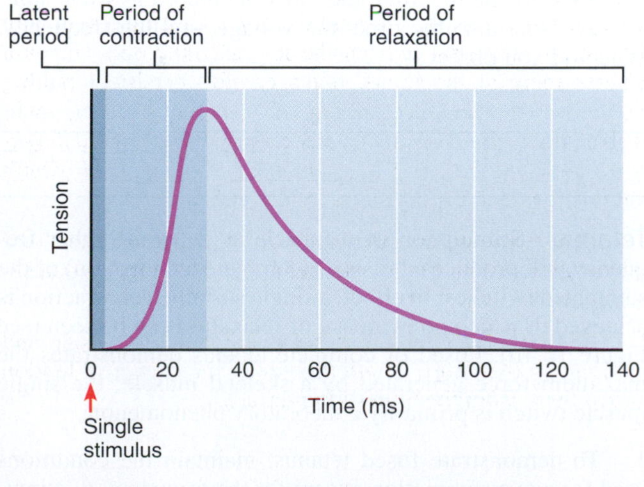
Latent period
Period of contraction
Period of relaxation
Tension
0
20
40
60
80
100
120
140
Time (ms)
Single stimulus

FIGURE 16.3 Tracing of a muscle twitch.

Determining the Threshold Stimulus

1. Assuming that you have already set up the recording apparatus, set the time marker to deliver one pulse per second and set the paper speed at a slow rate, approximately 0.1 cm per second.

2. Set the duration control on the stimulator between 7 and 15 milliseconds (msec), multiplier ×1. Set the voltage control at 0 V, multiplier ×1. Turn the sensitivity control knob of the stimulator fully clockwise (lowest value, greatest sensitivity).

3. Administer single stimuli to the muscle at 1-minute intervals, beginning with 0.1 V and increasing each successive stimulus by 0.1 V until a contraction is obtained (shown by a spike on the paper).

At what voltage did contraction occur? ______________ V

The voltage at which the first perceptible contractile response is obtained is called the **threshold stimulus.** All stimuli applied prior to this point are termed **subthreshold stimuli,** because at those voltages no response was elicited.

4. Stop the recording and mark the record to indicate the threshold stimulus, voltage, and time. _Do not remove the record from the recording surface;_ continue with the next experiment. _Remember: keep the muscle preparation moistened with Ringer's solution at all times._

Observing Graded Muscle Response to Increased Stimulus Intensity

1. Follow the previous setup instructions, but set the voltage control at the threshold voltage (as determined in the first experiment).

2. Deliver single stimuli at 1-minute intervals. Initially increase the voltage between shocks by 0.5 V; then increase the voltage by 1 to 2 V between shocks as the experiment continues, until contraction height increases no further. Stop the recording apparatus.

What voltage produced the highest spike (and thus the maximal strength of contraction)? ____________________ V

This voltage, called the **maximal stimulus** (for *your* muscle specimen), is the weakest stimulus at which all muscle cells are being stimulated.

3. Mark the record *maximal stimulus.* Record the maximal stimulus voltage and the time you completed the experiment.

4. What is happening to the muscle as the voltage is increased?

____

____

What is another name for this phenomenon?

____

5. Explain why the strength of contraction does not increase once the maximal stimulus is reached.

___

___

Timing the Muscle Twitch

1. Follow the previous setup directions, but set the voltage for the maximal stimulus (as determined in the preceding experiment) and set the paper advance or recording speed at maximum. Record the paper speed setting:

___ mm/sec

2. Determine the time required for the paper to advance 1 mm by using the formula

$$\frac{1 \text{ mm}}{\text{mm/sec (paper speed)}}$$

(Thus, if your paper speed is 25 mm/sec, each mm on the chart equals 0.04 sec.) Record the computed value:

1 mm = ___ sec

3. Deliver single stimuli at 1-minute intervals to obtain several "twitch" curves. Stop the recording.

4. Determine the duration of the latent, contraction, and relaxation phases of the twitches and record the data below.

Duration of latent period: ___ sec

Duration of contraction period: ___ sec

Duration of relaxation period: ___ sec

5. Label the record to indicate the point of stimulus, the beginning of contraction, the end of contraction, and the end of relaxation.

6. Allow the muscle to rest (but keep it moistened with Ringer's solution) before continuing with the next experiment.

Observing Graded Muscle Response to Increased Stimulus Frequency

Muscles subjected to frequent stimulation, without a chance to relax, exhibit two kinds of responses—wave summation and tetanus—depending on the level of stimulus frequency (Figure 16.4).

Wave Summation If a muscle is stimulated with a rapid series of stimuli of the same intensity before it has had a chance to relax completely, the response to the second and subsequent stimuli will be greater than to the first stimulus (see Figure 16.4a). This phenomenon, called **wave,** or **temporal, summation,** occurs because the muscle is already in a partially contracted state when subsequent stimuli are delivered.

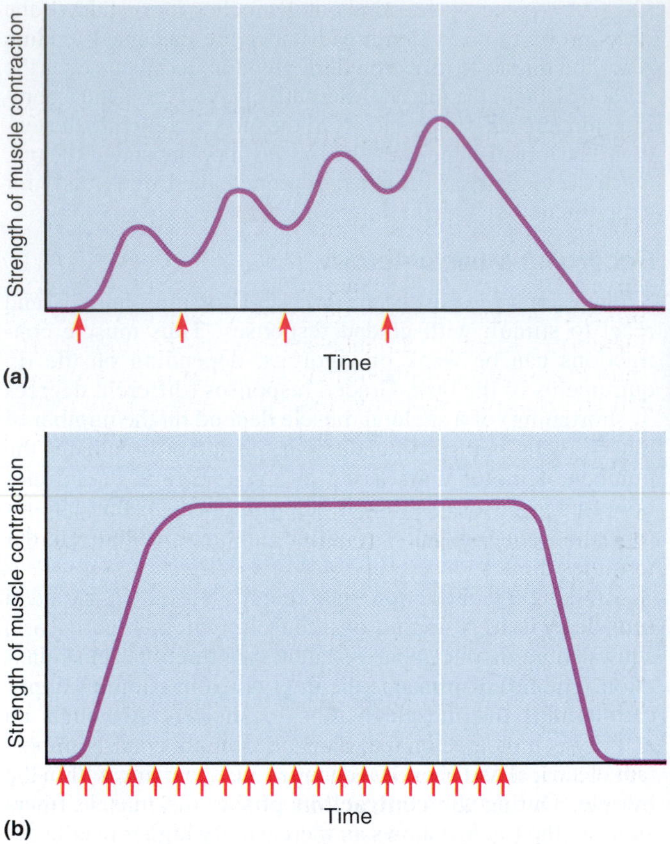
Strength of muscle contraction
Time
(a)
Strength of muscle contraction
Time
(b)

FIGURE 16.4 Muscle response to stimulation. Arrows represent stimuli **(a)** Wave summation at low frequency stimulation. **(b)** Fused tetanus occurs as stimulation rate is increased.

1. Set up the apparatus as in the previous experiment, setting the voltage to the maximal stimulus as determined earlier and the chart speed to maximum.

2. With the stimulator in single mode, deliver two successive stimuli as rapidly as possible.

3. Shut off the recorder and label the record as *wave summation.* Note also the time, the voltage, and the frequency. What did you observe?

___

___

Tetanus Stimulation of a muscle at an even higher frequency will produce a "fusion" (complete tetanization) of the summated twitches. In effect, a single sustained contraction is achieved in which no evidence of relaxation can be seen (see Figure 16.4b). Fused or complete tetanus demonstrates the maximum force generated by a skeletal muscle; the single muscle twitch is primarily a laboratory phenomenon.

1. To demonstrate fused tetanus, maintain the conditions used for wave summation except for the frequency of stimulation. Set the stimulator to deliver 60 stimuli per second.

2. As soon as you obtain a single smooth, sustained contraction (with no evidence of relaxation), discontinue stimulation and shut off the recorder.

3. Label the tracing with the conditions of experimentation, the time, and the area of *fused* or *complete tetanus*.

Inducing Muscle Fatigue

Muscle fatigue, the loss of the ability to contract, may be a result of the oxygen debt that occurs in the tissue after prolonged activity (through the accumulation of such waste products as lactic acid, as well as the depletion of ATP), depletion of nutrients, ion imbalances, or tissue damage. True muscle fatigue rarely occurs in the body because it is most often preceded by a subjective feeling of fatigue. Furthermore, fatigue of the neuromuscular junctions typically precedes fatigue of the muscle.

1. To demonstrate muscle fatigue, set up an experiment like the tetanus experiment but continue stimulation until the muscle completely relaxes and the contraction curve returns to the base line.

2. Measure the time interval between the beginning of complete tetanus and the beginning of fatigue (when the tracing begins its downward curve). Mark the record appropriately.

3. Determine the time required for complete fatigue to occur (the time interval from the beginning of fatigue until the return of the curve to the base line). Mark the record appropriately.

4. Allow the muscle to rest (keeping it moistened with Ringer's solution) for 10 minutes, and then repeat the experiment.

What was the effect of the rest period on the fatigued muscle?

____

____

What is the physiological basis for this reaction?

____

____

Determining the Effect of Load on Skeletal Muscle

When the fibers of a skeletal muscle are slightly stretched by a weight or tension, the muscle responds by contracting more forcibly and thus is capable of doing more work. When the actin and myosin barely overlap, sliding can occur along nearly the entire length of the actin filaments. If the load is increased beyond the optimum, the latent period becomes longer, contractile force decreases, and relaxation (fatigue) occurs more quickly. With excessive stretching, the muscle is unable to develop any active tension and no contraction occurs. Since the filaments no longer overlap at all with this degree of stretching, the sliding force cannot be generated.

If your equipment allows you to add more weights to the muscle specimen or to increase the tension on the muscle, perform the following experiment to determine the effect of loading on skeletal muscle and to develop a work curve for the frog's gastrocnemius muscle.

1. Set the stimulator to deliver the maximal voltage as previously determined.

2. Stimulate the unweighted muscle with single shocks at 1- to 2-second intervals to achieve three or four muscle twitches.

3. Stop the recording apparatus and add 10 g of weight or tension to the muscle. Restart and advance the recording about 1 cm, and then stimulate again to obtain three or four spikes.

4. Repeat the previous step seven more times, increasing the weight by 10 g each time until the total load on the muscle is 80 g or the muscle fails to respond. If the calcaneal tendon tears, the weight will drop, thus ending the trial. In such cases, you will need to prepare another frog's leg to continue the experiments and the maximal stimulus will have to be determined for the new muscle preparation.

5. When these "loading" experiments are completed, discontinue recording. Mark the curves on the record to indicate the load (in grams).

6. Measure the height of contraction (in millimeters) for each sequence of twitches obtained with each load, and insert this information on the chart on page 244.

7. Compute the work done by the muscle for each twitch (load) sequence.

Weight of load (g) × distance load lifted (mm) = work done

Enter these calculations into the chart in the column labeled Work done, Trial 1.

8. Allow the muscle to rest for 5 minutes. Then conduct a second trial in the same manner (i.e., repeat steps 2 through 7). Record this second set of measurements and calculations in the columns labeled Trial 2. Be sure to keep the muscle well moistened with Ringer's solution during the resting interval.

9. Using two different colors, plot a line graph of work done against the weight on the grid accompanying the chart for each trial. Label each plot appropriately.

10. Dismantle all apparatus and prepare the equipment for storage. Dispose of the frog remains in the appropriate container. Discard the gloves as instructed and wash and dry your hands.

11. Inspect your records of the experiments and make sure each is fully labeled with the experimental conditions, the date, and the names of those who conducted the experiments. For future reference, attach a tracing (or a copy of the tracing) for each experiment to this page. ■

	Distance load lifted (mm)		Work done	
Load (g)	Trial 1	Trial 2	Trial 1	Trial 2
0				
10				
20				
30				
40				
50				
60				
70				
80				

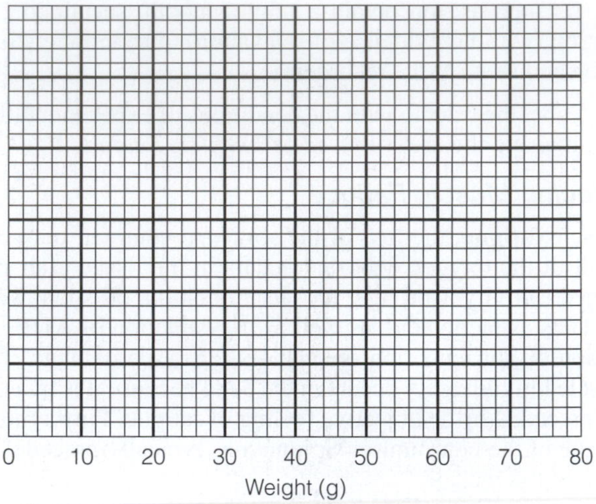
0 10 20 30 40 50 60 70 80
Weight (g)

ACTIVITY 4

Electromyography in a Human Subject Using BIOPAC®

Part 1: Temporal and Multiple Motor Unit Summation

This activity is an introduction to a procedure known as **electromyography,** the recording of skin-surface voltage manifestations of underlying skeletal muscle contraction. The actual visible recording of the resulting voltage waveforms is called an **electromyogram** (EMG).

A single skeletal muscle consists of numerous elongated *skeletal muscle cells,* also called *skeletal muscle fibers,* as discussed in Exercise 14. These muscle cells are excited by *motor neurons* of the central nervous system whose *axons* terminate at the muscle. An axon of a motor neuron branches profusely at the muscle; each branch produces multiple **axon terminals,** each of which innervates a single fiber. The number of muscle cells controlled by a single motor neuron can vary greatly, from five (for fine control needed in the hand) to 500 (for gross control, such as in the buttocks). The most important organizational concept in the physiology of muscle contraction is the *motor unit*, a single motor neuron and all of the cells within a muscle that it activates (see Figure 14.6 p. 192). Understanding gross muscular contraction depends upon realizing that a single muscle consists of multiple motor units, and that the gradual and coordinated activation of these motor units results in **graded contraction** of the whole muscle.

The nervous system controls muscle contraction by two mechanisms:

1. **Multiple motor unit summation (recruitment):** the gradual activation of more and more motor units

2. **Temporal (wave) summation:** an increase in the *frequency* of nerve impulses for each active motor unit

Thus, increasing the force of contraction of a muscle arises from gradually increasing the number of motor units being activated and increasing the frequency of nerve impulses delivered by those active motor units.

A final phenomenon, which is hardly noticeable except when performing electromyography, is **tonus,** a constant state of slight excitation of a muscle while it is in the relaxed state. Even while "at rest," a small number of motor units to a skeletal muscle remain slightly active to prepare the muscle for possible contraction.

Setting Up the Equipment

1. Connect the BIOPAC® unit to the computer.
2. Turn the computer **ON.**
3. Make sure the BIOPAC® unit is **OFF.**
4. Plug in the equipment as shown in Figure 16.5.
   - Electrode lead set—CH 3
   - Headphones—back of unit
5. Turn the BIOPAC® unit **ON.**
6. Attach three electrodes to the subject's dominant forearm as shown in Figure 16.6 and attach the electrode leads according to the colors indicated.
7. Start the BIOPAC® Student Lab program by double-clicking the icon on the desktop or by following your instructor's guidance.

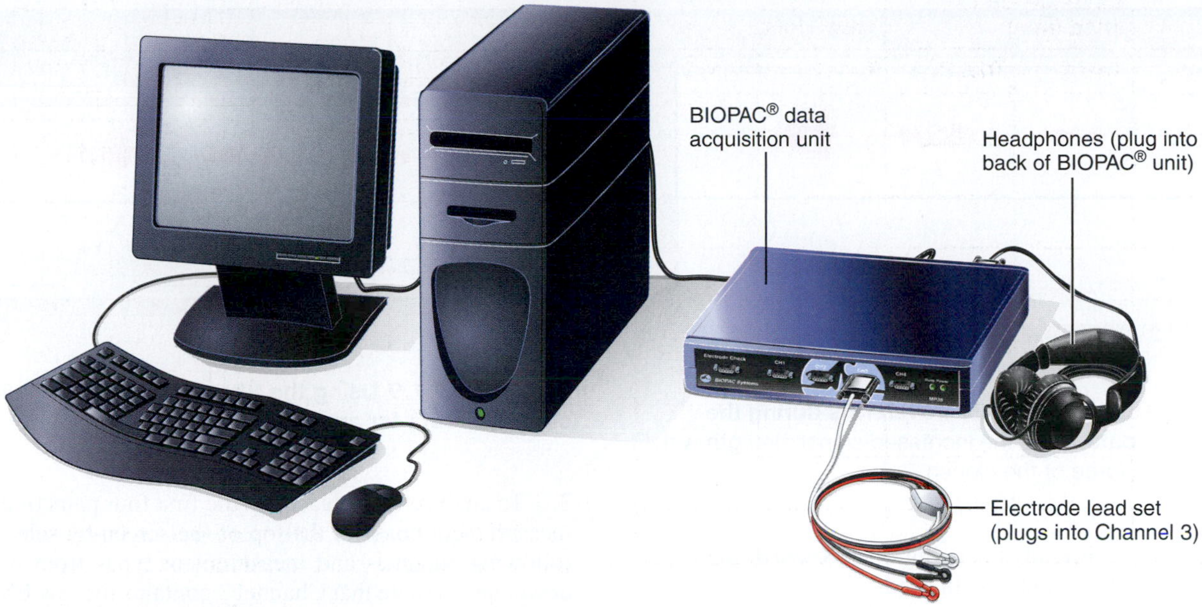
BIOPAC® data acquisition unit
Headphones (plug into back of BIOPAC® unit)
Electrode lead set (plugs into Channel 3)

FIGURE 16.5 Setting up the BIOPAC® equipment to observe temporal and multiple motor unit summation. Plug the headphones into the back of the data acquisition unit and the electrode lead set into Channel 3.

8. Select lesson **L01-EMG-1** from the menu and click **OK.**

9. Type in a filename that will save this subject's data on the computer hard drive. You may want to use the subject's last name followed by EMG-1 (for example, SmithEMG-1). Then click **OK.**

Calibrating the Equipment

1. With the subject in a still position, click **Calibrate.** This initiates the process by which the computer automatically establishes parameters to record the data properly for the subject.

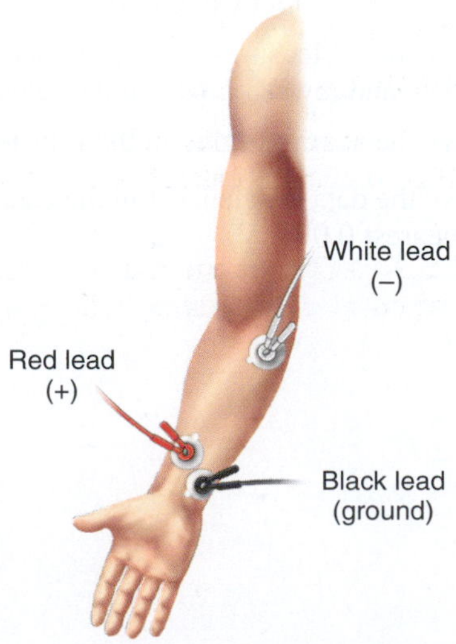
White lead
(−)
Red lead
(+)
Black lead
(ground)

FIGURE 16.6 Placement of electrodes and the appropriate attachment of electrode leads by color.

2. After you click **OK,** have the subject wait for 2 seconds, clench the fist tightly for 2 seconds, then release the fist and relax. The computer then automatically stops the recording.

3. Observe the recording of the calibration data, which should look like the waveform in Figure 16.7.

- If the data look very different, click **Redo Calibration** and repeat the steps above.
- If the data look similar, proceed to the next section.

Recording the Data

1. After you click **Record,** have the subject clench the fist softly for 2 seconds, then relax for 2 seconds, then clench harder for 2 seconds, and relax for 2 seconds, then clench even harder for 2 seconds, and relax for 2 seconds, and finally clench with maximum strength then relax. The result should be a series of four clenches of increasing intensity.

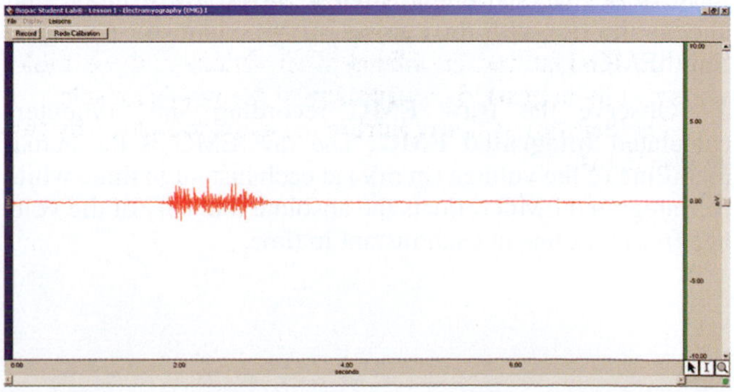

FIGURE 16.7 Example of waveform during the calibration procedure.

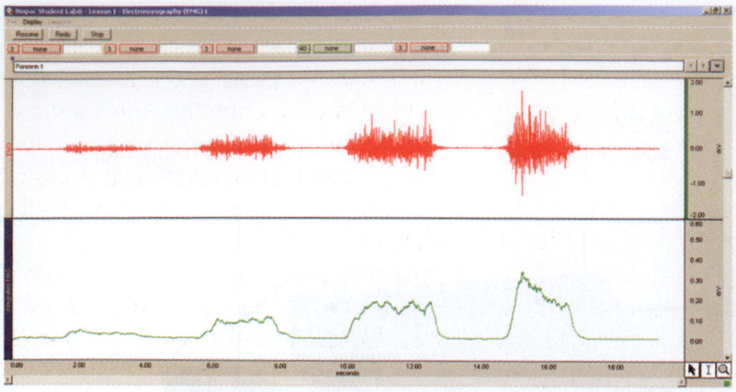

FIGURE 16.8 Example of waveforms during the recording of data. Note the increased signal strength with the increasing force of the clench.

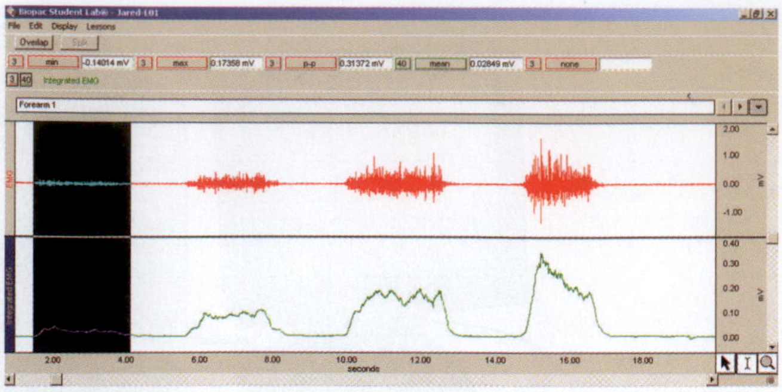

FIGURE 16.9 Using the I-beam cursor to highlight a cluster of data for analysis.

When the subject is ready to do this, click **Record;** then click **Suspend** when the subject is finished.

2. Observe the recording of the data, which should look like the waveforms in Figure 16.8.

- If the data look very different, click **Redo** and repeat the steps above.
- If the data look similar, click **STOP.** Click **YES** in response to the question, "Are you finished with both forearm recordings?"

Optional: Anyone can use the headphones to listen to an "auditory version" of the electrical activity of contraction by clicking **Listen** and having the subject clench and relax. Note that the frequency of the auditory signal corresponds with the frequency of action potentials stimulating the muscles. The signal will continue to run until you click **STOP.**

3. Click **Done,** and then remove all electrodes from the forearm.

- If you wish to record from another subject, choose the **Record from another subject** option and return to step 6 under Setting Up the Equipment.
- If you are finished recording, choose **Analyze current data file** and click **OK.** Proceed to Data Analysis, step 2.

Data Analysis

1. If you are just starting the BIOPAC® program to perform data analysis, enter **Review Saved Data** mode and choose the file with the subject's EMG data (for example, SmithEMG-1).

2. Observe the **Raw EMG** recording and computer-calculated **Integrated EMG.** The raw EMG is the actual recording of the voltage (in mV) at each instant in time, while the integrated EMG reflects the absolute intensity of the voltage from baseline at each instant in time.

3. To analyze the data, set up the first four pairs of channel/measurement boxes at the top of the screen by selecting the following channels and measurement types from the drop-down menus (note that Channel 3 contains the raw EMG data and Channel 40 contains the integrated EMG data):

Channel	Measurement
CH 3	min
CH 3	max
CH 3	p-p
CH 40	mean

4. Use the arrow cursor and click the I-beam cursor box at the lower right of the screen to activate the "area selection" function. Using the activated I-beam cursor, highlight the first EMG cluster, representing the first fist-clenching (Figure 16.9).

5. Notice that the computer automatically calculates the **min, max, p-p,** and **mean** for the selected area. These measurements, calculated from the data by the computer, represent the following:

min: Displays the *minimum* value in the selected area

max: Displays the *maximum* value in the selected area

p-p (peak-to-peak): Measures the difference in value between the highest and lowest values in the selected area

mean: Displays the average value in the selected area

6. Write down the data for clench 1 in the chart on page 247 (round to the nearest 0.01 mV).

7. Using the I-beam cursor, highlight the clusters for clenches 2, 3, and 4, and record the data in the chart below.

	Min	Max	P-P	Mean
Clench 1				
Clench 2				
Clench 3				
Clench 4				

From the data recorded in the chart, what trend do you observe for each of these measurements as the subject gradually increases the force of muscle contraction?

____

____

What is the relationship between maximum voltage for each clench and the number of motor units in the forearm that are being activated?

____

____

Part 2: Force Measurement and Fatigue

In this set of activities you will be observing graded muscle contractions and fatigue in a subject. **Graded muscle contractions,** which represent increasing levels of force generated by a muscle, depend upon: (1) the gradual activation of more motor units, and (2) increasing the frequency of motor neuron action potentials for each active motor unit. This permits a range of forces to be generated by any given muscle or group of muscles, all the way up to the maximum force.

For example, the biceps muscle will have more active motor units and exert more force when lifting a 10-kg object than when lifting a 2-kg object. In addition, the motor neuron of each active motor unit will increase the frequency of action potentials delivered to the motor units, resulting in tetanus. When all of the motor units of a muscle are activated and in a state of tetanus, the maximum force of that muscle is achieved. Recall that fatigue is a condition in which the muscle gradually loses some or all of its ability to contract after contracting for an extended period of time. **Physiological fatigue** occurs when there is a subsequent decrease in the amount of ATP available for contraction; increased levels of the anaerobic by-product, lactic acid; and ionic imbalances.

In this exercise, you will observe and measure graded contractions of the fist, and then observe fatigue, in both the dominant and nondominant arms. To measure the force generated during fist contraction, you will use a **hand dynamometer** (*dynamo* = force; *meter* = measure). The visual recording of force is called a **dynagram,** and the procedure of measuring the force itself is called **dynamometry.**

You will first record data from the subject's **dominant arm** (forearm 1) indicated by his or her "handedness," then repeat the procedures on the subject's **nondominant arm** (forearm 2) for comparison.

Setting Up the Equipment

1. Connect the BIOPAC® unit to the computer and turn the computer **ON.**

2. Make sure the BIOPAC® unit is **OFF.**

3. Plug in the equipment as shown in Figure 16.10.

- Hand dynamometer—CH 1
- Electrode lead set—CH 3
- Headphones—back of unit

4. Turn the BIOPAC® unit **ON.**

5. Attach three electrodes to the subject's *dominant* forearm (forearm 1) as shown in Figure 16.11, and attach the electrode leads according to the colors indicated.

6. Start the BIOPAC® Student Lab program on the computer by double-clicking the icon on the desktop or by following your instructor's guidance.

7. Select lesson **L02-EMG-2** from the menu and click **OK.**

8. Type in a filename that will save this subject's data on the computer hard drive. You may want to use subject's last name followed by EMG-2 (for example, SmithEMG-2). Then click **OK.**

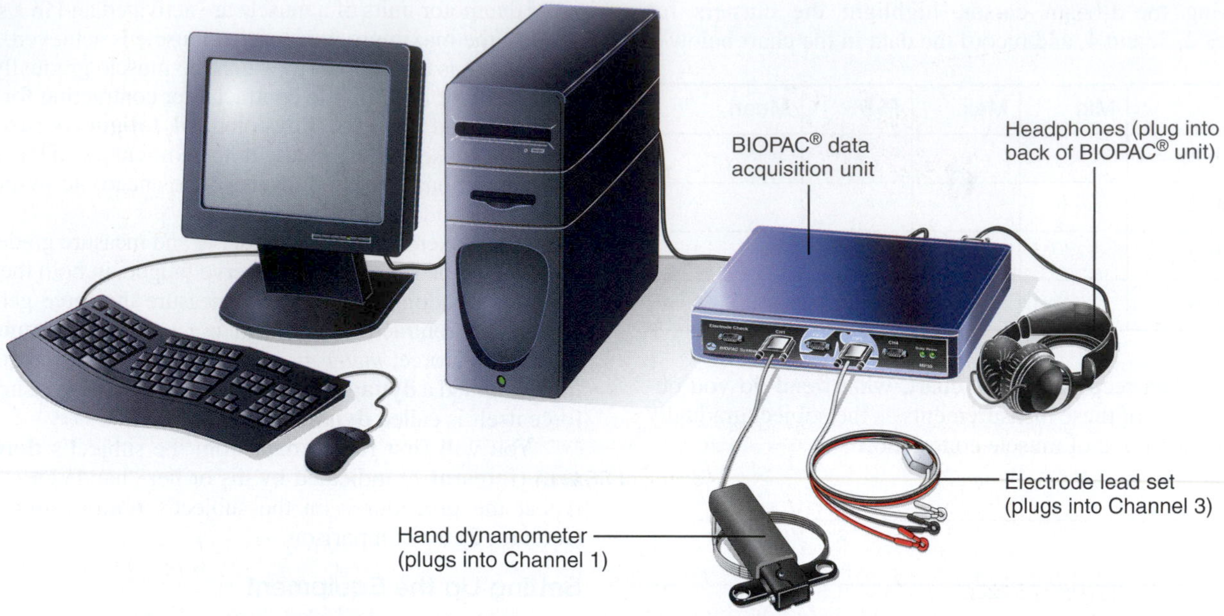
BIOPAC® data acquisition unit
Headphones (plug into back of BIOPAC® unit)
Electrode lead set (plugs into Channel 3)
Hand dynamometer (plugs into Channel 1)

FIGURE 16.10 Setting up the BIOPAC® equipment to observe graded muscle contractions and muscle fatigue. Plug the headphones into the back of the data acquisition unit, the hand dynamometer into Channel 1, and the electrode lead set into Channel 3.

Calibrating the Equipment

1. With the hand dynamometer at rest on the table, click **Calibrate.** This initiates the process by which the computer automatically establishes parameters to record the data properly for the subject.

2. A pop-up window prompts the subject to remove any grip force. This is to ensure that the dynamometer has been at rest on the table and that no force is being applied. When this is so, click **OK.**

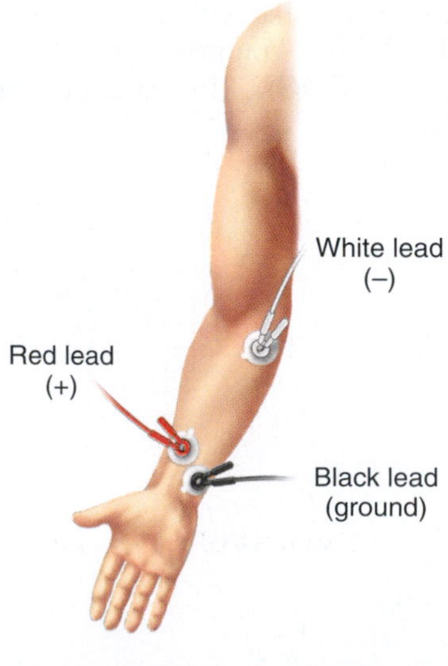
White lead (–)
Red lead (+)
Black lead (ground)

FIGURE 16.11 Placement of electrodes and the appropriate attachment of electrode leads by color.

3. As instructed by the pop-up window, the subject is to grasp the hand dynamometer with the dominant hand. With model SS25L, grasp as close to the crossbar as possible *without actually touching the crossbar,* as shown in Figure 16.12a. The crossbar should be held parallel to the floor, not at an angle. With model SS25LA, grasp the dynamometer with the palm of the hand against the short grip bar, as shown in Figure 16.12b. Hold model SS25LA vertically. When this is so, click **OK.** The instructions that follow apply to either model of dynamometer.

4. As instructed by the next pop-up window, have the subject wait 2 seconds, then squeeze the hand dynamometer as hard as possible for 2 seconds, and then relax. The computer will automatically stop the calibration.

5. When the subject is ready to proceed, click **OK** and follow the instructions in step 4. The calibration will stop automatically.

6. Observe the recording of the calibration data, which should look like the waveforms in Figure 16.13.

- If the data look very different, click **Redo Calibration** and repeat the steps above.
- If the data look similar, proceed to the next section.

Recording Incremental Force Data for the Forearm

1. Using the force data from the calibration procedure, estimate the **maximum force** that the subject generated (kg).

2. Divide that maximum force by four. In the next activity, the subject will gradually increase the force in approximately

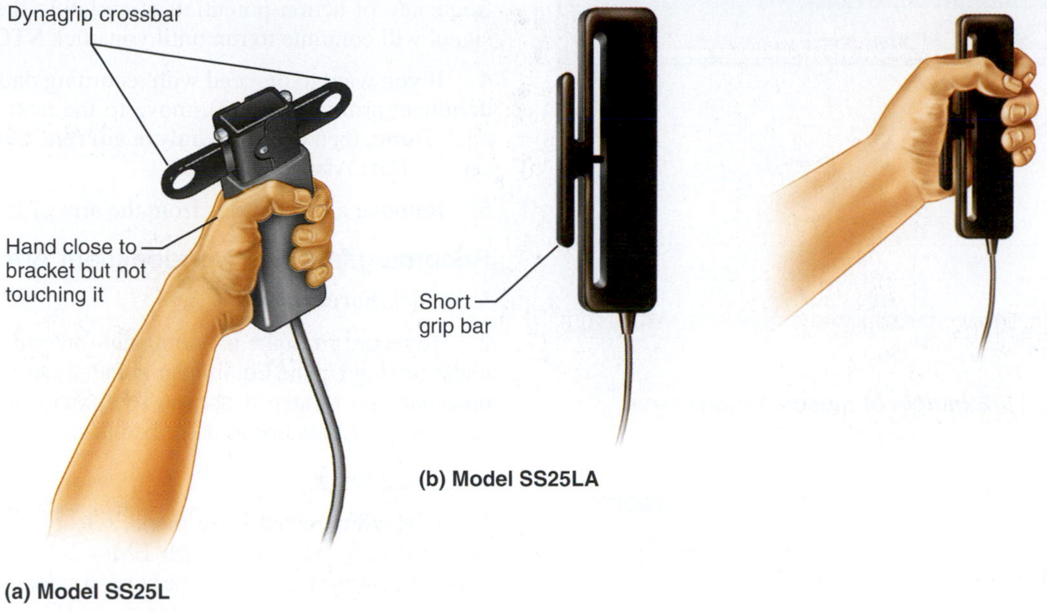
Dynagrip crossbar
Hand close to bracket but not touching it
(a) Model SS25L
Short grip bar
(b) Model SS25LA

FIGURE 16.12 Proper grasp of the hand dynamometer using either model SS25L or model SS25LA.

these increments. For example, if the maximum force generated was 20 kg, the increment will be 20/4 = 5 kg. The subject will grip at 5 kg, then 10 kg, then 15 kg, and then 20 kg. The subject should watch the tracing on the computer screen and compare it to the scale on the right to determine each target force.

3. After you click **Record,** have the subject wait 2 seconds, clench at the first force increment 2 seconds (for example, 5 kg), then relax 2 seconds, clench at the second force increment 2 seconds (10 kg), then relax 2 seconds, clench at the third force increment 2 seconds (15 kg), then relax 2 seconds, then clench with the maximum force 2 seconds (20 kg), and then relax. When the subject relaxes after the maximum clench, click **Suspend** to stop the recording.

4. Observe the recording of the data, which should look similar to the data in Figure 16.14.

- If the data look very different, click **Redo** and repeat the steps above.
- If the data look similar, proceed to observation and recording of muscle fatigue.

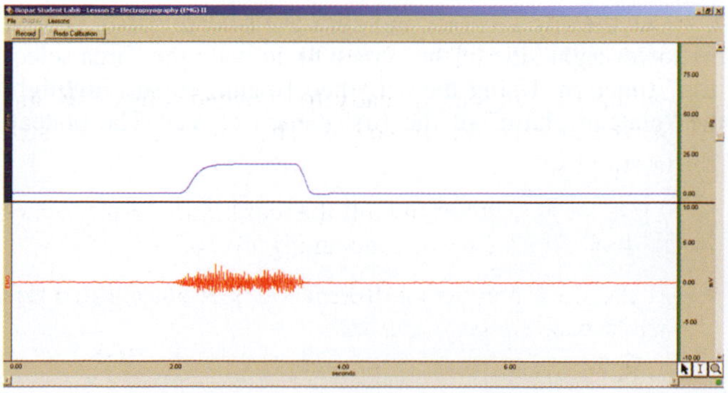

FIGURE 16.13 Example of calibration data. Force is measured in kilograms at the top and electromyography is measured in millivolts at the bottom.

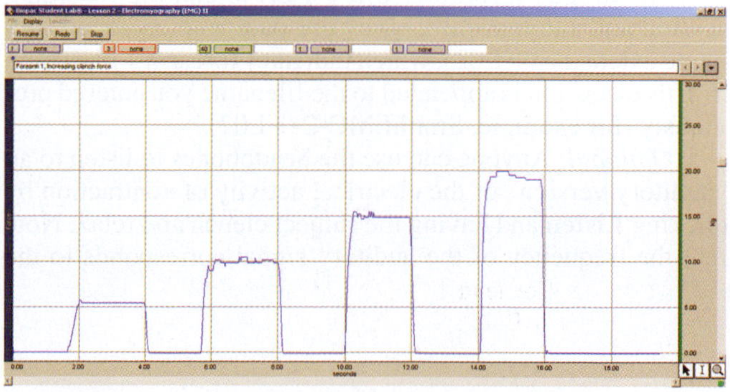

FIGURE 16.14 Example of incremental force data.

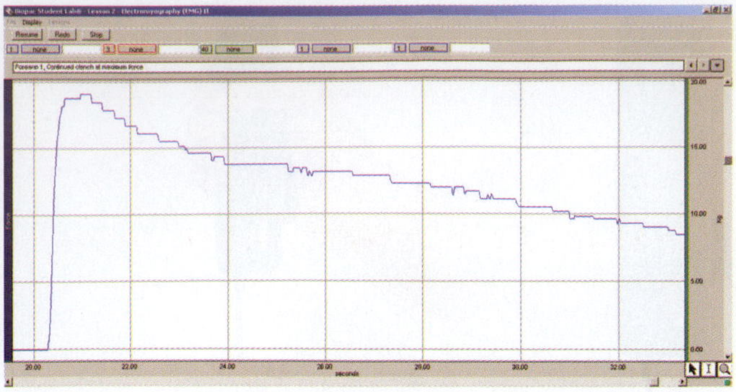

FIGURE 16.15 Example of muscle fatigue data.

Recording Muscle Fatigue Data for the Forearm

Continuing from the end of the incremental force recording, the subject will next record muscle fatigue.

1. After you click **Resume,** the recording will continue from where it stopped and the subject will clench the dynamometer with maximum force for as long as possible. A triangular "marker" will appear at the top of the data, denoting the beginning of this recording segment. When the subject's clench force falls below 50% of the maximum (for example, below 10 kg for a subject with 20 kg maximum force), click **Suspend.** The subject should not watch the screen during this procedure; those helping can inform the subject when it is time to relax.

2. Observe the recording of the data, which should look similar to the data in Figure 16.15.

- If the data look very different, click **Redo** and repeat the steps above.
- If the data look similar, click **STOP.**

3. When you click **STOP,** a dialog box comes up asking if you are sure you want to stop recording. Click **NO** to return to the **Resume** or **Stop** options, providing one last chance to redo the fatigue recording. Click **YES** to end the recording session and automatically save your data with an added filename extension (1-L02), which indicates forearm 1 for lesson 2. This extension is appended to the filename you entered previously (for example, SmithEMG-2-1-L02).

Optional: Anyone can use the headphones to listen to an "auditory version" of the electrical activity of contraction by clicking **Listen** and having the subject clench and relax. Note that the frequency of the auditory signal corresponds to the frequency of action potentials stimulating the muscles. The signal will continue to run until you click **STOP.**

4. If you want to proceed with recording data from the nondominant arm (forearm 2), move to the next section. If not, click **Done,** then choose **Analyze current data file** and proceed to Data Analysis, step 2.

5. Remove all electrodes from the arm of the subject.

Recording from the Nondominant Arm

1. Click **Forearm 2.**

2. To record from the nondominant forearm, return to step 5 under Setting Up the Equipment in Part 2, and repeat the entire procedure up to step 5 above. The extension 2-L02 will be added to your filename as described in step 3 above.

Data Analysis

1. In **Review Saved Data** mode, select the file that is to be analyzed (for example, Smith EMG-2-1-L02). If data was collected for both forearms, be sure to select the file with the proper extension (1-L02 dominant; 2-L02 nondominant).

2. Observe the recordings of the clench **Force** (kg), **Raw EMG** (mV), and computer-calculated **Integrated EMG** (mV). The force is the actual measurement of the strength of clench in kilograms at each instant in time. The raw EMG is the actual recording of the voltage (in mV) at each instant in time, and the integrated EMG indicates the absolute intensity of the voltage from baseline at each instant in time.

3. To analyze the data, set up the first three pairs of channel/measurement boxes at the top of the screen. Channel 1 contains the Force data, Channel 3 contains the Raw EMG data, and Channel 40 contains the Integrated EMG data. Select the following channels and measurement types:

Channel	Measurement
CH 1	mean
CH 3	p-p
CH 40	mean

4. Use the arrow cursor and click the I-beam cursor box on the lower right side of the screen to activate the "area selection" function. Using the activated I-beam cursor, highlight the "plateau phase" of the first clench cluster. The plateau

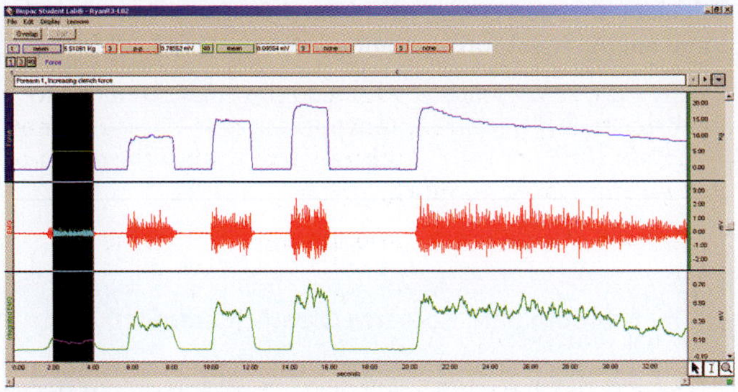

FIGURE 16.16 Highlighting the "plateau" of clench cluster 1.

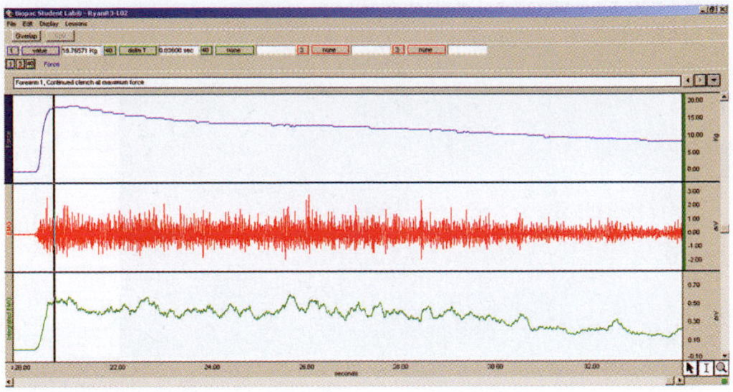

FIGURE 16.17 Selection of single point of maximum clench.

should be a relatively flat force in the middle of the cluster as shown in Figure 16.16.

5. Observe that the computer automatically calculates the **p-p** and **mean** values for the selected area. These measurements, calculated from the data by the computer, represent the following:

p-p (peak-to-peak): Measures the difference in value between the highest and lowest values in the selected area

mean: Displays the average value in the selected area

6. In the chart below, record the data for clench 1 (for example, 5-kg clench) to the nearest 0.01.

7. Using the I-beam cursor, highlight the clusters for the subsequent clenches and record the data in the chart.

Dominant Forearm Clench Increments

	Force at plateau Mean (kg)	Raw EMG P-P (msV)	Integrated EMG Mean (mV)
Clench 1			
Clench 2			
Clench 3			
Clench 4			

8. Scroll along the bottom of the data page to the segment that includes the recording of muscle fatigue (it should begin after the "marker" that appears at the top of the data).

9. Change the channel/measurement boxes so that the first two selected appear as follows (the third should be set to "none"):

Channel	Measurement
CH 1	value
CH 40	delta T

value: Measures the highest value in the selected area (CH 1 is force measured in kg)

delta T: Measures the time elapsed in the selected area (CH 40 is time measured in seconds)

10. Use the arrow cursor and click on the I-beam cursor box on the lower right side of the screen to activate the "area selection" function.

11. Using the activated I-beam cursor, select just the single point of maximum clench strength at the start of this data segment, as shown in Figure 16.17.

12. Note the maximum force measurement for this point (CH 1). Record this data in the next chart.

Dominant Forearm Fatigue Measurement

Maximum clench force (kg)	50% of the maximum clench force (divide maximum clench force by 2)	Time to fatigue (seconds)

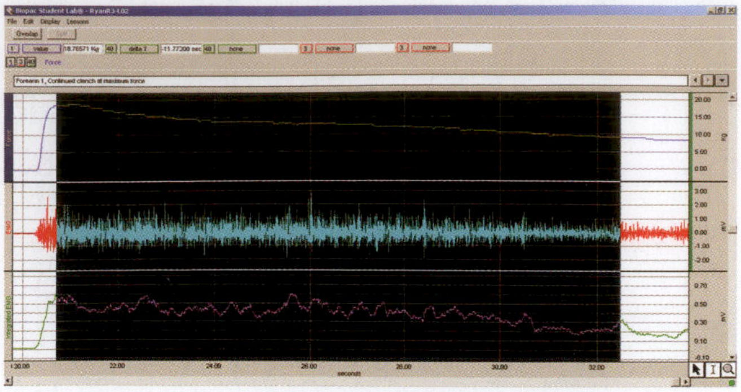

FIGURE 16.18 Highlighting to measure elapsed time to 50% of the maximum clench force.

13. Calculate the value of 50% of the maximum clench force, and record this in the data chart.

14. Using a metric tape measure, measure the circumference of the subject's dominant forearm at its greatest diameter:

_________ cm

15. Measure the amount of time that elapsed between the initial maximum force and the point at which the subject fatigued to 50% of this level. Using the activated I-beam cursor, highlight the area from the point of 50% clench force back to the point of maximal clench force, as shown in Figure 16.18.

16. Note the time it took the subject to reach this point of fatigue (CH 40 delta T) and record this data in the Dominant Forearm Fatigue Measurement chart.

Repeat Data Analysis for the Nondominant Forearm

1. Return to step 1 of the Data Analysis section and repeat the same measurements for the nondominant forearm (forearm 2).

2. Record your data in the two charts that follow; this data will be used for comparison.

3. Using a tape measure, measure the circumference of the subject's nondominant forearm at its greatest:

_________ cm

4. When finished, exit the program by going to the **File** menu at the top of the screen and clicking **Quit.**

Is there a difference in maximal force that was generated between the dominant and nondominant forearms? If so, how much?

____

Calculate the percentage difference in force between the dominant maximal force and nondominant maximal force.

____

____

Nondominant Forearm Clench Increments

	Force at plateau Mean (kg)	Raw EMG P-P (msV)	Integrated EMG Mean (mV)
Clench 1			
Clench 2			
Clench 3			
Clench 4			

Nondominant Forearm Fatigue Measurement

Maximum Clench Force (kg)	50% of the maximum clench force (divide maximum clench force by 2)	Time to fatigue (seconds)

Is there a difference in the circumference between the dominant and nondominant forearms? If so, how much?

____

If there is a difference in circumference, is this difference likely to be due to a difference in the *number* of muscle fibers in each forearm or in the *diameter* of each muscle fiber in the forearms? Explain.

____

____

Compare the time to fatigue between the two forearms.

____

____

NAME ______________________________

LAB TIME/DATE ______________________

REVIEW SHEET

EXERCISE 16

Skeletal Muscle Physiology: Frogs and Human Subjects

Muscle Activity

1. The following group of incomplete statements refers to a muscle cell in the resting state just before stimulation. Complete each statement by choosing the correct response from the key items.

Key:
- a. Na^+ diffuses out of the cell
- b. K^+ diffuses out of the cell
- c. Na^+ diffuses into the cell
- d. K^+ diffuses into the cell
- e. inside the cell
- f. outside the cell
- g. relative ionic concentrations on the two sides of the membrane
- h. electrical conditions
- i. activation of the sodium-potassium pump, which moves K^+ into the cell and Na^+ out of the cell
- j. activation of the sodium-potassium pump, which moves Na^+ into the cell and K^+ out of the cell

There is a greater concentration of Na^+ __________; there is a greater concentration of K^+ __________. When the stimulus is delivered, the permeability of the membrane at that point is changed; and __________, initiating the depolarization of the membrane. Almost as soon as the depolarization wave has begun, a repolarization wave follows it across the membrane. This occurs as __________. Repolarization restores the __________ of the resting cell membrane. The __________ is (are) reestablished by __________.

2. Number the following statements in the proper sequence to describe the contraction mechanism in a skeletal muscle cell. Number 1 has already been designated.

____1____ Depolarization occurs, and the action potential is generated.

__________ The muscle cell relaxes and lengthens.

__________ The calcium ion concentrations at the myofilaments increase; the myofilaments slide past one another, and the cell shortens.

__________ The action potential, carried deep into the cell by the T tubules, triggers the release of calcium ions from the sarcoplasmic reticulum.

__________ The concentration of the calcium ions at the myofilaments decreases as they are actively transported into the sarcoplasmic reticulum.

3. Refer to your observations of muscle fiber contraction on page 237–238 to answer the following questions.

a. Did your data support your hypothesis? __________

b. *Explain* your observations fully. __________

____

____

c. Draw a relaxed and a contracted sarcomere below.

Relaxed **Contracted**

Induction of Contraction in the Frog Gastrocnemius Muscle

4. Why is it important to destroy the brain and spinal cord of a frog before conducting physiological experiments on muscle contraction? ____

____

5. What kind of stimulus (electrical or chemical) travels from the motor neuron to skeletal muscle? ______________

What kind of stimulus (electrical or chemical) travels from the axon terminal to the sarcolemma? ______________

6. Give the name and duration of each of the three phases of the muscle twitch, and describe what is happening during each phase.

a. ______________, ______ msec, ______________________________

____

b. ______________, ______ msec, ______________________________

c. ______________, ______ msec, ______________________________

7. Use the items in the key to identify the conditions described.

Key:

a. maximal stimulus
b. multiple motor unit summation
c. subthreshold stimulus
d. tetanus
e. threshold stimulus
f. wave summation

__________ 1. sustained contraction without any evidence of relaxation

__________ 2. stimulus that results in no perceptible contraction

__________ 3. stimulus at which the muscle first contracts perceptibly

__________ 4. increasingly stronger contractions owing to stimulation at a rapid rate

__________ 5. increasingly stronger contractions owing to increased stimulus strength

__________ 6. weakest stimulus at which all muscle cells in the muscle are contracting

8. Complete the following statements by writing the appropriate words on the correspondingly numbered blanks at the right.

When a weak but smooth muscle contraction is desired, a few motor units are stimulated at a __1__ rate. If blue litmus paper is pressed to the cut surface of a fatigued muscle, the paper color changes to red, indicating low pH. This situation is caused by the accumulation of __2__ in the muscle. Within limits, as the load on a muscle is increased, the muscle contracts __3__ (more/less) strongly.

1. ______________________
2. ______________________
3. ______________________

9. During the frog experiment on muscle fatigue, how did the muscle contraction pattern change as the muscle began to fatigue?

____

____

How long was stimulation continued before fatigue was apparent? ______________________

If the sciatic nerve that stimulates the living frog's gastrocnemius muscle had been left attached to the muscle and the stimulus had been applied to the nerve rather than the muscle, would fatigue have become apparent sooner or later?

____

Explain your answer. ______________________

____

10. What will happen to a muscle in the body when its nerve supply is destroyed or badly damaged?

____

____

11. Explain the relationship between the load on a muscle and its strength of contraction. ______________________

____

____

12. The skeletal muscles are maintained in a slightly stretched condition for optimal contraction. How is this accomplished?

____

Why does stretching a muscle beyond its optimal length reduce its ability to contract? (Include an explanation of the events at the level of the myofilaments.) ______________________

____

____

13. If the length but not the tension of a muscle is changed, the contraction is called an isotonic contraction. In an isometric contraction, the tension is increased but the muscle does not shorten. Which type of contraction did you observe most often during the laboratory experiments? ______________________

Electromyography in a Human Subject Using BIOPAC®

14. If you were a physical therapist applying a constant voltage to the forearm, what might you observe if you gradually increased the *frequency* of stimulatory impulses, keeping the voltage constant each time?

____

____

15. Describe what is meant by the term *motor unit recruitment.* ____

____

16. Describe the physiological processes occurring in the muscle cells that account for the gradual onset of muscle fatigue.

____

____

____

17. Most subjects use their dominant forearm far more than their nondominant forearm. What does this indicate about degree of activation of motor units and these factors: muscle fiber diameter, maximum muscle fiber force, and time to muscle fatigue? (You may need to use your textbook for help with this one.)

____

____

____

____

18. Define *dynamometry.* ____

____

19. How might dynamometry be used to assess patients in a clinical setting? ____

____

____

EXERCISE 17

Histology of Nervous Tissue

MATERIALS

- ☐ Model of a "typical" neuron (if available)
- ☐ Compound microscope
- ☐ Immersion oil
- ☐ Prepared slides of an ox spinal cord smear and teased myelinated nerve fibers
- ☐ Prepared slides of Purkinje cells (cerebellum), pyramidal cells (cerebrum), and a dorsal root ganglion
- ☐ Prepared slide of a nerve (x.s.)

MasteringA&P™

Access practice quizzes and more in the Study Area at www.masteringaandp.com.

PAL

For access to anatomical models and more, check out Practice Anatomy Lab.

OBJECTIVES

1. To differentiate between the functions of neurons and neuroglia.
2. To list six types of neuroglia cells and indicate nervous system location of each.
3. To identify the important anatomical characteristics of a neuron on an appropriate diagram or projected slide.
4. To state the functions of axons, dendrites, axon terminals, neurofibrils, and myelin sheaths.
5. To explain how a nerve impulse is transmitted from one neuron to another.
6. To explain the role of Schwann cells in the formation of the myelin sheath.
7. To classify neurons according to structure and function.
8. To distinguish between a nerve and a tract and between a ganglion and a nucleus.
9. To describe the structure of a nerve, identifying the connective tissue coverings (endoneurium, perineurium, and epineurium) and citing their functions.

PRE-LAB QUIZZ

1. Circle the correct term. Nervous tissue is made up of two / three principal cell populations.
2. Neuroglia of the peripheral nervous system include
   a. ependymal cells and satellite cells
   b. oligodendrocytes and astrocytes
   c. satellite cells and Schwann cells
3. ___________________ are the basic functional units of nervous tissue.
4. These branching neuron processes serve as receptive regions and transmit electrical signals toward the cell body. They are:
   a. axons   c. dendrites
   b. collaterals   d. neuroglia
5. Circle True or False. Axons are the neuron processes that generate and conduct nerve impulses.
6. Most axons are covered with a fatty material called ___________________, which insulates the fibers and increases the speed of neurotransmission.
7. Circle the correct term. Neuron fibers (axons) running through the central nervous system form tracts / nerves of white matter.
8. Neurons can be classified according to structure. __________ neurons have many processes that issue from the cell body.
   a. Bipolar   b. Multipolar   c. Unipolar
9. Circle the correct term. Neurons can be classified according to function. Afferent / Efferent or motor neurons carry activating impulses from the central nervous system primarily to body muscles or glands.
10. Within a nerve, each fiber is surrounded by a delicate connective tissue sheath called the:
    a. endoneurium   b. epineurium   c. perineurium

The nervous system is the master integrating and coordinating system, continuously monitoring and processing sensory information both from the external environment and from within the body. Every thought, action, and sensation is a reflection of its activity. Like a computer, it processes and integrates new "inputs" with information previously fed into it ("programmed") to produce an appropriate response ("readout"). However, no computer can possibly compare in complexity and scope to the human nervous system.

Despite its complexity, nervous tissue is made up of just two principal cell populations: neurons and supporting cells referred to as **neuroglia** ("nerve glue"), or **glial cells.** The neuroglia in the central nervous system (CNS: brain and spinal cord) include *astrocytes, oligodendrocytes, microglia,* and *ependymal cells* (Figure 17.1). The most important glial cells in the peripheral nervous system (PNS), that is, in the neural structures outside the CNS, are *Schwann cells* and *satellite cells.*

Neuroglia serve the needs of the delicate neurons by bracing and protecting them. In addition, they act as phagocytes (microglia), myelinate the cytoplasmic extensions of the neurons (oligodendrocytes and Schwann cells), play a role in capillary-neuron exchanges, and control the chemical environment around neurons (astrocytes). Although neuroglia resemble neurons in some ways (they have fibrous cellular extensions), they are not capable of generating and transmitting nerve impulses, a capability that is highly developed in neurons. Our focus in this exercise is the highly excitable neurons.

Neuron Anatomy

Neurons are the basic functional units of nervous tissue. They are highly specialized to transmit messages (nerve impulses) from one part of the body to another. Although neurons differ structurally, they have many identifiable features in common (Figure 17.2a and c). All have a **cell body** from which slender processes extend. Although neuron cell bodies are typically found in the CNS in clusters called **nuclei,** occasionally they reside in **ganglia** (collections of neuron cell bodies outside the CNS). They make up the gray matter of the nervous system. Neuron fibers running through the CNS form **tracts** of white matter; in the PNS they form the peripheral **nerves.**

The neuron cell body contains a large round nucleus surrounded by cytoplasm *(neuroplasm).* The cytoplasm is riddled with neurofibrils and with darkly staining structures called Nissl bodies. **Neurofibrils,** the cytoskeletal elements of the neuron, have a support and intracellular transport function. **Nissl** (chromatophilic) **bodies,** an elaborate type of rough endoplasmic reticulum, are involved in the metabolic activities of the cell.

There are two types of neuron processes. **Dendrites** are *receptive regions* (they bear receptors for neurotransmitters released by other neurons) and **axons,** also called *nerve fibers,* generate and conduct *nerve impulses.* Neurons have only one axon (which may branch into **collaterals**) but may have many dendrites, depending on the neuron type.

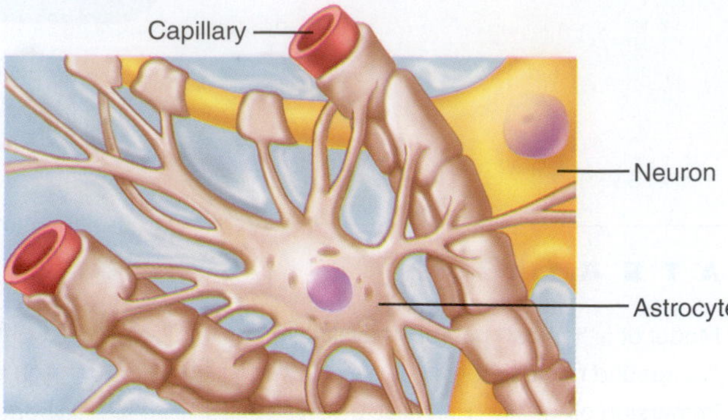

Capillary
Neuron
Astrocyte

(a) Astrocytes are the most abundant CNS neuroglia.

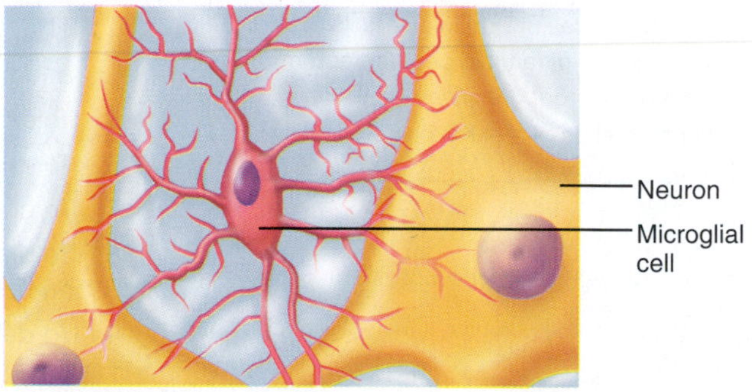

Neuron
Microglial cell

(b) Microglial cells are defensive cells in the CNS.

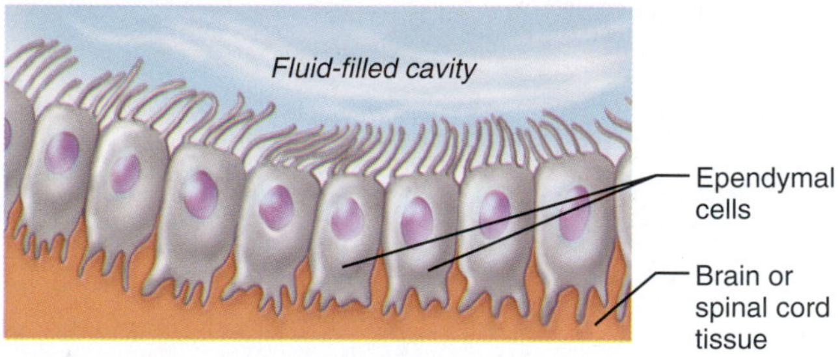

Fluid-filled cavity
Ependymal cells
Brain or spinal cord tissue

(c) Ependymal cells line cerebrospinal fluid–filled cavities.

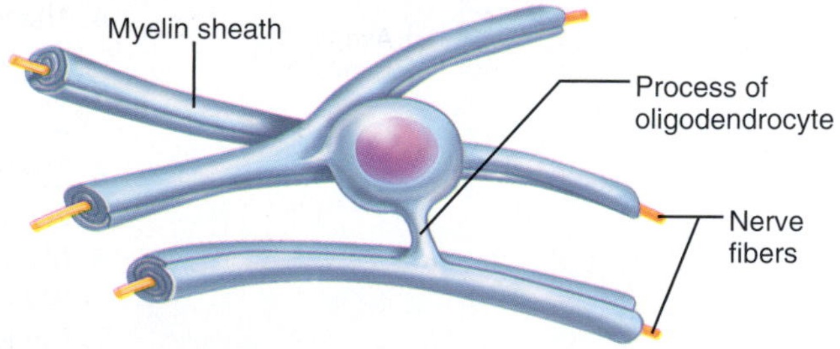

Myelin sheath
Process of oligodendrocyte
Nerve fibers

(d) Oligodendrocytes have processes that form myelin sheaths around CNS nerve fibers.

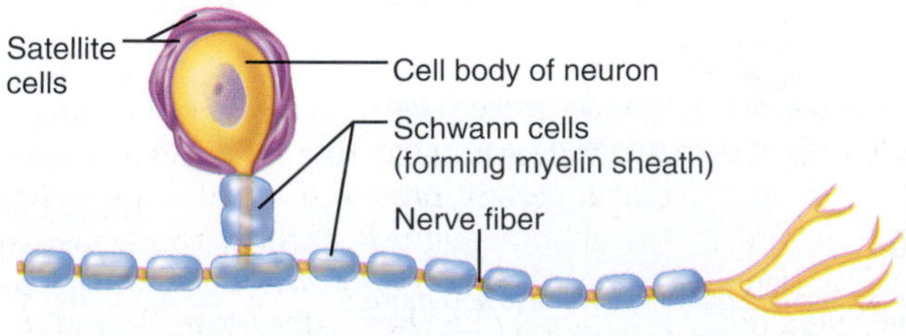

Satellite cells
Cell body of neuron
Schwann cells (forming myelin sheath)
Nerve fiber

(e) Satellite cells and Schwann cells (which form myelin) surround neurons in the PNS.

FIGURE 17.1 Neuroglia. (a–d) Supporting cells of the central nervous system. **(e)** Supporting cells of the peripheral nervous system.

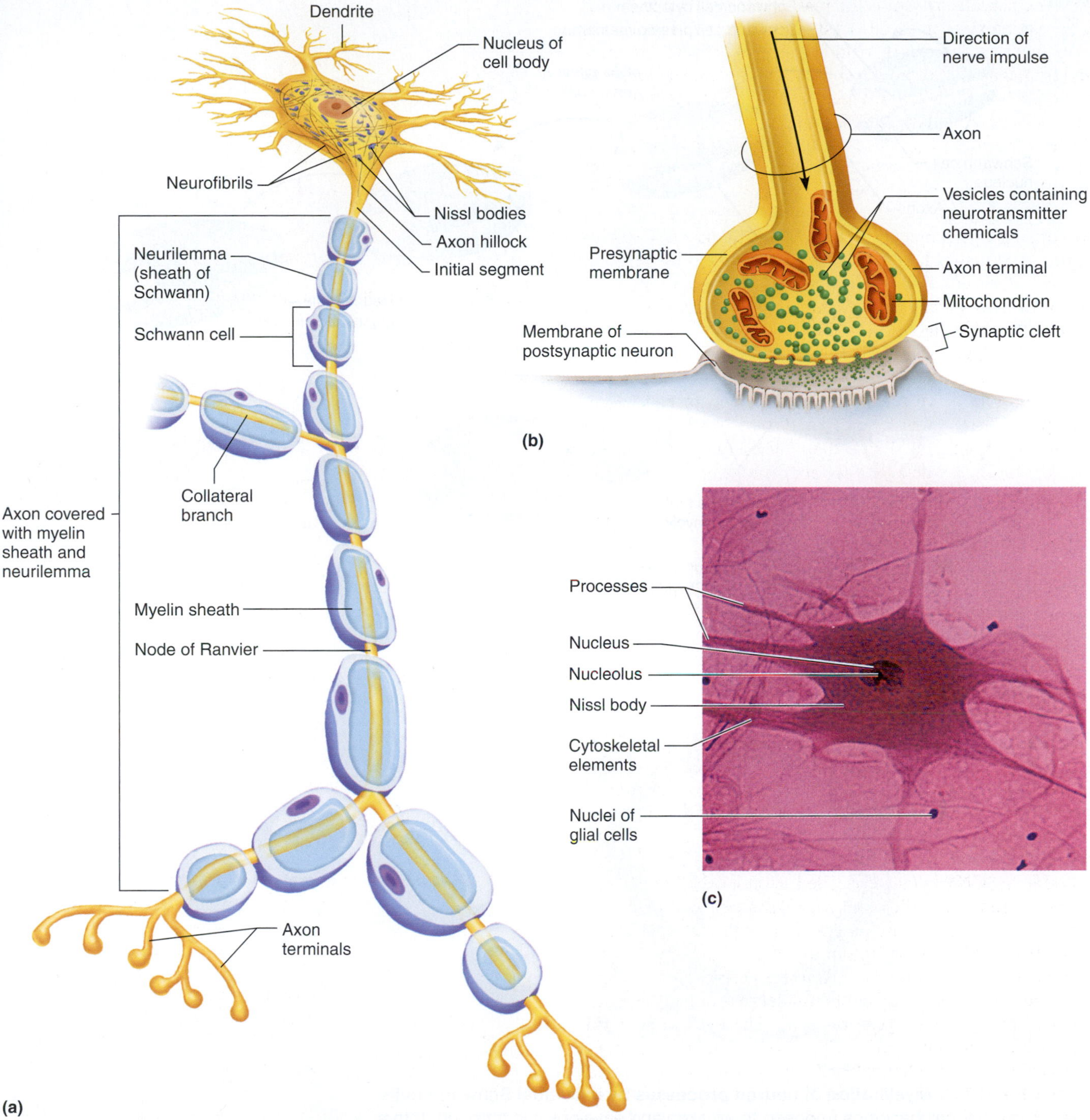
Dendrite
Nucleus of cell body
Neurofibrils
Nissl bodies
Axon hillock
Initial segment
Neurilemma (sheath of Schwann)
Schwann cell
Collateral branch
Axon covered with myelin sheath and neurilemma
Myelin sheath
Node of Ranvier
Axon terminals
(a)
Direction of nerve impulse
Axon
Vesicles containing neurotransmitter chemicals
Presynaptic membrane
Axon terminal
Mitochondrion
Membrane of postsynaptic neuron
Synaptic cleft
(b)
Processes
Nucleus
Nucleolus
Nissl body
Cytoskeletal elements
Nuclei of glial cells
(c)

FIGURE 17.2 Structure of a typical motor neuron. (a) Diagrammatic view. **(b)** An enlarged synapse. **(c)** Photomicrograph (250×).

In general, a neuron is excited by other neurons when their axons release neurotransmitters close to its dendrites or cell body. The electrical current produced travels across the cell body and if it is great enough, it elicits a regenerative electrical signal, an *impulse* or *action potential*, that travels down the axon. As Figure 17.2a shows, the axon (in motor neurons) begins just distal to a slightly enlarged cell body structure called the **axon hillock.** The point at which the axon hillock narrows to axon diameter is referred to as the *initial segment.* The axon ends in many small structures called **axon terminals,** or *synaptic knobs*, which form **synapses** or junctions with neurons or effector cells. These terminals store the neurotransmitter chemical in tiny vesicles. Each axon terminal is separated from the cell body or dendrites of the next (postsynaptic) neuron by a tiny gap called the **synaptic cleft** (Figure 17.2b). Thus, although they are close, there is no actual physical contact between neurons. When an impulse reaches the axon terminals, some of the *synaptic vesicles* rupture and release neurotransmitter into the synaptic cleft. The neurotransmitter then diffuses across the synaptic cleft to

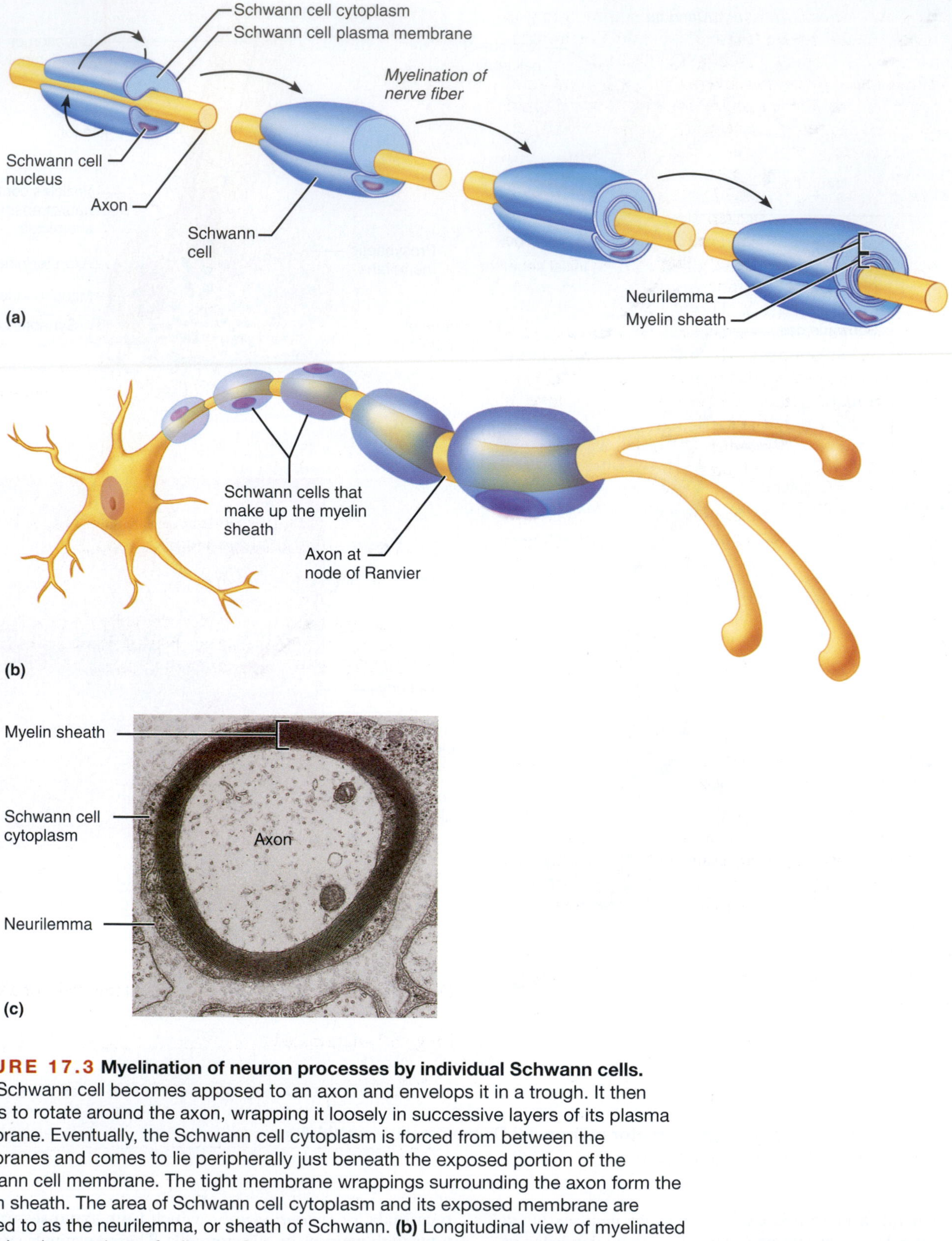
Schwann cell cytoplasm
Schwann cell plasma membrane
Myelination of nerve fiber
Schwann cell nucleus
Axon
Schwann cell
Neurilemma
Myelin sheath
(a)
Schwann cells that make up the myelin sheath
Axon at node of Ranvier
(b)
Myelin sheath
Schwann cell cytoplasm
Axon
Neurilemma
(c)

FIGURE 17.3 Myelination of neuron processes by individual Schwann cells. **(a)** A Schwann cell becomes apposed to an axon and envelops it in a trough. It then begins to rotate around the axon, wrapping it loosely in successive layers of its plasma membrane. Eventually, the Schwann cell cytoplasm is forced from between the membranes and comes to lie peripherally just beneath the exposed portion of the Schwann cell membrane. The tight membrane wrappings surrounding the axon form the myelin sheath. The area of Schwann cell cytoplasm and its exposed membrane are referred to as the neurilemma, or sheath of Schwann. **(b)** Longitudinal view of myelinated axon showing portions of adjacent Schwann cells and the node of Ranvier between them. **(c)** Electron micrograph of cross section through a myelinated axon (8,000×).

bind to membrane receptors on the next neuron, initiating an electrical current or *synaptic potential*. Specialized synapses between neurons and skeletal muscles are called neuromuscular junctions. They are discussed in Exercise 14.

Most long nerve fibers are covered with a fatty material called *myelin,* and such fibers are referred to as **myelinated fibers.** Axons in the peripheral nervous system are typically heavily myelinated by special supporting cells called **Schwann cells,** which wrap themselves tightly around the axon in jellyroll fashion (Figure 17.3). During the wrapping process, the cytoplasm is squeezed from between adjacent layers of the Schwann cell membranes, so that when the process is completed a tight core of plasma membrane material (protein-lipid material) encompasses the axon. This wrapping is the **myelin sheath.** The Schwann cell nucleus and the bulk of its cytoplasm end up just beneath the outermost portion of its plasma membrane. This peripheral part of the Schwann cell and its exposed plasma membrane is referred to as the **neurilemma,** or *sheath of Schwann.* Since the myelin sheath is formed by many individual Schwann cells, it is a discontinuous sheath. The gaps or indentations in the sheath are called **nodes of Ranvier** (see Figures 17.2 and 17.3).

Within the CNS, myelination is accomplished by glial cells called **oligodendrocytes** (see Figure 17.1d). These CNS sheaths do not exhibit the neurilemma seen in fibers myelinated by Schwann cells. Because of its chemical composition, myelin insulates the fibers and greatly increases the speed of neurotransmission by neuron fibers.

ACTIVITY 1

Identifying Parts of a Neuron

1. Study the typical motor neuron shown in Figure 17.2, noting the structural details described above, and then identify these structures on a neuron model.

2. Obtain a prepared slide of the ox spinal cord smear, which has large, easily identifiable neurons. Study one representative neuron under oil immersion and identify the cell body; the nucleus; the large, prominent "owl's eye" nucleolus; and the granular Nissl bodies. If possible, distinguish the axon from the many dendrites.

Sketch the cell in the space provided below, and label the important anatomical details you have observed. Compare your sketch to Figure 17.2c.

3. Obtain a prepared slide of teased myelinated nerve fibers. Using Figure 17.4 as a guide, identify the following: nodes of Ranvier, axon, Schwann cell nuclei, and myelin sheath.

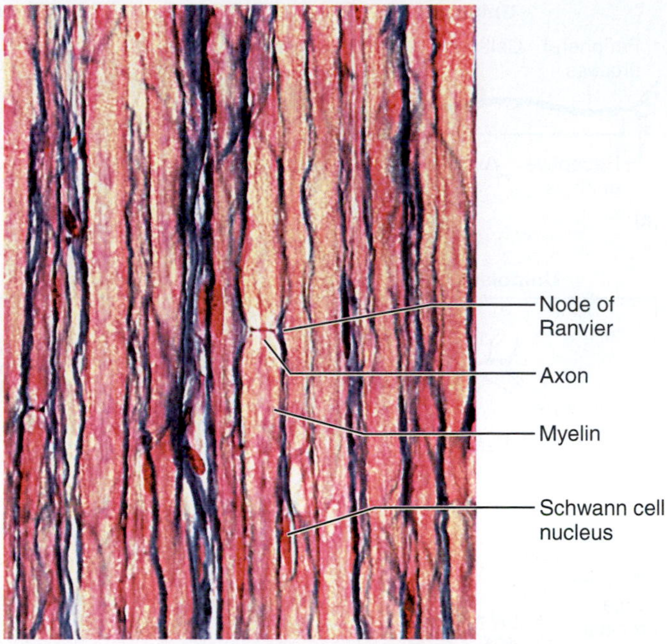
Node of Ranvier
Axon
Myelin
Schwann cell nucleus

FIGURE 17.4 Photomicrograph of a small portion of a peripheral nerve in longitudinal section (220×).

Sketch a portion of a myelinated nerve fiber in the space provided below, illustrating a node of Ranvier. Label the axon, myelin sheath, node, and neurilemma.

Do the nodes seem to occur at consistent intervals, or are they

irregularly distributed? ______________________________

Explain the functional significance of this finding: ________

____

____

Neuron Classification

Neurons may be classified on the basis of structure or of function.

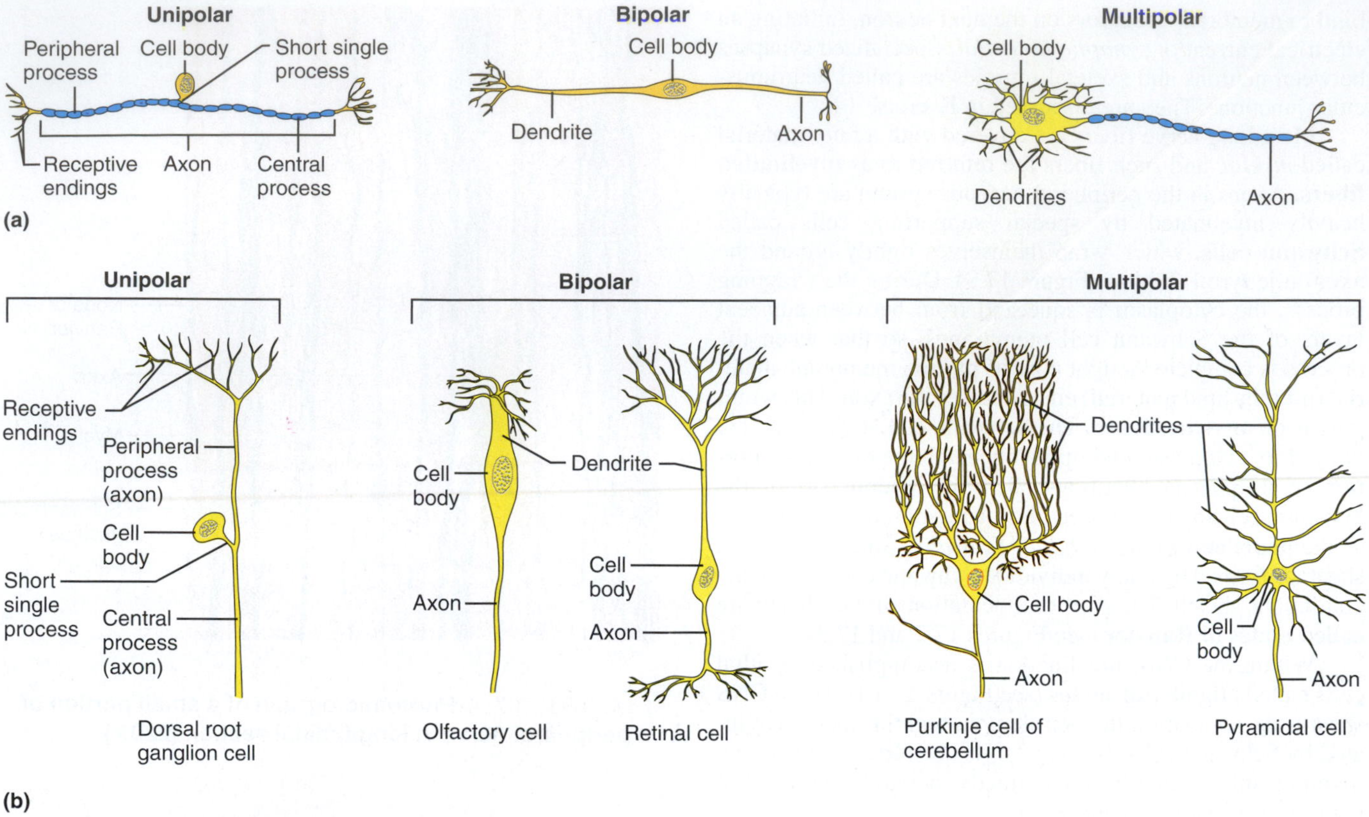
Unipolar
Peripheral process
Cell body
Short single process
Receptive endings
Axon
Central process
Bipolar
Cell body
Dendrite
Axon
Multipolar
Cell body
Dendrites
Axon
(a)
Unipolar
Receptive endings
Peripheral process (axon)
Cell body
Short single process
Central process (axon)
Dorsal root ganglion cell
Bipolar
Cell body
Dendrite
Axon
Cell body
Axon
Olfactory cell
Retinal cell
Multipolar
Dendrites
Cell body
Axon
Purkinje cell of cerebellum
Cell body
Axon
Pyramidal cell
(b)

FIGURE 17.5 Classification of neurons according to structure. **(a)** Classification of neurons based on structure (number of processes extending from the cell body). **(b)** Structural variations within the classes.

Classification by Structure

Structurally, neurons may be differentiated according to the number of processes attached to the cell body (Figure 17.5a). In **unipolar neurons,** one very short process, which divides into *peripheral* and *central processes,* extends from the cell body. Functionally, only the most distal portions of the peripheral process act as receptive endings; the rest acts as an axon along with the central process. Nearly all neurons that conduct impulses toward the CNS are unipolar.

Bipolar neurons have two processes attached to the cell body. This neuron type is quite rare, typically found only as part of the receptor apparatus of the eye, ear, and olfactory mucosa.

Many processes issue from the cell body of **multipolar neurons,** all classified as dendrites except for a single axon. Most neurons in the brain and spinal cord (CNS neurons) and those whose axons carry impulses away from the CNS fall into this last category.

ACTIVITY 2

Studying the Microscopic Structure of Selected Neurons

Obtain prepared slides of pyramidal cells of the cerebral cortex, Purkinje cells of the cerebellar cortex, and a dorsal root ganglion. As you observe them under the microscope, try to pick out the anatomical details depicted in Figure 17.5b and Figure 17.6. Notice that the neurons of the cerebral and cerebellar tissues (both brain tissues) are extensively branched; in contrast, the neurons of the dorsal root ganglion are more rounded. The many small nuclei visible surrounding the neurons are those of abutting glial cells.

Which of these neuron types would be classified as multipolar neurons?

____

Which as unipolar? ______________________________

Classification by Function

In general, neurons carrying impulses from sensory receptors in the internal organs (viscera), the skin, skeletal muscles, joints, or special sensory organs are termed **sensory,** or **afferent, neurons** (see Figure 17.7). The receptive endings of sensory neurons are often equipped with specialized receptors that are stimulated by specific changes in their immediate environment. The structure and function of these receptors arel considered separately in Exercise 23 (General Sensation). The cell bodies of sensory neurons are always found in

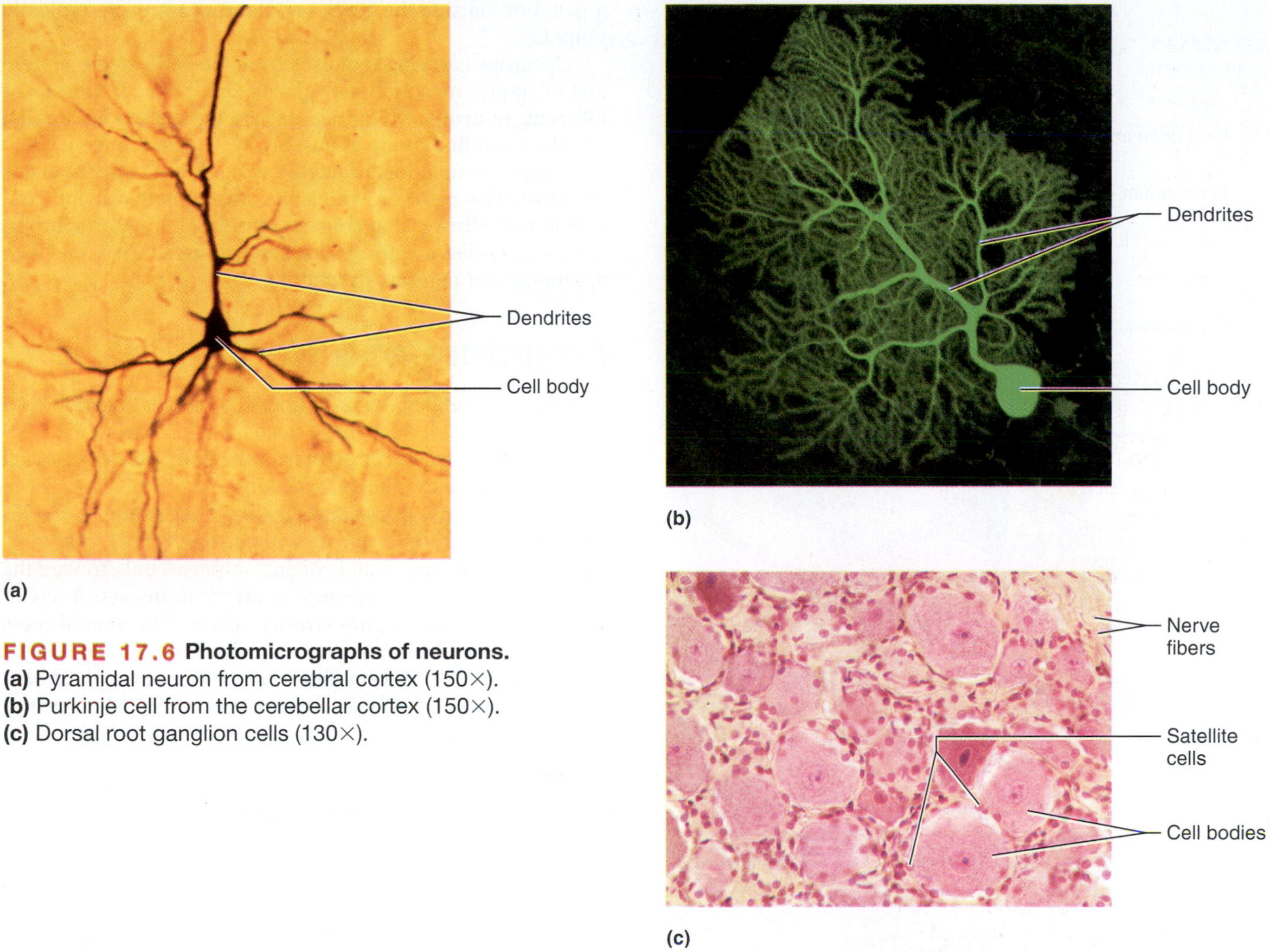
Dendrites
Cell body
(a)
Dendrites
Cell body
(b)
Nerve fibers
Satellite cells
Cell bodies
(c)

FIGURE 17.6 Photomicrographs of neurons.
(a) Pyramidal neuron from cerebral cortex (150×).
(b) Purkinje cell from the cerebellar cortex (150×).
(c) Dorsal root ganglion cells (130×).

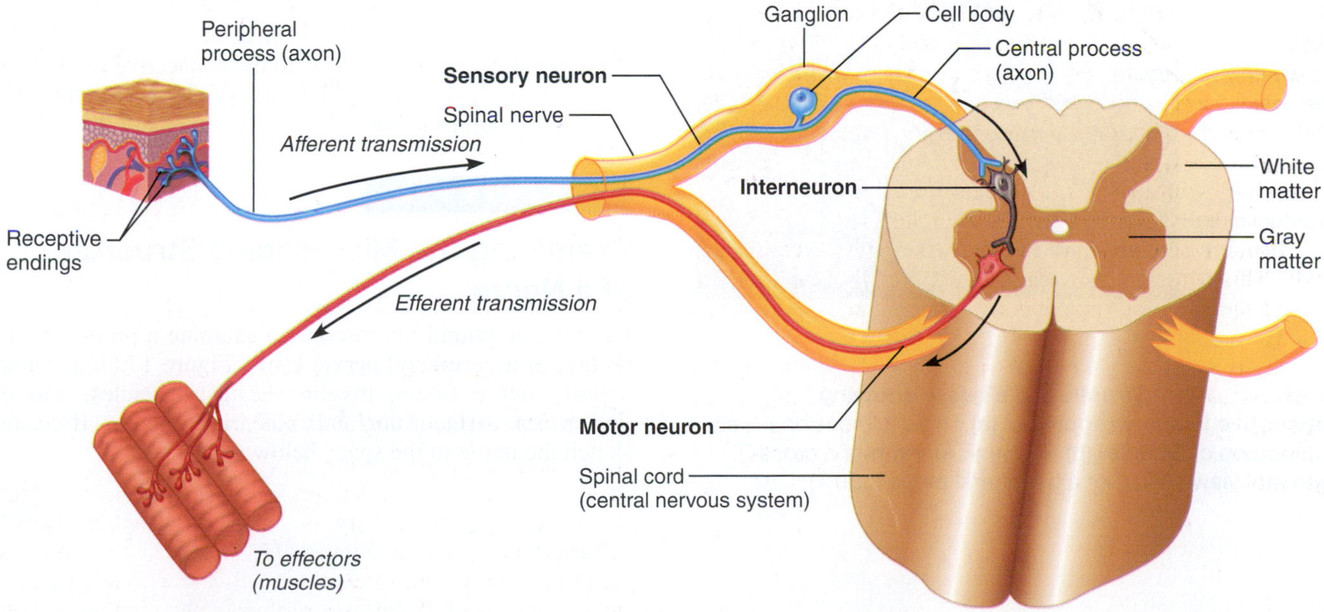
Peripheral process (axon)
Ganglion
Cell body
Central process (axon)
Sensory neuron
Spinal nerve
Afferent transmission
Receptive endings
Interneuron
White matter
Gray matter
Efferent transmission
Motor neuron
Spinal cord (central nervous system)
To effectors (muscles)

FIGURE 17.7 Classification of neurons on the basis of function. Sensory (afferent) neurons conduct impulses from the body's sensory receptors to the central nervous system; most are unipolar neurons with their nerve cell bodies in ganglia in the peripheral nervous system (PNS). Motor (efferent) neurons transmit impulses from the CNS to effectors (muscles). Interneurons (association neurons) complete the communication line between sensory and motor neurons. They are typically multipolar, and their cell bodies reside in the CNS.

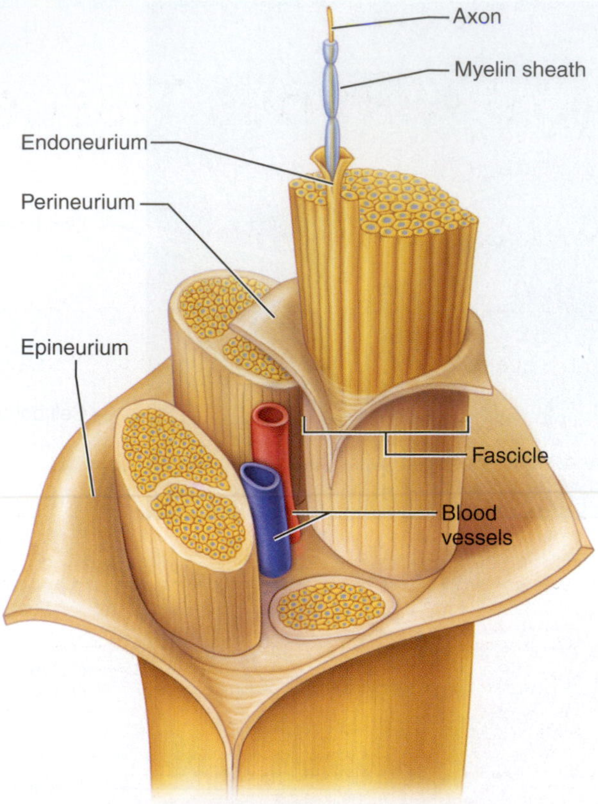
Axon
Myelin sheath
Endoneurium
Perineurium
Epineurium
Fascicle
Blood vessels
(a)

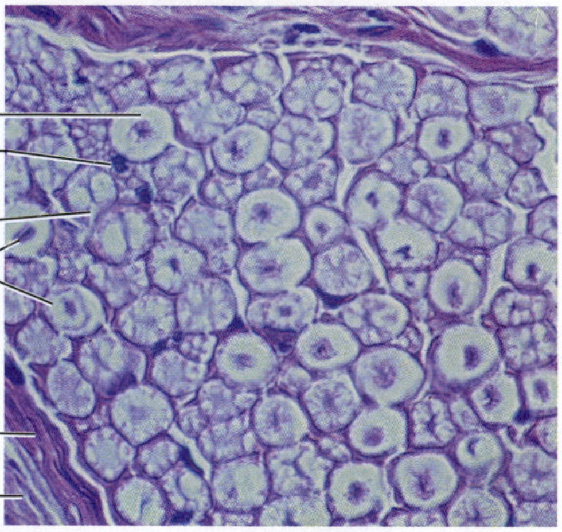
Myelin sheath
Nonmyelinated axon
Endoneurium
Heavily myelinated axons
Perineurium
Epineurium
(b)

FIGURE 17.8 Structure of a nerve showing connective tissue wrappings. (a) Three-dimensional view of a portion of a nerve. **(b)** Photomicrograph of a cross-sectional view of part of a peripheral nerve (330×).

a ganglion outside the CNS, and these neurons are typically unipolar.

Neurons carrying impulses from the CNS to the viscera and/or body muscles and glands are termed **motor,** or **efferent, neurons.** Motor neurons are most often multipolar, and their cell bodies are almost always located in the CNS.

The third functional category of neurons is **interneurons** or *association neurons,* which are situated between and contribute to pathways that connect sensory and motor neurons. Their cell bodies are always located within the CNS, and they are multipolar neurons structurally.

Structure of a Nerve

A nerve is a bundle of neuron fibers (axons) wrapped in connective tissue coverings that extends to and/or from the CNS and visceral organs or structures of the body periphery (such as skeletal muscles, glands, and skin).

Like neurons, nerves are classified according to the direction in which they transmit impulses. Nerves that carry only sensory processes and conduct impulses only toward the CNS are referred to as **sensory,** or **afferent, nerves.** A few of the cranial nerves are pure sensory nerves. The ventral roots of the spinal cord, which carry only motor fibers, can be considered **motor,** or **efferent, nerves.** Nerves carrying both sensory (afferent) and motor (efferent) fibers are called **mixed nerves;** most nerves of the body, including all spinal nerves, are mixed nerves.

Within a nerve, each fiber is surrounded by a delicate connective tissue sheath called an **endoneurium,** which insulates it from the other neuron processes adjacent to it. The endoneurium is often mistaken for the myelin sheath; it is instead an additional sheath that surrounds the myelin sheath. Groups of fibers are bound by a coarser connective tissue, called the **perineurium,** to form bundles of fibers called **fascicles.** Finally, all the fascicles are bound together by a white, fibrous connective tissue sheath called the **epineurium,** forming the cordlike nerve (Figure 17.8). In addition to the connective tissue wrappings, blood vessels and lymphatic vessels serving the fibers also travel within a nerve.

ACTIVITY 3

Examining the Microscopic Structure of a Nerve

Use the compound microscope to examine a prepared cross section of a peripheral nerve. Using Figure 17.8b as an aid, identify nerve fibers, myelin sheaths, fascicles, and endoneuriurn, perineurium, and epineurium sheaths. If desired, sketch the nerve in the space below. ■

NAME ______________________________

LAB TIME/DATE ______________________

REVIEW SHEET

EXERCISE 17

Histology of Nervous Tissue

1. The basic functional unit of the nervous system is the neuron. What is the major function of this cell type?

____

2. Name four types of neuroglia in the CNS, and list a function for each of these cells. (You will need to consult your textbook for this.)

Types	**Functions**
a. ______________________	a. ______________________
b. ______________________	b. ______________________
c. ______________________	c. ______________________
d. ______________________	d. ______________________

Name the PNS glial cell that forms myelin. ______________________

Name the PNS glial cell that surrounds dorsal root ganglion neurons. ______________________

3. Match each statement with a response chosen from the key.

Key:
a. afferent neuron
b. central nervous system
c. efferent neuron
d. ganglion
e. interneuron
f. neuroglia
g. neurotransmitters
h. nerve
i. nuclei
j. peripheral nervous system
k. synapse
l. tract

__________ 1. the brain and spinal cord collectively

__________ 2. specialized supporting cells in the CNS

__________ 3. junction or point of close contact between neurons

__________ 4. a bundle of nerve processes inside the CNS

__________ 5. neuron serving as part of the conduction pathway between sensory and motor neurons

__________ 6. ganglia and spinal and cranial nerves

__________ 7. collection of nerve cell bodies found outside the CNS

__________ 8. neuron that conducts impulses away from the CNS to muscles and glands

_________ 9. neuron that conducts impulses toward the CNS from the body periphery

_________ 10. chemicals released by neurons that stimulate or inhibit other neurons or effectors

Neuron Anatomy

4. Match the following anatomical terms (column B) with the appropriate description or function (column A).

Column A	Column B
_________ 1. region of the cell body from which the axon originates	a. axon
_________ 2. secretes neurotransmitters	b. axon terminal
_________ 3. receptive region of a neuron	c. axon hillock
_________ 4. insulates the nerve fibers	d. dendrite
_________ 5. site of the nucleus and most important metabolic area	e. myelin sheath
_________ 6. may be involved in the transport of substances within the neuron	f. neurofibril
_________ 7. essentially rough endoplasmic reticulum, important metabolically	g. neuronal cell body
_________ 8. impulse generator and transmitter	h. Nissl bodies

5. Draw a "typical" multipolar neuron in the space below. Include and label the following structures on your diagram: cell body, nucleus, nucleolus, Nissl bodies, dendrites, axon, axon collateral branch, myelin sheath, nodes of Ranvier, axon terminals, and neurofibrils.

6. What substance is found in synaptic vesicles of the axon terminal? _________________________

What role does this substance play in neurotransmission? _________________________

____

____

7. What anatomical characteristic determines whether a particular neuron is classified as unipolar, bipolar, or multipolar?

___

Make a simple line drawing of each type here.

Unipolar neuron **Bipolar neuron** **Multipolar neuron**

8. Correctly identify the sensory (afferent) neuron, interneuron (association neuron), and motor (efferent) neuron in the figure below.

Which of these neuron types is/are unipolar? ___

Which is/are most likely multipolar? ___

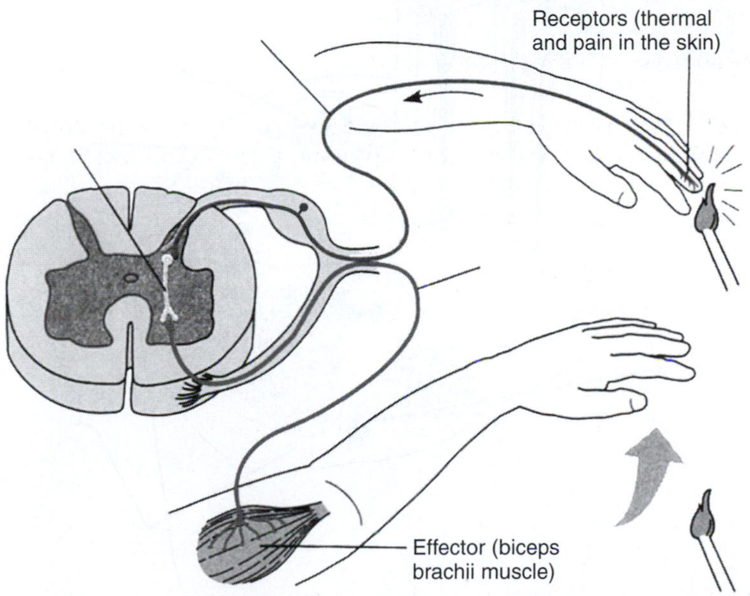
Receptors (thermal and pain in the skin)
Effector (biceps brachii muscle)

9. Describe how the Schwann cells form the myelin sheath and the neurilemma encasing the nerve processes.

___

___

Structure of a Nerve

10. What is a nerve? ___

___

11. State the location of each of the following connective tissue coverings.

endoneurium: ______________________________

perineurium: ______________________________

epineurium: ______________________________

12. What is the function of the connective tissue wrappings found in a nerve? ______________________________

____

13. Define *mixed nerve.* ______________________________

____

14. Identify all indicated parts of the nerve section.

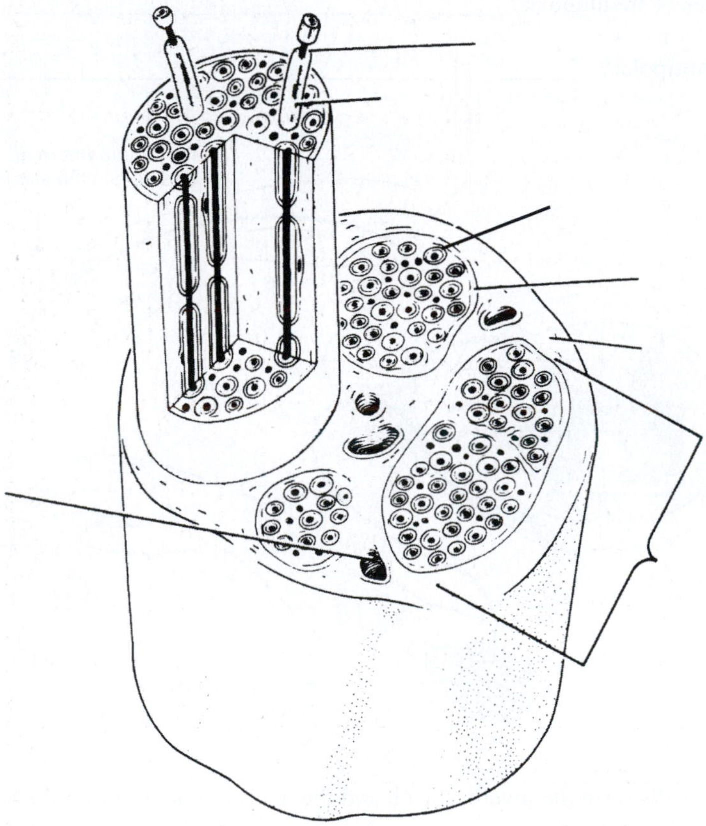

EXERCISE 18

Neurophysiology of Nerve Impulses

MATERIALS

- ☐ *Rana pipiens**
- ☐ Dissecting instruments and tray
- ☐ Disposable gloves
- ☐ Ringer's solution (frog) in dropper bottles, some at room temperature and some in an ice bath
- ☐ Thread
- ☐ Glass rods or probes
- ☐ Glass plates or slides
- ☐ Ring stand and clamp
- ☐ Stimulator; platinum electrodes
- ☐ Forceps
- ☐ Filter paper
- ☐ 0.01% hydrochloric acid (HCl) solution
- ☐ Sodium chloride (NaCl) crystals
- ☐ Heat-resistant mitts
- ☐ Bunsen burner
- ☐ Safety goggles
- ☐ Absorbent cotton
- ☐ Ether
- ☐ Pipettes
- ☐ 1-cc syringe with small-gauge needle
- ☐ 0.5% tubocurarine solution
- ☐ Frog board
- ☐ Oscilloscope
- ☐ Nerve chamber

*Instructor to provide freshly pithed frogs (*Rana pipiens*) for student experimentation.

MasteringA&P™

Access practice quizzes and more in the Study Area at www.masteringaandp.com.

PEx

This lab corresponds to PhysioEx Exercise 3. See p. PEx-35.

OBJECTIVES

1. To list the two major physiological properties of neurons.
2. To describe the polarized and depolarized states of the nerve cell membrane and to describe the events that lead to generation and conduction of a nerve impulse.
3. To explain how a nerve impulse is transmitted from one neuron to another.
4. To define *action potential, depolarization, repolarization, relative refractory period,* and *absolute refractory period.*
5. To list various substances and factors that can stimulate neurons.
6. To recognize that neurotransmitters may be either stimulatory or inhibitory in nature.
7. To state the site of action of the blocking agents ether and curare.

PRE-LAB QUIZ

1. Circle the correct term. Excitability / Conductivity is the ability to transmit nerve impulses to other neurons.
2. When a neuron is stimulated, the membrane becomes more permeable to Na^+ ions, which diffuse into the cell causing
   a. depolarization
   b. hyperpolarization
   c. repolarization
3. As an action potential progresses, the permeability to Na^+ decreases and the permeability to this ion increases:
   a. Ca^{2+}
   b. K^+
   c. Na^+
4. The period of time when the neuron is totally insensitive to further stimulation and cannot generate another action potential is
   a. absolute refractory period
   b. membrane potential
   c. repolarization
   d. threshold
5. What muscle and nerve will you need to isolate to study the physiology of nerve fibers?
   a. gastrocnemius and sciatic
   b. sartorius and femoral
   c. triceps brachii and radial
6. Circle True or False. Today you will use *alcohol* to inhibit the nerve impulse in your animal.
7. Circle the correct term. In the last portion of this experiment, you will electrically stimulate the frog's nerve and observe the compound action potential on a(n) oscilloscope / dissecting microscope.

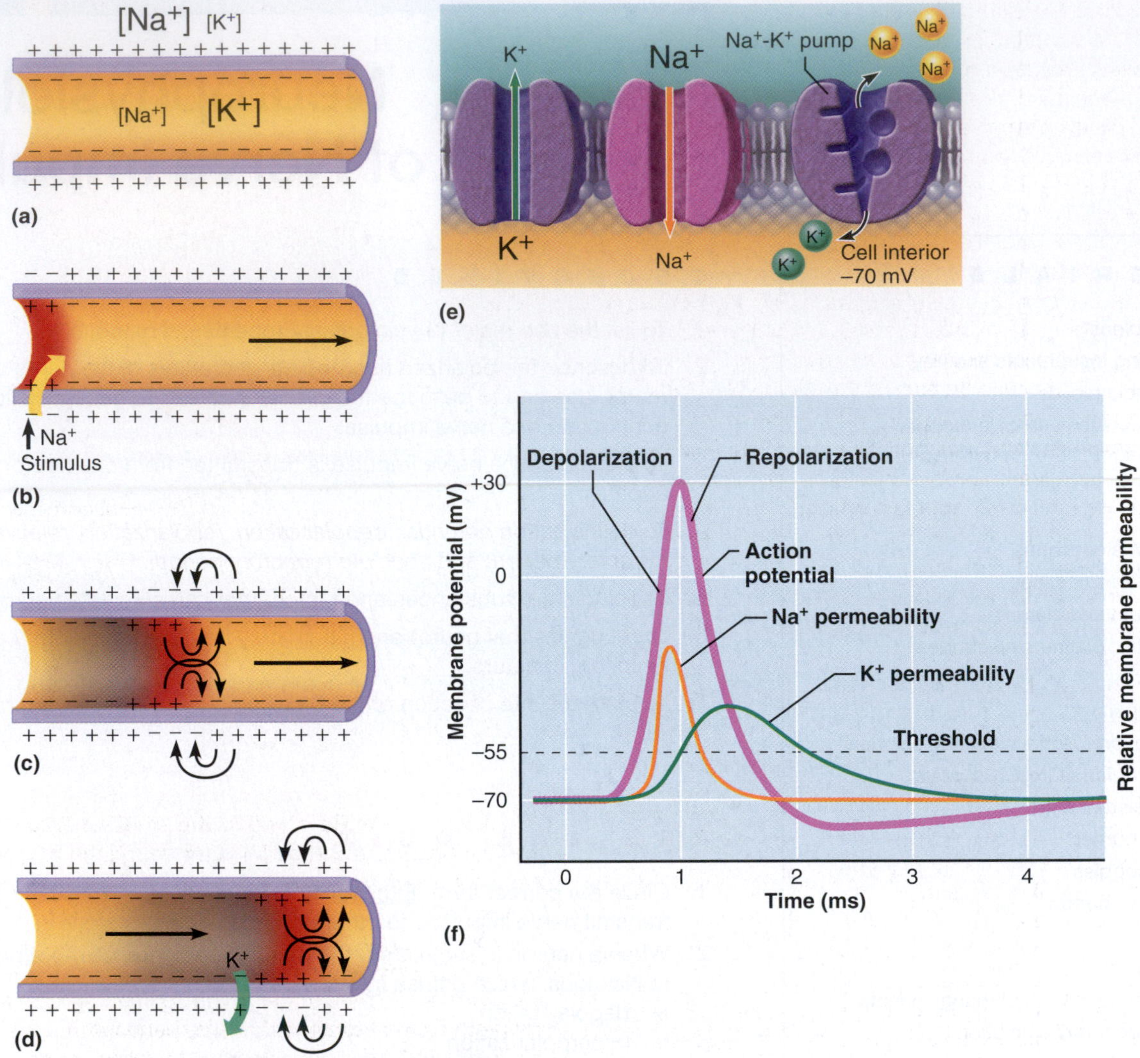
[Na+] [K+]
[Na+] [K+]
(a)
Na+
Stimulus
(b)
(c)
K+
(d)
K+
Na+
Na+-K+ pump
Na+
K+
Na+
Cell interior
–70 mV
(e)
Depolarization
Repolarization
Action potential
Na+ permeability
K+ permeability
Threshold
Membrane potential (mV)
+30
0
–55
–70
0
1
2
3
4
Time (ms)
Relative membrane permeability
(f)

FIGURE 18.1 The nerve impulse. (a) Resting membrane potential. There is an excess of positive ions at the external cell surface, with Na^+ the predominant extracellular fluid ion and K^+ the predominant intracellular ion. The plasma membrane has a low permeability to Na^+. **(b)** Depolarization—reversal of the resting potential. Application of a stimulus changes the membrane permeability, and Na^+ ions are allowed to diffuse rapidly into the cell. **(c)** Generation of the action potential or nerve impulse. If the stimulus is of adequate intensity, the depolarization wave spreads rapidly along the entire length of the membrane. **(d)** Repolarization—reestablishment of the resting potential. The negative charge on the internal plasma membrane surface and the positive charge on its external surface are reestablished by diffusion of K^+ ions out of the cell, proceeding in the same direction as in depolarization. **(e)** In the resting state, Na^+ ions leak into the cell and K^+ ions leak out. The resting membrane potential is maintained by the active sodium-potassium pump. **(f)** The action potential is caused by permeability changes in the plasma membrane.

Neurons are **excitable;** they respond to stimuli by producing an elecrical signal. Excited neurons communicate—they transmit electrical signals to neurons, muscles, glands and other tissues of the body, a property called **conductivity.** In a resting neuron, the interior of the cell membrane is slightly more negatively charged than the exterior surface as illustrated in Figure 18.1a. The difference in electrical charge produces a **resting membrane potential** across the membrane that is measured in millivolts. As in most cells, the predominant intracellular cation is K^+, and Na^+ is the predominant cation in the extracellular fluid. In a resting neuron, Na^+ leaks into the cell and K^+ leaks out. The resting membrane potential is maintained by the sodium-potassium pump, which transports Na^+ back out of the cell and K^+ back into the cell.

The Nerve Impulse

When a neuron receives an excitatory stimulus, the membrane becomes more permeable to sodium ions, and Na^+ diffuses down its electrochemical gradient into the cell. As a result, the interior of the membrane becomes less negative (Figure 18.1b), an event is called **depolarization.** If the stimulus is great enough to depolarize the initial segment of the axon to **threshold,** a **nerve impluse** or **action potential** is generated. The initial segment of the axon in multipolar neurons is at the axon hillock of the cell body. In peripheral sensory neurons, the initial segment is just proximal to the sensory receptor far from the cell body located in the dorsal root ganglion.

When the threshold voltage is reached, the membrane permeability to Na^+ increases rapidly (Figure 18.1f). As the neuron depolarizes, the polarity of the membrane reverses with the interior surface now becoming more positive than the exterior (Figure 18.1c). As the membrane permeability to Na^+ falls, the permeability to K^+ increases, and K^+ diffuses down its electrochemical gradient and out of the cell (Figure 18.1d). Once again the interior of the membrane becomes more negative than the exterior. This event is called **repolarization.**

The period of time when Na^+ permeability is rapidly changing and maximal, and the period immediately following when Na^+ permeability becomes restricted, together correspond to a time when the neuron is insensitive to further stimulation and cannot generate another action potential. This period is called the **absolute refractory period.** As Na^+ permeability is gradually restored to resting levels during repolarization, an especially strong stimulus to the neuron may provoke another action potential. This period of time is the **relative refractory period.** Restoration of the resting membrane potential restores the neuron's normal excitability.

Once generated, the action potential propagates along the entire length of the axon. It is never partially transmitted. Furthermore, it retains a constant amplitude and duration; the action potential is not small when a stimulus is small and large when a stimulus is large. Since the action potential of a given neuron is always the same, it is said to be an all-or-none response. When an action potential reaches the axon terminals, it causes neurotransmitter to be released. The neurotransmitter may be excitatory or inhibitory to the next cell in the transmission chain, depending on the receptor types on that cell. (The experiments in this exercise consider only excitatory neurotransmitters.)

The concentration of Na^+ and K^+ both inside and outside an active neuron change very little during a single action potential. Even when many action potentials are generated in a given area, the sodium-potassium pumps maintain the concentration differences across the membrane that are needed for the normal function of these excitable cells (Figure 18.1e).

Physiology of Nerve Fibers

The sciatic nerve is a bundle of axons that vary in diameter. An electrical signal recorded from a nerve represents the summed electrical activity of all the nerve fibers. This summed activity is called a **compound action potential**. Unlike an action potential in a single nerve fiber, the compound action potential varies in shape according to which fibers are producing action potentials. When a nerve is stimulated by external electrodes, as in our experiments, the largest fibers reach threshold first and generate action potentials. Higher-intensity stimuli are required to produce action potentials in smaller fibers.

In this laboratory session, you will investigate the functioning of nerve fibers by subjecting the sciatic nerve of a frog to various types of stimuli and blocking agents. Work in groups of two to four to lighten the workload.

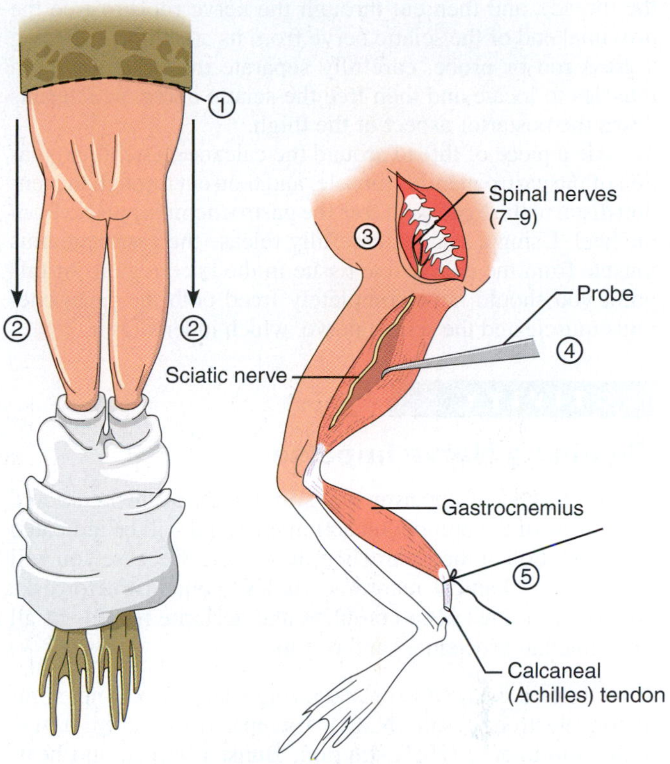

①
②
②
③
Spinal nerves (7–9)
Probe
④
Sciatic nerve
Gastrocnemius
⑤
Calcaneal (Achilles) tendon

FIGURE 18.2 Removal of the sciatic nerve and gastrocnemius muscle. (1) Cut through the frog's skin around the circumference of the trunk. **(2)** Pull the skin down over the trunk and legs. **(3)** Make a longitudinal cut through the abdominal musculature and expose the roots of the sciatic nerve (arising from spinal nerves 7–9). Ligate the nerve and cut the roots proximal to the ligature. **(4)** Use a glass probe to expose the sciatic nerve beneath the posterior thigh muscles. **(5)** Ligate the calcaneal (Achilles) tendon and cut it free distal to the ligature. Release the gastrocnemius muscle from the connective tissue of the knee region.

DISSECTION: Isolating the Gastrocnemius Muscle and Sciatic Nerve

1. Don gloves to protect yourself from any parasites the frogs might have. Obtain a pithed frog from your instructor, and bring it to your laboratory bench. Also obtain dissecting instruments, a tray, thread, two glass rods or probes, and frog Ringer's solution at room temperature from the supply area.

2. Prepare the sciatic nerve as illustrated in Figure 18.2. Place the pithed frog on the dissecting tray, dorsal side down. Make a cut through the skin around the circumference of the frog approximately halfway down the trunk, and then pull the skin down over the muscles of the legs. Open the abdominal cavity and push the abdominal organs to one side to expose the origin of the glistening white sciatic nerve, which arises from the last three spinal nerves. *Once the sciatic nerve has been exposed, it should be kept continually moist with room temperature Ringer's solution.*

3. Using a glass probe, slip a piece of thread moistened with Ringer's solution under the sciatic nerve close to its origin at the vertebral column. Make a single ligature (tie it firmly with

the thread), and then cut through the nerve roots to free the proximal end of the sciatic nerve from its attachments. Using a glass rod or probe, carefully separate the posterior thigh muscles to locate and then free the sciatic nerve, which runs down the posterior aspect of the thigh.

4. Tie a piece of thread around the calcaneal (Achilles) tendon of the gastrocnemius muscle, and then cut through the tendon distal to the ligature to free the gastrocnemius muscle from the heel. Using a scalpel, carefully release the gastrocnemius muscle from the connective tissue in the knee region. At this point you should have completely freed both the gastrocnemius muscle and the sciatic nerve, which innervates it.

ACTIVITY 1

Eliciting a Nerve Impulse

In this first set of experiments, stimulation of the nerve and generation of the compound action potential will be indicated by contraction of the gastrocnemius muscle. Because you will make no mechanical recording (unless your instructor asks you to), you must keep complete and accurate records of all experimental procedures and results.

1. Obtain a glass slide or plate, ring stand and clamp, stimulator, electrodes, salt (NaCl), forceps, filter paper, 0.01% hydrochloric acid (HCl) solution, Bunsen burner, and heat-resistant mitts. With glass rods, transfer the isolated muscle-nerve preparation to a glass plate or slide, and then attach the slide to a ring stand with a clamp. Allow the end of the sciatic nerve to hang over the free edge of the glass slide, so that it is easily accessible for stimulation. *Remember to keep the nerve moist at all times.*

2. You are now ready to investigate the response of the sciatic nerve to various stimuli, beginning with electrical stimulation. Using the stimulator and platinum electrodes, stimulate the sciatic nerve with single shocks, gradually increasing the intensity of the stimulus until the threshold stimulus is determined.

(The muscle as a whole will just barely contract at the threshold stimulus.) Record the voltage of this stimulus:

Threshold stimulus: ____________ V

Continue to increase the voltage until you find the point beyond which no further increase occurs in the strength of muscle contraction—that is, the point at which the maximal contraction of the muscle is obtained. Record this voltage below.

Maximal stimulus: ____________ V

Delivering multiple or repeated shocks to the sciatic nerve causes volleys of impulses in the nerve. Shock the nerve with multiple stimuli. Observe the response of the muscle. How does this response compare with the response to the single electrical shocks?

____

____

3. To investigate mechanical stimulation, pinch the free end of the nerve by firmly pressing it between two glass rods or by pinching it with forceps. What is the result?

____

4. Chemical stimulation can be tested by applying a small piece of filter paper saturated with HC1 solution (from the supply area) to the free end of the nerve. What is the result?

____

Drop a few grains of salt (NaCl) on the free end of the nerve. What is the result?

____

5. Now test thermal stimulation. Wearing the heat-resistant mitts, heat a glass rod for a few moments over a Bunsen burner. Then touch the rod to the free end of the nerve. What is the result?

____

What do these muscle reactions say about the irritability and conductivity of neurons?

____

____

Although most neurons within the body are stimulated to the greatest degree by a particular stimulus (in many cases, a chemical neurotransmitter), a variety of other stimuli may trigger nerve impulses, as illustrated by the experimental series just conducted. Generally, no matter what type of stimulus is present, if the affected part responds by becoming activated, it will always react in the same way. Familiar examples are the well-known phenomenon of "seeing stars" when you receive a blow to the head or press on your eyeball (try it), both of which trigger impulses in your optic nerves.

ACTIVITY 2

Inhibiting the Nerve Impulse

Numerous physical factors and chemical agents can impair the ability of nerve fibers to function. For example, deep pressure and cold temperature both block nerve impulse transmission by preventing the local blood supply from reaching the nerve fibers. Local anesthetics, alcohol, and numerous other chemicals are also very effective at blocking nerve transmission. Ether, one such chemical blocking agent, will be investigated first.

! Since ether is extremely volatile and explosive, perform this experiment in a vented hood. *Don safety goggles before beginning this procedure.*

1. Obtain another glass slide or plate, absorbent cotton. ether, and a pipette. Clamp the new glass slide to the ring stand slightly below the first slide of the apparatus setup for the previous experiment. With glass rods, gently position the sciatic nerve on this second slide, allowing a small portion of the nerve's distal end to extend over the edge. Place a piece of absorbent cotton soaked with ether under the midsection of the nerve on the slide, prodding it into position with a glass rod. Using a voltage slightly above the threshold stimulus, stimulate the distal end of the nerve at 2-minute intervals until the muscle fails to respond. (If the cotton dries before this, re-wet it with ether using a pipette.) How long did it take for anesthesia to occur?

__________ sec

2. Once anesthesia has occurred, stimulate the nerve beyond the anesthetized area, between the ether-soaked pad and the muscle. What is the result?

____

3. Remove the ether-soaked pad and flush the nerve fibers with Ringer's solution. Again stimulate the nerve at its distal end at 2-minute intervals. How long does it take for recovery?

____

Does ether exert its blocking effect on the nerve fibers *or* on the muscle cells?

______________________ Explain your reasoning.

____

____

If sufficient frogs are available and time allows, you may do the following classic experiment. In the 1800s Claude Bernard described an investigation into the effect of curare on nerve-muscle interaction. *Curare* was used by some South American Indian tribes to tip their arrows. Victims struck with these arrows were paralyzed, but the paralysis was not accompanied by loss of sensation.

1. Prepare another frog as described in steps 1 through 3 of the dissection instructions. However, in this case position the frog ventral side down on a frog board. In exposing the sciatic nerve, take care not to damage the blood vessels in the thigh region, as the success of the experiment depends on maintaining the blood supply to the muscles of the leg.

2. Expose and gently tie the left sciatic nerve so that it can be lifted away from the muscles of the leg for stimulation. Slip another length of thread under the nerve, and then tie the thread tightly around the thigh muscles to cut off circulation to the leg. The sciatic nerve should be above the thread and *not in* the ligated tissue. Expose and ligate the sciatic nerve of the *right* leg in the same manner, but this time do *not* ligate the thigh muscles.

3. Obtain a syringe and needle, and a vial of 0.5% tubocurarine. Slowly and carefully inject 1 cc of the tubocurarine into the dorsal lymph sac of the frog.* The dorsal lymph sacs are located dorsally at the level of the scapulae, so introduce the needle of the syringe just beneath the skin between the scapulae and toward one side of the spinal column. *Handle the tubocurarine very carefully, because it is extremely poisonous.* Do not get any on your skin.

4. Wait 15 minutes after injection of the tubocurarine to allow it to be distributed throughout the body in the blood and lymphatic stream. Then electrically stimulate the left sciatic nerve. Be careful not to touch any of the other tissues with the electrode. Gradually increasing the voltage, deliver single shocks until the threshold stimulus is determined for this specimen.

Threshold stimulus: __________ V

Now stimulate the right sciatic nerve with the same voltage intensity. Is there any difference in the reaction of the two muscles?

__________ If so, explain. ____________________

____

____

If you did not find any difference, wait an additional 10 to 15 minutes and restimulate both sciatic nerves.

What is the result? ____________________

____

5. To determine the site at which tubocurarine acts, directly stimulate each gastrocnemius muscle. What is the result?

____

____

____

Explain the difference between the responses of the right and left sciatic nerves.

____

____

____

*To obtain 1 cc of the tubocurarine, inject 1 cc of air into the vial through the rubber membrane, and then draw up 1 cc of the chemical into the syringe.

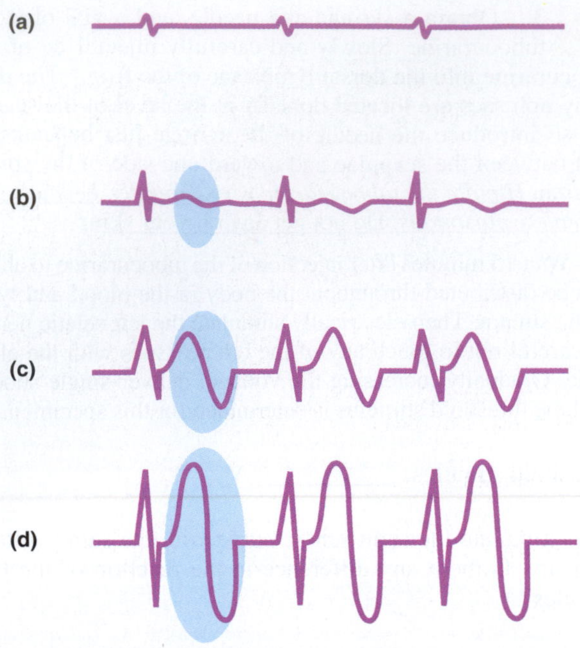
(a)
(b)
(c)
(d)

FIGURE 18.3 Recordings of the compound action potential from the sciatic nerve. The first compound action potential in each scan is circled. **(a)** Stimulus artifacts only **(b)–(d)** Increasing stimulas strengths reveal the graded nature of the compound action potential.

Explain the results when the muscles were stimulated directly.

___

___

___

At what site does tubocurarine (or curare) act?

___

The Oscilloscope: An Experimental Tool

In this exercise, a frog's sciatic nerve will be electrically stimulated, and the compound action potential generated will be observed on the oscilloscope. The **oscilloscope** is an instrument that visually displays the rapid changes in voltage that occur during an action potential. The dissected nerve will be placed in contact with two pairs of electrodes—*stimulating* and *recording*. The stimulating electrodes will be used to deliver a pulse of electricity to a point on the sciatic nerve. At another point on the nerve, a pair of recording electrodes connected to the oscilloscope will deliver the current to the oscilloscope, and the electrical pulse will be visible on the screen as a vertical deflection, or a *stimulus artifact* (Figure 18.3a). As the nerve is stimulated with increasingly higher voltage, the stimulus artifact increases in amplitude as well. When the stimulus voltage reaches a high enough level (threshold), a compound action potential will be generated by the nerve, and a *second* vertical deflection will appear on the screen, approximately 2 milliseconds after the stimulus artifact (Figure 18.3b–d). This second deflection reports the potential difference between the two recording electrodes—that is, between the first recording electrode that has already depolarized (and is in the process of repolarizing as the compound action potential travels along the nerve) and the second recording electrode.

ACTIVITY 3

Visualizing the Action Potential with an Oscilloscope

1. Obtain a nerve chamber, an oscilloscope, a stimulator, frog Ringer's solution (room temperature), a dissecting needle, and glass probes. Set up the experimental apparatus as illustrated in Figure 18.4. Connect the two stimulating electrodes to the output terminals of the stimulator and the two recording electrodes to the preamplifier of the oscilloscope.

2. Obtain another pithed frog, and prepare one of its sciatic nerves for experimentation as indicated on page 271 in steps 1 through 3 in the dissection instructions. While working, be careful not to touch the nerve with your fingers, and do not allow the nerve to touch the frog's skin.

3. When you have freed the sciatic nerve to the knee region with the glass probe, slip another thread length beneath that end of the nerve and make a ligature. Cut the nerve distal to this tied thread and then carefully lift the cut nerve away from the thigh of the frog by holding the threads at the nerve's proximal and distal ends. Place the nerve in the nerve chamber so that it rests across all four electrodes (the two stimulating and two recording electrodes) as shown in Figure 18.4. Flush the nerve with room temperature frog Ringer's solution.

4. Adjust the horizontal sweep according to the instructions given in the manual or by your instructor, and set the stimulator duration, frequency, and amplitude to their lowest settings.

5. Begin to stimulate the nerve with single stimuli, slowly increasing the voltage until a threshold stimulus is achieved. The compound action potential will appear as a small rounded "hump" immediately following the stimulus artifact. Record the voltage of the threshold stimulus:

Threshold stimulus: ____________ V

6. Flush the nerve with the Ringer's solution and continue to increase the voltage, watching as the vertical deflections produced by the compound action potential become diphasic (show both upward and downward vertical deflections). Record the voltage at which the compound action potential reaches its maximal amplitude; this is the maximal stimulus.

Maximal stimulus: ____________ V

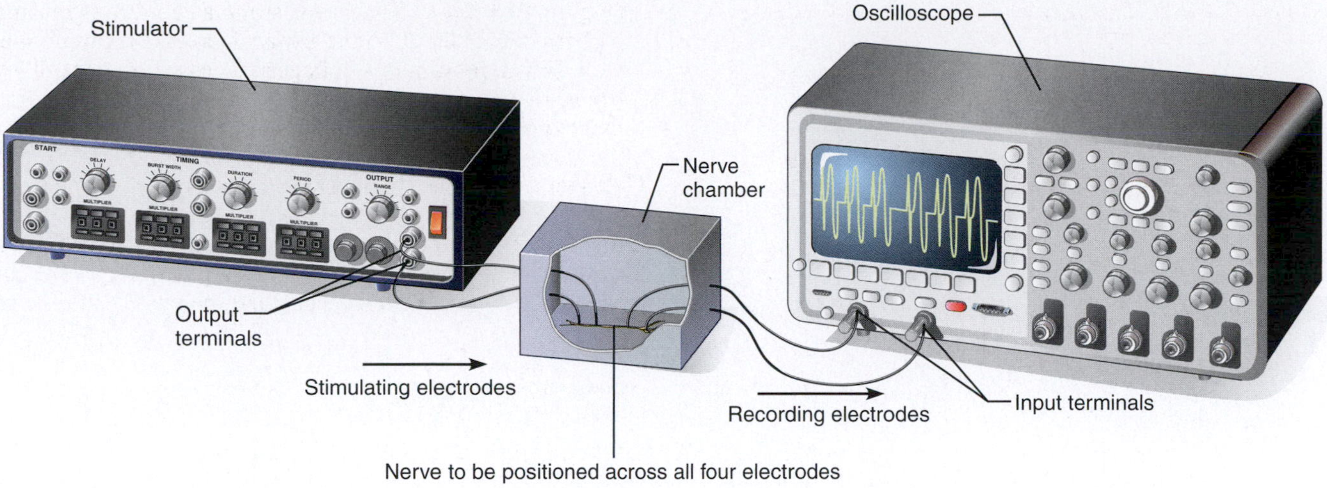
Stimulator
Oscilloscope
Nerve chamber
Output terminals
Stimulating electrodes
Recording electrodes
Input terminals
Nerve to be positioned across all four electrodes

FIGURE 18.4 Setup for oscilloscope visualization of action potentials in a nerve.

7. Set the stimulus voltage at a level just slightly lower than the maximal stimulus and gradually increase the frequency of stimulation. What is the effect on the size (amplitude) of the compound action potential?

____

8. Flush the nerve with room temperature Ringer's solution once again, and allow the nerve to sit for a few minutes while you obtain a bottle of Ringer's solution from the ice bath. Repeat steps 5 and 6 while your partner continues to flush the nerve preparation with the cold saline. Record the threshold and maximal stimuli, and watch the oscilloscope pattern carefully to detect any differences in the velocity or speed of conduction from what was seen previously.

Threshold stimulus: ____________ V

Maximal stimulus: ____________ V

9. Flush the nerve preparation with room temperature Ringer's solution again and then gently lift the nerve by its attached threads. Then turn the nerve around so that the end formerly resting on the stimulating electrodes now rests on the recording electrodes and vice versa. Stimulate the nerve. Is the impulse conducted in the opposite direction?

____

10. Dispose of the frog remains and gloves in the appropriate containers, clean the lab bench and equipment, and return your equipment to the proper supply area.

NAME ____________________________

LAB TIME/DATE ______________________

REVIEW SHEET

EXERCISE 18

Neurophysiology of Nerve Impulses

The Nerve Impulse

1. Match the terms in column B to the appropriate definition in column A.

Column A

_______ 1. period of depolarization of the neuron membrane during which it cannot respond to a second stimulus

_______ 2. reversal of the resting potential due to an influx of sodium ions

_______ 3. period during which potassium ions diffuse out of the neuron because of a change in membrane permeability

_______ 4. period of repolarization when only a strong stimulus will elicit an action potential

_______ 5. mechanism in which ATP is used to move sodium out of the cell and potassium into the cell; restores the resting membrane voltage and intracellular ionic concentrations

Column B

a. absolute refractory period

b. action potential

c. depolarization

d. relative refractory period

e. repolarization

f. sodium-potassium pump

2. Define the term *depolarization*. ____

____

How does an action potential differ from simple depolarization? ____________________________

____

3. Would a substance that decreases membrane permeability to sodium increase or decrease the probability of generating an action potential? Why?

____

____

4. The diagram here represents a section of an axon. Complete the figure by illustrating an area of resting membrane potential, an area of depolarization, and local current flow. Indicate the direction of the depolarization wave.

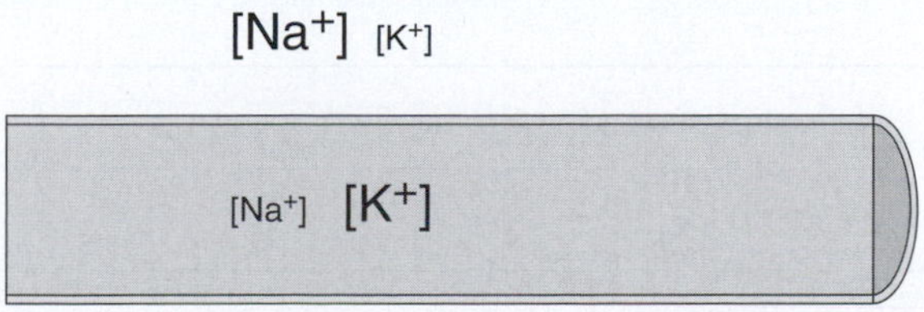
[Na+] [K+]
[Na+] [K+]

Physiology of Nerve Fibers: Eliciting and Inhibiting the Nerve Impulse

5. Respond appropriately to each question posed below. Insert your responses in the corresponding numbered blanks to the right.

1–3. Name three types of stimuli that resulted in action potential generation in the sciatic nerve of the frog.

4. Which of the stimuli resulted in the most effective nerve stimulation?

5. Which of the stimuli employed in that experiment might represent types of stimuli to which nerves in the human body are subjected?

6. What is the usual mode of stimulus transfer in neuron-to-neuron interactions?

7. Since the action potentials themselves were not visualized with an oscilloscope during this initial set of experiments, how did you recognize that impulses were being transmitted?

1. ______________________

2. ______________________

3. ______________________

4. ______________________

5. ______________________

6. ______________________

7. ______________________

6. How did the site of action of ether and tubocurarine differ? ______________________

____

In the curare experiment, why was one of the frog's legs ligated? ______________________

____

Visualizing the Action Potential with an Oscilloscope

7. Explain why the amplitude of the compound action potential recorded from the frog sciatic nerve increased when the voltage of the stimulus was increased above the threshold value. ______________________

____

8. What was the effect of cold temperature (flooding the nerve with iced Ringer's solution) on the functioning of the sciatic nerve tested? ______________________

____

9. When the nerve was reversed in position, was the impulse conducted in the opposite direction? ______________________

How can this result be reconciled with the concept of one-way conduction in neurons? ______________________

____

____

____

EXERCISE 19

Gross Anatomy of the Brain and Cranial Nerves

MATERIALS

- ☐ Human brain model (dissectible)
- ☐ Preserved human brain (if available)
- ☐ Three-dimensional model of ventricles
- ☐ Coronally sectioned human brain slice (if available)
- ☐ Materials as needed for cranial nerve testing (see Table 19.1): aromatic oils (e.g., vanilla and cloves); eye chart; ophthalmoscope penlight; safety pin; blunt probe (hot and cold); cotton; solutions of sugar, salt, vinegar, and quinine; ammonia; tuning fork; and tongue depressor
- ☐ Preserved sheep brain (meninges and cranial nerves intact)
- ☐ Dissecting instruments and tray
- ☐ Disposable gloves
- ☐ *The Human Nervous System: The Brain and Cranial Nerves* videotape

MasteringA&P™

Access practice quizzes and more in the Study Area at www.masteringaandp.com.

PAL

For access to anatomical models and more, check out Practice Anatomy Lab.

OBJECTIVES

1. To identify the following brain structures on a dissected specimen (or slices), human brain model, or appropriate diagram, and to state their functions:
   - *Cerebral hemisphere structures:* lobes, important fissures, lateral ventricles, basal ganglia, corpus callosum, fornix, septum pellucidum
   - *Diencephalon structures:* thalamus, intermediate mass, hypothalamus, optic chiasma, pituitary gland, mammillary bodies, pineal body, choroid plexus of the third ventricle, interventricular foramen
   - *Brain stem structures:* corpora quadrigemina, cerebral aqueduct, cerebral peduncles of the midbrain, pons, medulla, fourth ventricle
   - *Cerebellum structures:* cerebellar hemispheres, vermis, arbor vitae
2. To describe the composition of gray and white matter.
3. To locate the well-recognized functional areas of the human cerebral hemispheres.
4. To define *gyri, fissures,* and *sulci.*
5. To identify the three meningeal layers and state their function, and to locate the falx cerebri, falx cerebelli, and tentorium cerebelli.
6. To state the function of the arachnoid villi and dural sinuses.
7. To discuss the formation, circulation, and drainage of cerebrospinal fluid.
8. To identify at least four pertinent anatomical differences between the human brain and that of the sheep (or other mammal).
9. To identify the cranial nerves by number and name on a model or diagram, stating the origin and function of each.

PRE-LAB QUIZ

1. Circle the correct term. The central nervous system / peripheral nervous system consists of the brain and spinal cord.
2. Circle the correct term. The most superior portion of the brain is the cerebral hemispheres / brain stem.
3. Circle True or False. Deep grooves within the cerebral hemispheres are known as gyri.
4. On the ventral surface of the brain, you can observe the optic nerves and chiasma, the pituitary gland, and the mammillary bodies. These externally visible structures form the floor of the
   a. brain stem c. frontal lobe
   b. diencephalon d. occipital lobe
5. Circle the correct term. The lowest region of the brain stem, the medulla oblongata / cerebellum houses many vital autonomic centers involved in the control of heart rate, respiratory rhythm, and blood pressure.
6. Directly under the occipital lobes of the cerebrum is a large cauliflower-like structure known as the __________.
   a. brain stem b. cerebellum c. diencephalon

Text continues on next page.

7. Circle the correct term. The outer cortex of the brain contains the cell bodies of cerebral neurons and is known as white matter / gray matter.
8. The brain and spinal cord are covered and protected by three connective tissue layers called
   a. lobes
   b. meninges
   c. sulci
   d. ventricles
9. Circle True or False. Cerebrospinal fluid is produced by the frontal lobe of the cerebrum and is unlike any other body fluid.
10. How many pairs of cranial nerves are there?

    ____

When viewed alongside all nature's animals, humans are indeed unique, and the key to their uniqueness is found in the brain. Each of us is a composite reflection of our brain's experience. If all past sensory input could mysteriously and suddenly be "erased," we would be unable to walk, talk, or communicate in any manner. Spontaneous movement would occur, as in a fetus, but no voluntary integrated function of any type would be possible. Clearly we would cease to be the same individuals.

Because of the complexity of the nervous system, its anatomical structures are usually considered in terms of two principal divisions: the central nervous system and the peripheral nervous system. The **central nervous system (CNS)** consists of the brain and spinal cord, which primarily interpret incoming sensory information and issue instructions based on that information and on past experience. The **peripheral nervous system (PNS)** consists of the cranial and spinal nerves, ganglia, and sensory receptors. These structures serve as communication lines as they carry impulses—from the sensory receptors to the CNS and from the CNS to the appropriate glands, muscles, or other effector organs.

The PNS has two major subdivisions: the **sensory portion,** which consists of nerve fibers that conduct impulses toward the CNS, and the **motor portion,** which contains nerve fibers that conduct impulses away from the CNS. The motor portion, in turn, consists of the **somatic division** (sometimes called the *voluntary system*), which controls the skeletal muscles, and the **autonomic nervous system (ANS),** which controls smooth and cardiac muscles and glands. The ANS is often referred to as the *involuntary nervous system.* Its sympathetic and parasympathetic branches play a major role in maintaining homeostasis.

In this exercise both the brain (CNS) and cranial nerves (PNS) will be studied because of their close anatomical relationship.

The Human Brain

During embryonic development of all vertebrates, the CNS first makes its appearance as a simple tubelike structure, the **neural tube,** that extends down the dorsal median plane. By the fourth week, the human brain begins to form as an expansion of the anterior or rostral end of the neural tube (the end toward the head). Shortly thereafter, constrictions appear, dividing the developing brain into three major regions—**forebrain, midbrain,** and **hindbrain** (Figure 19.1). The remainder of the neural tube becomes the spinal cord.

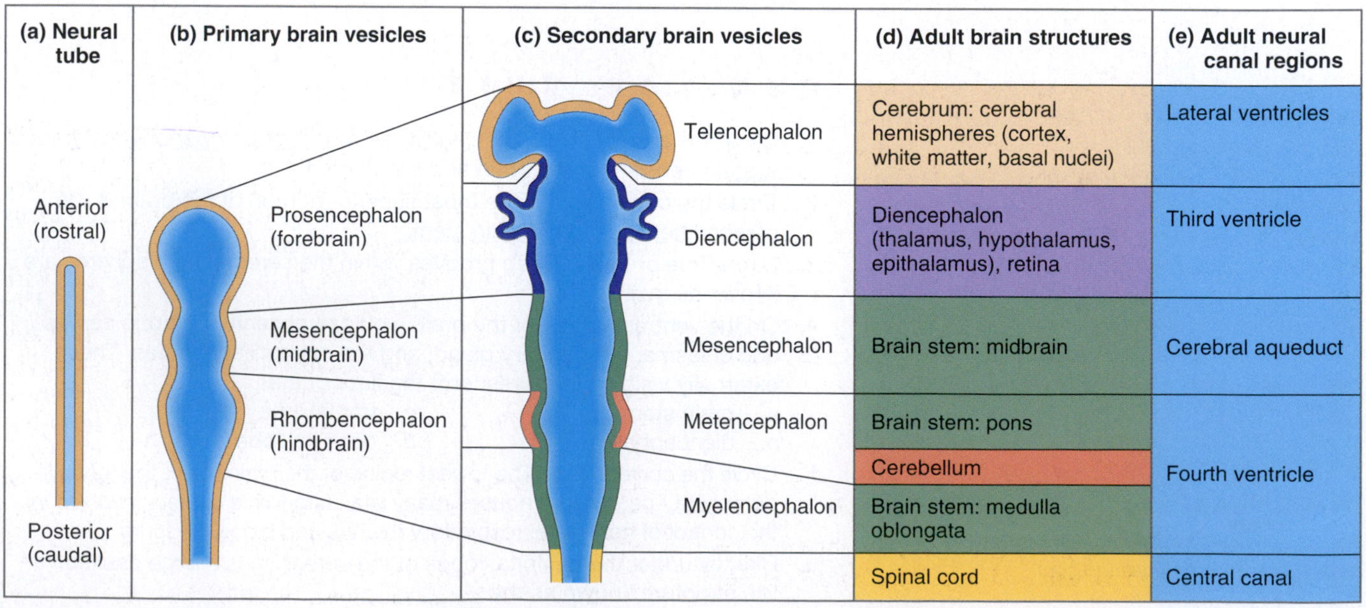
(a) Neural tube
(b) Primary brain vesicles
(c) Secondary brain vesicles
(d) Adult brain structures
(e) Adult neural canal regions
Anterior (rostral)
Posterior (caudal)
Prosencephalon (forebrain)
Mesencephalon (midbrain)
Rhombencephalon (hindbrain)
Telencephalon
Diencephalon
Mesencephalon
Metencephalon
Myelencephalon
Cerebrum: cerebral hemispheres (cortex, white matter, basal nuclei)
Diencephalon (thalamus, hypothalamus, epithalamus), retina
Brain stem: midbrain
Brain stem: pons
Cerebellum
Brain stem: medulla oblongata
Spinal cord
Lateral ventricles
Third ventricle
Cerebral aqueduct
Fourth ventricle
Central canal

FIGURE 19.1 Embryonic development of the human brain. (a) The neural tube subdivides into **(b)** the primary brain vesicles, which subsequently form **(c)** the secondary brain vesicles, which differentiate into **(d)** the adult brain structures. **(e)** The adult structures derived from the neural canal.

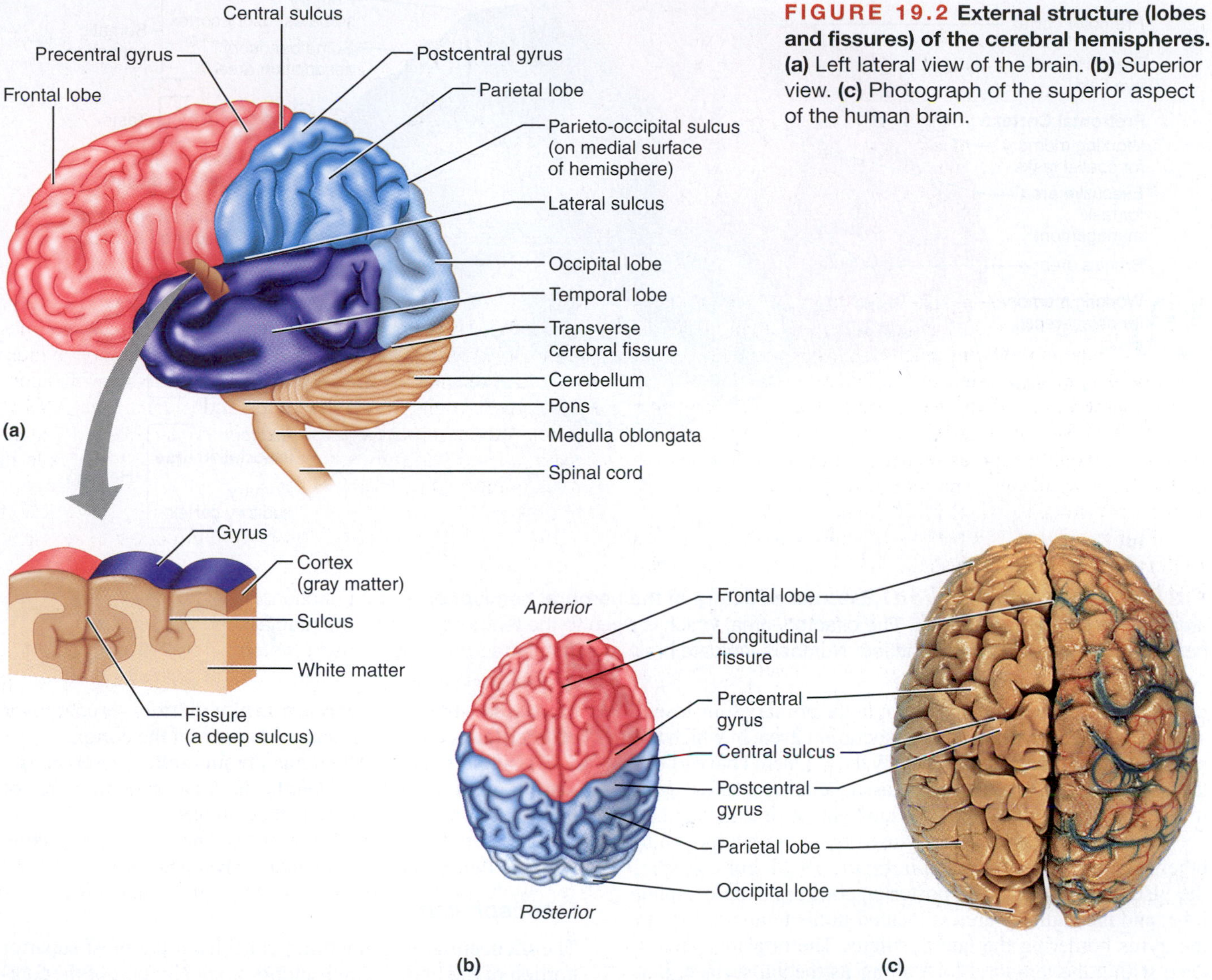

Central sulcus
Precentral gyrus
Postcentral gyrus
Frontal lobe
Parietal lobe
Parieto-occipital sulcus (on medial surface of hemisphere)
Lateral sulcus
Occipital lobe
Temporal lobe
Transverse cerebral fissure
Cerebellum
Pons
Medulla oblongata
Spinal cord
(a)
Gyrus
Cortex (gray matter)
Sulcus
White matter
Fissure (a deep sulcus)
Anterior
Posterior
Frontal lobe
Longitudinal fissure
Precentral gyrus
Central sulcus
Postcentral gyrus
Parietal lobe
Occipital lobe
(b)
(c)

FIGURE 19.2 External structure (lobes and fissures) of the cerebral hemispheres. (a) Left lateral view of the brain. **(b)** Superior view. **(c)** Photograph of the superior aspect of the human brain.

During fetal development, two anterior outpocketings extend from the forebrain and grow rapidly to form the cerebral hemispheres. Because of space restrictions imposed by the skull, the cerebral hemispheres are forced to grow posteriorly and inferiorly, and finally end up enveloping and obscuring the rest of the forebrain and most midbrain structures. Somewhat later in development, the dorsal hindbrain also enlarges to produce the cerebellum. The central canal of the neural tube, which remains continuous throughout the brain and cord, enlarges in four regions of the brain, forming chambers called **ventricles** (see Figure 19.8a and b, page 288).

ACTIVITY 1

Identifying External Brain Structures

Identify external brain structures using the figures cited. Also use a model of the human brain and other learning aids as they are mentioned.

Generally, the brain is studied in terms of four major regions: the cerebral hemispheres, diencephalon, brain stem, and cerebellum. The relationship between these four anatomical regions and the structures of the forebrain, midbrain, and hindbrain is also outlined in Figure 19.1.

Cerebral Hemispheres

The **cerebral hemispheres** are the most superior portion of the brain (Figure 19.2). Their entire surface is thrown into elevated ridges of tissue called **gyri** that are separated by shallow grooves called **sulci** or deeper grooves called **fissures.** Many of the fissures and gyri are important anatomical landmarks.

The cerebral hemispheres are divided by a single deep fissure, the **longitudinal fissure.** The **central sulcus** divides the **frontal lobe** from the **parietal lobe,** and the **lateral sulcus** separates the **temporal lobe** from the parietal lobe. The **parieto-occipital sulcus** on the medial surface of each hemisphere divides the **occipital lobe** from the parietal lobe. It is not visible externally. A fifth lobe of each cerebral hemisphere, the **insula,** is buried deep within the lateral sulcus, and is covered by portions of the temporal, parietal, and frontal lobes. Notice that most cerebral hemisphere lobes are named for the cranial bones that lie over them.

Some important functional areas of the cerebral hemispheres have also been located (Figure 19.2d). The **primary somatosensory cortex** is located in the **postcentral gyrus** of the parietal lobe. Impulses traveling from the body's sensory receptors (such as those for pressure, pain, and temperature) are localized in this area of the brain. ("This information is from

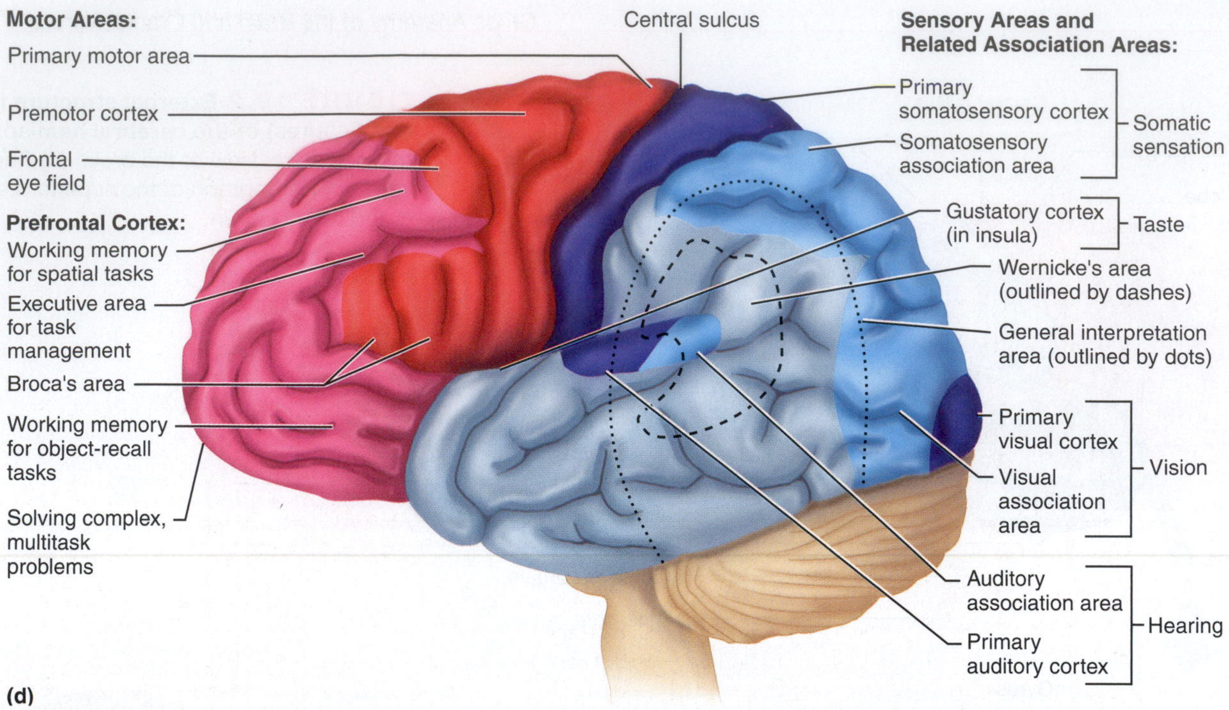

Motor Areas:
Primary motor area
Premotor cortex
Frontal eye field
Prefrontal Cortex:
Working memory for spatial tasks
Executive area for task management
Broca's area
Working memory for object-recall tasks
Solving complex, multitask problems
Central sulcus
Sensory Areas and Related Association Areas:
Primary somatosensory cortex
Somatosensory association area
Somatic sensation
Gustatory cortex (in insula)
Taste
Wernicke's area (outlined by dashes)
General interpretation area (outlined by dots)
Primary visual cortex
Visual association area
Vision
Auditory association area
Primary auditory cortex
Hearing
(d)

FIGURE 19.2 *(continued)* External structure of the cerebral hemispheres. (d) Functional areas of the left cerebral cortex. The olfactory area, which is deep to the temporal lobe on the medial hemispheric surface, is not identified. Numbers indicate brain regions plotted by the Brodmann system.

my big toe.") Immediately posterior to the primary somatosensory area is the **somatosensory association area,** in which the meaning of incoming stimuli is analyzed. ("Ouch! I have a *pain* there.") Thus, the somatosensory association area allows you to become aware of pain, coldness, a light touch, and the like.

Impulses from the special sense organs are interpreted in other specific areas also noted in Figure 19.2d. For example, the visual areas are in the posterior portion of the occipital lobe, and the auditory area is located in the temporal lobe in the gyrus bordering the lateral sulcus. The olfactory area is deep within the temporal lobe along its medial surface, in a region called the **uncus** (see Figure 19.4a, page 284).

The **primary motor area,** which is responsible for conscious or voluntary movement of the skeletal muscles, is located in the **precentral gyrus** of the frontal lobe. A specialized motor speech area called **Broca's area** is found at the base of the precentral gyrus just above the lateral sulcus. Damage to this area (which is located in only one cerebral hemisphere, usually the left) reduces or eliminates the ability to articulate words. Many areas involved in intellect, complex reasoning, and personality lie in the anterior portions of the frontal lobes, in a region called the **prefrontal cortex.**

A rather poorly defined region at the junction of the parietal and temporal lobes is **Wernicke's area,** an area in which unfamiliar words are sounded out. Like Broca's area, Wernicke's area is located in one cerebral hemisphere only, typically the left.

Although there are many similar functional areas in both cerebral hemispheres, each hemisphere is also a "specialist" in certain ways. For example, the left hemisphere is the "language brain" in most of us, because it houses centers associated with language skills and speech. The right hemisphere is more concerned with abstract, conceptual, or spatial processes—skills associated with artistic or creative pursuits.

The cell bodies of cerebral neurons involved in these functions are found only in the outermost gray matter of the cerebrum, the **cerebral cortex.** Most of the balance of cerebral tissue—the deeper **cerebral white matter**—is composed of fiber tracts carrying impulses to or from the cortex.

Using a model of the human brain (and a preserved human brain, if available), identify the areas and structures of the cerebral hemispheres described above.

Then continue using the model and preserved brain along with the figures as you read about other structures.

Diencephalon

The **diencephalon,** sometimes considered the most superior portion of the brain stem, is embryologically part of the forebrain, along with the cerebral hemispheres.

Turn the brain model so the ventral surface of the brain can be viewed. Using Figure 19.3 as a guide, start superiorly and identify the externally visible structures that mark the position of the floor of the diencephalon. These are the **olfactory bulbs** (synapse point of cranial nerve I) and **tracts, optic nerves** (cranial nerve II)**, optic chiasma** (where the fibers of the optic nerves partially cross over), **optic tracts, pituitary gland,** and **mammillary bodies.**

Brain Stem

Continue inferiorly to identify the **brain stem** structures—the **cerebral peduncles** (fiber tracts in the **midbrain** connecting the pons below with cerebrum above), the pons, and the medulla oblongata. *Pons* means "bridge," and the **pons** consists primarily of motor and sensory fiber tracts connecting the brain with lower CNS centers. The lowest brain stem region, the **medulla oblongata,** is also composed primarily of fiber tracts. You can see the **decussation of pyramids,** a crossover point for the major motor tracts (pyramidal tracts) descending from the motor areas of the cerebrum to the cord, on the medulla's surface. The medulla also houses many vital autonomic centers involved in the control of heart rate, respiratory rhythm, and blood pressure as well as involuntary centers involved in vomiting, swallowing, and so on.

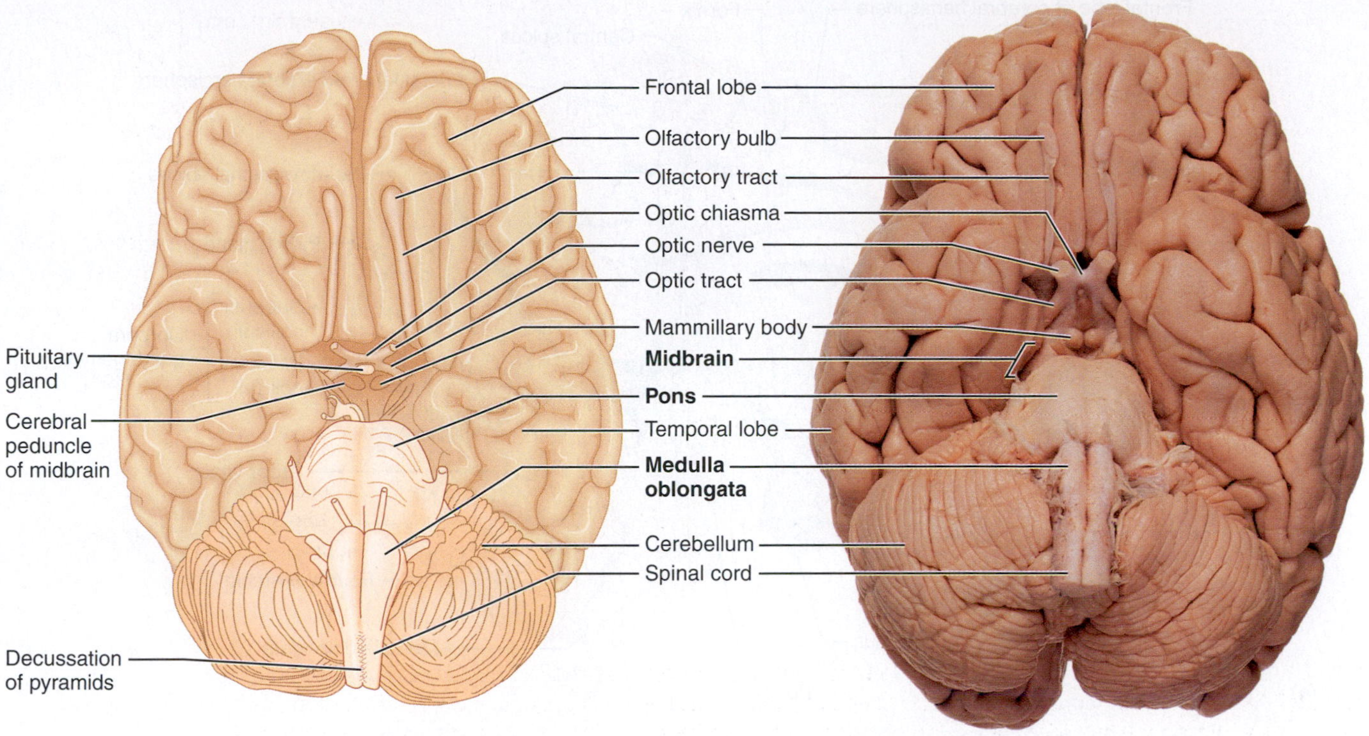

Frontal lobe
Olfactory bulb
Olfactory tract
Optic chiasma
Optic nerve
Optic tract
Mammillary body
Midbrain
Pons
Temporal lobe
Medulla oblongata
Cerebellum
Spinal cord
Pituitary gland
Cerebral peduncle of midbrain
Decussation of pyramids

FIGURE 19.3 Ventral (inferior) aspect of the human brain, showing the three regions of the brain stem. Only a small portion of the midbrain can be seen; the rest is surrounded by other brain regions.

Cerebellum

1. Turn the brain model so you can see the dorsal aspect. Identify the large cauliflower-like **cerebellum,** which projects dorsally from under the occipital lobes of the cerebrum. Notice that, like the cerebrum, the cerebellum has two major hemispheres and a convoluted surface (see Figure 19.6). It also has an outer cortex made up of gray matter with an inner region of white matter.

2. Remove the cerebellum to view the **corpora quadrigemina** (Figure 19.4), located on the posterior aspect of the midbrain, a brain stem structure. The two superior prominences are the **superior colliculi** (visual reflex centers); the two smaller inferior prominences are the **inferior colliculi** (auditory reflex centers).

ACTIVITY 2

Identifying Internal Brain Structures

The deeper structures of the brain have also been well mapped. Like the external structures, these can be studied in terms of the four major regions. As the internal brain areas are described, identify them on the figures cited. Also, use the brain model as indicated to help you in this study.

Cerebral Hemispheres

1. Take the brain model apart so you can see a median sagittal view of the internal brain structures (see Figure 19.4). Observe the model closely to see the extent of the outer cortex (gray matter), which contains the cell bodies of cerebral neurons. The pyramidal cells of the cerebral motor cortex (studied in Exercise 17, page 257) are representative of the neurons seen in the precentral gyrus.

2. Observe the deeper area of white matter, which is composed of fiber tracts. The fiber tracts found in the cerebral hemisphere white matter are called *association tracts* if they connect two portions of the same hemisphere, *projection tracts* if they run between the cerebral cortex and lower brain structures or spinal cord, and *commissures* if they run from one hemisphere to another. Observe the large **corpus callosum,** the major commissure connecting the cerebral hemispheres. The corpus callosum arches above the structures of the diencephalon and roofs over the lateral ventricles. Notice also the **fornix,** a bandlike fiber tract concerned with olfaction as well as limbic system functions, and the membranous **septum pellucidum,** which separates the lateral ventricles of the cerebral hemispheres.

3. In addition to the gray matter of the cerebral cortex, there are several "islands" of gray matter (clusters of neuron cell bodies) called **nuclei** buried deep within the white matter of the cerebral hemispheres. One important group of cerebral nuclei, called the **basal ganglia,*** flank the lateral and third ventricles. You can see these nuclei if you have a dissectible

*The historical term for these nuclei, *basal ganglia,* is a misleading term because ganglia are PNS structures. However, the name has been retained to differentiate these nuclei from the basal nuclei of the forebrain in the cerebrum.

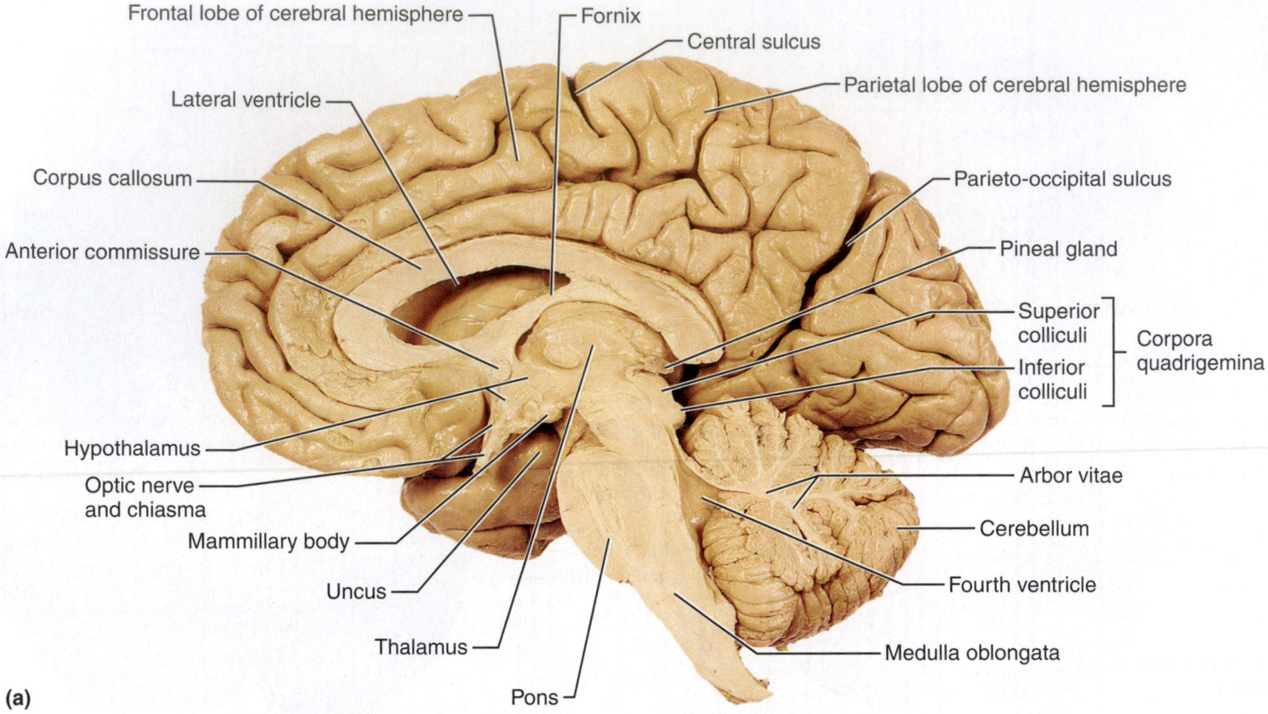
Frontal lobe of cerebral hemisphere
Fornix
Central sulcus
Parietal lobe of cerebral hemisphere
Lateral ventricle
Corpus callosum
Parieto-occipital sulcus
Anterior commissure
Pineal gland
Superior colliculi
Inferior colliculi
Corpora quadrigemina
Hypothalamus
Optic nerve and chiasma
Mammillary body
Uncus
Thalamus
Pons
Arbor vitae
Cerebellum
Fourth ventricle
Medulla oblongata

(a)

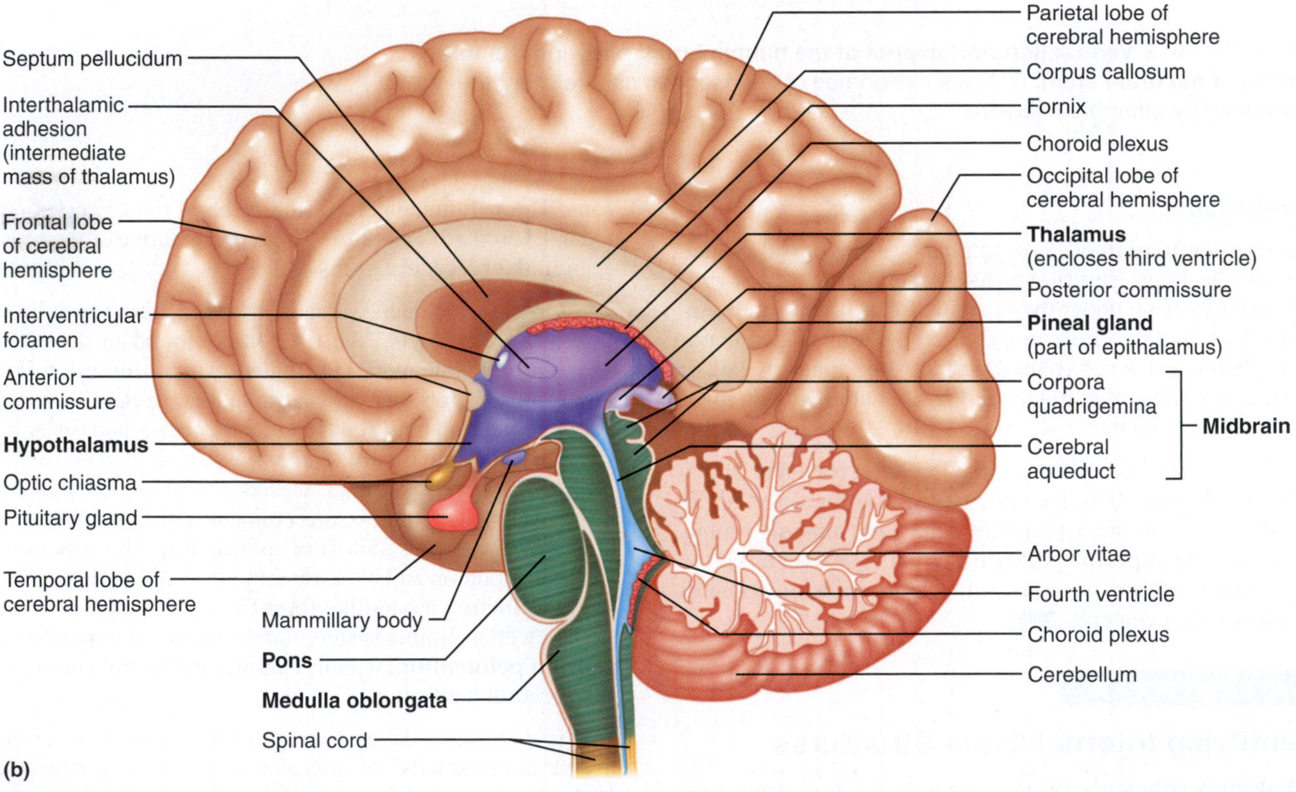
Septum pellucidum
Interthalamic adhesion (intermediate mass of thalamus)
Frontal lobe of cerebral hemisphere
Interventricular foramen
Anterior commissure
Hypothalamus
Optic chiasma
Pituitary gland
Temporal lobe of cerebral hemisphere
Mammillary body
Pons
Medulla oblongata
Spinal cord
Parietal lobe of cerebral hemisphere
Corpus callosum
Fornix
Choroid plexus
Occipital lobe of cerebral hemisphere
Thalamus (encloses third ventricle)
Posterior commissure
Pineal gland (part of epithalamus)
Corpora quadrigemina
Cerebral aqueduct
Midbrain
Arbor vitae
Fourth ventricle
Choroid plexus
Cerebellum

(b)

FIGURE 19.4 Diencephalon and brain stem structures as seen in a midsagittal section of the brain. (a) Photograph. **(b)** Diagrammatic view.

model or a coronally or cross-sectioned human brain slice. Otherwise, Figure 19.5 will suffice.

The basal ganglia, which are important subcortical motor nuclei (and part of the so-called *extrapyramidal system*), are involved in regulating voluntary motor activities. The most important of them are the arching, comma-shaped **caudate nucleus,** and the **lentiform nucleus,** which is composed of the **putamen** and **globus pallidus nuclei.** The closely associated *amygdaloid nucleus* (located at the tip of the caudate nucleus) is part of the *limbic system.*

The **corona radiata,** a spray of projection fibers coursing down from the precentral (motor) gyrus, combines with sensory fibers traveling to the sensory cortex to form a broad band of fibrous material called the **internal capsule.** The

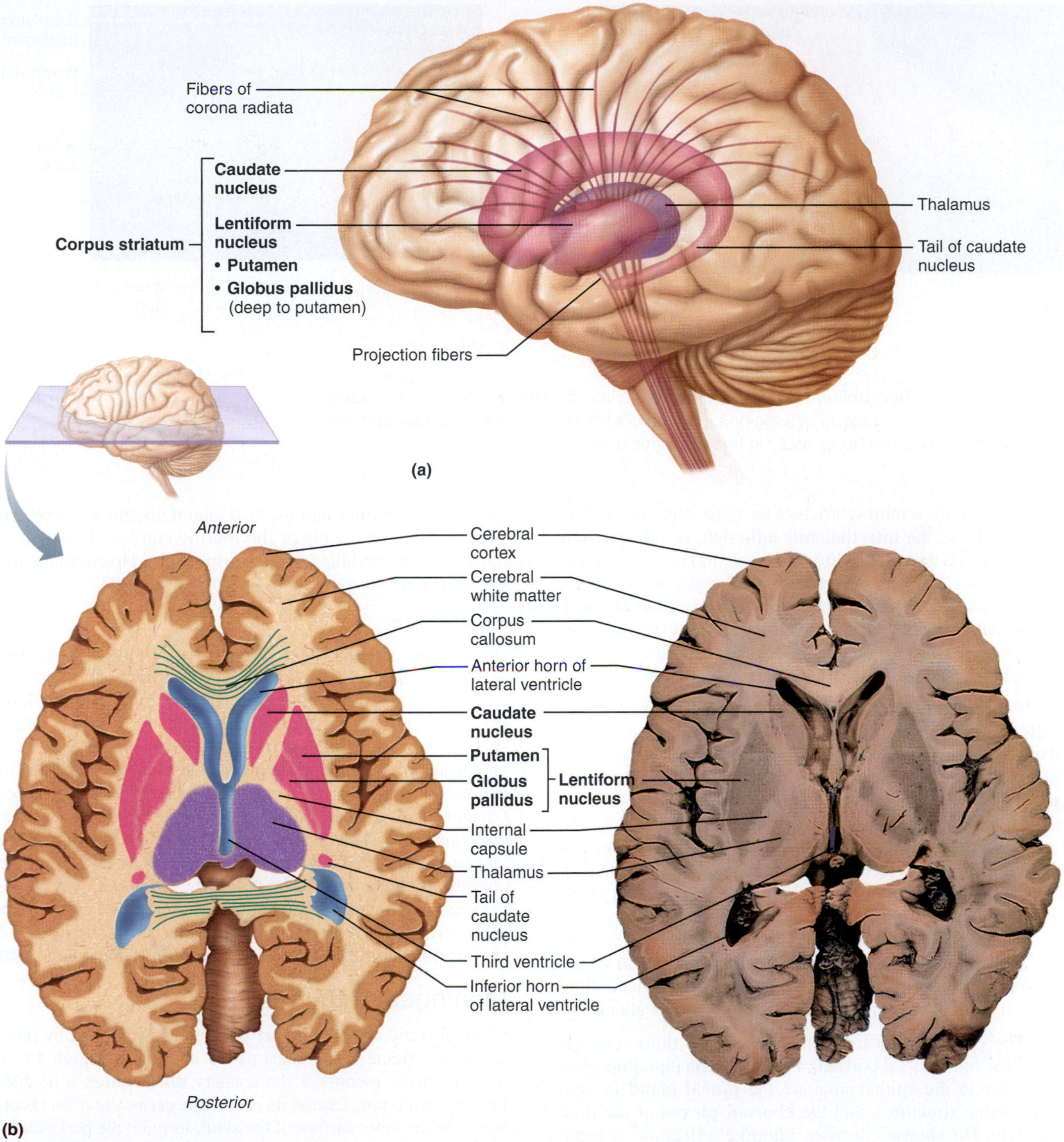

Fibers of corona radiata
Caudate nucleus
Lentiform nucleus
• Putamen
• Globus pallidus (deep to putamen)
Corpus striatum
Thalamus
Tail of caudate nucleus
Projection fibers
(a)
Anterior
Cerebral cortex
Cerebral white matter
Corpus callosum
Anterior horn of lateral ventricle
Caudate nucleus
Putamen
Globus pallidus
Lentiform nucleus
Internal capsule
Thalamus
Tail of caudate nucleus
Third ventricle
Inferior horn of lateral ventricle
Posterior
(b)

FIGURE 19.5 Basal ganglia. (a) Three-dimensional view of the basal ganglia showing their positions within the cerebrum. **(b)** A transverse section of the cerebrum and diencephalon showing the relationship of the basal ganglia to the thalamus and the lateral and third ventricles.

internal capsule passes between the diencephalon and the basal ganglia and through parts of the basal ganglia, giving them a striped appearance. This is why the caudate nucleus and the lentiform nucleus are sometimes referred to collectively as the **corpus striatum,** or "striped body" (Figure 19.5a).

4. Examine the relationship of the lateral ventricles and corpus callosum to the diencephalon structures; that is, thalamus and third ventricle—from the cross-sectional viewpoint (see Figure 19.5b).

Diencephalon

1. The major internal structures of the diencephalon are the thalamus, hypothalamus, and epithalamus (see Figure 19.4). The **thalamus** consists of two large lobes of gray matter that laterally

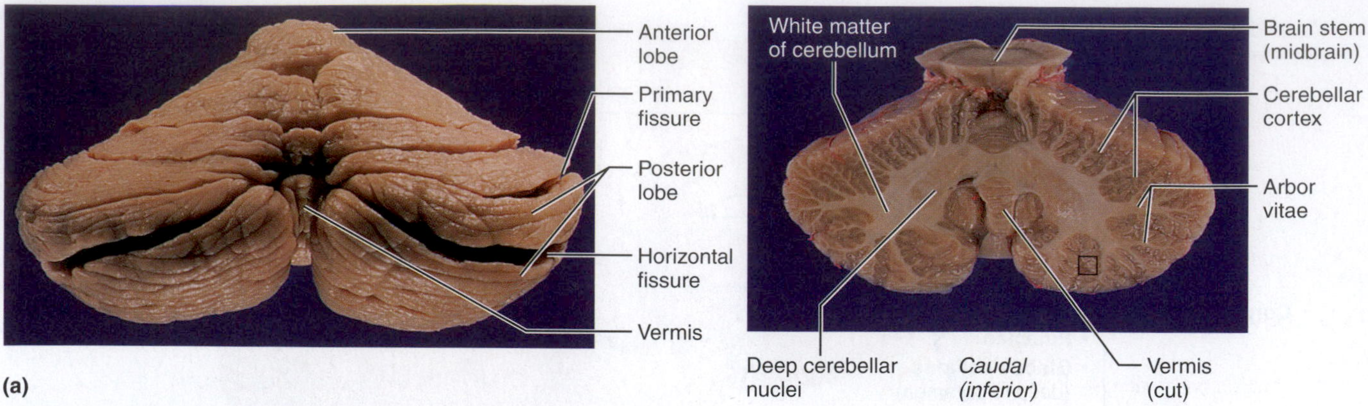
Anterior lobe
Primary fissure
Posterior lobe
Horizontal fissure
Vermis
(a)
White matter of cerebellum
Brain stem (midbrain)
Cerebellar cortex
Arbor vitae
Deep cerebellar nuclei
Caudal (inferior)
Vermis (cut)
(b)

FIGURE 19.6 Cerebellum. (a) Posterior (dorsal) view. **(b)** The cerebellum, sectioned to reveal its cortex and medullary regions. (The cerebellum is sectioned frontally and the brain stem is sectioned horizontally in this posterior view.)

enclose the shallow third ventricle of the brain. A slender stalk of thalamic tissue, the **interthalamic adhesion,** or **intermediate mass,** connects the two thalamic lobes and bridges the ventricle. The thalamus is a major integrating and relay station for sensory impulses passing upward to the cortical sensory areas for localization and interpretation. Locate also the **interventricular foramen** *(foramen of Monro),* a tiny orifice connecting the third ventricle with the lateral ventricle on the same side.

2. The **hypothalamus** makes up the floor and the inferolateral walls of the third ventricle. It is an important autonomic center involved in regulation of body temperature, water balance, and fat and carbohydrate metabolism as well as in many other activities and drives (sex, hunger, thirst). Locate again the pituitary gland, which hangs from the anterior floor of the hypothalamus by a slender stalk, the **infundibulum.** (The pituitary gland is usually not present in preserved brain specimens.) In life, the pituitary rests in the hypophyseal fossa of the sella turcica of the sphenoid bone. Its function is discussed in Exercise 27.

Anterior to the pituitary, identify the optic chiasma portion of the optic pathway to the brain. The **mammillary bodies,** relay stations for olfaction, bulge exteriorly from the floor of the hypothalamus just posterior to the pituitary gland.

3. The **epithalamus** forms the roof of the third ventricle and is the most dorsal portion of the diencephalon. Important structures in the epithalamus are the **pineal gland** (a neuroendocrine structure), and the **choroid plexus** of the third ventricle. The choroid plexuses, knotlike collections of capillaries within each ventricle, form the cerebrospinal fluid.

Brain Stem

1. Now trace the short midbrain from the mammillary bodies to the rounded pons below. Continue to refer to Figure 19.4. The **cerebral aqueduct** is a slender canal traveling through the midbrain; it connects the third ventricle to the fourth ventricle in the hindbrain below. The cerebral peduncles and the rounded corpora quadrigemina make up the midbrain tissue anterior and posterior (respectively) to the cerebral aqueduct.

2. Locate the hindbrain structures. Trace the rounded pons to the medulla oblongata below, and identify the fourth ventricle posterior to these structures. Attempt to identify the single median aperture and the two lateral apertures, three orifices found in the walls of the fourth ventricle. These apertures serve as conduits for cerebrospinal fluid to circulate into the subarachnoid space from the fourth ventricle.

Cerebellum

Examine the cerebellum. Notice that it is composed of two lateral hemispheres each with three lobes (*anterior, posterior,* and a deep *flocculonodular*) connected by a midline lobe called the **vermis** (Figure 19.6). As in the cerebral hemispheres, the cerebellum has an outer cortical area of gray matter and an inner area of white matter. The treelike branching of the cerebellar white matter is referred to as the **arbor vitae,** or "tree of life." The cerebellum is concerned with unconscious coordination of skeletal muscle activity and control of balance and equilibrium. Fibers converge on the cerebellum from the equilibrium apparatus of the inner ear, visual pathways, proprioceptors of tendons and skeletal muscles, and from many other areas. Thus the cerebellum remains constantly aware of the position and state of tension of the various body parts.

Meninges of the Brain

The brain (and spinal cord) are covered and protected by three connective tissue membranes called **meninges** (Figure 19.7). The outermost meninx is the leathery **dura mater,** a double-layered membrane. One of its layers (the *periosteal layer*) is attached to the inner surface of the skull, forming the periosteum. The other (the *meningeal layer*) forms the outermost brain covering and is continuous with the dura mater of the spinal cord.

The dural layers are fused together except in three places where the inner membrane extends inward to form a septum that secures the brain to structures inside the cranial cavity. One such extension, the **falx cerebri,** dips into the longitudinal fissure between the cerebral hemispheres to attach to the crista galli of the ethmoid bone of the skull (Figure 19.7a). The cavity created at this point is the large **superior sagittal sinus,** which collects blood draining from the brain tissue. The **falx cerebelli,** separating the two cerebellar hemispheres, and the **tentorium cerebelli,** separating the cerebrum from the cerebellum below, are two other important inward folds of the inner dural membrane.

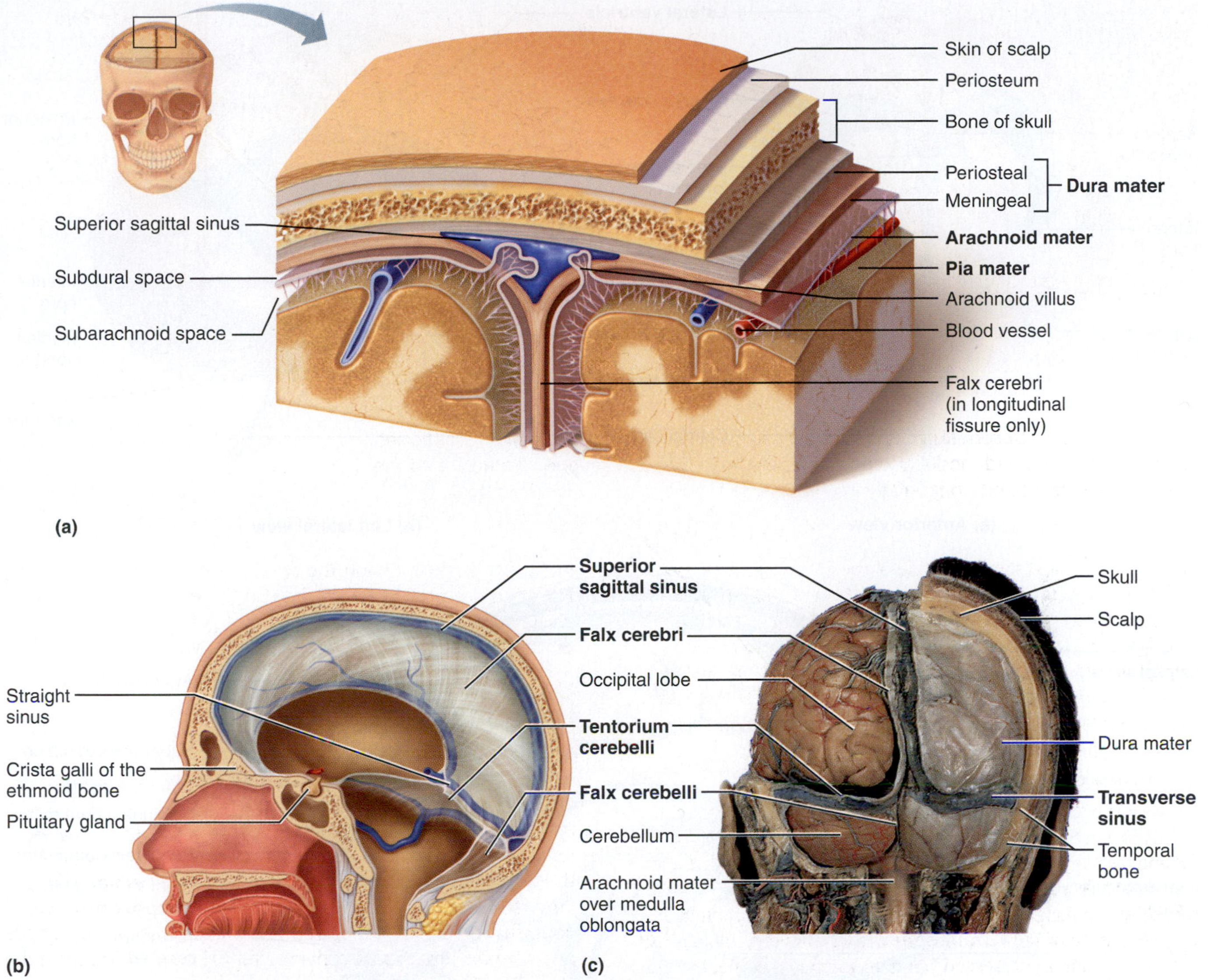

Skin of scalp
Periosteum
Bone of skull
Periosteal
Meningeal
Dura mater
Arachnoid mater
Pia mater
Arachnoid villus
Blood vessel
Falx cerebri
(in longitudinal
fissure only)
Superior sagittal sinus
Subdural space
Subarachnoid space
(a)
Straight
sinus
Crista galli of the
ethmoid bone
Pituitary gland
(b)
Superior
sagittal sinus
Falx cerebri
Occipital lobe
Tentorium
cerebelli
Falx cerebelli
Cerebellum
Arachnoid mater
over medulla
oblongata
(c)
Skull
Scalp
Dura mater
Transverse
sinus
Temporal
bone

FIGURE 19.7 Meninges of the brain. (a) Three-dimensional frontal section showing the relationship of the dura mater, arachnoid mater, and pia mater. The meningeal dura forms the falx cerebri fold, which extends into the longitudinal fissure and attaches the brain to the ethmoid bone of the skull. A dural sinus, the superior sagittal sinus, is enclosed by the dural membranes superiorly. Arachnoid villi, which return cerebrospinal fluid to the dural sinus, are also shown. **(b)** Position of the dural folds: the falx cerebri, tentorium cerebelli, and falx cerebelli. **(c)** Posterior view of the brain in place, surrounded by the dura mater. Sinuses between periosteal and meningeal dura contain venous blood.

The middle meninx, the weblike **arachnoid mater,** underlies the dura mater and is partially separated from it by the **subdural space.** Threadlike projections bridge the **subarachnoid space** to attach the arachnoid to the innermost meninx, the **pia mater.** The delicate pia mater is highly vascular and clings tenaciously to the surface of the brain, following its convolutions.

In life, the subarachnoid space is filled with cerebrospinal fluid. Specialized projections of the arachnoid tissue called **arachnoid villi** protrude through the dura mater to allow the cerebrospinal fluid to drain back into the venous circulation via the superior sagittal sinus and other dural sinuses.

Meningitis, inflammation of the meninges, is a serious threat to the brain because of the intimate association between the brain and meninges. Should infection spread to the neural tissue of the brain itself, life-threatening **encephalitis** may occur. Meningitis is often diagnosed by taking a sample of cerebrospinal fluid from the subarachnoid space. ■

Cerebrospinal Fluid

The cerebrospinal fluid, much like plasma in composition, is continually formed by the **choroid plexuses,** small capillary knots hanging from the roof of the ventricles of the brain. The

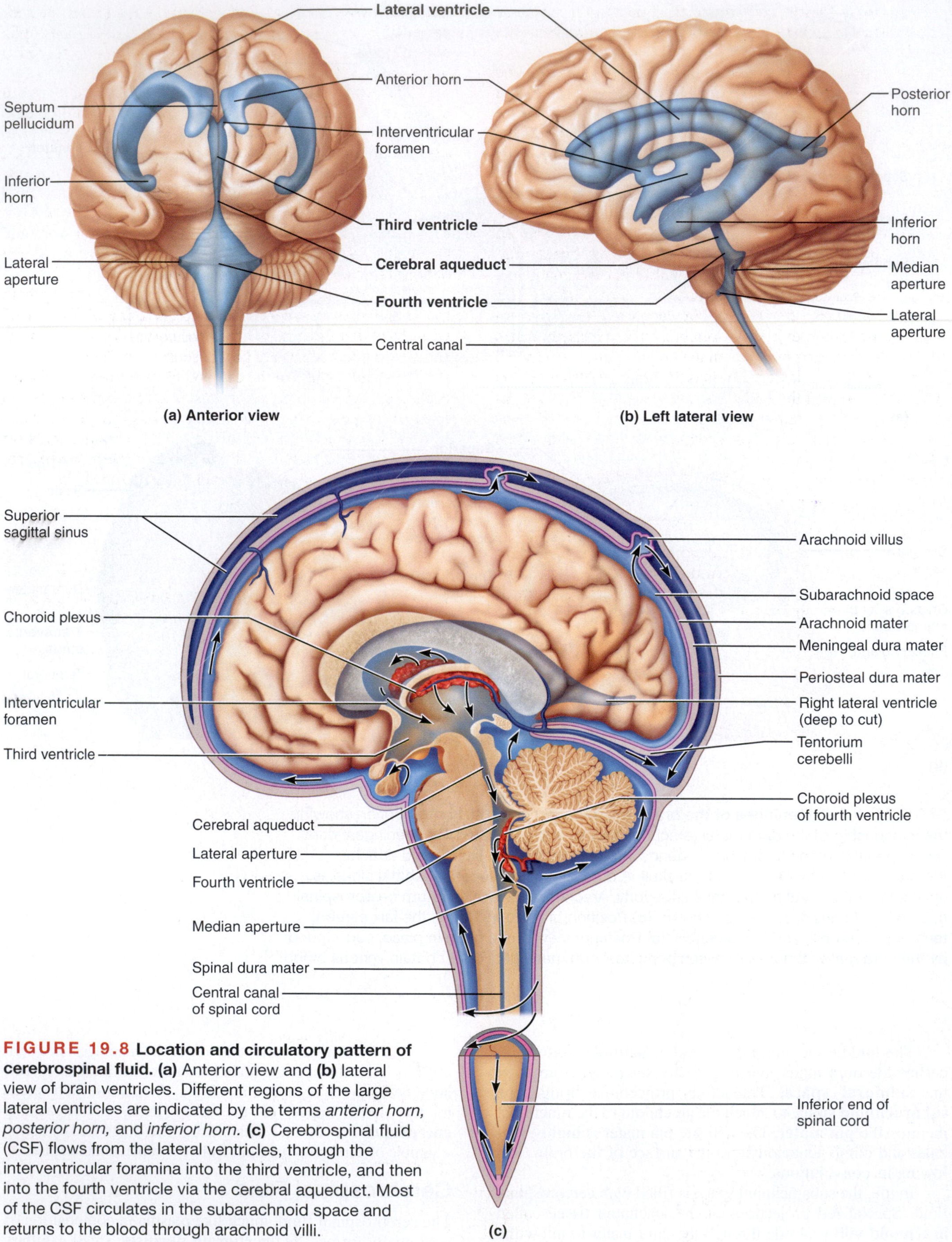

Lateral ventricle
Anterior horn
Septum pellucidum
Interventricular foramen
Inferior horn
Third ventricle
Cerebral aqueduct
Lateral aperture
Fourth ventricle
Central canal
(a) Anterior view
Posterior horn
Inferior horn
Median aperture
Lateral aperture
(b) Left lateral view
Superior sagittal sinus
Arachnoid villus
Subarachnoid space
Arachnoid mater
Meningeal dura mater
Choroid plexus
Periosteal dura mater
Interventricular foramen
Right lateral ventricle (deep to cut)
Third ventricle
Tentorium cerebelli
Choroid plexus of fourth ventricle
Cerebral aqueduct
Lateral aperture
Fourth ventricle
Median aperture
Spinal dura mater
Central canal of spinal cord
Inferior end of spinal cord
(c)

FIGURE 19.8 Location and circulatory pattern of cerebrospinal fluid. (a) Anterior view and **(b)** lateral view of brain ventricles. Different regions of the large lateral ventricles are indicated by the terms *anterior horn, posterior horn,* and *inferior horn.* **(c)** Cerebrospinal fluid (CSF) flows from the lateral ventricles, through the interventricular foramina into the third ventricle, and then into the fourth ventricle via the cerebral aqueduct. Most of the CSF circulates in the subarachnoid space and returns to the blood through arachnoid villi.

cerebrospinal fluid in and around the brain forms a watery cushion that protects the delicate brain tissue against blows to the head.

Within the brain, the cerebrospinal fluid circulates from the two lateral ventricles (in the cerebral hemispheres) into the third ventricle via the **interventricular foramina,** and then through the cerebral aqueduct of the midbrain into the fourth ventricle in the hindbrain (Figure 19.8). Some of the fluid reaching the fourth ventricle continues down the central canal of the spinal cord, but the bulk of it circulates into the subarachnoid space, exiting through the three foramina in the walls of the fourth ventricle (the two lateral and the single median apertures). The fluid returns to the blood in the dural sinuses via the arachnoid villi.

Ordinarily, cerebrospinal fluid forms and drains at a constant rate. However, under certain conditions—for example, obstructed drainage or circulation resulting from tumors or anatomical deviations—cerebrospinal fluid accumulates and exerts increasing pressure on the brain which, uncorrected, causes neurological damage in adults. In infants, **hydrocephalus** (literally, "water on the brain") is indicated by a gradually enlarging head. The infant's skull is still flexible and contains fontanels, so it can expand to accommodate the increasing size of the brain. ■

Cranial Nerves

The **cranial nerves** are part of the peripheral nervous system and not part of the brain proper, but they are most appropriately identified while studying brain anatomy. The 12 pairs of cranial nerves primarily serve the head and neck. Only one pair, the vagus nerves, extends into the thoracic and abdominal cavities. All but the first two pairs (olfactory and optic nerves) arise from the brain stem and pass through foramina in the base of the skull to reach their destination.

The cranial nerves are numbered consecutively, and in most cases their names reflect the major structures they control. The cranial nerves are described by name, number (Roman numeral), origin, course, and function in Table 19.1. This information should be committed to memory. A mnemonic device that might be helpful for remembering the cranial nerves in order is "*O*n *o*ccasion, *o*ur *t*rusty *t*ruck *a*cts *f*unny—*v*ery *g*ood *v*ehicle *a*ny*h*ow." The first letter of each word and the "a" and "h" of the final word "anyhow" will remind you of the first letter of the cranial nerve name.

Most cranial nerves are mixed nerves (containing both motor and sensory fibers). But close scrutiny of Table 19.1 will reveal that three pairs of cranial nerves (optic, olfactory, and vestibulocochlear) are purely sensory in function.

You may recall that the cell bodies of neurons are always located within the central nervous system (cortex or nuclei) or in specialized collections of cell bodies (ganglia) outside the CNS. Neuron cell bodies of the sensory cranial nerves are located in ganglia; those of the mixed cranial nerves are found both within the brain and in peripheral ganglia.

ACTIVITY 3

Identifying and Testing the Cranial Nerves

1. Observe the ventral surface of the brain model to identify the cranial nerves. Figure 19.9 may also aid you in this

Text continues on page 291.

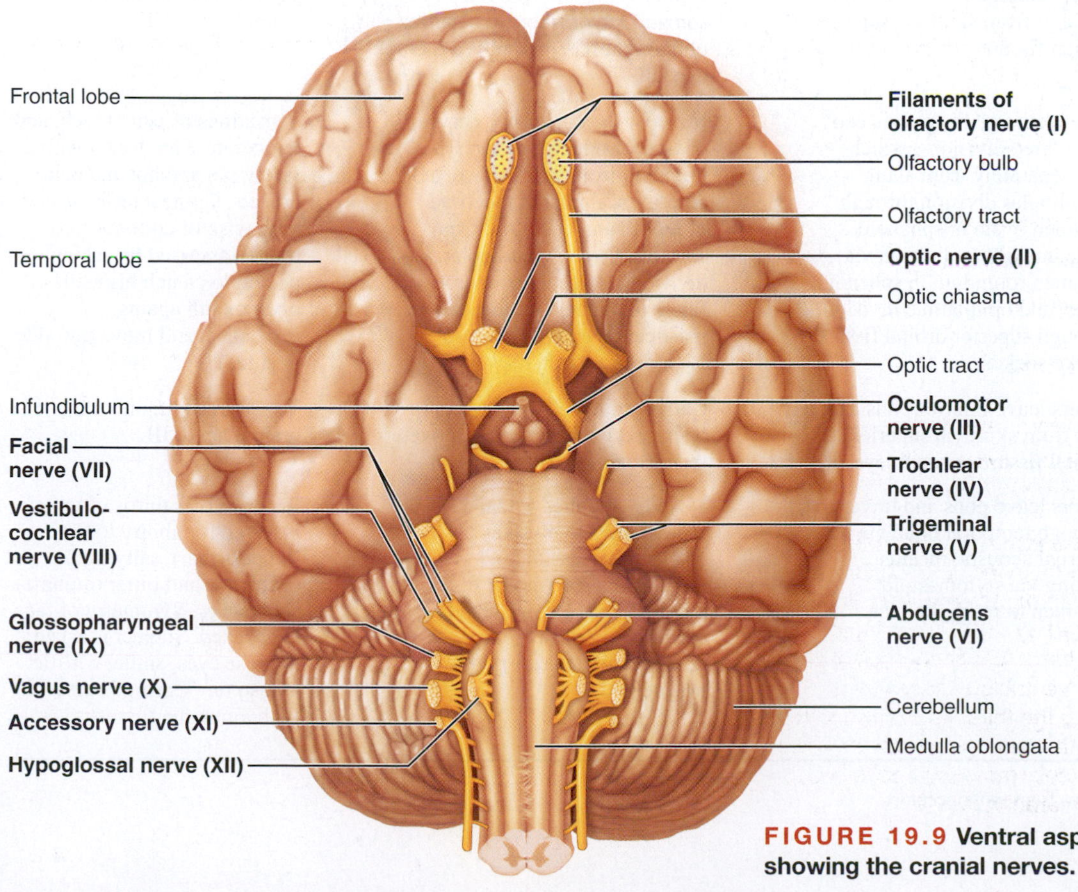
Frontal lobe
Temporal lobe
Infundibulum
Facial nerve (VII)
Vestibulo-cochlear nerve (VIII)
Glossopharyngeal nerve (IX)
Vagus nerve (X)
Accessory nerve (XI)
Hypoglossal nerve (XII)
Filaments of olfactory nerve (I)
Olfactory bulb
Olfactory tract
Optic nerve (II)
Optic chiasma
Optic tract
Oculomotor nerve (III)
Trochlear nerve (IV)
Trigeminal nerve (V)
Abducens nerve (VI)
Cerebellum
Medulla oblongata

FIGURE 19.9 Ventral aspect of the human brain, showing the cranial nerves. (See also Figure 19.3.)

TABLE 19.1 The Cranial Nerves (see Figure 19.9)

Number and name	Origin and course	Function*	Testing
I. Olfactory	Fibers arise from olfactory epithelium and run through cribriform plate of ethmoid bone to synapse in olfactory bulbs.	Purely sensory—carries afferent impulses associated with sense of smell.	Person is asked to sniff aromatic substances, such as oil of cloves and vanilla, and to identify each.
II. Optic	Fibers arise from retina of eye to form the optic nerve and pass through optic canal of orbit. Fibers partially cross over at the optic chiasma and continue on to the thalamus as the optic tracts. Final fibers of this pathway travel from the thalamus to the visual cortex as the optic radiation.	Purely sensory—carries afferent impulses associated with vision.	Vision and visual field are determined with eye chart and by testing the point at which the person first sees an object (finger) moving into the visual field. Fundus of eye viewed with ophthalmoscope to detect papilledema (swelling of optic disc, or point at which optic nerve leaves the eye) and to observe blood vessels.
III. Oculomotor	Fibers emerge from dorsal midbrain and course ventrally to enter the orbit. They exit from skull via superior orbital fissure.	Primarily motor—somatic motor fibers to inferior oblique and superior, inferior, and medial rectus muscles, which direct eyeball, and to levator palpebrae muscles of the superior eyelid; parasympathetic fibers to iris and smooth muscle controlling lens shape (reflex responses to varying light intensity and focusing of eye for near vision).	Pupils are examined for size, shape, and equality. Pupillary reflex is tested with penlight (pupils should constrict when illuminated). Convergence for near vision is tested, as is subject's ability to follow objects with the eyes.
IV. Trochlear	Fibers emerge from midbrain and exit from skull via superior orbital fissure.	Primarily motor—provides somatic motor fibers to superior oblique muscle that moves the eyeball.	Tested in common with cranial nerve III.
V. Trigeminal	Fibers emerge from pons and form three divisions, which exit separately from skull: mandibular division through foramen ovale in sphenoid bone, maxillary division via foramen rotundum in sphenoid bone, and ophthalmic division through superior orbital fissure of eye socket.	Mixed—major sensory nerve of face; conducts sensory impulses from skin of face and anterior scalp, from mucosae of mouth and nose, and from surface of eyes; mandibular division also contains motor fibers that innervate muscles of mastication and muscles of floor of mouth.	Sensations of pain, touch, and temperature are tested with safety pin and hot and cold objects. Corneal reflex tested with wisp of cotton. Motor branch assessed by asking person to clench his teeth, open mouth against resistance, and move jaw side to side.
VI. Abducens	Fibers leave inferior pons and exit from skull via superior orbital fissure to run to eye.	Carries somatic motor fibers to lateral rectus muscle that moves the eyeball.	Tested in common with cranial nerve III.
VII. Facial	Fibers leave pons and travel through temporal bone via internal acoustic meatus, exiting via stylomastoid foramen to reach the face.	Mixed—supplies somatic motor fibers to muscles of facial expression and parasympathetic motor fibers to lacrimal and salivary glands; carries sensory fibers from taste receptors of anterior portion of tongue.	Anterior two-thirds of tongue is tested for ability to taste sweet (sugar), salty, sour (vinegar), and bitter (quinine) substances. Symmetry of face is checked. Subject is asked to close eyes, smile, whistle, and so on. Tearing is assessed with ammonia fumes.

*Does not include sensory impulses from proprioceptors.

TABLE 19.1 *(continued)*

Number and name	Origin and course	Function*	Testing
VIII. Vestibulocochlear	Fibers run from inner-ear equilibrium and hearing apparatus, housed in temporal bone, through internal acoustic meatus to enter pons.	Purely sensory—vestibular branch transmits impulses associated with sense of equilibrium from vestibular apparatus and semicircular canals; cochlear branch transmits impulses associated with hearing from cochlea.	Hearing is checked by air and bone conduction using tuning fork.
IX. Glossopharyngeal	Fibers emerge from medulla and leave skull via jugular foramen to run to throat.	Mixed—somatic motor fibers serve pharyngeal muscles, and parasympathetic motor fibers serve salivary glands; sensory fibers carry impulses from pharynx, tonsils, posterior tongue (taste buds), and from chemoreceptors and pressure receptors of carotid artery.	A tongue depressor is used to check the position of the uvula. Gag and swallowing reflexes are checked. Subject is asked to speak and cough. Posterior third of tongue may be tested for taste.
X. Vagus	Fibers emerge from medulla and pass through jugular foramen and descend through neck region into thorax and abdomen.	Mixed—fibers carry somatic motor impulses to pharynx and larynx and sensory fibers from same structures; very large portion is composed of parasympathetic motor fibers, which supply heart and smooth muscles of abdominal visceral organs; transmits sensory impulses from viscera.	As for cranial nerve IX (IX and X are tested in common, since they both innervate muscles of throat and mouth).
XI. Accessory	Fibers arise from the superior aspect of spinal cord, enter the skull, and then travel through jugular foramen to reach muscles of neck and back.	Mixed (but primarily motor in function)—provides somatic motor fibers to sternocleidomastoid and trapezius muscles and to muscles of soft palate, pharynx, and larynx (spinal and medullary fibers respectively).	Sternocleidomastoid and trapezius muscles are checked for strength by asking person to rotate head and shrug shoulders against resistance.
XII. Hypoglossal	Fibers arise from medulla and exit from skull via hypoglossal canal to travel to tongue.	Mixed (but primarily motor in function)—carries somatic motor fibers to muscles of tongue.	Person is asked to protrude and retract tongue. Any deviations in position are noted.

*Does not include sensory impulses from proprioceptors.

study. Notice that the first (olfactory) cranial nerves are not visible on the model because they consist only of short axons that run from the nasal mucosa through the cribriform plate of the ethmoid bone. (However, the synapse points of the first cranial nerves, the *olfactory bulbs,* are visible on the model.)

2. The last column of Table 19.1 describes techniques for testing cranial nerves, which is an important part of any neurological examination. This information may help you understand cranial nerve function, especially as it pertains to some aspects of brain function. Conduct tests of cranial nerve function following directions given in the "testing" column of the table.

3. Several cranial nerve ganglia are named in the chart here. *Using your textbook or an appropriate reference,* fill in the chart by naming the cranial nerve the ganglion is associated with and stating its location.

Cranial nerve ganglion	Cranial nerve	Site of ganglion
Trigeminal		
Geniculate		
Inferior		
Superior		
Spiral		
Vestibular		

DISSECTION:
The Sheep Brain

The sheep brain is enough like the human brain to warrant comparison. Obtain a sheep brain, disposable gloves, dissecting tray, and instruments, and bring them to your laboratory bench.

1. Before beginning the dissection, turn your sheep brain so that you are viewing its left lateral aspect. Compare the various areas of the sheep brain (cerebrum, brain stem, cerebellum) to the photo of the human brain in Figure 19.10. Relatively speaking, which of these structures is obviously much larger in the human brain?

___

2. Place the intact sheep brain ventral surface down on the dissecting pan, and observe the dura mater. Feel its consistency and note its toughness. Cut through the dura mater along the line of the longitudinal fissure (which separates the cerebral hemispheres) to enter the superior sagittal sinus. Gently force the cerebral hemispheres apart laterally to expose the corpus callosum deep to the longitudinal fissure.

3. Carefully remove the dura mater and examine the superior surface of the brain. Notice that its surface, like that of the human brain, is thrown into convolutions (fissures and gyri). Locate the arachnoid mater, which appears on the brain surface as a delicate "cottony" material spanning the fissures. In contrast, the innermost meninx, the pia mater, closely follows the cerebral contours.

Ventral Structures

Figure 19.11a and b shows the important features of the ventral surface of the brain. Turn the brain so that its ventral surface is uppermost.

1. Look for the clublike olfactory bulbs anteriorly, on the inferior surface of the frontal lobes of the cerebral hemispheres. Axons of olfactory neurons run from the nasal mucosa through the perforated cribriform plate of the ethmoid bone to synapse with the olfactory bulbs.

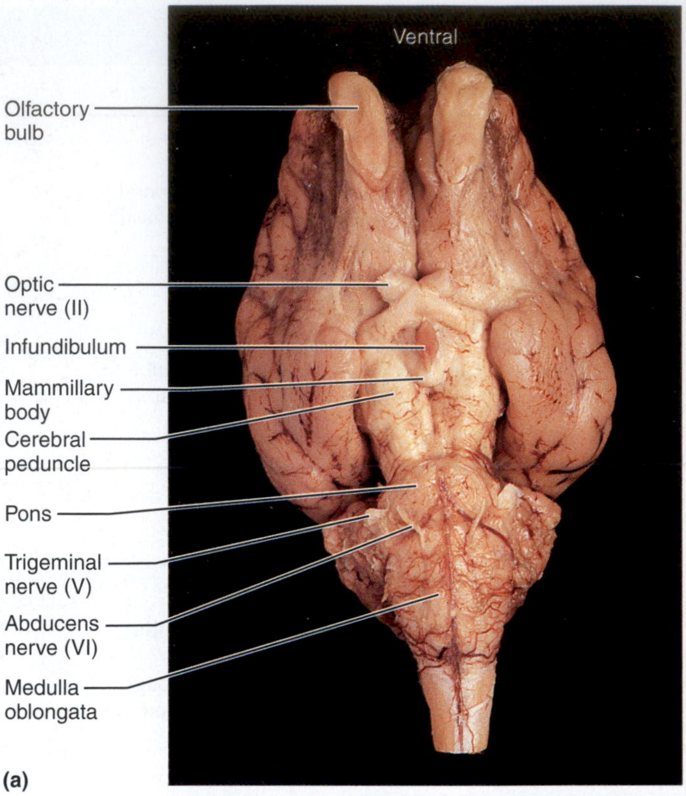
Ventral
Olfactory bulb
Optic nerve (II)
Infundibulum
Mammillary body
Cerebral peduncle
Pons
Trigeminal nerve (V)
Abducens nerve (VI)
Medulla oblongata
(a)

FIGURE 19.11 Intact sheep brain. (a) Photograph of ventral view.

How does the size of these olfactory bulbs compare with those of humans?

___

Is the sense of smell more important as a protective and a food-getting sense in sheep or in humans?

___

2. The optic nerve (II) carries sensory impulses from the retina of the eye. Thus this cranial nerve is involved in the sense of vision. Identify the optic nerves, optic chiasma, and optic tracts.

3. Posterior to the optic chiasma, two structures protrude from the ventral aspect of the hypothalamus—the infundibulum (stalk of the pituitary gland) immediately posterior to the optic chiasma and the mammillary body. Notice that the sheep's mammillary body is a single rounded eminence. In humans it is a double structure.

4. Identify the cerebral peduncles on the ventral aspect of the midbrain, just posterior to the mammillary body of the hypothalamus. The cerebral peduncles are fiber tracts connecting the cerebrum and medulla oblongata. Identify the large oculomotor nerves (III), which arise from the ventral midbrain surface, and

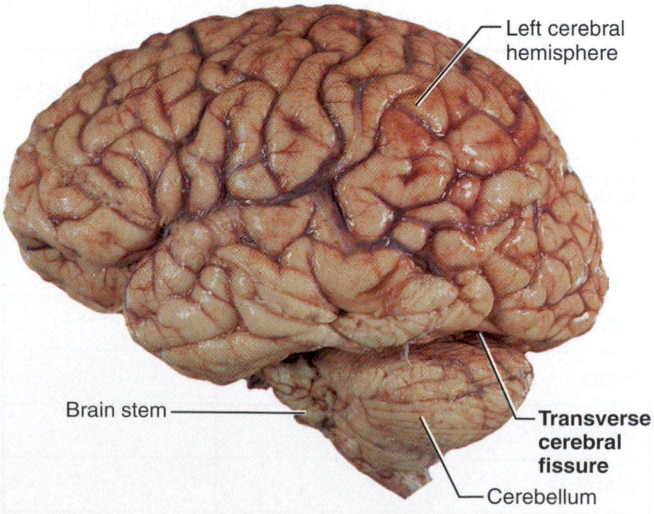
Left cerebral hemisphere
Brain stem
Transverse cerebral fissure
Cerebellum

FIGURE 19.10 Photo of lateral aspect of the human brain.

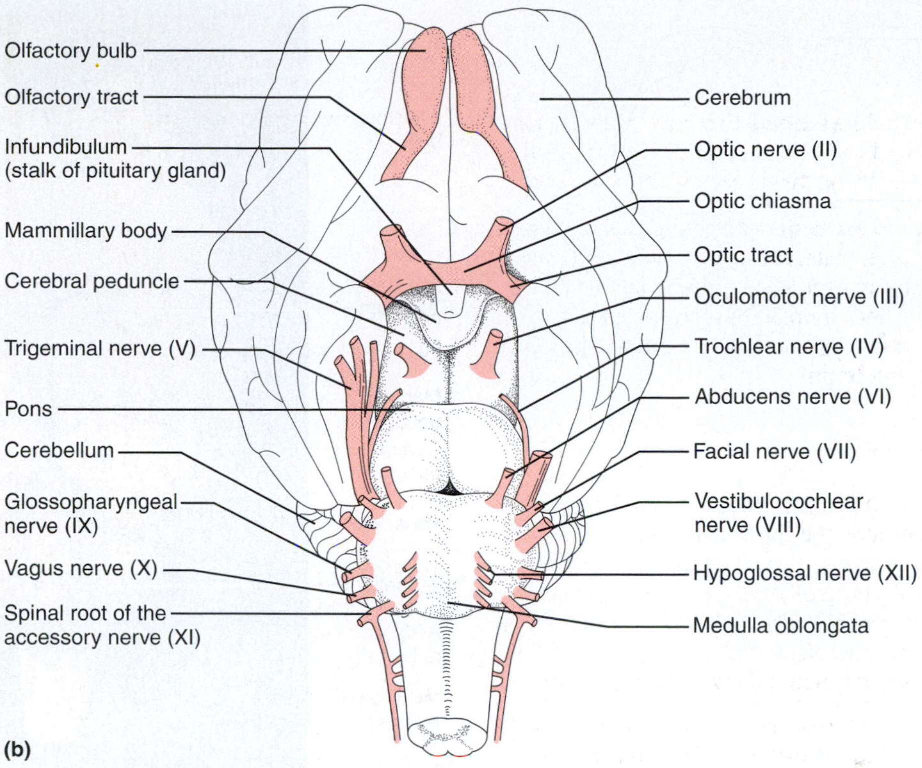
Olfactory bulb
Olfactory tract
Infundibulum (stalk of pituitary gland)
Mammillary body
Cerebral peduncle
Trigeminal nerve (V)
Pons
Cerebellum
Glossopharyngeal nerve (IX)
Vagus nerve (X)
Spinal root of the accessory nerve (XI)
Cerebrum
Optic nerve (II)
Optic chiasma
Optic tract
Oculomotor nerve (III)
Trochlear nerve (IV)
Abducens nerve (VI)
Facial nerve (VII)
Vestibulocochlear nerve (VIII)
Hypoglossal nerve (XII)
Medulla oblongata
(b)

FIGURE 19.11 ***(continued)*** **Intact sheep brain. (b)** Diagrammatic ventral view.

the tiny trochlear nerves (IV), which can be seen at the junction of the midbrain and pons. Both of these cranial nerves provide motor fibers to extrinsic muscles of the eyeball.

5. Move posteriorly from the midbrain to identify first the pons and then the medulla oblongata, both hindbrain structures composed primarily of ascending and descending fiber tracts.

6. Return to the junction of the pons and midbrain, and proceed posteriorly to identify the following cranial nerves, all arising from the pons:

- Trigeminal nerves (V), which are involved in chewing and sensations of the head and face.
- Abducens nerves (VI), which abduct the eye (and thus work in conjunction with cranial nerves III and IV)
- Facial nerves (VII), large nerves involved in taste sensation, gland function (salivary and lacrimal glands), and facial expression.

7. Continue posteriorly to identify:

- Vestibulocochlear nerves (VIII), purely sensory nerves that are involved with hearing and equilibrium.
- Glossopharyngeal nerves (IX), which contain motor fibers innervating throat structures and sensory fibers transmitting taste stimuli (in conjunction with cranial nerve VII).
- Vagus nerves (X), often called "wanderers," which serve many organs of the head, thorax, and abdominal cavity.
- Accessory nerves (XI), which serve muscles of the neck, larynx, and shoulder; actually arise from the spinal cord (C_1 through C_5) and travel superiorly to enter the skull before running to the muscles that they serve.
- Hypoglossal nerves (XII), which stimulate tongue and neck muscles.

It is likely that some of the cranial nerves will have been broken off during brain removal. If so, observe sheep brains of other students to identify those missing from your specimen.

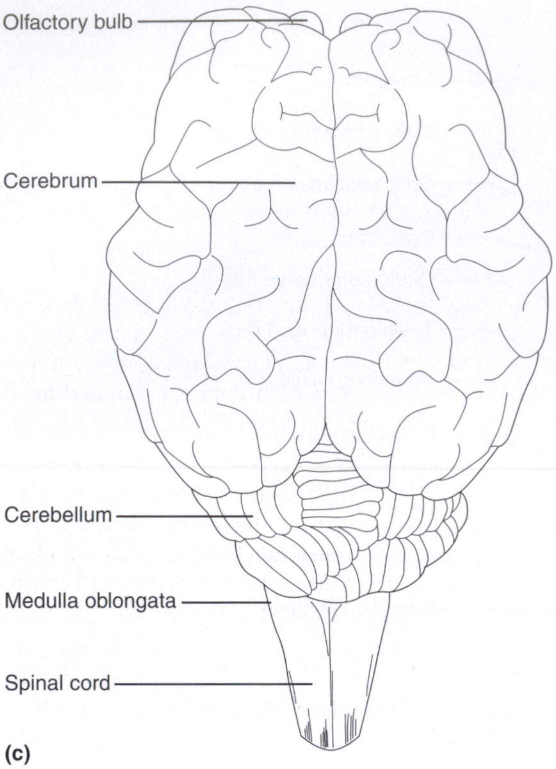
Olfactory bulb
Cerebrum
Cerebellum
Medulla oblongata
Spinal cord
(c)

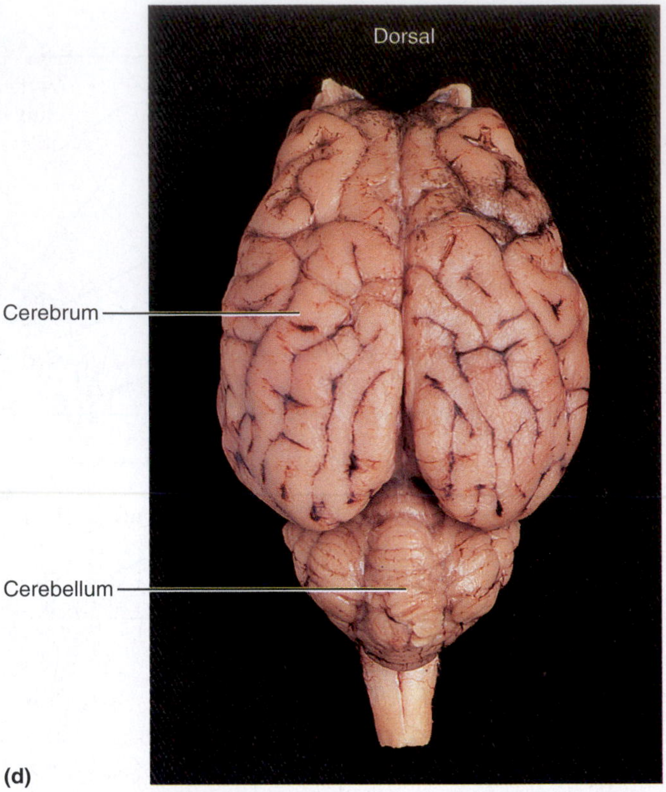
Dorsal
Cerebrum
Cerebellum
(d)

FIGURE 19.11 ***(continued)*** **Intact sheep brain. (c and d)** Diagrammatic and photographic views of the dorsal view, respectively.

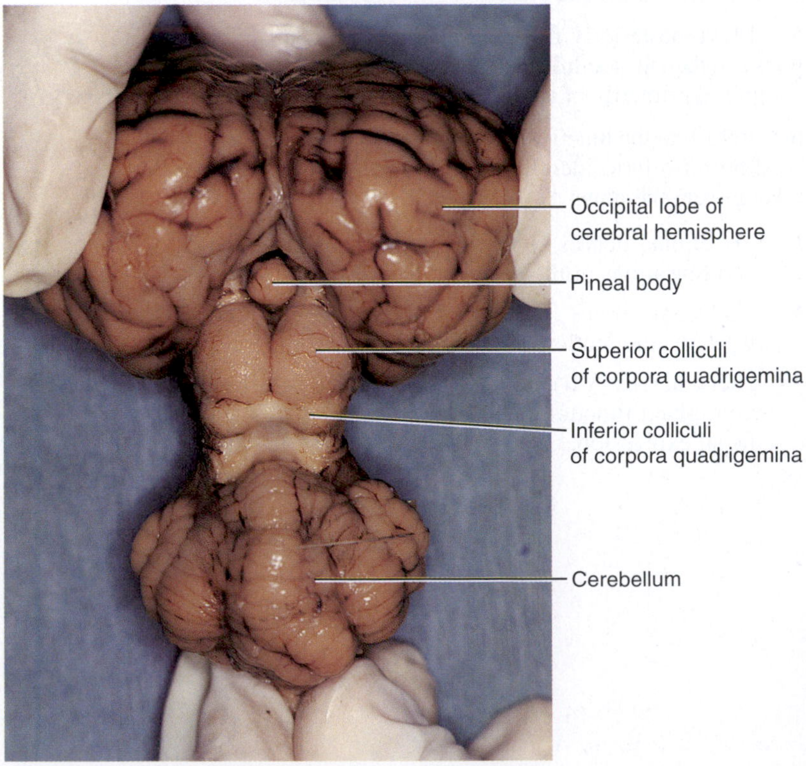
Occipital lobe of cerebral hemisphere
Pineal body
Superior colliculi of corpora quadrigemina
Inferior colliculi of corpora quadrigemina
Cerebellum

FIGURE 19.12 Means of exposing the dorsal midbrain structures of the sheep brain.

Dorsal Structures

1. Refer to Figure 19.11c and d as a guide in identifying the following structures. Reidentify the now exposed cerebral hemispheres. How does the depth of the fissures in the sheep's cerebral hemispheres compare to that of the fissures in the human brain?

____

2. Examine the cerebellum. Notice that, in contrast to the human cerebellum, it is not divided longitudinally, and that its fissures are oriented differently. What dural falx (falx cerebri or falx cerebelli) is missing that is present in humans?

____

3. Locate the three pairs of cerebellar peduncles, fiber tracts that connect the cerebellum to other brain structures, by lifting the cerebellum dorsally away from the brain stem. The most posterior pair, the inferior cerebellar peduncles, connect the cerebellum to the medulla. The middle cerebellar peduncles attach the cerebellum to the pons, and the superior cerebellar peduncles run from the cerebellum to the midbrain.

4. To expose the dorsal surface of the midbrain, gently separate the cerebrum and cerebellum as shown in Figure 19.12. Identify the corpora quadrigemina, which appear as four rounded prominences on the dorsal midbrain surface.

What is the function of the corpora quadrigemina?

____

____

Also locate the pineal gland, which appears as a small oval protrusion in the midline just anterior to the corpora quadrigemina.

Internal Structures

1. The internal structure of the brain can be examined only after further dissection. Place the brain ventral side down on the dissecting tray and make a cut completely through it in a superior to inferior direction. Cut through the longitudinal fissure, corpus callosum, and midline of the cerebellum. Refer to Figure 19.13 as you work.

2. A thin nervous tissue membrane immediately ventral to the corpus callosum that separates the lateral ventricles is the septum pellucidum. If it is still intact, pierce this membrane and probe the lateral ventricle cavity. The fiber tract ventral to the septum pellucidum and anterior to the third ventricle is the fornix.

How does the size of the fornix in this brain compare with the size of the human fornix?

____

____

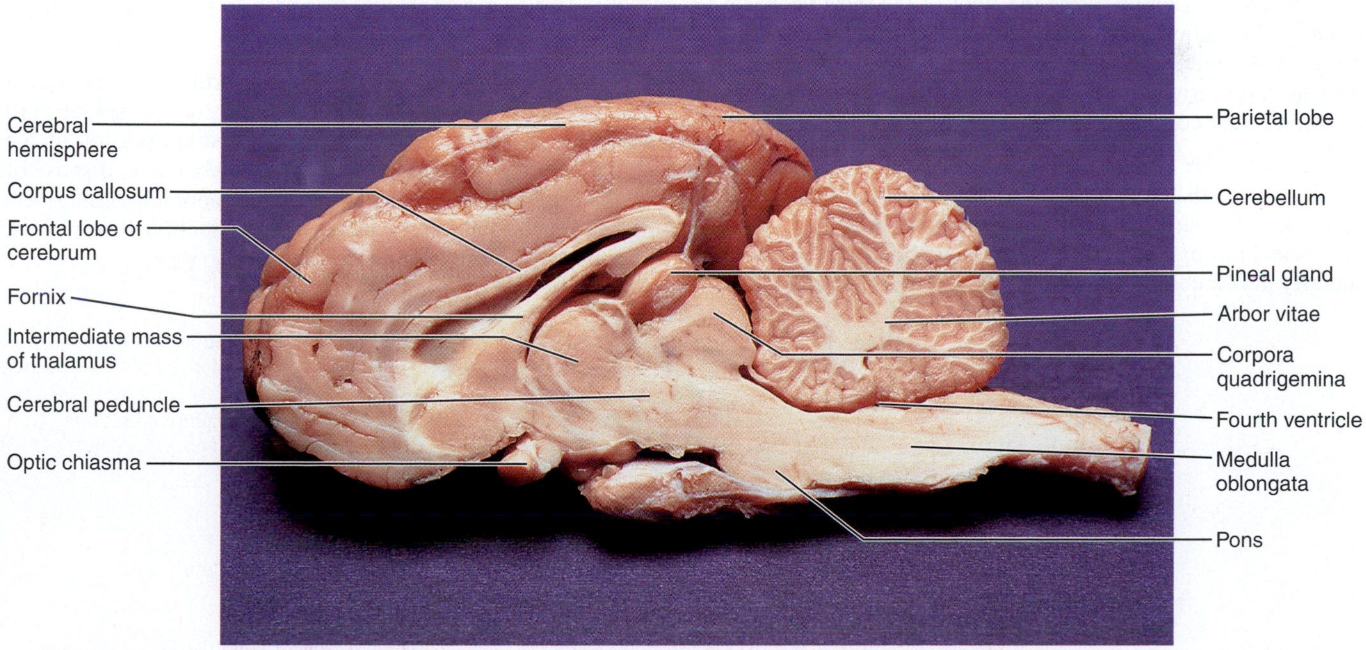
Cerebral hemisphere
Corpus callosum
Frontal lobe of cerebrum
Fornix
Intermediate mass of thalamus
Cerebral peduncle
Optic chiasma
Parietal lobe
Cerebellum
Pineal gland
Arbor vitae
Corpora quadrigemina
Fourth ventricle
Medulla oblongata
Pons

FIGURE 19.13 Photograph of sagittal section of the sheep brain showing internal structures.

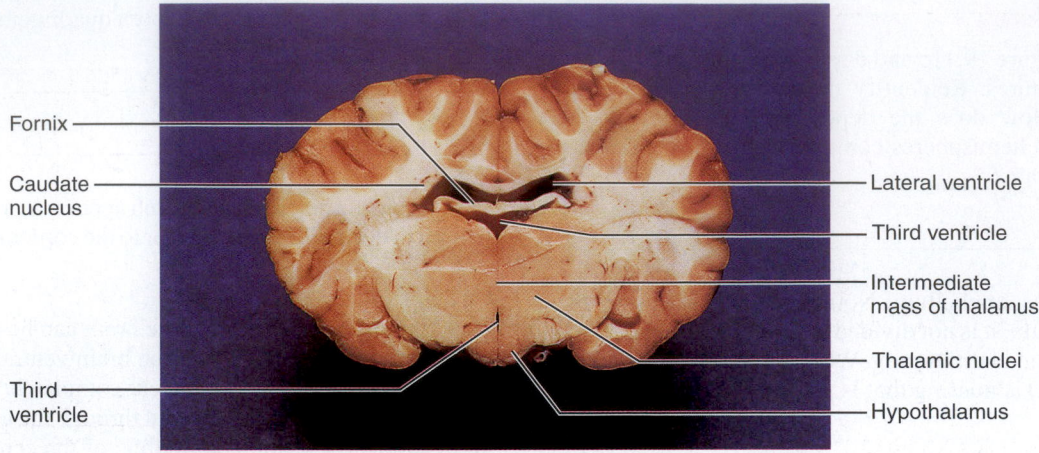

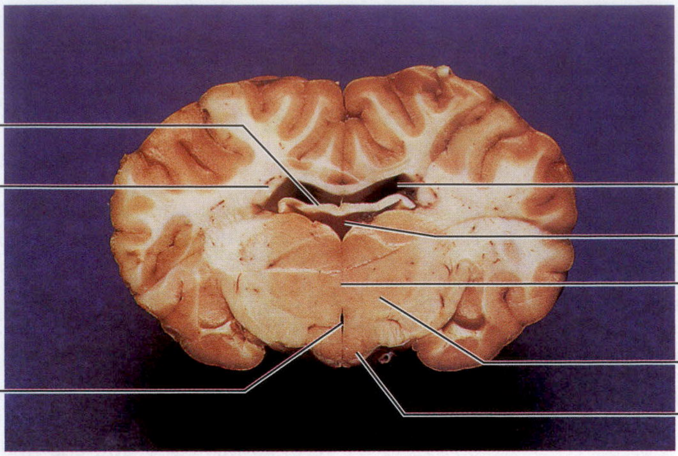
Fornix
Caudate nucleus
Third ventricle
Lateral ventricle
Third ventricle
Intermediate mass of thalamus
Thalamic nuclei
Hypothalamus

FIGURE 19.14 Frontal section of a sheep brain. Major structures include the thalamus, hypothalamus, and lateral and third ventricles.

Why do you suppose this is so? (Hint: What is the function of this band of fibers?)

---

3. Identify the thalamus, which forms the walls of the third ventricle and is located posterior and ventral to the fornix. The intermediate mass spanning the ventricular cavity appears as an oval protrusion of the thalamic wall. Anterior to the intermediate mass, locate the interventricular foramen, a canal connecting the lateral ventricle on the same side with the third ventricle.

4. The hypothalamus forms the floor of the third ventricle. Identify the optic chiasma, infundibulum, and mammillary body on its exterior surface. You can see the pineal gland at the superoposterior end of the third ventricle, just beneath the junction of the corpus callosum and fornix.

5. Locate the midbrain by identifying the corpora quadrigemina that form its dorsal roof. Follow the cerebral aqueduct (the narrow canal connecting the third and fourth ventricles) through the midbrain tissue to the fourth ventricle. Identify the cerebral peduncles, which form its anterior walls.

6. Identify the pons and medulla oblongata, which lie anterior to the fourth ventricle. The medulla continues into the spinal cord without any obvious anatomical change, but the point at which the fourth ventricle narrows to a small canal is generally accepted as the beginning of the spinal cord.

7. Identify the cerebellum posterior to the fourth ventricle. Notice its internal treelike arrangement of white matter, the arbor vitae.

8. If time allows, obtain another sheep brain and section it along the frontal plane so that the cut passes through the infundibulum. Compare your specimen with the photograph in Figure 19.14, and attempt to identify all the structures shown in the figure.

9. Check with your instructor to determine if cow spinal cord sections (preserved) are available for the spinal cord studies in Exercise 21. If not, save the small portion of the spinal cord from your brain specimen. Otherwise, dispose of all the organic debris in the appropriate laboratory containers and clean the laboratory bench, the dissection instruments, and the tray before leaving the laboratory.

EXERCISE 20

Electroencephalography

MATERIALS

- ☐ Oscilloscope and EEG lead-selector box or physiograph and high-gain preamplifier
- ☐ Cot (if available) or pillow
- ☐ Electrode gel
- ☐ EEG electrodes and leads
- ☐ Collodion gel or long elastic EEG straps
- ☐ BIOPAC® BIOPAC® MP36 (or MP35/30) data acquisition unit, PC or Mac computer, BIOPAC® Student Lab Software, electrode lead set, disposable vinyl electrodes, Lycra® swim cap (such as Speedo® brand) or supportive wrap (such as 3M Coban™ Self-adhering Support Wrap) to press electrodes against head for improved contact, and a cot or lab bench and pillow.

New versions of the BSL software (3.7.5 and higher for Windows, and 3.7.4 and higher for Mac) require slightly different channel settings and collection strategies. Instructions for using the newer software with the MP36/35 data acquisition units can be found on MasteringA&P. In addition, instructions for use of the NEW 2-channel data acquisition unit, the MP45, may also be found on MasteringA&P.

OBJECTIVES

1. To define *electroencephalogram,* and to discuss its clinical significance.
2. To describe or recognize typical tracings of the most common brain wave patterns (alpha, beta, theta, and delta waves), and to indicate the conditions under which each is most likely to be predominant.
3. To indicate the source of brain waves.
4. To define *alpha block.*
5. To monitor electroencephalography and recognize alpha rhythm.
6. To describe the effect of a sudden sound, mental concentration, and alkalosis on brain wave patterns.

PRE-LAB QUIZ

1. What does an electroencephalogram (EEG) actually measure?
   a. electrical activity of the brain
   b. electrical activity of the heart
   c. emotions
   d. physical activity of the subject
2. Circle the correct term. Alpha waves / Beta waves are typical of the attentive or awake state.
3. Circle True or False. Brain waves can change with age, sensory stimuli, and the chemical state of the body.
4. Where will you place the indifferent (ground) electrode on your subject?
   a. the earlobe
   b. the forehead
   c. over the occipital lobe
   d. over the temporal bone
5. During today's activity, students will instruct subjects to *hyperventilate.* What should the subjects do?
   a. breathe in a normal manner
   b. breathe rapidly
   c. breathe very slowly
   d. hold their breath until they almost pass out

MasteringA&P™

Access practice quizzes and more in the Study Area at www.masteringaandp.com.

Any physiological investigation of the brain can emphasize and expose only a very minute portion of its activity. Higher brain functions, such as consciousness and logical reasoning, are extremely difficult to investigate. It is obviously much easier to do experiments on the brain's input-output functions, some of which can be detected with appropriate recording equipment. Still, the ability to record brain activity does not necessarily guarantee an understanding of the brain.

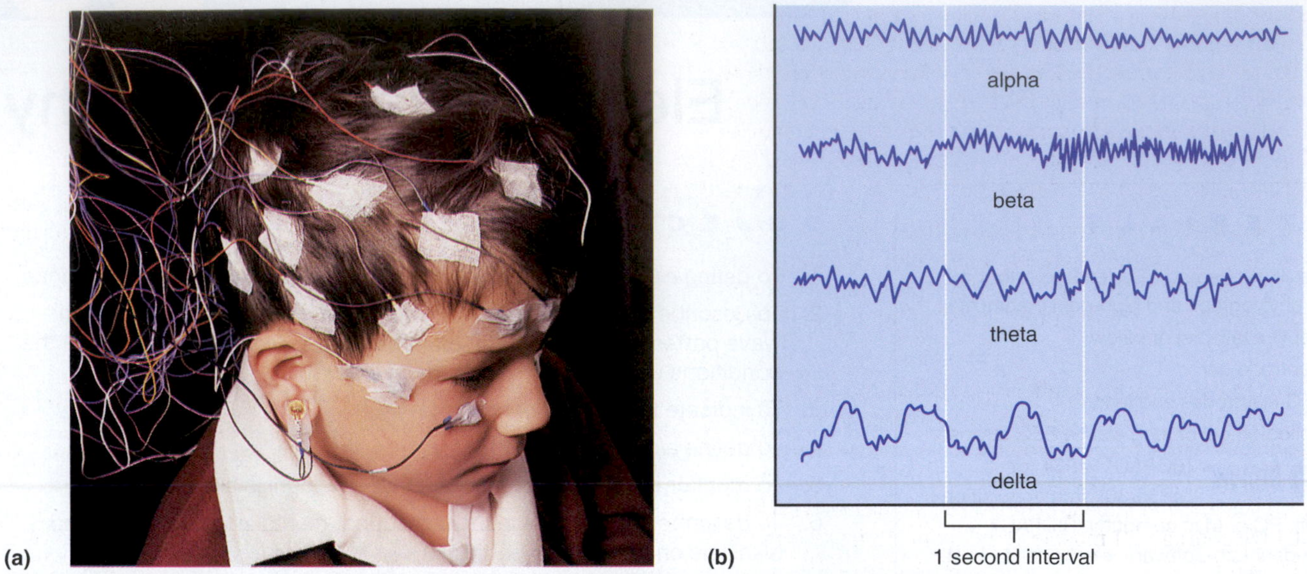
(a)
(b)
alpha
beta
theta
delta
1 second interval

FIGURE 20.1 Electroencephalography and brain waves. (a) To obtain a recording of brain wave activity (an EEG), electrodes are positioned on the patient's scalp and attached to a recording device called an electroencephalograph. **(b)** Typical EEGs. Alpha waves are typical of the awake but relaxed state; beta waves occur in the awake, alert state; theta waves are common in children but not in normal awake adults; delta waves occur during deep sleep.

Brain Wave Patterns and the Electroencephalogram

The **electroencephalogram (EEG),** a record of the electrical activity of the brain, can be obtained through electrodes placed at various points on the skin or scalp of the head. This electrical activity, which is recorded as waves (Figure 20.1), represents the summed synaptic activity of many neurons.

Certain characteristics of brain waves are known. They have a frequency of 1 to 30 hertz (Hz) or cycles per second, a dominant rhythm of 10 Hz, and an average amplitude (voltage) of 20 to 100 microvolts (μV). They vary in frequency in different brain areas, occipital waves having a lower frequency than those associated with the frontal and parietal lobes.

The first of the brain waves to be described by scientists were the alpha waves (or alpha rhythm). **Alpha waves** have an average frequency range of 8 to 13 Hz and are produced when the individual is in a relaxed state with the eyes closed. **Alpha block,** suppression of the alpha rhythm, occurs if the eyes are opened or if the individual begins to concentrate on some mental problem or visual stimulus. Under these conditions, the waves decrease in amplitude but increase in frequency. Under conditions of fright or excitement, the frequency increases still more.

Beta waves, closely related to alpha waves, are faster (14 to 30 Hz) and have a lower amplitude. They are typical of the attentive or alert state.

Very large (high-amplitude) waves with a frequency of 4 Hz or less that are seen in deep sleep are **delta waves. Theta waves** are large, abnormally contoured waves with a frequency of 4 to 7 Hz. Although theta waves are normal in children, they are abnormal in awake adults.

Brain waves change with age, sensory stimuli, brain pathology, and the chemical state of the body. (Glucose deprivation, oxygen poisoning, and sedatives all interfere with the rhythmic activity of brain output by disturbing the metabolism of neurons.) Sleeping individuals and patients in a coma have EEGs that are slower (lower frequency) than the alpha rhythm of normal adults. Fright, epileptic seizures, and various types of drug intoxication can be associated with comparatively faster cortical activity. Thus impairment of cortical function is indicated by neuronal activity that is either too fast or too slow; unconsciousness occurs at both extremes of the frequency range.

Because spontaneous brain waves are always present, even during unconsciousness and coma, the absence of brain waves (a "flat" EEG) is taken as clinical evidence of death. The EEG is used clinically to diagnose and localize many types of brain lesions, including epileptic foci, infections, abscesses, and tumors. ■

ACTIVITY 1

Observing Brain Wave Patterns Using an Oscilloscope or Physiograph

If one electrode (the *active electrode*) is placed over a particular cortical area and another (the *indifferent electrode*) is placed over an inactive part of the head, such as the earlobe, all of the activity of the cortex underlying the active electrode will, theoretically, be recorded. The inactive area provides a zero reference point, or a baseline, and the EEG represents the difference between "activities" occurring under the two electrodes.

1. Connect the EEG lead-selector box to the oscilloscope preamplifier, or connect the high-gain preamplifier to the physiograph channel amplifier. Adjust the horizontal sweep and sensitivity according to the directions given in the instrument manual or by your instructor.

2. Prepare the subject. The subject should lie undisturbed on a cot or on the lab bench with eyes closed in a quiet, dimly lit area.

(Someone who is able to relax easily makes a good subject.) Apply a small amount of electrode gel to the subject's forehead above the left eye and on the left earlobe. Press an electrode to each prepared area and secure each by (1) applying a film of collodion gel to the electrode surface and the adjacent skin or (2) using a long elastic EEG strap (knot tied at the back of the head). If collodion gel is used, allow it to dry before you continue.

3. Connect the active frontal lead (forehead) to the EEG lead-selector box outlet marked "L Frontal." Connect the lead from the indifferent electrode (earlobe) to the ground outlet (or to the appropriate input terminal on the high-gain preamplifier).

4. Turn the oscilloscope or physiograph on, and observe the EEG pattern of the relaxed subject for a period of 5 minutes. If the subject is truly relaxed, you should see a typical alpha-wave pattern. (If the subject is unable to relax and the alpha-wave pattern does not appear in this time interval, test another subject.) Discourage all muscle movement during the monitoring period.*

5. Abruptly and loudly clap your hands. The subject's eyes should open, and alpha block should occur. Observe the immediate brain wave pattern. How do the frequency and amplitude of the brain waves change?

____

____

Would you characterize this as beta rhythm? ____________

Why? ______________________________

____

6. Allow the subject about 5 minutes to achieve complete relaxation once again, then ask him or her to compute a number problem that requires concentration (for example, add 3 and 36, subtract 7, multiply by 2, add 50, etc.). Observe the brain wave pattern during the period of mental computation.

Observations: ______________________________

____

____

7. Once again allow the subject to relax until alpha rhythm resumes. Then, instruct him or her to hyperventilate for 3 minutes. *Be sure to tell the subject when to stop hyperventilating.* Hyperventilation rapidly flushes carbon dioxide out of the lungs, decreasing carbon dioxide levels in the blood and producing respiratory alkalosis.

Observe the changes in the rhythm and amplitude of the brain waves occurring during the period of hyperventilation.

*Note that 60-cycle "noise" (appearing as fast, regular, low amplitude waves superimposed on the more irregular brain waves) may interfere with the tracings being made, particularly if the laboratory has a lot of electrical equipment.

Observations: ______________________________

____

____

8. Think of other stimuli that might affect brain wave patterns. Test your hypotheses. Describe what stimuli you tested and what responses you observed.

____

____

____

____

ACTIVITY 2

Electroencephalography Using Biopac®

In this activity, the EEG of the subject will be recorded during a relaxed state, first with the eyes closed, then with the eyes open while silently counting to ten, and finally with the eyes closed again.

Setting Up the Equipment

1. Connect the BIOPAC® unit to the computer and turn the computer **ON.**

2. Make sure the BIOPAC® unit is **OFF.**

3. Plug in the equipment as shown in Figure 20.2.

- Electrode lead set—CH 1

4. Turn the BIOPAC® unit **ON.**

5. Attach three electrodes to the subject's scalp and ear as shown in Figure 20.3. Follow these important guidelines to assist in effective electrode placement:

- Select subjects with the easiest access to the scalp.
- Move as much hair out of the way as possible.
- Apply a dab of electrode gel to the spots where the electrodes will be attached.
- Apply pressure to the electrodes for 1 minute to ensure attachment.
- Use a swimcap or supportive wrap to maintain attachment.
- Do not touch the electrodes while recording.
- The earlobe electrode may be folded under the lobe itself.

6. When the electrodes are attached, the subject should lie down and relax with eyes closed for 5 minutes before recording.

7. Start the BIOPAC® Student Lab program on the computer by double-clicking the icon on the desktop or by following your instructor's guidance.

8. Select lesson **L03-EEG-1** from the menu, and click **OK.**

9. Type in a filename that will save this subject's data on the computer hard drive. You may want to use the subject's last name followed by EEG-1 (for example, SmithEEG-1), then click **OK.**

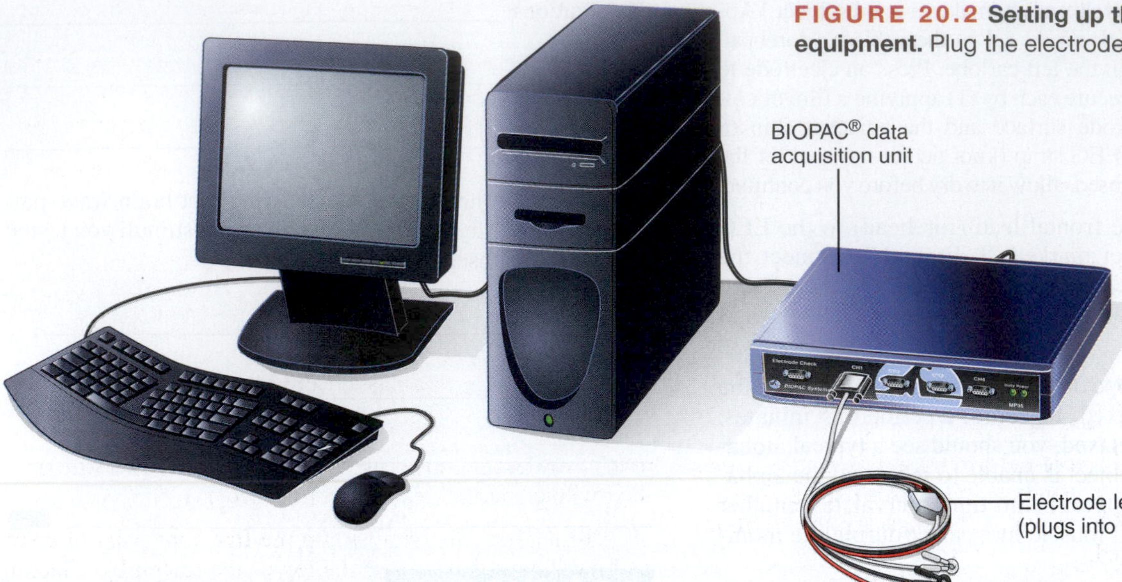
BIOPAC® data
acquisition unit
Electrode lead set
(plugs into Channel 1)

FIGURE 20.2 Setting up the BIOPAC® equipment. Plug the electrode set into Channel 1.

10. During this preparation, the subject should be very still and in a relaxed state with eyes closed. Allow the subject to relax with minimal stimuli.

Calibrating the Equipment

1. Make sure that the electrodes remain firmly attached to the surface of the scalp and earlobe. The subject should remain absolutely still and try to avoid movement of the body or face.
2. With the subject in a relaxed position, click **Calibrate.**
3. You will be prompted to check electrode attachment one final time. When ready, click **OK;** the computer will record for 15 seconds and stop automatically.
4. Observe the recording of the calibration data, which should look like Figure 20.4.

- If the data look very different, click **Redo Calibration** and repeat the steps above.
- If the data look similar, proceed to the next section.

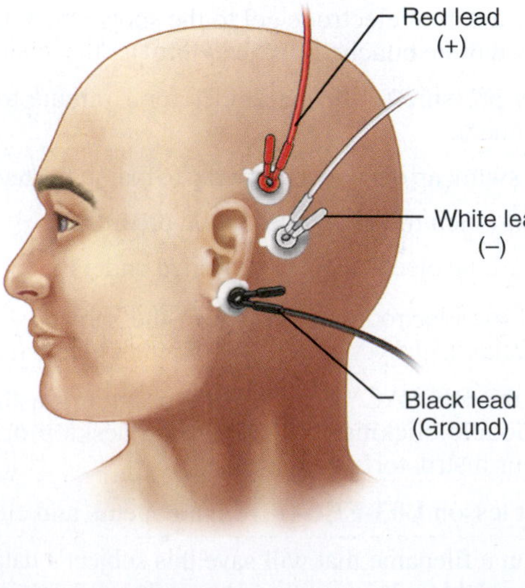
Red lead
(+)
White lead
(–)
Black lead
(Ground)

FIGURE 20.3 Placement of electrodes and the appropriate attachment of electrode leads by color.

Recording the Data

1. The subject should remain relaxed with eyes closed.
2. After clicking **Record,** the "director" will instruct the subject to keep his or her eyes closed for the first 10 seconds of recording, then open the eyes and *mentally* (not verbally) count to ten, then close the eyes again and relax for 10 seconds. The director will insert a marker by pressing the **F9** key (PC) or **ESC** key (Mac) when the command to open eyes is given, and another marker when the subject reaches the count of ten and closes the eyes. Click **Stop** 10 seconds after the subject recloses the eyes.
3. Observe the recording of the data, which should look similar to the data in Figure 20.5.

- If the data look very different, click **Redo** and repeat the steps above. If the subject moved too much during the recording, it is likely that artifact spikes will appear in the data.
- If the data look similar, proceed to the next step.

4. Click the buttons for each of the different EEG rhythms at the top of the data in the following order: **alpha, beta, delta,** and **theta.** When you click each button, the software will extract and display each of the specific frequency bands.

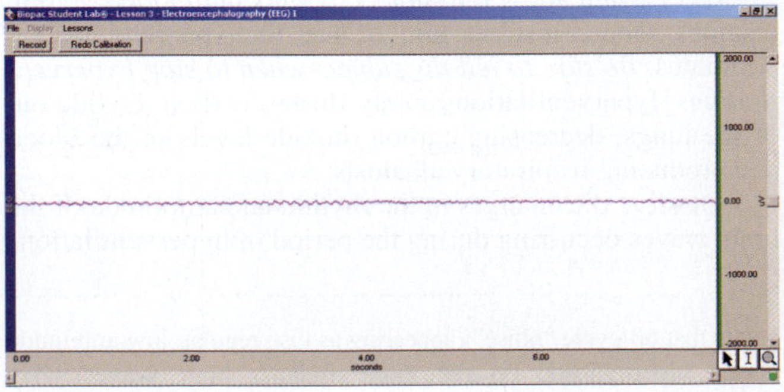

FIGURE 20.4 Example of calibration data.

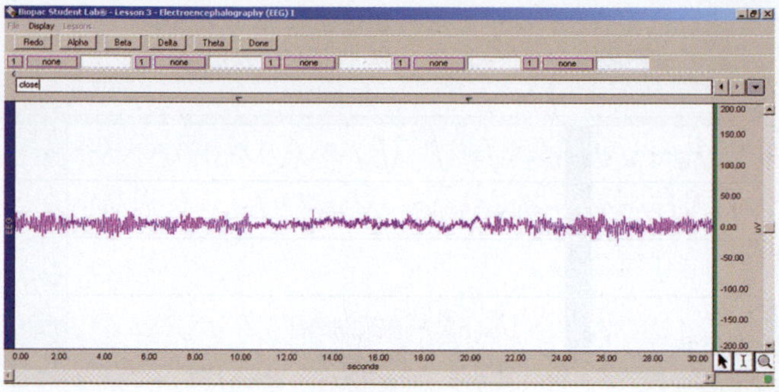

FIGURE 20.5 Example of EEG data.

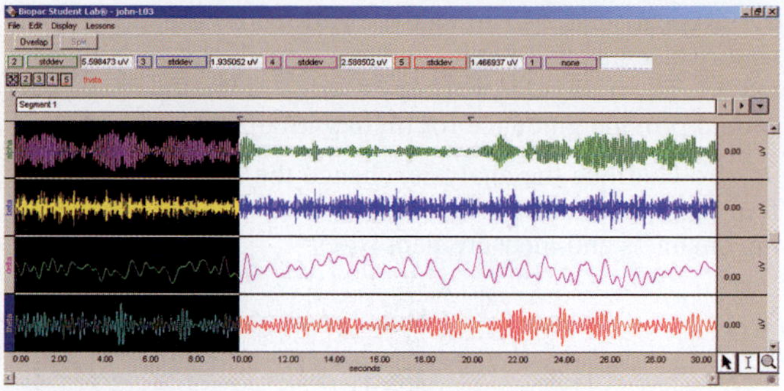

FIGURE 20.6 Highlighting the first data segment.

- To properly view each of the waveforms, you may have to click the **Display** menu and select **Autoscale Waveforms.** This function rescales the data for the rhythm band that is selected.

5. Observe the **alpha rhythm** band of data. The intensity of the alpha signal should decrease during the "eyes open" phase of the recording.

- If the data do not demonstrate this change, it is possible that the electrodes were not properly attached. You should redo the recording.

6. When finished, click **Done.** If you are certain you want to stop recording, click **YES.** Remove the electrodes from the subject's scalp.

7. A pop-up window will appear. To record from another subject select **Record from another subject** and return to step 5 under Setting Up the Equipment. If continuing to the Data Analysis section, select **Analyze current data file** and proceed to step 2.

Data Analysis

1. If just starting the BIOPAC® program to perform data analysis, enter **Review Saved Data** mode and choose the file with the subject's EEG data (for example, SmithEEG-1).

2. Observe the way the channel numbers are designated: CH 1—**raw EEG;** CH 2—**alpha;** CH 3—**beta;** CH 4—**delta;** and CH 5—**theta.**

3. To set up the display for optimal viewing, hide Channel 1. To do this, hold down the Ctrl key (PC) or Option key (Mac) while using the cursor to click channel box 1 (the small box with a 1 at the upper left of the screen).

4. To analyze the data, set up the first four pairs of channel/measurement boxes at the top of the screen by selecting the following channels and measurement types from the drop-down menus:

Channel	Measurement
CH 2	stdev
CH 3	stdev
CH 4	stdev
CH 5	stdev

stdev (standard deviation): This is a statistical calculation that estimates the variability of the data in the area highlighted by the I-beam cursor. This function minimizes the effects of extreme values and electrical artifacts that may unduly influence interpretation of the data.

5. Use the arrow cursor and click the I-beam cursor box at the lower right of the screen to activate the "area selection" function. Using the activated I-beam cursor, highlight the first 10-second segment of EEG data, which represents the subject at rest with eyes closed (Figure 20.6).

6. Observe that the computer automatically calculates the stdev for each of the channels of data (alpha, beta, delta, and theta).

7. Record the data for each rhythm in the chart below, rounding to the nearest 0.01 μV.

Standard Deviations (stdev) of Signals in Each Segment

Rhythm	Channel	Eyes closed Segment 1 Seconds 0–10	Eyes open Segment 2 Seconds 10–20	Eyes reclosed Segment 3 Seconds 20–30
Alpha	CH 2			
Beta	CH 3			
Delta	CH 4			
Theta	CH 5			

8. Repeat steps 5–7 to analyze and record the data for the next two segments of data, with eyes open, and with eyes reclosed. The triangular markers inserted at the top of the data should provide guidance for highlighting.

9. To continue the analysis, change the settings in the first four pairs of channel/measurement boxes. Select the following channels and measurement types:

Channel	Measurement
CH 2	Freq
CH 3	Freq
CH 4	Freq
CH 5	Freq

Freq (frequency): This gives the frequency in hertz (Hz) of an individual wave that is highlighted by the I-beam cursor.

10. To view an individual wave from among the high-frequency waveforms, you must use the zoom function. To activate the zoom function, use the cursor to click the magnifying glass at the lower-right corner of the screen (near the I-beam cursor box). The cursor will become a magnifying glass.

11. To examine individual waves within the **alpha** data, click that band with the magnifying glass until it is possible to observe the peaks and troughs of individual waves within **Segment 1.**

- To properly view each of the waveforms, you may have to click the **Display** menu and select **Autoscale Waveforms.** This function rescales the data for the rhythm band that is selected.

12. At this time, focus on alpha waves only. Reactivate the I-beam cursor by clicking its box in the lower-right corner. Highlight a *single* alpha wave from peak to peak as shown in Figure 20.7.

13. Read the calculated frequency (in Hz) in the measurement box for CH 2, and record this as the frequency of Wave 1 for alpha rhythm in the chart below.

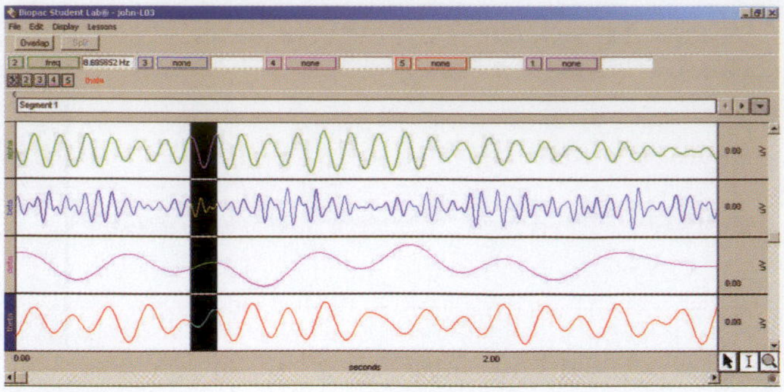
Biopac Student Lab® - john-L03
File Edit Display Lessons
Overlap
Split
freq
8.695652 Hz
none
Segment 1
0.00
seconds
2.00

FIGURE 20.7 Highlighting a single alpha wave from peak to peak.

14. Use the I-beam cursor to select two more individual **alpha** waves and record their frequencies in the chart.

15. You will now perform the same frequency measurements for three waves in each of the **beta** (CH 3), **delta** (CH 4), and **theta** (CH 5) data sets. Record these measurements in the chart.

16. Calculate the average of the three waves measured for each of the brain rhythms, and record the average in the chart.

17. When finished, answer the following questions and then exit the program by going to the **File** menu at the top of the page and clicking **Quit.**

Look at the waveforms you recorded and carefully examine all three segments of the alpha rhythm record. Is there a difference in electrical activity in this frequency range when the eyes are open versus closed? Describe your observations.

____

____

____

Carefully examine all three segments of the beta rhythm record. Is there a difference in electrical activity in this

Frequencies of Waves for Each Rhythm (Hz)

Rhythm	Channel	Wave 1	Wave 2	Wave 3	Average
Alpha	CH 2				
Beta	CH 3				
Delta	CH 4				
Theta	CH 5				

frequency range when the eyes are open versus closed? Describe what you observe.

____

____

____

This time, compare the intensity (height) of the alpha and beta waveforms throughout all three segments. Does the intensity of one signal appear more varied than the other in the record? Describe your observations.

____

____

____

Examine the data for the delta and theta rhythms. Is there any change in the waveform as the subject changes states? If so, describe the change observed.

____

____

____

The degree of variation in the intensity of the signal was estimated by calculating the standard deviation of the waves in each segment of data. In which time segment (eyes open, eyes closed, or eyes reclosed) is the difference in the standard deviations the greatest?

____

NAME ______________________________

LAB TIME/DATE ______________________

REVIEW SHEET

EXERCISE 20

Electroencephalography

Brain Wave Patterns and the Electroencephalogram

1. Define *EEG.* ______________________________

2. Identify the type of brain wave pattern described in each statement below.

_________ below 4 Hz; slow, large waves; normally seen during deep sleep

_________ rhythm generally apparent when an individual is in a relaxed, nonattentive state with the eyes closed

_________ correlated to the alert state; usually about 15 to 30 Hz

3. What is meant by the term *alpha block*? ______________________________

____

4. List at least four types of brain lesions that may be determined by EEG studies. ______________________________

____

5. What is the common result of hypoactivity or hyperactivity of the brain neurons? ______________________________

Observing Brain Wave Patterns

6. How was alpha block demonstrated in the laboratory experiment? ______________________________

____

7. What was the effect of mental concentration on the brain wave pattern? ______________________________

____

____

8. What effect on the brain wave pattern did hyperventilation have? ______________________________

Electroencephalography Using BIOPAC®

9. Observe the average frequency of the waves you measured for each rhythm. Did the calculated average for each fall within the specified range indicated in the introduction to encephalograms?

10. Suggest the possible advantages and disadvantages of using electroencephalography in a clinical setting.

EXERCISE 21

Spinal Cord, Spinal Nerves, and the Autonomic Nervous System

MATERIALS

- ☐ Spinal cord model (cross section)
- ☐ Three-dimensional laboratory charts or models of the spinal cord, spinal nerves, and sympathetic chain
- ☐ Red and blue pencils
- ☐ Preserved cow spinal cord sections with meninges and nerve roots intact (or spinal cord segment saved from the brain dissection in Exercise 19)
- ☐ Dissecting instruments and tray
- ☐ Disposable gloves
- ☐ Stereomicroscope
- ☐ Prepared slide of spinal cord (x.s.)
- ☐ Compound microscope
- ☐ BIOPAC® BIOPAC® MP36 (MP35/30) data acquisition unit, PC or Mac computer. BIOPAC® Student Lab Software, respiratory transducer belt, GSR finger leads, electrode lead set, disposable vinyl electrodes, conduction gel, and nine 8½ × 11 inch sheets of paper of different colors (white, black, red, blue, green, yellow, orange, brown, and purple) to be viewed in this sequence. New versions of the BSL software (3.7.5 and higher for Windows, and 3.7.4 and higher for Mac) require slightly different channel settings and collection strategies. Instructions for using the newer software with the MP36/35 data acquisition units can be found on MasteringA&P. In addition, instructions for use of the NEW 2-channel data acquisition unit, the MP45, may also be found on MasteringA&P.
- ☐ *The Human Nervous System: The Spinal Cord and Spinal Nerves* videotape

For instructions on animal dissections, see the dissection exercises starting on p. 697 in the cat, fetal pig, and rat editions of this manual.

MasteringA&P™

Access practice quizzes and more in the Study Area at www.masteringaandp.com.

PAL

For access to anatomical models and more, check out Practice Anatomy Lab.

OBJECTIVES

1. To identify important anatomical areas on a model or appropriate diagram of the spinal cord, and to name the neuron type found in these areas (where applicable).
2. To indicate two major areas where the spinal cord is enlarged, and to explain the reasons for the enlargement.
3. To define *conus medullaris, cauda equina,* and *filum terminale.*
4. To locate on a diagram the fiber tracts in the spinal cord, and to state their functional importance.
5. To list two major functions of the spinal cord.
6. To name the meningeal coverings of the spinal cord, and to state their function.
7. To describe the origin, fiber composition, and distribution of the spinal nerves, differentiating between roots, the spinal nerve proper, and rami, and to discuss the result of transecting these structures.
8. To discuss the distribution of the dorsal rami and ventral rami of the spinal nerves.
9. To identify the four major nerve plexuses, the major nerves of each, and their distribution.
10. To identify the site of origin and the function of the sympathetic and parasympathetic divisions of the autonomic nervous system, and to state how the autonomic nervous system differs from the somatic nervous system.
11. To record and analyze data associated with the galvanic skin response, a measure of changes that occur in the skin due to changes in autonomic stimulation.

PRE-LAB QUIZ

1. The spinal cord extends from the foramen magnum of the skull to the first or second lumbar vertebrae, where it terminates in the
   a. conus medullaris
   b. denticulate ligament
   c. filum terminale
   d. gray matter
2. How many pairs of spinal nerves do humans have?
   a. 10   c. 31
   b. 12   d. 47
3. Circle the correct term. In cross section, the gray / white matter of the spinal cord looks like a butterfly or the letter H.
4. Circle True or False. The cell bodies of sensory neurons are found in an enlarged area of the dorsal root called the gray commissure.

Text continues on next page.

5. Circle the correct term. Fiber tracts conducting impulses to the brain are called ascending or sensory / motor tracts.
6. Circle True or False. Because the spinal nerves arise from fusion of the ventral and dorsal roots of the spinal cord, and contain motor and sensory fibers, all spinal nerves are considered mixed nerves.
7. The ventral rami of all spinal nerves except for T_2 through T_{12} form complex networks of nerves known as _________.
   a. fissures c. plexuses
   b. ganglia d. sulci
8. Severe injuries to the _________ plexus cause weakness or paralysis of the entire upper limb.
   a. brachial c. lumbar
   b. cervical d. sacral
9. The _________ nervous system is the subdivision of the peripheral nervous system that regulates body activities that are generally not under conscious control.
   a. autonomic c. somatic
   b. peripheral d. vascular
10. The _________ division of the autonomic nervous system is responsible for the "fight-or-flight" response that readies the body to cope with situations that threaten homeostasis.
   a. parasympathetic c. somatic
   b. peripheral d. sympathetic

The cylindrical **spinal cord,** a continuation of the brain stem, is an association and communication center. It plays a major role in spinal reflex activity and provides neural pathways to and from higher nervous centers.

Anatomy of the Spinal Cord

Enclosed within the vertebral canal of the spinal column, the spinal cord extends from the foramen magnum of the skull to the first or second lumbar vertebra, where it terminates in the cone-shaped **conus medullaris** (Figure 21.1). Like the brain, the cord is cushioned and protected by meninges. The dura mater and arachnoid meningeal coverings extend beyond the conus medullaris, approximately to the level of S_2, and the **filum terminale,** a fibrous extension of the pia mater, extends even farther (into the coccygeal canal) to attach to the posterior coccyx. **Denticulate ligaments,** saw-toothed shelves of pia mater, secure the spinal cord to the bony wall of the vertebral column all along its length (see Figure 21.1c).

The cerebrospinal fluid-filled meninges extend well beyond the end of the spinal cord, providing an excellent site for removing cerebrospinal fluid for analysis (as when bacterial or viral infections of the spinal cord or the meninges are suspected) without endangering the delicate spinal cord. This procedure, called a *lumbar tap,* is usually performed below L_3. Additionally, "saddle block," or caudal anesthesia for childbirth, is normally administered (injected) between L_3 and L_5.

In humans, 31 pairs of spinal nerves arise from the spinal cord and pass through intervertebral foramina to serve the body area at their approximate level of emergence. The cord is about the size of a finger in circumference for most of its length, but there are obvious enlargements in the *cervical* and *lumbar* areas where the nerves serving the upper and lower limbs issue from the cord.

Because the spinal cord does not extend to the end of the vertebral column, the spinal nerves emerging from the inferior end of the cord must travel through the vertebral canal for some distance before exiting at the appropriate intervertebral foramina. This collection of spinal nerves traversing the inferior end of the vertebral canal is called the **cauda equina** (Figure 21.1a and d) because of its similarity to a horse's tail (the literal translation of *cauda equina*).

ACTIVITY 1

Identifying Structures of the Spinal Cord

Obtain a three-dimensional model or laboratory chart of a cross section of a spinal cord and identify its structures as they are described next.

Gray Matter

In cross section, the **gray matter** of the spinal cord looks like a butterfly or the letter H (Figure 21.2). The two dorsal projections are called the **dorsal (posterior) horns.** The two ventral projections are the **ventral (anterior) horns.** The tips of the ventral horns are broader and less tapered than those of the dorsal horns. In the thoracic and lumbar regions of the cord, there is also a lateral outpocketing of gray matter on each side referred to as the **lateral horn.** The central area of gray matter connecting the two vertical regions is the **gray commissure.** The gray commissure surrounds the **central canal** of the cord, which contains cerebrospinal fluid.

Neurons with specific functions can be localized in the gray matter. The dorsal horns contain interneurons and sensory fibers that enter the cord from the body periphery via the **dorsal root.** The cell bodies of these sensory neurons are found in an enlarged area of the dorsal root called the **dorsal root ganglion.** The ventral horns mainly contain cell bodies of motor neurons of the somatic nervous system (voluntary system), which send their axons out via the **ventral root** of the cord to enter the adjacent spinal nerve. The **spinal nerves** are formed from the fusion of the dorsal and ventral roots. The lateral horns, where present, contain nerve cell bodies of motor neurons of the autonomic nervous system (sympathetic division). Their axons also leave the cord via the ventral roots, along with those of the motor neurons of the ventral horns.

White Matter

The **white matter** of the spinal cord is nearly bisected by fissures (see Figure 21.2). The more open ventral fissure is the **ventral median fissure,** and the dorsal one is the shallow **dorsal median sulcus.** The white matter is composed of myelinated fibers—some running to higher centers, some

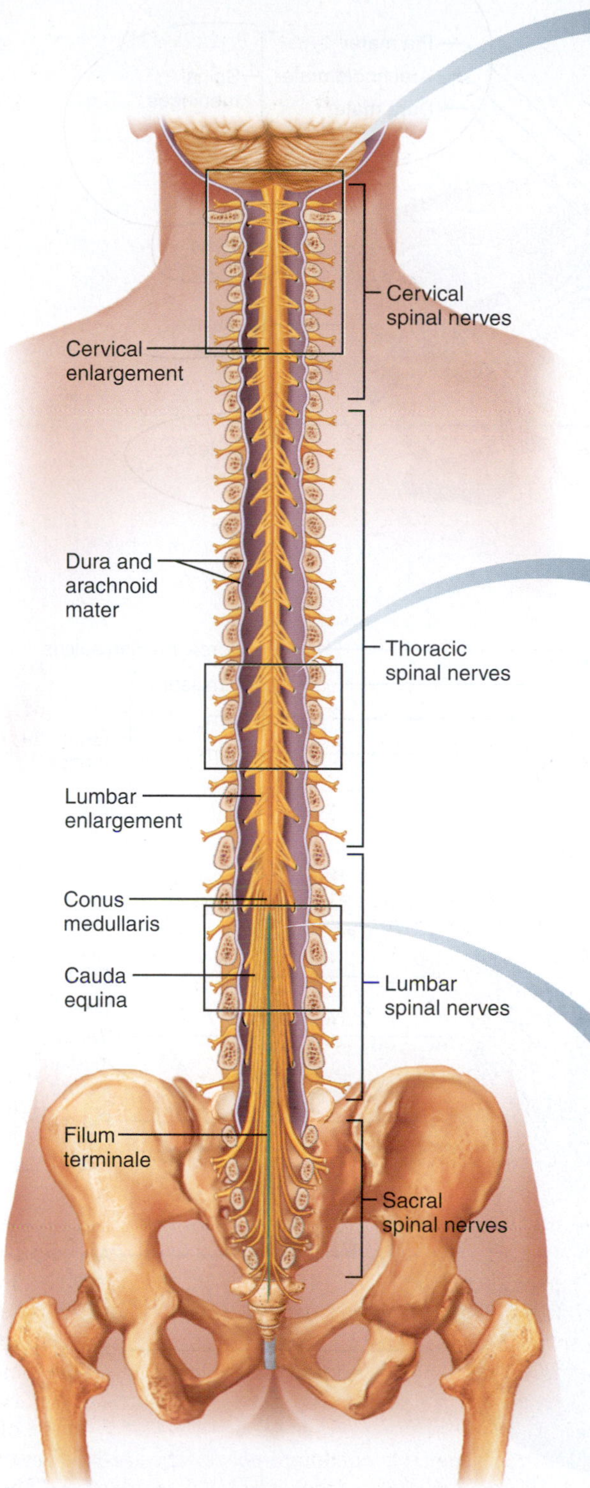
Cervical enlargement
Cervical spinal nerves
Dura and arachnoid mater
Thoracic spinal nerves
Lumbar enlargement
Conus medullaris
Cauda equina
Lumbar spinal nerves
Filum terminale
Sacral spinal nerves

(a) The spinal cord and its nerve roots, with the bony vertebral arches removed. The dura mater and arachnoid mater are cut open and reflected laterally.

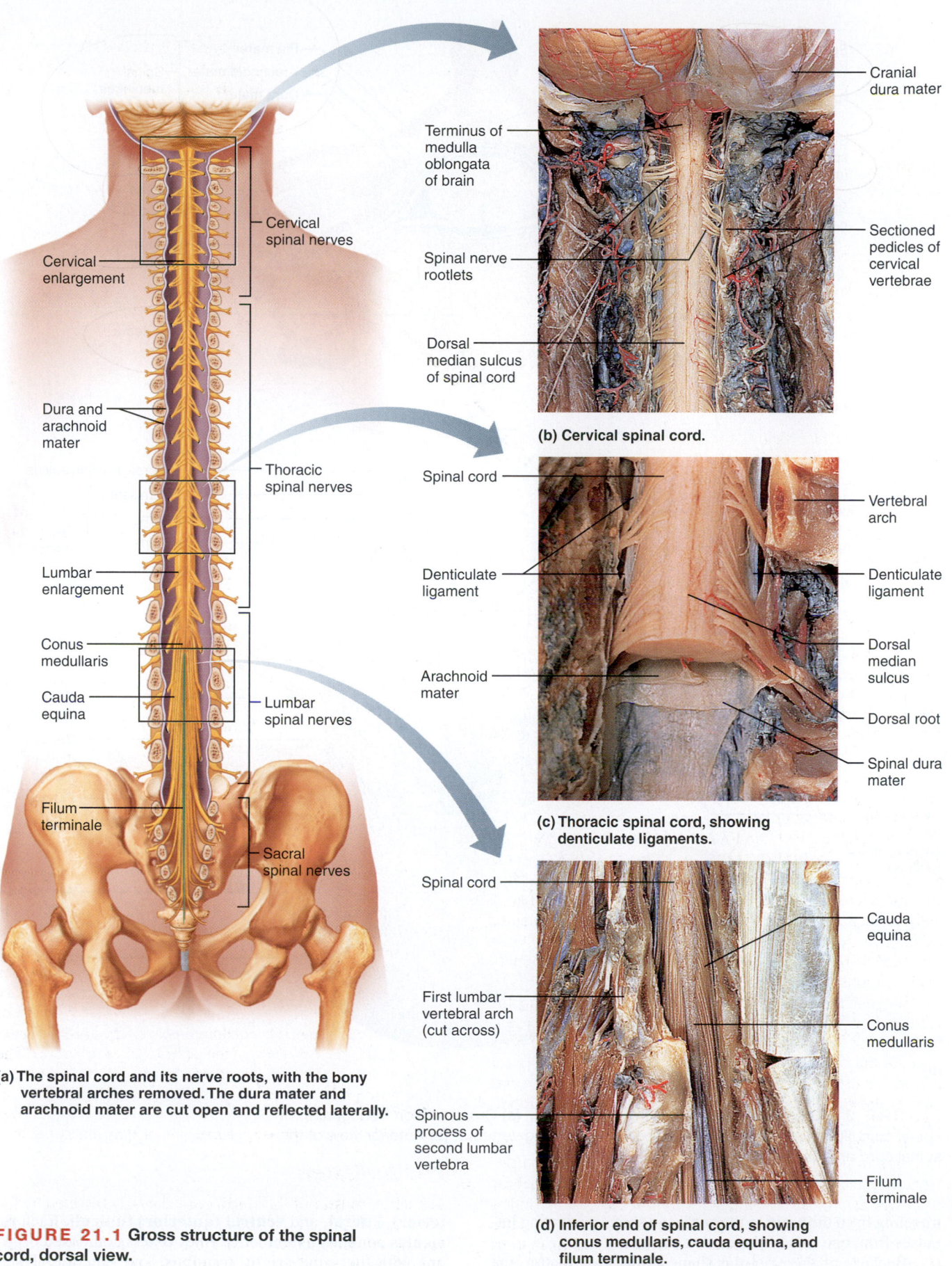

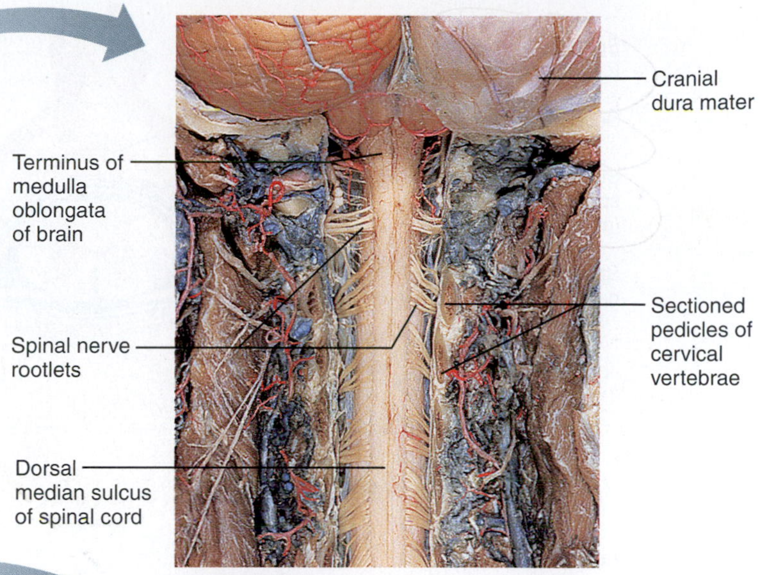
Cranial dura mater
Terminus of medulla oblongata of brain
Spinal nerve rootlets
Sectioned pedicles of cervical vertebrae
Dorsal median sulcus of spinal cord

(b) Cervical spinal cord.

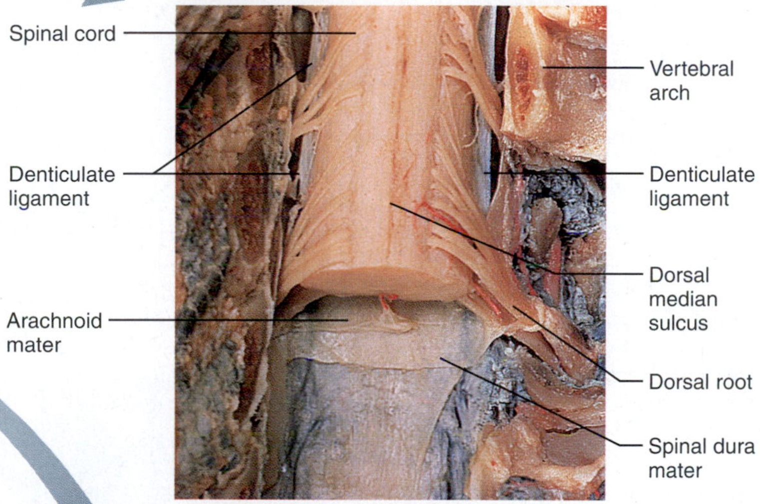
Spinal cord
Vertebral arch
Denticulate ligament
Denticulate ligament
Dorsal median sulcus
Arachnoid mater
Dorsal root
Spinal dura mater

(c) Thoracic spinal cord, showing denticulate ligaments.

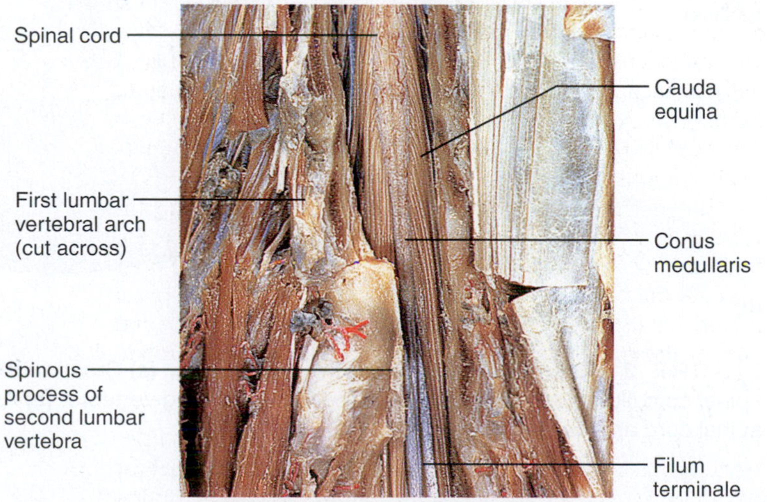
Spinal cord
Cauda equina
First lumbar vertebral arch (cut across)
Conus medullaris
Spinous process of second lumbar vertebra
Filum terminale

(d) Inferior end of spinal cord, showing conus medullaris, cauda equina, and filum terminale.

FIGURE 21.1 Gross structure of the spinal cord, dorsal view.

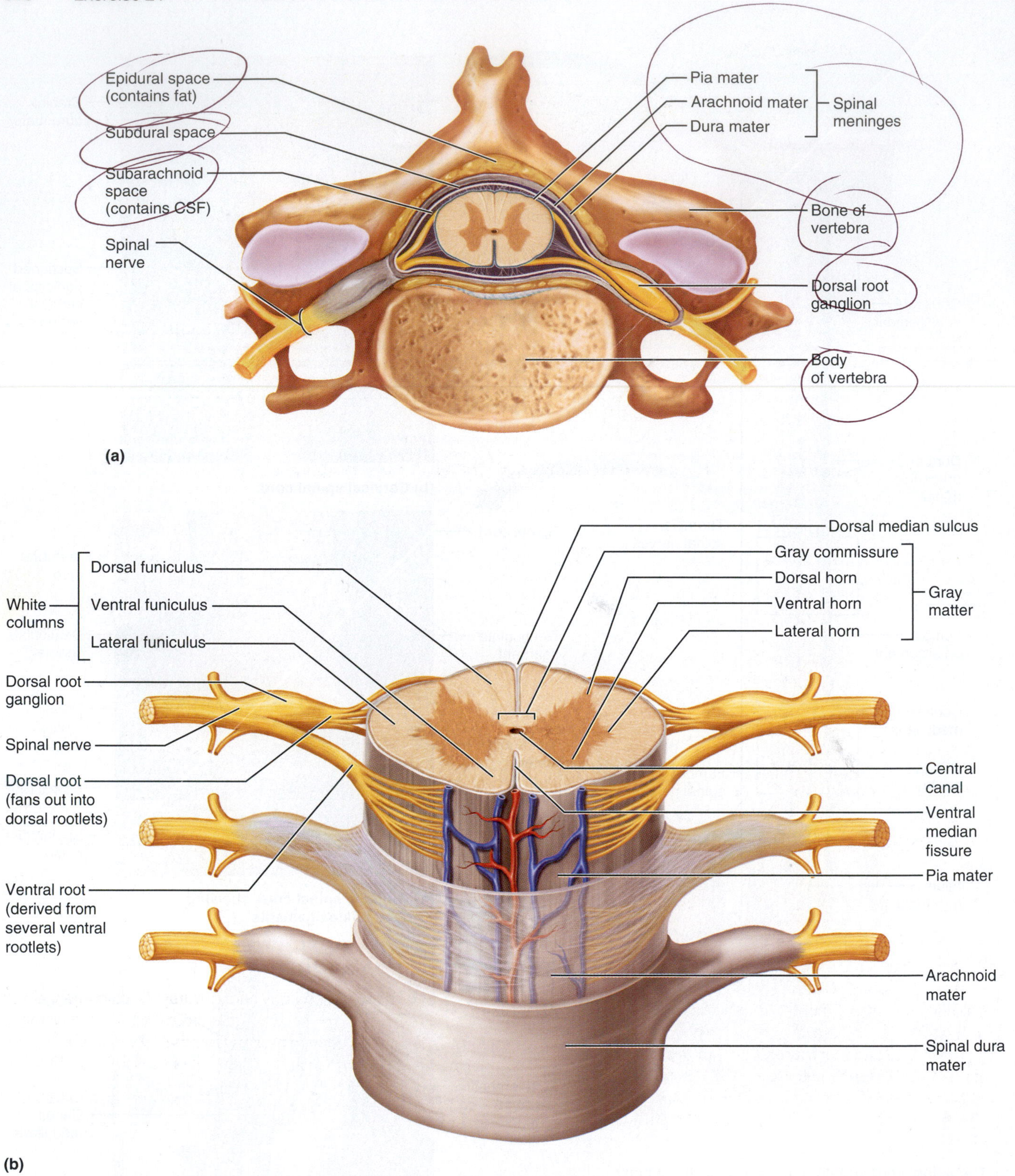
Epidural space (contains fat)
Subdural space
Subarachnoid space (contains CSF)
Spinal nerve
Pia mater
Arachnoid mater
Dura mater
Spinal meninges
Bone of vertebra
Dorsal root ganglion
Body of vertebra
(a)
Dorsal median sulcus
Gray commissure
Dorsal horn
Ventral horn
Lateral horn
Gray matter
White columns
Dorsal funiculus
Ventral funiculus
Lateral funiculus
Dorsal root ganglion
Spinal nerve
Dorsal root (fans out into dorsal rootlets)
Ventral root (derived from several ventral rootlets)
Central canal
Ventral median fissure
Pia mater
Arachnoid mater
Spinal dura mater
(b)

FIGURE 21.2 Anatomy of the human spinal cord. (a) Cross section through the spinal cord illustrating its relationship to the surrounding vertebra. **(b)** Anterior view of the spinal cord and its meningeal coverings.

traveling from the brain to the cord, and some conducting impulses from one side of the cord to the other.

Because of the irregular shape of the gray matter, the white matter on each side of the cord can be divided into three primary regions or **white columns:** the **dorsal (posterior), lateral,** and **ventral (anterior) funiculi.** Each funiculus contains a number of fiber **tracts** composed of axons with the same origin, terminus, and function. Tracts conducting sensory impulses to the brain are called *ascending,* or *sensory, tracts;* those carrying impulses

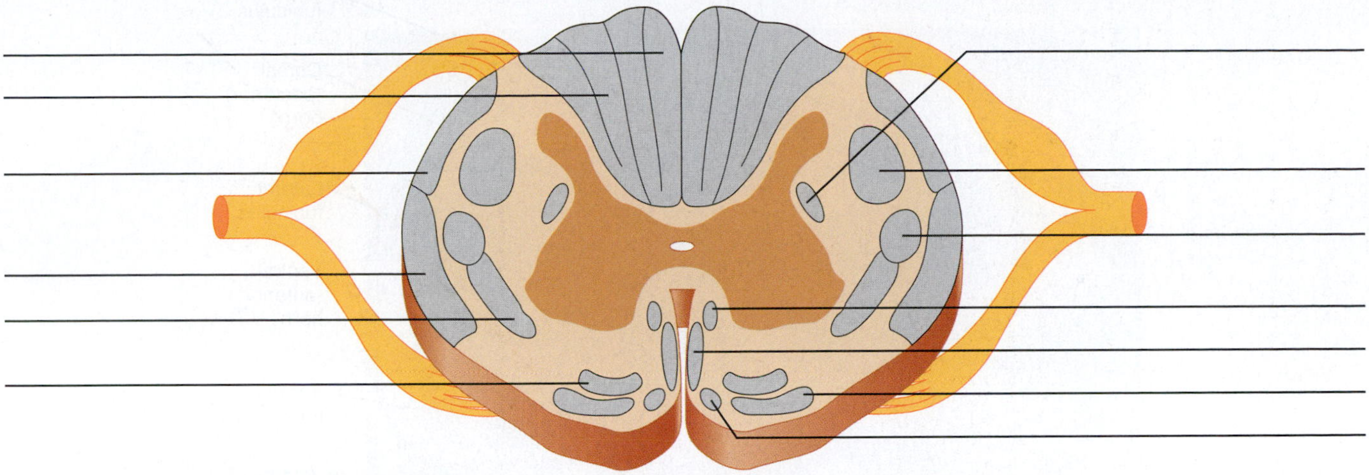
Ascending tracts
Descending tracts

FIGURE 21.3 Cross section of the spinal cord showing the relative positioning of its major tracts.

from the brain to the skeletal muscles are *descending,* or *motor, tracts.*

Because it serves as the transmission pathway between the brain and the body periphery, the spinal cord is an extremely important functional area. Even though it is protected by meninges and cerebrospinal fluid in the vertebral canal, it is highly vulnerable to traumatic injuries, such as might occur in an automobile accident.

When the cord is transected (or severely traumatized), both motor and sensory functions are lost in body areas normally served by that (and lower) regions of the spinal cord. Injury to certain spinal cord areas may even result in a permanent flaccid paralysis of both legs **(paraplegia)** or of all four limbs **(quadriplegia)**. ■

ACTIVITY 2

Identifying Spinal Cord Tracts

With the help of your textbook, label Figure 21.3 with the tract names that follow. Each tract is represented on both sides of the cord, but for clarity, label the motor tracts on the right side of the diagram and the sensory tracts on the left side of the diagram. *Color ascending tracts blue and descending tracts red.* Then fill in the functional importance of each tract beside its name below. As you work, try to be aware of how the naming of the tracts is related to their anatomical distribution.

Dorsal columns

Fasciculus gracilis ______________________

Fasciculus cuneatus ______________________

Dorsal spinocerebellar ______________________

Ventral spinocerebellar ______________________

Lateral spinothalamic ______________________

Ventral spinothalamic ______________________

Lateral corticospinal ______________________

Ventral corticospinal ______________________

Rubrospinal ______________________

Tectospinal ______________________

Vestibulospinal ______________________

Medial reticulospinal ______________________

Lateral reticulospinal ______________________ ■

DISSECTION:

Spinal Cord

1. Obtain a dissecting tray and instruments, disposable gloves, and a segment of preserved spinal cord (from a cow or saved from the brain specimen used in Exercise 19). Identify the tough outer meninx (dura mater) and the weblike arachnoid mater.

What name is given to the third meninx, and where is it found?

____

____

Peel back the dura mater and observe the fibers making up the dorsal and ventral roots. If possible, identify a dorsal root ganglion.

2. Cut a thin cross section of the cord and identify the ventral and dorsal horns of the gray matter with the naked eye or with the aid of a dissecting microscope.

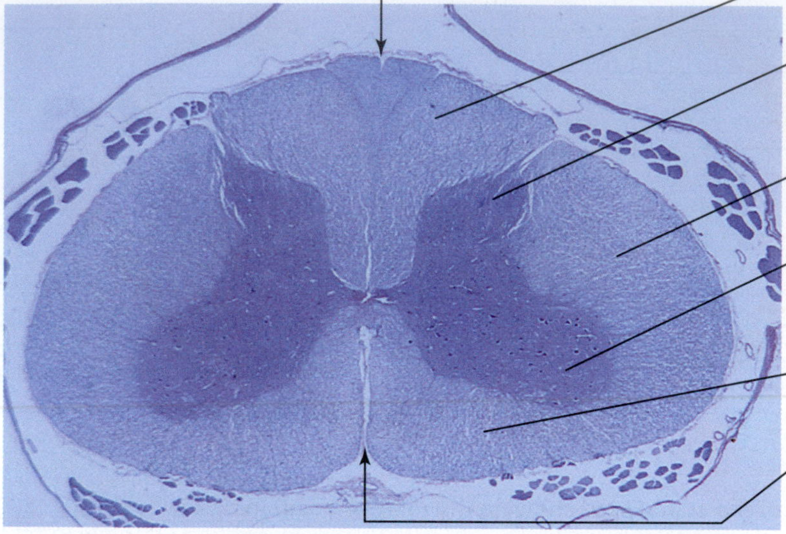
Dorsal median sulcus
Dorsal funiculus
Dorsal (posterior) horn
Lateral funiculus
Ventral (anterior) horn
Ventral funiculus
Ventral median fissure

FIGURE 21.4 Cross section of the spinal cord (10×).

How can you be certain that you are correctly identifying the ventral and dorsal horns?

___

___

Also identify the central canal, white matter, ventral median fissure, dorsal median sulcus, and dorsal, ventral, and lateral funiculi.

3. Obtain a prepared slide of the spinal cord (cross section) and a compound microscope. Refer to Figure 21.4 as you examine the slide carefully under low power. Observe the shape of the central canal.

Is it basically circular or oval? ___

Name the glial cell type that lines this canal. ___

What would you expect to find in this canal in the living animal?

___

Can any neuron cell bodies be seen? ___

Where? ___

___

What type of neurons would these most likely be—motor, sensory, or interneuron?

___

Spinal Nerves and Nerve Plexuses

The 31 pairs of human spinal nerves arise from the fusions of the ventral and dorsal roots of the spinal cord (see Figure 21.2a). Figure 21.5 shows how the nerves are named according to their point of issue. Because the ventral roots contain myelinated axons of motor neurons located in the cord and the dorsal roots carry sensory fibers entering the cord, all spinal nerves are **mixed nerves.** The first pair of spinal nerves leaves the vertebral canal between the base of the occiput and the atlas, but all the rest exit via the intervertebral foramina. The first through seventh pairs of cervical nerves emerge *above* the vertebra for which they are named; C_8 emerges between C_7 and T_1. (Notice that there are seven cervical vertebrae, but eight pairs of cervical nerves.) The remaining spinal nerve pairs emerge from the spinal cord *below* the same-numbered vertebra.

Almost immediately after emerging, each nerve divides into **dorsal** and **ventral rami.** (Thus each spinal nerve is only about 1 or 2 cm long.) The rami, like the spinal nerves, contain both motor and sensory fibers. The smaller dorsal rami serve the skin and musculature of the posterior body trunk at their approximate level of emergence. The ventral rami of spinal nerves T_2 through T_{12} pass anteriorly as the **intercostal nerves** to supply the muscles of intercostal spaces, and the skin and muscles of the anterior and lateral trunk. The ventral rami of all other spinal nerves form complex networks of nerves called **plexuses.** These plexuses serve the motor and sensory needs of the muscles and skin of the limbs. The fibers of the ventral rami unite in the plexuses (with a few rami supplying fibers to more than one plexus). From the plexuses the fibers diverge again to form peripheral nerves, each of which contains fibers from more than one spinal nerve. The four major nerve plexuses and their chief peripheral nerves are described in Tables 21.1–21.4 and

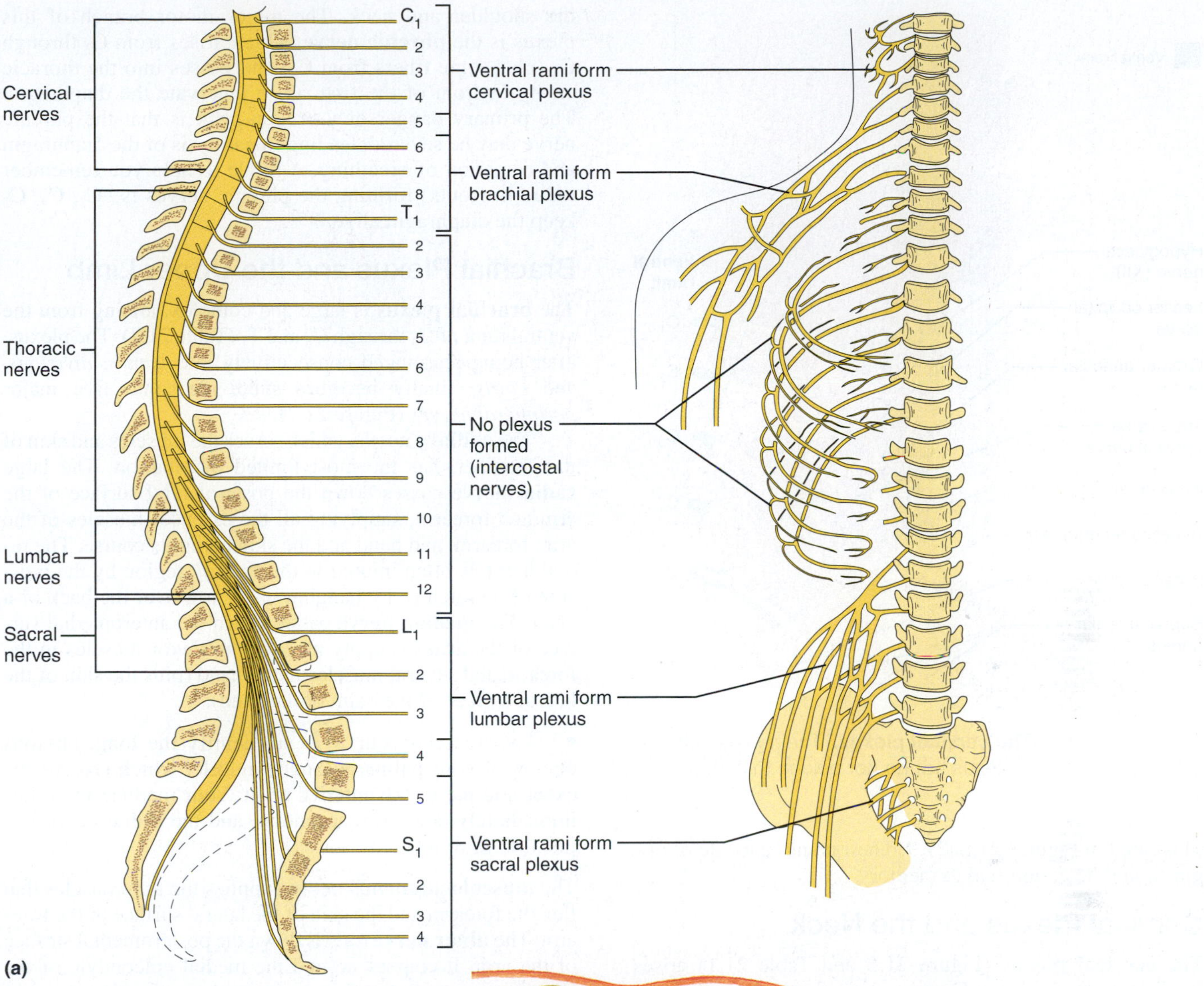
C1
2
3
4
5
6
7
8
T1
2
3
4
5
6
7
8
9
10
11
12
L1
2
3
4
5
S1
2
3
4
Cervical nerves
Thoracic nerves
Lumbar nerves
Sacral nerves
Ventral rami form cervical plexus
Ventral rami form brachial plexus
No plexus formed (intercostal nerves)
Ventral rami form lumbar plexus
Ventral rami form sacral plexus
(a)

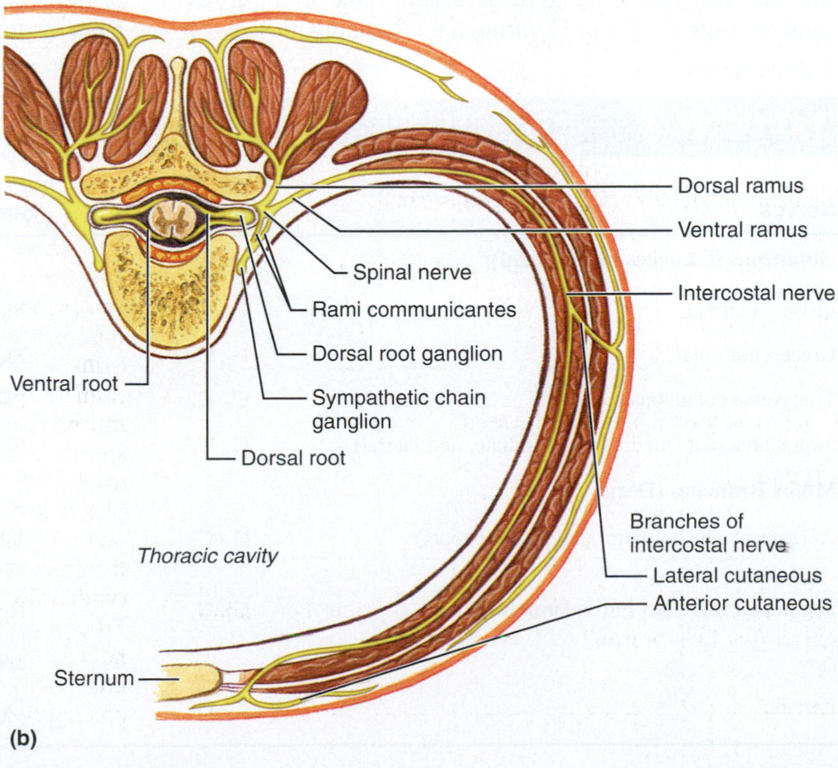
Dorsal ramus
Ventral ramus
Spinal nerve
Intercostal nerve
Rami communicantes
Dorsal root ganglion
Ventral root
Sympathetic chain ganglion
Dorsal root
Branches of intercostal nerve
Thoracic cavity
Lateral cutaneous
Anterior cutaneous
Sternum
(b)

FIGURE 21.5 Human spinal nerves. **(a)** Relationship of spinal nerves to vertebrae (areas of plexuses formed by the ventral rami are indicated). **(b)** Relative distribution of the ventral and dorsal rami of a spinal nerve (cross section of left trunk).

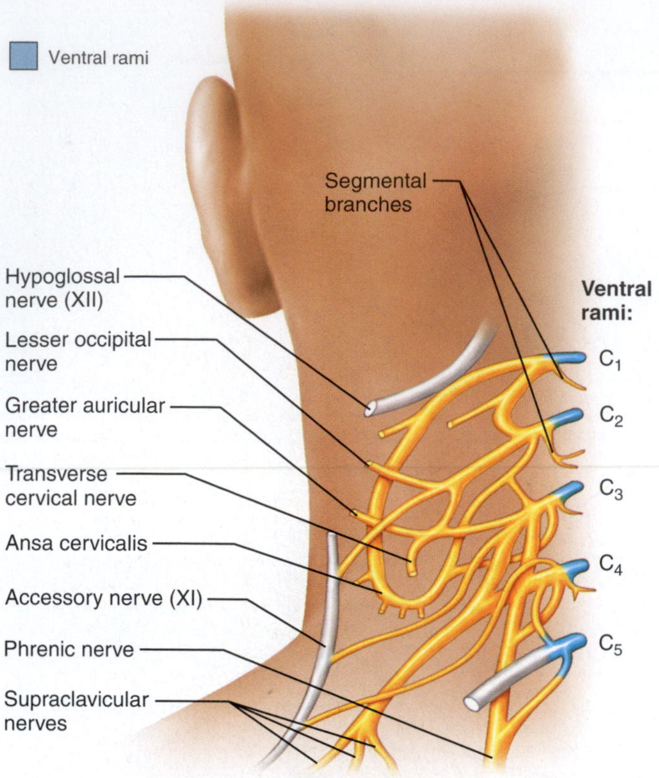
Ventral rami
Segmental branches
Hypoglossal nerve (XII)
Lesser occipital nerve
Greater auricular nerve
Transverse cervical nerve
Ansa cervicalis
Accessory nerve (XI)
Phrenic nerve
Supraclavicular nerves
Ventral rami:
C_1
C_2
C_3
C_4
C_5

FIGURE 21.6 The cervical plexus. The nerves colored gray connect to the plexus but do not belong to it. (See Table 21.1.)

illustrated in Figures 21.6–21.9. Their names and site of origin should be committed to memory.

Cervical Plexus and the Neck

The **cervical plexus** (Figure 21.6 and Table 21.1) arises from the ventral rami of C_1 through C_5 to supply muscles of the shoulder and neck. The major motor branch of this plexus is the **phrenic nerve,** which arises from C_3 through C_4 (plus some fibers from C_5) and passes into the thoracic cavity in front of the first rib to innervate the diaphragm. The primary danger of a broken neck is that the phrenic nerve may be severed, leading to paralysis of the diaphragm and cessation of breathing. A jingle to help you remember the rami (roots) forming the phrenic nerves is "C_3, C_4, C_5 keep the diaphragm alive."

Brachial Plexus and the Upper Limb

The **brachial plexus** is large and complex, arising from the ventral rami of C_5 through C_8 and T_1 (Table 21.2). The plexus, after being rearranged consecutively into *trunks, divisions,* and *cords,* finally becomes subdivided into five major *peripheral nerves* (Figure 21.7).

The **axillary nerve,** which serves the muscles and skin of the shoulder, has the most limited distribution. The large **radial nerve** passes down the posterolateral surface of the arm and forearm, supplying all the extensor muscles of the arm, forearm, and hand and the skin along its course. The radial nerve is often injured in the axillary region by the pressure of a crutch or by hanging one's arm over the back of a chair. The **median nerve** passes down the anteromedial surface of the arm to supply most of the flexor muscles in the forearm and several muscles in the hand (plus the skin of the lateral surface of the palm of the hand).

- Hyperextend your wrist to identify the long, obvious tendon of your palmaris longus muscle, which crosses the exact midline of the anterior wrist. Your median nerve lies immediately deep to that tendon, and the radial nerve lies just *lateral* to it.

The **musculocutaneous nerve** supplies the arm muscles that flex the forearm and the skin of the lateral surface of the forearm. The **ulnar nerve** travels down the posteromedial surface of the arm. It courses around the medial epicondyle of the humerus to supply the flexor carpi ulnaris, the ulnar head of

TABLE 21.1 Branches of the Cervical Plexus (See Figure 21.6)

Nerves	Spinal roots (ventral rami)	Structures served
Cutaneous Branches (Superficial)		
Lesser occipital	C_2 (C_3)	Skin on posterolateral aspect of neck
Greater auricular	C_2, C_3	Skin of ear, skin over parotid gland
Transverse cutaneous (cervical)	C_2, C_3	Skin on anterior and lateral aspect of neck
Supraclavicular (medial, intermediate, and lateral)	C_3, C_4	Skin of shoulder and anterior aspect of chest
Motor Branches (Deep)		
Ansa cervicalis (superior and inferior roots)	C_1–C_3	Infrahyoid muscles of neck (omohyoid, sternohyoid, and sternothyroid)
Segmental and other muscular branches	C_1–C_5	Deep muscles of neck (geniohyoid and thyrohyoid) and portions of scalenes, levator scapulae, trapezius, and sternocleidomastoid muscles
Phrenic	C_3–C_5	Diaphragm (sole motor nerve supply)

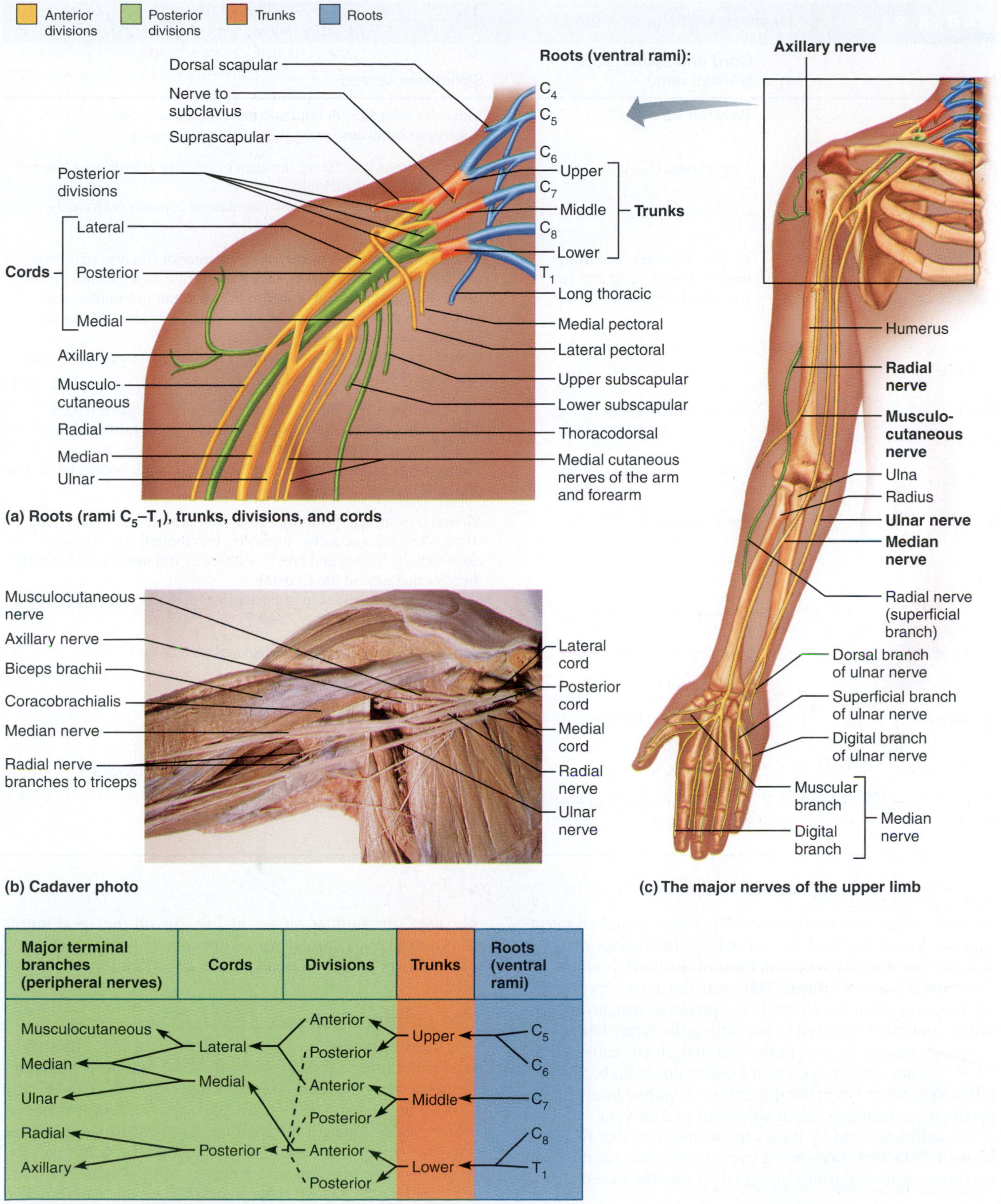
Anterior divisions
Posterior divisions
Trunks
Roots
Dorsal scapular
Nerve to subclavius
Suprascapular
Posterior divisions
Lateral
Cords
Posterior
Medial
Axillary
Musculo-cutaneous
Radial
Median
Ulnar
Roots (ventral rami):
C_4
C_5
C_6
Upper
C_7
Middle
Trunks
C_8
Lower
T_1
Long thoracic
Medial pectoral
Lateral pectoral
Upper subscapular
Lower subscapular
Thoracodorsal
Medial cutaneous nerves of the arm and forearm
Axillary nerve
Humerus
Radial nerve
Musculo-cutaneous nerve
Ulna
Radius
Ulnar nerve
Median nerve
Radial nerve (superficial branch)
Dorsal branch of ulnar nerve
Superficial branch of ulnar nerve
Digital branch of ulnar nerve
Muscular branch
Digital branch
Median nerve
Musculocutaneous nerve
Axillary nerve
Biceps brachii
Coracobrachialis
Median nerve
Radial nerve branches to triceps
Lateral cord
Posterior cord
Medial cord
Radial nerve
Ulnar nerve
Major terminal branches (peripheral nerves)
Cords
Divisions
Trunks
Roots (ventral rami)
Musculocutaneous
Median
Ulnar
Radial
Axillary
Lateral
Medial
Posterior
Anterior
Posterior
Anterior
Posterior
Anterior
Posterior
Upper
Middle
Lower
C_5
C_6
C_7
C_8
T_1

(a) Roots (rami C_5–T_1), trunks, divisions, and cords

(b) Cadaver photo

(c) The major nerves of the upper limb

(d) Flowchart summarizing relationships within the brachial plexus

FIGURE 21.7 The brachial plexus. (See Table 21.2.)

TABLE 21.2 Branches of the Brachial Plexus (See Figure 21.7)

Nerves	Cord and spinal roots (ventral rami)	Structures served
Axillary	Posterior cord (C_5, C_6)	Muscular branches: deltoid and teres minor muscles Cutaneous branches: some skin of shoulder region
Musculocutaneous	Lateral cord (C_5–C_7)	Muscular branches: flexor muscles in anterior arm (biceps brachii, brachialis, coracobrachialis) Cutaneous branches: skin on anterolateral forearm (extremely variable)
Median	By two branches, one from medial cord (C_8, T_1) and one from the lateral cord (C_5–C_7)	Muscular branches to flexor group of anterior forearm (palmaris longus, flexor carpi radialis, flexor digitorum superficialis, flexor pollicis longus, lateral half of flexor digitorum profundus, and pronator muscles); intrinsic muscles of lateral palm and digital branches to the fingers Cutaneous branches: skin of lateral two-thirds of hand, palm side and dorsum of fingers 2 and 3
Ulnar	Medial cord (C_8, T_1)	Muscular branches: flexor muscles in anterior forearm (flexor carpi ulnaris and medial half of flexor digitorum profundus); most intrinsic muscles of hand Cutaneous branches: skin of medial third of hand, both anterior and posterior aspects
Radial	Posterior cord (C_5–C_8, T_1)	Muscular branches: posterior muscles of arm, forearm, and hand (triceps brachii, anconeus, supinator, brachioradialis, extensors carpi radialis longus and brevis, extensor carpi ulnaris, and several muscles that extend the fingers) Cutaneous branches: skin of posterolateral surface of entire limb (except dorsum of fingers 2 and 3)
Dorsal scapular	Branches of C_5 rami	Rhomboid muscles and levator scapulae
Long thoracic	Branches of C_5–C_7 rami	Serratus anterior muscle
Subscapular	Posterior cord; branches of C_5 and C_6 rami	Teres major and subscapular muscles
Suprascapular	Upper trunk (C_5, C_6)	Shoulder joint; supraspinatus and infraspinatus muscles
Pectoral (lateral and medial)	Branches of lateral and medial cords (C_5–T_1)	Pectoralis major and minor muscles

the flexor digitorum profundus of the forearm, and all intrinsic muscles of the hand not served by the median nerve. It supplies the skin of the medial third of the hand, both the anterior and posterior surfaces. Trauma to the ulnar nerve, which often occurs when the elbow is hit, produces a smarting sensation commonly referred to as "hitting the funny bone."

Severe injuries to the brachial plexus cause weakness or paralysis of the entire upper limb. Such injuries may occur when the upper limb is pulled hard and the plexus is stretched (as when a football tackler yanks the arm of the halfback), and by blows to the shoulder that force the humerus inferiorly (as when a cyclist is pitched headfirst off his motorcycle and grinds his shoulder into the pavement). ■

Lumbosacral Plexus and the Lower Limb

The **lumbosacral plexus,** which serves the pelvic region of the trunk and the lower limbs, is actually a complex of two plexuses, the lumbar plexus and the sacral plexus (Figures 21.8 and 21.9). These plexuses interweave considerably and many fibers of the lumbar plexus contribute to the sacral plexus.

The Lumbar Plexus

The **lumbar plexus** arises from ventral rami of L_1 through L_4 (and sometimes T_{12}). Its nerves serve the lower abdominopelvic region and the anterior thigh (Table 21.3 and Figure 21.8). The largest nerve of this plexus is the **femoral nerve,** which passes beneath the inguinal ligament to innervate the anterior thigh muscles. The cutaneous branches of the femoral nerve (median and anterior femoral cutaneous and the saphenous nerves) supply the skin of the anteromedial surface of the entire lower limb.

TABLE 21.3 Branches of the Lumbar Plexus (See Figure 21.8)

Nerves	Spinal roots (ventral rami)	Structures served
Femoral	L_2–L_4	Skin of anterior and medial thigh via *anterior femoral cutaneous* branch; skin of medial leg and foot, hip and knee joints via *saphenous* branch; motor to anterior muscles (quadriceps and sartorius) of thigh; pectineus, iliacus
Obturator	L_2–L_4	Motor to adductor magnus (part), longus, and brevis muscles, gracilis muscle of medial thigh, obturator externus; sensory for skin of medial thigh and for hip and knee joints
Lateral femoral cutaneous	L_2, L_3	Skin of lateral thigh; some sensory branches to peritoneum
Iliohypogastric	L_1	Skin of lower abdomen, lower back, and hip; muscles of anterolateral abdominal wall (obliques and transversus) and pubic region
Ilioinguinal	L_1	Skin of external genitalia and proximal medial aspect of the thigh; inferior abdominal muscles
Genitofemoral	L_1, L_2	Skin of scrotum in males, of labia majora in females, and of anterior thigh inferior to middle portion of inguinal region; cremaster muscle in males

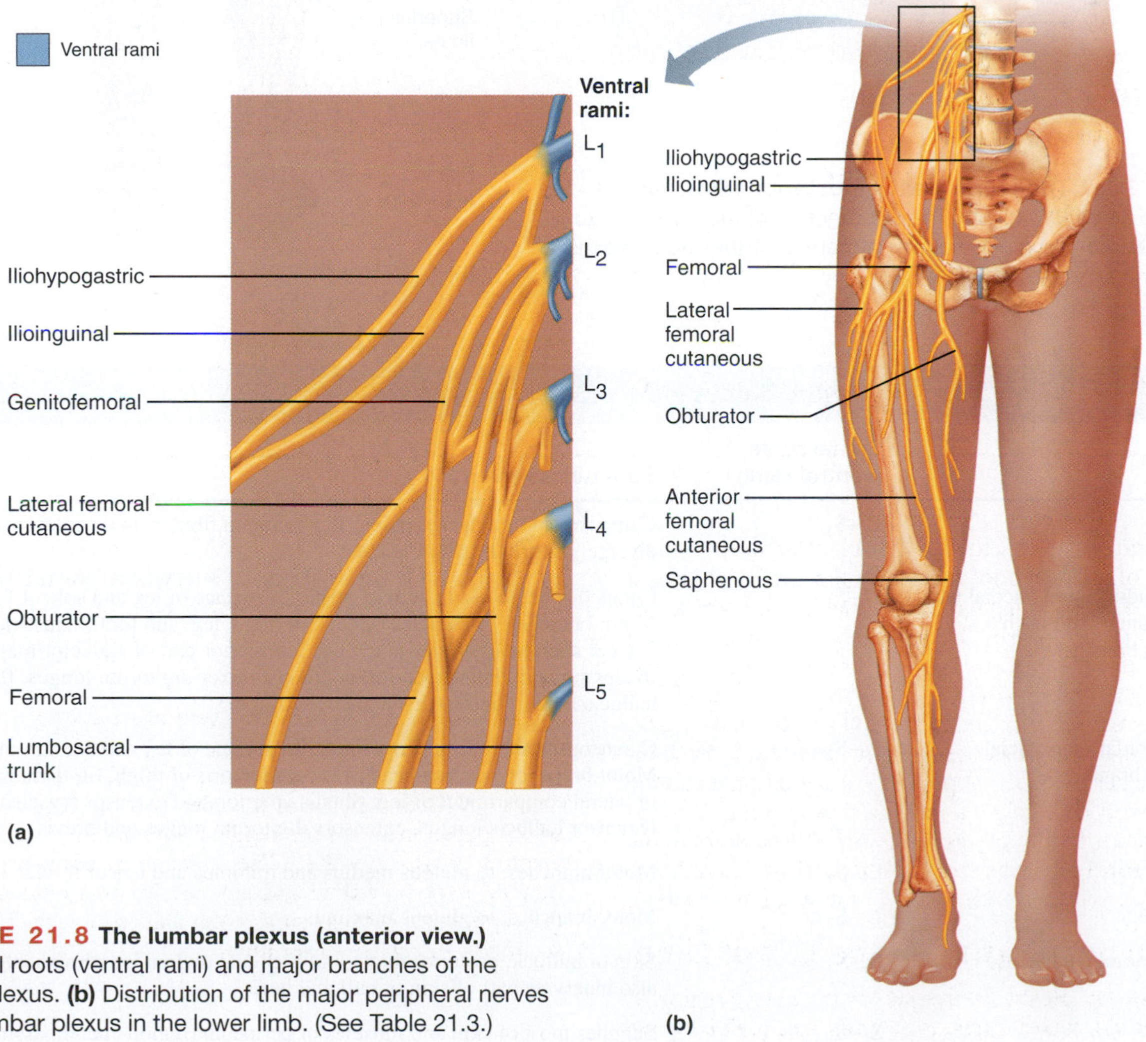
Ventral rami
Ventral rami:
L1
L2
L3
L4
L5
Iliohypogastric
Ilioinguinal
Genitofemoral
Lateral femoral cutaneous
Obturator
Femoral
Lumbosacral trunk
(a)
Iliohypogastric
Ilioinguinal
Femoral
Lateral femoral cutaneous
Obturator
Anterior femoral cutaneous
Saphenous
(b)

FIGURE 21.8 The lumbar plexus (anterior view.) **(a)** Spinal roots (ventral rami) and major branches of the lumbar plexus. **(b)** Distribution of the major peripheral nerves of the lumbar plexus in the lower limb. (See Table 21.3.)

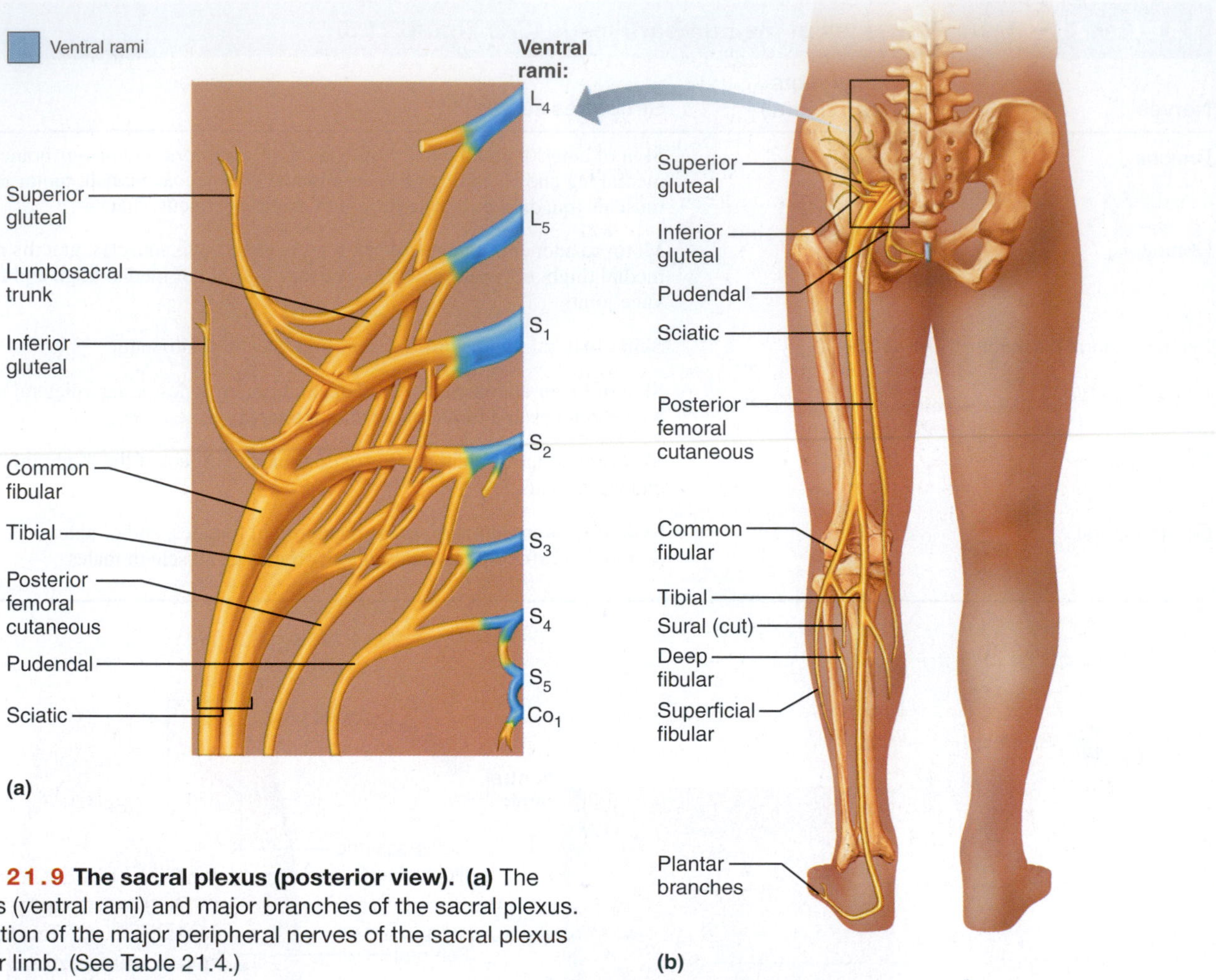
Ventral rami
Ventral rami:
L4
L5
S1
S2
S3
S4
S5
Co1
Superior gluteal
Lumbosacral trunk
Inferior gluteal
Common fibular
Tibial
Posterior femoral cutaneous
Pudendal
Sciatic
(a)
Superior gluteal
Inferior gluteal
Pudendal
Sciatic
Posterior femoral cutaneous
Common fibular
Tibial
Sural (cut)
Deep fibular
Superficial fibular
Plantar branches
(b)

FIGURE 21.9 The sacral plexus (posterior view). **(a)** The spinal roots (ventral rami) and major branches of the sacral plexus. **(b)** Distribution of the major peripheral nerves of the sacral plexus in the lower limb. (See Table 21.4.)

TABLE 21.4 Branches of the Sacral Plexus (See Figure 21.9)

Nerves	Spinal roots (ventral rami)	Structures served
Sciatic nerve	L_4–S_3	Composed of two nerves (tibial and common fibular) in a common sheath that diverge just proximal to the knee
Tibial (including sural, medial and lateral plantar, and medial calcaneal branches)	L_4–S_3	Cutaneous branches: to skin of posterior surface of leg and sole of foot Motor branches: to muscles of back of thigh, leg, and foot (hamstrings [except short head of biceps femoris], posterior part of adductor magnus, triceps surae, tibialis posterior, popliteus, flexor digitorum longus, flexor hallucis longus, and intrinsic muscles of foot)
Common fibular (superficial and deep branches)	L_4–S_2	Cutaneous branches: to skin of anterior surface of leg and dorsum of foot Motor branches: to short head of biceps femoris of thigh, fibularis muscles of lateral compartment of leg, tibialis anterior, and extensor muscles of toes (extensor hallucis longus, extensors digitorum longus and brevis)
Superior gluteal	L_4–S_1	Motor branches: to gluteus medius and minimus and tensor fasciae latae
Inferior gluteal	L_5–S_2	Motor branches: to gluteus maximus
Posterior femoral cutaneous	S_1–S_3	Skin of buttock, posterior thigh, and popliteal region; length variable; may also innervate part of skin of calf and heel
Pudendal	S_2–S_4	Supplies most of skin and muscles of perineum (region encompassing external genitalia and anus and including clitoris, labia, and vaginal mucosa in females, and scrotum and penis in males); external anal sphincter

The Sacral Plexus

Arising from L_4 through S_4, the nerves of the **sacral plexus** supply the buttock, the posterior surface of the thigh, and virtually all sensory and motor fibers of the leg and foot (Table 21.4 and Figure 21.9). The major peripheral nerve of this plexus is the **sciatic nerve,** the largest nerve in the body. The sciatic nerve leaves the pelvis through the greater sciatic notch and travels down the posterior thigh, serving its flexor muscles and skin. In the popliteal region, the sciatic nerve divides into the **common fibular nerve** and the **tibial nerve,** which together supply the balance of the leg muscles and skin, both directly and via several branches.

Injury to the proximal part of the sciatic nerve, as might follow a fall or disc herniation, results in a number of lower limb impairments. **Sciatica** (si-at'ĭ-kah), characterized by stabbing pain radiating over the course of the sciatic nerve, is common. When the sciatic nerve is completely severed, the leg is nearly useless. The leg cannot be flexed and the foot drops into plantar flexion (dangles), a condition called **footdrop.** ■

ACTIVITY 3

Identifying the Major Nerve Plexuses and Peripheral Nerves

Identify each of the four major nerve plexuses (and its major nerves) shown in Figures 21.6 to 21.9 on a large laboratory chart or model. Trace the courses of the nerves and relate those observations to the information provided in Tables 21.1 to 21.4. ■

The Autonomic Nervous System

The **autonomic nervous system (ANS)** is the subdivision of the peripheral nervous system (PNS) that regulates body activities that are generally not under conscious control. It is composed of a special group of motor neurons serving cardiac muscle (the heart), smooth muscle (found in the walls of the visceral organs and blood vessels), and internal glands. Because these structures typically function without conscious control, this system is often referred to as the *involuntary nervous system.*

There is a basic anatomical difference between the motor pathways of the **somatic** (voluntary) **nervous system,** which innervates the skeletal muscles, and those of the autonomic nervous system. In the somatic division, the cell bodies of the motor neurons reside in the CNS (brain stem or ventral horns of the spinal cord), and their axons, sheathed in cranial or spinal nerves, extend directly to the skeletal muscles they serve. However, the autonomic nervous system consists of chains of two motor neurons. The first motor neuron of each pair, called the *preganglionic neuron,* resides in the brain stem or lateral horn of the spinal cord. Its axon leaves the CNS to synapse with the second motor neuron *(ganglionic neuron),* whose cell body is located in a ganglion outside the CNS. The axon of the ganglionic neuron then extends to the organ it serves.

The autonomic nervous system has two major functional subdivisions (Figure 21.10). These, the sympathetic and parasympathetic divisions, serve most of the same organs but generally cause opposing or antagonistic effects.

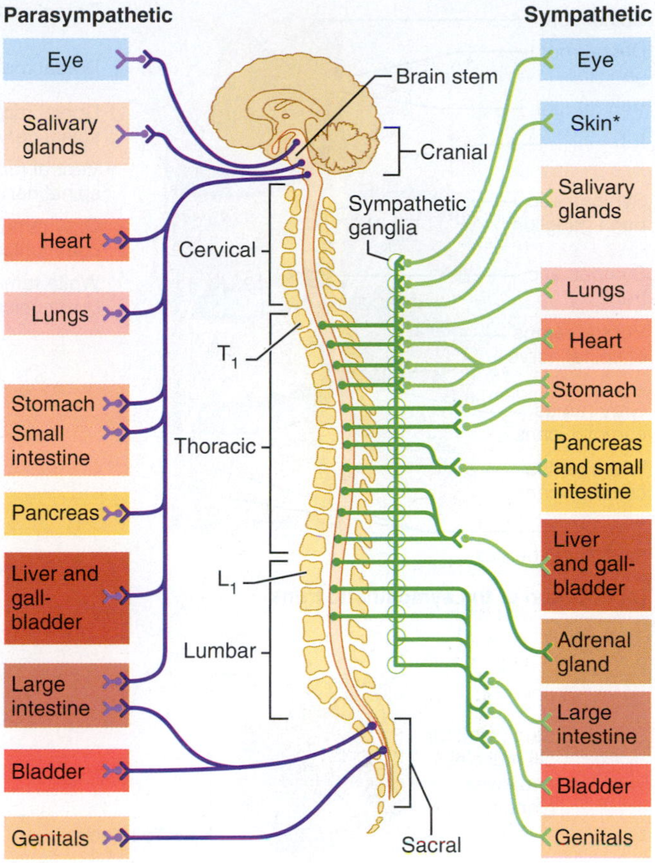
Parasympathetic
Sympathetic
Eye
Salivary glands
Heart
Lungs
Stomach
Small intestine
Pancreas
Liver and gall-bladder
Large intestine
Bladder
Genitals
Brain stem
Cranial
Sympathetic ganglia
Cervical
T_1
Thoracic
L_1
Lumbar
Sacral
Eye
Skin*
Salivary glands
Lungs
Heart
Stomach
Pancreas and small intestine
Liver and gall-bladder
Adrenal gland
Large intestine
Bladder
Genitals

FIGURE 21.10 Overview of the subdivisions of the autonomic nervous system. The parasympathetic and sympathetic divisions differ anatomically in (1) the sites of origin of their nerves, (2) the relative lengths of preganglionic and postganglionic fibers, and (3) the locations of their ganglia (synapse sites). Although sympathetic innervation to the skin(*) is shown only for the cervical area, all nerves to the periphery carry postganglionic sympathetic fibers.

Parasympathetic Division

The preganglionic neurons of the **parasympathetic,** or **craniosacral,** division are located in brain stem nuclei of cranial nerves III, VII, IX, X and in the S_2 through S_4 level of the spinal cord. The axons of the preganglionic neurons of the cranial region travel in their respective cranial nerves to the *immediate area* of the head and neck organs to be stimulated. There they synapse with the ganglionic neuron in a **terminal,** or **intramural** (literally, "within the walls"), **ganglion.** The ganglionic neuron then sends out a very short axon (postganglionic axon) to the organ it serves. In the sacral region, the preganglionic axons leave the ventral roots of the spinal cord and collectively form the **pelvic splanchnic nerves,** which travel to the pelvic cavity. In the pelvic cavity, the preganglionic axons synapse with the ganglionic neurons in ganglia located on or close to the organs served.

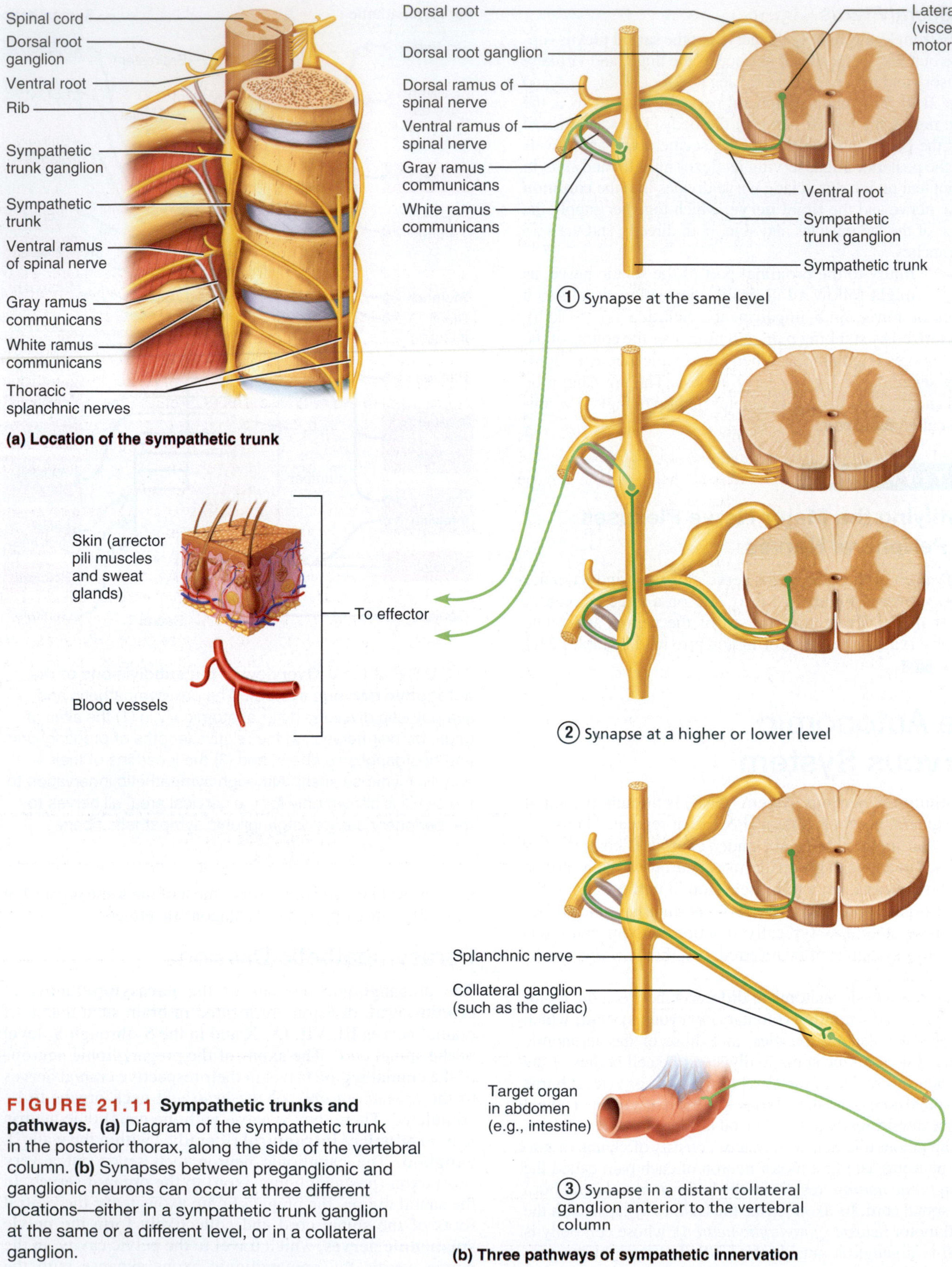
Spinal cord
Dorsal root ganglion
Ventral root
Rib
Sympathetic trunk ganglion
Sympathetic trunk
Ventral ramus of spinal nerve
Gray ramus communicans
White ramus communicans
Thoracic splanchnic nerves
(a) Location of the sympathetic trunk
Skin (arrector pili muscles and sweat glands)
Blood vessels
To effector
Dorsal root
Dorsal root ganglion
Dorsal ramus of spinal nerve
Ventral ramus of spinal nerve
Gray ramus communicans
White ramus communicans
Lateral horn (visceral motor zone)
Ventral root
Sympathetic trunk ganglion
Sympathetic trunk
1 Synapse at the same level
2 Synapse at a higher or lower level
Splanchnic nerve
Collateral ganglion (such as the celiac)
Target organ in abdomen (e.g., intestine)
3 Synapse in a distant collateral ganglion anterior to the vertebral column
(b) Three pathways of sympathetic innervation

FIGURE 21.11 Sympathetic trunks and pathways. (a) Diagram of the sympathetic trunk in the posterior thorax, along the side of the vertebral column. **(b)** Synapses between preganglionic and ganglionic neurons can occur at three different locations—either in a sympathetic trunk ganglion at the same or a different level, or in a collateral ganglion.

Sympathetic Division

The preganglionic neurons of the **sympathetic,** or **thoracolumbar,** division are in the lateral horns of the gray matter of the spinal cord from T_1 through L_2. The preganglionic axons leave the cord via the ventral root (in conjunction with the axons of the somatic motor neurons), enter the spinal nerve, and then travel briefly in the ventral ramus (Figure 21.11). From the ventral ramus, they pass through a small branch called the **white ramus communicans** to enter a **sympathetic trunk ganglion.** These two trunks **(chains)** lie alongside the vertebral column and are also called *paravertebral ganglia.*

Having reached the ganglion, a preganglionic axon may take one of three main courses (see Figure 21.11b). First, it may synapse with a ganglionic neuron in the sympathetic chain at that level. Second, the axon may travel upward or downward through the sympathetic chain to synapse with a ganglionic neuron at another level. In either of these two instances, the postganglionic axons then reenter the spinal nerve via a **gray ramus communicans** and travel in branches of a dorsal or ventral ramus to innervate skin structures (sweat glands, arrector pili muscles attached to hair follicles, and the smooth muscles of blood vessel walls) and thoracic organs. Third, the axon may pass through the ganglion without synapsing and form part of the **splanchnic nerves,** which travel to the viscera to synapse with a ganglionic neuron in a **collateral,** or **prevertebral, ganglion.** The major collateral ganglia—the *celiac, superior mesenteric, inferior mesenteric,* and *inferior hypogastric ganglia*—supply the abdominal and pelvic visceral organs. The postganglionic axon then leaves the ganglion and travels to a nearby visceral organ which it innervates.

ACTIVITY 4

Locating the Sympathetic Chain

Locate the sympathetic chain on the spinal nerve chart. ■

Autonomic Functioning

As noted earlier, most body organs served by the autonomic nervous system receive fibers from both the sympathetic and parasympathetic divisions. The only exceptions are the structures of the skin (sweat glands and arrector pili muscles attached to the hair follicles), the adrenal medulla, and essentially all blood vessels except those of the external genitalia, all of which receive sympathetic innervation only. When both divisions serve an organ, they have antagonistic effects. This is because their postganglionic axons release different neurotransmitters. The parasympathetic fibers, called **cholinergic fibers,** release acetylcholine; the sympathetic postganglionic fibers, called **adrenergic fibers,** release norepinephrine. (However, there are isolated examples of postganglionic sympathetic fibers, such as those serving sweat glands and some blood vessels that release acetylcholine.) The preganglionic fibers of both divisions release acetylcholine.

The parasympathetic division is often referred to as the housekeeping, or "resting and digesting," system because it maintains the visceral organs in a state most suitable for normal functions and internal homeostasis; that is, it promotes normal digestion and elimination. In contrast, activation of the sympathetic division is referred to as the "fight-or-flight" response because it readies the body to cope with situations that threaten homeostasis. Under such emergency conditions, the sympathetic nervous system induces an increase in heart rate and blood pressure, dilates the bronchioles of the lungs, increases blood sugar levels, and promotes many other effects that help the individual cope with a stressor.

Organ or function	Parasympathetic effect	Sympathetic effect
Heart		
Bronchioles of lungs		
Digestive tract activity		
Urinary bladder		
Iris of the eye		
Blood vessels (most)		
Penis/clitoris		
Sweat glands		
Adrenal medulla		
Pancreas		

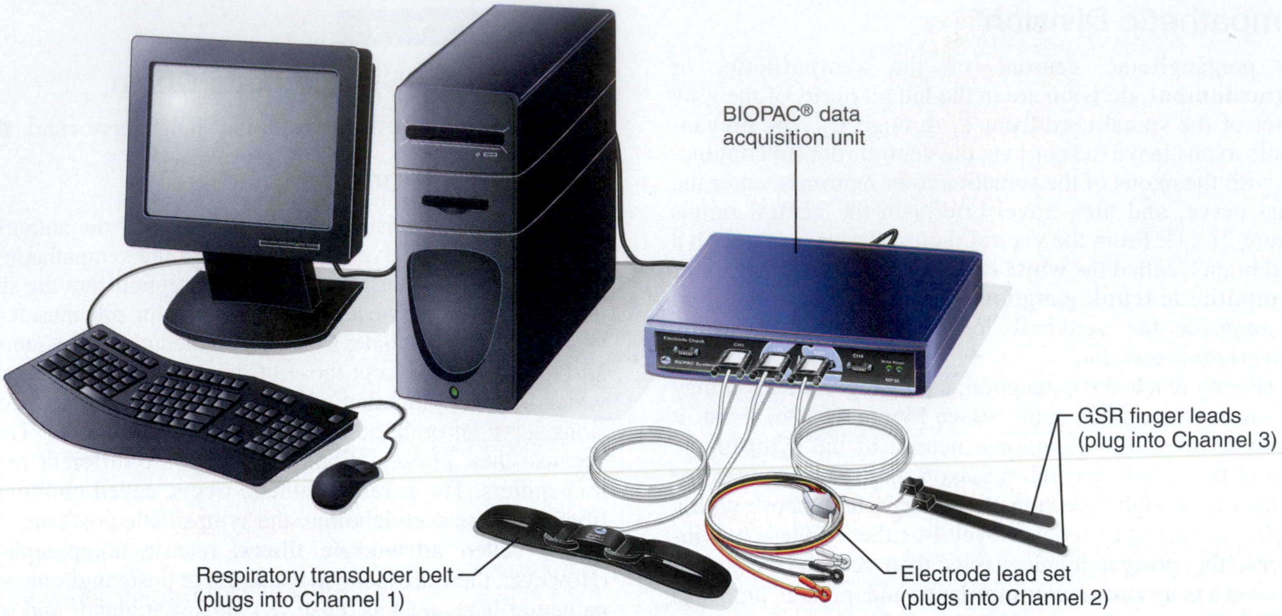
BIOPAC® data acquisition unit
GSR finger leads (plug into Channel 3)
Respiratory transducer belt (plugs into Channel 1)
Electrode lead set (plugs into Channel 2)

FIGURE 21.12 Setting up the BIOPAC® equipment. Plug the respiratory transducer belt into Channel 1, the electrode lead set into Channel 2, and the GSR finger leads into Channel 3.

As we grow older, our sympathetic nervous system gradually becomes less and less efficient, particularly in causing vasoconstriction of blood vessels. When elderly people stand up quickly after sitting or lying down, they often become light-headed or faint. This is because the sympathetic nervous system is not able to react quickly enough to counteract the pull of gravity by activating the vasoconstrictor fibers. So, blood pools in the feet. This condition, **orthostatic hypotension,** is a type of low blood pressure resulting from changes in body position as described. Orthostatic hypotension can be prevented to some degree if *slow* changes in position are made. This gives the sympathetic nervous system a little more time to react and adjust. ■

ACTIVITY 5

Comparing Sympathetic and Parasympathetic Effects

Several body organs are listed in the chart on the previous page. Using your textbook as a reference, list the effect of the sympathetic and parasympathetic divisions on each. ■

ACTIVITY 6

Exploring the Galvanic Skin Response within a Polygraph Using BIOPAC®

The autonomic nervous system is intimately integrated with the emotions, or affect, of an individual. A sad event, sharp pain, or simple stress can elicit measurable changes in autonomic regulation of heart rate, respiration, and blood pressure. In addition to these obvious physiological phenomena, more subtle autonomic changes can occur in the skin. Specifically, changes in autonomic tone in response to external circumstances can influence the rate of sweat gland secretion and blood flow to the skin that may not be overtly observable but nonetheless can be measured. The **galvanic skin response** is an electrophysiological measurement of changes that occur in the skin due to changes in autonomic stimulation.

The galvanic skin response is measured by recording the changes in **galvanic skin resistance (GSR)** and **galvanic skin potential (GSP).** Resistance, recorded in *ohms* (Ω), is a measure of the opposition of the flow of current from one electrode to another. Increasing resistance results in decreased current. Potential, measured in *volts* (V), is a measure of the amount of charge separation between two points. Increased sympathetic stimulation of sweat glands, in response to change in affect, decreases resistance on the skin because of increased water and electrolytes on the skin surface.

In this experiment you will record heart rate, respiration, and GSR while the subject is exposed to various conditions. Because "many" variables will be "recorded," this process is often referred to as a **polygraph.** The goal of this exercise is to record and analyze data to observe how this process works. This is not a "lie detector test," as its failure rate is far too high to provide true scientific or legal certainty. However, the polygraph can be used as an investigative tool.

Setting Up the Equipment

1. Connect the BIOPAC® unit to the computer and turn the computer **ON.**
2. Make sure the BIOPAC® unit is **OFF.**
3. Plug in the equipment as shown in Figure 21.12.

- Respiratory transducer belt—CH 1
- Electrode lead set—CH 2
- GSR finger leads—CH 3

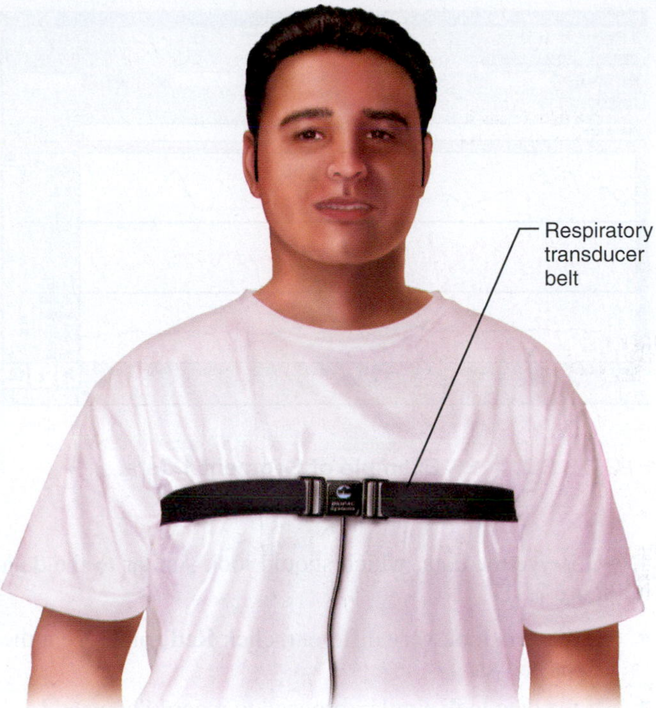
Respiratory
transducer
belt

FIGURE 21.13 Proper placement of the respiratory transducer belt around the subject's thorax.

4. Turn the BIOPAC unit **ON.**

5. Attach the respiratory transducer belt to the subject as shown in Figure 21.13. It should be fastened so that it is slightly tight even at the point of maximal expiration.

6. Fill both cavities of the GSR transducer with conduction gel, and attach the sensors to the subject's fingers as shown in Figure 21.14. The electrodes should be placed on

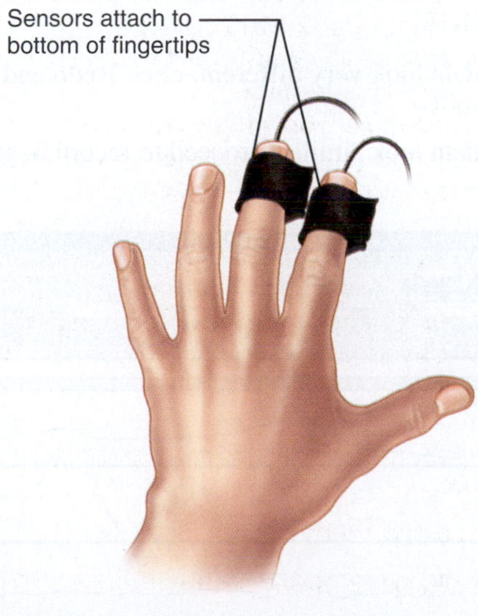
Sensors attach to
bottom of fingertips

FIGURE 21.14 Placement of the GSR finger leads to the fingers.

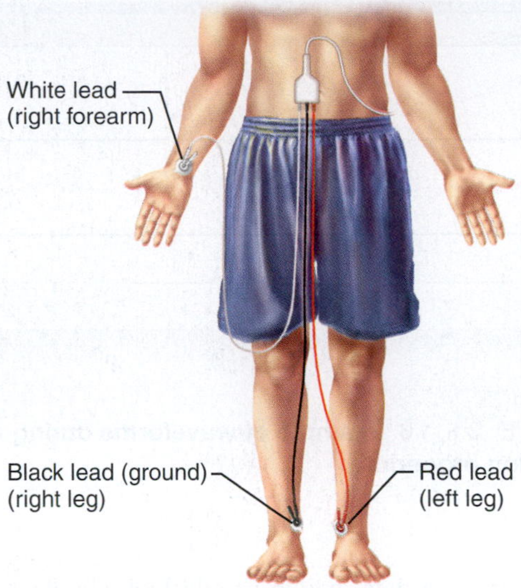
White lead
(right forearm)
Black lead (ground)
(right leg)
Red lead
(left leg)

FIGURE 21.15 Placement of electrodes and the appropriate attachment of electrode leads by color.

the middle and index fingers with the sensors on the skin, not the fingernail. They should fit snugly but not be so tight as to cut off circulation. To pick up a good GSR signal, it is important that the subject's hand have enough sweat (as it normally would). *The subject should not have freshly washed or cold hands.*

7. In order to record the heart rate, place the electrodes on the subject as shown in Figure 21.15. Place an electrode on the medial surface of each leg, just above the ankle. Place another electrode on the right anterior forearm just above the wrist.

8. Attach the electrode lead set to the electrodes according to the colors shown in Figure 21.15. Wait 5 minutes before starting the calibration procedure.

9. Start the BIOPAC® Student Lab program on the computer by double-clicking the icon on the desktop or by following your instructor's guidance.

10. Select lesson **L09-Poly-1** from the menu and click **OK.**

11. Type in a filename that will save this subject's data on the computer hard drive. You may want to use the subject's last name followed by Poly-1 (for example, SmithPoly-1), then click **OK.**

Calibrating the Equipment

1. Have the subject sit facing the director, but do not allow the subject to see the computer screen. The subject should remain immobile but be relaxed with legs and arms in a comfortable position.

2. When the subject is ready, click **Calibrate.** After 3 seconds, the subject will hear a beep and should inhale and exhale deeply for one breath.

3. Wait for the calibration to stop automatically after 10 seconds.

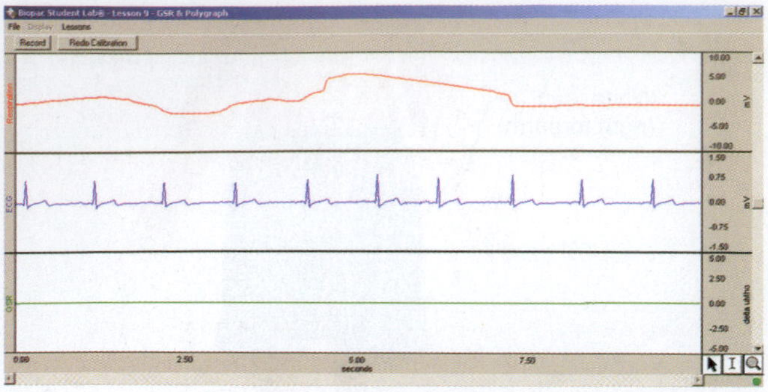

FIGURE 21.16 Example of waveforms during the calibration procedure.

4. Observe the data, which should look similar to that in Figure 21.16.

- If the data look very different, click **Redo Calibration** and repeat the steps above.
- If the data look similar, proceed to the next section.

Recording the Data

Hints to obtaining the best data:

- Do not let the subject see the data as it is being recorded.
- Conduct the exam in a quiet setting.
- Keep the subject as still as possible.
- Take care to have the subject move the mouth as little as possible when responding to questions.
- Make sure the subject is relaxed at resting heart rate before the exam begins.

The data will be recorded in three segments. The director must read through the directions for the entire segment before proceeding so that the subject can be prompted and questioned appropriately.

Segment 1: Baseline Data

1. When the subject and director are ready, click **Record.**
2. After waiting 5 seconds, the director will ask the subject to respond to the following questions and should remind the subject to minimize mouth movements when answering. Use the **F9** key (PC) or **ESC** key (Mac) to insert a marker after each response. Wait about 5 seconds between each answer.

- Quietly state your name.
- Slowly count down from ten to zero.
- Count backward from 30 by increasing odd numbers (subtract 1, then 3, then 5, etc.).
- Finally, the director lightly touches the subject on the cheek.

3. After the final, cheek-touching test, click **Suspend.**

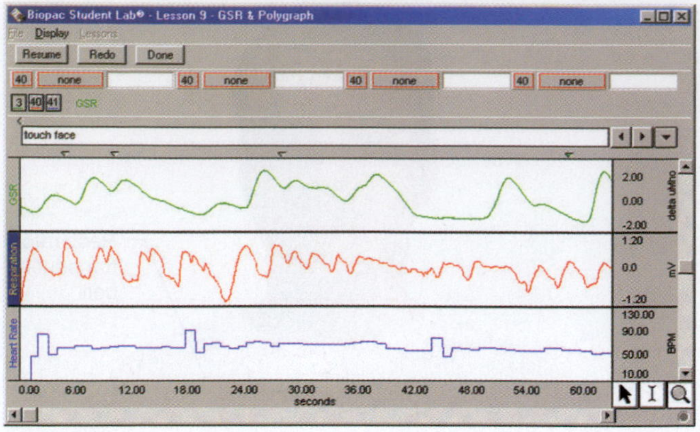
Biopac Student Lab® - Lesson 9 - GSR & Polygraph
Display
Resume
Redo
Done
GSR
touch face
Respiration
Heart Rate
seconds
delta uMho
mV
BPM

FIGURE 21.17 Example of Segment 1 data.

4. Observe the data, which should look similar to the data in Figure 21.17.

- If the data look very different, click **Redo** and repeat the steps above.
- If the data look similar, proceed to record Segment 2.

Segment 2: Response to Different Colors

1. When the subject and director are ready, click **Resume.**
2. The director will sequentially hold up nine differently colored paper squares about 2 feet in front of the subject's face. He or she will ask the subject to focus on the particular color for 10 seconds before moving to the next color in the sequence. The director will display the colors and insert a marker in the following order: white, black, red, blue, green, yellow, orange, brown, and purple. The director or assistant will use the **F9** key (PC) or **ESC** key (Mac) to insert a marker at the start of each color.
3. The subject will be asked to view the complete set of colors. After the color purple, click **Suspend.**
4. Observe the data, which should look similar to the data in Figure 21.18.

- If the data look very different, click **Redo** and repeat the steps above.
- If the data look similar, proceed to record Segment 3.

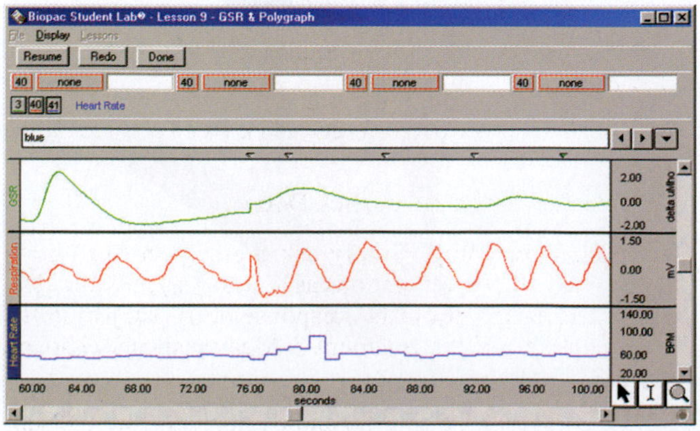
Biopac Student Lab® - Lesson 9 - GSR & Polygraph
Display
Resume
Redo
Done
Heart Rate
blue
GSR
Respiration
seconds
delta uMho
mV
BPM

FIGURE 21.18 Example of Segment 2 data.

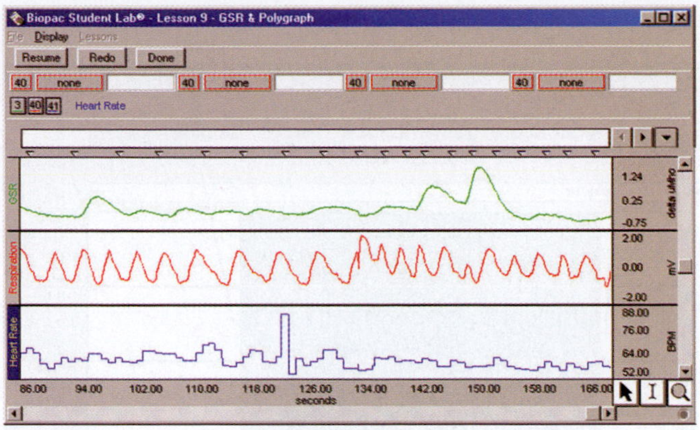
Biopac Student Lab® - Lesson 9 - GSR & Polygraph
File Display Lessons
Resume
Redo
Done
Heart Rate
GSR
Respiration
Heart Rate
seconds

FIGURE 21.19 Example of Segment 3 data.

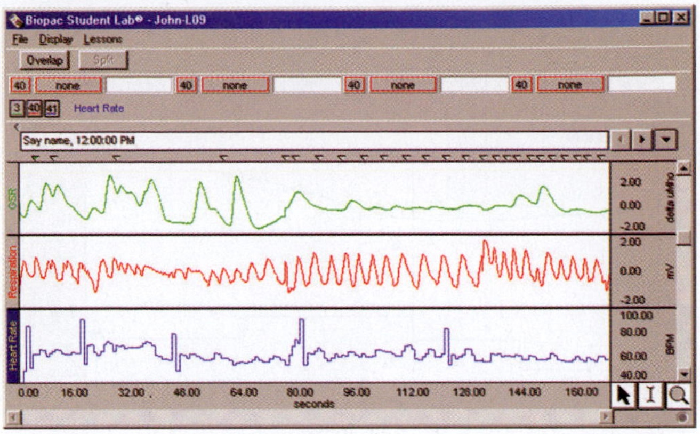
Biopac Student Lab® - John-L09
File Display Lessons
Overlap
Heart Rate
Say name, 12:00:00 PM
GSR
Respiration
Heart Rate
seconds

FIGURE 21.20 Example of polygraph recording with GSR, respiration, and heart rate.

Segment 3: Response to Different Questions

1. When the subject and director are ready, click **Resume.**

2. The director will ask the subject the ten questions in step 3 and note if the answer is Yes or No. In this segment, the recorder will use the **F9** key (PC) or **ESC** key (Mac) to insert a marker at the end of each question and the end of each answer. The director will circle the Yes or No response of the subject in the "Response" column of the Segment 3 Measurements chart on page 333.

3. The following questions are to be asked and answered either Yes or No:

- Are you currently a student?
- Are your eyes blue?
- Do you have any brothers?
- Did you earn an "A" on the last exam?
- Do you drive a motorcycle?
- Are you less than 25 years old?
- Have you ever traveled to another planet?
- Have aliens from another planet ever visited you?
- Do you watch *Sesame Street*?
- Have you answered all of the preceding questions truthfully?

4. After the last question is answered, click **Suspend.**

5. Observe the data, which should look similar to the data in Figure 21.19.

- If the data look very different, click **Redo** and repeat the steps above.
- If the data look similar, click **Done.**

6. Without recording, simply ask the subject to respond once again to all of the questions as honestly as possible. The director circles the Yes or No response of the subject in the "Truth" column of the Segment 3 Measurements chart on page 333.

7. Remove all of the sensors and equipment from the subject, and continue to Data Analysis.

Data Analysis

1. If just starting the BIOPAC® program to perform data analysis, enter **Review Saved Data** mode and choose the file with the subject's GSR data (for example, SmithPoly-1).

2. Observe how the channel numbers are designated, as shown in Figure 21.20: CH 3—**GSR;** CH 40—**Respiration;** CH 41—**Heart Rate.**

3. You may need to use the following tools to adjust the data in order to clearly view and analyze the first 5 seconds of the recording.

- Click the magnifying glass in the lower right corner of the screen (near the I-beam box) to activate the **zoom** function. Use the magnifying glass cursor to click on the very first waveforms until the first 5 seconds of data are represented (see horizontal time scale at the bottom of the screen).
- Select the **Display** menu at the top of the screen and click **Autoscale Waveforms** in the drop-down menu. This function will adjust the data for better viewing.

4. To analyze the data, set up the first three pairs of channel/measurement boxes at the top of the screen. (Each box activates a drop-down menu when you click it.) Select the following channels and measurement types:

Channel	Measurement
CH 41	value
CH 40	BPM
CH 3	value

Value: Displays the value of the measurement (for example, heart rate or GSR) at the point in time that is selected.

BPM: In this analysis, the BPM calculates breaths per minute when the area that is highlighted starts at the beginning of one inhalation and ends at the beginning of the next inhalation.

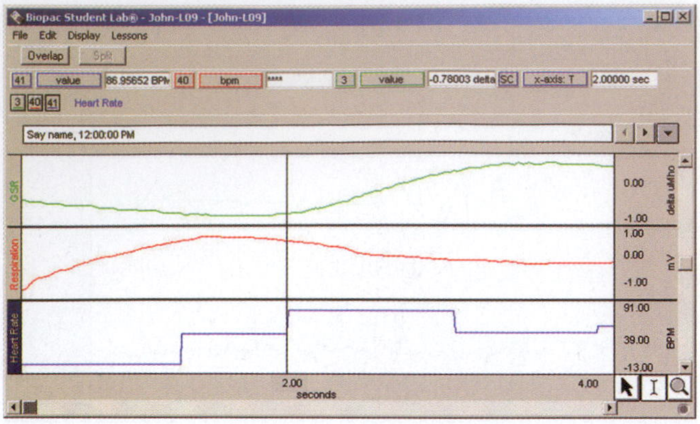

FIGURE 21.21 Selecting the two-second point for data analysis.

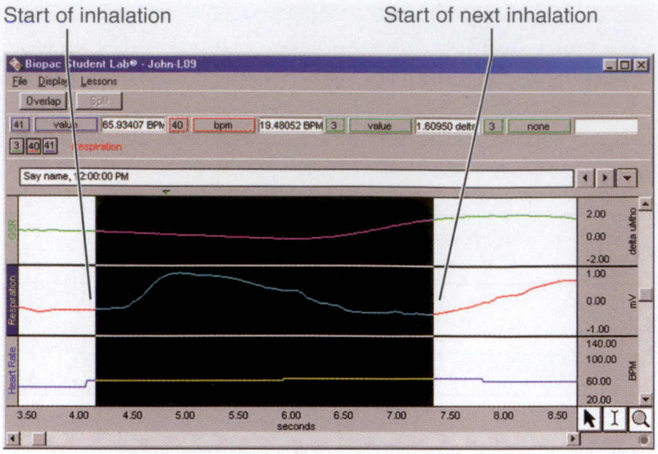
Start of inhalation
Start of next inhalation

FIGURE 21.22 Highlighting the waveforms from the start of one inhalation to the start of the next.

5. Use the arrow cursor and click the I-beam cursor box at the lower right side of the screen to activate the "area selection" function. Using the activated I-beam cursor, select the 2-second point on the data as shown in Figure 21.21. Record the heart rate and GSR values for Segment 1 data in the chart below. This point represents the resting or baseline data.

6. Using data from the first 6 seconds, use the I-beam cursor tool to highlight an area from the start of one inhalation to the start of the next inhalation, as shown in Figure 21.22. The start of an inhalation is indicated by the beginning of the ascension of the waveform. Record this as the baseline respiratory rate in the chart below showing measurements for Segment 1.

7. Using the markers as guides, scroll along the bottom scroll bar until the data from Segment 1 appears.

8. Analyze all parts of Segment 1. Using the tools described in steps 5 and 6, acquire the measurements for the heart rate, GSR, and respiration rate soon after each subject response. Use the maximum GSR value in that time frame as the point of measurement for GSR and heart rate. Use the beginning of two consecutive inhalations in that same time frame to measure respiration rate. Record these data in the chart showing measurements for Segment 1.

9. Repeat these same procedures to measure GSR, heart rate, and respiration rate for each color in Segment 2. Record these data in the chart of measurements for Segment 2.

Segment 1 Measurements

Procedure	Heart rate [CH 41 value]	Respiratory rate [CH 40 BPM]	GSR [CH 3 value]
Baseline			
Quietly say name			
Count from 10			
Count from 30			
Face is touched			

Segment 2 Measurements			
Color	**Heart rate [CH 41 value]**	**Respiratory rate [CH 40 BPM]**	**GSR [CH 3 value]**
White			
Black			
Red			
Blue			
Green			
Yellow			
Orange			
Brown			
Purple			

Segment 3 Measurements					
Question	**Response**	**Truth**	**Heart rate [CH 41 value]**	**Resp. rate [CH 40 BPM]**	**GSR [CH 3 value]**
Student?	Y N	Y N			
Blue eyes?	Y N	Y N			
Brothers?	Y N	Y N			
Earn “A”?	Y N	Y N			
Motorcycle?	Y N	Y N			
Under 25?	Y N	Y N			
Planet?	Y N	Y N			
Aliens?	Y N	Y N			
Sesame?	Y N	Y N			
Truthful?	Y N	Y N			

10. Repeat these same procedures to measure GSR, heart rate, and respiration rate for responses to each question in Segment 3. Record these data in the chart of measurements for Segment 3.

11. Examine GSR, heart rate, and respiration rate of the baseline data in the chart showing measurements for Segment 1.

12. For every condition to which the subject was exposed, write **H** if that value is higher than baseline, write **L** if the value is lower, and write **NC** if there is no significant change. Repeat this analysis for Segments 2 and 3.

Examine the data in the chart showing measurements for Segment 1. Is there any noticeable difference between the baseline GSR, heart rate, and respiration rate after each prompt? Under which prompts is the most significant change noted?

____

____

____

Examine the data in the chart showing measurements for Segment 2. Is there any noticeable difference between the baseline GSR, heart rate, and respiration rate after each color presentation? Under which colors is the most significant change noted?

____

____

____

Examine the data in the chart showing measurements for Segment 3. Is there any noticeable difference between the baseline GSR, heart rate, and respiration rate after each question? After which is the most significant change noted?

____

____

____

Speculate as to the reasons why a subject may demonstrate a change in GSR from baseline under different color conditions.

____

____

____

Speculate as to the reasons why a subject may demonstrate a change in GSR from baseline when a particular question is asked.

____

____

____

EXERCISE 22

Human Reflex Physiology

MATERIALS

- ☐ Reflex hammer
- ☐ Sharp pencils
- ☐ Cot (if available)
- ☐ Absorbent cotton (sterile)
- ☐ Tongue depressor
- ☐ Metric ruler
- ☐ Flashlight
- ☐ 100- or 250-ml beaker
- ☐ 10- or 25-ml graduated cylinder
- ☐ Lemon juice in dropper bottle
- ☐ Wide-range pH paper
- ☐ Large laboratory bucket containing freshly prepared 10% household bleach solution for saliva-soiled glassware
- ☐ Disposable autoclave bag
- ☐ Wash bottle containing 10% bleach solution
- ☐ Reaction time ruler (if available)

BIOPAC® BIOPAC® MP36 (MP35/30) data acquisition unit, PC or Mac computer. BIOPAC® Student Lab Software, hand switch, and headphones. New versions of the BSL software (3.7.5 and higher for Windows, and 3.7.4 and higher for Mac) require slightly different channel settings and collection strategies. Instructions for using the newer software with the MP36/35 data acquisition units can be found on MasteringA&P. In addition, instructions for use of the NEW 2-channel data acquisition unit, the MP45, may also be found on MasteringA&P.

Note: Instructions for using PowerLab® equipment can be found on MasteringA&P.

MasteringA&P™

Access practice quizzes and more in the Study Area at www.masteringaandp.com.

OBJECTIVES

1. To define *reflex* and *reflex arc.*
2. To name, identify, and describe the function of each element of a reflex arc.
3. To explain why reflex testing is an important part of every physical examination.
4. To describe and discuss several types of reflex activities as observed in the laboratory; to indicate the functional or clinical importance of each; and to categorize each as a somatic or autonomic reflex action.
5. To explain why cord-mediated reflexes are generally much faster than those involving input from the higher brain centers.
6. To investigate differences in reaction time of reflexes and learned responses.

PRE-LAB QUIZ

1. Define *reflex.* ____

   ____
2. Circle the correct term. Autonomic / Somatic reflexes include all those reflexes that involve stimulation of skeletal muscles.
3. In a reflex arc, the __________ transmits afferent impulses to the central nervous system.
   a. integration center
   b. motor neuron
   c. receptor
   d. sensory neuron
4. Circle True or False. Most reflexes are simple, two-neuron, monosynaptic reflex arcs.
5. Stretch reflexes are initiated by tapping a __________ which stretches the associated muscle.
   a. bone
   b. ligament
   c. tendon
6. An example of an autonomic reflex that you will be studying in today's lab is the __________ reflex.
   a. crossed-extensor
   b. gag
   c. plantar
   d. salivary
7. Name one of the pupillary reflexes you will be examining today.

   ____
8. Circle the correct term. The effectors of the salivary reflex are muscles / glands.
9. Circle True or False. Learned reflexes involve far fewer neural pathways and fewer types of higher intellectual activities, which shortens their response time.

Reflexes are rapid, predictable, involuntary motor responses to stimuli; they are mediated over neural pathways called reflex arcs.

Reflexes can be categorized into one of two large groups: autonomic reflexes and somatic reflexes. **Autonomic** (or visceral) **reflexes** are mediated through the autonomic nervous system, and we are not usually aware of them. These reflexes activate smooth muscles, cardiac muscle, and the glands of the body, and they regulate body functions such as digestion, elimination, blood pressure, salivation, and sweating. **Somatic reflexes** include all those reflexes that involve stimulation of skeletal muscles by the somatic division of the nervous system. An example of such a reflex is the rapid withdrawal of a hand from a hot object.

Reflex testing is an important diagnostic tool for assessing the condition of the nervous system. Distorted, exaggerated, or absent reflex responses may indicate degeneration or pathology of portions of the nervous system, often before other signs are apparent.

If the spinal cord is damaged, the easily performed reflex tests can help pinpoint the area (level) of spinal cord injury. Motor nerves above the injured area may be unaffected, whereas those at or below the lesion site may be unable to participate in normal reflex activity. ■

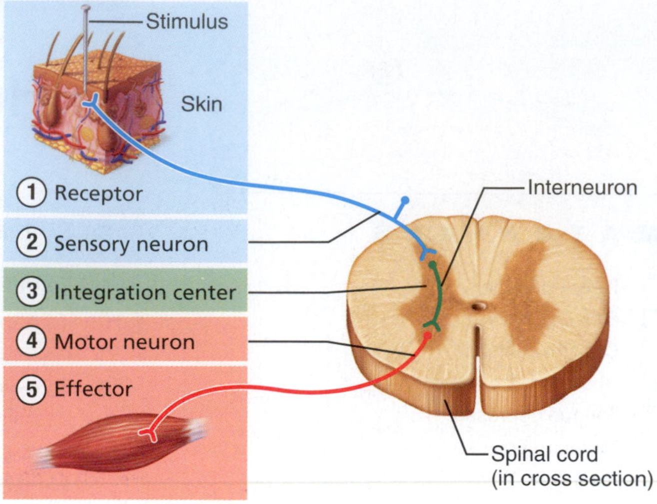
Stimulus
Skin
1 Receptor
2 Sensory neuron
3 Integration center
4 Motor neuron
5 Effector
Interneuron
Spinal cord
(in cross section)

FIGURE 22.1 The five basic components of reflex arcs. The reflex illustrated is polysynaptic.

Components of a Reflex Arc

Reflex arcs have five basic components (Figure 22.1):

1. The *receptor* is the site of stimulus action.
2. The *sensory neuron* transmits afferent impulses to the CNS.
3. The *integration center* consists of one or more neurons in the CNS.
4. The *motor neuron* conducts efferent impulses from the integration center to an effector organ.
5. The *effector,* muscle fibers or glands, responds to efferent impulses characteristically (by contracting or secreting, respectively).

The simple patellar or knee-jerk reflex shown in Figure 22.2a is an example of a simple, two-neuron, *monosynaptic*

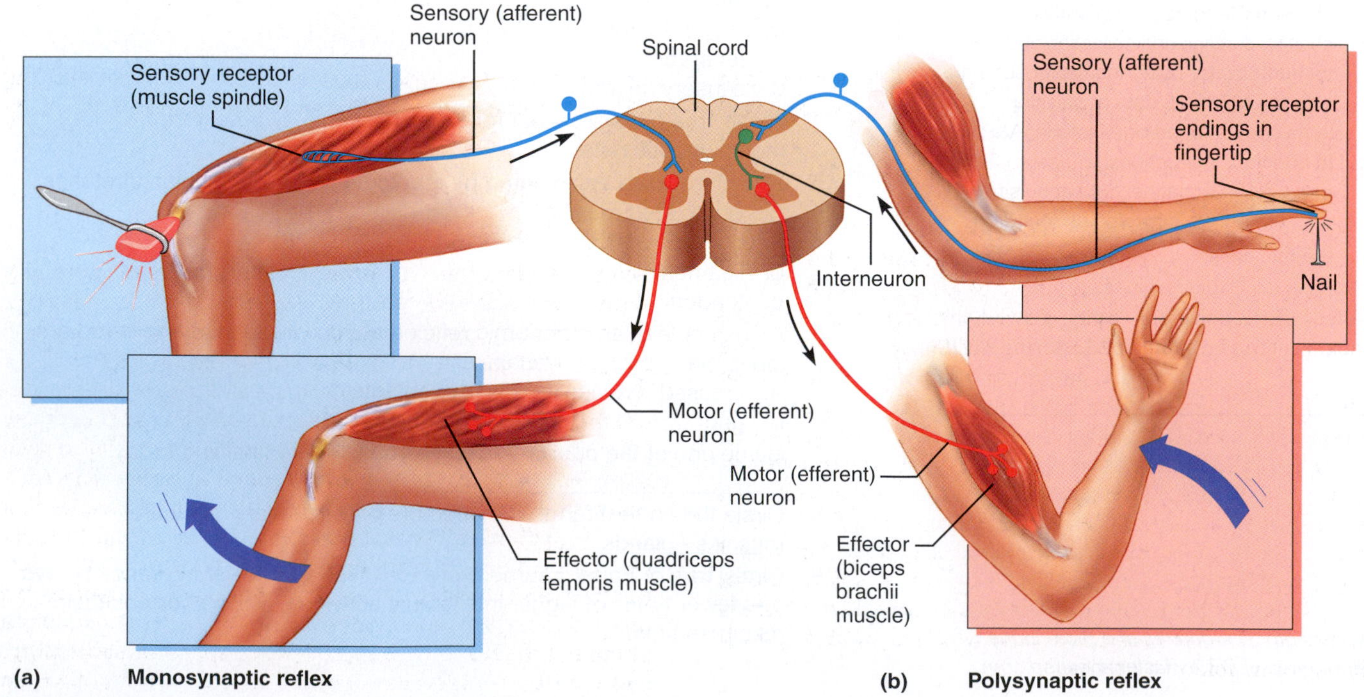
Sensory (afferent) neuron
Sensory receptor (muscle spindle)
Spinal cord
Sensory (afferent) neuron
Sensory receptor endings in fingertip
Interneuron
Nail
Motor (efferent) neuron
Motor (efferent) neuron
Effector (quadriceps femoris muscle)
Effector (biceps brachii muscle)
(a) Monosynaptic reflex
(b) Polysynaptic reflex

FIGURE 22.2 Monosynaptic and polysynaptic reflex arcs. The integration center is in the spinal cord, and in each example the receptor and effector are in the same limb. **(a)** The patellar reflex, a two-neuron monosynaptic reflex. **(b)** A flexor reflex, an example of a polysynaptic reflex.

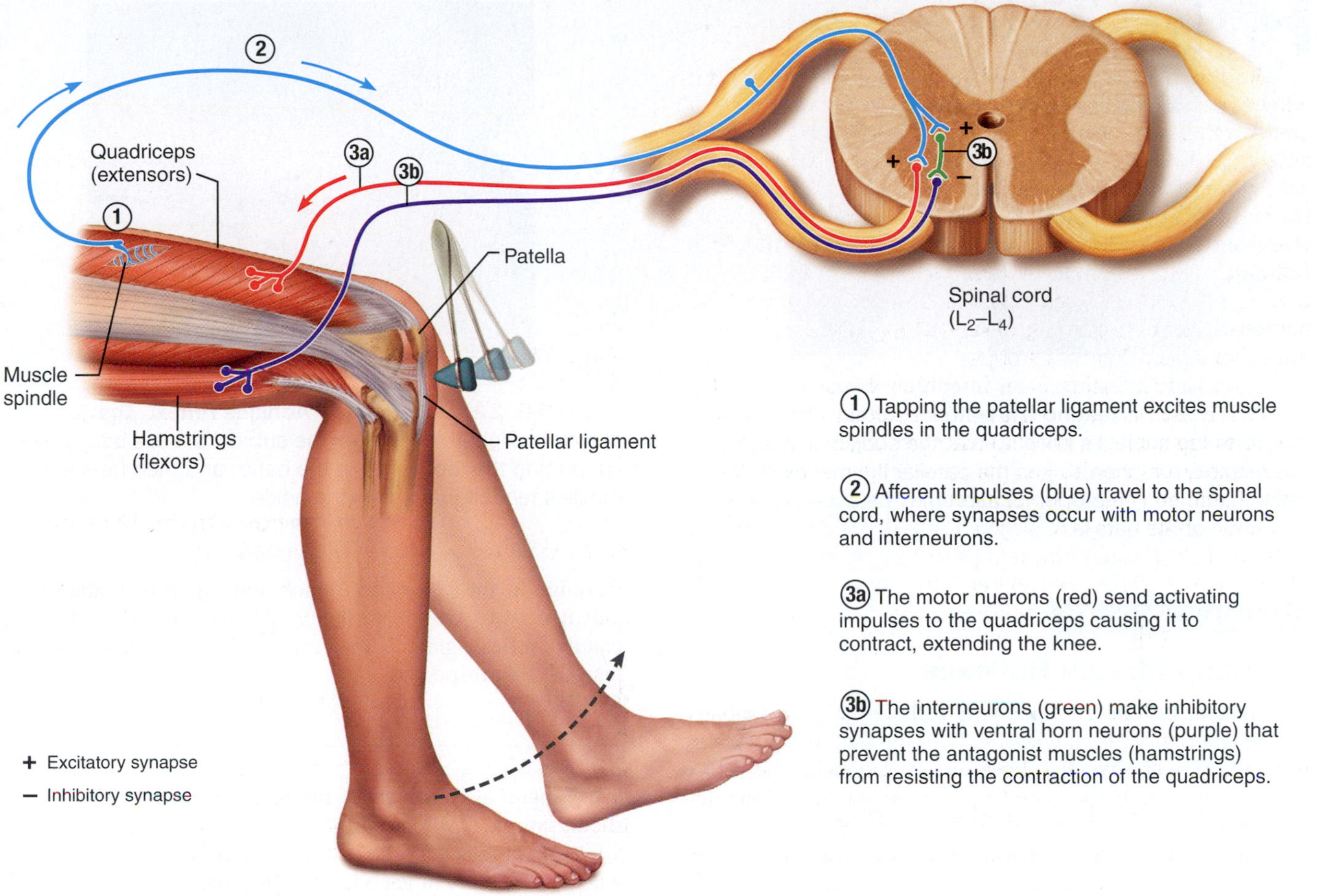
Quadriceps (extensors)
Muscle spindle
Hamstrings (flexors)
Patella
Patellar ligament
Spinal cord (L_2–L_4)
+ Excitatory synapse
– Inhibitory synapse
1 Tapping the patellar ligament excites muscle spindles in the quadriceps.
2 Afferent impulses (blue) travel to the spinal cord, where synapses occur with motor neurons and interneurons.
3a The motor nuerons (red) send activating impulses to the quadriceps causing it to contract, extending the knee.
3b The interneurons (green) make inhibitory synapses with ventral horn neurons (purple) that prevent the antagonist muscles (hamstrings) from resisting the contraction of the quadriceps.

FIGURE 22.3 The patellar (knee-jerk) reflex—a specific example of a stretch reflex.

(literally, "one synapse") reflex arc. It will be demonstrated in the laboratory. However, most reflexes are more complex and *polysynaptic,* involving the participation of one or more interneurons (association neurons) in the reflex arc pathway. A three-neuron reflex arc (flexor reflex) is diagrammed in Figure 22.2b. Since delay or inhibition of the reflex may occur at the synapses, the more synapses encountered in a reflex pathway, the more time is required to effect the reflex.

Reflexes of many types may be considered programmed into the neural anatomy. Many *spinal reflexes,* reflexes that are initiated and completed at the spinal cord level, occur without the involvement of higher brain centers. Generally these reflexes are present in animals whose brains have been destroyed, as long as the spinal cord is functional. Conversely, other reflexes require the involvement of the brain, since many different inputs must be evaluated before the appropriate reflex is determined. Superficial cord reflexes and pupillary responses to light are in this category. In addition, although many spinal reflexes do not require the involvement of higher centers, the brain is "advised" of spinal cord reflex activity and may alter it by facilitating or inhibiting the reflexes.

Somatic Reflexes

There are several types of somatic reflexes, including several that you will be eliciting during this laboratory session—the stretch, crossed extensor, superficial cord, corneal, and gag reflexes. Some require only spinal cord activity; others require brain involvement as well.

Spinal Reflexes

Stretch Reflexes

Stretch reflexes are important postural reflexes, normally acting to maintain posture, balance, and locomotion. Stretch reflexes are initiated by tapping a tendon, which stretches the muscle the tendon is attached to (Figure 22.3). This stimulates the muscle spindles and causes reflex contraction of the stretched muscle or muscles. Branches of the afferent fibers from the muscle spindles also synapse with interneurons controlling the antagonist muscles. The inhibition of those interneurons and the antagonist muscles that follows, called *reciprocal inhibition,* causes them to relax and prevents them from resisting (or reversing) the contraction of the stretched muscle. Additionally, impulses are relayed to higher brain centers (largely via the dorsal white columns) to advise of muscle length, speed of shortening, and the like—information needed to maintain muscle tone and posture. Stretch reflexes tend to be hypoactive or absent in cases of peripheral nerve damage or ventral horn disease and hyperactive in corticospinal tract lesions. They are absent in deep sedation and coma.

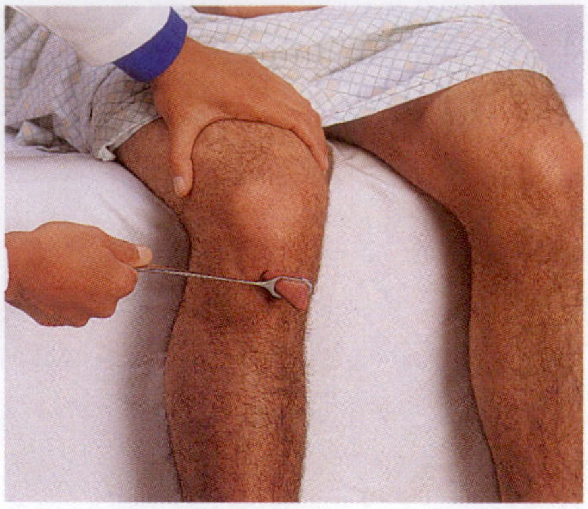

FIGURE 22.4 Testing the patellar reflex. The examiner supports the subject's knee so that the subject's muscles are relaxed, and then strikes the patellar ligament with the reflex hammer. The proper location may be ascertained by palpation of the patella.

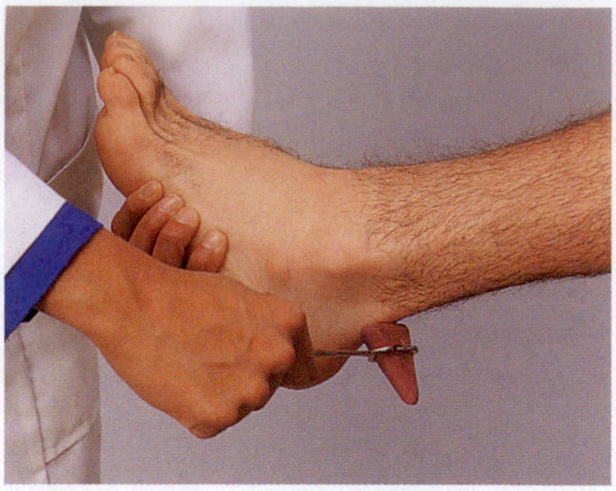

FIGURE 22.5 Testing the Achilles reflex. The examiner slightly dorsiflexes the subject's ankle by supporting the foot lightly in the hand, and then taps the Achilles tendon just above the ankle.

ACTIVITY 1

Initiating Stretch Reflexes

1. Test the **patellar,** or knee-jerk, **reflex** by seating a subject on the laboratory bench with legs hanging free (or with knees crossed). Tap the patellar ligament sharply with the reflex hammer just below the knee between the patella and the tibial tuberosity as shown in Figure 22.4. The knee-jerk response assesses the L_2–L_4 level of the spinal cord. Test both knees and record your observations. (Sometimes a reflex can be altered by your actions. If you encounter difficulty, consult your instructor for helpful hints.)

____

____

Which muscles contracted? ____________________

____

What nerve is carrying the afferent and efferent impulses?

____

2. Test the effect of mental distraction on the patellar reflex by having the subject add a column of three-digit numbers while you test the reflex again. Is the response more *or* less vigorous than the first response?

____

What are your conclusions about the effect of mental distraction on reflex activity?

____

3. Now test the effect of muscular activity occurring simultaneously in other areas of the body. Have the subject clasp the edge of the laboratory bench and vigorously attempt to pull it upward with both hands. At the same time, test the patellar reflex again. Is the response more or less vigorous than the first response?

____

4. Fatigue also influences the reflex response. The subject should jog in position until she or he is very fatigued (*really fatigued*—no slackers). Test the patellar reflex again, and record whether it is more or less vigorous than the first response.

____

Would you say that nervous system activity *or* muscle function is responsible for the changes you have just observed?

Explain your reasoning. ____________________

____

____

____

5. The **Achilles,** or ankle-jerk, **reflex** assesses the first two sacral segments of the spinal cord. With your shoe removed and your foot dorsiflexed slightly to increase the tension of the gastrocnemius muscle, have your partner sharply tap your calcaneal (Achilles) tendon with the broad side of the reflex hammer (Figure 22.5).

What is the result? ____________________

____

During walking, what is the action of the gastrocnemius at the ankle?

____

Crossed-Extensor Reflex

The **crossed-extensor reflex** is more complex than the stretch reflex. It consists of a flexor, or withdrawal, reflex followed by extension of the opposite limb.

This reflex is quite obvious when, for example, a stranger suddenly and strongly grips one's arm. The immediate response is to withdraw the clutched arm and push the intruder away with the other arm. The reflex is more difficult to demonstrate in a laboratory because it is anticipated, and under these conditions the extensor part of the reflex may be inhibited.

ACTIVITY 2

Initiating the Crossed-Extensor Reflex

The subject should sit with eyes closed and with the dorsum of one hand resting on the laboratory bench. Obtain a sharp pencil, and suddenly prick the subject's index finger. What are the results?

____

____

____

Did the extensor part of this reflex occur simultaneously or more slowly than the other reflexes you have observed?

____

What are the reasons for this? ________________________

____

____

The reflexes that have been demonstrated so far—the stretch and crossed-extensor reflexes—are examples of reflexes in which the reflex pathway is initiated and completed at the spinal cord level.

Superficial Cord Reflexes

The **superficial cord reflexes** (abdominal, cremaster, and plantar reflexes) result from pain and temperature changes. They are initiated by stimulation of receptors in the skin and mucosae. The superficial cord reflexes depend *both* on functional upper-motor pathways and on the cord-level reflex arc. Since only the plantar reflex can be tested conveniently in a laboratory setting, we will use this as our example.

The **plantar reflex,** an important neurological test, is elicited by stimulating the cutaneous receptors in the sole of the foot. In adults, stimulation of these receptors causes the toes to flex and move closer together. Damage to the pyramidal (or corticospinal) tract, however, produces *Babinski's sign,* an abnormal response in which the toes flare and the great toe moves in an upward direction. (In newborn infants, it is normal to see Babinski's sign due to incomplete myelination of the nervous system.)

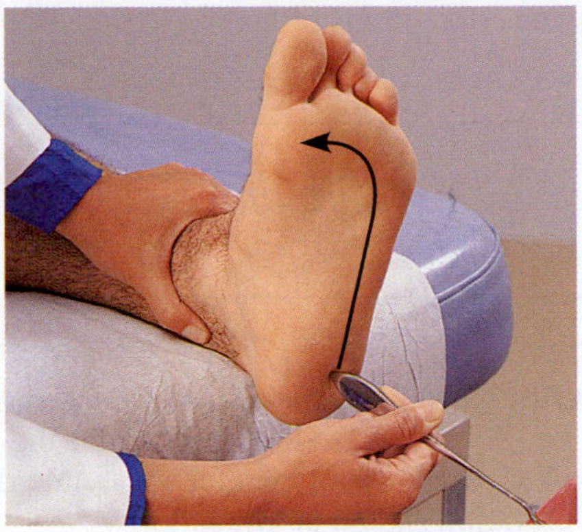

FIGURE 22.6 Testing the plantar reflex. Using a moderately sharp object, the examiner strokes the lateral border of the subject's sole, starting at the heel and continuing toward the big toe across the ball of the foot.

ACTIVITY 3

Initiating the Plantar Reflex

Have the subject remove a shoe and lie on the cot or laboratory bench with knees slightly bent and thighs rotated so that the posterolateral side of the foot rests on the cot. Alternatively, the subject may sit up and rest the lateral surface of the foot on a chair. Draw the handle of the reflex hammer firmly along the lateral side of the exposed sole from the heel to the base of the great toe (Figure 22.6).

What is the response? ____________________________

____

Is this a normal plantar reflex or a Babinski's sign?

____

Cranial Nerve Reflex Tests

In these experiments, you will be working with your lab partner to illustrate two somatic reflexes mediated by cranial nerves.

Corneal Reflex

The **corneal reflex** is mediated through the trigeminal nerve (cranial nerve V). The absence of this reflex is an ominous sign because it often indicates damage to the brain stem resulting from compression of the brain or other trauma.

ACTIVITY 4

Initiating the Corneal Reflex

Stand to one side of the subject; the subject should look away from you toward the opposite wall. Wait a few seconds and then quickly, *but gently,* touch the subject's cornea (on the side toward you) with a wisp of absorbent cotton. What reflexive reaction occurs when something touches the cornea?

____

What is the function of this reflex?

____

____

____

Gag Reflex

The **gag reflex** tests the somatic motor responses of cranial nerves IX and X. When the oral mucosa on the side of the uvula is stroked, each side of the mucosa should rise, and the amount of elevation should be equal.*

ACTIVITY 5

Initiating the Gag Reflex

For this experiment, select a subject who does not have a queasy stomach, because regurgitation is a possibility. Gently stroke the oral mucosa on each side of the subject's uvula with a tongue depressor. What happens?

____

____

! Discard the used tongue depressor in the disposable autoclave bag before continuing. *Do not* lay it on the laboratory bench at any time.

Autonomic Reflexes

The autonomic reflexes include the pupillary, ciliospinal, and salivary reflexes, as well as a multitude of other reflexes. Work with your partner to demonstrate the four autonomic reflexes described next.

Pupillary Reflexes

There are several types of pupillary reflexes. The **pupillary light reflex** and the **consensual reflex** will be examined here. In both of these pupillary reflexes, the retina of the eye is the receptor, the optic nerve (cranial nerve II) contains the afferent fibers, the oculomotor nerve (cranial nerve III) is responsible for conducting efferent impulses to the eye, and the smooth muscle of the iris is the effector. Many central nervous system centers are involved in the integration of these responses. Absence of normal pupillary reflexes is generally a late indication of severe trauma or deterioration of the vital brain stem tissue due to metabolic imbalance.

ACTIVITY 6

Initiating Pupillary Reflexes

1. Conduct the reflex testing in an area where the lighting is relatively dim. Before beginning, obtain a metric ruler and a flashlight. Measure and record the size of the subject's pupils as best you can.

Right pupil: _________ mm Left pupil: _________ mm

2. Stand to the left of the subject to conduct the testing. The subject should shield his or her right eye by holding a hand vertically between the eye and the right side of the nose.

3. Shine a flashlight into the subject's left eye. What is the pupillary response?

____

Measure the size of the left pupil: _________ mm

4. Without moving the flashlight, observe the right pupil. Has the same type of change (called a *consensual response*) occurred in the right eye?

____

Measure the size of the right pupil: _________ mm

The consensual response, or any reflex observed on one side of the body when the other side has been stimulated, is called a **contralateral response.** The pupillary light response, or any reflex occurring on the same side stimulated, is referred to as an **ipsilateral response.**

What does the occurrence of a contralateral response indicate about the pathways involved?

____

____

Was the sympathetic *or* the parasympathetic division of the autonomic nervous system active during the testing of these reflexes?

____

What is the function of these pupillary responses?

____

____

____

Ciliospinal Reflex

The **ciliospinal reflex** is another example of reflex activity in which pupillary responses can be observed. This response may initially seem a little bizarre, especially in view of the consensual reflex just demonstrated.

*The uvula is the fleshy tab hanging from the roof of the mouth just above the root of the tongue.

ACTIVITY 7

Initiating the Ciliospinal Reflex

1. While observing the subject's eyes, gently stroke the skin (or just the hairs) on the left side of the back of the subject's neck, close to the hairline.

What is the reaction of the left pupil? ________________

The reaction of the right pupil? ________________

2. If you see no reaction, repeat the test using a gentle pinch in the same area.

The response you should have noted—pupillary dilation—is consistent with the pupillary changes occurring when the sympathetic nervous system is stimulated. Such a response may also be elicited in a single pupil when more impulses from the sympathetic nervous system reach it for any reason. For example, when the left side of the subject's neck was stimulated, sympathetic impulses to the left iris increased, resulting in the ipsilateral reaction of the left pupil.

On the basis of your observations, would you say that the sympathetic innervation of the two irises is closely integrated?

________ Why or why not? ________________

____

____

Salivary Reflex

Unlike the other reflexes, in which the effectors were smooth or skeletal muscles, the effectors of the **salivary reflex** are glands. The salivary glands secrete varying amounts of saliva in response to reflex activation.

ACTIVITY 8

Initiating the Salivary Reflex

1. Obtain a small beaker, a graduated cylinder, lemon juice, and wide-range pH paper. After refraining from swallowing for 2 minutes, the subject is to expectorate (spit) the accumulated saliva into a small beaker. Using the graduated cylinder, measure the volume of the expectorated saliva and determine its pH.

Volume: ________________ cc pH: ________________

2. Now place 2 or 3 drops of lemon juice on the subject's tongue. Allow the lemon juice to mix with the saliva for 5 to 10 seconds, and then determine the pH of the subject's saliva by touching a piece of pH paper to the tip of the tongue.

pH: ________________

As before, the subject is to refrain from swallowing for 2 minutes. After the 2 minutes is up, again collect and measure the volume of the saliva and determine its pH.

Volume: ________________ cc pH: ________________

3. How does the volume of saliva collected after the application of the lemon juice compare with the volume of the first saliva sample?

____

How does the final saliva pH reading compare to the initial reading?

____

How does the final saliva pH reading compare to that obtained 10 seconds after the application of lemon juice?

____

What division of the autonomic nervous system mediates the reflex release of saliva?

____

Dispose of the saliva-containing beakers and the graduated cylinders in the laboratory bucket that contains bleach and put the used pH paper into the disposable autoclave bag. Wash the bench down with 10% bleach solution before continuing.

Reaction Time of Basic and Learned or Acquired Reflexes

The time required for reaction to a stimulus depends on many factors—sensitivity of the receptors, velocity of nerve conduction, the number of neurons and synapses involved, and the speed of effector activation, to name just a few. Some reflexes are *basic,* or inborn; others are *learned,* or *acquired,* reflexes resulting from practice or repetition. There is no clear-cut distinction between basic and learned reflexes, as most reflex actions are subject to modification by learning or conscious effort. In general, however, if the response involves a specific reflex arc, the synapses are facilitated and the response time is short. Learned reflexes involve a far larger number of neural pathways and many types of higher intellectual activities, including choice and decision making, which lengthens the response time.

There are various ways of testing reaction time of reflexes. The tests range from simple to ultrasophisticated. The following activities provide an opportunity to demonstrate the major time difference between simple and learned reflexes and to measure response time under various conditions.

ACTIVITY 9

Testing Reaction Time for Basic and Acquired Reflexes

1. Using a reflex hammer, elicit the patellar reflex in your partner. Note the relative reaction time needed for this basic reflex to occur.

2. Now test the reaction time for learned reflexes. The subject should hold a hand out, with the thumb and index finger extended. Hold a metric ruler so that its end is exactly 3 cm above the subject's outstretched hand. The ruler should be in the vertical position with the numbers reading from the bottom up. When the ruler is dropped, the subject should be able to grasp it between thumb and index finger as it passes, without having to change position. Have the subject catch the ruler five times, varying the time between trials. The relative speed of reaction can be determined by reading the number on the ruler at the point of the subject's fingertips.* (Thus if the number at the fingertips is 15 cm, the subject was unable to catch the ruler until 18 cm of length had passed through his or her fingers; 15 cm of ruler length plus 3 cm to account for the distance of the ruler above the hand.)† Record the number of centimeters that pass through the subject's fingertips (or the number of seconds required for reaction) for each trial:

Trial 1: ________ cm   Trial 4: ________ cm
________ sec   ________ sec
Trial 2: ________ cm   Trial 5: ________ cm
________ sec   ________ sec
Trial 3: ________ cm
________ sec

3. Perform the test again, but this time say a simple word each time you release the ruler. Designate a specific word as a signal for the subject to catch the ruler. On all other words, the subject is to allow the ruler to pass through his fingers. Trials in which the subject erroneously catches the ruler are to be disregarded. Record the distance the ruler travels (or the number of seconds required for reaction) in five *successful* trials:

Trial 1: ________ cm   Trial 4: ________ cm
________ sec   ________ sec
Trial 2: ________ cm   Trial 5: ________ cm
________ sec   ________ sec
Trial 3: ________ cm
________ sec

Did the addition of a specific word to the stimulus increase or decrease the reaction time?

____

4. Perform the testing once again to investigate the subject's reaction to word association. As you drop the ruler, say a word—for example, *hot.* The subject is to respond with a word he or she associates with the stimulus word—for example, *cold*—catching the ruler while responding. If unable to make a word association, the subject must allow the ruler to pass through his or her fingers. Record the distance the ruler travels (or the number of seconds required for reaction) in five successful trials, as well as the number of times the ruler is not caught by the subject.

Trial 1: ________ cm   Trial 4: ________ cm
________ sec   ________ sec
Trial 2: ________ cm   Trial 5: ________ cm
________ sec   ________ sec
Trial 3: ________ cm
________ sec

Number of times the subject did not catch the ruler:

____

You should have noticed quite a large variation in reaction time in this series of trials. Why is this so?

____

____

____

____

ACTIVITY 10

Measuring Reaction Time Using BIOPAC®

Setting Up the Equipment

1. Connect the BIOPAC® unit to the computer and turn the computer **ON.**

2. Make sure the BIOPAC® unit is **OFF.**

3. Plug in the equipment as shown in Figure 22.7.

- Hand switch—CH1
- Headphones—back of unit

4. Turn the BIOPAC® unit **ON.**

5. Start the BIOPAC® Student Lab program on the computer by double-clicking the icon on the desktop or by following your instructor's guidance.

6. Select lesson **L11-React-1** from the menu and click **OK.**

7. Type in a filename that will save this subject's data on the computer hard drive. You may want to use the subject's last name followed by React-1 (for example, SmithReact-1), then click **OK.**

Calibrating the Equipment

1. Seat the subject comfortably so that he or she cannot see the computer screen and keyboard.

2. Put the headphones on the subject and give the subject the hand switch to hold.

3. Inform the subject to push the hand switch button when a "click" is heard.

4. Click **Calibrate,** and then click **OK** when the subject is ready.

*Distance (d) can be converted to time (t) using the simple formula:

$$d \text{ (in cm)} = (1/2)(980 \text{ cm/sec}^2)t^2$$
$$t^2 = (d/490 \text{ cm/sec}^2)$$
$$t = 2\,\overline{(d/(490 \text{ cm/sec}^2)}$$

†An alternative would be to use a reaction time ruler, which converts distance to time (seconds).

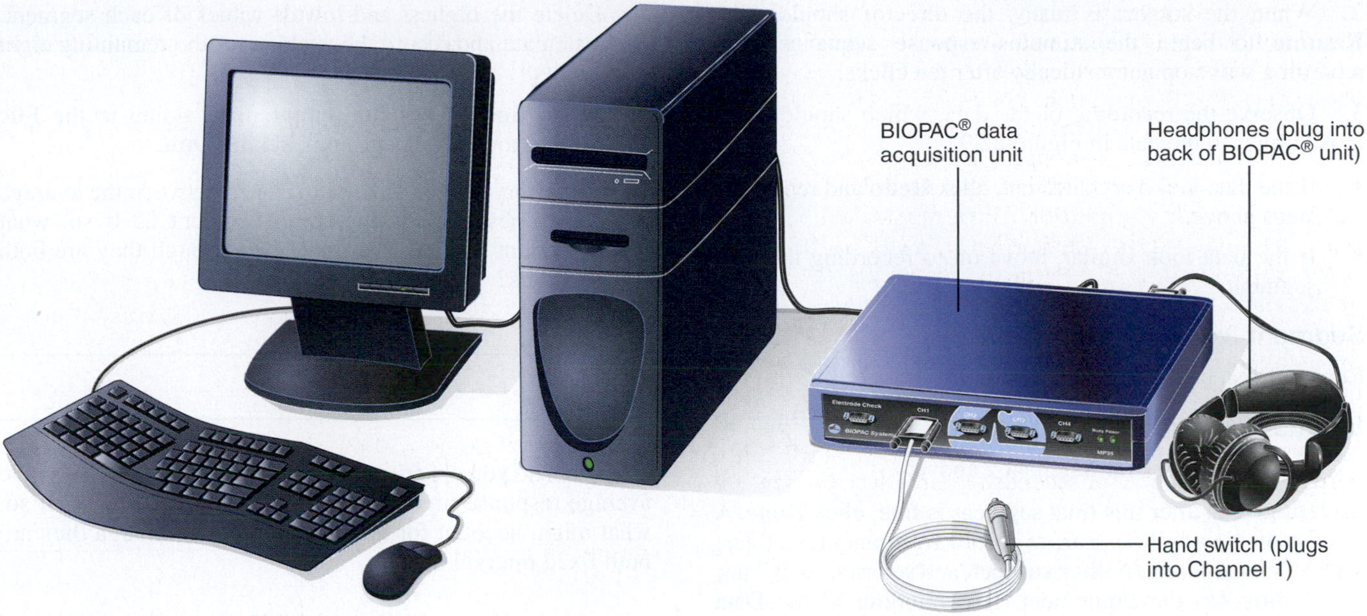
BIOPAC® data acquisition unit
Headphones (plug into back of BIOPAC® unit)
Hand switch (plugs into Channel 1)

FIGURE 22.7 Setting up the BIOPAC® equipment. Plug the headphones into the back of the unit and the hand switch into Channel 1.

5. Observe the recording of the calibration data, which should look like the waveforms in Figure 22.8.

- If the data look very different, click **Redo Calibration** and repeat the steps above.
- If the data look similar, proceed to the next section.

Recording the Data

In this experiment, you will record four different segments of data. In Segments 1 and 2, the subject will respond to random click stimuli. In Segments 3 and 4, the subject will respond to click stimuli at fixed intervals (about 4 seconds). The director will click **Record** to initiate the Segment 1 recording, and **Resume** to initiate Segments 2, 3, and 4. The subject should focus only on responding to the sound.

Segment 1: Random Trial 1

1. Each time a sound is heard, the subject should respond by pressing the button on the hand switch as quickly as possible.

2. When the subject is ready, the director should click **Record** to begin the stimulus-response sequence. The recording will stop automatically after ten clicks.

- A triangular marker will be inserted above the data each time a "click" stimulus occurs.
- An upward-pointing "pulse" will be inserted each time the subject responds to the stimulus.

3. Observe the recording of the data, which should look similar to the data in Figure 22.9.

- If the data look very different, click **Redo** and repeat the steps above.
- If the data look similar, move on to recording the next segment.

Segment 2: Random Trial 2

1. Each time a sound is heard, the subject should respond by pressing the button on the hand switch as quickly as possible.

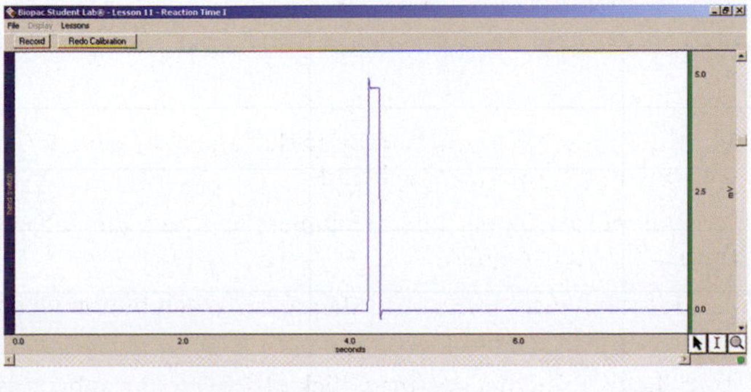

FIGURE 22.8 Example of waveforms during the calibration procedure.

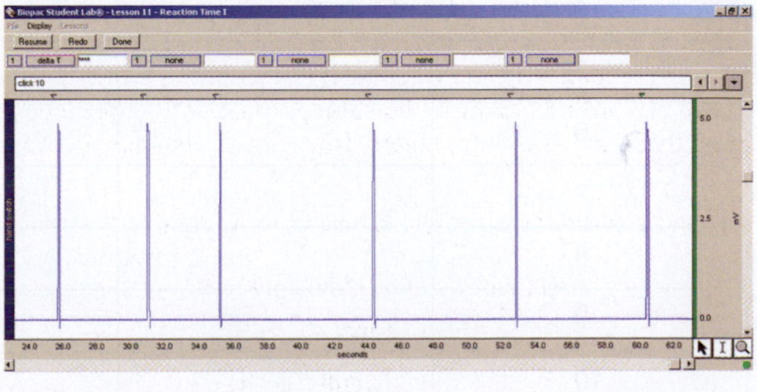

FIGURE 22.9 Example of waveforms during the recording of data.

2. When the subject is ready, the director should click **Resume** to begin the stimulus-response sequence. The recording will stop automatically after ten clicks.

3. Observe the recording of the data, which should again look similar to the data in Figure 22.9.

- If the data look very different, click **Redo** and repeat the steps above.
- If the data look similar, move on to recording the next segment.

Segment 3: Fixed Interval Trial 3

Repeat the steps for Segment 2 above.

Segment 4: Fixed Interval Trial 4

1. Repeat the steps for Segment 2 above.

2. If the data after this final segment is fine, click **Done.** A pop-up window will appear; to record from another subject select **Record from another subject,** and return to step 7 under Setting Up the Equipment. If continuing to the Data Analysis section, select **Analyze current data file** and proceed to step 2 in the Data Analysis section.

Data Analysis

1. If just starting the BIOPAC® program to perform data analysis, enter **Review Saved Data** mode and choose the file with the subject's reaction data (for example, SmithReact-1).

2. Observe that all ten reaction times are automatically calculated for each segment and are placed in the journal at the bottom of the computer screen.

3. Record the ten reaction times for each segment in the chart below.

4. Delete the highest and lowest values of each segment, then calculate and record the average for the remaining eight data points.

5. When finished, exit the program by going to the **File** menu at the top of the page and clicking **Quit.**

Do you observe a significant difference between the average response times of Segment 1 and Segment 2? If so, what might account for the difference, even though they are both random trials?

____

____

Likewise, do you observe a significant difference between the average response times of Segment 3 and Segment 4? If so, what might account for the difference, even though they are both fixed interval trials?

____

____

Optional Activity with BIOPAC® Reaction Time Measurement

To expand the experiment, choose another variable to test. Response to visual cues may be tested, or you may have the subject change the hand used when clicking the hand switch button. Devise the experiment, conduct the test, then record and analyze the data as described above. ■

Reaction Times (seconds)				
	Random		Fixed Interval	
Stimulus #	Segment 1	Segment 2	Segment 3	Segment 4
1				
2				
3				
4				
5				
6				
7				
8				
9				
10				
Average				

NAME ______________________________

LAB TIME/DATE ____________________

REVIEW SHEET

EXERCISE 22

Human Reflex Physiology

The Reflex Arc

1. Define *reflex.* ____

____

2. Name five essential components of a reflex arc: ____________________, ____________________,

____________________, ____________________, and ____________________

3. In general, what is the importance of reflex testing in a routine physical examination? ______________

____

____

Somatic and Autonomic Reflexes

4. Use the key terms to complete the statements given below.

Key: a. abdominal reflex
b. Achilles reflex
c. ciliospinal reflex
d. corneal reflex
e. crossed-extensor reflex
f. gag reflex
g. patellar reflex
h. plantar reflex
i. pupillary light reflex

Reflexes classified as somatic reflexes include a _______, _______, _______, _______, _______, _______, and _______.

Of these, the simple stretch reflexes are _______ and _______, and the superficial cord reflexes are _______ and _______.

Reflexes classified as autonomic reflexes include _______ and _______.

5. Name two cord-mediated reflexes. ____________________ and ____________________

____

Name two somatic reflexes in which the higher brain centers participate. ____________________

and ____

6. Can the stretch reflex be elicited in a pithed animal (that is, an animal in which the brain has been destroyed)? ________

Explain your answer. ____

____

7. Trace the reflex arc, naming efferent and afferent nerves, receptors, effectors, and integration centers, for the two reflexes listed. (Hint: Remember which nerve innervates the anterior thigh, and which nerve innervates the posterior thigh.)

patellar reflex: ______________________________

____

____

Achilles reflex: ______________________________

____

____

8. Three factors that influence the rapidity and effectiveness of reflex arcs were investigated in conjunction with patellar reflex testing—mental distraction, effect of simultaneous muscle activity in another body area, and fatigue.

Which of these factors increases the excitatory level of the spinal cord? ______________________________

Which factor decreases the excitatory level of the muscles? ______________________________

When the subject was concentrating on an arithmetic problem, did the change noted in the patellar reflex indicate that brain activity is necessary for the patellar reflex or only that it may modify it? ______________________________

____

9. Name the division of the autonomic nervous system responsible for each of the reflexes listed.

ciliospinal reflex: ______________________ salivary reflex: ______________________

pupillary light reflex: ______________________

10. The pupillary light reflex, the crossed-extensor reflex, and the corneal reflex illustrate the purposeful nature of reflex activity. Describe the protective aspect of each.

pupillary light reflex: ______________________________

corneal reflex: ______________________________

crossed-extensor reflex: ______________________________

____

____

____

11. Was the pupillary consensual response contralateral or ipsilateral? ______________________________

Why would such a response be of significant value in this particular reflex? ______________________________

____

12. Differentiate between the types of activities accomplished by somatic and autonomic reflexes. ______________

____

____

____

13. Several types of reflex activity were not investigated in this exercise. The most important of these are autonomic reflexes, which are difficult to illustrate in a laboratory situation. To rectify this omission, complete the following chart, using references as necessary.

Reflex	Organ involved	Receptors stimulated	Action
Micturition (urination)			
Hering-Breuer			
Defecation			
Carotid sinus			

Reaction Time of Basic and Learned or Acquired Reflexes

14. How do basic and learned or acquired reflexes differ? ______________________________

____

15. Name at least three factors that may modify reaction time to a stimulus. ______________________________

____

16. In general, how did the response time for the learned activity performed in the laboratory compare to that for the simple patelar reflex? ______________________________

17. Did the response time without verbal stimuli decrease with practice? __________ Explain the reason for this.

____

18. Explain, in detail, why response time increased when the subject had to react to a word stimulus.

____

____

19. When measuring reaction time in the BIOPAC® activity, was there a difference in reaction time when the stimulus was predictable versus unpredictable? Explain your answer.

____

____

EXERCISE 23

General Sensation

MATERIALS

- ☐ Compound microscope
- ☐ Immersion oil
- ☐ Prepared slides (longitudinal sections) of Pacinian corpuscles, Meissner's corpuscles, Golgi tendon organs, and muscle spindles
- ☐ Calipers or esthesiometer
- ☐ Small metric ruler
- ☐ Fine-point, felt-tipped markers (black, red, and blue)
- ☐ Blunt probes (obtainable from Carolina Biological Supply)
- ☐ Von Frey's hairs (or 1-in. lengths of horsehair glued to a matchstick) or sharp pencil
- ☐ Large beaker of ice water; chipped ice
- ☐ Hot water bath set at 45°C; laboratory thermometer
- ☐ Towel
- ☐ Four coins (nickels or quarters)
- ☐ Three large finger bowls or 1000-ml beakers

MasteringA&P™

Access practice quizzes and more in the Study Area at www.masteringaandp.com.

OBJECTIVES

1. To recognize various types of general sensory receptors as studied in the laboratory, and to describe the function and location of each type.
2. To define *exteroceptor, interoceptor,* and *proprioceptor*.
3. To explain the tactile two-point discrimination test, and to state its anatomical basis.
4. To define *tactile localization,* and to describe how this ability varies in different areas of the body.
5. To demonstrate and relate differences in relative density and distribution of tactile and thermoreceptors in the skin.
6. To define *adaptation* and *negative afterimage*.
7. To define *referred pain* and *projection.*

PRE-LAB QUIZ

1. Name one of the special senses. ____________________
2. Sensory receptors can be classified according to their source of stimulus. ________ are found close to the body surface and react to stimuli in the external environment.
   a. Exteroceptors
   b. Interoceptors
   c. Proprioceptors
   d. Visceroceptors
3. Circle True or False. General sensory receptors are widely distributed throughout the body and respond to, among other things, touch, pain, stretch, and changes in position.
4. Meissner's corpuscles respond to light touch. Where would you expect to find Meissner's corpuscles?
   a. deep within the dermal layer of hairy skin
   b. in the dermal papillae of hairless skin
   c. in the hypodermis of hairless skin
   d. in the uppermost portion of the epidermis
5. Pacinian corpuscles respond to
   a. deep pressure and vibrations
   b. light touch
   c. pain and temperature
6. Circle True or False. A map of the sensory receptors for touch, heat, cold, and pain shows that they are not evenly distributed throughout the body.
7. Circle the correct term. <u>Two-point threshold / Tactile localization</u> is the ability to determine which portion of the skin has been touched.
8. When a stimulus is applied for a prolonged period, the rate of receptor discharge slows, and conscious awareness of the stimulus declines. This phenomenon is known as
   a. accommodation
   b. adaptation
   c. adjustment
   d. discernment
9. You will test referred pain in this activity by immersing the subject's
   a. face in ice water to test the cranial nerve response
   b. elbow in ice water to test the ulnar nerve response
   c. hand in ice water to test the axillary nerve response
   d. leg in ice water to test the sciatic nerve response

People are irritable creatures. Hold a sizzling steak before them and their mouths water. Flash your high beams in their eyes on the highway and they cuss. Tickle them and they giggle. These "irritants" (the steak, the light, and the tickle) and many others are stimuli that continually assault us.

The body's **sensory receptors** react to stimuli or changes within the body and in the external environment. The tiny sensory receptors of the **general senses** react to touch, pressure, pain, heat, cold, stretch, vibration, and changes in position and are distributed throughout the body. In contrast to these widely distributed *general sensory receptors,* the receptors of the special senses are large, complex *sense organs* or small, localized groups of receptors. The **special senses** include sight, hearing, equilibrium, smell, and taste.

Sensory receptors may be classified according to the source of their stimulus. **Exteroceptors** react to stimuli in the external environment, and typically they are found close to the body surface. Exteroceptors include the simple cutaneous receptors in the skin and the highly specialized receptor structures of the special senses (the vision apparatus of the eye, and the hearing and equilibrium receptors of the ear, for example). **Interoceptors** or **visceroceptors** respond to stimuli arising within the body. Interoceptors are found in the internal visceral organs and include stretch receptors (in walls of hollow organs), chemoreceptors, and others. **Proprioceptors,** like interoceptors, respond to internal stimuli but are restricted to skeletal muscles, tendons, joints, ligaments, and connective tissue coverings of bones and muscles. They provide information on the position and degree of stretch of those structures.

The receptors of the special sense organs are complex and deserve considerable study. Thus the special senses (vision, hearing, equilibrium, taste, and smell) are covered separately in Exercises 24 through 26. Only the anatomically simpler **general sensory receptors**—cutaneous receptors and proprioceptors—will be studied in this exercise.

Structure of General Sensory Receptors

You cannot become aware of changes in the environment unless your sensory neurons and their receptors are operating properly. Sensory receptors are modified dendritic endings (or specialized cells associated with the dendrites) that are sensitive to specific environmental stimuli. They react to such stimuli by initiating a nerve impulse. Several histologically distinct types of general sensory receptors have been identified in the skin. Their structures are depicted in Figure 23.1.

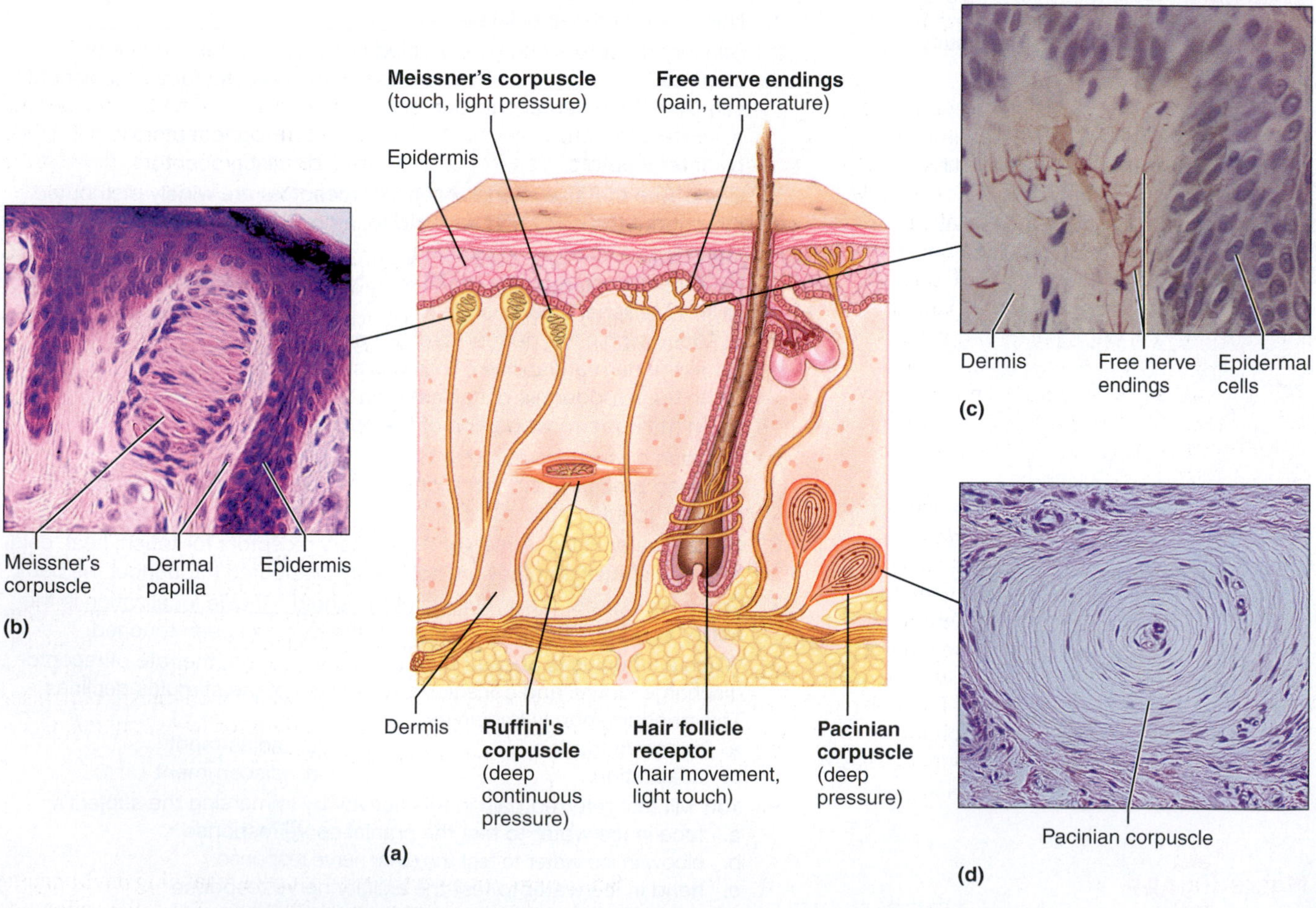
Meissner's corpuscle
(touch, light pressure)
Free nerve endings
(pain, temperature)
Epidermis
Dermis
Ruffini's corpuscle
(deep continuous pressure)
Hair follicle receptor
(hair movement, light touch)
Pacinian corpuscle
(deep pressure)
(a)
Meissner's corpuscle
Dermal papilla
Epidermis
(b)
Dermis
Free nerve endings
Epidermal cells
(c)
Pacinian corpuscle
(d)

FIGURE 23.1 Examples of cutaneous receptors. Drawing **(a)** and photomicrographs **(b–d)**. **(a)** Free nerve endings, hair follicle receptor, Meissner's corpuscles, Pacinian corpuscles, and Ruffini's corpuscle. Tactile (Merkel) discs are not illustrated. **(b)** Meissner's corpuscle in a dermal papilla (145×). **(c)** Free nerve endings at dermal-epidermal junction (185×). **(d)** Cross section of a Pacinian corpuscle in the dermis (100×).

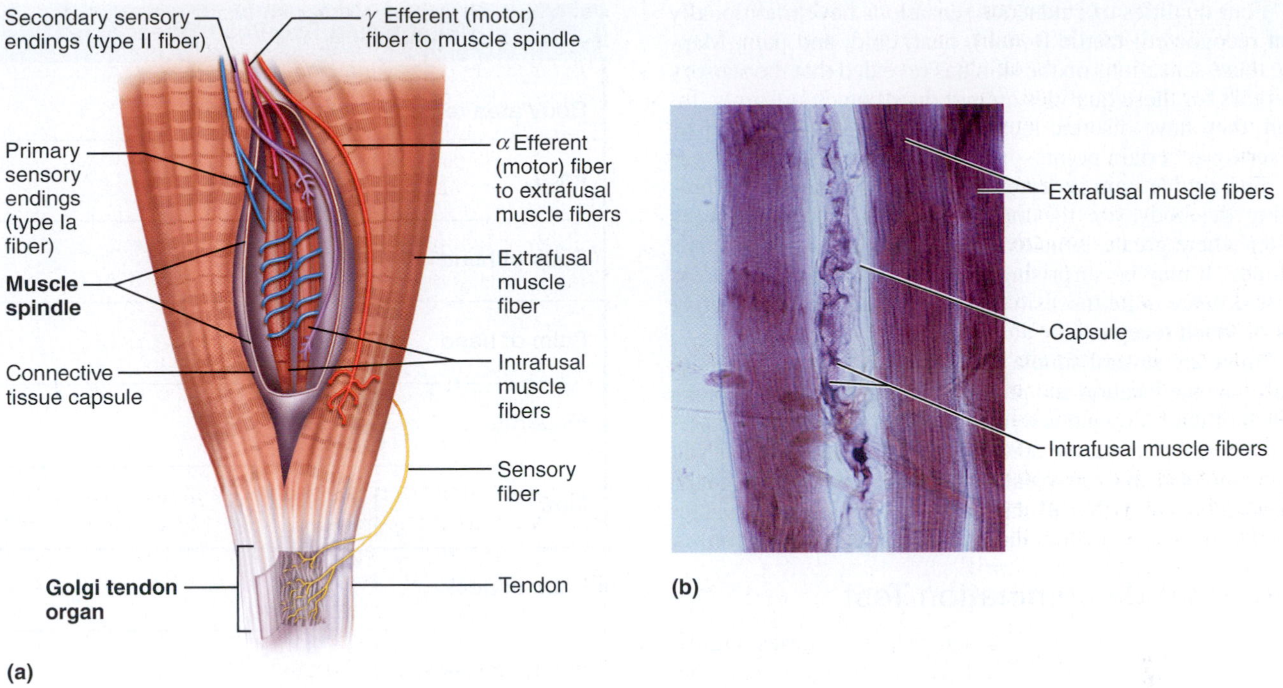

Secondary sensory endings (type II fiber)
γ Efferent (motor) fiber to muscle spindle
Primary sensory endings (type Ia fiber)
α Efferent (motor) fiber to extrafusal muscle fibers
Extrafusal muscle fiber
Muscle spindle
Connective tissue capsule
Intrafusal muscle fibers
Sensory fiber
Golgi tendon organ
Tendon
(a)
Extrafusal muscle fibers
Capsule
Intrafusal muscle fibers
(b)

FIGURE 23.2 Proprioceptors. (a) Diagrammatic view of a muscle spindle and Golgi tendon organ. Myelin has been omitted from all nerve fibers for clarity. **(b)** Photomicrograph of a muscle spindle (100×).

Many references link each type of receptor to specific stimuli, but there is still considerable controversy about the precise qualitative function of each receptor. Responses of all these receptors probably overlap considerably. Certainly, intense stimulation of any of them is always interpreted as pain.

The least specialized of the cutaneous receptors are the **free,** or **naked, nerve endings** of sensory neurons (Figure 23.1c), which respond chiefly to pain and temperature. The pain receptors are widespread in the skin and make up a sizable portion of the visceral interoceptors. Certain free nerve endings associate with specific epidermal cells to form **tactile (Merkel) discs,** or entwine in hair follicles to form **hair follicle receptors.** Both tactile discs and hair follicle receptors function as light touch receptors.

The other cutaneous receptors are a bit more complex, and the nerve endings are *encapsulated* by connective tissue cells. **Meissner's corpuscles,** also called **tactile corpuscles** because they respond to light touch, are located in the dermal papillae of hairless (glabrous) skin only (Figure 23.1b). **Ruffini's corpuscles** appear to respond to deep pressure and stretch stimuli. **Pacinian corpuscles,** also called **lamellated corpusles,** are anatomically more distinctive and lie deepest in the dermis (Figure 23.1c). Pacinian corpuscles respond only when deep pressure is first applied. They are best suited to monitor high-frequency vibrations.

ACTIVITY 1

Studying the Structure of Selected Sensory Receptors

1. Obtain a compound microscope and histologic slides of Pacinian and Meissner's corpuscles. Locate, under low power, a Meissner's corpuscle in the dermal layer of the skin. As mentioned above, these are usually found in the dermal papillae. Then switch to the oil immersion lens for a detailed study. Notice that the naked nerve fibers within the capsule are aligned parallel to the skin surface. Compare your observations to Figure 23.1b.

2. Next observe a Pacinian corpuscle located much deeper in the dermis. Try to identify the slender naked nerve ending in the center of the receptor and the heavy capsule of connective tissue surrounding it (which looks rather like an onion cut lengthwise). Also, notice how much larger the Pacinian corpuscles are than the Meissner's corpuscles. Compare your observations to Figure 23.1d.

3. Obtain slides of muscle spindles and Golgi tendon organs, the two major types of proprioceptors (Figure 23.2). In the slide of **muscle spindles,** note that minute extensions of the nerve endings of the sensory neurons coil around specialized slender skeletal muscle cells called **intrafusal cells,** or **fibers.** The **Golgi tendon organs** are composed of nerve endings that ramify through the tendon tissue close to the attachment between muscle and tendon. Stretching of muscles or tendons excites these receptors, which then transmit impulses that ultimately reach the cerebellum for interpretation. Compare your observations to Figure 23.2. ■

Receptor Physiology

Sensory receptors act as **transducers,** changing environmental stimuli into afferent nerve impulses. Since the action potential generated in all nerve fibers is essentially identical, the stimulus is identified entirely by the area of the brain's sensory cortex that is stimulated (which, of course, differs for the various afferent nerves).

Four qualities of cutaneous sensations have traditionally been recognized: tactile (touch), heat, cold, and pain. Mapping these sensations on the skin has revealed that the sensory receptors for these qualities are not distributed uniformly. Instead, they have discrete locations and are characterized by clustering at certain points—**punctate distribution.**

The simple pain receptors, extremely important in protecting the body, are the most numerous. Touch receptors cluster where greater sensitivity is desirable, as on the hands and face. It may be surprising to learn that rather large areas of the skin are quite insensitive to touch because of a relative lack of touch receptors.

There are several simple experiments you can conduct to investigate the location and physiology of cutaneous receptors. In each of the following activities, work in pairs with one person as the subject and the other as the experimenter. After you have completed an experiment, switch roles and go through the procedures again so that all class members obtain individual results. Keep an accurate account of each test that you perform.

Two-Point Discrimination Test

As noted, the density of the touch receptors varies significantly in different areas of the body. In general, areas that have the greatest density of tactile receptors have a heightened ability to "feel." These areas correspond to areas that receive the greatest motor innervation; thus they are also typically areas of fine motor control.

On the basis of this information, which areas of the body do you *predict* will have the greatest density of touch receptors?

____

ACTIVITY 2

Determining the Two-Point Threshold

1. Using calipers or an esthesiometer and a metric ruler, test the ability of the subject to differentiate two distinct sensations when the skin is touched simultaneously at two points. Beginning with the face, start with the caliper arms completely together. Gradually increase the distance between the arms, testing the subject's skin after each adjustment. Continue with this testing procedure until the subject reports that *two points* of contact can be felt. This measurement, the smallest distance at which two points of contact can be felt, is the **two-point threshold.**

2. Repeat this procedure on the back and palm of the hand, fingertips, lips, back of the neck, and ventral forearm. Record your results in the chart, Determining Two-Point Threshold.

3. Which area has the smallest two-point threshold?

____

Determining Two-Point Threshold

Body area tested	Two-point threshold (mm)
Face	
Back of hand	
Palm of hand	
Fingertip	
Lips	
Back of neck	
Ventral forearm	

Tactile Localization

Tactile localization is the ability to determine which portion of the skin has been touched. The tactile receptor field of the body periphery has a corresponding "touch" field in the brain's somatosensory cortex. Some body areas are well represented with touch receptors, allowing tactile stimuli to be localized with great accuracy, but touch-receptor density in other body areas allows only crude discrimination.

ACTIVITY 3

Testing Tactile Localization

1. The subject's eyes should be closed during the testing. The experimenter touches the palm of the subject's hand with a pointed black felt-tipped marker. The subject should then try to touch the exact point with his or her own marker, which should be of a different color. Measure the error of localization in millimeters.

2. Repeat the test in the same spot twice more, recording the error of localization for each test. Average the results of the three determinations, and record it in the chart Testing Tactile Location.

Testing Tactile Localization

Body area tested	Average error (mm)
Palm of hand	
Fingertip	
Ventral forearm	
Back of hand	
Back of neck	

Does the ability to localize the stimulus improve the second time? __________ The third time? __________

Explain. ______________________________

____

3. Repeat the preceding procedure on a fingertip, the ventral forearm, the back of a hand, and the back of the neck. Record the averaged results in the chart on the previous page.

4. Which area has the smallest error of localization?

____

Density and Location of Touch and Temperature Receptors

ACTIVITY 4

Plotting the Relative Density and Location of Touch and Temperature Receptors

1. Obtain two blunt probes, metric ruler, Von Frey's hair (or a sharp pencil), a towel, and black, red, and blue felt-tipped markers and bring them to your laboratory bench.

2. Place one blunt probe in a beaker of ice water and the other in a water bath controlled at 45°C. With a felt marker, draw a square (2 cm on each side) on the ventral surface of the subject's forearm. The square should not contain dots made previously. During the following tests, the subject's eyes are to remain closed. The subject should tell the examiner when a stimulus is detected.

3. Working in a systematic manner from one side of the marked square to the other, gently touch the Von Frey's hair or a sharp pencil tip to different points within the square. The *hairs* should be applied with a pressure that just causes them to bend; use the same pressure for each contact. Do not apply deep pressure. The goal is to stimulate only the more superficially located Meissner's corpuscles (as opposed to the Pacinian corpuscles located in the subcutaneous tissue). Mark with a *black dot* all points at which the touch is perceived.

4. Go to the area where the ice bath and 45°C water bath are set up. Remove the blunt probe from the ice water and, using a towel, quickly wipe it dry. Repeat the procedure outlined above, noting all points of cold perception with a *blue dot*.* Perception should be for temperature, not simply touch.

5. Remove the second blunt probe from the 45°C water bath and repeat the procedure once again, marking all points of heat perception with a *red dot*.*

6. After each student has acted as the subject, draw a "map" of your own and your partner's receptor areas in the squares provided here. Use the same color codes as on the skin.

*The blunt probes will have to be returned to the water baths approximately every 2 minutes to maintain the desired testing temperatures.

Student 1: __________ Student 2: __________

2 cm 2 cm

How does the density of the heat receptors correspond to that of the touch receptors?

____

To that of the cold receptors? __________

On the basis of your observations, which of these three types of receptors appears to be most abundant (at least in the area tested)?

____

Adaptation of Sensory Receptors

The number of impulses transmitted by sensory receptors often changes both with the intensity of the stimulus and with the length of time the stimulus is applied. In many cases, when a stimulus is applied for a prolonged period, the rate of receptor discharge slows and conscious awareness of the stimulus declines or is lost until some type of stimulus change occurs. This phenomenon is referred to as **adaptation.** The touch receptors adapt particularly rapidly, which is highly desirable. Who, for instance, would want to be continually aware of the pressure of clothing on their skin? The simple experiments to be conducted next allow you to investigate the phenomenon of adaptation.

ACTIVITY 5

Demonstrating Adaptation of Touch Receptors

1. The subject's eyes should be closed. Obtain four coins. Place one coin on the anterior surface of the subject's forearm, and determine how long the sensation persists for the subject. Duration of the sensation:

__________ sec

2. Repeat the test, placing the coin at a different forearm location. How long does the sensation persist at the second location?

__________ sec

3. After awareness of the sensation has been lost at the second site, stack three more coins atop the first one.

Does the pressure sensation return? __________

If so, for how long is the subject aware of the pressure in this instance?

______________ sec

Are the same receptors being stimulated when the four coins,

rather than the one coin, are used? _________

Explain. ____

____

4. To further illustrate the adaptation of touch receptors—in this case, the hair follicle receptors—gently and slowly bend one hair shaft with a pen or pencil until it springs back (away from the pencil) to its original position. Is the tactile sensation greater when the hair is being slowly bent or when it springs back?

____

Why is the adaptation of the touch receptors in the hair follicles particularly important to a woman who wears her hair in a ponytail? If the answer is not immediately apparent, consider the opposite phenomenon: what would happen, in terms of sensory input from her hair follicles, if these receptors did not exhibit adaptation?

____

____

ACTIVITY 6

Demonstrating Adaptation of Temperature Receptors

Adaptation of the temperature receptors can be tested using some very unsophisticated methods.

1. Obtain three large finger bowls or 1000-ml beakers and fill the first with 45°C water. Have the subject immerse her or his left hand in the water and report the sensation. Keep the left hand immersed for 1 minute and then also immerse the right hand in the same bowl.

What is the sensation of the left hand when it is first immersed?

____

What is the sensation of the left hand after 1 minute as compared to the sensation in the right hand just immersed?

____

Had adaptation occurred in the left hand? ______________

2. Rinse both hands in tap water, dry them, and wait 5 minutes before conducting the next test. Just before beginning the test, refill the finger bowl with fresh 45°C water, fill a second with ice water, and fill a third with water at room temperature.

3. Place the *left* hand in the ice water and the *right* hand in the 45°C water. What is the sensation in each hand after 2 minutes as compared to the sensation perceived when the hands were first immersed?

____

____

Which hand seemed to adapt more quickly?

____

4. After reporting these observations, the subject should then place both hands simultaneously into the finger bowl containing the water at room temperature. Record the sensation in the left hand: ____________________________

The right hand: ____________________________

The sensations that the subject experiences when both hands were put into room-temperature water are called **negative afterimages.** They are explained by the fact that sensations of heat and cold depend on the rapidity of heat loss or gain by the skin and differences in the temperature gradient.

Referred Pain

Experiments on pain receptor localization and adaptation are commonly conducted in the laboratory. However, there are certain problems with such experiments. Pain receptors are densely distributed in the skin, and they adapt very little, if at all. (This lack of adaptability is due to the protective function of the receptors. The sensation of pain often indicates tissue damage or trauma to body structures.) Thus no attempt will be made in this exercise to localize the pain receptors or to prove their nonadaptability, since both would cause needless discomfort to those of you acting as subjects and would not add any additional insight.

However, the phenomenon of referred pain is easily demonstrated in the laboratory, and such experiments provide information that may be useful in explaining common examples of this phenomenon. **Referred pain** is a sensory experience in which pain is perceived as arising in one area of the body when in fact another, often quite remote area is receiving the painful stimulus. Thus the pain is said to be "referred" to a different area. The phenomenon of **projection,** the process by which the brain refers sensations to their *usual* point of stimulation, provides the most simple explanation of such experiences. Many of us have experienced referred pain as a radiating pain in the forehead after quickly swallowing an ice-cold drink. Referred pain is important in many types of clinical diagnosis because damage to many visceral organs results in this phenomenon. For example, inadequate oxygenation of the heart muscle often results in pain being referred to the chest wall and left shoulder *(angina pectoris)*, and the reflux of gastric juice into the esophagus causes a sensation of intense discomfort in the thorax referred to as *heartburn.*

ACTIVITY 7

Demonstrating the Phenomenon of Referred Pain

Immerse the subject's elbow in a finger bowl containing ice water. In the chart below, record the quality (such as discomfort, tingling, or pain) and the quality progression of the sensations he or she reports for the intervals indicated. The elbow should be removed from ice water after the 2-minute reading. The last recording is to occur 3 minutes after removal of the subject's elbow from the ice water.

Also record the location of the perceived sensations. The ulnar nerve, which serves the medial third of the hand, is involved in the phenomenon of referred pain experienced during this test. How does the localization of this referred pain correspond to the areas served by the ulnar nerve?

____

Demonstrating Referred Pain

Time of observation	Quality of sensation	Localization of sensation
On immersion		
After 1 min		
After 2 min		
3 min after removal		

NAME ___________________________

LAB TIME/DATE ___________________

REVIEW SHEET

EXERCISE 23

General Sensation

Structure of General Sensory Receptors

1. Differentiate between interoceptors and exteroceptors relative to location and stimulus source.

interoceptor: ___________________________

exteroceptor: ___________________________

2. A number of activities and sensations are listed in the chart below. For each, check whether the receptors would be exteroceptors or interoceptors; and then name the specific receptor types. (Because visceral receptors were not described in detail in this exercise, you need only indicate that the receptor is a visceral receptor if it falls into that category.)

Activity or sensation	Exteroceptor	Interoceptor	Specific receptor type
Backing into a sun-heated iron railing			
Someone steps on your foot			
Reading a book			
Leaning on your elbows			
Doing sit-ups			
The "too full" sensation			
Seasickness			

Receptor Physiology

3. Explain how the sensory receptors act as transducers. ___________________________

____

4. Define *stimulus*. ___________________________

5. What was demonstrated by the two-point discrimination test? ___________________________

____

How well did your results correspond to your predictions? ___________________________

____

What is the relationship between the accuracy of the subject's tactile localization and the results of the two-point discrimination test? ___________________________

6. Define *punctate distribution.* ______________________________

____

7. Several questions regarding general sensation are posed below. Answer each by placing your response in the appropriately numbered blanks to the right.

1. Which cutaneous receptors are the most numerous?

2–3. Which two body areas tested were most sensitive to touch?

4–5. Which two body areas tested were least sensitive to touch?

6. Which appear to be more numerous—receptors that respond to cold or to heat?

7–9. Where would referred pain appear if the following organs were receiving painful stimuli—(7) gallbladder, (8) kidneys, and (9) appendix? (Use your textbook if necessary.)

10. Where was referred pain felt when the elbow was immersed in ice water during the laboratory experiment?

11. What region of the cerebrum interprets the kind and intensity of stimuli that cause cutaneous sensations?

1. ______________________________
2. ______________________________
3. ______________________________
4. ______________________________
5. ______________________________
6. ______________________________
7. ______________________________
8. ______________________________
9. ______________________________
10. ______________________________
11. ______________________________

8. Define *adaptation of sensory receptors.* ______________________________

____

9. Why is it advantageous to have pain receptors that are sensitive to all vigorous stimuli, whether heat, cold, or pressure?

____

Why is the nonadaptability of pain receptors important? ______________________________

____

10. Imagine yourself without any cutaneous sense organs. Why might this be very dangerous? ______________________________

____

____

11. Define *referred pain.* ______________________________

____

What is the probable explanation for referred pain? (Consult your textbook or an appropriate reference if necessary.)

____

____

EXERCISE 24

Special Senses: Vision

MATERIALS

- ☐ Chart of eye anatomy
- ☐ Dissectible eye model
- ☐ Prepared slide of longitudinal section of an eye showing retinal layers
- ☐ Compound microscope
- ☐ Preserved cow or sheep eye
- ☐ Dissecting instruments and tray
- ☐ Disposable gloves
- ☐ Metric ruler; meter stick
- ☐ Common (straight) pins
- ☐ Snellen eye chart (floor marked with chalk to indicate 20-ft. distance from posted Snellen chart)
- ☐ Ishihara's color plates
- ☐ Two pencils
- ☐ Test tubes
- ☐ Laboratory lamp or penlight
- ☐ Ophthalmoscope (if available)

MasteringA&P™

Access practice quizzes and more in the Study Area at www.masteringaandp.com.

PAL

For access to anatomical models and more, check out Practice Anatomy Lab.

OBJECTIVES

1. To describe the structure and function of the accessory visual structures.
2. To identify the structural components of the eye when provided with a model, an appropriate diagram, or a preserved sheep or cow eye, and to list the function(s) of each.
3. To describe the cellular makeup of the retina.
4. To discuss the mechanism of image formation on the retina.
5. To trace the visual pathway to the visual cortex, and to indicate the effects of damage to various parts of this pathway.
6. To define the following terms:

accommodation	convergence	hyperopia
astigmatism	emmetropia	myopia
cataract	glaucoma	refraction
conjunctivitis		

7. To discuss the importance of the pupillary and convergence reflexes.
8. To explain the difference between rods and cones with respect to visual perception and retinal localization.
9. To state the importance of an ophthalmoscopic examination.

PRE-LAB QUIZ

1. The anterior surface of each eye is protected by eyelids or __________.
   a. ducts
   b. lacrimal glands
   c. palpebrae
2. Name the mucous membrane that lines the internal surface of the eyelids and continues over the anterior surface of the eyeball.

   ____
3. How many extrinsic eye muscles are attached to the exterior surface of each eyeball?
   a. three
   b. four
   c. five
   d. six
4. The wall of the eye has three layers. The outermost fibrous layer is composed of the opaque, white sclera and the transparent __________.
   a. choroid
   b. ciliary gland
   c. cornea
   d. lacrima
5. Circle the correct term. Photoreceptor cells are distributed over the entire neural retina, except where the optic nerve leaves the eyeball. This site is called the macula lutea / optic disc.
6. Circle the correct term. The aqueous humor / vitreous humor is a clear, watery fluid that helps to maintain the intraocular pressure of the eye and provides nutrients for the avascular lens and cornea.
7. Photoreceptors of the eye include rods and cones. Which one is responsible for interpreting color, but can function only under conditions of high light intensity? ____________________

Text continues on next page.

8. Photoreceptors stimulate __________ neurons, which in turn stimulate ganglion cells.
   a. unipolar b. bipolar
   c. multipolar
9. Circle True or False. At the optic chiasma, the fibers from the medial side of each eye cross over to the opposite side.
10. The ability of the eye to focus differentially for objects of near vision is called
    a. accommodation
    b. adaptation
    c. astigmatism
    d. refraction

Anatomy of the Eye

External Anatomy and Accessory Structures

The adult human eye is a sphere measuring about 2.5 cm (1 inch) in diameter. Only about one-sixth of the eye's anterior surface is observable (Figure 24.1); the remainder is enclosed and protected by a cushion of fat and the walls of the bony orbit.

The **lacrimal apparatus** consists of the lacrimal gland, lacrimal canaliculi, lacrimal sac, and the nasolacrimal duct. The **lacrimal glands** are situated superior to the lateral aspect of each eye. They continually liberate a dilute salt solution (tears) that flows onto the anterior surface of the eyeball through several small ducts. The tears flush across the eyeball and through the **lacrimal puncta,** the tiny openings of the **lacrimal canaliculi** medially, then into the **lacrimal sac,** and finally into the **nasolacrimal duct,** which empties into the nasal cavity. The lacrimal secretion also contains **lysozyme,** an antibacterial enzyme. Because it constantly flushes the eyeball, the lacrimal fluid cleanses and protects the eye surface as it moistens and lubricates it. As we age, our eyes tend to become dry due to decreased lacrimation, and thus are more vulnerable to bacterial invasion and irritation.

The anterior surface of each eye is protected by the **eyelids,** or **palpebrae.** (See Figure 24.1.) The medial and lateral junctions of the upper and lower eyelids are referred to as the **medial** and **lateral commissures** *(canthi),* respectively. The **lacrimal caruncle,** a fleshy elevation at the medial commissure, produces a whitish oily secretion. A mucous membrane, the **conjunctiva,** lines the internal surface of the eyelids (as the *palpebral conjunctiva*) and continues over the anterior surface of the eyeball to its junction with the corneal

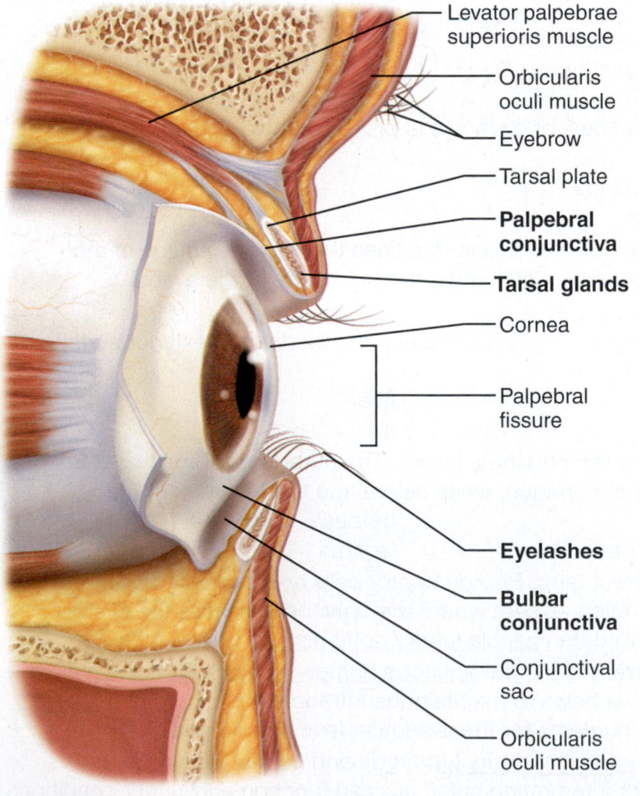

Levator palpebrae superioris muscle
Orbicularis oculi muscle
Eyebrow
Tarsal plate
Palpebral conjunctiva
Tarsal glands
Cornea
Palpebral fissure
Eyelashes
Bulbar conjunctiva
Conjunctival sac
Orbicularis oculi muscle
(a)

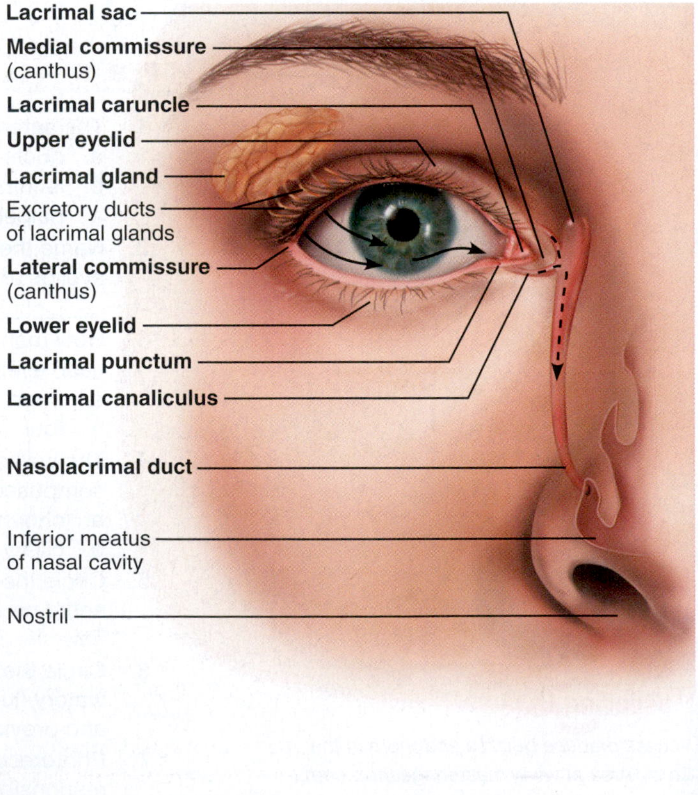

Lacrimal sac
Medial commissure (canthus)
Lacrimal caruncle
Upper eyelid
Lacrimal gland
Excretory ducts of lacrimal glands
Lateral commissure (canthus)
Lower eyelid
Lacrimal punctum
Lacrimal canaliculus
Nasolacrimal duct
Inferior meatus of nasal cavity
Nostril
(b)

FIGURE 24.1 External anatomy of the eye and accessory structures. (a) Lateral view; some structures shown in sagittal section. **(b)** Anterior view with lacrimal apparatus.

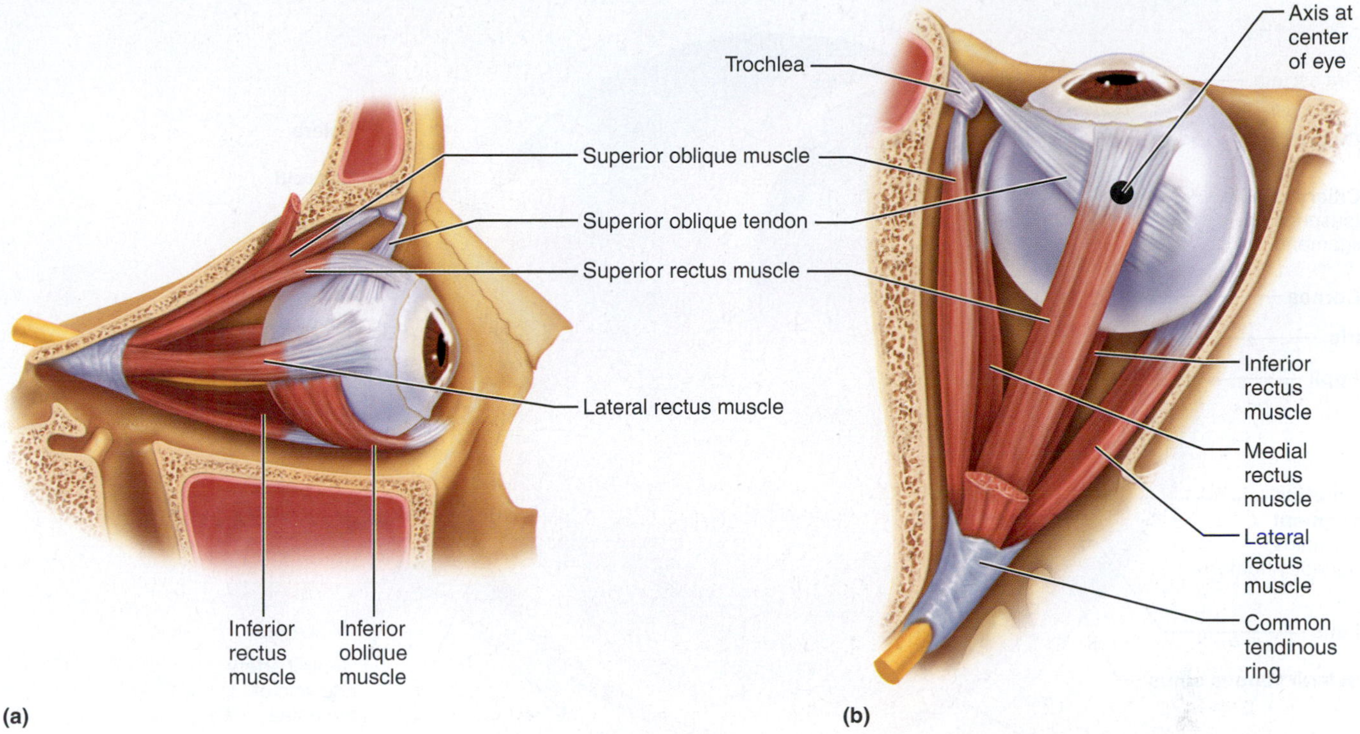
Trochlea
Superior oblique muscle
Superior oblique tendon
Superior rectus muscle
Lateral rectus muscle
Inferior rectus muscle
Inferior oblique muscle
(a)
Axis at center of eye
Inferior rectus muscle
Medial rectus muscle
Lateral rectus muscle
Common tendinous ring
(b)

Muscle	Action	Controlling cranial nerve
Lateral rectus	Moves eye laterally	VI (abducens)
Medial rectus	Moves eye medially	III (oculomotor)
Superior rectus	Elevates eye and turns it medially	III (oculomotor)
Inferior rectus	Depresses eye and turns it medially	III (oculomotor)
Inferior oblique	Elevates eye and turns it laterally	III (oculomotor)
Superior oblique	Depresses eye and turns it laterally	IV (trochlear)

(c)

FIGURE 24.2 Extrinsic muscles of the eye. (a) Lateral view of the right eye. **(b)** Superior view of the right eye. **(c)** Summary of actions of the extrinsic eye muscles and cranial nerves that control them.

epithelium (as the *bulbar conjunctiva*). The conjunctiva secretes mucus, which aids in lubricating the eyeball. Inflammation of the conjunctiva, often accompanied by redness of the eye, is called **conjunctivitis.**

Projecting from the border of each eyelid is a row of short hairs, the **eyelashes.** The **ciliary glands,** modified sweat glands, lie between the eyelash hair follicles and help lubricate the eyeball. Small sebaceous glands associated with the hair follicles and the larger **tarsal** *(meibomian)* **glands,** located posterior to the eyelashes, secrete an oily substance. An inflammation of one of the ciliary glands or a small oil gland is called a **sty.**

Six **extrinsic eye muscles** attached to the exterior surface of each eyeball control eye movement and make it possible for the eye to follow a moving object. The names and positioning of these extrinsic muscles are noted in Figure 24.2. Their actions are given in the chart accompanying that figure.

ACTIVITY 1

Identifying Accessory Eye Structures

Using Figure 24.1 or a chart of eye anatomy, observe the eyes of another student, and identify as many of the accessory structures as possible. Ask the student to look to the left. What extrinsic eye muscles are responsible for this action?

Right eye: ______________________________

Left eye: ______________________________

Internal Anatomy of the Eye

Anatomically, the wall of the eye is constructed of three layers (Figure 24.3). The outermost **fibrous layer** is a protective layer composed of dense avascular connective tissue. It has

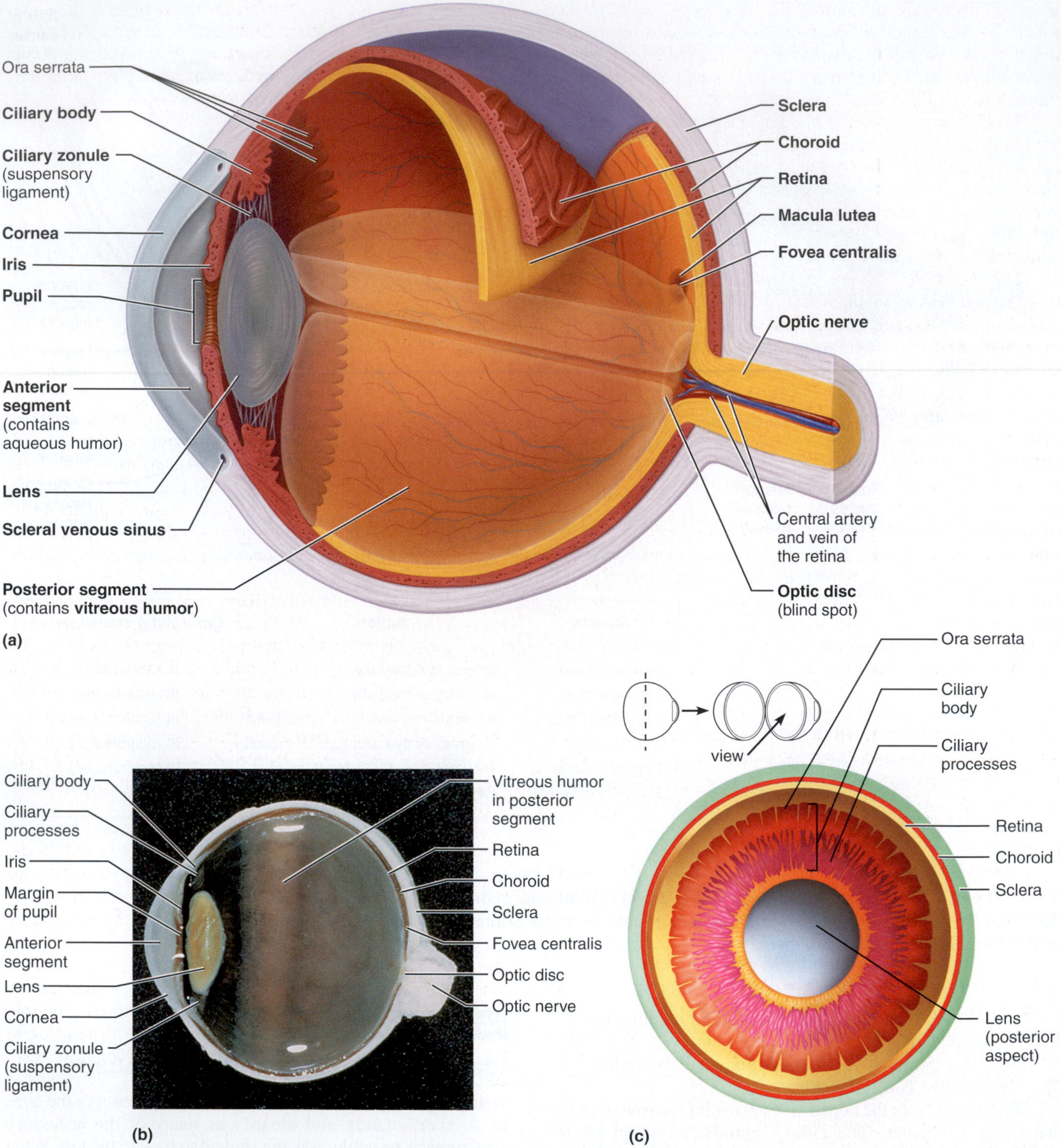

Ora serrata
Ciliary body
Ciliary zonule (suspensory ligament)
Cornea
Iris
Pupil
Anterior segment (contains aqueous humor)
Lens
Scleral venous sinus
Posterior segment (contains vitreous humor)
(a)
Sclera
Choroid
Retina
Macula lutea
Fovea centralis
Optic nerve
Central artery and vein of the retina
Optic disc (blind spot)
Ciliary body
Ciliary processes
Iris
Margin of pupil
Anterior segment
Lens
Cornea
Ciliary zonule (suspensory ligament)
Vitreous humor in posterior segment
Retina
Choroid
Sclera
Fovea centralis
Optic disc
Optic nerve
(b)
view
Ora serrata
Ciliary body
Ciliary processes
Retina
Choroid
Sclera
Lens (posterior aspect)
(c)

FIGURE 24.3 Internal anatomy of the eye. (a) Diagrammatic view of sagittal section of the eye. The vitreous humor is illustrated only in the bottom half of the eyeball. **(b)** Photograph of the human eye. **(c)** Posterior view of anterior half of the eye.

two obviously different regions: The opaque white **sclera** forms the bulk of the fibrous layer and is observable anteriorly as the "white of the eye." Its anteriormost portion is modified structurally to form the transparent **cornea,** through which light enters the eye.

The middle layer, called the **uvea,** is the **vascular layer.** Its posteriormost part, the **choroid,** is a blood-rich nutritive region containing a dark pigment that prevents light scattering within the eye. Anteriorly, the choroid is modified to form the **ciliary body,** which is chiefly composed of *ciliary* *muscles,* important in controlling lens shape, and *ciliary processes.* The ciliary processes secrete aqueous humor. The most anterior part of the uvea is the pigmented **iris.** The iris is incomplete, resulting in a rounded opening, the **pupil,** through which light passes.

The iris is composed of circularly and radially arranged smooth muscle fibers and acts as a reflexively activated diaphragm to regulate the amount of light entering the eye. In close vision and bright light, the circular muscles of the iris contract, and the pupil constricts. In distant vision and in dim light, the radial fibers contract, enlarging (dilating) the pupil and allowing more light to enter the eye.

The innermost **sensory layer** of the eye is the delicate, two-layered **retina** (Figures 24.3 and 24.4 on pages 366 and 368). The outer **pigmented epithelium** abuts the choroid and extends anteriorly to cover the ciliary body and the posterior side of the iris. The transparent inner **neural layer** extends anteriorly only to the ciliary body. It contains the photoreceptors, **rods** and **cones,** which begin the chain of electrical events that ultimately result in the transduction of light energy into nerve impulses that are transmitted to the optic cortex of the brain. Vision is the result. The photoreceptor cells are distributed over the entire neural retina, except where the optic nerve leaves the eyeball. This site is called the **optic disc,** or *blind spot.* Lateral to each blind spot, and directly posterior to the lens, is an area called the **macula lutea** ("yellow spot"), an area of high cone density. In its center is the **fovea centralis,** a minute pit about 0.4 mm in diameter, which contains mostly cones and is the area of greatest visual acuity. Focusing for discriminative vision occurs in the fovea centralis.

Light entering the eye is focused on the retina by the **lens,** a flexible crystalline structure held vertically in the eye's interior by the **suspensory ligament,** more specifically called the **ciliary zonule,** attached to the ciliary body. Activity of the ciliary muscle, which accounts for the bulk of ciliary body tissue, changes lens thickness to allow light to be properly focused on the retina.

In the elderly the lens becomes increasingly hard and opaque. **Cataracts,** which often result from this process, cause vision to become hazy or entirely obstructed. ■

The lens divides the eye into two segments: the **anterior segment** anterior to the lens, which contains a clear watery fluid called the **aqueous humor,** and the **posterior segment** behind the lens, filled with a gel-like substance, the **vitreous humor,** or **vitreous body.** The anterior segment is further divided into **anterior** and **posterior chambers,** located before and after the iris, respectively. The aqueous humor is continually formed by the capillaries of the **ciliary processes** of the ciliary body. It helps to maintain the intraocular pressure of the eye and provides nutrients for the avascular lens and cornea. The aqueous humor is reabsorbed into the **scleral venous sinus** *(canal of Schlemm).* The vitreous humor provides the major internal reinforcement of the posterior part of the eyeball, and helps to keep the retina pressed firmly against the wall of the eyeball. It is formed *only* before birth.

Anything that interferes with drainage of the aqueous fluid increases intraocular pressure. When intraocular pressure reaches dangerously high levels, the retina and optic nerve are compressed, resulting in pain and possible blindness, a condition called **glaucoma.** ■

ACTIVITY 2

Identifying Internal Structures of the Eye

Obtain a dissectible eye model and identify its internal structures described above. As you work, also refer to Figure 24.3. ■

Microscopic Anatomy of the Retina

As described above, the retina consists of two main types of cells: a pigmented *epithelium,* which abuts the choroid, and an inner cell layer composed of *neurons,* which is in contact with the vitreous humor (see Figure 24.4). The inner neural layer is composed of three major populations of cells. These are, from outer to inner aspect, the **photoreceptors** (rods and cones), the **bipolar cells,** and the **ganglion cells.**

The **rods** are the specialized receptors for dim light. Visual interpretation of their activity is in gray tones. The **cones** are color receptors that permit high levels of visual acuity, but they function only under conditions of high light intensity; thus, for example, no color vision is possible in moonlight. Mostly cones are found in the fovea centralis, and their number decreases as the retinal periphery is approached. By contrast, rods are most numerous in the periphery, and their density decreases as the macula is approached.

Light must pass through the ganglion cell layer and the bipolar neuron layer to reach and excite the rods and cones. As a result of a light stimulus, the photoreceptors undergo changes in their membrane potential that influence the bipolar neurons. These in turn stimulate the ganglion cells, whose axons leave the retina in the tight bundle of fibers known as the **optic nerve** (Figure 24.3). The retinal layer is thickest where the optic nerve attaches to the eyeball because an increasing number of ganglion cell axons converge at this point. It thins as it approaches the ciliary body. In addition to these three major cell types, the retina also contains horizontal cells and amacrine cells, which play a role in visual processing.

ACTIVITY 3

Studying the Microscopic Anatomy of the Retina

Use a compound microscope to examine a histologic slide of a longitudinal section of the eye. Identify the retinal layers by comparing your view to Figure 24.4. ■

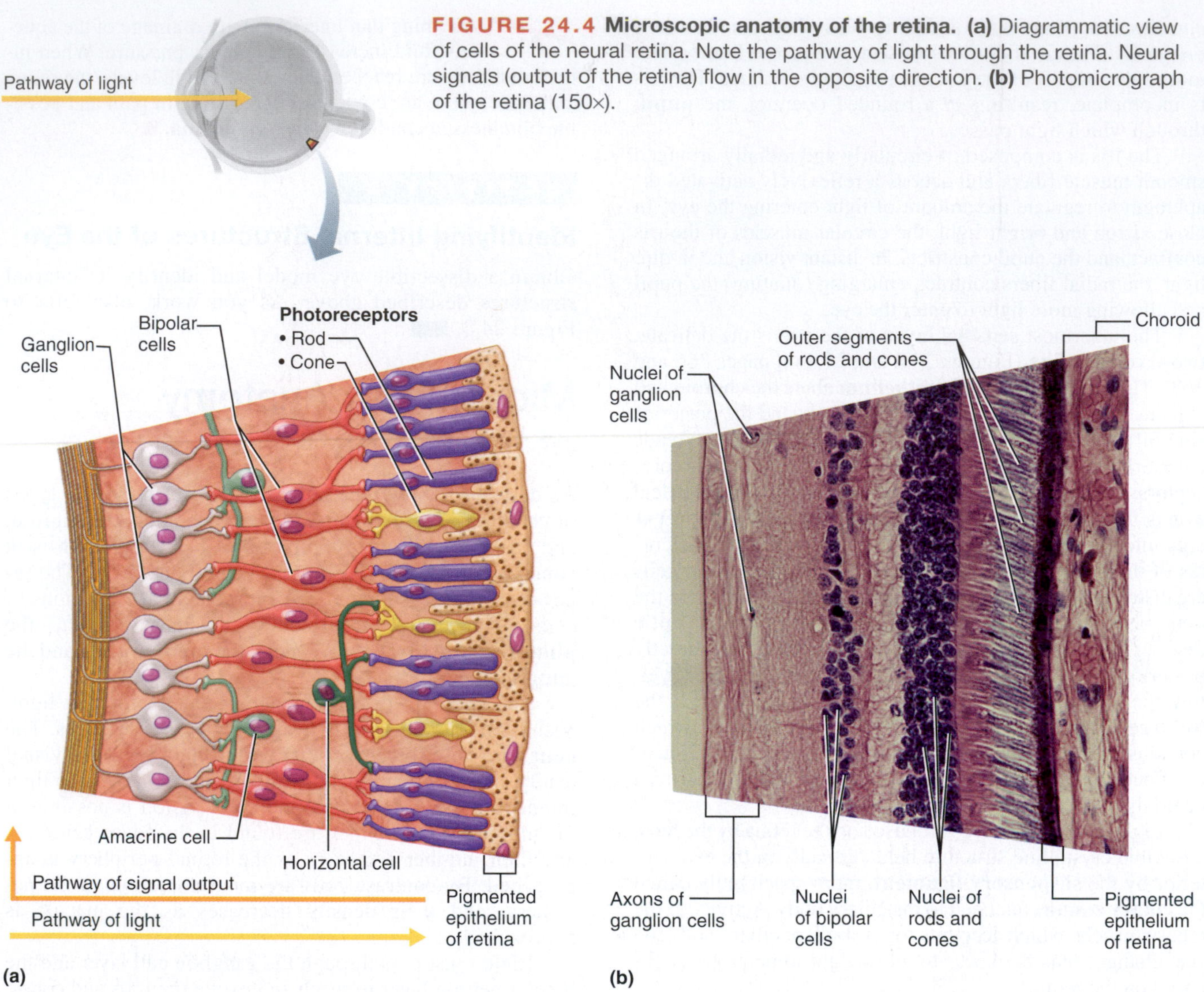

Pathway of light
Ganglion cells
Bipolar cells
Photoreceptors
• Rod
• Cone
Amacrine cell
Horizontal cell
Pathway of signal output
Pathway of light
Pigmented epithelium of retina
(a)
Choroid
Outer segments of rods and cones
Nuclei of ganglion cells
Axons of ganglion cells
Nuclei of bipolar cells
Nuclei of rods and cones
Pigmented epithelium of retina
(b)

FIGURE 24.4 Microscopic anatomy of the retina. (a) Diagrammatic view of cells of the neural retina. Note the pathway of light through the retina. Neural signals (output of the retina) flow in the opposite direction. **(b)** Photomicrograph of the retina (150×).

DISSECTION: The Cow (Sheep) Eye

1. Obtain a preserved cow or sheep eye, dissecting instruments, and a dissecting tray. Don disposable gloves.

2. Examine the external surface of the eye, noting the thick cushion of adipose tissue. Identify the optic nerve (cranial nerve II) as it leaves the eyeball, the remnants of the extrinsic eye muscles, the conjunctiva, the sclera, and the cornea. The normally transparent cornea is opalescent or opaque if the eye has been preserved. Refer to Figure 24.5 as you work.

3. Trim away most of the fat and connective tissue, but leave the optic nerve intact. Holding the eye with the cornea facing downward, carefully make an incision with a sharp scalpel into the sclera about 6 mm (¼ inch) above the cornea. (The sclera of the preserved eyeball is *very* tough, so you will have to apply substantial pressure to penetrate it.) Using scissors, complete the incision around the circumference of the eyeball paralleling the corneal edge.

4. Carefully lift the anterior part of the eyeball away from the posterior portion. Conditions being proper, the vitreous body should remain with the posterior part of the eyeball.

5. Examine the anterior part of the eye, and identify the following structures:

Ciliary body: Black pigmented body that appears to be a halo encircling the lens.

Lens: Biconvex structure that is opaque in preserved specimens.

Carefully remove the lens and identify the adjacent structures:

Iris: Anterior continuation of the ciliary body penetrated by the pupil.

Cornea: More convex anteriormost portion of the sclera; normally transparent but cloudy in preserved specimens.

6. Examine the posterior portion of the eyeball. Carefully remove the vitreous humor, and identify the following structures:

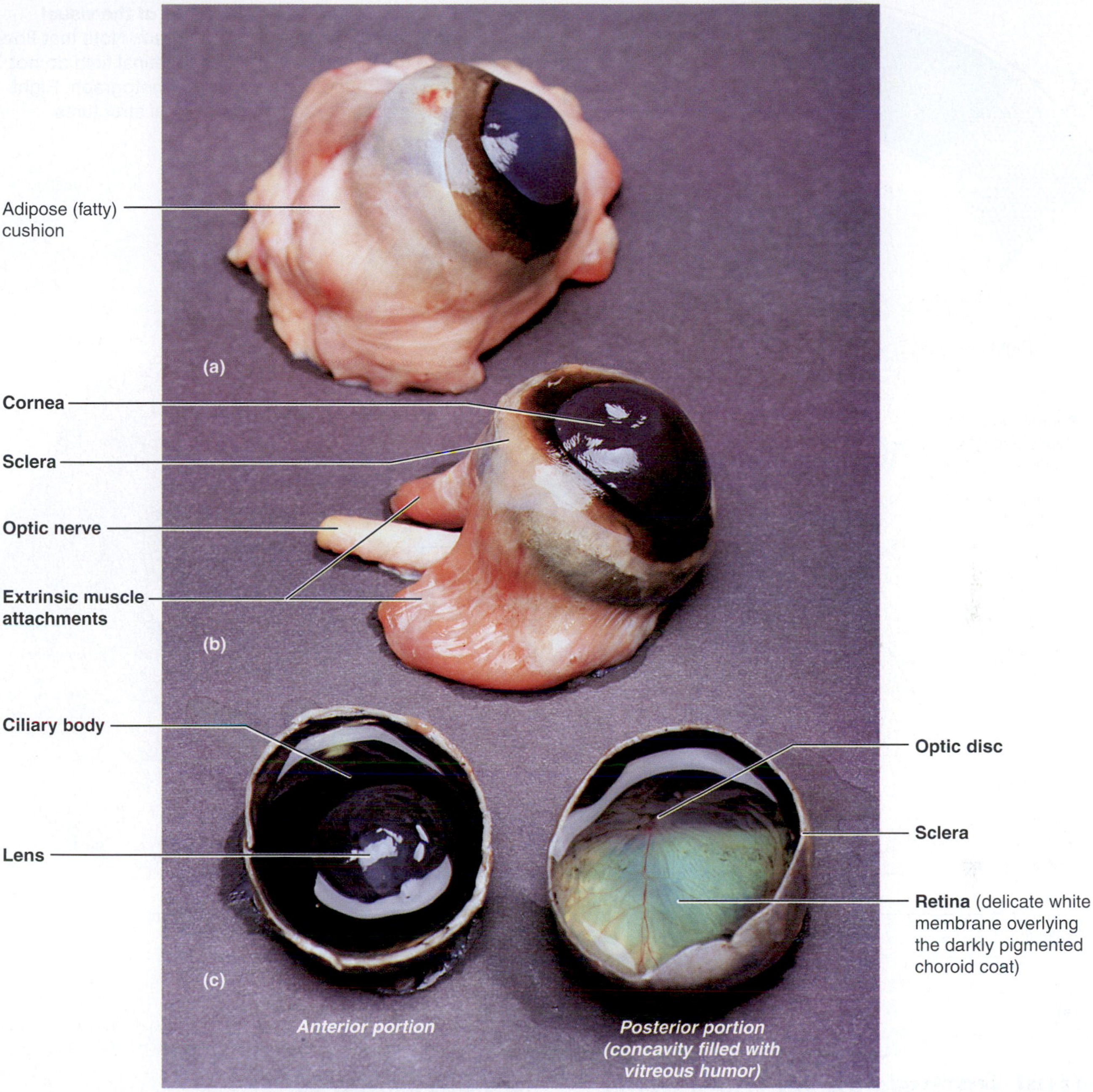

Adipose (fatty) cushion
(a)
Cornea
Sclera
Optic nerve
Extrinsic muscle attachments
(b)
Ciliary body
Lens
(c)
Anterior portion
Optic disc
Sclera
Retina (delicate white membrane overlying the darkly pigmented choroid coat)
Posterior portion (concavity filled with vitreous humor)

FIGURE 24.5 Anatomy of the cow eye. **(a)** Cow eye (entire) removed from orbit (notice the large amount of fat cushioning the eyeball). **(b)** Cow eye (entire) with fat removed to show the extrinsic muscle attachments and optic nerve. **(c)** Cow eye cut along the frontal plane to reveal internal structures.

Retina: The neural layer of the retina appears as a delicate tan, probably crumpled membrane that separates easily from the pigmented choroid.

Note its point of attachment. What is this point called?

____

Pigmented choroid coat: Appears iridescent in the cow or sheep eye owing to a special reflecting surface called the **tapetum lucidum.** This specialized surface reflects the light within the eye and is found in the eyes of animals that live under conditions of low-intensity light. It is not found in humans.

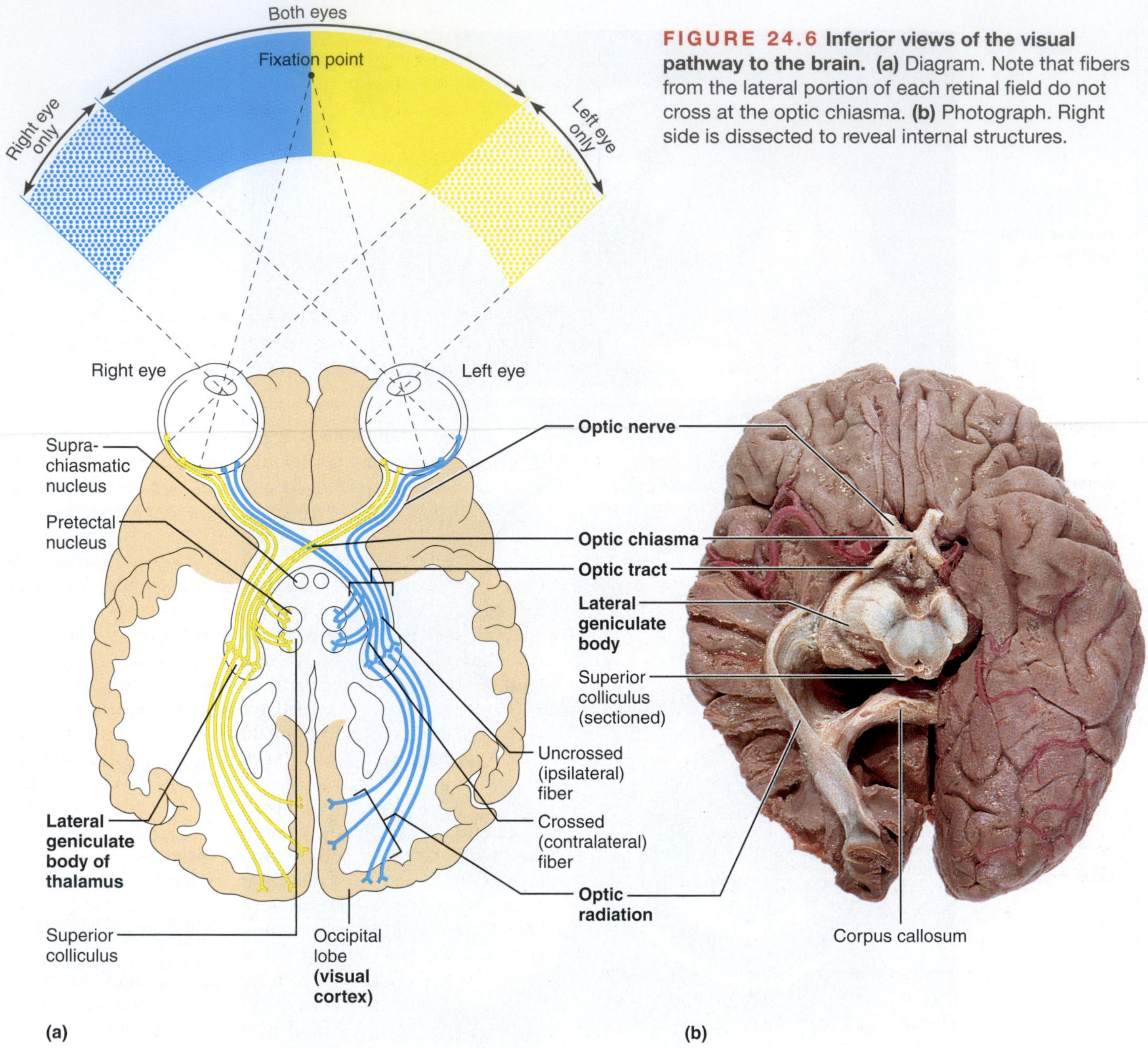
Both eyes
Fixation point
Right eye only
Left eye only
Right eye
Left eye
Supra-chiasmatic nucleus
Pretectal nucleus
Lateral geniculate body of thalamus
Superior colliculus
Occipital lobe (visual cortex)
Optic nerve
Optic chiasma
Optic tract
Lateral geniculate body
Superior colliculus (sectioned)
Uncrossed (ipsilateral) fiber
Crossed (contralateral) fiber
Optic radiation
Corpus callosum
(a)
(b)

FIGURE 24.6 Inferior views of the visual pathway to the brain. **(a)** Diagram. Note that fibers from the lateral portion of each retinal field do not cross at the optic chiasma. **(b)** Photograph. Right side is dissected to reveal internal structures.

Visual Pathways to the Brain

The axons of the ganglion cells of the retina converge at the posterior aspect of the eyeball and exit from the eye as the optic nerve. At the **optic chiasma,** the fibers from the medial side of each eye cross over to the opposite side (Figure 24.6). The fiber tracts thus formed are called the **optic tracts.** Each optic tract contains fibers from the lateral side of the eye on the same side and from the medial side of the opposite eye.

The optic tract fibers synapse with neurons in the **lateral geniculate body** of the thalamus, whose axons form the **optic radiation,** terminating in the **visual cortex** in the occipital lobe of the brain. Here they synapse with the cortical neurons, and visual interpretation occurs.

ACTIVITY 4

Predicting the Effects of Visual Pathway Lesions

After examining Figure 24.6, determine what effects lesions in the following areas would have on vision:

In the right optic nerve: ______________________________

____

Through the optic chiasma: ______________________________

____

In the left optic tract: ______________________________

____

X

FIGURE 24.7 Blind spot test figure.

In the right cerebral cortex (visual area): ______________

____

Visual Tests and Experiments

The Blind Spot

ACTIVITY 5

Demonstrating the Blind Spot

1. Hold Figure 24.7 about 46 cm (18 inches) from your eyes. Close your left eye, and focus your right eye on the X, which should be positioned so that it is directly in line with your right eye. Move the figure slowly toward your face- keeping your right eye focused on the X. When the dot focuses on the blind spot, which lacks photoreceptors, it will disappear.

2. Have your laboratory partner record in metric units the distance at which this occurs. The dot will reappear as the figure is moved closer. Distance at which the dot disappears:

Right eye ______________________________

Repeat the test for the left eye, this time closing the right eye and focusing the left eye on the dot. Record the distance at which the X disappears:

Left eye ______________________________

Refraction, Visual Acuity, and Astigmatism

When light rays pass from one medium to another, their velocity, or speed of transmission, changes, and the rays are bent, or **refracted.** Thus the light rays in the visual field are refracted as they encounter the cornea, lens, and vitreous humor of the eye.

The refractive index (bending power) of the cornea and vitreous humor are constant. But the lens's refractive index can be varied by changing the lens's shape—that is, by making it more or less convex so that the light is properly converged and focused on the retina. The greater the lens convexity, or bulge, the more the light will be bent and the stronger the lens. Conversely, the less the lens convexity (the flatter it is), the less it bends the light.

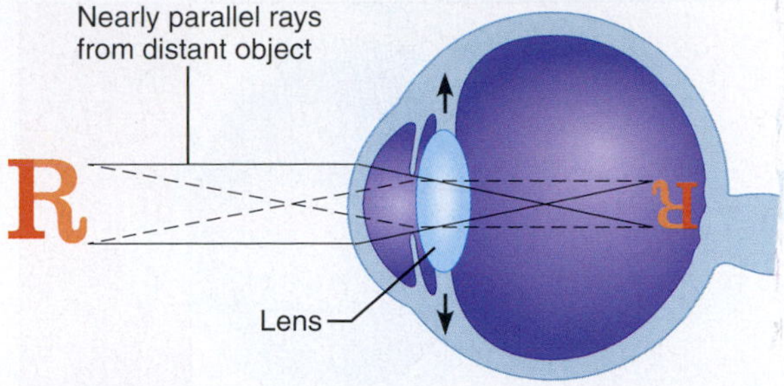
Nearly parallel rays from distant object
R
Lens

FIGURE 24.8 Refraction and real images. The refraction of light in the eye produces a real image (reversed, inverted, and reduced) on the retina.

In general, light from a distant source (over 6 m, or 20 feet) approaches the eye as parallel rays, and no change in lens convexity is necessary for it to focus properly on the retina. However, light from a close source tends to diverge, and the convexity of the lens must increase to make close vision possible. To achieve this, the ciliary muscle contracts, decreasing the tension on the suspensory ligament attached to the lens and allowing the elastic lens to "round up." Thus, a lens capable of bringing a *close* object into sharp focus is stronger (more convex) than a lens focusing on a more distant object. The ability of the eye to focus differentially for objects of near vision (less than 6 m, or 20 feet) is called **accommodation.** It should be noted that the image formed on the retina as a result of the refractory activity of the lens (Figure 24.8) is a **real image** (reversed from left to right, inverted, and smaller than the object).

The normal, or **emmetropic, eye** is able to accommodate properly (Figure 24.9a). However, visual problems may result (1) from lenses that are too strong or too "lazy" (overconverging and underconverging, respectively), (2) from structural problems such as an eyeball that is too long or too short to provide for proper focusing by the lens, or (3) from a cornea or lens with improper curvatures.

Individuals in whom the image normally focuses in front of the retina are said to have **myopia,** or nearsightedness (Figure 24.9b); they can see close objects without difficulty, but distant objects are blurred or seen indistinctly. Correction requires a concave lens, which causes the light reaching the eye to diverge.

If the image focuses behind the retina, the individual is said to have **hyperopia,** or farsightedness. Such persons have no problems with distant vision but need glasses with convex lenses to augment the converging power of the lens for close vision (Figure 24.9c).

Irregularities in the curvatures of the lens and/or the cornea lead to a blurred vision problem called **astigmatism.** Cylindrically ground lenses, which compensate for inequalities in the curvatures of the refracting surfaces, are prescribed to correct the condition. ■

Near-Point Accommodation

The elasticity of the lens decreases dramatically with age, resulting in difficulty in focusing for near or close vision. This condition is called **presbyopia**—literally, old vision. Lens elasticity can be tested by measuring the **near point of accommodation.** The near point of vision is about 10 cm from the eye in young adults. It is closer in children and farther in old age.

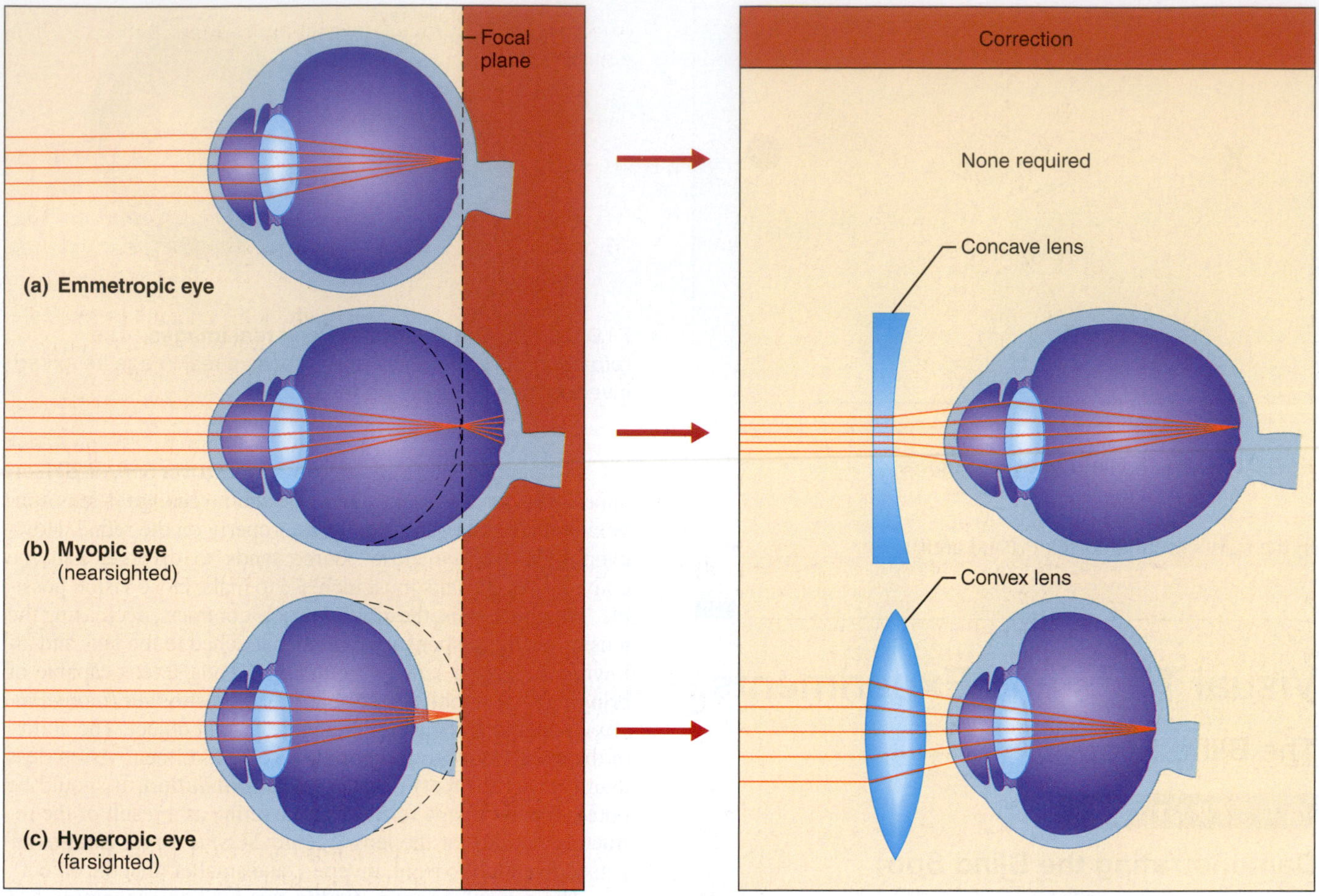
Focal plane
(a) Emmetropic eye
(b) Myopic eye (nearsighted)
(c) Hyperopic eye (farsighted)
Correction
None required
Concave lens
Convex lens

FIGURE 24.9 Problems of refraction. (a) In the emmetropic (normal) eye, light from both near and far objects is focused properly on the retina. **(b)** In a myopic eye, light from distant objects is brought to a focal point before reaching the retina. It then diverges. Applying a concave lens focuses objects properly on the retina. **(c)** In the hyperopic eye, light from a near object is brought to a focal point behind (past) the retina. Applying a convex lens focuses objects properly on the retina. The refractory effect of the cornea is ignored here.

ACTIVITY 6

Determining Near Point of Accommodation

To determine your near point of accommodation, hold a common straight pin at arm's length in front of one eye. (If desired, the text in the lab manual can be used rather than a pin.) Slowly move the pin toward that eye until the pin image becomes distorted. Have your lab partner use a metric ruler to measure the distance from your eye to the pin at this point, and record the distance below. Repeat the procedure for the other eye.

Near point for right eye: ______________________________

Near point for left eye: ______________________________

Visual Acuity

Visual acuity, or sharpness of vision, is generally tested with a Snellen eye chart, which consists of letters of various sizes printed on a white card. This test is based on the fact that letters of a certain size can be seen clearly by eyes with normal vision at a specific distance. The distance at which the normal, or emmetropic, eye can read a line of letters is printed at the end of that line.

ACTIVITY 7

Testing Visual Acuity

1. Have your partner stand 6 m (20 feet) from the posted Snellen eye chart and cover one eye with a card or hand. As your partner reads each consecutive line aloud, check for accuracy. If this individual wears glasses, give the test twice—first with glasses off and then with glasses on. *Do not remove contact lenses, but note that they were in place during the test.*

2. Record the number of the line with the smallest-sized letters read. If it is 20/20, the person's vision for that eye is normal. If it is 20/40, or any ratio with a value less than one, he or she has less than the normal visual acuity. (Such an individual is myopic.) If the visual acuity is 20/15, vision is better than normal, because this person can stand at 6 m (20 feet) from the

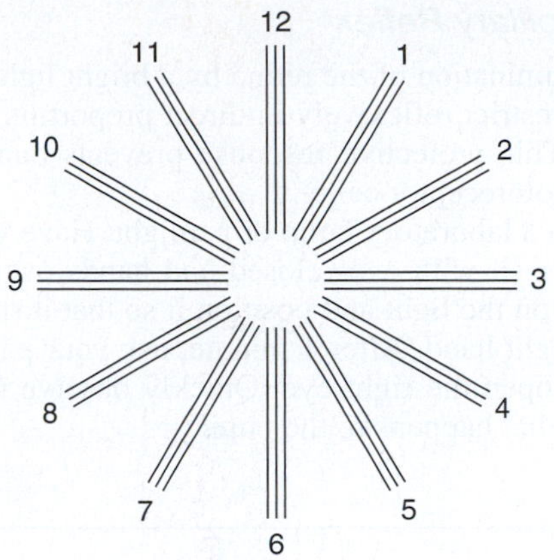
12
1
2
3
4
5
6
7
8
9
10
11

FIGURE 24.10 Astigmatism testing chart.

chart and read letters that are discernible by the normal eye only at 4.5 m (15 feet). Give your partner the number of the line corresponding to the smallest letters read, to record in step 4.

3. Repeat the process for the other eye.

4. Have your partner test and record your visual acuity. If you wear glasses, the test results *without* glasses should be recorded first.

Visual acuity, right eye without glasses: ____________

Visual acuity, right eye with glasses: ____________

Visual acuity, left eye without glasses: ____________

Visual acuity, left eye with glasses: ____________

ACTIVITY 8

Testing for Astigmatism

The astigmatism chart (Figure 24.10) is designed to test for defects in the refracting surface of the lens and/or cornea.

View the chart first with one eye and then with the other, focusing on the center of the chart. If all the radiating lines appear equally dark and distinct, there is no distortion of your refracting surfaces. If some of the lines are blurred or appear less dark than others, at least some degree of astigmatism is present.

Is astigmatism present in your left eye? ____________

Right eye? ____________

Color Blindness

Ishihara's color plates are designed to test for deficiencies in the cones or color photoreceptor cells. There are three cone types, each containing a different light-absorbing pigment. One type primarily absorbs the red wavelengths of the visible light spectrum, another the blue wavelengths, and a third the green wavelengths. Nerve impulses reaching the brain from these different photoreceptor types are then interpreted (seen) as red, blue, and green, respectively. Interpretation of the intermediate colors of the visible light spectrum is a result of overlapping input from more than one cone type.

ACTIVITY 9

Testing for Color Blindness

1. Find the interpretation table that accompanies the Ishihara color plates, and prepare a sheet to record data for the test. Note which plates are patterns rather than numbers.

2. View the color plates in bright light or sunlight while holding them about 0.8 m (30 inches) away and at right angles to your line of vision. Report to your laboratory partner what you see in each plate. Take no more than 3 seconds for each decision.

3. Your partner should record your responses and then check their accuracy with the correct answers provided in the color plate book. Is there any indication that you have some degree of color blindness? ____________ If so, what type?

____

____

Repeat the procedure to test your partner's color vision.

Binocular Vision

Humans, cats, predatory birds, and most primates are endowed with **binocular** (two-eyed) **vision.** Although both eyes look in approximately the same direction, they see slightly different views. Their visual fields, each about 170 degrees, overlap to a considerable extent; thus there is two-eyed vision at the overlap area (Figure 24.11).

In contrast, the eyes of many animals (rabbits, pigeons, and others) are more on the sides of their head. Such animals see in two different directions and thus have a panoramic field of view and **panoramic vision.** A mnemonic device to keep these straight is "Eyes in the front—likes to hunt. Eyes on the side—likes to hide."

Although both types of vision have their good points, binocular vision provides three-dimensional vision and an accurate means of locating objects in space. The slight differences

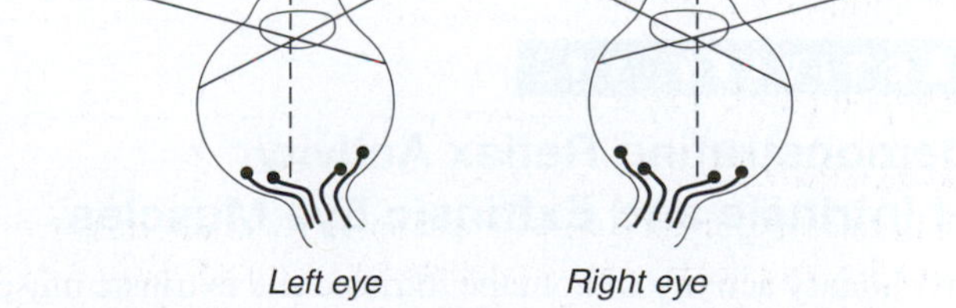
Area of overlap
Visual field of left eye
Visual field of right eye
Left eye
Right eye

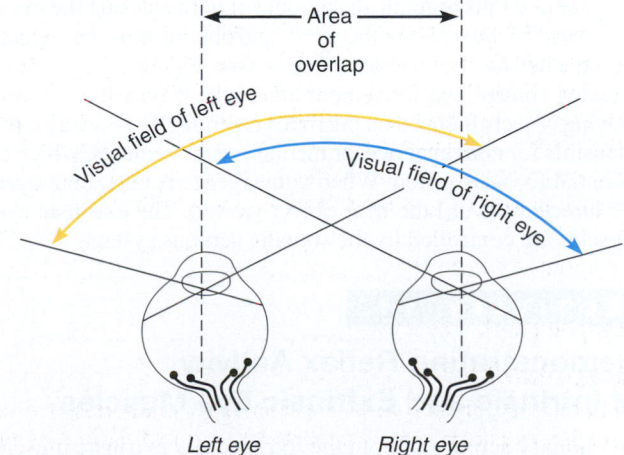

FIGURE 24.11 Overlapping of the visual fields.

between the views seen by the two eyes are fused by the higher centers of the visual cortex to give us *depth perception*. Because of the manner in which the visual cortex resolves these two different views into a single image, it is sometimes referred to as the "cyclopean eye of the binocular animal."

ACTIVITY 10

Testing for Binocular Vision

1. To demonstrate that a slightly different view is seen by each eye, perform the following simple experiment.

Close your left eye. Hold a pencil at arm's length directly in front of your right eye. Position another pencil directly beneath it and then move the lower pencil about half the distance toward you. As you move the lower pencil, make sure it remains in the *same plane* as the stationary pencil, so that the two pencils continually form a straight line. Then, without moving the pencils, close your right eye and open your left eye. Notice that with only the right eye open, the moving pencil stays in the same plane as the fixed pencil, but that when viewed with the left eye, the moving pencil is displaced laterally away from the plane of the fixed pencil.

2. To demonstrate the importance of two-eyed binocular vision for depth perception, perform this second simple experiment.

Have your laboratory partner hold a test tube erect about arm's length in front of you. With both eyes open, quickly insert a pencil into the test tube. Remove the pencil, bring it back close to your body, close one eye, and quickly and without hesitation insert the pencil into the test tube. (*Do not feel for the test tube with the pencil!*) Repeat with the other eye closed.

Was it as easy to dunk the pencil with one eye closed as with both eyes open?

____

Eye Reflexes

Both intrinsic (internal) and extrinsic (external) muscles are necessary for proper eye functioning. The *intrinsic muscles,* controlled by the autonomic nervous system, are those of the ciliary body (which alters the lens curvature in focusing) and the radial and circular muscles of the iris (which control pupillary size and thus regulate the amount of light entering the eye). The *extrinsic muscles* are the rectus and oblique muscles, which are attached to the eyeball exterior (see Figure 24.2). These muscles control eye movement and make it possible to keep moving objects focused on the fovea centralis. They are also responsible for **convergence,** or medial eye movements, which is essential for near vision. When convergence occurs, both eyes are directed toward the near object viewed. The extrinsic eye muscles are controlled by the somatic nervous system.

ACTIVITY 11

Demonstrating Reflex Activity of Intrinsic and Extrinsic Eye Muscles

Involuntary activity of both the intrinsic and extrinsic muscle types is brought about by reflex actions that can be observed in the following experiments.

Photopupillary Reflex

Sudden illumination of the retina by a bright light causes the pupil to constrict reflexively in direct proportion to the light intensity. This protective response prevents damage to the delicate photoreceptor cells.

Obtain a laboratory lamp or penlight. Have your laboratory partner sit with eyes closed and hands over his or her eyes. Turn on the light and position it so that it shines on the subject's right hand. After 1 minute, ask your partner to uncover and open the right eye. Quickly observe the pupil of that eye. What happens to the pupil?

____

Shut off the light and ask your partner to uncover and open the opposite eye. What are your observations of the pupil?

____

____

Accommodation Pupillary Reflex

Have your partner gaze for approximately 1 minute at a distant object in the lab—*not* toward the windows or another light source. Observe your partner's pupils. Then hold some printed material 15 to 25 cm (6 to 10 inches) from his or her face, and direct him or her to focus on it.

How does pupil size change as your partner focuses on the printed material?

____

Explain the value of this reflex. ____________

____

____

____

Convergence Reflex

Repeat the previous experiment, this time using a pen or pencil as the close object to be focused on. Note the position of your partner's eyeballs while he or she gazes at the distant object, and then at the close object. Do they change position as the object of focus is changed?

____________ In what way? ____________

____

____

____

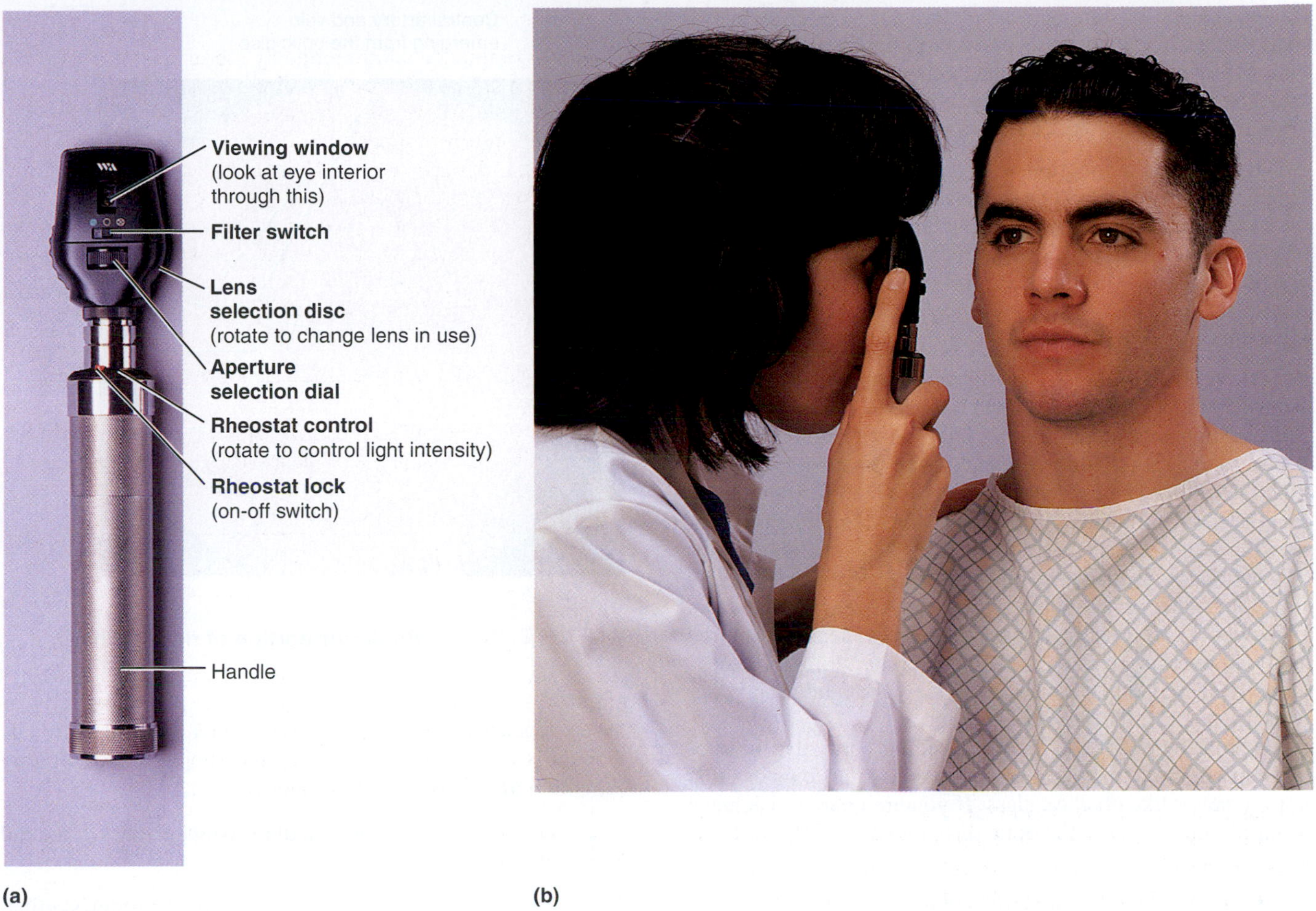
Viewing window
(look at eye interior through this)
Filter switch
Lens selection disc
(rotate to change lens in use)
Aperture selection dial
Rheostat control
(rotate to control light intensity)
Rheostat lock
(on-off switch)
Handle
(a)
(b)

FIGURE 24.12 Structure and use of an ophthalmoscope. (a) Structure of an ophthalmoscope. **(b)** Proper position for beginning to examine the right eye with an ophthalmoscope.

Explain the importance of the convergence reflex.

____

____

____

Ophthalmoscopic Examination of the Eye (Optional)

The ophthalmoscope is an instrument used to examine the *fundus,* or eyeball interior, to determine visually the condition of the retina, optic disc, and internal blood vessels. Certain pathologic conditions such as diabetes mellitus, arteriosclerosis, and degenerative changes of the optic nerve and retina can be detected by such an examination. The ophthalmoscope consists of a set of lenses mounted on a rotating disc (the **lens selection disc**), a light source regulated by a **rheostat control,** and a mirror that reflects the light so that the eye interior can be illuminated (Figure 24.12a).

The lens selection disc is positioned in a small slit in the mirror, and the examiner views the eye interior through this slit, appropriately called the **viewing window.** The focal length of each lens is indicated in diopters preceded by a plus (+) sign if the lens is convex and by a negative (−) sign if the lens is concave. When the zero (0) is seen in the **diopter window,** on the examiner side of the instrument, there is no lens positioned in the slit. The depth of focus for viewing the eye interior is changed by changing the lens.

The light is turned on by depressing the red **rheostat lock button** and then rotating the rheostat control in the clockwise direction. The **aperture selection dial** on the front of the instrument allows the nature of the light beam to be altered. The **filter switch,** also on the front, allows the choice of a green, unfiltered, or polarized light beam. Generally, green light allows for clearest viewing of the blood vessels in the eye interior and is most comfortable for the subject.

Once you have examined the ophthalmoscope and have become familiar with it, you are ready to conduct an eye examination.

ACTIVITY 12

Conducting an Ophthalmoscopic Examination

1. Conduct the examination in a dimly lit or darkened room with the subject comfortably seated and gazing straight ahead. To examine the right eye, sit face-to-face with the subject, hold

the instrument in your right hand, and use your right eye to view the eye interior (Figure 24.12b). You may want to steady yourself by resting your left hand on the subject's shoulder. To view the left eye, use your left eye, hold the instrument in your left hand, and steady yourself with your right hand.

2. Begin the examination with the 0 (no lens) in position. Grasp the instrument so that the lens disc may be rotated with the index finger. Holding the ophthalmoscope about 15 cm (6 inches) from the subject's eye, direct the light into the pupil at a slight angle—through the pupil edge rather than directly through its center. You will see a red circular area that is the illuminated eye interior.

3. Move in as close as possible to the subject's cornea (to within 5 cm, or 2 inches) as you continue to observe the area. Steady your instrument-holding hand on the subject's cheek if necessary. If both your eye and that of the subject are normal, the fundus can be viewed clearly without further adjustment of the ophthalmoscope. If the fundus cannot be focused, slowly rotate the lens disc counterclockwise until the fundus can be clearly seen. When the ophthalmoscope is correctly set, the fundus of the right eye should appear as shown in Figure 24.13. (**Note:** If a positive [convex] lens is required and your eyes are normal, the subject has hyperopia. If a negative [concave] lens is necessary to view the fundus and your eyes are normal, the subject is myopic.)

When the examination is proceeding correctly, the subject can often see images of retinal vessels in his own eye that appear rather like cracked glass. If you are unable to achieve a sharp focus or to see the optic disc, move medially or laterally and begin again.

4. Examine the optic disc for color, elevation, and sharpness of outline, and observe the blood vessels radiating from near its center. Locate the macula, lateral to the optic disc. It is a darker area in which blood vessels are absent, and the fovea appears to be a slightly lighter area in its center. The macula is most easily seen when the subject looks directly into the light of the ophthalmoscope.

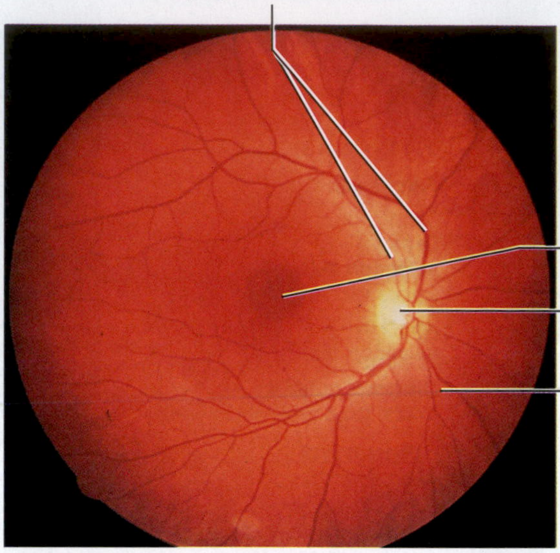
Central artery and vein emerging from the optic disc
Macula lutea
Optic disc
Retina

FIGURE 24.13 Posterior portion of right retina.

⚠ Do not examine the macula for longer than 1 second at a time.

5. When you have finished examining your partner's retina, shut off the ophthalmoscope. Change places with your partner (become the subject) and repeat steps 1–4.

NAME ______________________

LAB TIME/DATE ______________________

REVIEW SHEET

EXERCISE 24

Special Senses: Vision

Anatomy of the Eye

1. Name five accessory eye structures that contribute to the formation of tears and/or aid in lubrication of the eyeball, and then name the major secretory product of each. Indicate which has antibacterial properties by circling the correct secretory product.

Accessory structures	Product

2. The eyeball is wrapped in adipose tissue within the orbit. What is the function of the adipose tissue?

____

3. Why does one often have to blow one's nose after crying? ______________________

____

4. Identify the extrinsic eye muscle predominantly responsible for each action described below.

______________________ 1. turns the eye laterally

______________________ 2. turns the eye medially

______________________ 3. turns the eye up and laterally

______________________ 4. turns the eye inferiorly and medially

______________________ 5. turns the eye superiorly and medially

______________________ 6. turns the eye down and laterally

5. What is a sty? ______________________

____

Conjunctivitis? ______________________

6. Correctly identify each lettered structure in the diagram by writing the letter next to its name in the numbered list.

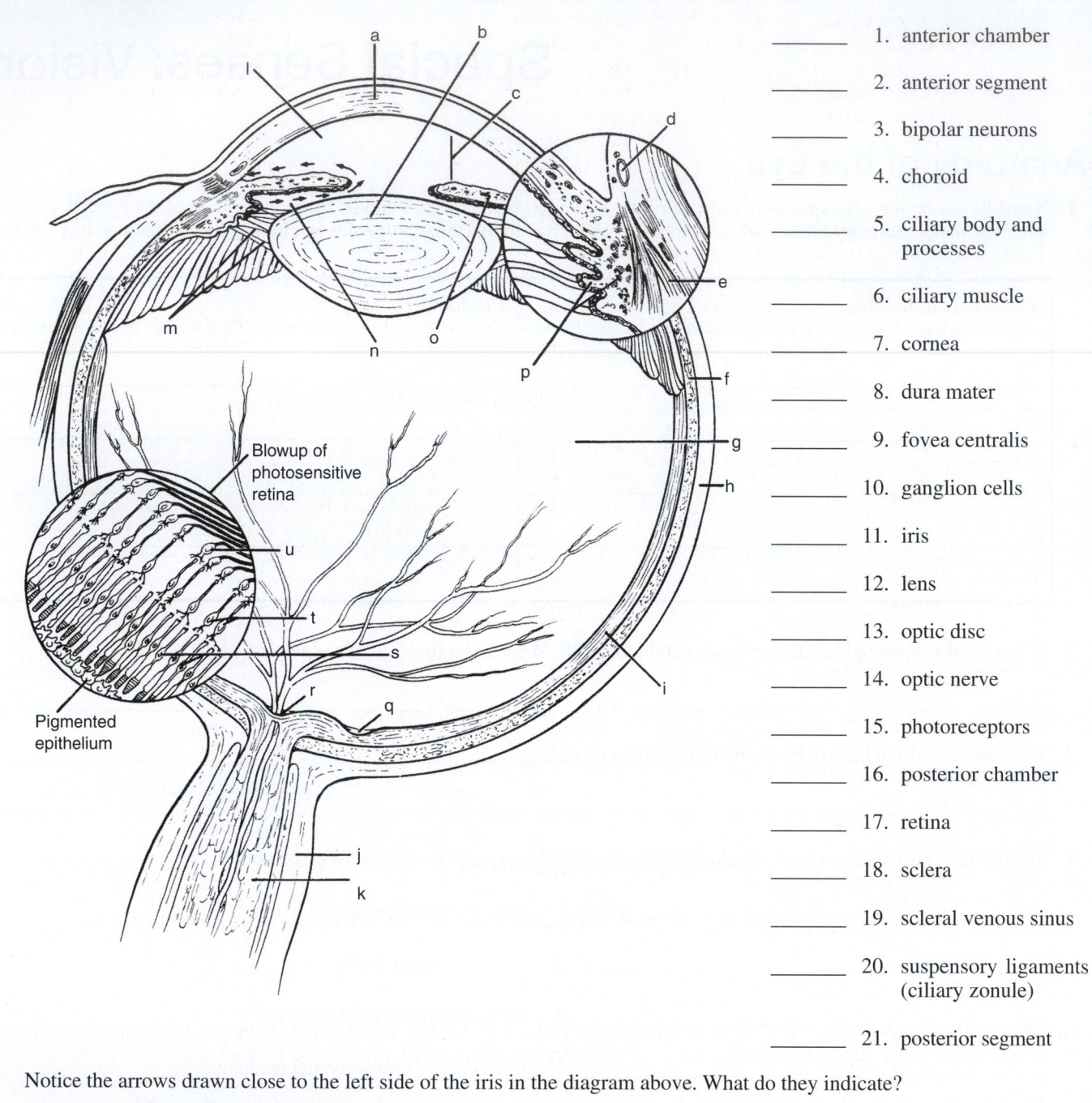

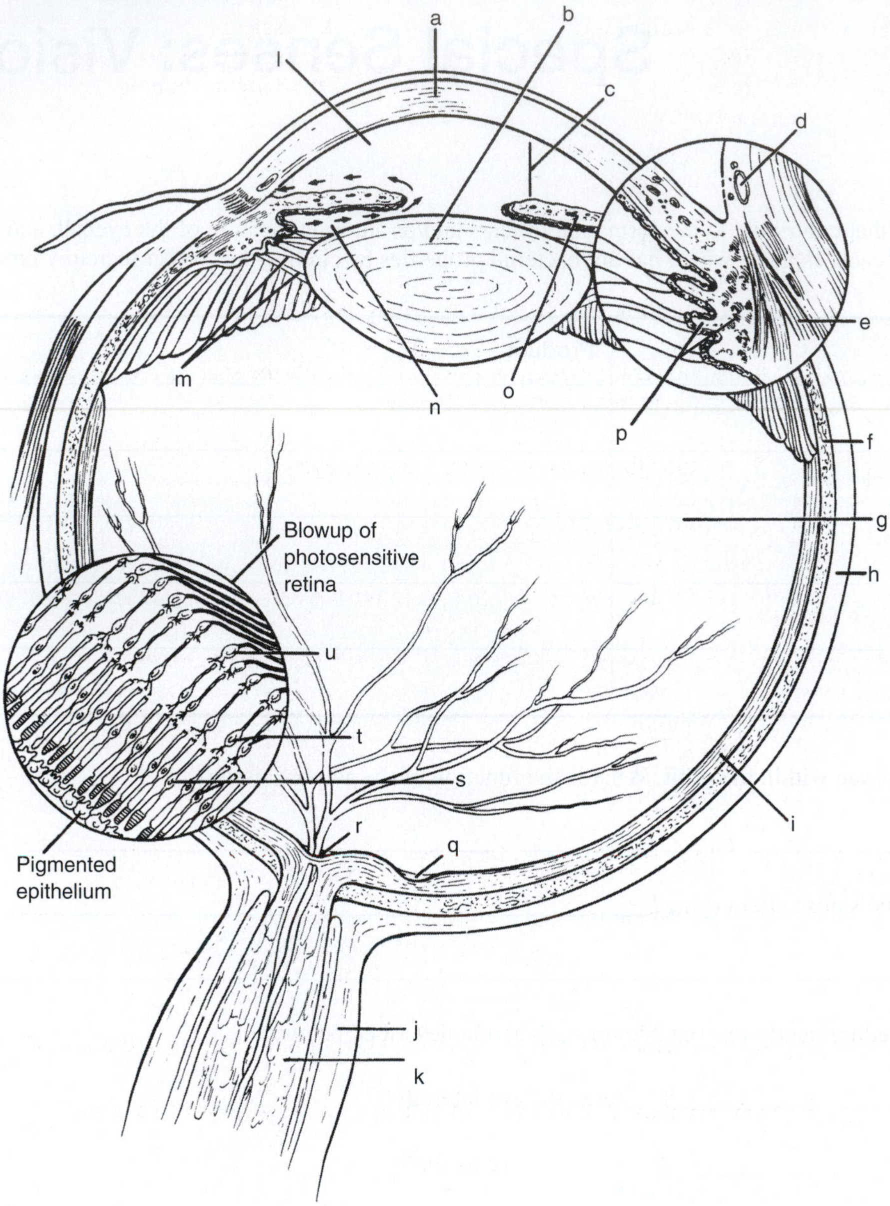
a
b
l
c
d
e
m
n
o
p
f
g
h
Blowup of
photosensitive
retina
u
t
s
r
q
i
Pigmented
epithelium
j
k

_______ 1. anterior chamber

_______ 2. anterior segment

_______ 3. bipolar neurons

_______ 4. choroid

_______ 5. ciliary body and processes

_______ 6. ciliary muscle

_______ 7. cornea

_______ 8. dura mater

_______ 9. fovea centralis

_______ 10. ganglion cells

_______ 11. iris

_______ 12. lens

_______ 13. optic disc

_______ 14. optic nerve

_______ 15. photoreceptors

_______ 16. posterior chamber

_______ 17. retina

_______ 18. sclera

_______ 19. scleral venous sinus

_______ 20. suspensory ligaments (ciliary zonule)

_______ 21. posterior segment

Notice the arrows drawn close to the left side of the iris in the diagram above. What do they indicate?

____

7. The iris is composed primarily of two smooth muscle layers, one arranged radially and the other circularly.

Which of these dilates the pupil? ____

8. You would expect the pupil to be dilated in which of the following circumstances? Circle the correct response(s).

a. in bright light b. in dim light c. focusing for near vision d. observing distant objects

9. The intrinsic eye muscles are controlled by (circle the correct response):

autonomic nervous system somatic nervous system

10. Match the key responses with the descriptive statements that follow.

Key: a. *aqueous humor*
b. choroid
c. ciliary body
d. ciliary processes of the ciliary body
e. cornea
f. fovea centralis
g. iris
h. lens
i. optic disc
j. retina
k. sclera
l. scleral venous sinus
m. vitreous humor

________________ 1. fluid filling the anterior segment of the eye

________________ 2. the "white" of the eye

________________ 3. part of the retina that lacks photoreceptors

________________ 4. modification of the choroid that controls the shape of the crystalline lens and contains the ciliary muscle

________________ 5. drains the aqueous humor from the eye

________________ 6. layer containing the rods and cones

________________ 7. substance occupying the posterior segment of the eyeball

________________ 8. forms the bulk of the heavily pigmented vascular layer

________, ________ 9. smooth muscle structures (2)

________________ 10. area of critical focusing and discriminatory vision

________________ 11. form (by filtration) the aqueous humor

________, ________ 12. light-bending media of the eye (4)

________, ________

________________ 13. anterior continuation of the sclera—your "window on the world"

________________ 14. composed of tough, white, opaque, fibrous connective tissue

Microscopic Anatomy of the Retina

11. The two major layers of the retina are the epithelial and neural layers. In the neural layer, the neuron populations are arranged as follows from the pigmented epithelial layer to the vitreous humor. (Circle the proper response.)

bipolar cells, ganglion cells, photoreceptors

photoreceptors, ganglion cells, bipolar cells

ganglion cells, bipolar cells, photoreceptors

photoreceptors, bipolar cells, ganglion cells

12. The axons of the ________________ cells form the optic nerve, which exits from the eyeball.

13. Complete the following statements by writing either *rods* or *cones* on each blank.

The dim light receptors are the ________. Only ________ are found in the fovea centralis, whereas mostly ________ are found in the periphery of the retina. ________ are the photoreceptors that operate best in bright light and allow for color vision.

Dissection of the Cow (Sheep) Eye

14. What modification of the choroid that is not present in humans is found in the cow eye? ____________

What is its function? ____________

____

15. What does the retina look like? ____________

At what point is it attached to the posterior aspect of the eyeball? ____________

Visual Pathways to the Brain

16. The visual pathway to the occipital lobe of the brain consists most simply of a chain of five cells. Beginning with the photoreceptor cell of the retina, name them and note their location in the pathway.

1. ____________
2. ____________
3. ____________
4. ____________
5. ____________

17. Visual field tests are done to reveal destruction along the visual pathway from the retina to the optic region of the brain. Note where the lesion is likely to be in the following cases.

Normal vision in left eye visual field; absence of vision in right eye visual field: ____________

Normal vision in both eyes for right half of the visual field; absence of vision in both eyes for left half of the visual field:

____

18. How is the right optic *tract* anatomically different from the right optic *nerve*? ____________

____

____

Visual Tests and Experiments

19. Match the terms in column B with the descriptions in column A.

	Column A	Column B
________	1. light bending	a. accommodation
________	2. ability to focus for close (less than 20 feet) vision	b. astigmatism
________	3. normal vision	c. convergence
________	4. inability to focus well on close objects (farsightedness)	d. emmetropia
________	5. nearsightedness	e. hyperopia
________	6. blurred vision due to unequal curvatures of the lens or cornea	f. myopia
________	7. medial movement of the eyes during focusing on close objects	g. refraction

20. Complete the following statements:

In farsightedness, the light is focused __1__ the retina. The lens required to treat myopia is a __2__ lens. The "near point" increases with age because the __3__ of the lens decreases as we get older. A convex lens, like that of the eye, produces an image that is upside down and reversed from left to right. Such an image is called a __4__ image.

1. ______________________
2. ______________________
3. ______________________
4. ______________________

21. Use terms from the key to complete the statements concerning near and distance vision.

Key: a. *contracted* b. *decreased* c. *increased* d. *relaxed* e. *taut*

During distance vision, the ciliary muscle is _______, the suspensory ligament is _______, the convexity of the lens is _______, and light refraction is _______. During close vision, the ciliary muscle is _______, the suspensory ligament is _______, lens convexity is _______, and light refraction is _______.

22. Explain why vision is lost when light hits the blind spot. ______________________

23. Using your Snellen eye test results, answer the following questions.

Is your visual acuity normal, less than normal, or better than normal? ______________________

Explain your answer. ______________________

____

Explain why each eye is tested separately when using the Snellen eye chart. ______________________

____

Explain 20/40 vision. ______________________

____

Explain 20/10 vision. ______________________

____

24. Define *astigmatism.* ______________________

____

How can it be corrected? ______________________

25. Define *presbyopia.* ______________________

____

What causes it? ______________________

26. To which wavelengths of light do the three cone types of the retina respond maximally?

_______________, _______________, and _______________

27. How can you explain the fact that we see a great range of colors even though only three cone types exist?

____

____

28. Explain the difference between binocular and panoramic vision. ______________________________

____

____

What is the advantage of binocular vision? ______________________________

What factor(s) are responsible for binocular vision? ______________________________

____

29. In the experiment on the convergence reflex, what happened to the position of the eyeballs as the object was moved closer to the subject's eyes? ______________________________

What extrinsic eye muscles control the movement of the eyes during this reflex? ______________________________

What is the value of this reflex? ______________________________

____

30. In the experiment on the photopupillary reflex, what happened to the pupil of the eye exposed to light?

_______________ What happened to the pupil of the nonilluminated eye? _______________

Explanation? ______________________________

____

31. Why is the ophthalmoscopic examination an important diagnostic tool? ______________________________

____

32. Many college students struggling through mountainous reading assignments are told that they need glasses for "eyestrain." Why is it more of a strain on the extrinsic and intrinsic eye muscles to look at close objects than at far objects?

____

____

EXERCISE 25

Special Senses: Hearing and Equilibrium

MATERIALS

- ☐ Three-dimensional dissectible ear model and/or chart of ear anatomy
- ☐ Otoscope (if available)
- ☐ Disposable otoscope tips (if available) and autoclave bag
- ☐ Alcohol swabs
- ☐ Compound microscope
- ☐ Prepared slides of the cochlea of the ear
- ☐ Absorbent cotton
- ☐ Pocket watch or clock that ticks
- ☐ Metric ruler
- ☐ Tuning forks (range of frequencies)
- ☐ Rubber mallet
- ☐ Audiometer
- ☐ Red and blue pencils
- ☐ Demonstration: Microscope focused on a slide of a crista ampullaris receptor of a semicircular canal
- ☐ Three coins of different sizes
- ☐ Rotating chair or stool
- ☐ Blackboard and chalk

MasteringA&P™

Access practice quizzes and more in the Study Area at www.masteringaandp.com.

PAL

For access to anatomical models and more, check out Practice Anatomy Lab.

OBJECTIVES

1. To identify, by appropriately labeling a diagram, the anatomical structures of the external, middle, and internal ear, and to explain their functions.
2. To describe the anatomy of the organ of hearing (spiral organ of Corti in the cochlea), and to explain its function in sound reception.
3. To describe the anatomy of the equilibrium organs of the internal ear (cristae ampullares and maculae), and to explain their relative function in maintaining equilibrium.
4. To explain how one is able to localize the source of sounds.
5. To define or explain *sensorineural deafness, conduction deafness,* and *nystagmus.*
6. To state the purpose of the Weber, Rinne, balance, Barany, and Romberg tests.
7. To describe the effects of acceleration on the semicircular canals.
8. To explain the role of vision in maintaining equilibrium.

PRE-LAB QUIZ

1. Circle the correct term. The ear is divided into three / four major areas.
2. The external ear is composed primarily of the __________ and the external acoustic meatus.
   a. auricle c. eardrum
   b. cochlea d. stapes
3. Circle the correct term. Sound waves that enter the external acoustic meatus eventually encounter the tympanic membrane / oval window, which then vibrates at the same frequency as the sound waves hitting it.
4. Three small bones found within the middle ear are the malleus, incus, and __________.
   a. auricle c. eardrum
   b. cochlea d. stapes
5. The snail-like __________, found in the inner ear, contains sensory receptors for hearing.
   a. cochlea c. semicircular canals
   b. lobule d. vestibule
6. Circle the correct term. Today you will use an ophthalmoscope / otoscope to examine the ear.
7. The __________ test is used for comparing bone and air-conduction hearing.
   a. Barany c. Weber
   b. Rinne
8. The equilibrium apparatus of the ear, the vestibular apparatus, is found in the
   a. external ear c. middle ear
   b. inner ear

Text continues on next page.

9. Circle the correct terms. The crista ampullaris / macula located in the semicircular duct / vestibule is essential for detecting static equilibrium.

10. Nystagmus is
   a. ability to hear only high frequency tones
   b. ability to hear only low frequency tones
   c. involuntary trailing of eyes in one direction, then rapid movement in the other
   d. sensation of dizziness

The ear is a complex structure containing sensory receptors for hearing and equilibrium. The ear is divided into three major areas: the *external ear,* the *middle ear,* and the *internal ear* (Figure 25.1). The external and middle ear structures serve the needs of the sense of hearing *only,* whereas internal ear structures function both in equilibrium and hearing reception.

Anatomy of the Ear

Gross Anatomy

ACTIVITY 1

Identifying Structures of the Ear

Obtain a dissectible ear model or chart of ear anatomy and identify the structures described below. Refer to Figure 25.1 as you work.

The **external,** or **outer, ear** is composed primarily of the auricle and the external acoustic meatus. The **auricle,** or **pinna,*** is the skin-covered cartilaginous structure encircling the auditory canal opening. In many animals, it collects and directs sound waves into the external auditory canal. In humans this function of the pinna is largely lost. The portion of the pinna lying inferior to the external auditory canal is the **lobule.**

The **external acoustic meatus,** or **external auditory canal,*** is a short, narrow (about 2.5 cm long by 0.6 cm wide) chamber carved into the temporal bone. In its skin-lined walls are wax-secreting glands called **ceruminous glands.** Sound waves that enter the external auditory meatus eventually encounter the **tympanic membrane,** or **eardrum,** which vibrates at exactly the same frequency as the sound wave(s) hitting it. The membranous eardrum separates the external from the middle ear.

The **middle ear** is essentially a small chamber—the **tympanic cavity**—found within the temporal bone. The cavity is spanned by three small bones, collectively called the **auditory ossicles** (**malleus, incus,** and **stapes**), which articulate to form a lever system that amplifies and transmits the vibratory motion of the eardrum to the fluids of the inner ear via the **oval window.** The ossicles are often referred to by their common names, that is, hammer, anvil, and stirrup, respectively.

Connecting the middle ear chamber with the nasopharynx is the **pharyngotympanic,** or **auditory, tube** (formerly known as the eustachian tube). Normally this tube is flattened and closed, but swallowing or yawning can cause it to open temporarily to equalize the pressure of the middle ear cavity

*Although the preferred anatomical terms for *pinna* and *external auditory canal* are *auricle* and *external acoustic meatus,* "pinna" and "external auditory canal" are heard often in clinical situations and will continue to be used here.

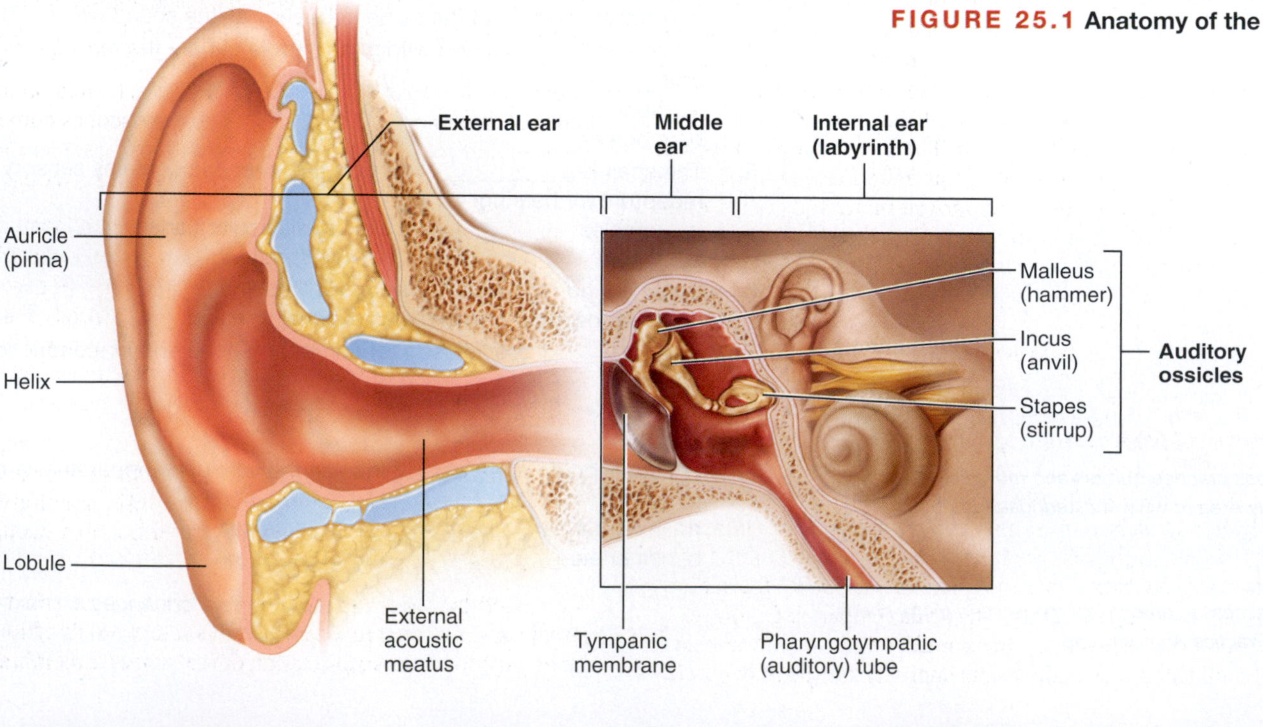
External ear
Middle ear
Internal ear (labyrinth)
Auricle (pinna)
Helix
Lobule
External acoustic meatus
Tympanic membrane
Pharyngotympanic (auditory) tube
Malleus (hammer)
Incus (anvil)
Stapes (stirrup)
Auditory ossicles

FIGURE 25.1 Anatomy of the ear.

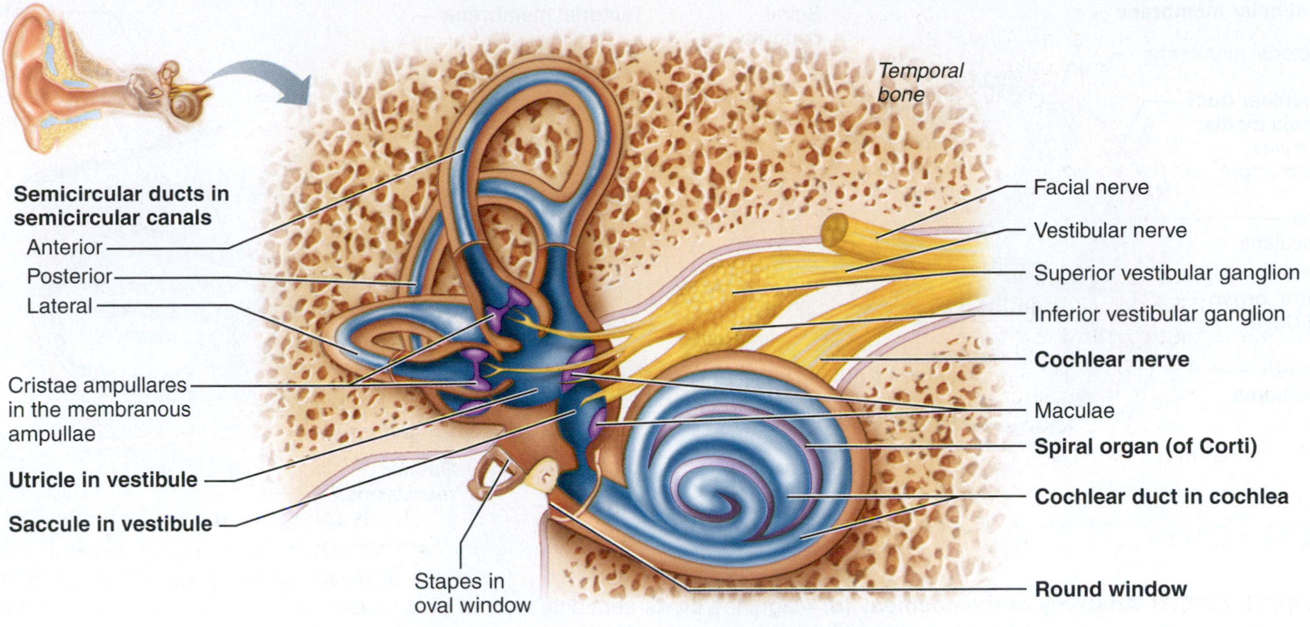
Temporal bone
Semicircular ducts in semicircular canals
Anterior
Posterior
Lateral
Cristae ampullares in the membranous ampullae
Utricle in vestibule
Saccule in vestibule
Stapes in oval window
Facial nerve
Vestibular nerve
Superior vestibular ganglion
Inferior vestibular ganglion
Cochlear nerve
Maculae
Spiral organ (of Corti)
Cochlear duct in cochlea
Round window

FIGURE 25.2 Internal ear. Right membranous labyrinth (blue) shown within the bony labyrinth (tan). The locations of sensory organs for hearing and equilibrium are shown in purple.

with external air pressure. This is an important function. The eardrum does not vibrate properly unless the pressure on both of its surfaces is the same.

Because the mucosal membranes of the middle ear cavity and nasopharynx are continuous through the pharyngotympanic tube, **otitis media,** or inflammation of the middle ear, is a fairly common condition, especially among youngsters prone to sore throats. In cases where large amounts of fluid or pus accumulate in the middle ear cavity, an emergency myringotomy (lancing of the eardrum) may be necessary to relieve the pressure. Frequently, tiny ventilating tubes are put in during the procedure. ■

The **internal,** or **inner, ear** consists of a system of bony and rather tortuous chambers called the **osseous,** or **bony, labyrinth,** which is filled with an aqueous fluid called **perilymph** (Figure 25.2). Suspended in the perilymph is the **membranous labyrinth,** a system that mostly follows the contours of the osseous labyrinth. The membranous labyrinth is filled with a more viscous fluid called **endolymph.** The three subdivisions of the bony labyrinth are the cochlea, the vestibule, and the semicircular canals, with the vestibule situated between the cochlea and semicircular canals. The **vestibule** and the **semicircular canals** are involved with equilibrium.

The snail-like **cochlea** (see Figures 25.2 and 25.3) contains the sensory receptors for hearing. The cochlear membranous labyrinth, the **cochlear duct,** is a soft wormlike tube about 3.8 cm long. It winds through the full two and three-quarter turns of the cochlea and separates the perilymph—containing cochlear cavity into upper and lower chambers, the **scala vestibuli** and **scala tympani** (the vestibular and tympanic ducts), respectively. The scala vestibuli terminates at the oval window, which "seats" the foot plate of the stirrup located laterally in the tympanic cavity. The scala tympani is bounded by a membranous area called the **round window.** The cochlear duct, itself filled with endolymph, supports the **spiral organ (of Corti),** which contains the receptors for hearing—the sensory hair cells and nerve endings of the **cochlear nerve,** a division of the vestibulocochlear nerve (VIII).

ACTIVITY 2

Examining the Ear with an Otoscope (Optional)

1. Obtain an otoscope and two alcohol swabs. Inspect your partner's external ear canal and then select the largest—*diameter* (not length!) speculum that will fit comfortably into his or her ear to permit full visibility. Clean the speculum thoroughly with an alcohol swab, and then attach the speculum to the battery-containing otoscope handle. Before beginning, check that the otoscope light beam is strong. (If not, obtain another otoscope or new batteries.) Some otoscopes come with disposable tips. Be sure to use a new tip for each ear examined. Dispose of these tips in an autoclave bag after use.

2. When you are ready to begin the examination, hold the lighted otoscope securely between your thumb and forefinger (like a pencil), and rest the little finger of the otoscope-holding hand against your partner's head. This maneuver forms a brace that allows the speculum to move as your partner moves and prevents the speculum from penetrating too deeply into the ear canal during unexpected movements.

3. Grasp the ear pinna firmly and pull it up, back, and slightly laterally. If your partner experiences pain or discomfort when the pinna is manipulated, an inflammation or infection of the external ear may be present. If this occurs, do not attempt to examine the ear canal.

4. Carefully insert the speculum of the otoscope into the external auditory canal in a downward and forward direction only far enough to permit examination of the tympanic membrane, or

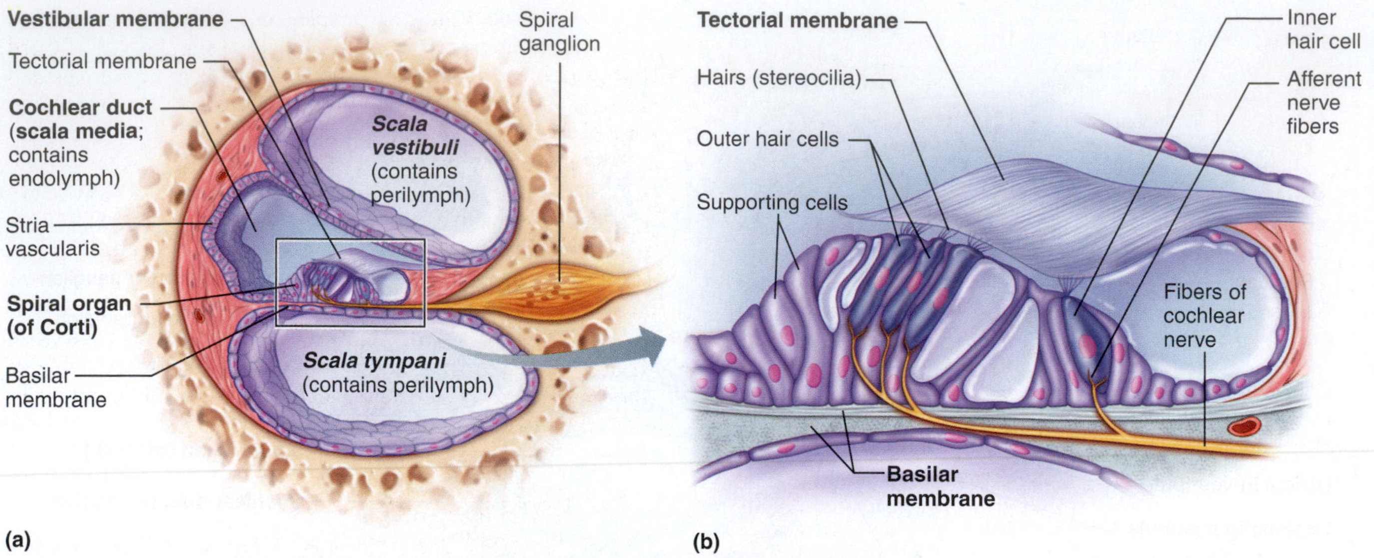
Vestibular membrane
Tectorial membrane
Cochlear duct (scala media; contains endolymph)
Stria vascularis
Spiral organ (of Corti)
Basilar membrane
Spiral ganglion
Scala vestibuli (contains perilymph)
Scala tympani (contains perilymph)
(a)
Tectorial membrane
Hairs (stereocilia)
Outer hair cells
Supporting cells
Inner hair cell
Afferent nerve fibers
Fibers of cochlear nerve
Basilar membrane
(b)

FIGURE 25.3 Anatomy of the cochlea. **(a)** Magnified cross-sectional view of one turn of the cochlea, showing the relationship of the three scalae. The scalae vestibuli and tympani contain perilymph; the cochlear duct (scala media) contains endolymph. **(b)** Detailed structure of the spiral organ (of Corti).

eardrum. Note its shape, color, and vascular network. The healthy tympanic membrane is pearly white. During the examination, notice if there is any discharge or redness in the canal and identify earwax.

5. After the examination, thoroughly clean the speculum with the second alcohol swab before returning the otoscope to the supply area.

Microscopic Anatomy of the Spiral Organ of Corti and the Mechanism of Hearing

The anatomical details of the spiral organ (of Corti) are shown in Figure 25.3. The hair (auditory receptor) cells rest on the **basilar membrane,** which forms the floor of the cochlear duct, and their "hairs" (stereocilia) project into a gelatinous membrane, the **tectorial membrane,** that overlies them. The roof of the cochlear duct is called the **vestibular membrane.** The endolymph-filled chamber of the cochlear duct is the **scala media.**

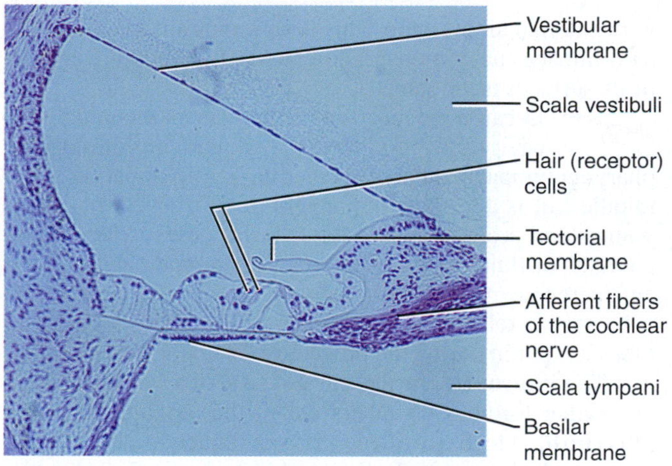
Vestibular membrane
Scala vestibuli
Hair (receptor) cells
Tectorial membrane
Afferent fibers of the cochlear nerve
Scala tympani
Basilar membrane

FIGURE 25.4 Histological image of the spiral organ (of Corti) (80×).

ACTIVITY 3

Examining the Microscopic Structure of the Cochlea

Obtain a compound microscope and a prepared microscope slide of the cochlea and identify the areas shown in Figure 25.4.

The mechanism of hearing begins as sound waves pass through the external auditory canal and through the middle ear into the internal ear, where the vibration eventually reaches the spiral organ (of Corti), which contains the receptors for hearing.

The popular "traveling wave" hypothesis of von Békésy suggests that vibration of the stirrup at the oval window initiates traveling waves that cause maximal displacements of the basilar membrane where they peak and stimulate the hair cells of the spiral organ in that region. Since the area at which the traveling waves peak is a high-pressure area, the vestibular membrane is compressed at this point and, in turn, compresses the endolymph and the basilar membrane of the cochlear duct. The resulting pressure on the perilymph in the scala tympani causes the membrane of the round window to bulge outward into the middle ear chamber, thus acting as a relief valve for the compressional wave. Georg von Békésy found that high-frequency waves (high-pitched sounds) peaked close to the oval window and that low-frequency waves (low-pitched sounds) peaked farther up the basilar membrane near the apex of the cochlea. The mechanism of sound reception by the spiral organ (of Corti) is complex. We do know that hair cells on the basilar membrane are

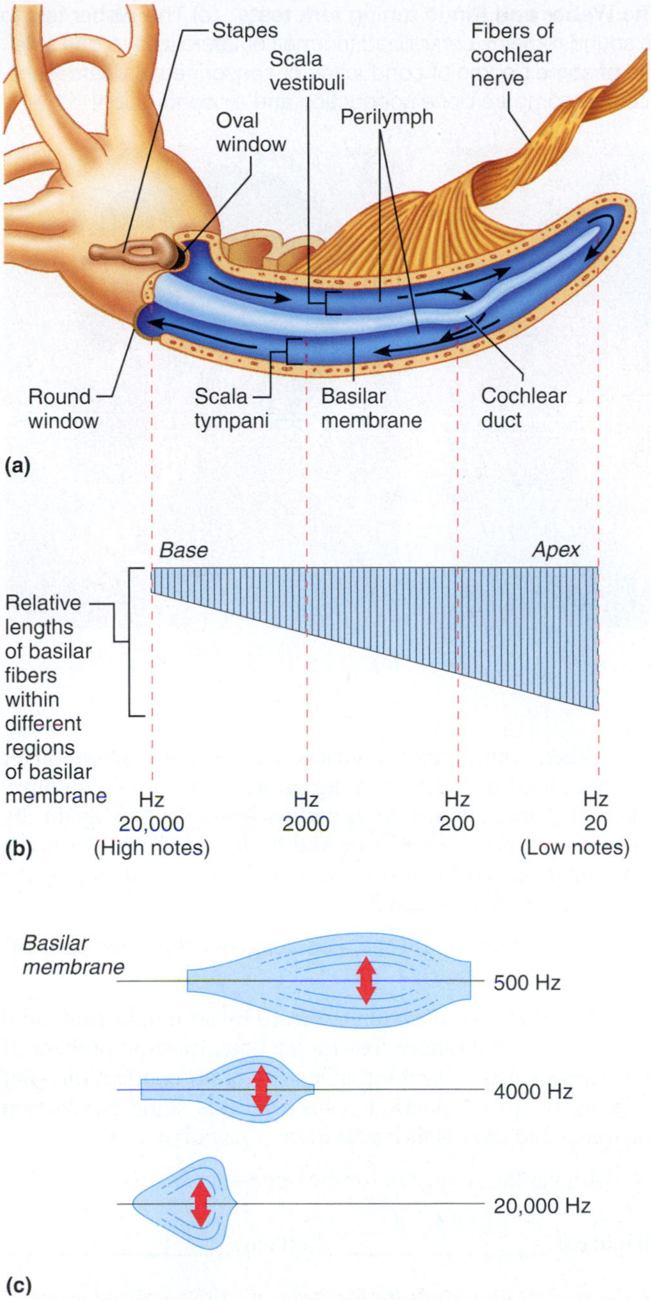
Stapes
Scala vestibuli
Fibers of cochlear nerve
Oval window
Perilymph
Round window
Scala tympani
Basilar membrane
Cochlear duct
(a)
Base
Apex
Relative lengths of basilar fibers within different regions of basilar membrane
Hz 20,000 (High notes)
Hz 2000
Hz 200
Hz 20 (Low notes)
(b)
Basilar membrane
500 Hz
4000 Hz
20,000 Hz
(c)

FIGURE 25.5 Resonance of the basilar membrane. The cochlea is depicted as if it has been uncoiled. **(a)** Fluid movement in the cochlea following the stirrup thrust at the oval window. Compressional wave thus created causes the round window to bulge into the middle ear. Pressure waves set up vibrations in the basilar membrane. **(b)** Fibers span the basilar membrane. The length of the fibers "tunes" specific regions to vibrate at specific frequencies. **(c)** Different frequencies of pressure waves in the cochlea stimulate particular hair cells and neurons.

uniquely stimulated by sounds of various frequencies and amplitude and once stimulated they depolarize and begin the chain of nervous impulses via the cochlear nerve to the auditory centers of the temporal lobe cortex. This series of events results in the phenomenon we call hearing (Figure 25.5).

By the time most people are in their sixties, a gradual deterioration and atrophy of the spiral organ of Corti begins and leads to a loss in the ability to hear high tones and speech sounds. This condition, **presbycusis,** is a type of sensorineural deafness. Because many elderly people refuse to accept their hearing loss and resist using hearing aids, they begin to rely more and more on their vision for clues as to what is going on around them and may be accused of ignoring people.

Although presbycusis is considered to be a disability of old age, it is becoming much more common in younger people as our world grows noisier. The damage (breakage of the "hairs" of the hair cells) caused by excessively loud sounds is progressive and cumulative. Each assault causes a bit more damage. Music played and listened to at deafening levels definitely contributes to the deterioration of hearing receptors. ■

ACTIVITY 4

Conducting Laboratory Tests of Hearing

Perform the following hearing tests in a quiet area. Test both the right and left ears.

Acuity Test

Have your lab partner pack one ear with cotton and sit quietly with eyes closed. Obtain a ticking clock or pocket watch and hold it very close to his or her *unpacked* ear. Then slowly move it away from the ear until your partner signals that the ticking is no longer audible. Record the distance in centimeters at which ticking is inaudible and then remove the cotton from the packed ear.

Right ear: ________________ Left ear: ________________

Is the threshold of audibility sharp or indefinite?

____

Sound Localization

Ask your partner to close both eyes. Hold the pocket watch at an audible distance (about 15 cm) from his or her ear, and move it to various locations (front, back, sides, and above his or her head). Have your partner locate the position by pointing in each instance. Can the sound be localized equally well at all positions?

____

If not, at what position(s) was the sound less easily located?

____

____

The ability to localize the source of a sound depends on two factors—the difference in the loudness of the sound reaching each ear and the time of arrival of the sound at each ear. How does this information help to explain your findings?

____

____

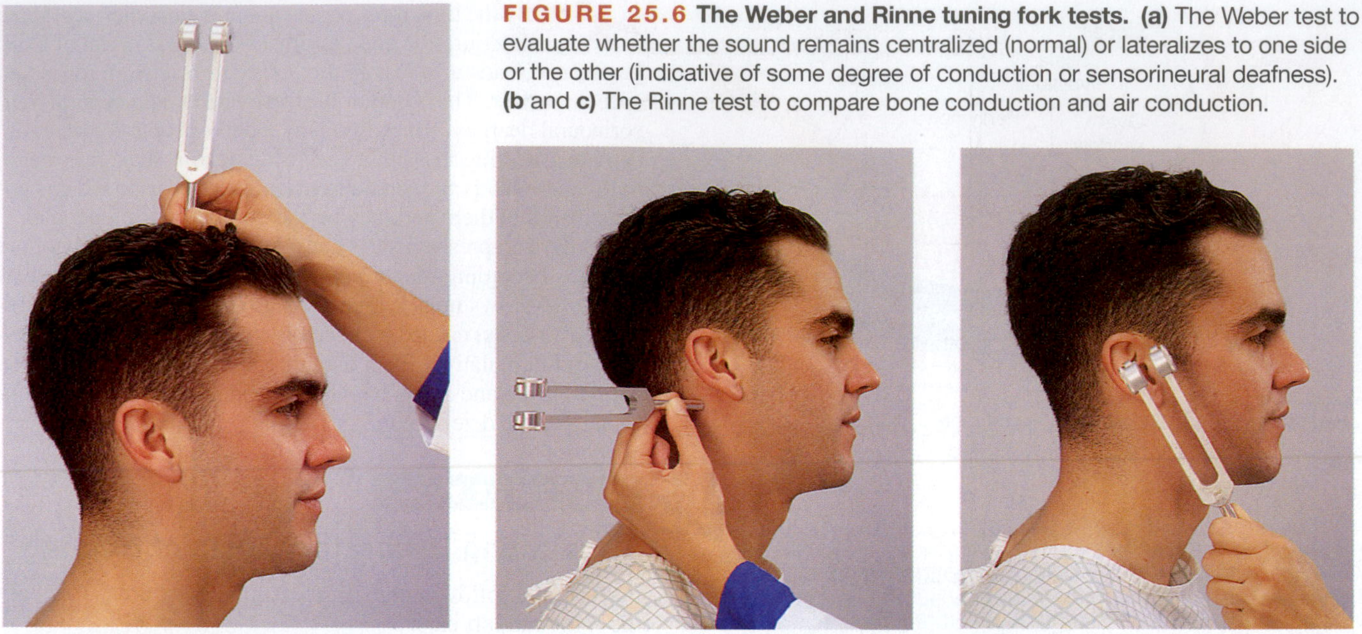

(a) (b) (c)

FIGURE 25.6 The Weber and Rinne tuning fork tests. (a) The Weber test to evaluate whether the sound remains centralized (normal) or lateralizes to one side or the other (indicative of some degree of conduction or sensorineural deafness). **(b and c)** The Rinne test to compare bone conduction and air conduction.

Frequency Range of Hearing

Obtain three tuning forks: one with a low frequency (75 to 100 Hz [cps]), one with a frequency of approximately 1000 Hz, and one with a frequency of 4000 to 5000 Hz. Strike the lowest-frequency fork on the heel of your hand or with a rubber mallet, and hold it close to your partner's ear. Repeat with the other two forks.

Which fork was heard most clearly and comfortably?

_______________ Hz

Which was heard least well? _______________ Hz

Weber Test to Determine Conduction and Sensorineural Deafness

Strike a tuning fork and place the handle of the tuning fork medially on your partner's head (see Figure 25.6a). Is the tone equally loud in both ears, or is it louder in one ear?

____

If it is equally loud in both ears, you have equal hearing or equal loss of hearing in both ears. If sensorineural deafness is present in one ear, the tone will be heard in the unaffected ear but not in the ear with sensorineural deafness. If conduction deafness is present, the sound will be heard more strongly in the ear in which there is a hearing loss due to sound conduction by the bone of the skull. Conduction deafness can be simulated by plugging one ear with cotton to interfere with the conduction of sound to the inner ear.

Rinne Test for Comparing Bone- and Air-Conduction Hearing

1. Strike the tuning fork, and place its handle on your partner's mastoid process (Figure 25.6b).

2. When your partner indicates that the sound is no longer audible, hold the still-vibrating prongs close to his auditory canal (Figure 25.6c). If your partner hears the fork again (by air conduction) when it is moved to that position, hearing is not impaired and the test result is to be recorded as positive (+). (Record below step 5.)

3. Repeat the test on the same ear, but this time test air-conduction hearing first.

4. After the tone is no longer heard by air conduction, hold the handle of the tuning fork on the bony mastoid process. If the subject hears the tone again by bone conduction after hearing by air conduction is lost, there is some conduction deafness and the result is recorded as negative (−).

5. Repeat the sequence for the opposite ear.

Right ear: _______________ Left ear: _______________

Does the subject hear better by bone or by air conduction?

____

Audiometry

When the simple tuning fork tests reveal a problem in hearing, audiometer testing is usually prescribed to determine the precise nature of the hearing deficit. An *audiometer* is an instrument (specifically, an electronic oscillator with earphones) used to determine hearing acuity by exposing each ear to sound stimuli of differing *frequencies* and *intensities.* The hearing range of human beings during youth is from 20 to 20,000 Hz, but hearing acuity declines with age, with reception for the high-frequency sounds lost first. Though this loss represents a major problem for some people, such as musicians, most of us tend to be fairly unconcerned until we begin to have problems hearing sounds in the range of 125 to 8000 Hz, the normal frequency range of speech.

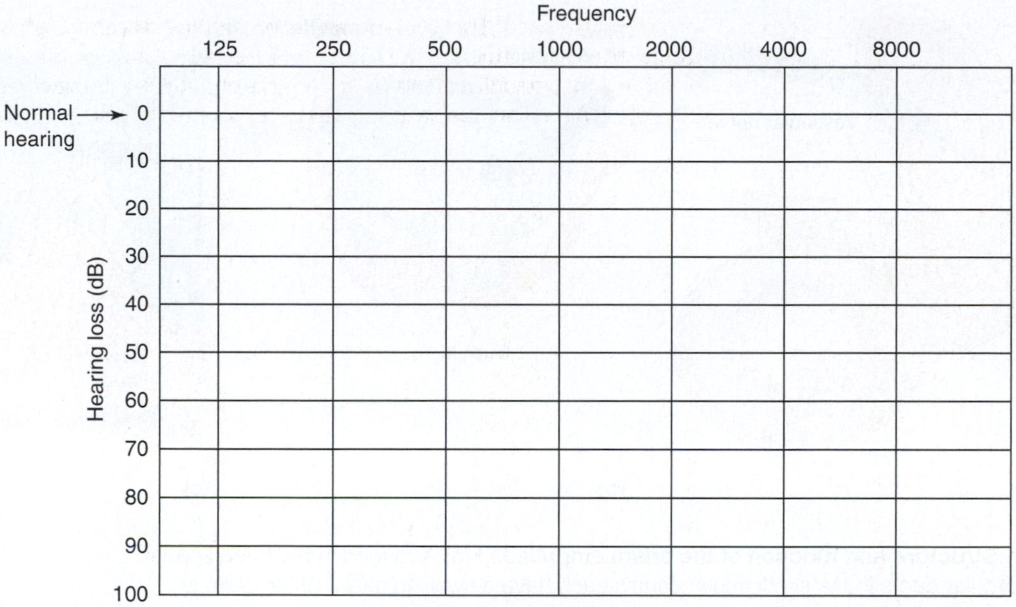
Frequency
125
250
500
1000
2000
4000
8000
Normal hearing
0
10
20
30
40
50
60
70
80
90
100
Hearing loss (dB)

The basic procedure of audiometry is to initially deliver tones of different frequencies to one ear of the subject at an intensity of 0 decibels (dB). (Zero decibels is not the complete absence of sound, but rather the softest sound intensity that can be heard by a person of normal hearing at each frequency.) If the subject cannot hear a particular frequency stimulus of 0 dB, the hearing threshold level control is adjusted until the subject reports that he or she can hear the tone. The number of decibels of intensity required above 0 dB is recorded as the hearing loss. For example, if the subject cannot hear a particular frequency tone until it is delivered at 30 dB intensity, then he or she has a hearing loss of 30 dB for that frequency.

ACTIVITY 5

Audiometry Testing

1. Obtain an audiometer and earphones, and a red and a blue pencil. Before beginning the tests, examine the audiometer to identify the two tone controls: one to regulate frequency and a second to regulate the intensity (loudness) of the sound stimulus. Identify the two output control switches that regulate the delivery of sound to one ear or the other (*red* to the right ear, *blue* to the left ear). Also find the *hearing threshold level control,* which is calibrated to deliver a basal tone of 0 dB to the subject's ears.

2. Place the earphones on the subject's head so that the red cord or ear-cushion is over the right ear and the blue cord or ear-cushion is over the left ear. Instruct the subject to raise one hand when he or she hears a tone.

3. Set the frequency control at 125 Hz and the intensity control at 0 dB. Press the red output switch to deliver a tone to the subject's right ear. If the subject does not respond, raise the sound intensity slowly by rotating the hearing level control counterclockwise until the subject reports (by raising a hand) that a tone is heard. Repeat this procedure for frequencies of 250, 500, 1000, 2000, 4000, and 8000.

4. Record the results in the grid above by marking a small red circle on the grid at each frequency-dB junction at which a tone was heard. Then connect the circles with a red line to produce a hearing acuity graph for the right ear.

5. Repeat steps 3 and 4 for the left (blue) ear, and record the results with blue circles and connecting lines on the grid.

Microscopic Anatomy of the Equilibrium Apparatus and Mechanisms of Equilibrium

The equilibrium apparatus of the inner ear, the **vestibular apparatus,** is in the vestibule and semicircular canals of the bony labyrinth. Their chambers are filled with perilymph, in which membranous labyrinth structures are suspended. The vestibule contains the saclike **utricle** and **saccule,** and the semicircular chambers contain **membranous semicircular ducts.** Like the cochlear duct, these membranes are filled with endolymph and contain receptor cells that are activated by the bending of their cilia.

Semicircular Canals

The semicircular canals are centrally involved in the **mechanism of dynamic equilibrium.** They are 1.2 cm in circumference and are oriented in three planes—horizontal, frontal, and sagittal. At the base of each semicircular duct is an enlarged region, the **ampulla,** which communicates with the utricle of the vestibule. Within each ampulla is a receptor region called a **crista ampullaris,** which consists of a tuft of hair cells covered

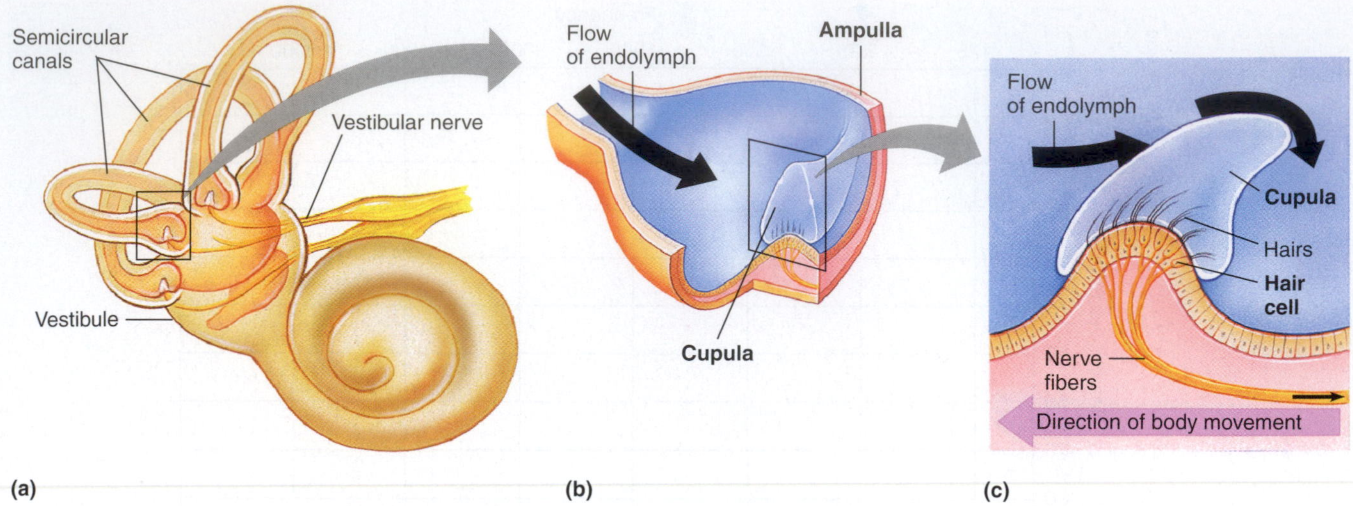
Semicircular canals
Vestibular nerve
Vestibule
Flow of endolymph
Ampulla
Cupula
Flow of endolymph
Cupula
Hairs
Hair cell
Nerve fibers
Direction of body movement
(a)
(b)
(c)

FIGURE 25.7 Structure and function of the crista ampullaris. (a) Arranged in the three spatial planes, the semicircular ducts in the semicircular canals each have a swelling called an ampulla at their base. **(b)** Each ampulla contains a crista ampullaris, a receptor that is essentially a cluster of hair cells with hairs projecting into a gelatinous cap called the cupula. **(c)** When the rate of rotation changes, inertia prevents the endolymph in the semicircular canals from moving with the head, so the fluid presses against the cupula, bending the hair cells in the opposite direction. The bending increases the frequency of action potentials in the sensory neurons in direct proportion to the amount of rotational acceleration. The mechanism adjusts quickly if rotation continues at a constant speed.

with a gelatinous cap, or **cupula** (Figure 25.7). When your head position changes in an angular direction, as when twirling on the dance floor or when taking a rough boat ride, the endolymph in the canal lags behind, pushing the cupula—like a swinging door—in a direction opposite to that of the angular motion (Figure 25.7c). This movement depolarizes the hair cells, resulting in enhanced impulse transmission up the vestibular division of the eighth cranial nerve to the brain. Likewise, when the angular motion stops suddenly, the inertia of the endolymph causes it to continue to move, pushing the cupula in the same direction as the previous body motion. This movement again initiates electrical changes in the hair cells (in this case, it hyperpolarizes them). (This phenomenon accounts for the reversed motion sensation you feel when you stop suddenly after twirling.) When you move at a constant rate of motion, the endolymph eventually comes to rest and the cupula gradually returns to its original position. The hair cells, no longer bent, send no new signals, and you lose the sensation of spinning. Thus the response of these dynamic equilibrium receptors is a reaction to *changes* in angular motion rather than to motion itself.

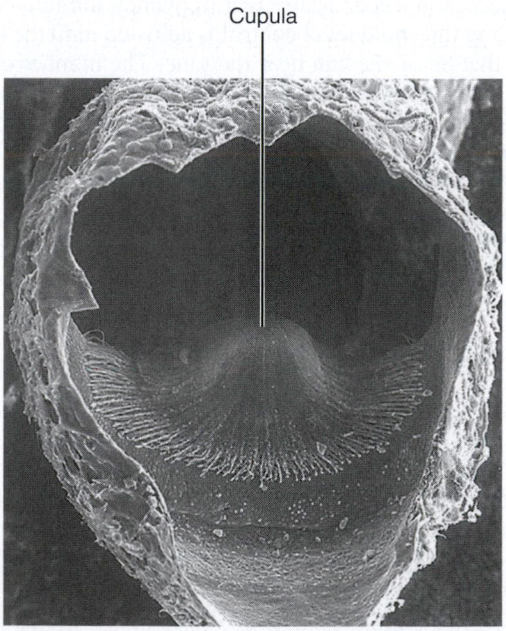
Cupula

FIGURE 25.8 Scanning electron micrograph of a crista ampullaris (290×).

ACTIVITY 6

Examining the Microscopic Structure of the Crista Ampullaris

Go to the demonstration area and examine the slide of a crista ampullaris. Identify the areas depicted in Figure 25.8 and Figure 25.7b and c. ■

Maculae

Maculae in the vestibule contain the **hair cells,** receptors that are essential to the **mechanism of static equilibrium.** The maculae respond to gravitational pull, thus providing information on which way is up or down, and to linear or straightforward changes in speed. They are located on the walls of the saccule and utricle. The hair cells in each macula are embedded in the **otolithic membrane,** a gelatinous material containing small grains of calcium carbonate **(otoliths).** When the head moves, the otoliths move in response to variations in gravitational pull. As they deflect different hair cells, they trigger hyperpolarization or depolarization of the hair cells and modify the rate of impulse transmission along the vestibular nerve (Figure 25.9).

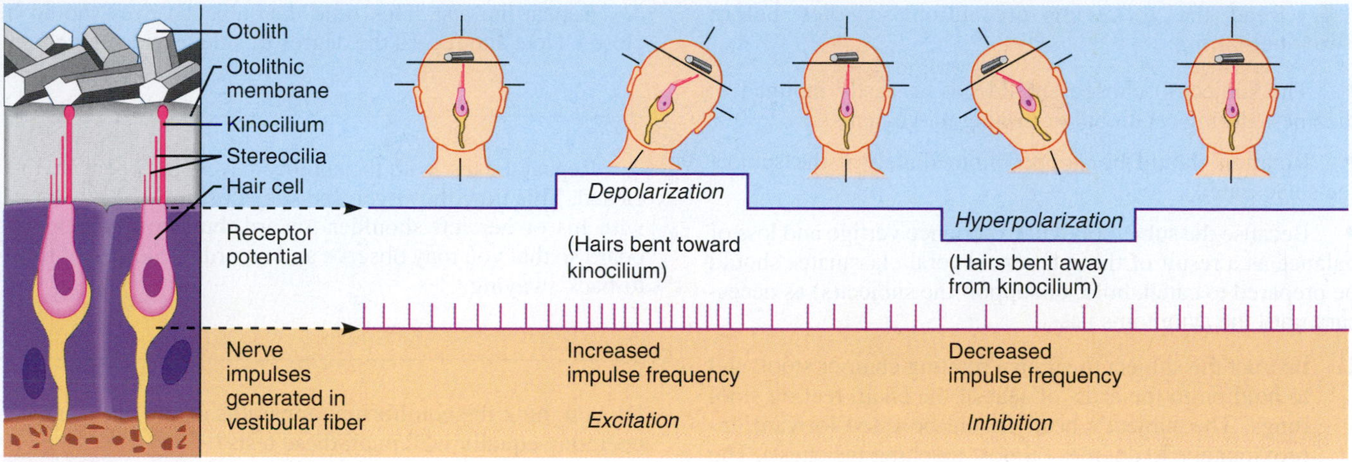
Otolith
Otolithic membrane
Kinocilium
Stereocilia
Hair cell
Receptor potential
Nerve impulses generated in vestibular fiber
Depolarization
(Hairs bent toward kinocilium)
Hyperpolarization
(Hairs bent away from kinocilium)
Increased impulse frequency
Excitation
Decreased impulse frequency
Inhibition

FIGURE 25.9 The effect of gravitational pull on a macula receptor in the utricle. When movement of the otolithic membrane bends the hair cells in the direction of the kinocilium, the vestibular fibers depolarize and generate action potentials more rapidly. When the hairs are bent in the direction away from the kinocilium, the hair cells become hyperpolarized, and the nerve fibers send impulses at a reduced rate (i.e., below the resting rate of discharge).

Although the receptors of the semicircular canals and the vestibule are responsible for dynamic and static equilibrium respectively, they rarely act independently. Complex interaction of many of the receptors is the rule. Processing is also complex and involves the brain stem and cerebellum as well as input from proprioceptors and the eyes.

ACTIVITY 7

Conducting Laboratory Tests on Equilibrium

The function of the semicircular canals and vestibule are not routinely tested in the laboratory, but the following simple tests illustrate normal equilibrium apparatus function as well as some of the complex processing interactions.

In the first balance test and the Barany test, you will look for **nystagmus,** which is the involuntary rolling of the eyes in any direction or the trailing of the eyes slowly in one direction, followed by their rapid movement in the opposite direction. It is normal during and after rotation; abnormal otherwise. The direction of nystagmus is that of its quick phase on acceleration.

Balance Tests

1. Have your partner walk a straight line, placing one foot directly in front of the other.

Is he or she able to walk without undue wobbling from side to side? ______________________

Did he or she experience any dizziness? ______________

The ability to walk with balance and without dizziness, unless subject to rotational forces, indicates normal function of the equilibrium apparatus.

Was nystagmus present? ______________________

2. Place three coins of different sizes on the floor. Ask your lab partner to pick up the coins, and carefully observe his or her muscle activity and coordination.

Did your lab partner have any difficulty locating and picking up the coins? ______________________

Describe your observations and your lab partner's observations during the test.

____

____

What kinds of interactions involving balance and coordination must occur for a person to move fluidly during this test?

____

____

3. If a person has a depressed nervous system, mental concentration may result in a loss of balance. Ask your lab partner to stand up and count backward from ten as rapidly as possible.

Did your lab partner lose balance? ______________

Barany Test (Induction of Nystagmus and Vertigo)*

This experiment evaluates the semicircular canals and should be conducted as a group effort to protect the test subject(s) from possible injury.

*__Vertigo__ is a sensation of dizziness and rotational movement when such movement is not occurring or has ceased.

⚠ Read the following precautionary notes before beginning:

- The subject(s) chosen should not be easily inclined to dizziness during rotational or turning movements.
- Rotation should be stopped immediately if the subject feels nauseated.
- Because the subject(s) will experience vertigo and loss of balance as a result of the rotation, several classmates should be prepared to catch, hold, or support the subject(s) as necessary until the symptoms pass.

1. Instruct the subject to sit on a rotating chair or stool, and to hold on to the arms or seat of the chair, feet on stool rungs. The subject's head should be tilted forward approximately 30 degrees (almost touching the chest). The horizontal (lateral) semicircular canal is stimulated when the head is in this position. The subject's eyes are to remain *open* during the test.
2. Four classmates should position themselves so that the subject is surrounded on all sides. The classmate posterior to the subject will rotate the chair.
3. Rotate the chair to the subject's right approximately 10 revolutions in 10 seconds, then suddenly stop the rotation.
4. Immediately note the direction of the subject's resultant nystagmus; and ask him or her to describe the feelings of movement, indicating speed and direction sensation. Record this information below.

____

____

____

If the semicircular canals are operating normally, the subject will experience a sensation that the stool is still rotating immediately after it has stopped and *will* demonstrate nystagmus.

When the subject is rotated to the right, the cupula will be bent to the left, causing nystagmus during rotation in which the eyes initially move slowly to the left and then quickly to the right. Nystagmus will continue until the cupula has returned to its initial position. Then, when rotation is stopped abruptly, the cupula will be bent to the right, producing nystagmus with its slow phase to the right and its rapid phase to the left. In many subjects, this will be accompanied by a feeling of vertigo and a tendency to fall to the right.

Romberg Test

The Romberg test determines the integrity of the dorsal white column of the spinal cord, which transmits impulses to the brain from the proprioceptors involved with posture.

1. Have your partner stand with his or her back to the blackboard.

2. Draw one line parallel to each side of your partner's body. He or she should stand erect, with eyes open and staring straight ahead for 2 minutes while you observe any movements. Did you see any gross swaying movements?

____

3. Repeat the test. This time the subject's eyes should be closed. Note and record the degree of side-to-side movement.

____

4. Repeat the test with the subject's eyes first open and then closed. This time, however, the subject should be positioned with his or her left shoulder toward, but not touching, the board so that you may observe and record the degree of front-to-back swaying.

____

Do you think the equilibrium apparatus of the inner ear was operating equally well in all these tests?

____

The proprioceptors? ______________________________

Why was the observed degree of swaying greater when the eyes were closed?

____

____

What conclusions can you draw regarding the factors necessary for maintaining body equilibrium and balance?

____

____

Role of Vision in Maintaining Equilibrium

To further demonstrate the role of vision in maintaining equilibrium, perform the following experiment. (Ask your lab partner to record observations and act as a "spotter.") Stand erect, with your eyes open. Raise your left foot approximately 30 cm off the floor, and hold it there for 1 minute.

Record the observations: ______________________________

____

____

Rest for 1 or 2 minutes; and then repeat the experiment with the same foot raised but with your eyes closed. Record the observations:

____

____

NAME ______________________________

LAB TIME/DATE ______________________

REVIEW SHEET

EXERCISE 25

Special Senses: Hearing and Equilibrium

Anatomy of the Ear

1. Select the terms from column B that apply to the column A descriptions. Some terms are used more than once.

Column A

_______, _______, _______, 1. structure composing the external ear

_______, _______, _______, 2. structures composing the internal ear

_______, _______, _______, 3. collectively called the ossicles

_______ 4. involved in equalizing the pressure in the middle ear with atmospheric pressure

_______ 5. vibrates at the same frequency as sound waves hitting it; transmits the vibrations to the ossicles

_______, _______ 6. contain receptors for the sense of balance

_______ 7. transmits the vibratory motion of the stirrup to the fluid in the scala vestibuli of the inner ear

_______ 8. acts as a pressure relief valve for the increased fluid pressure in the scala tympani; bulges into the tympanic cavity

_______ 9. passage between the throat and the tympanic cavity

_______ 10. fluid contained within the membranous labyrinth

_______ 11. fluid contained within the osseous labyrinth and bathing the membranous labyrinth

Column B

a. pharyngotympanic (auditory) tube

b. cochlea

c. endolymph

d. external auditory canal

e. incus (anvil)

f. malleus (hammer)

g. oval window

h. perilymph

i. pinna (auricle)

j. round window

k. semicircular canals

l. stapes (stirrup)

m. tympanic membrane

n. vestibule

2. Identify all indicated structures and ear regions in the following diagram.

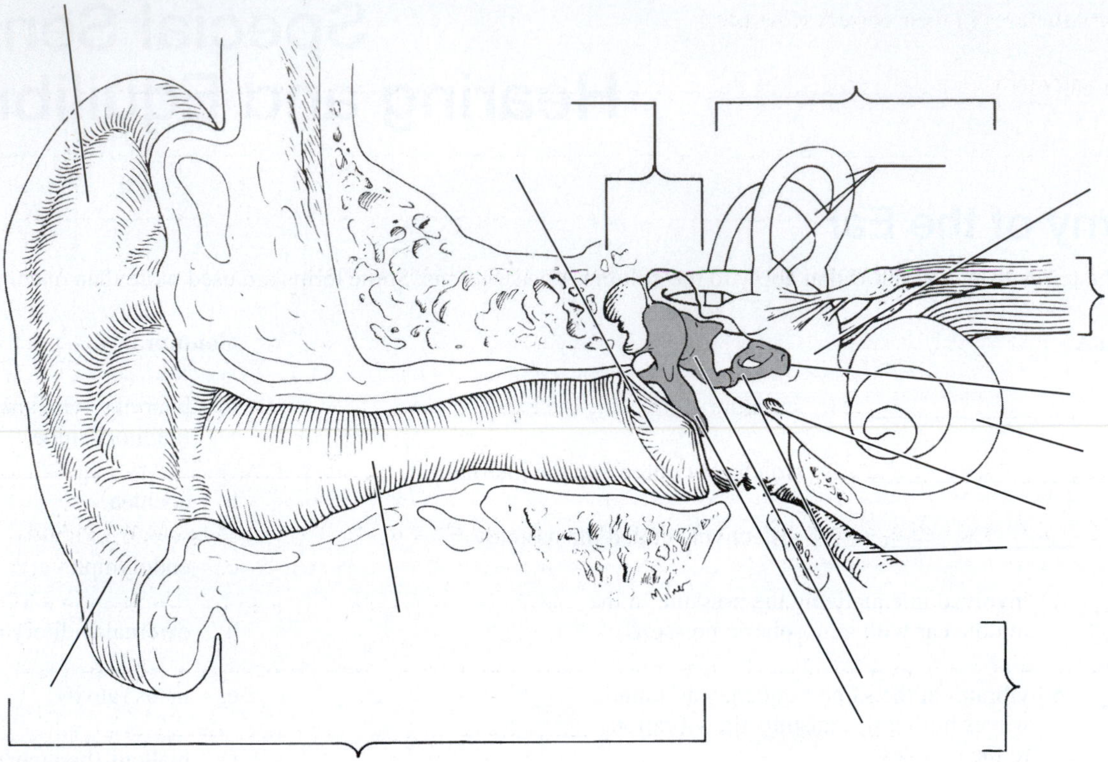

3. Match the membranous labyrinth structures listed in column B with the descriptive statements in column A.

Column A

_______, _______ 1. sacs found within the vestibule

_______ 2. contains the spiral organ (of Corti)

_______, _______ 3. sites of the maculae

_______ 4. positioned in all spatial planes

_______ 5. hair cells of spiral organ (of Corti) rest on this membrane

_______ 6. gelatinous membrane overlying the hair cells of the spiral organ (of Corti)

_______ 7. contains the crista ampullaris

_______, _______, _______, _______ 8. function in static equilibrium

_______, _______, _______, _______ 9. function in dynamic equilibrium

_______ 10. carries auditory information to the brain

_______ 11. gelatinous cap overlying hair cells of the crista ampullaris

_______ 12. grains of calcium carbonate in the maculae

Column B

a. ampulla

b. basilar membrane

c. cochlear duct

d. cochlear nerve

e. cupula

f. otoliths

g. saccule

h. semicircular ducts

i. tectorial membrane

j. utricle

k. vestibular nerve

4. Sound waves hitting the tympanic membrane (eardrum) initiate its vibratory motion. Trace the pathway through which vibrations and fluid currents are transmitted to finally stimulate the hair cells in the spiral organ (of Corti). (Name the appropriate ear structures in their correct sequence.)

Tympanic membrane → ______________________________

____

____

5. Describe how sounds of different frequency (pitch) are differentiated in the cochlea. ______________________________

____

____

____

6. Explain the role of the endolymph of the semicircular canals in activating the receptors during angular motion.

____

____

____

7. Explain the role of the otoliths in perception of static equilibrium (head position). ______________________________

____

____

____

Laboratory Tests

8. Was the auditory acuity measurement made during the experiment on page 387 the same or different for both ears?

______________________________ What factors might account for a difference in the acuity of the two ears?

____

____

9. During the sound localization experiment on page 387, note the position(s) in which the sound was least easily located.

How can this phenomenon be explained? ______________________________

____

10. In the frequency experiment on page 388, note which tuning fork was the most difficult to hear. ______________________________

What conclusion can you draw? ______________________________

____

11. When the tuning fork handle was pressed to your forehead during the Weber test, where did the sound seem to originate?

____

Where did it seem to originate when one ear was plugged with cotton? ____

How do sound waves reach the cochlea when conduction deafness is present? ____

____

12. Indicate whether the following conditions relate to conduction deafness (C) or sensorineural deafness (S).

_______ 1. can result from the fusion of the ossicles

_______ 2. can result from a lesion on the cochlear nerve

_______ 3. sound heard in one ear but not in the other during bone and air conduction

_______ 4. can result from otitis media

_______ 5. can result from impacted cerumen or a perforated eardrum

_______ 6. can result from a blood clot in the auditory cortex

13. The Rinne test evaluates an individual's ability to hear sounds conducted by air or bone. Which is more indicative of normal hearing? ____

14. Define *nystagmus.* ____

____

Define *vertigo.* ____

____

15. The Barany test investigated the effect that rotatory acceleration had on the semicircular canals. Explain *why* the subject still had the sensation of rotation immediately after being stopped. ____

____

16. What is the usual reason for conducting the Romberg test? ____

____

Was the degree of sway greater with the eyes open or closed? Why? ____

____

17. Normal balance, or equilibrium, depends on input from a number of sensory receptors. Name them.

____

18. What effect does alcohol consumption have on balance and equilibrium? Explain. ____

____

EXERCISE 26

Special Senses: Olfaction and Taste

MATERIALS

- ☐ Prepared slides: nasal olfactory epithelium (l.s.); the tongue showing taste buds (x.s.)
- ☐ Compound microscope
- ☐ Small mirror
- ☐ Paper towels
- ☐ Packets of granulated sugar
- ☐ Disposable autoclave bag
- ☐ Paper plates
- ☐ Equal-size food cubes of cheese, apple, raw potato, dried prunes, banana, raw carrot, and hard-cooked egg white (These prepared foods should be in an opaque container; a foil-lined egg carton would work well.)
- ☐ Toothpicks
- ☐ Disposable gloves
- ☐ Cotton-tipped swabs
- ☐ Paper cups
- ☐ Flask of distilled or tap water
- ☐ Prepared vials of oil of cloves, oil of peppermint, and oil of wintergreen or corresponding flavors found in the condiment section of a supermarket
- ☐ Chipped ice
- ☐ Five numbered vials containing common household substances with strong odors (herbs, spices, etc.)
- ☐ Nose clips
- ☐ Absorbent cotton

MasteringA&P™

Access practice quizzes and more in the Study Area at www.masteringaandp.com.

PAL

For access to anatomical models and more, check out Practice Anatomy Lab.

OBJECTIVES

1. To describe the location and cellular composition of the olfactory epithelium.
2. To describe the structure and function of the taste receptors.
3. To name the five basic qualities of taste sensation, and to list the chemical substances that elicit them.
4. To explain the interdependence between the senses of smell and taste.
5. To name two factors other than olfaction that influence taste appreciation of foods.
6. To define *olfactory adaptation.*

PRE-LAB QUIZ

1. Circle True or False. Receptors for olfaction and taste are classified as chemoreceptors because they respond to chemicals in solution.
2. The organ of smell is the __________, located in the roof of the nasal cavity.
   a. nares
   b. nostrils
   c. olfactory epithelium
   d. olfactory nerve
3. Circle the correct term. Olfactory receptor cells are bipolar / unipolar neurons whose olfactory cilia extend outward from the epithelium.
4. Most taste buds are located in __________, peglike projections of the tongue mucosa.
   a. cilia
   b. concha
   c. papillae
   d. supporting cells
5. Circle the correct term. Most taste buds are made of two / three types of modified epithelial cells.
6. There are five basic taste sensations. Name one. __________
7. Circle True or False. Texture, temperature, and smell have little or no effect on the sensation of taste.
8. You will use absorbent cotton and oil of wintergreen, peppermint, or cloves to test for olfactory
   a. accommodation
   b. adaptation
   c. identification
   d. recognition

The receptors for olfaction and taste are classified as **chemoreceptors** because they respond to chemicals in solution. Although five relatively specific types of taste receptors have been identified, the olfactory receptors are considered sensitive to a much wider range of chemical sensations. The sense of smell is the least understood of the special senses.

Location and Anatomy of the Olfactory Receptors

The **olfactory epithelium** (organ of smell) occupies an area of about 5 cm^2 in the roof of the nasal cavity (Figure 26.1a). Since the air entering the human nasal cavity must make a hairpin turn to enter the respiratory passages below, the nasal epithelium is in a rather poor position for performing its function. This is why sniffing, which brings more air into contact with the receptors, intensifies the sense of smell.

The specialized receptor cells in the olfactory epithelium are surrounded by **supporting cells,** non-sensory epithelial cells. The **olfactory receptor cells** are bipolar neurons whose **olfactory cilia** extend outward from the epithelium. Axonal nerve filaments emerging from their basal ends penetrate the cribriform plate of the ethmoid bone and proceed as the *olfactory nerves* to synapse in the olfactory bulbs lying on either side of the crista galli of the ethmoid bone. Impulses from neurons of the olfactory bulbs are then conveyed to the olfactory portion of the cortex (uncus) without synapsing in the thalamus.

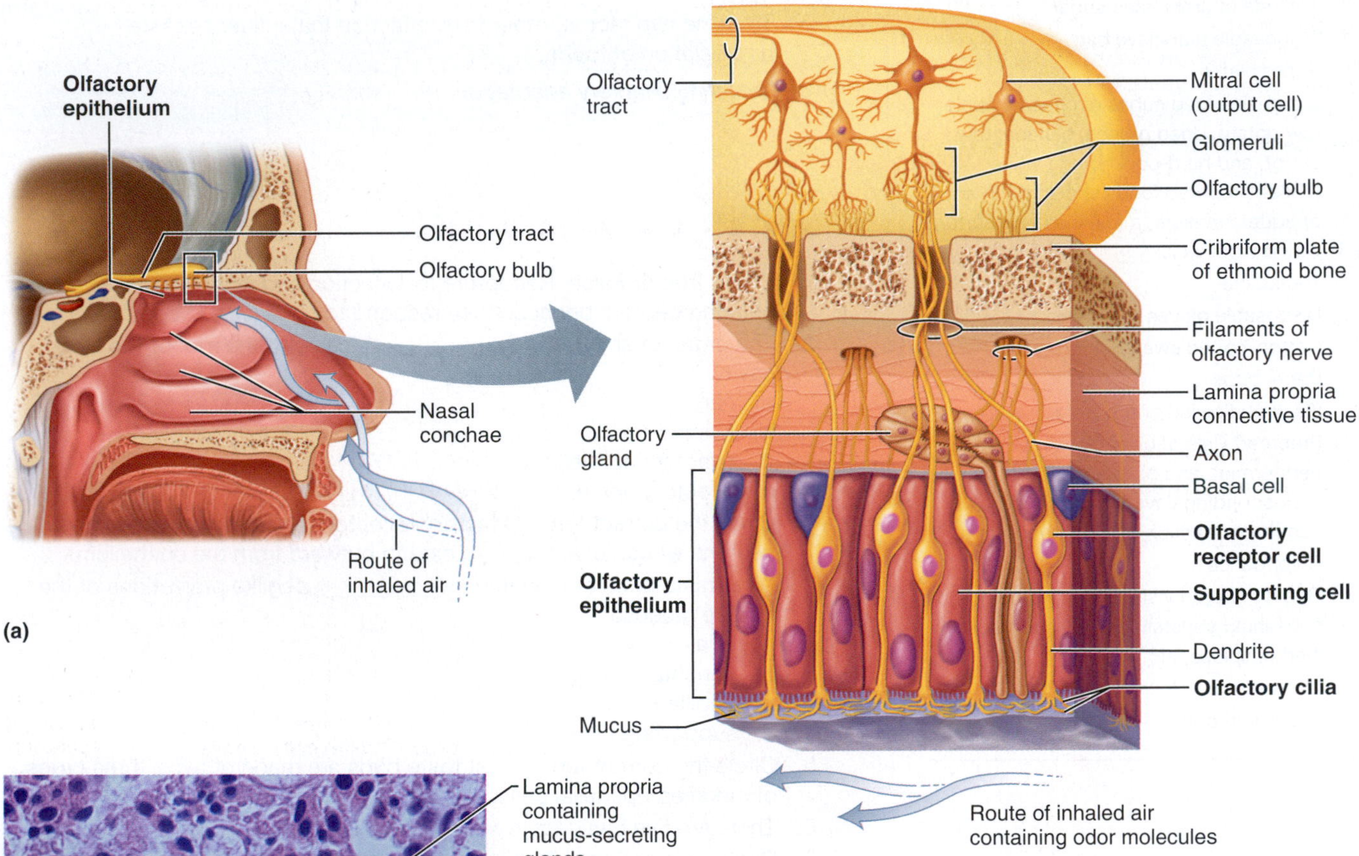
Olfactory epithelium
Olfactory tract
Olfactory bulb
Nasal conchae
Route of inhaled air
Olfactory tract
Mitral cell (output cell)
Glomeruli
Olfactory bulb
Cribriform plate of ethmoid bone
Filaments of olfactory nerve
Lamina propria connective tissue
Olfactory gland
Axon
Basal cell
Olfactory receptor cell
Olfactory epithelium
Supporting cell
Dendrite
Olfactory cilia
Mucus
Route of inhaled air containing odor molecules

(a)

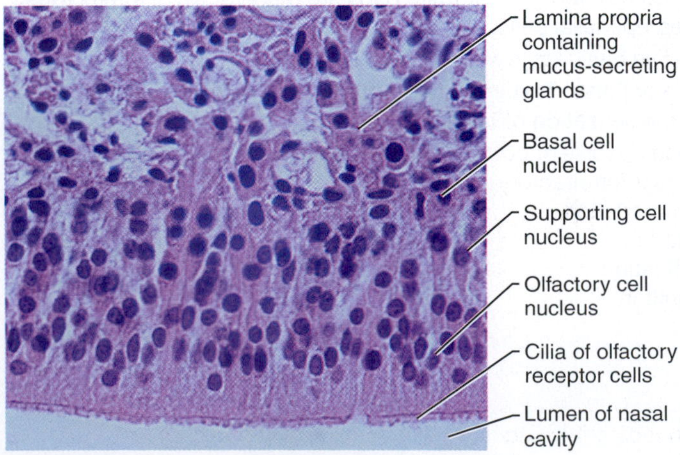
Lamina propria containing mucus-secreting glands
Basal cell nucleus
Supporting cell nucleus
Olfactory cell nucleus
Cilia of olfactory receptor cells
Lumen of nasal cavity

(b)

FIGURE 26.1 Location and cellular composition of olfactory epithelium. (a) Diagrammatic representation of olfactory epithelium. Enlarged view shows the course of the olfactory nerve filaments. **(b)** Histological image of the olfactory epithelium (220×).

ACTIVITY 1

Microscopic Examination of the Olfactory Epithelium

Obtain a longitudinal section of olfactory epithelium. Examine it closely using a compound microscope, comparing it to Figure 26.1b.

Location and Anatomy of Taste Buds

The **taste buds,** specific receptors for the sense of taste, are widely but not uniformly distributed in the oral cavity. Most are located in **papillae,** peglike projections of the mucosa, on the dorsal surface of the tongue (as described next). A few are found on the soft palate, epiglottis, pharynx, and inner surface of the cheeks.

Taste buds are located primarily on the sides of the large round **circumvallate papillae** (arranged in a V formation on the posterior surface of the tongue); in the side walls of the **foliate papillae;** and on the tops of the more numerous, mushroom-shaped **fungiform papillae.** (See Figure 26.2.)

- Use a mirror to examine your tongue. Which of the various papillae types can you pick out? ______________________

____

Each taste bud consists largely of a globular arrangement of two types of modified epithelial cells: the **gustatory,** or **taste cells,** which are the actual receptor cells, and basal cells. Several nerve fibers enter each taste bud and supply sensory nerve endings to each of the taste cells. The long microvilli of the receptor cells penetrate the epithelial surface through an opening called the **taste pore.** When these microvilli, called **gustatory hairs,** contact specific chemicals in the solution, the taste cells depolarize. The afferent fibers from the taste buds to the sensory cortex in the postcentral gyrus of the brain are carried in three cranial nerves: the *facial nerve (VII)* serves the anterior two-thirds of the tongue; the *glossopharyngeal nerve (IX)* serves the posterior third of the tongue; and the *vagus nerve (X)* carries a few fibers from the pharyngeal region.

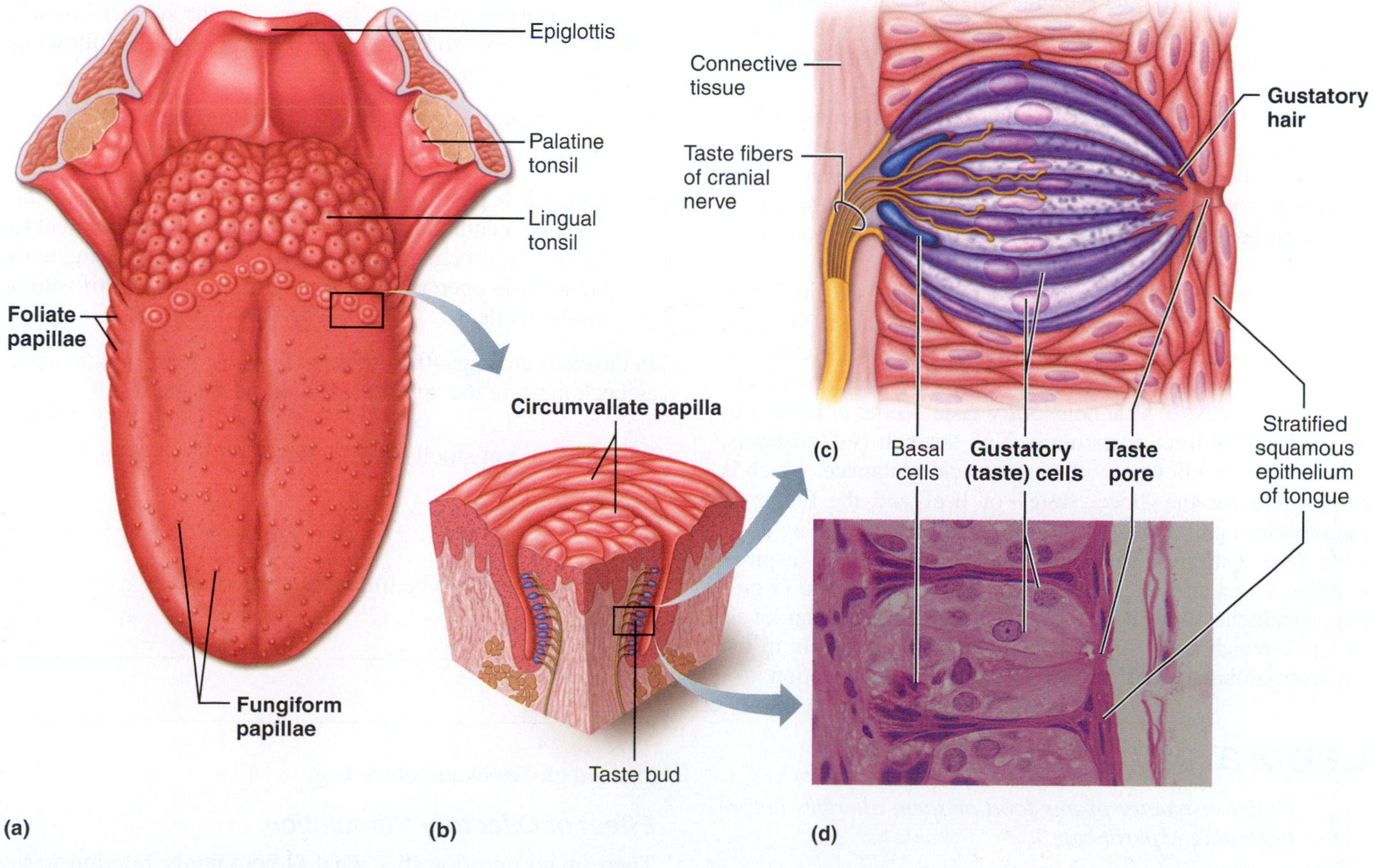
Epiglottis
Palatine tonsil
Lingual tonsil
Foliate papillae
Fungiform papillae
(a)
Circumvallate papilla
Taste bud
(b)
Connective tissue
Taste fibers of cranial nerve
Gustatory hair
(c)
Basal cells
Gustatory (taste) cells
Taste pore
Stratified squamous epithelium of tongue
(d)

FIGURE 26.2 Location and structure of taste buds. (a) Taste buds on the tongue are associated with papillae, projections of the tongue mucosa. **(b)** A sectioned circumvallate papilla shows the position of the taste buds in its lateral walls. **(c)** An enlarged view of a taste bud. **(d)** Photomicrograph of a taste bud (330×).

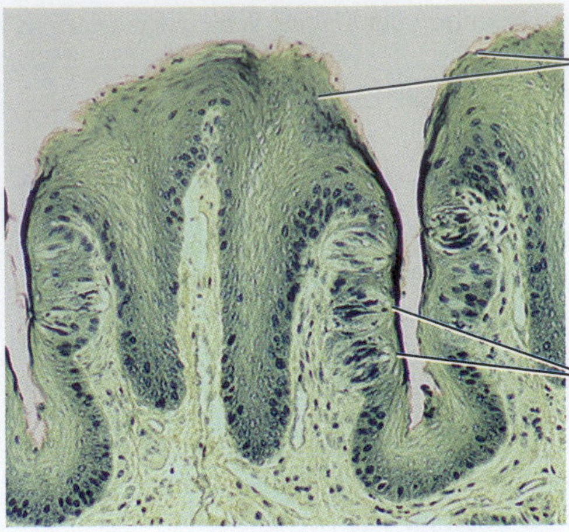
Foliate papillae
Taste buds

FIGURE 26.3 Taste buds on the lateral aspects of foliate papillae of the tongue (100×).

ACTIVITY 2

Microscopic Examination of Taste Buds

Obtain a microscope and a prepared slide of a tongue cross section. Use Figure 26.2b as a guide to aid you in locating the taste buds on the tongue papillae. Make a detailed study of one taste bud. Identify the taste pore and gustatory hairs if observed. Compare your observations to Figure 26.3.

When taste is tested with pure chemical compounds, most taste sensations can be grouped into one of five basic qualities—sweet, sour, bitter, salty, or umami (*u-mam'e;* "delicious"). Although all taste buds are believed to respond in some degree to all five classes of chemical stimuli, each type responds optimally to only one.

The *sweet* receptors respond to a number of seemingly unrelated compounds such as sugars (fructose, sucrose, glucose), saccharine, some lead salts, and some amino acids. Activation of *sour* receptors is mediated by hydrogen ions (H^+) and blockade of K^+ (or Na^+) channels. *Salty* taste seems to be due to influx of metal ions, particularly Na^+ through Na^+ channels, while *umami* is elicited by the amino acid glutamate, which is responsible for the "meat taste" of beef and the flavor of monosodium glutamate (MSG). *Bitter* taste is elicited by alkaloids (e.g. caffeine) and other substances such as aspirin. Umami, sweet, and bitter responses are mediated by a G protein, gustducin. Activation leads to increased intracellular levels of Ca^{2+}, which causes the opening of cation channels in the plasma membrance of the receptor cell and depolarization.

Laboratory Experiments

! *Notify instructor of any food or scent allergies before beginning experiments.*

ACTIVITY 3

Stimulating Taste Buds

1. Obtain several paper towels, a sugar packet, and a disposable autoclave bag and bring them to your bench.

2. With a paper towel, dry the dorsal surface of your tongue.

! Immediately dispose of the paper towel in the autoclave bag.

3. Tear off a corner of the sugar packet and shake a few sugar crystals on your dried tongue. Do *not* close your mouth.

Time how long it takes to taste the sugar. ____________ sec

Why couldn't you taste the sugar immediately?

____

ACTIVITY 4

Examining the Combined Effects of Smell, Texture, and Temperature on Taste

Effects of Smell and Texture

1. Ask the subject to sit with eyes closed and to pinch his or her nostrils shut.

2. Using a paper plate, obtain samples of the food items provided by your laboratory instructor. At no time should the subject be allowed to see the foods being tested. Wear plastic gloves and use toothpicks to handle food.

3. For each test, place a cube of food in the subject's mouth and ask him or her to identify the food by using the following sequence of activities:

- First, manipulate the food with the tongue.
- Second, chew the food.
- Third, if a positive identification is not made with the first two techniques and the taste sense, ask the subject to release the pinched nostrils and to continue chewing with the nostrils open to determine if a positive identification can be made.

In the chart on page 401, record the type of food, and then put a check mark in the appropriate column for the result.

Was the sense of smell equally important in all cases?

____

Where did it seem to be important and why?

____

____

Discard gloves in autoclave bag.

Effect of Olfactory Stimulation

There is no question that what is commonly referred to as taste depends heavily on stimulation of the olfactory receptors, particularly in the case of strongly odoriferous substances. The following experiment should illustrate this fact.

1. Obtain vials of oil of wintergreen, peppermint, and cloves, paper cup, flask of water, paper towels, and some

Identification by Texture and Smell				
Food tested	Texture only	Chewing with nostrils pinched	Chewing with nostrils open	Identification not made

fresh cotton-tipped swabs. Ask the subject to sit so that he or she cannot see which vial is being used, and to dry the tongue and close the nostrils.

2. Use a cotton swab to apply a drop of one of the oils to the subject's tongue. Can he or she distinguish the flavor?

____

! Put the used swab in the autoclave bag. *Do not redip the swab into the oil.*

3. Have the subject open the nostrils, and record the change in sensation he or she reports.

____

4. Have the subject rinse the mouth well and dry the tongue.

5. Prepare two swabs, each with one of the two remaining oils.

6. Hold one swab under the subject's open nostrils, while touching the second swab to the tongue.

Record the reported sensations. ______________________________

____

____

! 7. Dispose of the used swabs and paper towels in the autoclave bag before continuing.

Which sense, taste or smell, appears to be more important in the proper identification of a strongly flavored volatile substance?

____

Effect of Temperature

In addition to the effect that olfaction and food texture play in determining our taste sensations, the temperature of foods also helps determine if the food is appreciated or even tasted. To illustrate this, have your partner hold some chipped ice on the tongue for approximately a minute and then close his or her eyes. Immediately place any of the foods previously identified in his or her mouth and ask for an identification.

Results? ______________________________

____

____

Identification by Mouth and Nasal Inhalation			
Vial number	Identification with nose clips	Identification without nose clips	Other observations
1			
2			
3			
4			
5			

ACTIVITY 5

Assessing the Importance of Taste and Olfaction in Odor Identification

1. Go to the designated testing area. Close your nostrils with a nose clip, and breathe through your mouth. Breathing through your mouth only, attempt to identify the odors of common substances in the numbered vials at the testing area. Do not look at the substance in the container. Record your responses on the chart above.

2. Remove the nose clips, and repeat the tests using your nose to sniff the odors. Record your responses on the chart above.

3. Record any other observations you make as you conduct the tests.

4. Which method gave the best identification results?

____

What can you conclude about the effectiveness of the senses of taste and olfaction in identifying odors?

____

ACTIVITY 6

Demonstrating Olfactory Adaptation

Obtain some absorbent cotton and two of the following oils (oil of wintergreen, peppermint, or cloves). Place several drops of oil on the absorbent cotton. Press one nostril shut. Hold the cotton under the open nostril and exhale through the mouth. Record the time required for the odor to disappear (for olfactory adaptation to occur).

______________ sec

Repeat the procedure with the other nostril.

______________ sec

Immediately test another oil with the nostril that has just experienced olfactory adaptation. What are the results?

____

What conclusions can you draw? ______________________________

____

NAME ______________________________

LAB TIME/DATE ______________________

REVIEW SHEET

EXERCISE 26

Special Senses: Olfaction and Taste

Location and Anatomy of the Olfactory Receptors

1. Describe the location and cellular composition of the olfactory epithelium. ______________________________

____

____

2. How and why does sniffing improve your sense of smell? ______________________________

____

Location and Anatomy of Taste Buds

3. Name five sites where receptors for taste are found, and circle the predominant site.

______________________, ______________________, ______________________,

______________________, and ______________________

4. Describe the cellular makeup and arrangement of a taste bud. (Use a diagram, if helpful.) ______________________

____

Laboratory Experiments

5. Taste and smell receptors are both classified as ______________________________, because they both

respond to ____

6. Why is it impossible to taste substances with a dry tongue? ______________________________

____

7. The basic taste sensations are mediated by specific chemical substances or groups. Name them for the following taste modalities.

salt: ____________________ sour: ____________________ umami: ____________________

bitter: ____________________ sweet: ____________________

8. Name three factors that influence our appreciation of foods. Substantiate each choice with an example from the laboratory experience.

1. ____________ Substantiation: ______________________________

____

2. ____________ Substantiation: ______________________________

____

3. ____________ Substantiation: ______________________________

____

Which of the factors chosen is most important? ____________ Substantiate your choice with an example from everyday life. ______________________________

____

Expand on your explanation and choices by explaining why a cold, greasy hamburger is unappetizing to most people.

____

____

9. How palatable is food when you have a cold? ____________ Explain your answer. ____________

____

10. In your opinion, is olfactory adaptation desirable? ____________ Explain your answer.

____

____

EXERCISE 27

Functional Anatomy of the Endocrine Glands

MATERIALS

- ☐ Human torso model
- ☐ Anatomical chart of the human endocrine system
- ☐ Compound microscope
- ☐ Prepared slides of the anterior pituitary,* posterior pituitary, thyroid gland, parathyroid glands, adrenal gland, and pancreas*

For instructions on animal dissections, see the dissection exercises starting on page 697 in the cat, fetal pig, and rat editions of this manual.

*With differential staining if possible

OBJECTIVES

1. To identify and name the major endocrine glands and tissues of the body when provided with an appropriate diagram.
2. To list the hormones produced by the endocrine glands and discuss the general function of each.
3. To indicate the means by which hormones contribute to body homeostasis by giving appropriate examples of hormonal actions.
4. To cite mechanisms by which the endocrine glands are stimulated to release their hormones.
5. To describe the structural and functional relationship between the hypothalamus and the pituitary.
6. To describe a major pathological consequence of hypersecretion and hyposecretion of several of the hormones considered.
7. To correctly identify the histologic structure of the thyroid, parathyroid, pancreas, anterior and posterior pituitary, adrenal cortex, and adrenal medulla by microscopic inspection or when presented with an appropriate photomicrograph or diagram.
8. To name and point out the specialized hormone-secreting cells in the above tissues as studied in the laboratory.

PRE-LAB QUIZ

1. Define *hormone*. ______________________________
2. Circle the correct term. An endocrine / exocrine gland is a ductless gland that empties its hormone into the extracellular fluid, from which it enters the blood.
3. The pituitary gland, also known as the __________, is located in the sella turcica of the sphenoid bone.
   a. hypophysis
   b. hypothalamus
   c. thalamus
4. Circle True or False. The anterior pituitary gland is also referred to as the master endocrine gland because it controls the activity of many other endocrine glands.
5. The __________ gland composed of two lobes, is located in the throat, just inferior to the larynx.
   a. pancreas
   b. posterior pituitary
   c. thymus
   d. thyroid
6. The pancreas produces two hormones that are responsible for regulating blood sugar levels. Name the hormone that increases blood glucose levels. ____________________
7. Circle True or False. The gonads are considered to be both endocrine and exocrine glands.

Text continues on next page.

MasteringA&P™
Access practice quizzes and more in the Study Area at www.masteringaandp.com.

PAL
For access to anatomical models and more, check out Practice Anatomy Lab.

8. This gland is rather large in an infant, but begins to atrophy at puberty and is relatively inconspicuous by old age. It produces hormones that direct the maturation of T cells. It is the __________ gland.
   a. pineal
   b. testes
   c. thymus
   d. thyroid

9. Circle the correct term. Islets of Pancreatic islets / Acinar cells form the endocrine portion of the pancreas.

10. The outer cortex of the adrenal gland is divided into three zones (zonas). Which one produces aldosterone?
   a. zona fasciculata
   b. zona glomerulosa
   c. zona reticularis

The **endocrine system** is the second major control system of the body. Acting with the nervous system, it helps coordinate and integrate the activity of the body's cells. However, the nervous system employs electrochemical impulses to bring about rapid control, whereas the more slowly acting endocrine system employs chemical "messengers," or **hormones,** which are released into the blood to be transported throughout the body.

The term *hormone* comes from a Greek word meaning "to arouse." The body's hormones, which are steroids or amino acid–based molecules, arouse the body's tissues and cells by stimulating changes in their metabolic activity. These changes lead to growth and development and to the physiological homeostasis of many body systems. Although all hormones are bloodborne, a given hormone affects only the biochemical activity of a specific organ or organs. Organs that respond to a particular hormone are referred to as the **target organs** of that hormone. The ability of the target tissue to respond seems to depend on the ability of the hormone to bind with specific receptors (proteins) occurring on the cells' plasma membrane or within the cells.

Although the function of some hormone-producing glands (the anterior pituitary, thyroid, adrenals, parathyroids) is purely endocrine, the function of others (the pancreas and gonads) is mixed—both endocrine and exocrine. Both types of glands are derived from epithelium, but the endocrine, or ductless, glands release their product (always hormonal) directly into the extracellular fluid from which it enters blood or lymph. The exocrine glands release their products at the body's surface or upon an epithelial membrane via ducts. In addition, there are hormone-producing cells in the heart, the gastrointestinal tract, kidney, skin, adipose tissue, skeleton, and placenta, organs whose functions are primarily nonendocrine. Only the major endocrine organs, plus the pineal gland and the thymus, are considered here.

Gross Anatomy and Basic Function of the Endocrine Glands

Pituitary Gland (Hypophysis)

The *pituitary gland,* or *hypophysis,* is located in the hypophyseal fossa of the sella turcica of the sphenoid bone. It consists largely of two functional *lobes,* the **adenohypophysis,** or **anterior pituitary,** and the **neurohypophysis,** consisting mainly of the **posterior pituitary** (Figure 27.1). The pituitary gland is attached to the hypothalamus by a stalk called the **infundibulum.**

Anterior Pituitary Hormones

The anterior pituitary (lobe) secretes a number of hormones. Four of these are **tropic hormones.** A tropic hormone stimulates its target organ, which is also an endocrine gland, to secrete its hormones. Target organ hormones then exert their effects on other body organs and tissues. The anterior pituitary tropic hormones include:

- **Gonadotropins–follicle-stimulating hormone (FSH)** and **luteinizing hormone (LH)**—regulate gamete production and hormonal activity of the gonads (ovaries and testes). The precise roles of the gonadotropins are described in Exercise 43 along with other considerations of reproductive system physiology.
- **Adrenocorticotropic hormone (ACTH)** regulates the endocrine activity of the cortex portion of the adrenal gland.
- **Thyroid-stimulating hormone (TSH),** or **thyrotropin,** influences the growth and activity of the thyroid gland.

The two other important hormones produced by the anterior pituitary are not directly involved in the regulation of other endocrine glands of the body.

- **Growth hormone (GH)** is a general metabolic hormone that plays an important role in determining body size. It affects many tissues of the body; however, its major effects are exerted on the growth of muscle and the long bones of the body. Hyposecretion results in pituitary dwarfism in children. Hypersecretion causes gigantism in children and **acromegaly** (overgrowth of bones in hands, feet, and face) in adults. ■
- **Prolactin (PRL)** stimulates breast development and promotes and maintains lactation by the mammary glands after childbirth. It may stimulate testosterone production in males.

A less important secretory product of the anterior pituitary is pro-opiomelanocortin (POMC), a prohormone. POMC is split by enzymes into ACTH, enkephalin and beta endorphin (natural opiates), and melanocyte-stimulating hormone (MSH).

The anterior pituitary controls the activity of so many other endocrine glands that it has often been called the *master endocrine gland.* However, the anterior pituitary is controlled by neurosecretions, *releasing* or *inhibiting hormones,* produced by neurons of the ventral hypothalamus. These hypothalamic hormones are liberated into the **hypophyseal portal system** (Figure 27.1), and carried to cells of the anterior pituitary where they control release of anterior pituitary hormones.

Posterior Pituitary Hormones The posterior pituitary (lobe) is not an endocrine gland in a strict sense because it

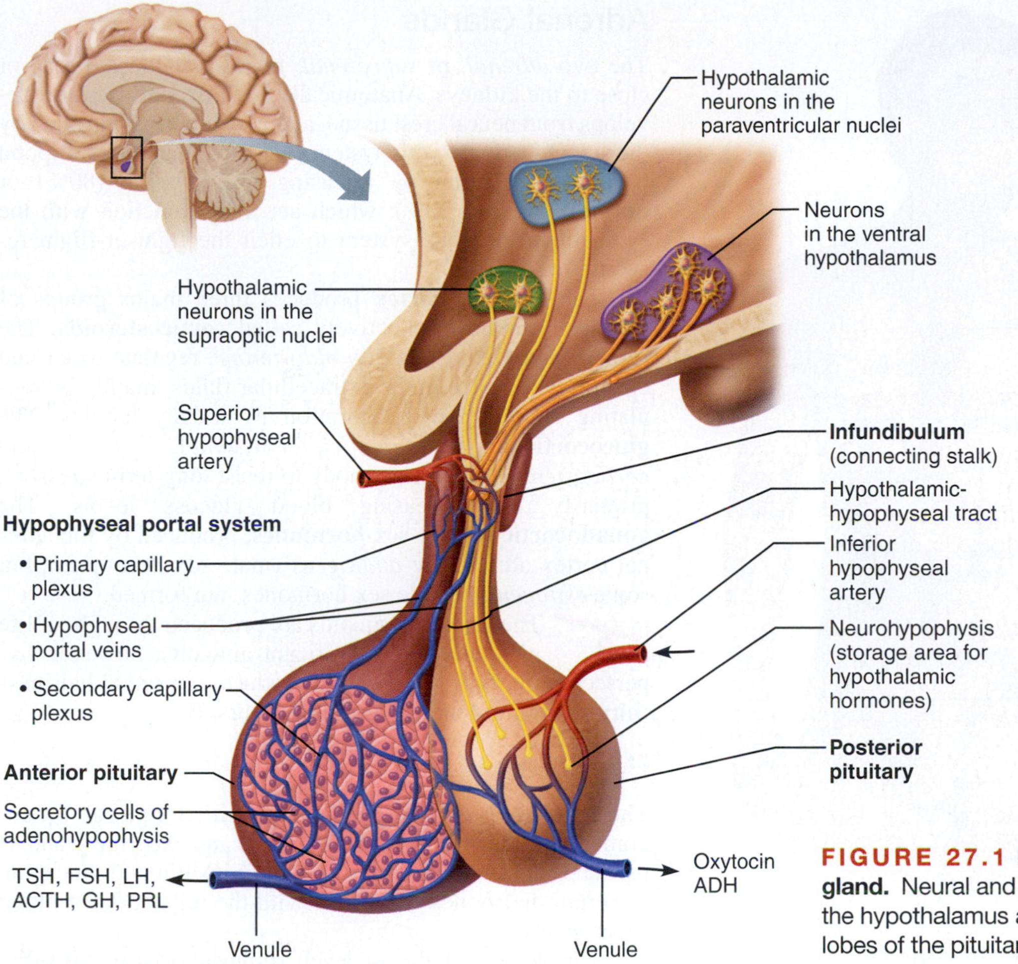
Hypothalamic neurons in the paraventricular nuclei
Neurons in the ventral hypothalamus
Hypothalamic neurons in the supraoptic nuclei
Superior hypophyseal artery
Infundibulum (connecting stalk)
Hypothalamic-hypophyseal tract
Inferior hypophyseal artery
Hypophyseal portal system
• Primary capillary plexus
• Hypophyseal portal veins
• Secondary capillary plexus
Neurohypophysis (storage area for hypothalamic hormones)
Posterior pituitary
Anterior pituitary
Secretory cells of adenohypophysis
TSH, FSH, LH, ACTH, GH, PRL
Oxytocin ADH
Venule
Venule

FIGURE 27.1 Hypothalamus and pituitary gland. Neural and vascular relationships between the hypothalamus and the anterior and posterior lobes of the pituitary are depicted.

does not synthesize the hormones it releases. Instead, it acts as a storage area for two hormones transported to it via the axons of neurons in the paraventricular and supraoptic nuclei of the hypothalamus. The hormones are released in response to nerve impulses from these neurons. The first of these hormones is **oxytocin,** which stimulates powerful uterine contractions during birth and coitus and also causes milk ejection in the lactating mother. The second, **antidiuretic hormone (ADH),** causes the distal and collecting tubules of the kidneys to reabsorb more water from the urinary filtrate, thereby reducing urine output and conserving body water.

Hyposecretion of ADH results in dehydration from excessive urine output, a condition called **diabetes insipidus.** Individuals with this condition experience an insatiable thirst. Hypersecretion results in edema, headache, and disorientation. ■

Pineal Gland

The *pineal gland* is a small cone-shaped gland located in the roof of the third ventricle of the brain. Its major endocrine product is **melatonin,** which exhibits a diurnal (daily) cycle. It peaks at night, making us drowsy, and is lowest around noon.

The endocrine role of the pineal body in humans is still controversial, but it is known to play a role in the biological rhythms (particularly mating and migratory behavior) of other animals. In humans, melatonin appears to exert some inhibitory effect on the reproductive system that prevents precocious sexual maturation.

Thyroid Gland

The *thyroid gland* is composed of two lobes joined by a central mass, or isthmus. It is located in the throat, just inferior to the larynx. It produces two major hormones, thyroid hormone and calcitonin.

Thyroid hormone (TH) is actually two physiologically active hormones known as **T_4 (thyroxine)** and **T_3 (triiodothyronine).** Because its primary function is to control the rate of body metabolism and cellular oxidation, TH affects virtually every cell in the body.

Hyposecretion of thyroxine leads to a condition of mental and physical sluggishness, which is called **myxedema** in the adult. Hypersecretion causes elevated metabolic rate, nervousness, weight loss, sweating, and irregular heartbeat. ■

Calcitonin decreases blood calcium levels by stimulating calcium salt deposit in the bones. Although it acts antagonistically to parathyroid hormone, the hormonal product of the parathyroid glands, calcitonin is not involved in day-to-day control of calcium homeostasis.

Parathyroid Glands

The *parathyroid glands* are found embedded in the posterior surface of the thyroid gland. Typically, there are two

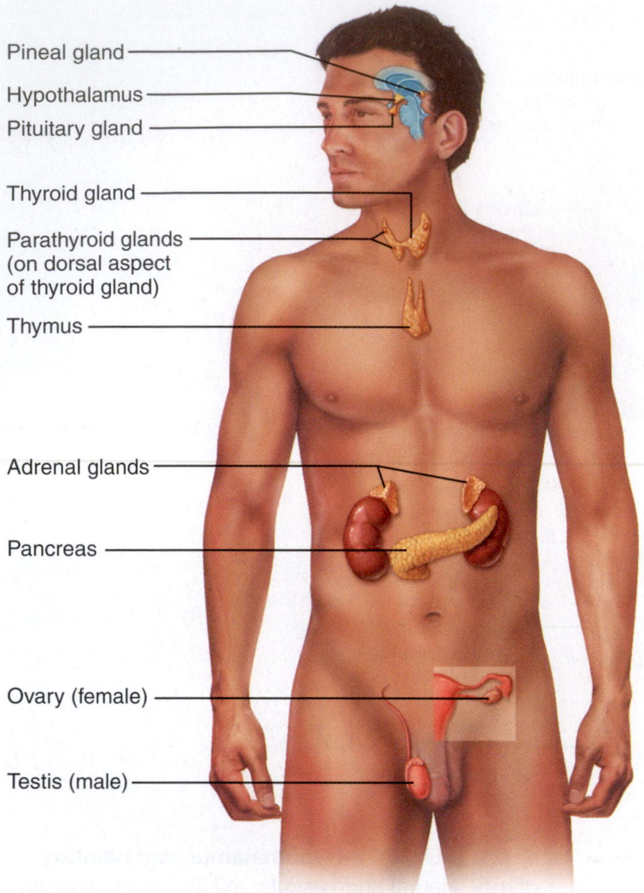
Pineal gland
Hypothalamus
Pituitary gland
Thyroid gland
Parathyroid glands (on dorsal aspect of thyroid gland)
Thymus
Adrenal glands
Pancreas
Ovary (female)
Testis (male)

FIGURE 27.2 Human endocrine organs.

small oval glands on each lobe, but there may be more and some may be located in other regions of the neck. They secrete **parathyroid hormone (PTH),** the most important regulator of calcium balance of the blood. When blood calcium levels decrease below a certain critical level, the parathyroids release PTH, which causes release of calcium from bone matrix and prods the kidney to reabsorb more calcium and less phosphate from the filtrate. PTH also stimulates the kidneys to convert vitamin D to its active D_3 form, *calcitriol.*

Hyposecretion increases neural excitability and may lead to **tetany,** prolonged muscle spasms that can result in respiratory paralysis and death. Hypersecretion of PTH results in loss of calcium from bones, causing deformation, softening, and spontaneous fractures. ■

Thymus

The *thymus* is a bilobed gland situated in the superior thorax, posterior to the sternum and anterior to the heart and lungs. Conspicuous in the infant, it begins to atrophy at puberty, and by old age it is relatively inconspicuous. The thymus produces several different families of hormones including **thymulin, thymosins,** and **thymopoietins.** These hormones are thought to be involved in the development of T lymphocytes and the immune response. Their role is poorly understood; they appear to act mainly locally as paracrines.

Adrenal Glands

The two *adrenal,* or *suprarenal, glands* are located atop or close to the kidneys. Anatomically, the **adrenal medulla** develops from neural crest tissue, and it is directly controlled by the sympathetic nervous system. The medullary cells respond to this stimulation by releasing **epinephrine** (80%) or **norepinephrine** (20%), which act in conjunction with the sympathetic nervous system to elicit the fight-or-flight response to stressors.

The **adrenal cortex** produces three major groups of steroid hormones, collectively called **corticosteroids.** The **mineralocorticoids,** chiefly *aldosterone,* regulate water and electrolyte balance in the extracellular fluids, mainly by regulating sodium ion reabsorption by kidney tubules. The **glucocorticoids** (*cortisol* [*hydrocortisone*], *cortisone,* and *corticosterone*) enable the body to resist long-term stressors, primarily by increasing blood glucose levels. The **gonadocorticoids,** or **sex hormones,** produced by the adrenal cortex are chiefly *androgens* (male sex hormones), but some *estrogens* (female sex hormones) are formed.

The gonadocorticoids are produced throughout life in relatively insignificant amounts; however, hypersecretion of these hormones produces abnormal hairiness **(hirsutism),** and masculinization occurs. ■

Pancreas

The *pancreas,* located partially behind the stomach in the abdomen, functions as both an endocrine and exocrine gland. It produces digestive enzymes as well as insulin and glucagon, important hormones concerned with the regulation of blood sugar levels.

Elevated blood glucose levels stimulate release of **insulin,** which decreases blood sugar levels, primarily by accelerating the transport of glucose into the body cells, where it is oxidized for energy or converted to glycogen or fat for storage.

Hyposecretion of insulin or some deficiency in the insulin receptors leads to **diabetes mellitus,** which is characterized by the inability of body cells to utilize glucose and the subsequent loss of glucose in the urine. Alterations of protein and fat metabolism also occur secondary to derangements in carbohydrate metabolism. Hypersecretion causes low blood sugar, or **hypoglycemia.** Symptoms include anxiety, nervousness, tremors, and weakness. ■

Glucagon acts antagonistically to insulin. When blood glucose levels are low, it stimulates the liver, its primary target organ, to break down glycogen stores to glucose and subsequently to release the glucose to the blood.

The Gonads

The *female gonads,* or *ovaries,* are paired, almond-sized organs located in the pelvic cavity. In addition to producing the female sex cells (ova), the ovaries produce two steroid hormone groups, the estrogens and progesterone. The endocrine and exocrine functions of the ovaries do not begin until the onset of puberty. The **estrogens** are responsible for the development of the secondary sex characteristics of the female at puberty (primarily maturation of the reproductive organs and development of the breasts) and act with progesterone to bring about cyclic changes of the uterine lining that occur during the menstrual cycle. The estrogens also help prepare the mammary glands for lactation. During pregnancy progesterone maintains the uterine musculature in a quiescent state and helps to prepare the breast tissue for lactation.

The paired oval *testes* of the male are suspended in a pouchlike sac, the scrotum, outside the pelvic cavity. In addition to the male sex cells (sperm), the testes produce the male sex hormone, **testosterone.** Testosterone promotes the maturation of the reproductive system accessory structures, brings about the development of the male secondary sex characteristics, and is responsible for sexual drive, or libido. Both the endocrine and exocrine functions of the testes begin at puberty. For a more detailed discussion of the function and histology of the ovaries and testes, see Exercises 42 and 43.

ACTIVITY 1

Identifying the Endocrine Organs

Locate the endocrine organs on Figure 27.2. Also locate these organs on the anatomical charts or torso.

Microscopic Anatomy of Selected Endocrine Glands

ACTIVITY 2

Examining the Microscopic Structure of Endocrine Glands

Obtain a microscope and one of each slide on the materials list. We will study only organs in which it is possible to identify the endocrine-producing cells. Compare your observations with the histology images in Figure 27.3a–f.

Thyroid Gland

1. Scan the thyroid under low power, noting the **follicles,** spherical sacs containing a pink-stained material *(colloid)*. Stored T_3 and T_4 are attached to the protein colloidal material stored in the follicles as **thyroglobulin** and are released gradually to the blood. Compare the tissue viewed to Figure 27.3a.

2. Observe the tissue under high power. Notice that the walls of the follicles are formed by simple cuboidal or squamous epithelial cells that synthesize the follicular products. The **parafollicular,** or **C, cells** you see between the follicles are responsible for calcitonin production.

When the thyroid gland is actively secreting, the follicles appear small, and the colloidal material has a ruffled border. When the thyroid is hypoactive or inactive, the follicles are large and plump, and the follicular epithelium appears to be squamouslike.

Parathyroid Glands

Observe the parathyroid tissue under low power to view its two major cell types, the chief cells and the oxyphil cells. Compare your observations to Figure 27.3b. The **chief cells,** which synthesize parathyroid hormone (PTH), are small and abundant, and arranged in thick branching cords. The function of the scattered, much larger **oxyphil cells** is unknown.

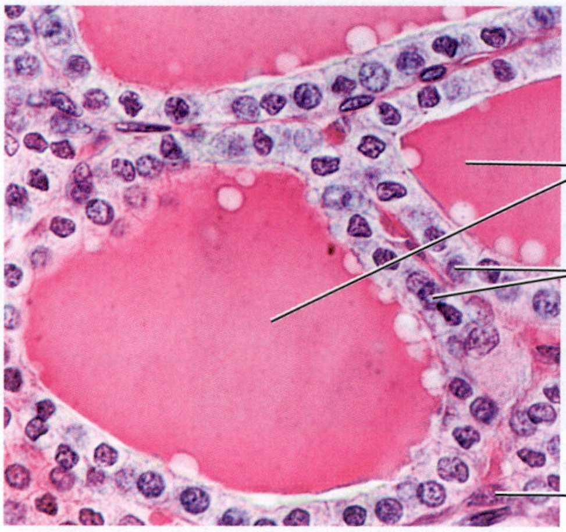
Colloid-filled follicles
Follicle cells
Blood vessel

(a) Thyroid gland (330×)

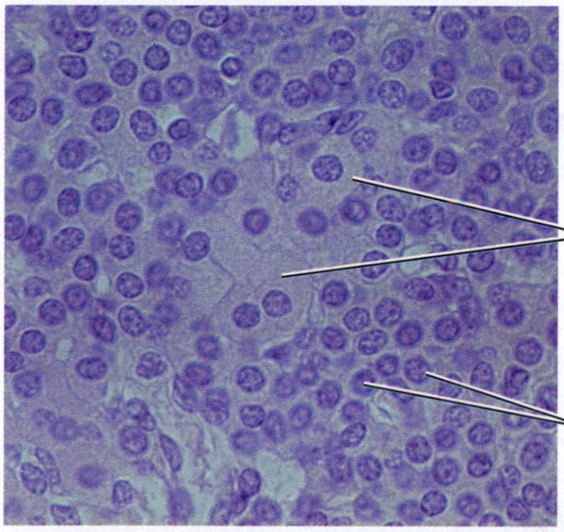
Oxyphil cells
Chief cells

(b) Parathyroid gland (330×)

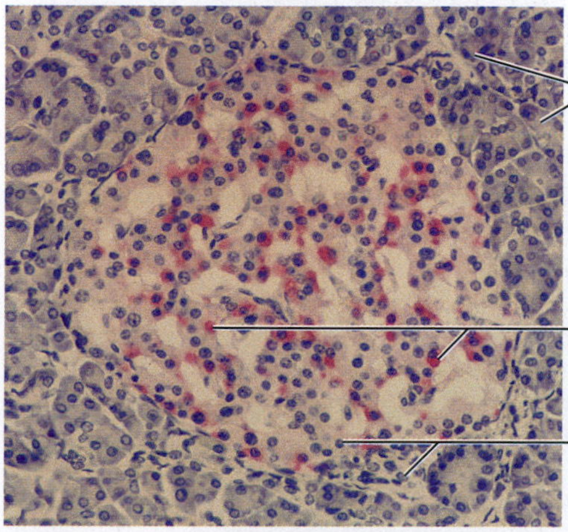
Exocrine (acinar) tissue of the pancreas
Alpha cells
Beta cells

(c) Pancreatic islet (170×)

FIGURE 27.3 Microscopic anatomy of selected endocrine organs.

Pancreas

1. Observe pancreas tissue under low power to identify the roughly circular **pancreatic islets (islets of Langerhans),** the endocrine portions of the pancreas. The islets are scattered

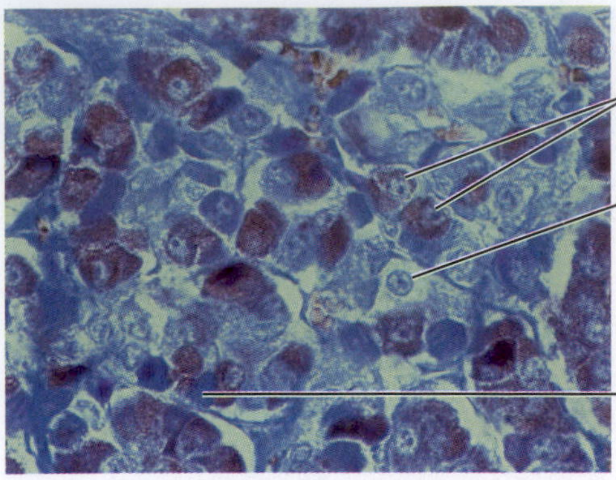
Acidophils
Chromophobe
Basophil

(d) Anterior pituitary (280×)

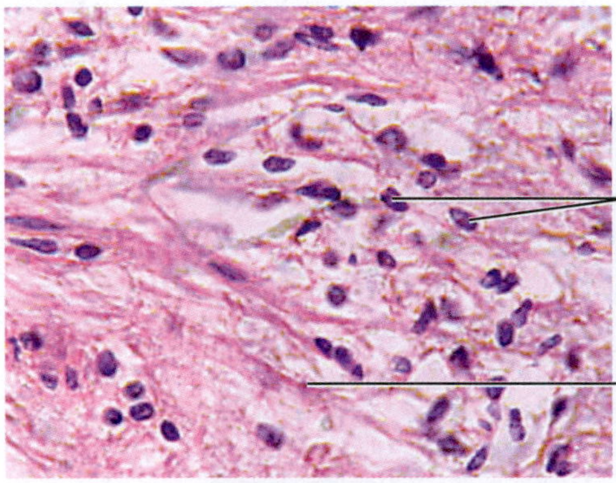
Pituicytes
Nerve fibers

(e) Posterior pituitary (465×)

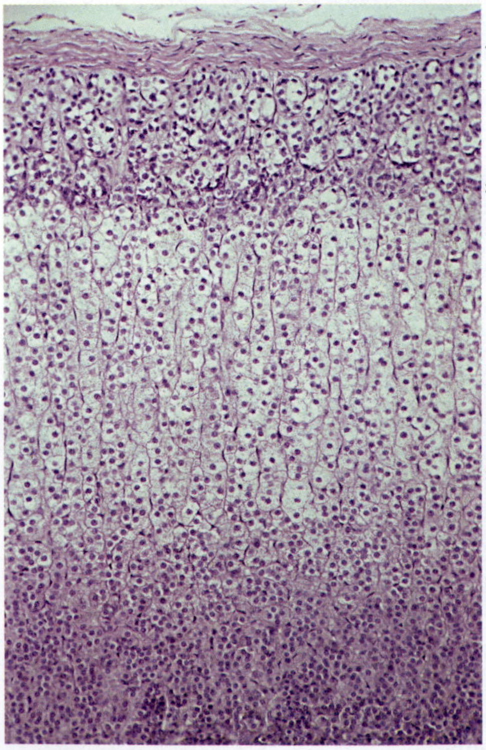
Zona glomerulosa
Zona fasciculata
Zona reticularis
Medulla cells

(f) Adrenal gland (110×)

FIGURE 27.3 *(continued)* Microscopic anatomy of selected endocrine organs.

amid the more numerous **acinar cells** and stain differently (usually lighter), which makes their identification possible. The deeper-staining acinar cells form the major portion of the pancreatic tissue. Acinar cells produce the exocrine secretion of hydrolytic enzymes that is released into the duodenum through the pancreatic duct. Alkaline fluid produced by duct cells accompanies the hydrolytic enzymes. See Figure 27.3c.

2. Focus on islet cells under high power. Notice that they are densely packed and have no definite arrangement. In contrast, the cuboidal acinar cells are arranged around secretory ducts. If special stains are used, it will be possible to distinguish the **alpha cells,** which tend to cluster at the periphery of the islets and produce glucagon, from the **beta cells,** which synthesize insulin. With these specific stains, the beta cells are larger and stain gray-blue, and the alpha cells are smaller and appear bright pink, as shown in Figure 27.3c.

Pituitary Gland

1. Observe the general structure of the pituitary gland under low power to differentiate between the glandular anterior pituitary and the neural posterior pituitary.

2. Using the high-power lens, focus on the nests of cells of the anterior pituitary. When differential stains are used it is possible to identify the specialized cell types that secrete the specific hormones. Using the photomicrograph in Figure 27.3d as a guide, locate the reddish brown–stained **acidophil cells,** which produce growth hormone and prolactin, and the **basophil cells,** whose deep-blue granules are responsible for the production of the tropic hormones (TSH, ACTH, FSH, and LH). **Chromophobes,** the third cellular population, do not take up the stain and appear rather dull and colorless. The role of the chromophobes is controversial, but they apparently are not directly involved in hormone production.

3. Switch your focus to the posterior pituitary where two hormones (oxytocin and antidiuretic hormone) synthesized by hypothalamic neurons are stored. Observe the axons of hypothalamic neurons that compose most of this portion of the pituitary. Also note the glial cells, or **pituicytes** (Figure 27.3e).

Adrenal Gland

1. Hold the slide of the adrenal gland up to the light to distinguish the outer cortex and inner medulla areas. Then scan the cortex under low power to distinguish the differences in cell appearance and arrangement in the three cortical areas. Refer to Figure 27.3f as you work. In the outermost **zona glomerulosa,** where most mineralocorticoid production occurs, the tightly packed cells are arranged in spherical clusters. The deeper intermediate **zona fasciculata** produces glucocorticoids. This is the thickest part of the cortex. Its cells are arranged in parallel cords. The innermost cortical zone, the **zona reticularis** produces sex hormones and some glucocorticoids. The cells here stain intensely and form a branching network.

2. Switch to higher power to view the large, lightly stained cells of the adrenal medulla, which produce epinephrine and norepinephrine. Notice their clumped arrangement. ■

NAME ______________________________

LAB TIME/DATE ______________________

REVIEW SHEET

EXERCISE 27

Functional Anatomy of the Endocrine Glands

Gross Anatomy and Basic Function of the Endocrine Glands

1. Both the endocrine and nervous systems are major regulating systems of the body; however, the nervous system has been compared to an airmail delivery system and the endocrine system to the Pony Express. Briefly explain this comparison.

____

____

____

2. Define *hormone.* ____

____

3. Chemically, hormones belong chiefly to two molecular groups, the ______________________________

and the ____.

4. Define *target organ.* ____

____

5. If hormones travel in the bloodstream, why don't all tissues respond to all hormones? ______________________

____

6. Identify the endocrine organ described by each of the following statements.

____________________ 1. located in the throat; bilobed gland connected by an isthmus

____________________ 2. found close to the kidney

____________________ 3. a mixed gland, located close to the stomach and small intestine

____________________ 4. paired glands suspended in the scrotum

____________________ 5. ride "horseback" on the thyroid gland

____________________ 6. found in the pelvic cavity of the female, concerned with ova and female hormone production

____________________ 7. found in the upper thorax overlying the heart; large during youth

____________________ 8. found in the roof of the third ventricle

7. The table below lists the functions of many of the hormones you have studied. From the keys below, fill in the hormones responsible for each function, and the endocrine glands that produce each hormone. Glands may be used more than once.

Hormones Key:

ACTH	estrogens	progesterone
ADH	FSH	prolactin
aldosterone	glucagon	PTH
calcitonin	insulin	T_3/T_4
cortisol	LH	testosterone
epinephrine	oxytocin	TSH

Glands Key:

adrenal cortex	pancreas
adrenal medulla	parathyroid glands
anterior pituitary	posterior pituitary
hypothalamus	testes
ovaries	thyroid gland

Function	*Hormone*(s)	*Gland*(s)
Regulate the function of another endocrine gland	1.	
	2.	
	3.	
	4.	
Maintenance of salt and water balance in the extracellular fluid	1.	
	2.	
Directly involved in milk production and ejection	1.	
	2.	
Controls the rate of body metabolism and cellular oxidation	1.	
Regulate blood calcium levels	1.	
	2.	
Regulate blood glucose levels; produced by the same "mixed" gland	1.	
	2.	
Released in response to stressors	1.	
	2.	
Drive development of secondary sex characteristics in males	1.	
Directly responsible for regulation of the menstrual cycle	1.	
	2.	

8. Although the pituitary gland is often referred to as the master gland of the body, the hypothalamus exerts some control over the pituitary gland. How does the hypothalamus control both anterior and posterior pituitary functioning?

___

___

9. Indicate whether the release of the hormones listed below is stimulated by (A) another hormone; (B) the nervous system (neurotransmitters, or neurosecretions); or (C) humoral factors (the concentration of specific nonhormonal substances in the blood or extracellular fluid). (Use your textbook as necessary.)

_______ 1. ACTH

_______ 2. calcitonin

_______ 3. estrogens

_______ 4. insulin

_______ 5. norepinephrine

_______ 6. parathyroid hormone

_______ 7. T_4/T_3

_______ 8. testosterone

_______ 9. TSH, FSH

10. Name the hormone(s) produced in *inadequate* amounts that directly result in the following conditions.

_______________________ 1. tetany

_______________________ 2. excessive diuresis without high blood glucose levels

_______________________ 3. loss of glucose in the urine

_______________________ 4. abnormally small stature, normal proportions

_______________________ 5. low BMR, mental and physical sluggishness

11. Name the hormone(s) produced in *excessive* amounts that directly result in the following conditions.

_______________________ 1. large hands and feet in the adult, large facial bones

_______________________ 2. nervousness, irregular pulse rate, sweating

_______________________ 3. demineralization of bones, spontaneous fractures

Microscopic Anatomy of Selected Endocrine Glands

12. Choose a response from the key below to name the hormone(s) produced by the cell types listed.

Key: a. *calcitonin*
b. GH, prolactin
c. glucagon
d. glucocorticoids
e. insulin
f. mineralocorticoids
g. PTH
h. T_4/T_3
i. TSH, ACTH, FSH, LH

_______ 1. parafollicular cells of the thyroid

_______ 2. follicular epithelial cells of the thyroid

_______ 3. beta cells of the pancreatic islets (islets of Langerhans)

_______ 4. alpha cells of the pancreatic islets (islets of Langerhans)

_______ 5. basophil cells of the anterior pituitary

_______ 6. zona fasciculata cells

_______ 7. zona glomerulosa cells

_______ 8. chief cells of the parathyroid

_______ 9. acidophil cells of the anterior pituitary

13. Six diagrams of the microscopic structures of the endocrine glands are presented here. Identify each and name all structures indicated by a leader line or bracket.

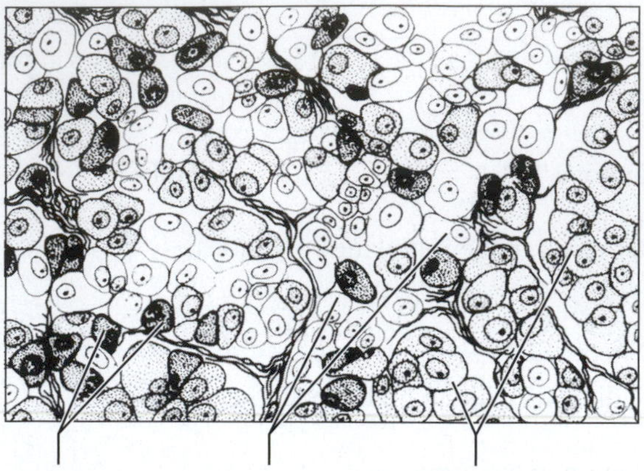

(a) ____________________

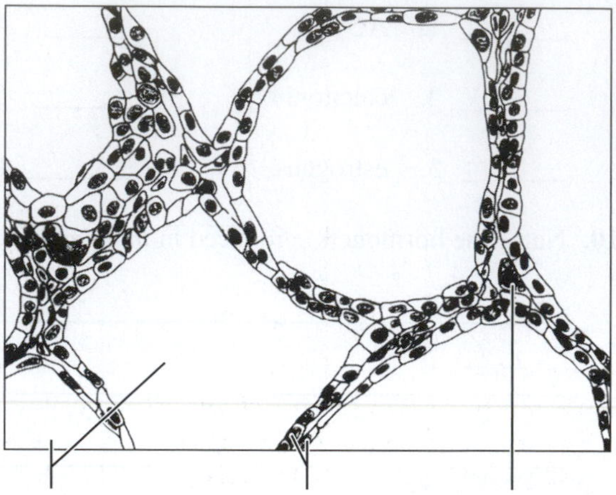

(d) ____________________

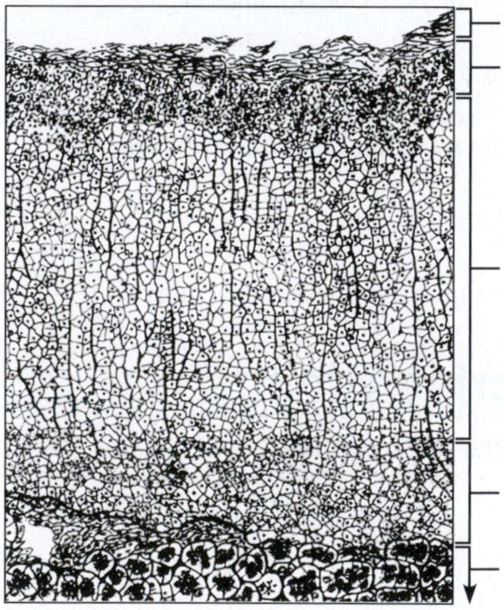

(b) ____________________

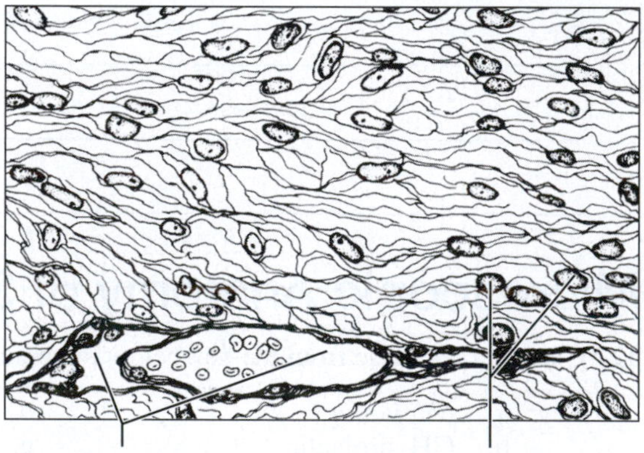

(e) ____________________

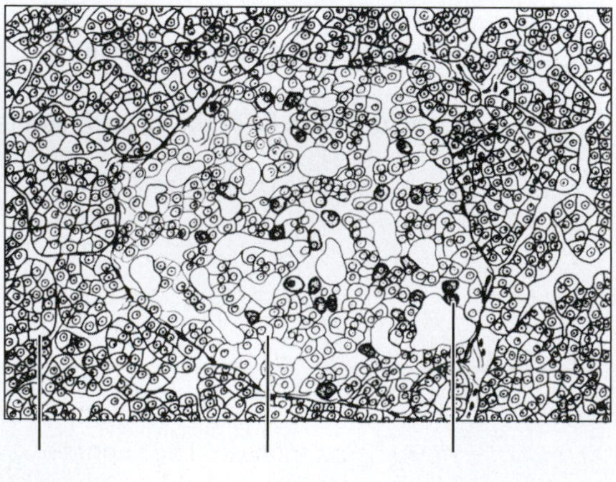

(c) ____________________

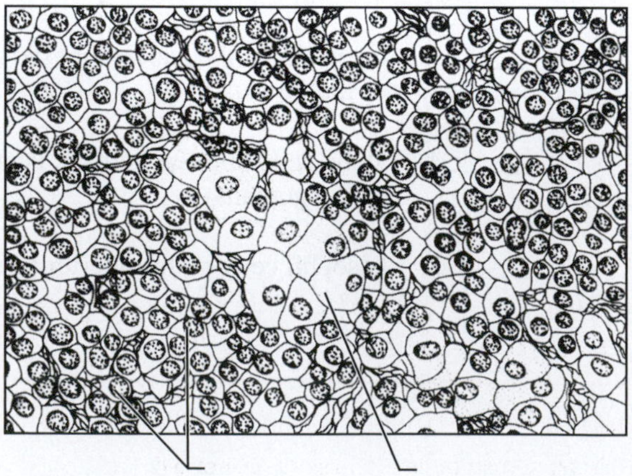

(f) ____________________

EXERCISE **28**

Role of Thyroid Hormone, Pituitary Hormone, Insulin, and Epinephrine

MATERIALS

Activity 1: Thyroid hormone and metabolic rate

- ☐ Glass desiccator, manometer, 20-ml glass syringe, two-hole rubber stopper, and T-valve (1 for every 3–4 students)
- ☐ Soda lime (desiccant)
- ☐ Hardware cloth squares
- ☐ Petrolatum
- ☐ Rubber tubing; tubing clamp; scissors
- ☐ 7.6-cm (3-in.) pieces of glass tubing
- ☐ Animal balances
- ☐ Heavy animal-handling gloves
- ☐ Young rats of the same sex, obtained 2 weeks prior to the laboratory session and treated as follows for 14 days:

  Group 1: control group—fed normal rat chow and water

  Group 2: experimental group A—fed normal rat chow and drinking water containing 0.02% propylthiouracil*

  Group 3: experimental group B—fed rat chow containing desiccated thyroid (2% by weight) and given normal drinking water
- ☐ Chart set up on chalkboard so that each student group can record its computed metabolic rate figures under the appropriate headings

*__Note to the Instructor:__ propylthiouracil (PTU) is degraded by light and should be stored in light-resistant containers or in the dark.

Text continues on next page.

MasteringA&P™

Access practice quizzes and more in the Study Area at www.masteringaandp.com.

PEx

This lab corresponds to PhysioEx Exercise 4. See p. PEx-59.

OBJECTIVES

1. To understand the physiological (and clinical) importance of metabolic rate measurement.
2. To investigate the effect of hypo-, hyper-, and euthyroid conditions on oxygen consumption and metabolic rate.
3. To assemble the necessary apparatus and properly use a manometer to obtain experimental results.
4. To calculate metabolic rate in terms of O_2 consumption.
5. To describe and explain the effect of pituitary hormones on the ovary.
6. To describe and explain the effects of hyperinsulinism.
7. To describe and explain the effect of epinephrine on the heart.

PRE-LAB QUIZ

1. Define *metabolism*. ______________________________

   ____
2. Circle the correct term. Catabolism / Anabolism is the process by which substances are broken down into simpler substances.
3. __________ is the single most important hormone responsible for influencing the rate of cellular metabolism and body heat production.
   a. Calcitonin
   b. Estrogen
   c. Insulin
   d. Thyroid hormone
4. Circle the correct term. Control / Experimental group B animals are assumed to have normal thyroid function and metabolic rate.
5. Basal metabolic rate (BMR) is:
   a. decreased in individuals with hyperthyroidism
   b. increased in individuals with hyperthyroidism
   c. increased in obese individuals
6. In this activity, __________ consumption will be measured with a respirator-manometer device.
   a. carbon dioxide
   b. food
   c. oxygen
   d. water
7. Circle True or False. Gonadotropins are produced by the anterior pituitary gland.
8. Circle the correct term. Many people with diabetes mellitus need injections of insulin / glucagon to maintain homeostasis.
9. What experiment will you do to observe the effects of epinephrine?
   a. Flush the heart of a dissected frog with epinephrine.
   b. Flush the heart of a dissected frog with insulin.
   c. Inject a rat with epinephrine.
   d. Inject a rat with follicle-stimulating hormone and epinephrine.

Activity 2: Pituitary hormone and ovary†

- ☐ Female frogs (*Rana pipiens*)
- ☐ Disposable gloves
- ☐ Battery jars
- ☐ Syringe (2-ml capacity)
- ☐ 20- to 25-gauge needle
- ☐ Frog pituitary extract
- ☐ Physiological saline
- ☐ Spring or pond water
- ☐ Wax marking pencils

Activity 3: Hyperinsulinism†

- ☐ 500- or 600-ml beakers
- ☐ 20% glucose solution
- ☐ Commercial insulin solution (400 Immunizing units [IU] per 100 ml of H_2O)
- ☐ Finger bowls
- ☐ Small (4–5 cm, or 1½–2 in.) freshwater fish (guppy, bluegill, or sunfish—listed in order of preference)
- ☐ Wax marking pencils

Activity 4: Epinephrine and heart†

- ☐ Frog (*Rana pipiens*)
- ☐ Dissecting instruments, tray, and pins
- ☐ Frog Ringer's solution in dropper bottle
- ☐ 1:1000 epinephrine (Adrenalin) solution in dropper bottles
- ☐ Disposable gloves

†*The Selected Actions of Hormones and Other Chemical Messengers* videotape (available to qualified adopters from Pearson Education) may be used in lieu of student participation in Activities 2–4 of Exercise 28.

The endocrine system exerts many complex and interrelated effects on the body as a whole, as well as on specific organs and tissues. Most scientific knowledge about this system is contemporary, and new information is constantly being presented. Many experiments on the endocrine system require relatively large laboratory animals; are time-consuming (requiring days to weeks of observation); and often involve technically difficult surgical procedures to remove the glands or parts of them, all of which makes it difficult to conduct more general types of laboratory experiments. Nevertheless, the four technically unsophisticated experiments presented here should illustrate how dramatically hormones affect body functioning.

To conserve laboratory specimens, experiments can be conducted by groups of four students. The use of larger working groups should not detract from benefits gained, since the major value of these experiments lies in observation.

ACTIVITY 1

Determining the Effect of Thyroid Hormone on Metabolic Rate

Metabolism is a broad term referring to all chemical reactions that are necessary to maintain life. It involves both *catabolism,* enzymatically controlled processes in which substances are broken down to simpler substances, and *anabolism,* processes in which larger molecules or structures are built from smaller ones. Most catabolic reactions in the body are accompanied by a net release of energy. Some of the liberated energy is captured to make ATP, the energy-rich molecule used by body cells to energize all their activities; the balance is lost in the form of thermal energy or heat. Maintaining body temperature is critically related to the heat-liberating aspects of metabolism.

Various foodstuffs make different contributions to the process of metabolism. For example, carbohydrates, particularly glucose, are generally broken down or oxidized to make ATP, whereas fats are utilized to form cell membranes and myelin sheaths, and to insulate the body with a fatty cushion. Fats are used secondarily for producing ATP, particularly when the diet is inadequate in carbohydrates. Proteins and amino acids tend to be conserved by body cells, and understandably so, since most structural elements of the body are proteinaceous in nature.

Thyroid hormone (collectively T_3 and T_4), produced by the thyroid gland, is the single most important hormone influencing the rate of cellular metabolism and body heat production.

Under conditions of excess thyroid hormone production (hyperthyroidism), an individual's basal metabolic rate (BMR), heat production, and oxygen consumption increase, and the individual tends to lose weight and become heat-intolerant and irritable. Conversely, hypothyroid individuals become mentally and physically sluggish and obese, and are cold-intolerant because of their low BMR. ■

Many factors other than thyroid hormone levels contribute to metabolic rate (for example, body size and weight, age, and activity level), but the focus of the following experiment is to investigate how differences in thyroid hormone concentration affect metabolism.

Three groups of laboratory rats will be used. *Control group* animals are assumed to have normal thyroid function (to be euthyroid) and to have normal metabolic rates for their relative body weights. *Experimental group A* animals have received water containing the chemical propylthiouracil, which counteracts or antagonizes the effects of thyroid hormone in the body. *Experimental group B* animals have been fed rat chow containing dried thyroid tissue, which contains thyroid hormone. The rates of oxygen consumption (an indirect means of determining metabolic rate) in the animals of the three groups will be measured and compared to investigate the effects of hyperthyroid, hypothyroid, and euthyroid conditions.

Oxygen consumption will be measured with a simple respirometer-manometer apparatus. Each animal will be placed in a closed chamber containing soda lime. As carbon dioxide is evolved and expired, it will be absorbed by the soda lime; therefore, the pressure changes observed will indicate the volume of oxygen consumed by the animal during the testing interval. Students will work in groups of three to four to assemble the apparatus, make preliminary weight measurements on the animals, and record the data.

Preparing the Respirometer-Manometer Apparatus

1. Obtain a desiccator, a two-hole rubber stopper, a 20-ml glass syringe, a hardware cloth square, a T-valve, scissors, rubber tubing, two short pieces of glass tubing, soda lime, a manometer, a clamp, and petrolatum, and bring them to your laboratory bench. The apparatus will be assembled as illustrated in Figure 28.1.

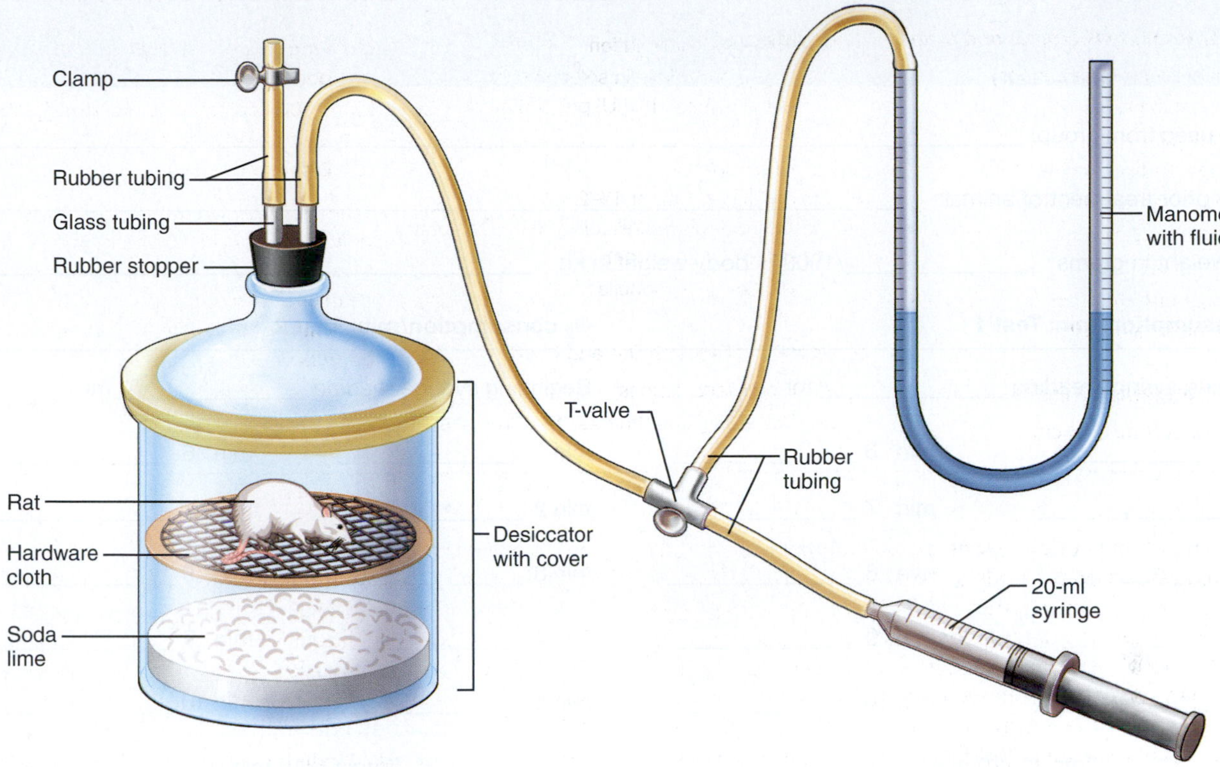
Clamp
Rubber tubing
Glass tubing
Rubber stopper
Rat
Hardware cloth
Soda lime
Desiccator with cover
T-valve
Rubber tubing
Manometer with fluid
20-ml syringe

FIGURE 28.1 Respirometer-manometer apparatus.

2. Shake soda lime into the bottom of the desiccator to thoroughly cover the glass bottom. Then place the hardware cloth on the ledge of the desiccator over the soda lime. The hardware cloth should be well above the soda lime, so that the animal will not be able to touch it. Soda lime is quite caustic and can cause chemical burns.

3. Lubricate the ends of the two pieces of glass tubing with petrolatum, and twist them into the holes in the rubber stopper until their distal ends protrude from the opposite side. *Do not plug the tubing with petrolatum.* Place the stopper into the desiccator cover, and set the cover on the desiccator temporarily.

4. Cut off a short (7.6-cm, or 3-in.) piece of rubber tubing, and attach it to the top of one piece of glass tubing extending from the stopper. Cut and attach a 30- to 35-cm (12- to 14-in.) piece of rubber tubing to the other glass tubing. Insert the T-valve stem into the distal end of the longer-length tubing.

5. Cut another short piece of rubber tubing; attach one end to the T-valve and the other to the nib of the 20-ml syringe. Remove the plunger of the syringe and grease its sides generously with petrolatum. Insert the plunger back into the syringe barrel and work it up and down to evenly disperse the petrolatum on the inner walls of the syringe, then pull the plunger out to the 20-ml marking.

6. Cut a piece of rubber tubing long enough to reach from the third arm of the T-valve to one arm of the manometer. (The manometer should be partially filled with water so that a U-shaped water column is seen.) Attach the tubing to the T-valve and the manometer arm.

7. Remove the desiccator cover and generously grease the cover's bottom edge with petrolatum. Place the cover back on the desiccator and firmly move it from side to side to spread the lubricant evenly.

8. Test the system for leaks as follows: Firmly clamp the *short* length of rubber tubing extending from the stopper. Now gently push in on the plunger of the syringe. If the system is properly sealed, the fluid in the manometer will move away from the rubber tubing attached to its arm. If there is an air leak, the manometer fluid level will not change, or it will change and then quickly return to its original level. If either of these events occurs, check all glass-to-glass or glass-to-rubber tubing connections. Smear additional petrolatum on suspect areas and test again. The apparatus must be airtight before experimentation can begin.

9. After ensuring that there are no leaks in the system, unclamp the short rubber tubing, and remove the desiccator cover.

Preparing the Animal

! 1. Put on the heavy animal-handling gloves, and obtain one of the animals as directed by your instructor. Handling it gently, weigh the animal to the nearest 0.1 g on the animal balance.

2. Carefully place the animal on the hardware cloth in the desiccator. The objective is to measure oxygen usage at basal levels, so you do not want to prod the rat into high levels of activity, which would produce errors in your measurements.

3. Record the animal's group (control or experimental group A or B) and its weight in kilograms (that is, weight in grams/1000) on the data sheet on page 418.

Metabolic Rate Data Sheet	
Animal used from group: ______________________	
14-day prior treatment of animal: ______________________	
Body weight in grams: ____________/1000 = body weight in kg: ____________	
O_2 consumption/min: Test 1	**O_2 consumption/min: Test 2**
Beginning syringe reading ____________ ml	Beginning syringe reading ____________ ml
min 1 ________ min 6 ________	min 1 ________ min 6 ________
min 2 ________ min 7 ________	min 2 ________ min 7 ________
min 3 ________ min 8 ________	min 3 ________ min 8 ________
min 4 ________ min 9 ________	min 4 ________ min 9 ________
min 5 ________ min 10 ________	min 5 ________ min 10 ________
________ Total O_2/10 min	________ Total O_2/10 min
Average ml O_2 consumed/10 min: ____________	
Milliliters O_2 consumed/hr: ____________	
Metabolic rate: ____________ ml O_2/kg/hr	
Averaged class results:	
Metabolic rate of control animals: ____________ ml O_2/kg/hr	
Metabolic rate of experimental group A animals (PTU-treated): ____________ ml O_2/kg/hr	
Metabolic rate of experimental group B animals (desiccated thyroid-treated): ____________ ml O_2/kg/hr	

Equilibrating the Chamber

1. Place the lid on the desiccator, and move it slightly from side to side to seal it firmly.

2. Leave the short tubing unclamped for 7 to 10 minutes to allow temperature equilibration in the chamber. (Since the animal's body heat will warm the air in the container, the air will expand initially. This must be allowed to occur before any measurements of oxygen consumption are taken. Otherwise, it would appear that the animal is generating oxygen rather than consuming it.)

Determining Oxygen Consumption of the Animal

1. Once again clamp the short rubber tubing extending from the desiccator lid stopper.

2. Check the manometer to make sure that the fluid level is the same in both arms. (If not, manipulate the syringe plunger to make the fluid levels even.) Record the time and the position of the bottom of the plunger (use ml marking) in the syringe.

3. Observe the manometer fluid levels at 1-minute intervals. Each time make the necessary adjustment to bring the fluid levels even in the manometer by carefully pushing the syringe plunger further into the barrel. Determine the amount of oxygen used per minute in each interval by computing the difference in air volumes within the syringe. For example, if after the first minute, you push the plunger from the 20- to the 17-ml marking, the oxygen consumption would be 3 ml/min. Then, if the plunger is pushed from 17 ml to 15 ml at the second minute reading, the oxygen usage during minute 2 would be 2 ml, and so on.

4. Continue taking readings (and recording oxygen consumption per minute interval on the data sheet) for 10 consecutive minutes or until the syringe plunger has been pushed nearly to the 0-ml mark. Then unclamp the short rubber

tubing, remove the desiccator cover, and allow the apparatus to stand open for 2 to 3 minutes to flush out the stale air.

5. Repeat the recording procedures for another 10-minute interval. *Make sure that you equilibrate the temperature within the chamber before beginning this second recording series.*

6. After you have recorded the animal's oxygen consumption for two 10-minute intervals, unclamp the short rubber tubing, remove the desiccator lid, and carefully return the rat to its cage.

Computing Metabolic Rate

Metabolic rate calculations are generally reported in terms of kcal/m²/hr and require that corrections be made to present the data in terms of standardized pressure and temperature conditions. These more complex calculations will not be used here, since the object is simply to arrive at some generalized conclusions concerning the effect of thyroid hormone on metabolic rate.

1. Obtain the average figure for milliliters of oxygen consumed per 10-minute interval by adding up the minute-interval consumption figures for each 10-minute testing series and dividing the total by 2.

$$\frac{\text{Total ml } O_2 \text{ in test series 1} + \text{total ml } O_2 \text{ in test series 2}}{2}$$

2. Determine oxygen consumption per hour using the following formula, and record the figure on the data sheet:

$$\frac{\text{Average ml } O_2 \text{ consumed}}{10 \text{ min}} \times \frac{60 \text{ min}}{\text{hr}} = \text{ml } O_2/\text{hr}$$

3. To determine the metabolic rate in milliliters of oxygen consumed per kilogram of body weight per hour so that the results of all experiments can be compared, divide the figure just obtained in step 2 by the animal's weight in kilograms (kg = lb ÷ 2.2).

$$\text{Metabolic rate} = \frac{\text{ml } O_2/\text{hr}}{\text{weight (kg)}} = \underline{\hspace{2cm}} \text{ ml } O_2/\text{kg/hr}$$

Record the metabolic rate on the data sheet and also in the appropriate space on the chart on the chalkboard.

4. Once all groups have recorded their final metabolic rate figures on the chalkboard, average the results of each animal grouping to obtain the mean for each experimental group. Also record this information on the data sheet.

ACTIVITY 2

Determining the Effect of Pituitary Hormones on the Ovary

As indicated in Exercise 43, anterior pituitary hormones called *gonadotropins,* specifically follicle-stimulating hormone (FSH) and luteinizing hormone (LH), regulate the ovarian cycles of the female. Although amphibians normally ovulate seasonally, many can be stimulated to ovulate "on demand" by injecting an extract of pituitary hormones.

1. Don disposable gloves, and obtain two frogs. Place them in separate battery jars to bring them to your laboratory bench. Also bring back a syringe and needle, a wax marking pencil, pond or spring water, and containers of pituitary extract and physiological saline.

2. Before beginning, examine each frog for the presence of eggs. Hold the frog firmly with one hand and exert pressure on its abdomen toward the cloaca (in the direction of the legs). If ovulation has occurred, any eggs present in the oviduct will be forced out and will appear at the cloacal opening. If no eggs are present, continue with step 3.

If eggs are expressed, return the animal to your instructor and obtain another frog for experimentation. Repeat the procedure for determining if eggs are present until two frogs that lack eggs have been obtained.

3. Aspirate 1 to 2 ml of the pituitary extract into a syringe. Inject the extract subcutaneously into the anterior abdominal (peritoneal) cavity of the frog you have selected to be the experimental animal. To inject into the peritoneal cavity, hold the frog with its ventral surface superiorly. Insert the needle through the skin and muscles of the abdominal wall in the lower quarter of the abdomen. Do not insert the needle far enough to damage any of the vital organs. With a wax marker, label its large battery jar "experimental," and place the frog in it. Add a small amount of pond water to the battery jar before continuing.

4. Aspirate 1 to 2 ml of physiological saline into a syringe and inject it into the peritoneal cavity of the second frog—this will be the control animal. (Make sure you inject the same volume of fluid into both frogs.) Place this frog into the second battery jar, marked "control." Allow the animals to remain undisturbed for 24 hours.

5. After 24 hours,* again check each frog for the presence of eggs in the cloaca. (See step 2.) If no eggs are present, make arrangements with your laboratory instructor to return to the lab on the next day (at 48 hours after injection) to check your frogs for the presence of eggs.

6. Return the frogs to the terrarium before leaving or continuing with the lab.

In which of the prepared frogs was ovulation induced?

____

Specifically, what hormone in the pituitary extract causes ovulation to occur?

____

ACTIVITY 3

Observing the Effects of Hyperinsulinism

Many people with diabetes mellitus need injections of insulin to maintain blood sugar (glucose) homeostasis. Adequate levels of blood glucose are essential for proper functioning of the nervous system; thus, the administration of insulin must be carefully controlled. If blood glucose levels fall precipitously, the patient will go into insulin shock.

*The student needs to inject the frog the day before the lab session or return to check results the day after the scheduled lab session.

A small fish will be used to demonstrate the effects of hyperinsulinism. Since the action of insulin on the fish parallels that in the human, this experiment should provide valid information concerning its administration to humans.

1. Prepare two finger bowls. Using a wax marking pencil, mark one A and the other B. To finger bowl A, add 100 ml of the commercial insulin solution. To finger bowl B, add 200 ml of 20% glucose solution.

2. Place a small fish in finger bowl A and observe its actions carefully as the insulin diffuses into its bloodstream through the capillary circulation of its gills.

Approximately how long did it take for the fish to become comatose?

____

What types of activity did you observe in the fish before it became comatose?

____

____

____

____

3. When the fish is comatose, carefully transfer it to finger bowl B and observe its actions. What happens to the fish after it is transferred?

____

____

____

____

Approximately how long did it take for this recovery?

____

4. After all observations have been made and recorded, carefully return the fish to the aquarium. ■

ACTIVITY 4

Testing the Effect of Epinephrine on the Heart

As noted in Exercise 27, the adrenal medulla and the sympathetic nervous system are closely interrelated, specifically because the cells of the adrenal medulla and the postganglionic axons of the sympathetic nervous system both release catecholamines. This experiment demonstrates the effects of epinephrine on the frog heart.

1. Obtain a frog, dissecting instruments and tray, disposable gloves, frog Ringer's solution, and dropper bottle of 1:1000 epinephrine solution. Bring them to your laboratory bench and don the gloves before beginning step 2.

2. Destroy the nervous system of the frog. (A frog used in the experiment on pituitary hormone effects may be used if your test results have already been obtained and are positive. Otherwise, obtain another frog.) Insert one blade of a scissors into its mouth as far as possible and quickly cut off the top of its head, posterior to the eyes. Then identify the spinal cavity and insert a dissecting needle into it to destroy the spinal cord.

3. Place the frog dorsal side down on a dissecting tray, and carefully open its ventral body cavity by making a vertical incision with the scissors.

4. Identify the beating heart, and carefully cut through the saclike pericardium to expose the heart tissue.

5. Visually count the heart rate for 1 minute, and record below. Keep the heart moistened with frog Ringer's solution during this interval.

Beats per minute: ______________________________

6. Flush the heart with epinephrine solution. Record the heartbeat rate per minute for 5 consecutive minutes.

minute 1 ______________ minute 4 ______________

minute 2 ______________ minute 5 ______________

minute 3 ______________

What was the effect of epinephrine on the heart rate?

____

Was the effect long-lived? ______________________________

7. Dispose of the frog in an appropriate container, and clean the dissecting tray and instruments before returning them to the supply area. ■

NAME ______________________________

LAB TIME/DATE ______________________

REVIEW SHEET

EXERCISE 28

Role of Thyroid Hormone, Pituitary Hormone, Insulin, and Epinephrine

Determining the Effect of Thyroid Hormone on Metabolic Rate

1. In the measurement of oxygen consumption in rats, which group had the highest metabolic rate?

______________________ Which group had the lowest metabolic rate? ______________________

Correlate these observations with the pretreatment these animals received. ______________________

Which group of rats was hyperthyroid? ______________________

Which euthyroid? ______________________ Which hypothyroid? ______________________

2. Since oxygen used = carbon dioxide generated, how do you know that what you measured was oxygen consumption?

3. What did changes in the fluid levels in the manometer arms indicate? ______________________

4. The techniques used in this set of laboratory experiments probably permitted several inaccuracies. One was the inability to control the activity of the rats. How would changes in their activity levels affect the results observed?

Another possible source of error was the lack of control over the amount of food consumed by the rats in the 14-day period preceding the laboratory session. If each of the rats had been force-fed equivalent amounts of food in that 14-day period, which group (do you think) would have gained the most weight?

______________________ Which the least? ______________________

Explain your answers. ______________________

5. TSH, produced by the anterior pituitary, causes the thyroid gland to release thyroid hormone to the blood. Which group of rats can be assumed to have the *highest* blood levels of TSH? ______________ Which the lowest? ______________

Explain your reasoning. ____

____

6. Use an appropriate reference to determine how each of the following factors modifies metabolic rate. Indicate increase by ↑ and decrease by ↓.

increased exercise ________ aging ________ infection/fever ________

small/slight stature ________ obesity ________ sex (♂ or ♀) ________

Determining the Effect of Pituitary Hormones on the Ovary

7. In the experiment on the effects of pituitary hormones, two anterior pituitary hormones caused ovulation to occur in the experimental animal. Which of these actually triggered ovulation or egg expulsion?

______________________ The normal function of the second hormone involved, ______________________,

is to ____.

8. Why was a second frog injected with saline? ________

Observing the Effects of Hyperinsulinism

9. Briefly explain what was happening within the fish's system when the fish was immersed in the insulin solution.

____

____

10. What is the mechanism of the recovery process observed? ______________________________

____

11. What would you do to help a friend who had inadvertently taken an overdose of insulin? ______________________

______________________ Why? ______________________

____

Testing the Effect of Epinephrine on the Heart

12. Based on your observations, what is the effect of epinephrine on the heart rate?

____

13. What is the role of this effect in the "fight-or-flight" response?

____

EXERCISE 29

Blood

MATERIALS

General supply area:*

- ☐ Disposable gloves
- ☐ Safety glasses (student-provided)
- ☐ Bucket or large beaker containing 10% household bleach solution for slide and glassware disposal
- ☐ Spray bottles containing 10% bleach solution
- ☐ Autoclave bag
- ☐ Designated lancet (sharps) disposal container
- ☐ Plasma (obtained from an animal hospital or prepared by centrifuging animal [for example, cattle or sheep] blood obtained from a biological supply house)
- ☐ Test tubes and test tube racks
- ☐ Wide-range pH paper
- ☐ Stained smears of human blood from a biological supply house or, if desired by the instructor, heparinized animal blood obtained from a biological supply house or an animal hospital (for example, dog blood), or EDTA-treated red cells (reference cells†) with blood type labels obscured (available from Immucor, Inc.)

*__Note to the Instructor:__ See directions for handling of soiled glassware and disposable items on page 424.

†The blood in these kits (each containing four blood cell types—A1, A2, B, and O—individually supplied in 10-ml vials) is used to calibrate cell counters and other automated clinical laboratory equipment. This blood has been carefully screened and can be safely used by students for blood typing and determining hematocrits. It is not usable for hemoglobin determinations or coagulation studies.

Text continues on next page.

MasteringA&P™

Access practice quizzes and more in the Study Area at www.masteringaandp.com.

PEx

This lab corresponds to PhysioEx Exercise 11. See p. PEx-161.

PAL

For access to anatomical models and more, check out Practice Anatomy Lab.

OBJECTIVES

1. To name the two major components of blood, and to state their average percentages in whole blood.
2. To describe the composition and functional importance of plasma.
3. To define *formed elements* and list the cell types composing them, cite their relative percentages, and describe their major functions.
4. To identify red blood cells, basophils, eosinophils, monocytes, lymphocytes, and neutrophils when provided with a microscopic preparation or appropriate diagram.
5. To provide the normal values for a total white blood cell count and a total red blood cell count, and to state the importance of these tests.
6. To conduct the following blood tests in the laboratory, and to state their norms and the importance of each.

   differential white blood cell count
   hematocrit
   hemoglobin determination
   clotting time
   ABO and Rh blood typing
   plasma cholesterol concentration
7. To discuss the reason for transfusion reactions resulting from the administration of mismatched blood.
8. To define *leukocytosis, leukopenia, leukemia, polycythemia,* and *anemia* and to cite a possible reason for each condition.

PRE-LAB QUIZ

1. Circle True or False. There are no special precautions that I need to observe when performing today's lab.
2. Three types of formed elements found in blood include erythrocytes, leucocytes, and ________.
   a. electrolytes c. platelets
   b. fibers d. sodium salts
3. Circle the correct term. Mature erythrocytes / leucocytes are the most numerous blood cells and do not have a nucleus.
4. The least numerous but largest of all agranulocytes is the
   a. basophil c. monocyte
   b. lymphocyte d. neutrophil
5. ________ are the leukocytes responsible for releasing histamine and other mediators of inflammation.
   a. Basophils c. Monocytes
   b. Eosinophils d. Neutrophils
6. Circle the correct term. When determining the hematocrit / hemoglobin, you will centrifuge whole blood in order to allow the formed elements to sink to the bottom of the sample.
7. Circle the correct term. Blood typing is based on the presence of proteins known as antigens / antibodies on the outer surface of the red blood cell plasma membrane.
8. Circle True or False. If an individual is transfused with the wrong type blood, the recipient's antibodies react with the donor's antigens, eventually clumping and hemolyzing the donated RBCs.

- ☐ Clean microscope slides
- ☐ Glass stirring rods
- ☐ Wright's stain in a dropper bottle
- ☐ Distilled water in a dropper bottle
- ☐ Sterile lancets
- ☐ Absorbent cotton balls
- ☐ Alcohol swabs (wipes)
- ☐ Paper towels
- ☐ Compound microscope
- ☐ Immersion oil
- ☐ Three-dimensional models (if available) and charts of blood cells
- ☐ Assorted slides of white blood count pathologies labeled "Unknown Sample ______"
- ☐ Timer

Because many blood tests are to be conducted in this exercise, it is advisable to set up a number of appropriately labeled supply areas for the various tests, as designated below. Some needed supplies are located in the general supply area.

Note: Artificial blood prepared by Ward's Natural Science can be used for differential counts, hematocrit, and blood typing.

Activity 4: Hematocrit

- ☐ Heparinized capillary tubes
- ☐ Microhematocrit centrifuge and reading gauge (if the reading gauge is not available, a millimeter ruler may be used)
- ☐ Capillary tube sealer or modeling clay

Activity 5: Hemoglobin determination

- ☐ Hemoglobinometer, hemolysis applicator, and lens paper; or Tallquist hemoglobin scale and test paper

Activity 6: Coagulation time

- ☐ Capillary tubes (nonheparinized)
- ☐ Fine triangular file

Activity 7: Blood typing

- ☐ Blood typing sera (anti-A, anti-B, and anti-Rh [anti-D])
- ☐ Rh typing box
- ☐ Wax marking pencil
- ☐ Toothpicks
- ☐ Medicine dropper
- ☐ Blood test cards or microscope slides

Activity 8: Demonstration

- ☐ Microscopes set up with prepared slides demonstrating the following bone (or bone marrow) conditions: macrocytic hypochromic anemia, microcytic hypochromic anemia, sickle cell anemia, lymphocytic leukemia (chronic), and eosinophilia

Activity 9: Cholesterol measurement

- ☐ Cholesterol test cards and color scale

In this exercise you will study plasma and formed elements of blood and conduct various hematologic tests. These tests are useful diagnostic tools for the physician because blood composition (number and types of blood cells, and chemical composition) reflects the status of many body functions and malfunctions.

ALERT: Special precautions when handling blood. This exercise provides information on blood from several sources: human, animal, human treated, and artificial blood. The decision to use animal blood for testing or to have students test their own blood will be made by the instructor in accordance with the educational goals of the student group. For example, for students in the nursing or laboratory technician curricula, learning how to safely handle human blood or other human wastes is essential. Whenever blood is being handled, special attention must be paid to safety precautions. Instructors who opt to use human blood are responsible for its safe handling. Precautions should be used regardless of the source of the blood. This will both teach good technique and ensure the safety of the students.

Follow exactly the safety precautions listed below.

1. Wear safety gloves at all times. Discard appropriately.
2. Wear safety glasses throughout the exercise.
3. Handle only your own, freshly let (human) blood.
4. Be sure you understand the instructions and have all supplies on hand before you begin any part of the exercise.
5. Do not reuse supplies and equipment once they have been exposed to blood.
6. Keep the lab area clean. Do not let anything that has come in contact with blood touch surfaces or other individuals in the lab. Pay attention to the location of any supplies and equipment that come into contact with blood.
7. Dispose of lancets immediately after use in a designated disposal container. Do not put them down on the lab bench, even temporarily.
8. Dispose of all used cotton balls, alcohol swabs, blotting paper, and so forth in autoclave bags and place all soiled glassware in containers of 10% bleach solution.
9. Wipe down the lab bench with 10% bleach solution when you finish.

Composition of Blood

Circulating blood is a rather viscous substance that varies from bright scarlet to a dull brick red, depending on the amount of oxygen it is carrying. The average volume of blood in the body is about 5–6 L in adult males and 4–5 L in adult females.

Blood is classified as a type of connective tissue because it consists of a nonliving fluid matrix (the **plasma**) in which living cells **(formed elements)** are suspended. The fibers typical of a connective tissue matrix become visible in blood only when clotting occurs. They then appear as fibrin threads, which form the structural basis for clot formation.

More than 100 different substances are dissolved or suspended in plasma (Figure 29.1), which is over 90% water. These include nutrients, gases, hormones, various wastes and metabolites, many types of proteins, and electrolytes. The composition of plasma varies continuously as cells remove or add substances to the blood.

Three types of formed elements are present in blood (Table 29.1). Most numerous are **erythrocytes,** or **red blood cells (RBCs),** which are literally sacs of hemoglobin

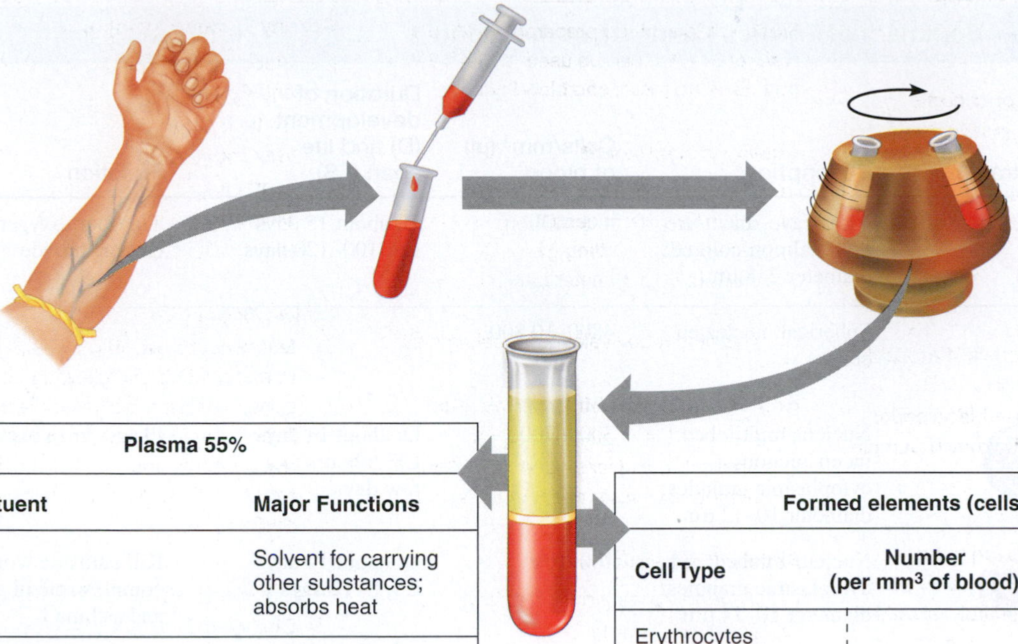

Plasma 55%	
Constituent	**Major Functions**
Water	Solvent for carrying other substances; absorbs heat
Salts (electrolytes) Sodium Potassium Calcium Magnesium Chloride Bicarbonate	Osmotic balance, pH buffering, regulation of membrane permeability
Plasma proteins Albumin	Osmotic balance, pH buffering
Fibrinogen	Clotting of blood
Globulins	Defense (antibodies) and lipid transport
Substances transported by blood Nutrients (glucose, fatty acids, amino acids, vitamins) Waste products of metabolism (urea, uric acid) Respiratory gases (O_2 and CO_2) Hormones	

Formed elements (cells) 45%		
Cell Type	**Number (per mm^3 of blood)**	**Functions**
Erythrocytes (red blood cells) 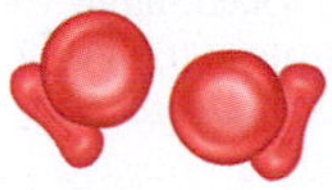	4 – 6 million	Transport oxygen and help transport carbon dioxide
Leukocytes (white blood cells)	4800 – 10,800	Defense and immunity
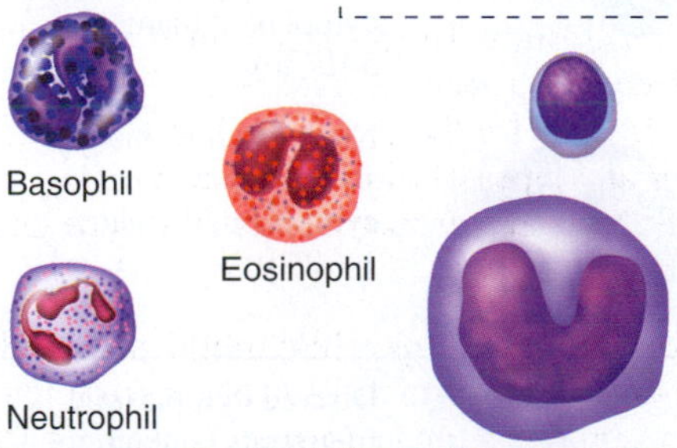 Basophil Eosinophil Lymphocyte Neutrophil Monocyte		
Platelets 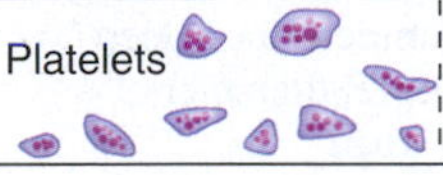	150,000 – 400,000	Blood clotting

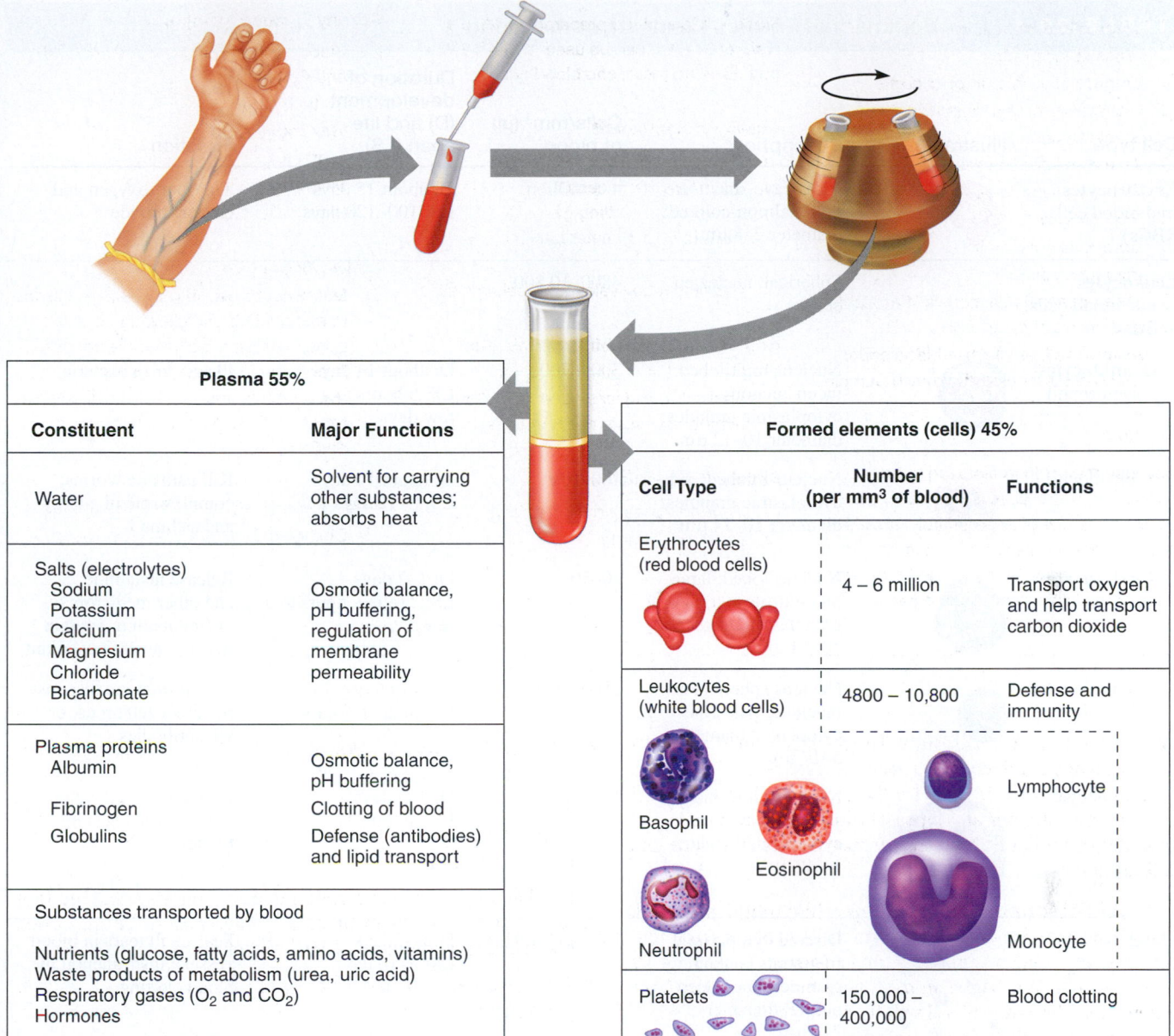

FIGURE 29.1 The composition of blood.

molecules that transport the bulk of the oxygen carried in the blood (and a small percentage of the carbon dioxide). **Leukocytes,** or **white blood cells (WBCs),** are part of the body's nonspecific defenses and the immune system, and **platelets** function in hemostasis (blood clot formation). Formed elements normally constitute 45% of whole blood; plasma accounts for the remaining 55%.

ACTIVITY 1

Determining the Physical Characteristics of Plasma

Go to the general supply area and carefully pour a few milliliters of plasma into a test tube. Also obtain some wide-range pH paper, and then return to your laboratory bench to make the following simple observations.

pH of Plasma

Test the pH of the plasma with wide-range pH paper. Record the pH observed. ______________________

Color and Clarity of Plasma

Hold the test tube up to a source of natural light. Note and record its color and degree of transparency. Is it clear, translucent, or opaque?

Color ______________________

Degree of transparency ______________________

TABLE 29.1 Summary of Formed Elements of the Blood

Cell type	Illustration	Description*	Cells/mm³ (μl) of blood	Duration of development (D) and life span (LS)	Function
Erythrocytes (red blood cells, RBCs)		Biconcave, anucleate disc; salmon-colored; diameter 7–8 μm	4–6 million	D: about 15 days LS: 100–120 days	Transport oxygen and carbon dioxide
Leukocytes (white blood cells, WBCs)		Spherical, nucleated cells	4800–10,800		
Granulocytes Neutrophil	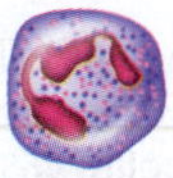	Nucleus multilobed; inconspicuous cytoplasmic granules; diameter 10–12 μm	3000–7000	D: about 14 days LS: 6 hours to a few days	Phagocytize bacteria
Eosinophil	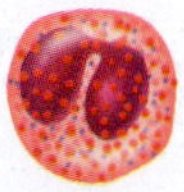	Nucleus bilobed; red cytoplasmic granules; diameter 10–14 μm	100–400	D: about 14 days LS: 8–12 days	Kill parasitic worms; complex role in allergy and asthma
Basophil	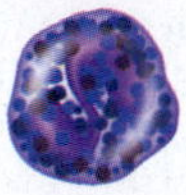	Nucleus lobed; large blue-purple cytoplasmic granules; diameter 10–14 μm	20–50	D: 1–7 days LS: ? (a few hours to a few days)	Release histamine and other mediators of in-flammation; contain heparin, an anticoagulant
Agranulocytes Lymphocyte		Nucleus spherical or indented; pale blue cytoplasm; diameter 5–17 μm	1500–3000	D: days to weeks LS: hours to years	Mount immune response by direct cell attack or via antibodies
Monocyte		Nucleus U- or kidney-shaped; gray-blue cytoplasm; diameter 14–24 μm	100–700	D: 2–3 days LS: months	Phagocytosis; develop into macrophages in tissues
Platelets	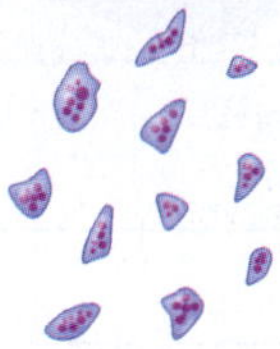	Discoid cytoplasmic fragments containing granules; stain deep purple; diameter 2–4 μm	150,000–400,000	D: 4–5 days LS: 5–10 days	Seal small tears in blood vessels; instrumental in blood clotting

*Appearance when stained with Wright's stain.

Consistency

While wearing gloves, dip your finger and thumb into plasma and then press them firmly together for a few seconds. Gently pull them apart. How would you describe the consistency of plasma (slippery, watery, sticky, granular)? Record your observations.

____

ACTIVITY 2

Examining the Formed Elements of Blood Microscopically

In this section, you will observe blood cells on an already prepared (purchased) blood slide or on a slide prepared from your own blood or blood provided by your instructor.

- Those using the purchased blood slide are to obtain a slide and begin their observations at step 6.
- Those testing blood provided by a biological supply source or an animal hospital are to obtain a tube of the supplied blood, disposable gloves, and the supplies listed in

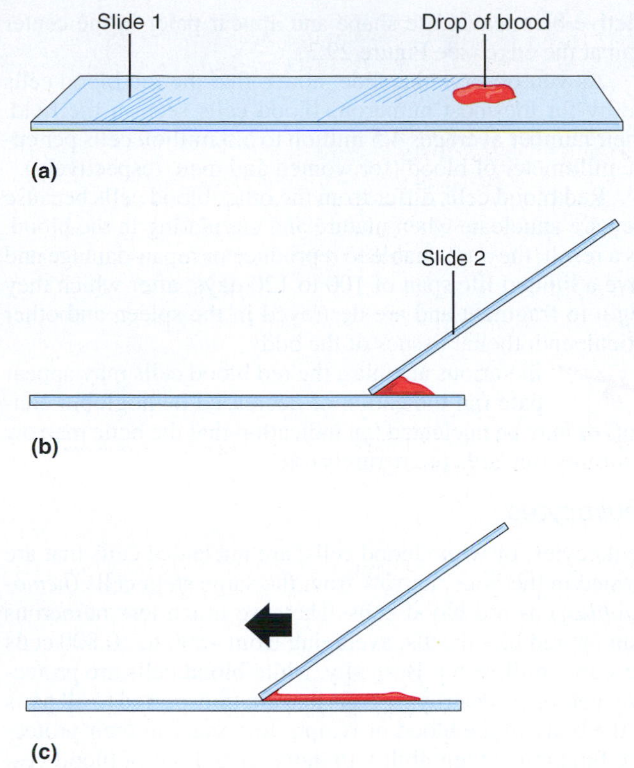
Slide 1
Drop of blood
(a)
Slide 2
(b)
(c)

FIGURE 29.2 Procedure for making a blood smear. (a) Place a drop of blood on slide 1 approximately ½ inch from one end. **(b)** Hold slide 2 at a 30° to 40° angle to slide 1 (it should touch the drop of blood) and allow blood to spread along entire bottom edge of angled slide. **(c)** Smoothly advance slide 2 to end of slide 1 (blood should run out before reaching the end of slide 1). Then lift slide 2 away from slide 1 and place it on a paper towel.

step 1, except for the lancets and alcohol swabs. After donning gloves, those students will go to step 3b to begin their observations.

- If you are examining your own blood, you will perform all the steps described below *except* step 3b.

1. Obtain two glass slides, a glass stirring rod, dropper bottles of Wright's stain and distilled water, two or three lancets, cotton balls, and alcohol swabs. Bring this equipment to the laboratory bench. Clean the slides thoroughly and dry them.
2. Open the alcohol swab packet and scrub your third or fourth finger with the swab. (Because the pricked finger may be a little sore later, it is better to prepare a finger on the nondominant hand.) Circumduct your hand (swing it in a cone-shaped path) for 10 to 15 seconds. This will dry the alcohol and cause your fingers to become engorged with blood. Then, open the lancet packet and grasp the lancet by its blunt end. Quickly jab the pointed end into the prepared finger to produce a free flow of blood. It is *not* a good idea to squeeze or "milk" the finger, as this forces out tissue fluid as well as blood. If the blood is not flowing freely, another puncture should be made.

! *Under no circumstances is a lancet to be used for more than one puncture.* Dispose of the lancets in the designated disposal container immediately after use.

3a. With a cotton ball, wipe away the first drop of blood; then allow another large drop of blood to form. Touch the blood to one of the cleaned slides approximately 1.3 cm, or ½ inch, from the end. Then quickly (to prevent clotting) use the second slide to form a blood smear as shown in Figure 29.2. When properly prepared, the blood smear is uniformly thin. If the blood smear appears streaked, the blood probably began to clot or coagulate before the smear was made, and another slide should be prepared. Continue at step 4.

3b. Dip a glass rod in the blood provided, and transfer a generous drop of blood to the end of a cleaned microscope slide. For the time being, lay the glass rod on a paper towel on the bench. Then, as described in step 3a and Figure 29.2, use the second slide to make your blood smear.

4. Dry the slide by waving it in the air. When it is completely dry, it will look dull. Place it on a paper towel, and flood it with Wright's stain. Count the number of drops of stain used. Allow the stain to remain on the slide for 3 to 4 minutes, and then flood the slide with an equal number of drops of distilled water. Allow the water and Wright's stain mixture to remain on the slide for 4 or 5 minutes or until a metallic green film or scum is apparent on the fluid surface. Blow on the slide gently every minute or so to keep the water and stain mixed during this interval.
5. Rinse the slide with a stream of distilled water. Then flood it with distilled water, and allow it to lie flat until the slide becomes translucent and takes on a pink cast. Then stand the slide on its long edge on the paper towel, and allow it to dry completely. Once the slide is dry, you can begin your observations.
6. Obtain a microscope and scan the slide under low power to find the area where the blood smear is the thinnest. After scanning the slide in low power to find the areas with the largest numbers of nucleated WBCs, read the following descriptions of cell types, and find each one on Figure 29.1 and Table 29.1. (The formed elements are also shown in Figure 29.3 and Figure 29.4.) Then, switch to

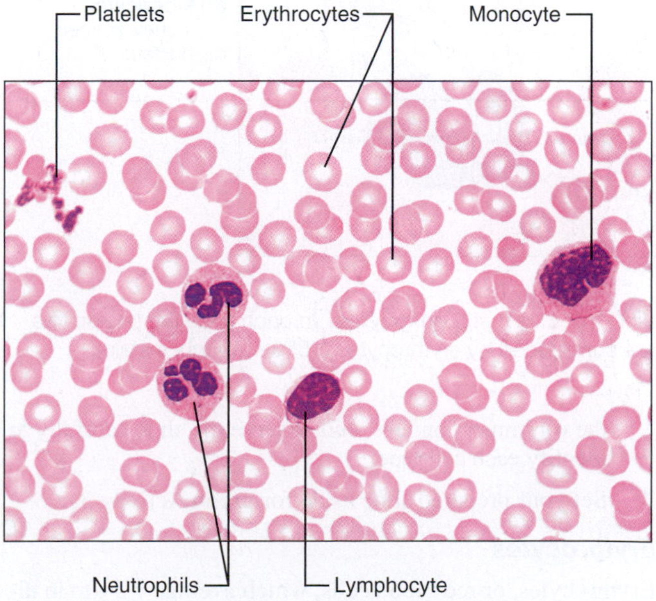
Platelets
Erythrocytes
Monocyte
Neutrophils
Lymphocyte

FIGURE 29.3 Photomicrograph of a human blood smear stained with Wright's stain (610×).

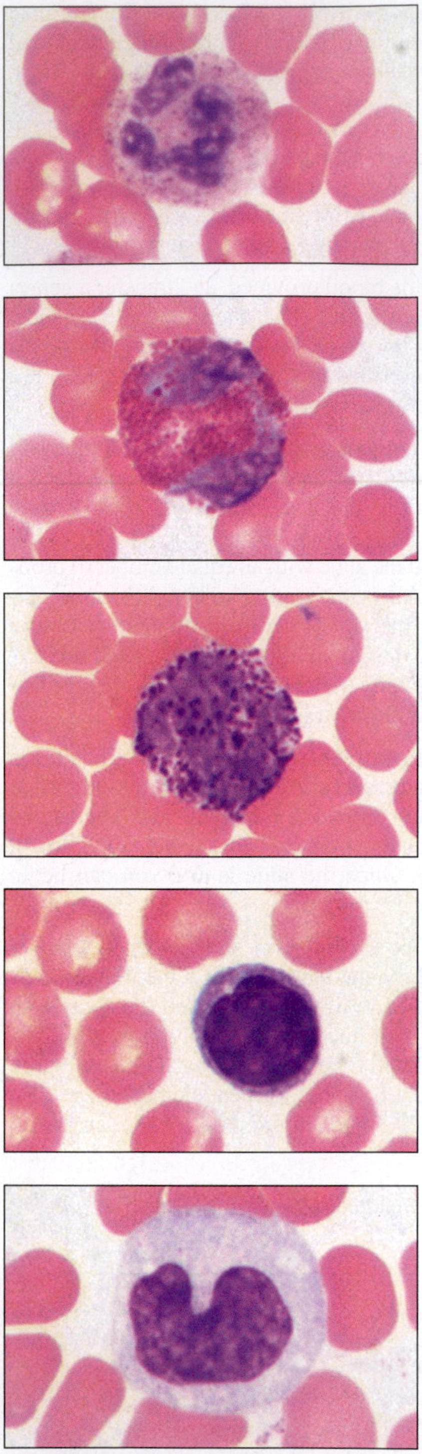

(a) Neutrophil; multilobed nucleus

(b) Eosinophil; bilobed nucleus, red cytoplasmic granules

(c) Basophil; bilobed nucleus, purplish-black cytoplasmic granules

(d) Small lymphocyte; large spherical nucleus

(e) Monocyte; kidney-shaped nucleus

FIGURE 29.4 Leukocytes. In each case the leukocytes are surrounded by erythrocytes (1850×, Wright's stain).

the oil immersion lens, and observe the slide carefully to identify each cell type.

7. Set your prepared slide aside for use in Activity 3.

Erythrocytes

Erythrocytes, or red blood cells, which average 7.5 μm in diameter, vary in color from a salmon red color to pale pink, depending on the effectiveness of the stain. They have a distinctive biconcave disc shape and appear paler in the center than at the edge (see Figure 29.3).

As you observe the slide, notice that the red blood cells are by far the most numerous blood cells seen in the field. Their number averages 4.5 million to 5.5 million cells per cubic millimeter of blood (for women and men, respectively).

Red blood cells differ from the other blood cells because they are anucleate when mature and circulating in the blood. As a result, they are unable to reproduce or repair damage and have a limited life span of 100 to 120 days, after which they begin to fragment and are destroyed in the spleen and other reticuloendothelial tissues of the body.

In various anemias, the red blood cells may appear pale (an indication of decreased hemoglobin content) or may be nucleated (an indication that the bone marrow is turning out cells prematurely). ■

Leukocytes

Leukocytes, or white blood cells, are nucleated cells that are formed in the bone marrow from the same stem cells (*hemocytoblast*) as red blood cells. They are much less numerous than the red blood cells, averaging from 4800 to 10,800 cells per cubic millimeter. Basically, white blood cells are protective, pathogen-destroying cells that are transported to all parts of the body in the blood or lymph. Important to their protective function is their ability to move in and out of blood vessels, a process called **diapedesis,** and to wander through body tissues by **amoeboid motion** to reach sites of inflammation or tissue destruction. They are classified into two major groups, depending on whether or not they contain conspicuous granules in their cytoplasm.

Granulocytes make up the first group. The granules in their cytoplasm stain differentially with Wright's stain, and they have peculiarly lobed nuclei, which often consist of expanded nuclear regions connected by thin strands of nucleoplasm. There are three types of granulocytes:

Neutrophil: The most abundant of the white blood cells (50% to 70% of the leukocyte population); nucleus consists of 3 to 6 lobes and the pale lilac cytoplasm contains fine cytoplasmic granules, which are generally indistinguishable and take up both the acidic (red) and basic (blue) dyes (*neutrophil* = neutral loving); functions as an active phagocyte. The number of neutrophils increases exponentially during acute infections. (See Figure 29.4a.)

Eosinophil: Represents 2% to 4% of the leukocyte population; nucleus is generally figure-8 or bilobed in shape; contains large cytoplasmic granules (elaborate lysosomes) that stain red-orange with the acid dyes in Wright's stain (see Figure 29.4b). Eosinophils are about the size of neutrophils and play a role in counterattacking parasitic worms. Eosinophils have complex roles in many other diseases, especially in allergy and asthma.

Basophil: Least abundant leukocyte type representing less than 1% of the population; large U- or S-shaped nucleus with two or more indentations. Cytoplasm contains coarse, sparse granules that are stained deep purple by the basic dyes in Wright's stain (see Figure 29.4c). The granules contain several chemicals, including histamine, a vasodilator that is discharged on exposure to antigens and helps mediate the inflammatory response. Basophils are about the size of neutrophils.

The second group, **agranulocytes,** or **agranular leukocytes,** contains no *visible* cytoplasmic granules. Although found in the bloodstream, they are much more abundant in lymphoid tissues. Their nuclei tend to be closer to the norm, that is, spherical, oval, or kidney-shaped. Specific characteristics of the two types of agranulocytes are listed below.

Lymphocyte: The smallest of the leukocytes, approximately the size of a red blood cell (see Figure 29.4d). The nucleus stains dark blue to purple, is generally spherical or slightly indented, and accounts for most of the cell mass. Sparse cytoplasm appears as a thin blue rim around the nucleus. Concerned with immunologic responses in the body; one population, the *B lymphocytes,* gives rise to *plasma cells* that produce antibodies released to blood. The second population, *T lymphocytes,* plays a regulatory role and destroys grafts, tumors, and virus-infected cells. Represents 25% or more of the WBC population.

Monocyte: The largest of the leukocytes; approximately twice the size of red blood cells (see Figure 29.4e). Represents 3% to 8% of the leukocyte population. Dark blue nucleus is generally kidney-shaped; abundant cytoplasm stains gray-blue. Once in the tissues, monocytes convert to macrophages, active phagocytes (the "long-term cleanup team"), increasing dramatically in number during chronic infections such as tuberculosis.

Students are often asked to list the leukocytes in order from the most abundant to the least abundant. The following silly phrase may help you with this task: *N*ever *l*et *m*onkeys *e*at *b*ananas (neutrophils, lymphocytes, monocytes, eosinophils, basophils).

Platelets

Platelets are cell fragments of large multinucleate cells **(megakaryocytes)** formed in the bone marrow. They appear as darkly staining, irregularly shaped bodies interspersed among the blood cells (see Figure 29.3). The normal platelet count in blood ranges from 150,000 to 400,000 per cubic millimeter. Platelets are instrumental in the clotting process that occurs in plasma when blood vessels are ruptured.

After you have identified these cell types on your slide, observe charts and three-dimensional models of blood cells if these are available. Do not dispose of your slide, as you will use it later for the differential white blood cell count. ■

Hematologic Tests

When someone enters a hospital as a patient, several hematologic tests are routinely done to determine general level of health as well as the presence of pathologic conditions. You will be conducting the most common of these tests in this exercise.

! Materials such as cotton balls, lancets, and alcohol swabs are used in nearly all of the following diagnostic tests. These supplies are at the general supply area and should be properly disposed of (glassware to the bleach bucket, lancets in a designated disposal container, and disposable items to the autoclave bag) immediately after use.

Other necessary supplies and equipment are at specific supply areas marked according to the test with which they are used. Since nearly all of the tests require a finger stab, if you will be using your own blood it might be wise to quickly read through the tests to determine in which instances more than one preparation can be done from the same finger stab. A little planning will save you the discomfort of a multiple-punctured finger.

An alternative to using blood obtained from the finger stab technique is using heparinized blood samples supplied by your instructor. The purpose of using heparinized tubes is to prevent the blood from clotting. Thus blood collected and stored in such tubes will be suitable for all tests except coagulation time testing.

Total White and Red Blood Cell Counts

A **total WBC count** or **total RBC count** determines the total number of that cell type per unit volume of blood. Total WBC and RBC counts are a routine part of any physical exam. Most clinical agencies use computers to conduct these counts. Since the hand counting technique typically done in college labs is rather outdated, total RBC and WBC counts will not be done here, but the importance of such counts (both normal and abnormal values) is briefly described below.

Total White Blood Cell Count

Since white blood cells are an important part of the body's defense system, it is essential to note any abnormalities in them.

Leukocytosis, an abnormally high WBC count, may indicate bacterial or viral infection, metabolic disease, hemorrhage, or poisoning by drugs or chemicals. A decrease in the white cell number below 4000/mm^3 **(leukopenia)** may indicate typhoid fever, measles, infectious hepatitis or cirrhosis, tuberculosis, or excessive antibiotic or X-ray therapy. A person with leukopenia lacks the usual protective mechanisms. **Leukemia,** a malignant disorder of the lymphoid tissues characterized by uncontrolled proliferation of abnormal WBCs accompanied by a reduction in the number of RBCs and platelets, is detectable not only by a total WBC count but also by a differential WBC count. ■

Total Red Blood Cell Count

Since RBCs are absolutely necessary for oxygen transport, a doctor typically investigates any excessive change in their number immediately.

An increase in the number of RBCs **(polycythemia)** may result from bone marrow cancer or from living at high altitudes where less oxygen is available. A decrease in the number of RBCs results in anemia. (The term **anemia** simply indicates a decreased oxygen-carrying capacity of blood that may result from a decrease in RBC number or size or a decreased hemoglobin content of the RBCs.) A decrease in RBCs may result suddenly from hemorrhage or more gradually from conditions that destroy RBCs or hinder RBC production. ■

Differential White Blood Cell Count

To make a **differential white blood cell count,** 100 WBCs are counted and classified according to type. Such a count is routine in a physical examination and in diagnosing illness, since any abnormality or significant elevation in percentages of WBC types may indicate a problem or the source of pathology.

ACTIVITY 3

Conducting a Differential WBC Count

1. Use the slide prepared for the identification of the blood cells in Activity 2. Begin at the edge of the smear and move the slide in a systematic manner on the microscope stage—either up and down or from side to side as indicated in Figure 29.5.

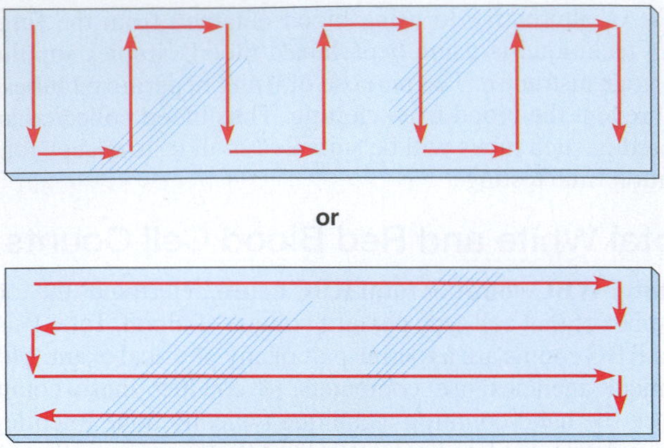
or

FIGURE 29.5 Alternative methods of moving the slide for a differential WBC count.

2. Record each type of white blood cell you observe by making a count in the first blank column of the chart below (for example, 卌 II = 7 cells) until you have observed and recorded a total of 100 WBCs. Using the following equation, compute the percentage of each WBC type counted, and record the percentages on the Hematologic Test Data Sheet on page 436.

$$\text{Percent (\%)} = \frac{\text{\# observed}}{\text{Total \# counted (100)}} \times 100$$

3. Select a slide marked "Unknown sample," record the slide number, and use the count chart below to conduct a differential count. Record the percentages on the data sheet on page 436.

How does the differential count from the unknown sample slide compare to a normal count?

Count of 100 WBCs

Cell type	Number observed: Student blood smear	Number observed: Unknown sample # ___
Neutrophils		
Eosinophils		
Basophils		
Lymphocytes		
Monocytes		

Using the text and other references, try to determine the blood pathology on the unknown slide. Defend your answer.

4. How does your differential white blood cell count correlate with the percentages given for each type on pages 428–429?

Hematocrit

The **hematocrit,** or **packed cell volume (PCV),** is routinely determined when anemia is suspected. Centrifuging whole blood spins the formed elements to the bottom of the tube, with plasma forming the top layer (see Figure 29.1). Since the blood cell population is primarily RBCs, the PCV is generally considered equivalent to the RBC volume, and this is the only value reported. However, the relative percentage of WBCs can be differentiated, and both WBC and plasma volume will be reported here. Normal hematocrit values for the male and female, respectively, are 47.0 ± 7 and 42.0 ± 5.

ACTIVITY 4

Determining the Hematocrit

The hematocrit is determined by the micromethod, so only a drop of blood is needed. If possible (and the centrifuge allows), all members of the class should prepare their capillary tubes at the same time so the centrifuge can be run only once.

1. Obtain two heparinized capillary tubes, capillary tube sealer or modeling clay, a lancet, alcohol swabs, and some cotton balls.

2. If you are using your own blood, cleanse a finger, and allow the blood to flow freely. Wipe away the first few drops and, holding the red-line-marked end of the capillary tube to the blood drop, allow the tube to fill at least three-fourths full by capillary action (Figure 29.6a). If the blood is not flowing freely, the end of the capillary tube will not be completely submerged in the blood during filling, air will enter, and you will have to prepare another sample.

If you are using instructor-provided blood, simply immerse the red-marked end of the capillary tube in the blood sample and fill it three-quarters full as just described.

3. Plug the blood-containing end by pressing it into the capillary tube sealer or clay (Figure 29.6b). Prepare a second tube in the same manner.

4. Place the prepared tubes opposite one another in the radial grooves of the microhematocrit centrifuge with the sealed ends abutting the rubber gasket at the centrifuge periphery (Figure 29.6c). This loading procedure balances the centrifuge and prevents blood from spraying everywhere by centrifugal force. *Make a note of the numbers of the grooves your tubes are in.* When all the tubes have been loaded, make sure the centrifuge is properly balanced, and secure the centrifuge cover. Turn the centrifuge on, and set the timer for 4 or 5 minutes.

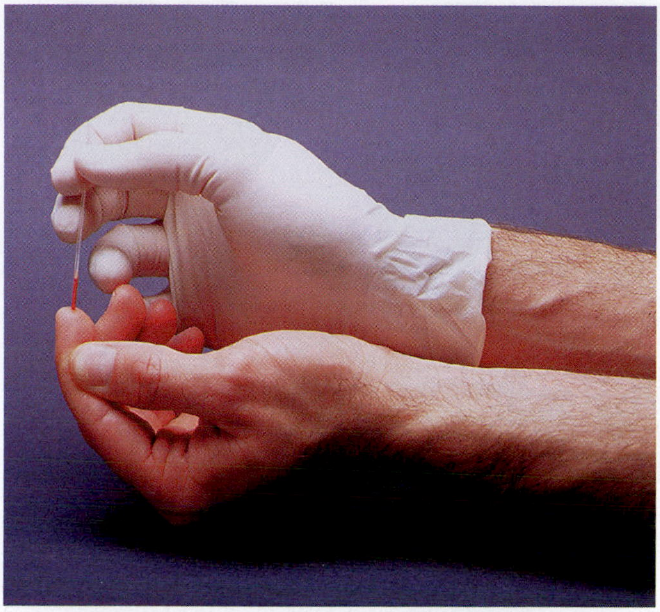

(a)

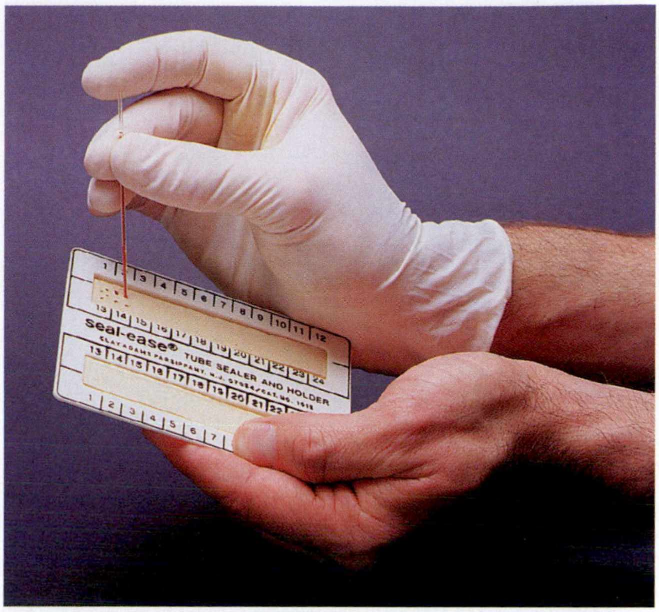
seal-ease® TUBE SEALER AND HOLDER

(b)

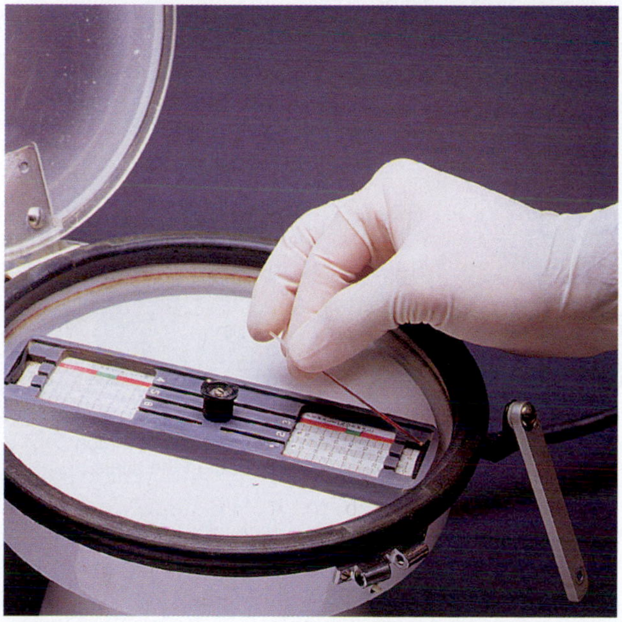

(c)

FIGURE 29.6 Steps in a hematocrit determination. **(a)** Load a heparinized capillary tube with blood. **(b)** Plug the blood-containing end of the tube with clay. **(c)** Place the tube in a microhematocrit centrifuge. (Centrifuge must be balanced.)

5. Determine the percentage of RBCs, WBCs, and plasma by using the microhematocrit reader. The RBCs are the bottom layer, the plasma is the top layer, and the WBCs are the buff-colored layer between the two. If the reader is not available, use a millimeter ruler to measure the length of the filled capillary tube occupied by each element, and compute its percentage by using the following formula:

$$\frac{\text{Height of the column composed of the element (mm)}}{\text{Height of the original column of whole blood (mm)}} \times 100$$

Record your calculations below and on the data sheet on page 436.

% RBC ________ % WBC ________ % plasma ________

Usually WBCs constitute 1% of the total blood volume. How do your blood values compare to this figure and to the normal percentages for RBCs and plasma? (See page 425.)

____

As a rule, a hematocrit is considered a more accurate test than the total RBC count for determining the RBC composition of the blood. A hematocrit within the normal range generally indicates a normal RBC number, whereas an abnormally high or low hematocrit is cause for concern. ■

Hemoglobin Concentration

As noted earlier, a person can be anemic even with a normal RBC count. Since hemoglobin (Hb) is the RBC protein responsible for oxygen transport, perhaps the most accurate way of measuring the oxygen-carrying capacity of the blood is to determine its hemoglobin content. Oxygen, which combines reversibly with the heme (iron-containing portion) of the hemoglobin molecule, is picked up by the blood cells in the lungs and unloaded in the tissues. Thus, the more hemoglobin molecules the RBCs contain, the more oxygen they will be able to transport. Normal blood contains 12 to 18 g of hemoglobin per 100 ml of blood. Hemoglobin content in men is slightly higher (13 to 18 g) than in women (12 to 16 g).

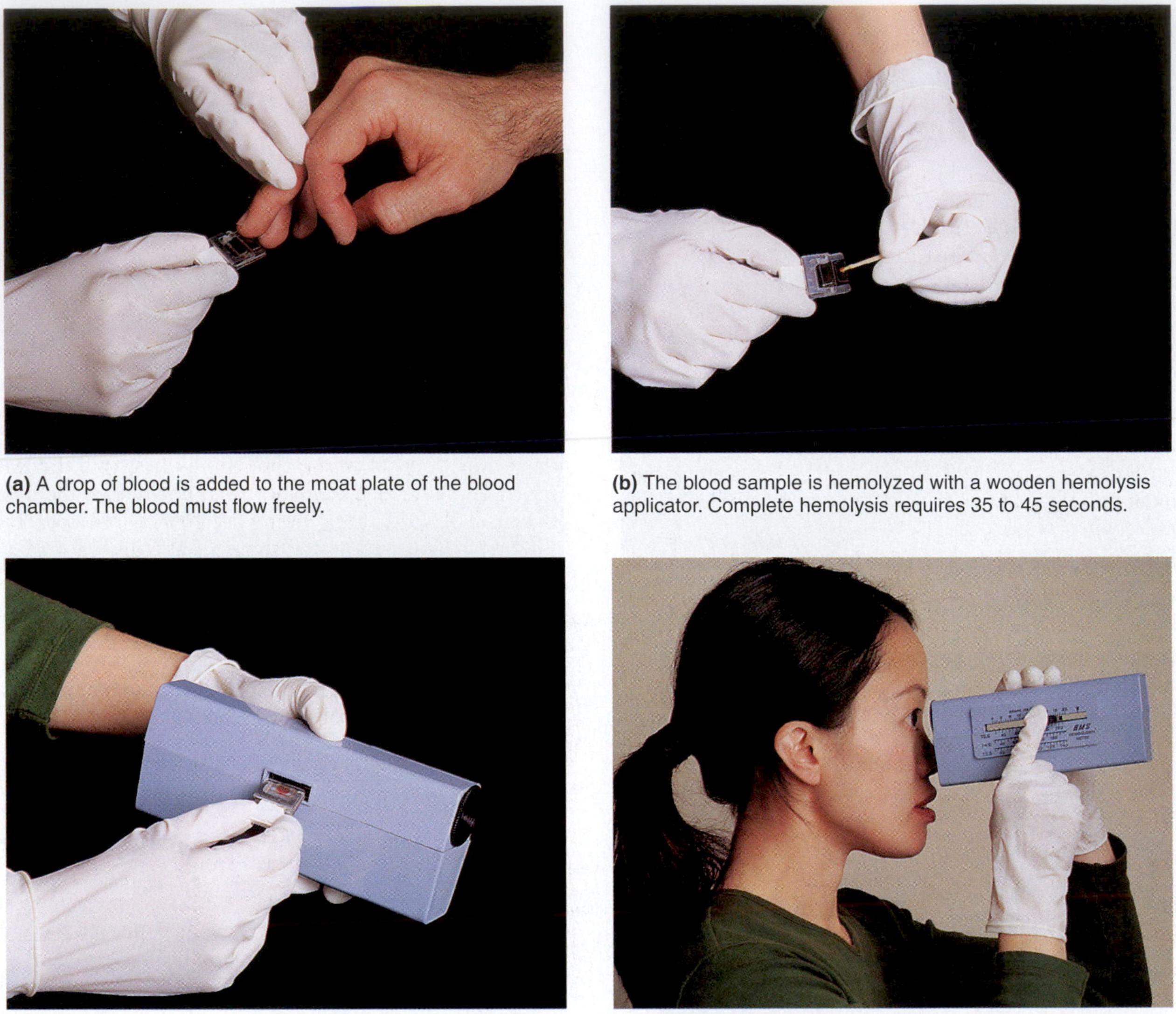

(a) A drop of blood is added to the moat plate of the blood chamber. The blood must flow freely.

(b) The blood sample is hemolyzed with a wooden hemolysis applicator. Complete hemolysis requires 35 to 45 seconds.

(c) The charged blood chamber is inserted into the slot on the side of the hemoglobinometer.

(d) The colors of the green split screen are found by moving the slide with the right index finger. When the two colors match in density, the grams/100 ml and % Hb are read on the scale.

FIGURE 29.7 Hemoglobin determination using a hemoglobinometer.

ACTIVITY 5

Determining Hemoglobin Concentration

Several techniques have been developed to estimate the hemoglobin content of blood, ranging from the old, rather inaccurate Tallquist method to expensive colorimeters, which are precisely calibrated and yield highly accurate results. Directions for both the Tallquist method and a hemoglobinometer are provided here.

Tallquist Method

1. Obtain a Tallquist hemoglobin scale, test paper, lancets, alcohol swabs, and cotton balls.

2. Use instructor-provided blood or prepare the finger as previously described. (For best results, make sure the alcohol evaporates before puncturing your finger.) Place one good-sized drop of blood on the special absorbent paper provided with the color scale. The blood stain should be larger than the holes on the color scale.

3. As soon as the blood has dried and loses its glossy appearance, match its color, under natural light, with the color standards by moving the specimen under the comparison scale so that the blood stain appears at all the various apertures. (The blood should not be allowed to dry to a brown color, as this will result in an inaccurate reading.) Because the colors on the scale represent 1% variations in hemoglobin content, it may be necessary to estimate the percentage if the color of your blood sample is intermediate between two color standards.

4. On the data sheet on page 436, record your results as the percentage of hemoglobin concentration and as grams per 100 ml of blood.

Hemoglobinometer Determination

1. Obtain a hemoglobinometer, hemolysis applicator, alcohol swab, and lens paper, and bring them to your bench. Test the hemoglobinometer light source to make sure it is working; if not, request new batteries before proceeding and test it again.

2. Remove the blood chamber from the slot in the side of the hemoglobinometer and disassemble the blood chamber by separating the glass plates from the metal clip. Notice as you do this that the larger glass plate has an H-shaped depression cut into it that acts as a moat to hold the blood, whereas the smaller glass piece is flat and serves as a coverslip.

3. Clean the glass plates with an alcohol swab, and then wipe them dry with lens paper. Hold the plates by their sides to prevent smearing during the wiping process.

4. Reassemble the blood chamber (remember: larger glass piece on the bottom with the moat up), but leave the moat plate about halfway out to provide adequate exposed surface to charge it with blood.

5. Obtain a drop of blood (from the provided sample or from your fingertip as before), and place it on the depressed area of the moat plate that is closest to you (Figure 29.7a).

6. Using the wooden hemolysis applicator, stir or agitate the blood to rupture (lyse) the RBCs (Figure 29.7b). This usually takes 35 to 45 seconds. Hemolysis is complete when the blood appears transparent rather than cloudy.

7. Push the blood-containing glass plate all the way into the metal clip and then firmly insert the charged blood chamber back into the slot on the side of the instrument (Figure 29.7c).

8. Hold the hemoglobinometer in your left hand with your left thumb resting on the light switch located on the underside of the instrument. Look into the eyepiece and notice that there is a green area divided into two halves (a split field).

9. With the index finger of your right hand, slowly move the slide on the right side of the hemoglobinometer back and forth until the two halves of the green field match (Figure 29.7d).

10. Note and record on the data sheet on page 436 the grams of Hb (hemoglobin)/100 ml of blood indicated on the uppermost scale by the index mark on the slide. Also record % Hb, indicated by one of the lower scales.

11. Disassemble the blood chamber once again, and carefully place its parts (glass plates and clip) into a bleach-containing beaker.

Generally speaking, the relationship between the PCV and grams of hemoglobin per 100 ml of blood is 3:1—for example, a PCV of 36 with 12 g of Hb per 100 ml of blood is a ratio of 3:1. How do your values compare?

____

Record on the data sheet (page 436) the value obtained from your data. ■

Bleeding Time

Normally a sharp prick of the finger or earlobe results in bleeding that lasts from 2 to 7 minutes (Ivy method) or 0 to 5 minutes (Duke method), although other factors such as altitude affect the time. How long the bleeding lasts is referred to as **bleeding time** and tests the ability of platelets to stop bleeding in capillaries and small vessels. Absence of some clotting factors may affect bleeding time, but prolonged bleeding time is most often associated with deficient or abnormal platelets.

Coagulation Time

Blood clotting, or **coagulation,** is a protective mechanism that minimizes blood loss when blood vessels are ruptured. This process requires the interaction of many substances normally present in the plasma (clotting factors, or procoagulants) as well as some released by platelets and injured tissues. Basically hemostasis proceeds as follows (Figure 29.8a): The injured tissues and platelets release **tissue factor (TF)** and **PF_3** respectively, which trigger the clotting mechanism, or cascade. Tissue factor and PF_3 interact with other blood protein clotting factors and calcium ions to form **prothrombin activator,** which in turn converts **prothrombin** (present in plasma) to **thrombin.** Thrombin then acts enzymatically to polymerize the soluble **fibrinogen** proteins (present in plasma) into insoluble **fibrin,** which forms a meshwork of strands that traps the RBCs and forms the basis of the clot (Figure 29.8b). Normally, blood removed from the body clots within 2 to 6 minutes.

ACTIVITY 6

Determining Coagulation Time

1. Obtain a *nonheparinized* capillary tube, a timer (or watch), a lancet, cotton balls, a triangular file, and alcohol swabs.

2. Clean and prick the finger to produce a free flow of blood. Discard the lancet in the disposal container.

3. Place one end of the capillary tube in the blood drop, and hold the opposite end at a lower level to collect the sample.

4. Lay the capillary tube on a paper towel.

Record the time. ______________________________

5. At 30-second intervals, make a small nick on the tube close to one end with the triangular file, and then carefully break the tube. Slowly separate the ends to see if a gel-like thread of fibrin spans the gap. When this occurs, record below and on the data sheet on page 436 the time for coagulation to occur. Are your results within the normal time range?

____

6. Put used supplies in the autoclave bag and broken capillary tubes into the sharps container. ■

Blood Typing

Blood typing is a system of blood classification based on the presence of specific glycoproteins on the outer surface of the RBC plasma membrane. Such proteins are called **antigens,** or **agglutinogens,** and are genetically determined. In many cases, these antigens are accompanied by plasma proteins, **antibodies** or **agglutinins,** that react with RBCs bearing different antigens, causing them to be clumped, agglutinated, and eventually hemolyzed. It is because of this phenomenon that a person's blood must be carefully typed before a whole blood or packed cell transfusion.

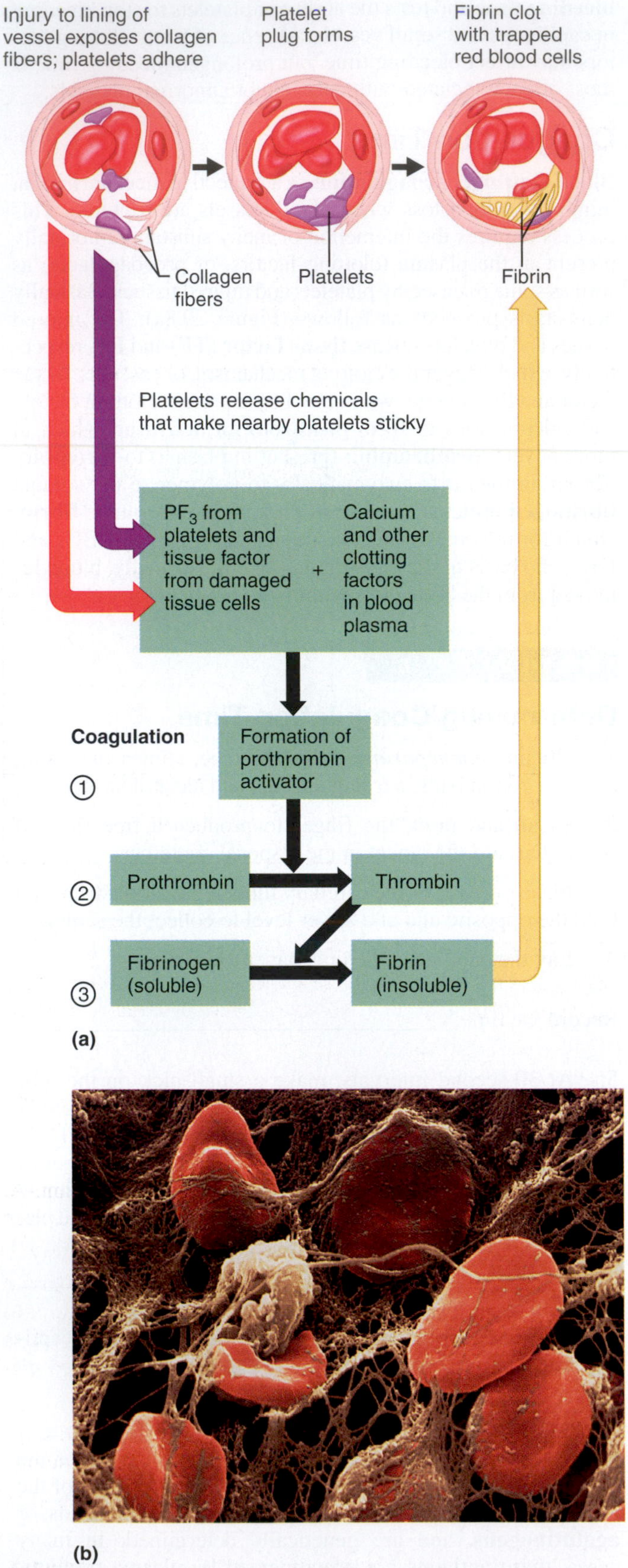
Injury to lining of vessel exposes collagen fibers; platelets adhere
Platelet plug forms
Fibrin clot with trapped red blood cells
Collagen fibers
Platelets
Fibrin
Platelets release chemicals that make nearby platelets sticky
PF3 from platelets and tissue factor from damaged tissue cells
+
Calcium and other clotting factors in blood plasma
Coagulation
Formation of prothrombin activator
1
Prothrombin
Thrombin
2
Fibrinogen (soluble)
Fibrin (insoluble)
3
(a)
(b)

FIGURE 29.8 Events of hemostasis and blood clotting. **(a)** Simple schematic of events. Steps numbered 1–3 represent the major events of coagulation. **(b)** Photomicrograph of RBCs trapped in a fibrin mesh (2700×).

Several blood typing systems exist, based on the various possible antigens, but the factors routinely typed for are antigens of the ABO and Rh blood groups which are most commonly involved in transfusion reactions. Other blood factors, such as Kell, Lewis, M, and N, are not routinely typed for unless the individual will require multiple transfusions. The basis of the ABO typing is shown in Table 29.2.

Individuals whose red blood cells carry the Rh antigen are Rh positive (approximately 85% of the U.S. population); those lacking the antigen are Rh negative. Unlike ABO blood groups, neither the blood of the Rh-positive (Rh^+) nor Rh-negative (Rh^-) individuals carries preformed anti-Rh antibodies. This is understandable in the case of the Rh-positive individual. However, Rh-negative persons who receive transfusions of Rh-positive blood become sensitized by the Rh antigens of the donor RBCs, and their systems begin to produce anti-Rh antibodies. On subsequent exposures to Rh-positive blood, typical transfusion reactions occur, resulting in the clumping and hemolysis of the donor blood cells.

Although the blood of dogs and other mammals does react with some of the human agglutinins (present in the antisera), the reaction is not as pronounced and varies with the animal blood used. Hence, the most accurate and predictable blood typing results are obtained with human blood. The artificial blood kit does not use any body fluids and produces results similar to but not identical to results for human blood.

ACTIVITY 7

Typing for ABO and Rh Blood Groups

Blood may be typed on glass slides or using blood test cards. Each method is described in this activity.

Typing Blood Using Glass Slides

1. Obtain two clean microscope slides, a wax marking pencil, anti-A, anti-B, and anti-Rh typing sera, toothpicks, lancets, alcohol swabs, medicine dropper, and the Rh typing box.

2. Divide slide 1 into halves with the wax marking pencil. Label the lower left-hand corner "anti-A" and the lower right-hand corner "anti-B." Mark the bottom of slide 2 "anti-Rh."

3. Place one drop of anti-A serum on the *left* side of slide 1. Place one drop of anti-B serum on the *right* side of slide 1. Place one drop of anti-Rh serum in the center of slide 2.

4. If you are using your own blood, cleanse your finger with an alcohol swab, pierce the finger with a lancet, and wipe away the first drop of blood. Obtain 3 drops of freely flowing blood, placing one drop on each side of slide 1 and a drop on slide 2. Immediately dispose of the lancet in a designated disposal container.

If using instructor-provided animal blood or EDTA-treated red cells, use a medicine dropper to place one drop of blood on each side of slide 1 and a drop of blood on slide 2.

! 5. Quickly mix each blood-antiserum sample with a *fresh* toothpick. Then dispose of the toothpicks and used alcohol swab in the autoclave bag.

6. Place slide 2 on the Rh typing box and rock gently back and forth. (A slightly higher temperature is required for precise Rh typing than for ABO typing.)

TABLE 29.2 ABO Blood Typing

ABO blood type	Antigens present on RBC membranes	Antibodies present in plasma	% of U.S. population White	Black	Asian
A	A	Anti-B	40	27	28
B	B	Anti-A	11	20	27
AB	A and B	None	4	4	5
O	Neither	Anti-A and anti-B	45	49	40

7. After 2 minutes, observe all three blood samples for evidence of clumping. The agglutination that occurs in the positive test for the Rh factor is very fine and difficult to perceive; thus if there is any question, observe the slide under the microscope. Record your observations in the Blood Typing chart.

8. Interpret your ABO results in light of the information in Figure 29.9. If clumping was observed on slide 2, you are Rh positive. If not, you are Rh negative.

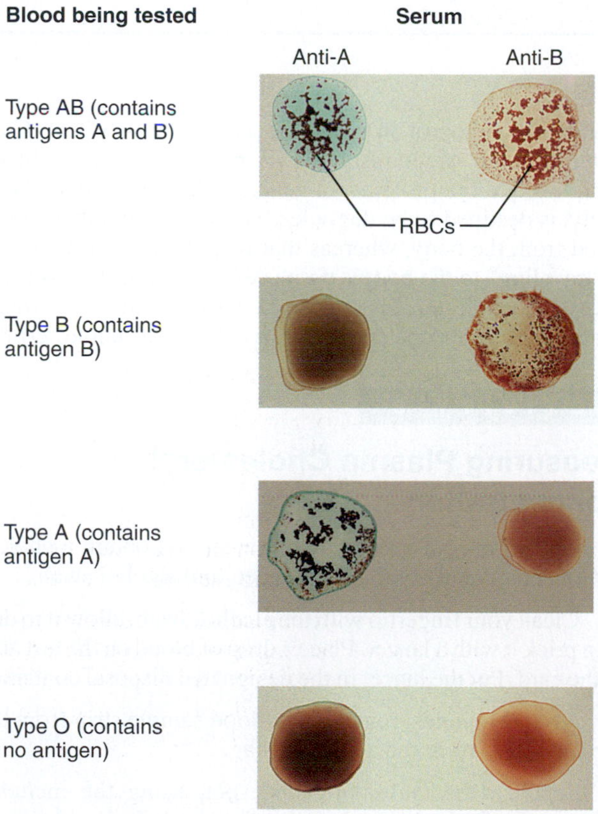
Blood being tested
Serum
Anti-A
Anti-B
Type AB (contains antigens A and B)
RBCs
Type B (contains antigen B)
Type A (contains antigen A)
Type O (contains no antigen)

FIGURE 29.9 Blood typing of ABO blood types. When serum containing anti-A or anti-B antibodies (agglutinins) is added to a blood sample, agglutination will occur between the antibody and the corresponding antigen (agglutinogen A or B). As illustrated, agglutination occurs with both sera in blood group AB, with anti-B serum in blood group B, with anti-A serum in blood group A, and with neither serum in blood group O.

Blood Typing

Result	Observed (+)	Not observed (−)
Presence of clumping with anti-A		
Presence of clumping with anti-B		
Presence of clumping with anti-Rh		

9. Record your blood type on the data sheet on page 436.

10. Put the used slides in the bleach-containing bucket at the general supply area; put disposable supplies in the autoclave bag.

Using Blood Typing Cards

1. Obtain a blood typing card marked A, B, and Rh, dropper bottles of anti-A serum, anti-B serum, and anti-Rh serum, toothpicks, lancets, and alcohol swabs.

2. Place a drop of anti-A serum in the spot marked anti-A, place a drop of anti-B serum on the spot marked anti-B, and place a drop of anti-Rh serum on the spot marked anti-Rh (or anti-D).

3. Carefully add a drop of blood to each of the spots marked "Blood" on the card. If you are using your own blood, refer to step 4 in the Activity 7 section Typing Blood Using Glass Slides. Immediately discard the lancet in the designated disposal container.

4. Using a new toothpick for each test, mix the blood sample with the antibody. Dispose of the toothpicks appropriately.

5. Gently rock the card to allow the blood and antibodies to mix.

6. After 2 minutes, observe the card for evidence of clumping. The Rh clumping is very fine and may be difficult to observe. Record your observations in the Blood Typing chart. Use Figure 29.9 to interpret your results.

7. Record your blood type on the chart on page 436, and discard the card in an autoclave bag. ■

Hematologic Test Data Sheet

Differential WBC count:

WBC	**Student blood smear**	**Unknown sample # ___**
% neutrophils	_________	_________
% eosinophils	_________	_________
% basophils	_________	_________
% monocytes	_________	_________
% lymphocytes	_________	_________

Hematocrit (PCV):

RBC _________ % of blood volume

WBC _________ % of blood volume   not generally reported

Plasma _________ % of blood

Hemoglobin (Hb) content:

Tallquist method: _________ g/100 ml of blood; _________ % Hb

Hemoglobinometer (type: _________________________________)

_________ g/100 ml of blood; _________ %Hb

Ratio (PCV to grams of Hb per 100 ml of blood): _________

Coagulation time _________

Blood typing:

ABO group _________ Rh factor _________

Cholesterol concentration _________ mg/dl of blood

ACTIVITY 8

Observing Demonstration Slides

Before continuing on to the cholesterol determination, take the time to look at the slides of *macrocytic hypochromic anemia, microcytic hypochromic anemia, sickle cell anemia, lymphocytic leukemia* (chronic), and *eosinophilia* that have been put on demonstration by your instructor. Record your observations in the appropriate section of the Exercise 29 Review Sheet. You can refer to your notes, the text, and other references later to respond to questions about the blood pathologies represented on the slides. ■

Cholesterol Concentration in Plasma

Atherosclerosis is the disease process in which the body's blood vessels become increasingly occluded by plaques. Because the plaques narrow the arteries, they can contribute to hypertensive heart disease. They also serve as focal points for the formation of blood clots (thrombi), which may break away and block smaller vessels farther downstream in the circulatory pathway and cause heart attacks or strokes.

Ever since medical clinicians discovered that cholesterol is a major component of the smooth muscle plaques formed during atherosclerosis, it has had a bad press. Today, virtually no physical examination of an adult is considered complete until cholesterol levels are assessed along with other lifestyle risk factors. A normal value for plasma cholesterol in adults ranges from 130 to 200 mg per 100 ml of plasma; you will use blood to make such a determination.

Although the total plasma cholesterol concentration is valuable information, it may be misleading, particularly if a person's high-density lipoprotein (HDL) level is high and low-density lipoprotein (LDL) level is relatively low. Cholesterol, being water insoluble, is transported in the blood complexed to lipoproteins. In general, cholesterol bound into HDLs is destined to be degraded by the liver and then eliminated from the body, whereas that forming part of the LDLs is "traveling" to the body's tissue cells. When LDL levels are excessive, cholesterol is deposited in the blood vessel walls; hence, LDLs are considered to carry the "bad" cholesterol.

ACTIVITY 9

Measuring Plasma Cholesterol Concentration

1. Go to the appropriate supply area, and obtain a cholesterol test card and color scale, lancet, and alcohol swab.

2. Clean your fingertip with the alcohol swab, allow it to dry, then prick it with a lancet. Place a drop of blood on the test area of the card. Put the lancet in the designated disposal container.

3. After 3 minutes, remove the blood sample strip from the card and discard in the autoclave bag.

4. Analyze the underlying test spot, using the included color scale. Record the cholesterol level below and on the data sheet above.

Cholesterol level _________ mg/dl

⚠ 5. Before leaving the laboratory, use the spray bottle of bleach solution and saturate a paper towel to thoroughly wash down your laboratory bench. ■

NAME ______________________________

LAB TIME/DATE ____________________

REVIEW SHEET

EXERCISE 29

Blood

Composition of Blood

1. What is the blood volume of an average-size adult male? __________ liters An average adult female? __________ liters

2. What determines whether blood is bright red or a dull brick-red? ______________________________

____

3. Use the key to identify the cell type(s) or blood elements that fit the following descriptive statements.

Key: a. red blood cell
b. megakaryocyte
c. eosinophil
d. basophil
e. monocyte
f. neutrophil
g. lymphocyte
h. formed elements
i. plasma

__________________ 1. most numerous leukocyte

__________________, __________________, and __________________ 2. granulocytes (3)

__________________ 3. also called an erythrocyte; anucleate formed element

__________________, __________________ 4. actively phagocytic leukocytes

__________________, __________________ 5. agranulocytes

__________________ 6. ancestral cell of platelets

__________________ 7. (a) through (g) are all examples of these

__________________ 8. number rises during parasite infections

__________________ 9. releases histamine; promotes inflammation

__________________ 10. many formed in lymphoid tissue

__________________ 11. transports oxygen

__________________ 12. primarily water, noncellular; the fluid matrix of blood

__________________ 13. increases in number during prolonged infections

__________________, __________________, __________________,

__________________, __________________ 14. the five types of white blood cells

4. List four classes of nutrients normally found in plasma. ______________________________,

______________________, ______________________, and ______________________

Name two gases. ______________________ and ______________________

Name three ions. ______________________, ______________________, and ______________________

5. Describe the consistency and color of the plasma you observed in the laboratory. ______________________

____

6. What is the average life span of a red blood cell? How does its anucleate condition affect this life span?

____

____

____

7. From memory, describe the structural characteristics of each of the following blood cell types as accurately as possible, and note the percentage of each in the total white blood cell population.

eosinophils: ____

____

neutrophils: ____

____

lymphocytes: ____

____

basophils: ____

____

monocytes: ____

____

8. Correctly identify the blood pathologies described in column A by matching them with selections from column B:

Column A		Column B
____________	1. abnormal increase in the number of WBCs	a. anemia
____________	2. abnormal increase in the number of RBCs	b. leukocytosis
____________	3. condition of too few RBCs or of RBCs with hemoglobin deficiencies	c. leukopenia
		d. polycythemia
____________	4. abnormal decrease in the number of WBCs	

Hematologic Tests

9. Broadly speaking, why are hematologic studies of blood so important in the diagnosis of disease?

___

___

10. In the chart below, record information from the blood tests you read about or conducted. Complete the chart by recording values for healthy male adults and indicating the significance of high or low values for each test.

Test	Student test results	Normal values (healthy male adults)	Significance	
			High values	Low values
Total WBC count	No data			
Total RBC count	No data			
Hematocrit				
Hemoglobin determination				
Bleeding time	No data			
Coagulation time				

11. Why is a differential WBC count more valuable than a total WBC count when trying to pin down the specific source of pathology? ___

___

___

12. What name is given to the process of RBC production? ___

What hormone acts as a stimulus for this process? ___

Why might patients with kidney disease suffer from anemia? ___

___

How can such patients be treated? ___

13. Discuss the effect of each of the following factors on RBC count. Consult an appropriate reference as necessary, and explain your reasoning.

long-term effect of athletic training (for example, running 4 to 5 miles per day over a period of six to nine months):

___

___

a permanent move from sea level to a high-altitude area: ____________________

____

____

14. Define *hematocrit.* ____________________

15. If you had a high hematocrit, would you expect your hemoglobin determination to be high or low? __________

Why? ____________________

____

16. What is an anticoagulant? ____________________

Name two anticoagulants used in conducting the hematologic tests. __________

and ____________________

What is the body's natural anticoagulant? ____________________

17. If your blood clumped with both anti-A and anti-B sera, your ABO blood type would be __________

To what ABO blood groups could you give blood? ____________________

From which ABO donor types could you receive blood? ____________________

Which ABO blood type is most common? __________ Least common? __________

18. What blood type is theoretically considered the universal donor? __________ Why? __________

____

19. Assume the blood of two patients has been typed for ABO blood type.

Typing results
Mr. Adams:

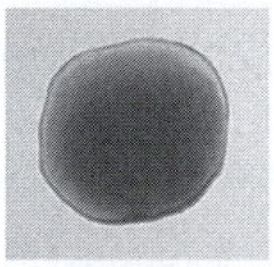

Blood drop and anti-A serum

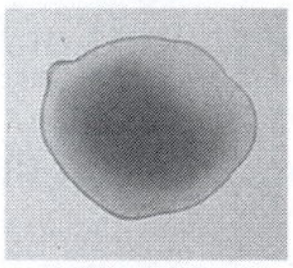

Blood drop and anti-B serum

Typing results
Mr. Calhoon:

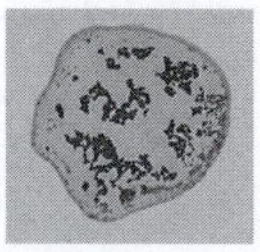

Blood drop and anti-A serum

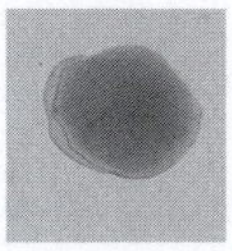

Blood drop and anti-B serum

On the basis of these results, Mr. Adams has type __________ blood, and Mr. Calhoon has type __________ blood.

20. Explain why an Rh-negative person does not have a transfusion reaction on the first exposure to Rh-positive blood but *does* have a reaction on the second exposure. ______________________________

____

What happens when an ABO blood type is mismatched for the first time? ______________________________

____

21. Record your observations of the five demonstration slides viewed.

a. Macrocytic hypochromic anemia: ______________________________

b. Microcytic hypochromic anemia: ______________________________

____

c. Sickle cell anemia: ______________________________

d. Lymphocytic leukemia (chronic): ______________________________

____

e. Eosinophilia: ______________________________

Which of the slides above (a through e) corresponds with the following conditions?

__________ 1. iron-deficient diet

__________ 2. a type of bone marrow cancer

__________ 3. genetic defect that causes hemoglobin to become sharp/spiky

__________ 4. lack of vitamin B_{12}

__________ 5. a tapeworm infestation in the body

__________ 6. a bleeding ulcer

22. Provide the normal, or at least "desirable," range for plasma cholesterol concentration.

______________________ mg/100 ml

23. Describe the relationship between high blood cholesterol levels and cardiovascular diseases such as hypertension, heart attacks, and strokes.

____

____

____

____

____

EXERCISE 30

Anatomy of the Heart

MATERIALS

- ☐ X ray of the human thorax for observation of the position of the heart in situ; X-ray viewing box
- ☐ Three-dimensional heart model and torso model or laboratory chart showing heart anatomy
- ☐ Red and blue pencils
- ☐ Highlighter
- ☐ Three-dimensional models of cardiac and skeletal muscle
- ☐ Compound microscope
- ☐ Prepared slides of cardiac muscle (l.s.)
- ☐ Preserved sheep heart, pericardial sacs intact (if possible)
- ☐ Dissecting instruments and tray
- ☐ Pointed glass rods or blunt probes
- ☐ Plastic rulers
- ☐ Disposable gloves
- ☐ Container for disposal of organic debris
- ☐ Laboratory detergent
- ☐ Spray bottle with 10% household bleach solution

MasteringA&P™

Access practice quizzes and more in the Study Area at www.masteringaandp.com.

PAL

For access to anatomical models and more, check out Practice Anatomy Lab.

OBJECTIVES

1. To describe the location of the heart.
2. To name and locate the major anatomical areas and structures of the heart when provided with an appropriate model, diagram, or dissected sheep heart, and to explain the function of each.
3. To trace the pathway of blood through the heart.
4. To explain why the heart is called a double pump, and to compare the pulmonary and systemic circuits.
5. To explain the operation of the atrioventricular and semilunar valves.
6. To trace the functional blood supply of the heart and name the associated blood vessels.
7. To describe the histology of cardiac muscle, and to note the importance of its intercalated discs and the spiral arrangement of its cells.

PRE-LAB QUIZ

1. The heart is enclosed in a double-walled sac called the
   a. apex c. pericardium
   b. mediastinum d. thorax
2. The heart is divided into __________ chambers.
   a. two c. four
   b. three d. five
3. What is the name of the two receiving chambers of the heart?
   ____
4. The left ventricle discharges blood into the __________, from which all systemic arteries of the body diverge to supply the body tissues.
   a. aorta c. pulmonary vein
   b. pulmonary artery d. vena cava
5. Circle True or False. Blood flows through the heart in one direction—from the atria to the ventricles.
6. Circle the correct term. The right atrioventricular valve, or tricuspid / mitral, prevents backflow into the right atrium when the right ventricle is contracting.
7. Circle the correct term. The heart serves as a double pump. The right / left side serves as the pulmonary circulation pump, shunting carbon dioxide–rich blood to the lungs.
8. The functional blood supply of the heart itself is provided by the
   a. aorta c. coronary arteries
   b. carotid arteries d. pulmonary trunk
9. Two microscopic features of cardiac cells that help distinguish them from other types of muscle cells are branching and
   a. intercalated discs c. sarcolemma
   b. myosin fibers d. striations
10. Circle the correct term. In the heart, the left / right ventricle has thicker walls and a basically circular cavity shape.

The major function of the **cardiovascular system** is transportation. Using blood as the transport vehicle, the system carries oxygen, digested foods, cell wastes, electrolytes, and many other substances vital to the body's homeostasis to and from the body cells. The system's propulsive force is the contracting heart, which can be compared to a muscular pump equipped with one-way valves. As the heart contracts, it forces blood into a closed system of large and small plumbing tubes (blood vessels) within which the blood is confined and circulated. This exercise deals with the structure of the heart, or circulatory pump. The anatomy of the blood vessels is considered separately in Exercise 32.

Gross Anatomy of the Human Heart

The **heart,** a cone-shaped organ approximately the size of a fist, is located within the mediastinum, or medial cavity, of the thorax. It is flanked laterally by the lungs, posteriorly by the vertebral column, and anteriorly by the sternum (Figure 30.1). Its more pointed **apex** extends slightly to the left and rests on the diaphragm, approximately at the level of the fifth intercostal space. Its broader **base,** from which the great vessels emerge, lies beneath the second rib and points toward the right shoulder. In situ, the right ventricle of the heart forms most of its anterior surface.

The apical pulse may be heard in the 5th intercostal space at the point of maximal intensity (PMI).

- If an X ray of a human thorax is available, verify the relationships described above; otherwise, Figure 30.1 should suffice.

The heart is enclosed within a double-walled fibroserous sac called the pericardium. The thin **epicardium,** or **visceral pericardium,** is closely applied to the heart muscle. It reflects downward at the base of the heart to form its companion serous membrane, the outer, loosely applied **parietal pericardium,** which is attached at the heart apex to the diaphragm. Serous fluid produced by these membranes allows the heart to beat in a relatively frictionless environment. The serous parietal pericardium, in turn, lines the loosely fitting superficial **fibrous pericardium** composed of dense connective tissue.

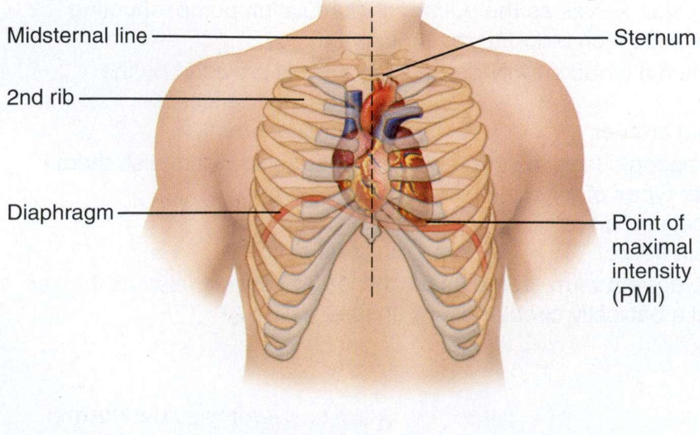
Midsternal line
Sternum
2nd rib
Diaphragm
Point of maximal intensity (PMI)

FIGURE 30.1 Location of the heart in the thorax.

Inflammation of the pericardium, **pericarditis,** causes painful adhesions between the serous pericardial layers. These adhesions interfere with heart movements. ■

The walls of the heart are composed primarily of cardiac muscle—the **myocardium**—which is reinforced internally by a dense fibrous connective tissue network. This network—the *fibrous skeleton of the heart*—is more elaborate and thicker in certain areas, for example, around the valves and at the base of the great vessels leaving the heart.

Figure 30.2 shows three views of the heart—external anterior and posterior views and a frontal section. As its anatomical areas are described in the text, consult the figure.

Heart Chambers

The heart is divided into four chambers: two superior **atria** (singular: *atrium*) and two inferior **ventricles,** each lined by thin serous endothelium called the **endocardium.** The septum that divides the heart longitudinally is referred to as the **interatrial** or **interventricular septum,** depending on which chambers it partitions. Functionally, the atria are receiving chambers and are relatively ineffective as pumps. Blood flows into the atria under low pressure from the veins of the body. The right atrium receives relatively oxygen-poor blood from the body via the **superior** and **inferior venae cavae** and the coronary sinus. Four **pulmonary veins** deliver oxygen-rich blood from the lungs to the left atrium.

The inferior thick-walled ventricles, which form the bulk of the heart, are the discharging chambers. They force blood out of the heart into the large arteries that emerge from its base. The right ventricle pumps blood into the **pulmonary trunk,** which routes blood to the lungs to be oxygenated. The left ventricle discharges blood into the **aorta,** from which all systemic arteries of the body diverge to supply the body tissues. Discussions of the heart's pumping action usually refer to ventricular activity.

Heart Valves

Four valves enforce a one-way blood flow through the heart chambers. The **atrioventricular (AV) valves,** located between the atrial and ventricular chambers on each side, prevent backflow into the atria when the ventricles are contracting. The left atrioventricular valve, also called the **mitral** or *bicuspid valve,* consists of two cusps, or flaps, of endocardium. The right atrioventricular valve, the **tricuspid valve,** has three cusps (Figure 30.3). Tiny white collagenic cords called the **chordae tendineae** (literally, heart strings) anchor the cusps to the ventricular walls. The chordae tendineae originate from small bundles of cardiac muscle, called **papillary muscles,** that project from the myocardial wall (see Figure 30.2b).

When blood is flowing passively into the atria and then into the ventricles during **diastole** (the period of ventricular filling), the AV valve flaps hang limply into the ventricular chambers and then are carried passively toward the atria by the accumulating blood. When the ventricles contract **(systole)** and compress the blood in their chambers, the intraventricular blood pressure rises, causing the valve flaps to be reflected superiorly, which closes the AV valves. The chordae tendineae, pulled taut by the contracting papillary muscles, anchor the flaps in a closed position that prevents backflow into the atria during ventricular contraction. If unanchored,

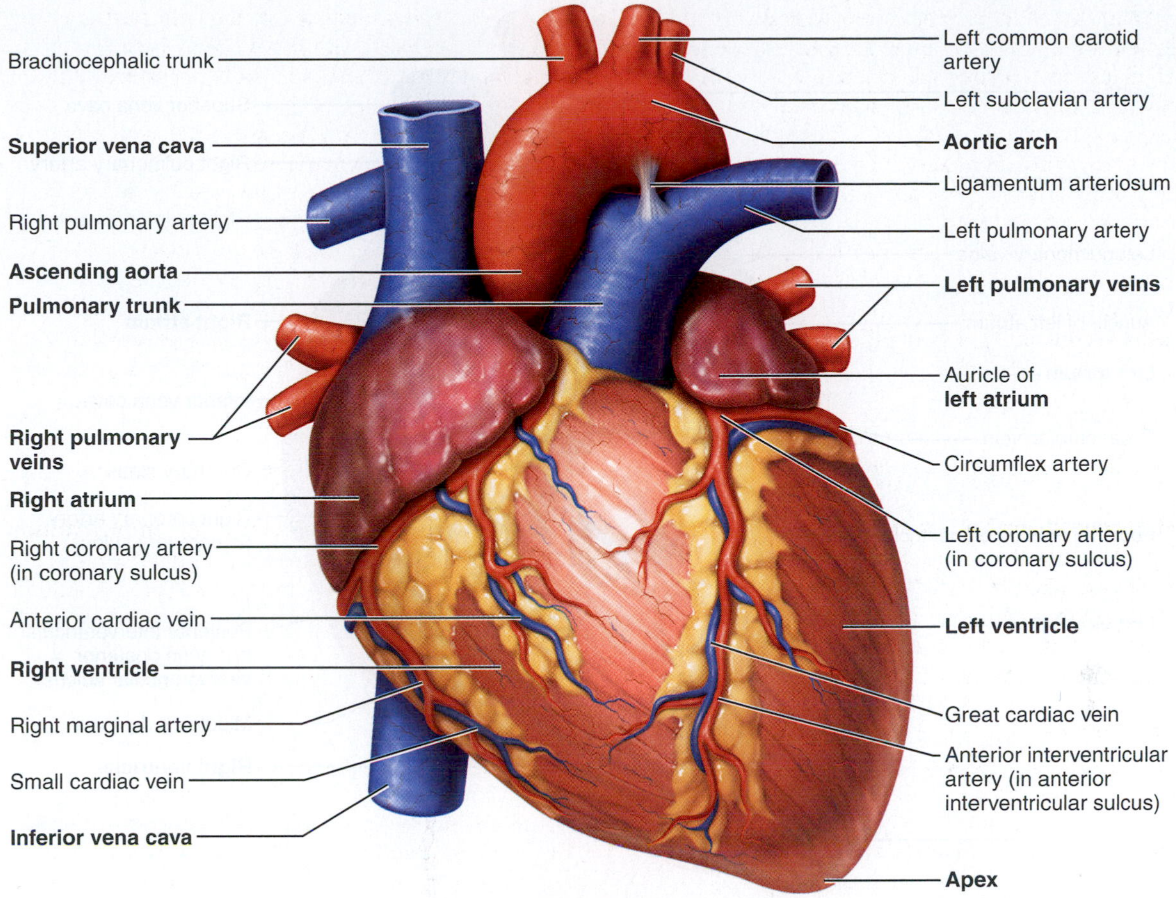
Brachiocephalic trunk
Superior vena cava
Right pulmonary artery
Ascending aorta
Pulmonary trunk
Right pulmonary veins
Right atrium
Right coronary artery (in coronary sulcus)
Anterior cardiac vein
Right ventricle
Right marginal artery
Small cardiac vein
Inferior vena cava
Left common carotid artery
Left subclavian artery
Aortic arch
Ligamentum arteriosum
Left pulmonary artery
Left pulmonary veins
Auricle of left atrium
Circumflex artery
Left coronary artery (in coronary sulcus)
Left ventricle
Great cardiac vein
Anterior interventricular artery (in anterior interventricular sulcus)
Apex

(a)

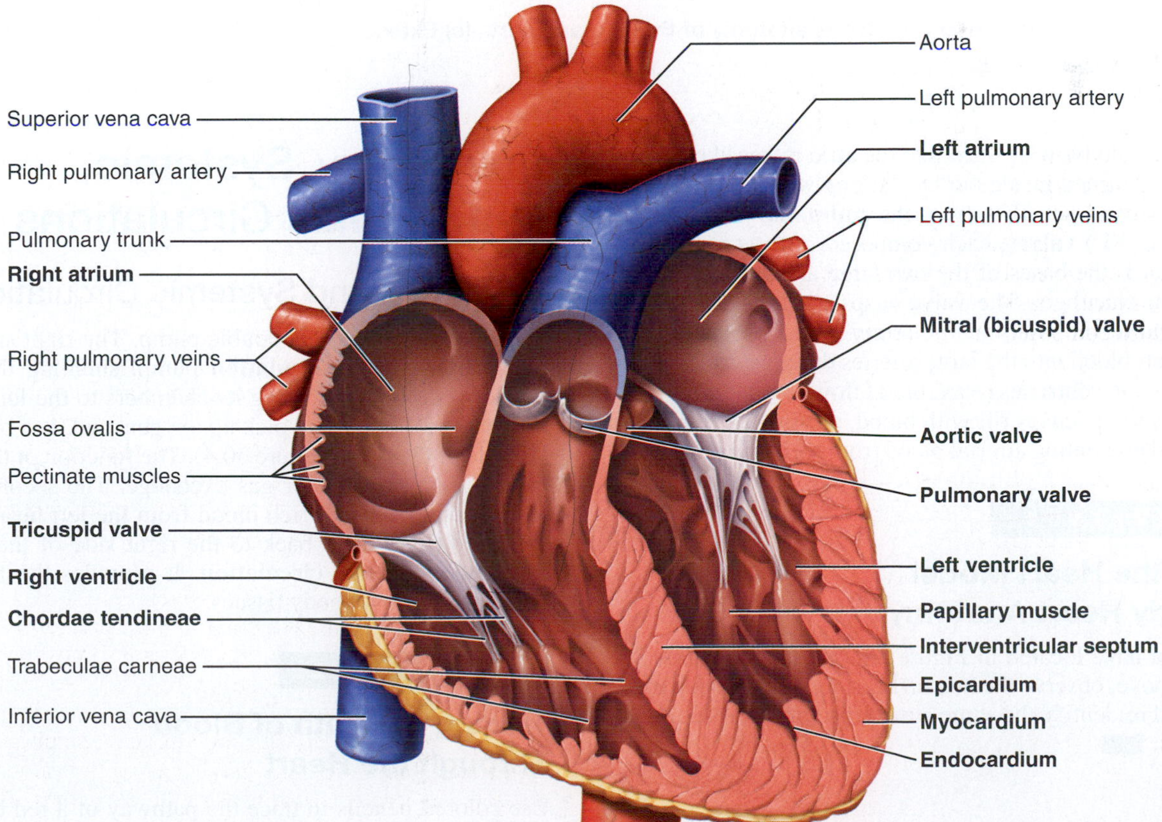
Superior vena cava
Right pulmonary artery
Pulmonary trunk
Right atrium
Right pulmonary veins
Fossa ovalis
Pectinate muscles
Tricuspid valve
Right ventricle
Chordae tendineae
Trabeculae carneae
Inferior vena cava
Aorta
Left pulmonary artery
Left atrium
Left pulmonary veins
Mitral (bicuspid) valve
Aortic valve
Pulmonary valve
Left ventricle
Papillary muscle
Interventricular septum
Epicardium
Myocardium
Endocardium

(b)

FIGURE 30.2 Gross anatomy of the human heart. (a) External anterior view. **(b)** Frontal section.

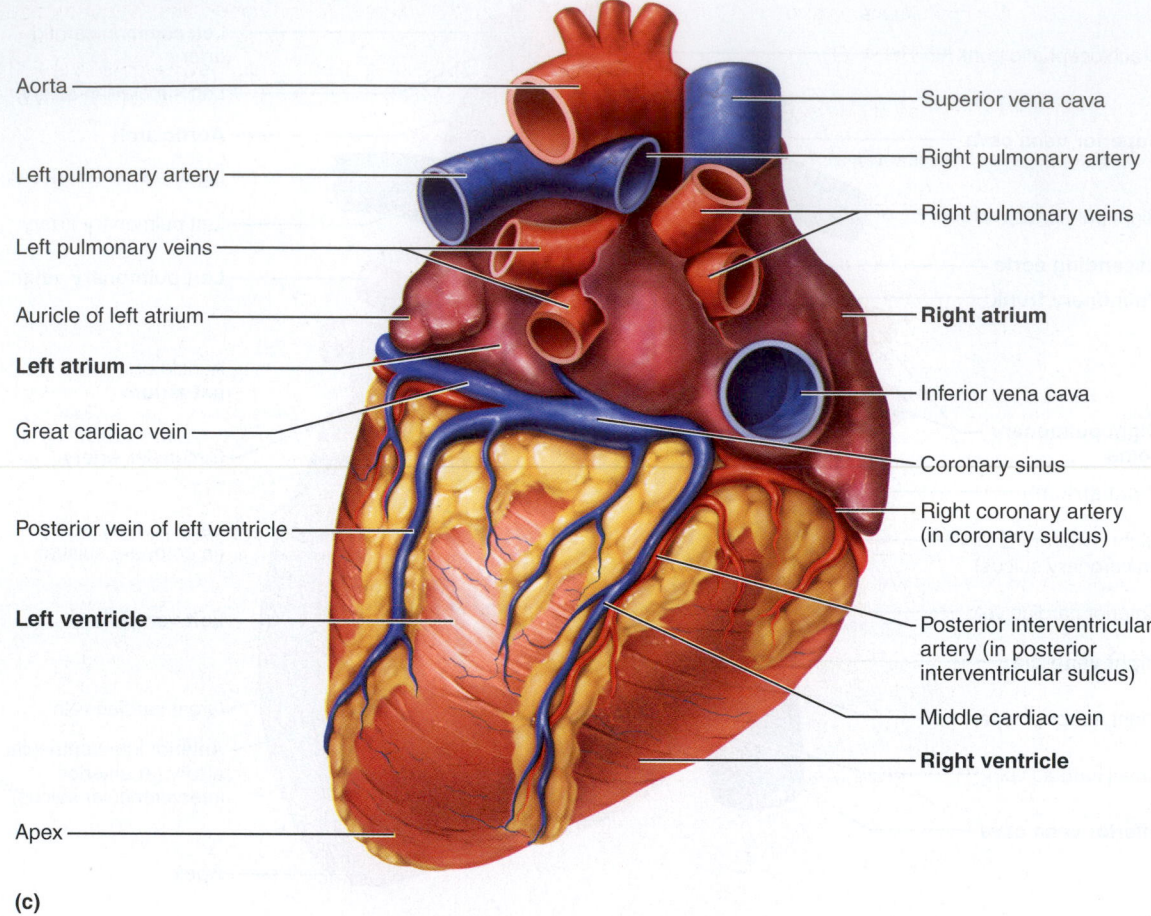

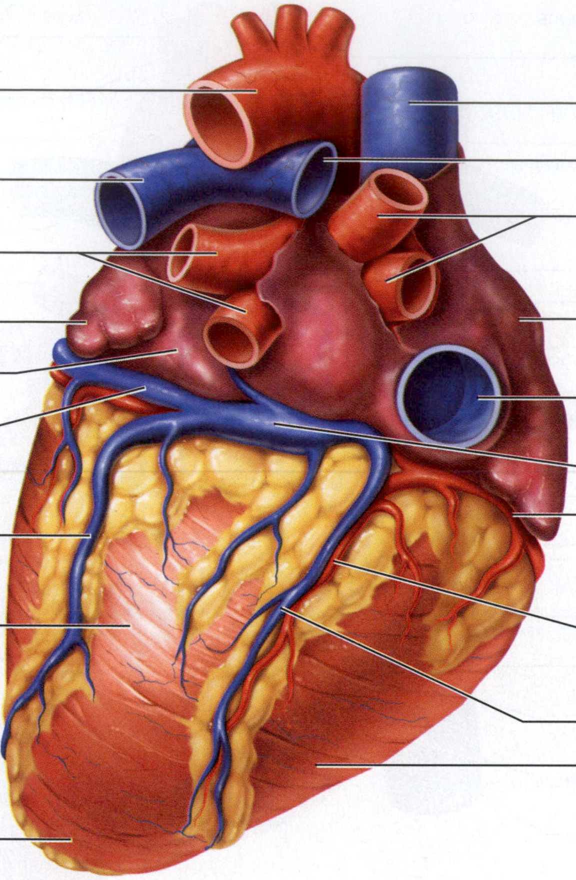
Aorta
Left pulmonary artery
Left pulmonary veins
Auricle of left atrium
Left atrium
Great cardiac vein
Posterior vein of left ventricle
Left ventricle
Apex
Superior vena cava
Right pulmonary artery
Right pulmonary veins
Right atrium
Inferior vena cava
Coronary sinus
Right coronary artery (in coronary sulcus)
Posterior interventricular artery (in posterior interventricular sulcus)
Middle cardiac vein
Right ventricle
(c)

FIGURE 30.2 *(continued)* Gross anatomy of the human heart. (c) Exterior posterior view.

the flaps would blow upward into the atria rather like an umbrella being turned inside out by a strong wind.

The second set of valves, the **pulmonary** and **aortic (semilunar, SL) valves,** each composed of three pocketlike cusps, guards the bases of the two large arteries leaving the ventricular chambers. The valve cusps are forced open and flatten against the walls of the artery as the ventricles discharge their blood into the large arteries during systole. However, when the ventricles relax, blood flows backward toward the heart and the cusps fill with blood, closing the semilunar valves and preventing arterial blood from reentering the heart.

ACTIVITY 1

Using the Heart Model to Study Heart Anatomy

When you have located in Figure 30.2 all the structures described above, observe the human heart model and laboratory charts and reidentify the same structures without referring to the figure.

Pulmonary, Systemic, and Cardiac Circulations

Pulmonary and Systemic Circulations

The heart functions as a double pump. The right side serves as the **pulmonary circulation** pump, shunting the carbon dioxide–rich blood entering its chambers to the lungs to unload carbon dioxide and pick up oxygen, and then back to the left side of the heart (Figure 30.4). The function of this circuit is strictly to provide for gas exchange. The second circuit, which carries oxygen-rich blood from the left heart through the body tissues and back to the right side of the heart, is called the **systemic circulation.** It provides the functional blood supply to all body tissues.

ACTIVITY 2

Tracing the Path of Blood Through the Heart

Use colored pencils to trace the pathway of a red blood cell through the heart by adding arrows to the frontal section diagram (Figure 30.2b). Use red arrows for the oxygen-rich blood and blue arrows for the less oxygen-rich blood.

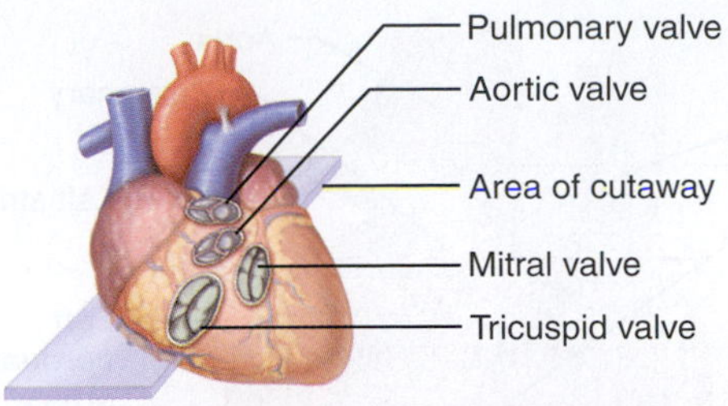
Pulmonary valve
Aortic valve
Area of cutaway
Mitral valve
Tricuspid valve

FIGURE 30.3 Heart valves. **(a)** Superior view of the two sets of heart valves (atria removed). **(b)** Photograph of the heart valves, superior view. **(c)** Photograph of the right AV valve. View begins in the right ventricle, looking toward the right atrium. **(d)** Coronal section of the heart.

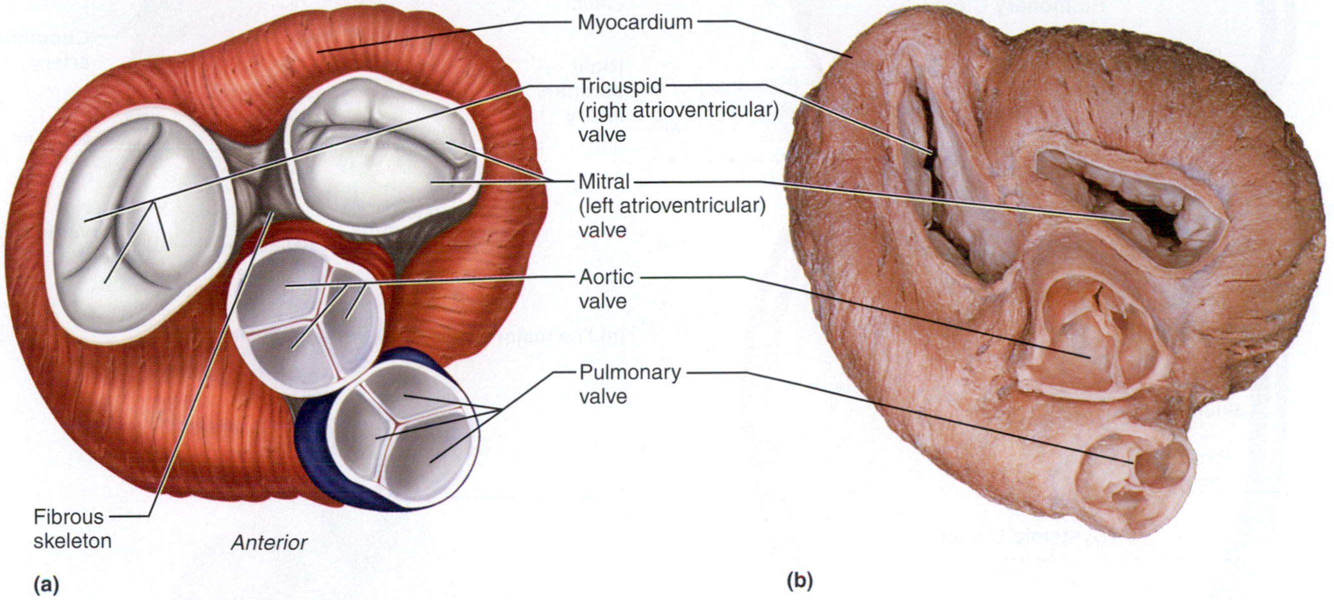
Myocardium
Tricuspid (right atrioventricular) valve
Mitral (left atrioventricular) valve
Aortic valve
Pulmonary valve
Fibrous skeleton
Anterior
(a)
(b)

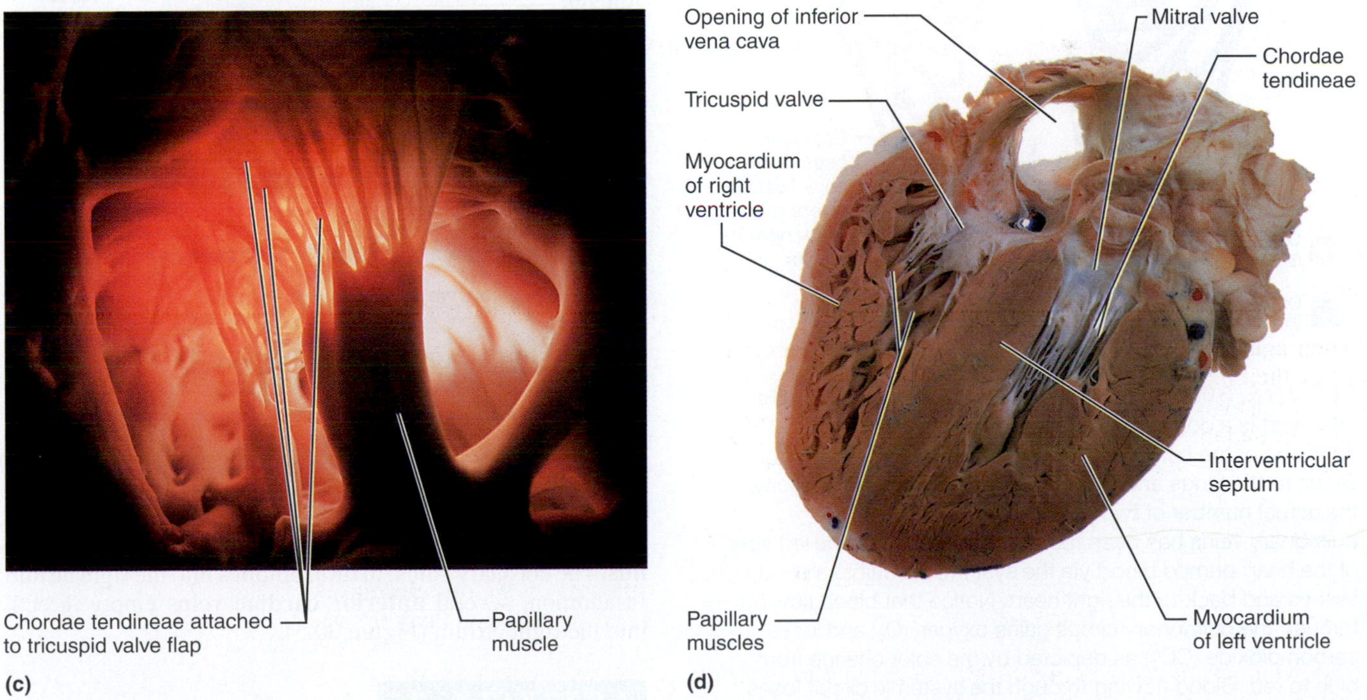
Chordae tendineae attached to tricuspid valve flap
Papillary muscle
(c)
Opening of inferior vena cava
Tricuspid valve
Myocardium of right ventricle
Mitral valve
Chordae tendineae
Interventricular septum
Papillary muscles
Myocardium of left ventricle
(d)

Coronary Circulation

Even though the heart chambers are almost continually bathed with blood, this contained blood does not nourish the myocardium. The functional blood supply of the heart is provided by the coronary arteries (see Figures 30.2 and 30.5). The **right** and **left coronary arteries** issue from the base of the aorta just above the aortic semilunar valve and encircle the heart in the **coronary sulcus** at the junction of the atria and ventricles. They then ramify over the heart's surface, the right coronary artery supplying the posterior surface of the ventricles and the lateral aspect of the right side of the heart, largely through

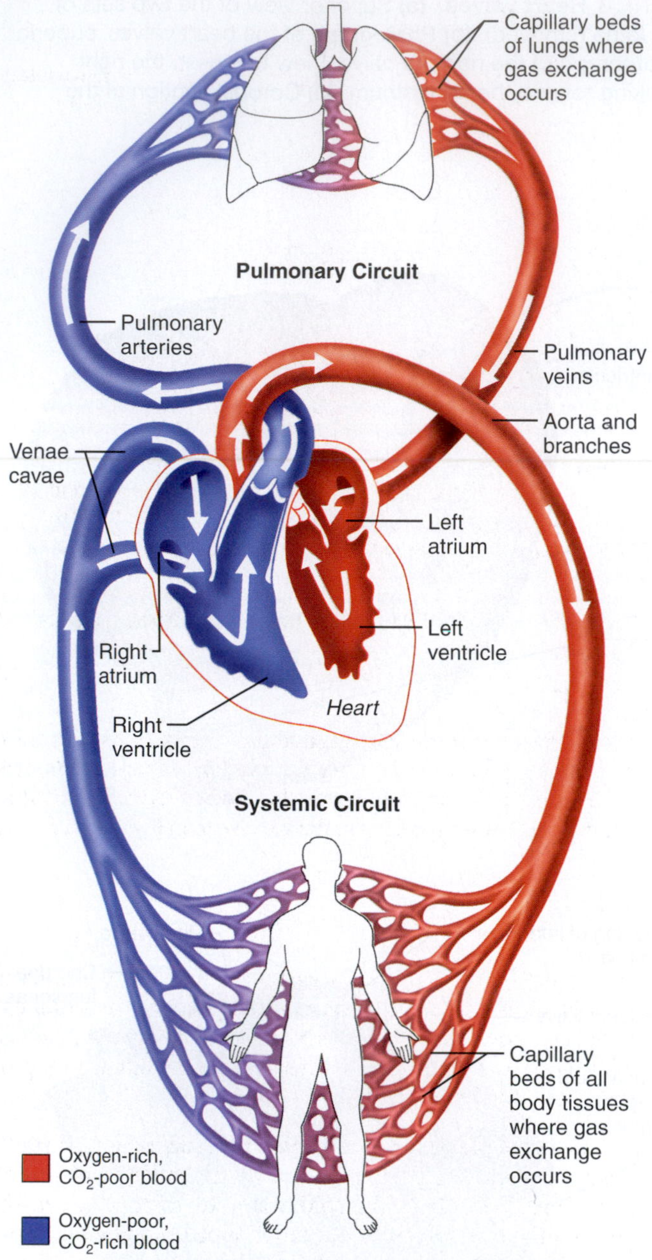
Capillary beds of lungs where gas exchange occurs
Pulmonary Circuit
Pulmonary arteries
Pulmonary veins
Aorta and branches
Venae cavae
Left atrium
Left ventricle
Right atrium
Right ventricle
Heart
Systemic Circuit
Capillary beds of all body tissues where gas exchange occurs
Oxygen-rich, CO_2-poor blood
Oxygen-poor, CO_2-rich blood

FIGURE 30.4 The systemic and pulmonary circuits. The heart is a double pump that serves two circulations. The right side of the heart pumps blood through the pulmonary circuit to the lungs and back to the left heart. (For simplicity, the actual number of two pulmonary arteries and four pulmonary veins has been reduced to one each.) The left side of the heart pumps blood via the systemic circuit to all body tissues and back to the right heart. Notice that blood flowing through the pulmonary circuit gains oxygen (O_2) and loses carbon dioxide (CO_2) as depicted by the color change from blue to red. Blood flowing through the systemic circuit loses oxygen and picks up carbon dioxide (red to blue color change).

its **posterior interventricular** and **right marginal artery** branches. The left coronary artery supplies the anterior ventricular walls and the laterodorsal part of the left side of the heart via its two major branches, the **anterior interventricular artery** and the **circumflex artery.** The coronary arteries and their branches are compressed during systole and fill when the heart is relaxed.

The myocardium is largely drained by the **great, middle,** and **small cardiac veins,** which empty into the **coronary sinus.** The coronary sinus, in turn, empties into the right atrium. In addition, several **anterior cardiac veins** empty directly into the right atrium (Figure 30.5).

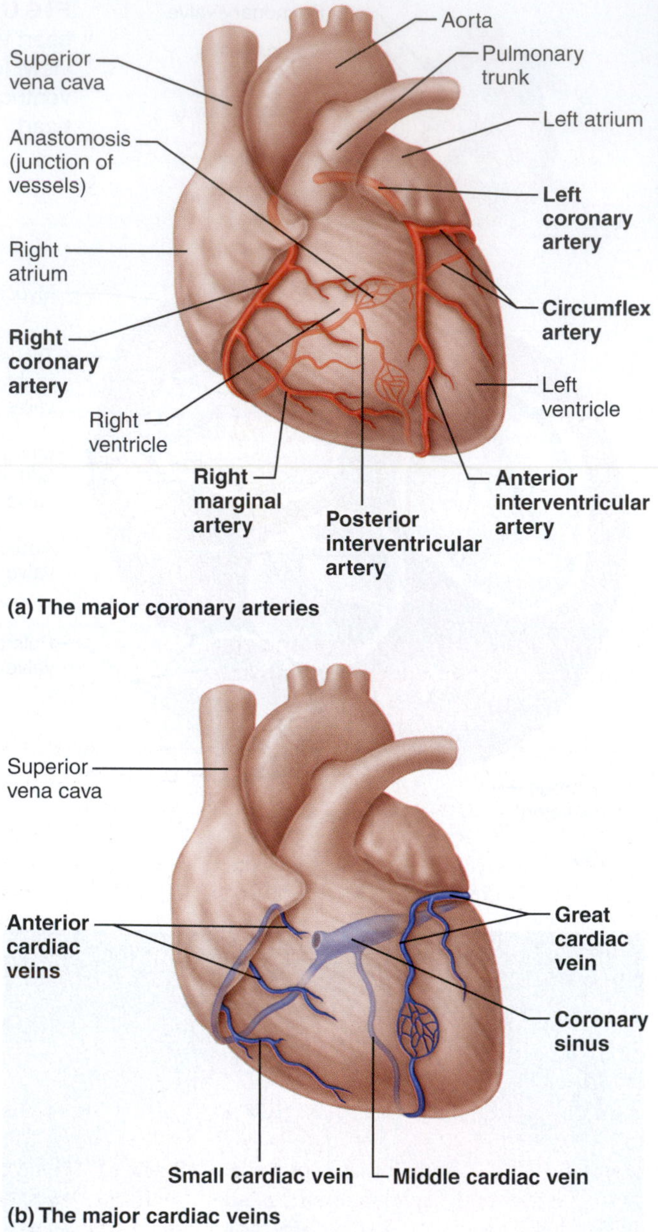
Aorta
Superior vena cava
Pulmonary trunk
Left atrium
Anastomosis (junction of vessels)
Left coronary artery
Right atrium
Circumflex artery
Right coronary artery
Left ventricle
Right ventricle
Right marginal artery
Posterior interventricular artery
Anterior interventricular artery
(a) The major coronary arteries
Superior vena cava
Anterior cardiac veins
Great cardiac vein
Coronary sinus
Small cardiac vein
Middle cardiac vein
(b) The major cardiac veins

FIGURE 30.5 Coronary circulation.

ACTIVITY 3

Using the Heart Model to Study Cardiac Circulation

1. Obtain a highlighter and highlight all the cardiac blood vessels in Figure 30.2a and c. Note how arteries and veins travel together.

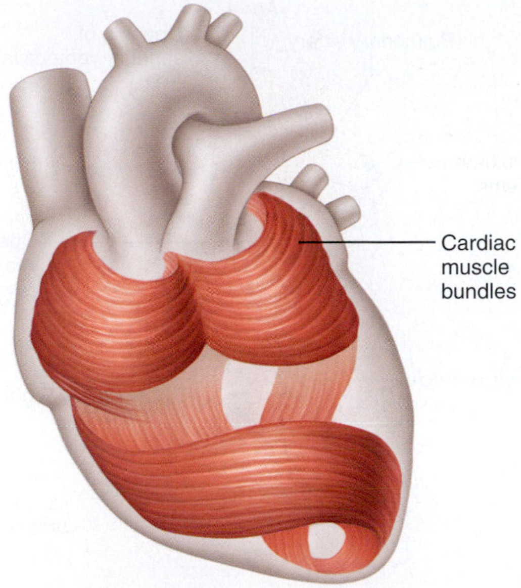
Cardiac
muscle
bundles

FIGURE 30.6 The circular and spiral arrangement of cardiac muscle bundles in the myocardium.

2. On a model of the heart, locate all the cardiac blood vessels shown in Figure 30.5. Use your finger to trace the pathway of blood from the right coronary artery to the lateral aspect of the right side of the heart and back to the right atrium. Name the arteries and veins along the pathway. Trace the pathway of blood from the left coronary artery to the anterior ventricular walls and back to the right atrium. Name the arteries and veins along the pathway. Note that there are multiple different pathways to distribute blood to these parts of the heart.

Microscopic Anatomy of Cardiac Muscle

Cardiac muscle is found in only one place—the heart. The heart acts as a vascular pump, propelling blood to all tissues of the body; cardiac muscle is thus very important to life. Cardiac muscle is involuntary, ensuring a constant blood supply.

The cardiac cells, only sparingly invested in connective tissue, are arranged in spiral or figure-8-shaped bundles (Figure 30.6). When the heart contracts, its internal chambers become smaller (or are temporarily obliterated), forcing the blood into the large arteries leaving the heart.

ACTIVITY 4

Examining Cardiac Muscle Tissue Anatomy

1. Observe the three-dimensional model of cardiac muscle, examining its branching cells and the areas where the cells interdigitate, the **intercalated discs.** These two structural features provide a continuity to cardiac muscle not seen in other muscle tissues and allow close coordination of heart activity.

2. Compare the model of cardiac muscle to the model of skeletal muscle. Note the similarities and differences between the two kinds of muscle tissue.

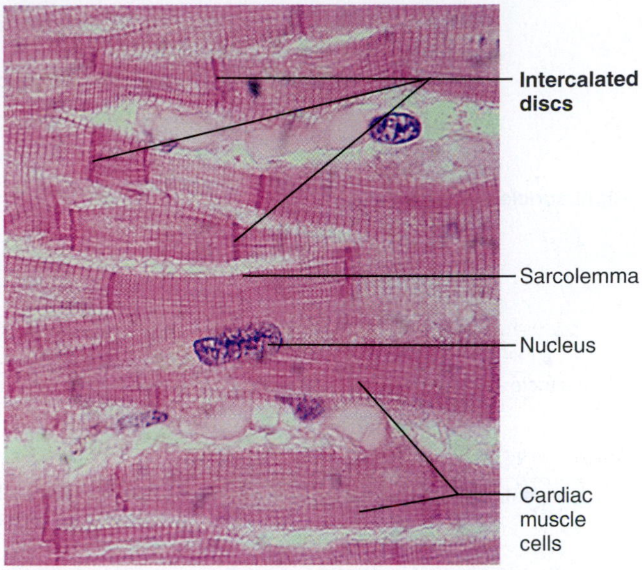
Intercalated
discs
Sarcolemma
Nucleus
Cardiac
muscle
cells

FIGURE 30.7 Photomicrograph of cardiac muscle (570×).

3. Obtain and observe a longitudinal section of cardiac muscle under high power. Identify the nucleus, striations, intercalated discs, and sarcolemma of the individual cells and then compare your observations to the view seen in Figure 30.7.

DISSECTION:

The Sheep Heart

Dissection of a sheep heart is valuable because it is similar in size and structure to the human heart. Also, a dissection experience allows you to view structures in a way not possible with models and diagrams. Refer to Figure 30.8 as you proceed with the dissection.

1. Obtain a preserved sheep heart, a dissecting tray, dissecting instruments, a glass probe, a plastic ruler, and gloves. Rinse the sheep heart in cold water to remove excessive preservatives and to flush out any trapped blood clots. Now you are ready to make your observations.

2. Observe the texture of the pericardium. Also, note its point of attachment to the heart. Where is it attached?

____

3. If the serous pericardial sac is still intact, slit open the parietal pericardium and cut it from its attachments. Observe the visceral pericardium (epicardium). Using a sharp scalpel, carefully pull a little of this serous membrane away from the myocardium. How do its position, thickness, and apposition to the heart differ from those of the parietal pericardium?

____

____

____

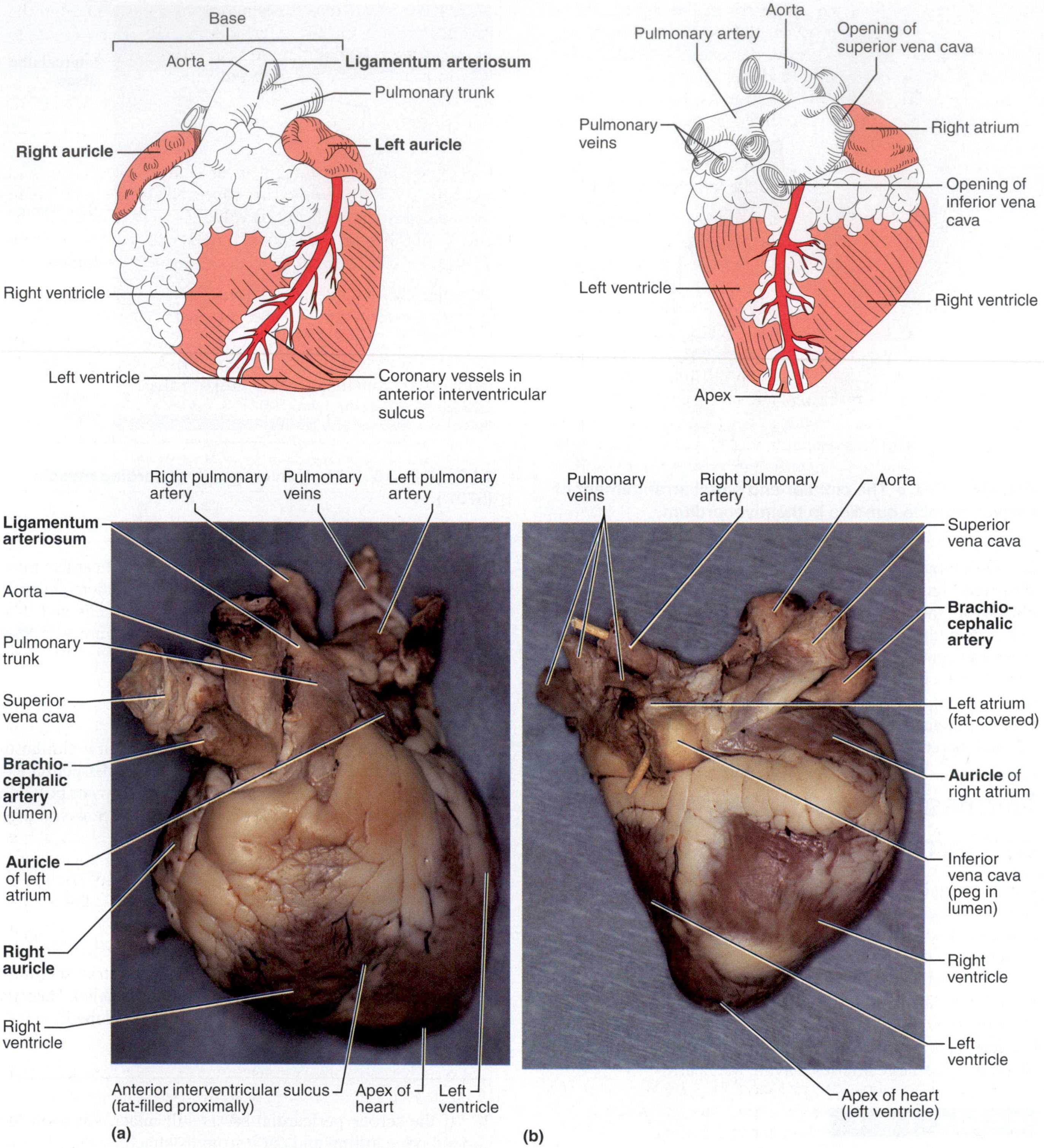
Base
Aorta
Ligamentum arteriosum
Pulmonary trunk
Right auricle
Left auricle
Right ventricle
Left ventricle
Coronary vessels in anterior interventricular sulcus
Aorta
Pulmonary artery
Opening of superior vena cava
Pulmonary veins
Right atrium
Opening of inferior vena cava
Left ventricle
Right ventricle
Apex
Right pulmonary artery
Pulmonary veins
Left pulmonary artery
Ligamentum arteriosum
Aorta
Pulmonary trunk
Superior vena cava
Brachio-cephalic artery (lumen)
Auricle of left atrium
Right auricle
Right ventricle
Anterior interventricular sulcus (fat-filled proximally)
Apex of heart
Left ventricle
(a)
Pulmonary veins
Right pulmonary artery
Aorta
Superior vena cava
Brachio-cephalic artery
Left atrium (fat-covered)
Auricle of right atrium
Inferior vena cava (peg in lumen)
Right ventricle
Left ventricle
Apex of heart (left ventricle)
(b)

FIGURE 30.8 Anatomy of the sheep heart. (a) Anterior view. **(b)** Posterior view. Diagrammatic views at top; photographs at bottom.

4. Examine the external surface of the heart. Notice the accumulation of adipose tissue, which in many cases marks the separation of the chambers and the location of the coronary arteries that nourish the myocardium. Carefully scrape away some of the fat with a scalpel to expose the coronary blood vessels.

5. Identify the base and apex of the heart, and then identify the two wrinkled **auricles,** earlike flaps of tissue projecting from the atrial chambers. The balance of the heart muscle is ventricular tissue. To identify the left ventricle, compress the ventricular chambers on each side of the longitudinal fissures carrying the coronary blood vessels. The side that feels

thicker and more solid is the left ventricle. The right ventricle feels much thinner and somewhat flabby when compressed. This difference reflects the greater demand placed on the left ventricle, which must pump blood through the much longer systemic circulation, a pathway with much higher resistance than the pulmonary circulation served by the right ventricle. Hold the heart in its anatomical position (Figure 30.8a), with the anterior surface uppermost. In this position the left ventricle composes the entire apex and the left side of the heart.

6. Identify the pulmonary trunk and the aorta extending from the superior aspect of the heart. The pulmonary trunk is more anterior, and you may see its division into the right and left pulmonary arteries if it has not been cut too closely to the heart. The thicker-walled aorta, which branches almost immediately, is located just beneath the pulmonary trunk. The first observable branch of the sheep aorta, the **brachiocephalic artery,** is identifiable unless the aorta has been cut immediately as it leaves the heart. The brachiocephalic artery splits to form the right carotid and subclavian arteries, which supply the right side of the head and right forelimb, respectively.

Gently pull on the aorta with your gloved fingers or forceps to stretch it. Repeat with the vena cava.

Which vessel is easier to stretch? ____________________

How does the elasticity of each vessel relate to its ability to withstand pressure?

____

Carefully clear away some of the fat between the pulmonary trunk and the aorta to expose the **ligamentum arteriosum,** a cordlike remnant of the **ductus arteriosus.** (In the fetus, the ductus arteriosus allows blood to pass directly from the pulmonary trunk to the aorta, thus bypassing the nonfunctional fetal lungs.)

7. Cut through the wall of the aorta until you see the aortic (semilunar) valve. Identify the two openings to the coronary arteries just above the valve. Insert a probe into one of these holes to see if you can follow the course of a coronary artery across the heart.

8. Turn the heart to view its posterior surface. The heart will appear as shown in Figure 30.8b. Notice that the right and left ventricles appear equal-sized in this view. Try to identify the four thin-walled pulmonary veins entering the left atrium. Identify the superior and inferior venae cavae entering the right atrium. Because of the way the heart is trimmed, the pulmonary veins and superior vena cava may be very short or missing. If possible, compare the approximate diameter of the superior vena cava with the diameter of the aorta.

Which is larger? ____________________

Which has thicker walls? ____________________

Why do you suppose these differences exist?

____

____

____

9. Insert a probe into the superior vena cava, through the right atrium, and out the inferior vena cava. Use scissors to cut along the probe so that you can view the interior of the right atrium. Observe the tricuspid valve.

How many flaps does it have? ____________________

Pour some water into the right atrium and allow it to flow into the ventricle. *Slowly and gently* squeeze the right ventricle to watch the closing action of this valve. (If you squeeze too vigorously, you'll get a face full of water!) Drain the water from the heart before continuing.

10. Return to the pulmonary trunk and cut through its anterior wall until you can see the pulmonary (semilunar) valve (Figure 30.9). Pour some water into the base of the pulmonary trunk to observe the closing action of this valve. How does its action differ from that of the tricuspid valve?

____

____

____

After observing pulmonary valve action, drain the heart once again. Extend the cut through the pulmonary trunk into the right ventricle. Cut down, around, and up through the tricuspid valve to make the cut continuous with the cut across the right atrium (see Figure 30.9).

11. Reflect the cut edges of the superior vena cava, right atrium, and right ventricle to obtain the view seen in Figure 30.9. Observe the comblike ridges of muscle throughout most of the right atrium. This is called **pectinate muscle** (*pectin* = comb). Identify, on the ventral atrial wall, the large opening of the inferior vena cava and follow it to its external opening with a probe. Notice that the atrial walls in the vicinity of the venae cavae are smooth and lack the roughened appearance (pectinate musculature) of the other regions of the atrial walls. Just below the inferior vena caval opening, identify the opening of the **coronary sinus,** which returns venous blood of the coronary circulation to the right atrium. Nearby, locate an oval depression, the **fossa ovalis,** in the interatrial septum. This depression marks the site of an opening in the fetal heart, the **foramen ovale,** which allows blood to pass from the right to the left atrium, thus bypassing the fetal lungs.

12. Identify the papillary muscles in the right ventricle, and follow their attached chordae tendineae to the flaps of the tricuspid valve. Notice the pitted and ridged appearance **(trabeculae carneae)** of the inner ventricular muscle.

13. Identify the **moderator band** (septomarginal band), a bundle of cardiac muscle fibers connecting the interventricular septum to anterior papillary muscles. It contains a branch of the atrioventricular bundle and helps coordinate contraction of the ventricle.

14. Make a longitudinal incision through the left atrium and continue it into the left ventricle. Notice how much thicker the myocardium of the left ventricle is than that of the right

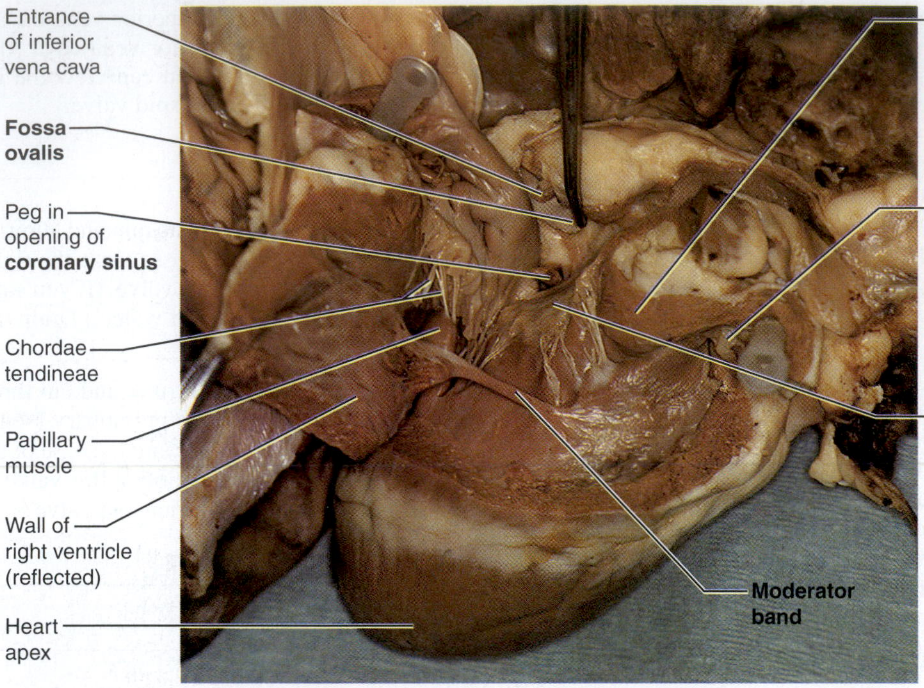
Entrance of inferior vena cava
Fossa ovalis
Peg in opening of coronary sinus
Chordae tendineae
Papillary muscle
Wall of right ventricle (reflected)
Heart apex
Cut surface of wall of right ventricle
Flap of pulmonary valve
Flap of tricuspid valve
Moderator band

FIGURE 30.9 Right side of the sheep heart opened and reflected to reveal internal structures.

ventricle. Measure the thickness of right and left ventricular walls and record the numbers.

____

How do your numbers compare with those of your classmates?

____

Compare the *shape* of the left ventricular cavity to the shape of the right ventricular cavity. (See Figure 30.10.)

____

____

Are the papillary muscles and chordae tendineae observed in the right ventricle also present in the left ventricle? ________

Count the number of cusps in the mitral valve. How does this compare with the number seen in the tricuspid valve?

____

How do the sheep valves compare with their human counterparts?

____

15. Reflect the cut edges of the atrial wall, and attempt to locate the entry points of the pulmonary veins into the left atrium. Follow the pulmonary veins, if present, to the heart exterior with a probe. Notice how thin-walled these vessels are.

16. Dispose of the organic debris in the designated container, clean the dissecting tray and instruments with detergent and water, and wash the lab bench with bleach solution before leaving the laboratory. ■

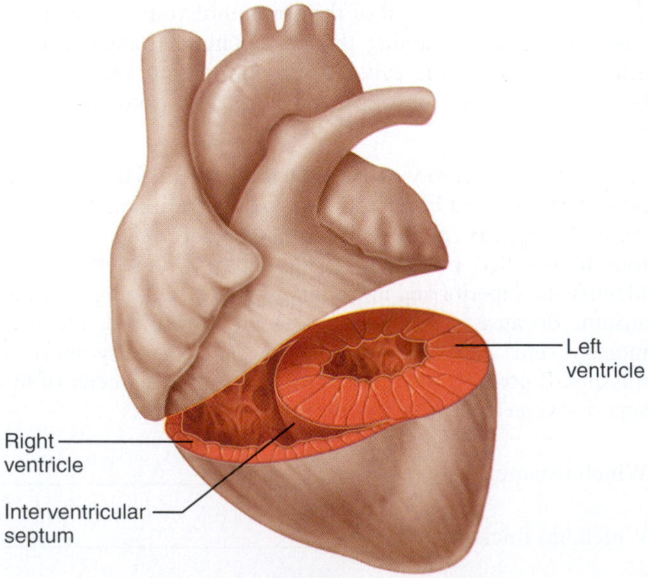
Left ventricle
Right ventricle
Interventricular septum

FIGURE 30.10 Anatomical differences between the right and left ventricles. The left ventricle has thicker walls, and its cavity is basically circular; by contrast, the right ventricle cavity is crescent-shaped and wraps around the left ventricle.

EXERCISE **31**

Conduction System of the Heart and Electrocardiography

MATERIALS

- ☐ Apparatus A or B:*

A: ECG recording apparatus, electrode paste, alcohol swabs, rubber straps or disposable electrodes

B: BIOPAC® BIOPAC® MP36 (or MP35/30) data acquisition unit, PC or Mac computer BIOPAC® Student Lab Software, electrode lead set, disposable vinyl electrodes

New versions of the BSL software (3.7.5 and higher for Windows, and 3.7.4 and higher for Mac) require slightly different channel settings and collection strategies. Instructions for using the newer software with the MP36/35 data acquisition units can be found on MasteringA&P. In addition, instructions for use of the NEW 2-channel data acquisition unit, the MP45, may also be found on MasteringA&P.

- ☐ Cot or lab table; pillow (optional)
- ☐ Millimeter ruler

*__Note:__ Instructions for using PowerLab® equipment can be found on MasteringA&P.

OBJECTIVES

1. To list and localize the elements of the intrinsic conduction, or nodal, system of the heart, and to describe how impulses are initiated and conducted through this system and the myocardium.
2. To interpret the ECG in terms of depolarization and repolarization events occurring in the myocardium; and to identify the P, QRS, and T waves on an ECG recording using an ECG recorder or BIOPAC®.
3. To calculate the heart rate, QRS interval, P–R interval, and Q–T interval from an ECG obtained during the laboratory period.
4. To define *tachycardia, bradycardia,* and *fibrillation.*

PRE-LAB QUIZ

1. Circle True or False. Cardiac muscle cells are electrically connected by gap junctions and behave as a single unit.
2. Because it sets the rate of depolarization for the normal heart, the __________ node is known as the pacemaker of the heart.
   a. atrioventricular
   b. Purkinje
   c. sinoatrial
3. Circle True or False. Activity of the nerves of the autonomic nervous system is essential for cardiac muscle to contract.
4. Today you will create a graphic recording of the electrical changes that occur during a cardiac cycle. This is known as an
   a. electrocardiogram
   b. electroencephalogram
   c. electromyogram
5. Circle the correct term. The typical ECG has three / six normally recognizable deflection waves.
6. In a typical ECG, the __________ wave signals the depolarization of the atria immediately before they contract.
   a. P   c. R
   b. Q   d. T
7. Circle True or False. The repolarization of the atria is usually masked by the large QRS complex.
8. Circle the correct term. A heart rate over 100 beats/minute is known as tachycardia / bradycardia.
9. How many electrodes will you place on your subject for today's activity if using a standard ECG apparatus?
   a. 3   c. 10
   b. 4   d. 12
10. Circle True or False. Electrical activity recorded by any lead depends on the location and orientation of the recording electrodes.

MasteringA&P™
Access practice quizzes and more in the Study Area at www.masteringaandp.com.

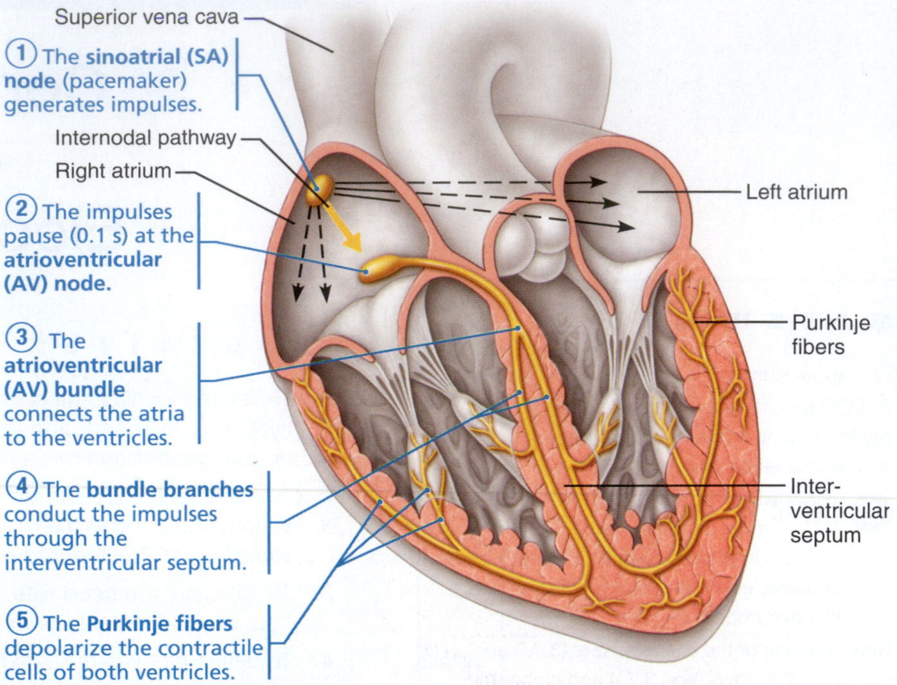
Superior vena cava
1 The sinoatrial (SA) node (pacemaker) generates impulses.
Internodal pathway
Right atrium
2 The impulses pause (0.1 s) at the atrioventricular (AV) node.
3 The atrioventricular (AV) bundle connects the atria to the ventricles.
4 The bundle branches conduct the impulses through the interventricular septum.
5 The Purkinje fibers depolarize the contractile cells of both ventricles.
Left atrium
Purkinje fibers
Inter-ventricular septum

FIGURE 31.1 The intrinsic conduction system of the heart. Dashed-line arrows indicate transmission of the impulse from the SA node through the atria. Solid yellow arrow indicates transmission of the impulse from the SA node to the AV node via the internodal pathway.

Heart contraction results from a series of electrical potential changes (depolarization waves) that travel through the heart preliminary to each beat. Because cardiac muscle cells are electrically connected by gap junctions, the entire myocardium behaves like a single unit, a **functional syncytium** (sin-sih′shum).

The Intrinsic Conduction System

The ability of cardiac muscle to beat is intrinsic—it does not depend on impulses from the nervous system to initiate its contraction and will continue to contract rhythmically even if all nerve connections are severed. However, two types of controlling systems exert their effects on heart activity. One of these involves nerves of the autonomic nervous system, which accelerate or decelerate the heartbeat rate depending on which division is activated. The second system is the **intrinsic conduction system,** or **nodal system,** of the heart, consisting of specialized noncontractile myocardial tissue. The intrinsic conduction system ensures that heart muscle depolarizes in an orderly and sequential manner (from atria to ventricles) and that the heart beats as a coordinated unit.

The components of the intrinsic conduction system include the **sinoatrial (SA) node,** located in the right atrium just inferior to the entrance to the superior vena cava; the **atrioventricular (AV) node** in the lower atrial septum at the junction of the atria and ventricles; the **AV bundle (bundle of His)** and right and left **bundle branches,** located in the interventricular septum; and the **Purkinje fibers,** essentially long strands of barrel-shaped cells called *Purkinje myocytes,* which ramify within the muscle bundles of the ventricular walls. The Purkinje fiber network is much denser and more elaborate in the left ventricle because of the larger size of this chamber (Figure 31.1).

The SA node, which has the highest rate of discharge, provides the stimulus for contraction. Because it sets the rate of depolarization for the heart as a whole, the SA node is often referred to as the *pacemaker.* From the SA node, the impulse spreads throughout the atria and to the AV node. This electrical wave is immediately followed by atrial contraction. At the AV node, the impulse is momentarily delayed (approximately 0.1 sec), allowing the atria to complete their contraction. It then passes through the AV bundle, the right and left bundle branches, and the Purkinje fibers, finally resulting in ventricular contraction. Note that the atria and ventricles are separated from one another by a region of electrically inert connective tissue, so the depolarization wave can be transmitted to the ventricles only via the tract between the AV node and AV bundle. Thus, any damage to the AV node-bundle pathway partially or totally insulates the ventricles from the influence of the SA node. Although autorhythmic cells are found throughout the heart, their rates of spontaneous depolarization differ. The nodal system increases the rate of heart depolarization and synchronizes heart activity.

Electrocardiography

The conduction of impulses through the heart generates electrical currents that eventually spread throughout the body. These impulses can be detected on the body's surface and recorded with an instrument called an *electrocardiograph.* The graphic recording of the electrical changes (depolarization followed by repolarization) occurring during the cardiac cycle is called an **electrocardiogram (ECG or EKG)** (Figure 31.2). For analysis, the ECG is divided into segments and intervals,

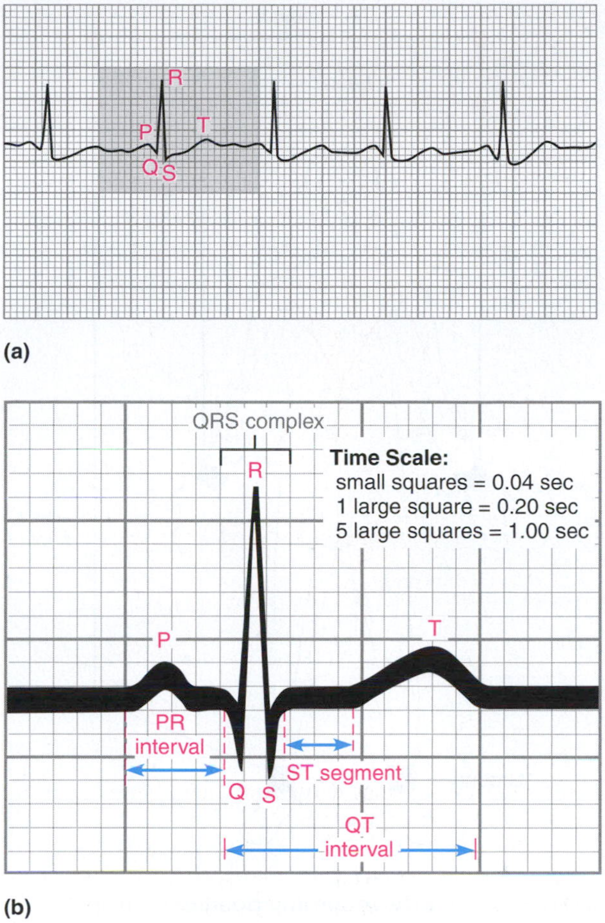
R
P
T
Q
S
(a)
QRS complex
R
Time Scale:
small squares = 0.04 sec
1 large square = 0.20 sec
5 large squares = 1.00 sec
P
T
PR
interval
ST segment
Q
S
QT
interval
(b)

FIGURE 31.2 The normal electrocardiogram. **(a)** Regular sinus rhythm. **(b)** Waves, segments, and intervals of a normal ECG.

which are defined in Table 31.1. The relationship between the deflection waves of an ECG and sequential excitation of the heart is shown in Figure 31.3.

TABLE 31.1 Boundaries of Each ECG Component

Feature	Boundaries
P wave	Start of P deflection to return to isoelectric line
P–R interval	Start of P deflection to start of Q deflection
P–R segment	End of P wave to start of Q deflection
QRS complex	Start of Q deflection to S return to isoelectric line
S–T segment	End of S deflection to start of T wave
Q–T interval	Start of Q deflection to end of T wave
T wave	Start of T deflection to return to isoelectric line
End T to next R	End of T wave to next R spike

It is important to understand what an ECG does *and does not* show: First, an ECG is a record of voltage and time—nothing else. Although we can and do infer that muscle contraction follows its excitation, sometimes it does not. Second, an ECG records electrical events occurring in relatively large amounts of muscle tissue (i.e., the bulk of the heart muscle), *not* the activity of nodal tissue which, like muscle contraction, can only be inferred. Nonetheless, abnormalities of the deflection waves and changes in the time intervals of the ECG are useful in detecting myocardial infarcts or problems with the conduction system of the heart. The P–R interval represents the time between the beginning of atrial depolarization and ventricular depolarization. Thus, it typically includes the period during which the depolarization wave passes to the AV node, atrial systole, and

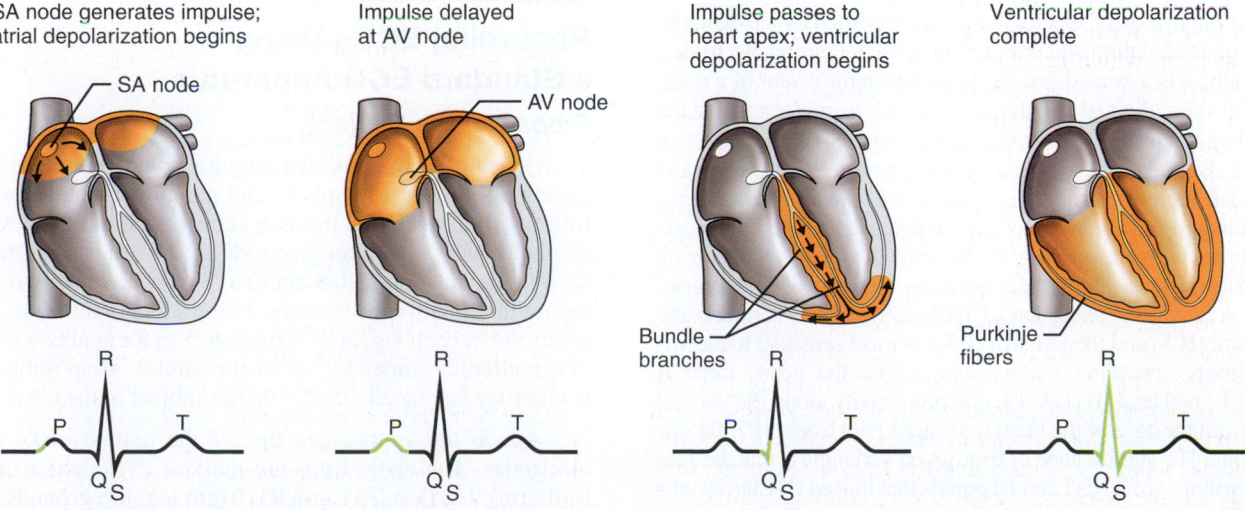

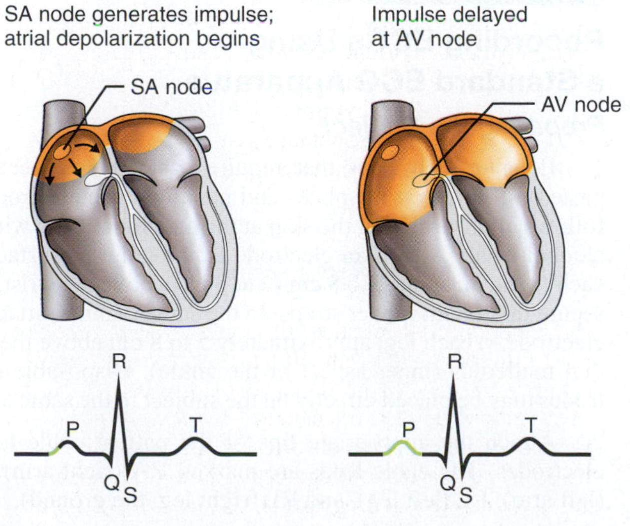
SA node generates impulse; atrial depolarization begins
SA node
Impulse delayed at AV node
AV node
Impulse passes to heart apex; ventricular depolarization begins
Bundle branches
Ventricular depolarization complete
Purkinje fibers
R
P
T
Q
S

FIGURE 31.3 The sequence of excitation of the heart related to the deflection waves of an ECG tracing.

the passage of the excitation wave to the balance of the conducting system. Generally, the P–R interval is about 0.16 to 0.18 sec. A longer interval may suggest a partial AV heart block caused by damage to the AV node. In total heart block, no impulses are transmitted through the AV node, and the atria and ventricles beat independently of one another—the atria at the SA node rate and the ventricles at their intrinsic rate, which is considerably slower.

A prolonged QRS complex (normally 0.08 sec) may indicate a right or left bundle branch block in which one ventricle is contracting later than the other. The Q–T interval is the period from the beginning of ventricular depolarization through repolarization and includes the time of ventricular contraction (the S–T segment). With a heart rate of 70 beats/min, this interval is normally 0.31 to 0.41 sec. As the rate increases, this interval becomes shorter; conversely, when the heart rate drops, the interval is longer. The repolarization of the atria, which occurs during the QRS interval, is generally obscured by the large QRS complex.

A heart rate over 100 beats/min is referred to as **tachycardia;** a rate below 60 beats/min is **bradycardia.** Although neither condition is pathological, prolonged tachycardia may progress to **fibrillation,** a condition of rapid uncoordinated heart contractions which makes the heart useless as a pump. Bradycardia in athletes is a positive finding; that is, it indicates an increased efficiency of cardiac functioning. Because *stroke volume* (the amount of blood ejected by a ventricle with each contraction) increases with physical conditioning, the heart can contract more slowly and still meet circulatory demands.

Twelve standard leads are used to record an ECG for diagnostic purposes. Three of these are bipolar leads that measure the voltage difference between the arms, or an arm and a leg, and nine are unipolar leads. Together the 12 leads provide a fairly comprehensive picture of the electrical activity of the heart.

For this investigation, four electrodes are used (Figure 31.4), and results are obtained from the three *standard limb leads* (also shown in Figure 31.4). Several types of physiographs or ECG recorders are available. Your instructor will provide specific directions on how to set up and use the available apparatus if standard ECG apparatus is used (Activity 1a). Instructions for use of BIOPAC® apparatus (Activity 1b) are provided on pages 462–466.

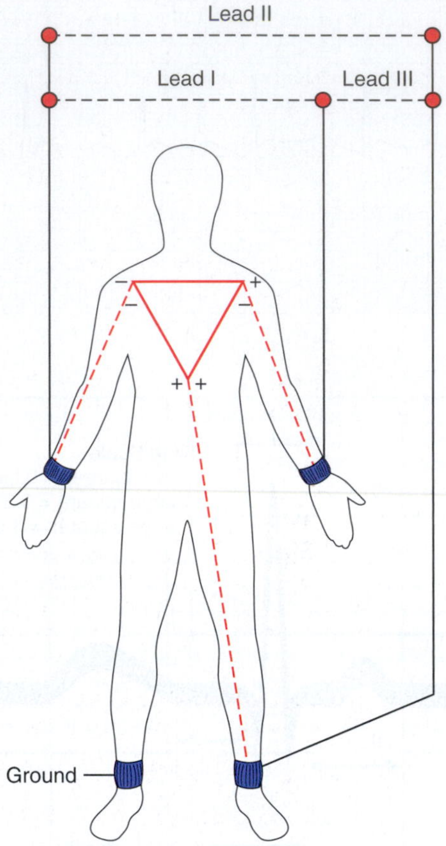
Lead II
Lead I
Lead III
Ground

FIGURE 31.4 ECG recording positions for the standard limb leads.

Understanding the Standard Limb Leads

As you might expect, electrical activity recorded by any lead depends on the location and orientation of the recording electrodes. Clinically, it is assumed that the heart lies in the center of a triangle with sides of equal lengths *(Einthoven's triangle)* and that the recording connections are made at the vertices (corners) of that triangle. But in practice, the electrodes connected to each arm and to the left leg are considered to connect to the triangle vertices. The standard limb leads record the voltages generated in the extracellular fluids surrounding the heart by the ion flows occurring simultaneously in many cells between any two of the connections. A recording using lead I (RA–LA), which connects the right arm (RA) and the left arm (LA), is most sensitive to electrical activity spreading horizontally across the heart. Lead II (RA–LL) and lead III (LA–LL) record activity along the vertical axis (from the base of the heart to its apex) but from different orientations. The significance of Einthoven's triangle is that the sum of the voltages of leads I and III equals that in lead II (Einthoven's law). Hence, if the voltages of two of the standard leads are recorded, that of the third lead can be determined mathematically.

Once the subject is prepared, the ECG will be recorded first under baseline (resting) conditions and then under conditions of fairly strenuous activity. Finally, recordings will be made while the subject holds his or her breath or carries out deep breathing. The activity recordings and those involving changes in respiratory rate or depth will be compared to the baseline recordings, and you will be asked to determine the reasons for the observed differences in the recordings.

ACTIVITY 1A

Recording ECGs Using a Standard ECG Apparatus

Preparing the Subject

1. If using electrodes that require paste, place electrode paste on four electrode plates and position each electrode as follows after scrubbing the skin at the attachment site with an alcohol swab. Attach an electrode to the anterior surface of each forearm, about 5 to 8 cm (2 to 3 in.) above the wrist, and secure them with rubber straps. In the same manner, attach an electrode to each leg, approximately 5 to 8 cm above the medial malleolus (inner aspect of the ankle). Disposable electrodes may be placed directly on the subject in the same areas.

2. Attach the appropriate tips of the patient cable to the electrodes. The cable leads are marked RA (right arm), LA (left arm), LL (left leg), and RL (right leg, the ground).

Making a Baseline Recording

1. Position the subject comfortably in a supine position on a cot (if available), or sitting relaxed on a laboratory chair.

2. Turn on the power switch and adjust the sensitivity knob to 1. Set the paper speed to 25 mm/sec and the lead selector to the position corresponding to recording from lead I (RA–LA).

3. Set the control knob at the **RUN** position and record the subject's at-rest ECG from lead I for 2 to 3 minutes or until the recording stabilizes. (You will need a tracing long enough to provide each student in your group with a representative segment.) The subject should try to relax and not move unnecessarily, because the skeletal muscle action potentials will also be picked up and recorded.

4. Stop the recording and mark it "lead I."

5. Repeat the recording procedure for leads II (RA–LL) and III (LA–LL).

6. Each student should take a representative segment of one of the lead recordings and label the record with the name of the subject and the lead used. Identify and label the P, QRS, and T waves. The calculations you perform for your recording should be based on the following information: Because the paper speed was 25 mm/sec, each millimeter of paper corresponds to a time interval of 0.04 sec. Thus, if an interval requires 4 mm of paper, its duration is 4 mm × 0.04 sec/mm = 0.16 sec.

7. Compute the heart rate. Obtain a millimeter ruler and measure the distance from the beginning of one QRS complex to the beginning of the next QRS complex. Enter this value into the following equation to find the time for one heartbeat.

_________ mm/beat × 0.04 sec/mm = _________ sec/beat

Now find the beats per minute, or heart rate, by using the figure just computed for seconds per beat in the following equation:

$$\frac{1}{\_\_\_\_\_\ \text{sec/beat}} \times 60\ \text{sec/min} = \_\_\_\_\_\_\_\_\ \text{beats/min}$$

Is the obtained value within normal limits? _______________

Measure the QRS complex, and compute its duration.

____

Measure the Q–T interval, and compute its duration.

____

Measure the P–R interval, and compute its duration.

____

Are the computed values within normal limits?

____

8. At the bottom of this page, attach segments of the ECG recordings from leads I through III. Make sure you indicate the paper speed, lead, and subject's name on each tracing. To the recording on which you based your previous computations, add your calculations for the duration of the QRS complex and the P–R and Q–T intervals above the respective area of tracing. Also record the heart rate on that tracing.

Recording the ECG for Running in Place

1. Make sure the electrodes are securely attached to prevent electrode movement while recording the ECG.

2. Set the paper speed to 25 mm/sec, and prepare to make the recording using lead I.

3. Record the ECG while the subject is running in place for 3 minutes. Then have the subject sit down, but continue to record the ECG for an additional 4 minutes. *Mark the recording* at the end of the 3 minutes of running and at 1 minute after cessation of activity.

4. Stop the recording. Compute the beats/min during the third minute of running, at 1 minute after exercise, and at 4 minutes after exercise. Record below:

_________ beats/min while running in place

_________ beats/min at 1 minute after exercise

_________ beats/min at 4 minutes after exercise

5. Compare this recording with the previous recording from lead I. Which intervals are shorter in the "running" recording?

____

Does the subject's heart rate return to resting level by 4 minutes after exercise?

____

Recording the ECG During Breath Holding

1. Position the subject comfortably in the sitting position.

2. Using lead I and a paper speed of 25 mm/sec, begin the recording. After approximately 10 seconds, instruct the subject to begin breath holding and mark the record to indicate the onset of the 1-minute breath-holding interval.

3. Stop the recording after 1 minute and remind the subject to breathe. Compute the beats/minute during the 1-minute experimental (breath-holding) period.

Beats/min during breath holding: _______________

4. Compare this recording with the lead I recording obtained under resting conditions.

What differences are seen? _______________

____

Attempt to explain the physiological reason for the differences you have seen. (Hint: A good place to start might be to check hypoventilation or the role of the respiratory system in acid-base balance of the blood.)

____

____

____

____

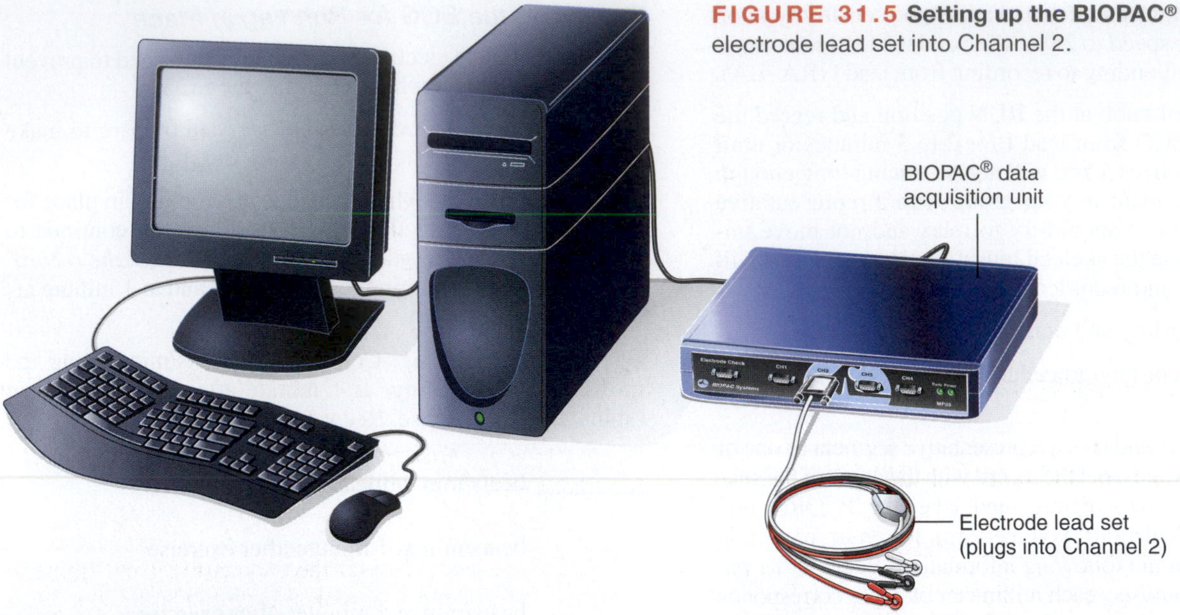
BIOPAC® data
acquisition unit
Electrode lead set
(plugs into Channel 2)

FIGURE 31.5 Setting up the BIOPAC® unit. Plug the electrode lead set into Channel 2.

ACTIVITY 1B

Electrocardiography Using BIOPAC®

In this activity, you will record the electrical activity of the heart under three different conditions: (1) while the subject is lying down, (2) after the subject sits up and breathes normally, and (3) after the subject has exercised and is breathing deeply.

Since the electrodes are not placed directly over the heart, artifacts can result from the recording of unwanted skeletal muscle activity. In order to obtain a clear ECG, it is important that the subject:

- Remain still during the recording.
- Refrain from laughing or talking during the recording.
- When in the sitting position, keep arms and legs steady and relaxed.

Setting Up the Equipment

1. Connect the BIOPAC® unit to the computer and turn the computer **ON.**
2. Make sure the BIOPAC® unit is **OFF.**
3. Plug in the equipment as shown in Figure 31.5:

- Electrode lead set—CH 2

4. Turn the BIOPAC® unit **ON.**
5. Place the three electrodes on the subject as shown in Figure 31.6, and attach the electrode leads according to the colors indicated. The electrodes should be placed on the medial surface of each leg, 5 to 8 cm (2 to 3 in.) superior to the ankle. The other electrode should be placed on the right anterior forearm 5 to 8 cm above the wrist.
6. The subject should lie down and relax in a comfortable position. A chair or place to sit up should be available nearby.
7. Start the BIOPAC® Student Lab program on the computer by double-clicking the icon on the desktop or by following your instructor's guidance.
8. Select lesson **L05-ECG-1** from the menu, and click **OK.**
9. Type in a filename that will save this subject's data on the computer hard drive. You may want to use the subject's last name followed by ECG-1 (for example, SmithECG-1), then click **OK.**

Calibrating the Equipment

- Examine the electrodes and the electrode leads to be certain they are properly attached.

1. The subject must remain still and relaxed. With the subject in a still position, click **Calibrate.** This will initiate the process whereby the computer will automatically establish parameters to record the data.
2. The calibration procedure will stop automatically after 8 seconds.
3. Observe the recording of the calibration data, which should look similar to the data in Figure 31.7.

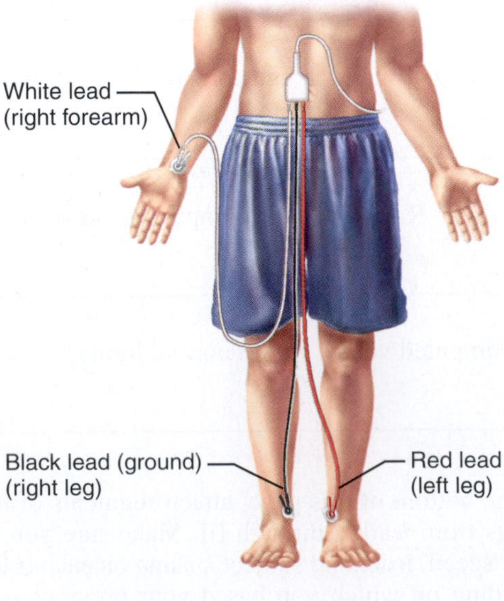
White lead
(right forearm)
Black lead (ground)
(right leg)
Red lead
(left leg)

FIGURE 31.6 Placement of electrodes and the appropriate attachment of electrode leads by color.

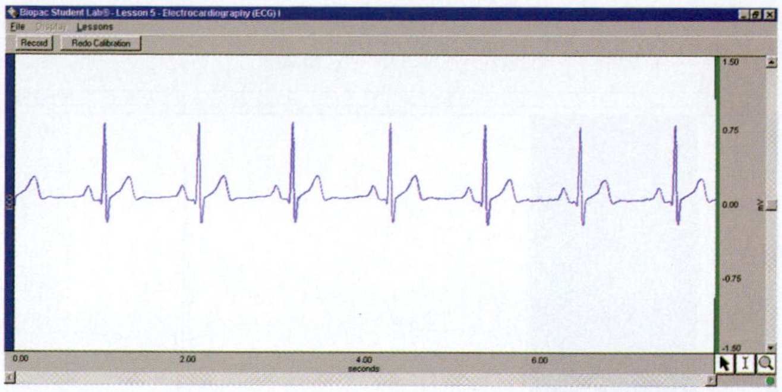
Record
Redo Calibration
1.50
0.75
0.00
-0.75
-1.50
0.00
2.00
4.00
seconds
6.00

FIGURE 31.7 **Example of calibration data.**

- If the data look very different, click **Redo Calibration** and repeat the steps above.
- If the data look similar, proceed to the next section.

Recording Segment 1: Subject Lying Down

1. To prepare for the recording, remind the subject to remain still and relaxed while lying down.

2. When prepared, click **Record** and gather data for 20 seconds. At the end of 20 seconds, click **Suspend.**

3. Observe the data, which should look similar to the data in Figure 31.8.

- If the data look very different, click **Redo** and repeat the steps above. Be certain to check attachment of the electrodes and leads, and remind the subject not to move, talk, or laugh.
- If the data look similar, move on to the next recording segment.

Recording Segment 2: After Subject Sits Up, With Normal Breathing

1. Tell the subject to be ready to sit up in the designated location. With the exception of breathing, the subject should try to remain motionless after assuming the seated position. *If the subject moves too much during recording after sitting up, unwanted skeletal muscle artifacts will affect the recording.*

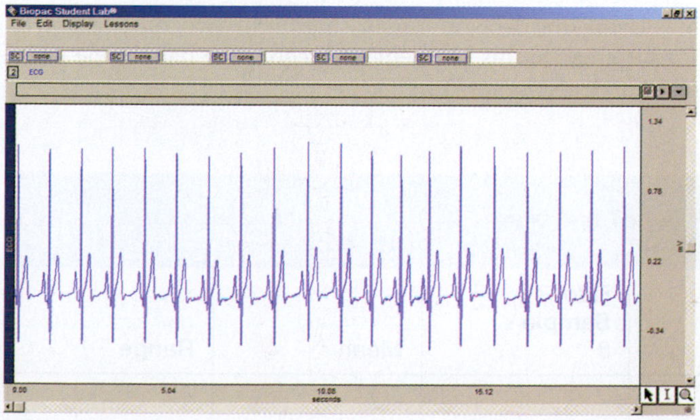

FIGURE 31.8 **Example of ECG data while the subject is lying down.**

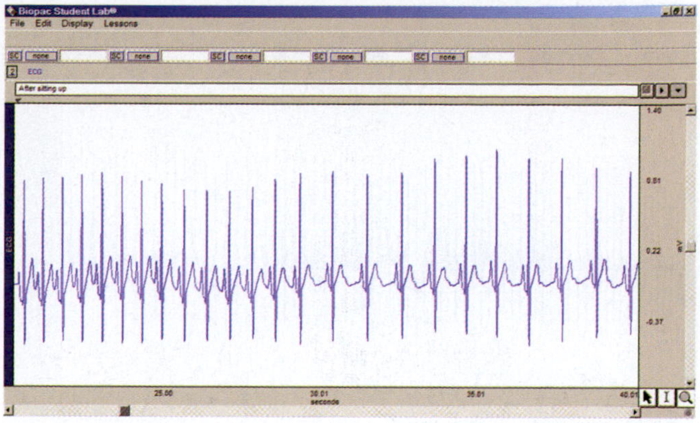

FIGURE 31.9 **Example of ECG data after the subject sits up and breathes normally.**

2. When prepared, instruct the subject to sit up. Immediately after the subject assumes a motionless state, click **Resume,** and the data will begin recording.

3. At the end of 20 seconds, click **Suspend** to stop recording.

4. Observe the data, which should look similar to the data in Figure 31.9.

- If the data look very different, click **Redo** and repeat the steps above. Be certain to check attachment of the electrodes, and do not click **Resume** until the subject is motionless.
- If the data look similar, move on to the next recording segment. ***Note:*** *For start of next segment, ignore the program instructions at the bottom of the computer screen.*

Recording Segment 3: After Subject Exercises, With Deep Breathing

1. Remove the electrode pinch connectors from the electrodes on the subject.

2. Have the subject do a brief round of exercise, such as jumping jacks or running in place for 1 minute, in order to elevate the heart rate.

3. As quickly as possible after the exercise, have the subject resume a motionless, seated position and reattach the pinch connectors. Once again, if the subject moves too much during recording, unwanted skeletal muscle artifacts will affect the data. After exercise, the subject is likely to be breathing deeply, but otherwise should remain as still as possible.

4. Immediately after the subject assumes a motionless, seated state, click **Resume,** and the data will begin recording. Record the ECG for 60 seconds in order to observe post-exercise recovery.

5. After 60 seconds, click **Suspend** to stop recording.

6. Observe the data, which should look similar to the data in Figure 31.10.

- If the data look very different, click **Redo** and repeat the steps above. Be certain to check attachment of the electrodes and leads, and remember not to click **Resume** until the subject is motionless.

7. When finished, click **Done.** Remove the electrodes from the subject.

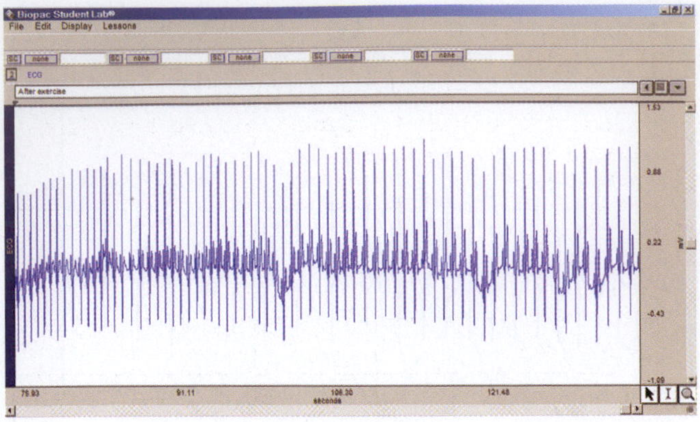

FIGURE 31.10 Example of ECG data after the subject exercises.

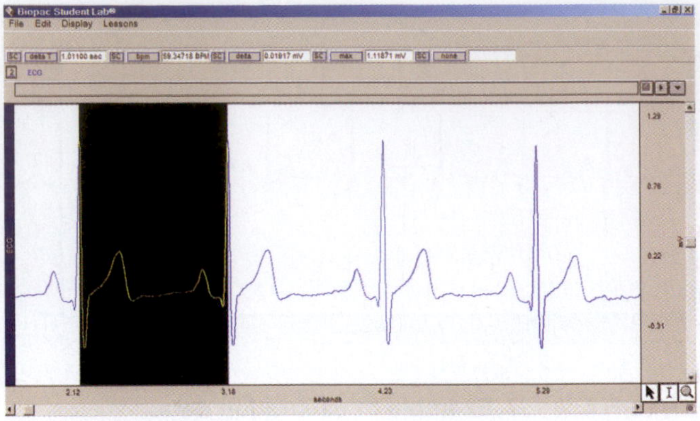

FIGURE 31.11 Example of highlighting from R-wave to R-wave.

8. A pop-up window will appear. To record from another subject, select **Record data from another subject** and return to step 5 under Setting Up the Equipment. If continuing to the Data Analysis section, select **Analyze current data file** and proceed to step 2 of the Data Analysis section.

Data Analysis

1. If just starting the BIOPAC® program to perform data analysis, enter **Review Saved Data** mode and choose the file with the subject's ECG data (for example, SmithECG-1).

2. Use the following tools to adjust the data in order to clearly view and analyze four consecutive cardiac cycles:

- Click the magnifying glass in the lower right corner of the screen (near the I-beam cursor box) to activate the **zoom** function. Use the magnifying glass cursor to click on the very first waveforms until there are about 4 seconds of data represented (see horizontal time scale at the bottom of the screen).
- Select the **Display** menu at the top of the screen, and click **Autoscale Waveforms.** This function will adjust the data for better viewing.
- Click the **Adjust Baseline** button. Two new buttons will appear; simply click these buttons to move the waveforms **Up** or **Down** so they appear clearly in the center of the screen. Once they are centered, click **Exit.**

3. Set up the first two pairs of channel/measurement boxes at the top of the screen by selecting the following channels and measurement types from the drop-down menus (note that Channel 2 contains the ECG data):

Channel	Measurement
CH 2	deltaT
CH 2	bpm

Analysis of Segment 1: Subject Lying Down

1. Use the arrow cursor and click the I-beam cursor box on the lower right side of the screen to activate the "area selection" function.

2. First measure **deltaT** and **bpm** in Segment 1 (approximately seconds 0–20). Using the I-beam cursor, highlight from the peak of one R-wave to the peak of the next R-wave, as shown in Figure 31.11.

3. Observe that the computer automatically calculates the **deltaT** and **bpm** for the selected area. These measurements represent the following:

deltaT (difference in time): Computes the elapsed time between the beginning and end of the highlighted area

bpm (beats per minute): Computes the beats per minute when highlighting from the R-wave of one cycle to the R-wave of another cycle

4. Record this data in the Segment 1 Samples chart below under R to R Sample 1 (round to the nearest 0.01 second and 0.1 beat per minute).

5. Using the I-beam cursor, highlight two other pairs of R to R areas in this segment and record the data in the same chart under Samples 2 and 3.

Segment 1 Samples for deltaT and bpm

Measure	Channel	R to R Sample 1	R to R Sample 2	R to R Sample 3	Mean	Range
deltaT	CH 2					
bpm	CH 2					

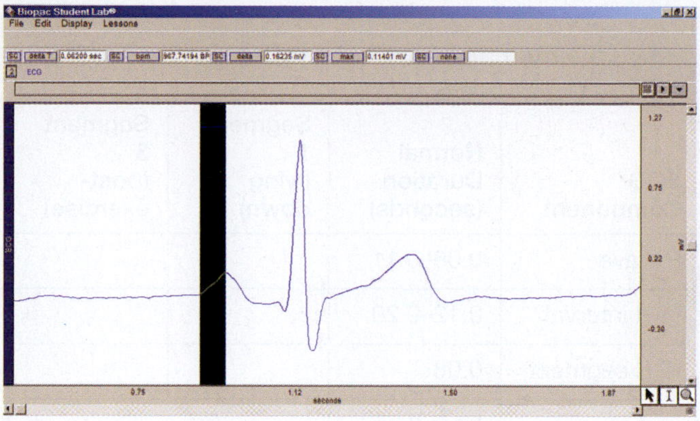

FIGURE 31.12 Example of a single ECG waveform with the first part of the P wave highlighted.

6. Calculate the means and ranges of the data in this chart.

7. Next, use the **zoom, Autoscale Waveforms,** and **Adjust Baseline** tools described above to focus in on one ECG waveform within Segment 1. See the example in Figure 31.12.

8. Once a single ECG waveform is centered for analysis, click the I-beam cursor box on the lower right side of the screen to activate the "area selection" function.

9. Using the highlighting function and **deltaT** computation, measure the duration of every component of the ECG waveform. Refer to Figure 31.2b and Table 31.1 for guidance in highlighting each component.

10. Highlight each component of one cycle. Observe the elapsed time, and record this data under Cycle 1 in the Segment 1 Elapsed Time chart.

11. Scroll along the horizontal axis at the bottom of the data to view and analyze two additional cycles in Segment 1. Record the elapsed time for every component of Cycle 2 and Cycle 3 in the Segment 1 Elapsed Time chart.

12. In the same chart, calculate the means for the three cycles of data and record.

Segment 1 Elapsed Time for ECG Components (seconds)

Component	Cycle 1	Cycle 2	Cycle 3	Mean
P wave				
P–R interval				
P–R segment				
QRS complex				
S–T segment				
Q–T interval				
T wave				
End T to next R				

Analysis of Segment 2: Subject Sitting Up and Breathing Normally

1. Scroll along the horizontal time bar until you reach the data for Segment 2 (approximately seconds 20–40). A marker with "After sitting up" should denote the beginning of this data.

2. As in the analysis of Segment 1, use the I-beam tool to highlight and measure the **deltaT** and **bpm** between three different pairs of R-waves in this segment, and record the data in the Segment 2 Samples chart below.

Analysis of Segment 3: After Exercise with Deep Breathing

1. Scroll along the horizontal time bar until you reach the data for Segment 3 (approximately seconds 40–60). A marker with "After exercise" should denote the beginning of this data.

Segment 2 Samples for deltaT and bpm

Measure	Channel	R to R Sample 1	R to R Sample 2	R to R Sample 3	Mean	Range
deltaT	CH 2					
bpm	CH 2					

Segment 3 Samples for deltaT and bpm

Measure	Channel	R to R Sample 1	R to R Sample 2	R to R Sample 3	Mean	Range
deltaT	CH 2					
bpm	CH 2					

Segment 3 Elapsed Time for ECG Components (seconds)				
Component	Cycle 1	Cycle 2	Cycle 3	Mean
P wave				
P–R interval				
P–R segment				
QRS complex				
S–T segment				
Q–T interval				
T wave				
End T to next R				

2. As before, use the I-beam tool to highlight and measure the **deltaT** and **bpm** between three pairs of R-waves in this segment, and record the data in the Segment 3 Samples chart on the previous page.

3. Using the instructions for steps 8 and 9 in the section Analysis of Segment 1, highlight and observe the elapsed time for each component of one cycle, and record these data under Cycle 1 in the Segment 3 Elapsed Time chart above.

4. Scroll along the horizontal axis at the bottom of the data to view and analyze two other cycles in Segment 3. Record the elapsed time for each component of Cycle 2 and Cycle 3 in the Segment 3 Elapsed Time chart above.

5. In the same chart, calculate the means for each component in Segment 3.

6. When finished, **Exit** the program.

Compare the average **deltaT** times and average **bpm** between the data in Segment 1 (lying down) and the data in Segment 3 (after exercise). Which is greater in each case?

____

____

What is the relationship between elapsed time (**deltaT**) between R-waves and the heart rate?

____

Is there a change in heart rate when the subject makes the transition from lying down (Segment 1) to a sitting position (Segment 2)?

____

Examine the average duration of each of the ECG components in Segment 1 and Segment 3. In the chart that follows, record the average values of each component. Draw a circle around those measures that fit within the normal range.

Average Duration for ECG Components			
ECG Component	Normal Duration (seconds)	Segment 1 (lying down)	Segment 3 (post-exercise)
P wave	0.06–0.11		
P–R interval	0.12–0.20		
P–R segment	0.08		
QRS complex	Less than 0.12		
S–T segment	0.12		
Q–T interval	0.31–0.41		
T wave	0.16		
End T to next R	varies		

Compare the Q–T intervals in the data while the subject is at rest versus after exercise; this interval corresponds closely to the duration of contraction of the ventricles. Describe and explain any difference.

____

____

Compare the duration in the period from the end of each T wave to the next R-spike while the subject is at rest versus after exercise. This interval corresponds closely to the duration of relaxation of the ventricles. Describe and explain any difference.

____

____

Which, if any, of the components in the Average Duration chart above demonstrates a significant difference from the normal value? Based on the events of the cardiac cycle and their representation in an ECG, what abnormality might the data suggest?

____

____

NAME ______________________

LAB TIME/DATE ______________________

REVIEW SHEET

EXERCISE 31

Conduction System of the Heart and Electrocardiography

The Intrinsic Conduction System

1. List the elements of the intrinsic conduction system in order, starting from the SA node.

SA node → ______________________ → ______________________ →

______________________ → ______________________

At what structure in the transmission sequence is the impulse temporarily delayed? ______________________

Why? ______________________

____

2. Even though cardiac muscle has an inherent ability to beat, the nodal system plays a critical role in heart physiology. What is that role? ______________________

____

Electrocardiography

3. Define *ECG.* ______________________

____

4. Draw an ECG wave form representing one heartbeat. Label the P, QRS, and T waves; the P–R interval; the S–T segment, and the Q–T interval.

5. Why does heart rate increase during running? ______________________

____

____

6. Describe what happens in the cardiac cycle in the following situations.

1. immediately before the P wave: ____________________

2. during the P wave: ____________________

3. immediately after the P wave (P–R segment): ____________________

4. during the QRS wave: ____________________

5. immediately after the QRS wave (S–T interval): ____________________

6. during the T wave: ____________________

7. Define the following terms.

1. *tachycardia:* ____________________

2. *bradycardia:* ____________________

3. *fibrillation:* ____________________

8. Which would be more serious, atrial or ventricular fibrillation? ____________________

Why? ____________________

____

9. Abnormalities of heart valves can be detected more accurately by auscultation than by electrocardiography. Why is this so?

____

____

____

EXERCISE 32

Anatomy of Blood Vessels

MATERIALS

- ☐ Compound microscope
- ☐ Prepared microscope slides showing cross sections of an artery and vein
- ☐ Anatomical charts of human arteries and veins (or a three-dimensional model of the human circulatory system)
- ☐ Anatomical charts of the following specialized circulations: pulmonary circulation, hepatic portal circulation, fetal circulation, arterial supply of the brain (or a brain model showing this circulation)

For instructions on animal dissections, see the dissection exercises starting on page 697 in the cat, fetal pig, and rat editions of this manual.

OBJECTIVES

1. To describe the tunics of blood vessel walls, and to state the function of each layer.
2. To correlate differences in artery, vein, and capillary structure with the functions of these vessels.
3. To recognize a cross-sectional view of an artery and vein when provided with a microscopic view or appropriate diagram.
4. To list and/or identify the major arteries arising from the aorta, and to indicate the body region supplied by each.
5. To list and/or identify the major veins draining into the superior and inferior venae cavae, and to indicate the body regions drained.
6. To point out and/or discuss the unique features of special circulations (pulmonary circulation, hepatic portal system, fetal circulation, cerebral arterial circle [circle of Willis]) in the body.

PRE-LAB QUIZ

1. Circle the correct term. Arteries / Veins drain tissues and return blood to the heart.
2. Circle True or False. It is through the walls of capillaries that actual gas exchange takes place between tissue cells and blood.
3. The __________ is the largest artery of the body.
   a. aorta
   b. carotid artery
   c. femoral artery
   d. subclavian artery
4. Circle the correct term. The largest branch of the abdominal aorta, the renal / superior mesenteric artery, supplies most of the small intestine and the first half of the large intestine.
5. The anterior tibial artery terminates with the __________ artery, which is often palpated in patients with circulatory problems to determine the circulatory efficiency of the lower limb.
   a. dorsalis pedis
   b. external iliac
   c. obturator
   d. tibial
6. Circle the correct term. Veins draining the head and upper extremities empty into the superior / inferior vena cava.
7. Located in the lower limb, the __________ is the longest vein in the body.
   a. external iliac
   b. fibular
   c. great saphenous
   d. internal iliac
8. Circle the correct term. The renal / hepatic veins drain the liver.
9. The function of the __________ is to drain the digestive viscera and carry dissolved nutrients to the liver for processing.
   a. fetal circulation
   b. hepatic portal circulation
   c. pulmonary circulation system
10. Circle the correct term. In the developing fetus, the umbilical artery / vein carries blood rich in nutrients and oxygen to the fetus.

MasteringA&P™

Access practice quizzes and more in the Study Area at www.masteringaandp.com.

PAL

For access to anatomical models and more, check out Practice Anatomy Lab.

The blood vessels constitute a closed transport system. As the heart contracts, blood is propelled into the large arteries leaving the heart. It moves into successively smaller arteries and then to the arterioles, which feed the capillary beds in the tissues. Capillary beds are drained by the venules, which in turn empty into veins that ultimately converge on the great veins entering the heart.

Arteries, carrying blood away from the heart, and veins, which drain the tissues and return blood to the heart, function simply as conducting vessels or conduits. Only the tiny capillaries that connect the arterioles and venules and ramify throughout the tissues directly serve the needs of the body's cells. It is through the capillary walls that exchanges are made between tissue cells and blood. Respiratory gases, nutrients, and wastes move along diffusion gradients. Thus, oxygen and nutrients diffuse from the blood to the tissue cells, and carbon dioxide and metabolic wastes move from the cells to the blood.

In this exercise you will examine the microscopic structure of blood vessels and identify the major arteries and veins of the systemic circulation and other special circulations.

Microscopic Structure of the Blood Vessels

Except for the microscopic capillaries, the walls of blood vessels are constructed of three coats, or *tunics* (Figure 32.1).

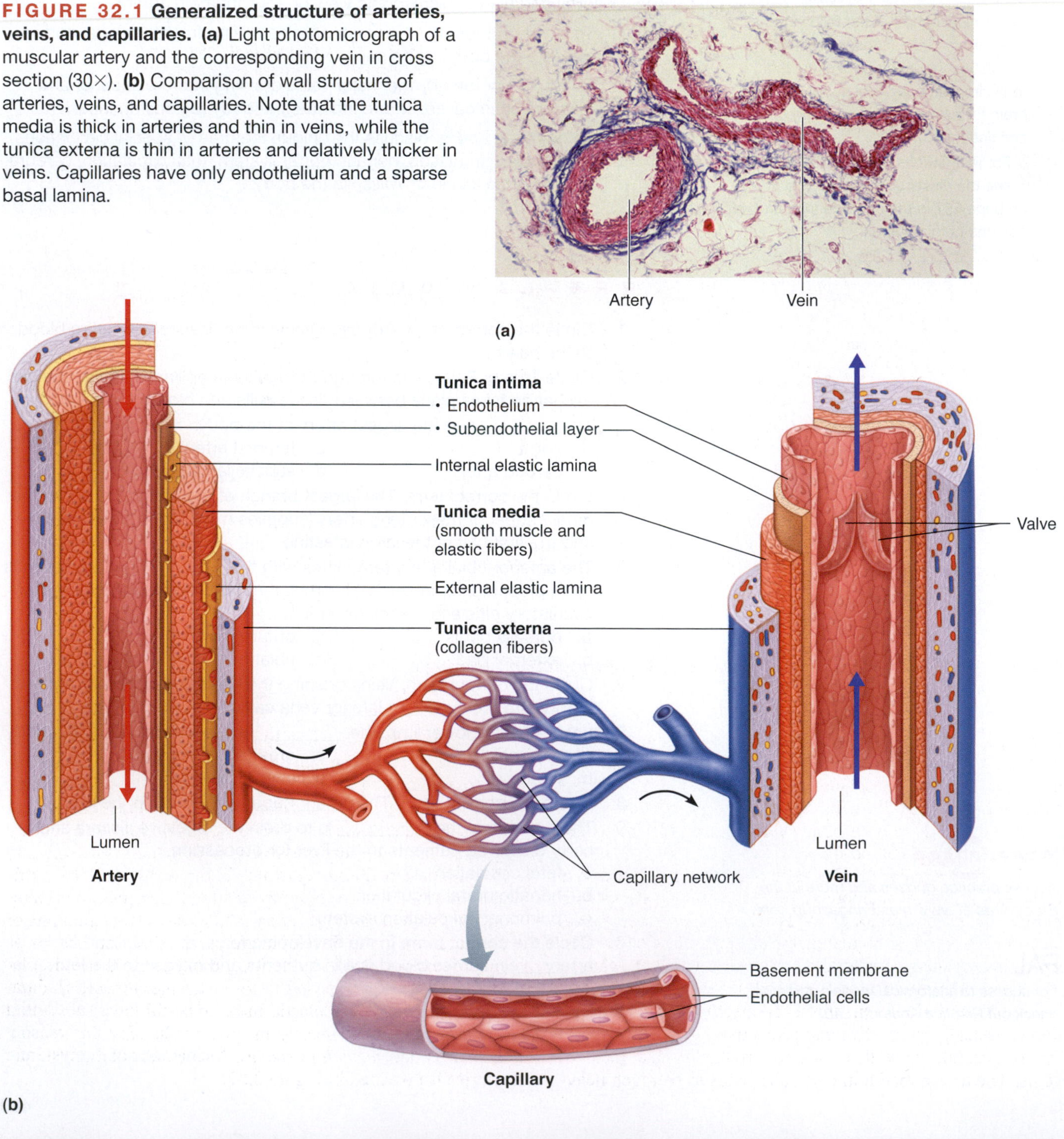
Artery
Vein
(a)
Tunica intima
• Endothelium
• Subendothelial layer
Internal elastic lamina
Tunica media
(smooth muscle and
elastic fibers)
External elastic lamina
Tunica externa
(collagen fibers)
Valve
Lumen
Artery
Capillary network
Lumen
Vein
Basement membrane
Endothelial cells
Capillary
(b)

FIGURE 32.1 Generalized structure of arteries, veins, and capillaries. (a) Light photomicrograph of a muscular artery and the corresponding vein in cross section (30×). **(b)** Comparison of wall structure of arteries, veins, and capillaries. Note that the tunica media is thick in arteries and thin in veins, while the tunica externa is thin in arteries and relatively thicker in veins. Capillaries have only endothelium and a sparse basal lamina.

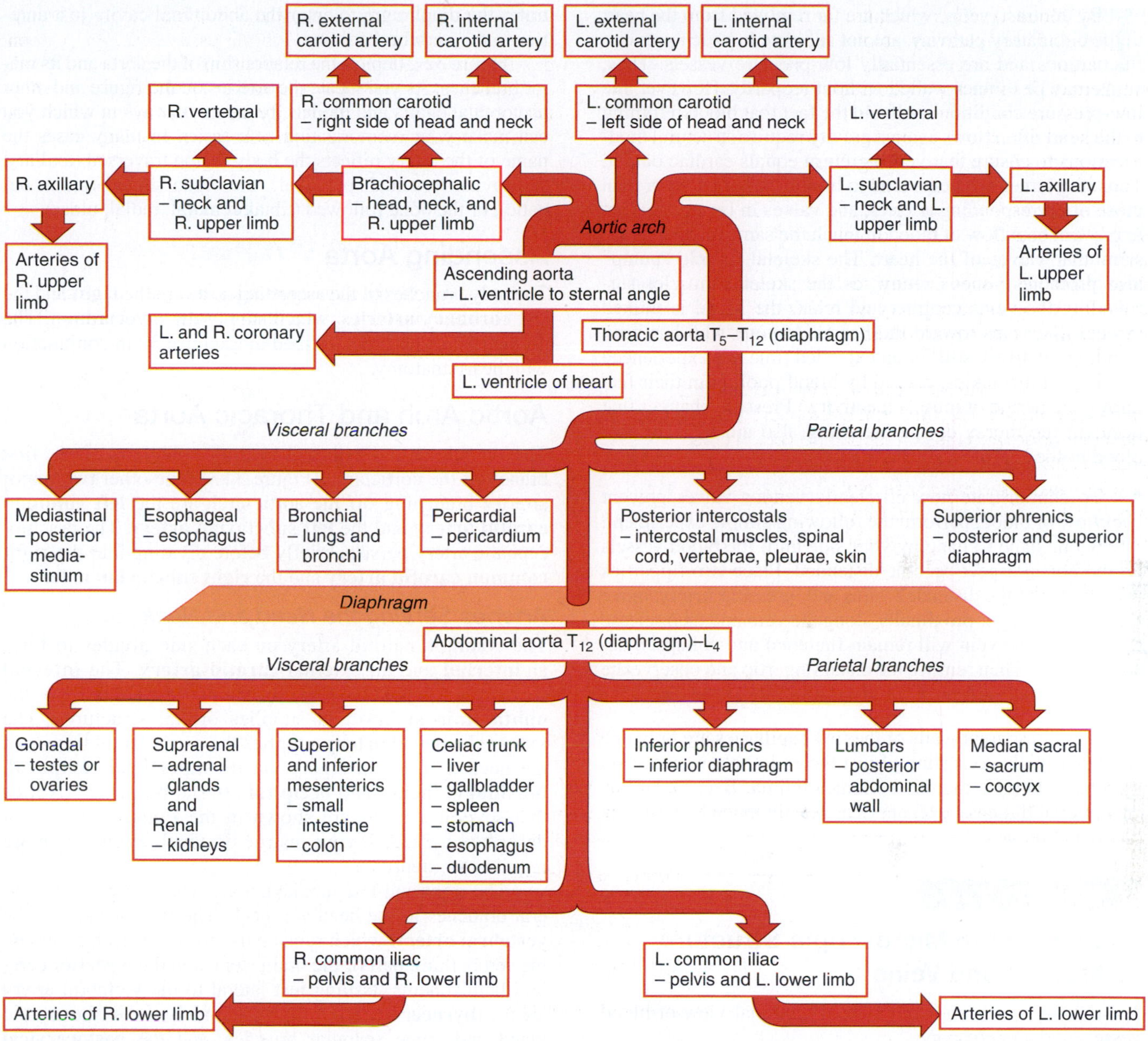
R. external carotid artery
R. internal carotid artery
L. external carotid artery
L. internal carotid artery
R. vertebral
R. common carotid – right side of head and neck
L. common carotid – left side of head and neck
L. vertebral
R. axillary
R. subclavian – neck and R. upper limb
Brachiocephalic – head, neck, and R. upper limb
Aortic arch
L. subclavian – neck and L. upper limb
L. axillary
Arteries of R. upper limb
Arteries of L. upper limb
Ascending aorta – L. ventricle to sternal angle
Thoracic aorta T_5–T_{12} (diaphragm)
L. and R. coronary arteries
L. ventricle of heart
Visceral branches
Parietal branches
Mediastinal – posterior mediastinum
Esophageal – esophagus
Bronchial – lungs and bronchi
Pericardial – pericardium
Posterior intercostals – intercostal muscles, spinal cord, vertebrae, pleurae, skin
Superior phrenics – posterior and superior diaphragm
Diaphragm
Abdominal aorta T_{12} (diaphragm)–L_4
Visceral branches
Parietal branches
Gonadal – testes or ovaries
Suprarenal – adrenal glands and Renal – kidneys
Superior and inferior mesenterics – small intestine – colon
Celiac trunk – liver – gallbladder – spleen – stomach – esophagus – duodenum
Inferior phrenics – inferior diaphragm
Lumbars – posterior abdominal wall
Median sacral – sacrum – coccyx
R. common iliac – pelvis and R. lower limb
L. common iliac – pelvis and L. lower limb
Arteries of R. lower limb
Arteries of L. lower limb

FIGURE 32.2 Schematic of the systemic arterial circulation. (R. = right, L. = left)

The **tunica intima,** or **interna,** which lines the lumen of a vessel, is a single thin layer of *endothelium* (squamous cells underlain by a scant basal lamina) that is continuous with the endocardium of the heart. Its cells fit closely together, forming an extremely smooth blood vessel lining that helps decrease resistance to blood flow.

The **tunica media** is the more bulky middle coat and is composed primarily of smooth muscle and elastin. The smooth muscle, under the control of the sympathetic nervous system, plays an active role in regulating the diameter of blood vessels, which in turn alters peripheral resistance and blood pressure.

The **tunica externa,** or **adventitia,** the outermost tunic, is composed of areolar or fibrous connective tissue. Its function is basically supportive and protective.

In general, the walls of arteries are thicker than those of veins. The tunica media in particular tends to be much heavier and contains substantially more smooth muscle and elastic tissue. This anatomical difference reflects a functional difference in the two types of vessels. Arteries, which are closer to the pumping action of the heart, must be able to expand as an increased volume of blood is propelled into them during systole and then recoil passively as the blood flows off into the circulation during diastole. Their walls must be sufficiently strong and resilient to withstand such pressure fluctuations. Since these larger arteries have such large amounts of elastic tissue in their media, they are often referred to as *elastic arteries.* Smaller arteries, further along in the circulatory pathway, are exposed to less extreme pressure fluctuations. They have less elastic tissue but still have substantial amounts of smooth muscle in their media. For this reason, they are called *muscular arteries.* A schematic of the systemic arteries is provided in Figure 32.2.

By contrast, veins, which are far removed from the heart in the circulatory pathway, are not subjected to such pressure fluctuations and are essentially low-pressure vessels. Thus, veins may be thinner-walled without jeopardy. However, the low-pressure condition itself and the fact that blood returning to the heart often flows against gravity require structural modifications to ensure that venous return equals cardiac output. Thus, the lumens of veins tend to be substantially larger than those of corresponding arteries, and valves in larger veins act to prevent backflow of blood in much the same manner as the semilunar valves of the heart. The skeletal muscle "pump" also promotes venous return; as the skeletal muscles surrounding the veins contract and relax, the blood is milked through the veins toward the heart. (Anyone who has been standing relatively still for an extended time has experienced swelling in the ankles, caused by blood pooling in their feet during the period of muscle inactivity.) Pressure changes that occur in the thorax during breathing also aid the return of blood to the heart.

• To demonstrate how efficiently venous valves prevent backflow of blood, perform the following simple experiment. Allow one hand to hang by your side until the blood vessels on the dorsal aspect become distended. Place two fingertips against one of the distended veins and, pressing firmly, move the superior finger proximally along the vein and then release this finger. The vein will remain flattened and collapsed despite gravity. Then remove the distal fingertip and observe the rapid filling of the vein.

The transparent walls of the tiny capillaries are only one cell layer thick, consisting of just the endothelium underlain by a basal lamina, that is, the tunica intima. Because of this exceptional thinness, exchanges are easily made between the blood and tissue cells.

ACTIVITY 1

Examining the Microscopic Structure of Arteries and Veins

1. Obtain a slide showing a cross-sectional view of blood vessels and a microscope.

2. Using Figure 32.1 as a guide, scan the section to identify a thick-walled artery. Very often, but not always, its lumen will appear scalloped due to the constriction of its walls by the elastic tissue of the media.

3. Identify a vein. Its lumen may be elongated or irregularly shaped and collapsed, and its walls will be considerably thinner. Notice the difference in the relative amount of elastic fibers in the media of the two vessels. Also, note the thinness of the intima layer, which is composed of flat squamous-type cells. ■

Major Systemic Arteries of the Body

The **aorta** is the largest artery of the body. Extending upward as the ascending aorta from the left ventricle, it arches posteriorly and to the left (aortic arch) and then courses downward as the descending aorta through the thoracic cavity. It penetrates the diaphragm to enter the abdominal cavity just anterior to the vertebral column.

Figure 32.2 depicts the relationship of the aorta and its major branches. As you locate the arteries on the figure and other anatomical charts and models, be aware of ways in which you can make your memorization task easier. In many cases the name of the artery reflects the body region traversed (axillary, subclavian, brachial, popliteal), the organ served (renal, hepatic), or the bone followed (tibial, femoral, radial, ulnar).

Ascending Aorta

The only branches of the ascending aorta are the **right** and the **left coronary arteries,** which supply the myocardium. The coronary arteries are described in Exercise 30 in conjunction with heart anatomy.

Aortic Arch and Thoracic Aorta

The **brachiocephalic** (literally, "arm-head") **trunk** is the first branch of the aortic arch (Figure 32.3). The other two major arteries branching off the aortic arch are the **left common carotid artery** and the **left subclavian artery.** The brachiocephalic artery persists briefly before dividing into the **right common carotid artery** and the **right subclavian artery.**

Arteries Serving the Head and Neck

The common carotid artery on each side divides to form an **internal** and an external **carotid artery**. The **internal carotid artery** serves the brain and gives rise to the **ophthalmic artery** that supplies orbital structures. The **external carotid artery** supplies the extracranial tissues of the neck and head, largely via its **superficial temporal, maxillary, facial,** and **occipital** arterial branches. (Notice that several arteries are shown in the figure that are not described here. Ask your instructor which arteries you are required to identify.)

The right and left subclavian arteries each give off several branches to the head and neck. The first of these is the **vertebral artery,** which runs up the posterior neck to supply the cerebellum, part of the brain stem, and the posterior cerebral hemispheres. Issuing just lateral to the vertebral artery are the **thyrocervical trunk,** which mainly serves the thyroid gland and some scapular muscles, and the **costocervical trunk,** which supplies deep neck muscles and some of the upper intercostal muscles. In the armpit, the subclavian artery becomes the axillary artery, which serves the upper limb.

Arteries Serving the Brain

A continuous blood supply to the brain is crucial because oxygen deprivation for even a few minutes causes irreparable damage to the delicate brain tissue. The brain is supplied by two pairs of arteries arising from the region of the aortic arch—the internal carotid arteries and the vertebral arteries. Figure 32.3a is a diagram of the brain's arterial supply.

Within the cranium, each internal carotid artery divides into **anterior** and **middle cerebral arteries,** which supply the bulk of the cerebrum. The right and left anterior cerebral arteries are connected by a short shunt called the **anterior communicating artery.** This shunt, along with shunts from each of the middle cerebral arteries, called the **posterior communicating arteries,** contribute to the formation of the **cerebral arterial circle (circle of Willis),** an arterial anastomosis at the base of the brain surrounding the pituitary gland and the optic chiasma.

FIGURE 32.3 Arteries of the head, neck, and brain. **(a)** Right aspect. **(b)** Drawing of the cerebral arteries. Cerebellum is not shown on the left side of the figure. **(c)** Cerebral arterial circle (circle of Willis) in a human brain.

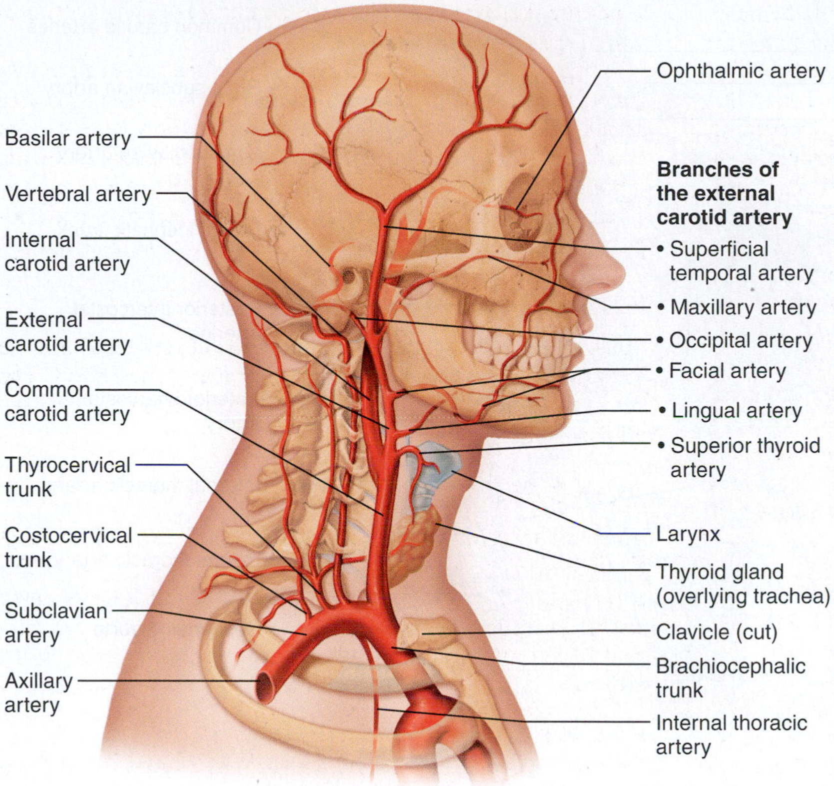
Ophthalmic artery
Basilar artery
Vertebral artery
Internal carotid artery
External carotid artery
Common carotid artery
Thyrocervical trunk
Costocervical trunk
Subclavian artery
Axillary artery
Branches of the external carotid artery
• Superficial temporal artery
• Maxillary artery
• Occipital artery
• Facial artery
• Lingual artery
• Superior thyroid artery
Larynx
Thyroid gland (overlying trachea)
Clavicle (cut)
Brachiocephalic trunk
Internal thoracic artery

(a)

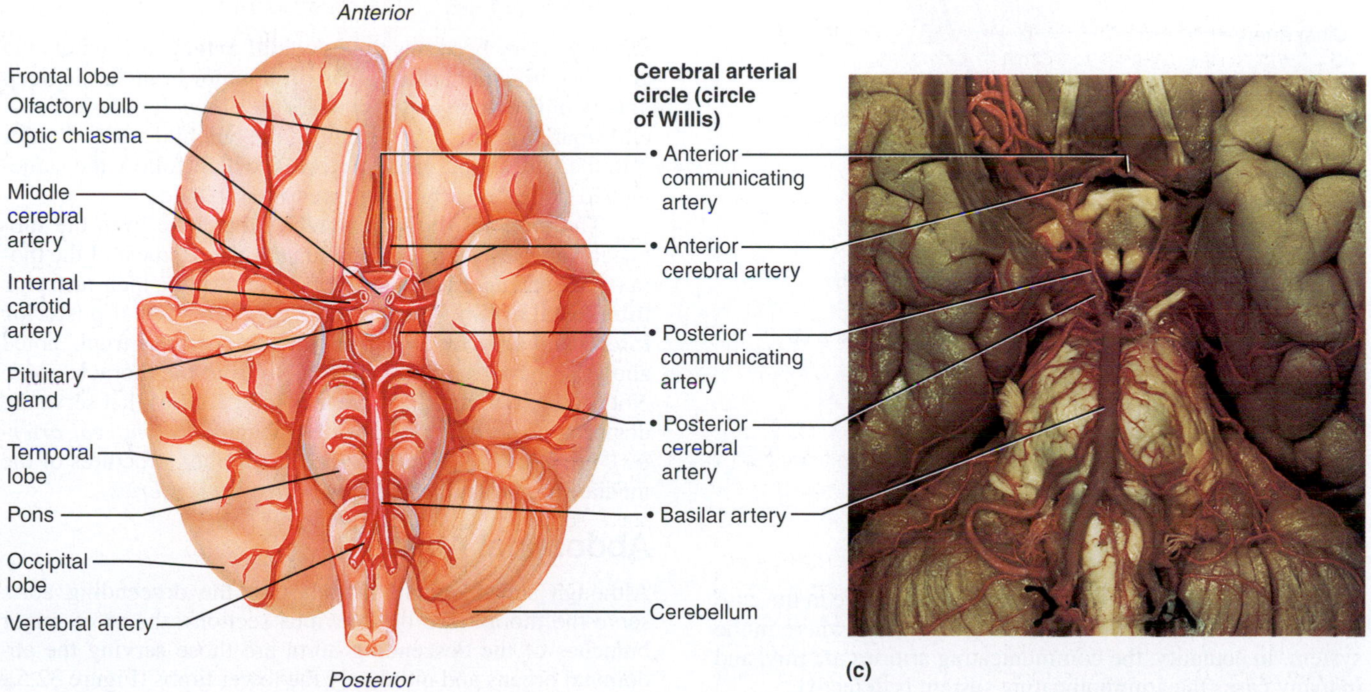
Anterior
Frontal lobe
Olfactory bulb
Optic chiasma
Middle cerebral artery
Internal carotid artery
Pituitary gland
Temporal lobe
Pons
Occipital lobe
Vertebral artery
Posterior
Cerebral arterial circle (circle of Willis)
• Anterior communicating artery
• Anterior cerebral artery
• Posterior communicating artery
• Posterior cerebral artery
• Basilar artery
Cerebellum

(b)

(c)

The paired **vertebral arteries** diverge from the subclavian arteries and pass superiorly through the foramina of the transverse process of the cervical vertebrae to enter the skull through the foramen magnum. Within the skull, the vertebral arteries unite to form a single **basilar artery,** which continues superiorly along the ventral aspect of the brain stem, giving off branches to the pons, cerebellum, and inner ear. At the base of the cerebrum, the basilar artery divides to form the posterior cerebral arteries. These supply portions of the temporal and occipital lobes of the cerebrum and complete the cerebral arterial circle posteriorly.

The uniting of the blood supply of the internal carotid arteries and the vertebral arteries via the cerebral arterial circle is a protective device that theoretically provides an alternative

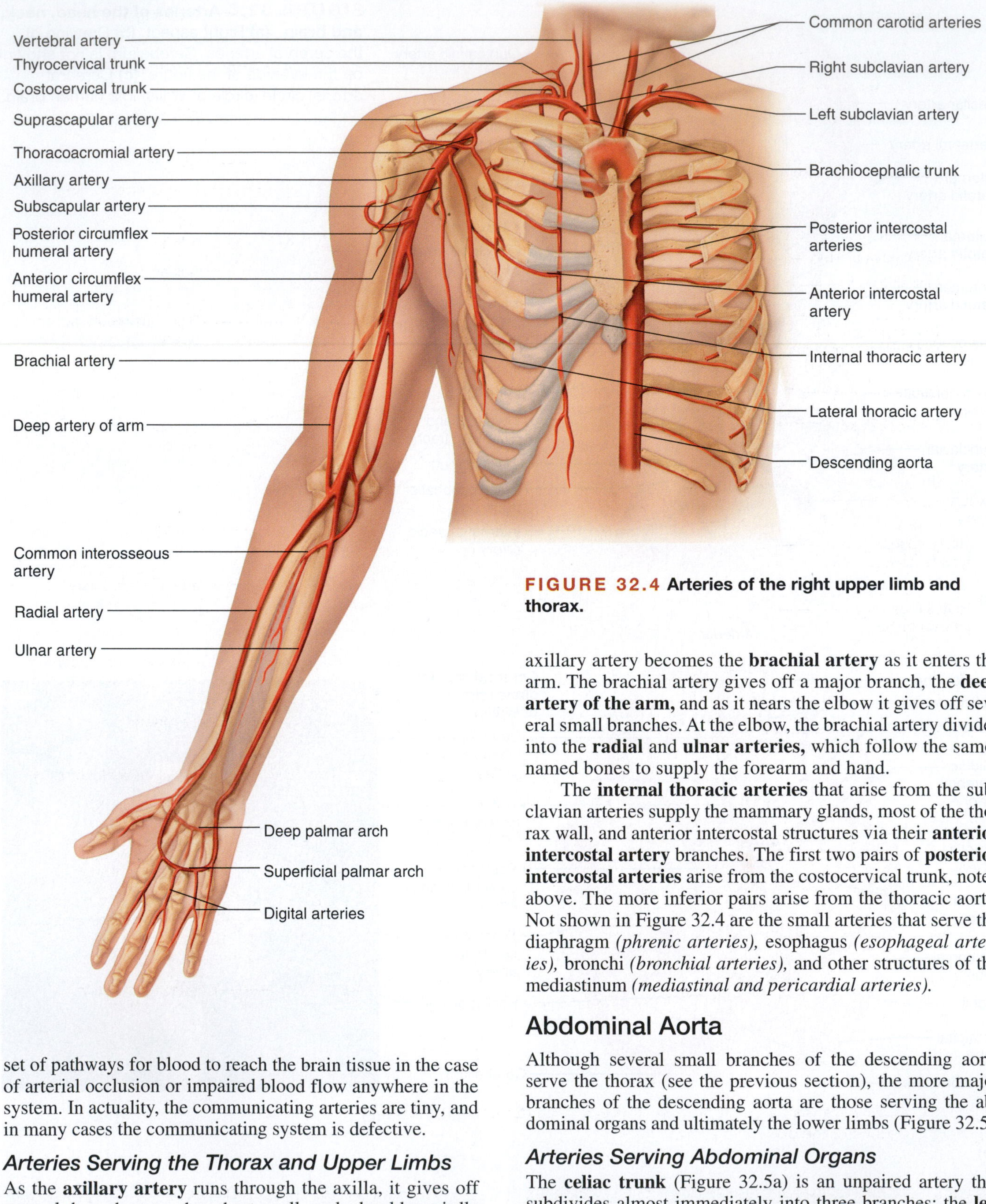

Vertebral artery
Thyrocervical trunk
Costocervical trunk
Suprascapular artery
Thoracoacromial artery
Axillary artery
Subscapular artery
Posterior circumflex humeral artery
Anterior circumflex humeral artery
Brachial artery
Deep artery of arm
Common interosseous artery
Radial artery
Ulnar artery
Deep palmar arch
Superficial palmar arch
Digital arteries
Common carotid arteries
Right subclavian artery
Left subclavian artery
Brachiocephalic trunk
Posterior intercostal arteries
Anterior intercostal artery
Internal thoracic artery
Lateral thoracic artery
Descending aorta

FIGURE 32.4 Arteries of the right upper limb and thorax.

set of pathways for blood to reach the brain tissue in the case of arterial occlusion or impaired blood flow anywhere in the system. In actuality, the communicating arteries are tiny, and in many cases the communicating system is defective.

Arteries Serving the Thorax and Upper Limbs

As the **axillary artery** runs through the axilla, it gives off several branches to the chest wall and shoulder girdle (Figure 32.4). These include the **thoracoacromial artery** (to shoulder and pectoral region), the **lateral thoracic artery** (lateral chest wall), the **subscapular artery** (to scapula and dorsal thorax), and the **anterior** and **posterior circumflex humeral arteries** (to the shoulder and the deltoid muscle). At the inferior edge of the teres major muscle, the axillary artery becomes the **brachial artery** as it enters the arm. The brachial artery gives off a major branch, the **deep artery of the arm,** and as it nears the elbow it gives off several small branches. At the elbow, the brachial artery divides into the **radial** and **ulnar arteries,** which follow the same-named bones to supply the forearm and hand.

The **internal thoracic arteries** that arise from the subclavian arteries supply the mammary glands, most of the thorax wall, and anterior intercostal structures via their **anterior intercostal artery** branches. The first two pairs of **posterior intercostal arteries** arise from the costocervical trunk, noted above. The more inferior pairs arise from the thoracic aorta. Not shown in Figure 32.4 are the small arteries that serve the diaphragm *(phrenic arteries),* esophagus *(esophageal arteries),* bronchi *(bronchial arteries),* and other structures of the mediastinum *(mediastinal and pericardial arteries).*

Abdominal Aorta

Although several small branches of the descending aorta serve the thorax (see the previous section), the more major branches of the descending aorta are those serving the abdominal organs and ultimately the lower limbs (Figure 32.5).

Arteries Serving Abdominal Organs

The **celiac trunk** (Figure 32.5a) is an unpaired artery that subdivides almost immediately into three branches: the **left gastric artery** supplying the stomach, the **splenic artery** supplying the spleen, and the **common hepatic artery,** which runs superiorly and gives off branches to the stomach **(right gastric artery),** duodenum, and pancreas. Where the **gastroduodenal artery** branches off, the common hepatic artery becomes the **hepatic artery proper,** which serves the liver. The **right** and **left gastroepiploic arteries,** branches of

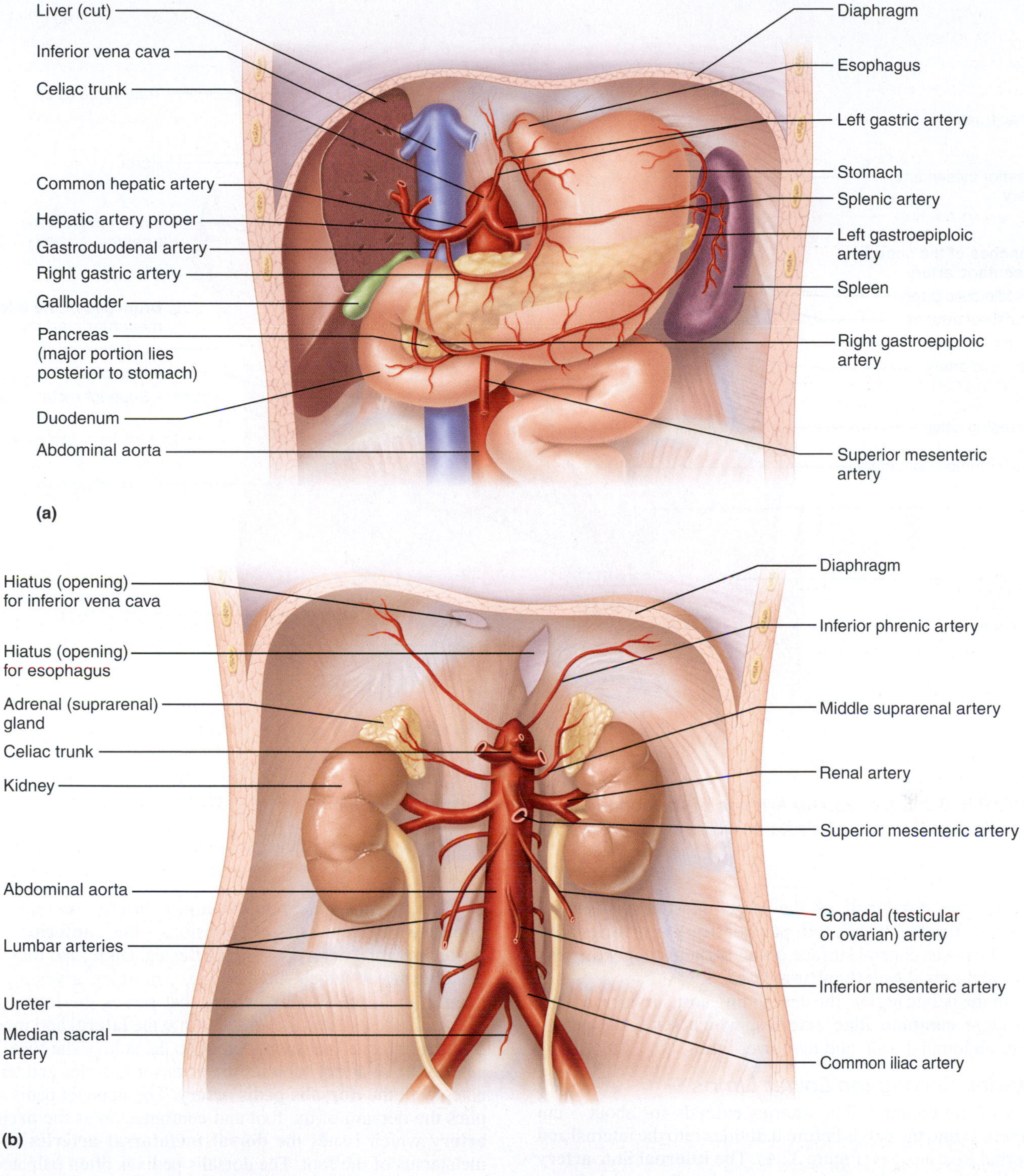
Liver (cut)
Inferior vena cava
Celiac trunk
Common hepatic artery
Hepatic artery proper
Gastroduodenal artery
Right gastric artery
Gallbladder
Pancreas (major portion lies posterior to stomach)
Duodenum
Abdominal aorta
Diaphragm
Esophagus
Left gastric artery
Stomach
Splenic artery
Left gastroepiploic artery
Spleen
Right gastroepiploic artery
Superior mesenteric artery
(a)
Hiatus (opening) for inferior vena cava
Hiatus (opening) for esophagus
Adrenal (suprarenal) gland
Celiac trunk
Kidney
Abdominal aorta
Lumbar arteries
Ureter
Median sacral artery
Diaphragm
Inferior phrenic artery
Middle suprarenal artery
Renal artery
Superior mesenteric artery
Gonadal (testicular or ovarian) artery
Inferior mesenteric artery
Common iliac artery
(b)

FIGURE 32.5 Arteries of the abdomen. **(a)** The celiac trunk and its major branches. **(b)** Major branches of the abdominal aorta.

the gastroduodenal and splenic arteries respectively, serve the left (greater) curvature of the stomach.

The largest branch of the abdominal aorta, the **superior mesenteric artery** (Figure 32.5b and 32.5c), supplies most of the small intestine (via the intestinal arteries) and the first half of the large intestine (via the ileocolic and colic arteries). Flanking the superior mesenteric artery on the left and right are the **middle suprarenal arteries** serving the adrenal glands that sit atop the kidneys.

The paired **renal arteries** (Figure 32.5b) supply the kidneys, and the **gonadal arteries,** arising from the ventral aortic surface just below the renal arteries, run inferiorly to serve the gonads. They are called **ovarian arteries** in the female and **testicular arteries** in the male. Since these vessels must travel through the inguinal canal to supply the testes, they are considerably longer in the male than the female.

The final major branch of the abdominal aorta is the **inferior mesenteric artery** (Figure 32.5b and 32.5c), which

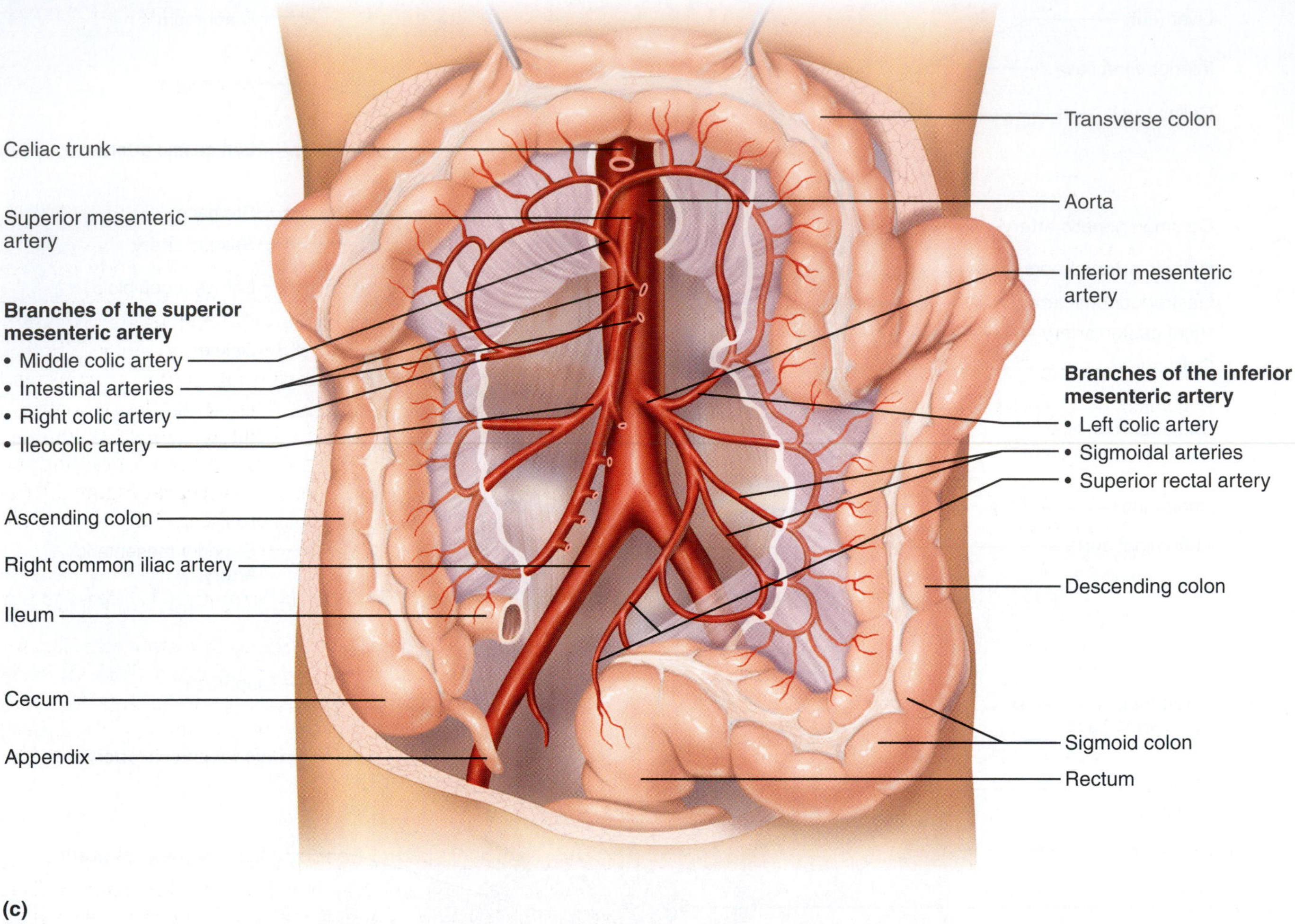
Celiac trunk
Superior mesenteric artery
Branches of the superior mesenteric artery
• Middle colic artery
• Intestinal arteries
• Right colic artery
• Ileocolic artery
Ascending colon
Right common iliac artery
Ileum
Cecum
Appendix
Transverse colon
Aorta
Inferior mesenteric artery
Branches of the inferior mesenteric artery
• Left colic artery
• Sigmoidal arteries
• Superior rectal artery
Descending colon
Sigmoid colon
Rectum
(c)

FIGURE 32.5 ***(continued)*** **Arteries of the abdomen. (c)** Distribution of the superior and inferior mesenteric arteries, transverse colon pulled superiorly.

supplies the distal half of the large intestine via several branches. Just below this, four pairs of **lumbar arteries** arise from the posterolateral surface of the aorta to supply the posterior abdominal wall (lumbar region).

In the pelvic region, the descending aorta divides into the two large **common iliac arteries,** which serve the pelvis, lower abdominal wall, and the lower limbs.

Arteries Serving the Lower Limbs

Each of the common iliac arteries extends for about 5 cm (2 inches) into the pelvis before it divides into the internal and external iliac arteries (Figure 32.6). The **internal iliac artery** supplies the gluteal muscles via the **superior** and **inferior gluteal arteries** and the adductor muscles of the medial thigh via the **obturator artery,** as well as the external genitalia and perineum (via the *internal pudendal artery,* not illustrated).

The **external iliac artery** supplies the anterior abdominal wall and the lower limb. As it continues into the thigh, its name changes to **femoral artery.** Proximal branches of the femoral artery, the **circumflex femoral arteries,** supply the head and neck of the femur and the hamstring muscles. The femoral artery gives off a deep branch, the **deep artery of the thigh** (also called the **deep femoral artery**), which is the main supply to the thigh muscles (hamstrings, quadriceps, and adductors). In the knee region, the femoral artery briefly becomes the **popliteal artery;** its subdivisions—the **anterior** and **posterior tibial arteries**—supply the leg, ankle, and foot. The posterior tibial, which supplies flexor muscles, gives off one main branch, the **fibular artery,** that serves the lateral calf (fibular muscles), and then divides into the **lateral** and **medial plantar arteries** that supply blood to the sole of the foot. The anterior tibial artery supplies the extensor muscles and terminates with the **dorsalis pedis artery.** The dorsalis pedis supplies the dorsum of the foot and continues on as the **arcuate artery** which issues the **dorsal metatarsal arteries** to the metatarsus of the foot. The dorsalis pedis is often palpated in patients with circulation problems of the leg to determine the circulatory efficiency to the limb as a whole.

- Palpate your own dorsalis pedis artery.

ACTIVITY 2

Locating Arteries on an Anatomical Chart or Model

Now that you have identified the arteries on Figures 32.2–32.6, attempt to locate and name them (without a reference) on a

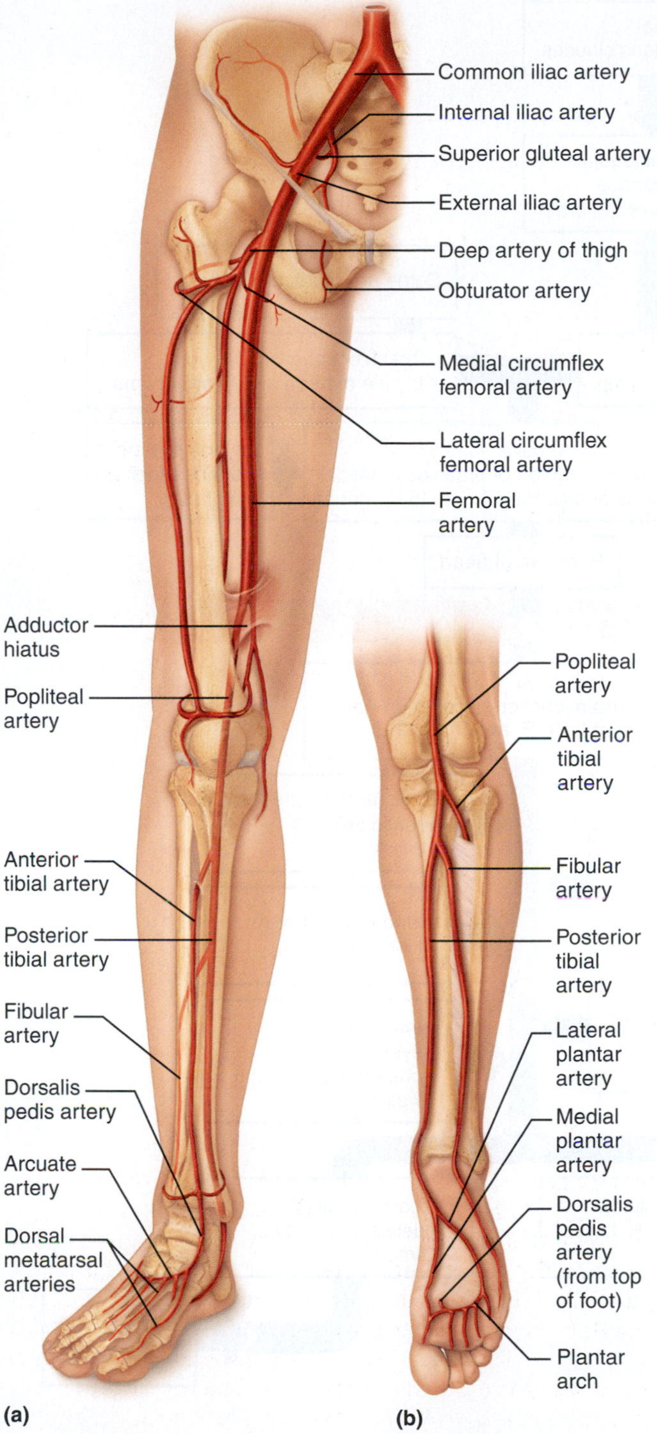
Common iliac artery
Internal iliac artery
Superior gluteal artery
External iliac artery
Deep artery of thigh
Obturator artery
Medial circumflex femoral artery
Lateral circumflex femoral artery
Femoral artery
Adductor hiatus
Popliteal artery
Anterior tibial artery
Posterior tibial artery
Fibular artery
Dorsalis pedis artery
Arcuate artery
Dorsal metatarsal arteries
(a)
Popliteal artery
Anterior tibial artery
Fibular artery
Posterior tibial artery
Lateral plantar artery
Medial plantar artery
Dorsalis pedis artery (from top of foot)
Plantar arch
(b)

FIGURE 32.6 Arteries of the right pelvis and lower limb. (a) Anterior view. **(b)** Posterior view.

large anatomical chart or three-dimensional model of the vascular system. ■

Major Systemic Veins of the Body

Arteries are generally located in deep, well-protected body areas. However, many veins follow a more superficial course and are often easily seen and palpated on the body surface. Most deep veins parallel the course of the major arteries, and in many cases the naming of the veins and arteries is identical except for the designation of the vessels as veins. Whereas the major systemic arteries branch off the aorta, the veins tend to converge on the venae cavae, which enter the right atrium of the heart. Veins draining the head and upper extremities empty into the **superior vena cava,** and those draining the lower body empty into the **inferior vena cava.** Figure 32.7 is a schematic of the systemic veins and their relationship to the venae cavae to get you started.

Veins Draining into the Inferior Vena Cava

The inferior vena cava, a much longer vessel than the superior vena cava, returns blood to the heart from all body regions below the diaphragm (see Figure 32.7). It begins in the lower abdominal region with the union of the paired **common iliac veins,** which drain venous blood from the legs and pelvis.

Veins of the Lower Limbs

Each common iliac vein is formed by the union of the **internal iliac vein,** draining the pelvis, and the **external iliac vein,** which receives venous blood from the lower limb (Figure 32.8). Veins of the leg include the **anterior** and **posterior tibial veins,** which serve the calf and foot. The anterior tibial vein is a superior continuation of the **dorsalis pedis vein** of the foot. The posterior tibial vein is formed by the union of the **medial** and **lateral plantar veins,** and ascends deep in the calf muscles. It receives the **fibular vein** in the calf and then joins with the anterior tibial vein at the knee to produce the **popliteal vein,** which crosses the back of the knee. The popliteal vein becomes the **femoral vein** in the thigh; the femoral vein in turn becomes the external iliac vein in the inguinal region.

The **great saphenous vein,** a superficial vein, is the longest vein in the body. Beginning in common with the **small saphenous vein** from the **dorsal venous arch,** it extends up the medial side of the leg, knee, and thigh to empty into the femoral vein. The small saphenous vein runs along the lateral aspect of the foot and through the calf muscle, which it drains, and then empties into the popliteal vein at the knee (Figure 32.8b).

Veins of the Abdomen

Moving superiorly in the abdominal cavity (Figure 32.9), the inferior vena cava receives blood from the posterior abdominal wall via several pairs of **lumbar veins,** and from the right ovary or testis via the **right gonadal vein.** (The **left gonadal [ovarian or testicular] vein** drains into the left renal vein superiorly.) The paired **renal veins** drain the kidneys. Just above the right renal vein, the **right suprarenal vein** (receiving blood from the adrenal gland on the same side) drains into the inferior vena cava, but its partner, the **left suprarenal**

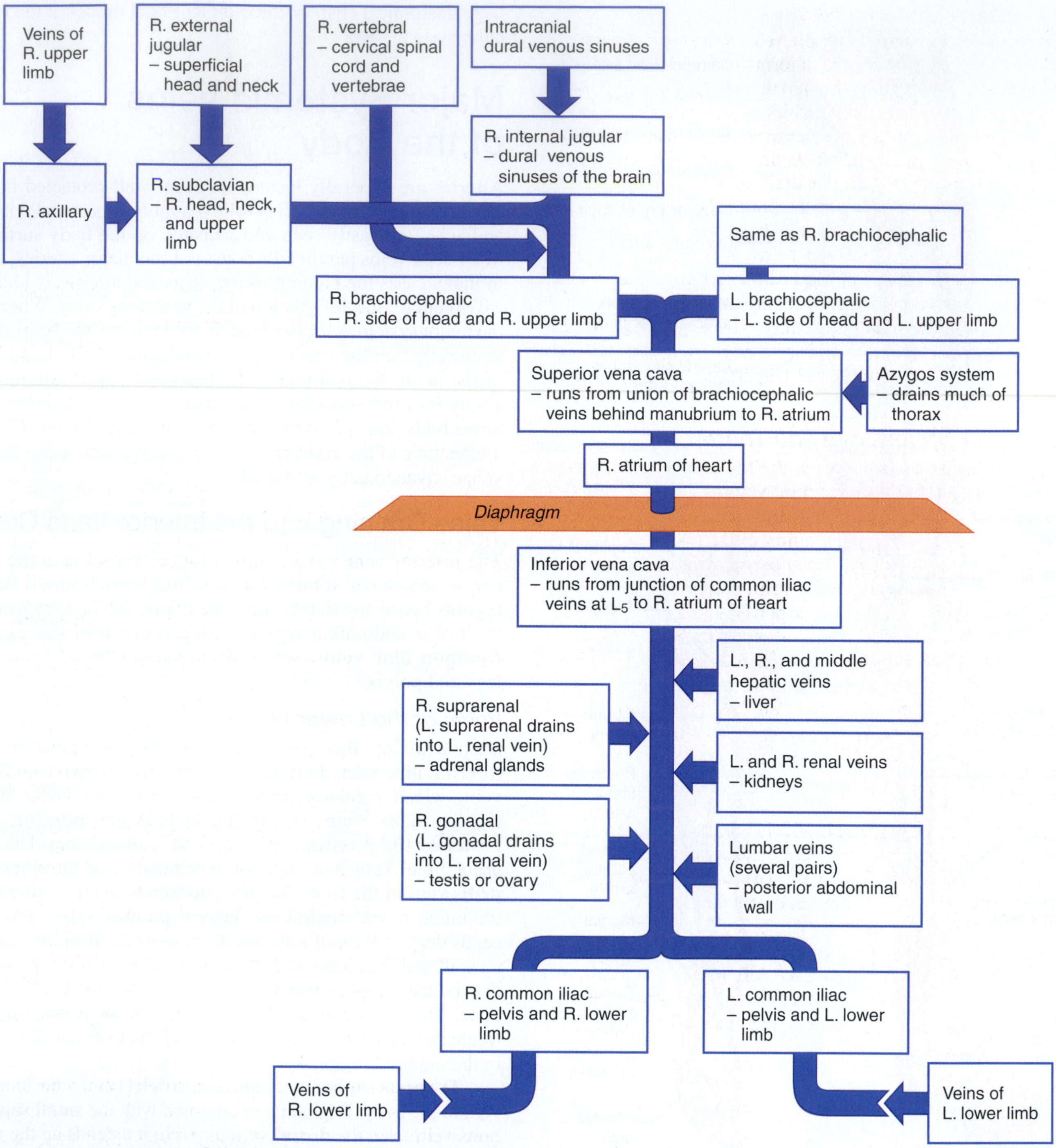
Veins of R. upper limb
R. external jugular – superficial head and neck
R. vertebral – cervical spinal cord and vertebrae
Intracranial dural venous sinuses
R. internal jugular – dural venous sinuses of the brain
R. axillary
R. subclavian – R. head, neck, and upper limb
Same as R. brachiocephalic
R. brachiocephalic – R. side of head and R. upper limb
L. brachiocephalic – L. side of head and L. upper limb
Superior vena cava – runs from union of brachiocephalic veins behind manubrium to R. atrium
Azygos system – drains much of thorax
R. atrium of heart
Diaphragm
Inferior vena cava – runs from junction of common iliac veins at L_5 to R. atrium of heart
L., R., and middle hepatic veins – liver
R. suprarenal (L. suprarenal drains into L. renal vein) – adrenal glands
L. and R. renal veins – kidneys
R. gonadal (L. gonadal drains into L. renal vein) – testis or ovary
Lumbar veins (several pairs) – posterior abdominal wall
R. common iliac – pelvis and R. lower limb
L. common iliac – pelvis and L. lower limb
Veins of R. lower limb
Veins of L. lower limb

FIGURE 32.7 Schematic of systemic venous circulation.

vein, empties into the left renal vein inferiorly. The **hepatic veins** drain the liver. The unpaired veins draining the digestive tract organs empty into a special vessel, the hepatic portal vein, which carries blood to the liver to be processed before it enters the systemic venous system. (The hepatic portal system is discussed separately on page 484.)

Veins Draining into the Superior Vena Cava

Veins draining into the superior vena cava are named from the superior vena cava distally, *but remember that the flow of blood is in the opposite direction*.

Veins of the Head and Neck

The **right** and **left brachiocephalic veins** drain the head, neck, and upper extremities and unite to form the superior vena cava (Figure 32.10). Notice that although there is only one brachiocephalic artery, there are two brachiocephalic veins.

Branches of the brachiocephalic veins include the internal jugular, vertebral, and subclavian veins. The **internal jugular veins** are large veins that drain the superior sagittal sinus and other **dural sinuses** of the brain. As they run inferiorly, they receive blood from the head and neck via the **superficial temporal** and **facial veins.** The **vertebral veins** drain the posterior aspect of the head including the cervical vertebrae and spinal cord. The **subclavian veins** receive venous blood from the upper extremity. The **external jugular vein** joins the subclavian vein near its origin to return the venous drainage of the extracranial (superficial) tissues of the head and neck.

Veins of the Upper Limb and Thorax

As the subclavian vein traverses the axilla, it becomes the **axillary vein** and then the **brachial vein** as it courses along the posterior aspect of the humerus (Figure 32.11). The brachial vein is formed by the union of the deep **radial** and **ulnar veins** of the forearm. The superficial venous drainage of the arm includes the **cephalic vein,** which courses along the lateral aspect of the arm and empties into the axillary vein; the **basilic vein,** found on the medial aspect of the arm and entering the brachial vein; and the **median cubital vein,** which runs between the cephalic and basilic veins in the anterior aspect of the elbow (this vein is often the site of choice for removing blood for testing purposes). The **median antebrachial vein** lies between the radial and ulnar veins, and terminates variably by entering the cephalic or basilic vein at the elbow.

The **azygos system** (Figure 32.11) drains the intercostal muscles of the thorax and provides an accessory venous system to drain the abdominal wall. The **azygos vein,** which drains the right side of the thorax, enters the dorsal aspect of the superior vena cava immediately before that vessel enters the right atrium. Also part of the azygos system are the **hemiazygos** (a continuation of the **left ascending lumbar vein** of the abdomen) and the **accessory hemiazygos veins,** which together drain the left side of the thorax and empty into the azygos vein.

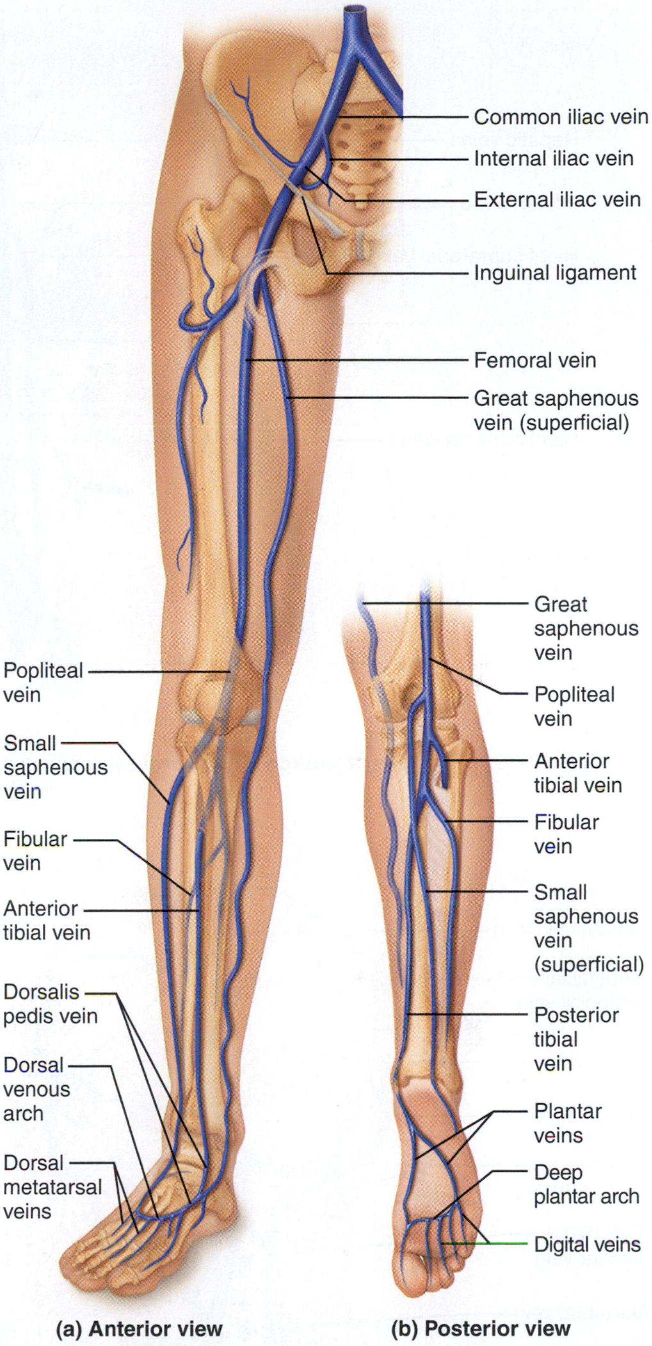
Common iliac vein
Internal iliac vein
External iliac vein
Inguinal ligament
Femoral vein
Great saphenous vein (superficial)
Popliteal vein
Small saphenous vein
Fibular vein
Anterior tibial vein
Dorsalis pedis vein
Dorsal venous arch
Dorsal metatarsal veins
Great saphenous vein
Popliteal vein
Anterior tibial vein
Fibular vein
Small saphenous vein (superficial)
Posterior tibial vein
Plantar veins
Deep plantar arch
Digital veins
(a) Anterior view
(b) Posterior view

FIGURE 32.8 Veins of the right pelvis and lower limb. **(a)** Anterior view. **(b)** Posterior view.

ACTIVITY 3

Identifying the Systemic Veins

Identify the important veins of the systemic circulation on the large anatomical chart or model without referring to the figures. ■

Special Circulations

Pulmonary Circulation

The pulmonary circulation (discussed previously in relation to heart anatomy on page 446) differs in many ways from systemic circulation because it does not serve the metabolic needs of the body tissues with which it is associated (in this case, lung tissue). It functions instead to bring the blood into close contact with the alveoli of the lungs to permit gas exchanges that rid the blood of excess carbon dioxide and replenish its supply of vital oxygen. The arteries of the pulmonary circulation are structurally much like veins, and they create a low-pressure bed in the lungs. (If the arterial pressure in the systemic circulation is 120/80, the pressure in the pulmonary artery is likely to be approximately 24/8.) The functional blood supply of the lungs is provided by the **bronchial arteries** (not shown), which diverge from the thoracic portion of the descending aorta.

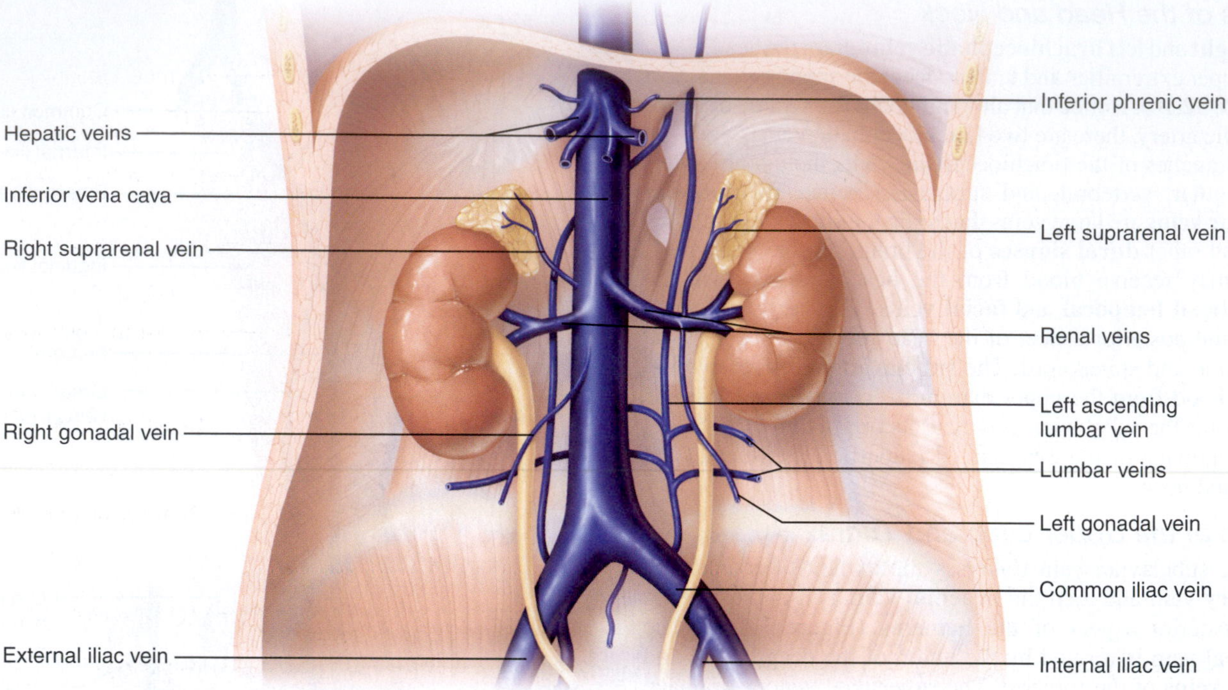
Inferior phrenic vein
Hepatic veins
Inferior vena cava
Right suprarenal vein
Left suprarenal vein
Renal veins
Left ascending lumbar vein
Right gonadal vein
Lumbar veins
Left gonadal vein
Common iliac vein
External iliac vein
Internal iliac vein

FIGURE 32.9 Venous drainage of abdominal organs not drained by the hepatic portal vein.

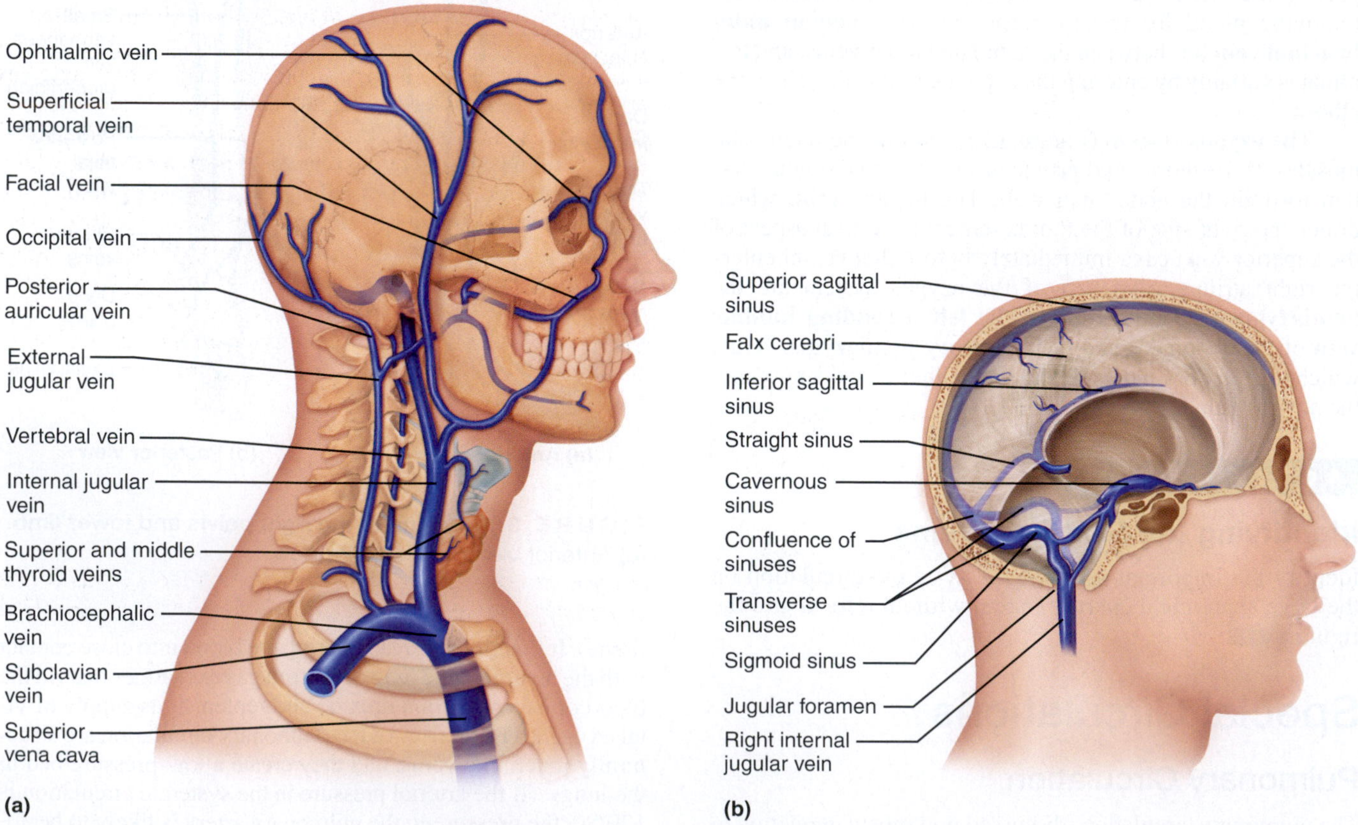
Ophthalmic vein
Superficial temporal vein
Facial vein
Occipital vein
Posterior auricular vein
External jugular vein
Vertebral vein
Internal jugular vein
Superior and middle thyroid veins
Brachiocephalic vein
Subclavian vein
Superior vena cava
(a)
Superior sagittal sinus
Falx cerebri
Inferior sagittal sinus
Straight sinus
Cavernous sinus
Confluence of sinuses
Transverse sinuses
Sigmoid sinus
Jugular foramen
Right internal jugular vein
(b)

FIGURE 32.10 Venous drainage of the head, neck, and brain. (a) Veins of the head and neck, right superficial aspect. **(b)** Dural sinuses of the brain, right aspect.

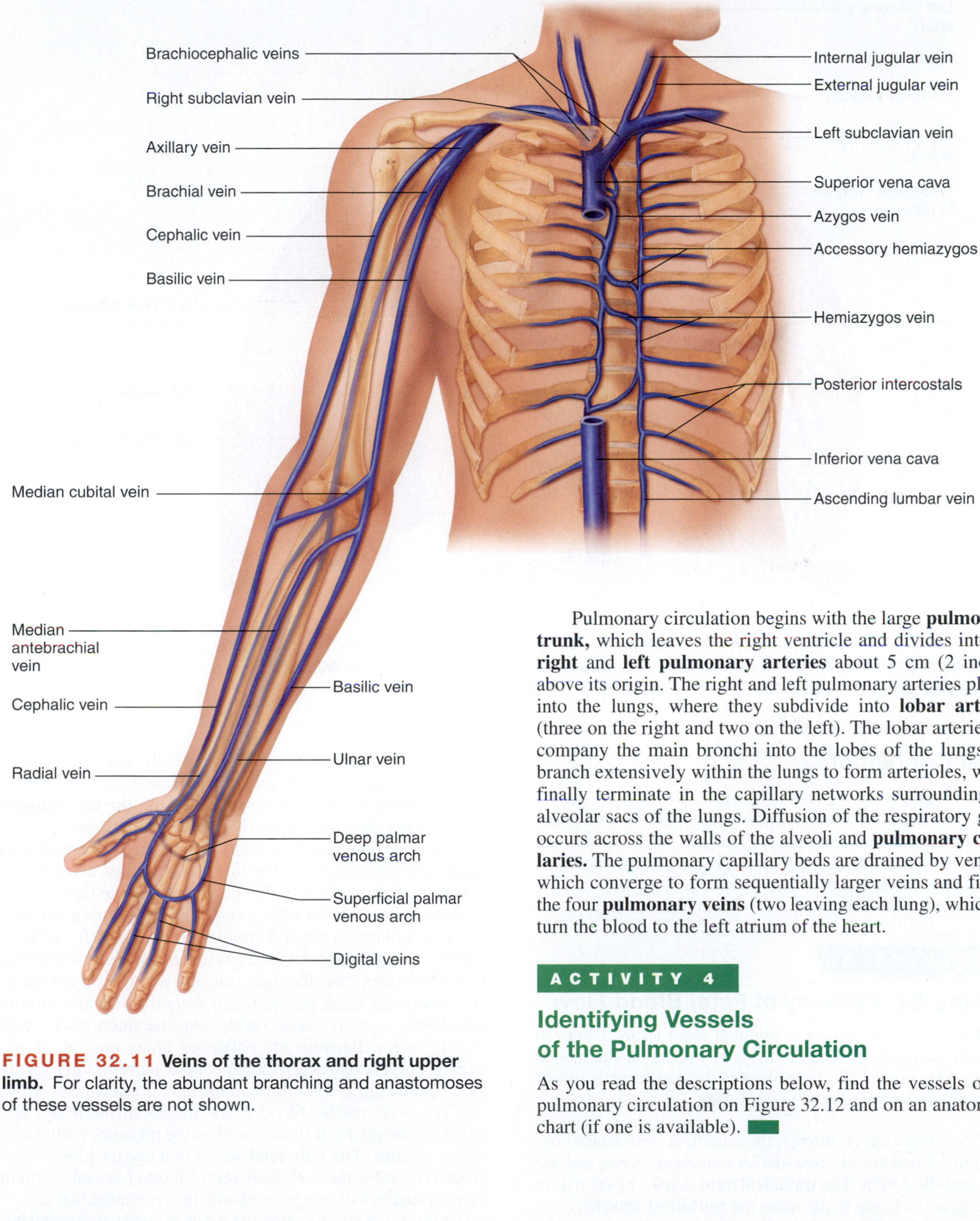
Brachiocephalic veins
Right subclavian vein
Axillary vein
Brachial vein
Cephalic vein
Basilic vein
Median cubital vein
Median antebrachial vein
Cephalic vein
Radial vein
Basilic vein
Ulnar vein
Deep palmar venous arch
Superficial palmar venous arch
Digital veins
Internal jugular vein
External jugular vein
Left subclavian vein
Superior vena cava
Azygos vein
Accessory hemiazygos vein
Hemiazygos vein
Posterior intercostals
Inferior vena cava
Ascending lumbar vein

FIGURE 32.11 Veins of the thorax and right upper limb. For clarity, the abundant branching and anastomoses of these vessels are not shown.

Pulmonary circulation begins with the large **pulmonary trunk,** which leaves the right ventricle and divides into the **right** and **left pulmonary arteries** about 5 cm (2 inches) above its origin. The right and left pulmonary arteries plunge into the lungs, where they subdivide into **lobar arteries** (three on the right and two on the left). The lobar arteries accompany the main bronchi into the lobes of the lungs and branch extensively within the lungs to form arterioles, which finally terminate in the capillary networks surrounding the alveolar sacs of the lungs. Diffusion of the respiratory gases occurs across the walls of the alveoli and **pulmonary capillaries.** The pulmonary capillary beds are drained by venules, which converge to form sequentially larger veins and finally the four **pulmonary veins** (two leaving each lung), which return the blood to the left atrium of the heart.

ACTIVITY 4

Identifying Vessels of the Pulmonary Circulation

As you read the descriptions below, find the vessels of the pulmonary circulation on Figure 32.12 and on an anatomical chart (if one is available).

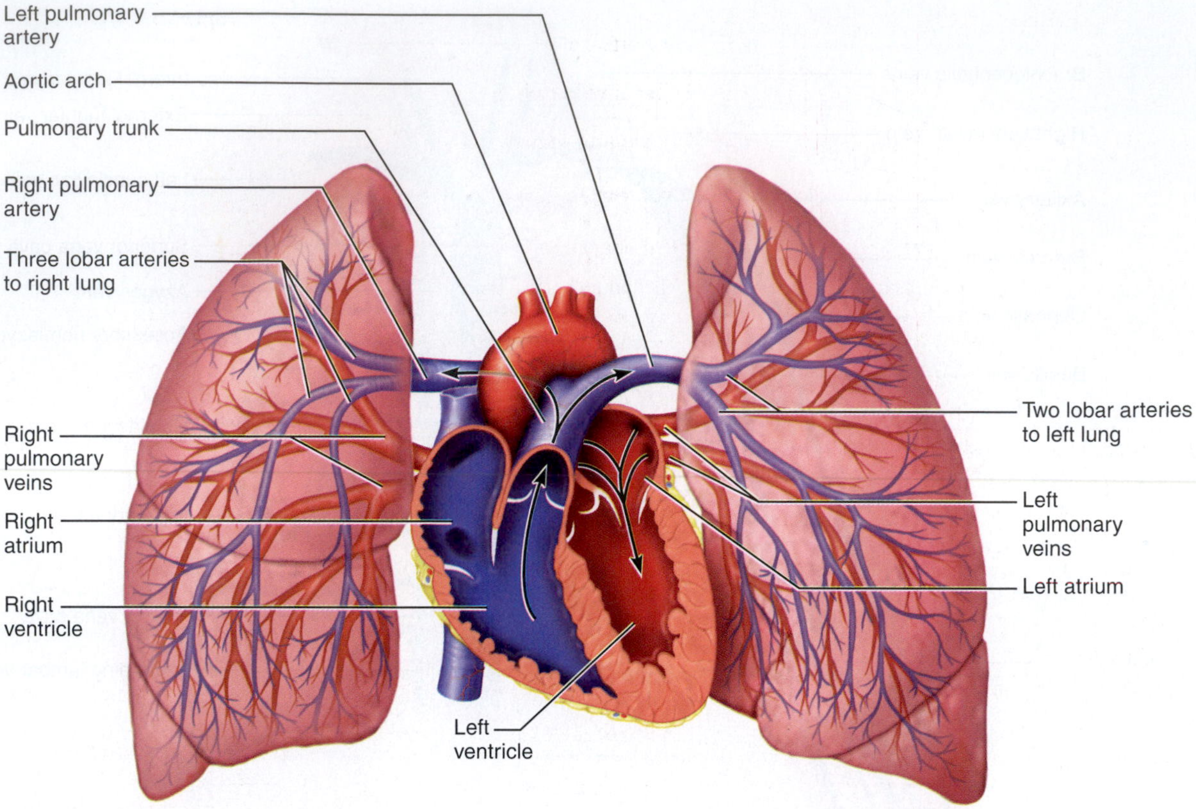
Left pulmonary artery
Aortic arch
Pulmonary trunk
Right pulmonary artery
Three lobar arteries to right lung
Right pulmonary veins
Right atrium
Right ventricle
Left ventricle
Two lobar arteries to left lung
Left pulmonary veins
Left atrium

FIGURE 32.12 The pulmonary circulation. The pulmonary arterial system is shown in blue to indicate that the blood carried is oxygen-poor. The pulmonary venous drainage is shown in red to indicate that the blood transported is oxygen-rich.

Fetal Circulation

In a developing fetus, the lungs and digestive system are not yet functional, and all nutrient, excretory, and gaseous exchanges occur through the placenta (see Figure 32.14a). Nutrients and oxygen move across placental barriers from the mother's blood into fetal blood, and carbon dioxide and other metabolic wastes move from the fetal blood supply to the mother's blood.

ACTIVITY 5

Tracing the Pathway of Fetal Blood Flow

Use Figure 32.13a and an anatomical chart (if available) to trace the pathway of fetal blood flow. Locate all the named vessels. Use Figure 32.13b to identify the named remnants of the foramen ovale and fetal vessels.

Fetal blood travels through the umbilical cord, which contains three blood vessels: two smaller umbilical arteries and one large umbilical vein. The **umbilical vein** carries blood rich in nutrients and oxygen to the fetus; the **umbilical arteries** carry carbon dioxide and waste-laden blood from the fetus to the placenta. The umbilical arteries, which transport blood away from the fetal heart, meet the umbilical vein at the *umbilicus* (navel, or belly button) and wrap around the vein within the cord en route to their placental attachments. Newly oxygenated blood flows in the umbilical vein superiorly toward the fetal heart. Some of this blood perfuses the liver, but the larger proportion is ducted through the relatively nonfunctional liver to the inferior vena cava via a shunt vessel called the **ductus venosus,** which carries the blood to the right atrium of the heart.

Because fetal lungs are nonfunctional and collapsed, two shunting mechanisms ensure that blood almost entirely bypasses the lungs. Much of the blood entering the right atrium is shunted into the left atrium through the **foramen ovale,** a flaplike opening in the interatrial septum. The left ventricle then pumps the blood out the aorta to the systemic circulation. Blood that does enter the right ventricle and is pumped out of the pulmonary trunk encounters a second shunt, the **ductus arteriosus,** a short vessel connecting the pulmonary trunk and the aorta. Because the collapsed lungs present an extremely high-resistance pathway, blood more readily enters the systemic circulation through the ductus arteriosus.

The aorta carries blood to the tissues of the body; this blood ultimately finds its way back to the placenta via the umbilical arteries. The only fetal vessel that carries highly oxygenated blood is the umbilical vein. All other vessels contain varying degrees of oxygenated and deoxygenated blood.

At birth, or shortly after, the foramen ovale closes and becomes the **fossa ovalis,** and the ductus arteriosus collapses and is converted to the fibrous **ligamentum arteriosum** (Figure 32.13b). Lack of blood flow through the umbilical vessels leads to their eventual obliteration, and the circulatory pattern becomes that of the adult. Remnants of the umbilical

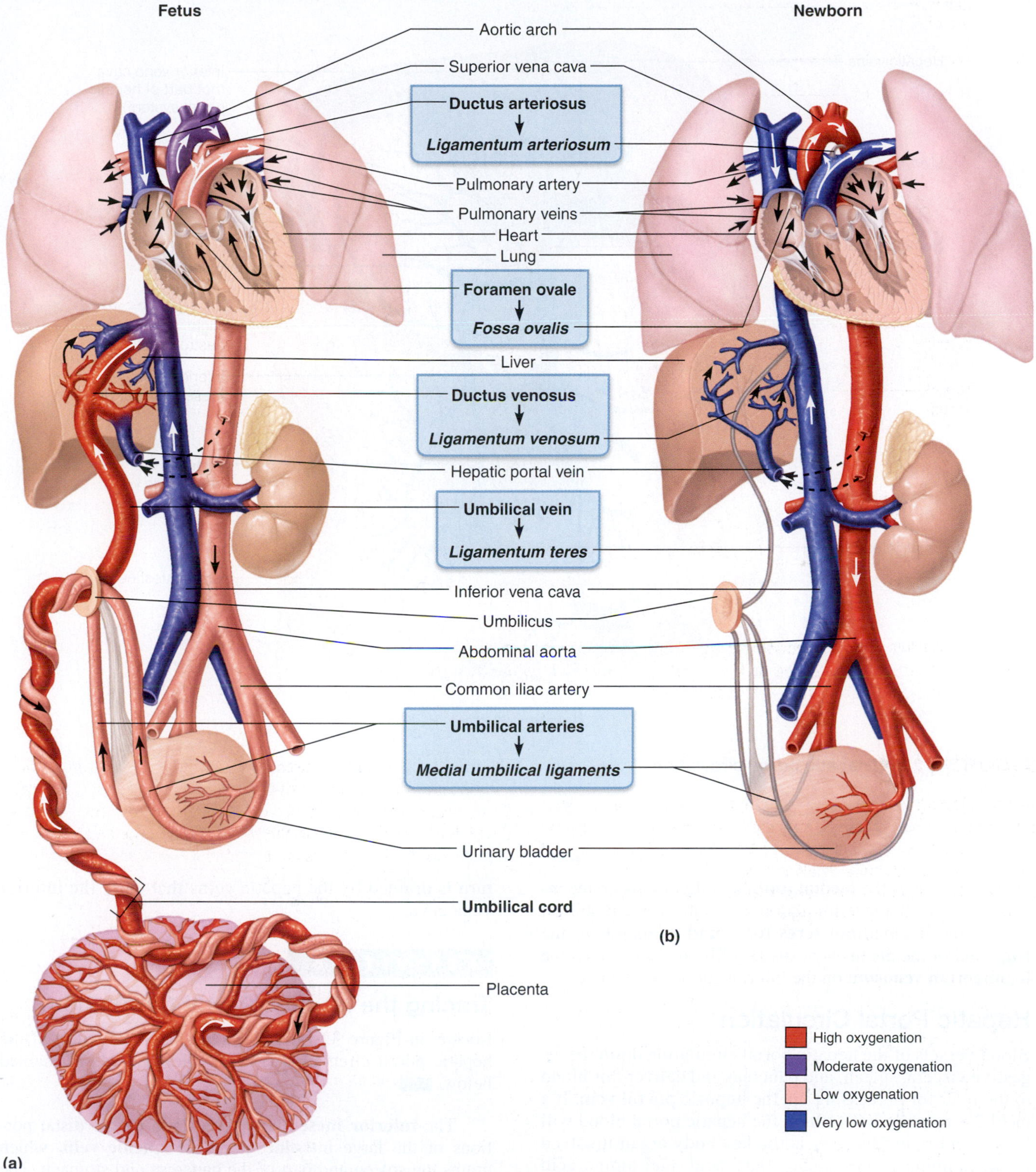

Fetus
Newborn
Aortic arch
Superior vena cava
Ductus arteriosus
Ligamentum arteriosum
Pulmonary artery
Pulmonary veins
Heart
Lung
Foramen ovale
Fossa ovalis
Liver
Ductus venosus
Ligamentum venosum
Hepatic portal vein
Umbilical vein
Ligamentum teres
Inferior vena cava
Umbilicus
Abdominal aorta
Common iliac artery
Umbilical arteries
Medial umbilical ligaments
Urinary bladder
Umbilical cord
Placenta
(a)
(b)
High oxygenation
Moderate oxygenation
Low oxygenation
Very low oxygenation

FIGURE 32.13 Circulation in fetus and newborn. Arrows indicate direction of blood flow. Arrows in the blue boxes go from the fetal structure to what it becomes after birth. **(a)** Special adaptations for embryonic and fetal life. The umbilical vein (red) carries oxygen- and nutrient-rich blood from the placenta to the fetus. The umbilical arteries (pink) carry waste-laden blood from the fetus to the placenta. **(b)** Changes in the cardiovascular system at birth. The umbilical vessels are occluded, as are the liver and lung bypasses (ductus venosus and arteriosus, and the foramen ovale).

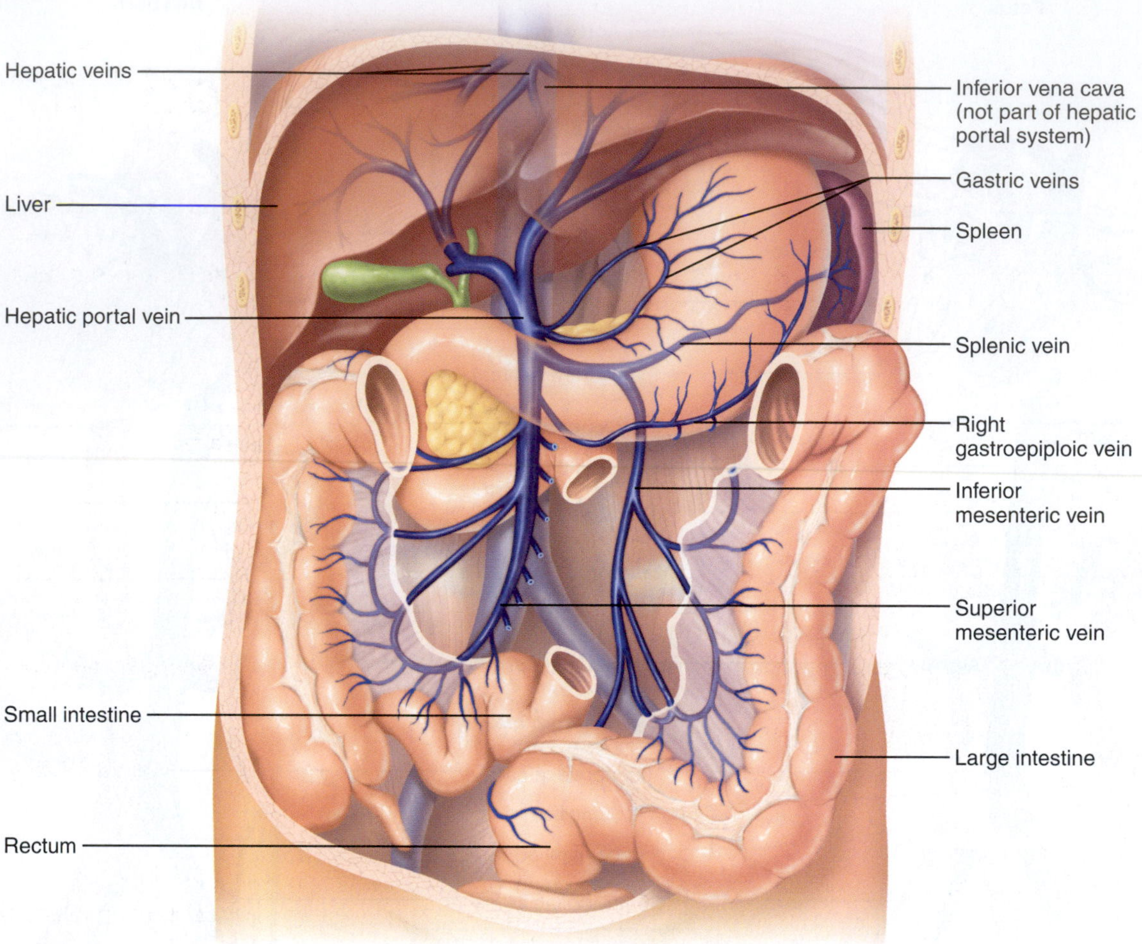
Hepatic veins
Liver
Hepatic portal vein
Small intestine
Rectum
Inferior vena cava (not part of hepatic portal system)
Gastric veins
Spleen
Splenic vein
Right gastroepiploic vein
Inferior mesenteric vein
Superior mesenteric vein
Large intestine

FIGURE 32.14 Hepatic portal circulation.

arteries persist as the **medial umbilical ligaments** on the inner surface of the anterior abdominal wall, of the umbilical vein as the **ligamentum teres** (or **round ligament**) of the liver, and of the ductus venosus as a fibrous band called the **ligamentum venosum** on the inferior surface of the liver.

Hepatic Portal Circulation

Blood vessels of the hepatic portal circulation drain the digestive viscera, spleen, and pancreas and deliver this blood to the liver for processing via the **hepatic portal vein.** If a meal has recently been eaten, the hepatic portal blood will be nutrient rich. The liver is the key body organ involved in maintaining proper sugar, fatty acid, and amino acid concentrations in the blood, and this system ensures that these substances pass through the liver before entering the systemic circulation. As blood percolates through the liver sinusoids, some of the nutrients are removed to be stored or processed in various ways for release to the general circulation. At the same time, the hepatocytes are detoxifying alcohol and other possibly harmful chemicals present in the blood, and the liver's macrophages are removing bacteria and other debris from the passing blood. The liver in turn is drained by the hepatic veins that enter the inferior vena cava.

ACTIVITY 6

Tracing the Hepatic Portal Circulation

Locate on Figure 32.14, and on an anatomical chart of the hepatic portal circulation (if available), the vessels named below.

The **inferior mesenteric vein,** draining the distal portions of the large intestine, joins the **splenic vein,** which drains the spleen and part of the pancreas and stomach. The splenic vein and the **superior mesenteric vein,** which receives blood from the small intestine and the ascending and transverse colon, unite to form the hepatic portal vein. The **left gastric vein,** which drains the lesser curvature of the stomach, drains directly into the hepatic portal vein.

For instructions on animal dissections, see the dissection exercises starting on page 697 in the cat, rat, and pig editions of this manual.

NAME ______________________

LAB TIME/DATE ______________________

REVIEW SHEET

EXERCISE 32

Anatomy of Blood Vessels

Microscopic Structure of the Blood Vessels

1. Cross-sectional views of an artery and of a vein are shown here. Identify each; on the lines to the sides, note the structural details that enabled you to make these identifications:

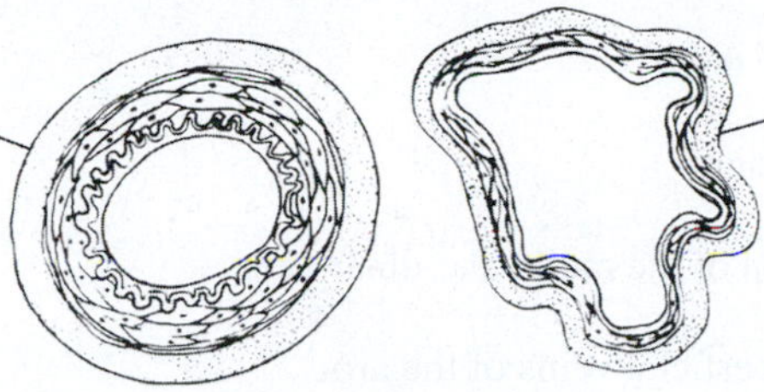

______________________ (vessel type)

______________________ (a)

______________________ (b)

______________________ (vessel type)

______________________ (a)

______________________ (b)

Now describe each tunic more fully by selecting its characteristics from the key below and placing the appropriate key letters on the answer lines.

Tunica intima __________ Tunica media __________ Tunica externa __________

Key:

a. innermost tunic
b. most superficial tunic
c. thin tunic of capillaries
d. especially thick in elastic arteries
e. contains smooth muscle and elastin
f. has a smooth surface to decrease resistance to blood flow

2. Why are valves present in veins but not in arteries? ______________________

____

____

3. Name two events *occurring within the body* that aid in venous return.

______________________ and ______________________

4. Why are the walls of arteries proportionately thicker than those of the corresponding veins? ______________________

____

____

Major Systemic Arteries and Veins of the Body

5. Use the key on the right to identify the arteries or veins described on the left.

__________ 1. the arterial system has one of these; the venous system has two

__________ 2. these arteries supply the myocardium

__________, __________ 3. two paired arteries serving the brain

__________ 4. longest vein in the lower limb

__________ 5. artery on the dorsum of the foot checked after leg surgery

__________ 6. serves the posterior thigh

__________ 7. supplies the diaphragm

__________ 8. formed by the union of the radial and ulnar veins

__________, __________ 9. two superficial veins of the arm

__________ 10. artery serving the kidney

__________ 11. veins draining the liver

__________ 12. artery that supplies the distal half of the large intestine

__________ 13. drains the pelvic organs

__________ 14. what the external iliac artery becomes on entry into the thigh

__________ 15. major artery serving the arm

__________ 16. supplies most of the small intestine

__________ 17. join to form the inferior vena cava

__________ 18. an arterial trunk that has three major branches, which run to the liver, spleen, and stomach

__________ 19. major artery serving the tissues external to the skull

__________, __________, __________ 20. three veins serving the leg

__________ 21. artery generally used to take the pulse at the wrist

Key:
a. anterior tibial
b. basilic
c. brachial
d. brachiocephalic
e. celiac trunk
f. cephalic
g. common carotid
h. common iliac
i. coronary
j. deep artery of the thigh
k. dorsalis pedis
l. external carotid
m. femoral
n. fibular
o. great saphenous
p. hepatic
q. inferior mesenteric
r. internal carotid
s. internal iliac
t. phrenic
u. posterior tibial
v. radial
w. renal
x. subclavian
y. superior mesenteric
z. vertebral

6. What is the function of the cerebral arterial circle (circle of Willis)?

____

7. The anterior and middle cerebral arteries arise from the ______________________________ artery.

They serve the ____________________ of the brain.

8. Trace the pathway of a drop of blood from the aorta to the left occipital lobe of the brain, noting all structures through which it flows. ____

____

9. The human arterial and venous systems are diagrammed on this page and the next. Identify all indicated blood vessels.

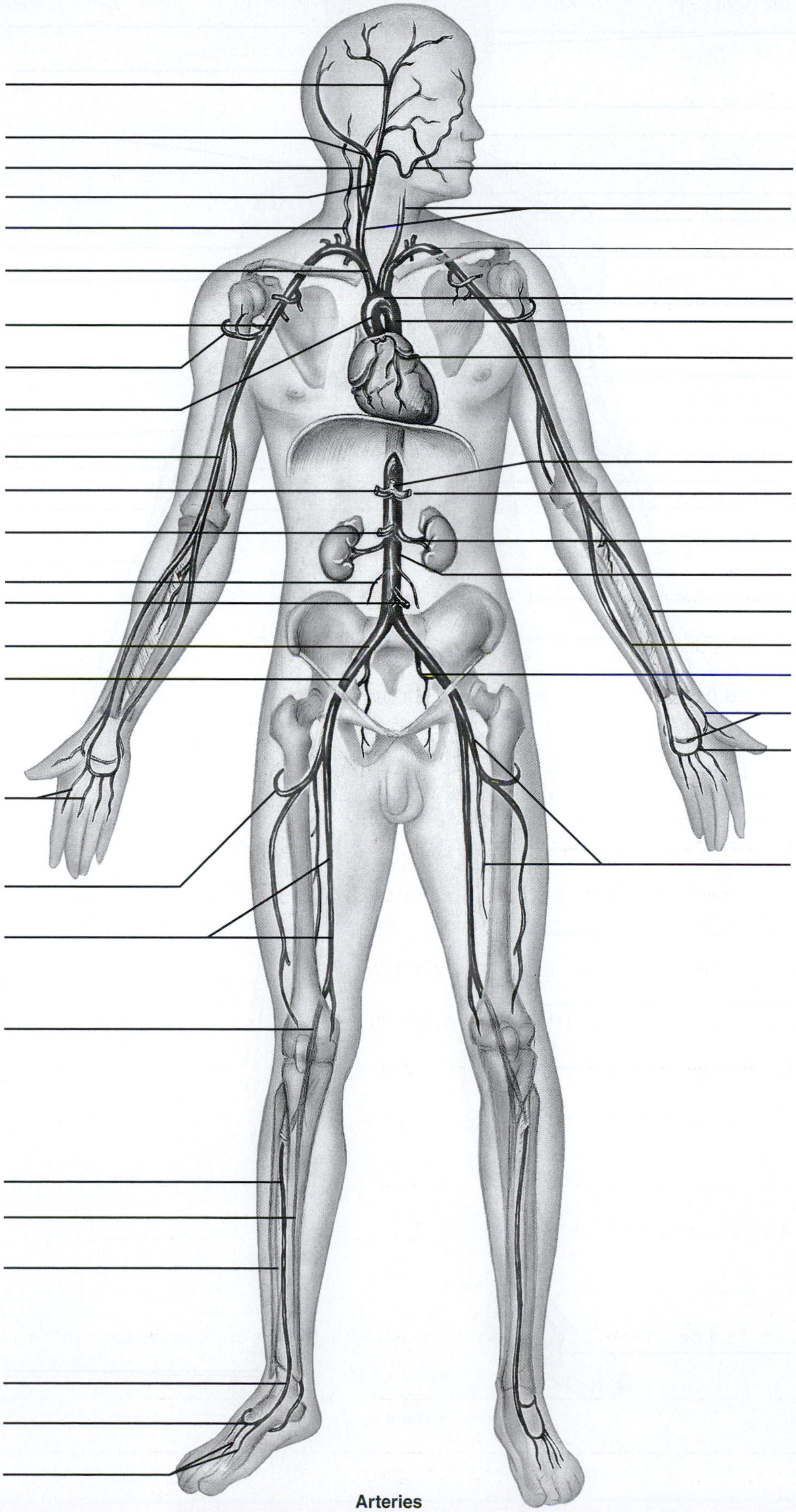

Arteries

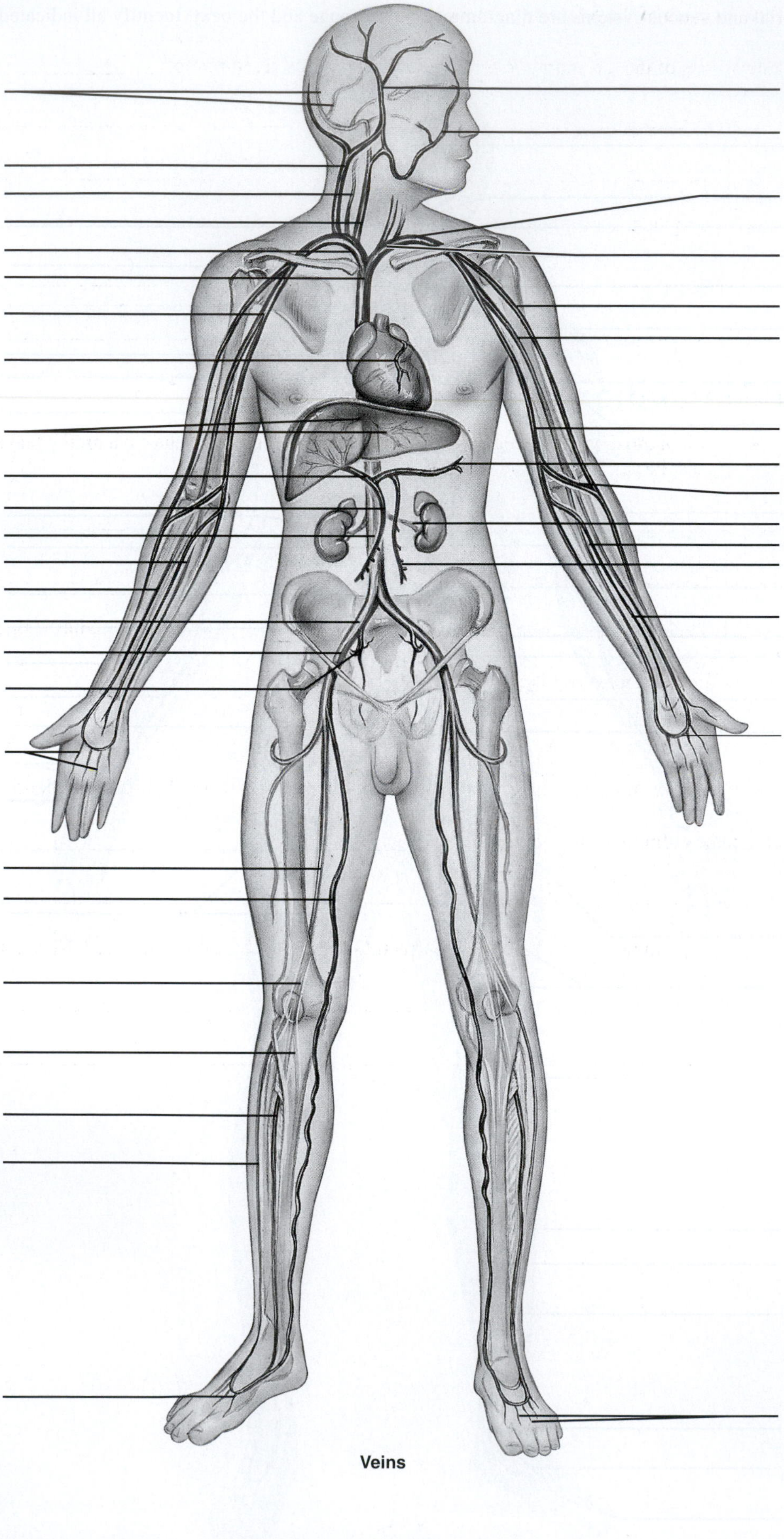

Veins

10. Trace the blood flow for each of the following situations.

a. from the capillary beds of the left thumb to the capillary beds of the right thumb: ______________________

____

____

b. from the mitral valve to the tricuspid valve by way of the great toe:

____

____

Pulmonary Circulation

11. Trace the pathway of a carbon dioxide gas molecule in the blood from the inferior vena cava until it leaves the bloodstream. Name all structures (vessels, heart chambers, and others) passed through en route.

____

____

12. Trace the pathway of oxygen gas molecules from an alveolus of the lung to the right ventricle of the heart. Name all structures through which it passes. Circle the areas of gas exchange. ______________________

____

13. Most arteries of the adult body carry oxygen-rich blood, and the veins carry carbon dioxide–rich blood.

How does this differ in the pulmonary arteries and veins? ______________________

____

14. How do the arteries of the pulmonary circulation differ structurally from the systemic arteries? What condition is indicated by this anatomical difference? ______________________

____

Hepatic Portal Circulation

15. What is the source of blood in the hepatic portal system? ______________________________

____

16. Why is this blood carried to the liver before it enters the systemic circulation? ______________________________

____

17. The hepatic portal vein is formed by the union of (a) ______________, which drains the ______________, ______________, ______________, ______________, and (b) ______________, which drains the ______________ and ______________. The ______________ vein, which drains the lesser curvature of the stomach, empties directly into the hepatic portal vein.

18. Trace the flow of a drop of blood from the small intestine to the right atrium of the heart, noting all structures encountered or passed through on the way. ______________________________

____

Fetal Circulation

19. For each of the following structures, first indicate its function in the fetus; and then note its fate (what happens to it or what it is converted to after birth). Circle the blood vessel that carries the most oxygen-rich blood.

Structure	Function in fetus	Fate
Umbilical artery		
Umbilical vein		
Ductus venosus		
Ductus arteriosus		
Foramen ovale		

20. What organ serves as a respiratory/digestive/excretory organ for the fetus? ______________________________

EXERCISE 33

Human Cardiovascular Physiology: Blood Pressure and Pulse Determinations

MATERIALS

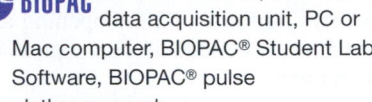

- ☐ Recording of "Interpreting Heart Sounds" (if available on free loan from the local chapters of the American Heart Association) or any of the suitable Internet sites on heart sounds
- ☐ Stethoscope
- ☐ Alcohol swabs
- ☐ Watch (or clock) with second hand
- BIOPAC® BIOPAC® MP36 (or MP35/30) data acquisition unit, PC or Mac computer, BIOPAC® Student Lab Software, BIOPAC® pulse plethysmograph

New versions of the BSL software (3.7.5 and higher for Windows, and 3.7.4 and higher for Mac) require slightly different channel settings and collection strategies. Instructions for using the newer software with the MP36/35 data acquisition units can be found on MasteringA&P. In addition, instructions for use of the NEW 2-channel data acquisition unit, the MP45, may also be found on MasteringA&P.

- ☐ Sphygmomanometer
- ☐ Felt marker
- ☐ Meter stick
- ☐ Cot (if available)
- ☐ Step stools (0.4 m [16 in.] and 0.5 m [20 in.] in height)
- ☐ Small basin suitable for the immersion of one hand
- ☐ Ice
- ☐ Laboratory thermometer

Note: Instructions for using PowerLab® equipment can be found on MasteringA&P.

MasteringA&P™

Access practice quizzes and more in the Study Area at www.masteringaandp.com.

PEx

This lab corresponds to PhysioEx Exercise 5. See p. PEx-75.

OBJECTIVES

1. To define *systole, diastole,* and *cardiac cycle.*
2. To indicate the normal length of the cardiac cycle, the relative pressure changes occurring within the atria and ventricles during the cycle, and the timing of valve closure.
3. To use the stethoscope to auscultate heart sounds, and to relate heart sounds to cardiac cycle events.
4. To describe the clinical significance of heart sounds and heart murmurs.
5. To demonstrate thoracic locations where the first and second heart sounds are most accurately auscultated.
6. To define *pulse, pulse deficit, blood pressure,* and *sounds of Korotkoff.*
7. To accurately determine a subject's apical and radial pulse.
8. To accurately determine a subject's blood pressure with a sphygmomanometer, and to relate systolic and diastolic pressures to events of the cardiac cycle.
9. To investigate the effects of exercise on blood pressure, pulse, and cardiovascular fitness.
10. To indicate factors affecting and/or determining blood flow and skin color.

PRE-LAB QUIZ

1. Circle the correct term. According to general usage, systole / diastole refers to ventricular relaxation.
2. A graph illustrating the pressure and volume changes during one heartbeat is called the
   a. blood pressure c. conduction system of the heart
   b. cardiac cycle d. electrical events of the heartbeat
3. Circle True or False. When ventricular systole begins, intraventricular pressure increases rapidly, closing the atrioventricular (AV) valves.
4. The average heart beats approximately ________ times per minute.
   a. 50 c. 100
   b. 75 d. 125
5. Circle the correct term. Abnormal heart sounds called murmurs / stroke can indicate valvular problems.
6. The term ________ refers to the alternating surges of pressure in an artery that occur with each contraction and relaxation of the left ventricle.
   a. diastole c. pulse
   b. murmur d. systole

Text continues on next page.

7. Circle the correct term. The pulse is most often taken at the lateral aspect of the wrist, above the thumb, compressing the popliteal / radial artery.
8. What device will you use today to measure your subject's blood pressure?

 ____

9. In reporting a blood pressure of 120/90, which number represents the *diastolic* pressure? ______________

10. The __________ are characteristic sounds that indicate the resumption of blood flow to the artery being occluded when taking blood pressure.
    a. ectopic heartbeats
    b. heart murmurs
    c. heart rhythms
    d. sounds of Korotkoff

Any comprehensive study of human cardiovascular physiology takes much more time than a single laboratory period. However, it is possible to conduct investigations of a few phenomena such as pulse, heart sounds, and blood pressure, all of which reflect the heart in action and the function of blood vessels. (The electrocardiogram is studied separately in Exercise 31.) A discussion of the cardiac cycle will provide a basis for understanding and interpreting the various physiological measurements taken.

Cardiac Cycle

In a healthy heart, the two atria contract simultaneously. As they begin to relax, the ventricles contract simultaneously. According to general usage, the terms **systole** and **diastole** refer to events of ventricular contraction and relaxation, respectively. The **cardiac cycle** is equivalent to one complete heartbeat—during which both atria and ventricles contract and then relax. It is marked by a succession of changes in blood volume and pressure within the heart.

Figure 33.1 is a graphic representation of the events of the cardiac cycle for the left side of the heart. Although pressure changes in the right side are less dramatic than those in the left, the same relationships apply.

We will begin the discussion of the cardiac cycle with the heart in complete relaxation (diastole). At this point, pressure in the heart is very low, blood is flowing passively from the pulmonary and systemic circulations into the atria and on through to the ventricles; the semilunar valves are closed, and the AV valves are open. Shortly, atrial contraction occurs and atrial pressure increases, forcing residual blood into the ventricles. Then ventricular systole begins and intraventricular pressure increases rapidly, closing the AV valves. When ventricular pressure exceeds that of the large arteries leaving the heart, the semilunar valves are forced open, and the blood in the ventricular chambers is expelled through the valves. During this phase, the aortic pressure reaches approximately 120 mm Hg in a healthy young adult. During ventricular systole, the atria relax and their chambers fill with blood, which results in gradually increasing atrial pressure. At the end of ventricular systole, the ventricles relax; the semilunar valves snap shut, preventing backflow, and momentarily, the ventricles are closed chambers. When the aortic semilunar valve snaps shut, a momentary increase in the aortic pressure results from the elastic recoil of the aorta after valve closure. This event results in the pressure fluctuation called the *dicrotic notch* (see Figure 33.1a). As the ventricles relax, the pressure within them begins to drop. When intraventricular pressure is again less than atrial pressure, the AV valves are forced open, and the ventricles again begin to fill with blood. Atrial and aortic pressures decrease, and the ventricles rapidly refill, completing the cycle.

The average heart beats approximately 75 beats per minute, and so the length of the cardiac cycle is about 0.8 second. Of this time period, atrial contraction occupies the first 0.1 second, which is followed by atrial relaxation and ventricular contraction for the next 0.3 second. The remaining 0.4 second is the quiescent, or ventricular relaxation, period. When the heart beats at a more rapid pace than normal, this last period decreases.

Notice that two different types of phenomena control the movement of blood through the heart: the alternate contraction and relaxation of the myocardium, and the opening and closing of valves, which is entirely dependent on the pressure changes within the heart chambers.

Study Figure 33.1 carefully to make sure you understand what has been discussed before continuing with the next portion of the exercise.

Heart Sounds

Sounds heard in the cardiovascular system result from turbulent blood flow. Two distinct sounds can be heard during each cardiac cycle. These heart sounds are commonly described by the monosyllables "lub" and "dup"; and the sequence is designated lub-dup, pause, lub-dup, pause, and so on. The first heart sound (lub) is referred to as S_1 and is associated with closure of the AV valves at the beginning of ventricular systole. The second heart sound (dup), called S_2, occurs as the semilunar valves close and corresponds with the end of systole. Figure 33.1a indicates the correlation of heart sounds with events of the cardiac cycle.

- Listen to the recording "Interpreting Heart Sounds" or another suitable recording so that you may hear both normal and abnormal heart sounds.

Abnormal heart sounds are called **murmurs** and often indicate valvular problems. In valves that do not close tightly, closure is followed by a swishing sound due to the backflow of blood (regurgitation). Distinct sounds, often described as high-pitched screeching, are associated with the tortuous flow of blood through constricted, or stenosed, valves. ■

ACTIVITY 1

Auscultating Heart Sounds

In the following procedure, you will auscultate your partner's heart sounds with an ordinary stethoscope. A number of more sophisticated heart-sound amplification systems are on the market, and your instructor may prefer to use one if it is available. If so, directions for the use of this apparatus will be provided by the instructor.

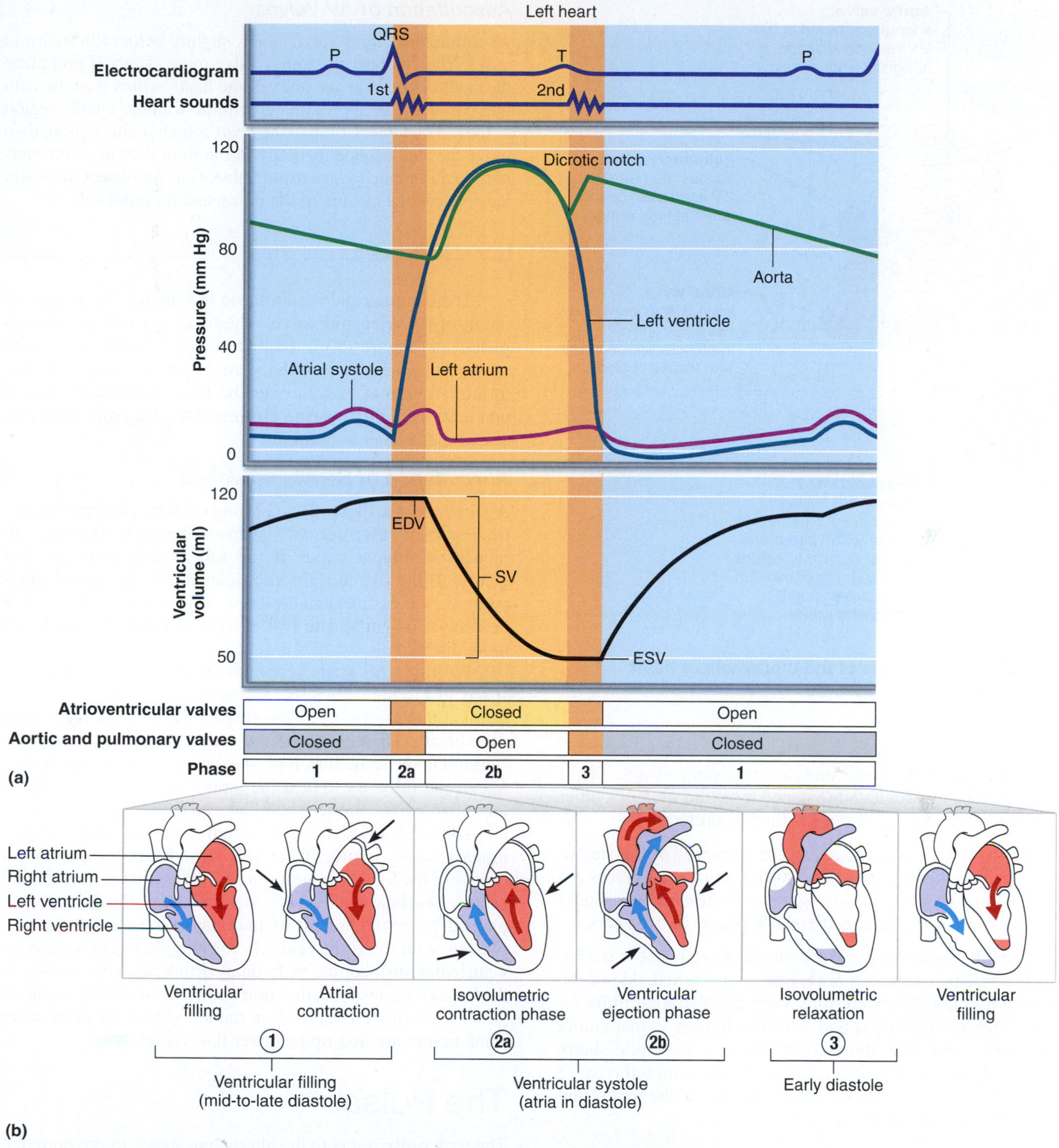
Left heart
QRS
P
T
P
Electrocardiogram
Heart sounds
1st
2nd
120
Dicrotic notch
80
Pressure (mm Hg)
Aorta
Left ventricle
40
Atrial systole
Left atrium
0
120
EDV
Ventricular volume (ml)
SV
50
ESV
Atrioventricular valves
Open
Closed
Open
Aortic and pulmonary valves
Closed
Open
Closed
Phase
1
2a
2b
3
1
(a)
Left atrium
Right atrium
Left ventricle
Right ventricle
Ventricular filling
Atrial contraction
Isovolumetric contraction phase
Ventricular ejection phase
Isovolumetric relaxation
Ventricular filling
1
2a
2b
3
Ventricular filling (mid-to-late diastole)
Ventricular systole (atria in diastole)
Early diastole
(b)

FIGURE 33.1 Summary of events occurring in the heart during the cardiac cycle. (a) An ECG tracing is superimposed on the graph (top) so that electrical events can be related to pressure and volume changes (center) in the left side of the heart. Pressures are lower in the right side of the heart. Time occurrence of heart sounds is also indicated. (EDV = end diastolic volume, SV = stroke volume, ESV = end systolic volume) **(b)** Events of phases 1 through 3 of the cardiac cycle are depicted in diagrammatic views of the heart.

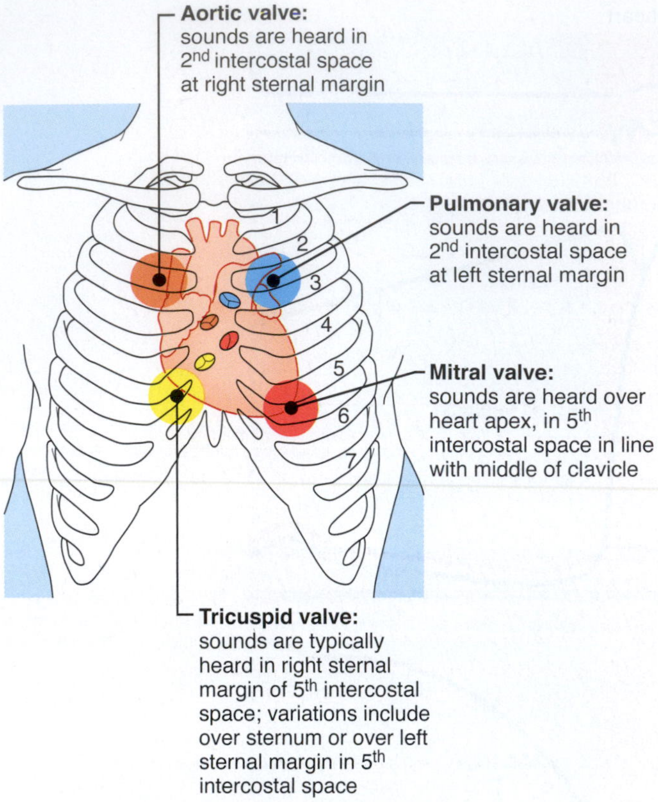
Aortic valve:
sounds are heard in 2nd intercostal space at right sternal margin
Pulmonary valve:
sounds are heard in 2nd intercostal space at left sternal margin
Mitral valve:
sounds are heard over heart apex, in 5th intercostal space in line with middle of clavicle
Tricuspid valve:
sounds are typically heard in right sternal margin of 5th intercostal space; variations include over sternum or over left sternal margin in 5th intercostal space
1
2
3
4
5
6
7

FIGURE 33.2 Areas of the thorax where heart sounds can best be detected.

1. Obtain a stethoscope and some alcohol swabs. Heart sounds are best auscultated (listened to) if the subject's outer clothing is removed, so a male subject is preferable.

2. With an alcohol swab, clean the earpieces of the stethoscope. Allow the alcohol to dry. Notice that the earpieces are angled. For comfort and best auscultation, the earpieces should be angled in a *forward* direction when placed into the ears.

3. Don the stethoscope. Place the diaphragm of the stethoscope on your partner's thorax, just to the sternal side of the left nipple at the fifth intercostal space, and listen carefully for heart sounds. The first sound will be a longer, louder (more booming) sound than the second, which is short and sharp. After listening for a couple of minutes, try to time the pause between the second sound of one heartbeat and the first sound of the subsequent heartbeat.

How long is this interval? ____________________ sec

How does it compare to the interval between the first and second sounds of a single heartbeat?

____

4. To differentiate individual valve sounds somewhat more precisely, auscultate the heart sounds over specific thoracic regions. Refer to Figure 33.2 for the positioning of the stethoscope.

Auscultation of AV Valves

As a rule, the mitral valve closes slightly before the tricuspid valve. You can hear the mitral valve more clearly if you place the stethoscope over the apex of the heart, which is at the fifth intercostal space, approximately in line with the middle region of the left clavicle. Listen to the heart sounds at this region; then move the stethoscope medially to the right margin of the sternum to auscultate the tricuspid valve. Can you detect the slight lag between the closure of the mitral and tricuspid valves?

____

There are normal variations in the site for "best" auscultation of the tricuspid valve. These range from the site depicted in Figure 33.2 (right sternal margin over fifth intercostal space) to over the sternal body in the same plane, to the left sternal margin over the fifth intercostal space. If you have difficulty hearing closure of the tricuspid valve, try one of these other locations.

Auscultation of Semilunar Valves

Again there is a slight dissynchrony of valve closure; the aortic semilunar valve normally snaps shut just ahead of the pulmonary semilunar valve. If the subject inhales deeply but gently, filling of the right ventricle (due to decreased intrapulmonary pressure) and closure of the pulmonary valve will be delayed slightly. The two sounds can therefore be heard more distinctly.

Position the stethoscope over the second intercostal space, just to the *right* of the sternum. The aortic valve is best heard at this position. As you listen, have your partner take a deep breath. Then move the stethoscope to the *left* side of the sternum in the same line, and auscultate the pulmonary valve. Listen carefully; try to hear the "split" between the closure of these two valves in the second heart sound.

Although at first it may seem a bit odd that the pulmonary valve issuing from the *right* heart is heard most clearly to the *left* of the sternum and the aortic valve of the left heart is best heard at the right sternal border, this is easily explained by reviewing heart anatomy. Because the heart is twisted, with the right ventricle forming most of the anterior ventricular surface, the pulmonary trunk actually crosses to the left as it issues from the right ventricle. Similarly, the aorta issues from the left ventricle at the left side of the pulmonary trunk before arching up and over that vessel.

The Pulse

The term **pulse** refers to the alternating surges of pressure (expansion and then recoil) in an artery that occur with each contraction and relaxation of the left ventricle. This difference between systolic and diastolic pressure is called the **pulse pressure**. (See page 498.) Normally the pulse rate (pressure surges per minute) equals the heart rate (beats per minute), and the pulse averages 70 to 76 beats per minute in the resting state.

Parameters other than pulse rate are also useful clinically. You may also assess the regularity (or rhythmicity) of the pulse, and its amplitude and/or tension—does the blood vessel expand and recoil (sometimes visibly) with the pressure waves? Can you feel it strongly, or is it difficult to detect? Is it regular like the ticking of a clock, or does it seem to skip beats?

ACTIVITY 2

Palpating Superficial Pulse Points

The pulse may be felt easily on any superficial artery when the artery is compressed over a bone or firm tissue. Palpate the following pulse or pressure points on your partner by placing the fingertips of the first two or three fingers of one hand over the artery. It helps to compress the artery firmly as you begin your palpation and then immediately ease up on the pressure slightly. In each case, notice the regularity of the pulse, and assess the degree of tension or amplitude. Figure 33.3 illustrates the superficial pulse points to be palpated.

Superficial temporal artery: Anterior to the ear, in the temple region.

Facial artery: Clench the teeth, and palpate the pulse just anterior to the masseter muscle on the mandible (in line with the corner of the mouth).

Common carotid artery: At the side of the neck.

Brachial artery: In the antecubital fossa, at the point where it bifurcates into the radial and ulnar arteries.

Radial artery: At the lateral aspect of the wrist, above the thumb.

Femoral artery: In the groin.

Popliteal artery: At the back of the knee.

Posterior tibial artery: Just above the medial malleolus.

Dorsalis pedis artery: On the dorsum of the foot.

Which pulse point had the greatest amplitude?

____

Which had the least? ___________________________________

Can you offer any explanation for this? ___________________

____

____

____

Because of its easy accessibility, the pulse is most often taken on the radial artery. With your partner sitting quietly, practice counting the radial pulse for 1 minute. Make three counts and average the results.

count 1 _______ count 2 _______

count 3 _______ average _______

Due to the elasticity of the arteries, blood pressure decreases and smooths out as blood moves farther away from the heart. A pulse, however, can still be felt in the fingers. A device called a plethysmograph or a piezoelectric pulse transducer can measure this pulse.

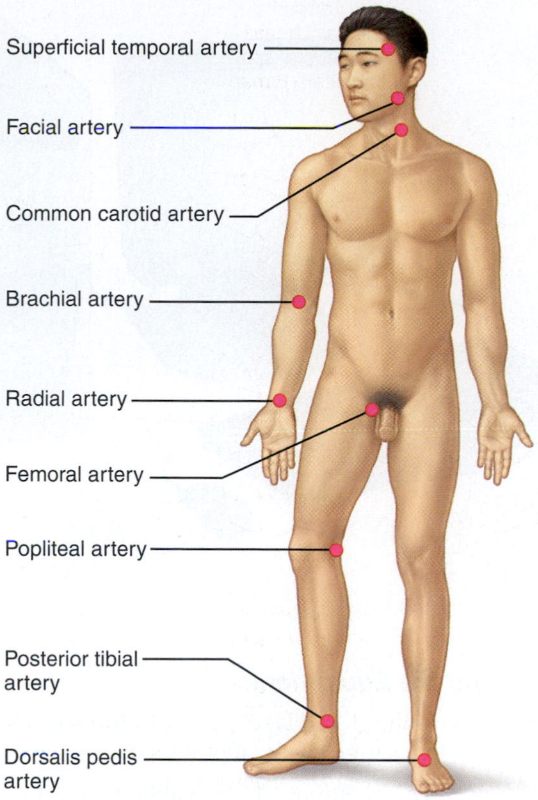
Superficial temporal artery
Facial artery
Common carotid artery
Brachial artery
Radial artery
Femoral artery
Popliteal artery
Posterior tibial artery
Dorsalis pedis artery

FIGURE 33.3 Body sites where the pulse is most easily palpated.

ACTIVITY 3

Measuring Pulse Using BIOPAC®

Setting Up the Equipment

1. Connect the BIOPAC® apparatus to the computer and turn the computer **ON.**
2. Make sure the BIOPAC® unit is **OFF.**
3. Plug in the equipment as shown in Figure 33.4:
   - Pulse transducer (plethysmograph)—CH 2
4. Turn the BIOPAC® unit **ON.**
5. Wrap the pulse transducer around the tip of the index finger as shown in Figure 33.5. Wrap the Velcro® around the finger gently (if wrapped too tight, it will reduce circulation to the finger and obscure the recording). *Do not wiggle the finger or move the plethysmograph cord during recording.*
6. Have the subject sit down with the forearms supported and relaxed.
7. Start the BIOPAC® Student Lab program on the computer by double-clicking the icon on the desktop or by following your instructor's guidance.
8. Select lesson **L07-ECG&P-1** from the menu, and click **OK.**
9. Type in a filename that will save this subject's data on the computer hard drive. You may want to use the subject's last name followed by Pulse (for example, SmithPulse-1), then click OK.

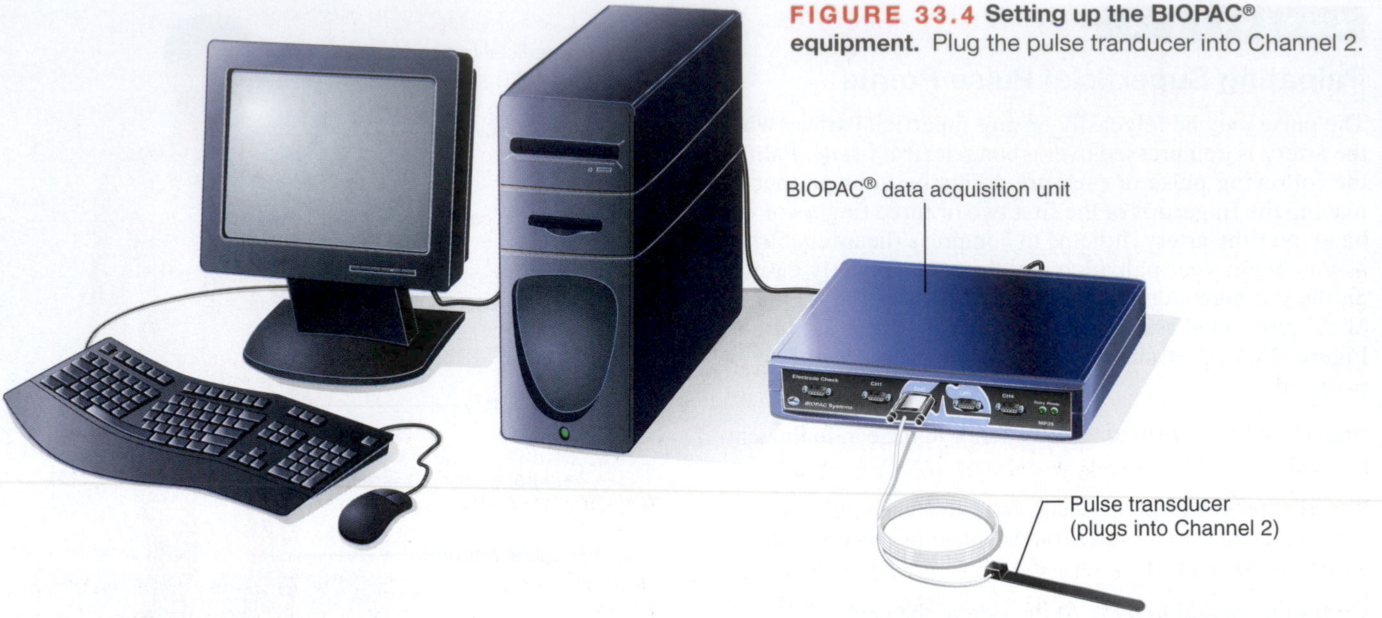
BIOPAC® data acquisition unit
Pulse transducer
(plugs into Channel 2)

FIGURE 33.4 Setting up the BIOPAC® equipment. Plug the pulse tranducer into Channel 2.

Calibrating the Equipment

1. When the subject is relaxed, click **Calibrate.** The calibration will stop automatically after 8 seconds.

2. Observe the recording of the calibration data, which should look like the waveforms in Figure 33.6.

- If the data look very different, click **Redo Calibration** and repeat the steps above. If there is no signal, you may need to loosen the Velcro® on the finger and check all attachments.
- If the data look similar, proceed to the next section.

Recording the Data

1. When the subject is ready, click **Record** to begin recording the pulse. After 30 seconds, click **Suspend.**

2. Observe the data, which should look similar to the pulse data in Figure 33.7.

- If the data look very different, click **Redo** and repeat the steps above.
- If the data look similar, go to the next step.

3. When you are finished, click **Done.** A pop-up window will appear; to record from another subject select **Record from another subject** and return to step 5 under Setting Up the Equipment. If continuing to the Data Analysis section, select **Analyze current data file** and proceed to step 2 in the Data Analysis section.

Data Analysis

1. If just starting the BIOPAC® program to perform data analysis, enter **Review Saved Data** mode and choose the file with the subject's pulse data (for example, SmithPulse-1).

2. Observe that pulse data is in the lower scale (ECG data was not recorded in this activity and will not be used for analysis).

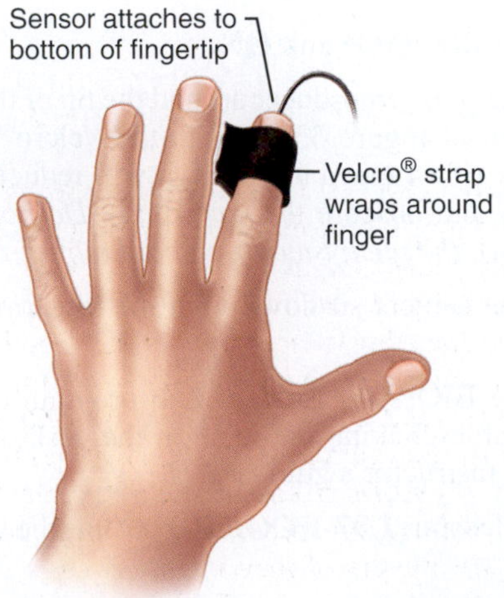
Sensor attaches to bottom of fingertip
Velcro® strap wraps around finger

FIGURE 33.5 Placement of the pulse transducer around the tip of the index finger.

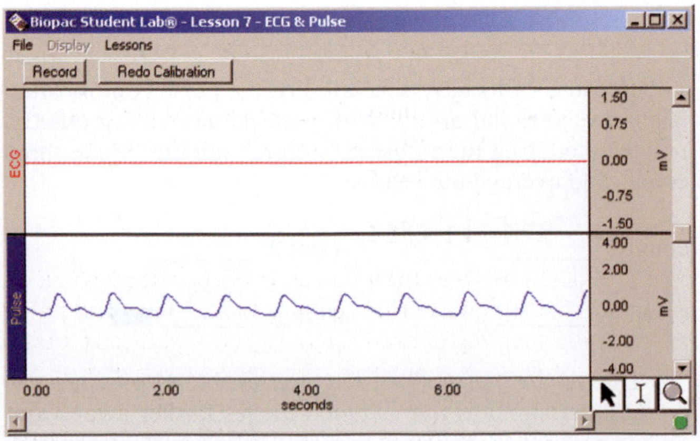
Biopac Student Lab® - Lesson 7 - ECG & Pulse
File
Display
Lessons
Record
Redo Calibration
ECG
Pulse
1.50
0.75
0.00
-0.75
-1.50
mV
4.00
2.00
0.00
-2.00
-4.00
mV
0.00
2.00
4.00
seconds
6.00

FIGURE 33.6 Example of waveforms during the calibration procedure.

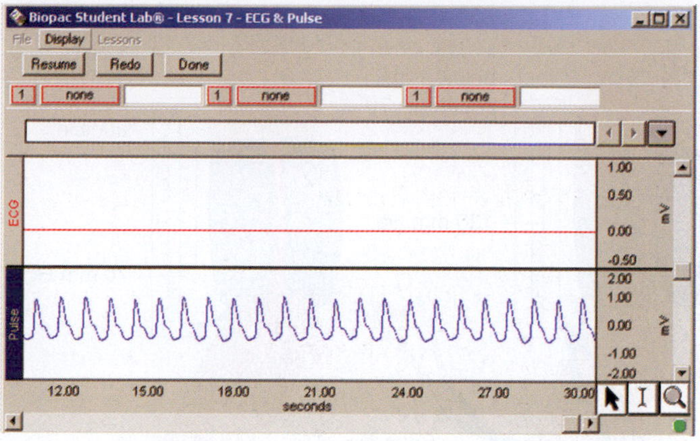

FIGURE 33.7 Example of waveforms during the recording of data.

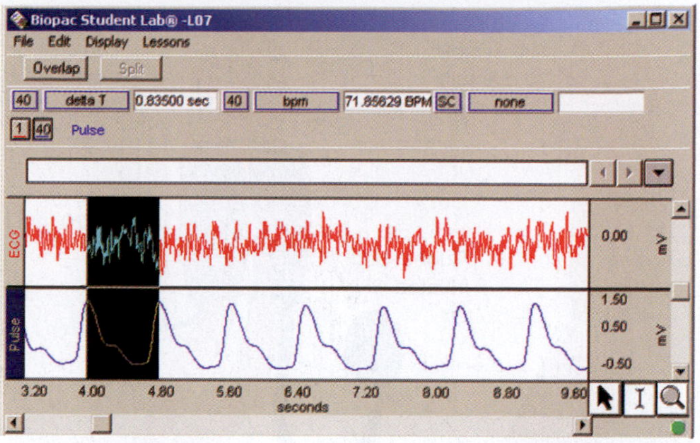

FIGURE 33.8 Using the I-beam cursor to highlight data for analysis.

3. To analyze the data, set up the first channel/measurement boxes at the top of the screen. Channel 40 contains the pulse data.

Channel	Measurement
CH 40	deltaT

deltaT: Measures the time elapsed in the selected area

4. Use the arrow cursor and click the I-beam cursor box on the lower right side of the screen to activate the "area selection" function. Using the activated I-beam cursor, highlight from the peak of one pulse to the peak of the next pulse as shown in Figure 33.8.

Observe the elapsed time between heartbeats and record here

(to the nearest 0.01 second): ______________________

5. Calculate the beats per minute by inserting the elapsed time into this formula:

(1 beat/ __________ sec) × (60 sec/min)

= __________ beats/min

Optional Activity with BIOPAC® Pulse Measurement

To expand the experiment, you can measure the effects of heat and cold on pulse rate. To do this, you can submerge the subject's hand (the hand without the plethysmograph!) in hot and/or ice water for 2 minutes, and then record the pulse. Alternatively, you can investigate change in pulse rate after brief exercise such as jogging in place.

Apical-Radial Pulse

The correlation between the apical and radial pulse rates can be determined by simultaneously counting them. The **apical pulse** (actually the counting of heartbeats) may be slightly faster than the radial because of a slight lag in time as the blood rushes from the heart into the large arteries where it can be palpated. However, any *large* difference between the values observed, referred to as a **pulse deficit,** may indicate cardiac impairment (a weakened heart that is unable to pump blood into the arterial tree to a normal extent), low cardiac output, or abnormal heart rhythms. In the case of atrial fibrillation or ectopic heartbeats, for instance, the second beat may follow the first so quickly that no second pulse is felt even though the apical pulse can still be heard (auscultated). Apical pulse counts are routinely ordered for those with cardiac decompensation.

ACTIVITY 4

Taking an Apical Pulse

With the subject sitting quietly, one student, using a stethoscope, should determine the apical pulse rate while another simultaneously counts the radial pulse rate. The stethoscope should be positioned over the fifth left intercostal space. The person taking the radial pulse should determine the starting point for the count and give the stop-count signal exactly 1 minute later. Record your values below.

apical count ____________________ beats/min

radial count ____________________ pulses/min

pulse deficit ____________________ pulses/min

Blood Pressure Determinations

Blood pressure (BP) is defined as the pressure the blood exerts against any unit area of the blood vessel walls, and it is generally measured in the arteries. Because the heart alternately contracts and relaxes, the resulting rhythmic flow of blood into the arteries causes the blood pressure to rise and fall during each beat. Thus you must take two blood pressure readings: the **systolic pressure,** which is the pressure in the arteries at the peak of ventricular ejection, and the **diastolic pressure,** which reflects the pressure during ventricular relaxation. Blood pressures are reported in millimeters of mercury (mm Hg), with the systolic pressure appearing first; 120/80 translates to 120 over 80, or a systolic pressure of 120 mm Hg and a diastolic pressure of 80 mm Hg. Normal blood pressure varies considerably from one person to another.

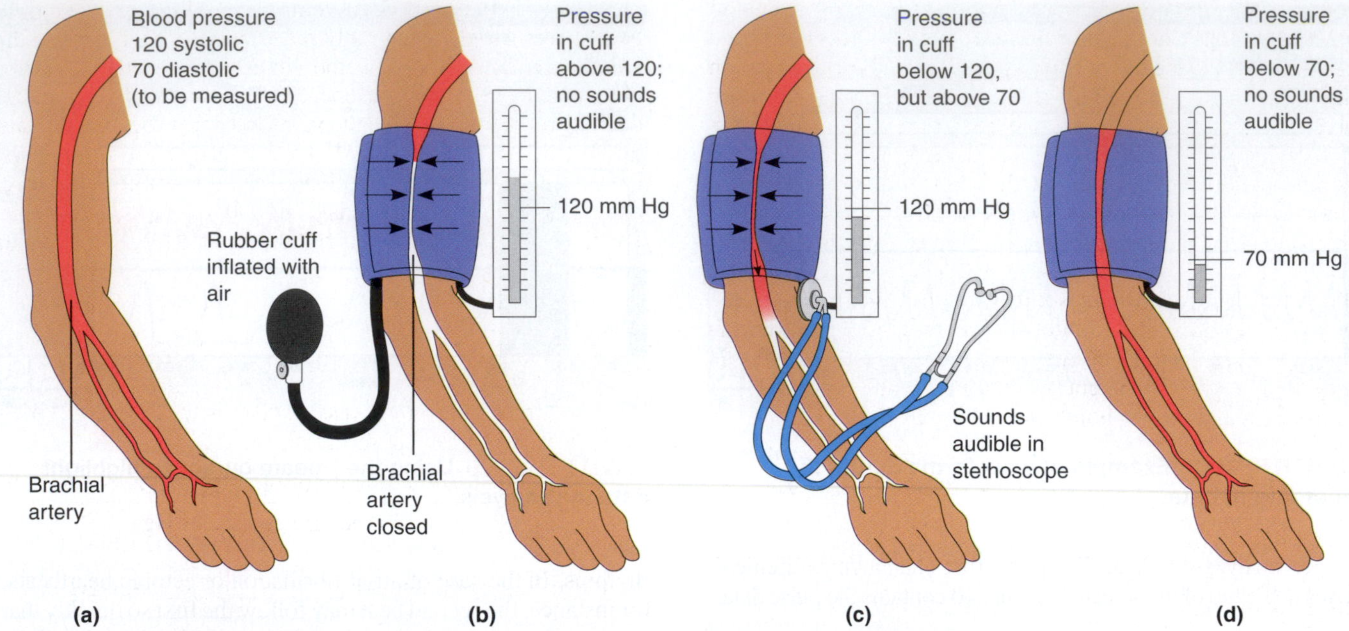

Blood pressure
120 systolic
70 diastolic
(to be measured)
Brachial
artery
(a)
Pressure
in cuff
above 120;
no sounds
audible
120 mm Hg
Rubber cuff
inflated with
air
Brachial
artery
closed
(b)
Pressure
in cuff
below 120,
but above 70
120 mm Hg
Sounds
audible in
stethoscope
(c)
Pressure
in cuff
below 70;
no sounds
audible
70 mm Hg
(d)

FIGURE 33.9 Procedure for measuring blood pressure. (a) The course of the brachial artery of the arm. Assume a blood pressure of 120/70. **(b)** The blood pressure cuff is wrapped snugly around the arm just above the elbow and inflated until blood flow into the forearm is stopped and no brachial pulse can be felt or heard. **(c)** Pressure in the cuff is gradually reduced while the examiner listens (auscultates) for sounds (of Korotkoff) in the brachial artery with a stethoscope. The pressure, read as the first soft tapping sounds are heard (the first point at which a small amount of blood is spurting through the constricted artery), is recorded as the systolic pressure. **(d)** As the pressure is reduced still further, the sounds become louder and more distinct, but when the artery is no longer restricted and blood flows freely, the sounds can no longer be heard. The pressure at which the sounds disappear is routinely recorded as the diastolic pressure.

In this procedure, you will measure arterial pressure by indirect means and under various conditions. You will investigate and demonstrate factors affecting blood pressure, and the rapidity of blood pressure changes.

ACTIVITY 5

Using a Sphygmomanometer to Measure Arterial Blood Pressure Indirectly

The **sphygmomanometer,** commonly called a *blood pressure cuff,* is an instrument used to obtain blood pressure readings by the auscultatory method (Figure 33.9). It consists of an inflatable cuff with an attached pressure gauge. The cuff is placed around the arm and inflated to a pressure higher than systolic pressure to occlude circulation to the forearm. As cuff pressure is gradually released, the examiner listens with a stethoscope for characteristic sounds called the **sounds of Korotkoff,** which indicate the resumption of blood flow into the forearm. The pressure at which the first soft tapping sounds can be detected is recorded as the systolic pressure. As the pressure is reduced further, blood flow becomes more turbulent, and the sounds become louder. As the pressure is reduced still further, below the diastolic pressure, the artery is no longer compressed; and blood flows freely and without turbulence. At this point, the sounds of Korotkoff can no longer be detected. The pressure at which the sounds disappear is recorded as the diastolic pressure.

1. Work in pairs to obtain radial artery blood pressure readings. Obtain a felt marker, stethoscope, alcohol swabs, and a sphygmomanometer. Clean the earpieces of the stethoscope with the alcohol swabs, and check the cuff for the presence of trapped air by compressing it against the laboratory table. (A partially inflated cuff will produce erroneous measurements.)

2. The subject should sit in a comfortable position with one arm resting on the laboratory table (approximately at heart level if possible). Wrap the cuff around the subject's arm, just above the elbow, with the inflatable area on the medial arm surface. The cuff may be marked with an arrow; if so, the arrow should be positioned over the brachial artery (Figure 33.9). Secure the cuff by tucking the distal end under the wrapped portion or by bringing the Velcro® areas together.

3. Palpate the brachial pulse, and lightly mark its position with a felt pen. Don the stethoscope, and place its diaphragm over the pulse point.

! *The cuff should not be kept inflated for more than 1 minute.* If you have any trouble obtaining a reading within this time, deflate the cuff, wait 1 or 2 minutes, and try again. (A prolonged interference with BP homeostasis can lead to fainting.)

4. Inflate the cuff to approximately 160 mm Hg pressure, and slowly release the pressure valve. Watch the pressure gauge as you listen carefully for the first soft thudding

sounds of the blood spurting through the partially occluded artery. Mentally note this pressure (systolic pressure), and continue to release the cuff pressure. You will notice first an increase, then a muffling, of the sound. For the diastolic pressure, note the pressure at which the sound becomes muffled or disappears. Controversy exists over which of the two points should be recorded as the diastolic pressure; so in some cases you may see readings such as 120/80/78, which indicates the systolic pressure followed by the *first* and *second diastolic end points.* The first diastolic end point is the pressure at which the sound muffles; the second is the pressure at which the sound disappears. It makes little difference here which of the two diastolic pressures is recorded, but be consistent. Make two blood pressure determinations, and record your results below.

First trial:	**Second trial:**
systolic pressure ________	systolic pressure ________
diastolic pressure ________	diastolic pressure ________

5. Compute the **pulse pressure** for each trial. The pulse pressure is the difference between the systolic and diastolic pressures, and it indicates the amount of blood forced from the heart during systole, or the actual "working" pressure. A narrowed pulse pressure (less than 30 mm Hg) may be a signal of severe aortic stenosis, constrictive pericarditis, or tachycardia. A widened pulse pressure (over 40 mm Hg) is common in hypertensive individuals.

Pulse pressure:

first trial ________ second trial ________

6. Compute the **mean arterial pressure (MAP)** for each trial using the following equation:

$$\text{MAP} = \text{diastolic pressure} + \frac{\text{pulse pressure}}{3}$$

first trial ________ second trial ________

ACTIVITY 6

Estimating Venous Pressure

It is not possible to measure venous pressure with the sphygmomanometer. The methods available for measuring it produce estimates at best, because venous pressures are so much lower than arterial pressures. The difference in pressure becomes obvious when these vessels are cut. If a vein is cut, the blood flows evenly from the cut. A lacerated artery produces rapid spurts of blood. In this activity, you will estimate venous pressures.

1. Obtain a meter stick, and ask your lab partner to stand with his or her right side toward the blackboard, arms hanging freely at the sides. On the board, mark the approximate level of the right atrium. (This will be just slightly higher than the point at which you ausculated the apical pulse.)

2. Observe the superficial veins on the dorsum of the right hand as the subject alternately raises and lowers it. Notice the collapsing and filling of the veins as internal pressures change. Have the subject repeat this action until you can determine the point at which the veins have just collapsed. Mark this hand level on the board. Then measure, in millimeters, the distance in the vertical plane from this point to the level of the right atrium (previously marked). Record this value. Distance of right arm from right atrium at point of venous collapse:

________________ mm

3. Compute the venous pressure (P_V), in millimeters of mercury, with the following formula:

$$P_V = \frac{1.056\ (\text{specific gravity of blood}) \times \text{mm (measured)}}{13.6\ (\text{specific gravity of Hg})}$$

Venous pressure computed:________________ mm Hg

Normal venous pressure varies from approximately 30 to 90 mm Hg. That of the hand ranges between 30 and 40 mm Hg. How does your computed value compare?

____

4. Because venous walls are so thin, pressure within them is readily affected by external factors such as muscle activity, deep pressure, and pressure changes occurring in the thorax during breathing. The Valsalva maneuver, which increases intrathoracic pressure, is used to demonstrate the effect of thoracic pressure changes on venous pressure.

To perform this maneuver take a deep breath, and then mimic the motions of exhaling forcibly, but without actually exhaling. In reaction to this, the glottis will close; and intrathoracic pressure will increase. (Most of us have performed this maneuver unknowingly in acts of defecation in which there is "straining at stool.") Have the same subject again stand next to the blackboard mark for the level of his or her right atrium. While the subject performs the Valsalva maneuver and raises and lowers one hand, determine the point of venous collapse and mark it on the board. Measure the distance of that mark from the right atrium level and record it below. Then compute the venous pressure and record it.

__________ mm Venous pressure: __________ mm Hg

How does this value compare with the venous pressure measurement computed for the relaxed state?

____

Explain: ____________________________________

____

____

Posture				
	Trial 1		Trial 2	
	BP	Pulse	BP	Pulse
Sitting quietly				
Reclining (after 2 to 3 min)				
Immediately on standing from the reclining position ("at attention" stance)				
After standing for 3 min				

ACTIVITY 7

Observing the Effect of Various Factors on Blood Pressure and Heart Rate

Arterial blood pressure is directly proportional to cardiac output (CO, amount of blood pumped out of the left ventricle per unit time) and peripheral resistance (PR) to blood flow, that is

$$BP = CO \times PR$$

Peripheral resistance is increased by blood vessel constriction (most importantly the arterioles), by an increase in blood viscosity, and by a loss of elasticity of the arteries (seen in arteriosclerosis). Any factor that increases either the cardiac output or the peripheral resistance causes an almost immediate reflex rise in blood pressure. A close examination of these relationships reveals that many factors—age, weight, time of day, exercise, body position, emotional state, and various drugs, for example—alter blood pressure. The influence of a few of these factors is investigated here.

The following tests are done most efficiently if one student acts as the subject; two are examiners (one taking the radial pulse and the other auscultating the brachial blood pressure); and a fourth student collects and records data. The sphygmomanometer cuff should be left on the subject's arm throughout the experiments (in a deflated state, of course) so that, at the proper times, the blood pressure can be taken quickly. In each case, take the measurements at least twice. For each of the following tests, students should formulate hypotheses, collect data, and write lab reports. (See Getting Started, page xv.) Conclusions should be shared with the class.

Posture

To monitor circulatory adjustments to changes in position, take blood pressure and pulse measurements under the conditions noted in the Posture chart above. Record your results on the chart.

Exercise

Blood pressure and pulse changes during and after exercise provide a good yardstick for measuring one's overall cardiovascular fitness. Although there are more sophisticated and more accurate tests that evaluate fitness according to a specific point system, the *Harvard step test* described here is a quick way to compare the relative fitness level of a group of people.

You will be working in groups of four, duties assigned as indicated above, except that student 4, in addition to recording the data, will act as the timer and call the cadence.

! Any student with a known heart problem should refuse to participate as the subject.

All four students may participate as the subject in turn, if desired, but the bench stepping is to be performed *at least twice* in each group—once with a well-conditioned person acting as the subject, and once with a poorly conditioned subject.

Bench stepping is the following series of movements repeated sequentially:

1. Place one foot on the step.
2. Step up with the other foot so that both feet are on the platform. Straighten the legs and the back.
3. Step down with one foot.
4. Bring the other foot down.

The pace for the stepping will be set by the "timer" (student 4), who will repeat "Up-2-3-4, up-2-3-4" at such a pace that each "up-2-3-4" sequence takes 2 sec (30 cycles/min).

1. Student 4 should obtain the step (0.5 m [20-in.] height for male subject or 0.4 m [16-in.] for a female subject) while baseline measurements are being obtained on the subject.

Exercise

Harvard step test for 5 min at 30/min	Baseline		Interval Following Test							
			Immediately		1 min		2 min		3 min	
	BP	P	BP	P	BP	P	BP	P	BP	P
Well-conditioned individual	____	____	____	____	____	____	____	____	____	____
Poorly conditioned individual	____	____	____	____	____	____	____	____	____	____

2. Once the baseline pulse and blood pressure measurements have been recorded on the Exercise chart above, the subject is to stand quietly at attention for 2 minutes to allow his or her blood pressure to stabilize before beginning to step.

3. The subject is to perform the bench stepping for as long as possible, up to a maximum of 5 minutes, according to the cadence called by the timer. The subject is to be watched for and warned against crouching (posture must remain erect). If he or she is unable to keep the pace for a span of 15 seconds, the test is to be terminated.

4. When the subject is stopped by the pacer for crouching, stops voluntarily because he or she is unable to continue, or has completed 5 minutes of bench stepping, he or she is to sit down. The duration of exercise (in seconds) is to be recorded, and the blood pressure and pulse are to be measured immediately and thereafter at 1-minute intervals for 3 minutes post-exercise.

Duration of exercise: ________ sec

5. The subject's *index of physical fitness* is to be calculated using the following formula:

$$\text{Index} = \frac{\text{duration of exercise in seconds} \times 100}{2 \times \text{sum of the three pulse counts in recovery}}$$

Scores are interpreted according to the following scale:

below 55	poor physical condition
55 to 62	low average
63 to 71	average
72 to 79	high average
80 to 89	good
90 and over	excellent

6. Record the test values on the Exercise chart above, and repeat the testing and recording procedure with the second subject.

When did you notice a greater elevation of blood pressure and pulse?

____

Explain:______________________________________

____

Was there a sizable difference between the after-exercise values for well-conditioned and poorly conditioned individuals?

__________ Explain:___________________________

____

Did the diastolic pressure also increase?______________

Explain:______________________________________

____

A Noxious Sensory Stimulus (Cold)

There is little question that blood pressure is affected by emotions and pain. This lability of blood pressure will be investigated through use of the **cold pressor test,** in which one hand will be immersed in unpleasantly (even painfully) cold water.

1. Measure the blood pressure and pulse of the subject as he or she sits quietly. Record these as the baseline values on the Noxious Sensory Stimulus chart (next page).

2. Obtain a basin and thermometer, fill the basin with ice cubes, and add water. When the temperature of the ice bath has reached 5°C, immerse the subject's other hand (the non-cuffed limb) in the ice water. With the hand still immersed, take blood pressure and pulse readings at 1-minute intervals for a period of 3 minutes, and record the values on the chart.

A Noxious Sensory Stimulus (Cold)							
Baseline		1 min		2 min		3 min	
BP	P	BP	P	BP	P	BP	P

How did the blood pressure change during cold exposure?

___

___

Was there any change in pulse?___

3. Subtract the respective baseline readings of systolic and diastolic blood pressure from the highest single reading of systolic and diastolic pressure obtained during cold immersion. (For example, if the highest experimental reading is 140/88 and the baseline reading is 120/70, then the differences in blood pressure would be systolic pressure, 20 mm Hg, and diastolic pressure, 18 mm Hg.) These differences are called the index of response. According to their index of response, subjects can be classified as follows:

Hyporeactors (stable blood pressure): Exhibit a rise of diastolic and/or systolic pressure ranging from 0 to 22 mm Hg or a drop in pressures

Hyperreactors (labile blood pressure): Exhibit a rise of 23 mm Hg or more in the diastolic and/or systolic blood pressure

Is the subject tested a hypo- or hyperreactor?

___

Skin Color as an Indicator of Local Circulatory Dynamics

Skin color reveals with surprising accuracy the state of the local circulation, and allows inferences concerning the larger blood vessels and the circulation as a whole. The Activity 8 experiments on local circulation illustrate a number of factors that affect blood flow to the tissues.

Clinical expertise often depends upon good observation skills, accurate recording of data, and logical interpretation of the findings. A single example will be given to demonstrate this statement: A massive hemorrhage may be internal and hidden (thus, not obvious) but will still threaten the blood delivery to the brain and other vital organs. One of the earliest compensatory responses of the body to such a threat is constriction of cutaneous blood vessels, which reduces blood flow to the skin and diverts it into the circulatory mainstream to serve other, more vital tissues. As a result, the skin of the face and particularly of the extremities becomes pale, cold, and eventually moist with perspiration. Therefore, pale, cold, clammy skin should immediately lead the careful diagnostician to suspect that the circulation is dangerously inefficient. Other conditions, such as local arterial obstruction and venous congestion, as well as certain pathologies of the heart and lungs, also alter skin texture, color, and circulation in characteristic ways.

ACTIVITY 8

Examining the Effect of Local Chemical and Physical Factors on Skin Color

The local blood supply to the skin (indeed, to any tissue) is influenced by (1) local metabolites, (2) oxygen supply, (3) local temperature, (4) autonomic nervous system impulses, (5) local vascular reflexes, (6) certain hormones, and (7) substances released by injured tissues. A number of these factors are examined in the simple experiments that follow. Each experiment should be conducted by students in groups of three or four. One student will act as the subject; the others will conduct the tests and make and record observations.

Vasodilation and Flushing of the Skin Due to Local Metabolites

1. Obtain a sphygmomanometer (blood pressure cuff) and stethoscope. You will also need a watch or clock with a second hand.

2. The subject should bare both arms by rolling up the sleeves as high as possible and then lay the forearms side by side on the bench top.

3. Observe the general color of the subject's forearm skin, and the normal contour and size of the veins. Notice whether skin color is bilaterally similar. Record your observations:

___

___

___

4. Apply the blood pressure cuff to one arm, and inflate it to 250 mm Hg. Keep it inflated for 1 minute. During this period, repeat the observations made above, and record the results:

____

____

5. Release the pressure in the cuff (leaving the deflated cuff in position), and again record the forearm skin color and the condition of the forearm veins. Make this observation immediately after deflation and then again 30 seconds later.

Immediately after deflation:______________________________

____

30 sec after deflation:______________________________

____

The above observations constitute your baseline information. Now conduct the following tests.

6. Instruct the subject to raise the cuffed arm above his or her head and to clench the fist as tightly as possible. While the hand and forearm muscles are tightly contracted, rapidly inflate the cuff to 240 mm Hg or more. This maneuver partially empties the hand and forearm of blood and stops most blood flow to the hand and forearm. Once the cuff has been inflated, the subject is to relax the fist and return the forearm to the bench top so that it can be compared to the other forearm.

7. Leave the cuff inflated for exactly 1 minute. During this interval, compare the skin color in the "ischemic" (blood-deprived) hand to that of the "normal" (non-cuffed-limb) hand. Quickly release the pressure immediately after 1 minute.

What are the subjective effects* of stopping blood flow to the arm and hand for 1 minute?

____

____

What are the objective effects (color of skin and condition of veins)?

____

____

How long does it take for the subject's ischemic hand to regain its normal color?

____

*Subjective effects are sensations—such as pain, coldness, warmth, tingling, and weakness—experienced by the subject. They are "symptoms" of a change in function.

Effects of Venous Congestion

1. Again, but with a different subject, observe and record the appearance of the skin and veins on the forearms resting on the bench top. This time, pay particular attention to the color of the fingers, particularly the distal phalanges, and the nail beds. Record this information:

____

____

2. Wrap the blood pressure cuff around one of the subject's arms, and inflate it to 40 mm Hg. Maintain this pressure for 5 minutes. Make a record of the subjective and objective findings just before the 5 minutes are up, and then again immediately after release of the pressure at the end of 5 minutes.

Subjective (arm cuffed):______________________________

____

Objective (arm cuffed):______________________________

____

Subjective (pressure released):______________________________

____

Objective (pressure released):______________________________

____

3. With still another subject, conduct the following simple experiment: Raise one arm above the head, and let the other hang by the side for 1 minute. After 1 minute, quickly lay both arms on the bench top, and compare their color.

Color of raised arm:______________________________

Color of dependent arm:______________________________

From this and the two preceding observations, analyze the factors that determine tint of color (pink or blue) and intensity of skin color (deep pink or blue as opposed to light pink or blue). Record your conclusions.

____

____

____

____

____

Collateral Blood Flow

In some diseases, blood flow to an organ through one or more arteries may be completely and irreversibly obstructed. Fortunately, in most cases a given body area is supplied both by one main artery and by anastomosing channels connecting the main artery with one or more neighboring blood vessels. Consequently, an organ may remain viable even though its main arterial supply is occluded, as long as the **collateral vessels** are still functional.

The effectiveness of collateral blood flow in preventing ischemia can be easily demonstrated.

1. Check the subject's hands to be sure they are *warm* to the touch. If not, choose another subject, or warm the subject's hands in 35°C water for 10 minutes before beginning.

2. Palpate the subject's radial and ulnar arteries approximately 2.5 cm (1 in.) above the wrist flexure, and mark their locations with a felt marker.

3. Instruct the subject to supinate one forearm and to hold it in a partially flexed (about a 30° angle) position, with the elbow resting on the bench top.

4. Face the subject and grasp his or her forearm with both of your hands, the thumb and fingers of one hand compressing the marked radial artery and the thumb and fingers of the other hand compressing the ulnar artery. Maintain the pressure for 5 minutes, noticing the progression of the subject's hand to total ischemia.

5. At the end of 5 minutes, release the pressure abruptly. Record the subject's sensations, as well as the intensity and duration of the flush in the previously occluded hand. (Use the other hand as a baseline for comparison.)

____

____

____

6. Allow the subject to relax for 5 minutes; then repeat the maneuver, but this time *compress only the radial artery.* Record your observations.

____

____

How do the results of the first test differ from those of the second test with respect to color changes during compression and to the intensity and duration of reactive hyperemia (redness of the skin)?

____

____

____

____

7. Once again allow the subject to relax for 5 minutes. Then repeat the maneuver, *with only the ulnar artery compressed.* Record your observations:

____

____

What can you conclude about the relative sizes of, and hand areas served by, the radial and ulnar arteries?

____

____

Effect of Mechanical Stimulation of Blood Vessels of the Skin

With moderate pressure, draw the blunt end of your pen across the skin of a subject's forearm. Wait 3 minutes to observe the effects, and then repeat with firmer pressure.

What changes in skin color do you observe with light-to-moderate pressure?

____

With heavy pressure?______________________________

The redness, or *flare,* observed after mechanical stimulation of the skin results from a local inflammatory response promoted by chemical mediators released by injured tissues. These mediators stimulate increased blood flow into the area and leaking of fluid (from the capillaries) into the local tissues. (**Note:** People differ considerably in skin sensitivity. Those most sensitive will show **dermatographism,** a condition in which the direct line of stimulation will swell quite obviously. This excessively swollen area is called a *wheal.*)

NAME ______________________________

LAB TIME/DATE ______________________

REVIEW SHEET

EXERCISE 33

Human Cardiovascular Physiology: Blood Pressure and Pulse Determinations

Cardiac Cycle

1. Using the grouped sets of terms to the right of the diagram, correctly identify each trace, valve closings and openings, and each time period of the cardiac cycle.

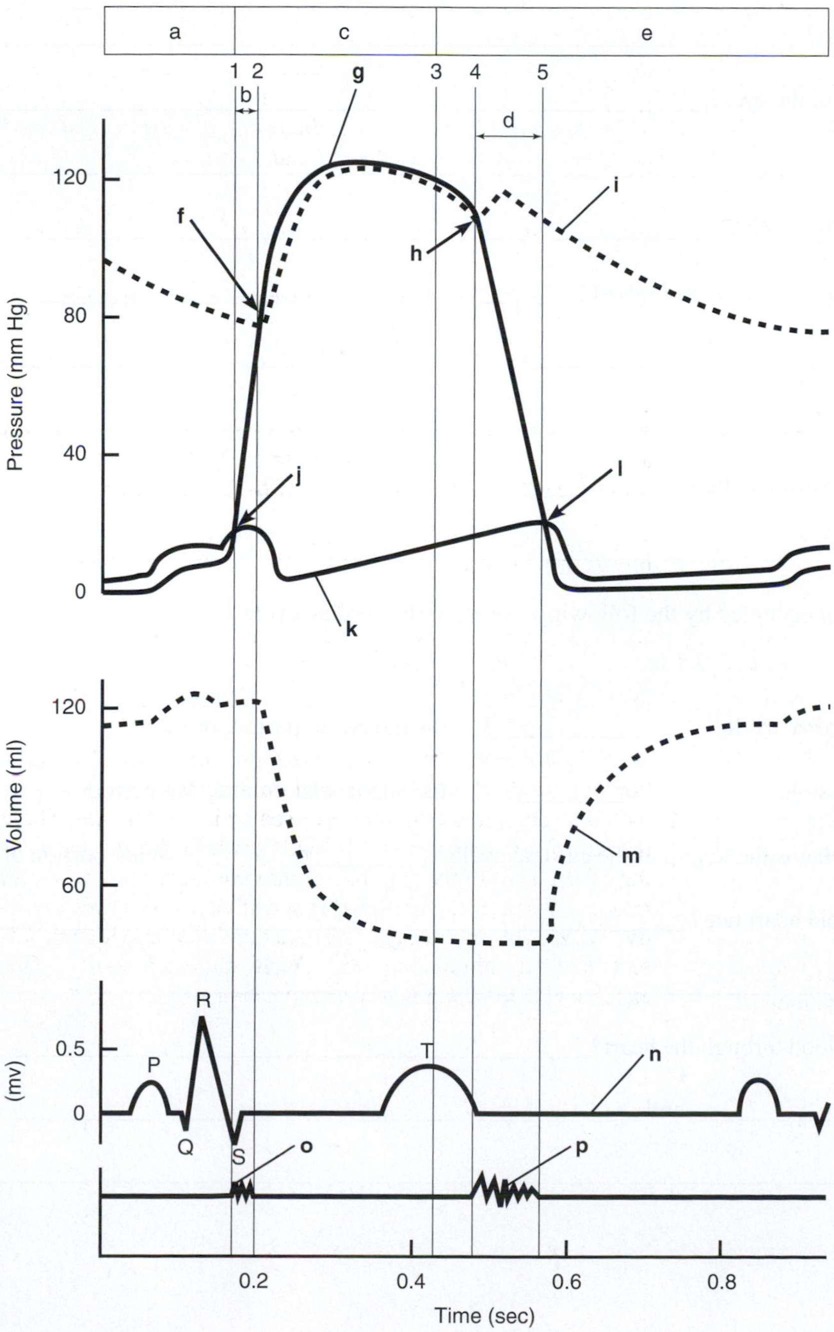

a
c
e
1 2
b
g
3 4 5
d
120
80
40
0
f
h
i
j
k
l
Pressure (mm Hg)
120
60
Volume (ml)
m
R
0.5
P
T
n
0
Q
S
(mv)
o
p
0.2
0.4
0.6
0.8
Time (sec)

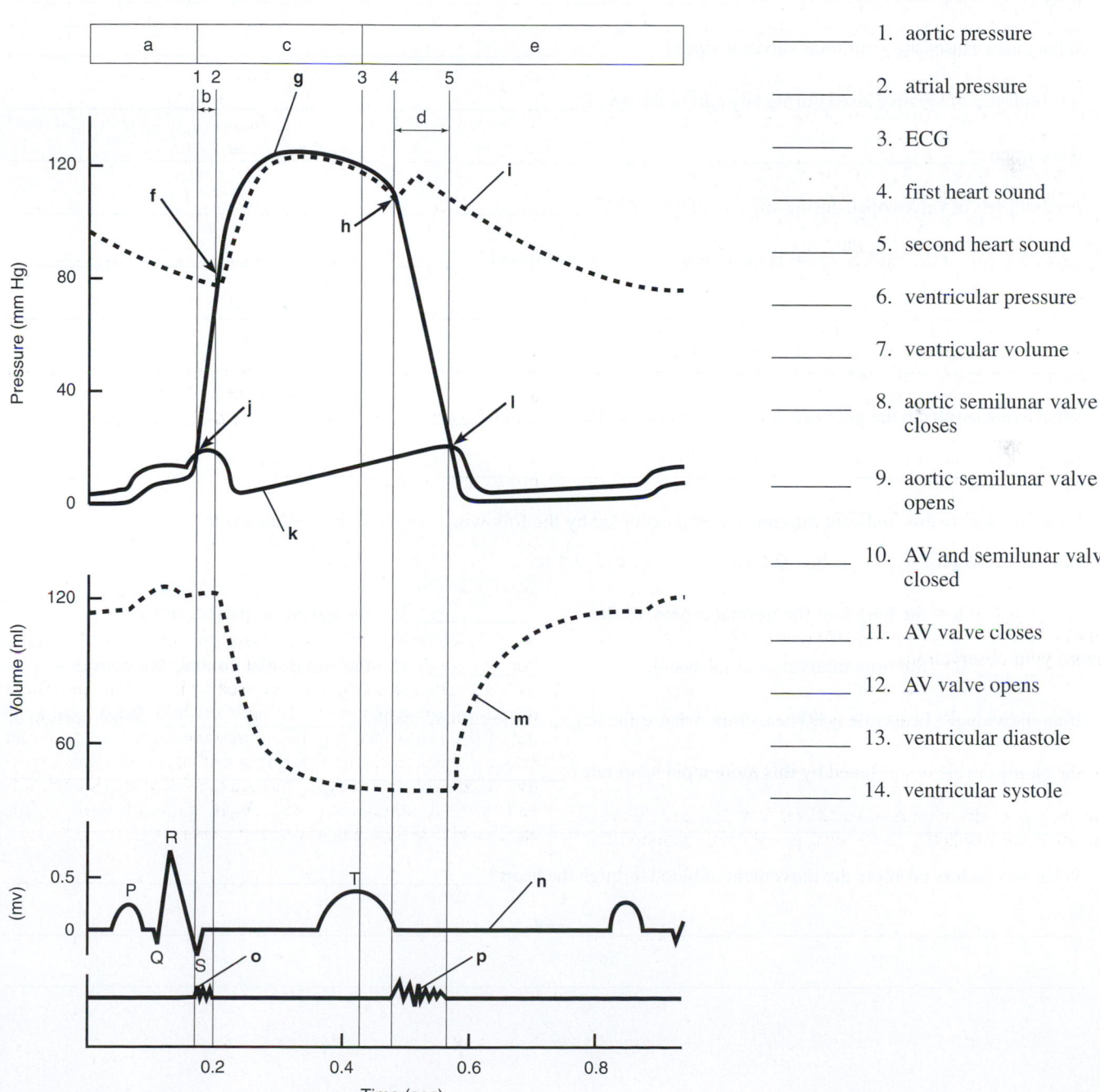

________ 1. aortic pressure

________ 2. atrial pressure

________ 3. ECG

________ 4. first heart sound

________ 5. second heart sound

________ 6. ventricular pressure

________ 7. ventricular volume

________ 8. aortic semilunar valve closes

________ 9. aortic semilunar valve opens

________ 10. AV and semilunar valves closed

________ 11. AV valve closes

________ 12. AV valve opens

________ 13. ventricular diastole

________ 14. ventricular systole

2. Define the following terms.

systole: ______

diastole: ______

cardiac cycle: ______

3. Answer the following questions concerning events of the cardiac cycle.

When are the AV valves closed? ______

What event within the heart causes the AV valves to open? ______

When are the semilunar valves closed? ______

What event causes the semilunar valves to open? ______

Are both sets of valves closed during any part of the cycle? ______

If so, when? ______

Are both sets of valves open during any part of the cycle? ______

At what point in the cardiac cycle is the pressure in the heart highest? ______

Lowest ______

What event results in the pressure deflection called the dicrotic notch? ______

4. Using the key below, indicate the time interval occupied by the following events of the cardiac cycle.

Key: a. *0.8 sec* b. *0.4 sec* c. *0.3 sec* d. *0.1 sec*

______ 1. the length of the normal cardiac cycle

______ 2. the time interval of atrial systole

______ 3. the quiescent period, or pause

______ 4. the ventricular contraction period

5. If an individual's heart rate is 80 beats/min, what is the length of the cardiac cycle? ______ What portion of the cardiac cycle is shortened by this more rapid heart rate? ______

6. What two factors promote the movement of blood through the heart? ______

______ and ______

Heart Sounds

7. Complete the following statements.

The monosyllables describing the heart sounds are __1__. The first heart sound is a result of closure of the __2__ valves, whereas the second is a result of closure of the __3__ valves. The heart chambers that have just been filled when you hear the first heart sound are the __4__, and the chambers that have just emptied are the __5__. Immediately after the second heart sound, both the __6__ and __7__ are filling with blood.

1. ______________________
2. ______________________
3. ______________________
4. ______________________
5. ______________________
6. ______________________
7. ______________________

8. As you listened to the heart sounds during the laboratory session, what differences in pitch, length, and amplitude (loudness) of the two sounds did you observe?______________________

____

____

9. In order to auscultate most accurately, indicate where you would place your stethoscope for the following sounds:

closure of the tricuspid valve:______________________

closure of the aortic semilunar valve:______________________

apical heartbeat:______________________

Which valve is heard most clearly when the apical heartbeat is auscultated?______________________

10. No one expects you to be a full-fledged physician on such short notice; but on the basis of what you have learned about heart sounds, how might abnormal sounds be used to diagnose heart problems?

____

____

____

The Pulse

11. Define *pulse*.______________________

12. Describe the procedure used to take the pulse.______________________

____

____

13. Identify the artery palpated at each of the pressure points listed.

at the wrist:______________________ on the dorsum of the foot:______________________

in front of the ear:______________________ at the side of the neck:______________________

14. When you were palpating the various pulse or pressure points, which appeared to have the greatest amplitude or tension?

______________________ Why do you think this was so?______________________

____

15. Assume someone has been injured in an auto accident and is hemorrhaging badly. What pressure point would you compress to help stop bleeding from each of the following areas?

the thigh:______________________ the calf:______________________

the forearm:______________________ the thumb:______________________

16. How could you tell by simple observation whether bleeding is arterial or venous?______________________

____

17. You may sometimes observe a slight difference between the value obtained from an apical pulse (beats/min) and that from an arterial pulse taken elsewhere on the body. What is this difference called?

____

Blood Pressure Determinations

18. Define *blood pressure.*______________________

19. Identify the phase of the cardiac cycle to which each of the following apply.

systolic pressure:______________________ diastolic pressure:______________________

20. What is the name of the instrument used to compress the artery and record pressures in the auscultatory method of determining blood pressure?______________________

21. What are the sounds of Korotkoff?______________________

____

What causes the systolic sound?______________________

____

What causes the disappearance of the sound?______________________

22. Interpret 145/85/82.______________________

____

____

23. Assume the following BP measurement was recorded for an elderly patient with severe arteriosclerosis: 170/110/–. Explain the inability to obtain the third reading.

____

____

24. Define *pulse pressure.* ____________________

Why is this measurement important? ____________________

____

25. How do venous pressures compare to arterial pressures? ____________________

Why? ____________________

26. What maneuver to increase the thoracic pressure illustrates the effect of external factors on venous pressure? ________

How is it performed? ____________________

____

27. What might an abnormal increase in venous pressure indicate? (Think!) ____________________

____

Observing the Effect of Various Factors on Blood Pressure and Heart Rate

28. What effect do the following have on blood pressure? (Indicate increase by ↑ and decrease by ↓.)

________ 1. increased diameter of the arterioles

________ 2. increased blood viscosity

________ 3. increased cardiac output

________ 4. hemorrhage

________ 5. arteriosclerosis

________ 6. increased pulse rate

29. In which position (sitting, reclining, or standing) is the blood pressure normally the highest?

____________________ The lowest? ____________________

What immediate changes in blood pressure did you observe when the subject stood up after being in the sitting or reclining position? ____________________

____

What changes in the blood vessels might account for the change? ____________________

____

____

After the subject stood for 3 minutes, what changes in blood pressure were observed? ____________________

____

How do you account for this change? ____________________

30. What was the effect of exercise on blood pressure?______________________________

On pulse rate?______________________ Do you think these effects reflect changes in cardiac output *or* in peripheral resistance?______________________________

Why are there normally no significant increases in diastolic pressure after exercise?______________________________

____

31. What effects of the following did you observe on blood pressure in the laboratory?

cold temperature:______________________

What do you think the effect of heat would be?______________________________

Why?______________________________

32. Differentiate between a hypo- and a hyperreactor relative to the cold pressor test.______________________________

____

Skin Color as an Indicator of Local Circulatory Dynamics

33. Describe normal skin color and the appearance of the veins in the subject's forearm before any testing was conducted.

____

34. What changes occurred when the subject emptied the forearm of blood (by raising the arm and making a fist) and the flow was occluded with the cuff?______________________________

What changes occurred during venous congestion?______________________________

____

35. What is the importance of collateral blood supplies?______________________________

____

36. Explain the mechanism by which mechanical stimulation of the skin produced a flare.______________________________

____

EXERCISE 34

Frog Cardiovascular Physiology

MATERIALS

- ☐ Dissecting instruments and tray
- ☐ Disposable gloves
- ☐ Petri dishes
- ☐ Medicine dropper
- ☐ Millimeter ruler
- ☐ Disposal container for organic debris
- ☐ Frog Ringer's solutions (at room temperature, 5°C, and 32°C)
- ☐ Frogs*
- ☐ Thread
- ☐ Large rubber bands
- ☐ Fine common pins
- ☐ Frog board
- ☐ Cotton balls
- ☐ Apparatus A or B:[†]
  A: Physiograph (polygraph), physiograph paper and ink, force transducer, transducer cable, transducer stand, stimulator output extension cable, electrodes

BIOPAC® BIOPAC® MP36 (or MP35/30) data acquisition unit, PC or Mac computer BIOPAC® BSL *PRO* Software, BIOPAC® HDW100A tension adjuster (or equivalent), BIOPAC® SS12LA force transducer with S-hook, small hook with thread, and transducer (or ring) stand

*Instructor will double-pith frogs as required for student experimentation.

[†]**Note:** Instructions for using PowerLab® equipment can be found on MasteringA&P.

Text continues on next page.

MasteringA&P™

Access practice quizzes and more in the Study Area at www.masteringaandp.com.

PEx

This lab corresponds to PhysioEx Exercise 6. See p. PEx-93.

OBJECTIVES

1. To list the properties of cardiac muscle as automaticity and rhythmicity, and define each property.
2. To explain the statement "Cardiac muscle has an intrinsic ability to beat."
3. To compare the intrinsic rate of contraction of the "pacemaker" of the frog heart (sinus venosus) to that of the atria and ventricle.
4. To compare the relative length of the refractory period of cardiac muscle with that of skeletal muscle, and to explain why it is not possible to tetanize cardiac muscle.
5. To define *extrasystole,* and to explain at what point in the cardiac cycle (and on an ECG tracing) an extrasystole can be induced.
6. To describe the effect of the following on heart rate: cold, heat, vagal stimulation, pilocarpine, atropine sulfate, epinephrine, digitalis, and potassium, sodium, and calcium ions.
7. To define *vagal escape* and discuss its value.
8. To define *ectopic pacemaker.*
9. To define *partial* and *total heart block.*
10. To describe how heart block was induced in the laboratory and explain the results, and to explain how heart block might occur in the human body.
11. To list the components of the microcirculatory system.
12. To identify an arteriole, venule, and capillaries in a frog's web, and to cite differences between relative size, rate of blood flow, and regulation of blood flow in these vessels.
13. To describe the effect of heat, cold, local irritation, and histamine on the rate of blood flow in the microcirculation, and to explain how these responses help maintain homeostasis.

PRE-LAB QUIZ

1. Circle True or False. Heart muscle can depolarize spontaneously in the absence of any external stimulation.
2. Spontaneous depolarization-repolarization events occur in a regular and continuous manner in cardiac muscle, a property known as
   a. automaticity b. rhythmicity c. synchronicity
3. How many chambers does the frog heart have?
   a. two b. three c. four
4. Circle True or False. Heart rate can be modified by extrinsic impulses from the autonomic nerves.
5. What is an extrasystole? ______________________________

   ____
6. Which chemical agent will you use to modify the frog heart rate?
   a. caffeine c. magnesium solution
   b. digitalis d. Ringer's solution

Text continues on next page.

New versions of the BSL software (3.7.5 and higher for Windows, and 3.7.4 and higher for Mac) require slightly different channel settings and collection strategies. Instructions for using the newer software with the MP36/35 data acquisition units can be found on MasteringA&P. In addition, instructions for use of the NEW 2-channel data acquisition unit, the MP45, may also be found on MasteringA&P.

- ☐ Dropper bottles of freshly prepared solutions (using frog Ringer's solution as the solvent) of the following:
  - 2.5% pilocarpine
  - 5% atropine sulfate
  - 1% epinephrine
  - 2% digitalis
  - 2% calcium chloride ($CaCl_2$)
  - 0.7% sodium chloride (NaCl)
  - 5% potassium chloride (KCl)
  - 0.01% histamine
  - 0.01 *N* HCl
- ☐ Dissecting pins
- ☐ Paper towels
- ☐ Compound microscope

7. The __________ nerve carries parasympathetic impulses to the heart.
   a. cardiac b. olfactory c. phrenic d. vagus
8. Circle the correct term. The phenomenon of vagal escape / heart block occurs when the heart stops momentarily then begins to beat again.
9. Circle True or False. The flow of blood through capillary beds is slow and intermittent.
10. __________ causes extensive vasodilation when applied to the frog web.
    a. Calcium c. Histamine
    b. Epinephrine d. Ringer's solution

Investigations of human cardiovascular physiology are very interesting, but many areas obviously do not lend themselves to experimentation. It would be tantamount to murder to inject a human subject with various drugs to observe their effects on heart activity or to expose the human heart in order to study the length of its refractory period. However, this type of investigation can be done on frogs or small laboratory animals and provides valuable data because the physiological mechanisms in these animals are similar, if not identical, to those in humans.

In this exercise, you will conduct the cardiac investigations just mentioned and others. In addition, you will observe the microcirculation in a frog's web and subject it to various chemical and thermal agents to demonstrate their influence on local blood flow.

Special Electrical Properties of Cardiac Muscle: Automaticity and Rhythmicity

Cardiac muscle differs from skeletal muscle both functionally and in its fine structure. Skeletal muscle must be electrically stimulated to contract. In contrast, heart muscle can and does depolarize spontaneously in the absence of external stimulation. This property, called **automaticity,** is due to plasma membranes that have reduced permeability to potassium ions but still allow sodium ions to slowly leak into the cells. This leakage causes the muscle cells to gradually depolarize until the action potential threshold is reached and *fast calcium channels* open, allowing Ca^{2+} entry from the extracellular fluid. Shortly thereafter, contraction occurs. Also, the spontaneous depolarization-repolarization events occur in a regular, continuous manner in cardiac muscle, a property called **rhythmicity.**

In the following experiment, you will observe these properties of cardiac muscle in vitro (that is, removed from the body). Work together in groups of three or four. (The instructor may choose to demonstrate this procedure if time or frogs are at a premium.)

ACTIVITY 1

Investigating the Automaticity and Rhythmicity of Heart Muscle

1. Obtain a dissecting tray and instruments, disposable gloves, two petri dishes, frog Ringer's solution, a metric ruler, and a medicine dropper, and bring them to your laboratory bench.

2. Don the gloves, and then request and obtain a doubly pithed frog from your instructor. Quickly open the thoracic cavity and observe the heart rate in situ (at the site or within the body).

Record the heart rate: ______________ beats/min

3. Dissect out the heart and the gastrocnemius muscle of the calf and place the removed organs in separate petri dishes containing frog Ringer's solution. (**Note:** Just in case you don't remember, the procedure for removing the gastrocnemius

muscle is provided on pages 240–241 in Exercise 16. The extreme care used in that procedure for the removal of the gastrocnemius muscle need not be exercised here.)

4. Observe the activity of the two organs for a few seconds.

Which is contracting? ______________________________

At what rate? ______________ beats/min

Is the contraction rhythmic? ______________

5. Sever the sinus venosus from the heart (see Figure 34.1). The **sinus venosus** of the frog's heart corresponds to the SA node of the human heart.

Does the sinus venosus continue to beat? ______________

If not, lightly touch it with a probe to stimulate it. Record its rate of contraction.

Rate: ______________ beats/min

6. Sever the right atrium from the heart; then remove the left atrium. Does each atrium continue to beat?

______________ Rate: ______________ beats/min

Does the ventricle continue to beat? ______________

Rate: ______________ beats/min

7. Notice in Figure 34.1 that frogs have a single ventricle. Fragment the ventricle to determine how small the ventricular fragments must be before the automaticity of ventricular muscle is abolished. Measure these fragments and record their approximate size.

________ mm × ________ mm × ________ mm

Which portion of the heart exhibited the most marked automaticity?

____

Which showed the least? ______________________________

8. Properly dispose of the frog and heart fragments in the organic debris container before continuing.

Baseline Frog Heart Activity

The heart's effectiveness as a pump is dependent both on intrinsic (within the heart) and extrinsic (external to the heart) controls. In this activity, you will investigate some of these factors.

The nodal system, in which the "pacemaker" imposes its depolarization rate on the rest of the heart, is one intrinsic factor that influences the heart's pumping action. If its impulses fail to reach the ventricles (as in heart block), the ventricles continue to beat but at their own inherent rate, which is much slower than that usually imposed on them. Although heart contraction does not depend on nerve impulses, its rate can be modified by extrinsic impulses reaching it through the autonomic nerves. Additionally, cardiac activity is modified by various chemicals, hormones, ions, and metabolites. The effects of several of these chemical factors are examined in the next experimental series.

The frog heart has two atria and a single, incompletely divided ventricle (see Figure 34.1). The pacemaker is

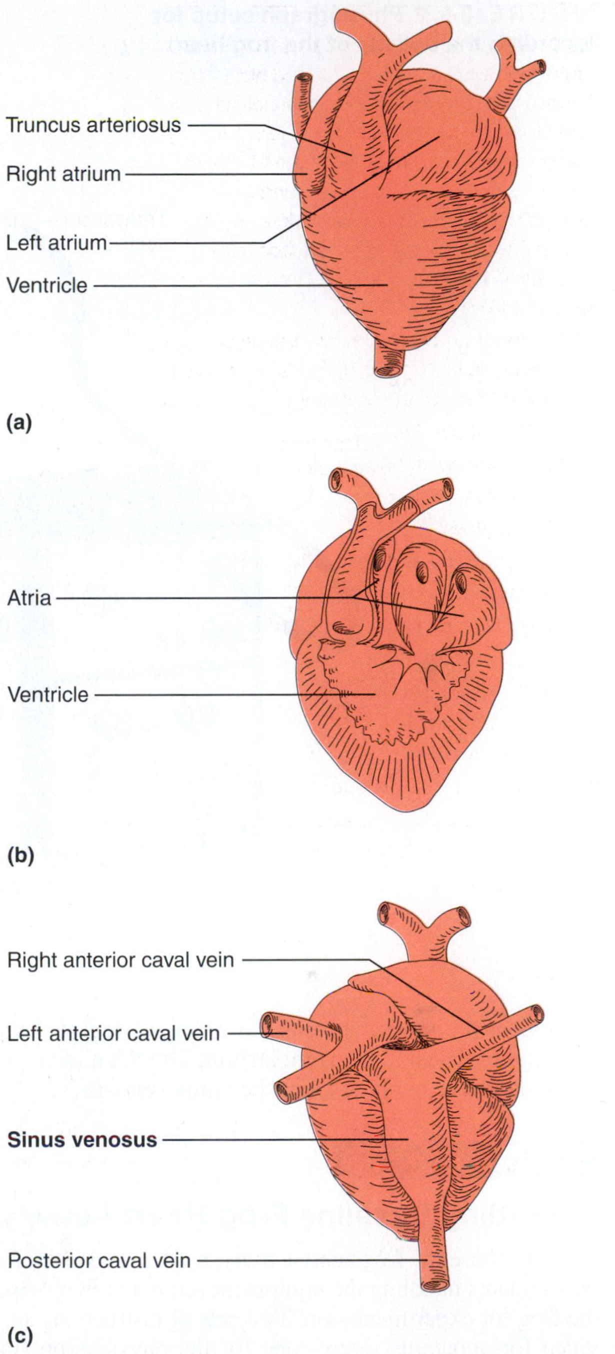
Truncus arteriosus
Right atrium
Left atrium
Ventricle
(a)
Atria
Ventricle
(b)
Right anterior caval vein
Left anterior caval vein
Sinus venosus
Posterior caval vein
(c)

FIGURE 34.1 Anatomy of the frog heart. (a) Ventral view showing the single truncus arteriosus leaving the undivided ventricle. **(b)** Longitudinal section showing the two atrial and single ventricular chambers. **(c)** Dorsal view showing the sinus venosus (pacemaker).

FIGURE 34.2 Physiograph setup for recording the activity of the frog heart.

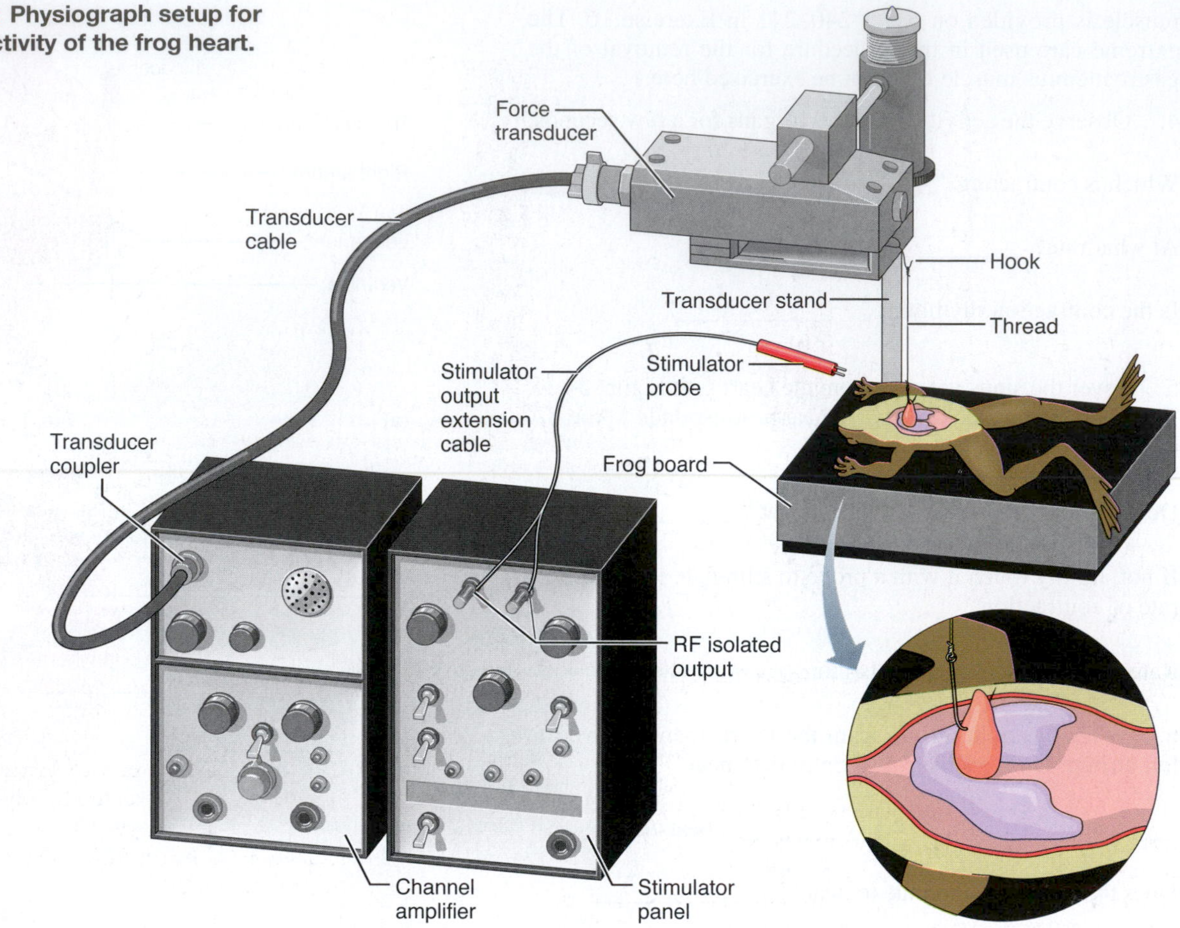
Force transducer
Transducer cable
Hook
Transducer stand
Thread
Stimulator output extension cable
Stimulator probe
Transducer coupler
Frog board
RF isolated output
Channel amplifier
Stimulator panel

located in the sinus venosus, an enlarged region between the venae cavae and the right atrium. The SA node of mammals may have evolved from the sinus venosus.

ACTIVITY 2

Recording Baseline Frog Heart Activity

To record baseline frog heart activity, work in groups of four—two students handling the equipment setup and two preparing the frog for experimentation. Two sets of instructions are provided for apparatus setup—one for the physiograph (Figure 34.2), the other for BIOPAC® (Figure 34.3). Follow the procedure outlined for the apparatus you will be using.

Apparatus Setups

Physiograph Apparatus Setup

1. Obtain a force transducer, transducer cable, and transducer stand, and bring them to the recording site.

2. Attach the force transducer to the transducer stand as shown in Figure 34.2.

3. Then attach the transducer cable to the transducer coupler (input) on the channel amplifier of the physiograph and to the force transducer.

4. Attach the stimulator output extension cable to output on the stimulator panel (red to red, black to black).

BIOPAC® Apparatus Setup

1. Connect the BIOPAC® apparatus to the computer and turn the computer **ON.**

2. Make sure the BIOPAC® unit is **OFF.**

3. Set up the equipment as shown in Figure 34.3.

4. Turn the BIOPAC® unit **ON.**

5. Launch the BIOPAC® BSL *PRO* software by clicking the icon on the desktop or by following your instructor's guidance.

6. Open the Frog Heart template by going to the **File** menu at the top of the screen and choosing **Open > Files of Type = Graph Template (GTL) > FrogHeart.gtl.**

7. Put the tension adjuster (BIOPAC® HDW100A, or equivalent) on the transducer stand, and attach the BIOPAC® SS12LA force transducer with the hook holes pointing down. Level the force transducer both horizontally and vertically.

8. Set the tension adjuster to approximately one quarter of its full range. (***Note:*** *Do not firmly tighten any of the thumbscrews at this stage.*) Select a force range of 0 to 50 grams for this experiment.

9. Select and attach the small S-hook to the force transducer.

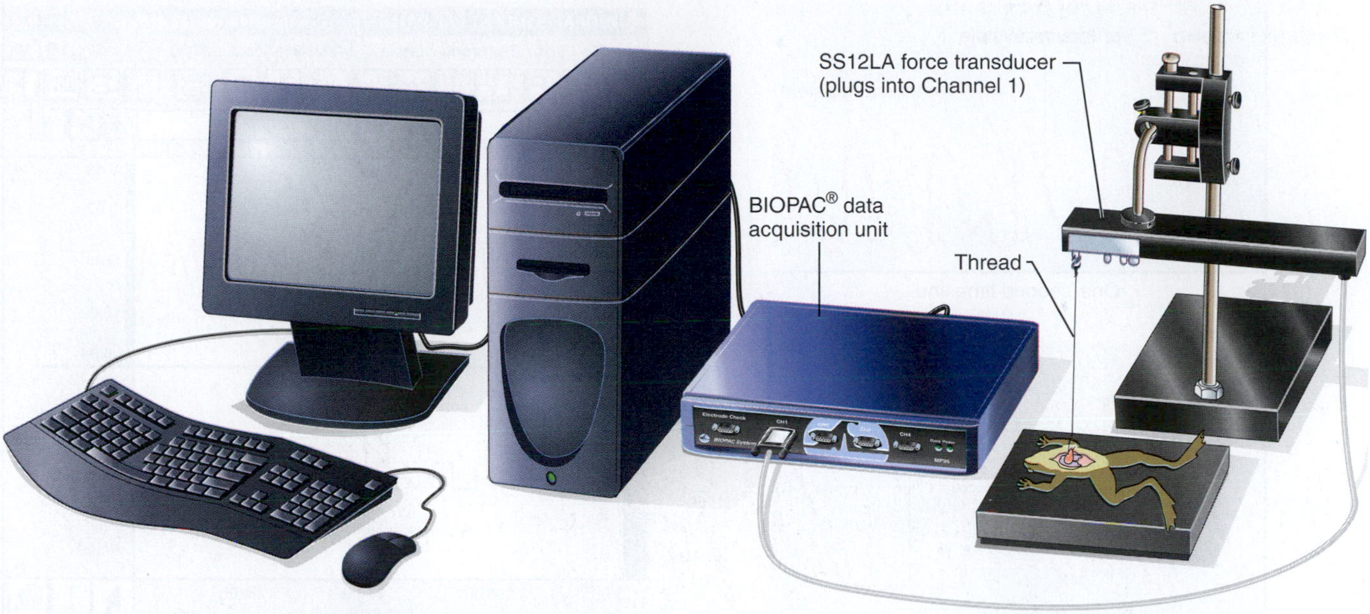
SS12LA force transducer
(plugs into Channel 1)
BIOPAC® data
acquisition unit
Thread

FIGURE 34.3 BIOPAC® setup for recording the activity of the frog heart.

Preparation of the Frog

1. Obtain room-temperature frog Ringer's solution, a medicine dropper, dissecting instruments and tray, disposable gloves, fine common pins (physiograph) or small hook (BIOPAC®), cotton ball, frog board, large rubber bands, and some thread, and bring them to your bench.

2. Don the gloves, and obtain a doubly pithed frog from your instructor.

3. Make a longitudinal incision through the abdominal and thoracic walls with scissors, and then cut through the sternum to expose the heart.

4. Grasp the pericardial sac with forceps, and cut it open so that the beating heart can be observed.

Is the sequence an atrial-ventricular one? ______________

5. Locate the vagus nerve, which runs down the lateral aspect of the neck and parallels the trachea and carotid artery. (In good light, it appears to be striated.) Slip an 18-inch length of thread under the vagus nerve so that it can later be lifted away from the surrounding tissues by the thread. Then place a Ringer's solution–soaked cotton ball over the nerve to keep it moistened until you are ready to stimulate it later in the procedure.

6. Using a medicine dropper, flush the heart with Ringer's solution. *From this point on the heart must be kept continually moistened with room-temperature Ringer's solution unless other solutions are being used for the experimentation.*

7. Attach the frog to the frog board using large rubber bands.

Physiograph Frog Heart Preparation

1. Bend a common pin to a 90° angle, and tie to its head a thread 0.46–0.5 m (18–20 inches) long. Force the pin through the apex of the heart (but do not penetrate the ventricular chamber) until the apex is well secured in the angle of the pin.

2. Tie the thread from the heart to the hook on the force transducer. Do not pull the thread too tightly. It should be taut enough to lift the heart apex upward, away from the thorax, but should *not* stretch the heart. Adjust the force transducer as necessary. (See Figure 34.2.)

BIOPAC® Frog Heart Preparation

1. Attach a small hook tied with thread to the frog heart, following the instructions in step 1 of the physiograph instructions above to insert it through the apex of the heart. Confirm that the prepared frog is firmly attached to the frog board, positioned below the ring stand with the line running vertically from the frog heart to the transducer.

2. Slide the tension adjuster/force transducer assembly down the ring stand until you can hang the loop loosely from the S-hook, then slide it back up until the line is taut but the heart muscle is not stretched (be careful not to tear the heart).

3. Position the tension adjuster and/or force transducer so that the top is level, with approximately 10 cm (4 inches) of line from the heart tendon to the S-hook. Adjust the assembly so that the thread line runs vertically; for a true reflection of the muscle's contractile force, the muscle must not be pulled at an angle.

4. Use the tension adjuster knob to make the line taut and tighten all thumbscrews to secure positioning of the assembly. Let the setup sit for a minute, then recheck the tension to make sure nothing has slipped or stretched.

5. Data may be distorted if the transducer line is not pulling directly vertical from the frog heart to the S-hook. Once again, align the frog as described above and make sure that the heart is not twisting the thread. If it is twisting, you will need to *carefully* remove the hook and repeat the setup.

6. BIOPAC® calculates *rate* data, which always trails the actual rate by one cycle. Data collection is a sensitive process and may display artifacts from table movement, heart movement (for instance, from breathing on the heart), and chemicals

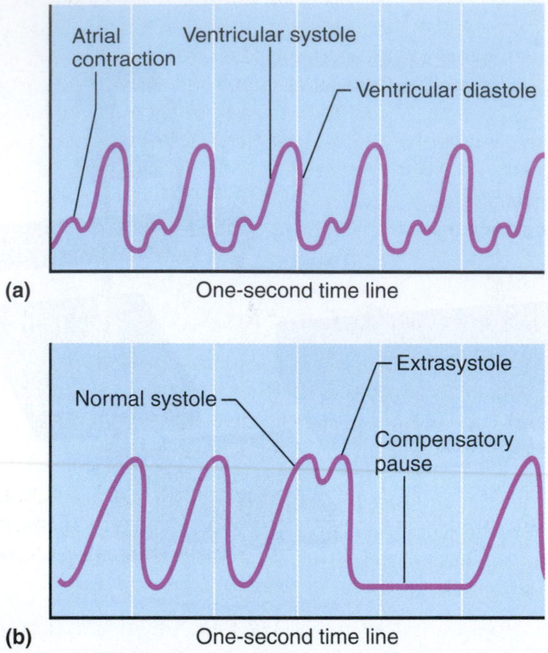
Atrial contraction
Ventricular systole
Ventricular diastole
(a)
One-second time line
Extrasystole
Normal systole
Compensatory pause
(b)
One-second time line

FIGURE 34.4 Physiograph recording of contractile activity of a frog heart. (a) Normal heartbeat. **(b)** Induction of an extrasystole.

touching the heart. To get the best data, keep the experimental area stable, clean, and clear of obstructions. Hints for obtaining best data:

- The SS12LA force transducer must be level on the horizontal and vertical planes.
- Set up the tension adjuster and force transducer in positions that minimize their movement when tension is applied. Keep the point of S-hook attachment as close as possible to the ring stand support.
- Position the tension adjuster so that you will not bump the cables or frog board when using the adjustment knob.
- Position and/or tape the force transducer cables where they will not be pulled or bumped easily.
- Make sure the frog board is on a stable surface.
- Make sure the frog is firmly attached to the frog board so it will not rise up when tension is applied.

Making the Baseline Recording

Using the Physiograph

1. Turn the amplifier on and balance the apparatus according to instructions provided by your instructor. Set the paper speed at 0.5 cm/sec. Press the record and paper advance buttons.

2. Set the signal magnet or time marker at 1/sec.

3. Record 12 to 15 normal heartbeats. Be sure you can distinguish atrial and ventricular contractions (see Figure 34.4). Then adjust the paper or scroll speed so that the peaks of ventricular contractions are approximately 2 cm apart. (Peaks indicate systole; troughs indicate diastole.) Pay attention to the relative force of heart contractions while recording.

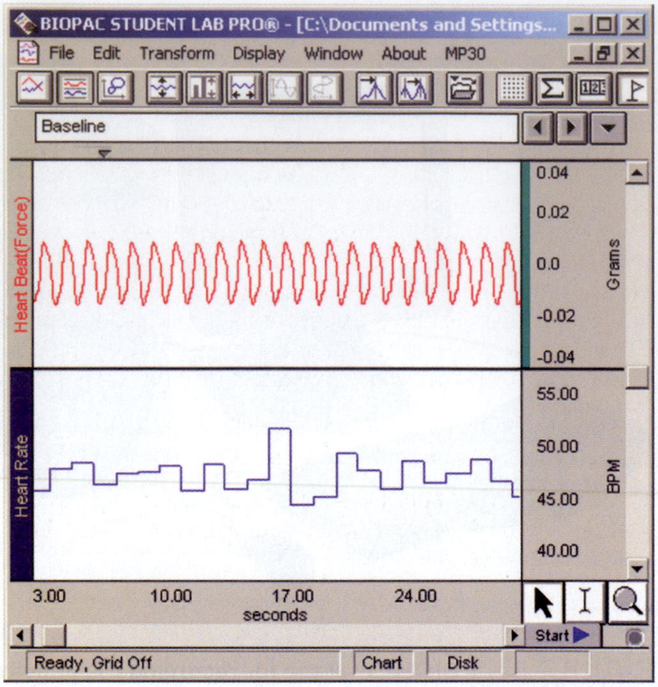
BIOPAC STUDENT LAB PRO® - [C:\Documents and Settings...
File Edit Transform Display Window About MP30
Baseline
Heart Beat(Force)
0.04
0.02
0.0
-0.02
-0.04
Grams
Heart Rate
55.00
50.00
45.00
40.00
BPM
3.00
10.00
17.00
seconds
24.00
Start
Ready, Grid Off
Chart
Disk

FIGURE 34.5 Example of baseline frog heart rate data.

Using BIOPAC®

1. Click **Start** to begin recording.

2. Observe at least 5 heart rate cycles, then click **Stop** to stop recording. Your data should look like that in Figure 34.5.

3. Choose **Save** from the File menu, and type in a filename to save the recorded data. You may want to save by your team's name followed by FrogHeart-1 (for example, Smith-FrogHeart-1).

Analyzing the Baseline Data

Count the number of ventricular contractions per minute from your physiograph or BIOPAC® data, and record:

______________ beats/min

Compute the A–V interval (period from the beginning of atrial contraction to the beginning of ventricular contraction).

______________ sec

How do the two tracings compare in time?

____

____

____

Mark the atrial and ventricular systoles on the record. *Remember to keep the heart moistened with Ringer's solution.*

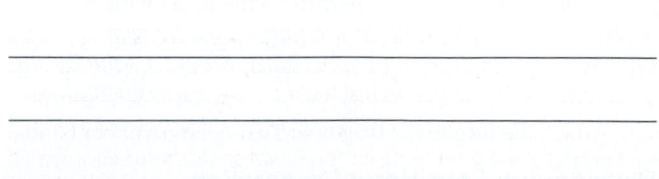

ACTIVITY 3

Investigating the Refractory Period of Cardiac Muscle Using the Physiograph

In conducting Exercise 16, you saw that repeated rapid stimuli could cause skeletal muscle to remain in a contracted state. In other words, the muscle could be tetanized. This was possible because of the relatively short refractory period of skeletal muscle. In this experiment, you will use the physiograph to investigate the refractory period of cardiac muscle and its response to stimulation.* During the procedure, one student should keep the stimulating electrodes in constant contact with the frog heart ventricle while another student operates the stimulator panel.

1. Using the physiograph, set the stimulator to deliver 20-V shocks of 2-msec duration, and begin recording.

2. Deliver single shocks at the beginning of ventricular contraction, the peak of ventricular contraction, and then later and later in the cardiac cycle.

3. Observe the recording for **extrasystoles,** which are extra beats that show up riding on the ventricular contraction peak. Also note the **compensatory pause,** which allows the heart to get back on schedule after an extrasystole. (See Figure 34.4b.)

During which portion of the cardiac cycle was it possible to induce an extrasystole?

____

____

4. Attempt to tetanize the heart by stimulating it at the rate of 20 to 30 impulses per second. What is the result?

____

____

Considering the function of the heart, why is it important that heart muscle cannot be tetanized?

____

____

ACTIVITY 4

Assessing Physical and Chemical Modifiers of Heart Rate

Now that you have observed normal frog heart activity, you will have an opportunity to investigate the effects of various factors that modify heart activity. In each case, record a few normal heartbeats before introducing the modifying factor. After removing the agent, allow the heart to return to its normal rate before continuing with the testing. On each record, indicate the point of introduction and removal of the modifying agent.

*BIOPAC® users may investigate the refractory period of cardiac muscle by using PhysioEx Exercise 6.

For each physical agent or solution that is applied:

If using the physiograph, increase the scroll or paper speed so that heartbeats appear as spikes 4 to 5 mm apart.

If using BIOPAC®, after applying each solution, click the **Start** button. When the effect is observed, record the effect for five cycles, then click **Stop.** Choose **Save** from the File menu to save your data.

(**Note:** Repeat these steps for each physical agent or chemical solution that is being applied.)

Temperature

1. Obtain 5°C and 32°C frog Ringer's solutions and medicine droppers.

2. Bathe the heart with 5°C Ringer's solution, and continue to record until the recording indicates a change in cardiac activity and five cardiac cycles have been recorded.

3. Stop recording, pipette off the cold Ringer's solution (remove the fluid by sucking it into the barrel of a medicine dropper), and flood the heart with room-temperature Ringer's solution.

4. Start recording again to determine the resumption of the normal heart rate. When this has been achieved, flood the heart with 32°C Ringer's solution, and again record five cardiac cycles after a change is noted.

5. Stop the recording, pipette off the warm Ringer's solution, and bathe the heart with room-temperature Ringer's solution once again.

What change occurred with the cold (5°C) Ringer's solution?

____

What change occurred with the warm (32°C) Ringer's solution?

____

6. Count the heart rate at the two temperatures, and record the data below.

________ beats/min at 5°C; ________ beats/min at 32°C

Chemical Agents

Pilocarpine Flood the heart with a 2.5% solution of pilocarpine. Record until a change in the pattern of the ECG is noticed. Pipette off the excess pilocarpine solution, and proceed immediately to the next test, which uses atropine as the testing solution. What happened when the heart was bathed in the pilocarpine solution?

____

Pilocarpine simulates the effect of parasympathetic (vagal) nerve stimulation by enhancing acetylcholine release; such drugs are called parasympathomimetic drugs.

Is pilocarpine an agonist or an antagonist of acetylcholine?

____

Atropine Sulfate Apply a few drops of atropine sulfate to the frog's heart, and observe the recording. If no changes are observed within 2 minutes, apply a few more drops. When you observe a response, pipette off the excess atropine sulfate and flood the heart with room-temperature Ringer's solution. What happens when the atropine sulfate is added?

____

____

Atropine is a drug that blocks the effect of the neurotransmitter acetylcholine, which is liberated by the parasympathetic nerve endings. Do your results accurately reflect this effect of atropine?

____

Is atropine an agonist or an antagonist of acetylcholine?

____

Epinephrine Flood the frog heart with epinephrine solution, and continue to record until a change in heart activity is noted.

What are the results? ______________________________

____

Which division of the autonomic nervous system does its effect mimic?

____

Digitalis Pipette off the excess epinephrine solution, and rinse the heart with room-temperature Ringer's solution. Continue recording, and when the heart rate returns to baseline values, bathe it in digitalis solution. What is the effect of digitalis on the heart?

____

____

____

Digitalis is a drug commonly prescribed for heart patients with congestive heart failure. It slows heart rate, providing more time for venous return and decreasing the work of the weakened heart. These effects are thought to be due to inhibition of the sodium-potassium pump and enhancement of Ca^{2+} entry into the myocardial fibers.

Various Ions To test the effect of various ions on the heart, apply the designated solution until you observe a change in heart rate or in strength of contraction. Pipette off the solution, flush with room-temperature Ringer's solution, and allow the heart to resume its normal rate before continuing. *Do not allow the heart to stop.* If the rate should decrease dramatically, flood the heart with room-temperature Ringer's solution.

Effect of Ca^{2+} (use 2% $CaCl_2$) ______________________________

____

Effect of Na^+ (use 0.7% NaCl) ______________________________

____

Effect of K^+ (use 5% KCl) ______________________________

____

Potassium ion concentration is normally higher within cells than in the extracellular fluid. *Hyperkalemia* decreases the resting potential of plasma membranes, thus decreasing the force of heart contraction. In some cases, the conduction rate of the heart is so depressed that **ectopic pacemakers** (pacemakers appearing erratically and at abnormal sites in the heart muscle) appear in the ventricles, and fibrillation may occur. Was there any evidence of premature beats in the recording of potassium ion effects? ______________

Was arrhythmia produced with any of the ions tested?

______________ If so, which? ______________________________

Vagus Nerve Stimulation

The vagus nerve carries parasympathetic impulses to the heart, which modify heart activity. If you are using the physiograph, you can test this by stimulating the vagus nerve.*

1. Remove the cotton placed over the vagus nerve. Using the previously tied thread, lift the nerve away from the tissues and place the nerve on the stimulating electrodes.

2. Using a duration of 0.5 msec at a voltage of 1 mV, stimulate the nerve at a rate of 50/sec. Continue stimulation until the heart stops momentarily and then begins to beat again **(vagal escape).** If no effect is observed, increase stimulus intensity and try again. If no effect is observed after a substantial increase in stimulus voltage, reexamine your "vagus nerve" to make sure that it is not simply strands of connective tissue.

3. Discontinue stimulation after you observe vagal escape, and flush the heart with room-temperature Ringer's solution until the normal heart rate resumes. What is the effect of vagal stimulation on heart rate?

____

The phenomenon of vagal escape demonstrates that many factors are involved in heart regulation and that any deleterious factor (in this case, excessive vagal stimulation) will be overcome, if possible, by other physiological mechanisms such as activation of the sympathetic division of the autonomic nervous system (ANS).

*BIOPAC® users may observe the effects of vagal stimulation by using PhysioEx Exercise 6.

Intrinsic Conduction System Disturbance (Heart Block)

1. Moisten a 25-cm (10-inch) length of thread and make a Stannius ligature (loop the thread around the heart at the junction of the atria and ventricle).

2. If using a physiograph, decrease the scroll or paper speed to achieve intervals of approximately 2 cm between the ventricular contractions, and record a few normal heartbeats.

3. Tighten the ligature in a stepwise manner while observing the atrial and ventricular contraction curves. As heart block occurs, the atria and ventricle will no longer show a 1:1 contraction ratio. Record a few beats each time you observe a different degree of heart block—a 2:1 ratio of atrial to ventricular contractions, 3:1, 4:1, and so on. As long as you can continue to count a whole number ratio between the two chamber types, the heart is in **partial heart block.** When you can no longer count a whole number ratio, the heart is in **total,** or **complete, heart block.**

4. When total heart block occurs, release the ligature to see if the normal A–V rhythm is reestablished. What is the result?

___

___

5. Attach properly labeled recordings (or copies of the recordings) made during this procedure to the last page of this exercise for future reference.

6. Dispose of the frog remains and gloves in appropriate containers, and dismantle the experimental apparatus before continuing. ■

The Microcirculation and Local Blood Flow

The thin web of a frog's foot provides an excellent opportunity to observe the flow of blood to, from, and within the capillary beds, where the real business of the circulatory system occurs. The collection of vessels involved in the exchange mechanism is referred to as the **microcirculation;** it consists of arterioles, venules, capillaries, and vascular shunts called *metarteriole–thoroughfare channels* (Figure 34.6).

The total cross-sectional area of the capillaries in the body is much greater than that of the veins and arteries combined. Thus, the velocity of flow through the capillary beds is quite slow. Capillary flow is also intermittent, because if all capillary beds were filled with blood at the same time, there would be no blood at all in the large vessels. The flow of blood into the capillary beds is regulated by the activity of muscular *terminal arterioles,* which feed the beds, and by *precapillary sphincters* at entrances to the true capillaries. The amount of blood flowing into the true capillaries of the bed is regulated most importantly by local chemical controls (local concentrations of carbon dioxide, histamine, pH). Thus a capillary bed may be flooded with blood or almost entirely bypassed depending on what is happening within the body or in a particular body region at any one time. You will investigate some of the local controls in the next group of experiments.

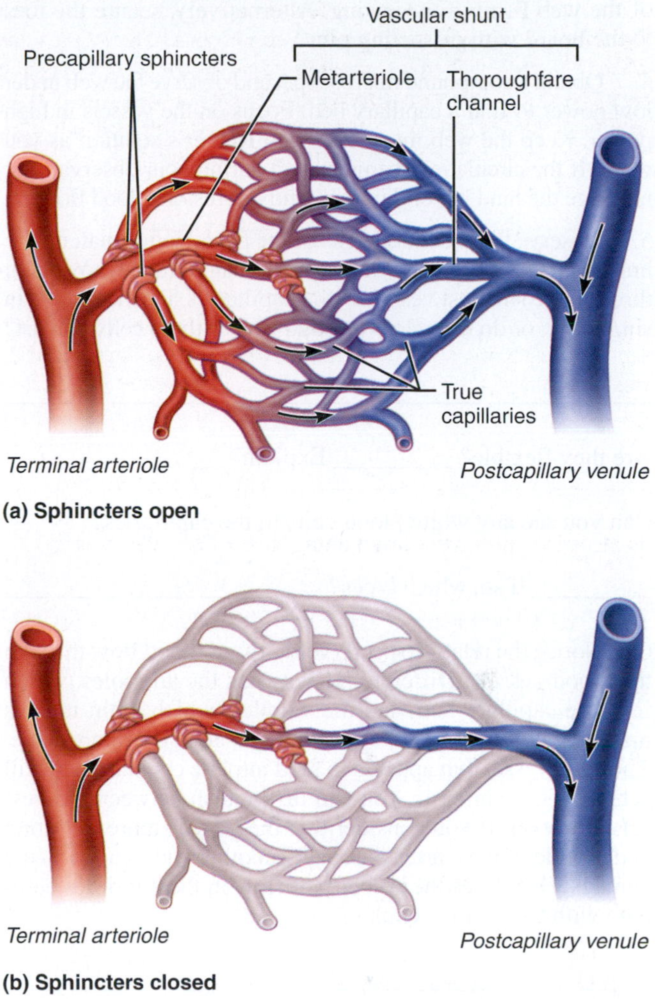
Vascular shunt
Precapillary sphincters
Metarteriole
Thoroughfare channel
True capillaries
Terminal arteriole
Postcapillary venule
(a) Sphincters open
Terminal arteriole
Postcapillary venule
(b) Sphincters closed

FIGURE 34.6 Anatomy of a capillary bed. The composite metarteriole–thoroughfare channels act as shunts to bypass the true capillaries when precapillary sphincters controlling blood entry into the true capillaries are constricted.

ACTIVITY 5

Investigating the Effect of Various Factors on the Microcirculation

1. Obtain a frog board (with a hole at one end), dissecting pins, disposable gloves, frog Ringer's solution (room temperature, 5°C, and 32°C), 0.01 *N* HCl, 0.01% histamine solution, 1% epinephrine solution, a large rubber band, and some paper towels.

2. Put on the gloves and obtain a frog (alive and hopping, *not* pithed). Moisten several paper towels with room-temperature Ringer's solution, and wrap the frog's body securely with them. One hind leg should be left unsecured and extending beyond the paper cocoon.

3. Attach the frog to the frog board (or other supporting structure) with a large rubber band and then carefully spread (but do not stretch) the web of the exposed hindfoot over the hole in the support. Have your partners hold the edges

of the web firmly for viewing. Alternatively, secure the toes to the board with dissecting pins.

4. Obtain a compound microscope, and observe the web under low power to find a capillary bed. Focus on the vessels in high power. Keep the web moistened with Ringer's solution as you work. If the circulation seems to stop during your observations, massage the hind leg of the frog gently to restore blood flow.

5. Observe the red blood cells of the frog. Notice that, unlike human RBCs, they are nucleated. Watch their movement through the smallest vessels—the capillaries. Do they move in single file, or do they flow through two or three cells abreast?

____

Are they flexible? _________ Explain. _________________

Can you see any white blood cells in the capillaries?

_________ If so, which types? _______________________

6. Notice the relative speed (velocity) of blood flow through the blood vessels. Differentiate between the arterioles, which feed the capillary bed, and the venules, which drain it. This may be tricky, because images are reversed in the microscope. Thus, the vessel that appears to feed into the capillary bed will actually be draining it. You can distinguish between the vessels, however, if you consider that the flow is more pulsating and turbulent in the arterioles and smoother and steadier in the venules. How does the velocity of flow in the arterioles compare with that in the venules?

____

In the capillaries? ____________________________

What is the relative difference in the diameter of the arterioles and capillaries?

____

Temperature

1. To investigate the effect of temperature on blood flow, flood the web with 5°C Ringer's solution two or three times to chill the entire area. Is a change in vessel diameter noticeable?

_________ Which vessels are affected? _______________

How? _______________________________________

2. Blot the web gently with a paper towel, and then bathe the web with warm (32°C) Ringer's solution. Record your observations.

____

Inflammation

1. Pipette 0.01 *N* HCl onto the frog's web. Hydrochloric acid will act as an irritant and cause a localized inflammatory response. Is there an increase or decrease in the blood flow into the capillary bed following the application of HCl?

____

What purpose do these local changes serve during a localized inflammatory response?

____

____

____

2. Flush the web with room-temperature Ringer's solution and blot.

Histamine

1. Histamine, which is released in large amounts during allergic responses, causes extensive vasodilation. Investigate this effect by adding a few drops of histamine solution to the frog web. What happens?

____

____

How does this response compare to that produced by HCl?

____

2. Blot the web and flood with 32°C Ringer's solution as before. Now add a few drops of 1% epinephrine solution, and observe the web. What are epinephrine's effects on the blood vessels?

____

Epinephrine is used clinically to reverse the vasodilation seen in severe allergic attacks (such as asthma) which are mediated by histamine and other vasoactive molecules.

3. Return the dropper bottles to the supply area and the frog to the terrarium. Properly clean your work area before leaving the lab. ■

NAME ______________________

LAB TIME/DATE ______________________

REVIEW SHEET

EXERCISE 34

Frog Cardiovascular Physiology

Special Electrical Properties of Cardiac Muscle: Automaticity and Rhythmicity

1. Define the following terms.

   automaticity: ______________________

   rhythmicity: ______________________

2. Discuss the anatomical differences between frog and human hearts. ______________________

   ____

   ____

3. Which region of the dissected frog heart had the highest intrinsic rate of contraction? ______________________

   The greatest automaticity? ______________________

   The greatest regularity or rhythmicity? ______________________ How do these properties correlate with the duties of a pacemaker? ______________________

   ____

   Is this region the pacemaker of the frog heart? ______________________

   Which region had the lowest intrinsic rate of contraction? ______________________

Investigating the Refractory Period of Cardiac Muscle

4. Define *extrasystole.* ______________________

5. Respond to the following questions if you used a physiograph. ______________________

   What was the effect of stimulation of the heart during ventricular contraction? ______________________

   During ventricular relaxation (first portion)? ______________________

   During the pause interval? ______________________

   What does this indicate about the refractory period of cardiac muscle? ______________________

Assessing Physical and Chemical Modifiers of Heart Rate

6. Describe the effect of thermal factors on the frog heart.

cold: ______________________ heat: ______________________

7. Once again refer to your recordings. Did the administration of the following produce any changes in force of contraction (shown by peaks of increasing or decreasing height)? If so, explain the mechanism.

epinephrine: ______________________

____

acetylcholine: ______________________

____

calcium ions: ______________________

____

8. Excessive amounts of each of the following ions would most likely interfere with normal heart activity. Note the type of changes caused in each case.

K^+: ______________________

Ca^{2+}: ______________________

Na^+: ______________________

9. Respond to the following questions if you used a physiograph. What was the effect of vagal stimulation on heart rate?

____

Which of the following factors cause the same (or very similar) heart rate–reducing effects: epinephrine, acetylcholine, atropine sulfate, pilocarpine, sympathetic nervous system activity, digitalis, potassium ions?

____

Which of the factors listed above would reverse or antagonize vagal effects? ______________________

____

10. What is vagal escape? ______________________

____

Why is vagal escape valuable in maintaining homeostasis? ______________________

____

11. How does the Stannius ligature used in the laboratory produce heart block? ______________________

____

____

12. Define *partial heart block,* and describe how it was recognized in the laboratory. ______

______

______

13. Define *total heart block,* and describe how it was recognized in the laboratory. ______

______

______

______

14. What do your heart block experiment results indicate about the spread of impulses from the atria to the ventricles?

______

______

Observing the Microcirculation Under Various Conditions

15. In what way are the red blood cells of the frog different from those of the human? ______

______

On the basis of this one factor, would you expect their life spans to be longer or shorter? ______

16. The following statements refer to your observation of one or more of the vessel types observed in the microcirculation in the frog's web. Characterize each statement by choosing one or more responses from the key.

Key: a. arteriole b. venule c. capillary

______ 1. smallest vessels observed

______ 2. vessel in which blood flow is rapid, pulsating

______ 3. vessel in which blood flow is least rapid

______ 4. red blood cells pass through these vessels in single file

______ 5. blood flow smooth and steady

______ 6. most numerous vessels

______ 7. vessels that deliver blood to the capillary bed

______ 8. vessels that serve the needs of the tissues via exchanges

______ 9. vessels that drain the capillary beds

17. Which of the vessel diameters changed most? ______

What division of the nervous system controls the vessels? ______

18. Discuss the effects of the following on blood vessel diameter (state specifically the blood vessels involved) and rate of blood flow. Then explain the importance of the reaction observed to the general well-being of the body.

local application of cold: ______________________________

____

local application of heat: ______________________________

____

inflammation (or application of HCl): ______________________________

____

histamine: ______________________________

____

EXERCISE 35

The Lymphatic System and Immune Response

MATERIALS

- ☐ Large anatomical chart of the human lymphatic system
- ☐ Prepared slides of lymph node, spleen, and tonsil
- ☐ Compound microscope
- ☐ Wax marking pencil
- ☐ Petri dish containing simple saline agar
- ☐ Medicine dropper
- ☐ Dropper bottles of red and green food color
- ☐ Dropper bottles of goat antibody to horse serum albumin, goat antibody to bovine serum albumin, goat antibody to swine serum albumin, horse serum albumin diluted to 20% with physiologic saline, unknown albumin sample diluted to 20% (prepared from horse, swine, and/or bovine albumin)
- ☐ Colored pencils

For instructions on animal dissections, see the dissection exercises starting on page 697 in the cat, fetal pig, and rat editions of this manual.

OBJECTIVES

1. To name the components of the lymphatic system.
2. To relate the function of the lymphatic system to that of the blood vascular system.
3. To describe the formation and composition of lymph, and to describe how it is transported through the lymphatic vessels.
4. To relate immunological memory, specificity, and differentiation of self from nonself to immune function.
5. To differentiate between the roles of B cells and T cells in the immune response.
6. To describe the structure and function of lymph nodes, and to indicate the localization of T cells, B cells, and macrophages in a typical lymph node.
7. To describe (or draw) the structure of the immunoglobulin monomer, and to name the five immunoglobulin subclasses.
8. To differentiate between antigen and antibody.
9. To understand the use of antigen-antibody reaction using the Ouchterlony test.

PRE-LAB QUIZ

1. Circle True or False. The lymphatic system protects the body by removing foreign material such as bacteria from the lymphatic stream.
2. Lymph is
   a. excess blood that has escaped from veins
   b. excess tissue fluid that has leaked out of capillaries
   c. excess tissue fluid that has escaped from arteries
3. Circle True or False. Lymphatic vessels have three tunics and are equipped with valves like veins.
4. __________, which serve as filters for the lymphatic system, occur at various points along the lymphatic vessels.
   a. Glands
   b. Lymph nodes
   c. Valves
5. Circle True or False. The immune response is a systemic response that occurs when the body recognizes a substance as foreign and acts to destroy or neutralize it.
6. Three characteristics of the immune response are the ability to distinguish self from nonself, memory, and
   a. autoimmunity
   b. specificity
   c. susceptibility
7. Circle the correct term. B cells / T cells differentiate in the thymus.
8. Circle the correct term. T cells mediate humoral / cellular immunity because they destroy cells infected with viruses and certain bacteria and parasites.
9. Circle True or False. Antibodies are produced by plasma cells in response to antigens and are found in all body secretions.

MasteringA&P™

Access practice quizzes and more in the Study Area at www.masteringaandp.com.

PEx

This lab corresponds to PhysioEx Exercise 12. See p. PEx-177.

PAL

For access to anatomical models and more, check out Practice Anatomy Lab.

The overall function of the lymphatic system is twofold: (1) it transports tissue fluid (lymph) to the blood vessels, and (2) it protects the body by removing foreign material such as bacteria from the lymphatic stream and by serving as a site for lymphocyte "policing" of body fluids and lymphocyte multiplication.

The Lymphatic System

The **lymphatic system** consists of a network of lymphatic vessels (lymphatics), lymphatic tissue, lymph nodes, and a number of other lymphoid organs, such as the tonsils, thymus, and spleen. We will focus on the lymphatic vessels and lymph nodes in this section. The white blood cells which are the central actors in body immunity are described later in this exercise.

Distribution and Function of Lymphatic Vessels and Lymph Nodes

As blood circulates through the body, the hydrostatic and osmotic pressures operating at the capillary beds result in fluid outflow at the arterial end of the bed and in its return at the

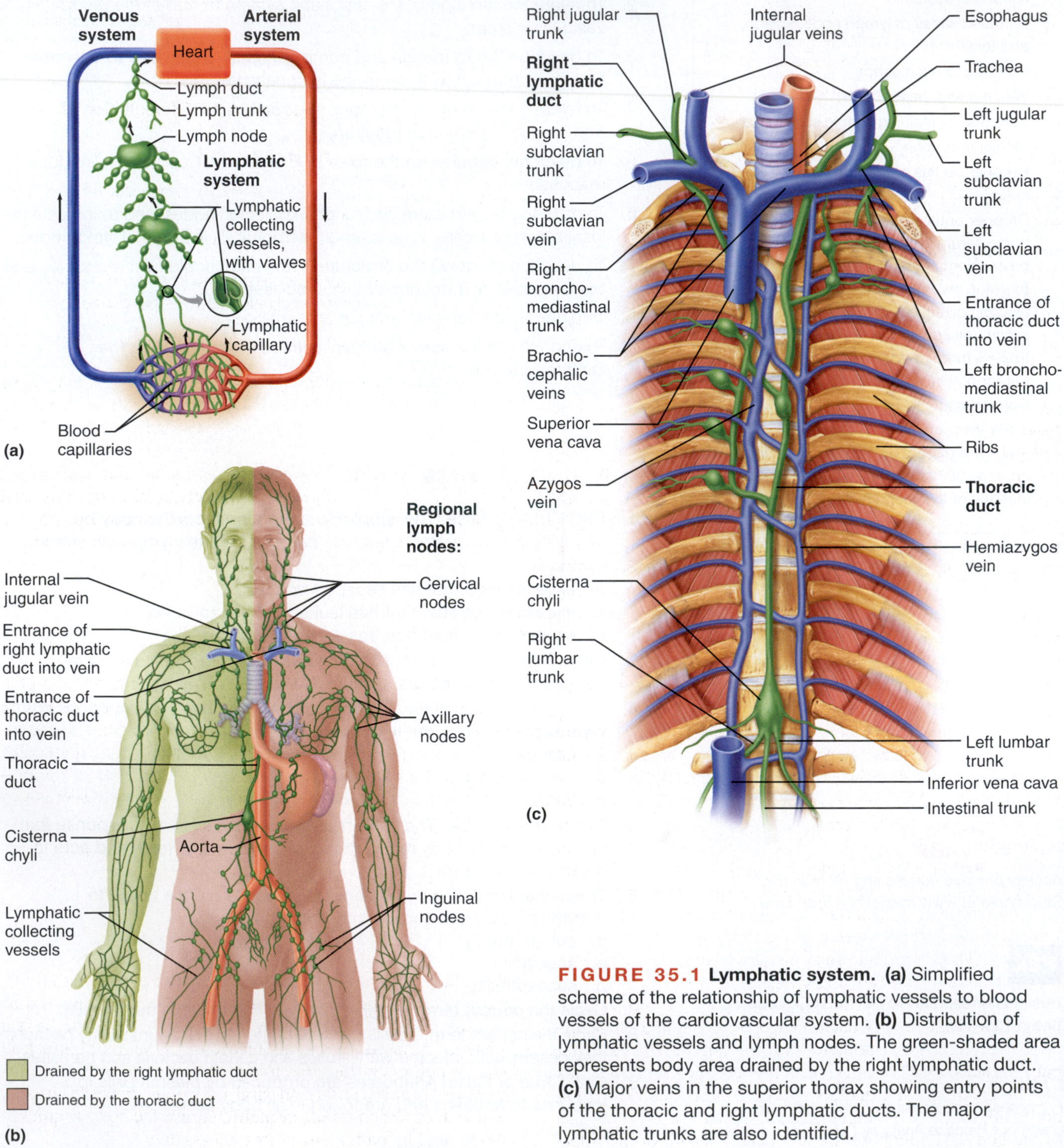
Venous system
Arterial system
Heart
Lymph duct
Lymph trunk
Lymph node
Lymphatic system
Lymphatic collecting vessels, with valves
Lymphatic capillary
Blood capillaries
(a)
Regional lymph nodes:
Internal jugular vein
Cervical nodes
Entrance of right lymphatic duct into vein
Entrance of thoracic duct into vein
Axillary nodes
Thoracic duct
Cisterna chyli
Aorta
Lymphatic collecting vessels
Inguinal nodes
Drained by the right lymphatic duct
Drained by the thoracic duct
(b)
Right jugular trunk
Internal jugular veins
Esophagus
Right lymphatic duct
Trachea
Left jugular trunk
Right subclavian trunk
Left subclavian trunk
Right subclavian vein
Left subclavian vein
Right broncho-mediastinal trunk
Entrance of thoracic duct into vein
Brachio-cephalic veins
Left broncho-mediastinal trunk
Superior vena cava
Ribs
Azygos vein
Thoracic duct
Hemiazygos vein
Cisterna chyli
Right lumbar trunk
Left lumbar trunk
Inferior vena cava
Intestinal trunk
(c)

FIGURE 35.1 Lymphatic system. (a) Simplified scheme of the relationship of lymphatic vessels to blood vessels of the cardiovascular system. **(b)** Distribution of lymphatic vessels and lymph nodes. The green-shaded area represents body area drained by the right lymphatic duct. **(c)** Major veins in the superior thorax showing entry points of the thoracic and right lymphatic ducts. The major lymphatic trunks are also identified.

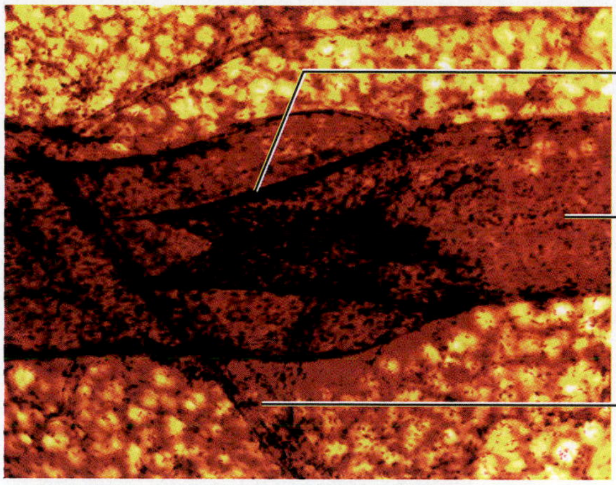
Valve leaflet
Lymph vessel
Blood capillary

FIGURE 35.2 Lymphatic vessel (550×).

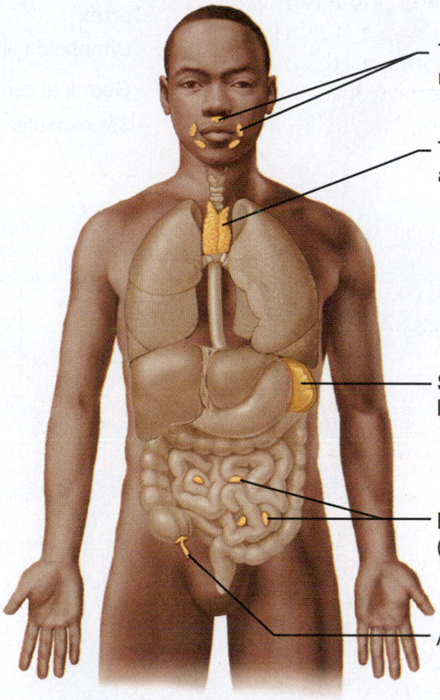
Tonsils (in pharyngeal region)
Thymus (in thorax; most active during youth)
Spleen (curves around left side of stomach)
Peyer's patches (in intestine)
Appendix

FIGURE 35.3 Lymphoid organs. Locations of the tonsils, thymus, spleen, appendix, and Peyer's patches.

venous end. However, not all of the lost fluid is returned to the bloodstream by this mechanism, and the fluid that lags behind in the tissue spaces must eventually return to the blood if the vascular system is to operate properly. (If it does not, fluid accumulates in the tissues, producing a condition called *edema.*) It is the microscopic, blind-ended **lymphatic capillaries** (Figure 35.1a), which ramify through nearly all the tissues of the body, that pick up this leaked fluid (primarily water and a small amount of dissolved proteins) and carry it through successively larger vessels—**lymphatic collecting vessels** to **lymphatic trunks**—until the lymph finally returns to the blood vascular system through one of the two large ducts in the thoracic region (Figure 35.1b). The **right lymphatic duct**, present in some but not all individuals, drains lymph from the right upper extremity, head, and thorax delivered by the jugular, subclavian, and bronchomediastinal trunks. In individuals without a right lymphatic duct, those trunks open directly into veins of the neck. The large **thoracic duct** receives lymph from the rest of the body (see Figure 35.1c). In humans, both ducts empty the lymph into the venous circulation at the junction of the internal jugular vein and the subclavian vein, on their respective sides of the body. Notice that the lymphatic system, lacking both a contractile "heart" and arteries, is a one-way system; it carries lymph only toward the heart.

Like veins of the blood vascular system, the lymphatic collecting vessels have three tunics and are equipped with valves (Figure 35.2). However, lymphatics tend to be thinner-walled, to have *more* valves, and to anastomose (form branching networks) more than veins. Since the lymphatic system is a pumpless system, lymph transport depends largely on the milking action of the skeletal muscles and on pressure changes within the thorax that occur during breathing.

As lymph is transported, it filters through bean-shaped **lymph nodes,** which cluster along the lymphatic vessels of the body. There are thousands of lymph nodes, but because they are usually embedded in connective tissue, they are not ordinarily seen. Within the lymph nodes are **macrophages,** phagocytes that destroy bacteria, cancer cells, and other foreign matter in the lymphatic stream, thus rendering many harmful substances or cells harmless before the lymph enters the bloodstream. Particularly large collections of lymph nodes are found in the inguinal, axillary, and cervical regions of the body. Although we are not usually aware of the filtering and protective nature of the lymph nodes, most of us have experienced "swollen glands" during an active infection. This swelling is a manifestation of the trapping function of the nodes.

Other lymphoid organs—the tonsils, thymus, and spleen (Figure 35.3)—resemble the lymph nodes histologically, and house similar cell populations (lymphocytes and macrophages).

ACTIVITY 1

Identifying the Organs of the Lymphatic System

Study the large anatomical chart to observe the general plan of the lymphatic system. Notice the distribution of lymph nodes, various lymphatics, the lymphatic trunks, and the location of the right lymphatic duct and the thoracic duct. Also identify the **cisterna chyli,** the enlarged terminus of the thoracic duct that receives lymph from the digestive viscera.

For instructions on animal dissections, see the dissection exercises starting on page 697 in the cat, rat, and fetal pig editions of this manual.

The Immune Response

The **adaptive immune system** is a functional system that recognizes something as foreign and acts to destroy or neutralize it. This response is known as the **immune response.** It is a systemic response and is not restricted to the initial infection site. When operating effectively, the immune response protects us from bacterial and viral infections, bacterial toxins, and cancer. When it fails or malfunctions, the body is quickly devastated by pathogens or its own assaults.

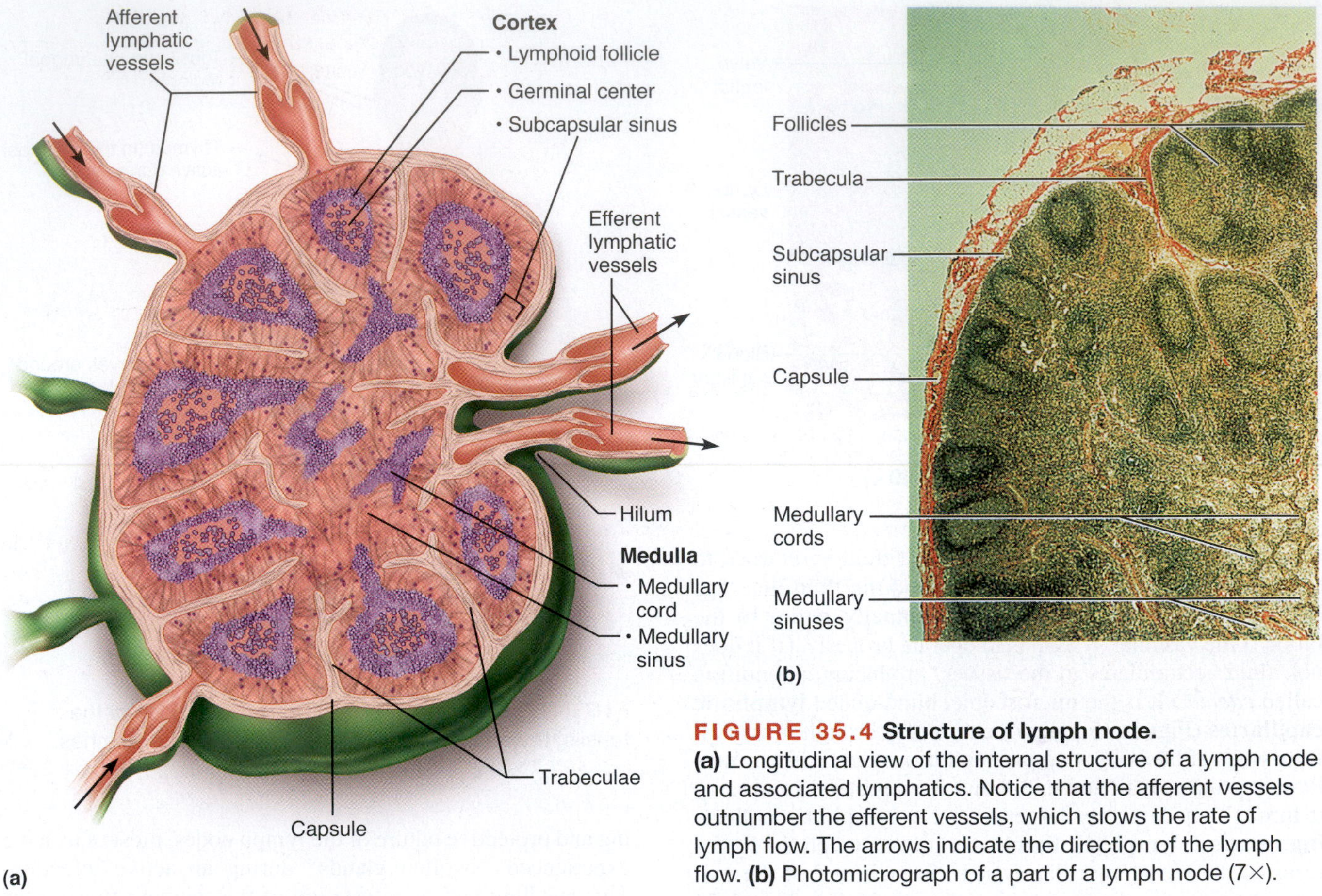
Afferent lymphatic vessels
Cortex
• Lymphoid follicle
• Germinal center
• Subcapsular sinus
Efferent lymphatic vessels
Hilum
Medulla
• Medullary cord
• Medullary sinus
Trabeculae
Capsule
(a)
Follicles
Trabecula
Subcapsular sinus
Capsule
Medullary cords
Medullary sinuses
(b)

FIGURE 35.4 Structure of lymph node. **(a)** Longitudinal view of the internal structure of a lymph node and associated lymphatics. Notice that the afferent vessels outnumber the efferent vessels, which slows the rate of lymph flow. The arrows indicate the direction of the lymph flow. **(b)** Photomicrograph of a part of a lymph node (7×).

Major Characteristics of the Immune Response

The most important characteristics of the immune response are its (1) **memory,** (2) **specificity,** and (3) **ability to differentiate self from nonself.** Not only does the immune system have a "memory" for previously encountered foreign antigens (the chicken pox virus for example), but this memory is also remarkably accurate and highly specific.

An almost limitless variety of macromolecules are *antigenic*—that is, capable of provoking an immune response and reacting with its products. Nearly all foreign proteins, many polysaccharides, and many small molecules (haptens), when linked to our own body proteins, exhibit this capability. The cells that recognize antigens and initiate the immune response are lymphocytes, the second most numerous members of the leukocyte, or white blood cell (WBC), population. Each immunocompetent lymphocyte is virtually monospecific; that is, it has receptors on its surface allowing it to bind with only one or a few very similar antigens.

As a rule, our own proteins are tolerated, a fact that reflects the ability of the immune system to distinguish our own tissues (self) from foreign antigens (nonself). Nevertheless, an inability to recognize self can and does occasionally happen, and our own tissues are attacked by the immune system. This phenomenon is called *autoimmunity.* Autoimmune diseases include multiple sclerosis (MS), myasthenia gravis, Graves' disease, glomerulonephritis, rheumatoid arthritis (RA), and type 1 (or insulin-dependent) diabetes mellitus.

Organs, Cells, and Cell Interactions of the Immune Response

The immune system utilizes as part of its arsenal the **lymphoid organs,** including the thymus, lymph nodes, spleen, tonsils, appendix, and bone marrow. Of these, the thymus and bone marrow are considered to be the *primary lymphoid organs.* The others are *secondary lymphoid areas.*

The stem cells that give rise to the immune system arise in the bone marrow. Their subsequent differentiation into one of the two populations of immunocompetent lymphocytes occurs in the primary lymphoid organs. The **B cells** (B lymphocytes) differentiate in bone marrow, and the **T cells** (T lymphocytes) differentiate in the thymus. While in their "programming organs" the lymphocytes become *immunocompetent,* an event indicated by the appearance of specific cell-surface proteins that enable the lymphocytes to respond (by binding) to a particular antigen.

After differentiation, the B and T cells leave the bone marrow and thymus, respectively; enter the bloodstream; and travel to peripheral (secondary) lymphoid organs, where clonal selection occurs. **Clonal selection** is triggered when an antigen binds to the specific cell-surface receptors of a T or B cell. This event causes the lymphocyte to proliferate rapidly, forming a clone of like cells, all bearing the same antigen-specific receptors. Then, in the presence of certain regulatory signals, the members of the clone specialize, or differentiate—some forming memory cells and others becoming effector or regulatory cells. Upon subsequent meetings with the same antigen, the immune response proceeds considerably faster because the troops are already mobilized and awaiting further orders, so to speak.

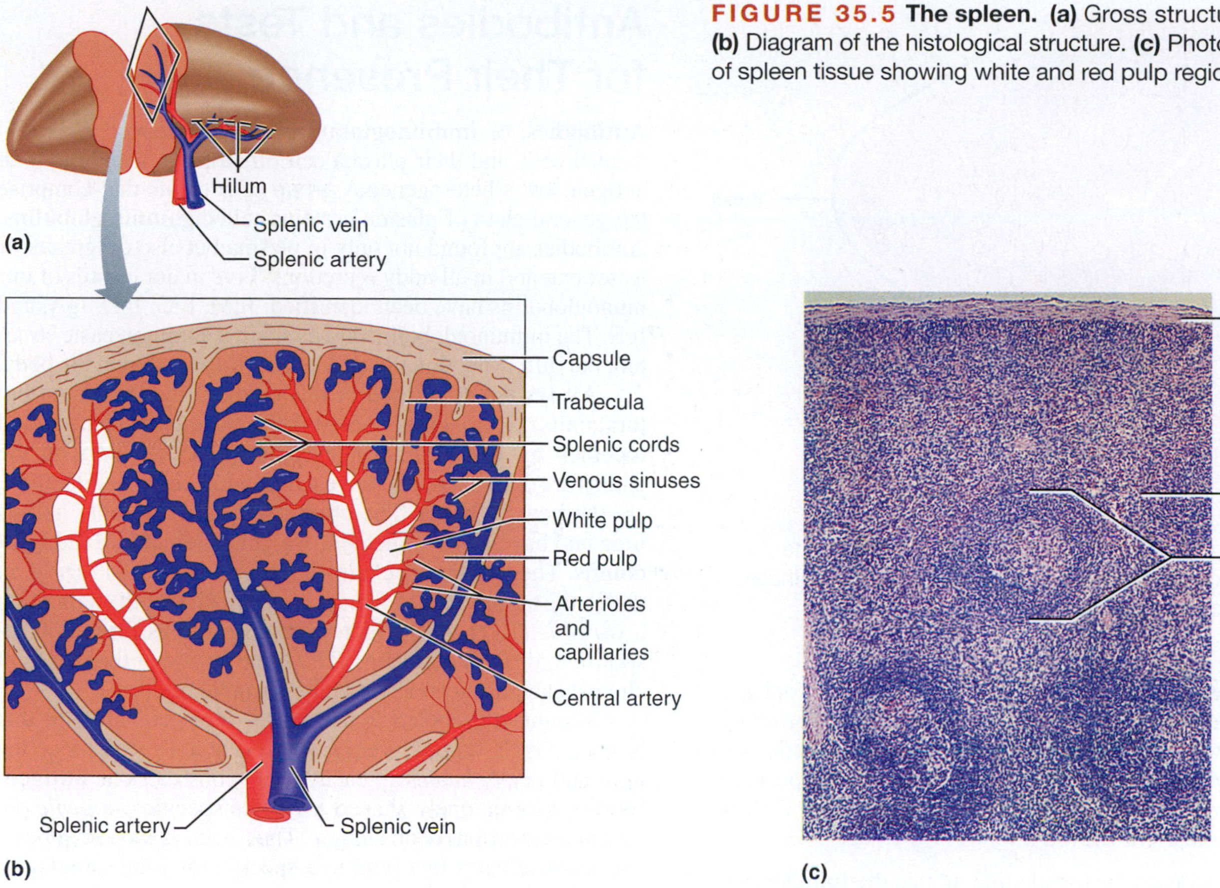
Hilum
Splenic vein
Splenic artery
(a)
Capsule
Trabecula
Splenic cords
Venous sinuses
White pulp
Red pulp
Arterioles and capillaries
Central artery
Splenic artery
Splenic vein
(b)
Capsule
Red pulp
White pulp
(c)

FIGURE 35.5 The spleen. (a) Gross structure. **(b)** Diagram of the histological structure. **(c)** Photomicrograph of spleen tissue showing white and red pulp regions (15×).

In the case of B cell clones, some become **memory B cells;** the others form antibody-producing **plasma cells**. Because the B cells act indirectly through the antibodies that their progeny release into the bloodstream (or other body fluids), they are said to provide **humoral immunity.** T cell clones are more diverse. Although all T cell clones also contain memory cells, some clones contain *cytotoxic T cells* (effector cells that directly attack virus-infected tissue cells). Others contain regulatory cells such as the *helper T cells* (that help activate the B cells and cytotoxic T cells) and still others contain *suppressor T cells* that can inhibit the immune response. Because certain T cells act directly to destroy cells infected with viruses, certain bacteria or parasites, and cancer cells, and to reject foreign grafts, T cells are said to mediate **cellular immunity.**

Absence or failure of thymic differentiation of T lymphocytes results in a marked depression of both antibody and cell-mediated immune functions. Additionally, the observation that the thymus naturally involutes with age has been correlated with the relatively immune-deficient status of elderly individuals. ■

All lymphoid tissues except the thymus and bone marrow contain both T and B cell–dependent regions.

ACTIVITY 2

Studying the Microscopic Anatomy of a Lymph Node, the Spleen, and a Tonsil

1. Obtain a compound microscope and prepared slides of a lymph node, spleen, and a tonsil. As you examine the lymph node slide, notice the following anatomical features, depicted in Figure 35.4. The node is enclosed within a fibrous **capsule,** from which connective tissue septa **(trabeculae)** extend inward to divide the node into several compartments. Very fine strands of reticular connective tissue issue from the trabeculae, forming the stroma of the gland within which cells are found.

In the outer region of the node, the **cortex,** some of the cells are arranged in globular masses, referred to as germinal centers. The **germinal centers** contain rapidly dividing B cells. The rest of the cortical cells are primarily T cells that circulate continuously, moving from the blood into the node and then exiting from the node in the lymphatic stream.

In the internal portion of the gland, the **medulla,** the cells are arranged in cordlike fashion. Most of the medullary cells are macrophages. Macrophages are important not only for their phagocytic function but also because they play an essential role in "presenting" the antigens to the T cells.

Lymph enters the node through a number of *afferent vessels,* circulates through *lymph sinuses* within the node, and leaves the node through *efferent vessels* at the **hilum.** Since each node has fewer efferent than afferent vessels, the lymph flow stagnates somewhat within the node. This allows time for the generation of an immune response and for the macrophages to remove debris from the lymph before it reenters the blood vascular system.

2. As you observe the slide of the spleen, look for the areas of lymphocytes suspended in reticular fibers, the **white pulp,** clustered around central arteries (Figure 35.5). The remaining tissue in the spleen is the **red pulp,** which is composed of

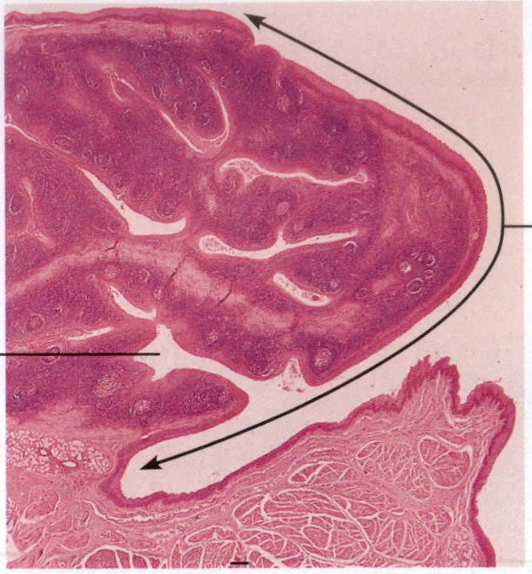
Tonsil
Crypt

FIGURE 35.6 Histology of a palatine tonsil. The luminal surface is covered with epithelium that invaginates deeply to form crypts (5×).

venous sinuses and areas of reticular tissue and macrophages called the **splenic cords.** The white pulp, composed primarily of lymphocytes, is responsible for the immune functions of the spleen. Macrophages remove worn-out red blood cells, debris, bacteria, viruses, and toxins from blood flowing through the sinuses of the red pulp.

3. As you examine the tonsil slide, notice the **follicles** containing **germinal centers** surrounded by scattered lymphocytes. The characteristic **crypts** (invaginations of the mucosal epithelium) of the tonsils trap bacteria and other foreign material (Figure 35.6). Eventually the bacteria work their way into the lymphoid tissue and are destroyed.

4. Compare and contrast the structure of the lymph node, spleen, and tonsils.

Antibodies and Tests for Their Presence

Antibodies, or **immunoglobulins (Igs),** produced by sensitized B cells and their plasma cell offspring in response to an antigen, are a heterogeneous group of proteins that comprise the general class of plasma proteins called **gamma globulins.** Antibodies are found not only in plasma but also (to greater or lesser extents) in all body secretions. Five major classes of immunoglobulins have been identified: IgM, IgG, IgD, IgA, and IgE. The immunoglobulin classes share a common basic structure but differ functionally and in their localization in the body.

All Igs are composed of one or more monomers (structural units). A monomer consists of four protein chains bound together by disulfide bridges (Figure 35.7). Two of the chains are quite large and have a high molecular weight; these are the **heavy chains.** The other two chains are only half as long and have a low molecular weight. These are called **light chains.** The two heavy chains have a *constant (C) region,* in which the amino acid sequence is identical in both chains, and a *variable (V) region,* which differs in the Igs formed in response to different antigens. The same is true of the two light chains; each has a constant and a variable region.

The intact Ig molecule has a three-dimensional shape that is generally Y-shaped. Together, the variable regions of the light and heavy chains in each "arm" construct one **antigen-binding site** uniquely shaped to "fit" a specific *antigenic determinant* (portion) of an antigen. Thus, each Ig monomer bears two identical sites that bind to a specific (and the same) antigen. Binding of the immunoglobulins to their complementary antigen(s) effectively immobilizes the antigens until they can be phagocytized or lysed by complement fixation.

Although the role of the immune system is to protect the body, symptoms of certain diseases involve excessively high antibody synthesis (as in multiple myeloma, a cancer of the bone marrow and adjacent bony structures) and/or the production of abnormal antibodies.

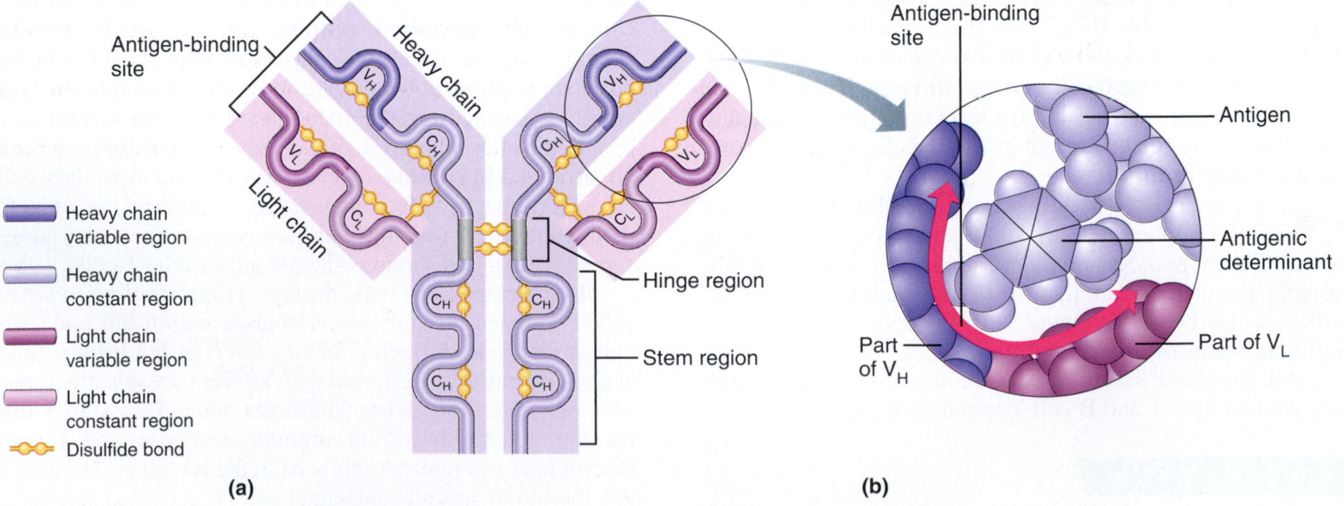
Antigen-binding site
Heavy chain
Light chain
V_H
C_H
V_L
C_L
Heavy chain variable region
Heavy chain constant region
Light chain variable region
Light chain constant region
Disulfide bond
Hinge region
Stem region
(a)
Antigen-binding site
Antigen
Antigenic determinant
Part of V_H
Part of V_L
(b)

FIGURE 35.7 Antibody structure. (a) Schematic antibody structure consists of two heavy chains and two light chains connected by disulfide bonds. **(b)** Enlargement of an antigen-binding site of an immunoglobulin.

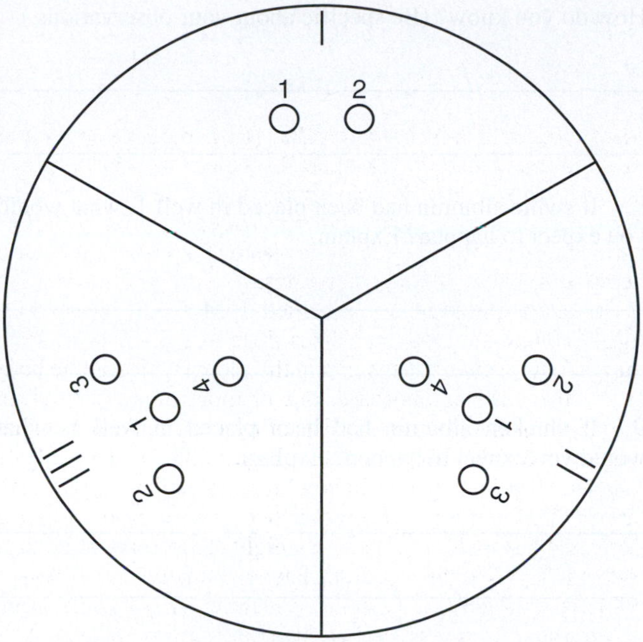
I
1
2
II
III
1
2
3
4
1
2
3
4

FIGURE 35.8 Template for well preparation for the Ouchterlony double-gel diffusion experiment.

The antigen-antibody reaction is used diagnostically in a variety of ways. One of the most familiar is blood typing. (See Exercise 29 for instructions on ABO and Rh blood typing.) Pregnancy tests also use the antigen-antibody reaction to test for the presence of human chorionic gonadotropin (hCG), a hormone produced early in pregnancy. Another technique for detecting antigens is the enzyme-linked immunoabsorbent assay (ELISA). Originally designed to measure antibody titer, ELISA has been modified for use in HIV-1 blood screening. The antigen-antibody test that we will use in this laboratory session is the Ouchterlony technique, which is used mainly for rapid screening of suspected antigens.

Ouchterlony Double-Gel Diffusion, an Immunological Technique

The Ouchterlony double-gel diffusion technique was developed in 1948 to detect the presence of particular antigens in sera or extracts. Antigens and antibodies are placed in wells in a gel and allowed to diffuse toward each other. If an antigen reacts with an antibody, a thin white line called a *precipitin line* forms. In the following activity, the double-gel diffusion technique will be used to identify antigens. Work in groups of no more than three.

ACTIVITY 3

Using the Ouchterlony Technique to Identify Antigens

1. Obtain one each of the materials for conducting the Ouchterlony test: petri dish containing saline agar; medicine dropper; wax marking pencil; and dropper bottles of red and green food dye, horse serum albumin, an unknown serum albumin sample, and antibodies to horse, bovine, and swine albumin. Put your initials and the number of the unknown albumin sample used on the bottom of the petri dish near the edge.

2. Use the wax marking pencil and the template (Figure 35.8) to divide the dish into three sections, and mark them I, II, and III.

3. Prepare sample wells, again using the template in Figure 35.8. Squeeze the medicine dropper bulb, and gently touch the tip to the surface of the agar. While releasing the bulb, push the tip down through the agar to the bottom of the dish. Lift the dropper vertically; this should leave a straight-walled well in the agar.

4. Repeat step 3 so that section I has two wells and sections II and III have four wells each.

5. To observe diffusion through the gel, nearly fill one well in section I with red dye and the other well with green dye. Be careful not to overfill the wells. Observe periodically for 30 to 45 minutes as the dyes diffuse through the agar. Draw your results as instructed in the Results section.

6. To demonstrate positive and negative results, fill the wells in section II as listed in Table 35.1. A precipitin line should form only between wells 1 and 2.

7. To test the unknown sample, fill the wells in section III as listed in Table 35.2.

8. Replace the cover on the petri dish, and incubate at room temperature for at least 16 hours. Make arrangements to observe the agar for precipitin lines after 16 hours. *The lines may begin to fade after 48 hours.* Draw the results as instructed in the following section, parts 1–3, indicating the location of all precipitin lines that form.

TABLE 35.1 Section II

Well	Solution
1	Horse serum albumin
2	Goat anti-horse albumin
3	Goat anti-bovine albumin
4	Goat anti-swine albumin

TABLE 35.2 Section III

Well	Solution
1	Unknown # ________
2	Goat anti-horse albumin
3	Goat anti-bovine albumin
4	Goat anti-swine albumin

Results

1. Demonstration of diffusion using dye. Draw the appearance of section I of your dish after 30 to 45 minutes on Figure 35.8. Use colored pencils.

2. Draw section II on Figure 35.8 as it appears after incubation, 16 to 48 hours. Be sure to number the wells.

3. Draw section III on Figure 35.8 as it appears after incubation, 16 to 48 hours. Be sure to number the wells.

Unknown # ________

4. What evidence for diffusion did you observe in section I?

____

____

5. Is there any evidence of a precipitate in section I?

____

6. Which of the sera functioned as an antigen in section II?

____

7. Which antibody reacted with the antigen in section II?

____

How do you know? (Be specific about your observations.)

____

____

8. If swine albumin had been placed in well 1, what would you expect to happen? Explain.

____

____

9. If chicken albumin had been placed in well 1, what would you expect to happen? Explain.

____

____

10. What antigens were present in the unknown solution?

____

How do you know? (Be specific about your observations.)

____

____

NAME ______________________________

LAB TIME/DATE ______________________

REVIEW SHEET

EXERCISE 35

The Lymphatic System and Immune Response

The Lymphatic System

1. Match the terms below with the correct letters on the diagram.

_______ 1. axillary lymph nodes

_______ 2. bone marrow

_______ 3. cervical lymph nodes

_______ 4. cisterna chyli

_______ 5. inguinal lymph nodes

_______ 6. lymphatic vessels

_______ 7. Peyer's patches (in intestine)

_______ 8. right lymphatic duct

_______ 9. spleen

_______ 10. thoracic duct

_______ 11. thymus

_______ 12. tonsils

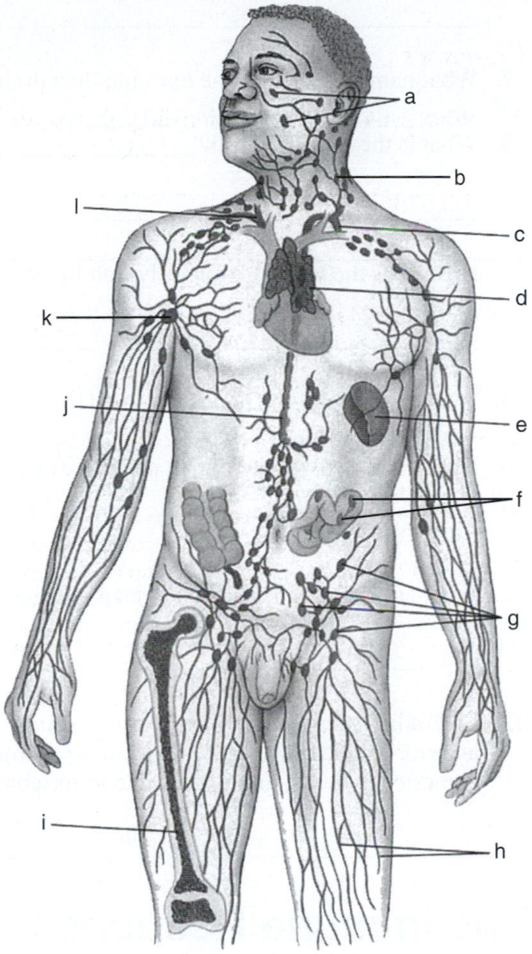

a
b
l
c
d
k
j
e
f
g
i
h

2. Explain why the lymphatic system is a one-way system, whereas the blood vascular system is a two-way system.

____

____

3. How do lymphatic vessels resemble veins? ______________________________________

How do lymphatic capillaries differ from blood capillaries? ______________________________

____

4. What is the function of the lymphatic vessels? ______________________________

____

5. What is lymph? ______________________________

6. What factors are involved in the flow of lymphatic fluid? ______________________________

____

7. What name is given to the terminal duct draining most of the body? ______________________________

8. What is the cisterna chyli? ______________________________

____

How does the composition of lymph in the cisterna chyli differ from that in the general lymphatic stream?

____

9. Which portion of the body is drained by the right lymphatic duct? ______________________________

10. Note three areas where lymph nodes are densely clustered: ______________________________,

______________________________, and ______________________________

11. What are the two major functions of the lymph nodes? ______________________________

and ______________________________

12. The radical mastectomy is an operation in which a cancerous breast, surrounding tissues, and the underlying muscles of the anterior thoracic wall, plus the axillary lymph nodes, are removed. After such an operation, the arm usually swells, or becomes edematous, and is very uncomfortable—sometimes for months. Why?

____

The Immune Response

13. What is the function of B cells in the immune response? ______________________________

____

14. What is the role of T cells? ______________________________

____

____

15. Define the following terms related to the operation of the immune system.

immunological memory: ______________________________

____

specificity: ______________________________

____

recognition of self from nonself: ______________________________

____

autoimmune disease: ______________________________

____

Studying the Microscopic Anatomy of a Lymph Node, the Spleen, and a Tonsil

16. In the space below, make a rough drawing of the structure of a lymph node. Identify the cortex area, germinal centers, and medulla. For each identified area, note the cell type (T cell, B cell, or macrophage) most likely to be found there.

17. What structural characteristic ensures a *slow* flow of lymph through a lymph node? ______________________________

____

Why is this desirable? ______________________________

____

18. What similarities in structure and function are found in the lymph nodes, spleen, and tonsils? ______________________________

____

____

Antibodies and Tests for Their Presence

19. Distinguish between antigen and antibody. ______________________________

____

20. Describe the structure of the immunoglobulin monomer, and label the diagram with the choices given in the key. ______

____

Key:

a. antigen-binding site
b. heavy chain
c. hinge region
d. light chain
e. stem region

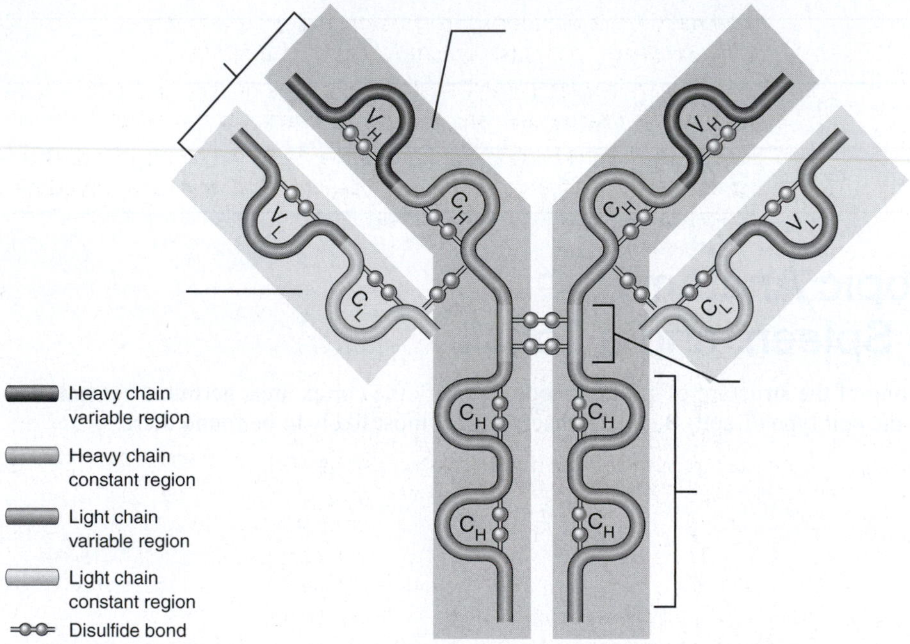

V_H
C_H
V_L
C_L
Heavy chain variable region
Heavy chain constant region
Light chain variable region
Light chain constant region
Disulfide bond

21. Are the genes coding for one antibody entirely different from those coding for a different antibody? ______________

Explain your answer. ______________________________

____

____

22. In the Ouchterlony test, what happened when the antibody to horse serum albumin mixed with horse serum albumin?

____

____

____

23. If the unknown antigen contained bovine and swine serum albumin, what would you expect to happen in the Ouchterlony test, and why? ______________________________

____

____

EXERCISE 36

Anatomy of the Respiratory System

MATERIALS

- ☐ Resin cast of the respiratory tree (if available)
- ☐ Human torso model
- ☐ Respiratory organ system model and/or chart of the respiratory system
- ☐ Larynx model (if available)
- ☐ Preserved inflatable lung preparation (obtained from a biological supply house) or sheep pluck fresh from the slaughterhouse
- ☐ Source of compressed air
- ☐ Dissecting tray
- ☐ Disposable gloves
- ☐ Disposable autoclave bag
- ☐ Prepared slides of the following (if available): trachea (cross section), lung tissue, both normal and pathological specimens (for example, sections taken from lung tissues exhibiting bronchitis, pneumonia, emphysema, or lung cancer)
- ☐ Compound and stereomicroscopes

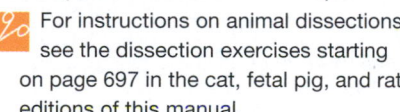

For instructions on animal dissections, see the dissection exercises starting on page 697 in the cat, fetal pig, and rat editions of this manual.

MasteringA&P™

Access practice quizzes and more in the Study Area at www.masteringaandp.com.

PAL

For access to anatomical models and more, check out Practice Anatomy Lab.

OBJECTIVES

1. To define the following terms: *respiratory system, pulmonary ventilation, external respiration,* and *internal respiration.*
2. To label the major respiratory system structures on a diagram (or identify them on a model), and to describe the function of each.
3. To recognize the histologic structure of the trachea (cross section) and lung tissue on prepared slides, and to describe the functions served by the observed structural modifications.

PRE-LAB QUIZ

1. The major role of the respiratory system is to
   a. dispose of waste products in a solid form
   b. permit the flow of nutrients through the body
   c. supply the body with carbon dioxide and dispose of oxygen
   d. supply the body with oxygen and dispose of carbon dioxide
2. Circle True or False. Four processes—pulmonary ventilation, external respiration, transport of respiratory gases, and internal respiration—must all occur in order for the respiratory system to function fully.
3. The upper respiratory structures include the nose, the larynx, and the
   a. epiglottis
   b. lungs
   c. pharynx
   d. trachea
4. Circle the correct term. The thyroid cartilage / oropharynx is the largest and most prominent of the laryngeal cartilages.
5. Circle True or False. The epiglottis forms a lid over the larynx when we swallow food: it closes off the respiratory passageway to incoming food or drink.
6. Air flows from the larynx to the trachea, and then enters the
   a. left and right lungs
   b. left and right main bronchi
   c. pharynx
   d. segmental bronchi
7. Circle the correct term. The lining of the trachea is pseudostratified ciliated columnar epithelium / transitional epithelium, which propels dust particles, bacteria, and other debris away from the lungs.
8. Circle True or False. All but the most minute branches of the respiratory tree have cartilaginous reinforcements in their walls.
9. ________________, tiny balloonlike expansions of the alveolar sacs, are composed of a single thin layer of squamous epithelium. They are the main structural and functional units of the lung and the actual sites of gas exchange.
10. Circle the correct term. Fissures divide the lungs into lobes, three on the right and two / three on the left.

Body cells require an abundant and continuous supply of oxygen. As the cells use oxygen, they release carbon dioxide, a waste product that the body must get rid of. These oxygen-using cellular processes, collectively referred to as *cellular respiration,* are more appropriately described in conjunction with the topic of cellular metabolism. The major role of the **respiratory system,** our focus in this exercise, is to supply the body with oxygen and dispose of carbon dioxide. To fulfill this role, at least four distinct processes, collectively referred to as **respiration,** must occur:

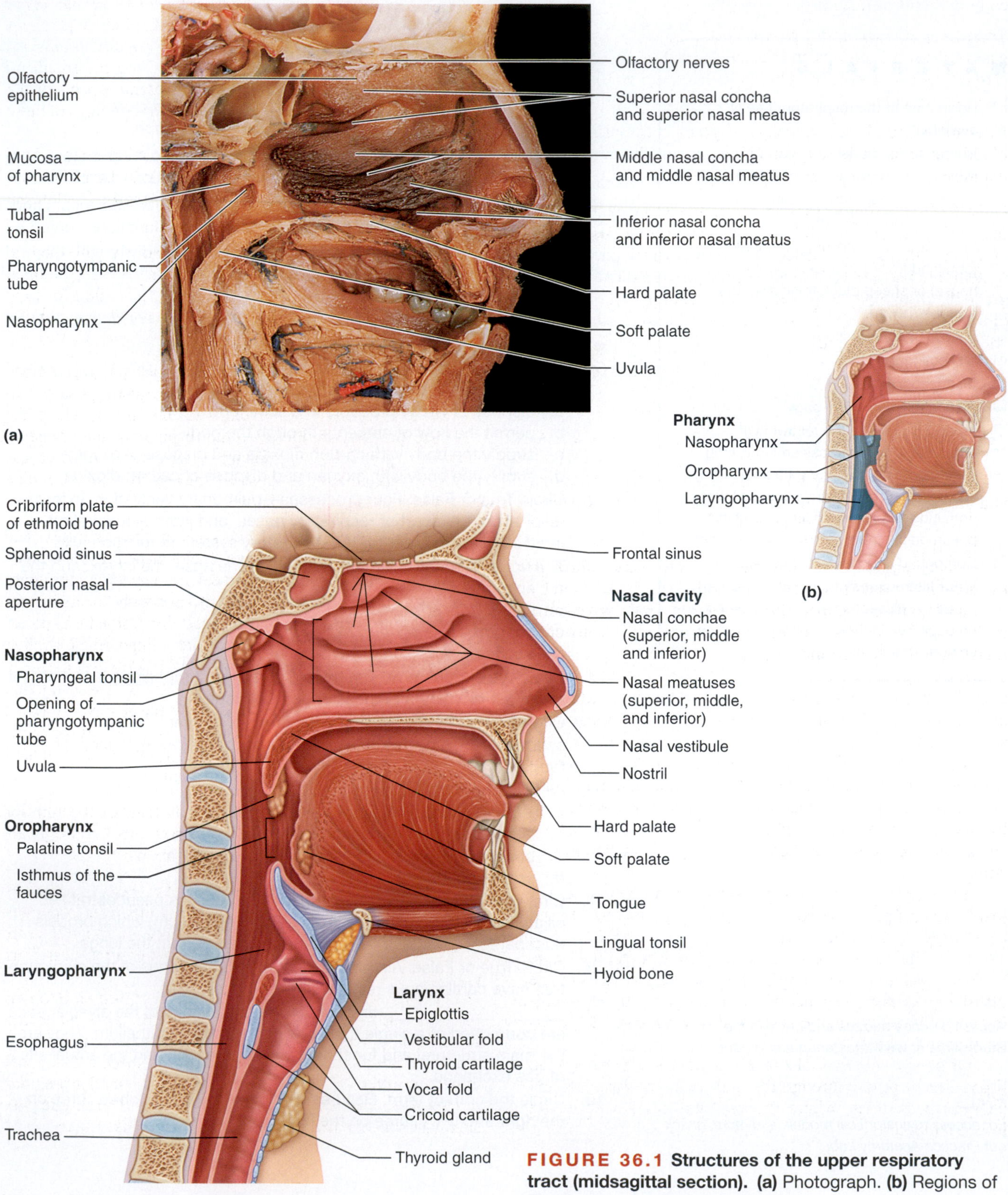
Olfactory epithelium
Mucosa of pharynx
Tubal tonsil
Pharyngotympanic tube
Nasopharynx
Olfactory nerves
Superior nasal concha and superior nasal meatus
Middle nasal concha and middle nasal meatus
Inferior nasal concha and inferior nasal meatus
Hard palate
Soft palate
Uvula
(a)
Pharynx
Nasopharynx
Oropharynx
Laryngopharynx
(b)
Cribriform plate of ethmoid bone
Sphenoid sinus
Posterior nasal aperture
Nasopharynx
Pharyngeal tonsil
Opening of pharyngotympanic tube
Uvula
Oropharynx
Palatine tonsil
Isthmus of the fauces
Laryngopharynx
Esophagus
Trachea
Frontal sinus
Nasal cavity
Nasal conchae (superior, middle and inferior)
Nasal meatuses (superior, middle, and inferior)
Nasal vestibule
Nostril
Hard palate
Soft palate
Tongue
Lingual tonsil
Hyoid bone
Larynx
Epiglottis
Vestibular fold
Thyroid cartilage
Vocal fold
Cricoid cartilage
Thyroid gland
(c)

FIGURE 36.1 Structures of the upper respiratory tract (midsagittal section). (a) Photograph. **(b)** Regions of the pharynx. **(c)** Diagrammatic view.

Pulmonary ventilation: The tidelike movement of air into and out of the lungs so that the gases in the alveoli are continuously changed and refreshed. Also more simply called *ventilation,* or *breathing*.

External respiration: The gas exchange between the blood and the air-filled chambers of the lungs (oxygen loading/carbon dioxide unloading).

Transport of respiratory gases: The transport of respiratory gases between the lungs and tissue cells of the body accomplished by the cardiovascular system, using blood as the transport vehicle.

Internal respiration: Exchange of gases between systemic blood and tissue cells (oxygen unloading and carbon dioxide loading).

Only the first two processes are the exclusive province of the respiratory system, but all four must occur for the respiratory system to "do its job." Hence, the respiratory and circulatory systems are irreversibly linked. If either system fails, cells begin to die from oxygen starvation and accumulation of carbon dioxide. Uncorrected, this situation soon causes death of the entire organism.

Upper Respiratory System Structures

The upper respiratory system structures—the nose, pharynx, and larynx—are shown in Figure 36.1 and described below. As you read through the descriptions, identify each structure in the figure.

Air generally passes into the respiratory tract through the **nostrils or nares,** and enters the **nasal cavity** (divided by the **nasal septum**). It then flows posteriorly over three pairs of lobelike structures, the **inferior, superior,** and **middle nasal conchae,** which increase the air turbulence. As the air passes through the nasal cavity, it is also warmed, moistened, and filtered by the nasal mucosa. The air that flows directly beneath the superior part of the nasal cavity may chemically stimulate the olfactory receptors located in the mucosa of that region. The nasal cavity is surrounded by the **paranasal sinuses** in the frontal, sphenoid, ethmoid, and maxillary bones. These sinuses, named for the bones in which they are located, act as resonance chambers in speech. Their mucosae, like that of the nasal cavity, warm and moisten the incoming air.

The nasal passages are separated from the oral cavity below by a partition composed anteriorly of the **hard palate** and posteriorly by the **soft palate.**

The genetic defect called **cleft palate** (failure of the palatine bones and/or the palatine processes of the maxillary bones to fuse medially) causes difficulty in breathing and oral cavity functions such as sucking and, later, mastication and speech. ■

Of course, air may also enter the body via the mouth. From there it passes through the oral cavity to move into the pharynx posteriorly, where the oral and nasal cavities are joined temporarily.

Commonly called the *throat,* the funnel-shaped **pharynx** connects the nasal and oral cavities to the larynx and esophagus inferiorly. It has three named parts (Figure 36.1):

1. The **nasopharynx** lies posterior to the nasal cavity and is continuous with it via the **posterior nasal aperture.** It lies above the soft palate; hence, it serves only as an air passage. High on its posterior wall is the *pharyngeal tonsil,* masses of lymphoid tissue that help to protect the respiratory passages from invading pathogens. The *pharyngotympanic (auditory) tubes,* which allow middle ear pressure to become equalized to atmospheric pressure, drain into the lateral aspects of the nasopharynx. The *tubal tonsils* surround the openings of these tubes into the nasopharynx.

   Because of the continuity of the middle ear and nasopharyngeal mucosae, nasal infections may invade the middle ear cavity and cause **otitis media** (middle ear inflammation), which is difficult to treat. ■

2. The **oropharynx** is continuous posteriorly with the oral cavity. Since it extends from the soft palate to the epiglottis of the larynx inferiorly, it serves as a common conduit for food and air. In its lateral walls are the *palatine tonsils.* The *lingual tonsil* covers the base of the tongue.

3. The **laryngopharynx**, like the oropharynx, accommodates both ingested food and air. It lies directly posterior to the upright epiglottis and extends to the larynx, where the common pathway divides into the respiratory and digestive channels. From the laryngopharynx, air enters the lower respiratory passageways by passing through the larynx (voice box) and into the trachea below.

The **larynx** (Figure 36.2) consists of nine cartilages. The two most prominent are the large shield-shaped **thyroid cartilage**, whose anterior medial laryngeal prominence is commonly referred to as *Adam's apple,* and the inferiorly located, ring-shaped **cricoid cartilage**, whose widest dimension faces posteriorly. All the laryngeal cartilages are composed of hyaline cartilage except the flaplike **epiglottis**, a flexible elastic cartilage located superior to the opening of the larynx. The epiglottis, sometimes referred to as the "guardian of the airways," forms a lid over the larynx when we swallow. This closes off the respiratory passageways to incoming food or drink, which is routed into the posterior esophagus, or food chute.

- Palpate your larynx by placing your hand on the anterior neck surface approximately halfway down its length. Swallow. Can you feel the cartilaginous larynx rising?

If anything other than air enters the larynx, a cough reflex attempts to expel the substance. Note that this reflex operates only when a person is conscious. Therefore, you should never try to feed or pour liquids down the throat of an unconscious person.

The mucous membrane of the larynx is thrown into two pairs of folds—the upper **vestibular folds,** also called the **false vocal cords,** and the lower **vocal folds,** or **true vocal cords,** which vibrate with expelled air for speech. The vocal cords are attached posterolaterally to the small triangular **arytenoid cartilages** by the *vocal ligaments.* The slitlike passageway between the folds is called the **glottis.**

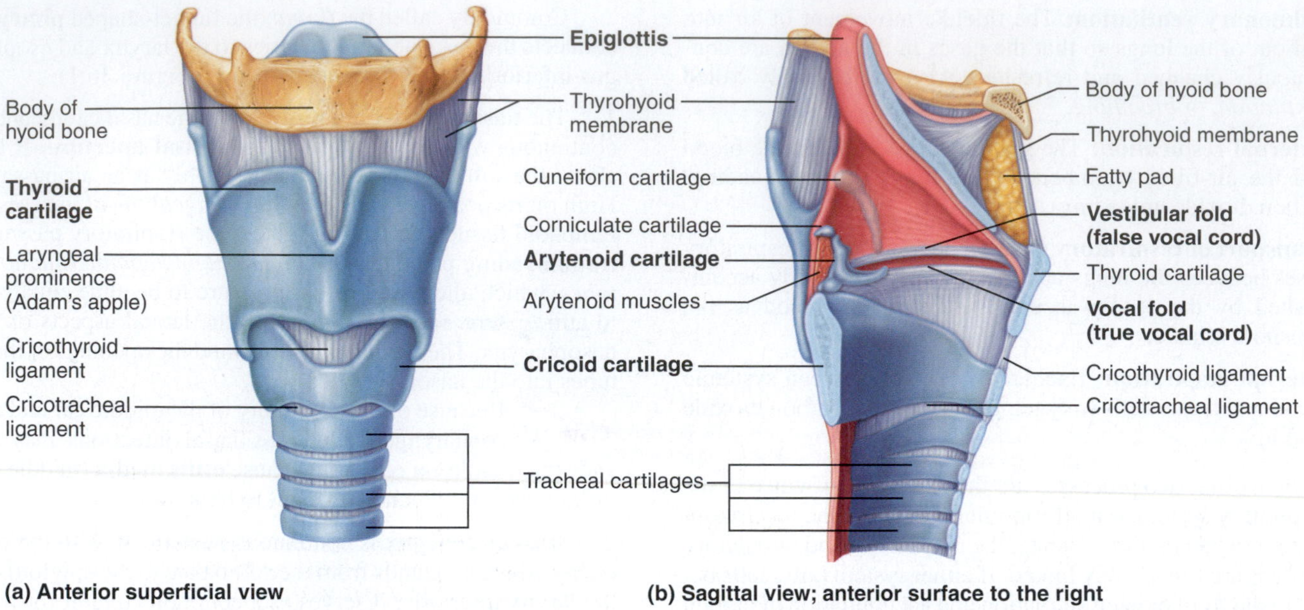
Epiglottis
Body of hyoid bone
Thyrohyoid membrane
Thyroid cartilage
Laryngeal prominence (Adam's apple)
Cricothyroid ligament
Cricotracheal ligament
Tracheal cartilages
Cuneiform cartilage
Corniculate cartilage
Arytenoid cartilage
Arytenoid muscles
Cricoid cartilage
Body of hyoid bone
Thyrohyoid membrane
Fatty pad
Vestibular fold (false vocal cord)
Thyroid cartilage
Vocal fold (true vocal cord)
Cricothyroid ligament
Cricotracheal ligament
(a) Anterior superficial view
(b) Sagittal view; anterior surface to the right

FIGURE 36.2 Structure of the larynx.

Lower Respiratory System Structures

Air entering the **trachea,** or windpipe, from the larynx travels down its length (about 11.0 cm or 4 inches) to the level of the *sternal angle* (or the disc between the fourth and fifth thoracic vertebrae). There the passageway divides into the right and left **main (primary) bronchi** (Figure 36.3), which plunge into their respective lungs at an indented area called the **hilum** (see Figure 36.5c). The right main bronchus is wider, shorter, and more vertical than the left, and foreign objects that enter the respiratory passageways are more likely to become lodged in it.

The trachea is lined with a ciliated, mucus-secreting, pseudostratified columnar epithelium, as are many of the other respiratory system passageways. The cilia propel mucus (produced by goblet cells) laden with dust particles, bacteria,

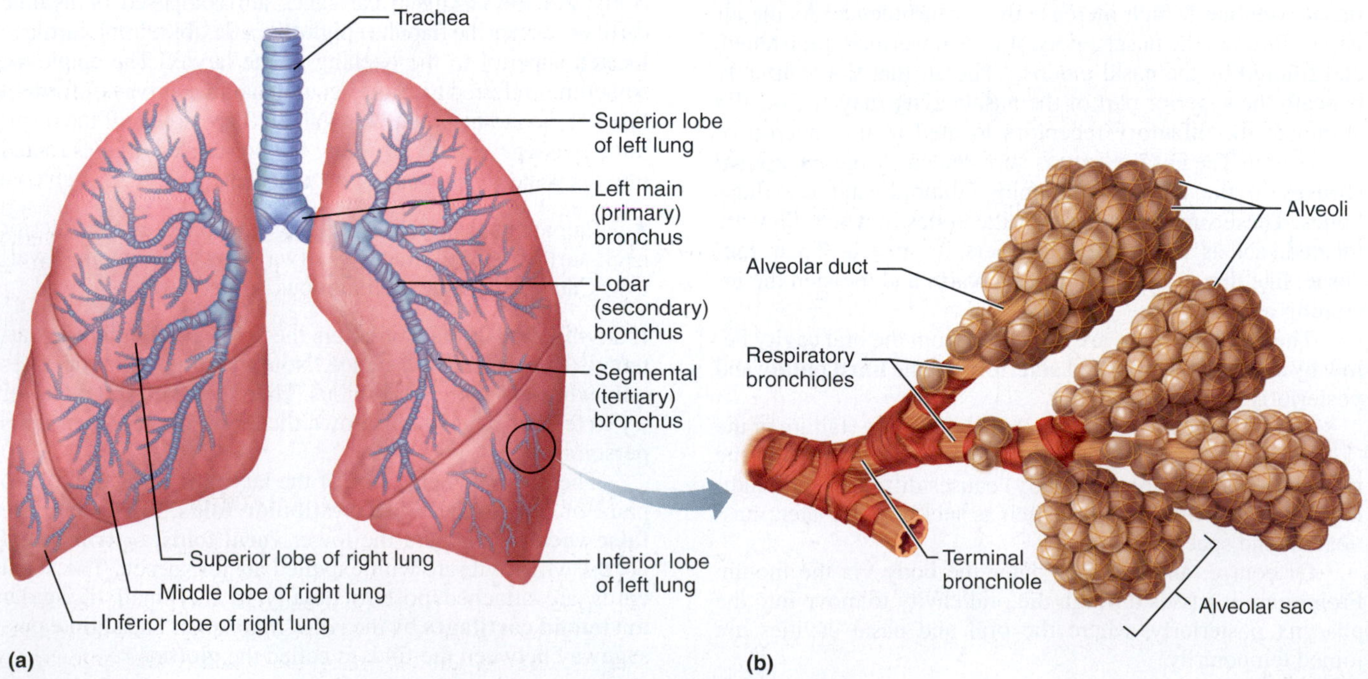
Trachea
Superior lobe of left lung
Left main (primary) bronchus
Lobar (secondary) bronchus
Segmental (tertiary) bronchus
Inferior lobe of left lung
Superior lobe of right lung
Middle lobe of right lung
Inferior lobe of right lung
(a)
Alveoli
Alveolar duct
Respiratory bronchioles
Terminal bronchiole
Alveolar sac
(b)

FIGURE 36.3 Structures of the lower respiratory tract. (a) Diagrammatic view. **(b)** Enlarged view of alveoli.

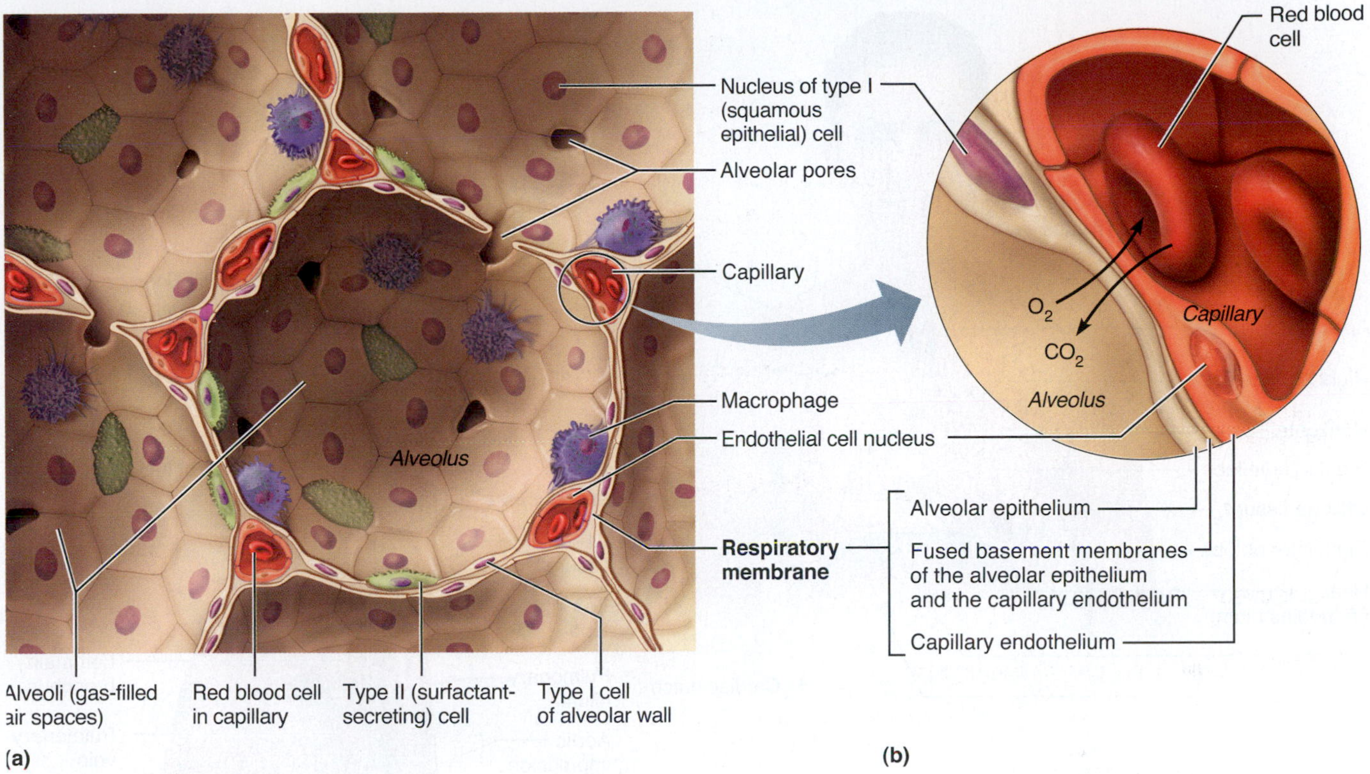
Nucleus of type I (squamous epithelial) cell
Alveolar pores
Capillary
Macrophage
Endothelial cell nucleus
Alveolus
Alveoli (gas-filled air spaces)
Red blood cell in capillary
Type II (surfactant-secreting) cell
Type I cell of alveolar wall
(a)
Red blood cell
O2
CO2
Capillary
Alveolus
Respiratory membrane
Alveolar epithelium
Fused basement membranes of the alveolar epithelium and the capillary endothelium
Capillary endothelium
(b)

FIGURE 36.4 Diagrammatic view of the relationship between the alveoli and pulmonary capillaries involved in gas exchange. (a) One alveolus surrounded by capillaries. **(b)** Enlargement of the respiratory membrane.

and other debris away from the lungs and toward the throat, where it can be expectorated or swallowed. The walls of the trachea are reinforced with C-shaped cartilaginous rings, the incomplete portion located posteriorly (see Figure 36.6, page 543). These C-shaped cartilages serve a double function: The incomplete parts allow the esophagus to expand anteriorly when a large food bolus is swallowed. The solid portions reinforce the trachea walls to maintain its open passageway regardless of the pressure changes that occur during breathing.

The primary bronchi further divide into smaller and smaller branches (the secondary, tertiary, on down), finally becoming the **bronchioles,** which have terminal branches called **respiratory bronchioles** (Figure 36.3b). All but the most minute branches have cartilaginous reinforcements in their walls, usually in the form of small plates of hyaline cartilage rather than cartilaginous rings. As the respiratory tubes get smaller and smaller, the relative amount of smooth muscle in their walls increases as the amount of cartilage declines and finally disappears. The complete layer of smooth muscle present in the bronchioles enables them to provide considerable resistance to air flow under certain conditions (asthma, hay fever, etc.). The continuous branching of the respiratory passageways in the lungs is often referred to as the **respiratory tree.** The comparison becomes much more meaningful in a resin cast of the respiratory passages.

- Observe a resin cast of respiratory passages if one is available.

The respiratory bronchioles in turn subdivide into several **alveolar ducts**, which terminate in alveolar sacs that rather resemble clusters of grapes. **Alveoli,** tiny balloonlike expansions along the alveolar sacs and occasionally found protruding from alveolar ducts and respiratory bronchioles, are composed of a single thin layer of squamous epithelium overlying a wispy basal lamina. The external surfaces of the alveoli are densely spiderwebbed with a network of pulmonary capillaries (Figure 36.4). Together, the alveolar and capillary walls and their fused basal laminae form the **respiratory membrane,** also called the *air-blood barrier.*

Because gas exchanges occur by simple diffusion across the respiratory membrane—oxygen passing from the alveolar air to the capillary blood and carbon dioxide leaving the capillary blood to enter the alveolar air—the alveolar sacs, alveolar ducts, and respiratory bronchioles are referred to collectively as **respiratory zone structures.** All other respiratory passageways (from the nasal cavity to the terminal bronchioles) simply serve as access or exit routes to and from these gas exchange chambers and are called **conducting zone structures.** Because the conducting zone structures have no exchange function, they are also referred to as *anatomical dead space.*

The Lungs and Their Pleural Coverings

The paired lungs are soft, spongy organs that occupy the entire thoracic cavity except for the *mediastinum,* which houses the heart, bronchi, esophagus, and other organs (Figure 36.5). Each lung is connected to the mediastinum by a **root** containing its vascular and bronchial attachments. The structures of the root enter (or leave) the lung via a medial indentation called the *hilum.* All structures distal to the primary bronchi are found within the lung substance. A lung's **apex,** the

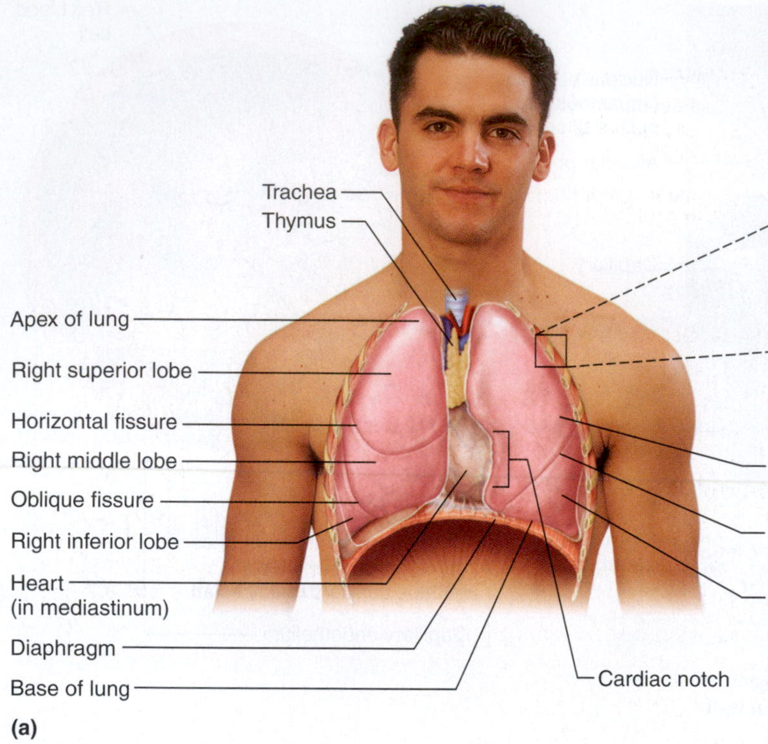
Trachea
Thymus
Apex of lung
Right superior lobe
Horizontal fissure
Right middle lobe
Oblique fissure
Right inferior lobe
Heart
(in mediastinum)
Diaphragm
Base of lung
Cardiac notch
Left superior lobe
Oblique fissure
Left inferior lobe
Intercostal muscle
Rib
Parietal pleura
Pleural cavity
Visceral pleura
Lung
(a)

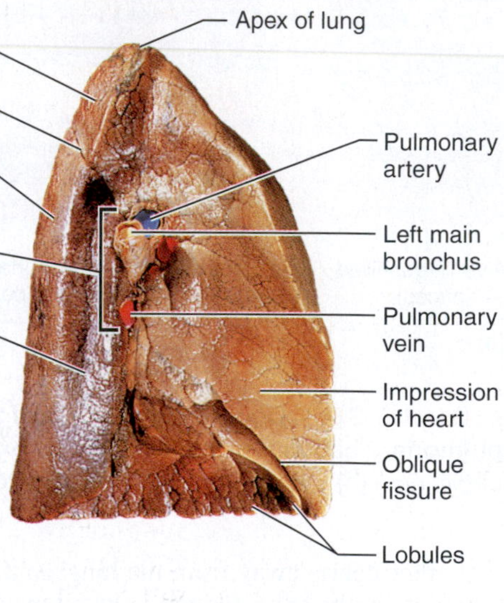
Apex of lung
Pulmonary artery
Left main bronchus
Pulmonary vein
Impression of heart
Oblique fissure
Lobules
Pulmonary hilum
Aortic impression
(b)

FIGURE 36.5 Anatomical relationships of organs in the thoracic cavity. (a) Anterior view of the thoracic organs. The lungs flank the central mediastinum. The inset at upper right depicts the pleurae and the pleural cavity. **(b)** Photograph of medial aspect of left lung. **(c)** Transverse section through the superior part of the thorax, showing the lungs and the main organs in the mediastinum.

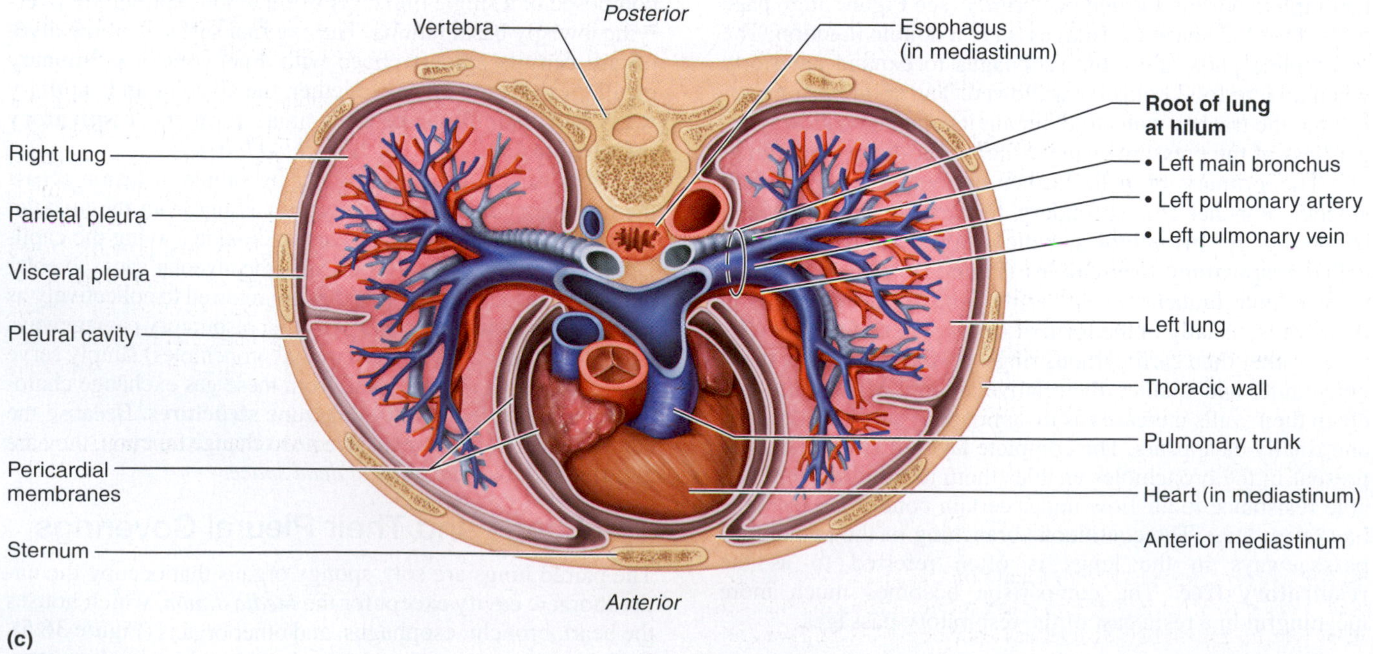
Posterior
Vertebra
Esophagus
(in mediastinum)
Root of lung
at hilum
• Left main bronchus
• Left pulmonary artery
• Left pulmonary vein
Right lung
Parietal pleura
Visceral pleura
Pleural cavity
Left lung
Thoracic wall
Pulmonary trunk
Pericardial
membranes
Heart (in mediastinum)
Anterior mediastinum
Sternum
Anterior
(c)

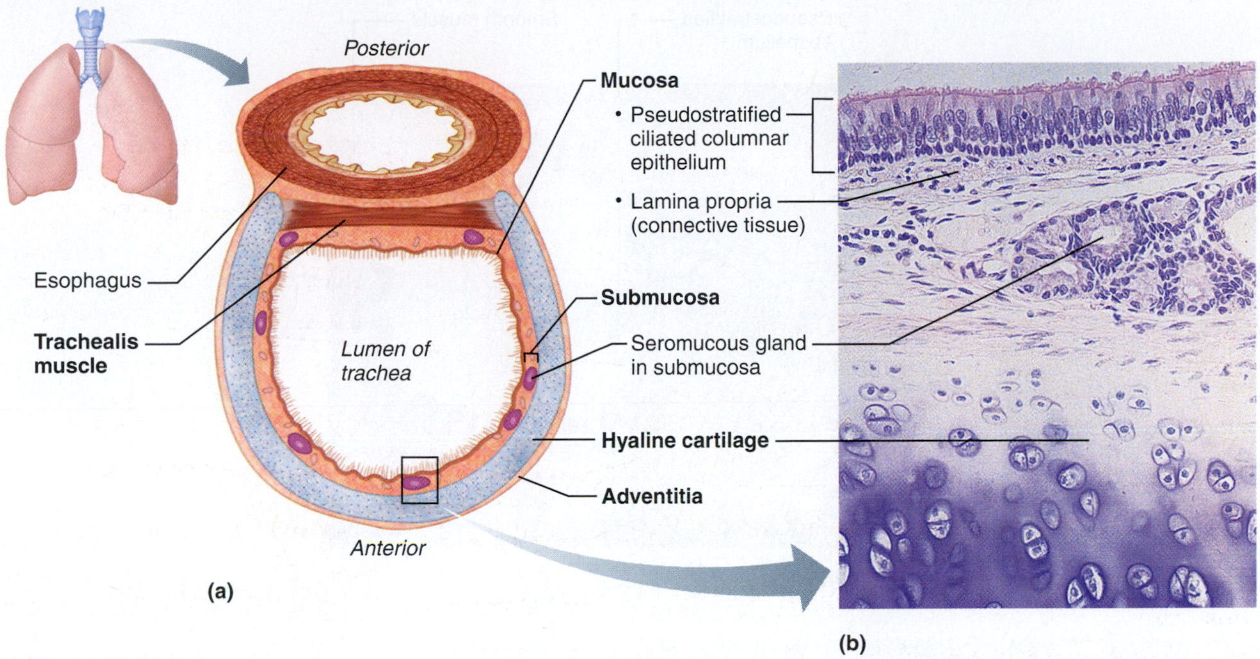
Posterior
Mucosa
• Pseudostratified ciliated columnar epithelium
• Lamina propria (connective tissue)
Esophagus
Submucosa
Trachealis muscle
Lumen of trachea
Seromucous gland in submucosa
Hyaline cartilage
Adventitia
Anterior
(a)
(b)

FIGURE 36.6 Tissue composition of the tracheal wall. (a) Cross-sectional view of the trachea. **(b)** Photomicrograph of a portion of the tracheal wall (110×).

narrower superior aspect, lies just deep to the clavicle, and its **base,** the inferior concave surface, rests on the diaphragm. Anterior, lateral, and posterior lung surfaces are in close contact with the ribs and, hence, are collectively called the **costal surface.** The medial surface of the left lung exhibits a concavity called the **cardiac notch,** which accommodates the heart where it extends left from the body midline. Fissures divide the lungs into a number of **lobes**—two in the left lung and three in the right. Other than the respiratory passageways and air spaces that make up the bulk of their volume, the lungs are mostly elastic connective tissue, which allows them to recoil passively during expiration.

Each lung is enclosed in a double-layered sac of serous membrane called the **pleura.** The outer layer, the **parietal pleura,** is attached to the thoracic walls and the **diaphragm;** the inner layer, covering the lung tissue, is the **visceral pleura.** The two pleural layers are separated by the **pleural cavity,** which is more of a potential space than an actual one. The pleural layers produce lubricating serous fluid that causes them to adhere closely to one another, holding the lungs to the thoracic wall and allowing them to move easily against one another during the movements of breathing.

ACTIVITY 1

Identifying Respiratory System Organs

Before proceeding, be sure to locate on the torso model, thoracic cavity structures model, larynx model, or an anatomical chart all the respiratory structures described—both upper and lower respiratory system organs. ■

For instructions on animal dissections, see the dissection exercises starting on page 697 in the cat, rat, and fetal pig editions of this manual.

ACTIVITY 2

Demonstrating Lung Inflation in a Sheep Pluck

A *sheep pluck* includes the larynx, trachea with attached lungs, the heart and pericardium, and portions of the major blood vessels found in the mediastinum (aorta, pulmonary artery and vein, venae cavae). If a sheep pluck is not available, a good substitute is a preserved inflatable pig lung.

⚠ Don disposable gloves, obtain a dissecting tray and a fresh sheep pluck (or a preserved pluck of another animal), and identify the lower respiratory system organs. Once you have completed your observations, insert a hose from an air compressor (vacuum pump) into the trachea and alternately allow air to flow in and out of the lungs. Notice how the lungs inflate. This observation is educational in a preserved pluck but it is a spectacular sight in a fresh one. Another advantage of using a fresh pluck is that the lung pluck changes color (becomes redder) as hemoglobin in trapped RBCs becomes loaded with oxygen.

⚠ Dispose of the gloves in the autoclave bag immediately after use. ■

ACTIVITY 3

Examining Prepared Slides of Trachea and Lung Tissue

1. Obtain a compound microscope and a slide of a cross section of the tracheal wall. Identify the smooth muscle layer, the hyaline cartilage supporting rings, and the pseudostratified ciliated epithelium. Use Figure 36.6 as a guide; also try

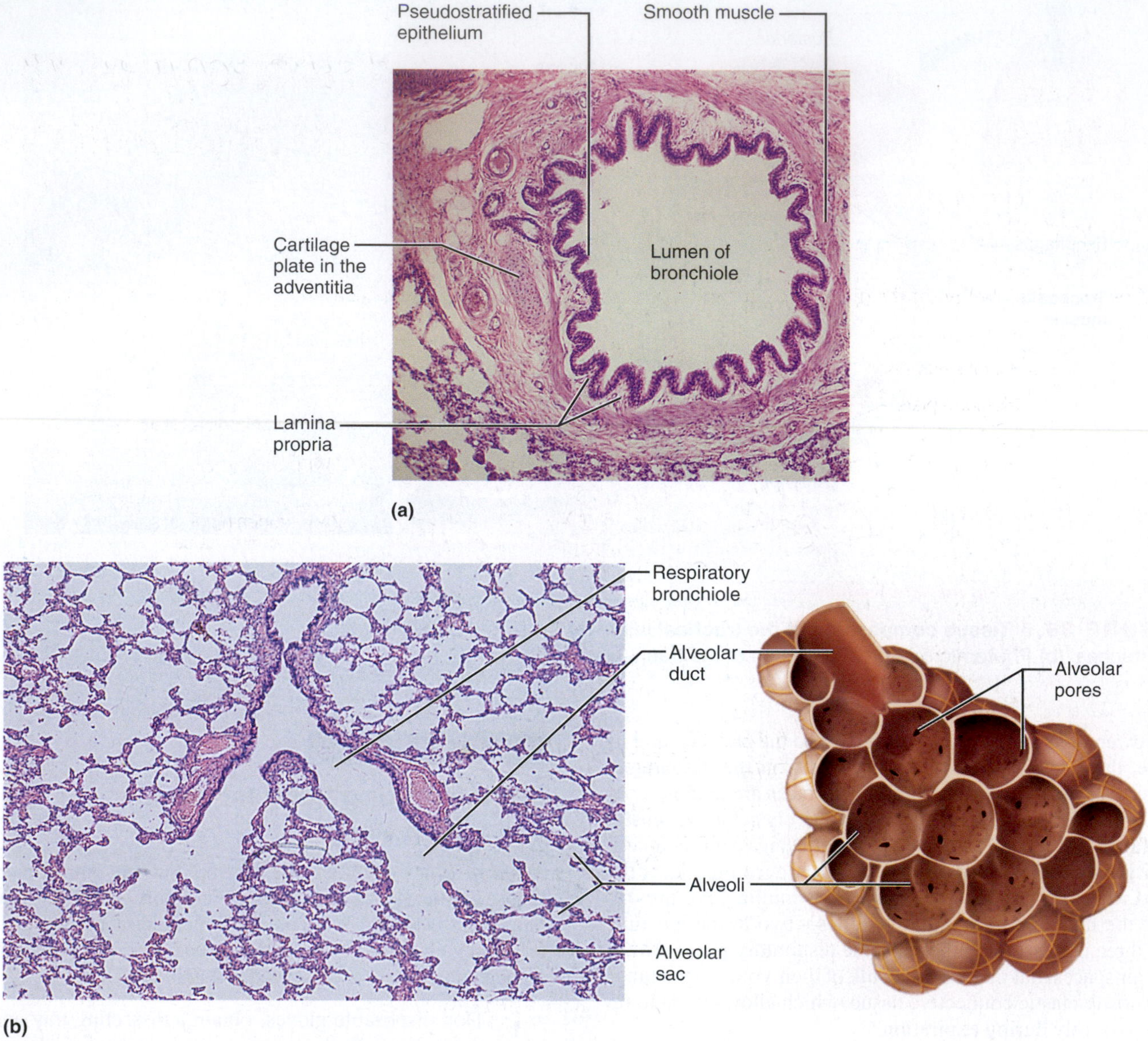
Pseudostratified epithelium
Smooth muscle
Cartilage plate in the adventitia
Lumen of bronchiole
Lamina propria
(a)
Respiratory bronchiole
Alveolar duct
Alveolar pores
Alveoli
Alveolar sac
(b)

FIGURE 36.7 Microscopic structure of a bronchiole and alveoli. **(a)** Photomicrograph of a section of a bronchiole (22×). **(b)** Photomicrograph and diagrammatic view of alveoli (70×).

to identify a few goblet cells in the epithelium (see Figure 6.3c and d on p. 71).

2. Obtain a slide of lung tissue for examination. The alveolus is the main structural and functional unit of the lung and is the actual site of gas exchange. Identify a bronchiole (Figure 36.7a) and the thin squamous epithelium of the alveolar walls (Figure 36.7b).

3. Examine slides of pathological lung tissues, and compare them to the normal lung specimens. Record your observations at the end of the Exercise 36 Review Sheet.

EXERCISE 37

Respiratory System Physiology

MATERIALS

- ☐ Model lung (bell jar demonstrator)
- ☐ Tape measure
- ☐ Stethoscope
- ☐ Alcohol swabs
- ☐ Apparatus A or B:

  A: Spirometer, disposable cardboard mouthpieces, nose clips, table (on chalkboard) for recording class data, disposable autoclave bag, battery jar containing 70% ethanol solution

  BIOPAC® BIOPAC® MP36 (or MP35/30) data acquisition unit, PC or Mac computer, BIOPAC® Student Lab Software, BIOPAC® airflow transducer, BIOPAC® calibration syringe, disposable mouthpiece, nose clip, bacteriological filter

  New versions of the BSL software (3.7.5 and higher for Windows, and 3.7.4 and higher for Mac) require slightly different channel settings and collection strategies. Instructions for using the newer software with the MP36/35 data acquisition units can be found on MasteringA&P. In addition, instructions for use of the NEW 2-channel data acquisition unit, the MP45, may also be found on MasteringA&P.
- ☐ Physiograph, pneumograph, and recording attachments for physiograph
- ☐ Paper bag
- ☐ 0.05 *M* NaOH
- ☐ Phenol red in a dropper bottle
- ☐ 100-ml beakers
- ☐ Straws

Text continues on next page.

MasteringA&P™

Access practice quizzes and more in the Study Area at www.masteringaandp.com.

PEx

This lab corresponds to PhysioEx Exercise 7. See p. PEx-105.

OBJECTIVES

1. To define the following (and be prepared to provide volume figures if applicable):

   inspiration
   expiration
   tidal volume
   vital capacity
   expiratory reserve volume
   inspiratory reserve volume
   minute respiratory volume
2. To explain the role of muscles and volume changes in the mechanical process of breathing.
3. To describe bronchial and vesicular breathing sounds.
4. To demonstrate proper usage of a spirometer.
5. To explain the relative importance of various mechanical and chemical factors in producing respiratory variations.
6. To explain the importance of the carbonic acid–bicarbonate buffer system in maintaining blood pH.

PRE-LAB QUIZ

1. Circle the correct term. Inspiration / Expiration is the phase of pulmonary ventilation when air passes out of the lungs.
2. Which of the following processes does *not* occur during inspiration?
   a. diaphragm moves to a flattened position
   b. gas pressure inside the lungs is lowered
   c. inspiratory muscles relax
   d. size of thoracic cavity increases
3. Circle True or False. Vesicular breathing sounds are produced by air rushing through the trachea and bronchi.
4. During normal quiet breathing, about _________ ml of air moves into and out of the lungs with each breath.
   a. 250
   b. 500
   c. 1000
   d. 2000
5. Circle the correct term. Tidal volume / Vital capacity is the maximum amount of air that can be exhaled after a maximal inspiration.
6. Circle True or False. The neural centers that control respiratory rhythm and maintain a rate of 12–18 respirations per minute are located in the medulla and thalamus.
7. Circle the correct term. Changes in pH and oxygen concentrations in the blood are monitored by chemoreceptor regions in the medulla / aortic and carotid bodies.
8. The carbonic acid–bicarbonate buffer system stabilizes arterial blood pH at:
   a. 2.0 ± 1.00
   b. 6.2 ± 0.07
   c. 7.4 ± 0.02
   d. 9.5 ±1.15

Text continues on next page.

- ☐ Concentrated HCl and NaOH in dropper bottles
- ☐ 250- and 50-ml beakers
- ☐ Plastic wash bottles containing distilled water
- ☐ Graduated cylinder (100 ml)
- ☐ Glass stirring rod
- ☐ Animal plasma
- ☐ pH meter (standardized with buffer of pH 7)
- ☐ Buffer solution (pH 7)
- ☐ 0.01 *M* HCl

Note: Instructions for using PowerLab® equipment can be found at MasteringA&P.

9. Circle the correct term. Acids / Bases released into the blood by the body cells tend to lower the pH of the blood and cause it to become acidic.
10. Circle True or False. Rate and depth of breathing, hyperventilation, and hypoventilation should have little or no effect on the acid-base balance of blood.

The body's trillions of cells require O_2 and give off CO_2 as a waste the body must get rid of. The **respiratory system** provides the link with the external environment for both taking in O_2 and eliminating CO_2, but it doesn't work alone. The cardiovascular system via its contained blood provides the watery medium for transporting O_2 and CO_2 in the body. Let's look into how the respiratory system carries out its role.

Mechanics of Respiration

Pulmonary ventilation, or **breathing,** consists of two phases: **inspiration,** during which air is taken into the lungs, and **expiration,** during which air passes out of the lungs. As the inspiratory muscles (external intercostals and diaphragm) contract during inspiration, the size of the thoracic cavity increases. The diaphragm moves from its relaxed dome shape to a flattened position, increasing the superoinferior volume. The external intercostals lift the rib cage, increasing the anteroposterior and lateral dimensions (Figure 37.1). Because the lungs adhere to the thoracic walls like flypaper owing to the presence of serous fluid in the pleural cavity, the intrapulmonary volume (volume within the lungs) also increases, lowering the air (gas) pressure inside the lungs. The gases then expand to fill the available space, creating a partial vacuum that causes air to flow into the lungs—constituting the act of inspiration. during expiration, the inspiratory muscles relax, and the natural tendency of the elastic lung tissue to recoil acts to decrease the intrathoracic and intrapulmonary volumes. As the gas molecules within the lungs are forced closer together, the intrapulmonary pressure rises to a point higher than atmospheric pressure. This causes gases to flow out of the lungs to equalize the pressure inside and outside the lungs—the act of expiration.

ACTIVITY 1

Operating the Model Lung

Observe the model lung, which demonstrates the principles involved in gas flows into and out of the lungs. It is a simple apparatus with a bottle "thorax," a rubber membrane "diaphragm," and balloon "lungs."

1. Go to the demonstration area and work the model lung by moving the rubber diaphragm up and down. The balloons will not fully inflate or deflate, but notice the *relative* changes in balloon (lung) size as the volume of the thoracic cavity is alternately increased and decreased.

2. Check the appropriate columns in the chart concerning these observations in the Exercise 37 Review Sheet at the end of this exercise.

3. Simulate a pneumothorax. Inflate the balloon lungs by pulling down on the diaphragm. Ask your lab partner to let air into the bottle "thorax" by loosening the rubber stopper.

What happens to the balloon lungs?

___

4. After observing the operation of the model lung, conduct the following tests on your lab partner. Use the tape measure to determine his or her chest circumference by placing the tape around the chest as high up under the armpits as possible. Record the measurements in inches in the appropriate space for each of the conditions on the following page.

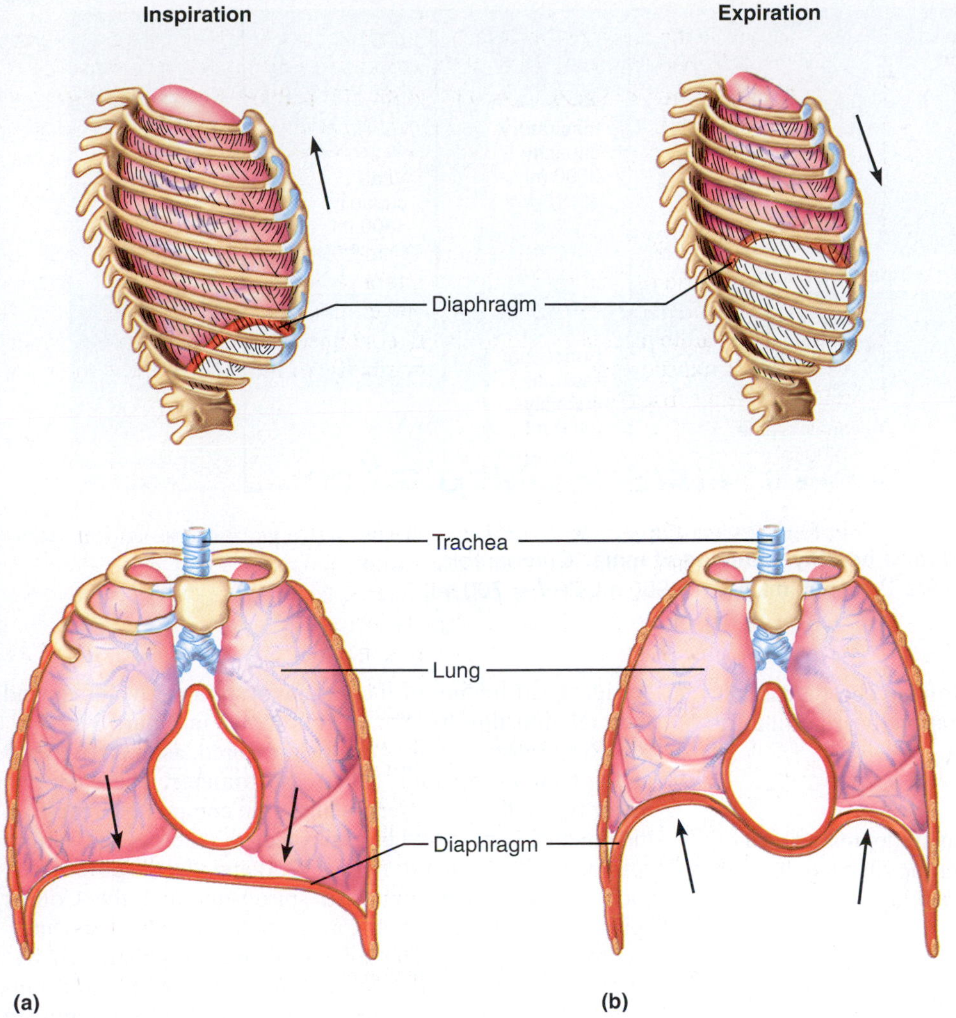
Inspiration
Expiration
Diaphragm
Trachea
Lung
Diaphragm
(a)
(b)

FIGURE 37.1 Rib cage and diaphragm positions during breathing. (a) At the end of a normal inspiration; chest expanded, diaphragm depressed. **(b)** At the end of a normal expiration; chest depressed, diaphragm elevated.

Quiet breathing:

Inspiration ______________ Expiration ______________

Forced breathing:

Inspiration ______________ Expiration ______________

Do the results coincide with what you expected on the basis of what you have learned thus far? _________

How does the structural relationship between the balloon-lungs and bottle-thorax differ from that seen in the human lungs and thorax?

____

____

____

Respiratory Sounds

As air flows in and out of the respiratory tree, it produces two characteristic sounds that can be picked up with a stethoscope (auscultated). The **bronchial sounds** are produced by air rushing through the large respiratory passageways (the trachea and the bronchi). The second sound type, **vesicular breathing sounds,** apparently results from air filling the alveolar sacs and resembles the sound of a rustling or muffled breeze.

ACTIVITY 2

Auscultating Respiratory Sounds

1. Obtain a stethoscope and clean the earpieces with an alcohol swab. Allow the alcohol to dry before donning the stethoscope.

2. Place the diaphragm of the stethoscope on the throat of the test subject just below the larynx. Listen for bronchial sounds on inspiration and expiration. Move the stethoscope down toward the bronchi until you can no longer hear sounds.

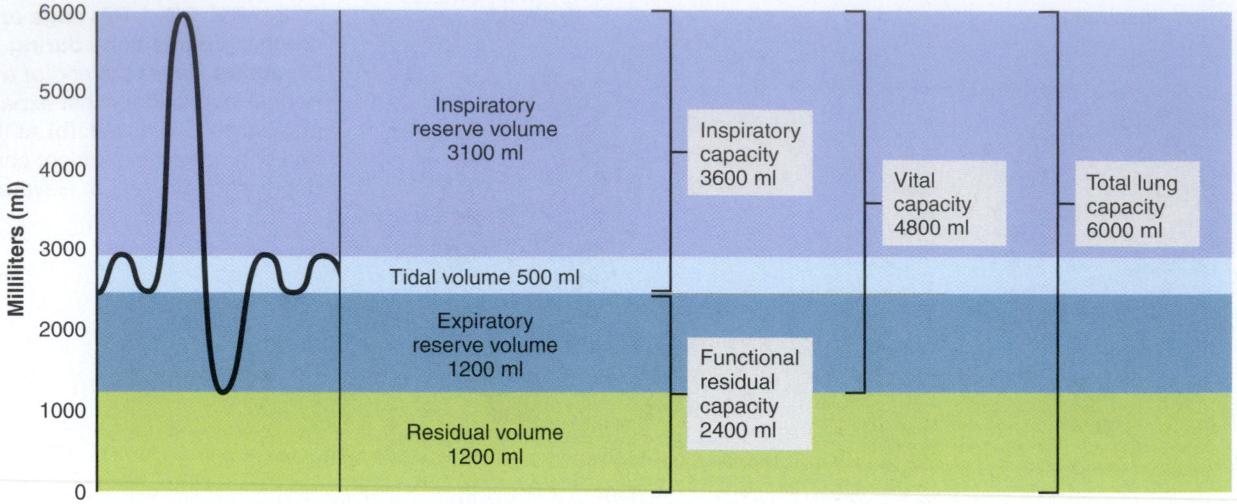
6000
5000
4000
3000
2000
1000
0
Milliliters (ml)
Inspiratory reserve volume 3100 ml
Tidal volume 500 ml
Expiratory reserve volume 1200 ml
Residual volume 1200 ml
Inspiratory capacity 3600 ml
Functional residual capacity 2400 ml
Vital capacity 4800 ml
Total lung capacity 6000 ml

FIGURE 37.2 Spirographic record for a healthy young adult male. Comparable values for a healthy young adult female are TV = 500 ml, IRV = 1900 ml, ERV = 700 ml, residual volume (RV) = 1100 ml.

3. Place the stethoscope over the following chest areas and listen for vesicular sounds during respiration (heard primarily during inspiration).

- At various intercostal spaces
- At the *triangle of auscultation* (a small depressed area of the back where the muscles fail to cover the rib cage; located just medial to the inferior part of the scapula)
- Under the clavicle

Diseased respiratory tissue, mucus, or pus can produce abnormal chest sounds such as rales (a rasping sound) and wheezing (a whistling sound).

Respiratory Volumes and Capacities—Spirometry

A person's size, sex, age, and physical condition produce variations in respiratory volumes. Normal quiet breathing moves about 500 ml of air in and out of the lungs with each breath. As you have seen in the first activity, a person can usually forcibly inhale or exhale much more air than is exchanged in normal quiet breathing. The terms given to the measurable respiratory volumes are defined next. You should memorize these terms and their normal values for an adult male.

Tidal volume (TV): Amount of air inhaled or exhaled with each breath under resting conditions (500 ml)

Inspiratory reserve volume (IRV): Amount of air that can be forcefully inhaled after a normal tidal volume inhalation (3100 ml)

Expiratory reserve volume (ERV): Amount of air that can be forcefully exhaled after a normal tidal volume exhalation (1200 ml)

Vital capacity (VC): Maximum amount of air that can be exhaled after a maximal inspiration (4800 ml)

$$VC = TV + IRV + ERV$$

An idealized tracing of the various respiratory volumes and their relationships to each other is shown in Figure 37.2.

Respiratory volumes can be measured, as in Activities 3 and 4, with an apparatus called a **spirometer.** There are two major types of spirometers, which give comparable results—the handheld dry, or wheel, spirometers (such as the Wright spirometer illustrated in Figure 37.3) and "wet" spirometers, such as the Phipps and Bird spirometer and the Collins spirometer (which is available in both recording and non-recording varieties). The somewhat more sophisticated wet spirometer consists of a plastic or metal *bell* within a rectangular or cylindrical tank that air can be added to or removed from (Figure 37.4). The outer tank contains water and has a tube running through it to carry air above the water level. The floating bottomless bell is inverted over the water-containing tank and connected to a volume indicator.

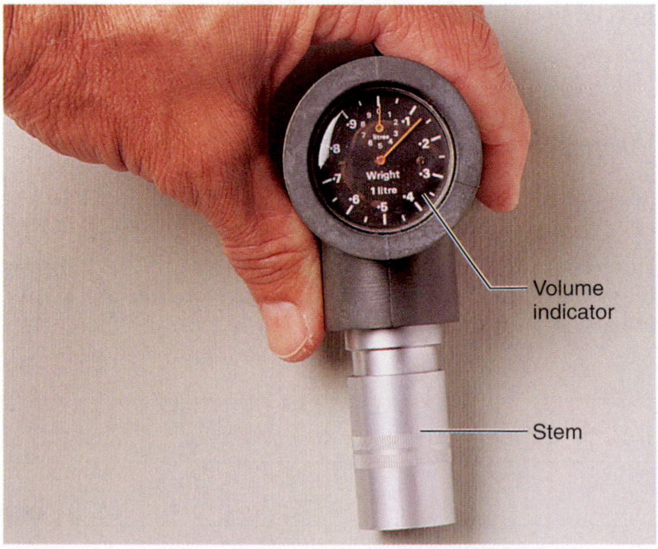
Wright
1 litre
Volume indicator
Stem

FIGURE 37.3 The Wright handheld dry spirometer. Reset to zero prior to each test.

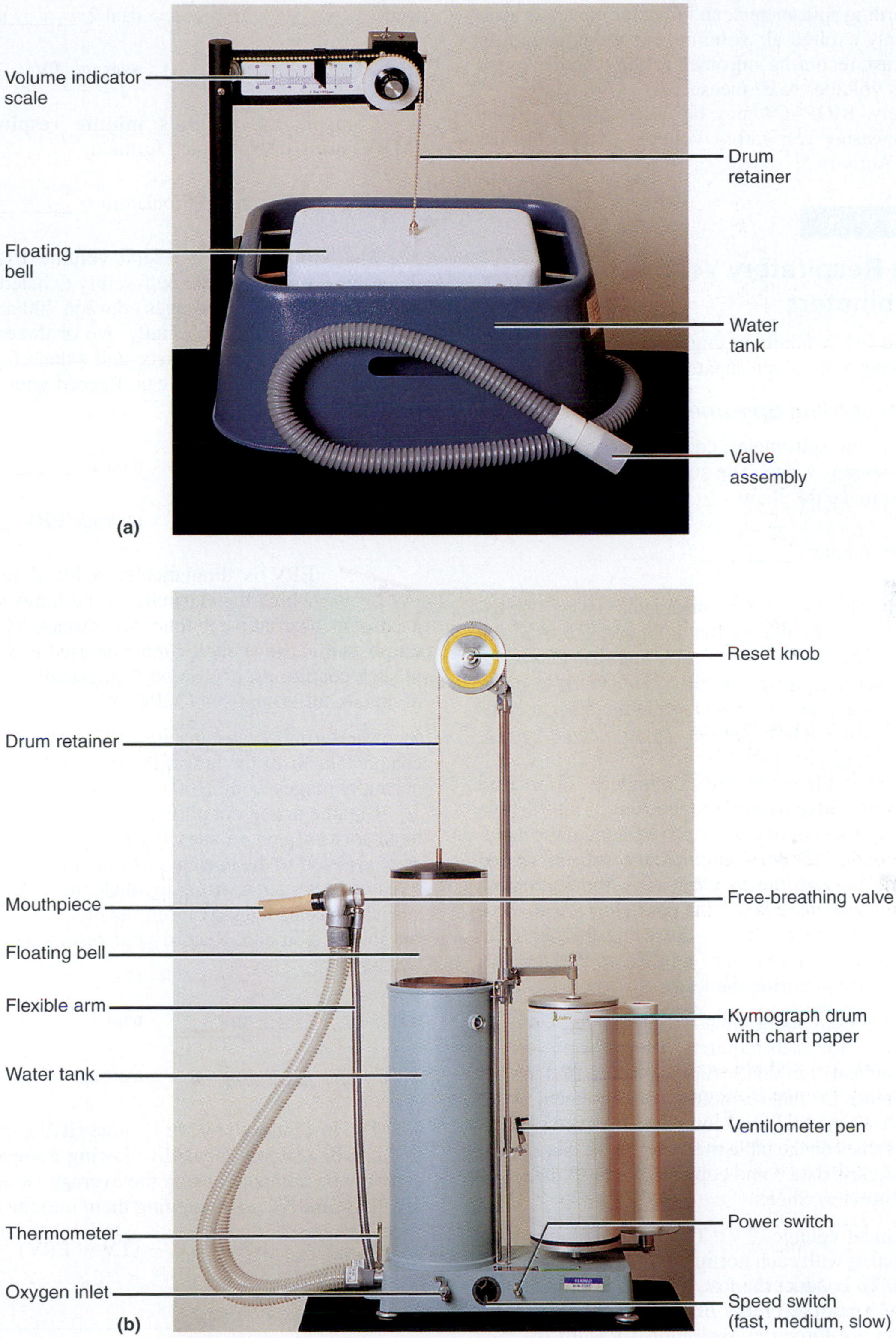
Volume indicator scale
Drum retainer
Floating bell
Water tank
Valve assembly
(a)
Reset knob
Drum retainer
Mouthpiece
Free-breathing valve
Floating bell
Flexible arm
Kymograph drum with chart paper
Water tank
Ventilometer pen
Power switch
Thermometer
Oxygen inlet
Speed switch (fast, medium, slow)
(b)

FIGURE 37.4 Wet spirometers. (a) The Phipps and Bird wet spirometer. **(b)** The Collins-9L wet recording spirometer.

In nonrecording spirometers, an indicator moves as air is *exhaled,* and only expired air volumes can be measured directly. By contrast, recording spirometers allow both inspired and expired gas volumes to be measured.

Alternatively, BIOPAC® may be used with an airflow transducer to measure respiratory volumes. Those instructions appear in Activity 5.

ACTIVITY 3

Measuring Respiratory Volumes Using Spirometers

The steps for using a nonrecording spirometer and a wet recording spirometer are given separately below.

Using a Nonrecording Spirometer

1. Before using the spirometer, count and record the subject's normal respiratory rate. The subject should face away from you as you make the count.

Respirations per minute: ________

Now identify the parts of the spirometer you will be using by comparing it to the illustration in Figure 37.3 or 37.4a. Examine the spirometer volume indicator *before beginning* to make sure you know how to read the scale. Work in pairs, with one person acting as the subject while the other records the data of the volume determinations. *Reset the indicator to zero before beginning each trial.*

Obtain a disposable cardboard mouthpiece. Insert it in the open end of the valve assembly (attached to the flexible tube) of the wet spirometer or over the fixed stem of the handheld dry spirometer. Before beginning, the subject should practice exhaling through the mouthpiece without exhaling through the nose, or prepare to use the nose clips (clean them first with an alcohol swab). If you are using the handheld spirometer, make sure its dial faces upward so that the volumes can be easily read during the tests.

2. The subject should stand erect during testing. Conduct the test three times for each required measurement. Record the data where indicated in this section, and then find the average volume figure for that respiratory measurement. After you have completed the trials and computed the averages, enter the average values on the table prepared on the chalkboard for tabulation of class data,* and copy all averaged data onto the Exercise 37 Review Sheet.

3. Measuring tidal volume (TV). The TV, or volume of air inhaled and exhaled with each normal respiration, is approximately 500 ml. To conduct the test, inhale a normal breath, and then exhale a normal breath of air into the spirometer mouthpiece. (Do not force the expiration!) Record the volume and repeat the test twice.

trial 1: ________ ml trial 2: ________ ml

trial 3: ________ ml average TV: ________ ml

4. Compute the subject's **minute respiratory volume (MRV)** using the following formula:

MRV = TV × respirations/min = ________ ml/min

5. Measuring expiratory reserve volume (ERV). The ERV is the volume of air that can be forcibly exhaled after a normal expiration. Normally it ranges between 700 and 1200 ml.

Inhale and exhale normally two or three times, then insert the spirometer mouthpiece and exhale forcibly as much of the additional air as you can. Record your results, and repeat the test twice again.

trial 1: ________ ml trial 2: ________ ml

trial 3: ________ ml average ERV: ________ ml

ERV is dramatically reduced in conditions in which the elasticity of the lungs is decreased by a chronic obstructive pulmonary disease (COPD) such as **emphysema.** Since energy must be used to *deflate* the lungs in such conditions, expiration is physically exhausting to individuals suffering from COPD. ■

6. Measuring vital capacity (VC). The VC, or total exchangeable air of the lungs (the sum of TV + IRV + ERV), normally ranges from 3100 ml to 4800 ml.

Breathe in and out normally two or three times, and then bend forward and exhale all the air possible. Then, as you raise yourself to the upright position, inhale as fully as possible. It is important to *strain* to inhale the maximum amount of air that you can. Quickly insert the mouthpiece, and exhale as forcibly as you can. Record your results and repeat the test twice again.

trial 1: ________ ml trial 2: ________ ml

trial 3: ________ ml average VC: ________ ml

7. The inspiratory reserve volume (IRV), or volume of air that can be forcibly inhaled following a normal inspiration, can now be computed using the average values obtained for TV, ERV, and VC and plugging them into the equation:

IRV = VC − (TV + ERV)

Record your average IRV: ____________________ ml

The normal IRV is substantial, ranging from 1900 to 3100 ml. How does your computed value compare?

____

Steps 8–10, which provide common directions for use of both nonrecording and recording spirometers, continue on page 556 after the wet recording spirometer directions.

*__Note to the Instructor:__ The format of class data tabulation can be similar to that shown here. However, it would be interesting to divide the class into smokers and nonsmokers and then compare the mean average VC and ERV for each group. Such a comparison might help to determine if smokers are handicapped in any way. It also might be a good opportunity for an informal discussion of the early warning signs of bronchitis and emphysema, which are primarily smokers' diseases.

Using a Recording Spirometer

1. In preparation for recording, familiarize yourself with the spirometer by comparing it to the equipment illustrated in Figure 37.4b.

2. Examine the chart paper, noting that its horizontal lines represent milliliter units. To apply the chart paper to the recording drum, first lift the drum retainer and then remove the kymograph drum. Wrap a sheet of chart paper around the drum, *making sure that the right edge overlaps the left.* Fasten it with tape, and then replace the kymograph drum and lower the drum retainer into its original position in the hole in the top of the drum.

3. Raise and lower the floating bell several times, noting as you do so that the *ventilometer pen* moves up and down on the drum. This pen, which writes in black ink, will be used for recording and should be adjusted so that it records in the approximate middle of the chart paper. This adjustment is made by repositioning the floating bell using the *reset knob* on the metal pulley at the top of the spirometer apparatus. The other pen, the respirometer pen, which records in red ink, will not be used for these tests and should be moved away from the drum's recording surface.

4. Record your normal respiratory rate. Clean the nose clips with an alcohol swab. While you wait for the alcohol to air dry, count and record your normal respiratory rate.

Respirations per minute: _________

5. Recording tidal volume. After the alcohol has air dried, apply the nose clips to your nose. This will enforce mouth breathing.

Open the *free-breathing valve.* Insert a disposable cardboard mouthpiece into the end (valve assembly) of the breathing tube, and then insert the mouthpiece into your mouth. Practice breathing for several breaths to get used to the apparatus. At this time, you are still breathing room air.

Set the spirometer switch to **SLOW** (32 mm/min). Close the free-breathing valve, and breathe in a normal manner for 2 minutes to record your tidal volume—the amount of air inspired or expired with each normal respiratory cycle. This recording should show a regular pattern of inspiration-expiration spikes and should gradually move upward on the chart paper. (A downward slope indicates that there is an air leak somewhere in the system—most likely at the mouthpiece.) Notice that on an apparatus using a counterweighted pen, such as the Collins-9L Ventilometer shown in Figure 37.4b, inspirations are recorded by upstrokes and expirations are recorded by downstrokes.*

6. Recording vital capacity. To record your vital capacity, take the deepest possible inspiration you can and then exhale to the greatest extent possible—really *push* the air out. The recording obtained should resemble that shown in Figure 37.5. Repeat the vital capacity measurement twice again. Then turn off the spirometer and remove the chart paper from the kymograph drum.

7. Determine and record your measured, averaged, and corrected respiratory volumes. Because the pressure and temperature inside the spirometer are influenced by room temperature and differ from those in the body, all measured values are to be multiplied by a **BTPS** (body temperature, atmospheric pressure, and water saturation) **factor.** At room temperature, the BTPS factor is typically 1.1 or very close to that value. Hence, you will multiply your average measured values by 1.1 to obtain your corrected respiratory volume values. Copy the averaged and corrected values onto the Exercise 37 Review Sheet.

- Tidal volume (TV). Select a typical resting tidal breath recording. Subtract the millimeter value of the trough (exhalation) from the millimeter value of the peak (inspiration). Record this value below as *measured TV 1.* Select two other TV tracings to determine the TV values for the TV 2 and TV 3 measurements. Then, determine your average TV and multiply it by 1.1 to obtain the BTPS-corrected average TV value.

measured TV 1: _______ ml average TV: _______ ml

measured TV 2: _______ ml corrected average TV:

measured TV 3: _______ ml _______ ml

Also compute your **minute respiratory volume (MRV)** using the following formula:

MRV = TV × respirations/min = _________ ml/min

- Inspiratory capacity (IC). In the first vital capacity recording, find the expiratory trough immediately preceding the maximal inspiratory peak achieved during vital capacity determination. Subtract the milliliter value of that expiration from the value corresponding to the peak of the maximal inspiration that immediately follows. For example, according to Figure 37.5, these values would be

$$6600 - 3650 = 2950 \text{ ml}$$

Record your computed value and the results of the two subsequent tests on the appropriate lines below. Then calculate the measured and corrected inspiratory capacity averages and record.

measured IC 1: _______ ml average IC: _______ ml

measured IC 2: _______ ml corrected average IC: _______ ml

measured IC 3: _______ ml

- Inspiratory reserve volume (IRV). Subtract the corrected average tidal volume from the corrected average for the inspiratory capacity and record below.

IRV = corrected average IC − corrected average TV

corrected average IRV _______ ml

- Expiratory reserve volume (ERV). Subtract the number of milliliters corresponding to the trough of the maximal expiration obtained during the vital capacity recording from milliliters corresponding to the last *normal* expiration

*If a Collins survey spirometer is used, the situation is exactly opposite: Upstrokes are expirations and downstrokes are inspirations.

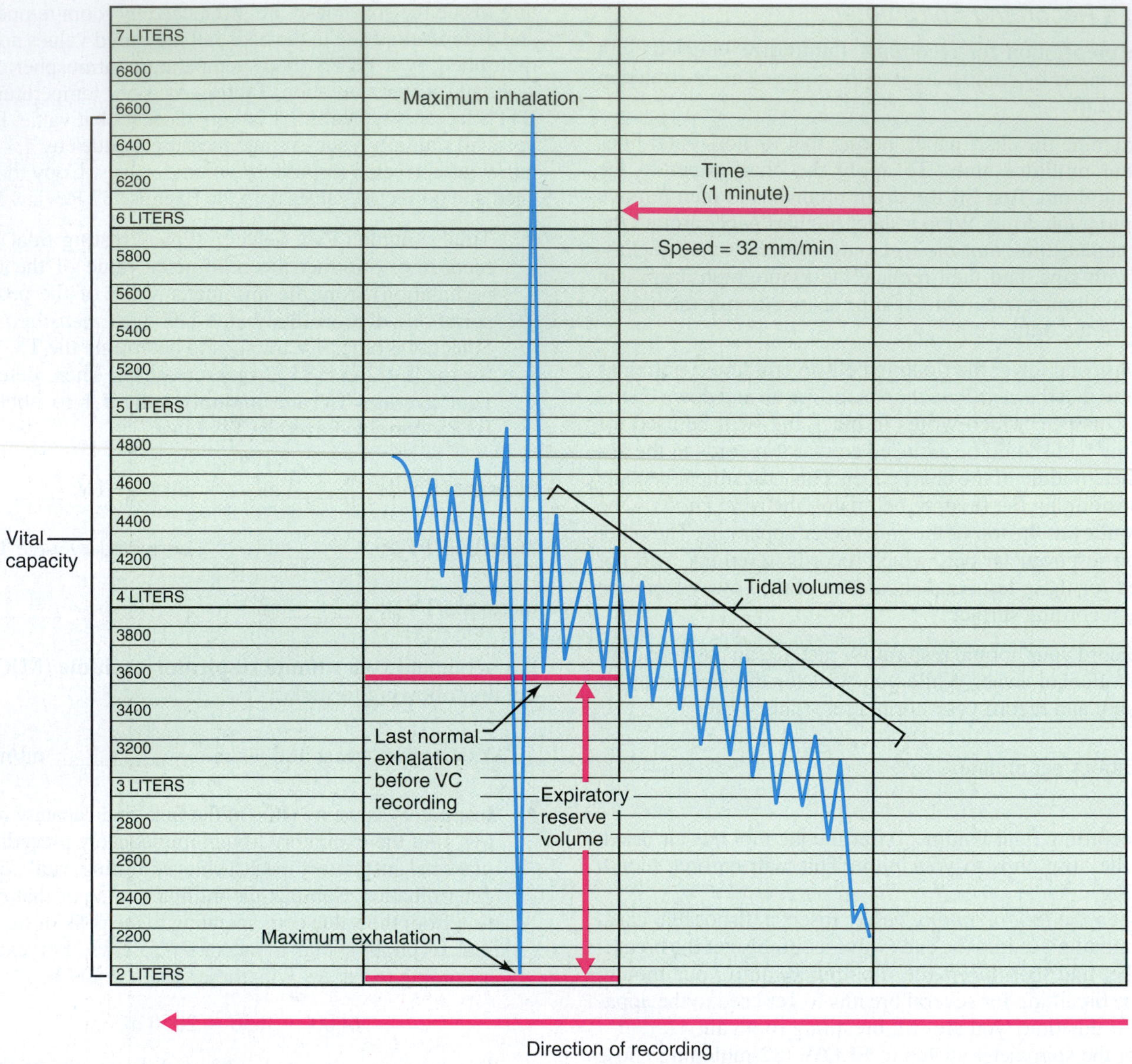
7 LITERS
6800
6600
6400
6200
6 LITERS
5800
5600
5400
5200
5 LITERS
4800
4600
4400
4200
4 LITERS
3800
3600
3400
3200
3 LITERS
2800
2600
2400
2200
2 LITERS
Maximum inhalation
Time
(1 minute)
Speed = 32 mm/min
Vital capacity
Tidal volumes
Last normal exhalation before VC recording
Expiratory reserve volume
Maximum exhalation
Direction of recording

FIGURE 37.5 A typical spirometry recording of tidal volume, inspiratory capacity, expiratory reserve volume, and vital capacity. At a drum speed of 32 mm/min, each vertical column of the chart represents a time interval of 1 minute. (Note that downstrokes represent exhalations and upstrokes represent inhalations.)

before the VC maneuver is performed. For example, according to Figure 37.5, these values would be

3650 ml − 2050 ml = 1600 ml

Record your measured and averaged values (three trials) below.

measured ERV 1: ______ ml average ERV: ______ ml

measured ERV 2: ______ ml corrected average ERV:

measured ERV 3: ______ ml ______ ml

- Vital capacity (VC). Add your corrected values for ERV and IC to obtain the corrected average VC. Record below and on the Exercise 37 Review Sheet.

corrected average VC: _________ ml

Now continue with step 8 (below) whether you are following the procedure for the nonrecording or recording spirometer.

8. Figure out how closely your measured average vital capacity volume compares with the *predicted values* for someone your age, sex, and height. Obtain the predicted figure either from Table 37.1 (male values) or Table 37.2 (female values). Notice that you will have to convert your height in inches to centimeters (cm) to find the corresponding value. This is easily done by multiplying your height in inches by 2.54.

Computed height: _________ cm

Predicted VC value (obtained from the appropriate table):

_________ ml

TABLE 37.1 Predicted Vital Capacities (ml) for Males

	Height in centimeters																								
Age	146	148	150	152	154	156	158	160	162	164	166	168	170	172	174	176	178	180	182	184	186	188	190	192	194
16	3765	3820	3870	3920	3975	4025	4075	4130	4180	4230	4285	4335	4385	4440	4490	4540	4590	4645	4695	4745	4800	4850	4900	4955	5005
18	3740	3790	3840	3890	3940	3995	4045	4095	4145	4200	4250	4300	4350	4405	4455	4505	4555	4610	4660	4710	4760	4815	4865	4915	4965
20	3710	3760	3810	3860	3910	3960	4015	4065	4115	4165	4215	4265	4320	4370	4420	4470	4520	4570	4625	4675	4725	4775	4825	4875	4930
22	3680	3730	3780	3830	3880	3930	3980	4030	4080	4135	4185	4235	4285	4335	4385	4435	4485	4535	4585	4635	4685	4735	4790	4840	4890
24	3635	3685	3735	3785	3835	3885	3935	3985	4035	4085	4135	4185	4235	4285	4330	4380	4430	4480	4530	4580	4630	4680	4730	4780	4830
26	3605	3655	3705	3755	3805	3855	3905	3955	4000	4050	4100	4150	4200	4250	4300	4350	4395	4445	4495	4545	4595	4645	4695	4740	4790
28	3575	3625	3675	3725	3775	3820	3870	3920	3970	4020	4070	4115	4165	4215	4265	4310	4360	4410	4460	4510	4555	4605	4655	4705	4755
30	3550	3595	3645	3695	3740	3790	3840	3890	3935	3985	4035	4080	4130	4180	4230	4275	4325	4375	4425	4470	4520	4570	4615	4665	4715
32	3520	3565	3615	3665	3710	3760	3810	3855	3905	3950	4000	4050	4095	4145	4195	4240	4290	4340	4385	4435	4485	4530	4580	4625	4675
34	3475	3525	3570	3620	3665	3715	3760	3810	3855	3905	3950	4000	4045	4095	4140	4190	4225	4285	4330	4380	4425	4475	4520	4570	4615
36	3445	3495	3540	3585	3635	3680	3730	3775	3825	3870	3920	3965	4010	4060	4105	4155	4200	4250	4295	4340	4390	4435	4485	4530	4580
38	3415	3465	3510	3555	3605	3650	3695	3745	3790	3840	3885	3930	3980	4025	4070	4120	4165	4210	4260	4305	4350	4400	4445	4495	4540
40	3385	3435	3480	3525	3575	3620	3665	3710	3760	3805	3850	3900	3945	3990	4035	4085	4130	4175	4220	4270	4315	4360	4410	4455	4500
42	3360	3405	3450	3495	3540	3590	3635	3680	3725	3770	3820	3865	3910	3955	4000	4050	4095	4140	4185	4230	4280	4325	4370	4415	4460
44	3315	3360	3405	3450	3495	3540	3585	3630	3675	3725	3770	3815	3860	3905	3950	3995	4040	4085	4130	4175	4220	4270	4315	4360	4405
46	3285	3330	3375	3420	3465	3510	3555	3600	3645	3690	3735	3780	3825	3870	3915	3960	4005	4050	4095	4140	4185	4230	4275	4320	4365
48	3255	3300	3345	3390	3435	3480	3525	3570	3615	3655	3700	3745	3790	3835	3880	3925	3970	4015	4060	4105	4150	4190	4235	4280	4325
50	3210	3255	3300	3345	3390	3430	3475	3520	3565	3610	3650	3695	3740	3785	3830	3870	3915	3960	4005	4050	4090	4135	4180	4225	4270
52	3185	3225	3270	3315	3355	3400	3445	3490	3530	3575	3620	3660	3705	3750	3795	3835	3880	3925	3970	4010	4055	4100	4140	4185	4230
54	3155	3195	3240	3285	3325	3370	3415	3455	3500	3540	3585	3630	3670	3715	3760	3800	3845	3890	3930	3975	4020	4060	4105	4145	4190
56	3125	3165	3210	3255	3295	3340	3380	3425	3465	3510	3550	3595	3640	3680	3725	3765	3810	3850	3895	3940	3980	4025	4065	4110	4150
58	3080	3125	3165	3210	3250	3290	3335	3375	3420	3460	3500	3545	3585	3630	3670	3715	3755	3800	3840	3880	3925	3965	4010	4050	4095
60	3050	3095	3135	3175	3220	3260	3300	3345	3385	3430	3470	3500	3555	3595	3635	3680	3720	3760	3805	3845	3885	3930	3970	4015	4055
62	3020	3060	3110	3150	3190	3230	3270	3310	3350	3390	3440	3480	3520	3560	3600	3640	3680	3730	3770	3810	3850	3890	3930	3970	4020
64	2990	3030	3080	3120	3160	3200	3240	3280	3320	3360	3400	3440	3490	3530	3570	3610	3650	3690	3730	3770	3810	3850	3900	3940	3980
66	2950	2990	3030	3070	3110	3150	3190	3230	3270	3310	3350	3390	3430	3470	3510	3550	3600	3640	3680	3720	3760	3800	3840	3880	3920
68	2920	2960	3000	3040	3080	3120	3160	3200	3240	3280	3320	3360	3400	3440	3480	3520	3560	3600	3640	3680	3720	3760	3800	3840	3880
70	2890	2930	2970	3010	3050	3090	3130	3170	3210	3250	3290	3330	3370	3410	3450	3480	3520	3560	3600	3640	3680	3720	3760	3800	3840
72	2860	2900	2940	2980	3020	3060	3100	3140	3180	3210	3250	3290	3330	3370	3410	3450	3490	3530	3570	3610	3650	3680	3720	3760	3800
74	2820	2860	2900	2930	2970	3010	3050	3090	3130	3170	3200	3240	3280	3320	3360	3400	3440	3470	3510	3550	3590	3630	3670	3710	3740

Courtesy of Warren E. Collins, Inc., Braintree, Mass.

TABLE 37.2 Predicted Vital Capacities (ml) for Females

	Height in centimeters																								
Age	146	148	150	152	154	156	158	160	162	164	166	168	170	172	174	176	178	180	182	184	186	188	190	192	194
16	2950	2990	3030	3070	3110	3150	3190	3230	3270	3310	3350	3390	3430	3470	3510	3550	3590	3630	3670	3715	3755	3800	3840	3880	3920
17	2935	2975	3015	3055	3095	3135	3175	3215	3255	3295	3335	3375	3415	3455	3495	3535	3575	3615	3655	3695	3740	3780	3820	3860	3900
18	2920	2960	3000	3040	3080	3120	3160	3200	3240	3280	3320	3360	3400	3440	3480	3520	3560	3600	3640	3680	3720	3760	3800	3840	3880
20	2890	2930	2970	3010	3050	3090	3130	3170	3210	3250	3290	3330	3370	3410	3450	3490	3525	3565	3605	3645	3695	3720	3760	3800	3840
22	2860	2900	2940	2980	3020	3060	3095	3135	3175	3215	3255	3290	3330	3370	3410	3450	3490	3530	3570	3610	3650	3685	3725	3765	3800
24	2830	2870	2910	2950	2985	3025	3065	3100	3140	3180	3220	3260	3300	3335	3375	3415	3455	3490	3530	3570	3610	3650	3685	3725	3765
26	2800	2840	2880	2920	2960	3000	3035	3070	3110	3150	3190	3230	3265	3300	3340	3380	3420	3455	3495	3530	3570	3610	3650	3685	3725
28	2775	2810	2850	2890	2930	2965	3000	3040	3070	3115	3155	3190	3230	3270	3305	3345	3380	3420	3460	3495	3535	3570	3610	3650	3685
30	2745	2780	2820	2860	2895	2935	2970	3010	3045	3085	3120	3160	3195	3235	3270	3310	3345	3385	3420	3460	3495	3535	3570	3610	3645
32	2715	2750	2790	2825	2865	2900	2940	2975	3015	3050	3090	3125	3160	3200	3235	3275	3310	3350	3385	3425	3460	3495	3535	3570	3610
34	2685	2725	2760	2795	2835	2870	2910	2945	2980	3020	3055	3090	3130	3165	3200	3240	3275	3310	3350	3385	3425	3460	3495	3535	3570
36	2655	2695	2730	2765	2805	2840	2875	2910	2950	2985	3020	3060	3095	3130	3165	3205	3240	3275	3310	3350	3385	3420	3460	3495	3530
38	2630	2665	2700	2735	2770	2810	2845	2880	2915	2950	2990	3025	3060	3095	3130	3170	3205	3240	3275	3310	3350	3385	3420	3455	3490
40	2600	2635	2670	2705	2740	2775	2810	2850	2885	2920	2955	2990	3025	3060	3095	3135	3170	3205	3240	3275	3310	3345	3380	3420	3455
42	2570	2605	2640	2675	2710	2745	2780	2815	2850	2885	2920	2955	2990	3025	3060	3100	3135	3170	3205	3240	3275	3310	3345	3380	3415
44	2540	2575	2610	2645	2680	2715	2750	2785	2820	2855	2890	2925	2960	2995	3030	3060	3095	3130	3165	3200	3235	3270	3305	3340	3375
46	2510	2545	2580	2615	2650	2685	2715	2750	2785	2820	2855	2890	2925	2960	2995	3030	3060	3095	3130	3165	3200	3235	3270	3305	3340
48	2480	2515	2550	2585	2620	2650	2685	2715	2750	2785	2820	2855	2890	2925	2960	2995	3030	3060	3095	3130	3160	3195	3230	3265	3300
50	2455	2485	2520	2555	2590	2625	2655	2690	2720	2755	2785	2820	2855	2890	2925	2955	2990	3025	3060	3090	3125	3155	3190	3225	3260
52	2425	2455	2490	2525	2555	2590	2625	2655	2690	2720	2755	2790	2820	2855	2890	2925	2955	2990	3020	3055	3090	3125	3155	3190	3220
54	2395	2425	2460	2495	2530	2560	2590	2625	2655	2690	2720	2755	2790	2820	2855	2885	2920	2950	2985	3020	3050	3085	3115	3150	3180
56	2365	2400	2430	2460	2495	2525	2560	2590	2625	2655	2690	2720	2755	2790	2820	2855	2885	2920	2950	2980	3015	3045	3080	3110	3145
58	2335	2370	2400	2430	2460	2495	2525	2560	2590	2625	2655	2690	2720	2750	2785	2815	2850	2880	2920	2945	2975	3010	3040	3075	3105
60	2305	2340	2370	2400	2430	2460	2495	2525	2560	2590	2625	2655	2685	2720	2750	2780	2810	2845	2875	2915	2940	2970	3000	3035	3065
62	2280	2310	2340	2370	2405	2435	2465	2495	2525	2560	2590	2620	2655	2685	2715	2745	2775	2810	2840	2870	2900	2935	2965	2995	3025
64	2250	2280	2310	2340	2370	2400	2430	2465	2495	2525	2555	2585	2620	2650	2680	2710	2740	2770	2805	2835	2865	2895	2925	2955	2990
66	2220	2250	2280	2310	2340	2370	2400	2430	2460	2495	2525	2555	2585	2615	2645	2675	2705	2735	2765	2800	2825	2860	2890	2920	2950
68	2190	2220	2250	2280	2310	2340	2370	2400	2430	2460	2490	2520	2550	2580	2610	2640	2670	2700	2730	2760	2795	2820	2850	2880	2910
70	2160	2190	2220	2250	2280	2310	2340	2370	2400	2425	2455	2485	2515	2545	2575	2605	2635	2665	2695	2725	2755	2780	2810	2840	2870
72	2130	2160	2190	2220	2250	2280	2310	2335	2365	2395	2425	2455	2480	2510	2540	2570	2600	2630	2660	2685	2715	2745	2775	2805	2830
74	2100	2130	2160	2190	2220	2245	2275	2305	2335	2360	2390	2420	2450	2475	2505	2535	2565	2590	2620	2650	2680	2710	2740	2765	2795

Courtesy of Warren E. Collins, Inc., Braintree, Mass.

Use the following equation to compute your VC as a percentage of the predicted VC value:

$$\% \text{ of predicted VC} = \frac{\text{average VC}}{\text{predicted value} \times 100}$$

% predicted VC value: ________ %

Figure 37.2 on page 552 is an idealized tracing of the respiratory volumes described and tested in this exercise. Examine it carefully. How closely do your test results compare to the values in the tracing?

____

9. Computing residual volume. A respiratory volume that cannot be experimentally demonstrated here is the residual volume (RV). RV is the amount of air remaining in the lungs after a maximal expiratory effort. The presence of residual air (usually about 1200 ml) that cannot be voluntarily flushed from the lungs is important because it allows gas exchange to go on continuously—even between breaths.

Although the residual volume cannot be measured directly, it can be approximated by using one of the following factors:

For ages 16–34	Factor = 0.250
For ages 35–49	Factor = 0.305
For ages 50–69	Factor = 0.445

Compute your predicted RV using the following equation:

$$\text{RV} = \text{VC} \times \text{factor}$$

! 10. Recording is finished for this subject. Before continuing with the next member of your group:

- Dispose of used cardboard mouthpieces in the autoclave bag.
- Swish the valve assembly (if removable) in the 70% ethanol solution, then rinse with tap water.
- Put a fresh mouthpiece into the valve assembly (or on the stem of the handheld spirometer). Using the procedures outlined above, measure and record the respiratory volumes for all members of your group. ■

Forced Expiratory Volume (FEV_T) Measurement

While not really diagnostic, pulmonary function tests can help the clinician distinguish between obstructive and restrictive pulmonary diseases. (In obstructive disorders, like chronic bronchitis and asthma, airway resistance is increased, whereas in restrictive diseases, such as polio and tuberculosis, total lung capacity declines.) Two highly useful pulmonary function tests used for this purpose are the FVC and the FEV_T (see Figure 37.6).

The **FVC** (forced vital capacity) measures the amount of gas expelled when the subject takes the deepest possible breath and then exhales forcefully and rapidly. This volume is reduced in those with restrictive pulmonary disease. The **FEV_T** (forced expiratory volume) involves the same basic testing procedure, but it specifically looks at the percentage of the vital capacity that is exhaled during specific time intervals of the FVC test. FEV_1, for instance, is the amount exhaled during the first second. Healthy individuals can expire 75% to 85% of their FVC in the first second. The FEV_1 is low in those with obstructive disease.

ACTIVITY 4

Measuring the FVC and FEV_1

Directions provided here for the FEV_T determination apply only to the recording spirometer.

1. Prepare to make your recording as described for the recording spirometer, steps 1–5 on page 555.

2. At a signal agreed upon by you and your lab partner, take the deepest inspiration possible and hold it for 1 to 2 seconds. As the inspiratory peak levels off, your partner is to change the drum speed to **FAST** (1920 mm/min) so that the distance between the vertical lines on the chart represents 1 second.

3. Once the drum speed is changed, exhale as much air as you can as rapidly and forcibly as possible.

4. When the tracing plateaus (bottoms out), stop recording and determine your FVC. Subtract the milliliter reading in the expiration trough (the bottom plateau) from the preceding inhalation peak (the top plateau). Record this value.

FVC: ________ ml

5. Prepare to calculate the FEV_1. Draw a vertical line intersecting with the spirogram tracing at the precise point that exhalation began. Identify this line as *line 1.* From line 1, measure 32 mm horizontally to the left, and draw a second vertical line. Label this as *line 2.* The distance between the two lines represents 1 second, and the volume exhaled in the first second is read where line 2 intersects the spirogram tracing. Subtract that milliliter value from the milliliter value of the inhalation peak (at the intersection of line 1), to determine the volume of gas expired in the first second. According to the values given in Figure 37.6, that figure would be 3400 ml (6800 ml − 3400 ml). Record your measured value below.

Milliliters of gas expired in second 1: ________ ml

6. To compute the FEV_1 use the following equation:

$$FEV_1 = \frac{\text{volume expired in second 1}}{\text{FVC volume}} \times 100\%$$

Record your calculated value below and on the Exercise 37 Review Sheet.

FEV_1: ________ % of FVC ■

ACTIVITY 5

Measuring Respiratory Volumes Using BIOPAC®

In this activity, you will measure respiratory volumes using the BIOPAC® airflow transducer. An example of these

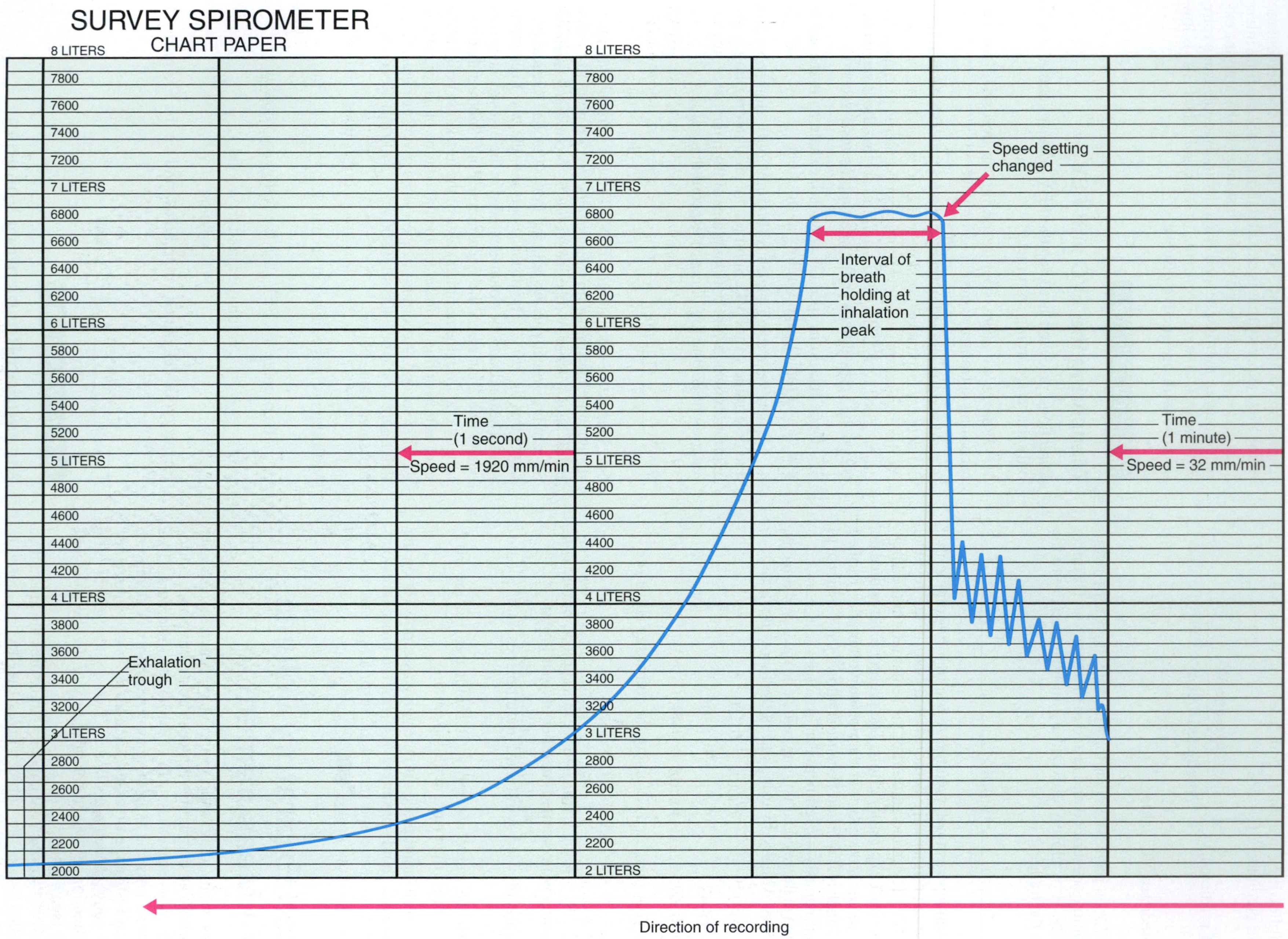
SURVEY SPIROMETER
CHART PAPER
8 LITERS
7800
7600
7400
7200
7 LITERS
6800
6600
6400
6200
6 LITERS
5800
5600
5400
5200
5 LITERS
4800
4600
4400
4200
4 LITERS
3800
3600
3400
3200
3 LITERS
2800
2600
2400
2200
2000
2 LITERS
Exhalation trough
Time (1 second)
Speed = 1920 mm/min
Interval of breath holding at inhalation peak
Speed setting changed
Time (1 minute)
Speed = 32 mm/min
Direction of recording

FIGURE 37.6 **A recording of the forced vital capacity (FVC) and forced expiratory volume (FEV) or timed vital capacity test.**

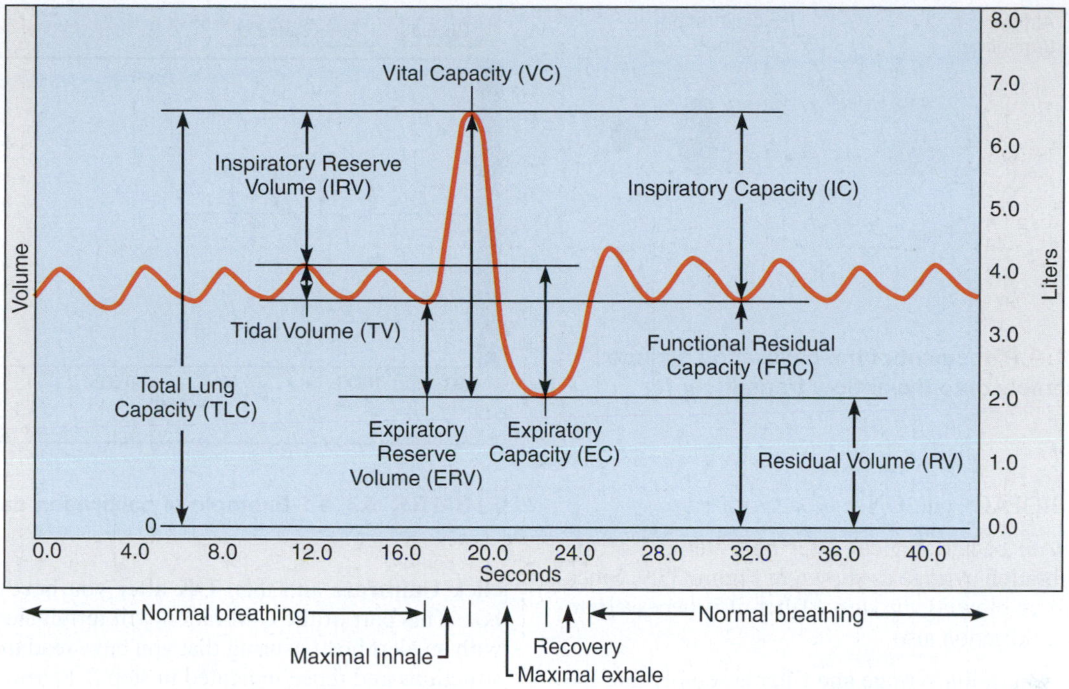
Vital Capacity (VC)
Inspiratory Reserve Volume (IRV)
Inspiratory Capacity (IC)
Tidal Volume (TV)
Functional Residual Capacity (FRC)
Total Lung Capacity (TLC)
Expiratory Reserve Volume (ERV)
Expiratory Capacity (EC)
Residual Volume (RV)
Volume
Liters
8.0
7.0
6.0
5.0
4.0
3.0
2.0
1.0
0.0
0
0.0
4.0
8.0
12.0
16.0
20.0
24.0
28.0
32.0
36.0
40.0
Seconds
Normal breathing
Maximal inhale
Recovery
Maximal exhale
Normal breathing

FIGURE 37.7 Example of a computer-generated spirogram.

volumes is demonstrated in the computer-generated spirogram in Figure 37.7. Since it is not possible to measure **residual volume (RV)** using the airflow transducer, assume that it is 1.0 liter for each subject, which is a reasonable estimation. It is also important to estimate the **predicted vital capacity** of the subject for comparison to the measured value. A rough estimate of the vital capacity in liters (VC) of a subject can be calculated using the following formulas based on height in centimeters (H) and age in years (A).

Male VC = (0.052)H − (0.022)A − 3.60

Female VC = (0.041)H − (0.018)A − 2.69

Because many factors besides height and age influence vital capacity, it should be assumed that measured values up to 20% above or below the calculated predicted value are normal.

Setting Up the Equipment

1. Connect the BIOPAC® unit to the computer and turn the computer **ON.**
2. Make sure the BIOPAC® unit is **OFF.**
3. Plug in the equipment as shown in Figure 37.8.

- Airflow transducer—CH 1

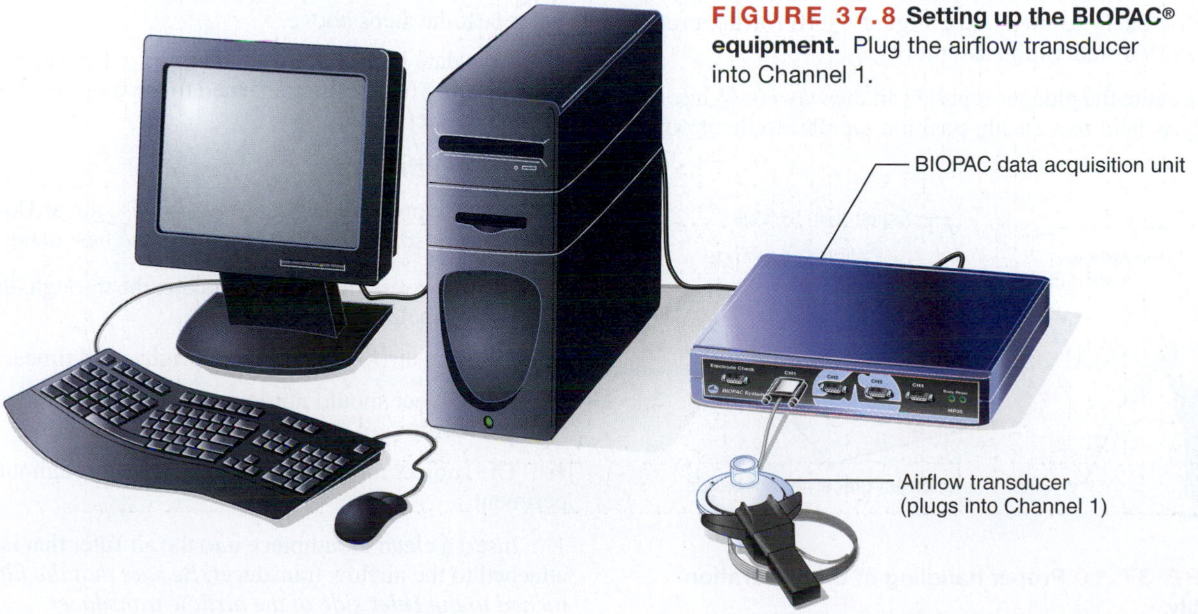
BIOPAC data acquisition unit
Airflow transducer (plugs into Channel 1)

FIGURE 37.8 Setting up the BIOPAC® equipment. Plug the airflow transducer into Channel 1.

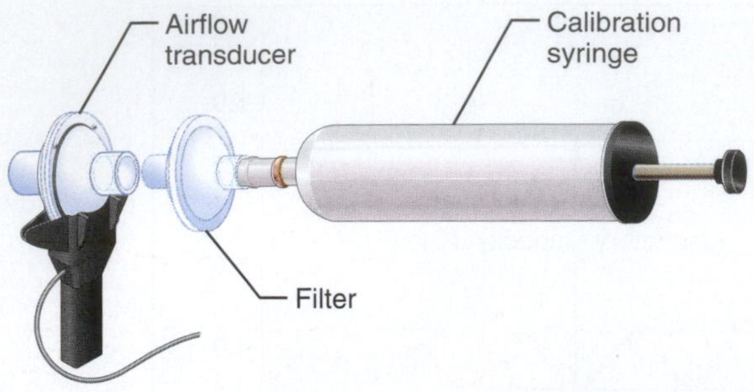
Airflow transducer
Calibration syringe
Filter

FIGURE 37.9 Placement of the calibration syringe and filter assembly onto the airflow transducer for calibration.

4. Turn the BIOPAC® unit **ON.**

5. Place a *clean* bacteriological filter onto the end of the BIOPAC® calibration syringe as shown in Figure 37.9. Since the subject will be blowing through a filter, it is necessary to use a filter for calibration also.

6. Insert the calibration syringe and filter assembly into the airflow transducer on the side labeled **Inlet.**

7. Start the BIOPAC® Student Lab program on the computer by double-clicking the icon on the desktop or by following your instructor's guidance.

8. Select lesson **L12-Lung-1** from the menu, and click **OK.**

9. Type in a filename that will save this subject's data on the computer hard drive. You may want to use the subject's last name followed by Lung-1 (for example, SmithLung-1), then click **OK.**

Calibrating the Equipment

Two precautions must be followed:

- The airflow transducer is sensitive to gravity, so it must be held directly parallel to the ground during calibration and recording.
- Do not hold onto the airflow transducer when it is attached to the calibration syringe and filter assembly—the syringe tip is likely to break. See Figure 37.10 for the proper handling of the calibration assembly.

1. Make sure the plunger is pulled all the way out. While the assembly is held in a steady position parallel to the ground, click **Calibrate** and then **OK** after you have read the alert box. This part of the calibration will terminate automatically with an alert box ensuring that you have read the onscreen instructions and those indicated in step 2, below.

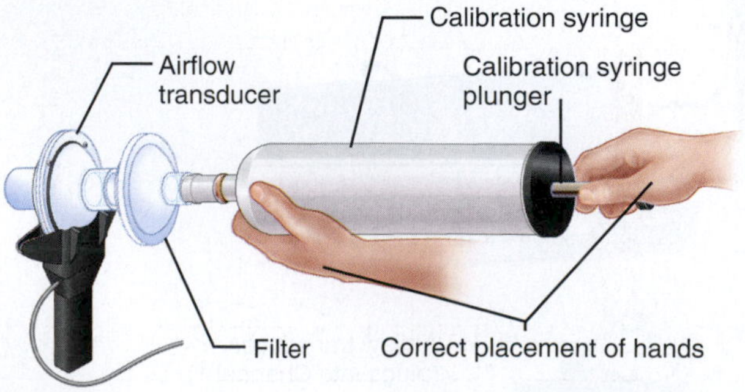
Calibration syringe
Airflow transducer
Calibration syringe plunger
Filter
Correct placement of hands

FIGURE 37.10 Proper handling of the calibration assembly.

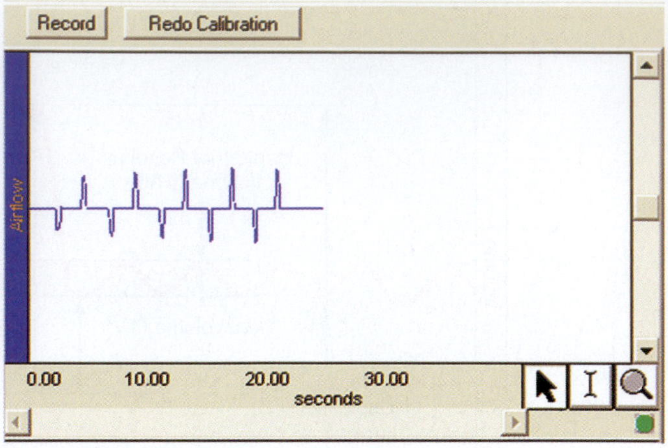
Record
Redo Calibration
Airflow
0.00
10.00
20.00
30.00
seconds

FIGURE 37.11 Example of calibration data.

2. The final part of the calibration involves simulating five breathing cycles using the calibration syringe. A single cycle consists of:

- Pushing the plunger in (taking 1 second for this stroke)
- Waiting for 2 seconds
- Pulling the plunger out (taking 1 second for this stroke)
- Waiting 2 seconds

Remember to hold the airflow transducer directly parallel to the ground during calibration and recording.

3. When ready to perform this second stage of the calibration, click **Yes.** After you have completed five cycles, click **End Calibration.**

4. Observe the data, which should look similar to that in Figure 37.11.

- If the data look very different, click **Redo Calibration** and repeat the steps above.
- If the data look similar, gently remove the calibration syringe, leaving the air filter attached to the transducer. Proceed to the next section.

Recording the data

Follow these procedures precisely, because the airflow transducer is very sensitive. Hints to obtain the best data:

- Always insert air filter on, and breathe through, the transducer side labeled **Inlet.**
- Keep the airflow transducer upright at all times.
- The subject should not look at the computer screen during the recording of data.
- The subject must keep a nose clip on throughout the experiment.

1. Insert a clean mouthpiece into the air filter that is already attached to the airflow transducer. *Be sure that the filter is attached to the* ***Inlet*** *side of the airflow transducer.*

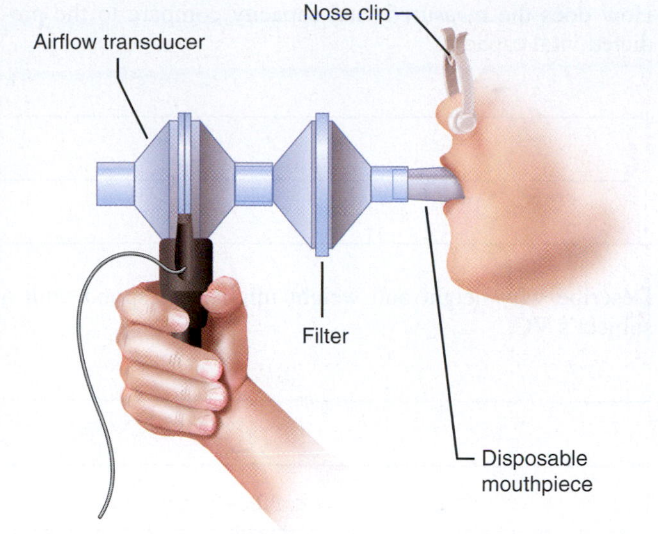
Nose clip
Airflow transducer
Filter
Disposable mouthpiece

FIGURE 37.12 Proper equipment setup for recording data.

2. Write the name of the subject on the mouthpiece and air filter. For safety purposes, each subject must use his or her own air filter and mouthpiece.

3. The subject should now place the nose clip on the nose (or hold the nose very tightly with finger pinch), wrap the lips tightly around the mouthpiece, and begin breathing normally through the airflow transducer, as shown in Figure 37.12.

4. When prepared, the subject will complete the following unbroken series with nose plugged and lips tightly sealed around the mouthpiece:

- Take five normal breaths (1 breath = inhale + exhale).
- Inhale as much air as possible.
- Exhale as much air as possible.
- Take five normal breaths.

5. When the subject is prepared to proceed, click **Record** on the first normal inhalation and proceed. When the subject finishes the last exhalation at the end of the series, click **Stop.**

6. Observe the data, which should look similar to that in Figure 37.13.

- If the data look very different, click **Redo** and repeat the steps above. Be certain that the lips are sealed around the mouthpiece, the nose is completely plugged, and the transducer is upright.
- If the data look similar, proceed to step 7.

7. When finished, click **Done.** A pop-up window will appear.

- Click **Yes** if you are done and want to stop recording.
- To record from another subject select **Record from another Subject** and return to step 1 under Recording the Data. You will not need to redo the calibration procedure for the second subject.
- If continuing to the data Analysis section, select **Analyze current data file** and proceed to step 2 of the Data Analysis section.

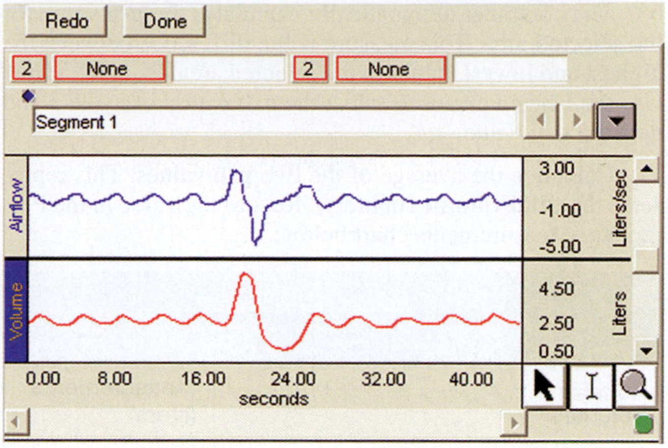
Redo
Done
2
None
2
None
Segment 1
Airflow
Volume
3.00
-1.00
-5.00
Liters/sec
4.50
2.50
0.50
Liters
0.00 8.00 16.00 24.00 32.00 40.00
seconds

FIGURE 37.13 Example of pulmonary data.

Data Analysis

1. If just starting the BIOPAC® program to perform data analysis, enter **Review Saved Data** mode and choose the file with the subject's Lung data (for example, SmithLung-1).

2. Observe how the channel numbers are designated: CH 1—Airflow; CH 2—Volume.

3. To set up the display for optimal viewing, hide CH 1—Airflow. To do this, hold down the Ctrl key (PC) or Option key (Mac) while using the cursor to click the Channel box 1 (the small box with a 1 at the upper left of the screen).

4. To analyze the data, set up the first pair of channel/measurement boxes at the top of the screen by selecting the following channel and measurement type from the drop-down menu:

Channel	Measurement
CH 2	p-p

5. Take two measures for an averaged TV calculation: Use the arrow cursor and click the I-beam cursor box on the lower right side of the screen to activate the "area selection" function. Using the activated I-beam cursor, highlight the inhalation of cycle 3 as shown in Figure 37.14.

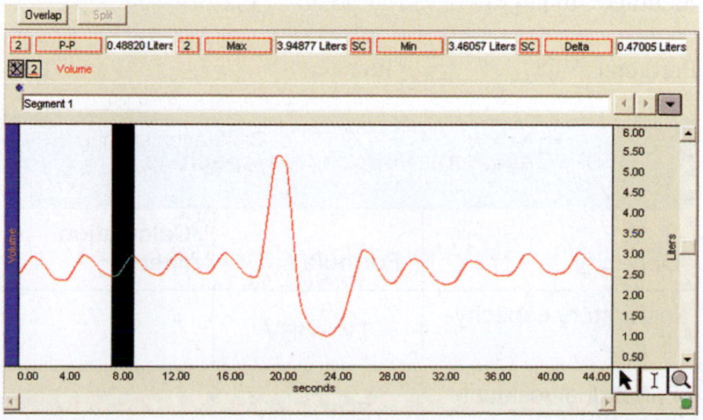
Overlap
Split
P-P
0.48820 Liters
Max
3.94877 Liters
Min
3.46057 Liters
Delta
0.47005 Liters
Volume
Segment 1
seconds
Liters

FIGURE 37.14 Highlighting data for the first three breaths.

6. The computer automatically calculates the **p-p** value for the selected area. This measure is the difference between the highest and lowest values in the selected area. Note the value. Use the I-beam cursor to select the exhalation of cycle 3 and note the **p-p** value.

7. Calculate the average of the two **p-p** values. This represents the **tidal volume** (in liters). Record the value in the Pulmonary Measurements chart below:

Pulmonary Measurements	
Volumes	**Measurements (liters)**
Tidal volume (TV)	
Inspiratory reserve volume (IRV)	
Expiratory reserve volume (ERV)	
Vital capacity (VC)	
Residual volume (RV)	1.00 (assumed)

8. Use the I-beam cursor to measure the IRV: Highlight from the peak of maximum inhalation to the peak of the last normal inhalation just before it (see Figure 37.7 for an example of IRV). Observe and record the **Δ (delta)** value in the chart (to the nearest 0.01 liter).

9. Use the I-beam cursor to measure the ERV: Highlight from the trough of maximum exhalation to the trough of the last normal exhalation just before it (see Figure 37.7 for an example of ERV). Observe and record the **Δ (delta)** value in the chart (to the nearest 0.01 liter).

10. Last, use the I-beam cursor to measure the VC: Highlight from the trough of maximum exhalation to the peak of maximum inhalation (see Figure 37.7 for an example of VC). Observe and record the **p-p** value in the chart (to the nearest 0.01 liter).

11. When finished, **Exit** the program.

Using the measured data, calculate the capacities listed in the Calculated Pulmonary Capacities chart.

Use the formula in the introduction of this activity (page 561) to calculate the predicted vital capacity of the subject based on height and age.

Predicted VC: _________ liters

Calculated Pulmonary Capacities		
Capacity	**Formula**	**Calculation (liters)**
Inspiratory capacity (IC)	= TV + IRV	
Functional residual capacity (FRC)	= ERV + RV	
Total lung capacity (TLC)	= TV + RV + IRV + ERV	

How does the measured vital capacity compare to the predicted vital capacity?

Describe why height and weight might correspond with a subject's VC.

What other factors might influence the VC of a subject?

Factors Influencing Rate and Depth of Respiration

The neural centers that control respiratory rhythm and maintain a rate of 12 to 18 respirations/min are located in the medulla and pons. On occasion, input from the stretch receptors in the lungs (via the vagus nerve to the medulla) modifies the respiratory rate, as in cases of extreme overinflation of the lungs (Hering-Breuer reflex).

Death occurs when medullary centers are completely suppressed, as from an overdose of sleeping pills or gross overindulgence in alcohol, and respiration ceases completely. ■

Although the nervous system centers initiate the basic rhythm of breathing, there is no question that physical phenomena such as talking, yawning, coughing, and exercise can modify the rate and depth of respiration. So, too, can chemical factors such as changes in oxygen or carbon dioxide concentrations in the blood or fluctuations in blood pH. Changes in carbon dioxide blood levels seem to act directly on the medulla control centers, whereas changes in pH and oxygen concentrations are monitored by chemoreceptor regions in the aortic and carotid bodies, which in turn send input to the medulla. The experimental sequence in Activity 6 is designed to test the relative importance of various physical and chemical factors in the process of respiration.

ACTIVITY 6

Visualizing Respiratory Variations Using the Physiograph-Pneumograph Apparatus*

The **pneumograph,** an apparatus that records variations in breathing patterns, is the best means of observing respiratory variations resulting from physical and chemical factors. The chest pneumograph is a coiled rubber hose that is attached around the thorax. As the subject breathes, chest movements produce pressure changes within the pneumograph that are transmitted to a recorder.

The instructor will demonstrate the method of setting up the recording equipment and discuss the interpretation of the results. Work in pairs so that one person can mark the record to identify the test for later interpretation. Ideally, the student being tested should face away from the recording apparatus to prevent voluntary modification of the record.

1. Attach the pneumograph tubing firmly, but not restrictively, around the thoracic cage at the level of the sixth rib, leaving room for chest expansion during testing. If the subject is female, position the tubing above the breasts to prevent slippage during testing. Set the pneumograph speed at 1 or 2, and the time signal at 10-second intervals. Record quiet breathing for 1 minute with the subject in a sitting position.

Record breaths per minute: ______________________

2. Make a vital capacity tracing: Record a maximal inhalation followed by a maximal exhalation. This should correlate to the vital capacity measurement obtained earlier and will provide a baseline for comparison during the rest of the pneumograph testing. Stop recording, and mark the graph appropriately to indicate tidal volume, expiratory reserve volume, inspiratory reserve volume, and vital capacity (the total of the three measurements). Also mark, with arrows, the direction the recording stylus moves during inspiration and during expiration.

Measure in mm the height of the vital capacity recording. divide the vital capacity measurement average recorded on the Exercise 37 Review Sheet by the millimeter figure to obtain the volume (in milliliters of air) represented by 1 mm on the recording. For example, if your vital capacity reading is 4000 ml and the vital capacity tracing occupies a vertical distance of 40 mm on the pneumograph recording, then a vertical distance of 1 mm equals 100 ml of air.

Record your computed value: __________ ml of air per mm

3. Record the subject's breathing as he or she performs activities from the following list. Make sure the record is marked accurately to identify each test conducted. Record your results on the Exercise 37 Review Sheet.

talking	swallowing water
yawning	coughing
laughing	lying down
standing	running in place
doing a math problem (concentrating)	

4. Without recording, have the subject breathe normally for 2 minutes, then inhale deeply and hold his or her breath for as long as he or she can.

Time the breath-holding interval: __________________ sec

As the subject exhales, turn on the recording apparatus and record the recovery period (time to return to normal breathing—usually slightly over 1 minute):

Time of recovery period: ________________________ sec

Did the subject have the urge to inspire *or* expire during breath holding? ______________________________

Without recording, repeat the above experiment, but this time exhale completely and forcefully *after* taking the deep breath.

What was observed this time? ______________________

____

Explain the results. (Hint: The vagus nerve is the sensory nerve of the lungs and plays a role here.)

____

____

____

5. Have the subject hyperventilate (breathe deeply and forcefully at the rate of 1 breath/4 sec) for about 30 sec.* Record both during and after hyperventilation. How does the pattern obtained during hyperventilation compare with that recorded during the vital capacity tracing?

____

Is the respiratory rate after hyperventilation faster *or* slower than during normal quiet breathing?

____

*A sensation of dizziness may develop. (As the carbon dioxide is washed out of the blood by hyperventilation, the blood pH increases, leading to a decrease in blood pressure and reduced cerebral circulation.) The subject may experience a lack of desire to breathe after forced breathing is stopped. If the period of breathing cessation—apnea—is extended, cyanosis of the lips may occur.

***Note to the Instructor:** This exercise may also be done without using the recording apparatus by simply having the students count the respiratory rate visually.

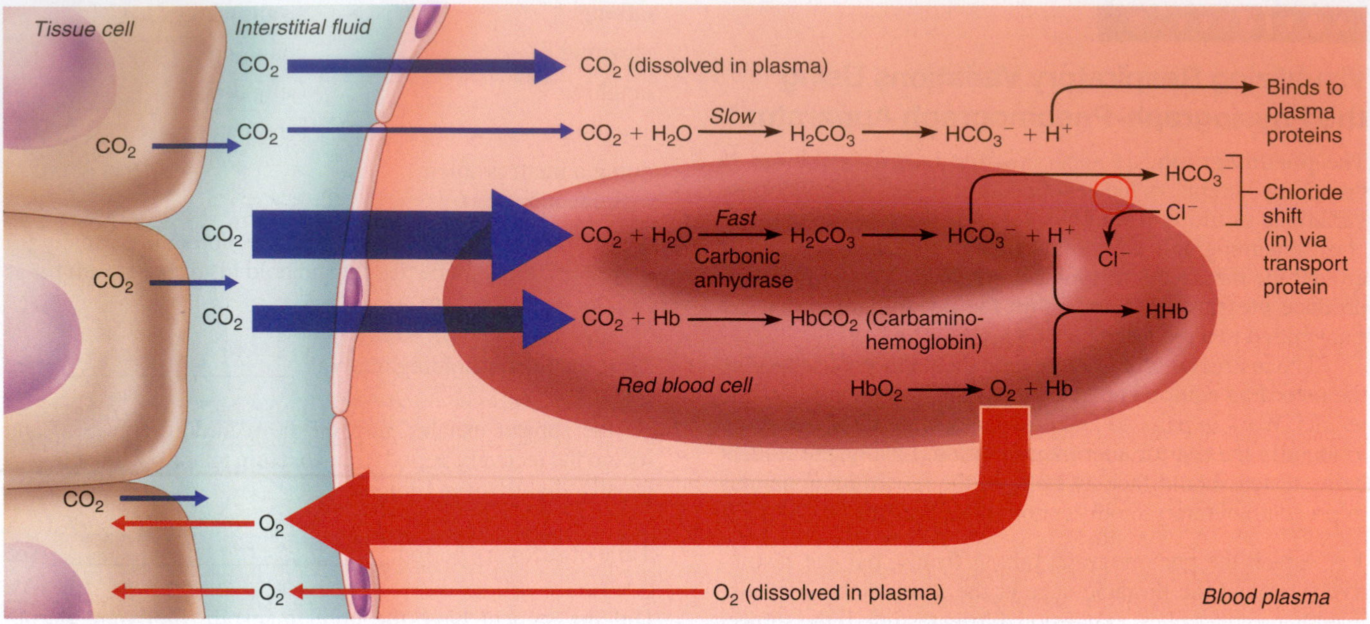

Tissue cell
Interstitial fluid
CO2
CO2 (dissolved in plasma)
CO2 + H2O Slow H2CO3 HCO3− + H+
Binds to plasma proteins
CO2 + H2O Fast Carbonic anhydrase H2CO3 HCO3− + H+
HCO3−
Cl−
Chloride shift (in) via transport protein
CO2 + Hb HbCO2 (Carbamino-hemoglobin)
HHb
Red blood cell
HbO2 O2 + Hb
O2
O2 (dissolved in plasma)
Blood plasma

FIGURE 37.15 Oxygen release and carbon dioxide pickup at the tissues.

6. Repeat the above test, but do not record until after hyperventilating. After hyperventilation, the subject is to hold his or her breath as long as he or she can. Can the breath be held for a longer or shorter time after hyperventilating?

___

7. Without recording, have the subject breathe into a paper bag for 3 minutes, then record his or her breathing movements.

During the bag-breathing exercise, the subject's partner should watch the subject carefully for any untoward reactions.

Is the breathing rate faster *or* slower than that recorded during normal quiet breathing?

___

After hyperventilating? ___

8. Run in place for 2 minutes, and then have your partner determine how long you can hold your breath.

Length of breath holding: ___ sec

9. To prove that respiration has a marked effect on circulation, conduct the following test. Have your lab partner record the rate and relative force of your radial pulse before you begin.

Rate: ___ beats/min Relative force: ___

Inspire forcibly. Immediately close your mouth and nose to retain the inhaled air, and then make a forceful and prolonged expiration. Your lab partner should observe and record the condition of the blood vessels of your neck and face, and again immediately palpate the radial pulse.

Observations: ___

Radial pulse: ___ beats/min Relative force: ___

Explain the changes observed. ___

___

___

Dispose of the paper bag in the autoclave bag. Keep the pneumograph records to interpret results and hand them in if requested by the instructor. Observation of the test results should enable you to determine which chemical factor, carbon dioxide or oxygen, has the greatest effect on modifying the respiratory rate and depth.

Role of the Respiratory System in Acid-Base Balance of Blood

As you have already learned, pulmonary ventilation is necessary for continuous oxygenation of the blood and removal of carbon dioxide (a waste product of cellular respiration) from the blood. Blood pH must be relatively constant for the cells of the body to function optimally. The carbonic acid–bicarbonate buffer system of the blood is extremely important because it helps stabilize arterial blood pH at 7.4 ± 0.02.

When carbon dioxide diffuses into the blood from the tissue cells, much of it enters the red blood cells, where it combines with water to form carbonic acid (Figure 37A.15):

$$H_2O + CO_2 \xrightarrow[\text{enzyme present in RBC}]{\text{carbonic anhydrase}} H_2CO_3$$

Some carbonic acid is also formed in the plasma, but that reaction is very slow because of the lack of the carbonic anhydrase enzyme. Shortly after it forms, carbonic acid dissociates to release bicarbonate (HCO_3^-) and hydrogen ions (H^+). The hydrogen ions that remain in the cells are neutralized, or buffered, when they combine with hemoglobin molecules. If they were not neutralized, the intracellular pH would become very acidic as H^+ ions accumulated. The bicarbonate ions diffuse out of the red blood cells into the plasma, where they become part of the carbonic acid–bicarbonate buffer system. As HCO_3^- follows its concentration gradient into the plasma, an electrical imbalance develops in the RBCs that draws Cl^- into them from the plasma. This exchange phenomenon is called the *chloride shift.*

Acids (more precisely, H^+) released into the blood by the body cells tend to lower the pH of the blood and to cause it to become acidic. On the other hand, basic substances that enter the blood tend to cause the blood to become more alkaline and the pH to rise. Both of these tendencies are resisted in large part by the carbonic acid–bicarbonate buffer system. If the H^+ concentration in the blood begins to increase, the H^+ ions combine with bicarbonate ions to form carbonic acid (a weak acid that does not tend to dissociate at physiological or acid pH) and are thus removed.

$$H^+ + HCO_3^- \rightarrow H_2CO_3$$

Likewise, as blood H^+ concentration drops below what is desirable and blood pH rises, H_2CO_3 dissociates to release bicarbonate ions and H^+ ions to the blood.

$$H_2CO_3 \rightarrow H^+ + HCO_3^-$$

The released H^+ lowers the pH again. The bicarbonate ions, being *weak* bases, are poorly functional under alkaline conditions and have little effect on blood pH unless and until blood pH drops toward acid levels.

In the case of excessively slow or shallow breathing (hypoventilation) or fast deep breathing (hyperventilation), the amount of carbonic acid in the blood can be greatly modified—increasing dramatically during hypoventilation and decreasing substantially during hyperventilation. In either situation, if the buffering ability of the blood is inadequate, respiratory acidosis or alkalosis can result. Therefore, maintaining the normal rate and depth of breathing is important for proper control of blood pH.

ACTIVITY 7

Demonstrating the Reaction Between Carbon dioxide (in Exhaled Air) and Water

1. Fill a beaker with 100 ml of distilled water.

2. Add 5 ml of 0.05 *M* NaOH and five drops of phenol red. Phenol red is a pH indicator that turns yellow in acidic solutions.

3. Blow through a straw into the solution.

What do you observe?

____

____

What chemical reaction is taking place in the beaker?

____

____

4. Discard the straw in the autoclave bag.

ACTIVITY 8

Observing the Operation of Standard Buffers

1. To observe the ability of a buffer system to stabilize the pH of a solution, obtain five 250-ml beakers and a wash bottle containing distilled water. Set up the following experimental samples:

Beaker 1:
(150 ml distilled water) pH ______

Beaker 2:
(150 ml distilled water and
1 drop concentrated HCl) pH ______

Beaker 3:
(150 ml distilled water and
1 drop concentrated NaOH) pH ______

Beaker 4: (150 ml standard buffer solution
[pH 7] and 1 drop concentrated HCl) pH ______

Beaker 5: (150 ml standard buffer solution
[pH 7] and 1 drop concentrated NaOH) pH ______

2. Using a pH meter standardized with a buffer solution of pH 7, determine the pH of the contents of each beaker and record above. After *each and every* pH recording, the pH meter switch should be turned to **STANDBY,** and the electrodes rinsed thoroughly with a stream of distilled water from the wash bottle.

3. Add 3 more drops of concentrated HCl to beaker 4, stir, and record the pH: ______

4. Add 3 more drops of concentrated NaOH to beaker 5, stir, and record the pH: ______

How successful was the buffer solution in resisting pH changes when a strong acid (HCl) or a strong base (NaOH) was added?

____

ACTIVITY 9

Exploring the Operation of the Carbonic Acid–Bicarbonate Buffer System

To observe the ability of the carbonic acid–bicarbonate buffer system of blood to resist pH changes, perform the following simple experiment.

1. Obtain two small beakers (50 ml), animal plasma, graduated cylinder, glass stirring rod, and a dropper bottle of 0.01 *M* HCl. Using the pH meter standardized with the buffer solution of pH 7.0, measure the pH of the animal plasma. Use only enough plasma to allow immersion of the electrodes and measure the volume used carefully.

pH of the animal plasma: ________

2. Add 2 drops of the 0.01 *M* HCl solution to the plasma; stir and measure the pH again.

pH of plasma plus 2 drops of HCl: ________

3. Turn the pH meter switch to **STANDBY,** rinse the electrodes, and then immerse them in a quantity of distilled water (pH 7) exactly equal to the amount of animal plasma used. Measure the pH of the distilled water.

pH of distilled water: ________

4. Add 2 drops of 0.01 *M* HCl, swirl, and measure the pH again.

pH of distilled water plus the two drops of HCl: ________

Is the plasma a good buffer? ________

What component of the plasma carbonic acid–bicarbonate buffer system was acting to counteract a change in pH when HCl was added?

____

NAME ____________________

LAB TIME/DATE ____________________

REVIEW SHEET

EXERCISE 37

Respiratory System Physiology

Mechanics of Respiration

1. For each of the following cases, check the column appropriate to your observations on the operation of the model lung.

	Diaphragm pushed up		Diaphragm pulled down	
Change	Increased	Decreased	Increased	Decreased
In internal volume of the bell jar (thoracic cage)				
In internal pressure				
In the size of the balloons (lungs)				
In direction of air flow	Into lungs	Out of lungs	Into lungs	Out of lungs

2. Base your answers to the following on your observations in question 1.

Under what internal conditions does air tend to flow into the lungs? ____________________

____

Under what internal conditions does air tend to flow out of the lungs? Explain why this is so. ____________________

____

3. Activation of the diaphragm and the external intercostal muscles begins the inspiratory process. What effect does contraction of these muscles have on thoracic volume, and how is this accomplished? ____________________

____

____

4. What was the approximate increase in diameter of chest circumference during a quiet inspiration? __________ inches

During forced inspiration? __________ inches

What temporary physiological advantage is created by the substantial increase in chest circumference during forced inspiration? ______________________________

____

5. The presence of a partial vacuum between the pleural membranes is integral to normal breathing movements. What would happen if an opening were made into the chest cavity, as with a puncture wound?

____

How is this condition treated medically? ______________________________

Respiratory Sounds

6. Which of the respiratory sounds is heard during both inspiration and expiration? ______________________________

Which is heard primarily during inspiration? ______________________________

7. Where did you best hear the vesicular respiratory sounds? ______________________________

Respiratory Volumes and Capacities—Spirometry or BIOPAC®

8. Write the respiratory volume term and the normal value that is described by the following statements.

Volume of air present in the lungs after a forceful expiration: ______________________________

Volume of air that can be expired forcibly after a normal expiration: ______________________________

Volume of air that is breathed in and out during a normal respiration: ______________________________

Volume of air that can be inspired forcibly after a normal inspiration: ______________________________

Volume of air corresponding to TV + IRV + ERV: ______________________________

9. Record experimental respiratory volumes as determined in the laboratory. (Corrected values are for the recording spirometer only.)

Average TV: ______________ ml

Corrected value for TV: ______________ ml

Average IRV: ______________ ml

Corrected value for IRV: ______________ ml

MRV: ______________ ml/min

Average ERV: ______________ ml

Corrected value for ERV: ______________ ml

Average VC: ______________ ml

Corrected value for VC: ______________ ml

% predicted VC: ______________ %

FEV_1: ______________ % FVC

10. Would your vital capacity measurement differ if you performed the test while standing? ________ While lying down?

________ Explain. ____

____

11. Which respiratory ailments can respiratory volume tests be used to detect?

____

____

12. Using an appropriate reference, complete the chart below.

		O_2	CO_2	N_2
% of composition of air	Inspired			
	Expired			

Factors Influencing Rate and Depth of Respiration

13. Where are the neural control centers of respiratory rhythm? ____________________ and ____________________

For questions 14–21, use your Activity 6 data (the pneumograph-physiograph recording or visual count).

14. In your data, what was the rate of quiet breathing?

Initial testing ________________________ breaths/min

Record observations of how the initial pneumograph recording was modified during the various testing procedures described below. Indicate the respiratory rate, and include comments on the relative depth of the respiratory peaks observed.

Test performed	Observations
Talking	
Yawning	
Laughing	
Standing	
Concentrating	
Swallowing water	
Coughing	
Lying down	
Running in place	

15. Record student data below.

Breath-holding interval after a deep inhalation: ________ sec length of recovery period: ________ sec

Breath-holding interval after a forceful expiration: ________ sec length of recovery period: ________ sec

After breathing quietly and taking a deep breath (which you held), was your urge to inspire *or* expire? ________

After exhaling and then holding one's breath, was the desire for inspiration *or* expiration? ________

Explain these results. (Hint: What reflex is involved here?) ________

____

____

16. Observations after hyperventilation: ________

____

17. Length of breath holding after hyperventilation: ________ sec

Why does hyperventilation produce apnea or a reduced respiratory rate? ________

____

18. Observations for rebreathing air: ________

____

Why does rebreathing air produce an increased respiratory rate? ________

____

19. What was the effect of running in place (exercise) on the duration of breath holding? ________

Explain this effect. ________

____

20. Record student data from the test illustrating the effect of respiration on circulation.

Radial pulse before beginning test: ________/min Radial pulse after testing: ________/min

Relative pulse force before beginning test: ________ Relative force of radial pulse after testing: ________

Condition of neck and facial veins after testing: ________

Explain these data. ________

____

21. Do the following factors generally increase (indicate with I) or decrease (indicate with D) the respiratory rate and depth?

increase in blood CO_2: ________ increase in blood pH: ________

decrease in blood O_2: ________ decrease in blood pH: ________

Did it appear that CO_2 or O_2 had a more marked effect on modifying the respiratory rate? ________

22. Where are sensory receptors sensitive to changes in blood pressure located? ________

____

23. Where are sensory receptors sensitive to changes in O_2 levels in the blood located? ________

____

24. What is the primary factor that initiates breathing in a newborn infant? ________

____

25. Blood CO_2 levels and blood pH are related. When blood CO_2 levels increase, does the pH increase or decrease?

________ Explain why. ________

____

26. Which, if any, of the measurable respiratory volumes would likely be exaggerated in a person who is cardiovascularly fit, such as a runner or a swimmer?

____

____

Which, if any, of the measurable respiratory volumes would likely be exaggerated in a person who has smoked a lot for over twenty years?

____

____

Role of the Respiratory System in Acid-Base Balance of Blood

27. Define *buffer.* ________

28. How successful was the laboratory buffer (pH 7) in resisting changes in pH when the acid was added? ________

____

When the base was added? ________

How successful was the buffer in resisting changes in pH when the additional aliquots (3 more drops) of the acid and base were added to the original samples? ________

29. What buffer system operates in blood plasma? ______________________________

Which member of the buffer system resists a *drop* in pH? ______________ Which resists a *rise* in pH? __________

30. Explain how the carbonic acid–bicarbonate buffer system of the blood operates. ______________________________

____

____

____

31. What happened when the carbon dioxide in exhaled air mixed with water? ______________________________

____

What role does exhalation of carbon dioxide play in maintaining relatively constant blood pH? ______________________________

____

____

EXERCISE 38

Anatomy of the Digestive System

MATERIALS

- ☐ Dissectible torso model
- ☐ Anatomical chart of the human digestive system
- ☐ Prepared slides of the liver and mixed salivary glands; of longitudinal sections of the gastroesophageal junction and a tooth; and of cross sections of the stomach, duodenum, and ileum
- ☐ Compound microscope
- ☐ Three-dimensional model of a villus (if available)
- ☐ Jaw model or human skull
- ☐ Three-dimensional model of liver lobules (if available)

For instructions on animal dissections, see the dissection exercises starting on page 697 in the cat, fetal pig, and rat editions of this manual.

MasteringA&P™

Access practice quizzes and more in the Study Area at www.masteringaandp.com.

PAL

For access to anatomical models and more, check out Practice Anatomy Lab.

OBJECTIVES

1. To state the overall function of the digestive system.
2. To identify on an appropriate diagram or torso model the organs of the alimentary canal, and to name their subdivisions if any.
3. To name and/or identify the accessory digestive organs.
4. To describe the general functions of the digestive system organs or structures.
5. To describe the general histologic structure of the alimentary canal wall and/or label a cross-sectional diagram of the wall with the following terms: mucosa, submucosa, muscularis externa, and serosa or adventitia.
6. To list and explain the specializations of the structure of the stomach and small intestine that contribute to their functional roles.
7. To list the major enzymes or enzyme groups produced by the salivary glands, stomach, small intestine, and pancreas.
8. To name human deciduous and permanent teeth, and to describe the anatomy of the generalized tooth.
9. To recognize (by microscopic inspection or by viewing an appropriate diagram or photomicrograph) the histologic structure of the following organs:

small intestine	tooth	liver
salivary glands	stomach	

PRE-LAB QUIZ

1. The digestive system
   a. eliminates undigested food
   b. provides the body with nutrients
   c. provides the body with water
   d. all of the above
2. Circle the correct term. Digestion / Absorption occurs when small molecules pass through epithelial cells into the blood for distribution to the body cells.
3. The __________ abuts the lumen of the alimentary canal and consists of epithelium, lamina propria, and muscularis mucosae.
   a. mucosa
   b. serosa
   c. submucosa
4. Circle the correct term. Approximately 25 cm long, the esophagus / alimentary canal conducts food from the pharynx to the stomach.
5. Wavelike contractions of the digestive tract that propel food along are called
   a. digestion
   b. elimination
   c. ingestion
   d. peristalsis

Text continues on next page.

6. The __________ is located on the left side of the abdominal cavity and is hidden by the liver and diaphragm.
   a. gallbladder
   b. large intestine
   c. small intestine
   d. stomach
7. Circle True or False. Nearly all nutrient absorption occurs in the small intestine.
8. Circle the correct term. The ascending colon / descending colon traverses down the left side of the abdominal cavity and becomes the sigmoid colon.
9. A tooth consists of two major regions, the crown and the
   a. dentin
   b. enamel
   c. gingiva
   d. root
10. Located inferior to the diaphragm, the __________ is the largest gland in the body, with four lobes.
   a. gallbladder
   b. liver
   c. pancreas
   d. thymus

The **digestive system** provides the body with the nutrients, water, and electrolytes essential for health. The organs of this system ingest, digest, and absorb food and eliminate the undigested remains as feces.

The digestive system consists of a hollow tube extending from the mouth to the anus, into which various accessory organs or glands empty their secretions (see Figure 38.2). Food material within this tube, the *alimentary canal,* is technically outside the body because it has contact only with the cells lining the tract. For ingested food to become available to the body cells, it must first be broken down *physically* (by chewing or churning) and *chemically* (by enzymatic hydrolysis) into its smaller diffusible molecules—a process called **digestion.** The digested end products can then pass through the epithelial cells lining the tract into the blood for distribution to the body cells—a process termed **absorption.** In one sense, the digestive tract can be viewed as a disassembly line, in which food is carried from one stage of its digestive processing to the next by muscular activity, and its nutrients are made available to the cells of the body en route.

The organs of the digestive system are traditionally separated into two major groups: the **alimentary canal,** or **gastrointestinal (GI) tract,** and the **accessory digestive organs.** The alimentary canal is approximately 9 meters long in a cadaver but is considerably shorter in a living person due to muscle tone. It consists of the mouth, pharynx, esophagus, stomach, and small and large intestines. The accessory structures include the teeth, which physically break down foods, and the salivary glands, gallbladder, liver, and pancreas, which secrete their products into the alimentary canal. These individual organs are described shortly.

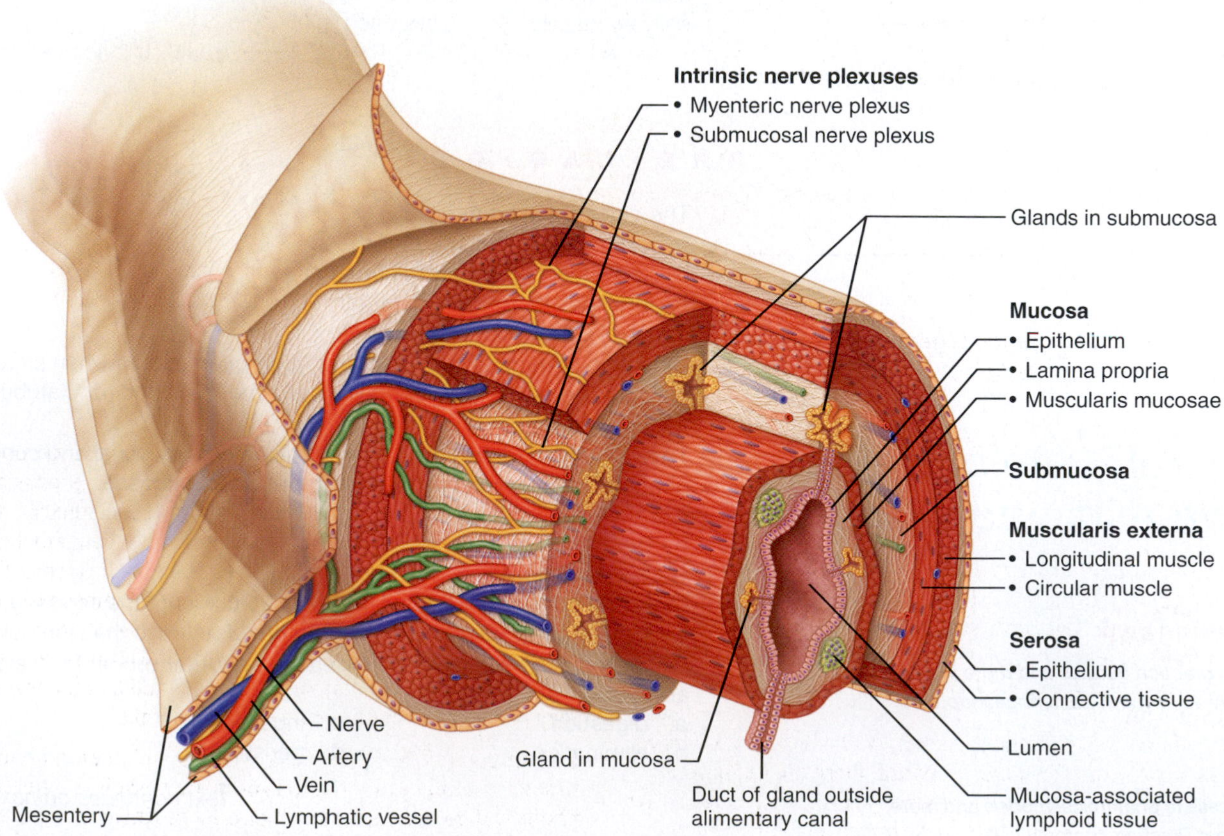
Intrinsic nerve plexuses
• Myenteric nerve plexus
• Submucosal nerve plexus
Glands in submucosa
Mucosa
• Epithelium
• Lamina propria
• Muscularis mucosae
Submucosa
Muscularis externa
• Longitudinal muscle
• Circular muscle
Serosa
• Epithelium
• Connective tissue
Lumen
Mucosa-associated lymphoid tissue
Duct of gland outside alimentary canal
Gland in mucosa
Nerve
Artery
Vein
Lymphatic vessel
Mesentery

FIGURE 38.1 Basic structural pattern of the alimentary canal wall.

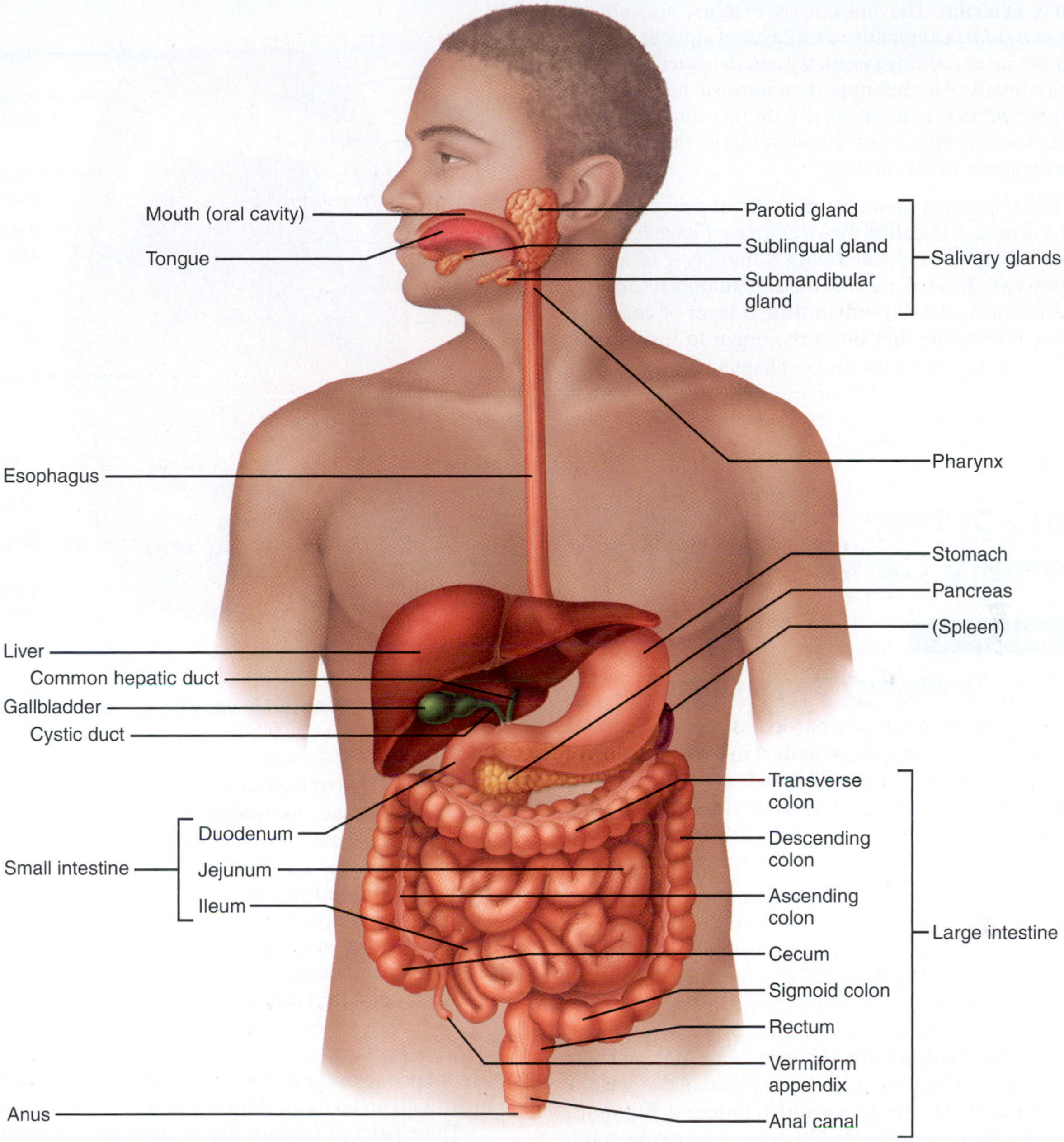

Mouth (oral cavity)
Tongue
Parotid gland
Sublingual gland
Submandibular gland
Salivary glands
Pharynx
Esophagus
Stomach
Pancreas
(Spleen)
Liver
Common hepatic duct
Gallbladder
Cystic duct
Transverse colon
Descending colon
Ascending colon
Cecum
Sigmoid colon
Rectum
Vermiform appendix
Anal canal
Large intestine
Small intestine
Duodenum
Jejunum
Ileum
Anus

FIGURE 38.2 The human digestive system: alimentary tube and accessory organs. (Liver and gallbladder are reflected superiorly and to the right.)

General Histological Plan of the Alimentary Canal

Because the alimentary canal has a shared basic structural plan (particularly from the esophagus to the anus), it makes sense to review that general structure as we begin studying this group of organs. Then our descriptions of individual organs can focus on their specializations for unique functions in the digestive process.

Essentially the alimentary canal walls have four basic **tunics** (layers). From the lumen outward, these are the *mucosa,* the *submucosa,* the *muscularis externa,* and either a *serosa* or *adventitia* (Figure 38.1). Each of these tunics has a predominant tissue type and a specific function in the digestive process.

Mucosa (mucous membrane): The mucosa is the wet epithelial membrane abutting the alimentary canal lumen. It consists of a surface *epithelium* (in most cases, a simple columnar), a *lamina propria* (areolar connective tissue on which the epithelial layer rests), and a *muscularis mucosae* (a scant layer of smooth muscle fibers that enable local movements of the mucosa). The major functions of the mucosa are secretion (of enzymes, mucus, hormones, etc.), absorption of digested foodstuffs, and protection (against bacterial invasion). A particular mucosal region may be involved in one or all three functions.

Submucosa: The submucosa is moderately dense connective tissue containing blood and lymphatic vessels, scattered lymphoid follicles, and nerve fibers. Its intrinsic nerve supply is called the *submucosal plexus.* Its major functions are nutrition and protection.

Muscularis externa: The muscularis externa, also simply called the *muscularis,* typically is a bilayer of smooth muscle, with the inner layer running circularly and the outer layer running longitudinally. Another important intrinsic nerve plexus, the *myenteric plexus,* is associated with this tunic. By controlling the smooth muscle of the muscularis, this plexus is the major regulator of GI motility.

Serosa: The outermost covering of most of the alimentary canal is the serosa, also called the *visceral peritoneum.* It consists of mesothelium associated with a thin layer of areolar connective tissue. In areas *outside* the abdominopelvic cavity, the serosa is replaced by an **adventitia,** a layer of coarse fibrous connective tissue that binds the organ to surrounding tissues. (This is the case with the esophagus.) The serosa reduces friction as the mobile digestive system organs work and slide across one another and the cavity walls. The adventitia anchors and protects the surrounded organ.

Organs of the Alimentary Canal

ACTIVITY 1

Identifying Alimentary Canal Organs

The sequential pathway and fate of food as it passes through the alimentary canal organs are described in the next sections. Identify each structure in Figure 38.2 and on the torso model or anatomical chart of the digestive system as you work.

Oral Cavity or Mouth

Food enters the digestive tract through the **oral cavity**, or **mouth** (Figure 38.3). Within this mucous membrane–lined cavity are the gums, teeth, tongue, and openings of the ducts of the salivary glands. The **lips (labia)** protect the opening of the chamber anteriorly, the **cheeks** form its lateral walls, and the **palate,** its roof. The anterior portion of the palate is referred to as the **hard palate** because bone (the palatine processes of the maxillae and horizontal plates of the palatine bones) underlies it. The posterior **soft palate** is a fibromuscular structure that is unsupported by bone. The **uvula,** a fingerlike projection of the soft palate, extends inferiorly from its posterior margin. The soft palate rises to close off the oral cavity from the nasal and pharyngeal passages during swallowing. The floor of the oral cavity is occupied by the muscular **tongue,** which is largely supported by the *mylohyoid muscle* (Figure 38.4) and attaches to the hyoid bone, mandible, styloid processes, and pharynx. A membrane called the **lingual frenulum** secures the inferior midline of the tongue to the floor of the mouth. The space between the lips and cheeks and the teeth is the **vestibule;** the area that lies within the teeth and gums (gingivae) is the **oral cavity** proper. (The teeth and gums are discussed in more detail on pages 586–587.)

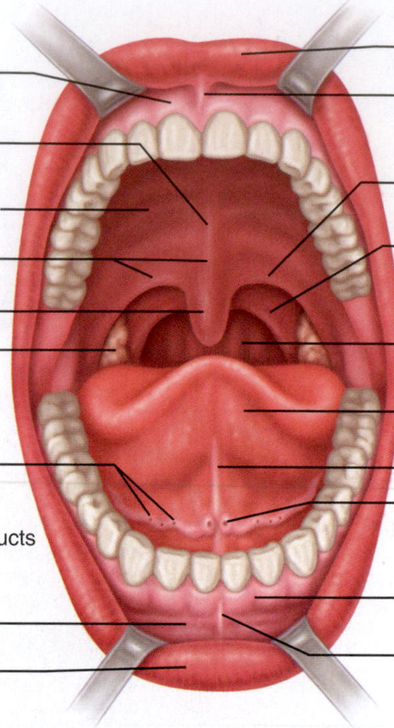

Gingivae (gums)
Palatine raphe
Hard palate
Soft palate
Uvula
Palatine tonsil
Sublingual fold with openings of sublingual ducts
Vestibule
Lower lip
Upper lip
Superior labial frenulum
Palatoglossal arch
Palatopharyngeal arch
Posterior wall of oropharynx
Tongue
Lingual frenulum
Opening of submandibular duct
Gingivae (gums)
Inferior labial frenulum

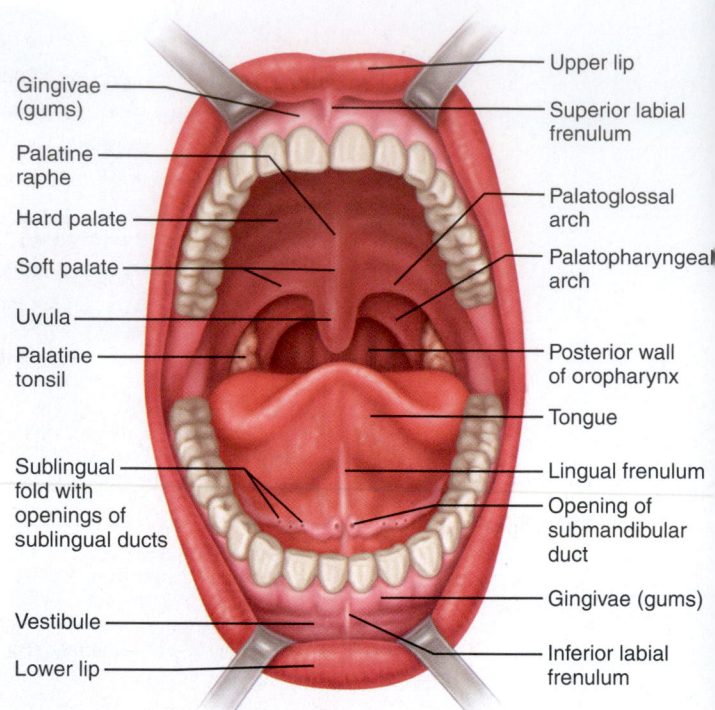

FIGURE 38.3 **Anterior view of the oral cavity.**

On each side of the mouth at its posterior end are masses of lymphoid tissue, the **palatine tonsils** (see Figure 38.3). Each lies in a concave area bounded anteriorly and posteriorly by membranes, the **palatoglossal arch** (anterior membrane) and the **palatopharyngeal arch** (posterior membrane). Another mass of lymphoid tissue, the **lingual tonsil** (see Figure 38.4), covers the base of the tongue, posterior to the oral cavity proper. The tonsils, in common with other lymphoid tissues, are part of the body's defense system.

Very often in young children, the palatine tonsils become inflamed and enlarge, partially blocking the entrance to the pharynx posteriorly and making swallowing difficult and painful. This condition is called **tonsillitis.**

Three pairs of salivary glands duct their secretion, saliva, into the oral cavity. One component of saliva, salivary amylase, begins the digestion of starchy foods within the oral cavity. (The salivary glands are discussed in more detail on page 587.)

As food enters the mouth, it is mixed with saliva and masticated (chewed). The cheeks and lips help hold the food between the teeth during mastication, and the highly mobile tongue manipulates the food during chewing and initiates swallowing. Thus the mechanical and chemical breakdown of food begins before the food has left the oral cavity. As noted in Exercise 26, the surface of the tongue is covered with papillae, many of which contain taste buds. Taste buds contain taste cells, receptors for taste sensation. So, in addition to its manipulative function, the tongue permits the enjoyment and appreciation of the food ingested.

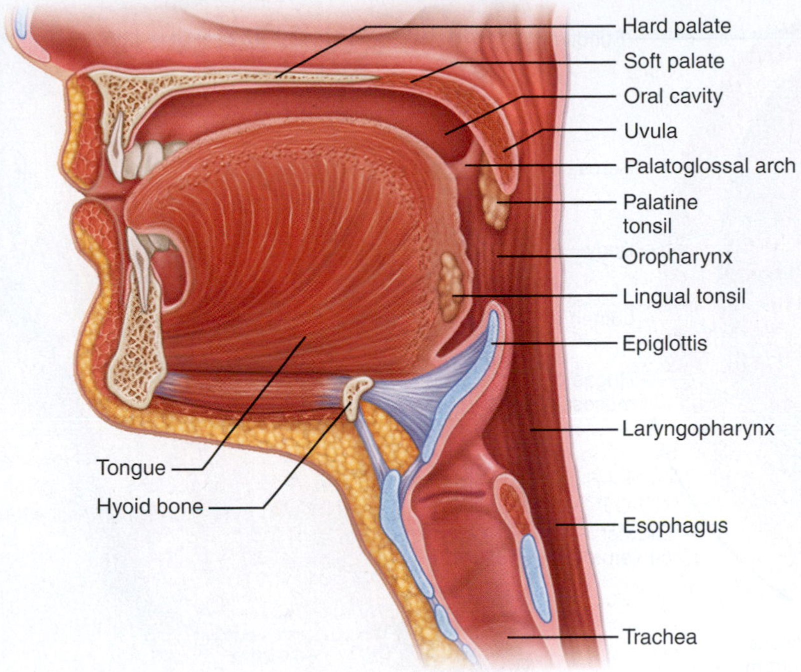

Hard palate
Soft palate
Oral cavity
Uvula
Palatoglossal arch
Palatine tonsil
Oropharynx
Lingual tonsil
Epiglottis
Laryngopharynx
Tongue
Hyoid bone
Esophagus
Trachea

FIGURE 38.4 Sagittal view of the head showing oral cavity and pharynx.

Pharynx

When the tongue initiates swallowing, the food passes posteriorly into the pharynx, a common passageway for food, fluid, and air (see Figure 38.4). The pharynx is subdivided anatomically into three parts—the **nasopharynx** (behind the nasal cavity), the **oropharynx** (behind the oral cavity extending from the soft palate to the epiglottis overlying the larynx), and the **laryngopharynx** (extending from the epiglottis to the base of the larynx), which is continuous with the esophagus.

The walls of the pharynx consist largely of two layers of skeletal muscles: an inner layer of longitudinal muscle (the levator muscles) and an outer layer of circular constrictor muscles, which initiate wavelike contractions that propel the food inferiorly into the esophagus. The mucosa of the oropharynx and laryngopharynx, like that of the oral cavity, contains a friction-resistant stratified squamous epithelium.

Esophagus

The **esophagus,** or gullet, extends from the pharynx through the diaphragm to the gastroesophageal sphincter in the superior aspect of the stomach. Approximately 25 cm long in humans, it is essentially a food passageway that conducts food to the stomach in a wavelike peristaltic motion. The esophagus has no digestive or absorptive function. The walls at its superior end contain skeletal muscle, which is replaced by smooth muscle in the area nearing the stomach. The **gastroesophageal sphincter,** a slight thickening of the smooth muscle layer at the esophagus-stomach junction, controls food passage into the stomach (see Figure 38.5). Since the esophagus is located in the thoracic rather than the abdominal cavity, its outermost layer is an *adventitia,* rather than the serosa.

Stomach

The **stomach** (Figures 38.2 and 38.5) is on the left side of the abdominal cavity and is hidden by the liver and diaphragm. Different regions of the saclike stomach are the **cardiac region** or **cardia** (the area surrounding the cardiac orifice through which food enters the stomach from the esophagus), the **fundus** (the dome-shaped portion of the stomach, superolateral to the cardiac region), the **body** (midportion of the stomach, inferior to the fundus), and the funnel-shaped **pyloric region** (consisting of the superiormost *pyloric antrum,* the more narrow *pyloric canal,* and the terminal *pylorus,* which is continuous with the small intestine through the **pyloric valve** or **sphincter**).

The concave medial surface of the stomach is called the **lesser curvature;** its convex lateral surface is the **greater curvature.** Extending from these curvatures are two mesenteries, called *omenta.* The **lesser omentum** extends from the liver to the lesser curvature of the stomach. The **greater omentum,** a saclike mesentery, extends from the greater curvature of the stomach, reflects downward over the abdominal contents to cover them in an apronlike fashion, and then blends with the **mesocolon** attaching the transverse colon to the posterior body wall. Figure 38.7 (page 582) illustrates the omenta as well as the other peritoneal attachments of the abdominal organs.

The stomach is a temporary storage region for food as well as a site for mechanical and chemical breakdown of food. It contains a third (innermost) *obliquely* oriented layer of smooth muscle in its muscularis externa that allows it to churn, mix, and pummel the food, physically reducing it to smaller fragments. **Gastric glands** of the mucosa secrete hydrochloric acid (HCl) and hydrolytic enzymes (primarily pepsinogen, the inactive form of *pepsin,* a protein-digesting enzyme), which begin the enzymatic, or chemical, breakdown of protein foods. The *mucosal glands* also secrete a viscous mucus that helps prevent the stomach itself from being digested by the proteolytic enzymes. Most digestive activity occurs in the pyloric region of the stomach. After the food is processed in the stomach, it resembles a creamy mass **(chyme),** which enters the small intestine through the pyloric sphincter.

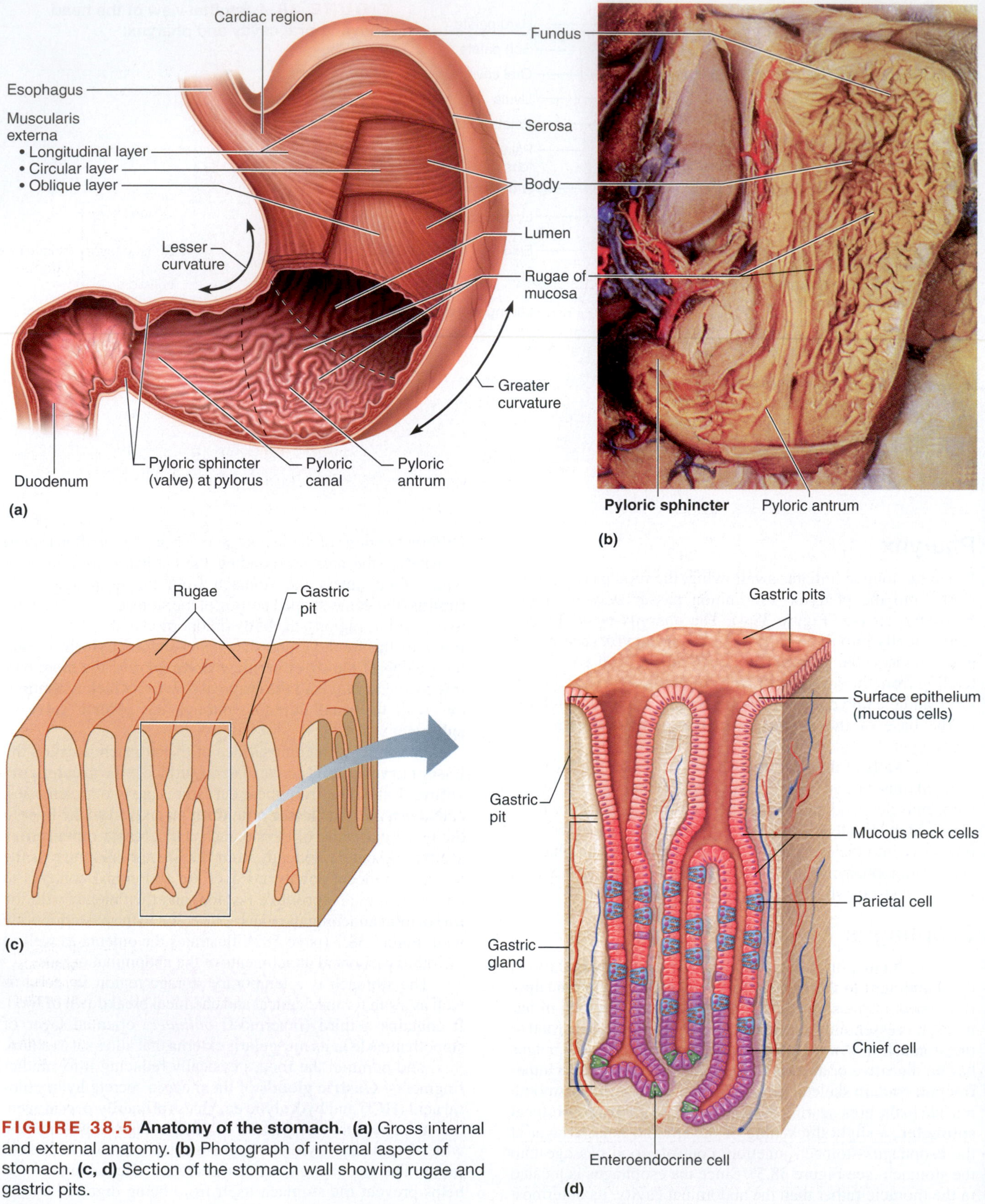
Cardiac region
Fundus
Esophagus
Muscularis externa
• Longitudinal layer
• Circular layer
• Oblique layer
Serosa
Body
Lumen
Lesser curvature
Rugae of mucosa
Greater curvature
Duodenum
Pyloric sphincter (valve) at pylorus
Pyloric canal
Pyloric antrum
(a)
Pyloric sphincter
Pyloric antrum
(b)
Rugae
Gastric pit
(c)
Gastric pits
Surface epithelium (mucous cells)
Gastric pit
Mucous neck cells
Parietal cell
Gastric gland
Chief cell
Enteroendocrine cell
(d)

FIGURE 38.5 Anatomy of the stomach. (a) Gross internal and external anatomy. **(b)** Photograph of internal aspect of stomach. **(c, d)** Section of the stomach wall showing rugae and gastric pits.

ACTIVITY 2

Studying the Histological Structure of Selected Digestive System Organs

To prepare for the histologic study you will be conducting now and later in the lab, obtain a microscope and the following slides: salivary glands (submandibular or sublingual); liver; cross sections of the duodenum, ileum, and stomach; and longitudinal sections of a tooth and the gastroesophageal junction.

1. **Stomach:** View the stomach slide first. Refer to Figure 38.6a as you scan the tissue under low power to locate the muscularis externa; then move to high power to more closely examine this layer. Try to pick out the three smooth muscle layers. How does the extra (oblique) layer of smooth muscle found in the stomach correlate with the stomach's churning movements?

___

___

___

Identify the gastric glands and the gastric pits (see Figures 38.5 and 38.6b). If the section is taken from the stomach fundus and is appropriately stained, you can identify, in the gastric glands, the blue-staining **chief** (or **zymogenic) cells,** which produce pepsinogen, and the red-staining **parietal cells,** which secrete HCl. The enteroendocrine cells that release hormones are indistinguishable. Draw a small section of the stomach wall, and label it appropriately.

2. **Gastroesophageal junction:** Scan the slide under low power to locate the mucosal junction between the end of the esophagus and the beginning of the stomach, the gastroesophageal junction. Compare your observations to Figure 38.6c. What is the functional importance of the epithelial differences seen in the two organs?

___

___

___

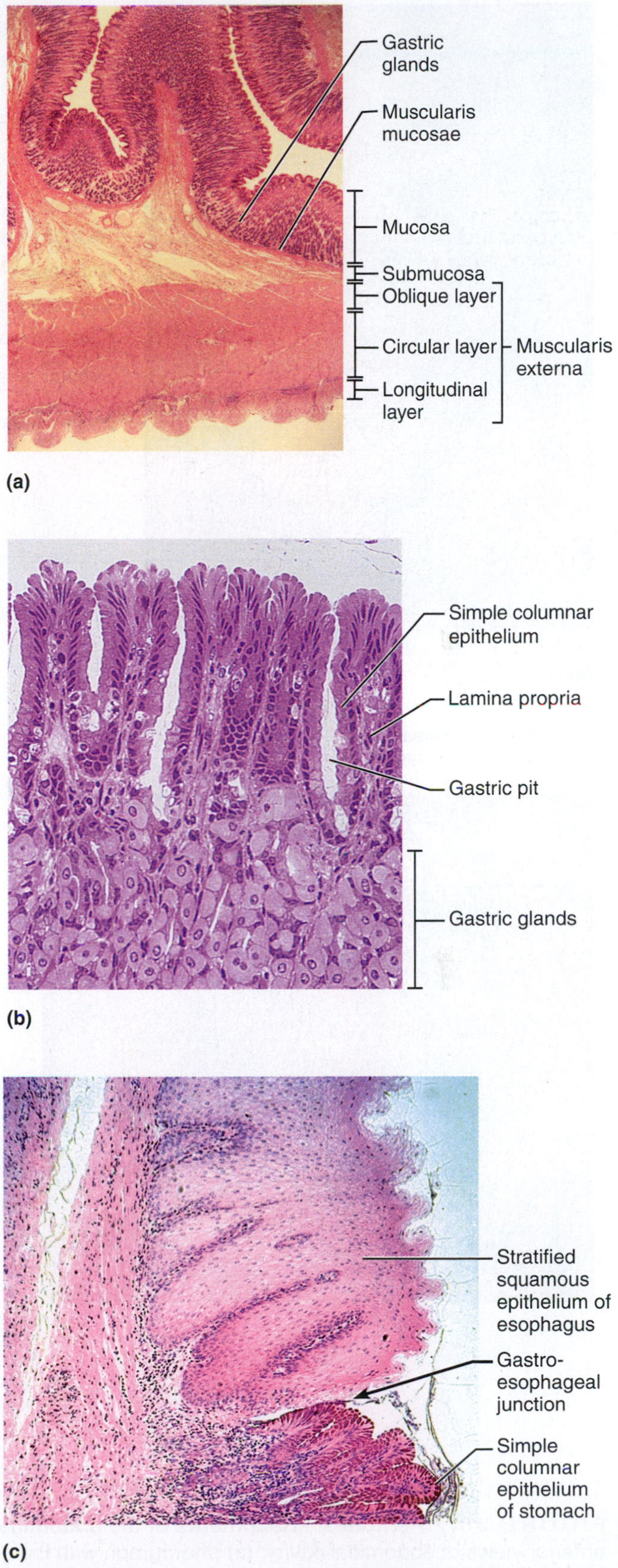
Gastric glands
Muscularis mucosae
Mucosa
Submucosa
Oblique layer
Circular layer
Longitudinal layer
Muscularis externa
(a)
Simple columnar epithelium
Lamina propria
Gastric pit
Gastric glands
(b)
Stratified squamous epithelium of esophagus
Gastro-esophageal junction
Simple columnar epithelium of stomach
(c)

FIGURE 38.6 Histology of selected regions of the stomach and gastroesophageal junction. **(a)** Stomach wall (30×). **(b)** Gastric pits and glands (120×). **(c)** Gastroesophageal junction, longitudinal section (70×).

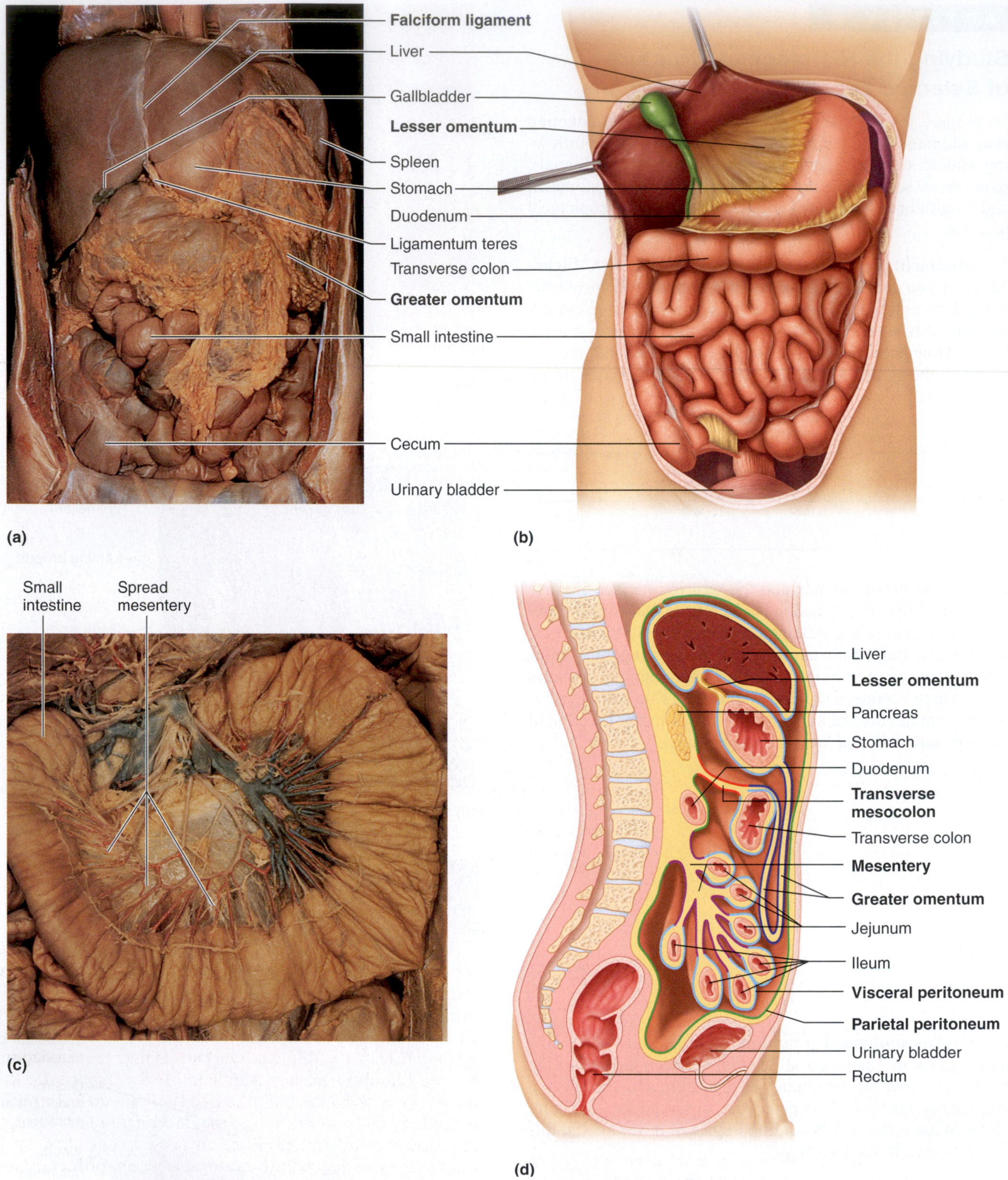

Falciform ligament
Liver
Gallbladder
Lesser omentum
Spleen
Stomach
Duodenum
Ligamentum teres
Transverse colon
Greater omentum
Small intestine
Cecum
Urinary bladder
(a)
(b)
Small intestine
Spread mesentery
(c)
Liver
Lesser omentum
Pancreas
Stomach
Duodenum
Transverse mesocolon
Transverse colon
Mesentery
Greater omentum
Jejunum
Ileum
Visceral peritoneum
Parietal peritoneum
Urinary bladder
Rectum
(d)

FIGURE 38.7 Peritoneal attachments of the abdominal organs. Superficial anterior views of abdominal cavity: **(a)** photograph with the greater omentum in place and **(b)** diagram showing greater omentum removed and liver and gallbladder reflected superiorly. **(c)** Mesentery of the small intestine. **(d)** Sagittal view of a male torso.

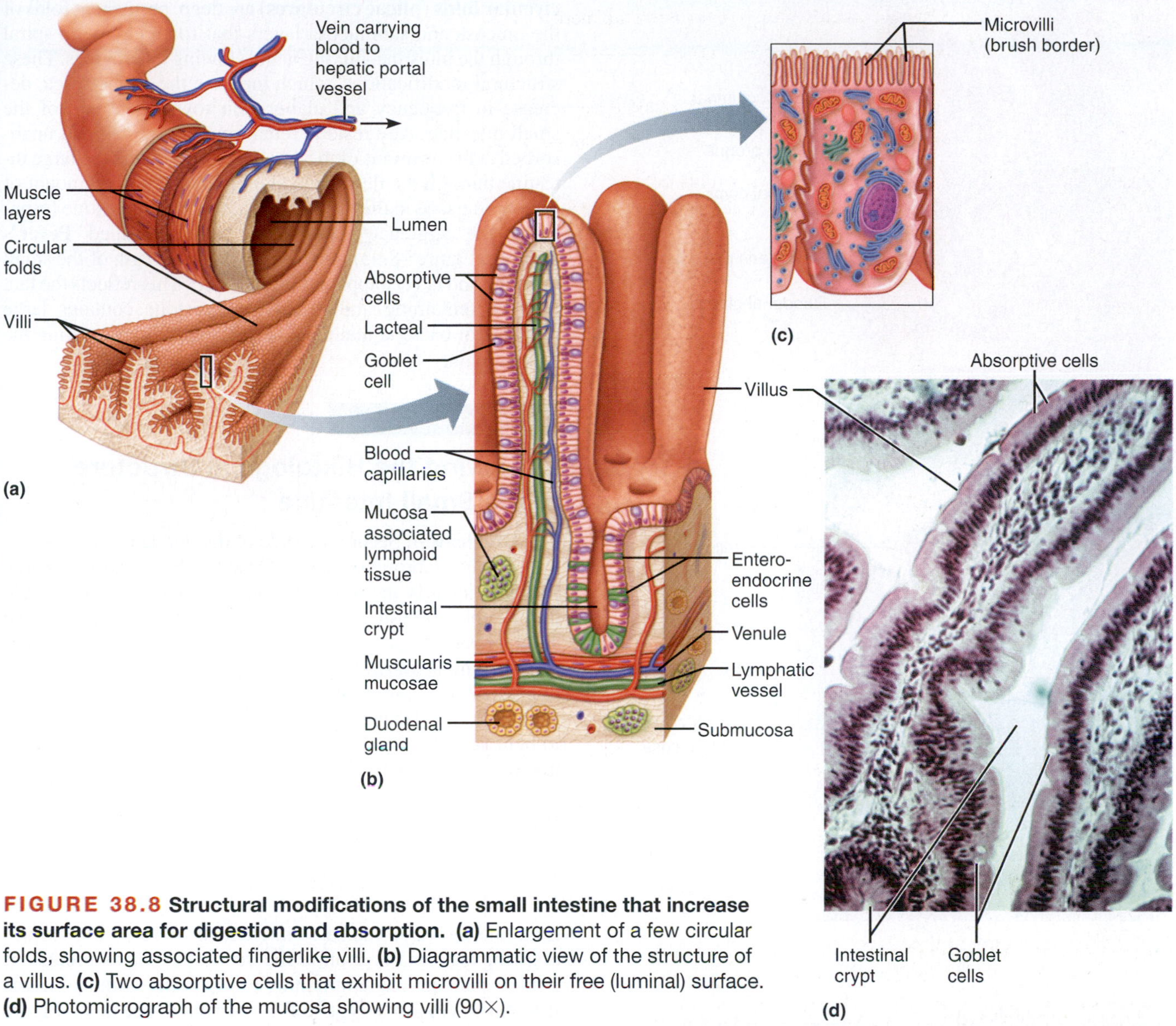

Vein carrying blood to hepatic portal vessel
Muscle layers
Lumen
Circular folds
Villi
(a)
Absorptive cells
Lacteal
Goblet cell
Blood capillaries
Mucosa associated lymphoid tissue
Intestinal crypt
Muscularis mucosae
Duodenal gland
(b)
Villus
Entero-endocrine cells
Venule
Lymphatic vessel
Submucosa
Microvilli (brush border)
(c)
Absorptive cells
Intestinal crypt
Goblet cells
(d)

FIGURE 38.8 Structural modifications of the small intestine that increase its surface area for digestion and absorption. **(a)** Enlargement of a few circular folds, showing associated fingerlike villi. **(b)** Diagrammatic view of the structure of a villus. **(c)** Two absorptive cells that exhibit microvilli on their free (luminal) surface. **(d)** Photomicrograph of the mucosa showing villi (90×).

Small Intestine

The **small intestine** is a convoluted tube, 6 to 7 meters (about 20 feet) long in a cadaver but only about 2 m (6 feet) long during life because of its muscle tone. It extends from the pyloric sphincter to the ileocecal valve. The small intestine is suspended by a double layer of peritoneum, the fan-shaped **mesentery,** from the posterior abdominal wall (see Figure 38.7), and it lies, framed laterally and superiorly by the large intestine, in the abdominal cavity. The small intestine has three subdivisions (see Figure 38.2): (1) the **duodenum** extends from the pyloric sphincter for about 25 cm (10 inches) and curves around the head of the pancreas; most of the duodenum lies in a retroperitoneal position. (2) The **jejunum,** continuous with the duodenum, extends for 2.5 m (about 8 feet). Most of the jejunum occupies the umbilical region of the abdominal cavity. (3) The **ileum**, the terminal portion of the small intestine, is about 3.6 m (12 feet) long and joins the large intestine at the **ileocecal valve.** It is located inferiorly and somewhat to the right in the abdominal cavity, but its major portion lies in the hypogastric region.

Brush border enzymes, hydrolytic enzymes bound to the microvilli of the columnar epithelial cells, and, more importantly, enzymes produced by the pancreas and ducted into the duodenum largely via the **main pancreatic duct** complete the enzymatic digestion process in the small intestine. Bile (formed in the liver) also enters the duodenum via the **bile duct** in the same area. At the duodenum, the ducts join to form the bulblike **hepatopancreatic ampulla** and empty their products into the duodenal lumen through the **major duodenal papilla**, an orifice controlled by a muscular valve called the **hepatopancreatic sphincter** (see Figure 38.15 on page 589).

Nearly all nutrient absorption occurs in the small intestine, where three structural modifications that increase the mucosa absorptive area appear—the microvilli, villi, and circular folds (Figure 38.8). **Microvilli** are minute projections of the surface plasma membrane of the columnar epithelial lining cells of the mucosa. **Villi** are the fingerlike projections of the mucosa tunic that give it a velvety appearance and texture. The

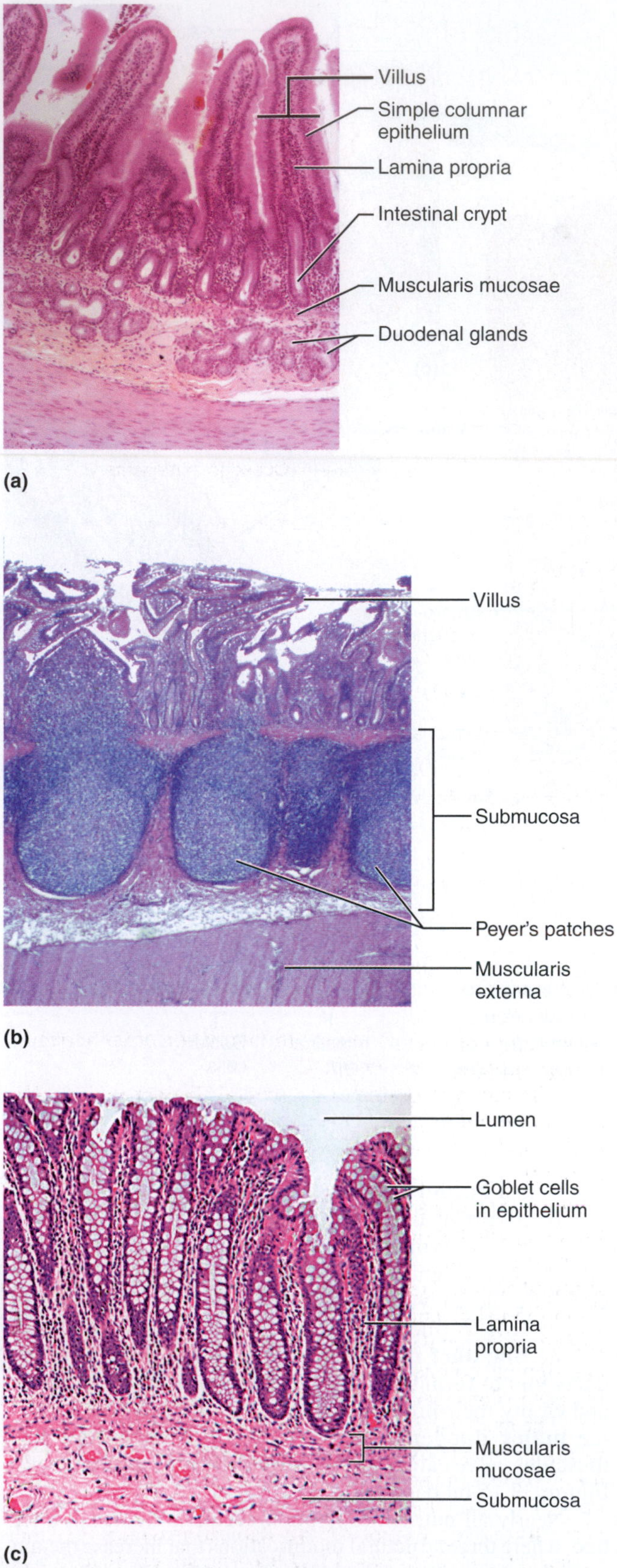
Villus
Simple columnar epithelium
Lamina propria
Intestinal crypt
Muscularis mucosae
Duodenal glands
(a)
Villus
Submucosa
Peyer's patches
Muscularis externa
(b)
Lumen
Goblet cells in epithelium
Lamina propria
Muscularis mucosae
Submucosa
(c)

FIGURE 38.9 Histology of selected regions of the small and large intestines. Cross-sectional views. **(a)** Duodenum of the small intestine (15×). **(b)** Ileum of the small intestine (5×). **(c)** Large intestine (30×).

circular folds (plicae circulares) are deep, permanent folds of the mucosa and submucosa layers that force chyme to spiral through the intestine, mixing it and slowing its progress. These structural modifications, which increase the surface area, decrease in frequency and elaboration toward the end of the small intestine. Any residue remaining undigested and unabsorbed at the terminus of the small intestine enters the large intestine through the ileocecal valve. In contrast, the amount of lymphoid tissue in the submucosa of the small intestine (especially the aggregated lymphoid follicles called **Peyer's patches,** Figure 38.9b) increases along the length of the small intestine and is very apparent in the ileum. This reflects the fact that the remaining undigested food residue contains large numbers of bacteria that must be prevented from entering the bloodstream.

ACTIVITY 3

Observing the Histological Structure of the Small Intestine

1. **Duodenum:** Secure the slide of the duodenum (cross section) to the microscope stage. Observe the tissue under low power to identify the four basic tunics of the intestinal wall—that is, the **mucosa** (the lining and its three sublayers), the **submucosa** (areolar connective tissue layer deep to the mucosa), the **muscularis externa** (composed of circular and longitudinal smooth muscle layers), and the **serosa** (the outermost layer, also called the *visceral peritoneum*). Consult Figure 38.9a to help you identify the scattered mucus-producing **duodenal glands** (Brunner's glands) in the submucosa.

What type of epithelium do you see here? ______________

____

Examine the large leaflike *villi,* which increase the surface area for absorption. Notice the scattered mucus-producing goblet cells in the epithelium of the villi. Note also the **intestinal crypts** (crypts of Lieberkühn, see also Figure 38.8), invaginated areas of the mucosa between the villi containing the cells that produce intestinal juice, a watery mucus-containing mixture that serves as a carrier fluid for absorption of nutrients from the chyme. Sketch and label a small section of the duodenal wall, showing all layers and villi.

2. **Ileum:** The structure of the ileum resembles that of the duodenum, except that the villi are less elaborate (most of the absorption has occurred by the time the ileum is reached).

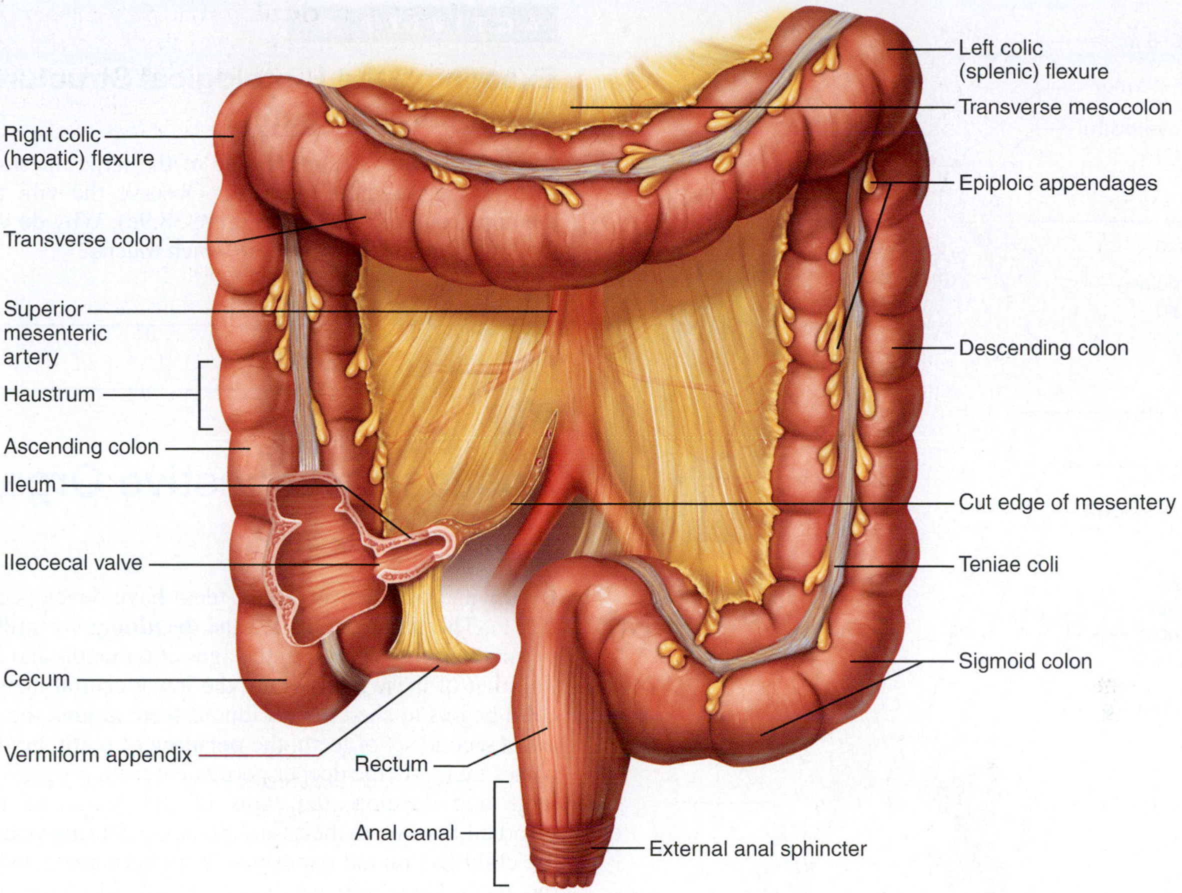
Left colic
(splenic) flexure
Transverse mesocolon
Right colic
(hepatic) flexure
Epiploic appendages
Transverse colon
Superior
mesenteric
artery
Descending colon
Haustrum
Ascending colon
Ileum
Cut edge of mesentery
Ileocecal valve
Teniae coli
Sigmoid colon
Cecum
Vermiform appendix
Rectum
Anal canal
External anal sphincter

FIGURE 38.10 The large intestine. (Section of the cecum removed to show the ileocecal valve.)

Secure a slide of the ileum to the microscope stage for viewing. Observe the villi, and identify the four layers of the wall and the large, generally spherical Peyer's patches (Figure 38.9b). What tissue composes Peyer's patches?

____

3. If a villus model is available, identify the following cells or regions before continuing: absorptive epithelium, goblet cells, lamina propria, slips of the muscularis mucosae, capillary bed, and lacteal. If possible, also identify the intestinal crypts. ■

Large Intestine

The **large intestine** (Figure 38.10) is about 1.5 m (5 feet) long and extends from the ileocecal valve to the anus. It encircles the small intestine on three sides and consists of the following subdivisions: **cecum, vermiform appendix, colon, rectum,** and **anal canal.**

The blind tubelike appendix, which hangs from the cecum, is a trouble spot in the large intestine. Since it is generally twisted, it provides an ideal location for bacteria to accumulate and multiply. Inflammation of the appendix, or appendicitis, is the result. ■

The colon is divided into several distinct regions. The **ascending colon** travels up the right side of the abdominal cavity and makes a right-angle turn at the **right colic (hepatic) flexure** to cross the abdominal cavity as the **transverse colon.** It then turns at the **left colic (splenic) flexure** and continues down the left side of the abdominal cavity as the **descending colon,** where it takes an S-shaped course as the **sigmoid colon.** The sigmoid colon, rectum, and the anal canal lie in the pelvis anterior to the sacrum and thus are not considered abdominal cavity structures. Except for the transverse and sigmoid colons, which are secured to the dorsal body wall by mesocolons (see Figure 38.7), the colon is retroperitoneal.

The anal canal terminates in the **anus,** the opening to the exterior of the body. The anus, which has an external sphincter of skeletal muscle (the voluntary sphincter) and an internal sphincter of smooth muscle (the involuntary sphincter), is normally closed except during defecation when the undigested remains of the food and bacteria are eliminated from the body as feces.

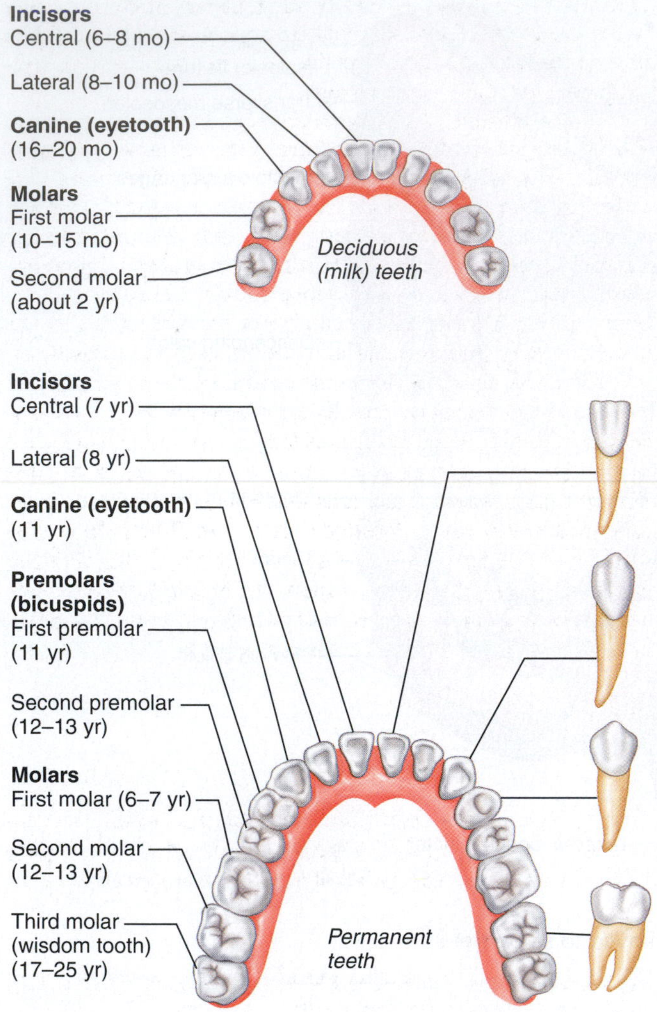
Incisors
Central (6–8 mo)
Lateral (8–10 mo)
Canine (eyetooth)
(16–20 mo)
Molars
First molar
(10–15 mo)
Second molar
(about 2 yr)
Deciduous
(milk) teeth
Incisors
Central (7 yr)
Lateral (8 yr)
Canine (eyetooth)
(11 yr)
Premolars
(bicuspids)
First premolar
(11 yr)
Second premolar
(12–13 yr)
Molars
First molar (6–7 yr)
Second molar
(12–13 yr)
Third molar
(wisdom tooth)
(17–25 yr)
Permanent
teeth

FIGURE 38.11 Human deciduous teeth and permanent teeth. (Approximate time of teeth eruption shown in parentheses.)

In the large intestine, the longitudinal muscle layer of the muscularis externa is reduced to three longitudinal muscle bands called the **teniae coli.** Since these bands are shorter than the rest of the wall of the large intestine, they cause the wall to pucker into small pocketlike sacs called **haustra.** Fat-filled pouches of visceral peritoneum, called *epiploic appendages,* hang from the colon's surface.

The major function of the large intestine is to consolidate and propel the unusable fecal matter toward the anus and eliminate it from the body. While it does that chore, it (1) provides a site for the manufacture, by intestinal bacteria, of some vitamins (B and K), which it then absorbs into the bloodstream; and (2) reclaims most of the remaining water from undigested food, thus conserving body water.

Watery stools, or **diarrhea,** result from any condition that rushes undigested food residue through the large intestine before it has had sufficient time to absorb the water (as in irritation of the colon by bacteria). Conversely, when food residue remains in the large intestine for extended periods (as with atonic colon or failure of the defecation reflex), excessive water is absorbed and the stool becomes hard and difficult to pass causing **constipation.** ■

ACTIVITY 4

Examining the Histological Structure of the Large Intestine

Large intestine: Secure a slide of the large intestine to the microscope stage for viewing. Observe the villi and note the numerous goblet cells (Figure 38.9c). Why do you think the large intestine produces so much mucus?

____

____ ■

Accessory Digestive Organs

Teeth

By the age of 21, two sets of teeth have developed (Figure 38.11). The initial set, called the **deciduous** (or **milk**) **teeth,** normally appears between the ages of 6 months and 2½ years. The first of these to erupt are the lower central incisors. The child begins to shed the deciduous teeth around the age of 6, and a second set of teeth, the **permanent teeth,** gradually replaces them. As the deeper permanent teeth progressively enlarge and develop, the roots of the deciduous teeth are resorbed, leading to their final shedding. During years 6 to 12, the child has mixed dentition—both permanent and deciduous teeth. Generally, by the age of 12, all of the deciduous teeth have been shed, or exfoliated.

Teeth are classified as **incisors, canines** *(eye teeth),* **premolars** *(bicuspids),* and **molars.** Teeth names reflect differences in relative structure and function. The incisors are chisel-shaped and exert a shearing action used in biting. Canines are cone-shaped or fanglike, the latter description being much more applicable to the canines of animals whose teeth are used for the tearing of food. Incisors, canines, and premolars typically have single roots, though the first upper premolars may have two. The lower molars have two roots, but the upper molars usually have three. The premolars have two *cusps* (grinding surfaces); the molars have broad crowns with rounded cusps specialized for the fine grinding of food.

Dentition is described by means of a **dental formula,** which designates the numbers, types, and position of the teeth in one side of the jaw. (Because tooth arrangement is bilaterally symmetrical, it is only necessary to designate one side of the jaw.) The complete dental formula for the deciduous teeth from the medial aspect of each jaw and proceeding posteriorly is as follows:

$$\frac{\text{Upper teeth: 2 incisors, 1 canine, 0 premolars, 2 molars}}{\text{Lower teeth: 2 incisors, 1 canine, 0 premolars, 2 molars}} \times 2$$

This formula is generally abbreviated to read as follows:

$$\frac{2,1,0,2}{2,1,0,2} \times 2 = 20 \text{ (number of deciduous teeth)}$$

The 32 permanent teeth are then described by the following dental formula:

$$\frac{2,1,2,3}{2,1,2,3} \times 2 = 32 \text{ (number of permanent teeth)}$$

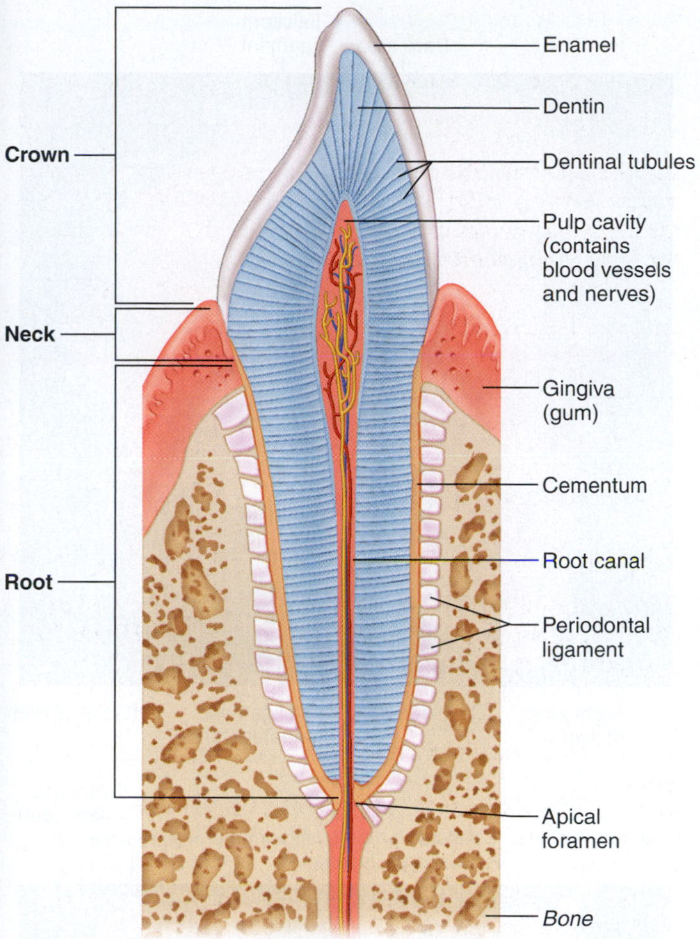
Crown
Neck
Root
Enamel
Dentin
Dentinal tubules
Pulp cavity (contains blood vessels and nerves)
Gingiva (gum)
Cementum
Root canal
Periodontal ligament
Apical foramen
Bone

FIGURE 38.12 Longitudinal section of human canine tooth within its bony alveolus.

Although 32 is designated as the normal number of permanent teeth, not everyone develops a full complement. In many people, the third molars, commonly called *wisdom teeth,* never erupt.

ACTIVITY 5

Identifying Types of Teeth

Identify the four types of teeth (incisors, canines, premolars, and molars) on the jaw model or human skull. ■

A tooth consists of two major regions, the **crown** and the **root.** A longitudinal section made through a tooth shows the following basic anatomical plan (Figure 38.12). The crown is the superior portion of the tooth. The portion of the crown visible above the **gingiva,** or **gum,** is referred to as the *clinical crown.* The entire area covered by **enamel** is called the *anatomical crown.* Enamel is the hardest substance in the body and is fairly brittle. It consists of 95% to 97% inorganic calcium salts and thus is heavily mineralized. The crevice between the end of the anatomical crown and the upper margin of the gingiva is referred to as the *gingival sulcus* and its apical border is the *gingival margin.*

That portion of the tooth embedded in the alveolar portion of the jaw is the root, and the root and crown are connected by a slight constriction, the **neck.** The outermost surface of the root is covered by **cementum,** which is similar to bone in composition and less brittle than enamel. The cementum attaches the tooth to the **periodontal ligament,** which holds the tooth in the alveolar socket and exerts a cushioning effect. **Dentin,** which composes the bulk of the tooth, is the bonelike material medial to the enamel and cementum.

The **pulp cavity** occupies the central portion of the tooth. **Pulp,** connective tissue liberally supplied with blood vessels, nerves, and lymphatics, occupies this cavity and provides for tooth sensation and supplies nutrients to the tooth tissues. **Odontoblasts,** specialized cells that reside in the outer margins of the pulp cavity, produce the dentin. The pulp cavity extends into distal portions of the root and becomes the **root canal.** An opening at the root apex, the **apical foramen,** provides a route of entry into the tooth for blood vessels, nerves, and other structures from the tissues beneath.

ACTIVITY 6

Studying Microscopic Tooth Anatomy

Observe a slide of a longitudinal section of a tooth, and compare your observations with the structures detailed in Figure 38.12. Identify as many of these structures as possible. ■

Salivary Glands

Three pairs of major **salivary glands** (see Figure 38.2) empty their secretions into the oral cavity.

Parotid glands: Large glands located anterior to the ear and ducting into the mouth over the second upper molar through the parotid duct.

Submandibular glands: Located along the medial aspect of the mandibular body in the floor of the mouth, and ducting under the tongue to the base of the lingual frenulum.

Sublingual glands: Small glands located most anteriorly in the floor of the mouth and emptying under the tongue via several small ducts.

Food in the mouth and mechanical pressure (even chewing rubber bands or wax) stimulate the salivary glands to secrete saliva. Saliva consists primarily of *mucin* (a viscous glycoprotein), which moistens the food and helps to bind it together into a mass called a **bolus,** and a clear serous fluid containing the enzyme *salivary amylase.* Salivary amylase begins the digestion of starch (a large polysaccharide), breaking it down into disaccharides, or double sugars, and glucose. Parotid gland secretion is mainly serous; the submandibular is a mixed gland that produces both mucin and serous components; and the sublingual gland produces mostly mucin.

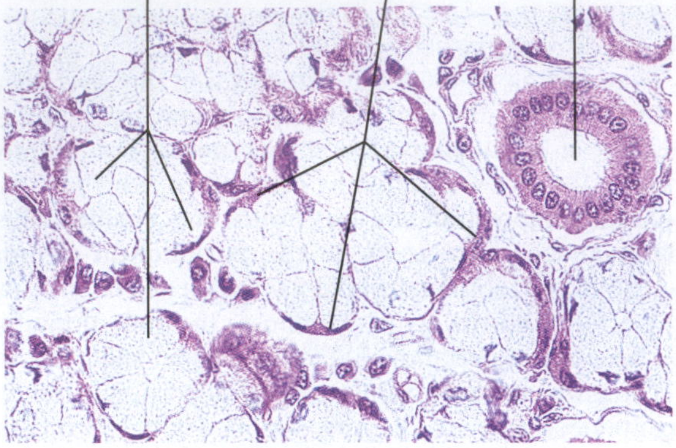
Mucous cells
Serous demilunes
Duct

FIGURE 38.13 Histology of a mixed salivary gland. Sublingual gland (180×).

ACTIVITY 7

Examining Salivary Gland Tissue

Examine salivary gland tissue under low power and then high power to become familiar with the appearance of a glandular tissue. Notice the clustered arrangement of the cells around their ducts. The cells are basically triangular, with their pointed ends facing the duct orifice. If possible, differentiate between mucus-producing cells, which look hollow or have a clear cytoplasm, and serous cells, which produce the clear, enzyme-containing fluid and have granules in their cytoplasm. The serous cells often form *demilunes* (caps) around the more central mucous cells. Figure 38.13 may be helpful in this task.

Liver and Gallbladder

The **liver** (see Figure 38.2), the largest gland in the body, is located inferior to the diaphragm, more to the right than the left side of the body. As noted earlier, it hides the stomach from view in a superficial observation of abdominal contents. The human liver has four lobes and is suspended from the diaphragm and anterior abdominal wall by the **falciform ligament** (see Figure 38.14).

The liver is one of the body's most important organs, and it performs many metabolic roles. However, its digestive function is to produce bile, which leaves the liver through the **common hepatic duct** and then enters the duodenum through the **bile duct** (Figure 38.15). Bile has no enzymatic action but emulsifies fats (breaks up large fat particles into smaller ones), thus creating a larger surface area for more efficient lipase activity. Without bile, very little fat digestion or absorption occurs.

When digestive activity is not occurring in the digestive tract, bile backs up into the **cystic duct** and enters the **gallbladder,** a small, green sac on the inferior surface of the liver. It is stored there until needed for the digestive process. While in the gallbladder, bile is concentrated by the removal of water and some ions. When fat-rich food enters the duodenum, a hormonal stimulus causes the gallbladder to contract, releasing the stored bile and making it available to the duodenum.

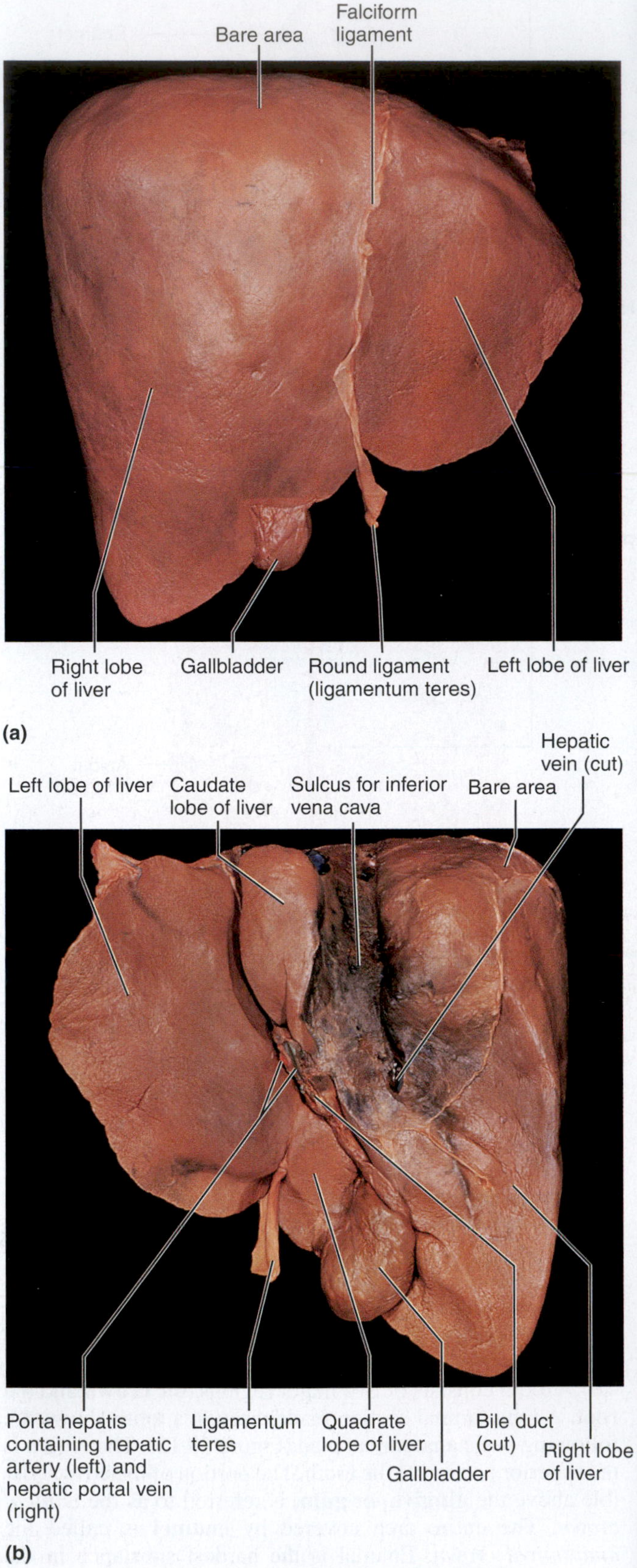
Bare area
Falciform ligament
Right lobe of liver
Gallbladder
Round ligament (ligamentum teres)
Left lobe of liver
(a)
Left lobe of liver
Caudate lobe of liver
Sulcus for inferior vena cava
Bare area
Hepatic vein (cut)
Porta hepatis containing hepatic artery (left) and hepatic portal vein (right)
Ligamentum teres
Quadrate lobe of liver
Gallbladder
Bile duct (cut)
Right lobe of liver
(b)

FIGURE 38.14 Gross anatomy of the human liver. **(a)** Anterior view. **(b)** Posteroinferior aspect. The four liver lobes are separated by a group of fissures in this view.

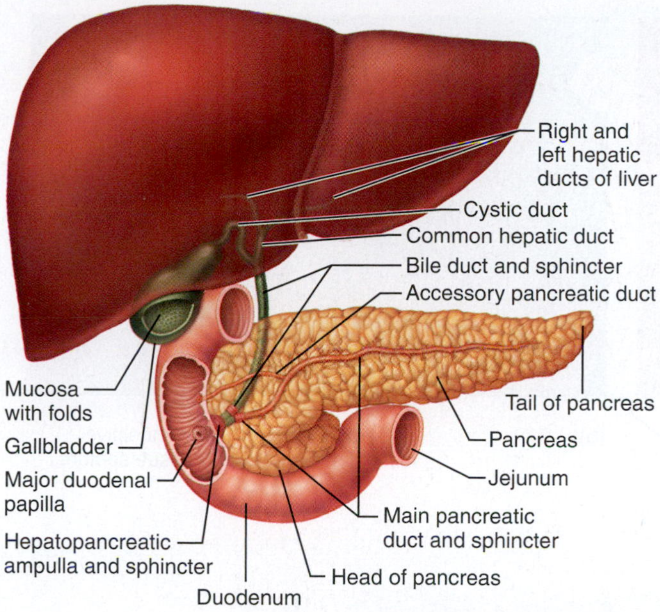
Right and
left hepatic
ducts of liver
Cystic duct
Common hepatic duct
Bile duct and sphincter
Accessory pancreatic duct
Mucosa
with folds
Tail of pancreas
Gallbladder
Pancreas
Major duodenal
papilla
Jejunum
Main pancreatic
duct and sphincter
Hepatopancreatic
ampulla and sphincter
Head of pancreas
Duodenum

FIGURE 38.15 Ducts of accessory digestive organs.

If the common hepatic or bile duct is blocked (for example, by wedged gallstones), bile is prevented from entering the small intestine, accumulates, and eventually backs up into the liver. This exerts pressure on the liver cells, and bile begins to enter the bloodstream. As the bile circulates through the body, the tissues become yellow, or jaundiced.

Blockage of the ducts is just one cause of jaundice. More often it results from actual liver problems such as **hepatitis** (an inflammation of the liver) or **cirrhosis,** a condition in which the liver is severely damaged and becomes hard and fibrous. Cirrhosis is almost guaranteed in those who drink excessive alcohol for many years. ■

As demonstrated by its highly organized anatomy, the liver (Figure 38.16) is very important in the initial processing of the nutrient-rich blood draining the digestive organs. Its structural and functional units are called **lobules.** Each lobule is a basically cylindrical structure consisting of cordlike arrays of **hepatocytes** or *liver cells,* which radiate outward from a central vein running upward in the longitudinal axis of the lobule. At each of the six corners of the lobule is a **portal triad** (portal tract), so named because three basic structures are always present there: a *portal arteriole* (a branch of the *hepatic artery,* the functional blood supply of the liver), a *portal venule* (a branch of the *hepatic portal vein* carrying nutrient-rich blood from the digestive viscera), and a *bile duct.* Between the liver cells are blood-filled spaces, or **sinusoids,** through which blood from the hepatic portal vein and hepatic artery percolates. Star-shaped **hepatic macrophages,** special phagocytic cells, also called **Kupffer cells,** line the sinusoids and remove debris such as bacteria from the blood as it flows past, while the hepatocytes pick up oxygen and nutrients. Much of the glucose transported to the liver from the digestive system is stored as glycogen in the liver for later use, and amino acids are taken from the blood by the liver cells and utilized to make plasma proteins. The sinusoids empty into the central vein, and the blood ultimately drains from the liver via the *hepatic veins.*

Bile is continuously being made by the hepatocytes. It flows through tiny canals, the **bile canaliculi,** which run between adjacent cells toward the bile duct branches in the triad regions, where the bile eventually leaves the liver. Notice that the directions of blood and bile flow in the liver lobule are exactly opposite.

ACTIVITY 8

Examining the Histology of the Liver

Examine a slide of liver tissue and identify as many as possible of the structural features illustrated in Figure 38.16. Also examine a three-dimensional model of liver lobules if this is available. Reproduce a small pie-shaped section of a liver lobule in the space below. Label the hepatocytes, the Kupffer cells, sinusoids, a portal triad, and a central vein. ■

Pancreas

The **pancreas** is a soft, triangular gland that extends horizontally across the posterior abdominal wall from the spleen to the duodenum (see Figure 38.2). Like the duodenum, it is a retroperitoneal organ (see Figure 38.7). As noted in Exercise 27, the pancreas has both an endocrine function (it produces the hormones insulin and glucagon) and an exocrine (enzyme-producing) function. It produces a whole spectrum of hydrolytic enzymes, which it secretes in an alkaline fluid into the duodenum through the pancreatic duct. Pancreatic juice is very alkaline. Its high concentration of bicarbonate ion (HCO_3^-) neutralizes the acidic chyme entering the duodenum from the stomach, enabling the pancreatic and intestinal enzymes to operate at their optimal pH. (Optimal pH for digestive activity to occur in the stomach is very acidic and results from the presence of HCl; that for the small intestine is slightly alkaline.) See Figure 27.3c.

For instructions on animal dissections, see the dissection exercises starting on page 697 in the cat, rat, and fetal pig editions of this manual.

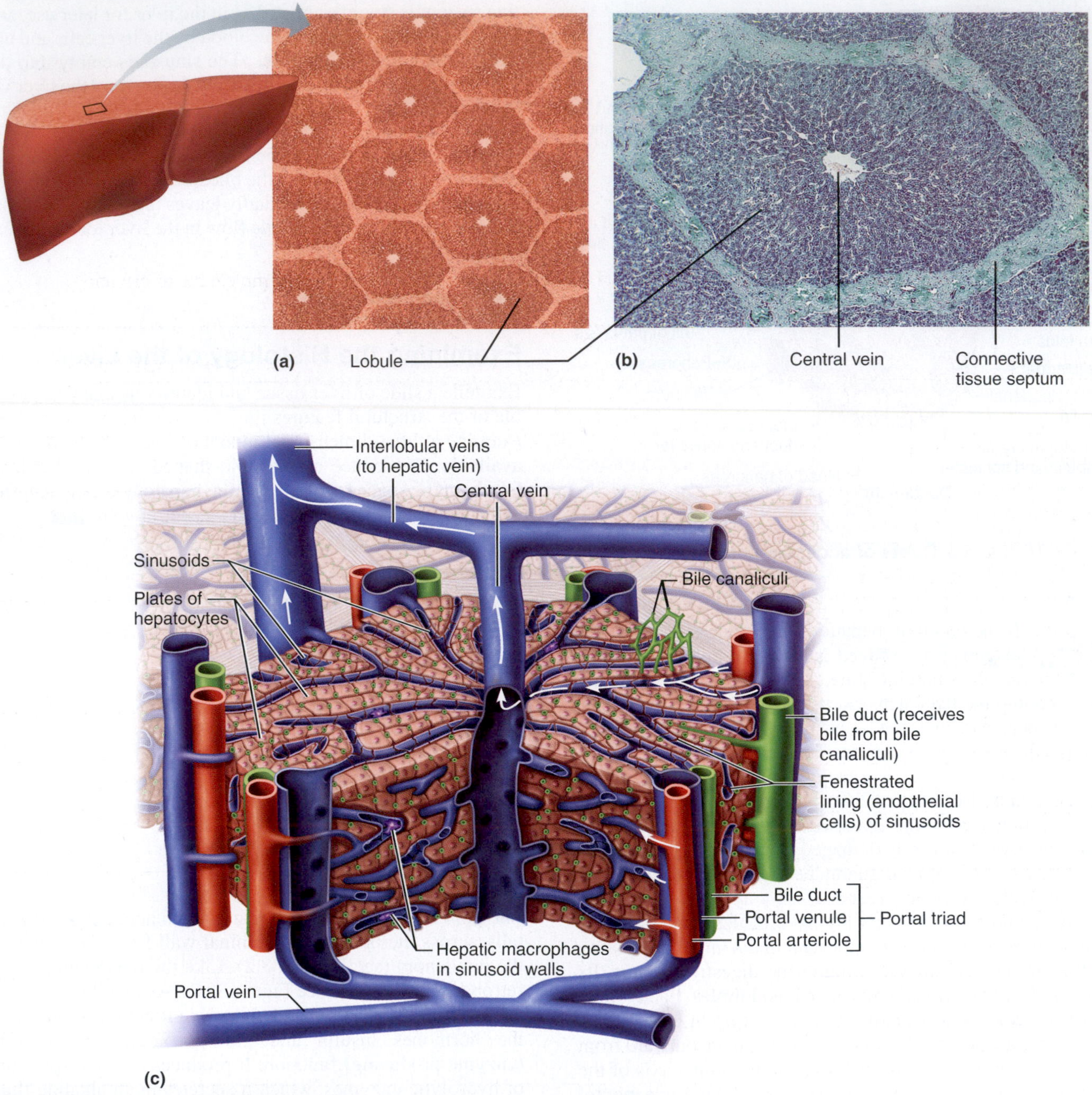
(a)
Lobule
(b)
Central vein
Connective tissue septum
Interlobular veins (to hepatic vein)
Central vein
Sinusoids
Plates of hepatocytes
Bile canaliculi
Bile duct (receives bile from bile canaliculi)
Fenestrated lining (endothelial cells) of sinusoids
Bile duct
Portal venule
Portal arteriole
Portal triad
Hepatic macrophages in sinusoid walls
Portal vein
(c)

FIGURE 38.16 Microscopic anatomy of the liver, diagrammatic view. **(a)** Schematic view of the cut surface of the liver showing the hexagonal nature of its lobules. **(b)** Photomicrograph of one liver lobule (40X). **(c)** Enlarged three-dimensional view of one liver lobule. Arrows show direction of blood flow. Bile flows in the opposite direction toward the bile ducts.

NAME ______________________

LAB TIME/DATE ______________

REVIEW SHEET

EXERCISE 38

Anatomy of the Digestive System

General Histological Plan of the Alimentary Canal

1. The general anatomical features of the alimentary canal are listed below. Fill in the table to complete the information.

Wall layer	Subdivisions of the layer (if applicable)	Major functions
mucosa		
submucosa		
muscularis externa		
serosa or adventitia		

Organs of the Alimentary Canal

2. The tubelike digestive system canal that extends from the mouth to the anus is known as the ______________ canal or the ______________ tract.

3. How is the muscularis externa of the stomach modified? ______________

____

How does this modification relate to the function of the stomach? ______________

____

4. What transition in epithelial type exists at the gastroesophageal junction? ______________

____

How do the epithelia of these two organs relate to their specific functions? ______________

____

5. Differentiate between the colon and the large intestine. ______________

____

6. Match the items in column B with the descriptive statements in column A.

Column A

_______ 1. structure that suspends the small intestine from the posterior body wall

_______ 2. fingerlike extensions of the intestinal mucosa that increase the surface area for absorption

_______ 3. large collections of lymphoid tissue found in the submucosa of the small intestine

_______ 4. deep folds of the mucosa and submucosa that extend completely or partially around the circumference of the small intestine

_______, _______ 5. regions that break down foodstuffs mechanically

_______ 6. mobile organ that manipulates food in the mouth and initiates swallowing

_______ 7. conduit for both air and food

_______, _______, _______ 8. three structures continuous with and representing modifications of the peritoneum

_______ 9. the "gullet"; no digestive/absorptive function

_______ 10. folds of the gastric mucosa

_______ 11. sacculations of the large intestine

_______ 12. projections of the plasma membrane of a mucosal epithelial cell

_______ 13. valve at the junction of the small and large intestines

_______ 14. primary region of food and water absorption

_______ 15. membrane securing the tongue to the floor of the mouth

_______ 16. absorbs water and forms feces

_______ 17. area between the teeth and lips/cheeks

_______ 18. wormlike sac that outpockets from the cecum

_______ 19. initiates protein digestion

_______ 20. structure attached to the lesser curvature of the stomach

_______ 21. organ distal to the stomach

_______ 22. valve controlling food movement from the stomach into the duodenum

_______ 23. posterosuperior boundary of the oral cavity

_______ 24. location of the hepatopancreatic sphincter through which pancreatic secretions and bile pass

_______ 25. serous lining of the abdominal cavity wall

_______ 26. principal site for the synthesis of vitamin K by microorganisms

_______ 27. region containing two sphincters through which feces are expelled from the body

_______ 28. bone-supported anterosuperior boundary of the oral cavity

Column B

a. anus

b. appendix

c. circular folds

d. esophagus

e. frenulum

f. greater omentum

g. hard palate

h. haustra

i. ileocecal valve

j. large intestine

k. lesser omentum

l. mesentery

m. microvilli

n. oral cavity

o. parietal peritoneum

p. Peyer's patches

q. pharynx

r. pyloric valve

s. rugae

t. small intestine

u. soft palate

v. stomach

w. tongue

x. vestibule

y. villi

z. visceral peritoneum

7. Correctly identify all organs depicted in the diagram below.

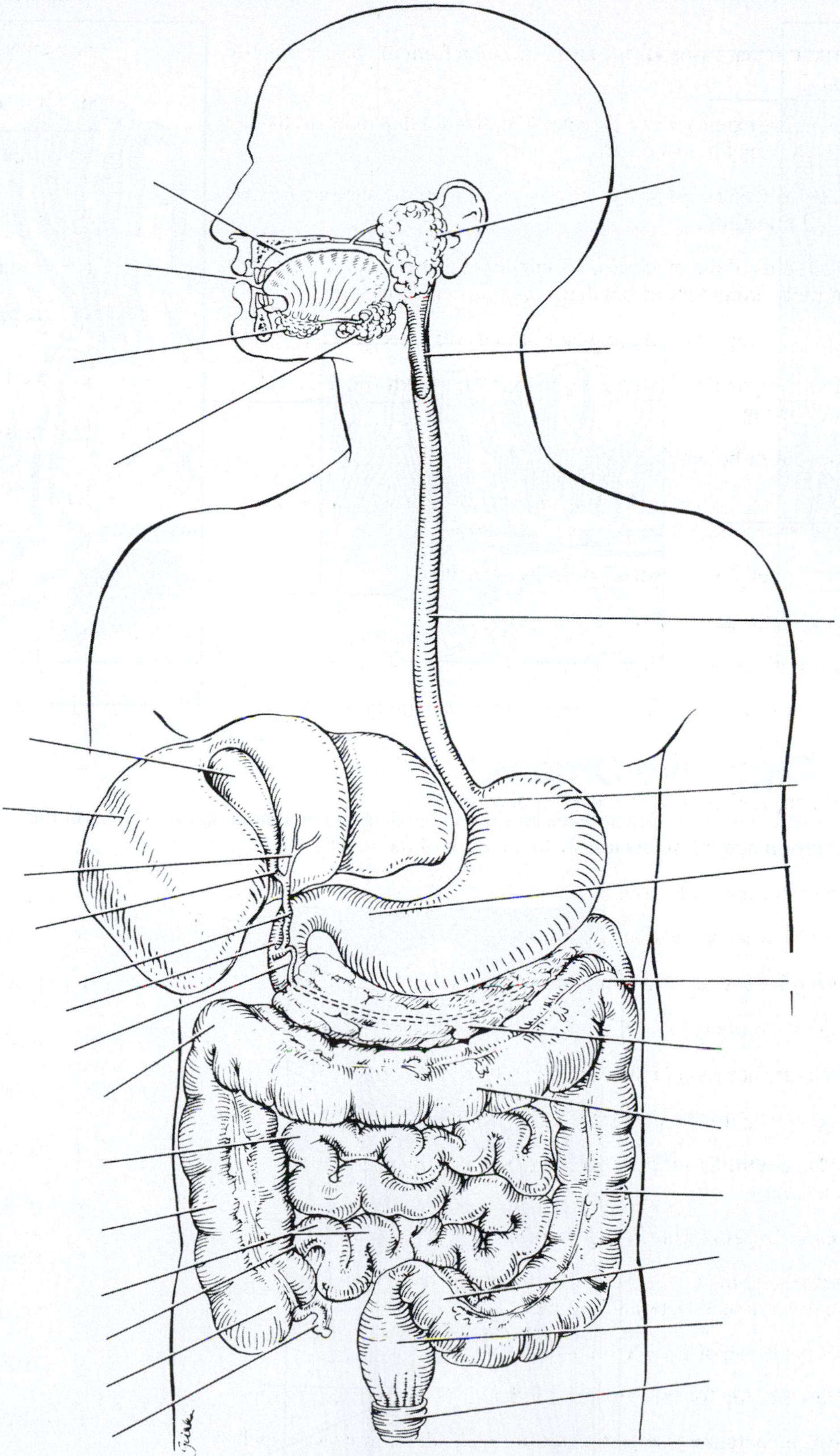

8. You have studied the histological structure of a number of organs in this laboratory. Three of these are diagrammed below. Identify and correctly label each.

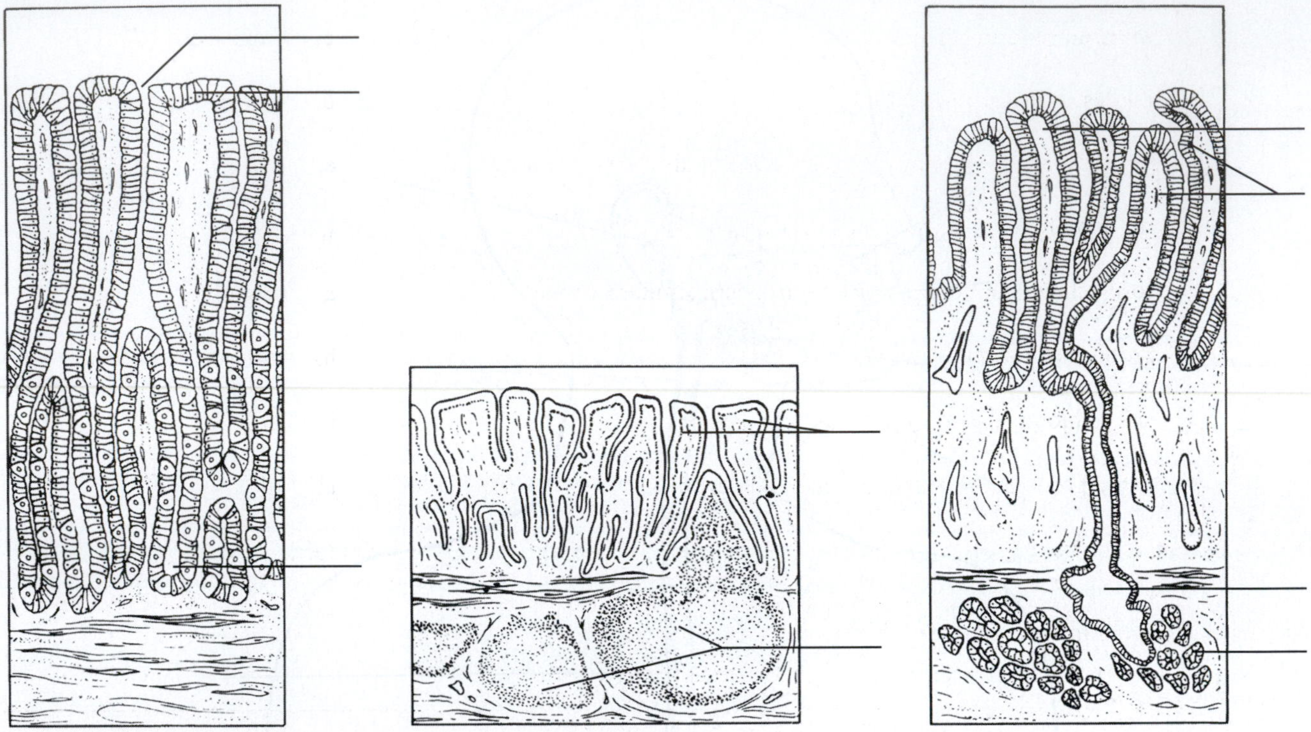

(a) ______________ (b) ______________ (c) ______________

Accessory Digestive Organs

9. Correctly label all structures provided with leader lines in the diagram of a molar below. (Note: Some of the terms in the key for question 10 may be helpful in this task.)

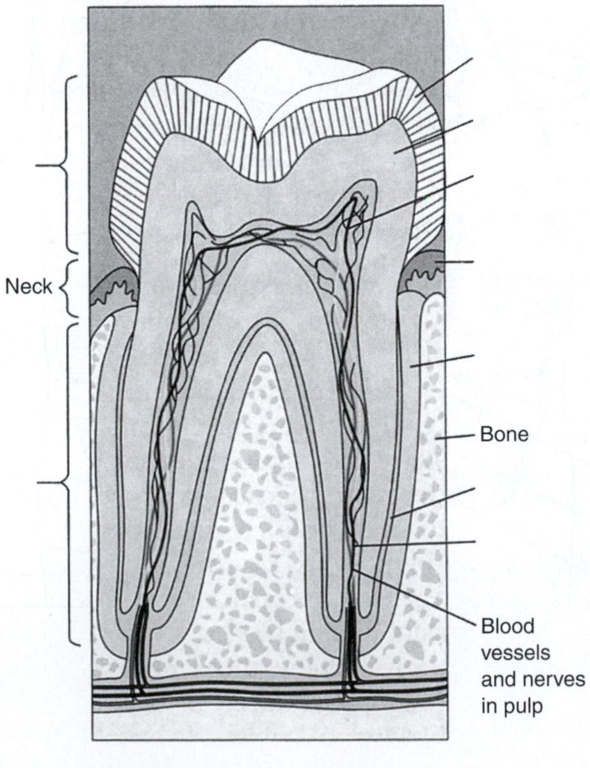
Neck
Bone
Blood vessels and nerves in pulp

10. Use the key to identify each tooth area described below.

			Key:
_______	1. visible portion of the tooth in situ	a.	anatomical crown
_______	2. material covering the tooth root	b.	cementum
_______	3. hardest substance in the body	c.	clinical crown
_______	4. attaches the tooth to bone and surrounding alveolar structures	d.	dentin
_______	5. portion of the tooth embedded in bone	e.	enamel
_______	6. forms the major portion of tooth structure; similar to bone	f.	gingiva
_______	7. produces the dentin	g.	odontoblast
_______	8. site of blood vessels, nerves, and lymphatics	h.	periodontal ligament
_______	9. entire portion of the tooth covered with enamel	i.	pulp
		j.	root

11. In the human, the number of deciduous teeth is ______________; the number of permanent teeth is ______________.

12. The dental formula for permanent teeth is $\frac{2,1,2,3}{2,1,2,3} \times 2$

Explain what this means. ____

____

What is the dental formula for the deciduous teeth? ______________ × ______________ = ______________

13. Which teeth are the "wisdom teeth"? ____

14. Various types of glands form a part of the alimentary tube wall or duct their secretions into it. Match the glands listed in column B with the function/locations described in column A.

	Column A		Column B
_______	1. produce(s) mucus; found in the submucosa of the small intestine	a.	duodenal glands
_______	2. produce(s) a product containing amylase that begins starch breakdown in the mouth	b.	gastric glands
_______	3. produce(s) a whole spectrum of enzymes and an alkaline fluid that is secreted into the duodenum	c.	intestinal crypts
_______	4. produce(s) bile that it secretes into the duodenum via the bile duct	d.	liver
_______	5. produce(s) HCl and pepsinogen	e.	pancreas
_______	6. found in the mucosa of the small intestine; produce(s) intestinal juice	f.	salivary glands

15. Which of the salivary glands produces a secretion that is mainly serous? ____

16. What is the role of the gallbladder? ______________________________

17. Name three structures always found in the portal triad regions of the liver. ______________________________,

______________________________, and ______________________________

18. Where would you expect to find the Kupffer cells of the liver? ______________________________

What is their function? ______________________________

19. Why is the liver so dark red in the living animal? ______________________________

____

20. The pancreas has two major populations of secretory cells—those in the islets and the acinar cells. Which population serves

the digestive process? ______________________________

EXERCISE 39

Chemical and Physical Processes of Digestion

MATERIALS

Part I: Enzyme Action

General Supply Area

- ☐ Hot plates
- ☐ 250-ml beakers
- ☐ Boiling chips
- ☐ Test tubes and test tube rack
- ☐ Wax markers
- ☐ Water bath set at 37°C (if not available, incubate at room temperature and double the time)
- ☐ Ice water bath
- ☐ Chart on chalkboard for recording class results

Activity 1: Starch Digestion

- ☐ Dropper bottle of distilled water
- ☐ Dropper bottles of the following:
  - 1% alpha-amylase solution*
  - 1% boiled starch solution, freshly prepared†
  - 1% maltose solution
  - Lugol's iodine solution (IKI)
  - Benedict's solution
- ☐ Spot plate

Activity 2: Protein Digestion

- ☐ Dropper bottles of 1% trypsin and 0.01% BAPNA solution

*The alpha-amylase must be a low-maltose preparation for good results.

†Prepare by adding 1 g starch to 100 ml distilled water; boil and cool; add a pinch of salt (NaCl). Prepare fresh daily.

Text continues on next page.

OBJECTIVES

1. To list the digestive system enzymes involved in the digestion of proteins, fats, and carbohydrates; to state their site of origin; and to summarize the environmental conditions promoting their optimal functioning.
2. To recognize the variation between different types of enzyme assays.
3. To name the end products of protein, fat, and carbohydrate digestion.
4. To perform the appropriate chemical tests to determine if digestion of a particular foodstuff has occurred.
5. To cite the function(s) of bile in the digestive process.
6. To discuss the possible role of temperature and pH in the regulation of enzyme activity.
7. To define *enzyme, catalyst, control, substrate,* and *hydrolase.*
8. To explain why swallowing is both a voluntary and a reflex activity.
9. To discuss the role of the tongue, larynx, and gastroesophageal sphincter in swallowing.
10. To compare and contrast segmentation and peristalsis as mechanisms of propulsion.

PRE-LAB QUIZ

1. Circle the correct term. Enzymes are <u>catalysts / substrates</u> that increase the rate of chemical reactions without becoming a part of the product.
2. A(n) __________ is a specimen or standard against which all experimental samples are compared.
   a. assay   b. control   c. substrate   d. trial
3. One enzyme that you will be studying today, produced by the salivary glands and secreted into the mouth, hydrolyzes starch to maltose. It is ____________________________.
4. Circle True or False. When you use iodine to test for starch, a color change to blue-black indicates a positive starch test.
5. If Benedict's test in the starch assay produces a __________ precipitate, then your test will be recorded as positive for maltose.
   a. blue to black   b. green to orangec. white
6. The enzyme __________, produced by the pancreas, is responsible for breaking down proteins.
   a. amylase   b. kinase   c. lipase   d. trypsin
7. Circle the correct term. The enzyme <u>pancreatic lipase / pepsin</u> hydrolyzes neutral fats to their component monoglycerides and fatty acids.
8. Circle True or False. Both smooth and skeletal muscles are involved in the physical processes of digestion.
9. __________ movements are local contractions that mix foodstuffs with digestive juices and increase the rate of absorption.
   a. Deglutition   b. Elimination   c. Peristaltic   d. Segmental

MasteringA&P™

Access practice quizzes and more in the Study Area at www.masteringaandp.com.

PEx

This lab corresponds to PhysioEx Exercise 8. See p. PEx-119.

Activity 3: Bile Action and Fat Digestion

- ☐ Dropper bottles of 1% pancreatin solution, litmus cream (fresh cream to which powdered litmus is added to achieve a deep blue color), 0.1 *N* HCl, and vegetable oil
- ☐ Bile salts (sodium taurocholate)
- ☐ Parafilm (small squares to cover the test tubes)

Part II: Physical Processes

Activity 5: Observing Digestion

- ☐ Water pitcher
- ☐ Paper cups
- ☐ Stethoscope
- ☐ Alcohol swab
- ☐ Disposable autoclave bag
- ☐ Watch, clock, or timer

Activity 6: Videotape

- ☐ Television and VCR for independent viewing of videocassette by student
- ☐ *Interactive Physiology*®, Digestive System

Because nutrients can be absorbed only when broken down to their monomers, food digestion is a prerequisite to food absorption. You have already studied mechanisms of passive and active absorption in Exercise 5 and/or 5B. Before proceeding, review that material.

Chemical Digestion of Foodstuffs: Enzymatic Action

Enzymes are large protein molecules produced by body cells. They are biological **catalysts,** meaning that they increase the rate of a chemical reaction without themselves becoming part of the product. The digestive enzymes are hydrolytic enzymes, or **hydrolases.** Their **substrates,** or the molecules on which they act, are organic food molecules which they break down by adding water to the molecular bonds, thus cleaving the bonds between the subunits or monomers.

The various hydrolytic enzymes are highly specific in their action. Each enzyme hydrolyzes only one or a small group of substrate molecules, and specific environmental conditions are necessary for it to function optimally. Since digestive enzymes actually function outside the body cells in the digestive tract, their hydrolytic activity can also be studied in a test tube. Such an in vitro study provides a convenient laboratory environment for investigating the effect of such variations on enzymatic activity.

Figure 39.1 is a flowchart of the progressive digestion of carbohydrates, proteins, fats, and nucleic acids. It summarizes the specific enzymes involved, their site of formation, and their site of action. Acquaint yourself with the flowchart before beginning this experiment, and refer to it as necessary during the laboratory session.

General Instructions for Activities 1–3

Work in groups of four, with each group taking responsibility for setting up and conducting one of the following experiments. In each of the digestive procedures being studied (starch, protein, and fat digestion) and in the amylase assay, you are directed to boil the contents of one or more test tubes. To do this, obtain a 250-ml beaker, boiling chips, and a hot plate from the general supply area. Place a few boiling chips into the beaker, add about 125 ml of water, and bring to a boil. Place the test tube for each specimen in the water for the number of minutes specified in the directions. You will also be using a 37°C and an ice water bath for parts of these experiments. You will need to use your time very efficiently in order to set up and perform each test properly.

Upon completion of the experiments, each group should communicate its results to the rest of the class by recording them in a chart on the chalkboard. All members of the class should observe the **controls** (the specimens or standards against which experimental samples are compared) as well as the positive and negative examples of all experimental results. Additionally, all members of the class should be able to explain the tests used and the results observed and anticipated for each experiment. Note that water baths and hot plates are at the general supply area.

ACTIVITY 1

Assessing Starch Digestion by Salivary Amylase

1. From the general supply area, obtain a test tube rack, 10 test tubes, and a wax marking pencil. From the Activity 1 supply area, obtain a dropper bottle of distilled water and dropper bottles of maltose, amylase, and starch solutions.

2. In this experiment you will investigate the hydrolysis of starch to maltose by **salivary amylase** (the enzyme produced by the salivary glands and secreted into the mouth), so you will need to be able to identify the presence of starch and maltose to determine to what extent the enzymatic activity has occurred. Thus controls must be prepared to provide a known standard against which comparisons can be made. Starch decreases and sugar increases as digestion occurs, according to the following formula:

$$\text{Starch} + \text{water} \xrightarrow{\text{amylase}} \text{maltose}$$

Two students should prepare the controls (tubes 1A to 3A) while the other two prepare the experimental samples (tubes 4A to 6A).

- Mark each tube with a wax pencil and load the tubes as indicated in the Salivary Amylase chart on page 600, using 3 drops (gtt) of each indicated substance.
- Place all tubes in a rack in the 37°C water bath for approximately 1 hour. Shake the rack gently from time to time to keep the contents evenly mixed.
- At the end of the hour, perform the amylase assay described below.
- While these tubes are incubating, proceed to Physical Processes: Mechanisms of Food Propulsion and Mixing (page 603). Be sure to monitor the time so as to complete this activity as needed.

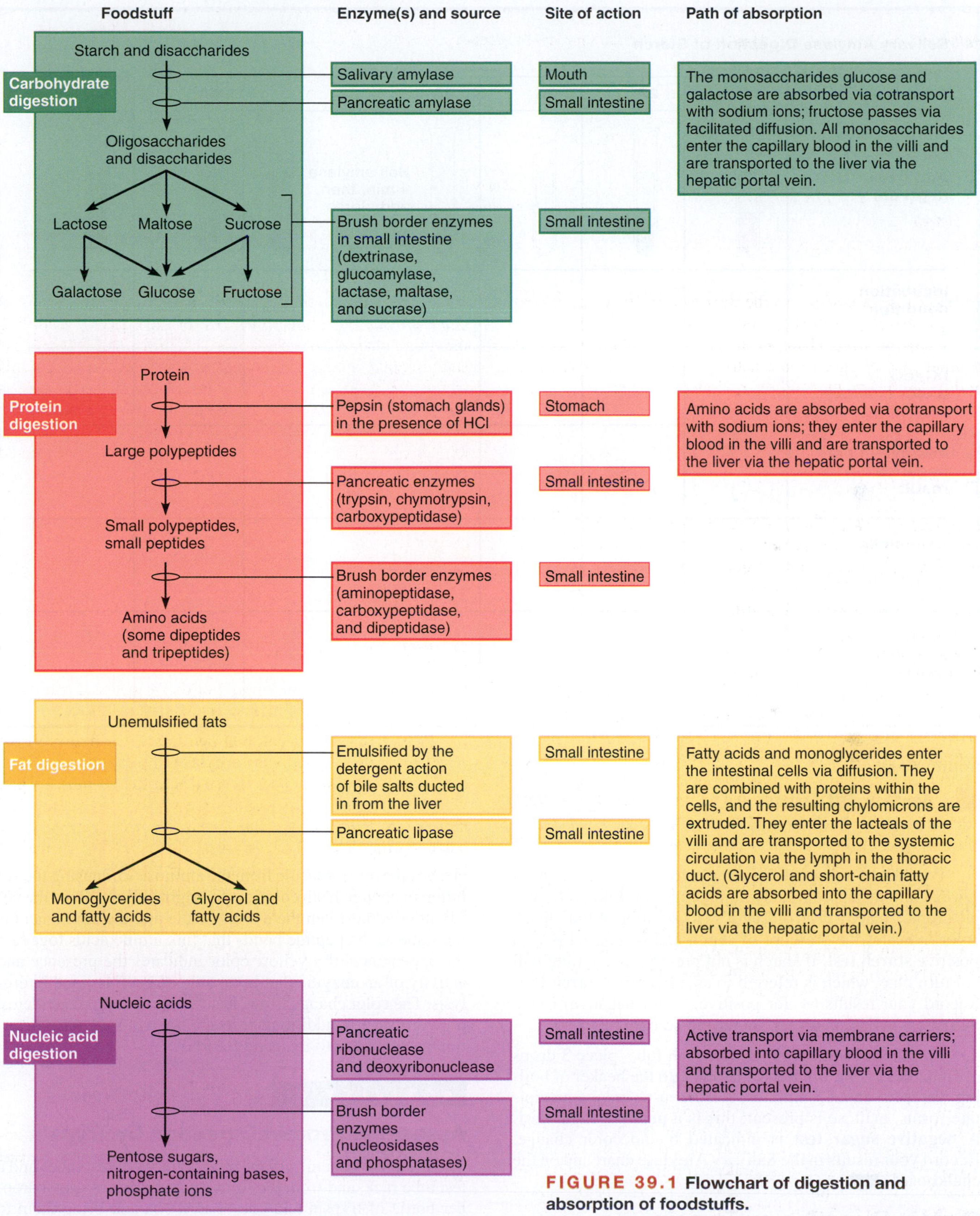
Foodstuff
Enzyme(s) and source
Site of action
Path of absorption
Carbohydrate digestion
Starch and disaccharides
Salivary amylase
Mouth
Pancreatic amylase
Small intestine
Oligosaccharides and disaccharides
Lactose
Maltose
Sucrose
Galactose
Glucose
Fructose
Brush border enzymes in small intestine (dextrinase, glucoamylase, lactase, maltase, and sucrase)
Small intestine
The monosaccharides glucose and galactose are absorbed via cotransport with sodium ions; fructose passes via facilitated diffusion. All monosaccharides enter the capillary blood in the villi and are transported to the liver via the hepatic portal vein.
Protein digestion
Protein
Pepsin (stomach glands) in the presence of HCl
Stomach
Large polypeptides
Pancreatic enzymes (trypsin, chymotrypsin, carboxypeptidase)
Small intestine
Small polypeptides, small peptides
Brush border enzymes (aminopeptidase, carboxypeptidase, and dipeptidase)
Small intestine
Amino acids (some dipeptides and tripeptides)
Amino acids are absorbed via cotransport with sodium ions; they enter the capillary blood in the villi and are transported to the liver via the hepatic portal vein.
Fat digestion
Unemulsified fats
Emulsified by the detergent action of bile salts ducted in from the liver
Small intestine
Pancreatic lipase
Small intestine
Monoglycerides and fatty acids
Glycerol and fatty acids
Fatty acids and monoglycerides enter the intestinal cells via diffusion. They are combined with proteins within the cells, and the resulting chylomicrons are extruded. They enter the lacteals of the villi and are transported to the systemic circulation via the lymph in the thoracic duct. (Glycerol and short-chain fatty acids are absorbed into the capillary blood in the villi and transported to the liver via the hepatic portal vein.)
Nucleic acid digestion
Nucleic acids
Pancreatic ribonuclease and deoxyribonuclease
Small intestine
Brush border enzymes (nucleosidases and phosphatases)
Small intestine
Pentose sugars, nitrogen-containing bases, phosphate ions
Active transport via membrane carriers; absorbed into capillary blood in the villi and transported to the liver via the hepatic portal vein.

FIGURE 39.1 Flowchart of digestion and absorption of foodstuffs.

Amylase Assay

1. After one hour, obtain a spot plate and dropper bottles of Lugol's solution (for the IKI, or iodine, test) and Benedict's solution from the Activity 1 supply area. Set up your boiling water bath using a hot plate, boiling chips, and a 250-ml beaker obtained from the general supply area.

2. While the water is heating, mark six depressions of the spot plate 1A–6A (*A* for amylase) for sample identification.

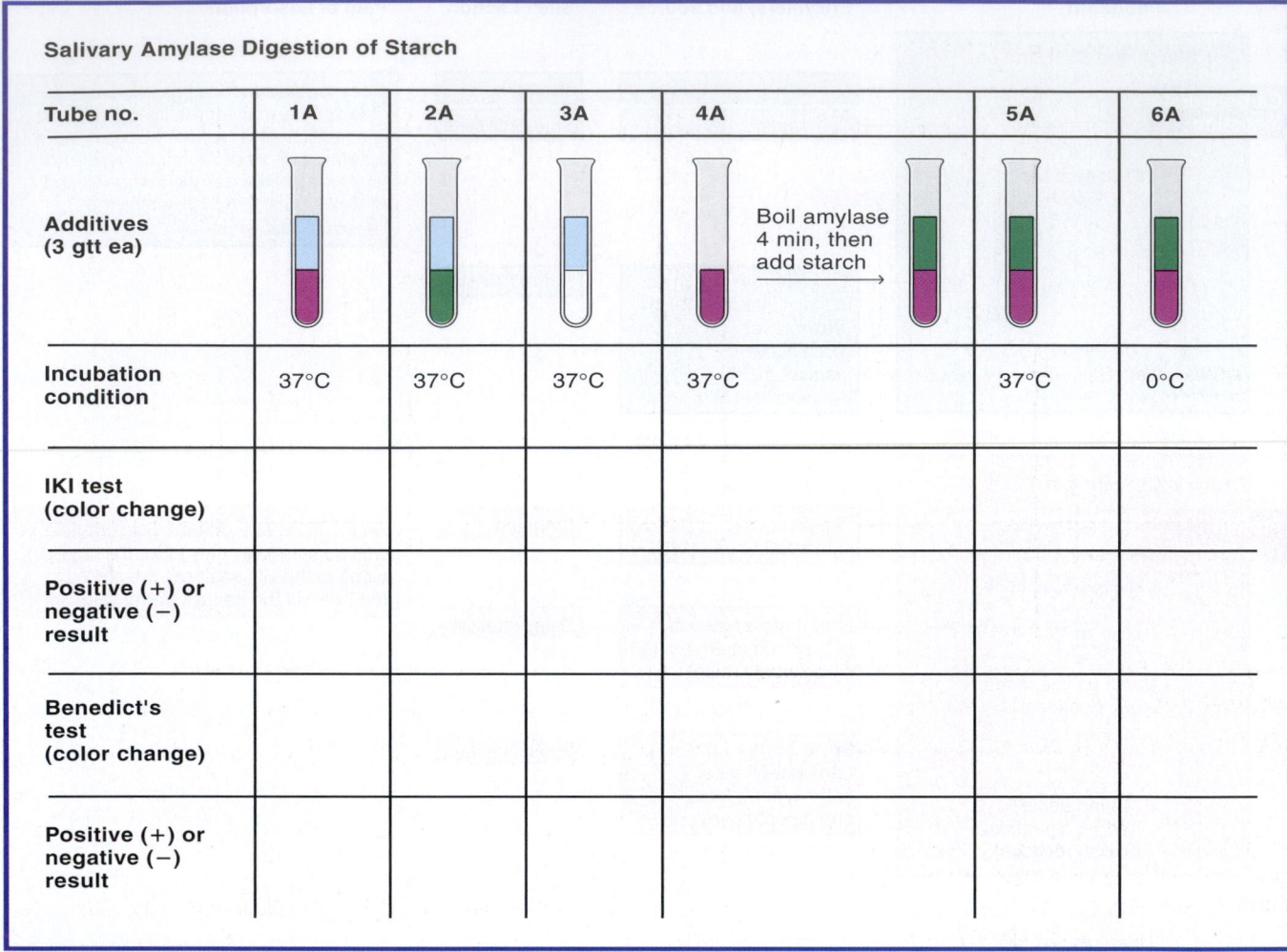

Salivary Amylase Digestion of Starch

Tube no.	1A	2A	3A	4A	5A	6A
Additives (3 gtt ea)		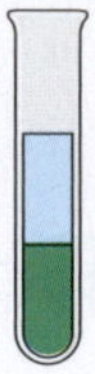		 Boil amylase 4 min, then add starch → 		
Incubation condition	37°C	37°C	37°C	37°C	37°C	0°C
IKI test (color change)						
Positive (+) or negative (−) result						
Benedict's test (color change)						
Positive (+) or negative (−) result						

Additive key:

■ = Amylase ■ = Starch □ = Maltose □ = Water

3. Pour about a drop of the sample from each of the tubes 1A–6A into the appropriately numbered spot. Into each sample droplet, place a drop of Lugol's IKI solution. A blue-black color indicates the presence of starch and is referred to as a **positive starch test.** If starch is not present, the mixture will not turn blue, which is referred to as a **negative starch test.** Record your results (+ for positive, − for negative) in the Salivary Amylase chart and on the chalkboard.

4. Into the remaining mixture in each tube, place 3 drops of Benedict's solution. Put each tube into the beaker of boiling water for about 5 minutes. If a green-to-orange precipitate forms, maltose is present; this is a **positive sugar test.** A **negative sugar test** is indicated by no color change. Record your results in the Salivary Amylase chart and on the chalkboard.

Protein Digestion by Trypsin

Trypsin, an enzyme produced by the pancreas, hydrolyzes proteins to small fragments (proteoses, peptones, and peptides). BAPNA (*N*-alpha-benzoyl-L-arginine-*p*-nitroanilide) is a synthetic trypsin substrate consisting of a dye covalently bound to an amino acid. Trypsin hydrolysis of BAPNA cleaves the dye molecule from the amino acid, causing the solution to change from colorless to bright yellow. Since the covalent bond between the dye molecule and the amino acid is the same as the peptide bonds that link amino acids together, the appearance of a yellow color indicates the presence and activity of an enzyme that is capable of peptide bond hydrolysis. The color change from clear to yellow is direct evidence of hydrolysis, so additional tests are not required when determining trypsin activity using BAPNA.

ACTIVITY 2

Assessing Protein Digestion by Trypsin

1. From the general supply area, obtain five test tubes and a test tube rack, and from the Activity 2 supply area get a dropper bottle of trypsin and one of BAPNA and bring them to your bench.

2. Two students should prepare the controls (tubes 1T and 2T) while the other two prepare the experimental samples (tubes 3T to 5T).

Trypsin Digestion of Protein

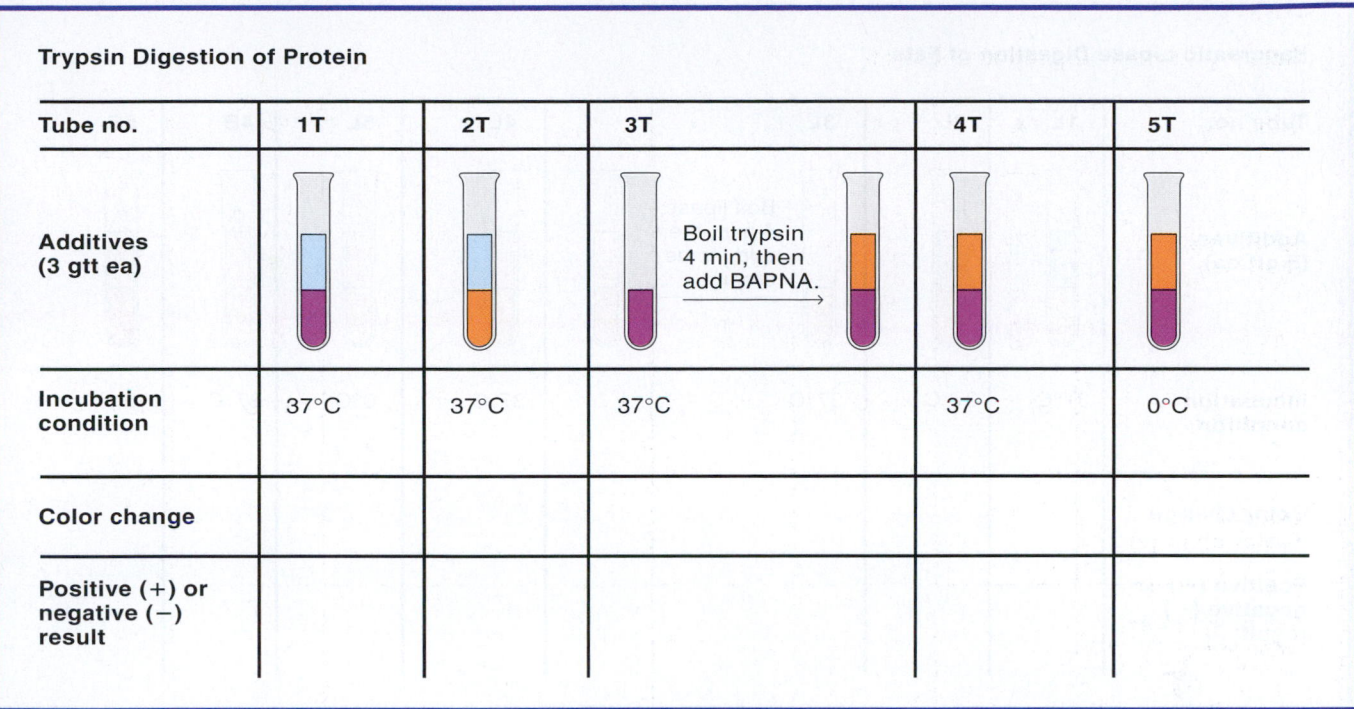

Tube no.	**1T**	**2T**	**3T**	**4T**	**5T**
Additives (3 gtt ea)	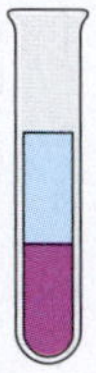	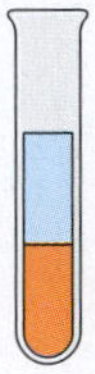	 Boil trypsin 4 min, then add BAPNA. → 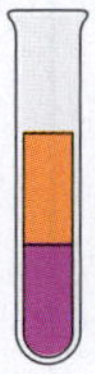	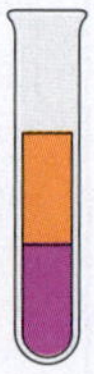	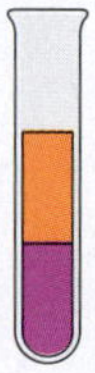
Incubation condition	37°C	37°C	37°C	37°C	0°C
Color change					
Positive (+) or negative (−) result					

Additive key:

= Trypsin = BAPNA = Water

- Mark each tube with a wax pencil and load the tubes as indicated in the Trypsin chart, using 3 drops (gtt) of each indicated substance.
- Place all tubes in a rack in the appropriate water bath for approximately 1 hour. Shake the rack occasionally to keep the contents well mixed.
- At the end of the hour, examine the tubes for the results of the trypsin assay (detailed below).
- While these tubes are incubating, proceed to Physical Processes: Mechanisms of Food Propulsion and Mixing (page 603).

Trypsin Assay

Since BAPNA is a synthetic colorigenic (color-producing) substrate, the presence of yellow color indicates a **positive hydrolysis test;** the dye molecule has been cleaved from the amino acid. If the sample mixture remains clear, a **negative hydrolysis test** has occurred.

Record the results in the Trypsin chart and on the chalkboard.

Pancreatic Lipase Digestion of Fats and the Action of Bile

The treatment that fats and oils go through during digestion in the small intestine is a bit more complicated than that of carbohydrates or proteins—pretreatment with bile to physically emulsify the fats is required. Hence, two sets of reactions occur.

First:

$$\text{Fats/oils} \xrightarrow[\text{(emulsification)}]{\text{bile}} \text{minute fat/oil droplets}$$

Then:

$$\text{Fat/oil droplets} \xrightarrow[\text{(digestion)}]{\text{lipase}} \text{monoglycerides and fatty acids}$$

The term **pancreatin** describes the enzymatic product of the pancreas, which includes enzymes that digest proteins, carbohydrates, nucleic acids, and fats. It is used here to investigate the properties of **pancreatic lipase,** which hydrolyzes fats and oils to their component monoglycerides and two fatty acids (and occasionally to glycerol and three fatty acids).

The fact that some of the end products of fat digestion (fatty acids) are organic acids that decrease the pH provides an easy way to recognize that digestion is ongoing or completed. You will be using a pH indicator called *litmus blue* to follow these changes; it changes from blue to pink as the test tube contents become acid.

ACTIVITY 3

Demonstrating the Emulsification Action of Bile and Assessing Fat Digestion by Lipase

1. From the general supply area, obtain nine test tubes and a test tube rack, plus one dropper bottle of each of the solutions in the Activity 3 supply area.

2. Although *bile,* a secretory product of the liver, is not an enzyme, it is important to fat digestion because of its emulsifying

Pancreatic Lipase Digestion of Fats

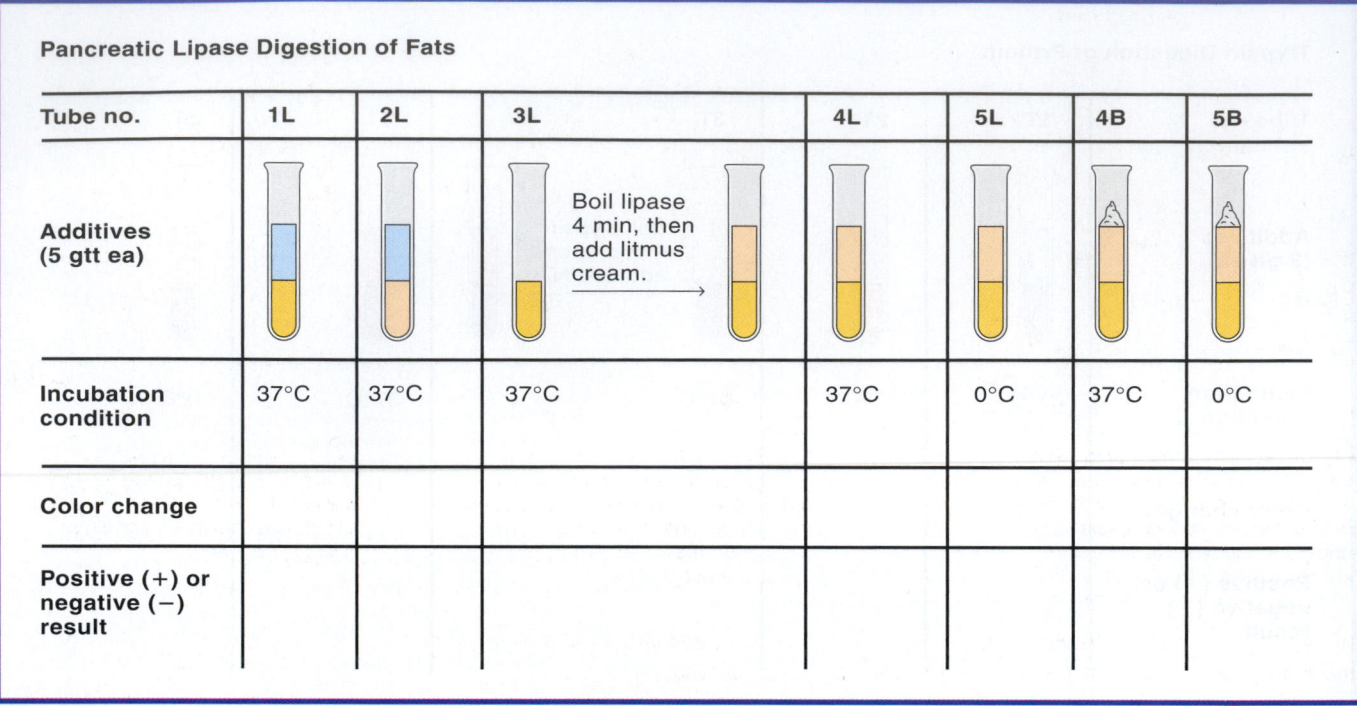

Tube no.	1L	2L	3L		4L	5L	4B	5B
Additives (5 gtt ea)	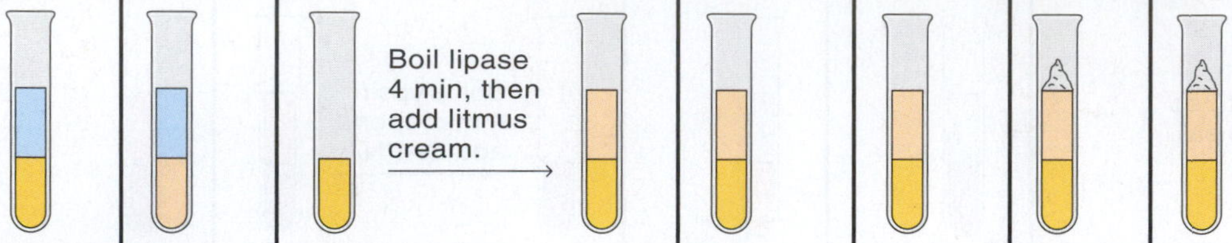			Boil lipase 4 min, then add litmus cream. →				
Incubation condition	37°C	37°C	37°C		37°C	0°C	37°C	0°C
Color change								
Positive (+) or negative (−) result								

Additive key:

= Lipase   = Litmus cream   = Water   = Pinch bile salts

action (the physical breakdown of larger particles into smaller ones) on fats. Emulsified fats provide a larger surface area for enzymatic activity. To demonstrate the action of bile on fats, prepare two test tubes and mark them 1E and 2E (*E* for emulsified fats).

- To tube 1E, add 10 drops of water and 2 drops of vegetable oil.
- To tube 2E, add 10 drops of water, 2 drops of vegetable oil, and a pinch of bile salts.
- Cover each tube with a small square of Parafilm, shake vigorously, and allow the tubes to stand at room temperature.

After 10 to 15 minutes, observe both tubes. If emulsification has not occurred, the oil will be floating on the surface of the water. If emulsification has occurred, the fat droplets will be suspended throughout the water, forming an emulsion.

In which tube has emulsification occurred? ______________

3. Two students should prepare the controls (1L and 2L, *L* for lipase), while the other two students in the group set up the experimental samples (3L to 5L, 4B, and 5B, where *B* is for bile) as illustrated in the Pancreatic Lipase chart.

- Mark each tube with a wax pencil and load the tubes using 5 drops (gtt) of each indicated solution.
- Place a pinch of bile salts in tubes 4B and 5B.
- Cover each tube with a small square of Parafilm, and shake to mix the contents of the tube.
- Remove the Parafilm, and place all tubes in a rack in the appropriate water bath for approximately 1 hour. Shake the test tube rack from time to time to keep the contents well mixed.
- At the end of the hour, perform the lipase assay (below).
- While these tubes are incubating proceed to Physical Processes: Mechanisms of Food Propulsion and Mixing (p. 603). Be sure to monitor the time so as to complete this activity when needed.

Lipase Assay

The basis of this assay is a pH change that is detected by a litmus powder indicator. Alkaline or neutral solutions containing litmus are blue but will turn reddish in the presence of acid. Since fats are digested to fatty acids (organic acids) during hydrolysis, they lower the pH of the sample they are in. Litmus cream (fresh cream providing the fat substrate to which litmus powder was added) will turn from blue to pink if the solution is acid. Because the effect of hydrolysis is directly seen, additional assay reagents are not necessary.

1. To prepare a color control, add 0.1 *N* HCl drop by drop to tubes 1L and 2L (covering the tubes with a square of Parafilm after each addition and shaking to mix) until the cream turns pink.

2. Record the color of the tubes in the Pancreatic Lipase chart and on the chalkboard. ■

ACTIVITY 4

Reporting Results and Conclusions

1. Share your results with the class as directed in the General Instructions on page 598.

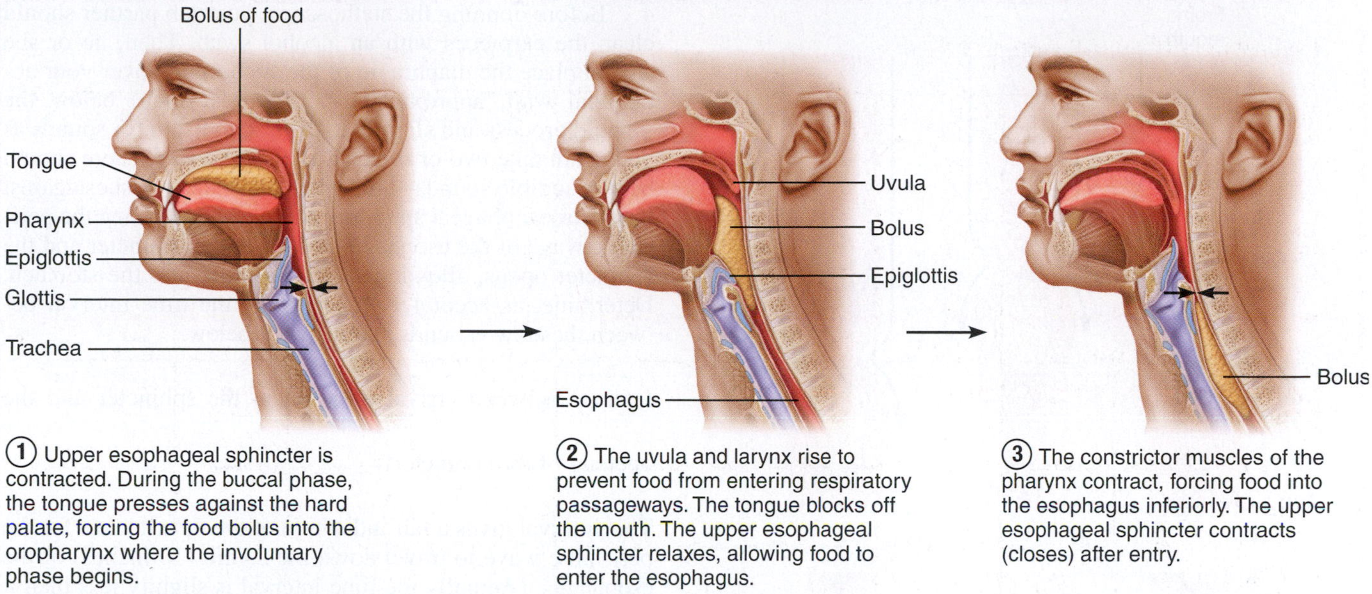
Bolus of food
Tongue
Pharynx
Epiglottis
Glottis
Trachea
Uvula
Bolus
Epiglottis
Esophagus
Bolus
① Upper esophageal sphincter is contracted. During the buccal phase, the tongue presses against the hard palate, forcing the food bolus into the oropharynx where the involuntary phase begins.
② The uvula and larynx rise to prevent food from entering respiratory passageways. The tongue blocks off the mouth. The upper esophageal sphincter relaxes, allowing food to enter the esophagus.
③ The constrictor muscles of the pharynx contract, forcing food into the esophagus inferiorly. The upper esophageal sphincter contracts (closes) after entry.

FIGURE 39.2 Swallowing. The process of swallowing consists of voluntary (buccal) (step ①) and involuntary (pharyngeal-esophageal) phases (steps ②–③).

2. Suggest additional experiments, and carry out experiments if time permits.

3. Prepare a lab report for the experiments on digestion. See Getting Started on page xv.

Physical Processes: Mechanisms of Food Propulsion and Mixing

Although enzyme activity is a very important part of the overall digestion process, foods must also be processed physically (by chewing and churning) and moved by mechanical means along the tract if digestion and absorption are to be completed. Just about any time organs exhibit mobility, muscles are involved, and movements of and in the gastrointestinal tract are no exception. Although we tend to think only of smooth muscles when visceral activities are involved, both skeletal and smooth muscles are involved in digestion. This fact is amply demonstrated by the simple activities that follow.

Deglutition (Swallowing)

Swallowing, or **deglutition,** which is largely the result of skeletal muscle activity, occurs in two phases: *buccal* (mouth) and *pharyngeal-esophageal.* The initial phase—the buccal (Figure 39.2 step ①)—is voluntarily controlled and initiated by the tongue. Once begun, the process continues involuntarily in the pharynx and esophagus, through peristalsis, resulting in the delivery of the swallowed contents to the stomach (Figure 39.2 steps ②–③).

ACTIVITY 5

Observing Movements and Sounds of Digestion

1. Obtain a pitcher of water, a stethoscope, a paper cup, an alcohol swab, and an autoclave bag in preparation for making the following observations.

2. While swallowing a mouthful of water, consciously note the movement of your tongue during the process. Record your observations.

____

____

____

3. Repeat the swallowing process while your laboratory partner watches the externally visible movements of your larynx. (This movement is more obvious in a male, since males have a larger Adam's apple.) Record your observations.

____

____

What do these movements accomplish? __________________

____

____

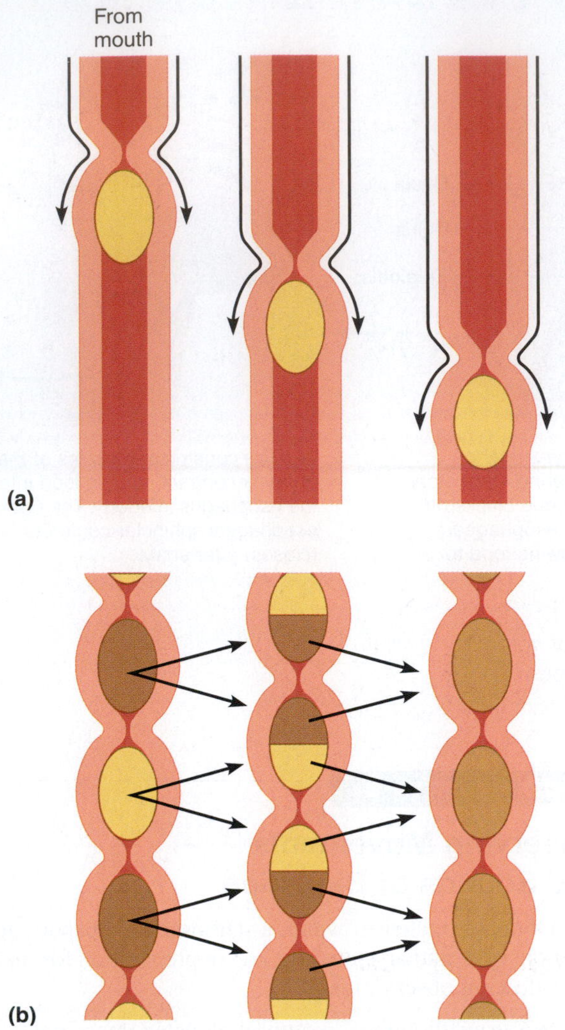

From mouth
(a)
(b)

FIGURE 39.3 Peristaltic and segmental movements of the digestive tract. (a) Peristalsis: neighboring segments of the intestine alternately contract and relax, moving food along the tract. **(b)** Segmentation: single segments of intestine alternately contract and relax. Because inactive segments exist between active segments, food mixing occurs to a greater degree than food movement. Peristalsis is superimposed on segmentation during digestion.

4. Before donning the stethoscope, your lab partner should clean the earpieces with an alcohol swab. Then, he or she should place the diaphragm of the stethoscope over your abdominal wall, approximately 2.5 cm (1 inch) below the xiphoid process and slightly to the left, to listen for sounds as you again take two or three swallows of water. There should be two audible sounds—one when the water splashes against the gastroesophageal sphincter and the second when the peristaltic wave of the esophagus arrives at the sphincter and the sphincter opens, allowing water to gurgle into the stomach. Determine, as accurately as possible, the time interval between these two sounds and record it below.

Interval between arrival of water at the sphincter and the

opening of the sphincter: ________ sec

This interval gives a fair indication of the time it takes for the peristaltic wave to travel down the 25 cm (10 inches) of the esophagus. (Actually the time interval is slightly less than it seems, because pressure causes the sphincter to relax before the peristaltic wave reaches it.)

! Dispose of the used paper cup in the autoclave bag. ■

Segmentation and Peristalsis

Although several types of movements occur in the digestive tract organs, peristalsis and segmentation are most important as mixing and propulsive mechanisms (Figure 39.3).

Peristaltic movements are the major means of propelling food through most of the digestive viscera. Essentially they are waves of contraction followed by waves of relaxation that squeeze foodstuffs through the alimentary canal, and they are superimposed on segmental movements.

Segmental movements are local constrictions of the organ wall that occur rhythmically. They serve mainly to mix the foodstuffs with digestive juices and to increase the rate of absorption by continually moving different portions of the chyme over adjacent regions of the intestinal wall. However, segmentation is also an important means of food propulsion in the small intestine, and slow segmenting movements called haustral contractions are frequently seen in the large intestine.

ACTIVITY 6

Viewing Segmental and Peristaltic Movements

If a videotape showing some of the propulsive movements is available, go to a viewing station to view it before leaving the laboratory. Alternatively, use the *Interactive Physiology*® module on the Digestive System to observe gut motility. ■

NAME ______________________________

LAB TIME/DATE ______________________

REVIEW SHEET

EXERCISE 39

Chemical and Physical Processes of Digestion

Chemical Digestion of Foodstuffs: Enzymatic Action

1. Match the following definitions with the proper choices from the key.

Key: a. catalyst b. control c. enzyme d. substrate

__________ 1. substance on which a catalyst works

__________ 2. biologic catalyst; protein in nature

__________ 3. increases the rate of a chemical reaction without becoming part of the product

__________ 4. provides a standard of comparison for test results

2. List the three characteristics of enzymes. ______________________________

____

____

3. The enzymes of the digestive system are classified as hydrolases. What does this mean?

____

____

4. Fill in the following chart about the various digestive system enzymes encountered in this exercise.

Enzyme	Organ producing it	Site of action	Substrate(s)	Optimal pH
Salivary amylase				
Trypsin				
Lipase (pancreatic)				

5. Name the end products of digestion for the following types of foods.

proteins: ______________________ carbohydrates: ______________________

fats: ______________________ and ______________________

6. You used several indicators or tests in the laboratory to determine the presence or absence of certain substances. Choose the correct test or indicator from the key to correspond to the condition described below.

Key: a. Lugol's iodine (IKI) b. Benedict's solution c. litmus d. BAPNA

_______ 1. used to test for protein hydrolysis, which was indicated by a yellow color

_______ 2. used to test for the presence of starch, which was indicated by a blue-black color

_______ 3. used to test for the presence of fatty acids, which was evidenced by a color change from blue to pink

_______ 4. used to test for the presence of reducing sugars (maltose, sucrose, glucose) as indicated by a blue to green or orange color change

7. What conclusions can you draw when an experimental sample gives both a positive starch test and a positive maltose test after incubation? ______________________________

Why was 37°C the optimal incubation temperature? ______________________________

Why did very little, if any, starch digestion occur in test tube 4A? ______________________________

____

When starch was incubated with amylase at 0°C, did you see any starch digestion? ______________________________

Why or why not? ______________________________

____

Assume you have made the statement to a group of your peers that amylase is capable of starch hydrolysis to maltose. If you had not done control tube 1A, what objection to your statement could be raised? ______________________________

____

What if you had not done tube 2A? ______________________________

8. In the exercise concerning trypsin function, why was an enzyme assay like Benedict's or Lugol's IKI (which test for the presence of a reaction product) not necessary? ______________________________

Why was tube 1T necessary? ______________________________

Why was tube 2T necessary? ______________________________

Trypsin is a protease similar to pepsin, the protein-digesting enzyme in the stomach. Would trypsin work well in the stomach? ____________ Why? ______________________________

9. In the procedure concerning pancreatic lipase digestion of fats and the action of bile salts, how did the appearance of tubes 1E and 2E differ? ______________________________

____

Explain the reason for the difference. ______________________________

____

Why did the litmus indicator change from blue to pink during fat hydrolysis? ______________________

____

Why is bile not considered an enzyme? ______________________________________

How did the tubes containing bile compare with those not containing bile? ______________________

What role does bile play in fat digestion? ____________________________________

____

10. The three-dimensional structure of a functional protein is altered by intense heat or nonphysiological pH even though peptide bonds may not break. Such inactivation is called denaturation, and denatured enzymes are nonfunctional. Explain why.

____

____

What specific experimental conditions resulted in denatured enzymes? ______________________

____

11. Pancreatic and intestinal enzymes operate optimally at a pH that is slightly alkaline, yet the chyme entering the duodenum from the stomach is very acid. How is the proper pH for the functioning of the pancreatic-intestinal enzymes ensured?

____

____

12. Assume you have been chewing a piece of bread for 5 or 6 minutes. How would you expect its taste to change during this interval? ____

Why? ____

13. Note the mechanism of absorption (passive or active transport) of the following food breakdown products, and indicate by a check mark (✓) whether the absorption would result in their movement into the blood capillaries or the lymph capillaries (lacteals).

Substance	Mechanism of absorption	Blood	Lymph
Monosaccharides			
Fatty acids and glycerol			
Amino acids			
Water			
Na^+, Cl^-, Ca^{2+}			

14. People on a strict diet to lose weight begin to metabolize stored fats at an accelerated rate. How does this condition affect blood pH? ____

15. Using a flowchart, trace the pathway of a ham sandwich (ham = protein and fat; bread = starch) from the mouth to the site of absorption of its breakdown products, noting where digestion occurs and what specific enzymes are involved.

16. Some of the digestive organs have groups of secretory cells that liberate hormones into the blood. These exert an effect on the digestive process by acting on other cells or structures and causing them to release digestive enzymes, expel bile, or increase the motility of the digestive tract. For each hormone below, note the organ producing the hormone and its effects on the digestive process. Include the target organs affected.

Hormone	Produced by	Target organ(s) and effects
Secretin		
Gastrin		
Cholecystokinin		

Physical Processes: Mechanisms of Food Propulsion and Mixing

17. Complete the following statements.

Swallowing, or __1__, occurs in two phases—the __2__ and __3__. One of these phases, the __4__ phase, is voluntary. During the voluntary phase, the __5__ is used to push the food into the back of the throat. During swallowing, the __6__ rises to ensure that its passageway is covered by the epiglottis so that the ingested substances don't enter the respiratory passageways. It is possible to swallow water while standing on your head because the water is carried along the esophagus involuntarily by the process of __7__. The pressure exerted by the foodstuffs on the __8__ sphincter causes it to open, allowing the foodstuffs to enter the stomach.

The two major types of propulsive movements that occur in the small intestine are __9__ and __10__. One of these movements, __11__, acts to continually mix the foods and to increase the absorption rate by moving different parts of the chyme mass over the intestinal mucosa, but it has less of a role in moving foods along the digestive tract.

1. ____________________
2. ____________________
3. ____________________
4. ____________________
5. ____________________
6. ____________________
7. ____________________
8. ____________________
9. ____________________
10. ____________________
11. ____________________

EXERCISE 40

Anatomy of the Urinary System

MATERIALS

- ☐ Human dissectible torso model, three-dimensional model of the urinary system, and/or anatomical chart of the human urinary system
- ☐ Dissecting instruments and tray
- ☐ Pig or sheep kidney, doubly or triply injected
- ☐ Disposable gloves
- ☐ Three-dimensional models of the cut kidney and of a nephron (if available)
- ☐ Compound microscope
- ☐ Prepared slides of a longitudinal section of kidney and cross sections of the bladder

For instructions on animal dissections, see the dissection exercises starting on page 697 in the cat, fetal pig, and rat editions of this manual.

OBJECTIVES

1. To describe the function of the urinary system.
2. To identify, on an appropriate diagram or torso model, the urinary system organs and to describe the general function of each.
3. To compare the course and length of the urethra in males and females.
4. To identify these regions of the dissected kidney (longitudinal section): hilum, cortex, medulla, medullary pyramids, major and minor calyces, pelvis, renal columns, and fibrous and perirenal fat capsules.
5. To trace the blood supply of the kidney from the renal artery to the renal vein.
6. To define the nephron as the physiological unit of the kidney, and to describe its anatomy.
7. To define *glomerular filtration, tubular reabsorption,* and *tubular secretion,* and to indicate the nephron areas involved in these processes.
8. To define *micturition,* and to explain pertinent differences in the control of the two bladder sphincters (internal and external).
9. To recognize microscopic or diagrammatic views of the histologic structure of the kidney and bladder.

PRE-LAB QUIZZ

1. Circle the correct term. In its excretory role, the urinary system is primarily concerned with the removal of carbon-containing / nitrogenous wastes from the body.
2. The __________ perform(s) the excretory and homeostatic functions of the urinary system.
   a. kidneys
   b. ureters
   c. urinary bladder
   d. all of the above
3. Circle the correct term. The cortex / medulla of the kidney is segregated into triangular regions with a striped appearance.
4. Circle the correct term. As the renal artery approaches a kidney, it is divided into branches known as the segmental arteries / afferent arterioles.
5. What do we call the anatomical units responsible for the formation of urine?
   ____
6. This knot of coiled capillaries, found in the kidneys, forms the filtrate. It is the
   a. arteriole
   b. glomerulus
   c. podocyte
   d. tubule
7. The section of the renal tubule closest to the glomerular capsule is the
   a. collecting duct
   b. distal convoluted tubule
   c. loop of Henle
   d. proximal convoluted tubule
8. Circle the correct term. The afferent / efferent arteriole drains the glomerular capillary bed.
9. Circle True or False. During tubular reabsorption, components of the filtrate move from the bloodstream into the tubule.
10. Circle the correct term. The internal / external urethral sphincter consists of skeletal muscle and is voluntarily controlled.

MasteringA&P™

Access practice quizzes and more in the Study Area at www.masteringaandp.com.

PAL

For access to anatomical models and more, check out Practice Anatomy Lab.

Metabolism of nutrients by the body produces wastes (carbon dioxide, nitrogenous wastes, ammonia, and so on) that must be eliminated from the body if normal function is to continue. Although excretory processes involve several organ systems (the lungs excrete carbon dioxide and skin glands excrete salts and water), it is the **urinary system** that is primarily concerned with the removal of nitrogenous wastes from the body. In addition to this purely excretory function, the kidney maintains the electrolyte, acid-base, and fluid balances of the blood and is thus a major, if not *the* major, homeostatic organ of the body.

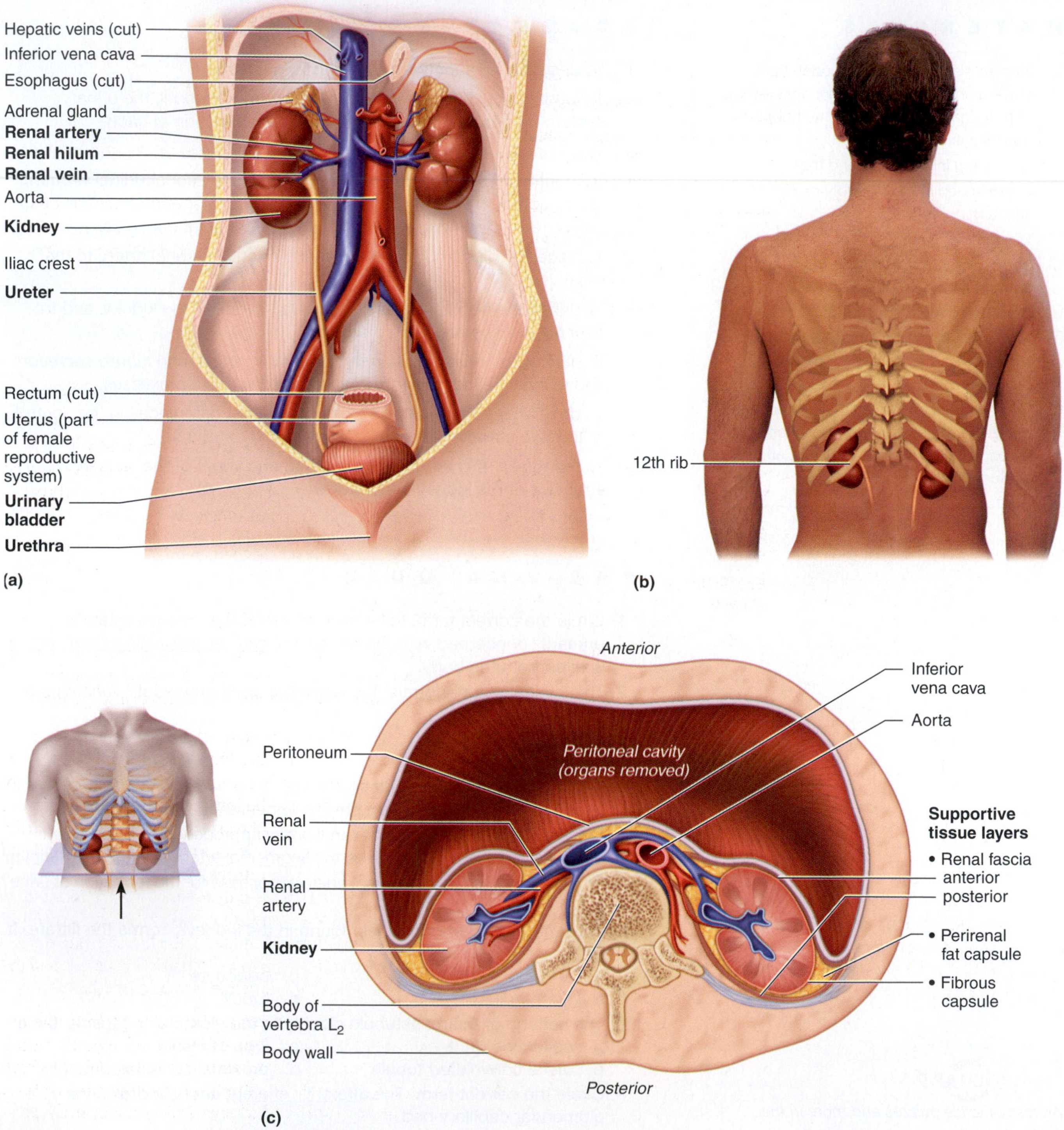
Hepatic veins (cut)
Inferior vena cava
Esophagus (cut)
Adrenal gland
Renal artery
Renal hilum
Renal vein
Aorta
Kidney
Iliac crest
Ureter
Rectum (cut)
Uterus (part of female reproductive system)
Urinary bladder
Urethra
(a)
12th rib
(b)
Anterior
Inferior vena cava
Aorta
Peritoneum
Peritoneal cavity (organs removed)
Renal vein
Renal artery
Kidney
Body of vertebra L_2
Body wall
Supportive tissue layers
• Renal fascia
anterior
posterior
• Perirenal fat capsule
• Fibrous capsule
Posterior
(c)

FIGURE 40.1 Organs of the urinary system. **(a)** Anterior view of the female urinary organs. Most unrelated abdominal organs have been removed. **(b)** Posterior in situ view showing the position of the kidneys relative to the twelfth rib pair. **(c)** Cross section of the abdomen viewed from inferior direction. Note the retroperitoneal position and supportive tissue layers of the kidneys.

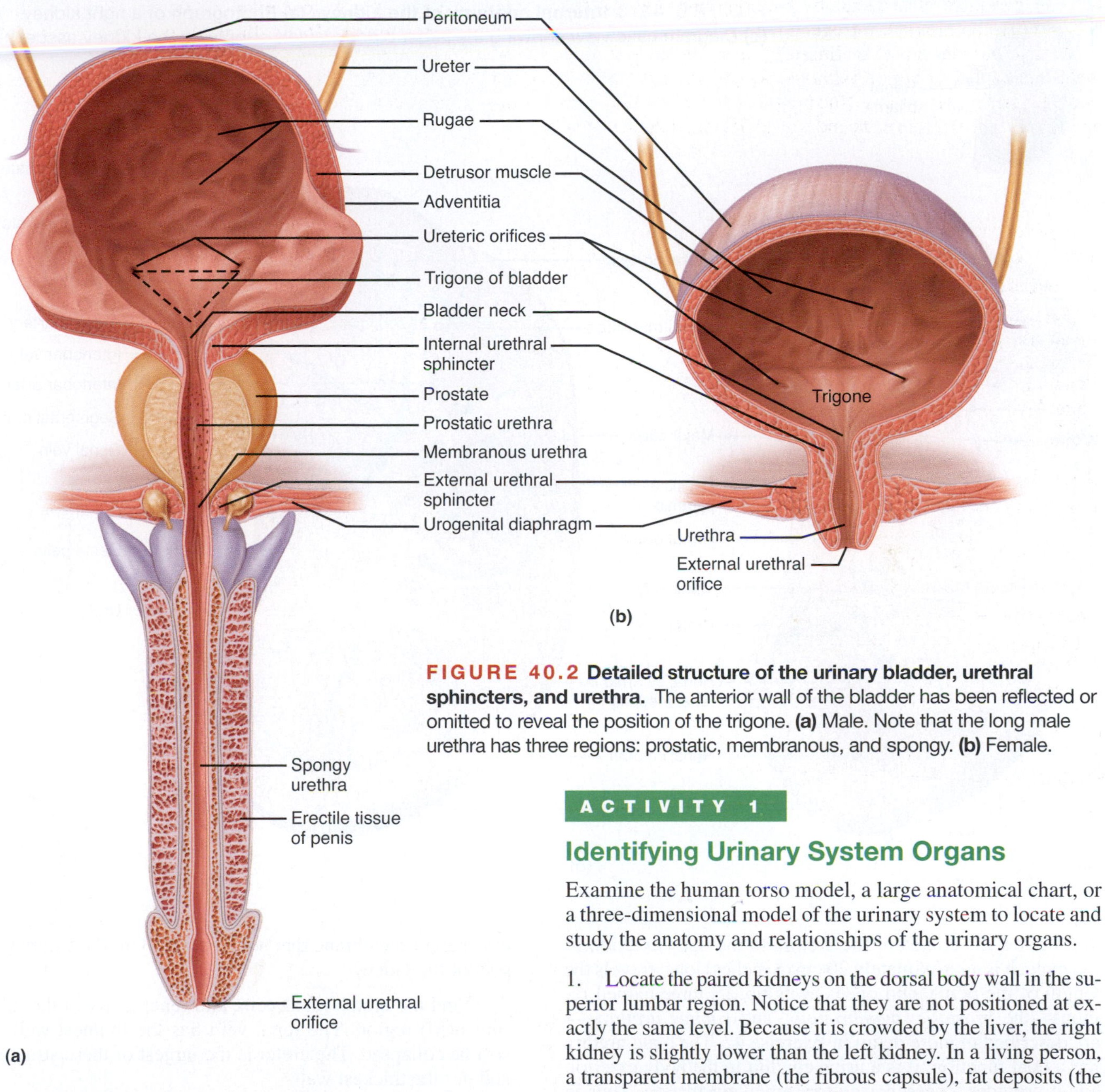
Peritoneum
Ureter
Rugae
Detrusor muscle
Adventitia
Ureteric orifices
Trigone of bladder
Bladder neck
Internal urethral sphincter
Prostate
Prostatic urethra
Membranous urethra
External urethral sphincter
Urogenital diaphragm
Spongy urethra
Erectile tissue of penis
External urethral orifice
(a)
Trigone
Urethra
External urethral orifice
(b)

FIGURE 40.2 Detailed structure of the urinary bladder, urethral sphincters, and urethra. The anterior wall of the bladder has been reflected or omitted to reveal the position of the trigone. **(a)** Male. Note that the long male urethra has three regions: prostatic, membranous, and spongy. **(b)** Female.

To perform its functions, the kidney acts first as a blood filter, and then as a blood processor. It allows toxins, metabolic wastes, and excess ions to leave the body in the urine, while retaining needed substances and returning them to the blood. Malfunction of the urinary system, particularly of the kidneys, leads to a failure in homeostasis which, unless corrected, is fatal.

Gross Anatomy of the Human Urinary System

The urinary system (Figure 40.1) consists of the paired kidneys and ureters and the single urinary bladder and urethra. The **kidneys** perform the functions described above and manufacture urine in the process. The remaining organs of the system provide temporary storage reservoirs or transportation channels for urine.

ACTIVITY 1

Identifying Urinary System Organs

Examine the human torso model, a large anatomical chart, or a three-dimensional model of the urinary system to locate and study the anatomy and relationships of the urinary organs.

1. Locate the paired kidneys on the dorsal body wall in the superior lumbar region. Notice that they are not positioned at exactly the same level. Because it is crowded by the liver, the right kidney is slightly lower than the left kidney. In a living person, a transparent membrane (the fibrous capsule), fat deposits (the *perirenal fat capsules*), and the fibrous renal fascia surround and hold the kidneys in place in a retroperitoneal position.

When the fatty material surrounding the kidneys is reduced or too meager in amount (in cases of rapid weight loss or in very thin individuals), the kidneys are less securely anchored and may drop to a more inferior position in the abdominal cavity. This phenomenon is called **ptosis.** ■

2. Observe the **renal arteries** as they diverge from the descending aorta and plunge into the indented medial region **(hilum)** of each kidney. Note also the **renal veins,** which drain the kidneys (circulatory drainage) and the two **ureters,** which drain urine from the kidneys and conduct it by peristalsis to the bladder for temporary storage.

3. Locate the **urinary bladder,** and observe the point of entry of the two ureters into this organ. Also locate the single **urethra,** which drains the bladder. The triangular region of the bladder delineated by these three openings (two ureteral and one urethral orifice) is referred to as the **trigone** (Figure 40.2).

FIGURE 40.3 Internal anatomy of the kidney. (a) Photograph of a right kidney. **(b)** Diagrammatic view showing the larger blood vessels supplying the kidney tissue.

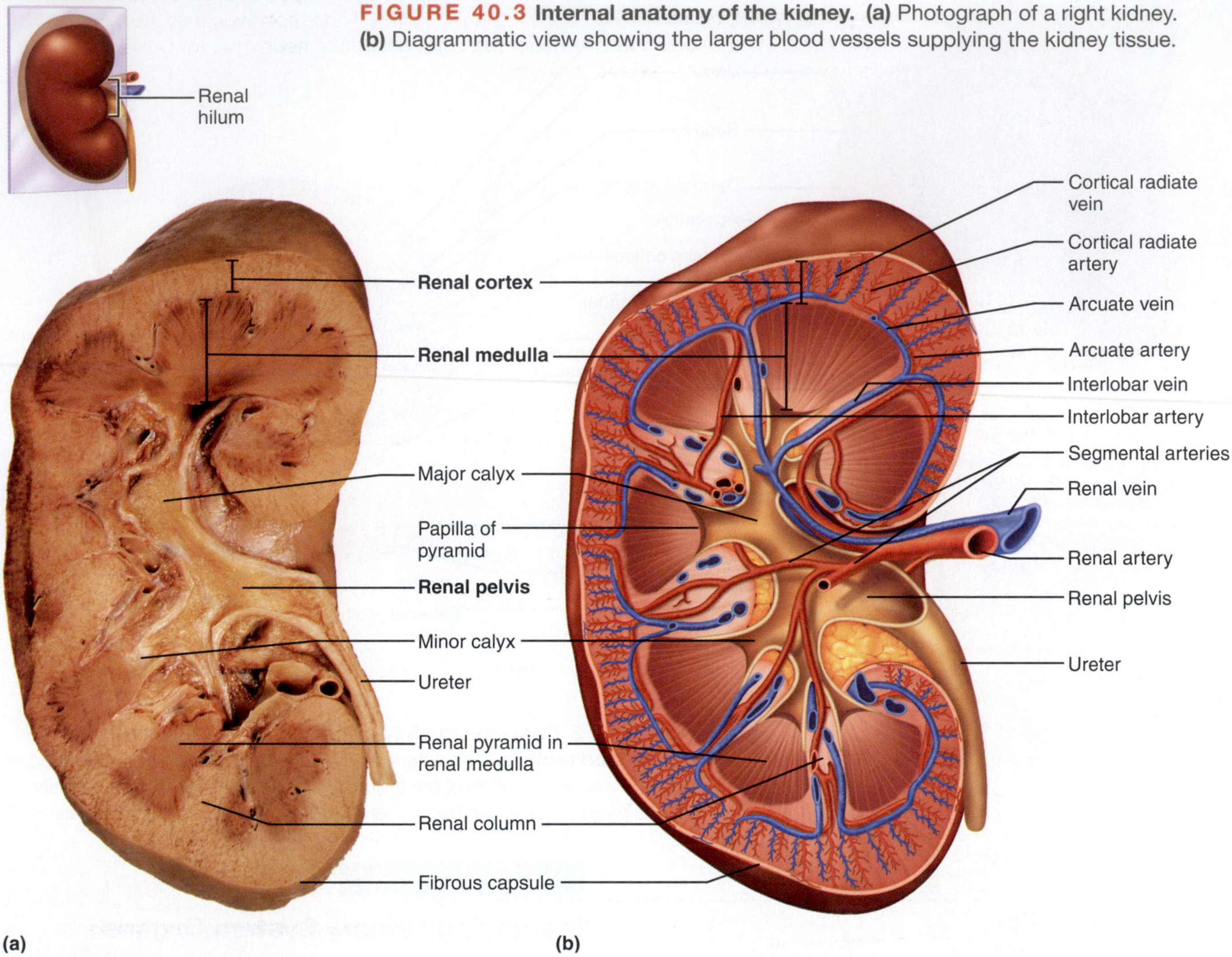

Renal hilum
Renal cortex
Renal medulla
Major calyx
Papilla of pyramid
Renal pelvis
Minor calyx
Ureter
Renal pyramid in renal medulla
Renal column
Fibrous capsule
Cortical radiate vein
Cortical radiate artery
Arcuate vein
Arcuate artery
Interlobar vein
Interlobar artery
Segmental arteries
Renal vein
Renal artery
Renal pelvis
Ureter
(a)
(b)

4. Follow the course of the urethra to the body exterior. In the male, it is approximately 20 cm (8 inches) long, travels the length of the **penis,** and opens at its tip. Its three named regions—the *prostatic, membranous,* and *spongy urethrae*—are described in more detail in Exercise 42. The male urethra has a dual function: it is a urine conduit to the body exterior, and it provides a passageway for semen ejaculation. Thus in the male, the urethra is part of both the urinary and reproductive systems. In females, the urethra is very short, approximately 4 cm (1½ inches) long (see Figure 40.2). There are no common urinary-reproductive pathways in the female, and the female's urethra serves only to transport urine to the body exterior. Its external opening, the **external urethral orifice,** lies anterior to the vaginal opening.

DISSECTION: Gross Internal Anatomy of the Pig or Sheep Kidney

1. In preparation for dissection, don gloves. Obtain a preserved sheep or pig kidney, dissecting tray, and instruments. Observe the kidney to identify the **fibrous capsule,** a smooth, transparent membrane that adheres tightly to the external aspect of the kidney.

2. Find the ureter, renal vein, and renal artery at the hilum (indented) region. The renal vein has the thinnest wall and will be collapsed. The ureter is the largest of these structures and has the thickest wall.

3. Make a cut through the longitudinal axis (frontal section) of the kidney and locate the anatomical areas described below and depicted in Figure 40.3.

Kidney cortex: The superficial kidney region, which is lighter in color. If the kidney is doubly injected with latex, you will see a predominance of red and blue latex specks in this region indicating its rich vascular supply.

Medullary region: Deep to the cortex; a darker, reddish-brown color. The medulla is segregated into triangular regions that have a striped, or striated, appearance—the **medullary (renal) pyramids.** The base of each pyramid faces toward the cortex. Its more pointed *papilla,* or *apex,* points to the innermost kidney region.

Renal columns: Areas of tissue that are more like the cortex in appearance, which segregate and dip inward between the pyramids.

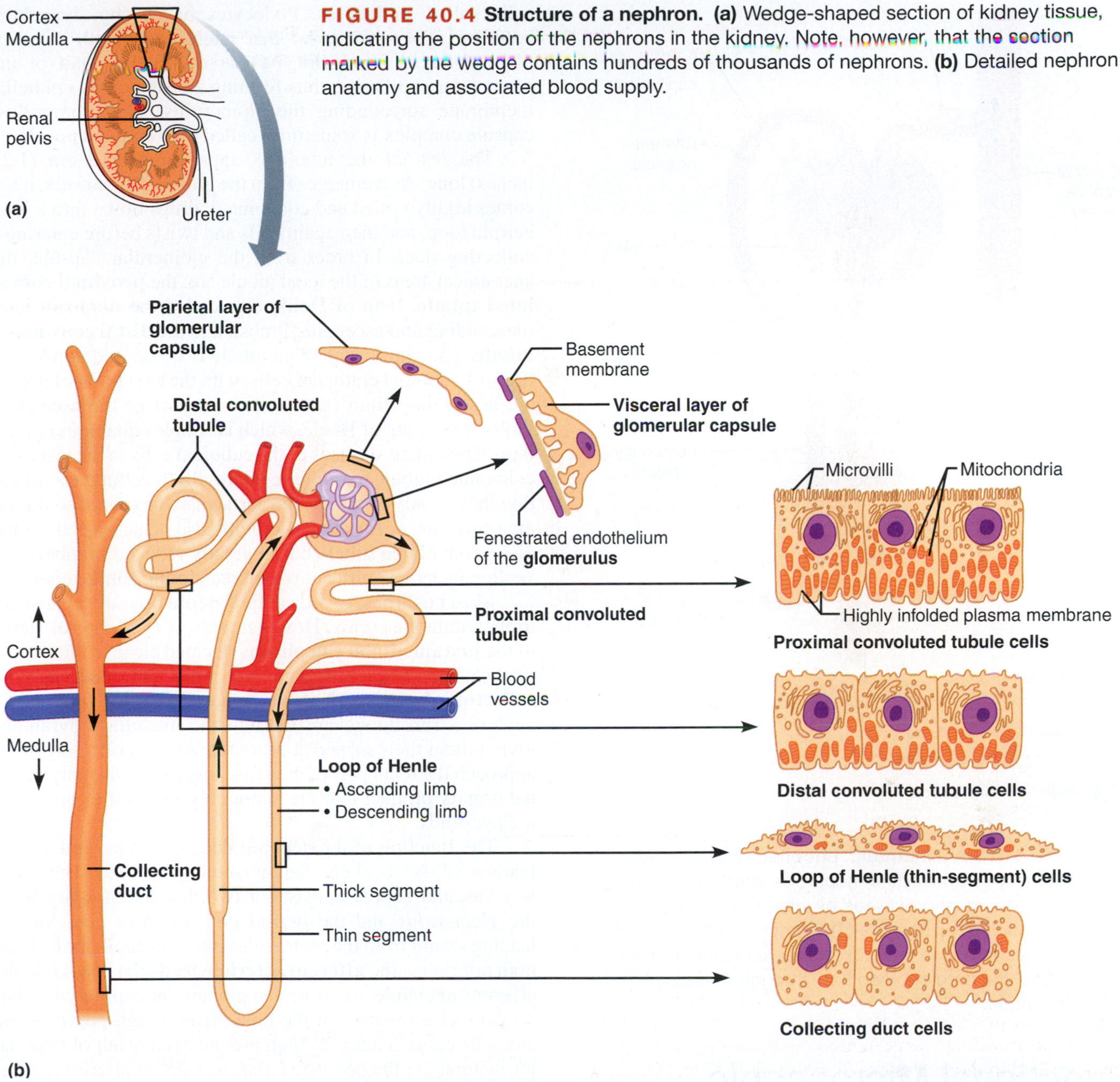
Cortex
Medulla
Renal pelvis
(a)
Ureter
Parietal layer of glomerular capsule
Basement membrane
Distal convoluted tubule
Visceral layer of glomerular capsule
Microvilli
Mitochondria
Fenestrated endothelium of the glomerulus
Highly infolded plasma membrane
Proximal convoluted tubule
Proximal convoluted tubule cells
Cortex
Blood vessels
Medulla
Distal convoluted tubule cells
Loop of Henle
• Ascending limb
• Descending limb
Loop of Henle (thin-segment) cells
Collecting duct
Thick segment
Thin segment
Collecting duct cells
(b)

FIGURE 40.4 Structure of a nephron. (a) Wedge-shaped section of kidney tissue, indicating the position of the nephrons in the kidney. Note, however, that the section marked by the wedge contains hundreds of thousands of nephrons. **(b)** Detailed nephron anatomy and associated blood supply.

Renal pelvis: Extending inward from the hilum; a relatively flat, basinlike cavity that is continuous with the **ureter,** which exits from the hilum region. Fingerlike extensions of the pelvis should be visible. The larger, or primary, extensions are called the **major calyces** (singular: *calyx*); subdivisions of the major calyces are the **minor calyces.** Notice that the minor calyces terminate in cuplike areas that enclose the apexes of the medullary pyramids and collect urine draining from the pyramidal tips into the pelvis.

4. If the preserved kidney is doubly or triply injected, follow the renal blood supply from the renal artery to the **glomeruli.** The glomeruli appear as little red and blue specks in the cortex region. (See Figure 40.4.)

Approximately a fourth of the total blood flow of the body is delivered to the kidneys each minute by the large **renal arteries.** As a renal artery approaches the kidney, it breaks up into branches called **segmental arteries,** which enter the hilum. Each segmental artery, in turn, divides into several **interlobar arteries,** which ascend toward the cortex in the renal column areas. At the top of the medullary region, these arteries give off arching branches, the **arcuate arteries,** which curve over the bases of the medullary pyramids. Small **cortical radiate arteries** branch off the arcuate arteries and ascend into the cortex, giving off the individual **afferent arterioles,** which provide the capillary networks (glomeruli and peritubular capillary beds) that supply the nephrons, or functional units, of the kidney. Blood draining from the nephron capillary networks in the cortex enters the **cortical radiate veins** and then drains through the **arcuate veins** and the **interlobar veins** to finally enter the **renal vein** in the pelvis region. (There are no segmental veins.)

Dispose of the kidney specimen as your instructor specifies.

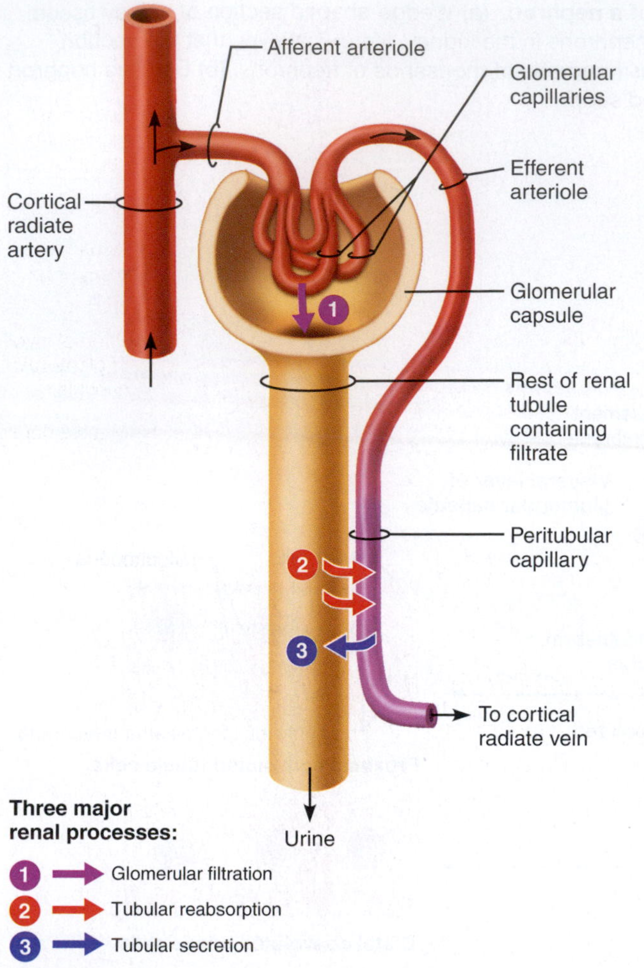
Afferent arteriole
Glomerular capillaries
Efferent arteriole
Cortical radiate artery
Glomerular capsule
Rest of renal tubule containing filtrate
Peritubular capillary
To cortical radiate vein
Urine
Three major renal processes:
Glomerular filtration
Tubular reabsorption
Tubular secretion

FIGURE 40.5 A schematic, uncoiled nephron. A kidney actually has millions of nephrons acting in parallel. The three major renal processes by which the kidneys adjust the composition of plasma are **(1)** glomerular filtration, **(2)** tubular reabsorption, and **(3)** tubular secretion. Black arrows show the path of blood flow through the renal microcirculation.

Functional Microscopic Anatomy of the Kidney and Bladder

Kidney

Each kidney contains over a million nephrons, which are the anatomical units responsible for forming urine. Figure 40.4 depicts the detailed structure and the relative positioning of the nephrons in the kidney.

Each nephron consists of two major structures: a **glomerulus** (a capillary knot) and a renal tubule. During embryologic development, each **renal tubule** begins as a blind-ended tubule that gradually encloses an adjacent capillary cluster, or glomerulus. The enlarged end of the tubule encasing the glomerulus is the **glomerular (Bowman's) capsule,** and its inner, or visceral, wall consists of highly specialized cells called **podocytes.** Podocytes have long, branching processes *(foot processes)* that interdigitate with those of other podocytes and cling to the endothelial wall of the glomerular capillaries, thus forming a very porous epithelial membrane surrounding the glomerulus. The glomerulus-capsule complex is sometimes called the **renal corpuscle.**

The rest of the tubule is approximately 3 cm (1.25 inches) long. As it emerges from the glomerular capsule, it becomes highly coiled and convoluted, drops down into a long hairpin loop, and then again coils and twists before entering a collecting duct. In order from the glomerular capsule, the anatomical areas of the renal tubule are: the **proximal convoluted tubule, loop of Henle** also called the **nephron loop** (descending and ascending limbs), and the **distal convoluted tubule.** The wall of the renal tubule is composed almost entirely of cuboidal epithelial cells, with the exception of part of the descending limb (and sometimes part of the ascending limb) of the loop of Henle, which is simple squamous epithelium. The lumen surfaces of the cuboidal cells in the proximal convoluted tubule have dense microvilli (a cellular modification that greatly increases the surface area exposed to the lumen contents, or filtrate). Microvilli also occur on cells of the distal convoluted tubule but in greatly reduced numbers, revealing its less significant role in reclaiming filtrate contents.

Most nephrons, called **cortical nephrons,** are located entirely within the cortex. However, parts of the loops of Henle of the **juxtamedullary nephrons** (located close to the cortex-medulla junction) penetrate well into the medulla. The **collecting ducts,** each of which receives urine from many nephrons, run downward through the medullary pyramids, giving them their striped appearance. As the collecting ducts approach the renal pelvis, they fuse together and empty the final urinary product into the minor calyces via the papillae of the pyramids.

The function of the nephron depends on several unique features of the renal circulation (see Figure 40.5). The capillary vascular supply consists of two distinct capillary beds, the *glomerulus* and the *peritubular capillary bed.* Vessels leading to and from the glomerulus, the first capillary bed, are both arterioles: the **afferent arteriole** feeds the bed while the **efferent arteriole** drains it. The glomerular capillary bed has no parallel elsewhere in the body. It is a high-pressure bed along its entire length. Its high pressure is a result of two major factors: (1) the bed is *fed and drained* by arterioles (arterioles are high-resistance vessels as opposed to venules, which are low-resistance vessels), and (2) the afferent feeder arteriole is larger in diameter than the efferent arteriole draining the bed. The high hydrostatic pressure created by these two anatomical features forces out fluid and blood components smaller than proteins from the glomerulus into the glomerular capsule. That is, it forms the filtrate that is processed by the nephron tubule.

The **peritubular capillary bed** arises from the efferent arteriole draining the glomerulus. This set of capillaries clings intimately to the renal tubule and empties into the cortical radiate veins that leave the cortex. The peritubular capillaries are *low-pressure*, porous capillaries adapted for absorption rather than filtration and readily take up the solutes and water reabsorbed from the filtrate by the tubule cells. The juxtamedullary nephrons have additional looping vessels, called the **vasa recta** (straight vessels), that parallel the long loops of Henle in the medulla. Hence, the two capillary beds of the nephron have very different, but complementary, roles: the glomerulus produces the filtrate, and the peritubular capillaries reclaim most of that filtrate.

Additionally, each nephron has a region called a **juxtaglomerular apparatus (JGA),** which plays important roles in regulating the rate of filtration and in systemic blood pressure. The JGA consists of (1) *juxtaglomerular (JG) cells* (also called *granular cells*), which function as blood pressure sensors in the walls of the afferent arterioles near the glomerulus, and (2) a *macula densa,* a specialized group of columnar chemoreceptor cells in the distal convoluted tubule abutting the JG cells (Figure 40.6a).

Urine formation is a result of three processes: *filtration, reabsorption,* and *secretion* (see Figure 40.5). **Filtration,** the role of the glomerulus, is largely a passive process in which a portion of the blood passes from the glomerular bed into the glomerular capsule. This filtrate then enters the proximal convoluted tubule where tubular reabsorption and secretion begin. During **tubular reabsorption,** many of the filtrate components move through the tubule cells and return to the blood in the peritubular capillaries. Some of this reabsorption is passive, such as that of water which passes by osmosis, but the reabsorption of most substances depends on active transport processes and is highly selective. Which substances are reabsorbed at a particular time depends on the composition of the blood and needs of the body at that time. Substances that are almost entirely reabsorbed from the filtrate include water, glucose, and amino acids. Various ions are selectively reabsorbed or allowed to go out in the urine according to what is required to maintain appropriate blood pH and electrolyte composition. Waste products (urea, creatinine, uric acid, and drug metabolites) are reabsorbed to a much lesser degree or not at all. Most (75% to 80%) of tubular reabsorption occurs in the proximal convoluted tubule. The rest occurs in other areas, especially the distal convoluted tubules and collecting ducts.

Tubular secretion is essentially the reverse process of tubular reabsorption. Substances such as hydrogen and potassium ions and creatinine move either from the blood of the peritubular capillaries through the tubular cells or from the tubular cells into the filtrate to be disposed of in the urine. This process is particularly important for the disposal of substances not already in the filtrate (such as drug metabolites), and as a device for controlling blood pH.

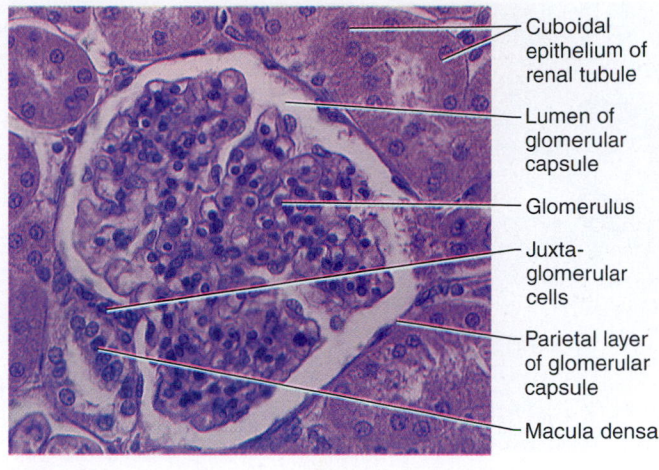

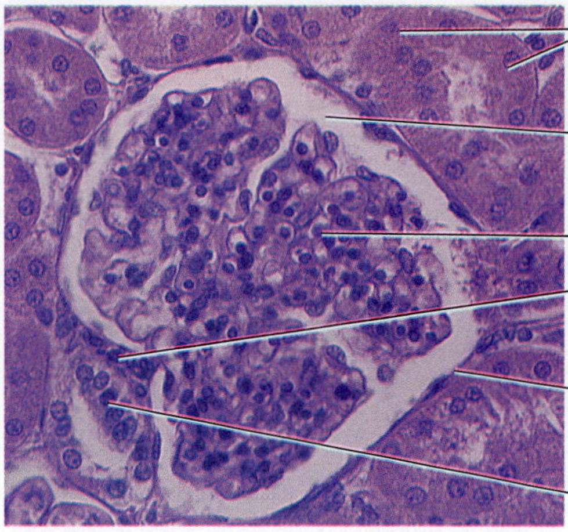
Cuboidal epithelium of renal tubule
Lumen of glomerular capsule
Glomerulus
Juxta-glomerular cells
Parietal layer of glomerular capsule
Macula densa

(a)

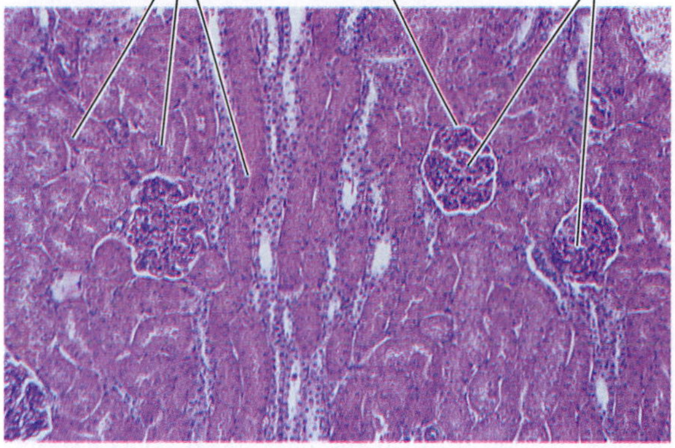
Renal tubules
Lumen of the glomerular capsule
Glomeruli

(b)

FIGURE 40.6 Microscopic structure of kidney tissue. **(a)** Detailed structure of the glomerulus (170×). **(b)** Low-power view of the renal cortex (50×).

ACTIVITY 2

Studying Nephron Structure

1. Begin your study of nephron structure by identifying the glomerular capsule, proximal and distal convoluted tubule regions, and the loop of Henle on a model of the nephron. Then, obtain a compound microscope and a prepared slide of kidney tissue to continue with the microscope study of the kidney.

2. Hold the longitudinal section of the kidney up to the light to identify cortical and medullary areas. Then secure the slide on the microscope stage, and scan the slide under low power.

3. Move the slide so that you can see the cortical area. Identify a glomerulus, which appears as a ball of tightly packed material containing many small nuclei (Figures 40.6). It is usually delineated by a vacant-appearing region (corresponding to the space between the visceral and parietal layers of the glomerular capsule) that surrounds it.

4. Notice that the renal tubules are cut at various angles. Try to differentiate between the fuzzy cuboidal epithelium of the proximal convoluted tubule, which has dense microvilli, and that of the distal convoluted tubule with sparse microvilli. Also identify the thin-walled loop of Henle. ■

Bladder

Although urine production by the kidney is a continuous process, urine is usually removed from the body when voiding is convenient. In the meantime the **urinary bladder,** which receives urine via the ureters and discharges it via the urethra, stores it temporarily.

Voiding, or **micturition,** is the process in which urine empties from the bladder. Two sphincter muscles or valves (see Figure 40.2), the **internal urethral sphincter** (more superiorly located) and the **external urethral sphincter** (more inferiorly located) control the outflow of urine from the bladder. Ordinarily, the bladder continues to collect urine until about 200 ml have accumulated, at which time the stretching of the bladder wall activates stretch receptors. Impulses transmitted to the central nervous system subsequently produce reflex contractions of the bladder wall through parasympathetic

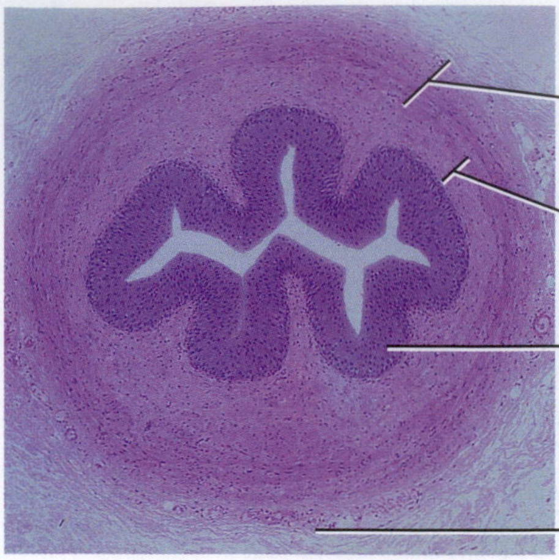
Circular layer
Longitudinal layer
Smooth muscle
Transitional epithelium
Adventitia

FIGURE 40.7 Structure of the ureter wall. Cross section of ureter (15×).

nervous system pathways (in other words, via the pelvic splanchnic nerves). As contractions increase in force and frequency, stored urine is forced past the internal sphincter, which is a smooth muscle involuntary sphincter, into the superior part of the urethra. It is then that a person feels the urge to void. The inferior external sphincter consists of skeletal muscle and is voluntarily controlled. If it is not convenient to void, the opening of this sphincter can be inhibited. Conversely, if the time is convenient, the sphincter may be relaxed and the stored urine flushed from the body. If voiding is inhibited, the reflex contractions of the bladder cease temporarily and urine continues to accumulate in the bladder. After another 200 to 300 ml of urine have been collected, the *micturition reflex* will again be initiated.

Lack of voluntary control over the external sphincter is referred to as **incontinence.** Incontinence is normal in children 2 years old or younger, as they have not yet gained control over the voluntary sphincter. In adults and older children, incontinence is generally a result of spinal cord injury, emotional problems, bladder irritability, or some other pathology of the urinary tract. ■

ACTIVITY 3

Studying Bladder Structure

1. Return the kidney slide to the supply area, and obtain a slide of bladder tissue. Scan the bladder tissue. Identify its three layers: mucosa, muscular layer, and fibrous adventitia.

2. Study the mucosa with its highly specialized transitional epithelium. The plump, transitional epithelial cells have the ability to slide over one another, thus decreasing the thickness of the mucosa layer as the bladder fills and stretches to accommodate the increased urine volume. Depending on the degree of stretching of the bladder, the mucosa may be three to eight cell layers thick. Compare the transitional epithelium of the mucosa to that shown in Figure 6.3h (page 73).

3. Examine the heavy muscular wall (detrusor muscle), which consists of three irregularly arranged muscular layers. The innermost and outermost muscle layers are arranged longitudinally; the middle layer is arranged circularly. Attempt to differentiate the three muscle layers.

4. Draw a small section of the bladder wall, and label all regions or tissue areas.

5. Compare your sketch of the bladder wall to the structure of the ureter wall shown in Figure 40.7. How are the two organs similar histologically?

___

___

___

What is/are the most obvious differences?

___

___

For instructions on animal dissections, see the dissection exercises starting on page 697 in the cat, rat, and fetal pig editions of this manual.

NAME ______________________

LAB TIME/DATE ______________________

REVIEW SHEET

EXERCISE 40

Anatomy of the Urinary System

Gross Anatomy of the Human Urinary System

1. Complete the following statements.

The kidney is referred to as an excretory organ because it excretes __1__ wastes. It is also a major homeostatic organ because it maintains the electrolyte, __2__, and __3__ balance of the blood.

Urine is continuously formed by the __4__ and is routed down the __5__ by the mechanism of __6__ to a storage organ called the __7__. Eventually, the urine is conducted to the body __8__ by the urethra. In the male, the urethra is __9__ centimeters long and transports both urine and __10__. The female urethra is __11__ centimeters long and transports only urine.

Voiding or emptying the bladder is called __12__. Voiding has both voluntary and involuntary components. The voluntary sphincter is the __13__ sphincter. An inability to control this sphincter is referred to as __14__.

1. ______________________
2. ______________________
3. ______________________
4. ______________________
5. ______________________
6. ______________________
7. ______________________
8. ______________________
9. ______________________
10. ______________________
11. ______________________
12. ______________________
13. ______________________
14. ______________________

2. What is the function of the fat cushion that surrounds the kidneys in life? ______________________

____

3. Define *ptosis.* ______________________

____

4. Why is incontinence a normal phenomenon in the child under 1½ to 2 years old? ______________________

____

What events may lead to its occurrence in the adult? ______________________

____

____

5. Complete the labeling of the diagram to correctly identify the urinary system organs.

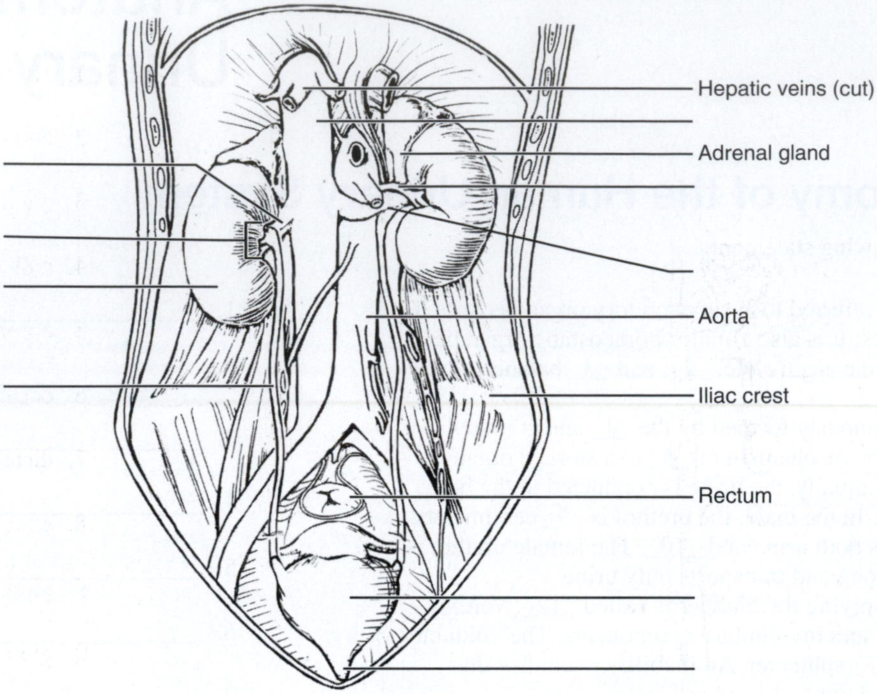
Hepatic veins (cut)
Adrenal gland
Aorta
Iliac crest
Rectum

Gross Internal Anatomy of the Pig or Sheep Kidney

6. Match the appropriate structure in column B to its description in column A.

Column A	Column B
_______ 1. smooth membrane, tightly adherent to the kidney surface	a. cortex
_______ 2. portion of the kidney containing mostly collecting ducts	b. medulla
_______ 3. portion of the kidney containing the bulk of the nephron structures	c. minor calyx
_______ 4. superficial region of kidney tissue	d. fibrous capsule
_______ 5. basinlike area of the kidney, continuous with the ureter	e. renal column
_______ 6. a cup-shaped extension of the pelvis that encircles the apex of a pyramid	f. renal pelvis
_______ 7. area of cortical tissue running between the medullary pyramids	

Functional Microscopic Anatomy of the Kidney and Bladder

7. Match each lettered structure in the diagram of the nephron (and associated renal blood supply) with the correct name in the numbered list.

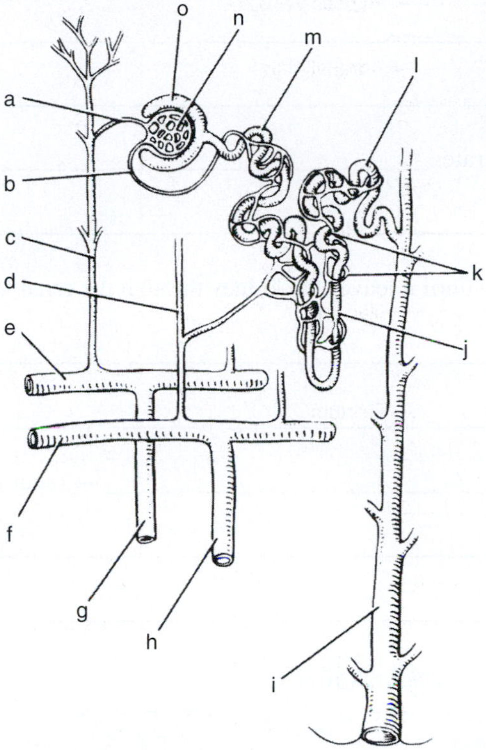

o
n
m
l
a
b
c
d
e
k
j
f
g
h
i

_______ 1. afferent arteriole

_______ 2. arcuate artery

_______ 3. arcuate vein

_______ 4. collecting duct

_______ 5. cortical radiate artery

_______ 6. cortical radiate vein

_______ 7. distal convoluted tubule

_______ 8. efferent arteriole

_______ 9. glomerular capsule

_______ 10. glomerulus

_______ 11. interlobar artery

_______ 12. interlobar vein

_______ 13. loop of Henle

_______ 14. peritubular capillaries

_______ 15. proximal convoluted tubule

8. Use the terms provided in question 7 to identify the following descriptions.

_______________ 1. site of filtrate formation

_______________ 2. primary site of tubular reabsorption

_______________ 3. secondarily important site of tubular reabsorption

_______________ 4. structure that conveys the processed filtrate (urine) to the renal pelvis

_______________ 5. blood supply that directly receives substances from the tubular cells

_______________ 6. its inner (visceral) membrane forms part of the filtration membrane

9. Explain *why* the glomerulus is such a high-pressure capillary bed. _______________

____

How does its high-pressure condition aid its function of filtrate formation? _______________

____

10. What structural modification of certain tubule cells enhances their ability to reabsorb substances from the filtrate?

____

11. Explain the mechanism of tubular secretion, and explain its importance in the urine-formation process. ____________

____

____

12. Compare and contrast the composition of blood plasma and glomerular filtrate. ____________

____

13. Trace a drop of blood from the time it enters the kidney via the renal artery until it leaves the kidney through the renal vein.

Renal artery → ____

____

____ → renal vein

14. Define *juxtaglomerular apparatus.* ____

____

15. Label the figure using the key letters of the correct terms.

Key: a. *juxtaglomerular cells*

b. cuboidal epithelium

c. macula densa

d. glomerular capsule (parietal layer)

e. distal convoluted tubule

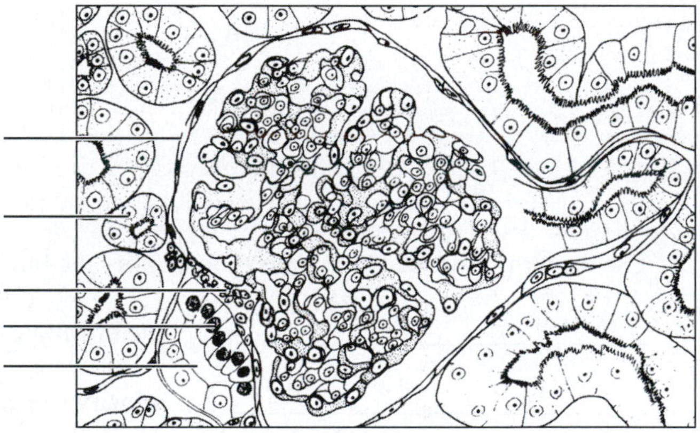

16. Trace the anatomical pathway of a molecule of creatinine (metabolic waste) from the glomerular capsule to the urethra. Note each microscopic and/or gross structure it passes through in its travels. Name the subdivisions of the renal tubule.

Glomerular capsule → ____

____

____ → urethra

17. What is important functionally about the specialized epithelium (transitional epithelium) in the bladder?

____

EXERCISE 41

Urinalysis

MATERIALS

- ☐ Disposable gloves
- ☐ Student urine samples collected at the beginning of the laboratory or "normal" artificial urine provided by the instructor*
- ☐ Numbered "pathological" urine specimens provided by the instructor*
- ☐ Wide-range pH paper
- ☐ Dipsticks: individual (Clinistix, Ketostix, Albustix, Hemastix) or combination (Chemstrip or Multistix)
- ☐ Urinometer
- ☐ Test tubes, test tube rack, and test tube holders
- ☐ 10-cc graduated cylinders
- ☐ Test reagents for sulfates: 10% barium chloride solution, dilute HCl (hydrochloric acid)
- ☐ Hot plate
- ☐ 500-ml beaker
- ☐ Test reagent for phosphates: dilute nitric acid, dilute ammonium molybdate
- ☐ Glass stirring rod
- ☐ Test reagent for chloride: 3.0% silver nitrate solution ($AgNO_3$), freshly prepared
- ☐ Clean microscope slide and coverslip
- ☐ Compound microscope
- ☐ Test reagent for urea: concentrated nitric acid in dropper bottles
- ☐ Test reagent for glucose (Clinitest tablets) and Clinitest color chart
- ☐ Medicine droppers

Text continues on next page.

OBJECTIVES

1. To list the physical characteristics of urine, and to indicate the normal pH and specific gravity ranges.
2. To list substances that are normal urinary constituents.
3. To conduct various urinalysis tests and procedures, and to use them to determine the substances present in a urine specimen.
4. To define the following urinary conditions:

calculi	ketonuria	pyuria
glycosuria	hematuria	casts
albuminuria	hemoglobinuria	

5. To explain the implications and possible causes of conditions listed in objective 4.

PRE-LAB QUIZ

1. Normal urine is usually pale yellow to amber in color, due to the presence of
   a. hemochrome
   b. melanin
   c. urochrome
2. Circle the correct term. The average pH value of urine is <u>6.0 / 11.0</u>.
3. Circle True or False. Glucose can usually be found in all normal urine.
4. __________, like other blood proteins, is /are too large to pass through the glomerular filtration membrane and is/are normally not found in urine.
   a. Albumin c. Nitrates
   b. Chloride d. Sulfate
5. Circle the correct term. <u>Hematuria / Ketonuria</u>, the appearance of red blood cells in the urine, almost always indicates pathology of the urinary system.
6. The appearance of bile pigments in the urine, a condition known as __________, can be an indication of liver disease.
   a. albuminuria c. ketonuria
   b. bilirubinuria d. pyuria
7. Circle the correct term. <u>Casts / Calculi</u> are hardened cell fragments formed in the distal convoluted tubules and collecting ducts and flushed out of the urinary tract.
8. When determining the presence of inorganic constituents such as sulfates, phosphates, and chlorides, you will be looking for the "formation of a precipitate." What is a precipitate? ____________________

____

MasteringA&P™

Access practice quizzes and more in the Study Area at www.masteringaandp.com.

PEx

This lab corresponds to PhysioEx Exercise 9. See p. PEx-131.

Blood composition depends on three major factors: diet, cellular metabolism, and urinary output. In 24 hours, the kidneys' 2 million nephrons filter 150 to 180 liters of blood plasma through their glomeruli into the tubules, where it is selectively processed by tubular reabsorption and secretion. In the same period, urinary output, which contains by-products of metabolism and excess ions, is 1.0 to 1.8 liters. In healthy individuals, the kidneys can maintain blood constancy despite wide variations in diet and metabolic activity. With certain pathological conditions, urine composition often changes dramatically.

*Directions for making artificial urine are provided in the Instructor's Guide for this manual.

- ☐ Timer (watch or clock with a second hand)
- ☐ Ictotest reagent tablets and test mat
- ☐ Flasks and laboratory buckets containing 10% bleach solution
- ☐ Disposable autoclave bags
- ☐ *Demonstration:* Instructor-prepared specimen of urine sediment set up for microscopic analysis

Characteristics of Urine

To be valuable as a diagnostic tool, a urinalysis must be done within 30 minutes after the urine is voided or on refrigerated urine. Freshly voided urine is generally clear and pale yellow to amber in color. This normal yellow color is due to *urochrome,* a pigment metabolite arising from the body's destruction of hemoglobin (via bilirubin or bile pigments). As a rule, color variations from pale yellow to deeper amber indicate the relative concentration of solutes to water in the urine. The greater the solute concentration, the deeper the color. Abnormal urine color may be due to certain foods, such as beets, various drugs, bile, or blood.

The odor of freshly voided urine is slightly aromatic, but bacterial action gives it an ammonia-like odor when left standing. Some drugs, vegetables (such as asparagus), and various disease processes (such as diabetes mellitus) alter the characteristic odor of urine. For example, the urine of a person with uncontrolled diabetes mellitus (and elevated levels of ketones) smells fruity or acetone-like.

The pH of urine ranges from 4.5 to 8.0, but its average value, 6.0, is slightly acidic. Diet may markedly influence the pH of the urine. For example, a diet high in protein (meat, eggs, cheese) and whole wheat products increases the acidity of urine. Such foods are called *acid ash foods.* On the other hand, a vegetarian diet *(alkaline ash diet)* increases the alkalinity of the urine.* A bacterial infection of the urinary tract may also result in urine with a high pH.

Specific gravity is the relative weight of a specific volume of liquid compared with an equal volume of distilled water. The specific gravity of distilled water is 1.000, because 1 ml weighs 1 g. Since urine contains dissolved solutes, it weighs more than water, and its customary specific gravity ranges from 1.001 to 1.030. Urine with a specific gravity of 1.001 contains few solutes and is considered very dilute. Dilute urine commonly results when a person drinks excessive amounts of water, uses diuretics, or suffers from diabetes insipidus or chronic renal failure. Conditions that produce urine with a high specific gravity include limited fluid intake, fever, and kidney inflammation, called *pyelonephritis.* If urine becomes excessively concentrated, some of the substances normally held in solution begin to precipitate or crystallize, forming **kidney stones,** or **renal calculi.**

*The naming of foods as acid or alkaline *ash* derives from the fact that if an acid ash food is burned to ash, the pH of the ash is acidic. The ash of alkaline ash foods is alkaline.

Normal constituents of urine (in order of decreasing concentration) include water; urea;* sodium,† potassium, phosphate, and sulfate ions; creatinine;* and uric acid.* Much smaller but highly variable amounts of calcium, magnesium, and bicarbonate ions are also found in the urine. Abnormally high concentrations of any of these urinary constituents may indicate a pathological condition.

Abnormal Urinary Constituents

Abnormal urinary constituents are substances not normally present in the urine when the body is operating properly. ■

Glucose

The presence of glucose in the urine, a condition called **glycosuria,** indicates abnormally high blood sugar levels. Normally, blood sugar levels are maintained between 80 and 100 mg/100 ml of blood. At this level all glucose in the filtrate is reabsorbed by the tubular cells and returned to the blood. Glycosuria may result from carbohydrate intake so excessive that normal physiological and hormonal mechanisms cannot clear it from the blood quickly enough. In such cases, the active transport reabsorption mechanisms of the renal tubules for glucose are exceeded—but only temporarily.

Pathological glycosuria occurs in conditions such as uncontrolled diabetes mellitus, in which the body cells are unable to absorb glucose from the blood because the pancreatic islet cells produce inadequate amounts of the hormone insulin, or there is some abnormality of the insulin receptors. Under such circumstances, the body cells increase their metabolism of fats, and the excess and unusable glucose spills out in the urine.

Albumin

Albuminuria, or the presence of albumin in urine, is an abnormal finding. Albumin is the single most abundant blood protein and is very important in maintaining the osmotic pressure of the blood. Albumin, like other blood proteins, is too large to pass through the glomerular filtration membrane. Thus, albuminuria is generally indicative of abnormally increased permeability of the glomerular membrane. Certain nonpathological conditions, such as excessive exertion, pregnancy, or overabundant protein intake, can temporarily increase the membrane permeability, leading to **physiological albuminuria.** Pathological conditions resulting in albuminuria include events that damage the glomerular membrane, such as kidney trauma due to blows, the ingestion of poisons or heavy metals, bacterial toxins, glomerulonephritis, and hypertension.

*Urea, uric acid, and creatinine are the most important nitrogenous wastes found in urine. Urea is an end product of protein breakdown; uric acid is a metabolite of purine breakdown; and creatinine is associated with muscle metabolism of creatine phosphate.

†Sodium ions appear in relatively high concentration in the urine because of reduced urine volume, not because large amounts are being excreted. Sodium is the major positive ion in the plasma; under normal circumstances, most of it is actively reabsorbed.

Ketone Bodies

Ketone bodies (acetoacetic acid, beta-hydroxybutyric acid, and acetone) normally appear in the urine in very small amounts. **Ketonuria,** the presence of these intermediate products of fat metabolism in excessive amounts, usually indicates that abnormal metabolic processes are occurring. The result may be *acidosis* and its complications. Ketonuria is an expected finding during starvation, or diets very low in carbohydrates, when inadequate food intake forces the body to use its fat stores. Ketonuria coupled with a finding of glycosuria is generally diagnostic for diabetes mellitus.

Red Blood Cells

Hematuria, the appearance of red blood cells, or erythrocytes, in the urine, almost always indicates pathology of the urinary tract, because erythrocytes are too large to pass through the glomerular pores. Possible causes include irritation of the urinary tract organs by calculi (kidney stones), which produces frank bleeding; infection or tumors of the urinary tract; or physical trauma to the urinary organs. In healthy menstruating females, it may reflect accidental contamination of the urine sample with the menstrual flow.

Hemoglobin

Hemoglobinuria, the presence of hemoglobin in the urine, is a result of the fragmentation, or hemolysis, of red blood cells. As a result, hemoglobin is liberated into the plasma and subsequently appears in the kidney filtrate. Hemoglobinuria indicates various pathological conditions including hemolytic anemias, transfusion reactions, burns, poisonous snake bites, or renal disease.

Nitrites

The presence of urinary nitrites might indicate a bacterial infection, particularly *E. coli* or other gram-negative rods. Nitrites are valuable for early detection of bladder infections.

Bile Pigments

Bilirubinuria, the appearance of bilirubin (bile pigments) in urine, is an abnormal finding and usually indicates liver pathology, such as hepatitis, cirrhosis, or bile duct blockage. Bilirubinuria is signaled by a yellow foam that forms when the urine sample is shaken.

Urobilinogen is produced in the intestine from bilirubin and gives feces a brown color. Some urobilinogen is reabsorbed into the blood and either excreted back into the intestine by the liver or excreted by the kidneys in the urine. Complete absence of urobilinogen may indicate renal disease or obstruction of bile flow in the liver. Increased levels may indicate hepatitis A, cirrhosis, or biliary disease.

White Blood Cells

Pyuria is the presence of white blood cells or other pus constituents in the urine. It indicates inflammation of the urinary tract.

Casts

Any complete discussion of the varieties and implications of casts is beyond the scope of this exercise. However, because they always represent a pathological condition of the kidney or urinary tract, they should at least be mentioned. **Casts** are hardened cell fragments, usually cylindrical, which are formed in the distal convoluted tubules and collecting ducts and then flushed out of the urinary tract. Hyaline casts are formed from a mucoprotein secreted by tubule cells. These casts form when the filtrate flow rate is slow, the pH is low, or the salt concentration is high, all conditions which cause protein to denature. Red blood cell casts are typical in glomerulonephritis, as red blood cells leak through the filtration membrane and stick together in the tubules. White blood cell casts form when the kidney is inflamed, which is typically a result of pyelonephritis but sometimes occurs with glomerulonephritis. Degenerated renal tubule cells form granular or waxy casts. Broad waxy casts may indicate end-stage renal disease. Two cast types are shown in Figure 41.1 (page 626).

ACTIVITY 1

Analyzing Urine Samples

In this part of the exercise, you will use prepared dipsticks and perform chemical tests to determine the characteristics of normal urine as well as to identify abnormal urinary components. You will investigate two or more urine samples. The first, designated as the *standard urine specimen* in the chart on page 625, will be either yours or a "normal" sample provided by your instructor. The second will be an unknown urine specimen provided by your instructor. Make the following determinations on both samples, and record your results by circling the appropriate item or description or by adding data to complete the chart. If you have more than one unknown sample, accurately identify each sample by number.

! *Obtain and wear disposable gloves throughout this laboratory session.* Although the instructor-provided urine samples are actually artificial urine (concocted in the laboratory to resemble real urine), you should still observe the techniques of safe handling of body fluids as part of your learning process. When you have completed the laboratory procedures: (1) dispose of the gloves, used pH paper strips, and dipsticks in the autoclave bag; (2) put used glassware in the bleach-containing laboratory bucket; (3) wash the lab bench down with 10% bleach solution.

Determination of the Physical Characteristics of Urine

1. Determine the color, transparency, and odor of your "normal" sample and one of the numbered pathological samples, and circle the appropriate descriptions in the Urinalysis Results chart.

2. Obtain a roll of wide-range pH paper to determine the pH of each sample. Use a fresh piece of paper for each test, and dip the strip into the urine to be tested two or three times before comparing the color obtained with the chart on the dispenser. Record your results in the chart. (If you will be using one of the combination dipsticks—Chemstrip or Multistix—this pH determination can be done later.)

3. To determine specific gravity, obtain a urinometer cylinder and float. Mix the urine well, and fill the urinometer cylinder about two-thirds full with urine.

4. Examine the urinometer float to determine how to read its markings. In most cases, the scale has numbered lines separated by a series of unnumbered lines. The numbered lines

give the reading for the first two decimal places. You must determine the third decimal place by reading the lower edge of the meniscus—the curved surface representing the urine-air junction—on the stem of the float.

5. Carefully lower the urinometer float into the urine. Make sure it is floating freely before attempting to take the reading. Record the specific gravity of both samples in the chart. *Do not dispose of this urine if the samples that you have are less than 200 ml in volume* because you will need to make several more determinations.

Determination of Inorganic Constituents in Urine

Sulfates Using a 10-cc graduated cylinder, add 5 ml of urine to a test tube, and then add a few drops of dilute hydrochloric acid and 2 ml of 10% barium chloride solution. The appearance of a white precipitate (barium sulfate) indicates the presence of sulfates in the sample. Clean the graduated cylinder and the test tubes well after use. Record your results.

Phosphates Obtain a hot plate and a 500-ml beaker. To prepare the hot water bath, half fill the beaker with tap water and heat it on the hot plate. Add 5 ml of urine to a test tube, and then add three or four drops of dilute nitric acid and 3 ml of ammonium molybdate. Mix well with a glass stirring rod, and then heat gently in a hot water bath. Formation of a yellow precipitate indicates the presence of phosphates in the sample. Record your results.

Chlorides Place 5 ml of urine in a test tube, and add several drops of silver nitrate ($AgNO_3$). The appearance of a white precipitate (silver chloride) is a positive test for chlorides. Record your results.

Nitrites Use a combination dipstick to test for nitrites. Record your results.

Determination of Organic Constituents in Urine

Individual dipsticks or combination dipsticks (Chemstrip or Multistix) may be used for many of the tests in this section. If combination dipsticks are used, be prepared to take the readings on several factors (pH, protein [albumin], glucose, ketones, blood/hemoglobin, leukocytes, urobilinogen, bilirubin, and nitrites) at the same time. Generally speaking, results for all of these tests may be read *during* the second minute after immersion, but readings taken after 2 minutes have passed should be considered invalid. Pay careful attention to the directions for method and time of immersion and disposal of excess urine from the strip, regardless of the dipstick used. If you are testing your own urine and get an unanticipated result, it is helpful to know that most of the combination dipsticks produce false positive or negative results for certain solutes when the subject is taking vitamin C, aspirin, or certain drugs.

Urea Put two drops of urine on a clean microscope slide and *carefully* add one drop of concentrated nitric acid to the urine. Slowly warm the mixture on a hot plate until it begins to dry at the edges, but do not allow it to boil or to evaporate to dryness. When the slide has cooled, examine the edges of the preparation under low power to identify the rhombic or hexagonal crystals of urea nitrate, which form when urea and nitric acid react chemically. Keep the light low for best contrast. Record your results.

Glucose Use a combination dipstick or obtain a vial of Clinistix, and conduct the dipstick test according to the instructions on the vial. Record your results in the Urinalysis Results chart.

Because the Clinitest reagent is routinely used in clinical agencies for glucose determinations in pediatric patients (children), it is worthwhile to conduct this test as well. Obtain the Clinitest tablets and the associated color chart. You will need a timer (watch or clock with a second hand) for this test. Using a medicine dropper, put 5 drops of urine into a test tube; then rinse the dropper and add 10 drops of water to the tube. Add a Clinitest tablet. Wait 15 seconds and then compare the color obtained to the color chart. Record your results.

Albumin Use a combination dipstick or obtain the Albustix dipsticks, and conduct the determinations as indicated on the vial. Record your results.

Ketones Use a combination dipstick or obtain the Ketostix dipsticks. Conduct the determinations as indicated on the vial. Record your results.

Blood/Hemoglobin Test your urine samples for the presence of hemoglobin by using a Hemastix dipstick or a combination dipstick according to the directions on the vial. Usually a short drying period is required before making the reading, so read the directions carefully. Record your results.

Bilirubin Using a combination dipstick, determine if there is any bilirubin in your urine samples. Record your results.

Also conduct the Ictotest for the presence of bilirubin. Using a medicine dropper, place one drop of urine in the center of one of the special test mats provided with the Ictotest reagent tablets. Place one of the reagent tablets over the drop of urine, and then add two drops of water directly to the tablet. If the mixture turns purple when you add water, bilirubin is present. Record your results.

Leukocytes Use a combination dipstick to test for leukocytes. Record your results.

Urobilinogen Use a combination dipstick to test for urobilinogen. Record your results.

Clean up your area following the procedures on page 623.

Urinalysis Results*

Observation or test	Normal values	Standard urine specimen	Unknown specimen (# ______)
Physical Characteristics			
Color	Pale yellow	Yellow: pale medium dark other ______	Yellow: pale medium dark other ______
Transparency	Transparent	Clear Slightly cloudy Cloudy	Clear Slightly cloudy Cloudy
Odor	Characteristic	Describe: ______	Describe: ______
pH	4.5–8.0	______	______
Specific gravity	1.001–1.030	______	______
Inorganic Components			
Sulfates	Present	Present Absent	Present Absent
Phosphates	Present	Present Absent	Present Absent
Chlorides	Present	Present Absent	Present Absent
Nitrites	Absent	Present Absent	Present Absent
Organic Components			
Urea	Present	Present Absent	Present Absent
Glucose Dipstick: ______	Negative	Record results: ______	Record results: ______
Clinitest	Negative	______	______
Albumin Dipstick: ______	Negative	______	______
Ketone bodies Dipstick: ______	Negative	______	______
RBCs/hemoglobin Dipstick: ______	Negative	______	______
Bilirubin Dipstick: ______	Negative	______	______
Ictotest	Negative (no color change)	Negative Positive (purple)	Negative Positive (purple)
Leukocytes	Absent	Present Absent	Present Absent
Urobilinogen	Present	Present Absent	Present Absent

*Circle the appropriate description if provided; otherwise, record the results you observed. Identify dipsticks used.

ACTIVITY 2

Analyzing Urine Sediment Microscopically (Optional)

If your instructor so indicates, conduct a microscopic analysis of urine sediment (of "real" urine). The urine sample to be analyzed microscopically has been centrifuged to spin the more dense urine components to the bottom of a tube, and some of the sediment (mounted on a slide) has been stained with Sedi-stain to make the components more visible.

Go to the demonstration microscope to conduct this study. Using the lowest light source possible, examine the slide under low power to determine if any of the sediments illustrated in Figure 41.1 can be seen.

Unorganized sediments: Chemical substances that form crystals or precipitate from solution; for example, calcium oxalates, carbonates, and phosphates; uric acid; ammonium ureates; and cholesterol. Also, if one has been taking antibiotics or certain drugs such as sulfa drugs, these may be detectable in the urine in crystalline form. Normal urine contains very small amounts of crystals, but conditions such as urinary retention or urinary tract infection may cause the appearance of much larger amounts (and their possible consolidation into calculi). The high-power lens may be needed to view the various crystals, which tend to be much more minute than the organized (cellular) sediments.

Organized sediments: Include epithelial cells (rarely of any pathological significance), pus cells (white blood cells), red blood cells, and casts. Urine is normally negative for organized sediments, and the presence of the last three categories mentioned, other than trace amounts, always indicates kidney pathology.

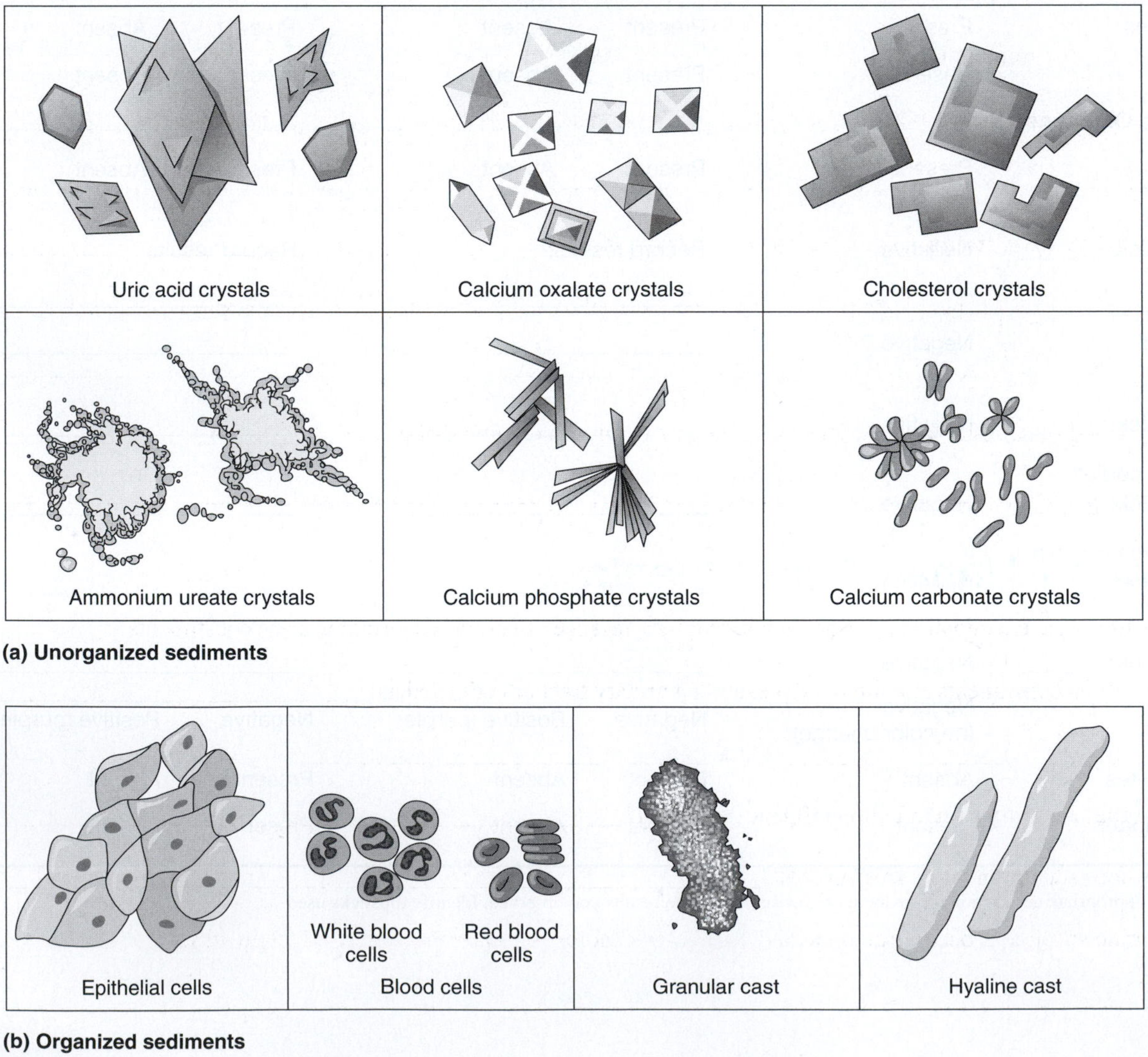
Uric acid crystals
Calcium oxalate crystals
Cholesterol crystals
Ammonium ureate crystals
Calcium phosphate crystals
Calcium carbonate crystals
(a) Unorganized sediments
Epithelial cells
White blood cells
Red blood cells
Blood cells
Granular cast
Hyaline cast
(b) Organized sediments

FIGURE 41.1 Examples of sediments. (a) Unorganized sediments. **(b)** Organized sediments.

NAME ______________________________

LAB TIME/DATE ______________________

REVIEW SHEET

EXERCISE 41

Urinalysis

Characteristics of Urine

1. What is the normal volume of urine excreted in a 24-hour period? ________

2. Assuming normal conditions, note whether each of the following substances would be (a) in greater relative concentration in the urine than in the glomerular filtrate, (b) in lesser concentration in the urine than in the glomerular filtrate, or (c) absent from both the urine and the glomerular filtrate.

________ 1. water	________ 6. amino acids	________ 11. uric acid
________ 2. phosphate ions	________ 7. glucose	________ 12. creatinine
________ 3. sulfate ions	________ 8. albumin	________ 13. pus (WBCs)
________ 4. potassium ions	________ 9. red blood cells	________ 14. nitrites
________ 5. sodium ions	________ 10. urea	

3. Explain why urinalysis is a routine part of any good physical examination. ______________________________

____

4. What substance is responsible for the normal yellow color of urine? ______________________________

5. Which has a greater specific gravity: 1 ml of urine or 1 ml of distilled water? ________ Explain your answer. __________

____

6. Explain the relationship between the color, specific gravity, and volume of urine. ______________________________

____

Abnormal Urinary Constituents

7. A microscopic examination of urine may reveal the presence of certain abnormal urinary constituents.

Name three constituents that might be present if a urinary tract infection exists. ______________________________,

______________________________, and ______________________________

8. How does a urinary tract infection influence urine pH? ______________________________

How does starvation influence urine pH? ______________________________

9. All urine specimens become alkaline and cloudy on standing at room temperature. Explain why. __________________

____

10. Several specific terms have been used to indicate the presence of abnormal urine constituents. Identify each of the abnormalities described below by inserting a term from the key at the right that names the condition.

______________ 1. presence of erythrocytes in the urine

______________ 2. presence of hemoglobin in the urine

______________ 3. presence of glucose in the urine

______________ 4. presence of albumin in the urine

______________ 5. presence of ketone bodies (acetone and others) in the urine

______________ 6. presence of pus (white blood cells) in the urine

Key:

a. albuminuria
b. glycosuria
c. hematuria
d. hemoglobinuria
e. ketonuria
f. pyuria

11. What are renal calculi, and what conditions favor their formation? ______________

____

12. Glucose and albumin are both normally absent in the urine, but the reason for their exclusion differs. Explain the reason for the absence of glucose. ______________

Explain the reason for the absence of albumin. ______________

13. The presence of abnormal constituents or conditions in urine may be associated with diseases, disorders, or other causes listed in the key. Select and list all conditions associated with each numbered item.

______ 1. low specific gravity

______ 2. high specific gravity

______ 3. glucose

______ 4. albumin

______ 5. blood cells

______ 6. hemoglobin

______ 7. bilirubin

______ 8. ketone bodies

______ 9. casts

______ 10. pus

Key:

a. cystitis (inflammation of the bladder)
b. diabetes insipidus
c. diabetes mellitus
d. eating a 5-lb box of sweets for lunch
e. glomerulonephritis
f. gonorrhea
g. hemolytic anemias
h. hepatitis, cirrhosis of the liver
i. kidney stones
j. pregnancy, exertion
k. pyelonephritis
l. starvation

14. Name the three major nitrogenous wastes found in the urine. ______________,

______________, and ______________

15. Explain the difference between organized and unorganized sediments. ______________

____

____

EXERCISE 42

Anatomy of the Reproductive System

MATERIALS

- ☐ Three-dimensional models or large laboratory charts of the male and female reproductive tracts
- ☐ Prepared slides of cross sections of the penis, seminal vesicles, epididymis, uterus showing endometrium (proliferative phase), and uterine tube
- ☐ Compound microscope

For instructions on animal dissections, see the dissection exercises starting on page 697 in the cat, fetal pig, and rat editions of this manual.

OBJECTIVES

1. To discuss the general function of the reproductive system.
2. To identify and name the structures of the male and female reproductive systems when provided with an appropriate model or diagram, and to discuss the general function of each.
3. To define *semen,* discuss its composition, and name the organs involved in its production.
4. To trace the pathway followed by a sperm from its site of formation to the external environment.
5. To name the exocrine and endocrine products of the testes and ovaries, indicating the cell types or structures responsible for the production of each.
6. To discuss the microscopic structure of the penis, seminal vesicles, epididymis, uterine (fallopian) tube, and uterus, and to relate structure to function.
7. To define *gonad, ejaculation,* and *erection*.
8. To discuss the function of the fimbriae and ciliated epithelium of the uterine tubes.
9. To identify the fundus, body, and cervical regions of the uterus.
10. To define *endometrium, myometrium*, and *ovulation*.
11. To discuss the reproduction-related function of female mammary glands.

PRE-LAB QUIZ

1. The essential organs of reproduction are the __________, which produce the germ cells.
   a. accessory male glands c. seminal vesicles
   b. gonads d. uterus
2. Circle the correct term. The paired oval testes lie in the scrotum / prostate outside the abdominopelvic cavity, where they are kept slightly cooler than body temperature.
3. After sperm are produced, they enter the first part of the duct system, the __________.
   a. ductus deferens c. epididymis
   b. ejaculatory duct d. urethra
4. The prostate, seminal vesicles, and bulbourethral glands produce __________, the liquid medium in which sperm leaves the body.
   a. seminal fluid c. urine
   b. testosterone d. water
5. Circle the correct term. The interstitial cells / seminiferous tubules produce testosterone, the hormonal product of the testis.
6. The endocrine products of the ovaries are estrogen and
   a. luteinizing hormone c. prolactin
   b. progesterone d. testosterone
7. Circle the correct term. The labia majora / clitoris are/is homologous to the penis.
8. The __________ is a pear-shaped organ that houses the embryo or fetus during its development.
   a. bladder c. uterus
   b. cervix d. vagina

Text continues on next page.

MasteringA&P™

Access practice quizzes and more in the Study Area at www.masteringaandp.com.

PAL

For access to anatomical models and more, check out Practice Anatomy Lab.

9. Circle the correct term. The endometrium / myometrium, the thick mucosal lining of the uterus, has a superficial layer that sloughs off periodically.

10. Circle the correct term. A developing egg is ejected from the ovary at the appropriate stage of maturity in an event known as menstruation / ovulation.

Other organ systems of the body function primarily to sustain the existing individual, but the reproductive system is unique. Most simply stated, the biological function of the **reproductive system** is to perpetuate the species.

The essential organs of reproduction are the **gonads,** the testes and the ovaries, which produce the germ cells. The reproductive role of the male is to manufacture sperm and to deliver them to the female reproductive tract. The female, in turn, produces eggs. If the time is suitable, the combination of sperm and egg produces a fertilized egg, which is the first cell of a new individual. Once fertilization has occurred, the female uterus provides a nurturing, protective environment in which the embryo, later called the fetus, develops until birth.

Gross Anatomy of the Human Male Reproductive System

The primary reproductive organs of the male are the **testes,** the male gonads, which have both an exocrine (sperm production) and an endocrine (testosterone production) function. All other reproductive structures are conduits or sources of secretions, which aid in the safe delivery of the sperm to the body exterior or female reproductive tract.

ACTIVITY 1

Identifying Male Reproductive Organs

As the following organs and structures are described, locate them on Figure 42.1, and then identify them on a three-dimensional model of the male reproductive system or on a large laboratory chart.

The paired oval testes lie in the **scrotum** outside the abdominopelvic cavity. The temperature there (approximately 94°F, or 34°C) is slightly lower than body temperature, a requirement for producing viable sperm.

The accessory structures forming the *duct system* are the epididymis, the ductus deferens, the ejaculatory duct, and the urethra. The **epididymis** is an elongated structure running up

FIGURE 42.1 Reproductive organs of the human male. (a) Sagittal view.

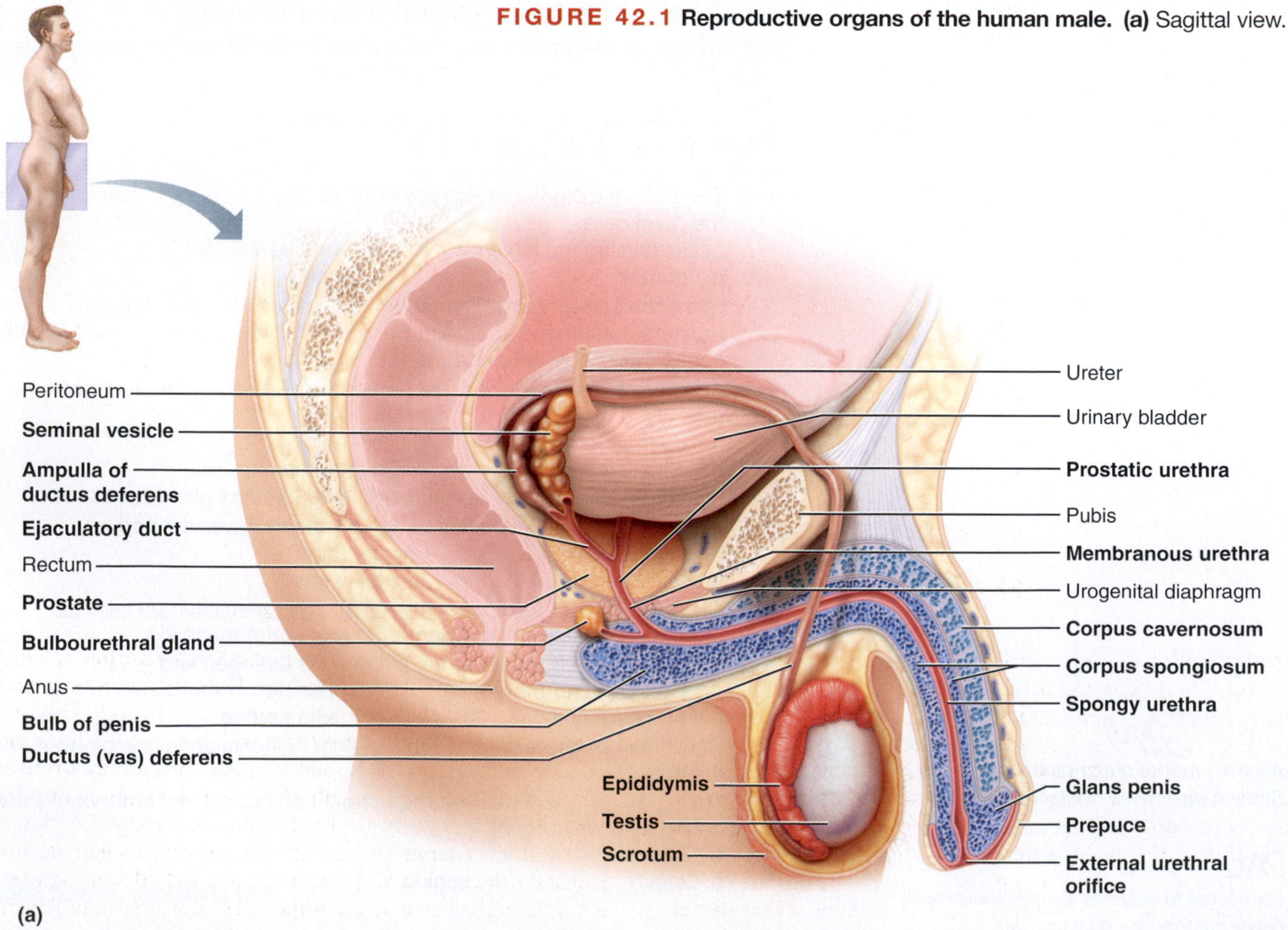
Peritoneum
Seminal vesicle
Ampulla of ductus deferens
Ejaculatory duct
Rectum
Prostate
Bulbourethral gland
Anus
Bulb of penis
Ductus (vas) deferens
Epididymis
Testis
Scrotum
Ureter
Urinary bladder
Prostatic urethra
Pubis
Membranous urethra
Urogenital diaphragm
Corpus cavernosum
Corpus spongiosum
Spongy urethra
Glans penis
Prepuce
External urethral orifice
(a)

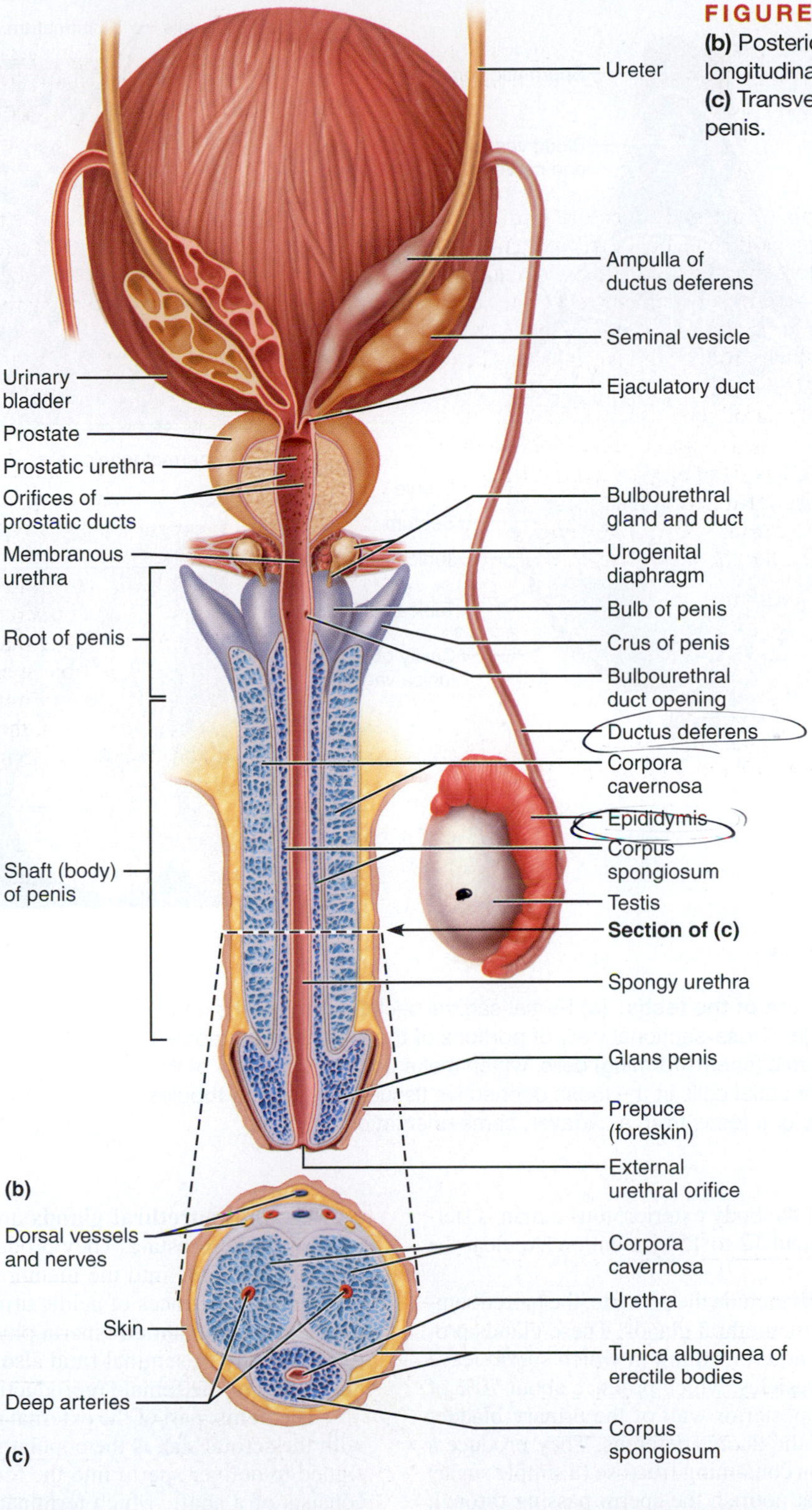

Ureter
Ampulla of ductus deferens
Seminal vesicle
Ejaculatory duct
Urinary bladder
Prostate
Prostatic urethra
Orifices of prostatic ducts
Membranous urethra
Root of penis
Bulbourethral gland and duct
Urogenital diaphragm
Bulb of penis
Crus of penis
Bulbourethral duct opening
Ductus deferens
Corpora cavernosa
Epididymis
Corpus spongiosum
Shaft (body) of penis
Testis
Section of (c)
Spongy urethra
Glans penis
Prepuce (foreskin)
External urethral orifice
(b)
Dorsal vessels and nerves
Corpora cavernosa
Urethra
Skin
Tunica albuginea of erectile bodies
Deep arteries
Corpus spongiosum
(c)

FIGURE 42.1 *(continued)* **(b)** Posterior view showing a longitudinal section of the penis. **(c)** Transverse section of the penis.

the posterolateral aspect of the testis and capping its superior aspect. The epididymis forms the first portion of the duct system and provides a site for immature sperm entering it from the testis to complete their maturation process. The **ductus deferens,** or **vas deferens** (sperm duct), arches superiorly from the epididymis, passes through the inguinal canal into the pelvic cavity, and courses over the superior aspect of the urinary bladder. In life, the ductus deferens is enclosed along with blood vessels and nerves in a connective tissue sheath called the **spermatic cord** (see Figure 42.2). The terminus of the ductus deferens enlarges to form the region called the **ampulla,** which empties into the **ejaculatory duct.** During **ejaculation,** contraction of the ejaculatory duct propels the sperm through the prostate to the **prostatic urethra,** which in turn empties into the **membranous urethra** and then into the **spongy urethra,** which runs through the length of the penis to the body exterior.

The spermatic cord is easily palpated through the skin of the scrotum. When a *vasectomy* is performed, a small incision is made in each side of the scrotum, and each ductus deferens is cut through or cauterized. Although sperm are still produced,

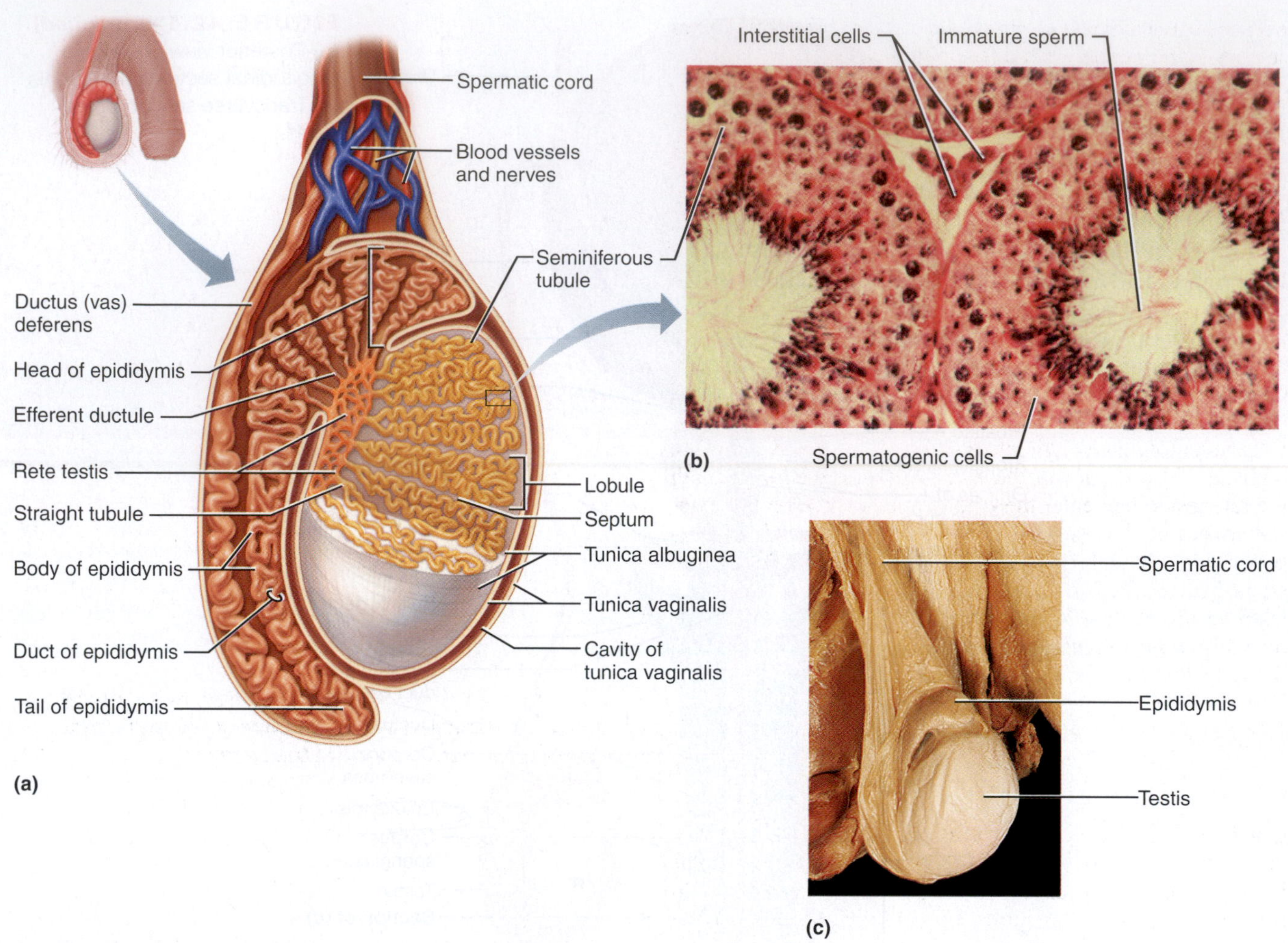
Spermatic cord
Blood vessels and nerves
Seminiferous tubule
Ductus (vas) deferens
Head of epididymis
Efferent ductule
Rete testis
Straight tubule
Body of epididymis
Duct of epididymis
Tail of epididymis
Lobule
Septum
Tunica albuginea
Tunica vaginalis
Cavity of tunica vaginalis
(a)
Interstitial cells
Immature sperm
Spermatogenic cells
(b)
Spermatic cord
Epididymis
Testis
(c)

FIGURE 42.2 Structure of the testis. (a) Partial sagittal section of the testis and associated epididymis. **(b)** Cross-sectional view of portions of the seminiferous tubules, showing the spermatogenic (sperm-forming) cells, which make up the epithelium of the tubule walls, and the interstitial cells in the loose connective tissue between the tubules (160×). **(c)** External view of a testis from a cadaver; same orientation as in (a).

they can no longer reach the body exterior; thus a man is sterile after this procedure (and 12 to 15 ejaculations to clear the conducting tubules).

The *accessory glands* include the prostate, the paired seminal vesicles, and the bulbourethral glands. These glands produce **seminal fluid,** the liquid medium in which sperm leave the body. The **seminal vesicles,** which produce about 70% of seminal fluid, lie at the posterior wall of the urinary bladder close to the terminus of the ductus deferens. They produce a viscous alkaline secretion containing fructose (a simple sugar) and other substances that nourish the sperm passing through the tract or that promote the fertilizing capability of sperm in some way. The duct of each seminal vesicle merges with a ductus deferens to form the ejaculatory duct (mentioned above); thus sperm and seminal fluid enter the urethra together.

The **prostate** encircles the urethra just inferior to the bladder. It secretes a milky fluid into the urethra, which plays a role in activating the sperm.

Hypertrophy of the prostate, a troublesome condition commonly seen in elderly men, constricts the urethra so that urination is difficult. ■

The **bulbourethral glands** are tiny, pea-shaped glands inferior to the prostate. They produce a thick, clear, alkaline mucus that drains into the membranous urethra. This secretion neutralizes traces of acidic urine in the urethra just prior to ejaculation of **semen** (sperm plus seminal fluid). The relative alkalinity of seminal fluid also buffers the sperm against the acidity of the female reproductive tract.

The **penis,** part of the external genitalia of the male along with the scrotal sac, is the copulatory organ of the male. Designed to deliver sperm into the female reproductive tract, it consists of a shaft, which terminates in an enlarged tip, the **glans penis** (see Figure 42.1a and b). The skin covering the penis is loosely applied, and it reflects downward to form a circular fold of skin, the **prepuce,** or **foreskin,** around the proximal end of the glans. (The foreskin is removed in the surgical procedure called *circumcision.*) Internally, the penis consists primarily of three elongated cylinders of erectile tissue, which engorge with blood during sexual excitement. This causes the penis to become rigid and enlarged so that it may more adequately serve as a penetrating device. This event is called **erection.** The paired dorsal cylinders are the corpora cavernosa. The single ventral **corpus spongiosum** surrounds the penile urethra (see Figure 42.1c).

Microscopic Anatomy of Selected Male Reproductive Organs

Each **testis** is covered by a dense connective tissue capsule called the **tunica albuginea** (literally, "white tunic"). Extensions of this sheath enter the testis, dividing it into a number of lobes, each of which houses one to four highly coiled **seminiferous tubules,** the sperm-forming factories (Figure 42.2). The seminiferous tubules of each lobe converge to empty the sperm into another set of tubules, the **rete testis,** at the posterior of the testis. Sperm traveling through the rete testis then enter the epididymis, located on the exterior aspect of the testis, as previously described. Lying between the seminiferous tubules and softly padded with connective tissue are the **interstitial cells,** which produce testosterone, the hormonal product of the testis. Unless directed by your instructor to conduct a microscopic study of the testis here, you will conduct that study in Exercise 43.

ACTIVITY 2

Penis

Obtain a slide of a cross section of the penis. Scan the tissue under low power to identify the urethra and the cavernous bodies. Compare your observations to Figure 42.1c and Figure 42.3. Observe the lumen of the urethra carefully. What type of epithelium do you see?

___

Explain the function of this type of epithelium.

___

___

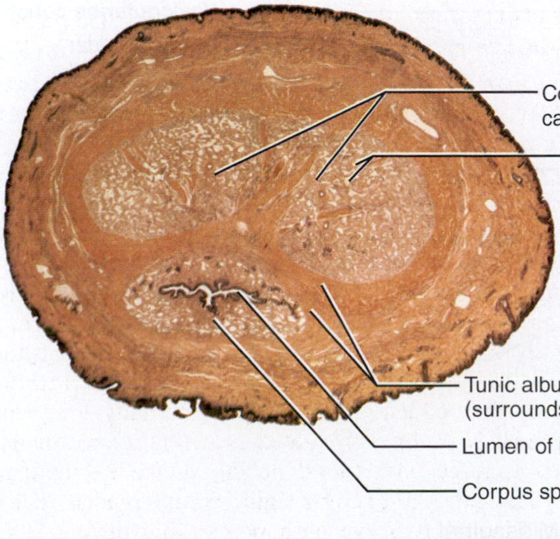
Corpora cavernosae
Venous cavities
Tunic albuginea (surrounds corpora)
Lumen of urethra
Corpus spongiosum

FIGURE 42.3 Transverse section of the penis (3×).

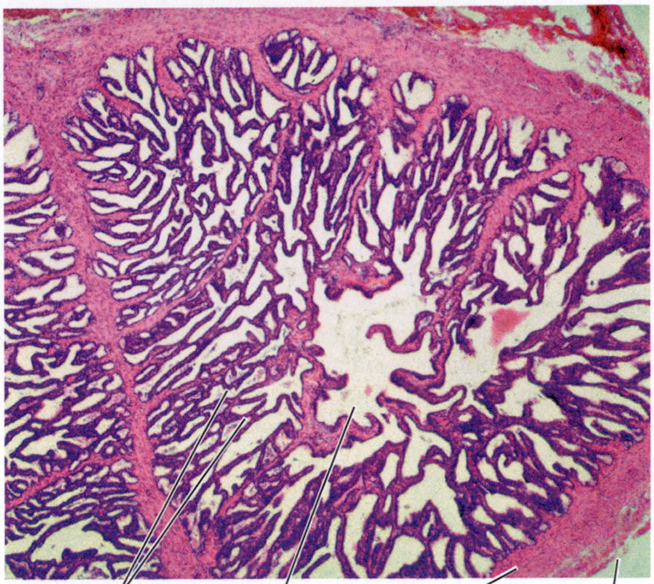
Mucosal folds
Lumen of seminal tubule
Muscular wall
Connective tissue

FIGURE 42.4 Cross-sectional view of a seminal vesicle with its elaborate network of mucosal folds. Vesicular secretion is seen in the lumen (7×).

ACTIVITY 3

Seminal Vesicle

Obtain a slide showing a cross-sectional view of the seminal vesicle. Examine the slide at low magnification to get an overall view of the highly folded mucosa of this gland. Switch to higher magnification, and notice that the folds of the vesicle protrude into the lumen where they divide further, giving the lumen a honeycomb look (Figure 42.4). Notice that the loose connective tissue lamina propria is underlain by smooth muscle fibers—first a circular layer, and then a longitudinal layer. Identify the vesicular secretion in the lumen, a viscous substance that is rich in sugar and prostaglandins.

ACTIVITY 4

Epididymis

Obtain a slide of a cross section of the epididymis. Notice the abundant tubule cross sections resulting from the fact that the coiling epididymis tubule has been cut through many times in the specimen. Look for sperm in the lumen of the tubule. Examine the composition of the tubule wall carefully. Identify the *stereocilia* (not shown) of the pseudostratified columnar epithelial lining. These nonmotile microvilli absorb excess fluid and pass nutrients to the sperm in the lumen. Now identify the smooth muscle layer. What do you think the function of the smooth muscle is?

___

___

Gross Anatomy of the Human Female Reproductive System

The **ovaries** (female gonads) are the primary reproductive organs of the female. Like the testes of the male, the ovaries produce both an exocrine product (eggs, or ova) and endocrine products (estrogens and progesterone). The other accessory structures of the female reproductive system transport, house, nurture, or otherwise serve the needs of the reproductive cells and/or the developing fetus.

The reproductive structures of the female are generally considered in terms of internal organs and external organs, or external genitalia.

ACTIVITY 5

Identifying Female Reproductive Organs

As you read the descriptions of these structures, locate them on Figures 42.5 and 42.6 and then on the female reproductive system model or large laboratory chart.

External Genitalia

The **external genitalia (vulva)** consist of the mons pubis, the labia majora and minora, the clitoris, the urethral and vaginal orifices, the hymen, and the greater vestibular glands. The **mons pubis** is a rounded fatty eminence overlying the pubic symphysis. Running inferiorly and posteriorly from the mons pubis are two elongated, pigmented, hair-covered skin folds, the **labia majora,** which are homologous to the scrotum. These enclose two smaller hair-free folds, the **labia minora.** (Terms indicating only one of the two folds in each case are *labium majus* and *minus,* respectively.) The labia minora, in turn, enclose a region called the **vestibule,** which contains many structures—

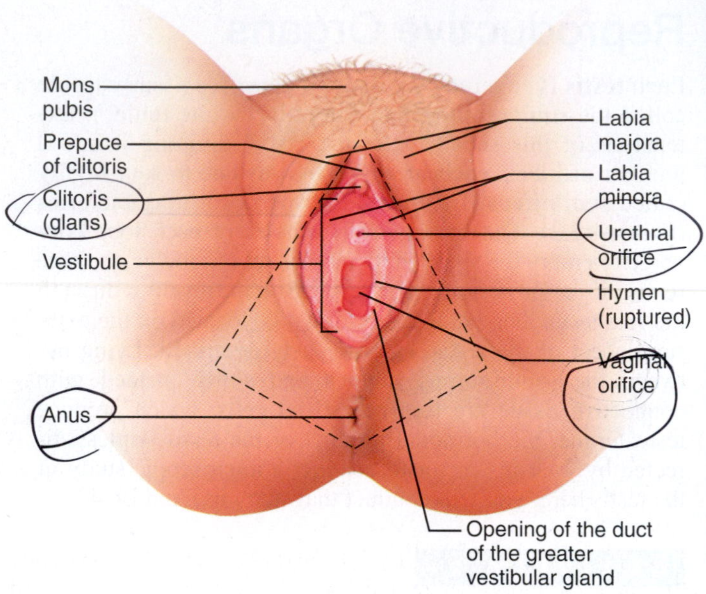

Mons pubis
Prepuce of clitoris
Clitoris (glans)
Vestibule
Anus
Labia majora
Labia minora
Urethral orifice
Hymen (ruptured)
Vaginal orifice
Opening of the duct of the greater vestibular gland

FIGURE 42.5 External genitalia of the human female. The region enclosed by dashed lines is the perineum.

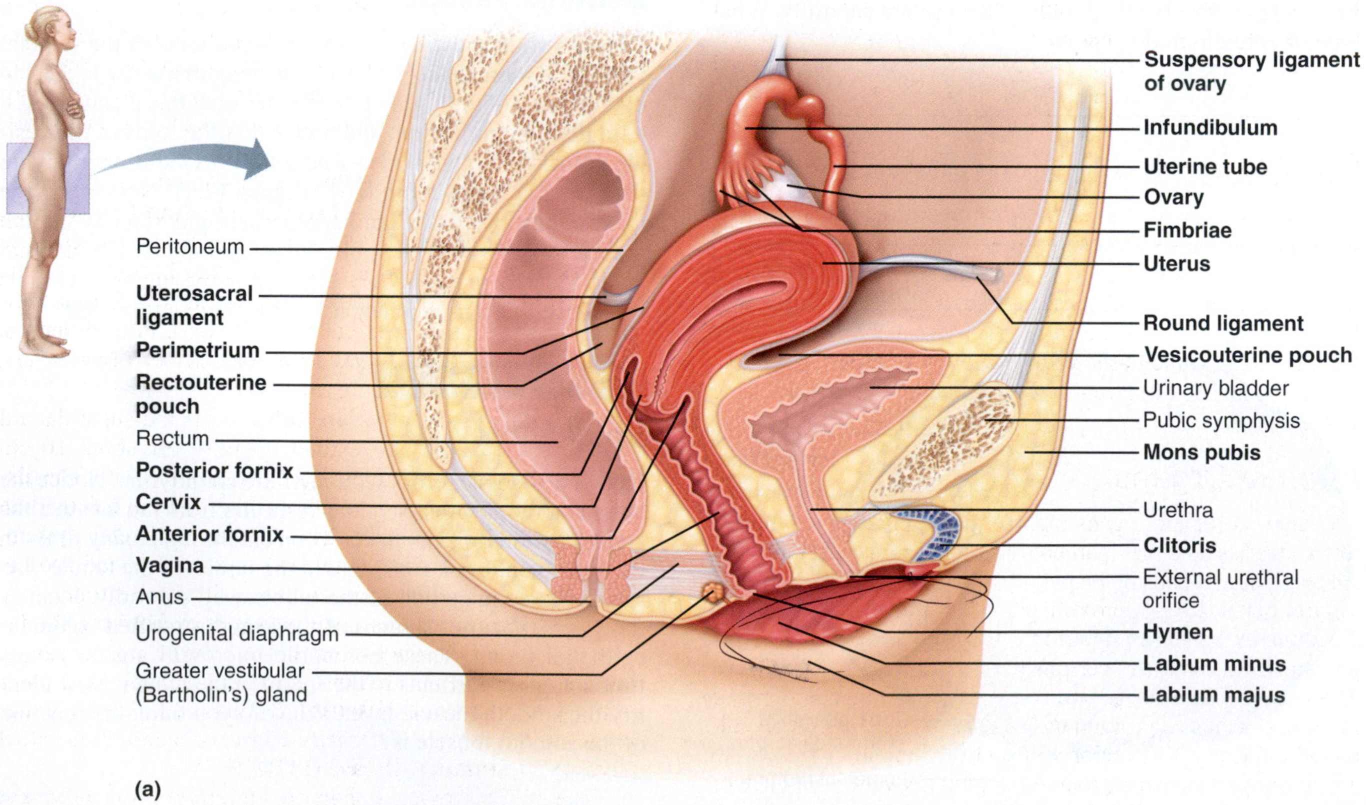

Peritoneum
Uterosacral ligament
Perimetrium
Rectouterine pouch
Rectum
Posterior fornix
Cervix
Anterior fornix
Vagina
Anus
Urogenital diaphragm
Greater vestibular (Bartholin's) gland
Suspensory ligament of ovary
Infundibulum
Uterine tube
Ovary
Fimbriae
Uterus
Round ligament
Vesicouterine pouch
Urinary bladder
Pubic symphysis
Mons pubis
Urethra
Clitoris
External urethral orifice
Hymen
Labium minus
Labium majus
(a)

FIGURE 42.6 Internal reproductive organs of the human female. (a) Midsagittal section of the human female reproductive system.

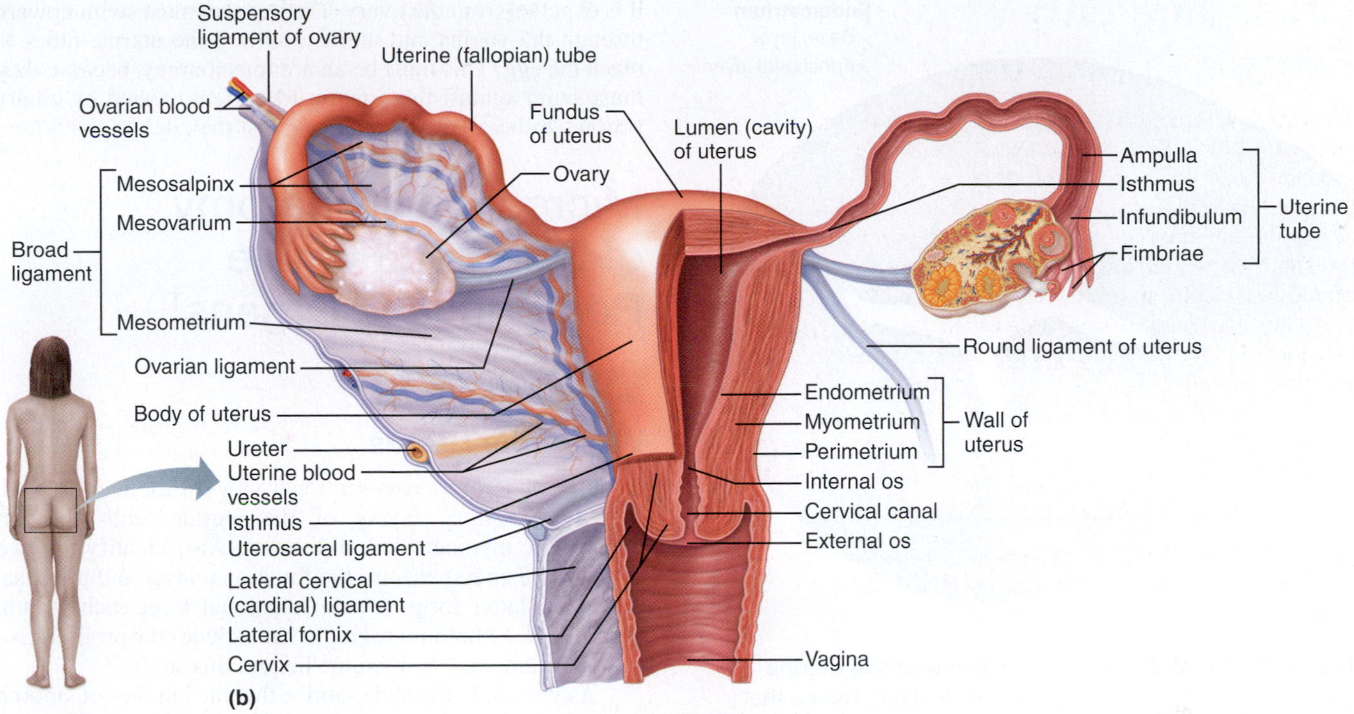

Suspensory ligament of ovary
Uterine (fallopian) tube
Ovarian blood vessels
Fundus of uterus
Lumen (cavity) of uterus
Ovary
Ampulla
Isthmus
Infundibulum
Fimbriae
Uterine tube
Broad ligament
Mesosalpinx
Mesovarium
Mesometrium
Ovarian ligament
Body of uterus
Round ligament of uterus
Endometrium
Myometrium
Perimetrium
Wall of uterus
Ureter
Uterine blood vessels
Internal os
Isthmus
Cervical canal
Uterosacral ligament
External os
Lateral cervical (cardinal) ligament
Lateral fornix
Cervix
Vagina
(b)

FIGURE 42.6 ***(continued)*** **(b)** Posterior view. The posterior walls of the vagina, uterus, and uterine tubes, and the broad ligament have been removed on the right side to reveal the shape of the lumen of these organs.

the clitoris, most anteriorly, followed by the urethral orifice and the vaginal orifice. The diamond-shaped region between the anterior end of the labial folds, the ischial tuberosities laterally, and the anus posteriorly is called the **perineum.**

The **clitoris** is a small protruding structure, homologous to the penis. Like its counterpart, it is composed of highly sensitive, erectile tissue. It is hooded by skin folds of the anterior labia minora, referred to as the **prepuce of the clitoris.** The urethral orifice, which lies posterior to the clitoris, is the outlet for the urinary system and has no reproductive function in the female. The vaginal opening is partially closed by a thin fold of mucous membrane called the **hymen** and is flanked by the pea-sized, mucus-secreting **greater vestibular glands.** These glands (see Figure 42.5) lubricate the distal end of the vagina during coitus.

Internal Organs

The internal female organs include the vagina, uterus, uterine tubes, ovaries, and the ligaments and supporting structures that suspend these organs in the pelvic cavity (see Figure 42.6). The **vagina** extends for approximately 10 cm (4 inches) from the vestibule to the uterus superiorly. It serves as a copulatory organ and birth canal and permits passage of the menstrual flow. The pear-shaped **uterus,** situated between the bladder and the rectum, is a muscular organ with its narrow end, the **cervix,** directed inferiorly. The major portion of the uterus is referred to as the **body;** its superior rounded region above the entrance of the uterine tubes is called the **fundus.** A fertilized egg is implanted in the uterus, which houses the embryo or fetus during its development.

In some cases, the fertilized egg may implant in a uterine tube or even on the abdominal viscera, creating an **ectopic pregnancy.** Such implantations are usually unsuccessful and may even endanger the mother's life because the uterine tubes cannot accommodate the increasing size of the fetus. ■

The **endometrium,** the thick mucosal lining of the uterus, has a superficial **functional layer,** or **stratum functionalis,** that sloughs off periodically (about every 28 days) in response to cyclic changes in the levels of ovarian hormones in the woman's blood. This sloughing-off process, which is accompanied by bleeding, is referred to as **menstruation,** or **menses.** The deeper **basal layer,** or **stratum basalis** (see Figure 42.7), forms a new functionalis after menstruation ends.

The **uterine,** or **fallopian, tubes** enter the superolateral region of the uterus and extend laterally for about 10 cm (4 inches) toward the ovaries in the peritoneal cavity. The distal ends of the tubes are funnel-shaped and have fingerlike projections called **fimbriae.** Unlike in the male duct system, there is no actual contact between the female gonad and the initial part of the female duct system—the uterine tube.

Because of this open passageway between the female reproductive organs and the peritoneal cavity, reproductive system infections, such as gonorrhea and other **sexually transmitted diseases (STDs),** can cause widespread inflammations of the pelvic viscera, a condition called **pelvic inflammatory disease (PID).** ■

The internal female organs are all retroperitoneal, except the ovaries. They are supported and suspended somewhat freely by ligamentous folds of peritoneum. The peritoneum takes an undulating course. From the pelvic cavity floor it

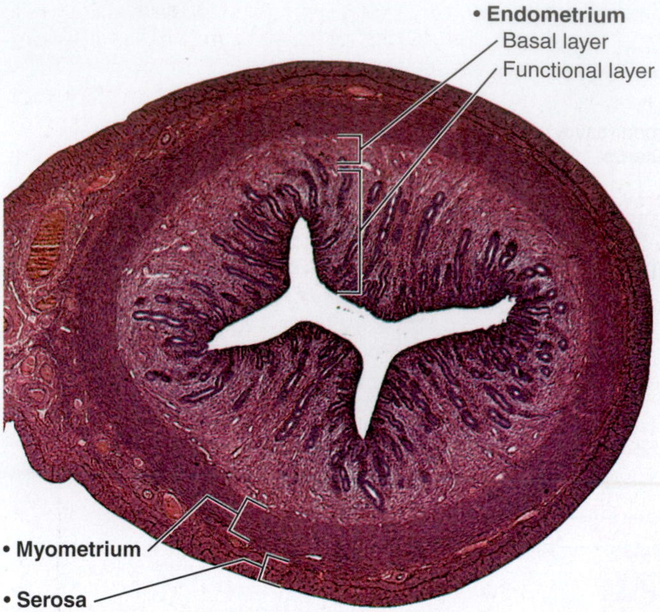
• Endometrium
Basal layer
Functional layer
• Myometrium
• Serosa

FIGURE 42.7 Cross-sectional view of the uterine wall. The mucosa is in the proliferative stage. Notice that the basal layer has virtually no glands compared to the functional layer (5×).

moves superiorly over the top of the bladder, reflects over the anterior and posterior surfaces of the uterus, and then over the rectum, and up the posterior body wall. The fold that encloses the uterine tubes and uterus and secures them to the lateral body walls is the **broad ligament** (Figure 42.6b). The part of the broad ligament specifically anchoring the uterus is called the **mesometrium** and that anchoring the uterine tubes, the **mesosalpinx.** The **round ligaments,** fibrous cords that run from the uterus to the labia majora, and the **uterosacral ligaments,** which course posteriorly to the sacrum, also help attach the uterus to the body wall. The ovaries are supported medially by the **ovarian ligament** (extending from the uterus to the ovary), laterally by the **suspensory ligaments,** and posteriorly by a fold of the broad ligament, the **mesovarium.**

Within the ovaries, the female gametes (eggs) begin their development in saclike structures called *follicles.* The growing follicles also produce *estrogens.* When a developing egg has reached the appropriate stage of maturity, it is ejected from the ovary in an event called **ovulation.** The ruptured follicle is then converted to a second type of endocrine gland, called a *corpus luteum,* which secretes progesterone (and some estrogens).

The flattened almond-shaped ovaries lie adjacent to the uterine tubes but are not connected to them; consequently, an ovulated egg* enters the pelvic cavity. The waving fimbriae of the uterine tubes create fluid currents that, if successful, draw the egg into the lumen of the uterine tube, where it begins its passage to the uterus, propelled by the cilia of the tubule walls. The usual and most desirable site of fertilization is the uterine tube, because the journey to the uterus takes about 3 to 4 days and an egg is viable for up to 24 hours after it is expelled from the ovary. Thus, sperm must swim upward through the vagina and uterus and into the uterine tubes to reach the egg. This must be an arduous journey, because they must swim against the downward current created by ciliary action—rather like swimming against the tide!

*To simplify this discussion, the ovulated cell is called an egg. What is actually expelled from the ovary is an earlier stage of development called a secondary oocyte. These matters are explained in more detail in Exercise 43.

Microscopic Anatomy of Selected Female Reproductive Organs†

ACTIVITY 6

Wall of the Uterus

Obtain a slide of a cross-sectional view of the uterine wall. Identify the three layers of the uterine wall—the endometrium, myometrium, and serosa. Also identify the two strata of the endometrium: the functional layer and the basal layer. The latter forms a new functional layer each month. Figure 42.7, a photomicrograph that includes the proliferative endometrium, may be of some help in this study.

As you study the slide, notice that the bundles of smooth muscle are oriented in several different directions. What is the function of the **myometrium** (smooth muscle layer) during the birth process?

____

____

ACTIVITY 7

Uterine Tube

Obtain a slide of a cross-sectional view of a uterine tube for examination. Using Figure 42.8 as a guide, notice the highly folded mucosa (the folds nearly fill the tubule lumen). Then switch to high power to examine the ciliated secretory epithelium.

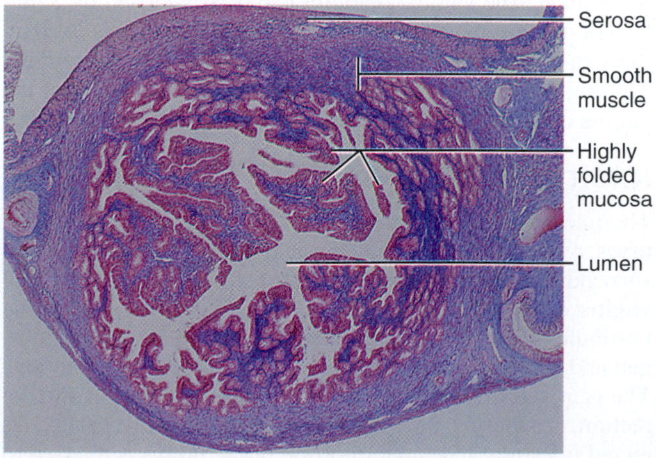
Serosa
Smooth muscle
Highly folded mucosa
Lumen

FIGURE 42.8 Cross-sectional view of the uterine tube (7×).

†A microscopic study of the ovary is described in Exercise 43.

The Mammary Glands

The **mammary glands** exist within the breasts, of course, in both sexes, but they normally have a reproduction-related function only in females. Since the function of the mammary glands is to produce milk to nourish the newborn infant, their importance is more closely associated with events that occur when reproduction has already been accomplished. Periodic stimulation by the female sex hormones, especially estrogens, increases the size of the female mammary glands at puberty. During this period, the duct system becomes more elaborate, and fat is deposited—fat deposition being the more important contributor to increased breast size.

The rounded, skin-covered mammary glands lie anterior to the pectoral muscles of the thorax, attached to them by connective tissue. Slightly below the center of each breast is a pigmented area, the **areola,** which surrounds a centrally protruding **nipple** (Figure 42.9.)

Internally each mammary gland consists of 15 to 25 **lobes** which radiate around the nipple and are separated by fibrous connective tisse and adipose, or fatty, tissue. Within each lobe are smaller chambers called **lobules,** containing the glandular **alveoli** that produce milk during lactation. The alveoli of each lobule pass the milk into a number of **lactiferous ducts,** which join to form an expanded storage chamber, the **lactiferous sinus,** as they approach the nipple. The sinuses open to the outside at the nipple.

For instructions on animal dissections, see the dissection exercises starting on page 697 in the cat, fetal pig, and rat editions of this manual.

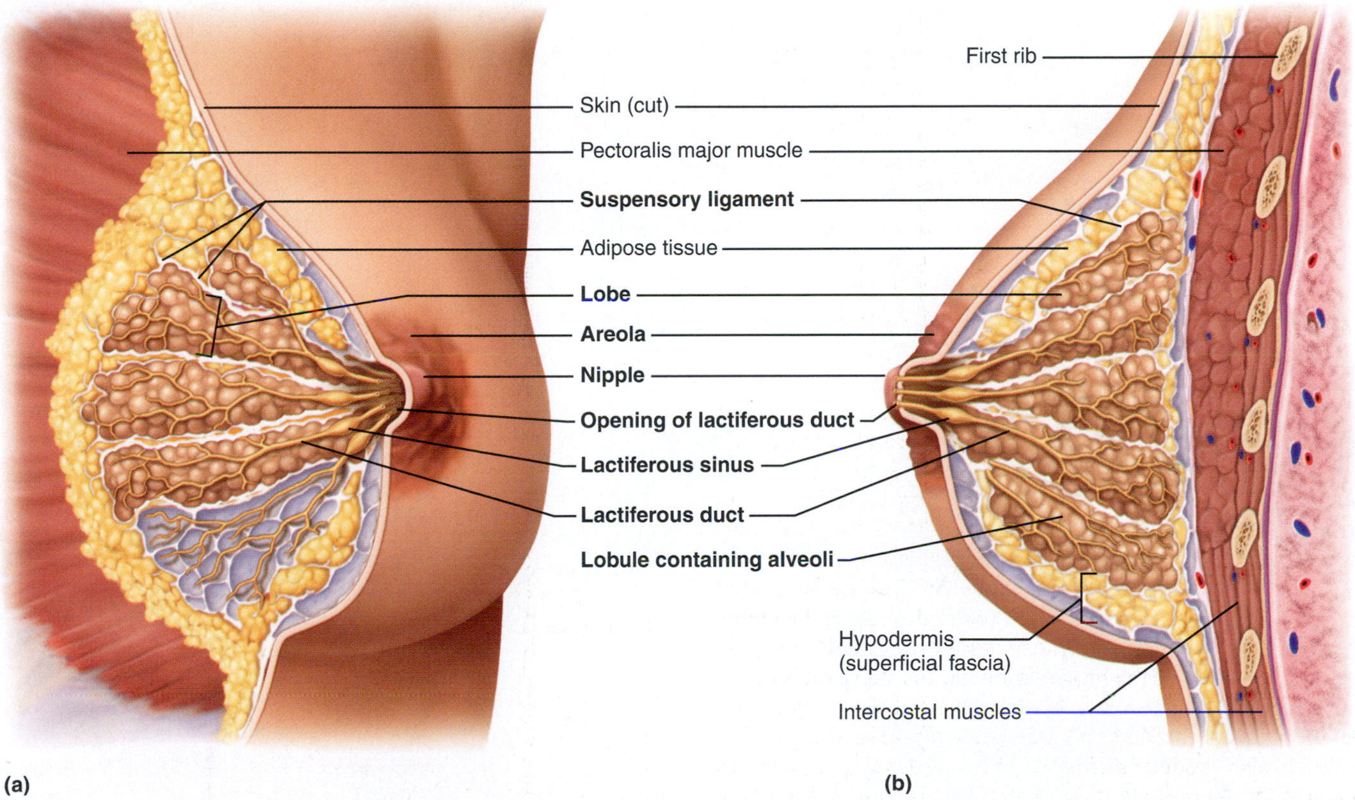
First rib
Skin (cut)
Pectoralis major muscle
Suspensory ligament
Adipose tissue
Lobe
Areola
Nipple
Opening of lactiferous duct
Lactiferous sinus
Lactiferous duct
Lobule containing alveoli
Hypodermis (superficial fascia)
Intercostal muscles
(a)
(b)

FIGURE 42.9 Anatomy of lactating mammary gland. (a) Anterior view of partially dissected breast. **(b)** Sagittal section of the breast.

EXERCISE 43

Physiology of Reproduction: Gametogenesis and the Female Cycles

MATERIALS

- ☐ Three-dimensional models illustrating meiosis, spermatogenesis, and oogenesis
- ☐ Sets of "pop it" beads in two colors with magnetic centromeres*
- ☐ Compound microscope
- ☐ Prepared slides of testis and human sperm
- ☐ *Demonstration:* microscopes set up to demonstrate the stages of oogenesis in *Ascaris megalocephala* listed below
  - Slide 1: Primary oocyte with fertilization membrane, sperm nucleus, and aligned tetrads apparent
  - Slide 2: Formation of the first polar body
  - Slide 3: Secondary oocyte with dyads aligned
  - Slide 4: Formation of the ovum and second polar body
  - Slide 5: Fusion of the male and female pronuclei to form the fertilized egg
- ☐ Prepared slides of ovary and uterine endometrium (showing menstrual, proliferative, and secretory stages)

*Chromosome Simulation Lab Activity available from Ward's Natural Science

OBJECTIVES

1. To define *meiosis, gametogenesis, oogenesis, spermatogenesis, spermiogenesis, synapsis, haploid, diploid,* and *menses*.
2. To relate the stages of spermatogenesis to the cross-sectional structure of the seminiferous tubule.
3. To discuss the microscopic structure of the ovary; to be prepared to identify primary, secondary, and vesicular follicles and the corpus luteum; and to state the hormonal products of the last two structures.
4. To relate the stages of oogenesis to follicle development in the ovary.
5. To cite similarities and differences between mitosis and meiosis and between spermatogenesis and oogenesis.
6. To describe sperm anatomy, and to relate it to function.
7. To discuss the phases and control of the menstrual cycle.
8. To discuss the effect of FSH and LH on the ovary, and to describe the feedback relationship between anterior pituitary gonadotropins and ovarian hormones.
9. To describe the effect of FSH and LH on testicular function.

PRE-LAB QUIZ

1. Human gametes contain _________ chromosomes.
   a. 13 b. 23 c. 36 d. 46
2. The end product of meiosis is
   a. two diploid daughter cells
   b. two haploid daughter cells
   c. four diploid daughter cells
   d. four haploid daughter cells
3. Circle the correct terms. A grouping of four chromatids, known as a dyad / tetrad, occurs only during mitosis / meiosis.
4. _________ cells extend inward from the periphery of the seminiferous tubule and provide nourishment to the spermatids as they begin their transformation into sperm.
   a. Interstitial
   b. Leydig
   c. Sustentacular
5. Circle the correct term. The acrosome / midpiece of the sperm contains enzymes involved in the penetration of the egg.
6. Circle the correct term. Within each ovary, the immature ovum develops in a saclike structure called a corpus / follicle.
7. As the primordial follicle grows and its epithelium changes from squamous to cuboidal cells, it becomes a(n) _________ and begins to produce estrogens.
   a. oocyte
   b. oogonia
   c. primary follicle
   d. primary ovum
8. Circle True or False. A sudden release of luteinizing hormone by the anterior pituitary triggers ovulation.

Text continues on next page.

MasteringA&P™

Access practice quizzes and more in the Study Area at www.masteringaandp.com.

9. Circle the correct term. The corpus luteum / corpus albicans is a solid glandular structure with a scalloped lumen that develops from a ruptured follicle.

10. The __________ phase of the female cycle occurs from days 1–5 and is signaled by the sloughing off of the thick functional layer of the endometrium.
   a. endometrial
   b. menstrual
   c. proliferative
   d. secretory

Every human being, so far, has developed from the union of gametes. Gametes, produced only in the testis or ovary, are unique cells. They have only half the normal chromosome number (designated as ***n,*** or the **haploid complement**) seen in all other body cells.

Meiosis

In humans, gametes have 23 chromosomes instead of the 46 present in other tissue cells. Theoretically, every gamete has a full set of genetic instructions, a conclusion borne out by the observation that some animals can develop from an egg that is artificially stimulated, as by a pinprick, rather than by sperm entry. (This is less true of the sperm, because some have an incomplete sex chromosome.)

Gametogenesis, the process of gamete formation, involves reduction of the chromosome number by half. This event is important to maintain the characteristic chromosomal number of the species generation after generation. Otherwise there would be a doubling of chromosome number with each succeeding generation, and the cells would become so chock-full of genetic material there would be little room for anything else.

Egg and sperm chromosomes that carry genes for the same traits are called **homologous chromosomes.** When the sperm and egg fuse to form the **zygote,** or fertilized egg, it is said to contain 23 pairs of homologous chromosomes, or the **diploid (2*n*)** chromosome number of 46. The zygote, once formed, then divides to produce the cells needed to construct the multicellular human body. All cells of the developing human body have a chromosome content exactly identical in quality and quantity to that of the fertilized egg. This is assured by the nuclear division process called **mitosis.** (Mitosis was considered in depth in Exercise 4. You may want to review it at this time.)

To produce gametes with the reduced (haploid) chromosomal number, **meiosis,** a specialized type of nuclear division, occurs in the ovaries and testes during gametogenesis. Before meiosis begins, the chromosomes are replicated in the *mother cell* or stem cell just as they are before mitosis. As a result, the mother cell briefly has double the normal diploid genetic complement. The stem cell then undergoes two consecutive nuclear divisions, termed *meiosis I* and *II,* or the first and second maturation divisions, *without replicating the chromosomes before the second division.* Cytokinesis produces four haploid daughter cells, rather than the two diploid daughter cells resulting from mitotic division.

The entire process of meiosis is complex and is dealt with here only to the extent necessary to reveal important differences between this type of nuclear division and mitosis. Essentially, each meiotic division involves the same phases and events seen in mitosis (prophase, metaphase, anaphase, and telophase), but during the first maturation division (meiosis I) an event not seen in mitosis occurs during prophase. The homologous chromosomes, each now a duplicated structure, begin to pair so that they become closely aligned along their entire length. This pairing is called **synapsis.** As a result, 23 **tetrads** (groupings of four chromatids) form, become attached to the spindle fibers, and begin to align themselves on the spindle equator. While in synapsis, the "arms" of adjacent homologous chromosomes coil around each other, forming many points of **crossover,** or **chiasmata.** (Perhaps this could be called the conjugal bed of the cell!) When anaphase of meiosis I begins, the homologues separate from one another, breaking and exchanging parts at points of crossover, and move apart toward opposite poles of the cell. The centromeres holding the "sister" *chromatids* (threads of chromatin), or **dyads,** together do *not* break at this point (Figure 43.1).

During the second maturation division, events parallel those in mitosis, except that the daughter cells do *not* replicate their chromosomes before this division, and each daughter cell has only half of the homologous chromosomes rather than a complete set. The crossover events and the way in which the homologues align on the spindle equator during the

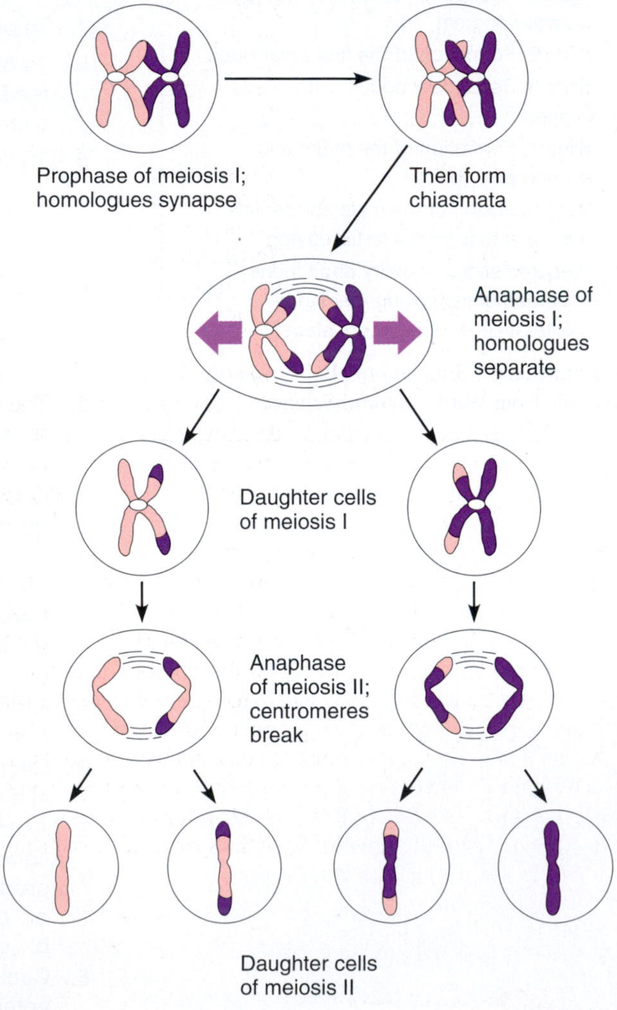

Prophase of meiosis I; homologues synapse
Then form chiasmata
Anaphase of meiosis I; homologues separate
Daughter cells of meiosis I
Anaphase of meiosis II; centromeres break
Daughter cells of meiosis II

FIGURE 43.1 Events of meiosis involving one pair of homologous chromosomes. Male homologue is purple; female homologue is pink.

first maturation division introduce an immense variability in the resulting gametes, which explains why, barring identical twins, we are all genetically unique.

ACTIVITY 1

Identifying Meiotic Phases and Structures

1. Obtain a model depicting the events of meiosis, and follow the sequence of events during the first and second maturation divisions. Identify prophase, metaphase, anaphase, and telophase in each. Also identify tetrads and chiasmata during the first maturation division and dyads (groupings of two chromatids connected by centromeres) in the second maturation division. Note ways in which the daughter cells resulting from meiosis I differ from the mother cell and how the gametes differ from both cell populations. (Use the key on the model, your textbook, or an appropriate reference as necessary to aid you in these observations.)

2. Using strings of colored "pop it" beads with magnetic centromeres, demonstrate the phases of meiosis, including crossing over, for a cell with a diploid (2*n*) number of 4. Use one bead color for the male chromosomes and another color for the female chromosomes.

3. Ask your instructor to verify the accuracy of your "creation" before returning the beads to the supply area.

Spermatogenesis

Human sperm production, or **spermatogenesis,** begins at puberty and continues without interruption throughout life. The average male ejaculation contains about a quarter billion sperm. Because only one sperm fertilizes an ovum, it seems that nature has tried to ensure that the perpetuation of the species will not be endangered for lack of sperm.

As explained in Exercise 42, spermatogenesis (the process of gametogenesis in males, Figure 43.2) occurs in the seminiferous tubules of the testes. The primitive stem cells or **spermatogonia,** found at the tubule periphery, divide extensively to build up the stem cell line. Before puberty, all divisions are mitotic divisions that produce more spermatogonia. At puberty, however, under the influence of follicle-stimulating hormone (FSH) secreted by the anterior pituitary gland, each mitotic division of a spermatogonium produces one spermatogonium and one **primary spermatocyte,** which is destined to undergo meiosis. As meiosis occurs, the dividing cells approach the lumen of the tubule. Thus the progression of meiotic events can be followed from the tubule periphery to the lumen. It is important to recognize that **spermatids,** haploid cells that are the actual product of meiosis, are not functional gametes. They are nonmotile cells and have too much excess baggage to function well in a reproductive capacity. Another process, called **spermiogenesis,** which follows meiosis, strips away the extraneous cytoplasm from the spermatid, converting it to a motile, streamlined **sperm.**

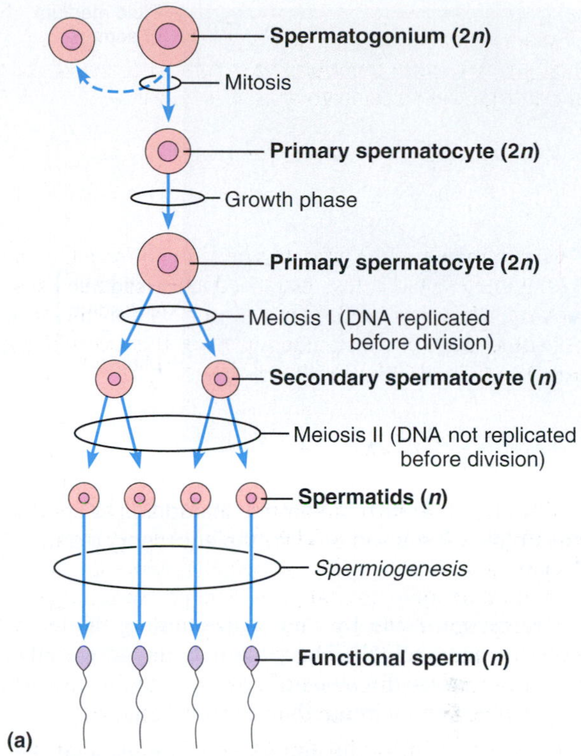

Spermatogonium (2n)
Mitosis
Primary spermatocyte (2n)
Growth phase
Primary spermatocyte (2n)
Meiosis I (DNA replicated before division)
Secondary spermatocyte (n)
Meiosis II (DNA not replicated before division)
Spermatids (n)
Spermiogenesis
Functional sperm (n)

(a)

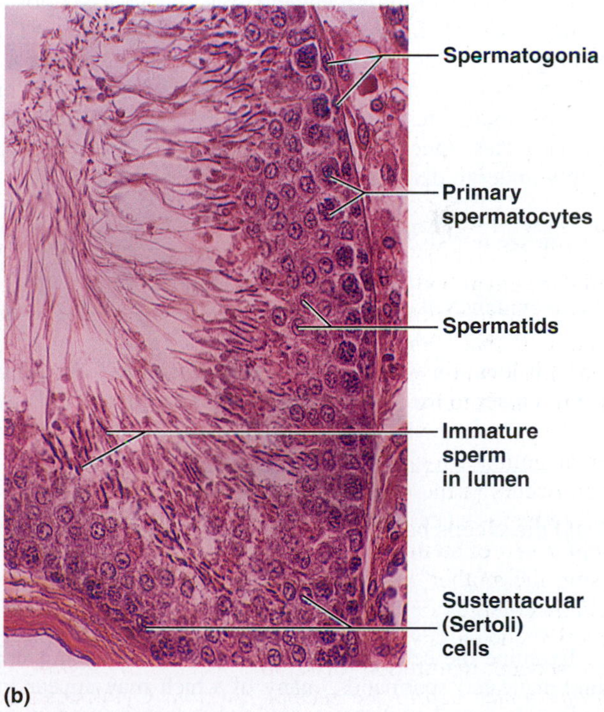

Spermatogonia
Primary spermatocytes
Spermatids
Immature sperm in lumen
Sustentacular (Sertoli) cells

(b)

FIGURE 43.2 Spermatogenesis. (a) Flowchart of meiotic events and spermiogenesis. **(b)** Micrograph of an active seminiferous tubule (230×).

ACTIVITY 2

Examining Events of Spermatogenesis

1. Obtain a slide of the testis and a microscope. Examine the slide under low power to identify the cross-sectional views of the cut seminiferous tubules. Then rotate the high power lens into position and observe the wall of one of the cut tubules. As you work, refer to Figure 43.2 to make the following identifications.

2. Scrutinize the cells at the periphery of the tubule. The cells in this area are the spermatogonia. About half of these will form primary spermatocytes, which begin meiosis.

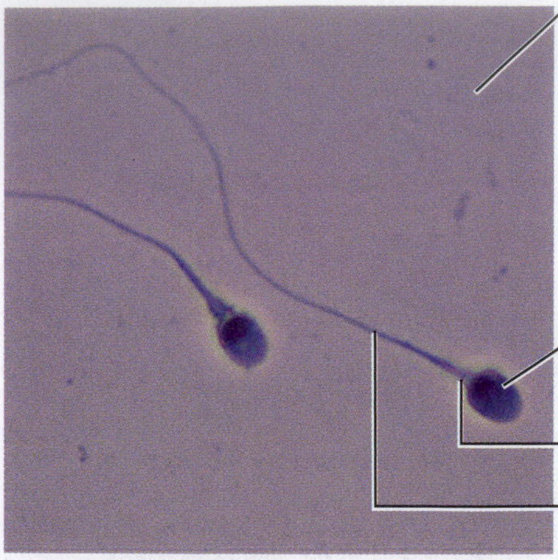
Fluid medium of semen
Head with acrosome
Midpiece
Tail
Sperm

FIGURE 43.3 Sperm in semen. In addition to sperm, semen contains fluids secreted by the accessory glands (1500×).

These are recognizable by their pale-staining nuclei with centrally located nucleoli. The remaining daughter cells resulting from mitotic divisions of spermatogonia stay at the tubule periphery to maintain the germ cell line.

3. Observe the cells in the middle of the tubule wall. There you should see a large number of cells (spermatocytes) that are obviously undergoing a nuclear division process. Look for coarse clumps of chromatin or threadlike chromosomes (visible only during nuclear division) that have the appearance of coiled springs. Attempt to differentiate between the larger primary spermatocytes and the somewhat smaller secondary spermatocytes. Once formed, the secondary spermatocytes quickly undergo division and so are more difficult to find.

Can you see tetrads? ______________________________

Is there evidence of crossover? ______________________

In which location would you expect to see cells containing tetrads, closer to the spermatogonia or closer to the lumen?

____

Would these cells be primary or secondary spermatocytes?

____

4. Examine the cells at the tubule lumen. Identify the small round-nucleated spermatids, many of which may appear lopsided and look as though they are starting to lose their cytoplasm. See if you can find a spermatid embedded in an elongated cell type—a **sustentacular,** or **Sertoli, cell**—which extends inward from the periphery of the tubule. The sustentacular cells nourish the spermatids as they begin their transformation into sperm. Also in the adluminal area (area toward the lumen), locate immature sperm, which can be identified by their tails. The sperm develop directly from the spermatids by the loss of extraneous cytoplasm and the development of a propulsive tail.

5. Identify the **interstitial cells,** or **Leydig cells,** lying external to and between the seminiferous tubules. LH (luteinizing hormone), also called *interstitial cell–stimulating hormone* in males, prompts these cells to produce testosterone, which acts synergistically with FSH to stimulate sperm production.

In the next stage of sperm development, spermiogenesis, all the superficial cytoplasm is sloughed off, and the remaining cell organelles are compacted into the three regions of the mature sperm. At the risk of oversimplifying, these anatomical regions are the *head,* the *midpiece,* and the *tail,* which correspond roughly to the activating and genetic region, the metabolic region, and the locomotor region, respectively. The mature sperm is a streamlined cell equipped with an organ of locomotion and a high rate of metabolism that enable it to move long distances quickly to get to the egg. It is a prime example of the correlation of form and function.

The pointed sperm head contains the DNA, or genetic material, of the chromosomes. Essentially it is the nucleus of the spermatid. Anterior to the nucleus is the **acrosome,** which contains enzymes involved in sperm penetration of the egg.

In the midpiece of the sperm is a centriole which gives rise to the filaments that structure the sperm tail. Wrapped tightly around the centriole are mitochondria that provide the ATP needed for contractile activity of the tail.

The tail is a typical flagellum produced by a centriole. When powered by ATP, the tail propels the sperm.

6. Obtain a prepared slide of human sperm, and view it with the oil immersion lens. Compare what you see to the photograph of sperm in Figure 43.3. Identify the head, acrosome, and tail regions. Deformed sperm, for example sperm with multiple heads or tails, are sometimes present in such preparations. Did you observe any?

______________ If so, describe them. ______________

____

7. Examine the model of spermatogenesis to identify the spermatogonia, the primary and secondary spermatocytes, the spermatids, and the functional sperm. ■

Demonstration of Oogenesis in *Ascaris* (Optional)

Oogenesis (the process of producing an egg) in mammals is difficult to demonstrate. However, the process may be studied rather easily in the transparent eggs of *Ascaris megalocephala*, an invertebrate roundworm parasite found in the intestine of mammals. Since its diploid chromosome number is 4, the chromosomes are easily counted.

ACTIVITY 3

Examining Meiotic Events Microscopically

Go to the demonstration area where the slides are set up, and make the following observations:

1. Scan the first demonstration slide to identify a *primary oocyte,* the cell type that begins the meiotic process. It will have what appears to be a relatively thick cell membrane; this is the *fertilization membrane* that the oocyte produces after sperm penetration. Find and study a primary oocyte that is undergoing the first maturation division. Look for a barrel-shaped spindle with two tetrads (two groups of four beadlike

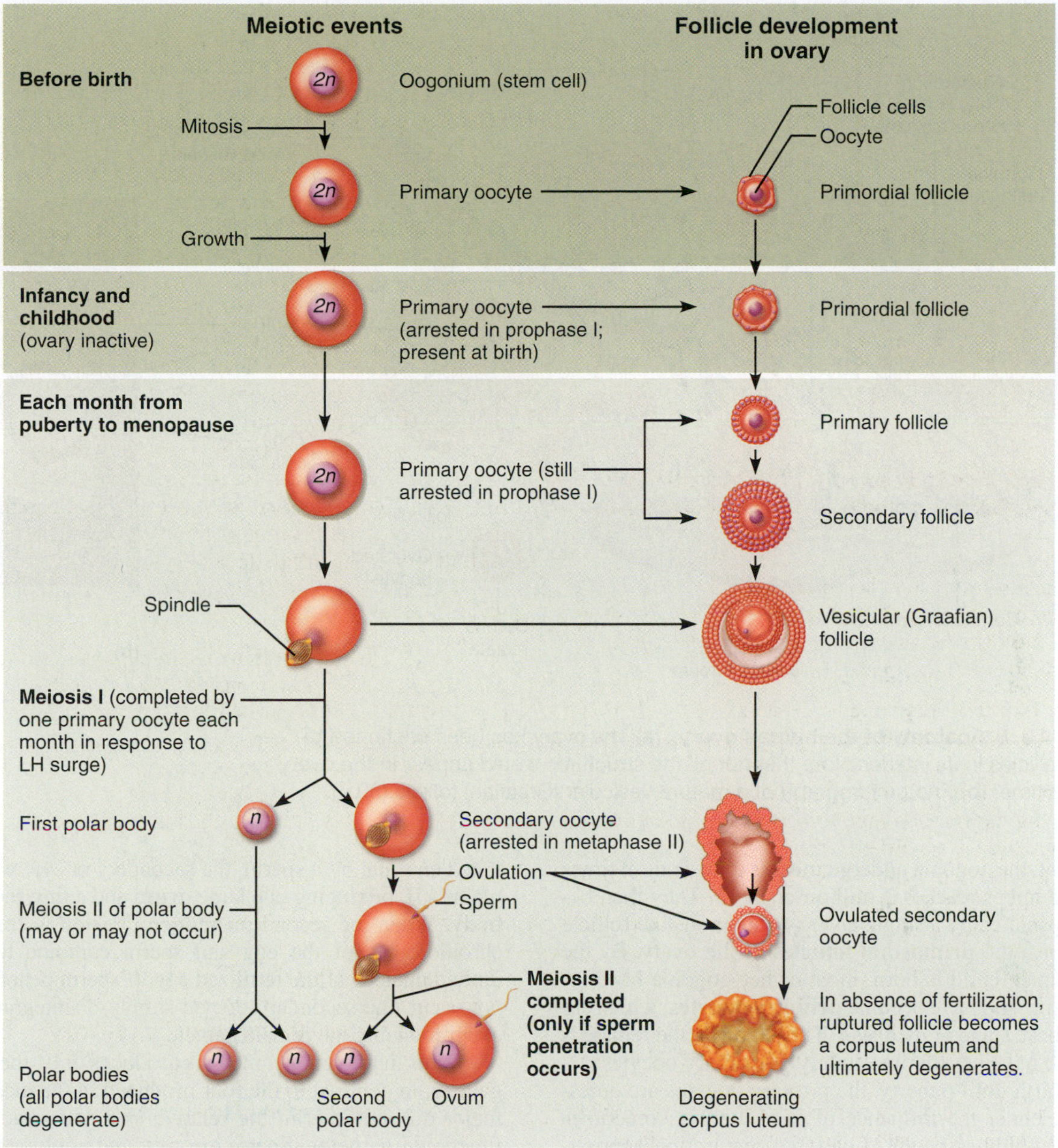
Meiotic events
Follicle development in ovary
Before birth
Oogonium (stem cell)
Mitosis
Follicle cells
Oocyte
Primary oocyte
Primordial follicle
Growth
Infancy and childhood (ovary inactive)
Primary oocyte (arrested in prophase I; present at birth)
Primordial follicle
Each month from puberty to menopause
Primary follicle
Primary oocyte (still arrested in prophase I)
Secondary follicle
Spindle
Vesicular (Graafian) follicle
Meiosis I (completed by one primary oocyte each month in response to LH surge)
First polar body
Secondary oocyte (arrested in metaphase II)
Ovulation
Sperm
Meiosis II of polar body (may or may not occur)
Ovulated secondary oocyte
Meiosis II completed (only if sperm penetration occurs)
In absence of fertilization, ruptured follicle becomes a corpus luteum and ultimately degenerates.
Polar bodies (all polar bodies degenerate)
Second polar body
Ovum
Degenerating corpus luteum
2n
n

FIGURE 43.4 Oogenesis. Left, flowchart of meiotic events. Right, correlation with follicular development and ovulation in the ovary.

chromosomes) in it. Most often the spindle is located at the periphery of the cell. (The sperm nucleus may or may not be seen, depending on how the cell was cut.)

2. Observe slide 2. Locate a cell in which half of each tetrad (a dyad) is being extruded from the cell surface into a smaller cell called the *first polar body*.

3. On slide 3, attempt to locate a *secondary oocyte* (a daughter cell produced during meiosis I) undergoing the second maturation division. In this view, you should see two dyads (two groups of two beadlike chromosomes) on the spindle.

4. On slide 4, locate a cell in which the *second polar body* is being formed. In this case, both it and the ovum will now contain two chromosomes, the haploid number for *Ascaris*.

5. On slide 5, identify a *fertilized egg* or a cell in which the sperm and ovum nuclei (actually *pronuclei*) are fusing to form a single nucleus containing four chromosomes.

Human Oogenesis and the Ovarian Cycle

Once the adult ovarian cycle is established, gonadotropic hormones produced by the anterior pituitary influence the development of ova in the ovaries and their cyclic production of female sex hormones. Within an ovary, each immature ovum develops within a saclike structure called a *follicle,* where it is encased by one or more layers of smaller cells called **follicle cells** (when one layer is present) or **granulosa cells** (when there is more than one layer).

The process of **oogenesis,** or female gamete formation, which occurs in the ovary, is similar to spermatogenesis occurring in the testis, but there are some important differences. The process, schematically outlined in Figure 43.4, begins with primitive stem cells called **oogonia,** located in the ovarian cortices of the developing female fetus. During fetal

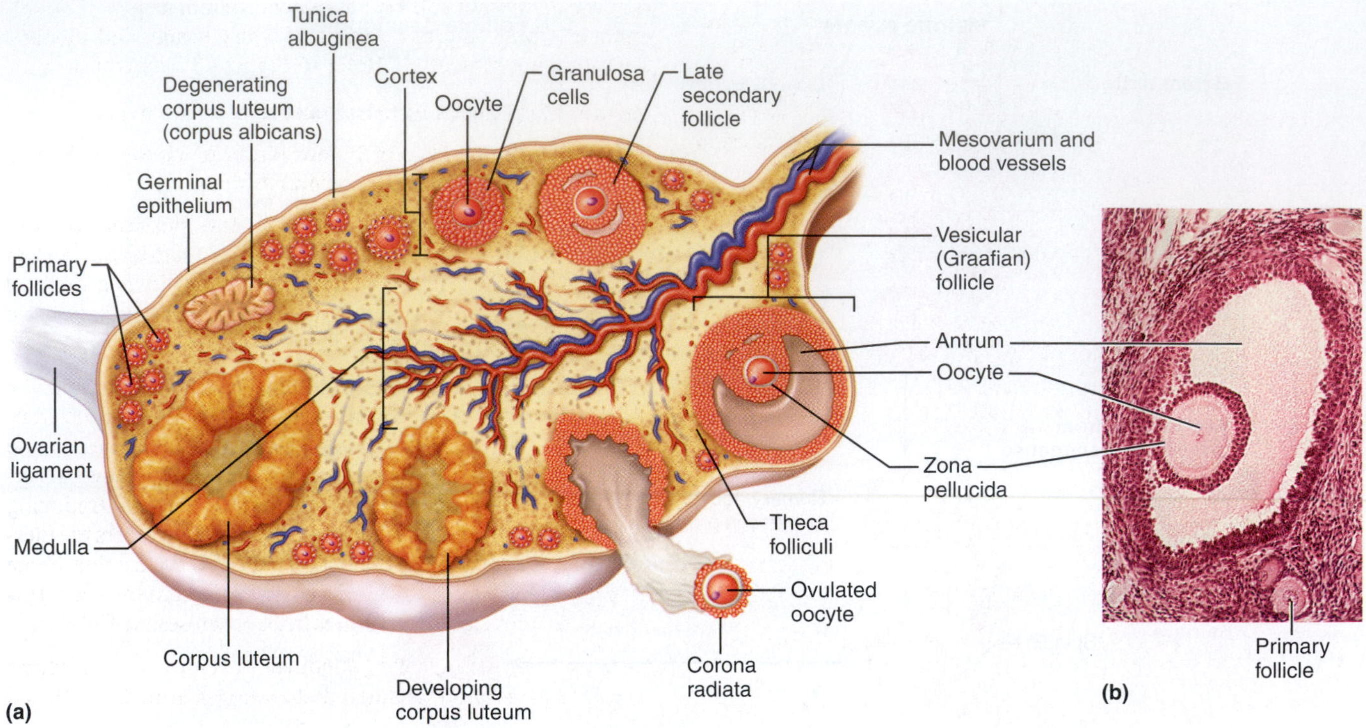
Tunica albuginea
Degenerating corpus luteum (corpus albicans)
Cortex
Oocyte
Granulosa cells
Late secondary follicle
Mesovarium and blood vessels
Germinal epithelium
Primary follicles
Vesicular (Graafian) follicle
Antrum
Oocyte
Zona pellucida
Theca folliculi
Ovarian ligament
Medulla
Ovulated oocyte
Corona radiata
Corpus luteum
Developing corpus luteum
(a)
Primary follicle
(b)

FIGURE 43.5 Anatomy of the human ovary. (a) The ovary has been sectioned to reveal the follicles in its interior. Note that not all the structures would appear in the ovary at the same time. **(b)** Photomicrograph of a mature vesicular (Graafian) follicle (30×).

development, the oogonia undergo mitosis thousands of times until their number reaches 2 million or more. They then become encapsulated by a single layer of squamouslike follicle cells and form the **primordial follicles** of the ovary. By the time the female child is born, most of her oogonia have increased in size and have become **primary oocytes,** which are in the prophase stage of meiosis I. Thus at birth, the female is presumed to have her lifetime supply of primary oocytes.

From birth until puberty, the primary oocytes are quiescent. Then, under the influence of FSH, one or sometimes more of the follicles begin to undergo maturation approximately every 28 days.

As a follicle grows, its epithelium changes from squamous to cuboidal cells and it comes to be called a **primary follicle** (see also Figure 43.5). The primary follicle begins to produce estrogens, and the primary oocyte completes its first maturation division, producing two haploid daughter cells that are very disproportionate in size. One of these is the **secondary oocyte,** which contains nearly all of the cytoplasm in the primary oocyte. The other is the tiny **first polar body.** The first polar body may then complete the second maturation division, producing two more polar bodies. These eventually disintegrate for lack of sustaining cytoplasm.

As the follicle containing the secondary oocyte continues to enlarge, blood levels of estrogens rise. Initially, estrogen exerts a negative feedback influence on the release of gonadotropins by the anterior pituitary. However, approximately in the middle of the 28-day cycle, as the follicle reaches the mature **vesicular,** or **Graafian, follicle** stage, rising estrogen levels become highly stimulatory and a sudden burstlike release of LH (and, to a lesser extent, FSH) by the anterior pituitary triggers ovulation. The secondary oocyte is extruded and begins its journey along the uterine tube to the uterus. If penetrated en route by a sperm, the secondary oocyte will undergo meiosis II, producing one large **ovum** and a tiny **second polar body.** When the second maturation division is complete, the chromosomes of the egg and sperm combine to form the diploid nucleus of the fertilized egg. If sperm penetration does not occur, the secondary oocyte simply disintegrates without ever producing the female gamete.

Thus in the female, meiosis produces only one functional gamete, in contrast to the four produced in the male. Another major difference is in the relative size and structure of the functional gametes. Sperm are tiny and equipped with tails for locomotion. They have few organelles and virtually no nutrient-containing cytoplasm; hence the nutrients contained in semen are essential to their survival. In contrast, the egg is a relatively large nonmotile cell, well stocked with cytoplasmic reserves that nourish the developing embryo until implantation can be accomplished. Essentially all the zygote's organelles are "delivered" by the egg.

Once the secondary oocyte has been expelled from the ovary, LH transforms the ruptured follicle into a **corpus luteum,** which begins producing progesterone and estrogen. Rising blood levels of the two ovarian hormones inhibit FSH release by the anterior pituitary. As FSH declines, its stimulatory effect on follicular production of estrogens ends, and estrogen blood levels begin to decline. Since rising estrogen levels triggered LH release by the anterior pituitary, falling estrogen levels result in declining levels of LH in the blood. Corpus luteum secretory function is maintained by high blood levels of LH. Thus as LH blood levels begin to drop toward the end of the 28-day cycle, progesterone production ends and the corpus luteum begins to degenerate and is replaced by scar tissue **(corpus albicans).** The graphs in Figure 43.7 depict the hormone relationships described here.

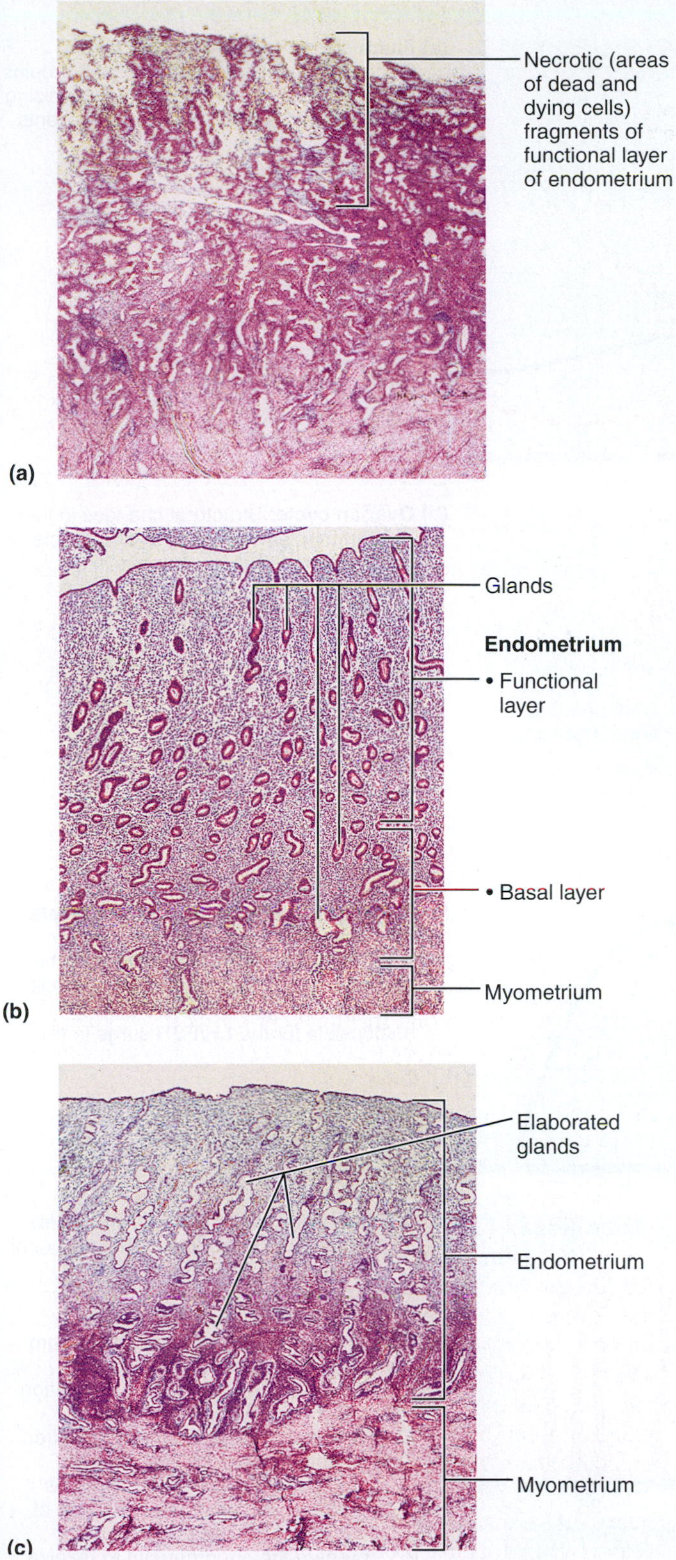
Necrotic (areas of dead and dying cells) fragments of functional layer of endometrium
(a)
Glands
Endometrium
• Functional layer
• Basal layer
Myometrium
(b)
Elaborated glands
Endometrium
Myometrium
(c)

FIGURE 43.6 Endometrial changes during the menstrual cycle. (a) Onset of menstruation (20×). **(b)** Early proliferative phase (20×). **(c)** Early secretory phase (all 20×).

ACTIVITY 4

Examining Oogenesis in the Ovary

Because many different stages of ovarian development exist within the ovary at any one time, a single microscopic preparation will contain follicles at many different stages of development. Obtain a cross section of ovary tissue, and identify the following structures. Refer to Figure 43.5 as you work.

Germinal epithelium: Outermost layer of the ovary.

Primary follicle: One or a few layers of cuboidal follicle cells surrounding the larger central developing ovum.

Secondary (growing) follicles: Follicles consisting of several layers of follicle (granulosa) cells surrounding the central developing ovum, and beginning to show evidence of fluid accumulation and **antrum** (central cavity) formation. Follicle development may take more than one cycle.

Vesicular (Graafian) follicle: At this stage of development, the follicle has a large antrum containing fluid produced by the granulosa cells. The developing secondary oocyte is pushed to one side of the follicle and is surrounded by a capsule of several layers of granulosa cells called the **corona radiata** (radiating crown). When the secondary oocyte is released, it enters the uterine tubes with its corona radiata intact. The connective tissue stroma (background tissue) adjacent to the mature follicle forms a capsule, called the **theca folliculi,** that encloses the follicle.

Corpus luteum: A solid glandular structure or a structure containing a scalloped lumen that develops from the ruptured follicle. ■

ACTIVITY 5

Comparing and Contrasting Oogenesis and Spermatogenesis

Examine the model of oogenesis, and compare it with the spermatogenesis model. Note differences in the size and structure of the functional gametes. ■

The Menstrual Cycle

The **uterine cycle,** or **menstrual cycle,** is hormonally controlled by estrogens and progesterone secreted by the ovary. It is normally divided into three stages: menstrual, proliferative, and secretory. The endometrial changes are described in Figure 43.7 and shown in Figure 43.6. Figure 43.7 shows how they correlate with hormonal and ovarian changes.

If fertilization has occurred, the embryo will produce a hormone much like LH, which will maintain the function of the corpus luteum. Otherwise, as the corpus luteum begins to deteriorate, lack of ovarian hormones in the blood causes blood vessels supplying the endometrium to kink and become spastic, setting the stage for menses to begin by the 28th day.

Although Figure 43.7d assumes a classic 28-day cycle, the length of the menstrual cycle is highly variable, sometimes as short as 21 days or as long as 38.

ACTIVITY 6

Observing Histological Changes in the Endometrium During the Menstrual Cycle

Obtain slides showing the menstrual, secretory, and proliferative phases of the uterine endometrium. Observe each carefully, comparing their relative thicknesses and vascularity. As you work, refer to the corresponding photomicrographs in Figure 43.6. ■

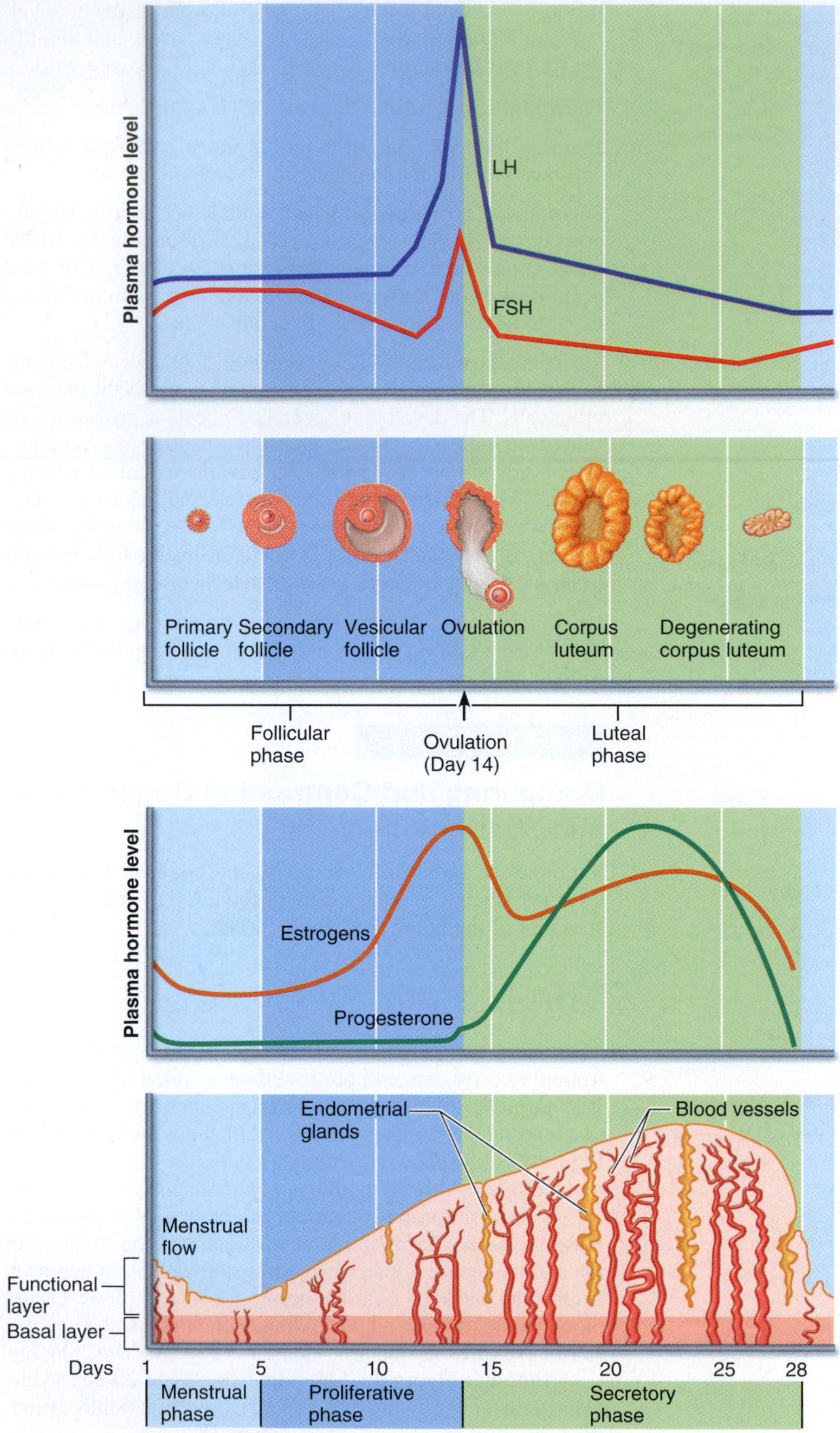
Plasma hormone level
LH
FSH
Primary follicle
Secondary follicle
Vesicular follicle
Ovulation
Corpus luteum
Degenerating corpus luteum
Follicular phase
Ovulation (Day 14)
Luteal phase
Plasma hormone level
Estrogens
Progesterone
Endometrial glands
Blood vessels
Menstrual flow
Functional layer
Basal layer
Days 1 5 10 15 20 25 28
Menstrual phase
Proliferative phase
Secretory phase

(a) Fluctuation of gonadotropin levels: Fluctuating levels of pituitary gonadotropins (follicle-stimulating hormone and luteinizing hormone) in the blood regulate the events of the ovarian cycle.

(b) Ovarian cycle: Structural changes in the ovarian follicles during the ovarian cycle are correlated with (d) changes in the endometrium of the uterus during the uterine cycle.

(c) Fluctuation of ovarian hormone levels: Fluctuating levels of ovarian hormones (estrogens and progesterone) cause the endometrial changes of the uterine cycle. The high estrogen levels are also responsible for the LH/FSH surge in (a).

(d) The three phases of the uterine cycle:
- Menstrual: Shedding of the functional layer of the endometrium. Approximately days 1–5.
- Proliferative: Rebuilding of the functional layer of the endometrium under influence of estrogens. Approximately days 6–14. Ovulation occurs at the end of this phase.
- Secretory: Begins immediately after ovulation under the influence of progesterone. The blood supply is enriched and glandular secretion of nutrients increases, events that prepare the endometrium to receive an embryo.

FIGURE 43.7 Correlation of anterior pituitary and ovarian hormones with structural changes in the ovary and uterus. The time bar applies to all parts of the figure.

NAME ______________________________

LAB TIME/DATE ____________________

REVIEW SHEET

EXERCISE 43

Physiology of Reproduction: Gametogenesis and the Female Cycles

Meiosis

1. The following statements refer to events occurring during mitosis and/or meiosis. For each statement, decide if the event occurs in (a) mitosis only, (b) meiosis only, or (c) both mitosis and meiosis.

_______ 1. dyads are visible

_______ 2. tetrads are visible

_______ 3. product is two diploid daughter cells genetically identical to the mother cell

_______ 4. product is four haploid daughter cells quantitatively and qualitatively different from the mother cell

_______ 5. involves the phases prophase, metaphase, anaphase, and telophase

_______ 6. occurs throughout the body

_______ 7. occurs only in the ovaries and testes

_______ 8. provides cells for growth and repair

_______ 9. homologues synapse; chiasmata are seen

_______ 10. chromosomes are replicated before the division process begins

_______ 11. provides cells for perpetuation of the species

_______ 12. consists of two consecutive nuclear divisions, without chromosomal replication occurring before the second division

2. Describe the process of synapsis. ____

____

3. How does crossover introduce variability in the daughter cells? ______________________

____

4. Define *homologous chromosomes*. ____

____

Spermatogenesis

5. The cell types seen in the seminiferous tubules are listed in the key. Match the correct cell type(s) with the descriptions given below.

Key: a. primary spermatocyte
b. secondary spermatocyte
c. spermatogonium
d. sustentacular cell
e. spermatid
f. sperm

_______________ 1. primitive stem cell

_______________ 2. haploid

_______________ 3. provides nutrients to developing sperm

_______________ 4. products of meiosis II

_______________ 5. product of spermiogenesis

_______________ 6. product of meiosis I

6. Why are spermatids not considered functional gametes? _______________

____

7. Differentiate between *spermatogenesis* and *spermiogenesis*. _______________

____

____

8. Draw a sperm below, and identify the acrosome, head, midpiece, and tail. Then beside each label, note the composition and function of each of these sperm structures.

9. The life span of a sperm is very short. What anatomical characteristics might lead you to suspect this even if you didn't know its life span? _______________

Oogenesis, the Ovarian Cycle, and the Menstrual Cycle

10. The sequence of events leading to germ cell formation in the female begins during fetal development. By the time the child is born, all viable oogonia have been converted to _______________.

In view of this fact, how does the total germ cell potential of the female compare to that of the male?

____

11. The female gametes develop in structures called *follicles*. What is a follicle? _______________

____

How are primary and vesicular follicles anatomically different? ______________________________

____

____

____

What is a corpus luteum? ______________________________

____

12. What is the major hormone produced by the vesicular follicle? ______________________________

By the corpus luteum? ______________________________

13. Use the key to identify the cell type you would expect to find in the following structures.

Key: a. oogonium b. primary oocyte c. secondary oocyte d. ovum

_______ 1. forming part of the primary follicle in the ovary

_______ 2. in the uterine tube before fertilization

_______ 3. in the mature vesicular follicle of the ovary

_______ 4. in the uterine tube shortly after sperm penetration

14. The cellular product of spermatogenesis is four ______________________________; the final product of oogenesis is one ______________________________ and three ______________________________. What is the function of this unequal cytoplasmic division seen during oogenesis in the female? ______________________________

____

____

What is the fate of the three tiny cells produced during oogenesis? ______________________________

Why? ______________________________

15. The following statements deal with anterior pituitary and ovarian hormones and with hormonal interrelationships. Name the hormone(s) described in each statement.

______________________ 1. produced by primary follicles in the ovary

______________________ 2. ovulation occurs after its burstlike release

______________________ and ______________________ 3. exert negative feedback on the anterior pituitary relative to FSH secretion

______________________ 4. stimulates LH release by the anterior pituitary

______________________ 5. stimulates the corpus luteum to produce progesterone and estrogen

______________________ 6. maintains the hormonal production of the corpus luteum in a nonpregnant woman

16. Why does the corpus luteum deteriorate toward the end of the ovarian cycle? ______________________________

____

17. For each statement below dealing with hormonal blood levels during the female ovarian and menstrual cycles, decide whether the condition in column A is usually (a) greater than, (b) less than, or (c) essentially equal to the condition in column B.

	Column A	Column B
_______	1. amount of LH in the blood during menses	amount of LH in the blood at ovulation
_______	2. amount of FSH in the blood on day 6 of the cycle	amount of FSH in the blood on day 20 of the cycle
_______	3. amount of estrogen in the blood during menses	amount of estrogen in the blood at ovulation
_______	4. amount of progesterone in the blood on day 14	amount of progesterone in the blood on day 23
_______	5. amount of estrogen in the blood on day 10	amount of progesterone in the blood on day 10

18. What uterine tissue undergoes dramatic changes during the menstrual cycle? _______________

19. When during the female menstrual cycle would fertilization be unlikely? Explain why. _______________

____

20. Assume that a woman could be an "on demand" ovulator like the rabbit, in which copulation stimulates the hypothalamic–anterior pituitary axis and causes LH release, and an oocyte was ovulated and fertilized on day 26 of her 28-day cycle. Why would a successful pregnancy be unlikely at this time?

____

____

____

21. The menstrual cycle depends on events within the female ovary. The stages of the menstrual cycle are listed below. For each, note its approximate time span and the related events in the uterus; and then to the right, record the ovarian events occurring simultaneously. Pay particular attention to hormonal events.

Menstrual cycle stage	Uterine events	Ovarian events
Menstruation		
Proliferative		
Secretory		

EXERCISE 44

Survey of Embryonic Development

MATERIALS

- ☐ Prepared slides of sea urchin development (zygote through larval stages)
- ☐ Compound microscope
- ☐ Three-dimensional human development models or plaques (if available)
- ☐ *Demonstration:* Phases of human development in *Life Before Birth: Normal Fetal Development,* Second Edition (by Marjorie A. England, 1996, Mosby-Wolfe)
- ☐ Disposable gloves
- ☐ *Demonstration:* Pregnant cat, rat, or pig uterus (one per laboratory session) with uterine wall dissected to allow student examination
- ☐ Three-dimensional model of pregnant human torso
- ☐ *Demonstration:* Fresh or formalin-preserved human placenta (obtained from a clinical agency)
- ☐ Prepared slide of placenta tissue

MasteringA&P™
Access practice quizzes and more in the Study Area at www.masteringaandp.com.

OBJECTIVES

1. To define *fertilization* and *zygote*.
2. To define and discuss the function of *cleavage* and *gastrulation*.
3. To name the three primary germ layers, and to discuss the importance of each.
4. To differentiate between the blastula and gastrula forms of the sea urchin and human when provided with appropriate models or diagrams.
5. To define *blastocyst* and/or *chorionic vesicle*.
6. To identify the following structures of a human chorionic vesicle when provided with an appropriate diagram, and to state the function of each.

inner cell mass	chorionic villi	yolk sac
trophoblast	amnion	allantois

7. To describe the process and timing of implantation in the human.
8. To define *decidua basalis* and *decidua capsularis*.
9. To state the germ-layer origin of several body organs and organ systems of the human.
10. To describe developmental direction.
11. To describe the gross anatomy and general function of the human placenta.

PRE-LAB QUIZ

1. Circle the correct term. The fertilized egg, or zygote / embryo, appears as a single cell surrounded by a fertilization membrane and a jellylike membrane.
2. The uniting of the egg and sperm nuclei is known as
   a. embryogensis
   b. fertilization
   c. implantation
   d. mitosis
3. Circle True or False. Cleavage is a series of mitotic divisions without any intervening growth periods that results in a multicellular embryonic body.
4. Circle the correct term. As a result of gastrulation, a three / four-layered embryo forms, with each layer corresponding to a primary germ layer.
5. The __________ gives rise to the epidermis of the skin and the nervous system.
   a. ectoderm
   b. endoderm
   c. mesoderm
6. By the ninth week of development, the embryo is referred to as a
   a. blastocyst
   b. blastomere
   c. fetus
   d. gastrula
7. Circle True or False. The placenta is composed solely of embryonic membranes.
8. Circle the correct term. The allantois / amnion encases the young embryonic body in a fluid-filled chamber that acts to protect the developing embryo against trauma.
9. What is the function of the placenta? ______________________________
____

Because reproduction is such a familiar event, we tend to lose sight of the wonder of the process. One part of that process, the development of the embryo, is the concern of embryologists who study the changes in structure that occur from the time of fertilization until the time of birth.

Early development in all animals involves three basic types of activities, which are integrated to ensure the formation of a viable offspring: (1) an increase in cell number and subsequent cell growth; (2) cellular specialization; and (3) morphogenesis, the formation of functioning organ systems. This exercise first provides a rather broad overview of the changes in structure that take place during embryonic development in sea urchins. The pattern of changes in this marine animal provides a basis of comparison with developmental events in the human.

Developmental Stages of Sea Urchins and Humans

ACTIVITY 1

Microscopic Study of Sea Urchin Development

1. Obtain a compound microscope and a set of slides depicting embryonic development of the sea urchin. Draw simple diagrams of your observations as you work.

2. Observe the fertilized egg, or **zygote,** which appears as a single cell immediately surrounded by a fertilization membrane and a jellylike membrane. After an egg is penetrated by a sperm, the egg and the sperm nuclei fuse to form a single nucleus. This process is called **fertilization.** Within 2 to 5 minutes after sperm penetration, a fertilization membrane forms beneath the jelly coat to prevent the entry of additional sperm. Draw the zygote and label the fertilization and jelly membranes.

Zygote

3. Observe the cleavage stages. Once fertilization has occurred, the zygote begins to divide, forming a mass of successively smaller and smaller cells, called **blastomeres.** This series of mitotic divisions without intervening growth periods is referred to as **cleavage,** and it results in a multicellular embryonic body. As the division process continues, a solid ball of cells forms. (At the 32-cell stage, it is called the **morula,** and the embryo resembles a raspberry in form.) Then the cell mass hollows out to become the embryonic form called the **blastula,** which is a ball of cells surrounding a central cavity. The blastula is the final product of cleavage.

The cleavage stage of embryonic development provides a large number of building blocks (cells) with which to fashion the forming body. (If this is a little difficult to understand, consider trying to build a structure with a huge block of granite rather than with small bricks.)

In the spaces below, diagram the 2-, 4-, 8-, and 16-cell stages as you observe them.

2-cell stage 4-cell stage

8-cell stage 16-cell stage

Identify and sketch the blastula stage of cleavage—a ball of cells with an apparently lighter center due to the presence of the central cavity.

Blastula

4. Identify the **early gastrula** form (which follows the blastula in the developmental sequence). The gastrula looks as if one end of the blastula has been indented or pushed into the central cavity, forming a two-layered embryo. In time, a third layer of cells appears between the initial two cell layers. Thus, as a result of **gastrulation,** a three-layered embryo forms, each layer corresponding to a **primary germ layer** from which all body tissues develop. The innermost layer, the **endoderm,** and the middle layer, the **mesoderm,** form the internal organs; the outermost layer, the **ectoderm,** forms the surface tissues of the body.

Draw a gastrula below. Label the ectoderm and endoderm. If you can see the third layer of cells, the mesoderm, budding off between the other two layers, label that also.

Gastrula

5. Gastrulation in the sea urchin is followed by the appearance of the free-swimming larval form, in which the three germ layers have differentiated into the various tissues and organs of the animal's body.

The larvae exist for a few days in the unattached form and then settle to the ocean bottom to attach and develop into the sessile adult form. If time allows, observe the larval form on the prepared slides. Record your observations in a simple drawing below.

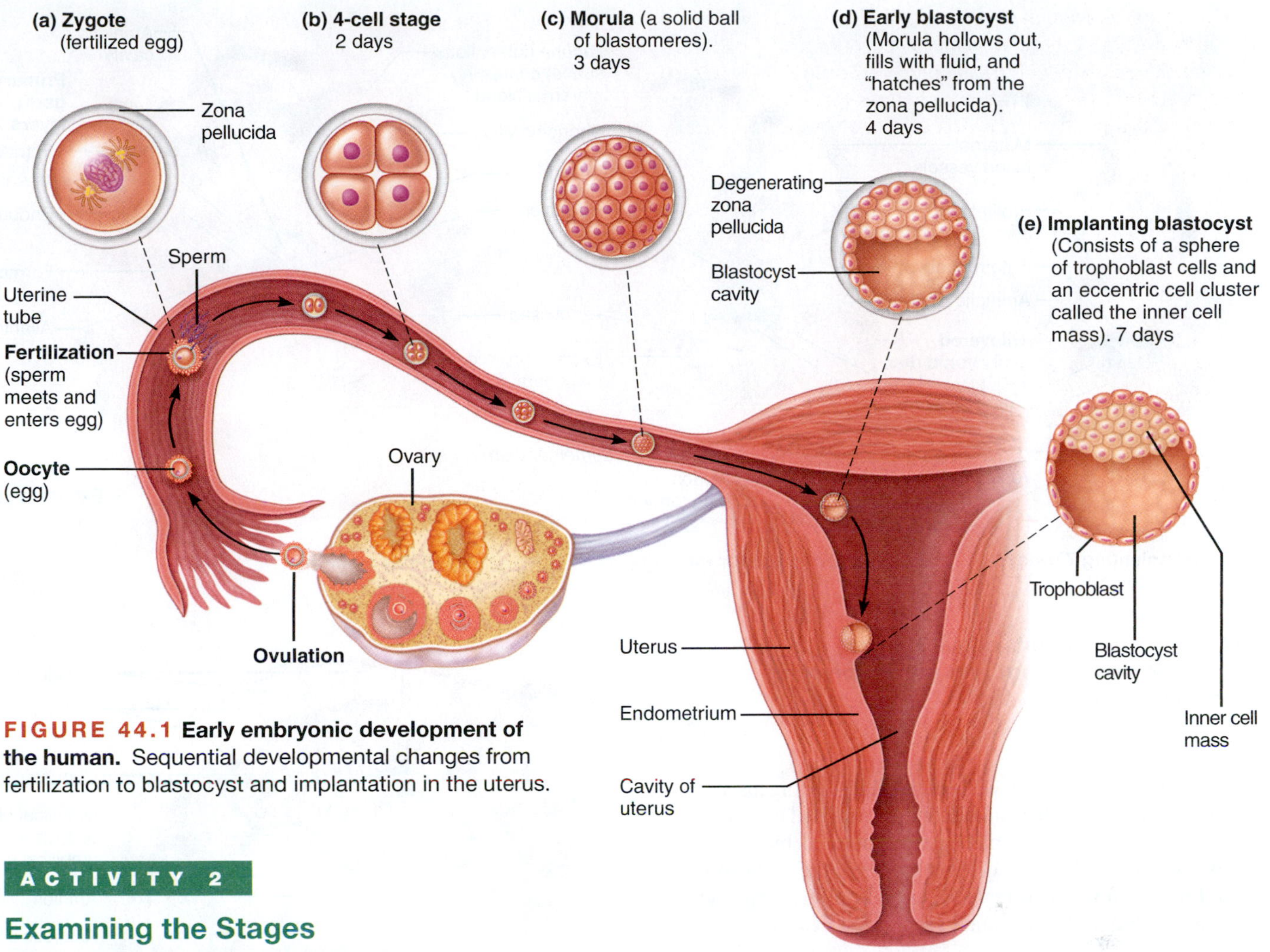
(a) Zygote
(fertilized egg)
Zona pellucida
(b) 4-cell stage
2 days
(c) Morula (a solid ball of blastomeres).
3 days
(d) Early blastocyst
(Morula hollows out, fills with fluid, and "hatches" from the zona pellucida).
4 days
Degenerating zona pellucida
Blastocyst cavity
(e) Implanting blastocyst
(Consists of a sphere of trophoblast cells and an eccentric cell cluster called the inner cell mass). 7 days
Sperm
Uterine tube
Fertilization
(sperm meets and enters egg)
Oocyte
(egg)
Ovary
Ovulation
Uterus
Endometrium
Cavity of uterus
Trophoblast
Blastocyst cavity
Inner cell mass

FIGURE 44.1 Early embryonic development of the human. Sequential developmental changes from fertilization to blastocyst and implantation in the uterus.

ACTIVITY 2

Examining the Stages of Human Development

Examine the models or plaques of human development that are on display. If these are not available, use Figures 44.1 and 44.2 for this study. Observe the models to identify the various stages of human development and respond to the questions posed below.

1. Observe the fertilized egg, or zygote, which appears as a single cell immediately surrounded by a jellylike *zona pellucida* and then a crown of granulosa cells (the *corona radiata*).

2. Next, observe the cleavage stages.

Is the human cleavage process similar to that in the sea urchin?

____

Why do you suppose this is so? ______________________

____

Observe the blastula, the final product of cleavage, which is called the **blastocyst** or **chorionic vesicle** in the human. Unlike the sea urchin, only a portion of the blastula cells in the human contribute to the formation of the embryonic body—those seen at one side of the blastocyst (Figure 44.1e) forming the **inner cell mass (ICM).** The rest of the blastocyst—that enclosing the central cavity and overriding the ICM—is referred to as the **trophoblast.** The trophoblast becomes an extraembryonic membrane called the **chorion,** which forms the fetal portion of the **placenta.**

3. Observe the *implanting* blastocyst shown on the model or in the figures. By approximately the seventh day after ovulation, a developing human embryo (blastocyst) is floating free in the uterine cavity. About that time, it adheres to the uterine wall over the ICM area, and implantation begins. The trophoblast cells secrete enzymes that erode the uterine mucosa at the point of attachment to reach the vascular supply in the submucosa. By the fourteenth day after ovulation, implantation is completed and the uterine mucosa has grown over the burrowed-in embryo. The portion of the uterine wall beneath the ICM, destined to take part in placenta formation, is called the **decidua basalis** and that surrounding the rest of the blastocyst is called the **decidua capsularis.** Identify these regions.

By the time implantation has been completed, embryonic development has progressed to the **gastrula stage,** and the three primary germ layers are present and are beginning to differentiate (Figure 44.2). Within the next 6 weeks, virtually all of the body organ systems will have been laid down at least in rudimentary form by the germ layers. The outermost layer, *ectoderm,* gives rise to the epidermis of the skin and the nervous system. The deepest layer, the *endoderm,* forms the mucosa of the digestive and respiratory tracts and associated glands. *Mesoderm,* the middle layer, forms virtually everything lying

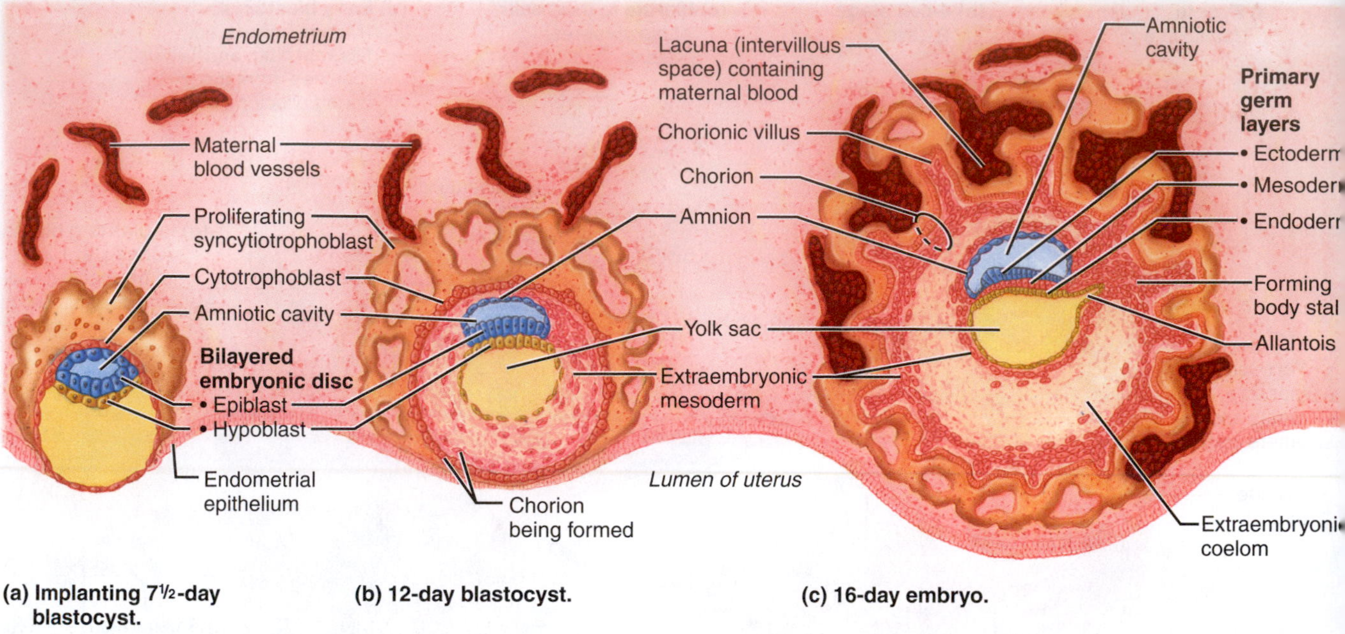
Endometrium
Maternal blood vessels
Proliferating syncytiotrophoblast
Cytotrophoblast
Amniotic cavity
Bilayered embryonic disc
• Epiblast
• Hypoblast
Endometrial epithelium
Chorion being formed
Lacuna (intervillous space) containing maternal blood
Chorionic villus
Chorion
Amnion
Yolk sac
Extraembryonic mesoderm
Lumen of uterus
Amniotic cavity
Primary germ layers
• Ectoderm
• Mesoderm
• Endoderm
Forming body stal
Allantois
Extraembryoni
coelom
(a) Implanting 7½-day blastocyst.
(b) 12-day blastocyst.
(c) 16-day embryo.

FIGURE 44.2 Events of placentation, early embryonic development, and extraembryonic membrane formation.

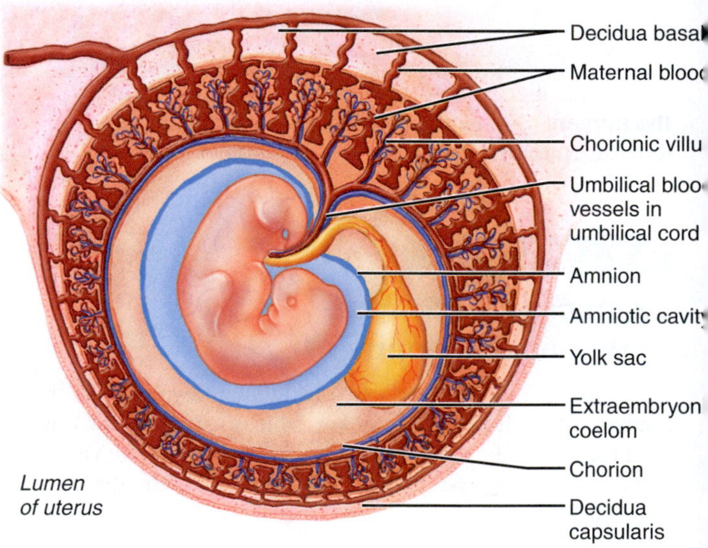
Decidua basal
Maternal blood
Chorionic villu
Umbilical bloo vessels in umbilical cord
Amnion
Amniotic cavit
Yolk sac
Extraembryon coelom
Chorion
Decidua capsularis
Lumen of uterus

(d) 4½-week embryo.

between the two (skeleton, walls of the digestive organs, urinary system, skeletal muscles, circulatory system, and others).

All the groundwork has been completed by the eighth week. By the ninth week of development, the embryo is referred to as a **fetus,** and from this point on, the major activities are growth and tissue and organ specialization.

4. Again observe the blastocyst to follow the formation of the embryonic membranes and the placenta (see Figure 44.2). Notice the villus extensions of the trophoblast. By the time implantation is complete, the trophoblast has differentiated into the chorion, and its large elaborate villi are lying in the blood-filled sinusoids in the uterine tissue. This composite of uterine tissue and **chorionic villi** is called the **placenta** (Figure 44.3), and all exchanges to and from the embryo occur through the chorionic membranes.

Three embryonic membranes (originating in the ICM) have also formed by this time—the amnion, the allantois, and the yolk sac (Figure 44.2). Attempt to identify each.

- The **amnion** encases the young embryonic body in a fluid-filled chamber that protects the embryo against mechanical trauma and temperature extremes and prevents adhesions during rapid embryonic growth.
- The **yolk sac** in humans has lost its original function, which was to pass nutrients to the embryo after digesting the yolk mass. The placenta has taken over that task; also, the human egg has very little yolk. However, the yolk sac is not totally useless. The embryo's first blood cells originate here, and the primordial germ cells migrate from it into the embryo's body to seed the gonadal tissue.
- The **allantois,** which protrudes from the posterior end of the yolk sac, is also largely redundant in humans because of the placenta. In birds and reptiles, it is a repository for embryonic wastes. In humans, it is the structural basis on which the mesoderm migrates to form the body stalk, or **umbilical cord,** which attaches the embryo to the placenta.

5. Go to the demonstration area to view the photographic series, *Life Before Birth: Normal Fetal Development*. These photographs illustrate human development in a way you will long remember. After viewing them, respond to the following questions.

In your own words, what do the chorionic villi look like?

___

___

___

What organs or organ systems appear *very* early in embryonic development?

____

____

Does development occur in a rostral to caudal (head to toe) direction, or vice versa?

____

Does development occur in a distal to proximal direction, or vice versa?

____

Does spontaneous movement occur in utero? ______________

How does the mother recognize this? ______________

____

The very young embryo has been described as resembling "an astronaut suspended and floating in space." Do you think this definition is appropriate?

__________ Why or why not? ______________________

____

____

____

What is vernix caseosa? ______________________

____

____

What is lanugo? ______________________________

____

____

In Utero Development

ACTIVITY 3

Identifying Fetal Structures

1. Put on disposable gloves. Go to the appropriate demonstration area and observe the fetuses in the Y-shaped animal (cat, rat, or pig) uterus. Identify the following fetal or fetal-related structures:

- **Placenta** (a composite structure formed from the uterine mucosa and the fetal chorion).

Describe its appearance. ______________________

____

____

____

- **Umbilical cord.** Describe its relationship to the placenta and fetus.

____

____

____

- **Amniotic sac.** Identify the transparent amnion surrounding a fetus. Open one amniotic sac and note the amount, color, and consistency of the fluid.

____

____

____

Remove a fetus and observe the degree of development of the head, body, and extremities. Is the skin thick or thin?

____

What is the basis for your response? ______________

____

____

2. Observe the model of a pregnant human torso. Identify the placenta. How does it differ in shape from the animal placenta observed?

____

____

____

Identify the umbilical cord. In what region of the uterus does implantation usually occur, as indicated by the position of the placenta?

____

____

What might be the consequence if it occurred lower?

___

___

___

Why would a feet-first position (breech presentation) be less desirable than the positioning of the model?

___

___

___

Gross and Microscopic Anatomy of the Placenta

The placenta is a remarkable temporary organ. Composed of maternal and fetal tissues, it is responsible for providing nutrients and oxygen to the embryo and fetus while removing carbon dioxide and metabolic wastes.

ACTIVITY 4

Studying Placental Structure

1. Notice that the human placenta on display has two very different-appearing surfaces—one smooth and the other spongy, roughened, and torn-looking.

Which is the fetal side? ___

What is the basis for your conclusion? ___

___

___

Identify the umbilical cord. Within the cord, identify the umbilical vein and two umbilical arteries. What is the function of the umbilical vein?

___

The umbilical arteries? ___

___

2. Obtain a microscope slide of placental tissue. Observe the tissue carefully, comparing it to Figure 44.3. Identify the *intervillous spaces* (maternal sinusoids), which are blood-filled in life. Identify the villi, and notice their rich vascular supply. Draw a small representative diagram of your observations below and label it appropriately.

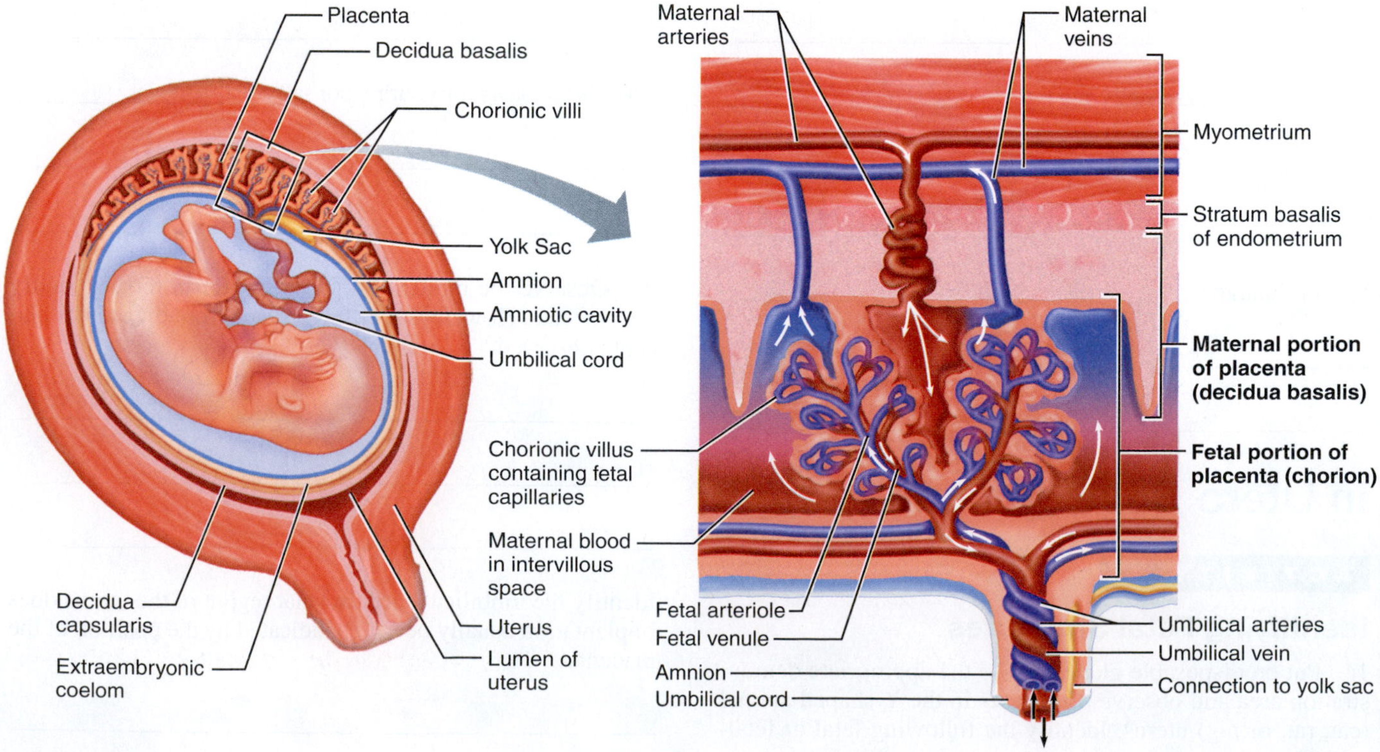
Placenta
Decidua basalis
Chorionic villi
Yolk Sac
Amnion
Amniotic cavity
Umbilical cord
Chorionic villus containing fetal capillaries
Maternal blood in intervillous space
Uterus
Lumen of uterus
Decidua capsularis
Extraembryonic coelom
Maternal arteries
Maternal veins
Myometrium
Stratum basalis of endometrium
Maternal portion of placenta (decidua basalis)
Fetal portion of placenta (chorion)
Fetal arteriole
Fetal venule
Amnion
Umbilical cord
Umbilical arteries
Umbilical vein
Connection to yolk sac

FIGURE 44.3 Diagrammatic representation of the structure of the placenta for a 13-week fetus.

NAME ____________________

LAB TIME/DATE ____________________

REVIEW SHEET

EXERCISE 44

Survey of Embryonic Development

Developmental Stages of Sea Urchins and Humans

1. Define *zygote*. ____________________

2. Describe how you were able to tell by observation when a sea urchin egg was fertilized. ____________________

____

3. Use the key choices to identify the embryonic stage or process described below.

Key: a. blastula b. cleavage c. fertilization d. gastrulation e. morula f. zygote

_______ 1. fusion of male and female pronuclei

_______ 2. solid ball of embryonic cells

_______ 3. process of rapid mitotic cell division without intervening growth periods

_______ 4. combination of egg and sperm

_______ 5. process involving cell rearrangements to form the three primary germ layers

_______ 6. embryonic stage in which the embryo consists of a hollow ball of cells

4. What is the importance of cleavage in embryonic development? ____________________

____

How is cleavage different from mitotic cell division, which occurs later in life? ____________________

____

____

5. The cells of the human blastula (blastocyst) have various fates. Which blastocyst structures have the following fates?

____________ 1. produces the embryonic body

____________ 2. becomes the chorion and cooperates with uterine tissues to form the placenta

____________ 3. produces the amnion, yolk sac, and allantois

____________ 4. produces the primordial germ cells

____________ 5. an embryonic membrane that provides the structural basis for the body stalk or umbilical cord

6. Using the letters on the diagram, correctly identify each of the following maternal or embryonic structures.

______ amnion	______ chorion	______ decidua basalis	______ endoderm
______ body stalk	______ chorionic villi	______ decidua capsularis	______ mesoderm
		______ ectoderm	______ uterine cavity

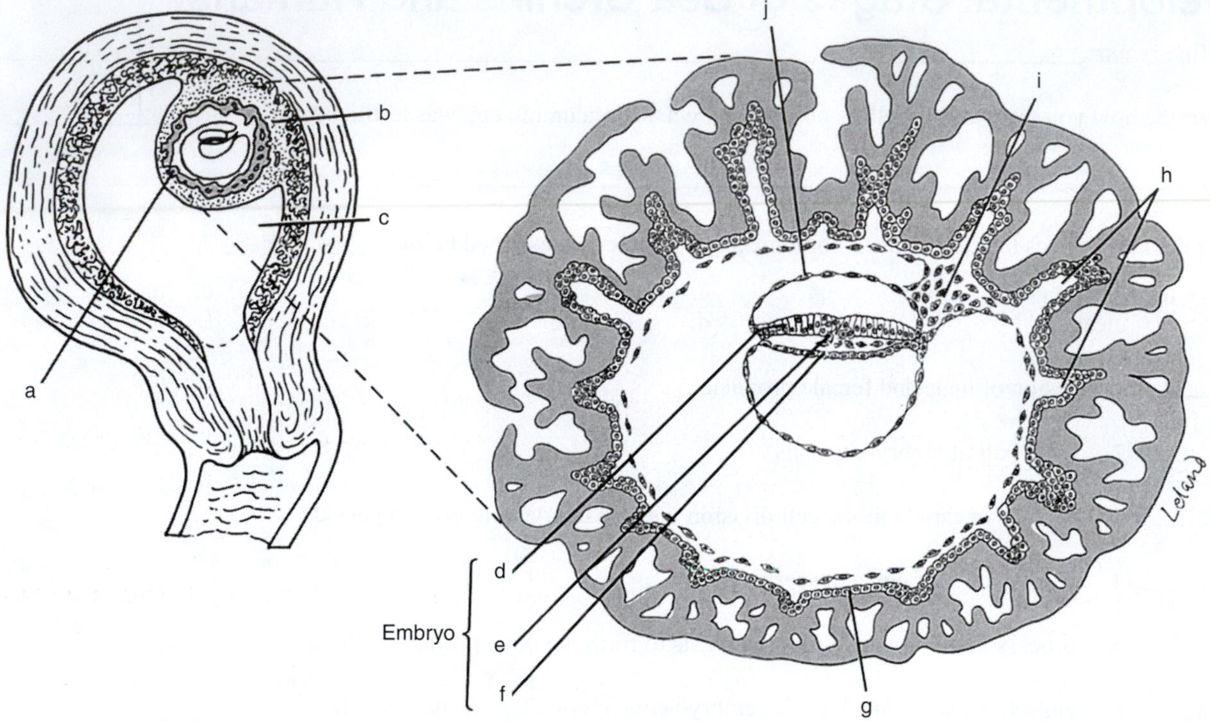
a
b
c
d
e
f
g
h
i
j
Embryo
Leland

7. Explain the process and importance of gastrulation. ____

____

____

8. What is the function of the amnion and the amniotic fluid? ____

____

____

9. Describe the process of implantation, noting the role of the trophoblast cells. ____

____

____

10. How many days after fertilization is implantation generally completed? _______ What event in the female menstrual cycle ordinarily occurs just about this time if implantation does not occur? ____

____

11. What name is given to the part of the uterine wall directly under the implanting embryo? ____________________

That surrounding the rest of the embryonic structure? ____________________

12. Using an appropriate reference, find out what *decidua* means and state the definition. ____________________

____

How is this terminology applicable to the deciduas of pregnancy? ____________________

____

13. Referring to the illustrations and text of *Life Before Birth: Normal Fetal Development,* answer the following:

Which two organ systems are extensively developed in the *very young* embryo?

____________________ and ____________________

Describe the direction of development by circling the correct descriptions below:

proximal-distal   distal-proximal   caudal-rostral   rostral-caudal

Does bodily control during infancy develop in the same directions? Think! Can an infant pick up a common pin (pincer grasp) or wave his arms earlier? Is arm-hand or leg-foot control achieved earlier?

____

14. Note whether each of the following organs or organ systems develops from the (a) ectoderm, (b) endoderm, or (c) mesoderm. Use an appropriate reference as necessary.

______ 1. skeletal muscle	______ 4. respiratory mucosa	______ 7. nervous system
______ 2. skeleton	______ 5. circulatory system	______ 8. serosa membrane
______ 3. lining of gut	______ 6. epidermis of skin	______ 9. liver, pancreas

In Utero Development

15. Make the following comparisons between a human and the pregnant dissected animal structures.

Comparison object	Human	Dissected animal
Shape of the placenta		
Shape of the uterus		

16. Where in the human uterus do implantation and placentation ordinarily occur? ____________________

____

17. Describe the function(s) of the placenta. ____________________

____

What embryonic membranes has the placenta more or less "put out of business"? ____________________

____

18. When does the human embryo come to be called a fetus? ____________________

19. What is the usual and most desirable fetal position in utero? ____________________

Why is this the most desirable position? ____________________

____

____

____

Gross and Microscopic Anatomy of the Placenta

20. Describe fully the gross structure of the human placenta as observed in the laboratory. ____________________

____

____

21. What is the tissue origin of the placenta: fetal, maternal, or both? ____________________

22. What placental barriers must be crossed to exchange materials? ____________________

____

EXERCISE 45

Principles of Heredity

MATERIALS

- ☐ Pennies (for coin tossing)
- ☐ PTC (phenylthiocarbamide) taste strips
- ☐ Sodium benzoate taste strips
- ☐ Chart drawn on chalkboard for tabulation of class results of human phenotype/genotype determinations

Activity 5: Blood typing

- ☐ Anti-A and Anti-B sera
- ☐ Clean microscope slides
- ☐ Toothpicks
- ☐ Wax pencils
- ☐ Sterile lancets
- ☐ Alcohol swabs
- ☐ Beaker containing 10% bleach solution
- ☐ Disposable autoclave bag

Activity 6: Hemoglobin phenotyping

- ☐ Electrophoresis equipment and power supply
- ☐ 1.2% agarose gels
- ☐ 1X TBE (Tris-Borate/EDTA) buffer pH 8.4
- ☐ Micropipette or variable automicropipette with tips
- ☐ Marking pen
- ☐ Safety goggles (student-provided)
- ☐ Metric ruler
- ☐ Plastic baggies
- ☐ Disposable gloves
- ☐ Coomassie protein stain solution
- ☐ Coomassie de-stain solution
- ☐ Distilled water
- ☐ Staining tray
- ☐ 100-ml gradulated cylinder
- ☐ Hemoglobin samples dissolved in TBE solubilizing buffer with bromophenol blue:
  HbA, labeled
  HbS, labeled
  HbA + HbS, labeled
  Unknown samples of each of the above

MasteringA&P™
Access practice quizzes and more in the Study Area at www.masteringaandp.com.

OBJECTIVES

1. To define *allele, heterozygous, homozygous, dominance, recessiveness, genotype, phenotype,* and *incomplete dominance*.
2. To gain practice working out simple genetics problems using a Punnett square.
3. To become familiar with basic laws of probability.
4. To observe selected human phenotypes, and to determine their genotype basis.
5. To separate variants of hemoglobin using agarose gel electrophoresis.

PRE-LAB QUIZ

1. Circle the correct term. Genes that code for the same genetic trait are alleles / sister chromatids.
2. Circle True or False. A heterozygous individual will have two of the same alleles in a chromosome pair.
3. The allele with less potency, which is present but not expressed, is the __________ allele.
   a. dominant
   b. genotypic
   c. homozygous
   d. recessive
4. Circle the correct term. The physical appearance of an individual's genetic makeup, the characteristics that we can see, is that person's gentotype / phenotype.
5. A __________ is used to demonstrate the genotypes of potential offspring that might result from mating.
   a. karyotype
   b. phenotype cross
   c. Punnett square
6. A condition known as __________ can result when heterozygous individuals exhibit a phenotype intermediate between homozygous individuals, and both alleles are expressed in the offspring.
   a. incomplete dominance
   b. sex-linked inheritance
   c. total dominance
   d. total heterozygosity
7. Circle the correct term. Possession of the X / Y chromosome determines maleness.
8. Circle True or False. The Y chromosome is only about one-third the size of the X chromosome, and it lacks many of the genes that are found on the X.
9. Circle True or False. Males are more likely to suffer from hemophilia because it is a result of receiving the sex-linked gene from the father.

The field of genetics is bristling with excitement. Complex gene-splicing techniques have allowed researchers to precisely isolate genes coding for specific proteins and then to use those genes to harvest large amounts of specific proteins and even to cure some dreaded human diseases. At present, growth hormone, insulin, erythropoietin, and interferon produced by these genetic engineering techniques are available for clinical use, and the list is growing daily.

Comprehending the genetics involved in such studies requires arduous training. However, anyone can gain a basic understanding and appreciation of how genes regulate our various traits (dimples and hair color, for example). The focus of this exercise is to provide a "genetics sampler" or relatively simple introduction to the principles of heredity.

Introduction to the Language of Genetics

In humans all cells, except eggs and sperm, contain 46 chromosomes, that is, the diploid number. This number is established when fertilization occurs and the egg and sperm fuse, combining the 23 chromosomes (or haploid complement) each is carrying. The diploid chromosomal number is maintained throughout life in nearly all cells of the body by the precise process of mitosis. As explained in Exercise 43, the diploid chromosomal number actually represents two complete (or nearly complete) sets of genetic instructions—one from the egg and the other from the sperm—or 23 pairs of *homologous chromosomes*.

Genes coding for the same traits on each pair of homologous chromosomes are called **alleles.** The alleles may be identical or different in their influence. For example, the pair of alleles coding for hairline shape on your forehead may specify either straight across or widow's peak. When both alleles in a homologous chromosome pair have the same expression, the individual is **homozygous** for that trait. When the alleles differ in their expression, the individual is **heterozygous** for the given trait; and often only one of the alleles, called the **dominant gene,** will exert its effects. The allele with less potency, the **recessive gene,** will be present but suppressed. Whereas dominant genes, or alleles, exert their effects in both homozygous and heterozygous conditions, as a rule recessive alleles *must* be present in double dose (homozygosity) to exert their influence.

An individual's actual genetic makeup, that is, whether he is homozygous or heterozygous for the various alleles, is called **genotype.** The manner in which genotype is expressed (for example, the presence of a widow's peak [see Figure 45.2 on page 672] or not, blue versus brown eyes) is referred to as **phenotype.**

The complete story of heredity is much more complex than just outlined, and in actuality the expression of many traits (for example, eye color) is determined by the interaction of many allele pairs. However, our emphasis here will be to investigate only the less complex aspects of genetics.

Dominant-Recessive Inheritance

One of the best ways to master the terminology and learn the principles of heredity is to work out the solutions to some genetic crosses in much the same manner Gregor Mendel did in his classic experiments on pea plants. (Mendel, an Austrian monk of the mid-1800s, found evidence in these experiments that each gamete contributes just one allele to each pair in the zygote.)

To work out the various simple monohybrid (one pair of alleles) crosses in this exercise, you will be given the genotype of the parents. You will then determine the possible genotypes of their offspring by using a grid called the *Punnett square,* and you will record the percentages of both genotype and phenotype. To illustrate the procedure, an example of one of Mendel's pea plant crosses is outlined next.

Alleles: *T* (determines *tallness;* dominant)
t (determines *dwarfism;* recessive)

Genotypes of parents: *TT* (♂) × *tt* (♀)

Phenotypes of parents: Tall × dwarf

To use the Punnett, or checkerboard, square, write the alleles (actually gametes) of one parent across the top and the gametes of the other parent down the left side. Then combine the gametes across and down to achieve all possible combinations, as follows:

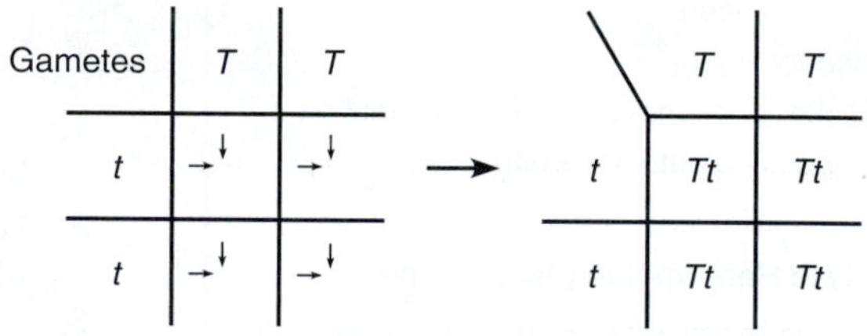
Gametes
T
T
t
t
T
T
t
Tt
Tt
t
Tt
Tt

Results: Genotypes 100% *Tt* (all heterozygous)
Phenotypes 100% tall (because *T*, which determines tallness, is dominant and all contain the *T* allele)

ACTIVITY 1

Working Out Crosses Involving Dominant and Recessive Genes

For each of the following crosses, draw your own Punnett square and use the technique outlined above to determine the genotypes and phenotypes of the offspring.

1. Genotypes of parents: *Tt* (♂) × *tt* (♀)

% of each genotype: ______________________

% of each phenotype: ________% tall, ________% dwarf

2. Genotypes of parents: *Tt* (♂) × *Tt* (♀)

% of each genotype: ______________________

% of each phenotype: ________% tall, ________% dwarf

3. Genotypes of parents: *TT* (♂) × *Tt* (♀)

% of each genotype: ______________________

% of each phenotype: ________% tall, ________% dwarf

Incomplete Dominance

The concepts of dominance and recessiveness are somewhat arbitrary and artificial in some instances because so-called dominant genes may be expressed differently in homozygous and heterozygous individuals. This produces a condition called **incomplete dominance,** or *intermediate inheritance*. In such cases, both alleles express themselves in the offspring. The crosses are worked out in the same manner as indicated previously, but heterozygous offspring exhibit a phenotype intermediate between that of the homozygous individuals. Some examples follow.

ACTIVITY 2

Working Out Crosses Involving Incomplete Dominance

1. The inheritance of flower color in snapdragons illustrates the principle of incomplete dominance. The genotype *RR* is expressed as a red flower, *Rr* yields pink flowers, and *rr* produces white flowers. Work out the following crosses to determine the expected phenotypes and both genotypic and phenotypic percentages.

a. Genotypes of parents: *RR* × *rr*

% of each genotype: ______________________________

% of each phenotype: ______________________________

b. Genotypes of parents: *Rr* × *rr*

% of each genotype: ______________________________

% of each phenotype: ______________________________

____

c. Genotypes of parents: *Rr* × *Rr*

% of each genotype: ______________________________

% of each phenotype: ______________________________

____

2. In humans, the inheritance of sickle cell anemia/trait is determined by a single pair of alleles that exhibit incomplete dominance. Individuals homozygous for the sickling gene (*s*) have *sickle cell anemia* (SCA). In double dose (*ss*), the sickling gene causes production of a very abnormal hemoglobin, which crystallizes and becomes sharp and spiky under conditions of oxygen deficit. This, in turn, leads to clumping and hemolysis of red blood cells in the circulation, which causes a great deal of pain and can be fatal. Heterozygous individuals (*Ss*) have the *sickle cell trait* (SCT); they make both normal and sickling hemoglobin. Usually these individuals are healthy, but prolonged decreases in blood oxygen levels can lead to a sickle cell crisis. Individuals with the genotype *SS* form normal hemoglobin. Work out the following crosses:

a. Parental genotypes: *SS* × *ss*

% of each genotype: ______________________________

% of each phenotype: ______________________________

b. Parental genotypes: *Ss* × *Ss*

% of each genotype: ______________________________

% of each phenotype: ______________________________

____

c. Parental genotypes: *ss* × *Ss*

% of each genotype: ______________________________

% of each phenotype: ______________________________

Sex-Linked Inheritance

Of the 23 pairs of homologous chromosomes, 22 pairs are referred to as **autosomes.** Autosomes contain genes that determine most body (somatic) characteristics. The 23rd pair, the **sex chromosomes,** determine the sex of an individual, that is, whether an individual will be male or female. Normal females possess two sex chromosomes that look alike, the X chromosomes. Males possess two dissimilar sex chromosomes, referred to as X and Y. Possession of the Y chromosome determines maleness. A photomicrograph of a male's chromosome complement (male karyotype) is shown in Figure 45.1. The Y sex chromosome is only

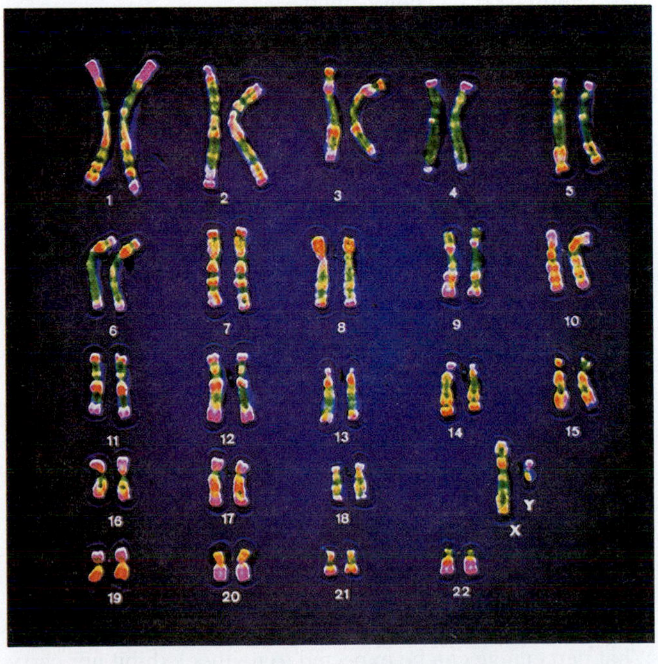
1
2
3
4
5
6
7
8
9
10
11
12
13
14
15
16
17
18
Y
X
19
20
21
22

FIGURE 45.1 Karyotype (chromosomal complement) of human male. Each pair of homologous chromosomes is numbered except the sex chromosomes, which are identified by their letters, X and Y.

about a third the size of the X sex chromosome, and it lacks many of the genes (directing characteristics other than sex) that are found on the X.

Genes present *only* on the X sex chromosome are called *sex-linked* (or X-linked) genes. Some examples of sex-linked genes include those that determine normal color vision (or, conversely, color blindness), and normal clotting ability (as opposed to hemophilia, or bleeder's disease). The alleles that determine color blindness and hemophilia are recessive alleles. In females, *both* X chromosomes must carry the recessive alleles for a woman to express either of these conditions, and thus they tend to be infrequently seen. However, should a male receive a sex-linked recessive allele for these conditions, he will exhibit the recessive phenotype because his Y chromosome does not contain alleles for that gene.

The critical point to understand about sex-linked inheritance is the *absence* of male to male (that is, father to son) transmission of sex-linked genes. The X of the father *will* pass to each of his daughters but to none of his sons. Males always inherit sex-linked conditions from their mothers (via the X chromosome); they may inherit holandric conditions (from alleles on the Y chromosome) from their fathers.

ACTIVITY 3

Working Out Crosses Involving Sex-Linked Inheritance

1. A heterozygous woman carrying the recessive gene for color blindness marries a man who is color-blind. Assume the dominant gene is X^C (allele for normal color vision) and the recessive gene is X^c (determines color blindness). The mother's genotype is X^CX^c and the father's X^cY. Do a Punnett square to determine the answers to the following questions.

According to the laws of probability, what percentage of all their children will be color-blind?

________%

What is the percentage of color-blind individuals by sex?

________% males; ________% females

What percentage will be carriers? ____________________%

What is the sex of the carriers? ____________________

2. A heterozygous woman carrying the recessive gene for hemophilia marries a man who is not a hemophiliac. Assume the dominant gene is X^H and the recessive gene is X^h. The woman's genotype is X^HX^h and her husband's genotype is X^HY. What is the potential percentage and sex of their offspring who will be hemophiliacs?

________% males; ________% females

What percentage can be expected to neither exhibit nor carry the allele for hemophilia?

________% males; ________% females

What is the anticipated sex and percentage of individuals who will be carriers for hemophilia?

________%; ________sex

Probability

Segregation (or parceling out) of chromosomes to daughter cells (gametes) during meiosis and the combination of egg and sperm are random or chance events. Hence, the possibility that certain genomes will arise and be expressed is based on the laws of probability. The randomness of gene recombination from each parent determines individual uniqueness and explains why siblings, however similar, never have totally corresponding traits (unless, of course, they are identical twins). The Punnett square method that you have been using to work out the genetics problems actually provides information on the *probability* of the appearance of certain genotypes considering all possible events. Probability (*P*) is defined as:

$$P = \frac{\text{number of specific events or cases}}{\text{total number of events or cases}}$$

If an event is certain to happen, its probability is 1. If it happens one out of every two times, its probability is ½; if one out of four times, its probability is ¼, and so on.

When figuring the probability of separate events occurring together (or consecutively), the probability of each event must be multiplied together to get the final probability figure. For example, the probability of a penny coming up "heads" in each toss is ½ (because it has two sides—heads and tails). But the probability of a tossed penny coming up heads four times in a row is $\frac{1}{2} \times \frac{1}{2} \times \frac{1}{2} \times \frac{1}{2} = \frac{1}{16}$.

ACTIVITY 4

Exploring Probability

1. Obtain two pennies and perform the following simple experiment to explore the laws of probability.

a. Toss one penny into the air ten times, and record the number of heads (H) and tails (T) observed.

____________ heads ____________ tails

Probability: ________/10 tails; ________/10 heads

b. Now simultaneously toss two pennies into the air for 24 tosses, and record the results of each toss below. In each case, report the probability in the lowest fractional terms.

# of HH: ________ Probability: ____________

# of HT: ________ Probability: ____________

# of TT: ________ Probability: ____________

Does the first toss have any influence on the second?

____

Does the third toss have any influence on the fourth?

____

c. Do a Punnett square using HT for the "alleles" of one "parental" coin and HT for the "alleles" of the other.

Probability of HH: ________

Probability of HT: ________

Probability of TT: ________

How closely do your coin-tossing results correlate with the percentages obtained from the Punnett square results?

____

____

2. Determine the probability of having a boy or girl offspring for each conception.

Parental genotypes: XY × XX

Probability of males: ________%

Probability of females: ________%

3. Dad wants a baseball team! What are the chances of his having nine sons in a row?

________ (Sorry, Dad! Let the girls play.)

Genetic Determination of Selected Human Characteristics

Most human traits are determined by multiple alleles or the interaction of several gene pairs. However, a few visible human traits or phenotypes can be traced to a single gene pair. It is some of these that will be investigated here.

ACTIVITY 5

Using Phenotype to Determine Genotype

For each of the characteristics described here, determine (as best you can) both your own phenotype and genotype, and record this information on the following chart. Since it is impossible to know if you are homozygous or heterozygous for a nondetrimental trait when you exhibit its dominant expression, you are to record your genotype as *A*—(or *B*—, and so on, depending on the letter used to indicate the alleles) in such cases. As you will see, all the traits examined here are nonharmful characteristics. Generally speaking, in humans dominant gene disorders are presumed to be heterozygous (one dominant and one recessive allele), because having two such defective mutated alleles is usually not compatible with life.

Record of Human Genotypes/Phenotypes

Characteristic	Phenotype	Genotype
PTC taste *(P,p)*		
Sodium benzoate taste *(S,s)*		
Sex *(X,Y)*		
Dimples *(D,d)*		
Widow's peak *(W,w)*		
Proximal finger hair *(H,h)*		
Freckles *(F,f)*		
Blaze *(B,b)*		
ABO blood type (I^A, I^B, i)		

On the other hand, if you exhibit the recessive trait, you are homozygous for the recessive allele and should record it accordingly as *aa* (*bb*, *cc*, and so on). When you have completed your observations, also record your data on the chalkboard chart for tabulation of class results.

PTC taste: Obtain a PTC taste strip. PTC, or phenylthiocarbamide, is a harmless chemical that some people can taste and others find tasteless. Chew the strip. If it tastes slightly bitter, you are a "taster" and possess the dominant gene *(P)* for this trait. If you cannot taste anything, you are a nontaster and are homozygous recessive *(pp)* for the trait. Approximately 70% of the people in the United States are tasters.

Sodium benzoate taste: Obtain a sodium benzoate taste strip and chew it. A different pair of alleles (from that determining PTC taste) determines the ability to taste sodium benzoate. If you can taste it, you have at least one of the dominant alleles *(S)*. If not, you are homozygous recessive *(ss)* for the trait. Also record whether sodium benzoate tastes salty, bitter, or sweet to you (if a taster). Even though PTC and sodium benzoate taste are inherited independently, they interact to determine a person's taste sensations. Individuals who find PTC bitter and sodium benzoate salty tend to like sauerkraut, buttermilk, spinach, and other slightly bitter or salty foods.

Sex: The genotype XX determines the female phenotype, whereas XY determines the male phenotype.

Dimpled cheeks: The presence of dimples in one or both cheeks is due to a dominant gene *(D)*. Absence of dimples indicates the homozygous recessive condition *(dd)*.

Widow's peak: A distinct downward V-shaped hairline at the middle of the forehead is referred to as a widow's peak. It is determined by a dominant allele *(W)*, whereas the straight or continuous forehead hairline is determined by the homozygous recessive condition *(ww)* (see Figure 45.2).

Dominant traits
Recessive traits
Widow's peak
Straight hairline
Finger hair
No finger hair
Freckles
No freckles

FIGURE 45.2 Selected examples of human phenotypes.

Proximal finger hair: Critically examine the dorsum of the proximal segment (phalanx) of fingers 3 and 4. If no hair is obvious, you are recessive *(hh)* for this condition. If hair is seen, you have the dominant gene *(H)* for this trait (which, however, is determined by multigene inheritance) (see Figure 45.2).

Freckles: Freckles are the result of a dominant gene. Use *F* as the dominant allele and *f* as the recessive allele (see Figure 45.2).

Blaze: A lock of hair different in color from the rest of scalp hair is called a blaze; it is determined by a dominant gene. Use *B* for the dominant gene and *b* for the recessive gene.

Blood type: Inheritance of the ABO blood type is based on the existence of three alleles designated as I^A, I^B, and *i*. Both I^A and I^B are dominant over *i*, but neither is dominant over the other. Thus the possession of I^A and I^B will yield type AB blood, whereas the possession of the I^A and *i* alleles will yield type A blood, and so on as explained in Exercise 29A. There

TABLE 45.1 Blood Groups

ABO blood group	Genotype
A	I^AI^A or I^Ai
B	I^BI^B or I^Bi
AB	I^AI^B
O	ii

are four ABO blood groups or phenotypes, A, B, AB, and O, and their correlation to genotype is indicated in Table 45.1.

Assuming you have previously typed your blood, record your phenotype and genotype in the chart on page 671. If not, type your blood following the instructions on pages 434–435, and then enter your results in the table.

Dispose of any blood-soiled supplies by placing the glassware in the bleach-containing beaker and all other items in the autoclave bag.

Once class data have been tabulated, scrutinize the results. Is there a single trait that is expressed in an identical manner by all members of the class?

Because all human beings have 23 pairs of homologues and each pair segregates independently at meiosis, the number of possible combinations at segregation is more than 8 million! On the basis of this information, what would you guess are the chances of any two individuals in the class having identical phenotypes for all 14 traits investigated?

Hemoglobin Phenotype Identification Using Agarose Gel Electrophoresis

Agarose gel electrophoresis separates molecules based on charge. In the appropriate buffer with an alkaline pH, hemoglobin molecules will move toward the anode of the apparatus at different speeds, based on the number of negative charges on the molecules. Agarose gel provides a medium for travel and slows the migration down a bit.

Sickle cell anemia slides were observed in Exercise 29A and the trait/disease is discussed earlier in this exercise. The beta chains of hemoglobin S (HbS) contain a base substitution where a valine replaces glutamic acid. As a result of the substitution, HbS has fewer negative charges than the predominant form of adult hemoglobin (HbA) and can be separated from HbA using agarose gel electrophoresis.

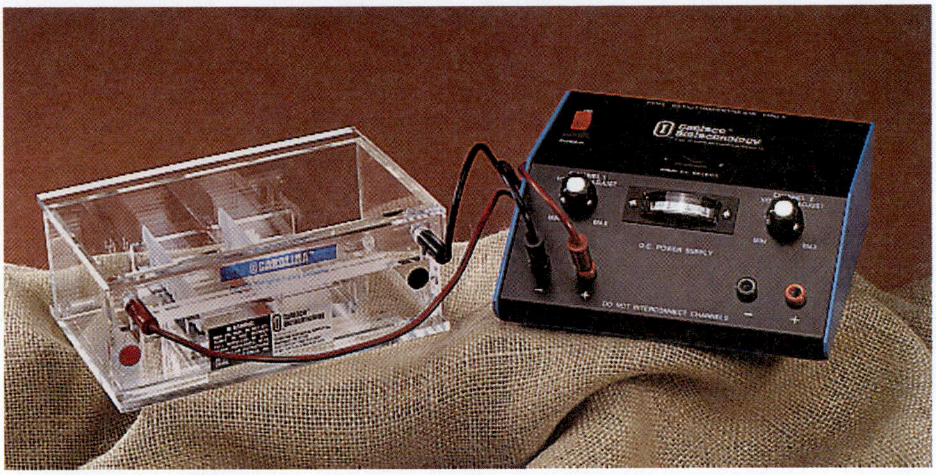

FIGURE 45.3 **Agarose gel electrophoresis equipment and power supply.**

ACTIVITY 6

Using Agarose Gel Electrophoresis to Identify Normal Hemoglobin, Sickle Cell Anemia, and Sickle Cell Trait

1. You will need an electrophoresis unit and power supply (Figure 45.3), a 1.2% agarose gel with eight wells, 1X TBE buffer, micropipettes or a variable automatic micropipette (2–20 μl) with tips, samples of hemoglobin (marked AA, AS, SS, and unknown #_________) dissolved in TBE solubilizing buffer containing bromophenol blue, a marking pen, safety goggles, metric ruler, plastic baggie, and disposable gloves.

2. Record the number of your unknown sample._________

3. Place the agarose gel into the electrophoresis unit.

4. Using a micropipette, carefully add 15 μl of hemoglobin sample AA to wells number 1 and 5, AS to wells 2 and 6, SS to wells 3 and 7, and the unknown samples to wells 4 and 8.

5. Slowly add electrophoresis buffer until the gels are covered with about 0.25 cm of buffer.

6. ! The electrophoresis unit runs with high voltage. Do not attempt to open it while the power supply is attached. Close and lock the electrophoresis unit, and connect the unit to the power source, red to red and black to black.

7. Run the unit about 50 minutes at 120 volts until the bromophenol blue is about 0.25 cm from the anode.

8. Turn the power supply **OFF**. Disconnect the cables.

9. Open the electrophoresis unit, carefully remove the gel, and slide the gel into a plastic baggie. You should be able to see the hemoglobin bands on the gel. Mark each of the bands on the gel with the marking pen.

Alternatively the gels may be stained with Coomassie blue. Obtain a flask of Coomassie blue stain, a flask of de-staining solution, a flask of distilled water, a staining tray, and a 100-ml graduated cylinder, and do the following:

a. Carefully remove the gel from the plastic plate, and place the gel into a staining dish.

b. Add about 30 ml of stain (enough stain to cover the gel), and be sure the agarose is not stuck to the dish.

c. Allow the gel to remain in the stain for at least an hour (more time might be necessary) and then remove the stain. Rinse the gel and dish with distilled water.

d. Add about 100 ml of de-staining solution. Change the solution after a day. If the background stain has been reduced enough to see the bands, place the staining dish over a light source, and observe the bands. If the stain is still too dark, repeat the de-staining process until the bands can be observed.

e. To store the gels, refrigerate in a baggie with a small amount of de-stain solution or dry on a glass plate.

10. Draw the banding patterns for samples 1 through 8 in the figure for question 16 in the Review Sheet, page 678. Based on the banding patterns of the known samples, what are the genotypes of your unknown samples? Record on the chart.

11. Rinse the electrophoresis unit in distilled or de-ionized water, and clean the glass plates with soap and water. ■

NAME ______________________

LAB TIME/DATE ______________

REVIEW SHEET

EXERCISE 45

Principles of Heredity

Introduction to the Language of Genetics

1. Match the key choices with the definitions given below.

Key: a. alleles
b. autosomes
c. dominant
d. genotype
e. heterozygous
f. homozygous
g. phenotype
h. recessive
i. sex chromosomes

__________ 1. actual genetic makeup

__________ 2. chromosomes determining maleness/femaleness

__________ 3. situation in which an individual has identical alleles for a particular trait

__________ 4. genes not expressed unless they are present in homozygous condition

__________ 5. expression of a genetic trait

__________ 6. situation in which an individual has different alleles making up his genotype for a particular trait

__________ 7. genes for the same trait that may have different expressions

__________ 8. chromosomes regulating most body characteristics

__________ 9. the more potent gene allele; masks the expression of the less potent allele

Dominant-Recessive Inheritance

2. In humans, farsightedness is inherited by possession of a dominant allele *(A)*. If a man who is homozygous for normal vision *(aa)* marries a woman who is heterozygous for farsightedness *(Aa)*, what proportion of their children would be expected to be

farsighted? ________%

3. A metabolic disorder called phenylketonuria (PKU) is due to an abnormal recessive gene *(p)*. Only homozygous recessive individuals exhibit this disorder. What percentage of the offspring will be anticipated to have PKU if the parents are *Pp* and

pp? ________%

4. A man obtained 32 spotted and 10 solid-color rabbits from a mating of two spotted rabbits.

Which trait is dominant? ______________________ Recessive? ______________________

What is the probable genotype of the rabbit parents? ________ × ________

5. Assume that the allele controlling brown eyes (*B*) is dominant over that controlling blue eyes (*b*) in human beings. (In actuality, eye color in humans is an example of polygenic inheritance, which is much more complex than this.) A blue-eyed man marries a brown-eyed woman, and they have six children, all brown-eyed. What is the most likely genotype of the father?

_______ Of the mother? _______ If the seventh child had *blue* eyes, what could you conclude about the parents' genotypes?

____

____

Incomplete Dominance

6. Tail length on a bobcat is controlled by incomplete dominance. The alleles are *T* for normal tail length and *t* for tail-less.

What name could/would you give to the tails of heterozygous (*Tt*) cats? ___________________

How would their tail length compare with that of *TT* or *tt* bobcats? ___________________

7. If curly-haired individuals are genotypically *CC*, straight-haired individuals are *cc*, and wavy-haired individuals are heterozygotes (*Cc*), what percentage of the various phenotypes would be anticipated from a cross between a *CC* woman and a *cc* man?

_________% curly _________% wavy _________% straight

Sex-Linked Inheritance

8. What does it mean when someone says a particular characteristic is sex-linked? ___________________

____

9. You are a male, and you have been told that hemophilia "runs in your genes." Whose ancestors, your mother's or your father's, should you investigate? ___________________ Why? ___________________

____

____

10. An X^CX^c female marries an X^CY man. Do a Punnett square for this match.

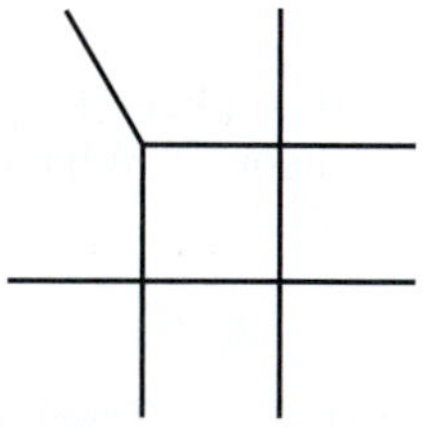

What is the probability of producing a color-blind son? _________

A color-blind daughter? _________

A daughter who is a carrier for the color-blind allele? _________

11. Why are consanguineous marriages (marriages between blood relatives) prohibited in most cultures?

____

____

Probability

12. What is the probability of having three daughters in a row? ____________

13. A man and a woman, each of seemingly normal intellect, marry. Although neither is aware of the fact, each is a heterozygote for the allele for mental retardation. Is the allele for mental retardation dominant or recessive? ____________

What are the chances of their having one mentally retarded child? ____________

What are the chances that all their children (they plan a family of four) will be mentally retarded? ____________

Genetic Determination of Selected Human Characteristics

14. Look back at your data to complete this section. For each of the situations described here, determine if an offspring with the characteristics noted is possible with the parental genotypes listed. Check (✓) the appropriate column.

		Possibility	
Parental genotypes	**Phenotype of child**	**Yes**	**No**
$Ff \times ff$	Freckles		
$DD \times dd$	Dimples		
$HH \times Hh$	Proximal finger hair		
$I^Ai \times I^Bi$	Type O blood		
$I^AI^B \times ii$	Type B blood		

15. You have dimples, and you would like to know if you are homozygous or heterozygous for this trait. You have six brothers and sisters. By observing your siblings, how could you tell, with some degree of certainty, that you are a heterozygote?

____

____

____

Using Agarose Gel Electrophoresis to Identify Hemoglobin Phenotypes

16. Draw the banding patterns you obtained on the figure below. Indicate the genotype of each band.

Sample genotype	Well	Banding pattern
1. ____________	1. ☐	
2. ____________	2. ☐	
3. ____________	3. ☐	
4. ____________	4. ☐	
5. ____________	5. ☐	
6. ____________	6. ☐	
7. ____________	7. ☐	
8. ____________	8. ☐	

17. What is the genotype of sickle cell anemia? _________ Sickle cell trait? _________

18. Why does sickle cell hemoglobin behave differently from normal hemoglobin during agarose gel electrophoresis?

____

____

EXERCISE 46

Surface Anatomy Roundup

MATERIALS

- ☐ Articulated skeletons
- ☐ Three-dimensional models or charts of the skeletal muscles of the body
- ☐ Hand mirror
- ☐ Stethoscope
- ☐ Alcohol swabs
- ☐ Washable markers

OBJECTIVES

1. To define *surface anatomy* and explain why it is an important field of study, and to define *palpation*.
2. To describe and palpate the major surface features of the cranium, face, and neck.
3. To describe the easily palpated bony and muscular landmarks of the back, and to locate the vertebral spines on the living body.
4. To list the bony surface landmarks of the thoracic cage, explain how they relate to the major soft organs of the thorax, and explain how to find the second to eleventh ribs.
5. To name and palpate the important surface features on the anterior abdominal wall, and to explain how to palpate a full bladder.
6. To define and explain the following: *linea alba, umbilical hernia,* examination for an inguinal hernia, *linea semilunaris,* and *McBurney's point*.
7. To locate and palpate the main surface features of the upper limb.
8. To explain the significance of the antecubital fossa, pulse points in the distal forearm, and the anatomical snuff box.
9. To describe and palpate the surface landmarks of the lower limb.
10. To explain exactly where to administer an injection in the gluteal region and in the other major sites of intramuscular injection.

PRE-LAB QUIZ

1. Why is it useful to study surface anatomy?
   a. You can easily locate deep muscle insertions.
   b. You can relate external surface landmarks to the location of internal organs.
   c. You can study cadavers more easily.
   d. You really can't learn that much by studying surface anatomy; it's a gimmick.
2. Circle the correct term. Palpation / Dissection allows you to feel internal structures through the skin.
3. The galea aponeurotica binds to the subcutaneous tissue of the cranium to form the
   a. mastoid process
   b. occipital protuberance
   c. true scalp
   d. xiphoid process
4. The __________ is the most prominent neck muscle and also the neck's most important landmark.
   a. buccinator
   b. epicranius
   c. masseter
   d. sternocleidomastoid
5. The three boundaries of the __________ are the trapezius medially, the latissimus dorsi inferiorly, and the scapula laterally.
   a. torso triangle
   b. triangle of ausculation
   c. triangle of back muscles
   d. triangle of Burney
6. Circle True or False. The lungs do not fill the inferior region of the pleural cavity.

Text continues on next page.

MasteringA&P™
Access practice quizzes and more in the Study Area at www.masteringaandp.com.

7. Circle True or False. With the exception of a full bladder, most internal pelvic organs are not easily palpated through the skin of the body surface.
8. On the dorsum of your hand is a grouping of superficial veins known as the __________, which provides a site for drawing blood and inserting intravenous catheters.
   a. anatomical snuff box
   b. dorsal venous network
   c. radial and ulnar veins
   d. palmar arches
9. Circle True or False. To avoid harming major nerves and blood vessels, clinicians who administer intramuscular injections in the gluteal region of adults use the gluteus medius muscle.
10. The large femoral artery and vein descend vertically through the __________, formed by the border of the inguinal ligament, the medial border of the adductor longus muscle, and the medial border of the sartorius muscle.
    a. femoral triangle
    b. lateral condyle
    c. medial condyle
    d. quadriceps

Surface anatomy is a valuable branch of anatomical and medical science. True to its name, **surface anatomy** does indeed study the *external surface* of the body, but more importantly, it also studies *internal* organs as they relate to external surface landmarks and as they are seen and felt through the skin. Feeling internal structures through the skin with the fingers is called **palpation** (literally, "touching").

Surface anatomy is living anatomy, better studied in live people than in cadavers. It can provide a great deal of information about the living skeleton (almost all bones can be palpated) and about the muscles and blood vessels that lie near the body surface. Furthermore, a skilled examiner can learn a good deal about the heart, lungs, and other deep organs by performing a surface assessment. Thus, surface anatomy serves as the basis of the standard physical examination. For those planning a career in the health sciences or physical education, a study of surface anatomy will show you where to take pulses, where to insert tubes and needles, where to locate broken bones and inflamed muscles, and where to listen for the sounds of the lungs, heart, and intestines.

We will take a regional approach to surface anatomy, exploring the head first and proceeding to the trunk and the limbs. You will be observing and palpating your own body as you work through the exercise, because your body is the best learning tool of all. To aid your exploration of living anatomy, skeletons and muscle models or charts are provided around the lab so that you can review the bones and muscles you will encounter. For skin sites you are asked to mark that you cannot reach on your own body, it probably would be best to choose a male student as a subject.

ACTIVITY 1

Palpating Landmarks of the Head

The head (Figures 46.1 and 46.2) is divided into the cranium and the face.

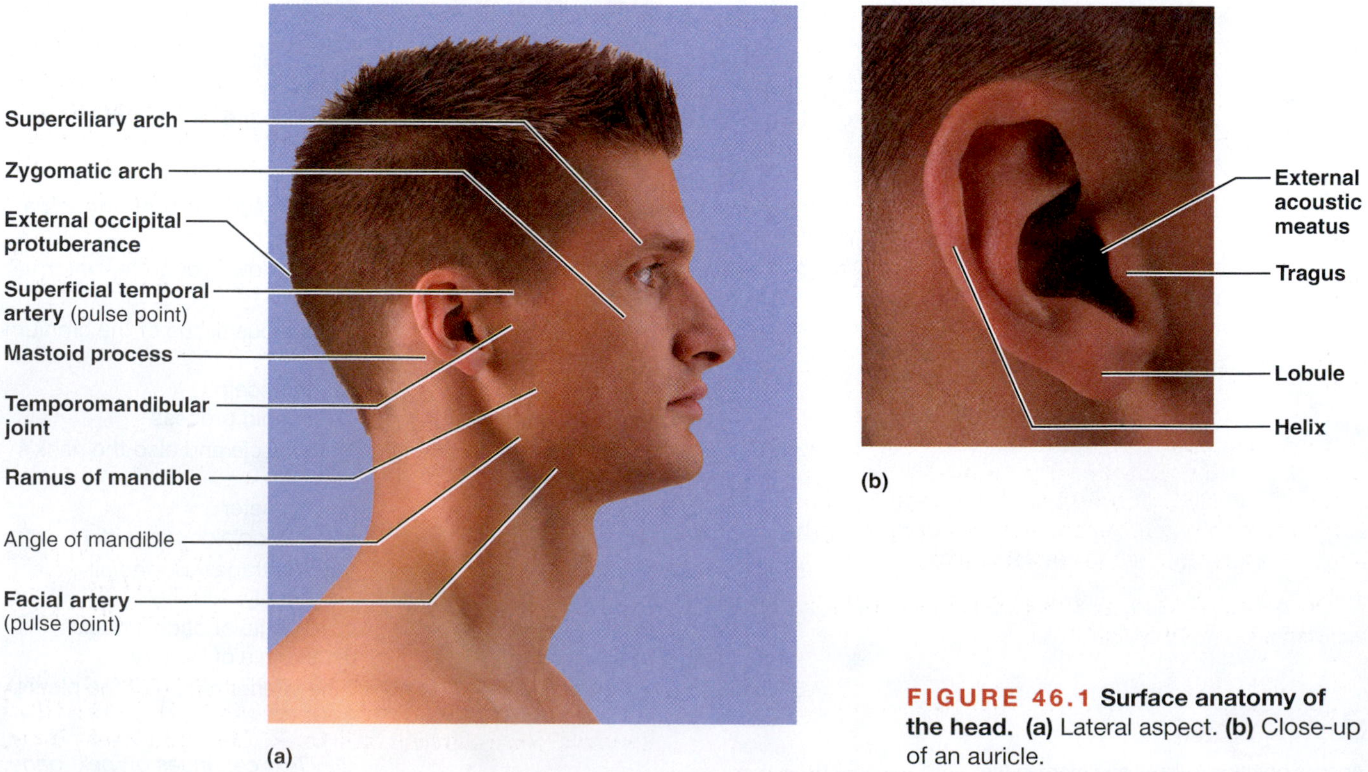
Superciliary arch
Zygomatic arch
External occipital protuberance
Superficial temporal artery (pulse point)
Mastoid process
Temporomandibular joint
Ramus of mandible
Angle of mandible
Facial artery (pulse point)
(a)
External acoustic meatus
Tragus
Lobule
Helix
(b)

FIGURE 46.1 Surface anatomy of the head. (a) Lateral aspect. **(b)** Close-up of an auricle.

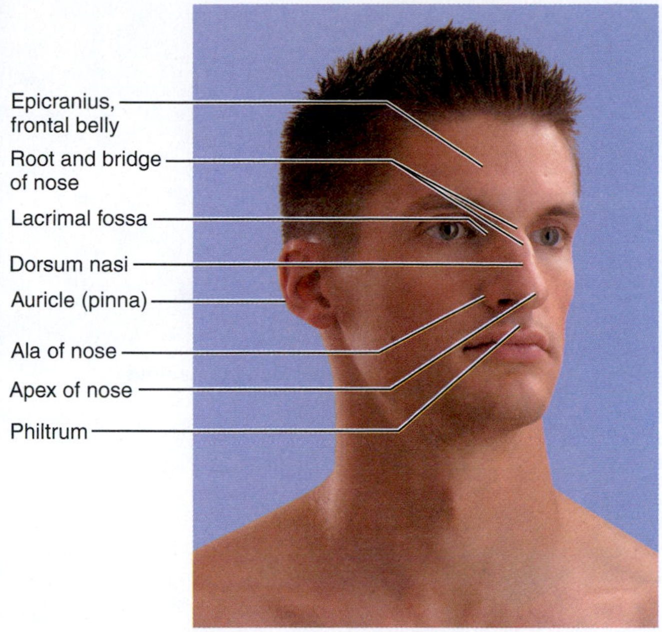

Epicranius, frontal belly
Root and bridge of nose
Lacrimal fossa
Dorsum nasi
Auricle (pinna)
Ala of nose
Apex of nose
Philtrum

FIGURE 46.2 Surface structures of the face.

Cranium

1. Run your fingers over the superior surface of your head. Notice that the underlying cranial bones lie very near the surface. Proceed to your forehead and palpate the **superciliary arches** (brow ridges) directly superior to your orbits (see Figure 46.1).

2. Move your hand to the posterior surface of your skull, where you can feel the knoblike **external occipital protuberance.** Run your finger directly laterally from this projection to feel the ridgelike *superior nuchal line* on the occipital bone. This line, which marks the superior extent of the muscles of the posterior neck, serves as the boundary between the head and the neck. Now feel the prominent **mastoid process** on each side of the cranium just posterior to your ear.

3. The **frontal belly** of the epicranius (see Figure 46.2) inserts superiorly onto the broad aponeurosis called the *galea aponeurotica* (Table 15.1, pages 202 and 204) that covers the superior surface of the cranium. This aponeurosis binds tightly to the overlying subcutaneous tissue and skin to form the true **scalp.** Push on your scalp, and confirm that it slides freely over the underlying cranial bones. Because the scalp is only loosely bound to the skull, people can easily be "scalped" (in industrial accidents, for example). The scalp is richly vascularized by a large number of arteries running through its subcutaneous tissue. Most arteries of the body constrict and close after they are cut or torn, but those in the scalp are unable to do so because they are held open by the dense connective tissue surrounding them.

What do these facts suggest about the amount of bleeding that accompanies scalp wounds?

____

____

Face

The surface of the face is divided into many different regions, including the *orbital, nasal, oral* (mouth), and *auricular* (ear) areas.

1. Trace a finger around the entire margin of the bony orbit. The **lacrimal fossa,** which contains the tear-gathering lacrimal sac, may be felt on the medial side of the eye socket.

2. Touch the most superior part of your nose, its **root,** which lies between the eyebrows (see Figure 46.2). Just inferior to this, between your eyes, is the **bridge** of the nose formed by the nasal bones. Continue your finger's progress inferiorly along the nose's anterior margin, the **dorsum nasi,** to its tip, the **apex.** Place one finger in a nostril and another finger on the flared winglike **ala** that defines the nostril's lateral border. Then feel the **philtrum,** the shallow vertical groove on the upper lip below the nose.

3. Grasp your **auricle,** the shell-like part of the external ear that surrounds the opening of the **external acoustic meatus** (Figure 46.1). Now trace the ear's outer rim, or **helix,** to the **lobule** (earlobe) inferiorly. The lobule is easily pierced, and since it is not highly sensitive to pain, it provides a convenient place to hang an earring or obtain a drop of blood for clinical blood analysis. Feel the **tragus,** the stiff projection just anterior to the external acoustic meatus. Next, place a finger on your temple just anterior to the auricle. There, you will be able to feel the pulsations of the **superficial temporal artery,** which ascends to supply the scalp (Figure 46.1).

4. Run your hand anteriorly from your ear toward the orbit, and feel the **zygomatic arch** just deep to the skin. This bony arch is easily broken by blows to the face. Next, place your fingers on the skin of your face, and feel it bunch and stretch as you contort your face into smiles, frowns, and grimaces. You are now monitoring the action of several of the subcutaneous **muscles of facial expression** (Table 15.1, pages 202 and 204).

5. On your lower jaw, palpate the parts of the bony **mandible:** its anterior body and its posterior ascending **ramus.** Press on the skin over the mandibular ramus, and feel the **masseter muscle** bulge when you clench your teeth. Palpate the anterior border of the masseter, and trace it to the mandible's inferior margin. At this point, you will be able to detect the pulse of your **facial artery** (Figure 46.1). Finally, to feel the **temporomandibular joint,** place a finger directly anterior to the external acoustic meatus of your ear, and open and close your mouth several times. The bony structure you feel moving is the *head of the mandible.* ■

ACTIVITY 2

Palpating Landmarks of the Neck

Bony Landmarks

1. Run your fingers inferiorly along the back of your neck, in the posterior midline, to feel the *spinous processes* of the cervical vertebrae. The spine of C_7, the *vertebra prominens,* is especially prominent.

2. Now, beginning at your chin, run a finger inferiorly along the anterior midline of your neck (Figure 46.3). The first hard structure you encounter will be the U-shaped **hyoid bone,**

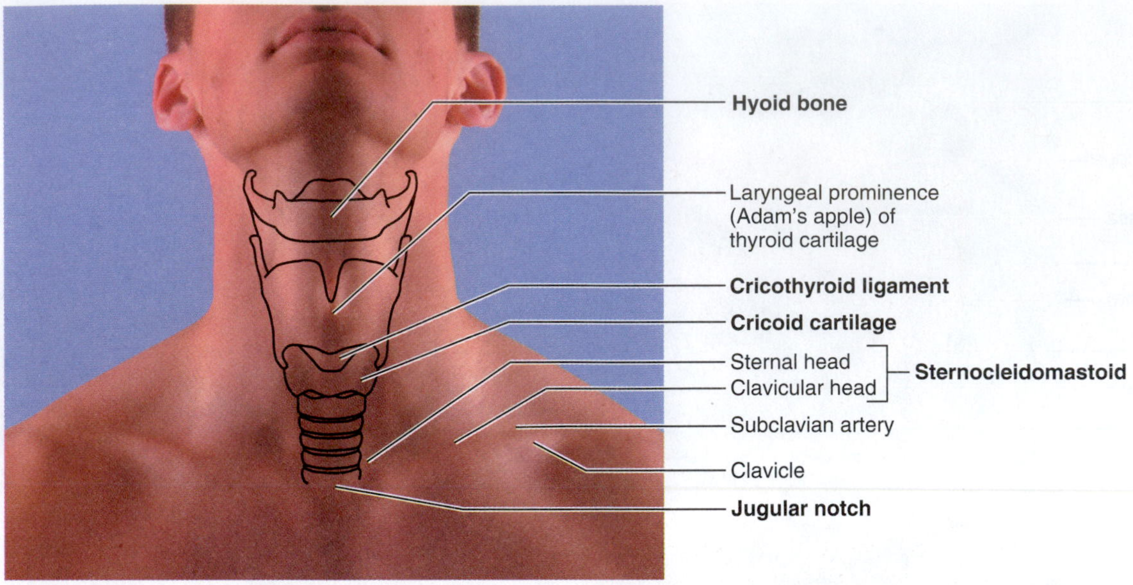
Hyoid bone
Laryngeal prominence (Adam's apple) of thyroid cartilage
Cricothyroid ligament
Cricoid cartilage
Sternal head
Clavicular head
Sternocleidomastoid
Subclavian artery
Clavicle
Jugular notch

FIGURE 46.3 Anterior surface of the neck. A diagram of the underlying skeleton of the larynx is superimposed on a photograph of the neck.

which lies in the angle between the floor of the mouth and the vertical part of the neck. Directly inferior to this, you will feel the **laryngeal prominence** (Adam's apple) of the thyroid cartilage. Just inferior to the laryngeal prominence, your finger will sink into a soft depression (formed by the **cricothyroid ligament**) before proceeding onto the rounded surface of the **cricoid cartilage.** Now swallow several times, and feel the whole larynx move up and down.

3. Continue inferiorly to the trachea. Attempt to palpate the *isthmus of the thyroid gland,* which feels like a spongy cushion over the second to fourth tracheal rings (see Figure 46.3). Then, try to palpate the two soft lateral *lobes* of your thyroid gland along the sides of the trachea.

4. Move your finger all the way inferiorly to the root of the neck, and rest it in the **jugular notch,** the depression in the superior part of the sternum between the two clavicles. By pushing deeply at this point, you can feel the cartilage rings of the trachea.

Muscles

The **sternocleidomastoid** is the most prominent muscle in the neck and the neck's most important surface landmark. You can best see and feel it when you turn your head to the side.

Obtain a hand mirror, hold it in front of your face, and turn your head sharply from right to left several times. You will be able to see both heads of this muscle, the **sternal head** medially and the **clavicular head** laterally (Figure 46.3). Several important structures lie beside or beneath the sternocleidomastoid:

- The *cervical lymph nodes* lie both superficial and deep to this muscle. (Swollen cervical nodes provide evidence of infections or cancer of the head and neck.)
- The *common carotid artery* and *internal jugular vein* lie just deep to the sternocleidomastoid, a relatively superficial location that exposes these vessels to danger in slashing wounds to the neck.
- Just lateral to the inferior part of the sternocleidomastoid is the large **subclavian artery** on its way to supply the upper limb. By pushing on the subclavian artery at this point, one can stop the bleeding from a wound anywhere in the associated limb.
- Just anterior to the sternocleidomastoid, superior to the level of your larynx, you can feel a carotid pulse—the pulsations of the **external carotid artery** (Figure 46.4).
- The *external jugular vein* descends vertically, just superficial to the sternocleidomastoid and deep to the skin (Figure 46.3). To make this vein "appear" on your neck, stand before the mirror, and gently compress the skin superior to your clavicle with your fingers.

Triangles of the Neck

The sternocleidomastoid muscles divide each side of the neck into the posterior and anterior triangles (Figure 46.4a).

1. The **posterior triangle** is defined by the sternocleidomastoid anteriorly, the trapezius posteriorly, and the clavicle inferiorly. Palpate the borders of the posterior triangle.

The **anterior triangle** is defined by the inferior margin of the mandible superiorly, the midline of the neck anteriorly, and the sternocleidomastoid posteriorly.

2. The contents of these two triangles are shown in Figure 46.4b. The posterior triangle contains many important nerves and blood vessels, including the **accessory nerve** (cranial nerve XI), most of the **cervical plexus,** and the **phrenic nerve.** In the inferior part of the triangle are the **external jugular vein,** the trunks of the **brachial plexus,** and the **subclavian artery.** These structures are relatively superficial and are easily cut or injured by wounds to the neck.

In the neck's anterior triangle, important structures include the **submandibular gland,** the **suprahyoid** and **infrahyoid muscles,** and parts of the **carotid arteries** and **jugular veins** that lie superior to the sternocleidomastoid.

- Palpate your carotid pulse.

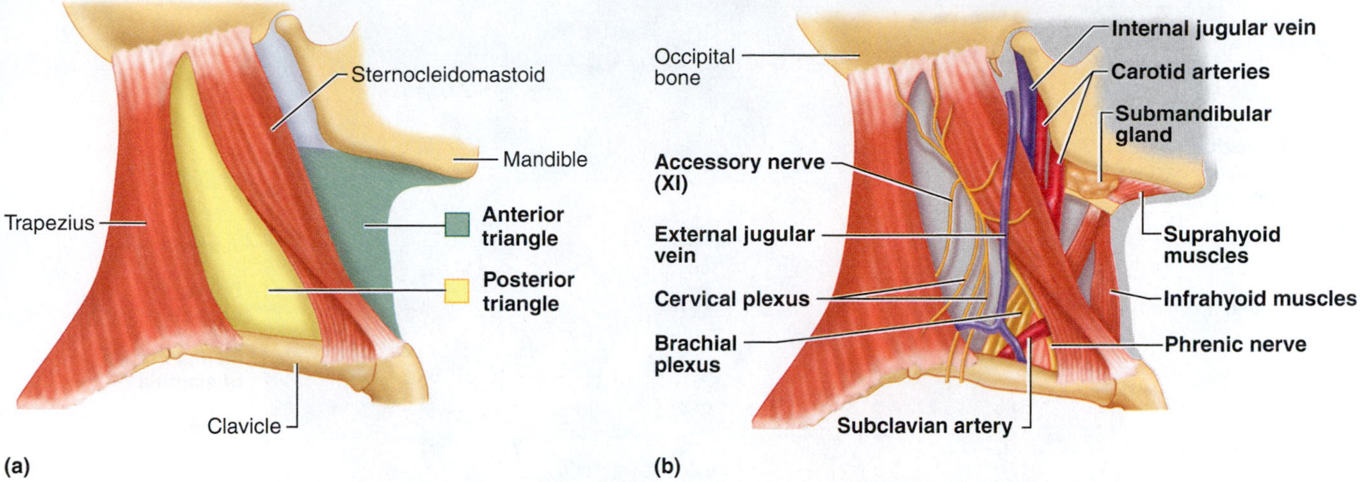
Sternocleidomastoid
Mandible
Trapezius
Anterior triangle
Posterior triangle
Clavicle
(a)
Occipital bone
Internal jugular vein
Carotid arteries
Submandibular gland
Accessory nerve (XI)
External jugular vein
Suprahyoid muscles
Cervical plexus
Infrahyoid muscles
Brachial plexus
Phrenic nerve
Subclavian artery
(b)

FIGURE 46.4 Anterior and posterior triangles of the neck. (a) Boundaries of the triangles. **(b)** Some contents of the triangles.

A wound to the posterior triangle of the neck can lead to long-term loss of sensation in the skin of the neck and shoulder, as well as partial paralysis of the sternocleidomastoid and trapezius muscles. Explain these effects. ■

ACTIVITY 3

Palpating Landmarks of the Trunk

The trunk of the body consists of the thorax, abdomen, pelvis, and perineum. The *back* includes parts of all of these regions, but for convenience it is treated separately.

The Back

Bones

1. The vertical groove in the center of the back is called the **posterior median furrow** (Figure 46.5). The *spinous processes* of the vertebrae are visible in the furrow when the spinal column is flexed.

- Palpate a few of these processes on your partner's back (C_7 and T_1 are the most prominent and the easiest to find).
- Also palpate the posterior parts of some ribs, as well as the prominent **spine of the scapula** and the scapula's long **medial border.**

  The scapula lies superficial to ribs 2 to 7; its **inferior angle** is at the level of the spinous process of vertebra T_7. The medial end of the scapular spine lies opposite the T_3 spinous process.

2. Now feel the **iliac crests** (superior margins of the iliac bones) in your own lower back. You can find these crests effortlessly by resting your hands on your hips. Locate the most superior point of each crest, a point that lies roughly halfway between the posterior median furrow and the lateral side of the body (see Figure 46.5). A horizontal line through these two superior points, the **supracristal line,** intersects L_4, providing a simple way to locate that vertebra. The ability to locate L_4 is essential for performing a *lumbar puncture*, a procedure in which the clinician inserts a needle into the vertebral canal of the spinal column directly superior or inferior to L_4 and withdraws cerebrospinal fluid.

3. The *sacrum* is easy to palpate just superior to the cleft in the buttocks. You can feel the *coccyx* in the extreme inferior part of that cleft, just posterior to the anus.

Muscles The largest superficial muscles of the back are the **trapezius** superiorly and **latissimus dorsi** inferiorly (Figure 46.5). Furthermore, the deeper **erector spinae** muscles are very evident in the lower back, flanking the vertebral column like thick vertical cords.

1. Shrug your shoulders to feel the trapezius contracting just deep to the skin.

2. Feel your partner's erector spinae muscles contract and bulge as he straightens his spine from a slightly bent-over position.

The superficial muscles of the back fail to cover a small area of the rib cage called the **triangle of auscultation** (see Figure 46.5). This triangle lies just medial to the inferior part of the scapula. Its three boundaries are formed by the trapezius medially, the latissimus dorsi inferiorly, and the scapula laterally. The physician places a stethoscope over the skin of this triangle to listen for lung sounds (*auscultation* = listening). To hear the lungs clearly, the doctor first asks the patient to fold the arms together in front of the chest and then flex the trunk.

What do you think is the precise reason for having the patient take this action?

---

3. Have your partner assume the position just described. After cleaning the earpieces with an alcohol swab, use the stethoscope to auscultate the lung sounds. Compare the clarity of the

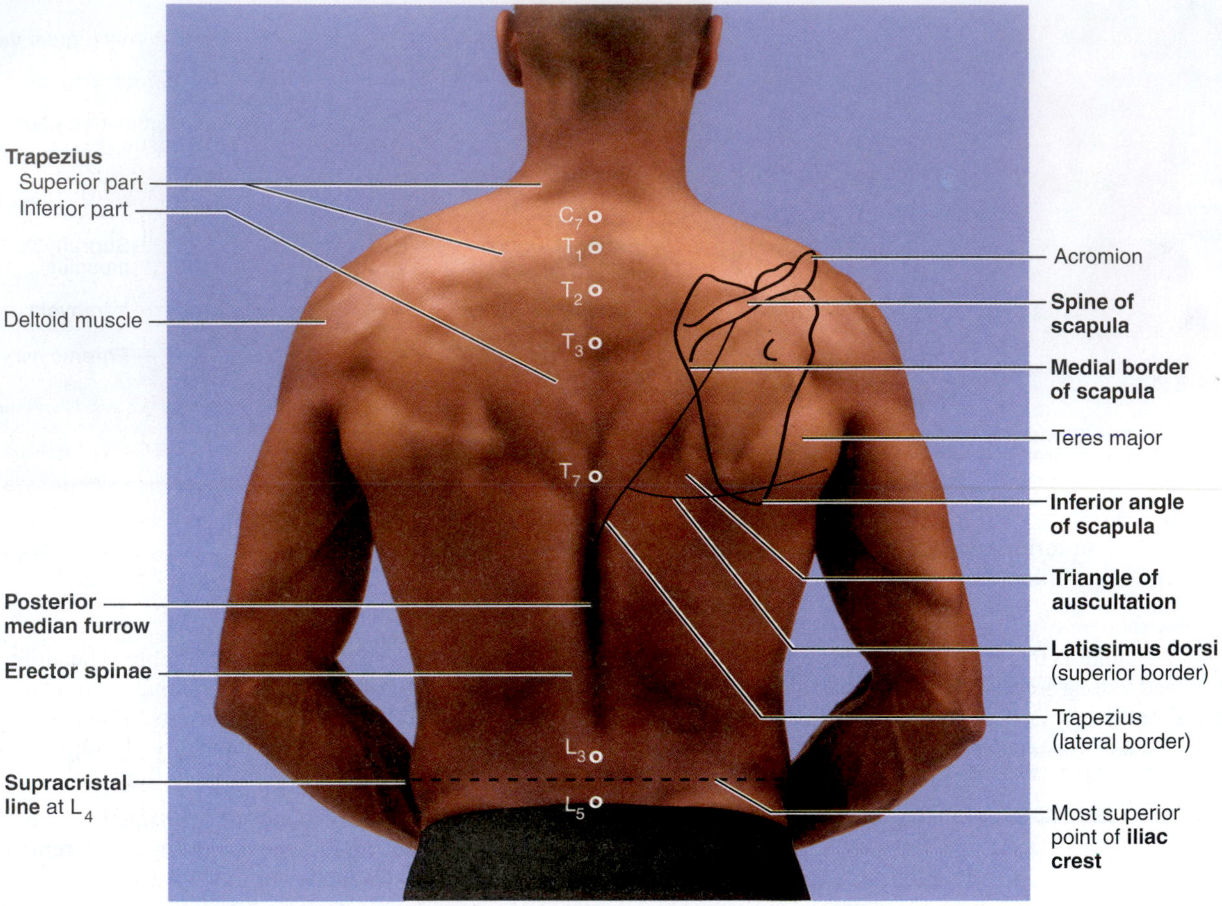
Trapezius
Superior part
Inferior part
Deltoid muscle
Posterior median furrow
Erector spinae
Supracristal line at L4
C7
T1
T2
T3
T7
L3
L5
Acromion
Spine of scapula
Medial border of scapula
Teres major
Inferior angle of scapula
Triangle of auscultation
Latissimus dorsi (superior border)
Trapezius (lateral border)
Most superior point of iliac crest

FIGURE 46.5 Surface anatomy of the back.

lung sounds heard over the triangle of auscultation to that over other areas of the back.

The Thorax

Bones

1. Start exploring the anterior surface of your partner's bony *thoracic cage* (Figures 46.6 and 46.7) by defining the extent of the *sternum*. Use a finger to trace the sternum's triangular *manubrium* inferior to the jugular notch, its flat *body,* and the tongue-shaped **xiphoid process.** Now palpate the ridgelike **sternal angle,** where the manubrium meets the body of the sternum. Locating the sternal angle is important because it directs you to the second ribs (which attach to it). Once you find the second rib, you can count down to identify every other rib in the thorax (except the first and sometimes the twelfth rib, which lie too deep to be palpated). The sternal angle is a highly reliable landmark—it is easy to locate, even in overweight people.

2. By locating the individual ribs, you can mentally "draw" a series of horizontal lines of "latitude" that you can use to map and locate the underlying visceral organs of the thoracic cavity. Such mapping also requires lines of "longitude," so let us construct some vertical lines on the wall of your partner's trunk. As he lifts an arm straight up in the air, extend a line inferiorly from the center of the axilla onto his lateral thoracic wall. This is the **midaxillary line** (see Figure 46.6a). Now estimate the midpoint of his **clavicle,** and run a vertical line inferiorly from that point toward the groin. This is the **midclavicular line,** and it will pass about 1 cm medial to the nipple.

3. Next, feel along the V-shaped inferior edge of the rib cage, the **costal margin.** At the **infrasternal angle,** the superior angle of the costal margin, lies the **xiphisternal joint.** Deep to the xiphisternal joint, the heart lies on the diaphragm.

4. The thoracic cage provides many valuable landmarks for locating the vital organs of the thoracic and abdominal cavities. On the anterior thoracic wall, ribs 2–6 define the superior-to-inferior extent of the female breast, and the fourth intercostal space indicates the location of the **nipple** in men, children, and small-breasted women. The right costal margin runs across the anterior surface of the liver and gallbladder. Surgeons must be aware of the inferior margin of the *pleural cavities* because if they accidentally cut into one of these cavities, a lung collapses. The inferior pleural margin lies adjacent to vertebra T_{12} near the posterior midline (see Figure 46.6b) and runs horizontally across the back to reach rib 10 at the midaxillary line. From there, the pleural margin ascends to rib 8 in the midclavicular line (see Figure 46.6a) and to the level of the xiphisternal joint near the anterior midline. The *lungs* do not fill the inferior region of the pleural cavity. Instead, their inferior borders run at a level that is

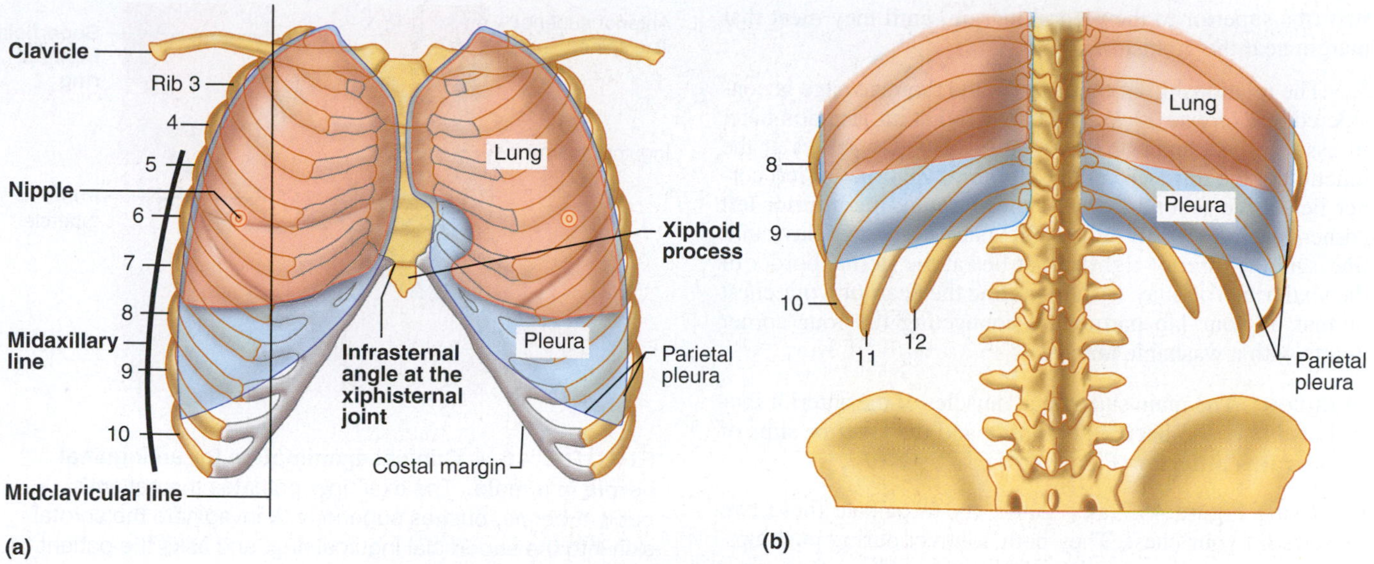

Clavicle
Rib 3
4
5
Nipple
6
7
8
Midaxillary line
9
10
Midclavicular line
Lung
Xiphoid process
Pleura
Infrasternal angle at the xiphisternal joint
Parietal pleura
Costal margin
(a)
Lung
Pleura
8
9
10
11
12
Parietal pleura
(b)

FIGURE 46.6 The bony rib cage as it relates to the underlying lungs and pleural cavities. Both the pleural cavities (blue) and the lungs (pink) are outlined. **(a)** Anterior view. **(b)** Posterior view.

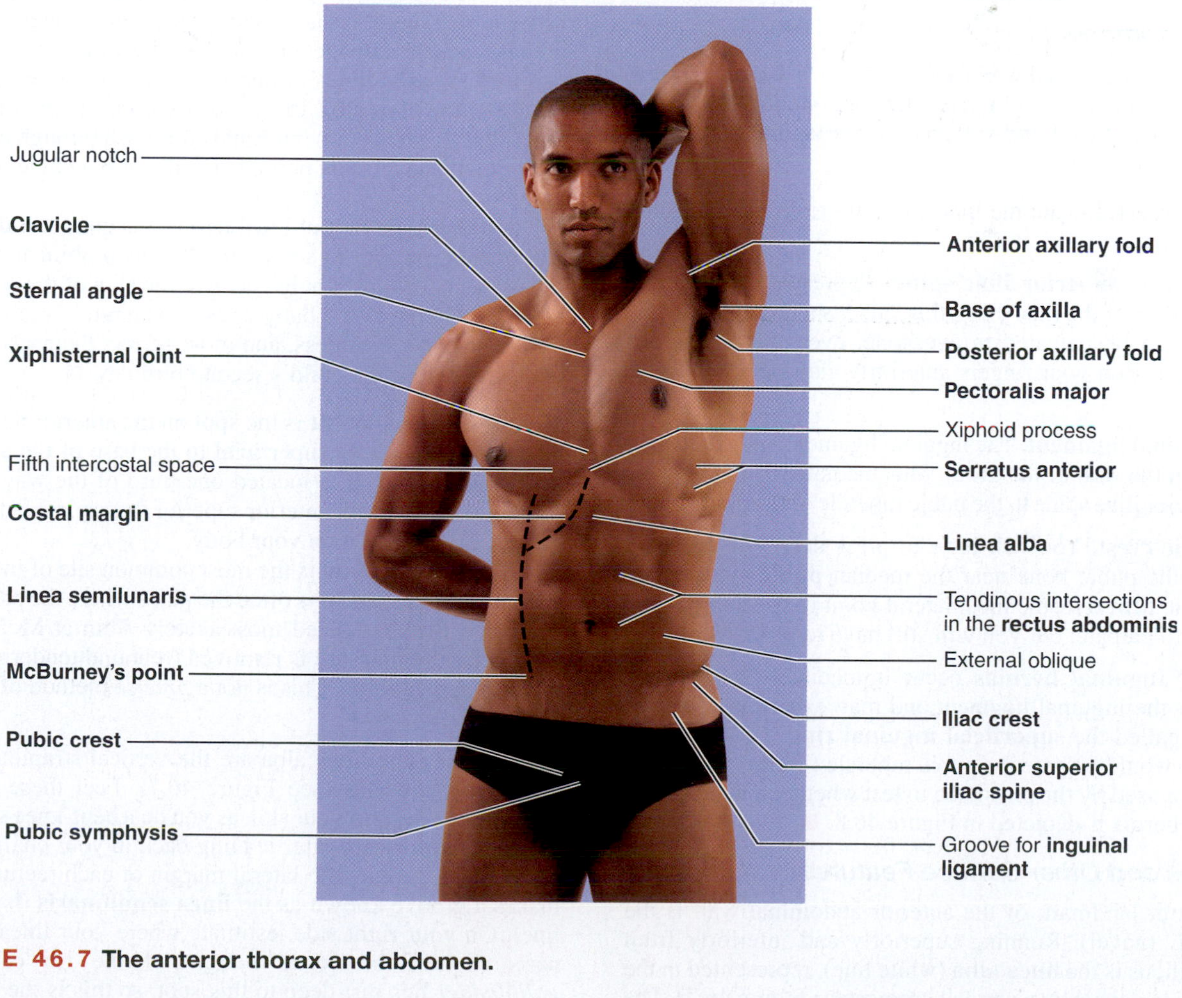

Jugular notch
Clavicle
Sternal angle
Xiphisternal joint
Fifth intercostal space
Costal margin
Linea semilunaris
McBurney's point
Pubic crest
Pubic symphysis
Anterior axillary fold
Base of axilla
Posterior axillary fold
Pectoralis major
Xiphoid process
Serratus anterior
Linea alba
Tendinous intersections in the rectus abdominis
External oblique
Iliac crest
Anterior superior iliac spine
Groove for inguinal ligament

FIGURE 46.7 The anterior thorax and abdomen.

two ribs superior to the pleural margin, until they meet that margin near the xiphisternal joint.

5. The relationship of the *heart* to the thoracic cage is considered in Exercise 30. We will review that information here. In essence, the superior right corner of the heart lies at the junction of the third rib and the sternum; the superior left corner lies at the second rib, near the sternum; the inferior left corner lies in the fifth intercostal space in the midclavicular line; and the inferior right corner lies at the sternal border of the sixth rib. You may wish to outline the heart on your chest or that of your lab partner by connecting the four corner points with a washable marker.

Muscles The main superficial muscles of the anterior thoracic wall are the **pectoralis major** and the anterior slips of the **serratus anterior** (Figure 46.7).

- Using Figure 46.7 as a guide, try to palpate these two muscles on your chest. They both contract during push-ups, and you can confirm this by pushing yourself up from your desk with one arm while palpating the muscles with your opposite hand. ■

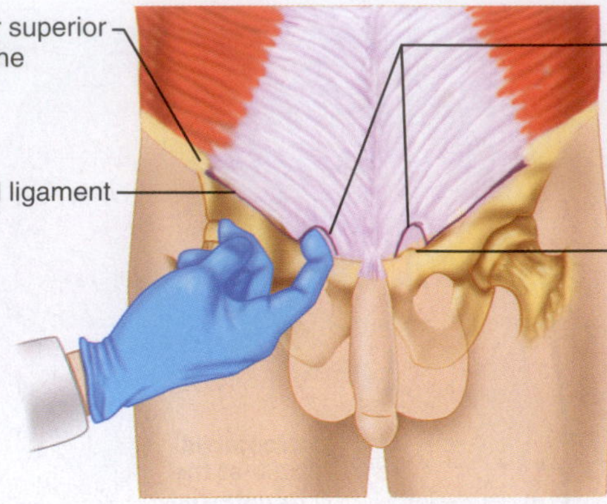
Anterior superior iliac spine
Superficial inguinal ring
Inguinal ligament
Pubic tubercle

FIGURE 46.8 Clinical examination for an inguinal hernia in a male. The examiner palpates the patient's pubic tubercle, pushes superiorly to invaginate the scrotal skin into the superficial inguinal ring, and asks the patient to cough. If an inguinal hernia exists, it will push inferiorly and touch the examiner's fingertip.

ACTIVITY 4

Palpating Landmarks of the Abdomen

Bony Landmarks

The anterior abdominal wall (see Figure 46.7) extends inferiorly from the costal margin to an inferior boundary that is defined by several landmarks. Palpate these landmarks as they are described below.

1. **Iliac crest.** Locate the iliac crests by resting your hands on your hips.

2. **Anterior superior iliac spine.** Representing the most anterior point of the iliac crest, this spine is a prominent landmark. It can be palpated in everyone, even those who are overweight. Run your fingers anteriorly along the iliac crest to its end.

3. **Inguinal ligament.** The inguinal ligament, indicated by a groove on the skin of the groin, runs medially from the anterior superior iliac spine to the pubic tubercle of the pubic bone.

4. **Pubic crest.** You will have to press deeply to feel this crest on the pubic bone near the median **pubic symphysis.** The **pubic tubercle,** the most lateral point of the pubic crest, is easier to palpate, but you will still have to push deeply.

Inguinal hernias occur immediately superior to the inguinal ligament and may exit from a medial opening called the **superficial inguinal ring.** To locate this ring, one would palpate the pubic tubercle (Figure 46.8). The procedure used by the physician to test whether a male has an inguinal hernia is depicted in Figure 46.8. ■

Muscles and Other Surface Features

The central landmark of the anterior abdominal wall is the *umbilicus* (navel). Running superiorly and inferiorly from the umbilicus is the **linea alba** (white line), represented in the skin of lean people by a vertical groove (see Figure 46.7). The linea alba is a tendinous seam that extends from the xiphoid process to the pubic symphysis, just medial to the rectus abdominis muscles (Table 15.3, pages 207–208). The linea alba is a favored site for surgical entry into the abdominal cavity because the surgeon can make a long cut through this line with no muscle damage and minimal bleeding.

Several kinds of hernias involve the umbilicus and the linea alba. In an **acquired umbilical hernia,** the linea alba weakens until intestinal coils push through it just superior to the navel. The herniated coils form a bulge just deep to the skin.

Another type of umbilical hernia is a **congenital umbilical hernia,** present in some infants: The umbilical hernia is seen as a cherry-sized bulge deep to the skin of the navel that enlarges whenever the baby cries. Congenital umbilical hernias are usually harmless, and most correct themselves automatically before the child's second birthday. ■

1. **McBurney's point** is the spot on the anterior abdominal skin that lies directly superficial to the base of the appendix (see Figure 46.7). It is located one-third of the way along a line between the right anterior superior iliac spine and the umbilicus. Try to find it on your body.

McBurney's point is the most common site of incision in appendectomies, and it is often the place where the pain of appendicitis is experienced most acutely. Pain at McBurney's point after the pressure is removed (rebound tenderness) can indicate appendicitis. This is not a *precise* method of diagnosis, however.

2. Flanking the linea alba are the vertical straplike **rectus abdominis** muscles (see Figure 46.7). Feel these muscles contract just deep to your skin as you do a bent-knee sit-up (or as you bend forward after leaning back in your chair). In the skin of lean people, the lateral margin of each rectus muscle makes a groove known as the **linea semilunaris** (half-moon line). On your right side, estimate where your linea semilunaris crosses the costal margin of the rib cage. The *gallbladder* lies just deep to this spot, so this is the standard point of incision for gallbladder surgery. In muscular people, three horizontal grooves can be seen in the skin covering the

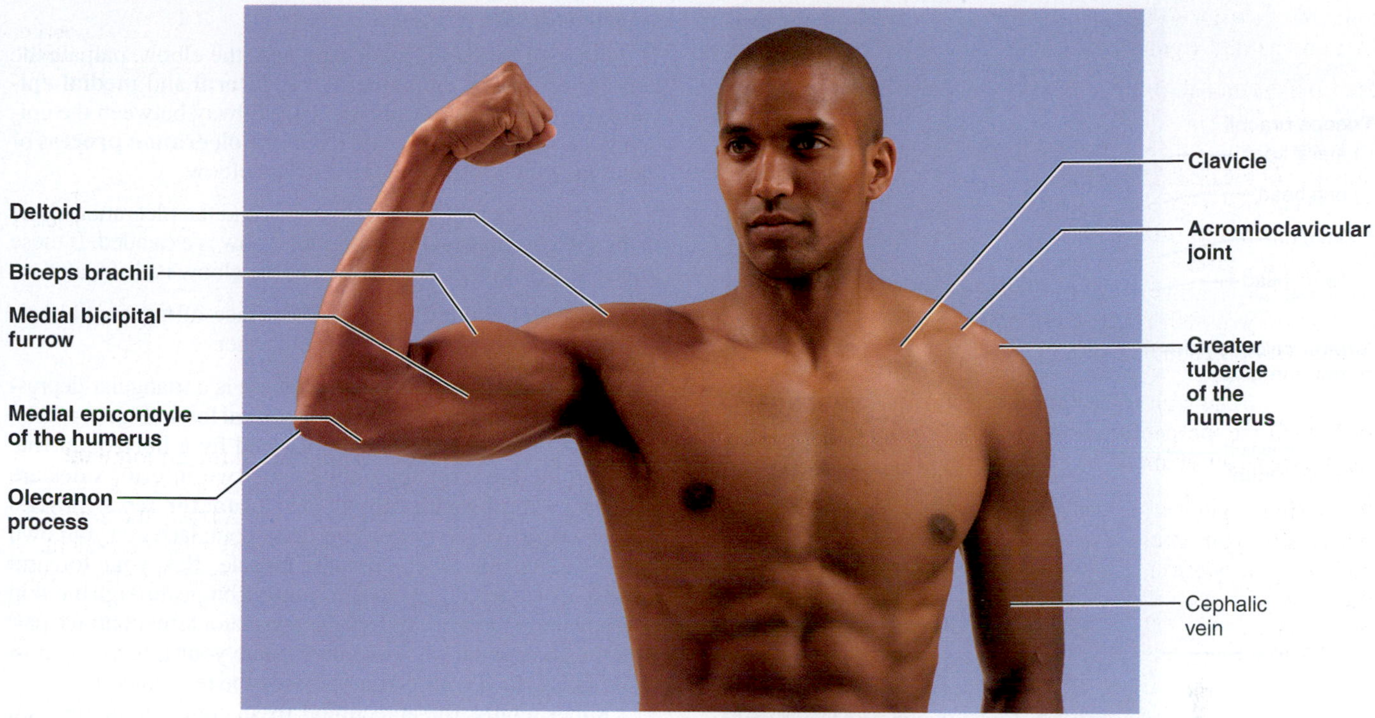
Deltoid
Biceps brachii
Medial bicipital furrow
Medial epicondyle of the humerus
Olecranon process
Clavicle
Acromioclavicular joint
Greater tubercle of the humerus
Cephalic vein

FIGURE 46.9 Shoulder and arm.

rectus abdominis. These grooves represent the **tendinous insertions** (or **intersections** or **inscriptions**), fibrous bands that subdivide the rectus muscle. Because of these subdivisions, each rectus abdominis muscle presents four distinct bulges. Try to identify these insertions on yourself or your partner.

3. The only other major muscles that can be seen or felt through the anterior abdominal wall are the lateral **external obliques**. Feel these muscles contract as you cough, strain, or raise your intra-abdominal pressure in some other way.

4. Recall that the anterior abdominal wall can be divided into four quadrants (see Figure 1.7a). A clinician listening to a patient's **bowel sounds** places the stethoscope over each of the four abdominal quadrants, one after another. Normal bowel sounds, which result as peristalsis moves air and fluid through the intestine, are high-pitched gurgles that occur every 5 to 15 seconds.

- Use the stethoscope to listen to your own or your partner's bowel sounds.

Abnormal bowel sounds can indicate intestinal disorders. Absence of bowel sounds indicates a halt in intestinal activity, which follows long-term obstruction of the intestine, surgical handling of the intestine, peritonitis, or other conditions. Loud tinkling or splashing sounds, by contrast, indicate an increase in intestinal activity. Such loud sounds may accompany gastroenteritis (inflammation and upset of the GI tract) or a partly obstructed intestine. ■

The Pelvis and Perineum

The bony surface features of the *pelvis* are considered with the bony landmarks of the abdomen (page 686) and the gluteal region (page 690). Most *internal* pelvic organs are not palpable through the skin of the body surface. A full *bladder,* however, becomes firm and can be felt through the abdominal wall just superior to the pubic symphysis. A bladder that can be palpated more than a few centimeters above this symphysis is retaining urine and dangerously full, and it should be drained by catheterization. ■

ACTIVITY 5

Palpating Landmarks of the Upper Limb

Axilla

The **base of the axilla** is the groove in which the underarm hair grows (see Figure 46.7). Deep to this base lie the axillary *lymph nodes* (which swell and can be palpated in breast cancer), the large *axillary vessels* serving the upper limb, and much of the brachial plexus. The base of the axilla forms a "valley" between two thick, rounded ridges, the **axillary folds.** Just anterior to the base, clutch your **anterior axillary fold,** formed by the pectoralis major muscle. Then grasp your **posterior axillary fold.** This fold is formed by the latissimus dorsi and teres major muscles of the back as they course toward their insertions on the humerus.

Shoulder

1. Again locate the prominent spine of the scapula posteriorly. Follow the spine to its lateral end, the flattened **acromion** on the shoulder's summit. Then, palpate the **clavicle** anteriorly, tracing this bone from the sternum to the shoulder. Notice the clavicle's curved shape.

2. Now locate the junction between the clavicle and the acromion on the superolateral surface of your shoulder, at the **acromioclavicular joint.** To find this joint, thrust your arm anteriorly repeatedly until you can palpate the precise point of pivoting action.

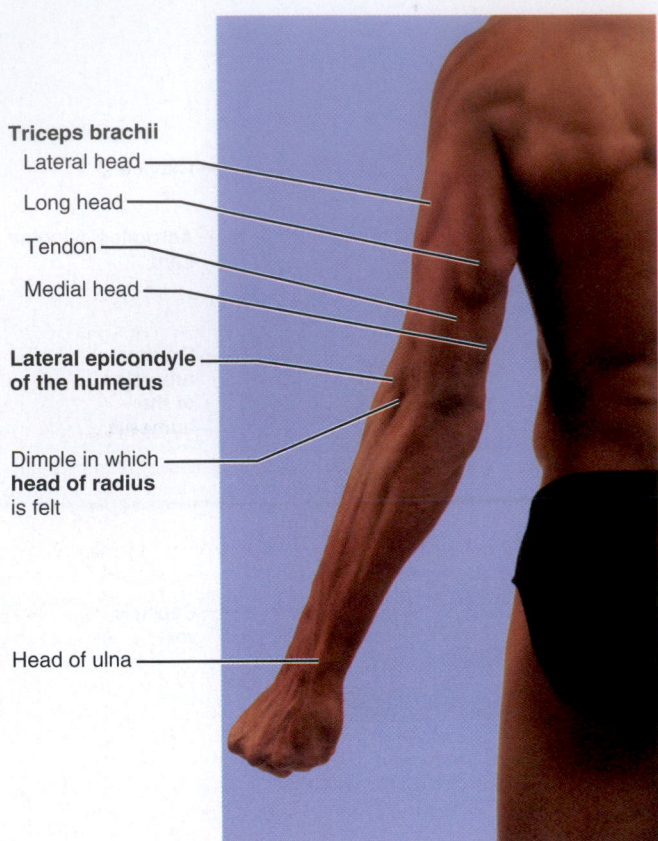

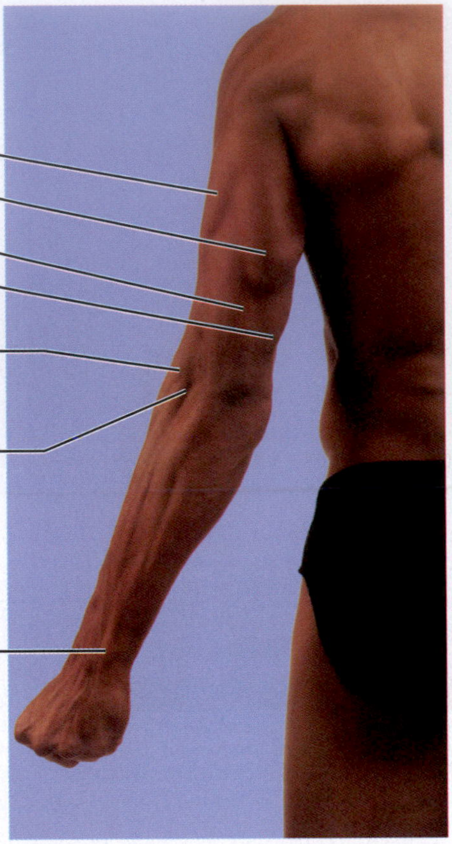
Triceps brachii
Lateral head
Long head
Tendon
Medial head
Lateral epicondyle of the humerus
Dimple in which head of radius is felt
Head of ulna

FIGURE 46.10 Surface anatomy of the upper limb, posterior view.

3. Next, place your fingers on the **greater tubercle** of the humerus. This is the most lateral bony landmark on the superior surface of the shoulder. It is covered by the thick **deltoid muscle,** which forms the rounded superior part of the shoulder. Intramuscular injections are often given into the deltoid, about 5 cm (2 inches) inferior to the greater tubercle (refer to Figure 46.17a, page 691).

Arm

Remember, according to anatomists, the arm runs only from the shoulder to the elbow, and not beyond.

1. In the arm, palpate the humerus along its entire length, especially along its medial and lateral sides.

2. Feel the **biceps brachii** muscle contract on your anterior arm when you flex your forearm against resistance. The medial boundary of the biceps is represented by the **medial bicipital furrow** (see Figure 46.9). This groove contains the large *brachial artery,* and by pressing on it with your fingertips you can feel your *brachial pulse.* Recall that the brachial artery is the artery routinely used in measuring blood pressure with a sphygmomanometer.

3. All three heads of the **triceps brachii** muscle (lateral, long, and medial) are visible through the skin of a muscular person (Figure 46.10).

Elbow Region

1. In the distal part of your arm, near the elbow, palpate the two projections of the humerus, the **lateral** and **medial epicondyles** (Figures 46.9 and 46.10). Midway between the epicondyles, on the posterior side, feel the **olecranon process** of the ulna, which forms the point of the elbow.

2. Confirm that the two epicondyles and the olecranon all lie in the same horizontal line when the elbow is extended. If these three bony processes do not line up, the elbow is dislocated.

3. Now feel along the posterior surface of the medial epicondyle. You are palpating your ulnar nerve.

4. On the anterior surface of the elbow is a triangular depression called the **antecubital fossa,** or **cubital fossa** (Figure 46.11). The triangle's superior *base* is formed by a horizontal line between the humeral epicondyles; its two inferior sides are defined by the **brachioradialis** and **pronator teres** muscles (see Figure 46.11b). Try to define these boundaries on your own limb. To find the brachioradialis muscle, flex your forearm against resistance, and watch this muscle bulge through the skin of your lateral forearm. To feel your pronator teres contract, palpate the antecubital fossa as you pronate your forearm against resistance. (Have your partner provide the resistance.)

Superficially, the antecubital fossa contains the **median cubital vein** (see Figure 46.11a). Clinicians often draw blood from this superficial vein and insert intravenous (IV) catheters into it to administer medications, transfused blood, and nutrient fluids. The large **brachial artery** lies just deep to the median cubital vein (see Figure 46.11b), so a needle must be inserted into the vein from a shallow angle (almost parallel to the skin) to avoid puncturing the artery. Other structures that lie deep in the fossa are also shown in Figure 46.11b.

5. The median cubital vein interconnects the larger **cephalic** and **basilic veins** of the upper limb. These veins are visible through the skin of lean people (see Figure 46.11a). Examine your arm to see if your cephalic and basilic veins are visible.

Forearm and Hand

The two parallel bones of the forearm are the medial *ulna* and the lateral *radius.*

1. Feel the ulna along its entire length as a sharp ridge on the posterior forearm (confirm that this ridge runs inferiorly from the olecranon process). As for the radius, you can feel its distal half, but most of its proximal half is covered by muscle. You can, however, feel the rotating **head** of the radius. To do this, extend your forearm, and note that a dimple forms on the posterior lateral surface of the elbow region (see Figure 46.10). Press three fingers into this dimple, and rotate your free hand as if you were turning a doorknob. You will feel the head of the radius rotate as you perform this action.

2. Both the radius and ulna have a knoblike **styloid process** at their distal ends. Figure 46.12 shows a way to locate these processes. Do not confuse the ulna's styloid process with the conspicuous **head of the ulna,** from which the styloid process stems. Confirm that the styloid process of the radius lies about 1 cm (0.4 inch) distal to that of the ulna.

Colles' fracture of the wrist is an impacted fracture in which the distal end of the radius is pushed proximally into the shaft of the radius. This sometimes occurs

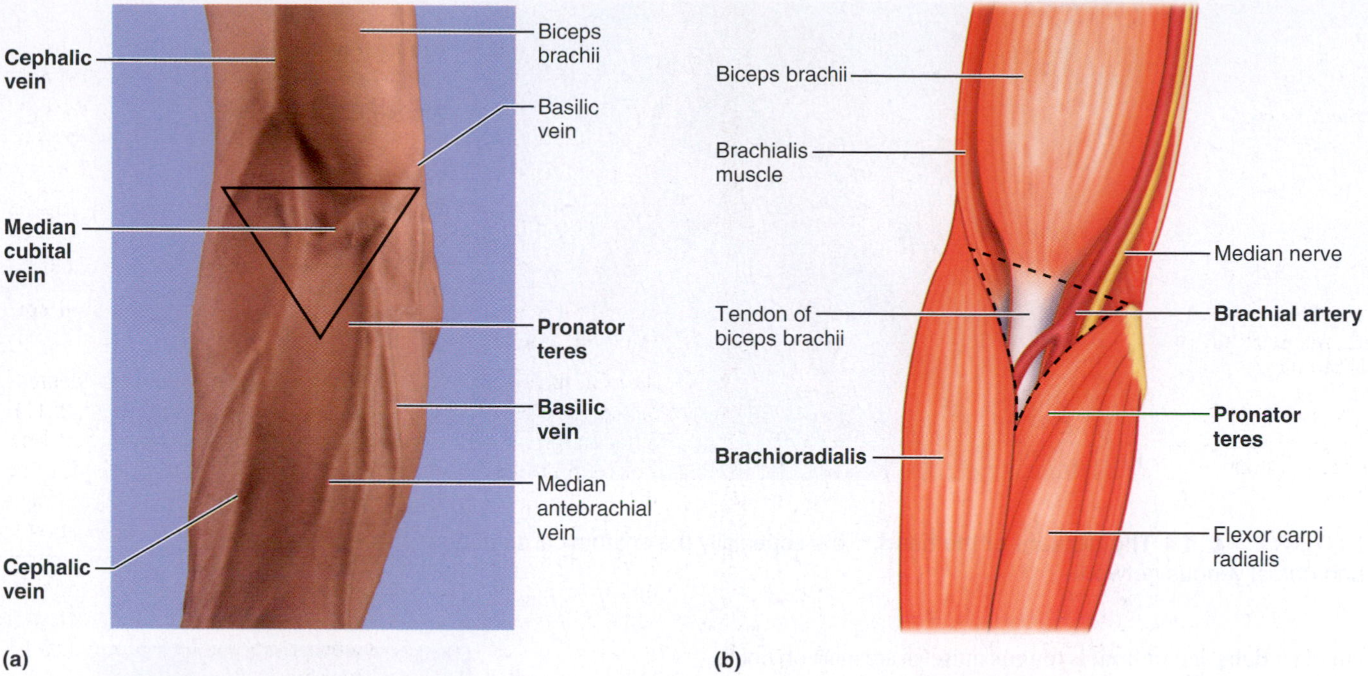
Cephalic vein
Biceps brachii
Basilic vein
Median cubital vein
Pronator teres
Basilic vein
Median antebrachial vein
Cephalic vein
(a)
Biceps brachii
Brachialis muscle
Median nerve
Tendon of biceps brachii
Brachial artery
Pronator teres
Brachioradialis
Flexor carpi radialis
(b)

FIGURE 46.11 The antecubital (cubital) fossa on the anterior surface of the right elbow (outlined by the triangle). (a) Photograph. **(b)** Diagram of deeper structures in the fossa.

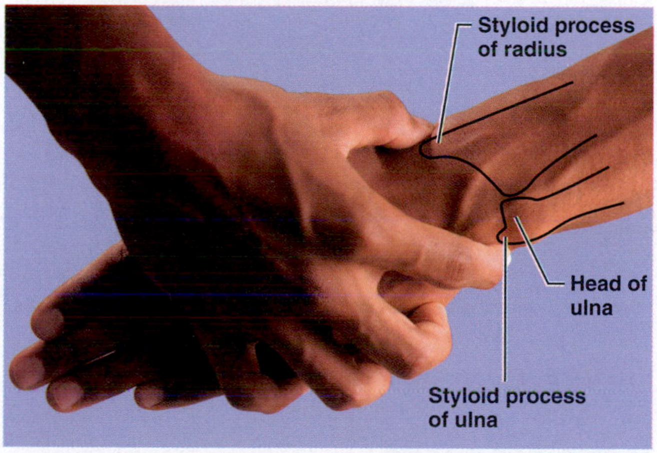
Styloid process of radius
Head of ulna
Styloid process of ulna

FIGURE 46.12 A way to locate the styloid processes of the ulna and radius. The right hand is palpating the left hand in this picture. Note that the head of the ulna is not the same as its styloid process. The styloid process of the radius lies about 1 cm distal to the styloid process of the ulna.

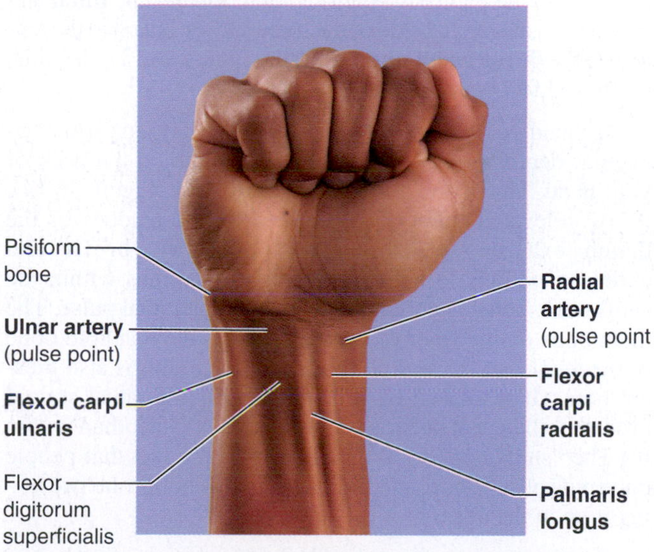
Pisiform bone
Ulnar artery (pulse point)
Flexor carpi ulnaris
Flexor digitorum superficialis
Radial artery (pulse point
Flexor carpi radialis
Palmaris longus

FIGURE 46.13 The anterior surface of the distal forearm and fist. The tendons of the flexor muscles guide the clinician to several sites for pulse taking.

when someone falls on outstretched hands, and it most often happens to elderly women with osteoporosis. Colles' fracture bends the wrist into curves that resemble those on a fork. ■

Can you deduce how physicians use palpation to diagnose a Colles' fracture?

____

____

3. Next, feel the major groups of muscles within your forearm. Flex your hand and fingers against resistance, and feel the anterior *flexor muscles* contract. Then extend your hand at the wrist, and feel the tightening of the posterior *extensor muscles*.

4. Near the wrist, the anterior surface of the forearm reveals many significant features (Figure 46.13). Flex your fist against resistance; the tendons of the main wrist flexors will bulge the skin of the distal forearm. The tendons of the **flexor**

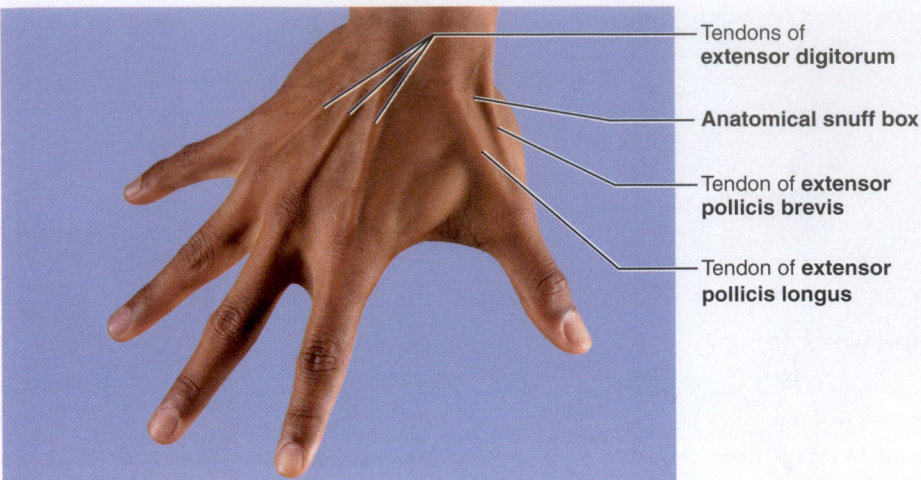

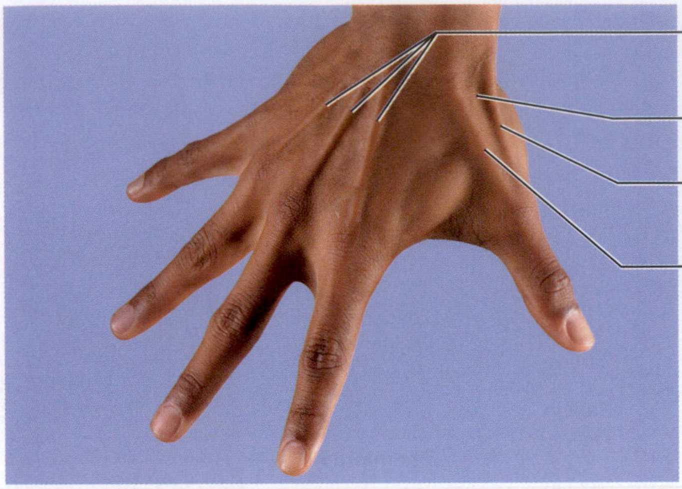
Tendons of extensor digitorum
Anatomical snuff box
Tendon of extensor pollicis brevis
Tendon of extensor pollicis longus

FIGURE 46.14 The dorsum of the hand. Note especially the anatomical snuff box and dorsal venous network.

carpi radialis and **palmaris longus** muscles are most obvious. (The palmaris longus, however, is absent from at least one arm in 30% of all people, so your forearm may exhibit just one prominent tendon instead of two.) The **radial artery** lies just lateral to (on the thumb side of) the flexor carpi radialis tendon, where the pulse is easily detected (Figure 46.13). Feel your radial pulse here. The *median nerve* (which innervates the thumb) lies deep to the palmaris longus tendon. Finally, the **ulnar artery** lies on the medial side of the forearm, just lateral to the tendon of the **flexor carpi ulnaris.** Using Figure 46.13 as a guide, locate and feel your ulnar arterial pulse.

5. Extend your thumb and point it posteriorly to form a triangular depression in the base of the thumb on the back of your hand. This is the **anatomical snuff box** (Figure 46.14). Its two elevated borders are defined by the tendons of the thumb extensor muscles, **extensor pollicis brevis** and **extensor pollicis longus.** The radial artery runs within the snuff box, so this is another site for taking a radial pulse. The main bone on the floor of the snuff box is the scaphoid bone of the wrist, but the styloid process of the radius is also present here. (If displaced by a bone fracture, the radial styloid process will be felt outside of the snuff box rather than within it.) The "snuff box" took its name from the fact that people once put snuff (tobacco for sniffing) in this hollow before lifting it up to the nose.

6. On the dorsum of your hand, observe the superficial veins just deep to the skin. This is the **dorsal venous network,** which drains superiorly into the cephalic vein. This venous network provides a site for drawing blood and inserting intravenous catheters and is preferred over the median cubital vein for these purposes. Next, extend your hand and fingers, and observe the tendons of the **extensor digitorum** muscle.

7. The anterior surface of the hand also contains some features of interest (Figure 46.15). These features include the *epidermal ridges* (fingerprints) and many **flexion creases** in the skin. Grasp your **thenar eminence** (the bulge on the palm that contains the thumb muscles) and your **hypothenar eminence** (the bulge on the medial palm that contains muscles that move the little finger).

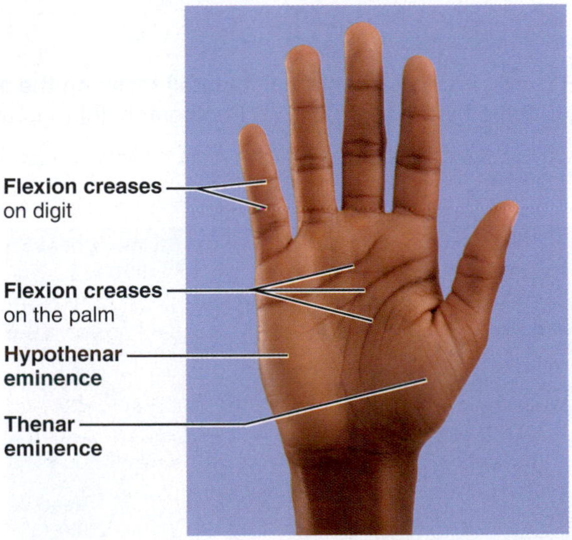
Flexion creases on digit
Flexion creases on the palm
Hypothenar eminence
Thenar eminence

FIGURE 46.15 The palmar surface of the hand.

ACTIVITY 6

Palpating Landmarks of the Lower Limb

Gluteal Region

Dominating the gluteal region are the two *prominences* (cheeks) of the buttocks. These are formed by subcutaneous fat and by the thick **gluteus maximus** muscles. The midline groove between the two prominences is called the **natal cleft** (*natal* = rump) or **gluteal cleft.** The inferior margin of each prominence is the horizontal **gluteal fold,** which roughly corresponds to the inferior margin of the gluteus maximus.

1. Try to palpate your **ischial tuberosity** just above the medial side of each gluteal fold (it will be easier to feel if you sit down or flex your thigh first). The ischial tuberosities are the robust inferior parts of the ischial bones, and they support the body's weight during sitting.

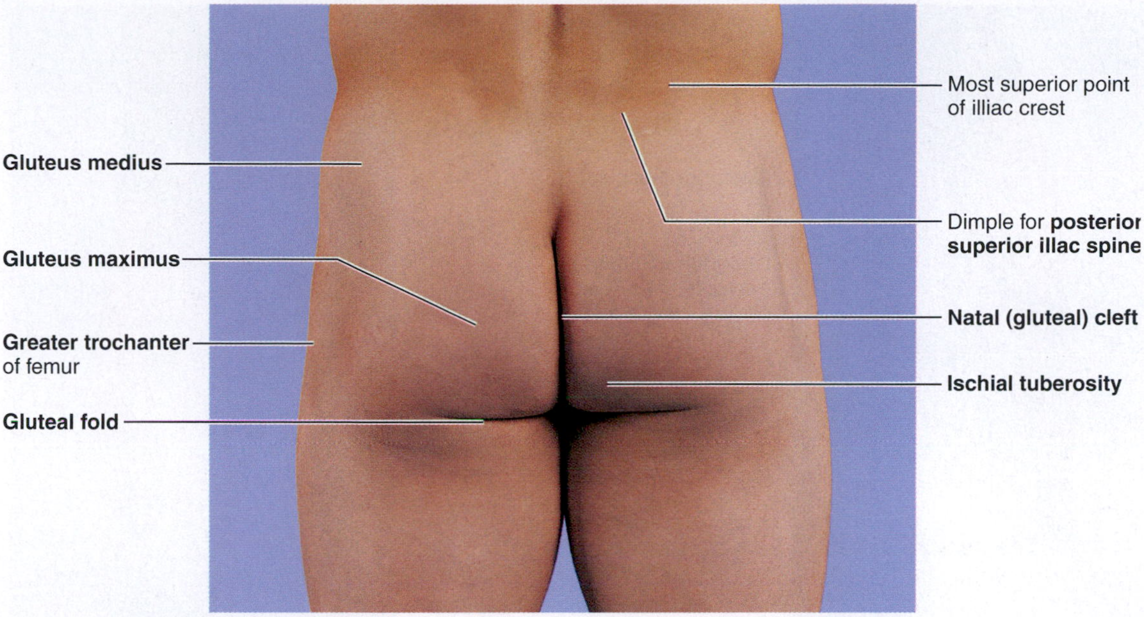
Most superior point of illiac crest
Gluteus medius
Dimple for posterior superior illac spine
Gluteus maximus
Natal (gluteal) cleft
Greater trochanter of femur
Ischial tuberosity
Gluteal fold

FIGURE 46.16 The gluteal region. The region extends from the iliac crests superiorly to the gluteal folds inferiorly. Therefore, it includes more than just the prominences of the buttock.

2. Next, palpate the **greater trochanter** of the femur on the lateral side of your hip (Figure 46.16). This trochanter lies just anterior to a hollow and about 10 cm (one hand's breadth, or 4 inches) inferior to the iliac crest. To confirm that you have found the greater trochanter, alternately flex and extend your thigh. Because this trochanter is the most superior point on the lateral femur, it moves with the femur as you perform this movement.

3. To palpate the sharp **posterior superior iliac spine** (see Figure 46.16), locate your iliac crests again, and trace each to its most posterior point. You may have difficulty feeling this spine, but it is indicated by a distinct dimple in the skin that is easy to find. This dimple lies two to three finger breadths lateral to the midline of the back. The dimple also indicates the position of the *sacroiliac joint,* where the hip bone attaches to the sacrum of the spinal column. (You can check *your* "dimples" out in the privacy of your home.)

The gluteal region is a major site for administering intramuscular injections. When giving such injections, extreme care must be taken to avoid piercing a major nerve that lies just deep to the gluteus maximus muscle. Can you guess what nerve this is?

It is the thick *sciatic nerve,* which innervates much of the lower limb. Furthermore, the needle must avoid the gluteal nerves and gluteal blood vessels, which also lie deep to the gluteus maximus.

To avoid harming these structures, the injections are most often applied to the **gluteus *medius*** (not maximus) muscle superior to the cheeks of the buttocks, in a safe area called the **ventral gluteal site** (Figure 46.17b). To locate this site, mentally

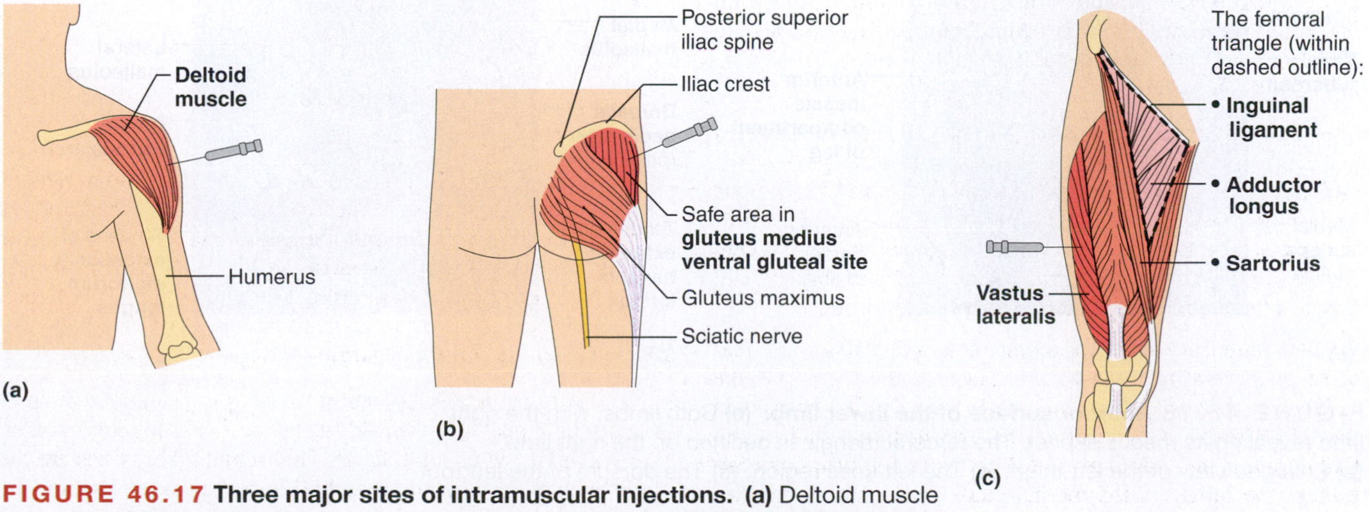
Deltoid muscle
Humerus
(a)
Posterior superior iliac spine
Iliac crest
Safe area in gluteus medius ventral gluteal site
Gluteus maximus
Sciatic nerve
(b)
The femoral triangle (within dashed outline):
• Inguinal ligament
• Adductor longus
• Sartorius
Vastus lateralis
(c)

FIGURE 46.17 Three major sites of intramuscular injections. (a) Deltoid muscle of the arm. **(b)** Ventral gluteal site (gluteus medius). **(c)** Vastus lateralis in the lateral thigh. The femoral triangle is also shown.

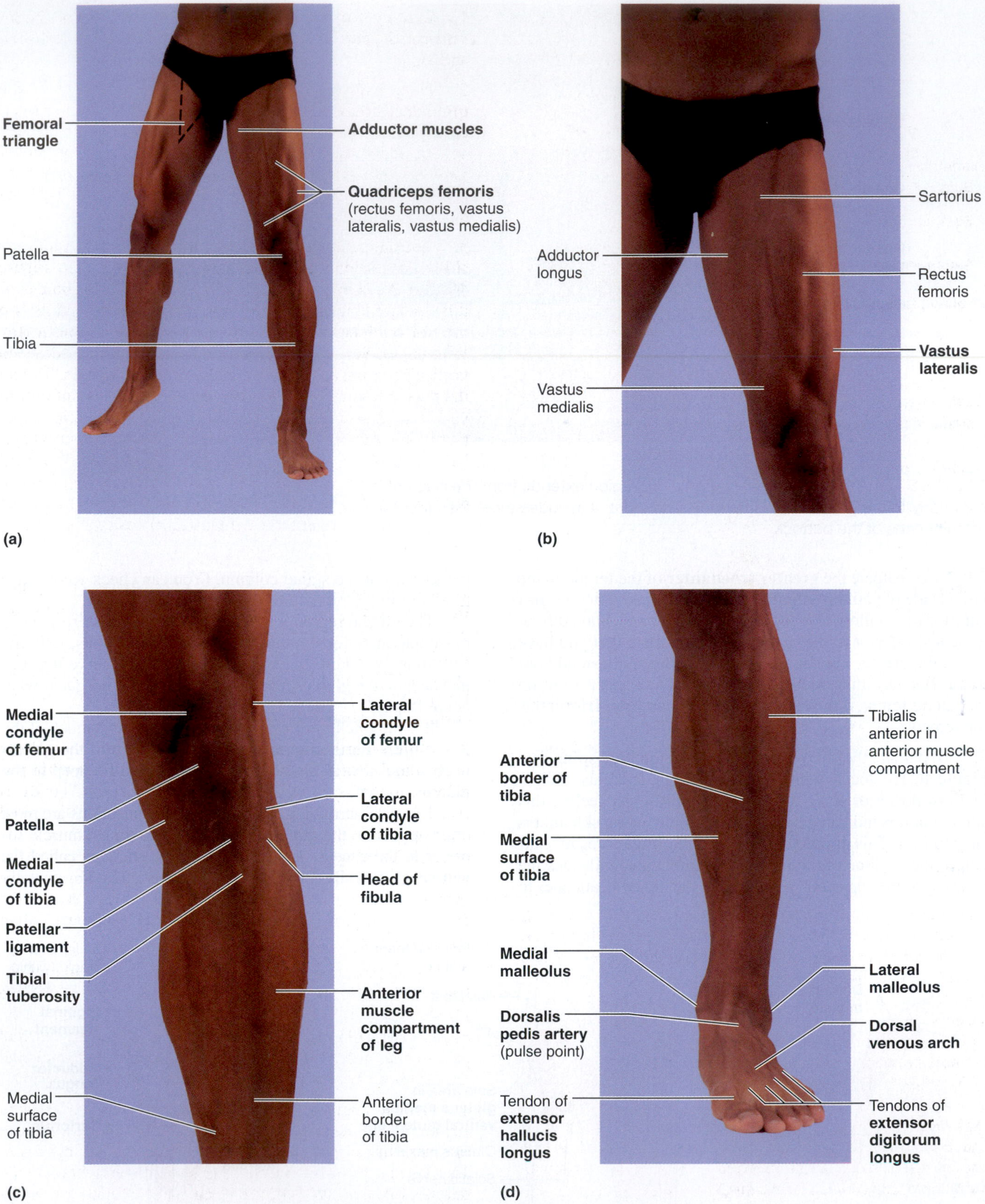
Femoral triangle
Adductor muscles
Quadriceps femoris (rectus femoris, vastus lateralis, vastus medialis)
Patella
Tibia
(a)
Sartorius
Adductor longus
Rectus femoris
Vastus lateralis
Vastus medialis
(b)
Medial condyle of femur
Lateral condyle of femur
Patella
Lateral condyle of tibia
Medial condyle of tibia
Head of fibula
Patellar ligament
Tibial tuberosity
Anterior muscle compartment of leg
Medial surface of tibia
Anterior border of tibia
(c)
Tibialis anterior in anterior muscle compartment
Anterior border of tibia
Medial surface of tibia
Medial malleolus
Lateral malleolus
Dorsalis pedis artery (pulse point)
Dorsal venous arch
Tendon of extensor hallucis longus
Tendons of extensor digitorum longus
(d)

FIGURE 46.18 Anterior surface of the lower limb. (a) Both limbs, with the right limb revealing its medial aspect. The femoral triangle is outlined on the right limb. **(b)** Enlarged view of the left thigh. **(c)** The left knee region. **(d)** The dorsum of the left foot.

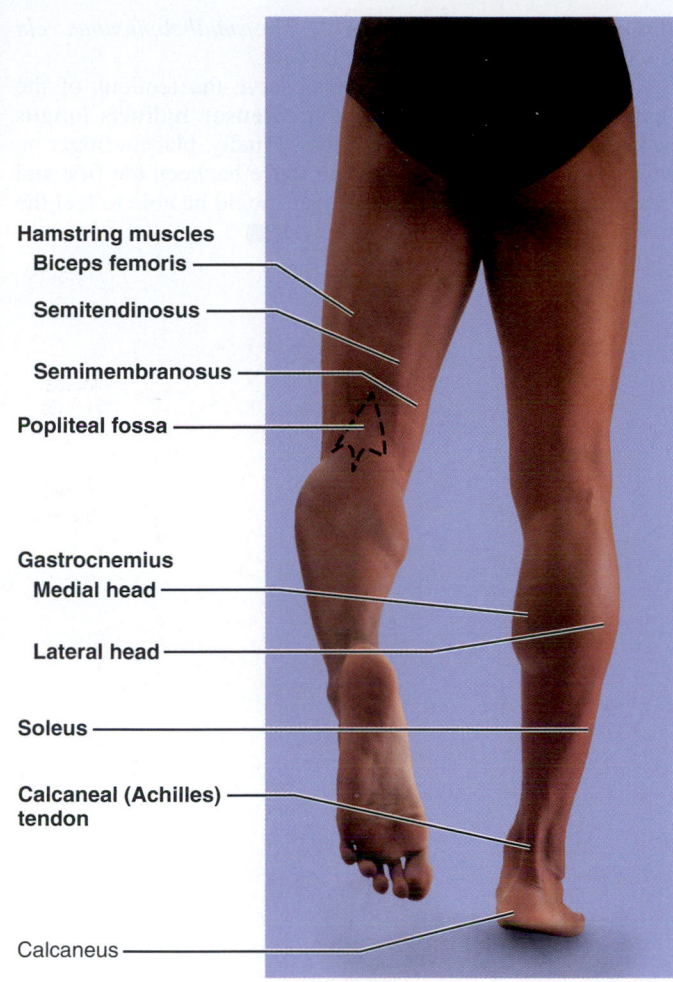

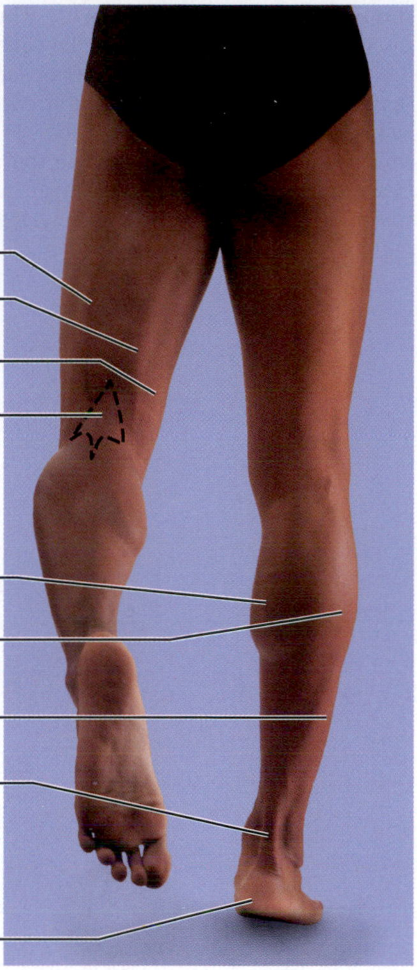
Hamstring muscles
Biceps femoris
Semitendinosus
Semimembranosus
Popliteal fossa
Gastrocnemius
Medial head
Lateral head
Soleus
Calcaneal (Achilles) tendon
Calcaneus

FIGURE 46.19 Posterior surface of the lower limb. Notice the diamond-shaped popliteal fossa posterior to the knee.

draw a line laterally from the posterior superior iliac spine (dimple) to the greater trochanter; the injection would be given 5 cm (2 inches) superior to the midpoint of that line. Another safe way to locate the ventral gluteal site is to approach the lateral side of the patient's left hip with your extended right hand (or the right hip with your left hand). Then, place your thumb on the anterior superior iliac spine and your index finger as far posteriorly on the iliac crest as it can reach. The heel of your hand comes to lie on the greater trochanter, and the needle is inserted in the angle of the V formed between your thumb and index finger about 4 cm (1.5 inches) inferior to the iliac crest.

Gluteal injections are not given to small children because their "safe area" is too small to locate with certainty and because the gluteal muscles are thin at this age. Instead, infants and toddlers receive intramuscular shots in the prominent **vastus lateralis** muscle of the thigh.

Thigh

The thigh is pictured in Figures 46.18 and 46.19. Much of the femur is clothed by thick muscles, so the thigh has few palpable bony landmarks.

1. Distally, feel the **medial** and **lateral condyles of the femur** and the **patella** anterior to the condyles (see Figure 46.18c and a).

2. Next, palpate your three groups of thigh muscles—the **quadriceps femoris muscles** anteriorly, the **adductor muscles** medially, and the **hamstrings** posteriorly (see Figures 46.18a and b and 46.19). The **vastus lateralis,** the lateral muscle of the quadriceps group, is a site for intramuscular injections. Such injections are administered about halfway down the length of this muscle (see Figure 46.17c).

3. The anterosuperior surface of the thigh exhibits a three-sided depression called the **femoral triangle** (see Figure 46.18a). As shown in Figure 46.17c, the superior border of this triangle is formed by the **inguinal ligament,** and its two inferior borders are defined by the **sartorius** and **adductor longus** muscles. The large *femoral artery* and *vein* descend vertically through the center of the femoral triangle. To feel the pulse of your femoral artery, press inward just inferior to your midinguinal point (halfway between the anterior superior iliac spine and the pubic tubercle). Be sure to push hard, because the artery lies somewhat deep. By pressing very hard on this point, one can stop the bleeding from a hemorrhage in the lower limb. The femoral triangle also contains most of the *inguinal lymph nodes* (which are easily palpated if swollen).

Leg and Foot

1. Locate your patella again, then follow the thick **patellar ligament** inferiorly from the patella to its insertion on the superior tibia (see Figure 46.18c). Here you can feel a rough projection, the **tibial tuberosity.** Continue running your fingers inferiorly along the tibia's sharp **anterior border** and its flat **medial surface**—bony landmarks that lie very near the surface throughout their length.

2. Now, return to the superior part of your leg, and palpate the expanded **lateral** and **medial condyles of the tibia** just inferior to the knee. (You can distinguish the tibial condyles from the femoral condyles because you can feel the tibial condyles move with the tibia during knee flexion.) Feel the bulbous **head of the fibula** in the superolateral region of the leg (see Figure 46.18c). Try to feel the *common fibular nerve* (nerve to the anterior leg and foot) where it wraps around the fibula's *neck* just inferior to its head. This nerve is often bumped against the bone here and damaged.

3. In the most distal part of the leg, feel the **lateral malleolus** of the fibula as the lateral prominence of the ankle (see Figure 46.18d). Notice that this lies slightly inferior to the **medial malleolus** of the tibia, which forms the ankle's medial prominence. Place your finger just posterior to the medial malleolus to feel the pulse of your *posterior tibial artery.*

4. On the posterior aspect of the knee is a diamond-shaped hollow called the **popliteal fossa** (see Figure 46.19). Palpate the large muscles that define the four borders of this fossa: The **biceps femoris** forming the superolateral border, the **semitendinosus** and **semimembranosus** defining the superomedial border, and the two heads of the **gastrocnemius** forming the inferior border. The *popliteal artery* and *vein* (main vessels to the leg) lie deep within this fossa. To feel a popliteal pulse, flex your leg at the knee and push your fingers firmly into the popliteal fossa. If a physician is unable to

feel a patient's popliteal pulse, the femoral artery may be narrowed by atherosclerosis.

5. Observe the dorsum (superior surface) of your foot. You may see the superficial **dorsal venous arch** overlying the proximal part of the metatarsal bones (Figure 46.18d). This arch gives rise to both saphenous veins (the main superficial veins of the lower limb). Visible in lean people, the *great saphenous vein* ascends along the medial side of the entire limb (see Figure 32.8, page 479). The *small saphenous vein* ascends through the center of the calf.

As you extend your toes, observe the tendons of the **extensor digitorum longus** and **extensor hallucis longus** muscles on the dorsum of the foot. Finally, place a finger on the extreme proximal part of the space between the first and second metatarsal bones. Here you should be able to feel the pulse of the **dorsalis pedis artery.** ■

NAME ______________________________

LAB TIME/DATE ____________________

REVIEW SHEET

EXERCISE 46

Surface Anatomy Roundup

__________ **1.** A blow to the cheek is most likely to break what superficial bone or bone part? (a) superciliary arches, (b) the philtrum, (c) zygomatic arch, (d) the tragus

__________ **2.** Rebound tenderness (a) occurs in appendicitis, (b) is whiplash of the neck, (c) is a sore foot from playing basketball, (d) occurs when the larynx falls back into place after swallowing.

__________ **3.** The anatomical snuff box (a) is in the nose, (b) contains the styloid process of the radius, (c) is defined by tendons of the flexor carpi radialis and palmaris longus, (d) cannot really hold snuff.

__________ **4.** Some landmarks on the body surface can be seen or felt, but others are abstractions that you must construct by drawing imaginary lines. Which of the following pairs of structures is abstract and invisible? (a) umbilicus and costal margin, (b) anterior superior iliac spine and natal cleft, (c) linea alba and linea semilunaris, (d) McBurney's point and midaxillary line, (e) philtrum and sternocleidomastoid

__________ **5.** Many pelvic organs can be palpated by placing a finger in the rectum or the vagina, but only one pelvic organ is readily palpated through the skin. This is the (a) nonpregnant uterus, (b) prostate gland, (c) full bladder, (d) ovaries, (e) rectum.

__________ **6.** A muscle that contributes to the posterior axillary fold is the (a) pectoralis major, (b) latissimus dorsi, (c) trapezius, (d) infraspinatus, (e) pectoralis minor, (f) a and e.

__________ **7.** Which of the following is not a pulse point? (a) anatomical snuff box, (b) inferior margin of mandible anterior to masseter muscle, (c) center of distal forearm at palmaris longus tendon, (d) medial bicipital furrow on arm, (e) dorsum of foot between the first two metatarsals

__________ **8.** Which pair of ribs inserts on the sternum at the sternal angle? (a) first, (b) second, (c) third, (d) fourth, (e) fifth

__________ **9.** The inferior angle of the scapula is at the same level as the spinous process of which vertebra? (a) C_5, (b) C_7, (c) T_3, (d) T_7, (e) L_4

__________ **10.** An important bony landmark that can be recognized by a distinct dimple in the skin is the (a) posterior superior iliac spine, (b) styloid process of the ulna, (c) shaft of the radius, (d) acromion.

__________ **11.** A nurse missed a patient's median cubital vein while trying to withdraw blood and then inserted the needle far too deeply into the cubital fossa. This error could cause any of the following problems, except this one: (a) paralysis of the ulnar nerve, (b) paralysis of the median nerve, (c) bruising the insertion tendon of the biceps brachii muscle, (d) blood spurting from the brachial artery.

__________ **12.** Which of these organs is almost impossible to study with surface anatomy techniques? (a) heart, (b) lungs, (c) brain, (d) nose

__________ **13.** A preferred site for inserting an intravenous medication line into a blood vessel is the (a) medial bicipital furrow on arm, (b) external carotid artery, (c) dorsal venous arch of hand, (d) popliteal fossa.

__________ **14.** One listens for bowel sounds with a stethoscope placed (a) on the four quadrants of the abdominal wall; (b) in the triangle of auscultation; (c) in the right and left midaxillary line, just superior to the iliac crests; (d) inside the patient's bowels (intestines), on the tip of an endoscope.

DISSECTION EXERCISE 1

Dissection and Identification of Fetal Pig Muscles

MATERIALS

- ☐ Preserved and injected fetal pig (one for every two to four students)
- ☐ Dissecting instruments and trays
- ☐ Twine
- ☐ Metric ruler
- ☐ Name tag and large plastic bag
- ☐ Disposable gloves
- ☐ Embalming fluid
- ☐ Paper towels
- ☐ Organic debris container

OBJECTIVES

1. To name and locate the muscles on a dissection animal.
2. To recognize similarities and differences between human and fetal pig musculature.

The skeletal muscles of all mammals are named in a similar fashion. However, some muscles that are separate in lower animals are fused in humans, and some muscles present in lower animals are lacking in humans.

This exercise involves dissection of the pig musculature, in conjunction with the study of human muscles, to enhance your knowledge of the human muscular system. Since the aim is to become familiar with the muscles of the human body, you should pay particular attention to the similarities between pig and human muscles. However, pertinent differences will be pointed out as you encounter them. See Exercise 15 of this manual for a discussion of the anatomy of the human muscular system.

ACTIVITY 1

Examining the Surface Anatomy of the Fetal Pig

The preserved laboratory animals purchased for dissection have been embalmed with a solution that prevents deterioration of the tissues. The animals generally are delivered in plastic bags containing a small amount of embalming fluid. *Do not dispose of this fluid when removing the pig*. It is important to keep the pig's tissues moist, as this animal will be used for dissection exercises until the end of the course. The embalming fluid may cause your eyes to smart and may dry your skin, but these small irritants are more desirable than working with a dissection specimen that has become hard and odoriferous because of bacterial action. Don disposable gloves before beginning dissection.

Obtain a fetal pig, dissecting tray, dissecting instruments, and a name tag. Use a pencil to mark the name tag with the names of the members of your group, and set it aside. You will attach the name tag to the plastic bag at the end of the laboratory session so that you can identify your animal in subsequent laboratory sessions.

1. Before actually beginning to dissect the fetal pig, it is worth taking a few minutes to become familiar with the surface anatomy of your dissection animal. Place your specimen on its side on the dissection tray and identify its four major body regions—head, neck, trunk, and tail. Refer to Figure D1.1 as you conduct the following observations.

Head: On the head identify the following structures.

- **Mouth,** bounded by fleshy lips.
- **External nares (nostrils)** at the end of the rostrum, or snout.
- **Auricle (pinna)** of the external ear, which surrounds the external auditory canal that leads to the eardrum.
- **Eyes,** bounded by the eyelids.
- **Nictitating membrane**—Pull the upper and lower eyelids apart to view the nictitating membrane (which helps to keep the eyeball clean by sweeping across it) at the medial corner of the eye.

PAL
For access to anatomical models and more, check out Practice Anatomy Lab.

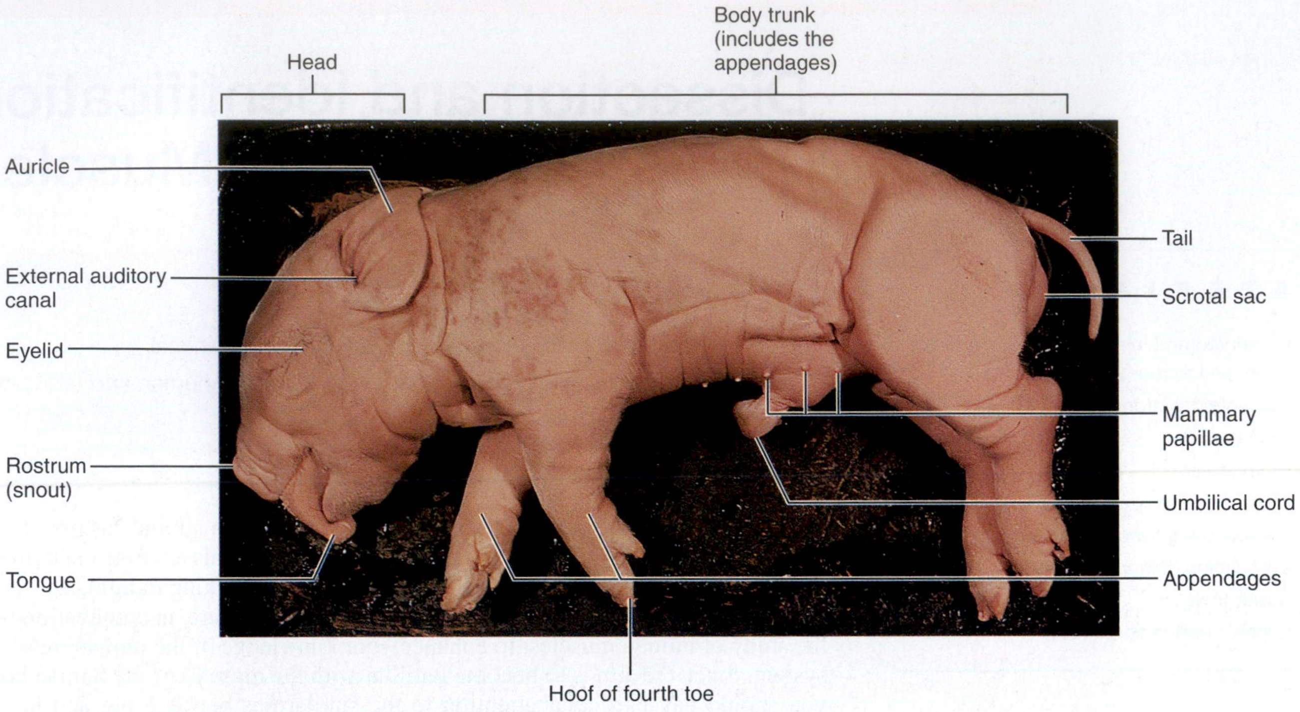
Head
Body trunk (includes the appendages)
Auricle
External auditory canal
Eyelid
Rostrum (snout)
Tongue
Tail
Scrotal sac
Mammary papillae
Umbilical cord
Appendages
Hoof of fourth toe

FIGURE D1.1 External features of a fetal pig, lateral aspect.

Neck: Corresponds to the cervical region of the spine, located between the head and body trunk.

Trunk: Accounts for most of the body and includes the appendages (fore- and hindlimbs).

Identify the following surface anatomy landmarks on the trunk of your specimen.

- **Mammary papillae** (teats), five to seven on each side of the ventral trunk surface. These papillae, which are associated with the mammary glands, allow the mature females to suckle their young.
- **Thorax,** the anterior trunk region encased by the rib cage.
- **Abdomen,** the posterior trunk region from which the **umbilical cord** projects ventrally. (The umbilical cord attaches the developing pig to the placenta.)
- **Anus,** the posterior external opening of the digestive tract.

While examining the posterior abdominal surface, determine the sex of your fetal pig. In females, there is a common opening for the urogenital organs positioned just ventral to the anus in a protrusion called the **genital papilla.** If the animal is a male, the opening of the penis, called the **preputial orifice,** will be seen just posterior to the umbilical cord, and the **scrotal sacs,** the paired skin sacs that contain the testes in the adult animal, will appear as swollen regions between the hind legs.

- **Appendages**—Examine the forelimbs and hindlimbs. Notice that the digits (toes) end in **hooves** and that the first toe is absent. Like other ungulates, pigs walk on their toes; the third and fourth toes bear most of the weight and are the largest digits.
- **Tail:** The posteriormost projection from the body trunk.

2. The normal gestation period, or period of development, is about 115 days for pigs. A fetal pig that is 18 mm long has completed about one-third of its developmental period (about 36 days); one 40 mm long has completed about half (56 days), a 220-mm specimen has gestated for about 100 days, and one that is 300 mm long is full term. Determine the approximate age of your fetal pig specimen by measuring the length of its body. Record below.

______________ mm length

______________ approximate gestation age

ACTIVITY 2

Preparing the Fetal Pig for Dissection

1. Now you are ready to begin your dissection. Place the pig dorsal side down on the dissecting tray. To secure the animal to the dissecting tray, make a loop knot with twine around one upper limb, carry the twine under the inferior surface of the tray, and secure the opposing limb. Repeat for the lower extremities.

2. Make a short, *shallow* midventral incision at the base of the throat to just penetrate the skin. From this point on, use scissors. Continue the cut the length of the ventral body surface to the umbilicus. Cut laterally around the umbilicus on each side, and continue with two incisions (flanking the median line) to the pelvic region (Figure D1.2).

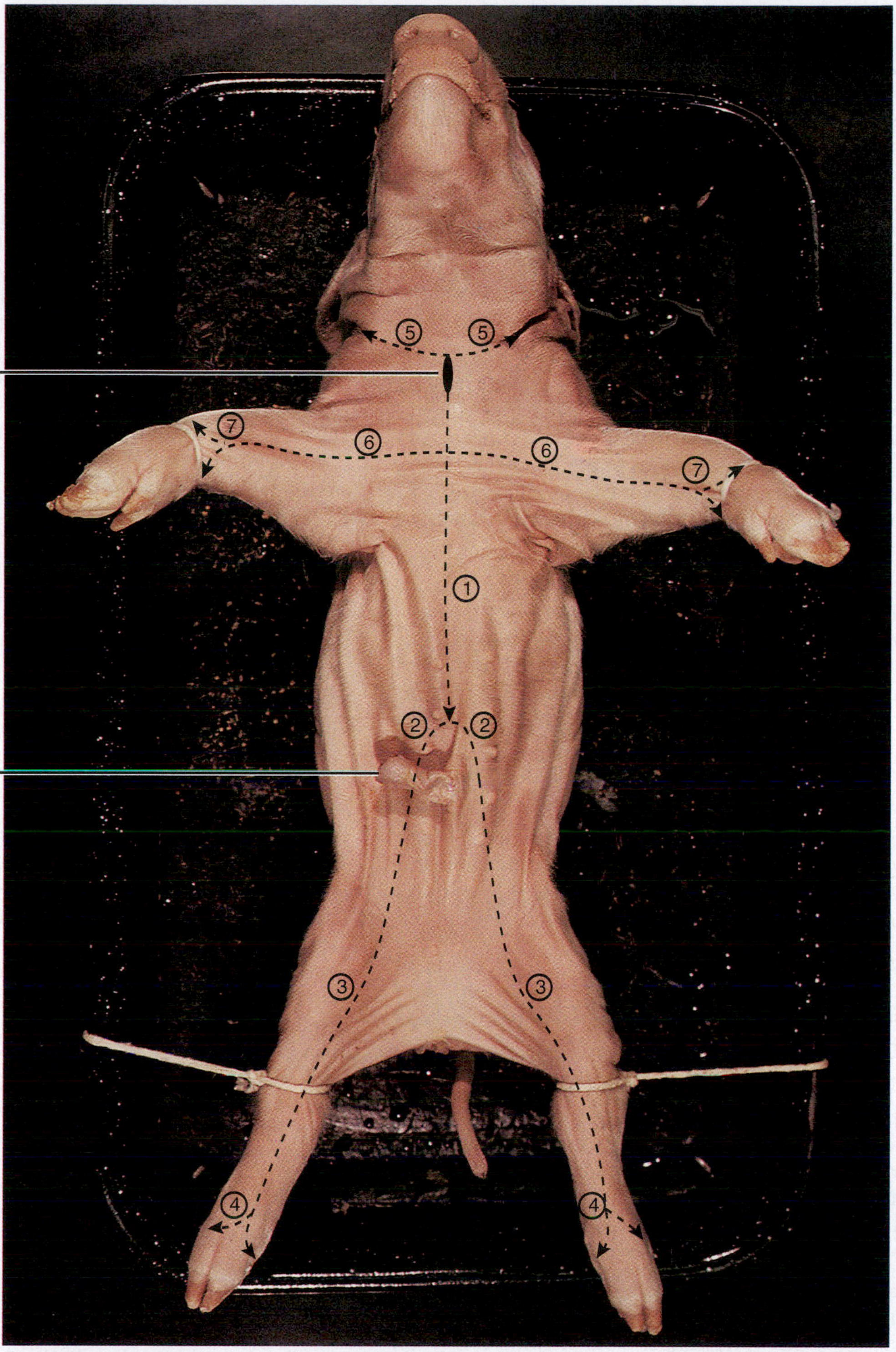
Initial incision
Umbilical cord
1
2
2
3
3
4
4
5
5
6
6
7
7

FIGURE D1.2 Incisions to be made in skinning pig. Numbers indicate sequence.

3. Continue the incisions from the pelvic region down the medial surface of each leg to the hoof, and cut the skin completely around the ankles.

4. Return to the neck. Make an incision completely around the neck. Then cut down each foreleg to the wrist. Completely cut the skin around the wrists.

5. Now free the skin from the loose connective tissue (superficial fascia) that binds it to the underlying structures. With one hand, grasp the skin on one side of the midline ventral incision. Using your fingers, the closed scissor tips, or a blunt probe, break through the "cottony" connective tissue fibers to release the skin from the muscle beneath. Work toward the dorsal surface and then upward toward the neck region. As you

pull the skin from the body, you should see small, white, cordlike structures that extend from it to the muscles at fairly regular intervals. These are the cutaneous nerves, which serve the skin. You will also see (particularly on the ventral surface) that a thin layer of muscle fibers adheres to the skin. These are the cutaneous muscles, which enable the pig to move or twitch the skin to get rid of irritants such as insects. Do not remove the skin from the head, since the muscles of the pig are not sufficiently similar to human head muscles to merit study.

6. Complete the skinning process by releasing the skin from the forelimbs, lower torso, and hindlimbs in the same manner. Set the skin to one side; it will be used to wrap the skinned specimen for storage.

7. Note the neck incision where the pig's blood vessels were injected with latex (Figure D1.2). If time allows, begin on the side *opposite* the incision, using forceps or fingers, and carefully remove as much gelatinous material, fat, and fascia from the surface of the muscles as possible. Since you will dissect only one side of the animal's musculature, you need to perform this clearing process only on one side.

8. Inspect your skinned pig and compare to the view in Figure D1.3b. Notice that the cleavage lines between the muscles are indistinct because of the overlying connective tissue. Additionaly, a fetal pig's muscles are incompletely developed (remember, it's a *fetal* pig) and are easily torn. Fetal pigs are excellent for other types of anatomy studies, but they are less than ideal for a study of the muscular system. Nonetheless, your examination of the fetal pig musculature should help you understand the structure of human muscles.

9. If the muscle dissection exercises are to be done at a later laboratory session, follow the cleanup instructions noted in the following box. *Prepare your pig for storage in this way every time the pig is used.*

Preparing the Dissection Animal for Storage

1. To prevent internal organs from drying out, dampen a layer of folded paper towels with embalming fluid, and wrap them snugly around the animal's torso. (Do not use *water-soaked* paper towels, which encourage the growth of mold.) Make sure the dissected areas are completely enveloped. Return the animal's skin flaps to their normal position over the ventral cavity body organs.

2. Place the animal in a plastic storage bag. Add more embalming fluid if necessary, press out excess air, and securely close the bag with a rubber band or twine.

3. Make sure your name tag is securely attached, and place the animal in the designated storage container.

4. Clean all dissecting equipment with soapy water, rinse, and dry it for return to the storage area. Wash down the lab bench and properly dispose of organic debris and your gloves before leaving the laboratory.

ACTIVITY 3

Dissecting Trunk and Neck Muscles

The proper dissection of muscles involves carefully separating one muscle from another so that they can be identified and transecting the superficial muscles to study the deeper layers. In general, when you want to transect a muscle, you should completely free it from all adhering connective tissue and then cut through it about halfway between its origin and insertion attachments.

Before you begin dissecting, observe your skinned pig. If you look carefully, you will see changes in the direction of muscle fibers, which will help you locate the muscle borders. As a rule, all the fibers of one muscle are held together by a connective tissue sheath (epimysium) and run in the same general direction. Pulling in slightly different directions on two adjacent muscles will usually allow you to expose the normal cleavage line between them.

Once you have identified the cleavage lines, use a blunt probe to break the connective tissue connections between them and separate the muscles from one another. If the muscles separate as clean, distinct bundles, your procedure is probably correct. If you observe a ragged or "chewed-up" appearance, you are probably tearing a muscle apart rather than separating it from the adjacent muscles. Only those muscles most easily identified and separated out will be identified in this exercise because of the immaturity of the dissection specimen and time considerations.

Anterior Neck Muscles

1. Using Figures D1.3 and D1.4 as guides, examine the anterior neck surface of the pig and identify the following superficial neck muscles. (The platysma belongs in this group but was probably removed during the skinning process.) The **sternomastoid** (corresponding to part of the sternocleidomastoid muscle in humans) is a thick but narrow muscle extending obliquely from the mastoid process to the sternum along the side of the neck. The large external jugular vein, which drains the head, should be obvious, crossing the anterior aspect of this muscle. The **mylohyoid** muscle parallels the bottom aspect of the chin, and the **digastric** muscle abuts the mylohyoid laterally (see Figure D1.3a). Although it is not a neck muscle, at this time you can also identify the fleshy **masseter** muscle (see Figure D1.5), which flanks the digastric muscle laterally. The **sternohyoid** is a narrow straplike muscle united to its partner, which runs along the midventral aspect of the neck. Notice the large masses of thymus gland tissue lying deep to these muscles.

2. The deeper muscles of the anterior neck of the pig are small and straplike and hardly worth the effort of the dissection; however, you can see one of these deeper muscles with a minimum of effort. Transect the sternomastoid and the sternohyoid approximately at midbelly (halfway between the origin and the insertion). Reflect the cut ends back to identify the **sternothyroid** (not illustrated), which runs downward on the anterior surface of the throat just deep and lateral to the sternohyoid muscle. Notice that after it originates from the sternum, this muscle separates into two parts that insert into the lateral and ventral surfaces of the thyroid cartilage of the larynx.

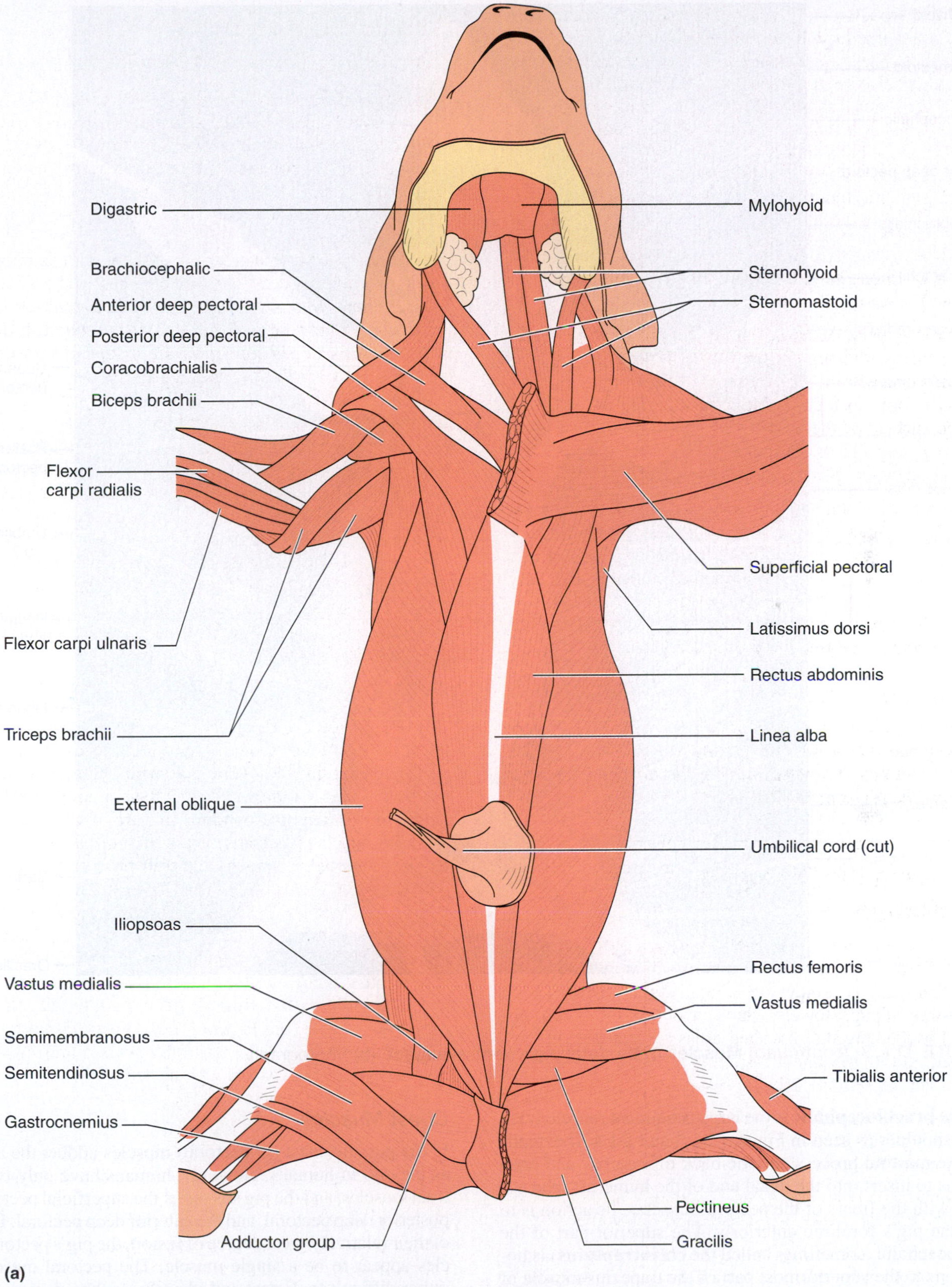
Digastric
Mylohyoid
Brachiocephalic
Sternohyoid
Anterior deep pectoral
Sternomastoid
Posterior deep pectoral
Coracobrachialis
Biceps brachii
Flexor carpi radialis
Superficial pectoral
Latissimus dorsi
Flexor carpi ulnaris
Rectus abdominis
Triceps brachii
Linea alba
External oblique
Umbilical cord (cut)
Iliopsoas
Rectus femoris
Vastus medialis
Vastus medialis
Semimembranosus
Tibialis anterior
Semitendinosus
Gastrocnemius
Pectineus
Adductor group
Gracilis
(a)

FIGURE D1.3 Muscles of the fetal pig's ventral aspect. (a) Diagrammatic view. Superficial muscles are shown on the left side of the body; deeper muscles are shown on the right. (Sartorius muscles of the thigh have been removed.) **(b)** Photograph on next page.

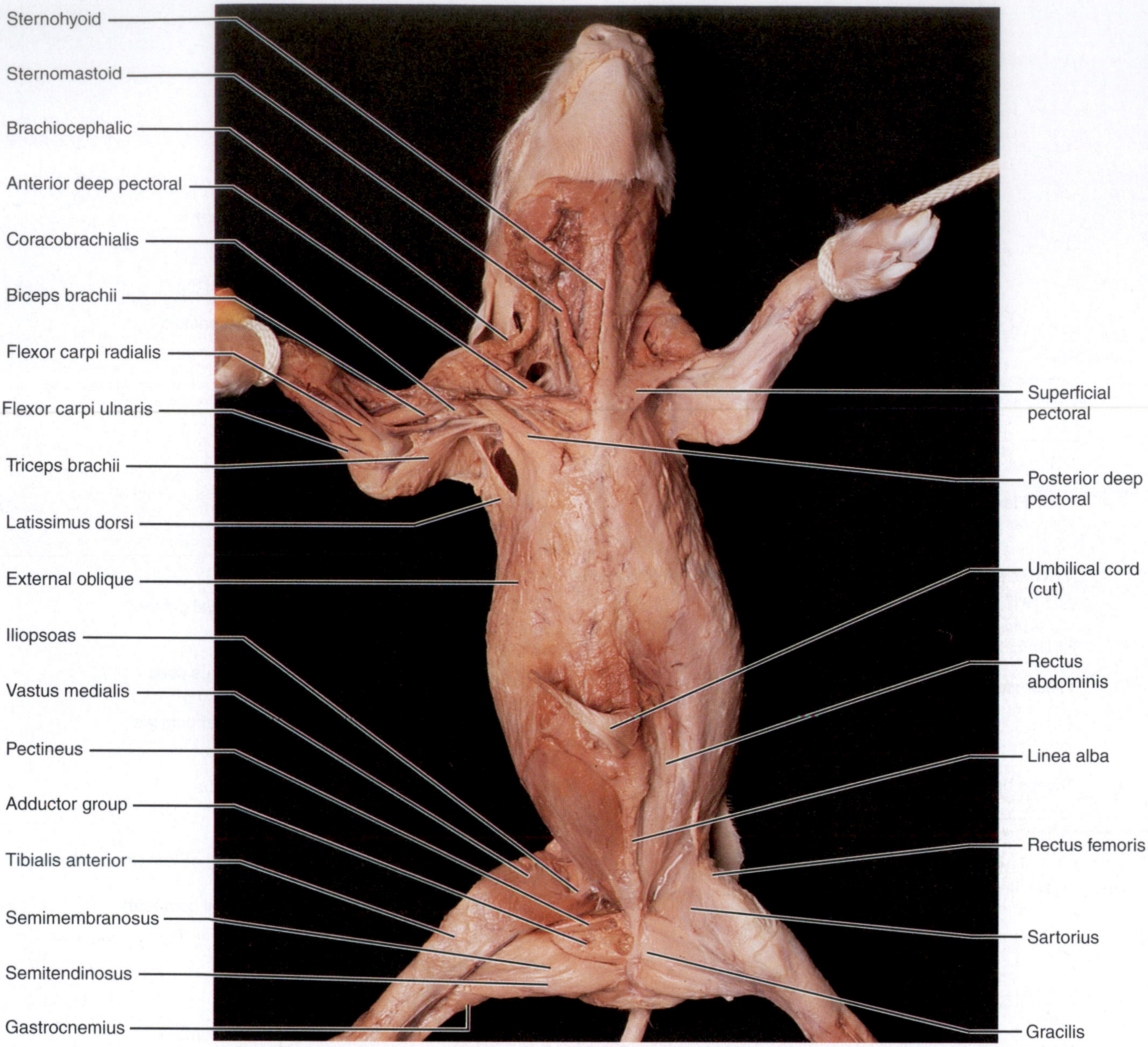
Sternohyoid
Sternomastoid
Brachiocephalic
Anterior deep pectoral
Coracobrachialis
Biceps brachii
Flexor carpi radialis
Flexor carpi ulnaris
Triceps brachii
Latissimus dorsi
External oblique
Iliopsoas
Vastus medialis
Pectineus
Adductor group
Tibialis anterior
Semimembranosus
Semitendinosus
Gastrocnemius
Superficial pectoral
Posterior deep pectoral
Umbilical cord (cut)
Rectus abdominis
Linea alba
Rectus femoris
Sartorius
Gracilis

(b)

FIGURE D1.3 ***(continued)*** **Muscles of the fetal pig's ventral aspect. (b)** Photograph.

3. The **brachiocephalic** is the large prominent muscle lying on the shoulder as seen in Figures D1.3 and D1.4. It extends from the mastoid process and the back of the neck and head and runs to insert into the distal end of the humerus, where it blends with the fibers of the pectoral muscles. Its action is to move the pig's forelimb anteriorly. The superior part of the brachiocephalic (sometimes called the **clavotrapezius**) is homologous to the superiormost part of the trapezius muscle of humans, which inserts into the clavicle. The pig, however, lacks a clavicle, so the fibers of this muscle fuse with those of the clavobrachialis running to the arm, forming the composite brachiocephalic muscle.

Chest Muscles

In the pig, the chest (or pectoral) muscles adduct the arm just as they do in humans. However, humans have only two pectoral muscles, and the pig has three: the superficial pectoral, the posterior deep pectoral, and the anterior deep pectoral. Because of their relatively great degree of fusion, the pig's pectoral muscles appear to be a single muscle. The pectoral muscles are rather difficult to dissect and identify, as they do not separate from one another easily. Refer to Figure D1.4 as you work.

The **superficial pectoral** is a thin, fan-shaped band of muscle that arises from the sternum just anterior to the posterior deep pectoral (discussed below) and inserts along nearly the entire length of the humerus and on the forearm fascia.

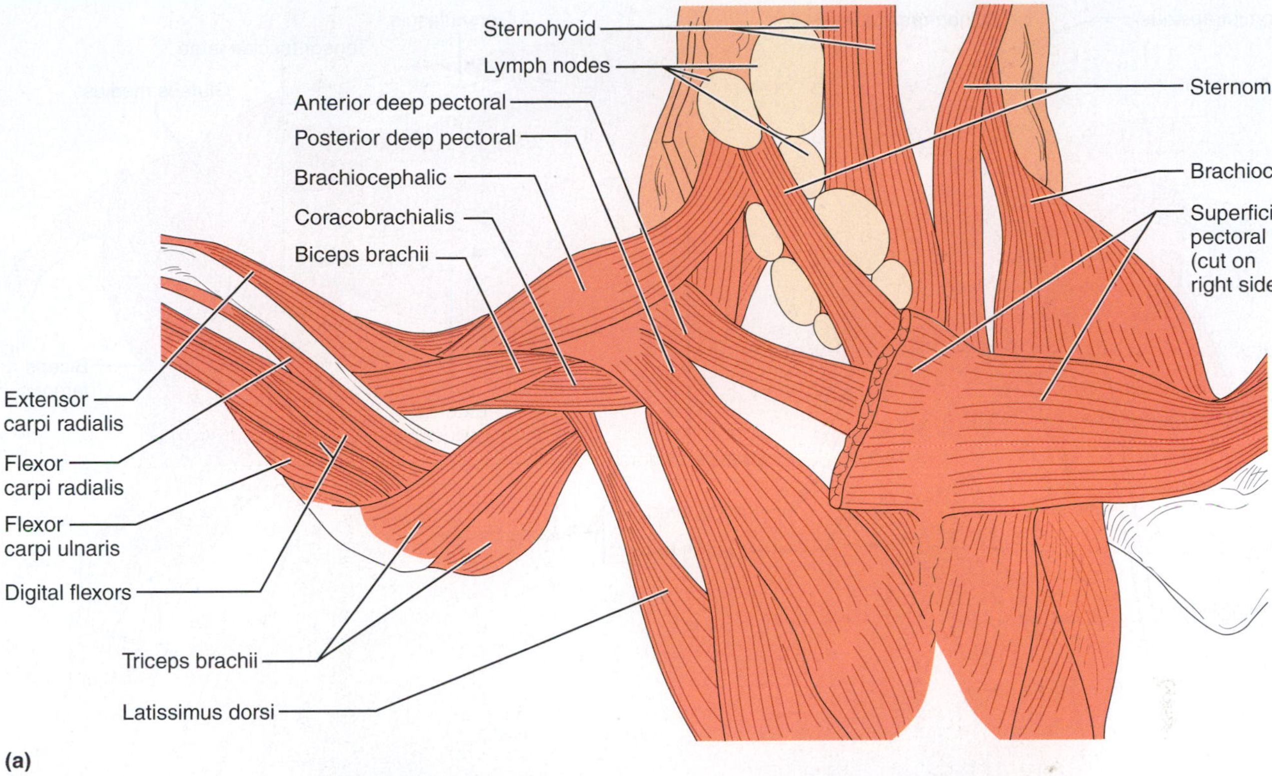
Sternohyoid
Lymph nodes
Anterior deep pectoral
Posterior deep pectoral
Brachiocephalic
Coracobrachialis
Biceps brachii
Sternomastoid
Brachiocephalic
Superficial pectoral (cut on right side)
Extensor carpi radialis
Flexor carpi radialis
Flexor carpi ulnaris
Digital flexors
Triceps brachii
Latissimus dorsi
(a)

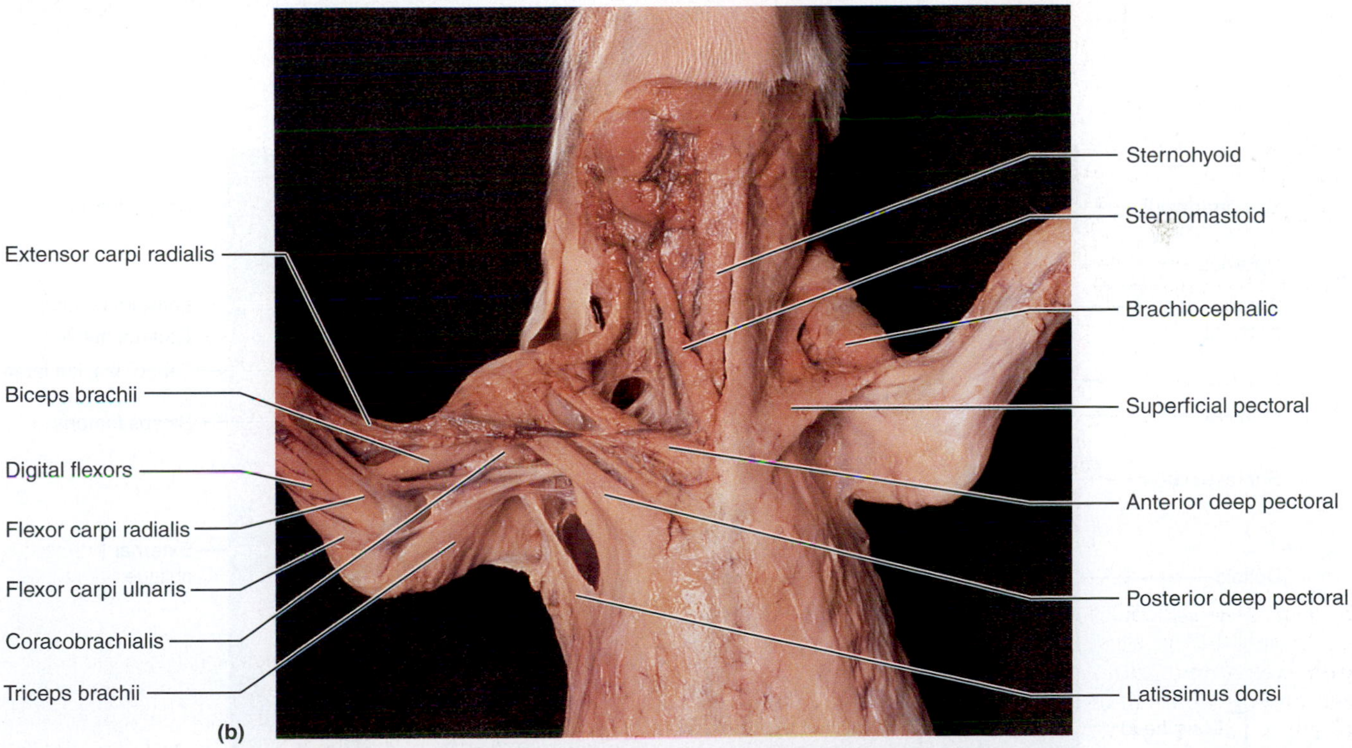
Sternohyoid
Sternomastoid
Brachiocephalic
Superficial pectoral
Anterior deep pectoral
Posterior deep pectoral
Latissimus dorsi
Extensor carpi radialis
Biceps brachii
Digital flexors
Flexor carpi radialis
Flexor carpi ulnaris
Coracobrachialis
Triceps brachii
(b)

FIGURE D1.4 Muscles of the fetal pig's ventral trunk and neck. (a) Diagrammatic view. **(b)** Photograph.

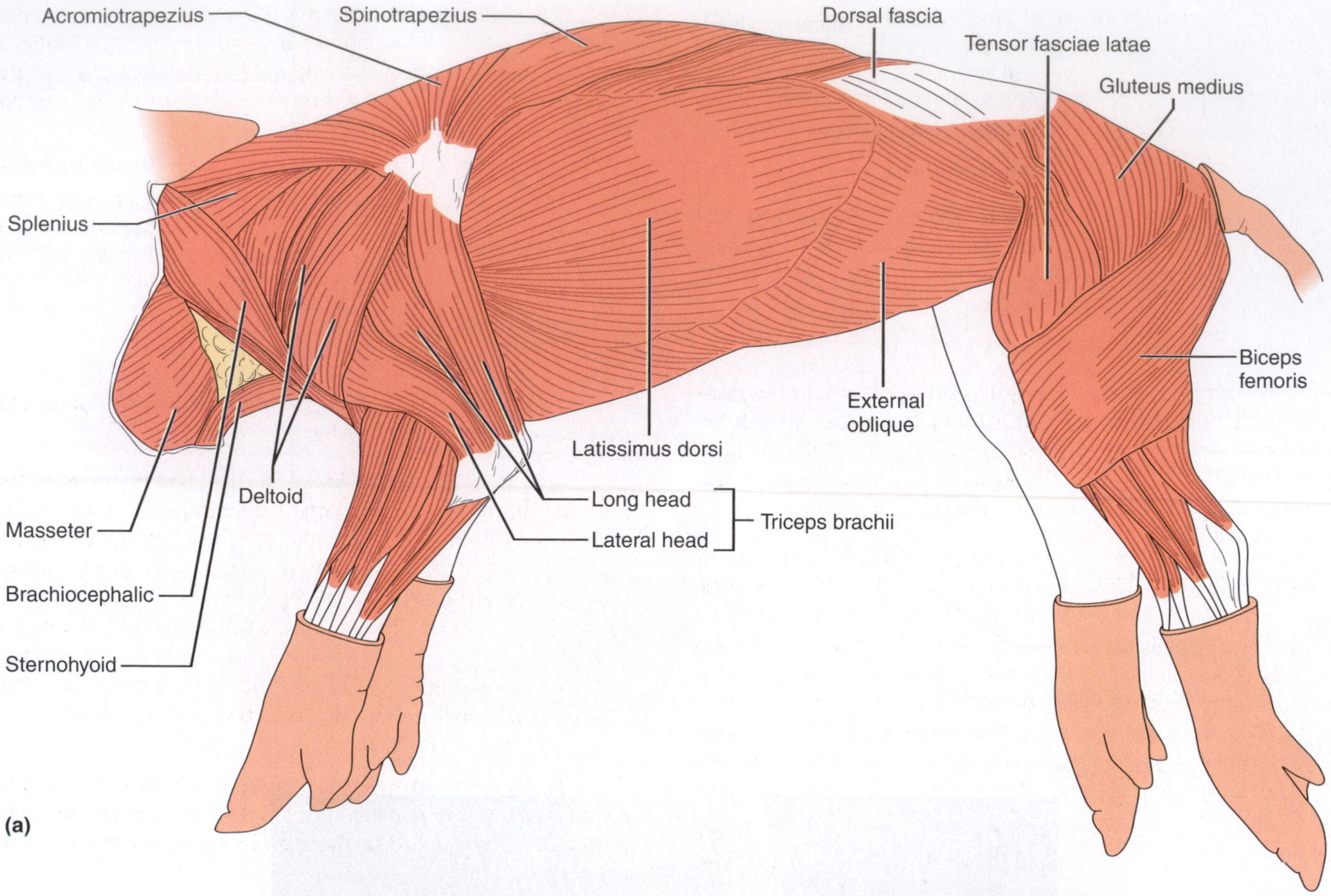
Acromiotrapezius
Spinotrapezius
Dorsal fascia
Tensor fasciae latae
Gluteus medius
Splenius
Biceps femoris
External oblique
Latissimus dorsi
Deltoid
Long head
Lateral head
Triceps brachii
Masseter
Brachiocephalic
Sternohyoid

(a)

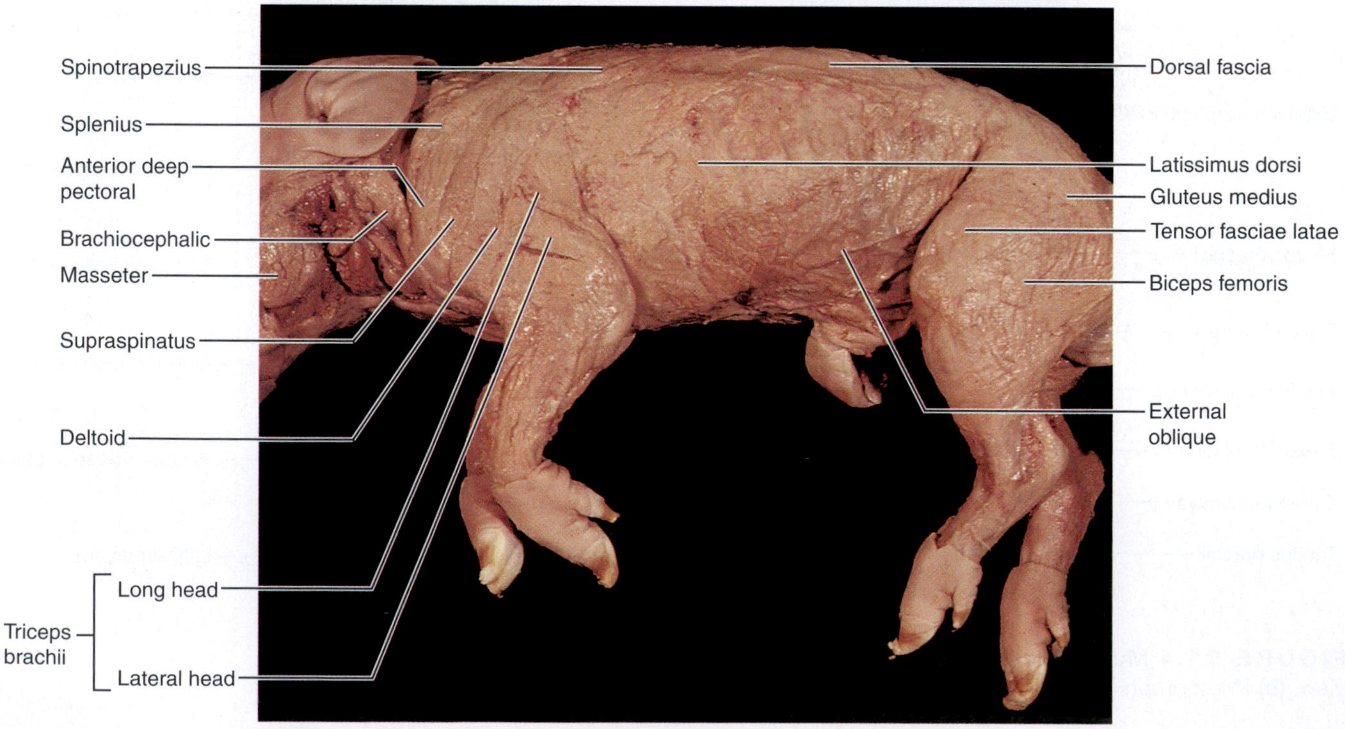
Spinotrapezius
Splenius
Anterior deep pectoral
Brachiocephalic
Masseter
Supraspinatus
Deltoid
Triceps brachii
Long head
Lateral head
Dorsal fascia
Latissimus dorsi
Gluteus medius
Tensor fasciae latae
Biceps femoris
External oblique

(b)

FIGURE D1.5 Superficial musculature of the fetal pig, lateral aspect. **(a)** Diagrammatic view. **(b)** Photograph.

The diagonally oriented fibers of the **posterior deep pectoral** can be seen inferior to the superficial pectoral. The muscle originates on the ventral sternum and costal cartilages of the fourth to ninth ribs and runs obliquely to insert on the proximal humerus. Slip the handle of your scalpel between the superficial pectoral and the posterior deep pectoral muscles to separate them.

The **anterior deep pectoral** is a thin straplike muscle deep to the superficial pectoral muscle. It extends superiorly from the posterior deep pectoral to run over the shoulder to insert on the dorsal scapular region.

Muscles of the Abdominal Wall

Now identify and trace the origins and insertions of the muscles of the abdominal wall, all of which are similar in function to the same-named muscles in humans. Refer back to Figure D1.3 for these muscle dissections. Avoid the umbilical area and the external genitalia if the pig is a male.

1. The **rectus abdominis** is a superficial longitudinal band of muscle running immediately lateral to the midline of the body on the abdominal surface from the pubis to the rib cage. To expose this muscle, slit the aponeurosis (of the oblique and transversus muscles) that encloses it. Note that the four transverse tendinous intersections of the rectus abdominis are absent or difficult to identify in the pig. Identify the **linea alba,** which separates the rectus abdominis muscles. Note the relationship of the rectus abdominis to the other abdominal muscles and their fascia.

2. The **external oblique** is a sheetlike muscle immediately lateral to (and running beneath) the rectus abdominis. (See also Figure D1.5.) Transect the external oblique to reveal the anterior attachment of the rectus abdominis and the **internal oblique** (not illustrated). Reflect the external oblique, and observe the deeper muscle. Notice that the fiber direction of the internal oblique runs perpendicular to that of the external oblique.

3. Transect the internal oblique muscle to reveal the fibers of the **transversus abdominis** (not illustrated), with its fibers running transversely across the abdomen (dorsoventrally in the animal). The transversus is very thin and may be difficult to separate from the internal oblique.

Superficial Muscles of the Posterior Trunk and Neck

Refer to Figure D1.5 as you dissect the following superficial muscles of the posterior surface.

1. Turn your pig onto its lateral surface and begin your dissection. Humans have one large single trapezius muscle, but the pig has three separate muscles: the clavotrapezius, the acromiotrapezius, and the spinotrapezius. As noted earlier, the most superior muscle of the group, the clavotrapezius (part of the brachiocephalic muscle in Figure D1.5a; also see Figure D1.6b), is homologous to that part of the human trapezius that inserts into the clavicle. Slip a probe under this muscle and follow it to its apparent origin. Release the clavotrapezius muscle from the adjoining muscles.

2. The **acromiotrapezius** is a short fan-shaped muscle posterior to the clavotrapezius. It originates from the cervical vertebral spines and inserts on the scapular spine by a broad, easily

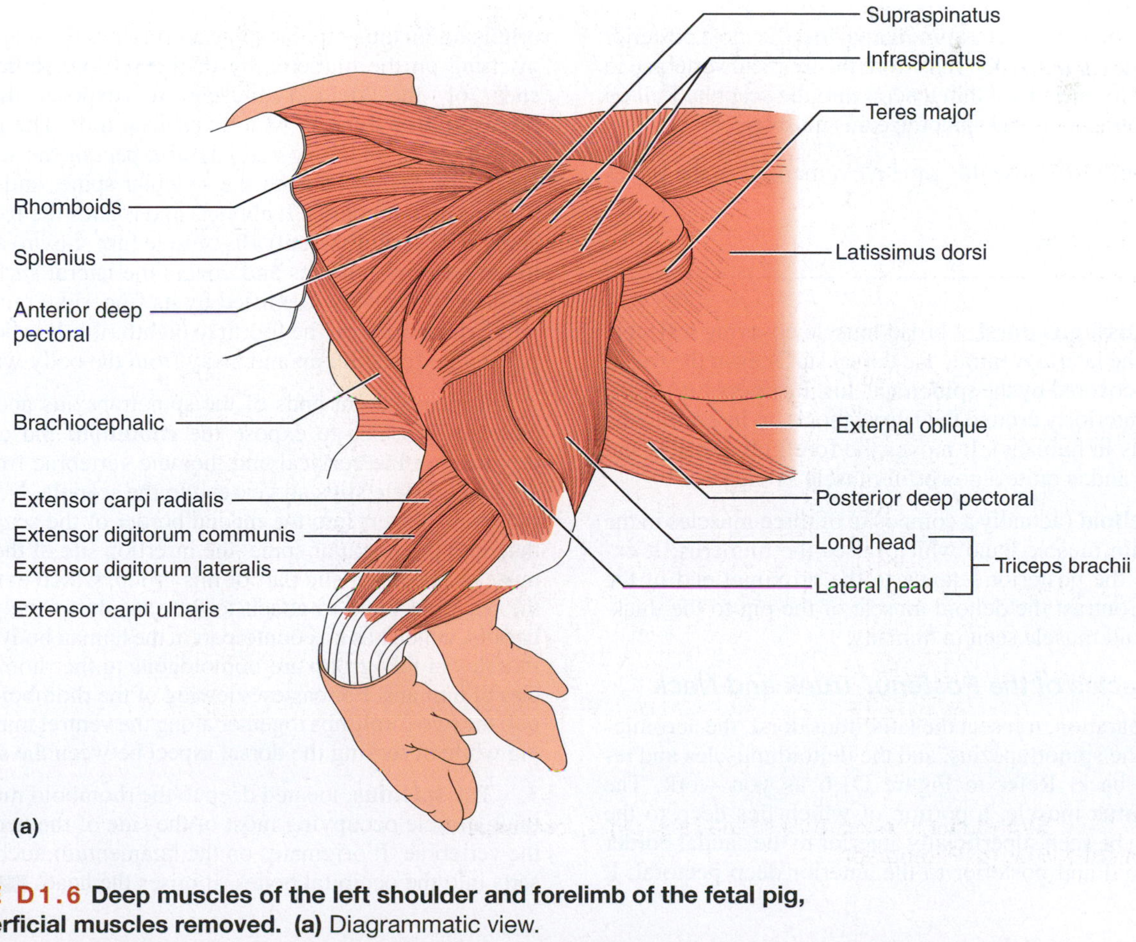
Supraspinatus
Infraspinatus
Teres major
Rhomboids
Splenius
Anterior deep pectoral
Latissimus dorsi
Brachiocephalic
External oblique
Extensor carpi radialis
Extensor digitorum communis
Extensor digitorum lateralis
Extensor carpi ulnaris
Posterior deep pectoral
Long head
Lateral head
Triceps brachii
(a)

FIGURE D1.6 Deep muscles of the left shoulder and forelimb of the fetal pig, with superficial muscles removed. (a) Diagrammatic view.

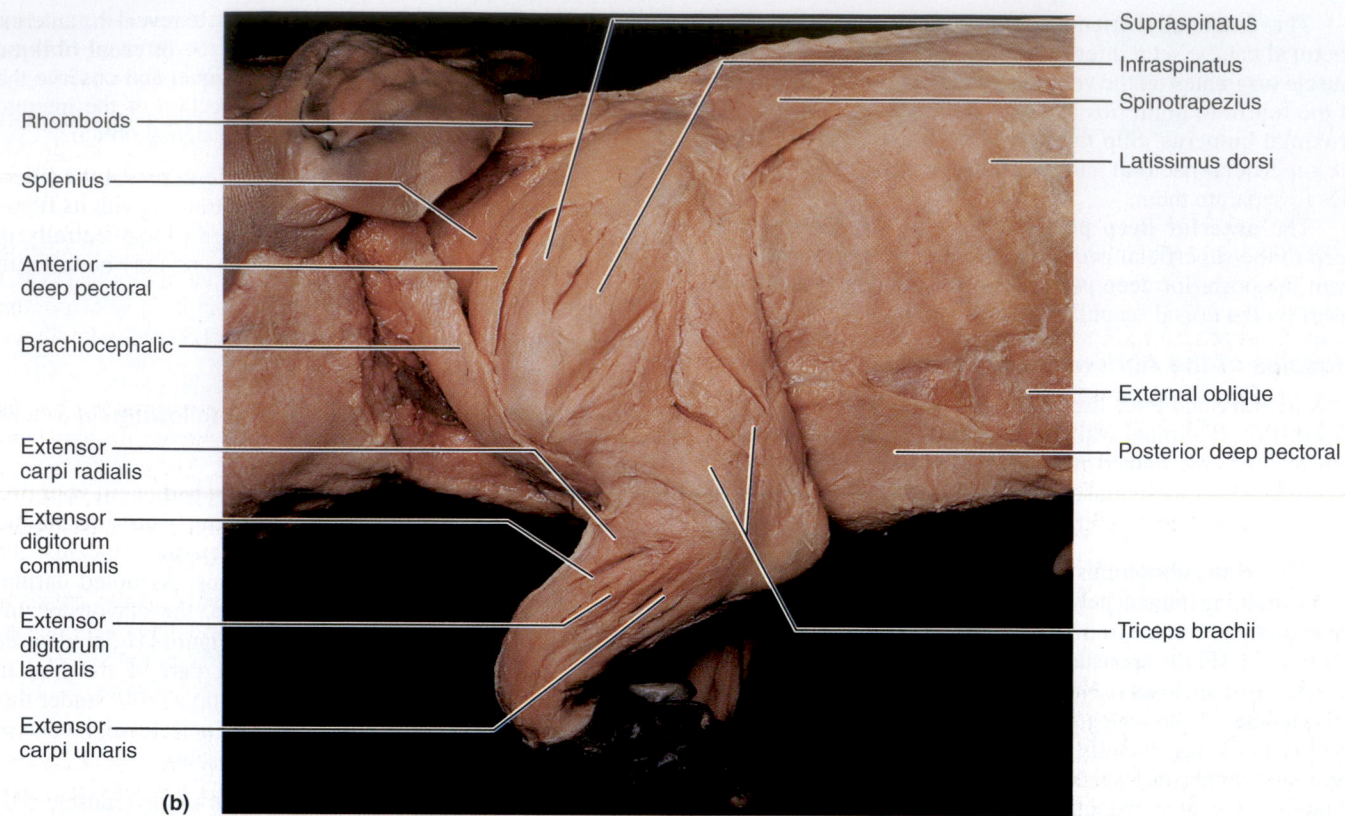

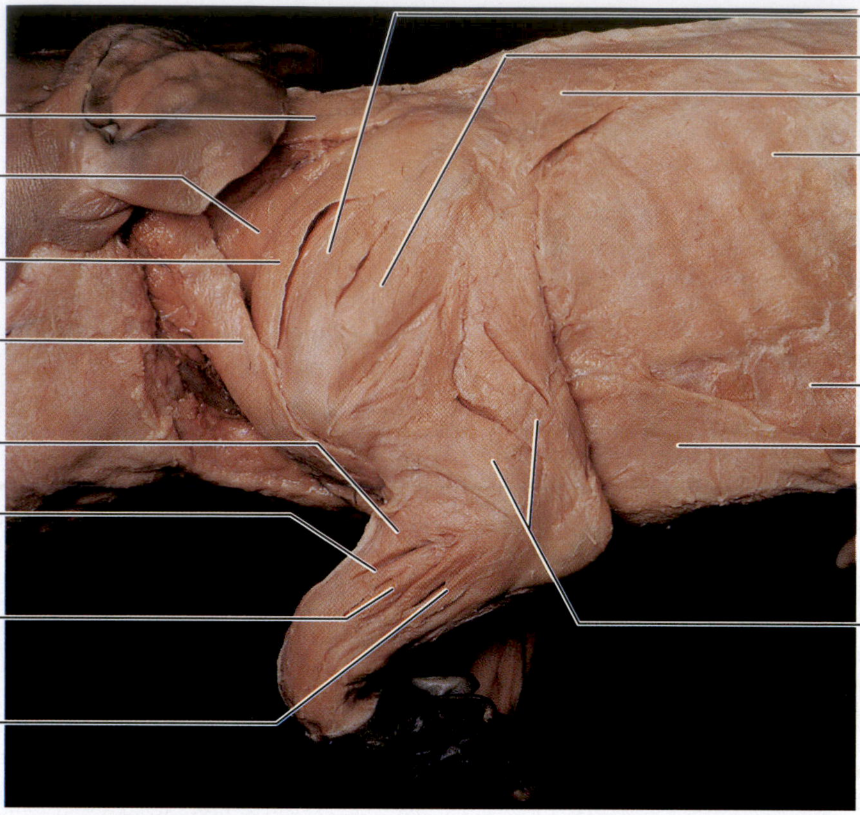
Rhomboids
Splenius
Anterior deep pectoral
Brachiocephalic
Extensor carpi radialis
Extensor digitorum communis
Extensor digitorum lateralis
Extensor carpi ulnaris
Supraspinatus
Infraspinatus
Spinotrapezius
Latissimus dorsi
External oblique
Posterior deep pectoral
Triceps brachii

(b)

FIGURE D1.6 *(continued)* Deep muscles of the left shoulder and forelimb of the fetal pig, with superficial muscles removed. (b) Photograph.

identified aponeurosis. The **spinotrapezius,** the most posterior trapezius muscle in the pig, runs from the thoracic vertebrae to the scapula to insert by a thin tendon into the scapula. Pull on the acromiotrapezius and spinotrapezius muscles.

Do they appear to cause the same movement(s) in the pig as in humans?

____

3. The **latissimus dorsi,** a broad muscle covering a sizable portion of the lateral, ventral, and dorsal surfaces of the trunk, is partially covered by the spinotrapezius. Its fibers run downward and anteriorly around the lateral thorax to insert into the humerus (as in humans). It moves the forelimb dorsally and posteriorly and is quite a powerful muscle in pigs.

4. The **deltoid** (actually a composite of three muscles in the pig) is a thin muscle band which raises the humerus. It extends from the posterior scapula to the proximal end of the humerus. Contrast the deltoid muscle of the pig to the thick, fleshy deltoid muscle seen in humans.

Deep Muscles of the Posterior Trunk and Neck

1. In preparation, transect the latissimus dorsi, the acromiotrapezius, the spinotrapezius, and the deltoid muscles and reflect them back. Refer to Figure D1.6 as you work. The **supraspinatus** muscle, a portion of which lies deep to the deltoid, can be seen superficially anterior to the caudal border of the deltoid and posterior to the anterior deep pectoral. It originates on the scapular spine and clothes the scapula before inserting on the humerus by dual insertion. Reflect the cut ends of the deltoid muscle to expose the thicker **infraspinatus** muscle, which lies deep to it. The infraspinatus originates on the posterolateral aspect of the scapula, occupies the region beneath the scapular spine, and inserts on the proximal humerus. It abducts and rotates the forelimb laterally. The **serratus ventralis** muscle (not shown) arises deep to the pectoral muscles and covers the lateral surface of the rib cage. It is easily identified by its fingerlike muscular origins, which arise on the fourth to eighth ribs. It is best seen by raising the forelimb up and away from the body wall.

2. Reflect the cut ends of the spinotrapezius and acromiotrapezius muscles to expose the **rhomboid** muscles. These originate on the cervical and thoracic vertebrae from several separate muscle slips and extend to the scapula. Note that the rhomboids insert into the medial border of the scapula rather than into the scapular spine, the insertion site of the trapezius muscles. The cephalic part of this group, which extends from the occipital bone to attach to the scapula, is the **rhomboid capitis,** which has no counterpart in the human body. The other muscles in this group are homologous to the rhomboid muscles of humans. For easiest viewing of the rhomboid muscles, pull the two forelimbs together along the ventral midline of the pig while observing the dorsal aspect between the scapulae.

3. The **splenius,** located deep to the rhomboid muscles, is a thick muscle occupying most of the side of the neck close to the vertebrae. It originates on the ligamentum nuchae and inserts into the occipital bones. It raises the head.

ACTIVITY 4

Dissecting Forelimb Muscles

Refer to Figures D1.4 and D1.6 as you study these muscles.

Upper Forelimb Muscles

1. The triceps muscle of the pig **(triceps brachii)** can be easily identified if the pig is placed on its side (Figure D1.6). It is a large fleshy muscle covering the posterior aspect and much of the side of the humerus. As in humans, this muscle arises from three heads, which originate from the humerus and scapula and insert jointly into the olecranon process of the ulna. Remove the fascia from the superior region of the lateral arm surface to identify the lateral and long heads of the triceps. The long head is medial to the lateral head on the posterior surface of the arm. The deep medial head can be exposed by transecting and reflecting the lateral head.

How does the function of this muscle in the pig compare to its function in humans?

____

2. The **brachialis** (not shown) is located anterior to the lateral head of the triceps muscle. Identify its origin on the humerus, and trace its course as it crosses the elbow laterally and inserts on the ulna. It flexes the foreleg of the pig.

3. In the pig, the **biceps brachii** muscle (Figure D1.4) is a small, slender, spindle-shaped muscle that is covered almost completely by the clavobrachialis portion of the brachiocephalic muscle. The biceps brachii in the pig arises via a single tendon from the scapula in the area anterior to the glenoid cavity and inserts into the proximal radius and ulna. It flexes the forelimb of the pig. Follow this muscle to its origin.

Does the biceps have two heads in the pig? ______________

4. The **coracobrachialis** of the pig is insignificant and can be seen as a very small muscle crossing the ventral aspect of the shoulder joint. It runs beneath the biceps brachii to insert on the humerus and has the same function as its homologue in humans.

Lower Forelimb Muscles

Identification of the lower forelimb muscles is difficult because of the tough fascia sheath that encases them.

1. Remove as much of this connective tissue as possible, and cut through the ligaments that secure the tendons at the wrist (transverse carpal ligaments) so that you will be able to follow the various muscles to their insertions. Only selected forelimb muscles will be identified in the pig.

2. Begin your identification of the forearm muscles by examining the lateral surface of the forelimb, which mostly houses extensor muscles of the carpals, metacarpals, and digits. The muscles of this region are very much alike in appearance and are difficult to identify unless a definite order is followed. Thus you will begin with the most posterior muscles and proceed to the anterior aspect. Remember to check the tendons of insertion to verify your muscle identification in each case. These muscles are illustrated in Figure D1.6.

3. Follow the **extensor carpi ulnaris** muscle from the lateral epicondyle of the humerus to the ulnar side of the fifth metacarpal. Often this muscle has a shiny insertion tendon that helps in its identification.

4. The **extensor digitorum lateralis,** or **longus,** muscle is similar in appearance to the extensor carpi ulnaris and lies just anterior to it. The extensor digitorum lateralis arises from the distal humerus and inserts by a divided tendon into the digits. This muscle has no human counterpart.

5. You can see the **extensor digitorum communis** for its entire length along the lateral surface of the forelimb just anterior to the extensor digitorum lateralis. Trace it to its four tendons, which insert on the second to the fifth digits. The extensor digitorum communis and the two extensor muscles described above act together to extend the digits.

6. The **extensor carpi radialis** is anterior to the extensor digitorum communis and is sometimes partially covered by it. It originates from the distal end of the humerus and inserts into the distal end of the radius. It rotates the foot of the pig.

7. The ribbonlike muscle seen on the lateral surface of the humerus and passing down the foreleg to insert on the styloid process of the radius is the **brachioradialis** (not shown). It rotates the forelimb. (If your removal of the fascia was not very careful, this muscle may have been removed in the clearing process.)

8. Turn the pig so that you can easily observe the ventral forelimb muscles, mostly flexors and pronators. For the following series of muscle identifications refer to Figure D1.4. Your examination of these muscles will be cursory, so you need not attempt to trace the origins and insertions of these muscles. In the pig, as in humans, most of these muscles arise from the medial epicondyle of the humerus. Identify the following muscles or muscle groups by beginning at the posterior surface of the medial forearm and working anteriorly. The **digital flexor** muscles are two thin muscles seen adjacent to the flexor carpi ulnaris. The **flexor carpi radialis** is located anterior to the **flexor carpi ulnaris** and acts with it to flex the wrist.

ACTIVITY 5

Dissecting Hindlimb Muscles

With the pig on its side, remove the fat and fascia from all the thigh surfaces, but do not cut through or remove the **fascia lata,** which is a tough white aponeurosis covering the anterolateral surface of the thigh from the hip to the leg.

Muscles of the Posterolateral Hindlimb

Referring to Figures D1.3, D1.5, and D1.7, identify the following muscles of the hip, thigh, and posterior shank.

1. The **tensor fasciae latae** is wide and thin at its superior end, where it originates on the iliac crest and narrows to a long tendon that inserts into the fascia lata. As its name indicates, it tenses or tightens the fascia lata. Additionally, it flexes the hip and extends the knee. Transect this muscle. The

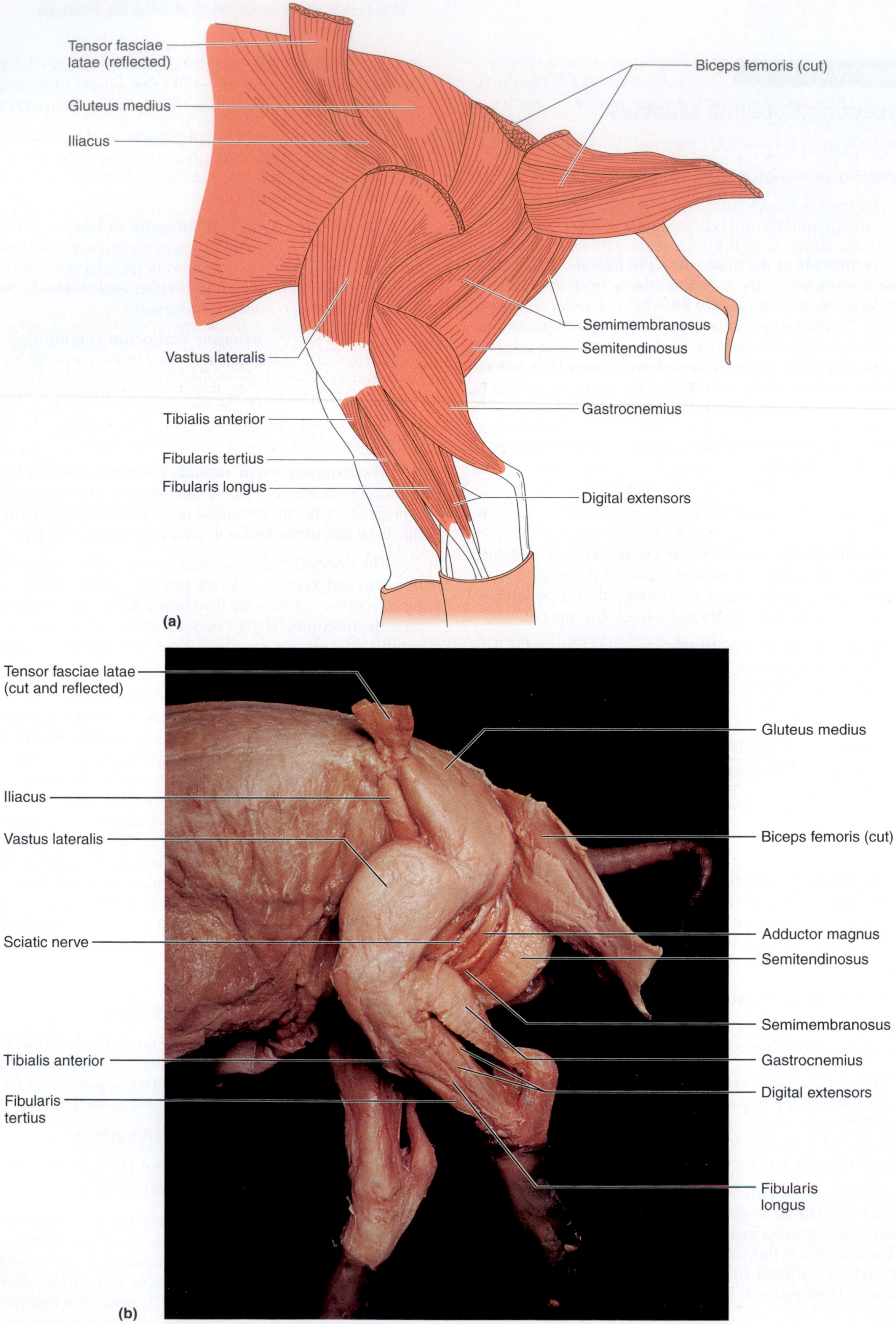
Tensor fasciae
latae (reflected)
Gluteus medius
Iliacus
Biceps femoris (cut)
Semimembranosus
Semitendinosus
Vastus lateralis
Gastrocnemius
Tibialis anterior
Fibularis tertius
Fibularis longus
Digital extensors
(a)
Tensor fasciae latae
(cut and reflected)
Gluteus medius
Iliacus
Biceps femoris (cut)
Vastus lateralis
Adductor magnus
Sciatic nerve
Semitendinosus
Semimembranosus
Gastrocnemius
Tibialis anterior
Digital extensors
Fibularis
tertius
Fibularis
longus
(b)

FIGURE D1.7 Muscles of the lateral aspect of the left thigh and leg of the fetal pig. (a) Diagrammatic view. **(b)** Photograph.

gluteus medius is a thick muscle lying posterior and deep to the tensor fasciae latae. It originates from the lumbodorsal and gluteal fascias and inserts into the greater trochanter of the femur. It abducts the thigh.

2. The **gluteus maximus** muscle in the pig is very thin. Located posterior to the tensor fasciae latae, it overlies the anterior portion of the gluteus medius muscle, with which it acts synergistically. No attempt will be made to distinguish the limits of the individual gluteal muscles.

3. The **hamstring muscles,** a group in the hindlimb, consist of the biceps femoris, the semitendinosus, and the semimembranosus muscles. The **biceps femoris** is a large triangular muscle of the thigh that covers most of its posterolateral surface. Trace it from its origin on the ischium to its insertion on the distal femur and proximal tibia. It abducts and extends the thigh and flexes the shank. Transect this muscle and reflect its cut ends to identify the following muscles lying deep to it. The **semitendinosus** is a thick, narrow muscle band lying deep to the posterior border of the biceps femoris. Contrary to what its name implies ("half-tendon"), this muscle is muscular and fleshy except at its insertion. Trace this muscle from its origin on the pelvic girdle to its insertion on the proximal tibia. It bends the knee. The **semimembranosus,** a large muscle lying medial to the semitendinosus and largely obscured by it, is the most posteriorly placed muscle of the thigh. The distal end of this broad muscle overlies and is closely associated with the adductor magnus muscle. Trace it from its origin on the pelvic girdle to its insertion on the proximal tibia. Before identifying the muscles of the shank, identify the large **vastus lateralis** muscle, which lies anterior to the semimembranosus muscle and is part of the quadriceps muscle group to be dealt with later.

How does the semimembranosus compare with its human homologue?

____

4. Remove the heavy fascia covering the lateral surface of the shank and proceed from posterior to anterior aspect to identify the following muscles on the posterolateral shank (leg). Reflect the lower portion of the biceps femoris muscle to see the origin of the following muscles. The **triceps surae** is a large composite muscle of the calf, which is homologous to the same-named muscle in humans. The **gastrocnemius** part of the triceps surae is the largest muscle on the shank. As in humans, it has two heads and inserts via the calcaneal (Achilles) tendon into the calcaneus. Run a probe beneath this muscle and then transect it to reveal the **soleus,** which is deep to the gastrocnemius (see Figure D1.8).

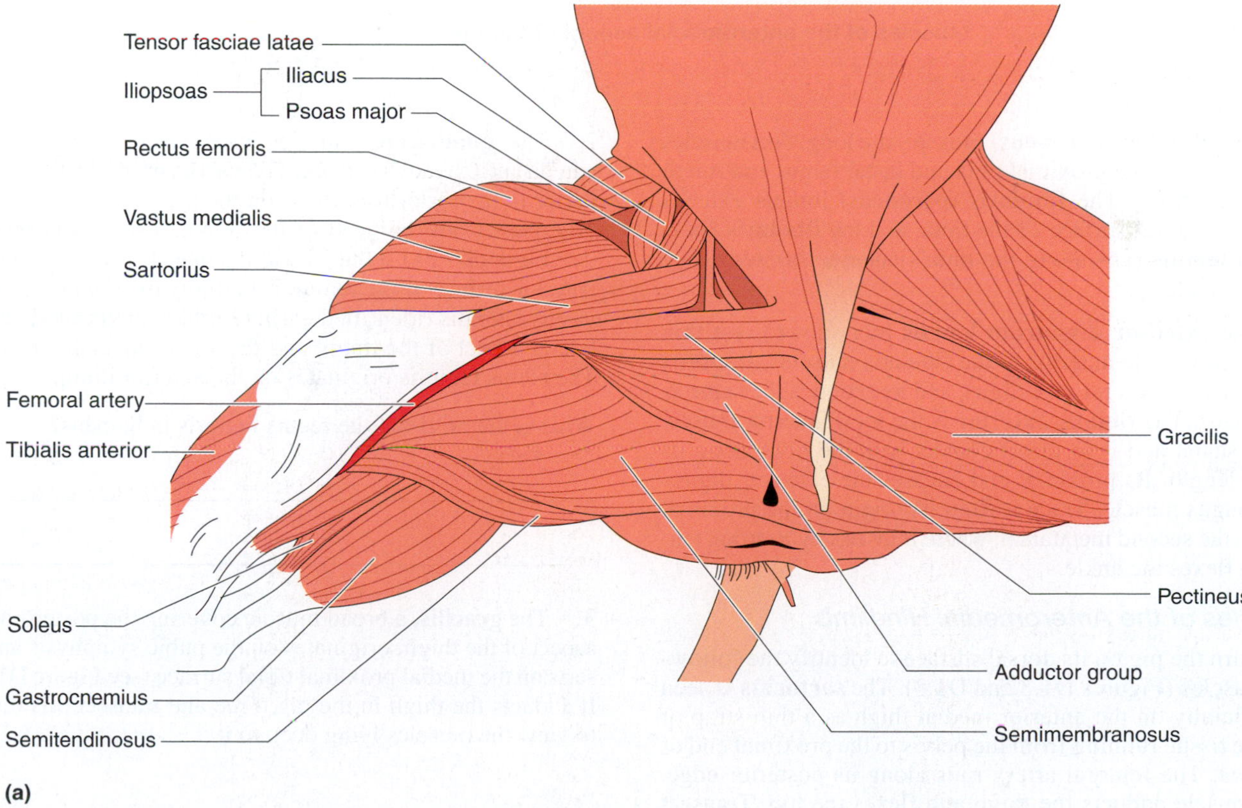
Tensor fasciae latae
Iliopsoas
Iliacus
Psoas major
Rectus femoris
Vastus medialis
Sartorius
Femoral artery
Tibialis anterior
Soleus
Gastrocnemius
Semitendinosus
Gracilis
Pectineus
Adductor group
Semimembranosus
(a)

FIGURE D1.8 Muscles of the anteromedial aspect of the right hindlimb.
(a) Diagrammatic view. (Gracilis muscle has been removed from the right thigh.)

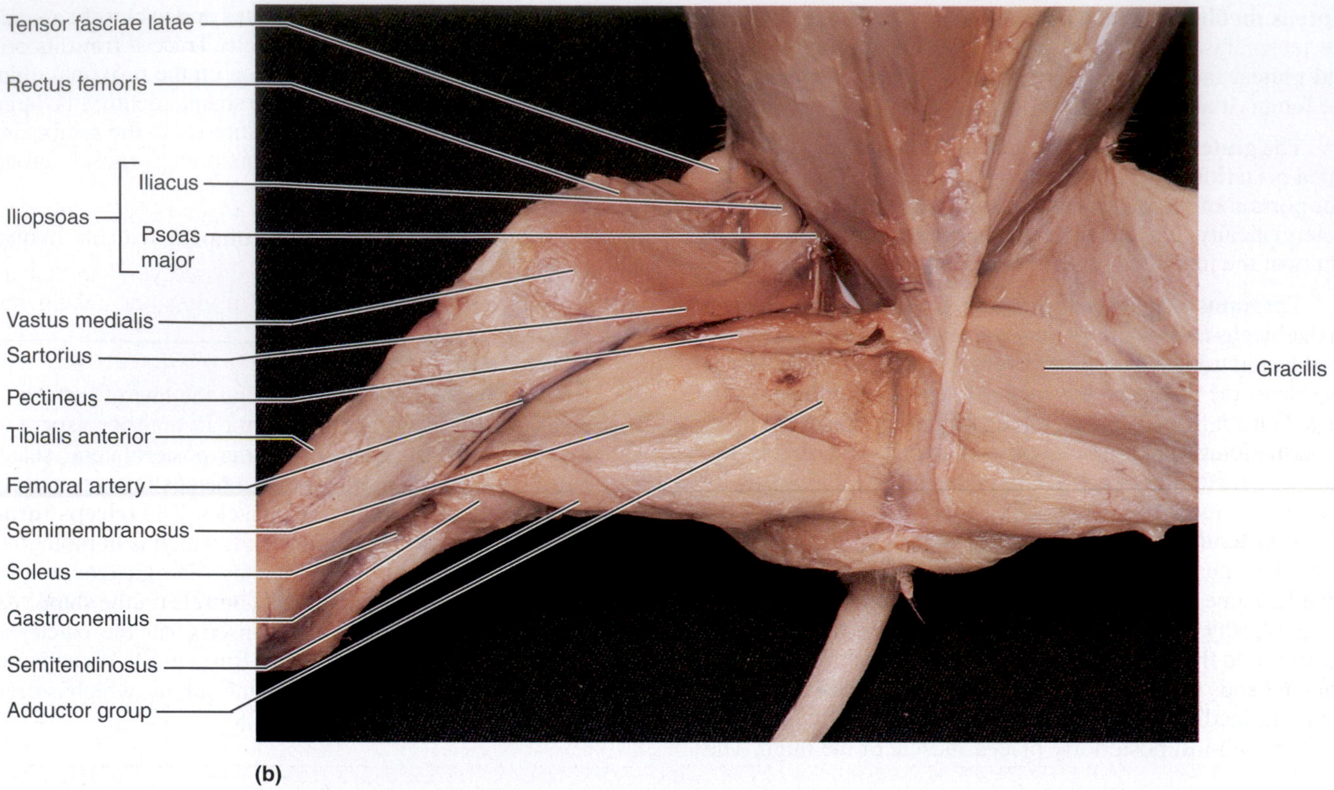
Tensor fasciae latae
Rectus femoris
Iliacus
Iliopsoas
Psoas major
Vastus medialis
Sartorius
Pectineus
Tibialis anterior
Femoral artery
Semimembranosus
Soleus
Gastrocnemius
Semitendinosus
Adductor group
Gracilis
(b)

FIGURE D1.8 ***(continued)*** **Muscles of the anteromedial aspect of the right hindlimb. (b)** Photograph.

5. The **fibularis (peroneus) muscles** are long, thin muscles that arise from the proximal tibia and distal femur and run to tarsal insertions. The **fibularis (peroneus) longus** extends along the lateral surface of the tibia, and the **fibularis (peroneus) tertius** runs along the tibia's anterior surface. They flex the ankle.

6. The **extensor digitorum longus** (not shown) is a thin elongated muscle anterior to the fibularis muscles. It inserts by divided tendon into the digits, and its action is the same as in humans. The **tibialis anterior** is the most anterior muscle of the shank and is in direct contact with the tibia along its whole length. Its proximal end encases the extensor digitorum longus muscle. Trace it from its origin on the proximal tibia to the second metatarsal, where it inserts by a strong tendon. It flexes the ankle.

Muscles of the Anteromedial Hindlimb

1. Turn the pig on its dorsal surface to identify the following muscles (Figures D1.3 and D1.8). The **sartorius** is seen superficially on the anterior medial thigh as a thin strap of muscle tissue running from the pelvis to the proximal end of the tibia. The femoral artery runs along its posterior edge. This muscle adducts the thigh and flexes the hip. Transect this muscle.

2. The **quadriceps** muscles insert in common into the patella and extend the shank. The **vastus medialis** lies just superior to the sartorius muscle on the medial thigh surface. It originates from the head of the femur. The **rectus femoris** (see Figure D1.3a), a thick bandlike muscle running along the anterior surface of the femur, is partially overlain by both the vastus lateralis (identified earlier), which clothes the anterolateral aspect of the thigh, and the vastus medialis medially. The rectus femoris originates on the anterior ilium.

What is the origin of the rectus femoris in humans?

____

____

3. The **gracilis,** a broad muscle covering the posteromedial aspect of the thigh, originates on the pubic symphysis and inserts on the medial proximal tibial surface (see Figure D1.3a). It adducts the thigh in the pig. Free and transect this muscle to view the muscles lying deep to it.

How does this compare with the human gracilis?

____

____

4. The **adductor magnus** lies deep to the gracilis. Separate this muscle from the semimembranosus muscle, which lies inferior to it and with which it is closely associated.

5. Examine the superior margin of the adductor magnus to locate the long slender **pectineus** muscle. It is normally covered by the sartorius muscle distally. It originates on the pubic bone and inserts on the medial aspect of the femur. This muscle is similar in action to its human homologue and acts to adduct the thigh.

6. Just superior to the pectineus muscle, you can see a small portion of the **iliopsoas,** a composite muscle whose action is similar to that of its counterpart in humans.

7. The dissection of the remaining musculature of the anterior shank in the pig is included on pages 709–710 and shown in Figure D1.7.

8. Properly clean your work area and instruments with soapy water, then dry and dispose of your gloves and organic debris in the assigned containers. Prepare your animal for storage as instructed in the box on page 700 before leaving the laboratory.

DISSECTION REVIEW

Many human muscles are modified from those of the pig (or any quadruped) as a result of the requirements of an upright posture. The following questions refer to these differences.

1. How does the human trapezius muscle differ from the pig's? ____

____

____

2. How does the deltoid differ? ____

____

____

3. How do the size and orientation of the human sartorius muscle differ than that in the pig?

____

____

4. Explain these differences in terms of differences in function. ____

____

____

5. The human rectus abdominis is definitely divided by four transverse tendons (tendinous intersections). These tendons are absent or difficult to identify in the pig. How do these tendons affect the human upright posture?

____

____

DISSECTION EXERCISE 2

Dissection of the Spinal Cord and Spinal Nerves of the Fetal Pig

MATERIALS

- ☐ Disposable gloves
- ☐ Dissecting instruments and tray
- ☐ Animal specimen from previous dissection
- ☐ Bone cutters
- ☐ Embalming fluid
- ☐ Paper towels

OBJECTIVES

1. To identify on a dissected animal the spinal cord, spinal nerves and roots, dorsal root ganglia, and the spinal dura mater.
2. To identify on a dissected animal the sympathetic trunk; the radial, median, and ulnar nerves of the forelimb; and the femoral, sciatic, and internal and external popliteal nerves of the hindlimb.

As in humans, the spinal cord of the fetal pig is a basically cylindrical mass of nervous tissue lying within the vertebral canal of the spinal column. It is enlarged in the cervical and lumbar regions where the spinal nerves serving the limbs arise. The pig has 33 pairs of spinal nerves (as compared to 31 in humans). Of these, 8 are cervical, 14 thoracic, 7 lumbar, and 4 sacral.

A complete dissection of the spinal cord and spinal nerves of the pig would be extraordinarily time-consuming and exacting and is not warranted in a basic anatomy and physiology course. However, it is desirable to have some dissection work to complement your study of the anatomical charts. To provide some hands-on experience you will carry out a partial dissection of the spinal cord and of the brachial and lumbar plexuses and identify some of the major nerves. See Exercise 21 of this manual for a discussion of human spinal nerves.

ACTIVITY 1

Exposing and Examining the Spinal Cord

The photograph in Figure D2.1 shows the dorsal aspect of nearly the entire spinal cord of the fetal pig as well as spinal nerves issuing from the cord. If a prosected (previously dissected) animal is provided for these observations, this is the view you will see. If you will be dissecting your pig, only a section of the cord need be exposed and studied during this laboratory session. If a prosected specimen is used, begin your observations at step 5. (Note that the dura mater may already be slit and pinned.)

1. Don gloves and obtain dissecting tools and tray and your animal specimen.
2. Place your pig on the dissecting tray ventral side down and carefully remove the muscles overlying approximately 10 cm (4 inches) of the vertebral column. Although this muscle clearing can be done anywhere along the length of the spinal cord, it is most easily done in the thoracic region.
3. Using bone cutters, cut across the pedicles of the exposed vertebrae and remove the vertebral arches to reveal the spinal cord lying snugly in the vertebral canal. Continue to chip away carefully at the bone and remove the epidural fat until you have exposed several spinal nerves at the lateral aspects of the cord.
4. Observe the tough, glistening dura mater covering the cord and spinal roots. Notice that it is not fused with the periosteum lining the vertebral canal (as is the dura mater covering the brain in the cranial cavity).

PAL

For access to anatomical models and more, check out Practice Anatomy Lab.

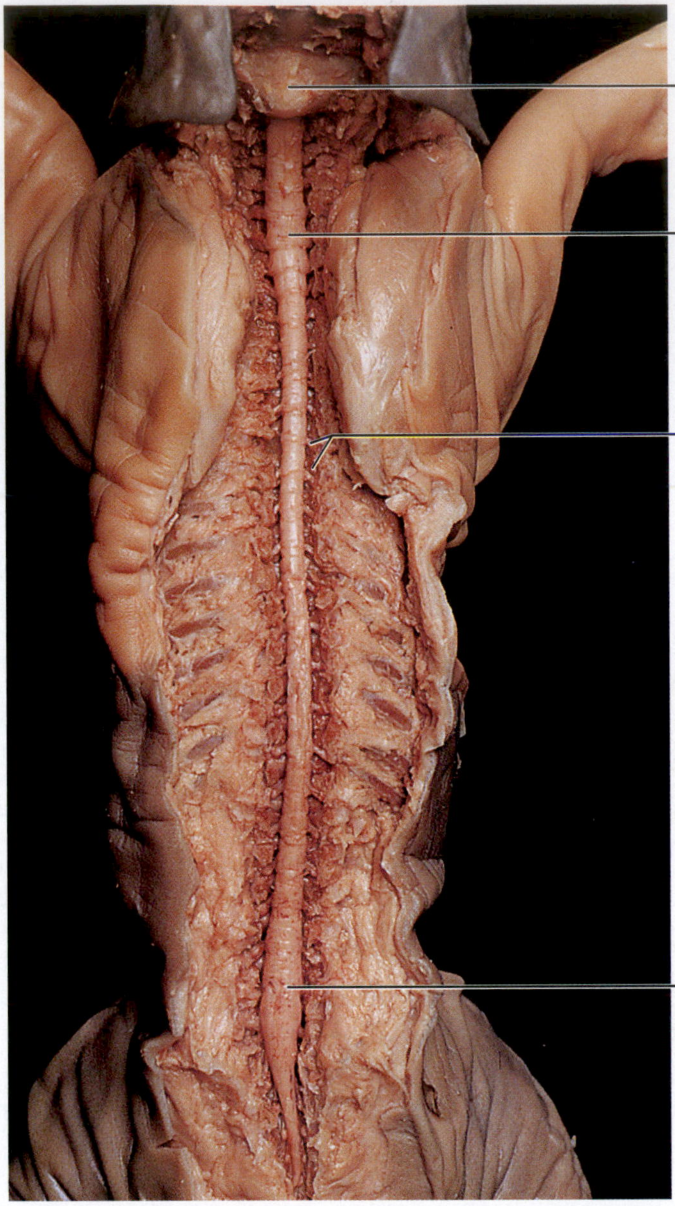
Brain
Cervical enlargement of spinal cord
Spinal nerves
Lumbar enlargement of spinal cord

FIGURE D2.1 Spinal cord of the fetal pig, dorsal aspect. The spinal nerves and cervical and lumbar enlargements of the spinal cord are obvious in this view.

5. Slit open the dura mater, and observe that each **spinal nerve** arises from two roots—the **dorsal (sensory)** and **ventral (motor) roots**—which, in turn, originate from several rootlets issuing from the cord.

6. Identify the **dorsal root ganglia,** small round enlargements on the dorsal roots that are the "residences" of the cell bodies of sensory neurons serving the body periphery.

7. Trace a spinal nerve laterally to see it divide into **dorsal** and **ventral rami.** The dorsal rami supply structures of the dorsal body wall, whereas ventral rami distribute to the anterior and lateral body trunk and (via plexuses) to the limbs.

8. If you have exposed a region of the cord other than the thoracic region, you may be able to identify other anatomical features of the spinal cord, such as the **cervical** or **lumbar enlargements,** or the conus medullaris and cauda equina at the caudal end of the cord. ■

ACTIVITY 2

Identifying Nerves of the Brachial Plexus

1. If a prosected specimen is provided, begin your observations of the brachial plexus at step 2. Otherwise, place your pig dorsal side down on the dissecting tray to make these initial observations on the nerves of the brachial plexus. Reflect the cut ends of the left pectoralis muscles to expose the large brachial plexus in the axillary region. The nerves appear as glistening white cords. Carefully clean the exposed nerves as far back toward their points of origin as possible. The **brachial plexus,** which consists of tough interconnected nerves emerging from the last three cervical and first thoracic vertebrae, should now be visible. Refer to Figure D2.2 as you work.

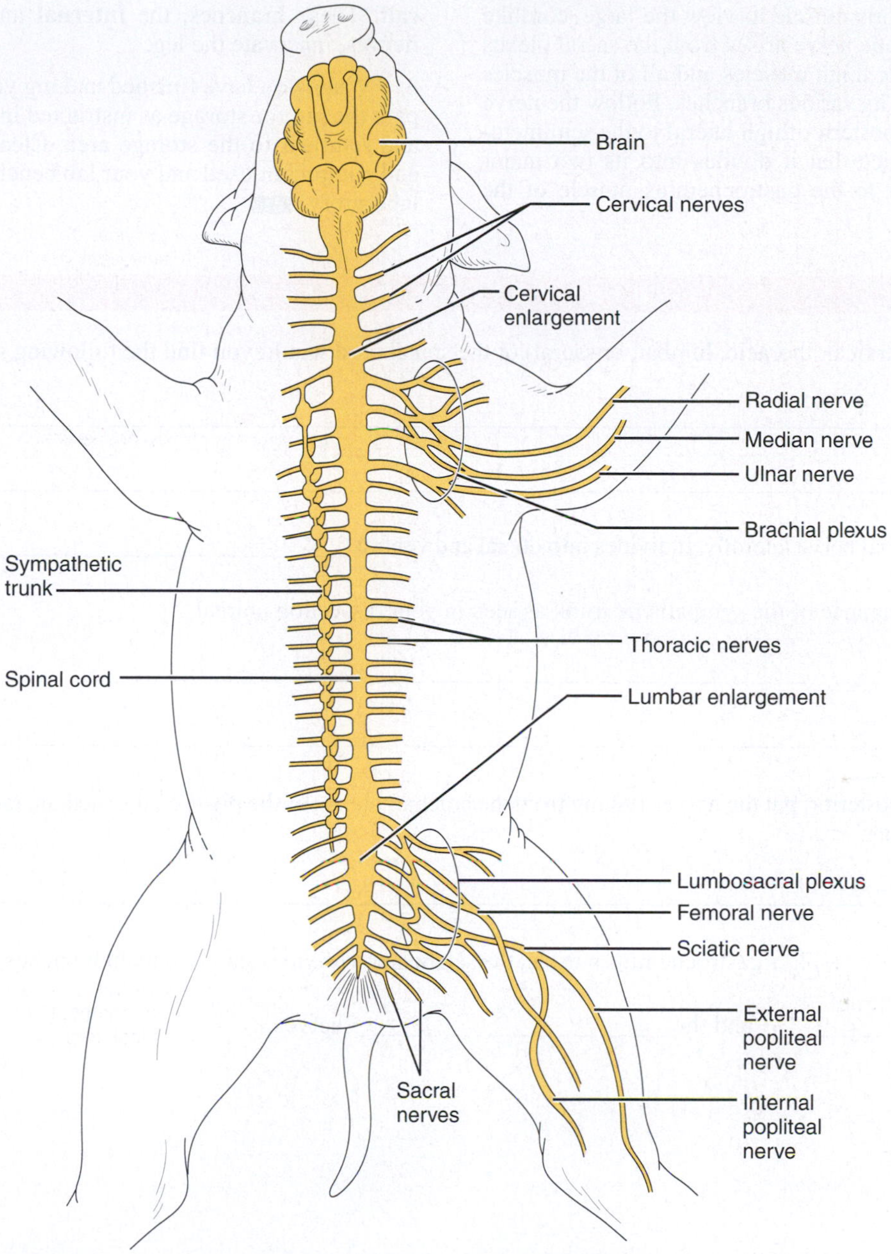
Brain
Cervical nerves
Cervical enlargement
Radial nerve
Median nerve
Ulnar nerve
Brachial plexus
Sympathetic trunk
Thoracic nerves
Spinal cord
Lumbar enlargement
Lumbosacral plexus
Femoral nerve
Sciatic nerve
External popliteal nerve
Sacral nerves
Internal popliteal nerve

FIGURE D2.2 Brachial plexus and lumbosacral plexus of the fetal pig, ventral aspect.

2. The **median nerve** is closely associated with the brachial artery and vein, which course through the arm. It supplies most of the ventral muscles of the forearm. The **radial nerve** is a large nerve seen just superior to the median nerve. Follow it into the triceps brachii muscle of the arm. The **ulnar nerve** is the most posterior of the large brachial plexus nerves. It supplies forearm and hand muscles that are not innervated by the median nerve.

3. Although not actually part of the brachial plexus, the **sympathetic trunk** (chain of sympathetic nervous system ganglia) may be easily identified in the thoracic region. Pull the heart and lungs ventrally to expose the descending aorta. Carefully examine the dorsolateral surface of the aorta as it travels through the thorax. The sympathetic trunk, lying alongside the aorta, appears as a slender white cord with segmental enlargements (the ganglia) in it. ■

ACTIVITY 3

Identifying Nerves of the Lumbosacral Plexus

Now we are ready to examine the nerves of the posterior **lumbosacral plexus,** composed of branches of the last three lumbar and first sacral nerves.

1. To locate the **femoral nerve** (see Figure D2.2) as it emerges from the pelvic cavity to enter the ventromedial aspect of the thigh, first identify the femoral artery and vein, with which it is closely associated. Follow it into the muscles and skin of the anterior thigh, which it supplies.

2. Turn the pig ventral side down so that you can see the posterior aspect of the lower limb. Reflect the ends of the

transected biceps femoris muscle to view the large, cordlike **sciatic nerve.** The sciatic nerve arises from the sacral plexus and serves the posterior thigh muscles and all of the muscles of the leg and foot via its various branches. Follow the nerve as it travels down the posterior thigh lateral to the semimembranosus muscle. Notice that it divides into its two major branches just superior to the gastrocnemius muscle of the calf. These branches, the **internal** and **external popliteal nerves,** innervate the leg.

3. When you have finished making your observations, prepare the pig for storage as instructed in the box on page 700, and return it to the storage area. Clean all dissecting tools and equipment used and your lab bench before you leave the laboratory. ■

DISSECTION REVIEW

1. In what region (cervical, thoracic, lumbar, or sacral) of the spinal cord would you find the following special features?

Enlargements: ______________________________

Cauda equina: ______________________________

2. As you trace a spinal nerve laterally, it divides into dorsal and ventral ______________________ (rami/roots).

3. Describe the appearance of the sympathetic trunk as seen in your dissection animal.

____

____

4. From anterior to posterior, put the nerves issuing from the brachial plexus of the pig (i.e., the median, radial, and ulnar nerves) in their proper order.

____

5. Just superior to the fetal pig's gastrocnemius muscle, the sciatic nerve divides into two main branches, the ______________

______________________ and the ______________________ nerves.

DISSECTION EXERCISE 3

Identification of Selected Endocrine Organs of the Fetal Pig

MATERIALS

- ☐ Disposable gloves
- ☐ Dissecting instruments and tray
- ☐ Animal specimen from previous dissections
- ☐ Bone cutters
- ☐ Embalming fluid
- ☐ Paper towels

OBJECTIVES

1. To prepare the fetal pig for observation by opening the ventral body cavity.
2. To identify and name the major endocrine organs on a dissected fetal pig.

ACTIVITY 1

Opening the Ventral Body Cavity

1. Don gloves. Obtain your dissection animal, and place it on the dissecting tray, ventral side up. Using scissors, make two longitudinal incisions through the ventral body wall, beginning just superior and lateral to the midline of the pubic bone. Cut to and around each side of the umbilical cord until the cuts meet medially. Continue anteriorly to the rib cage with a single midline incision. Check the incision guide in the Figure D3.1 inset as you work.

2. Using bone cutters, angle slightly (1.3 cm, or ½ inch) to the right or left of the sternum, and continue the cut through the rib cartilages, just lateral to the body midline, to the base of the throat.

3. Again using scissors, make two lateral cuts on both sides of the ventral body surface anterior and posterior to the diaphragm, which separates the thoracic and abdominal parts of the ventral body cavity. Leave the diaphragm intact to maintain the separation.

4. Spread the skin flaps laterally to expose the thoracic and abdominal organs.

5. If the body cavity contains a dark fluid, flush it out with tap water before continuing. ■

ACTIVITY 2

Identifying Organs

A helpful adjunct to identifying selected endocrine organs of the fetal pig is a general overview of ventral body cavity organs, as shown in Figure D3.1. You will study the organ systems housed in the ventral body cavity in later units, so the objective here is simply to identify the most important organs and those that will help you to locate the desired endocrine organs (marked *). A schematic and photographs showing the relative positioning of several of the animal's endocrine organs is provided in Figure D3.2. See Exercises 27 and 28A of this manual for a discussion of the human endocrine system.

Neck and Thoracic Cavity Organs

Trachea: The windpipe; runs down the midline of the throat and then divides just anterior to the lungs to form the bronchi, which plunge into the lungs on either side.

***Thymus:** Glandular structure overlying the trachea and superior to and partly covering the heart (and thyroid gland). The thymus is intimately involved (via its

PAL

For access to anatomical models and more, check out Practice Anatomy Lab.

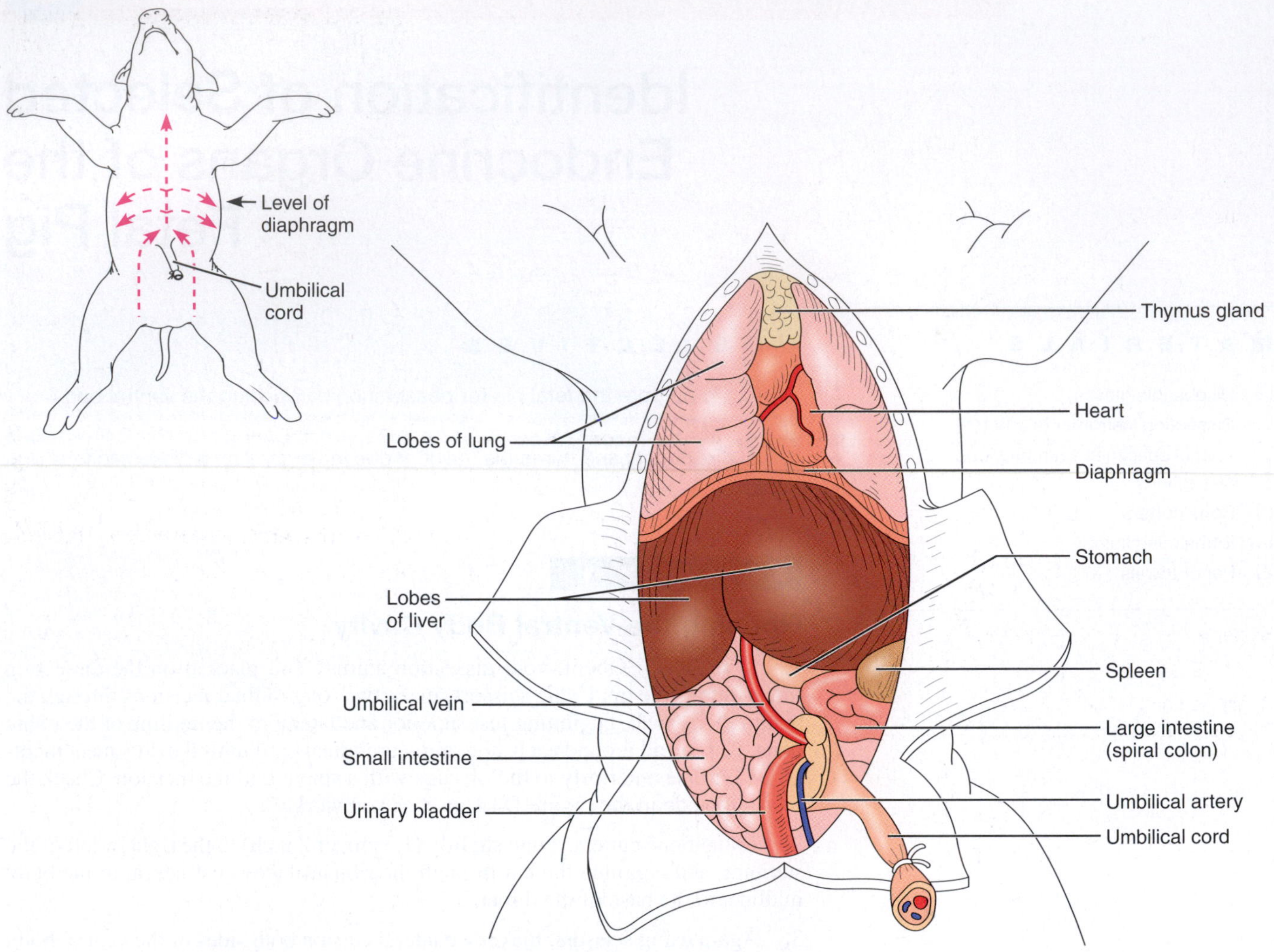
Level of
diaphragm
Umbilical
cord
Thymus gland
Heart
Lobes of lung
Diaphragm
Stomach
Lobes
of liver
Spleen
Umbilical vein
Large intestine
(spiral colon)
Small intestine
Urinary bladder
Umbilical artery
Umbilical cord

FIGURE D3.1 Ventral body cavity organs of the fetal pig, greater omentum removed. Inset indicates incision lines for opening the ventral body cavity.

hormones) in programming the immune system. Because your dissection animal is a fetus, the thymus will be quite large (see Figure D3.2).

***Thyroid:** Probably overlain by part of the thymus gland, it overlies the trachea at the level of the third or fourth tracheal ring. This endocrine organ's hormones are the main hormones regulating the body's metabolic rate (see Figure D3.2).

Heart: In the mediastinum, enclosed by the pericardium.

Lungs: Paired organs flanking the heart.

Abdominal Cavity Organs

Liver: Large multilobed organ lying under the umbrella of the diaphragm.

- Lift the large drapelike, fat-infiltrated greater omentum covering the abdominal organs to expose these organs:

Stomach: Dorsally located sac to the left side of the liver.

Spleen: Flattened brown organ curving around the lateral aspect of the stomach.

Small intestine: Tubelike organ continuing posteriorly from the stomach.

***Pancreas:** Diffuse gland lying deep to and between the small intestine and stomach. Lift the stomach with your forceps; you should see the pancreas situated in the delicate mesentery behind the stomach (see Figures D3.2 and D6.2, pages 719 and 738). This gland is extremely important in regulating blood sugar levels.

Large intestine: Forms a tight coil in the abdomen and ultimately terminates in the rectum.

- Push the intestines to one side with a probe to reveal the deeper organs in the abdominal cavity.

Kidneys: Bean-shaped organs located toward the dorsal body wall surface and behind the peritoneum.

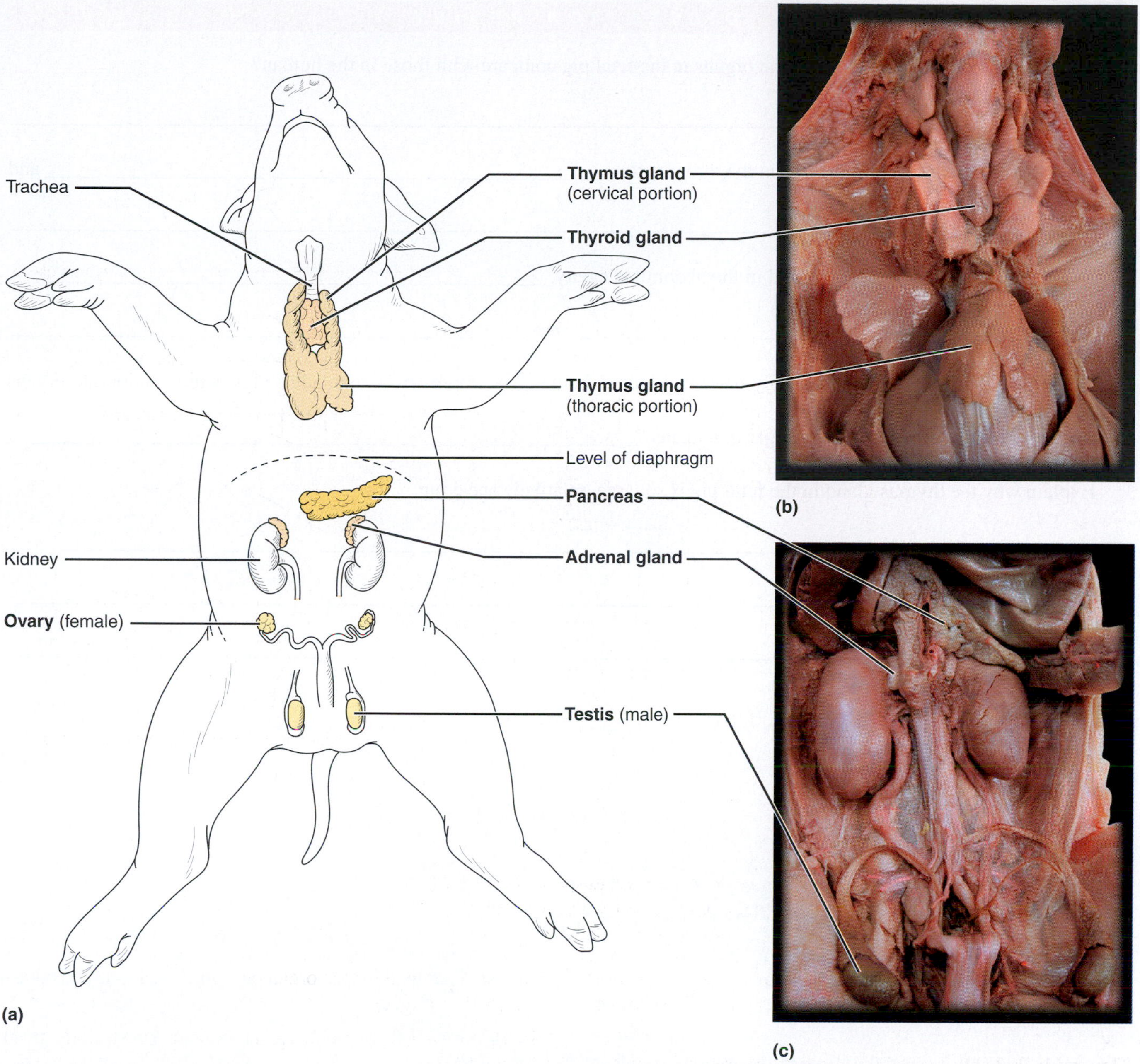
Trachea
Thymus gland (cervical portion)
Thyroid gland
Thymus gland (thoracic portion)
Level of diaphragm
Pancreas
Kidney
Adrenal gland
Ovary (female)
Testis (male)
(a)
(b)
(c)

FIGURE D3.2 Endocrine organs in the fetal pig. (a) Drawing. **(b)** and **(c)** Photographs.

***Adrenal glands:** Seen above and medial to each kidney, embedded in fatty tissue. These small glands produce corticosteroids important in the stress reponse, and in preventing abnormalities of water and electrolyte balance in the body.

***Gonads (ovaries or testes):** Sex organs producing sex hormones. The location of the gonads is illustrated in Figure D3.2, but their identification is deferred until the reproductive system organs are considered (Dissection Exercise 8).

ACTIVITY 3

Preparing the Animal for Storage

Before leaving the lab, prepare your animal for storage as instructed in the box on page 700 and return it to the storage container.

Clean and dry all dissecting equipment and your lab bench. Properly dispose of your gloves before leaving the laboratory.

DISSECTION REVIEW

1. How do the locations of the endocrine organs in the fetal pig compare with those in the human?

____

2. Name two endocrine organs located in the throat region. ______________________________ and ______________________________

3. Name three endocrine organs located in the abdominal cavity. ______________________________, ______________________________, and ______________________________

4. Given the assumption (not necessarily true) that human beings have more stress than adult pigs, which endocrine organs would you expect to be relatively larger in humans? ______________________________

5. Explain why the thymus gland in the fetal pig is so large, relatively speaking.

____

____

____

DISSECTION EXERCISE 4

Dissection of Blood Vessels and Main Lymphatic Ducts of the Fetal Pig

MATERIALS

- ☐ Disposable gloves
- ☐ Dissecting instruments and tray
- ☐ Animal specimen from previous dissections
- ☐ Bone cutters
- ☐ Twine
- ☐ Embalming fluid
- ☐ Paper towels

PAL

For access to anatomical models and more, check out Practice Anatomy Lab.

OBJECTIVES

1. To identify some of the most important blood vessels of the fetal pig.
2. To identify the thoracic and right lymphatic ducts of the fetal pig.
3. To point out anatomical differences between the vascular system of the human and the fetal pig.

To identify the blood vessels of the circulatory system and main lymphatic ducts of the pig, it is necessary to open the ventral body cavity. If Dissection Exercise 3 was conducted, you have already opened your animal's ventral body cavity and identified many of its organs. In such a case, begin this exercise with Activity 2 (Preparing to Identify the Blood Vessels). See Exercise 32 of this manual for a discussion of the human anatomy of the blood vessels, and Exercise 35A for a discussion of the lymphatic system.

ACTIVITY 1

Opening the Ventral Body Cavity

1. Don gloves. Place the pig dorsal side down on the dissecting tray and secure its forelimbs and hindlimbs to the dissecting tray with twine. Using scissors, make two longitudinal incisions through the ventral body wall beginning just superior and lateral to the midline of the pubic bone. The outline drawing provided in Figure D4.1a indicates where to make the incisions. Cut to and then around each side of the umbilical cord until the cuts meet medially. Continue anteriorly to the rib cage with a single midline incision.

2. Angle bone cutters slightly (1.3 cm, or ½ inch) to the right or left of the sternum, and continue the cut through the rib cartilages just lateral to the body midline, to the base of the neck.

3. Make two lateral cuts on both sides of the ventral body surface, anterior and posterior to the dome-shaped muscular diaphragm. Leave the diaphragm intact. Spread the thoracic walls laterally to expose the thoracic organs.

4. If the body cavity contains a dark fluid, flush it out with tap water before continuing.

5. Using Figure D4.1b as a guide, and a blunt probe as your isolating instrument, identify the organs of the thoracic and abdominal cavities.

Neck and Thoracic Cavity Organs

Thyroid: An oval, reddish organ (gland) at the base of the throat.

Heart: In the mediastinum enclosed by the pericardium.

Thymus: A large, brownish elongated mass of tissue extending over the heart and superior thoracic organs.

Lungs: Flank the heart.

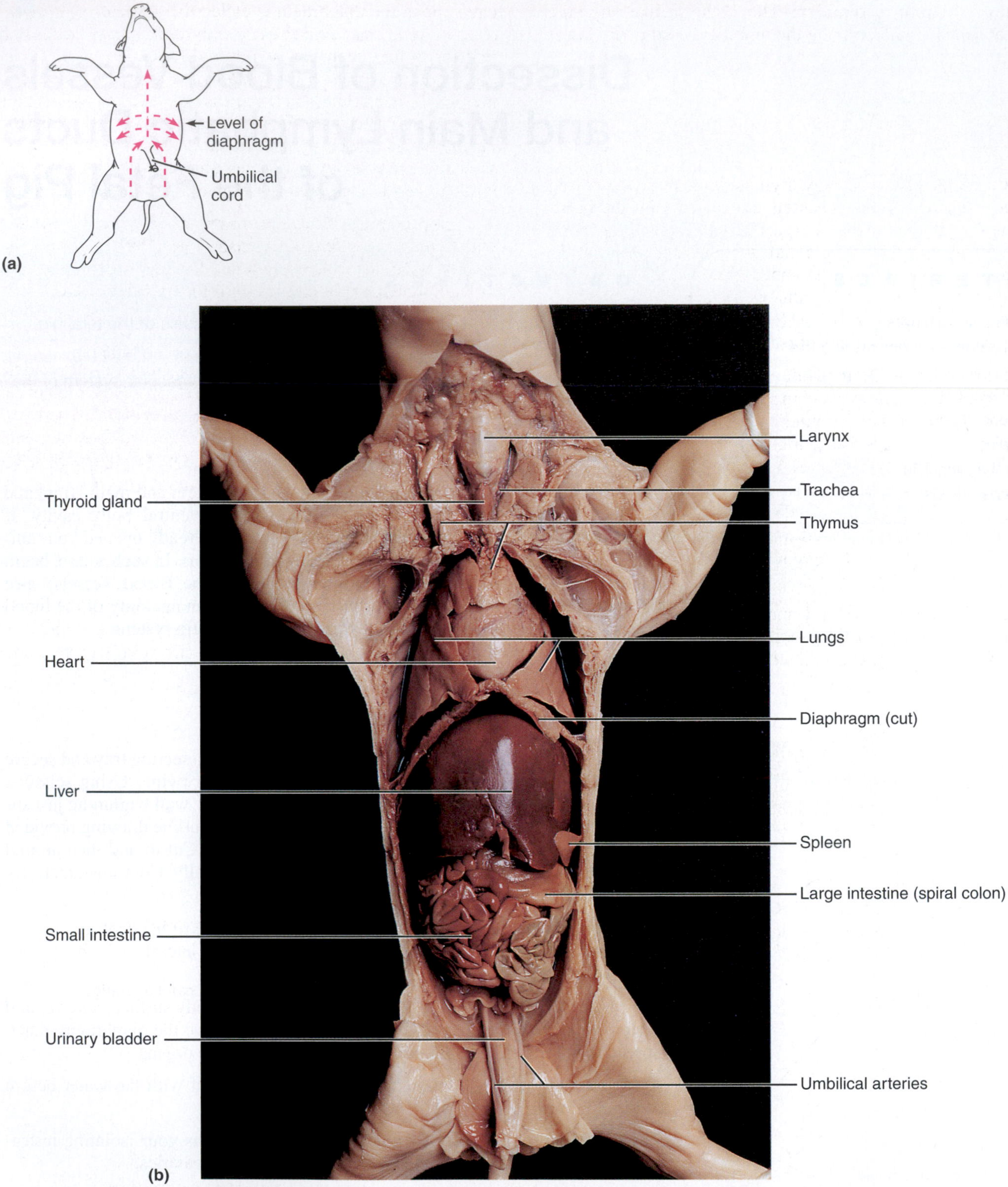
Level of
diaphragm
Umbilical
cord
(a)
Larynx
Trachea
Thyroid gland
Thymus
Lungs
Heart
Diaphragm (cut)
Liver
Spleen
Large intestine (spiral colon)
Small intestine
Urinary bladder
Umbilical arteries
(b)

FIGURE D4.1 Ventral body cavity organs of the fetal pig. (a) Incision guide for opening the ventral body cavity. **(b)** Photograph of opened ventral cavity.

Abdominal Cavity Organs

Liver: A large, brown, multilobed organ posterior to the diaphragm that dominates the abdominal cavity.

Stomach: To the left of and nearly covered by the liver.

Spleen: A brown organ curving around the lateral aspect of the stomach.

Small intestine: Continues posteriorly from the stomach.

Large intestine: Also called the **spiral colon,** a large compact mass of dark coils lying within the coils of the small intestine.

Urinary (allantoic) bladder: The large saclike structure seen attached to and entering the umbilical cord in the lower abdominal wall.

ACTIVITY 2

Preparing to Identify the Blood Vessels

1. Carefully clear away any thymus tissue or fat obscuring the heart and large blood vessels associated with the heart. Before identifying the blood vessels, try to locate the *phrenic nerve* (from the cervical plexus), which innervates the diaphragm. The phrenic nerves can be seen ventral to the root of the lung on each side, passing to the diaphragm. Also attempt to locate the *vagus nerve* (cranial nerve X) passing laterally along the trachea and dorsal to the lung.

2. Slit the parietal pericardium and reflect it superiorly; then cut it away from its heart attachments. Review the structures of the heart. Notice its pointed inferior end (apex) and its broader superior base. Identify the two atria, which appear darker in color than the inferior ventricles. Identify the *coronary arteries* in the sulcus on the ventral surface of the heart; these should be injected with red latex. (As an aid to blood vessel identification, laboratory specimens prepared for dissection have the arteries injected with red latex and the veins injected with blue latex. Exceptions to this will be noted as you encounter them.)

3. Identify the two large venae cavae entering the right atrium. The venae cavae, depending on the source, have been given various names. For example, that portion of the vena cava above the diaphragm is called the *superior vena cava, anterior vena cava, cranial vena cava,* and *precava,* while that inferior to the diaphragm is referred to as the *inferior vena cava, posterior vena cava, caudal vena cava,* or *postcava* in the pig. To keep things as simple as possible, we will use the terms used for the homologous vessels in humans—**superior vena cava** and **inferior vena cava.** The caval veins drain the same relative body areas as in humans. The superior vena cava is the largest dark-colored vessel entering the base of the heart.

4. Also identify the **pulmonary trunk** (usually injected with blue latex), extending anteriorly from the right ventricle, and its two divisions, the **right** and **left pulmonary arteries.** Trace the pulmonary arteries until they enter the lungs. Follow the **pulmonary veins** as they leave the lungs until they enter the left atrium.

5. Identify the **ascending aorta** arising from the left ventricle and running dorsal to the superior vena cava and to the left of the body midline. Also identify the **ductus arteriosus,** a short vessel connecting the aorta to the pulmonary trunk at the point where the pulmonary trunk divides to form the pulmonary arteries. This fetal structure shunts blood from the pulmonary artery to the aorta, bypassing the nonfunctional fetal lungs. After birth it is occluded and becomes the fibrous ligamentum arteriosum.

6. While you are in the chest, take time to identify the major lymphatic ducts that return leaked plasma to the blood vascular system. Because lymphatic vessels are extremely thin-walled, it is difficult to locate them in a dissection unless the animal has been triply injected (with yellow or green for

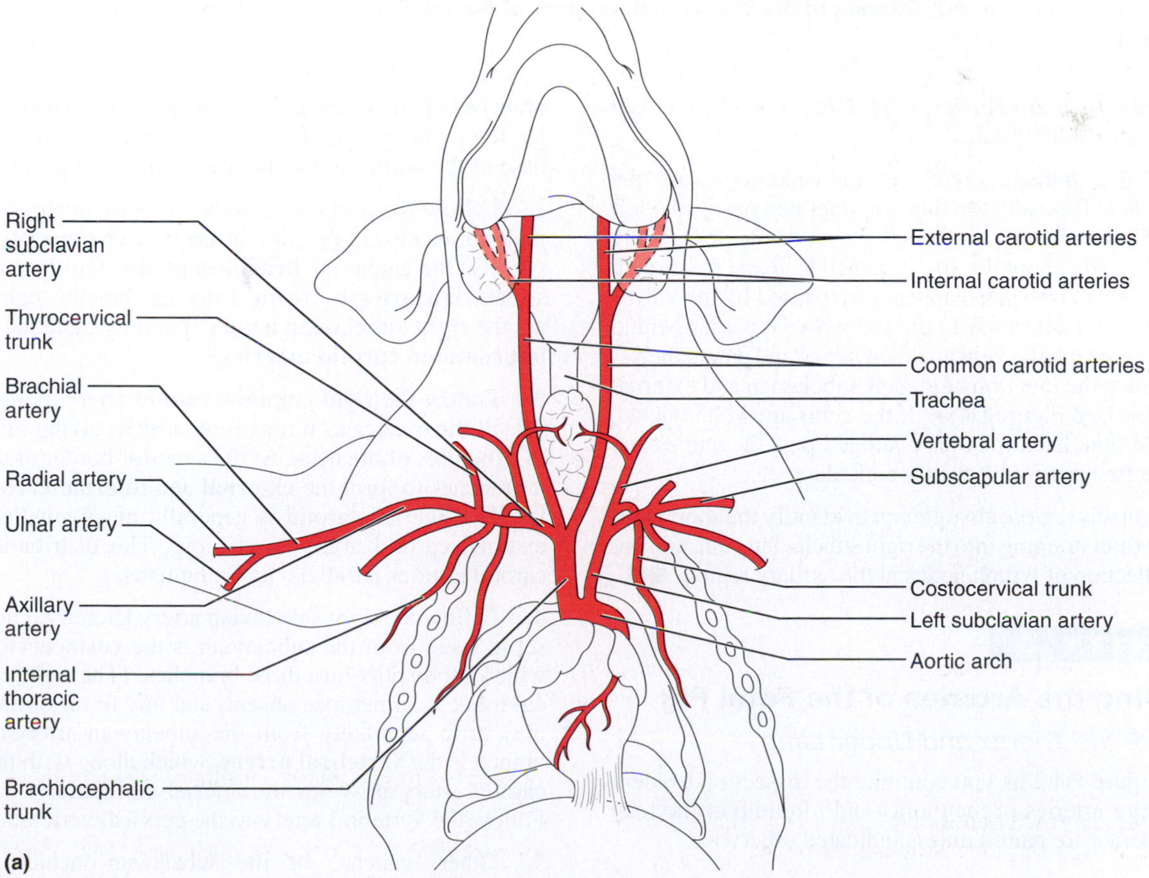

Right subclavian artery
Thyrocervical trunk
Brachial artery
Radial artery
Ulnar artery
Axillary artery
Internal thoracic artery
Brachiocephalic trunk
(a)
External carotid arteries
Internal carotid arteries
Common carotid arteries
Trachea
Vertebral artery
Subscapular artery
Costocervical trunk
Left subclavian artery
Aortic arch

FIGURE D4.2 Arteries of the thorax and forelimb of the fetal pig. (a) Diagrammatic view.

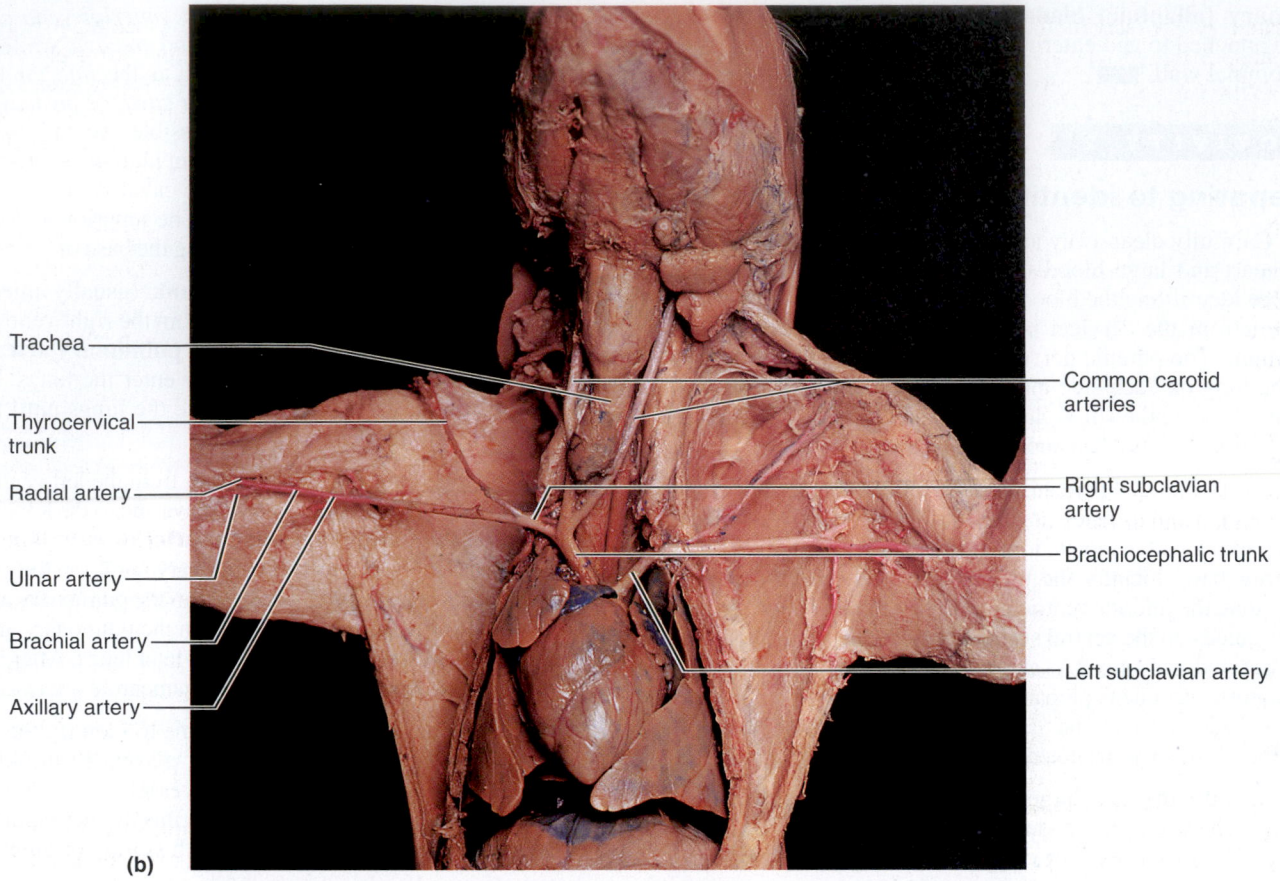

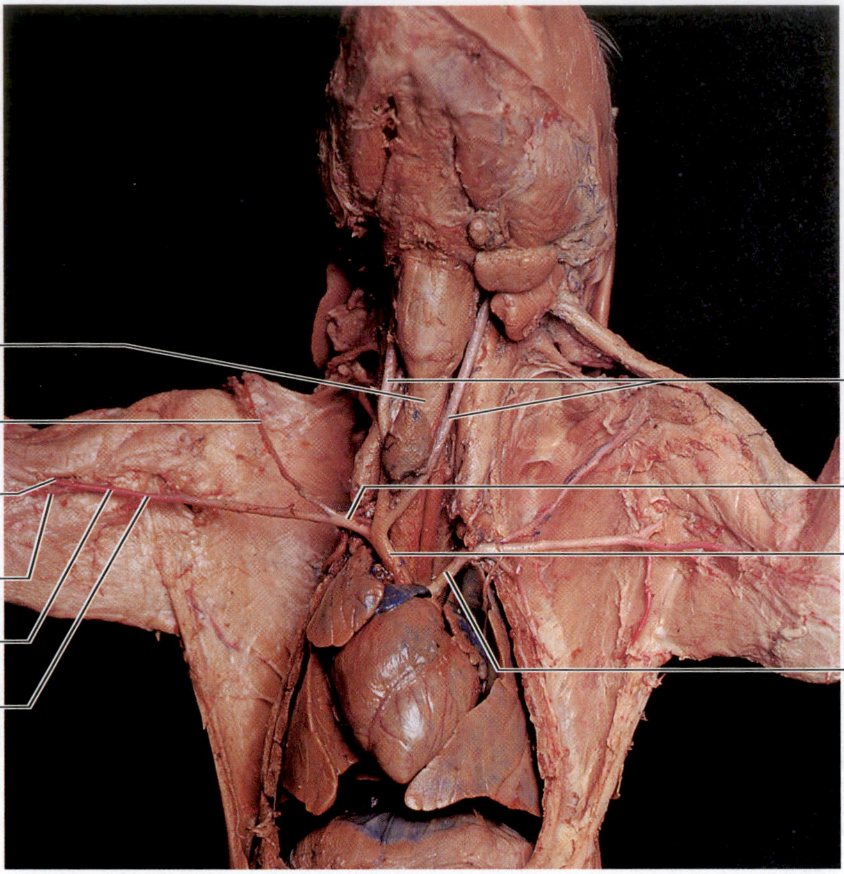
Trachea
Thyrocervical trunk
Radial artery
Ulnar artery
Brachial artery
Axillary artery
Common carotid arteries
Right subclavian artery
Brachiocephalic trunk
Left subclavian artery
(b)

FIGURE D4.2 *(continued)* Arteries of the thorax and forelimb of the fetal pig. **(b)** Photograph.

the lymphatic system). However, the large thoracic duct can be localized and identified.

7. Move the thoracic organs to the side to locate the **thoracic duct.** Typically the thoracic duct lies just to the left of the midline and abuts the dorsal aspect of the descending aorta. It is usually about the size of pencil lead and red-brown with a segmented or beaded appearance caused by the valves within it. Trace it anteriorly to the site where it passes behind the left brachiocephalic vein and then bends and enters the venous system at the junction of the left subclavian and external jugular veins (see Figure D4.4). If the veins are well injected, some of the blue latex may have slipped past the valves and entered the first portion of the thoracic duct.

8. While in this region, also attempt to identify the short **right lymphatic duct** draining into the right subclavian vein, and notice the collection of lymph nodes in the axillary region.

ACTIVITY 3

Identifying the Arteries of the Fetal Pig

Arteries of the Thorax and Upper Limb

Refer to Figure D4.2 as you continue the dissection to identify the major arteries of the thorax and forelimb of the fetal pig. All arteries are paired unless indicated otherwise.

1. Reidentify the aorta as it emerges from the left ventricle. As noted in the dissection of the sheep heart, the first branches of the aorta are the **coronary arteries,** which supply the myocardium. The coronary arteries emerge from the base of the aorta and can be seen on the surface of the heart.

2. Follow the aorta as it arches **(aortic arch).** Identify its major branches. In the pig, the aortic arch gives off two large vessels, the unpaired **brachiocephalic trunk** and the **left subclavian artery.** Arising from the brachiocephalic trunk are the **right subclavian artery** (laterally), and the **right** and **left common carotid arteries.**

3. Follow the right common carotid artery along the right side of the trachea as it moves anteriorly, giving off branches to structures of the neck. At the superior border of the larynx, it branches to form the **external** and **internal carotid arteries.** The internal carotid is generally much smaller than the external carotid artery in the pig. The distribution of the carotid arteries parallels that in humans.

4. Follow the right subclavian artery laterally. The first vessel to issue from the subclavian is the **costocervical trunk,** which soon splits into three branches. (The right costocervical trunk is sometimes absent, and any or all of its branches may arise separately from the subclavian artery.) The first branch is the **vertebral artery,** which along with the internal carotid artery provides the arterial circulation of the brain. Follow the vertebral artery to the cervical vertebrae.

5. Other branches of the subclavian include the large **thyrocervical trunk** (to the deep muscles of the neck and shoulder and the thyroid and parotid glands) and the **internal**

thoracic artery (serving the ventral thoracic wall). These vessels arise from the subclavian approximately opposite one another, distal to the costocervical trunk. As the subclavian artery passes in front of the first rib, it splits into the **axillary artery,** which provides several branches to the trunk and shoulder muscles, and the large **subscapular artery,** which serves the subscapular muscle region.

6. Follow the course of the axillary artery into the arm, where it becomes the **brachial artery** and travels with the median nerve down the length of the humerus. At the elbow, the brachial artery branches to produce the two major arteries serving the forearm and hand, the **radial** and **ulnar arteries.**

7. Return to the thorax, lift the left lung, and follow the course of the **descending aorta** through the thoracic cavity. The esophagus overlies it along its course. Notice the eight or nine pairs of intercostal arteries that branch laterally from the aorta in the thoracic region and the several arteries supplying the esophagus.

Arteries of the Abdomen and Lower Limb

1. Follow the aorta through the diaphragm into the abdominal cavity. Carefully pull the peritoneum away from its ventral surface to allow identification of the following vessels (Figure D4.3):

Celiac trunk (artery): The first abdominal branch of the aorta diverging from the aorta immediately as it enters the abdominal cavity. This unpaired artery is highly variable in the pig. Its branches supply the stomach, liver, gallbladder, pancreas, and spleen. Trace as many of its branches to these organs as possible.

Superior (or **anterior) mesenteric artery:** Emerges from the ventral surface of the abdominal aorta approximately 1.3 cm (½ inch) posterior to the celiac trunk; supplies the small intestine and most of the large intestine. Spread the mesenteric attachment of the small intestine to observe the branches of this artery as they run to supply the small intestine.

Adrenolumbar arteries: Paired arteries diverging from the aorta slightly posterior to the anterior mesenteric artery; supply the muscles of the body wall and the adrenal glands.

Renal arteries: Large paired arteries arising from the aorta about 2.5 cm (1 inch) posterior to the adrenolumbar arteries and coursing to the kidneys, which they supply. Generally, the right renal artery is superiorly located. Branching of the renal arteries frequently occurs as they approach and enter the kidneys.

Gonadal arteries (testicular or **ovarian arteries):** Slender paired arteries serving the gonads.

Inferior (or **posterior) mesenteric artery:** Thin unpaired vessel arising from the ventral surface of the aorta posterior to the gonadal arteries; supplies the second half of the large intestine and rectum.

External iliac arteries: Large arteries that issue from the lateral surface of the aorta and continue through the body wall, passing under the inguinal ligament to enter the thigh. The small **circumflex iliac artery,** a branch of the external iliac, serves the muscles of the abdominal wall.

2. The descending aorta ends posteriorly by dividing into three arteries, the two lateral **umbilical arteries** and the midline **median sacral artery,** which continues posteriorly

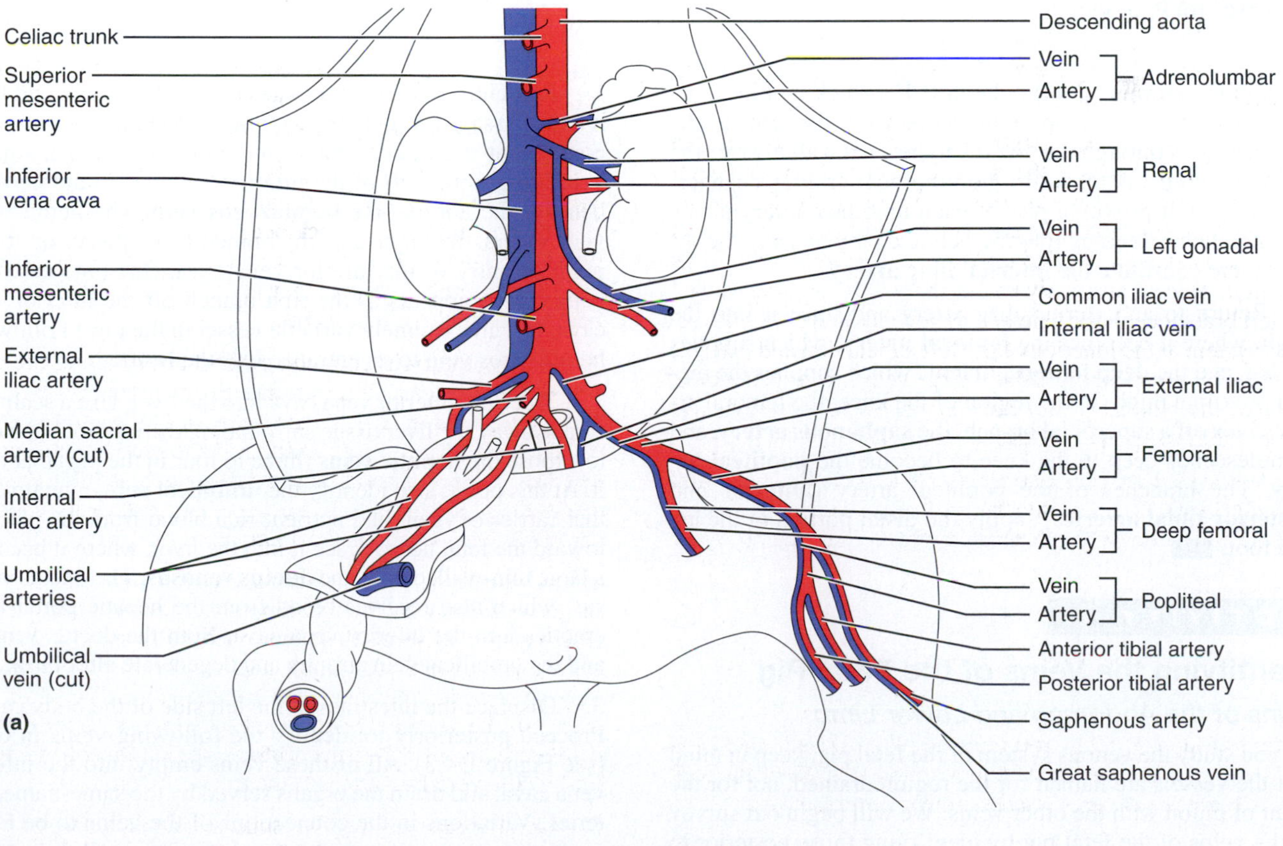
Celiac trunk
Superior mesenteric artery
Inferior vena cava
Inferior mesenteric artery
External iliac artery
Median sacral artery (cut)
Internal iliac artery
Umbilical arteries
Umbilical vein (cut)
(a)
Descending aorta
Vein
Artery
Adrenolumbar
Vein
Artery
Renal
Vein
Artery
Left gonadal
Common iliac vein
Internal iliac vein
Vein
Artery
External iliac
Vein
Artery
Femoral
Vein
Artery
Deep femoral
Vein
Artery
Popliteal
Anterior tibial artery
Posterior tibial artery
Saphenous artery
Great saphenous vein

FIGURE D4.3 Arteries and veins of the abdomen and hindlimb of the fetal pig.
(a) Diagrammatic view.

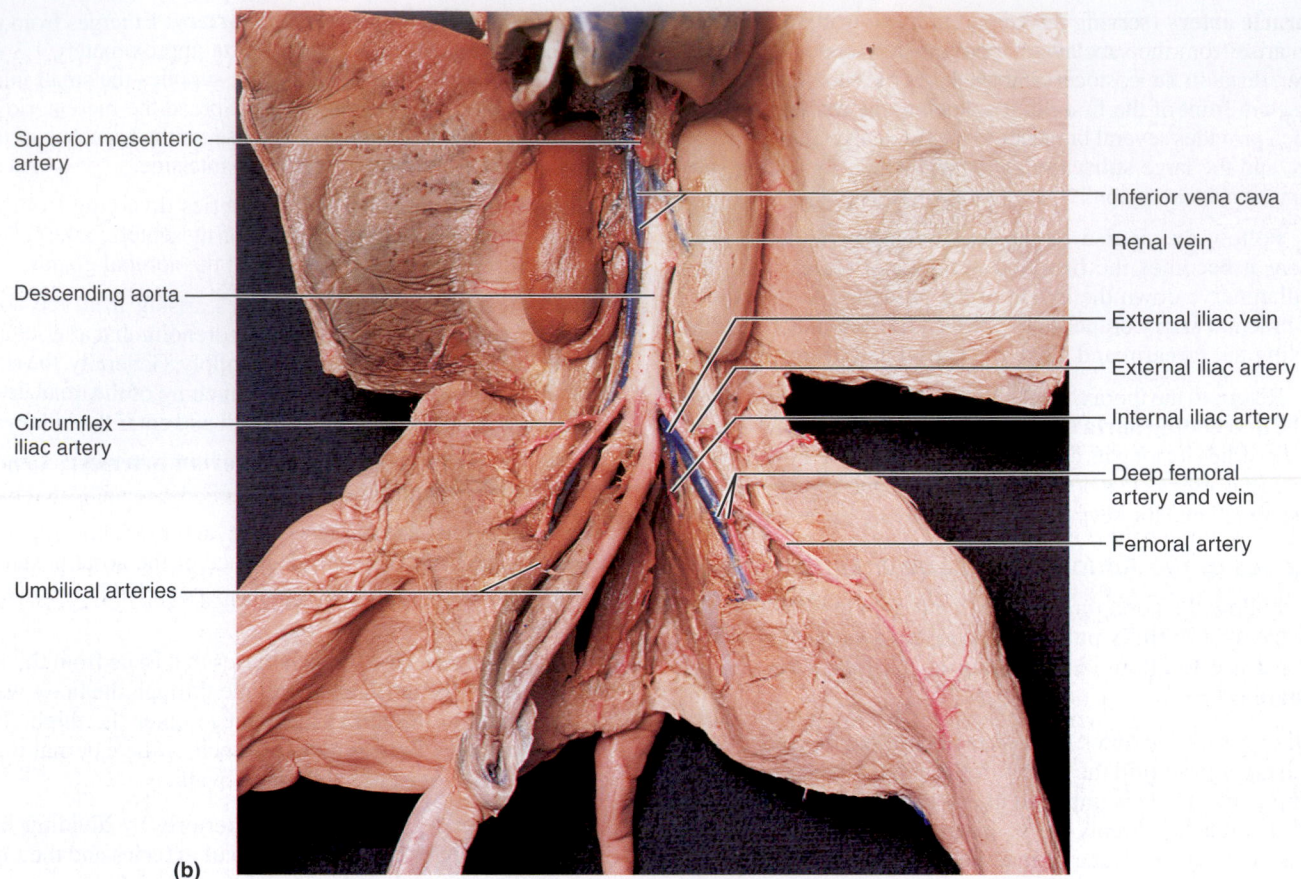

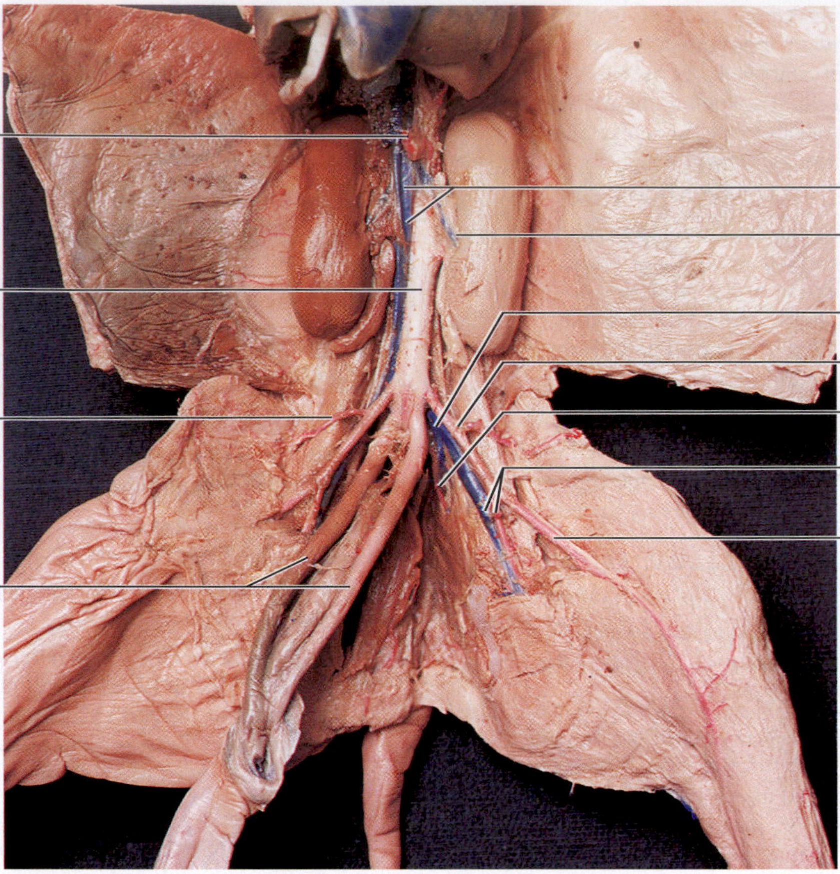
Superior mesenteric artery
Descending aorta
Circumflex iliac artery
Umbilical arteries
Inferior vena cava
Renal vein
External iliac vein
External iliac artery
Internal iliac artery
Deep femoral artery and vein
Femoral artery
(b)

FIGURE D4.3 ***(continued)*** **Arteries and veins of the abdomen and hindlimb of the fetal pig. (b)** Photograph.

through the sacrum to serve the tail. The thick umbilical arteries, which carry oxygen-poor blood from the fetus to the placenta, pass through the ventral abdominal wall to enter the umbilical cord. (After birth, the umbilical arteries degenerate.) The short proximal part of each umbilical artery plus a small branch extending into the pelvic cavity to serve the organs there constitute the **internal iliac artery.**

3. Return to an external iliac artery and trace it into the thigh, where it becomes the **femoral artery,** which supplies the leg, and the **deep femoral artery,** which supplies the medial proximal thigh. In the region of the knee, the femoral artery gives off a superficial branch, the **saphenous artery,** and then descends deep to the knee to become the **popliteal artery.** The branches of the popliteal artery (**anterior** and **posterior tibial arteries**) supply the distal portion of the leg and foot. ■

ACTIVITY 4

Identifying the Veins of the Fetal Pig

Veins of the Abdomen and Lower Limb

As you study the venous system of the fetal pig, keep in mind that the vessels are named for the region drained, not for the point of union with the other veins. We will begin our survey of the veins of the fetal pig by identifying those posterior to the heart.

1. Reidentify the inferior vena cava in the thorax and trace it to its passage through the diaphragm. Note as you follow its course that the **intercostal veins** drain into a much smaller vein lying to the left of the inferior vena cava (and partially beneath the aorta)—the **hemiazygos vein.** The hemiazygos vein enters the right atrium, immediately posterior to the point of entry of the inferior vena cava. (In some cases the hemiazygos appears as the first branch off the inferior vena cava; it is an extremely variable vessel in the pig.) Follow the hemiazygos vein to its entrance into the heart.

2. Trace the inferior vena cava into the liver. Use a scalpel to scrape away the liver tissue surrounding the inferior vena cava to expose the **hepatic veins** (three to four in the pig) that enter it. At this point, also identify the **umbilical vein,** a fetal vessel that carries oxygen- and nutrient-rich blood from the placenta toward the fetal heart. Trace it into the liver, where it becomes a large thin-walled duct, the **ductus venosus.** The ductus venosus, which also receives blood from the hepatic portal vein, empties into the inferior vena cava. Both the ductus venosus and the umbilical vein atrophy and degenerate after birth.

3. Displace the intestines to the left side of the body cavity. Proceed posteriorly to identify the following veins in order (see Figure D4.3). All of these veins empty into the inferior vena cava, and drain the organs served by the same-named arteries. Variations in the connections of the veins to be localized are common. If you observe deviations, call them to the attention of your instructor.

Adrenolumbar veins: From the adrenal glands and body wall. (Generally the right adrenolumbar vein drains into the inferior vena cava, and the left drains into the left renal vein posteriorly.)

Renal veins: From the kidneys. (It is common to find two renal veins on the right side.)

Gonadal veins (testicular or **ovarian veins):** These veins are small and generally poorly injected. The right member drains into the inferior vena cava, and the left gonadal vein empties into the left renal vein.

Common iliac veins: Form the inferior vena cava by their union.

4. The common iliac veins are formed in turn by the union of the **internal iliac** and **external iliac veins.** The more medial internal iliac veins receive branches from the pelvic organs and gluteal region. The external iliac vein receives blood from the lower extremity. As the external iliac vein enters the thigh by running beneath the inguinal ligament, it receives the **deep femoral vein,** which drains the thigh and the external genital region, and then becomes the femoral vein, which receives blood from the thigh, leg, and foot.

5. Follow the **femoral vein** down the thigh to identify the **great saphenous vein,** a superficial vein that courses along the medial aspect of the hindlimb (accompanied by the saphenous artery and nerve) to enter the femoral vein. The femoral vein is formed by the union of this vein and the **popliteal vein,** which is found deep in the thigh.

Veins of the Thorax and Upper Limb

1. Now we are ready to identify the veins anterior to the heart (Figure D4.4). Reidentify the superior vena cava as it enters the right atrium. Trace it anteriorly to identify its initial branches.

Costocervical trunks: Paired veins entering the superior vena cava just before it enters the heart; receive blood from three veins. The most important of these is the **vertebral vein,** which drains the cervical vertebral area.

Internal thoracic (mammary) veins: Enter the superior vena cava at the level of the third rib; drain the ribs and mammary glands.

Right and **left brachiocephalic veins:** Form the superior vena cava by their union.

2. Reflect the pectoral muscles, and trace the brachiocephalic vein laterally. Identify the three large veins that typically join to form it.

Internal jugular vein: Drains the deeper regions of the head; parallels and lies adjacent to the trachea, common carotid artery, and vagus nerve in the neck region.

External jugular vein: Lateral to the internal jugular vein; drains the superficial tissues of the head. It is formed by the union of the external and internal maxillary veins in the general area of the submaxillary gland. Occasionally the jugular veins unite before entering the brachiocephalic vein.

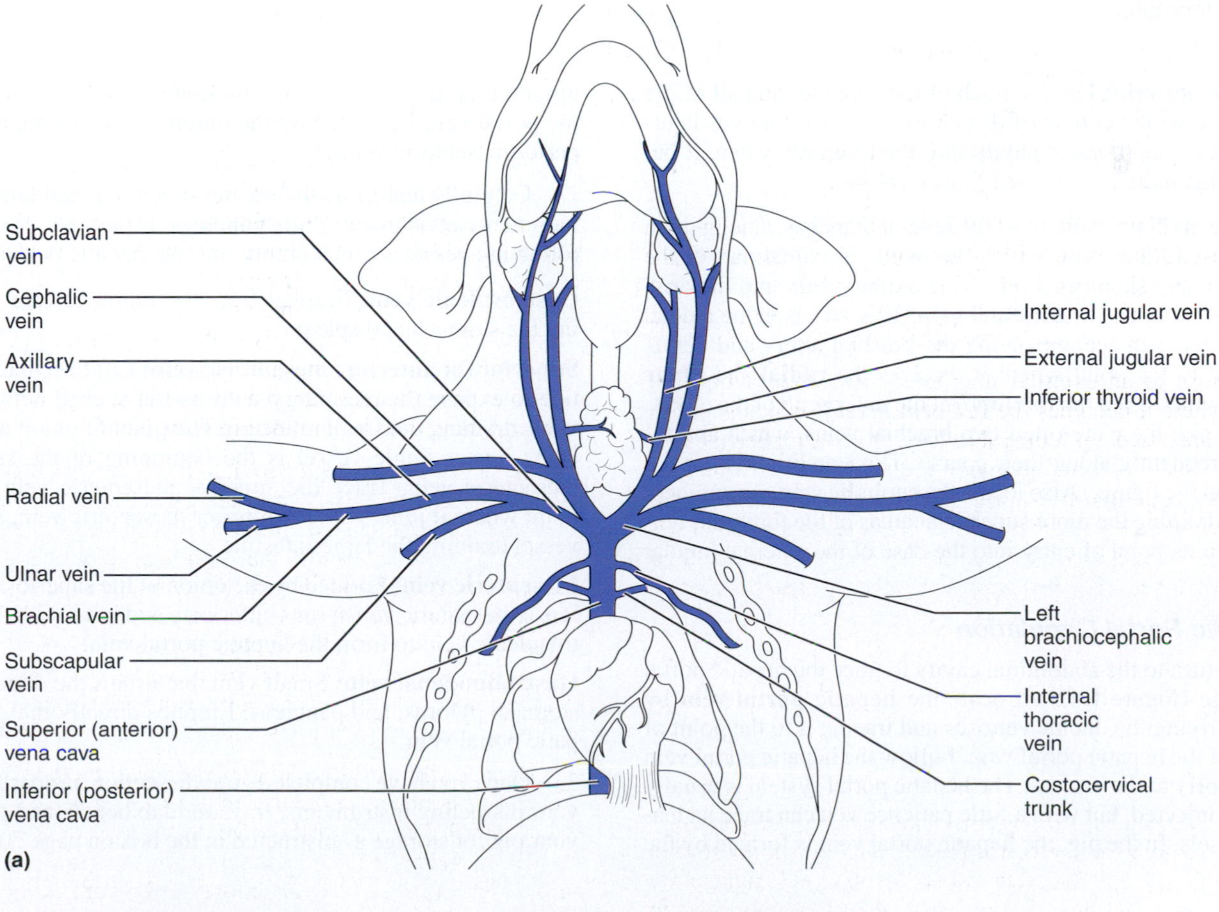

Subclavian vein
Cephalic vein
Axillary vein
Radial vein
Ulnar vein
Brachial vein
Subscapular vein
Superior (anterior) vena cava
Inferior (posterior) vena cava
(a)
Internal jugular vein
External jugular vein
Inferior thyroid vein
Left brachiocephalic vein
Internal thoracic vein
Costocervical trunk

FIGURE D4.4 Veins of the thorax and forelimb of the fetal pig. (a) Diagrammatic view.

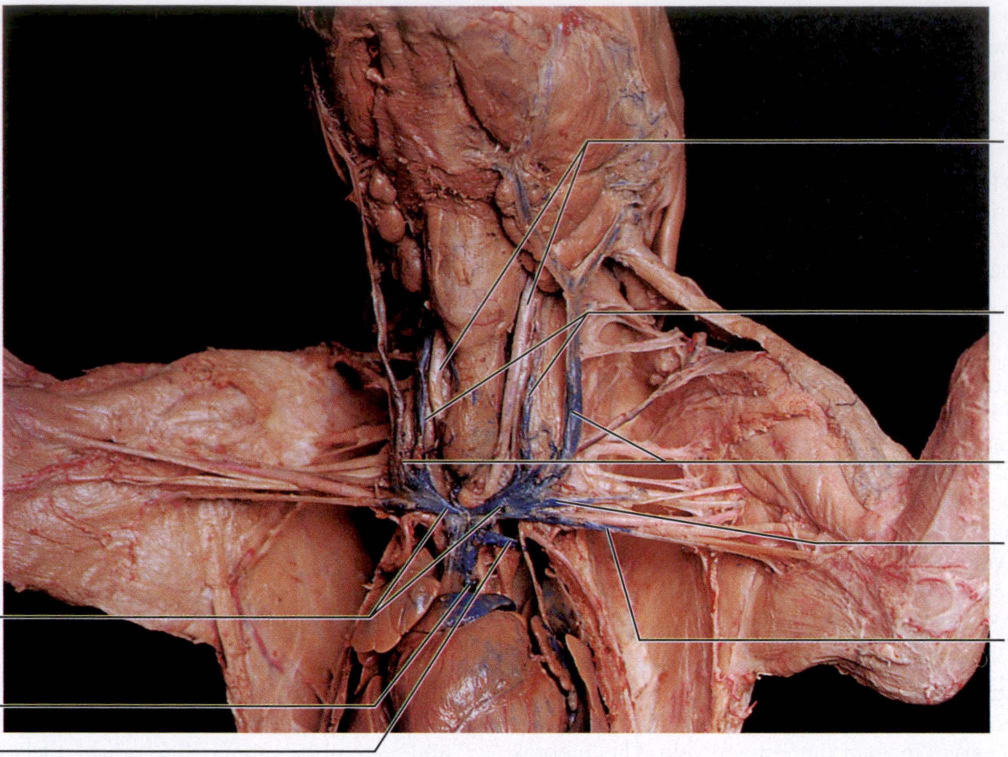
Common carotid arteries
Internal jugular veins
External jugular veins
Left subclavian vein
Left axillary vein
Brachiocephalic veins
Superior vena cava
Costocervical trunk
(b)

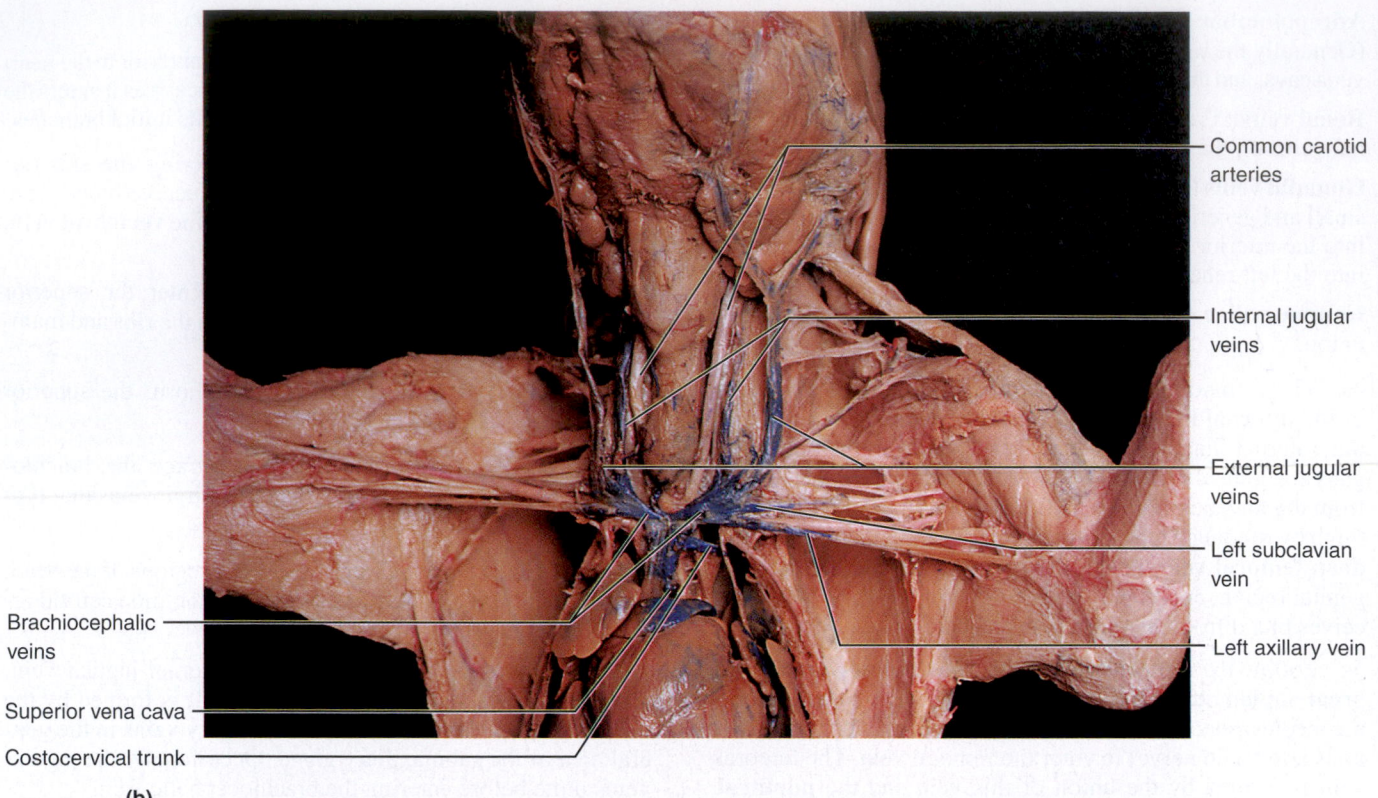

FIGURE D4.4 ***(continued)*** **Veins of the thorax and forelimb of the fetal pig.** **(b)** Photograph.

Subclavian vein: Drains much of the shoulder and all of the arm. Follow the course of this short vessel as it moves laterally out of the thoracic cavity into the forelimb, where it becomes the axillary vein (see Figure D4.4b).

3. The **axillary vein** gives off several branches, among them the **subscapular vein,** which drains the proximal part of the forelimb and shoulder. Follow the axillary vein into the arm, where it becomes the **brachial vein.** This vein is in the medial side of the arm accompanying the brachial artery and nerve. Trace it to the point where it receives the **radial** and **ulnar veins** (which drain the forelimb) at the inner bend of the elbow. (In the pig, there are often two brachial veins, which anastomose frequently along their course. This condition is not depicted in the figure.) Also locate the **cephalic vein,** a prominent vessel draining the more superficial areas of the forelimb. Follow it to its point of entry into the base of the external jugular vein.

Hepatic Portal Circulation

1. Return to the abdominal cavity to trace the hepatic portal drainage (Figure D4.5). Locate the **hepatic portal vein** by reidentifying the ductus venosus and tracing it to the point of entry of the hepatic portal vein. Follow the hepatic portal vein posteriorly to the viscera. The hepatic portal system is usually poorly injected, but with a little patience you can trace its major vessels. In the pig, the hepatic portal vein is formed by the union of the gastrosplenic and mesenteric veins. (In humans, the portal vein is formed by the union of the splenic and superior mesenteric veins.)

2. Carefully tear away the greater omentum, and loosen the mesenteric attachment of the pancreas. If possible, locate the following vessels, which empty into the hepatic portal vein.

Gastrosplenic vein: Formed from the union of veins draining the stomach and spleen.

Superior (or **anterior) mesenteric vein:** Lift the small intestine to expose the mesentery with its fan-shaped network of veins draining the small intestine. The point of union of these veins into a single vessel is the beginning of the superior mesenteric vein. Trace the superior mesenteric vein to the point where it joins with the inferior mesenteric vein, a small vessel draining the large intestine.

Mesenteric vein: Formed by the union of the superior and inferior mesenteric veins; runs anteriorly and joins with the gastrosplenic vein to form the hepatic portal vein.

Gastroduodenal vein: Small vein that drains the duodenum, stomach pylorus, and pancreas. Empties directly into the hepatic portal vein.

3. Once you have completed your dissection, properly clean your dissecting instruments, tray, and lab bench, and prepare your pig for storage as instructed in the box on page 700.

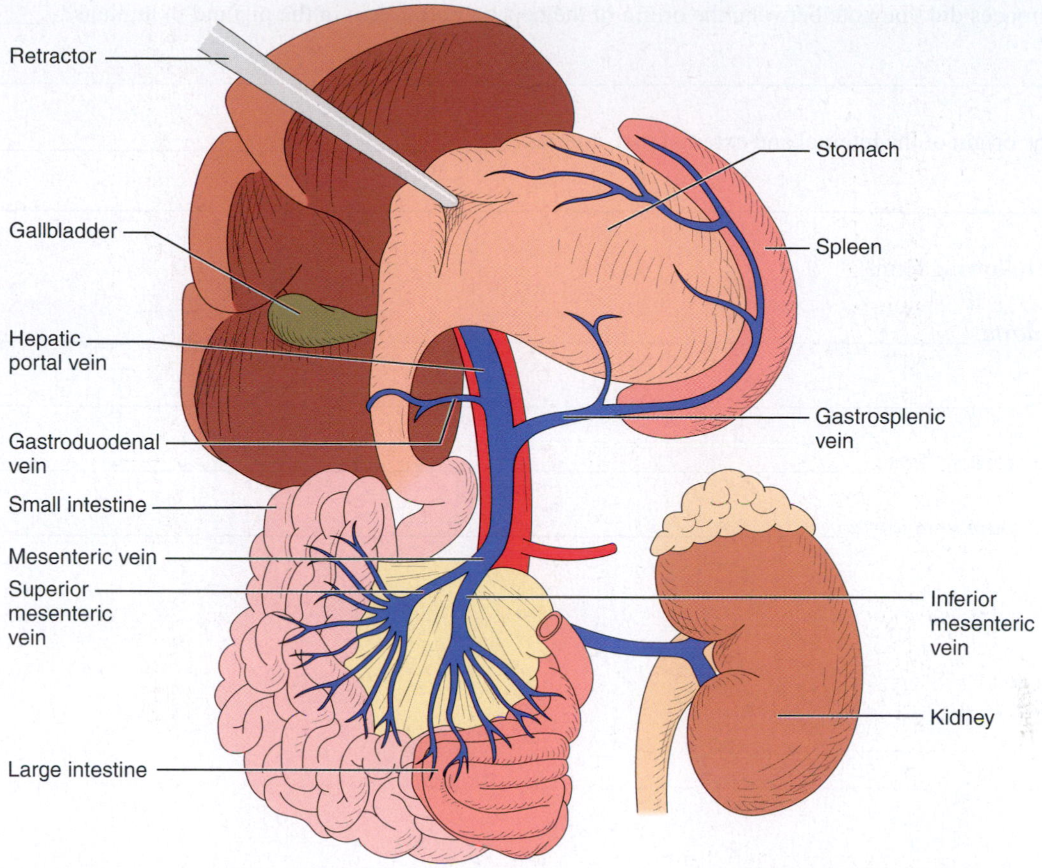
Retractor
Gallbladder
Hepatic portal vein
Gastroduodenal vein
Small intestine
Mesenteric vein
Superior mesenteric vein
Large intestine
Stomach
Spleen
Gastrosplenic vein
Inferior mesenteric vein
Kidney

FIGURE D4.5 Hepatic portal circulation of the fetal pig.

DISSECTION REVIEW

1. Is the fetal pig's lymphatic drainage pattern basically similar or dissimilar to that of humans? ______________________

2. What is the role of the following?

   a. thoracic duct ____

   b. right lymphatic duct ____

3. What differences did you observe between the origin of the common carotid arteries in the pig and in the human?

   ____

   ____

4. How do the relative sizes of the external and internal jugular veins differ in the human and the pig? ______________

   ____

5. How do the brachial veins of the pig differ from those of humans? ______________________

   ____

6. What differences did you note between the origin of the hepatic portal vein in the pig and in humans? ______________

____

Between the origin of the internal and external iliac arteries? ______________

____

7. Define the following terms.

ascending aorta: ______________

aortic arch: ______________

descending thoracic aorta: ______________

descending abdominal aorta: ______________

DISSECTION EXERCISE 6

Dissection of the Digestive System of the Fetal Pig

MATERIALS

- ☐ Disposable gloves
- ☐ Dissecting instruments and tray
- ☐ Animal specimen from previous dissections
- ☐ Twine
- ☐ Bone cutters
- ☐ Hand lens
- ☐ Embalming fluid
- ☐ Paper towels

PAL

For access to anatomical models and more, check out Practice Anatomy Lab.

OBJECTIVES

1. To identify on a dissected animal the organs composing the alimentary canal, and to name their subdivisions if any.
2. To name and identify the accessory digestive organs of the dissection animal.

Obtain your pig. Don gloves and secure it to the dissecting tray with the pig's lateral surface down for the initial portion of this dissection procedure. Obtain all necessary dissecting instruments. If you have completed the dissection of the circulatory and respiratory systems, the abdominal cavity is already exposed and many of the digestive system structures have been previously identified. However, duplication of effort generally provides a good learning experience, so you will trace and identify all the digestive system structures in this exercise. See Exercise 38 of this manual for a discussion of the human digestive system.

ACTIVITY 1

Identifying Oral Cavity Structures

1. To expose and identify the **salivary glands,** cut through the skin in a diagonal plane from an initial incision below the ear to the mouth. Continue the cut posteriorly to the anterior aspect of the forelimb. This will produce a triangular flap of skin that can be folded back to identify the salivary glands. Once the skin has been reflected back, carefully remove the underlying connective tissue and observe the tissues to differentiate between *muscle tissue,* appearing mostly as small parallel bundles of fibers in this region, and the *glandular tissue,* which is quite different in texture (bunches of nodular-looking tissue). Many lymph nodes are present in this area and should be removed if they obscure the salivary glands (Figure D6.1). Locate the large, thin, triangular-shaped **parotid gland** on the cheek, just inferior to the ear. In the fetal pig, this gland may be poorly developed and somewhat diffuse. Follow its duct over the surface of the masseter muscle to the angle of the mouth. The small, compact **submandibular gland** is anterior to the parotid gland and partially underlies it, and is easily mistaken for a lymph node. The **sublingual gland** is small and located at the base of the tongue anterior to the submandibular gland. The ducts of these two salivary glands run deep and parallel to each other and empty on the side of the frenulum of the tongue. These need not be identified in the pig.

2. To expose and identify the structures of the oral cavity, use bone cutters to cut through the mandible just anterior to the angle to free the lower jaw from the maxilla. Observe the teeth of the pig. The dental formula for the fetal pig is as follows:

$$\frac{3,1,4,0}{3,1,4,0} \times 2 = 32$$

Notice that many of the fetal teeth are incompletely developed, or may not have emerged through the gums yet. If you have a very young fetus with no teeth in evidence, apply pressure to the gums to feel if there are any "bumps" indicating unerupted teeth or cut into the gum tissue to determine whether the deciduous teeth are present. Identify the **hard** and **soft palates** (see Figure D5.1 on page 732.) and use a probe to trace the hard palate to its posterior limits. Does the pig have a uvula?

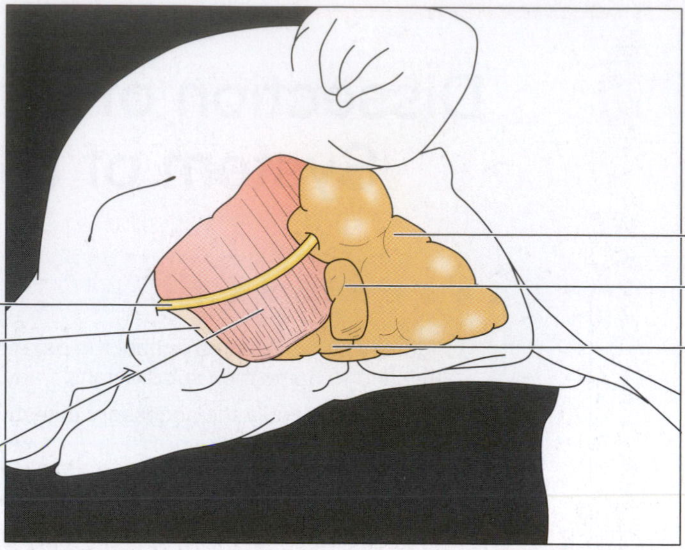
Parotid gland
Submandibular gland
Sublingual gland
Facial nerve
Duct of parotid gland
Masseter muscle
(a)

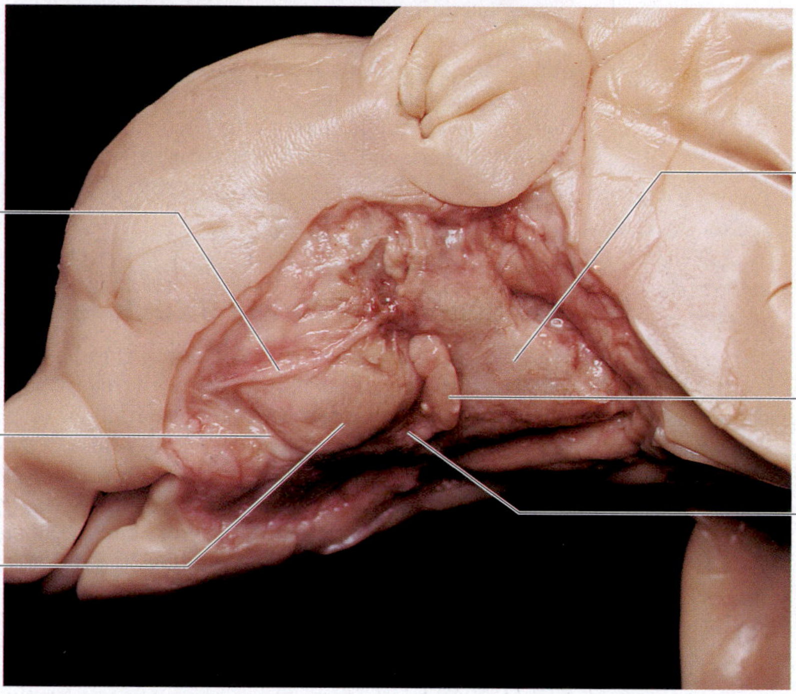
Parotid gland
Facial nerve
Submandibular gland
Duct of parotid gland
Sublingual gland
Masseter muscle
(b)

FIGURE D6.1 Salivary glands of the fetal pig. (a) Diagrammatic view. **(b)** Photograph.

3. Identify the **oropharynx** beneath the soft palate at the rear of the oral cavity, and the *nasopharynx,* the continuation of the nasal cavities, superior to it. Identify the **tongue** and rub your fingers across its surface to feel the papillae. As in humans, the tongue plays a role in the manipulation of food in the mouth, and its papillae house the taste buds.

4. Locate the **lingual frenulum** that attaches the tongue to the floor of the mouth. Trace the tongue posteriorly until you locate the **epiglottis,** which appears as a small white tissue tab and which closes the respiratory passageway when swallowing occurs. Also identify the **esophagus** posterior to the epiglottis and trace it to the diaphragm, which it penetrates.

ACTIVITY 2

Identifying Digestive Organs in the Abdominal Cavity

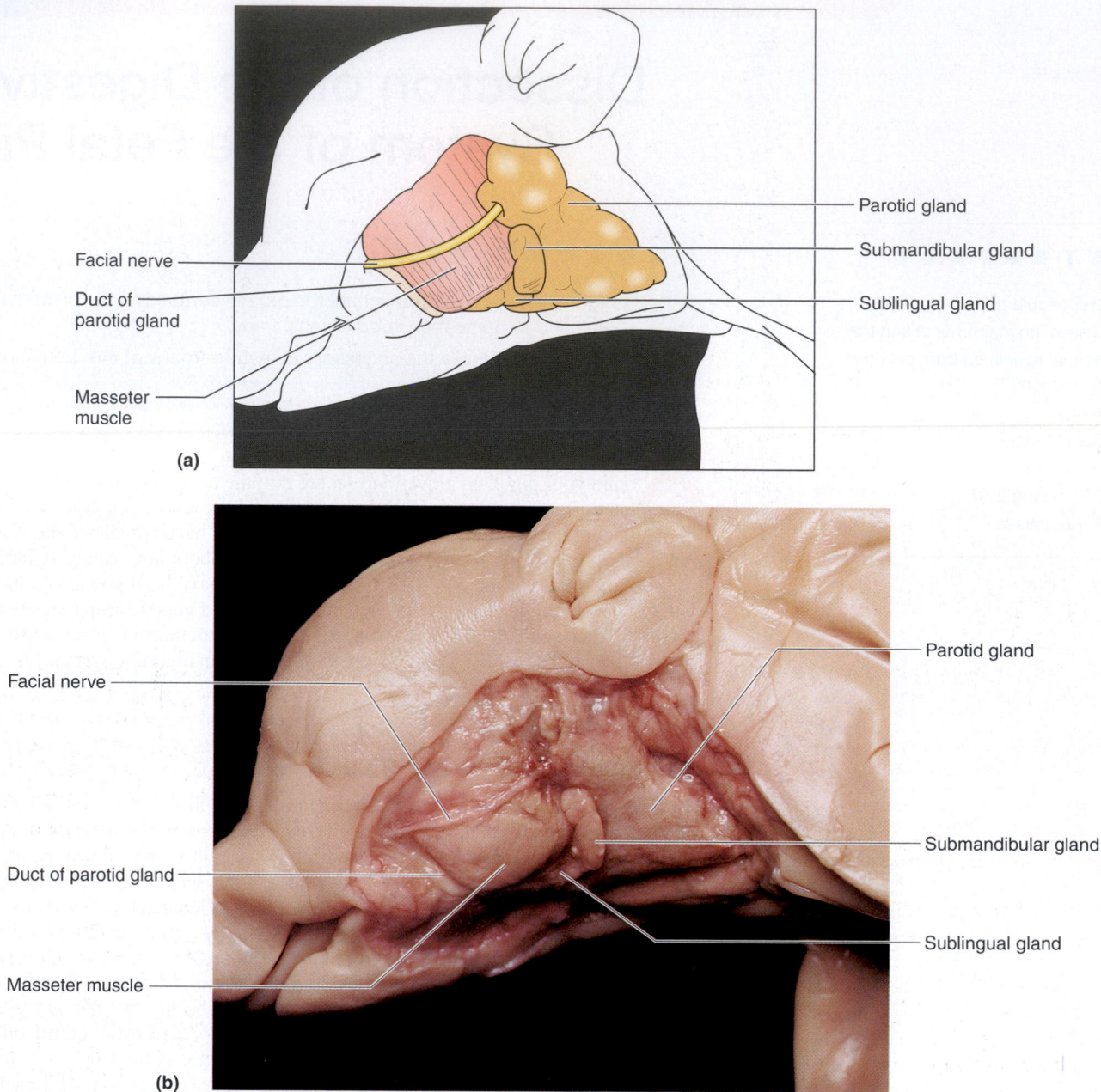

1. Using Figure D6.2 as a guide, locate the abdominal alimentary canal structures. If you have not opened the abdominal cavity previously, make a midline incision from the rib cage to the umbilical cord and then make two lateral cuts to encircle the umbilicus. Continue these cuts, staying lateral to the body midline, posteriorly until you reach the pubic bone. Make four lateral cuts, two parallel to the rib cage and two at the inferior

DISSECTION EXERCISE 7

Dissection of the Urinary System of the Fetal Pig

MATERIALS

- ☐ Disposable gloves
- ☐ Dissecting instruments and tray
- ☐ Animal specimen from previous dissections
- ☐ Twine or pins
- ☐ Embalming fluid
- ☐ Paper towels

OBJECTIVE

To identify on a dissection specimen the urinary system organs, and to describe the general function of each.

The structures of the reproductive and urinary systems are often considered together as the *urogenital system,* since they have common embryologic origins. However, the emphasis in this dissection is on identifying the structures of the urinary tract (Figures D7.1 and D7.2) with only a few references to contiguous reproductive structures. (Dissection Exercise 8 is a study of the anatomy of the reproductive system.) See Exercise 40 of this manual for a discussion of the human urinary system.

ACTIVITY

Identifying Urinary Organs of the Fetal Pig

1. Don gloves. Obtain your dissection specimen, and pin or tie its limbs to the dissection tray. Reflect the abdominal viscera (especially the small intestine) to locate the **kidneys** high on the dorsal body wall. Notice that the kidneys in the pig, as in the human, are retroperitoneal (behind the peritoneum).

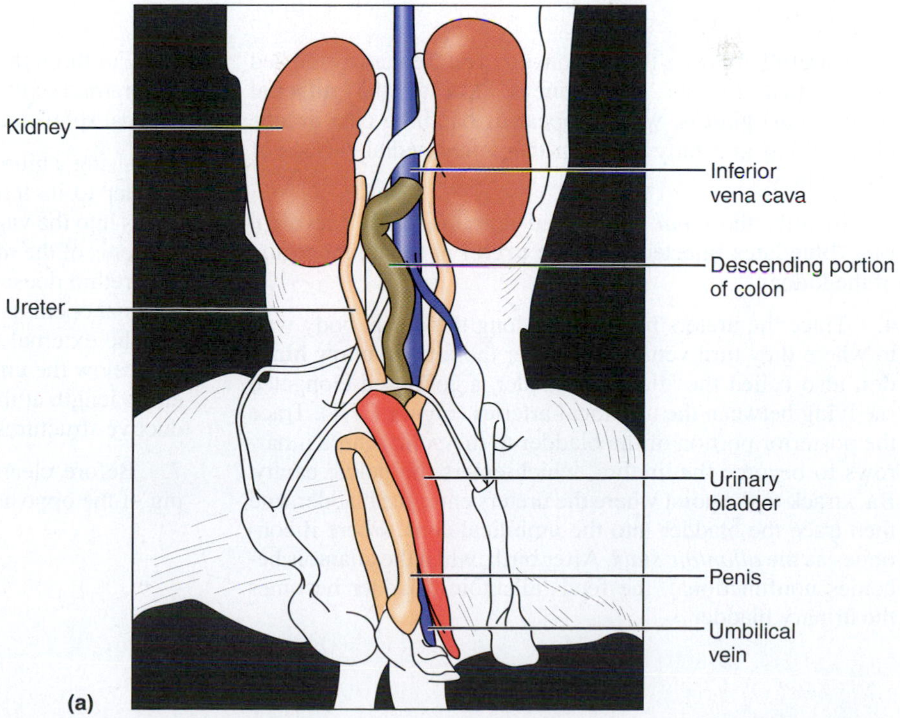

Kidney
Ureter
Inferior vena cava
Descending portion of colon
Urinary bladder
Penis
Umbilical vein
(a)

FIGURE D7.1 Urinary system of the male fetal pig. (a) Diagrammatic view.

PAL

For access to anatomical models and more, check out Practice Anatomy Lab.

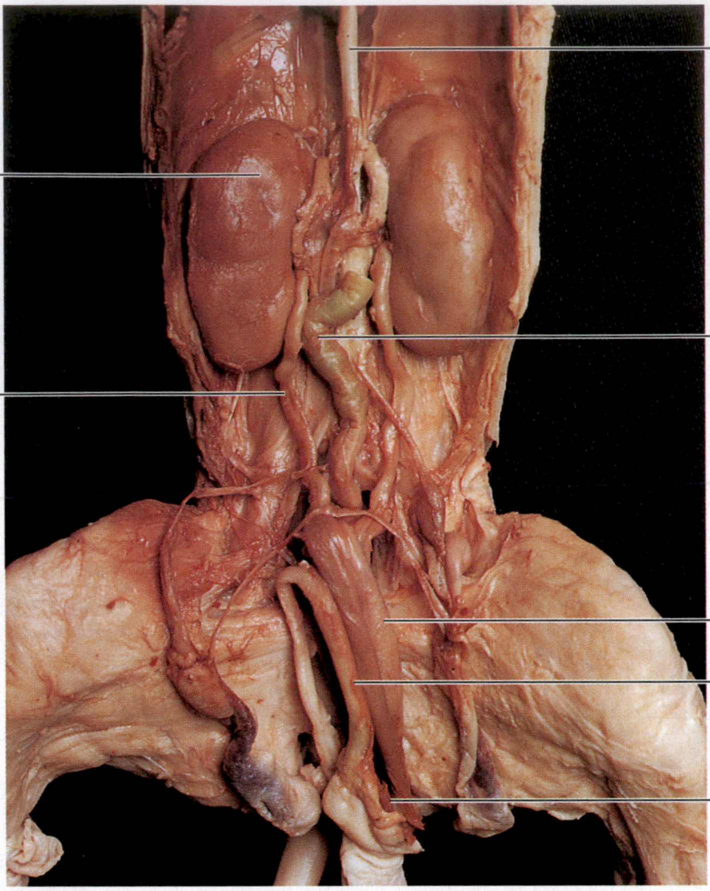

Inferior vena cava
Kidney
Descending portion of colon
Ureter
Urinary bladder
Penis
Umbilical vein
(b)

FIGURE D7.1 *(continued)* Urinary system of the male fetal pig. (b) Photograph. Some reproductive structures are shown.

2. Carefully remove the peritoneum, and clear away the bed of fat that invests the kidneys. Locate the **adrenal** *(suprarenal)* **glands,** which appear as bandlike pale orange glands lying in a fatty mass on the anteromedial surface of each kidney.

3. Identify the *renal artery* (red latex injected), the *renal vein* (blue latex injected), and the **ureter** at the hilum region of the kidney.

4. Trace the ureters posteriorly along the dorsal body wall to where they turn ventrally to enter the fetal **urinary bladder,** also called the allantoic bladder, a collapsed elongated sac lying between the umbilical arteries (Figure D7.1). Trace the posterior portion of the bladder to the point where it narrows to become the urethra, which enters the pelvic cavity. Backtrack to the point where the ureters enter the bladder, and then trace the bladder into the umbilical cord, where it continues as the *allantoic stalk.* After birth, when the allantois becomes nonfunctional, the fetal (allantoic) bladder becomes the urinary bladder.

5. Cut through the bladder wall and examine the region of the urethral exit to see if you can locate any evidence of the internal sphincter.

6. Using a blunt probe, trace the urethra as it exits from the bladder to its terminus in the **urogenital sinus,** which also opens into the vagina in the female pig (Figure D7.2) and into the penis of the male (see Figure D7.1). In the human female, the urethra does not empty into the vagina but has a separate external opening located anterior to the vaginal orifice. Identify the external urogenital opening in the male pig located just below the umbilicus. Do not expose the urethra along its entire length at this time because you may damage the reproductive structures, which will be studied later.

7. Before cleaning up the dissection materials, observe a pig of the opposite sex.

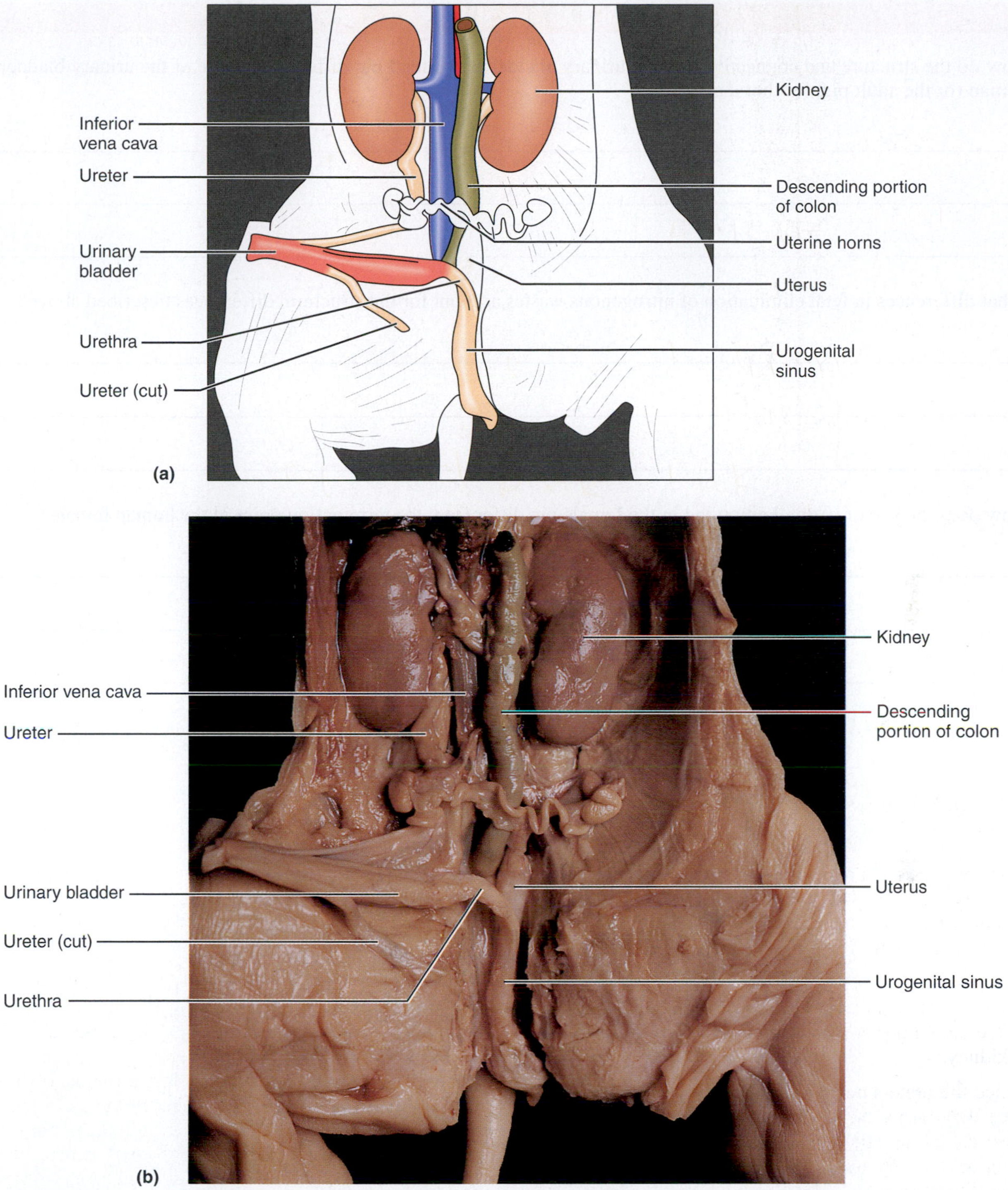
Kidney
Inferior vena cava
Ureter
Descending portion of colon
Uterine horns
Urinary bladder
Uterus
Urethra
Urogenital sinus
Ureter (cut)
(a)
Kidney
Inferior vena cava
Descending portion of colon
Ureter
Urinary bladder
Uterus
Ureter (cut)
Urogenital sinus
Urethra
(b)

FIGURE D7.2 Urinary system of the female fetal pig. (a) Diagrammatic view. **(b)** Photograph. Some reproductive structures also indicated.

DISSECTION REVIEW

1. How do the structure and connectivity of the urinary bladder of the fetal pig differ from those of the urinary bladder of the human (or the adult pig, for that matter)?

____

____

____

2. What differences in fetal elimination of nitrogenous wastes account for the structural differences described above?

____

____

____

3. How does the site of urethral emptying in the female pig differ from the termination point in the human female?

____

____

____

DISSECTION EXERCISE 8

Dissection of the Reproductive System of the Fetal Pig

MATERIALS

- ☐ Disposable gloves
- ☐ Dissecting instruments and tray
- ☐ Animal specimen from previous dissections
- ☐ Bone cutters
- ☐ Small metric ruler (for female pigs)
- ☐ Embalming fluid
- ☐ Paper towels

PAL

For access to anatomical models and more, check out Practice Anatomy Lab.

OBJECTIVES

1. To identify the major reproductive structures of the male and female dissection animals.
2. To recognize and discuss pertinent differences between the reproductive structures of the human and the dissection animal.

Obtain your pig, a dissecting tray, the necessary dissection instruments, and disposable gloves. After you have completed the study of the reproductive structures of your specimen, observe a pig of the opposite sex. (The following instructions assume that the abdominal cavity has been opened in previous dissection exercises.) See Exercise 42 of this manual for a discussion of the human reproductive system.

ACTIVITY 1

Identifying Male Reproductive System Organs

Refer to Figure D8.1 as you identify the structures.

1. Identify the **preputial orifice** on the ventral body surface just posterior to the umbilical cord; this is the external urethral orifice. Carefully cut through the skin of the orifice and identify the distal end of the **penis,** which lies within a fold of skin (the prepuce). Notice that there is a major difference between the human and the fetal pig penis that has to do with location. In human males the penis is an external structure; in pigs it lies internal to the external urethral orifice on the ventral body surface. Reflect the cut midventral strip of the body wall to identify the thin cord-like penis shaft, which lies immediately posterior to the urinary bladder. Cross section the penis to observe the relative positioning of the cavernous bodies.

2. Using bone cutters, carefully cut through the pubic symphysis. Follow the penis posteriorly to the point where the **urethra** enters it. As in the human male, the male pig urethra serves as both a urine and a sperm conduit. Also identify the paired **bulbourethral glands,** which flank the penile-urethral junction laterally.

3. Continue to follow the urethra superiorly until you can see the point where the right and left **ductus deferens** (each encapsulated in a spermatic cord) loop over the ureters to enter the dorsal aspect of the urethra. (This positioning of the spermatic cord reflects the fact that initially the testis is in the same relative position as the ovary is in the female. In the fetus, the testis descends laterally and ventrally to the ureters to enter the scrotal sac.)

4. Identify the small paired **seminal vesicles** on the dorsal aspect of the urethra adjacent to the points of ductus deferens entry. Also, attempt to find the **prostate** at the bladder-urethra junction by carefully dissecting the tissue between the seminal vesicles. In the adult pig, the prostate can be found in this position, but it is generally difficult to identify in the fetal pig.

5. Trace the **spermatic cord** (see Figure D8.1) posteriorly through the inguinal canal into the scrotum. Identify the **scrotal sac** exteriorly on the ventral body surface anterior to the anus. Make a shallow incision into the scrotum to expose the

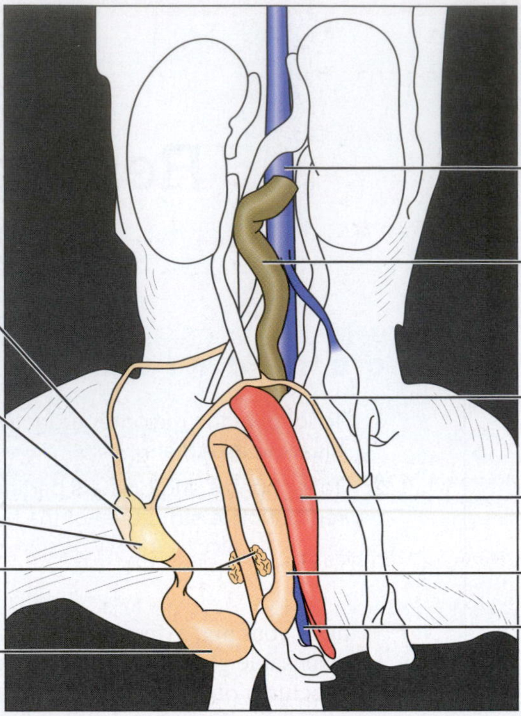

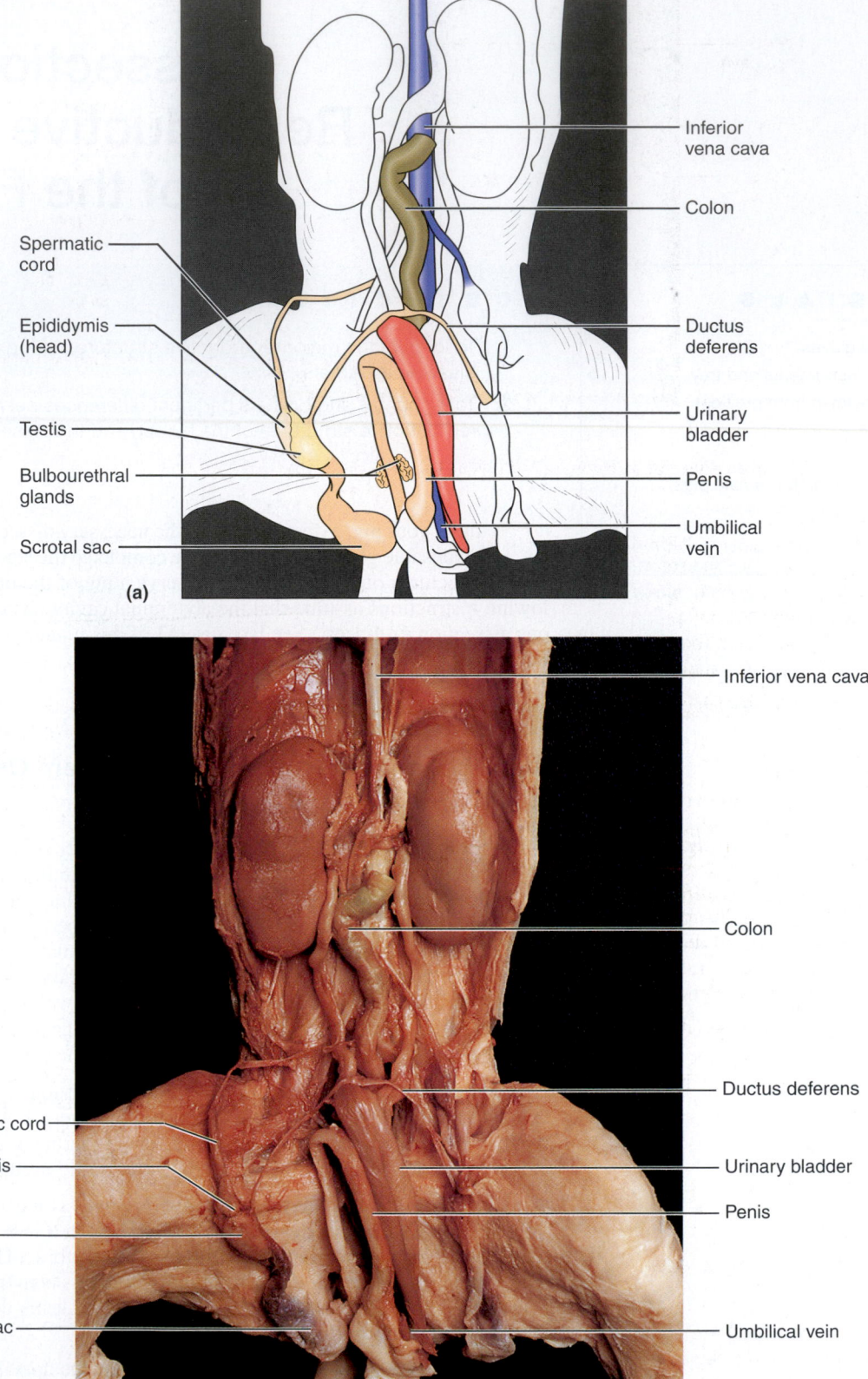

Inferior vena cava
Colon
Spermatic cord
Epididymis (head)
Ductus deferens
Urinary bladder
Testis
Bulbourethral glands
Penis
Umbilical vein
Scrotal sac
(a)

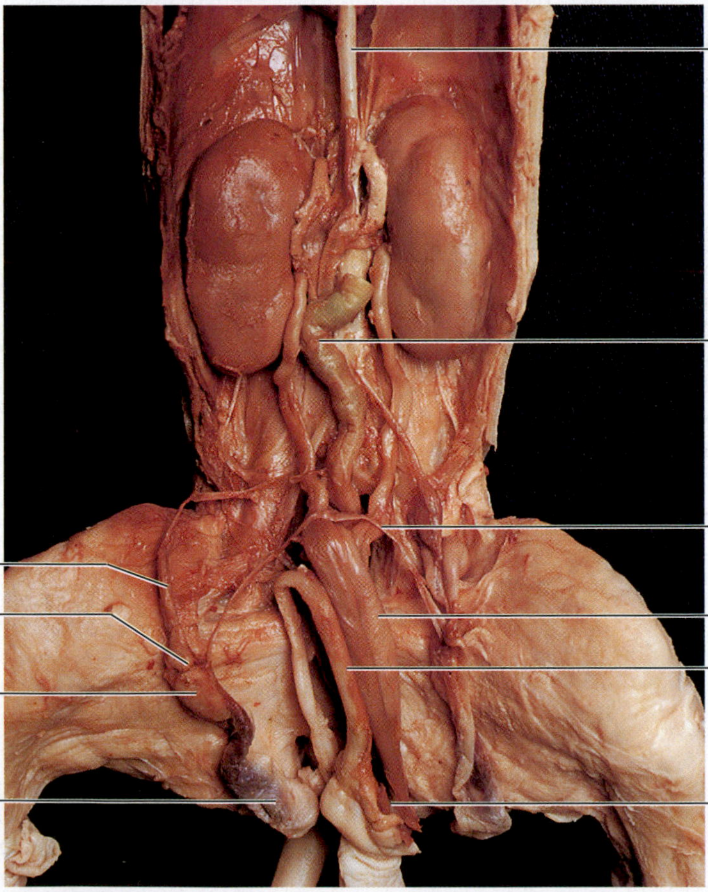

Inferior vena cava
Colon
Ductus deferens
Spermatic cord
Epididymis (head)
Urinary bladder
Penis
Testis
Scrotal sac
Umbilical vein
(b)

FIGURE D8.1 Reproductive system of the male fetal pig. (a) Diagrammatic view. **(b)** Photograph.

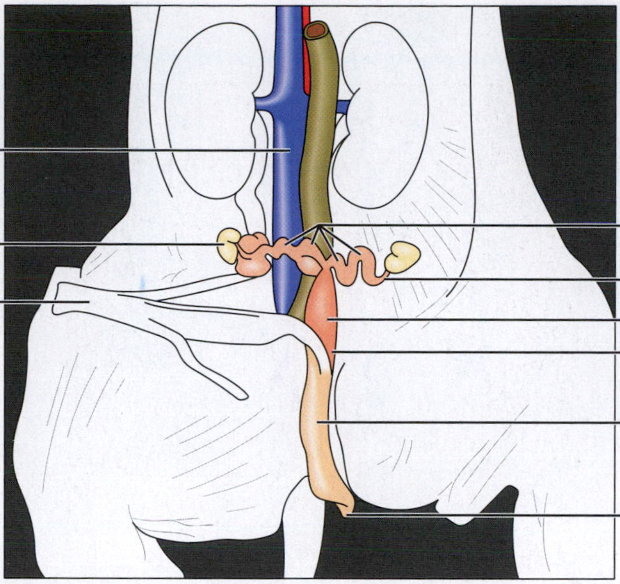
Inferior vena cava
Ovary
Urinary bladder
Uterine horns
Uterine tube
Body of uterus
Vagina
Urogenital sinus
Genital papilla
(a)

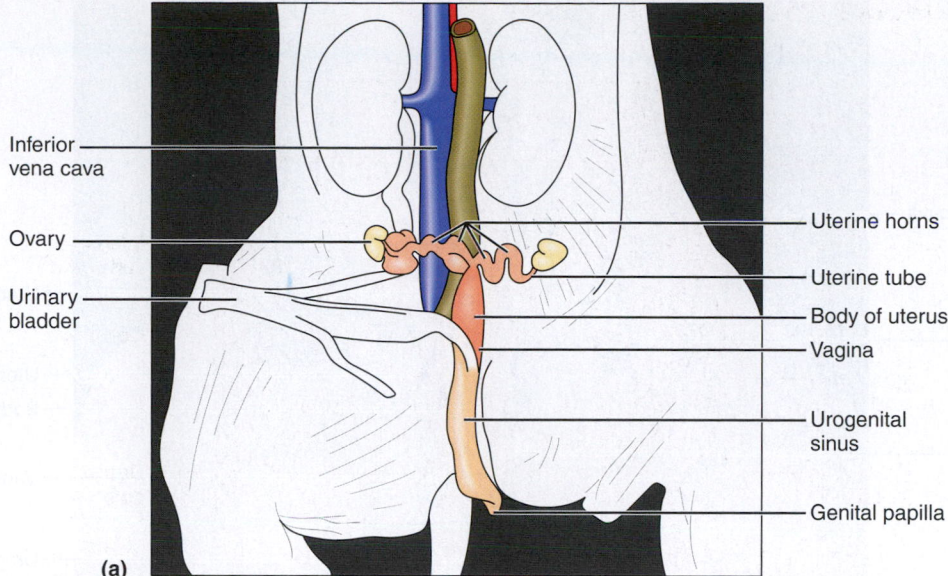

FIGURE D8.2 Reproductive system of the female fetal pig. (a) Diagrammatic view.

testis lying in a membranous sac (inguinal sac). (It may be necessary to flush the interior of the scrotum with tap water to remove coagulated blood, which tends to accumulate there.) Gently grasp the inguinal sac and pull it up and out of the scrotum. As you do so, notice that the spermatic cord moves more posteriorly into the inguinal canal.

6. Carefully cut open the inguinal sac and identify the following structures.

Testis: The male gonad.

Epididymis: A fine convoluted tube running along the posterior, lateral, and anterior surfaces of the testis; conveys sperm from the testis to the ductus deferens.

Gubernaculum: A tough cord extending from the posterior end of the epididymis to the inguinal pouch and scrotum. Its attachments are established early in development, and its failure to grow as rapidly as other body structures in the area results in its "pulling" the testis into the scrotum, thus causing its descent.

7. Make a longitudinal cut through the testis and epididymis. Can you see the tubular nature of the epididymis and rete testis portion of the testis with the naked eye?

ACTIVITY 2

Identifying Organs of the Female Reproductive System

Refer to Figure D8.2 as you identify the structures described next.

1. Unlike the pear-shaped simplex, or one-part, uterus of the human, the uterus of the pig is Y-shaped (bipartite, or bicornuate) and consists of a **uterine body** from which two **uterine horns** (cornua) diverge. Such an enlarged uterus enables the animal to produce litters. Examine the abdominal cavity and identify the urinary bladder and the body of the uterus lying just dorsal to it.

2. Follow one of the uterine horns as it travels superiorly in the body cavity, noting the *broad ligament,* a thin mesentery that helps anchor it and the other reproductive structures to the body wall. Approximately halfway up the length of the uterine horn, it should be possible to identify the more important *round ligament.* This cord of connective tissue extends laterally and posteriorly from the uterine horn to the region of the body wall that corresponds to the inguinal region of the male.

3. Just caudal to the kidney, identify the **uterine tube** (also called the **oviduct** in pigs) and **ovary** at the distal end of the uterine horn. Observe how the funnel-shaped end of the uterine tube curves around the ovary. As in the human, the distal end of the tube is fimbriated, or fringed, and the tube is lined with ciliated epithelium. Gently dislodge the ovarian end of the uterine tube and examine it to identify the *ostium,* the fimbriae-surrounded opening to the lumen of the uterine tube. The uterine tubes of the pig are tiny and relatively much shorter than in the human. Identify the *ovarian ligament,* a short thick cord that anchors the ovary to the uterus. You may also be able to identify the *suspensory ligament* that anchors the ovary to the body wall. Also observe the *ovarian artery and vein,* passing through the mesentery to the ovary and uterine structures.

4. Return to the body of the uterus and follow it caudad to the bony pelvis. Use bone cutters to cut through the median line of the pelvis (pubic symphysis), cutting carefully so that the underlying urethra is not damaged. Expose the pelvic region by pressing the thighs dorsally. Follow the uterine body caudally to the vagina and note the point where the urethra draining the bladder and the **vagina** enter a common chamber, the **urogenital sinus.** How does this anatomical arrangement compare to that seen in the human female?

5. On the outer surface of the pig, observe the **vulva,** which is similar to the human vulva. Identify the external urogenital orifice, which is hooded by the **genital papilla.**

6. Insert one blade of a scissors into the lateral aspect of the urogenital orifice and cut anteriorly along the length of the urogenital sinus and vagina. Reflect the cut edges and identify the following structures.

Clitoris: A small elevation on the ventral surface of the sinus, just anterior to the genital papilla.

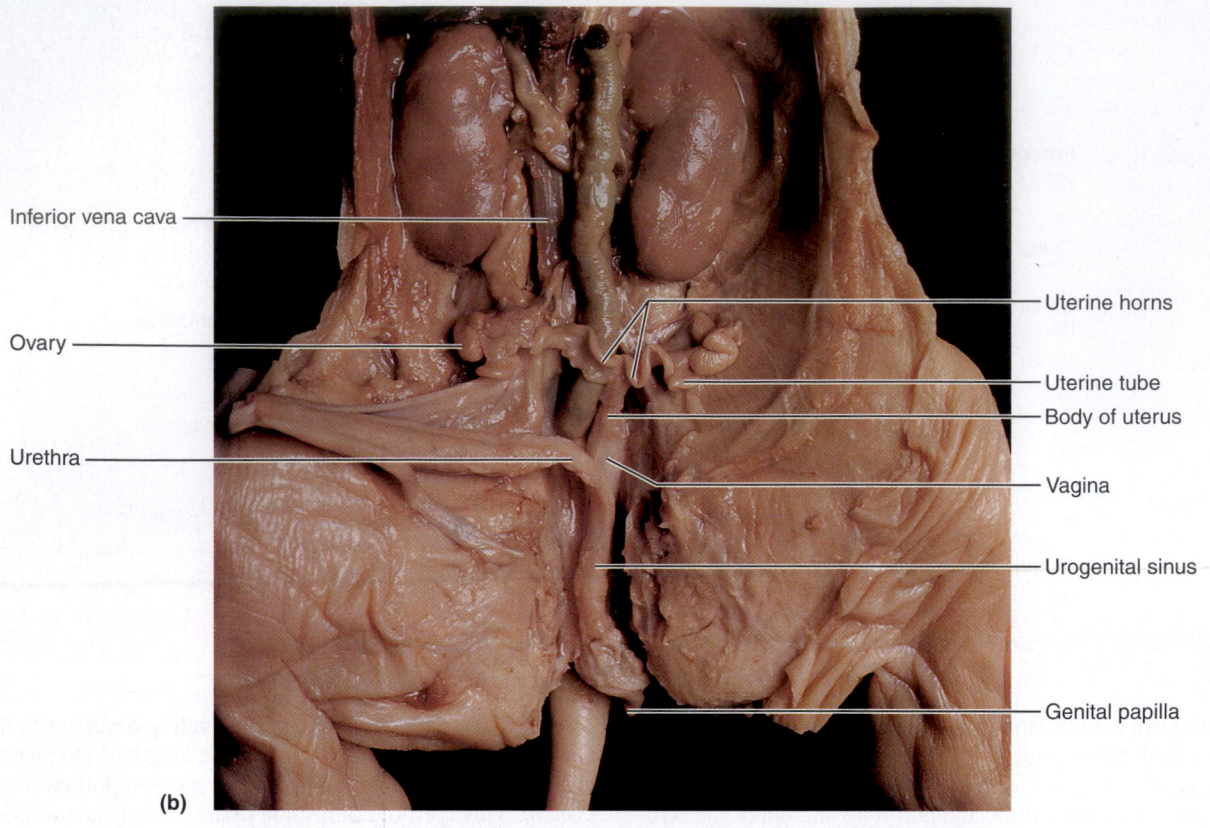

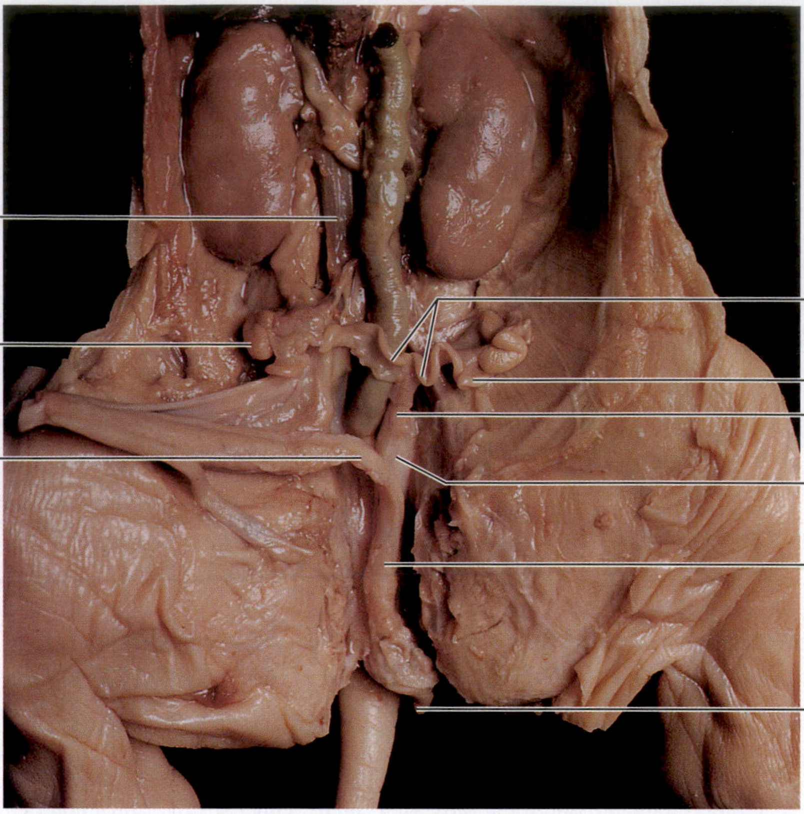
Inferior vena cava
Ovary
Urethra
Uterine horns
Uterine tube
Body of uterus
Vagina
Urogenital sinus
Genital papilla
(b)

FIGURE D8.2 ***(continued)*** **Reproductive system of the female fetal pig. (b)** Photograph.

Urethral opening into the urogenital sinus.

Cervix: The muscular ring separating the vagina from the uterus. Notice that the vagina is comparatively much shorter in the pig than in the human. Its limits are demarcated by the cervix anteriorly and the urogenital sinus posteriorly. Measure the approximate length of the pig's vagina and record below.

_______________ mm

7. When you have completed your observations of the reproductive system of male and female pigs, clean your dissecting instruments and tray and your lab bench, and properly prepare the pig for storage as instructed in the box on page 700.

DISSECTION REVIEW

1. The female pig has a _______________ uterus; that of the human female is _______________. Explain the difference in structure of these two uterine types. _______________

____

2. What reproductive advantage is conferred by the pig's uterine type?

____

____

3. Cite differences noted between the pig and the human relative to the following structures:

uterine tubes or oviducts _______________

____

urethral and vaginal openings in the female _______________

____

PhysioEx™ 9.0

PhysioEx™ 9.0 by
Peter Zao, North Idaho College
Timothy Stabler, Indiana University Northwest
Lori Smith, American River College
Andrew Lokuta, University of Wisconsin–Madison
Edwin Griff, University of Cincinnati

PhysioEx™ 9.0

EXERCISE 1

Cell Transport Mechanisms and Permeability

PRE-LAB QUIZ

1. Circle the correct term. A passive process, diffusion / osmosis is the movement of solute molecules from an area of greater concentration to an area of lesser concentration.
2. A solution surrounding a cell is *hypertonic* if:
   a. it contains fewer nonpenetrating solute particles than the interior of the cell.
   b. it contains more nonpenetrating solute particles than the interior of the cell.
   c. it contains the same amount of nonpenetrating solute particles as the interior of the cell.
3. Which of the following would require an input of energy?
   a. diffusion
   b. filtration
   c. osmosis
   d. vesicular transport
4. Circle the correct term. In pinocytosis / phagocytosis, parts of the plasma membrane and cytoplasm expand and flow around a relatively large or solid material and engulf it.
5. Circle the correct term. In active / passive processes, the cell provides energy in the form of ATP to power the transport process.

Exercise Overview

The molecular composition of the plasma membrane allows it to be selective about what passes through it. It allows nutrients and appropriate amounts of ions to enter the cell and keeps out undesirable substances. For that reason, we say the plasma membrane is **selectively permeable.** Valuable cell proteins and other substances are kept within the cell, and metabolic wastes pass to the exterior.

Transport through the plasma membrane occurs in two basic ways: either passively or actively. In **passive processes,** the transport process is driven by concentration or pressure differences *(gradients)* between the interior and exterior of the cell. In **active processes,** the cell provides energy (ATP) to power the transport.

Two key passive processes of membrane transport are **diffusion** and **filtration.** Diffusion is an important transport process for every cell in the body. **Simple diffusion** occurs without the assistance of membrane proteins, and **facilitated diffusion** requires a membrane-bound carrier protein that assists in the transport.

In both simple and facilitated diffusion, the substance being transported moves *with* (or *along* or *down*) the *concentration gradient* of the solute (from a region of its higher concentration to a region of its lower concentration). The process does not require energy from the cell. Instead, energy in the form of **kinetic energy** comes from the constant motion of the molecules. The movement of solutes continues until the solutes are evenly dispersed throughout the solution. At this point, the solution has reached **equilibrium.**

A special type of diffusion across a membrane is **osmosis.** In osmosis, water moves with its concentration gradient, from a higher concentration of water to a lower concentration of water. It moves in response to a higher concentration of solutes on the other side of a membrane.

In the body, the other key passive process, **filtration,** usually occurs only across capillary walls. Filtration depends upon a *pressure gradient* as its driving force. It is not a selective process. It is dependent upon the size of the pores in the filter.

The two key active processes (recall that active processes require energy) are **active transport** and **vesicular transport.** Like facilitated diffusion, active transport uses a membrane-bound carrier protein. Active transport differs from facilitated diffusion because the solutes move *against* their concentration gradient and because ATP is used to power the transport. Vesicular transport includes phagocytosis, endocytosis, pinocytosis, and exocytosis. These processes are not covered in this exercise. The activities in this exercise will explore the cell transport mechanisms individually.

ACTIVITY 1

Simulating Dialysis (Simple Diffusion)

OBJECTIVES

1. To understand that diffusion is a passive process dependent upon a solute concentration gradient.
2. To understand the relationship between molecular weight and molecular size.
3. To understand how solute concentration affects the rate of diffusion.
4. To understand how molecular weight affects the rate of diffusion.

Introduction

Recall that all molecules possess *kinetic energy* and are in constant motion. As molecules move about randomly at high speeds, they collide and bounce off one another, changing direction with each collision. For a given temperature, all matter has about the same average kinetic energy. Smaller molecules tend to move faster than larger molecules because kinetic energy is directly related to both mass and velocity ($KE = \frac{1}{2}mv^2$).

When a **concentration gradient** (difference in concentration) exists, the net effect of this random molecular movement is that the molecules eventually become evenly distributed throughout the environment—in other words, diffusion occurs. **Diffusion** is the movement of molecules from a region of their higher concentration to a region of their lower concentration. The driving force behind diffusion is the kinetic energy of the molecules themselves.

The diffusion of particles into and out of cells is modified by the plasma membrane, which is a physical barrier. In general, molecules diffuse passively through the plasma membrane if they are small enough to pass through its pores (and are aided by an electrical and/or concentration gradient) or if they can dissolve in the lipid portion of the membrane (as in the case of CO_2 and O_2). A membrane is called *selectively permeable, differentially permeable,* or *semipermeable* if it allows some solute particles (molecules) to pass but not others.

The diffusion of *solute particles* dissolved in water through a selectively permeable membrane is called **simple diffusion.** The diffusion of *water* through a differentially permeable membrane is called **osmosis.** Both simple diffusion and osmosis involve movement of a substance from an area of its higher concentration to an area of its lower concentration, that is, *with* (or *along* or *down*) its concentration gradient.

This activity provides information on the passage of water and solutes through selectively permeable membranes. You can apply what you learn to the study of transport mechanisms in living, membrane-bounded cells. The dialysis membranes used each have a different *molecular weight cutoff (MWCO),* indicated by the number below it. You can think of MWCO in terms of pore size: the larger the MWCO number, the larger the pores in the membrane. The molecular weight of a solute is the number of grams per mole, where a mole is the constant Avogadro's number 6.02×10^{23} molecules/mole. The larger the molecular weight, the larger the mass of the molecule. The term molecular mass is sometimes used instead of molecular weight.

EQUIPMENT USED The following equipment will be depicted on-screen: left and right beakers—used for diffusion of solutes; dialysis membranes with various molecular weight cutoffs (MWCOs).

Experiment Instructions

Go to the home page in the PhysioEx software and click **Exercise 1: Cell Transport Mechanisms and Permeability.** Click **Activity 1: Simulating Dialysis (Simple Diffusion),** and take the online **Pre-lab Quiz** for Activity 1.

After you take the online Pre-lab Quiz, click the **Experiment** tab and begin the experiment. The experiment instructions are reprinted here for your reference. The opening screen for the experiment is shown below.

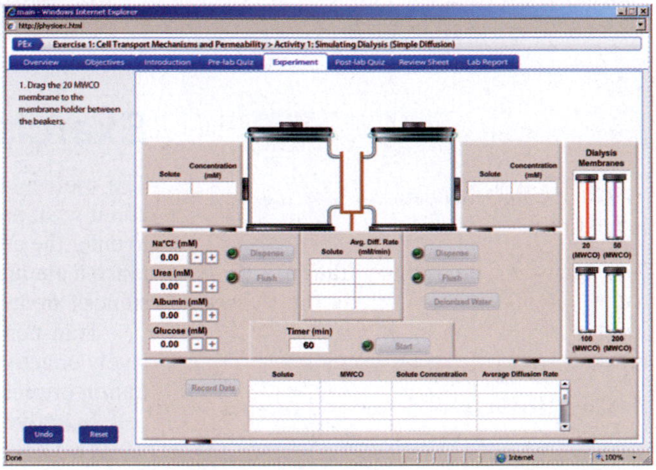
main - Windows Internet Explorer
http://physioex.html
PEx
Exercise 1: Cell Transport Mechanisms and Permeability > Activity 1: Simulating Dialysis (Simple Diffusion)
Overview
Objectives
Introduction
Pre-lab Quiz
Experiment
Post-lab Quiz
Review Sheet
Lab Report
1. Drag the 20 MWCO membrane to the membrane holder between the beakers.
Solute
Concentration (mM)
Dialysis Membranes
Na+Cl- (mM)
0.00
Urea (mM)
0.00
Albumin (mM)
0.00
Glucose (mM)
0.00
Dispense
Flush
Avg. Diff. Rate (mM/min)
Deionized Water
Timer (min)
60
Start
20 (MWCO)
50 (MWCO)
100 (MWCO)
200 (MWCO)
Record Data
MWCO
Solute Concentration
Average Diffusion Rate
Undo
Reset
Done
Internet
100%

1. Drag the 20 MWCO membrane to the membrane holder between the beakers.

2. Increase the Na^+Cl^- concentration to be dispensed to the left beaker to 9.00 m*M* by clicking the + button beside the Na^+Cl^- display. Click **Dispense** to fill the left beaker with 9.00 m*M* Na^+Cl^- solution.

3. Note that the concentration of Na^+Cl^- in the left beaker is displayed in the concentration window to the left of the beaker. Click **Deionized Water** and then click **Dispense** to fill the right beaker with deionized water.

4. After you start the run, the barrier between the beakers will descend, allowing the solutions in each beaker to have access to the dialysis membrane separating them. You will be

able to determine the amount of solute that passes through the membrane by observing the concentration display to the side of each beaker. A level above zero in Na^+Cl^- concentration in the right beaker indicates that Na^+ and Cl^- ions are diffusing from the left beaker into the right beaker through the selectively permeable dialysis membrane. Note that the timer is set to 60 minutes. The simulation compresses the 60-minute time period into 10 seconds of real time. Click **Start** to start the run and watch the concentration display to the side of each beaker for any activity.

5. Click **Record Data** to display your results in the grid (and record your results in Chart 1).

CHART 1 Dialysis Results (average diffusion rate in m*M*/min)

	Membrane MWCO			
Solute	**20**	**50**	**100**	**200**
Na^+Cl^-				
Urea				
Albumin				
Glucose				

PREDICT Question 1
The molecular weight of urea is 60.07. Do you think urea will diffuse through the 20 MWCO membrane?

____

6. Click **Flush** beneath each of the beakers to prepare for the next run.

7. Increase the urea concentration to be dispensed to the left beaker to 9.00 m*M* by clicking the + button beside the urea display. Click **Dispense** to fill the left beaker with 9.00 m*M* urea solution.

8. Click **Deionized Water** and then click **Dispense** to fill the right beaker with deionized water.

9. Click **Start** to start the run and watch the concentration display to the side of each beaker for any activity.

10. Click **Record Data** to display your results in the grid (and record your results in Chart 1).

11. Click the 20 MWCO membrane in the membrane holder to automatically return it to the membrane cabinet and then click **Flush** beneath each beaker to prepare for the next run.

12. Drag the 50 MWCO membrane to the membrane holder between the beakers. Increase the Na^+Cl^- concentration to be dispensed to the left beaker to 9.00 m*M*. Click **Dispense** to fill the left beaker with 9.00 m*M* Na^+Cl^- solution.

13. Click **Deionized Water** and then click **Dispense** to fill the right beaker with deionized water.

14. Click **Start** to start the run and watch the concentration display to the side of each beaker for any activity.

15. Click **Record Data** to display your results in the grid (and record your results in Chart 1).

16. Click **Flush** beneath each of the beakers to prepare for the next run.

17. Increase the Na^+Cl^- concentration to be dispensed to the left beaker to 18.00 m*M*. Click **Dispense** to fill the left beaker with 18.00 m*M* Na^+Cl^- solution.

18. Click **Deionized Water** and then click **Dispense** to fill the right beaker with deionized water.

19. Click **Start** to start the run and watch the concentration display to the side of each beaker for any activity.

20. Click **Record Data** to display your results in the grid (and record your results in Chart 1).

21. Click the 50 MWCO membrane in the membrane holder to automatically return it to the membrane cabinet and then click **Flush** beneath each beaker to prepare for the next run.

22. Drag the 100 MWCO membrane to the membrane holder between the beakers. Increase the Na^+Cl^- concentration to be dispensed to the left beaker to 9.00 m*M*. Click **Dispense** to fill the left beaker with 9.00 m*M* Na^+Cl^- solution.

23. Click **Deionized Water** and then click **Dispense** to fill the right beaker with deionized water.

24. Click **Start** to start the run and watch the concentration display to the side of each beaker for any activity.

25. Click **Record Data** to display your results in the grid (and record your results in Chart 1).

26. Click **Flush** beneath each of the beakers to prepare for the next run.

27. Increase the urea concentration to be dispensed to the left beaker to 9.00 m*M*. Click **Dispense** to fill the left beaker with 9.00 m*M* urea solution.

28. Click **Deionized Water** and then click **Dispense** to fill the right beaker with deionized water.

29. Click **Start** to start the run and watch the concentration display to the side of each beaker for any activity.

30. Click **Record Data** to display your results in the grid (and record your results in Chart 1).

31. Click the 100 MWCO membrane in the membrane holder to automatically return it to the membrane cabinet and then click **Flush** beneath each beaker to prepare for the next run.

PREDICT Question 2
Recall that glucose is a monosaccharide, albumin is a protein with 607 amino acids, and the average molecular weight of a single amino acid is 135 g/mole. Will glucose or albumin be able to diffuse through the 200 MWCO membrane?

____

32. Drag the 200 MWCO membrane to the membrane holder between the beakers. Increase the glucose concentration to be dispensed to the left beaker to 9.00 m*M*. Click **Dispense** to fill the left beaker with 9.00 m*M* glucose solution.

33. Click **Deionized Water** and then click **Dispense** to fill the right beaker with deionized water.

34. Click **Start** to start the run and watch the concentration display to the side of each beaker for any activity.

35. Click **Record Data** to display your results in the grid (and record your results in Chart 1).

36. Click **Flush** beneath each of the beakers to prepare for the next run.

37. Increase the albumin concentration to be dispensed to the left beaker to 9.00 m*M*. Click **Dispense** to fill the left beaker with 9.00 m*M* albumin solution.

38. Click **Deionized Water** and then click **Dispense** to fill the right beaker with deionized water.

39. Click **Start** to start the run and watch the concentration display to the side of each beaker for any activity.

40. Click **Record Data** to display your results in the grid (and record your results in Chart 1).

After you complete the experiment, take the online **Post-lab Quiz** for Activity 1.

Activity Questions

1. Did any solutes move through the 20 MWCO membrane? Why or why not?

____

____

2. Did Na^+Cl^- move through the 50 MWCO membrane?

____

____

3. Describe how the size of a molecule (molecular weight) affects its rate of diffusion.

____

____

4. What happened to the rate of diffusion when you increased the Na^+Cl^- solute concentration?

____

____

ACTIVITY 2

Simulated Facilitated Diffusion

OBJECTIVES

1. To understand that some solutes require a carrier protein to pass through a membrane because of size or solubility limitations.
2. To observe how the concentration of solutes affects the rate of facilitated diffusion.
3. To observe how the number of transport proteins affects the rate of facilitated diffusion.
4. To understand how transport proteins can become saturated.

Introduction

Some molecules are lipid insoluble or too large to pass through pores in the cell's plasma membrane. Instead, they pass through the membrane by a passive transport process called **facilitated diffusion.** For example, sugars, amino acids, and ions are transported by facilitated diffusion. In this form of transport, solutes combine with carrier-protein molecules in the membrane and are then transported *with* (or *along* or *down)* their concentration gradient. The carrier-protein molecules in the membrane might have to change shape slightly to accommodate the solute, but the cell does not have to expend the energy of ATP.

Because facilitated diffusion relies on carrier proteins, solute transport varies with the number of available carrier-protein molecules in the membrane. The carrier proteins can become saturated if too much solute is present and the maximum transport rate is reached. The carrier proteins are embedded in the plasma membrane and act like a shield, protecting the hydrophilic solute from the lipid portions of the membrane.

Facilitated diffusion typically occurs in one direction for a given solute. The greater the concentration difference between one side of the membrane and the other, the greater the rate of facilitated diffusion.

EQUIPMENT USED The following equipment will be depicted on-screen: left and right beakers—used for diffusion of solutes; dialysis membranes with various molecular weight cutoffs (MWCOs); membrane builder—used to build membranes with different numbers of glucose protein carriers.

Experiment Instructions

Go to the home page in the PhysioEx software and click **Exercise 2: Cell Transport Mechanisms and Permeability.** Click **Activity 2: Simulated Facilitated Diffusion,** and take the online **Pre-lab Quiz** for Activity 2.

After you take the online Pre-lab Quiz, click the **Experiment** tab and begin the experiment. The experiment instructions are reprinted here for your reference. The opening screen for the experiment is shown on the following page.

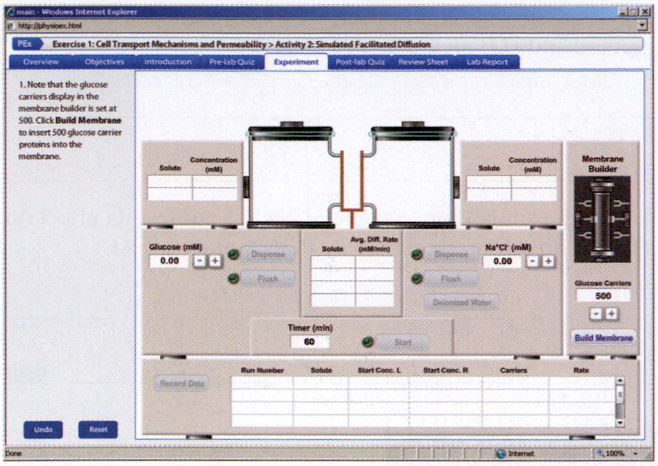
main - Windows Internet Explorer
http://physioex.html
PEx
Exercise 1: Cell Transport Mechanisms and Permeability > Activity 2: Simulated Facilitated Diffusion
Overview
Objectives
Introduction
Pre-lab Quiz
Experiment
Post-lab Quiz
Review Sheet
Lab Report
1. Note that the glucose carriers display in the membrane builder is set at 500. Click Build Membrane to insert 500 glucose carrier proteins into the membrane.
Solute
Concentration (mM)
Membrane Builder
Glucose (mM)
0.00
Dispense
Flush
Avg. Diff. Rate (mM/min)
Na+Cl- (mM)
Deionized Water
Glucose Carriers
500
Build Membrane
Timer (min)
60
Start
Record Data
Run Number
Start Conc. L
Start Conc. R
Carriers
Rate
Undo
Reset
Done
Internet
100%

1. Note that the glucose carriers display in the membrane builder is set at 500. Click **Build Membrane** to insert 500 glucose carrier proteins into the membrane.

2. Drag the membrane to the membrane holder between the beakers.

3. Increase the glucose concentration to be dispensed to the left beaker to 2.00 m*M* by clicking the + button beside the glucose display. Click **Dispense** to fill the left beaker with 2.00 m*M* glucose solution.

4. Note that the concentration of glucose in the left beaker is displayed in the concentration window to the left of the beaker. Click **Deionized Water** and then click **Dispense** to fill the right beaker with deionized water.

5. After you start the run, the barrier between the beakers will descend, allowing the solutions in each beaker to have access to the dialysis membrane separating them. You will be able to determine the amount of solute that passes through the membrane by observing the concentration display to the side of each beaker. A level above zero in glucose concentration in the right beaker indicates that glucose is diffusing from the left beaker into the right beaker through the selectively permeable dialysis membrane. Note that the timer is set to 60 minutes. The simulation compresses the 60-minute time period into 10 seconds of real time. Click **Start** to start the run and watch the concentration display to the side of each beaker for any activity.

6. Click **Record Data** to display your results in the grid (and record your results in Chart 2).

CHART 2 Facilitated Diffusion Results (glucose transport rate, m*M*/min)

	Number of glucose carrier proteins		
Glucose concentration	500	700	100
2 m*M*			
8 m*M*			
10 m*M*			
2 m*M* w/ 2.00 m*M* $Na^{+}Cl^{-}$			

7. Click **Flush** beneath each of the beakers to prepare for the next run.

8. Increase the glucose concentration to be dispensed to the left beaker to 8.00 m*M* by clicking the + button beside the glucose display. Click **Dispense** to fill the left beaker with 8.00 m*M* glucose solution.

9. Click **Deionized Water** and then click **Dispense** to fill the right beaker with deionized water.

10. Click **Start** to start the run and watch the concentration display to the side of each beaker for any activity.

11. Click **Record Data** to display your results in the grid (and record your results in Chart 2).

12. Click the membrane in the membrane holder to automatically return it to the membrane builder and then click **Flush** beneath each beaker to prepare for the next run.

? PREDICT Question 1
What effect do you think increasing the number of protein carriers will have on the glucose transport rate?

____

13. Increase the number of glucose carriers to 700 by clicking the + button beneath the glucose carriers display. Click **Build Membrane** to insert 700 glucose carrier proteins into the membrane.

14. Drag the membrane to the membrane holder between the beakers. Increase the glucose concentration to be dispensed to the left beaker to 2.00 m*M*. Click **Dispense** to fill the left beaker with 2.00 m*M* glucose solution.

15. Click **Deionized Water** and then click **Dispense** to fill the right beaker with deionized water.

16. Click **Start** to start the run and watch the concentration display to the side of each beaker for any activity.

17. Click **Record Data** to display your results in the grid (and record your results in Chart 2).

18. Click **Flush** beneath each of the beakers to prepare for the next run.

19. Increase the glucose concentration to be dispensed to the left beaker to 8.00 m*M*. Click **Dispense** to fill the left beaker with 8.00 m*M* glucose solution.

20. Click **Deionized Water** and then click **Dispense** to fill the right beaker with deionized water.

21. Click **Start** to start the run and watch the concentration display to the side of each beaker for any activity.

22. Click **Record Data** to display your results in the grid (and record your results in Chart 2).

23. Click the membrane in the membrane holder to automatically return it to the membrane builder and then click **Flush** beneath each beaker to prepare for the next run.

24. Decrease the number of glucose carriers to 100 by clicking the – button beneath the glucose carriers display. Click **Build Membrane** to insert 100 glucose carrier proteins into the membrane.

25. Drag the membrane to the membrane holder between the beakers. Increase the glucose concentration to be dispensed to the left beaker to 10.00 m*M*. Click **Dispense** to fill the left beaker with 10.00 m*M* glucose solution.

26. Click **Deionized Water** and then click **Dispense** to fill the right beaker with deionized water.

27. Click **Start** to start the run and watch the concentration display to the side of each beaker for any activity.

28. Click **Record Data** to display your results in the grid (and record your results in Chart 2).

29. Click the membrane in the membrane holder to automatically return it to the membrane builder and then click **Flush** beneath each beaker to prepare for the next run.

30. Increase the number of glucose carriers to 700. Click **Build Membrane** to insert 700 glucose carrier proteins into the membrane.

PREDICT Question 2
What effect do you think adding Na^+Cl^- will have on the glucose transport rate?

____

31. Increase the glucose concentration to be dispensed to the left beaker to 2.00 m*M*. Click **Dispense** to fill the left beaker with 2.00 m*M* glucose solution.

32. Increase the Na^+Cl^- concentration to be dispensed to the right beaker to 2.00 m*M*. Click **Dispense** button to fill the right beaker with 2.00 m*M* Na^+Cl^- solution.

33. Click **Start** to start the run and watch the concentration display to the side of each beaker for any activity.

34. Click **Record Data** to display the results in the grid (and record your results in Chart 2).

After you complete the experiment, take the online **Post-lab Quiz** for Activity 2.

Activity Questions

1. Are the solutes moving with or against their concentration gradient in facilitated diffusion?

____

____

2. What happened to the rate of facilitated diffusion when the number of carrier proteins was increased?

____

____

3. Explain why equilibrium was not reached with 10 m*M* glucose and 100 membrane carriers.

____

____

4. In the simulation you added Na^+Cl^- to test its effect on glucose diffusion. Explain why there was no effect.

____

____

ACTIVITY 3

Simulating Osmotic Pressure

OBJECTIVES

1. To explain how osmosis is a special type of diffusion.
2. To understand that osmosis is a passive process that depends upon the concentration gradient of water.
3. To explain how tonicity of a solution relates to changes in cell volume.
4. To understand conditions that affect osmotic pressure.

Introduction

A special form of diffusion, called **osmosis,** is the diffusion of water through a selectively permeable membrane. (A membrane is called *selectively permeable, differentially permeable,* or *semipermeable* if it allows some molecules to pass but not others.) Because water can pass through the pores of most membranes, it can move from one side of a membrane to the other relatively freely. Osmosis takes place whenever there is a difference in water concentration between the two sides of a membrane.

If we place distilled water on both sides of a membrane, *net* movement of water does not occur. Remember, however, that water molecules would still move between the two sides of the membrane. In such a situation, we would say that there is no *net* osmosis.

The concentration of water in a solution depends on the number of solute particles present. For this reason, increasing the solute concentration coincides with decreasing the water concentration. Because water moves down its concentration gradient (from an area of its higher concentration to an area of its lower concentration), it always moves *toward* the solution with the highest concentration of solutes. Similarly, solutes also move down their concentration gradients.

If we position a *fully* permeable membrane (permeable to solutes and water) between two solutions of differing concentrations, then all substances—solutes and water—diffuse freely, and an equilibrium will be reached between the two sides of the membrane. However, if we use a selectively permeable membrane that is impermeable to the solutes, then we have established a condition where water moves but solutes do not. Consequently, water moves toward the more concentrated solution, resulting in a *volume increase* on that side of the membrane.

By applying this concept to a closed system where volumes cannot change, we can predict that the *pressure* in the more concentrated solution will rise. The force that would need to be applied to oppose the osmosis in a closed system is the **osmotic pressure.** Osmotic pressure is measured in *millimeters of mercury (mm Hg)*. In general, the more impermeable the solutes, the higher the osmotic pressure.

Osmotic changes can affect the volume of a cell when it is placed in various solutions. The concept of **tonicity** refers to the way a solution affects the volume of a cell. The tonicity of a solution tells us whether or not a cell will shrink or swell. If the concentration of impermeable solutes is the *same* inside and outside of the cell, the solution is **isotonic.** If there is a *higher* concentration of impermeable solutes *outside* the cell than in the cell's interior, the solution is **hypertonic.** Because the net movement of water would be out of the cell, the cell would *shrink* in a hypertonic solution. Conversely, if the concentration of impermeable solutes is *lower* outside of the cell than in the cell's interior, then the solution is **hypotonic.** The net movement of water would be into the cell, and the cell would *swell* and possibly burst.

EQUIPMENT USED The following equipment will be depicted on-screen: left and right beakers—used for diffusion of solutes; dialysis membranes with various molecular weight cutoffs (MWCOs).

Experiment Instructions

Go to the home page in the PhysioEx software and click **Exercise 1: Cell Transport Mechanisms and Permeability.** Click **Activity 3: Simulating Osmotic Pressure,** and take the online **Pre-lab Quiz** for Activity 3.

After you take the online Pre-lab Quiz, click the **Experiment** tab and begin the experiment. The experiment instructions are reprinted here for your reference. The opening screen for the experiment is shown below.

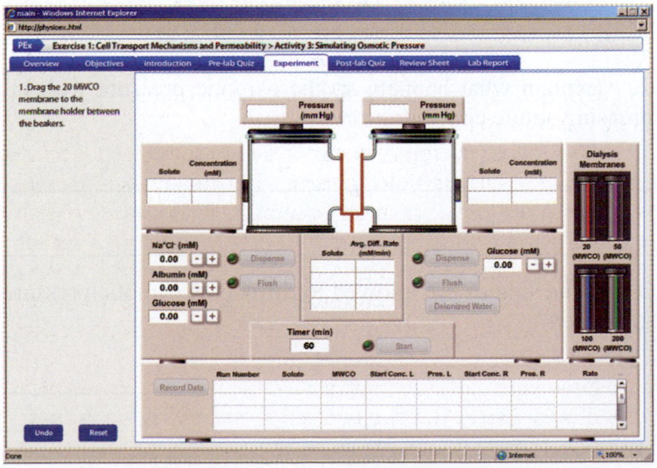
main - Windows Internet Explorer
Exercise 1: Cell Transport Mechanisms and Permeability > Activity 3: Simulating Osmotic Pressure
Overview
Objectives
Introduction
Pre-lab Quiz
Experiment
Post-lab Quiz
Review Sheet
Lab Report
1. Drag the 20 MWCO membrane to the membrane holder between the beakers.
Pressure (mm Hg)
Pressure (mm Hg)
Solute
Concentration (mM)
Dialysis Membranes
Albumin (mM)
Glucose (mM)
Dispense
Flush
Avg. Diff. Rate (mM/min)
Deionized Water
Timer (min)
60
Start
Run Number
MWCO
Start Conc. L
Pres. L
Start Conc. R
Pres. R
Rate
Record Data
Undo
Reset

1. Drag the 20 MWCO membrane to the membrane holder between the beakers.

2. Increase the Na^+Cl^- concentration to be dispensed to the left beaker to 5.00 m*M* by clicking the + button beside the Na^+Cl^- display. Click **Dispense** to fill the left beaker with 5.00 m*M* Na^+Cl^- solution.

3. Note that the concentration of Na^+Cl^- in the left beaker is displayed in the concentration window to the left of the beaker. Click **Deionized Water** and then click **Dispense** to fill the right beaker with deionized water.

4. After you start the run, the barrier between the beakers will descend, allowing the solutions in each beaker to have access to the dialysis membrane separating them. You can observe the changes in pressure in the two beakers by watching the pressure display above each beaker. You will also be able to determine the amount of solute that passes through the membrane by observing the concentration display to the side of each beaker. A level above zero in Na^+Cl^- concentration in the right beaker indicates that Na^+ and Cl^- ions are diffusing from the left beaker into the right beaker through the selectively permeable dialysis membrane. Note that the timer is set to 60 minutes. The simulation compresses the 60-minute time period into 10 seconds of real time. Click **Start** to start the run and watch the pressure display above each beaker for any activity.

5. Click **Record Data** to display your results in the grid (and record your results in Chart 3).

CHART 3 Osmosis Results

Solute	Membrane (MWCO)	Pressure on left (mm Hg)	Diffusion rate (m*M*/min)
Na^+Cl^-			
Na^+Cl^-			
Na^+Cl^-			
Glucose			
Glucose			
Glucose			
Albumin w/glucose			

6. Click **Flush** beneath each of the beakers to prepare for the next run.

7. Increase the Na^+Cl^- concentration to be dispensed to the left beaker to 10.00 m*M* by clicking the + button beside the Na^+Cl^- display. Click **Dispense** to fill the left beaker with 10.00 m*M* Na^+Cl^- solution.

8. Click **Deionized Water** and then click **Dispense** to fill the right beaker with deionized water.

PREDICT Question 1
What effect do you think increasing the Na^+Cl^- concentration will have?

____

9. Click **Start** to start the run and watch the pressure display above each beaker for any activity.

10. Click **Record Data** to display your results in the grid (and record your results in Chart 3).

11. Click the 20 MWCO membrane in the membrane holder to automatically return it to the membrane cabinet and then click **Flush** beneath each beaker to prepare for the next run.

12. Drag the 50 MWCO membrane to the membrane holder between the beakers. Increase the Na^+Cl^- concentration to be dispensed to the left beaker to 10.00 m*M*. Click **Dispense** to fill the left beaker with 10.00 m*M* Na^+Cl^- solution.

13. Click **Deionized Water** and then click **Dispense** to fill the right beaker with deionized water.

14. Click **Start** to start the run and watch the pressure display above each beaker for any activity.

15. Click **Record Data** to display your results in the grid (and record your results in Chart 3).

16. Click the 50 MWCO membrane in the membrane holder to automatically return it to the membrane cabinet and then click **Flush** beneath each beaker to prepare for the next run.

17. Drag the 100 MWCO membrane to the membrane holder between the beakers. Increase the glucose concentration to be dispensed to the left beaker to 8.00 m*M* by clicking the + button beside the glucose display beneath the left beaker. Click **Dispense** to fill the left beaker with 8.00 m*M* glucose solution.

18. Click **Deionized Water** and then click **Dispense** to fill the right beaker with deionized water.

19. Click **Start** to start the run and watch the pressure display above each beaker for any activity.

20. Click **Record Data** to display your results in the grid (and record your results in Chart 3).

21. Click **Flush** beneath each of the beakers to prepare for the next run.

22. Increase the glucose concentration to be dispensed to the left beaker to 8.00 m*M*. Click **Dispense** to fill the left beaker with 8.00 m*M* glucose solution.

23. Increase the glucose concentration to be dispensed to the right beaker to 8.00 m*M* by clicking the + button beside the glucose display beneath the right beaker. Click **Dispense** to fill the right beaker with 8.00 m*M* glucose solution.

24. Click **Start** to start the run and watch the pressure display above each beaker for any activity.

25. Click **Record Data** to display your results in the grid (and record your results in Chart 3).

26. Click the 100 MWCO membrane in the membrane holder to automatically return it to the membrane cabinet and then click **Flush** beneath each beaker to prepare for the next run.

27. Drag the 200 MWCO membrane to the membrane holder between the beakers. Increase the glucose concentration to be dispensed to the left beaker to 8.00 m*M*. Click **Dispense** to fill the left beaker with 8.00 m*M* glucose solution.

28. Click **Deionized Water** and then click **Dispense** to fill the right beaker with deionized water.

29. Click **Start** to start the run and watch the pressure display above each beaker for any activity.

30. Click **Record Data** to display your results in the grid (and record your results in Chart 3).

31. Click **Flush** beneath each of the beakers to prepare for the next run.

32. Increase the albumin concentration to be dispensed to the left beaker to 9.00 m*M*. Click **Dispense** to fill the left beaker with 9.00 m*M* albumin solution.

33. Increase the glucose concentration to be dispensed to right beaker to 10.00 m*M*. Click **Dispense** to fill the right beaker with 10.00 m*M* glucose solution.

? PREDICT Question 2
What do you think will be the pressure result of the current experimental conditions?

____

34. Click **Start** to start the run and watch the pressure display above each beaker for any activity.

35. Click **Record Data** to display your results in the grid (and record your results in Chart 3).

After you complete the experiment, take the online **Post-lab Quiz** for Activity 3.

Activity Questions

1. Which membrane resulted in the greatest pressure with Na^+Cl^- as the solute? Why?

____

____

2. Explain what happens to the osmotic pressure with increasing solute concentration.

____

____

3. If the solutes are allowed to diffuse, is osmotic pressure generated?

____

____

4. If the solute concentrations are equal, is osmotic pressure generated? Why or why not?

____

____

ACTIVITY 4

Simulating Filtration

OBJECTIVES

1. To understand that filtration is a passive process dependent upon a pressure gradient.
2. To understand that filtration is not a selective process.
3. To explain that the size of the membrane pores will determine what passes through.
4. To explain the effect that increasing the hydrostatic pressure has on the filtration rate and how this correlates to events in the body.
5. To understand the relationship between molecular weight and molecular size.

Introduction

Filtration is the process by which water and solutes pass through a membrane (such as a dialysis membrane) from an area of higher hydrostatic (fluid) pressure into an area of lower hydrostatic pressure. Like diffusion, filtration is a passive process. For example, fluids and solutes filter out of the capillaries in the kidneys into the kidney tubules because blood pressure in the capillaries is greater than the fluid pressure in the tubules. So, if blood pressure increases, the rate of filtration increases.

Filtration is not a selective process. The amount of *filtrate*—the fluids and solutes that pass through the membrane—depends almost entirely on the *pressure gradient* (the difference in pressure between the solutions on the two sides of the membrane) and on the *size* of the *membrane pores*. Solutes that are too large to pass through are retained by the capillaries. These solutes usually include blood cells and proteins. Ions and smaller molecules, such as glucose and urea, can pass through.

In this activity the pore size is measured as a *molecular weight cutoff (MWCO)*, which is indicated by the number below the filtration membrane. You can think of MWCO in terms of pore size: the larger the MWCO number, the larger the pores in the filtration membrane. The molecular weight of a solute is the number of grams per mole, where a mole is the constant Avogadro's number 6.02×10^{23} molecules/mole. You will also analyze the filtration membrane for the presence or absence of solutes that might be left sticking to the membrane.

EQUIPMENT USED The following equipment will be depicted on-screen: top and bottom beakers—used for filtration of solutes; dialysis membranes with various molecular weight cutoffs (MWCOs); membrane residue analysis station—used to analyze the filtration membrane.

Experiment Instructions

Go to the home page in the PhysioEx software and click **Exercise 1: Cell Transport Mechanisms and Permeability.** Click **Activity 4: Simulating Filtration** and take the online **Pre-lab Quiz** for Activity 4.

After you take the online Pre-lab Quiz, click the **Experiment** tab and begin the experiment. The experiment instructions are reprinted here for your reference. The opening screen for the experiment is shown above.

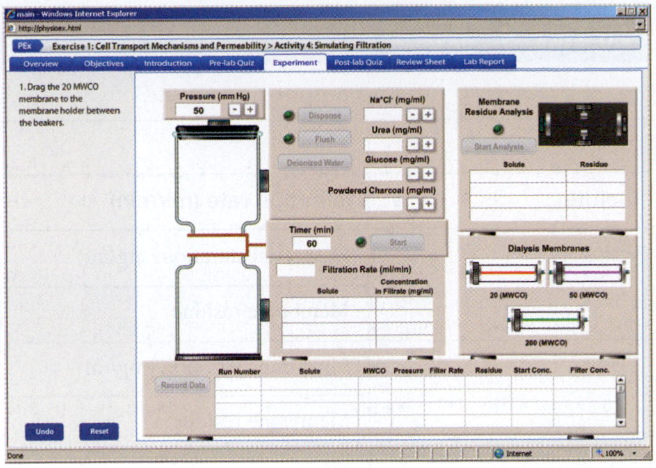
Exercise 1: Cell Transport Mechanisms and Permeability > Activity 4: Simulating Filtration
Overview
Objectives
Introduction
Pre-lab Quiz
Experiment
Post-lab Quiz
Review Sheet
Lab Report
1. Drag the 20 MWCO membrane to the membrane holder between the beakers.
Pressure (mm Hg)
50
Dispense
Flush
Deionized Water
Na⁺Cl⁻ (mg/ml)
Urea (mg/ml)
Glucose (mg/ml)
Powdered Charcoal (mg/ml)
Membrane Residue Analysis
Start Analysis
Solute
Residue
Timer (min)
60
Start
Filtration Rate (ml/min)
Concentration in Filtrate (mg/ml)
Dialysis Membranes
20 (MWCO)
50 (MWCO)
200 (MWCO)
Record Data
Run Number
MWCO
Pressure
Filter Rate
Start Conc.
Filter Conc.
Undo
Reset

1. Drag the 20 MWCO membrane to the membrane holder between the beakers.

2. Increase the concentration of Na^+Cl^-, urea, glucose, and powdered charcoal to be dispensed to 5.00 mg/ml by clicking the + button beside the display for each solute. Click **Dispense** to fill the top beaker.

3. After you start the run, the membrane holder below the top beaker retracts, and the solution will filter through the membrane into the beaker below. You will be able to determine whether solute particles are moving through the filtration membrane by observing the concentration displays beside the bottom beaker. A rise in detected solute concentration indicates that the solute particles are moving through the filtration membrane. Note that the pressure is set at 50 mm Hg and the timer is set to 60 minutes. The simulation compresses the 60-minute time period into 10 seconds of real time. Click **Start** to start the run and watch the concentration displays beside the bottom beaker for any activity.

4. Drag the 20 MWCO membrane to the holder in the membrane residue analysis unit. Click **Start Analysis** to begin analysis (and cleaning) of the membrane.

5. Click **Record Data** to display your results in the grid (and record your results in Chart 4).

6. Click the 20 MWCO membrane in the membrane holder to automatically return it to the membrane cabinet and then click **Flush** to prepare for the next run.

? **PREDICT Question 1**
What effect will increasing the pore size of the filter have on the filtration rate?

____

7. Drag the 50 MWCO membrane to the membrane holder between the beakers. With the concentration of Na^+Cl^-, urea, glucose, and powdered charcoal still set to 5.00 mg/ml, click **Dispense** to fill the top beaker.

8. Click **Start** to start the run and watch the concentration displays beside the bottom beaker for any activity.

CHART 4 Filtration Results

		Membrane (MWCO)			
		20	50	200	200
Solute	Filtration rate (ml/min)				
Na^+Cl^-	Filter concentration (mg/ml)				
	Membrane residue				
Urea	Filter concentration (mg/ml)				
	Membrane residue				
Glucose	Filter concentration (mg/ml)				
	Membrane residue				
Powdered charcoal	Filter concentration (mg/ml)				
	Membrane residue				

9. Drag the 50 MWCO membrane to the holder in the membrane residue analysis unit. Click **Start Analysis** to begin analysis (and cleaning) of the membrane.

10. Click **Record Data** to display your results in the grid (and record your results in Chart 4).

11. Click the 50 MWCO membrane in the membrane holder to automatically return it to the membrane cabinet and then click **Flush** to prepare for the next run.

12. Drag the 200 MWCO membrane to the membrane holder between the beakers. With the concentration of Na^+Cl^-, urea, glucose, and powdered charcoal still set to 5.00 mg/ml, click **Dispense** to fill the top beaker.

13. Click **Start** to start the run and watch the concentration displays beside the bottom beaker for any activity.

14. Drag the 200 MWCO membrane to the holder in the membrane residue analysis unit. Click **Start Analysis** to begin analysis (and cleaning) of the membrane.

15. Click **Record Data** to display your results in the grid (and record your results in Chart 4).

16. Click the 200 MWCO membrane in the membrane holder to automatically return it to the membrane cabinet and then click **Flush** to prepare for the next run.

? PREDICT Question 2
What will happen if you increase the pressure above the beaker (the driving pressure)?

____

17. Increase the pressure to 100 mm Hg by clicking on the + button beside the pressure display above the top beaker.

18. Drag the 200 MWCO membrane to the membrane holder between the beakers. With the concentration of Na^+Cl^-, urea, glucose, and powdered charcoal still set to 5.00 mg/ml, click **Dispense** to fill the top beaker.

19. Click **Start** to start the run and watch the concentration displays beside the bottom beaker for any activity.

20. Drag the 200 MWCO membrane to the holder in the membrane residue analysis unit. Click **Start Analysis** to begin analysis (and cleaning) of the membrane.

21. Click **Record Data** to display your results in the grid (and record your results in Chart 4).

After you complete the Experiment, take the online **Post-lab Quiz** for Activity 4.

Activity Questions

1. Explain your results with the 20 MWCO filter. Why weren't any of the solutes present in the filtrate?

____

____

2. Describe two variables that affected the rate of filtration in your experiments.

____

____

3. Explain how you can increase the filtration rate through living membranes.

4. Judging from the filtration results, indicate which solute has the largest molecular weight.

ACTIVITY 5

Simulating Active Transport

OBJECTIVES

1. To understand that active transport requires cellular energy in the form of ATP.
2. To explain how the balance of sodium and potassium is maintained by the Na^+-K^+ pump, which moves both ions against their concentration gradients.
3. To understand coupled transport and be able to explain how the movement of sodium and potassium is independent of other solutes, such as glucose.

Introduction

Whenever a cell uses cellular energy (ATP) to move substances across its membrane, the process is an *active transport process.* Substances moved across cell membranes by an active transport process are generally unable to pass by diffusion. There are several reasons why a substance might not be able to pass through a membrane by diffusion: it might be too large to pass through the membrane pores, it might not be lipid soluble, or it might have to move *against*, rather than with, a concentration gradient.

In one type of active transport, substances move across the membrane by combining with a carrier-protein molecule. This kind of process resembles an enzyme-substrate interaction. ATP hydrolysis provides the driving force, and, in many cases, the substances move *against* concentration gradients or electrochemical gradients or both. The carrier proteins are commonly called **solute pumps.** Substances that are moved into cells by solute pumps include amino acids and some sugars. Both of these kinds of solutes are necessary for the life of the cell, but they are lipid insoluble and too large to pass through membrane pores.

In contrast, sodium ions (Na^+) are ejected from the cells by active transport. There is more Na^+ outside the cell than inside the cell, so Na^+ tends to remain in the cell unless actively transported out. In the body, the most common type of solute pump is the Na^+-K^+ (sodium-potassium) pump, which moves Na^+ and K^+ in opposite directions across cellular membranes. Three Na^+ ions are ejected from the cell for every two K^+ ions entering the cell. Note that there is more K^+ inside the cell than outside the cell, so K^+ tends to remain outside the cell unless actively transported in.

Membrane carrier proteins that move more than one substance, such as the Na^+-K^+ pump, participate in *coupled transport*. If the solutes move in the same direction, the carrier is a *symporter*. If the solutes move in opposite directions, the carrier is an *antiporter*. A carrier that transports only a single solute is a *uniporter*.

EQUIPMENT USED The following equipment will be depicted on-screen: Simulated cell inside a large beaker.

Experiment Instructions

Go to the home page in the PhysioEx software and click **Exercise 1: Cell Transport Mechanisms and Permeability.** Click **Activity 5: Simulating Active Transport** and take the online **Pre-lab Quiz** for Activity 5.

After you take the online Pre-lab Quiz, click the **Experiment** tab and begin the experiment. The experiment instructions are reprinted here for your reference. The opening screen for the experiment is shown below.

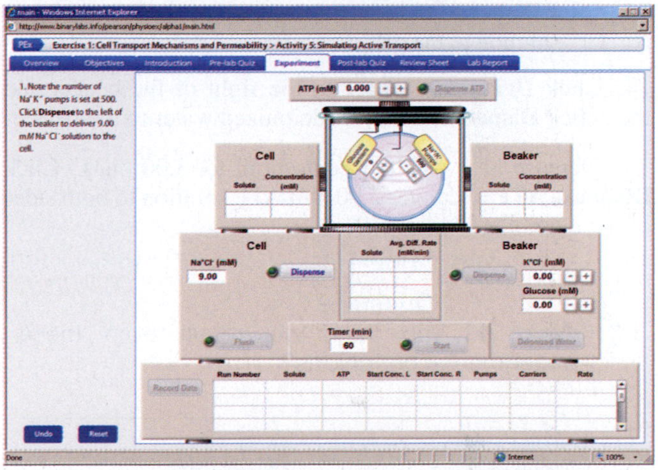
Exercise 1: Cell Transport Mechanisms and Permeability > Activity 5: Simulating Active Transport
Cell
Beaker
Timer (min)
60
Undo
Reset

1. Note the number of Na^+-K^+ pumps is set at 500. Click **Dispense** to the left of the beaker to deliver 9.00 m*M* Na^+Cl^- solution to the cell.

2. Increase the K^+Cl^- concentration to be delivered to the beaker to 6.00 m*M* by clicking the + button beside the K^+Cl^-display. Click **Dispense** to the right of the beaker to deliver 6.00 m*M* K^+Cl^- solution to the beaker.

3. Increase the ATP concentration to 1.00 m*M* by clicking the + button beside the ATP display above the beaker. Click **Dispense ATP** to deliver 1.00 m*M* ATP solution to both sides of the membrane.

4. After you start the run, the solutes will move across the cell membrane, simulating active transport. You will be able to determine the amount of solute that is transported across the membrane by observing the concentration displays on both sides of the beaker (the display on the left shows the concentrations inside the cell and the display on the right shows the concentrations inside the beaker). Note that the timer is set to 60 minutes. The simulation compresses the 60-minute time period into 10 seconds of real time. Click **Start** to start the run and watch the concentration displays on both sides of the beaker for any activity.

5. Click **Record Data** to display your results in the grid.

6. Click **Flush** to reset the beaker and simulated cell.

7. Click **Dispense** to the left of the beaker to deliver 9.00 m*M* Na^+Cl^- solution to the cell.

8. Increase the K^+Cl^- concentration to be delivered to the beaker to 6.00 m*M* by clicking the + button beside the K^+Cl^- display. Click **Dispense** to the right of the beaker to deliver 6.00 m*M* K^+Cl^- solution to the beaker.

9. Increase the ATP concentration to 3.00 m*M* by clicking the + button beside the ATP display above the beaker. Click **Dispense ATP** to deliver 3.00 m*M* ATP solution to both sides of the membrane.

10. Click **Start** to start the run and watch the concentration displays on both sides of the beaker for any activity.

11. Click **Record Data** to display your results in the grid.

12. Click **Flush** to reset the beaker and simulated cell.

13. Click **Dispense** to the left of the beaker to deliver 9.00 m*M* Na^+Cl^- solution to the cell.

14. Click **Deionized Water** to the right of the beaker and then click **Dispense** to deliver deionized water to the beaker.

15. Increase the ATP concentration to 3.00 m*M*. Click **Dispense ATP** to deliver 3.00 m*M* ATP solution to both sides of the membrane.

? PREDICT Question 1
What do you think will result from these experimental conditions?

____

16. Click **Start** to start the run and watch the concentration displays on both sides of the beaker for any activity.

17. Click **Record Data** to display your results in the grid.

18. Click **Flush** to reset the beaker and simulated cell.

19. Increase the number of Na^+-K^+ pumps to 800 by clicking the + button beneath the Na^+-K^+ pump display. Click **Dispense** to the left of the beaker to deliver 9.00 m*M* Na^+Cl^- solution to the cell.

20. Increase the K^+Cl^- concentration to be delivered to the beaker to 6.00 m*M*. Click **Dispense** to the right of the beaker to deliver 6.00 m*M* K^+Cl^- solution to the beaker.

21. Increase the ATP concentration to 3.00 m*M*. Click **Dispense ATP** to deliver 3.00 m*M* ATP solution to both sides of the membrane.

22. Click **Start** to start the run and watch the concentration displays on both sides of the beaker for any activity.

23. Click **Record Data** to display your results in the grid.

24. Click **Flush** to reset the beaker and simulated cell.

25. With the number of Na^+-K^+ pumps still set to 800, increase the number of glucose carriers to 400 by clicking the + button beneath the glucose carriers display. Click **Dispense** to the left of the beaker to deliver 9.00 m*M* Na^+Cl^- solution to the cell.

? PREDICT Question 2
Do you think the addition of glucose carriers will affect the transport of sodium or potassium?

____

26. Increase the K^+Cl^- concentration to be delivered to the beaker to 6.00 m*M*. Increase the glucose concentration to be delivered to the beaker to 10.00 m*M*. Click **Dispense** to the right of the beaker to deliver 6.00 m*M* K^+Cl^- and 10.00 m*M* glucose solution to the beaker.

27. Increase the ATP concentration to 3.00 m*M*. Click **Dispense ATP** to deliver 3.00 m*M* ATP solution to both sides of the membrane.

28. Click **Start** to start the run and watch the concentration displays on both sides of the beaker for any activity.

29. Click **Record Data** to display your results in the grid.

After you complete the experiment, take the online **Post-lab Quiz** for Activity 5.

Activity Questions

1. In the initial trial the number of Na^+-K^+ pumps is set to 500, the Na^+Cl^- concentration is set to 9.00 m*M*, the K^+Cl^- concentration is set to 6.00 m*M*, and the ATP concentration is set to 1.00 m*M*. Explain what happened and why. What would happen if no ATP had been dispensed?

____

____

2. Why was there no transport when you dispensed only Na^+Cl^-, even though ATP was present?

____

____

3. What happens to the rate of transport of Na^+ and K^+ when you increase the number of Na^+-K^+ pumps?

____

____

4. Explain why the Na^+ and K^+ transports were unaffected by the addition of glucose.

____

____

PhysioEx 9.0

REVIEW SHEET

EXERCISE 1

NAME ______________________

LAB TIME/DATE ______________________

Cell Transport Mechanisms and Permeability

ACTIVITY 1 Simulating Dialysis (Simple Diffusion)

1. Describe two variables that affect the rate of diffusion. ______________________

____

2. Why do you think the urea was not able to diffuse through the 20 MWCO membrane? How well did the results compare with your prediction? ______________________

3. Describe the results of the attempts to diffuse glucose and albumin through the 200 MWCO membrane. How well did the results compare with your prediction? ______________________

____

4. Put the following in order from smallest to largest molecular weight: glucose, sodium chloride, albumin, and urea. ______

____

ACTIVITY 2 Simulated Facilitated Diffusion

1. Explain one way in which facilitated diffusion is the same as simple diffusion and one way in which it differs. ______

____

2. The larger value obtained when more glucose carriers were present corresponds to an increase in the rate of glucose transport. Explain why the rate increased. How well did the results compare with your prediction? ______

____

3. Explain your prediction for the effect Na^+Cl^- might have on glucose transport. In other words, explain why you picked the choice that you did. How well did the results compare with your prediction? ______

ACTIVITY 3 Simulating Osmotic Pressure

1. Explain the effect that increasing the Na^+Cl^- concentration had on osmotic pressure and why it has this effect. How well did the results compare with your prediction? ______________________

____

2. Describe one way in which osmosis is similar to simple diffusion and one way in which it is different. ______

______

3. Solutes are sometimes measured in milliosmoles. Explain the statement, "Water chases milliosmoles." ______

______

4. The conditions were 9 m*M* albumin in the left beaker and 10 m*M* glucose in the right beaker with the 200 MWCO membrane in place. Explain the results. How well did the results compare with your prediction? ______

______

ACTIVITY 4 Simulating Filtration

1. Explain in your own words why increasing the pore size increased the filtration rate. Use an analogy to support your statement. How well did the results compare with your prediction? ______

______

2. Which solute did not appear in the filtrate using any of the membranes? Explain why. ______

______

3. Why did increasing the pressure increase the filtration rate but not the concentration of solutes? How well did the results compare with your prediction? ______

______

ACTIVITY 5 Simulating Active Transport

1. Describe the significance of using 9 m*M* sodium chloride inside the cell and 6 m*M* potassium chloride outside the cell, instead of other concentration ratios. ______

______

2. Explain why there was no sodium transport even though ATP was present. How well did the results compare with your prediction? ______

3. Explain why the addition of glucose carriers had no effect on sodium or potassium transport. How well did the results compare with your prediction? ______

4. Do you think glucose is being actively transported or transported by facilitated diffusion in this experiment? Explain your answer. ______

______

PhysioEx™ 9.0

EXERCISE 2

Skeletal Muscle Physiology

PRE-LAB QUIZ

1. Circle the correct term. The potential difference / resistance across the plasma membrane is the result of the different ion permeabilities.
2. Depolarization of skeletal muscle is caused by a(n)
   a. influx of Ca^{2+}
   b. influx of K^+
   c. influx of Na^+
   d. release of Na^+
3. The ________ wave follows the depolarization wave across the sarcolemma.
   a. hyperpolarization
   b. refraction
   c. repolarization
4. Circle the correct term. Contraction of a skeletal muscle that results from a single stimulus is called a twitch / tetanus.
5. Circle True or False. The voltage at which the first noticeable contractile response is obtained is called the threshold stimulus.
6. Circle True or False. A single muscle is made up of many motor units, and the gradual activation of these motor units results in a graded contraction of the whole muscle.
7. A muscle contraction that is smooth and sustained as a result of high-frequency stimulation is
   a. tetanus
   b. tonus
   c. a twitch
8. Muscle fatigue, the loss of the ability to contract, may be a result of
   a. oxygen deficit in the tissue after prolonged activity
   b. potassium buildup in the tissue after prolonged activity
   c. sodium deficit in the tissue after prolonged activity
   d. the accumulation of ATP in the tissue after prolonged activity

Exercise Overview

Humans make voluntary decisions to walk, talk, stand up, and sit down. Skeletal muscles, which are usually attached to the skeleton, make these actions possible (view Figure 2.1). Skeletal muscles characteristically span two joints and attach to the skeleton via **tendons,** which attach to the periosteum of a bone. Skeletal muscles are composed of hundreds to thousands of individual cells called **muscle fibers,** which produce **muscle tension** (also referred to as **muscle force**). Skeletal muscles are remarkable machines. They provide us with the manual dexterity to create magnificent works of art and can generate the brute force needed to lift a 45-kilogram sack of concrete.

When a skeletal muscle is isolated from an experimental animal and mounted on a **force transducer,** you can generate **muscle contractions** with controlled **electrical stimulation.** Importantly, the contractions of this isolated muscle are known to mimic those of working muscles in the body. That is, in vitro experiments reproduce in vivo functions. Therefore, the activities you perform in this exercise will give you valuable insight into skeletal muscle physiology.

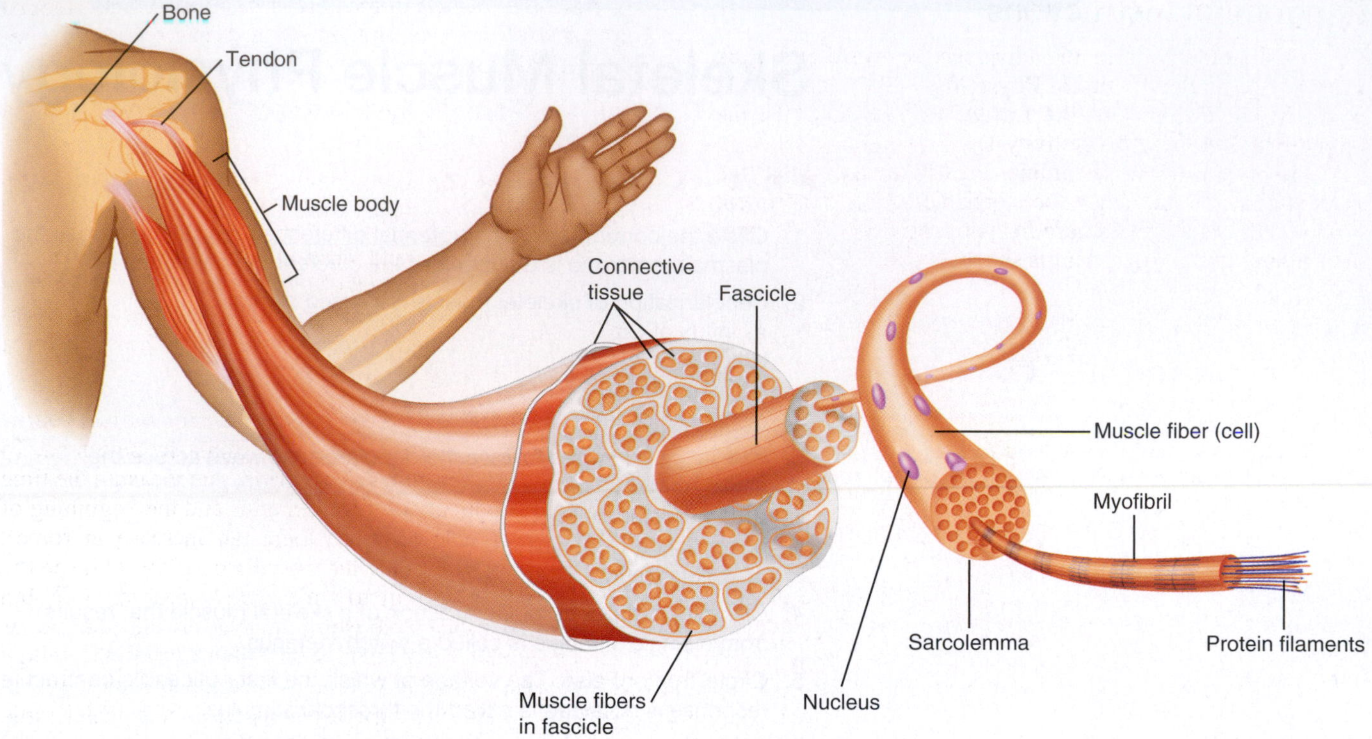
Bone
Tendon
Muscle body
Connective tissue
Fascicle
Muscle fiber (cell)
Myofibril
Sarcolemma
Protein filaments
Muscle fibers in fascicle
Nucleus

FIGURE 2.1 **Structure of a skeletal muscle.**

ACTIVITY 1

The Muscle Twitch and the Latent Period

OBJECTIVES

1. To understand the terms *excitation-contraction coupling, electrical stimulus, muscle twitch, latent period, contraction phase,* and *relaxation phase.*
2. To initiate muscle twitches with electrical stimuli of varying intensity.
3. To identify and measure the duration of the latent period.

Introduction

A **motor unit** consists of a **motor neuron** and all of the **muscle fibers** it innervates. The motor neuron and a muscle fiber intersect at the **neuromuscular junction.** Specifically, the neuromuscular junction is the location where the axon terminal of the neuron meets a specialized region of the muscle fiber's plasma membrane. This specialized region is called the **motor end plate.**

An action potential in a motor neuron triggers the release of acetylcholine from its terminal. Acetylcholine then diffuses onto the muscle fiber's plasma membrane (or **sarcolemma**) and binds to receptors in the motor end plate, initiating a change in ion permeability that results in a *graded depolarization* of the muscle plasma membrane (the end-plate potential). The events that occur at the neuromuscular junction lead to the **end-plate potential.** The end-plate potential triggers a series of events that results in the contraction of a muscle cell. This entire process is called **excitation-contraction coupling.**

You will be simulating excitation-contraction coupling in this and subsequent activities, but you will be using electrical pulses, rather than acetylcholine, to trigger action potentials. The pulses will be administered by an electrical stimulator that can be set for the precise voltage, frequency, and duration of shock desired. When applied to a muscle that has been surgically removed from an animal, a single electrical stimulus will result in a **muscle twitch**—the mechanical response to a single action potential. A muscle twitch has three phases: the *latent period,* the *contraction phase,* and the *relaxation phase.*

1. The **latent period** is the period of time that elapses between the generation of an action potential in a muscle cell and the start of muscle contraction. Although no force is generated during the latent period, chemical changes (including the release of calcium from the sarcoplasmic reticulum) occur intracellularly in preparation for contraction.

2. The **contraction phase** starts at the end of the latent period and ends when muscle tension peaks.

3. The **relaxation phase** is the period of time from peak tension until the end of the muscle contraction

EQUIPMENT USED The following equipment will be depicted on-screen: intact, viable skeletal muscle dissected off the leg of a frog; electrical stimulator—delivers the desired amount and duration of stimulating voltage to the muscle via electrodes resting on the muscle; mounting stand—includes a force transducer to measure the amount of force, or tension, developed by the muscle; oscilloscope—displays the stimulated muscle twitch and the amount of active, passive, and total force developed by the muscle.

Experiment Instructions

Go to the home page in the PhysioEx software and click **Exercise 2: Skeletal Muscle Physiology.** Click **Activity 1: The Muscle Twitch and the Latent Period,** and take the online **Pre-lab Quiz** for Activity 1.

After you take the online Pre-lab Quiz, click the **Experiment** tab and begin the experiment. The experiment instructions are reprinted here for your reference. The opening screen for the experiment is shown below.

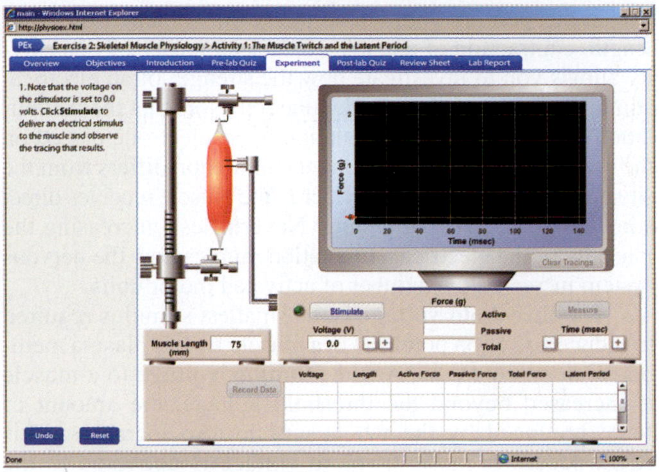
main - Windows Internet Explorer
http://physioex.html
PEx
Exercise 2: Skeletal Muscle Physiology > Activity 1: The Muscle Twitch and the Latent Period
Overview
Objectives
Introduction
Pre-lab Quiz
Experiment
Post-lab Quiz
Review Sheet
Lab Report
1. Note that the voltage on the stimulator is set to 0.0 volts. Click Stimulate to deliver an electrical stimulus to the muscle and observe the tracing that results.
Force (g)
Time (msec)
Clear Tracings
Stimulate
Voltage (V)
0.0
Force (g)
Active
Passive
Total
Measure
Time (msec)
Muscle Length (mm)
75
Voltage
Length
Active Force
Passive Force
Total Force
Latent Period
Record Data
Undo
Reset
Done
Internet
100%

1. Note that the voltage on the stimulator is set to 0.0 volts. Click **Stimulate** to deliver an electrical stimulus to the muscle and observe the tracing that results.

2. The tracing on the oscilloscope indicates active muscle force. Note that no muscle force developed because the voltage was set to zero. Click **Record Data** to display your results in the grid (and record your results in Chart 1).

CHART 1 Latent Period Results

Voltage	Active force (g)	Latent period (msec)

3. Increase the voltage to 3.0 volts by clicking the + button beside the voltage display.

4. Click **Stimulate** and observe the tracing that results.

5. Note the muscle force that developed. Click **Record Data** to display your results in the grid (and record your results in Chart 1).

6. Click **Clear Tracings** to remove the tracings from the oscilloscope.

7. Increase the voltage to 4.0 volts by clicking the + button beside the voltage display.

8. Click **Stimulate** and observe the tracing that results. Note that the trace starts at the left side of the screen and stays flat for a short period of time. Remember that the X-axis displays elapsed time in milliseconds. Also note how the force during the twitch also changes.

9. Click **Measure** on the stimulator. A thin, vertical yellow line appears at the far left side of the oscilloscope screen. To measure the length of the latent period, you measure the time between the application of the stimulus and the beginning of the first observable response (here, an increase in force). Click the + button beside the time display. You will see the vertical yellow line start to move across the screen. Watch what happens in the time (msec) display as the line moves across the screen. Keep clicking the + button until the yellow line reaches the point in the tracing where the graph stops being a flat line and begins to rise (this is the point at which muscle tension starts to develop). If the yellow line moves past the desired point, click the − button to move it backward.

When the yellow line is positioned correctly, click **Record Data** to display the latent period in the grid (and record your results in Chart 1).

10. Click **Clear Tracings** to remove the tracings from the oscilloscope.

> **?** **PREDICT Question 1**
> Will changes to the stimulus voltage alter the duration of the latent period? Explain.
>
> ______________________________

11. You will now gradually increase the voltage to observe how changes to the stimulus voltage alter the duration of the latent period.

- Increase the voltage by 2.0 volts.
- Click **Stimulate** and observe the tracing that results.
- Click **Measure** on the stimulator and then click the + button until the yellow line reaches the point in the tracing where the graph stops being a flat line and begins to rise.
- Click **Record Data** (and record your results in Chart 1).

Repeat this step until you reach 10.0 volts.

After you complete the experiment, take the online **Post-lab Quiz** for Activity 1.

Activity Questions

1. Draw a graph that depicts a single skeletal muscle twitch, placing time on the X-axis and force on the Y-axis. Label the phases of this muscle twitch and describe what is happening in the muscle during each phase.

2. During the latent period of a skeletal muscle twitch, there is an apparent lack of muscle activity. Describe the electrical and chemical changes that occur in the muscle during this period.

___

___

___

ACTIVITY 2

The Effect of Stimulus Voltage on Skeletal Muscle Contraction

OBJECTIVES

1. To understand the terms *motor neuron, muscle twitch, motor unit, recruitment, stimulus voltage, threshold stimulus,* and *maximal stimulus.*
2. To understand how motor unit recruitment can increase the tension a whole muscle develops.
3. To identify a threshold stimulus voltage.
4. To observe the effect of increases in stimulus voltage on a whole muscle.
5. To understand how increasing stimulus voltage to an isolated muscle in an experiment mimics motor unit recruitment in the body.

Introduction

A skeletal muscle produces **tension** (also known as **muscle force**) when nervous or electrical stimulation is applied. The force generated by a whole muscle reflects the number of active **motor units** at a given moment. A strong muscle contraction implies that many motor units are activated, with each unit developing its maximal tension, or force. A weak muscle contraction implies that fewer motor units are activated, but each motor unit still develops its maximal tension. By increasing the number of active motor units, we can produce a steady increase in muscle force, a process called **motor unit recruitment.**

Regardless of the number of **motor units** activated, a single stimulated contraction of whole skeletal muscle is called a **muscle twitch.** A tracing of a muscle twitch is divided into three phases: the latent period, the contraction phase, and the relaxation phase. The latent period is a short period between the time of muscle stimulation and the beginning of a muscle response. Although no force is generated during this interval, chemical changes occur intracellularly in preparation for contraction (including the release of calcium from the sarcoplasmic reticulum). During the contraction phase, the myofilaments utilize the cross-bridge cycle and the muscle develops tension. Relaxation takes place when the contraction has ended and the muscle returns to its normal resting state and length.

In this activity you will stimulate an isometric, or fixed-length, contraction of an isolated skeletal muscle. This activity allows you to investigate how the strength of an electrical stimulus affects whole-muscle function. Note that these simulations involve indirect stimulation by an electrode placed on the surface of the muscle. Indirect stimulation differs from the situation in vivo, where each fiber in the muscle receives direct stimulation via a nerve ending. Nevertheless, increasing the intensity of the electrical stimulation mimics how the nervous system increases the number of activated motor units.

The **threshold voltage** is the smallest stimulus required to induce an action potential in a muscle fiber's plasma membrane, or sarcolemma. As the **stimulus voltage** to a muscle is increased beyond the threshold voltage, the amount of force produced by the whole muscle also increases. This result occurs because, as more voltage is delivered to the whole muscle, more muscle fibers are activated and, thus, the total force produced by the muscle increases. Maximal tension in the whole muscle occurs when all the muscle fibers have been activated by a sufficiently strong stimulus (referred to as the **maximal voltage**). Stimulation with voltages greater than the maximal voltage will not increase the force of contraction. This experiment is analogous to, and accurately mimics, muscle activity in vivo, where the recruitment of additional motor units increases the total muscle force produced. This phenomenon is called *motor unit recruitment.*

EQUIPMENT USED The following equipment will be depicted on-screen: intact, viable skeletal muscle dissected off the leg of a frog; electrical stimulator—delivers the desired amount and duration of stimulating voltage to the muscle via electrodes resting on the muscle; mounting stand—includes a force transducer to measure the amount of force, or tension, developed by the muscle; oscilloscope—displays the stimulated muscle twitch and the amount of active, passive, and total force developed by the muscle.

Experiment Instructions

Go to the home page in the PhysioEx software and click **Exercise 2: Skeletal Muscle Physiology.** Click **Activity 2: The Effect of Stimulus Voltage on Skeletal Muscle Contraction,** and take the online **Pre-lab Quiz** for Activity 2.

After you take the online Pre-lab Quiz, click the **Experiment** tab and begin the experiment. The experiment instructions are reprinted here for your reference. The opening screen for the experiment is shown on the following page.

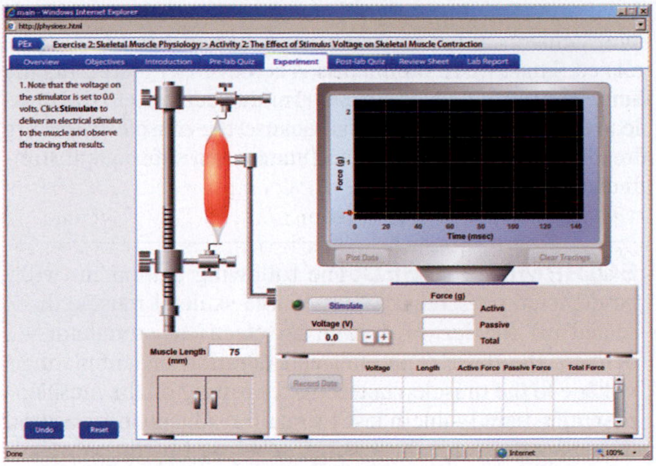
main - Windows Internet Explorer
http://physioex.html
PEx
Exercise 2: Skeletal Muscle Physiology > Activity 2: The Effect of Stimulus Voltage on Skeletal Muscle Contraction
Overview
Objectives
Introduction
Pre-lab Quiz
Experiment
Post-lab Quiz
Review Sheet
Lab Report
1. Note that the voltage on the stimulator is set to 0.0 volts. Click Stimulate to deliver an electrical stimulus to the muscle and observe the tracing that results.
Force (g)
Time (msec)
Plot Data
Clear Tracings
Stimulate
Voltage (V)
0.0
Force (g)
Active
Passive
Total
Muscle Length (mm)
75
Record Data
Voltage
Length
Active Force
Passive Force
Total Force
Undo
Reset
Done
Internet
100%

1. Note that the voltage on the stimulator is set to 0.0 volts. Click **Stimulate** to deliver an electrical stimulus to the muscle and observe the tracing that results.

2. Note the active force display and then click **Record Data** to display your results in the grid (and record your results in Chart 2).

3. Increase the voltage to 0.2 volts by clicking the + button beside the voltage display. Click **Stimulate** to deliver an electrical stimulus to the muscle and observe the tracing that results.

4. Note the active force display and then click **Record Data** to display your results in the grid (and record your results in Chart 2).

5. You will now gradually increase the voltage and stimulate the muscle to determine the minimum voltage required to generate active force.

   - Increase the voltage by 0.1 volts and then click **Stimulate.**
   - If no active force is generated, increase the voltage by 0.1 volts and stimulate the muscle again. When active force is generated, click **Record Data** to display your results in the grid (and record your results in Chart 2).

6. Enter the threshold voltage for this experiment in the field below and then click **Submit** to record your answer in the lab report. ________ volts

7. Click **Clear Tracings** to clear the tracings on the oscilloscope.

> **? PREDICT Question 1**
> As the stimulus voltage is increased from 1.0 volt up to 10 volts, what will happen to the amount of active force generated with each stimulus?
> ____

8. Increase the voltage on the stimulator to 1.0 volt and then click **Stimulate.**

9. Note the active force display and then click **Record Data** to display your results in the grid (and record your results in Chart 2).

CHART 2 Effect of Stimulus Voltage on Skeletal Muscle Contraction

Voltage	Active force (g)

10. You will now gradually increase the voltage and stimulate the muscle to determine the maximal voltage.

   - Increase the voltage by 0.5 volts.
   - Click **Stimulate** and observe the tracing that results.
   - Note the active force display and then click **Record Data** to display your results in the grid (and record your results in Chart 2).

   Repeat this step until you reach 10.0 volts.

11. Click **Plot Data** to view a summary of your data on a plotted grid. Click **Submit** to record your plot in the lab report.

12. Enter the maximal voltage for this experiment in the field below and then click **Submit** to record your answer in the lab report. ________ volts

After you complete the experiment, take the online **Post-lab Quiz** for Activity 2.

Activity Questions

1. For a single skeletal muscle twitch, explain the effect of increasing stimulus voltage.

____

____

2. How is this effect achieved in vivo?

____

____

ACTIVITY 3

The Effect of Stimulus Frequency on Skeletal Muscle Contraction

OBJECTIVES

1. To understand the terms *stimulus frequency, wave summation,* and *treppe.*
2. To observe the effect of an increasing stimulus frequency on the force developed by an isolated skeletal muscle.
3. To understand how increasing stimulus frequency to an isolated skeletal muscle induces the summation of twitch force.

Introduction

As demonstrated in Activity 2, increasing the stimulus voltage to an isolated skeletal muscle (up to a maximal value) results in an increase of force produced by the whole muscle. This experimental result is analogous to motor unit recruitment in the body. Importantly, this result relies on being able to increase the single stimulus intensity in the experiment. You will now explore another way to increase the force produced by an isolated skeletal muscle.

When a muscle first contracts, the force it is able to produce is less than the force it is able to produce with subsequent stimulations within a relatively short time span. **Treppe** is the progressive increase in force generated when a muscle is stimulated in succession, such that muscle twitches follow one another closely, with each successive twitch peaking slightly higher than the one before. This step-like increase in force is why treppe is also known as the staircase effect. For the first few twitches, each successive twitch produces slightly more force than the previous twitch as long as the muscle is allowed to fully relax between stimuli and the stimuli are delivered relatively close together.

When a skeletal muscle is stimulated repeatedly, such that the stimuli arrive one after another within a short period of time, muscle twitches can overlap with each other and result in a stronger muscle contraction than a stand-alone twitch. This phenomenon is known as wave summation. **Wave summation** occurs when muscle fibers that are developing tension are stimulated again before the fibers have relaxed. Thus, wave summation is achieved by increasing the **stimulus frequency,** or rate of stimulus delivery to the muscle. Wave summation occurs because the muscle fibers are already in a partially contracted state when subsequent stimuli are delivered.

EQUIPMENT USED The following equipment will be depicted on-screen: intact, viable skeletal muscle dissected off the leg of a frog; an electrical stimulator—delivers the desired amount and duration of stimulating voltage to the muscle via electrodes resting on the muscle; mounting stand—includes a force transducer to measure the amount of force, or tension, developed by the muscle; oscilloscope—displays the stimulated muscle twitch and the amount of active, passive, and total force developed by the muscle.

Experiment Instructions

Go to the home page in the PhysioEx software and click **Exercise 2: Skeletal Muscle Physiology.** Click **Activity 3: The Effect of Stimulus Frequency on Skeletal Muscle Contraction,** and take the online **Pre-lab Quiz** for Activity 3.

After you take the online Pre-lab Quiz, click the **Experiment** tab and begin the experiment. The experiment instructions are reprinted here for your reference. The opening screen for the experiment is shown below.

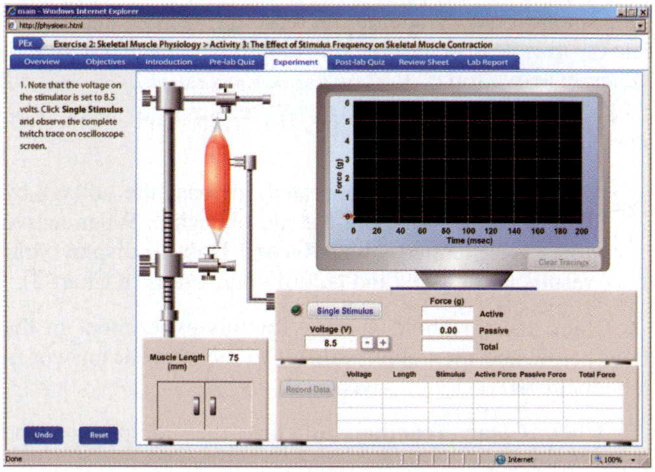
main - Windows Internet Explorer
http://physioex.html
PEx
Exercise 2: Skeletal Muscle Physiology > Activity 3: The Effect of Stimulus Frequency on Skeletal Muscle Contraction
Overview
Objectives
Introduction
Pre-lab Quiz
Experiment
Post-lab Quiz
Review Sheet
Lab Report
1. Note that the voltage on the stimulator is set to 8.5 volts. Click Single Stimulus and observe the complete twitch trace on oscilloscope screen.
Force (g)
Time (msec)
Clear Tracings
Single Stimulus
Voltage (V)
8.5
Force (g)
Active
0.00
Passive
Total
Muscle Length (mm)
75
Voltage
Length
Stimulus
Active Force
Passive Force
Total Force
Record Data
Undo
Reset
Done
Internet
100%

1. Note that the voltage on the stimulator is set to 8.5 volts. Click **Single Stimulus** and observe the tracing that results on the oscilloscope.

2. Note the active force display and then click **Record Data** to display your results in the grid (and record your results in Chart 3).

3. Click **Single Stimulus** and allow the trace to rise and completely fall. *Immediately after* the trace has returned to baseline, click **Single Stimulus** again.

CHART 3 **Effect of Stimulus Frequency on Skeletal Muscle Contraction**

Voltage	Stimulus	Active force (g)

4. Note the active force for the second muscle twitch and click **Record Data** to display your results in the grid (and record your results in Chart 3).

5. You should have observed an increase in active force generated by the muscle with the immediate second stimulus. This increase demonstrates the phenomenon of treppe. Click **Clear Tracings** to clear the tracings on the oscilloscope.

6. You will now investigate the process of wave summation. Click **Single Stimulus** and watch the trace rise and begin to fall. *Before* the trace falls completely back to the baseline, click **Single Stimulus** again. (You can simply click **Single Stimulus** twice in quick succession in order to achieve this.)

7. Note the active force for the second muscle twitch and click **Record Data** to display your results in the grid (and record your results in Chart 3).

? PREDICT Question 1
As the stimulus frequency increases, what will happen to the muscle force generated with each successive stimulus? Will there be a limit to this response?

____

____

____

8. Now stimulate the muscle at a higher frequency by clicking **Single Stimulus** four times in rapid succession.

9. Note the active force display and then click **Record Data** to display your results in the grid (and record your results in Chart 3).

10. Click **Clear Tracings** to clear the tracings on the oscilloscope.

? PREDICT Question 2
In order to produce sustained muscle contractions with an active force value of 5.2 grams, do you think you need to increase the stimulus voltage?

____

11. Increase the voltage to 10.0 volts by clicking the + button beside the voltage display. After setting the voltage, click **Single Stimulus** four times in rapid succession.

12. Note the active force display and then click **Record Data** to display your results in the grid (and record your results in Chart 3).

13. Click **Clear Tracings** to clear the tracings on the oscilloscope.

14. Return the voltage to 8.5 volts by clicking the − button beside the voltage display. After setting the voltage, click **Single Stimulus** as many times as you can in rapid succession. Note the active force display. If you did not achieve an active force of 5.2 grams, click **Clear Tracings** and then click **Single Stimulus** even more rapidly. Repeat this step until you achieve an active force of 5.2 grams.

When you achieve an active force of 5.2 grams, click **Record Data** to display your results in the grid (and record your results in Chart 3).

After you complete the experiment, take the online **Post-lab Quiz** for Activity 3.

Activity Questions

1. Why is treppe also known as the staircase effect?

____

____

2. What changes are thought to occur in the skeletal muscle to allow treppe to be observed?

____

____

3. How does the frequency of stimulation affect the amount of force generated by a skeletal muscle?

____

____

4. Explain how wave summation is achieved in vivo.

____

____

ACTIVITY 4

Tetanus in Isolated Skeletal Muscle

OBJECTIVES

1. To understand the terms *stimulus frequency, unfused tetanus, fused tetanus,* and *maximal tetanic tension.*
2. To observe the effect of an increasing stimulus frequency on an isolated skeletal muscle.
3. To understand how increasing the stimulus frequency to an isolated skeletal muscle leads to unfused or fused tetanus.

Introduction

As demonstrated in Activity 3, increasing the **stimulus frequency** to an isolated skeletal muscle results in an increase in force produced by the whole muscle. Specifically, you observed that, if electrical stimuli are applied to a skeletal muscle in quick succession, the overlapping twitches generated more force with each successive stimulus. However, if stimuli continue to be applied frequently to a muscle over a prolonged period of time, the maximum possible muscle force from each stimulus will eventually reach a plateau—a state known as **unfused tetanus.** If stimuli are then applied with even greater frequency, the twitches will begin to fuse so that the peaks and valleys of each twitch become indistinguishable from one another—this state is known as **complete (fused) tetanus.** When the stimulus frequency reaches a value beyond which no further increases in force are generated by the muscle, the muscle has reached its **maximal tetanic tension.**

EQUIPMENT USED The following equipment will be depicted on-screen: intact, viable skeletal muscle dissected off the leg of a frog; electrical stimulator—delivers the desired amount and duration of stimulating voltage to the muscle via electrodes resting on the muscle; mounting stand—includes a force transducer to measure the amount of force, or tension, developed by the muscle; oscilloscope—displays the stimulated muscle twitch and the amount of active, passive, and total force developed by the muscle.

Experiment Instructions

Go to the home page in the PhysioEx software and click **Exercise 2: Skeletal Muscle Physiology.** Click **Activity 4: Tetanus in Isolated Skeletal Muscle** and take the online **Pre-lab Quiz** for Activity 4.

After you take the online Pre-lab Quiz, click the **Experiment** tab and begin the experiment. The experiment instructions are reprinted here for your reference. The opening screen for the experiment is shown above.

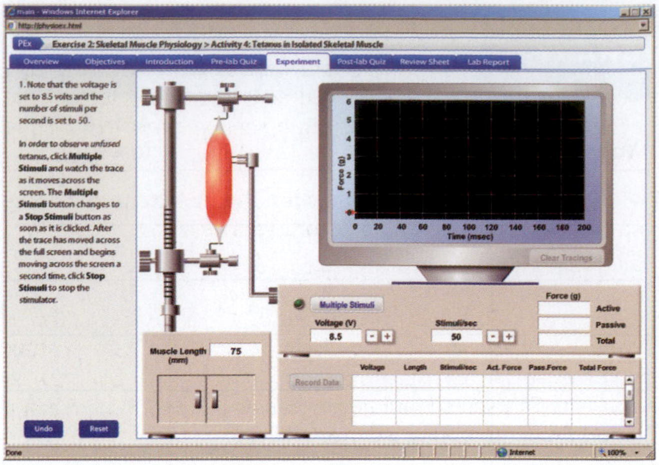
Exercise 2: Skeletal Muscle Physiology > Activity 4: Tetanus in Isolated Skeletal Muscle
Overview
Objectives
Introduction
Pre-lab Quiz
Experiment
Post-lab Quiz
Review Sheet
Lab Report
1. Note that the voltage is set to 8.5 volts and the number of stimuli per second is set to 50.
In order to observe unfused tetanus, click Multiple Stimuli and watch the trace as it moves across the screen. The Multiple Stimuli button changes to a Stop Stimuli button as soon as it is clicked. After the trace has moved across the full screen and begins moving across the screen a second time, click Stop Stimuli to stop the stimulator.
Force (g)
Time (msec)
Clear Tracings
Multiple Stimuli
Voltage (V)
8.5
Stimuli/sec
50
Active
Passive
Total
Muscle Length (mm)
75
Record Data
Voltage
Length
Stimuli/sec
Act. Force
Pass.Force
Total Force
Undo
Reset

1. Note that the voltage is set to 8.5 volts and the number of stimuli per second is set to 50. To observe *unfused* tetanus, click **Multiple Stimuli** and watch the trace as it moves across the screen. The **Multiple Stimuli** button changes to a **Stop Stimuli** button after it is clicked. After the trace has moved across the full screen and begins moving across the screen a second time, click **Stop Stimuli** to stop the stimulator.

2. Click **Record Data** to display your results in the grid (and record your results in Chart 4).

CHART 4 Tetanus in Isolated Skeletal Muscle

Stimuli/second	Active force (g)

? PREDICT Question 1
As the stimulus frequency increases further, what will happen to the muscle tension and twitch appearance with each successive stimulus? Will there be a limit to this response?

____

____

3. In order to observe *fused* tetanus, increase the stimuli/sec setting to 130 by clicking the + button beside the stimuli/sec display. Click **Multiple Stimuli** and observe the resulting trace. After the trace has moved across the full screen and begins moving across the screen a second time, click **Stop Stimuli.**

4. Note the fused tetanus and click **Record Data** to display your results in the grid (and record your results in Chart 4).

5. Click **Clear Tracings** to clear the oscilloscope screen.

6. Increase the stimuli/sec setting to 140 by clicking the + button beside the stimuli/sec display. Click **Multiple Stimuli** and observe the resulting trace. After the trace has moved across the full screen and begins moving across the screen a second time, click **Stop Stimuli.**

7. Note the fused tetanus and click **Record Data** to display your results in the grid (and record your results in Chart 4).

8. Click **Clear Tracings** to clear the oscilloscope screen.

9. You will now observe the effect of incremental increases in the number of stimuli per second above 140 stimuli per second.

- Increase the stimuli/sec setting by 2.
- Click **Multiple Stimuli** and observe the resulting trace. After the trace has moved across the full screen and begins moving across the screen a second time, click **Stop Stimuli.**
- Click **Record Data** to display your results in the grid (and record your results in Chart 4).
- Click **Clear Tracings** to clear the oscilloscope screen.

Repeat this step until you reach 150 stimuli per second.

After you complete the experiment, take the online **Post-lab Quiz** for Activity 4.

Activity Questions

1. Explain what you think is being summated in the skeletal muscle to allow a high stimulus frequency to induce a smooth, continuous skeletal muscle contraction.

____

____

2. Why do many toddlers receive a tetanus shot (and then subsequent booster shots, as needed, later in life)? How does the condition known as "lockjaw" relate to tetanus shots?

____

____

ACTIVITY 5

Fatigue in Isolated Skeletal Muscle

OBJECTIVES

1. To understand the terms *stimulus frequency, complete (fused) tetanus, fatigue,* and *rest period.*
2. To observe the development of skeletal muscle fatigue.
3. To understand how the length of intervening rest periods determines the onset of fatigue.

Introduction

As demonstrated in Activities 3 and 4, increasing the stimulus frequency to an isolated skeletal muscle induces an increase of force produced by the whole muscle. Specifically, if voltage stimuli are applied to a muscle frequently in quick succession, the skeletal muscle generates more force with each successive stimulus.

However, if stimuli continue to be applied frequently to a muscle over a prolonged period of time, the maximum force of each twitch eventually reaches a plateau—a state known as *unfused tetanus*. If stimuli are then applied with even greater frequency, the twitches begin to fuse so that the peaks and valleys of each twitch become indistinguishable from one another—this state is known as **complete (fused) tetanus**. When the **stimulus frequency** reaches a value beyond which no further increase in force is generated by the muscle, the muscle has reached its **maximal tetanic tension.**

In this activity you will observe the phenomena of skeletal muscle *fatigue*. Fatigue refers to a decline in a skeletal muscle's ability to maintain a constant level of force, or tension, after prolonged, repetitive stimulation. You will also demonstrate how intervening **rest periods** alter the onset of fatigue in skeletal muscle. The causes of fatigue are still being investigated and multiple molecular events are thought to be involved, though the accumulations of lactic acid, ADP, and P_i in muscles are thought to be the major factors causing fatigue in the case of high-intensity exercise.

Common definitions for **fatigue** are:

- The failure of a muscle fiber to produce tension because of previous contractile activity.
- A decline in the muscle's ability to maintain a constant force of contraction after prolonged, repetitive stimulation.

EQUIPMENT USED The following equipment will be depicted on-screen: intact, viable skeletal muscle dissected off the leg of a frog; electrical stimulator—delivers the desired amount and duration of stimulating voltage to the muscle via electrodes resting on the muscle; mounting stand—includes a force transducer to measure the amount of force, or tension, developed by the muscle; oscilloscope—displays the stimulated muscle twitch and the amount of active, passive, and total force developed by the muscle.

Experiment Instructions

Go to the home page in the PhysioEx software and click **Exercise 2: Skeletal Muscle Physiology.** Click **Activity 5: Fatigue in Isolated Skeletal Muscle,** and take the online **Pre-lab Quiz** for Activity 5.

After you take the online Pre-lab Quiz, click the **Experiment** tab and begin the experiment. The experiment instructions are reprinted here for your reference. The opening screen for the experiment is shown on the following page.

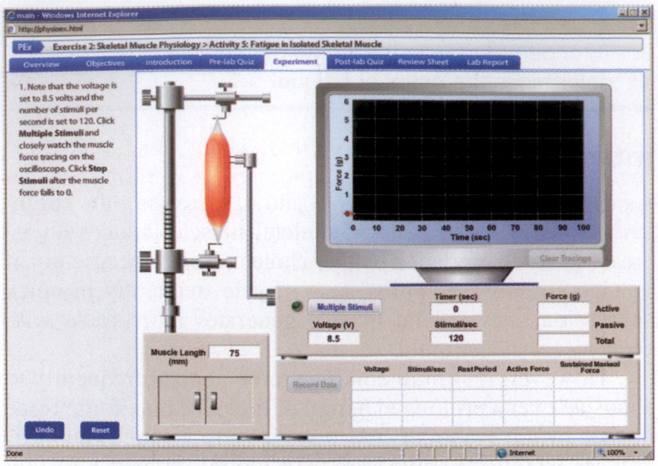
main - Windows Internet Explorer
Exercise 2: Skeletal Muscle Physiology > Activity 5: Fatigue in Isolated Skeletal Muscle
Overview
Objectives
Introduction
Pre-lab Quiz
Experiment
Post-lab Quiz
Review Sheet
Lab Report
1. Note that the voltage is set to 8.5 volts and the number of stimuli per second is set to 120. Click Multiple Stimuli and closely watch the muscle force tracing on the oscilloscope. Click Stop Stimuli after the muscle force falls to 0.
Force (g)
Time (sec)
Clear Tracings
Multiple Stimuli
Timer (sec)
0
Force (g)
Active
Passive
Total
Voltage (V)
8.5
Stimuli/sec
120
Muscle Length (mm)
75
Record Data
Voltage
Stimuli/sec
Rest Period
Active Force
Sustained Maximal Force
Undo
Reset
Done
Internet
100%

1. Note that the voltage is set to 8.5 volts and the number of stimuli per second is set to 120. Click **Multiple Stimuli** and closely watch the muscle force tracing on the oscilloscope. Click **Stop Stimuli** after the muscle force falls to 0.

2. Click **Record Data** to display your results in the grid (and record your results in Chart 5).

CHART 5 Fatigue Results

Rest period (sec)	Active force (g)	Sustained maximal force (sec)

3. Click **Clear Tracings** to clear the oscilloscope screen.

> **? PREDICT Question 1**
> If the stimulator is briefly turned off for defined periods of time, what will happen to the length of time that the muscle is able to sustain maximal developed tension when the stimulator is turned on again?
>
> ______________________________
>
> ______________________________

4. To demonstrate the onset of fatigue after a variable rest period, you will be clicking the **Multiple Stimuli** button on and off three times. Read through the steps below before proceeding. Watch the timer closely to help you determine when to turn the stimulator back on.

- Click **Multiple Stimuli.**
- After the muscle force falls to 0, click **Stop Stimuli** to turn off the stimulator.
- Wait 10 seconds, then click **Multiple Stimuli** to turn the stimulator back on.
- Click **Stop Stimuli** after the muscle force falls to 0.
- Wait 20 seconds, then click **Multiple Stimuli** to turn the stimulator back on.
- Click **Stop Stimuli** after the muscle force falls to 0.

5. Click **Record Data** to display your results in the grid (and record your results in Chart 5).

After you complete the experiment, take the online **Post-lab Quiz** for Activity 5.

Activity Questions

1. What proposed mechanisms most likely explain why fatigue develops?

____

____

2. What would you recommend to an interested friend as the best ways to delay the onset of fatigue?

____

____

ACTIVITY 6

The Skeletal Muscle Length-Tension Relationship

OBJECTIVES

1. To understand the terms *isometric contraction, active force, passive force, total force,* and *length-tension relationship.*
2. To understand the effect that resting muscle length has on tension development when the muscle is maximally stimulated in an isometric experiment.
3. To explain the molecular basis of the skeletal muscle length-tension relationship.

Introduction

Skeletal muscle contractions are either isometric or isotonic. When a muscle attempts to move a load that is equal to the force generated by the muscle, the muscle contracts isometrically. During an **isometric** contraction, the muscle stays at a fixed length (*isometric* means "same length"). An example of isometric muscle contraction is when you stand in a doorway and push on the doorframe. The load that you are attempting to move (the doorframe) can easily equal the force generated by your muscles, so your muscles do not shorten even though they are actively contracting.

Isometric contractions are accomplished experimentally by keeping both ends of the muscle in a fixed position while electrically stimulating the muscle. Resting length (the length of the muscle before stimulation) is an important factor in determining the amount of force that a muscle can develop when stimulated. **Passive force** is generated by stretching the muscle and results from the elastic recoil of the tissue itself. This passive force is largely caused by the protein titin, which acts as a molecular bungee cord. **Active force** is generated when myosin thick filaments bind to actin thin filaments, thus engaging the cross bridge cycle and ATP hydrolysis. Think of the skeletal muscle as having two force properties: it exerts passive force when it is stretched (like a rubber band exerts passive force) and active force when it is stimulated. **Total force** is the sum of passive and active forces.

This activity allows you to set and hold constant the length of the isolated skeletal muscle and subsequently stimulate it with individual maximal voltage stimuli. A graph relating the three forces generated and the fixed length of the muscle will be automatically plotted after you stimulate the muscle. In muscle physiology this graph is known as the **isometric length-tension relationship.** The results of this simulation can be applied to human muscles to understand how optimum resting length will result in maximum force production.

To understand why muscle tissue behaves as it does, you must understand tension at the cellular level. If you have difficulty understanding the results of this activity, review the sliding filament model of muscle contraction. Think of the length-tension relationship in terms of those sarcomeres that are too short, those that are too long, and those that have the ideal amount of thick and thin filament overlap.

EQUIPMENT USED The following equipment will be depicted on-screen: intact, viable skeletal muscle dissected off the leg of a frog; electrical stimulator—delivers the desired amount and duration of stimulating voltage to the muscle via electrodes resting on the muscle; mounting stand—includes (1) a force transducer to measure the amount of force, or tension, developed by the muscle and (2) a gearing system that allows the hook through the muscle's lower tendon to be moved up or down, thus altering the fixed length of the muscle; oscilloscope—displays the stimulated muscle twitch and the amount of active, passive, and total force developed by the muscle.

Experiment Instructions

Go to the home page in the PhysioEx software and click **Exercise 2, Skeletal Muscle Physiology.** Click **Activity 6, The Skeletal Muscle Length-Tension Relationship,** and take the online **Pre-lab Quiz** for Activity 6.

After you take the online Pre-lab Quiz, click the **Experiment** tab and begin the experiment. The experiment instructions are reprinted here for your reference. The opening screen for the experiment is shown below.

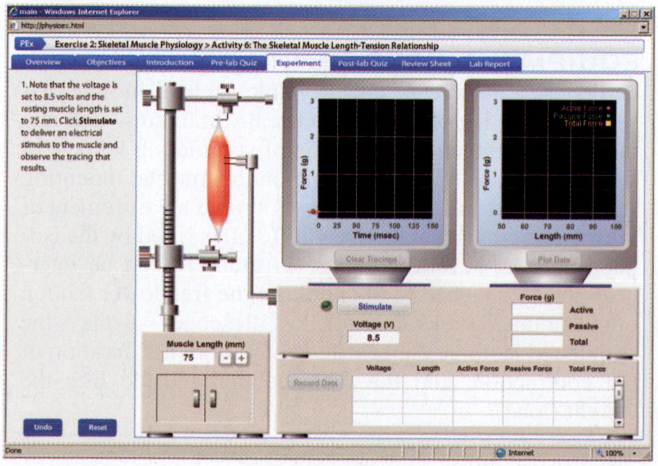
main - Windows Internet Explorer
http://physioex.html
PEx
Exercise 2: Skeletal Muscle Physiology > Activity 6: The Skeletal Muscle Length-Tension Relationship
Overview
Objectives
Introduction
Pre-lab Quiz
Experiment
Post-lab Quiz
Review Sheet
Lab Report
1. Note that the voltage is set to 8.5 volts and the resting muscle length is set to 75 mm. Click Stimulate to deliver an electrical stimulus to the muscle and observe the tracing that results.
Force (g)
Time (msec)
Length (mm)
Clear Tracings
Plot Data
Stimulate
Voltage (V)
8.5
Force (g)
Active
Passive
Total
Muscle Length (mm)
75
Record Data
Voltage
Length
Active Force
Passive Force
Total Force
Undo
Reset
Done
Internet
100%

1. Note that the voltage is set to 8.5 volts and the resting muscle length is set to 75 mm. Click **Stimulate** to deliver an electrical stimulus to the muscle and observe the tracing that results.

2. You should see a single muscle twitch tracing on the left oscilloscope display and three data points (representing active, passive, and total force generated during this twitch) plotted on the right display. The yellow box represents the total force, the red dot contained within the yellow box represents the active force, and the green square represents the passive force. Click **Record Data** to display your results in the grid (and record your results in Chart 6).

CHART 6 Skeletal Muscle Length-Tension Relationship

Length (mm)	Active force (g)	Passive force (g)	Total force (g)

PREDICT Question 1
As the resting length of the muscle is changed, what will happen to the amount of total force the muscle generates during the stimulated twitch?

____

____

3. You will now gradually shorten the muscle to determine the effect of muscle length on active, passive, and total force.

- Shorten the muscle by 5 mm by clicking the – button beside the muscle length display.
- Click **Stimulate** to deliver an electrical stimulus to the muscle and note the values of the total, active, and passive forces relative to those observed at the original 75 mm.
- Click **Record Data** to display your results in the grid (and record your results in Chart 6).

Repeat these steps until you reach a muscle length of 50 mm.

4. Click **Clear Tracings** to clear the left oscilloscope display.

5. Lengthen the muscle to 80 mm by clicking the + button beside the muscle length display. Click **Stimulate** to deliver an electrical stimulus to the muscle and note the values of the total, active, and passive forces relative to those observed at the original 75 mm.

6. Click **Record Data** to display your results in the grid (and record your results in Chart 6).

7. You will now gradually lengthen the muscle to determine the effect of muscle length on active, passive, and total force.

- Lengthen the muscle by 10 mm by clicking the + button beside the muscle length display.
- Click **Stimulate** to deliver an electrical stimulus to the muscle and note the values of the total, active, and passive forces relative to those observed at the original 75 mm.
- Click **Record Data** to display your results in the grid (and record your results in Chart 6).

Repeat these steps until you reach a muscle length of 100 mm.

8. Click **Plot Data** to view a summary of your data on a plotted grid. Click **Submit** to record your plot in the lab report.

After you complete the experiment, take the online **Post-lab Quiz** for Activity 6.

Activity Questions

1. Explain what happens in the skeletal muscle sarcomere to result in the changes in active, passive, and total force when the resting muscle length is changed.

____

____

2. Explain the dip in the total force curve as the muscle was stretched to longer lengths. (Hint: Keep in mind that you are measuring the sum of active and passive forces.)

____

____

ACTIVITY 7

Isotonic Contractions and the Load-Velocity Relationship

OBJECTIVES

1. To understand the terms *isotonic concentric contraction, load, latent period, shortening velocity,* and *load-velocity relationship.*
2. To understand the effect that increasing load (that is, weight) has on an isolated skeletal muscle when the muscle is stimulated in an isotonic contraction experiment.
3. To understand the load-velocity relationship in isolated skeletal muscle.

Introduction

Skeletal muscle contractions can be described as either isometric or isotonic. When a muscle attempts to move an object (the **load**) that is equal in weight to the force generated by the muscle, the muscle is observed to contract isometrically. In an isometric contraction, the muscle stays at a fixed length (*isometric* means "same length").

During an **isotonic contraction,** the skeletal muscle length changes and, thus, the load moves a measurable distance. If the muscle length shortens as the load moves, the contraction is called an **isotonic *concentric* contraction.** An isotonic concentric contraction occurs when a muscle generates a force greater than the load attached to the muscle's end. In this type of contraction, there is a **latent period** during which there is a rise in muscle tension but no observable movement of the weight. After the muscle tension exceeds the weight of the load, an isotonic concentric contraction can begin. Thus, the latent period gets longer as the weight of the load gets larger. When the building muscle force exceeds the load, the muscle shortens and the weight moves. Eventually, the force of the muscle contraction will decrease as the muscle twitch begins the relaxation phase, and the load will therefore start to return to its original position.

An isotonic twitch is not an all-or-nothing event. If the load is increased, the muscle must generate more force to move it and the latent period will therefore get longer because it will take more time for the necessary force to be generated by the muscle. The speed of the contraction (muscle **shortening velocity**) also depends on the load that the muscle is attempting to move. Maximal shortening velocity is attained with minimal load attached to the muscle. Conversely, the heavier the load, the slower the muscle twitch. You can think of lifting an object from the floor as an example. A light object can be lifted quickly (high velocity), whereas a heavier object will be lifted with a slower velocity for a shorter duration.

In an isotonic muscle contraction experiment, one end of the muscle remains free (unlike in an isometric contraction experiment, where both ends of the muscle are held in a fixed position). Different weights (loads) can then be attached to the free end of the isolated muscle, while the other end is held in a fixed position by the force transducer. If the weight (the load) is less than the tension generated by the whole muscle, then the muscle will be able to lift it with a measurable distance, velocity, and duration. In this activity, you will change the weight (load) that the muscle will try to move as it shortens.

EQUIPMENT USED The following equipment will be depicted on-screen: intact, viable skeletal muscle dissected off the leg of a frog; electrical stimulator—delivers the desired amount and duration of stimulating voltage to the muscle via electrodes resting on the muscle; mounting stand—includes a ruler that allows a rapid measurement of the distance (cm) that the weight (load) is lifted by the isolated muscle; several weights (in grams)—can be interchangeably attached to the hook on the free lower tendon of the mounted skeletal muscle; oscilloscope—displays the stimulated isotonic concentric contraction, the duration of the contraction, and the distance that muscle lifts the weight (load).

Experiment Instructions

Go to the home page in the PhysioEx software and click **Exercise 2: Skeletal Muscle Physiology.** Click **Activity 7: Isotonic Contractions and the Load-Velocity Relationship,** and take the online **Pre-lab Quiz** for Activity 7.

After you take the online Pre-lab Quiz, click the **Experiment** tab and begin the experiment. The experiment instructions are reprinted here for your reference. The opening screen for the experiment is shown below.

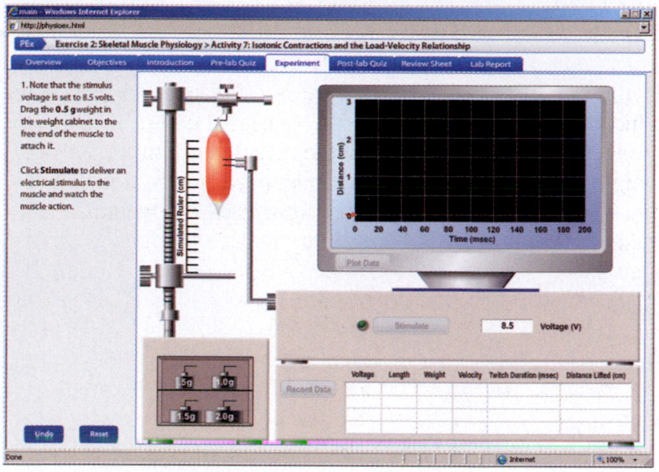
main - Windows Internet Explorer
http://physioex.html
PEx
Exercise 2: Skeletal Muscle Physiology > Activity 7: Isotonic Contractions and the Load-Velocity Relationship
Overview
Objectives
Introduction
Pre-lab Quiz
Experiment
Post-lab Quiz
Review Sheet
Lab Report
1. Note that the stimulus voltage is set to 8.5 volts. Drag the 0.5 g weight in the weight cabinet to the free end of the muscle to attach it.
Click Stimulate to deliver an electrical stimulus to the muscle and watch the muscle action.
Simulated Ruler (cm)
Distance (cm)
Time (msec)
Plot Data
Stimulate
8.5
Voltage (V)
Record Data
Voltage
Length
Weight
Velocity
Twitch Duration (msec)
Distance Lifted (cm)
.5g
1.0g
1.5g
2.0g
Undo
Reset
Done
Internet
100%

1. Note that the stimulus voltage is set to 8.5 volts. Drag the 0.5-g weight in the weight cabinet to the free end of the muscle to attach it. Click **Stimulate** to deliver an electrical stimulus to the muscle and watch the muscle action.

2. Observe that, as the muscle shortens in length, it lifts the weight off the platform. The muscle then lengthens as it relaxes and lowers the weight back down to the platform. Click **Stimulate** again and try to watch both the muscle and the oscilloscope screen at the same time.

3. Click **Record Data** to display your results in the grid (and record your results in Chart 7).

CHART 7 Isotonic Contraction Results

Weight (g)	Velocity (cm/sec)	Twitch duration (msec)	Distance lifted (cm)

? PREDICT Question 1
As the load on the muscle *increases*, what will happen to the latent period, the shortening velocity, the distance that the weight moved, and the contraction duration?

____

____

4. Remove the 0.5-g weight by dragging it back to the weight cabinet. Drag the 1.0-g weight to the free end of the muscle to attach it. Click **Stimulate** and observe the muscle and the oscilloscope screen.

5. Click **Record Data** to display your results in the grid (and record your results in Chart 7).

6. Remove the 1.0-g weight by dragging it back to the weight cabinet. Drag the 1.5-g weight to the free end of the muscle to attach it. Click **Stimulate** and observe the muscle and the oscilloscope screen.

7. Click **Record Data** to display your results in the grid (and record your results in Chart 7).

8. Remove the 1.5-g weight by dragging it back to the weight cabinet. Drag the 2.0-g weight to the free end of the muscle to attach it. Click **Stimulate** and observe the muscle and the oscilloscope screen.

9. Click **Record Data** to display your results in the grid (and record your results in Chart 7).

10. Click **Plot Data** to generate a muscle load-velocity relationship. Watch the display carefully as the program animates the development of a load-velocity relationship for the data you have collected. Click **Submit** to record your plot in the lab report.

After you complete the experiment, take the online **Post-lab Quiz** for Activity 7.

Activity Questions

1. Explain the relationship between the load attached to a skeletal muscle and the initial velocity of skeletal muscle shortening.

____

____

____

2. Explain why it will take you longer to perform ten repetitions lifting a 20-pound weight than it would to perform the same number of repetitions with a 5-pound weight.

____

____

____

PhysioEx 9.0

REVIEW SHEET

EXERCISE 2

NAME ____________________

LAB TIME/DATE ____________________

Skeletal Muscle Physiology

ACTIVITY 1 The Muscle Twitch and the Latent Period

1. Define the terms *skeletal muscle fiber, motor unit, skeletal muscle twitch, electrical stimulus,* and *latent period.* ____________________

2. What is the role of acetylcholine in a skeletal muscle contraction? ____________________

3. Describe the process of excitation-contraction coupling in skeletal muscle fibers. ____________________

4. Describe the three phases of a skeletal muscle twitch. ____________________

5. Does the duration of the latent period change with different stimulus voltages? How well did the results compare with your prediction? ____________________

6. At the threshold stimulus, do sodium ions start to move into or out of the cell to bring about the membrane depolarization?

ACTIVITY 2 The Effect of Stimulus Voltage on Skeletal Muscle Contraction

1. Describe the effect of increasing stimulus voltage on isolated skeletal muscle. Specifically, what happened to the muscle force generated with stronger electrical stimulations and why did this change occur? How well did the results compare with your prediction? ______________________

____

2. How is this change in whole-muscle force achieved in vivo? ______________________

____

3. What happened in the isolated skeletal muscle when the maximal voltage was applied? ______________________

____

ACTIVITY 3 The Effect of Stimulus Frequency on Skeletal Muscle Contraction

1. What is the difference between stimulus intensity and stimulus frequency? ______________________

2. In this experiment you observed the effect of stimulating the isolated skeletal muscle multiple times in a short period with complete relaxation between the stimuli. Describe the force of contraction with each subsequent stimulus. Are these results called treppe or wave summation? ______________________

____

3. How did the frequency of stimulation affect the amount of force generated by the isolated skeletal muscle when the frequency of stimulation was increased such that the muscle twitches did not fully relax between subsequent stimuli? Are these results called treppe or wave summation? How well did the results compare with your prediction? ______________________

____

____

____

4. To achieve an active force of 5.2 g, did you have to increase the stimulus voltage above 8.5 volts? If not, how did you achieve an active force of 5.2 g? How well did the results compare with your prediction? ______________________

____

____

5. Compare and contrast frequency-dependent wave summation with motor unit recruitment (previously observed by increasing the stimulus voltage). How are they similar? How was each achieved in the experiment? Explain how each is achieved in vivo. ______________________

____

ACTIVITY 4 Tetanus in Isolated Skeletal Muscle

1. Describe how increasing the stimulus frequency affected the force developed by the isolated whole skeletal muscle in this activity. How well did the results compare with your prediction? ______________________

____

2. Indicate what type of force was developed by the isolated skeletal muscle in this activity at the following stimulus frequencies: at 50 stimuli/sec, at 140 stimuli/sec, and above 146 stimuli/sec. ______________________

____

3. Beyond what stimulus frequency is there no further increase in the peak force? What is the muscle tension called at this frequency? ______________________

ACTIVITY 5 Fatigue in Isolated Skeletal Muscle

1. When a skeletal muscle fatigues, what happens to the contractile force over time? ______________________

____

2. What are some proposed causes of skeletal muscle fatigue? ______________________

____

3. Turning the stimulator off allows a small measure of muscle recovery. Thus, the muscle will produce more force for a longer time period if the stimulator is briefly turned off than if the stimuli were allowed to continue without interruption. Explain why this might occur. How well did the results compare with your prediction? ______________________

____

____

____

4. List a few ways that humans could delay the onset of fatigue when they are vigorously using their skeletal muscles. _____

____

ACTIVITY 6 The Skeletal Muscle Length-Tension Relationship

1. What happens to the amount of total force the muscle generates during the stimulated twitch? How well did the results compare with your prediction? ______________________

____

____

2. What is the key variable in an isometric contraction of a skeletal muscle? ______________________

3. Based on the unique arrangement of myosin and actin in skeletal muscle sarcomeres, explain why active force varies with changes in the muscle's resting length. ____

____

____

4. What skeletal muscle lengths generated passive force? (Provide a range.) ________________

5. If you were curling a 7-kg dumbbell, when would your bicep muscles be contracting isometrically? ________________

ACTIVITY 7 Isotonic Contractions and the Load-Velocity Relationship

1. If you were using your bicep muscles to curl a 7-kg dumbbell, when would your muscles be contracting isotonically?

____

2. Explain why the latent period became longer as the load became heavier in the experiment. How well did the results compare with your prediction? ____

____

____

3. Explain why the shortening velocity became slower as the load became heavier in this experiment. How well did the results compare with your prediction? ____

____

____

4. Describe how the shortening distance changed as the load became heavier in this experiment. How well did the results compare with your prediction? ____

____

____

5. Explain why it would take you longer to perform 10 repetitions lifting a 10-kg weight than it would to perform the same number of repetitions with a 5-kg weight. ____

____

6. Describe what would happen in the following experiment: A 2.5-g weight is attached to the end of the isolated whole skeletal muscle used in these experiments. Simultaneously, the muscle is maximally stimulated by 8.5 volts and the platform supporting the weight is removed. Will the muscle generate force? Will the muscle change length? What is the name for this type of contraction? ____

____

PhysioEx™ 9.0

EXERCISE 3

Neurophysiology of Nerve Impulses

PRE-LAB QUIZ

1. Circle the correct term. Excitability / Conductivity is the ability to transmit nerve impulses to other neurons.
2. When a neuron is stimulated, the membrane becomes more permeable to Na^+ ions, which diffuse into the cell, causing
   a. depolarization
   b. hyperpolarization
   c. repolarization
3. As an action potential progresses, the permeability to Na^+ decreases and the permeability to ________ increases.
   a. Ca^{2+}
   b. K^+
   c. Na^+
4. The period of time when the neuron is totally insensitive to further stimulation and cannot generate another action potential is
   a. absolute refractory period
   b. membrane potential
   c. repolarization
   d. threshold
5. What muscle and nerve will you need to isolate to study the physiology of nerve fibers?
   a. gastrocnemius and sciatic
   b. sartorius and femoral
   c. triceps brachii and radial

Exercise Overview

The nervous system contains two general types of cells: **neurons** and neuroglia (or glial cells). This exercise focuses on neurons. Neurons respond to their local environment by generating an electrical signal. For example, sensory neurons in the nose generate a signal (called a **receptor potential**) when odor molecules interact with receptor proteins on the membrane of these olfactory sensory neurons. Thus, sensory neurons can respond directly to sensory stimuli. The receptor potential can trigger another electrical signal (called an **action potential**), which travels along the membrane of the sensory neuron's axon to the brain—you could say that the action potential is conducted to the brain.

The action potential causes the release of **chemical neurotransmitters** onto neurons in olfactory regions of the brain. These chemical neurotransmitters bind to receptor proteins on the membrane of these brain **interneurons.** In general, interneurons respond to chemical neurotransmitters released by other neurons. In the nose the odor molecules are sensed by sensory neurons. In the brain the odor is perceived by the activity of interneurons responding to neurotransmitters. Any resulting action or behavior is caused by the subsequent activity of **motor neurons,** which can stimulate muscles to contract (see Exercise 2).

In general each neuron has three functional regions for signal transmission: a receiving region, a conducting region, and an output region, or secretory region. Sensory neurons often have a receptive ending specialized to detect a specific sensory stimulus, such as odor, light, sound, or touch. The **cell body** and **dendrites** of interneurons receive stimulation by neurotransmitters at structures called **chemical synapses** and produce **synaptic potentials.** The conducting region is

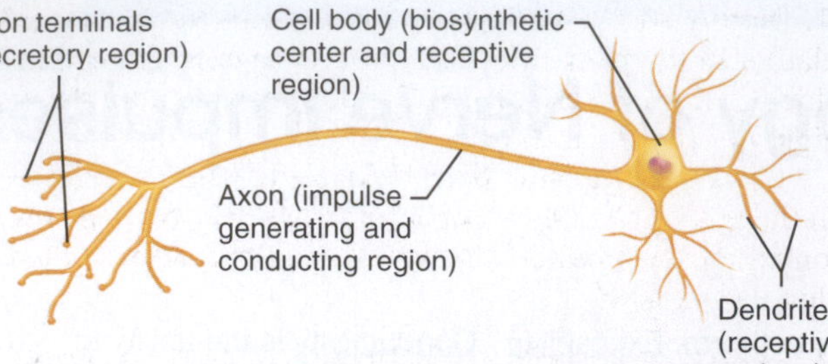
Axon terminals
(secretory region)
Cell body (biosynthetic center and receptive region)
Axon (impulse generating and conducting region)
Dendrites
(receptive regions)

FIGURE 3.1 A neuron with functional areas identified.

usually an **axon,** which ends in an output region (the axon terminal) where neurotransmitter is released (view Figure 3.1).

Although the neuron is a single cell surrounded by a continuous plasma membrane, each region contains distinct membrane proteins that provide the basis for the functional differences. Thus, the receiving end has receptor proteins and proteins that generate the receptor potential, the conducting region has proteins that generate and conduct action potentials, and the output region has proteins to package and release neurotransmitters. Membrane proteins are found throughout the neuronal membrane—many of these proteins transport ions (see Exercise 1).

The signals generated and conducted by neurons are electrical. In ordinary household devices, electric current is carried by electrons. In biological systems, currents are carried by positively or negatively charged **ions.** Like charges repel each other and opposite charges attract. In general, ions cannot easily pass through the lipid bilayer of the plasma membrane and must pass through **ion channels** formed by integral membrane proteins. Some channels are usually open (leak channels) and others are gated, meaning that the channel can be in an open or closed configuration. Channels can also be selective for which ions are allowed to pass. For example, sodium channels are mostly permeable to sodium ions when open, and potassium channels are mostly permeable to potassium ions when open. The term **conductance** is often used to describe **permeability.** In general, ions will flow through an open channel from a region of higher concentration to a region of lower concentration (see Exercise 1). In this exercise you will explore some of these characteristics applied to neurons.

Although it is possible to measure the ionic currents through the membrane (even the currents passing through single ion channels), it is more common to measure the potential difference, or voltage, across the membrane. This membrane voltage is usually called the **membrane potential,** and the units are **millivolts (mV).** One can think of the membrane as a battery, a device that separates and stores charge. A typical household battery has a positive and a negative pole so that when it is connected, for example through a lightbulb in a flashlight, current flows through the bulb. Similarly, the plasma membrane can store charge and has a relatively positive side and a relatively negative side. Thus, the membrane is said to be **polarized.** When these two sides (intracellular and extracellular) are connected through open ion channels, current in the form of ions can flow in or out across the membrane and thus change the membrane voltage.

ACTIVITY 1

The Resting Membrane Potential

OBJECTIVES

1. To define the term *resting membrane potential*.
2. To measure the resting membrane potential in different parts of a neuron.
3. To determine how the resting membrane potential depends on the concentrations of potassium and sodium.
4. To understand the ion conductances/ion channels involved in the resting membrane potential.

Introduction

The receptor potential, synaptic potentials, and action potentials are important signals in the nervous system. These potentials refer to changes in the membrane potential from its resting level. In this activity you will explore the nature of the resting potential. The **resting membrane potential** is really a potential difference between the inside of the cell (intracellular) and the outside of the cell (extracellular) across the membrane. It is a steady-state condition that depends on the resting permeability of the membrane to ions and on the intracellular and extracellular concentrations of those ions to which the membrane is permeable.

For many neurons, Na^+ and K^+ are the most important ions, and the concentrations of these ions are established by transport proteins, such as the Na^+-K^+ pump, so that the intracellular Na^+ concentration is low and the intracellular K^+ concentration is high. Inside a typical cell, the concentration of K^+ is ~150 m*M* and the concentration of Na^+ is ~5 m*M*. Outside a typical cell, the concentration of K^+ is ~5 m*M* and the concentration of Na^+ is ~150 m*M*. If the membrane is permeable to a particular ion, that ion will diffuse down its concentration gradient from a region of higher concentration to a region of lower concentration. In the generation of the resting membrane potential, K^+ ions diffuse out across the membrane, leaving behind a net negative charge—large anions that cannot cross the membrane.

The membrane potential can be measured with an amplifier. In the experiment the extracellular solution is connected to a ground (literally, the earth) which is defined as 0 mV. To record the voltage across the membrane, a microelectrode is inserted through the membrane without significantly damaging it. Typically, the microelectrode is made by pulling a thin glass pipette to a fine hollow point and filling the pulled pipette with a salt solution. The salt solution conducts electricity like a wire, and the glass insulates it. Only the tip of the microelectrode is inserted through the membrane, and the filled tip of the microelectrode makes electrical contact with the intracellular solution. A wire connects the microelectrode to the input of the amplifier so that the amplifier records the membrane potential, the voltage across the membrane between the intracellular and grounded extracellular solutions.

The membrane potential and the various signals can be observed on an oscilloscope. An electron beam is pulled up or down according to the voltage as it sweeps across a phosphorescent screen. Voltages below 0 mV are negative and voltages above 0 mV are positive. For this first activity, the time of the sweep is set for 1 second per division, and the sensitivity is set to 10 mV per division; a division is the distance between gridlines on the oscilloscope.

EQUIPMENT USED The following equipment will be depicted on-screen: neuron (in vitro)—a large, dissociated (or cultured) neuron; three extracellular solutions—control, high potassium, and low sodium; microelectrode—a probe with a very small tip that can impale a single neuron (In an actual wet lab, a microelectrode manipulator is used to position the microelectrode. For simplicity, the microelectrode manipulator will not be depicted in this activity.); microelectrode manipulator controller—controls movement of the manipulator; microelectrode amplifier—used to measure the voltage between the microelectrode and a reference; oscilloscope—used to observe voltage changes.

Experiment Instructions

Go to the home page in the PhysioEx software, and click **Exercise 3: Neurophysiology of Nerve Impulses.** Click **Activity 1: The Resting Membrane Potential,** and take the online **Pre-lab Quiz** for Activity 1.

After you take the online Pre-lab Quiz, click the **Experiment** tab and begin the experiment. The experiment instructions are reprinted here for your reference. The opening screen for the experiment is shown below.

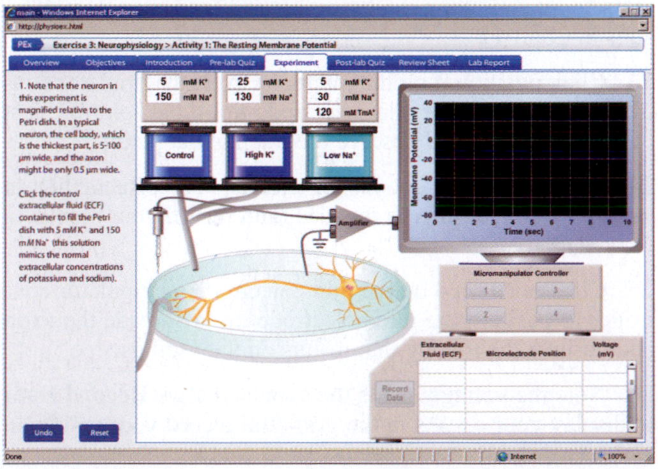
Exercise 3: Neurophysiology > Activity 1: The Resting Membrane Potential
Overview
Objectives
Introduction
Pre-lab Quiz
Experiment
Post-lab Quiz
Review Sheet
Lab Report
Control
High K+
Low Na+
Amplifier
Membrane Potential (mV)
Time (sec)
Micromanipulator Controller
Extracellular Fluid (ECF)
Microelectrode Position
Voltage (mV)
Record Data
Undo
Reset

1. Note that the neuron in this experiment is magnified relative to the petri dish. In a typical neuron, the cell body, which is the thickest part, is 5–100 μm wide, and the axon might be only 0.5 μm wide.

   Click the ***control*** **extracellular fluid (ECF)** container to fill the petri dish with 5 m*M* K^+ and 150 m*M* Na^+ (this solution mimics the normal extracellular concentrations of potassium and sodium).

2. Note that a reference electrode is already positioned in the petri dish. This reference electrode is connected to ground through the amplifier.

   Click position **1** on the microelectrode manipulator controller to position the microelectrode tip in the solution, just outside the cell body, and observe the tracing that results on the oscilloscope.

3. Note the oscilloscope tracing of the voltage outside the cell body and click **Record Data** to display your results in the grid (and record your results in Chart 1).

4. Click position **2** on the microelectrode manipulator controller to position the microelectrode tip just inside the cell body and observe the tracing that results.

5. Note the oscilloscope tracing of the voltage inside the cell body and click **Record Data** to display your results in the grid (and record your results in Chart 1). This is the resting membrane potential; that is, the potential difference between intracellular and extracellular membrane voltages. By convention, the extracellular resting membrane voltage is taken to be 0 mV.

6. Click position **3** on the microelectrode manipulator controller to position the microelectrode tip in the solution, just outside the axon, and observe the tracing that results.

7. Note the oscilloscope tracing of the voltage outside the axon and click **Record Data** to display your results in the grid (and record your results in Chart 1).

CHART 1 Resting Membrane Potential

Extracellular fluid (ECF)	Microelectrode position	Voltage (mV)

8. Click position **4** on the microelectrode manipulator controller to position the microelectrode tip just inside the axon and observe the tracing that results.

9. Note the oscilloscope tracing of the voltage inside the axon and click **Record Data** to display your results in the grid (and record your results in Chart 1).

PREDICT Question 1
Predict what will happen to the resting membrane potential if the extracellular K^+ concentration is increased.

____

____

10. You will now change the concentrations of the ions in the extracellular fluid to determine which ions contribute most to the separation of charge across the membrane. The extracellular potassium concentration is normally low, so you will first increase the extracellular potassium concentration.

In the high K^+ ECF solution the K^+ concentration has been increased fivefold, from 5 to 25 m*M*. To keep the number of positive charges in the extracellular solution constant, the Na^+ concentration has been reduced by 20 m*M*, from 150 to 130 m*M*. As you will see, this relatively small decrease in Na^+ will not by itself change the membrane potential. Note that in this activity, the generation of the action potential (which is covered in Activities 3–9) is blocked with a toxin. Click the **high K^+ ECF** container to change the solution in the petri dish to 25 m*M* K^+ and 130 m*M* Na^+.

11. Note the voltage inside the axon and click **Record Data** to display your results in the grid (and record your results in Chart 1).

12. Click position **3** on the microelectrode manipulator controller to position the microelectrode tip in the solution, just outside the axon, and observe the tracing that results.

13. Note the voltage outside the axon and click **Record Data** to display your results in the grid (and record your results in Chart 1).

14. Click position **1** on the microelectrode manipulator controller to position the microelectrode tip in the solution, just outside the cell body, and observe the tracing that results.

15. Note the voltage outside the cell body and click **Record Data** to display your results in the grid (and record your results in Chart 1).

16. Click position **2** on the microelectrode manipulator controller to position the microelectrode tip just inside the cell body and observe the tracing that results on the oscilloscope.

17. Note the voltage inside the cell body and click **Record Data** to display your results in the grid (and record your results in Chart 1).

18. Click the *control* ECF container to change back to the normal K^+ concentration and note the change in voltage inside the cell body.

19. You will now decrease the extracellular Na^+ concentration (the extracellular Na^+ concentration is normally high).

The extracellular sodium concentration in the low Na^+ solution has been decreased fivefold, from 150 m*M* to 30 m*M*. To keep the number of positive charges constant in the extracellular solution, the Na^+ has been replaced by the same amount of a large monovalent cation. Note that the extracellular Na^+ concentration, even in the low Na^+ ECF, is higher than the intracellular Na^+ concentration. Click the **low Na^+ ECF** container to change the solution in the petri dish to 5 m*M* K^+ and 30 m*M* Na^+.

20. Note the voltage inside the cell body and click **Record Data** to display your results in the grid (and record your results in Chart 1).

21. Click position **1** on the microelectrode manipulator controller to position the microelectrode tip in the solution, just outside the cell body, and observe the tracing that results.

22. Note the voltage outside the cell body and click **Record Data** to display your results in the grid (and record your results in Chart 1).

23. Click position **3** on the microelectrode manipulator controller to position the microelectrode tip in the solution, just outside the axon, and observe the tracing that results.

24. Note the voltage outside the axon and click **Record Data** to display your results in the grid (and record your results in Chart 1).

25. Click position **4** on the microelectrode manipulator controller to position the microelectrode tip just inside the axon and observe the tracing that results on the oscilloscope.

26. Note the voltage inside the axon and click **Record Data** to display your results in the grid (and record your results in Chart 1).

After you complete the experiment, take the online **Post-lab Quiz** for Activity 1.

Activity Questions

1. Explain why the resting membrane potential had the same value in the cell body and in the axon.

____

____

2. Describe what would happen to a resting membrane potential if the sodium-potassium transport pump was blocked.

____

____

3. Describe what would happen to a resting membrane potential if the concentration of large intracellular anions that are unable to cross the membrane is experimentally increased.

____

____

ACTIVITY 2

Receptor Potential

OBJECTIVES

1. To define the terms *sensory receptor, receptor potential, sensory transduction, stimulus modality,* and *depolarization.*
2. To determine the *adequate stimulus* for different sensory receptors.
3. To demonstrate that the receptor potential amplitude increases with stimulus intensity.

Introduction

The receiving end of a sensory neuron, the **sensory receptor,** has receptor proteins (as well as other membrane proteins) that can generate a signal called the **receptor potential** when the sensory neuron is stimulated by an appropriate, adequate stimulus. In this activity you will use the same recording instruments and microelectrode that you used in Activity 1. However, in this activity, you will record from the sensory receptor of three different sensory neurons and examine how these neurons respond to sensory stimuli of different modalities.

The sensory region will be shown disconnected from the rest of the neuron so that you can record the receptor potential in isolation. Similar results can sometimes be obtained by treating a whole neuron with chemicals that block the responses generated by the axon. The molecules localized to the sensory receptor ending are able to generate a receptor potential when an adequate stimulus is applied. The energy in the stimulus (for example, chemical, physical, or heat) is changed into an electrical response that involves the opening or closing of membrane ion channels. The general process that produces this change is called **sensory transduction,** which occurs at the receptor ending of the sensory neuron. Sensory transduction can be thought of as a type of signal transduction where the signal is the sensory stimulus.

You will observe that, with an appropriate stimulus, the amplitude of the receptor potential increases with stimulus intensity. Such a response is an example of a potential that is graded with stimulus intensity. These responses are sometimes referred to as *graded potentials,* or *local potentials.* Thus, the receptor potential is a graded, or local, potential. If the response (receptor potential) is a change in membrane potential from the negative resting potential to a less negative level, the membrane becomes less polarized and the change is called **depolarization.**

EQUIPMENT USED The following equipment will be depicted on-screen: three sensory receptors—Pacinian (lamellar) corpuscle, olfactory receptor, and free nerve ending; microelectrode—a probe with a very small tip that can impale a single neuron (In an actual wet lab, a microelectrode manipulator is used to position the microelectrodes. For simplicity, the microelectrode manipulator will not be depicted in this activity.); microelectrode amplifier—used to measure the voltage between the microelectrode and a reference; stimulator—used to select the stimulus modality (pressure, chemical, heat, or light) and intensity (low, moderate, or high); oscilloscope—used to observe voltage changes.

Experiment Instructions

Go to the home page in the PhysioEx software, and click **Exercise 3: Neurophysiology of Nerve Impulses.** Click **Activity 2: Receptor Potential,** and take the online **Pre-lab Quiz** for Activity 2.

After you take the online Pre-lab Quiz, click the **Experiment** tab and begin the experiment. The experiment instructions are reprinted here for your reference. The opening screen for the experiment is shown below.

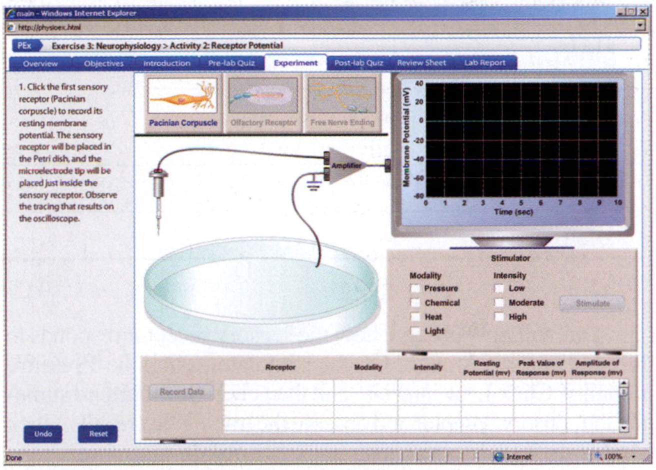
main - Windows Internet Explorer
http://physioex.html
PEx
Exercise 3: Neurophysiology > Activity 2: Receptor Potential
Overview
Objectives
Introduction
Pre-lab Quiz
Experiment
Post-lab Quiz
Review Sheet
Lab Report
1. Click the first sensory receptor (Pacinian corpuscle) to record its resting membrane potential. The sensory receptor will be placed in the Petri dish, and the microelectrode tip will be placed just inside the sensory receptor. Observe the tracing that results on the oscilloscope.
Pacinian Corpuscle
Olfactory Receptor
Free Nerve Ending
Amplifier
Membrane Potential (mV)
Time (sec)
Stimulator
Modality
Pressure
Chemical
Heat
Light
Intensity
Low
Moderate
High
Stimulate
Receptor
Modality
Intensity
Resting Potential (mv)
Peak Value of Response (mv)
Amplitude of Response (mv)
Record Data
Undo
Reset
Done
Internet
100%

1. Note that the timescale on the oscilloscope has been changed from 1 second per division to 10 milliseconds per division, so that you can observe the responses recorded in the sensory receptors more clearly. Click the first sensory receptor (Pacinian corpuscle) to record its resting membrane potential. The sensory receptor will be placed in the petri dish, and the microelectrode tip will be placed just inside the sensory receptor. Observe the tracing that results on the oscilloscope.

2. Note the voltage inside the sensory receptor and click **Record Data** to display your results in the grid (and record your results in Chart 2).

PREDICT Question 1
The adequate stimulus for a Pacinian corpuscle is pressure or vibration on the skin. For a Pacinian corpuscle, which modality will induce a receptor potential of the largest amplitude?

____

____

CHART 2 Receptor Potential

	Receptor potential (mV)		
Stimulus modality	Pacinian (lamellar) corpuscle	Olfactory receptor	Free nerve ending
None			
Pressure			
Low			
Moderate			
High			
Chemical			
Low			
Moderate			
High			
Heat			
Low			
Moderate			
High			
Light			
Low			
Moderate			
High			

3. You will now observe how the sensory receptor responds to different sensory stimuli. On the stimulator, click the **Pressure** modality. Click **Low** intensity and then click **Stimulate** to stimulate the sensory receptor and observe the tracing that results. Click **Moderate** intensity and then click **Stimulate** and observe the tracing that results. Click **High** intensity and then click **Stimulate** and observe the tracing that results. Click **Record Data** to display your results in the grid (and record your results in Chart 2).

4. On the stimulator, click the **Chemical** (odor) modality. Click **Low** intensity and then click **Stimulate** to stimulate the sensory receptor and observe the tracing that results. Click **Moderate** intensity and then click **Stimulate** and observe the tracing that results. Click **High** intensity and then click **Stimulate** and observe the tracing that results. Click **Record Data** to display your results in the grid (and record your results in Chart 2).

5. On the stimulator, click the **Heat** modality. Click **Low** intensity and then click **Stimulate** to stimulate the sensory receptor and observe the tracing that results. Click **Moderate** intensity and then click **Stimulate** and observe the tracing that results. Click **High** intensity and then click **Stimulate** and observe the tracing that results. Click **Record Data** to display your results in the grid (and record your results in Chart 2).

6. On the stimulator, click the **Light** modality. Click **Low** intensity and then click **Stimulate** to stimulate the sensory receptor and observe the tracing that results. Click **Moderate** intensity and then click **Stimulate** and observe the tracing that results. Click **High** intensity and then click **Stimulate** and observe the tracing that results. Click **Record Data** to display your results in the grid (and record your results in Chart 2).

? PREDICT Question 2
The adequate stimuli for olfactory receptors are chemicals, typically odorant molecules. For an olfactory receptor, which modality will induce a receptor potential of the largest amplitude?

____

7–12. Repeat steps 1–6 with the next sensory receptor: olfactory receptor.

13–18. Repeat steps 1–6 with the next sensory receptor: free nerve ending.

After you complete the experiment, take the online **Post-lab Quiz** for Activity 2.

Activity Questions

1. Are graded receptor potentials always depolarizing? Do graded receptor potentials always make it easier to induce action potentials?

____

____

2. Based on the definition of membrane depolarization in this activity, define membrane *hyperpolarization*.

____

____

3. What do you think is the adequate stimulus for sensory receptors in the ear? Can you think of a stimulus that would inappropriately activate the sensory receptors in the ear if the stimulus had enough intensity?

____

____

ACTIVITY 3

The Action Potential: Threshold

OBJECTIVES

1. To define the terms *action potential, nerve, axon hillock, trigger zone,* and *threshold.*
2. To predict how an increase in extracellular K^+ could trigger an action potential.

Introduction

In this activity you will explore changes in potential that occur in the axon. Axons are long, thin structures that conduct a signal called the **action potential.** A **nerve** is a bundle of axons.

Axons are typically studied in a nerve chamber. In this activity the axon will be draped over wires that make electrical contact with the axon and can therefore record the electrical activity in the axon. Because the axon is so thin, it is very difficult to insert an electrode across the membrane into the axon. However, some of the charge (ions) that crosses the membrane to generate the action potential can be recorded from outside the membrane (extracellular recording), as you will do in this activity. The molecular mechanisms underlying the action potential were explored more than 50 years ago with intracellular recording using the giant axons of the squid, which are about 1 millimeter in diameter.

In this activity the axon will be artificially disconnected from the cell body and dendrites. In a typical multipolar neuron (view Figure 3.1 in the Exercise Overview), the axon extends from the cell body at a region called the **axon hillock.** In a myelinated axon, this first region is called the initial segment. An action potential is usually initiated at the junction of the axon hillock and the initial segment; therefore, this region is also referred to as the **trigger zone.**

You will use an electrical stimulator to explore the properties of the action potential. Current passes from the stimulator to one of the stimulation wires, then across the axon, and then back to the stimulator through a second wire. This current will depolarize the axon. Normally, in a sensory neuron, the depolarizing receptor potential spreads passively to the axon hillock and produces the depolarization needed to evoke the action potential. Once an action potential is generated, it is regenerated down the membrane of the axon. In other words, the action potential is **propagated,** or *conducted,* down the axon (see Activity 6).

You will now generate an action potential at one end of the axon by stimulating it electrically and record the action potential that is propagated down the axon. The extracellular action potential that you record is similar to one that would be recorded across the membrane with an intracellular microelectrode, but much smaller. For simplicity, only one axon is depicted in this activity.

EQUIPMENT USED The following equipment will be depicted on-screen: nerve chamber; axon; oscilloscope—used to observe timing of stimuli and voltage changes in the axon; stimulator—used to set the stimulus voltage and to deliver pulses that depolarize the axon; stimulation wires (S); recording electrodes (wires R1 and R2)—used to record voltage changes in the axon. (The first set of recording electrodes, R1, is 2 centimeters from the stimulation wires, and the second set of recording electrodes, R2, is 2 centimeters from R1.)

Experiment Instructions

Go to the home page in the PhysioEx software, and click **Exercise 3: Neurophysiology of Nerve Impulses.** Click **Activity 3: The Action Potential: Threshold,** and take the online **Pre-lab Quiz** for Activity 3.

After you take the online Pre-lab Quiz, click the **Experiment** tab and begin the experiment. The experiment instructions are reprinted here for your reference. The opening screen for the experiment is shown below.

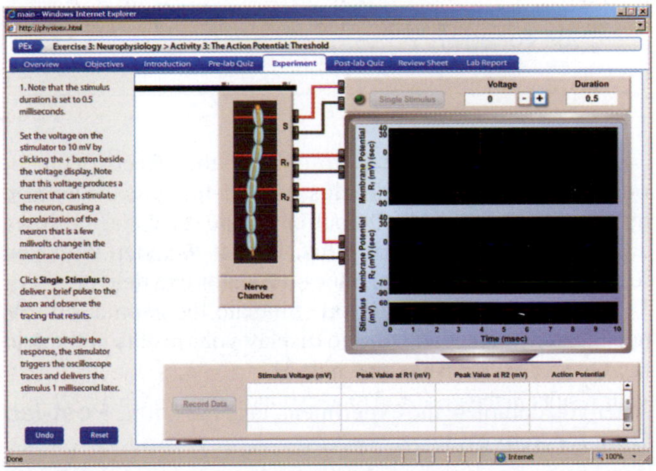
main - Windows Internet Explorer
PEx
Exercise 3: Neurophysiology > Activity 3: The Action Potential: Threshold
Overview
Objectives
Introduction
Pre-lab Quiz
Experiment
Post-lab Quiz
Review Sheet
Lab Report
1. Note that the stimulus duration is set to 0.5 milliseconds.
Set the voltage on the stimulator to 10 mV by clicking the + button beside the voltage display. Note that this voltage produces a current that can stimulate the neuron, causing a depolarization of the neuron that is a few millivolts change in the membrane potential
Click Single Stimulus to deliver a brief pulse to the axon and observe the tracing that results.
In order to display the response, the stimulator triggers the oscilloscope traces and delivers the stimulus 1 millisecond later.
Undo
Reset
S
R₁
R₂
Nerve Chamber
Single Stimulus
Voltage
0
Duration
0.5
Membrane Potential R₁ (mV) (sec)
Membrane Potential R₂ (mV) (sec)
Stimulus (mV)
Time (msec)
Record Data
Stimulus Voltage (mV)
Peak Value at R1 (mV)
Peak Value at R2 (mV)
Action Potential
Done
Internet
100%

1. Note that the stimulus duration is set to 0.5 milliseconds. Set the voltage on the stimulator to 10 mV by clicking the + button beside the voltage display. Note that this voltage produces a current that can stimulate the neuron, causing a depolarization of the neuron that is a change of a few millivolts in the membrane potential.

Click **Single Stimulus** to deliver a brief pulse to the axon and observe the tracing that results. In order to display the response, the stimulator triggers the oscilloscope traces and delivers the stimulus 1 millisecond later.

2. Note that the recording electrodes R1 and R2 record the extracellular voltage, rather than the actual membrane potential. The 10 mV depolarization at the site of stimulation only occurs locally at that site and is not recorded farther down the axon. At this initial stimulus voltage, there was no action potential. Click **Record Data** to display your results in the grid (and record your results in Chart 3).

CHART 3 Threshold			
Stimulus voltage (mV)	Peak value at R1 (μV)	Peak value at R2 (μV)	Action potential

3. You will increase the stimulus voltage until you observe an action potential at recording electrode 1 (R1). Increase the voltage by 10 mV by clicking the + button beside the voltage display and then click **Single Stimulus.** The voltage at which you first observe an action potential is the **threshold voltage.** Note that the action potential recorded extracellularly is quite small. Intracellularly, the membrane potential would change from −70 mV to about +30 mV. Click **Record Data** to display your results in the grid (and record your results in Chart 3).

? **PREDICT Question 1**
How will the action potential at R1 (or R2) change as you continue to increase the stimulus voltage?

____

4. You will now continue to observe the effects of incremental increases of the stimulus voltage. Increase the voltage by 10 mV by clicking the + button beside the voltage display and then click **Single Stimulus.** Repeat this step until you reach the maximum voltage the stimulator can deliver.

Repeat this step until you stimulate the axon at 50 mV and then click **Record Data** to display your results in the grid (and record your results in Chart 3).

After you complete the experiment, take the online **Post-lab Quiz** for Activity 3.

Activity Questions

1. Explain why the threshold voltage is not always the same value (between axons and within an axon).

____

____

2. Describe how the action potential is regenerated by local ion flux at each location on the axon.

____

____

3. Why doesn't the peak value of the action potential increase with stronger stimuli?

____

____

ACTIVITY 4

The Action Potential: Importance of Voltage-Gated Na^+ Channels

OBJECTIVES

1. To define the term *voltage-gated channel.*
2. To describe the effect of tetrodotoxin on the voltage-gated Na^+ channel.
3. To describe the effect of lidocaine on the voltage-gated Na^+ channel.
4. To examine the effects of tetrodotoxin and lidocaine on the action potential.
5. To predict the effect of lidocaine on pain perception and to predict the site of action in the sensory neurons (nociceptors) that sense pain.

Introduction

The action potential (as seen in Activity 3) is generated when voltage-gated sodium channels open in sufficient numbers. **Voltage-gated sodium channels** open when the membrane depolarizes. Each sodium channel that opens allows Na^+ ions to diffuse into the cell down their electrochemical gradient. When enough sodium channels open so that the amount of sodium ions that enters via these voltage-gated channels overcomes the leak of potassium ions (recall that the potassium leak via passive channels establishes and maintains the negative resting membrane potential), threshold for the action potential is reached, and an action potential is generated.

In this activity you will observe what happens when these voltage-gated sodium channels are blocked with chemicals. One such chemical is tetrodotoxin (TTX), a toxin found in puffer fish, which is extremely poisonous. Another such chemical is lidocaine, which is typically used to block pain in dentistry and minor surgery.

EQUIPMENT USED The following equipment will be depicted on-screen: nerve chamber; axon; oscilloscope—used to observe timing of stimuli and voltage changes in the axon; stimulator—used to set the stimulus voltage and the interval between stimuli and to deliver pulses that depolarize the axon; stimulation wires (S); recording electrodes (wires R1 and R2)—used to record voltage changes in the axon (The first set of recording electrodes, R1, is 2 centimeters from the stimulation wires, and the second set of recording electrodes, R2, is 2 centimeters from R1.); tetrodotoxin (TTX); lidocaine.

Experiment Instructions

Go to the home page in the PhysioEx software and click **Exercise 3: Neurophysiology of Nerve Impulses.** Click **Activity 4: The Action Potential: Importance of Voltage-Gated Na^+ Channels,** and take the online **Pre-lab Quiz** for Activity 4.

After you take the online Pre-lab Quiz, click the **Experiment** tab and begin the experiment. The experiment instructions are reprinted here for your reference. The opening screen for the experiment is shown below.

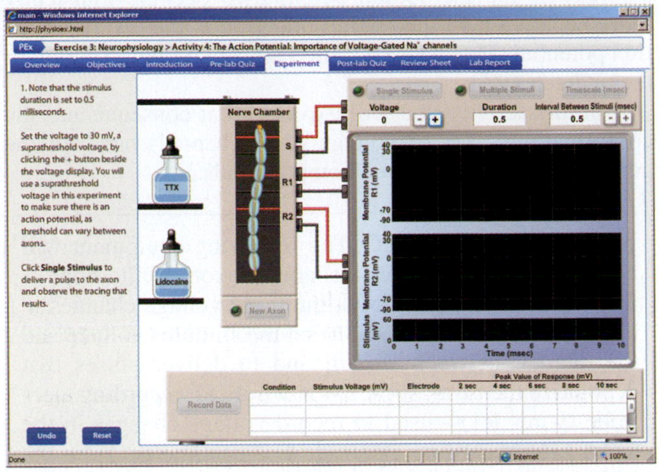
Exercise 3: Neurophysiology > Activity 4: The Action Potential: Importance of Voltage-Gated Na+ channels
Overview
Objectives
Introduction
Pre-lab Quiz
Experiment
Post-lab Quiz
Review Sheet
Lab Report
Single Stimulus
Multiple Stimuli
Timescale (msec)
Voltage
Duration
Interval Between Stimuli (msec)
Nerve Chamber
TTX
Lidocaine
New Axon
Time (msec)
Record Data
Condition
Stimulus Voltage (mV)
Electrode
Peak Value of Response (mV)
Undo
Reset

1. Note that the stimulus duration is set to 0.5 milliseconds. Set the voltage to 30 mV, a suprathreshold voltage, by clicking the + button beside the voltage display. You will use a suprathreshold voltage in this experiment to make sure there is an action potential, as threshold can vary between axons. Click **Single Stimulus** to deliver a pulse to the axon and observe the tracing that results.

2. Enter the peak value of the response at R1 and R2 in the field below and then click **Submit** to record your answer in the lab report. _______________ μV

3. Click **Timescale** on the stimulator to change the timescale on the oscilloscope from milliseconds to seconds.

4. You will now deliver successive stimuli separated by 2.0-second intervals to observe what the control action potentials look like at this timescale. Set the interval between stimuli to 2.0 seconds by clicking the + button beside the "Interval between Stimuli" display. Click **Multiple Stimuli** to deliver pulses to the axon every 2 seconds. The stimuli will be stopped after 10 seconds.

5. Note the peak values of the responses at R1 and R2 and click **Record Data** to display your results in the grid (and record your results in Chart 4).

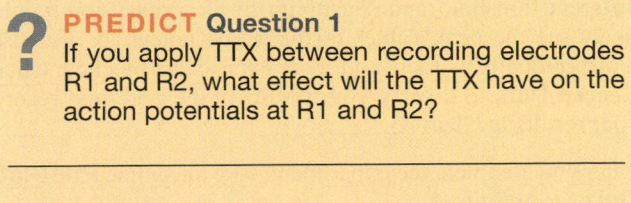

? **PREDICT Question 1**
If you apply TTX between recording electrodes R1 and R2, what effect will the TTX have on the action potentials at R1 and R2?

____

____

6. Drag the dropper cap of the TTX bottle to the axon between recording electrodes R1 and R2 to apply a drop of TTX to the axon.

7. Click **Multiple Stimuli** to deliver pulses to the axon every 2 seconds. The stimuli will be stopped after 10 seconds.

8. Note the peak values of the responses at R1 and R2 and click **Record Data** to display your results in the grid (and record your results in Chart 4).

9. Click **New Axon** to select a new axon. TTX is irreversible and there is no known antidote for TTX poisoning.

? **PREDICT Question 2**
If you apply lidocaine between recording electrodes R1 and R2, what effect will the lidocaine have on the action potentials at R1 and R2?

____

____

10. Drag the dropper cap of the lidocaine bottle to the axon between recording electrodes R1 and R2 to apply a drop of lidocaine to the axon.

11. Set the interval between stimuli to 2.0 seconds by clicking the + button beside the "Interval between Stimuli" display. Click **Multiple Stimuli** to deliver pulses to the axon every 2 seconds. The stimuli will be stopped after 10 seconds.

CHART 4 Effects of Tetrodotoxin and Lidocaine

			Peak value of response (μV)				
Condition	Stimulus voltage (mV)	Electrodes	2 sec	4 sec	6 sec	8 sec	10 sec

12. Note the peak values of the responses at R1 and R2. For simplicity, this experiment was performed on a single axon, where the action potential is an "all-or-none" event. If you had treated a bundle of axons (a nerve), each with a slightly different threshold and sensitivity to the drugs, you would likely see the peak values of the action potentials decrease more gradually as more and more axons were blocked. Click **Record Data** to display your results in the grid (and record your results in Chart 4).

After you complete the experiment, take the online **Post-lab Quiz** for Activity 4.

Activity Questions

1. If depolarizing membrane potentials open voltage-gated sodium channels, what closes them?

____

2. Why must a sushi chef go through years of training to prepare puffer fish for human consumption?

____

3. For action potential generation and propagation, are there any other cation channels that could substitute for the voltage-gated sodium channels if the sodium channels were blocked?

____

ACTIVITY 5

The Action Potential: Measuring Its Absolute and Relative Refractory Periods

OBJECTIVES

1. To define *inactivation* as it applies to a voltage-gated sodium channel.
2. To define the *absolute refractory period* and *relative refractory period* of an action potential.
3. To define the relationship between stimulus frequency and the generation of action potentials.

Introduction

Voltage-gated sodium channels in the plasma membrane of an excitable cell open when the membrane depolarizes. About 1–2 milliseconds later, these same channels inactivate, meaning they no longer allow sodium to go through the channel. These inactivated channels cannot be reopened by depolarization for an additional period of time (usually many milliseconds). Thus, during this time, fewer sodium channels can be opened. There are also voltage-gated potassium channels that open during the action potential. These potassium channels open more slowly. They contribute to the repolarization of the action potential from its peak, as more potassium flows out through this second type of potassium channel (recall there are also passive potassium channels that let potassium leak out, and these leak channels are always open). The flux through extra voltage-gated potassium channels opposes the depolarization of the membrane to threshold, and it also causes the membrane potential to become transiently more negative than the resting potential at the end of an action potential. This phase is called after-hyperpolarization, or the undershoot.

In this activity you will explore what consequences the conformation states of voltage-gated channels have for the generation of subsequent action potentials.

EQUIPMENT USED The following equipment will be depicted on-screen: nerve chamber; axon; oscilloscope—used to observe timing of stimuli and voltage changes in the axon; stimulator—used to set the stimulus voltage and the interval between stimuli and to deliver pulses that depolarize the axon; stimulation wires (S); recording electrode (wires R1)—used to record voltage changes in the axon. (The recording electrode is 2 centimeters from the stimulation wires.)

Experiment Instructions

Go to the home page in the PhysioEx software and click **Exercise 3: Neurophysiology of Nerve Impulses.** Click **Activity 5: The Action Potential: Measuring Its Absolute and Relative Refractory Periods,** and take the online **Pre-lab Quiz** for Activity 5.

After you take the online Pre-lab Quiz, click the **Experiment** tab and begin the experiment. The experiment instructions are reprinted here for your reference. The opening screen for the experiment is shown below.

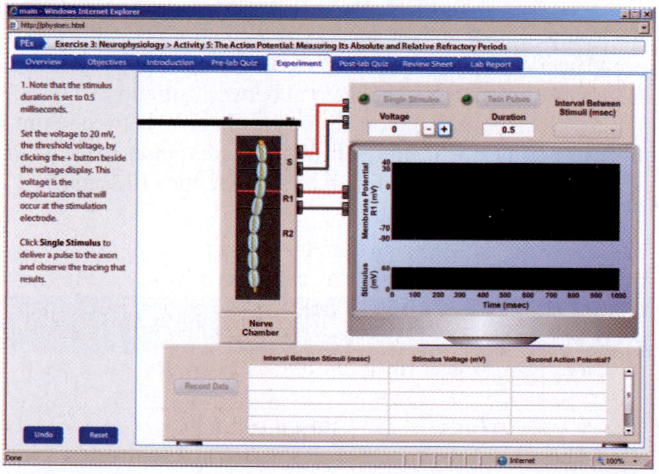
main - Windows Internet Explorer
PEx
Exercise 3: Neurophysiology > Activity 5: The Action Potential: Measuring Its Absolute and Relative Refractory Periods
Overview
Objectives
Introduction
Pre-lab Quiz
Experiment
Post-lab Quiz
Review Sheet
Lab Report
1. Note that the stimulus duration is set to 0.5 milliseconds.
Set the voltage to 20 mV, the threshold voltage, by clicking the + button beside the voltage display. This voltage is the depolarization that will occur at the stimulation electrode.
Click Single Stimulus to deliver a pulse to the axon and observe the tracing that results.
Single Stimulus
Twin Pulses
Voltage
Duration
Interval Between Stimuli (msec)
S
R1
R2
Nerve Chamber
Membrane Potential R1 (mV)
Stimulus (mV)
Time (msec)
Interval Between Stimuli (msec)
Stimulus Voltage (mV)
Second Action Potential?
Record Data
Undo
Reset

1. Note that the stimulus duration is set to 0.5 milliseconds. Set the voltage to 20 mV, the threshold voltage, by clicking the + button beside the voltage display. This voltage is the depolarization that will occur at the stimulation electrode. Click **Single Stimulus** to deliver a pulse to observe an action potential at this timescale.

2. You will now deliver two successive stimuli separated by 250 milliseconds. Set the interval between stimuli to 250 milliseconds by selecting 250 in the "Interval between Stimuli"

pull-down menu. Click **Twin Pulses** to deliver two pulses to the axon and observe the tracing that results. Click **Record Data** to display your results in the grid (and record your results in Chart 5).

CHART 5 Absolute and Relative Refractory Periods

Interval between stimuli (msec)	Stimulus voltage (mV)	Second action potential?

3. Decrease the interval between stimuli to 125 milliseconds by selecting 125 in the "Interval between Stimuli" pull-down menu. Click **Twin Pulses** to deliver two pulses to the axon and observe the tracing that results. Click **Record Data** to display your results in the grid (and record your results in Chart 5).

4. Decrease the interval between stimuli to 60 milliseconds by selecting 60 in the "Interval between Stimuli" pull-down menu. Click **Twin Pulses** to deliver two pulses to the axon and observe the tracing that results.

Note that, at this stimulus interval, the second stimulus did not generate an action potential. Click **Record Data** to display your results in the grid (and record your results in Chart 5).

5. A second action potential can be generated at this stimulus interval, but the stimulus intensity must be increased. This interval is part of the relative refractory period, the time after an action potential when a second action potential can be generated if the stimulus intensity is increased.

Increase the stimulus intensity by 5 mV by clicking the + button beside the voltage display and then click **Twin Pulses** to deliver two pulses to the axon. Repeat this step until you generate a second action potential. After you generate a second action potential, click **Record Data** to display your results in the grid (and record your results in Chart 5).

? PREDICT Question 1
If you further decrease the interval between the stimuli, will the threshold for the second action potential change?

____

____

6. You will now decrease the interval until the second action potential fails again. (So that you can clearly observe two action potentials at the shorter interval between stimuli, the timescale on the oscilloscope has been set to 10 msec per division.) Decrease the interval between stimuli by 50% and then click **Twin Pulses** to deliver two pulses to the axon. When the second action potential fails, click **Record Data** to display your results in the grid (and record your results in Chart 5).

7. You will now increase the stimulus intensity until a second action potential is generated again. Increase the stimulus intensity by 5 mV by clicking the + button beside the voltage display and then click **Twin Pulses** to deliver two pulses to the axon. Repeat this step until you generate a second action potential. After you generate a second action potential, click **Record Data** to display your results in the grid (and record your results in Chart 5).

8. You will now determine the interval between stimuli at which a second action potential cannot be generated, no matter how intense the stimulus. Increase the stimulus intensity to 60 mV (the highest voltage on the stimulator). Decrease the interval between stimuli by 50% and then click **Twin Pulses** to deliver two pulses to the axon. Repeat this step until the second action potential fails.

The interval at which the second action potential fails is the **absolute refractory period,** the time after an action potential when the neuron cannot fire a second action potential, no matter how intense the stimulus. Click **Record Data** to display your results in the grid (and record your results in Chart 5).

After you complete the experiment, take the online **Post-lab Quiz** for Activity 5.

Activity Questions

1. Explain how the absolute refractory period ensures directionality of action potential propagation.

____

____

2. Some tissues (for example, cardiac muscle) have long absolute refractory periods. Why would this be beneficial?

____

____

3. What do you think is the benefit of a relative refractory period in an axon of a sensory neuron?

____

____

ACTIVITY 6

The Action Potential: Coding for Stimulus Intensity

OBJECTIVES

1. To observe the response of axons to longer periods of stimulation.
2. To examine the relationship between stimulus intensity and the frequency of action potentials.

Introduction

As seen in Activity 3, the action potential has a constant amplitude, regardless of the stimulus intensity—it is an "all-or-none" event. As seen in Activity 5, the absolute refractory period is the time after an action potential when the neuron cannot fire a second action potential, no matter how intense the stimulus, and the relative refractory period is the time after an action potential when a second action potential can be generated if the stimulus intensity is increased.

In this activity you will use these concepts to begin to explore how the axon codes the stimulus intensity as *frequency,* the number of events (in this case, action potentials) per unit time. To demonstrate this phenomenon you will use longer periods of stimulation that are more representative of real-life stimuli. For example, when you encounter an odor, the odor is normally present for seconds (or longer), unlike the very brief stimuli used in Activities 3–5. These longer stimuli allow the axon of the neuron to generate additional action potentials as soon as it has recovered from the first. As seen in Activity 5, the length of this recovery period changes depending on the stimulus intensity. For example, at threshold, a second action potential can occur only after the axon has recovered from the absolute refractory period and the entire relative refractory period.

We will not consider the phenomenon of adaptation, which is a decrease in the response amplitude that often occurs with prolonged stimuli. For example, with most odors, after many seconds, you no longer smell the odor, even though it is still present. This decrease in response is due to adaptation.

EQUIPMENT USED The following equipment will be depicted on-screen: nerve chamber; axon; oscilloscope—used to observe timing of stimuli and voltage changes in the axon; stimulator—used to set the voltage and duration of stimuli and to deliver pulses that depolarize the axon; stimulation wires (S); recording electrode (wires R1)—used to record voltage changes in the axon (The recording electrode is 2 centimeters from the stimulation wires).

Experiment Instructions

Go to the home page in the PhysioEx software and click **Exercise 3: Neurophysiology of Nerve Impulses.** Click **Activity 6: The Action Potential: Coding for Stimulus Intensity,** and take the online **Pre-lab Quiz** for Activity 6.

After you take the online Pre-lab Quiz, click the **Experiment** tab and begin the experiment. The experiment instructions are reprinted here for your reference. The opening screen for the experiment is shown below.

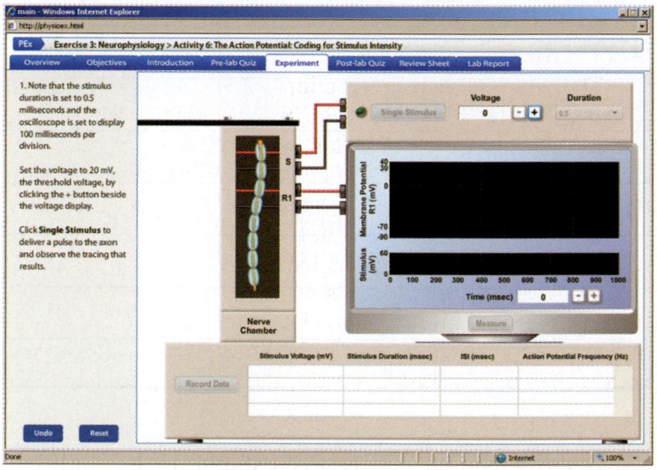
main - Windows Internet Explorer
http://physioex.html
PEx
Exercise 3: Neurophysiology > Activity 6: The Action Potential: Coding for Stimulus Intensity
Overview
Objectives
Introduction
Pre-lab Quiz
Experiment
Post-lab Quiz
Review Sheet
Lab Report
1. Note that the stimulus duration is set to 0.5 milliseconds and the oscilloscope is set to display 100 milliseconds per division.
Set the voltage to 20 mV, the threshold voltage, by clicking the + button beside the voltage display.
Click Single Stimulus to deliver a pulse to the axon and observe the tracing that results.
Single Stimulus
Voltage
Duration
S
R1
Membrane Potential R1 (mV)
Stimulus (mV)
Time (msec)
Measure
Nerve Chamber
Stimulus Voltage (mV)
Stimulus Duration (msec)
ISI (msec)
Action Potential Frequency (Hz)
Record Data
Undo
Reset
Done
Internet
100%

1. Note that the stimulus duration is set to 0.5 milliseconds and the oscilloscope is set to display 100 milliseconds per division. Set the voltage to 20 mV, the threshold voltage, by clicking the + button beside the voltage display. Click **Single Stimulus** to deliver a pulse to the axon and observe the tracing that results.

2. Note how the action potential looks at this timescale and click **Record Data** to display your results in the grid (and record your results in Chart 6).

CHART 6 Frequency of Action Potentials

Stimulus voltage (mV)	Stimulus duration (msec)	ISI (msec)	Action potential frequency (Hz)

3. Increase the stimulus duration to 500 milliseconds by selecting 500 from the duration pull-down menu. Click **Single Stimulus** to deliver a pulse to the axon and observe the tracing that results. The stimulus is delivered after a delay of 100 milliseconds so that you can easily see the timing of the stimulus.

4. At the site of stimulation, the stimulus keeps the membrane of the axon at threshold for a long time, but this depolarization does not spread to the recording electrode. After one action potential has been generated and the axon has fully recovered from its absolute and relative refractory periods, the stimulus is still present to generate another action potential.

Measure the time (in milliseconds) between action potentials. This interval should be a bit longer than the relative refractory period (measured in Activity 5). Click **Measure** to help determine the time between action potentials. A thin, vertical yellow line appears at the far left side of the oscilloscope screen. You can move the line in 10-millisecond increments by clicking the + and – buttons beside the time display, which shows the time at the line. Click **Submit** to display your answer in the data table (and record your results in Chart 6).

5. The interval between action potentials is sometimes called the interspike interval (ISI). Action potentials are sometimes referred to as spikes because of their rapid time course. From the ISI, you can calculate the action potential frequency. The frequency is the reciprocal of the interval and is usually expressed in hertz (Hz), which is events (action potentials) per second. From the ISI you entered, calculate the frequency of action potentials with a prolonged (500 msec) threshold stimulus intensity. Frequency = 1/ISI. Click **Submit** to display your answer in the data table (and record your results in Chart 6).

6. A stimulus intensity of 30 mV was able to generate a second action potential toward the end of the relative refractory period in Activity 5. With this stronger stimulus, the second action potential can occur after a shorter time. Increase the stimulus intensity to 30 mV by clicking the + button beside the voltage display. Click **Single Stimulus** to deliver this stronger stimulus and observe the tracing that results.

7. Click **Submit** to display your answer in the data table (and record your results in Chart 6). Click **Measure** to help determine the time between action potentials. A thin, vertical yellow line appears at the far left side of the oscilloscope screen. You can move the line in 10-millisecond increments by clicking the + and – buttons beside the time display, which shows the time at the line.

8. From the ISI you entered, calculate the frequency of action potentials with a prolonged (500 msec) 30-mV stimulus intensity. Frequency = 1/ISI. Click **Submit** to display your answer in the data table (and record your results in Chart 6).

9. A stimulus intensity of 45 mV was able to generate a second action potential in the middle of the relative refractory period in Activity 5. With this even stronger stimulus, the second action potential can occur after an even shorter time. Increase the stimulus intensity to 45 mV.

? PREDICT Question 1
What effect will the increased stimulus intensity have on the frequency of action potentials?

____

____

10. Click **Single Stimulus** to deliver the stronger, 45-mV stimulus and observe the tracing that results.

11. Click **Submit** to display your answer in the data table (and record your results in Chart 6). Click **Measure** to help determine the time between action potentials. A thin, vertical yellow line appears at the far left side of the oscilloscope screen. You can move the line in 10-millisecond increments by clicking the + and – buttons beside the time display, which shows the time at the line.

12. From the ISI you entered, calculate the frequency of action potentials with a prolonged (500 msec) 45-mV stimulus intensity. Frequency = 1/ISI. Click **Submit** to display your answer in the data table (and record your results in Chart 6).

After you complete the experiment, take the online **Post-lab Quiz** for Activity 6.

Activity Questions

1. Compare the action potential frequency in a temperature-sensitive sensory neuron exposed to warm water and then hot water.

____

____

2. When a long-duration stimulus is applied, what two determinants of an action potential refractory period are being overcome?

____

____

3. Suggest several ways to pharmacologically overcome a neuron's refractory period and thereby increase the action potential frequency.

____

____

ACTIVITY 7

The Action Potential: Conduction Velocity

OBJECTIVES

1. To define and measure *conduction velocity* for an action potential.
2. To examine the effect of myelination on conduction velocity.
3. To examine the effect of axon diameter on conduction velocity.

Introduction

Once generated, the action potential is propagated, or conducted, down the axon. In other words, all-or-none action potentials are regenerated along the entire length of the axon. This propagation ensures that the amplitude of the action potential does not diminish as it is conducted along the axon. In some cases, such as the sensory neuron traveling from your toe to the spinal cord, the axon can be quite long (in this case, up to a 1 meter). Propagation/conduction occurs because there are voltage-gated sodium and potassium channels located along the axon and because the large depolarization that constitutes the action potential (once generated at the trigger zone) easily brings the next region of the axon to threshold. The **conduction velocity** can be easily calculated by knowing both the distance the action potential travels and the amount of time it takes. Velocity has the units of distance per time, typically meters/second. An experimental stimulus artifact (see Activity 3) provides a convenient marker of the stimulus time because it travels very quickly (for our purposes, instantaneously) along the axon.

Several parameters influence the conduction velocity in an axon, including the axon diameter and the amount of myelination. **Myelination** refers to a special wrapping of the membrane from glial cells (or neuroglia) around the axon. In the central nervous system, oligodendrocytes are the glia that wrap around the axon. In the peripheral nervous system, the Schwann cells are the glia that wrap around the axon. Many glial cells along the axon contribute a myelin sheath, and the myelin sheaths are separated by gaps called nodes of Ranvier.

In this activity you will compare the conduction velocities of three axons: (1) a large-diameter, heavily myelinated axon, often called an A fiber (the terms axon and fiber are synonymous), (2) a medium-diameter, lightly myelinated axon (called the B fiber), and (3) a thin, unmyelinated fiber (called the C fiber). Examples of these axon types in the body include the axon of the sensory Pacinian corpuscle (an A fiber), the axon of both the olfactory sensory neuron and a free nerve ending (C fibers), and a visceral sensory fiber (a B fiber).

EQUIPMENT USED The following equipment will be depicted on-screen: nerve chamber; three axons—A fiber, B fiber, and C fiber; oscilloscope—used to observe timing of stimuli and voltage changes in the axon; stimulator—used to set the stimulus voltage and to deliver pulses that depolarize the axon; stimulation wires (S); recording electrodes (wires R1 and R2)—used to record voltage changes in the axon. (The first set of recording electrodes, R1, is 2 centimeters from the stimulation wires, and the second set of recording electrodes, R2, is 2 centimeters from R1.)

Experiment Instructions

Go to the home page in the PhysioEx software and click **Exercise 3: Neurophysiology of Nerve Impulses.** Click **Activity 7: The Action Potential: Conduction Velocity,** and take the online **Pre-lab Quiz** for Activity 7.

After you take the online Pre-lab Quiz, click the **Experiment** tab and begin the experiment. The experiment instructions are reprinted here for your reference. The opening screen for the experiment is shown below.

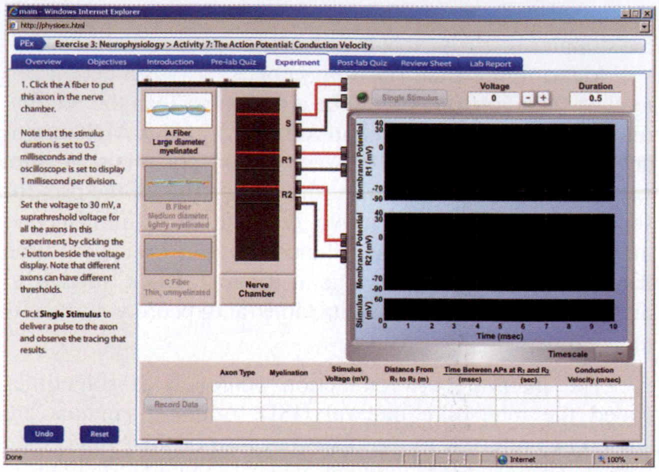
Exercise 3: Neurophysiology > Activity 7: The Action Potential: Conduction Velocity
Overview
Objectives
Introduction
Pre-lab Quiz
Experiment
Post-lab Quiz
Review Sheet
Lab Report
1. Click the A fiber to put this axon in the nerve chamber.
A Fiber
Large diameter
myelinated
B Fiber
C Fiber
Nerve Chamber
Single Stimulus
Voltage
Duration
Record Data
Undo
Reset

1. Click the A fiber to put this axon in the nerve chamber. Note that the stimulus duration is set to 0.5 milliseconds and the oscilloscope is set to display 1 millisecond per division.

Set the voltage to 30 mV, a suprathreshold voltage for all the axons in this experiment, by clicking the + button beside the voltage display. Note that different axons can have different thresholds. Click **Single Stimulus** to deliver a pulse to the axon and observe the tracing that results.

2. Click **Record Data** to display your results in the grid (and record your results in Chart 7).

3. Note the difference in time between the action potential recorded at R1 and the action potential recorded at R2. The distance between these sets of recording electrodes is 10 centimeters (0.1 m). Convert the time from milliseconds to seconds and then click **Submit** to display your results in the grid (and record your results in Chart 7).

4. Calculate the conduction velocity in meters/second by dividing the distance between R1 and R2 (0.1 m) by the time it took for the action potential to travel from R1 to R2. Click

CHART 7 Conduction Velocity

Axon type	Myelination	Stimulus voltage (mV)	Distance from R1 to R2 (m)	Time between action potentials at R1 and R2		Conduction velocity (m/sec)
				(msec)	(sec)	

Submit to display your results in the grid (and record your results in Chart 7).

> **PREDICT Question 1**
> How will the conduction velocity in the B fiber compare with that in the A fiber?
>
> ______________________________________
>
> ______________________________________

5. Click the **B fiber** to put this axon in the nerve chamber. Set the timescale on the oscilloscope to 10 milliseconds per division by selecting 10 in the timescale pull-down menu. Click **Single Stimulus** to deliver a pulse to the axon and observe the tracing that results.

6–8. Repeat steps 2–4 with the B fiber (and record your results in Chart 7).

> **PREDICT Question 2**
> How will the conduction velocity in the C fiber compare with that in the B fiber?
>
> ______________________________________
>
> ______________________________________

9. Click the **C fiber** to put this axon in the nerve chamber. Set the timescale on the oscilloscope to 50 milliseconds per division by selecting 50 in the timescale pull-down menu. Click **Single Stimulus** to deliver a pulse to the axon and observe the tracing that results.

10–12. Repeat steps 2–4 with the C fiber (and record your results in Chart 7).

After you complete the experiment, take the online **Post-lab Quiz** for Activity 7.

Activity Questions

1. The squid utilizes a very large-diameter, unmyelinated axon to execute a rapid escape response when it perceives danger. How is this possible, given that the axon is unmyelinated?

____

____

2. When you burn your finger on a hot stove, you feel sharp, immediate pain, which later becomes slow, throbbing pain. These two types of pain are carried by different pain axons. Speculate on the axonal diameter and extent of myelination of these axons.

____

____

3. Why do humans possess a mixture of axons, some large-diameter, heavily myelinated axons and some small-diameter, relatively unmyelinated axons?

____

____

ACTIVITY 8

Chemical Synaptic Transmission and Neurotransmitter Release

OBJECTIVES

1. To define *neurotransmitter, chemical synapse, synaptic vesicle,* and *postsynaptic potential.*
2. To determine the role of calcium ions in neurotransmitter release.

Introduction

A major function of the nervous system is communication. The axon conducts the action potential from one place to another. Often, the axon has branches so that the action potential is conducted to several places at about the same time. At the end of each branch, there is a region called the axon terminal that is specialized to release packets of chemical neurotransmitters from small (~30-nm diameter) intracellular membrane-bound vesicles, called **synaptic vesicles. Neurotransmitters** are extracellular signal molecules that act on local targets as paracrine agents, on the neuron releasing the chemical as autocrine agents, and sometimes as hormones (endocrine agents) that reach their target(s) via the circulation. These chemicals are released by exocytosis and diffuse across a small extracellular space (called the synaptic gap, or synaptic cleft) to the target (most often the receiving end of another neuron or a muscle or gland). The neurotransmitter molecules often bind to membrane receptor proteins on the target, setting in motion a sequence of molecular events that can open or close membrane ion channels and cause the membrane potential in the target cell to change. This region where the neurotransmitter is released from one neuron and binds to a receptor on a target cell is called a **chemical synapse,** and the change in membrane potential of the target is called a synaptic potential, or **postsynaptic potential.**

In this activity you will explore some of the steps in neurotransmitter release from the axon terminal. Exocytosis of synaptic vesicles is normally triggered by an increase in calcium ions in the axon terminal. The calcium enters from outside the cell through membrane calcium channels that are opened by the depolarization of the action potential. The axon terminal has been greatly magnified in this activity so that you can visualize the release of neurotransmitter. Different from the other activities in this exercise, however, this procedure of directly seeing neurotransmitter release is not easily done in the lab; rather, neurotransmitter is usually detected by the postsynaptic potentials it triggers or by collecting and analyzing chemicals at the synapse after robust stimulation of the neurons.

EQUIPMENT USED The following equipment will be depicted on-screen: neuron (in vitro)—a large, dissociated (or cultured) neuron with magnified axon terminals; four extracellular solutions—control Ca^{2+}, no Ca^{2+}, low Ca^{2+}, and Mg^{2+}.

Experiment Instructions

Go to the home page in the PhysioEx software and click **Exercise 3: Neurophysiology of Nerve Impulses.** Click **Activity 8: Chemical Synaptic Transmission and Neurotransmitter Release,** and take the online **Pre-lab Quiz** for Activity 8.

After you take the online Pre-lab Quiz, click the **Experiment** tab and begin the experiment. The experiment instructions are reprinted here for your reference. The opening screen for the experiment is shown below.

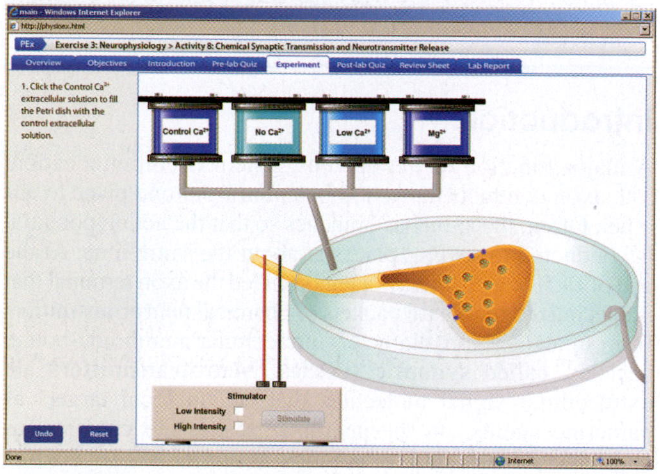
main - Windows Internet Explorer
http://physioex.html
PEx
Exercise 3: Neurophysiology > Activity 8: Chemical Synaptic Transmission and Neurotransmitter Release
Overview
Objectives
Introduction
Pre-lab Quiz
Experiment
Post-lab Quiz
Review Sheet
Lab Report
1. Click the Control Ca2+ extracellular solution to fill the Petri dish with the control extracellular solution.
Control Ca2+
No Ca2+
Low Ca2+
Mg2+
Stimulator
Low Intensity
High Intensity
Stimulate
Undo
Reset

1. Click the *control Ca^{2+}* extracellular solution to fill the petri dish with the control extracellular solution.

2. Click **Low Intensity** on the stimulator and then click **Stimulate** to stimulate the neuron (axon) with a threshold stimulus that generates a low frequency of action potentials. Observe the release of neurotransmitter.

3. Click **High Intensity** on the stimulator and then click **Stimulate** to stimulate the neuron with a longer, more intense stimulus to generate a burst of action potentials. Observe the release of neurotransmitter.

? PREDICT Question 1
You have just observed that each action potential in a burst can trigger additional neurotransmitter release. If calcium ions are removed from the extracellular solution, what will happen to neurotransmitter release at the nerve terminal?

____

____

4–6. Repeat steps 1–3 with the *no Ca^{2+}* extracellular solution.

? PREDICT Question 2
What will happen to the amount of neurotransmitter release when low amounts of calcium are added back to the extracellular solution?

____

____

7–9. Repeat steps 1–3 with the *low Ca^{2+}* extracellular solution.

? PREDICT Question 3
What will happen to neurotransmitter release when magnesium is added to the extracellular solution?

____

____

10–12. Repeat steps 1–3 with the *Mg^{2+}* extracellular solution.

After you complete the experiment, take the online **Post-lab Quiz** for Activity 8.

Activity Questions

1. If you added more sodium to the extracellular solution, could the sodium substitute for the missing calcium?

____

____

2. How does botulinum toxin block synaptic transmission? Why is it used for cosmetic procedures?

____

____

ACTIVITY 9

The Action Potential: Putting It All Together

OBJECTIVES

1. To identify the functional areas (for example, the sensory ending, axon, and postsynaptic membrane) of a two-neuron circuit.
2. To predict and test the responses in each functional area to a very weak, subthreshold stimulus.
3. To predict and test the responses in each functional area to a moderate stimulus.
4. To predict and test the responses in each functional area to an intense stimulus.

Introduction

In the nervous system, sensory neurons respond to adequate sensory stimuli, generating action potentials in the axon if the stimulus is strong enough to reach threshold (the action potential is an "all-or-nothing" event). Via chemical synapses, these sensory neurons communicate with interneurons that process the information. Interneurons also communicate with motor neurons that stimulate muscles and glands, again, usually via chemical synapses.

After performing Activities 1–8, you should have a better understanding of how neurons function by generating changes from their resting membrane potential. If threshold is reached, an action potential is generated and propagated. If the stimulus is more intense, then action potentials are generated at a higher frequency, causing the release of more neurotransmitter at the next synapse. At an excitatory synapse the chemical neurotransmitter binds to receptors at the receiving end of the next cell (usually the cell body or dendrites of an interneuron), causing ion channels to open, resulting in a depolarization toward threshold for an action potential in the interneuron's axon. This depolarizing synaptic potential (called an excitatory postsynaptic potential) is graded in amplitude, depending on the amount of neurotransmitter and the number of channels that open. In the axon, the amplitude of this synaptic potential is coded as the frequency of action potentials. Neurotransmitters can also cause inhibition, which will not be covered in this activity.

In this activity you will stimulate a sensory neuron, predict the response of that cell and its target, and then test those predictions.

EQUIPMENT USED The following equipment will be depicted on-screen: neuron (in vitro)—a large, dissociated (or cultured) neuron; interneuron (in vitro)—a large, dissociated (or cultured) interneuron; microelectrodes—small probes with very small tips that can impale a single neuron (In an actual wet lab, a microelectrode manipulator is used to position the microelectrodes. For simplicity, the microelectrode manipulator will not be depicted in this activity.); microelectrode amplifier—used to measure the voltage between the microelectrodes and a reference; oscilloscope—used to observe the changes in voltage across the membrane of the neuron and interneuron; stimulator —used to set the stimulus intensity (low or high) and to deliver pulses to the neuron.

Experiment Instructions

Go to the home page in the PhysioEx software and click **Exercise 3: Neurophysiology of Nerve Impulses.** Click **Activity 9: The Action Potential: Putting It All Together,** and take the online **Pre-lab Quiz** for Activity 9.

After you take the online Pre-lab Quiz, click the **Experiment** tab and begin the experiment. The experiment instructions are reprinted here for your reference. The opening screen for the experiment is shown below.

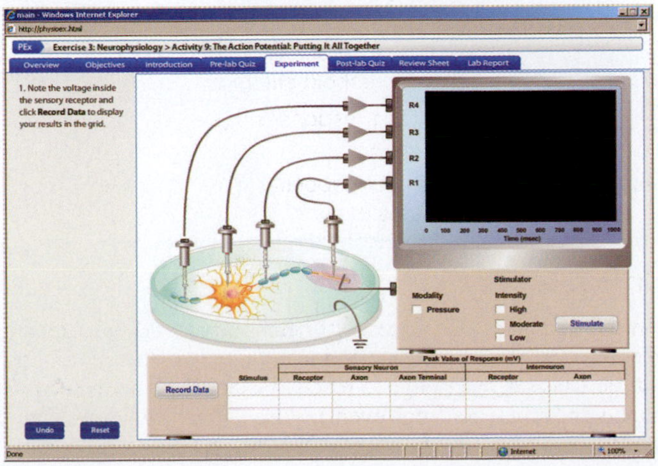
Exercise 3: Neurophysiology > Activity 9: The Action Potential: Putting It All Together
Overview
Objectives
Introduction
Pre-lab Quiz
Experiment
Post-lab Quiz
Review Sheet
Lab Report
Stimulate
Record Data
Undo
Reset

1. Note the membrane potential at the sensory receptor and the receiving end of the interneuron and click **Record Data** to display your results in the grid (and record your results in Chart 9).

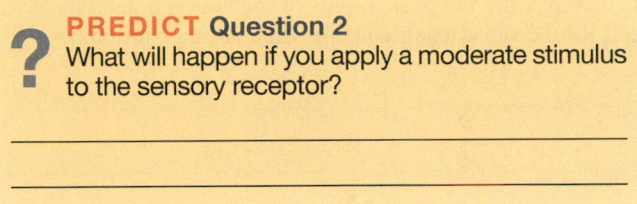

> **PREDICT Question 1**
> What will happen if you apply a very weak, subthreshold stimulus to the sensory receptor?

2. Click **Very Weak** intensity on the stimulator and then click **Stimulate** to stimulate the receiving end of the sensory neuron and observe the tracing that results.

3. Click **Record Data** to display your results in the grid (and record your results in Chart 9). The stimulus lasts 500 msec.

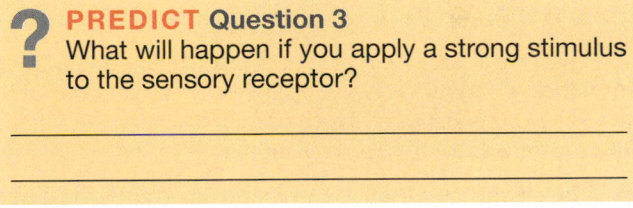

> **PREDICT Question 2**
> What will happen if you apply a moderate stimulus to the sensory receptor?

4. Click **Moderate** intensity on the stimulator and then click **Stimulate** to stimulate the sensory receptor and observe the tracing that results.

5. Click **Record Data** to display your results in the grid (and record your results in Chart 9).

> **PREDICT Question 3**
> What will happen if you apply a strong stimulus to the sensory receptor?

6. Click **Strong** intensity on the stimulator and then click **Stimulate** to stimulate the sensory receptor and observe the tracing that results.

7. Click **Record Data** to display your results in the grid (and record your results in Chart 9).

After you complete the experiment, take the online **Post-lab Quiz** for Activity 9.

CHART 9 Putting It All Together

Stimulus	Sensory neuron			Interneuron	
	Membrane Potential (mV) Receptor	AP frequency (Hz) in axon	Vesicles released from axon terminal	Membrane potential (mV) receiving end	AP frequency (Hz) in axon
None					
Weak					
Moderate					
Strong					

Activity Questions

1. Why were the peak values of the action potentials at R2 and R4 the same when you applied a strong stimulus?

2. If the axons were unmyelinated, would the peak value of the action potential at R4 change relative to that at R2?

PhysioEx 9.0

REVIEW SHEET

EXERCISE 3

Neurophysiology of Nerve Impulses

NAME ______________________

LAB TIME/DATE ______________________

ACTIVITY 1 The Resting Membrane Potential

1. Explain why increasing extracellular K^+ reduces the net diffusion of K^+ out of the neuron through the K^+ leak channels.

____

2. Explain why increasing extracellular K^+ causes the membrane potential to change to a less negative value. How well did the results compare with your prediction? ______________________

____

____

3. Explain why a change in extracellular Na^+ did not alter the membrane potential in the resting neuron. ______________________

____

4. Discuss the relative permeability of the membrane to Na^+ and K^+ in a resting neuron. ______________________

____

____

5. Discuss how a change in Na^+ or K^+ conductance would affect the resting membrane potential. ______________________

____

ACTIVITY 2 Receptor Potential

1. Sensory neurons have a resting potential based on the efflux of potassium ions (as demonstrated in Activity 1). What passive channels are likely found in the membrane of the olfactory receptor, in the membrane of the Pacinian corpuscle, and in the membrane of the free nerve ending? ______________________

____

____

2. What is meant by the term *graded potential*? ______________________

____

____

3. Identify which of the stimulus modalities induced the largest amplitude receptor potential in the Pacinian corpuscle. How well did the results compare with your prediction? ______________________________

4. Identify which of the stimulus modalities induced the largest-amplitude receptor potential in the olfactory receptors. How well did the results compare with your prediction? ______________________________

5. The olfactory receptor also contains a membrane protein that recognizes isoamyl acetate and, via several other molecules, transduces the odor stimulus into a receptor potential. Does the Pacinian corpuscle likely have this isoamyl acetate receptor protein? Does the free nerve ending likely have this isoamyl acetate receptor protein? ______________________________

6. What type of sensory neuron would likely respond to a green light? ______________________________

ACTIVITY 3 The Action Potential: Threshold

1. Define the term *threshold* as it applies to an action potential. ______________________________

2. What change in membrane potential (depolarization or hyperpolarization) triggers an action potential? ______________________________

3. How did the action potential at R1 (or R2) change as you increased the stimulus voltage above the threshold voltage? How well did the results compare with your prediction? ______________________________

4. An action potential is an "all-or-nothing" event. Explain what is meant by this phrase. ______________________________

5. What part of a neuron was investigated in this activity? ______________________________

ACTIVITY 4 The Action Potential: Importance of Voltage-Gated Na^+ Channels

1. What does TTX do to voltage-gated Na^+ channels? ______

2. What does lidocaine do to voltage-gated Na^+ channels? How does the effect of lidocaine differ from the effect of TTX? ______

3. A nerve is a bundle of axons, and some nerves are less sensitive to lidocaine. If a nerve, rather than an axon, had been used in the lidocaine experiment, the responses recorded at R1 and R2 would be the sum of all the action potentials (called a compound action potential). Would the response at R2 after lidocaine application necessarily be zero? Why or why not? ______

4. Why are fewer action potentials recorded at R2 when TTX is applied between R1 and R2? How well did the results compare with your prediction? ______

5. Why are fewer action potentials recorded at R2 when lidocaine is applied between R1 and R2? How well did the results compare with your prediction? ______

6. Pain-sensitive neurons (called nociceptors) conduct action potentials from the skin or teeth to sites in the brain involved in pain perception. Where should a dentist inject the lidocaine to block pain perception? ______

ACTIVITY 5 The Action Potential: Measuring Its Absolute and Relative Refractory Periods

1. Define *inactivation* as it applies to a voltage-gated sodium channel. ______

2. Define the *absolute refractory period.* ______________________________

____

____

3. How did the threshold for the second action potential change as you further decreased the interval between the stimuli? How well did the results compare with your prediction? ______________________________

____

4. Why is it harder to generate a second action potential during the relative refractory period? ______________________________

____

____

ACTIVITY 6 The Action Potential: Coding for Stimulus Intensity

1. Why are multiple action potentials generated in response to a long stimulus that is above threshold? ______________________________

____

____

2. Why does the frequency of action potentials increase when the stimulus intensity increases? How well did the results compare with your prediction? ______________________________

____

3. How does threshold change during the relative refractory period? ______________________________

____

____

4. What is the relationship between the interspike interval and the frequency of action potentials? ______________________________

____

____

ACTIVITY 7 The Action Potential: Conduction Velocity

1. How did the conduction velocity in the B fiber compare with that in the A fiber? How well did the results compare with your prediction? ______________________________

____

____

2. How did the conduction velocity in the C fiber compare with that in the B fiber? How well did the results compare with your prediction? ____

3. What is the effect of axon diameter on conduction velocity? ____

4. What is the effect of the amount of myelination on conduction velocity? ____

5. Why did the time between the stimulation and the action potential at R1 differ for each axon? ____

6. Why did you need to change the timescale on the oscilloscope for each axon? ____

ACTIVITY 8 Chemical Synaptic Transmission and Neurotransmitter Release

1. When the stimulus intensity is increased, what changes: the number of synaptic vesicles released or the amount of neurotransmitter per vesicle? ____

2. What happened to the amount of neurotransmitter release when you switched from the control extracellular fluid to the extracellular fluid with no Ca^{2+}? How well did the results compare with your prediction? ____

3. What happened to the amount of neurotransmitter release when you switched from the extracellular fluid with no Ca^{2+} to the extracellular fluid with low Ca^{2+}? How well did the results compare with your prediction? ____

4. How did neurotransmitter release in the Mg^{2+} extracellular fluid compare to that in the control extracellular fluid? How well did the result compare with your prediction? ___

___

___

5. How does Mg^{2+} block the effect of extracellular calcium on neurotransmitter release? ___

___

___

ACTIVITY 9 The Action Potential: Putting It All Together

1. Why is the resting membrane potential the same value in both the sensory neuron and the interneuron? ___

___

___

2. Describe what happened when you applied a very weak stimulus to the sensory receptor. How well did the results compare with your prediction? ___

___

___

3. Describe what happened when you applied a moderate stimulus to the sensory receptor. How well did the results compare with your prediction? ___

___

___

4. Identify the type of membrane potential (graded receptor potential or action potential) that occurred at R1, R2, R3, and R4 when you applied a moderate stimulus. (View the response to the stimulus.) ___

___

___

5. Describe what happened when you applied a strong stimulus to the sensory receptor. How well did the results compare with your prediction? ___

___

___

PhysioEx™ 9.0

EXERCISE 4

Endocrine System Physiology

PRE-LAB QUIZ

1. Define *metabolism.* ______________________________
2. Circle the correct term. Catabolism / Anabolism is the process by which substances are broken down into simpler substances.
3. ________ is the single most important hormone responsible for influencing the rate of cellular metabolism and body heat production.
   a. Calcitonin
   b. Estrogen
   c. Insulin
   d. Thyroid hormone
4. Circle the correct term. Control / Experimental group animals are assumed to have normal thyroid function and metabolic rate.
5. Basal metabolic rate (BMR) is:
   a. decreased in individuals with hyperthyroidism
   b. increased in individuals with hyperthyroidism
   c. increased in obese individuals
6. In this exercise, ________ consumption will be measured with a respirator-manometer device.
   a. carbon dioxide
   b. food
   c. oxygen
   d. water
7. Circle True or False. Gonadotropins are produced by the anterior pituitary gland.
8. Circle the correct term. Many people with diabetes mellitus need injections of insulin / glucagon to maintain homeostasis.

Exercise Overview

In the human body the **endocrine system** (in addition to the nervous system) coordinates and integrates the functions of different physiological systems (view Figure 4.1). Thus, the endocrine system plays a critical role in maintaining **homeostasis.** This role begins with chemicals, called **hormones,** secreted from ductless **endocrine glands,** which are tissues that have an epithelial origin. Endocrine glands secrete hormones into the extracellular fluid compartments. More specifically, the blood usually carries hormones (sometimes attached to specific plasma proteins) to their **target cells.** Target cells can be very close to, or very far from, the source of the hormone.

Hormones bind to high-affinity **receptors** located on the target cell's surface, in its cytosol, or in its nucleus. These hormone receptors have remarkable **sensitivity,** as the hormone concentration in the blood can range from 10^{-9} to 10^{-12} molar! A hormone-receptor complex forms and can then exert a **biological action** through signal-transduction cascades and alteration of gene transcription at the target cell. The physiological response to hormones can vary from seconds to hours to days, depending on the chemical nature of the hormone and its receptor location in the target cell.

The chemical structure of the hormone is important in determining how it will interact with target cells. *Peptide* and *catecholamine hormones* are fast-acting hormones that attach to a plasma-membrane receptor and cause a second-messenger cascade in the cytoplasm of the target cell. For example, a chemical called cAMP (cyclic adenosine monophosphate) is synthesized from a molecule of ATP. The

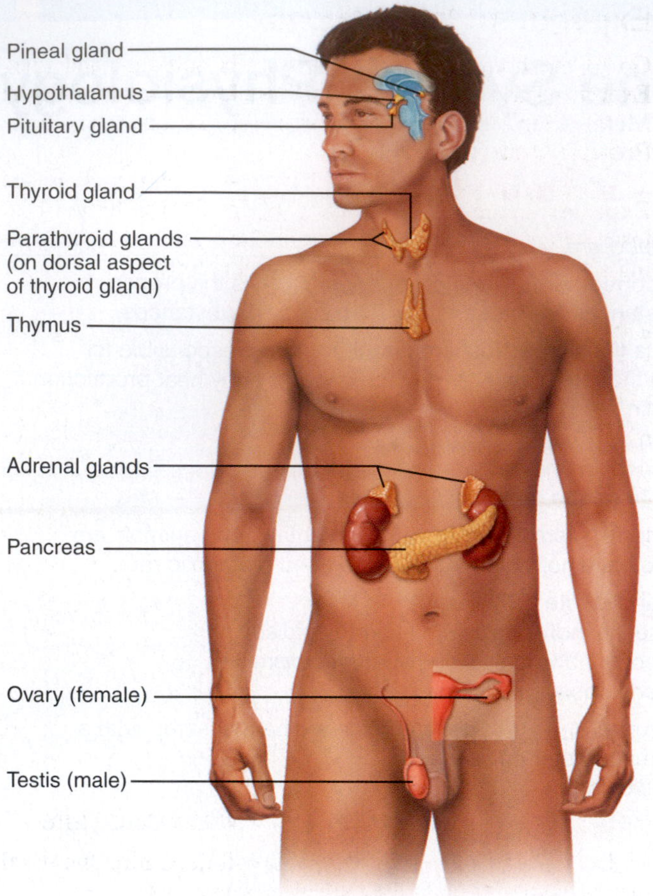
Pineal gland
Hypothalamus
Pituitary gland
Thyroid gland
Parathyroid glands (on dorsal aspect of thyroid gland)
Thymus
Adrenal glands
Pancreas
Ovary (female)
Testis (male)

FIGURE 4.1 Selected endocrine organs of the body.

synthesis of this chemical makes the cell more metabolically active and, therefore, more able to respond to a stimulus.

Steroid hormones and *thyroxine* (thyroid hormone) are slow-acting hormones that enter the target cell and interact with the nucleus to affect the transcription of various proteins that the cell can synthesize. The hormones enter the nucleus and attach at specific points on the DNA. Each attachment causes the production of a specific mRNA, which is then moved to the cytoplasm, where ribosomes can translate the mRNA into a protein.

Keep in mind that the organs of the endocrine system do not function independently. The activities of one endocrine gland are often coordinated with the activities of other glands. No one system functions independently of any other system. For this reason, we will be stressing feedback mechanisms and how we can use them to predict, explain, and understand hormone effects.

Given the powerful influence that hormones have on homeostasis, **negative feedback mechanisms** are important in regulating hormone secretion, synthesis, and effectiveness at target cells. Negative feedback ensures that if the body needs a particular hormone, that hormone will be produced until there is too much of it. When there is too much of the hormone, its release will be inhibited.

Rarely, the body regulates hormones via a *positive feedback mechanism.* The release of *oxytocin* from the posterior pituitary is one of these rare instances. Oxytocin is a hormone that causes the muscle layer of the uterus, called the *myometrium,* to contract during childbirth. This contraction of the myometrium causes additional oxytocin to be released, allowing stronger contractions. Unlike what happens in negative feedback mechanisms, the increase in circulating levels of oxytocin does not inhibit oxytocin secretion.

Many experimental methods can be used to study the functions of an endocrine gland. These methods include removing the gland from an animal and then injecting, implanting, or feeding glandular extracts into a normal animal or an animal deprived of the gland being studied. In this exercise you will use these methods to gain a deeper understanding of the *function* and *regulation* of some of the endocrine glands.

ACTIVITY 1

Metabolism and Thyroid Hormone

OBJECTIVES

1. To understand the terms *basal metabolic rate (BMR), thyroid-stimulating hormone (TSH), thyroxine, goiter, hypothyroidism, hyperthyroidism, thyroidectomized*, and *hypophysectomized.*
2. To observe how negative feedback mechanisms regulate hormone release.
3. To understand thyroxine's role in maintaining the basal metabolic rate.
4. To understand the effect of TSH on the basal metabolic rate.
5. To understand the role of the hypothalamus in regulating the secretion of thyroxine and TSH.

Introduction

Metabolism is the broad range of biochemical reactions occurring in the body. Metabolism includes *anabolism* and *catabolism.* Anabolism is the building up of small molecules into larger, more complex molecules via enzymatic reactions. Energy is stored in the chemical bonds formed when larger, more complex molecules are formed.

Catabolism is the breakdown of large, complex molecules into smaller molecules via enzymatic reactions. The breaking of chemical bonds in catabolism releases energy that the cell can use to perform various activities, such as forming ATP. The cell does not use all the energy released by bond breaking. Much of the energy is released as heat to maintain a fixed body temperature, especially in humans. Humans are *homeothermic* organisms that need to maintain a fixed body temperature to maintain the activity of the various metabolic pathways in the body.

The most important hormone for maintaining metabolism and body heat is **thyroxine** (thyroid hormone), also known as *tetraiodothyronine,* or T_4. Thyroxine is secreted by the thyroid gland, located in the neck.

The production of thyroxine is controlled by the pituitary gland, or hypophysis, which secretes **thyroid-stimulating hormone (TSH).** The blood carries TSH to its target tissue, the thyroid gland. TSH causes the thyroid gland to increase in size and secrete thyroxine into the general circulation. If TSH levels are too high, the thyroid gland enlarges. The resulting glandular swelling in the neck is called a **goiter**.

The **hypothalamus** in the brain is also a vital participant in thyroxine and TSH production. It is a primary endocrine gland that secretes several hormones that affect the pituitary gland, or hypophysis, which is also located in the brain.

Thyrotropin-releasing hormone (TRH) is directly linked to thyroxine and TSH secretion. TRH from the hypothalamus stimulates the anterior pituitary to produce TSH, which then stimulates the thyroid to produce thyroxine.

These events are part of a classic negative feedback mechanism. When circulation levels of thyroxine are low, the hypothalamus secretes more TRH to stimulate the pituitary gland to secrete more TSH. The increase in TSH further stimulates the secretion of thyroxine from the thyroid gland. The increased levels of thyroxine will then influence the hypothalamus to reduce its production of TRH.

TRH travels from the hypothalamus to the pituitary gland via the **hypothalamic-pituitary portal system.** This specialized arrangement of blood vessels consists of a single **portal vein** that connects two capillary beds. The hypothalamic-pituitary portal system transports many other hormones from the hypothalamus to the pituitary gland. The hypothalamus primarily secretes *tropic* hormones, which stimulate the secretion of other hormones. TRH is an example of a tropic hormone because it stimulates the release of TSH from the pituitary gland. TSH itself is also an example of a tropic hormone because it stimulates production of thyroxine.

In this activity you will investigate the effects of thyroxine and TSH on a rat's metabolic rate. The metabolic rate will be indicated by the amount of oxygen the rat consumes per time per body mass. You will perform four experiments on three rats: a normal rat, a thyroidectomized rat (a rat whose thyroid gland has been surgically removed), and a hypophysectomized rat (a rat whose pituitary gland has been surgically removed). You will determine (1) the rat's basal metabolic rate, (2) its metabolic rate after it has been injected with thyroxine, (3) its metabolic rate after it has been injected with TSH, and (4) its metabolic rate after it has been injected with propylthiouracil, a drug that inhibits the production of thyroxine.

EQUIPMENT USED The following equipment will be depicted on-screen: three refillable syringes—used to inject the rats with propylthiouracil (a drug that inhibits the production of thyroxine by blocking the incorporation of iodine into the hormone precursor molecule), thyroid-stimulating hormone (TSH), and thyroxine; airtight, glass animal chamber—provides an isolated, sealed system in which to measure the amount of oxygen consumed by the rat in a specified amount of time (Opening the clamp on the left tube allows outside air into the chamber, and closing the clamp will create a closed, airtight system. The T-connector on the right tube allows you to connect the chamber to the manometer or to connect the fluid-filled manometer to the syringe filled with air.); soda lime (found at the bottom of the glass chamber)—absorbs the carbon dioxide given off by the rat; manometer—U-shaped tube containing fluid (As the rat consumes oxygen in the isolated, sealed system, this fluid will rise in the left side of the U-shaped tube and fall in the right side of the tube.); syringe—used to inject air into the tube and thus measure the amount of air that is needed to return the fluid columns in the manometer to their original levels; animal scale—used to measure body weight; three white rats—a *normal* rat, a *thyroidectomized* (Tx) rat (a rat whose thyroid gland has been surgically removed), and a *hypophysectomized* (Hypox) rat (a rat whose pituitary gland has been surgically removed).

Experiment Instructions

Go to the home page in the PhysioEx software and click **Exercise 4: Endocrine System Physiology.** Click **Activity 1: Metabolism and Thyroid Hormone,** and take the online **Pre-lab Quiz** for Activity 1.

After you take the online Pre-lab Quiz, click the **Experiment** tab and begin the experiment. The experiment instructions are reprinted here for your reference. The opening screen for the experiment is shown below.

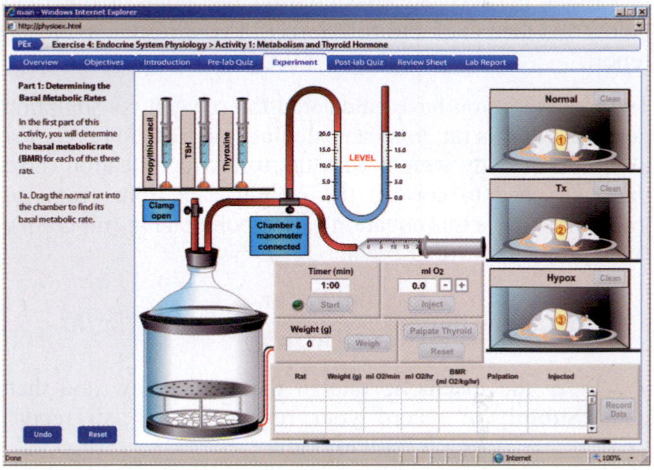
Exercise 4: Endocrine System Physiology > Activity 1: Metabolism and Thyroid Hormone
Part 1: Determining the Basal Metabolic Rates
Propylthiouracil
TSH
Thyroxine
LEVEL
Clamp open
Chamber & manometer connected
Normal
Tx
Hypox
Timer (min)
ml O2
Weight (g)
Start
Inject
Weigh
Palpate Thyroid
Reset
Undo

Part 1: Determining the Basal Metabolic Rates

In the first part of this activity, you will determine the basal metabolic rate (BMR) for each of the three rats.

1a. Drag the *normal* rat into the chamber to find its BMR.

1b. Click **Weigh** to determine the rat's weight.

1c. Click the clamp on the left tube (top of the chamber) to close it. This will prevent any outside air from entering the chamber and ensure that the only oxygen the rat is breathing is the oxygen inside the closed system.

1d. Note that the timer is set to one minute. Click **Start** beneath the timer to measure the amount of oxygen consumed by the rat in one minute in the sealed chamber. Note what happens to the water levels in the manometer as time progresses.

1e. Click the T-connector knob to connect the manometer and syringe.

1f. Click the clamp on the left tube (top of the chamber) to open it so the rat can breathe outside air.

1g. Observe the difference between the level in the left and right arms of the manometer. Estimate the volume of O_2 that you will need to inject to make the levels equal by counting the divisions on both sides. This volume is equivalent to the amount of oxygen that the rat consumed during the minute in the sealed chamber. Click the + button under the ml O_2 display until you reach the estimated volume. Then click **Inject** and watch what happens to the fluid in the two arms. When the volume levels are equalized, the word "Level" will appear and stay on the screen.

- If you have not injected enough oxygen, the word "Level" will not appear. Click the + to increase the volume and then click **Inject** again.

- If you have injected too much oxygen, the word "Level" will flash and then disappear. Click the button to decrease the volume and then click **Inject** again. Click **Record Data** when the levels are equalized.

1h. Calculate the oxygen consumption per hour for this rat using the following equation:

$$\frac{\text{ml } O_2 \text{ consumed}}{1 \text{ minute}} \times \frac{60 \text{ minutes}}{1 \text{ hr}} = \text{ml } O_2/\text{hr}$$

Enter the oxygen consumption per hour in the field below and then click **Submit** to record your results in the lab report. ____________________ ml O_2/hr

1i. Now that you have calculated the oxygen consumption per hour for this rat, you can calculate the metabolic rate per kilogram of body weight with the following equation (note that you need to convert the weight data from grams to kilograms to use this equation): Metabolic rate = (ml O_2/hr)/(weight in kg) = ml O_2/kg/hr.

$$\text{Metabolic rate} = \frac{\text{ml } O_2/\text{hr}}{\text{weight in kg}} = \text{ml } O_2/\text{kg/hr}$$

Enter the metabolic rate in the field below and then click **Submit** to record your results in the lab report. ____________________ ml O_2/kg/hr

1j. Click **Palpate Thyroid** to manually check the size of the thyroid and, thus, whether a goiter is present. After reviewing the findings, click **Submit** to record your results in the lab report.

1k. Drag the rat from the chamber back to its cage and then click **Restore** (beneath **Palpate Thyroid**) to restore the apparatus to its initial state.

> **? PREDICT Question 1**
> Make a prediction about the basal metabolic rate (BMR) of the remaining rats compared with the BMR of the normal rat you just measured.
>
> ____
>
> ____

2a.–2k. Repeat steps 1a–1k for the *thyroidectomized (Tx)* rat.

3a.–3k. Repeat steps 1a–1k for the *hypophysectomized (Hypox)* rat.

> **? PREDICT Question 2**
> What do you think will happen to the metabolic rates of the rats after you inject them with thyroxine?
>
> ____
>
> ____

Part 2: Determining the Effect of Thyroxine on Metabolic Rate

In this part of the activity, you will investigate the effects of thyroxine injections on the metabolic rates of all three rats.

4a. Drag the syringe filled with *thyroxine* to the *normal* rat's hindquarters. Release the mouse button to inject thyroxine into the rat. (In this experiment, the effects of the injection are immediate. In a wet lab, you would have to inject the rats daily with thyroxine for 1–2 weeks).

4b. In this part of the activity, the rat's weight, the amount of oxygen consumed by the rat in one minute, the rat's oxygen consumption per hour, the rat's metabolic rate, and the result of the thyroid palpation will be generated automatically after you drag the rat into the chamber.

Drag the injected rat into the chamber and note the results (and record your results in Chart 1).

4c. Drag the rat from the chamber back to its cage and then click **Clean** to clear all traces of thyroxine from the rat and clean the syringe. (In this experiment, the thyroxine is removed instantly. In a wet lab, clearance would take weeks or require that a different rat be used.)

5a.–5c, Repeat steps 4a–4c with the *thyroidectomized (Tx)* rat (and record your results in Chart 1).

6a.–6c. Repeat steps 4a–4c with the *hypophysectomized (Hypox)* rat (and record your results in Chart 1).

> **? PREDICT Question 3**
> What do you think will happen to the metabolic rates of the rats after you inject them with TSH?
>
> ____
>
> ____

Part 3: Determining the Effect of TSH on Metabolic Rate

In this part of the activity, you will investigate the effects of TSH injections on the metabolic rates of all three rats.

7a. Drag the syringe filled with *TSH* to the *normal* rat's hindquarters. Release the mouse button to inject TSH into the rat. (In this experiment, the effects of the injection are immediate. In a wet lab, you would have to inject the rats daily with TSH for 1–2 weeks.)

7b. In this part of the activity, the rat's weight, the amount of oxygen consumed by the rat in one minute, the rat's oxygen consumption per hour, the rat's metabolic rate, and the result of the thyroid palpation will be generated automatically after you drag the rat into the chamber.

Drag the injected rat into the chamber and note the results (and record your results in Chart 1).

7c. Drag the rat from the chamber back to its cage and then click **Clean** to clear all traces of TSH from the rat and clean the syringe. (In this experiment, the TSH is removed instantly. In a wet lab, clearance would take weeks or require that a different rat be used.)

8a.–8c. Repeat steps 7a–7c with the *thyroidectomized (Tx)* rat (and record your results in Chart 1).

9a.–9c. Repeat steps 7a–7c with the *hypophysectomized (Hypox)* rat (and record your results in Chart 1).

CHART 1 Effects of Hormones on Metabolic Rate

	Normal rat	Thyroidectomized rat	Hypophysectomized rat
Baseline			
Weight	____ grams	____ grams	____ grams
ml O_2 used in 1 minute	____ ml	____ ml	____ ml
ml O_2 used per hour	____ ml	____ ml	____ ml
Metabolic rate	____ ml O_2/kg/hr	____ ml O_2/kg/hr	____ ml O_2/kg/hr
Palpation results	____	____	____
With thyroxine			
Weight	____ grams	____ grams	____ grams
ml O_2 used in 1 minute	____ ml	____ ml	____ ml
ml O_2 used per hour	____ ml	____ ml	____ ml
Metabolic rate	____ ml O_2/kg/hr	____ ml O_2/kg/hr	____ ml O_2/kg/hr
Palpation results	____	____	____
With TSH			
Weight	____ grams	____ grams	____ grams
ml O_2 used in 1 minute	____ ml	____ ml	____ ml
ml O_2 used per hour	____ ml	____ ml	____ ml
Metabolic rate	____ ml O_2/kg/hr	____ ml O_2/kg/hr	____ ml O_2/kg/hr
Palpation results	____	____	____
With propylthiouracil			
Weight	____ grams	____ grams	____ grams
ml O_2 used in 1 minute	____ ml	____ ml	____ ml
ml O_2 used per hour	____ ml	____ ml	____ ml
Metabolic rate	____ ml O_2/kg/hr	____ ml O_2/kg/hr	____ ml O_2/kg/hr
Palpation results	____	____	____

? PREDICT Question 4

Propylthiouracil (PTU) is a drug that inhibits the production of thyroxine by blocking the attachment of iodine to tyrosine residues in the follicle cells of the thyroid gland (iodinated tyrosines are linked together to form thyroxine). What do you think will happen to the metabolic rates of the rats after you inject them with PTU?

____

____

Part 4: Determining the Effect of Propylthiouracil on Metabolic Rate

In this part of the activity, you will investigate the effects of propylthiouracil injections on the metabolic rates of all three rats.

10a. Drag the syringe filled with *propylthiouracil* to the *normal* rat's hindquarters. Release the mouse button to inject propylthiouracil into the rat. (In this experiment, the effects of the injection are immediate. In a wet lab, you would have to inject the rats daily with propylthiouracil for 1–2 weeks).

10b. In this part of the activity, the rat's weight, the amount of oxygen consumed by the rat in one minute, the rat's oxygen consumption per hour, the rat's metabolic rate, and the result of the thyroid palpation will be generated automatically after you drag the rat into the chamber.

Drag the injected rat into the chamber and note the results (and record your results in Chart 1).

10c. Drag the rat from the chamber back to its cage and then click **Clean** to clear all traces of propylthiouracil from the rat and clean the syringe. (In this experiment, the propylthiouracil is removed instantly. In a wet lab, clearance would take weeks or require that a different rat be used.)

11a.–11c. Repeat steps 10a–10c with the *thyroidectomized (Tx)* rat (and record your results in Chart 1).

12a.–12c. Repeat steps 10a–10c with the *hypophysectomized (Hypox)* rat (and record your results in Chart 1).

After you complete the experiment, take the online **Post-lab Quiz** for Activity 1.

Activity Questions

1. Using a water-filled manometer, you observed the amount of oxygen consumed by rats in a sealed chamber. What happened to the carbon dioxide the rat produced while in the sealed chamber?

2. What would happen to the fluid levels of the manometer (and, thus, the results of the metabolism experiment) if the rats in the sealed chamber were engaged in physical activity (such as running in a wheel)?

3. Describe the role of the hypothalamus in the production of thyroxine.

4. What does it mean if a hormone is a *tropic* hormone?

5. How could you treat a thyroidectomized rat so that it functions like a "normal" rat? How would you verify that your treatments were safe and effective?

6. What is the role of the hypothalamus in the production of thyroid-stimulating hormone (TSH)?

7. How does thyrotropin-releasing hormone (TRH) travel from the hypothalamus to the pituitary gland?

8. Why didn't the administration of TSH have any effect on the metabolic rate of the thyroidectomized rat?

9. Why didn't the administration of propylthiouracil have any effect on the metabolic rate of either the thyroidectomized rat or the hypophysectomized rat?

10. Propylthiouracil inhibits the production of thyroxine by blocking the attachment of iodine to the amino acid tyrosine. What naturally occurring problem in some parts of the world does this drug mimic?

ACTIVITY 2

Plasma Glucose, Insulin, and Diabetes Mellitus

OBJECTIVES

1. To understand the use of the terms *insulin, type 1 diabetes mellitus, type 2 diabetes mellitus,* and *glucose standard curve.*
2. To understand how fasting plasma glucose levels are used to diagnose diabetes mellitus.
3. To understand the assay that is used to measure plasma glucose.

Introduction

Insulin is a hormone produced by the beta cells of the endocrine portion of the pancreas. This hormone is vital to the regulation of **plasma glucose** levels, or "blood sugar," because the hormone enables our cells to absorb glucose from the bloodstream. Glucose absorbed from the blood is either used as fuel for metabolism or stored as glycogen (also known as animal starch), which is most notable in liver and muscle cells. About 75% of glucose consumed during a meal is stored as glycogen. As humans do not feed continuously (we are considered "discontinuous feeders"), the production of glycogen from a meal ensures that a supply of glucose will be available for several hours after a meal.

Furthermore, the body has to maintain a certain level of plasma glucose to continuously serve nerve cells because these cell types use only glucose for metabolic fuel. When glucose levels in the plasma fall below a certain value, the alpha cells of the pancreas are stimulated to release the hormone **glucagon.** Glucagon stimulates the breakdown of stored glycogen into glucose, which is then released back into the blood.

When the pancreas does not produce enough insulin, **type 1 diabetes mellitus** results. When the pancreas produces sufficient insulin but the body fails to respond to it, **type 2 diabetes mellitus** results. In either case, glucose remains in the bloodstream, and the body's cells are unable to take it up to serve as the primary fuel for metabolism. The kidneys then filter the excess glucose out of the plasma. Because the reabsorption of filtered glucose involves a finite number of transporters in kidney tubule cells, some of the excess glucose is not reabsorbed into the circulation. Instead, it passes out of the body in urine (hence *sweet urine,* as the name **diabetes mellitus** suggests).

The inability of body cells to take up glucose from the blood also results in skeletal muscle cells undergoing protein catabolism to free up amino acids to be used in forming glucose in the liver. This action puts the body into a negative nitrogen balance from the resulting protein depletion and tissue wasting. Other associated problems include poor wound healing and poor resistance to infections.

This activity is divided into two parts. In Part 1, you will generate a **glucose standard curve,** which will be explained in the experiment. In Part 2, you will use the glucose standard curve to measure the fasting plasma glucose levels from several patients to diagnose the presence or absence of diabetes mellitus. A patient with FPG values greater than or equal to 126 mg/dl in two FPG tests is diagnosed with diabetes. FPG values between 110 and 126 mg/dl indicate impairment or borderline impairment of insulin-mediated glucose uptake by cells. FPG values less than 110 mg/dl are considered normal.

EQUIPMENT USED The following equipment will be depicted on-screen: deionized water—used to adjust the volume so that it is the same for each reaction; glucose standard; enzyme color reagent; barium hydroxide; heparin; blood samples from five patients; test tubes—used as reaction vessels for the various tests; test tube incubation unit—used to incubate, mix, and centrifuge the samples; spectrophotometer—used to measure the amount of light absorbed or transmitted by a pigmented solution.

Experiment Instructions

Go to the home page in the PhysioEx software and click **Exercise 4: Endocrine System Physiology.** Click **Activity 2: Plasma Glucose, Insulin, and Diabetes Mellitus,** and take the online **Pre-lab Quiz** for Activity 2.

After you take the online Pre-lab Quiz, click the **Experiment** tab and begin the experiment. The experiment instructions are reprinted here for your reference. The opening screen for the experiment is shown below.

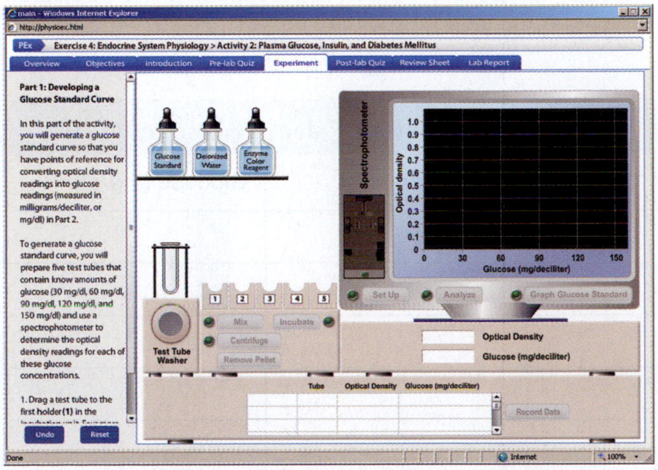
main - Windows Internet Explorer
http://physioex.html
PEx
Exercise 4: Endocrine System Physiology > Activity 2: Plasma Glucose, Insulin, and Diabetes Mellitus
Overview
Objectives
Introduction
Pre-lab Quiz
Experiment
Post-lab Quiz
Review Sheet
Lab Report
Part 1: Developing a Glucose Standard Curve
In this part of the activity, you will generate a glucose standard curve so that you have points of reference for converting optical density readings into glucose readings (measured in milligrams/deciliter, or mg/dl) in Part 2.
To generate a glucose standard curve, you will prepare five test tubes that contain know amounts of glucose (30 mg/dl, 60 mg/dl, 90 mg/dl, 120 mg/dl, and 150 mg/dl) and use a spectrophotometer to determine the optical density readings for each of these glucose concentrations.
1. Drag a test tube to the first holder (1) in the
Undo
Reset
Glucose Standard
Deionized Water
Enzyme Color Reagent
Spectrophotometer
Optical density
Glucose (mg/deciliter)
Set Up
Analyze
Graph Glucose Standard
Test Tube Washer
Mix
Incubate
Centrifuge
Remove Pellet
Optical Density
Glucose (mg/deciliter)
Tube
Optical Density
Glucose (mg/deciliter)
Record Data
Done
Internet
100%

Part 1: Developing a Glucose Standard Curve

In this part of the activity, you will generate a glucose standard curve so that you have points of reference for converting optical density readings into glucose readings (measured in milligrams/deciliter, or mg/dl) in Part 2.

To generate a glucose standard curve, you will prepare five test tubes that contain known amounts of glucose (30 mg/dl, 60 mg/dl, 90 mg/dl, 120 mg/dl, and 150 mg/dl) and use a spectrophotometer to determine the optical density readings for each of these glucose concentrations.

1. Drag a test tube to the first holder (**1**) in the incubation unit. Four more test tubes will automatically be placed in the incubation unit.

2. Drag the dropper cap of the glucose standard bottle to the first tube in the incubation unit to dispense one drop of glucose standard solution into the tube. The dropper will automatically move across and dispense glucose standard to the remaining tubes. Note that each tube receives one additional drop of glucose standard (tube 2 receives 2 drops, tube 3 receives 3 drops, tube 4 receives 4 drops, and tube 5 receives 5 drops).

3. Drag the dropper cap of the deionized water bottle to the first tube in the incubation unit to dispense four drops of deionized water into the tube. The dropper will automatically move across and dispense deionized water to the remaining tubes. Note that each tube receives one less drop of deionized water (tube 2 receives 3 drops, tube 3 receives 2 drops, tube 4 receives 1 drop, and tube 5 does not receive any drops).

4. Click **Mix** to mix the contents of the tubes.

5. Click **Centrifuge** to centrifuge the contents of the tubes. After the centrifugation process, the tubes will automatically rise.

6. Click **Remove Pellet** to remove any pellets formed during the centrifugation process. Pellets can contain reagent precipitates and debris from the laboratory environment.

7. Drag the dropper cap of the enzyme color reagent bottle to the first tube in the incubation unit to dispense five drops of enzyme color reagent into each tube.

8. Click **Incubate** to incubate the contents of the tubes. The incubation unit will gently agitate the test tube rack, evenly mixing the contents of all test tubes throughout the incubation.

9. Click **Set Up** on the spectrophotometer to warm up the instrument and get it ready for your sample readings.

10. Drag tube 1 to the spectrophotometer.

11. Click **Analyze** to analyze the sample. A data point will appear on the monitor to show the optical density and the glucose concentration of the sample. These values will also appear in the optical density and glucose displays.

12. Click **Record Data** to display your results in the grid (**and** record your results in Chart 2.1). The tube will automatically be placed in the test tube washer.

CHART 2.1 Glucose Standard Curve Results

Tube	Optical density	Glucose (mg/dl)
1		
2		
3		
4		
5		

13. You will now analyze the samples in the remaining tubes.
    - Drag the next tube into the spectrophotometer.
    - Click **Analyze** to analyze the sample. A data point will appear on the monitor to show the optical density and the glucose concentration of the sample. These values will also appear in the optical density and glucose displays.
    - Click **Record Data** to display your results in the grid (and record your results in Chart 2.1). The tube will automatically be placed in the test tube washer.

    Repeat this step until you analyze all five tubes.
14. Click **Graph Glucose Standard** to generate the glucose standard curve on the monitor. You will use this graph in Part 2.

> **? PREDICT Question 1**
> How would you measure the amount of plasma glucose in a patient sample?
>
> ____
>
> ____

Part 2: Measure Fasting Plasma Glucose Levels

In this part of the activity, you will use the glucose standard curve you generated in Part 1 to measure the fasting plasma glucose levels from five patients to diagnose the presence or absence of diabetes mellitus. Note the addition of two reagent bottles (barium hydroxide and heparin) and blood samples from the five patients. To undergo the fasting plasma glucose (FPG) test, patients must fast for a minimum of 8 hours prior to the blood draw.

A patient with FPG values greater than or equal to 126 mg/dl in two FPG tests is diagnosed with diabetes. FPG values between 110 and 126 mg/dl indicate impairment or borderline impairment of insulin-mediated glucose uptake by cells. FPG values less than 110 mg/dl are considered normal.

15. Drag a test tube to the first holder **(1)** in the incubation unit. Four more test tubes will automatically be placed in the incubation unit.
16. Drag the dropper cap of the first patient blood sample to the first tube in the incubation unit to dispense three drops of the sample. Three drops from each sample will automatically be dispensed into a separate tube.
17. Drag the dropper cap of the deionized water bottle to the first tube in the incubation unit to dispense five drops of deionized water into each tube.
18. Barium hydroxide dissolves and thus clears both proteins and cell membranes (so that clear glucose readings can be obtained). Drag the dropper cap of the barium hydroxide bottle to the first tube in the incubation unit to dispense five drops of barium hydroxide into each tube.
19. Drag the dropper cap of the heparin bottle to the first tube in the incubation unit to dispense a drop of heparin into each tube. Heparin prevents blood clots, which would interfere with clear glucose readings.
20. Click **Mix** to mix the contents of the tubes.
21. Click **Centrifuge** to centrifuge the contents of the tubes. After the centrifugation process, the tubes will automatically rise.
22. Click **Remove Pellet** to remove any pellets formed during the centrifugation process. Pellets can contain reagent precipitates and debris from the laboratory environment.
23. Drag the dropper cap of the enzyme color reagent bottle to the first tube in the incubation unit to dispense five drops of enzyme color reagent into each tube.
24. Click **Incubate** to incubate the contents of the tubes. The incubation unit will gently agitate the test tube rack, evenly mixing the contents of all test tubes throughout the incubation.
25. Click **Set Up** on the spectrophotometer to warm up the instrument and get it ready for your sample readings.
26. Click **Graph Glucose Standard** to display the glucose standard curve you generated in Part 1 on the monitor.
27. Drag tube 1 to the spectrophotometer.
28. Click **Analyze** to analyze the sample. A horizontal line will appear on the monitor to show the optical density of the sample. The optical density will also appear in the optical density display.
29. Drag the movable ruler (the vertical red line on the right side of the monitor) to the intersection of the horizontal yellow line (the optical density of the sample) and the glucose standard curve. Note the change in the glucose display as you move the line. The glucose concentration where the lines intersect is the fasting plasma glucose for this patient. Click **Record Data** to display your results in the grid (and record your results in Chart 2.2). The tube will automatically be placed in the test tube washer, and the monitor will be cleared (except for the glucose standard curve).

CHART 2.2 Fasting Plasma Glucose Results

Sample	Optical density	Glucose (mg/dl)
1		
2		
3		
4		
5		

30. You will now analyze the samples in the remaining tubes.
    - Drag the next tube into the spectrophotometer.
    - Click **Analyze** to analyze the sample. A data point will appear on the monitor to show the optical density and the glucose concentration of the sample. These values will also appear in the optical density and glucose displays.

• Click **Record Data** to display your results in the grid. The tube will automatically be placed in the test tube washer (and record your results in Chart 2.2).

Repeat this step until you analyze all five tubes.

After you complete the experiment, take the online **Post-lab Quiz** for Activity 2.

Activity Questions

1. How would you know if your glucose standard curve was aberrant and thus inappropriate for patient diagnostics?

____

____

2. What are potential sources of variability when generating a glucose standard curve?

____

____

3. What recommendations would you make to a patient with fasting plasma glucose levels in the impaired/borderline-impaired range who was in the impaired/borderline-impaired range for the oral glucose tolerance test?

____

____

4. The amount of corn syrup in the American diet has been described as alarmingly high (especially in the foods that children eat). In the context of this activity, predict the likely trends in the fasting plasma glucose levels of our children as they mature.

____

____

ACTIVITY 3

Hormone Replacement Therapy

OBJECTIVES

1. To understand the terms *hormone replacement therapy, follicle-stimulating hormone (FSH), estrogen, calcitonin, osteoporosis, ovariectomized,* and *T score.*
2. To understand how estrogen levels affect bone density.
3. To understand the potential benefits of hormone replacement therapy.

Introduction

Follicle-stimulating hormone (FSH) is an anterior pituitary peptide hormone that stimulates ovarian follicle growth. Developing ovarian follicles then produce and secrete a steroid hormone called **estrogen** into the plasma. Estrogen has numerous effects on the female body and homeostasis, including the stimulation of bone growth and protection against **osteoporosis** (a reduction in the quantity of bone characterized by decreased bone mass and increased susceptibility to fractures).

After menopause, the ovaries stop producing and secreting estrogen. One of the effects and potential health problems of menopause is a loss of bone density that can result in osteoporosis and bone fractures. For this reason, postmenopausal treatments to prevent osteoporosis often include hormone replacement therapy. Estrogen can be administered to increase bone density. Calcitonin (secreted by C cells in the thyroid gland) is another peptide hormone that can be administered to counteract the development of osteoporosis. Calcitonin inhibits osteoclast activity and stimulates calcium uptake and deposition in long bones.

In this activity you will use three **ovariectomized** rats that are no longer producing estrogen because their ovaries have been surgically removed. A **T score** is a quantitative measurement of the mineral content of bone, used as an indicator of the structural strength of the bone and as a screen for osteoporosis. The three rats were chosen because each has a baseline T score of 2.61, indicating osteoporosis. T scores are interpreted as follows: normal = +1 to −0.99; osteopenia (bone thinning) = −1.0 to −2.49; osteoporosis = −2.5 and below.

You will administer either estrogen therapy or calcitonin therapy to these rats, representing two types of **hormone replacement therapy.** The third rat will serve as an untreated control and receive daily injections of saline. The vertebral bone density (VBD) of each rat will be measured with dual X-ray absorptiometry (DXA) to obtain its T score after treatment.

EQUIPMENT USED The following equipment will be depicted on-screen: three ovariectomized rats (Note that if this were an actual wet lab, the ovariectomies would have been performed on the rats a month before the experiment to ensure that no residual hormones remained in the rats' systems.); saline; estrogen; calcitonin; reusable syringe—used to inject the rats; anesthesia—used to immobilize the rats for the X-ray scanning; dual X-ray absorptiometry bone-density scanner (DXA)—used to measure vertebral bone density of the rats.

Experiment Instructions

Go to the home page in the PhysioEx software and click **Exercise 4: Endocrine System Physiology.** Click **Activity 3: Hormone Replacement Therapy,** and take the online **Pre-lab Quiz** for Activity 3.

After you take the online Pre-lab Quiz, click the **Experiment** tab and begin the experiment. The experiment instructions are reprinted here for your reference. The opening screen for the experiment is shown on the following page.

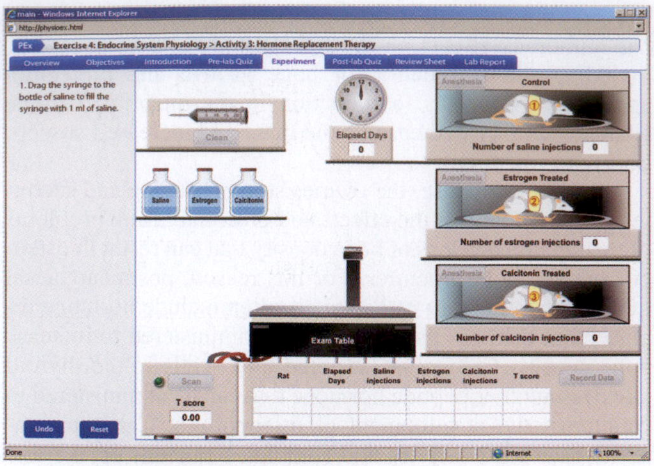
main - Windows Internet Explorer
http://physioex.html
PEx
Exercise 4: Endocrine System Physiology > Activity 3: Hormone Replacement Therapy
Overview
Objectives
Introduction
Pre-lab Quiz
Experiment
Post-lab Quiz
Review Sheet
Lab Report
1. Drag the syringe to the bottle of saline to fill the syringe with 1 ml of saline.
Clean
Elapsed Days
0
Anesthesia
Control
Number of saline injections
0
Saline
Estrogen
Calcitonin
Estrogen Treated
Number of estrogen injections
Calcitonin Treated
Number of calcitonin injections
Exam Table
Scan
T score
0.00
Rat
Elapsed Days
Saline injections
Estrogen injections
Calcitonin injections
T score
Record Data
Undo
Reset
Done
Internet
100%

1. Drag the syringe to the bottle of saline to fill the syringe with 1 ml of saline.

2. Drag the syringe to the *control* rat, placing the tip of the needle in the rat's lower abdominal area. Injections into this area are considered *intraperitoneal* and will quickly be circulated by the abdominal blood vessels.

3. Click **Clean** beneath the syringe holder to clean the syringe of all residues.

4. Drag the syringe to the bottle of estrogen to fill the syringe with 1 ml of estrogen.

5. Drag the syringe to the *estrogen-treated* rat, placing the tip of the needle in the rat's lower abdominal area.

6. Click **Clean** beneath the syringe holder to clean the syringe of all residues.

7. Drag the syringe to the bottle of calcitonin to fill the syringe with 1 ml of calcitonin.

8. Drag the syringe to the *calcitonin-treated* rat, placing the tip of the needle in the rat's lower abdominal area.

9. Click **Clean** beneath the syringe holder to clean the syringe of all residues.

10. Click the clock face to advance one day (24 hours).

11. Each rat must receive seven injections over the course of seven days (one injection per day). The remaining injections will be automated. Click the clock face to repeat the series of injections until you have injected each of the rats seven times.

? PREDICT Question 1
What effect will the saline injections have on the control rat's vertebral bone density?

____

? PREDICT Question 2
What effect will the estrogen injections have on the estrogen-treated rat's vertebral bone density?

____

? PREDICT Question 3
What effect will the calcitonin injections have on the calcitonin-treated rat's vertebral bone density?

____

12. Click **Anesthesia** above the *control* rat's cage to immobilize the control rat with a gaseous anesthetic for X-ray scanning.

13. Drag the anesthetized rat to exam table for X-ray scanning.

14. Click **Scan** to activate the scanner. The T score will appear in the T score display. Click **Record Data** to record your results in the grid (and record your results in Chart 3). The control rat will be automatically returned to its cage.

CHART 3 Hormone Replacement Therapy Results

Rat	T score

15. You will now obtain the T scores for the remaining rats. Perform these steps to obtain the T score for the *estrogen-treated* rat, then repeat these steps to obtain the T score for the *calcitonin-treated* rat.

- Click **Anesthesia** above the rat's cage to immobilize the rat with a gaseous anesthetic for X-ray scanning.
- Drag the anesthetized rat to exam table for X-ray scanning.
- Click **Scan** to activate the scanner. The T score will appear in the T score display.
- Click **Record Data** to record your results in the grid (and record your results in Chart 3). The rat will be automatically returned to its cage.

After you complete the experiment, take the online **Post-lab Quiz** for Activity 3.

Activity Questions

1. Recently, hormone replacement therapy has been prominent in the popular press. Describe a hormone replacement therapy that you have seen in the news, and highlight its benefits, its potential risks, the reasons to continue and the reasons to discontinue its use.

____

____

2. In hormone replacement therapy, how is the hormone dose determined by the prescribing physician?

ACTIVITY 4

Measuring Cortisol and Adrenocorticotropic Hormone

OBJECTIVES

1. To understand the terms *cortisol, adrenocorticotropic hormone (ACTH), corticotropin-releasing hormone (CRH), Cushing's syndrome, iatrogenic, Cushing's disease,* and *Addison's disease.*
2. To understand how CRH controls ACTH secretion and ACTH controls cortisol secretion.
3. To understand how negative feedback mechanisms influence the levels of tropic CRH and ACTH.
4. To measure the blood levels of cortisol and ACTH in five patients and correlate these readings with symptoms and diagnoses.
5. To distinguish between Cushing's syndrome and Cushing's disease.

Introduction

Cortisol, a hormone secreted by the *adrenal cortex,* is important in the body's response to many kinds of stress. Cortisol release is stimulated by **adrenocorticotropic hormone (ACTH),** a tropic hormone released by the anterior pituitary. A *tropic* hormone stimulates the secretion of another hormone. ACTH release, in turn, is stimulated by **corticotropin-releasing hormone (CRH),** a tropic hormone from the hypothalamus. Increased levels of cortisol negatively feed back to inhibit the release of both ACTH and CRH.

Increased cortisol in the blood, or *hypercortisolism,* is referred to as **Cushing's syndrome** if the increase is caused by an adrenal gland tumor. Cushing's syndrome can also be **iatrogenic** (that is, physician induced). For example, physician-induced Cushing's syndrome can occur when glucocorticoid hormones, such as prednisone, are administered to treat rheumatoid arthritis, asthma, or lupus. Cushing's syndrome is often referred to as "steroid diabetes" because it results in hyperglycemia. In contrast, **Cushing's disease** is hypercortisolism caused by an anterior pituitary tumor. People with Cushing's disease exhibit increased levels of ACTH and cortisol.

Decreased cortisol in the blood, or *hypocortisolism,* can occur because of adrenal insufficiency. In primary adrenal insufficiency, also known as **Addison's disease,** the low cortisol is directly caused by gradual destruction of the adrenal cortex and ACTH levels are typically elevated as a compensatory effect. Secondary adrenal insufficiency also results in low levels of cortisol, usually caused by damage to the anterior pituitary. Therefore, the levels of ACTH are also low in secondary adrenal insufficiency.

As you can see, a variety of endocrine disorders can be related to both high and low levels of cortisol and ACTH. Table 4.1 summarizes these endocrine disorders.

TABLE 4.1 Cortisol and ACTH Disorders

	Cortisol level	ACTH level
Cushing's syndrome (primary hypercortisolism)	High	Low
Iatrogenic Cushing's syndrome	High	Low
Cushing's disease (secondary hypercortisolism)	High	High
Addison's disease (primary adrenal insufficiency)	Low	High
Secondary adrenal insufficiency (hypopituitarism)	Low	Low

EQUIPMENT USED The following equipment will be depicted on-screen: plasma samples from five patients; HPLC (high-performance liquid chromatography) column—used to quantitatively measure the amount of cortisol and ACTH in the patient samples; HPLC detector—provides the hormone concentration in the patient sample; reusable syringe—used to inject the patient samples into the HPLC injection port; HPLC injection port—used to inject the patient samples into the HPLC column.

Experiment Instructions

Go to the home page in the PhysioEx software and click **Exercise 4: Endocrine System Physiology.** Click **Activity 4: Measuring Cortisol and Adrenocorticotropic Hormone,** and take the online **Pre-lab Quiz** for Activity 4.

After you take the online Pre-lab Quiz, click the **Experiment** tab and begin the experiment. The experiment instructions are reprinted here for your reference. The opening screen for the experiment is shown below.

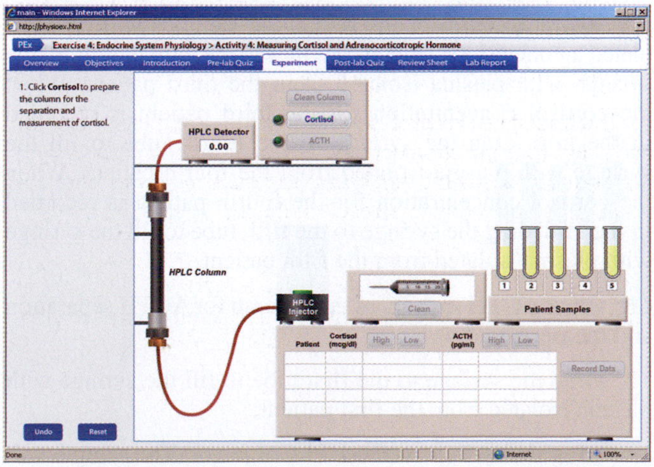
Exercise 4: Endocrine System Physiology > Activity 4: Measuring Cortisol and Adrenocorticotropic Hormone
Overview
Objectives
Introduction
Pre-lab Quiz
Experiment
Post-lab Quiz
Review Sheet
Lab Report
1. Click Cortisol to prepare the column for the separation and measurement of cortisol.
Clean Column
Cortisol
ACTH
HPLC Detector
0.00
HPLC Column
HPLC Injector
Clean
Patient Samples
Record Data
Undo
Reset

1. Click **Cortisol** to prepare the column for the separation and measurement of cortisol.

2. Drag the syringe to the first tube to fill the syringe with plasma isolated from the first patient.

3. Drag the syringe to the HPLC injector. The sample will enter the tubing and flow through the column. The cortisol concentration in the patient sample will appear in the HPLC detector display.

4. Click **Record Data** to display your results in the grid (and record your results in Chart 4).

CHART 4 Measurement of Cortisol

Patient	Cortisol (mcg/dl)	Cortisol level	ACTH (pg/ml)	ACTH level
1				
2				
3				
4				
5				

5. Click **Clean** beneath the syringe to prepare it for the next sample. Click **Clean Column** to remove residual cortisol from the column.

6. Drag the syringe to the second tube to fill the syringe with plasma isolated from the second patient.

7. Drag the syringe to the HPLC injector. The sample will enter the tubing and flow through the column. The cortisol concentration in the patient sample will appear in the HPLC detector display.

8. Click **Record Data** to display your results in the grid (and record your results in Chart 4).

9. Click **Clean** beneath the syringe to prepare it for the next sample. Click **Clean Column** to remove residual cortisol from the column.

10. The procedure for the remaining samples will be completed automatically. Drag the syringe to the third tube to fill the syringe with plasma isolated from the third patient. When the cortisol concentration for the third patient is recorded in the grid, drag the syringe to the fourth tube to fill the syringe with plasma isolated from the fourth patient. When the cortisol concentration for the fourth patient is recorded in the grid, drag the syringe to the fifth tube to fill the syringe with plasma isolated from the fifth patient.

11. Click **ACTH** to prepare the column for ACTH separation and measurement.

12. Drag the syringe to the first tube to fill the syringe with plasma isolated from the first patient.

13. Drag the syringe to the HPLC injector. The sample will enter the tubing and flow through the column. The ACTH concentration in the patient sample will appear in the HPLC detector display.

14. Click **Record Data** to display your results in the grid.

15. Click **Clean** beneath the syringe to prepare it for the next sample. Click **Clean Column** to remove residual ACTH from the column.

16. Drag the syringe to the second tube to fill the syringe with plasma isolated from the second patient.

17. Drag the syringe to the HPLC injector. The sample will enter the tubing and flow through the column. The ACTH concentration in the patient sample will appear in the HPLC detector display.

18. Click **Record Data** to display your results in the grid (and record your results in Chart 4).

19. Click **Clean** beneath the syringe to prepare it for the next sample. Click **Clean Column** to remove residual ACTH from the column.

20. The procedure for the remaining samples will be completed automatically. Drag the syringe to the third tube to fill the syringe with plasma isolated from the third patient. When the ACTH concentration for the third patient is recorded in the grid, drag the syringe to the fourth tube to fill the syringe with plasma isolated from the fourth patient. When the ACTH concentration for the fourth patient is recorded in the grid, drag the syringe to the fifth tube to fill the syringe with plasma isolated from the fifth patient.

21. Indicate whether the cortisol and ACTH concentrations (levels) for each patient are high or low using the breakpoints shown in Table 4.2. Click the row of the patient and then click **High** or **Low** next to cortisol and ACTH.4.2.

TABLE 4.2 Abnormal Morning Cortisol and ACTH Levels

ACTH level	High	Low
Cortisol	≥23 mcg/dl	<5 mcg/dl
ACTH	≥80 pg/ml	<20 pg/ml

Note: 1 mcg = 1 µg = 1 microgram

After you complete the experiment, take the online **Post-lab Quiz** for Activity 4.

Activity Questions

1. Discuss the benefits and drawbacks of giving glucocorticoids to young children that have significant allergy-induced asthma.

____

____

2. Explain the difference between Cushing's syndrome and Cushing's disease.

____

____

PhysioEx™ 9.0

REVIEW SHEET

EXERCISE 4

NAME ________________

LAB TIME/DATE ________________

Endocrine System Physiology

ACTIVITY 1 Metabolism and Thyroid Hormone

Part 1

1. Which rat had the fastest basal metabolic rate (BMR)? ________________

2. Why did the metabolic rates differ between the normal rat and the surgically altered rats? How well did the results compare with your prediction? ________________

3. If an animal has been thyroidectomized, what hormone(s) would be missing in its blood? ________________

4. If an animal has been hypophysectomized, what effect would you expect to see in the hormone levels in its body? ________________

Part 2

5. What was the effect of thyroxine injections on the normal rat's BMR? ________________

6. What was the effect of thyroxine injections on the thyroidectomized rat's BMR? How does the BMR in this case compare with the normal rat's BMR? Was the dose of thyroxine in the syringe too large, too small, or just right? ________________

7. What was the effect of thyroxine injections on the hypophysectomized rat's BMR? How does the BMR in this case compare with the normal rat's BMR? Was the dose of thyroxine in the syringe too large, too small, or just right? ______

______

______

Part 3

8. What was the effect of thyroid-stimulating hormone (TSH) injections on the normal rat's BMR? ______

______

______

9. What was the effect of TSH injections on the thyroidectomized rat's BMR? How does the BMR in this case compare with the normal rat's BMR? Why was this effect observed? ______

______

10. What was the effect of TSH injections on the hypophysectomized rat's BMR? How does the BMR in this case compare with the normal rat's BMR? Was the dose of TSH in the syringe too large, too small, or just right? ______

______

______

Part 4

11. What was the effect of propylthiouracil (PTU) injections on the normal rat's BMR? Why did this rat develop a palpable goiter?

______

______

12. What was the effect of PTU injections on the thyroidectomized rat's BMR? How does the BMR in this case compare with the normal rat's BMR? Why was this effect observed? ______

______

13. What was the effect of PTU injections on the hypophysectomized rat's BMR? How does the BMR in this case compare with the normal rat's BMR? Why was this effect observed? ______

______

______

ACTIVITY 2 Plasma Glucose, Insulin, and Diabetes Mellitus

1. What is a glucose standard curve, and why did you need to obtain one for this experiment? Did you correctly predict how you would measure the amount of plasma glucose in a patient sample using the glucose standard curve? ______________

____

____

2. Which patient(s) had glucose reading(s) in the diabetic range? Can you say with certainty whether each of these patients has type 1 or type 2 diabetes? Why or why not? ______________

____

____

3. Describe the diagnosis for patient 3, who was also pregnant at the time of this assay. ______________

____

____

4. Which patient(s) had normal glucose reading(s)? ______________

____

5. What are some lifestyle choices these patients with normal plasma glucose readings might recommend to the borderline impaired patients? ______________

____

____

ACTIVITY 3 Hormone Replacement Therapy

1. Why were ovariectomized rats used in this experiment? How does the fact that the rats are ovariectomized explain their baseline T scores? ______________

____

____

2. What effect did the administration of saline injections have on the control rat? How well did the results compare with your prediction? ______________

____

____

3. What effect did the administration of estrogen injections have on the estrogen-treated rat? How well did the results compare with your prediction? ______________________________

____

____

4. What effect did the administration of calcitonin injections have on the calcitonin-treated rat? How well did the results compare with your prediction? ______________________________

____

____

5. What are some health risks that postmenopausal women must consider when contemplating estrogen hormone replacement therapy? ______________________________

____

____

ACTIVITY 4 Measuring Cortisol and Adrenocorticotropic Hormone

1. Which patient would most likely be diagnosed with Cushing's disease? Why? ______________________________

____

____

2. Which two patients have hormone levels characteristic of Cushing's syndrome? ______________________________

____

____

3. Patient 2 is being treated for rheumatoid arthritis with prednisone. How does this information change the diagnosis? ______

____

____

4. Which patient would most likely be diagnosed with Addison's disease? Why? ______________________________

____

____

PhysioEx 9.0

EXERCISE 5

Cardiovascular Dynamics

PRE-LAB QUIZ

1. Circle the correct term. According to general usage, systole / diastole refers to ventricular relaxation.
2. A graph illustrating the pressure and volume changes during one heartbeat is called the
   a. blood pressure
   b. cardiac cycle
   c. conduction system of the heart
   d. electrical events of the heartbeat
3. Circle True or False. When ventricular systole begins, intraventricular pressure increases rapidly, closing the atrioventricular (AV) valves.
4. The average heart beats approximately ________ times per minute.
   a. 50 c. 100
   b. 75 d. 125
5. Circle the correct term. Abnormal heart sounds called murmurs / strokes can indicate valvular problems.
6. The term ________ refers to the alternating surges of pressure in an artery that occur with each contraction and relaxation of the left ventricle.
   a. diastole c. pulse
   b. murmur d. systole
7. Circle the correct term. The pulse is most often taken at the lateral aspect of the wrist, above the thumb, compressing the popliteal / radial artery.
8. What device will you use today to measure your subject's blood pressure? ________
9. In reporting a blood pressure of 120/90, which number represents the *diastolic* pressure? ________
10. The ________ are characteristic sounds that indicate the resumption of blood flow to the artery being occluded when taking blood pressure.
    a. ectopic heartbeats
    b. heart murmurs
    c. heart rhythms
    d. sounds of Korotkoff

Exercise Overview

The cardiovascular system is composed of a pump—the heart—and blood vessels that distribute blood containing oxygen and nutrients to every cell of the body. The principles governing blood flow are the same physical laws that apply to the flow of liquid through a system of pipes. For example, one very basic law in fluid mechanics is that the flow rate of a liquid through a pipe is directly proportional to the difference between the pressures at the two ends of the pipe (the **pressure gradient**) and inversely proportional to the pipe's **resistance** (a measure of the degree to which the pipe hinders, or resists, the flow of the liquid).

$$\text{Flow} = \text{pressure gradient}/\text{resistance} = \Delta P/R$$

This basic law also applies to blood flow. The "liquid" is blood, and the "pipes" are blood vessels. The pressure gradient is the difference between the pressure in arteries and the pressure in veins that results when blood is pumped into arteries. Blood flow rate is directly proportional to the pressure gradient and inversely proportional to resistance.

Blood flow is the amount of blood moving through a body area or the entire cardiovascular system in a given amount of time. Total blood flow is proportional to **cardiac output** (the amount of blood the heart is able to pump per minute). Blood flow to specific body areas can vary dramatically in a given time period. Organs differ in their requirements from moment to moment, and blood vessels have different sized diameters in their lumen (opening) to regulate local blood flow to various areas in response to the tissues' immediate needs. Consequently, blood flow can increase to some areas and decrease to other areas at the same time.

Resistance is a measure of the degree to which the blood vessel hinders, or resists, the flow of blood. The main factors that affect resistance are (1) blood vessel *radius,* (2) blood vessel *length,* and (3) blood *viscosity.*

Radius

The smaller the blood vessel radius, the greater the resistance, because of frictional drag between the blood and the vessel walls. Contraction of smooth muscle of the blood vessel, or **vasoconstriction,** results in a decrease in the blood vessel radius. Lipid deposits can also cause the radius of an artery to decrease, preventing blood from reaching the coronary tissue, which frequently leads to a heart attack. Alternately, relaxation of smooth muscle of the blood vessel, or **vasodilation,** causes an increase in the blood vessel radius. Blood vessel radius is the single most important factor in determining blood flow resistance.

Length

The longer the vessel length, the greater the resistance—again, because of friction between the blood and vessel walls. The length of a person's blood vessels change only as a person grows. Otherwise, the length generally remains constant.

Viscosity

Viscosity is blood "thickness," determined primarily by **hematocrit**—the fractional contribution of red blood cells to total blood volume. The higher the hematocrit, the greater the viscosity. Under most physiological conditions, hematocrit does not vary much and blood viscosity remains more or less constant.

The Effect of Blood Pressure and Vessel Resistance on Blood Flow

Blood flow is directly proportional to blood pressure because the pressure difference (ΔP) between the two ends of a vessel is the driving force for blood flow. Peripheral resistance is the friction that opposes blood flow through a blood vessel. This relationship is represented in the following equation:

$$\text{Blood flow (ml/min)} = \frac{\Delta P}{\text{peripheral resistance}}$$

Three factors that contribute to peripheral resistance are blood viscosity (η), blood vessel length (L), and the radius of the blood vessel (r). These relationships are expressed in the following equation:

$$\text{Peripheral resistance} = \frac{8L\eta}{\pi r^4}$$

From this equation you can see that the viscosity of the blood and the length of the blood vessel are directly proportional to peripheral resistance. The peripheral resistance is inversely proportional to the fourth power of the vessel radius. If you combine the two equations, you get the following result:

$$\text{Blood flow (ml/min)} = \frac{\Delta P \pi r^4}{8L\eta}$$

From this combination you can see that blood flow is directly proportional to the fourth power of vessel radius, which means that small changes in vessel radius result in dramatic changes in blood flow.

ACTIVITY 1

Studying the Effect of Blood Vessel Radius on Blood Flow Rate

OBJECTIVES

1. To understand how blood vessel radius affects blood flow rate.
2. To understand how vessel radius is changed in the body.
3. To understand how to interpret a graph of blood vessel radius versus blood flow rate.

Introduction

Controlling **blood vessel radius** (one-half of the diameter) is the principal method of controlling blood flow. Controlling blood vessel radius is accomplished by contracting or relaxing the smooth muscle within the blood vessel walls (vasoconstriction or vasodilation).

To understand why radius has such a pronounced effect on blood flow, consider the physical relationship between blood and the vessel wall. Blood in direct contact with the vessel wall flows relatively slowly because of the friction, or drag, between the blood and the lining of the vessel. In contrast, blood in the center of the vessel flows more freely because it is not rubbing against the vessel wall. The free-flowing blood in the middle of the vessel is called the **laminar flow.** Now picture a fully constricted (small-radius) vessel and a fully dilated (large-radius) vessel. In the fully constricted vessel, proportionately more blood is in contact with the vessel wall and there is less laminar flow, significantly impeding the rate of blood flow in the fully constricted vessel relative to that in the fully dilated vessel.

In this activity you will study the effect of blood vessel radius on blood flow. The experiment includes two glass beakers and a tube connecting them. Imagine that the left beaker is your heart, the tube is an artery, and the right beaker is a destination in your body, such as another organ.

EQUIPMENT USED The following equipment will be depicted on-screen: left beaker—simulates blood flowing from the heart; flow tube between the left and right beaker—simulates an artery; right beaker—simulates another organ (for example, the biceps brachii muscle).

Experiment Instructions

Go to the home page in the PhysioEx software and click **Exercise 5: Cardiovascular Dynamics.** Click **Activity 1: Studying the Effect of Blood Vessel Radius on Blood Flow Rate,** and take the online **Pre-lab Quiz** for Activity 1.

After you take the online Pre-lab Quiz, click the **Experiment** tab and begin the experiment. The experiment instructions are reprinted here for your reference. The opening screen for the experiment is shown below.

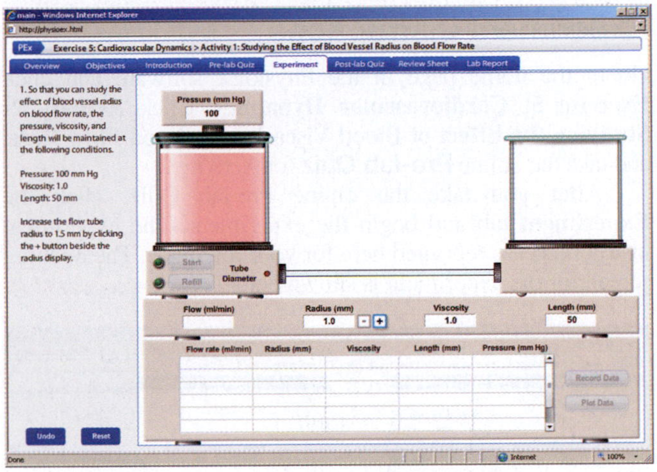
Exercise 5: Cardiovascular Dynamics > Activity 1: Studying the Effect of Blood Vessel Radius on Blood Flow Rate
Overview
Objectives
Introduction
Pre-lab Quiz
Experiment
Post-lab Quiz
Review Sheet
Lab Report
Pressure (mm Hg)
100
Tube Diameter
Flow (ml/min)
Radius (mm)
1.0
Viscosity
1.0
Length (mm)
50
Record Data
Plot Data
Undo
Reset

1. So that you can study the effect of blood vessel radius on blood flow rate, the pressure, viscosity, and length will be maintained at the following conditions:

Pressure: 100 mm Hg

Viscosity: 1.0

Length: 50 mm

Increase the flow tube radius to 1.5 mm by clicking the + button beside the radius display.

2. Click **Start** and then watch the fluid move into the right beaker. (Fluid moves slowly under some conditions—be patient!) Pressure propels fluid from the left beaker to the right beaker through the flow tube. The flow rate is shown in the flow rate display after the left beaker has finished draining.

3. Click **Record Data** to display your results in the grid (and record your results in Chart 1).

4. Click **Refill** to replenish the left beaker.

PREDICT Question 1
What do you think will happen to the flow rate if the radius is increased by 0.5 mm?

____

CHART 1 Effect of Blood Vessel Radius on Blood Flow Rate

Flow (ml/min)	Radius (mm)

5. Increase the flow tube radius to 2.0 mm by clicking the + button beside the radius display. Click **Start** and watch the fluid move into the right beaker.

6. Click **Record Data** to display your results in the grid (and record your results in Chart 1).

7. Click **Refill** to replenish the left beaker.

8. You will now observe the effect of incremental increases in flow tube radius.

- Increase the flow tube radius by 0.5 mm.
- Click **Start** and then watch the fluid move into the right beaker.
- Click **Record Data** to display your results in the grid (and record your results in Chart 1).
- Click **Refill** to replenish the left beaker.

Repeat this step until you reach a flow tube radius of 5.0 mm.

PREDICT Question 2
Do you think a graph plotted with radius on the X-axis and flow rate on the Y-axis will be linear (a straight line)?

____

9. Click **Plot Data** to view a summary of your data on a plotted grid. Radius will be displayed on the X-axis and flow rate will be displayed on the Y-axis. Click **Submit** to record your plot in the lab report.

After you complete the experiment, take the online **Post-lab Quiz** for Activity 1.

Activity Questions

1. Describe the relationship between vessel radius and blood flow rate.

____

____

2. In this activity you altered the radius of the flow tube by clicking the + and − buttons. Explain how and why the radius of blood vessels is altered in the human body.

____

____

3. Describe the appearance of your plot of blood vessel radius versus blood flow rate and relate the plot to the relationship between these two variables.

____

____

4. Describe an advantage of slower blood velocity in some areas of the body, for example, in the capillaries of our fingers.

____

____

ACTIVITY 2

Studying the Effect of Blood Viscosity on Blood Flow Rate

OBJECTIVES

1. To understand how blood viscosity affects blood flow rate.
2. To list the components in the blood that contribute to blood viscosity.
3. To explain conditions that might lead to viscosity changes in the blood.
4. To understand how to interpret a graph of viscosity versus blood flow.

Introduction

Viscosity is the thickness, or "stickiness," of a fluid. The more viscous a fluid, the more resistance to flow. Therefore, the flow rate will be slower for a more viscous solution. For example, consider how much more slowly maple syrup pours out of a container than milk does.

The viscosity of blood is due to the presence of plasma proteins and formed elements, which include white blood cells (leukocytes), red blood cells (erythrocytes), and platelets (thrombocytes). Formed elements and plasma proteins in the blood slide past one another, increasing the resistance to flow. With a viscosity of 3–5, blood is much more viscous than water (usually given a viscosity value of 1).

A body in homeostatic balance has a relatively stable blood consistency. Nevertheless, it is useful to examine the effects of blood viscosity on blood flow to predict what might occur in the human cardiovascular system when homeostatic imbalances occur. Factors such as dehydration and altered blood cell numbers do alter blood viscosity. For example, polycythemia is a condition in which excess red blood cells are present, and certain types of anemia result in fewer red blood cells. Increasing the number of red blood cells increases blood viscosity, and decreasing the number of red blood cells decreases blood viscosity.

In this activity you will examine the effects of blood viscosity on blood flow rate. The experiment includes two glass beakers and a tube connecting them. Imagine that the left beaker is your heart, the tube is an artery, and the right beaker is a destination in your body, such as another organ.

EQUIPMENT USED The following equipment will be depicted on-screen: left beaker—simulates blood flowing from the heart; flow tube between the left and right beaker—simulates an artery; right beaker—simulates another organ (for example, the biceps brachii muscle).

Experiment Instructions

Go to the home page in the PhysioEx software and click **Exercise 5: Cardiovascular Dynamics.** Click **Activity 2: Studying the Effect of Blood Viscosity on Blood Flow Rate,** and take the online **Pre-lab Quiz** for Activity 2.

After you take the online Pre-lab Quiz, click the **Experiment** tab and begin the experiment. The experiment instructions are reprinted here for your reference. The opening screen for the experiment is shown below.

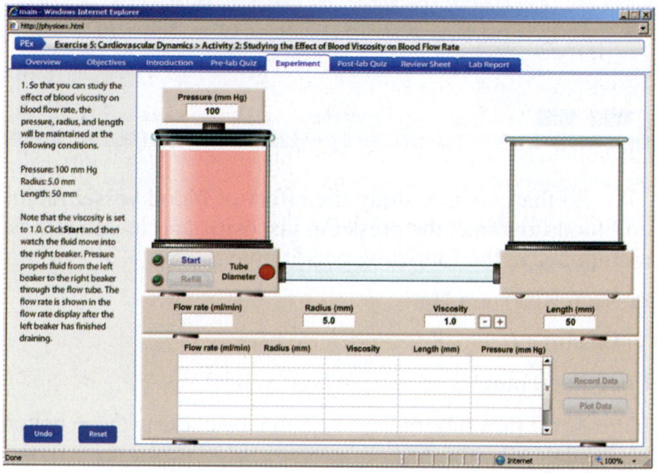
main - Windows Internet Explorer
http://physioex.html
PEx
Exercise 5: Cardiovascular Dynamics > Activity 2: Studying the Effect of Blood Viscosity on Blood Flow Rate
Overview
Objectives
Introduction
Pre-lab Quiz
Experiment
Post-lab Quiz
Review Sheet
Lab Report
1. So that you can study the effect of blood viscosity on blood flow rate, the pressure, radius, and length will be maintained at the following conditions.
Pressure: 100 mm Hg
Radius: 5.0 mm
Length: 50 mm
Note that the viscosity is set to 1.0. Click Start and then watch the fluid move into the right beaker. Pressure propels fluid from the left beaker to the right beaker through the flow tube. The flow rate is shown in the flow rate display after the left beaker has finished draining.
Pressure (mm Hg)
100
Start
Refill
Tube Diameter
Flow rate (ml/min)
Radius (mm)
5.0
Viscosity
1.0
Length (mm)
50
Flow rate (ml/min)
Radius (mm)
Viscosity
Length (mm)
Pressure (mm Hg)
Record Data
Plot Data
Undo
Reset
Done
Internet
100%

1. So that you can study the effect of blood viscosity on blood flow rate, the pressure, radius, and length will be maintained at the following conditions:

Pressure: 100 mm Hg

Radius: 5.0 mm

Length: 50 mm

Note that the viscosity is set to 1.0. Click **Start** and then watch the fluid move into the right beaker. Pressure propels fluid from the left beaker to the right beaker through the flow tube. The flow rate is shown in the flow rate display after the left beaker has finished draining.

2. Click **Record Data** to display your results in the grid (and record your results in Chart 2).

CHART 2	Effect of Blood Viscosity on Blood Flow Rate
Flow (ml/min)	Viscosity

3. Click **Refill** to replenish the left beaker.

PREDICT Question 1
What effect do you think increasing the viscosity will have on the fluid flow rate?

___

4. Increase the fluid viscosity to 2.0 by clicking the + button beside the viscosity display. Click **Start** and then watch the fluid move into the right beaker.

5. Click **Record Data** to display your results in the grid (and record your results in Chart 2).

6. Click **Refill** to replenish the left beaker.

7. You will now observe the effect of incremental increases in viscosity.

- Increase the viscosity by 1.0.
- Click **Start** and then watch the fluid move into the right beaker.
- Click **Record Data** to display your results in the grid (and record your results in Chart 2).
- Click **Refill** to replenish the left beaker.

Repeat this step until you reach a viscosity of 8.0.

8. Click **Plot Data** to view a summary of your data on a plotted grid. Viscosity will be displayed on the X-axis and flow rate will be displayed on the Y-axis. Click **Submit** to record your plot in the lab report.

After you complete the experiment, take the online **Post-lab Quiz** for Activity 2.

Activity Questions

1. Describe the effect on blood flow rate when blood viscosity was increased.

___

___

2. Explain why the relationship between viscosity and blood flow rate is inversely proportional.

___

___

3. What might happen to blood flow if you increased the number of blood cells?

___

___

ACTIVITY 3

Studying the Effect of Blood Vessel Length on Blood Flow Rate

OBJECTIVES

1. To understand how blood vessel length affects blood flow rate.
2. To explain conditions that can lead to blood vessel length changes in the body.
3. To compare the effect of blood vessel length changes with the effect of blood vessel radius changes on blood flow rate.

Introduction

Blood vessel lengths increase as we grow to maturity. The longer the vessel, the greater the resistance to blood flow through the blood vessel because there is a larger surface area in contact with the blood cells. Therefore, when blood vessel length increases, friction increases. Our blood vessel lengths stay fairly constant in adulthood, unless we gain or lose weight. If we gain weight, blood vessel lengths can increase, and if we lose weight, blood vessel lengths can decrease.

In this activity you will study the physical relationship between blood vessel length and blood flow. Specifically, you will study how blood flow changes in blood vessels of constant radius but different lengths. The experiment includes two glass beakers and a tube connecting them. Imagine that the left beaker is your heart, the tube is an artery, and the right beaker is a destination in your body, such as another organ.

EQUIPMENT USED The following equipment will be depicted on-screen: left beaker—simulates blood flowing from the heart; flow tube between the left and right beaker—simulates an artery; right beaker—simulates another organ (for example, the biceps brachii muscle).

Experiment Instructions

Go to the home page in the PhysioEx software and click **Exercise 5: Cardiovascular Dynamics.** Click **Activity 3: Studying the Effect of Blood Vessel Length on Blood Flow Rate,** and take the online **Pre-lab Quiz** for Activity 3.

After you take the online Pre-lab Quiz, click the **Experiment** tab and begin the experiment. The experiment instructions are reprinted here for your reference. The opening screen for the experiment is shown on the following page.

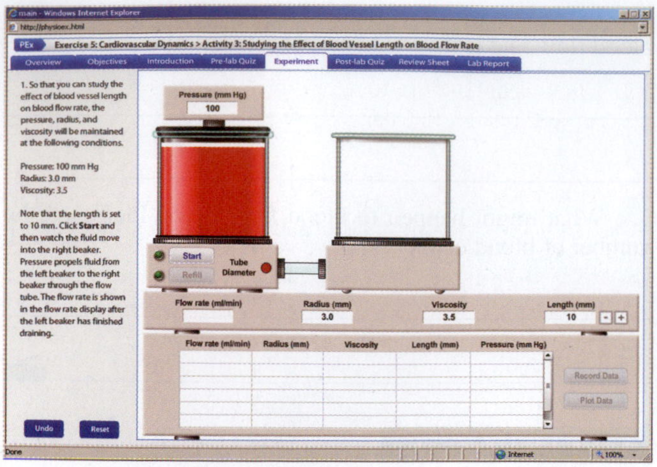
Exercise 5: Cardiovascular Dynamics > Activity 3: Studying the Effect of Blood Vessel Length on Blood Flow Rate
Overview
Objectives
Introduction
Pre-lab Quiz
Experiment
Post-lab Quiz
Review Sheet
Lab Report
Pressure (mm Hg)
100
Start
Refill
Tube Diameter
Flow rate (ml/min)
Radius (mm)
3.0
Viscosity
3.5
Length (mm)
10
Record Data
Plot Data
Undo
Reset

1. So that you can study the effect of blood vessel length on blood flow rate, the pressure, radius, and viscosity will be maintained at the following conditions:

Pressure: 100 mm Hg

Radius: 3.0 mm

Viscosity: 3.5

Note that the length is set to 10 mm. Click **Start** and then watch the fluid move into the right beaker. Pressure propels fluid from the left beaker to the right beaker through the flow tube. The flow rate is shown in the flow rate display after the left beaker has finished draining.

2. Click **Record Data** to display your results in the grid (and record your results in Chart 3).

CHART 3 Effect of Blood Vessel Length on Blood Flow Rate

Flow (ml/min)	Flow Tube length (mm)

3. Click **Refill** to replenish the left beaker.

? PREDICT Question 1
What effect do you think increasing the flow tube length will have on the fluid flow rate?

____

4. Increase the flow tube length to 15 mm by clicking the + button beside the length display. Click **Start** and then watch the fluid move into the right beaker.

5. Click **Record Data** to display your results in the grid (and record your results in Chart 3).

6. Click **Refill** to replenish the left beaker.

7. You will now observe the effect of incremental increases in flow tube length.

- Increase the flow tube length by 5 mm.
- Click **Start** and then watch the fluid move into the right beaker.
- Click **Record Data** to display your results in the grid (and record your results in Chart 3).
- Click **Refill** to replenish the left beaker.

Repeat this step until you reach a flow tube length of 40 mm.

8. Click **Plot Data** to view a summary of your data on a plotted grid. Length will be displayed on the X-axis and flow rate will be displayed on the Y-axis. Click **Submit** to record your plot in the lab report.

After you complete the experiment, take the online **Post-lab Quiz** for Activity 3.

Activity Questions

1. Is the relationship between blood vessel length and blood flow rate directly proportional or inversely proportional? Why?

____

____

2. Which of the following can vary in size more quickly: blood vessel diameter or blood vessel length?

____

____

3. Describe what happens to resistance when blood vessel length increases.

____

____

ACTIVITY 4

Studying the Effect of Blood Pressure on Blood Flow Rate

OBJECTIVES

1. To understand how blood pressure affects blood flow rate.
2. To understand what structure produces blood pressure in the human body.
3. To compare the plot generated for pressure versus blood flow to those generated for radius, viscosity, and length.

Introduction

The pressure difference between the two ends of a blood vessel is the driving force behind blood flow. This pressure difference is referred to as a pressure gradient. In the cardiovascular system, the force of contraction of the heart provides the initial pressure and vascular resistance contributes to the pressure gradient. If the heart changes its force of contraction, the blood vessels need to be able to respond to the change in force. Large arteries close to the heart have more elastic tissue in their tunics in order to accommodate these changes.

In this activity you will look at the effect of pressure changes on blood flow (recall from the blood flow equation that a change in blood flow is directly proportional to the pressure gradient). The experiment includes two glass beakers and a tube connecting them. Imagine that the left beaker is your heart, the tube is an artery, and the right beaker is a destination in your body, such as another organ.

EQUIPMENT USED The following equipment will be depicted on-screen: left beaker—simulates blood flowing from the heart; flow tube between the left and right beaker—simulates an artery; right beaker—simulates another organ (for example, the biceps brachii muscle).

Experiment Instructions

Go to the home page in the PhysioEx software and click **Exercise 5: Cardiovascular Dynamics.** Click **Activity 4: Studying the Effect of Blood Pressure on Blood Flow Rate,** and take the online **Pre-lab Quiz** for Activity 4.

After you take the online Pre-lab Quiz, click the **Experiment** tab and begin the experiment. The experiment instructions are reprinted here for your reference. The opening screen for the experiment is shown below.

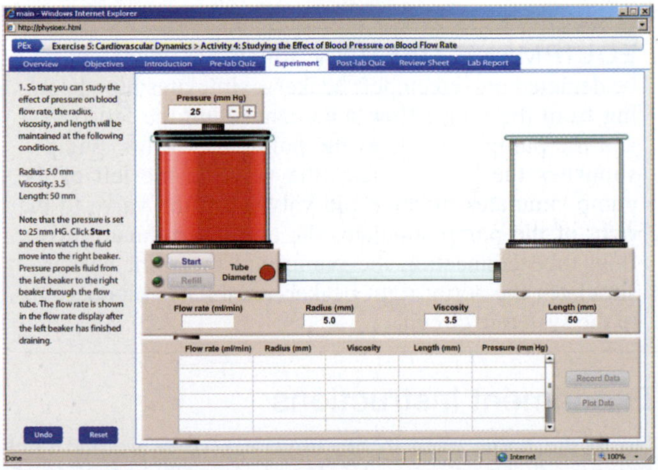
main - Windows Internet Explorer
http://physioex.html
PEx
Exercise 5: Cardiovascular Dynamics > Activity 4: Studying the Effect of Blood Pressure on Blood Flow Rate
Overview
Objectives
Introduction
Pre-lab Quiz
Experiment
Post-lab Quiz
Review Sheet
Lab Report
1. So that you can study the effect of pressure on blood flow rate, the radius, viscosity, and length will be maintained at the following conditions.
Radius: 5.0 mm
Viscosity: 3.5
Length: 50 mm
Note that the pressure is set to 25 mm HG. Click Start and then watch the fluid move into the right beaker. Pressure propels fluid from the left beaker to the right beaker through the flow tube. The flow rate is shown in the flow rate display after the left beaker has finished draining.
Pressure (mm Hg)
25
Start
Refill
Tube Diameter
Flow rate (ml/min)
Radius (mm)
5.0
Viscosity
3.5
Length (mm)
50
Flow rate (ml/min)
Radius (mm)
Viscosity
Length (mm)
Pressure (mm Hg)
Record Data
Plot Data
Undo
Reset
Done
Internet
100%

1. So that you can study the effect of pressure on blood flow rate, the radius, viscosity, and length will be maintained at the following conditions:

Radius: 5.0 mm

Viscosity: 3.5

Length: 50 mm

Note that the pressure is set to 25 mm Hg. Click **Start** and then watch the fluid move into the right beaker. Pressure propels fluid from the left beaker to the right beaker through the flow tube. The flow rate is shown in the flow rate display after the left beaker has finished draining.

2. Click **Record Data** to record your results in the grid (and record your results in Chart 4).

CHART 4 Effect of Blood Pressure on Blood Flow Rate

Flow (ml/min)	Pressure (mm Hg)

3. Click **Refill** to replenish the left beaker.

PREDICT Question 1
What effect do you think increasing the pressure will have on the fluid flow rate?

____

4. Increase the pressure to 50 mm Hg by clicking the + button beside the pressure display. Click **Start** and then watch the fluid move into the right beaker.

5. Click **Record Data** to record your results in the grid (and record your results in Chart 4).

6. Click **Refill** to replenish the left beaker.

7. You will now observe the effect of incremental increases in pressure.

- Increase the pressure by 25 mm Hg.
- Click **Start** and then watch the fluid move into the right beaker.
- Click **Record Data** to display your results in the grid (and record your results in Chart 4).
- Click **Refill** to replenish the left beaker.

Repeat this step until you reach a pressure of 200 mm Hg.

PREDICT Question 2
Do you think a graph plotted with pressure on the X-axis and flow rate on the Y-axis will be linear (a straight line)?

____

8. Click **Plot Data** to view a summary of your data on a plotted grid. Pressure will be displayed on the X-axis and flow rate will be displayed on the Y-axis. Click **Submit** to record your plot in the lab report.

After you complete the experiment, take the online **Post-lab Quiz** for Activity 4.

Activity Questions

1. How does increasing the driving pressure affect the blood flow rate?

____

____

2. Is the relationship between blood pressure and blood flow rate directly proportional or inversely proportional? Why?

____

____

3. How does the cardiovascular system increase pressure?

____

____

4. Although changing blood pressure can be used to alter the blood flow rate, this approach causes problems if it continues indefinitely. Explain why.

____

____

ACTIVITY 5

Studying the Effect of Blood Vessel Radius on Pump Activity

OBJECTIVES

1. To understand the terms *systole* and *diastole.*
2. To predict how a change in blood vessel radius will affect flow rate.
3. To predict how a change in blood vessel radius will affect heart rate.
4. To observe the compensatory mechanisms for maintaining blood pressure.

Introduction

In the human body, the heart beats approximately 70 strokes each minute. Each heartbeat consists of a filling interval, when blood moves into the chambers of the heart, and an ejection period, when blood is actively pumped into the aorta and the pulmonary trunk.

The pumping activity of the heart can be described in terms of the phases of the cardiac cycle. Heart chambers fill during **diastole** (relaxation of the heart) and pump blood out during **systole** (contraction of the heart). As you can imagine, the length of time the heart is relaxed is one factor that determines the amount of blood within the heart at the end of the filling interval. Up to a point, increasing ventricular filling time results in a corresponding increase in ventricular volume. The volume in the ventricles at the end of diastole, just before cardiac contraction, is called the **end diastolic volume,** or **EDV.** The volume ejected by a single ventricular contraction is the **stroke volume,** and the volume remaining in the ventricle after contraction is the **end systolic volume,** or **ESV.**

The human heart is a complex, four-chambered organ consisting of two individual pumps (the right and left sides). The right side of the heart pumps blood through the lungs into the left side of the heart. The left side of the heart, in turn, delivers blood to the systems of the body. Blood then returns to the right side of the heart to complete the circuit.

Recall that cardiac output **(CO)** is equal to blood flow. To determine CO, you multiply heart rate (HR) by stroke volume (SV): CO = HR × SV. From the equation for flow (flow $= \Delta P/R$), you can determine the equation for blood pressure: $\Delta P =$ flow $\times R$. Substituting CO in the equation for flow, you get: $\Delta P = \text{HR} \times \text{SV} \times R$.

Therefore, to maintain blood pressure, the cardiovascular system can alter heart rate, stroke volume, or resistance. For example, if resistance decreases, heart rate can increase to maintain the pressure difference.

In this activity you will explore the operation of a simple, one-chambered pump and apply the physical concepts in the experiment to the operation of either of the two pumps of the human heart. The stroke volume and the difference in pressure will remain constant. You will explore the effect that a change in resistance has on heart rate and the compensatory mechanisms that the cardiovascular system uses to maintain blood pressure.

EQUIPMENT USED The following equipment will be depicted on-screen: left beaker—simulates blood coming from the lungs; flow tube connecting the left beaker and the pump—simulates the pulmonary veins; pump—simulates the left ventricle (the valve to the left of the pump simulates the bicuspid valve, and the valve to the right of the pump simulates the aortic semilunar valve); flow tube connecting the pump and the right beaker—simulates the aorta; right beaker—simulates blood going to the systemic circuit.

Experiment Instructions

Go to the home page in the PhysioEx software and click **Exercise 5: Cardiovascular Dynamics.** Click **Activity 5: Compensation: Studying the Effect of Blood Vessel Radius on Pump Activity** and take the online **Pre-lab Quiz** for Activity 5.

After you take the online Pre-lab Quiz, click the **Experiment** tab and begin the experiment. The experiment instructions are reprinted here for your reference. The opening screen for the experiment is shown on the following page.

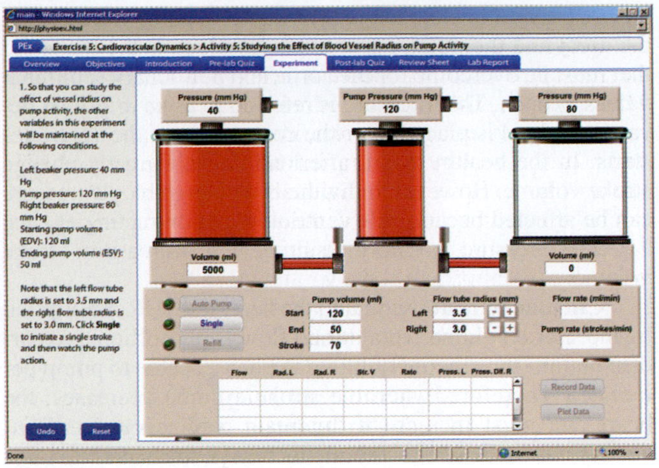
main - Windows Internet Explorer
http://physioex.html
PEx
Exercise 5: Cardiovascular Dynamics > Activity 5: Studying the Effect of Blood Vessel Radius on Pump Activity
Overview
Objectives
Introduction
Pre-lab Quiz
Experiment
Post-lab Quiz
Review Sheet
Lab Report
1. So that you can study the effect of vessel radius on pump activity, the other variables in this experiment will be maintained at the following conditions.
Left beaker pressure: 40 mm Hg
Pump pressure: 120 mm Hg
Right beaker pressure: 80 mm Hg
Starting pump volume (EDV): 120 ml
Ending pump volume (ESV): 50 ml
Note that the left flow tube radius is set to 3.5 mm and the right flow tube radius is set to 3.0 mm. Click Single to initiate a single stroke and then watch the pump action.
Pressure (mm Hg)
40
Pump Pressure (mm Hg)
120
Pressure (mm Hg)
80
Volume (ml)
5000
Volume (ml)
0
Auto Pump
Single
Refill
Pump volume (ml)
Start 120
End 50
Stroke 70
Flow tube radius (mm)
Left 3.5
Right 3.0
Flow rate (ml/min)
Pump rate (strokes/min)
Flow
Rad. L
Rad. R
Str. V
Rate
Press. L
Press. Dif. R
Record Data
Plot Data
Undo
Reset
Done
Internet
100%

1. So that you can study the effect of vessel radius on pump activity, the other variables in this experiment will be maintained at the following conditions:

Left beaker pressure: 40 mm Hg

Pump pressure: 120 mm Hg

Right beaker pressure: 80 mm Hg

Starting pump volume (EDV): 120 ml

Ending pump volume (ESV): 50 ml

Note that the left flow tube radius is set to 3.5 mm and the right flow tube radius is set to 3.0 mm. Click **Single** to initiate a single stroke and then watch the pump action.

2. Click **Auto Pump** to initiate 10 strokes and then watch the pump action. The flow rate is shown in the flow rate display and the pump rate is shown in the pump rate display after the left beaker has finished draining.

3. Click **Record Data** to display your results in the grid (and record your results in Chart 5).

CHART 5 **Effect of Blood Vessel Radius on Pump Activity**

Flow rate (ml/min)	Right radius (mm)	Pump rate (strokes/min)

4. Click **Refill** to replenish the left beaker.

PREDICT Question 1
If you increase the flow tube radius, what will happen to the pump rate to maintain constant pressure?

____

5. Increase the right flow tube radius to 3.5 mm by clicking the + button beside the right flow tube radius display. Click **Auto Pump** to initiate 10 strokes and then watch the pump action.

6. Click **Record Data** to display your results in the grid (and record your results in Chart 5).

7. Click **Refill** to replenish the left beaker.

8. You will now observe the effect of incremental increases in the right flow tube radius.

- Increase the right flow tube radius by 0.5 mm.
- Click **Auto Pump** to initiate 10 strokes and then watch the pump action.
- Click **Record Data** to display your results in the grid (and record your results in Chart 5).
- Click **Refill** to replenish the left beaker.

Repeat this step until you reach a right flow tube radius of 5.0 mm.

9. Click **Plot Data** to view a summary of your data on a plotted grid. Right flow tube radius will be displayed on the X-axis and flow rate will be displayed on the Y-axis. Click **Submit** to record your plot in the lab report.

After you complete the experiment, take the online **Post-lab Quiz** for Activity 5.

Activity Questions

1. Describe the position of the pump during diastole.

____

____

2. Describe the position of the pump during systole.

____

____

3. Describe what happened to the flow rate when the blood vessel radius was increased.

____

____

4. Explain what happened to the resistance and the pump rate to maintain pressure when the radius was increased.

____

____

ACTIVITY 6

Studying the Effect of Stroke Volume on Pump Activity

OBJECTIVES

1. To understand the effect a change in venous return has on stroke volume.
2. To explain how stroke volume is changed in the heart.
3. To explain the Frank-Starling law of the heart.
4. To define *preload, contractility,* and *afterload.*
5. To distinguish between intrinsic and extrinsic control of contractility of the heart.
6. To explore how heart rate and stroke volume contribute to cardiac output and blood flow.

Introduction

In a normal individual, 60% of the blood contained within the heart is ejected from the heart during ventricular systole, leaving 40% of the blood behind. The blood ejected by the heart—the **stroke volume**—is the difference between the **end diastolic volume (EDV),** the volume in the ventricles at the end of diastole, just before cardiac contraction, and **end systolic volume (ESV),** the volume remaining in the ventricle after contraction. That is, stroke volume = EDV − ESV. Many factors affect stroke volume, the most important of which include *preload, contractility,* and *afterload.* We will look at these defining factors and how they relate to stroke volume.

The Frank-Starling law of the heart states that, when more than the normal volume of blood is returned to the heart by the venous system, the heart muscle will be stretched, resulting in a more forceful contraction of the ventricles. This, in turn, will cause more than normal blood to be ejected by the heart, raising the stroke volume. The degree to which the ventricles are stretched by the end diastolic volume (EDV) is referred to as the **preload.** Thus, the preload results from the amount of ventricular filling between strokes, or the magnitude of the EDV. Ventricular filling could increase when the heart rate is slow because there will be more time for the ventricles to fill. Exercise increases venous return and, therefore, EDV. Factors such as severe blood loss and dehydration decrease venous return and EDV.

The **contractility** of the heart refers to strength of the cardiac muscle contraction (usually the ventricles) and its ability to generate force. A number of extrinsic mechanisms, including the sympathetic nervous system and hormones, control the force of cardiac muscle contraction, but they are not the focus of this activity. The focus of this activity will be the intrinsic controls of contractility (those that reside entirely within the heart). When the end diastolic volume increases, the cardiac muscle fibers of the ventricles stretch and lengthen. As the length of the cardiac sarcomere increases, so does the force of contraction. Cardiac muscle, like skeletal muscle, demonstrates a **length-tension relationship.** At rest, cardiac muscles are at a less than optimum overlap length for maximum tension production in the healthy heart. Therefore, when the heart experiences an increase in stretch with an increase in venous return and, therefore, EDV, it can respond by increasing the force of contraction, yielding a corresponding increase in stroke volume.

Afterload is the back pressure generated by the blood in the aorta and the pulmonary trunk. Afterload is the threshold that must be overcome for the aortic and pulmonary semilunar values to open. This pressure is referred to as an *after*load because the load is placed after the contraction of the ventricles starts. In the healthy heart, afterload doesn't greatly change stroke volume. However, individuals with high blood pressure can be affected because the ventricles are contracting against a greater pressure, possibly resulting in a decrease in stroke volume.

Cardiac output is equal to the heart rate (HR) multiplied by the stroke volume. Total blood flow is proportional to cardiac output (the amount of blood the heart is able to pump per minute). Therefore, when the stroke volume decreases, the heart rate must increase to maintain cardiac output. Conversely, when the stroke volume increases, the heart rate must decrease to maintain cardiac output.

Even though our simple pump in this experiment does not work exactly like the human heart, you can apply the concepts presented to basic cardiac function. In this activity you will examine how the activity of the pump is affected by changing the starting (EDV) and ending volumes (ESV).

EQUIPMENT USED The following equipment will be depicted on-screen: left beaker—simulates blood coming from the lungs; flow tube connecting the left beaker and the pump—simulates the pulmonary veins; pump—simulates the left ventricle (the valve to the left of the pump simulates the bicuspid valve, and the valve to the right of the pump simulates the aortic semilunar valve); flow tube connecting the pump and the right beaker—simulates the aorta; right beaker—simulates blood going to the systemic circuit.

Experiment Instructions

Go to the home page in the PhysioEx software and click **Exercise 5: Cardiovascular Dynamics.** Click **Activity 6: Studying the Effect of Stroke Volume on Pump Activity** and take the online **Pre-lab Quiz** for Activity 6.

After you take the online Pre-lab Quiz, click the **Experiment** tab and begin the experiment. The experiment instructions are reprinted here for your reference. The opening screen for the experiment is shown below.

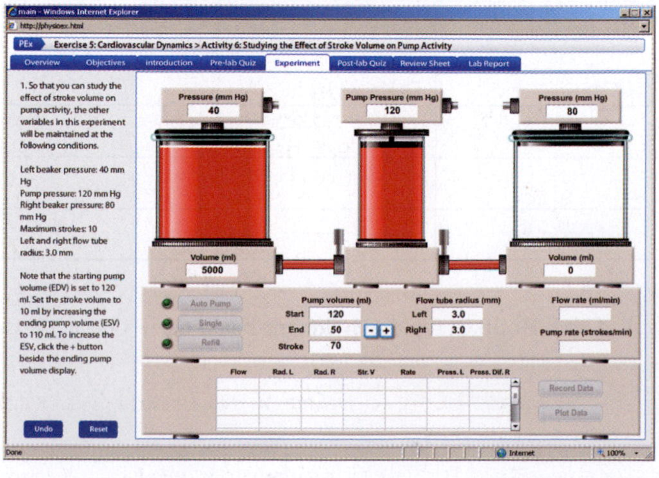
main - Windows Internet Explorer
http://physioex.html
PEx
Exercise 5: Cardiovascular Dynamics > Activity 6: Studying the Effect of Stroke Volume on Pump Activity
Overview
Objectives
introduction
Pre-lab Quiz
Experiment
Post-lab Quiz
Review Sheet
Lab Report
1. So that you can study the effect of stroke volume on pump activity, the other variables in this experiment will be maintained at the following conditions.
Left beaker pressure: 40 mm Hg
Pump pressure: 120 mm Hg
Right beaker pressure: 80 mm Hg
Maximum strokes: 10
Left and right flow tube radius: 3.0 mm
Note that the starting pump volume (EDV) is set to 120 ml. Set the stroke volume to 10 ml by increasing the ending pump volume (ESV) to 110 ml. To increase the ESV, click the + button beside the ending pump volume display.
Pressure (mm Hg)
40
Pump Pressure (mm Hg)
120
Pressure (mm Hg)
80
Volume (ml)
5000
Volume (ml)
0
Auto Pump
Single
Refill
Pump volume (ml)
Start
120
End
50
Stroke
70
Flow tube radius (mm)
Left
3.0
Right
3.0
Flow rate (ml/min)
Pump rate (strokes/min)
Flow
Rad. L
Rad. R
Str. V
Rate
Press. L
Press. Dif. R
Record Data
Plot Data
Undo
Reset
Done
Internet
100%

1. So that you can study the effect of stroke volume on pump activity, the other variables in this experiment will be maintained at the following conditions:

Left beaker pressure: 40 mm Hg

Pump pressure: 120 mm Hg

Right beaker pressure: 80 mm Hg

Maximum strokes: 10

Left and right flow tube radius: 3.0 mm

Note that the starting pump volume (EDV) is set to 120 ml. Set the stroke volume to 10 ml by increasing the ending pump volume (ESV) to 110 ml. To increase the ESV, click the + button beside the ending pump volume display.

2. Click **Auto Pump** to initiate 10 strokes and then watch the pump action. The flow rate is shown in the flow rate display and the pump rate is shown in the pump rate display after the left beaker has finished draining.

3. Click **Record Data** to display your results in the grid (and record your results in Chart 6).

CHART 6 Effect of Stroke Volume on Pump Activity

Flow rate (ml/min)	Stroke volume (ml)	Pump rate (strokes/min)

4. Click **Refill** to replenish the left beaker.

? PREDICT Question 1
If the pump rate is analogous to the heart rate, what do you think will happen to the rate when you increase the stroke volume?

____

5. Increase the stroke volume to 20 ml by decreasing the ESV. To decrease the ending pump volume, click the − button beside the ending pump volume display.

6. Click **Auto Pump** to initiate 10 strokes and then watch the pump action.

7. Click **Record Data** to display your results in the grid (and record your results in Chart 6).

8. Click **Refill** to replenish the left beaker.

9. You will now observe the effect of incremental increases in the stroke volume.

- Increase the stroke volume by 10 ml by decreasing the ending pump volume (ESV).
- Click **Auto Pump** to initiate 10 strokes and then watch the pump action.
- Click **Record Data** to display your results in the grid (and record your results in Chart 6).
- Click **Replenish** to refill the left beaker.

Repeat this step until you reach a stroke volume of 60 ml.

10. Increase the stroke volume by 20 ml by decreasing the ending pump volume (ESV). Click **Auto Pump** to initiate 10 strokes and then watch the pump action.

11. Click **Record Data** to display your results in the grid (and record your results in Chart 6).

12. Click **Refill** to replenish the left beaker.

13. Increase the stroke volume by 20 ml by decreasing the ending pump volume (ESV). Click **Auto Pump** to initiate 10 strokes and then watch the pump action.

14. Click **Record Data** to display your results in the grid (and record your results in Chart 6).

15. Click **Plot Data** to view a summary of your data on a plotted grid. Stroke volume will be displayed on the X-axis and flow rate will be displayed on the Y-axis. Click **Submit** to record your plot in the lab report.

After you complete the Experiment, take the online **Post-lab Quiz** for Activity 6.

Activity Questions

1. Describe how the heart responds to an increase in end diastolic volume (include the terms *preload* and *contractility* in your explanation).

____

____

2. Explain what happened to the pump rate when the stroke volume increased. Why?

____

____

3. Judging from the simulation results, explain why an athlete's resting heart rate might be lower than that of an average person.

____

____

ACTIVITY 7

Compensation in Pathological Cardiovascular Conditions

OBJECTIVES

1. To understand how aortic stenosis affects flow of blood through the heart.
2. To explain ways in which the cardiovascular system might compensate for changes in peripheral resistance.
3. To understand how the heart compensates for changes in afterload.
4. To explain how valves affect the flow of blood through the heart.

Introduction

If a blood vessel is compromised, your cardiovascular system can compensate to some degree. Aortic valve stenosis is a condition where there is a partial blockage of the aortic semilunar valve, increasing resistance to blood flow and left ventricular **afterload.** Therefore, the pressure that must be reached to open the aortic valve increases. The heart could compensate for a change in afterload by increasing contractility, the force of contraction. Increasing contractility will increase cardiac output by increasing stroke volume. To increase contractility, the myocardium becomes thicker. Athletes similarly improve their hearts through cardiovascular conditioning. That is, the thickness of the myocardium increases in diseased hearts with aortic valve stenosis and in athletes' hearts (though the chamber volume increases in athletes' hearts and decreases in diseased hearts).

Valves are important in the heart because they ensure that blood flows in one direction through the heart. The valves in the activity will ensure that blood moves in a single direction. Because the right flow tube represents the aorta (which is actually on the left side of the heart), decreasing the right flow tube radius simulates stenosis, or narrowing of the aortic valve.

Plaques in the arteries, known as **atherosclerosis,** can similarly cause an increase in resistance. An increase in peripheral resistance results in a decreased flow rate. Atherosclerosis is a type of **arteriosclerosis** in which the arteries have lost their elasticity. Atherosclerosis is one of the conditions that leads to heart disease.

In this activity you will test three different compensation mechanisms and predict which mechanism will make the best improvement in flow rate. The three mechanisms include (1) increasing the left flow tube radius (that is, increasing preload), (2) increasing the pump's pressure (that is, increasing contractility), and (3) decreasing the pressure in the right beaker (that is, decreasing afterload).

EQUIPMENT USED The following equipment will be depicted on-screen: left beaker—simulates blood coming from the lungs; flow tube connecting the left beaker and the pump—simulates the pulmonary veins; pump—simulates the left ventricle (the valve to the left of the pump simulates the bicuspid valve, and the valve to the right of the pump simulates the aortic semilunar valve); flow tube connecting the pump and the right beaker—simulates the aorta; right beaker—simulates blood going to the systemic circuit.

Experiment Instructions

Go to the home page in the PhysioEx software and click **Exercise 5: Cardiovascular Dynamics.** Click **Activity 7: Compensation in Pathological Cardiovascular Conditions** and take the online **Pre-lab Quiz** for Activity 7.

After you take the online Pre-lab Quiz, click the **Experiment** tab and begin the experiment. The experiment instructions are reprinted here for your reference. The opening screen for the experiment is shown below.

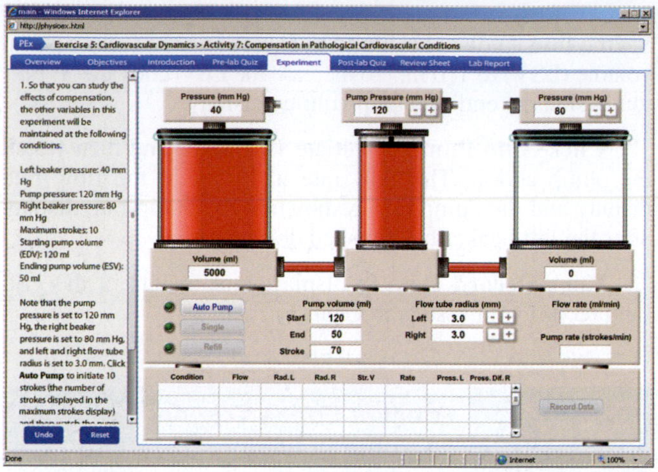
Exercise 5: Cardiovascular Dynamics > Activity 7: Compensation in Pathological Cardiovascular Conditions
Pressure (mm Hg)
Pump Pressure (mm Hg)
Volume (ml)
Auto Pump
Single
Refill
Pump volume (ml)
Flow tube radius (mm)
Flow rate (ml/min)
Pump rate (strokes/min)
Record Data
Undo
Reset

1. So that you can study the effects of compensation, the other variables in this experiment will be maintained at the following conditions:

Left beaker pressure: 40 mm Hg

Maximum strokes: 10

Starting pump volume (EDV): 120 ml

Ending pump volume (ESV): 50 ml

Note that the pump pressure is set to 120 mm Hg, the right beaker pressure is set to 80 mm Hg, and left and right flow tube radius is set to 3.0 mm. Click **Auto Pump** to initiate 10 strokes (the number of strokes displayed in the maximum strokes display) and then watch the pump action. The flow rate is shown in the flow rate display and the pump rate is shown in the pump rate display after the left beaker has finished draining.

2. Click **Record Data** to display your results in the grid (and record your results in Chart 7). This will be your baseline, or "normal," data point for flow rate.

3. Click **Refill** to replenish the left beaker.

4. Decrease the right flow tube radius to 2.5 mm by clicking the – button beside the right flow tube radius display. Click **Auto Pump** to initiate 10 strokes and then watch the pump action.

5. Click **Record Data** to display your results in the grid (and record your results in Chart 7).

6. Click **Refill** to replenish the left beaker.

CHART 7 Compensation Results

Condition	Flow rate (ml/min)	Left radius (mm)	Right radius (mm)	Pump rate (strokes/ min)	Pump pressure (mm Hg)	Right beaker pressure (mm Hg)

PREDICT Question 1
You will now test three mechanisms to compensate for the decrease in flow rate caused by the decreased flow tube radius. Which mechanism do you think will have the greatest compensatory effect?

____

7. Increase the left flow tube radius to 3.5 mm by clicking the + button beside the left flow tube radius display. Click **Auto Pump** to initiate 10 strokes and then watch the pump action.

8. Click **Record Data** to display your results in the grid (and record your results in Chart 7).

9. Click **Refill** to replenish the left beaker.

10. Increase the left flow tube radius to 4.0 mm. Click **Auto Pump** to initiate 10 strokes and then watch the pump action.

11. Click **Record Data** to display your results in the grid (and record your results in Chart 7).

12. Click **Refill** to replenish the left beaker.

13. Increase the left flow tube radius to 4.5 mm. Click **Auto Pump** to initiate 10 strokes and then watch the pump action.

14. Click **Record Data** to display your results in the grid (and record your results in Chart 7).

15. Click **Refill** to replenish the left beaker.

16. Decrease the left flow tube radius to 3.0 mm by clicking the − button beside the left flow tube radius display and increase the pump pressure to 130 mm Hg by clicking the + button beside the pump pressure display. Click **Auto Pump** to initiate 10 strokes and then watch the pump action.

17. Click **Record Data** to display your results in the grid (and record your results in Chart 7).

18. Click **Refill** to replenish the left beaker.

19. Increase the pump pressure to 140 mm Hg by clicking the + button beside the pump pressure display. Click **Auto Pump** to initiate 10 strokes and then watch the pump action.

20. Click **Record Data** to display your results in the grid (and record your results in Chart 7).

21. Click **Refill** to replenish the left beaker.

22. Increase the pump pressure to 150 mm Hg. Click **Auto Pump** to initiate 10 strokes and then watch the pump action.

23. Click **Record Data** to display your results in the grid (and record your results in Chart 7).

24. Click **Refill** to replenish the left beaker.

25. Decrease the pump pressure to 120 mm Hg by clicking the − button beside the pump pressure display and decrease the right (destination) beaker pressure to 70 mm Hg by clicking the − button beside the right beaker pressure display. Click **Auto Pump** to initiate 10 strokes and then watch the pump action.

26. Click **Record Data** to display your results in the grid (and record your results in Chart 7).

27. Click **Refill** to replenish the left beaker.

28. Decrease the right (destination) beaker pressure to 60 mm Hg by clicking the − button beside the right beaker pressure display. Click **Auto Pump** to initiate 10 strokes and then watch the pump action.

29. Click **Record Data** to display your results in the grid (and record your results in Chart 7).

30. Click **Refill** to replenish the left beaker.

31. Decrease the right (destination) beaker pressure to 50 mm Hg. Click **Auto Pump** to initiate 10 strokes and then watch the pump action.

32. Click **Record Data** to display your results in the grid (and record your results in Chart 7).

33. Click **Refill** to replenish the left beaker.

? PREDICT Question 2
What do you think will happen if the pump pressure and the beaker pressure are the same?

____

34. Increase the right (destination) beaker pressure to 120 mm Hg by clicking the + button beside the right beaker pressure display. Click **Auto Pump** to initiate 10 strokes and then watch the pump action.

After you complete the experiment, take the online **Post-lab Quiz** for Activity 7.

Activity Questions

1. Explain why a thicker myocardium is seen in both the athlete's heart and the diseased heart.

____

____

2. Describe what the term *afterload* means.

____

____

3. Explain which mechanism in the simulation had the greatest compensatory effect.

____

____

4. Describe the mechanism used in the human heart to compensate for aortic stenosis.

____

____

PhysioEx 9.0

REVIEW SHEET

EXERCISE 5

NAME ____________________

LAB TIME/DATE ____________________

Cardiovascular Dynamics

ACTIVITY 1 Studying the Effect of Blood Vessel Radius on Blood Flow Rate

1. Explain how the body establishes a pressure gradient for fluid flow. ____________________

2. Explain the effect that the flow tube radius change had on flow rate. How well did the results compare with your prediction?

3. Describe the effect that radius changes have on the laminar flow of a fluid. ____________________

4. Why do you think the plot was not linear? (Hint: Look at the relationship of the variables in the equation.) How well did the results compare with your prediction? ____________________

ACTIVITY 2 Studying the Effect of Blood Viscosity on Blood Flow Rate

1. Describe the components in the blood that affect viscosity. ____________________

2. Explain the effect that the viscosity change had on flow rate. How well did the results compare with your prediction?

3. Describe the graph of flow versus viscosity. ____________________

4. Discuss the effect that polycythemia would have on viscosity and on blood flow. ____________________

ACTIVITY 3 Studying the Effect of Blood Vessel Length on Blood Flow Rate

1. Which is more likely to occur, a change in blood vessel radius or a change in blood vessel length? Explain why.

2. Explain the effect that the change in blood vessel length had on flow rate. How well did the results compare with your prediction?

3. Explain why you think blood vessel radius can have a larger effect on the body than changes in blood vessel length (use the blood flow equation).

4. Describe the effect that obesity would have on blood flow and why.

ACTIVITY 4 Studying the Effect of Blood Pressure on Blood Flow Rate

1. Explain the effect that pressure changes had on flow rate. How well did the results compare with your prediction?

2. How does the plot differ from the plots for tube radius, viscosity, and tube length? How well did the results compare with your prediction?

3. Explain why pressure changes are not the best way to control blood flow.

4. Use your data to calculate the increase in flow rate in ml/min/mm Hg.

ACTIVITY 5 Studying the Effect of Blood Vessel Radius on Pump Activity

1. Explain the effect of increasing the right flow tube radius on the flow rate, resistance, and pump rate.

2. Describe what the left and right beakers in the experiment correspond to in the human heart. ______________________________

____

3. Briefly describe how the human heart could compensate for flow rate changes to maintain blood pressure. ______________

____

ACTIVITY 6 Studying the Effect of Stroke Volume on Pump Activity

1. Describe the Frank-Starling law in the heart. ____

____

2. Explain what happened to the pump rate when you increased the stroke volume. Why do you think this occurred? How well did the results compare with your prediction? ____

____

3. Describe how the heart alters stroke volume. ____

____

4. Describe the intrinsic factors that control stroke volume. ____________________________________

____

ACTIVITY 7 Compensation in Pathological Cardiovascular Conditions

1. Explain how the heart could compensate for changes in peripheral resistance. ______________________

____

2. Which mechanism had the greatest compensatory effect? How well did the results compare with your prediction? ________

____

3. Explain what happened when the pump pressure and the beaker pressure were the same. How well did the results compare with your prediction? ____

____

4. Explain whether it would be better to adjust heart rate or blood vessel diameter to achieve blood flow changes at a local level (for example, in just the digestive system). ____

____

PhysioEx™ 9.0

EXERCISE 6

Cardiovascular Physiology

PRE-LAB QUIZ

1. Circle True or False. Heart muscle can depolarize spontaneously in the absence of any external stimulation.
2. Spontaneous depolarization-repolarization events occur in a regular and continuous manner in cardiac muscle, a property known as
   a. automaticity
   b. rhythmicity
   c. synchronicity
3. How many chambers does the frog heart have?
   a. two
   b. three
   c. four
4. Circle True or False. Heart rate can be modified by extrinsic impulses from the autonomic nerves.
5. What is an extrasystole? ______________________________

   ____
6. Which chemical agent will you use to modify the frog heart rate?
   a. caffeine
   b. digitalis
   c. magnesium solution
   d. Ringer's solution
7. The ________ nerve carries parasympathetic impulses to the heart.
   a. cardiac
   b. olfactory
   c. phrenic
   d. vagus
8. Circle the correct term. The phenomenon of vagal escape / heart block occurs when the heart stops momentarily, then begins to beat again.
9. Circle True or False. The flow of blood through capillary beds is slow and intermittent.
10. ________ causes extensive vasodilation when applied to the frog web.
    a. Calcium
    b. Epinephrine
    c. Histamine
    d. Ringer's solution

Exercise Overview

Cardiac muscle and some types of smooth muscle contract spontaneously, without any external stimuli. Skeletal muscle is unique in that it requires depolarizing signals from the nervous system to contract. The heart's ability to trigger its own contractions is called **autorhythmicity**.

If you isolate cardiac pacemaker muscle cells, place them into cell culture, and observe them under a microscope, you can see the cells contract. Autorhythmicity occurs because the plasma membrane in cardiac pacemaker muscle cells has reduced permeability to potassium ions but still allows sodium and calcium ions to slowly leak into the cells. This leakage causes the muscle cells to slowly depolarize until the action potential threshold is reached and L-type calcium channels open, allowing Ca^{2+} entry from the extracellular fluid. Shortly thereafter, contraction of the remaining cardiac muscle occurs prior to potassium-dependent repolarization. The spontaneous

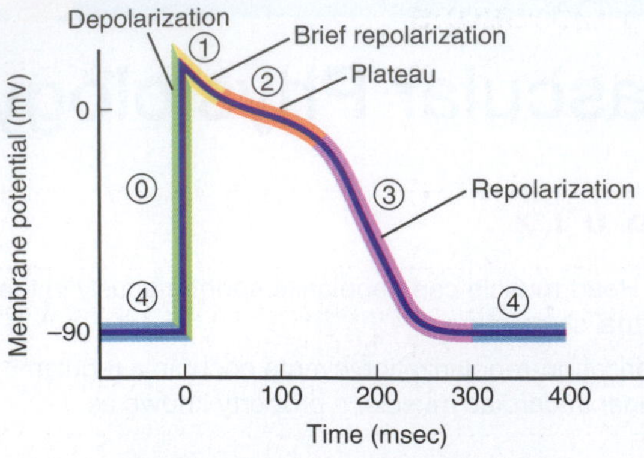
Depolarization
Brief repolarization
Plateau
Repolarization
①
②
③
④
⓪
Membrane potential (mV)
0
–90
0
100
200
300
400
Time (msec)

FIGURE 6.1 The cardiac action potential.

depolarization-repolarization events occur in a regular and continuous manner in cardiac pacemaker muscle cells, leading to **cardiac action potentials** in the majority of cardiac muscle.

There are five main phases of membrane polarization in a cardiac action potential (view Figure 6.1).

- **Phase 0** is similar to depolarization in the neuronal action potential. Depolarization causes voltage-gated sodium channels in the cell membrane to open, increasing the flow of sodium ions into the cell and increasing the membrane potential.
- In **phase 1**, the open sodium channels begin to inactivate, decreasing the flow of sodium ions into the cell and causing the membrane potential to fall slightly. At the same time, voltage-gated potassium channels close and voltage-gated calcium channels open. The subsequent decrease in the flow of potassium out of the cell and increase in the flow of calcium into the cell act to depolarize the membrane and curb the fall in membrane potential caused by the inactivation of sodium channels.
- In **phase 2**, known as the **plateau phase**, the membrane remains in a depolarized state. Potassium channels stay closed, and long-lasting (L-type) calcium channels stay open. This plateau lasts about 0.2 seconds, or 200 milliseconds.
- In **phase 3**, the membrane potential gradually falls to more negative values when a second set of potassium channels that began opening in phases 1 and 2 allows significant amounts of potassium to flow out of the cell. The falling membrane potential causes calcium channels to close, reducing the flow of calcium into the cell and repolarizing the membrane until the resting potential is reached.
- In **phase 4**, the resting membrane potential is again established in cardiac muscle cells and is maintained until the next depolarization arrives from neighboring cardiac pacemaker cells.

The total cardiac action potential lasts 250–300 milliseconds.

ACTIVITY 1

Investigating the Refractory Period of Cardiac Muscle

OBJECTIVES

1. To observe the autorhythmicity of the heart.
2. To understand the phases of the cardiac action potential.
3. To observe and induce extrasystoles on an ECG tracing from an isolated, intact heart.
4. To relate the refractory period to wave summation and tetanus.

Introduction

Recall that **wave summation** occurs when a skeletal muscle is stimulated with such frequency that muscle twitches overlap and result in a stronger contraction than a single muscle twitch. When the stimulations are frequent enough, the muscle reaches a state of fused tetanus, during which the individual muscle twitches cannot be distinguished. Tetanus occurs in skeletal muscle because skeletal muscle has a relatively short **absolute refractory period** (a period during which action potentials cannot be generated no matter how strong the stimulus).

Unlike skeletal muscle, cardiac muscle has a relatively long refractory period and is thus incapable of wave summation. In fact, cardiac muscle is incapable of reacting to *any* stimulus before approximately the middle of phase 3, and will not respond to a normal cardiac stimulus before phase 4. The period of time between the beginning of the cardiac action potential and the approximate middle of phase 3 is the **absolute refractory period**. The period of time between the absolute refractory period and phase 4 is the **relative refractory period**. The total refractory period of cardiac muscle is 200–250 milliseconds—almost as long as the contraction of the cardiac muscle.

In this activity you will use external stimulation to better understand the refractory period of cardiac muscle. You will use a frog heart, which is anatomically similar to the human heart. The frog heart has two atria and a single, incompletely divided ventricle.

EQUIPMENT USED The following equipment will be depicted on-screen: oscilloscope display—displays the ECG tracing from the frog heart; electrical stimulator—used to a apply electrical shocks to the frog heart; electrode holder—locks electrodes in place for stimulation; external stimulation electrode; apparatus for sustaining an isolated frog heart—includes 23°C Ringer's solution; frog heart.

Experiment Instructions

Go to the home page in the PhysioEx software and click **Exercise 6: Cardiovascular Physiology.** Click **Activity 1: Investigating the Refractory Period of Cardiac Muscle,** and take the online **Pre-lab Quiz** for Activity 1.

After you take the online Pre-lab Quiz, click the **Experiment** tab and begin the experiment. The experiment instructions are reprinted here for your reference. The opening screen for the experiment is shown on the following page.

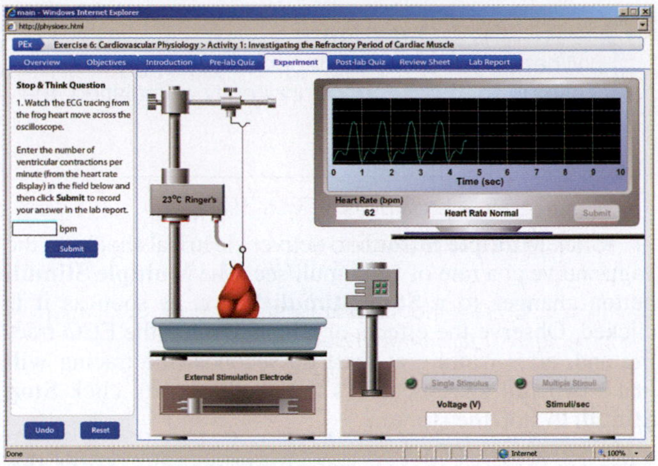
main - Windows Internet Explorer
http://physioex.html
PEx
Exercise 6: Cardiovascular Physiology > Activity 1: Investigating the Refractory Period of Cardiac Muscle
Overview
Objectives
Introduction
Pre-lab Quiz
Experiment
Post-lab Quiz
Review Sheet
Lab Report
Stop & Think Question
1. Watch the ECG tracing from the frog heart move across the oscilloscope.
Enter the number of ventricular contractions per minute (from the heart rate display) in the field below and then click Submit to record your answer in the lab report.
bpm
Submit
Undo
Reset
23°C Ringer's
0 1 2 3 4 5 6 7 8 9 10
Time (sec)
Heart Rate (bpm)
62
Heart Rate Normal
Submit
External Stimulation Electrode
Single Stimulus
Multiple Stimuli
Voltage (V)
Stimuli/sec
Done
Internet
100%

1. Watch the ECG tracing from the frog heart move across the oscilloscope. Enter the number of ventricular contractions per minute (from the heart rate display) in the field below and then click **Submit** to record your answer in the lab report.

_____ beats/min

2. Drag the external stimulation electrode to the electrode holder to the right of the frog heart. The electrode will touch the ventricular muscle tissue.

> **? PREDICT Question 1**
> When you increase the frequency of the stimulation, what do you think will happen to the amplitude (height) of the ventricular systole wave?
>
> ______________________________

3. Deliver single shocks in succession by clicking **Single Stimulus** rapidly. You might need to practice to acquire the correct technique. You should see a "doublet," or double peak, which contains an **extrasystole,** or extra contraction of the ventricles, and then a compensatory pause, which allows the heart to get back on schedule after the extrasystole. When you see a doublet, click **Submit** to record the tracing in the lab report.

> **? PREDICT Question 2**
> If you deliver multiple stimuli (20 stimuli per second) to the heart, what do you think will happen?
>
> ______________________________

4. Click **Multiple Stimuli** to deliver electrical shocks to the heart at a rate of 20 stimuli/sec. The **Multiple Stimuli** button changes to a **Stop Stimuli** button as soon as it is clicked. Observe the effects of stimulation on the ECG tracing and, after a few seconds, click **Stop Stimuli** to stop the stimuli.

After you complete the experiment, take the online **Post-lab Quiz** for Activity 1.

Activity Questions

1. Describe how the frog heart and human heart differ anatomically.

____

____

2. What does an extrasystole correspond to? How did you induce an extrasystole on the ECG tracing?

____

____

3. Explain why it is important that wave summation and tetanus do not occur in the cardiac muscle.

____

____

ACTIVITY 2

Examining the Effect of Vagus Nerve Stimulation

OBJECTIVES

1. To understand the role that the sympathetic and parasympathetic nervous systems have on heart activity.
2. To explain the consequences of vagal stimulation and vagal escape.
3. To explain the functionality of the sinoatrial node.

Introduction

The autonomic nervous system has two branches: the **sympathetic** nervous system ("fight or flight") and **parasympathetic** nervous system ("resting and digesting"). At rest both the sympathetic and parasympathetic nervous systems are working but the parasympathetic branch is more active. The sympathetic nervous system becomes more active when needed, for example, during exercise and when confronting danger.

Both the parasympathetic and sympathetic nervous systems supply nerve impulses to the heart. Stimulation of the sympathetic nervous system increases the rate and force of contraction of the heart. Stimulation of the parasympathetic nervous system decreases the heart rate without directly changing the force of contraction. The vagus nerve (cranial nerve X) carries the signal to the heart. If stimulation of the vagus nerve (vagal stimulation) is excessive, the heart will stop beating. After a short time, the ventricles will begin to beat again. The resumption of the heartbeat is referred to as **vagal escape** and can be the result of sympathetic reflexes or initiation of a rhythm by the Purkinje fibers.

The **sinoatrial node (SA node)** is a cluster of autorhythmic cardiac cells found in the right atrial wall in the human heart. The SA node has the fastest rate of spontaneous depolarization, and, for that reason, it determines the heart rate and is therefore referred to as the heart's "**pacemaker.**" In the absence of parasympathetic stimulation, sympathetic stimulation, and hormonal controls, the SA node generates action potentials 100 times per minute.

EQUIPMENT USED The following equipment will be depicted on-screen: oscilloscope display—displays the ECG tracing from the frog heart; electrical stimulator—used to apply electrical shocks to the frog heart; electrode holder—locks electrodes in place for stimulation; vagus nerve stimulation electrode; apparatus for sustaining an isolated, intact frog heart—includes 23°C Ringer's solution; frog heart with vagus nerve (thin, white strand to the right).

Experiment Instructions

Go to the home page in the PhysioEx software and click **Exercise 6: Cardiovascular Physiology.** Click **Activity 2: Examining the Effect of Vagus Nerve Stimulation,** and take the online **Pre-lab Quiz** for Activity 2.

After you take the online Pre-lab Quiz, click the **Experiment** tab and begin the experiment. The experiment instructions are reprinted here for your reference. The opening screen for the experiment is shown below.

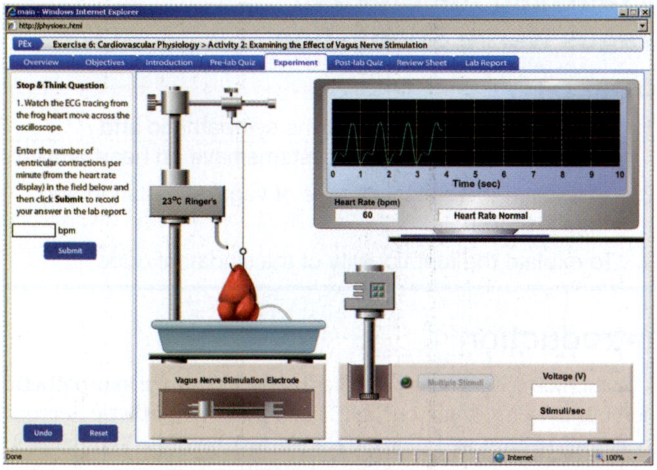
Exercise 6: Cardiovascular Physiology > Activity 2: Examining the Effect of Vagus Nerve Stimulation
Overview
Objectives
Introduction
Pre-lab Quiz
Experiment
Post-lab Quiz
Review Sheet
Lab Report
Stop & Think Question
1. Watch the ECG tracing from the frog heart move across the oscilloscope.
Enter the number of ventricular contractions per minute (from the heart rate display) in the field below and then click Submit to record your answer in the lab report.
bpm
Submit
23°C Ringer's
Time (sec)
Heart Rate (bpm)
60
Heart Rate Normal
Vagus Nerve Stimulation Electrode
Multiple Stimuli
Voltage (V)
Stimuli/sec
Undo
Reset

1. Watch the ECG tracing from the frog heart move across the oscilloscope. Enter the number of ventricular contractions per minute (from the heart rate display) in the field below and then click **Submit** to record your answer in the lab report.

_____ beats/min

2. Drag the vagus nerve stimulation electrode to the electrode holder to the right of the heart. Note that, when the electrode locks in place, the vagus nerve is draped over the electrode. Stimuli will go directly to the vagus nerve and indirectly to the heart.

3. Enter the number of ventricular contractions per minute (from the heart rate display) in the field below and then and click **Submit** to record your answer in the lab report.

_____ beats/min

? PREDICT Question 1
What do you think will happen if you apply multiple stimuli to the heart by indirectly stimulating the vagus nerve?

____

4. Click **Multiple Stimuli** to deliver electrical shocks to the vagus nerve at a rate of 50 stimuli/sec. The **Multiple Stimuli** button changes to a **Stop Stimuli** button as soon as it is clicked. Observe the effects of stimulation on the ECG tracing and, after waiting at least 20 seconds (the tracing will make two full sweeps across the oscilloscope), click **Stop Stimuli** to stop the stimuli.

After you complete the experiment, take the online **Post-lab Quiz** for Activity 2.

Activity Questions

1. Describe how stimulation of the vagus nerves affects the heart rate.

____

____

2. How does the sympathetic nervous system affect heart rate and the force of contraction?

____

____

3. Describe the mechanism of vagal escape.

____

____

4. What would happen to the heart rate if the vagus nerve were cut?

____

____

ACTIVITY 3

Examining the Effect of Temperature on Heart Rate

OBJECTIVES

1. To define the terms *hyperthermia* and *hypothermia.*
2. To contrast the terms *homeothermic* and *poikilothermic.*
3. To understand the effect that temperature has on the frog heart.
4. To understand the effect that temperature could have on the human heart.

Introduction

Humans are **homeothermic**, which means that the human body maintains an internal body temperature within the 35.8–38.2°C range even though the external temperature is changing. When the external temperature is elevated, the hypothalamus is signaled to activate heat-releasing mechanisms, such as sweating and vasodilation, to maintain the body's internal temperature. During extreme external temperature conditions, the body might not be able to maintain homeostasis and either **hyperthermia** (elevated body temperature) or **hypothermia** (low body temperature) could result. In contrast, the frog is a **poikilothermic** animal. Its internal body temperature changes depending on the temperature of its external environment because it lacks internal homeostatic regulatory mechanisms.

Ringer's solution, also known as Ringer's irrigation, consists of essential electrolytes (chloride, sodium, potassium, calcium, and magnesium) in a physiological solution and is required to keep the isolated, intact heart viable. In this activity you will explore the effect of temperature on heart rate using a Ringer's solution incubated at different temperatures.

EQUIPMENT USED The following equipment will be depicted on-screen: oscilloscope display—displays the ECG tracing from the frog heart; electrical stimulator—used to a apply electrical shocks to the frog heart; electrode holder—locks electrodes in place for stimulation; external stimulation electrode; apparatus for sustaining an isolated, intact frog heart—includes 5°C, 23°C, and 32°C Ringer's solution; frog heart.

Experiment Instructions

Go to the home page in the PhysioEx software and click **Exercise 6: Cardiovascular Physiology.** Click **Activity 3: Examining the Effect of Temperature on Heart Rate,** and take the online **Pre-lab Quiz** for Activity 3.

After you take the online Pre-lab Quiz, click the **Experiment** tab and begin the experiment. The experiment instructions are reprinted here for your reference. The opening screen for the experiment is shown below.

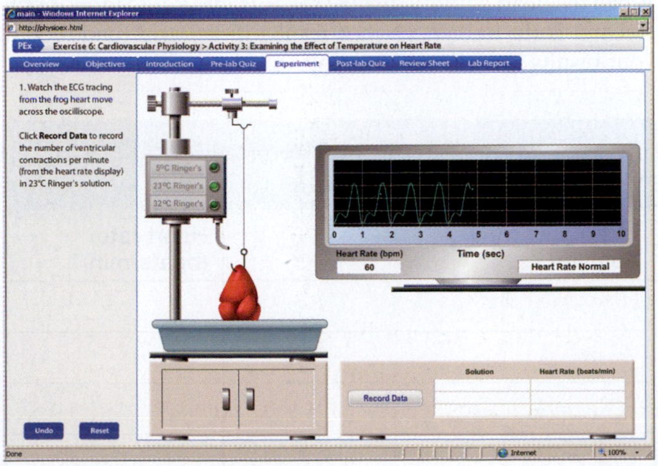
Heart Rate (bpm)
Time (sec)
Heart Rate Normal
Record Data
Solution
Heart Rate (beats/min)
Undo
Reset

1. Watch the ECG tracing from the frog heart move across the oscilloscope. Click **Record Data** to record the number of ventricular contractions per minute (from the heart rate display) in 23°C Ringer's solution.

? PREDICT Question 1
What effect will decreasing the temperature of the Ringer's solution have on the heart rate of the frog?

____

2. Click **5°C Ringer's** to observe the effects of lowering the temperature.

3. When the heart activity display reads *Heart Rate Stable*, click **Record Data** to display your results in the grid (and record your results in Chart 3).

CHART 3 Effect of Temperature on Heart Rate

Solution	Heart rate (beats/min)

4. Click **23°C Ringer's** to bathe the heart and return it to room temperature. When the heart activity display reads *Heart Rate Normal*, you can proceed.

? PREDICT Question 2
What effect will increasing the temperature of the Ringer's solution have on the heart rate of the frog?

____

5. Click **32°C Ringer's** to observe the effects of increasing the temperature.

6. When the heart activity display reads *Heart Rate Stable*, click **Record Data** to display your results in the grid (and record your results in Chart 3).

After you complete the experiment, take the online **Post-lab Quiz** for Activity 3.

Activity Questions

1. Explain the importance of Ringer's solution (essential electrolytes in physiological saline) in maintaining the autorhythmicity of the heart.

____

____

2. Describe the effect of lower temperature on heart rate.

____

____

3. Explain the effect that fever would have on heart rate. Explain why.

____

____

ACTIVITY 4

Examining the Effects of Chemical Modifiers on Heart Rate

OBJECTIVES

1. To distinguish between cholinergic and adrenergic modifiers of heart rate.
2. To define agonist and antagonist modifiers of heart rate.
3. To observe the effects of epinephrine, pilocarpine, atropine, and digitalis on heart rate.
4. To relate chemical modifiers of the heart rate to sympathetic and parasympathetic activation.

Introduction

Although the heart does not need external stimulation to beat, it can be affected by extrinsic controls, most notably, the autonomic nervous system. The sympathetic nervous system is activated in times of "fight or flight," and sympathetic nerve fibers release **norepinephrine** (also known as **noradrenaline**) and **epinephrine** (also known as **adrenaline**) at their cardiac synapses.

Norepinephrine and epinephrine increase the frequency of action potentials by binding to β_1 adrenergic receptors embedded in the plasma membrane of **sinoatrial (SA) node** (pacemaker) cells. Working through a cAMP second-messenger mechanism, binding of the ligand opens sodium and calcium channels, increasing the rate of depolarization and shortening the period of repolarization, thus increasing the heart rate.

The parasympathetic nervous system, our "resting and digesting branch," usually dominates, and parasympathetic nerve fibers release **acetylcholine** at their cardiac synapses. Acetylcholine decreases the frequency of action potentials by binding to muscarinic cholinergic receptors embedded in the plasma membrane of the SA node cells. Acetylcholine indirectly opens potassium channels and closes calcium and sodium channels, decreasing the rate of depolarization and, thus, decreasing heart rate.

Chemical modifiers that inhibit, mimic, or enhance the action of acetylcholine in the body are labeled **cholinergic**. Chemical modifiers that inhibit, mimic, or enhance the action of epinephrine in the body are **adrenergic**. If the modifier works in the same fashion as the neurotransmitter (acetylcholine or norepinephrine), it is an **agonist**. If the modifier works in opposition to the neurotransmitter, it is an **antagonist**. In this activity you will explore the effects of pilocarpine, atropine, epinephrine, and digitalis on heart rate.

EQUIPMENT USED The following equipment will be depicted on-screen: oscilloscope display—displays the ECG tracing; apparatus for sustaining an isolated intact frog heart—includes 23°C Ringer's solution; pilocarpine; atropine; epinephrine; digitalis; frog heart.

Experiment Instructions

Go to the home page in the PhysioEx software and click **Exercise 6: Cardiovascular Physiology.** Click **Activity 4: Examining the Effects of Chemical Modifiers on Heart Rate,** and take the online **Pre-lab Quiz** for Activity 4.

After you take the online Pre-lab Quiz, click the **Experiment** tab and begin the experiment. The experiment instructions are reprinted here for your reference. The opening screen for the experiment is shown below.

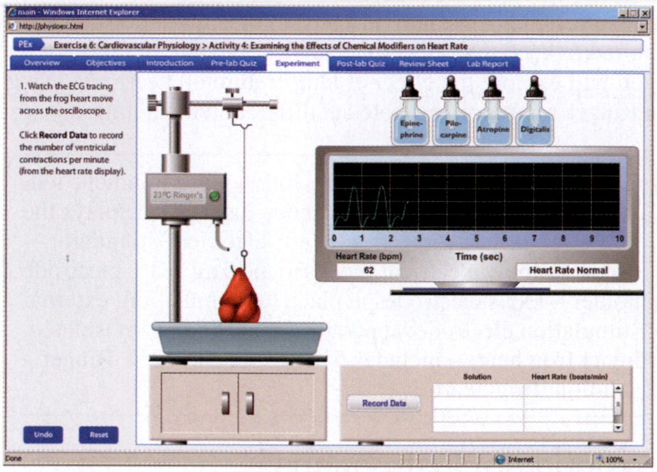
main - Windows Internet Explorer
http://physioex.html
PEx
Exercise 6: Cardiovascular Physiology > Activity 4: Examining the Effects of Chemical Modifiers on Heart Rate
Overview
Objectives
Introduction
Pre-lab Quiz
Experiment
Post-lab Quiz
Review Sheet
Lab Report
1. Watch the ECG tracing from the frog heart move across the oscilloscope.
Click Record Data to record the number of ventricular contractions per minute (from the heart rate display).
23°C Ringer's
Epine-phrine
Pilo-carpine
Atropine
Digitalis
0 1 2 3 4 5 6 7 8 9 10
Heart Rate (bpm)
62
Time (sec)
Heart Rate Normal
Solution
Heart Rate (beats/min)
Record Data
Undo
Reset
Done
Internet
100%

1. Watch the ECG tracing from the frog heart move across the oscilloscope. Click **Record Data** to record the number of ventricular contractions per minute (from the heart rate display) and record your results in Chart 4.

2. Drag the dropper cap of the epinephrine bottle to the frog heart to release epinephrine onto the heart.

3. Observe the ECG tracing and the heart activity display. When the heart activity display reads *Heart Rate Stable*, click **Record Data** to display your results in the grid (and record your results in Chart 4).

CHART 4 Effects of Chemical Modifiers on Heart Rate

Solution	Heart rate (beats/min)

4. Click **23°C Ringer's** (room temperature) to bathe the heart and flush out the epinephrine. When the heart activity display reads *Heart Rate Normal*, you can proceed.

PREDICT Question 1
Pilocarpine is a cholinergic drug, an acetylcholine agonist. Predict the effect that pilocarpine will have on heart rate.

____

5. Drag the dropper cap of the pilocarpine bottle to the frog heart to release pilocarpine onto the heart.

6. Observe the ECG tracing and the heart activity display. When the heart activity display reads *Heart Rate Stable*, click **Record Data** to display your results in the grid (and record your results in Chart 4).

7. Click **23°C Ringer's** (room temperature) to bathe the heart and flush out the pilocarpine. When the heart activity display reads *Heart Rate Normal*, you can proceed.

PREDICT Question 2
Atropine is another cholinergic drug, an acetylcholine antagonist. Predict the effect that atropine will have on heart rate.

____

8. Drag the dropper cap of the atropine bottle to the frog heart to release atropine onto the heart.

9. Observe the ECG tracing and the heart activity display. When the heart activity display reads *Heart Rate Stable*, click **Record Data** to display your results in the grid (and record your results in Chart 4).

10. Click **23°C Ringer's** (room temperature) to bathe the heart and flush out the atropine. When the heart activity display reads *Heart Rate Normal*, you can proceed.

11. Drag the dropper cap of the digitalis bottle to the frog heart to release digitalis onto the heart.

12. Observe the ECG tracing and the heart activity display. When the heart activity display reads *Heart Rate Stable*, click **Record Data** to display your results in the grid (and record your results in Chart 4).

After you complete the experiment, take the online **Post-lab Quiz** for Activity 4.

Activity Questions

1. Define *agonist* and *antagonist.* Clearly distinguish between the two and give examples used in this activity.

____

____

2. Describe the effect of epinephrine on heart rate and force of contraction.

____

____

3. What is the effect of atropine on heart rate?

____

____

4. Describe the effect of digitalis on heart rate and force of contraction.

____

____

ACTIVITY 5

Examining the Effects of Various Ions on Heart Rate

OBJECTIVES

1. To understand the movement of ions that occurs during the cardiac action potential.
2. To describe the potential effect of potassium, sodium, and calcium ions on heart rate.
3. To explain how calcium channel blockers might be used pharmaceutically to treat heart patients.
4. To define the terms *inotropic* and *chronotropic.*

Introduction

In cardiac muscle cells, action potentials are caused by changes in permeability to ions due to the opening and closing of ion channels. The permeability changes that occur for the cardiac muscle cell involve potassium, sodium, and calcium ions. The concentration of potassium is greater inside the cardiac muscle cell than outside the cell. Sodium and calcium are present in larger quantities outside the cell than inside the cell.

The resting cell membrane favors the movement of potassium more than sodium or calcium. Therefore, the resting membrane potential of cardiac cells is determined mainly by the ratio of extracellular and intracellular concentrations of potassium. View Table 6.1 for a summary of the phases of the cardiac action potential and ion movement during each phase.

Calcium channel blockers are used to treat high blood pressure and abnormal heart rates. They block the movement of calcium through its channels throughout all phases of the cardiac action potentials. Consequently, because less calcium gets through, both the rate of depolarization and the force of the contraction are reduced. Modifiers that affect heart rate are **chronotropic**, and modifiers that affect the force of contraction are **inotropic**. Modifiers that lower heart rate are negative chronotropic, and modifiers that increase heart rate are positive chronotropic. The same adjectives describe inotropic modifiers. Therefore, negative inotropic drugs decrease the force of contraction of the heart and positive inotropic drugs increase the force of contraction of the heart.

TABLE 6.1

Phase of cardiac action potential	Ion movement
Phase 0 (rapid depolarization)	Sodium moves in
Phase 1 (small repolarization)	Sodium movement decreases
Phase 2 (plateau)	Potassium movement out decreases Calcium moves in
Phase 3 (repolarization)	Potassium moves out Calcium movement decreases
Phase 4 (resting potential)	Potassium moves out Little sodium or calcium moves in

EQUIPMENT USED The following equipment will be depicted on-screen: oscilloscope display—displays the ECG tracing from the frog heart; apparatus for sustaining frog heart—includes 23°C Ringer's solution; calcium ions; sodium ions; potassium ions; frog heart.

Experiment Instructions

Go to the home page in the PhysioEx software and click **Exercise 6: Cardiovascular Physiology.** Click **Activity 5: Examining the Effects of Various Ions on Heart Rate,** and take the online **Pre-lab Quiz** for Activity 5.

After you take the online Pre-lab Quiz, click the **Experiment** tab and begin the experiment. The experiment instructions are reprinted here for your reference. The opening screen for the experiment is shown below.

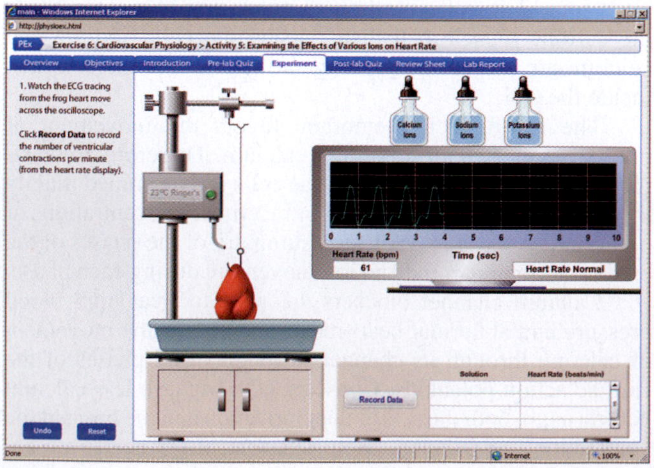
main - Windows Internet Explorer
http://physioex.html
PEx
Exercise 6: Cardiovascular Physiology > Activity 5: Examining the Effects of Various Ions on Heart Rate
Overview
Objectives
Introduction
Pre-lab Quiz
Experiment
Post-lab Quiz
Review Sheet
Lab Report
1. Watch the ECG tracing from the frog heart move across the oscilloscope.
Click Record Data to record the number of ventricular contractions per minute (from the heart rate display).
23°C Ringer's
Calcium Ions
Sodium Ions
Potassium Ions
0 1 2 3 4 5 6 7 8 9 10
Heart Rate (bpm)
61
Time (sec)
Heart Rate Normal
Solution
Heart Rate (beats/min)
Record Data
Undo
Reset
Done
Internet
100%

1. Watch the ECG tracing from the frog heart move across the oscilloscope. Click **Record Data** to record the number of ventricular contractions per minute (from the heart rate display).

> **PREDICT Question 1**
> Because calcium channel blockers are negative chronotropic and negative inotropic, what effect do you think increasing the concentration of calcium will have on heart rate?
>
> ______________________________

2. Drag the dropper cap of the calcium ions bottle to the frog heart to release calcium ions onto the heart. Note the change in heart rate after you drop the calcium ions onto the heart.

3. When the heart activity display reads *Heart Rate Stable*, click **Record Data** to display your results in the grid (and record your results in Chart 5).

CHART 5 Effects of Various Ions on Heart Rate

Solution	Heart rate (beats/min)

4. Click **23°C Ringer's** (room temperature) to bathe the heart and flush out the calcium. When the heart activity display reads *Heart Rate Normal*, you can proceed.

5. Drag the dropper cap of the sodium ions bottle to the frog heart to release sodium ions onto the heart. Note the immediate change in the heart rate and the change in heart rate over time after you drop the sodium ions onto the heart.

6. After waiting at least 20 seconds (the tracing will make two full sweeps across the oscilloscope), click **Record Data** to display your results in the grid (and record your results in Chart 5).

7. Click **23°C Ringer's** (room temperature) to bathe the heart and flush out the sodium. When the heart activity display reads *Heart Rate Normal*, you can proceed.

> **PREDICT Question 2**
> Excess potassium outside of the cardiac cell decreases the resting potential of the plasma membrane, thus decreasing the force of contraction. What effect (if any) do you think it will *initially* have on heart rate?
>
> ______________________________

8. Drag the dropper cap of the potassium ions bottle to the frog heart to release potassium ions onto the heart. Note the immediate change in heart rate and the change in heart rate over time after you drop the potassium ions onto the heart.

9. After waiting at least 20 seconds (the tracing will make two full sweeps across the oscilloscope), click **Record Data** to display your results in the grid (and record your results in Chart 5).

After you complete the experiment, take the online **Post-lab Quiz** for Activity 5.

Activity Questions

1. Define chronotropic and inotropic effects on the heart.

____

____

2. Describe the effect of adding calcium ions to the frog heart.

____

____

3. Calcium channel blockers are often used to treat high blood pressure. Explain how their effects would benefit individuals with high blood pressure.

____

____

4. Describe the initial effect of adding potassium ions to the frog heart.

____

____

PhysioEx 9.0

REVIEW SHEET

EXERCISE 6

NAME ____________________

LAB TIME/DATE ____________________

Cardiovascular Physiology

ACTIVITY 1 Investigating the Refractory Period of Cardiac Muscle

1. Explain why the larger waves seen on the oscilloscope represent the stimulus required for the ventricles to contract.

2. Explain why the amplitude of the wave did not change when you increased the frequency of the stimulation. (Hint: Relate your response to the cardiac muscle refractory period.) How well did the results compare with your prediction?

3. Why is it only possible to induce an extrasystole during relaxation?

4. Explain why wave summation and tetanus are not possible in cardiac muscle tissue. How well did the results compare with your prediction?

ACTIVITY 2 Examining the Effect of Vagus Nerve Stimulation

1. Explain the effect that extreme vagus nerve stimulation had on the heart. How well did the results compare with your prediction?

2. Explain two ways that the heart can overcome excessive vagal stimulation.

3. Describe how the sympathetic and parasympathetic nervous systems work together to regulate heart rate.

4. What do you think would happen to the heart rate if the vagus nerve was cut?

ACTIVITY 3 Examining the Effect of Temperature on Heart Rate

1. Explain the effect that decreasing the temperature had on the frog heart. How do you think the human heart would respond? How well did the results compare with your prediction? ______

______

2. Describe why Ringer's solution is required to maintain heart contractions. ______

______

3. Explain the effect that increasing the temperature had on the frog heart. How do you think the human heart would respond? How well did the results compare with your prediction? ______

______

ACTIVITY 4 Examining the Effects of Chemical Modifiers on Heart Rate

1. Describe the effect that pilocarpine had on the heart and why it had this effect. How well did the results compare with your prediction? ______

______

2. Atropine is an acetylcholine antagonist. Does atropine inhibit or enhance the effects of acetylcholine? Describe your results and how they correlate with how the drug works. How well did the results compare with your prediction? ______

______

3. Describe the benefits of administering digitalis. ______

______

4. Distinguish between cholinergic and adrenergic chemical modifiers. Include examples of each in your discussion. ______

______

ACTIVITY 5 Examining the Effects of Various Ions on Heart Rate

1. Describe the effect that increasing the calcium ions had on the heart. How well did the results compare with your prediction?

______

2. Describe the effect that increasing the potassium ions initially had on the heart in this activity. Relate this to the resting membrane potential of the cardiac muscle cell. How well did the results compare with your prediction? ______

______

3. Describe how calcium channel blockers are used to treat patients and why. ______

______

PhysioEx™ 9.0

EXERCISE 7

Respiratory System Mechanics

PRE-LAB QUIZ

1. Circle the correct term. Inspiration / Expiration is the phase of pulmonary ventilation when air passes out of the lungs.
2. Which of the following processes does *not* occur during inspiration?
   a. diaphragm moves to a flattened position
   b. gas pressure inside the lungs is lowered
   c. inspiratory muscles relax
   d. size of thoracic cavity increases
3. Circle True or False. Vesicular breathing sounds are produced by air rushing through the trachea and bronchi.
4. During normal quiet breathing, about ________ ml of air moves into and out of the lungs with each breath.
   a. 250   c. 1000
   b. 500   d. 2000
5. Circle the correct term. Tidal volume / Vital capacity is the maximum amount of air that can be exhaled after a maximal inspiration.
6. Circle True or False. The neural centers that control respiratory rhythm and maintain a rate of 12–18 respirations per minute are located in the medulla and thalamus.
7. Circle the correct term. Changes in pH and oxygen concentrations in the blood are monitored by chemoreceptor regions in the medulla / aortic and carotid bodies.
8. The carbonic acid–bicarbonate buffer system stabilizes arterial blood pH at:
   a. 2.0 ± 1.00 c. 7.4 ± 0.02
   b. 6.2 ± 0.07 d. 9.5 ± 1.15
9. Circle the correct term. Acids / Bases released into the blood by the body cells tend to lower the pH of the blood and cause it to become acidic.
10. Circle True or False. Rate and depth of breathing, hyperventilation, and hypoventilation should have little or no effect on the acid-base balance of blood.

Exercise Overview

The physiological function of the respiratory system is essential to life. If problems develop in most other physiological systems, we can survive for some time without addressing them. But if a persistent problem develops within the respiratory system (or the circulatory system), death can occur in minutes.

The primary role of the respiratory system is to distribute oxygen to, and remove carbon dioxide from, *all* the cells of the body. The respiratory system works together with the circulatory system to achieve this. **Respiration** includes **ventilation,** or the movement of air into and out of the lungs (breathing), and the transport (via blood) of oxygen and carbon dioxide between the lungs and body cells (view Figure 7.1). The heart pumps deoxygenated blood to pulmonary capillaries, where gas exchange occurs between blood and **alveoli** (air sacs in the lungs), thus oxygenating the blood. The heart then pumps the oxygenated blood to body tissues, where oxygen is used for cell metabolism. At the same time, carbon dioxide (a waste product of metabolism) from body tissues diffuses into the blood. This carbon dioxide–enriched, oxygen-reduced blood then returns to the heart, completing the circuit.

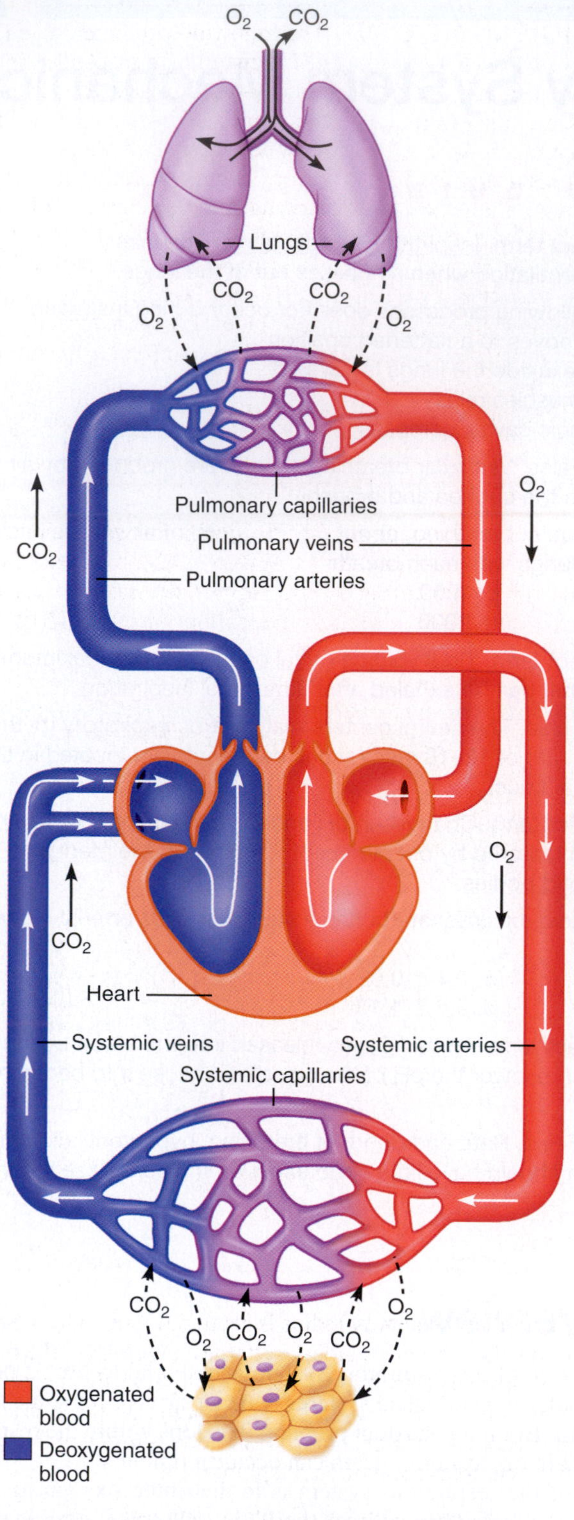

O_2
CO_2
Lungs
CO_2
CO_2
O_2
O_2
Pulmonary capillaries
O_2
CO_2
Pulmonary veins
Pulmonary arteries
O_2
CO_2
Heart
Systemic veins
Systemic arteries
Systemic capillaries
CO_2
O_2
O_2
CO_2
O_2
CO_2
Oxygenated blood
Deoxygenated blood

FIGURE 7.1 Relationship between external respiration and internal respiration.

Ventilation is the result of skeletal muscle contraction. When the **diaphragm**—a dome-shaped muscle that divides the thoracic and abdominal cavities—and the **external intercostal muscles** contract, the volume in the thoracic cavity increases. This increase in thoracic volume reduces the pressure in the thoracic cavity, allowing atmospheric gas to enter the lungs (a process called **inspiration**). When the diaphragm and the external intercostals relax, the pressure in the thoracic cavity increases as the volume decreases, forcing air out of the lungs (a process called **expiration**). Inspiration is considered an *active* process because muscle contraction requires the use of ATP, whereas expiration is usually considered a *passive* process because the muscles relax, rather than contract. When a person is running, however, expiration becomes an active process, resulting from the contraction of **internal intercostal muscles** and **abdominal muscles.** In this case, both inspiration and expiration are considered *active* processes because muscle contraction is needed for both.

The amount of air that flows into and out of the lungs in 1 minute is the pulmonary **minute ventilation,** which is calculated by multiplying the **frequency of breathing** by the volume of each breath (the **tidal volume**). Ventilation must be regulated at all times to maintain oxygen in arterial blood and carbon dioxide in venous blood at their normal levels—that is, at their normal **partial pressures.** The *partial pressure* of a gas is the proportion of pressure that the gas exerts in a mixture. For example, in the atmosphere at sea level, the total pressure is 760 mm Hg. Oxygen makes up 21% of the total atmosphere and, therefore, has a partial pressure (P_{O_2}) of 160 mm Hg (760 mm Hg $\times$ 0.21).

Oxygen and carbon dioxide diffuse down their partial pressure gradients, from high partial pressures to low partial pressures. Oxygen diffuses from the alveoli of the lungs into the blood, where it can dissolve in plasma and attach to hemoglobin, and then diffuses from the blood into the tissues. Carbon dioxide (produced by the metabolic reactions of the tissues) diffuses from the tissues into the blood and then diffuses from the blood into the alveoli for export from the body.

In this exercise you will investigate the basic mechanics and regulation of the respiratory system. The concepts you will explore with a simulated lung will help you understand the operation of the human respiratory system in better detail.

ACTIVITY 1

Measuring Respiratory Volumes and Calculating Capacities

OBJECTIVES

1. To understand the use of the terms *ventilation, inspiration, expiration, diaphragm, external intercostals, internal intercostals, abdominal-wall muscles, expiratory reserve volume (ERV), forced vital capacity (FVC), tidal volume (TV), inspiratory reserve volume (IRV), residual volume (RV),* and *forced expiratory volume in one second (FEV_1).*
2. To understand the roles of skeletal muscles in the mechanics of breathing.
3. To understand the volume and pressure changes in the thoracic cavity during ventilation of the lungs.
4. To understand the effects of airway radius and, thus, resistance on airflow.

Introduction

The two phases of **ventilation,** or breathing, are (1) **inspiration,** during which air is taken into the lungs, and (2) **expiration,** during which air is expelled from the lungs. Inspiration occurs as the **external intercostal muscles** and the **diaphragm** contract. The diaphragm, normally a dome-shaped muscle, flattens as it moves inferiorly while the external intercostal muscles, situated between the ribs, lift the rib cage. These cooperative actions increase the thoracic volume. Air rushes into the lungs because this increase in thoracic volume creates a partial vacuum.

During quiet expiration, the inspiratory muscles relax, causing the diaphragm to rise superiorly and the chest wall to move inward. Thus, the **thorax** returns to its normal shape because of the elastic properties of the lung and thoracic wall. As in a deflating balloon, the pressure in the lungs rises, forcing air out of the lungs and airways. Although expiration is normally a *passive* process, **abdominal-wall muscles** and the **internal intercostal muscles** can also contract during expiration to force additional air from the lungs. Such forced expiration occurs, for example, when you exercise, blow up a balloon, cough, or sneeze.

Normal, quiet breathing moves about 500 ml (0.5 liter) of air (the **tidal volume**) into and out of the lungs with each breath, but this amount can vary due to a person's size, sex, age, physical condition, and immediate respiratory needs. In this activity you will measure the following respiratory volumes (the values given for the normal adult male and female are approximate).

Tidal volume (TV): Amount of air inspired and then expired with each breath under resting conditions (500 ml)

Inspiratory reserve volume (IRV): Amount of air that can be forcefully inspired after a normal tidal volume inspiration (male, 3100 ml; female, 1900 ml)

Expiratory reserve volume (ERV): Amount of air that can be forcefully expired after a normal tidal volume expiration (male, 1200 ml; female, 700 ml)

Residual volume (RV): Amount of air remaining in the lungs after forceful and complete expiration (male, 1200 ml; female, 1100 ml)

Respiratory capacities are calculated from the respiratory volumes. In this activity you will calculate the following respiratory capacities.

Total lung capacity (TLC): Maximum amount of air contained in lungs after a maximum inspiratory effort: TLC = TV + IRV + ERV + RV (male, 6000 ml; female, 4200 ml)

Vital capacity (VC): Maximum amount of air that can be inspired and then expired with maximal effort: VC = TV + IRV + ERV (male, 4800 ml; female 3100 ml)

You will also perform two pulmonary function tests in this activity.

Forced vital capacity (FVC): Amount of air that can be expelled when the subject takes the deepest possible inspiration and forcefully expires as completely and rapidly as possible

Forced expiratory volume (FEV_1): Measures the percentage of the vital capacity that is expired during 1 second of the FVC test (normally 75%–85% of the vital capacity)

EQUIPMENT USED The following equipment will be depicted on-screen: simulated human lungs suspended in a glass bell jar; rubber diaphragm—used to seal the jar and change the volume and, thus, pressure in the jar (As the diaphragm moves inferiorly, the volume in the bell jar increases and the pressure drops slightly, creating a partial vacuum in the bell jar. This partial vacuum causes air to be sucked into the tube at the top of the bell jar and then into the simulated lungs. As the diaphragm moves up, the decreasing volume and rising pressure within the bell jar forces air out of the lungs.); adjustable airflow tube—connects the lungs to the atmosphere; oscilloscope; three different breathing patterns: normal tidal volumes, expiratory reserve volume (ERV), and forced vital capacity (FVC).

Experiment Instructions

Go to the home page in the PhysioEx software and click **Exercise 7: Respiratory System Mechanics.** Click **Activity 1: Measuring Respiratory Volumes and Calculating Capacities,** and take the online **Pre-lab Quiz** for Activity 1.

After you take the online Pre-lab Quiz, click the **Experiment** tab and begin the experiment. The experiment instructions are reprinted here for your reference. The opening screen for the experiment is shown below.

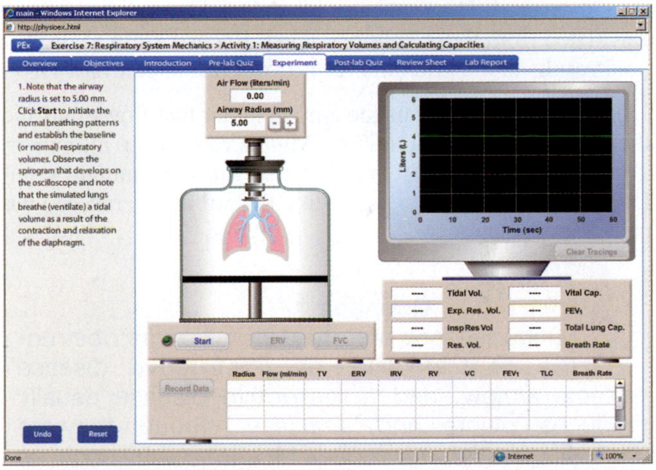
main - Windows Internet Explorer
Exercise 7: Respiratory System Mechanics > Activity 1: Measuring Respiratory Volumes and Calculating Capacities
Overview
Objectives
Introduction
Pre-lab Quiz
Experiment
Post-lab Quiz
Review Sheet
Lab Report
Air Flow (liters/min)
Airway Radius (mm)
Start
ERV
FVC
Record Data
Undo
Reset
Tidal Vol.
Exp. Res. Vol.
Insp Res Vol
Res. Vol.
Vital Cap.
Total Lung Cap.
Breath Rate
Time (sec)
Clear Tracings

1. Note that the airway radius is set to 5.00 mm. Click **Start** to initiate the normal breathing patterns and establish the baseline (or normal) respiratory volumes. Observe the spirogram that develops on the oscilloscope and note that the simulated lungs breathe (ventilate) a tidal volume as a result of the contraction and relaxation of the diaphragm.

2. Click **Record Data** to display your results in the grid (and record your results in Chart 1).

3. Click **Clear Tracings** to clear the spirogram on the oscilloscope.

4. You will now complete the measurement of respiratory volumes and determine the respiratory capacities. First, click **Start** to initiate the normal breathing pattern. After 10 seconds, click **ERV.** Wait another 10 seconds and then click **FVC** to

CHART 1 Respiratory Volumes and Capacities

Radius (mm)	Flow (ml/min)	TV (ml)	ERV (ml)	IRV (ml)	RV (ml)	VC (ml)	FEV_1 (ml)	TLC (ml)

complete the measurement of respiratory volumes. When you click ERV, the program will simulate forced expiration using the contraction of the internal intercostal muscles and abdominal-wall muscles. When you click FVC, the lungs will first inspire maximally and then expire fully to demonstrate forced vital capacity.

5. Note that, in addition to the tidal volume, the expiratory reserve volume, inspiratory reserve volume, and residual volume were measured. The vital capacity and total lung capacity were calculated from those volumes. Click **Record Data** to display your results in the grid (and record your results in Chart 1).

6. Minute ventilation is the amount of air that flows into and then out of the lungs in a minute. Minute ventilation (ml/min) = TV (ml/breath) × BPM (breaths/min). Enter the minute ventilation in the field below and then click **Submit** to record your answer in the lab report. ________ ml/min

? PREDICT Question 1
Lung diseases are often classified as obstructive or restrictive. An **obstructive** disease affects *airflow*, and a **restrictive** disease usually reduces *volumes and capacities*. Although they are not diagnostic, pulmonary function tests such as forced expiratory volume (FEV_1) can help a clinician determine the difference between obstructive and restrictive diseases. Specifically, an FEV_1 is the forced volume expired in 1 second.

In obstructive diseases such as chronic bronchitis and asthma, airway radius is decreased. Thus, FEV_1 will:

________________________________.

7. You will now explore what effect changing the airway radius has on pulmonary function. Decrease the airway radius to 4.50 mm by clicking the − button beneath the airway radius display.

8. Click **Start** to initiate the normal breathing pattern. After 10 seconds, click **ERV.** Wait another 10 seconds and then click **FVC.** The FEV_1 will appear in the FEV_1 display beneath the oscilloscope.

9. Click **Record Data** to display your results in the grid (and record your results in Chart 1).

10. You will now gradually decrease the airway radius.

- Decrease the airway radius by 0.50 mm by clicking the − button beneath the airway radius display.
- Click **Start** to initiate the normal breathing pattern. After 10 seconds, click **ERV.** Wait another 10 seconds and then click **FVC.** The FEV_1 will appear in the FEV_1 display beneath the oscilloscope.
- Click **Record Data** to display your results in the grid (and record your results in Chart 1).

Repeat this step until you reach an airway radius of 3.00 mm.

11. A useful way to express FEV_1 is as a percentage of the forced vital capacity (FVC). Using the FEV_1 and FVC values from the data grid, calculate the FEV_1 (%) by dividing the FEV_1 volume by the FVC volume (in this case, the VC is equal to the FVC) and multiply by 100%. Enter the FEV_1 (%) for an airway radius of 5.0 mm in the field below and then click **Submit** to record your answer in the lab report.

FEV_1 (%) for an airway radius of 5.0 (mm): ____________

12. Enter the FEV_1 (%) for an airway radius of 3.00 mm in the field below and then click **Submit** to record your answer in the lab report.

FEV_1 (%) for an airway radius of 3.00 (mm): ____________

After you complete the experiment, take the online **Post-lab Quiz** for Activity 1.

Activity Questions

1. When you forcefully exhale your entire expiratory reserve volume, any air remaining in your lungs is called the residual volume (RV). Why is it impossible to further exhale the RV (that is, *where* is this air volume trapped, and *why* is it trapped)?

____

____

2. How do you measure a person's RV in a laboratory?

____

____

3. Draw a spirogram that depicts a person's volumes and capacities before and during a significant cough.

ACTIVITY 2

Comparative Spirometry

OBJECTIVES

1. To understand the terms *spirometry, spirogram, emphysema, asthma, inhaler, moderate exercise, heavy exercise, tidal volume (TV), expiratory reserve volume (ERV), inspiratory reserve volume (IRV), residual volume (RV), vital capacity (VC), total lung capacity (TLC), forced vital capacity (FVC),* and *forced expiratory volume in one second (FEV_1).*
2. To observe and compare spirograms collected from resting, healthy patients to those taken from an emphysema patient.
3. To observe and compare spirograms collected from resting, healthy patients to those taken from a patient suffering an acute asthma attack.
4. To observe and compare the spirogram collected from an asthmatic patient *while* suffering an acute asthma attack to that taken after the patient uses an inhaler for relief.
5. To observe and compare spirograms collected from volunteers engaged in moderate exercise and heavy exercise.

Introduction

In this activity you will explore the changes to normal respiratory volumes and capacities when pathophysiology develops and during aerobic exercise by recruiting volunteers to breathe into a water-filled spirometer. The spirometer is a device that measures the volume of air inspired and expired by the lungs over a specified period of time. Several lung capacities and flow rates can be calculated from this data to assess pulmonary function. With your knowledge of respiratory mechanics, you can predict, document, and explain changes to the volumes and capacities in each state.

Emphysema breathing: With emphysema, there is a significant loss of elastic recoil in the lung tissue. This loss of elastic recoil occurs as the disease destroys the walls of the alveoli. Airway resistance is also increased as the lung tissue in general becomes more flimsy and exerts less anchoring on the surrounding airways. Thus, the lung becomes overly compliant and expands easily. Conversely, a great effort is required to expire because the lungs can no longer passively recoil and deflate. Each expiration requires a noticeable and exhausting muscular effort, and a person with emphysema expires slowly.

Acute asthma attack breathing: During an acute asthma attack, bronchiole smooth muscle spasms and, thus, the airways become constricted (that is, reduced in diameter). They also become clogged with thick mucus secretions. These changes lead to significantly increased airway resistance.

Underlying these symptoms is an airway inflammatory response brought on by triggers such as allergens (for example, dust and pollen), extreme temperature changes, and even exercise. Like with emphysema, the airways collapse and pinch closed before a forced expiration is completed. Thus, the volumes and peak flow rates are significantly reduced during an asthma attack. Unlike with emphysema, the elastic recoil is not diminished in an acute asthma attack.

When an acute asthma attack occurs, many people seek to relieve symptoms with an inhaler, which atomizes the medication and allows for direct application onto the afflicted airways. Usually, the medication includes a smooth muscle relaxant (for example, a β_2 agonist or an acetylcholine antagonist) that relieves the bronchospasms and induces bronchiole dilation. The medication can also contain an anti-inflammatory agent, such as a corticosteroid, that suppresses the inflammatory response. The use of the inhaler reduces airway resistance.

Breathing during exercise: During *moderate* aerobic exercise, the human body has an increased metabolic demand, which is met, in part, by changes in respiration. Specifically, both the rate of breathing and the tidal volume increase. These two respiratory variables do not increase by the same amount. The increase in the tidal volume is greater than the increase in the rate of breathing. During *heavy* exercise, further changes in respiration are required to meet the extreme metabolic demands of the body. In this case both the rate of breathing and the tidal volume increase to their maximum tolerable limits.

EQUIPMENT USED The following equipment will be depicted on-screen: a classic water-filled spirometer with an attached rotating drum that records the analog spirogram in real time; breathing patterns from a variety of patients: unforced breathing and forced vital capacity for a "normal" patient, a patient with emphysema, and a patient with asthma (during an attack and after using an inhaler); and the breathing patterns from a patient during moderate and heavy exercise.

Experiment Instructions

Go to the home page in the PhysioEx software and click **Exercise 7: Respiratory System Mechanics.** Click **Activity 2: Comparative Spirometry,** and take the online **Pre-lab Quiz** for Activity 2.

After you take the online Pre-lab Quiz, click the **Experiment** tab and begin the experiment. The experiment instructions are reprinted here for your reference. The opening screen for the experiment is shown below.

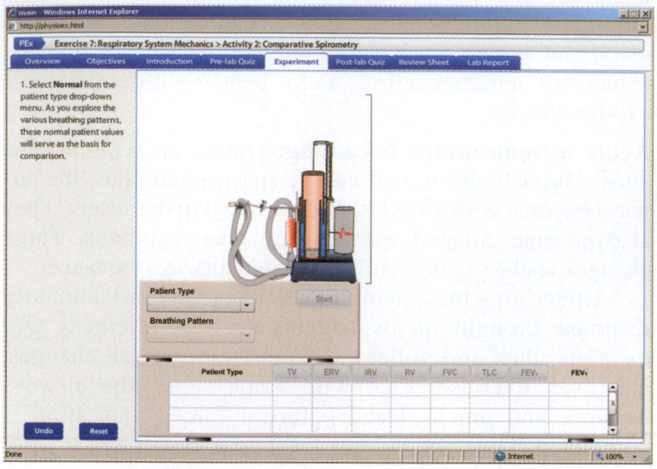
main - Windows Internet Explorer
Exercise 7: Respiratory System Mechanics > Activity 2: Comparative Spirometry
Overview
Objectives
Introduction
Pre-lab Quiz
Experiment
Post-lab Quiz
Review Sheet
Lab Report
1. Select Normal from the patient type drop-down menu. As you explore the various breathing patterns, these normal patient values will serve as the basis for comparison.
Patient Type
Breathing Pattern
Start
Patient Type
TV
ERV
IRV
RV
FVC
TLC
Undo
Reset

1. Select **Normal** from the patient type drop-down menu. As you explore the various breathing patterns, these normal patient values will serve as the basis for comparison.

2. Select **Unforced Breathing** from the breathing pattern drop-down menu.

3. Click **Start** to record the patient's unforced breathing pattern and watch as the drum starts turning and the spirogram develops on the paper rolling off the drum.

4. Note the volume levels (in milliliters) on the Y-axis of the spirogram. When half the screen is filled with unforced tidal volumes and the spirogram has paused, select **Forced Vital Capacity** from the breathing pattern drop-down menu.

5. Click **Start** to record the patient's forced vital capacity. The spirogram ends as the paper rolls to the right edge of the screen.

6. Click on each of the buttons in the data recorder to measure respiratory volumes and capacities. Start with tidal volume (TV) and work your way to the right. When you measure each volume or capacity, (1) a bracket appears on the spirogram to indicate where that measurement originates and (2) the value (in milliliters) displays in the grid. After you complete all the measurements, the FEV_1 (%) ratio will automatically be calculated. The FEV_1 (%) = (FEV_1/FVC) × 100%. Record your results in Chart 2.

PREDICT Question 1

With emphysema, there is a significant loss of elastic recoil in the lung tissue and a noticeable, exhausting muscular effort is required for each expiration. Inspiration actually becomes easier because the lung is now overly compliant. What lung values will change (from those of the normal patient) in the spirogram when the patient with emphysema is selected?

____

7. Select **Emphysema** from the patient type drop-down menu.

8. Select **Unforced Breathing** from the breathing pattern drop-down menu.

9. Click **Start** to record the patient's unforced breathing pattern and watch as the drum starts turning and the spirogram develops on the paper rolling off the drum.

10. Note the volume levels on the Y-axis of the spirogram. When half the screen is filled with unforced tidal volumes and the spirogram has paused, select **Forced Vital Capacity** from the breathing pattern drop-down menu.

11. Click **Start** to record the patient's forced vital capacity. The spirogram ends as the paper rolls to the right edge of the screen.

12. Click on each of the buttons in the data recorder to measure respiratory volumes and capacities. Start with tidal volume (TV) and work your way to the right. Record your results in Chart 2.

CHART 2 Spirometry Results

Patient type	TV (ml)	ERV (ml)	IRV (ml)	RV (ml)	FVC (ml)	TLC (ml)	FEV_1 (ml)	FEV_1 (%)
Normal								
Emphysema								
Acute asthma attack								
Plus inhaler								
Moderate exercise								
Heavy exercise								

PREDICT Question 2
During an acute asthma attack, airway resistance is significantly increased by (1) increased thick mucous secretions and (2) airway smooth muscle spasms. What lung values will change (from those of the normal patient) in the spirogram for a patient suffering an acute asthma attack?

___

13. Select **Acute Asthma Attack** from the patient type drop-down menu.

14. Select **Unforced Breathing** from the breathing pattern drop-down menu.

15. Click **Start** to record the patient's uforced breathing pattern and watch as the drum starts turning and the spirogram develops on the paper rolling off the drum.

16. Note the volume levels on the Y-axis of the spirogram. When half the screen is filled with unforced tidal volumes and the spirogram has paused, select **Forced Vital Capacity** from the breathing pattern drop-down menu.

17. Click **Start** to record the patient's forced vital capacity. The spirogram ends as the paper rolls to the right edge of the screen.

18. Click on each of the buttons in the data recorder to measure respiratory volumes and capacities. Start with tidal volume (TV) and work your way to the right. Record your results in Chart 2.

PREDICT Question 3
When an acute asthma attack occurs, many people seek relief from the increased airway resistance by using an inhaler. This device atomizes the medication and induces bronchiole dilation (though it can also contain an anti-inflammatory agent). What lung values will change *back* to those of the normal patient in the spirogram after the asthma patient uses an inhaler?

___

19. Select **Plus Inhaler** from the patient type drop-down menu.

20. Select **Unforced Breathing** from the breathing pattern drop-down menu.

21. Click **Start** to record the patient's unforced breathing pattern and watch as the drum starts turning and the spirogram develops on the paper rolling off the drum.

22. Note the volume levels on the Y-axis of the spirogram. When half the screen is filled with unforced tidal volumes and the spirogram has paused, select **Forced Vital Capacity** from the breathing pattern drop-down menu.

23. Click **Start** to record the patient's forced vital capacity. The spirogram ends as the paper rolls to the right edge of the screen.

24. Click on each of the buttons in the data recorder to measure respiratory volumes and capacities. Start with tidal volume (TV) and work your way to the right. Record your results in Chart 2.

PREDICT Question 4
During moderate aerobic exercise, the human body will change its respiratory cycle in order to meet increased metabolic demands. During heavy exercise, further changes in respiration are required to meet the extreme metabolic demands of the body. Which lung value will change more during moderate exercise, the ERV or the IRV?

___

25. Select **Moderate Exercise** from the patient type drop-down menu. Note that the selection of a breathing pattern is not applicable because our central nervous system automatically adjusts and maintains the depth and frequency of breathing to meet the increased metabolic demands while we exercise. We do not normally alter this pattern with conscious intervention.

26. Click **Start** to record the patient's breathing pattern and watch as the drum starts turning and the spirogram develops on the paper rolling off the drum.

27. Click on each of the buttons in the data recorder to measure respiratory volumes and capacities. Start with tidal volume (TV) and work your way to the right. *ND* indicates this measurement or calculation was not done. Record your results in Chart 2.

28. Select **Heavy Exercise** from the patient type drop-down menu.

29. Click **Start** to record the patient's breathing pattern and watch as the drum starts turning and the spirogram develops on the paper rolling off the drum.

30. Click on each of the buttons in the data recorder to measure respiratory volumes and capacities. Start with tidal volume (TV) and work your way to the right. Record your results in Chart 2.

After you complete the experiment, take the online **Post-lab Quiz** for Activity 2.

Activity Questions

1. Why is residual volume (RV) above normal in a patient with emphysema?

___

___

2. Why did the asthmatic patient's inhaler medication fail to return all volumes and capacities to normal values right away?

___

___

3. Looking at the spirograms generated in this activity, state an easy way to determine whether a person's exercising effort is moderate or heavy.

____

____

ACTIVITY 3

Effect of Surfactant and Intrapleural Pressure on Respiration

OBJECTIVES

1. To understand the terms *surfactant, surface tension, intrapleural space, intrapleural pressure, pneumothorax,* and *atelectasis.*
2. To understand the effect of surfactant on surface tension and lung function.
3. To understand how negative intrapleural pressure prevents lung collapse.

Introduction

At any gas-liquid boundary, the molecules of the liquid are attracted more strongly to each other than they are to the gas molecules. This unequal attraction produces tension at the liquid surface, called **surface tension.** Because surface tension resists any force that tends to increase surface area of the gas-liquid boundary, it acts to decrease the size of hollow spaces, such as the alveoli, or microscopic air spaces within the lungs.

If the film lining the air spaces in the lung were pure water, it would be very difficult, if not impossible, to inflate the lungs. However, the aqueous film covering the alveolar surfaces contains **surfactant,** a detergent-like mixture of lipids and proteins that decreases surface tension by reducing the attraction of water molecules to each other. You will explore the importance of surfactant in this activity.

Between breaths, the pressure in the pleural cavity, the **intrapleural pressure,** is less than the pressure in the alveoli. Two forces cause this negative pressure condition: (1) the tendency of the lung to recoil because of its elastic properties and the surface tension of the alveolar fluid and (2) the tendency of the compressed chest wall to recoil and expand outward. These two forces pull the lungs away from the thoracic wall, creating a partial vacuum in the pleural cavity.

Because the pressure in the intrapleural space is lower than atmospheric pressure, any opening created in the pleural membranes equalizes the intrapleural pressure with atmospheric pressure by allowing air to enter the pleural cavity, a condition called **pneumothorax.** A pneumothorax can then lead to lung collapse, a condition called **atelectasis.** In this activity, the **intrapleural space** is the space between the wall of the glass bell jar and the outer wall of the lung it contains.

EQUIPMENT USED The following equipment will be depicted on-screen: simulated human lungs suspended in a glass bell jar; rubber diaphragm—used to seal the jar and change the volume and, thus, pressure in the jar (As the diaphragm moves inferiorly, the volume in the bell jar increases and the pressure drops slightly, creating a partial vacuum in the bell jar. This partial vacuum causes air to be sucked into the tube at the top of the bell jar and then into the simulated lungs. As the diaphragm moves up, the decreasing volume and rising pressure within the bell jar forces air out of the lungs.); valve—allows intrapleural pressure in the left side of bell jar to equalize with atmospheric pressure; surfactant—amphipathic lipids (dipalmitoylphosphatidylcholine, phosphatidylglycerol, and palmitic acid) and short, synthetic peptides in a mixture that mimics the surfactant found in human lungs (surfactant molecules reduce surface tension in alveoli by adsorbing to the air-water interface, with their hydrophilic parts in the water and their hydrophobic parts facing toward the air); oscilloscope.

Experiment Instructions

Go to the home page in the PhysioEx software and click **Exercise 7: Respiratory System Mechanics.** Click **Activity 3: Effect of Surfactant and Intrapleural Pressure on Respiration,** and take the online **Pre-lab Quiz** for Activity 3.

After you take the online Pre-lab Quiz, click the **Experiment** tab and begin the experiment. The experiment instructions are reprinted here for your reference. The opening screen for the experiment is shown below.

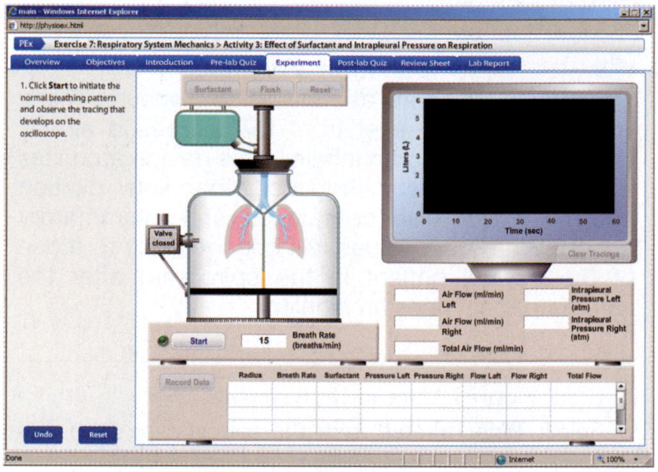
main - Windows Internet Explorer
http://physioex.html
PEx
Exercise 7: Respiratory System Mechanics > Activity 3: Effect of Surfactant and Intrapleural Pressure on Respiration
Overview
Objectives
Introduction
Pre-lab Quiz
Experiment
Post-lab Quiz
Review Sheet
Lab Report
1. Click Start to initiate the normal breathing pattern and observe the tracing that develops on the oscilloscope.
Surfactant
Flush
Reset
Valve closed
Liters (L)
Time (sec)
Clear Tracings
Start
15
Breath Rate (breaths/min)
Air Flow (ml/min) Left
Air Flow (ml/min) Right
Total Air Flow (ml/min)
Intrapleural Pressure Left (atm)
Intrapleural Pressure Right (atm)
Record Data
Radius
Breath Rate
Surfactant
Pressure Left
Pressure Right
Flow Left
Flow Right
Total Flow
Undo
Reset
Done
Internet
100%

1. Click **Start** to initiate the normal breathing pattern and observe the tracing that develops on the oscilloscope.

2. Click **Record Data** to display your results in the grid (and record your results in Chart 3). This data represents breathing in the absence of surfactant.

3. Click **Surfactant** twice to dispense two aliquots of the synthetic lipids and peptides onto the interior lining of the lungs.

4. Click **Start** to initiate breathing in the presence of surfactant and observe the tracing that develops.

5. Click **Record Data** to display your results in the grid (and record your results in Chart 3).

CHART 3 Effect of Surfactant and Intrapleural Pressure on Respiration

Surfactant	Intrapleural pressure left (atm)	Intrapleural pressure right (atm)	Airflow left (ml/min)	Airflow right (ml/min)	Total airflow (ml/min)

> **PREDICT Question 1**
> What effect will adding more surfactant have on these lungs?
>
> ______________________________________

6. Click **Surfactant** twice to dispense two more aliquots of the synthetic lipids and proteins onto the interior lining of the lungs.

7. Click **Start** to initiate breathing in the presence of additional surfactant and observe the tracing that develops.

8. Click **Record Data** to display your results in the grid (and record your results in Chart 3).

9. Click **Clear Tracings** to clear the tracing on the oscilloscope.

10. Click **Flush** to clear the lungs of surfactant from the previous run.

11. Click **Start** to initiate breathing and observe the tracing that develops. Notice the negative pressure condition displayed below the oscilloscope when the lungs inflate.

12. Click **Record Data** to display your results in the grid (and record your results in Chart 3).

13. Click the valve on the left side of the glass bell jar to open it.

14. Click **Start** to initiate breathing and observe the tracing that develops.

15. Click **Record Data** to display your results in the grid (and record your results in Chart 3).

> **PREDICT Question 2**
> What will happen to the collapsed lung in the left side of the glass bell jar if you close the valve?
>
> ______________________________________

16. Click the valve on the left side of the glass bell jar to close it.

17. Click **Start** to initiate breathing and observe the tracing that develops.

18. Click **Record Data** to display your results in the grid (and record your results in Chart 3).

19. Click the **Reset** button above the glass bell jar to draw the air out of the intrapleural space and return the lung to its normal resting condition.

20. Click **Start** to initiate breathing and observe the tracing that develops.

21. Click **Record Data** to display your results in the grid (and record your results in Chart 3).

After you complete the experiment, take the online **Post-lab Quiz** for Activity 3.

Activity Questions

1. Why is normal quiet breathing so difficult for premature infants?

____

____

2. Why does a pneumothorax frequently lead to atelectasis?

____

____

PhysioEx 9.0

REVIEW SHEET

EXERCISE 7

NAME ______________________

LAB TIME/DATE ______________________

Respiratory System Mechanics

ACTIVITY 1 Measuring Respiratory Volumes and Calculating Capacities

1. What would be an example of an everyday respiratory event the ERV button simulates? ______________________

2. What additional skeletal muscles are utilized in an ERV activity?

____

3. What was the FEV_1 (%) at the initial radius of 5.00 mm?

____

4. What happened to the FEV_1 (%) as the radius of the airways decreased? How well did the results compare with your prediction?

____

5. Explain why the results from the experiment suggest that there is an obstructive, rather than a restrictive, pulmonary problem.

____

ACTIVITY 2 Comparative Spirometry

1. What lung values changed (from those of the normal patient) in the spirogram when the patient with emphysema was selected? Why did these values change as they did? How well did the results compare with your prediction?

____

____

2. Which of these two parameters changed more for the patient with emphysema, the FVC or the FEV_1? ______________________

3. What lung values changed (from those of the normal patient) in the spirogram when the patient experiencing an acute asthma attack was selected? Why did these values change as they did? How well did the results compare with your prediction?

____

____

____

4. How is having an acute asthma attack similar to having emphysema? How is it different? ______________________

____

5. Describe the effect that the inhaler medication had on the asthmatic patient. Did all the spirogram values return to "normal"? Why do you think some values did not return all the way to normal? How well did the results compare with your prediction?

____

____

____

6. How much of an increase in FEV_1 do you think is required for it to be considered significantly improved by the medication?

____

7. With moderate aerobic exercise, which changed more from normal breathing, the ERV or the IRV? How well did the results compare with your prediction?

____

____

8. Compare the breathing rates during normal breathing, moderate exercise, and heavy exercise. ______________________________

____

ACTIVITY 3 Effect of Surfactant and Intrapleural Pressure on Respiration

1. What effect does the addition of surfactant have on the airflow? How well did the results compare with your prediction?

____

____

2. Why does surfactant affect airflow in this manner? ______________________________

____

3. What effect did opening the valve have on the left lung? Why does this happen?

____

____

4. What effect on the collapsed lung in the left side of the glass bell jar did you observe when you closed the valve? How well did the results compare with your prediction?

____

____

5. What emergency medical condition does opening the left valve simulate?

____

6. In the last part of this activity, you clicked the Reset button to draw the air out of the intrapleural space and return the lung to its normal resting condition. What emergency procedure would be used to achieve this result if these were the lungs in a living person?

____

____

7. What do you think would happen when the valve is opened if the two lungs were in a single large cavity rather than separate cavities? ____

____

PhysioEx™ 9.0

EXERCISE 8

Chemical and Physical Processes of Digestion

PRE-LAB QUIZ

1. Circle the correct term. Enzymes are catalysts / substrates that increase the rate of chemical reactions without becoming a part of the product.
2. A(n) ________ is a specimen or standard against which all experimental samples are compared.
   a. assay
   b. control
   c. substrate
   d. trial
3. One enzyme that you will be studying today, produced by the salivary glands and secreted into the mouth, hydrolyzes starch to maltose. It is ________.
4. Circle True or False. When you use iodine to test for starch, a color change to blue-black indicates a positive starch test.
5. If Benedict's test in the starch assay produces a ________ precipitate, then your test will be recorded as positive for maltose.
   a. blue to black
   b. green to orange
   c. white
6. The enzyme ________, produced by the pancreas, is responsible for breaking down proteins.
   a. amylase
   b. kinase
   c. lipase
   d. trypsin
7. Circle the correct term. The enzyme pancreatic lipase / pepsin hydrolyzes neutral fats to their component monoglycerides and fatty acids.
8. Circle True or False. Both smooth and skeletal muscles are involved in the physical processes of digestion.
9. ________ movements are local contractions that mix foodstuffs with digestive juices and increase the rate of absorption.
   a. Deglutition
   b. Elimination
   c. Peristaltic
   d. Segmental

Exercise Overview

The **digestive system,** also called the gastrointestinal system, consists of the digestive tract (also called the gastrointestinal tract, or GI tract) and accessory glands that secrete enzymes and fluids needed for digestion. The digestive tract includes the mouth, pharynx, esophagus, stomach, small intestine, colon, rectum, and anus. The major functions of the digestive system are to ingest food, to break food down to its simplest components, to extract nutrients from these components for absorption into the body, and to eliminate wastes.

Most of the food we consume cannot be absorbed into our bloodstream without first being broken down into smaller subunits. **Digestion** is the process of

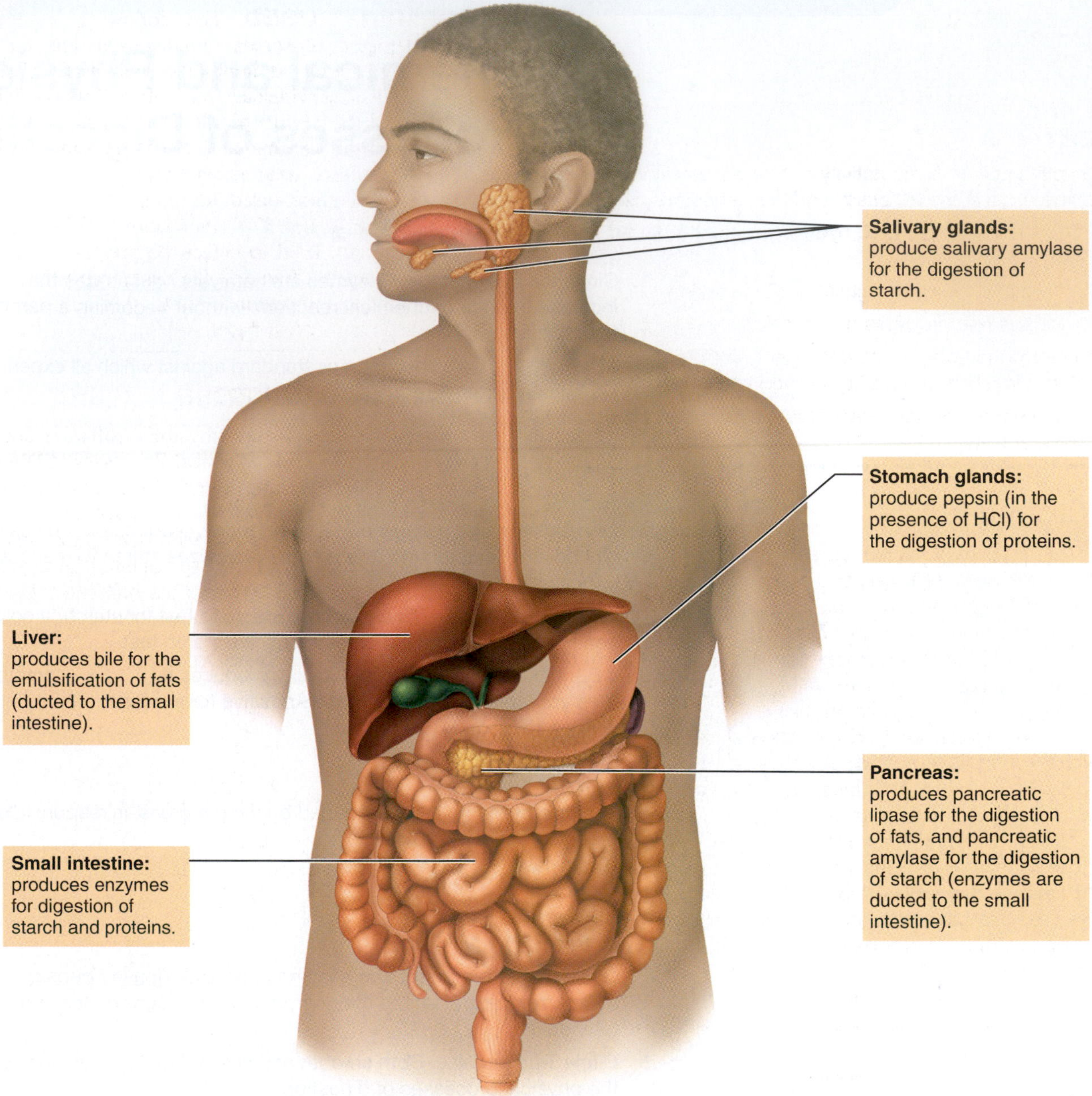
Salivary glands:
produce salivary amylase for the digestion of starch.
Stomach glands:
produce pepsin (in the presence of HCl) for the digestion of proteins.
Liver:
produces bile for the emulsification of fats (ducted to the small intestine).
Pancreas:
produces pancreatic lipase for the digestion of fats, and pancreatic amylase for the digestion of starch (enzymes are ducted to the small intestine).
Small intestine:
produces enzymes for digestion of starch and proteins.

FIGURE 8.1 The human digestive system. A few sites of chemical digestion and the organs that produce the enzymes of chemical digestion.

breaking down food molecules into smaller molecules with the aid of enzymes in the digestive tract. **Enzymes** are large protein molecules produced by body cells. They are biological catalysts that increase the rate of a chemical reaction without becoming part of the product. The digestive enzymes are hydrolytic enzymes, or **hydrolases,** which break down organic food molecules, or **substrates,** by adding water to the molecular bonds, thus cleaving the bonds between the subunits, or monomers.

A hydrolytic enzyme is highly specific in its action. Each enzyme hydrolyzes one substrate molecule or, at most, a small group of substrate molecules. Specific environmental conditions are necessary for an enzyme to function optimally. For example, in extreme environments, such as high temperature, an enzyme can unravel, or denature, because of the effect that temperature has on the three-dimensional structure of the protein.

Because digestive enzymes actually function outside the body cells in the digestive tract lumen, their hydrolytic activity can also be studied in vitro in a test tube. Such in vitro studies provide a convenient laboratory environment for investigating the effect of various factors on enzymatic activity. View Figure 8.1 for an overview of chemical digestion sites in the body.

ACTIVITY 1

Assessing Starch Digestion by Salivary Amylase

OBJECTIVES

1. To explain how enzyme activity can be assessed with enzyme assays: the IKI assay and the Benedict's assay.
2. To define enzyme, catalyst, hydrolase, substrate, and control.
3. To understand the specificity of amylase action.
4. To name the end products of carbohydrate digestion.
5. To perform the appropriate chemical tests to determine whether digestion of a particular food has occurred.
6. To discuss the possible effect of temperature and pH on amylase activity.

Introduction

In this activity you will investigate the hydrolysis of starch to maltose by **salivary amylase,** the enzyme produced by the salivary glands and secreted into the mouth. For you to be able to detect whether or not enzymatic action has occurred, you need to be able to identify the presence of the substrate and the product to determine to what extent hydrolysis has occurred. Thus, **controls** must be prepared to provide a known standard against which comparisons can be made. With positive controls, all of the required substances are included and a positive result is expected. Sometimes negative controls are included. With negative controls, a negative result is expected. Negative results with negative controls validate the experiment. Negative controls are used to determine whether there are any contaminating substances in the reagents. So, when a positive result is produced but a negative result is expected, one or more contaminating substances are present to cause the change.

With amylase activity, starch decreases and maltose increases as digestion proceeds according to the following equation.

$$\text{Starch} + \text{water} \xrightarrow{\text{amylase}} \text{maltose}$$

Because the chemical changes that occur as starch is digested to maltose cannot be seen by the naked eye, you need to conduct an **enzyme assay,** the chemical method of detecting the presence of digested substances. You will perform two enzyme assays on each sample. The IKI assay detects the presence of starch, and the Benedict's assay tests for the presence of reducing sugars, such as glucose or maltose, which are the digestion products of starch. Normally a caramel-colored solution, IKI turns blue-black in the presence of starch. Benedict's reagent is a bright blue solution that changes to green to orange to reddish brown with increasing amounts of maltose. It is important to understand that enzyme assays only indicate the presence or absence of substances. It is up to you to analyze the results of the experiments to decide whether enzymatic hydrolysis has occurred.

EQUIPMENT USED The following equipment will be depicted on-screen: amylase—an enzyme that digests starch; starch—a complex carbohydrate substrate; maltose—a disaccharide substrate; pH buffers—solutions used to adjust the pH of the solution; deionized water—used to adjust the volume so that it is the same for each reaction; test tubes—used as reaction vessels for the various tests; incubators—used for temperature treatments (boiling, freezing, and 37°C incubation); IKI—found in the assay cabinet; used to detect the presence of starch; Benedict's reagent—found in the assay cabinet; used to detect the products of starch digestion (this includes the reducing sugars maltose and glucose).

Experiment Instructions

Go to the home page in the PhysioEx software and click **Exercise 8: Chemical and Physical Processes of Digestion.** Click **Activity 1: Assessing Starch Digestion by Salivary Amylase,** and take the online **Pre-lab Quiz** for Activity 1.

After you take the online Pre-lab Quiz, click the **Experiment** tab and begin the experiment. The experiment instructions are reprinted here for your reference. The opening screen for the experiment is shown below.

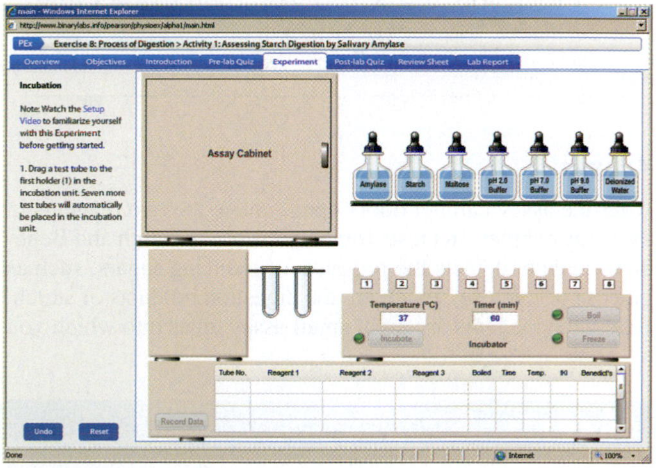
Exercise 8: Process of Digestion > Activity 1: Assessing Starch Digestion by Salivary Amylase
Overview
Objectives
Introduction
Pre-lab Quiz
Experiment
Post-lab Quiz
Review Sheet
Lab Report
Incubation
Assay Cabinet
Amylase
Starch
Maltose
Temperature (°C)
Timer (min)
Incubator
Undo
Reset

Incubation

1. Drag a test tube to the first holder **(1)** in the incubation unit. Seven more test tubes will automatically be placed in the incubation unit.

2. Add the substances indicated below to tubes 1 through 7.

Tube 1: amylase, starch, pH 7.0 buffer

Tube 2: amylase, starch, pH 7.0 buffer

Tube 3: amylase, starch, pH 7.0 buffer

Tube 4: amylase, deionized water, pH 7.0 buffer

Tube 5: deionized water, starch, pH 7.0 buffer

Tube 6: deionized water, maltose, pH 7.0 buffer

Tube 7: amylase, starch, pH 2.0 buffer

Tube 8: amylase, starch, pH 9.0 buffer

To add a substance to a test tube, drag the dropper cap of the bottle on the solutions shelf to the top of the test tube.

3. Click the number **(1)** under the first test tube. The tube will descend into the incubation unit. All other tubes should remain in the raised position.

4. Click **Boil** to boil tube 1. After boiling for a few moments, the tube will automatically rise.

5. Click the number **(2)** under the second test tube. The tube will descend into the incubation unit. All other tubes should remain in the raised position.

6. Click **Freeze** to freeze tube 2. After freezing for a few moments, the tube will automatically rise.

7. Click **Incubate** to start the run. Note that the incubation temperature is set at 37°C and the timer is set at 60 min. The incubation unit will gently agitate the test tube rack, evenly mixing the contents of all test tubes throughout the incubation. The simulation compresses the 60-minute time period into 10 seconds of real time, so what would be a 60-minute incubation in real time will take only 10 seconds in the simulation. When the incubation time elapses, the test tube rack will automatically rise, and the doors to the assay cabinet will open.

? PREDICT Question 1
What effect do you think boiling and freezing will have on the activity of the amylase enzyme?

____

____

Assays

After the assay cabinet doors open, notice the two reagents in the assay cabinet. IKI tests for the presence of starch and Benedict's reagent detects the presence of reducing sugars, such as glucose or maltose, which are the digestion products of starch. Below the reagents are eight small assay tubes into which you will dispense a small amount of test solution from the incubated samples in the incubation unit, plus a drop of IKI.

8. Drag the first tube in the incubation unit to the first small assay tube on the left side of the assay cabinet to decant approximately half of the contents in the test tube into the assay tube. The decanting step will automatically repeat for the remaining tubes in the incubation unit.

9. Drag the IKI dropper cap to the first assay tube to dispense a drop of IKI into the assay tube. The dropper will automatically move across and dispense IKI to the remaining tubes.

10. Inspect the tubes for color change. A blue-black color indicates a positive starch test. If starch is not present, the mixture will look like diluted IKI, a negative starch test. Intermediate starch amounts result in a pale-gray color. Click **Record Data** to display your results in the grid (and record your results in Chart 1).

11. Drag the Benedict's reagent dropper cap to the test tube in the first holder **(1)** in the incubation unit to dispense five drops of Benedict's reagent into the tube. The dropper will automatically move across and dispense Benedict's reagent to the remaining tubes.

12. Click **Boil.** The entire tube rack will descend into the incubation unit and automatically boil the tube contents for a few moments.

13. Inspect the tubes for color change. A green-to-reddish color indicates that a reducing sugar is present; this is a positive sugar test. An orange-colored sample contains more sugar than a green sample. A reddish-brown color indicates even more sugar. A negative sugar test is indicated by no color change from the original bright blue. Click **Record Data** to display your results in the grid (and record your results in Chart 1).

After you complete the experiment, take the online **Post-lab Quiz** for Activity 1.

CHART 1 Salivary Amylase Digestion of Starch

Tube No.	1	2	3	4	5	6	7	8
Additives	Amylase Starch pH 7.0 buffer	Amylase Starch pH 7.0 buffer	Amylase Starch pH 7.0 buffer	Amylase Deionized water pH 7.0 buffer	Deionized water Starch pH 7.0 buffer	Deionized water Maltose pH 7.0 buffer	Amylase Starch pH 2.0 buffer	Amylase Starch pH 9.0 buffer
Incubation condition	Boil first, then incubate at 37°C for 60 minutes	Freeze first, then incubate at 37°C for 60 minutes	37°C 60 minutes	37°C 60 minutes	37°C 60 minutes	37°C 60 minutes	37°C 60 minutes	37°C 60 minutes
IKI test								
Benedict's test								

Activity Questions

1. Describe the effect that boiling had on the activity of amylase. Why did boiling have this effect? How does the effect of freezing differ from the effect of boiling?

___

___

2. What is the purpose for including tube 3 and what can you conclude from the result?

___

___

3. Describe how you determined the optimal pH for amylase activity.

___

___

4. Judging from what you learned in this activity, suggest a reason why salivary amylase would be much less active in the stomach.

___

___

ACTIVITY 2

Exploring Amylase Substrate Specificity

OBJECTIVES

1. Explain how hydrolytic enzyme activity can be assessed with the IKI assay and the Benedict's assay.
2. Understand the specificity that enzymes have for their substrate.
3. Understand the difference between the substrates starch and cellulose.
4. Explain what would be the substrate specificity of peptidase.
5. Explain how bacteria might aid in digestion.

Introduction

In this activity you will investigate the specificity that enzymes have for their substrates. To do this you will hydrolyze starch to maltose and maltotriose using **salivary amylase,** the enzyme produced by the salivary glands and secreted into the mouth. To detect whether or not enzymatic action has occurred, you need to be able to identify the presence of the substrate and the product to determine to what extent hydrolysis has occurred. The **substrate** is the substance that the enzyme acts on. The enzyme has a pocket called the **active site,** which the substrate or substrates must fit into temporarily for catalysis to occur. The substrate is often held in the active site by non-covalent bonds (weak bonds), such as ionic bonds and hydrogen bonds.

With amylase activity, starch decreases and sugar increases as digestion proceeds according to the following equation.

$$\text{Starch} + \text{water} \xrightarrow{\text{amylase}} \text{maltose} + \text{maltotriose} + \text{starch}$$

Because the chemical changes that occur as starch is digested to maltose cannot be seen by the naked eye, you need to conduct an **enzyme assay,** the chemical method of detecting the presence of digested substances. You will perform two enzyme assays on each sample. The IKI assay detects the presence of starch or cellulose and the Benedict's assay tests for the presence of reducing sugars, such as glucose or maltose, which are the digestion products of starch. Normally a caramel-colored solution, IKI turns blue-black in the presence of starch or cellulose. Benedict's reagent is a bright blue solution that changes to green to orange to reddish brown with increasing amounts of maltose. It is important to understand that enzyme assays only indicate the presence or absence of substances. It is up to you to analyze the results of the experiments to decide whether enzymatic hydrolysis has occurred.

Starch is a polysaccharide found in plants, where it is used to store energy. Plants also have the polysaccharide **cellulose,** which provides rigidity to their cell walls. Both polysaccharides are polymers of glucose, but the glucose molecules are linked differently. You will be testing salivary amylase to determine whether it digests cellulose. Also, you will investigate to see whether a bacterial suspension can digest cellulose and whether **peptidase,** a pancreatic enzyme that digests peptides, can break down starch.

EQUIPMENT USED The following equipment will be depicted on-screen: amylase—an enzyme that digests starch; starch—a polysaccharide; pH 7.0 buffer—a solution used to set the pH of the test tube solution; deionized water—used to adjust the test tube solution volume so it is the same for each reaction; glucose—a reducing sugar that is the monosaccharide subunit of both starch and cellulose; cellulose—a complex carbohydrate found in the cell wall of plants; peptidase—a pancreatic enzyme that breaks down peptides; bacteria—a suspension of live bacteria; test tubes—used as reaction vessels for the various tests; incubators—used for temperature treatments (37°C incubation); IKI—found in the assay cabinet; used to detect the presence of starch or cellulose; Benedict's reagent—found in the assay cabinet; used to detect the products of starch and cellulose digestion.

Experiment Instructions

Go to the home page in the PhysioEx software and click **Exercise 8: Chemical and Physical Processes of Digestion.** Click **Activity 2: Exploring Amylase Substrate Specificity,** and take the online **Pre-lab Quiz** for Activity 2.

After you take the online Pre-lab Quiz, click the **Experiment** tab and begin the experiment. The experiment instructions are reprinted here for your reference. The opening screen for the experiment is shown on the following page.

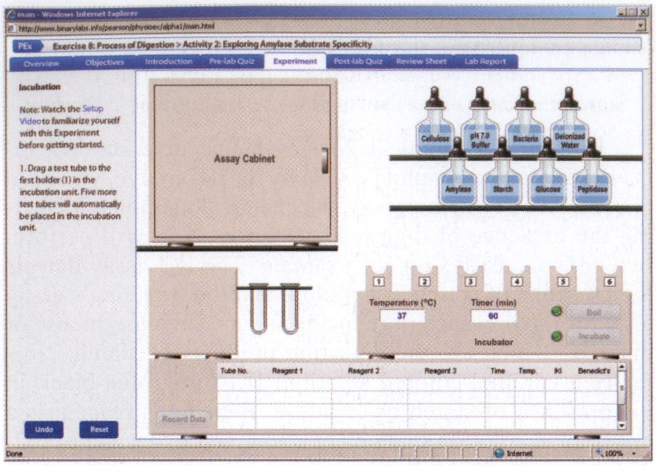
Exercise 8: Process of Digestion > Activity 2: Exploring Amylase Substrate Specificity
Overview
Objectives
Introduction
Pre-lab Quiz
Experiment
Post-lab Quiz
Review Sheet
Lab Report
Incubation
Note: Watch the Setup Video to familiarize yourself with this Experiment before getting started.
1. Drag a test tube to the first holder (1) in the incubation unit. Five more test tubes will automatically be placed in the incubation unit.
Assay Cabinet
Cellulose
pH 7.0 Buffer
Bacteria
Deionized Water
Amylase
Starch
Glucose
Peptidase
Temperature (°C)
37
Timer (min)
60
Incubator
Boil
Incubate
Tube No.
Reagent 1
Reagent 2
Reagent 3
Time
Temp.
IKI
Benedict's
Record Data
Undo
Reset

Incubation

1. Drag a test tube to the first holder **(1)** in the incubation unit. Five more test tubes will automatically be placed in the incubation unit.

2. Add the substances indicated below to tubes 1 through 6.

Tube 1: amylase, starch, pH 7.0 buffer

Tube 2: amylase, glucose, pH 7.0 buffer

Tube 3: amylase, cellulose, pH 7.0 buffer

Tube 4: cellulose, pH 7.0 buffer, deionized water

Tube 5: peptidase, starch, pH 7.0 buffer

Tube 6: bacteria, cellulose, pH 7.0 buffer

To add a substance to a test tube, drag the dropper cap of the bottle on the solutions shelf to the test tube.

3. Click **Incubate** to start the run. Note that the incubation temperature is set at 37°C and the timer is set at 60 min. The incubation unit will gently agitate the test tube rack, evenly mixing the contents of all test tubes throughout the incubation. The simulation compresses the 60-minute time period into 10 seconds of real time, so what would be a 60-minute incubation in real life will take only 10 seconds in the simulation. When the incubation time elapses, the test tube rack will automatically rise, and the doors to the assay cabinet will open.

PREDICT Question 1
Do you think test Tube 3 will show a positive Benedict's test?

____

____

Assays

After the assay cabinet doors open, notice the two reagents in the assay cabinet. IKI tests for the presence of starch and Benedict's reagent detects the presence of reducing sugars, such as glucose or maltose, which are the digestion products of starch. Below the reagents are seven small assay tubes into which you will dispense a portion of the incubated samples, plus a drop of IKI.

4. Drag the first tube in the incubation unit to the first small assay tube on the left side of the assay cabinet to decant approximately half of the contents in the test tube into the assay tube. The decanting step will automatically repeat for the remaining tubes in the incubation unit.

5. Drag the IKI dropper cap to the first assay tube to dispense a drop of IKI into the assay tube. The dropper will automatically dispense IKI into the remaining tubes.

6. Inspect the tubes for color change. A blue-black color indicates a positive starch test. If starch is not present, the mixture will look like diluted IKI, a negative starch test. Intermediate starch amounts result in a pale-gray color. Click **Record Data** to display your results in the grid (and record your results in Chart 2).

7. Drag the Benedict's reagent dropper cap to the test tube in the first holder **(1)** in the incubation unit to dispense five drops of Benedict's reagent into the tube. The dropper will automatically move across and dispense Benedict's reagent to the remaining tubes.

8. Click **Boil**. The entire tube rack will descend into the incubation unit and automatically boil the tube contents for a few moments.

9. Inspect the tubes for color change. A green-to-reddish color indicates that a reducing sugar is present; this is a positive sugar test. An orange-colored sample contains more sugar than a green sample. A reddish-brown color indicates even more sugar. A negative sugar test is indicated by no color change from the original bright blue. Click **Record Data** to display your results in the grid (and record your results in Chart 2).

CHART 2 Enzyme Digestion of Starch and Cellulose

Tube No.	1	2	3	4	5	6
Additives	Amylase Starch pH 7.0 buffer	Amylase Glucose pH 7.0 buffer	Amylase Cellulose pH 7.0 buffer	Deionized water Cellulose pH 7.0 buffer	Peptidase Starch pH 7.0 buffer	Bacteria Cellulose pH 7.0 buffer
Incubation condition	37°C 60 minutes	37°C 60 minutes	37°C 60 minutes	37°C 60 minutes	37°C 60 minutes	37°C 60 minutes
IKI test						
Benedict's test						

After you complete the Experiment, take the online **Post-lab Quiz** for Activity 2.

Activity Questions

1. Does amylase use cellulose as a substrate?

2. What effect did the addition of bacteria have on the digestion of cellulose?

3. What effect did the addition of peptidase to the starch have? Why?

4. What is the smallest subunit into which starch can be broken down?

ACTIVITY 3

Assessing Pepsin Digestion of Protein

OBJECTIVES

1. Explain how the enzyme activity of pepsin can be assessed with the BAPNA assay.
2. Identify the substrate specificity of pepsin.
3. Discuss the effects of temperature and pH on pepsin activity.
4. Understand the pH specificity of enzyme activity and how it relates to human physiology.

Introduction

In this activity, you will explore the digestion of protein **(peptides).** Peptides are two or more **amino acids** linked together by a peptide bond. A peptide chain containing 10 to 100 amino acids is typically called a **polypeptide. Proteins** can consist of a large peptide chain (more than 100 amino acids) or even multiple peptide chains.

During digestion, **chief cells** of the stomach glands secrete a protein-digesting enzyme called **pepsin.** Pepsin **hydrolyzes** peptide bonds. This activity breaks up ingested proteins and polypeptides into smaller peptide chains and free amino acids. In this activity, you will use **BAPNA** as a **substrate** to assess pepsin activity. BAPNA is a synthetic "peptide" that releases a yellow dye **product** when hydrolyzed. BAPNA solutions turn yellow in the presence of an active peptidase, such as pepsin, but otherwise remain colorless.

To quantify the pepsin activity in each test solution, you will use a **spectrophotometer** to measure the amount of yellow dye produced. A spectrophotometer shines light through the sample and then measures how much light is absorbed. The fraction of light absorbed is expressed as the sample's **optical density.** Yellow solutions, where BAPNA has been hydrolyzed, will have optical densities greater than zero. The greater the optical density, the more hydrolysis has occurred. Colorless solutions, in contrast, do not absorb light and will have an optical density near zero.

Some negative controls are included in this activity. With negative controls, a negative result is expected. Negative results with negative controls validate the experiment. Negative controls are used to determine whether there are any contaminating substances in the reagents. So, when a positive result is produced but a negative result is expected, one or more contaminating substances are present to cause the change.

EQUIPMENT USED The following equipment will be depicted on-screen: pepsin—an enzyme that digests peptides; BAPNA—a synthetic "peptide"; pH buffers—solutions used to set the pH of the test tube solution; deionized water—used to adjust the test tube solution volume so it is the same for each reaction; test tubes—used as reaction vessels for the various tests; incubators—used for temperature treatments (boiling and 37°C incubation); spectrophotometer—found in the assay cabinet; used to measure the optical density of solutions.

Experiment Instructions

Go to the home page in the PhysioEx software and click **Exercise 8: Chemical and Physical Processes of Digestion.** Click **Activity 3: Assessing Pepsin Digestion of Protein,** and take the online **Pre-lab Quiz** for Activity 3.

After you take the online Pre-lab Quiz, click the **Experiment** tab and begin the experiment. The experiment instructions are reprinted here for your reference. The opening screen for the experiment is shown below.

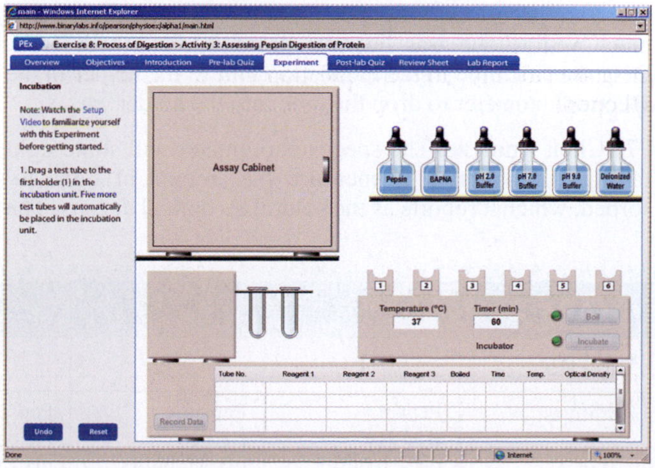
Assay Cabinet
Pepsin
BAPNA
pH 2.0 Buffer
pH 7.0 Buffer
pH 9.0 Buffer
Deionized Water
Temperature (°C)
Timer (min)
Incubator
Undo
Reset

Incubation

1. Drag a test tube to the first holder **(1)** in the incubation unit. Five more test tubes will automatically be placed in the incubation unit.

2. Add the substances indicated below to tubes 1 through 6.

Tube 1: pepsin, BAPNA, pH 2.0 buffer

Tube 2: pepsin, BAPNA, pH 2.0 buffer

Tube 3: pepsin, deionized water, pH 2.0 buffer

Tube 4: deionized water, BAPNA, pH 2.0 buffer

Tube 5: pepsin, BAPNA, pH 7.0 buffer

Tube 6: pepsin, BAPNA, pH 9.0 buffer

To add a substance to a test tube, drag the dropper cap of the bottle on the solutions shelf to the test tube.

3. Click the number (1) under the first test tube. The tube will descend into the incubation unit. All other tubes should remain in the raised position.

4. Click **Boil** to boil tube 1. After boiling for a few moments, the tube will automatically rise.

5. Click **Incubate** to start the run. Note that the incubation temperature is set at 37°C and the timer is set at 60 min. The incubation unit will gently agitate the test tube rack, evenly mixing the contents of all test tubes throughout the incubation. The simulation compresses the 60-minute time period into 10 seconds of real time, so what would be a 60-minute incubation in real life will take only 10 seconds in the simulation. When the incubation time elapses, the test tube rack will automatically rise, and the doors to the assay cabinet will open. The spectrophotometer is in the assay cabinet.

? PREDICT Question 1
At which pH do you think pepsin will have the highest activity?

____

____

Assays

6. You will now use the spectrophotometer to measure how much yellow dye was liberated from BAPNA hydrolysis. Drag the first tube in the incubation unit to the holder in the spectrophotometer to drop the tube into the holder.

7. Click **Analyze.** The spectrophotometer will shine light through the solution to measure the amount of light absorbed, which it reports as the solution's optical density. The optical density of the sample is shown in the optical density display.

8. Click **Record Data** to display your results in the grid (and record your results in Chart 3).

9. Drag the tube to its original position in the incubation unit.

10. Analyze the remaining five tubes by repeating the following steps for each tube.

- Drag the tube to the holder in the spectrophotometer to drop the tube into the holder.
- Click **Analyze.**
- Drag the tube to its original position in the incubation unit.

After you have analyzed all five tubes, click **Record Data** to display your results in the grid. (and record your results in Chart 3).

After you complete the experiment, take the online **Post-lab Quiz** for Activity 3.

Activity Questions

1. Describe the significance of the optimum pH for pepsin observed in the simulation and the secretion of pepsin by the chief cells of the gastric glands.

____

____

2. Would pepsin be active in the mouth? Explain your answer.

____

____

3. What are the subunit products of peptide digestion?

____

____

4. Describe the reason for including control tube 4.

____

____

CHART 3 Pepsin Digestion of Protein

Tube No.	1	2	3	4	5	6
Additives	Pepsin BAPNA pH 2.0 buffer	Pepsin BAPNA pH 2.0 buffer	Pepsin Deionized water pH 2.0 buffer	Deionized water BAPNA pH 2.0 buffer	Pepsin BAPNA pH 7.0 buffer	Pepsin BAPNA pH 9.0 buffer
Incubation condition	Boil first, then incubate at 37°C for 60 minutes	37°C 60 minutes	37°C 60 minutes	37°C 60 minutes	37°C 60 minutes	37°C 60 minutes
Optical density						

ACTIVITY 4

Assessing Lipase Digestion of Fat

OBJECTIVES

1. Explain how the enzyme activity of pancreatic lipase can be assessed with a pH-based measurement.
2. Identify the hydrolysis products of fat digestion.
3. Understand the role that bile plays in fat digestion.
4. Understand the significance of pH specificity of lipase activity and how it relates to human physiology.
5. Discuss the difficulty of using pH to measure digestion when comparing the activity of lipase at various pHs.

Introduction

Fats and oils belong to a diverse class of molecules called lipids. **Triglycerides,** a type of lipid, make up both fats and oils. At room temperature, fats are solid and oils are liquid. Both are poorly soluble in water. This insolubility of triglycerides presents a challenge during digestion because they tend to clump together, leaving only the surface molecules exposed to **lipase** enzymes. To overcome this difficulty, **bile salts** are secreted into the small intestine during digestion to physically emulsify lipids. Bile salts act like a detergent, separating the lipid clumps and increasing the surface area accessible to lipase enzymes.

As a result, two reactions must occur. First,

$$\text{Triglyceride clumps} \xrightarrow[\text{(emulsification)}]{\text{bile}} \text{minute triglyceride droplets}$$

Then,

$$\text{Triglyceride} \xrightarrow{\text{lipase}} \text{monoglyceride} + \text{two fatty acids}$$

Lipase hydrolyzes each triglyceride to a monoglyceride and two fatty acids. In addition to the **pancreatic lipase** secreted into the small intestine, **lingual lipase** and **gastric lipase** are also secreted. Even though bile salts are not secreted in the mouth or the stomach, small amounts of lipids are digested by these other lipases.

Because some of the end products of fat digestion are acidic (that is, fatty acids), lipase activity can be easily measured by monitoring the solution's **pH.** A solution containing fatty acids liberated by lipase activity will have a lower pH than a solution without such fatty acid production. You will record pH in this activity with a **pH meter.**

EQUIPMENT USED The following equipment will be depicted on-screen: lipase—an enzyme that digests triglycerides; vegetable oil—a mixture of triglycerides; bile salts—a solution that physically separates fats into smaller droplets; pH buffers—solutions used to set the pH of the test tube solution; deionized water—used to adjust the test tube solution volume so it is the same for each reaction; test tubes—used as reaction vessels for the various tests; incubators—used for temperature treatments (boiling and 37°C incubation); pH meter—found in the assay cabinet; used to measure pH.

Experiment Instructions

Go to the home page in the PhysioEx software and click **Exercise 8: Chemical and Physical Processes of Digestion.** Click **Activity 4: Assessing Lipase Digestion of Fat,** and take the online **Pre-lab Quiz** for Activity 4.

After you take the online Pre-lab Quiz, click the **Experiment** tab and begin the experiment. The experiment instructions are reprinted here for your reference. The opening screen for the experiment is shown below.

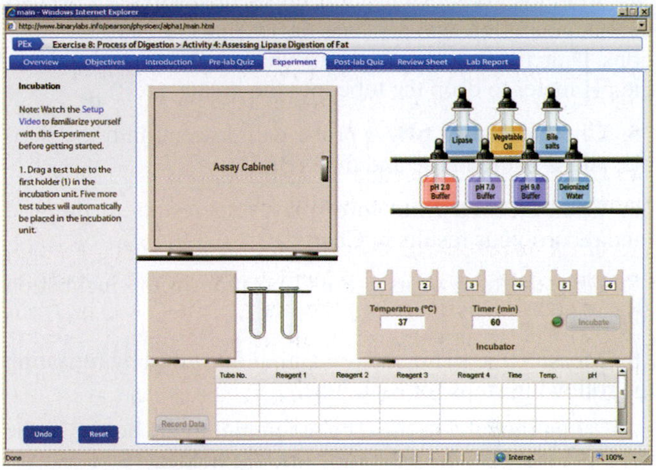
Exercise 8: Process of Digestion > Activity 4: Assessing Lipase Digestion of Fat
Overview
Objectives
Introduction
Pre-lab Quiz
Experiment
Post-lab Quiz
Review Sheet
Lab Report
Incubation
Assay Cabinet
Lipase
Vegetable Oil
Bile salts
pH 2.0 Buffer
pH 7.0 Buffer
pH 9.0 Buffer
Deionized Water
Temperature (°C)
Timer (min)
Incubate
Incubator
Tube No.
Reagent 1
Reagent 2
Reagent 3
Reagent 4
Time
Temp.
pH
Record Data
Undo
Reset

Incubation

1. Drag a test tube to the first holder **(1)** in the incubation unit. Five more test tubes will automatically be placed in the incubation unit.
2. Add the substances indicated below to tubes 1 through 6.

Tube 1: lipase, vegetable oil, bile salts, pH 7.0 buffer

Tube 2: lipase, vegetable oil, deionized water, pH 7.0 buffer

Tube 3: lipase, deionized water, bile salts, pH 9.0 buffer

Tube 4: deionized water, vegetable oil, bile salts, pH 7.0 buffer

Tube 5: lipase, vegetable oil, bile salts, pH 2.0 buffer

Tube 6: lipase, vegetable oil, bile salts, pH 9.0 buffer

To add a substance to a test tube, drag the dropper cap of the bottle on the solutions shelf to the test tube.

3. Click **Incubate** to start the run. Note that the incubation temperature is set at 37°C and the timer is set at 60 min. The incubation unit will gently agitate the test tube rack, evenly mixing the contents of all test tubes throughout the incubation. The simulation compresses the 60-minute time period into 10 seconds of real time, so what would be a 60-minute incubation in real life will take only 10 seconds in the simulation. When the incubation time elapses, the test tube rack will automatically rise, and the doors to the assay cabinet will open.

PREDICT Question 1
Which tube do you think will have the highest lipase activity?

___

___

Assays

4. After the assay cabinet doors open, you will see a pH meter that you will use to measure the final pH of your test solutions. Drag the first tube in the incubation unit to the holder in the pH meter to drop the tube into the holder.

5. Click **Measure pH.** A probe will descend into the sample, take a pH reading, and then retract.

6. Click **Record Data** to display your results in the grid (and record your results in Chart 4).

7. Drag the tube to its original position in the incubation unit.

8. Measure the pH in the remaining five tubes by repeating the following steps for each tube.

- Drag the tube in the incubation unit to the holder in the pH meter to drop the tube into the holder.
- Click **Measure pH.**
- Drag the tube to its original position in the incubation unit.

After you have measured the pH in all five tubes, click **Record Data** to display your results in the grid (and record your results in Chart 4).

After you complete the experiment, take the online **Post-lab Quiz** for Activity 4.

Activity Questions

1. Describe how lipase activity is measured in the simulation.

___

___

2. Can you determine if fat hydrolysis occurred in tube 5? Why or why not?

___

___

3. Would pancreatic lipase be active in the mouth? Why or why not?

___

___

4. Describe the physical separation of fats by bile salts.

___

___

CHART 4 Pancreatic Lipase Digestion of Triglycerides and the Action of Bile

Tube No.	1	2	3	4	5	6
Additives	Lipase Vegetable oil Bile salts pH 7.0 buffer	Lipase Vegetable oil Deionized water pH 7.0 buffer	Lipase Deionized water Bile salts pH 9.0 buffer	Deionized water Vegetable oil Bile salts pH 7.0 buffer	Lipase Vegetable oil Bile salts pH 2.0 buffer	Lipase Vegetable oil Bile salts pH 9.0 buffer
Incubation condition	37°C 60 minutes	37°C 60 minutes	37°C 60 minutes	37°C 60 minutes	37°C 60 minutes	37°C 60 minutes
pH						

PhysioEx 9.0

REVIEW SHEET

EXERCISE 8

Chemical and Physical Processes of Digestion

NAME ______________________

LAB TIME/DATE ______________________

ACTIVITY 1 Assessing Starch Digestion by Salivary Amylase

1. List the substrate and the subunit product of amylase. ______________________

____

2. What effect did boiling and freezing have on enzyme activity? Why? How well did the results compare with your prediction?

____

____

3. At what pH was the amylase most active? Describe the significance of this result. ______________________

____

4. Briefly describe the need for controls and give an example used in this activity. ______________________

____

5. Describe the significance of using a 37°C incubation temperature to test salivary amylase activity. ______________________

____

ACTIVITY 2 Exploring Amylase Substrate Specificity

1. Describe why the results in tube 1 and tube 2 are the same. ______________________

____

2. Describe the result in tube 3. How well did the results compare with your prediction? ______________________

____

3. Describe the usual substrate for peptidase. ______________________

____

4. Explain how bacteria can aid in digestion. ______________________

____

ACTIVITY 3 Assessing Pepsin Digestion of Protein

1. Describe the effect that boiling had on pepsin and how you could tell that it had that effect.

2. Was your prediction correct about the optimal pH for pepsin activity? Discuss the physiological correlation behind your results.

3. What do you think would happen if you reduced the incubation time to 30 minutes for tube 5?

ACTIVITY 4 Assessing Lipase Digestion of Fat

1. Explain why you can't fully test the lipase activity in tube 5.

2. Which tube had the highest lipase activity? How well did the results compare with your prediction? Discuss possible reasons why it may or may not have matched.

3. Explain why pancreatic lipase would be active in both the mouth and the intestine.

4. Describe the process of bile emulsification of lipids and how it improves lipase activity.

PhysioEx™ 9.0

EXERCISE 9

Renal System Physiology

PRE-LAB QUIZ

1. Normal urine is usually pale yellow to amber in color, due to the presence of
   a. hemochrome
   b. melanin
   c. urochrome
2. Circle the correct term. The average pH value of urine is 6.0 / 11.0.
3. Circle True or False. Glucose can usually be found in all normal urine.
4. ________, like other blood proteins, is/are too large to pass through the glomerular filtration membrane and is/are normally not found in urine.
   a. Albumin
   b. Chloride
   c. Nitrates
   d. Sulfate
5. Circle the correct term. Hematuria / Ketonuria, the appearance of red blood cells in the urine, almost always indicates pathology of the urinary system.
6. The appearance of bile pigments in the urine, a condition known as ________, can be an indication of liver disease.
   a. albuminuria
   b. bilirubinuria
   c. ketonuria
   d. pyuria
7. Circle the correct term. Casts / Calculi are hardened cell fragments formed in the distal convoluted tubules and collecting ducts and flushed out of the urinary tract.
8. When determining the presence of inorganic constituents such as sulfates, phosphates, and chlorides, you will be looking for the "formation of a precipitate." What is a precipitate?

   ____

Exercise Overview

The **kidney** is *both* an excretory and a regulatory organ. By filtering the water and solutes in the blood, the kidneys are able to *excrete* excess water, waste products, and even foreign materials from the body. However, the kidneys also *regulate* (1) plasma osmolarity (the concentration of a solution expressed as osmoles of solute per liter of solvent), (2) plasma volume, (3) the body's acid-base balance, and (4) the body's electrolyte balance. All these activities are extremely important for maintaining homeostasis in the body.

The paired kidneys are located between the posterior abdominal wall and the abdominal peritoneum. The right kidney is slightly lower than the left kidney. Each human kidney contains approximately one million **nephrons,** the functional units of the kidney.

Each nephron is composed of a **renal corpuscle** and a **renal tubule.** The renal corpuscle consists of a "ball" of capillaries, called the *glomerulus,* which is enclosed by a fluid-filled capsule, called *Bowman's capsule,* or the glomerular capsule. An **afferent arteriole** supplies blood to the glomerulus. As blood flows through the glomerular capillaries, protein-free plasma filters into the Bowman's capsule, a process called **glomerular filtration.** An **efferent arteriole** then drains the glomerulus of the remaining blood (view Figure 9.1).

The filtrate flows from Bowman's capsule into the start of the renal tubule, called the **proximal convoluted tubule,** then into the **loop of Henle,** a U-shaped hairpin loop, and, finally, into the **distal convoluted tubule** before emptying into

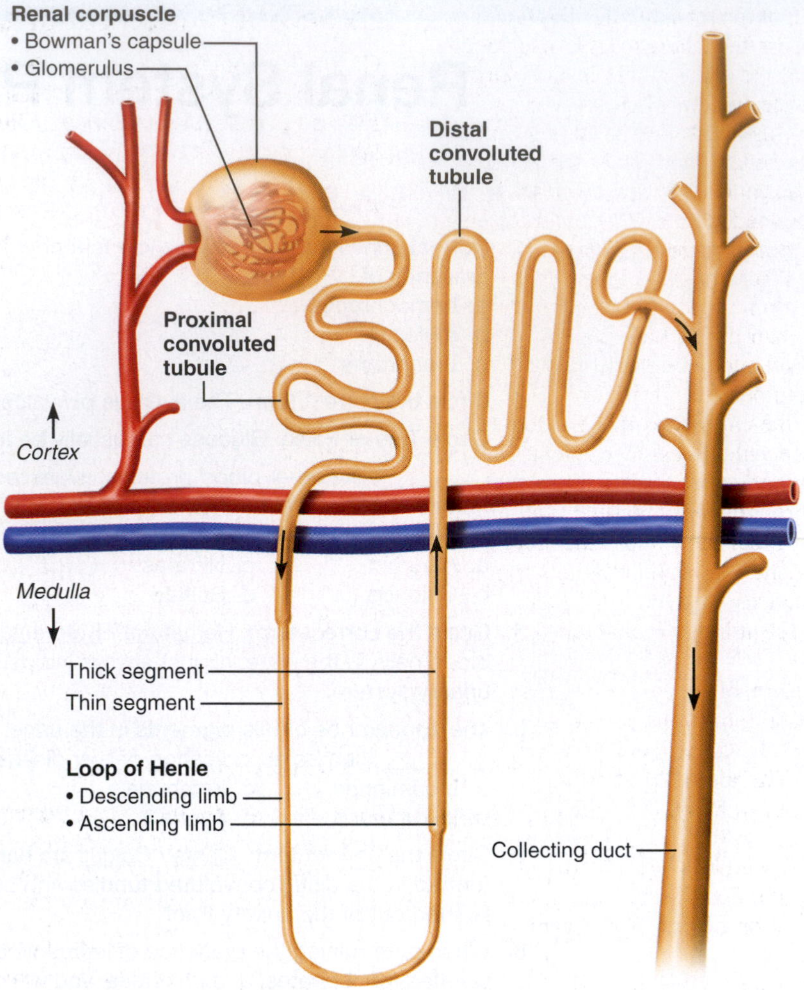
Renal corpuscle
• Bowman's capsule
• Glomerulus
Distal convoluted tubule
Proximal convoluted tubule
Cortex
Medulla
Thick segment
Thin segment
Loop of Henle
• Descending limb
• Ascending limb
Collecting duct

FIGURE 9.1 **Location and structure of nephrons.**

a **collecting duct**. From the collecting duct, the filtrate flows into, and collects in, the minor calyces.

The nephron performs three important functions that process blood into filtrate and urine: (1) glomerular filtration, (2) tubular reabsorption, and (3) tubular secretion. **Glomerular filtration** is a passive process in which fluid passes from the lumen of the glomerular capillary into the glomerular capsule of the renal tubule. **Tubular reabsorption** moves most of the filtrate back into the blood, leaving mainly salt water and the wastes in the lumen of the tubule. Some of the desirable, or needed, solutes are actively reabsorbed, and others move passively from the lumen of the tubule into the interstitial spaces. **Tubular secretion** is essentially the reverse of tubular reabsorption and is a process by which the kidneys can rid the blood of additional unwanted substances, such as creatinine and ammonia.

The reabsorbed solutes and water that move into the interstitial space between the nephrons need to be returned to the blood, or the kidneys will rapidly swell like balloons. The **peritubular capillaries** surrounding the renal tubule reclaim the reabsorbed substances and return them to general circulation. Peritubular capillaries arise from the efferent arteriole exiting the glomerulus and empty into the renal veins leaving the kidney.

ACTIVITY 1

The Effect of Arteriole Radius on Glomerular Filtration

OBJECTIVES

1. To understand the terms *nephron, glomerulus, glomerular capillaries, renal tubule, filtrate, Bowman's capsule, renal corpuscle, afferent arteriole, efferent arteriole, glomerular capillary pressure,* and *glomerular filtration rate.*
2. To understand how changes in afferent arteriole radius impact glomerular capillary pressure and filtration.
3. To understand how changes in efferent arteriole radius impact glomerular capillary pressure and filtration.

Introduction

Each of the million **nephrons** in each kidney contains two major parts: (1) a tubular component, the **renal tubule,** and (2) a vascular component, the **renal corpuscle** (view Figure 9.1). The **glomerulus** is a tangled capillary knot that filters fluid from the blood into the lumen of the renal tubule. The function of the renal tubule is to process the filtered fluid, also called the **filtrate.** The beginning of the renal tubule is an enlarged end called **Bowman's capsule** (or the glomerular capsule), which surrounds the glomerulus and serves to funnel the filtrate into the rest of the renal tubule. Collectively, the glomerulus and Bowman's capsule are called the renal corpuscle.

Two arterioles are associated with each glomerulus: an **afferent arteriole** feeds the **glomerular capillary** bed and an **efferent arteriole** drains it. These arterioles are responsible for blood flow through the glomerulus. The diameter of the efferent arteriole is smaller than the diameter of the afferent arteriole, restricting blood flow out of the glomerulus. Consequently, the pressure in the glomerular capillaries forces fluid through the endothelium of the capillaries into the lumen of the surrounding Bowman's capsule. In essence, everything in the blood except for the blood cells (red and white) and plasma proteins is filtered through the glomerular wall. From the Bowman's capsule, the filtrate moves into the rest of the renal tubule for processing. The job of the tubule is to reabsorb all the beneficial substances from its lumen and allow the wastes to travel down the tubule for elimination from the body.

During glomerular filtration, blood enters the glomerulus from the afferent arteriole and protein-free plasma flows from the blood across the walls of the glomerular capillaries and into the Bowman's capsule. The **glomerular filtration rate** is an index of kidney function. In humans, the filtration rate ranges from 80 to 140 ml/min, so that, in 24 hours, as much as 180 liters of filtrate is produced by the glomeruli. The filtrate formed is devoid of cellular debris, is essentially protein free, and contains a concentration of salts and organic molecules similar to that in blood.

The glomerular filtration rate can be altered by changing arteriole resistance or arteriole hydrostatic pressure. In this activity, you will explore the effect of arteriole radius on glomerular capillary pressure and filtration in a single nephron. You can apply the concepts you learn by studying a single nephron to understand the function of the kidney as a whole.

EQUIPMENT USED The following equipment will be depicted on-screen: source beaker for blood (first beaker on left side of screen)—simulates blood flow and pressure (mm Hg) from general circulation to the nephron; drain beaker for blood (second beaker on left side of screen)—simulates the renal vein; flow tube with adjustable radius—simulates the afferent arteriole and connects the blood supply to the glomerular capillaries; second flow tube with adjustable radius—simulates the efferent arteriole and drains the glomerular capillaries into the peritubular capillaries, which ultimately drain into the renal vein (drain beaker); simulated nephron (The filtrate forms in Bowman's capsule, flows through the renal tubule—the tubular components—and empties into a collecting duct, which in turn drains into the urinary bladder.); nephron tank; glomerulus—"ball" of capillaries that forms part of the filtration membrane; glomerular (Bowman's) capsule—forms part of the filtration membrane and a capsular space where the filtrate initially forms; proximal convoluted tubule; loop of Henle; distal convoluted tubule; collecting duct; drain beaker for filtrate (beaker on right side of screen)—simulates the urinary bladder.

Experiment Instructions

Go to the home page in the PhysioEx software and click **Exercise 9: Renal System Physiology.** Click **Activity 1: The Effect of Arteriole Radius on Glomerular Filtration,** and take the online **Pre-lab Quiz** for Activity 1.

After you take the online Pre-lab Quiz, click the **Experiment** tab and begin the experiment. The experiment instructions are reprinted here for your reference. The opening screen for the experiment is shown below.

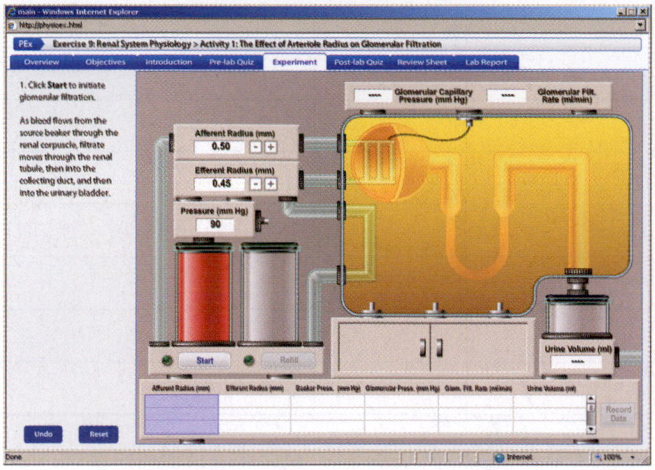
Exercise 9: Renal System Physiology > Activity 1: The Effect of Arteriole Radius on Glomerular Filtration
Overview
Objectives
Introduction
Pre-lab Quiz
Experiment
Post-lab Quiz
Review Sheet
Lab Report
1. Click Start to initiate glomerular filtration.
As blood flows from the source beaker through the renal corpuscle, filtrate moves through the renal tubule, then into the collecting duct, and then into the urinary bladder.
Glomerular Capillary Pressure (mm Hg)
Glomerular Filt. Rate (ml/min)
Afferent Radius (mm)
0.50
Efferent Radius (mm)
0.45
Pressure (mm Hg)
90
Start
Refill
Urine Volume (ml)
Undo
Reset
Record Data

1. Click **Start** to initiate glomerular filtration. As blood flows from the source beaker through the renal corpuscle, filtrate moves through the renal tubule, then into the collecting duct, and then into the urinary bladder.

2. The glomerular capillary pressure display shows the hydrostatic blood pressure in the glomerular capillaries that promotes filtration, and the filtration rate display shows the flow rate of the fluid moving from the lumen of the glomerular capillaries into the lumen of Bowman's capsule. Click **Record Data** to display your results in the grid (and record your results in Chart 1).

3. Click **Refill** to replenish the source beaker and prepare the nephron for the next run.

PREDICT Question 1
What will happen to the glomerular capillary pressure and filtration rate if you decrease the radius of the afferent arteriole?

____

4. Decrease the radius of the afferent arteriole to 0.45 mm by clicking the – button beside the afferent radius display. Click **Start** to initiate glomerular filtration.

5. Note the glomerular capillary pressure and glomerular filtration rate displays and click **Record Data** to display your results in the grid (and record your results in Chart 1).

6. Click **Refill** to replenish the source beaker and prepare the nephron for the next run.

CHART 1 Effect of Arteriole Radius on Glomerular Filtration

Afferent arteriole radius (mm)	Efferent arteriole radius (mm)	Glomerular capillary pressure (mm Hg)	Glomerular filtration rate (ml/min)

7. You will now observe the effect of incremental decreases in the radius of the afferent arteriole.

- Decrease the radius of the afferent arteriole by 0.05 mm by clicking the – button beside the afferent radius display.
- Click **Start** to initiate glomerular filtration.
- Note the glomerular capillary pressure and glomerular filtration rate displays and click **Record Data** to display your results in the grid (and record your results in Chart 1).
- Click **Refill** to replenish the source beaker and prepare the nephron for the next run.

Repeat this step until you reach an afferent arteriole radius of 0.35 mm.

? PREDICT Question 2
What will happen to the glomerular capillary pressure and filtration rate if you increase the radius of the afferent arteriole?

____

8. Increase the radius of the afferent arteriole to 0.55 mm by clicking the + button beside the afferent radius display. Click **Start** to initiate glomerular filtration.

9. Note the glomerular capillary pressure and glomerular filtration rate displays and click **Record Data** to display your results in the grid (and record your results in Chart 1).

10. Click **Refill** to replenish the source beaker and prepare the nephron for the next run.

11. Increase the radius of the afferent arteriole to 0.60 mm. Click **Start** to initiate glomerular filtration.

12. Note the glomerular capillary pressure and glomerular filtration rate displays and click **Record Data** to display your results in the grid (and record your results in Chart 1).

13. Click **Refill** to replenish the source beaker and prepare the nephron for the next run.

? PREDICT Question 3
What will happen to the glomerular capillary pressure and filtration rate if you decrease the radius of the efferent arteriole?

____

14. Decrease the radius of the afferent arteriole to 0.50 mm by clicking the – button beside the afferent radius display. Click **Start** to initiate glomerular filtration.

15. Note the glomerular capillary pressure and glomerular filtration rate displays and click **Record Data** to display your results in the grid (and record your results in Chart 1).

16. Click **Refill** to replenish the source beaker and prepare the nephron for the next run.

17. You will now observe the effect of incremental decreases in the radius of the efferent arteriole.

- Decrease the radius of the efferent arteriole by 0.05 mm by clicking the – button beside the efferent radius display.
- Click **Start** to initiate glomerular filtration.

- Note the glomerular capillary pressure and glomerular filtration rate displays and click **Record Data** to display your results in the grid (and record your results in Chart 1).
- Click **Refill** to replenish the source beaker and prepare the nephron for the next run.

Repeat this step until you reach an efferent arteriole radius of 0.30 mm.

After you complete the experiment, take the online **Post-lab Quiz** for Activity 1.

Activity Questions

1. Activation of sympathetic nerves that innervate the kidney leads to a decreased urine production. Knowing that fact, what do you think the sympathetic nerves do to the afferent arteriole?

___

___

2. How is this effect of the sympathetic nervous system beneficial? Could this effect become harmful if it goes on too long?

___

___

ACTIVITY 2

The Effect of Pressure on Glomerular Filtration

OBJECTIVES

1. To understand the terms *glomerulus, glomerular capillaries, renal tubule, filtrate, Starling forces, Bowman's capsule, renal corpuscle, afferent arteriole, efferent arteriole, glomerular capillary pressure*, and *glomerular filtration rate.*
2. To understand how changes in glomerular capillary pressure affect glomerular filtration rate.
3. To understand how changes in renal tubule pressure affect glomerular filtration rate.

Introduction

Cellular metabolism produces a complex mixture of waste products that must be eliminated from the body. This excretory function is performed by a combination of organs, most importantly, the paired kidneys. Each kidney consists of approximately one million nephrons, which carry out three crucial processes: (1) glomerular filtration, (2) tubular reabsorption, and (3) tubular secretion.

Both the blood pressure in the **glomerular capillaries** and the **filtrate** pressure in the **renal tubule** can have a significant impact on the **glomerular filtration rate.** During glomerular filtration, blood enters the **glomerulus** from the **afferent arteriole. Starling forces** (hydrostatic and osmotic pressure gradients) drive protein-free fluid between the blood in the glomerular capillaries and the filtrate in **Bowman's capsule.** The glomerular filtration rate is an index of kidney function. In humans, the filtration rate ranges from 80 to 140 ml/min, so that, in 24 hours, as much as 180 liters of filtrate is produced by the glomerular capillaries. The filtrate formed is devoid of blood cells, is essentially protein free, and contains a concentration of salts and organic molecules similar to that in blood.

Approximately 20% of the blood that enters the glomerular capillaries is normally filtered into Bowman's capsule, where it is then referred to as filtrate. The unusually high hydrostatic blood pressure in the glomerular capillaries promotes this filtration. Thus, the glomerular filtration rate can be altered by changing the afferent arteriole resistance (and, therefore, the hydrostatic pressure). In this activity you will explore the effect of blood pressure on the glomerular filtration rate in a single nephron. You can apply the concepts you learn by studying a single nephron to understand the function of the kidney as a whole.

EQUIPMENT USED The following equipment will be depicted on-screen: left source beaker (first beaker on left side of screen)—simulates blood flow and pressure (mm Hg) from general circulation to the nephron; drain beaker for blood (second beaker on left side of screen)—simulates the renal vein; flow tube with adjustable radius—simulates the afferent arteriole and connects the blood supply to the glomerular capillaries; second flow tube with adjustable radius—simulates the efferent arteriole and drains the glomerular capillaries into the peritubular capillaries, which ultimately drain into the renal vein (drain beaker); simulated nephron (The filtrate forms in Bowman's capsule, flows through the renal tubule—the tubular components—and empties into a collecting duct, which in turn drains into the urinary bladder.); nephron tank; glomerulus—"ball" of capillaries that forms part of the filtration membrane; glomerular (Bowman's) capsule—forms part of the filtration membrane and a capsular space where the filtrate initially forms; proximal convoluted tubule; loop of Henle; distal convoluted tubule; collecting duct; one-way valve between end of collecting tube (duct) and urinary bladder—used to restrict the flow of filtrate into the urinary bladder, increasing the volume and pressure in the renal tubule; drain beaker for filtrate (beaker on right side of screen)—simulates the urinary bladder.

Experiment Instructions

Go to the home page in the PhysioEx software and click **Exercise 9: Renal System Physiology.** Click **Activity 2: The Effect of Pressure on Glomerular Filtration,** and take the online **Pre-lab Quiz** for Activity 2.

After you take the online Pre-lab Quiz, click the **Experiment** tab and begin the experiment. The experiment instructions are reprinted here for your reference. The opening screen for the experiment is shown on the following page.

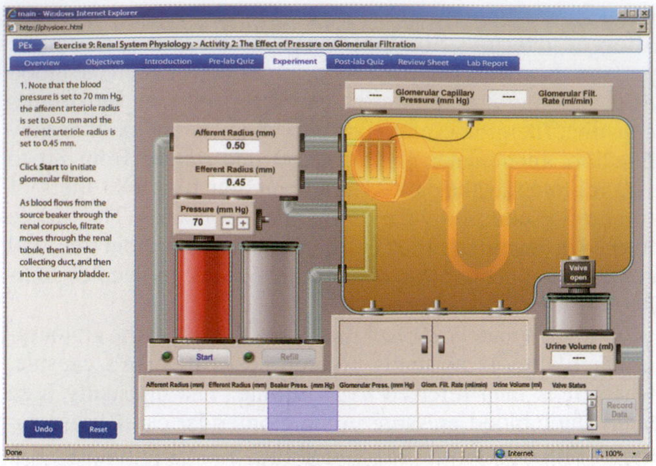
main - Windows Internet Explorer
http://physioex.html
PEx
Exercise 9: Renal System Physiology > Activity 2: The Effect of Pressure on Glomerular Filtration
Overview
Objectives
Introduction
Pre-lab Quiz
Experiment
Post-lab Quiz
Review Sheet
Lab Report
1. Note that the blood pressure is set to 70 mm Hg, the afferent arteriole radius is set to 0.50 mm and the efferent arteriole radius is set to 0.45 mm.
Click Start to initiate glomerular filtration.
As blood flows from the source beaker through the renal corpuscle, filtrate moves through the renal tubule, then into the collecting duct, and then into the urinary bladder.
Glomerular Capillary Pressure (mm Hg)
Glomerular Filt. Rate (ml/min)
Afferent Radius (mm)
0.50
Efferent Radius (mm)
0.45
Pressure (mm Hg)
70
Valve open
Start
Refill
Urine Volume (ml)
Afferent Radius (mm)
Efferent Radius (mm)
Beaker Press. (mm Hg)
Glomerular Press. (mm Hg)
Glom. Filt. Rate (ml/min)
Urine Volume (ml)
Valve Status
Record Data
Undo
Reset
Done
Internet
100%

1. Note that the blood pressure is set to 70 mm Hg, the afferent arteriole radius is set to 0.50 mm, and the efferent arteriole radius is set to 0.45 mm. Click **Start** to initiate glomerular filtration. As blood flows from the source beaker through the renal corpuscle, filtrate moves through the renal tubule, then into the collecting duct, and then into the urinary bladder.

2. The glomerular capillary pressure display shows the hydrostatic blood pressure in the glomerular capillaries that promotes filtration, and the filtration rate display shows the flow rate of the fluid moving from the lumen of the glomerular capillaries into the lumen of Bowman's capsule. Click **Record Data** to display your results in the grid (and record your results in Chart 2.)

3. Click **Refill** to replenish the source beaker and prepare the nephron for the next run.

PREDICT Question 1
What will happen to the glomerular capillary pressure and filtration rate if you increase the blood pressure in the left source beaker?

____

4. Increase the blood pressure to 80 mm Hg by clicking the + button beside the pressure display. Click **Start** to initiate glomerular filtration.

5. Note the glomerular capillary pressure and glomerular filtration rate displays and click **Record Data** to display your results in the grid (and record your results in Chart 2).

6. Click **Refill** to replenish the source beaker and prepare the nephron for the next run.

7. You will now observe the effect of further incremental increases in blood pressure.

- Increase the blood pressure by 10 mm Hg by clicking the + button beside the pressure display.
- Click **Start** to initiate glomerular filtration.
- Note the glomerular capillary pressure and glomerular filtration rate displays and click **Record Data** to display your results in the grid (and record your results in Chart 2).
- Click **Refill** to replenish the source beaker and prepare the nephron for the next run.

Repeat this step until you reach a blood pressure of 100 mm Hg.

PREDICT Question 2
What will happen to the filtrate pressure in Bowman's capsule (not directly measured in this experiment) and the filtration rate if you close the one-way valve between the collecting duct and the urinary bladder?

____

8. Note that the valve between the collecting duct and the urinary bladder is open. Decrease the blood pressure to 70 mm Hg by clicking the button beside the pressure display. Click **Start** to initiate glomerular filtration.

CHART 2 Effect of Arteriole Radius on Glomerular Filtration

Blood pressure (mm Hg)	Valve (open or closed)	Glomerular capillary pressure (mm Hg)	Glomerular filtration rate (ml/min)	Urine volume (ml)

9. Note the glomerular capillary pressure and glomerular filtration rate displays and click **Record Data** to display your results in the grid (and record your results in Chart 2).

10. Click **Refill** to replenish the source beaker and prepare the nephron for the next run.

11. Click the valve between the collecting duct and the urinary bladder to close it. Click **Start** to initiate glomerular filtration.

12. Note the glomerular capillary pressure and glomerular filtration rate displays and click **Record Data** to display your results in the grid (and record your results in Chart 2).

13. Click **Refill** to replenish the source beaker and prepare the nephron for the next run.

14. Increase the blood pressure to 100 mm Hg. Click **Start** to initiate glomerular filtration.

15. Note the glomerular capillary pressure and glomerular filtration rate displays and click **Record Data** to display your results in the grid (and record your results in Chart 2).

16. Click **Refill** to replenish the source beaker and prepare the nephron for the next run.

17. Click the valve between the collecting duct and the urinary bladder to open it. Click **Start** to initiate glomerular filtration.

18. Note the glomerular capillary pressure and glomerular filtration rate displays and click **Record Data** to display your results in the grid (and record your results in Chart 2).

After you complete the experiment, take the online **Post-lab Quiz** for Activity 2.

Activity Questions

1. Judging from the results in this laboratory activity, what *should be* the effect of blood pressure on glomerular filtration?

____

____

2. Persistent high blood pressure with inadequate glomerular filtration is now a frequent problem in Western cultures. Using the concepts in this activity, explain this health problem.

____

____

ACTIVITY 3

Renal Response to Altered Blood Pressure

OBJECTIVES

1. To understand the terms *nephron, renal tubule, filtrate, Bowman's capsule, blood pressure, afferent arteriole, efferent arteriole, glomerulus, glomerular filtration rate,* and *glomerular capillary pressure.*
2. To understand how blood pressure affects glomerular capillary pressure and glomerular filtration.
3. To observe which is more effective: changes in afferent or efferent arteriole radius when changes in blood pressure occur.

Introduction

In humans approximately 180 liters of filtrate flows into the **renal tubules** every day. As demonstrated in Activity 2, the **blood pressure** supplying the **nephron** can have a substantial impact on the **glomerular capillary pressure** and **glomerular filtration.** However, under most circumstances, glomerular capillary pressure and glomerular filtration remain relatively constant despite changes in blood pressure because the nephron has the capacity to alter its **afferent** and **efferent arteriole** radii.

During glomerular filtration, blood enters the **glomerulus** from the afferent arteriole. **Starling forces** (primarily hydrostatic pressure gradients) drive protein-free fluid out of the glomerular capillaries and into **Bowman's capsule.** Importantly for our body's homeostasis, a relatively constant glomerular filtration rate of 125 ml/min is maintained despite a wide range of blood pressures that occur throughout the day for an average human.

Activities 1 and 2 explored the independent effects of arteriole radii and blood pressure on glomerular capillary pressure and glomerular filtration. In the human body, these effects occur simultaneously. Therefore, in this activity, you will alter both variables to explore their combined effects on glomerular filtration and observe how changes in one variable can compensate for changes in the other to maintain an adequate glomerular filtration rate.

EQUIPMENT USED The following equipment will be depicted on-screen: left source beaker (first beaker on left side of screen)—simulates blood flow and pressure (mm Hg) from general circulation to the nephron; drain beaker for blood (second beaker on left side of screen)—simulates the renal vein; flow tube with adjustable radius—simulates the afferent arteriole and connects the blood supply to the glomerular capillaries; second flow tube with adjustable radius—simulates the efferent arteriole and drains the glomerular capillaries into the peritubular capillaries, which ultimately drain into the renal vein (drain beaker); simulated nephron (The filtrate forms in Bowman's capsule, flows through the renal tubule—the tubular components—and empties into a collecting duct, which in turn drains into the urinary bladder.); nephron tank; glomerulus—"ball" of capillaries that forms part of the filtration membrane; glomerular (Bowman's) capsule—forms part of the filtration membrane and a capsular space where the filtrate initially forms; proximal convoluted tubule; loop of Henle; distal convoluted tubule; collecting duct; one-way valve between end of collecting tube (duct) and urinary bladder—used to restrict the flow of filtrate into the urinary bladder, increasing the volume and pressure in the renal tubule; drain beaker for filtrate (beaker on right side of screen)—simulates the urinary bladder.

Experiment Instructions

Go to the home page in the PhysioEx software and click **Exercise 9: Renal System Physiology.** Click **Activity 3: Renal Response to Altered Blood Pressure,** and take the online **Pre-lab Quiz** for Activity 3.

After you take the online Pre-lab Quiz, click the **Experiment** tab and begin the experiment. The experiment instructions are reprinted here for your reference. The opening screen for the experiment is shown below.

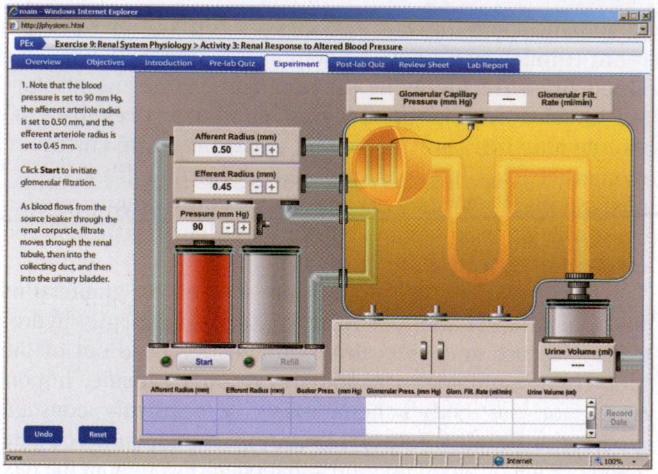
main - Windows Internet Explorer
http://physioex.html
PEx
Exercise 9: Renal System Physiology > Activity 3: Renal Response to Altered Blood Pressure
Overview
Objectives
Introduction
Pre-lab Quiz
Experiment
Post-lab Quiz
Review Sheet
Lab Report
1. Note that the blood pressure is set to 90 mm Hg, the afferent arteriole radius is set to 0.50 mm, and the efferent arteriole radius is set to 0.45 mm.
Click Start to initiate glomerular filtration.
As blood flows from the source beaker through the renal corpuscle, filtrate moves through the renal tubule, then into the collecting duct, and then into the urinary bladder.
Glomerular Capillary Pressure (mm Hg)
Glomerular Filt. Rate (ml/min)
Afferent Radius (mm)
0.50
Efferent Radius (mm)
0.45
Pressure (mm Hg)
90
Start
Refill
Urine Volume (ml)
Afferent Radius (mm)
Efferent Radius (mm)
Beaker Press. (mm Hg)
Glomerular Press. (mm Hg)
Glom. Filt. Rate (ml/min)
Urine Volume (ml)
Record Data
Undo
Reset
Done
Internet
100%

1. Note that the blood pressure is set to 90 mm Hg, the afferent arteriole radius is set to 0.50 mm, and the efferent arteriole radius is set to 0.45 mm. Click **Start** to initiate glomerular filtration. As blood flows from the source beaker through the renal corpuscle, filtrate moves through the renal tubule, then into the collecting duct, and then into the urinary bladder.

2. The glomerular capillary pressure display shows the hydrostatic blood pressure in the glomerular capillaries that promotes filtration, and the filtration rate display shows the flow rate of the fluid moving from the lumen of the glomerular capillaries into the lumen of Bowman's capsule. Click **Record Data** to display your results in the grid (and record your results in Chart 3).

3. Click **Refill** to replenish the source beaker and prepare the nephron for the next run.

4. You will now observe how the nephron might operate to keep the glomerular filtration rate relatively constant despite a large drop in blood pressure. Decrease the blood pressure to 70 mm Hg by clicking the – button beside the pressure display. Click **Start** to initiate glomerular filtration.

5. Note the glomerular capillary pressure and glomerular filtration rate displays and click **Record Data** to display your results in the grid (and record your results in Chart 3).

6. Click **Refill** to replenish the source beaker and prepare the nephron for the next run.

7. Increase the afferent arteriole radius to 0.60 mm by clicking the + button beside the afferent radius display. Click **Start** to initiate glomerular filtration.

8. Note the glomerular capillary pressure and glomerular filtration rate displays and click **Record Data** to display your results in the grid (and record your results in Chart 3).

9. Click **Refill** to replenish the source beaker and prepare the nephron for the next run.

10. Return the afferent arteriole radius to 0.50 mm by clicking the – button beside the afferent radius display and decrease the efferent radius to 0.35 mm by clicking the button beside the efferent radius display. Click **Start** to initiate glomerular filtration.

11. Note the glomerular capillary pressure and glomerular filtration rate displays and click **Record Data** to display your results in the grid (and record your results in Chart 3).

12. Click **Refill** to replenish the source beaker and prepare the nephron for the next run.

? PREDICT Question 1
What will happen to the glomerular capillary pressure and glomerular filtration rate if both of these arteriole radii changes are implemented simultaneously with the low blood pressure condition?

____

CHART 3 Renal Response to Altered Blood Pressure

Afferent arteriole radius (mm)	Efferent arteriole radius (mm)	Blood pressure (mm Hg)	Glomerular capillary pressure (mm Hg)	Glomerular filtration rate (ml/min)

13. Set the afferent arteriole radius to 0.60 mm and keep the efferent arteriole radius at 0.35 mm. Click **Start** to initiate glomerular filtration.

14. Note the glomerular capillary pressure and glomerular filtration rate displays and click **Record Data** to display your results in the grid (and record your results in Chart 3).

After you complete the experiment, take the online **Post-lab Quiz** for Activity 3.

Activity Questions

1. How could an increased urine volume be viewed as beneficial to the body?

___

___

2. Diuretics are frequently given to people with persistent high blood pressure. Why?

___

___

ACTIVITY 4

Solute Gradients and Their Impact on Urine Concentration

OBJECTIVES

1. To understand the terms *antidiuretic hormone (ADH), reabsorption, loop of Henle, collecting duct, tubule lumen, interstitial space,* and *peritubular capillaries.*
2. To explain the process of water reabsorption in specific regions of the nephron.
3. To understand the role of ADH in water reabsorption by the nephron.
4. To describe how the kidneys can produce urine that is four times more concentrated than the blood.

Introduction

As filtrate moves through the tubules of a nephron, solutes and water move *from* the **tubule lumen** *into* the **interstitial spaces** of the nephron. This movement of solutes and water relies on the total solute concentration gradient in the interstitial spaces surrounding the tubule lumen. The interstitial fluid is comprised mostly of NaCl and urea. When the nephron is permeable to solutes or water, equilibrium will be reached between the interstitial fluid and the tubular fluid contents.

Antidiuretic hormone (ADH) increases the water permeability of the **collecting duct,** allowing water to flow to areas of higher solute concentration, from the tubule lumen into the surrounding interstitial spaces. **Reabsorption** describes this movement of filtered solutes and water from the lumen of the renal tubules back into the plasma. The reabsorbed solutes and water that move into the interstitial space need to be returned to the blood, or the kidneys will rapidly swell like balloons. The **peritubular capillaries** surrounding the renal tubule reclaim the reabsorbed substances and return them to general circulation. Peritubular capillaries arise from the efferent arteriole exiting the glomerulus and empty into the renal veins leaving the kidney.

Without reabsorption, we would excrete the solutes and water that our bodies need to maintain homeostasis. In this activity you will examine the process of passive reabsorption that occurs while filtrate travels through a nephron and urine is formed. While completing the experiment, assume that when ADH is present, the conditions favor the formation of the most concentrated urine possible.

EQUIPMENT USED The following equipment will be depicted on-screen: simulated nephron surrounded by interstitial space between the nephron and peritubular capillaries (Reabsorbed solutes, such as glucose, will move from the lumen of the tubule into the interstitial space, and then into the peritubular capillaries that branch out from the efferent arteriole.); drain beaker for filtrate—simulates the urinary bladder; antidiuretic hormone (ADH).

Experiment Instructions

Go to the home page in the PhysioEx software and click **Exercise 9: Renal System Physiology.** Click **Activity 4: Solute Gradients and Their Impact on Urine Concentration,** and take the online **Pre-lab Quiz** for Activity 4.

After you take the online Pre-lab Quiz, click the **Experiment** tab and begin the experiment. The experiment instructions are reprinted here for your reference. The opening screen for the experiment is shown below.

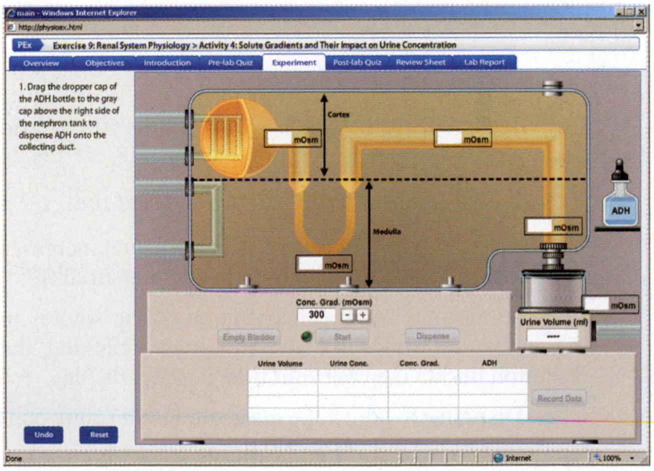
main - Windows Internet Explorer
http://physioex.html
PEx
Exercise 9: Renal System Physiology > Activity 4: Solute Gradients and Their Impact on Urine Concentration
Overview
Objectives
Introduction
Pre-lab Quiz
Experiment
Post-lab Quiz
Review Sheet
Lab Report
1. Drag the dropper cap of the ADH bottle to the gray cap above the right side of the nephron tank to dispense ADH onto the collecting duct.
Cortex
Medulla
mOsm
ADH
Conc. Grad. (mOsm)
300
Empty Bladder
Start
Dispense
Urine Volume (ml)
Urine Volume
Urine Conc.
Conc. Grad.
Record Data
Undo
Reset
Done
Internet
100%

1. Drag the dropper cap of the ADH bottle to the gray cap above the right side of the nephron tank to dispense ADH onto the collecting duct.

2. Click **Dispense** beneath the concentration gradient display to adjust the maximum total solute concentration in the interstitial fluid to 300 mOsm. Because the blood solute concentration is also 300 mOsm, there is no osmotic difference between the lumen of the tubule and the surrounding interstitial fluid.

3. Click **Start** to initiate filtration. Filtrate will flow through the nephron, and solutes and water will move out of the tubules into the interstitial space. Fluid will also move back

into the peritubular capillaries, thus completing the process of reabsorption.

4. Click **Record Data** to display your results in the grid (and record your results in Chart 4).

CHART 4 Solute Gradients and Their Impact on Urine Concentration

Urine volume (ml)	Urine concentration (mOsm)	Concentration gradient (mOsm)

5. Click **Empty Bladder** to prepare for the next run.

PREDICT Question 1
What will happen to the urine volume and concentration as the solute gradient in the interstitial space is increased?

____

6. Increase the maximum concentration of the solutes in the interstitial space to 600 mOsm by clicking the + button beside the concentration gradient display. Click **Dispense** to adjust the maximum total solute concentration in the interstitial fluid.

7. Click **Start** to initiate filtration.

8. Click **Record Data** to display your results in the grid (and record your results in Chart 4).

9. Click **Empty Bladder** to prepare for the next run.

10. You will now observe the effect of incremental increases in maximum total solute concentration in the interstitial fluid.

- Increase the maximum concentration of the solutes in the interstitial space by 300 mOsm by clicking the + button beside the concentration gradient display.
- Click **Dispense** to adjust the maximum total solute concentration in the interstitial fluid.
- Click **Start** to initiate filtration.
- Click **Record Data** to display your results in the grid (and record your results in Chart 4).
- Click **Empty Bladder** to prepare for the next run.

Repeat this step until you reach the maximum total solute concentration in the interstitial fluid of 1200 mOsm.

After you complete the experiment, take the online **Post-lab Quiz** for Activity 4.

Activity Questions

1. From what you learned in this activity, speculate on ways that desert rats are able to concentrate their urine significantly more than humans.

____

____

2. Judging from this activity, what would be a reasonable mechanism for diuretics?

____

____

ACTIVITY 5

Reabsorption of Glucose via Carrier Proteins

OBJECTIVES

1. To understand the terms *reabsorption, carrier proteins, apical membrane, secondary active transport, facilitated diffusion,* and *basolateral membrane.*
2. To understand the role that glucose carrier proteins play in removing glucose from the filtrate.
3. To understand the concept of a glucose carrier transport maximum and why glucose is not normally present in the urine.

Introduction

Reabsorption is the movement of filtered solutes and water from the lumen of the renal tubules back into the plasma. Without reabsorption, we would excrete the solutes and water that our bodies require for homeostasis.

Glucose is not very large and is therefore easily filtered out of the plasma into Bowman's capsule as part of the filtrate. To ensure that glucose is reabsorbed into the body so that it can fuel cellular metabolism, glucose **carrier proteins** are present in the proximal tubule cells of the nephron. There are a finite number of these glucose carriers in each renal tubule cell. Therefore, if too much glucose is present in the filtrate, it will not all be reabsorbed and glucose will be inappropriately excreted into the urine.

Glucose is first absorbed by **secondary active transport** at the **apical membrane** of proximal tubule cells and then it leaves the tubule cell via **facilitated diffusion** along the **basolateral membrane.** Both types of carrier proteins that transport these molecules across the tubule membranes are transmembrane proteins. Because carrier proteins are needed to move glucose from the lumen of the nephron into the interstitial spaces, there is a limit to the amount of glucose that can be reabsorbed. When all glucose carriers are bound with the glucose they are transporting, excess glucose in the filtrate is eliminated in urine.

In this activity, you will examine the effect of varying the number of glucose transport proteins in the *proximal convoluted tubule*. It is important to note that, normally, the number of glucose carriers is constant in a human kidney and that it is the plasma glucose that varies during the day. Plasma glucose will be held constant in this activity, and the number of glucose carriers will be varied.

EQUIPMENT USED The following equipment will be depicted on-screen: simulated nephron surrounded by interstitial space between the nephron and peritubular capillaries (Reabsorbed solutes, such as glucose, will move from the lumen of the tubule into the interstitial space, and then into the peritubular capillaries that branch out from the efferent arteriole.); drain beaker for filtrate—simulates the urinary bladder; glucose carrier protein control box—used to adjust the number of glucose carriers that will be inserted into the proximal tubule.

Experiment Instructions

Go to the home page in the PhysioEx software and click **Exercise 9: Renal System Physiology.** Click **Activity 5: Reabsorption of Glucose via Carrier Proteins,** and take the online **Pre-lab Quiz** for Activity 5.

After you take the online Pre-lab Quiz, click the **Experiment** tab and begin the experiment. The experiment instructions are reprinted here for your reference. The opening screen for the experiment is shown below.

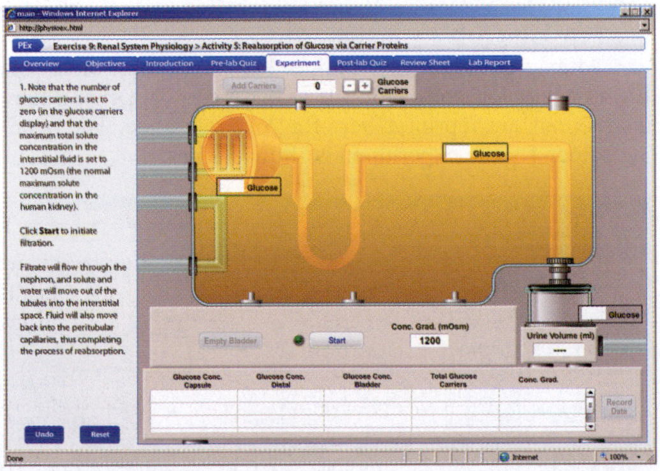

1. Note that the number of glucose carriers is set to zero (in the glucose carriers display) and that the maximum total solute concentration in the interstitial fluid is set to 1200 mOsm (the normal maximum solute concentration in the human kidney). Click **Start** to initiate filtration. Filtrate will flow through the nephron, and solute and water will move out of the tubules into the interstitial space. Fluid will also move back into the peritubular capillaries, thus completing the process of reabsorption.

2. Click **Record Data** to display your results in the grid (and record your results in Chart 5). The concentrations of glucose in Bowman's capsule, the distal convoluted tubule, and the urinary bladder will be displayed in the grid.

CHART 5 Reabsorption of Glucose via Carrier Proteins

Glucose concentration (m*M*)			
Bowman's capsule	Distal convoluted tubule	Urinary bladder	Glucose carriers

3. Click **Empty Bladder** to prepare the nephron for the next run.

PREDICT Question 1
What will happen to the glucose concentration in the urinary bladder as glucose carriers are added to the proximal tubule?

____

4. Increase the number of glucose carriers to 100 (an arbitrary number) by clicking the + button beside the glucose carriers display. Click **Add Carriers** to insert the specified number of glucose carrier proteins per unit area into the membrane of the proximal tubule.

5. Click **Start** to initiate filtration.

6. Click **Record Data** to display your results in the grid (and record your results in Chart 5).

7. Click **Empty Bladder** to prepare the nephron for the next run.

8. You will now observe the effect of incremental increases in the number of glucose carriers.

- Increase the number of glucose carriers by 100 by clicking the + button beside the glucose carriers display.
- Click **Add Carriers** to insert the specified number of glucose carrier proteins per unit area into the membrane of the proximal tubule.
- Click **Start** to initiate filtration.
- Click **Record Data** to display your results in the grid (and record your results in Chart 5).
- Click **Empty Bladder** to prepare the nephron for the next run.

Repeat this step until you have inserted 400 glucose carrier proteins per unit area into the membrane of the proximal tubule.

After you complete the experiment, take the online **Post-lab Quiz** for Activity 5.

Activity Questions

1. Why would your family physician at the turn of the twentieth century taste your urine?

____

____

ACTIVITY 6

The Effect of Hormones on Urine Formation

OBJECTIVES

1. To understand the terms *antidiuretic hormone (ADH), aldosterone, reabsorption, loop of Henle, distal convoluted tubule, collecting duct, tubule lumen,* and *interstitial space.*
2. To understand how the hormones aldosterone and ADH affect renal processes in a human kidney.
3. To understand the role of ADH in water reabsorption by the nephron.
4. To understand the role of aldosterone in solute reabsorption and secretion by the nephron.

Introduction

The concentration and volume of urine excreted by our kidneys will change depending on what our body needs for homeostasis. For example, if a person consumes a large quantity of water, the excess water will be eliminated as a large volume of dilute urine. On the other hand, when dehydration occurs, there is a clear benefit in being able to produce a small volume of concentrated urine to retain water. Activity 4 demonstrated how the total solute concentration gradient in the interstitial spaces surrounding the tubule lumen makes it possible to excrete concentrated urine.

Aldosterone is a hormone produced by the adrenal cortex under the control of the body's *renin-angiotensin system.* A decrease in blood pressure is detected by cells in the afferent arteriole, triggering the release of renin. Renin acts as a proteolytic enzyme, causing angiotensinogen to be converted into angiotensin I. Endothelial cells throughout the body possess a *converting enzyme* that converts angiotensin I into angiotensin II. Angiotensin II signals the adrenal cortex to secrete aldosterone. Aldosterone acts on the distal convoluted tubule cells in the nephron to promote the reabsorption of sodium from filtrate *into* the body and the secretion of potassium *from* the body. This electrolyte shift, coupled with the addition of **antidiuretic hormone (ADH),** also causes more water to be reabsorbed into the blood, resulting in increased blood pressure.

ADH is manufactured by the hypothalamus and stored in the posterior pituitary gland. ADH levels are influenced by the osmolality of body fluids and the volume and pressure of the cardiovascular system. A 1% change in body osmolality will cause this hormone to be secreted. The primary action of this hormone is to increase the permeability of the collecting duct to water so that more water is reabsorbed into the body by inserting aquaporins, or water channels, in the apical membrane. Without this water reabsorption, the body would quickly dehydrate.

Thus, our kidneys tightly regulate the amount of water and solutes excreted to maintain water balance in the body. If water intake is down, or if there has been a fluid loss from the body, the kidneys work to conserve water by making the urine very hyperosmotic (having a relatively high solute concentration) to the blood. If there has been a large intake of fluid, the urine is more hypo-osmotic. In the normal individual, urine osmolarity varies from 50 to 1200 milliosmoles/kg of water.

EQUIPMENT USED The following equipment will be depicted on-screen: simulated nephron surrounded by interstitial space between the nephron and peritubular capillaries (Reabsorbed solutes, such as glucose, will move from the lumen of the tubule into the interstitial space, and then into the peritubular capillaries that branch out from the efferent arteriole.); drain beaker for filtrate—simulates the urinary bladder; aldosterone; antidiuretic hormone (ADH).

Experiment Instructions

Go to the home page in the PhysioEx software and click **Exercise 9: Renal System Physiology.** Click **Activity 6: The Effect of Hormones on Urine Formation,** and take the online **Pre-lab Quiz** for Activity 6.

After you take the online Pre-lab Quiz, click the **Experiment** tab and begin the experiment. The experiment instructions are reprinted here for your reference. The opening screen for the experiment is shown below.

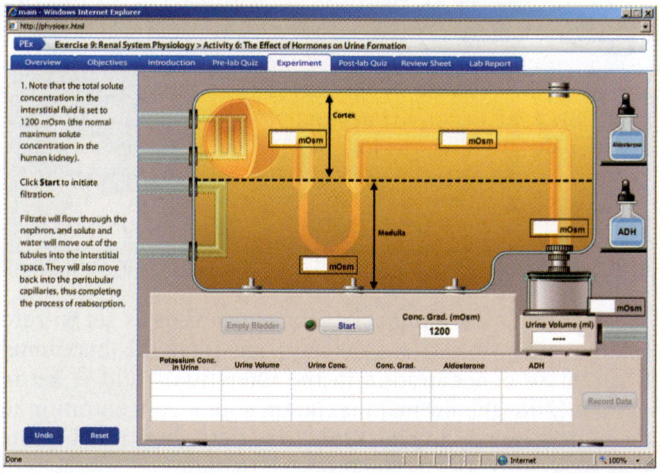
main - Windows Internet Explorer
Exercise 9: Renal System Physiology > Activity 6: The Effect of Hormones on Urine Formation
Overview
Objectives
Introduction
Pre-lab Quiz
Experiment
Post-lab Quiz
Review Sheet
Lab Report
1. Note that the total solute concentration in the interstitial fluid is set to 1200 mOsm (the normal maximum solute concentration in the human kidney).
Click Start to initiate filtration.
Filtrate will flow through the nephron, and solute and water will move out of the tubules into the interstitial space. They will also move back into the peritubular capillaries, thus completing the process of reabsorption.
Cortex
Medulla
mOsm
ADH
Empty Bladder
Start
Conc. Grad. (mOsm)
1200
Urine Volume (ml)
Potassium Conc. in Urine
Urine Volume
Urine Conc.
Conc. Grad.
Aldosterone
Record Data
Undo
Reset
Done
Internet

1. Note that the total solute concentration in the interstitial fluid is set to 1200 mOsm (the normal maximum solute concentration in the human kidney). Click **Start** to initiate filtration. Filtrate will flow through the nephron, and solute and water will move out of the tubules into the interstitial space. They will also move back into the peritubular capillaries, thus completing the process of reabsorption.

2. Click **Record Data** to display your results in the grid (and record your results in Chart 6). You will use this baseline data to compare the conditions of the filtrate and urine volume in the presence of the hormones aldosterone and ADH.

CHART 6 The Effect of Hormones on Urine Formation

Potassium concentration (m*M*)	Urine volume (ml)	Urine concentration (mOsm)	Aldosterone	ADH

3. Click **Empty Bladder** to prepare the nephron for the next run.

PREDICT Question 1
What will happen to the urine volume (compared with baseline) when aldosterone is added to the distal tubule?

____

4. Drag the dropper cap of the aldosterone bottle to the gray cap above the right side of the nephron tank to dispense aldosterone into the tank surrounding the distal tubule and the collecting duct.

5. Click **Start** to initiate filtration.

6. Click **Record Data** to display your results in the grid (and record your results in Chart 6).

7. Click **Empty Bladder** to prepare the nephron for the next run.

8. Drag the dropper cap of the ADH bottle to the gray cap at the top right side of the nephron tank to dispense ADH into the tank surrounding the distal tubule and the collecting duct.

PREDICT Question 2
What will happen to the urine volume (compared with baseline) when ADH is added to the collecting duct?

____

9. Click **Start** to initiate filtration.

10. Click **Record Data** to display your results in the grid (and record your results in Chart 6).

11. Click **Empty Bladder** to prepare the nephron for the next run.

PREDICT Question 3
What will happen to the urine volume and the urine concentration (compared with baseline) in the presence of both aldosterone and ADH?

____

12. Drag the dropper cap of the aldosterone bottle and then the dropper cap of the ADH bottle to the gray cap above the right side of the nephron tank to dispense aldosterone and ADH into the tank surrounding the distal convoluted tubule and the collecting duct.

13. Click **Start** to initiate filtration.

14. Click **Record Data** to display your results in the grid (and record your results in Chart 6).

After you complete the experiment, take the online **Post-lab Quiz** for Activity 6.

Activity Questions

1. Why does ethanol consumption lead to a dramatic increase in urine production?

____

____

2. Why do angiotensin converting enzyme (ACE) inhibitors given to people with hypertension lead to increased urine production?

____

____

PhysioEx 9.0

REVIEW SHEET

EXERCISE 9

NAME ______________________

LAB TIME/DATE ______________

Renal System Physiology

ACTIVITY 1 The Effect of Arteriole Radius on Glomerular Filtration

1. What are two primary functions of the kidney? ____

____

2. What are the components of the renal corpuscle? ____

____

3. Starting at the renal corpuscle, list the components of the renal tubule as they are encountered by filtrate. ____________

____

____

4. Describe the effect of decreasing the afferent arteriole radius on glomerular capillary pressure and filtration rate. How well did the results compare with your prediction? ____

____

5. Describe the effect of increasing the afferent arteriole radius on glomerular capillary pressure and filtration rate. How well did the results compare with your prediction? ____

____

6. Describe the effect of decreasing the efferent arteriole radius on glomerular capillary pressure and filtration rate. How well did the results compare with your prediction? ____

____

7. Describe the effect of increasing the efferent radius on glomerular capillary pressure and filtration rate. ____________

____

ACTIVITY 2 The Effect of Pressure on Glomerular Filtration

1. As blood pressure increased, what happened to the glomerular capillary pressure and the glomerular filtration rate? How well did the results compare with your prediction? ____

____

2. Compare the urine volume in your baseline data with the urine volume as you increased the blood pressure. How did the urine volume change? ______________________________

____

3. How could the change in urine volume with the increase in blood pressure be viewed as being beneficial to the body?

____

____

4. When the one-way valve between the collecting duct and the urinary bladder was closed, what happened to the filtrate pressure in Bowman's capsule (this is not directly measured in this experiment) and the glomerular filtration rate? How well did the results compare with your prediction? ______________________________

____

5. How did increasing the blood pressure alter the results when the valve was closed? ______________________________

____

ACTIVITY 3 Renal Response to Altered Blood Pressure

1. List the several mechanisms you have explored that change the glomerular filtration rate. How does each mechanism specifically alter the glomerular filtration rate? ______________________________

____

2. Describe and explain what happened to the glomerular capillary pressure and glomerular filtration rate when *both* arteriole radii changes were implemented simultaneously with the low blood pressure condition. How well did the results compare with your prediction? ______________________________

____

____

3. How could you adjust the afferent or efferent radius to compensate for the effect of reduced blood pressure on the glomerular filtration rate? ______________________________

____

4. Which arteriole radius adjustment was more effective at compensating for the effect of low blood pressure on the glomerular filtration rate? Explain why you think this difference occurs. ______________________________

____

5. In the body, how does a nephron maintain a near-constant glomerular filtration rate despite a constantly fluctuating blood pressure? ______________________________

____

ACTIVITY 4 Solute Gradients and Their Impact on Urine Concentration

1. What happened to the urine concentration as the solute concentration in the interstitial space was increased? How well did the results compare to your prediction? ______________________________

2. What happened to the volume of urine as the solute concentration in the interstitial space was increased? How well did the results compare to your prediction? ______________________________

3. What do you think would happen to urine volume if you did not add ADH to the collecting duct? ______________________________

4. Is most of the tubule filtrate reabsorbed into the body or excreted in urine? Explain. ______________________________

5. Can the reabsorption of solutes influence water reabsorption from the tubule fluid? Explain. ______________________________

ACTIVITY 5 Reabsorption of Glucose via Carrier Proteins

1. What happens to the concentration of glucose in the urinary bladder as the number of glucose carriers increases? ______________________________

2. What types of transport are utilized during glucose reabsorption and where do they occur? ______________________________

3. Why does the glucose concentration in the urinary bladder become zero in these experiments? ______________________________

4. A person with type 1 diabetes cannot make insulin in the pancreas, and a person with untreated type 2 diabetes does not respond to the insulin that is made in the pancreas. In either case, why would you expect to find glucose in the person's urine?

____

ACTIVITY 6 The Effect of Hormones on Urine Formation

1. How did the addition of aldosterone affect urine volume (compared with baseline)? Can the reabsorption of solutes influence water reabsorption in the nephron? Explain. How well did the results compare with your prediction? ______

______

______

2. How did the addition of ADH affect urine volume (compared with baseline)? How well did the results compare with your prediction? Why did the addition of ADH also affect the concentration of potassium in the urine (compared with baseline)?

______

______

3. What is the principal determinant for the release of aldosterone from the adrenal cortex? ______

______

4. How did the addition of both aldosterone and ADH affect urine volume (compared with baseline)? How well did the results compare with your prediction? ______

______

______

5. What is the principal determinant for the release of ADH from the posterior pituitary gland? Does ADH favor the formation of dilute or concentrated urine? Explain why. ______

______

______

6. Which hormone (aldosterone or ADH) has the greater effect on urine volume? Why? ______

______

______

7. If ADH is not available, can the urine concentration still vary? Explain your answer. ______

______

______

8. Consider this situation: you want to reabsorb sodium ions but you do not want to increase the volume of the blood by reabsorbing large amounts of water from the filtrate. Assuming that aldosterone and ADH are both present, how would you adjust the hormones to accomplish the task? ______

______

PhysioEx 9.0

EXERCISE 10

Acid-Base Balance

PRE-LAB QUIZ

1. Circle the correct term. A substance that dissolves in water to release hydrogen ions is a(n) acid / base.
2. The measurement of the inverse logarithm of the hydrogen ion concentration is ________.
   a. buffering capacity
   b. pH
   c. neutralization
   d. acidity
3. Which of the following is *not* a regulatory mechanism for acid/base balance in the body?
   a. the kidneys
   b. protein buffers
   c. the respiratory system
   d. the digestive system
4. Circle True or False. The normal pH range of the blood is alkaline.
5. Name the gas that we exhale to maintain blood pH homeostasis. ________
6. Circle the correct term. The weak acid that dissociates into H^+ and HCO_3^- is sulfuric acid / carbonic acid.
7. When the pH of the blood is <7.35, we refer to this condition as ________.
   a. alkalosis
   b. acidosis
   c. homeostasis
   d. metabolism
8. Circle True or False. The kidneys excrete carbon dioxide gas to return pH levels to a normal range.
9. Acid-base imbalances can have metabolic or ________ causes.

Exercise Overview

pH denotes the hydrogen ion concentration, $[H^+]$, in a solution (such as body fluids). The reciprocal relationship between pH and $[H^+]$ is defined by the following equation.

$$pH = \log(1/[H^+])$$

Because the relationship is reciprocal, $[H^+]$ is higher at *lower* pH values (indicating higher acid levels) and lower at *higher* pH values (indicating lower acid levels).

The pH of a body's fluid is also referred to as its **acid-base balance.** An **acid** is a substance that releases H^+ in solution. A **base,** often a hydroxyl ion (OH^-) or bicarbonate ion (HCO_3^-), is a substance that binds, or buffers, the H^+. A **strong acid** completely dissociates in solution, releasing all of its hydrogen ions and, thus, lowering the solution's pH. A **weak acid** dissociates incompletely and does not release all of its hydrogen ions in solution, producing a lesser effect on the solution's pH. A **strong base** has a strong tendency to bind to H^+, raising the solution's pH. A **weak base** binds less of the H^+, producing a lesser effect on the solution's pH.

The pH of body fluids is very tightly regulated. Blood and tissue fluids normally have a pH between 7.35 and 7.45. Under pathological conditions, blood pH as low as 6.9 or as high as 7.8 has been recorded, but a higher or lower pH cannot sustain human life. The narrow range from 7.35 to 7.45 is remarkable when you consider the vast number of biochemical reactions that take place in the body. The human body normally produces a large amount of H^+ as the result of metabolic processes; ingested acids; and the products of fat, sugar, and amino acid metabolism. The regulation of a relatively constant internal pH is one of the major physiological functions of the body's organ systems.

To maintain pH homeostasis, the body utilizes both *chemical* and *physiological* buffering systems. Chemical buffers are composed of a mixture of weak acids and weak bases. They help regulate the body's pH levels by binding H^+ and removing it from solution as its concentration begins to rise or by releasing H^+ into solution as its concentration begins to fall. The body's three major chemical buffering systems are the *bicarbonate, phosphate,* and *protein buffer systems.* We will not focus on chemical buffering systems in this exercise, but keep in mind that chemical buffers are the fastest form of compensation and can return pH to normal within a fraction of a second.

The body's two major physiological buffering systems are the **renal system** and the **respiratory system.** The renal system is the slower of the two, taking hours to days to do its work. The respiratory system usually works within minutes, but cannot handle the amount of pH change that the renal system can. These physiological buffer systems help regulate body pH by controlling the output of acids, bases, or carbon dioxide (CO_2) from the body. For example, if there is too much acid in the body, the renal system may respond by excreting more H^+ from the body in urine. Similarly, if there is too much carbon dioxide in the blood, the respiratory system may respond by increasing ventilation to expel the excess carbon dioxide. Carbon dioxide levels have a direct effect on pH because the addition of carbon dioxide to the blood results in the generation of more H^+. The following equation shows what happens when carbon dioxide combines with water in the blood, producing carbonic acid.

$$H_2O + CO_2 \rightleftarrows \underset{\text{carbonic acid}}{H_2CO_3} \rightleftarrows H^+ + \underset{\text{bicarbonate ion}}{HCO_3^-}$$

ACTIVITY 1

Hyperventilation

OBJECTIVES

1. To introduce pH homeostasis in the body.
2. To understand the normal ranges for pH and P_{CO_2}.
3. To recognize respiratory alkalosis and its causes.
4. To interpret an oscilloscope tracing for hyperventilation and compare it with a tracing for normal breathing.

Introduction

Acid-base imbalances can have respiratory and metabolic causes. When diagnosing these disorders, two key signs are evaluated: the pH and the partial pressure of carbon dioxide in the blood (**P_{CO_2}**). The normal range for pH is between 7.35 and 7.45, and the normal range for P_{CO_2} is between 35 and 45 mm Hg. When the pH falls below 7.35, the body is said to be in a state of **acidosis.** When the pH rises above 7.45, the body is said to be in a state of **alkalosis.**

Respiratory alkalosis is the condition of too little carbon dioxide in the blood. Respiratory alkalosis commonly results from traveling to high altitude (where the air contains less oxygen) or hyperventilation, which can be brought on by fever, panic attack, or anxiety. Hyperventilation, defined as an increase in the rate and depth of breathing, removes carbon dioxide from the blood faster than it is being produced by the cells of the body, reducing the amount of H^+ in the blood and, thus, increasing the blood's pH. The following equation shows the shift in the equilibrium that results in the increase in blood pH due to less carbon dioxide in the blood.

$$H_2O + \mathbf{CO_2} \leftarrow \underset{\text{carbonic acid}}{H_2CO_3} \leftarrow \mathbf{H^+} + \underset{\text{bicarbonate ion}}{HCO_3^-}$$

The renal system can compensate for alkalosis by retaining H^+ and excreting bicarbonate ions to lower the blood pH levels back to the normal range.

EQUIPMENT USED The following equipment will be depicted on-screen: simulated lung chamber; pH meter; oscilloscope; two breathing patterns: normal and hyperventilation.

Experiment Instructions

Go to the home page in the PhysioEx software and click **Exercise 10: Acid-Base Balance.** Click **Activity 1: Hyperventilation,** and take the online **Pre-lab Quiz** for Activity 1.

After you take the online Pre-lab Quiz, click the **Experiment** tab and begin the experiment. The experiment instructions are reprinted here for your reference. The opening screen for the experiment is shown below.

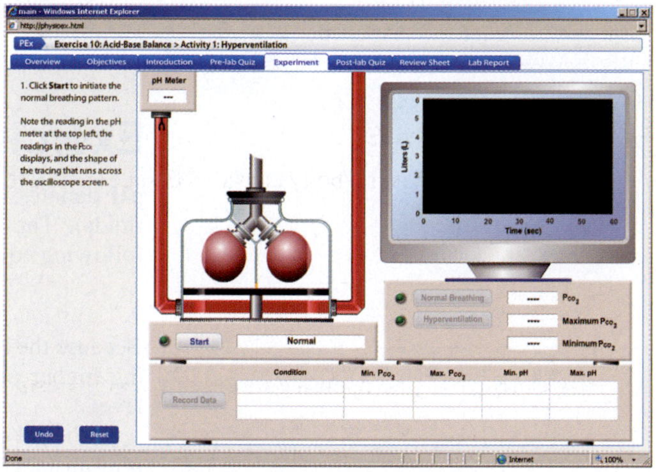
main - Windows Internet Explorer
Exercise 10: Acid-Base Balance > Activity 1: Hyperventilation
Overview
Objectives
Introduction
Pre-lab Quiz
Experiment
Post-lab Quiz
Review Sheet
Lab Report
1. Click Start to initiate the normal breathing pattern.
Note the reading in the pH meter at the top left, the readings in the Pco2 displays, and the shape of the tracing that runs across the oscilloscope screen.
pH Meter
Liters (L)
Time (sec)
Normal Breathing
Hyperventilation
Pco2
Maximum Pco2
Minimum Pco2
Start
Normal
Condition
Min. Pco2
Max. Pco2
Min. pH
Max. pH
Record Data
Undo
Reset

1. Click **Start** to initiate the normal breathing pattern. Note the reading in the pH meter at the top left, the readings in the P_{CO_2} displays, and the shape of the tracing that runs across the oscilloscope screen.

2. Click **Record Data** to display your results in the grid (and record your results in Chart 1).

PREDICT Question 1
What do you think will happen to the pH and P_{CO_2} levels with hyperventilation?

____

CHART 1 Hyperventilation Breathing Patterns

Condition	Minimum P_{CO_2}	Maximum P_{CO_2}	Minimum pH	Maximum pH

3. Click **Start** to initiate the normal breathing pattern. After the normal breathing tracing runs for 10 seconds, click **Hyperventilation** to initiate the hyperventilation breathing pattern. Note the reading in the pH meter at the top left, the readings in the P_{CO_2} displays, and the shape of the tracing that runs across the oscilloscope screen.

4. Click **Record Data** to display your results in the grid (and record your results in Chart 1).

5. Click **Start** to initiate the normal breathing pattern. After the normal breathing tracing runs for 10 seconds, click **Hyperventilation** to initiate the hyperventilation breathing pattern. After the hyperventilation tracing runs for 10 seconds, click **Normal Breathing** to return to the normal breathing pattern. Note the reading in the pH meter at the top left, the readings in the P_{CO_2} displays, and the shape of the tracing that runs across the oscilloscope screen.

6. Click **Record Data** to display your results in the grid (and record your results in Chart 1).

After you complete the experiment, take the online **Post-lab Quiz** for Activity 1.

Activity Questions

1. At what pH range is the body considered to be in a state of respiratory alkalosis?

____

____

2. How can the body compensate for respiratory alkalosis?

____

____

3. How did the tidal volume change with hyperventilation?

____

____

4. What might cause a person to hyperventilate?

____

____

ACTIVITY 2

Rebreathing

OBJECTIVES

1. To understand how rebreathing can simulate hypoventilation.
2. To observe the results of respiratory acidosis.
3. To describe the causes of respiratory acidosis.

Introduction

The body is said to be in a state of **acidosis** when the pH of the blood falls below 7.35 (although a pH of 7.35 is technically not acidic). Respiratory acidosis is the result of impaired respiration, or *hypoventilation*, which leads to the accumulation of too much carbon dioxide in the blood. The causes of impaired respiration include airway obstruction, depression of the respiratory center in the brain stem, lung disease (such as emphysema and chronic bronchitis), and drug overdose.

Recall that carbon dioxide contributes to the formation of carbonic acid when it combines with water through a reversible reaction catalyzed by carbonic anhydrase. The carbonic acid then dissociates into hydrogen ions and bicarbonate ions. Because hypoventilation results in elevated carbon dioxide levels in the blood, the equilibrium shifts, the H^+ levels increase, and the pH value of the blood decreases.

$$H_2O + \mathbf{CO_2} \rightarrow \underset{\text{carbonic acid}}{H_2CO_3} \rightarrow \mathbf{H^+} + \underset{\text{bicarbonate ion}}{HCO_3^-}$$

Rebreathing is the action of breathing in air that was just expelled from the lungs. Rebreathing results in the accumulation of carbon dioxide in the blood. Breathing into a paper bag is an example of rebreathing. (Note that breathing into a paper bag can deplete the body of oxygen and is therefore not the best therapy for hyperventilation because it can mask other life-threatening emergencies, such as a heart attack or asthma.) In this activity, you will observe what happens to pH and carbon dioxide levels in the blood during rebreathing. In the body, the kidneys regulate the acid-base balance by altering the amount of H^+ and HCO_3^- excreted in the urine.

EQUIPMENT USED The following equipment will be depicted on-screen: simulated lung chamber; pH meter; oscilloscope; two breathing patterns: normal and rebreathing.

Experiment Instructions

Go to the home page in the PhysioEx software and click **Exercise 10: Acid-Base Balance.** Click **Activity 2: Rebreathing,** and take the online **Pre-lab Quiz** for Activity 2.

After you take the online Pre-lab Quiz, click the **Experiment** tab and begin the experiment. The experiment instructions are reprinted here for your reference. The opening screen for the experiment is shown below.

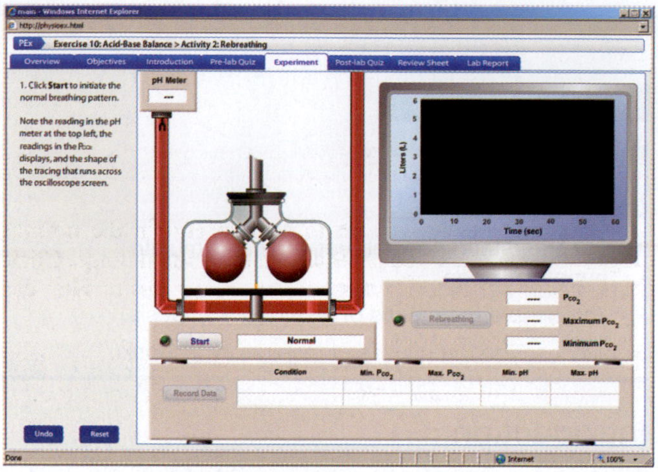
Exercise 10: Acid-Base Balance > Activity 2: Rebreathing
Overview
Objectives
Introduction
Pre-lab Quiz
Experiment
Post-lab Quiz
Review Sheet
Lab Report
1. Click Start to initiate the normal breathing pattern.
Note the reading in the pH meter at the top left, the readings in the P_{CO_2} displays, and the shape of the tracing that runs across the oscilloscope screen.
pH Meter
Liters (L)
Time (sec)
P_{CO_2}
Maximum P_{CO_2}
Minimum P_{CO_2}
Rebreathing
Start
Normal
Condition
Min. P_{CO_2}
Max. P_{CO_2}
Min. pH
Max. pH
Record Data
Undo
Reset

1. Click **Start** to initiate the normal breathing pattern. Note the reading in the pH meter at the top left, the readings in the P_{CO_2} displays, and the shape of the tracing that runs across the oscilloscope screen.
2. Click **Record Data** to display your results in the grid (and record your results in Chart 2).

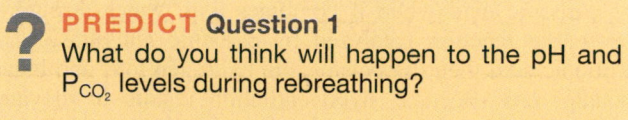

> **? PREDICT Question 1**
> What do you think will happen to the pH and P_{CO_2} levels during rebreathing?
>
> ____

3. Click **Start** to initiate the normal breathing pattern. After the normal breathing tracing runs for 10 seconds, click **Rebreathing** to initiate the rebreathing pattern. Note the reading in the pH meter at the top left, the readings in the P_{CO_2} displays, and the shape of the tracing that runs across the oscilloscope screen.
4. Click **Record Data** to display your results in the grid (and record your results in Chart 2).

After you complete the experiment, take the online **Post-lab Quiz** for Activity 1.

Activity Questions

1. Did the pH level of the blood change at all with rebreathing? If so, how did it change?

____

____

2. What happens to the pH level of the blood when there is too much carbon dioxide remaining in the blood?

____

____

3. How did the tidal volumes change with rebreathing?

____

____

4. Describe two ways in which too much carbon dioxide might remain in the blood.

____

____

ACTIVITY 3

Renal Responses to Respiratory Acidosis and Respiratory Alkalosis

OBJECTIVES

1. To understand renal compensation mechanisms for respiratory acidosis and respiratory alkalosis.
2. To explore the functional unit of the kidneys that responds to acid-base balance.
3. To observe the changes in ion concentrations that occur with renal compensation.

Introduction

The kidneys play a major role in maintaining fluid, electrolyte, and acid-base balance in the body's internal environment. By regulating the amount of water lost in the urine, the kidneys defend the body against excessive hydration or dehydration. By regulating the acidity of urine and the rate of electrolyte excretion, the kidneys maintain plasma pH and electrolyte levels within normal limits.

CHART 2 Normal Breathing Patterns

Condition	Minimum P_{CO_2}	Maximum P_{CO_2}	Minimum pH	Maximum pH

Renal compensation is the body's primary method of compensating for conditions of respiratory acidosis or respiratory alkalosis. The kidneys regulate the acid-base balance by altering the amount of H^+ and HCO_3^- excreted in the urine. If we revisit the equation for the dissociation of carbonic acid, a weak acid, we see that the conservation of bicarbonate ion (base) has the same net effect as the loss of acid, H^+.

$$H_2O + CO_2 \rightleftarrows \underset{\text{carbonic acid}}{H_2CO_3} \rightleftarrows H^+ + \underset{\text{bicarbonate ion}}{HCO_3^-}$$

In this activity you will examine how the renal system compensates for respiratory acidosis or respiratory alkalosis. Respiratory acidosis is generally caused by the accumulation of carbon dioxide in the blood from hypoventilation, but it can also be caused by rebreathing. Acidosis results in a lower-than-normal blood pH. Respiratory alkalosis is caused by a depletion of carbon dioxide, often caused by an episode of hyperventilation, and results in an elevated blood pH.

You will primarily be working with the variable P_{CO_2}. Recall that the normal range for pH is between 7.35 and 7.45 and the normal range for P_{CO_2} is between 35 and 45 mm Hg. You will observe how increases and decreases in P_{CO_2} affect the levels of H^+ and HCO_3^- that the kidneys excrete in urine. The functional unit for adjusting the plasma composition is the **nephron.** Remember that although the renal system can partially compensate for pH imbalances with a respiratory cause, the kidneys cannot fully compensate if respirations have not returned to normal because the carbon dioxide levels will still be abnormal.

EQUIPMENT USED The following equipment will be depicted on-screen: source beaker for blood (first beaker on left side of screen); drain beaker for blood (second beaker on left side of screen); simulated nephron (The filtrate forms in Bowman's capsule and flows through the renal tubule—the tubular components, and empties into a collecting duct which, in turn drains into the urinary bladder.); nephron tank; glomerulus—"ball" of capillaries that forms part of the filtration membrane; glomerular (Bowman's) capsule—forms part of the filtration membrane and a capsular space where the filtrate initially forms; proximal convoluted tubule; loop of Henle; distal convoluted tubule; collecting duct; drain beaker for filtrate (beaker on right side of screen)—simulates the urinary bladder.

Experiment Instructions

Go to the home page in the PhysioEx software and click **Exercise 10: Acid-Base Balance.** Click **Activity 3: Renal Responses to Respiratory Acidosis and Respiratory Alkalosis** and take the online **Pre-lab Quiz** for Activity 3.

After you take the online Pre-lab Quiz, click the **Experiment** tab and begin the experiment. The experiment instructions are reprinted here for your reference. The opening screen for the experiment is shown below.

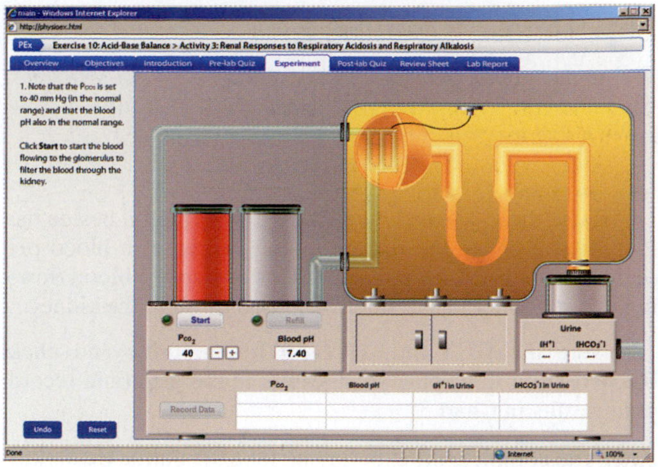
Exercise 10: Acid-Base Balance > Activity 3: Renal Responses to Respiratory Acidosis and Respiratory Alkalosis
Overview
Objectives
Introduction
Pre-lab Quiz
Experiment
Post-lab Quiz
Review Sheet
Lab Report
Start
Refill
Blood pH
7.40
40
Urine
Record Data
Undo
Reset

1. Note that the P_{CO_2} is set to 40 mm Hg (in the normal range) and that the blood pH is also in the normal range. Click **Start** to start the blood flowing to the glomerulus to filter the blood through the kidney.

2. Note the $[H^+]$ and $[HCO_3^-]$ in the urine and click **Record Data** to display your results in the grid (and record your results in Chart 3).

CHART 3 Renal Responses to Respiratory Acidosis and Respiratory Alkalosis

P_{CO_2}	Blood pH	$[H^+]$ in urine	$[HCO_3^-]$ in urine

3. Click **Refill** to replenish the source beaker.

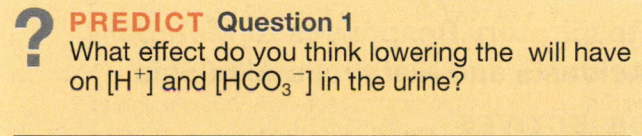

PREDICT Question 1
What effect do you think lowering the will have on $[H^+]$ and $[HCO_3^-]$ in the urine?

____

4. Lower the P_{CO_2} to 30 by clicking the – button beside the P_{CO_2} display. Note the corresponding increase in blood pH (above the normal range). Click **Start** to start the blood flowing to the glomerulus to filter the blood through the kidney.

5. Note the $[H^+]$ and $[HCO_3^-]$ in the urine and click **Record Data** to display your results in the grid (and record your results in Chart 3).

6. Click **Refill** to replenish the source beaker.

PREDICT Question 2
What effect do you think raising the P_{CO_2} will have on $[H^+]$ and $[HCO_3^-]$ in the urine?

7. Raise the P_{CO_2} to 60 by clicking the + button beside the P_{CO_2} display. Note the corresponding decrease in blood pH (below the normal range). Click **Start** to start the blood flowing to the glomerulus to filter the blood through the kidney.

8. Note the $[H^+]$ and $[HCO_3^-]$ in the urine and click **Record Data** to display your results in the grid (and record your results in Chart 3).

After you complete the experiment, take the online **Post-lab Quiz** for Activity 3.

Activity Questions

1. Describe how the kidneys respond to respiratory acidosis.

2. What P_{CO_2} corresponded to respiratory acidosis?

3. Describe how the kidneys respond to respiratory alkalosis.

4. What P_{CO_2} corresponded to respiratory alkalosis?

ACTIVITY 4

Respiratory Responses to Metabolic Acidosis and Metabolic Alkalosis

OBJECTIVES

1. To understand the causes of metabolic acidosis and metabolic alkalosis.
2. To observe the physiological changes that occur with an increase and decrease in metabolic rate.
3. To explain how the respiratory system compensates for metabolic acidosis and alkalosis.

Introduction

Conditions of acidosis and alkalosis that do not have respiratory causes are termed *metabolic acidosis and metabolic alkalosis.* **Metabolic acidosis** is characterized by low plasma HCO_3^- and pH. The causes of metabolic acidosis include:

- **Ketoacidosis,** a buildup of keto acids that can result from diabetes mellitus
- **Salicylate poisoning,** a toxic condition resulting from ingestion of too much aspirin or oil of wintergreen (a substance often found in laboratories)
- The ingestion of too much alcohol, which metabolizes into acetic acid
- Diarrhea, which results in the loss of bicarbonate with the elimination of intestinal contents
- Strenuous exercise, which can cause a buildup of lactic acid from anaerobic muscle metabolism

Metabolic alkalosis is characterized by elevated plasma HCO_3^- and pH. The causes of metabolic alkalosis include:

- Ingestion of alkali, such as antacids or bicarbonate
- Vomiting, which can result in the loss of too much H^+
- Constipation, which may result in significant reabsorption of HCO_3^-

Increases or decreases in the body's normal metabolic rate can also result in metabolic acidosis or alkalosis. Recall that carbon dioxide—a waste product of metabolism—mixes with water in plasma to form carbonic acid, which in turn forms H^+.

$$H_2O + CO_2 \rightleftarrows \underset{\text{carbonic acid}}{H_2CO_3} \rightleftarrows H^+ + \underset{\text{bicarbonate ion}}{HCO_3^-}$$

An increase in the normal metabolic rate causes more carbon dioxide to form as a metabolic waste product, resulting in the formation of more H^+ and, therefore, lower plasma pH, potentially causing acidosis. Other acids that are also normal metabolic waste products (such as ketone bodies and phosphoric, uric, and lactic acids) would likewise accumulate with an increase in metabolic rate.

Conversely, a decrease in the normal metabolic rate causes less carbon dioxide to form as a metabolic waste product, resulting in the formation of less H^+ and, therefore, higher plasma pH, potentially causing alkalosis. Many factors can affect the rate of cell metabolism. For example, fever, stress, or the ingestion of food all cause the rate of cell metabolism to *increase*. Conversely, a fall in body temperature or a decrease in food intake causes the rate of cell metabolism to *decrease*.

The respiratory system compensates for metabolic acidosis or alkalosis by expelling or retaining carbon dioxide in the blood. During metabolic acidosis, respiration increases to expel carbon dioxide from the blood, thus decreasing $[H^+]$ and raising the pH. During metabolic alkalosis, respiration decreases to promote the accumulation of carbon dioxide in the blood, thus increasing $[H^+]$ and decreasing the pH.

The renal system also compensates for metabolic acidosis and alkalosis by conserving or excreting bicarbonate ions. Nevertheless, in this activity, you will focus on respiratory compensation of metabolic acidosis and alkalosis.

EQUIPMENT USED The following equipment will be depicted on-screen: simulated heart pump; simulated lung chamber; oscilloscope.

Experiment Instructions

Go to the home page in the PhysioEx software and click **Exercise 10: Acid-Base Balance.** Click **Activity 4: Respiratory Responses to Metabolic Acidosis and Metabolic Alkalosis,** and take the online **Pre-lab Quiz** for Activity 4.

After you take the online Pre-lab Quiz, click the **Experiment** tab and begin the experiment. The experiment instructions are reprinted here for your reference. The opening screen for the experiment is shown below.

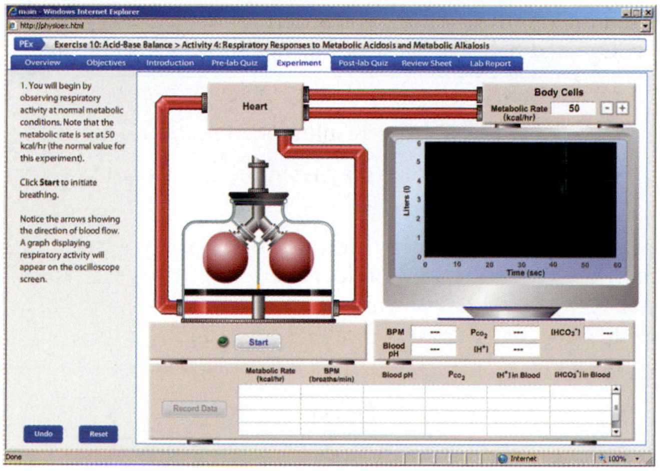
Exercise 10: Acid-Base Balance > Activity 4: Respiratory Responses to Metabolic Acidosis and Metabolic Alkalosis
Overview
Objectives
Introduction
Pre-lab Quiz
Experiment
Post-lab Quiz
Review Sheet
Lab Report
Heart
Body Cells
Metabolic Rate (kcal/hr)
50
Start
Record Data
Undo
Reset

1. You will begin by observing respiratory activity at normal metabolic conditions. Note that the metabolic rate is set at 50 kcal/hr (the normal value for this experiment). Click **Start** to initiate breathing and blood flow. Notice the arrows showing the direction of blood flow. A graph displaying respiratory activity will appear on the oscilloscope screen.

2. Note the data in the displays below the oscilloscope screen and click **Record Data** to display your results in the grid (and record your results in Chart 4).

3. Increase the metabolic rate to 60 kcal/hr by clicking the + button beside the metabolic rate display. Click **Start** to initiate breathing and blood flow.

4. Note the data in the displays below the oscilloscope screen and click **Record Data** to display your results in the grid (and record your results in Chart 4).

5. Click **Clear Tracings** to clear the tracings on the oscilloscope.

? PREDICT Question 1
What do you think will happen when the metabolic rate is increased to 80 kcal/hr?

____

6. Increase the metabolic rate to 80 kcal/hr by clicking the + button beside the metabolic rate display. Click **Start** to initiate breathing and blood flow.

7. Note the data in the displays below the oscilloscope screen and click **Record Data** to display your results in the grid (and record your results in Chart 4).

8. Click **Clear Tracings** to clear the tracings on the oscilloscope.

9. Decrease the metabolic rate to 40 kcal/hr by clicking the − button beside the metabolic rate display. Click **Start** to initiate breathing and blood flow.

10. Note the data in the displays below the oscilloscope screen and click **Record Data** to display your results in the grid (and record your results in Chart 4).

11. Click **Clear Tracings** to clear the tracings on the oscilloscope.

? PREDICT Question 2
What do you think will happen when the metabolic rate is decreased to 20 kcal/hr?

____

CHART 4 Respiratory Responses to Metabolic Acidosis and Metabolic Alkalosis

Metabolic rate	BPM (breaths/min)	Blood pH	P_{CO_2}	$[H^+]$ in blood	$[HCO_3^-]$ in blood

12. Decrease the metabolic rate to 20 kcal/hr by clicking the button beside the metabolic rate display. Click **Start** to initiate breathing and blood flow.

13. Note the data in the displays below the oscilloscope screen and click **Record Data** to display your results in the grid (and record your results in Chart 4).

After you complete the experiment, take the online **Post-lab Quiz** for Activity 4.

Activity Questions

1. Describe what happens to carbon dioxide and pH with increased metabolism.

2. Describe the respiratory response to metabolic acidosis.

3. When the respiratory system compensates for the metabolic acidosis, does the pH increase or decrease in value?

4. Describe the respiratory response to metabolic alkalosis.

PhysioEx 9.0

REVIEW SHEET 10

EXERCISE

NAME ______________________

LAB TIME/DATE ______________________

Acid-Base Balance

ACTIVITY 1 Hyperventilation

1. Describe the normal ranges for pH and carbon dioxide in the blood. ______________________

____

____

2. Describe what happened to the pH and the carbon dioxide levels with hyperventilation. How well did the results compare with your prediction? ______________________

____

____

3. Explain how returning to normal breathing after hyperventilation differed from hyperventilation without returning to normal breathing. ______________________

____

____

4. Describe some possible causes of respiratory alkalosis. ______________________

____

____

ACTIVITY 2 Rebreathing

1. Describe what happened to the pH and the carbon dioxide levels during rebreathing. How well did the results compare with your prediction? ______________________

____

____

2. Describe some possible causes of respiratory acidosis. ______________________

____

____

3. Explain how the renal system would compensate for respiratory acidosis. ______

______

______

ACTIVITY 3 Renal Responses to Respiratory Acidosis and Respiratory Alkalosis

1. Describe what happened to the concentration of ions in the urine when the P_{CO_2} was lowered. How well did the results compare with your prediction? ______

______

______

2. What condition was simulated when the P_{CO_2} was lowered? ______

______

______

3. Describe what happened to the concentration of ions in the urine when the P_{CO_2} was raised. How well did the results compare with your prediction? ______

______

______

4. What condition was simulated when the P_{CO_2} was raised? ______

______

______

ACTIVITY 4 Respiratory Responses to Metabolic Acidosis and Metabolic Alkalosis

1. Describe what happened to the blood pH when the metabolic rate was increased to 80 kcal/hr. What body system was compensating? How well did the results compare with your prediction? ______

______

______

2. List and describe some possible causes of metabolic acidosis. ______

______

______

3. Describe what happened to the blood pH when the metabolic rate was decreased to 20 kcal/hr. What body system was compensating? How well did the results compare with your prediction? ______________________________

____

____

4. List and describe some possible causes of metabolic alkalosis. ______________________________

____

____

PhysioEx 9.0 EXERCISE 11

Blood Analysis

PRE-LAB QUIZ

1. Three types of formed elements found in blood include erythrocytes, leukocytes, and ________.
   a. electrolytes
   b. fibers
   c. platelets
   d. sodium salts
2. Circle the correct term. Mature erythrocytes / leukocytes are the most numerous blood cells and do not have a nucleus.
3. The least numerous but largest of all agranulocytes is the
   a. basophil
   b. lymphocyte
   c. monocyte
   d. neutrophil
4. ________ are the leukocytes responsible for releasing histamine and other mediators of inflammation.
   a. Basophils
   b. Eosinophils
   c. Monocytes
   d. Neutrophils
5. Circle the correct term. When determining the hematocrit / hemoglobin, you will centrifuge whole blood in order to allow the formed elements to sink to the bottom of the sample.
6. Circle the correct term. Blood typing is based on the presence of proteins known as antigens / antibodies on the outer surface of the red blood cell plasma membrane.
7. Circle True or False. If an individual is transfused with the wrong type blood, the recipient's antibodies react with the donor's antigens, eventually clumping and hemolyzing the donated RBCs.

Exercise Overview

Blood transports soluble substances to and from all cells of the body. Laboratory analysis of our blood can reveal important information about how well this function is being achieved. The five activities in this exercise simulate common laboratory tests performed on blood: (1) *hematocrit* determination, (2) *erythrocyte sedimentation rate*, (3) *hemoglobin* determination, (4) *blood typing,* and (5) total *cholesterol* determination.

Hematocrit refers to the percentage of red blood cells (RBCs), or erythrocytes, in a sample of whole blood. A hematocrit of 48 means that 48% of the volume of blood consists of RBCs. RBCs transport oxygen to the cells of the body. Therefore, the higher the hematocrit, the more RBCs are present in the blood and the greater the oxygen-carrying potential of the blood. Males usually have higher hematocrit levels than females because males have higher levels of testosterone. In addition to promoting the male sex characteristics, testosterone is responsible for stimulating the release of erythropoietin from the kidneys. Erythropoietin (EPO) is a hormone that stimulates the synthesis of RBCs. Therefore, higher levels of testosterone lead to more EPO secretion and, thus, higher hematocrit levels.

The **erythrocyte sedimentation rate (ESR)** measures the settling of RBCs in a vertical, stationary tube of blood during one hour. In a healthy individual, RBCs do not settle very much in an hour. In some disease conditions, increased production of fibrinogen and immunoglobulins causes the RBCs to clump together, stack up, and form a column (called a *rouleaux formation*). RBCs in a rouleaux formation are heavier and settle faster (that is, they display an increase in the sedimentation rate.)

Hemoglobin (Hb), a protein found in RBCs, is necessary for the transport of oxygen from the lungs to the cells of the body. Four polypeptide chains of amino acids comprise the globin part of the molecule. Each polypeptide chain has a heme unit—a group of atoms that includes an atom of iron to which a molecule of oxygen binds. Each polypeptide chain, if it folds correctly, can bind a molecule of oxygen. Therefore, each hemoglobin molecule can carry four molecules of oxygen. Oxygen combined with hemoglobin forms oxyhemoglobin, which has a bright red color.

All of the cells in the human body, including RBCs, are surrounded by a plasma membrane that contains genetically determined glycoproteins, called antigens. On RBC membranes, there are certain antigens, called **agglutinogens,** that determine a person's blood type. Blood typing is used to identify the **ABO blood groups,** which are determined by the presence or absence of two antigens: **type A** and **type B.** Because these antigens are genetically determined, a person has two copies (alleles) of the gene for these antigens, one copy from each parent.

Cholesterol is a lipid substance that is essential for life—it is an important component of all cell membranes and is the base molecule of steroid hormones, vitamin D, and bile salts. Cholesterol is produced in the human liver and is present in some foods of animal origin, such as milk, meat, and eggs. Because cholesterol is a hydrophobic lipid, it needs to be wrapped in protein packages, called **lipoproteins,** to travel in the blood (which is mostly water) from the liver and digestive organs to the cells of the body.

ACTIVITY 1

Hematocrit Determination

OBJECTIVES

1. To understand the terms *hematocrit, red blood cells, hemoglobin, buffy coat, anemia,* and *polycythemia.*
2. To understand how the hematocrit (packed red blood cell volume) is determined.
3. To understand the implications of elevated or decreased hematocrit.
4. To understand the importance of proper disposal of laboratory material that comes in contact with blood.

Introduction

Hematocrit refers to the percentage of **red blood cells (RBCs),** or erythrocytes, in a sample of whole blood. A hematocrit of 48 means that 48% of the volume of blood consists of RBCs. RBCs transport oxygen to the cells of the body. Therefore, the higher the hematocrit, the more RBCs are present in the blood and the higher the oxygen-carrying potential of the blood. Hematocrit values are determined by spinning a microcapillary tube filled with a sample of whole blood in a special microhematocrit centrifuge. This procedure separates the blood cells from the blood plasma. A **buffy coat** layer of white blood cells (WBCs) appears as a thin, white layer *between* the heavier RBC layer and the lighter, yellow plasma.

The hematocrit is determined after centrifuging by measuring the height of the RBC layer (in millimeters) and dividing that by the height of the total blood sample (in millimeters). This calculation gives the percentage of the total blood volume consisting of RBCs. The average hematocrit for males is 42–52%, and the average hematocrit for females is 37–47%. A lower-than-normal hematocrit indicates **anemia,** and a higher-than-normal hematocrit indicates **polycythemia.**

Anemia is a condition in which insufficient oxygen is transported to the body's cells. There are many possible causes for anemia, including inadequate numbers of RBCs, a decreased amount of the oxygen-carrying pigment **hemoglobin** in the RBCs, and abnormally shaped hemoglobin. The heme portion of a hemoglobin molecule contains an atom of iron to which a molecule of oxygen can bind. If adequate iron is not available, the body cannot manufacture hemoglobin, resulting in the condition *iron-deficiency anemia. Aplastic anemia* results from the failure of the bone marrow to produce adequate red blood cell numbers. *Sickle cell anemia* is an inherited condition in which the protein portion of hemoglobin molecules folds incorrectly when oxygen levels are low. As a result, oxygen molecules cannot bind to the misshapen hemoglobin, the RBCs develop a sickle shape, and anemia results. Regardless of the underlying cause, anemia causes a reduction in the blood's ability to transport oxygen to the cells of the body.

Polycythemia refers to an increase in RBCs, resulting in a higher-than-normal hematocrit. There are many possible causes of polycythemia, including living at high altitudes, strenuous athletic training, and tumors in the bone marrow. In this activity you will simulate the blood test used to determine hematocrit.

EQUIPMENT USED The following equipment will be depicted on-screen: six heparinized capillary tubes (heparin keeps blood from clotting); blood samples from six individuals: sample 1: a healthy male living in Boston, sample 2: a healthy female living in Boston, sample 3: a healthy male living in Denver, sample 4: a healthy female living in Denver, sample 5: a male with aplastic anemia, sample 6: a female with iron-deficiency anemia; capillary tube sealer—a clay material (shown as an orange-yellow substance) used to seal the capillary tubes on one end so the blood sample can be centrifuged without having the blood spray out of the tube; microhematocrit centrifuge—used to centrifuge the samples (rotates at 14,500 revolutions per minute); metric ruler; biohazardous waste disposal—used to properly dispose of equipment that comes in contact with blood.

Experiment Instructions

Go to the home page in the PhysioEx software and click **Exercise 11: Blood Analysis.** Click **Activity 1: Hematocrit Determination,** and take the online **Pre-lab Quiz** for Activity 1.

After you take the online Pre-lab Quiz, click the **Experiment** tab and begin the experiment. The experiment instructions are reprinted here for your reference. The opening screen for the experiment is shown below.

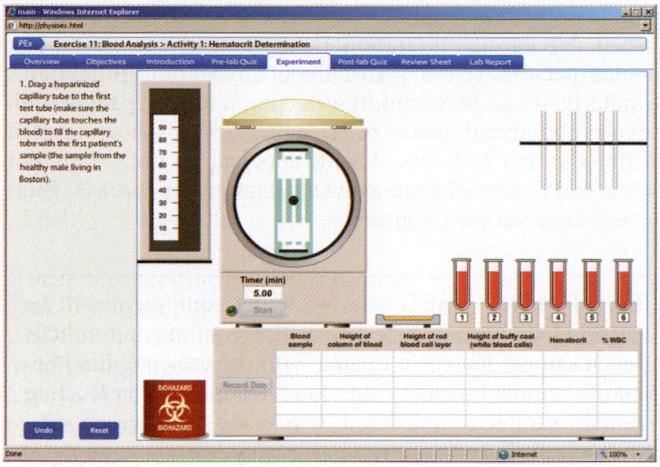
main - Windows Internet Explorer
http://physioex.html
PEx
Exercise 11: Blood Analysis > Activity 1: Hematocrit Determination
Overview
Objectives
Introduction
Pre-lab Quiz
Experiment
Post-lab Quiz
Review Sheet
Lab Report
1. Drag a heparinized capillary tube to the first test tube (make sure the capillary tube touches the blood) to fill the capillary tube with the first patient's sample (the sample from the healthy male living in Boston).
90
80
70
60
50
40
30
20
10
Timer (min)
5.00
Start
1
2
3
4
5
6
Blood sample
Height of column of blood
Height of red blood cell layer
Height of buffy coat (white blood cells)
Hematocrit
% WBC
Record Data
BIOHAZARD
BIOHAZARD
Undo
Reset
Done
Internet
100%

1. Drag a heparinized capillary tube to the first test tube (make sure the capillary tube touches the blood) to fill the capillary tube with the first patient's sample (the sample from the healthy male living in Boston).

2. Drag the capillary tube containing sample 1 to the container of capillary tube sealer to seal one end of the tube.

3. Drag the capillary tube to the microhematocrit centrifuge. The remaining samples will automatically be prepared for centrifugation.

4. Note that the timer is set to 5 minutes. Click **Start** to centrifuge the samples for 5 minutes at 14,500 revolutions per minute. The simulation compresses the 5-minute time period into 5 seconds of real time.

5. Drag capillary tube 1 from the centrifuge to the metric ruler to measure the height of the column of blood and the height of each layer.

6. Click **Record Data** to display your results in the grid (and record your results in Chart 1).

7. Drag capillary tube 1 to the biohazardous waste disposal.

> **?** **PREDICT Question 1**
> Predict how the hematocrits of the patients living in Denver, Colorado (approximately one mile above sea level), will compare with the hematocrit levels of the patients living in Boston, Massachusetts (at sea level).
>
> ____

8. You will now measure the column and layer heights of the remaining samples.

- Drag the next capillary tube from the centrifuge to the metric ruler.
- Click **Record data** to display your results in the grid (and record your results in Chart 1). The tube will automatically be placed in the biohazardous waste disposal.

Repeat this step for each of the remaining samples.

After you complete the experiment, take the online **Post-lab Quiz** for Activity 1.

CHART 1 Hematocrit Determination

	Total height of column of blood (mm)	Height of red blood cell layer (mm)	Height of buffy coat (mm)	Hematocrit	% WBC
Sample 1 (healthy male living in Boston)					
Sample 2 (healthy female living in Boston)					
Sample 3 (healthy male living in Denver)					
Sample 4 (healthy female living in Denver)					
Sample 5 (male with aplastic anemia)					
Sample 6 (female with iron-deficiency anemia)					

Activity Questions

1. How do you calculate the hematocrit after you centrifuge the total blood sample? What does the result of this calculation indicate?

____

____

2. What is the significance of the "buffy coat" after you centrifuge the total blood sample?

____

____

3. As noted in the Exercise Overview, the average hematocrit for males is 42–52%, the average hematocrit for females is 37–47%, and erythropoietin is a hormone that is responsible for the synthesis of RBCs. Given this information, explain how a female could have a consistent hematocrit of 48, large, well-defined skeletal muscles, and an abnormally deep voice.

____

____

____

ACTIVITY 2

Erythrocyte Sedimentation Rate

OBJECTIVES

1. To understand *erythrocyte sedimentation rate (ESR), red blood cells (RBCs),* and *rouleaux formation.*
2. To learn how to perform an erythrocyte sedimentation rate blood test.
3. To understand the results (and their implications) from an erythrocyte sedimentation rate blood test.
4. To understand the importance of proper disposal of laboratory material that comes in contact with blood.

Introduction

The **erythrocyte sedimentation rate (ESR)** measures the settling of **red blood cells (RBCs)** in a vertical, stationary tube of whole blood during one hour. In a healthy individual, red blood cells do not settle very much in an hour. In some disease conditions, increased production of fibrinogen and immunoglobulins cause the RBCs to clump together, stack up, and form a dark red column (called a **rouleaux formation**). RBCs in a rouleaux formation are heavier and settle faster (that is, they exhibit an increase in the settling rate).

The ESR is neither very specific nor diagnostic, but it can be used to follow the progression of certain diseases, including sickle cell anemia, some cancers, and inflammatory diseases, such as rheumatoid arthritis. When the disease worsens, the ESR increases. When the disease improves, the ESR decreases.

The ESR can be elevated in iron-deficiency anemia, and menstruating females sometimes develop anemia and show an increase in ESR. The ESR can also be used to evaluate a patient with chest pains because the ESR is elevated in established myocardial infarction (heart attack) but normal in angina pectoris (chest pain without myocardial infarction). Similarly, it can be useful in screening a female patient with severe abdominal pains because the ESR is not elevated within the first 24 hours of acute appendicitis but is elevated in the early stage of acute pelvic inflammatory disease (PID) or ruptured ectopic pregnancy.

EQUIPMENT USED The following equipment will be depicted on-screen: blood samples from six individuals (each sample has been treated with the anticoagulant heparin): sample 1: healthy individual, sample 2: menstruating female, sample 3: individual with sickle cell anemia, sample 4: individual with iron-deficiency anemia, sample 5: individual suffering a myocardial infarction, sample 6: individual with angina pectoris; sodium citrate—used to bind with calcium and prevent the blood samples from clotting so they can be easily poured into the narrow sedimentation rate tubes; test tubes—used as reaction vessels for the tests; sedimentation tubes (contained in cabinet); magnifying chamber—used to help read the millimeter markings on the sedimentation tubes; biohazardous waste disposal—used to properly dispose of equipment that comes in contact with blood.

Experiment Instructions

Go to the home page in the PhysioEx software and click **Exercise 11: Blood Analysis.** Click **Activity 2: Erythrocyte Sedimentation Rate,** and take the online **Pre-lab Quiz** for Activity 2.

After you take the online Pre-lab Quiz, click the **Experiment** tab and begin the experiment. The experiment instructions are reprinted here for your reference. The opening screen for the experiment is shown below.

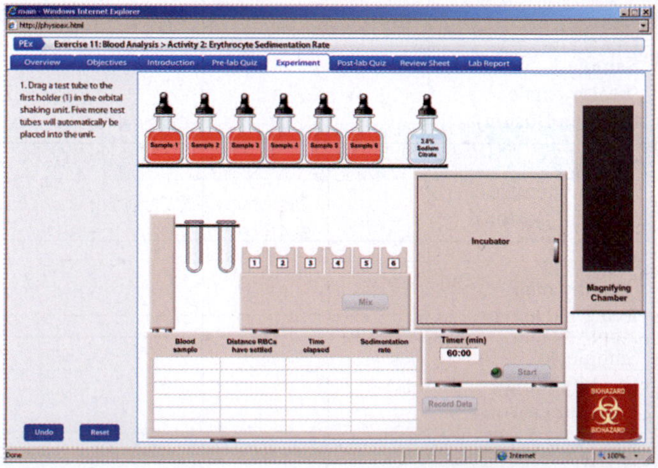
main - Windows Internet Explorer
PEx
Exercise 11: Blood Analysis > Activity 2: Erythrocyte Sedimentation Rate
Overview
Objectives
Introduction
Pre-lab Quiz
Experiment
Post-lab Quiz
Review Sheet
Lab Report
1. Drag a test tube to the first holder (1) in the orbital shaking unit. Five more test tubes will automatically be placed into the unit.
Sample 1
Sample 2
Sample 3
Sample 4
Sample 5
Sample 6
3.8% Sodium Citrate
Incubator
Magnifying Chamber
Mix
Blood sample
Distance RBCs have settled
Time elapsed
Sedimentation rate
Timer (min)
60:00
Start
Record Data
BIOHAZARD
Undo
Reset

1. Drag a test tube to the first holder (1) in the orbital shaking unit. Five more test tubes will automatically be placed into the unit.

2. Drag the dropper cap of the sample 1 bottle (the sample from the healthy individual) to the first test tube (1) in the orbital shaking unit to dispense one milliliter of blood into the tube. The remaining five samples will be automatically dispensed.

3. Drag the dropper cap of the 3.8% sodium citrate bottle to the first test tube to dispense 0.5 milliliters of sodium citrate into each of the tubes.

4. Click **Mix** to mix the samples.

5. Drag the first test tube to the first sedimentation tube in the incubator to pour the contents of the test tube into the sedimentation tube.

6. Drag the now empty test tube to the biohazardous waste disposal. The contents of the remaining test tubes will automatically be poured into the sedimentation tubes, and the empty tubes will automatically be placed in the biohazardous waste disposal.

7. Note that the timer is set to 60 minutes. Click **Start** to incubate the sedimentation tubes for 60 minutes. The simulation compresses the 60-minute time period into 6 seconds of real time.

8. Drag the first sedimentation tube to the magnifying chamber to examine the tube. The tube is marked in millimeters (the distance between two marks is 5 mm).

9. Click **Record Data** to display your results in the grid (and record your results in Chart 2).

10. Drag the sedimentation tube to the biohazardous waste disposal.

? PREDICT Question 1
How will the sedimentation rate for sample 6 (unhealthy individual) compare with the sedimentation rate for sample 1 (healthy individual)?

____

11. You will now measure the sedimentation rate for the remaining samples.

- Drag the next sedimentation tube to the magnifying chamber to examine the tube.
- Click **Record Data** to display your results in the grid (and record your results in Chart 2). The tube will automatically be placed in the biohazardous waste disposal.

Repeat this step for each of the remaining samples.

After you complete the experiment, take the online **Post-lab Quiz** for Activity 2.

Activity Questions

1. Why is ESR useful, even though it is neither specific nor sensitive?

____

____

2. Describe the physical process underlying an accelerated erythrocyte sedimentation rate.

____

____

ACTIVITY 3

Hemoglobin Determination

OBJECTIVES

1. To understand the terms *hemoglobin (Hb), anemia, heme, oxyhemoglobin,* and *hemoglobinometer.*
2. To learn how to determine the amount of hemoglobin in a blood sample.
3. To understand the results and their implications when examining the amounts of hemoglobin present in a blood sample.
4. To understand the importance of proper disposal of laboratory material that comes in contact with blood.

CHART 2 Erythrocyte Sedimentation Rate

Blood sample	Distance RBCs have settled (mm)	Elapsed time	Sedimentation rate
Sample 1 (healthy individual)			
Sample 2 (menstruating female)			
Sample 3 (individual with sickle cell anemia)			
Sample 4 (individual with iron-deficiency anemia)			
Sample 5 (individual suffering a myocardial infarction)			
Sample 6 (individual with angina pectoris)			

Introduction

Hemoglobin (Hb), a protein found in red blood cells, is necessary for the transport of oxygen from the lungs to the cells of the body. Four polypeptide chains of amino acids comprise the globin part of the molecule. Each polypeptide chain has a **heme** unit—a group of atoms that includes an atom of iron to which a molecule of oxygen binds. Each polypeptide chain, if it folds correctly, can bind a molecule of oxygen. Therefore, each hemoglobin molecule can carry four molecules of oxygen. Oxygen combined with hemoglobin forms **oxyhemoglobin,** which has a bright-red color. Anemia results when insufficient oxygen is carried in the blood.

A quantitative hemoglobin measurement is used to determine the classification and possible causes of anemia and also gives useful information on some other disease conditions. For example, a person can have anemia with a normal red blood cell count if there is inadequate hemoglobin in the red blood cells. Normal blood contains an average of 12–18 grams of hemoglobin per 100 milliliters of blood. A healthy male has 13.5–18 g/100 ml and a healthy female has 12–16 g/100 ml. Hemoglobin levels increase in patients with polycythemia, congestive heart failure, and chronic obstructive pulmonary disease (COPD). Hemoglobin levels also increase when dwelling at high altitudes. Hemoglobin levels decrease in patients with anemia, hyperthyroidism, cirrhosis of the liver, renal disease, systemic lupus erythematosus, and severe hemorrhage.

The hemoglobin level of a blood sample is determined by stirring the blood with a wooden stick to rupture, or lyse, the red blood cells. The color intensity of the hemolyzed blood reflects the amount of hemoglobin present. A **hemoglobinometer** transmits green light through the hemolyzed blood sample and then compares the amount of light that passes through the sample to standard color intensities to determine the hemoglobin content of the sample.

EQUIPMENT USED The following equipment will be depicted on-screen: blood samples from five individuals: sample 1: healthy male, sample 2: healthy female, sample 3: female with iron-deficiency anemia, sample 4: male with polycythemia, sample 5: female Olympic athlete; hemolysis sticks—used to stir the blood samples to lyse the red blood cells, thereby releasing their hemoglobin; blood chamber dispenser—used to dispense a blood chamber slide with a depression for the blood sample; hemoglobinometer—used to analyze the hemoglobin level in each sample; biohazardous waste disposal—used to properly dispose of equipment that comes in contact with blood.

Experiment Instructions

Go to the home page in the PhysioEx software and click **Exercise 11: Blood Analysis.** Click **Activity 3: Hemoglobin Determination,** and take the online **Pre-lab Quiz** for Activity 3.

After you take the online Pre-lab Quiz, click the **Experiment** tab and begin the experiment. The experiment instructions are reprinted here for your reference. The opening screen for the experiment is shown below.

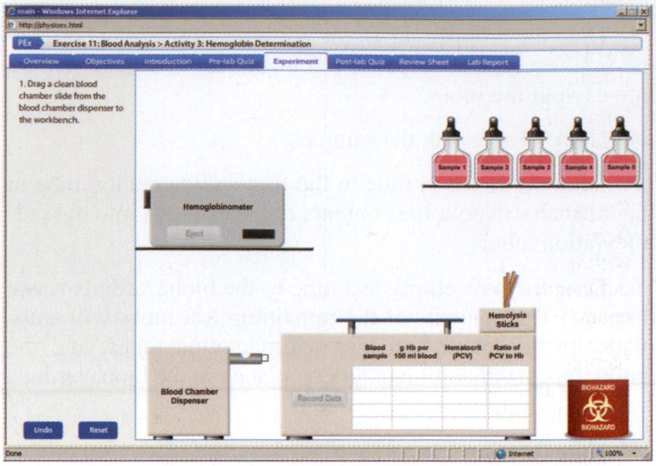
Exercise 11: Blood Analysis > Activity 3: Hemoglobin Determination
Overview
Objectives
Introduction
Pre-lab Quiz
Experiment
Post-lab Quiz
Review Sheet
Lab Report
1. Drag a clean blood chamber slide from the blood chamber dispenser to the workbench.
Sample 1
Sample 2
Sample 3
Sample 4
Sample 5
Hemoglobinometer
Eject
Hemolysis Sticks
Blood sample
g Hb per 100 ml blood
Hematocrit (PCV)
Ratio of PCV to Hb
Blood Chamber Dispenser
Record Data
BIOHAZARD
Undo
Reset

1. Drag a clean blood chamber slide from the blood chamber dispenser to the workbench.

2. Drag the bottle cap from the sample 1 bottle (the sample from the healthy male) to the depression in the blood chamber slide to dispense a drop of blood into the depression.

3. Drag a hemolysis stick to the drop of blood in the chamber to stir the blood sample for 45 seconds, lysing the red blood cells and releasing their hemoglobin.

4. Drag the hemolysis stick to the biohazardous waste disposal.

5. Drag the blood chamber slide to the dark rectangular slot on the hemoglobinometer to analyze the sample. After you insert the blood chamber slide into the hemoglobinometer, you will see a blowup of the inside of the hemoglobinometer.

6. The left half of the circular field shows the intensity of green light transmitted by blood sample 1. The right half of the circular field shows the intensity of green light for known levels of hemoglobin present in blood. Drag the lever on the right side of the hemoglobinometer down until the shade of green in the right half of the field matches the shade of green in the left half of the field and then click **Record Data** to display your results in the grid (and record your results in Chart 3).

7. Click **Eject** to remove the blood chamber slide from the hemoglobinometer.

8. Drag the blood chamber slide from the hemoglobinometer to the biohazardous waste disposal.

? PREDICT Question 1
How will the hemoglobin levels for the female Olympic athlete (sample 5) compare with the hemoglobin levels for the healthy female (sample 2)?

____

CHART 3 Hemoglobin Determination

Blood sample	Hb in grams per 100 ml of blood	Hematocrit (PCV)	Ratio of PCV to Hb
Sample 1 (healthy male)		48	
Sample 2 (healthy female)		44	
Sample 3 (female with iron-deficiency anemia)		40	
Sample 4 (male with polycythemia)		60	
Sample 5 (female Olympic athlete)		60	

9. You will now measure the hemoglobin levels for each of the remaining samples.

- Drag a blood chamber slide to the workbench.
- Drag the bottle cap from the next sample bottle to the depression in the slide.
- Drag a hemolysis stick to the drop of blood in the chamber (after stirring the sample, the hemolysis stick will automatically be placed in the biohazardous waste disposal).
- Drag the blood chamber slide to the dark rectangular slot on the hemoglobinometer.
- Drag the lever on the right side of the hemoglobinometer down until the shade of green in the right half of the field matches the shade of green in the left half of the field and then click **Record Data** to display your results in the grid (and record your results in Chart 3).
- Click **Eject** to remove the blood chamber slide from the hemoglobinometer (the slide will automatically be placed in the biohazardous waste disposal).

Repeat this step until you analyze all five samples.

After you complete the experiment, take the online **Post-lab Quiz** for Activity 3.

Activity Questions

1. As mentioned in the introduction to this activity, hemoglobin levels increase for people living at high altitudes. Given that the atmospheric pressure of oxygen significantly declines as you ascend to higher elevations, why do you think hemoglobin levels would increase for those living at high altitudes?

____

____

2. Just by looking at the color of a freshly drawn blood sample, how could you distinguish between blood that is well oxygenated and blood that is poorly oxygenated?

____

____

ACTIVITY 4

Blood Typing

OBJECTIVES

1. To understand the terms *antigens, agglutinogens, ABO antigens, Rh antigens*, and *agglutinins*.
2. To learn how to perform a blood-typing assay.
3. To understand the results and their implications when examining agglutination reactions.
4. To understand the importance of proper disposal of laboratory material that comes in contact with blood.

Introduction

All of the cells in the human body, including red blood cells, are surrounded by a plasma membrane that contains genetically determined glycoproteins, called **antigens.** On red blood cell membranes, there are certain antigens, called **agglutinogens,** that determine a person's blood type. If a blood transfusion recipient has antibodies (called **agglutinins**) that react with the antigens present on the transfused cells, the red blood cells will become clumped together, or agglutinated, and then lysed, resulting in a potentially life-threatening blood transfusion reaction. It is therefore important to determine an individual's blood type before performing blood transfusions to avoid mixing incompatible blood. Although many different antigens are present on red blood cell membranes, the **ABO** and **Rh antigens** cause the most vigorous and potentially fatal transfusion reactions.

The ABO blood groups are determined by the presence or absence of two antigens: type A and type B. Because these antigens are genetically determined, a person has two copies (alleles) of the gene for these proteins, one copy from each parent. The presence of these antigens is due to a dominant allele, and their absence is due to a recessive allele.

- A person with type A blood can have two alleles for the type A antigen or one allele for the type A antigen and one allele for the absence of either the type A or type B antigen.
- A person with type B blood can have two alleles for the type B antigen or one allele for the type B antigen and one allele for the absence of either the type A or type B antigen.
- A person with type AB blood has one allele for the type A antigen and one allele for the type B antigen.
- A person with type O blood has two recessive alleles and has neither the type A nor type B antigen.

TABLE 11.1 ABO Blood Types

Blood type	Antigens on RBCs	Antibodies present in plasma
A	A	anti-B
B	B	anti-A
AB	A and B	none
O	none	anti-A and anti-B

Antibodies against the A and B antigens are found preformed in the blood plasma. A person has antibodies only for the antigens not on his or her red blood cells, so a person with type A blood will have anti-B antibodies. View Table 11.1 for a summary of the antigens on red blood cells and the antibodies in the plasma for each blood type.

The Rh factor is another genetically determined protein that can be present on red blood cell membranes. Approximately 85% of the population is Rh positive (Rh^+), and their red blood cells have this protein on their surface. Antibodies against the Rh factor are not found preformed in the plasma. They are produced by an Rh negative (Rh^-) individual only after exposure to blood cells from someone who is Rh^+. Such exposure can occur during pregnancy when Rh^+ blood cells from the baby cross the placenta and expose the mother to the antigen.

To determine an individual's blood type, drops of an individual's blood sample are mixed separately with antiserum containing antibodies to either type A antigens, type B antigens, or Rh antigens. An agglutination reaction (showing clumping) indicates the presence of the agglutinogen.

EQUIPMENT USED The following equipment will be depicted on-screen: blood samples from six individuals with different blood types; anti-A serum (blue bottle), anti-B serum (yellow bottle), and anti-Rh serum (white bottle), containing antibodies to the A antigen, B antigen, and Rh antigen, respectively; blood-typing slide dispenser; color-coded stirring sticks—used to mix the blood sample and the serum (blue: used with anti-A serum, yellow: used with the anti-B serum, white: used with the anti-Rh serum); light box—used to view the blood type samples; biohazardous waste disposal—used to properly dispose of equipment that comes in contact with blood.

Experiment Instructions

Go to the home page in the PhysioEx software and click **Exercise 11, Blood Analysis.** Click **Activity 4, Blood Typing,** and take the online **Pre-lab Quiz** for Activity 4.

After you take the online Pre-lab Quiz, click the **Experiment** tab and begin the experiment. The experiment instructions are reprinted here for your reference. The opening screen for the experiment is shown above.

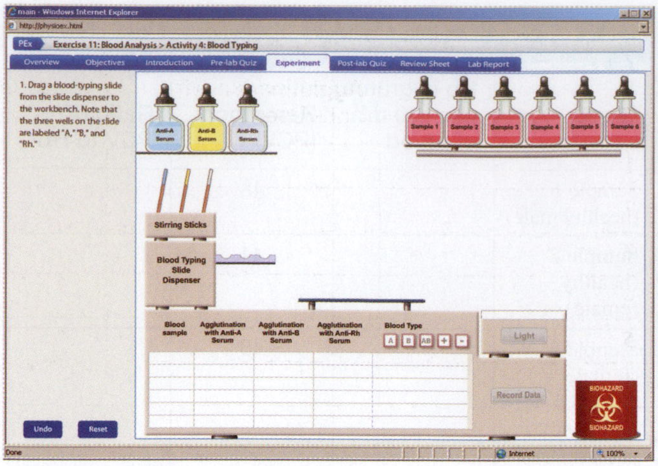
main - Windows Internet Explorer
Exercise 11: Blood Analysis > Activity 4: Blood Typing
Overview
Objectives
Introduction
Pre-lab Quiz
Experiment
Post-lab Quiz
Review Sheet
Lab Report
1. Drag a blood-typing slide from the slide dispenser to the workbench. Note that the three wells on the slide are labeled "A," "B," and "Rh."
Anti-A Serum
Anti-B Serum
Anti-Rh Serum
Sample 1
Sample 2
Sample 3
Sample 4
Sample 5
Sample 6
Stirring Sticks
Blood Typing Slide Dispenser
Blood sample
Agglutination with Anti-A Serum
Agglutination with Anti-B Serum
Agglutination with Anti-Rh Serum
Blood Type
Light
Record Data
BIOHAZARD
Undo
Reset
Done
Internet
100%

1. Drag a blood-typing slide from the slide dispenser to the workbench. Note that the three wells on the slide are labeled "A," "B," and "Rh."
2. Drag the dropper cap of the sample 1 bottle to well A on the blood-typing slide to dispense a drop of blood into each well.
3. Drag the dropper cap of the anti-A serum bottle to well A on the blood-typing slide to dispense a drop of anti-A serum into the well.
4. Drag the dropper cap of the anti-B serum bottle to well B on the blood-typing slide to dispense a drop of anti-B serum into the well.
5. Drag the dropper cap of the anti-Rh serum bottle to well Rh on the blood-typing slide to dispense a drop of anti-Rh serum into the well.
6. Drag a blue-tipped stirring stick to well A to mix the blood and anti-A serum.
7. Drag the stirring stick to the biohazardous waste disposal.
8. Drag a yellow-tipped stirring stick to well B to mix the blood and anti-B serum.
9. Drag the stirring stick to the biohazardous waste disposal.
10. Drag a white-tipped stirring stick to well Rh to mix the blood and anti-Rh serum.
11. Drag the stirring stick to the biohazardous waste disposal.
12. Drag the blood-typing slide to the light box and then click **Light** to analyze the slide.
13. Under each of the wells, click **Positive** if agglutination occurred (the sample shows clumping) or click **Negative** if agglutination did not occur (the sample looks smooth).

CHART 4 Blood Typing Results

Blood sample	Agglutination with anti-A serum	Agglutination with anti-B serum	Agglutination with anti-Rh serum	Blood type
1				
2				
3				
4				
5				
6				

14. Click **Record Data** to display your results in the grid (and record your results in Chart 4).

15. Drag the blood-typing slide to the biohazardous waste disposal.

> **PREDICT Question 1**
> If the patient's blood type is AB^-, what would be the appearance of the A, B, and Rh samples?
>
> ______________________________

16. You will now analyze the remaining samples.

- Drag a blood-typing slide from the slide dispenser to the workbench. The next sample will be added to each well on the slide, the appropriate antiserum will be added to each well, the sample and antisera will be mixed, and the slide will be placed in the light box.
- Under each of the wells, click **Positive** if agglutination occurred (the sample shows clumping) or click **Negative** if agglutination did not occur (the sample looks smooth).
- Click **Record Data** to display your results in the grid (and record your results in Chart 4).

Repeat this step until you analyze all six samples.

17. You will now indicate the blood type for each sample and indicate whether the sample is Rh positive or Rh negative.

- Click the row for the sample in the grid (and record your results in Chart 4).
- Click A, B, AB, or O above the blood type column to indicate the blood type.
- Click the − button or the + button above the blood type column to indicate whether the sample is Rh negative or Rh positive.

Repeat this step for all six samples. Record your results in Chart 4.

After [illegible] lete the experiment, take the online **Post-lab** [illegible] 4.

[illegible] stions

[illegible] nst the A and B antigens are found in the [illegible] has antibodies only for the antigens that are not present on their red blood cells. Using this information, list the antigens found on red blood cells and the antibodies in the plasma for blood types 1) AB−, 2) O+, 3) B−, and 4) A+.

____

____

2. If an individual receives a bone marrow transplant from someone with a different ABO blood type, what happens to the recipient's ABO blood type?

____

____

ACTIVITY 5

Blood Cholesterol

OBJECTIVES

1. To understand the terms *cholesterol, lipoproteins, low-density lipoprotein (LDL), hypocholesterolemia, hypercholesterolemia,* and *atherosclerosis.*
2. To learn how to test for total blood cholesterol using a colorimetric assay.
3. To understand the results and their implications when examining total blood cholesterol.
4. To understand the importance of proper disposal of laboratory material that comes in contact with blood.

Introduction

Cholesterol is a lipid substance that is essential for life—it is an important component of all cell membranes and is the base molecule of steroid hormones, vitamin D, and bile salts. Cholesterol is produced in the human liver and is present in some foods of animal origin, such as milk, meat, and eggs. Because cholesterol is a water-insoluble lipid, it needs to be wrapped in protein packages, called **lipoproteins,** to travel in the blood (which is mostly water) from the liver and digestive organs to the cells of the body.

One type of lipoprotein package, called **low-density lipoprotein (LDL),** has been identified as a potential source of damage to the interior of arteries. LDLs can contribute to **atherosclerosis,** the buildup of plaque, in these blood vessels.

A total blood cholesterol determination does not measure the level of LDLs, but it does provide valuable information about the total amount of cholesterol in the blood.

Less than 200 milligrams of total cholesterol per deciliter of blood is considered desirable. Between 200 and 239 mg/dl is considered borderline high cholesterol. Over 240 mg/dl is considered high blood cholesterol **(hypercholesterolemia)** and is associated with an increased risk of cardiovascular disease. Abnormally low blood cholesterol levels (total cholesterol lower than 100 mg/dl) can also suggest a problem. Low levels may indicate hyperthyroidism (overactive thyroid gland), liver disease, inadequate absorption of nutrients from the intestine, or malnutrition. Other reports link **hypocholesterolemia** (low blood cholesterol) to depression, anxiety, and mood disturbances, which are thought to be controlled by the level of available serotonin, a neurotransmitter. There is evidence of a relationship between low levels of blood cholesterol and low levels of serotonin in the brain.

In this test for total blood cholesterol, a sample of blood is mixed with enzymes that produce a colored reaction with cholesterol. The intensity of the color indicates the amount of cholesterol present. The cholesterol tester compares the color of the sample to the colors of known levels of cholesterol (standard values).

EQUIPMENT USED The following equipment will be depicted on-screen: lancets—sharp, needlelike instruments used to prick the finger to obtain a drop of blood; four patients (represented by an extended finger); alcohol wipes—used to cleanse the patient's fingertip before it is punctured with the lancet; color wheel—divided into shades of green that correspond to total cholesterol levels; cholesterol strips—contain chemicals that convert, by a series of reactions, the cholesterol in the blood sample into a green-colored solution; biohazardous waste disposal—used to properly dispose of equipment that comes in contact with blood.

Experiment Instructions

Go to the home page in the PhysioEx software and click **Exercise 11: Blood Analysis.** Click **Activity 5: Blood Cholesterol,** and take the online **Pre-lab Quiz** for Activity 5.

After you take the online Pre-lab Quiz, click the **Experiment** tab and begin the experiment. The experiment instructions are reprinted here for your reference. The opening screen for the experiment is shown below.

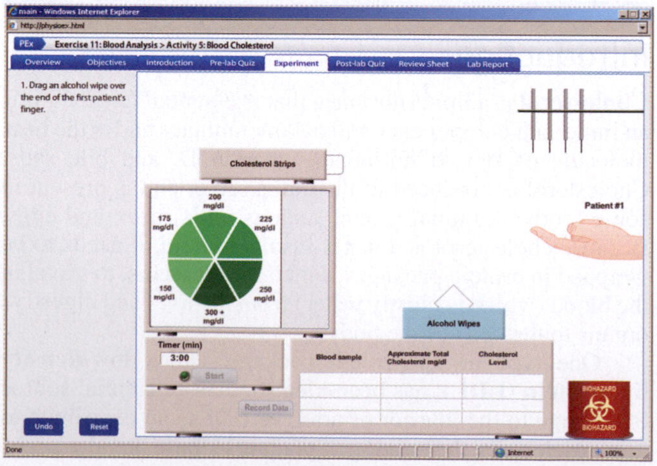

Exercise 11: Blood Analysis > Activity 5: Blood Cholesterol
Overview
Objectives
Introduction
Pre-lab Quiz
Experiment
Post-lab Quiz
Review Sheet
Lab Report
1. Drag an alcohol wipe over the end of the first patient's finger.
Cholesterol Strips
Patient #1
Alcohol Wipes
Timer (min)
3:00
Start
Record Data
Blood sample
Approximate Total Cholesterol mg/dl
Cholesterol Level
Undo
Reset

1. Drag an alcohol wipe over the end of the first patient's finger.

2. Drag the alcohol wipe to the biohazardous waste disposal.

3. Drag a lancet to the tip of the patient's finger to prick the finger and obtain a drop of blood.

4. Drag the lancet to the biohazardous waste disposal.

5. Drag a cholesterol strip to the finger to transfer a drop of blood from the patient's finger to the strip.

6. Drag the cholesterol strip to the rectangular box to the right of the color wheel.

7. Click **Start** to start the timer. It takes three minutes for the chemicals in the cholesterol strip to react with the blood. The simulation compresses the 3-minute time period into 3 seconds of real time.

8. Click the color on the color wheel that most closely matches the color on the cholesterol strip.

9. Click **Record Data** to display your results in the grid (and record your results in Chart 5).

CHART 5 Total Cholesterol Determination

Blood sample	Approximate total cholesterol (mg/dl)	Cholesterol level
1		
2		
3		
4		

10. Drag the cholesterol test strip to the biohazardous waste disposal.

PREDICT Question 1
Patient 4 prefers to cook all his meat in lard or bacon grease. Knowing this dietary preference, you anticipate his total cholesterol level to be:

____

11. You will now test the total cholesterol levels for the remaining patients.

- Drag an alcohol wipe over the end of the patient's finger. The alcohol wipe will automatically be placed in the biohazardous waste disposal.
- Drag a lancet to the tip of the patient's finger to prick the finger and obtain a drop of blood. The lancet will automatically be placed in the biohazardous waste disposa
- Drag a cholesterol strip to the finger to transfer of blood from the patient's finger to the strip

- Drag the cholesterol strip to the rectangular box to the right of the color wheel. The timer will automatically run for three minutes to allow the chemicals in the cholesterol strip to react with the blood.
- Click the color on the color wheel that most closely matches the color on the cholesterol strip.
- Click **Record Data** to display your results in the grid (and record your results in Chart 5). The cholesterol strip will automatically be placed in the biohazardous waste disposal.

Repeat this step until you determine the total cholesterol levels for all four patients.

After you complete the experiment, take the online **Post-lab Quiz** for Activity 5.

Activity Questions

1. Why do cholesterol plaques occur in arteries and not veins?

___

___

2. Phytosterols can alter absorption of certain molecules by the intestinal tract. Why would they be a beneficial dietary supplement for people with high LDL levels?

___

___

PhysioEx™ 9.0

REVIEW SHEET

EXERCISE 11

NAME ______________________________

LAB TIME/DATE ______________________

Blood Analysis

ACTIVITY 1 Hematocrit Determination

1. List the hematocrits for the healthy male (sample 1) and female (sample 2) living in Boston (at sea level) and indicate whether they are normal or whether they indicate anemia or polycythemia.

____

____

2. Describe the difference between the hematocrits for the male and female living in Boston. Why does this difference between the sexes exist?

____

____

3. List the hematocrits for the healthy male and female living in Denver (approximately one mile above sea level) and indicate whether they are normal or whether they indicate anemia or polycythemia.

____

____

4. How did the hematocrit levels of the Denver residents differ from those of the Boston residents? Why? How well did the results compare with your prediction?

____

____

5. Describe how the kidneys respond to a chronic decrease in oxygen and what effect this has on hematocrit levels.

____

____

6. List the hematocrit for the male with aplastic anemia (sample 5) and indicate whether it is normal or abnormal. Explain your response.

____

____

7. List the hematocrit for the female with iron-deficiency anemia (sample 6) and indicate whether it is normal or abnormal. Explain your response.

____

____

ACTIVITY 2 Erythrocyte Sedimentation Rate

1. Describe the effect that sickle cell anemia has on the sedimentation rate (sample 3). Why do you think that it has this effect?

____

____

2. How did the sedimentation rate for the menstruating female (sample 2) compare with the sedimentation rate for the healthy individual (sample 1)? Why do you think this occurs?

____

____

3. How did the sedimentation rate for the individual with angina pectoris (sample 6) compare with the sedimentation rate for the healthy individual (sample 1)? Why? How well did the results compare with your prediction?

____

____

4. What effect does iron-deficiency anemia (sample 4) have on the sedimentation rate?

____

____

5. Compare the sedimentation rate for the individual suffering a myocardial infarction (sample 5) with the sedimentation rate for the individual with angina pectoris (sample 6). Explain how you might use this data to monitor heart conditions.

____

____

ACTIVITY 3 Hemoglobin Determination

1. Is the male with polycythemia (sample 4) deficient in hemoglobin? Why?

____

____

2. How did the hemoglobin levels for the female Olympic athlete (sample 5) compare with the hemoglobin levels for the healthy female (sample 2)? Is either person *deficient* in hemoglobin? How well did the results compare with your prediction?

____

____

3. List conditions in which hemoglobin levels would be expected to decrease. Provide reasons for the change when possible.

4. List conditions in which hemoglobin levels would be expected to increase. Provide reasons for the change when possible.

5. Describe the ratio of hematocrit to hemoglobin for the healthy male (sample 1) and female (sample 2). (A normal ratio of hematocrit to grams of hemoglobin is approximately 3:1.) Discuss any differences between the two individuals.

6. Describe the ratio of hematocrit to hemoglobin for the female with iron-deficiency anemia (sample 3) and the female Olympic athlete (sample 5). (A normal ratio of hematocrit to grams of hemoglobin is approximately 3:1.) Discuss any differences between the two individuals.

ACTIVITY 4 Blood Typing

1. How did the appearance of the A, B, and Rh samples for the patient with AB^- blood type compare with your prediction?

2. Which blood sample contained the rarest blood type?

3. Which blood sample contained the universal donor?

4. Which blood sample contained the universal recipient?

5. Which blood sample did not agglutinate with any of the antibodies tested? Why?

6. What antibodies would be found in the plasma of blood sample 1?

7. When transfusing an individual with blood that is compatible but not the same type, it is important to separate packed cells from the plasma and administer only the packed cells. Why do you think this is done? (Hint: Think about what is *in plasma* versus what is *on RBCs*.)

8. List the blood samples in this activity that represent people who could donate blood to a person with type B^+ blood.

ACTIVITY 5 Blood Cholesterol

1. Which patient(s) had desirable cholesterol level(s)?

2. Which patient(s) had elevated cholesterol level(s)?

3. Describe the risks for the patient(s) you identified in question 2.

4. Was the cholesterol level for patient 4 low, desirable, or high? How well did the results compare with your prediction? What advice about diet and exercise would you give to this patient? Why?

5. Describe some reasons why a patient might have abnormally low blood cholesterol.

PhysioEx™ 9.0 EXERCISE 12

Serological Testing

PRE-LAB QUIZ

1. Circle True or False. The lymphatic system protects the body by removing foreign material such as bacteria from the lymphatic stream.
2. Lymph is
   a. excess blood that has escaped from veins
   b. excess tissue fluid that has filtered out of capillaries
   c. excess tissue fluid that has escaped from arteries
3. Circle True or False. Lymphatic vessels have three tunics and are equipped with valves like veins.
4. ________, which serve as filters for the lymphatic system, occur at various points along the lymphatic vessels.
   a. Glands
   b. Lymph nodes
   c. Valves
5. Circle True or False. The immune response is a systemic response that occurs when the body recognizes a substance as foreign and acts to destroy or neutralize it.
6. Three characteristics of the immune response are the ability to distinguish self from nonself, memory, and
   a. autoimmunity
   b. specificity
   c. susceptibility
7. Circle the correct term. B cells / T cells differentiate in the thymus.
8. Circle the correct term. T cells mediate humoral / cellular immunity because they destroy cells infected with viruses and certain bacteria and parasites.
9. Circle True or False. Antibodies are produced by plasma cells in response to antigens and are found in all body secretions.

Exercise Overview

Immunology, the study of the immune system, focuses on chemical interactions that are difficult to observe. A number of chemical techniques have been developed to visually represent antibodies and antigens in the **serum,** the fluid portion of the blood with the clotting factors removed. The study and use of these techniques is referred to as **serology.** These techniques are performed in vitro, outside of the body, and are primarily used as diagnostic tools to detect disease. Other applications include pregnancy testing and drug testing. These immunological techniques depend upon the principle that an antibody binds only to specific, corresponding antigens. The tests are relatively expensive to perform, so these activities will allow you to perform them without the sometimes cost-prohibitive supplies.

Antigens and Antibodies

The word **antigen** is derived from two words: *anti*body and *gen*erator. Antigens do not produce antibodies, but early scientists noted that when antigens were present, antibodies appeared. Plasma cells actually produce antibodies.

Antigens include proteins, polysaccharides, and various small molecules that stimulate antibody production. Antigens are often molecules that are described as **nonself,** or foreign to the body. There are also self-antigens that act as identifier tags, such as the proteins found on the surface of red blood cells.

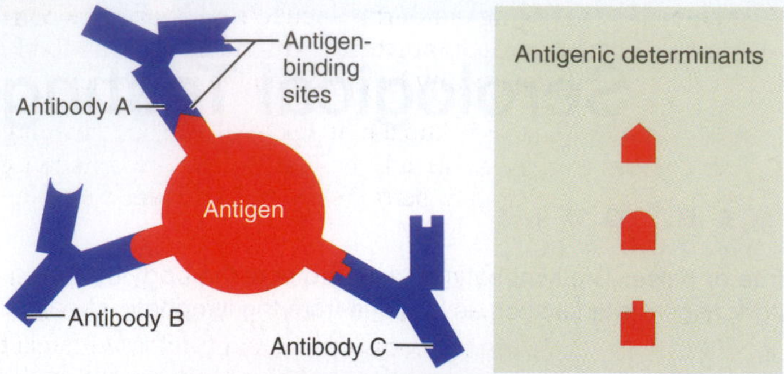
Antigen-binding sites
Antibody A
Antigen
Antibody B
Antibody C
Antigenic determinants

FIGURE 12.1 Antigen-antibody interaction with antigenic determinants.

Most often, antigens are a portion of an infectious agent, such as a bacterium or a virus, and the body produces antibodies in response to the presence of the infectious agent.

Antigens are often large and have multiple antigenic sites—locations that can bind to antibodies. We refer to these sites as **antigenic determinants,** or **epitopes.** The antibody has a corresponding antigen-binding site that has a "lock-and-key" recognition for the antigenic determinant on the antigen (view Figure 12.1). All of the simulated tests presented in this exercise take advantage of antigen-antibody specificity. These tests include direct fluorescent antibody technique, Ouchterlony technique, ELISA (enzyme-linked immunosorbent assay), and Western blotting technique.

Nonspecific Binding

The lock-and-key recognition that antigen and antibody have for each other is much like the specificity that an enzyme and its substrate have for one another. However, with antigen and antibody, **nonspecific binding** sometimes occurs. For this reason you will perform a number of washing steps in this exercise to remove any nonspecific binding.

Positive and Negative Controls

You will also use **positive** and **negative controls** to ensure that the test is working accurately. Positive controls include a substance that is known to react positively, thus giving you a standard against which to base your results. Negative controls include substances that should not react. A positive result with a negative control is a "false positive," which would invalidate all other results. Likewise, a negative result with a positive control is a "false negative," which would also invalidate your results.

ACTIVITY 1

Using Direct Fluorescent Antibody Technique to Test for Chlamydia

OBJECTIVES

1. To understand how fluorescent antibodies can be used diagnostically to detect the presence of a specific antigen.
2. To observe how to test for the sexually transmitted disease chlamydia.
3. To distinguish between antigens and antibodies.
4. To understand the terms *epitope* and *antigenic determinant.*
5. To observe nonspecific binding that can result between antigen and antibody.

Introduction

The direct fluorescent antibody technique uses antibodies to directly detect the presence of antigen. A fluorescent dye molecule attached to these antibodies acts as a visual signal for a positive result. This technique is typically used to test for antigens from infectious agents, such as bacteria or viruses. In this activity you will test for the presence of *Chlamydia trachomatis* (a bacterium that invades the cells of its host) using fluorescently labeled antibodies to detect the presence of the antigen and, therefore, the bacterium. *Chlamydia trachomatis* is an important infectious agent because it causes the sexually transmitted disease **chlamydia.** Left untreated, chlamydia can lead to sterility in men and women.

Chlamydia trachomatis is an obligate, intracellular bacterium, which means that it can only survive inside a host cell. The life cycle of the bacterium has two cellular types. The infectious cell type is the small, dense **elementary body,** which is capable of attaching to the host cell. The **reticulate body** is a larger, less-dense cell, which divides actively once inside the host cell. The reticulate body is also referred to as the vegetative form. The life cycle of *Chlamydia* begins when the elementary body enters the host cell and continues as the elementary body changes inside the host cell into a reticulate body. The reticulate body divides into more reticulate bodies and converts back to the elementary body form for release to infect other cells.

In this activity you will test three patient samples and two control samples for the *Chlamydia* infection. An epithelial scraping from the male urethra or from the cervix of the uterus is performed to collect squamous cells from the surface. The elementary bodies are measured by reacting antigen-specific antibodies to infected cells. The fluorescent dye attached to the antigen-specific antibodies makes the complex detectable. The sample is viewed with a fluorescent microscope. The presence of ten or more elementary bodies in a field of view with a diameter of 5 millimeters is considered a positive result. The elementary bodies will be stained green inside red host cells.

EQUIPMENT USED The following equipment will be depicted on-screen: five samples: patient A, patient B, patient C, a positive control, and a negative control; incubator; fluorescent microscope; 95% ethyl alcohol—used for fixing the sample to the microscope slide; chlamydia fluorescent antibody (Chlamydia FA)—antibodies specific for the *Chlamydia* antigen with a fluorescent dye attached; fluorescent antibody mounting media (FA mounting)—used to mount the prepared sample to the slide when ready for viewing under the microscope; phosphate buffered saline (PBS)—used to wash off excess antibodies and prevent nonspecific binding of the antigen and antibody; fluorescent antibody buffer (FA buffer)—used to remove excess ethyl alcohol; petri dishes—used for incubation of the slides to keep them moist; microscope slides—an incubation vessel where the antigen and antibody react; cotton-tipped applicators—used for application and mixing of the antibodies with the samples; filter paper—used to keep the samples moist in the petri dishes; biohazardous waste disposal.

Experiment Instructions

Go to the home page in the PhysioEx software and click **Exercise 12: Serological Testing.** Click **Activity 1: Using Direct Fluorescent Antibody Technique to Test for Chlamydia,** and take the online **Pre-lab Quiz** for Activity 1.

After you take the online Pre-lab Quiz, click the **Experiment** tab and begin the experiment. The experiment instructions are reprinted here for your reference. The opening screen for the experiment is shown below.

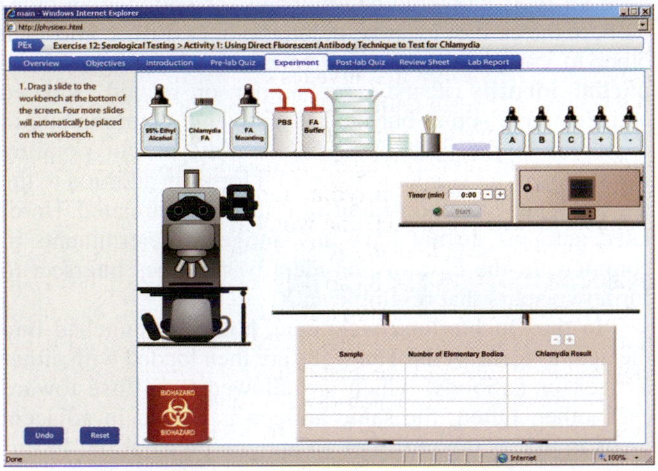

Exercise 12: Serological Testing > Activity 1: Using Direct Fluorescent Antibody Technique to Test for Chlamydia
Overview
Objectives
Introduction
Pre-lab Quiz
Experiment
Post-lab Quiz
Review Sheet
Lab Report
1. Drag a slide to the workbench at the bottom of the screen. Four more slides will automatically be placed on the workbench.
PBS
Sample
Number of Elementary Bodies
Chlamydia Result
Undo
Reset

1. Drag a slide to the workbench at the bottom of the screen. Four more slides will automatically be placed on the workbench.

2. The patient samples have been suspended in a small amount of buffer and placed in dropper bottles for ease of dispensing. Drag the dropper cap of the patient A sample bottle to the first slide on the workbench to dispense a drop of the sample onto the slide. A drop from each sample will be placed on a separate slide.

3. Drag the dropper cap of the 95% ethyl alcohol bottle to the first slide on the workbench to dispense three drops of ethyl alcohol onto each slide.

4. Set the timer to 5 minutes by clicking the + button beside the timer display. Click **Start** to start the timer and allow the ethyl alcohol to fix the sample to the slide and prevent the sample from being washed off in the subsequent washing steps. The simulation compresses the 5-minute time period into 5 seconds of real time.

5. Drag the fluorescent antibody (FA) buffer squirt bottle to the first slide to rinse all five slides and remove excess ethyl alcohol.

6. Drag an applicator stick to the chlamydia fluorescent antibody (FA) bottle to soak its cotton tip with antibodies that are specific for *Chlamydia* and labeled with a fluorescent tag.

7. Drag the applicator stick to the first slide to apply the chlamydia fluorescent antibody. Separate applicator sticks will automatically be soaked in chlamydia fluorescent antibody and applied to each slide. Each applicator will automatically be placed in the biohazardous waste disposal.

8. Drag a petri dish to the workbench. A piece of filter paper will be placed into the petri dish. The filter paper has been moistened with fluorescent antibody buffer to keep the samples from drying out during incubation. Four more petri dishes (and filter paper moistened with fluorescent antibody buffer) will automatically be placed on the workbench.

9. Drag the first slide into the first petri dish. The remaining four slides will automatically be placed into the remaining petri dishes and all five petri dishes will be loaded into the incubator.

10. Set the timer to 20 minutes by clicking the + button next to the timer display. Click **Start** to incubate the samples at 25°C. During incubation the antibodies will react with the corresponding antigens if they are present in the sample. The petri dishes will automatically be removed from the incubator when the time is complete. The simulation compresses the 20-minute incubation time period into 10 seconds of real time.

11. Drag the phosphate buffered saline (PBS) squirt bottle to the first petri dish to wash off excess antibodies and prevent nonspecific binding of the antigen and antibody. The timer will count down 10 minutes for a thorough washing.

12. Click the first petri dish to open the dish and remove the slide. The slides will automatically be removed from the remaining petri dishes.

13. Drag the first petri dish to the biohazardous waste disposal. The remaining petri dishes will automatically be placed in the biohazardous waste disposal.

14. Drag the dropper cap of the fluorescent antibody (FA) mounting media to the first slide to dispense a drop of mounting media onto each slide to mount the sample to the slide.

15. Drag the first slide (patient A) to the fluorescent microscope. Count the number of elementary bodies you see through the microscope (recall that elementary bodies stain green). Click **Submit** to display your results in the grid (and record your results in Chart 1). After you click **Submit,** the slide will automatically be placed in the biohazardous waste disposal.

CHART 1 Direct Fluorescent Antibody Technique Results

Sample	Number of elementary bodies	Chlamydia result
Patient A		
Patient B		
Patient C		
Positive control		
Negative control		

16. Repeat step 15 for Patient B.

17. Repeat step 15 for Patient C.

18. Repeat step 15 for the Positive Control.

19. Repeat step 15 for the Negative Control.

20. You will now indicate whether each sample is negative or positive for *Chlamydia.* Click the row for the sample in the grid and then click the − button or the + button above the Chlamydia result column to indicate whether the sample is negative or positive for *Chlamydia.* Repeat this step for all five samples. Record your results in Chart 1.

After you complete the experiment, take the online **Post-lab Quiz** for Activity 1.

Activity Questions

1. With this technique, is the antigen or antibody found on the patient sample? Explain how you know this.

____

____

2. Explain the difference between an antigen and an epitope (antigenic determinant).

____

____

3. When a sample has a small number of elementary bodies but not enough to be a positive result, there appears to have been some nonspecific binding that was not removed by the washing steps. Which sample displayed this property?

____

____

ACTIVITY 2

Comparing Samples with Ouchterlony Double Diffusion

OBJECTIVES

1. To observe the precipitation reaction between antigen and antibody.
2. To distinguish between *epitope* and *antigen.*
3. To understand the specificity that antibodies have for their epitopes.
4. To observe how related proteins might share epitopes in common.

Introduction

The Ouchterlony technique is also known as double diffusion. In this technique antigen and antibody diffuse toward each other in a semisolid medium made up of clear, clarified agar. When the antigen and antibody are in optimal proportions, cross-linking of the antigen and antibody occurs, forming an insoluble precipitate, called a **precipitin line.** These lines can then be used to visually identify similarities between antigens. If optimum proportions have not been met—for example, if there is excess antigen or excess antibody—then no visible precipitate will form. This technique provides easily visible evidence of the binding between antigen and antibody, and sophisticated equipment is not needed to observe the antigen-antibody reaction.

The Ouchterlony technique is designed to determine whether antigens are identical, related, or unrelated. Antigens have **identity** if they are identical. Identical antigens have all their antigenic determinants, or epitopes, in common. In the case of identity, precipitin lines diffuse into each other to completely fuse and form an arc. Antigens have **partial identity** if they are similar or related. Related antigens have some, but not all, antigenic determinants in common. In the case of partial identity, a spur pointing toward the more similar antigen well forms in addition to the arc. Antigens have **non-identity** if they are unrelated. Unrelated antigens do not have any antigenic determinants in common. In the case of non-identity, the lines intersect to form two spurs that resemble an X.

In the Ouchterlony technique, holes are punched into the agar to form wells. The wells are then loaded with either antigen or antibody, which are allowed to diffuse toward each other. Often, the same antigen is placed in adjacent wells to assess the purity of an antigen preparation. In this case a smooth arc with no spurs should be seen, as the antigens are identical. Multiple antibodies can also be placed in a center well. The antibodies will diffuse out in all directions and react with the antigens that are placed in the surrounding wells.

In this activity you will use human and bovine (from cows) albumin as the antigens, and the antibodies will be made in goats against albumin from either humans or cows. The goals are to identify an unknown antigen and to observe the patterns produced by the various relationships: identity, partial identity, and non-identity.

EQUIPMENT USED The following equipment will be depicted on-screen: goat anti–human albumin (Goat A-H)—an antiserum containing antibodies produced by goats against human albumin; goat anti–bovine albumin (Goat A-B)—an antiserum containing antibodies produced by goats against bovine (cow) albumin; bovine serum albumin (BSA); human serum albumin (HSA); unknown antigen; petri dishes filled with clear agar; well cutter.

Experiment Instructions

Go to the home page in the PhysioEx software and click **Exercise 12: Serological Testing.** Click **Activity 2: Comparing Samples with Ouchterlony Double Diffusion:** and take the online **Pre-lab Quiz** for Activity 2.

After you take the online Pre-lab Quiz, click the **Experiment** tab and begin the experiment. The experiment instructions are reprinted here for your reference. The opening screen for the experiment is shown on the following page.

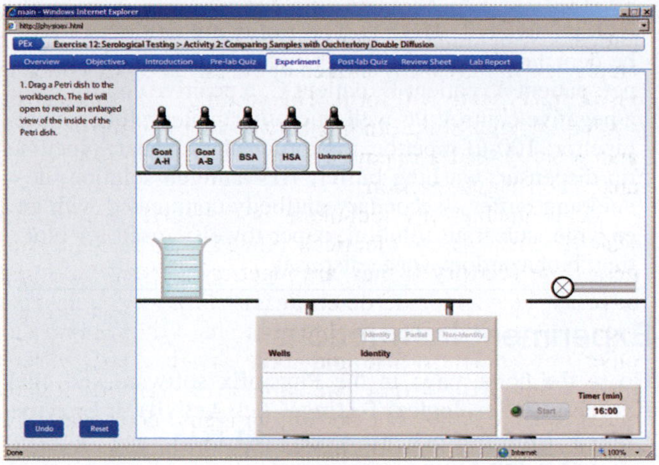
main - Windows Internet Explorer
PEx
Exercise 12: Serological Testing > Activity 2: Comparing Samples with Ouchterlony Double Diffusion
Overview
Objectives
Introduction
Pre-lab Quiz
Experiment
Post-lab Quiz
Review Sheet
Lab Report
1. Drag a Petri dish to the workbench. The lid will open to reveal an enlarged view of the inside of the Petri dish.
Goat A-H
Goat A-B
BSA
HSA
Unknown
Identity
Partial
Non-Identity
Wells
Identity
Timer (min)
Start
16:00
Undo
Reset
Done
Internet
100%

1. Drag a petri dish to the workbench. The lid will open to reveal an enlarged view of the inside of the petri dish.

2. Drag the well cutter to the middle of the enlarged view of the petri dish to punch a hole in the agar in the middle of the petri dish. Drag the well cutter to the upper left, upper right, lower left, and lower right of the petri dish to punch four more holes in the agar. After you punch all five wells into the agar, the wells will be labeled 1–5.

3. Drag the dropper cap of the goat anti–human albumin (Goat A-H) bottle to well 1 to fill it with a sample.

4. Drag the dropper cap of the goat anti–bovine albumin (Goat A-B) bottle to well 1 to fill it with a sample.

5. Drag the dropper cap of the bovine serum albumin (BSA) bottle to well 2 to fill it with a sample.

6. Drag the dropper cap of the bovine serum albumin (BSA) bottle to well 3 to fill it with a sample.

7. Drag the dropper cap of the human serum albumin (HSA) bottle to well 4 to fill it with a sample.

8. Drag the dropper cap of the unknown antigen bottle to well 5 to fill it with a sample.

> **? PREDICT Question 1**
> How do you think human serum albumin and bovine serum albumin will compare?
>
> ______________________________

9. Note that the timer is set to 16 hours. Click **Start** to start the timer. The antigen and antibodies will diffuse toward each other and form a precipitate, detected as a precipitin line. The simulation compresses the 16-hour time period into 10 seconds of real time.

10. You will now examine the precipitin lines that formed and indicate the relationship between each pair of antigens. Click the row for the wells containing the antigens in the grid and then click **Identity, Partial,** or **Non-Identity** above the identity column to indicate whether the antigens have identity, partial identity, or non-identity. Repeat this step for the four pairs of antigens. Record your results in Chart 2.

CHART 2 Ouchterlony Double Diffusion Results

Wells	Identity
2 and 5	
2 and 3	
3 and 4	
4 and 5	

After you complete the Experiment, take the online **Post-lab Quiz** for Activity 2.

Activity Questions

1. Which type of identity was present between the samples in this activity? Describe this type of identity.

____

____

2. Describe the importance of what you place in the center well.

____

____

3. Why do you think it is important for the agar to be clear and clarified?

____

____

4. Describe the role that albumin plays in the blood.

____

____

ACTIVITY 3

Indirect Enzyme-Linked Immunosorbent Assay (ELISA)

OBJECTIVES

1. To understand how the enzyme-linked immunosorbent assay (ELISA) is used as a diagnostic test.
2. To distinguish between the direct and the indirect ELISA.
3. To describe the basic structure of antibodies.
4. To define *seroconversion.*
5. To understand how the indirect ELISA is used to detect antibodies against HIV.

Introduction

The **enzyme-linked immunosorbent assay (ELISA)** is used to test for the presence of an antigen or antibody. The assay is considered enzyme linked because an enzyme is chemically linked to an antibody in both the direct and indirect versions of the test. Immunosorbent refers to the fact that either antigens or antibodies are being adsorbed (stuck) to plastic. If the test is designed to detect an antigen or antigens, it is a **direct ELISA** because it is directly looking for the foreign substance. An **indirect ELISA** is designed to detect antibodies that the patient has made against the antigen. A positive result with the indirect ELISA requires **seroconversion.** Seroconversion occurs when a patient goes from testing negative for a specific antibody to testing positive for the same antibody.

In the direct ELISA, a 96-well microtiter plate is coated with homologous antibodies made against the antigen of interest. The number of wells makes it easy to test many samples at the same time. The patient serum sample is added to the plate to test for the presence of the antigen that binds to the antibody coating on the plate. ELISA takes advantage of the fact that protein sticks well to plastic. A secondary antibody is added to the plate after the patient serum sample is added. If the antigen is present, a "sandwich" of antibody, antigen, and secondary antibody will form. The secondary antibody is chemically linked to an enzyme. When the substrate is added, the enzyme converts the substrate from a colorless compound to a colored compound. The amount of color produced will be proportional to the amount of antigen binding to the antibodies and thus indicates whether the patient is positive for the antigen. If the antigen is not present, the secondary (enzyme-linked) antibodies will be rinsed away with the washing steps and the substrate will not be converted and will remain colorless. A common use of the direct ELISA is a home pregnancy test, which detects human chorionic gonadotropin (hCG), a hormone present in the urine of pregnant women.

In the indirect ELISA, a 96-well microtiter plate is coated with antigens. The patient serum sample is added to test for the presence of antibodies that bind to the antigens on the plate. The secondary antibody that is added has an enzyme linked to it that binds to the **constant region** of the primary antibody if it is present in the patient sample. The constant region of an antibody has the same sequence of amino acids within a class of antibodies (for example, all IgG antibodies have the same constant region). The **variable region** of an antibody provides the diversity of antibodies and is the site to which the antigen binds. The configuration that forms in the indirect ELISA is antigen, primary antibody, and secondary antibody. Just as in the direct ELISA, the addition of substrate is used to determine whether the sample is positive for the presence of antibody.

In this activity you will use the indirect ELISA to test for the presence of antibodies made against human immunodeficiency virus (HIV). You will use positive and negative controls to verify the results. You will note that an indeterminate result can be obtained if there is not enough color produced to warrant a positive result. The cause of an indeterminate result could be either nonspecific binding or that the individual has been recently infected and has not yet produced enough antibodies for a positive result. In either case, the individual would be retested.

EQUIPMENT USED The following equipment will be depicted on-screen: five samples in the samples cabinet: patient A, patient B, patient C, a positive control, and a negative control; 96-well microtiter plate; multichannel pipettor; 100-μl pipettor; microtiter plate reader; pipettor tip dispenser; washing buffer; HIV antigen solution; developing buffer—secondary antibody conjugated with an enzyme; substrate solution; paper towels—used for blotting; biohazardous waste disposal.

Experiment Instructions

Go to the home page in the PhysioEx software and click **Exercise 12: Serological Testing.** Click **Activity 3: Enzyme-Linked Immunosorbent Assay (ELISA),** and take the online **Pre-lab Quiz** for Activity 3.

After you take the online Pre-lab Quiz, click the **Experiment** tab and begin the experiment. The experiment instructions are reprinted here for your reference. The opening screen for the experiment is shown below.

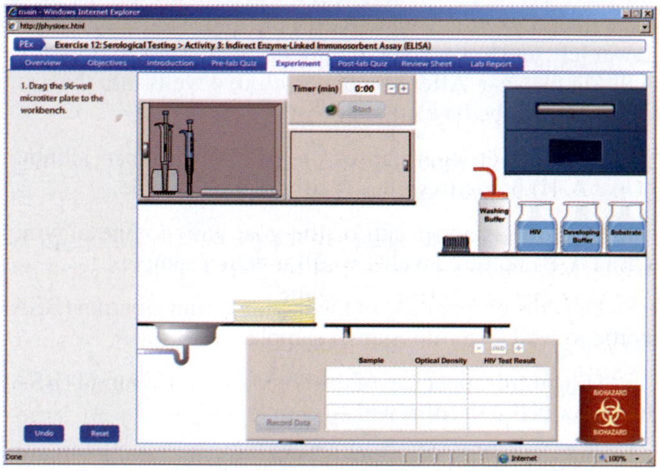
Exercise 12: Serological Testing > Activity 3: Indirect Enzyme-Linked Immunosorbent Assay (ELISA)
1. Drag the 96-well microtiter plate to the workbench.
Timer (min)
0:00
Start
Washing Buffer
HIV
Developing Buffer
Substrate
Sample
Optical Density
HIV Test Result
Record Data
Undo
Reset
BIOHAZARD

1. Drag the 96-well microtiter plate to the workbench.

2. Drag the multichannel pipettor to the pipette tip dispenser to insert the tips.

3. Drag the multichannel pipettor to the HIV antigens bottle to draw the antigen solution into the tips.

4. Drag the multichannel pipettor directly over the microtiter plate to dispense the liquid into the wells in one row of the plate.

5. Drag the multichannel pipettor to the biohazardous waste disposal for removal and disposal of the tips.

6. Set the timer to 14 hours by clicking the + button beside the timer display. This incubation time allows the antigens to stick to the plastic wells of the microtiter plate. Click **Start** to start the timer. The simulation compresses the 14-hour time period into 10 seconds of real time.

7. Drag the washing buffer squeeze bottle to the microtiter plate to remove excess antigens that are not adsorbed (stuck) to the plate.

8. Drag the microtiter plate to the sink to dump the contents of the tray into the sink to remove the washing buffer and excess antigens that are not stuck to the plastic.

9. Drag the microtiter plate to the paper towels. The plate will be pressed to the surface of the paper towels to remove the remaining liquid from the wells. In a typical ELISA, you would perform multiple washing steps to reduce any nonspecific binding. The number of washing steps in this simulation has been reduced for simplicity.

10. Drag the 100-μl pipettor to the tip dispenser to place a tip onto the pipettor.

11. Drag the 100-μl pipettor to the test tube containing the positive control sample (+) to draw the sample into the tip.

12. Drag the 100-μl pipettor to the microtiter plate to dispense the sample into the wells of the plate. The tip will automatically be removed and disposed of in the biohazardous waste disposal. Each of the remaining samples will automatically be dispensed into plate.

13. Set the timer to 1 hour by clicking the + button beside the timer display. This incubation time allows the antigens stuck to the plastic to bind to the antibodies present in the sample. Click **Start** to start the timer. The simulation compresses the 1-hour time period into 10 seconds of real time.

14. Drag the washing buffer squeeze bottle to the microtiter plate to wash off excess antibodies and prevent nonspecific binding of the antigen and antibody.

15. Drag the microtiter plate to the sink to dump washing buffer and unbound antibodies into the sink.

16. Drag the microtiter plate to the paper towels. The plate will be pressed to the surface of the paper towels to remove the remaining liquid from the wells.

17. Drag the multichannel pipettor to the pipette tip dispenser to insert the tips.

18. Drag the multichannel pipettor to the developing buffer bottle to draw the developing buffer into the tips. The developing buffer contains the conjugated secondary antibody.

19. Drag the multichannel pipettor to the microtiter plate to dispense the solution into the wells. The tips will automatically be removed and disposed of in the biohazardous waste disposal.

20. Set the timer to 1 hour and then click **Start** to start the timer and allow the conjugated secondary antibody to bind to the primary antibody if it is present in the sample.

21. Drag the washing buffer squeeze bottle to the microtiter plate to remove any nonspecific binding that occurred.

22. Drag the microtiter plate to the sink to dump the contents of the tray into the sink.

23. Drag the microtiter plate to the paper towels. The plate will be pressed to the surface of the paper towels to remove the remaining liquid from the wells.

24. Drag the multichannel pipettor to the pipette tip dispenser to insert the tips.

25. Drag the multichannel pipettor to the substrate bottle to draw the substrate into the tips.

26. Drag the multichannel pipettor to the microtiter plate to dispense the solution into the wells. The tips will automatically be removed and disposed of in the biohazardous waste disposal.

27. An enlargement of the wells will appear. The development will progress over time. To determine the optical density for each sample (the samples are in the first row, from top to bottom, of the microtiter plate):

- Click the well and the optical density will appear in the window of the microtiter plate reader.
- Click **Record Data** to display your results in the grid (and record your results in Chart 3).

CHART 3 Indirect ELISA Results

Sample	Optical density	HIV test result
Patient A		
Patient B		
Patient C		
Positive control		
Negative control		

28. You will now indicate whether the result for each sample is negative, indeterminate, or positive for HIV.

- A result of <0.300 is read as negative for HIV-1.
- A result of 0.300–0.499 is read as indeterminate (need to retest).
- A result of >0.500 is read as positive for HIV-1.

Click the row for the sample in the grid and then click the − button, **IND,** or the + button above the HIV test result column to indicate whether the result for the sample is positive, indeterminate, or negative for HIV. Repeat this step for all five samples. Record your results in Chart 3.

After you complete the Experiment, take the online **Post-lab Quiz** for Activity 3.

Activity Questions

1. Describe how you can tell that this test is the indirect ELISA rather than the direct ELISA.

____

____

2. Describe what the secondary antibody binds to in this activity and why.

____

____

3. Define *seroconversion.* How can you tell that a sample has seroconverted?

____

____

ACTIVITY 4

Western Blotting Technique

OBJECTIVES

1. To compare the Western blotting technique to the ELISA.
2. To observe the use of the Western blotting technique to test for HIV.
3. To distinguish between antigens and antibodies.

Introduction

Southern blotting was developed by Ed Southern in 1975 to identify DNA. A variation of this technique, developed to identify RNA, was named Northern blotting, thus continuing the directional theme. Western blotting, another variation that identifies proteins, is named by the same convention.

Western blotting uses an electrical current to separate proteins on the basis of their size and charge. This technique uses **gel electrophoresis** to separate the proteins in a gel matrix. Because the resulting gel is fragile and would be difficult to use in further tests, the proteins are then transferred to a **nitrocellulose membrane.** The original Western blotting technique used blotting (diffusion) to transfer the proteins, but electricity is also used now for the transfer of the proteins to nitrocellulose strips. These strips are commercially available, eliminating the need for the electrophoresis and transfer equipment. In this activity you will begin the procedure after the HIV (human immunodeficiency virus) antigens have already been transferred to nitrocellulose and cut into strips.

Western blotting is also known as **immunoblotting** because the proteins that are transferred, or blotted, onto the membrane are later treated with antibodies—the same procedure used in the **indirect enzyme-linked immunosorbent assay (ELISA).** The ELISA is considered enzyme linked because an enzyme is chemically linked to an antibody in both the direct and indirect versions of the test. Immunosorbent refers to the fact that either antigens or antibodies are being adsorbed (stuck) to plastic. If the test is designed to detect an antigen or antigens, it is a **direct ELISA** because it is directly looking for the foreign substance. An **indirect ELISA** is designed to detect antibodies that the patient has made against the antigen.

Similar to the secondary antibodies used in the indirect ELISA technique, the secondary antibodies in the Western blot have an enzyme attached to them, allowing for the use of color to detect a particular protein. The secondary antibody binds to the constant region of the primary antibody found in the patient's sample. The main difference between these techniques is that the ELISA technique uses a well that corresponds to a mixture of antigens, and the Western blot has a discrete protein band that represents the specific antigen that the antibody is recognizing. Like HIV, Lyme disease can also be detected with the Western blot technique.

The initial test for HIV is the ELISA, which is less expensive and easier to perform than the Western blot. The Western blot is used as a confirmatory test after a positive ELISA because the ELISA is prone to false-positive results. The bands from a positive Western blot are from antibodies binding to specific proteins and glycoproteins from the human immunodeficiency virus. A positive result from the Western blot is determined by the presence of particular protein bands (view Table 12.1).

EQUIPMENT USED The following equipment will be depicted on-screen: washing buffer; developing buffer—secondary antibody conjugated with an enzyme; substrate solution; five samples in the samples cabinet: patient A, patient B, patient C, positive control, and negative control; rocking apparatus; nitrocellulose strips; troughs; tray; biohazardous waste disposal.

Experiment Instructions

Go to the home page in the PhysioEx software and click **Exercise 12: Serological Testing.** Click **Activity 4: Western Blotting Technique,** and take the online **Pre-lab Quiz** for Activity 4.

After you take the online Pre-lab Quiz, click the **Experiment** tab and begin the experiment. The experiment instructions are reprinted here for your reference. The opening screen for the experiment is shown below.

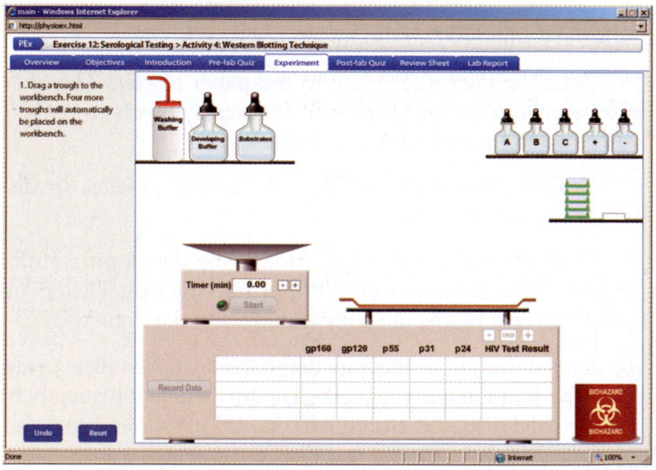
main - Windows Internet Explorer
http://physioex.html
PEx
Exercise 12: Serological Testing > Activity 4: Western Blotting Technique
Overview
Objectives
Introduction
Pre-lab Quiz
Experiment
Post-lab Quiz
Review Sheet
Lab Report
1. Drag a trough to the workbench. Four more troughs will automatically be placed on the workbench.
Washing Buffer
Developing Buffer
Substrates
A
B
C
+
-
Timer (min)
0.00
Start
gp160
gp120
p55
p31
p24
HIV Test Result
Record Data
Undo
Reset
BIOHAZARD
Done
Internet
100%

1. Drag a trough to the tray on the workbench. Four more troughs will automatically be placed on the tray.

2. Click the stack of nitrocellulose strips to place a nitrocellulose strip in each trough.

3. Drag the dropper cap of the patient A sample bottle to the first trough to dispense the antiserum from patient A to the nitrocellulose strip. A drop of antiserum for each patient will be dispensed into a separate trough.

4. Drag the tray holding the five troughs to the rocking apparatus.

5. Set the timer to 60 minutes by clicking the + button beside the timer display. Click **Start** to gently rock the samples and allow the antibodies to react with the antigens bound to the nitrocellulose. The tray will automatically be returned to the workbench and each trough will be drained into the biohazardous waste disposal when the time is complete. The simulation compresses the 60-minute time period into 10 seconds of real time.

6. Drag the washing buffer squirt bottle to the first trough to dispense washing buffer in each trough. Each trough will be automatically drained into the biohazardous waste container. The washing step removes any nonspecific binding of antibodies that occurred.

7. Drag the dropper cap of the developing buffer bottle to the first trough to dispense developing buffer to each trough.

8. Drag the tray holding the five troughs to the rocking apparatus.

9. Set the timer to 60 minutes by clicking the + button beside the timer display. Click **Start** to gently rock the samples and allow the antibodies to react with the antibodies bound to the nitrocellulose. The tray will automatically be returned to the workbench and each trough will be drained into the biohazardous waste disposal when the time is complete.

10. Drag the washing buffer squirt bottle to the first trough to add washing buffer to each trough. Each trough will be automatically drained into the biohazardous waste container. The washing step removes any nonspecific binding of secondary conjugated antibodies. Excess secondary conjugated antibodies could react erroneously with the substrate and give a false-positive result.

11. Drag the dropper cap of the substrates bottle to the first trough to dispense the substrates (tetramethyl benzidine and hydrogen peroxide) into each trough. The substrates are the chemicals that are being changed by the enzyme that is linked to the antibody.

12. Drag the tray holding the five troughs to the rocking apparatus.

13. Set the timer to 10 minutes by clicking the + button beside the timer display. Click **Start** to gently rock the samples and allow the enzyme to react with the substrates. The tray will automatically be returned to the workbench when the time is complete. The simulation compresses the 10-minute time period into 10 seconds of real time.

14. To determine the antigens present for each sample:

- Click the nitrocellulose strip inside the trough to visualize the results.
- Click **Record Data** to display your results in the grid (and record your results in Chart 4). The bands present on the nitrocellulose strip represent the antibodies present in the sample that have reacted with the antigens (bands) on the strip (view Table 12.1).

Repeat this step for all five samples.

TABLE 12.1 HIV Antigens

Abbreviation	Description
gp160	Glycoprotein 160, a viral envelope precursor
gp120	Glycoprotein 120, a viral envelope protein that binds to CD4
p55	A precursor to the viral core protein p24
gp41	A final envelope glycoprotein
p31	Reverse transcriptase
p24	A viral core protein

15. You will now indicate whether the result for each sample is negative, indeterminate, or positive for HIV. The criteria for reporting a positive result varies slightly from agency to agency. The Centers for Disease Control and Prevention recommend the following criteria:

- If no bands are present, the result is negative.
- If bands are present but they do not match the criteria for a positive result, the result is indeterminate. Patients that are deemed indeterminate after multiple tests should be monitored and tested again at a later date.
- If either p31 or p24 is present *and* gp160 or gp120 is present, the result is positive. Click the row for the sample in the grid and then click the − button, **IND,** or the + button above the HIV test result column to indicate whether the result for the sample is positive, indeterminate, or negative for HIV.

Repeat this step for all five samples. Record your results in Chart 4.

After you complete the experiment, take the online **Post-lab Quiz** for Activity 4.

CHART 4 Western Blot Results

Sample	gp160	gp120	p55	p31	p24	HIV test result
Patient A						
Patient B						
Patient C						
Positive control						
Negative control						

Activity Questions

1. Describe how gel electrophoresis is used to separate proteins.

____

____

2. In a patient sample that is positive for HIV, would antibodies or antigens be present when using the Western blot technique? How do you know?

____

____

PhysioEx 9.0

REVIEW SHEET

EXERCISE 12

NAME ______________________

LAB TIME/DATE ______________________

Serological Testing

ACTIVITY 1 Using Direct Fluorescent Antibody Technique to Test for Chlamydia

1. Describe the importance of the washing steps in the direct antibody fluorescence test. ______________________

2. Explain where the epitope (antigenic determinant) is located. ______________________

3. Describe how a positive result is detected in this serological test. ______________________

4. How would the results be affected if a negative control gave a positive result? ______________________

ACTIVITY 2 Comparing Samples with Ouchterlony Double Diffusion

1. Describe how you were able to determine what antigen is in the unknown well. ______________________

2. Why does the precipitin line form? ______________________

3. Did you think human serum albumin and bovine serum albumin would have epitopes in common? How well did the results compare with your prediction? ______________________________

ACTIVITY 3 Indirect Enzyme-Linked Immunosorbent Assay (ELISA)

1. Describe how the direct and indirect ELISA are different. ______________________________

2. Discuss why a patient might test indeterminate. ______________________________

3. How would your results have been affected if your negative control had given an indeterminate result? ______________________________

4. Briefly describe the basic structure of antibodies. ______________________________

ACTIVITY 4 Western Blotting Technique

1. Describe why the HIV Western blot is a more specific test than the indirect ELISA for HIV. ______________________________

2. Explain the procedure for a patient with an indeterminate HIV Western blot result. ______________________________

3. Briefly describe how the nitrocellulose strips were prepared before the patient samples were added to them. ______________________________

4. Describe the importance of the washing steps in the procedure. ______________________________

The Metric System

Measurement	Unit and abbreviation	Metric equivalent	Metric to English conversion factor	English to metric conversion factor
Length	1 kilometer (km)	= 1000 (10^3) meters	1 km = 0.62 mile	1 mile = 1.61 km 1 yard = 0.914 m 1 foot = 0.305 m
	1 meter (m)	= 100 (10^2) centimeters = 1000 millimeters	1 m = 1.09 yards 1 m = 3.28 feet 1 m = 39.37 inches	
	1 centimeter (cm)	= 0.01 (10^{-2}) meter	1 cm = 0.394 inch	1 foot = 30.5 cm 1 inch = 2.54 cm
	1 millimeter (mm)	= 0.001 (10^{-3}) meter	1 mm = 0.039 inch	
	1 micrometer (μm) [formerly micron (μ)]	= 0.000001 (10^{-6}) meter		
	1 nanometer (nm) [formerly millimicron (mμ)]	= 0.000000001 (10^{-9}) meter		
	1 angstrom (Å)	= 0.0000000001 (10^{-10}) meter		
Area	1 square meter (m^2)	= 10,000 square centimeters	1 m^2 = 1.1960 square yards 1 m^2 = 10.764 square feet	1 square yard = 0.8361 m^2 1 square foot = 0.0929 m^2
	1 square centimeter (cm^2)	= 100 square millimeters	1 cm^2 = 0.155 square inch	1 square inch = 6.4516 cm^2
Mass	1 metric ton (t)	= 1000 kilograms	1 t = 1.103 ton	1 ton = 0.907 t
	1 kilogram (kg)	= 1000 grams	1 kg = 2.205 pounds	1 pound = 0.4536 kg
	1 gram (g)	= 1000 milligrams	1 g = 0.0353 ounce 1 g = 15.432 grains	1 ounce = 28.35 g
	1 milligram (mg)	= 0.001 gram	1 mg = approx. 0.015 grain	
	1 microgram (μg)	= 0.000001 gram		
Volume (solids)	1 cubic meter (m^3)	= 1,000,000 cubic centimeters	1 m^3 = 1.3080 cubic yards 1 m^3 = 35.315 cubic feet	1 cubic yard = 0.7646 m^3 1 cubic foot = 0.0283 m^3
	1 cubic centimeter (cm^3 or cc)	= 0.000001 cubic meter = 1 milliliter	1 cm^3 = 0.0610 cubic inch	1 cubic inch = 16.387 cm^3
	1 cubic millimeter (mm^3)	= 0.000000001 cubic meter		
Volume (liquids and gases)	1 kiloliter (kl or kL)	= 1000 liters	1 kL = 264.17 gallons	1 gallon = 3.785 L
	1 liter (l or L)	= 1000 milliliters	1 L = 0.264 gallons 1 L = 1.057 quarts	1 quart = 0.946 L
	1 milliliter (ml or mL)	= 0.001 liter = 1 cubic centimeter	1 ml = 0.034 fluid ounce 1 ml = approx. $\frac{1}{4}$ teaspoon 1 ml = approx. 15–16 drops (gtt.)	1 quart = 946 ml 1 pint = 473 ml 1 fluid ounce = 29.57 ml 1 teaspoon = approx. 5 ml
	1 microliter (μl or μL)	= 0.000001 liter		
Time	1 second (s or sec)	= $\frac{1}{60}$ minute		
	1 millisecond (ms or msec)	= 0.001 second		
Temperature	Degrees Celsius (°C)		°F = $\frac{9}{5}$ (°C) + 32	°C = $\frac{5}{9}$ (°F − 32)

Credits

ILLUSTRATIONS

All illustrations are by Imagineering STA Media Services unless otherwise noted.

Exercise 1
1.1: Imagineering STA Media Services/Precision Graphics. 1.2, 1.4: Precision Graphics. 1.7: Adapted from Marieb and Mallatt, *Human Anatomy,* 3e, F1.10, © Benjamin Cummings, 2003.

Exercise 3
3.2–3.4, Activity 3: Precision Graphics.

Exercise 4
4.3: Tomo Narashima.

Exercise 5
5.1: Precision Graphics.

Exercise 6
6.2: Precision Graphics.

Exercise 7
7.1, 7.2, 7.7: Electronic Publishing Services, Inc.

Exercise 10
10.1–10.6, 10.8: Nadine Sokol.

Exercise 11
Table 11.1: Laurie O'Keefe.

Exercise 12
12.2a,b: Nadine Sokol.

Exercise 13
13.5: Precision Graphics.

Exercise 14
14.5: Electronic Publishing Services, Inc.

Exercise 15
15.1: Imagineering/Adapted from Martini, *Fundamentals of Anatomy & Physiology,* 4e, F11.1, Upper Saddle River, NJ: Prentice-Hall, © Frederic H. Martini, 1998.

Exercise 16
16.1–16.3: Precision Graphics. 16.7–16.9, 16.13–16.18, 16.U1: Biopac Systems.

Exercise 17
17.1: Imagineering STA Media Services/Precision Graphics. 17.2, 17.5: Precision Graphics. 17.8a: Electronic Publishing Services, Inc.

Exercise 18
18.2–18.4: Precision Graphics.

Exercise 19
19.1, 19.5, 19.7–19.9: Electronic Publishing Services, Inc. 19.11b,c, 19.13a: Precision Graphics.

Exercise 20
20.4–20.7: Biopac Systems.

Exercise 21
21.2b: Electronic Publishing Services, Inc. 21.16–21.22: Biopac Systems.

Exercise 22
22.1: Electronic Publishing Services, Inc. 22.8, 22.9: Biopac Systems.

Exercise 24
24.1, 24.3a, 24.4a: Electronic Publishing Services, Inc. 24.7: Shirley Bortoli. 24.10, 24.11: Precision Graphics.

Exercise 25
25.1–25.3: Electronic Publishing Services, Inc./Precision Graphics.

Exercise 26
26.1, 26.2: Electronic Publishing Services, Inc.

Exercise 29
29.2, 29.5: Precision Graphics.

Exercise 30
30.1: Electronic Publishing Services, Inc./Precision Graphics. 30.2, 30.3a, 30.6: Electronic Publishing Services, Inc. 30.8: Precision Graphics.

Exercise 31
31.1: Electronic Publishing Services, Inc. 31.2, 31.4: Precision Graphics. 31.7–31.12: Biopac Systems.

Exercise 32
32.1: Adapted from Tortora and Grabowski, *Principles of Anatomy and Physiology,* 9e, F21.1, New York: Wiley, © Biological Sciences Textbooks and Sandra Reynolds Grabowski, 2000. 32.2, 32.3a, 32.4–32.12, 32.14: Electronic Publishing Services, Inc. 32.3b,c: Kristin Mount.

Exercise 33
33.2: Precision Graphics. 33.6–33.8: Biopac Systems. 33.9: Precision Graphics.

Exercise 34
34.1, 34.2, 34.4: Precision Graphics. 34.5: Biopac Systems. 34.6a,b: Electronic Publishing Services, Inc.

Exercise 35
35.8: Precision Graphics.

Exercise 36
36.1–36.5, 36.7: Electronic Publishing Services, Inc.

Exercise 37
37.5, 37.6: Precision Graphics. 37.11, 37.13, 37.14: Biopac Systems.

Exercise 38
38.1–38.5, 38.8–38.10, 38.15: Electronic Publishing Services, Inc. 38.16: Electronic Publishing Services, Inc./Precision Graphics.

Exercise 39
Tables: Precision Graphics.

Exercise 40
40.1a, 40.2: Electronic Publishing Services, Inc.

Exercise 41
41.1: Precision Graphics.

Exercise 42
42.1, 42.2a, 42.6: Electronic Publishing Services, Inc.

Exercise 43
43.1: Precision Graphics. 43.2: Electronic Publishing Services, Inc.

Exercise 44
44.1: Electronic Publishing Services, Inc.

Exercise 46
46.17: Precision Graphics.

Cat Dissection Exercises
CD1.1, CD3.1, CD3.3, CD7.1: Precision Graphics. CD2.1a, CD2.2a, CD3.2, CD4.2, CD4.3, CD6.1, CD7.2a, CD8.2a, CD9.1a: Kristin Mount.

Fetal Pig Dissection Exercises
Kristen Mount.

Rat Dissection Exercises
Imagineering STA Media Services.

PhysioEx Exercises
Screenshots: BinaryLabs, Inc. Art: F2.1: Precision Graphics/Adapted from Stanfield, *Principles of Human Physiology,* 4e. F12.1, Upper Saddle River, NJ, © Pearson Science. 6.1: Precision Graphics/Adapted from Stanfield, *Principles of Human Physiology,* 4e. F13.13, Upper Saddle River, NJ, © Pearson Science. 7.1: Precision Graphics/Adapted from Stanfield, *Principles of Human Physiology,* 4e. F16.1, Upper Saddle River, NJ, © Pearson Science. All other figures are adapted from Human *Anatomy & Physiology,* 8/e by Elaine Marieb & Katja Hoehn.

PHOTOGRAPHS

Exercise 1
1.3 top: Jenny Thomas, Pearson Science. 1.3a: Howard Sochurek. 1.3b: James Cavallini/Photo Researchers. 1.3c: CNRI/Science Photo Library/Photo Researchers. 1.6b: Custom Medical Stock Photography.

Exercise 2

2.1–2.4a, 2.5b,c: Elena Dorfman, Pearson Science. 2.4b,c, 2.5a, 2.6a–c: David L. Bassett. 2.7: Carolina Biological Supply/Phototake.

Exercise 3

3.1: Leica. 3.5: Victor P. Eroschenko, Pearson Science.

Exercise 4

4.1: Don Fawcett and Science Source/Photo Researchers. 4.4a–f: Ed Reschke.

Exercise 5

5.2a–c: Richard Megna/Fundamental Photographs. 5.3a–c: David M. Philips/Photo Researchers.

Exercise 6

6.3a: G. W. Willis/Visuals Unlimited. 6.3b,f,h, 6.5d,j,k: Allen Bell, University of New England. Pearson Science. 6.3c,e, 6.5c,f,i: Nina Zanetti, Pearson Science. 6.3d, 6.5a,b,g,h,l, 6.7b: Ed Reschke. 6.3g: R. G. Kessel and R. H. Kardon/Visuals Unlimited. 6.5e: Ed Reschke/Peter Arnold. 6.6: Biophoto Associates/Photo Researchers. 6.7a: Eric Graves/Photo Researchers. 6.7c: Victor P. Eroschenko.

Exercise 7

7.2a: Ed Reschke/Peter Arnold. 7.3: Pearson Science. 7.5b: Carolina Biological Supply/Phototake. 7.5d: Manfred Kage/Peter Arnold. 7.6a: Victor P. Eroschenko. 7.6b: From *Gray's Anatomy* by Henry Gray, © Churchill Livingstone, UK. 7.7a: Cabisco/Visuals Unlimited. 7.7b: John D. Cunningham/Visuals Unlimited. 7.RS.2: Marian Rice.

Exercise 8

8.2a–c: Steve Downing. 8.3: Pearson Science.

Exercise 9

9.4c: Ed Reschke/Peter Arnold. 9.5: Ed Reschke. 9.RS.2: Alan Bell, Pearson Science.

Exercise 10

10.4a,b, 10.5: Ralph T. Hutchings. 10.6c: Elena Dorfman, Pearson Science. 10.9c: From the David Bassett *Atlas of Human Anatomy*. 10.17b: Dissection by Shawn Miller, photography by Mark Nielsen and Alexa Doig. 10.18c: Pearson Science.

Exercise 11

11.5b: National Library of Medicine. Table 11.1: From *A Stereoscopic Atlas of Human Anatomy* by David L. Bassett.

Exercise 12

12.1: Jack Scanlon, Holyoke Community College, MA. 12.2c,d: R. T. Hutchings.

Exercise 13

13.6c: From *A Stereoscopic Atlas of Human Anatomy* by David L. Bassett. 13.7d: L. Bassett/Visuals Unlimited. 13.8a: Mark Nielsen, University of Utah; Pearson Science. 13.8c: Video Surgery/Photo Researchers.

Exercise 14

14.1a: Marian Rice. 14.3, 14.6: Victor P. Eroschenko, Pearson Science. 14.4: John D. Cunningham/Visuals Unlimited.

Exercise 15

15.4b: From *A Stereoscopic Atlas of Human Anatomy* by David L. Bassett. 15.5a, 15.8b, 15.9a: Dissection by Shawn Miller, photography by Mark Nielsen and Alexa Doig. 15.11f, L: Bassett/Visuals Unlimited. 15.13b: From the David Bassett *Atlas of Human Anatomy*.

Exercise 17

17.2c: Triarch/Visuals Unlimited. 17.3c: Don Fawcett/Photo Researchers. 17.4: Eroschenko's Interactive Histology. 17.6a: Carolina Biological Supply/Phototake. 17.6b: Thomas Deerinck, NCMIR/SPL/Photo Researchers. 17.6c: Nina Zanetti, Pearson Science. 17.8b: Victor P. Eroschenko, Pearson Science.

Exercise 19

19.2c: Robert A. Chase. 19.3: Ralph T. Hutchings/Visuals Unlimited. 19.4a: Ralph T. Hutchings. 19.5b: Pat Lynch/Photo Researchers. 19.6a,b, 19.7c: From *A Stereoscopic Atlas of Human Anatomy* by David L. Bassett. 19.10: A. Glauberman/Photo Researchers. 19.11a,d: Sharon Cummings, University of California, Davis; Pearson Science. 19.12, 19.13b, 19.14: Elena Dorfman, Pearson Science.

Exercise 20

20.1a: Alexander Tsiaras/Science Source/Photo Researchers.

Exercise 21

21.1b: From *A Stereoscopic Atlas of Human Anatomy* by David L. Bassett. 21.1c,d: L. Bassett/Visuals Unlimited. 21.4: Victor P. Eroschenko, Pearson Science. 21.7b: Ralph T. Hutchings.

Exercise 22

22.4–22.6: Richard Tauber, Pearson Science.

Exercise 23

23.1b: Kilgore College Biology Dept., Kilgore, Texas. 23.1c,d, 23.2b: Victor P. Eroschenko, Pearson Science.

Exercise 24

24.3b: From *A Stereoscopic Atlas of Human Anatomy* by David L Bassett. 24.4b: Ed Reschke/Peter Arnold. 24.5: Elena Dorfman, Pearson Science. 24.6b: Stephen Spector; courtesy of Charles Thomas, Kansas University Medical Center; Pearson Science. 24.12a,b: Richard Tauber, Pearson Science. 24.13: A. L. Blum/Visuals Unlimited.

Exercise 25

25.4: Victor P. Eroschenko, Pearson Science. 25.6a–c: Richard Tauber, Pearson Science. 25.8: I. M. Hunter-Duvar, Department of Otolaryngology, The Hospital for Sick Children, Toronto.

Exercise 26

26.1b, 26.3: Victor P. Eroschenko, Pearson Science. 26.2d: Carolina Biological Supply/Phototake.

Exercise 27

27.3a: Michael Ross/Photo Researchers. 27.3b,d: Victor P. Eroschenko, Pearson Science. 27.3c: Carolina Biological Supply/Phototake. 27.3e: Benjamin Widrevitz, Natural Sciences Division, College of DuPage, Glen Ellyn, IL. 27.3f: Ed Reschke/Peter Arnold.

Exercise 29

29.3: Ed Reschke/Peter Arnold. 29.4a–e: Nina Zanetti, Pearson Science. 29.6a-c, 29.7a–d: Elena Dorfman, Pearson Science. 29.8: Meckes and Ottawa/Photo Researchers. 29.9: Pearson Science.

Exercise 30

30.3b, 30.3d: From A *Stereoscopic Atlas of Human Anatomy* by David L. Bassett. 30.3c: Lennart Nilsson, *The Body Victorious*, New York: Dell, © Boehringer Ingelheim International GmbH. 30.7: Ed Reschke. 30.8a,b, 30.9: Wally Cash, Kansas State University; Pearson Science. 30.RS.3: Ed Reschke.

Exercise 32

32.1a: Gladden Willis/Visuals Unlimited. 32.3c: David L. Bassett.

Exercise 35

35.2: Ed Reschke/Peter Arnold. 35.4b: Biophoto Associates/Photo Researchers. 35.5c: LUMEN Histology, Loyola University Medical Education Network. 35.6: John Cunningham/Visuals Unlimited.

Exercise 36

36.1a, 36.5b: From *A Stereoscopic Atlas of Human Anatomy* by David L. Bassett. 36.5a: Richard Tauber, Pearson Science. 36.6b: Victor P. Eroschenko, Pearson Science. 36.7a: Ed Reschke/Peter Arnold. 36.7b: Carolina Biological Supply/Phototake.

Exercise 37

37.3, 37.4a,b: Elena Dorfman, Pearson Science.

Exercise 38

38.5b: From *Color Atlas of Histology* by Leslie P. Garner and James L. Hiatt,

© Williams and Wilkins, 1990. 38.6a, 38.9a: Nina Zanetti, Pearson Science. 38.6b, 38.13: Victor P. Eroschenko, Pearson Science. 38.6c: Roger C. Wagner, Dept. of Biological Sciences, University of Delaware. 38.7a,c, 38.14a,b, 38.16b: From *A Stereoscopic Atlas of Human Anatomy* by David L. Bassett. 38.8d: LUMEN Histology, Loyola University Medical Education Network. 38.9b: Steve Downing. 38.9c: University of Kansas Medical Center, Department of Anatomy and Cell Biology.

Exercise 40

40.1b: Richard Tauber, Pearson Science. 40.3a: Ralph T. Hutchings. 40.6a,b, 40.7: Victor P. Eroschenko, Pearson Science.

Exercise 42

42.2b: Ed Reschke. 42.2c: From *A Stereoscopic Atlas of Human Anatomy* by David L. Bassett. 42.3: Harry Plymale. 42.4: Roger C. Wagner, University of Delaware. 42.7: Biodisc/Visuals Unlimited. 42.8: Victor P. Eroschenko, Pearson Science.

Exercise 43

43.2.b: Pearson Science. 43.3: M. Abbey/Visuals Unlimited. 43.5b: Ed Reschke. 43.6a–c: Victor P. Eroschenko, Pearson Science.

Exercise 45

45.1: CNRI/SPL/Photo Researchers. 45.2.1: Llewellyn/Uniphoto Picture Agency. 45.2.2: photos.com. 45.2.3, 45.2.4: Ogust/Image Works. 45.2.5: Boisvieux/Explorer/Photo Researchers. 45.2.6: Anthony Loveday, Pearson Science. 45.3: Carolina Biological Supply.

Exercise 46

46.1a,b, 46.3, 46.5, 46.7, 46.9–46.16, 46.18a–d, 46.19: John Wilson White, Pearson Science. 46.2: Jenny Thomas, Pearson Science.

Cat Dissection Exercises

CD1.2–CD1.13, CD2.3b CD4.1, CD4.4b, CD4.5, CD6.2, CD6.3, CD7.3, CD7.4, CD7.5b, CD8.1b, CD9.2b: Shawn Miller (dissection) and Mark Nielsen (photography), Pearson Science. CD2.1b, CD9.1b: Paul Waring, Pearson Science. CD2.2b, CD7.2b, CD8.2b: Elena Dorfman, Pearson Science. CD3.3b,c: Yvonne Baptiste-Szymanski, Pearson Science.

Fetal Pig Dissection Exercises

PD1.1, PD1.2: Jack Scanlon, Holyoke Community College, Pearson Science. PD1.3b, PD1.4b, PD1.5b, PD1.6b, PD1.7b, PD1.8b, PD2.1, PD4.1b, PD4.2b, PD4.3b, PD4.4b, PD5.1, PD5.2b, PD6.1b, PD6.2b, PD7.1b, PD7.2b, PD8.1b, PD8.2b: Elena Dorfman, Pearson Science. PD3.2b,c: Charles J. Venglarik, Pearson Science.

Rat Dissection Exercises

Robert J. Sullivan, Pearson Science.

TRADEMARK ACKNOWLEDGMENTS

3M is a trademark of 3M.
Adrenaline is a registered trademark of King Pharmaceuticals.
Albustix, Clinistix, Clinitest, Hemastix, Ictotest, Ketostix, and Multistix are registered trademarks of Bayer.
Betadine is a registered trademark of Purdue Products L. P.
Chemstrip is a registered trademark of Roche Diagnostics.
Coban is a trademark of 3M.
Harleco is a registered trademark of EMD Chemicals Inc.
Hefty® Baggies® is a registered trademark of Pactiv Corporation.
Landau is a registered trademark of Landau Uniforms.
Lycra is a registered trademark of INVISTA.
Macintosh, Power Macintosh and Mac OS X are registered trademarks of Apple Computer, Inc.
Novocain is a registered trademark of Sterling Drug, Inc.
Parafilm is a registered trademark of Pechiney Incorporated.
Sedi-stain is a registered trademark of Becton, Dickinson and Company.
Speedo is a registered trademark of Speedo International.
VELCRO® is a registered trademark of VELCRO Industries B. V.
Wampole is a registered trademark of Wampole Laboratories.
Windows is either a registered trademark or trademark of Microsoft Corporation in the United States and/or other countries.

Index

NOTE: Page numbers in **boldface** indicate a definition. A *t* following a page number indicates tabular material and an *f* indicates an illustration.